UNITED STATES DEPARTMENT OF AGRICULTURE
MISCELLANEOUS PUBLICATION NO. 401

Washington, D. C. Issued June 1942

GEOGRAPHICAL GUIDE
to
FLORAS OF THE WORLD

An annotated list with special reference to useful plants and common plant names

---

PART I

Africa, Australia, North America, South America, and Islands of the Atlantic, Pacific, and Indian Oceans

---

By

S. F. BLAKE
Senior Botanist
Division of Plant Exploration and Introduction
Bureau of Plant Industry

and

ALICE C. ATWOOD
Botanical Bibliographer
Library

UNITED STATES
GOVERNMENT PRINTING OFFICE
WASHINGTON: 1942

For sale by the Superintendent of Documents, Washington, D. C. - - - - Price 75 cents

FLORAS OF THE WORLD

An annotated selected list of floras and floristic works relating to vascular plants, including bibliographies and publications dealing with useful plants and vernacular names.

Part I

Africa, Australasia, Insular Floras, North America, and South America

By S. F. Blake, *senior botanist, Division of Plant Exploration and Introduction, Bureau of Plant Industry,* and Alice C. Atwood, *botanical bibliographer, Library*[1]

CONTENTS

[1] The authors wish to express their thanks to the authorities and librarians of the Gray Herbarium, the Arnold Arboretum, and the New York Botanical Garden for permission to consult their libraries freely. Several botanists have examined the lists of titles of regions with which they are familiar and have suggested additional papers. Dr. John Hendley Barnhart, of the New York Botanical Garden, has rendered much assistance by furnishing full names and dates of some authors from his extensive card catalog of botanists, now including about 35,000 cards.

INTRODUCTION

The primary purpose of this publication is usefulness, not bibliographical completeness. Its aim is to furnish an annotated catalog of all the now useful floras and floristic works, including those in periodical literature, that list or describe the complete vascular flora (or the phanerogams only) of any region or locality, and to include as well all publications dealing on the same scale with useful and medicinal plants, vernacular names, and botanical bibliography. In general, only the later works relating to a given region or subject are included, although earlier publications are usually mentioned under the titles of works that have replaced them. Publications dealing with only a part of the flora, such as a single family or larger group of the flowering plants or the pteridophytes only, or a single group of useful plants, such as those producing dyes, resins, or forage, are excluded, except for complete papers on edible, medicinal, or woody plants. Occasional titles relating to weeds and poisonous plants, chiefly from the botanical point of view, have been included, although no attempt has been made to include all papers on these subjects. In the case of little-known regions without complete floras, some publications of an incomplete nature, such as lists of species obtained by individual collectors or expeditions, have been inserted, as have also some miscellaneous papers not falling strictly within the scope of the list as defined. In general, no attempt has been made to evaluate the titles listed, but references to critical reviews are given when available. Papers that are primarily ecological have been excluded unless they contain fairly complete lists of the plants of the areas discussed. Works of the "popular" type, intended for beginners in botany, have been omitted, as well as works that are now only of historical value. The list is intended to be complete through 1939, and no later titles have been inserted.

The selective nature of this work has left the authors free to omit a few papers dealing with areas so small that they can be disregarded by all but the local botanist dealing in a very special way with the regions concerned, and also those few lists that are so full of errors or so incomplete that their inclusion would serve no useful purpose. These intentional omissions aside, this publication can be taken as providing an annotated summary of the available nominally complete publications on the vascular flora and the economic botany of the world. Unintentional omissions will, of course, be found by users, and additions will be gratefully received by the authors.

No list of the scope of this one has ever been published. Daydon Jackson's Guide to the Literature of Botany,[2] now over 60 years old, dealt almost entirely with independently published works, and its few annotations were mainly bibliographical. A 12-page publication by Prof. G. L. Goodale,[3] of about the same date, listed geographically and chronologically the more important independently published

---

[2] JACKSON, B. D. GUIDE TO THE LITERATURE OF BOTANY; BEING A CLASSIFIED SELECTION OF BOTANICAL WORKS, INCLUDING NEARLY 6,000 TITLES NOT GIVEN IN PRITZEL'S THESAURUS. xl, 626 p. 22 cm. London, 1881. (Index Society Publications, viii.)

[3] GOODALE, GEORGE L. THE FLORAS OF DIFFERENT COUNTRIES. Bibl. Contrib. Libr. Harvard Univ. 9. 12 p 1879. (Reprinted from Bul. Libr. Harvard Univ. no. 10, 11, and 12.)—Partly annotated list, geographically grouped, then arranged in chronological order; includes for the most part only separately published works, many of which are now only of historical value. Almost none of the smaller local floras are included.

floras of the world, and similar but briefer lists have appeared in various textbooks on systematic or other phases of botany, but none of these represented any real search of the voluminous available literature.

The part of this list now published, covering all the world except Europe and Asia and the islands closely associated with them (namely, those north of 40° north latitude in the Atlantic Ocean and north of 30° north latitude in the Pacific), includes about 2,597 primary titles and 428 subsidiary titles (supplements, reviews, etc., as well as the few works listed as not available for examination). At least as many more papers have been examined and rejected as not fulfilling the promise of their titles.

The list is based in the first instance on the classified subject catalog of botany in the library of the United States Department of Agriculture, begun in 1906 under the direction of the late Dr. Frederick V. Coville by Marjorie F. Warner and Alice C. Atwood, and continued by them and their assistants. This catalog provides a fairly complete index to the systematic and economic publications of the world for the last third of a century, as well as to much of the earlier literature. Practically all the pertinent works on the shelves of the libraries of the Gray Herbarium, the Arnold Arboretum, and the New York Botanical Garden have been examined independently by one of the authors, and the titles given in the bibliographies of the papers cited in the present list have been checked. The geographically arranged lists in the Bradley Bibliography,[4] the Catalogue of the Library of the Arnold Arboretum (vol. 2), and the publications of the Lloyd Library [5] have also been utilized, as well as Bay's Bibliographies of Botany.[6] Certain series of periodicals have been gone through, but no complete examination of the botanical periodicals of the world has been attempted, nor have the unclassified catalogs of the libraries of the Kew Herbarium and the British Museum been searched, although they would, no doubt, provide some additional items.

After the bulk of the list was essentially completed, a search through the principal periodical bibliographical publications added a considerable number of mostly minor titles, which previously had been overlooked or rejected. The publications examined for this purpose include: Biological Abstracts (vol. 1–13, pt. 8), Botanical Abstracts (vol. 1–15), Bohnensieg's Repertorium annuum literaturae botanicae periodicae (vol. 1–8), Botanisches Centralblatt (vol. 1–175, heft 6) and its Beihefte (vol. 1–9), Bulletin de la Société Botanique de France (vol. 1–86, pt. 2), Engler's Botanische Jahrbücher (vol. 1–70, pt. 3), Fortschritte der Botanik (vol. 1–8), International Catalogue of Scientific Literature (vol. 1–14), and Just's Botanischer Jahresbericht (vol. 1–57, heft 1).

---

[4] REHDER, ALFRED. THE BRADLEY BIBLIOGRAPHY. A GUIDE TO THE LITERATURE OF THE WOODY PLANTS OF THE WORLD PUBLISHED BEFORE THE BEGINNING OF THE TWENTIETH CENTURY. 5 v. 30 cm. Cambridge [Mass.], 1911–1918. (Pub. Arnold Arboretum no. 3.)—Phytography (floras, etc.) in v. 1.

[5] HOLDEN, WILLIAM, and WYCOFF, EDITH. BIBLIOGRAPHICAL CONTRIBUTIONS FROM THE LLOYD LIBRARY, CINCINNATI, OHIO. v. 1, no. 2–13. 1911–1914.—Each number a bibliography of the floristic literature of a country or region, alphabetically arranged by authors. Those referring to regions in the present list are: No. 9 (North America and West Indies), no. 10 (South America and Antarctic regions), no. 12 (Oceanica, including Australasia), and no. 13 (Africa).

[6] BAY, J. C. BIBLIOGRAPHIES OF BOTANY. A CONTRIBUTION TOWARD A BIBLIOTHECA BIBLIOGRAPHICA . . . Prog. Rei Bot. 3: 331–456. 1910.—Includes, among other material, a list of "general and comprehensive bibliographies" and one of "national (regional) bibliographies."

A few general works and textbooks dealing with the flora or the economic plants of the whole world, for the most part familiar to every working botanist, have not been included in this publication, but may be mentioned here. The most important ones dealing primarily with the flora of the world are De Candolle's Prodromus and its continuation, the Monographiae Phanerogamarum, Bentham and Hooker's Genera Plantarum, Baillon's Histoire des Plantes, Engler and Prantl's Die Natürlichen Pflanzenfamilien, Dalla Torre and Harms' Genera Siphonogamarum, the extensive series of volumes forming Das Pflanzenreich, the Index Kewensis, and the Index Londinensis. Some outstanding works dealing with vernacular names are Gerth van Wijk,[7] Lyons[8] (which also gives uses), Bedevian,[9] and Meyer.[10]

Numerous works, mostly textbooks, treating of the economic and medicinal plants of the world are available, among which the following may be mentioned: For economic plants in general, Clute's Useful Plants of the World,[11] Hill's Economic Botany,[12] Holland's Overseas Plant Products,[13] Rosenthal's Synopsis Plantarum Diaphoricarum,[14] Wiesner's Die Rohstoffe des Pflanzenreichs,[15] and Wehmer's Die Pflanzenstoffe[16]; for edible plants, Bois' Les Plantes Alimentaires,[17]

---

[7] GERTH VAN WIJK, H. L. A DICTIONARY OF PLANT NAMES. 2 v. 29.5 cm. The Hague, 1911–1916.—Vol. 1, bibliography; alphabetical list of botanical names, with English, French, German and Dutch vernacular names; vol. 2, index to vernacular names.

[8] LYONS, A. B. PLANT NAMES SCIENTIFIC AND POPULAR, INCLUDING IN THE CASE OF EACH PLANT THE CORRECT BOTANICAL NAME IN ACCORDANCE WITH THE REFORMED NOMENCLATURE, TOGETHER WITH BOTANICAL AND POPULAR SYNONYMS . . . ed. 2. 630 p. 20.5 cm. Detroit, 1907.—Alphabetical list of 2,327 genera, each with one or more species, with vernacular names and principal uses; index of vernacular names.

[9] BEDEVIAN, A. K. ILLUSTRATED POLYGLOTTIC DICTIONARY OF PLANT NAMES IN LATIN, ARABIC, ARMENIAN, ENGLISH, FRENCH, GERMAN, ITALIAN AND TURKISH LANGUAGES, INCLUDING ECONOMIC, MEDICINAL, POISONOUS AND ORNAMENTAL PLANTS AND COMMON WEEDS . . . xv, 644, 455 p. illus., port. 21 cm. Cairo, 1936.—Alphabetical list of 3,657 vascular and cellular plants, with vernacular names in eight languages; notes on transliteration and pronunciation; bibliography; separate lists of names in each language, with references to the first part of the work.

[10] MEYER, HANS. BUCH DER HOLZNAMEN . . . xviii, 564 p. 25.5 cm. Hannover, 1933–36.—Alphabetical list of vernacular names of trees in all languages, with botanical equivalent, family name, and habitat.

[11] CLUTE, W. N. THE USEFUL PLANTS OF THE WORLD. 2d ed. 220 p. 22 cm. Indianapolis, 1932. (1st ed., 1928, reprinted from Amer. Bot. suppl. v. 33–34, 1927–28.)—Popular treatment, with lists of principal species grouped by uses.

[12] HILL, A. F. ECONOMIC BOTANY. A TEXTBOOK OF USEFUL PLANTS AND PLANT PRODUCTS. x, 592 p. 225 fig. 24 cm. N. Y., 1937.—Discussion of useful plants, grouped by uses; systematic list of species; bibliography, composed chiefly of publications in English.

[13] HOLLAND, J. H. OVERSEAS PLANT PRODUCTS. vii, 279 p. 18.5 cm. London, 1937.—Briefly annotated alphabetical list of trade names, with botanical names and localities, of "all the natural products of vegetable origin, imported on a commercial scale into the docks under the control of the Port of London Authority and into other ports, for landing and delivery to the consignees in the markets of the United Kingdom. Others of economic value in the countries of production, including many that have been sent to Kew from time to time for identification, are also included." Has bibliography.

[14] ROSENTHAL, D. A. SYNOPSIS PLANTARUM DIAPHORICARUM. SYSTEMATISCHE UEBERSICHT DER HEIL-, NUTZ- UND GIFTPFLANZEN ALLER LÄNDER. xxvi, 1359 p. 22.5 cm. Erlangen, 1861 (1861–62).—Systematic list of medicinal, useful, and poisonous plants, with German vernacular names. officinal names, range, brief account of uses.

[15] WIESNER, JULIUS VON. DIE ROHSTOFFE DES PFLANZENREICHS. 4. aufl. . . . Herausgegeben von Paul Krais und Wilhelm von Brehmer. 2 v. (2,253 p.). illus. 25 cm. Leipzig, 1927–28. (1st ed., 1873.)—Encyclopedic work with especial emphasis on chemical features, the plants grouped under such headings as Alkaloide, Farbstoffe, Hölzer, etc., with extensive bibliographies and annotated systematic lists of species.

[16] WEHMER, CARL. DIE PFLANZENSTOFFE BOTANISCH-SYSTEMATISCH BEARBEITET. BESTANDTEILE UND ZUSAMMENSETZUNG DER EINZELNEN PFLANZEN UND DEREN PRODUKTE. PHANEROGAMEN. 2., neubearb. und verm. aufl. 2 v. (1,511 p.). Jena, 1929–31 (1st ed., 1911.)—Ergänzungsband zur zweiten auflage: nachträge aus den jahren 1930–1934, mitbearb. von Magdalene Hadders. Aus dem nachlass hrsg. von Hans Amelung . . . iv, 244 p. port. 25.5 cm. Jena, 1935.—Systematic list of 4,493 spermatophytes, with habitat, products, chemical properties, references.

[17] BOIS, DÉSIRÉ. LES PLANTES ALIMENTAIRES CHEZ TOUS LES PEUPLES ET À TRAVERS LES AGES. HISTOIRE, UTILISATION, CULTURE. 4 v. 698 fig. 25.5 cm. Paris. 1927–37.—Descriptive account of wild and cultivated edible plants, with history, uses, vernacular names, and references. Vol. 1, vegetables; v. 2, fruits; v. 3, spices and condiments; v. 4, sources of drinks. "May be regarded as the fourth edition of 'A. Paillieux and D. Bois, Le potager d'un curieux . . . '" 1885.

and Sturtevant's Notes on Edible Plants[18]; for medicinal plants, Baillon's Traité de Botanique Médicale Phanérogamique,[19] Bentley and Trimen's Medicinal Plants,[20] Dragendorff's Die Heilpflanzen,[21] and Dujardin-Beaumetz and Égasse's Les Plantes Médicinales;[22] for woody plants, in addition to the Bradley Bibliography[23] and Meyer's Buch der Holznamen,[24] Ahern and Newton's Bibliography,[25] Howard's Manual of the Timbers of the World,[26] and Record and Mell's[27] Timbers of Tropical America.

The more strictly bibliographical part of the present work has been the chief concern of Miss Atwood, the selection and annotation of the titles that of S. F. Blake. For the purpose of this publication, it has not seemed worth while to enter into the niceties of citation so dear to the heart of the confirmed bibliographer. Intercalated unnumbered pages, variant cover and title-page titles, unpaged errata lists and the like have in practically all cases been passed over, but no effort has been spared to secure accuracy in all the bibliographical matter actually presented. Authors' given names, when more than one, are cited only as initials in the body of the work, but full names and dates, wherever available, are given in the author index (prepared by Miss Atwood).

Subdivisions under the principal geographic units have been dispensed with as far as possible, to avoid multiplication of subheadings, but under each geographic heading works covering the whole region are listed first (under the heading *General*), followed by the local and partial titles (under the heading *Local*). Wherever it seems desirable, the geographical information given in the title has

---

[18] STURTEVANT, E. L. STURTEVANT'S NOTES ON EDIBLE PLANTS. Edited by U. P. Hedrick. vii, 686 p. port. 31 cm. Albany, 1919. (New York Dept. of agriculture. 27th Annual report. v. 2, pt. 2; also Report of the New York agricultural experiment station, 1919, II.)—Alphabetical list of 2,897 plants, with habitat, uses, principal English names, and references; bibliography. Edible fungi are not included.

[19] BAILLON, H. E. TRAITÉ DE BOTANIQUE MÉDICALE PHANÉROGAMIQUE. 1499 p. 3487 fig. 24 cm. Paris, 1883–84.—General account of organography and physiology; systematic account of medicinal plants of the world, with vernacular names and uses, and brief descriptions of families and genera and of the more important species.

[20] BENTLEY, ROBERT, and TRIMEN, HENRY. MEDICINAL PLANTS, BEING DESCRIPTIONS WITH ORIGINAL FIGURES OF THE PRINCIPAL PLANTS EMPLOYED IN MEDICINE AND AN ACCOUNT OF THE CHARACTERS, PROPERTIES, AND USES OF THEIR PARTS AND PRODUCTS OF MEDICINAL VALUE. 4 v. 306 col. pl. 24 cm. London, 1880 (1875–80).—Systematic account with descriptions, official names, composition, uses, references; 306 species figured, many other mentioned. Designed as an illustrated botanical guide to official British, Indian, and United States pharmacopoeias.

[21] DRAGENDORFF, GEORG. DIE HEILPFLANZEN DER VERSCHIEDENEN VÖLKER UND ZEITEN. IHRE ANWENDUNG, WESENTLICHEN BESTANDTHEILE UND GESCHICHTE . . . vi, 884 p. 24.5 cm. Stuttgart, 1898.—Briefly annotated systematic list, mostly without references and with few vernacular names.

[22] DUJARDIN-BEAUMETZ, GEORGES, and ÉGASSE, ED. LES PLANTES MÉDICINALES INDIGÈNES ET EXOTIQUES, LEURS USAGES THÉRAPEUTIQUES PHARMACEUTIQUES ET INDUSTRIELS vii, 845 p. 1034 fig., 40 col. pl. 28.5 cm. Paris, 1889.—Descriptive account of more significant medicinal plants of the world, alphabetically arranged, with account of properties and uses; index of botanical and vernacular names; list of species classified by uses.

[23] See footnote 4, p. 4.

[24] See footnote 10, p. 5.

[25] AHERN, G. P., and NEWTON, H. K. A BIBLIOGRAPHY ON WOODS OF THE WORLD, EXCLUSIVE OF THE TEMPERATE REGION OF NORTH AMERICA AND WITH EMPHASIS ON TROPICAL WOODS. Scient. Contrib. Trop. Plant Res. Found. no. 10. 77 p. 1928.—List of 1,530 titles, geographically arranged, with translations of non-English titles; subject index. Includes chiefly items dealing with forestry and commercial aspects of woods; floras and purely taxonomic works are mostly omitted.

[26] HOWARD, A. L. A MANUAL OF THE TIMBERS OF THE WORLD, THEIR CHARACTERISTICS AND USES. Rev. ed. to which is appended an index of vernacular names. xxiii, 672 p. illus. 18.5 cm. London, 1934. (1st ed., 1920.)—Annotated list, alphabetically arranged by vernacular names, with range vernacular names, uses, character of wood; bibliography, geographical lists of timbers; indices of botanical and vernacular names.

[27] RECORD, S. J., and MELL, C. D. TIMBERS OF TROPICAL AMERICA. xviii, 610 p. 50 pl. 26 cm. New Haven, 1924.—General account of forests in whole of West Indies, Mexico, Central America, and South America; systematic account of families and more important genera and species of trees, with vernacular names, uses, description of wood and wood anatomy.

been supplemented in the annotation, usually in the form of a parenthetical statement. The figures of longitude given are based on the Greenwich meridian. The annotations in all cases are to be regarded as supplementary to the title.

Each publication is listed in full only once, but cross-references are given immediately following the pertinent geographic heading. It has not seemed necessary to provide cross-references from smaller geographical units to the larger units of which they form a part, and the latter should always be consulted as well. Except in a very few cases, all of which are so indicated in the annotation, all the works listed have been examined by one or both of the authors. The comparatively few works that, from their titles, appear to be of value, but which have not been examined or found reviewed, are listed directly after the appropriate geographic heading.

The arrangement adopted is an alphabetical-geographical one, summarized in the table of contents. The Times Survey Atlas of the World (London, 1922) has been followed in general for geographic boundaries. No great difficulty has been found in arranging the titles under readily intelligible geographic headings except in the cases of Africa and Insular Floras. The difficulty in Africa arises not so much from the numerous changes of name following the World War, which were accompanied by relatively few and mostly unimportant changes of boundaries, as from the great area covered by some of the principal floras and from the vagueness of many of the geographical divisions of Africa that do not correspond with political boundaries. The residue of works that could not be definitely placed under any of the recognized political divisions was finally listed under the headings Sahara, South Africa (in general, the region south of the Tropic of Capricorn), and West Africa (in general, the region from Mauritania to Angola). Cross-references to these are provided under the separate countries concerned, but reciprocal references under these three nonpolitical divisions have not been considered necessary.

The arrangement of the very numerous insular floras has offered particular difficulty. Some papers deal with the flora of the whole or only a part of definite island groups, others with that of a single island of a group or with several scattered islands or groups of islands; and the boundaries of such commonly used geographic terms as Melanesia, Micronesia, and Polynesia differ somewhat in different atlases and are usually not clearly marked on actual maps. The method finally adopted is that of listing under separate headings papers dealing with an island group as a whole and those dealing individually with any of its constituent islands when these are of any considerable size. Papers treating of the flora of several islands or groups of islands are almost always listed in full under the first island mentioned. Cross-references are provided in all cases. All the islands north of 40° north latitude in the Atlantic and 30° north latitude in the Pacific are placed under the continents and countries to which they belong (Greenland under North America; Iceland, Spitzbergen, Faroe Islands, etc., under Europe; Aleutian Islands under Alaska, and so on); the islands stretching from the southern tip of Kamchatka to and including Formosa, also Hainan and the Bonin Islands, as well as Ceylon, under Asia; New Zealand (with Stewart Island), Tasmania, and the small islands immediately off Australia under Australasia; Cedros, Guadalupe, Revillagigedo, Socorro, Tres Marías, and a few smaller islands under Mexico; Fernando Po, São Thomé (St. Thomas) with Principe, and Zanzibar under Africa; West Indies under North America. The remaining islands, forming the section Insular Floras of this list, are divided somewhat arbitrarily into three groups: Atlantic Ocean islands, between 70° west and 20° east longitude (Greenwich meridian); Indian Ocean islands (including Arabian Sea and Bay of Bengal), between 20° and about 95° east longitude, terminating with the Nicobar Islands (but including Christmas Island and Keeling or Cocos Islands); and the Pacific Ocean islands, from about 95° east to 70° west longitude, including all the Dutch East Indies from Sumatra eastward, with the small islands immediately south of them. The islands in the Mediterranean Sea, except those belonging to Africa, will be treated in the second part of this work.

The examination and annotation of several thousand publications has brought to attention so many common faults and omissions in the preparation of floristic papers that some comments will not be out of place. The essential features of even the barest list of plants include a title accurately describing its contents; the most definite possible statement of the geographical area covered, with reference, in the case of small or obscure localities, to the distance and direction from some locality that can be found on any reasonably detailed map, supplemented, particularly in the case of oceanic islands, by details of latitude and longitude; and a statement of the material on which the list is based, whether the author's collections or observations alone or all published or unpublished information. Any list that attempts to represent the known flora of a region should include also a statistical summary, not necessarily detailed, of the number of species included, preferably divided among the vascular cryptogams, gymnosperms, monocotyledons, and dicotyledons, with similar figures for at least the larger families, and a bibliography in which the titles and references are given accurately and in full, not merely to the volume of a work or periodical or the first page of a paper.

Additional items that add greatly to the value of a flora include accounts of the topography, hydrography, climate, geology, and soils; botanical explorations and list of collectors, with full names and some biographical information; list of herbaria in which specimens are deposited; general and special features of vegetation, ecology, phytogeography, life zones, endemic species; notes on native and cultivated useful plants; local vernacular names; lists of doubtful and excluded species, with reference to previous publications; a gazetteer of localities, particularly those not to be found in ordinary atlases, supplemented, if possible, by a map; a list of botanical names first published in the work; and an index. An estimate of the relative completeness of the work may well be added, and, if circumstances warrant, some account of conditions affecting botanical collecting, such as methods of transportation, available accommodation, and so on. The inclusion or omission of keys, descriptions, references, and synonymy depends on the purpose for which the publication is intended and on the availability of such information in other publications. Authors of works dealing with regions not covered by descriptive floras, for instance Alaska or most of the Central American countries, should remember that the utility of such publications is increased manyfold by the inclusion of keys, supplemented if possible by brief descriptions, and that, if choice must be made, some of the space taken up by detailed citation of exsiccatae and various other less essential matter can be much more profitably employed for this purpose. Authors of supplements to published material should give a full reference to the title and place of publication of the original paper.

Despite its occasional omissions, the present publication is sufficiently complete to afford data for a summary of the knowledge of the floras of more than half the world; all, that is, except Europe and Asia and their associated islands. The extent of the lack of detailed knowledge of the floras of much of this part of the world, with the honorable exceptions of Greenland, Australasia, and a considerable proportion of the islands, is brought out in the following paragraphs, which necessarily pass over the numerous strictly local lists or partial floras.

Africa.—Of the 49 geographical divisions here employed (excluding the Sahara, South Africa, and West Africa), there are more or less complete floras, containing descriptions or keys, of seven, as follows: Algeria (1888–1910), Anglo-Egyptian Sudan (1929), Angola (1937—; in process of publication), Egypt (1912, plus list of 220 additional forms in 1930), Swaziland and Transvaal together (1926—; in process of publication), and Tunisia (1902). Of these, only the still unfinished floras of Angola and of Transvaal and Swaziland can really be regarded as satisfactory. The known flora of Algeria has been greatly increased since 1910; Broun and Massey's Flora of the Sudan is rather a field book than a sufficient flora; Muschler's Flora of Egypt is presumably as unreliable as his other publications; and the keyed list of the flora of Tunis (with Algeria) published by Battandier and Trabut in 1902, with its predecessor, Bonnet and Barratte's Catalogue of 1896, is long out of date.

Some 19 additional countries or colonies are represented by more or less extensively annotated lists, without keys or descriptions except of the new species: Belgian Congo (1909–21), Eritrea (1903–07; never completed), Gambia (1907, 1922), Gold Coast (1914, 1930, 1937), Italian Somaliland (1929–36; extensive, but not a comprehensive flora), Kenya (1922, 1936), Liberia (1937), Libya (1910), Morocco (1931–34), Natal (1921), Nyasaland (1898), Rio de Oro (1914), São Thomé (1886–92), Sierra Leone (1916; not seen), Southern Rhodesia (1916), South-West Africa (1917–28), Tanganyika (1929—; not yet finished), Uganda (1902), and possibly Zanzibar (ca. 1890).

The following 23 countries, colonies, and protectorates are entirely without general floras or lists limited to their area, although all but 2 (Bechuanaland, Ruanda and Urundi) have been the subject of partial papers cited in this bibliography: Basutoland, Bechuanaland, Bechuanaland Protectorate, British Somaliland, Cameroons, Cape of Good Hope, Dahomey, Ethiopia (except for Richard's ancient flora), French Equatorial Africa, French Guinea, French Somaliland, French West Africa, Ivory Coast, Mauritania, Mozambique, Nigeria (except for Hooker and Bentham's ancient Niger Flora), Northern Rhodesia,

Orange Free State, Portuguese Guinea, Ruanda and Urundi, Senegal, Spanish Guinea, and Togo. All of these, however, are covered by the general floras of Africa—Flora Capensis, Flora of Tropical Africa, Flora of West Tropical Africa—so far as their botany was known at the time the works were written. The lists of species given by Engler in Die Pflanzenwelt Afrikas should always be consulted for information on the flora of these areas, as well as Chevalier's Énumération (1920) for the west coast region from Senegal to Calabar.

On the whole, the principal countries in Africa most insufficiently known or inadequately covered by published works are, to the north of the Equator, *Rio de Oro,[28] Mauritania, Senegal, Portuguese Guinea, French Guinea, Sierra Leone, Ivory Coast, Togo, parts of French West Africa and French Equatorial Africa, Spanish Guinea, Uganda, Kenya, *Eritrea, *Ethiopia, *French and *British Somaliland; and south of the Equator, *Northern Rhodesia, Nyasaland, *Mozambique, *Bechuanaland, and Orange Free State. A single comprehensive summary of the very extensive work done on the flora of the Belgian Congo since the Durands' Sylloge of 1909 is also greatly to be desired, as well as similar summaries for Algeria, Tunisia, Libya, Nyasaland, and Uganda.

AUSTRALASIA.[29]—The only general flora of Australia, that of Bentham, is now over 60 years old. All the principal subdivisions of Australasia, however, are provided with comparatively modern lists or descriptive floras: New South Wales (Maiden and Betche's Census, 1916, and Moore and Betche's descriptive Handbook, 1893); Northern Territory (Ewart and Davies' keyed Flora, 1917); Queensland (Bailey's descriptive Queensland flora, 1899–1902, and his Comprehensive Catalogue, 1913); South Australia (Black's descriptive Flora, 1922–29, as well as Tate's Census, 1889–95, and his keyed Handbook, 1890); Victoria (Ewart's descriptive Flora, 1930, and the Field Naturalists' Club's Census, 1928–31); Western Australia (Gardner's unannotated Enumeratio, 1931); New Zealand (Cheeseman's descriptive Manual, 1925); Tasmania (Rodway's briefly descriptive Tasmanian Flora, 1903). There are also a large number of substantial local lists, and the flora as a whole can be regarded as more satisfactorily covered by published works than that of any equally extensive division of the earth's surface except Europe.

INSULAR FLORAS.—Insular floras, from their compactness and aloofness, seem to have had a peculiar attraction for botanists.

Of the 17 Atlantic islands or groups of islands included in this list, all are more or less well covered by existing lists, although only one group (Bermuda Islands) has a modern descriptive flora.

Of the 21 islands or groups of islands listed in the Indian Ocean, the smaller islands as a rule are pretty well covered. Some of the larger floras are comparatively old. Kurz's report on the Andaman Islands dates from 1870 (with a subsequent nominal list by Rogers in 1903), Baker's descriptive Flora of Mauritius from 1877 (with a list of additions by Johnston in 1895), Balfour's Botany of Rodriguez from 1879, Jacob de Cordemoy's Flora of Réunion from 1895, and Kurz's Sketch of the Vegetation of the Nicobar Islands from 1876.

---

[28] Those marked with an asterisk are especially in need of floras.

[29] The term Australasia, as here used refers only to Australia, New Zealand, and Tasmania, the minor islands being listed under Insular Floras.

The greatest need in this field is a satisfactory flora of Madagascar. Baron's Compendium dates from 1901 to 1906, Palacký's Catalogus from 1905 to 1907, and both are merely lists, as is also the catalog now being published in parts by the Académie Malgache. The excellent descriptive Flore de Madagascar edited by Humbert has so far covered only 6 of the 189 families represented in the flora, and the prospects of its completion under present disturbed world conditions are not favorable. The large proportion of endemics in this flora of over 5,000 species and the extent to which the descriptions are scattered in botanical literature make the publication of a descriptive flora of the island one of the greatest desiderata in the botanical literature of the world.

The 23 islands or groups of islands in the Pacific Ocean are very unequally covered. The chief ones of which knowledge of the flora, however incomplete, is fairly well organized are the Australian Islands (Antipodes, Auckland, Bounty, Campbell, Chatham, Kermadec, Lord Howe, Macquarie, Norfolk, Snares), Borneo (Merrill's Bibliographic Enumeration, 1921, with additions to 1926), Cocos Island, Fiji Islands (Seemann, 1865–73, with extensive lists of additions by several recent authors), Galapagos Islands, Java (Koorders' Exkursionsflora, 1911–37), Juan Fernandez Islands (Skottsberg, 1921), some islands or groups in Melanesia (New Hebrides, Guillaumin, 1928 and subsequently; New Caledonia, listed separately beyond; and, mostly in less complete form, the flora of several smaller islands), Micronesia (Kanehira's Enumeration, 1935; also Guam, Merrill, 1914, and several of the smaller islands), New Caledonia (Guillaumin, 1911, with a bewildering array of subsequent additions), Philippine Islands (Merrill, 1923–26, with subsequent additions to 1930), parts of Polynesia (Marquesas, Tuamotu, and Austral Islands, etc., Brown and Brown, 1931–35; Samoan and Society Islands, listed separately beyond; and numerous single islands or small groups), Samoan Islands (Reinecke, 1896–98, with additions; Christophersen, 1935–38), San Ambrosio and San Felix (Skottsberg, 1937), and Society Islands (Drake del Castillo, 1892, and Setchell's recent papers).

Most of these islands, so far as modern floras go, are covered by lists only, with the principal exception of Koorders' Exkursionsflora von Java and Drake del Castillo's Flore de la Polynésie Française, so that much research is necessary for the identification of specimens. Sumatra and New Guinea are very inadequately provided with floras, the former having Miquel's Prodromus of 1860 and various later lists covering individual collections; the latter various lists of separate collections, but no general flora since its treatment in Miquel's Flora Indiae Batavae in 1855–59. The Hawaiian Islands have Hillebrand's Flora of 1888, now far from satisfactory, and Degener's Flora Hawaiiensis, a well-illustrated current work marred by its piecemeal publication in loose-leaf form, a method of publication which is unique among floras at present, and, it is to be hoped, forever. Celebes has among modern floras only Koorders' Verslag of 1898, covering the Minahasa Peninsula, with a set of supplements extending to 1922. Many of the smaller groups or single islands possess fairly complete floras. The extent to which the known flora of the smaller islands in Polynesia can be increased by careful collecting is indicated by the statistical table recently given by St. John [30] of the results of the Mangarevan Expedition.

NORTH AMERICA.—*Alaska.*—Hultén's Flora of the Aleutian Islands and Westernmost Alaska Peninsula (1937), Porsild's Contributions to the Flora of Alaska (1939), and several fairly modern local lists are available, but there is no general flora except Rothrock's compiled list of 1872, now of little more than historical value. A flora of Alaska by Eric Hultén is in course of publication.

Canada.—The only general list, by John Macoun, dates from 1883–90, with supplements by James M. Macoun extending to 1906. Of the 15 Provinces and districts here recognized (including Newfoundland, politically not a part of Canada), only one-third possess a modern list: Franklin (Simmons, 1913), Manitoba (Jackson and others, 1922), Prince Edward Island (Hurst, 1933, an unannotated list), Quebec (Louis-Marie, 1931; also Marie-Victorin's Flore Laurentienne, 1935, covering the southern part), Saskatchewan (Fraser and Russell, 1937). There are older lists for Labrador (Macoun, 1897), New Brunswick (Fowler, 1885, with additions to 1905), and Nova Scotia (Lindsay, 1877), none of which adequately represents present knowledge, and a fair number of local lists.

Central America—Of the eight Central American countries (including Mexico), only three possess modern floras, all written by Paul C. Standley alone or in cooperation with other botanists (British Honduras, 1936; Costa Rica, 1937–38; Salvador, 1925). The first contains brief descriptions of the woody plants, and the second of the dicotyledons; the third is an annotated list only. Guatemala, Honduras, and Nicaragua are quite without real floras, although two of them are the subject of publications so miscalled; the two last are the least known of Central American countries, but a few local lists can be cited. Panama possesses, in addition to Seemann's now ancient Flora (1852–54), Standley's Flora of Barro Colorado Island (1933) and his Flora of the Panama Canal Zone (1928), the latter with keys and often brief descriptions. Mexico, by far the largest and richest in species of Central American republics, has had no complete flora since the general one of Hemsley (Biologia Centrali-Americana . . . Botany, 1879–88), which covered also the other Central American countries and is now very far from representing the known vascular plants. The gap is well filled, as far as woody plants are concerned, by Standley's Trees and Shrubs of Mexico (1920–26), but this includes probably less than half the total flora. The only complete flora of any Mexican State is Standley's Flora of Yucatan (1930), but there are also floras by Reiche of localities in the vicinity of Mexico City and others of several islands by other authors as well as Brandegee's papers on Baja California.

Greenland.—The by no means extensive flora of Greenland (scarcely 400 species) probably can boast of more pages of botanical literature per species than that of any other region in this list. The last complete list with localities and annotations is that of Lange, as revised by Rosenvinge in 1892, but there are a comparatively large number of carefully prepared lists of later date for different parts of the island.

---

[30] Bul. Bernice P. Bishop Mus. 133: 56. 1935.

United States.—Of the 49 geographical units (48 States plus the District of Columbia), 7 (Arizona, Delaware, Florida, Idaho, Oregon, South Carolina, and Virginia [31]) have never had a separate flora, and 4 others (Georgia in 1849; Kentucky, 1853; Louisiana, 1852; Massachusetts, 1835) have not had a published flora for around a century. Of the remaining 38, some 28 possess State lists that are more or less up to date: Alabama, 1901; Arkansas, 1891, with additions to 1926; California, 1925; Colorado, 1906; Connecticut, 1910, with additions to 1930; District of Columbia, 1919, with additions to 1930; Indiana, 1900, with additions to 1935; Iowa, 1933; Kansas, 1893; Michigan, 1904, with additions to 1908; Mississippi, 1921; Missouri, 1935; Montana, 1900, with additions to 1905; Nebraska, 1936; Nevada, 1925 (but so completely combined with that of Utah that it is a regional rather than a State flora); New Jersey, 1890; New Mexico, 1915; New York, 1924; North Dakota, 1918; Ohio, 1932, with additions to 1939; Oklahoma, 1930; Pennsylvania, 1903; South Dakota, 1932; Texas, 1938; Utah, 1925 (see remark under Nevada); Vermont, 1937; Washington, 1906; West Virginia, 1913, with additions to 1936. Of these 28 State floras, 18 were published or were brought approximately up to date by supplements within the last 20 years (that is, subsequent to 1919).[32] The published floras of the remaining 10 States cannot be regarded as at all adequately reflecting existing knowledge (Illinois, 1876; Maine, 1892, with additions to 1897; Maryland, 1910; Minnesota, 1884, with additions to 1887; New Hampshire, 1874; North Carolina, 1899; Rhode Island, 1920, an unreliable list; Tennessee, 1901; Wisconsin, 1883; Wyoming, 1896). A severe critique might reduce from 28 to not more than about 13 the number of States with moderately satisfactory floras.

One or more local floras covering larger or smaller areas are cited in this list for all the 48 States except Nevada. California, Indiana, Iowa, Massachusetts, Michigan, New York, Pennsylvania, and Washington are particularly well supplied with them. Insofar as the identification of vascular plants is concerned, all the States are covered by local or regional floras with the exception of Arizona, for which only Gray's unfinished Synoptical Flora, now more than 40 years old, is available; two floras of the State, however, are in an advanced state of preparation.

West Indies.—Good modern floras are available of the Bahama Islands (Britton and Millspaugh, 1920), Hispaniola (Urban, 1920–21), Jamaica (Fawcett and Rendle, 1910—; not yet finished), and Puerto Rico (Britton and Wilson, 1923–30; also Urban, 1903–11), all of which except those by Urban are descriptive floras. In the Lesser Antilles about half the islands are the subject of lists or descriptive floras; in order from north to south, Anegada, American

[31] Gronovius' Flora Virginica, although of great historical importance, can hardly be regarded as a real flora of the State, as it was based on the collections of a single botanist made in only a part of the whole area, gives almost no specific localities, and, moreover, uses the old polynomial nomenclature, although its second edition (1762) was published after the Species Plantarum.

[32] One more flora has been published, too late for inclusion in this List, bringing the total of more modern State floras to 19. GATES, F. C. ANNOTATED LIST OF THE PLANTS OF KANSAS: FERNS AND FLOWERING PLANTS (cover title, FLORA OF KANSAS). 266 p. maps. Topeka, 1940. (Publication of Agricultural Experiment Station, Kansas State College of Agriculture and Applied Science.)—In addition, a very full State flora of Indiana and an unannotated check list of the flora of West Virginia have appeared in 1940. DEAM, C. C. FLORA OF INDIANA. 1236 p. illus. (incl. maps). Indianapolis, 1940. (Publication of Department of Conservation, Division of Forestry, Indiana.)—CORE, E. L. A CATALOGUE OF THE VASCULAR PLANTS OF WEST VIRGINIA. Castanea 5: 31–73. 1940.

Virgin Islands (formerly Danish West Indies; St. Thomas, St. John, and St. Croix), Anguilla, Dutch West Indies (St. Martin, St. Eustatius, Saba), Guadeloupe, Martinique, and St. Vincent. There are also comparatively ancient lists of the plants of St. Bartholomew (1826) and Barbados (1848). The others, including such sizable islands as Barbuda, St. Christopher, Nevis, Antigua, Montserrat, Dominica, St. Lucia, and Grenada, are covered only by Grisebach's inclusive and now rather out-of-date Flora of the British West Indian Islands (1859–64) and his list of plants of the Lesser Antilles (1857). The greatest gap in the botanical literature of the West Indies, however, is the lack of any flora of Cuba since the relatively ancient lists of Grisebach (1866) and Sauvalle (1873), both based primarily on Charles Wright's collections.

South America.—Of the 14 units (13 countries and Patagonia, treated separately for convenience), only 4 possess recent and essentially complete lists of vascular plants—Patagonia (Macloskie, 1903–14, an annotated list with keys and brief descriptions), Surinam (list by Pulle, 1906, as well as his descriptive flora, still in process of publication), Uruguay (list by Herter, 1930, with subsequent additions), and Venezuela (list by Knuth, 1926–28). Reiche's descriptive Flora de Chile (1896–1911), in itself not too trustworthy, was left incomplete before reaching the end of the dicotyledons, and Gay's earlier flora (1845–54) is now completely out of date. A descriptive flora of Peru, by Macbride and collaborators, is now in course of publication. Martius' Flora Brasiliensis (1840–1906), the greatest of all published floras both in actual size and in number of species described, befitting the country whose flora is the richest in the world, is in great part too old to be of much more than historical value.

None of the seven remaining countries has even a comprehensive list of species, with the exception of that for British Guiana, now almost a century old. For Argentina, the chief sources are Grisebach's two lists of 1874 and 1879, based on the collections of Lorentz, Hieronymus, and Schickendantz, and the catalog by Hauman and others, which reaches only to the Droseraceae (Engler and Prantl system) and of which no new part has appeared since 1923; there are, however, numerous local lists and papers on medicinal and woody plants. Bolivia has only the lists of plants collected by Rusby, Bang, Herzog, Buchtien, and R. S. Williams. Our knowledge of the general flora of British Guiana rests almost entirely on the collections of Robert and Richard Schomburgk, made about a century ago and worked up principally by Bentham, then listed by Richard Schomburgk in 1848 in a form not easy to consult; there are several papers on Mount Roraima and a recent flora by Graham of the Kartabo region, with keys and brief descriptions.

Colombia and Ecuador, of all South American countries, are the most incompletely represented in floristic literature. Both are highly mountainous, with great numbers of endemic species, but the only available general floras, in both cases never completed, date from the 1860's, and there are no real local floras. French Guiana has had no general flora, although the woods and useful plants are well covered, and the chief sources for knowledge of the flora are Sagot's very fragmentary list and Benoist's list of his own collections. For Paraguay, the principal basis is Chodat's account of Hassler's collections, with Morong and Britton's list of the plants collected by Morong around Asuncion.

It has seemed worth while to attempt to select a list of 100 floras and floristic works of greatest general utility. In this list Africa, Insular Floras, and North America are represented by approximately the same number of titles, whereas Australasia and South America are assigned a smaller number; the former because it can be adequately covered thus, the latter because few works of broad scope are available. The list follows the alphabetical arrangement of titles in the body of the work. An asterisk indicates works in process of publication, a dagger those that were never completed.

Africa [33] (25 titles).—†Engler, Die Pflanzenwelt Afrikas; Battandier and Trabut, Flore de l'Algérie; *Carrisso, Conspectus Florae Angolensis; Durand and Durand, Sylloge Florae Congolanae; Muschler, Manual Flora of Egypt; †Pirotta, Flora della Colonia Eritrea; Chiovenda, Flora Somala; Dinklage, Verzeichnis der Flora von Liberia; Durand and Barratte, Florae Libycae Prodromus; Pampanini, Plantae Tripolitanae, and his Prodromo della Flora Cirenaica; Jahandiez and Maire, Catalogue des Plantes du Maroc; Bews, Introduction to the Flora of Natal and Zululand; Burkill, List of the Known Plants Occurring in British Central Africa, Nyasaland, and the British Territory North of the Zambesi; Maire, Études sur la Flore et la Végétation du Sahara central; Harvey and Sonder, Flora Capensis; Marloth, Flora of South Africa; Eyles, Record of Plants Collected in Southern Rhodesia; Dinter, Index der aus Deutsch-Südwestafrika bis zum Jahre 1917 bekannt gewordenen Pflanzenarten; Engler, Die Pflanzenwelt Ost-Afrikas und der Nachbargebiete; *Peter, Flora von Deutsch-Ostafrika; *Burtt Davy, Manual of the Flowering Plants and Ferns of the Transvaal with Swaziland; Bonnet and Barratte, Catalogue Raisonné des Plantes Vasculaires de la Tunisie; Hutchinson and Dalziel, Flora of West Tropical Africa; *Oliver, Flora of Tropical Africa.

Australasia (11 titles).—Bentham, Flora Australiensis; Maiden and Betche, Census of New South Wales Plants; Moore and Betche, Handbook of the Flora of New South Wales; Ewart and Davies, Flora of the Northern Territory; Bailey, Comprehensive Catalogue of Queensland Plants, and his Queensland Flora; Black, Flora of South Australia; Ewart, Flora of Victoria; Gardner, Enumeratio Plantarum Australiae Occidentalis; Cheeseman, Manual of the New Zealand Flora; Rodway, Tasmanian Flora.

Insular floras (25 titles).—Hemsley, Report on the Scientific Results of the Voyage of H. M. S. *Challenger* . . . Botany; Trelease, Botanical Observations on the Azores; Britton, Flora of Bermuda; Pitard and Proust, Les Iles Canaries, Flore de l'Archipel; Chevalier, Les Iles du Cap Vert; Schenck, Vergleichende Darstellung der Pflanzengeographie der subantarktischen Inseln; Skottsberg, Botanical Survey of the Falkland Islands; Menezes, Flora do Archipelago da Madeira; *Académie Malgache, Catalogue des Plantes de Madagascar; Palacký, Catalogus Plantarum Madagascariensium; Baker, Flora of Mauritius and the Seychelles; Jacob de Cordemoy, Flore de

[33] Chevalier, Flore vivante de l'Afrique occidentale française, should probably be included in this list, but the work has not been available for examination.

l'île de la Réunion; Balfour, Botany of Sokotra and Abd-el-Kuri; Merrill, Bibliographic Enumeration of Bornean Plants; Miquel, Flora Indiae Batavae (including the supplement, Prodromus Florae Sumatranae); Stewart, Botanical Survey of the Galapagos Islands; Hillebrand, Flora of the Hawaiian Islands; Koorders, Exkursions-flora von Java; Skottsberg, Phanerogams of the Juan Fernandez Islands; Kanehira, Enumeration of Micronesian Plants; Guillaumin, Catalogue des Plantes Phanérogames de la Nouvelle-Calédonie; Schumann and Lauterbach, Die Flora der Deutschen Schutzgebiete in der Südsee; Merrill, Enumeration of Philippine Flowering Plants; Drake, Illustrationes Florae Insularum Maris Pacifici; Brown and Brown, Flora of Southeastern Polynesia.

North America (30 titles).—*New York Botanical Garden, North American Flora; Macoun, Catalogue of Canadian Plants; Simmons, Survey of the Phytogeography of the Arctic American Archipelago; Marie-Victorin, Flore Laurentienne; Hemsley, Biologia Centrali-Americana . . . Botany; Standley and Record, Forests and Flora of British Honduras; Standley, Flora of Costa Rica, his Trees and Shrubs of Mexico, and his Flora of the Panama Canal Zone; Standley and Calderón, Lista Preliminar de las Plantas de El Salvador; Lange, Conspectus Florae Groenlandicae; †Gray, Synoptical Flora of North America; Britton and Brown, Illustrated Flora of the Northern United States, Canada and the British Possessions; Gray's New Manual of Botany; Small, Flora of the Southeastern United States, and his Manual of the Southeastern Flora; *Abrams, Illustrated Flora of the Pacific States; Howell, Flora of Northwest America; Rydberg, Flora of the Rocky Mountains and Adjacent Plains, and his Flora of the Prairies and Plains of Central North America; Jepson, Manual of the Flowering Plants of California; Tidestrom, Flora of Utah and Nevada; Wooton and Standley, Flora of New Mexico; Coulter, Botany of Western Texas; Piper, Flora of the State of Washington; Grisebach, Flora of the British West Indian Islands, and his Catalogus Plantarum Cubensium; Urban, Flora Domingensis; *Fawcett and Rendle, Flora of Jamaica; Britton and Wilson, Botany of Porto Rico and the Virgin Islands.

South America (9 titles).—Grisebach, Symbolae ad Floram Argentinam; Martius, Flora Brasiliensis; †Reiche, Flora de Chile; Chodat, Plantae Hasslerianae; Macloskie, Flora Patagonica; *Macbride, Flora of Peru; *Pulle, Flora of Surinam; Herter, Enumeratio Plantarum Vascularium Sponte Nascentium Republica Orientali Uruguayensi; Knuth, Initia Florae Venezuelensis.

ANNOTATED LIST OF FLORAS AND FLORISTIC WORKS

Africa

In addition to the division by countries and colonies, the headings Sahara, South Africa, and West Africa are used for titles not falling readily under the names of political units.

Durand, Théophile, and Schinz, Hans. Conspectus florae Africae, ou énumération des plantes d'Afrique. v. 1, pt. 2, v. 5. 25 cm. Bruxelles [etc.], 1895–98.
List of vascular plants, with references, synonyms, local and general range. Incomplete: v. 1, pt. 2, Ranunculaceae—Frankeniaceae (Bentham and Hooker system); v. 5, Monocotyledoneae and Gymnospermae. No more published.

Engler, Adolf. Die pflanzenwelt Afrikas insbesondere seiner tropischen gebiete. Grundzüge der pflanzenverbreitung in Afrika und die charakterpflanzen

Afrikas. v. 1–3, v. 5, h. 1. illus., plates, maps. 25.5 cm. Leipzig, 1908–25. (Die vegetation der erde . . . hrsg. von A. Engler und O. Drude, v. 9.)

Vol. 1: Phytogeographical; lists of principal collectors, with their localities and dates, and map of routes (p. xvi–xxviii); general features of flora, plant formations, etc., of the 68 phytogeographical regions recognized (including Cape Verde, Canary, and Madeira Islands), with numerous lists of species and references to principal publications; climate, soils, plant formations, etc., of Africa as a whole; vol. 2, running account of pteridophytes, gymnosperms, and monocotyledons, with keys to genera and mention of more important species; vol. 3, the same, Casuarinaceae—Umbelliferae; vol. 4, to cover Sympetalae and cellular plants (not yet published); vol. 5, pt. 1, ecology; vol. 5, pt. 2–5, to cover botanical explorations and bibliography (not yet published). In several cases the lists of species given in this work are the fullest available.

Rochebrune, A. T. de. Toxicologie africaine. Étude botanique, historique, ethnographique, chimique, physiologique, thérapeutique, pharmacologique, posologique, etc., sur les végétaux toxiques et suspects propres au continent africain et aux îles adjacentes. 2 v. 631 fig. 25.5 cm. Paris, 1897 (1896)–99.

Detailed account of poisonous vascular plants, systematically arranged, with synonymy, vernacular names, range, description, history, chemistry, physiology, therapeutics, pharmacology, etc. Incomplete; Ranunculaceae—Mimosaceae. Originally appeared in Bul. Soc. Hist. Nat. Autun v. 8–12. 1895–99.

Shantz, H. L., and Marbut, C. F. The vegetation and soils of Africa. x, 263 p. (incl. map). plates. 21 cm. N. Y., 1923. (American Geographical Society. Research series no. 13.)

Cited here for its bibliography (p. 85–112), consisting chiefly of works on travel, alphabetically arranged by authors. Very few strictly botanical works are listed.

Thonner, Franz. The flowering plants of Africa. An analytical key to the genera of African phanerogams. xvi, 647 p. 150 pl. on 75, map. 25.5 cm. London, 1915. (German edition, 1908, Die blütenpflanzen Afrikas.)

Bibliography, keys to families and genera, somewhat annotated; statistics, glossary, brief list of vernacular names.

ALGERIA

See also Egypt (Issa Bey; Schweinfurth, Arabische pflanzennamen); Sahara.

The following publication has not been available for examination:

Doumergue, François. Herborisations oranaises. Première partie: Oran. Bul. Soc. Géog. et d'Arch. Oran v. 33, fasc. 136. 76 p. 1913. (Cited from Bul. Soc. Bot. France **62:** 90. 1915.)

General

Battandier, J. A., and Trabut, Louis. Algérie. Plantes médicinales, essences et parfums. 39 p. 24 cm. Alger, 1889.

Annotated list of wild and cultivated species. Much the same material appears in Battandier, J. A. Algérie. Plantes médicinales. 61 p. 21 cm. Alger-Mustapha, 1900. (Exposition universelle de 1900.)

——— and **Trabut, Louis.** Flore analytique et synoptique de l'Algérie et de la Tunisie. 406 p. 22.5 cm. Alger, 1902 [1904].

Flora of 3,316 vascular plants, in form of keys, with very brief indication of range.

——— and **Trabut, Louis.** Flore de l'Algérie contenant la description de toutes les plantes signalées jusqu'à ce jour comme spontanées en Algérie et catalogue des plantes du Maroc. 2 v. 6 pl. 25.5 cm. Alger, 1888–95.—Supplément aux phanérogames. 90 p. 24 cm. Paris, Alger, 1910.

Descriptive flora of Algeria, with keys; additional species recorded by Ball from Morocco are listed, but not described. Replaces (as to Monocotyledones) their "Flore d'Alger" (1884). The supplement includes also a list of herbaria consulted and bibliography. The first volume is by Battandier alone.

Bouquet, J. Matière médicale indigène de l'Afrique du Nord. Trav. Off. Natl. Mat. Prem. Vég. (Paris) Notice 8. 29 p. 1921.

Includes annotated list of medicinal plants from region of Algeria, Tunis, Sahara, etc.

Cosson, Ernest. Compendium florae atlanticae, seu expositio methodica plantarum omnium in Algeria necnon in Regno Tunetano et Imperio Maroccano hucusque notarum, ou flore des États Barbaresques, Algérie, Tunisie et Maroc. 2 v. 2 maps. 24.5 cm. Paris, 1881–87.

(1) Alphabetical lists of collectors, well annotated; gazetteer, bibliography, phytogeographical regions, map showing itineraries of principal collecting expeditions, list of principal localities with names of collectors; (2) list of botanists supplementary to annotated list of collectors; descriptive flora of vascular plants (in Latin), with references and vernacular names. Incomplete (Ranunculaceae—Cruciferae). A second edition of the "Répertoire alphabétique des principales localités . . . ," with map, separately published in 1882 (142 p.).

Ducellier, Léon, and **Maire, René.** Végétaux adventices observés dans l'Afrique du Nord. Bul. Soc. Hist. Nat. Afrique Nord **14:** 304–325. 1923; **16:** 126–131. 1925.

Annotated list of vascular plants; sources of origin, etc.

Foureau, Fernand. Essai de catalogue des noms arabes et berbères de quelques plantes, arbustes et arbres algériens et sahariens ou introduits et cultivés en Algérie. 48 p. 28 cm. Paris, 1896.

Alphabetical list of Arab and Touareg or Berber vernacular names, with botanical equivalents.

France. Comité régional d'Algérie. Répertoire des plantes médicinales de l'Afrique du Nord. 62 p. 24 cm. Alger, 1924.

Directions for collecting commercial material of drug plants; annotated list of species, alphabetically arranged by vernacular names. Originally published in Revue Hort. Algérie v. 27. 1923.

France. Ministère de la guerre—Direction des affaires de l'Algérie. Catalogue explicatif et raisonné de l'Exposition permanente des produits de l'Algérie . . . suivi du catalogue méthodique des produits algériens à l'Exposition universelle de Paris en 1855. 207 p. 22.5 cm. Paris, 1855.

Includes classified and annotated lists of useful plants.

Lapie, Georges, and **Maige, Albert.** Flore forestière illustrée comprenant toutes les espèces ligneuses de l'Algérie et les espèces ligneuses les plus répandues en Tunisie, au Maroc et dans le midi de la France. . . . viii, 357 p. 881 fig., map. 22 cm. Paris, [1915?].

A popular handbook of forestry for Algeria; includes illustrated keys and alphabetically arranged list of 482 species; vernacular names.

Maire, René. Contributions à l'étude de la flore de l'Afrique du Nord 1–28. Bul. Soc. Hist. Nat. Afrique Nord **9:** 172—183. 1918; **12:** 42–52, 180(bis)–187. 1921; **13:** 37–44, 209–220. 1922; **14:** 118–159. 1923; **15:** 70–92, 95–106, 380–395. 1 fig. 1924; **17:** 104–126. 1926. —Mém. Soc. Sci. Nat. Maroc no. 15. 58 p. 1926.—Bul. Soc. Hist. Nat. Afrique Nord **19:** 29–66. 1928.—Bul. Soc. Sci. Nat. Maroc **8:** 128–143. 1928.—Bul. Soc. Hist. Nat. Afrique Nord **20:** 12–42. 1929.—Mém. Soc. Sci. Nat. Maroc no. 21. 19 p. 1929.—Bul. Soc. Hist. Nat. Afrique Nord **20:** 171–220. 1929; **22:** 30–72, 275–330. 3 fig. 1931; **23:** 163–222. 1932; **24:** 194–232. 1933.—Bul. Soc. Sci. Nat. Maroc **13:** 263–275. 1933.—Bul. Soc. Hist. Nat. Afrique Nord **25:** 286–326. 1934; **26:** 184–234. 1935; **27:** 203–238, 241–270. pl. 5–16. 1936; **28:** 332–388. pl. 29–40. 1937; **29:** 403–458. pl. 19. 1938; **30:** 255–314, 327–370. pl. 17, 19–20. 1939.

New species, new records, and critical notes on a large number of species from Morocco, Algeria, Libya, and the Sahara. The series includes 3,101 separate numbered notes (through 1939). No. 27 (1939) by Maire and M. Weiller.

——— Les progrès des connaissances botaniques en Algérie depuis 1830. 229 p. 8 port. Paris, 1931.

Chronologically arranged account of botanical explorations, with biographical notes on collectors and included bibliography.

Trabut, Louis. Flore du nord de l'Afrique. Répertoire des noms indigènes des plantes spontanées, cultivées et utilisées dans le nord de l'Afrique. 355 p. 24 cm. Alger, 1935. (1830–1930. Collection du centenaire de l'Algérie. Études scientifiques.)

Previous work on vernacular names; alphabetical list of botanical names, with vernacular equivalents; alphabetical list of vernacular names, with references to generic names; similar list of French vernacular names.

Local

Bonnet, Edmond, and Maury, Paul. D'Aïn-Sefra à Djenien-Bou-Resq. Voyage botanique dans le Sud-Oranais. Jour. de Bot. **2**: 277–301, 312–322. 1888.

Botanical explorations of authors, with lists of species of different localities; list of vascular plants, Characeae, and mushrooms collected, with localities.

Clary, L. R. Catalogue des plantes observées à Daya (Algérie). Bul. Soc. Hist. Nat. Toulouse **22**: 115–172. 1888.

Physiography, general features of flora, botanical explorations; list of vascular plants, with localities.

Debeaux, Odon. Catalogue des plantes observées dans le territoire de Boghar (Algérie). Actes Soc. Linn. Bordeaux **23** (III, 3): 159–208, 247–317. 1860. (Reprinted, 121 p. Bordeaux, 1861.)

Physiography, geology, etc.; annotated list of 730 vascular plants.

——— Flore de la Kabylie du Djurjura (Algérie). Revue Bot. (Toulouse) **11**: 1–416d. 1893.—Additions et corrections . . . l. c. **12**: 113–160. 1894. (Reprinted, 468 p. Paris, 1894.)

Physiography, botanical explorations, bibliography; list of 1,710 vascular plants with references and localities; phytogeography.

Doumergue, François. Les hauts plateaux oranais de l'ouest au point de vue botanique. Compt. Rend. Assoc. Franç. Avanc. Sci. **25** (2): 374–403. 1897.

Itineraries, topography; list of vascular plants and Characeae collected by author, with localities; critical notes on several species.

Ducellier, Léon. Étude phytogéographique des dunes de la baie d'Alger. Revue Gén. Bot. **23**: 273–308, 321–340. 8 fig. (incl. 2 maps), pl. 6–7. 1911.

Soil, climate, plant associations; unannotated list of vascular plants. The dunes occupy a stretch about 15 km. long.

Fitting, Hans, and Littmann, Enno. Arabische pflanzennamen aus der umgegend von Biskra (Algerien). Zeitschr. Deut. Morgenländ. Gesell. **65**: 336–348. 1911.

Annotated alphabetical list of vernacular names, with botanical equivalents.

Flahault, Charles. Rapport sur les herborisations de la Société. Bul. Soc. Bot. France **53** (Sess. Extr.): lxxxviii–clxxix. pl. 15–38, tab. 1907.

Running account of topography and vegetation, with numerous lists of species observed at different localities in Oran by the Société Botanique de France; brief bibliography.

Gay, Hippolyte. Florule de Blida. Revue Bot. (Toulouse) **7**: 87–155. 1888–89. (Reprint, 68 p.)

Cultivated plants, plant zones; annotated list of vascular plants; also list of mosses and hepatics (by L. Corbière).

Hanoteau, Adolphe, and Letourneux, Aristide. Flore de la Kabylie. *In* their La Kabylie . . . 2. éd. rev. et augm. v. 1, p. 107–208. Paris, 1893. (1st ed., 1868.)

Botanical explorations, general features of flora; list of vascular (1,619 species) and cellular plants, with localities, prepared with the aid of E. Cosson.

Hochreutiner, B. P. G. Le Sud-oranais. Études floristiques et phytogéographiques faites au cours d'une exploration dans le sud-ouest de l'Algérie en 1901. Annuaire Conserv. et Jard. Bot. Genève **7/8**: 22–276. 1 fig., 24 pl. 1904.

Itinerary, plant habitats with lists of species, origin of flora; list of vascular plants collected by author, with localities; lists of cellular plants, galls, and insects collected (by other authors).

Julien, Adr. Flore de la région de Constantine; comprenant la description des charactères botaniques des plantes de la contrée, de leurs propriétés et leurs usages chez les européens et chez les indigènes. 332 p. 21.5 cm. Constantine, 1894.

General features of flora, botanical stations with lists of species; annotated catalog of vascular and cellular plants (over 1,400 species), with notes on uses.

Lapie, Georges. Étude phytogéographique de la Kabylie du Djurjura. Revue Géog. **3**: 1–154. 12 fig. (incl. 2 maps), 3 maps. 1909.

Includes numerous lists of vascular and cellular plants in the account of different plant formations, but no complete list; bibliography.

Lefranc, Edmond. La Calle. Topographie, botanique et climatologie. 26 p. 24 cm. Paris, 1867. (Reprinted with some additions from Bul. Soc. Bot. France v. 9 and v. 12.)

Botanical explorations, physiography, plant zones; annotated list of vascular plants collected by the author in 1859–61.

——— Catalogue des plantes récoltées dans la subdivision de Sidi-bel-Abbès, pendant les années 1863 et 1864. Bul. Soc. Bot. France **13:** 26–31, 45–51, 65–71. 1866.

List of vascular plants and Characeae, with localities.—See also his Sidi-bel-Abbès. Topographie, climatologie et botanique. l. c. **12:** 383–395. 1865.

Maire, René. Contribution à la étude de la flore des montagnes de Numidie. Bul. Soc. Hist. Nat. Afrique Nord **18:** 71–76. 1927.

List of species collected by the author in 1924–25, mostly new to the region.

——— Contribution à l'étude de la flore du Djurdjura. Bul. Soc. Hist. Nat. Afrique Nord **5:** 235–238. 1913; **7:** 49–61. 1916.

Plant associations, phytogeography; records of species new or rare in the region.

——— La végétation des montagnes du Sud Oranais. Bul. Soc. Hist. Nat. Afrique Nord **7** [bis]: 210–292. pl. 4–17. 1916.

Itinerary; physiography; botanical explorations and plant formations of different localities; short list of Arabic names; list of new or noteworthy vascular and cellular plants; phytogeography, bibliography.

——— and **Wilczek, Ernest.** Florule des Iles Habibas. Bul. Soc. Hist. Nat. Afrique Nord **26** [bis]: 61–75. 2 fig. (maps), 2 pl. 1936.

Topography, geology, general features of flora, plant associations, origin of flora; list of cellular and vascular plants, list of cultivated plants. Based on collections of authors and Faure in 1934. (Off coast of Oran.)

Reboud, Victor. Catalogue des plantes les plus intéressantes observées dans le cercle de Bou-Saâda (Algérie), de septembre 1864 à juillet 1865. Bul. Soc. Bot. France **14:** 130–136. 1867.

List of vascular plants, with localities.—See also **Nicholas, G.** Liste des plantes récoltées à Bou-Saâda et observations sur quelques unes d'entre elles. Bul. Soc. Hist. Nat. Afrique Nord **6:** 139–148. pl. 1. 1914.

——— Catalogue des plantes phanérogames croissant spontanément dans le cercle de Souk-Ahras. Bul. Acad. Hippone **14:** 12–53. 1879. (Preprinted, 44 p. Bône, 1878.)

General features of flora, physiography, etc.; list of 944 vascular plants, briefly annotated as to locality.

Sériziat, Charles. Catalogue des plantes de Tébessa. *In* his Études sur Tébessa et ses environs. Bul. Acad. Hippone **22:** 186–215. 1887.

Annotated list of 320 vascular and cellular plants.—See also **Desorthés.** Supplément . . . l. c. **24:** 120–129. 1891.

Société Botanique de France. Liste des plantes observées aux environs de Biskra et dans l'Aurès . . . Session de Biskra (avril 1892). 26 p. map. 24.5 cm. Alger, 1892.

Unannotated lists of plants from several localities.

Trabut, Louis. D'Oran à Mécheria. Notes botaniques et catalogue des plantes remarquables. 36 p. 24 cm. Alger, 1887.

General features of flora; catalog of species.

ANGLO-EGYPTIAN SUDAN

See also Egypt (Issa Bey; Schweinfurth, Beitrag); Ethiopia (Chiovenda, Plantae novae); Northern Rhodesia (Fries); West Africa (Gürke; Oliver).

General

Broun, A. F., and **Massey, R. E.** Flora of the Sudan. 502 p. 19 cm. London, 1929.

Key to families (by J. Hutchinson); list of spermatophytes, with vernacular names, brief descriptions, localities, uses.

Local

Chipp, T. F. The Imatong Mountains, Sudan. Bul. Misc. Inform. Kew **1929:** 177–197. pl. 8–11, map. 1929.

Geography, topography, general features of flora, phytogeography; annotated list, by M. B. Moss, of cellular and vascular plants collected by Chipp. The Imatong Mountains lie just north of the boundary between Anglo-Egyptian Sudan and Uganda.

Crowfoot, G. M. Flowering plants of the northern and central Sudan . . . Forming a companion volume to the List of the Sudan flora. xxv p. 1 fig. (map), plates. 22.5 cm. Leominster, [1928].

General features of flora; note on certain species of Acacia known as "samr" and "seyal"; figures of about 175 plants, with very brief descriptions (color of flowers, etc.), and vernacular names. The "List of the Sudan flora" referred to is Broun and Massey's Flora.

Grant, J. A. The botany of the Speke and Grant expedition, an enumeration of the plants collected during the journey of the late Captain J. H. Speke and Captain (now Lieut.-Col.) J. A. Grant from Zanzibar to Egypt . . . Trans. Linn. Soc. Bot. v. 29. 190 p. 136 pl., map. 1872–75.

Itinerary; annotated list of native names, with uses; annotated list of 702 vascular plants. (Dicotyledones, Monocotyledones-Glumiferae, by D. Oliver; Monocotyledones-Petaloideae, Filices, etc., by J. G. Baker.) (From landing on edge of Kingam River opposite Zanzibar to Cairo, via west side of Victoria Nyanza and the Nile; does not include Zanzibar.)

Lester-Garland, L. V. Some plants from Jebel Marra, Darfur. Jour. Bot. **59:** 46–48. 1921.

List of vascular plants collected by Captain H. Lynes in 1920, with altitudes and general range.—See also **Norman, Cecil.** Plants from Jebel Marra, Darfur. Jour. Bot. **62:** 134–139. 1924.

Muriel, C. E. Appendix to Report on the Sudan forests. 2d ed. 11 p. Cairo, 1901.

List of plants mainly from the White Nile, the Blue Nile, and Kordofan, alphabetically arranged by vernacular names, with Arabic and Hameg vernacular names, habitat, remarks on uses.

Rein, G. K. Die im englischen Sudan, in Uganda und dem nördlichen Kongostaate wild und halbwild wachsenden nutzpflanzen. Tropenpflanzer **13:** 374–379, 532–539. 1909; **15:** 217–220, 387–393. 1911.

Annotated list of principal useful wild plants, grouped by uses.

Schweinfurth, Georg. Bericht über die botanischen ergebnisse der ersten Niam-Niam-reise, Januar-Juli 1870. Bot. Zeit. **29:** 301–317, 324–341, 351–366, 372–376. illus. 1871.

Statistics, general features of flora, running systematic account of principal plants (usually with generic names only), notes on principal cultivated plants.

——— Flora des Soturba an der nubischen küste. Verh. K. K. Zool.-Bot. Gesell. Wien **15:** 537–560. 1865.

General features of flora; annotated list of 265 vascular plants, with vernacular names, based on the author's collections in 1864–65; list of 71 vascular plants from Makaur Island; list of 62 vascular plants from Wady Gadireh, Egypt (on the Red Sea, lat. 25° N.).

——— Pflanzen-namen der Bega-sprache zwischen Suakin und Berber . . . Zeitschr. Gesell. Erdk. Berlin **4:** 334–346. 1869.

Systematic list of 146 plants, with vernacular names; short lists of species collected at various localities on the Blue Nile and the White Nile.

——— Vegetations-skizzen vom Bachr-el-Gasal. Bot. Zeit. **28:** 81–88. 1870.

Brief sketch of flora, with mention of economic plants; no list of species.

Shaw, W. B. K. The flora of the Libyan Desert. Bul. Misc. Inform. Kew **1934:** 281–289. 1934.

General features of flora; list of spermatophytes collected by author on two trips extending from Cairo as far as Tekro and northern Dar Fur. For account of his first trip see, under same title, l. c. **1931:** 161–166.

ANGOLA (PORTUGUESE WEST AFRICA)

See also Ethiopia (Engler, Ueber die hochgebirgsflora); West Africa (Gürke, Oliver).

General

Carrisso, L. W., ed. Conspectus florae angolensis. v. 1, fasc. 1. (Ranunculaceae–Malvaceae por A. W. Exell e F. A. Mendonça.) (ix-xxiii, 176 p. map.) 24.5 cm. Lisboa, 1937.
Topography, plant zones, etc. (in Portuguese and English); list of vascular plants, with keys, references, synonymy, citation of specimens (with indication of herbaria), habit, habitat, flowering dates, general range.

Ficalho, F. M. C. de M., *conde de.* Plantas uteis da Africa Portugueza. 279 p. 23 cm. Lisboa, 1884.—Indice para a obra Plantas uteis . . . por A. S. Barjona de Freitas. xvii p. Lisboa, 1908.
History of cultivated plants; annotated systematic list of phanerogams, with vernacular names. Includes only Dicotyledones and Gymnospermae. Refers chiefly to Angola, but includes also the other Portuguese possessions in Africa. A corrected and somewhat enlarged revision of his Nomes vulgares de algumas plantas africanas, principalmente angolenses, published in Bol. Soc. Geog. Lisboa **2**: 601–619, 707–716. 1881; **3**: 479–492, 612–623. 1882.

Gossweiler, John. Bibliografia das viagens feitas em Angola por exploradores de diversas nacionalidades, cujos relatórios se referem à flora, fauna e agricultura. Bol. Serv. Agr. e Com., Colon. e Flor. Angola **7** (28/31, Ja./D. 1935): 121–125. 1937.
Unannotated list of 55 titles, alphabetically arranged by authors.

——— Elementos para a história de exploração botânica de Angola. Itinerário e relacão dos viajantes e exploradores que fizeram colecções botânicas em Angola. Bol. Soc. Broteriana II, **13**: 283–305. map. [1939?]
Brief account of botanical explorations; alphabetical list of collectors, with biographical data and account of work in Angola.

——— Relação dos viajantes que coligiram plantas em Angola actualmente depositadas nos institutos botânicos de Londres, Berlim, Paris, Montpellier, Zurich, Lisboa, Coimbra, Cabo de Boa Esperanca e Washington onde estão sendo estudadas desde 1868. Bol. Serv. Agr. e Com., Colon. e Flor. Angola **7** (28/31, Ja./D. 1935): 127–130. 1937.
Alphabetical list of collectors, with dates and localities visited.

Welwitsch, Frederico. Synopse explicativa das amostras de madeiras e drogas medicinaes e de outros objectos mormente ethnographicos colligidos na provincia de Angola enviados á Exposição Internacional de Londres em 1862, incluindo os que foram offerecidos ao gabinete pharmacologico da Escola Medico-Cirurgica de Lisboa. 56 p. 21 cm. Lisboa, 1862.
Annotated lists of 149 timbers and medicinal and other useful plants, with vernacular names and uses.

Local

Exell, A. W., Good, R. D'O., and others. Mr. John Gossweiler's plants from Angola and Portuguese Congo. Jour. Bot. **64** (Suppl.): 1–24, 1–40. 1 fig. 1926; **65** (Suppl.): 25–80, 41–72. 1 fig. 1927; **66** (Suppl.): 81–176. 1 fig. 1928; **67** (Suppl.): 177–203, 73–120. 1 fig. 1929; **68** (Suppl.): 121–144. 1930; **69** (Suppl.): 145–152, 153–160, 161–168. 2 fig. 1931; **70** (Suppl.): 205–212, 213–220, 221–228. fig. 1–2. 1932; **71** (Suppl.): 229–236, 237–243, 169–175. fig. 3. 1933.—Dicotyledones. Addendum. l. c. **73** (Suppl.): 1–12. 1935; **74** (Suppl.): 13–24. 1936.
List, with localities and collector's numbers, general range, and frequent field notes. Polypetalae and Gamopetalae separately paged.—See also Pteridophyta, by A. H. G. Alston, l. c. **72** (Suppl.). 11 p. 1934. Index to Polypetalae **71** (Suppl.): 237–243, Gamopetalae, p. 171–175.

Hiern, W. P., Rendle, A. B., and others. Catalogue of the African plants collected by Dr. Friedrich Welwitsch in 1853–61. 2 v. in 6. port. 22 cm. London, 1896–1901.
Biography and bibliography of Welwitsch; annotated list of vascular plants, with often ample field notes or descriptions. Vol. [1], 4 pts., Dicotyledons by Hiern; v. 2, pt. 1, Monocotyledons and Gymnosperms, by A. B. Rendle; v. 2, pt. 2, Cryptogamia, by various authors. Refers entirely to Angola.

Sousa, A. de Figuèiredo Gomes e. Contribuição para o estudo da flora do distrito de Mossâmedes. Bol. Agência Geral Colón. (Lisboa) **6**: 41–64. 15 fig. 1930.

General features of flora; annotated tabular list of phanerogams collected, the families arranged alphabetically. The same text, with fewer illustrations and two families added at the end of the list, is given in Bol. Serv. Agr. e Com. Angola **3**: 76–87. 1933.

Warburg, Otto. Kunene-Sambesi-expedition, H. Baum, 1903. xi, 593 p. 106 fig., 13 pl. (1 col.), map. 25 cm. Berlin, 1903.

Includes list by numerous authors of cellular and vascular plants (p. 155–427) collected in southern Angola, with localities, general range, and frequent annotations; tabular list by G. Hegi, showing general range of plants collected; phytogeography by Warburg; annotated list of useful plants, classified by uses, by Warburg.

BASUTOLAND

See also South Africa.

Local

Phillips, E. P. A contribution to the flora of the Leribe Plateau and environs: with a discussion on the relationships of the floras of Basutoland, the Kalahari, and the southeastern regions. Annals So. African Mus. **16**: 1–379. 3 fig. (maps), 7 pl. (incl. map). 1917.

Geology, general features of flora, alien flora, statistics; annotated list of vascular plants, with extralimital range, native names and uses.

BECHUANALAND

See South Africa.

BECHUANALAND PROTECTORATE

See also South Africa; West Africa (Oliver).

Local

Bremekamp, C. E. B., and **Obermeÿer, A. A.** Sertum kalahariense, a list of the plants collected. *In* Scientific results of the Vernay-Lang Kalahari expedition, March to September, 1930. Annals Transvaal Mus. **16**: 399–442. 1935.

List of 326 vascular plants, with data and general range, collected on expedition from Gaberones (Bechuanaland Protectorate) to Ghanzi, across Ngamiland to the junction of the Zambesi and Chobe (Kwando) Rivers, then through Southern Rhodesia, ending near Makarikari Lake. Of the species collected in the central Kalahari, 204 are additions to N. E. Brown's list of 1909.

Lugard, E. J., and **Brown, N. E.** The flora of Ngamiland. Bul. Misc. Inform. Kew **1909**: 81–146. map. 1909.

Physiography, climate, general features of flora (by Lugard); annotated list of 374 vascular plants, of which 92 are described as new species.

Seiner, Franz. Pflanzengeographische beobachtungen in der Mittel-Kalahari. *In* **Engler, A.** Beiträge zur flora von Afrika XXXVIII. Bot. Jahrb. Engler **46**: 1–50. 4 pl. 1911.

Account of plant formations, with included running list of species observed by author, between Serue (Serowe) and Rietfontein-Nord via Tsau.

BELGIAN CONGO

See also Anglo-Egyptian Sudan (Rein); Cameroons (Mildbraed, Botanik); Ethiopia (Engler, Ueber die hochgebirgsflora); Tanganyika (Mildbraed); West Africa (Gürke; Oliver).

General

Dewèvre, Alfred. Les plantes utiles du Congo. 2. éd., rev. et corr. 65 p. 22.5 cm. Bruxelles, Paris, 1894.

General considerations; annotated list of cultivated and wild plants, classified by uses.

Duchesne, Fl. Les essences forestières du Congo Belge III. Leurs dénominations indigènes. 265 p. 27.5 cm. Ixelles-Bruxelles, 1938.
Alphabetical list of vernacular names with botanical equivalents, with region where used, dialect, and source of names; systematic list of botanical names, with vernacular equivalents.

Durand, Théophile. Les explorations botaniques au Congo Belge et leurs résultats. Bul. Cl. Sci. Acad. Roy. Belg. **1909**: 1347–1374. 1909.
Chronological account of botanical explorations, with bibliography.

——— and **Durand, Hélène.** Sylloge florae congolanae (Phanerogamae). 716 p. 26.5 cm. Bruxelles, 1909. (Also issued as Bul. Jard. Bot. État Bruxelles v. 2. 1910.)
List of botanical collectors, statistics; list of 3,546 phanerogams, with references, synonyms, localities, and collectors; statistics for botanical districts. Summarizes knowledge of the flora through 1908.—See also **Wildeman, É. de.** Additions à la flore du Congo. Bul. Jard. Bot. État Bruxelles **4**: 1–241. 1914; **5**: 117–268, 269–364, 365–412. 1916–19; **6**: 1–64, 65–129. 35 pl. 1919; **7**: 1–88, 271–316. 1920–21. (Lists of additions to Sylloge, with localities and collectors' numbers. The number of species is said to have reached 6,300 in 1915.)

Staner, Pierre. Plantes congolaises à fruits comestibles. Pub. Inst. Natl. Étude Agron. Congo Belge Sér. Sci. no. 4. 56 p. 9 fig. [1935.]
Annotated systematic list of native plants with edible fruits, with vernacular names and description of fruit; key to the genera, based on the fruit; index of vernacular names.

——— and **Boutique, R.** Matériaux pour l'étude des plantes médicinales indigènes du Congo Belge. Mém. Sect. Sci. Nat. et Méd. Inst. Roy. Colon. Belge Collect. in-8°. v. 5, fasc. 6. 228 p. 17 fig. (partly on plates). 1937.
Annotated systematic list of plants, with range and uses; bibliography.

Wildeman, Émile de. A propos de médicaments indigènes congolais. Mém. Sect. Sci. Nat. et Méd. Inst. Roy. Colon. Belge Collect. in-8°. v. 3, fasc. 3. 127 p. 1935.
General considerations, plants used for particular diseases; list of vernacular and botanical names.

——— Documents pour l'étude de l'alimentation végétale de l'indigène du Congo Belge. Mém. Sect. Sci. Nat. et Méd. Inst. Roy. Colon. Belge Collect. in-8°. v. 2, fasc. 4. 263 p. 1934.
General considerations; alphabetical list of vernacular and botanical names, with brief annotations under the latter; bibliography.

——— Les forêts congolaises et leurs principales essences économiques. 214 p. map. 25 cm. Bruxelles, 1926. (Bibliothèque-Congo no. 19.)
Includes (p. 117–207) briefly annotated list of timber trees, alphabetically arranged by botanical names.

——— Notices sur des plantes utiles ou intéressantes de la flore du Congo. 2 v. illus., plates. 23 cm. Bruxelles, 1903–08. (Publication de l'État indépendant du Congo.)
Discusses various species at considerable length, and includes annotated alphabetical list of vernacular names, with botanical equivalents, where known, of the woods of the Congo and neighboring regions, with corresponding list of botanical names (v. 1, p. 353–385).

——— Remarques à propos de la forêt équatoriale congolaise. Mém. Sect. Sci. Nat. et Méd. Inst. Roy. Colon. Belge Collect. in-8°. v. 2, fasc. 2. 120 p. 3 maps. 1934.
Cited here for its bibliography (p. 112–120) of publications relating to the forests of the Belgian Congo.

Local

Burtt, B. D. A botanical reconnaissance in the Virunga volcanoes of Kigezi Ruanda, Kivu. Bul. Misc. Inform. Kew **1934**: 145–165. 1 fig. (map), pl. 7. 1934.
Running account of author's trip in 1930 with mention of plants collected. The Virunga or Mufumbiro volcanoes are a particularly isolated range of peaks that lie about 120 miles south of the Ruwenzori ice cap, in the great divide between the east tropical African savannas and the forests of the Congo basin.

They are partly in Belgian Congo (Parc National Albert) and partly on the boundary between Uganda and the mandated territory of Ruanda and Urundi and the boundary between the latter and Belgian Congo.

Cotton, A. D. The arborescent Senecios of the Virunga Mountains. Bul. Misc. Inform. Kew **1931:** 289–297. 1 fig. (map). 1931.

Cited here for its annotated list of works on the flora of the Virunga Mountains.

Delevoy, Gaston. Les essences forestières du Katanga. His La question forestière au Katanga, v. 2. xv, 525 p. illus. 23.5 cm. Bruxelles, 1929.

Alphabetical list of woody plants, with references, vernacular names, general and local range, description, uses.

Engler, Adolf. Verzeichniss der auf der Graf v. Götzen'schen expedition bei der besteigung des Kirunga gesammelten pflanzen. *In* **Götzen, G. A.** von. Durch Afrika von ost nach west . . . p. 374–384. Berlin, 1895.

List of 79 species, collected at 2,000 meters elevation and above on Kirunga. Mount Kirunga is in the Mufumbiro group.

Fishlock, C. W. L., and Hancock, G. L. R. Notes on the flora and fauna of Ruwenzori, with special reference to the Bujuku Valley. Jour. East Africa and Uganda Nat. Hist. Soc. no. 44. p. 205–229. 12 pl. 1932.

Explorations, plant zones with lists of plants collected by authors in 1931; notes on fauna, bibliography.

Ghesquière, Jean. Note sur la forêt de Lonkala (Sankuru). Bul. Agr. Congo Belge **16:** 549–555. fig. 218–219. 1925.

Includes systematic list of timbers, with vernacular names and brief description of wood.

Gillet, Just., and Paque, Égide. Plantes principales de la région de Kisantu. Leur nom indigène, leur nom scientifique, leurs usages. (Notes botaniques sur la région du Bas- et Moyen-Congo, fasc. 1.) Annales Mus. Congo Bot. V, [v. 4], fasc. 1. 120 p. 22 fig. 1910.

Alphabetical annotated list of vernacular names, with Portuguese names, scientific names and uses; cross index of scientific names.

Hauman, Lucien. Esquisse de la végétation des hautes altitudes sur le Ruwenzori (Résultats botaniques de la Mission belge pour l'exploration scientifique du Ruwenzori, 1932). Bul. Cl. Sci. Acad. Roy. Belg. V, **19:** 602–616, 702–717, 900–917. illus., map. 1933.

Plant associations; list of 75 vascular plants known from above 3,800 meters altitude; climate, soil, fauna, ecology, phytogeography.

——— La flore. *In* **Grunne, X.** de, and others. Vers les glaciers de l'équateur. Le Ruwenzori. Mission scientifique belge 1932. p. 251–259. fig. 123. Bruxelles, 1937.

Physiography, general features of flora; annotated list of 72 vascular plants now known from 3,800–5,000 meters elevation; bibliography.

Lebrun, Jean. Les essences forestières des régions montagneuses du Congo oriental. Pub. Inst. Natl. Étude Agron. Congo Belge Sér. Scient. no. 1. 263 p. (incl. 28 fig., 18 pl. (pl. 1, map)). 1935. (Les essences forestières du Congo Belge II.)

Plant zones, bibliography; descriptive account of trees, with keys, references, vernacular names, citation of exsiccatae, uses.

——— Rapport sur un voyage d'études botaniques dans le district du Kibali-Ituri. Bul. Agr. Congo Belge **25:** 386–437. fig. 62–75 (incl. map). 1934.

Includes several short lists of vascular plants collected by author, the longest being one from Mount Ruwenzori.

Rendle, A. B., Baker, E. G., and Moore, S. LeM. An account of the plants collected on Mount Ruwenzori by Dr. A. G. F. Wollaston. Jour. Linn. Soc. Bot. **38:** 228–279. pl. 16–19. 1908.

Botanical explorations, general features of flora; annotated list of phanerogams.

Robyns, Walter. La colonisation végétale des laves récentes du volcan Rumoka (laves de Kateruzi). Mém. Sect. Sci. Nat. et Méd. Inst. Roy. Colon. Belge Collect. in-8°. v. 1, no. 1. 32 p. 10 pl., fold. map. 1932. (Contribution à l'étude de la végétation du Parc National Albert.)

Includes list of 27 plants (26 vascular) found on Volcan Rumoka, north of Lake Kivu, after the volcanic eruption of 1912 which completely destroyed the vegetation.

Staner, Pierre. Contributions à l'étude de la flore du Parc National Albert et du Kivu. I–II. Plantes récoltées par M. le Dr. Scaetta. Revue Zool. et Bot. Afric. **23**: 208–230. 1933; **24**: 213–224. 1933.

Lists of vascular plants, with data. The Parc National Albert is directly northeast of Lake Kivu, partly in Belgian Congo, partly in Ruanda and Urundi.

Vermoesen, Camille. Manuel des essences forestières du Congo Belge (région équatoriale et Mayombe). xii, 282 p. illus., col. plates. 18.5 cm. Bruxelles, 1923.

Pocket manual of principal useful trees, with vernacular names, range, description, and uses. This is pt. 1 of Les essences forestières du Congo Belge.

Wildeman, Émile de. Campagnie du Kasai. Mission permanente d'études scientifiques. Résultats de ses recherches botaniques et agronomiques. viii, 463 p. 45 pl. (i. e. 46), 2 maps. 28 cm. Bruxelles, 1910.

Includes (p. 229–440) list of vascular plants collected in the Bassin du Kasai, with occasional annotations; physiography, etc.

——— Contribution à l'étude de la flore du Katanga. cxliv, 264 p. (incl. 19 pl.). 25 cm. Bruxelles, 1921.—Supplément I–V. 26 cm. Bruxelles, 1927–32.

Botanical distribution, with florulas of Albertville-Toa, Nyassa, Tanganyika, Victoria Nyanza, Lukafu, Elisabethville, Kundelungu, etc.; notes on useful trees, statistics; list of known vascular plants, with localities, collectors' names, and vernacular names (indexed). Including the first three supplements, the known flora amounted to 2,230 species.

——— Documents pour l'étude de la géo-botanique congolaise. Bul. Soc. Roy. Bot. Belg. v. 51, fasc. 3. 406 p. 117 pl. (incl. maps). 1913.

General features of flora, phytogeography; lists of species, with remarks on vegetation, from coast region (Moanda, Banana-Nemloa, Boma), Mayombe (Bingila), Bas-Congo (Matadi-Vivi, Kisantu, Kimuenza, Léopoldville), the tropical forest region (region of the Equator, Nouvelle-Anvers, Beni, Irumu, Mawambi, Avakubi, Yambuya, Buta-Bima, Imese, Likimi, Lake Léopold II, Bena-Dibele, Kondue), Haut-Ubangi and Uele (Fort-Sibu-Rapides de Bangui), Kasai (Popokabaka, Katola, Madibi, Luano, Kikwite, Atènes-Illongonga, Dilolo-Kimpuki), Moyen-Katanga or Haut-Congo (Nyangwe-Kasongo), Lake Albert Edward and Ruwenzori (Ninagongo, Rutschuru, west slope of Ruwenzori (Butago)), region of Lakes Kivu and Tanganyika (Albertville-Toa), Haut-Katanga (Lukafu, Elisabethville, Kundelungu, Manika), Nile region.

——— Études de systématique et de géographie botaniques sur la flore du Bas- et du Moyen-Congo. Annales Mus. Congo Bot. V, v. 1–3. 230 pl. 1903–12.

Vol. 1, list of cellular and vascular plants, with references, localities, collectors' numbers, and occasional annotations; based on specimens from various collectors examined by the author since 1903; vol. 2, same since 1906; vol. 3, a later series of lists.

——— Études sur la flore des districts du Bangala et de l'Ubangi (Congo Belge). xvii, 464 p. col. front., 52 fig., 20 pl., map. 27.5 cm. Bruxelles, 1911. (Plantae Thonnerianae congolenses ser. 2.)

Includes (p. 267–403) list of vascular plants known from the region.

——— Mission Émile Laurent (1903–1904). 2 v. illus., 138 fig., 185 pl., port., map. 29 cm. Bruxelles, 1905–07.

Includes partly annotated list of vascular plants collected, with references and localities; list of vernacular names of laticiferous plants, with localities and botanical names (p. 527–534). (Lower Congo and Kasai River region.)

——— Mission forestière et agricole du Comte Jacques de Briey . . . au Mayumbe (Congo Belge). xiv, 468 p. 62 fig., 15 pl., port., 2 fold. tab. 23 cm. Bruxelles, 1920.

Includes (p. 235–266) list of plants collected; also much information on woods and other useful plants.

——— Plantae bequaertianae. Études sur les récoltes botaniques du Dr. J. Bequaert, chargé de missions au Congo Belge (1913–1915). v. 1–6, fasc. 1. 22 cm. Gand, Paris, 1921–32.

Lists of vascular plants, with data; families not in strictly systematic order. The collections were made chiefly in the regions of Mayombe, Semliki, Ruwenzori, and the valley of the Rutshuru near Lake Albert.

Wildeman, Émile de. Sur des plantes médicinales ou utiles du Mayumbe (Congo Belge) d'après des notes du R. P. Wellens (1891–1924). Mém. Sect. Sci. Nat. et Méd. Inst. Roy. Colon. Belge Collect. in-8°. v. 6, fasc. 4. 97 p. 1938.
Alphabetical annotated list of botanical and vernacular names.

BRITISH SOMALILAND PROTECTORATE

See also Ethiopia (Engler, Ueber die hochgebirgsflora); Italian Somaliland (Chiovenda, Flora Somala); West Africa (Gürke; Oliver).

General

Drake-Brockman, R. E. British Somaliland. xvi, 334 p. illus., plates, map. 23 cm. London, 1912.
Includes chapter on "the flora and fauna" (p. 221–238), with general account of vegetation; on "products of commercial value" (p. 239–270), with account of principal plant and animal products; Appendix III, "List of trees and plants the fruit or leaves of which are eaten by the Somalis" (p. 281–283), an annotated list of vernacular names, with botanical names when known; Appendix IV, "A collection of the flora of British Somaliland communicated to Kew during the years 1905–11, together with their vernacular names" (p. 285–294), an unannotated list of 318 plants, not in systematic order; Appendix VI, "Some notes on the myrrh, frankincense, bdelliums, and gum resins of British Somaliland" (p. 301–323).

——— The economic resources of British Somaliland. Trans. Internatl. Cong. Trop. Agr. 3d, London, 1914. **2**: 488–500. 1917.
Includes section on wild and cultivated vegetable products of economic importance.

Local

Franchet, A. R. Sertulum somalense. *In* **Revoil, Georges.** Fauna et flore des pays Çomalis. [pt. 9.] 70 p. 6 pl. Paris, 1882.
Bibliography; list of 144 vascular plants, with localities.

Oliver, Daniel. Flora of Somali-Land. *In* **James, F. L.** The unknown horn of Africa. p. 317–323. 4 pl. London, 1888.
List of nearly 150 vascular plants collected by James and Thrupp, with localities. The specimens were collected in British Somaliland and Ethiopia.

CAMEROONS (INCLUDING CAMEROUN)

See also Ethiopia (Engler, Ueber die hochgebirgsflora); French West Africa (Chevalier); West Africa.

General

Bertin, André. Les bois du Cameroun. 312 p. plates, maps. 25 cm. Paris, 1920. (Mission d'études forestières envoyée dans les colonies françaises par les Ministères de la guerre, de l'armement et des colonies, t. 4.)
Includes (p. 31–96) lists of scientific and vernacular names. Refers mostly to the French mandate.

Local

Braun, Johannes. Botanischer bericht über die flora von Kamerun. Mitt. Forschungsreis. u. Gelehr. Deut. Schutzgeb. **2**: 141–176. 1889.
Annotated list of fungi and vascular plants, collected by the author in 1887–1889, in Kamerun, Gross Batanga, and Malimba.

Gilg, Ernst. Die bis jetzt bekannten hohen bäume Kameruns, welche wertvolles holz geben oder als nutzhölzer in frage kommen könnten. Notizbl. K. Bot. Gart. Berlin **5**: 123–131. 1909.
List with heights, vernacular names, and occasional annotations.

Hédin, Louis. Étude sur la forêt et les bois du Cameroun sous mandat français. 230 p. plates, map. 25.5 cm. Paris, 1930.
Includes (p. 41–200) annotated list of trees of Cameroun, with vernacular names, notes on uses, and description of wood characters; bibliography.

Hooker, J. D. On the plants of the temperate regions of the Cameroons Mountains and islands in the Bight of Benin; collected by Mr. Gustav Mann . . . Jour. Linn. Soc. Bot. **7**: 171–240. 1864.

Itinerary, phytogeography; list of vascular and cellular plants, with localities and general range, from Cameroon Mountain and the islands of Fernando Po, São Thomé, and Principe.

Mildbraed, Johannes. Botanik. *In* Wissenschaftliche ergebnisse der zweiten deutschen Zentral-Afrika-expedition 1910–1911 unter führung Adolf Friedrichs, Herzogs zu Mecklenburg. v. 2. 202 p. 90 pl. Leipzig, 1922.

Includes general account of vegetation and lists of species collected from Belgian Congo (Kimuenza, Stanley Pool to Molundu), Kamerun (Molundu, Jukaduma, Kadei, Lomie, Ebolowa, and Sangmelima, Kribi), and the islands of Annobon (p. 146–164) and Fernando Po (p. 164–195), the last two lists including also species recorded in the Flora of tropical Africa.

——— Probeflächen-aufnahmen aus dem Kameruner regenwald. Notizbl Bot. Gart. Berlin **10**: 951–976. 1930.

Includes partly annotated running list of vascular plants collected by author at Likomba, with some vernacular names.

——— Von den Bulus genutzte wildwachsende pflanzen des Südkameruner waldlandes. Notizbl. K. Bot. Gard. Berlin App. 27. 43 p. 1913.

Annotated list, classified by uses, with descriptions, vernacular names and uses; systematic list of plants mentioned, with vernacular names.

——— Zur kenntnis der vegetationsverhältnisse Nord-Kameruns. Bot. Jahrb. Engler **65**: 1–52. 1932.

General features of flora between Bongor and Garua; annotated list of vascular plants collected or observed by author and others; phytogeography.

Santesson, C. G. Einige drogen aus dem Kamerun-gebiete und ihre einheimische verwendung. Arkiv Bot. v. 20A, no. 8. 34 p. 1926.

Annotated list of 42 species, with vernacular names and uses; list classified by uses; list of scientific and vernacular names.

CAPE OF GOOD HOPE

See also South Africa.

General

Sim, T. R. The forests and forest flora of the Colony of the Cape of Good Hope. vii, 361 p. 160 pl., port., map. 32.5 cm. Aberdeen, Scotland, 1907.

Forests, economic aspects, etc.; keys to genera; annotated descriptive flora of woody plants.

Local

Adamson, R. S. The vegetation and flora of Robben Island. Trans. Roy. Soc. So. Africa **22**: 279–296. pl. 13–14. 1934.

History, climate, geology, soils, vegetation; tabular list of spermatophytes, lists of cultivated plants. Robben Island lies off the entrance to Table Bay, 4 miles west of Blaauwberg.

Bolus, Frank, and Bolus, Louisa. Key to the flora of the Cape Peninsula. Annals Bolus Herb. **1**: 22–35, 116–125, 173–184. 1914–15.

Keys to families, genera, and species, intended to supplement Bolus and Wolley-Dod's "List of the flowering plants and ferns . . ." Key to genera and species incomplete (Ranunculaceae—Anacardiaceae).

Bolus, Harry, and Wolley-Dod, A. H. A list of the flowering plants and ferns of the Cape Peninsula, with notes on some of the critical species. Trans. So. African Phil. Soc. **14**: 207–373. 1903.

Physiography, general features of flora, plant associations, vertical range, statistics, phytogeography, effect of fires, etc.; annotated list of 2,351 vascular plants and Characeae. (Small peninsula on which Cape Town is situated.)

Compton, R. H. The flora of the Whitehill district. Trans. Roy. Soc. So. Africa **19**: 269–329. 1931.

Topography, climate, etc.; annotated list of over 700 vascular plants. The area covered is from Bantams on the west to Baviaans on the east and from the summit of Ngaap Kop on the north to the crest of the Witteberg Range on the south; the total area being about 40 square miles.

Duthie, A. V. List of vascular cryptogams and flowering plants of the Stellenbosch Flats. Annals Univ. Stellenbosch v. 8, sect. A, no. 4. 52 p. 1930.

Annotated list of vascular plants.

Dyer, R. A. The vegetation of the divisions of Albany and Bathurst. Mem. Bot. Survey So. Africa no. 17. 138 p. illus. (maps), plates. 1937.

Mainly ecological; includes sketch of botanical explorations, topography, geology, economic plants, bibliography.

Galpin, E. E. A contribution to the knowledge of the flora of the Drakensberg. Rep. So. African Assoc. Adv. Sci. 6 (1908) : 209–229. 1909.

Author's itinerary in 1904; list of vascular plants collected, with data. This was the first botanical collection from the Cape Colony part of the Drakensberg.

Levyns, M. R. A guide to the flora of the Cape Peninsula. xv, 284 p. 199 fig., 3 col. pl. 22.5 cm. Cape Town, Johannesburg [1929].

Glossary, keys to families and genera; briefly descriptive catalog of genera, with indication of number of species and mention of occasional species. Covers essentially same area as Bolus and Wolley-Dod's "List of the flowering plants and ferns of the Cape Peninsula."

Miller, O. B. A list of some native names of trees, shrubs, etc., in use in the Transkeian Territories. Bul. Forest Dept. Union So. Africa no. 8. 23 p. 1923.

List of vernacular names, with botanical equivalents; alphabetical list of botanical names, with Xosa vernacular names and notes on uses, etc. In both English and Dutch.

Muir, John. The vegetation of the Riversdale area, Cape Province. Mem. Bot. Survey So. Africa no. 13. 82 p. map. 1929.

Mainly ecological, but includes account of botanical explorations, geology, etc.

Pearson, H. H. W., and others. List of the plants collected in the Percy Sladen memorial expeditions, 1908–9, 1910–11. Annals So. African Mus. **9:** 30–59, 61–90, 129–192, 193–272, 355–448. 16 (i. e. 17) fig., 8 pl. 1912–17.

Enumeration of vascular and cellular plants, by various authors; not in systematic order, but indexed (p. 431–448). The itineraries of the various expeditions are summarized on pages 359 to 361. The collections are chiefly from western Cape of Good Hope, Bushmanland, Namaqualand, and Great Namaqualand.

Pegler, Alice. On the flora of Kentani. Annals Bolus Herb. **2:** 1–14, 112–128, 163–184. 1916–18.

Physiography, climate, phenology; annotated list of vascular plants. (On southeast coast, 32°30′ S. Lat., 28°18′ E. Long.)

Phillips, E. P. Some notes on a collecting trip to French Hoek. African Jour. Sci. **16:** 450–478. 1919.

Topography, geology, ecology, statistics of flora; annotated list of 335 spermatophytes collected by author in 1913.

Salisbury, F. S. Naturalised plants of Albany and Bathurst. Rec. Albany Mus. **3:** 163–177. 1919.

Annotated list of vascular plants.

Schönland, Selmar. List of the flowering plants found in the districts of Albany and Bathurst, Cape Colony. Rec. Albany Mus. **2:** 44–64, 97–136. 1907.

Physiography, etc.; annotated list of phanerogams. Incomplete (Cycadaceae-Anacardiaceae).

——— Phanerogamic flora of the divisions of Uitenhage and Port Elizabeth. Mem. Bot. Survey So. Africa no. 1. 118 p. 1 fig. (map). 1919.

Topography, geology, climate, statistics, plant formations, phytogeography; list of 2,332 phanerogams, with localities.

Sim, T. R. Sketch and check list of the flora of Kaffraria. 92 p. 19 cm. Cape Town, 1894.

Unannotated list (2,449 species) of vascular plants, with occasional vernacular names; general remarks on physiography and vegetation.

DAHOMEY

See also West Africa.

General

Bertherand, E. L. Acclimatation. Flore médicale de l'Afrique occidentale. 16 p. 25 cm. [Alger,] 1886.

Annotated systematic list of medicinal and other useful plants of Dahomey, based on the observations and collections of Abbé Cordioux.

Local

Hua, Henri. Une collection botanique du haut Dahomey et de la vallée du Niger moyen, récoltée par M. de Gironcourt, en 1908–1910. Bul. Mus. Hist. Nat. Paris **22**: 330–335. 1916.

Lists of angiosperms, with localities, collected at about 10° N. Lat. in Dahomey, and in "Niger moyen."

EGYPT

See also Anglo-Egyptian Sudan (Grant); West Africa (Oliver). E. Boissier, Flora Orientalis (which includes Egypt south to the first cataract), and G. E. Post, Flora of Syria, Palestine, and Sinai, will be given in the second part of this work. Papers referring particularly to the Sinai Peninsula, geographically a part of Asia but politically a part of Egypt, are included under Egypt.

General

Ascherson, Paul, and Schweinfurth, Georg. Illustration de la flore d'Egypte. Mém. Inst. Égyptien **2**: 25–260. 1889 [1887].—Supplément. l. c. 745–821. 1889.

Botanical explorations, phytogeographical divisions; list of 1,316 spontaneous vascular plants (including supplement) with vernacular names and indication of distribution by districts; cultivated plants listed but not numbered; index of vernacular names. The supplement includes a Florula Rhinocoluraea (list of 305 vascular and cellular plants known from the vicinity of El-Arîch) and a Florula Sirbonis (list of 59 vascular and cellular plants collected at Sebakh-el-Berdâwil), both by Ascherson only. Both localities are in the Sinai Peninsula.

Hartmann, Fernande. L'agriculture dans l'ancienne Égypte. 332 p. 76 fig. 24.5 cm. Paris, 1923.

Includes chapter (p. 17–70) on wild and cultivated economic plants, classified by uses.

Issa Bey, Ahmed. Dictionnaire des noms des plantes en latin, français, anglais et arabe. xiv, 227 p. port. 28 cm. Le Caire, 1930.

Bibliography; alphabetical list of botanical names, with vernacular names in the languages mentioned; indices of vernacular names in each language, including that in Arabic (with Arabic numbering of pages).

Loret, Victor. La flore pharaonique d'après les documents hiéroglyphiques et les spécimens découverts dans les tombes. 2. éd., rev. et augm. 145 p. 25 cm. Paris, 1892. (1st ed., 1887.)

Bibliography; annotated list of 202 vascular and cellular plants; vernacular names in French, Hebrew, Arabic, Coptic, and hieroglyphic.

Muschler, Reno. A manual flora of Egypt . . . 2 v. 23 cm. Berlin, 1912.

Descriptive flora of 1,503 vascular plants, with references, localities, and Egyptian names; botanical explorations, with bibliographical references; phytogeography, geology, climate; tables showing local distribution of all the species in Egypt, and their range in the Mediterranean Basin and elsewhere; list of cultivated plants; glossary; index of Arabian names.—See also **Simpson, N. D.** Some supplementary records to Muschler's Manual flora of Egypt. Including many species collected by G. W. Murray. Bul. Tech. and Sci. Serv. Min. Agr. Egypt no. 93. 59 p. 1930. (Records of 400 cellular and vascular plants, including 220 species and varieties not in Muschler's Flora.)

Ramis, A. I. Bestimmungstabellen zur flora von Aegypten. iv, 221 p. 25.5 cm. Jena, 1929.

Flora of vascular plants, in form of keys, briefly annotated.

Schweinfurth, Georg. Arabische pflanzennamen aus Aegypten, Algerien und Jemen. xxiv, 232 p. 32 cm. Berlin, 1912.
Includes Latin-Arabic and Arabic-Latin lists from Egypt; from Jemen (Yemen), after Forskål, and from Jemen and southern Arabia; from Biskra, and the coast and Tal Mountain region of northwest Algeria; also similar list of the parts of the date palm.

——— Beitrag zur flora Aethiopiens. 1. abth. xii, 311 p. 4 pl. 33.5 cm. Berlin, 1867.
Annotated list of 1,051 vascular and cellular plants of the Egyptian region, based on collections of C. G. Ehrenberg, Cienkowski, Schimper, and others, with localities and vernacular names; bibliography; compiled list of 3,947 (3,958) vascular plants (including cultivated plants, not all the latter numbered) known from the Nile region, with brief indication of range. No more published. The second list, by Schweinfurth and P. Ascherson, bears the title: "Aufzählung sämmtlichler zur zeit bekannten phanerogamen- und gefässkryptogamen-arten aus dem gesammtgebiete der Nil-Länder." It covers Egypt proper, Nubia proper, Kordofan, Sennaar, Abyssinia, Bahr-el-Abiad, and the Ukereve region.

Sherborn, C. D. Bibliography of scientific and technical literature relating to Egypt, 1800–1900. ii, 155 p. 27 cm. Cairo, 1915. (Reprint of prel. ed., 1910.)
Unannotated list, arranged alphabetically by authors, with key in form of subject index.

Société horticole d'Alexandrie. Liste des plantes cultivées en Égypte. ii, 190 p. 22 cm. Alexandrie, 1901.
Briefly annotated tabular lists, the species divided by habit, etc.

Woenig, Franz. Die pflanzen in alten Aegypten. Ihre heimat, geschichte, kultur und ihre mannigfache verwendung im sozialen leben, in kultus, sitten, gebräuchen, medizin, kunst. Nach den eigenen bildlichen darstellungen der alten Aegypter, pflanzenresten aus gräberfunden, zeugnissen alter schriftsteller und den ergebnissen der neuen forschungen. 2. aufl. 425 p. 177 fig. 22 cm. Leipzig, 1897. (1st ed., 1886.)
Deals mainly with useful and medicinal plants.

Local

Christensen, Carl. Index to Pehr Forsskål: Flora aegyptiaco-arabica 1775 with a revision of Herbarium Forsskålii contained in the Botanical Museum of the University of Copenhagen. Dansk Bot. Arkiv v. 4, no. 3. 54 p. 1922.
History of Forsskål's herbarium, bibliography, Forsskål's itinerary, list of new genera described by Forsskål, with synonymy, list of species listed in Flora Aegyptiaco-arabica, with identifications, similar list of new species described by Rottbøll and Vahl from Forsskål's plants, index.

Decaisne, Joseph. Enumération des plantes recueillies par M. Bové dans les deux Arabies, la Palestine, la Syrie et l'Egypte. Ann. Sci. Nat. II, Bot. 2: 5–18, 239–270. 1834; 3: 257–291. pl. 7. 1835.
List of 283 cellular and vascular plants from Sinai collected mostly by Bové, with localities. Reprinted with the title Florula Sinaica, a subtitle in the original.

Drar, Mohammed. Enumeration of the plants collected at Gebel Elba during two expeditions. Bull Tech. and Sci. Serv. Min. Agr. Egypt no. 149. vii, 123 p. 29 pl. 1937.
Botanical explorations; annotated list of vascular plants, with localities. (Southeast corner of Egypt just north of 22° N. Lat.)

Ducros, A. H. Essai sur le droguier populaire arabe de l'Inspectorat des pharmacies du Caire. Mém. Inst. Égypte v. 15. viii, 165 p. 9 pl. 1930.
Annotated list of 233 species, with vernacular names (French and Arabic), brief description and uses.

Hart, H. C. Some account of the fauna and flora of Sinai, Petra, and Wâdy 'Arabah. x, 225 p. 12 pl., 2 maps. 29.5 cm. London, 1891.
Includes (p. 79–120) list of vascular and cellular plants collected by author, with localities, also (p. 121–172) "An analysis of the flora of Sinai and general remarks on its botany, and that of the Dead Sea basin," containing tabular

list of 509 vascular plants of the Sinaitic Peninsula, with remarks on phytogeography and a list of the tropical flora of the basin, etc. Pages 1–120 reprinted from Trans. Roy. Irish Acad. **28** (Science) : 373–452. 1885.

Hume, W. F. List of plants noted in the Sinai Peninsula. *In* his The topography and geology of the peninsula of Sinai (southeastern portion). p. 207–233. Cairo, 1906.

List of 650 plants (532 vascular) with Arabic name, localities, and name of collector. An account of botanical explorations and general features of flora is given on pages 96 to 111.

Hurst, H. A. List of desert-plants collected at Ramleh, near Alexandria, Egypt, from September 1875 to April 1876. Mem. Lit. and Phil. Soc. Manchester III, **6**: 141–156. 1879.

Botanical explorations, etc; unannotated list of phanerogams.

Klunzinger, C. B. Die vegetation der egyptisch-arabischen wüste bei Koseir. Zeitschr. Gesell. Erdk. Berlin **13**: 432–462. 1878.

General features of the area and of its flora, plant zones; running account of principal plants, grouped by habitats. Koseir (Kosseir, Qosêr) is on the Red Sea coast, about 26° N. Lat.

Kotschy, Theodor. Die vegetation und der canal auf dem Isthmus von Suez. Oesterr. Bot. Zeitschr. **8**: 41–54. 1858.

General features of flora; lists of species from different localities.

Muschler, Reno. Beitrag zur kenntnis der flora von "el-Tor." (Sinai-halbinsel.) Verhandl. Bot. Ver. Prov. Brandenb. **49**: 66–146. 1908.

Topography, geology, climate; annotated list of 204 spermatophytes, with references, local and extralimital range, vernacular names; botanical explorations, general features of flora, phytogeography, alphabetical list of vernacular names, index.

Range, Paul. Die flora der Isthmuswüste. Auf grund des vom verfasser in den jahren 1915/16 gesammelten, von . . . Dr. Georg Schweinfurth bestimmten materials unter benutzung der einschlägigen literatur bearbeitet. Veröffentl. Gesell, Palästina-Forsch. 7. 44 p. map. 1921.

Botanical explorations, geology, climate, general features of flora, annotated list of 240 spermatophytes collected by the author in 1915–16, with data; tabular list of 479 spermatophytes known from the area, with indication of occurrence in neighboring regions (Sinai, Egypt, Palestine) ; briefly annotated bibliography.

Rikli, Martin, and Rübel, Eduard. Zur kenntnis von flora und vegetationsverhältnissen der Libyschen Wüste. Vierteljahrsschr. Naturf. Gesell. Zürich **73** (Beibl. 15, Festschrift Hans Schinz) : 190–232. 1 fig. (map), pl. 7–8. 1928.

Climate, botanical regions; annotated lists of species from Egyptian Marmara, the Libyan Desert in Egypt, Siwa Oasis, and the Nile Valley from Assuan to Wadi Halfa.

Schneider, Oscar. Ueber die flora der wüste von Ramleh. Sitzungs-Ber. Naturw. Gesell. Isis Dresden **1871**: 153–161. 1871.

Plant zones; annotated list of plants collected by author.

Schweinfurth, Georg. The flora of the desert surrounding Helwân (east). *In* **May, W. P.** Helwân and the Egyptian desert. 2d ed. p. 16–38. plates. London, 1904.

General features of flora; list of 191 spermatophytes known from desert east of Helwân.

———, **Ascherson, Paul,** and **Taubert, Paul.** Primitiae florae Marmaricae. Bul. Herb. Boissier **1**: 433–449, 584–603, 644–682. pl. 20. 1893.

Geography, topography, geology, general features of flora, botanical investigations (with footnote references) ; annotated list of 401 plants (378 vascular). The term Marmarica refers to the coast of Cyrenaica and Egypt, between 23° and 29° E. Long.—See also **Stapf, Otto.** Additions to the florula Marmarica. Bul. Misc. Inform. Kew **1907**: 365–369. 1907.—Also **Pampanini, R.,** and **Zanon, V.** Un manipolo di piante della Marmarica. Bul. Soc. Bot. Ital. **1919**: 46–48. 1919.—Also **Gauba, Erwin.** Neue und seltene pflanzen aus der ägyptischen Marmarika. Bot. Jahrb. Engler **67**: 275–289. 1935.

Zohary, Michael. Die phytogeographische gliederung der flora der Halbinsel Sinai. Beih. Bot. Centralbl. abt. B, **52**: 549–621. 3 fig. (incl. map). 1935.

Topography, botanical explorations, general features of flora, phytogeography; tabular list of 942 vascular plants, with indication of biological form, distribution in the area, and phytogeographical relationships; bibliography.

ERITREA

See also Ethiopia (Chiovenda, Plantae novae); Libya (Cortesi); West Africa (Oliver). The following publication has not been available:

Cuoco, L. Fauna e flora medica e industriale della colonia Eritrea. 109 p. 12°. Napoli, 1898. (Cited from Naturae Novitates **20**: 272. 1898.)

General

Chiovenda, Emilio. Vegetali utilizzati nella medicina indigena dell'Eritrea, Somalia, e regioni vicine. Atti Cong. Studi Colon. 1., Firenze, 1931. **7**: 351–376. 1931.

Briefly annotated list, grouped by uses, with vernacular names.

Fiori, Adriano. Boschi e piante legnose dell'Eritrea. 428 p. 177 fig. (incl. pl.). 25 cm. Firenze, 1909–12.

Forestry regulations, general features of flora, plant formations, phytogeography, bibliography; annotated descriptive flora of woody plants; key to genera; table of specific gravities of woods; bibliography; index of vernacular names.

Pirotta, Romualdo. Flora della colonia Eritrea. Annuario R. Ist. Bot. Roma v. 8. 464 p. 12 pl. 1903–07.

Botanical explorations (with footnote references); annotated list of vascular plants with references, localities, and names of collectors; not in systematic order. Incomplete; the manuscript of fasc. 4, prepared by Chiovenda to complete the work, was lost by the printing establishment to which it was sent (Chiovenda, Atti Cong. Studi Colon. 1., Firenze, 1931. **3**: 50. 1931).

Rovesti, Paolo. Medicamenti, aromi e droghe nei mercati indigeni dell'Eritrea. Riv. Ital. Essenz., Profumi e Piante Off. **15**: 19–29. illus. 1933.

Briefly annotated lists, alphabetically arranged by vernacular names.

Schweinfurth, Georg. Le piante utili dell'Eritrea. Bol. Soc. Africana Italia **10**: 233–286. 1891. (Reprinted.)

Annotated list of useful plants, classified by uses, with vernacular names.

Tancredi, A. M. Notizie e studi sulla colonia Eritrea. viii, 276 p. 25 cm. Roma, 1913.

Includes annotated list (p. 109–117) of useful plants, with vernacular names, as well as general account of plant zones, cultivated plants, and woody vegetation (p. 96–107).

Local

Chiovenda, Emilio. Contributo alla conoscenza della flora del Paese dei Cunama (Eritrea). Agricolt. Colon. **18**: 342–352. 1924.

List of vascular and cellular plants from region of Cunama and Baria collected by Corni, Calciatai, and Braccini.

——— Le piante raccolte dai prof. G. Dainelli e O. Marinelli in Eritrea nel 1905–06. Nuovo Gior. Bot. Ital. n. s., **26**: 147–168. 1919.

List of vascular plants and Hepaticae collected in Assaorta.

——— Le piante raccolte dal Dr. Nello Beccari in Eritrea nel 1905. Nuovo Gior. Bot. Ital. n. s., **26**: 89–114. 3 fig. 1919.

List of vascular plants collected mainly in the high basin of the Barca River, with localities.

Courbon, Alfred. Flore de l'île de Dissée (Mer Rouge). Annales Sci. Nat. IV, Bot. **18**: 130–158. pl. 7. 1862.

Physiography, etc.; annotated list of 68 vascular plants.

Fiori, Adriano. Piante raccolte in nella colonia Eritrea nel 1909. Nuovo Gior. Bot. Ital. n. s., **19**: 412–462. 1912; **20**: 345–394. 1913.

Annotated list of 970 vascular plants.

——— Pteridophytae et Phanerogamae erythraeae a cl. I. Baldrati lectae et communicatae. Nuovo Gior. Bot. Ital. n. s., **23**: 466–494. 1916.

List of localities; list of vascular plants, with localities.

Lanza, Domenico, and Mattei, G. E. Plantae erythraeae a L. Senni annis 1905–07 lectae. Bol. R. Orto Bot. e Giard. Colon. Palermo **8**: 73–123, 267–278. pl. 3–10. 1909; **9**: 3–75. pl. 1–4. 1910.

List of 550 vascular plants, with synonymy, data, and frequent annotations.

Martelli, Ugolini. Contribuzione alla flora di Massaua. Nuovo Gior. Bot. Ital. 20: 359–371. 1888.
List of phanerogams, with data, collected by O. Beccari in 1870 and Dr. Arcadipane in 1886; short list of plants collected at Assab by Beccari in 1870.

Pasquale, G. A. Cenni sulla flora di Assab. Atti R. Accad. Sci. Fis. e Mat. (Napoli) s. 2, v. 1, no. 12. 12 p. 1888 (1885).
List of 61 plants (60 angiosperms), with localities and occasional vernacular names, collected by G. B. Licata.

Schweinfurth, Georg. Sammlung arabisch-aethiopischer pflanzen. Ergebnisse von reisen in den jahren 1881, 88, 89, 91 und 92. Bul. Herb. Boissier v. 2, app. II. p. 1–113. 1894; v. 4, app. II. p. 115–266. 1896; v. 7, app. II. p. 267–340. 1899.
Annotated list of 870 vascular plants with data, frequent vernacular names and annotations, from Eritrea, also from Jemen and other parts of Arabia. Never completed; ends with Balsaminaceae (Engler and Prantl system).

Terracciano, Achille. Escursione botanica alle terre degli Habab. Bol. Soc. Geog. Ital. III, 5: 774–786, 987–991. map. 1892; 6: 30–40, 121–137, 214–228. 1893.
Account of author's explorations in 1892, with included mention of plants collected and observed.

——— Florula di Anfilah. Contribuzione alla conoscenza della flora dell'Africa orientale II. Annuario R. Ist. Bot. Roma 5: 89–121. pl. 7. 1894.
Phytogeography; list of 141 vascular plants, with localities; table showing local distribution of species listed.

ETHIOPIA

See also British Somaliland (Oliver): Egypt (Schweinfurth, Beitrag); Eritrea (Chiovenda, Vegetali utilizzati nella medicina); Italian Somaliland (Chiovenda, Flora Somala); Libya (Cortesi); West Africa (Gürke; Oliver).

General

Hemsley, W. B. On the vegetable productions of Abyssinia. Jour. Trav. and Nat. Hist. 1: 309–318. 1868.
General features of flora; running account of principal useful plants, grouped by families.

Richard, Achille. Tentamen florae abyssinicae, seu enumeratio plantarum hucusque in plerisque Abyssiniae provinciis detectarum et praecipue a beatis doctoribus Richard Quartin Dillon et Antonio Petit (annis 1838–1843) lectarum. 2 v. 23.5 cm., and atlas of 102 (i. e. 103) pl. 49.5 cm. Parisiis, [1847–51]. (Voyage en Abyssinie . . . exécuté par . . . A. Lefebvre . . . Histoire naturelle, Botanique.)
Annotated descriptive flora of phanerogams.

Rovesti, Paolo. Indagini sui prodotti erboristici dei mercati abissini in Etiopia. Riv. Ital. Essen., Profumi e Piante Off. 15: 179–191. 12 fig. 1933.
Includes annotated list, alphabetically arranged by vernacular names, with botanical names and uses.

Local

Avetta, Carlo. Prima—[quinta] contribuzione alla flora dello Scioa. Nuovo Gior. Bot. Ital. 21: 303–311, 332–339, 344–352. 1889; 22: 234–239, 242–247. 1890.
Lists of vascular plants, with data, mostly collected by Dr. Vincenzo Ragazzi in vicinity of Let-Marefia. The "Prima contribuzione" follows II–III in paging. The lists were combined and issued with title: Materiali per la flora dello Scioa, in Annuario R. Ist. Bot. Roma 6: 44–66. pl. 3–4. 1897.

Braun, Alexander. Beitrag zur kenntniss der abyssinischen culturpflanzen. Flora 31: 89–98. 1848.
Annotated list of 36 species.

Chiovenda, Emilio. La collezione botanica fatta dall'Ing. Edgard Taschdjian nell'Impero Etiopico nel 1935–36. Malpighia 34: 485–539. 1937.
List of 291 vascular plants collected in area south and west of Lago Tana, with data, vernacular names, and uses.

Chiovenda, Emilio. Etiopia. Osservazioni botaniche, agrarie ed industriali fatte nell' Abissinia settentrionale nell'anno 1909. Monog. e Rap. Colon. Min. Colon. (Italy) no. 24. 132 p. 1912.
Itinerary; annotated lists of cultivated and wild useful plants; index of vernacular names.

——— Plantae novae vel minus notae e regione aethiopica. Annali Bot. **9:** 51–85, 125–152, 315–322. 1911; **10:** 383–415. 1912; **13:** 371–410. 1915.—Bull. Soc. Bot. Ital. **1917:** 21–27, 38–39, 53–61, 69–70. 1917.—Lav. R. Ist. Bot. Catania no. 1. 32 p. 1928.
Records of 230 species and varieties, in great part new, from Ethiopia, Eritrea, Nubia, and Italian Somaliland.

Engler, Adolf. Ueber die hochgebirgsflora des tropischen Afrika. Abhandl. K. Akad. Wiss. Berlin Phys.-Math. Cl. 1891, Abh. II. 461 p. 1892.
Botanical explorations, with bibliographical references included; lists of plants of various elevated areas in Africa, with discussion; systematic list of vascular plants, with references, localities (comparatively detailed, with altitudes and collectors), ecological classification, and general range. Covers in general the region from Ethiopia (Abyssinia) and Somalia to Mozambique, and on the west coast Angola, Kamerun, São Thomé, and Fernando Po.

——— Ueber die vegetationsverhältnisse von Harar und des Gallahochlandes auf grund der expedition von Freiherrn von Erlanger und Hrn. Oscar Neumann. Sitzungsber. K. Preuss. Akad. Wiss. **1906** (2): 726–747. 1906.
Includes lists of species from various localities and habitats.

Martelli, Ugolino. Florula bogosensis. Enumerazione delle piante dei Bogos raccolte dal Dott. O. Beccari nell'anno 1870 . . . vii, 169 p. 1 pl. 23.5 cm. Firenze, 1886.
List of vascular and cellular plants, with references and localities, and general range.

Paillieux, Auguste, and **Bois, Désiré.** De quelques plantes alimentaires de l'Abyssinie. Revue Sci. Nat. Appl. Soc. Natl. Acclim. France **37:** 803–809. 1890.
Annotated list of 11 species, with vernacular names.

Pax, Ferdinand. Die von Felix Rosen in Abyssinien gesammelten pflanzen. *In* **Engler, A.** Beiträge zur flora von Afrika XXX. Bot. Jahrb. Engler **39:** 602–662. 1907.
List of 350 vascular and cellular plants, with reference and data.

Penzig, Otto. Piante raccolte in un viaggio botanico fra i Bogos ed i Mensa, nell' Abissinia settentrionale. Atti Cong. Bot. Internatl. Genova, 1892. p. 310–368. 1893.
Author's itinerary, lists of species from various localities; list of 664 vascular plants, with localities.

Raffeneau-Delile, Alire. Énumération des plantes recueillies par MM. Ferret et Galinier. *In* **Ferret, A.,** and **Galinier, J. G.** Voyage en Abyssinie dans les provinces du Tigre, du Samen et de l'Ahmara . . . v. 3, p. 85–163. pl. 6–15. Paris, 1847.
Annotated list of 144 species (118 phanerogams).

Schweinfurth, Georg. Abyssinische pflanzennamen. Eine alphabetische aufzählung von namen einheimischer gewächse in Tigrinja sowie in anderen semitischen und hamitischen sprachen von Abyssinien, unter beifügung der botanischen artbezeichnung. Abhandl. K. Akad. Wiss. Berlin 1893, Phys. Abh. II. 84 p. 1893.
Sources, etc.: list of vernacular names in native characters, with phonetic rendering and botanical name; alphabetical lists of vernacular and botanical names, with equivalents.

FRENCH EQUATORIAL AFRICA

See also Anglo-Egyptian Sudan (Schweinfurth, Bericht); French West Africa (Chevalier); West Africa.

General

Jumelle, Henri. Catalogue descriptif des collections botaniques du Musée colonial de Marseille: Afrique équatoriale française. Annales Mus. Colon. Marseille v. 31 (IV, 1), fasc. 1. 80 p. 1923.
Partly annotated list, mainly composed of useful plants, these classified by uses.

Local

Baudon, Alfred. Notes sur la flore et les plantes économiques du Bas-Congo. Annales Mus. Colon. Marseille **17** (II, 7) : 361–410. 5 fig. 1909.

Deforestation, etc.; list of vascular plants collected by author, with localities, vernacular names and notes on uses. The author's studies were made in Moyen-Congo, at Brazzaville and north to the Alima River.

Bertin, A. J. Les bois du Gabon. 2. éd., rev. et corr. 304 p. map. 25.5 cm. Paris, 1929. (1st ed., 1918.) (Mission d'études forestières envoyée dans les colonies françaises par les Ministères de la guerre, de l'armement et des colonies. t. 2.)

Includes (p. 160–304) annotated alphabetical list of vernacular names of trees, with botanical equivalents, and alphabetical index to botanical names treated in Chevalier's "La forêt et les bois du Gabon," with page references.

Chevalier, Auguste. Études sur la flore de l'Afrique centrale française (bassins de l'Oubangui et du Chari). Tome 1. Enumération des plantes récoltées. xii, 451 p. port. 25 cm. Paris, 1913. (Mission Chari–Lac Tchad (1902–1904).)

Botanical explorations; list of vascular and cellular plants collected, with data and occasional annotations. (Regions of Moyen- and Haut-Oubangui, Lac Tchad, Bas-, Moyen- and Haut-Chari.)

——— La forêt et les bois du Gabon. vii, 468 p. 31 fig., 28 pl. 25 cm. Paris, 1917. (Les végétaux utiles de l'Afrique tropicale française fasc. 9.)

Annotated list of collectors, annotated systematic list of trees, with descriptive notes, vernacular names, uses, and habitat; list of trees classed by uses, etc.

Jardin, Édelestan. Apercu sur la flore du Gabon avec quelques observations sur les plantes les plus importantes. Bul. Soc. Linn. Normandie IV, **4**: 135–203. 1890. (Reprint 71 p., Paris, 1891.)

Geography, botanical explorations; unannotated list of known cellular and vascular plants, with vernacular names; annotated list of vernacular names of undetermined plants; notes upon numerous species.

Louis, Xavier, and Monod, Théodore. Notes botaniques sur la région nord de la subdivision d'Ati (colonie du Tchad) . . . Bul. Agence Écon. Colon. Auton. et Terr. Afr. (France) **27**: 593–637. pl. 1934.

Climate, physiography, transcription of native names; annotated list of 198 vascular plants, with vernacular names, local range.

Maire, René. Contribution à l'étude de la flore du Tibesti. (Mission au Tibesti. 2e vol. Botanique.) Mém. Acad. Sci. Inst. France II, **62** (Mém.) : 1–39. 1936.

Botanical explorations, geology, general features of flora, bibliography; list of cellular and vascular plants known from the region (159 native species of vascular plants), with localities, general range, and vernacular names; list of cultivated plants, list of Toubou vernacular names with botanical equivalents.

Payeur-Didelot, J. F. Notes et mémoires sur le Gabon-Congo et la côte occidentale d'Afrique. IX. Coup d'oeil sur les espèces utilisables de la flore gabonaise. Bul. Soc. Géog. de l'Est **19**: 177–208, 397–422. 1897.

Sketch of principal woods and useful plants, grouped by uses.

Pellegrin, François. Collections botaniques rapportées par la Mission Tilho de la région Niger-Tchad. Bul. Mus. Hist. Nat. Paris **17**: 459–466, 566–571. 1911; **18**: 46–50. 1912.

List of vascular plants, with localities; general range given. For notes on useful plants, with a revised list of plants collected, see his Les collections botaniques rapportées par la Mission Tilho avec indications sur les plantes utiles. France. Ministère des colonies. Documents scientifiques de la Mission Tilho (1906–1909). v. 3. p. 459–484. 4 fig. Paris, 1914. (Reprinted.)

——— De quelques bois utiles du Gabon. Bul. Soc. Bot. France **84**: 639–645. 1938.

Unannotated systematic list, with vernacular names.

——— La flore du Mayombe d'après les récoltes de M. Georges Le Testu. Mém. Soc. Linn. Normandie v. 26 (II, 10), fasc. 2. 125 p. 7 fig., pl. 4–11. 1924; n.s., Sect. Bot. v. 1, fasc. 3. 83 p. 13 fig., 6 pl., map. 1928; fasc. 4. 114, viii p. 8 pl. 1938.

Annotated list of vascular plants with references and citation of specimens, with data. (Pt. 1) Dicotyledones Polypetalae; (pt. 2) Gamopetalae except

Rubiaceae; (pt. 3) Rubiaceae, Monocotyledones and Pteridophyta, also "Notes sur la végétation . . ." by G. Le Testu. (Valleys of the Nyanga and Nguonyé (Ngunie) Rivers, Gabon.)

FRENCH GUINEA

See also French West Africa (Perrot); Senegal; West Africa.

General

Pobéguin, Henri. Essai sur la flore de la Guinée française. Produits forestiers, agricoles et industriels. 392 p. 80 pl., map. 25 cm. Paris, 1906.
General features of flora, useful plants, sketch of indigenous plants by families, vernacular names (indexed).

——— Les plantes médicinales de la Guinée. 85 p. 25.5 cm. Paris, 1912. (Bibliothèque d'agriculture coloniale.)
Annotated list, with vernacular names and uses, alphabetically arranged by scientific names; list classified by uses.

Local

Corre, Armand. Esquisse de la flore et de la faune médicales et économiques du Rio-Nunez. Arch. Méd. Navale **26**: 14–40. 1876.
Account of wild and cultivated economic and medicinal plants, with vernacular names.

FRENCH SOMALILAND

See also West Africa (Oliver).
The following publication has not been available:
Aubert de la Rüe, Edgar. La Somalie française. Paris, 1938. (Cited from Chevalier, A. Compt. Rend. Acad. Sci. Paris **209**: 73–74. 1939.)

General

Chevalier, Auguste. La flore de la Somalie française et la forêt-relique du Mont Goudah. Compt. Rend. Acad. Sci. Paris **209**: 73–76. 1939.
Botanical explorations; general features of flora, with mention of some species.

——— La Somalie française. Sa flore et ses productions végétales. Revue Bot. Appl. et Agr. Trop. **19**: 663–687 (incl. fig. 2–3, pl. 17–19). 1939.
Geology, climate, botanical explorations, principal woody plants, grasses, notes on useful plants.

Local

Franchet, Adrien. Plantes du voyage au golfe de Tadjourah recueillies par M. L. Faurot. Jour. de Bot. **1**: 117–123, 134–136. 1887.
List of 105 spermatophytes, with localities and vernacular names, collected by Dr. Louis Faurot in 1885–86 at Obock and Tadjourah (a few from Ile de Kamarane, Arabia).

FRENCH WEST AFRICA

See also Dahomey (Hua); French Equatorial Africa (Pellegrin, Collections); Mauritania; Sahara; Senegal (Chevalier); West Africa.
The following publication has not been available:
Chevalier, Auguste. Flore vivante de l'Afrique occidentale française. v. 1 (360 p.). Paris, 1938.—Vol. 1 covers Gymnosperms and early dicotyledonous families of French West Africa, Togo, North Cameroons, Ubangi-Shari-Chad, and French Sahara. (Not seen; cited from Fortschr. Bot. **8**: 83, 85. 1939.)

General

Jumelle, Henri. Catalogue descriptif des collections botaniques du Musée colonial de Marseille: Afrique occidentale française. Annales Mus. Colon. Marseille v. 25 (III, 5), fasc. 1. 95 p. 1917.
Annotated list of useful plants, classified by uses, with vernacular names.

Perrot, Émile. Sur les productions végétales indigènes ou cultivées de l'Afrique occidentale française (Sahara, Soudan nigérien, Haute-Volta, Guinée).

Trav. Off. Natl. Mat. Prem. Vég. (Paris) Notice 31. viii, 468 p. 23 pl., 3 maps. 1929.
Includes running account of principal wild and cultivated economic plants.

Local

Bonnet, Edmond. Énumération des plantes recueillies par M. R. Chudeau, dans la région de Tombouctou et du Moyen-Niger. Mém. Soc. Bot. France no. 20. 19 p. 1911.
List of 228 vascular plants collected between Araouan and Djenné.

Chevalier, Auguste. La végétation de la région de Tombouctou. Actes Cong. Internatl. Bot. 1, Paris, 1900. p. 248–275 (incl. pl. 10–11). 1900.
Botanical explorations, plant zones, biology of plants, general features of flora, effects of cultivation; annotated list of 52 wild and cultivated woody plants, with localities and vernacular names; bibliography. The area covered is in the western Sudan, between 5° and 10° W. Long., from 17° N. Lat. to the southern edge of the Sahara Desert.

Hagerup, Olaf. Étude des types biologiques de Raunkiaer dans la flore autour Tombouctou. Biol. Meddel. K. Danske Vidensk. Selsk. v. 9, no. 4. 116 p. 5 fig., 5 pl. 1930.
Includes list of 138 phaneragams collected by author, the families arranged alphabetically.

GAMBIA

See also Senegal (Rançon) ; West Africa.

General

Williams, F. N. Florula gambica, une contribution à la flore de la colonie brittanique. Bul. Herb. Boiss. II, 7: 81–96, 193–208, 369–386. 1907.
Geography, botanical explorations, published floras; annotated list of 285 angiosperms.

Local

Dawe, M. T. List of plants collected in the Gambia. 11 p. 21.5 cm. Bathurst, 1922.
Annotated list of vascular and cellular plants collected by author, with localities, vernacular names, and uses.

Kew. Royal Gardens. Botany of Gambia delimitation commission. Bul. Misc. Inform. Kew 1891: 268–275. 1891.
Annotated list of vascular and cellular plants, with vernacular names and uses, collected in 1890–91 by Dr. Brown Lester in vicinity of River Gambia. — See l. c. 1892: 45–48, for an abstract of the report of the French members of the commission.

GOLD COAST

See also West Africa.

General

Burtt Davy, Joseph, and Hoyle, A. C., eds. Draft of first descriptive check-list of the Gold Coast. Compiled by members of the Gold Coast Forest Department . . . 152, vii p. 25 cm. Oxford, 1937. (Check-lists of the forest trees and shrubs of the British Empire, no. 3.)
List of woody plants, alphabetically arranged by families, with brief descriptions, vernacular names, citation of exsiccatae. The area covered includes Gold Coast colony, with Ashanti, Northern Territories, and the part of Togoland under British mandate. Processed publication, cover-title only.

Chipp, T. F. The forest officers' handbook of the Gold Coast, Ashanti and the Northern Territories. 148 p. (incl. 20 pl.). 3 diagr., 5 maps. 25.5 cm. London, [1922?].
Includes (p. 55–69) classified list of economic plants, and index of vernacular names of principal forest trees.

Chipp, T. F. A list of the herbaceous plants and undershrubs of the Gold Coast, Ashanti, and the Northern Territories. 55 p. 25 cm. London, 1914.
Annotated list, with localities and collectors' numbers; classified list of useful plants.

Irvine, F. R. Plants of the Gold Coast. lxxix, 521 p. front., 69 fig. (pt. on plates). 20.5 cm. London, 1930.
Native botany and herbalism, native uses of plants, general features of flora, classified lists of useful plants; annotated alphabetical list of about 1,200 phanerogams, with vernacular names in several native languages (indexed), brief descriptions, uses; no systematic list of species.

Local

Cummins, H. A. Botany of Ashanti Expedition. Bul. Misc. Inform. Kew **1898**: 65–82. 1898.
Topography, climate, general features of flora, agriculture, economic products; list of vascular plants and bryophytes collected by author between Cape Coast Castle and the Moinsi Hills, chiefly at Assin Yan Kumassi, with general range indicated.

ITALIAN SOMALILAND

See also Eritrea (Chiovenda, Vegetali utilizzati nella medicina); Ethiopia (Chiovenda, Plantae novae; Engler Ueber die hochgebirgsflora); Libya (Cortesi); West Africa (Gürke; Oliver).

General

Chiovenda, Emilio. Flora Somala. xvi, 436 p. 50 pl. (incl. map). 24.5 cm. Roma, 1929.—Flora Somala II. xvi, 482 p. 247 fig. 25 cm. Modena, 1932. (Lavori eseguiti presso l'Istituto botanico della R. Università di Modena 3.)—Flora Somala III. Raccolte somale dei proff. G. Pollacci, L. Maffei, R. Ciferri e N. Puccioni fatte negli anni 1934 e 1935. Atti Ist. Bot. Giov. Briosi R. Univ. Pavia IV, **7**: 117–160. 12 pl. 1936.—Nuovo contributo alla flora della Somalia Italiana. Atti Soc. Nat. e Mat. Modena **66**: 1–19. 1935.
Geology, climate, general features and affinities of flora, phytogeography, briefly annotated list of useful native plants, with vernacular names; list of vascular and cellular plants, with references, citation of exsiccatae, vernacular names. The 1929 publication is based on specimens collected by Profs. G. Stefanini and G. Paoli in British and Italian Somaliland and in eastern Ethiopia east of a line drawn from Dolo to Zeila, the other publications on later collections. Not a complete flora.

Local

Chiovenda, Emilio. Le collezioni botaniche della Missione Stefanini-Paoli nella Somalia Italiana. 241 p. 6 fig. (incl. map), 24 pl. 28.5 cm. Firenze, 1916. (Resultati scientifici della Missione Stefanini-Paoli . . . v. 1.)
Botanical explorations; annotated list of vascular and cellular plants: appendix containing annotated list of species collected by Mangano, Scassellati, Mazzocchi, and Provenzale; vernacular names (indexed).

——— Contributo alla conoscenza della flora Somala Transjubense. Piante raccolte dal Dr. Pompeo Gorini nel 1925. Agricolt. Colon. **20**: 42–49, 103–109. 1926.
List of spermatophytes, with occasional vernacular names and notes.

Mattei, G. E. Contribuzioni alla flora della Somalia Italiana. Bol. R. Orto Bot. e Giard. Colon. Palermo **7**: 85–112, 165–191. 1908.
Two centuries of angiosperms, with references, data, and frequent notes and vernacular names. Specimens collected by Dr. Cesare Macaluso in regions of Goscia, Brava, Merca, and Mogadiscio.

Paoli, Guido. Cenno sulle piante utili e utilizzabili raccolte dalla Missione. *In* **Stefanini, Giuseppe, and Paoli, G.** Ricerche idrogeologiche, botaniche ed entomologiche fatte nella Somalia Italiana meridionale. Relaz. e. Monog. Agr.-Colon. Ist. Agr. Colon. Ital. **7**: 225–250. 1916.
Annotated lists of useful plants, classified by uses.

IVORY COAST

See also West Africa.

General

Aubréville, André. La flore forestière de la Côte d'Ivoire. 3 v. 351 pl., maps. 28.5 cm. Paris, 1936.
Forest formations, key to families; systematic descriptive flora of woody plants, with vernacular names.

Bégué, Louis. Contribution à l'étude de la végétation forestière de la Haute-Côte d'Ivoire. Pub. Com. Études Hist. et Scient. Afrique Occid. Franç. sér. B, no. 4. 126 p. 19 pl., map. 1937.
Mainly ecological, with numerous short lists of species from different localities; tabular list of principal woody plants with vernacular names.

Bertin, André. Les bois de la Côte d'Ivoire. 176 p. 8 pl., map. 25.5 cm. Paris, 1918. (Mission d'études forestières envoyée dans les colonies françaises par les Ministères de la guerre, de l'armement et des colonies. I.)
Includes lists of scientific and vernacular names, with references to Chevalier's "Les bois de la Côte d'Ivoire."

Chevalier, Auguste. Première étude sur les bois de la Côte d'Ivoire. 314 p. 1 fig., map. 24.5 cm. Paris, 1909. (Les végétaux utiles de l'Afrique tropicale française. fasc. 5.)
Includes (p. 109–274) annotated list of trees, systematically arranged, with vernacular names and frequent descriptions.

KENYA

See also Ethiopia (Engler, Ueber die hochgebirgsflora); Tanganyika (Engler, Die pflanzenwelt Ost-Afrikas; Greenway); West Africa (Oliver).
The following publication has not been available:
Chiovenda, Emilio. Raccolte botaniche (Embryophyta diploidalia). Fatte dai missionare della consolata nel Kenya. [Lavori Ist. Bot. R. Univ. Modena Mem. 6.] 160 p. 1935. (Title cited from Bot. Centralbl. n. f., **31** (v. **173**, Lit.): 107. 1938.)

General

Battiscombe, Edward. Trees and shrubs of Kenya Colony. A revision and enlargement [by I. R. Dale] of "A descriptive catalogue of some of the common trees and woody plants of Kenya Colony, by E. Battiscombe." 201 p. 24.5 cm. Nairobi, 1936.
Native tribes, forest types; annotated list of woody plants, with vernacular names, local range, brief descriptions. A few herbaceous species are included.—See also **Dale, I. R.** Additions and corrections . . . Inst. Paper Imp. Forestry Inst. Oxford **18**: 20–28. 1939.

Kenya. Colony and protectorate of Kenya—Forestry department. Preliminary catalogue: trees, shrubs, and herbs. 34 p. 23 cm. Nairobi, 1922.
Systematic list "of the more common plants" with vernacular names.

Local

Fries, T. C. E., and **Fries, Robert.** Beiträge zur kenntnis der flora des Kenia, Mt. Aberdare und Mt. Elgon. Notizbl. Bot. Gart. Berlin **8**: 389–423, 475–504, 547–573, 661–704. [16] fig. 1923–24; **9**: 16–36, 173–189, 299–333, 485–522, 1103–1135. fig. 15–17. 1924–27; **10**: 69–101, 266–272, 594–621, 897–917. 1927–30; **11**: 20–47. 1 fig. 1930.
Annotated list of species collected by the Swedish expeditions of 1920 and 1921–22; the families not in systematic order. Treatments of various families contributed by other authors.

Gregory, J. W. The great rift valley, being the narrative of a journey to Mount Kenya and Lake Baringo . . . xxi, 422 p. 23 fig. (incl. maps), 20 (i. e. 21) pl., 2 maps. 24 cm. London, 1896.
Includes (p. 389–400) list of vascular plants collected, with localities, by E. G. Baker, J. Britten, A. B. Rendle, and W. Carruthers.

LIBERIA

See also West Africa.

General

Cooper, G. P., and Record, S. J. The evergreen forests of Liberia. Bul. Yale Univ. School Forestry no. 31. xi, 153 p. 15 pl. on 8, map. 1931.

History, topography, climate, general features of flora; list of 286 woody plants, with vernacular names, brief descriptions, citations of collections, and descriptions of woods; composition of sample plots (as to species, etc.); uses of woods; alphabetical list of vernacular names, with botanical equivalents.

Dinklage, Max. Verzeichnis der flora von Liberia. Repert. Spec. Nov. Fedde **41**: 235–271. 1937.

Botanical explorations; list of known vascular plants, with citation of exsiccatae but no localities or annotations.

Stapf, Otto. The known plants of Liberia, mainly from materials in the herbarium of the Royal Botanic Gardens, Kew. *In* **Johnston, Harry.** Liberia. v. 2, p. 570–669. fig. 246–266. London, 1906.

Botanical explorations; list of vascular plants, with brief descriptions, localities, collectors' names, and occasional annotations.

LIBYA

See also Algeria (Maire, Contributions à l'étude); Egypt (Issa Bey; Schweinfurth, Ascherson, and Taubert).

The following publications have not been available:

Corti, Roberto. Prodromo della flora e vegetazione del Fezzân (Resultati scientifici delle Missioni nel Sahara Italiano). fasc. 1–3 (250 p.). Firenze, 1938. (Ed. provvisoria.) (Cited from Nuovo Gior. Bot. Ital. **45**: ccii. 1938.)

Nannizzi, Arturo. Le piante coltivate ed utili della Libia. viii, 192 p. 21 cm. Siena, 1913. (Cited from International catalogue of scientific literature, 13th annual issue M, Botany. p. 206.)

General

Ascherson, Paul. Die aus dem mittlern Nordafrika, dem gebiete der Rohlfs'schen expedition nach Kufra bekannt gewordenen pflanzen. *In* **Rohlfs, Gerhard.** Kufra. Reise von Tripolis nach der oase Kufra . . . p. 386–559. Leipzig, 1881.

Botanical explorations, with bibliographical references; annotated lists of vascular and cellular plants known from Tripolitania (437 species), Fezzan (200 species), Kufra (39 species), Audjilagruppe (Aujila: 48 species), and Cyrenaica, including Turkish Marmara (493 species), with localities, vernacular names, and references. Cultivated species are included and numbered in the lists. Under the title "Plantarum Africae septentrionalis mediae hucusque cognitarum conspectus," the list of 775 plants (759 vascular) was reprinted in Bot. Centralbl. **8**: 278–287. 1881.

[**Borzi, Antonino.**] Elenco alfabetico degli autori che si occuparono della Libia sotto l'aspetto botanico ed agrario e della loro pubblicazioni . . . Monog. e Rap. Colon. Min. Colon. (Italy) no. 9. 32 p. 1912; no. 11. 30 p. 1912.

Partly annotated list, alphabetically arranged by authors; includes also collectors, with brief biographical notes.

Cortesi, Fabrizio. Piante officinali e della medicina popolare delle Colonie italiane d'Africa e regioni limitrofe. Rassegna Econ. Colon. (Italy) **24**: 71–126. 1936.

Bibliography: systematic list of vascular plants, with vernacular names and brief account of uses, summarizing available information; lists of botanical and vernacular names, without references to pages. Covers Libya, Eritrea, Italian Somaliland, and Ethiopia.

Durand, Ernest, and Barratte, Gustave. Florae libycae prodromus, ou catalogue raisonné des plantes de Tripolitaine . . . avec la collaboration de P. Ascherson, W. Barbey et R. Muschler . . . cxxvii, 330 p. 20 pl., map. 35 cm. Genève, 1910.

Physiography, biographical sketches of botanical collectors, synoptical tables showing distribution, phytogeography, list of cultivated plants, bibliography; geology (by S. Meunier); annotated list of 1,156 plants (1,026 vascular), with references, local and extralimital range. Area covered includes Tripolitaine, the Fezzan, Cyrenaica, and Turkish Marmara.

Trotter, Alessandro. Flora economica della Libia. Statistica illustrata delle piante coltivate e delle spontanee utili delle regione. Seguita da una appendice intorno alle droghe del commercio locale. 375 p. 145 pl., 2 profiles. 19.5 cm. Roma, 1915.

Bibliography, physiography, plant formations, climate, etc.; annotated list of 379 useful plants, with vernacular names and uses; classified lists of useful plants; list of drugs; indices.

Local

Ascherson, Paul. Florula der oasengruppe Kufra nach den sammlungen und beobachtungen von G. Rohlfs. Verhandl. Bot. Ver. Prov. Brandenb. **23** (Sitzungsber.): 27–33. 1882.

General features of flora, cultivated plants; annotated list of 39 vascular plants.—See also **Lusina, G.** Aggiunte alla flora delle oasi di Cufra. Annali Bot. **19**: 423–425. 1932.

Béguinot, Augusto, and Vaccari, Antonio. Contributo alla flora della Libia in base a piante raccolte dall'ottobre 1911 al luglio 1912. Monog. e Rap. Colon. Min. Colon. (Italy) no. 16. 72 p. 10 pl. 1912.—Secondo contributo . . . con notizie sulle piante utili osservate. l. c. 1913, no. 7. 40 p. 1913. (Also in Mem. R. Accad. Sci. Lett. ed Arti Modena III, **10** (2): 553–584. 1913.)—Terzo—[quarto] contributo . . . Annali Bot. **12**: 87–150. 1913; **13**: 9–34. 1914.

(1) Localities, reduction of species described by Viviani in his "Florae libycae specimen," phenology, life forms, phytogeography; annotated list of 402 vascular plants. (2) Annotated list of 172 species; notes on useful plants, with vernacular names. (3) Annotated list of 395 species; table of 596 species mentioned in the three contributions, with indication of locality and occasional notes. (4) Annotated list of 231 species.

Corti, Roberto. Le raccolte botaniche nel sud Cirenaico del Prof. L. di Caporiacco (1933—Spediz. Marchesi) e del Prof. U. Mónterin (1934–R. Soc. Geogr. Italiana) e la florula delle Oasi di Cufra e del Gebél Auenát. Nuovo Gior. Bot. Ital. **45**: ccii-ccxl. 3 fig. 1938.

Botanical explorations, with footnote references; list of spermatophytes known from Oasis of Kufra and Gebél Auenát, with citation of specimens and general range; phytogeography, etc., with lists of species from numerous localities in the region.

Letourneux, A. H. Note sur un voyage botanique à Tripoli de Barbarie. Bul. Soc. Bot. France **36**: 91–99. 1889.

Includes list of about 300 vascular plants and Characeae from the Oasis of Tripoli; list of about 100 cultivated plants, with French and Arabic vernacular names.

Lusina, Giuseppe. La flora. *In* R. Società geografica italiana. Resultati scientifici della missione alla Oasi di Giarabùb (1926–1927). fasc. 4, p. 551–558. Roma, 1931.

Alphabetical list of 65 vascular plants; bibliography, phytogeography. (Tripolitania.)

Maire, René, and Weiller, Marc. Contributions à l'étude de la flore de la Libye. *In* **Maire, R.** Contributions à l'étude de la flore de l'Afrique du Nord fasc. 27. Bul. Soc. Hist. Nat. Afrique Nord **30**: 255–314. pl. 17. 1939.

Annotated list of cellular and vascular plants, new or rare in Libya, collected by authors in April 1938.

Pampanini, Renato. Piante di Bengasi e del suo territorio raccolte dal Rev. P. D. Vito Zanon della Missione dei PP. Giuseppini al Fuehat. Nuovo Gior. Bot. Ital. n. s., **23**: 260–293. 1916; **24**: 113–171. 1917.

Botanical explorations, with biblographical footnotes; special localities; two lists of vascular and cellular plants, with localities, collectors, and vernacular names.—See also **Pampanini, R.**, and Zanon, V. Nuovi contributi alla conoscenza della flora della Cirenaica. l. c. **26**: 205–220. 1919.—Nuovi contributi . . . II. Bull. Soc. Bot. Ital. **1922**: 6–7. 1922.

——— Plantae tripolitanae ab auctore anno 1913 lectae et repertorium florae vascularis Tripolitaniae. xiv, 334 p. 1 fig., 9 pl., map. 25 cm. Firenze, 1914. (La missione Franchetti in Tripolitania (Il Gebèl), appendice I.)

New records, bibliography; annotated list of vascular plants known from region, with references, localities, and collectors. The appendix (p. 287–328) contains lists of cellular plants, by various authors.

Pampanini, Renato. Prodromo della flora cirenaica. xxxviii, 577 p. 2 fig., 6 pl. 24.5 cm. Forlì, 1930.—Aggiunte al "Prodromo della flora cirenaica" dalle mie raccolte in Cirenaica negli anni 1933–34. Rend. Sem. Fac. Sci. R. Univ. Cagliari **8:** 53–79. 1938.

Botanical explorations, list of botanical collectors with biographical notes, chronological list of collectors, statistics of collectors, etc.; list of 1,617 cellular and vascular plants (1,053 vascular), with references, localities, and collectors; gazetteer, bibliography. The 1938 additions bring the total to 1,197 species of vascular plants, plus 529 varieties and forms.

Trotter, Alessandro. Nuovi materiali per una flora della Tripolitania. Nuovo Gior. Bot. Ital. n. s. **22:** 319–364. 5 fig., pl. 14–18. 1915.

Rare or noteworthy species collected by author in 1914.

——— Osservazioni e notizie intorno alla flora ed alla vegetazione dell'altopiano della Cirenaica. Annali R. Scuola Sup. Agr. Portici II, v. 18, [no. 4]. 23 pp 1923.

Localities, etc.; list of spermatophytes collected in 1919 by Prof. E. De Cillis.

MAURITANIA

See also French West Africa; West Africa.

Local

Benoist, Raymond. Plantes récoltées par M. Mond en Mauritanie. Bul. Mus. Hist. Nat. Paris **33:** 188–192. 1927.

List of phanerogams, with localities. (Along coast, from St. Louis to Cape Blanco.)

Bonnet, Edmond. Partie botanique A. Étude systématique. *In* **Gruvel, A.,** and **Chudeau, R.** Mission en Mauritanie occidentale. Actes Soc. Linn. Bordeaux **63:** 15–37. pl. 2. 1909.

General features of flora; list of 119 phanerogams, with localities and general range.

——— and **Pellegrin, François.** Énumération des plantes recueillies par M. R. Chudeau dans le nord-ouest de la Mauritanie. Compt. Rend. Assoc. Franç. Avanc. Sci. **43** (2): 463–469. 1915.

List of species previously recorded by the authors, with bibliography; list of additional species, with localities.

Jumelle, Henri. Plantes de Mauritanie récoltées par le Lieutenant Boëry. Annales Mus. Colon. Marseille v. 36 (IV, 6), no. 1. 43 p. 3 pl. 1928.

Climate, geology, geography, plant formations; list of 104 vascular plants, with localities and extralimital range; based on a collection made in the region of Kiffa in 1924–1926.

Maire, René. Florule du Zemmour. His Contributions à l'étude de la flore du Sahara occidental, fasc. 6. Bul. Soc. Hist. Nat. Afrique Nord **26:** 148–162. 1935.

Topography, etc.; annotated list of 69 vascular plants collected by Luthereau in 1934, with vernacular names and uses. Zemmour is on the border between Rio de Oro and northern Mauritania, between 25° and 26° N. Lat.

Zolotarevsky, B., and **Murat, M.** Rapport scientifique sur les recherches de la Mission d'études de la biologie des acridiens en Mauritanie (A. O. F.). Bul. Soc. Hist. Nat. Afrique Nord **29:** 29–103. pl. 2–8, map. 1938.

Includes sketch of vegetation (plant zones, etc., p. 50–75) and list of plants cited, with vernacular names (p. 96–101), also bibliography. (Western and northern Mauritania.)

MOROCCO (INCLUDING SPANISH MOROCCO)

See also Algeria (Battandier and Trabut, Flore de l'Algérie; Cosson; Ducellier and Maire; France, Comité regional d'Algérie; Lapis and Maige; Maire, Contributions à l'étude de la flore de l'Afrique du nord; Trabut, Flore du nord de l'Afrique); Egypt (Issa Bey).

General

Ball, John. Spicilegium florae maroccanae. Jour. Linn. Soc. Bot. **16:** 281–742. pl. 9–28. 1877–78.

Botanical explorations, phytogeography, list of collectors; annotated list of vascular (and a few cellular) plants known from the region, with collectors, localities, and general range.

Emberger, Louis. Les arbres du Maroc et comment les reconnaître. 317 p. 17 cm. Paris, 1938.
Semipopular work, with keys, description, uses, Arabic and Berber vernacular names, etc.

——— and **Maire, René.** Tableau phytogéographique du Maroc. 1ère partie. Mém. Soc. Sci. Nat. Maroc no. 38. viii, 187 p. 16 fig. (incl. maps, diagrs.), 16 pl. 1934.
Relates mainly to soil and climate; included here for its account of botanical explorations (p. 5–35), with footnote references.

Gattefossé, Jean. Les plantes dans la thérapeutique indigène au Maroc. *In* **Perrot, E.,** and **Gentil, L.** Sur les productions végétales du Maroc. Trav. Off. Natl. Mat. Prem. Vég. (Paris) Notice **10**: 72–123. 1921.
Bibliography; annotated list, with vernacular names (indexed).

——— and **Jahandiez, Émile.** Essai de bibliographie botanique marocaine. Bul. Soc. Sci. Nat. Maroc **2**: 71–86. 1922.
List of 6 titles on vernacular names and 182 (chronologically arranged) on geographical and economic botany.

Jahandiez, Émile, and **Maire, René.** Catalogue des plantes du Maroc (spermatophytes et ptéridophytes) . . . Avec la collaboration de: J. A. Battandier . . . L. Ducellier . . . L. Emberger . . . P. Font-Quer . . . 3 v. (xl, 913 p.). 25.5 cm. Alger, 1931–34.
Geographical divisions, bibliography of 415 titles; annotated list of vascular plants, with localities and extralimital range. Includes Spanish as well as French Morocco.

Local

Boitel, M. Liste des plantes recueillies autour de Rabat et de Salé (Maroc). Bul. Soc. Sci. Nat. Maroc **1**: 41–50. 1921.
Annotated list of phanerogams.

Bonnet, Edmond. Florule de Dar-el-Beïda (Maroc). Naturaliste **11**: 194–195, 203. 1889.
Unannotated list of plants collected by Alphonse Mellerio in 1887–88. Dar-el-Beïda is also known as Casablanca.

——— Florule d'El Araich (Maroc). Naturaliste **7/9**: 403–404. 1887.
List of 155 vascular plants collected in 1886 by Alphonse Mellerio. El Araich, also known as Larache, is on the west coast, 87 km. southwest of Tangiers. (Spanish Morocco.)

Braun-Blanquet, Josias. Zur kenntnis der vegetationsverhältnisse des Grossen Atlas. Vierteljahrsschr. Naturf. Gesell. Zürich **73** (Beibl. 15, Festschrift Hans Schinz): 334–357. fig. 5–7, pl. 12–13. 1928.
Plant formations, etc.; list of species not previously recorded.

——— and **Maire, René.** Études sur la végétation et la flore marocaines. Comptes-rendus des herborisations de la Société Botanique de France, session du Maroc, 1921. Mém. Soc. Sci. Nat. Maroc **8**: 5–244. 10 pl., map. 1924.
Plant geography and associations; list of cellular and vascular plants, with localities. (From Mogador to Tanant and Demnat, north to Mazagan and Casablanca, then east through Fez to beyond Tlemsen.)

Caballero, Arturo. Datos botanicos del territorio de Ifni. Trab. Mus. Nac. Cien. Nat. y Jard. Bot. Madrid, Ser. Bot. no. 28. 36 p. 3 pl. 1935; no. 30. 33 p. 7 pl. 1935.
Botanical explorations of author in 1934; annotated list of vascular plants collected. Unfinished; includes Sympetalae, part of Choripetalae. (Spanish possession on west coast, about 29° N Lat.)

——— Enumeración de las plantes herborizadas en el Rif. Mem. R. Soc. Españ. Hist. Nat. **8**: 241–292. 1914.
Itinerary; list of 395 vascular plants collected by the author in 1912, with localities and occasional annotations. (Spanish Morocco.)

——— Excursión botánica a Melilla en 1915. Trab. Mus. Nac. Cien. Nat. Madrid, Ser. Bot. no. 11. 39 p. 2 pl. 1917.

List of vascular plants collected by author, with localities and frequent annotations. (Spanish Morocco.)

Dubard, Marcel, and Cayla, Victor. Liste de quelques plantes utiles du Maroc d'après les documents rapportés par M. de Gironcourt. Agr. Prat. Pays Chauds **9** (1) : 95–106. 1909.

Annotated list of 134 species, with Arabic names.

Emberger, Louis. Notes sur la végétation des montagnes du Tadla autour de Ksiba. Bul. Soc. Sci. Nat. Maroc **8**: 216–219. 1928.

General features of flora; list of vascular plants collected by author in 1927. (Moyen-Atlas.)

——— Recherches botaniques et phytogéographiques dans le Grand-Atlas oriental (massifs du Ghat et du Mgoun). Mém. Soc. Sci. Nat. Maroc no. 33. 49 p. 2 pl. 1932.

Botanical explorations, general features of flora, plant formations; list of more important vascular plants collected, with data.

——— and **Maire, René.** Spicilegium rifanum. Mém. Soc. Sci. Nat. Maroc no. 17. 59 p. 7 pl. 1927.

General features of flora; list of cellular and vascular plants collected by authors in 1926 in the eastern "Atlas rifain."

Font y Quer, Pío. Nota sobre la flora subalpina de la cumbre del Lexhab (Marruecos). Mem. Acad. Cien. y Artes Barcelona **22**: 335–352. 1931.

Author's expeditions in 1929–30, general features of flora; list of species collected, arranged in ecological groups.

Gandoger, Michel. Florule de Ceuta (Maroc). Bul. Soc. Bot. France **54**: 77–81. 1907.

Unannotated list of vascular plants collected by author. (Spanish Morocco.)

Hooker, J. D., and Ball, John. Journal of a tour in Marocco and the Great Atlas. xvi, 489 p. illus., plates, map. 22.5 cm. London, 1878.

Includes chapter by J. D. Hooker "On some of the economic plants of Marocco" (p. 386–404), and by J. Ball "On the mountain flora of two valleys in the Great Atlas of Marocco" (p. 423–446) with table showing distribution.

Humbert, Henri. Végétation de l'Atlas saharien occidental et additions à l'étude botanique de l'Ari Ayachi. Bul. Soc. Hist. Nat. Afrique Nord **19**: 204–240. pl. 20–27. 1928.

General features of flora, with lists of principal vascular plants of different localities; list of rarer species, with localities; list of vernacular names.

——— Végétation du Grand Atlas marocain oriental. Exploration botanique de l'Ari Ayachi. Bul. Soc. Hist. Nat. Afrique Nord **15**: 147–234. pl. 2–14. 1924.

Geography, geology, plant formations; list of alpine and subalpine species collected by author in Ayachi in 1923, with localities and frequent annotations.

Jahandiez, Émile. Contributions à l'étude de la flore du Maroc. Missions de l'Institut scientifique cherifien en 1921. I. Recherches botaniques dans le Tadla. II. Herborisations dans le Moyen Atlas. III. Recherches botaniques dans le Grand Atlas. IV. Liste méthodique des espèces recueillies. Mém. Soc. Sci. Nat. Maroc [no. 4] (v. 3, no. 1). 123 p. 9 pl. 1923.

Includes (chapter 4) list of vascular and cellular plants collected, with references and localities.

——— and **Weiller, Marc.** Herborisations dans le sud-ouest marocain et l'Anti-Atlas. Cavanillesia **5**: 5–23. 1932.

Itinerary; list of vascular plants collected in 1931 between Mogador and Agadir, in the Province of Sous, and in the Anti-Atlas, with localities.

Joly, Alexandre. Liste des espèces végétales récoltées à Tétuan (Maroc) pendant l'hiver et le printemps de l'année 1905. Bul. Soc. Hist. Nat. Afrique Nord **4**: 108–112. 1912.

List of vascular plants collected, arranged by habitats. (Spanish Morocco.)

——— La végétation dans les Beni Znassen (Maroc). Compt. Rend. Assoc. Franç. Avanc. Sci. **39** (2) : 86–93. 1911.

General features of flora; list of vascular plants collected by author in 1908.

Lester-Garland, L. V. The botany of the Maroccan Middle Atlas. Jour. Bot. **58**: 97–101. 1930.

List of species collected in 1919 by Capt. Hubert Lynes near Azrou at 4,000 to 6,000 feet altitude.

Litardière, René de, and Maire, René. Contributions à l'étude de la flore du Grand Atlas. Mém. Soc. Sci. Nat. Maroc v. 4, no. 1 (i. e. no. 6). 32 p. 6 fig. 1924.—Contributions à l'étude de la flore du Maroc. l. c. no. 26. 56 p. 6 fig. 1930.

List of 205 vascular plants, rare or new in the area, with data and frequent annotations.

Lowe, R. T. A list of plants observed or collected at Mogador and in its immediate environs during a few days' visit to the place in April 1859; with notes and observations. Jour. Linn. Soc. Bot. 5: 26–45. 1861 (1860).

General features of flora; list of 177 vascular plants, with notes on many of the species.

Maire, René. Études sur la végétation et la flore du Grand Atlas et du Moyen Atlas marocains. Mém. Soc. Sci. Nat. Maroc no. 7. 220 p. 16 pl. 1924.

Botanical explorations (with references), physiography, plant associations, phytogeography; list of cellular and vascular plants collected by author, with localities.

——— Note sur la flore du Siroua. Bul. Soc. Hist. Nat. Afrique Nord **15:** 52–53. 1924.

Brief list of phanerogams collected by Prof. L. Gentile in 1923, with localities. This is the first information on the flora of this mountain mass, which connects the Grand Atlas and the Anti-Atlas.

——— La végétation de l'Anti-Atlas oriental et du Djebel Bani. Bul. Soc. Sci. Nat. Maroc **13:** 156–164. 1933.

Explorations of author in 1932; general features of flora, with included lists of species.

——— La végétation des montagnes des Glaoua. Bul. Soc. Sci. Nat. Maroc **14:** 140–146. 1934.

Running account of vegetation, with enumeration of principal species. The Glaoua Mountains are between the central and the eastern Grand Atlas.

Mas y Guindal, Joaquín. Materiales para la flora de Marruecos. Cavanillesia **1:** 97–102, 131–133. 1928; **2:** 164–166. 1930; **3:** 97–99. 1931; **4:** 81–87. 1931; **5:** 178–180. 1932; **6:** 29–32. 1933.

Includes unannotated lists of vascular plants from (v. 1) Yebala, Beni Ider and Ben Karrik, (v. 2) Larache, (v. 3) Castillejos, (v. 4) Rincón de Medik, (v. 5) Akarrat, El Mesin and Dra-el-Asef, (v. 6) Ilguersa, Zinat, and other localities. (Spanish Morocco.)

——— Notes sur la flore de Tetouan et de ses environs. Trav. Off. Natl. Mat. Prem. Vég. (Paris) Notice 37. 46 p. 1932. (Reprinted from L'Hygiène sociale no. 77.)

Botanical explorations, geography, general features of flora; list of vascular plants collected by author, with localities; bibliography. (Spanish Morocco.)

——— Plantas medicinales del Protectorado Español de Marruecos. Ceuta, 1929.

Annotated list, with range, uses, vernacular names, etc. Not seen; notice from Cavanillesia **3:** 70. 1930. (Spanish Morocco.)

Moreau. Contribution à l'étude de la flore de la Chaouia. Arch. Méd. et Pharm. Mil. **62:** 625–633. 1913.

General features of vegetation; unannotated list of about 500 vascular plants collected by author, nearly all in vicinity of Casablanca.

Murbeck, Svante. Contributions à la connaissance de la flore du Maroc I–II. Lunds Univ. Årsskr. n.f., avd. 2, v. 18, no. 3. 76 p. 4 fig., 12 pl. 1922; v. 19, no. 1. 68 p. 8 fig., 7 pl. 1923.

List of vascular plants collected by author in 1921, mostly in vicinity of Marrakech and in the Grand Atlas, with data and frequent annotations.

Pau, Carlos. Plantas del norte de Yebala (Marruecos). Mem. R. Soc. Españ. Hist. Nat. **12:** 263–401. 1924.

Author's itinerary, bibliography; annotated list of vascular plants. (Spanish Morocco.)

Pitard, C.-J. Contribution à l'étude de la flore du Maroc. 80 p. 25.5 cm. Paris, 1931.

List of new or rare species collected by author in 1911–13, as well as by Lieut. Mouret in 1913. Includes sections on mosses, fungi, and lichens by other authors.

Sennen, *frère,* and **Mauricio,** *frère.* Catálogo de la flora del Rif Oriental y principalmente de las cabilas limítrofes con Melilla. xv, 159 p. map. 25.5 cm. Melilla, [1933?]

Annotated list of vascular plants. (Spanish Morocco.)

Sietti, Henri. Nouvelle contribution à l'histoire naturelle de l'île d'Alborán. Bul. Soc. Sci. Nat. Maroc **13:** 10–22. 3 pl. 1933.

Geography, scientific explorations, physiography, geology, etc.; list of lichens and phanerogams (about 10 species); zoology. Alborán is 83 km. south of Point Guardas Viejas (Spain) and 54 km. north of Cape Tres Forcas. (Spanish Morocco.)

Vidal y López, Manuel. Flórula del campamento legionario de Dar Riffien (Cabila de Anyhera). Bol. Soc. Ibérica Cien. Nat. **25:** 190–205. 1926.

Unannotated list of phanerogams collected by author and others in 1921. (Spanish Morocco.)

MOZAMBIQUE

See also Angola (Ficalho); Ethiopia (Engler, Ueber die hochgebirgsflora); South Africa; Tanganyika (Engler, Die pflanzenwelt Ost-Afrikas); West Africa (Oliver).

General

Gomes de Almeida, A. Plantas venenosas e medicinas dos indígenas de Moçambique. Bol. Agr. e Pec. Moçambique **1930** (1/2): 9–29. 1930.

Annotated list, with vernacular names; bibliography.

Sim, T. R. Forest flora and forest resources of Portuguese East Africa. vi, 166 p. 100 pl. 25×30.5 cm. Aberdeen, 1909.

Physiography, general features of flora and agriculture, forests, etc.; briefly descriptive flora of woody plants, with vernacular names; timbers, barks, oils, etc.

Local

Maugham, R. C. F. Zambesia: a general description of the valley of the Zambesi River, from its delta to the river Aroangwa, with its history, agriculture, fauna, and ethnography. xiv, 408 p. plates, port., map. 22.5 cm. London, 1910.

Includes (p. 163–195) chapter on "Zambesian flora," with general account of vegetation and of cultivated plants, and unannotated "List of plants, etc., observed in Zambesia" (p. 188–195), the families arranged alphabetically, the botanical names without authorities.

Peters, W. C. H., and others. Naturwissenschaftliche reise nach Mossambique . . . in den jahren 1842 bis 1848 ausgeführt. Botanik. 2 v. 60 (i. e. 61) pl. 51 cm. Berlin, 1862–64.

List of vascular plants, with localities and vernacular names, by various botanists.

Pires de Lima, Américo. Subsídios para o estudo da flora de Moçambique. Espermáfitas do litoral-norte. Brotéria Sér. Bot. **19:** 107–143. 1921: **20:** 5–11. 1922. —2a série . Bol. Soc. Broteriana II, **2:** 126–152. 1924.

Annotated list of 150 vascular plants, with descriptions of most species.

Rendle, A. B., Baker, E. G., Moore, S. Le M., and **Gepp, A.** A contribution to our knowledge of the flora of Gazaland: being an account of collections made by C. F. M. Swynnerton . . . Jour. Linn. Soc. Bot. **40:** 1–245. 7 pl. (incl. map). 1911.

Account of expedition (by Swynnerton); annotated list of vascular and cellular plants collected "chiefly in the high country which forms the boundary between eastern Rhodesia and Portuguese territory."—The notes on useful plants are brought together in **Jackson, J. R.** Notes on the economic flora of Gazaland. Gard. Chron. III, **51:** 283–284, 296–297, 321–322, 332–333. 1912.

Schinz, Hans. Plantae Menyharthianae. Ein beitrag zur kenntnis der flora des unteren Sambesi. Denkschr. Math.-Naturw. Kl. K. Akad. Wiss. Wien **78:** 367–445. 1905. (Reprinted, 79 p.)

Biographical sketch and itinerary of Menyharth, notes on cultivated plants, climate (by J. Fényi); list of cellular and vascular plants, with localities, based

241306°—42——4

on collections by Menyharth made mainly in Boroma, and on another collection by the Haynald Observatory at Zumba, Boroma, and Quelimane.

Schinz, Hans, and Junod, H. A. Zur kenntnis der pflanzenwelt der Delagoa-Bay. Bul. Herb. Boissier **7**: 869–892. 1899.—Mém. Herb. Boissier **10**: 25–75. 1900.—I. nachtrag. Bul. Herb. Boissier II, **3**: 653–662. 1903.

Climate, phenology, general features of flora, bibliography; list of vascular plants, with localities and vernacular names. Forms nos. XI, XII, XV of Schinz's "Beiträge zur kenntnis der Afrikanischen flora" and nos. IX, XI, XX of Mitteilungen aus dem Botanischen Museum der Universität Zürich.

Sousa, A. de Figuèiredo Gomes e. Flora de Moçambique. Lista de algumas plantas classificadas do herbário da Direcção de agricultura. Bol. Agr. e Pec. Moçambique **1932**: 119-129. 1933.

Annotated list of phanerogams; obviously very far from a complete flora.

NATAL

See also Orange Free State (Markötter); South Africa.

General

Bews, J. W. An introduction to the flora of Natal and Zululand. vi, 248 p. 21 cm. Pietermaritzburg, 1921.

Botanical explorations, bibliography, keys to families and genera, with briefly annotated list of 3,786 spermatophytes; vernacular names (indexed).

Henkel, J. S. The woody plants of Natal and Zululand. A field book of the woody plants . . . xii, 252 p. 2 pl. 22 cm. Durban and Pietermaritzburg, 1934.

Glossary; descriptive flora of woody plants in form of artificial key, with localities; no systematic list; bibliography.

Wood, J. M. A handbook to the flora of Natal. 202 p. 21.5 cm. Durban, 1907.

Keys to families and genera and (pteridophytes only) to species; enumeration of species (3,447 plus 83 varieties) with indication of altitudinal range.

——— Indigenous food plants. Rpt. Colon. Herb. (Natal) **1900**: 12–24. 1901.

Annotated list, the plants classified by uses.—See also his notes on medicinal plants, l. c. **1894**: 4–9. 1895. (List alphabeted by vernacular names, with uses.)

——— Natal plants. 6 v. 600 pl. 28 cm. Durban, 1898–1912.

Plates with descriptive letterpress. M. S. Evans joint author of vol. **1**.

——— Revised list of the flora of Natal. Trans. So. African Phil. Soc. **18**: 121–280. 1908.—Supplement containing Apocynaceae, Asclepiadeae, corrections and additions. Trans. Roy. Soc. So. Africa **1**: 453–472. 1910.—Addendum . . . l. c. **3**: 47–60. 1913.

Plant zones; annotated list of 3,308 species and 83 varieties of vascular plants (in first part).

Local

Bews, J. W. The plant ecology of the Drakensberg Range. Annals Natal Mus. **3**: 511–565. pl. 36–39. 1917.

Includes lists of species grouped by habitat.

Bryant, A. T. Zulu medicine and medicine-men. Annals Natal Govt. Mus. **2**: 1–103. 1909.

Includes annotated list (p. 77–103) of plants discussed, with vernacular names, uses, etc.

Forbes, H. M. L. An account of the flora of the Malvern district. So. African Jour. Nat. Hist. **2**: 195–208. 1920.

Physiography, etc.; running account of habitats, with lists of species (not in systematic order). (A suburb of Durban.)

——— The flora of Isipingo. So. African Jour. Sci. **18**: 348–358. 1922.

Includes running lists of species, classed by habitats. (Near Durban.)

——— A note on the flora of Salisbury Island (Durban Bay). So. African Jour. Sci. **20**: 304–308. pl. 7–8. 1923.

Plant formations; list of species collected, grouped by habit.

Wood, J. M. Flora of Durban and vicinity. List of native plants observed within a radius of five miles from Durban. Rpt. Natal Bot. Gard. **1887** (Rpt. Colon. Herb.): 1–15. 1888.—Additional list . . . l. c. **1888**: 19–20. 1889.—Additions . . . Ann. Rpt. Colon. Herb. **1892**: 5–8. [1893.]
List of dicotyledons and gymnosperms, with localities. The 1893 paper lists the monocotyledons and vascular cryptogams.

NIGERIA

See also West Africa.
The following publication has not been available:
Anonymous. Vocabulary of Nigerian names of trees, shrubs, and herbs. Lagos, 1936. (Cited from Dalziel, J. M., The useful plants of west tropical Africa, p. xi. 1937.)

General

Ainslie, J. R. A list of plants used in native medicine in Nigeria. Inst. Paper Imp. Forestry Inst. Oxford no. 7. 92, 9, 7 p. 1937.
Annotated list of 366 species, with vernacular names, uses, and methods of use; list of diseases, with references to species used as remedies. (Processed publication.)

Holland, J. H. The useful plants of Nigeria . . . Bul. Misc. Inform. Kew, Add. Ser. 9. 4 pts. (vi, 963 p.) 2 plans, map. 1908–22.
History of colony, physiography, hydrography, climate, list of botanical collectors, botanical gardens, agriculture, etc.; annotated systematic list of useful vascular plants, with references and vernacular names (indexed); bibliography of "books and papers of general interest for west Africa."

Hooker, J. D., and Bentham, George. Flora nigritiana; or, a catalogue of the plants of the river Niger, the island of Fernando Po, and adjacent parts of western tropical Africa; from the collections of Dr. Th. Vogel: to which are added those of Mr. G. Don and other travellers. *In* **Hooker, W. J.,** ed. Niger flora . . . p. 199–577. pl. 17–50. London, 1849.
Annotated list of 974 phanerogams, with localities and general range.

Local

Dalziel, J. M. A Hausa botanical vocabulary. 119 p. 23 cm. London, 1916.
Alphabetical list of vernacular names for plants, foods, etc., with botanical equivalents and notes; index of botanical genera.

——— Notes on the botanical resources of Yola Province, Northern Nigeria. Bul. Misc. Inform. Kew **1910**: 133–142. 1910.
Running account of useful plants, with vernacular names.

——— Vegetable products of Kontagora Province, Northern Nigeria. Bul. Imp. Inst. **5**: 255–266. 1907.
Running account of principal useful wild plants, with vernacular names and uses.

Dawodu, T. B. A provisional list of the indigenous and naturalised flowering plants of the town and island of Lagos and Etube Metta district. 22, ix p. 17.5 cm. Etube Metta, 1902.
Unannotated systematic list, with vernacular names.

Foster, E. W. Notes on Nigerian trees and plants. 69 p. 21.5 cm. Guildford [Eng.], [1914].
Annotated systematic list, with vernacular names and uses; refers mostly to woody plants.

Hutchinson, John. A contribution to the flora of Northern Nigeria. Plants collected on the Bauchi Plateau by Mr. H. V. Lely. Bul. Misc. Inform. Kew **1921**: 353–407. 10 fig. 1921.
Phytogeography; annotated list of about 400 vascular plants; general range given.

Kennedy, J. D. Forest flora of Southern Nigeria. 242 p. 25 cm. Lagos, 1936.
Forest types; annotated systematic list, with vernacular names, brief descriptions, uses.

Kew. Royal Gardens. Indigenous plants of Yoruba-land. Bul. Misc. Inform. Kew **1891**: 206–219. 1891.

Tabular list of 147 species, with Yoruba names (and meanings), localities, and notes on uses; based on collections by Alvan Millson.

——— The timbers of Southern Nigeria. Bul. Misc. Inform. Kew **1908**: 189–195. 1908.

Annotated list, mainly by H. N. Thompson.

Lely, H. V. The useful trees of Northern Nigeria. xii, 128 p. (incl. illus., pl.). 27.5 cm. London, 1925.

Descriptive account of 120 trees, alphabetically arranged by botanical names, with vernacular names and uses; figure of leaves and flowers of each species; table showing flowering season of each species.

Rendle, A. B., Baker, E. G., Wernham, H. F., and **Moore, S. LeM.** Catalogue of the plants collected by Mr. and Mrs. P. A. Talbot in the Oban district, South Nigeria. x, 157 p. 17 pl. 22.5 cm. London, 1913.

Physiography, geology, general features of flora; descriptions of new species; systematic list of species, with collectors' numbers and general range.

NORTHERN RHODESIA

See also Nyasaland (Burkill); West Africa (Oliver).

Local

Fries, R. E. Botanische untersuchungen. viii, 354, xi p. 40 fig., 22 pl., map. 32.5 cm. Stockholm, 1914–16.—Ergänzungsheft. 135 p. 18 fig., 16 pl. 32.5 cm. Stockholm, 1921. (Wissenschaftliche ergebnisse der schwedischen Rhodesia-Kongo-expedition 1911–1912 unter leitung von Eric graf von Rosen. bd. 1.)

Annotated list of vascular plants. The "Ergänzungsheft" includes notes on general features of vegetation, etc., and (p. 71–78) a "Florula bangweolensis" (Lake Bangweolo region, Northern Rhodesia). The expedition traveled from south of Lake Bangweolo along the east side of the long chain of lakes to Gondokoro in the Sudan.

NYASALAND

See also West Africa (Oliver).

General

Burkill, I. H. List of the known plants occurring in British Central Africa, Nyasaland, and the British territory north of the Zambesi. *In* **Johnston, H. H.** British Central Africa . . . 2d ed. London, 1898. p. 233–284.—A supplementary list . . . p. 284a–284l.

Botanical explorations; list of vascular and cellular plants, with localities and names of collectors; nearly 2,100 species, of which nearly 1,900 are phanerogams. (Nyasaland and Northern Rhodesia.)

Burtt Davy, Joseph, and **Hoyle, A. C.,** eds. Nyasaland Protectorate. Compiled by the Imperial Forestry Institute in collaboration with the Forest Department of Nyasaland. 111 p. 22 cm. Oxford, 1936. (Check-lists of the forest trees and shrubs of the British Empire. no. 2.)

Forest types; list of woody plants, with vernacular names (indexed), the families arranged alphabetically; list of exotics.

Local

Britten, James, and others. The plants of Milanji, Nyasa-land, collected by Mr. Alexander Whyte . . . Trans. Linn. Soc. Bot. II, 4: 1–67. pl. 1–10, map. 1894.

Physiography, etc. (by W. Carruthers); list of vascular and cellular plants, with references, localities, and extralimital range.

ORANGE FREE STATE

See also South Africa.

Local

Markötter, E. I. 'n Plantegeografiese skets en die flora van Witzieshoek, O. V. S.; Oliviershoekpas, Natal; en Koolhoek, O. V. S. Annals Univ. Stellenbosch v. 8, sect. A, no. 1. 50 p. 1 fig. (map). 1930.

Topography, climate, plant formations, ecology; annotated list of vascular plants, with citation of exsiccatae.

PORTUGUESE GUINEA

See also Angola (Ficalho); West Africa.

General

Sousa, A. de Figuèiredo Gomes e. Notas sôbre a flora da Guiné portuguesa. Bol. Agência Geral Colôn. Portugal **5** (44): 99–139. illus. 1929.

Physiography, soils, climate, general features of flora; annotated list of useful wild and cultivated plants, with vernacular names.

RIO DE ORO

General

Almonte, Enrique d'. Ensayo de una breve descripción del Sahara español. Bol. R. Soc. Geog. (Madrid) **56**: 129–347. plates, 3 maps. 1914.

Sketch of geography, hydrography, geology, climate, fauna, bibliography, etc.; includes (p. 216–235) a chapter "Indicaciones relativas á la flora," with list of known vascular plants, with localities and vernacular names, and remarks on general features of vegetation. Covers the colony of Rio de Oro (proper), the protectorate, and occupied territory, reaching from 20° 46′ to 29° N. Lat.

RUANDA AND URUNDI

See Belgian Congo (Burtt; Cotton; Staner); Northern Rhodesia (Fries); Tanganyika (Meyer; Mildbraed); West Africa (Oliver).

SAHARA

See also French West Africa.

Bruno, Alessandro. I prodotti utili della vegetazione sahariana. Bol. Soc. Africana Italia **25**: 265–280. tab. 1906.

Annotated list of 57 useful plants, with vernacular names (on chart).

Chevalier, Auguste. Les productions végétales du Sahara et de ses confins nord et sud . . . Revue Bot. Appl. et Agr. Trop. **12**: 669–924. 1932.

Includes (p. 825–898) "Liste des plantes cultivées ou à cultiver ou spontanées et utilisées par les indigènes dans le Sahara et sur ses confins nord et sud," an annotated systematic list, with vernacular names, localities, and uses, preceded by chapters on general features of flora and on edible plants, woods, and various other useful plants, with bibliographies.

Diels, Ludwig. Beiträge zur flora der Zentral-Sahara und ihrer pflanzengeographie. Nach der sammelausbeute des freiherrn Hans Gyer von Schwappenburg. Bot. Jahrb. Engler **54** (Beibl. **120**): 51–155. pl. 1 (map). 1917.

Itinerary, annotated bibliography; annotated list of Characeae and vascular plants; plant geography, climate, general features of vegetation, elements of flora. The list itself incorporated into Maire, Études sur la flore et la végétation du Sahara central (1933). Covers in general the area from 2° to 12° E. Long., 21° to 29° N. Lat.

Dürkop, Erich. Die nutzpflanzen der Sahara. Beih. Tropenpfl. **4**: 157–204. 1903. (Preprinted as Inaug.-diss., with title: Die wirtschafts- und handelsgeographischen provinzen der Sahara, begründet durch nützliche pflanzen. 56 p. 1902.)

Geography, climate, general features of vegetation; running account of principal and minor useful plants, geographically grouped, with footnote references.

Gram, Karl (i. e. **Kai**). Karplantevegetationen i Moydir (Emmidir) i Centralsahara. 168 p. 45 fig. København, 1935. (Dissertation.)
Mainly ecological; includes annotated list of 180 vascular plants collected by author in 1922–23. (Not seen; cited from Bot. Centralbl. n. f., **28** (v. **170**) : 376. 1937.) (Near Ahaggar, about 25° 20′ N. Lat., 4° 20′ E. Long.)

Maire, René. [1.]–7. contribution à l'étude de la flore du Sahara occidental. Bul. Soc. Hist. Nat. Afrique Nord **13**: 24–28. 1922; **14**: 159–160. 1923; **16**: 87–97. 1925; **18**: 9–11. 1927; **25**: 10–20. 1934; **26**: 148–162. 1935; **27**: 344–354. 1936.
Lists of plants from various localities, with data and vernacular names. [No. 1] has title: Plantes récoltées par l'expédition Augiéras . . . (1920–1921); No. 6 (1935) is a "Florule du Zemmour" (see under Mauritania).

——— Études sur la flore et la végétation du Sahara central. I–II. Mém. Soc. Hist. Nat. Afrique Nord no. 3. 272 p. 22 fig., 36 pl., 2 col. pl., 2 maps. 1933. (Mission du Hoggar II.)
Botanical explorations, bibliography; annotated list of cellular and vascular plants known from the central Sahara, with localities, general range, notes on uses; lists of botanical names, with Tamachek (Touareg) equivalents, and of Tamachek vernacular names. Covers in general the area between 21° and 29° N. Lat. and 2° and 12° E. Long., with the southern boundary of Algeria as its southern limit.

Monod, Théodore. Mission saharienne Augiéras-Draper, 1927–1928. Phanérogames. (Liste des récoltes classées par ordre géographique.) Bul. Mus. Hist. Nat. Paris II, **4**: 756–774. 1 fig. 1932.
Phytogeographical regions, climate, etc.; unannotated lists of phanerogams collected by author at 15 different localities between Atakor–n–Ahaggar and the Niger River.

SÃO THOMÉ AND PRINCIPE

See also Angola (Ficalho); Cameroons (Hooker); Ethiopia (Engler, Ueber die hochgebirgsflora); West Africa.

General

Cardoso, João, jr. Plantas medicinaes das ilhas de S. Thomé e Principe. *In* his Subsidios para a materia medica e thérapeutica das possessoes ultramarinas portuguezas. v. 1, p. 141–142. Lisboa, 1902.
Unannotated alphabetical list of medicinal plants.

Local

Henriques, J. A. Contribuição para o estudo da flora d'Africa. Catalogo da flora de ilha de S. Thomé. Bol. Soc. Broteriana **4**: 129–221. illus., 8 pl. 1886; **5**: 196–220. 6 pl. 1887; **10**: 97–165. 1892.
Botanical explorations, physiography, general features of flora, cultivated plants; list of vascular and cellular plants, with localities, collectors, and general range. Musci by C. Müller, Hepaticae by F. Stephani, Fungi by G. Winter, Lichenes by Nylander. Subtitle of first part Flora de S. Thomé.—See also his A ilha de S. Tomé sob o ponto de vista historico-natural e agricola. l. c. **27**: 1–197. 83 fig. (incl. plates, maps), map. 1917. (List of cellular and vascular plants, p. 165–197, based on his earlier list.)

Moller, A. F. Einige medizinische pflanzen von S. Thomé (Westafrika). Ber. Deut. Pharm. Gesell. **7**: 352–355, 491–501. 1897.
Annotated list, with vernacular names and uses.

SENEGAL

See also French Guiana (Lanessan, Les plantes utiles, p. 335–350, 792–841); Gambia (Kew, Royal Gardens); West Africa.

General

Chevalier, Auguste. Nos connaissances actuelles sur la géographie botanique et la flore économique du Sénégal et du Soudan. *In* Lasnet, Alexandre, and others. Une mission au Sénégal . . . p. 197–267. 7 fig. Paris, 1900. (Exposition universelle de 1900—Les colonies françaises.) (Reprinted.)

Plant zones; running account of native and cultivated useful plants; bibliography.

Guillemin, J. A., Perrottet, Samuel, and Richard, Achille. Florae Senegambiae tentamen, seu historia plantarum in diversis Senegambiae regionibus a peregrinatoribus Perrottet et Leprieur detectarum. xi, 316 p. 72 pl. 25.5 cm. Parisiis, 1830–33.
Botanical explorations; annotated flora with vernacular names and diagnoses or descriptions. Only one volume was issued (Ranunculaceae—Myrtaceae). Covers Senegal and French Guinea.

Sambuc, Camille. Contribution à l'étude de la flore et de la matière médicale de la Sénégambie. 102 p. 26.5 cm. Montpellier, 1887. (Thèse.)
Botanical explorations, geography, languages, phytogeography; annotated list of medicinal plants, arranged by uses.

Sébire, Albert. Les plantes utiles du Sénégal: plantes indigènes, plantes exotiques. lxx, 341 p. 38 fig., 6 pl. 18.5 cm. Paris, [1899].
Agriculture; lists of vernacular names, classified by uses; running descriptive account of cultivated and wild useful plants, with vernacular names.

Vallot, Joseph. Études sur la flore du Sénégal. Bul. Soc. Bot. France **29**: 168–239. map. 1882. (Reprinted, 80 p. Paris, 1883.)
Annotated alphabetical list of 114 collectors, with bibliographies; general bibliography; list of works cited; gazetteer; list of explorers who died in tropical Africa; annotated list of vascular plants, with localities, exsiccatae, general range. Unfinished; includes only Ranunculaceae—Capparidaceae.

Local

Noury, M. F. Contribution à l'étude de la flore de la Sénégambie et du nord du Foutah-Djallon. Arch. Méd. Navale **51**: 199–210, 305–317, 366–384. 1889.
Author's itinerary, botanical explorations, soil, etc.; annotated list of plants collected, with frequent descriptions and notes on uses. The area referred to lies between Gambia and Rio-Nuñez.

Rançon, André. La flore utile du bassin de la Gambie. Bul. Soc. Géog. Com. Bordeaux II, **18**: 324–338, 353–382, 385–410, 417–442, 467–477, 496–508, 518–541, 545–558. map. 1895. (Reprinted with title: Étude botanique exotique. La flore utile . . . 160 p. 1895.)
Running account of wild and cultivated useful plants, grouped by uses, with vernacular names.

SIERRA LEONE

See also West Africa.
The following publication has not been available:
Lane-Poole, C. E. The trees, shrubs, herbs, and climbers of Sierra Leone. 159 p. Freetown, 1916. (Cited from G. P. Ahern and H. K. Newton, A bibliography on woods of the world. Scient. Contrib. Trop. Plant Res. Found, **10**: 59. 1928.)

General

Elliot, G. F. S. Sierra Leone. Reports on botany and geology. Colon. Rep. Misc. Gt. Brit. no. 3. 78 p. 1893.
Geology, climate, etc.; annotated classified list of useful plants, with vernacular names (indexed).

Sabine, Joseph. Some account of the edible fruits of Sierra Leone . . . from the journal and personal communication of Mr. George Don, A. L. S. Trans. Hort. Soc. London **5**: 439–466. col. pl. 1824.
Annotated list based on George Don's expedition in 1822; list of fruits recorded by Afzelius; list of cultivated fruits.

SOUTH AFRICA

South Africa, a term used in various senses, is here taken as the area south of the middle course of the Zambesi River, including South-West Africa, Bechuanaland Protectorate, Southern Rhodesia, southern Mozambique, and the countries south of them. The floras listed under this head are only those referring to all or most of this region, those relating to definite smaller areas being cited under their respective geographical subdivisions.

Goester, L. E. De geneesmiddelen van Groot Nederland. Overzicht der in Zuid-Afrika meest gebruikte plantaardige geneesmiddelen. Phar. Weekbl.

Nederland. **51:** 1019–1024, 1035–1041, 1107–1117, 1129–1140, 1200–1204, 1240–1249, 1265–1275. 1914.

Annotated systematic list, with vernacular names and uses, of principal medicinal plants of Cape of Good Hope, Natal, Orange Free State, and Transvaal.

Harvey, W. H., Sonder, W., and others. Flora capensis: being a systematic description of the plants of Cape Colony, Caffraria and Port Natal (and neighboring territories). 7 v. 23 cm. Dublin, 1859–65; London, 1896–1933.

Descriptive flora of spermatophytes, with keys, references, and citation of specimens. Vol. 1–3 (1859–65) by Harvey and Sonder; the remaining volumes (1896–1925) were edited by Dyer, and the limits were extended northward to the Tropic of Capricorn. The botanical regions adopted in the later volumes are discussed in the preface to v. 6. Vol. 5, sect. 2, supplement (1933), by A. W. Hill, contains the gymnosperms. The history of the work is discussed by Dyer in Kew Bull. **1925:** 289–293.

Macowan, Peter, and Bolus, Harry. Catalogue of printed books and papers relating to South Africa. Pt. 1. Botany. Trans. So. African Phil. Soc. **2:** 111–187. 1882.

List of 551 papers and books, arranged alphabetically by authors, with authors' dates and bibliographical annotations.—See also **Schönland, S.** South African botanical literature. Mem. Bot. Survey So. Africa **4:** 69–85. 1922. (Classified bibliography, mainly supplementary to Macowan and Bolus' "Catalogue.")

Marloth, Rudolf. Dictionary of the common names of plants with list of foreign plants cultivated in the open. 175 p. 22 cm. Cape Town, 1917.

Annotated alphabetical list of about 2,000 vernacular names (Dutch, English, and Kafir) with botanical equivalents; systematic list of scientific names, with vernacular equivalents; list of mixed vernacular and scientific names of foreign plants cultivated in the open in South Africa; bibliography. "Supplement" to his Flora of South Africa.

——— The flora of South Africa, with synoptical tables of the genera of the higher plants. 4 v. in 6. illus., plates (pt. col.). 29.5 cm. Capetown, 1913–32.

Semipopular account, with excellent illustrations, keys to families and genera, and mention of principal species; not a systematic flora. Contains much information on biology and adaptations of plants. The northern boundary of the area covered follows the Tropic of Capricorn from Walfisch Bay to the Limpopo River, then the latter to its mouth.

Phillips, E. P. A brief historical sketch of the development of botanical science in South Africa and the contribution of South Africa to botany. So. African Jour. Sci. **27:** 39–80. 1930.

Sketch of botanical explorations and publications, with much biographical information.

——— Economic plants of South Africa. Off. Year Book Union So. Africa **8** (1910–25): 47–53. map. 1927.

Annotated systematic list of 64 phanerogams, with vernacular names and uses.

——— The genera of South African flowering plants. 702 p. 25 cm. Cape Town, 1926. (Botanical survey of South Africa. Memoir no. 10.)

Synopsis of classification; systematic descriptions of families and genera, with keys and brief account of geographical distribution, with especial reference to South Africa. Covers the Union of South Africa and South-West Africa. Replaces Harvey, W. H. The genera of South African plants. ed. 2. 1868.—For additions see his Newly described and newly recorded genera . . . So. African Jour. Sci. **30:** 220–221. 1933.

Smith, Andrew. A contribution to South African materia medica, chiefly from plants in use among the natives. 3d ed., largely extended. viii, 238 p. 22 cm. Cape Town, [etc., 1895]. (1st ed., 18—; 2d ed., 1888.)

Annotated lists, classified by uses, etc., with vernacular names.

Steyn, D. G. The toxicology of plants in South Africa; together with a consideration of poisonous foodstuffs and fungi. xii, 631 p. 135 fig. 24 cm. [Johannesburg,] 1934. (South African agricultural series, vol. 13.)

General account of poisonous plants; annotated systematic list of poisonous plants, with vernacular names, range, toxicity, etc.; extensive bibliography.

Watt, J. M., and Breyer-Brandwijk, M. G. The medicinal and poisonous plants of southern Africa, being an account of their medicinal uses, chemical composition, pharmacological effects, and toxicology in man and animal. xx, 314 p. plates (pt. col.). 26 cm. Edinburgh, 1932.

Annotated systematic list of vascular and cellular plants, with vernacular names, uses, references, etc.; bibliography.

SOUTHERN RHODESIA

See also Bechuanaland Protectorate (Bremekamp and Obermeÿer); South Africa; West Africa (Oliver).

General

Eyles, Frederick. A record of plants collected in Southern Rhodesia. Trans. Roy. Soc. So. Africa **5:** 273–564. 1916.

List of 2,397 species of cellular and vascular plants, with references, localities, and collectors.

Local

Gibbs, L. S. A contribution to the botany of Southern Rhodesia. Jour. Linn. Soc. Bot. **37:** 425–494. pl. 17–20. 1906.

General features of flora; annotated list of vascular and cellular plants collected by the author in Matopo Hills and at Victoria Falls, Zambesi, Southern Rhodesia, in Aug.–Oct. 1905; bibliography.

Gilliland, H. B. Notes on the flora of Rhodesian Manicaland. I. Jour. So. African Bot. **4:** 143–156. pl. 41–45. 1938.

Annotated list of bryophytes and vascular plants not previously recorded from Manicaland, partly in collaboration with other authors; bibliography. Pt. 1 covers bryophytes, pteridophytes, and gymnosperms.

Monro, C. F. H. Some indigenous trees of Southern Rhodesia. Proc. Rhodesia Scient. Assoc. v. 8, pt. 2. 123 p. 1908.

Orthography of vernacular names; alphabetical list of vernacular names, with botanical equivalents and authorities; descriptive list of trees referred to.

Norlindh, Tycho, and Weimarck, Henning. Beiträge zur kenntnis der flora von Süd-Rhodesia. Bot. Notiser **1932:** 1–21. 6 fig. (incl. map). 1932; **1934;** 64–110. 10 fig. 1934; **1935:** 357–408. 21 fig. 1935; **1936:** 1–50. 13 fig. 1936; **1937:** 161–201, 419–446. 12 fig. 1937.

Itinerary, general features of flora; list of plants collected by the Swedish expedition to South Africa and Southern Rhodesia in 1930–31, with data, the families not in systematic order.

Steedman, E. C. A description of some trees, shrubs and lianes of Southern Rhodesia. xix, 191 p. (incl. 85 pl.). 7 pl. 25 cm. Bulawayo, 1933.

General features of flora, systematic list of species included; descriptive partial flora, with vernacular names and occasional notes on uses.

SOUTH-WEST AFRICA

See also Cape of Good Hope (Pearson); South Africa; West Africa (Oliver).

General

Dinter, Kurt. Index der aus Deutsch-Südwestafrika bis zum jahre 1917 bekannt gewordenen pflanzenarten. Repert. Spec. Nov. Fedde **15:** 77–92, 340–355, 426–433. 1917–19; **16:** 167–174, 239–247, 335–344. 1919–20; **17:** 185–192, 258–265, 303–311. 1921; **18:** 13–16, 248–256, 423–444. 1922; **19:** 93–96, 187–192, 235–240, 315–320. 1923–24; **20:** 314–316. 1924; **22:** 108–112, 375–383. 1925–26; **23:** 130–137, 227–236, 363–371. 1926–27; **24:** 13–16, 302–304, 366–368. 1927–28; **25:** 49–51, 209–210. 1928.

Alphabetical list of 2,708 vascular and cellular plants, with references to original description and localities and collectors' numbers.

———. Die vegetabilische veldkost Deutsch-Südwest-Afrikas. 47 p. 13 pl. 23.5 cm. Okahandja, 1912.

Annotated, briefly descriptive list of edible plants grouped by parts used, with vernacular names. The South African word "veldkost" covers all edible wild plants and animal products.

Schinz, Hans. Die pflanzenwelt Deutsch-Südwest-Afrikas. Bul. Herb. Boissier **4** (App. 3): 1–57. 1896; **5** (App. 3): 59–101. tab. 1. 1897; Mem. Herb. Boissier **1**: 103–131. 1900.

List of cellular and vascular plants, with citation of exsiccatae and localities and extralimital range; intended to include all known up to August 30, 1896. Incomplete; reached Crotalaria (Engler-Prantl system).

Local

Bolus, Frank, Bolus, Louisa and Glover, Ruth. Flowering plants and ferns collected on the Great Karasberg by the Percy Sladen Memorial Expedition, 1912–1913. Annals Bolus Herb. **1**: 9–19, 72–75, 97–114. pl. 1–6, 8, 13–14. 1914–15.

List with localities and collecting numbers. Preceded (p. 1–8) by an account of the topography, climate, etc., by H. H. W. Pearson. (In Great Namaqualand.)

Range, Paul. Die flora des Namalandes I–XIV. Repert. Spec. Nov. Fedde **30**: 129–158. pl. 129 (map). 1932; **33**: 1–22. 1933; **35**: 35–42. 1934; **36**: 1–19, 97–109, 241–264. 1934; **38**: 122–130, 256–280. 1935; **39**: 55–66, 283–287. 1935–36; **43**: 251–256, 262–271. 1938; **44**: 101–114. 1938; **45**: 320. pl. 241–244 (maps). 1938.

Botanical explorations, geology, climate, plant regions, general features of flora, bibliography; systematic list of vascular plants (including cultivated species), with references, citation of exsiccatae, extralimital range, occasional vernacular names and uses; short conspectus of cellular plants, fossil flora, statistics (1,646 species), alphabetical list of collectors with brief biographical information, list of localities. Page 320 (vol. 45) is table of contents.

Seiner, Franz. Ergebnisse einer bereisung der Omaheke in den jahren 1910–1912. Mitt. Deut. Schutzgeb. Wiss. Beih. Deut. Kolonialbl. **26**: 225–316. 5 pl., 4 maps. 1913.

Includes (p. 238–255) a considerable list of vascular plants, partly annotated, grouped into 73 plant formations. The area covered is northeastern South-West Africa.

SPANISH GUINEA

See also Cameroons (Hooker; Mildbraed, Botanik); Ethiopia (Engler, Ueber die hochgebirgsflora); West Africa.

Local

Hooker, J. D. On the vegetation of Clarence Peak, Fernando Po; with descriptions of the plants collected by Mr. Gustav Mann on the higher parts of that mountain. Jour. Linn. Soc. Bot. **6**: 1–23. 1862 (1861).

Phytogeography; list of vascular plants, with localities and general range.

Mildbraed, Johannes. Neue arten von der Insel Annobon. Notizbl. Bot. Gart. Berlin **13**: 697–705. 1937.

Descriptions and reductions to synonymy of the undescribed new species given in his list of the plants of Annobon in Wissenschaftliche ergebnisse der zweiten deutschen Zentral-Afrika-expedition 1910–11, bd. 2, Botanik. 1922. p. 157–164 (see under Cameroons).

SWAZILAND

See South Africa; Transvaal (Burtt Davy and Pott, A first check-list; Burtt Davy, A manual).

TANGANYIKA

See also Anglo-Egyptian Sudan (Grant); Ethiopia (Engler, Ueber die hochgebirgsflora); Northern Rhodesia (Fries); West Africa (Gürke; Oliver); Zanzibar (Sacleux).

General

Bally, P. R. O. Heil- und giftpflanzen der eingeborenen von Tanganyika. Beih. Repert. Spec. Nov. Fedde v. 102. 87 p. 19 pl. 1938.

General account of native medicine, arrow poisons, fish and insect poisons, drinks and stimulants; annotated systematic list of 345 vascular plants, with native names and uses; list classified by uses; indices.

Bally, P. R. O. Native medicinal and poisonous plants of East Africa. Bul. Misc. Inform. Kew **1937:** 10–26. 1937.
Briefly annotated list of vascular plants, with vernacular names and uses. Refers principally to Tanganyika Territory.

Braun, Karl. Gewürze und aromatika der völker des früheren Deutsch-Ostafrika. Heil- u. Gewürz-Pflanz. **11:** 55–86, 113–141. 1928.
Annotated list of 80 species, with vernacular names (indexed) and uses, alphabetically arranged by scientific names; bibliography.

——— Pflanzen aus Deutsch-Ostafrika, ihre namen und verwendung bei den eingeborenen. Arch. Pharm. **262:** 368–381. 1924; **263:** 123–139. 1925; **265:** 45–61. 1927.
Annotated list of 91 species, alphabetically arranged, with vernacular names, notes on uses and references.

Busse, Walter. Ueber heil- und nutzpflanzen Deutsch-Ostafrikas. Ber. Deut. Pharm. Gesell. **14:** 187–207. 2 pl. 1904.
Discussion of "hunger-foods," Strychnos, etc.; running account of medicinal and useful plants.

Engler, Adolf, ed. Die pflanzenwelt Ost-Afrikas und der nachbargebiete. Theil B. Die nutzpflanzen Ost-Afrikas. 535 p. 24 fig. 27 cm. Berlin, 1895.
Chapters on palms, grasses, bananas, edible and other useful plants, medicinal and ornamental plants, by various authors.

——— and others. Die pflanzenwelt Ost-Afrikas und der nachbargebiete. Theil C. Verzeichniss der bis jetzt aus Ost-Afrika bekannt gewordenen pflanzen. 433 p. 45 pl. 27.5 cm. Berlin, 1895.
Geographical regions (39), with mention of most important collectors; systematic list of cellular and vascular plants by various authors, with areal occurrence, extralimital range, and occasional annotations. Covers primarily Tanganyika (German East Africa), southern British East Africa, and Portuguese East Africa south to the Zambesi River, but indicates distribution of included species in other areas of Africa.

Greenway, P. J. A Swahili dictionary of plant names. xvi, 112 p. 16 cm. Dar es Salaam, 1937.
Bibliography; briefly annotated alphabetical list of vernacular names, with botanical equivalents (not indexed) and uses. "The Swahili language is spoken by the mixed races of Arabs, negroes and others who inhabit the east coast of Africa."

Peter, Albert. Flora von Deutsch-Ostafrika. Zusammenstellung der in Deutsch-Ostafrika beobachteten farnartigen gewächse und blütenpflanzen mit literatur-nachweisen, angabe der verbreitung auf der erde und bestimmungstabellen. Beih. Repert. Spec. Nov. Fedde v. 40, pt. 1–2, lfg. 3. plates. 1929–38.
Author's itineraries; annotated list of vascular plants, with keys, references, citation of localities, general range, occasional vernacular names and uses. Not yet completed. Lieferung 3 (pt. 2) reaches Basella; each part accompanied by separately paged Anhang containing full descriptions of the new species.

Sadebeck, Richard. Die tropischen nutzpflanzen Ostafrikas, ihre anzucht und ihre ev. plantagenbetrieb. Jahrb. Hamburg. Wiss. Anst. **9:** 203–228. 1891.
Account of the most important species.

Local

Engler, Adolf, and others. Die von W. Goetze am Rukwa-see und Nyassa-see sowie in den zwischen beiden seen gelegenen gebirgsländern, insbesondere dem Kinga-gebirge gesammelten pflanzen, nebst einigen nachträgen (durch * bezeichnet) zu Bericht III. Berichte über die botanischen ergebnisse der Nyassa-see- und Kinga-gebirgs-expedition . . . IV. *In* his Beiträge zur flora von Afrika XXII. Bot. Jahrb. Engler **30:** 239–445. 7 fig., pl. 4–22. 1901.
List of cellular and vascular plants, with data and occasional vernacular names.

——— and others. Die von W. Goetze und Dr. Stuhlmann im Ulugurugebirge, sowie die von W. Goetze in der Kisaki- und Khutu-steppe und in Uhehe gesammelten pflanzen. Berichte über die botanischen ergebnisse der Nyassa-see- und Kinga-gebirgs-expedition der Hermann- und Elise- geb. Heckmann-Wentzel-stiftung. III. *In* his Beiträge zur flora von Afrika XX. Bot. Jahrb. Engler **28:** 332–510. pl. 5–10. 1900.

Botanical explorations; list of cellular and vascular plants, with localities, vernacular names, and uses.

Engler, Adolf. Ueber die flora des gebirgslandes von Usambara, auf grund der von Herrn Carl Holst daselbst gemachten sammlungen. *In* his Beiträge zur flora von Afrika V. Bot. Jahrb. Engler **17**: 156–168. 1893.
Plant formations, with lists of species; no general list.

Heering, Wilhelm. Dr. E. Obst's botanische sammlung aus dem abflusslosen rumpfschollenland des nordöstlichen Deutsch-Ostafrika. Mitt. Geog. Gesell. Hamburg **29**: 205–216. 1915.
Annotated list of cellular and vascular plants, collected from Wakindiga-land and on the summit of Mount Ufiomi, both regions previously unknown botanically. Most of the specimens collected were lost.

Jaeger, Fritz. Das hochland der Riesenkrater und die umliegenden hochländer Deutsch-Ostafrikas. Ergebnisse einer amtlichen forschungsreise ins abflusslose gebiet des nördlichen Deutsch-Ostafrika 1906–07. Mitt. Deut. Schutzgeb. Wiss. Beih. Deut. Kolonialbl. Ergänzungsh. no. 4. iv, 133 p. 8 fig. 3 pl., 3 maps. 1911.
Includes (p. 88–94) unannotated list of 522 vascular plants collected by the expedition, grouped by localities and plant formations. The identifications were made at the Berlin Herbarium.

Merker, Moritz. Pflanzen der Massai-steppe im süden, westen und südwesten des Kilimandscharo und Meru, welche bei den Massais teils als medizinalpflanzen, teils als anderweitig nützliche oder schädliche pflanzen beachtung finden. Notizbl. K. Bot. Gart. Berlin **3**: 194–197. 1902.
Unannotated list of 85 vascular plants with Massai and botanical names.

Meyer, Hans. Ergebisse einer reise durch das zwischenseengebiet Ostafrikas 1911. Mitt. Deut. Schutzgeb. Wiss. Beih. Deut. Kolonialbl. Ergänzungsh. no. 6. 127 p. illus. (incl. maps), 8 pl., 3 maps. 1913.
Includes (p. 93–101) list of species from several localities in northern German East Africa collected by author and identified at the Berlin Herbarium.

Mildbraed, Johannes. Wissenschaftliche ergebnisse der Deutschen Zentral-Afrika-expedition 1907–1908 unter führung Adolph Friedrichs, Herzogs zu Mecklenburg. bd. II. Botanik. 718 p. illus., 78 pl. 27 cm. Leipzig, 1910–14.
Includes list of vascular and cellular plants, with localities, general range, and occasional annotations, by G. Brause, G. Hieronymus, R. Pilger, and others; not in systematic order. Part 7 is ecological and phytogeographical. The expedition worked in northwestern Tanganyika, Ruanda and Urundi, northeastern Belgian Congo, and Uganda.

Schweinfurth, Georg. Phanerogams. *In* **Höhnel, Ludwig von.** Discovery of Lakes Rudolf and Stefanie. v. 2, p. 350–368. London and N. Y., 1894.
List of 171 vascular plants, mostly from Kilimanjaro, with locality and altitude; lists of mosses, liverworts, and lichens by other authors.

Trotha, Hellmuth von. Der wald um Tabora in Deutsch-Ostafrika. Notizbl. K. Bot. Gart. Berlin **5**: 212–231. 1911.
Annotated list of 184 useful plants, with vernacular names.

TOGO

See also French West Africa (Chevalier); Gold Coast (Burtt Davy and Hoyle); West Africa.

General

Volkens, Georg. Die nutzpflanzen Togos. Notizbl. K. Bot. Gart. Berlin App. 22. 119 p. 60 fig. 1909–10.
Annotated list, classified by uses, with vernacular names.

Local

Büttner, Richard. Zur flora des Togolandes. Verzeichniss der von Dr. R. Büttner in den jahren 1890 und 1891 im Togohinterlande, zumeist in der nähe der Forschungsstation Bismarckburg gesammelten pflanzen. Mitt. Forschungsreis. u. Gelehr. Deut. Schutzgeb. **6**: 231–236. 1893.
Unannotated list of vascular and cellular plants from Adeliland.

Gaisser, Karl. Die produktion der eingeborenen des bizirkes Sokode-Bassari (Schutzgebiet Togo). Mitt. Deut. Schutzgeb. Wiss. Beih. Deut. Kolonialbl. **25**: 239–318. illus. 1912.

Includes account of wild and cultivated plants and animals of economic importance and alphabetical list of vernacular names of plants in the Tschaudjo language, with botanical equivalents and names in other native languages.

TRANSVAAL

See also South Africa; West Africa (Oliver).

General

Burtt Davy, Joseph. Alien plants spontaneous in the Transvaal. Rpt. So. African Assoc. Adv. Sci. **2**: 252–299. 1904.

Annotated list of 141 species; economic aspects, phytogeography.

——— A manual of the flowering plants and ferns of the Transvaal with Swaziland, South Africa. pt. 1–2 (xxxv, 529 p., incl. 80 fig., map.) 19.5 cm. London, 1926–32.

Glossary, key to families, etc.; annotated briefly descriptive flora of vascular plants, with vernacular names and citation of collectors' numbers. To contain about 4,500 species; pt. 1 contains Marattiaceae–Bombacaceae (Hutchinsonian sequence of families), pt. 2, Malvaceae–Umbelliferae.

——— Native trees of the Transvaal. Transvaal Agr. Jour. **5**: 413–433. pl. 69–75. 1907.—Additions and corrections . . . Rpt. Transvaal Dept. Agr. **1906/07**: 148–154. 1908.

Life zones, with lists of species; list of 269 trees, with localities and vernacular names.—See also his Notes on some Transvaal trees and shrubs. Bul. Misc. Inform. Kew **1908**: 145–175. 1908.

——— and **Pott, Mrs. Reno.** A first check-list of the flowering plants and ferns of the Transvaal and Swaziland. Annals Transvaal Mus. **3**: 119–182. 1912.—Additions and corrections to the recorded flora . . . So. African Jour. Sci. **9**: 343–356. 1913; **15**: 570–571. 1919.—Addendum to the First check-list . . . Annals Transvaal Mus. **6**: 119–135. 1920.

Unannotated list of 3,298 vascular plants (including the "Additions and corrections"). Mrs. Pott is joint author of "A first check-list . . ." and sole author of the "Addendum . . ."—See also **Bremekamp, C. E. B.** New or otherwise noteworthy plants from the northern Transvaal. l. c. **15**: 233–264. 1933.

Verdoorn, I. C. Edible wild fruits of the Transvaal. Bul. Dept. Agr. and Forestry Union So. Africa no. 185 (Plant Ind. Ser. no. 29). 55 p. 14 fig., 7 col. pl. 1938.

Annotated systematic list, with vernacular names, keys, and brief descriptions.

Local

Galpin, E. E. Botanical survey of the Springbok flats, Transvaal. Mem. Bot. Survey So. Africa no. 12. 100 p. plates. 1927.

Mainly ecological; includes (p. 58–79) briefly annotated list of 584 phanerogams.

Glover, P. E. A contribution to the ecology of the highveld flora. So. African Jour. Sci. **34**: 224–259. 16 fig. 1937.

Includes annotated list of vascular plants from Frankenwald, 12 miles from Johannesburg.

Obermeÿer, A. A. A preliminary list of the plants found in the Kruger National Park. Annals Transvaal Mus. **17**: 185–227. map. 1937.

Geography, climate, general features of flora; list of over 300 vascular plants, mostly collected by Land in 1932, with data and general range.

——— **Schweickerdt, H. G.,** and **Verdoorn, I. C.** An enumeration of plants collected in the northern Transvaal. Bothalia **3**: 223–258. 1 fig. (map). 1937.

List of vascular plants, with data, collected chiefly in vicinity of the farm "Zoutpan" at the western end of the Zoutpansberg Range.

Smuts, J. C., and **Hutchinson, John.** The vegetation of Schoemanskloof, eastern Transvaal. Bul. Misc. Inform. Kew **1933**: 417–427. 1933.

General features of flora (by Smuts); list of vascular plants collected by Smuts in June 1932, with occasional annotations (by Hutchinson).

Verdoorn, I. C. Notes on the vegetation of the Fountains Valley, Pretoria. So. African Jour. Sci. **26:** 190–194. 1929.

Geology, etc.; running partial list of vascular plants collected by author, arranged according to season.

TUNISIA

See also Algeria (Battandier and Trabut; Bouquet; Cosson; Ducellier and Maire; France, Comité regional d'Algérie; Lapie and Maige; Trabut).

General

Bocquillon-Limousin, Henri. Les plantes utiles de la Tunisie. Monde des Plant. **4:** 241–244, 260, 276–279, 288–289, 305–312. 1895.

Briefly annotated systematic lists of medicinal and other useful plants.

Bonnet, Edmond, and Barratte, Gustave. Catalogue raisonné des plantes vasculaires de la Tunisie. xlix, 519 p. 25 cm. Paris, 1896. (Exploration scientifique de la Tunisie.)—**Bonnet, E.** Additions et corrections . . . Jour. de Bot. **13:** 83–86. 1899.

Botanical explorations, physiography, geology, climate, cultivated plants, affinities of flora (by Doumet-Adanson); annotated list of vascular plants, with references, localities, and extralimital range.

Cuénod, Auguste. Les plantes médicinales de Tunisie I. Distribution géographique . . . *In* Comité régional des plantes médicinales de Tunisie. Les plantes médicinales de Tunisie. p. 9–35. Tunis, 1920.

Annotated list of principal wild species, alphabetically arranged by vernacular names.

Local

Bardin, L., *père.* Catalogue méthodique et raisonné des plantes vasculares de Carthage et de ses environs, avec indication des plantes médicinales indigènes. Revue Tunis. Inst. Carthage **5:** 95–110, 183–203, 372–378, 509–516. 1898; **6:** 96–106, 189–203, 353–359, 472–485. 1899.

Annotated list of 694 vascular plants based on author's collections, with notes on medicinal plants.

Burollet, P. A. La Sahel de Sousse, monographie phytogéographique. Annales Serv. Bot. Tunisie v. 4, fasc. 2. 270 p. 8 pl., map. 1927.

Botanical explorations, physiography, climate, geology, water, etc.; mostly unannotated list of 1,038 species and varieties of vascular and cellular plants (754 species of vascular plants) known from the region; ecology, plant formations, weeds, cultivated and useful plants, history of vegetation, bibliography. The term Sahel de Sousse, as used by the author, refers to the coastal region for an extent of about 100 km. about Sousse (Susa).

Chabrolin, Charles. Note sur la flore des îles de la Galite et du Galiton. Bul. Soc. Hist. Nat. Afrique Nord **24:** 233–242. 1933.

Geology, botanical explorations, etc.; unannotated lists of phanerogams of Iles de La Galite and Galiton, Cape Serrat, with additional records for Ile Plane and Ile Cani. These islands are about 40 km. north of Cape Serrat, Tunisia.

Cosson, E. S. Exploration de la Kroumirie centrale. Bul. Soc. Bot. France **32:** 296–324. 1885.

General features of flora, with lists of species of different localities; list of vascular plants and mosses collected by the author and others in 1883, with localities.

Gillot, F. X. Notes botaniques. *In* **Chaignon, H. de.** Contributions à l'histoire naturelle de la Tunisie. Bul. Soc. Hist. Nat. Autun **17:** 116–142. 1904.

List of vascular plants, mosses, and lichens collected by Vicomte H. de Chaignon in 1902–1904 at Bir M'Cherga, Aïn-el-Aker, and Zaghouan, with notes on more important species.

——— Une journée d'herborisation à Souk-el-Khemis (Tunisie). Bul. Soc. Bot. France **47:** 289–296. 1900.

List of vascular plants collected by author, with localities.

Joly, Alexandre. Note sur la flore du sud Tunisien (Jefara et Diebel Labiod). Compt. Rend. Assoc. Franç. Avanc. Sci. **38** (1909): 569–577. 1910.

General features of flora; list of spermatophytes collected by author, with localities.

Lindberg, Harald. Itinera mediterranea. Ein beitrag zur kenntnis der westmediterranen flora auf grund eines materials von gefässpflanzen, gesammelt in Tunesien und Sizilien im jahre 1924 und in Spanien und Marokko im jahre 1926. Acta Soc. Scient. Fenn. n. s., B, v. 1, no. 2. 178 p. 32 pl. 1932.
Author's itinerary; list of about 2,100 vascular plants, with localities.

Murbeck, Svante. Contributions à la connaissance des renonculacées [-polypodiacées] de la flore du nord-ouest de l'Afrique et plus spécialement de la Tunisie. Lunds Univ. Årsskr. afd. 2, v. 33, no. 12. 126 p. pl. 1–6. 1897; v. 34, no. 7. 41 p. pl. 7–9. 1898; v. 35, no. 3. 30 p. 2 fig., pl. 10–12. 1899; v. 36, no. 1. 34 p. 9 fig., pl. 13–15. 1900.—Contributions . . . 2. sér. l. c. n. f., afd. 2, v. 1, no. 4. 40 p. pl. 1–7. 1905; v. 2, no. 1. p. 41–83. pl. 8–20. 1905.
List of vascular plants collected mainly by the author, with localities and frequent annotations. Ser. 2 (no. 1) includes a florula of Djebel Bargou, a branch of the Atlas Mountains, in the northeast part of central Tunisia, arranged by habitats.

Pitard, C. J. Rapports sur les herborisations faites par la Société pendant la session de Tunisie. Bul. Soc. Bot. France **56** (Sess. Extr.): cxi–cxcviii. pl. 2–19. 1910.
Physiography, soils, human influence on flora; lists of plants collected by members of the Société Botanique de France in 1909, grouped by localities and habitats; phytogeography. (Central Tunisia, in region of Gabès, Gafra, and Tozeur.)

UGANDA

See also Anglo-Egyptian Sudan (Grant; Rein); Belgian Congo (Burtt; Cotton); Ethiopia (Engler, Ueber die hochgebirgsflora); Kenya (Fries); Northern Rhodesia (Fries); Tanganyika (Engler, Die pflanzenwelt Ost-Afrikas; Mildbraed); West Africa (Oliver).
All papers on Mount Ruwenzori are placed under Belgian Congo.

General

Burtt Davy, Joseph, and Bolton, Florence. Uganda protectorate. 132 p. 21.5 cm. Oxford, 1935. (Check-lists of the forest trees and shrubs of the British Empire, no. 1.)
Botanical explorations, plant zones, annotated list of more important trees, bibliography; annotated list of 1,146 species and varieties of native woody plants, with vernacular names (not indexed) and citation of exsiccatae; list of cultivated woody plants.

Wright, C. H. List of the plants occurring in the Uganda Protectorate. *In* **Johnston, H. H.** The Uganda Protectorate. p. 329–351. New York, London, 1902.
Botanical explorations; list of vascular and cellular plants, with localities and collectors.

Local

Baker, E. G., Moore, S. LeM., and Rendle, A. B. The botany of the Anglo-German Uganda boundary commission. Jour. Linn. Soc. Bot. **37**: 116–227. pl. 1–4. 1905.
Phytogeography; list of phanerogams collected by Dr. A. G. Bagshawe, with localities, vernacular names, uses and general range. The area covered was along the boundary between the present Uganda and Tanganyika west of Lake Victoria Nyanza and the island of Buvúma in the lake.

Dawe, M. T. Report on a botanical mission through the forest districts of Buddu and the Western and Nile Provinces of the Uganda Protectorate. 63 p. 5 pl., map. 33 cm. London, 1906.
Physiography, general features of flora, etc.; list of vascular and cellular plants collected, with localities and occasional vernacular names and uses.—See also **Stapf, Otto, and others.** Plantae novae Daweanae in Uganda lectae . . . with an appendix by M. T. Dawe. Jour. Linn. Soc. Bot. **37**: 495–544. pl. 21–23 (map). 1906.

Lugard, E. J., and Bullock, A. A. The flora of Mount Elgon. Bul. Misc. Inform. Kew **1933**: 49–106. 1 fig. (map). 1933.
Physiography, climate, natives, fauna, plant zones, bibliography (by Lugard); list of 649 vascular plants collected in 1930–31 by E. J. and Mrs. C. Lugard, with data and general range.

WEST AFRICA

Under this heading are included general works covering all or much of the region from Mauritania to Angola, inclusive.

Ascherson, Paul. Botanisch-ethnographische notizen aus Guinea. Aus den aufzeichnungen von Thonning in Schumacher's Beskrivelse af Guineiske planter. Zeitschr. Ethnol. **11**: 231–258. 1879.

Annotated list of 166 wild and cultivated plants, mostly economic, with vernacular names (indexed). Refers to the region of "Upper Guinea."

Chevalier, Auguste. Énumération des plantes cultivées par les indigènes en Afrique tropicale et des espèces naturalisées dans le même pays et ayant probablement été cultivées a une époque plus ou moins reculée. Bul. Soc. Natl. Acclim. France **59**: 65–79, 104–110, 133–138, 239–242, 312–318, 341–346, 386–392. 1912.

Systematic list of about 293 species and varieties, with brief notes on origin and area where cultivated. Covers region from Mauritania to Congo.

——— Enumération des plantes récoltées dans les régions suivantes: Mauritanie, Oasis sahariennes, Sénégal, Casamance, Guinée française, Soudan français, Côte d'Ivoire, Dahomey et en outre Gold-Coast, Sierra-Leone, Lagos, Old-Calabar. xiii, 798 p. port., map. 25 cm. Paris, 1920. (Exploration botanique de l'Afrique occidentale française. t. 1.)

List of vascular and cellular plants collected by the author and associates, with localities, collectors' numbers, vernacular names, and occasional annotations.

Dalziel, J. M. The useful plants of west tropical Africa . . . Being an appendix to the Flora of west tropical Africa . . . xi, 612 p. 25.5 cm. London, 1937.

Bibliography; annotated systematic list of spermatophytes, with vernacular names and their localities, uses, and references. Covers same area as Hutchinson and Dalziel, Flora of west tropical Africa. The information and references in Holland's The useful plants of Nigeria are not repeated, but references to his work are given under each species.

Gürke, Max. Uebersicht über die gebiete des tropischen Afrika, in welchen deutsche reisende ihre im Berliner Botanischen Museum niedergelegten sammlungen zusammen brachten, mit angabe der wichtigsten, über ihre reisen und deren ergebnisse veröffentlichten aufsätze. *In* **Engler, A.** Beiträge zur flora von Afrika I. Bot. Jahrb. Engler **14**: 279–292. 1891.

Briefly annotated list of collectors and their itineraries, with bibliographies.

Hua, Henri. Les explorations botaniques dans les colonies françaises de l'Afrique tropicale, d'aprés les collections conservées au Muséum d'histoire naturelle de Paris. Actes Cong. Internatl. Bot. 1er Paris, 1900. p. 239–247. 1900.

Brief account of principal collectors in Senegal, French Guinea, French Sudan, Ivory Coast, Dahomey, and French Equatorial Africa (French Congo).

Hutchinson, John, and Dalziel, J. M. Flora of west tropical Africa. The British west African colonies, British Cameroons, the French and Portuguese colonies south of the Tropic of Cancer to Lake Chad, and Fernando Po. 2 v. 381 fig., 2 maps. 26 cm. London, 1927–36.

General features of flora, botanical explorations, bibliography, glossary, key to families; systematic list of spermatophytes, with keys, note on habit, references, citation of localities and exsiccatae. Covers area along Tropic of Cancer from coast east to the western boundary of Chad Territory (about 15° E.), then south along the eastern border of Nigeria to the coast, including Cameroon Mountain, Fernando Po, and the smaller islands off the coast, the whole nearly equivalent to the "Upper Guinea" of the Flora of Tropical Africa.

Moller, A. F. Medizinische pflanzen Westafrikas. Ber. Deut. Pharm. Gesell. **8**: 46–53, 93–100, 240–243. 1898.

Briefly annotated list of medicinal species of "Ober- und Unter-Guinea," in part systematically arranged, with occasional vernacular names.

Moloney, Alfred. Sketch of the forestry of West Africa, with particular reference to its present principal commercial products. vi, 533 p. 20 cm. London, 1887.

Includes (p. 269–458) briefly annotated "list of economic plants of western Africa," by J. M. Hillier. Covers principally the regions of Upper Guinea and Lower Guinea, as defined in Oliver, Flora of Tropical Africa.

Oliver, Daniel, and others. Flora of tropical Africa. v. 1–10, pt. 1. 22.5 cm. London, 1868–1937.

Descriptive flora of spermatophytes, with keys, synonymy, local and general range, citation of exsiccatae. Specimens cited are grouped under the following geographical regions as defined in vol. 7: Upper Guinea (west coast region, from the mouth of the Senegal River to Lake Chad, south to the southern boundary of Cameroons, including Fernando Po); North Central (the Sahara north to the Tropic of Cancer, eastward to about 26° E. Long., south to border of Upper Guinea and Belgian Congo); Nile Land (the Nile Basin from the 26th meridian to the coast, south to border of Belgian Congo and Tanganyika); Lower Guinea (west coast region from southern boundary of Cameroons to Tropic of Capricorn, bounded on the east by Belgian Congo, Rio Kwango, and the 20th meridian); South Central (Belgian Congo, and Angola east of the 20th meridian); Mozambique (Tanganyika, Mozambique, Rhodesia, etc., south to the Tropic of Capricorn). Vol. 1–3 were written by Oliver, with the assistance of other botanists at Kew, especially W. P. Hiern and J. G. Baker; the remaining volumes, written mostly by the botanists at Kew, were edited by W. T. T. Dyer (v. 4–6, pt. 1, v. 7–8), David Prain (v. 6, pt. 2, v. 9), and A. W. Hill (v. 10, pt. 1). Not yet completed.

Unwin, A. H. West African forests and forestry. 527 p. 110 fig. on plates. 26 cm. London, 1920.

Includes annotated lists, with vernacular names and uses, of the trees of Gambia, Sierra Leone, Liberia, Ivory Coast, Gold Coast, Togo, Nigeria, Cameroons; briefly annotated bibliography.

ZANZIBAR

See also West Africa (Gürke; Oliver).

General

Sacleux, Charles, *père.* Essai de catalogue des plantes de Zanzibar, Pemba, Mombassa, Amou et de la grande terre en face de ces îles jusqu'à la ligne de partage des eaux entre la côte et les grands lacs. xxxvi p. 17 cm. [Zanzibar, 18–?] (Extrait de Dictionnaire français Swahili. Zanzibar, 1888–91.)

Systematic list, with localities and vernacular names.

Werth, Emil. Die vegetation der insel Sansibar. 97 p. 6 fig., map. 24.5 cm. Berlin, 1901. (Inaug.-diss., Bern.—Reprinted from Mitt. Seminars Orient. Sprach. Univ. Berlin v. 4, abt. 3. 1901.)

Mainly ecological; includes sections on cultivated plants and weeds; no list of species.

AUSTRALASIA

Here used in a restricted sense to include only Australia, New Zealand, and Tasmania, the other islands commonly included in Australasia being given under Insular Floras.

AUSTRALIA

See also Tasmania (Hooker).

Bailey, F. M. Concise history of Australian botany. Proc. Roy. Soc. Queensland **8** (2): xvii–xli, xlv–xlvii. 1891.

Annotated list of collectors, chronologically arranged.

Bentham, George. Flora australiensis: a description of the plants of the Australian territory. 7 v. 22 cm. London, 1863–78.

Collections examined; descriptive flora of vascular plants, with keys, synonymy, local and general range; includes the plants of Tasmania. Title page says "by George Bentham . . . assisted by Ferdinand Mueller . . . ," but in preface (v. 1, p. 12*) Bentham states "I alone am therefore responsible for the details of this work, for the limitation given to genera and species, for their characters and description."

Cheel, Edwin. A review of the flora of the arid and semi-arid regions of Australia. Rpt. Austral. and New Zeal. Assoc. Adv. Sci. **23**: 307–337. 1937.

Brief account of botanical explorations, forage plants, erosion control, bibliography; mostly unannotated list of plants recorded from the region.

Guilfoyle, W. R. Australian plants suitable for gardens, parks, timber reserves, etc. 478 p. (incl. plates). 23 cm. Melbourne, etc., [1911?].
Alphabetical, briefly annotated list of native plants suitable for cultivation; lists of common names of eucalypts and of other plants. The plates are mostly habit photographs of trees and shrubs.

Maiden, J. H. Australian indigenous plants providing human foods and food-adjuncts. Proc. Linn. Soc. N. S. Wales II, **3**: 481–556. 1888.
Annotated list of 212 plants, alphabetically arranged by scientific names, with uses and vernacular names.

——— Australian vegetation. *In* The commonwealth of Australia: federal handbook . . . p. 163–209. Melbourne, [1914].
General sketch of Australian flora.

——— A bibliography of Australian economic botany. Pt. 1. v, 61 p. 24 cm. Sydney, 1892. (Technical education series, no. 10. Department of Public Instruction [N. S. Wales].)
Annotated bibliography, the titles classified by subjects; indices of authors and subjects. "I have omitted all papers and works which consist of 'pure' botany."

——— Indigenous vegetable drugs. Agr. Gaz. N. S. Wales **9**: 1106–1127. 1898; **10**: 40–53, 131–141. 1899. (Part II reprinted as Misc. Pub. Dept. Agr. N. S. Wales no. 256.)
Annotated systematic list.

——— Native food plants. Agr. Gaz. N. S. Wales **10**: 117–130, 279–290, 618–628, 730–740. 1899. (Reprinted as Misc. Pub. Dept. Agr. N. S. Wales no. 282. 65 p.)
Annotated systematic list, with vernacular names, uses, and localities. Covers Australia generally.

——— Records of Australian botanists—(a) General, (b) New South Wales. Jour. and Proc. Roy. Soc. N. S. Wales **42**: 60–132. 1908.—1st suppl. Rpt. Australas. Assoc. Adv. Sci. **13** (1911): 224–243. pl. 20–21 (ports.). 1912.—2d suppl. Jour. and Proc. Roy. Soc. N. S. Wales **55**: 150–169. pl. 8–9 (ports.). 1921.
Bibliography; alphabetical lists of botanists, with biographical data. Index in first supplement covers the whole series of similar papers on different States to that date.

——— Records of the earlier French botanists as regards Australian plants. Jour. and Proc. Roy. Soc. N. S. Wales **44**: 123–155. pl. 3–13 (ports.). 1910.
Bibliographic and biographical data.

——— Useful Australian plants no. 1–108. *In* Agr. Gaz. N. S. Wales v. 4–17, 19–20, 22–25. plates. 1893–1914.
Accounts of various species, in no regular order; descriptions, range, uses.

——— The useful native plants of Australia (including Tasmania). xii, 696 p. 22 cm. Sydney, 1889.
Annotated list, with vernacular names and uses, the plants classified under 11 headings and arranged alphabetically by scientific names; indices of vernacular and botanical names.

Mueller, Ferdinand von. Second systematic census of Australian plants, with chronologic, literary and geographic annotations. Pt. I.—Vasculares. 244 p. 21 x 28 cm. Melbourne, 1889.
Systematic list of vascular plants, with references to place of original publication and to Bentham's "Flora Australiensis" and the author's "Fragmenta" and indication of range.

Sutton, C. S. Bibliography of the Australian floristic and ecological plant-geography. Rpt. Australas. Assoc. Adv. Sci. **16** (1923): 110–118. 1924.
Unannotated list, arranged alphabetically by authors.

NEW SOUTH WALES

See also Queensland (Francis); South Australia (Wood, Floristics and ecology of the mallee); Victoria (Stirling, Notes on a census).

General

Anderson, R. H. The naturalised flora of New South Wales (excluding Gramineae). Contrib. N. S. Wales Natl. Herb. **1**: 16–33. 1939.
Systematic list of 415 angiosperms, with indication of country of origin.

Dixon, W. A. The plants of New South Wales; an analytical key to the flowering plants, except grasses and rushes, and ferns of the state, set out in an original method, with an up-to-date list of native and introduced flora. xxiv, 322 p. 3 (i. e. 6) pl. 17 cm. Sydney, 1906.

Keys to vascular plants.

Maiden, J. H. The forest flora of New South Wales. 8 v. 295 pl. 32 cm. Sydney, 1902–25.

Descriptions, vernacular names, uses, timber, habitat, and localities; the species not in systematic order.

——— Some reputed medicinal plants of New South Wales. Proc. Linn. Soc. N. S. Wales II, **3**: 355–393. 1888.

Annotated list of 70 native species, alphabetically arranged by scientific names.

——— and **Betche, Ernst.** A census of New South Wales plants. xx, 216 p. 24.5 cm. Sydney, 1916.

Systematic list of vascular plants with references to original descriptions and to Bentham's Flora Australiensis. There is a Supplement, listing Algae only.

Moore, Charles, and Betche, Ernst. Handbook of the flora of New South Wales. A description of the flowering plants and ferns indigenous to New South Wales. xxxix, 582 p. 22 cm. Sydney, 1893.—**Moore, Charles.** Plants with their habitats, discovered to be indigenous to this colony since the publication of the Handbook . . . Jour. and Proc. Roy. Soc. N. S. Wales **27**: 84–85. 1893.

Botanical explorations, glossary; descriptive flora, with keys; list of plants from Lord Howe and Norfolk Islands not included in the descriptive flora; list of naturalized plants, with indication of native range; glossary of specific names.

Rietmann, Othmar. Die nutzpflanzen von Neu-Südwales. Ber. St. Gall. Naturw. Gesell. **1864/65**: 105–118. 1865.

Running account of principal useful plants.

Local

Baker, R. T. On the botany of Rylstone and the Goulburn River districts. Pt. I. Proc. Linn. Soc. N. S. Wales **21**: 427–466. 1896.

Physiography, geology; annotated list of about 350 vascular plants. (About 100 miles northwest of Sydney.)

Barwick, A. C. The botany of the "clears" and "basalt masses," County of Hunter, N. S. Wales. Proc. Linn. Soc. N. S. Wales **28**: 932–943. 1904.

General features of flora; annotated list of vascular plants. The "clears" are areas of volcanic soil, characterized by the absence of underbrush. (Central eastern New South Wales.)

Cambage, R. H. Notes on the native flora of New South Wales. X. The Federal Capital Territory. Proc. Linn. Soc. N. S. Wales **43**: 673–711. 2 fig. (incl. map), pl. 71–74. 1919.

Botanical explorations, topography, geology, climate, soil, general features of flora; list of vascular plants (and Marchantiaceae) observed.

Chisholm, E. C. The Comboyne plateau. Its general conformation and flora. Proc. Linn. Soc. N. S. Wales **50**: 284–293. 1 fig. (map). 1925.—Additional flora . . . l. c. **52**: 378–379. 1927.—Further additions . . . l. c. **59**: 143–155. 1934.—Final additions . . . l. c. **62**: 65–72. 1937.

Physiography, general features of flora; unannotated list of vascular plants. (Near the coast southwest of Port Macquarie.)

——— The flora of Marrangaroo, County of Cook. Austral. Nat. **5**: 122–129. 1 fig. (map). 1923.

Topography, climate, distribution of some trees; unannotated list of vascular plants. (About 100 miles from Sydney.)

Cleland, J. B. List of plants growing upon Milson Island, Hawkesbury River. Rpt. Govt. Bur. Microbiol. N. S. Wales **3**: 218–225. 1914.

Annotated list of 234 native and 55 introduced vascular plants. (Near Brooklyn, north of Sydney.)

Davis, Consett, Day, M. F., and Waterhouse, D. F. Notes on the terrestrial ecology of the Five Islands. I. Proc. Linn. Soc. N. S. Wales **63**: 357–388. 8 fig. (incl. maps), pl. 15–19. 1939.

Includes annotated list (p. 380–383) of vascular plants. (Near Port Kembla.)

Dwyer, J. W. A floral survey of the south western slopes of N. S. Wales round about Temora and Barmedman. Austral. Nat. **4**: 212–224. 1921.

Soils; annotated list of cellular and vascular plants collected by author from Stockinbingal to Wyalong. (Southeastern New South Wales.)

Hamilton, A. A. Topographical and ecological notes on the flora of the Blue Mountains. Proc. Linn. Soc. N. S. Wales **40**: 386–413. 1915.

Botanical explorations (with included bibliography), general features of flora, etc.; annotated list of vascular plants. (About 70 miles west of Sydney.)

Hamilton, A. G. A list of the indigenous plants of the Mudgee district. Proc. Linn. Soc. N. S. Wales II, **2**: 259–306. 1887.

Geology, climate, etc.; annotated list of 1,208 vascular plants, with notes on uses. (About 40 miles east of Wellington.)

——— On the flora of Mt. Wilson. Proc. Linn. Soc. N. S. Wales **24**: 346–372. 1899.

Physiography, climate, general features of flora; list of 545 vascular plants; bibliography. Mount Wilson is on the boundary between New South Wales and Queensland, about 60 miles from the coast.

Haviland, F. E. Notes on the indigenous plants in the Cobar district. Proc. Linn. Soc. N. S. Wales **36**: 507–540. 1 fig. (map). 1912; **38**: 639–655. 1914.

Geology, general features of flora, climate; annotated list of 504 vascular and cellular plants. (North-central New South Wales.)

Maiden, J. H. The botany of Howell (Bora Creek): a tin-granite flora. Proc. Linn. Soc. N. S. Wales **31**: 63–72. 1906.

Physiography, etc.; list of vascular plants. (Northeastern New South Wales, on western slope of New England Range.)

——— A contribution towards a flora of Mount Kosciusko. Agr. Gaz. N. S. Wales **9**: 720–740. 1898.—A second contribution . . . l. c. **10**: 1001–1042. pl. 1899. (Reprinted as Misc. Pub. Dept. Agr. N. S. Wales 241, 331.)

General features of flora, bibliography; partly annotated lists of vascular plants found at different elevations. Mount Kosciusko (7,328 ft.) is the highest mountain in Australia. The second contribution is similarly arranged and has also notes on the general characteristics of alpine plants and on the phytogeographic relation of Mount Kosciusko and Tasmania. (Southeastern New South Wales.)

——— The Dorrigo forest reserve. Agr. Gaz. N. S. Wales **5**: 218–223, 599–633. map. 1894; **7**: 38–39. 1897; **8**: 523. 1897. (1st pt. reprinted 41 p.)

Includes annotated list of vascular plants collected by the author in 1893. (In County of Fitzroy, on the north coast.)

——— A list of plants collected by Mr. Richard Helms in the Australian Alps, February, 1893. Agr. Gaz. N. S. Wales **5**: 836–841. 1894.

List of phanerogams, with localities and altitudes, collected in the neighborhood of Mount Kosciusko, in southeastern New South Wales.

——— A list of plants collected in the vicinity of the Jenolan Caves, by W. F. Blakely and J. C. Wiburd. Agr. Gaz. N. S. Wales **12**: 1390–1394. 1901.

Unannotated list of vascular plants. (About 70 miles almost due west of Sydney.)

Ross, W. J. C. Notes on the flora of Bathurst and its connection with the geology of the district. Rpt. Australas. Assoc. Adv. Sci. **7**: 467–481. 1898.

Physiography, general features of flora; tabular list of spermatophytes, with indication of geological formations on which they occur. (About 100 miles northwest of Sydney.)

Turner, Frederick. Botany of north-eastern New South Wales. Proc. Linn. Soc. N. S. Wales **31**: 365–392. 1906.

Physiography, climate, general features of flora (including useful plants); no list of species (1,767 known). (Area between 32° S. Lat. and Queensland boundary, west to 152°20′ and 151° E. Long.)

——— Botany of north-western New South Wales. Proc. Linn. Soc. N. S. Wales **30**: 32–91. map. 1905.

Physiography, climate, useful plants of aborigines, general features of flora; unannotated list of 1,137 vascular plants. The area covered lies between the northern boundary of New South Wales and 32° S. Lat. and between 147° and 151°20′ E. Long.

Turner, Frederick. Botany of south-western New South Wales. Proc. Linn. Soc. N. S. Wales **29**: 132–181. 1904.

Climate, general features of flora, with notes on economic plants and vernacular names; unannotated list of vascular plants. The area covered extends from 33° S. Lat. to the Murray (or Hume) River, and from 141° to 147° E. Long.

——— Botany of the Darling, New South Wales. Proc. Linn. Soc. N. S. Wales **28**: 406–442. 1903.

Physiography, climate, general features of flora, botanical explorations; list of 760 vascular plants from vicinity of Darling River, between 29° and 33° S. Lat. and 141° to 147° E. Long.

——— The vegetation of New England, New South Wales. Proc. Linn. Soc. N. S. Wales **28**: 276–311. 1903.

Geology, physiography, climate, general features of flora; list of 708 vascular plants, with localities. The area covered is that portion of New South Wales extending "northwards along the Dividing Range from a little south of Armidale to the Queensland border."

Woolls, William. Plants indigenous and naturalised in the neighbourhood of Sydney . . . [2d ed.] 71 p. 21.5 cm. Sydney, 1891. (1st ed., 1880.)

Climate, etc.; unannotated lists of native and naturalized vascular plants and Characeae (1,465 species, of which 1,290 are native).

NORTHERN TERRITORY

See also South Australia (Tate, A census, and A handbook).
Northern Territory was a part of South Australia from 1863 through 1910.

General

Ewart, A. J., and **Davies, O. B.** The flora of the Northern Territory . . . with appendices by J. H. Maiden . . . and by A. A. Hamilton and Edwin Cheel . . . viii, 387 p. (incl. 27 pl., 14 tab.). map. 24 cm. Melbourne, 1917.

Partly annotated list of vascular plants, with keys or brief diagnoses, localities, uses; alphabetical list of vernacular names; classified lists of useful and poisonous plants.—See also **Ewart** and others. Contributions to the flora of Australia no. 30–34. Proc. Roy. Soc. Victoria n. s., **38**: 77–87, 164–182. 4 fig. 1926; **39**: 1–12, 154–161. 5 fig. 1926–27; **40**: 81–87. 1928. (Reprinted.)

Local

Black, J. M. Botany. *In* Scientific notes on an expedition into the interior of Australia carried out by Capt. S. A. White . . . from July to October, 1913. Trans. and Proc. Roy. Soc. So. Austral. **38**: 460–471. pl. 38–39. 1914.

Botanical explorations; annotated list of 178 vascular plants from South Australia and Northern Territory, mostly from the latter.

Cleland, J. B., and **Johnston, T. H.** Aboriginal names and uses of plants at the Granites, Central Australia. Trans. Roy. Soc. So. Austral. **63**: 22–26. 1939.

Systematic list, with vernacular names and uses. (Nearly 400 miles northwest of Alice Springs.)

——— and **Johnston, T. H.** The ecology of the aborigines of Central Australia; botanical notes. Trans. and Proc. Roy. Soc. So. Austral. **57**: 113–124. pl. 5. 1933.

Account of principal useful plants, grouped by uses; annotated systematic list of useful plants, with vernacular names. (Vicinity of Mount Liebig, MacDonnell Range.)

Holtze, Maurice. Narrative of an exploring tour across Melville Island, with notes on its botany. Trans. and Proc. Roy. Soc. So. Austral. **15**: 114–120. 1892.

Itinerary of author in October 1887, with list of vascular plants collected. (40 miles north of Port Darwin.)

Kempe, Hermann. Plants indigenous to the neighbourhood of Hermannsburg, on the river Finke, central Australia. Trans. and Proc. Roy. Soc. So. Austral. **3**: 129–137. 1880; **5**: 19–23. 1882.

Annotated list of 287 vascular plants. Also cited as "Indigenous plants of the MacDonnell Range." (Southwestern part of Territory.)

Mueller, Ferdinand von. A catalogue of plants collected during Mr. Alexander Forrest's geographical exploration of north-west Australia in 1879. Jour. and Proc. Roy. Soc. N. S. Wales **14**: 81–95. map. 1881.

List of vascular plants, with localities, collected between King's Sound and Port Darwin.

——— and **Tate, Ralph.** List of plants collected during Mr. Tietkens' expedition into central Australia, 1889. Trans. and Proc. Roy. Soc. So. Austral. **13**: 94–109. 1890.—Supplemental notes to the list of plants . . . l. c. **13**: 170–171. 1890.

Localities; list of 250 vascular plants, with localities.

Rendle, A. B., Moore, S. LeM., Baker, E. G., and **Gepp, A.** Capt. G. H. Wilkins's Groote Eylandt plants. Jour. Bot. **64**: 89–99. 1926.

Unannotated list of species collected on Groote Eylandt in 1925 and Crocodile Islands in 1924.

Tate, Ralph. Botany. *In* Report on the work of the Horn Scientific Expedition to central Australia. v. 3, p. 117–194. London, Melbourne, 1896.

Botanical regions, general features of flora, phytogeography, botanical explorations; annotated list of 614 vascular plants known from the Larapintine region; list of plants new or rare in the central Eremian region. The Larapintine region of the author is the southwestern part of Northern Territory, from the MacDonnell Range south to Engoordina (Mount Encurdina, 25°10′ S. Lat., 134°16′ E. Long.) and west to Lake Amadeus. His central Eremian region is the low plains around Lake Eyre in South Australia.

QUEENSLAND

General

Bailey, F. M. Comprehensive catalogue of Queensland plants, both indigenous and naturalised . . . 879 p. illus. (incl. 976 numb. fig.), 16 col. pl. 25 cm. Brisbane, [1913].

Systematic list of vascular and cellular plants with frequent annotations, especially on economic plants. Replaces his "Catalogue of the indigenous and naturalised plants of Queensland" (1890).

——— Medicinal plants of Queensland. Proc. Linn. Soc. N. S. Wales **5**: 1–29. 1880.

Annotated systematic list of vascular and cellular plants, with uses.

——— The Queensland flora. 6 v. (2,015 p.). 88 pl. 24.5 cm. Brisbane, 1899–1902.—General index. 66 p. 1905.

Descriptive flora of vascular plants, with keys, notes on wood and uses, vernacular names (separate indices), local and general range. The "General index" includes only scientific names.

——— Queensland woods, with a brief popular description of the trees, their distribution, qualities, uses of timber, etc., etc. [ed. 3.] vi, 13–139 p. 20 cm. London, 1899. (1st ed., 1886.)

Systematic list of 601 species, with brief botanical descriptions, some vernacular names, and notes on wood.

——— A sketch of the economic plants of Queensland. 73 p. 21 cm. Brisbane, 1888. (Queensland Commission. Centennial International Exhibition, Melbourne, 1888.)

Annotated list, alphabetically arranged, including wild and cultivated plants.

——— The weeds and suspected poisonous plants of Queensland . . . 245 p. 408 fig. 24.5 cm. Brisbane, 1906 (1907).

Annotated descriptive treatment of vascular and cellular plants.

——— and others. Contributions to the Queensland flora [no. 17–22]. Bot. Bul. Dept. Agr. and Stock Queensl. no. 17–22. 14, 16, 14, 20, 23, 36 p. illus. 1913–20.

Additional or new species and critical notes. No. 1–15 of this series preceded Bailey's "Queensland flora"; no. 16 dealt mainly with fungi; no. 18–19, by J. F. Bailey and C. T. White; no. 20–21, by C. T. White; no. 22, by C. T. White and W. D. Francis.

Domin, Karel. Beiträge zur flora und pflanzengeographie Australiens. teil 1, abt. 1–2 (951 p.); abt. 3, lfg. 1–8 (sep. paged). 183 fig., 36 pl. 32 cm. Stuttgart, 1915–29. (Bibliotheca Botanica heft 85, 89.)

Annotated list with synonymy, local and general range (with citation of collectors), and critical remarks. Refers primarily to Queensland, and is essentially a flora of that province. Reaches Solanum (2,800 species).

Maiden, J. H. Records of Queensland botanists. Rpt. Australas. Assoc. Adv. Sci. 12 (1909): 373–384. 2 port. 1910.
Bibliography; alphabetical list of botanists and collectors, with biographical information.

Local

Bailey, F. M. Botany of the Bellenden-Ker expedition. *In* **Meston, A.** Report of the government scientific expedition to Bellenden-Ker Range . . . p. 29–30. Brisbane, 1889.
List of vascular plants collected from Mulgrave River to summit of Bellenden-Ker, and at Bartle Frere and Harvey's Creek, with localities and references.—See also his report *in* **Meston, Archibald.** Expedition to the Bellenden-Ker Range. p. 9–17. 1904, for annotated list of species collected on 1904 expedition. (Northeastern Queensland.)

——— A contribution towards a flora of Mount Perry. Proc. Roy. Soc. Queensland 1: 61–76. 1884.—Pt. II, by **James Keys.** l. c. 2: 41–55. 1885.
(1) Unannotated list of vascular plants with vernacular names, based on Keys' collections. (2) Physiography, soil, general features of flora, partly annotated list of additional vascular plants and of cellular plants. The area treated includes the Burnett, Boolboonda, and Normanby Ranges, and is about 70 miles northwest of Maryborough.

——— A few words about the flora of the islands of Torres Straits and the mainland about Somerset. Rpt. Australas. Assoc. Adv. Sci. 7 (1898): 423–447. 1898.
General features of flora; annotated list of plants known from Thursday Island.

——— and **Woods, J. E. T.** A census of the flora of Brisbane. Proc. Linn. Soc. N. S. Wales 4: 137–204. 1879.
List of 1,228 vascular plants, with localities and indication of general range. The area covered includes Moreton Island and the country within 25 miles of Brisbane.

Francis, W. D. Australian rain forest trees, excluding the species confined to the tropics. xi, 347 p. 226 fig., pl., map. 24.5 cm. Brisbane, 1929.
Distribution and characters of rain forests in Australia, anatomical modifications, key to species; descriptive systematic list, with keys to genera and range; uses and references given for more important species. Covers east coast region from Cape York to Cape Otway (Queensland, New South Wales, Victoria).

Gibbs, L. S. A contribution to the phyto-geography of Bellenden-Ker. Jour. Bot. 55: 297–310. 1917.
General features of flora; list of species collected by the author, with localities; general range given. The list is by S. LeM. Moore, except palms (Beccari), orchids (Rendle), ferns (Gepp), and fungi (Ramsbottom). (Northeastern Queensland.)

Longman, H. A. The plants of Mast-head Island. Proc. Roy. Soc. Queensland 25: 17–23. 1914.
General features of flora; list of 26 phanerogams. (Islet in the Capricorn group off the coast northeast of Gladstone.)

MacGillivray, W. D. K., and **Rodway, F. A.** Plants on islands of the Bunker and Capricorn groups. Rpt. Great Barrier Reef Com. 3: 58–63. 1931.
General features of flora of Lady Musgrave, Hoskyn, and Fairfax Islands (Bunker group), North-west, Tryon, Heron, Masthead, Wilson, and One-tree Islands (Capricorn group); list of angiosperms collected by authors in 1927. (Off coast of Queensland on Tropic of Capricorn, 152° E. Long.)

• **Palmer, Edward.** On plants used by the natives of north Queensland, Flinders and Mitchell Rivers, for food, medicine, etc., etc. Jour. and Proc. Roy. Soc. N. S. Wales 17: 93–113. 1884.
Annotated list of 106 vascular plants classified by uses, including a few from other parts of Australia.

Shirley, John. A preliminary list of plants of the National Park, Macpherson Range. Proc. Roy. Soc. Queensland **31**: 21–27. 1920.
List of trees, vines, and ferns. (Southeastern Queensland, on border of New South Wales.)

Wedd, Joseph. Excursion to Newmarket, 8th October, 1910. Report on botany. Queensland Nat. **1**: 177–180. 1911.
General features of flora; unannotated list of 166 vascular plants observed by author.

——— and **White, C. T.** List of plants collected at Glasshouse Mountains, September 11th to 13th, 1909. Queensland Nat. **1**: 119–120. 1910.
Unannotated list of vascular plants.

White, C. T. The biology of North-West Islet, Capricorn group. (H.) Botany. List of plants collected . . . with notes by Dr. W. Macgillivray. Austral. Zool. **4**: 251–252. 1926.
Annotated list of phanerogams. (Off the coast on the Tropic of Capricorn.)

——— Flora of Moreton Island. Queensland Nat. **4**: 86–91. 1924.
Botanical explorations, ecology, with running list of vascular plants. (Off the coast near Brisbane.)

——— Flora of the Bunya Mountains. Queensland Agr. Jour. n. s., **13**: 25–31. 1920.
General features of flora; unannotated list of vascular plants collected (mainly by the author) above 2,000 feet elevation. (Southeastern Queensland.)

——— Ligneous plants collected for the Arnold Arboretum in north Queensland by S. F. Kajewski in 1929. Contrib. Arnold Arboretum no. 4. 113 p. 9 pl. 1933.
Soils, general features of flora, etc.; annotated list of woody plants collected in rain forest region on Atherton Tableland, Mount Bartle Frere and Daintree River.

SOUTH AUSTRALIA

See also Northern Territory (Black; Tate); Western Australia (Mueller and Tate).

General

Black, J. M. Flora of South Australia. 746 p. (incl. 336 fig., 53 pl., map). 25 cm. Adelaide, 1922–29. (Handbooks of the flora and fauna of South Australia, issued by the British Science Guild, South Australian Branch.)
Botanical explorations, glossary; descriptive flora of 2,430 vascular plants (2,046 indigenous), with keys. The treatment of Orchidaceae is by R. S. Rogers.—See also his Additions to the flora . . . [1]–37. *In* Trans. and Proc. Roy. Soc. So. Austral. v. 33, 35–37, 39–62. 1909–1938.

——— The naturalised flora of South Australia. 192 p. 206 fig. 18 cm. Adelaide, 1909.
Descriptive flora of 368 phanerogams, with keys; figure of one species in each genus.

Maiden, J. H. A century of botanical endeavor in South Australia. Rpt. Australas. Assoc. Adv. Sci. **11** (1907): 158–199. 1908.
Explorations, biographical notices, etc., with mention of publications.

Tate, Ralph. A census of the indigenous flowering plants and vascular cryptogams of extra-tropical South Australia. Trans. and Proc. Roy. Soc. So. Austral. **12**: 67–128. pl. 5. 1889.—A supplement . . . l. c. **19**: 79–83. 1895.
Phytogeographical regions and districts; systematic list of 1,935 species with distribution indicated by districts. An earlier census of the same region was published by the author in 1880–82 (including additions). Covers South Australia and Northern Territory as far north as the Tropic of Capricorn.

——— A handbook of the flora of extratropical South Australia, containing the flowering plants and ferns. vi, 303 p. map. 19 cm. Adelaide, 1890.
Flora of 1,935 vascular plants, in form of keys; systematic list, with indication of distribution; glossaries. Covers South Australia and Northern Territory as far north as the Tropic of Capricorn. Northern Territory was a part of South Australia from 1863 until January 1, 1911, when it became a separate territory.

Wood, J. G. The vegetation of South Australia. 164 p. 58 fig. (incl. maps), map. Adelaide, 1937. (Handbooks of the flora and fauna of South Australia, issued by the South Australian branch of the British Science Guild.)
Ecological, with bibliographies; includes (p. 162–164) list of vernacular names of common species.

Local

Adamson, R. S., and **Osborn, T. G. B.** On the ecology of the Ooldea district. Trans. and Proc. Roy. Soc. So. Austral. **46:** 539–564. pl. 32–36. 1922.
Includes (p. 559–563) tabular list of known spermatophytes, with habitat indicated. Ooldea is 427 miles west of Port Augusta, on the eastern edge of the Nullarbor Plain, about 30° S. Lat. and 132° E. Long.

Black, J. M. Botany. *In* Results of the South Australian Museum expedition to Strzelecki and Cooper Creeks. Trans. and Proc. Roy. Soc. So. Austral. **41:** 631–653. pl. 38–43. 1917.
Botanical explorations; annotated list of 153 vascular plants. (About 28° S. Lat., 140° E. Long.)

——— Botany. *In* Scientific notes on an expedition into the northwestern regions of South Australia. Trans. and Proc. Roy. Soc. So. Austral. **39:** 823–842. pl. 69–70. 1915.
Botanical explorations, localities; annotated list of 200 vascular plants.

Cleland, J. B. The botanical features between Oodnadatta and Ernabella in the Musgrave Ranges, with a locality list of plants from the north-west of South Australia identified by Mr. J. M. Black . . . Trans. and Proc. Roy. Soc. So. Austral. **60:** 114–126. 1936.
General features of flora; list of vascular plants, with localities, mostly based on plants collected by the "Seventh expedition for the study of the Australian aborigine." The Musgrave Ranges are in about 26° S. Lat., 132° E. Long.

——— List of plants occurring between Outer Harbour and Sellick's Beach. So. Austral. Nat. **15:** 9–13, 33–40. 1933–34.
List of 459 species and 2 varieties of vascular plants, with localities; 155 of the species are introduced.

——— Notes on the flora north-west of Port Augusta between Lake Torrens and Tarcoola. Trans. and Proc. Roy. Soc. So. Austral. **54:** 140–147. 1 fig. (map), pl. 5–7. 1930.
Notes on commercial sandalwood (*Eucarya spicata*); list of 149 vascular plants, with localities. (South-central South Australia.)

——— and **Black, J. M.** An enumeration of the vascular plants of Kangaroo Island. Trans. and Proc. Roy. Soc. So. Austral. **51:** 24–61. 1927.
Botanical explorations, with included bibliography; list of 725 species of vascular plants, of which 653 are native, with localities and occasional annotations. (Near the coast southeast of Adelaide.)

——— and **Black, J. M.** The plants of the Encounter Bay district. So. Austral. Nat. **6:** 22–30. 1925.—Additional records. l. c. **8:** 39–41. 1927.
Mostly unannotated list of 603 species and 10 varieties of vascular plants. (About 40 miles south of Adelaide.)

——— **Black, J. M.,** and **Reese, L.** The flora of the north-east corner of South Australia, north of Cooper's Creek. Trans. and Proc. Roy. Soc. So. Austral. **49:** 103–120. 1 fig. (map). 1925.—Additions to the flora of Minnie Downs. So. Austral. Nat. **8:** 28–29. 1927.
Botanical explorations, soils, general features of flora; list of 275 vascular plants (including additions), with localities. (Coopers Creek is about 28° S. Lat., 140° E. Long.)

——— and **Johnston, T. H.** Aboriginal names and uses of plants in the northern Flinders Ranges. Trans. Roy. Soc. So. Austral. **63:** 172–179. 1939.
List of vascular and cellular plants, with vernacular names and notes on uses, collected by authors in May, 1937, in region of Nepabunna; also additional names and uses recorded by Hale and Tindale.

Fawcett, S. G. M., and **Vance, C. E.** List of vascular plants. *In* The Sir Joseph Banks Islands. Reports of the expedition of the McCoy Society . . . Proc. Roy. Soc. Victoria n. s., **50:** 306–311. 1938.
Bibliography; tabular list of phanerogams, showing range on islands, habitat, and habit. (In Spencer Gulf.)

Koch, Max. A list of plants collected on Mt. Lyndurst Run, S. Australia. Trans. and Proc. Roy. Soc. So. Austral. **22:** 101–118. 1898.—Supplementary list . . . l. c. **24:** 81–85. 1900.

Annotated list of vascular plants, with vernacular names and notes on uses; list of introduced plants. In the two papers 390 indigenous and about 20 introduced species are recorded. (About 30° S. Lat., 138° E. Long.)

Maiden, J. H. A contribution to the botany of South Australia. Trans. and Proc. Roy. Soc. So. Austral. **32:** 252–286. 1908.

Lists of species from Kangaroo Island, Flinders Island, Williams Island, Neptune Isles, Wedge Island, Thistle Island, Taylor Island, Revesby Island in Sir Joseph Banks Group, Althorpe Isles, and other localities, collected in 1907 by the author and others. The islands are near the mouth of Spencer Gulf, in southeastern South Australia.

Osborn, T. G. B. The ecology of Pearson Islands. (The flora and fauna of Nuyts Archipelago and the Investigator Group, no. 8.) Trans. and Proc. Roy. Soc. So. Austral. **47:** 97–118. 1 fig. (map), pl. 4–9. 1923.

Physiography, climate, ecology, plant associations, soils; annotated tabular list of 52 vascular plants. (Southeast of Flinders Island, about 34° S. Lat., 134° E. Long.)

——— Notes on the vegetation of Flinders Island. (The flora and fauna of Nuyts Archipelago and the Investigator Group, no. 18.) Trans. and Proc. Roy. Soc. So. Austral. **49:** 276–289. pl. 21–23. 1925.

Exploration, physiography, climate, soil, plant associations, general features of flora; tabular list of 108 vascular plants, with notes on habitat, etc.; bibliography. (33°43′ S. Lat., 134°30′ E. Long.)

——— A sketch of the ecology of Franklin Islands. (Flora and fauna of Nuyt's Archipelago no. 3.) Trans. and Proc. Roy. Soc. So. Austral. **46:** 194–206. pl. 8–11. 1922.

Physiography, climate, general features of flora; list of phanerogams. The Franklin Islands are in 32°37′ S. Lat., 133°39′ E. Long.

Tate, Ralph. The botany of Kangaroo Island. Trans. and Proc. Roy. Soc. So. Austral. **6:** 116–171. map. 1883.—A revision of the flora of Kangaroo Island . . . l. c. **12:** 62–66. 1889.

History, physiography, geology, soils, climate, botanical explorations, artificial influences on vegetation, general botanical features, phytogeography, etc.; annotated list of 415 vascular plants. The "Revision" contains additions and corrections bringing the total to 513 species.

——— List of some plants inhabiting the north-eastern part of the Lake Torrens basin. Trans. and Proc. Roy. Soc. So. Austral. **6:** 100–106. 1883.

Geology, soils; list of vascular plants, with localities, collected by the author in 1883.

——— The natural history of the country around the head of the Great Australian Bight. Trans. and Proc. Phil. Soc. Adelaide, So. Austral. **1878/79:** 94–128. pl. 4. 1879.

Includes (p. 118–124) general features of flora, annotated list of phanerogams. Refers mainly to Bunda Plateau, a name used by the author for the region in extreme southwestern South Australia from the head of the Great Australian Bight (about 131° E. Long.) to Wilson Bluff, on the boundary of Western Australia, north to Ooldea Water and Boundary Dam (about 29°20′ S. Lat.).

——— On the geological and botanical features of southern Yorke-Peninsula, South Australia. Trans. and Proc. Roy. Soc. So. Austral. **13:** 112–120. 1890.

Geology, general features of flora; list of 232 vascular plants; phytogeography.

——— Plants of the Lake Eyre basin. Trans. and Proc. Roy. Soc. So. Austral. **11:** 85–100. 1889.

Plant geography, botanical explorations; list of 388 vascular plants, with localities.

Tepper, J. G. O. Die flora von Clarendon und umgegend (Süd-Australien). Bot. Centralbl. **63:** 1–9, 33–39. 1895.

Topography, geology, climate, general features of flora; list of 446 vascular plants collected by author, with flowering dates. (Southeastern part of South Australia, 12 miles south of Adelaide.)

Tepper, J. G. O. On the characteristics and distribution of the native and naturalised plants about Ardrossan, Yorke's Peninsula. Trans. and Proc. Roy. Soc. So. Austral. **3**: 25–45. pl. 4 (fig. 1). 1880.—Appendix to the list of plants about Ardrossan. l. c. 175–179. 1880.

Physiography, climate, general features of flora; list of 371 native and naturalized vascular plants, with local distribution indicated.

Wood, J. G. An analysis of the vegetation of Kangaroo Island and the adjacent peninsulas. Trans. and Proc. Roy. Soc. So. Austral. **54**: 105–139. 3 fig. (maps). 1930.

Mainly ecological; includes tabular list of species of Kangaroo Island and Fleurieu, Eyre, and Yorke Peninsulas, with indication of localities; bibliography.

——— Floristics and ecology of the mallee. Trans. and Proc. Roy. Soc. So. Austral. **53**: 359–378. 2 fig. (incl. map). 1929.

Climate, ecology and growth forms, bibliography; tabular list of about 600 spermatophytes, with indication of range. The "mallee" is a scrub formation; the area covered in this paper is mainly in southern South Australia, with adjacent parts of Victoria and New South Wales.

VICTORIA

See also Queensland (Francis); South Australia (Wood, Floristics and ecology of the mallee).

General

Ewart, A. J. Flora of Victoria. 1,257 p. 349 fig., col. pl. 22.5 cm. [Melbourne,] 1930.—Flora . . . additions and alterations. Victorian Nat. **53**: 60–63. 1936.

Geology, climate, general features of flora, naturalized plants; annotated descriptive flora of vascular plants, with keys, local and general range.

——— and **Sutton, C. S.** Vernacular names of Victorian plants. Jour. Dept. Agr. Victoria **9**: 383–390, 532–541. 1911; **10**: 443–448, 567–572. 1912; **12**: 82–91. 1914; **13**: 487–493. 1915; **14**: 55–58, 180–186, 240–243, 499–502. 1916. (Reprinted.)

Systematic tabular list, with "popular" names and brief notes on general appearance and uses.

——— and **Tovey, J. R.** The weeds, poison plants, and naturalized aliens of Victoria. viii, 110 p. 33 col. pl. 24.5 cm. Melbourne, 1909.

Annotated list of weeds and noxious plants; annotated alphabetical tabular list of naturalized plants.—See also **Audas, J. W.**, and **Morris, P. F.** Supplement . . . 19 p. Melbourne, 1925.

Field Naturalists' Club of Victoria. A census of the plants of Victoria with their regional distribution and the vernacular names as adopted by the Plant Names Committee . . . 2d ed. xvi, 93 p. 2 maps. 17.5 cm. Melbourne, 1928. (1st ed., 1923.)—Supplement 1[-4]. Victorian Nat. **40**: 147, 245. 1923–24; **42**: 130–132. 1925; **48**: 142–148. 1931.

Systematic list of vascular plants with English names and brief indication of local range; list of species erroneously recorded in Mueller's "Key"; alphabetical list of naturalized plants.

Maiden, J. H. Records of Victorian botanists. Victorian Nat. **25**: 101–117. port. 1908.

Bibliography; alphabetical list of botanists and collectors with biographical information.

Mueller, Ferdinand von. Key to the system of Victorian plants . . . 2 v. 152 (i. e. 210) fig., map. 18 cm. Melbourne, 1885–88.

Vol. 1 (1887–88): descriptive flora of vascular plants in form of keys. Vol. 2 (1885): systematic list of species, with indication of local range; full-page figures with dissections.—See also his Supplement to the enumeration of Victorian plants, comprising the species added since Part II of the Key to the system of our native vegetation was published . . . Victorian Nat. **5**: 14–16. 1888.

——— and **Becker, Ludwig.** Indigenous vegetable productions. *In* Report on the resources of the colony of Victoria. Trans. Phil. Inst. Victoria v. 4, app. 1, p. 21–30. 1860.

Running account of principal economic plants of Victoria.

Local

Adcock, G. H. Census of indigenous plants of the Geelong district. Geelong Nat. **6**: 31–36. 1896. (Reprinted, 14 p.)

Unannoted list of vascular plants. (Southwestern coast.)

——— List of plants from the Cape Otway forest. Arranged according to Baron von Mueller's "Key to the system of Victorian plants." Geelong Nat. **5** (1): 13–19. 1895.—Supplementary list . . . l. c. **5** (4): 18–19. 1896. (Reprinted 15, [2] p.)

Unannotated list of vascular plants. (Southwestern coast.)

Daley, Charles. Around Noojee. Victorian Nat. **41**: 166–183. 1925.

General features of flora; list of vascular plants collected in October 1924. (North of Warragul, about 60 miles southeast of Melbourne.)

D'Alton, St. Eloy. The botany of the "Little Desert," Wimmera, Victoria. Victorian Nat. **30**: 65–78. pl. 4 (map). 1913.

General features of flora; list of vascular plants, prepared by C. S. Sutton. (Northwestern Victoria.)

Ewart, A. J. Biological survey of Wilson's Promontory. Victorian Nat. **25**: 142–149. pl. (map). 1909; **26**: 129–137. 1910; **27**: 178–180. 1911.

Includes alphabetical list of 600 vascular plants (including additions in 2d and 3d reports). The papers include "Botanical reports" by J. W. Audas.—See also **Audas, J. W.** Recent botanical investigations at Wilson's Promontory. l. c. **27**: 208–214. 1911. (About 120 miles southeast of Melbourne.)

——— The flora of the Victorian Alps. Victorian Nat. **27**: 104–120. pl. 11 (map). 1910.

Physiography, bibliography; unannotated alphabetical list of 334 vascular plants, with reference to collector or recorder; general features of flora (by J. W. Audas).—See also **Tadgell, A. J.** A contribution to "The flora of the Victorian Alps." l. c. **38**: 105–118. 1922. Includes 130 additions. (Southeastern Victoria.)

Hardy, A. D. The Mallee: Ouyen to Pinnaroo. Victorian Nat. **30**: 148–167, 173–174. 2 pl., map. 1914.

Topography, general features of flora; unannotated list of vascular plants; bibliography. The Mallee country is defined as the northwest corner of Victoria bounded on the south by a line drawn from Swan Hill to the border of South Australia; it includes also adjacent New South Wales and South Australia.

Paton, D. J. The plants of the "Whipstick" scrub, Bendigo. Victorian Nat. **40**: 189–204. map. 1924.

General features of flora; list of angiosperms, with flowering dates. "Whipstick" scrub is a name applied to a region with dense thickets of shrubby eucalypts entangled with vines.

Patton, R. T. List of vascular plants. *In* Lady Julia Percy Island. Reports of the expedition of the McCoy Society . . . Proc. Roy. Soc. Victoria **49**: 362. 1937.

Annotated list of 32 species. Lady Julia Percy Island is a small volcanic island in Bass Strait, between Victoria and Tasmania.

Spencer, Baldwin, and French, Charles. Trip to Croajingolong. Victorian Nat. **6**: 1–38. 5 pl., map. 1889.

Includes, besides a running account of plants seen, general notes on flora and list of rarer species collected by authors. (Southeasternmost Victoria.)

Stirling, James. Notes on a census of the flora of the Australian Alps. Trans. and Proc. Bot. Soc. Edinb. **22**: 319–395. 3 pl., map. 1904.

Physiography, general features of flora with notes on economic plants; list of 1,019 vascular and cellular plants, with local range, mention of States other than Victoria in which the species occur, and altitude reached. (Southeastern Victoria and adjacent New South Wales.)

——— Notes on the flora of Mount Hotham. Victorian Nat. **4**: 72–78. 1887.

History, topography, geology, general features of flora; list of spermatophytes collected by author. (In the Australian Alps, about 70 miles north of Bairnsdale.)

——— Notes on the physiography of the western portion of the County of Croajingolong. Proc. Roy. Soc. Victoria n. s., **1**: 80–108. pl. 1889.

Includes (p. 93–103) list of vascular plants, with localities. (Southeasternmost Victoria.)

Stirling, James. The phanerogamia of the Mitta Mitta source basin and their habitats. Trans. and Proc. Roy. Soc. Victoria **19**: 1–23. 1883; **21**: 29–51. 1885.
Physiography, etc.; annotated list of spermatophytes. (In the Australian Alps not far from New South Wales.)

Sullivan, David. A complete census of the flora of the Grampians and Pyrenees. Rpt. Australas. Assoc. Adv. Sci. **2** (1890): 497–525. 1890.
General features of flora; list of 708 vascular and cellular plants, with localities.—See also **Campbell, A. G.** A census of Grampian plants. Victorian Nat. **28**: 105–111. 1911. (Botanical explorations, physiography, geology, general features of flora; list of 136 species additional to Sullivan's list.) (Southwestern Victoria.)

Sutton, C. S. Notes on the Sandringham flora. Victorian Nat. **28**: 5–20. 1911.—Supplementary notes . . . l. c. **29**: 79–96. map. 1912.
General features of flora, plant associations; list of 460 vascular plants (including supplement), with indication of local habitat. (About 15 miles south of Melbourne.)

——— A sketch of the Keilor Plains flora. Victorian Nat. **33**: 112–123, 128–143. 1 fig. (map). 1916–1917.
Geology, climate, soil, plant associations, endemic species; partly annotated list of 444 vascular plants. (The region covered has a radius of about 30 miles around Melbourne.)

Tadgell, A. J. Mount Bogong and its flora. Victorian Nat. **41**: 56–80. 1 fig. (map). 1924.
Includes (p. 72–79) alphabetical list of about 269 vascular plants, with altitudes; also list of 24 cellular plants. (In the Australian Alps, about 50 miles from the New South Wales line.)

Tisdall, H. T. List of plants growing in the neighborhood of Walhalla, with remarks. *In* **Herman, Hyman.** Report on the Walhalla gold-field. (Special reports. Dept. of mines, Victoria.) p. 59–64. Melbourne, 1901.
Partly annotated list of spermatophytes, without authorities for botanical names. (Southern part of Victoria, about 80 miles east of Melbourne.)

Tovey, J. R. Some notes on Coode Island and its flora. Victorian Nat. **28**: 57–61. 1911.
Running account of vascular plants observed by author. (Mouth of Yarra River, near Melbourne.)

Zimmer, W. J. The flora of the far north-west of Victoria. Its distribution in relation to soil types, and its value in the prevention of soil erosion. 86 p. 6 pl. 26 cm. [Melbourne,] 1937.
Mainly ecological; includes (p. 73–86) unannotated list of 402 vascular plants in area between the Murray River on the north, the South Australian border on the west, a line east from the South Australian border to Hattah, and thence easterly to the Murray River at Wemen. (Processed publication.)

WESTERN AUSTRALIA

See also Northern Territory (Mueller).

General

Alexander, W. B., Lane-Poole, C. E., and Herbert, D. A. Lists of the principal indigenous West Australian plants of economic importance and of naturalised aliens and weeds established in the state, with their vernacular names. Jour. and Proc. Roy. Soc. West. Austral. **6**: 41–46. 1920.
Tabular list of botanical names, with vernacular names and brief notes on uses; list of naturalized plants.

Fitzgerald, W. V. Trees of Western Australia, with notes on their uses and distribution. Jour. and Proc. Mueller Bot. Soc. West. Austral. **1** (11): 1–78. 1903.
Life zones; annotated list, with brief descriptions, and notes on timber and other uses.

Gardner, C. A. Enumeratio plantarum Australiae occidentalis. A systematic census of the plants occurring in Western Australia. iv, 150 p. 24.5 cm. Perth, 1931.

Unannotated list of vascular plants, with references to original descriptions; includes naturalized plants. Originally started as supplement to Jour. Dept. Agr. West. Austral. s. 2, v. 7.

Maiden, J. H. Records of Western Australian botanists. Jour. West. Austral. Nat. Hist. Soc. [2] (6) : 5–27. 1909.

Bibliography; alphabetical list of botanists, with biographical information.

Local

Diels, Ludwig. Die pflanzenwelt von West-Australien südlich des Wendekreises. Mit einer einleitung über die pflanzenwelt gesamt-Australiens in grundzügen. xii, 413 p. 82 fig. (incl. maps), 34 pl., map. 25.5 cm. Leipzig, 1906. (Die vegetation der erde, v. 7.)

Mainly ecological and phytogeographical; includes (p. 41–73) sketch of botanical explorations and bibliography.

——— and **Pritzel, Ernst.** Fragmenta phytographiae Australiae occidentalis. Beiträge zur kenntnis der pflanzen Westaustraliens, ihrer verbreitung und ihrer lebens-verhältnisse. Bot. Jahrb. Engler **35:** 55–662. 70 fig. 1904–05.

Records and critical notes on many species, mainly from the authors' collections in extratropical Western Australia in 1900–01.

Domin, Karel. Additions to the flora of Western and north-western Australia. Jour. Linn. Soc. Bot. **41:** 245–283. pl. 10–13. 1912.—New additions to the flora of Western Australia. Věstník K. České Společn. Náuk Třída Mat.-Přír. (Mém. Soc. Roy. Sci. Bohème Cl. Sci.) 1921/22. no. 2. 125 p. 1923.

(1912) New species and critical notes covering Pteridophyta, Monocotyledones, and Casuarinaceae, based mainly on collections by Dr. E. Clement and Capt. A. A. Dorrien-Smith; bibliography. (1923) List of additions covering the Dicotyledons, with references and critical notes. (Southwestern part of Western Australia and region between Ashburton and De Gray Rivers.)

Fitzgerald, W. V. The botany of the Kimberleys, north-west Australia. Jour. and Proc. Roy. Soc. West. Austral. **3:** 102–224. 1918.

Physiography; annotated list of 676 vascular plants. (About 26° S. Lat., 120° E. Long.)

Gardner, C. A. Botanical notes. Kimberley division of Western Australia. Forests Dept. Bul. West. Austral. 32. 105 p. 18 pl., map. 1923.

Physiography, climate, general features of vegetation, plant formations; annotated list of vascular plants collected on Kimberley Exploration Expedition of 1921. (Northwestern Western Australia.)

——— List of the naturalised plants of extra-tropical Western Australia. Jour. and Proc. Roy. Soc. West. Austral. **11:** 69–80. 1925.

Partly annotated systematic list of 237 plants, with native habitat; refers only to the region lying south of 26° S. Lat.

Maiden, J. H. Notes on Acacia, no. II. Tropical Western Australia. Jour. and Proc. Roy. Soc. N. S. Wales **51:** 71–124. pl. 1–4. 1917.

Consists of an annotated list of 23 papers or collections bearing on botany of the "Nor-West," each so extensively annotated with reference to the acacias dealt with that the bibliography is difficult to follow. (Coastal region from about Tropic of Capricorn north to boundary of Northern Territory.)

Moore, S. Le M. The botanical results of a journey into the interior of Western Australia; with some observations on the nature and relations of the desert flora. Jour. Linn. Soc. Bot. **34:** 171–261. 1899.

General features of flora; annotated list of vascular plants collected by author in 1894–95, with localities; statistics of Western Australian desert flora (said to number 867 species), phytogeography, homoplasy. (South-central Western Australia, mostly between Southern Cross and Mount Margaret, near Lake Carey.)

Morrison, Alexander. The vegetation of Western Australia. West. Austral. Yearbook **1900/01:** 298–341. 1902.

General features of flora; unannotated list of extratropical Western Australian plants. List based on one published by Mueller in 1896, brought up to date by Morrison.

Mueller, Ferdinand von. Observations on plants, collected during Mr. Joseph Bradshaw's expedition to the Prince Regent's River. Proc. Linn. Soc. N. S. Wales II, **6:** 457–478. 1892.

List of vascular plants, with localities and occasional annotations; based on material collected by W. T. Allen between Cambridge Gulf and Prince Regent's River, in northernmost Western Australia.

Mueller, Ferdinand von. The plants indigenous around Sharks Bay and its vicinity, chiefly from the collections of the Honorable John Forrest. 24 p. 33.5 cm. Perth, 1883.

Botanical explorations; list of vascular plants, with localities and occasional annotations. (On the west coast at about 26° S. Lat.)

——— Plants of north-western Australia. 19 p. 33.5 cm. Perth. 1881.

Botanical explorations; list of phanerogams, with localities and occasional annotations, based on plants collected by John Forrest's trigonometrical survey of the Nickol Bay district in 1878, and his expedition in 1879 between Nickol Bay and King's Sound. (Nickol Bay is in 20°35′ S. Lat., 116°50′ E. Long., King's Sound 16°50′ S. Lat., 123°20′ E. Long.)

——— and **Tate, Ralph.** Botany [of the Elder exploring expedition] (phanerogams and vascular cryptogams). Trans. and Proc. Roy. Soc. So. Austral. **16**: 333–383. 1896.

List of about 700 species, with localities, collected between Warrina, South Australia, and the Murchison district of Western Australia. (Central and northern South Australia and southern Western Australia.)

Ostenfeld, Carl. Additions and notes to the flora of extra-tropical W. Australia. His Contributions to West Australian botany III. Biol. Meddel. K. Danske Vidensk. Selsk. v. 3, no. 2. 144 p. 19 fig., 12 pl. 1921.

Partly annotated list of vascular plants collected by author in 1914.

Tepper, J. G. O. The flora of Roebuck Bay, West Australia. Trans. and Proc. Roy. Soc. So. Austral. **17**: 13–20. 1893.

Physiography, climate, etc.; list of 143 vascular plants collected by J. W. O. Tepper in 1889–91. Roebuck Bay, also known as Port Broome, is in about 18° S. Lat., 122° E. Long.—German translation in Bot. Centralbl. **54**: 257–261. 1893.

NEW ZEALAND

Allan, H. H. New Zealand trees and shrubs and how to identify them. x, 188 p. plates. 19 cm. Auckland, [1928].

Popular treatment in form of artificial key with brief descriptions, the species not in systematic order.

Andersen, J. C. Popular names of New Zealand plants. Trans. and Proc. New Zeal. Inst. **56**: 659–714. 1926; **57**: 905–977. 1927.

Principles of popular nomenclature; list of Maori and English vernacular names, under alphabetically arranged scientific names; (pt. 2) alphabetical list of vernacular names with botanical equivalents and reference to authorities for use; bibliography.

Cheeseman, T. F. Catalogue of the plants of New Zealand, including both indigenous and naturalised species. 33 p. 24.5 cm. Wellington, 1906.

List of 1,571 native and 527 naturalized vascular plants, with general distribution in area indicated; based on 1906 edition of the author's Manual.

——— Manual of the New Zealand flora. 2d ed., rev. and enl. by the author. Ed. by W. R. B. Oliver. xliv, 1163 p. 24 cm. Wellington, 1925. (1st ed., 1906.)

History of botanical explorations, with mention of chief publications (continued for 1905–24 by Oliver); list of publications of Cheeseman; descriptive flora of vascular plants (1,763 species), with keys, references, synonymy, local and general range; annotated list of 576 naturalized plants, list of Maori vernacular names, glossary.

——— and **Hemsley, W. B.** Illustrations of the New Zealand flora. 2 v. 250 (i. e., 251) pl. 32 cm. Wellington, 1914.

Excellent plates of characteristic plants; the text relates mainly to history, range, and habitat of the species described. The work is intended to accompany Cheeseman's Manual. A systematic list of previously published illustrations of the New Zealand flora is given in vol. 2.

Cockayne, Leonard. Report on the dune-areas of New Zealand, their geology, botany, and reclamation. 76 p. 14 pl. 33.5 cm. Wellington, 1911.

Mainly ecological and geological; includes (p. 33–34) table showing distribution, range, vernacular names, etc., of 147 vascular plants, followed by list of naturalized plants; bibliography.

Cockayne, Leonard. The vegetation of New Zealand. 2d ed., almost entirely rewritten, thoroughly rev., and enl. xxvi, 456 p. 87 pl., 3 maps. 25 cm. Leipzig, 1928. (Die vegetation der erde, v. 14.) (1st ed., 1921.)

Botanical explorations, with mention of principal publications; bibliography; physiography, ecology, agriculture, phytogeography, statistics, etc.

——— and **Allan, H. H.** An annotated list of groups of wild hybrids in the New Zealand flora. Annals Bot. **48:** 1–55. 1934.

List of 491 known or supposed hybrids; bibliography.

——— and **Turner, E. P.** The trees of New Zealand. 171 p. 118 fig. 21 cm. Wellington, 1928.

General matter; descriptions and photographs of 106 trees with vernacular names, alphabetically arranged by scientific name; additional descriptive list of shrubby plants; annotated list of timbers.

Colenso, William. On the vegetable food of the ancient New Zealanders before Cook's visit. Trans. and Proc. New Zeal. Inst. **13:** 3–38. 1881.

Includes descriptive account of useful plants, both wild and cultivated.

Hamilton, Augustus. List of papers and notices on botanical subjects in the "Transactions of the New Zealand Institute" and other publications. Trans. and Proc. New Zeal. Inst. **36:** 342–372. 1904.

Alphabetically arranged by authors, the papers of each author in chronological order.

Hooker, J. D. Flora Novae-Zelandiae. 2 v. 130 col. pl. 32.5 cm. London, 1853(1852)–55. (The botany of the Antarctic voyage of H. M. discovery ships *Erebus* and *Terror*, in the years 1839–1843 . . . pt. II.)

Botanical explorations, origin of flora, phytogeography; annotated descriptive flora, with vernacular names. Vol. 1, phanerogams; vol. 2, cryptogams, supplement, index.

——— Handbook of the New Zealand flora; a systematic description of the native plants of New Zealand and the Chatham, Kermadec's, Lord Auckland's, Campbell's and Macquarrie's Islands. 2 v. 21.5 cm. London, 1864–67.

Descriptive flora, with local and general range, and uses. Vol. 1 contains vascular plants, v. 2 cellular.

Kirk, Thomas. The forest flora of New Zealand. xv, 345 p. 142 pl. 34 cm. Wellington, 1889.

Descriptive account of economic trees and shrubs, with history, range, uses, Maori names, classified list of most important woods; synopsis of classification; glossaries.

——— The students' flora of New Zealand and the outlying islands. vi, 408 p. 25 cm. Wellington, [1899].

Descriptive flora, with keys, synonymy, local and general range; English vernacular names, glossary. Incomplete (Ranunculaceae—Compositae).

Laing, R. M., and **Blackwell, E. W.** Plants of New Zealand. 3d ed., rev. and enl. xv, 468 p. (incl. front., 173 [i. e. 174] fig.). 22.5 cm. Auckland [etc.], 1927.

General features of flora, phytogeography, etc.; key to families, descriptive flora (very incomplete); glossary. The illustrations are practically all from photographs of living specimens.

Thomson, G. M. The naturalisation of animals and plants in New Zealand. x, 607 p. map. 24.5 cm. Cambridge, Engl., 1922.

Includes (p. 363–501) annotated list of introduced vascular plants (over 600 of which have become naturalized), with details of first records, etc.; also chapter on "Alteration in flora since European occupation of New Zealand" (p. 517–535).

NORTH ISLAND

See also South Island (Buchanan, Botanical notes).

Adams, James. On the botany of Hikurangi Mountain. Trans. and Proc. New Zeal. Inst. **30:** 414–433. 1898.

General features of flora; list of vascular plants collected by D. Petrie and the author on Mount Hikurangi, in the Raukumara Range, and in the East Cape district, with localities.

——— On the botany of Te Aroha Mountain. Trans. and Proc. New Zeal. Inst. **17:** 275–287. 1885.

Physiography, general features of flora; annotated list of vascular plants. (Provincial district of Auckland.)

Allan, H. H., and **Dalrymple, K. W.** Ferns and flowering-plants of Mayor Island, N. Z. Trans. and Proc. New Zeal. Inst. **56**: 34–36. pl. 5. 1926.

Physiography, general features of flora, etc.; list of 98 species and varieties of vascular plants. (Off the southeast coast of the provincial district of Auckland.)

Aston, B. C. Botanical notes made on a journey across the Tararuas. Trans. and Proc. New Zeal. Inst. **42**: 13–25. 1 fig. (map), pl. 1–7. 1910.

Physiography, general features of flora; unannotated list of plants seen in the valley of the Pakuratahi; list of additions to Petrie's flora of the Tararuas (1908), with localities. (About 40 miles northeast of Wellington.)

——— List of phanerogamic plants indigenous in the Wellington Province. Trans. and Proc. New Zeal. Inst. **43**: 225–247. 1911.

Annotated list of about 680 phanerogams, showing altitudinal range.

——— Notes on the phanerogamic flora of the Ruahine Mountain-chain, with a list of the plants observed thereon. Trans. and Proc. New Zeal. Inst. **46**: 40–54. 1 fig. (map). 1914.

General features of flora; annotated list of phanerogams, showing altitudinal range; brief bibliography.—For additions, see his Indigenous plants of the Ruahine and Kaimanawa Mountains. l. c. **47**: 68–70. 1915. (Along the boundary between the provincial districts of Wellington and Hawkes Bay.)

——— The vegetation of the Tarawera Mountains, New Zealand. Trans. and Proc. New Zeal. Inst. **48**: 304–314. pl. 19–22. 1916.

Includes list of plants collected in 1913 and later on Tarawera Mountains; bibliography. The vegetation of that region was nearly or quite completely destroyed by a volcanic eruption in 1886. (On eastern side of Lake Tarawera, in provincial district of Auckland, about 38°12′ S Lat., 176°26′ E. Long.)

Buchanan, John. List of plants found on Miramar Peninsula, Wellington Harbour. Trans. and Proc. New Zeal. Inst. **5**: 349–352. 1873.

General features of flora; list of species, arranged in ecological groups.

——— Notes on the botany of Mount Egmont and neighbourhood, New Zealand, February, 1867. Jour. Linn. Soc. Bot. **10**: 57–62. 1869 (1867). (Reprinted in his Botanical notes on the Kaikoura Mountains and Mount Egmont. p. 10–16. 1867.)

General features of flora, etc.; unannotated list of vascular plants. (Provincial district of Taranaki, on southwest coast.)

——— On the botany of Kawau Island: physical features and causes influencing the distribution of species. Trans. and Proc. New Zeal. Inst. **9**: 503–527. 1877.

Physiography, geology, general features of flora, plant formations; unannotated lists of indigenous and introduced vascular plants. (In Hauraki Gulf, 27 miles north of Auckland.)

——— and **Kirk, Thomas.** List of plants found in the northern district of the Province of Auckland. Trans. and Proc. New Zeal. Inst. **2**: 239–247. 1870.

General features of flora; list of vascular plants, with indication of local distribution.—See also **Kirk, T.** On the botany of the northern part of the Province of Auckland. l. c. **3**: 166–177. 1871.

Carse, Harry. On the flora of the Mangonui County. Trans. and Proc. New Zeal. Inst. **43**: 194–224. 1 fig. (map). 1911.—On some additions . . . l. c. **45**: 276–277. 1913.—The ferns and fern allies . . . l. c. **47**: 76–93. 1915.—Some further additions . . . l. c. **48**: 237–243. 1916.

Botanical explorations, physiography, plant formations; annotated list of 538 vascular plants. (Near the north end of North Island.)

——— On the flora of the Mauku district. Trans. and Proc. New Zeal. Inst. **34**: 362–386. 1902.

General features of flora; list of 405 vascular plants, with localities. (Lower middle part of peninsula of Auckland.)

Cheeseman, T. F. Notes on the Three Kings Islands. Trans. and Proc. New Zeal. Inst. **20**: 141–150. 1888.—Further notes . . . l. c. **23**: 408–424. pl. 37–38 (37, map). 1891.

(1888) Physiography, general features of flora, history; unannotated list of 82 vascular plants. (1891) Similar notes, including ornithology; unannotated list of 143 vascular plants. (Northwest of North Cape, in 34°9′ S. Lat., 172°8′ E. Long.)

Cheeseman, T. F. On the botany of the Titirangi district of the Province of Auckland. Trans. and Proc. New Zeal. Inst. **4:** 270–284. 1872.

General features of flora; critical notes on several species; unannotated lists of about 460 native and 110 naturalized vascular plants. (About 10 miles southwest of Auckland.)

——— On the flora of the North Cape district. Trans. and Proc. New Zeal. Inst. **29:** 333–385. pl. 23 (map). 1897.

Botanical explorations (with included bibliography), physiography, general features of flora; annotated list of vascular plants observed between Mangonui (Mongonui) and North Cape.

Cockayne, Leonard. Notes on a brief botanical visit to the Poor Knights Islands. Trans. and Proc. New Zeal. Inst. **38:** 351–360. 1906.

Physiography, plant formations; list of vascular plants. (Off the east coast of Auckland provincial district, 35°29′ S. Lat., 174°23′ E. Long.)

——— Report on a botanical survey of Kapiti Island. 23 p. 8 pl., map. 33 cm. Wellington, 1907.

Physiography, climate, plant formations, affiinities of flora; tabular list of indigenous vascular plants, showing vernacular names, general and local range; list of introduced species.—For additions see **Aston, B. C.** Plant habitats hitherto unrecorded. Trans. and Proc. New Zeal. Inst. **47:** 67–68. 1915. (Off the west coast about 30 miles north of Wellington.)

——— Report on a botanical survey of the Tongariro National Park. 42 p. 16 pl., map. 33 cm. Wellington, 1908.

History, climate, geology, ecology, fauna, etc.; table of 260 indigenous vascular plants, showing vernacular names, local and general range; bibliography. (Extreme northern Wellington provincial district.)

——— Report on a botanical survey of the Waipoua kauri forest. 44 p. 10 pl., map. 33 cm. Wellington, 1908.

Physiography, climate, plant formations, ecology, etc.; table of indigenous vascular plants, showing vernacular names, local and general range, and "life-form"; bibliography. (Northwest part of North Island, near Hokianga Harbor.)

Colenso, William. On the botany, geographic and oeconomic, of the North Island of the New Zealand group. Trans. and Proc. New Zeal. Inst. v. 1, pt. 3, [art. 3]. 58 p. 1869.

Includes (p. 29–58) sketch of economic plants with uses and vernacular names. Half title: On the botany of the North Island of New Zealand.

Cranwell, L. M., and **Moore, L. B.** Botanical notes on the Hen and Chicken Islands. Rec. Auckland Inst. and Mus. **1:** 301–318. 2 fig. (maps), pl. 73. 1935.

Topography, geology, climate, animal life, plant communities, phytogeography; list of cellular and vascular plants (263 vascular); bibliography. (Off the northeast coast of Auckland, 35°58′ S. Lat., 174°45′ E. Long.)

——— and **Moore, L. B.** The vegetation of Maungapohatu. Rec. Auckland Inst. and Mus. **1:** 71–80. 1931.

Physiography, general features of flora; annotated list of cellular and vascular plants collected in March 1930. Maungapohatu is a mountain, not previously studied botanically, in the Urewera country, East Cape botanical district.

Hamilton, W. M. The Little Barrier Island. Hauturu. New Zeal. Jour. Sci. and Technol. **17:** 465–495, 717–749. illus. (incl. maps). 1935–36; **18:** 557–578. illus. 1936. (Reprinted as Bul. Dept. Scient. and Ind. Res. New Zeal. no. 54. 1937.)

History, geology, soils, climate, plant associations, bibliography, etc.; tabular annotated list of native (351 species) and introduced spermatophytes, also bryophytes. (West of Great Barrier Island, near the mouth of Hauraki Gulf.)

Kirk, Thomas. Notes on the botany of the East Cape district. Trans. and Proc. New Zeal. Inst. **29:** 509–532. 1897.

Botanical explorations, physiography, general features of flora; partly annotated list of about 500 vascular plants.

——— Notes on the botany of Waiheke, Rangitoto, and other islands in the Hauraki Gulf. Trans. and Proc. New Zeal. Inst. **11:** 444–454. 1879.

Physiography, general features of flora; list of vascular plants from Rangitoto Island.

Kirk, Thomas. On the botany of the Great Barrier Island. Trans. and Proc. New Zeal. Inst. **1**: 144–157. 1869.

General features of flora; critical notes on some species; unannotated lists of native and naturalized vascular plants (about 400 species); also lists of plants from Little Barrier Island and Kororariki, Bay of Islands. (Off the central and northern part of the east coast, provincial district of Auckland.)

——— On the botany of the Thames gold-fields. Trans. and Proc. New Zeal. Inst. **2**: 89–100. 1870.

General features of flora; list of 397 vascular plants, arranged by ecological groups.—For additions see **Adams, J.** (same title) l. c. **16**: 385–393. 1884. (Central eastern coast of Auckland provincial district.)

——— On the flora of the Isthmus of Auckland and the Takapuna district. Pt. 1. Trans. and Proc. New Zeal. Inst. **3**: 148–161. 1871.

General features of flora; unannotated lists of native (about 800 species) and naturalized vascular plants. Pt. 2 (l. c. **4**: 228–238. 1872) relates to cellular plants. (Vicinity of Auckland city and the Takapuna region immediately north of it across Waitemata Harbor.)

——— On the naturalized plants of Port Nicholson and the adjacent district. Trans. and Proc. New Zeal. Inst. **10**: 362–378. 1878.

General factors of introduction; annotated list. (Province of Wellington.)

Oliver, W. R. B. Vegetation of Poor Knights Islands. New Zeal. Jour. Sci. and Techn. **7**: 376–384. 3 fig. 1925.

Geology, destruction of flora by pigs, general features of flora, with mention of many species; no list.—See also **Cranwell, L. M.** New plant records from the Poor Knights Islands . . . Rec. Auckland Inst. and Mus. **2**: 101–110. 2 fig. (maps), pl. 25–26. 1937. (Off the east coast of Auckland provincial district, 35°29′ S. Lat., 174°23′ E. Long.)

——— The vegetation of White Island, New Zealand. Jour. Linn. Soc. Bot. **43**: 41–47. 2 fig., pl. 2–3. 1915.

Physiography, geology, climate, bird life, plant formations; annotated list of 12 vascular plants. White Island is a small volcanic cone, with an atmosphere charged with hydrochloric acid from Whakari Volcano.

Petrie, Donald. Account of a visit to Mount Hector, a high peak of the Tararuas, with list of flowering-plants. Trans. and Proc. New Zeal. Inst. **40**: 289–304. 1908.

Physiography, general features of flora; list of phanerogams "hitherto observed on the higher parts of the Tararuas" with altitudes. (About 30 miles northeast of Wellington.)

Sladden, Bernard. Karewa: an island sanctuary. New Zeal. Jour. Sci. and Technol. **7**: 182–187. 3 fig. 1924.

Topography, fauna; running account of plants, with vernacular names. (In entrance of Tauranga Harbor, Province of Auckland.)

——— Tuhua, or Mayor Island. New Zeal. Jour. Sci. and Technol. **8**: 193–210. 5 fig. (incl. map). 1926.

History, topography, general features of flora; list of vascular plants; notes on fauna. (Off the southeast coast of the provincial district of Auckland.)

——— and **Falla, R. A.** Alderman Islands. A general description, with notes on the flora and fauna. New Zeal. Jour. Sci. and Technol. **9**: 193–205. 4 fig. (incl. map). 1927.

History, physiography, plant formations; list of vascular plants. The second part (p. 282–290) is entirely zoological. (Off east coast of Auckland provincial district, 36°58′ S. Lat., 176°5′ E. Long.)

Wall, Arnold, and **Cranwell, L. M.** The botany of Auckland. 47 p. illus. (incl. map). 23 cm. Auckland, 1936.

Vegetation of special localities, etc.; tabular list of 559 vascular plants, with habitat, vernacular names, and occasional localities. Reprinted from the New Zealand Herald. (Vicinity of city of Auckland.)

Zotov, V. D., and others. An outline of the vegetation and flora of the Tararua Mountains. Trans. and Proc. Roy. Soc. New Zeal. **68**: 259–324. pl. 27–33, 2 maps, tab. 1938.

Mainly ecological; includes (p. 290–310) annotated list of vascular plants by V. D. Zotov, N. L. Elder, and A. D. Beddie, also lists of mosses and hepatics; bibliography. (About 40 miles northeast of Wellington.)

SOUTH ISLAND

The following publication has not been available:

Dunedin Field Club. Catalogues of the indigenous and introduced flowering plants, ferns and seaweeds, occurring in the Dunedin district. Dunedin, 1896. (Cited from Cockayne, L. The vegetation of New Zealand. 1921. p. 15.)—See also **Aston, B. C.** List of plants supplementary to the Dunedin Field Club's Catalogue . . . Trans. and Proc. New Zeal. Inst. **31:** 747. 1898.

Anonymous. [List of trees and shrubs found in the Province of Otago.] *In* New Zealand Exhibition, 1865. Reports and awards . . . p. 267–275. Dunedin, 1866.

Briefly annotated list with vernacular names, the plants classified by uses.

Armstrong, J. B. A short sketch of the flora of the Province of Canterbury, with catalogue of species. Trans. and Proc. New Zeal. Inst. **12:** 325–353. 1880.

Botanical districts, forage plants, weeds, economic plants, etc.; list of 857 vascular plants, with general distribution in area indicated; also cellular plants, and list of naturalized plants additional to that in same journal, v. 4, p. 284.

Armstrong, J. F. On the naturalized plants of the Province of Canterbury. Trans. and Proc. New Zeal. Inst. **4:** 284–290. 1872.

General features of occurrence of introduced plants; list of 171 species. For additions see vol. **12**, p. 353.

Aston, B. C. The botany of Tapu-ae-nuku, New Zealand. Bul. Misc. Inform. Kew **1916:** 182–184. 1916.

Account of an ascent of Mount Tapu-ae-nuku, the highest point in the northern part of South Island (9,467 ft.), with mention of plants found. (Province of Marlborough.)

Buchanan, John. Notes on the botany of the Province of Marlborough, made during a visit there in the months of November, December, and January, 1866–67. Jour. Linn. Soc. Bot. **10:** 63–68. 1869 (1867–68). (Reprinted in his Botanical notes on the Kaikoura Mountains and Mount Egmont. 1867. p. 3–10.)

General features of flora, etc.; unannotated list of vascular plants.

——— Sketch of the botany of Otago. Trans. and Proc. New Zeal. Inst. **1** (3): 22–53. 1869.

General features of flora; annotated list of more conspicuous or useful plants; tabular list of vascular plants, with vernacular names and general features of distribution.

Cheeseman, T. F. Contributions to a flora of the Nelson Provincial district. Trans. and Proc. New Zeal. Inst. **14:** 301–329. 1882.

Annotated list of 666 vascular plants.—For annotated supplementary list, see **Kirk, T.** Additional contributions to the flora . . . l. c. **18:** 318–324. 1886. (Northwestern part of South Island.)

Cockayne, Leonard. List of plants from the Franz Josef Glacier to the coastline. *In* **Bell, J. M.** A geographical report on the Franz Josef Glacier . . . p. 11–14. Wellington, 1910.

Unannotated list of vascular plants (356 species including the additions).—See also his Additional species for the neighbourhood of the Franz Josef Glacier from the sea to about 1,200 m. altitude, with his List of species of Omeroa Saddle (Some hitherto-unrecorded plant habitats VII). Trans. and Proc. New Zeal. Inst. **44:** 55–59. 1912. (In Southern Alps, near the center of eastern boundary of Westland.)

——— Notes on the vegetation of the Open Bay Islands. Trans. and Proc. New Zeal. Inst. **37:** 368–377. pl. 23. 1905.

Physiography, soil, plant zones; list of vascular plants. The Open Bay Islands are two small islets in the Tasman Sea, 3 miles from the shore of south Westland, nearly opposite the mouth of the Okuru River.

——— On a collection of plants from the Solanders. Trans. and Proc. New Zeal. Inst. **41:** 404–405. 1909.

List of 19 vascular plants collected by Capt. J. Bollons in 1908. (About 23 miles south of southwest end of South Island.)

——— Report of a botanical survey of Stewart Island. 68 p. 9 pl., map. 33 cm. Wellington, 1909.

History, botanical explorations, physiography, climate, ecology, plant formations, phytogeography, bird life, etc.; table showing local and general range, vernacular names, etc., of indigenous vascular plants; list of introduced plants; bibliography. (Large island just south of South Island.)

Hamilton, Augustus. List of plants collected in the district of Okarito, Westland. Trans. and Proc. New Zeal. Inst. **11**: 435–438. 1879.

Unannotated list of vascular plants collected by author in 1877.—See also **Kirk, T.** Notes on Mr. Hamilton's collection . . . l. c. 439–444.

Kirk, Thomas. On the botany of the Bluff Hill. Trans. and Proc. New Zeal. Inst. **10**: 400–406. 1878.

General features of flora; unannotated list of vascular plants observed by author on January 1, 1877. (Southeastern coast of Southland.)

——— On the flowering plants of Stewart Island. Trans. and Proc. New Zeal. Inst. **17**: 213–228. 1885.

Botanical explorations, physiography, geology, general features of flora, special notes on endemic or rarer species; no list of flora (said to be about 450 species). The paper is followed by "On the ferns and fern allies of Stewart Island," p. 228–234, by the same author, an annotated list. (Large island just south of South Island.)

Laing, R. M. Some notes on the botany of the Spenser Mountains, with a list of the species collected. Trans. and Proc. New Zeal. Inst. **44**: 60–75. 1912.

History, topography, general features of flora, etc.; unannotated list of vascular plants. (In Southern Alps, on border between provincial districts of Canterbury and Nelson.)

——— The vegetation of Banks Peninsula, with a list of species (flowering-plants and ferns). Trans. and Proc. New Zeal. Inst. **51**: 355–408. 1 fig. (map). 1919.—Supplement 1. l. c. **55**: 438–444. 1 fig. 1924.

Physiography, climate, agents altering plant-distribution, plant associations, phytogeography, bibliography; annotated list of indigenous vascular plants. (Southeast of Christchurch.)

——— and **Oliver, W. R. B.** Vegetation of the upper Bealey River basin, with a list of the species. Trans. and Proc. New Zeal. Inst. **59**: 715–730. 1929.—Vegetation of the Bealey River basin. Supplement to list of species (1929), with notes. l. c. **64**: 1–10. pl. 1. 1934.

General features of flora, plant formations; annotated list of vascular plants. H. W. Gourlay, joint author of supplement. (In vicinity of Mount Rolleston in Southern Alps, Province of Canterbury.)

Lindsay, W. L. Contributions to New Zealand botany. 102 p. 28.5 cm. London and Edinburgh, 1868.

Botanical explorations, climate, physiography, geology, etc.; unannotated list of vascular and cellular plants; critical notes on various species; index of Maori names, etc. The area covered is the Province of Otago.

Petrie, Donald. List of the flowering plants indigenous to Otago, with indications of their distribution and range in altitude. Trans. and Proc. Nek Zeal. Inst. **28**: 540–591. 1896.—Supplement . . . l. c. **29**: 421–422. 1897.

Annotated list of phanerogams collected by the author in Otago and on Stewart Island; list of species reported by others.—For additions and corrections, with sketch of botanical explorations, see **Kirk, T.** On the history of botany in Otago. l. c. **29**: 532–540. 1897; also **Aston, B. C.** l. c. **31**: 747. 1899.

——— A visit to Stewart Island, with notes on its flora. Trans. and Proc. New Zeal. Inst. **13**: 323–332. 1881.

Physiography, general features of flora; list of phanerogams collected by author and G. M. Thomson in January 1880. The paper refers only to the region of Paterson's Inlet and near Port Pegasus. (Large island just south of South Island.)

Poppelwell, D. L. Botanical results of an excursion to the Upper Makarora Valley and the Haast Pass, supported by a list of the species observed. Trans. and Proc. New Zeal. Inst. **49**: 161–166. 1917.

Unannotated list of vascular plants collected by author and others in December 1915. (Along border between provincial districts of Westland and Otago.)

——— Notes of a botanical excursion to Bunkers Island (Stewart Island). Trans. and Proc. New Zeal. Inst. **50**: 154–157. 1918.

Geology, plant formations; unannotated list of 75 vascular plants. (Off northeast coast of Stewart Island.)

Poppelwell, D. L. Notes of a botanical excursion to Long Island, near Stewart Island, including a list of species. Trans. and Proc. New Zeal. Inst. **49**: 167–171. 1917.

Physiography, ecology, plant formations; list of 75 vascular plants. (Off southwest coast of Stewart Island.)

——— Notes of a botanical excursion to northern portion of the Eyre Mountains. Trans. and Proc. New Zeal. Inst. **45**: 288–293. 1913.

Annotated list of 177 vascular plants collected by author in April 1912. (Near Queenstown, Province of Otago.)

——— Notes of a botanical visit to Herekopere Island, Stewart Island. Trans. and Proc. New Zeal. Inst. **47**: 142–144. 1915.

Running account of plants observed by author in November 1913. (In Foveaux Strait, 5 or 6 miles from Halfmoon Bay.)

——— Notes on a botanical visit to Coll or Bench Island (Stewart Island). Trans. and Proc. New Zeal. Inst. **50**: 158–159. 1918.

General features of flora; unannotated list of 54 vascular plants. (Off northeast coast of Stewart Island.)

——— Notes on the botany of Routeburn Valley and Lake Harris Saddle. Trans. and Proc. New Zeal. Inst. **46**: 22–29. 1914.

General features of flora; annotated list of 205 vascular plants. (Along northwestern border of Southland and Otago divisions, provincial district of Otago.)

——— Notes on the indigenous vegetation of Ben Lomond, with a list of species. Trans. and Proc. New Zeal. Inst. **52**: 248–252. 1920.

Annotated list of 164 vascular plants. (Provincial district of Otago, near Queenstown.)

——— Notes on the indigenous vegetation of the northeastern portion of the Hokonui Hills, with a list of species. Trans. and Proc. New Zeal. Inst. **52**: 239–247. 1920.

Annotated list of 285 vascular plants. (Southeastern part of Southland portion of Otago, between Gore, Winton, and Lumsden.)

——— Notes on the plant covering of Codfish Island and the Rugged Islands. Trans. and Proc. New Zeal. Inst. **44**: 76–87. pl. 9. 1912.

Topography, plant associations; annotated list of 116 vascular plants. (Off northeast coast of Stewart Island.)

——— Notes on the plant-covering of Pukeokaoka, Stewart Island. Trans. and Proc. New Zeal. Inst. **48**: 244–245. 1916.

Running account of plants observed by the author in January 1915. (Small island in Foveaux Strait, between Motunui and Herekopere Islands.)

——— Notes on the plant-covering of the Breaksea Islands, Stewart Island. Trans. and Proc. New Zeal. Inst. **48**: 246–252. 1916.

Physiography, plant formations, introduced plants; list of 69 indigenous vascular plants. (Group of 6 small islands east of Stewart Island, between Port Adventure and Lord's River, 47°6′ S. Lat., 168°15′ E. Long.)

——— Notes on the plant covering of the Garvie Mountains, with a list of species. Trans. and Proc. New Zeal. Inst. **47**: 120–142. 1915.

Botanical explorations, topography, plant associations; annotated list of 360 vascular plants. (On border of Otago and Southland portions of the provincial district of Otago.)

——— and **Thomson, W. A.** Notes of a botanical visit to Hollyford Valley and Martin's Bay, with a list of indigenous plants. Trans. and Proc. New Zeal. Inst. **50**: 146–154. 1918.

Topography, plant associations; annotated list of vascular plants collected by authors and others in 1916–17. (Northwestern part of Otago proper.)

Simpson, George, and Thomson, J. S. Results of a brief botanical excursion to Rough Peaks Range. New Zeal. Jour. Sci. and Technol. **8**: 372–378. 4 fig. 1926.

Lists of vascular plants from different altitudes collected by authors in March 1926. The range had not been botanized before. (Near Lake Wakatipu, Province of Otago.)

Smith, J. C. List of phanerogamic plants indigenous in the Southland district. Trans. and Proc. New Zeal. Inst. **46**: 220–246. 1 fig. (map). 1914.

Annotated list of vascular plants. (Southwestern part of South Island.)

——— Notes on the botany of Lake Hauroko district. Trans. and Proc. New Zeal. Inst. **43**: 248–253. 1911.

General features of flora; annotated list of 225 vascular plants. (Southland region, provincial district of Otago.)

Smith, J. C. The vegetation of Awarua Plain. Trans. and Proc. New Zeal. Inst. **58**: 55–56. pl. 1. 1927.

General features of flora; list of vascular plants. The Awarua Plain is a swamp bog covering many thousand acres. (Southland region, provincial district of Otago.)

Smith, W. W. Plants naturalised in the County of Ashburton. Trans. and Proc. New Zeal. Inst. **36**: 203–225. 1904.

General account of introductions, etc.; list of 368 angiosperms, with country of origin. (Central coastal region, Province of Canterbury.)

Speight, Robert, Cockayne, Leonard, and Laing, R. M. The **Mount Arrowsmith** district; a study in physiography and plant ecology. Trans. and Proc. New Zeal. Inst. **43**: 313–378. 3 fig. (incl. 2 maps), pl. 3–7. 1911.

Topography, hydrography, glaciers, plant formations, etc.; annotated list of 357 vascular plants; bibliography. (In Southern Alps, on border between Provinces of Canterbury and Westland.)

———**Wall, Arnold, and Laing, R. M.**, eds. Natural history of Canterbury. Issued by the Philosophical Institute of Canterbury . . . ix, 299 p. plates, ports., maps. 22 cm. Christchurch, 1927.

Includes among other contributions dealing with botany, "The history of botanical research in Canterbury" (p. 97–103, including bibliography), by R. M. Laing; "The vegetation and flora of the Canterbury Plain" (p. 115–144), by L. Cockayne, with annotated tabular list of 341 vascular plants; "Alpine and sub-alpine flora" (p. 145–159), by A. Wall, with lists of characteristic species growing at the higher altitudes.

Townson, William. On the vegetation of the Westport district. Trans. and Proc. New Zeal. Inst. **39**: 380–433. pl. 14–14a (incl. map). 1907.

Topography, geology, general features of flora; annotated list of 789 vascular plants.—See also **Petrie, D.** Some additions to the flora of the Westport district. l. c. **46**: 30–31. 1914. (Provincial district of Nelson.)

Wall, Arnold. The botany of Christchurch. 41 p. Christchurch, [1922].

"The paper gives details of plant stations, with a complete list of species." (Not seen; data from Bot. Abstr. **14**: 366. 1925.) (Province of Canterbury.)

——— The flora of Mount Cook. 55 p. illus. Christchurch, 1925.

Plant communities, growth forms, etc.; list of vascular plants. (Not seen; data from Bot. Abstr. **15**: 529. 1926.) (Province of Canterbury.)

TASMANIA

See also Australia (general works).

General

Archer, William. Notes on the vegetable products of Tasmania . . . *In* **Whiting, George.** The products and resources of Tasmania, as illustrated in the International exhibition, 1862. 2d ed. enl. and corr. p. 39–42. Hobart Town, 1862. (Reprinted in Technologist **3**: 1–5. 1863.)

Annotated list, mainly of useful trees.

Hooker, J. D. Flora Tasmaniae. 2 v. 200 col. pl. 33 cm. London, 1860 (1855–60). (The botany of the Antarctic voyage of H. M. discovery ships *Erebus* and *Terror*, in the years 1839–1843 . . . pt. III.)

General features of floras of Australia, Tasmania, etc.; phytogeography; annotated list of edible plants of Australia; botanical explorations in Australia; annotated descriptive flora of Tasmania, with key to families, local and general range. Vol. 1, Dicotyledons and Coniferae; vol. 2, Monocotyledons and cryptogams (the cellular cryptogams by other authors).

Maiden, J. H. Records of Tasmanian botanists. Papers and Proc. Roy. Soc. Tasmania **1909**: 9–29. [1910.]

Bibliography; alphabetical list of botanists and collectors with biographical information.

Rodway, Leonard. The Tasmanian flora. xix, 320 p. plates. 24.5 cm. Hobart, 1903.

Flora of vascular plants, with keys, brief descriptions, local and general range; glossary, including vernacular names (not given in text).

Spicer, W. W. A handbook of the plants of Tasmania. xxiv, 160 p. 4 pl. 19 cm. Hobart Town, 1878.

Annotated list of 1,096 vascular plants, with keys; glossary. Intended for popular use.

Local

Gibbs, L. S. Notes on the phytogeography and flora of the mountain summit plateaux of Tasmania. Jour. Ecology **8**: 1–17, 89–117. 1920.

Geology, physiography, climate, plant formations, descriptions of mountains, plant associations of mountain summit plateaus, phytogeography (all with footnote references); annotated list of cellular and vascular plants collected by the author in 1914–15, with general range indicated.

Mueller, Ferdinand von. Deal Island. Papers and Proc. Roy. Soc. Tasmania **1883**: 282–283. 1884.

Unannotated list of vascular plants. (In Kent's Group in Bass Strait, 39° 28′ S. Lat., 147° 22′ E. Long.)

——— Flora of King Island. List of plants collected . . . Victorian Nat. **4**: 140–146. 1888.

Unannotated lists of vascular plants and of introduced species. (Between Tasmania and Australia, 40° S. Lat., 144° E. Long.)

Sutton, C. S. A sketch of the vegetation of the Cradle Mountain, Tasmania, and a census of its plants. Papers and Proc. Roy. Soc. Tasmania **1928**: 132–159. 1929.

Mainly ecological; includes tabular list of vascular plants, with ecological occurrence indicated. (In west-central Tasmania, 41° 42′ S. Lat., 145° 58′ E. Long.)

INSULAR FLORAS

ATLANTIC OCEAN

ASCENSION ISLAND

General

Hemsley, W. B. Ascension. *In* his Report on the botany of the Bermudas and various other islands . . . 2d pt. Report on the scientific results of the voyage of H. M. S. *Challenger* during the years 1873–76. Botany. v. 1, pt. 2, p. 31–48. pl. 16–17. London, 1885 (1884).

Physiography, general features of flora, etc.; annotated list of plants (including cellular cryptogams by other authors). Includes 12 phanerogams and 14 vascular cryptogams.

Local

Rudmose Brown, R. N. Contributions towards the botany of Ascension. Trans. and Proc. Bot. Soc. Edinb. **23**: 199–204. 1906. (Reprinted in Report on the scientific results of the voyage of S. Y. *Scotia* . . . v. 3, Botany. 1912.)

Physiography, etc.; annotated list of vascular (10 species) and cellular plants collected by the author.

Watson, Sereno. Notes upon a collection of plants from the island of Ascension. Proc. Amer. Acad. Arts and Sci. **26**: 161–163. 1891. (Reprinted as Contributions to American botany, Gray Herbarium, no. 18, 4.)

List of 16 vascular plants (plus 6 cellular), collected on United States Eclipse Expedition of 1889; adds 9 vascular plants to Hemsley's flora (1884).

AZORES

General

Seubert, Moritz. Flora azorica quam ex collectionibus schedisque Hochstetteri patris et filii elaboravit . . . vi, 50 p. 15 pl. 28 cm. Bonnae, 1844.

Annotated list of 400 (or 391) species, of which 67 are cellular plants; 50 species endemic; geography, physiography, zones of vegetation.

Trelease, William. Botanical observations on the Azores. Ann. Rpt. Missouri Bot. Gard. **8:** 77–220. pl. 12–66. 1897.

Annotated list of all plants known from the Azores, with references to descriptions; geography, meteorology, general features of vegetation; bibliography.—See also **Tutin, T. G.,** and **Warburg, E. F.** Notes on the flora of the Azores. (Contributions from the University herbarium, Cambridge.) Jour. Bot. **70:** 7–13, 38–46. 1932.

Watson, H. C. Botany of the Azores. *In* **Godman, F. DuC.** Natural history of the Azores . . . p. 113–288. London, 1870.

Annotated list of 478 vascular plants; history of botanical exploration, with critical remarks on previous lists; geographical affinities of flora, with tabular list.

BERMUDA ISLANDS

General

Britton, N. L. Flora of Bermuda. xi, 585 p. illus., col. pl. 23.5 cm. New York, 1918.

Complete flora of vascular and cellular plants, including cultivated as well as native and introduced species, with keys, synonymy, description, extralimital range, and text figure of all native and naturalized species above the thallophytes; brief description of region and statistics of distribution; annotated bibliography; glossary; list of principal collections made in Bermuda. Musci by E. G. Britton, Hepaticae by A. W. Evans, Lichenes by L. W. Riddle, Fungi by F. J. Seaver. Algae by M. A. Howe.

Hemsley, W. B. The Bermudas. *In* his Report on the botany of the Bermudas and various other islands. . . . Report on the scientific results of the voyage of H. M. S. *Challenger* during the years 1873–76. Botany. v. 1, pt. 1. 135 p. pl. 1–13. London, 1885 (1884).

Physiography, climate, general features of flora, botanical explorations, statistics; systematic list of vascular plants (also cellular plants, by various authors) with localities, names of collectors, general range. Includes 326 vascular plants, of which 144 are probably indigenous.

CANARY ISLANDS

General

Lindinger, Leonhard. Beiträge zur kenntnis von vegetation und flora der Kanarischen Inseln. 2. teil. Flora der Kanarischen Inseln. Berichtigungen und nachträge zu J. Pitard et L. Proust, Les îles Canaries. Flore de l'archipel. Paris, 1909. Abhandl. Gebiet Auslandsk. Hamburg. Univ. **21** (reihe C, Naturw. bd. **8**): 135–350. 1926.

Bibliography; list of vascular plants (alphabetically arranged by families) with citation of all records additional to Pitard and Proust.

Pitard, Joseph, and **Proust, Louis.** Les Iles Canaries. Flore de l'archipel. 502 p. 19 pl. 25 cm. Paris, [1908].

Botanical collections, topography, climate, soils, zones of vegetation, phytogeography, endemic forms; annotated list of vascular plants (about 1,352 species), Musci, and Hepaticae, with pertinent synonymy, localities, and general range.

Sauer, Fritz. Catalogus plantarum in Canariensibus insulis sponte et subsponte crescentium. 78 p. 21 cm. Halis Saxonum, 1880. (Dissertation.)

Bibliography; tabular list, showing general distribution; annotations, statistics, phytogeography.

Webb, P. B., and **Berthelot, Sabin.** Phytographia canariensis. *In* their Histoire naturelle des Iles Canaries. t. 3, 2. ptie. 4 sect. in 6 v. 288 pl. (pt. col.). 35 cm. Paris, 1836–1850.

Descriptive flora, in Latin, with synonyms, vernacular names, local and extralimital range, of vascular and cellular plants (the latter by C. Montagne). For dates of publication see Jour. Soc. Bibl. Nat. Hist. **1:** 55–58. 1937.

Local

Bolle, Carl. Florula insularum olim Purpurariarum, nunc Lanzarote et Fuerteventura cum minoribus Isleta de Lobos et la Graciosa in archipelago canariensi. Bot. Jahrb. Engler **14:** 230–257. 1891.

Annotated list of 400 vascular plants. The general features of flora are discussed in his "Botanische rückblicke auf der Inseln Lanzarote und Fuertaventura," l. c. **16:** 224–261. 1892. A single unnumbered page "Omissa et addenda ad Florulam . . .," raising the total to 416, was issued with Bot. Jahrb. Engler v. 15 in 1892 (date according to Bot. Centralbl. **51:** 89).

Knoche, Herman. Die Kanarischen Inseln. His Vagandi mos. Reiseskizzen eines botanikers, I. 304 p. front., illus. (incl. maps), 23 pl. 25.5 cm. Strasbourg, 1923.

Ecology of different islands, etc.; bibliography (p. 40–41); list of vascular plants (and lichens) collected by the author (p. 211–279).

Masferrer y Arquimbau, Ramón. Recuerdos botánicos de Tenerife; ó sea, datos para el estudio de la flora canaria. Anales Soc. Españ. Hist. Nat. **9:** 309–369. 1880; **10:** 139–230. 1881; **11:** 307–398. 1882.

Annotated list of spermatophytes of Tenerife, with notes on those of the Canary Islands in general, Madeira, and Azores. Never completed; Plantago is last genus treated.

Schenck, Heinrich. Beiträge zur kenntnis der vegetation der Canarischen Inseln . . . *In* Wissenschaftliche ergebnisse der Deutschen tiefsee-expedition auf dem dampfer "*Valdivia*" 1898–1899. v. 2, teil 1, p. 227–406. 71 fig. (incl. maps), pl. 16–27. Jena, 1907.

Botanical explorations, bibliography, physiography, climate, plant formations, lists of species of different localities, etc.

CAPE VERDE ISLANDS

General

Cardoso, João, jr. Agrupamentos para a classificação thérapeutica das plantas medicinaes do Archipelago de Cabo Verde. *In* his Subsidios para a materia medica e thérapeutica das possessões ultramarinas portuguezas. v. 1, p. 1–12. Lisboa, 1902.

Unannotated list of medicinal plants, grouped by uses.—On p. 97–100, unannotated alphabetical "Catalogo de plantas medicinaes do Archipelago de Cabo Verde."

Chevalier, Auguste. Les Iles du Cap Vert. Géographie, biogéographie, agriculture, flore de l'archipel. Revue Bot. Appl. et Agr. Trop. **15:** 733–1090. fig. 22–32 (incl. map), 16 pl. 1935.

Author's itinerary and bibliography, prehistory, discovery, colonization, geography, climate, geology, soil, general features of fauna, population, bibliography; botanical explorations, statistics, phytogeography, analysis of flora, zones of vegetation, plant associations, weeds, agriculture, chronological bibliography of flora; annotated list of 636 vascular plants known from the group, including cultivated plants (about 300 species considered native), with local and general range and vernacular names (indexed); lists of cellular plants collected by the author.

Schmidt, J. A. Beiträge zur flora der Cap Verdischen Inseln. Mit berücksichtigung aller bis jetzt daselbst bekannten wildwachsenden und kultivirten pflanzen . . . viii, 356 p. 22 cm. Heidelberg, 1852.

History, climate, cultivated plants, phytogeography, plant formations, statistics; list of wild and cultivated vascular and cellular plants (435 vascular), with localities, collectors' names, and general distribution.

Local

Coutinho, A. X. P. Herbarii gorgonei universitatis olisiponensis catalogus. Arq. Univ. Lisboa **1:** 265–334. 1914.—Supplementum . . . l. c. **2:** 27–59. 1915.

List of 311 vascular plants, with references and localities, in the herbarium of Lisbon University, collected principally by Cardoso and Borjona in Cape Verde Islands.

Krause, E. H. L. Flora der insel St. Vincent in der Capverdengruppe. Bot. Jahrb. Engler **14:** 394–425. 1891.

Physiography, climate, etc.; annotated list of 183 vascular plants; plant formations, phytogeography, origin of flora.

COCKBURN ISLAND

General

Schenck, Heinrich. Vergleichende darstellung der pflanzengeographie der subantarktischen inseln . . . VI. Antarktisches polargebiet. *In* Wissenschaftliche ergebnisse der Deutschen tiefsee-expedition auf dem dampfer "*Valdivia*" 1898–1899. v. 2, teil 1, p. 161–178. Jena, 1905.

Botanical explorations, bibliography, climate; separate tabular lists of cellular plants from Cockburn Island, Victoria Land, and Gerlache Strait. The only vascular plant known from the area is *Deschampsia antarctica* (see p. 162). Cockburn is a small island off the east coast of James Ross Island, Louis-Philippe Land, West Antarctica.

FALKLAND ISLANDS

General

Schenck, Heinrich. Vergleichende darstellung der pflanzengeographie der subantarktischen inseln . . . III. Falkland-inseln. *In* Wissenschaftliche ergebnisse der Deutschen tiefsee-expedition auf dem dampfer "*Valdivia*" 1898–1899. v. 2, teil 1, p. 96–106. Jena, 1905.

Botanical collections, bibliography, physiography, climate, general features of flora (no complete list), plant formations.

Skottsberg, Carl. A botanical survey of the Falkland Islands. (Botanische ergebnisse der Schwedischen expedition nach Patagonien und dem Feuerlande 1907–09. III.) K. Svenska Vetensk.-Akad. Handl. v. 50, no. 3. 129 p. 10 fig., 14 pl., map. 1913.

Botanical explorations, annotated list of all known vascular plants, with localities and extralimital range; phytogeography, growth forms, phenology, plant associations, bibliography.—See also his Bemerkungen zu einigen von M. Gandoger neuerdings von den Falkland-inseln beschriebenen pflanzen. Bot. Jahrb. Engler **50** (Beibl. **112**): 13–17. 1913, for minor corrections and for reduction of numerous new species described by Gandoger.

——— Studien über das pflanzenleben der Falkland-inseln. *In* Wissenschaftliche ergebnisse der Schwedischen südpolar-expedition 1901–1903. v. 4, pt. 10. 58 p. 5 fig. Stockholm, 1909.

Physiography, climate, plant associations, periodicity, bibliography, etc.; annotated list of vascular plants (including all previous records); plant geography (139 native species, of which 10 are endemic).

Wright, C. H. Flora of the Falkland Islands. Jour. Linn. Soc. Bot. **39**: 313–339. 1911.

Annotated list of vascular plants, with references.

Local

Birger, Selim. Die vegetation bei Port Stanley auf den Falklandsinseln. Bot. Jahrb. Engler **39**: 275–305. 1 fig., 2 pl. 1906.

Botanical explorations with bibliographical footnotes, climate, plant associations, phenology; includes list of species new to the islands.

Skottsberg, Carl. C. V. Marquand's "Additions to the flora of the Falkland Islands." Acta Horti Gothoburg. **4**: 303–306. 1929.

Critical notes on the paper listed (Bull. Misc. Inform. Kew **1923**: 369–371. 1923), as well as on the nomenclature of certain plants in his own "Botanical survey" of 1913.

Vallentin, Mrs. E. F., and Cotton, Mrs. E. M. Illustrations of the flowering plants and ferns of the Falkland Islands. xii p., 65 l. 64 col. pl. 25.5 cm. London, 1921.

Colored plates with dissections, accompanied by descriptions and brief annotations.

FERNANDO NORONHA

General

Hemsley, W. B. Fernando-Noronha and contiguous islets. *In* his Report on the botany of the Bermudas and various other islands . . . Report on the scientific results of the voyage of H. M. S. *Challenger* during the years 1873–76. Botany. v. 1, pt. 2, p. 9–30. pl. 14–15, 47. London, 1885 (1884).

Physiography, climate, general features of flora; systematic list of plants (vascular and cellular), with statement of general range. Includes 58 phanerogams and no vascular cryptogams.

Ridley, H. N. Notes on the botany of Fernando Noronha. Jour. Linn. Soc. Bot. **27**: 1–95. pl. 1–4. 1890.

Physiography, history, origin of flora, etc.; annotated list of vascular and cellular plants, the latter contributed by other authors; geology.

GRAHAM LAND

General

Skottsberg, Carl. Einige bemerkungen über die vegetationsverhältnisse des Graham-Landes. *In* Wissenschaftliche ergebnisse der Schwedischen südpolar-expedition 1901–1903. v. 4, pt. 13. 16 p. 2 fig. (incl. map), 3 pl. Stockholm, 1912.

Climate, etc.; lists of plants of Graham Land, adjacent small islands, and South Shetland Islands, collected by this expedition; bibliography. Only one vascular plant listed (*Aira antarctica* [*Deschampsia antarctica*]); *Colobanthus crassifolius* (d'Urv.) Hook. f. also is known from Graham Land (Charcot's first expedition).

——— Some remarks upon the geographical distribution of vegetation in the colder southern hemisphere. Ymer **25**: 402–427. 6 fig., pl. 7–8 (maps). 1905.

Mainly phytogeographical, with summary of knowledge of flora of Antarctic islands; bibliography. Cited here for its sketch (p. 410–414) of the flora of Graham Land.

MADEIRA ISLANDS

General

Lowe, R. T. A manual flora of Madeira and the adjacent islands of Porto Santo and the Desertas. 2 v. (v. 1–2, pt. 1). 19.5 cm. London, 1868 (1857–72).

Plant zones; annotated descriptive flora of vascular plants. Incomplete (Thalamiflorae, Calyciflorae, Corolliflorae).—See also **Menezes, C. A. de.** Notice sur les phanérogames de Madère et Porto Santo, non indiquées dans la flore de ces îles de R. T. Lowe. 22 p. 23 cm. Funchal, 1899. Includes 390 species of spermatophytes, with brief annotations and vernacular names; list of special localities with local species; bibliography. Nearly the same material, lacking list of special localities and bibliography, was published as his "Catalogo das phanerogamicas de Madeira e do Porto Santo não indicadas na flora d'estas ilhas do R. T. Lowe." 1894.

Menezes, C. A. de. Flora do archipelago da Madeira (phanerogamicas e cryptogamicas vasculares). 282 p. 22 cm. Funchal, 1914.

Annotated list of 951 vascular plants (both wild and cultivated), with localities, pertinent references, vernacular names, and usually brief descriptions; climate, zones of vegetation, annotated list of botanists, bibliography.—For supplement, see his "Subsidios para o estudio da flora . . ." Brotéria Sér. Bot. **20**: 113–119. 1922 and "Novos subsidios . . ." l. c. **22**: 20–27. 1926.—Also his Mais algumas notas sobre a flora madeirense. l. c. **23**: 73–77. 1927; Nota acêrca de algumas plantas de flora do arquipélago da Madeira. Jor. Sci. Mat., Fís. e Nat. Lisboa **24** (III, 5): 183–187. 1927.

Porto da Cruz, *visconde do.* A flora madeirense na medicina popular. Brotéria Ciên. Nat. **4**: 35–46, 71–78, 139–144, 145–154. 3 fig. (incl. map). 1935.

Running account of medicinal plants, with vernacular and botanical names, grouped under the diseases for which they are used.

Vahl, Martin. Madeiras vegetation. Geografisk monografi. 172 p. 1 pl. 23.5 cm. København og Kristiania, 1904.

Physiography, climate, ecology, phytogeography, etc.; cited here mainly for its bibliography.

Local

Menezes, C. A. de. A flora dos mais altos picos de Ilha da Madeira. Brotéria Sér. Bot. **22**: 124–132. 1926.

List of 44 vascular plants, with localities.

PALMER LAND

General

Hooker, J. D. Flora antarctica. 2 v. 187 pl. 30.5 cm. London, 1844–48. (The botany of the Antarctic voyage of H. M. discovery ships *Erebus* and *Terror*, in the years 1839–1843 . . . pt. I.)

Vol. 1: Lord Auckland's Group and Campbell's Island. Physiography, etc.; annotated list of vascular and cellular plants. Vol. 2: Antarctic regions (exclusive of Lord Auckland's Group and Campbell's Island). Annotated list of vascular and cellular plants from Fuegia, Falkland Islands, Palmer Land, South Shetland Islands, South Georgia, Tristan d'Acunha, and Kerguelen's Land. All the regions mentioned, except Palmer Land, are covered by separate later floras.

ST. HELENA

General

Hemsley, W. B. St. Helena. *In* his Report on the botany of the Bermudas and various other islands . . . 2d pt. Report on the scientific results of the voyage of H. M. S. *Challenger* during the years 1873–76. Botany. v. 1, pt. 2, p. 49–122. pl. 18–22, 48–51. London, 1885 (1884).

Physiography, climate, historical sketch of the vegetation, geographical affinities, etc.; annotated list of plants (vascular and cellular), with extra-limital range. About 94 species are considered probably indigenous.

Melliss, J. C. St. Helena: a physical, historical, and topographical description of the island, including its geology, fauna, flora, and meteorology. xiv, 426 p. plates (partly col.). 26.5 cm. London, 1875.

(Botany, p. 221–383, col. pl. 25–56.) Annotated list of 1,058 plants (of which 907 are vascular) growing on the island, a very large number of which are known only in cultivation. Nearly all the endemic plants are illustrated.

SALVAGE ISLANDS

General

Lowe, R. T. Florulae salvagicae tentamen; or a list of plants collected in the Salvages or Salvage Islands by Sr. Constantino Cabral de Noronha . . . 24 p. 18 cm. London, 1869.

Physiography, etc.; annotated list of 74 plants (64 vascular), with extra-limital range; general features of flora.—See also **Menezes, C. A. de.** Subsidios para o conhecimento da flora das Ilhas Selvagens. Jor. Sci. Math., Fís. e Nat. Lisboa **23:** 187–194. 1924.

SOUTH GEORGIA

See also **Australian Islands** (Cheeseman, On the systematic botany).

General

Schenck, Heinrich. Vergleichende darstellung der pflanzengeographie der subantarktischen inseln . . . II. Südgeorgien, nebst Sandwich-inseln, Bouvet-insel. *In* Wissenschaftliche ergebnisse der Deutschen tiefsee-expedition auf dem dampfer "*Valdivia*" 1898–1899. v. 2, teil 1, p. 82–96. Jena, 1905.

Botanical collections, bibliography, geography, physiography, climate; tabular list of 18 vascular plants; formations, etc. No botanical collections have yet been made on the Sandwich Islands (56°–59° S., 26°–28° W.) or Bouvet Island (54°26′ S., 3°24′ E.).

Skottsberg, Carl. Die gefässpflanzen Südgeorgiens. *In* Wissenschaftliche ergebnisse der Schwedischen südpolar-expedition 1901–1903. v. 4, pt. 3. 12 p. 2 pl., map. Stockholm, 1905.

Annotated list of 19 vascular plants; phytogeographical relationships.—See also his The vegetation of South Georgia. *In* Wissenschaftliche ergebnisse der Schwedischen südpolar-expedition 1901–1903. v. 4, pt. 12. 36 p. 4 fig., 6 pl., map. Stockholm, 1912. (Botanical explorations, physiography, etc.; list of species; origin of flora, plant formations, etc.)—For additions, see **Taylor, Norman.** Plants collected on the South Georgia expedition. Sci. Bul. Mus. Brooklyn Inst. Arts and Sci. **2:** 60–63. 1914.

SOUTH ORKNEY ISLANDS

General

Rudmose Brown, R. N., Wright, C. H., and **Darbishire, O. V.** The botany of the South Orkneys. Trans. and Proc. Bot. Soc. Edinb. **23:** 101–110. pl. 3. 1905. (Reprinted with additions and with omission of portion on mosses in Report of the scientific results of the voyage of S. Y. *Scotia* . . . v. 3, Botany. 1912.)

History, topography, climate; lists of mosses (Wright) and lichens (Darbishire); no vascular plants known.

TRINIDAD ISLAND

General

Hemsley, W. B. South Trinidad. *In* his Report on the botany of the Bermudas and various other islands . . . 2d pt. Report on the scientific results of the voyage of H. M. S. *Challenger* during the years 1873–76. Botany. v. 1, pt. 2, p. 123–132. pl. 23–24. London, 1885 (1884).

Physiography, etc.; annotated list of plants (vascular and cellular); general range stated. Includes 9 phanerogams and 4 ferns. Trinidad or South Trinidad is in 20°30′ S. Lat., 29°22′ W. Long.

TRISTAN DA CUNHA ISLANDS

General

Christophersen, Erling, and others. Plants of Tristan da Cunha. 19 p. 27.5 cm. Oslo, 1937. (Scientific results of the Norwegian Antarctic expeditions 1927–1928 et sqq., no. 16.)

Botanical explorations, phytogeography; annotated list of plants collected by Einar Siggeson in 1934; unannotated list of cellular and vascular (104 species) plants known from the group, not including Gough Island; bibliography.

Hemsley, W. B. The Tristan da Cunha group. *In* his Report on the botany of the Bermudas and various other islands . . . 2d pt. Report on the scientific results of the voyage of H. M. S. *Challenger* during the years 1873–76. Botany. v. 1, pt. 2, p. 133–185. pl. 25–39, 45–46. London, 1885 (1884).

Physiography, botanical explorations, statistical table of distribution; annotated list of vascular and cellular plants (65 vascular). The group includes Tristan da Cunha, Inaccessible, and Nightingale Islands.

Local

Christophersen, Erling. Plants of Gough Island (Diego Alvarez). 16 p. 3 fig. 27.5 cm. Oslo, 1934. (Scientific results of the Norwegian Antarctic expeditions 1927–1928 et sqq., no. 13.)

Phytogeography; list of plants collected by Lars Christensen in 1933; unannotated list of cellular and vascular (38) plants recorded from the island; bibliography.

Dyer, R. A. The flora of Tristan da Cunha: H. M. S. *Carlisle* expedition, 1937. Bothalia **3:** 589–612. 7 fig. 1939.

Botanical explorations, affinities, history, and general features of flora; list of cellular and vascular plants collected by author. Adds 6 phanerogams, all presumably introduced.

Phillips, E. P. A list of the phanerogams and ferns collected by Mr. P. C. Keytel on the island of Tristan da Cunha, 1908–1909. Annals So. African Mus. **9:** 96–103. 1913.

List of plants collected, with general range: list of additional species recorded from the island. Includes 37 species not previously recorded.

Rudmose Brown, R. N. The botany of Gough Island.—I. Phanerogams and ferns. Jour. Linn. Soc. Bot. **37:** 238–249. 1 fig., pl. 7–9. 1905. (Reprinted in Report on the scientific results of the voyage of S. Y. *Scotia* . . . v. 3, Botany. 1912.)

Annotated list of 27 vascular plants (2 endemic), with general range; physiography, general features of vegetation, bibliography.

Wilkins, G. H. Gough Island. Jour. Bot. **63**: 65–70. 1925.
General features of flora; annotated list of vascular and cellular plants (phanerogams by E. G. Baker and Wilkins; cryptogams by A. Gepp, H. N. Dixon, and R. Paulson).

INDIAN OCEAN

ALDABRA

See also Seychelles (Hemsley; Sommerhayes).

General

Hemsley, W. B., and others. Flora of Aldabra: with notes on the flora of the neighbouring islands. Bul. Misc. Inform. Kew **1919**: 108–153. 1919.
Physiography, climate, botanical explorations, plant associations, systematic list of vascular plants (about 171 apparently native species) with collectors' names and general range; geographical relationships of flora; bibliography. Includes also lists of plants from other atolls in this general area—Assumption (p. 140–142), Cosmoledo (p. 142–143), Astove (p. 143), Farquhar (p. 143), Providence (p. 143–144), St. Pierre (p. 144–145), Gloriosa (p. 145), Amirante group (p. 145), Coëtivy (p. 146), Agalega (p. 146), Cargados (p. 147), Chagos Archipelago (p. 148).

ANDAMAN ISLANDS

General

Kurz, Sulpiz. Report on the vegetation of the Andaman Islands, accompanied by a report on the forests. 75 p. map. 33 cm. Calcutta, 1870. (Reprint with additions and corrections of an earlier report (27, xxv, 13 p.), 1870.)
Annotated list of vascular plants (520 species indigenous, 76 introduced); general floral features; report on forests; vernacular names; no index. An 1867 edition is listed by B. D. Jackson.

Parkinson, C. E. A forest flora of the Andaman Islands. An account of the trees, shrubs and principal climbers of the islands. v, v, xiii, 325 p. front., 6 pl. 18 cm. Simla, 1923.
Annotated descriptive flora of woody plants, with keys (about 650 species, of which 540 indigenous); vernacular names (indexed); lists of species; geography, physiography, climate, general features of vegetation; history of botanical exploration; bibliography; glossary.

Prain, David. The non-indigenous species of the Andaman flora. Jour. Asiatic Soc. Bengal **59** (2): 235–261. 1891. (Reprinted in his Memoirs and memoranda, chiefly botanical. 1894.)
Annotated lists of cultivated and introduced species, as observed in 1866 and 1889–90; general study of subject of introduced plants.

Rogers, C. G. A preliminary list of the plants of the Andaman Islands. ii, 51 p. 23.5 cm. Port Blair, Andaman Isl., 1903.
Nominal list, unannotated, "Drawn up for me by Mr. J. S. Gamble . . . chiefly from the 'Flora of British India.' "

Local

Kloss, C. B. In the Andamans and Nicobars. The narrative of a cruise in the schooner *Terrapin*, with notices of the islands, their fauna, ethnology, etc. xvi, 373 p. illus., plates. 24.5 cm. London, 1903.
Contains, in addition to passing notes in the text, a list of "Principal forest trees of the Andamans" (p. 336–338) with vernacular names and uses (p. 339–341), and "Principal flora of the Nicobars" (p. 345–349).

Prain, David. On a botanical visit to Little Andaman and the Nicobars. Proc. Asiatic Soc. Bengal **1891**: 156–175. 1892. (Reprinted in his Memoirs and memoranda, chiefly botanical. 1894.)
Physiography, general features of flora; separate lists for Little Andaman (61 species, of which 56 are vascular plants) and Car Nicobar and Batti Malv (113 species, of which 112 are vascular plants).

CHAGOS ARCHIPELAGO

See also Aldabra (Hemsley); Maldive Islands (Willis and Gardiner).

General

Willis, J. C., and Gardiner, J. S. Flora of the Chagos Archipelago. Trans. Linn. Soc. Zool. **19**: 301–306. 1931.

Botanical explorations, etc.; annotated list of vascular plants (98 native or escaped species).

Local

Hemsley, W. B. Report on the vegetation of Diego Garcia. Jour. Linn. Soc. Bot. **22**: 332–340. 1886.

Botanical collections, physiography, etc.; tabular list of 43 probably indigenous vascular plants, with extralimital distribution indicated.

CHRISTMAS ISLAND

General

Andrews, C. W. A monograph of Christmas Island (Indian Ocean). Physical features and geology . . . with descriptions of the fauna and flora by numerous contributors. xiii, 337 p. illus., 21 pl., map. 22.5 cm. London, 1900.

Contains (p. 171–195, pl. 17–18) an annotated list of 152 vascular plants by E. G. Baker, A. B. Rendle, and A. Gepp.

Ridley, H. N. An expedition to Christmas Island. Jour. Straits Branch Roy. Asiatic Soc. **45**: 137–271. 1906.—Christmas Island flora—additional notes. l. c. **48**: 107–108. 1907.

Botanical explorations; critical notes on vernacular names; list of introduced plants, arranged by date of first record; list of endemic species; annotated list of cultivated plants; note on timbers; annotated list of vascular and cellular plants (about 151 indigenous vascular plants, of which 29 are endemic); methods of dissemination.

COCO ISLANDS

General

Prain, David. The vegetation of the Coco group. Jour. Asiatic Soc. Bengal **60** (2): 283–406. 1892. (Reprinted in his Memoirs and memoranda, chiefly botanical. 1894.)

Physiography, general features of flora, annotated list of 358 species (of which 307 are vascular plants), with extralimital distribution; statistical study of distribution and probable origin of flora. The islands are in the Bay of Bengal, not far north of the Andaman Islands.

COMORO ISLANDS

See also Madagascar (Jumelle).

General

Voeltzkow, Alfred. Flora und fauna der Comoren. *In* his Reise in Ostafrika in den jahren 1903–1905 . . . v. 3, p. 429–480. Stuttgart, 1917.

Flora (p. 430–454): botanical explorations; list of cellular and vascular plants with localities and extralimital range, prepared with assistance of G. Schellenberg; bibliography.

CROZET ISLANDS

See also Kerguelen Island (Schenck); Macdonald Islands (Hemsley).

General

Hemsley, W. B. The Crozets. *In* his Report on the botany of the Bermudas and various other islands . . . 2d pt. Report on the scientific results of the voyage of H. M. S. *Challenger* during the years 1873–76. Botany. v. 1, pt. 2, p. 207–210. pl. 40. London, 1885 (1884).

Physiography, etc.; annotated list of plants (vascular and cellular). Includes 5 phanerogams and 2 ferns.

Schenck, Heinrich. Die gefässpflanzen der Deutschen südpolar-expedition 1901–1903 gesammelt auf der Possession-insel (Crozet-gruppe), Kerguelen, Heard-insel, St. Paul und Neu-Amsterdam. *In* Deutsche südpolar-expedition . . . 8. bd. Botanik. p. 97–123. 10 fig. Berlin, 1906.

Annotated lists of vascular plants of Possession Island (18 species), Kerguelen (several lists), Heard Island (7 species), St. Paul and New Amsterdam (4 and 8 species, respectively).

GLORIOSA ISLANDS

See also Aldabra.

Local

Hemsley, W. B. Islands of the Indian Ocean. List of plants in the Kew herbarium collected in some of the smaller tropical islands by Dr. Coppinger. *In* his Report on present state of knowledge of various insular floras . . . Report on the scientific results of the voyage of H. M. S. *Challenger* during the years 1873–76. Botany. v. 1, pt. 1, p. 16–17. London, 1885.

List of vascular plants from Du Lise and Gloriosa Islands, of the Gloriosa group, off Madagascar; Bird Island, in the Seychelles; Eagle Island, Isle des Roches, and Poivre Island, of the Amirante group; Providence and Alphonse Islands and Cerf Islets, Providence Reef, northeast of Madagascar.

KEELING ISLANDS

See also Molucca Islands (Hemsley).

General

Forbes, H. O. List of the Keeling Atoll plants. *In* his A naturalist's wanderings in the eastern archipelago . . . from 1878 to 1883. p. 42–43. London, 1885.

Unannotated tabular list of vascular and cellular plants collected by Charles Darwin and by the author.

Henslow, J. S. Florula keelingensis. An account of the native plants of the Keeling Islands. Annals Nat. Hist. **1:** 337–347. 1838.

Annotated list of 21 vascular and cellular plants collected by Darwin on the voyage of the *Beagle.*

Local

Guppy, H. B. The dispersal of plants as illustrated by the flora of the Keeling or Cocos Islands. Jour. Trans. Victoria Inst. (London) **24:** 267–301. 1890.

Mainly phytogeographical; contains (p. 272–273) a list of 16 phanerogams not recorded from the islands previous to 1888.

KERGUELEN ISLAND

See also Crozet Islands (Schenck); Prince Edward Islands (Moseley); Australian Islands (Cheeseman, On the systematic botany).

General

Hemsley, W. B. Kerguelen Island. *In* his Report on the botany of the Bermudas and various other islands . . . 2d pt. Report on the scientific results of the voyage of H. M. S. *Challenger* during the years 1873–76. Botany. v. 1, pt. 2, p. 211–243. pl. 40. London, 1885 (1884).

Physiography, general features of flora; annotated list of plants (vascular and cellular). Includes 21 phanerogams and 6 pteridophytes.

Hooker, J. D. Botany [of Kerguelen Island]. *In* An account of the petrological, botanical, and zoological collections made in Kerguelen's Land and Rodriguez during the Transit of Venus expeditions . . . 1874–75. Phil. Trans. Roy. Soc. **168** (extra vol.): 9–23. pl. 1–2. 1879.

Sketch of botanical explorations and general features of flora; annotated list of 21 phanerogams, 6 pteridophytes, and 1 Nitella.

Schenck, Heinrich. Vergleichende darstellung der pflanzengeographie der subantarktischen inseln insbesondere über flora und vegetation von Kerguelen. I. Kerguelenbezirk (Kerguelen-, Prinz Eduard-, Crozet-, Macdonald-inseln). *In* Wissenschaftliche ergebnisse der Deutschen tiefsee-expedition auf dem dampfer "*Valdivia*" 1898–1899. v. 2, teil 1, p. 9–82. 33 fig., pl. 1–10, map (in text). Jena, 1905.

Botanical explorations, bibliography, geography, physiography, climate; tabular list of all vascular plants recorded from Kerguelen, Prince Edward (Marion), Crozet (Possession), and Heard Islands (35 species, of which 5 are introduced), with extralimital range; plant formations, phenology, anatomy, etc. Each island treated separately as to botanical history, bibliography, etc.

LACCADIVE ISLANDS

See also Maldive Islands (Willis and Gardiner).

General

Prain, David. Botany of the Laccadives. Jour. Bombay Nat. Hist. Soc. **7:** 268–295, 460–486. 1893; **8:** 57–86, 488. 1893–94. (Reprinted in his Memoirs and memoranda, chiefly botanical. 1894.)

Geography, physiography, botanical collections, bibliography; methods of introduction, etc.; annotated list of 206 plants, of which 194 are vascular.

Local

Willis, J. C. Notes on the flora of Minikoi. Annals Roy. Bot. Gard. Peradeniya **1:** 39–43. 1901.

Bibliography; unannotated list of 134 vascular plants, with vernacular names.

MACDONALD ISLANDS

See also Crozet Islands (Schenck); Kerguelen Island (Schenck); Prince Edward Islands (Moseley).

Local

Hemsley, W. B. MacDonald group. Heard Island. *In* his Report on the botany of the Bermudas and various other islands . . . 2d pt. Report on the scientific results of the voyage of H. M. S. *Challenger* during the years 1873–76. Botany. v. 1, pt. 2, p. 245–248. pl. 53. London, 1885 (1884).

Physiography, etc.; botanical explorations and composition of flora of Heard Island, Kerguelen, the Prince Edward group (Marion Island), and the Crozets, with table; annotated list of 17 plants of Heard Island (5 vascular).

MADAGASCAR

See also French Guiana (Lanessan, Les plantes utiles, p. 873–888, for useful plants of the small island of Nossi Bé).

General

Académie malgache. Catalogue des plantes de Madagascar. 23 nos. 25 cm. Tananarive, 1931–35.

Lists of species, usually arranged alphabetically, with synonymy, vernacular names, citations of localities, geographic area (in Madagascar), and general range; bibliographies. Each family or group of families published separately. Still in course of publication.

Baillon, H. E., and **Drake del Castillo, Emmanuel.** Histoire naturelle des plantes. 6 v. plates. 33 cm. Paris, 1886–1903. (**Grandidier, Alfred.** Histoire physique, naturelle et politique de Madagascar. v. 28–30, 34–36.)

Vol. 2–6 (v. 28–29, 34–36 of whole series) consist of plates only (Hibbertia—Epallage); Baillon's name appears on title page of all but vol. 6, where it is replaced by that of Drake del Castillo. Vol. 1 (v. 30) by Drake del Castillo, published in 1902, is a descriptive flora, with keys, synonymy, vernacular names, uses, local and general range, and citation of exsiccatae; incomplete (208 p.: Ranunculaceae—Leguminosae, Proteaceae), corresponding to only 46 of the plates. No index. Vol. 4 (v. 34) is vol. 2, pt. 4 of the atlas. Vol. 39 (1898–1915) contains the mosses, by F. Renauld and J. Cardot.

Baron, Richard. Compendium des plantes malgaches. Revue de Madagascar **3** (2): 741–760, 856–865, 936–950. 1901: **4** (1): 225–240, 534–542; **4** (2): 39–48, 225–242. 1902; **5** (1): 435–441; **5** (2): 44–54, 134–144, 312–322, 429–437. 1903; **6** (1): 346–353. 1904; **7** (1): 60–67, 247–258; **7** (2): 246–257. 1905; **8**: 350–364, 530–549, 708–726, 813–841, 915–936. 1906.
Systematic enumeration, with synonyms, uses, vernacular names.

——— The flora of Madagascar. Jour. Linn. Soc. Bot. **25**: 246–294. 1 fig. (map). 1889.
General features of flora, statistics, list of endemic genera, description of three floral regions recognized with lists of genera and species; list of introduced species.

——— Notes on the economic plants of Madagascar. Bul. Misc. Inform. Kew **1890**: 203–215. 1890.
Annotated systematic list of 109 species, with vernacular names.

Dandouau, André. Catalogue alphabétique des noms malgaches de végétaux. Bul. Écon. Madagascar **10** (Suppl.): i–xlii. 1910; **11** (Suppl.): xliii–cxxvi. 1911.
Alphabetical list of vernacular names, compiled from previous works and collected by the author himself, with botanical equivalents.

Heckel, Édouard. Catalogue alphabétique raisonné des plantes médicales et toxiques de Madagascar avec leur emploi indigène. Annales Mus. Colon. Marseille v. 11 (s. 2, v. 1), fasc. 2, p. 59–204. 3 fig. 1903. (Reprinted with title, Les plantes médicinales et toxiques de Madagascar. 148 p.)
Extensively annotated catalog, alphabeted by vernacular names; index of scientific names, with vernacular equivalents.

——— Les plantes utiles de Madagascar. Catalogue alphabétique des plantes utiles et en particulier des plantes médicinales et toxiques de Madagascar avec leurs noms malgaches et leurs emplois. Annales Mus. Colon. Marseille **18** (s. 2, v. 8): 5–372. 72 fig. 1910. (Reprinted.)
Annotated catalog, alphabeted by vernacular names; index of scientific names with vernacular equivalents.

Humbert, Henri, ed. Flore de Madagascar (plantes vasculaires). Publiée sous les auspices du gouvernement général de Madagascar . . . 23ᵉ, 29ᵉ, 32ᵉ, 37ᵉ, 40ᵉ, 49ᵉ famille[s]. illus. 24, 26 cm. Tananarive, 1936–39.
Descriptive flora, with keys, synonymy, citation of exsiccatae with localities, vernacular names, extralimital range. Each family a separately paged monograph, with cover title only. Six families so far published (out of 189): *Aponogetonaceae, Cyperaceae, Lemnaceae, Commelinaceae, Liliaceae, Orchidaceae* (the last not seen).

Jumelle, Henri. Madagascar et Comores. *In* his Catalogue descriptif des collections botaniques . . . Annales Mus. Colon. Marseille v. 24 (s. 3, v. 4), fasc. 1, p. 1–79. 1916.
Annotated lists of economic plants, classified according to uses, with vernacular and scientific names (both indexed).

Olsen, Aksel. Madagassiske plantenavne. Nyt Mag. Naturvidensk. **54**: 57–147. 1916.
Alphabetical index of 1,930 vernacular names, with botanical equivalents and brief indication of habit and often of uses.

Palacký, Jan. Catalogus plantarum madagascariensium. 5 fasc. (55, 38, 89, 60, 57 p.). 24 cm. Pragae, 1906–07 (1905–07).—Filices madagascarienses. 32 p. 24 cm. Pragae, 1906.
List of scientific names, with abbreviated references and indication if endemic, of "plusque 5000 esp." Families not in strictly systematic order; no index; fasc. 3 was the last to be published. Baron's Compendium was not available to the author. The list of pteridophytes is alphabetically arranged, with references to localities or collectors.

Perrier de la Bâthie, Henri. Les plantes introduites à Madagascar. Revue Bot. Appl. et Agr. Trop. **11**: 719–729, 833–837, 920–923, 991–999. 1931; **12**: 48–52, 128–133, 213–220, 296–301, 372–383, 462–468, 530–543. 1932.
Annotated lists of cultivated, escaped, and naturalized plants; altogether about 900 species.

Local

Alleizette, Charles d', and **Poisson, Henri.** Contribution à l'étude de la végétation des environs de Tananarive. Bul. Mus. Hist. Nat. Paris **17**: 171–189. 1911.

General features of vegetation, plant zones; annotated tabular list of vascular plants, grouped by habitat, with vernacular names, based on observations made in 1905 and 1906.

Jeannot, Eugène. Les productions végétales de la région des Betsimisaraka-Betanimena (Madagascar). Revue Cult. Colon. **8**: 38–44, 73–77, 97–102, 134–142, 168–176, 201–206, 231–233. 1901. (Reprinted 39 p.)

Native tribes, climate, plant zones; annotated list of useful plants, with vernacular names and botanical names when known, grouped by uses. (Central eastern Madagascar, from Antongile to Mananjary, west to the east ridge of the basin of Mangoro and the mountains of Tanala.)

Lecomte, Henri. Madagascar. Les bois de la forêt d'Analamazaotra . . . avec la collaboration de M. Paul Danguy . . . 188 p. 56 pl. 32.5 cm. Paris, 1922.

Mainly concerned with anatomical features of the woods, but contains illustrations of many species, with vernacular names and uses.

Perrier de la Bâthie, Henri. Le Tsaratanana, l'Ankaratra et l'Andringitra. Mém. Acad. Malgache 3. 68, [3] p. map. 1927.

Includes (p. 53–66) lists of endemic plants of these three mountain groups.

Poisson, Henri. Recherches sur la flore méridionale de Madagascar. 230 p. 30 fig., 16 pl. (incl. 3 maps). 25 cm. Paris, 1912.

General summary of physiography, climate, etc.; study of xerophytic plants, etc.; short list of vernacular names with botanical identifications, bibliography.

Viguier, René, and **Humbert, Henri.** Plantes récoltées à Madagascar en 1912. Bul. Soc. Linn. Normandie VII, **3**: 325–356. 1921; **4**: 37–52. 1922; **5**: 125–140. 1923; **6**: 169–200. 1924; **7**: 193–208. 1925.

List of spermatophytes collected by authors, with data, vernacular names, and frequent annotations. Unfinished; reaches Orchidaceae (part).

MALDIVE ISLANDS

General

Willis, J. C., and **Gardiner, J. S.** The botany of the Maldive Islands. Annals Roy. Bot. Gard. Peradeniya **1**: 45–164. pl. 2. 1901.

Botanical explorations, physiography, etc.; annotated list of all vascular plants known from the Maldives, Laccadives, and Chagos, with vernacular names, uses, etc.; notes on vegetation of various atolls; index of Maldivian vernacular names; economic plants, classified by uses; origin of flora.

MASCARENE ISLANDS

See also French Guiana (Lanessan, Les plantes utiles, p. 188–213, 528–582, for useful plants of Réunion).

Local

Baker, J. G. Flora of Mauritius and the Seychelles: a description of the flowering plants and ferns of those islands. 19, L, 557 p. 22 cm. London, 1877.

Botanical collections, physiography, statistics of flora (1,327 species); descriptive flora; vernacular names (separate index).—See also **Johnston, H. H.** Additions to the flora of Mauritius as recorded in Baker's "Flora of Mauritius and the Seychelles." Trans. and Proc. Bot. Soc. Edinb. **20**: 391–407. 1895. (Annotated list of 103 species, including cellular cryptogams, not recorded in Baker's Flora.)—Also **Vaughan, R. E.** Contributions to the flora of Mauritius.—I. An account of the naturalized flowering plants recorded from Mauritius since the publication of Baker's "Flora of Mauritius and the Seychelles" (1877). Jour. Linn. Soc. Bot. **51**: 285–308. 1937. (Discussion of methods of introduction; annotated list, with descriptions; bibliography.)

Balfour, I. B. Botany [of Rodriguez]. *In* An account of the petrological, botanical, and zoological collections made in Kerguelen's Land and Rodriguez during the Transit of Venus expeditions . . . 1874–75. Phil. Trans. Roy. Soc. **168** (extra vol.); 302–387. pl. 19–36. 1879.

General features of flora, with geographic relationships; annotated list of species (297 phanerogams, of which 108 are introduced, and 25 vascular cryptogams); vernacular names.

Bouton, Louis. Plantes médicinales de Maurice. 2. éd. v, 147 p. 23 cm. Port Louis, 1864. (1st ed., 1857.)
Systematic list of medicinal plants (native and cultivated) with French, English, and native names and uses.

Daruty, Clément. Plantes médicinales de l'île Maurice et des pays intertropicaux . . . xiii, lxii, 123, xii p. 23.5 cm. Maurice, 1886.
Includes tabular alphabetical list of Creole names of medicinal plants of Mauritius, with botanical, Tamoul, and Hindu names, and uses.

Horne, John. Notes on the flora of Flat Island. Trans. Roy. Soc. Mauritius **19:** 116–151. 1887. (Preprinted 30 p. 1886.)
General features of flora, naturalized plants; annotated list of 122 vascular plants (of which 69 are indigenous), with vernacular names.

Jacob de Cordemoy, Eugène. Flore de l'île de la Réunion. xxvii, 574 p. 24.5 cm. Paris, 1895.
Physiography, climate, general features of flora, history of botanical explorations; annotated list of cellular and vascular plants, with vernacular names, uses, brief synonymy, and often descriptions.

Johnston, H. H. Report on the flora of Ile des Aigrettes, Mauritius. Trans. and Proc. Bot. Soc. Edinb. **20:** 317–331. 1895.
Physiography; annotated list of 55 vascular plants.

——— Report on the flora of Les Bénitiers, Mauritius. Trans. and Proc. Bot. Soc. Edinb. **20:** 331–335. 1895.
Physiography; annotated list of plants from Les Bénitiers, two small rocky islets very close to Mauritius.

——— Report on the flora of Round Island, Mauritius. Trans. and Proc. Bot. Soc. Edinb. **20:** 237–264. 1894.
Geography, physiography, relationships and general features of flora, short bibliography: annotated list of 52 plants (of which 41 are vascular), including those found by previous collectors.

——— Report on the flora of the outlying islands in Mahébourg Bay, Mauritius. Trans. and Proc. Bot. Soc. Edinb. **20:** 353–374. 1895.
Physiography, etc.; annotated list of 53 plants from Ile de la Passe, Ile Vakois, Ile aux Fouquets, Ile aux Fous, Ile Marianne, Rocher des Oiseaux.

Jumelle, Henri. Réunion. *In* his Catalogue descriptif des collections botaniques . . . Annales Mus. Colon. Marseille v. 24 (s. 3, v. 4), fasc. 1, p. 81–112. 1916.
Annotated lists of economic plants, classified according to uses, with vernacular and scientific names (both indexed).

LeClerc, Jules. Des plantes médicinales de l'île de la Réunion et de leur application à la thérapeutique. 84 p. 22.5 cm. St. Denis (Réunion), 1864.
Running account of medicinal plants, classified by uses; index of vernacular names only.

NARCONDAM ISLAND

General

Prain, David. On the flora of Narcondam and Barren Island. Jour. Asiatic Soc. Bengal **62** (2): 17–86. pl. 3–4 (maps). 1893. (Reprinted in his Memoirs and memoranda, chiefly botanical. 1894.)
Physiography, general features of flora, annotated list of 174 species (of which 161 are vascular plants); discussion of nature and origin of flora.

NEW AMSTERDAM AND ST. PAUL ISLANDS

See also Crozet Islands (Schenck).

General

Hemsley, W. B. Amsterdam and St. Paul Islands. *In* his Report on the botany of the Bermudas and various other islands . . . 2d pt. Report on the scientific results of the voyage of H. M. S. *Challenger* during the years 1873–76. Botany, v. 1, pt. 2, p. 259–281. pl. 39, 41–45, 52. London, 1885 (1884).

Botanical explorations, composition of flora, with table; annotated list of plants (vascular and cellular). Includes 19 phanerogams and 19 pteridophytes.

Reichardt, H. W. Ueber die flora der insel St. Paul im Indischen Ocean. Verhandl. K. K. Zool.-Bot. Gesell. Wien **21**: 3–36. 1871.

Physiography, botanical investigations, general features of flora, etc. (with bibliographical footnotes); annotated list of cellular and vascular plants, with local and general range. Includes notes on the neighboring Amsterdam Island with list of its flora (p. 9–11).

Schenck, Heinrich. Ueber flora und vegetation von St. Paul und Neu-Amsterdam. *In* Wissenschaftliche ergebnisse der Deutschen tiefsee-expedition auf dem dampfer "*Valdivia*" 1898–1899. v. 2, teil 1, p. 179–218. 14 fig., pl. 11–15. Jena, 1905.

Botanical investigations, physiography, climate, general features of vegetation, phenology, etc.; tabular list of 36 native vascular plants and supplementary lists of introduced ones; geographical relationships.

NICOBAR ISLANDS

See also Andaman Islands (Kloss; Prain, On a botanical visit).

General

Kurz, Sulpiz. A sketch of the vegetation of the Nicobar Islands. Jour. Asiatic Soc. Bengal **45** (2): 105–164. pl. 12–13. 1876.

Geography, geology, soils, plant formations; annotated list of 624 vascular plants.

PRINCE EDWARD ISLANDS

See also Kerguelen Island (Schenck).

Local

Hemsley, W. B. Prince Edward group. Marion Island. *In* his Report on the botany of the Bermudas and various other islands . . . 2d pt. Report on the scientific results of the voyage of H. M. S. *Challenger* during the years 1873–76. Botany. v. 1, pt. 2, p. 187–206. pl. 53. London, 1885 (1884).

Physiography, general features of vegetation; annotated list of plants (vascular and cellular); extralimital range given. Includes 8 phanerogams and 6 pteridophytes.

Moseley, H. N. On the botany of Marion Island, Kerguelen's Land, and Yong Island of the Heard group. Jour. Linn. Soc. Bot. **14**: 387–388. 1874.

List of vascular plants of Marion Island and Yong Island. Followed by a tabular list by **D. Oliver,** List of plants collected by H. N. Moseley . . . on Kerguelen's Land, Marion Island, and Yong Island, p. 389–390.—For general features of flora, physiography, etc., see **Moseley, H. N.** Notes on the flora of Marion Island. l. c. **15**: 481–486. 1877.

SEYCHELLES

See also Gloriosa Islands; Mascarene Islands (Baker).

General

Summerhayes, V. S. An enumeration of the angiosperms of the Seychelles Archipelago. Trans. Linn. Soc. Zool. **19**: 261–299. table. 1931.

Botanical explorations, general features of flora, statistics, phytogeography; list of angiosperms (480 species), with references, localities, and collectors' numbers, and general range.—For the Pteridophyta, see **Christensen, C.** On the ferns of the Seychelles and the Aldabra group. Trans. Linn. Soc. Bot. II, **7**: 409–425. pl. 45. 1912.

Local

Diels, Ludwig. Beiträge zur kenntnis der vegetation und flora der Seychellen . . . *In* Wissenschaftliche ergebnisse der Deutschen tiefsee-expedition auf dem dampfer "*Valdivia*" 1898–1899. v. 2, teil 1, p. 407–466. 35 fig., pl. 28–44, map. Jena. 1922.

Bibliography, general description of vegetation; annotated list of vascular plants collected by Deutsche Tiefsee Expedition of 1899.

Hemsley, W. B. Flora of Seychelles and Aldabra: new phanerogamia, chiefly of the Percy Sladen Trust expedition, with some emendations in synonymy. Jour. Bot. v. 54, suppl. II. 24 p. 1916; **54:** 361–363. 1916.

Critical notes and descriptions of new species; never completed.

SOCOTRA

General

Balfour, I. B. Botany of Socotra. Trans. Roy. Soc. Edinb. v. 21. xix, 446 p. 100 pl., map. 1888.

Botanical explorations, physiography, economic plants, phytogeography; annotated list of vascular (584 species) and cellular plants.

——— Botany of Sokotra and Abd-el-Kuri. Angiospermae [and] Pteridophyta. *In* **Forbes, H. O.**, ed. The natural history of Sokotra and Abd-el-Kuri . . . p. 447–542. illus., pl. 26A and B (col.). Liverpool, 1903.

List of collectors; annotated list of 618 angiosperms now known from Sokotra and 61 known from Abd-el-Kuri; annotated list of ferns, mosses, and liverworts of the two areas. Lists of fungi, lichens, seaweeds, and diatoms, in part by other authors, are also given.

Local

Vierhapper, Fritz. Beiträge zur kenntnis der flora Südarabiens und der inseln Sokótra, Sémha und 'Abd el Kûri. Bearbeitung der von Dr. St. Paulay und Professor Dr. O. Simony auf der expedition der Kaiserlichen akademie der wissenschaften . . . von December 1898 bis mitte März 1899 gesammelten gefässpflanzen. I. theil. Gefässpflanzen der inseln Sokótra, Sémha und 'Abd el Kûri. Denkschr. Math.-Naturw. Kl. K. Akad. Wiss. Wien **71:** 321–490. 35 fig., 17 pl. 1907.

Annotated list of 221 vascular plants, with references, localities, and citations from Balfour and Forbes.

PACIFIC OCEAN

AUSTRALIAN ISLANDS

(Including Antipodes, Auckland, Bounty, Campbell, Chatham, Kermadec, Lord Howe, Macquarie, Norfolk, Snares.)

See also New Zealand (Cheeseman, Manual; Hooker, Handbook; Kirk, The students' flora); Polynesia (Merrill).

Local

Buchanan, John. On the flowering plants and ferns of the Chatham Islands. Trans. and Proc. New Zeal. Inst. **7:** 333–341. pl. 12–14. 1875.

List of 205 vascular plants, partly annotated, based principally on the collections of H. H. Travers in 1866 and 1871.

Cheeseman, T. F. On the flora of the Kermadec Islands . . . Trans. and Proc. New Zeal. Inst. **20:** 151–181. 1888.

Physiography, general features of flora, phytogeography, zoology; annotated list of 115 vascular plants; table showing general distribution; list of naturalized plants.

——— On the systematic botany of the islands to the south of New Zealand. *In* **Chilton, C.** The subantarctic islands of New Zealand. v. 2, p. 389–471. Wellington, 1909.

Refers to Snares, Lord Auckland, Campbell, Antipodes, Bounty, and Macquarie Islands. Geography, botanical explorations; annotated list of vascular plants, with localities; list of naturalized plants; table showing local and general distribution of each species; phytogeography (including table showing local distribution of the 37 vascular plants known from Kerguelen and South Georgia Islands, p. 457), with annotated list of the 53 endemic species. The grasses, omitted in this paper, are treated in the following one, "Gramina of the subantarctic islands . . ." by **Donald Petrie**, p. 472–481.

——— The vascular flora of Macquarie Island. 63 p. map. 31 cm. Sydney, 1919. (Australasian Antarctic expedition, 1911–14. Scientific reports. Ser. C. v. 7, pt. 3.)

Physiography, climate, botanical explorations, extensively annotated list of vascular plants (34 native) with references, localities, and general range; affinities and origin of flora; bibliography.

Downing, C. T. On Norfolk Island, its character and productions. Papers and Proc. Roy. Soc. Tasmania **3**: 195–212. 1859. (Reprinted in Edinb. New Phil. Jour. n. s., **16**: 322–328. 1862.)

Includes (p. 202–207, 210–212) notes on useful plants, wild and cultivated.

Endlicher, S. F. L. Prodromus florae norfolkicae sive catalogus stirpium quae in insula Norfolk annis 1804 et 1805 a Ferdinando Bauer collectae et depictae nunc in Museo caesareo palatino rerum naturalium Vindobonae servantur. viii, 100 p. 23.5 cm. Vindobonae, 1833.

List of 152 species, of which 139 are vascular; descriptive flora, with references and frequent annotations.

Hemsley, W. B. The flora of Lord Howe Island. Annals Bot. **10**: 221–284. 1896.

History, physiography, botanical explorations, general features of flora; list of 209 vascular plants, with references, general range, and occasional annotations; tables showing general range of genera and species; phytogeography; bibliography.

Hooker, J. D. On the botany of Raoul Island, one of the Kermadec group in the South Pacific Ocean. Jour. Linn. Soc. Bot. **1**: 125–129. 1857 (1856).

Geography, general features of flora; list of 42 plants collected by McGillivray on the voyage of H. M. S. *Herald.*

Kirk, Thomas. On the botany of Antipodes Island. Trans. and Proc. New Zeal. Inst. **23**: 436–441. 1891.

Physiography, zoology, general features of flora; unannotated list of 55 vascular plants.

——— On the botany of the Snares. Trans. and Proc. New Zeal. Inst. **23**: 426–431. 1891.

Physiography, zoology, general features of flora, with special notes on more important species; unannotated list of vascular plants.

Laing, R. M. A revised list of the Norfolk Island flora, with some notes on the species. Trans. and Proc. New Zeal. Inst. **47**: 1–39. 1915.

Botanical explorations, present condition of flora, tests for distinguishing indigenous species, plant associations, vegetation of Phillip Island; phytogeography, bibliography; annotated list of 175 vascular plants, with local and general range; includes the plants of Phillip Island.

Maiden, J. H. The flora of Norfolk Island. Proc. Linn. Soc. N. S. Wales **28**: 692–785. pl. 38. 1904.

Botanical explorations; annotated list of vascular and cellular plants; introduced plants; early accounts of vegetation; bibliography; notes on flora of Phillip Island and Nepean Island.

——— Observations on the vegetation of Lord Howe Island. Proc. Linn. Soc. N. S. Wales **23**: 112–158. pl. 1–4. 1898.—Some further observations on the vegetation of Lord Howe Island. l. c. **24**: 381–384. pl. 32–33. 1899; **26**: 156–158. 1901; **27**: 347–351. pl. 15. 1902; **39**: 377–384. pl. 28. 1914; **45**: 564–566. 1921.

The first paper contains notes on author's visit in 1898 (?), climate, etc.; annotated bibliography; botanical explorations; list of vascular plants, mainly additions to Hemsley's list (1896); introduced plants, etc.; brings flora up to 217 indigenous and 20 introduced species. In vol. 39, p. 379, is a list of plants collected on Admiralty Islets (near Lord Howe Island) by Charles Hedley in September 1908. Titles of various numbers differ.

Mueller, Ferdinand von. The vegetation of the Chatham-Islands. 86 p. 7 pl. 24.5 cm. Melbourne, 1864.

Botanical explorations, phytogeography; annotated list of 87 vascular plants, with references and vernacular names.

Oliver, W. R. B. The vegetation and flora of Lord Howe Island. Trans. and Proc. New Zeal. Inst. **49**: 94–161. 3 fig., pl. 10–16. 1917.

Physiography, climate, plant formations, origin of flora, phytogeography; annotated list of 209 vascular plants, with reference to records, local and general range; list of introduced plants; annotated bibliography; excluded species.

Oliver, W. R. B. The vegetation of the Kermadec Islands. Trans. and Proc. New Zeal. Inst. **42**: 118–175. 1 fig. (map), pl. 12–23. 1910.

Phytogeography, botanical explorations, geology, climate, introduced plants and animals, plant formations, geographical distribution; annotated list of indigenous vascular plants, with general range given; list of introduced plants; bibliography.

Schenck, Heinrich. Vergleichende darstellung der pflanzengeographie der subantarktischen inseln . . . V. Inseln südlich von Neuseeland [Die Snares, Antipoden-, Lord Auckland-, Campbell- und Macquarie-inseln]. *In* Wissenschaftliche ergebnisse der Deutschen tiefsee-expedition auf dem dampfer "*Valdivia*" 1898–1899. v. 2, teil 1, p. 130–161. Jena, 1905.

Botanical collectors, bibliography; tabular list of 182 native vascular plants, showing distribution inside and outside the region, with additional lists of introduced species; climate, etc.; general features of flora of each group discussed separately.

BORNEO

See also Dutch East Indies (general works; also Steenis, On the origin).

General

Beccari, Odoardo. Note botaniche e botanico industriali bornensi. Webbia **5**: 451–581. 6 fig. 1923.

Annotated lists of useful plants of Borneo, classified by uses; vernacular names indexed.

Merrill, E. D. A bibliographic enumeration of Bornean plants. Jour. Straits Branch Roy, Asiatic Soc. spec, no. 637 p. 1921.

Systematic list of Spermatophyta (4,924 species, considered to be probably only 50 to 60 percent of real total), with references and synonymy, citation of exsiccatae, general range; general features of flora, history of botanical collections.—See also his New or noteworthy Bornean plants I–III. Jour. Straits Branch Roy. Asiatic Soc. **85**: 157–201. 1922; **86**: 312–342. 1922; Jour. Malay. Branch Roy. Asiatic Soc. **1**: 22–45. 1923.—Also his Additions to our knowledge of the Bornean flora, [I]–II. Philippine Jour. Sci. **21**: 515–534. 1922; **30**: 79–87. 1926.

——— A contribution to the bibliography of the botany of Borneo. Sarawak Mus. Jour. **2**: 99–136. 1915.

Bibliography of 479 titles.—Additional titles in his "Bibliographic enumeration . . .," p. 2–6. 1921.

Local

Endert, F. H. Lijst van nuttige planten met korte aanteekeningen over het gebruik. *In* Indisch comite voor wetenschappelijke onderzoekingen. Midden-Oost-Borneo expeditie, 1925. p. 292–312. Weltevreden, 1927.

Annotated list, the plants classified by uses.

Gibbs, L. S. A contribution to the flora and plant formations of Mount Kinabalu and the highlands of British North Borneo. Jour. Linn. Soc. Bot. **42**: 1–240. 8 fig., 8 pl. (incl. map). 1914.

Annotated list of plants (vascular and cellular) collected in the Provinces of Dent and Keppel British North Borneo, January to March 1910, by the author; general range given; botanical exploration, orography, meteorology, plant formations; bibliography.

Irmscher, Edgar, ed. Beiträge zur kenntnis der flora von Borneo. Mitt. Inst. Allg. Bot. Hamburg v. 7. 310 p. illus., 10 pl., 2 maps. 1927–37.

Itinerary (by H. Winkler); list of vascular and cellular plants collected by Prof. Hans Winkler in central Borneo along the Melawi and upper Kapuas Rivers in 1924–25, with data; families not in systematic order. Still in process of publication.

Merrill, E. D. The flora of Banguey Island. Philippine Jour. Sci. **29**: 341–427. 1 fig. (map). 1926.

Briefly annotated list of about 410 vascular plants from Banguey Island, north of Borneo; general range given; general features of island.

Merrill, E. D. Plantae elmerianae borneenses. Univ. Calif. Pub. Bot. v. 15. 316 p. 1929.

Annotated list of about 945 vascular plants collected by A. D. E. Elmer in British North Borneo in 1921–23.

Moulton, J. C. An account of the various expeditions to Mt. Kinabalu. Sarawak Mus. Jour. **2**: 137–176. 1915.

Extensively annotated list of expeditions to Mount Kinabalu, 1851–1913; general account of region and methods of approach.

Ridley, H. N. Additions to the flora of Borneo and other Malay islands: [I]–XIV. Bul. Misc. Inform. Kew **1930**: 364–373. 1930; **1931**: 33–39, 493–499. 1931; **1933**: 190–202. 1933; **1936**: 17–21. 1936; **1938**: 110–123, 173–175, 175–176, 221–242, 275–285, 306–307. 1938; **1939**: 29–31, 275–290, 504–509, 509–517. 1939.

Descriptions of new species and new records, grouped by families, but not in systematic order. Pt. 8 is by S. J. van Ooststroom, pt. 12–14 by H. K. Airy Shaw.

Stapf, Otto. On the flora of Mount Kinabalu, in North Borneo. Trans. Linn. Soc. Bot. II, **4**: 69–263. pl. 11–20. 1894.

Annotated list of vascular plants (also Muscineae and Hepaticae; 342 phanerogams, of which 199 are endemic); botanical exploration, orography, geology, climate, botanical zones and formations, geographical affinities of flora, tables of distribution.

CELEBES

See also Dutch East Indies (general works; also Koorders-Schumacher; Steenis, On the origin); Molucca Islands (Rant).

Local

Koorders, S. H. Verslag eener botanische dienstreis door de Minahasa, tevens eerste overzicht der flora van n. o. Celebes uit een wetenschappelijk en praktisch oogpunt. Meded. Lands Plant. Buitenz. no. 19. xxvi, 716 p. 17 pl. (incl. maps). 1898.—[1.]–3. nachtrag zu meiner enumeratio specierum phanerogamarum Minahassae. Natuurk. Tijdschr. Nederland.-Indië **61**: 250–261. 1902; **63**: 76–89, 90–99. 1904.—Supplement . . . deel 1, afl. 1. 30 p. 10 pl. 26 cm. Batavia, 1918.—Supplement . . . deel 1, afl. 2. Bul. Jard. Bot. Buitenz. III, **2**: 242–260. pl. 11–13. 1920. (Reprinted as p. 31–50.)—Supplement . . . deel 2–3. Uitgegeven door Mevr. de wed. A. Koorders-Schumacher. 2 v. 127 pl. 29 cm. Buitenzorg, 1922.

Annotated list (p. 253) of 1,375 phanerogams of Minahasa Peninsula, Celebes; general range, vernacular names (indexed), uses; lists of vernacular names and remarks on uses, etc.; geography, physiography (p. viii). The supplements contain mostly new or recently described species. [1. nachtrag] has title Eenige aanvullingen en verbeteringen van mijn Verslag . . .

CLIPPERTON ISLAND

General

Snodgrass, R. E., and Heller, Edmund. The birds of Clipperton and Cocos Islands. Proc. Washington Acad. Sci. **4**: 501–520. 1902.

On p. 516 it is stated that Clipperton Island is "absolutely destitute of vegetation."

COCOS ISLAND

See also Galapagos Islands (Svenson).

General

Robinson, B. L. Plants collected by Messrs. Snodgrass and Heller on Cocos Island of the Pacific. *In* his Flora of the Galapagos Islands. Proc. Amer. Acad. Arts and Sci. **38**: 261–263. 1902.

Unannotated list of cellular and vascular plants.

Stewart, Alban. Notes on the botany of Cocos Island. Proc. Calif. Acad. Sci. IV, **1**: 375–404. pl. 31–34. 1912. (Expedition of the California Academy of Sciences to the Galapagos Islands, 1905–06. V.)

Physiography, plant geography, etc.; annotated list of vascular plants (and mosses) based chiefly on collections made by the author in 1905.

DUTCH EAST INDIES

See also Borneo; Celebes; Java; Molucca Islands; New Guinea; Sumatra.

General

Boerlage, J. G. Handleiding tot de kennis der flora van Nederlandsch Indië. Beschrijving van de families en geslachten der Nederl. Indische phanerogamen. 3 v. 22.5 cm. Leiden, 1890–1900.

Descriptive only as to families and genera; species listed at end of each family with statement of range, but without references. No vernacular names. Vol. 1–3, pt. 1, complete the Dicotyledons; no more published.

Cammerloher, Hermann, and others. Contributions à l'étude de la flore des Indes Néerlandaises I–XXXIV. *In* Bul. Jard. Bot. Buitenz. s. 3, v. 5–15. 1923–38.

Revisionary treatments of miscellaneous families, not in systematic order, by various botanists.

Clercq, F. S. A. de. Nieuw plantkundig woordenboek voor Nederlandsch Indië. Met korte aanwijzingen van het nuttig gebruik der planten en hare beteekenis in het volksleven en met registers der inlandsche en wetenschappelijke benamingen . . . xx, 395 p. 28 cm. Amsterdam, 1909.

Alphabetical list of vernacular names (dialects stated in each case) with cross-references to the alphabetically arranged list of scientific names, which is briefly annotated as to uses; additional index of vernacular names of the parts of the plants (fruits, flowers, etc.) ; systematic list of families and genera mentioned.

Duyster, Marinus. Giftige indische planten en plantenbestanddeelen. 143, xiv p. 24.5 cm. Bandoeng, 1927.

Annotated systematic list of poisonous plants, with vernacular names, chemical properties, etc.

Eeden, F. W. van. Houtsoorten van Nederlandsch Oost-Indië. Tevens beschrijving der meest bekende boomen van den Nederlandsch-Indischen Archipel en hunne waarde voor de huishouding. 3. vermeerderde uitgave bewerkt door J. J. Duyfjes. 341 p. 19.5 cm. Haarlem, 1905 (1906). (1st ed., 1872; 2d, 1886.)

Annotated list of 815 species of trees of the Dutch East Indies, arranged by families, with vernacular names.

Filet, G. J. Plantkundig woordenboek voor Nederlandsch-Indië; met korte aanwijzingen van het geneeskundig- en huishoudelijk gebruik der planten, en vermelding der verschillende inlandsche en wetenschappelijke benamingen. 2. ed. x, 348 p. 22.5 cm. Amsterdam, 1888. (1st ed., 1876.)

Alphabetical list of 9,283 vernacular names (plus additions), with botanical equivalents, indication of locality where used, and notes on uses; index of scientific names. The first edition contained 8,860 names.

Grevelink, A. H. B. Planten van Nederlandsch-Indië bruikbaar voor handel, nijverheid en geneeskunde. vi, xlviii, 871 p. 8 fig. 24 cm. Amsterdam, 1883.

Extensively annotated systematic list of useful plants, both wild and cultivated; brief descriptions, vernacular names, uses. Much less complete than Heyne's work.

Heyne, K. De nuttige planten van Nederlandsch Indië. 2e herziene en vermeerderde druk. 3 v. 27.5 cm. Batavia, 1927. (1st ed., 4 v., 1913–17; rev. ed. (herdruk) vol. 1, 1922.)

Extensively annotated systematic list of useful plants, wild and cultivated, with vernacular names, brief descriptions, uses, commercial details of the more important species; bibliography; synoptical list of species mentioned, classified by uses; indices of scientific and vernacular names.

Internationale circumpacifische onderzoek-commissie. A short history and the present position of botanical investigation in the Dutch East Indies. 22 p. illus. 27 cm. Amsterdam, 1923.

History of botanical exploration and investigation, laboratories, etc.; unannotated list of principal botanical publications, list of periodicals.

Koorders, S. H. Lijst der phanerogamen-geslachten van den Maleischen Archipel (Index generum phanerogamorum Archipelagi Indici). Natuurk. Tijdschr. Nederland.-Indië **55**: 312–352. 1896.

Systematic tabular list of genera, with indication of habit.

Miquel, F. A. W. Flora van Nederlandsch Indië (Flora Indiae batavae). 3 v. in 4. 41 pl., 2 maps. 24.5 cm. Amsterdam, 1855–59.

Descriptive flora (in Latin, habitat notes in Dutch) covering the spermatophytes; synonymy, references, vernacular names. The area covered extends from the Andaman Islands to northwestern Australia and western and southern New Guinea. Supplement 1 is a flora of Sumatra, here treated separately.

Ochse, J. J., and Bakhuizen van den Brink, R. C. Vegetables of the Dutch East Indies (edible tubers, bulbs, rhizomes and spices included). Survey of the indigenous and foreign plants serving as pot-herbs and side-dishes. xxxvi, 1005 p. 463 fig., col. pl. 27.5 cm. Buitenzorg, 1931.

Annotated list, the families and their species arranged alphabetically, with vernacular names, description, use, references, and figure of each species; index of native vernacular names used in Dutch East Indies outside of Java and Madoera (Madura) Island. An entirely revised and much enlarged edition of Ochse's Tropische groenten (1925).

Prillieux, Édouard. Sur les productions agricoles et forestières des possessions hollandaises des Indes orientales. Bul. Soc. Acclim. France III, 1 [v. 21]: 359–389. 1874. (Reprinted.)

Systematic list of 247 woods exhibited at the Paris Exposition of 1867, with vernacular names and brief annotations.

Pulle, A. A. The vegetation. *In* **Rutten, L. M. R.**, ed. Science in the Netherlands East Indies. p. 164–191 (incl. 2 fig. (ports.), plates). Amsterdam, 1929].

Rather detailed account of botanical investigations, with included bibliography.

Sirks, M. J. Indische natuuronderzoek . . . Meded. Kolon. Inst. Amsterdam no. 6, Afd. Handelsmus. no. 2. xi, 303 p. ports. 1915.

Sketch of the development of natural history in the Dutch colonies (Old and New World) with bibliographical footnotes and much biographical information; index to personal names.

Stürler, F. A. von. De vruchten van Nederlandsch Oost-Indië. 83 p. 22 fig., 10 pl. 25 cm. Tiel, 1907.

Descriptions of the principal cultivated and wild fruits; bibliography.

Local

Britten, James, and others. Prodromus florae timorensis; compiled in the Botanical department of the British Museum. *In* **Forbes, H. O.** A naturalist's wanderings in the eastern archipelago . . . from 1878 to 1883. p. 497–523. London, 1885.

Botanical explorations, Forbes' itinerary; unannotated list of vascular plants. Polypetalae by J. Britten, Gamopetalae and Apetalae by W. Fawcett, Monocotyledones by H. N. Ridley, Filices by W. Carruthers.

Cardoso, João, jr. Plantas medicinaes de ilha de Timor. *In* his Subsidios para a materia medica e thérapeutica das possessões ultramarinas portuguezas. v. 1, p. 233–236. Lisboa, 1902.

Unannotated alphabetical list of medicinal plants.

Decaisne, Joseph. Description d'un herbier de l'île de Timor. Nouv. Annales Mus. Hist. Nat. Paris **3**: 333–501. pl. 16–21. 1834. (Reprinted with title: Herbarii timorensis descriptio . . . 173 p. Parisiis, 1835.)

List of about 550 species of vascular plants, with Latin descriptions; collections, physiography, etc.

Docters van Leeuwen, W. M. Botanical results of a trip to the Salajar Islands. Blumea **2**: 239–277. I fig. (map). 1937.—Addenda et corrigenda. l. c. **3**: 236–237. 1939.

General features of flora, botanical explorations, geology, climate; unannotated lists of vascular plants of islands of Salajar, Kajoeadi, Bonèrate, Pasitaloe, Kalao, Kalaotoa, and Djampéa, south of Celebes, collected by the author in 1913; bibliography.

——— Krakatau, 1883 to 1933. A. Botany. Annales Jard. Bot. Buitenz. v. 46/47. xii, 506 p. 1 fig. (map), 37 pl., map. 1936.

History, changes in vegetation after eruption, comparison with similar areas elsewhere, botanical explorations, soil, means of transport, associations, etc.; annotated lists of galls, mosses, hepatics, and 324 vascular plants; bibliography. Replaces the earlier works of the same author, Ernst, and others.

Docters van Leeuwen, W. M. Die vegetation der insel Toppers Hoedje in der Sunda-Strasse. Natuurk. Tijdschr. Nederland.-Indië **94**: 149–169. 1934.
Geology, general features of flora; annotated list of 50 vascular plants collected by the author in 1928 and 1932. (Small island between Sumatra and Java.)

——— The vegetation of the island of Sebesy, situated in the Sunda-Strait, near the islands of the Krakatau-group; in the year 1921. Annales Jard. Bot. Buitenz. **32**: 135–192. pl. 36–37 (maps). 1923.
History after eruption of 1883, etc.; briefly annotated list of 359 vascular plants from island of Sebesy, and comparative list of plants of Krakatau (including Verlaten—259 species) and Sebesy; bibliography.

Henderson, M. R. The "padang" flora of Jemala, in the Anamba Islands, N. E. I. Gard. Bul. Straits Settlem. **5**: 234–240. 1931.
General features of flora; annotated list of plants collected by the author in April 1928.

Junghuhn, Franz. Plantae junghuhnianae. Enumeratio plantarum, quas in insulis Java et Sumatra detexit Fr. Junghuhn. 522 p. 23.5 cm. Lugduni-Batavorum, 1853–55 [1851–57].
List of vascular and cellular plants collected by Junghuhn, with description of new species and various critical remarks and annotations. Issued in 4 parts; descriptions by Miquel and others.

Koorders-Schumacher, Anna. Systematisches verzeichnis der zum herbar Koorders gehörenden, in Niederländisch-Ostindien, besonders in den jahren 1888–1903 gesammelten phanerogamen und pteridophyten nach den original-einsammlungsnotizen und bestimmungs-etiketten, unter der leitung von Dr. S. H. Koorders zusammengestellt. II.—V. Abt. Sumatra, Celebes, Lombok . . . 62, 160 p. 24.5 cm. Buitenzorg, 1910–14.
Several systematic lists, in part annotated, of plants of the Koorders herbarium, as follows: Sumatra (several lists for different regions; phanerogams only); Celebes (vascular plants); Lombok and other islands (mainly references to other lists; no proper list of species).

Kurz, Sulpiz. Korte schets der vegetatie van het eiland Bangka. Natuurk. Tijdschr. Nederland.-Indië **27**: 142–258. 1864.
Plant formations, climate, etc.; annotated list of phanerogams (959 species); separate list of vernacular names. German translation by J. K. Hasskarl of introductory pages (143–156) in Bot. Zeit. **23**: 6–8, 15–19. 1865.

Lam, H. J. Miangas (Palmas). Scattered annotations . . . 66 p. 20 fig. and 3 maps on 11 pl., 1 tab. 26.5 cm. Batavia, 1932. (Indisch comité voor wetenschappelijke onderzoekingen Batavia VI.)
Topography, geology, general features of flora and fauna, history, etc.; list of 88 vascular plants collected by the author in 1926, with vernacular names and general range; bibliography. Miangas or Palmas Island, the northernmost of the Talaud (Talaur) Islands, is in about 5°33′ N. Lat., 126°34′ E. Long.

Malm, Jacob von. Die phanerogamenflora der Kleinen Sunda-inseln und ihre beziehungen. Repert. Spec. Nov. Fedde **34**: 255–307. 1934.
Geology, climate, plant formations; partial list of spermatophytes collected by the Rensch expedition in 1927 and the Elbert expedition in 1909 on Lombok, Sumbawa, and Flores (the species of wide range omitted); phytogeography, bibliography.

Rendle, A. B., and others. Dr. H. O. Forbes's Malayan plants. Jour. Bot. **62** (Suppl.): 1–48. 1924; **63** (Suppl.): 49–136. 3 fig. 1925; **64** (Suppl.): 137–149. 1 fig. 1926.
List of vascular and cellular plants collected by H. O. Forbes in Java and Sumatra in 1878–83, with localities, collecting numbers, and frequent annotations. Polypetalae by E. G. Baker, Gamopetalae and Monochlamydeae by S. LeM. Moore. Monocotyledones by H. N. Ridley, Pteridophyta by A. Gepp (also Bryophyta and Lichenes by other authors).

Spanoghe, J. B. Prodromus florae timorensis. Linnaea **15**: 161–208, 314–350, 476–480. 1841.
List of 914 phanerogams, with references and occasional localities. Gramineae, Cyperaceae, and some other families are omitted.

Steenis, C. G. G. J. van. Botanical results of a trip to the Anambas and Natoena Islands . . . with notes on the vegetation of Djemadja by M. R. Henderson. Bul. Jard. Bot. Buitenz. III, **12:** 151–211. 11 fig. (incl. map), pl. 1932.

Geography, geology, climate, botanical explorations, plant formations, phytogeography, classified list of useful wild and cultivated plants; no systematic list of flora. These islands form the Poelau Toedjoeh and are situated in the center of the Sunda Shelf in the South China Sea between 3° and 4° N. and 106° and 109° E.

——— On the origin of the Malaysian mountain flora. Bul. Jard. Bot. Buitenz. III, **13:** 135–262, 289–417. 10 fig. (incl. maps), tab., 2 maps. 1934–35; **14:** 56–72. 1936.

Mainly phytogeographical; includes (p. 155–260) list of Malaysian mountain species of temperate genera of mosses, hepatics, and vascular plants, with list of localities and altitudinal range, the families arranged alphabetically. Covers Malay Peninsula, islands from Sumatra to New Guinea, and Philippine Islands.

Teysmann, J. E. Verslag eener botanische reis over Timor en de daaronder ressorteerende eilanden Samauw, Alor, Solor, Floris en Soemba. Natuurk. Tijdschr, Nederland.-Indië **34:** 348–517. 1874.

Includes (p. 486–515) list of vernacular names, with botanical equivalents and brief account of useful plants.

FIJI ISLANDS

See also Polynesia (general works).

General

Horne, John. List of plants found in Fiji. *In* his A year in Fiji . . . p. 256–284. London, 1881.

Alphabetical lists of spermatophytes and pteridophytes known from Fiji, with indication if endemic or introduced. For descriptions and reductions of some of the many new species indicated, see **Baker, J. G.** Recent additions to our knowledge of the flora of Fiji. Jour. Linn. Soc. Bot. **20:** 358–373. 1883.

Seemann, Berthold. Flora vitiensis: a description of the plants of the Viti or Fiji Islands, with an account of their history, uses, and properties. xxxiii, 453 p. 100 pl. (incl. map, pl. 3–100 col.). 32 cm. London, 1865–73.

Descriptive flora of vascular and cellular plants, including cultivated species (descriptions in Latin, annotations in English); vernacular names (not indexed); botanical exploration, physiography, general features of flora, list of endemic forms, etc.—For descriptions of many new species from the islands, see **Gillespie, J. W.** New plants from Fiji I–III. Bul. Bern. P. Bishop Mus. 74. 99 p. 57 fig., pl. 1930; 83. 72 p. 40 fig., pl. 1931; 91. 81 p. 43 fig. 1932.—Also **Smith, A. C.** Fijian plant studies. l. c. 141. 166 p. 83 fig. 1936.

——— Viti: an account of a government mission to the Vitian or Fijian Islands in the years 1860–61. xv, 447 p. illus., 4 pl., map. 22.5 cm. Cambridge, 1862.

Includes chapters (p. 274–380) dealing with useful plants.

Wright, C. H. A list of Fijian plant names. Bul. Dept. Agr. Fiji no. 9 (i. e. 10). 10 p. 1918.—Errata and addenda. Agr. Cir. Dept. Agr. Fiji **1:** 59–60. 1920.

Alphabetical list of vernacular names with botanical equivalents, including all those previously recorded.—See also **Parham, [H. B. R.].** Names of a few Fijian plants and their botanical equivalents. 13 p. Suva, [1935].

Local

Gibbs, L. S. A contribution to the montane flora of Fiji (including cryptogams), with ecological notes. Jour. Linn. Soc. Bot. **39:** 130–210. 1 fig. (map), pl. 11–16. 1909.

Annotated list of plants collected by author in 1907, mostly at Nadarivatu, Mount Victoria Range, island of Viti Levu; general range given; physiography and climatology of Viti Levu, botanical explorations, affinities of flora, bibliography.

Turrill, W. B. A contribution to flora of Fiji. Jour. Linn. Soc. Bot. **43:** 15–39. 1915.

Annotated list of phanerogams collected by Sir Everard im Thurn in 1905–07.

GALAPAGOS ISLANDS

General

Robinson, B. L. Flora of the Galapagos Islands. Proc. Amer. Acad. Arts and Sci. **38**: 77–269. 3 pl. 1902. (Contr. Gray Herb. n. s., no. 24.)

Botanical collections, bibliography; systematic list, with pertinent synonymy, exsiccatae, extralimital range, frequent annotations; tables of botanical collections, distribution of species by islands (vascular plants); general features of flora, botanical features of each island, origin of flora. Treatment of cellular cryptogams by W. G. Farlow (except Hepaticae, by A. W. Evans).

Stewart, Alban. A botanical survey of the Galapagos Islands. Proc. Calif. Acad. Sci. IV, **1**: 7–252. pl. 1–19 (incl. map). 1911.

Similar in plan to the flora by Robinson and supplementary to it; vascular plants only. The plates are mostly of cacti. The general features of the vegetation on each island are discussed by the author in "Some observations concerning the botanical conditions on the Galapagos Islands." Trans. Wisc. Acad. Sci. **18** (1): 270–340. 1915.

Local

Christophersen, Erling. A collection of plants from the Galapagos Islands. Nyt Mag. Naturvidensk. **70**: 67–95. pl. 1931 (1932).

Recent botanical explorations; adds about 19 species to the flora of the islands, based principally on a collection made by Miss Borghild Rorud (Mrs. Rambech) in 1926–27.

Svenson, H. K. Plants of the Astor expedition, 1930 (Galapagos and Cocos Islands). Amer. Jour. Bot. **22**: 208–277 (incl. 9 pl.). 1935. (Brooklyn Botanic Garden Contribution no. 69.)

General features of flora, etc.; annotated list of vascular plants collected, including several new to the Galapagos Islands and two new to Cocos Island.

HAWAIIAN ISLANDS

See also Polynesia.

The following publication has not been available:

Kaaiakamanu, D. M., and **Akina, J. K.** Hawaiian herbs of medicinal value, found among the mountains and elsewhere in the Hawaiian Islands, and known to the Hawaiians to possess curative and palliative properties most effective in removing physical ailments. Translated by Akaiko Akana . . . 74 p. 8°. Honolulu, 1922. (Not seen; cited from Merrill, E. D. Polynesian botanical bibliography, p. 110.)

General

Bryan, E. H., jr. Hawaiian nature notes. 285 p. illus. (incl. maps). 22.5 cm. Honolulu, 1933.

Includes (p. 81–87) chapter on "Food plants of the ancient Hawaiians," with bibliography.

Bryan, W. A. Natural history of Hawaii; being an account of the Hawaiian people, the geology and geography of the islands, and the native and introduced plants and animals of the group. 596 p. (incl. 117 pl.). Honolulu, 1915.

General sketch of the flora, p. 189–230, with small and very crowded photographs of many species; botanical bibliography, p. 508.

Crawford, D. L. Hawaii's crop parade. A review of useful products derived from the soil in the Hawaiian Islands, past and present. 305 p. 24 cm. Honolulu, 1937.

Annotated list of wild and cultivated plant products, alphabetically arranged by vernacular names.

Degener, Otto. Flora hawaiiensis, or the new illustrated flora of the Hawaiian Islands. Book I–III. illus. (pt. col.). 24 cm. [Honolulu,] 1932–38.

Descriptive flora, with vernacular names, synonymy, type locality, local and general range, uses. Issued in loose-leaf form, unpaged; each species occupying a separate sheet (dated), with illustration on verso. In process of publication.

Forbes, C. N. Notes on the naturalized flora of the Hawaiian Islands. Occas. Papers Bern. P. Bishop Mus. **4**: 323–334 (no. 5, p. 23–34). 1911.

Annotated list of introduced vascular plants, most of which are not mentioned in Hillebrand's Flora (1888).

Fosberg, F. R. A key to the families of monocotyledons in the Hawaiian Islands. Occas. Papers Univ. Hawaii no. 18. 8 p. 1934.—**St. John, Harold,** and **Fosberg, F. R.** Identification of Hawaiian plants. A key to the families of dicotyledons . . . l. c. no. 36. 53 p. [1938.]

Keys to families; alphabetical list of families, each with botanical characters and alphabetical list of genera.

Heller, A. A. Observations on the ferns and flowering plants of the Hawaiian Islands. Minn. Bot. Studies **1** (Bul. 9, Geol. and Nat. Hist. Survey Minn. Bot. Ser. II): 760–922. pl. 42–69 (incl. 2 maps). 1897.

Geography, physiography; annotated list of species, with many critical notes. To be taken as supplementary to Hillebrand's Flora.

Hillebrand, William. Flora of the Hawaiian Islands; a description of their phanerogams and vascular cryptogams. xcvi, 673 p. pl., map. 22 cm. London and New York, 1888.

Geography, physiography, life zones; outline of descriptive botany; descriptive flora of vascular plants (999 species, of which 653 are endemic), with synonymy, references, and vernacular names.

Judd, A. F. Trees and plants. *In* Ancient Hawaiian civilization. A series of lectures delivered at the Kamehameha schools. p. 273–281. Honolulu, [1933].

Tabular list of useful plants, alphabetically arranged by vernacular names, with botanical names and uses.

MacCaughey, Vaughan. History of botanical exploration in Hawaii. Hawaiian Forester and Agr. **15**: 388–396, 417–429, 508–510. 1918; **16**: 25–28, 49–54. 1919.

Chronological sketch of botanical exploration and botanists, brief biographical sketches, bibliography.

——— A survey of the Hawaiian land flora. Bot. Gaz. **64**: 90–114. 5 fig. (incl. map). 1917.

A brief but useful summary of the flora.

Neal, M. C. Plants used medicinally. *In* **Handy, E. S. C.,** and others. Outline of Hawaiian therapeutics. Bul. Bern. P. Bishop Mus. **126**: 39–49. 1934.

Unannotated list, arranged alphabetically by vernacular names, with botanical equivalents and references; bibliography.

Pope, W. T. Manual of wayside plants of Hawaii. Including illustrations, descriptions, habits, uses and methods of control of such plants as have a wild nature of growth, exclusive of ferns. 289 p. 160 fig. 23.5 cm. Honolulu, 1929.

Annotated, descriptive systematic list of Hawaiian weeds; bibliography.

Rock, J. F. C. The indigenous trees of the Hawaiian Islands. 518 p. 215 fig. 26.5 cm. Honolulu, 1913.

Life zones, forest regions, floral aspects; descriptive flora of woody plants, with synonymy, references, extensive annotations, and photographs, mostly from fresh or living specimens; vernacular names (separately indexed).

——— List of Hawaiian names of plants. Bot. Bul. Bd. Agr. and Forestry Hawaii no. 2. 20 p. 1913.

Alphabetical lists of vernacular names of wild plants, introduced plants, Algae, etc., with botanical equivalents.

——— The poisonous plants of Hawaii. Hawaiian Forester and Agr. **17**: 59–62, 97–101. 1920.

Annotated list.—See also **Zschokke, T. C.** Poisonous plants now found in the Hawaiian Islands. Agr. Notes Agr. Ext. Serv. Univ. Hawaii no. 49. 4 p. 1933. (Processed publication.)

Wilder, G. P. Fruits of the Hawaiian Islands. rev. ed. 247 p. (incl. 121 pl.). 26.5 cm. Honolulu, 1911.

Photographic plates, with brief descriptions and remarks, of 121 native and naturalized fruits of the Hawaiian Islands.

Local

Bitter, Georg. Ergebnisse einer reise nach dem Pacific (Prof. Dr. Schauinsland 1896/97). Die phanerogamische pflanzenwelt der insel Laysan. Abhandl. Naturw. Ver. Bremen **16**: 430–439. pl. 4. 1900.

Annotated list of 26 vascular plants; phytogeographical relationships.—See also **Schauinsland, H. H.** Drei monate auf einer koralleninsel (Laysan). 104 p. 23 cm. Bremen, 1899. (Annotated list of 27 vascular plants, p. 93–100.)

Caum, E. L. Notes on the flora and fauna of Lehua and Kaula Islands. Occas. Papers Bern. P. Bishop Mus. v. 11, no. 21. 17 p. (incl. 1 fig., 3 pl.). 1936.

General features of flora; annotated list of 35 vascular plants; also list of birds.

——— Notes on the flora of Molokini. Occas. Papers Bern. P. Bishop Mus. 9 (1): 15–18. 1930.

Botanical explorations; list of vascular plants; changes in vegetation.

Christophersen, Erling. Vascular plants of Johnston and Wake Islands. Occas. Papers Bern. P. Bishop Mus. v. 9, no. 13, 20 p. (incl. 3 fig. (maps), 5 pl.). 1931. (Tanager expedition publication no. 6.)

Physiography, general features of flora; annotated list of vascular plants.

——— and **Caum, E. L.** Vascular plants of the Leeward Islands, Hawaii. Bul. Bern. P. Bishop Mus. 81. 41 p. 3 fig. (maps), 16 pl. 1931. (Tanager expedition publication no. 7.)

Physiography, general features of flora of each island with list of its species; annotated list of vascular plants, with localities and collectors' numbers; bibliography.—See also **St. John, H.** Additions to the flora of Midway Islands. (Hawaiian plant studies 3.) Occas. Papers Bern. P. Bishop Mus. v. 11, no. 14. 4 p. 1935.

Degener, Otto. Illustrated guide to the more common or noteworthy ferns and flowering plants of Hawaii National Park, with descriptions of ancient Hawaiian customs and an introduction to the geologic history of the islands. xv, 312 p. 95 fig. 23.5 cm. Honolulu, 1930.

Not a flora, but contains running accounts of the more interesting or striking plants with ethnological and other information and vernacular names. The park is on the islands of Hawaii and Maui.

Forbes, C. N. An enumeration of Niihau plants. Occas. Papers Bern. P. Bishop Mus. **5**: 99–108 (no. 3, p. 17–26). 3 pl., map. 1913.

List of vascular plants collected by the author in 1913.—See also **St. John, H.** Additions to the flora of Niihau. l. c. v. 9, no. 14. 11 p. (incl. 3 pl.). 1931.

——— Notes on the flora of Kahoolawe and Molokini. Occas. Papers Bern. P. Bishop Mus. **5**: 85–92 (no. 3, p. 3–10). 5 pl. (incl. 2 maps). 1913.

List of vascular plants collected by the author in 1913.

Hosaka, E. Y. Ecological and floristic studies in Kipapa Gulch, Oahu. Occas. Papers Bern. P. Bishop Mus. **13**: 175–232. 18 fig. (incl. map). 1937.

Includes list of bryophytes and vascular plants; bibliography. This locality is in the Koolau Range, near Pearl Harbor.

Kraebel, C. J. Mauna Kea plant list. Hawaiian Forester and Agr. **19**: 2–4. pl. 1925.

Alphabetical list of botanical names, with English and Hawaiian vernacular names, covering the plants seen from Kuaiau to about 10,000 feet altitude, with indication of habit, etc.

St. John, Harold, and **Hosaka, E. Y.** Weeds of the pineapple fields of the Hawaiian Islands. Res. Pub. Univ. Hawaii no. 6. 196 p. illus. 1932.

Descriptive systematic list, with figures of each species.

Skottsberg, Carl. Remarks on the flora of the high Hawaiian volcanoes. Acta Horti Gothoburg. **6**: 47–65. 1931.

Includes annotated list of alpine and subalpine species.

——— Vascular plants from the Hawaiian Islands. I–II. Acta Horti Gothoburg. **2**: 185–284. 9 fig. 1926; **10**: 97–193. 23 fig. 1936.

Annotated lists of species collected by the author, with numerous critical notes.

JAVA

See also Dutch East Indies (general works; also Junghuhn; Rendle; Steenis, On the origin).

The following publication has not been available:

Koorders, S. H. Flora van Tjibodas, umfassend die blütenpflanzen, welche in der botanischen Tjibodas-waldreserve und oberhalb derselben auf den West-Javanischen vulkanen Pangerango und Gede wildwachsend vorkommen. 5 lfg. Batavia, 1918–19. (Cited from Tectona **13**: 478. 1920.)

General

Backer, C. A. Handboek voor de flora van Java. v. 1–3. 27 cm. Batavia, 1924–28.

Descriptive flora of vascular plants, with keys, synonyms, vernacular names; the treatment of the species in form of keys. Pt. 3 (issued in 1924) takes the work through Musaceae–Heliconia (Engler-Prantl system).

——— Plantes exotiques naturalisées dans Java. Annales Jard. Bot. Buitenz. Suppl. **3** (1): 393–420. 1910.

Annotated alphabetical list of 153 species.

——— Schoolflora voor Java. (Ranunculaceae—Myrtaceae.) clxxix, 676 p. 12 pl. 23.5 cm. Weltevreden, 1911.

A flora in the form of ample keys, with vernacular names (indexed) and glossaries.

Boldingh, Isaac. Zakflora voor de landbouwstreken op Java. xvii, 204 p. 21.5 cm. Batavia, 1916.

Pocket flora in the form of keys, treating the vascular plants of the cultivated regions of Java.

Junghuhn, Franz. Die pflanzendecke. Das pflanzenreich Java's vom physiognomischen und physischgeographischen gesichtspunkte betrachtet. *In* his Java, seine gestalt, pflanzendecke und innere bauart. Nach der 2. verb. aufl. des holländischen original in's deutsche übertragen von J. K. Hasskarl. v. 1, p. 133–483. Leipzig, 1857.

General features of vegetation, plant formations, etc., with partly annotated bibliography (p. 133–148).

Koorders, S. H. Exkursionsflora von Java, umfassend die blütenpflanzen mit besonderer berücksichtigung der im hochgebirge wildwachsenden arten. 3 v. 139 fig., 19 pl., 4 maps and atlas (v. 4) of 1313 (i. e. 1110) fig. 26 cm. Jena, 1911–37.

Descriptive flora of Spermatophyta, with descriptions mostly in form of keys, local and extralimital range, vernacular names. The atlas, never completed, covers Cycadaceae–Ranunculaceae, and Leguminosae (part); Abt. 6 not published.—For Pteridophyta, see **Backer, C. A.**, and **Posthumus, O.** Varenflora voor Java. 370 p. 81 fig. Buitenzorg, 1939. (Not seen.)

——— Plantkundig woordenboek voor de boomen van Java med korte aanteekeningen over de bruikbaarheid van het hout. Meded. Lands Plant. Buitenz. no. 12. xx, 173 p. 1894.

Systematic list of trees, with vernacular names (separately indexed) and brief annotations.

——— and **Valeton, Theodoric.** Bijdrage no. 1 [–10] tot de kennis der boomsoorten van (op) Java. (Additamenta ad cognitionem florae arboreae javanicae.) Meded. Lands Plant. Buitenz. no. 11. 363 p. 1894; no. 14. 228 p. 1895; no. 16. 320 p. 1896; no. 17. 328 p. 1896; no. 33. 464 p. 1900; no. 40. [201] p. 1900; no. 42. 231 p. 1900; no. 59. 285 p. 1902; no. 61. 407 p. 1903; no. 68. 287 p. 1904.—no. 11–13. Meded. Dep. Landb. [Nederl.-Indië] no. 2. 277, 12 p. 1906; no. 10. 782 p. 1910; no. 18. 286 p. 1914.—Atlas der baumarten von Java im anschluss an die "Bijdragen tot de kennis der boomsoorten van Java." 4 v. (16 lfg.). 800 pl. 26 cm. Leiden, 1913–18.

Revisionary treatments of families of woody plants, in Dutch and Latin; general index to vernacular and botanical names in pt. 13.

Vorderman, A. G. Javaansche geneesmiddelen. Geneesk. Tijdschr. Nederland.-Indië **34**: 269–343. 1894.

Bibliography; briefly annotated list of 130 drug plants, alphabetically arranged by vernacular names, with botanical and Chinese names and references to literature; indices.

Local

Backer, C. A. Flora van Batavia. Deel I. Dicotyledones dialypetalae (Thalamiflorae en Disciflorae). xvi, 405, 25, 18 p. 26.5 cm. Batavia, 1907. (Mededeelingen uitgaande van het Departement van Landbouw no. 4.)

Descriptive flora of the vicinity of Batavia, Java, with keys and vernacular names (separately indexed).

——— Onkruidflora der javasche suikerrietgronden. afl. 1–5 (8, lxxvii, 907 p.). map and atlas, afl. 1–12 (pl. 1–384). 25 cm. Soerabaia, 1928–38. (Handboek ten dienste van de suikerrietcultuur en de rietsuiker-fabricago op Java, 7. deel.)

Full descriptions, with vernacular names and keys to genera and species, of the weeds of Javan sugar fields; habit figures, with occasional details, of each species. Text completed, but plates still in process of publication.

Backer, C. A., and **Slooten, D. F. van.** Geïllustreerd handboek der javaansche theeonkruiden en hunne beteekenis voor de cultuur. 47 p. 240 pl., map. 24 cm. Batavia, 1924.

Annotated descriptive account of weeds of teafields, with vernacular names, keys, and with plate of each species. Includes most of the widespread weeds of the Tropics of the Old World.

Buitenzorg. Jardin botanique. Flore de Buitenzorg. v. 1–6. illus., plates. 26 cm. Leiden, 1898–1922.

Descriptive flora, with keys, in German. Of the 6 parts so far published, the only ones relating to vascular plants are vol. 1, Pteridophyta, by M. Raciborski, and vol. 6 (in 2 parts, text and plates), Orchidaceae, by J. J. Smith.

Clason, E. W. The vegetation of the Upper-Badak region of Mount Kelut (East Java). Bul. Jard. Bot. Buitenz. III, **13**: 509–518. pl. 5–6. 1935.

Topography, volcanic history, general features of vegetation; list of vascular and cellular plants, annotated by symbols, and grouped by size and habit.

Docters van Leeuwen, W. M. Beitrag zur kenntnis der gipfelvegetation der in Mittel-Java gelegenen vulkane Soembing und Sindoro. Bul. Jard. Bot. Buitenz. III, **11**: 28–56. 2 fig., pl. 4–13. 1930.

Includes annotated list of vascular and cellular plants occurring above 2,400 meters elevation.

——— Biology of plants and animals occurring in the higher parts of Mount Pangrango-Gedeh in West-Java. Verhandl. K. Akad. Wetensch. Amsterdam Afd. Natuurk., 2. Sect., v. 31. 278 p. 66 fig., 30 pl., fold. tab. 1933.

Includes account of botanical explorations, with footnote references (p. 1–22), and annotated systematic lists of plants from the higher parts of Mount Pangrango--Gedeh (p. 143–250) and from other mountain tops and the vicinity of Tjibodas (p. 251–268).

——— Über die erneuerung der verbrannten alpinen flora des Merbaboegebirges in Zentral-Java. Ber. Deut. Bot. Gesell. **31**: 151–157. 3 fig. 1913.

Includes lists of alpine species collected before and after the big fire of 1912.

Elbert, J. E. W. Ueber die zonare verbreitung der vegetation auf den Lawuvulkan Mittel-Javas. Meded. Rijks Herb. Leiden no. 12. 31 p. 1912.

List by H. Hallier (p. 7–31) of vascular plants and 1 lichen collected by Elbert on Mount Lawu, with altitudes and general range.

Koorders, S. H. Floristischer überblick über die blütenpflanzen des urwaldes von Tjibodas auf dem Vulkan Gede in West-Java nebst einer nummerliste und einer systematischen übersicht der dort für botanische untersuchungen von mir numerierten waldbäume. Bot. Jahrb. Engler **50** (Suppl.): 278–303. 1914.

General features of flora, etc.; unannotated systematic list of trees.

——— Verslag van eene dienstreis naar de Karimon-djawa-eilanden. Natuurk. Tijdschr. Nederland.-Indië **48**: 20–132. pl., map. 1889.

Bibliography, topography, general features of botany and zoology, annotated list of useful plants collected on the different islands of the group, zoology, geology. The islands lie north of Djapara, Java.

——— Versuch einer arten-aufzählung der hochgebirgsflora von Tosari und Ngadisari. His Notizen über die phanerogamenflora von Java. II–III, V. Natuurk. Tijdschr. Nederland.-Indië **60**: 241–280, 370–374. 1901; **62**: 213–266. 1903.

Botanical explorations; annotated list of 342 vascular plants, with vernacular names and brief descriptions, the families arranged alphabetically.

Koorders-Schumacher, Anna. Systematisches verzeichnis der zum herbar Koorders gehörenden, in Niederländisch-Ostindien, besonders in den jahren 1888–1903 gesammelten phanerogamen und pteridophyten nach den original-einsammlungsnotizen und bestimmungs-etiketten, unter der leitung von Dr. S. H. Koorders zusammengestellt. I. Abt. Java. v. p. 24.5 cm. Buitenzorg, 1910–13.

Systematic list of vascular plants collected by Koorders, with localities, collecting numbers and occasional annotations. Numerous separate pagings, mostly by families.

Penzig, Otto. I prodotti vegetali del mercato di Buitenzorg (Giava). Atti Soc. Ligust. Sci. Nat. e Geog. **9**: 405–429. 1899.

Descriptive account of useful plants sold in the bazar of Buitenzorg.

Pijl, Leendert van der. The re-establishment of vegetation on Mt. Goentoer (Java). Annales Jard. Bot. Buitenz. **48**: 129–152. pl. 19–20. 1938.
History of eruptions which destroyed the flora and the process of revegetation, etc.; tabular list of known cellular and vascular plants; bibliography.

Rock, J. F. The forest of Mt. Gedah, West Java. Hawaiian Plant. Rec. **22**: 67–104. illus. 1920.
General features of flora, life zones, very brief synoptical list of more important plant families and their principal representatives.

Schmucker, Theodor. Beiträge zur kenntnis der hochgebirgsflora Javas und zur theorie der pflanzenausbreitung. Beih. Bot. Centralbl. 2. Abt. **43**: 34–68. 1926.
Mainly phytogeographic; contains (p. 65–67) list of species known in Java above 2,500 meters elevation; bibliography.

Smith, J. J. Een botanische reis naar de Duizend-eilanden. Teysmannia **18**: 450–456. 1907.
Annotated alphabetical lists of vascular plants collected by author in December 1906, on Poeloe Paniki, Poeloe Pendjaliran, and Poeloe Doea. (Thousand Islands, north of Java near its western end.)

Steenis, C. G. G. J. van. Schets van de flora van het eiland Dapoer (Duizend Eilanden). Trop. Natuur **24**: 31–34. 6 fig. (incl. map). 1935
Running account of plants observed.

Vorderman, A. G. Inlandsche namen van eenige Madoereesche planten en simplicia . . . Natuurk. Tijdschr. Nederland.-Indië **59**: 140–197. 1900.
Madoera (Madura) language (by H. N. Kiliaan), botanical glossary; annotated alphabetical list of 408 vernacular names; index of botanical names.

JUAN FERNANDEZ ISLANDS

See also Polynesia (Merrill).

General

Hemsley, W. B. Juan Fernandez and Masafuera. *In* his Report on the botany of Juan Fernandez . . . Report on the scientific results of the voyage of H. M. S. *Challenger* during the years 1873–76. Botany. v. 1, pt. 3, p. 1–96. pl. 54–63. London, 1885 (1884).
Physiography, botanical exploration, general features of vegetation, tabular analysis of flora, extinct species, etc.; annotated list of plants (vascular and cellular). Includes 124 phanerogams and 44 pteridophytes.

Johow, Friedrich. Estudios sobre la flora de las islas de Juan Fernandez. xi, 287 p. 8 fig., 18 pl., 2 maps. 32.5 cm. Santiago de Chile, 1896.
Geography and geology (by Roberto Pöhlmann), botanical explorations, bibliography; annotated list of vascular and cellular plants; evolution of the flora, statistical tables, plant formations, etc.

Skottsberg, Carl. The phanerogams of the Juan Fernandez Islands. *In* his The natural history of Juan Fernandez and Easter Island. v. 2, Botany, pt. 2, no. 7. p. 95–240. 39 fig., pl. 10–20 (1 col.). Uppsala, 1922. (Preprint, 1921.)
Enumeration of the indigenous (142 species) and introduced (130 species) phanerogams, with references, localities, general range, and critical notes; general features of flora; bibliography.—See also **Christensen, C.,** and **Skottsberg, C.** The Pteridophyta of the Juan Fernandez Islands. l. c. pt. 1, no. 1. p. 1–46. 7 fig., pl. 1–5. 1920. (List of 51 Pteridophyta, with references, localities, and critical notes.)—For additions, see his Notes on some recent collections made in the islands of Juan Fernandez. Acta Horti Gothoburg. **4**: 155–171. illus. 1929.

MELANESIA

See also New Caledonia; New Guinea; Polynesia (Endlicher; Guppy; Jouan; Merrill).

Local

Beck-Mannagetta, Günther. Flora des Stewart-atolls im Stillen Ocean. Annalen K. K. Naturhist. Hofmus. Wien **3**: 251–256. 1888.
General features of flora; list of 17 plants (14 vascular) collected on the voyage of the *Novara* in 1858. The atoll, also known as Sikéiana (Sikai Anna),

8°22′ S. Lat., 162°58′ E. Long., is just east of the Solomon Islands. The specimens were collected on Faule or Small Island, one of the 4 composing the atoll.

Bradtke, W. The flora of a small tropical island . . . With a foreword by C. T. White. Queensland Nat. **2**: 133–135. 1921.

List of 22 vascular plants collected on a small coral island in the Duke of York Islands, Bismarck Archipelago, with vernacular names.

Burkill, I. H. On a collection of plants from New Britain (Neu Pommern). Proc. Cambridge Phil. Soc. **9**: 90–97. 1896.

Botanical explorations, etc.; list of vascular plants collected in 1875 by Baron A. von Hügel; bibliography. Known flora is about 160 species.

Guillaumin, André. Contribution à la flore des Nouvelles-Hébrides I–III. Bul. Soc. Bot. France **66**: 267–277. 1920; **74**: 693–712. 1928; **76**: 298–303. 1929.

Part I consists of a list of 111 vascular plants collected on the island of Efate (Vaté or Sandwich) by Levat in 1883, with localities and vernacular names; the number previously known was 13. Part II is a list of those now known from the New Hebrides, including Banks and Torres Islands (312 species and varieties), with localities and names of collectors; bibliography, briefly annotated list of collectors. Part III. Supplément aux plantes recueillies par M. Levat.—For additions see his Contribution to the flora of the New Hebrides. Plants collected by S. F. Kajewski in 1928 and 1929. Jour. Arnold Arboretum **12**: 221–264. 3 fig. (incl. map). 1931; **13**: 1–29, 81–126. 3 fig., pl. 43. 1932; **14**: 53–61. 1933.—Also his Contribution à la flore des Nouvelles-Hébrides. Plantes recueillies par M. et Mme. Aubert de la Rüe en 1934 (phanérogames). Bul. Soc. Bot. France **82**: 346–354. 1 fig. (map). 1935.

——— A florula of the island of Espiritu Santo, one of the New Hebrides. Jour. Linn. Soc. Bot. **51**: 547–566. 1938.

Climate, botanical explorations, list of plants previously collected; list of vascular plants collected by Ina Baker and Mrs. Zita Baker in 1933–34; bibliography.

Guppy, H. B. The Solomon Islands and their natives. xvi, 384 p. plates (incl. map). 25.5 cm. London. 1887.

Contains, p. 294–304, an annotated "List of plants collected in the islands of Bougainville Straits, Solomon group, during 1884," with vernacular names.

Hemsley, W. B. The Admiralty Islands. *In* his Report on the botany of Juan Fernandez . . . Report on the scientific results of the voyage of H. M. S. *Challenger* during the years 1873–76. Botany. v. 1, pt. 3, p. 227–275. London, 1885.

Physiography, general features of flora, table of distribution; annotated list of vascular and cellular plants (69 phanerogams).

Kew. Royal Gardens. Flora of the Solomon Islands. Bul. Misc. Inform. Kew **1894**: 211–215. 1894; **1895**: 132–139. 1895.

Annotated list of vascular plants, based on a collection made by the officers of H. M. S. *Penguin* in 1894; (second paper) records and descriptions of plants additional to those mentioned in several earlier papers cited in this publication.

Krämer, Augustin. Beiträge zu einer monographie der Hermit-inseln (Luf-archipel). *In* Forschungsreise S. M. S. *Planet* 1906/07. Herausgegeben von Reichs-Marine-Amt. **5**: 57–122. fig. 17–49 (incl. map). Berlin, 1909.

Includes (p. 117–122) general features of flora and briefly annotated list of vascular plants collected by author, alphabetically arranged by vernacular names, with botanical names where known. These were the first plants collected on these islands. (West of Admiralty Islands, 1°15′ S. Lat., 145°45′ E. Long.)

Lauterbach, Carl. Beiträge zur flora von Neu-Mecklenburg. Bot. Jahrb. Engler **45**: 354–365. 1911.

Includes species not recorded by Schumann and Lauterbach, Die flora der Deutschen schutzgebiete in der Südsee, and its Nachträge (see under New Guinea).

Rechinger, Karl. Bearbeitung der Musci, Pteridophytae und Siphonogamae des Neu-Guinea-archipels, der Pteridophytae und Siphonogamae von Ceylon, Hawaii und Hongkong . . . endlich nachträge und berichtungen zu den vorhergehenden teilen. *In* his Botanische und zoologische ergebnisse einer wissenschaftlichen forschungsreise nach den Samoainseln, dem Neuguinea-archipel und den Salomonsinseln von März bis Dezember 1905. V. teil. Denkschr. Math.-Naturw. Kl. K. Akad. Wiss. Wien **89**: 443–672, 699–708. 30 fig., 8 pl. 1914.

Includes (p. 462–463) unannotated list of 60 vascular and cellular plants from the Shortland Islands, previously unknown botanically (Poperang Island and "Taubeninsel"); also (p. 468–621) list of vascular plants collected by author in Neu-Pommern, Solomon Islands, and Shortland Islands, with data. The Shortland Islands are south of the east end of Bougainville Island, one of the Solomon Islands.

Schumann, K. M., and others. Die flora von Neu-Pommern. Notizbl. K. Bot. Gart. Berlin **2**: 59–158. map. 1898.

Botanical investigations, physiography, plant formations; annotated list of cellular and vascular plants by various authors.

MICRONESIA

See also New Guinea (Schumann and Lauterbach); Polynesia (Endlicher; Guppy; Merrill).

General

Kanehira, Ryôzô. An enumeration of Micronesian plants. Jour. Dept. Agr. Kyushu Imp. Univ. **4**: 237–464. pl. 2 (map). 1935.

Botanical explorations; list of 1,219 vascular plants, with local and general distribution. The area covered includes the Marianne, Caroline, and Marshall groups. The principal islands are Saipan, Tinian, Rota, Yap, Palau, Truk, Ponape, Kusai, Jaluit, and Guam.

——— Flora micronesica. 8, 468, 37 p. 211 fig., 21 pl., map. 26 cm. Tokyo, 1933.

Includes a general sketch of the flowering plants, with footnote bibliography; the trees and shrubs (including descriptions and figures); list of vascular plants known from Micronesia. Entirely in Japanese, except for botanical names.

Local

Bryan, E. H., jr. The plants of Guam, pt. I–XXVI. Appeared in Guam Recorder, v. 13, no. 8–v. 16, no. 4, 6, 8. Nov. 1936–Nov. 1939. I–IV, reprinted.)

Botanical explorations and publications, plant associations; descriptive flora of vascular plants, with keys, uses, and vernacular names. Includes cultivated as well as wild species; total number given as 588. Not yet completed. Information in part furnished by the author.

Bryan, W. A. A monograph of Marcus Island. Occas. Papers Bern. P. Bishop Mus. **2**: 77–124. 8 fig., map. 1903.

Includes (p. 122–124) annotated list of 10 indigenous plants (only partly named botanically), also cultivated plants.

Christian, F. W. Ponapean trees, plants and shrubs. *In* his The Caroline Islands . . . p. 328–352. London, 1899.

Annotated alphabetical list of vernacular names used in Ponape, with botanical identification when known.

Diels, Ludwig, ed. Beiträge zur flora von Mikronesien und Polynesien . . . Ser. I–V. Bot. Jahrb. Engler **52**: 1–18. 2 fig. 1914; **56**: 429–577. 7 fig. 1921; **59**: 1–29. 1 fig. 1924; **63**: 271–323. 1930; **69**: 395–400. 1938.

Treatments of various families or of single novelties, not in systematic order, contributed by specialists; no approach to a systematic flora. Ser. I (Beiträge zur flora von Mikronesien) edited by G. Volkens.

Hosokawa, Takehide. An enumeration of the plants collected from Ponape. Kudoa **5**: 41–55, 79–96. 1937.

List of plants collected by author, with collecting numbers, data, and vernacular names. First installment not seen; second reaches Ulmaceae (Engler and Prantl system). (Processed publication.)

——— Phytogeographical relationship between the Bonin and the Marianne Islands laying stress upon the distribution of the families, genera and special species of their vernacular [i. e. indigenous] plants. Jour. Soc. Trop. Agr. (Formosa) **6**: 201–209, 657–670. 1 fig. (map). 1934.

Included here for its unannotated list of vascular plants of the Marianne Islands (p. 662–669), with mention of islands on which each occurs.

Hosokawa, Takehide. A preliminary account of the phytogeographical study on Truk, Caroline. Bul. Bio-geog. Soc. Japan **7**: 171–255. 1937.

Includes annotated list of 217 plants known from the island. Text in Japanese. (Not seen; data from Bot. Jahrb. Engler **69** (Lit.) : 70. 1938.)

——— Preliminary account of the vegetation of the Marianne Islands group. Bul. Bio-geog. Soc. Japan **5**: 124–172. 9 fig., pl. 10–14 (incl. map). 1934.

Includes (p. 129–151) list of vascular plants, with mention of islands on which each occurs and indication of endemic species; also (in Japanese) botanical explorations, topography, climate, phytogeography, bibliography, etc.; brief English summary.

Kanehira, Ryôzô. On the flora of Rota Island. Syokubutu oyobi dôbutu (Bot. and Zool.) **4**: 63–70. 12 fig. 1936.

Contains lists of woody plants, including some found also on other islands of the group. (In Japanese.)

Koidzumi, Gen-ichi. The vegetation of Jaluit Island. Bot. Mag. Tokyo **29**: 242–257. 3 fig. (incl. map). 1915.

List of 59 spontaneous vascular plants, with vernacular names and general range; geography, meteorology, plant formations, bibliography.

Merrill, E. D. An enumeration of the plants of Guam. Philippine Jour. Sci. C, Bot. **9**: 17–155. 1914.—Additions to the flora of Guam. l. c. **15**: 539–544. 1919 (1920).

Annotated list of vascular plants (545 species, of which only 225 are truly indigenous and about 61 endemic), as well as cellular cryptogams, with general range given, citation of exsiccatae, vernacular names (not indexed); general features of flora, with especial consideration of the origin of the weeds. The "Additions" comprise 14 species.

Safford, W. E. The useful plants of the island of Guam. Contrib. U. S. Natl. Herb. v. 9. 416 p. 69 pl., map. 1905.

Geography, history, scientific explorations, physiography, climate, plant associations, agriculture, etc.; bibliography; annotated alphabetical list of plants (wild and cultivated), with brief descriptions, uses, and vernacular names.

Tuyama, Takasi. Plants of Marcus Island. Jour. Jap. Bot. **14**: 425–426. 1938.

List of 11 vascular plants collected by Y. Yabe, with Japanese vernacular names. Text is in Japanese.

Volkens, Georg. Die flora der Marshallinseln. Notizbl. K. Bot. Gart. Berlin **4**: 83–91. 1903.

List of species, with vernacular names and names of collectors, and occasional annotations; includes also the species known from the Gilbert Islands.

——— Die vegetation der Karolinen, mit besonderer berücksichtigung der von Yap. Bot. Jahrb. Engler **31**: 412–477. pl. 11–14. 1901.

Briefly annotated list of vascular and cellular plants of the island of Yap; vernacular names; general features of physiography, etc., of the Carolines; geography, geology, climate, plant formations, and cultivated plants of Yap.

MOLUCCA ISLANDS

See also Dutch East Indies (general works; also Steenis, On the origin); New Guinea (Lauterbach; Warburg).

General

Warburg, Otto. Die botanische erforschung der Molukken seit Rumph's zeiten. *In* Rumphius gedenboek 1702–1902. Uitgegeven door het Koloniaal museum te Haarlem. p. 63–78. Haarlem, 1902.

Sketch of botanical explorations, with itineraries and references to publications; list of islands with names of collectors.

Local

Forbes, H. O. List of plants from Timor-Laut. *In* his A naturalist's wanderings in the eastern archipelago . . . from 1878 to 1883. p. 354–355. London, 1885.

Unannotated list of vascular plants collected by the author.

Hemsley, W. B. The south-eastern Moluccas. *In* his Report on the botany of Juan Fernandez . . . Report on the scientific results of the voyage of

H. M. S. *Challenger* during the years 1873–76. Botany. v. 1, pt. 3, p. 101–226. pl. 64–65. London, 1885.

General features of vegetation; table of distribution, with discussion, including lists of plants from Keeling (or Cocos) Islands (in the Indian Ocean), Diego Garcia (Chagos Islands), Rarotonga (Hervey or Cook group), and various Pacific islands (Palmerston Island, Cook group; Enderbury, Gardner, Hull, Mary, Phoenix, and Sydney Islands, in the Phoenix Islands; Fanning Island; Starbuck Island; Suwarrow Island; Ducie Island); annotated list of plants (vascular and cellular) from the islands of Arrou (Aru), Babar, Dammar, Ki (Kei), Little Ki, Lakor, Larat, Letti, Moa, Timor Laut, and Wetter, including 367 phanerogams.

Merrill, E. D. An interpretation of Rumphius's Herbarium amboinense. 595 p. 2 maps. 23 cm. Manila, 1917.

Account of Rumphius and his work, etc.; annotated systematic list of plants listed by Rumphius, with synonymy, and citation of Amboina specimens; list of doubtful species; list of names in Rumphius' work, with botanical equivalents.

Rant, Anton. Twee botanische reizen naar Ambon. Natuurk. Tijdschr. Nederland.-Indië **94**: 100–133. 1 fig., 4 pl. 1934.

Includes unannotated list of cellular and vascular plants collected by author in Amboina (Ambon) and Ceram, also similar list from Celebes and Boeroe (Leksoela); bibliography.

Thistleton-Dyer, W. T. Report on the botany of Hr. H. O. Forbes's expedition to Timor-Laut. Jour. Linn. Soc. Bot. **21**: 370–374. 1885.

Geography, general features of flora; unannotated list of species, lacking a few of the species given in Forbes' own list (1885).

NEW CALEDONIA

See also Polynesia (Jouan; Merrill); French Guiana (Lanessan, Les plantes utiles, p. 233–279, 663–701).

General

Guillaumin, André. Catalogue des plantes phanérogames de la Nouvelle-Calédonie et dépendances (Iles des Pins et Loyalty). Annales Mus. Colon. Marseille **19** (II, **9**): 77–290. fold. map. 1911.

Botanical explorations, annotated list of collectors; systematic list of spermatophytes with localities and collectors; index of vernacular names.

——— Contribution à la flore de la Nouvelle-Calédonie [I–II]. Annales Mus. Colon, Marseille II, **9**: 55–75. 1911.—III–VI. *In* Lecomte, H. Notulae systematicae **2**: 41–46, 99–105. 1911.—VII–L. Bul. Mus. Hist. Nat. Paris **17**: 349–357, 453–459, 558–566. 1911; **18**: 39–46, 91–101, 166–176, 324–331, 373–384, 466–468. 1912; **19**: 376–383, 509–524. 1913; **20**: 93–96. 1914; **25**: 213–217, 288–295, 372–378, 499–505, 645–652. 1919; **26**: 77–84, 174–179, 254–261, 361–368, 434–435. 1920; **27**: 119–125, 257–259, 558–562. 1921; **28**: 103–108, 196–198, 545–546. 1922; **29**: 112–118. 1923; **31**: 100–103, 209–212, 480–481, 482–484. 1925; **32**: 229–232. 1926; **33**: 110–114, 272–276. 1927.—LI–LIII. Bul. Mus. Hist. Nat. Paris II, **1**: 117–123, 216–218. 1929.—LIV. Candollea **5**: 148–152. 1934 (1932).—LV–LXIV. Bul. Mus. Hist. Nat. Paris II, **2**: 165–171. 1930; **4**: 688–704. 1932; **5**: 242–249, 322–327. 1933; **6**: 198–203, 302–307. 1934.—LXV–LXVI. Notulae Syst. Mus. Hist. Nat. Paris **5**: 13–16, 131–134. 1935–36.—LXVII–LXX. Bul. Mus. Hist. Nat. Paris II, **10**: 433–434, 518–521, 623–627. 1938; **11**: 412–416. 1939.

Annotated lists of specimens of different collectors. Nos. 1 and 2 published under title: Contribution á la flore de Bourail (Nouvelle-Calédonie).

——— Matériaux pour la flore de la Nouvelle-Calédonie. I–V. *In* Lecomte, H. Notulae systematicae **3**: 5–65, 160–165, 260–263. 1914–16.—VI–XXII. Bul. Soc. Bot. France **66**: 310–313. 1920; **67**: 27–29, 47–54, 64–66, 121–124, 345–347. 1920–21; **69**: 31–34, 65–69, 507–509. 1922–23; **71**: 939–946, 1101–1112. 1925; **72**: 89–92. 1925; **73**: 102–107, 429–433, 437–442. 1926.—XXIII. Arch. de Bot. (Caen) Bul. Mens. **1**: 73–77. 1927.—XXIV–XXV. Bul. Soc. Bot. France **74**: 924–930. 1928; **75**: 288–294. 1928.—XXVI–XXVII. Arch. de Bot. (Caen) Mém. v. 2, no. 3. 47 p. [1929]; v. 3, no. 5. 47 p. 1929.—XXVIII–LV. Bul. Soc. Bot. France **79**: 225–226, 335–341, 515–516, 689–691. 1932; **80**: 35–38, 476–480. 1933; **81**: 3–17, 242–246, 283–285, 454–455. 1934; **82**: 47–48, 274–283. 1935; **83**: 294–315, 485–488, 577–578. 1936–37; **84**: 54–61, 98–100, 159–161, 255–

257, 462–463, 1937–38; **85**: 19–21, 37–47, 202–203, 294–305, 626–653. 1938–1939; **86**: 174, 275–278. 1939.

Revisionary treatments of miscellaneous families and genera of spermatophytes.

Vieillard, Eugène. Plantes utiles de la Nouvelle–Calédonie. Annales Sci. Nat. IV, Bot. **16**: 28–76. 1862.

Annotated list of some useful plants, with vernacular names and uses.

Local

Beauvisage, Georges. Genera montrouzierana plantarum Novae Caledoniae. Annales Soc. Bot. Lyon **26**: 1–96. 1901.

Critical discussion of the numerous new genera described by Montrouzier in 1860.

Däniker, A. U. Katalog der Pteridophyta und Embryophyta siphonogama. (Ergebnisse der reise von Dr. A. U. Däniker nach Neu-Caledonien und den Loyalty-inseln (1924/6) 4.) Vierteljahrsschr. Naturf. Gesell. Zürich v. 77–78, Beibl. 19. p. 1–114, 115–235, 237–338, 339–395. 1932–33. (Mitteilungen aus dem Botanischen museum der Universität Zürich CXLII.)

Geology, etc.; list of vascular plants, with data, collected by the author in 1924–26, with a few by other collectors; occasional critical notes. Pt. 4 goes through Asclepiadaceae (Engler and Prantl system).

Guillaumin, André, and Beauvisage, Georges. Species montrouzieranae seu enumeratic plantarum in Nova Caledonia terrisque adjacentibus a R. P. Montrouzier lectarum. Annales Soc. Bot. Lyon **38**: 76–130. 3 fig. (incl. map). 1914.

Enumeration of all known specimens collected by Montrouzier from New Caledonia and nearby regions, with reductions to synonymy; references to earlier papers on same subject.

Montrouzier, Xavier, *père*. Flore de l'île Art (près de la Nouvelle Calédonie). Mém. Cl. Sci. Acad. Imp. Sci., Belles-lettr. et Arts Lyon n. s., **10**: 173–254. 1860.

Annotated list of vascular plants, most of which are described as new; mention of principal edible and textile plants.

Rendle, A. B., Baker, E. G., and Moore, S. LeM. A systematic account of the plants collected in New Caledonia and the Isle of Pines by Prof. R. H. Compton, M. A., in 1914. Part I. Flowering plants (angiosperms). Jour. Linn. Soc. Bot. **45**: 245–417. pl. 13–24. 1921.

Annotated list of 830 angiosperms, of which 230 are described as new.—For Part II. **Compton, R. H.** Gymnosperms, Pteridophyta, see l. c. 421–462. pl. 26–27. 1922.

Sarasin, Fritz, and Roux, Jean. Nova Caledonia. Forschungen in Neu-Caledonien und auf den Loyalty-inseln. B. Botanik. Redaktion Hans Schinz und A. Guillaumin. v. 1, lfg. 1–3. plates. 28.5 cm. Berlin und Wiesbaden, 1914–21.

Annotated lists of miscellaneous families of vascular and cellular plants collected by Sarasin, by various authors; localities, general range, synonymy; geographic botany, floral statistics (p. 256–293).

Schlechter, Rudolf. Beiträge zur kenntnis der flora von Neu-Kaledonien. Bot. Jahrb. Engler **39**: 1–274. 23 fig. 1906.

List of vascular plants collected by author. Treatment of various families contributed by other authors.

White, C. T. Ligneous plants collected in New Caledonia by C. T. White in 1923. Jour. Arnold Arboretum **7**: 74–103. 1926.

Annotated list of woody plants collected by the author. Gymospermae by E. H. Wilson, Angiospermae by A. Guillaumin.

NEW GUINEA

See also Dutch East Indies (general works; also Steenis, On the origin).

General

Lam, H. J. Materials towards a study of the flora of the island of New Guinea. Blumea **1**: 115–159. 2 fig. (maps). 1934.

General review of the subject; plant associations, statistics, list of endemic or nearly endemic genera, phytogeography, list of principal collectors, bibliography (including index of families treated in Lauterbach's "Beiträge zur flora von Papuasien" and in "Nova Guinea. Résultats des expéditions scientifiques à la Nouvelle Guinée," vol. 8, 12, and 14).

Lauterbach, Carl. Beiträge zur flora von Papuasien I–XXIV. Bot. Jahrb. Engler **49**: 1–169. illus. 1912; **40**: 1–80, 81–170, 288–383. illus. 1913; **52**: 19–176, 177–220. illus. 1914–15; **54**: 69–96, 97–128, 129–261. illus. 1919; **55**: 19–144, 145–312. illus. 1917–18; **56**: 31–160, 161–272, 273–400, 401–414. 1920–21; **57**: 321–400, 401–528, 529–608. 1922; **58**: 1–96, 97–181, 244–304, 305–448, 449–495. 1923; **59**: 41–128, 505–567. illus. 1924–25; **60**: 1–174. pl. 1. 1925; **61**: 1–64, 65–222. 1927; **62**: 284–304, 338–416, 417–501, 550–569. pl. 25. 1928–29; **63**: 1–28, 419–476. 1929–30; **66**: 161–215. 1934: **67**: 143–236. 1935; **68**: 244–247, 248–268. 1937; **69**: 252–294. 1938; **70**: 95–148. 1939.

Treatments of miscellaneous families and genera by various authors; continued by L. Diels (no. XXIII on) after Lauterbach's death. Some of the articles are merely descriptions of new species, others revisionary treatments with keys. The species listed in nos. I–VI are indexed in Bot. Jahrb. Engler **55**: 295–312. 1918. For index to families treated in parts I–XX (1912–1934), see Lam, H. J., Blumea **1**: 147–150. 1934.

Lorentz, H. A., and others, eds. Nova Guinea. Résultats de l'expédition scientifique néerlandaise à la Nouvelle-Guinée . . . Botanique. v. 8, v. 12, v. 14, v. 18. plates. 31.5 cm. Leiden, 1909–36.

Revisionary treatments of miscellaneous families by various authors. For index to families (v. 8, 12, 14), see Lam, H. J., Blumea **1**: 150–155. 1934. Publication discontinued; replaced by Nova Guinea, a journal of botany, zoology [etc.] of the Papuan region. n. s., v. 1, 1937—(no botanical articles to date).

Warburg, Otto. Das pflanzenkleid und die nutzpflanzen Neu-Guineas. *In* **Krieger, M.** Neu-Guinea. p. 36–72. pl. 2–13. Berlin, [1899].

General features of flora, plant formations, condensed account of useful plants.

Local

Beccari, Odoardo. Catalogue of the plants of the Fly River, 1877. *In* **Albertis, L. M. d'.** New Guinea . . . v. 2, p. 396–400. London, 1880.

Unannotated list of 314 vascular plants collected by d'Albertis; many of the species are named only to genus.

Diels, Ludwig. Beiträge zur flora des Saruwaged-gebirges. Bot. Jahrb. Engler **62**: 452–501. 1929.

Botanical explorations, with bibliographical references; list of cellular and vascular plants collected by Chr. Keysser in 1912 and 1916. These mountains are in northeastern New Guinea.

Gibbs, L. S. Dutch N. W. New Guinea. A contribution to the phytogeography and flora of the Arfak Mountains, etc. iv, 226 p. 16 fig., 4 pl. 26.5 cm. London, 1917.

Botanical explorations, topography, plant associations, phytogeography, bibliography; annotated list of vascular and cellular plants collected by the author at the Angi Lakes, altitude 7,000–9,000 feet, in the Arfak Mountains in December 1913.

Kärnbach, Ludwig. Ueber die nutzpflanzen der eingeborenen in Kaiser-Wilhelmsland. Bot. Jahrb. Engler **16** (Beibl. **37**) : 10–19. 1892.

Annotated systematic list of wild and cultivated vascular plants, with uses, but almost no vernacular names.

Kew. Royal Gardens. Flora of British New Guinea. Bul. Misc. Inform. Kew **1899**: 95–133. 1899.

List of about 130 vascular plants (and a few cellular) from Mount Scratchley, with some from other localities; by various authors.

Mueller, Ferdinand von. Descriptive notes on Papuan plants, I–IX. 2 vol. 21 cm. Melbourne, 1875–[90].

Consist of nine parts, each containing one or more annotated lists of species. In vol. 1, p. 37–40, a list is given of 144 plants (not Papuan) to which Linnaeus gave the same binomial that had been used by pre-Linnaean authors.

Rendle, A. B., and others. Dr. H. O. Forbes's New Guinea plants. Jour. Bot. **61** (Suppl.): 1–64. 1923; **62**: 53–55. 1924.

List of New Guinea plants collected by H. O. Forbes in 1885–86, with localities and frequent annotations. Polypetalae by E. G. Baker, Gamopetalae and Monochlamydeae by S. L. Moore, Monocotyledons by A. B. Rendle, Filices by A. Gepp; Musci and Lichenes by other authors.

Ridley, H. N. Report on the botany of the Wollaston expedition to Dutch New Guinea, 1912–13 . . . assisted by Messrs. E. G. Baker, S. Moore, H. F. Wernham, C. H. Wright, and others. Trans. Linn. Soc. Bot. II, **9**: 1–284. pl. 1–6. 1916.

Itinerary, botanical regions; annotated list of vascular and cellular plants, with references and general range.

Scheffer, R. H. C. C. Énumération des plantes de la Nouvelle-Guinée, avec description de espèces nouvelles. Annales Jard. Bot. Buitenz. **1**: 1–60. 1876.—Epilogue . . . l. c. 178–181. 1876.

List of phanerogams collected by Teysmann in 1871, mostly without localities except in the case of new species.

Schumann, K. M. Plantae bammlerianae. Notizbl. K. Bot. Gart. Berlin **1**: 44–57. 1895.

Annotated list of cellular and vascular plants collected by Bammler on the Tami Islands.

——— and **Hollrung, Max.** Die flora von Kaiser Wilhelms Land. v, 137 p. 25.5 cm. Berlin, [1889]. (Beiheft zu den Nachrichten über Kaiser Wilhelms Land und den Bismarck-archipel. 1889.)

Systematic list of cellular and vascular plants, with localities, collectors' names, general range, occasional annotations.

——— and **Lauterbach, Carl.** Die flora der deutschen schutzgebiete in der Südsee. xvi, 613 p. 22 pl., map. 28.5 cm. Leipzig, 1901 (1900). —Nachträge . . . (mit ausschluss Samoa's und der Karolinen). 446 p. 14 pl., port. 28 cm. Leipzig, 1905.

Botanical collections, bibliography; list of cellular and vascular plants (2,208 species, of which 1,560 are vascular), with references, citation of collections, and general range, but no vernacular names. The area covered includes Kaiser Wilhelmsland (northeast quarter of New Guinea), Bismarck Archipelago, Admiralty Islands, western half of the Solomon Islands, Marshall Islands, Caroline Islands, Palau (Pelew) Islands, and the Ladrones (Marianne Islands). The supplement includes an index to the genera and species of both parts.

Valeton, Theodoric. Plantae papuanae. Bul. Dépt. Agr. Indes Néerland. no. 10, iii, 70 p. 1907.

Systematic list, with frequent annotations, of spermatophytes collected on the expedition of A. Wichmann to the north coast in 1903 and on that of P. Meyes and Rochfort to the south and southwest coast in 1903–04; bibliography.

Warburg, Otto. Beiträge zur kenntnis der papuanischen flora. Bot. Jahrb. Engler **13**: 230–455. 1891.

Phytogeography; annotated list of spermatophytes collected by author in New Guinea, the Moluccas, the Bismarck Archipelago, and various small neighboring islands; general range stated. The author defines Papuasia as the region between 130° and 154° E. Long. and 0° and 11° S. Lat., including New Guinea and adjacent small islands, the Bismarck Archipelago, the Aru Islands, the Kei (Key) Islands, the Admiralty Islands, and the Louisiade Archipelago. The term is used by some to include the Solomon Islands also.

——— Bergpflanzen aus Kaiser Wilhelms-land, gesammelt auf der Zöller'schen expedition im Finisterregebirge von F. Hellwig. Bot. Jahrb. Engler **16**: 1–32. pl. 1. 1892.

Itinerary, general features of flora; annotated list of cellular and vascular plants collected at or above 1,000 meters elevation.

White, C. T. A contribution to our knowledge of the flora of Papua (British New Guinea). Proc. Roy. Soc. Queensland **34**: 5–65. 1923.

General features of flora, history of botanical exploration; list of vascular plants (about 800 species), nearly all collected by the author; bibliography.

——— Ligneous plants collected in the territory of Papua (British New Guinea) in 1925–26 by L. J. Brass. Jour. Arnold Arboretum **10**: 197–274. 1929.

General features of flora; annotated list of woody spermatophytes; short list of herbaceous plants collected.—See also **Rehder, A.** A supplement to C. T. White, "Ligneous plants . . ." l. c. 14: 62–67. 1933.

PARACEL ISLANDS

General

Gagnepain, François. Quelques plantes des îlots de la Mer de Chine. Bul. Mus. Hist. Nat. Paris II, **6**: 286. 1934.—Deuxième note . . . Notulae Syst. Mus. Hist. Nat. Paris **6**: 35–37. 1 fig. (map). 1937.

List, with localities and dates, of plants collected on the Paracel Islands (off Annam) and North Danger, Loaita, Itu-Aba, and Spratly Islands to the south of the Paracels.

Sawada, Kaneyoshi, and **Sasaki, Syun'iti.** [Plants of the Paracel Archipelago.] Trans. Nat. Hist. Soc. Formosa **12**: 89–90. 1922.

List of 28 plants collected by Y. Ono on Drummond, Money, Lovell, Duncan, and Lincoln Islands in 1922. (In Japanese.) Title and annotation from Merrill, E. D., and Walker, E. H. A bibliography of eastern Asiatic botany. p. 433. 1938.

PHILIPPINE ISLANDS

See also Dutch East Indies (Steenis, On the origin).

General

Brown, W. H., ed. Minor products of Philippine forests. 3 v. illus. 23.5 cm. Manila, 1920–21. (Bulletin no. 22, Bureau of Forestry.)

Illustrated treatments of various forest products other than timber, by several authors. Vol. 1 includes palms and palm products, bamboos, fiber plants; vol. 2, resins, oils, etc., wild food plants, natural dyes; vol. 3, wild ornamental plants (herbs), soap substitutes, official medicinal plants, poisonous plants, miscellaneous useful wild plants, edible fungi, medical uses of Philippine plants. Vernacular names are given; complete index in vol. 3.

Merrill, E. D. A dictionary of the native plant names of the Philippine Islands. [Pub.] Bur. Govt. Lab. Philippine Isl. no. 8. 193 p. Manila, 1903.

Alphabetical list of about 5,000 vernacular names, with botanical equivalents; corresponding list of scientific names with collected vernacular names and brief annotations on habit, uses, etc.

——— An enumeration of Philippine flowering plants. 4 v. 6 maps. 24 cm. Manila, 1922 (1923)–26.

Annotated list of vascular plants (8,120 species, of which 5,532 are endemic), with synonymy, exsiccatae, local and extralimital range, vernacular names (v. 1–3). Vol. 4 contains geography and geology, Philippine alphabet and vernacular names, botanical history, ecology, floral relationships, bibliography, additions to enumeration, and general indices of vernacular and scientific names.—Pages 1–239 (and 6 maps) of v. 4 reprinted with title: A discussion and bibliography of Philippine flowering plants, as Pop. Bul. Bur. Sci. Philippine Isl. no. 2. 1926.—See also his Additions to our knowledge of the Philippine flora, I–III. Philippine Jour. Sci. **26**: 447–496. 1925; **29**: 475–496. 1926; **30**: 389–430. 1926.—Also **Quisumbing, Eduardo.** New or interesting Philippine plants, I. l. c. **41**: 315–371. 28 fig., 3 pl. 1930.

——— Species blancoanae. A critical revision of the Philippine species of plants described by Blanco and by Llanos. 423 p. 24 cm. Manila, 1918.

Description of editions, etc.; annotated list of species described by Blanco and by Llanos, with identifications. Replaces his A review of the identifications of the species described in Blanco's Flora de Filipinas. [Pub.] Bur. Govt. Lab. Philippine Isl. no. 27. 132 p. 1905; also his Additional identifications . . . Philippine Jour. Sci. C, Bot. **2**: 429–436. 1907.

Pardo de Tavera, T. H. Plantas medicinales de Filipinas. 339 p. 22 cm. Madrid, 1892. (English translation and revision, Medicinal plants of the Philippines, by **J. B. Thomas, jr.** 269 p. Philadelphia, 1901.)

Annotated systematic list, with vernacular names, uses, description, and habitat; index of species classified by uses and by diseases.

Philippine Islands. Bureau of forestry. A dictionary of names applied to trees of the first, second, and third groups. Bul. Bur. Forestry Philippine Isl. no. 23. 40 p. 1923.

List of officially recognized names, with botanical equivalents; alphabetical list of other vernacular names for these, with cross-references.

Reyes, L. J. Philippine woods. Tech. Bul. Dept. Agr. and Comm. Philippines no. 7. 536 p. 88 pl. 1938.
Bibliography, principal distinctive characters used in wood anatomy; systematic list of 264 important and about 200 minor species, with vernacular names, distribution, size, general and structural characteristics of wood; classification of Philippine woods into 4 groups, weight, durability, etc., of woods.

Scribner, F. L. List of Philippine agricultural products and fiber plants. Bul. Bur. Agr. Philippine Isl. no. 5. 47 p. 1904.
Briefly annotated alphabetical list of fruits and vegetables, with vernacular names; annotated tabular list of fiber plants.

Wester, P. J. The food plants of the Philippines. Bul. Bur. Agr. Philippine Isl. no. 39. 3d ed. rev. x, 236 p. 67 pl., map. 1924. (1st ed. in Philippine Agr. Rev. v. 9. 1916; 2d ed. l. c. v. 14. 1921 (1922).)
Topography, climate, etc.; commercial statistics; annotated list of species (wild and cultivated) alphabeted by vernacular names; index of scientific names with vernacular equivalents; bibliography.

Whitford, H. N. The principal forest trees. (The forests of the Philippines. Pt. II.) Bul. Bur. Forestry Philippine Isl. no. 10. 113 p. 103 pl. 1911.
Systematic list, with scientific, official, and usual trade names; key to principal timber trees; descriptive account of habit, bark, wood, uses, localities, vernacular names of each species, systematically arranged.

Local

Brown, W. H., Merrill, E. D., and Yates, H. S. The revegetation of Volcano Island, Luzon, Philippine Islands, since the eruption of Taal Volcano in 1911. Philippine Jour. Sci. C, Bot. **12**: 177–248. 2 fig. (maps), pl. 4–16 (incl. map). 1917.
Topography, former vegetation, revegetation, plant associations, etc.; annotated list of vascular plants found since 1914.

Kanehira, Ryôzô. A list of plants collected at Davao, P. I. Bult. Sci. Fak. Terkult. Kjuŝu Imp. Univ. Fukuoka, Japan **7**: 228–238. 1936.
List of vascular plants collected by the author, with localities and general range. (Text in Japanese.) (Mindanao.)

Merrill, E. D. A flora of Manila. 490 p. 24 cm. Manila, 1912. [Philippine Islands. Bureau of science. Publication no. 5.]
Descriptive flora of the wild and most of the cultivated vascular plants (1,007 species), with keys, local and extralimital range, vernacular names; glossary. The area covered is only about 100 sq. km. around Manila, but the flora is nearly complete for most inhabited regions at low altitudes in all parts of the Philippine Islands.—See also his Notes on the flora of Manila, with special reference to the introduced element. Philippine Jour. Sci. C, Bot. **7**: 145–208. 1912. (Statistical tables of distribution of the species treated in his "Flora of Manila," with discussion; lists of introduced species.)—Also his Nomenclatural and systematic notes on the flora of Manila. l. c. 227–251. 1912.

——— The flora of Mount Halcon, Mindoro. Philippine Jour. Sci. C, Bot. **2**: 251–309. 1907.
Botanical collections, plant formations, etc.; annotated list of spermatophytes with general range and occasional annotations. Contains references to papers on Pteridophyta, Orchidaceae, and Bryophyta of same locality.

——— The flora of the Lamao Forest Reserve. Philippine Jour. Sci. **1** (Suppl. 1): 1–141. 1906.
Topography, general features of flora, statistics; annotated list of hepatics, mosses, and vascular plants; general range indicated. The reserve is on Mount Mariveles, Province of Bataan, Luzon.

——— Hugh Cuming's letters to Sir William J. Hooker. Philippine Jour. Sci. **30**: 153–184. port. 1926.
Includes (p. 174–176) explanation of erroneous labeling of some of Cuming's "Philippine" plants, with list of collecting numbers and localities.

——— On a collection of plants from the Batanes and Babuyanes Islands. Philippine Jour. Sci. C, Bot. **3**: 385–442. 1909.
Geography, phytogeography; list of about 415 vascular plants, with localities. The Batanes (or Bashi) and the Babuyanes Islands form a group extending from near the north coast of Luzon to within about 160 km. of southern Formosa.

Merrill, E. D., and Merritt, M. L. The flora of Mount Pulog. Philippine Jour. Sci. C, Bot. **5**: 287–403. 4 pl., tab., map. 1910.

Geology, types of vegetation, statistical table of distribution, etc.; list of 528 cellular and vascular plants, of which 224 are endemic. Mount Pulog is the highest peak in Luzon.

Robinson, C. B. Botanical notes upon the island of Polillo. Philippine Jour. Sci. C, Bot. **6**: 185–228. 1911.

Geography, phytogeography, etc.; list of vascular plants. Polillo is an island off eastern Luzon.

POLYNESIA

See also Fiji Islands; Hawaiian Islands; Micronesia (Diels); Molucca Islands; Samoan Islands; Society Islands.

The following publication has not been available:

Seurat, L. G. Observations sur la structure, la faune et la flore de l'île Marutea du Sud. Papeete, 1904.

General

Drake del Castillo, Emmanuel. Illustrationes florae insularum Maris Pacifici. 458 p. 50 pl. 33 cm. Paris, 1886–1892.

The introduction summarizes the physiography, plant distribution, general features of flora, principal botanical explorations. Pages 103 to end contain a systematic list of the vascular plants of Polynesia (2,189 species), with synonyms, localities, collectors, and general range. The area covered includes Viti (Fiji), Tonga, Samoa, Society, Paumoto (Tuamotu), Marquesas, and the Hawaiian Islands. The exquisite plates illustrate characteristic species of various families.

Endlicher, Stephan. Bemerkungen über die flora der Südseeinseln. Annalen Wiener Mus. Naturgesch. **1**: 129–190. pl. 13–16. 1836.

Synoptical list of 1,672 species (of which 1,382 are vascular plants) with localities and authorities for records. The area included is Polynesia, Micronesia, and eastern Melanesia. The introduction consists mainly of geographical discussion.

Guppy, H. B. The Polynesians and their plant names. Jour. Trans. Victoria Inst. (London) **29**: 135–170. 1897.

Includes alphabetical list of important plants with vernacular names in Indian Archipelago, Malagasy, West Pacific, Fiji, West and East Polynesia, Micronesia, and elsewhere.—See also **Safford, W. E.** Cultivated plants of Polynesia and their vernacular names, an index to the origin and migration of the Polynesians. Proc. Pan-Pacific Scient. Conf. 1st, 1920 (Spec. Pub. Bern. P. Bishop Mus. **7**): 183–187. 1921.

Jouan, Henri. Les plantes alimentaires de l'Océanie. Mém. Soc. Natl. Sci. Nat. Cherbourg **19** (II, **9**): 33–83. 1875.

Extensively annotated list of food plants, with vernacular names. Covers the Society, Marquesas, Sandwich (Hawaiian), Fiji, and Tonga Islands, New Caledonia, Wallis I., etc.

——— Les plantes industrielles de l'Océanie. Mém. Soc. Natl. Sci. Nat. Cherbourg **20** (II, **10**): 145–240. 1876.

Annotated list of useful plants (excluding edible plants), classified by uses; vernacular names given, but mostly not indexed. Covers the Society, Marquesas, Sandwich, Fiji, and Tonga Islands, New Caledonia, Wallis I., etc.

Merrill, E. D. Polynesian botanical bibliography, 1773–1935. Bul. Bern. P. Bishop Mus. 144. 194 p. 1937.

List of full titles of periodicals cited; briefly annotated list of about 2,600 titles, arranged alphabetically by authors. Replaces his "Bibliography of Polynesian botany," l. c. 13. 68 p. 1924. The region covered is essentially the Pacific basin between 30° N. and 30° S. Lat., excluding the Bonin Islands, east to Juan Fernandez. New Caledonia, New Hebrides, Norfolk, and the Kermadec Islands are included. The larger archipelagos near New Guinea, such as the Louisiades, Solomons, and Bismarck Archipelago, are excluded.

Local

Beaglehole, Ernest, and Beaglehole, Pearl. Ethnology of Pukapuka. Bul. Bern. P. Bishop Mus. 150. v, 419 p. 55 fig. (incl. maps), 6 pl. 1938.

Contains (p. 25–26) section on "Plant life," including list of "all plants growing in Pukapuka, as identified by Marie C. Neal, E. H. Bryan, Jr., and F. Raymond Fosberg," alphabetically arranged by vernacular names, with notes on uses. (Pukapuka or Danger Islands, an atoll northeast of Samoan Islands, 165°50′ W. Long., 10°55′ S. Lat.)

Brown, F. B. H., and Brown, E. D. W. Flora of southeastern Polynesia. Bul. Bern. P. Bishop Mus. 84. 194 p. 18 fig. (incl. maps), 35 pl. 1931; 89. 123 p. 19 fig. (incl. maps), 21 pl. 1931; 130. 386 p. 70 fig., 9 pl. 1935.

Botanical explorations, topography, climate, native agriculture, etc.; annotated descriptive flora of vascular plants, with vernacular names and uses; bibliographies. The total indigenous flora includes 453 species and varieties. Pt. I, Monocotyledons; pt. II, Pteridophytes; pt. III, Dicotyledons. Covers the Marquesas, Tuamotu, and Austral Islands, Rapa Island, and the Mangareva-Henderson Island group.

Burkill, I. H., and Crosby, C. S. The flora of Vavau, one of the Tonga Islands. Jour. Linn. Soc. Bot. **35**: 20–65. 1901.

List of vascular plants, including 83 species additional to Hemsley's "Flora of the Tonga or Friendly Islands" (1894); general range given; botanical explorations, general features of vegetation, statistical table.

Cheeseman, T. F. The flora of Rarotonga, the chief island of the Cook group. Trans. Linn. Soc. Bot. II, **6**: 261–313. pl. 31–35. 1903.

Annotated list of 334 vascular plants, with vernacular names and general range; physiography, general features of vegetation.

Christophersen, Erling. Vegetation of Pacific equatorial islands. Bul. Bern. P. Bishop Mus. 44. 79 p. 7 pl. 1927. (Whippoorwill expedition, Pub. no. 2.)

Separate annotated lists of vascular plants of Christmas, Jarvis, Fanning, Washington, Palmyra, Howland, and Baker Islands, all atolls of coral origin; geography, meteorology, soil, plant formations; bibliography.

Cranwell, L. M. Flora of Manihiki, Cook group. Rec. Auckland Inst. and Mus. **1**: 169–171. 1933.

Geography, general features of flora; annotated list of cellular and vascular plants, with vernacular names; bibliography.

Drake del Castillo, Emmanuel. Flore de la Polynésie française. Description des plantes vasculaires qui croissent spontanément ou qui sont généralement cultivées aux îles de la Société, Marquises, Pomotou, Gambier et Wallis. xxiv, 352 p. map. 26 cm. Paris, 1892.

Physiography geographical affinities, botanical collectors, annotated bibliography; descriptive flora, the vernacular names given but not indexed. Includes 588 species, of which 161 are endemic.

Fosberg, F. R. Vegetation of Vostok Island, central Pacific. Proc. Hawaiian Acad. Sci. **11** (Spec. Pub. Bern. P. Bishop Mus. 30): 19. 1937.

List of 1 cellular and 2 vascular plants collected by Capt. W. J. Anderson. Vostok Island is a tiny isolated coral island in 10°66′ S. Lat., 152°23′ W. Long.

Fuentes, Francisco. Reseña botánica sobre la Isla de Pascua. Bol. Mus. Nac. Chile **5**: 320–337 1913.

Geology, climate, history, notes on introduced and cultivated plants; list of 124 vascular and cellular plants (including cultivated species), with general range and occasional annotations. (Easter Island.)

Guillaumin, André, Camus, Aimée, and Tardieu-Blot, Mme. Plantes vasculaires récoltées à l'Ile de Pâques par la Mission Franco-Belge. Bul. Mus. Hist. Nat. Paris II, **8**: 552–556. 1936.

Botanical explorations, with bibliographical references; list of vascular plants collected in 1934–35, including additions to the known flora. (Easter Island.)

Hedley, Charles. General account of the Atoll of Funafuti. Mem. Austral. Mus. Sydney **3**: 1–71. 3 fig. 1896.

Includes account of vegetation (p. 20–41), with mention of uses of plants. Funafuti is one of the Ellice Islands.

Hemsley, W. B. Elizabeth Island . . . *In* his Report on present state of knowledge of various insular floras . . . Report on the scientific results of the voyage of H. M. S. *Challenger* during the years 1873–76. Botany. v.1, Introd. p. 15. London, 1885.

List of 4 phanerogams. Elizabeth or Henderson Island lies between Pitcairn Island and Ducie Island, southeast of the Tuamotu Islands.

Hemsley, W. B. The flora of the Tonga or Friendly Islands, with descriptions of and notes on some new or remarkable plants, partly from the Solomon Islands. Jour. Linn. Soc. Bot. **30**: 158–217. pl. 9–11. 1894.

Geography, geology, botanical explorations; systematic list of vascular plants (290 species), with names of collectors and general range; table of distribution of the species.

——— Maldon Island . . . *In* his Report on present state of knowledge of various insular floras . . . Report on the scientific results of the voyage of H. M. S. *Challenger* during the years 1873–76. Botany. v. 1, Introd. p. 17–18. London, 1885.

List of phanerogams collected by James Macrae. Maldon or Malden Island is in 3° 58′ S. Lat., 155° W. Long.

Jardin, Édelestan. Essai sur l'histoire naturelle de l'archipel de Mendana ou des Marquises. 2. ptie: Botanique. Mém. Soc. Imp. Sci. Nat. Cherbourg **5**: 289–331. 1857. (Reprinted with title: Essai d'une flore de l'archipel des Marquises. 43 p. 1858.)

List of vascular (221 species) and cellular plants collected or observed by the author in 1853–54 in the northwest group of the archipelago, with vernacular names and some annotations; notes on vegetation and knowledge of plants by natives.

Linton, A. M. Notes on the vegetation of Penrhyn and Manihiki Islands. Jour. Polynesian Soc. **42**: 300–307. pl. 1933.

General notes on agriculture; annotated list of useful plants, with mention of others, bibliography. Penrhyn and Manihiki are two of the Cook Islands.

Macgregor, Gordon. Ethnology of Tokelau Islands. Bul. Bern. P. Bishop Mus. 146. iv, 183 p. 25 fig. (incl. map), 10 pl., tab. 1937.

Includes (p. 11) list of known plants, alphabetically arranged by vernacular names. The Tokelau Islands are four atolls north of Samoa, also known as Union Group or Union Islands, about 8°–11° S. Lat., 171°–173° W. Long.

Maiden, J. H. The botany of Funafuti, Ellice group. Proc. Linn. Soc. N. S. Wales **29**: 539–556. 1904.

Annotated list of vascular plants and one lichen, with vernacular names and uses.

——— Notes on the botany of Pitcairn Island. Rpt. Australas. Assoc. Adv. Sci. **8**: 262–271. 1901.

Botanical explorations, bibliography; partly annotated list of vascular plants, with vernacular names.

Riley, L. A. M. Notes on the flora of Rapa. ("St. George" Pacific expedition, 1924–5.) Bul. Misc. Inform. Kew **1926**: 51–56. 1926.

General features of flora: list of 17 vascular plants. Rapa or Oparu Island is not far southeast of the Tubuai or Austral Islands.

Rock, J. F. C. Palmyra Island with a description of its flora . . . with the cooperation of O. Beccari, A. Zahlbruckner, U. Martelli, H. L. Lyon, and M. A. Howe. Bul. Col. Hawaii 4. 53 p. (incl. 20 pl.). 1916.

History, physiography, etc.; annotated list of 15 vascular plants (and 23 cellular).

St. John, Harold, and Fosberg, F. R. Vegetation of Flint Island, central Pacific. Occas. Papers Bern. P. Bishop Mus. v. 12, no. 24. 4 p. 1937.

General features of flora; annotated list of vascular plants, with authors' collecting numbers [based on a visit to the island Oct. 16, 1934]. Flint Island is a tiny coral island in the south central Pacific, 11° 25′ 43″ S. Lat., 151° 48′ W. Long.

Seurat, L. G. Flore économique de la Polynésie française. Bul. Soc. Natl. Acclim. France **52**: 310–326, 355–359, 369–376. 1905.

Climate, general features of flora; running account of useful plants, grouped by uses.

——— Observations sur la structure de l'île Timoe (Crescent). 8 p. 8°. 1903.

Contains short lists of fauna and flora. (Not seen; title and annotation supplied by H. St. John.) (Southeastern part of Tuamotu Islands.)

Skottsberg, Carl. The phanerogams of Easter Island. *In* his The natural history of Juan Fernandez and Easter Island. v. 2, Botany, pt. 2, no. 5. p. 61–84. 2 fig., pl. 6–9. Uppsala, 1922.

Enumeration of phanerograms (30 native species, with supplementary lists of introduced species), with localities, general range, and critical notes; bibliography.—See also **Christensen, C.**, and **Skottsberg, C.** The ferns of Easter Island. l. c. pt. 1, no. 2, p. 49–53. 3 fig. 1920.

Trelease, William. Report on the collection of plants made by Dr. [W. S.] Dixon, U. S. N. *In* Report of the eclipse expedition to Caroline Island, May, 1883. Mem. Natl. Acad. Sci. **2**: 88–90. 1884.

Annotated list of wild and cultivated vascular and cellular plants. The list of wild vascular species is repeated by Hemsley in Report on the scientific results of the voyage of H. M. S. *Challenger* during the years 1873–76. Botany. v. 1, Introd. p. 18. 1885. Caroline Island (not to be confused with the Caroline Islands, in Micronesia) is in 9° 57′ S. Lat., 150° 75′ W. Long.

Wilder, G. P. The flora of Makatea. Bul. Bern. P. Bishop Mus. 120. 49 p. 1 fig. (map), 5 pl. 1934.

Topography, climate, phosphate industry, general features of animal and plant life; annotated list of vascular plants, with vernacular names and brief descriptions. Makatea is one of the Tuamotu Islands, also known as Paumotu Islands or Low Archipelago.

——— Flora of Rarotonga. Bul. Bern. P. Bishop Mus. 86. 113 p. 3 fig., 8 pl. 1931.

Topography, history, etc.; annotated list of vascular plants, with vernacular names and brief descriptions. Rarotonga is one of the Cook Islands.

SAMOAN ISLANDS

See also Polynesia (general works).

General

Bülow, Werner von. Die Samoa-inseln und ihre einheimischen nutzpflanzen. Gartenfl. **45**: 412–415, 452–454, 518–520, 543–544, 574–575, 628–633. 1896.

Brief account, with vernacular names, the plants classified by uses.

Christophersen, Erling. Flowering plants of Samoa. Bul. Bern. P. Bishop Mus. 128. 221 p. 32 fig. 1935; 154. 77 p. 21 fig., 3 pl. 1938.

Botanical explorations; annotated list of angiosperms; vernacular names, bibliography. Apparently based mostly on specimens in Bernice P. Bishop Museum (collected from 1921 on). Pt. 2 contains new records, etc.

Krämer, Augustin. Die Samoa-inseln. v. 2. illus., plates. 31 cm. Stuttgart, 1903.

The chapter on "Flora" (p. 359–388) contains an annotated alphabetical list of vernacular names with botanical identifications and uses.

Lauterbach, Carl. Beiträge zur flora der Samoa-inseln. Bot. Jahrb. Engler **41**: 215–238. 1908.

Annotated list of species supplementary to Reinecke's and Krämer's papers. Essentially the same material published in Jahresber. Schles. Gesell. Vaterl. Cult. **86** (2. abt. Zool.-Bot. Sekt.): 17–26. 1909.

Reinecke, Franz. Die flora der Samoa-inseln. Bot. Jahrb. Engler **23**: 237–368. 7 fig., pl. 4–5. 1896; **25**: 578–708. 1 fig., pl. 8–13. 1898. (Reprinted.)

Botanical collections, geography, physiography, etc.; annotated list of cellular and vascular plants, with vernacular names (indexed) and extralimital range.

——— Die nutzpflanzen Samoas und ihre verwendung. Jahresber. Schles. Gesell. Vaterl. Cult. **73**, Abt. II (Naturw. C): 22–46. 1896.

Annotated list of native and cultivated useful plants, classified by uses.

——— Samoa. 312 p. illus., map. 22 cm. Berlin, [1902].

General account of Samoa and its inhabitants; contains (p. 302–312) an annotated list of edible plants.

Setchell, W. A. American Samoa: pt. I. Vegetation of Tutuila Island. Pt. II. Ethnobotany of the Samoans. Pt. III. Vegetation of Rose Atoll. vi, 275 p. 57 fig., 37 pl. 25.5 cm. Washington, 1924. (Carnegie Inst. Washington, Pub. 341—Dept. Mar. Biol. v. 20.)

(I) Physiography, climate, plant formations, statistics of flora, etc.; annotated list of vascular and cellular plants with vernacular names and general range; bibliography. (II) Bibliography; running account of useful plants, classified by uses. (III) Geography, history, geology, climate, fauna, ecology, etc.; annotated list of vascular and cellular plants. Rose Atoll is in 14° 32′ S. Lat., 168° 11′ W. Long.

Local

Horne, John. List of plants found in Upolu, while on a short visit to the Samoan, or Navigator Islands. *In* his A year in Fiji . . . p. 284–286. London, 1881.

Unannotated list of vascular plants.

SAN AMBROSIO AND SAN FELIX

General

Hemsley, W. B. On the vegetation of San Ambrosio and San Felix. *In* his Report on the botany of Juan Fernandez . . . Report on the scientific results of the voyage of H. M. S. *Challenger* during the years 1873–76. Botany. v. 1, pt. 3, p. 97–100. London, 1885.

Physiography, etc.; annotated list of species (8 phanerogams).

Philippi, Federico. La flora de las islas San Ambrosio i San Félix. Anales Univ. Chile **47** (Secc. 1): 185–194. 1875.

Annotated list of 12 phanerograms.

Skottsberg, Carl. Die flora der Desventuradas-inseln (San Felix und San Ambrosio) nach den sammlungen F. Johows und mit einfügung seiner hinterlassenen schriften herausgegeben und ergänzt . . . Göteborgs K. Vetensk.-o. Vitterhets-Samh. Handl. V, B, v. 5, no. 6. 87 p. 46 fig. (incl. map). 1937.

Topography, botanical investigations; critically annotated list of vascular and cellular plants; phytogeography, ecology; bibliography.

Local

Johnston, I. M. The flora of San Felix Island. Jour. Arnold Arboretum **16**: 440–447. pl. 165. 1935.

Physiography, general features of flora, botanical explorations; annotated list of 7 vascular plants.

SOCIETY ISLANDS

See also Polynesia (general works; also Drake del Castillo; Seurat); French Guiana (Lanessan, Les plantes utiles, p. 842–872).

General

Guillemin, J. B. A. Zephyritis taitensis. Énumération des plantes découvertes par les voyageurs, dans les Iles de la Société, principalement dans celle de Taïti. Annales Sci. Nat. II, Bot. **6**: 297–320. 1836; **7**: 177–192, 241–255, 340–370. 1837.—Supplément . . . par **E. Jardin.** Mém. Soc. Imp. Sci. Nat. Cherbourg **7**: 239–244. 1860.

Botanical explorations, phytogeography, with special lists; list of cellular and vascular plants (363 species) known from the Society Islands, with references and localities, and many descriptions quoted from Forster's manuscript. The supplement adds 123 species of cellular and vascular plants.

Henry, Teuira. Flora of the Society and Tuamotu Islands. *In* her Ancient Tahiti. Bul. Bern. P. Bishop Mus. **48**: 33–68. 1928.

Extensively annotated list of wild and cultivated useful plants, with vernacular names.

Seurat, L. G. Vocabulaire des termes d'histoire naturelle (animaux et plantes), dans les dialectes tahitien, tuamotu, mangarévien et marquisien. 28 p. 22.5 cm. [Chalon-s.-Saône, 19–?]

Alphabetical list of vernacular names, with scientific equivalents; bibliography.

Local

Butteaud, Édouard. Flore tahitienne. 128 p. 23.5 cm. Papeete, 1891.

Partly annotated list of vascular and cellular plants; alphabetical list of Tahitian names, with botanical and French equivalents.

Cuzent, Gilbert. Iles de la Société. Tahiti. Considérations géologiques, météorologiques et botaniques sur l'île . . . 275 p. 1 pl., 2 maps. 25 cm. Rochefort, 1860.

Physiography, anthropology, etc.; economic plants; annotated list of 532 vascular plants, of which 248 are introduced, prepared with the assistance of M. Pancher.

Nadeaud, Jean. Énumération des plantes indigènes de l'île de Tahiti . . . v, 86 p. 24 cm. Paris, 1873.

General features of flora; annotated list of 508 plants (417 vascular) collected by author in 1856–59, with vernacular names (indexed).

——— Plantes usuelles des Tahitiens. 52 p. 25 cm. Montpellier, 1864. (Thèse.)

Extensively annotated list of useful plants of Tahiti, with vernacular names.

Setchell, W. A. Phytogeographical notes on Tahiti. I. Land vegetation. Univ. Calif. Pub. Bot. **12:** 241–290. 1926.

Plant formations, geographical relationships, statistics, etc.; bibliography.

——— Tahitian spermatophytes collected by W. A. Setchell, C. B. Setchell and H. E. Parks. Univ. Calif. Pub. Bot. **12:** 143–213. pl. 23–36. 1926.

Annotated list of species, with vernacular names; references to literature.

SUMATRA

See also Dutch East Indies (general works; also Junghuhn; Koorders-Schumacher; Rendle; Steenis, On the origin).

General

Miquel, F. A. W. Prodromus florae sumatranae. xx, 656 p. 4 pl. 23.5 cm. Amsterdam, 1860. (His Flora Indiae batavae. Supplementum primum.) (German ed., 1862: Sumatra, seine pflanzenwelt und deren erzeugnisse.)

Botanical explorations, bibliography, physiography, climate, plant formations, economic plants; list of spermatophytes (2,642 species) with references to descriptions in vol. 1–3 of his Flora Indiae Batavae and descriptions of plants not included in those volumes.

Local

Bartlett, H. H. Sumatran plants collected in Asahan and Karoland, with notes on their vernacular names. Papers Mich. Acad. Sci. **6:** 1–66. 1927.

Two annotated lists; vernacular names, uses; notes on spelling of vernacular names, with bibliography of subject.—See also his The Batak lands of north Sumatra . . . Nat. and Appl. Sci. Bul. Univ. Philippines **4:** 211–323. 2 maps. 1935. (Includes account of collections made in Oostkust, with footnote references; gazetteer.)

Bruijn Kops, G. F. de. Lijst van boomen en planten, gevonden wordende in de residentie Riouw, op de oostkust van Sumatra en omliggende landen. Tijdschr. Indische Taal-, Land- en Volkenk. **2:** 271–317. 1854.

Alphabetical list of Malayan names, with botanical equivalents and uses.

Hasselt, A. L. van, and Boerlage, J. G. Bijdragen tot de kennis der flora van Midden-Sumatra. 50 p. 8 pl. 27.5 cm. Leiden, 1884. (Midden-Sumatra. Reizen en onderzoekingen der Sumatra-expeditie . . . 1877–79, beschreven . . . van P. J. Veth. 4. deel, natuurlijke historie, 2. gedeelte, Flora.)

Annotated list of vascular and cellular plants, collected on the Aardrijkskundig Genootschap Sumatra expedition of 1877–79.

Merrill, E. D. An enumeration of plants collected in Sumatra by W. N. and C. M. Bangham. Contrib. Arnold Arboretum no. 8. 178 p. 14 pl. 1934.

Geographical affinities of flora, botanical explorations (with partial bibliography); list of about 484 vascular plants collected in 1931–32 in northern Sumatra.

——— New Sumatran plants. I–IV. Papers Mich. Acad. Sci. **19:** 149–203. pl. 16–35. 1934; **20:** 95–112. 1935; **23:** 177–202. 1938; **24** (1): 63–92. 1939.

Part 1: descriptions of new species based on the collections of H. H. Bartlett, H. S. Yates, and others; pt. 2: records of species new to Sumatra and range extensions in Sumatra; pt. 3–4: new species and new records.

Ridley, H. N. A botanical excursion to northern Sumatra. Jour. Malay. Branch Roy. Asiatic Soc. **1:** 46–113. 1923.

Annotated list of vascular plants collected by the author and others in 1921.

Ridley, H. N. The flora of the Mentawi Islands. Bul. Misc. Inform. Kew **1926**: 57–94. 1926.

General features of flora; list of 365 species collected by the expedition of C. Boden Kloss in 1914. (West of Sumatra.)

——— Plants from Bencoolen, Sumatra. Collected by Mr. C. J. Brooks. Bul. Misc. Inform. Kew **1925**: 76–94. 1925.

List of phanerogams (also mosses, by H. N. Dixon). (West coast of Sumatra.)

——— Spermatophyta and Pterydophyta [sic]. *In* Results of an expedition to Korinchi Peak, Sumatra. Pt. IV. Jour. Fed. Malay States Mus. **8**: 13–135. 1917.

List of 778 vascular plants, with locality, general range, and frequent annotations.

Steenis, C. G. G. J. van. Report of a botanical trip to the Ranau region, south Sumatra. Bul. Jard. Bot. Buitenz. III, **13**: 1–56. 11 fig. (incl. pl., maps). 1933.

Topography, geology, botanical explorations (with footnote references), plant associations, phytogeography; no complete list of species.

North America

New York Botanical Garden. North American flora. New York, 1905–1939.

Complete but unannotated descriptive systematic treatment of the flora of the North American continent, Greenland, and the West Indies; Trinidad, Tobago, and Curaçao are not included. The work is planned to occupy 34 volumes, of which v. 16–34 are to be devoted to vascular plants; none of the latter are completed.

ALASKA

See also Canada (Macoun); Mackenzie (Macoun and Holm); Yukon (Macoun, List of plants collected during 1911 and 1912); Greenland (Hooker; Ostenfeld, Flora Arctica); United States (general works).

General

Dall, W. H. Alaska and its resources. xii, 627 p. illus., plates, map. 23.5 cm. Boston, 1870.

Includes (p. 589–594) unannotated list of useful plants, with localities; also bibliography of Alaska.

Rothrock, J. T. Sketch of the flora of Alaska. Ann. Rpt. Smithsn. Inst. **1867**: 433–463. 1872.

General features of flora; compiled list of 732 vascular and cellular plants, with localities.

Taylor, R. F. Pocket guide to Alaska trees. Misc. Pub. U. S. Dept. Agr. 55. 39 p. illus., map. 1929.

General features of forests; annotated briefly descriptive list of 28 trees, with local range; a popular treatment.

Local

Anderson, J. P. Notes on the flora of Sitka, Alaska. Proc. Iowa Acad. Sci. **23**: 427–482. pl. 16–31. [1917.]

Topography, climate, botanical explorations, life zones, plant associations, economic plants, annotated list of fungi.

——— Plants of southeastern Alaska. Proc. Iowa Acad. Sci. **25**: 427–449. [1919].—Supplemental list . . . l. c. **26**: 327–331. [1920.]

Localities visited; annotated list of vascular plants, mainly collected at Sitka and Juneau.

——— Plants used by the Eskimo of the northern Bering Sea and Arctic regions of Alaska. Amer. Jour. Bot. **26**: 714–716. 1939.

Annotated list of 40 vascular plants, with vernacular names; used mostly for food.

Bongard, H. G. Observations sur la végétation de l'île de Sitcha. Mém. Acad. Sci. St. Pétersb. VI. Sci. Math., Phys., et Nat. **2**: 119–177. pl. 1–6. 1832 [1833].

Climate, soil, etc.; list of 222 vascular plants collected by Mertens, with technical annotations.

Cooper, W. S. The seed-plants and ferns of the Glacier Bay National Monument, Alaska. Bul. Torrey Club **57**: 327–338. 1930.—Additions to the flora . . . 1935–1936. l. c. **66**: 453–456. 1939.

Recent history of region, plant communities; annotated list of vascular plants.

Coville, F. V. Flora of the region. *In* **Brooks, A. H.** The Mount McKinley region, Alaska . . . U. S. Geol. Survey Prof. Paper **70**: 208–211. 1911.

List of vascular plants collected by L. M. Prindle, with localities.

——— List of plants, collected in Seward Peninsula by Arthur J. Collier. *In* **Brooks, A. H.** Reconnaissances in the Cape Nome and Norton Bay regions, Alaska, in 1900. p. 167–174. Washington, 1901.

Annotated list of vascular plants, with localities; also cellular plants. A publication of the U. S. Geological Survey.

——— and **Wight, W. F.** List of plants collected in northern Alaska by W. L. Poto in 1901. *In* **Mendenhall, W. C.** Reconnaissance from Fort Hamlin to Kotzebue Sound, Alaska . . . U. S. Geol. Survey Prof. Paper **10**: 58–65. 1902.

Annotated list of vascular plants.

——— **Wight, W. F.,** and **Merrill, E. D.** Arctic plants collected. *In* **Schrader, F. C.** A reconnaissance in northern Alaska across the Rocky Mountains, along Koyukuk, John, Anaktuvuk, and Colville Rivers, and the Arctic coast to Cape Lisburne, in 1901. U. S. Geol. Survey Prof. Paper **20**: 130–134. 1904.

List of vascular plants, with localities, collected from Nulato north to Point Barrow.

Eastwood, Alice. A descriptive list of the plants collected by Dr. F. E. Blaisdell at Nome City, Alaska. Bot. Gaz. **33**: 126–149, 199–213, 284–299. 10 fig. (incl. map). 1902.

Physiography, etc.; list of 170 vascular plants, with descriptions of all above Juncaceae.

Evans, W. H. The agricultural outlook of the coast region of Alaska. U. S. Dept. Agr. Yearbook **1897**: 553–576. pl. 32–35. 1898.

Includes (p. 561–566) notes on forests, grasses, berries, and other useful native plants.

Funston, Frederick, and **Coville, F. V.** Botany of Yakutat Bay, Alaska. Contrib. U. S. Natl. Herb. **3**: 325–353. 1896.

Annotated list (137 species and varieties of vascular plants, 27 of cellular plants) by Coville; field report by Funston.

Gorman, M. W. Economic botany of southeastern Alaska. Pittonia **3**: 64–85. 1896.

Annotated list of useful plants.

Gray, Asa. Plants. *In* Report of the International Polar Expedition to Point Barrow, Alaska. p. 191–192. Washington, 1885.

Annotated list of vascular and cellular plants, the latter based on a report by Farlow.

Hultén, Eric. Flora of the Aleutian Islands and westernmost Alaska Peninsula with notes on the flora of Commander Islands. 397 p. 6 fig. (incl. maps), 16 pl. 26 cm. Stockholm, 1937.

Botanical explorations, geology, climate, plant associations, phytogeography; annotated list of 477 species of vascular plants, with citation of exsiccatae and map of distribution of each species in the region (plus 2 species added in "Additions," p. 341–342); bibliography. The region covered includes the westernmost tip of the Alaska Peninsula (from 162° westward), the Aleutian Islands, and the Commander Islands.

Kjellman, F. R. Fanerogamer från Vest-Eskimåernas land. *In* **Nordenskiöld, A. E.** Vega-expeditionens vetenskapliga iakttagelser. v. 2, p. 25–60. pl. 3. Stockholm, 1883.

Annotated list of phanerogams collected by the Vega Expedition in 1879 at Port Clarence, containing about 45 species not in Seemann's Flora.

——— Fanerogamfloran på S:t Lawrence-ön. *In* **Nordenskiöld, A. E.** Vega-expeditionens vetenskapliga iakttagelser. v. 2, p. 1–23. pl. 1–2. Stockholm, 1883.

Previous botanical explorations, plant geography, etc.; unannotated list of 96 phanerogams collected on St. Lawrence Island by the Vega Expedition in 1879; list of species recorded but not found by the author.

Knowlton, F. H. List of plants collected by Mr. Charles L. McKay at Nushagak, Alaska, in 1881, for the United States National Museum. Proc. U. S. Natl. Mus. **8** (1885) : 213–221. 1886.

Annotated list of vascular and cellular plants.

Kurtz, Fritz. Die flora des Chilcatgebietes im südöstlichen Alaska, nach den sammlungen der gebrüder Krause. Bot. Jahrb. Engler **19**: 327–431. 1894.

Physiography, general features of the flora, life zones, useful plants, list of vernacular names (of Tlinkit Indians), bibliography; list of 498 vascular plants with localities, also cellular plants.

Macoun, J. M. A list of plants of the Pribilof Islands, Bering Sea, with notes on their distribution. *In* **Jordan, D. S.** The fur seals and fur-seal islands of the North Pacific Ocean. pt. 3, p. 559–587. pl. 87–94. Washington, 1899.

Botanical explorations, general features of flora; annotated list of 184 vascular plants (also cellular plants) known from the area; table of distribution; bibliography.—See also **McAtee, W. L.** Additions to the flora . . . Torreya **22**: 67. 1922.

Merriam, C. H. Plants of the Pribilof Islands, Bering Sea, with critical notes by J. N. Rose. Proc. Biol. Soc. Washington **7**: 133–150. 1892.

Physiography, climate, general features of flora; annotated list of vascular and cellular plants, mainly from specimens collected by the author in 1891.

Mertie, J. B. The Chandalar-Sheenjek district, Alaska. U. S. Geol. Survey Bul. **810**: 87–139. 2 fig. (maps), 2 fold. maps (in pocket). 1930.

Includes general account of woody plants and list of vascular plants collected by author and identified by P. C. Standley (p. 105–109).

Muir, John. The cruise of the *Corwin*. Journal of the Arctic expedition of 1881 in search of DeLong and the *Jeannette*. Edited by W. F. Badè. xxxi, 278 p. plates, map. 21.5 cm. Boston, 1917.

The "Botanical notes" (p. 259–272) include general features of vegetation, physiography, etc., of Unalaska, St. Lawrence Island, St. Michael, Golofnin Bay, Kotzebue Sound, Cape Thompson, Cape Prince of Wales, 20 miles east of Cape Lisburne, Cape Wankarem (Siberia), Plover Bay (Siberia), Herald Island (Siberia), Wrangell Land (Siberia); all but the first two with lists of species collected. The "Botanical notes" are reprinted in Torreya **18**: 197–210. 1918.

Osgood, W. H. Natural history of the Cook Inlet region, Alaska. North Amer. Fauna **21**: 51–81. pl. 6–7. 1901.

Includes (p. 53–56) general features of the flora and list of woody plants collected or observed.

Porsild, A. E. Contributions to the flora of Alaska. Rhodora **41**: 143–183, 199–254, 262–301. 1 fig. (map), pl. 551–554. 1939.

Botanical collections examined, author's itinerary in 1926, with general features of flora at various localities and list of principal collecting localities; list of vascular plants collected (648 species and principal subspecies and varieties, of which 93 are new to Alaska) with frequent annotations.

——— Flora of Little Diomede Island in Bering Strait. Proc. and Trans. Roy. Soc. Canada III, **32** (V, Biol. Sci.): 21–38. 1938.

Topography, geology, climate, fauna, general features of flora; annotated list of vascular plants collected by author in 1926; bibliography. Also known as Krusenstern Island.

Seemann, Berthold. Flora of western Eskimaux-land. *In* his The botany of the voyage of H. M. S. *Herald* . . . during the years 1845–51. p. 9–56. pl. 1–11 (incl. map). London, 1852.

History, geography, soil, climate, general features of flora; list of 315 plants (249 vascular), with references, localities, and occasional annotations. (Coast from Norton Sound to Franklin Point, also St. Lawrence Island.)

Steffen, Hans. Ein beitrag zur flora von Alaska. Beih. Bot. Centralbl. Abt. B, **54**: 547–556. 3 fig. 1936.

General features of vegetation; list of 70 angiosperms collected in vicinity of Circle Springs, near Fairbanks, by Joseph Ulmer.—See also his Verstreute beiträge zur flora der Arktis I. Zur flora von Alaska. l. c. **58**: 100–104. 2 fig. 1938. (Adds 39 species.)

Tatewaki, Misao, and Kobayashi, Yoshio. A contribution to the flora of the Aleutian Islands. Jour. Faculty Agr. Hokkaido Imp. Univ. **36**: 1–119. 5 fig., 8 pl., map. 1934.

Geography, plant communities; annotated list of vascular plants collected by the authors.

Turner, L. M. Contributions to the natural history of Alaska. Results of investigations made chiefly in the Yukon district and the Aleutian Islands. 226 p. 26 pl. (pt. col.). 30 cm. Washington, 1886. (Arctic series of publications issued in connection with the Signal Service, U. S. Army, no. II.)

Includes (p. 61–85) a partly annotated list of vascular and cellular plants, with localities, based on Rothrock's list of 1872.

CANADA

See also United States (general works, also Northeastern and Western).

The boundaries of the Canadian Provinces have undergone considerable change during their recent history. The maps in the Times Survey Atlas of the World (1922) represent the present boundaries (except that Labrador now extends much farther west into Quebec), and have been followed as a guide. The three districts into which the Northwest Territories are divided are here separated, for convenience, as follows: Mackenzie, the mainland between Yukon Territory and 102° W. Long.; Keewatin, mainland (exclusive of Boothia and Melville Peninsulas) between 102° W. Long. and Hudson Bay, together with islands (Southampton, Coats, and Mansel Islands, and those south of them) in Hudson and James Bays; Franklin, Boothia and Melville Peninsulas and all islands in Hudson Strait and Arctic waters except those adjacent to the Yukon coast.

Adams, John. A bibliography of Canadian plant geography to the end of the year 1920 [–35]. Trans. Roy. Canad. Inst. **16**: 293–355. 1928; **17**: 103–145. 1929; **17**: 227–265, [for 1921–25] 267–295. 1930; [for 1926–30] **18**: 343–373. 1932; [for 1931–35] **21**: 95–134. 1936.

Unannotated list of floristic papers, grouped chronologically and arranged alphabetically by authors; 3,399 Canadian titles, plus 197 on Greenland in the "Appendix on the flora of Greenland" in the 1936 paper. M. H. Norwell was joint author in 1932 and 1936.

Fyles, Faith. Principal poisonous plants of Canada. Bul. Expt. Farms Canada 2d ser., no. 39. xi, 112 p. illus., 8 col. pl. 1920.

Annotated account, with descriptions, properties, symptoms, and remedies; bibliography.

MacKay, A. H. Bibliography of Canadian botany, 1900[–15]. Proc. and Trans. Roy. Soc. Canada II, **7** (sect. IV): 141–142. 1901; **8** (sect. IV): 157–160. 1902; **9** (sect. IV): 169–172. 1903; **10** (sect. IV): 153–160. 1905; **11** (sect. IV): 143–152. 1906; **12** (sect. IV): 33–48. 1906; III, **4** (sect. IV): 121–153. 1911; **5** (sect. IV): 177–189. 1912; **6** (sect. IV): 129–137. 1913; **7** (sect. IV): 175–185. 1914. Trans. Roy. Soc. Canada III, **8** (sect. IV): 25–35. 1914; **9** (sect. IV): 251–261. 1916; **10** (sect. IV): 189–199. 1917.

Annual bibliographies, alphabeted by authors. Earlier numbers have title Botanical bibliography of Canada.

Macoun, John. Catalogue of Canadian plants. 3 v. 24.5 cm. Montreal, 1883–1902.

Annotated list of vascular and cellular plants with localities; Alaskan and Newfoundland plants are included. The vascular plants (v. 1–2, 1883–1890) number 3,209 species (including additions in v. 2).—For additions see **Macoun, J. M.** Contributions to Canadian botany I–XVIII. Canad. Rec. Sci. **6**: 23–27, 76–88, 141–153, 198–210, 264–276, 318–329, 405–415, 459–469. 1894–95; **7**: 39–50, 267–286, 463–477. 1896–97. Ottawa Nat. **12**: 161–172. 1898; **13**: 159–169. 1899; **15**: 71–79, 267–275. 1901–02; **16**: 211–223. 1903; **20**: 135–141, 162–171. 1906. (Reprinted.)

Morton, B. R., and **Lewis, R. G.** Native trees of Canada. Bul. Forest Branch Dept. Interior Canada no. 61. 233 p. (incl. illus. (maps), plates). 1917.

Descriptions, range, uses, maps showing local range of each species; synoptical descriptive tables, figures of leaves, fruits, and often twigs, habit photographs.

Penhallow, D. P. A review of Canadian botany from the first settlement of New France to the nineteenth century. Proc. and Trans. Roy. Soc. Canada **5** (sect. IV): 45–61. 1887.—Review . . . from 1800 to 1895. l. c. **II, 3** (sect. IV): 3–56. 1897.

Botanical explorers, botanical gardens and societies; bibliographies.

Provancher, Léon, *abbé.* Flore canadienne ou descriptions de toutes les plantes des forêts, champs, jardins et eaux du Canada . . . 2 v. (cont. paged). 420 fig. 22.5 cm. Québec, 1862.

Descriptive flora, with vernacular names (largely artificial) and uses; includes cultivated plants. Of little value as a flora.

ALBERTA

See also Mackenzie (Preble); Manitoba (Tyrrell); United States, Western (Kirkwood; Rydberg, Flora of the Rocky Mountains).

Local

Farr, E. M. Contributions to a catalogue of the flora of the Canadian Rocky Mountains and the Selkirk Range. Contrib. Bot. Lab. Univ. Penn. **3**: 1–88. map. 1907.

General features of flora; list of vascular plants, with localities, based mainly on the author's collections in 1904–05. The region covered is in southern British Columbia and southern Alberta, between Banff and Glacier.

Raup, H. M. Botanical investigations in Wood Buffalo Park. Bul. Natl. Mus. Canada no. 74 (Biol. Ser. 20). 174 p. illus., maps. 1935.

Botanical investigations, soils, geology, physiography, climate, types of vegetation, plant associations; annotated list of vascular plants (449 species and 12 additional varieties); bibliography. The region is in northeastern Alberta and southeastern Mackenzie, south of Great Slave Lake.

——— Phytogeographic studies in the Athabaska–Great Slave Lake region. I. Catalogue of vascular plants. Jour. Arnold Arboretum **17**: 180–240, 241–315. pl. 190–200, fold. map. 1936.

Botanical explorations, geography (with gazetteer); annotated list of 739 vascular plants (696 species), with citation of exsiccatae. Covers northeastern Alberta, northwestern Saskatchewan, and southeastern Mackenzie (about 105°–119° W. Long., 55°–63° N. Lat.).

Standley, P. C. Plants of the Alpine Club expedition to the Mount Robson region. Canad. Alpine Jour., special no. 1912. p. 76–97. 5 pl. 1912 (1913).

Annotated list of 147 species collected in the Mount Robson region of eastern British Columbia and western Alberta in 1911 by N. Hollister and J. H. Riley.

BRITISH COLUMBIA

See also Alberta (Farr; Standley); Yukon (Macoun, List of plants collected by Dr. G. M. Dawson); United States, Western (Frye and Rigg; Kirkwood; McMinn and Maino; Rydberg, Flora of the Rocky Mountains).

General

Anderson, J. R. Trees and shrubs, food, medicinal and poisonous plants of British Columbia. 165 p. illus., map. 25.5 cm. Victoria, B. C., 1925.

Descriptions of woody plants; annotated lists of food plants, with native methods of cooking, and of medicinal and poisonous plants.

Griffith, B. G. A pocket guide to the trees and shrubs of British Columbia. 69 p. 23 cm. Victoria, 1934.

Systematic list, with keys and descriptions.

Local

Brown, Robert (of Campster). On the vegetable products, used by the northwest American Indians as food and medicine, in the arts and in superstitious rites. Trans. Bot. Soc. Edinb. **9**: 378–396. 1868. (Also in Pharm. Jour. and Trans. II, **10**: 89–94, 168–174. 1868.)

Descriptive account, the plants classified by uses. Refers mainly to Oregon, Washington, and British Columbia.

Carter, W. R., and Newcombe, C. F. A preliminary catalogue of the flora of Vancouver and Queen Charlotte Islands. 86 p. 26 cm. Victoria, 1921.

Annotated list of vascular plants.—For additions (1921–26) see Rpt. Prov. Mus. Nat. Hist. Brit. Columbia **1921**: 16; **1922**: 15; **1923**: 15; **1924**: 23; **1925**: 17; **1926**: 10.—Additions since 1926, l. c. **1937**: 11.

Davidson, John. Contribution to the flora of British Columbia (Garibaldi Mountain region). Ann. Rpt. (2d, 1914) Bot. Off. Brit. Columbia **1**: 58–74. 4 pl. 1915.

Annotated list of vascular plants.

——— List of plants found during the botanical exploration of the Skagit River basin. Ann. Rpt. (3d, 1915) Bot. Off. Brit. Columbia **1**: 111–134. 1916.

List of 456 vascular plants, with localities.

Glendenning, Reginald. An annotated check list of the flowering plants and ferns, native and introduced, growing without cultivation in the Cowichan district, Vancouver Island, B. C. Cowichan Field Nat. Club Bul. no. 1. 20 p. 1918.

Annotated list of vascular plants. The area covered includes the valleys of the Chemainus, Cowichan, and Koksilah Rivers, from Chemainus south to Shawnigan and west to Cowichan Lake. Publication has only cover title.

Graham, Roy. Notes on the flora of the Peace River. Rpt. Prov. Mus. Nat. Hist. Brit. Columbia **1933**: B13–B24. 1934.

List of about 250 vascular plants; general features of flora; useful and poisonous plants.

Henry, J. K. Flora of southern British Columbia and Vancouver Island, with many references to Alaska and northern species. xiii, 363 p. 21.5 cm. Toronto, 1915.

Descriptive flora of vascular plants, with keys; brief account of floral regions; glossary.

Osgood, W. H. Natural history of the Queen Charlotte Islands, British Columbia. North Amer. Fauna **21**: 7–50. 1 fig., 5 pl. (incl. map). 1901.

Includes (p. 11–16) general features of the flora and list of cellular and vascular plants collected.

Raup, H. M. Phytogeographic studies in the Peace and upper Liard River regions, Canada, with a catalogue of the vascular plants. Contrib. Arnold Arboretum no. 6. 230 p. 6 fig., 9 pl., map. 1934.

History, geology, climate, phytogeography, formations, bibliography; annotated list of 754 species and 30 additional varieties of vascular plants. (Northeastern British Columbia and adjacent Mackenzie.)

Rosendahl, C. O. Observations on plant distribution in Renfrew district of Vancouver Island. Postelsia **1906**: 1–132. pl. 1–11. 1906.

Topography, climate, plant formations, etc.; annotated list of 253 vascular plants.

Smith, H. I. Materia medica of the Bella Coola and neighbouring tribes of British Columbia. Bul. Natl. Mus. Canada **56**: 47–68. 1929.

Systematic list of medicinal plants, with methods of use. (Bella Coola tribe of Bella Coola valley, Gitksan of Skeena River, the Carrier who live in the territory behind both these tribes, and the Sikani of the headwaters of Peace River.)

Steedman, E. V. Ethnobotany of the Thompson Indians of British Columbia. Based on field notes by James A. Teit. Ann. Rpt. Bur. Amer. Ethnol. **45** (1927/28): 441–522. 1930.

Principles of plant names used by these Indians; annotated classified lists of medicinal, edible, and other useful plants, with vernacular and botanical names; list of botanical names. (Southwestern British Columbia.)

Ulke, Titus. A flora of Yoho Park, British Columbia. Contrib. Biol. Lab. Catholic Univ. Amer. no. 14. 89 p. 10 pl. 1934.

Botanical explorations, geology; annotated list of 565 vascular plants, with keys; bibliography. (On eastern edge of British Columbia, about 51°20′ N. Lat.)

——— List of the vascular plants of the Horsethief Creek-Purcell Range, B. C. Canad. Field-Nat. **49**: 49–55, 71–76. 1935.

Annotated list of 405 vascular plants collected by the author in 1928. (Extreme southeastern British Columbia.)

FRANKLIN

See also Greenland (Holm; Hooker; Ostenfeld, Flora Arctica; Rikli); Mackenzie (Porsild; Preble; Stefánsson); Manitoba (Tyrrell).

General

Simmons, H. G. A survey of the phytogeography of the Arctic American archipelago with some notes about its exploration. Lunds Univ. Årsskr. n. f., afd. 2, v. 9, no. 19. 183 p. 2 maps. 1913.

Botanical explorations, with table showing their results in number of species; geography, geology, temperature; list of islands, and localities where collections have been made, with position; annotated list of 204 vascular plants, with localities; phytogeography, origin of flora, bibliography. The region covered is bounded by Banks Land, Victoria Land, Boothia Felix, Melville Peninsula, Southampton and adjacent islands, Baffin Land, Ellesmere Land, and Prince Patrick Island.

Local

Ambronn, Hermann. Phanerogamen und gefäss-kryptogamen vom Kinguafjord. *In* Die internationale polarforschung 1882–83. Die Deutsche expeditionen . . . v. 2, p. 75–92. Berlin, 1890.

Annotated list of vascular plants. (Baffin Land.)

Fernald, M. L. Baffin Land plants collected by the Macmillan expedition, 1922. Rhodora **25**: 111–114. 1923.

List of vascular plants, with localities.

Greely, A. W. Botany. *In* his Report on the proceedings of the United States expedition to Lady Franklin Bay, Grinnell Land. v. 2, p. 11–18. Washington, 1888. (International polar expedition.)

Includes annotated list of 69 vascular plants collected in vicinity of Fort Conger, Grinnell Land (81°44′ N., 64°45′ W.); also list of mosses and lichens.—Essentially the same list is given by **G. Vasey,** Bot. Gaz. **10**: 364–366. 1885. (Ellesmere Land.)

Johansen, Frits. Vascular plants collected during the Canadian Hudson Strait expedition in 1927. Canad. Field-Nat. **48**: 126–131. 1934.

List of about 110 vascular plants collected by the author in 1927 at Port Burwell, Wakeham Bay, and Eric Cove, Ungava; Lake Harbour, at south end of Baffin Island; and Nottingham Island, at west end of Hudson Strait.

Macoun, J. M. List of plants collected by Mr. J. G. McMillan on Melville Island, in the autumn of 1908 and early summer of 1909. *In* **Bernier, J. E.** Report on the Dominion of Canada government expedition to the Arctic islands and Hudson Strait on board the D. G. S. *Arctic*. p. 489–491. Ottawa, 1910.

Unannotated list of vascular plants (and 2 mosses and lichens).

Ostenfeld, C. H. Vascular plants collected in Arctic North America (King William Land, King Point and Herschell Isl.) by the *Gjöa* expedition under Captain Roald Amundsen 1904–1906. Skr. Vidensk.-Selsk. Christiania I. Math.-Naturv. Kl. 1909. no. 8. vi, 73 p. 3 pl. 1910.

Separate annotated lists for King William Land (63 species) and for King Point and Herschell Island (110 species); references to earlier publications. (King William Land, Franklin; King Point and Herschell Island, Yukon.)

Polunin, Nicholas. The flora of Akpatok Island, Hudson Strait. Jour. Bot. **72**: 197–204. 1934.

General features of flora; list of 391 cellular and vascular plants. There are 123 angiosperms, including 22 species, forms, and hybrids of Draba.

Simmons, H. G. Stray contributions to the botany of North Devon and some other islands, visited in 1900–1902. With . . . an alphabetical index to no. 2, Flora of Ellesmereland, no. 16, Plants of N. W. Greenland, and this paper. Report of the 2d Norwegian Arctic expedition in the *Fram* 1898–1902. no. 19. 36 p. 1 fig. (map). Kristiania, 1909.

Includes list of vascular plants from various localities in North Devon, Castle Island and Devil's Isle, North Kent, Graham and Buckingham Islands, Heiberg Land, Schei Island and Big Island, Rignes Lands.

——— The vascular plants in the flora of Ellesmereland. Report of the 2d Norwegian Arctic expedition in the *Fram* 1898–1902. no. 2. 197 p. 5 fig., 10 pl., map (in text). Kristiania, 1906.

Geography, botanical explorations, phytogeography, gazetteer; annotated list of 115 species; bibliography.

KEEWATIN

See also Franklin (Simmons, A survey); Mackenzie (Porsild); Manitoba (Tyrrell); Greenland (Hooker; Ostenfeld, Flora Arctica).

Local

Polunin, Nicholas. The flora of Southampton Island, Hudson Bay. Jour. Bot. **76**: 93–103. 1938.

Botanical explorations; list of 151 species and 20 minor forms of vascular plants now known, with names of collectors; bibliography.

Polunin, Nicholas. Vascular plants from Mansel (Mansfield) Island, N. W. T. Canad Field-Nat. **52**: 5–9. 1938.
History, botanical explorations; list of 45 species and 4 minor forms of angiosperms; bibliography.

Raup, H. M. Pteridophyta and Spermatophyta of Southampton Island. Mem. Carnegie Mus. Pittsburgh **12** (3): 17–30. 1936.
Botanical explorations; annotated list of 109 vascular plants, of which 66 had not been recorded previously.

LABRADOR

See also Newfoundland (Fernald, A botanical expedition; Waghorne).

General

Macoun, J. M. List of the plants known to occur on the coast and in the interior of the Labrador Peninsula. Ann. Rpt. Geol. Survey Canada n. s., **8**: 353L–366L. 1897.
Table of vascular plants, showing range in Labrador.

Macoun, John. Catalogue of the plants reported by various travellers and others as growing on the coast of Labrador. *In* **Packard, A. S.** The Labrador coast. p. 451–474. New York, London, 1891.
Annotated list of vascular plants.

Meyer, E. H. F. De plantis labradoricis libri tres. xxii, 218 p. 19.5 cm. Lipsiae, 1830.
Bibliography of works on flora of Labrador and other northern regions; annotated list of 198 plants (168 vascular) from Labrador, with extralimital range; climate, phytogeography, etc., of Labrador and other northern regions.

Schlechtendal, D. F. L. von. Ueber die flora von Labrador. Linnaea **10**: 76–114. 1836 [1835?].
Partly annotated list of 32 cellular and 206 vascular plants then known from Labrador, with references to E. Meyer's paper (1830) and Hooker's Flora Boreali-Americana; general features of flora.

Local

Abbe, E. C. Botanical results of the Grenfell-Forbes northern Labrador expedition, 1931. Rhodora **38**: 102–161. 7 fig. (maps), pl. 408–411. 1936. (Contrib. Gray Herb. 111.)
Geology, botanical explorations with bibliographical footnotes, author's itinerary, general features of flora, phytogeography; annotated list of vascular plants. The region covered is on the northeast coast from Aillik to Ikordlearsuk.

Ascherson, P. F. A. Nachtrag zur flora von Labrador. Flora **43**: 369–370. 1860.
List of vascular plants collected by Lundberg at Nain in 1847 and not recorded by previous writers.

Butler, S. R. Labrador plants. Canad. Nat. n. s., **5**: 350–353. 1870.
Annotated list of vascular plants collected by the author, mainly at Amour Point, Forteau Bay, and Caribou Island; includes also several plants collected by Miss Macfarlane. (Extreme southern Labrador.)

Delabarre, E. B. Report of the Brown-Harvard expedition to Nachvak, Labrador, in the year 1900. Bul. Geog. Soc. Phila. **3**: 65–212. plates, maps. 1902.
Includes chapter on botany (p. 167–201); general features of flora, annotated bibliography, list of vascular and cellular plants collected, with localities. For corrections, see footnote to his chapter on Flora *in* **Grenfell, W. T.** Labrador, the country and the people. New ed. p. 392–393. N. Y., 1913. (1st ed., 1909.)

Fernald, M. L., and Sornborger, J. D. Some recent additions to the Labrador flora. Ottawa Nat. **13**: 89–107. 1899.
Annotated list of rarer species collected by the Bowdoin College expedition of 1891 and by J. D. Sornborger.

Hantzsch, Bernhard. Verzeichnis einer anzahl im nordöstlichsten Labrador gesammelter pflanzenarten. *In* his Beiträge zur kenntnis des nordöstlichsten Labradors. Mitt. Ver. Erdk. Dresden **1** (8): 220–222. 1909. (English transl.

List of a number of species of plants collected in the most north-eastern part of Labrador, in Canad. Nat. **45**: 173–174. 1931.)

List of plants collected by the author and others north of 59° 30′ N. Lat., with a few Eskimo names and uses.

Kew. Royal Gardens. Plants from Labrador. Bul. Misc. Inform. Kew **1907**: 76–88. 1907.

List of vascular and cellular plants collected in 1905 and 1906 by Sir William MacGregor on the Chidley Peninsula at the north tip of Labrador and at Anse Sablon on the south coast.

MacKay, A. H. Labrador plants. (Collected by W. H. Prest on the Labrador coast north of Hamilton Inlet, from the 25th of June to the 12th of August, 1901.) Proc. and Trans. Nova Scot. Inst. Sci. **10**: 507–508. 1903.

Unannotated list of phanerogams.

St. John, Harold. Notes sur la flore du Labrador. Naturaliste Canad. **51**: 195–201. 1925.

Critical notes on a list of Labrador plants published by V. A. Huard in 1923, in the same journal (**49**: 183–186), which was based on observations by Abbé P. Lamay.

Wetmore, R. H. Plants of the Hamilton Inlet and Lake Melville region, Labrador. Rhodora **25**: 4–12. 1923.

List of vascular plants collected by the author in 1921, with localities.

Woodworth, R. H. Interesting plants of northern Labrador. Rhodora **29**: 54–57. 1927.

Annotated list of rarer species collected in northernmost Labrador by the author in 1926.

MACKENZIE

See also Alberta (Raup, Botanical investigations in Wood Buffalo Park, and Phytogeographic studies); British Columbia (Raup); Manitoba (Tyrrell); Greenland (Hooker; Ostenfeld, Flora Arctica).

Local

Macoun, J. M., and Holm, Theodore. The vascular plants of the Arctic coast of America west of the 100th meridian, collected by the Canadian Arctic expedition 1913–18. Report of the Canadian Arctic expedition 1913–18. v. 5, Botany. pt. A. 24 p. 13 pl., map. Ottawa, 1921.

List of 230 species, with localities, from Alaska, Yukon, and Mackenzie.

Porsild, A. E. Edible roots and berries of northern Canada. 17 p. 13 fig. 24.5 cm. Ottawa, 1937.

Annotated list of about 20 edible vascular plants of Arctic and sub-Arctic Canada, as well as some lichens.

Preble, E. A. Trees and shrubs of Athabaska-Mackenzie region. *In* his A biological investigation of the Athabaska-Mackenzie region. North Amer. Fauna **27**: 515–534. fig. 16, pl. 24–25. 1908.

Annotated list, with notes on Indian uses. Covers Mackenzie and islands north of it, and the portions of Alberta and Saskatchewan north of 55° N. Lat.

Stefánsson, Vilhjálmur. My life with the Eskimo. ix, 538 p. plates, 2 maps. 23 cm. N. Y., 1913.

Contains (p. 445–448) notes on useful plants, with lists of species collected at Coronation Gulf (Mackenzie), southwestern Victoria Island (Franklin), Cape Bathurst (Mackenzie), and King Point (Yukon).

MANITOBA

See also United States, Western (Rydberg, Flora of the prairies and plains).

General

Burman, W. A. The Phanerogamia and Pteridophyta of Manitoba. *In* A handbook to Winnipeg and the Province of Manitoba prepared for the 79th annual meeting of the British Association . . . 1909. p. 157–182. map. Winnipeg, 1909.

Botanical districts; unannotated list of vascular plants.

Jackson, V. W., Higham, J. F., and Groh, Herbert. Check list of Manitoba flora (with notes). 35 p. illus. (incl. map). 24 cm. [Winnipeg,] 1922.

Briefly annotated list of 1,029 vascular plants. The map shows floral zones. In 1925 [?] the senior author issued an unpaged, unannotated, multigraphed list of vascular plants (Check list of Manitoba flora revised to December, 1924).

Local

Christy, R. M. Notes on the botany of Manitoba. Jour. Bot. **25**: 271–276, 290–301. 1887.

General features of flora; annotated list of vascular plants (and 1 moss) collected by the author in 1883, mostly near Carberry.

Lowe, C. W. The flora of Warrens Landing, Lake Winnipeg, Man. Canad. Field-Nat. **34**: 26–30. 2 fig. (incl. map). 1920.

General features of flora; unannotated list of vascular plants. (North end of Lake Winnipeg.)

Størmer, Per. Plants collected by Frits Johansen in 1929, at Hudson Bay Railway and Port Churchill in Arctic Canada. Nyt Mag. Naturvidensk. **73**: 259–272. 1933.

Annotated list of vascular plants.

Tyrrell, J. W. Classified list of plants. *In* his Across the sub-arctics of Canada. p. 251–272. Toronto, 1897.

List of 233 plants (196 vascular), with localities, collected by the author in 1893 along route between Lake Athabasca and west coast of Hudson Bay, together with a collection made in 1885 at Ashe Inlet on north shore of Hudson Strait and a collection made by Marjorie Lofthouse at Fort Churchill. Identifications by J. Macoun.

NEW BRUNSWICK

General

Fowler, James. Preliminary list of the plants of New Brunswick. Bul. Nat. Hist. Soc. New Brunswick **1** (4): 8–84. 1885. (Reprinted 82 p. 1885.)

Previous publications; annotated list of 942 vascular plants, with additional list of ballast plants. See also Report of the Committee on botany [list of plants new to the Province . . .] l. c. **3** (11): 48–50. 1893.—Also **Hay, G. U.** Additions to plants of New Brunswick. l. c. **5**: 358–363. 1905.

Local

Fowler, James. Report on the flora of St. Andrews, N. B. Contrib. Canad. Biol. Mar. Biol. Sta. Canada **1901**: 41–48. 1901.

General features of flora; unannotated list of 331 plants (299 vascular) collected by the author in June-August, 1900.—Also appeared with slight changes as A visit to St. Andrew's, N. B., with a catalogue of plants collected in its vicinity. Proc. Nat. Hist. Assoc. Miramichi **2**: 21–28. 1901.

Klugh, A. B. Notes on the flora of the St. Croix River valley and Passamaquoddy region, New Brunswick. Contrib. Canad. Biol. Mar. Biol. Sta. Canada **1906/10**: 265–276. pl. 27–29. 1912.

Annotated list of vascular and some cellular plants collected by the author in July to September 1909, arranged in ecological groups.

Vroom, James. A list of flowering plants and ferns found in Charlotte County, New Brunswick. 12 p. 23.5 cm. St. Stephen, N. B., 1887.

Briefly annotated list of vascular plants.

NEWFOUNDLAND (INCLUDING ST. PIERRE ET MIQUELON)

General

Reeks, Henry. A list of the flowering plants and ferns of Newfoundland, with meteorological observations. 30 p. 21 cm. Newbury [England], 1873.

List, p. 2–7, of 371 species identified by the author in 1866–68. Many of the identifications are erroneous, and the list is long out of date.

Waghorne, A. C. The flora of Newfoundland, Labrador and St. Pierre et Miquelon. Proc. and Trans. Nova Scot. Inst. Sci. **8**: 359–373. 1893; **9**: 83–100, 361–401. 1896–98.
Bibliography; list of 538 species (Polypetalae and Gamopetalae), with localities; largely compiled.

Local

Arsène, Louis, *frère.* Contribution to the flora of the islands of St. Pierre et Miquelon. Rhodora **29**: 117–133, 144–158, 173–191, 204–221. 1927.
Previous botanical work, with bibliographical footnotes, author's explorations, soil, phytogeography; annotated list of newly recorded or otherwise noteworthy species; table of known vascular plants (487 species, of which 391 are native), showing names of collectors.

Delamare, Ernest, Renauld, F., and **Cardot, J.** Flora miquelonensis. Florule de l'île Miquelon (Amérique du Nord) . . . Annales Soc. Bot. Lyon **15**: 65–143. 1888. (Reprinted 78 p.)
Botanical explorations, topography, geology, climate, phytogeography; annotated list of vascular plants, mosses, hepatics, lichens, algae.

Eames, E. H. Notes upon the flora of Newfoundland. Rhodora **11**: 85–99. 1909.
Records of plants from the western coast, collected in 1908 by E. H. Eames and C. C. Godfrey.

Fernald, M. L. A botanical expedition to Newfoundland and southern Labrador. Rhodora **13**: 109–162. pl. 86–91. 1911. (Contrib. Gray Herb. 40.)
Previous botanical explorations, with footnote references to publications; itinerary of author's expedition, with mention of more important plants collected; geographic origin of flora. Known vascular plants are "scarcely 1,000 species."

——— Recent discoveries in the Newfoundland flora. Rhodora **35**: 1–16, 47–63, 80–107, 120–140, 161–185, 203–223, 230–247, 265–283, 298–315, 327–346, 364–386, 395–403. 29 fig. (maps), pl. 232–273. 1933. (Contrib. Gray Herb. 101.)
Itinerary, etc.; phytogeography; range extensions, critical notes, and revisionary treatments of numerous species of plants.

——— Two summers of botanizing in Newfoundland. Rhodora **28**: 49–63, 74–87, 89–111, 115–129, 145–155, 161–178, 181–204, 210–225, 234–241. pl. 153–155. 1926–27. (Contrib. Gray Herb. 76.)
Running account of the author's expeditions in 1924 and 1925; annotated list of noteworthy species collected.

Robinson, B. L., and **Schrenk, Hermann von.** Notes upon the flora of Newfoundland. Canad. Rec. Sci. **7**: 3–31. 1896. (Reprinted.)
Previous explorations, general features of flora; annotated list of species collected by the authors in 1894. (Avalon Peninsula.)

NOVA SCOTIA

General

Lindsay, A. W. H. A catalogue of the flora of Nova Scotia arranged according to Gray's Manual of botany . . . Proc. and Trans. Nova Scot. Inst. Nat. Sci. **4**: 184–222. 1877.
Tabular list of vascular and cellular plants, with localities.—See also **How, Henry.** Additions . . . l. c. p. 312–319. 1877.

Prest, W. H. Edible wild plants of Nova Scotia. Proc. and Trans. Nova Scot. Inst. Sci. **11**: 387–416. 1908.
Annotated list of 77 plants classified according to parts used.

Local

Bailey, L. W. Notes on the geology and botany of Digby Neck. Proc. and Trans. Nova Scot. Inst. Sci. **9**: 69–82. pl. 4–6. 1896.
Includes unannotated partial list of spermatophytes. (On peninsula in southwestern Nova Scotia.)

Campbell, G. G. List of plants collected in the neighbourhood of Truro, Nova Scotia, during the summer of 1883 and 1884. Proc. and Trans. Nova Scot. Inst. Nat. Sci. **6**: 209–225. 1885.—Supplementary list . . . l. c. **6**: 283–285. 1886.
Annotated list of vascular plants; Aster and Solidago omitted.

Cox, G. H. List of plants collected in and around the town of Shelburne. Proc. and Trans. Nova Scot. Inst. Sci. **8**: 439–444. 1895.
Mostly unannotated list of vascular plants.

Fernald, M. L. The Gray Herbarium expedition to Nova Scotia, 1920. Rhodora **23**: 89–111, 130–132, 153–171, 184–195, 223–245, 257–278, 284–300. pl. 130. 1921. (Contrib. Gray Herb. **63**.)
Journal of expedition, with comments on various plants; annotated list of noteworthy species collected, including 110 new to Canada and 122 more first recorded from Nova Scotia.

——— Notes on the flora of western Nova Scotia, 1921. Rhodora **24**: 157–164, 165–180, 201–207. 1922. (Contrib. Gray Herb. 66.)
Itinerary; annotated list of noteworthy species collected, including 37 new to Canada and 25 more new to Nova Scotia.

Fowler, James. Report on the flora of Canso, Nova Scotia. Further Contrib. Canad. Biol. Mar. Biol. Sta. Canada **1902/05**: 59–70. 1907.
General features of flora; unannotated list of 298 plants (244 vascular).

Perry, L. M. The vascular flora of St. Paul Island, Nova Scotia. Rhodora **33**: 105–126. 1 fig. (map). 1931. (Contrib. Gray Herb. 94.)
Physiography, general features of flora; annotated list of about 300 vascular plants collected by the author and Dr. Muriel V. Roscoe in 1929.

Robinson, C. B. Contributions to a flora of Nova Scotia. I. Plants collected in eastern Nova Scotia in August, 1906. Bul. Pictou Acad. Scient. Assoc. **1**: 30–44. 1907.
List of cellular and vascular plants collected by the author, with localities and occasional annotations.

Rousseau, Jacques. Notes floristiques sur l'est de la Nouvelle-Écosse. Naturaliste Canad. **65**: 285–315, 317–335. 8 fig. 1938.
Ecology; annotated list of vascular plants collected by author in 1930, mostly from vicinity of Guysborough; bibliographical footnotes.

St. John, Harold. Sable Island, with a catalogue of its vascular plants. Proc. Boston Soc. Nat. Hist. **36**: 1–103. pl. 1–2. 1921. (Contrib. Gray Herb. 62.)
History, physiographic changes, animals, botanical explorations, plant habitats, phytogeography, etc.; annotated list of vascular plants (147 native species, varieties and forms; 51 adventive plants).

ONTARIO

See also Quebec (D'Urban; Potter); Michigan (Dodge, Flora of St. Clair County); Minnesota (Densmore); New York (Zenkert).

General

White, J. H. The forest trees of Ontario and the more commonly planted foreign trees. A guide to their identification . . . 80, v p. plates. 24.5 cm. Toronto, 1925.
Brief descriptions of the species, with illustrations of leaves, fruit, and twigs.

Local

Bailey, L. H. Plants collected or observed on Hunters' Island, British America, July 26 and 27, 1886. Bul. Geol. and Nat. Hist. Survey Minn. **3**: 55–56. 1887.
Mostly unannotated list of vascular plants. (Southwestern Ontario, 48°15′ N. Lat., 91°30′ W. Long.)

Bell, John. On the plants of the Manitoulin Islands. Rpt. Prog. Geol. Survey Canada **1866/69**: 449–469. 1870.
Topography, general features of flora; list of vascular plants, mosses, hepatics, lichens, and Characeae.

Billings, Braddish, jr. List of plants observed growing principally within four miles of Prescott, C. W., and for the most part in 1860. Annals Bot. Soc. Canada **1**: 114–140. [1862.]
Annotated list of vascular plants, mosses, hepatics, lichens, and Characeae. Appears to consolidate his two earlier lists with similar titles in Canad. Nat. and Geol. **3**: 39–50. 1858; **5**: 14–24. 1860.

Dewey, L. H. The plants of islands 74 and 85 in the French River, Ontario. Canad. Field-Nat. **53:** 127–130. 1939.

Topography, geology, climate, general features of flora; list of vascular plants collected by the author in 1939.

Dickson, J. M., and **Alexander, Andrew.** Flora of Hamilton district. Jour. and Proc. Hamilton Assoc. **13:** 95–127. 1897.

Bibliography; unannotated list of vascular plants and Marchantiaceae.

Dodge, C. K. Annotated list of flowering plants and ferns of Point Pelee, Ont., and neighbouring districts. Mem. Canada Geol. Survey **54** (Biol. Ser. 2). 131 p. 1914.

General features of flora; annotated list of 623 vascular plants. Also issued in a French edition, 138 p., 1917.

—— The flowering plants, ferns and fern allies growing without cultivation in Lambton County, Ontario. Rpt. Mich. Acad. Sci. **16:** 132–200. 1914.

Physiography, habitats, etc.; annotated list of vascular plants.

Fletcher, James. Flora ottawaensis. 101 p. 23.5 cm. [Ottawa, 1888–99.]

Botanical collections, etc.; annotated list of phanerogams. Incomplete; breaks off in middle of treatment of *Scirpus pungens*. Pages 1–13 originally issued in Trans. Ottawa Field Nat. Club, v. 4 (Ottawa Nat., v. 2), p. 28–32, 61–64, 77–80. 1888, later reprinted and with the remaining pages issued as Suppl. to the same journal, v. 4–6 (Ottawa Nat., v. 2–4). Pages 62–69 are numbered 121–128.

Gibson, John, and **Macoun, John.** The plants of the eastern coast of Lake Huron, and their distribution through the northern and western portions of British North America. Canad. Jour. Sci., Lit. and Hist. n. s., **14:** 635–657. 1875. (Reprinted 23 p. [1876?])

Tabular list of vascular plants of eastern shore of Lake Huron and southern and western shores of Georgian Bay, with indication of distribution in eastern Ontario, Lake Superior, and westward.—For ecological discussion of this region see their Botany of the eastern coast of Lake Huron, l. c. 467–478.

Holmes, E. M. Medicinal plants used by the Cree Indians, Hudson's Bay Territory. Pharm. Jour. and Trans. III, **15:** 302–304. 1884. (Reprinted in Amer. Jour. Pharm. **56:** 617–621. 1884.)

Annotated list.—See also **Haydon, Walton.** Medicinal plants used by the Cree Indians. Therapeutic Gaz. (Detroit) **8:** 398–399. 1884. (Same material as in Holmes' paper.)—Also **Strath, R.** Materia medica, pharmacy and therapeutics of the Cree Indians of the Hudson Bay territory. St. Paul Med. Jour. **5:** 735–746. 1903.

Hosie, R. C. Botanical investigations in Batchawana Bay region, Lake Superior . . . with A catalogue of the vascular plants, by T. M. C. Taylor . . . Bul. Natl. Mus. Canada 88, Biol. Ser. no. 23. v, 152 p. 34 fig. (incl. map). 1938.

Mainly ecological; includes annotated list of vascular plants. Batchawana Bay is about 20 miles north of Sault Ste. Marie, Ontario.

Kellerman, W. A. Flora of Hen and Chickens Islands, 1903. Ohio Nat. **4:** 190–191. 1904.

Lists of vascular plants from these islands.—See also **Schaffner, J. H.** The flora of Little Chicken Island. l. c. **3:** 331–332. 1902. (In Canadian waters at west end of Lake Erie.)

Kirkconnell, Watson. Botanical survey of South Victoria. A local check-list of plants found in the six southern townships of Victoria County, Ontario. 15 p. 21 cm. Lindsay, Ontario, 1926.

Topography, soils, climate; unannotated list of 620 higher fungi, bryophytes, and vascular plants.

Macoun, J. M. List of plants collected on the Rupert and Moose Rivers, along the shores of James' Bay, and on the islands in James' Bay, during the summers of 1885 and 1887 . . . *In* **Low, A. P.** Report on explorations in James' Bay and country east of Hudson Bay . . . Ann. Rpt. Geol. and Nat. Hist. Survey Canada n. s., **3** (2, Rpt. J): 63–74. 1888.

Tabular list of vascular plants.

Macoun, John. Catalogue . . . of the plants collected by Dr. Robert Bell along the Michipicoten River and in the southern part of the basin of Moose River. *In* **Bell, Robert.** Reports . . . on the geology of the basin of Moose River . . . Rpt. Prog. Geol. and Nat. Hist. Survey Canada **1880/82** (C): 17C–28C. 1883.

General features of the flora; tabular list of 320 vascular and cellular plants, with locality indicated.

Macoun, John, and Gibson, John. The rarer plants of the Province of Ontario. Trans. and Proc. Bot. Soc. Edinb. **12:** 300–307. 1875.

Physiography, geology, etc.; list of rarer phanerogams, with synonymy, descriptions, and local range.

Saunders, William. List of plants collected chiefly in the immediate neighbourhood of London, C. W. Canad. Jour. Sci., Lit. and Hist. n. s., **8:** 219–238. 1863. (Reprinted 20 p. [1863?].)

Annotated list of vascular plants, mosses, hepatics, and lichens.

Scott, William. The seed plants of Toronto and vicinity. *In* **Faull, J. H.** The natural history of the Toronto region. p. 100–140. Toronto, 1913.

Partly annotated list of 798 spermatophytes. An annotated list of pteridophytes, by Ivey, T. J., is in same volume (p. 141–149).

Spotton, H. B. List of plants collected in the vicinity of the town of Barrie. Canad. Jour. Sci., Lit. and Hist. n.s., **15:** 46–50. 1876.

Unannotated list of vascular plants.

PRINCE EDWARD ISLAND

General

Hurst, Blythe. Flowering plants and ferns of Prince Edward Island. Trans. Roy. Canad. Inst. **19:** 251–273. 1933.

Previous publications; unannotated list of 594 species and varieties of vascular plants.—See also **Adams, John.** Some additions to the flora of Prince Edward Island. Canad. Field-Nat. **51:** 105–107. 1937.

QUEBEC

See also Franklin (Johansen); Mackenzie (Porsild); Ontario (Macoun, J. M., List of plants collected on the Rupert and Moose Rivers).

General

Adams, John. Medicinal plants of Quebec. Ann. Rpt. Quebec Soc. Protect. Plants **7:** 50–54. 1915.

Briefly annotated list of 60 species, classified according to parts used.

Louis-Marie, *pere.* Flore-manuel de la province de Québec. 319 p. illus. 21.5 cm. Montréal, 1931. (Contribution no. 23 Institut agricole d'Oka.)

Elementary morphology, etc.; briefly annotated elementary flora of vascular plants, in form of keys; vernacular names, glossary.

Local

Baum, H. E. A scanty flora. Plant World **4:** 145–146. pl. 10. 1901.

List of 3 phanerogams collected in 1900 by Miss E. E. Leach on Bird Rock, at the northern tip of the Magdalen Islands, a famous rookery of gannets and other water birds.

Bowman, P. W. Notes on the flora of the Matemek River district, "North Shore," Quebec, Canada. Rhodora **34:** 48–55. 1932.

Annotated list of vascular plants collected by the author in 1927.

Campbell, Robert. The flora of Montreal Island. Canad. Rec. Sci. **5:** 208–234. 1892; **6:** 397–405. 1895; **7:** 146–151. 1896; **8:** 11–24, 349–365. 1899–1901.

Annotated list of vascular plants, with four supplements (including the mosses). First contribution in vol. 8 has title: The Gramineae, Cyperaceae and Juncaceae of Montreal Island, by H. B. Cushing and Robert Campbell.—See also **Emberson, F. C.** The trees of Montreal Island. l. c. **9:** 78–83. 1903. (Unannotated list.)

——— Notes on the flora of Cap-à-l'Aigle. Canad. Rec. Sci. **4:** 54–68. 1890.—Supplemental notes . . . **5:** 38–40. 1892.—Additional notes . . . **8:** 281–295. 1901.

Annotated list of vascular plants collected on the Murray and Loutre Rivers, Charlevoix County, Quebec.

241306°—42——10

Carrier, J. C. Le flore de l'île de Montréal, Canada. Bul. Acad. Internatl. Géog. Bot. **10:** 166–173, 228–232, 281–283. 1901; **11:** 22, 79–80, 188. 1902; **12:** 55–56, 209–210. 1903; **13:** 268–281. 1904; **20:** 53–56. 1910.
Annotated list of 633 vascular plants.

D'Urban, W. S. M. Catalogue of plants collected in the counties of Argenteuil and Ottawa, in 1858. Canad. Nat. and Geol. **6:** 120–137. 1861.
Annotated list of 362 vascular and cellular plants collected by the author. Argenteuil County is in Quebec, Ottawa in Ontario.

Lewis, H. F. An annotated list of vascular plants collected on the north shore of the Gulf of St. Lawrence, 1927–1930. Canad. Field-Nat. **45:** 129–135, 174–179, 199–204, 225–228. 1931; **46:** 12–18, 36–40, 64–66, 89–95. 1932.
Geology; list of anomalous occurrences of calcicolous plants; annotated list of 567 vascular plants (including minor forms). Covers essentially same area as St. John's "Botanical exploration of the north shore of the Gulf of St. Lawrence . . ." (1922).

Macoun, J. M. List of plants collected at Lake Mistassini, Rupert River and Rupert House . . . 1885. *In* **Low, A. P.** Report of the Mistassini expedition 1884–5. Ann. Rpt. Geol. and Nat. Hist. Survey Canada n. s., **1** (Rpt. D): 36D–44D. 1885.
Tabular list of 306 vascular plants, with locality indicated.

Macoun, John. List and notes . . . of plants collected by Mr. Richardson on the Magdalen Islands. *In* **Richardson, James.** Report of a geological exploration of the Magdalen Islands. Rpt. Prog. Geol. Survey Canada **1879–80:** 12G–15G. 1881.
Tabular list of 79 vascular plants; general features of flora.—See also **Provancher, L. A.** Liste des plantes . . . *In* his Un naturaliste aux îles de la Madeleine. Naturaliste Canad. **19:** 246–247. 1890.

Marie-Victorin, *frère.* Études floristiques sur la région du lac Saint-Jean. Contrib. Lab. Bot. Univ. Montreal no. 4. 174 p. 28 fig. 1925.
Physiography, etc.; botanical explorations; author's itinerary, with critical notes on various species; list of cellular and vascular plants collected, with localities; plant associations, etc.

——— La flore du Témiscouata. Rapport sur une nouvelle exploration botanique de ce comté de la province de Québec. Naturaliste Canad. **41:** 99–108, 115–119, 132–138, 148–155, 165–170, 181–188. illus. (incl. map). 1915; **42:** 6–12, 18–29, 34–44, 51–59, 68–79, 106–111, 121–126, 136–143, 153–158, 168–175, 181–187. illus. (incl. map). 1915–16.
Bibiography, climate, geology, plant formations; annotated alphabetical lists of cellular and vascular plants.

——— Flore laurentienne. 917 p. 324 fig., maps (fig. A-K and fold. map). 29 cm. Montréal, 1935.
Botanical explorations, bibliography, physiography, geology, climate, phytogeography, evolution; annotated descriptive flora of 1,568 vascular plants, with keys, notes on biology; glossary. Covers the southern part of Quebec, north to about the level of the Saguenay River, Lake St. John, and the upper Ottawa River.

——— and **Meilleur, René.** La florule de la Grosse-Ile. Naturaliste Canad. **66:** 107–122. 13 fig. 1939.
Topography, geology, etc.; list of 255 vascular plants; phytogeography. (In the St. Lawrence River, 40 miles downstream from Quebec.)

Northrop, J. I. Plant notes from Termiscouata [sic] County, Canada. Bul. Torrey Club **14:** 230–238. 1887. (Reprinted as Contrib. Herb. Columbia Col. no. 3.—Also in Osborn, H. F. A naturalist in the Bahamas. 1910.)
General features of flora; systematic list of vascular plants collected by author mostly at Notre Dame du Lac, on west shore of Lake Témiscouata.—See also **Northrop, J. I. and A. B.** Plant notes from Tadousac and Temiscouata County, Canada. Bul. Torrey Club **17:** 27–32. 1890.

Penhallow, D. P. Notes on the flora of Cacouna, P. Q. Canad. Rec. Sci. **4:** 432–460. 1891.
Geology, general features of flora; annotated list of 212 vascular plants collected by the author in August 1891.

——— Notes on the flora of St. Helen's Island, Montreal. Canad. Rec. Sci. **4:** 369–372. 1891.
General features of woody flora; alphabetical list of plants observed by author.

Polunin, Nicholas. Vascular plants from Diana Bay, Hudson Strait. Canad. Field-Nat. **51:** 111–114. 1937.

List of 75 vascular plants. Diana Bay is on the south shore of Hudson Strait, 61° N. Lat., 70° W. Long.

Potter, David. Plants collected in the southern region of James Bay. Rhodora **36:** 274–284. 1 fig. (map). 1934.

Tabular list of vascular plants collected by the author in 1929, with indication of localities.

Rousseau, Jacques. Études floristiques sur la région de Matapédia (Québec). Bul. Mus. Natl. Canada **66** (Sér. Biol. no. 17) : 1–25. 1 fig. (map), 2 pl. 1931.

General features of flora; annotated list of vascular plants collected by the author in 1929; bibliography.

Saint-Cyr, D. N. List of plants gathered by D. N. Saint-Cyr, on the north shore, from St. Paul's Bay to Ouatchechou, and in the islands of Mingan, Anticosti and Grand Mecatina, during the summer of 1882 and the month of July 1885, during . . . his two trips to the lower St. Lawrence and the Gulf. *In* [Sessional papers, Prov. of Quebec] v. 20, pt. 3, no. 17B, p. 94–114. 1887.

List of vascular and cellular plants, with localities, identified by J. Macoun.

St. John, Harold. A botanical exploration of the north shore of the Gulf of St. Lawrence including an annotated list of the species of vascular plants. Mem. Canada Dept. Mines 126 (Biol. Ser. 4). iii, 130 p. 9 fig., pl., 2 maps. 1922.

Geology, physiography, botanical explorations, relations between chemical nature of soils and plant distribution, bibliographies; annotated list (614 species and varieties, of which 574 are native).

Schmitt, Joseph. Monographie de l'île d'Anticosti (Golfe Saint-Laurent). vi, 370 p. plates, maps. 25.5 cm. Paris, 1904.

Annotated list of vascular plants (identifications revised by Macoun), p. 59–234.—For additions, see **Adams, John.** Some additions to the vascular flora of Anticosti Island. Canad. Field-Nat. **48:** 63–65. 1934.

SASKATCHEWAN

See also Alberta (Raup, Phytogeographic studies); Mackenzie (Preble); Manitoba (Tyrrell); United States, Western (Rydberg, Flora of the Rocky Mountains, and Flora of the prairies and plains).

General

Fraser, W. P., and **Russell, R. C.** List of the flowering plants, ferns and fern allies of Saskatchewan. 46 p. 25.5 cm. Saskatoon, 1937.—Additions . . . 7 p. 28 cm. [Saskatoon,] 1938.

Annotated list of vascular plants, with localities for the rarer species. The list of Additions processed.

YUKON

See also Franklin (Ostenfeld); Mackenzie (Macoun and Holm; Porsild; Stefánsson); Greenland (Hooker; Ostenfeld, Flora arctica); United States, Western (Kirkwood).

Local

Britton, N. L., and **Rydberg, P. A.** An enumeration of the flowering plants collected by R. S. Williams and by J. B. Tarleton. Bul. N. Y. Bot. Gard. **2:** 149–187. 1901. (Contributions to the botany of the Yukon Territory, 4.)

List with localities and occasional annotations. The pteridophytes of the same collectors are listed by L. M. Underwood, l. c. 148–149. (Klondike region.)

Macoun, J. M. [List of plants collected during 1911 and 1912.] *In* Cairnes, **D. D.** The Yukon-Alaska international boundary, between Porcupine and Yukon Rivers. Mem. Canada Geol. Survey **67** (Geol. Ser. **49**) : 13–18. 1914.

List of vascular and cellular plants, preceded (p. 10–12) by notes on general features of flora by D. D. Cairnes.

Macoun, John. List of plants collected by Dr. G. M. Dawson in the Yukon district and adjacent northern portion of British Columbia in 1887. *In* Dawson, **G. M.** Report on an exploration in the Yukon district . . . Montreal, 1888.

p. 215B–228B. (With Ann. Rpt. Geol. and Nat. Hist. Survey Canada n. s., v. 3, pt. 1. 1889.)

List of 201 plants (198 vascular) with localities.

Macoun, John. List of plants collected by Mr. J. B. Tyrrell in the Klondike region in 1899. Ottawa Nat. 13: 209–218. 1899.

List of vascular and cellular plants.

CENTRAL AMERICA (INCLUDING MEXICO)

Hemsley, W. B. Biologia centrali-americana; or, contributions to the knowledge of the fauna and flora of Mexico and Central America . . . Botany. 4 v. 110 (i. e. 111) pl. (pt. col., and incl. map). 31 cm. London, 1879–88.

The fundamental modern work. Systematic list of vascular plants of Mexico (excluding Lower California) and Central America, with synonymy, citation of exsiccatae, and statement of extralimital range, based chiefly on the material in Kew Herbarium; phytogeography; sketch of botanical exploration, in form of an annotated list of collectors; summary and analysis of flora; bibliography; separate lists of plants from Cozumel Island collected in 1885 by G. F. Gaumer, and from Holbox, Mugeres, Cozumel, and Ruatan Islands collected in 1886 by Gaumer. Includes 12,233 species, of which 11,626 are Phanerogamae; 8,193 of the latter are endemic.

BRITISH HONDURAS

General

Sprague, T. A., and **Riley, L. A. M.** Materials for a flora of British Honduras. Bul. Misc. Inform. Kew **1924:** 1–20. 1924.

Systematic list with citation of exsiccatae, vernacular names, and occasional annotations; geography and physiography; plant formations; botanical explorations; short bibliography. Incomplete: Ranunculaceae–Elaeocarpaceae (Bentham and Hooker system). Total number of phanerogams estimated at 4,000; this enumeration not expected to contain over 800.

Standley, P. C., and **Record, S. J.** The forests and flora of British Honduras. Pub. Field Mus. Nat. Hist. 350, Bot. Ser. v. 12. 432 p. 16 pl. 1936.

Geography, geology, soils, climate, agriculture, forests, timbers, economic trees, bibliography, relationships of flora, botanical explorations; partly annotated list of 2,125 vascular plants, with vernacular names and uses, and keys and brief descriptions of woody plants.

COSTA RICA

For Cocos Island, see under Insular Floras.

General

Alfaro, Anastasio. Lista de las plantas encontradas hasta ahora en Costa Rica y en los territorios limítrofes, extractada de la "Biologia centrali-americana." Anales Mus. Nac. Costa Rica **1:** 1–101. 1888.

General features of flora; list of 3,386 vascular plants of Costa Rica and other Central American countries, of which 1,218 are definitely Costa Rican.

Durand, Théophile, and **Pittier, H. F.** Primitiae florae costaricensis. 3 v. 22 cm. Bruxelles, 1891–96; San José, 1898–1901.

List of species, with citation of exsiccatae and statement of extralimital range, prepared with the aid of many collaborators; sketch of plant geography; botanical exploration; gazetteer (1(1): 46–48). Durand's name appears as co-author only of vol. 1. Incomplete; includes (in vascular plants) Pteridophyta, Araceae, Iridaceae, Piperaceae, most Polypetalae, Euphorbiaceae, Gamopetalae except Solanaceae and Rubiaceae. The pagination of vol. 1 is involved. Vol. 1 (in 3 fasc.) 208, 191, 297 p. 1891–96; v. 2 (in 7 fasc.) 405 p. 1898–1900; v. 3 (fasc. 1). 69 p. 1901. Vol. 1 reprinted with changed pagination from Bul. Soc. Roy. Bot. Belg. **30:** 7–97, 196–305. 1891; **31:** 119–215. 1892; **32:** 122–201. 1893; **35:** 151–297. 1896; vol. 2 from Anales Inst. Fís.-Geog. Costa Rica **9:** 77–223. 1898.

Pittier, H. F. Ensayo sobre las plantas usuales de Costa Rica. xi, 176 p. 31 pl. 25 cm. Washington, 1908.

Extensively annotated alphabetical list of vernacular (Spanish) names of wild and cultivated plants; cross-reference index of scientific names, separate index of vernacular names used by Indian tribes; bibliography of flora of Costa Rica; sketch of botanical exploration; plant distribution; discussion of economic plants, classified by uses; origin of vernacular names.

Polakowsky, Hellmuth. Die pflanzenwelt von Costa-Rica; ein beitrag zur kenntnis der pflanzen-geographie und der flora von Central-Amerika. Jahresber. Ver. Erdk. Dresden **16** (Wiss. th.): 25–124. map. 1879. (Spanish translation by Manuel Carazo Peralta, with annotations by H. Pittier, as "La flora de Costa Rica. Contribución al estudio de la fitogeografía centro-americana." Anales Inst. Fís.-Geog. Nac. Costa Rica **2**: 177–201. 1890.)

Geography, plant geography (indicated on the map), systematic list of plants (cellular and vascular) known from Costa Rica, with localities and collectors' names.

Standley, P. C. Flora of Costa Rica. Pub. Field Mus. Nat. Hist. 391, 392, 420, 429, Bot. Ser. v. 18. 1616 p. map. 1937–38.

Physiography, plant zones, botanical explorations, bibliography; list of 6,085 spermatophytes with vernacular names, local and extralimital range, citation of type collection (if from Costa Rica) and often of other local collections; brief descriptions of genera and species in the case of the Dicotyledons. Orchidaceae by Oakes Ames. Index, p. 1573–1616, issued [1939]. A translation with slight changes has begun to appear as Museo Nacional de Costa Rica, Serie Botánica v. 1 (pt. 1, no. 1–3, p. 1–146. 1937–39).

GUATEMALA

See also Mexico (Bukasov).

General

Guérin, René, González, Darío, and García Salas M., Jorge. Catalogue des produits présentés par la république de Guatémala à l'Exposition universelle de Paris, 1900. 117 p. 23 cm. Guatémala, 1900.

Annotated, classified lists of plant products; general summary of geographic and other information about Guatemala.

Niederlein, Gustav. The Republic of Guatemala. 63 p. 23.5 cm. Philadelphia, 1898.

Includes (p. 19–30) chapter on vegetation with unannotated classified lists of principal economic plants, with vernacular and scientific names, also alphabetical lists of native names of woods and medicinal plants, without botanical equivalents.

Smith, J. D. Enumeratio plantarum guatemalensium necnon salvadorensium hondurensium nicaraguensium costaricensium. 8 v. plates. 23 cm. Oquawkae, 1889–1907. (Title varies, that of vol. 1 is: Enumeratio plantarum guatemalensium imprimis a H. de Tuerckheim collectarum; v. 2–3: Enumeratio plantarum guatemalensium.)

Each volume (1–7) consists of copies of the data of specimens distributed by J. D. Smith from Guatemala (and in the later volumes other Central American republics) during successive periods of years, arranged in systematic order. Vol. 8 contains an alphabetical list of all the species so distributed with references to their enumeration in earlier volumes. Lists 3,736 species, of which 1,189 are not given by Hemsley. Vol. 1–5 include also his Undescribed plants from Guatemala I–XXI, reprinted from Bot. Gaz. v. 12–16, 18–21, 23–25, 26.

[Tejada, R.?] Catálogo de plantas reputadas medicinales en la República de Guatemala. 71 p. 26 cm. Guatemala, [1913?]. (At head of title: Fiestas de Minerva de 1913. Exposición nacional.)

Alphabetical list of vernacular names (and botanical names when known), with localities and brief notes on medicinal uses. The "segunda edición" (55 p., Guatemala, 1929) seems to be identical with the first, except for the omission of the short preface.

Local

Blake, S. F. Native names and uses of some plants of eastern Guatemala and Honduras. Contrib. U. S. Natl. Herb. **24**: 87–100. pl. 29–33. 1922.

Briefly annotated list of vernacular and botanical names with local uses.

Guatemala. Ministerio de agricultura. Flora guatemalteca. Herbario nacional. Plantas clasificadas científicamente por el Prof. Paul C. Standley . . . y revisadas por el Prof. Ulises Rojas. v. 1 (19 p.). 26 cm. Guatemala, 1929.
Alphabetical list of 583 vernacular names with botanical equivalents.

Johnston, J. R. Lista de plantas de Guatemala en el herbario de la Escuela nacional de agricultura, Chimaltenango. Revista Agr. (Guatemala) **15:** 203–206, 249–260, 314–321. 1938; **16:** 129–133. 1939. (Reprinted with title: Catálogo de plantas de Guatemala. 52 p.)
Mostly unannotated list of plants, collected principally by the author, with some vernacular names.

Lundell, C. L. Plants probably utilized by the Old Empire Maya of Petén and adjacent lowlands. Papers Mich. Acad. Sci. **24:** 37–56. 1939.
Topography, climate, anciently cultivated trees; annotated list of useful plants, grouped by uses; bibliography.

——— The vegetation of Petén. With an appendix: Studies of Mexican and Central American plants I. ix, 244 p. 39 pl. (incl. map). 29.5 cm. Washington, D. C., 1937. (Carnegie Institution of Washington Publication no. 478.)
Botanical explorations, geology, climate, soils, history of vegetation, ecology, etc.; annotated lists of vascular plants known from northern Petén (785 species), central Petén (835 species), and southern Petén (191 species), with vernacular names.

Recinos, Adrián. Monografía del Departamento de Huehuetenango, República de Guatemala. xiv, 269 p. illus. (incl. maps). 25 cm. Guatemala, 1913.
Includes chapter (p. 50–86) giving general features of topography, life zones, and annotated list of wild and cultivated economic plants, classified by uses, with vernacular names.

Record, S. J., and Kuylen, Henry. Trees of the lower Río Motagua valley, Guatemala. Trop. Woods, Yale Univ. School Forestry **7:** 10–29. 1926.
General features of tree flora; annotated systematic list of trees, with vernacular names (indexed).

HONDURAS

See also Central America (Hemsley, for list including the plants of Ruatan Island); Guatemala (Smith; Blake).

General

Record, S. J. Trees of Honduras. Trop. Woods, Yale Univ. School Forestry **10:** 10–47. 1927.
Annotated list, with vernacular names and uses; index of vernacular names.—See also **Standley, P. C.** A second list of the trees of Honduras. l. c. **21:** 9–41. 1930.—Additions . . . l. c. **37:** 27–39. 1934.

Local

Conzemius, Édouard. Economic plants of the Bay Islands (Honduras). Gard. Chron. III, **81:** 50–51, 69–70, 81, 117–118, 133, 180–181, 217, 270, 305, 413–414. fig. 42. 1927.
Geography, physiography, climate, natives, etc.; annotated list of wild and cultivated useful plants.

Popenoe, Wilson. The useful plants of Copan. Amer. Anthropol. n. s., **21:** 125–138. 1919.
Climate, soil, etc.; annotated list of useful plants, wild and cultivated, classified by uses, with vernacular names. Refers to the valley of the Copán River in western Honduras.

Standley, P. C. Flora of the Lancetilla Valley, Honduras. Pub. Field Mus. Nat. Hist. 283, Bot. Ser. v. 10. 418 p. 68 pl. 1931.
Geology, climate, general features of flora, useful plants, botanical explorations; list of cellular plants; briefly descriptive flora of vascular plants, with vernacular names and uses. (Near Tela, Department of Atlántida.)

——— The woody plants of Siguatepeque, Honduras. Jour. Arnold Arboretum **11:** 15–46. 1930.
General features of flora; annotated list of woody phanerogams collected by the author, with vernacular names. (Department of Comayagua.)

Yuncker, T. G. A contribution to the flora of Honduras. Pub. Field Mus. Nat. Hist. 405, Bot. Ser. **17**: 285–407. 18 pl. 1938.
Botanical explorations, general features of flora around Siguatepeque; annotated list of cellular and vascular plants collected by the author in 1934 and 1936 near Siguatepeque (Department of Comayagua), Potrerillas and Lake Yojoa (Department Cortés), including also species from the Lancetilla-Tela region not recorded by Standley.

MEXICO

See also Central America (Hemsley, for special lists of the plants of Cozumel, Holbox, and Mugeres Islands, Yucatan).

General

Alcocer, G. V. Catálogo de los frutos comestibles mexicanos. Anales Mus. Nac. México II, **2**: 413–488. 1905.
Annotated systematic list of wild and cultivated fruits, with vernacular names. Apparently never finished.

Bukasov, S. M. Vozdelyvaemye rastenii͡a Meksiki, Gvatemaly i Kolumbii (The cultivated plants of Mexico, Guatemala, and Colombia). Prilozh. Trudy Prikl. Bot., Gen. i Selek. (Suppl. Bul. Appl. Bot., Gen. and Plant-Breed. Leningr.) no. 47 553, [16,] xxxvii p. 365 fig. (incl. maps), 2 col. pl., fold. tab., fold. map. 1930
Geography, climate, centers of origin, general features of agricultural regions, native and foreign cultivated plants. In Russian, with fairly full English summary. Supplementary articles on various plants by several authors. Work based on the expedition of the Institute of Applied Botany in 1925–26.

Conzatti, Cassiano, and Smith, L. C. Flora sinóptica mexicana. pt. 2–3 (23, 225 p.). port., 9 tab. 22.5 cm. Oaxaca, 1895–97. (2d ed., pt. 3. 335 p. tab. México, 1910.)
Descriptive flora, with keys; in most genera only a part of the species are described, the others merely listed with locality; very few vernacular names. Incomplete; includes only Corolliflorae (Asclepiadales—Bignoniales), 2,505 species and 75 varieties. Pt. 1 never published.

Flores, Leopoldo. Manuel terapéutico de plantas mexicanas. 102 p. 21.5 cm. México, 1909.
Annotated list of medicinal plants, arranged alphabetically by vernacular names; supplementary list of same classified by uses.

Léon, Nicolás. Biblioteca botánico-mexicana. Catálogo bibliográfico, biográfico y crítico de autores y escritos referentes á vegetales de México y sus aplicaciones, desde la conquista hasta el presente. 372 p. 23 cm. México, 1895.
Alphabetical list of 805 authors (with additions) and their works; botanical explorations, collectors.

Martínez, Maximino. Catálogo de nombres vulgares y científicos de plantas mexicanas. 551 p. 19 cm. México, 1937.
Alphabetical list of over 13,000 vernacular names, with indication of habit, color of flower, etc., botanical name and locality; includes cultivated plants; bibliography; no index of botanical names. Replaces his Catálogo alfabético de nombres vulgares y científicos de plantas que existen en México. 1923–[29?], which in turn replaced Ramírez and Alcocer's Sinonimia vulgar y científica de las plantas mexicanas. 1902.

——— Las plantas medicinales de México. 644 p. illus., col. plates. 20 cm. México, 1933.
Annotated lists of (1) identified plants whose properties have been studied, (2) identified plants with imperfectly known properties, (3) unidentified plants, (4) plants extracted from the "Libro del Judio" published at Mérida, Yucatán, in 1834. Vernacular and botanical names, localities, uses, chemistry, properties, references, etc.

——— Las plantas mas utiles que existen en la República mexicana. 381 p. illus. 23 cm. México, 1928.
List of about 100 more important species, arranged alphabetically by vernacular names, with uses and bibliography under each species; indices.

Ramírez, José. La vegetación de México; recopilación y análisis de las principales clasificaciones propuestas. 271 p. maps. 23 cm. México, 1899.
Discussion of phytogeographical regions of Mexico proposed by various writers, including the author. Listed here on account of its tables (p. 148–248) of Mexican localities with altitudes, arranged alphabetically by States.

Rose, J. N. Notes on useful plants of Mexico. Contrib. U. S. Natl. Herb. **5**: 209–259. pl. 28–64. 1899.
Annotated list of economic plants, classified by uses.

Standley, P. C. Trees and shrubs of Mexico. Contrib. U. S. Natl. Herb. v. 23. 1721 p. 1920–26.
Complete flora of the native woody vascular plants (about 5,700 species) with keys, brief descriptions, synonymy, statement of local and extralimital range, extensive account of vernacular names, and local uses; brief history of early explorations. Treatment of several families contributed by specialists.

Suc, Louis. Les plantes médicinales du Mexique. 236 p. 24 cm. Toulouse, 1912. (Université de Toulouse. Faculté mixte de médicine et de pharmacie. 1911–12. no. 15.)
Brief sketch of botanical explorations, life zones, etc.; systematic list of medicinal plants, with vernacular names, uses, composition, etc., with footnote references; bibliography of 20 titles.

Urbina, Manuel. Catálogo de plantas mexicanas (fanerógamas). vi, 487 p. 23.5 cm. México, 1897.
Systematic list, with synonymy, citation of exsiccatae with localities, and vernacular names, based on the material in the Museo Nacional of Mexico. Includes only about 3,000 species, and so is very far from representing the total flora.

Local

Batalla, M. A., and Ramírez Cantú, Débora. Contribución al estudio floristico del Valle de México. Estudio cualitativo de las sinecias que habitan los Cerros del Norte del Valle de México. Anales Inst. Biol. México **10**: 227–267. 20 fig. 1939.
Geology, geography, climate, ecology; list of 220 vascular plants from the Sierra de Guadalupe, with localities; bibliography. (State of Mexico and (?) Federal District.)

Brandegee, T. S. A collection of plants from Baja California, 1889. Proc. Calif. Acad. Sci. II, **2**: 117–216. pl. 2–11, map. 1889.
Annotated list of vascular plants collected by the author in Baja California (from Margarita Island northward) in 1889; brief sketch of physiography and useful plants.

——— Flora of the Cape region of Baja California. Proc. Calif. Acad. Sci. II, **3**: 108–182, 218–227. 1891–92.—Additions . . . Zoe **4**: 398–408. pl. 31. 1894.
List of 679 species (of which 674 are vascular plants), annotated as to locality, from region "south of a line drawn along the northern base of the mountains from Todos Santos to La Paz"; brief account of physiography. The additions bring the total number of vascular plants to 795.

Bravo Hollis, Helia. Observaciones florísticas y geobotánicas en el valle de Mezquital, Hidalgo. Anales Inst. Biol. México **8**: 3–82 (incl. 12 pl.). 9 fold. tab. 1937.
Mainly ecological; includes list (p. 57–82) of cellular and vascular plants, with localities; bibliography.

Cuevas, Benjamin. Plantas medicinales de Yucatán y guía médica práctica doméstica. 278 p. 37 col. pl., port. 23 cm. Mérida, Yucatán, 1913.
Alphabetical list of vernacular names (with botanical equivalents when known), brief descriptions, medicinal and other uses. The second part of the work (p. 129–278) is a popular medical guide, with separate title page dated 1908. The colored figures (3 on each plate) are tiny and unsatisfactory.

Eastwood, Alice. List of plants recorded from Cedros Island, Mexico. Proc. Calif. Acad. Sci. IV, **18**: 420–441. 1929.
Botanical explorations, list of species first described from the island, etc.; annotated list of 176 vascular plants.

Eastwood, Alice. A list of plants recorded from the Tres Marías Islands, Mexico. Proc. Calif. Acad. Sci. IV, **18**: 442–468. 1929.
Botanical explorations, list of species first described from the island; list of 324 vascular plants with citation of exsiccatae.

——— List of the plants recorded from Guadalupe Island, Mexico. Proc. Calif. Acad. Sci. IV, **18**: 394–420. pl. 33–34. 1929.
Botanical explorations, list of species originally described from the island, etc.; annotated list of 161 vascular plants.

Ferris, R. S. Preliminary report on the flora of the Tres Marías Islands. Contrib. Dudley Herb. Stanford Univ. **1**: 63–81. 4 pl. 1927.
General features of flora, etc.; annotated list of vascular plants collected by the author in 1925, composed chiefly of those not hitherto recorded. Total flora said to consist of about 285 species.

Goldman, E. A. Plant records of an expedition to Lower California. Contrib. U. S. Natl. Herb. **16**: 309–371. pl. 104–133, map. 1916.
Annotated list of vascular plants (and 1 lichen) collected by E. W. Nelson and E. A. Goldman in 1905–06 on a trip covering the whole extent of the peninsula, with vernacular names and notes on uses.

Greene, E. L. The botany of Cedros Island. Pittonia **1**: 194–208. 1888.—A supplementary list of Cedros Island plants. l. c. **1**: 266–269. 1889.
Geography, etc.; annotated list of 91 vascular plants.

——— Botany of the Coronados Islands. West Amer. Sci. **1**: 69–71. 1885.
Topography, etc.; mostly unannotated list of nearly 70 vascular plants. (Off northwesternmost Lower California, 32°22′ N. Lat., 117°30′ W. Long.)

——— A catalogue of the flowering plants and ferns of Guadalupe Island. Bul. Calif. Acad. Sci. **1**: 220–228. 1885.
Adds 13 species to Watson's list; notes on other species. Preceded (p. 214–220) by "Notes on Guadalupe Island"; general features of flora, climate, etc.

——— The vegetation of the San Benito Islands. Pittonia **1**: 261–266. 1889.
General features of flora; list of 24 vascular plants, of which 5 are endemic.

Heilprin, Angelo. The temperate and alpine floras of the giant volcanoes of Mexico. (Being a report from the Committee on the Michaux legacy.) Proc. Amer. Phil. Soc. **30**: 4–22. 1892.
Includes list of species reaching 10,000 feet or higher on Orizaba, Popocatepetl, Ixtaccihuatl, and Nevada de Toluca, with discussion.

Johnston, I. M. Expedition of the California Academy of Sciences to the Gulf of California in 1921. The botany (the vascular plants). Proc. Calif. Acad. Sci. IV, **12**: 951–1218. map. 1924.
Annotated list of 456 vascular plants (and 20 fungi and lichens) collected in Lower California, Sonora, and islands in the Gulf of California; physiography, phytogeography, and history of botanical exploration.

——— The flora of the Revillagigedo Islands. Proc. Calif. Acad. Sci. IV, **20**: 9–104. 1931.
Botanical explorations; physiography, etc., of each island with tabular list of species recorded and remarks on distribution, origin of flora; annotated list of 120 vascular plants.

Loesener, Theodor. Ueber Maya-namen und nutzanwendung yucatekischer pflanzen. *In* Festschrift Eduard Seler . . . herausgegeben von W. Lehmann. p. 321–343. Stuttgart, 1922.
Bibliography; systematic list of 348 plants, with Maya names (and translations) and notes on uses. (Yucatan.)

Millspaugh, C. F. Plantae yucatanae (regionis antillanae). Plants of the insular, coastal and plain regions of the peninsula of Yucatan, Mexico. Pub. Field Columbian Mus. 69, 92, Bot. Ser. **3**: 1–84, 85–151. illus., plates, map. 1903–04.
Descriptive flora, with keys based primarily on fruiting characters; citation of exsiccatae; vernacular names, uses; text figure of each species (usually only fruiting details). Incomplete: fasc. 1, Polypodiaceae and Schizaeaceae (by Millspaugh), Gramineae and Cyperaceae (by Millspaugh and Agnes Chase); fasc. 2, Compositae (Millspaugh and Chase). Never completed.

——— Vegetation of Alacran Reef. Pub. Field Mus. Nat. Hist. 187, Bot. Ser. **2**: 421–431. 1 fig., 2 pl., map. 1916.
Topography, etc.; annotated list of 18 angiosperms. (In Campeche Bank, 75 miles north of Yucatan, about 22°33′ N. Lat., 89°41′ W. Long.)

Ortega, J. G. Informe sobre las plantas de las Islas Marías. Bol. Pro-Cult. Region. S. C. L. Mazatlán, Sinaloa **2** (2) : 5–9. 1936.

Unannotated list of vascular plants of Tres Marías Islands, with vernacular names.

Ramírez Laguna, Antonio. Nota acerca del aprovechamiento de algunas plantas de importancia económica en la región del valle de Mezquital, Hgo. Anales Inst. Biol. México **8**: 83–115. 10 fig. 1937.

Annotated list, alphabetically arranged by botanical names, with vernacular names and uses.

Reiche, K. F. Flora excursoria en el valle central de México. Claves analíticas y descripciones de las familias y géneros fanerógamicos. 303 p. 22 cm. México, 1926.

Keys to and descriptions of families and genera, with mention of more important species; vernacular names. (State of Mexico.)

——— La vegetación en los alrededores de la capital de México. 143 p. 27 fig., map. 24 cm. México, 1914.

Bibliography, physiography, plant formations, etc.; unannotated list of vascular plants, mosses, lichens, and fungi. Reprinted with some changes in Bot. Jahrb. Engler **58** (Beibl. **129**) : 1–116. 1922.

Reko, B. P. De los nombres botánicos aztecas (tomado de la obra inédita "Sinonimia vulgar y científica de la flora oaxaqueña"). México Antiguo **1**: 113–157. 1919.

Etymology, botanical terminology; annotated list of Aztec names, with botanical identifications.

Riley, L. A. M. Contribution to the flora of Sinaloa. Bul. Misc. Inform. Kew **1923**: 103–115, 163–175, 333–346, 388–401. 1923; **1924**: 206–222. 1924.

Systematic list, with citation of exsiccatae, vernacular names, and occasional critical notations; brief sketch of geography and collectors, particularly Seemann. Incomplete (5 parts) ; includes only Polypetalae (Bentham and Hooker system).

Rose, J. N. List of plants collected by Dr. Edward Palmer in 1890 on Carmen Island. Contrib. U. S. Natl. Herb. **1**: 129–134. pl. 12–14. 1892.

Physiography, etc.; annotated list of 68 phanerogams.

——— Plants of the Tres Marias Islands. North Amer. Fauna **14**: 77–91. 2 fig. 1899.

Annotated list of vascular plants collected by E. W. Nelson in 1897.

Roys, R. L. The ethno-botany of the Maya. Pub. Middle Amer. Res. Ser. Tulane Univ. Louisiana no. 2. xxiv, 359 p. 1931.

Classified list of diseases, with remedies; annotated list of Maya plant names, with botanical identification and medicinal uses; alphabetical list of botanical names, with vernacular equivalents; annotated list of Maya zoological names; climate and food supply; bibliography. Refers to Yucatan only.

Sprague, T. A. Sessé and Mociño's Plantae Novae Hispaniae and Flora Mexicana. Bul. Misc. Inform. Kew **1926**: 417–425. 1926.

History of these works; gazetteer of localities mentioned in them; bibliography.

Standley, P. C. Flora of Yucatan. Pub. Field Mus. Nat. Hist. 279, Bot. Ser. **3**: 157–492. 1930.

Geology, climate, general features of vegetation, botanical explorations, Maya vernacular names, bibliography; annotated list of 1,263 plants (nearly all vascular), with vernacular names (Maya, Spanish, and English), uses, and very brief descriptions. The area covered includes Campeche, Yucatan, and the territory of Quintana Roo. The information on useful plants in this work has been extracted and classified in a paper by R. M. Harper, Useful plants of Yucatan, Bul. Torrey Club **59**: 279–288. 1932.

Vasey, George, and Rose, J. N. Plants collected in 1889 at Socorro and Clarion Islands, Pacific Ocean. Proc. U. S. Natl. Mus. **13**: 145–149. 1890. (Scientific results of explorations by the U. S. Fish Commission steamer *Albatross*, no. 16.)

List of 26 species (18 from Socorro, 12 from Clarion Island).

Watson, Sereno. List of plants collected by Dr. Edward Palmer in southwestern Chihuahua, Mexico, in 1885. (His Contributions to American botany [no. 13, pt. 1].) Proc. Amer. Acad. Arts and Sci. **21**: 414–445. 1886.

Localities, etc.; systematic list of vascular plants, with localities, collector's numbers, and frequent annotations.

Watson, Sereno. On the flora of Guadalupe Island, Lower California. [and] List of a collection of plants from Guadalupe Island, made by Dr. Edward Palmer, with his notes upon them. Proc. Amer. Acad. Arts and Sci. **11:** 105–112, 112–121. 1876.

Climate, general features of flora, etc.; annotated list of 119 plants (of which 105 are vascular).

Williams, Llewelyn. Arboles y arbustos del Istmo de Tehuantepec, México. Lilloa **4:** 137–171. 7 fig. 1939.

Plant formations; list of woody and herbaceous plants, alphabeted by vernacular names, with botanical equivalents. (Vera Cruz and Oaxaca.)

NICARAGUA

See also Guatemala (Smith).

General

Lévy, Pablo. Notas geográficas y económicas sobre la República de Nicaragua; su historia, topografía, clima, producciones y riquezas, poblacion y costumbres, gobierno, agricultura, industria, comercio, etc. . . . xvi, 627 p. fold. map. 26.5 cm. Paris, 1873.

Contains sections (p. 166–191) on woods and other useful plants, classified by uses, with vernacular names.

Ramírez Goyena, Miguel. Flora nicaragüense. 2 v. (1064 p.). 29 cm. [Managua,] 1909–11.

Descriptive flora of vascular plants (plus *Chara*); vernacular names (separate index); detailed account of medicinal plants. Not of value as a flora.

Local

Record, S. J. Trees and shrubs collected by F. C. Englesing in northeastern Nicaragua. Trop. Woods, Yale Univ. School Forestry **17:** 18–38. 1929.

Annotated list, with vernacular names (indexed).

PANAMA

For Cocos Island, see Insular Floras.

General

Seemann, Berthold. Flora of the Isthmus of Panama. *In* his The botany of the voyage of H. M. S. *Herald.* p. 57–254. pl. 12–50 (incl. map). London, 1852–54.

Annotated list of 1,204 species (of which 1,164 are vascular plants); botanical exploration, physiography, economic plants, agriculture. (Includes not only the present country of Panama but also the former Territory of Darien; i. e., the coastal region of Colombia from the Isthmus of Panama south to Río San Juan, about 4°10′ N. Lat.)

Local

Johansen, Holger. A handbook of the principal trees and shrubs of the Ancon and Balboa districts, Panama Canal Zone. 97 p. (incl. front., 32 pl., map). 25.5 cm. [Washington,] 1925.

Annotated list including both wild and cultivated species, with location of the living specimens.

Record, S. J. Contributions to the arborescent flora of western Panama. Trop. Woods, Yale Univ. School Forestry **16:** 9–35. 1928.

Systematic list of trees and a few herbaceous plants collected by G. P. Cooper in 1926–28, with vernacular names (indexed).

Standley, P. C. The flora of Barro Colorado Island, Panama. Contrib. Arnold Arboretum no. 5. 178 p. 21 pl., map. 1933.

Physiography, botanical explorations, partly annotated lists of cellular and vascular plants, with brief diagnostic characters of many of the species; bibliography. Replaces the author's work of the same title published in Smithsonian Misc. Coll. v. 78, no. 8. 1927, and its two supplements (Pub. Field Mus. Nat. Hist. 258, Bot. Ser. **4:** 143–158. 1929; Jour. Arnold Arboretum **11:** 119–129. 1930); first supplement by L. A. Kenoyer and Standley.

Standley, P. C. Flora of the Panama Canal Zone. Contrib. U. S. Natl. Herb. v. 27. x, 416 p. 7 fig., 67 pl. 1928.

A condensed flora of the spermatophytes, with keys, vernacular names and uses, and often brief descriptions. The introduction discusses the physiography, composition of the flora, agriculture, and history of botanical exploration.

SALVADOR

General

Guzmán, D. J. Especies utiles de la flora salvadoreña médico-industrial con aplicación a la medicina, farmacia, agricultura, artes, industrias y comercio. 2 v. in 1 (678 p.). 24.5 cm. San Salvador, 1924–26.

Annotated list, with vernacular names, brief descriptions, and uses (largely medicinal); appendix on timbers, etc.

Salvador. Ministero de instrucción pública. Flora salvadoreña. 5 v. 500 pl. (part col.). 18.5 x 28.5 cm. [San Salvador] 1926–[32?].

Photographs, with brief descriptive text, of both wild and cultivated plants, especially those which are useful or ornamental; vernacular names. The plates represent specimens in the herbarium of Félix Choussy.

Standley, P. C., and Calderón, Salvador. Lista preliminar de las plantas de El Salvador. 274 p. 23.5 cm. San Salvador, [1925].—2. suplemento. p. 275–302. 12 pl. San Salvador, 1927.

Annotated list of about 2,070 native and cultivated species, including nonvascular plants; vernacular names. The first supplement is a part of the original list (p. 273–274).

GREENLAND

See also Canada (Adams); United States, general (Gray; Watson).

General

Holm, Theodore. Contributions to the morphology, synonymy, and geographical distribution of Arctic plants. Report of the Canadian Arctic expedition 1913–18. v. 5, Botany, pt. B. 139 p. illus. 1922.

Morphological and systematic studies, geographical distribution, composition of Arctic flora; tables of distribution (including Old World), with discussion; extensive bibliography of Arctic floras. The geographical table of species covers Greenland, Arctic American archipelago, Spitzbergen, Nova Zembla, and Arctic Scandinavia, Russia, and Siberia, and includes also records of the same species from other northern or elevated regions.

Hooker, J. D. Outlines of the distribution of Arctic plants. Trans. Linn. Soc. Bot. **23**: 251–348. pl. 32 (map). 1861.

Plant geography; table showing world-wide distribution of Arctic vascular plants (805 species); critical systematic notes. The table shows general distribution of all vascular plants known from north of 66°30′ N. Lat. in both hemispheres.

Lange, Johan, and others. Conspectus florae groenlandicae. Meddel. om Grønland. v. 3 in 3 pts. (L, 1,016 p.). 57 fig., 2 pl., 3 maps. 1880–94.

Pt. 1: Botanical explorations; bibliography; notes on Greenland (by A. Kornerup); annotated list of 378 vascular plants; index of vernacular names, with supplementary list. A French translation of the Danish introductory matter is given (p. 215–229). [Pt. 1 is practically a reprint of the first issue of 1880 (Meddel. om Grønland v. 3. xxxvi, 231 p. 3 maps).] Pt. 2: Additional collections, chronological table of collections made in Greenland, 1813–1886; additions and corrections to annotated list, bringing total to 395 species; treatment of mosses (by Lange and C. Jensen). Pt. 3: Cellular cryptogams (by various authors); list of vascular plants, reducing total number to 374 (by L. K. Rosenvinge); index of vascular plants; gazetteer of localities mentioned. Rosenvinge's list, "Andet tillaeg til Grønlands fanerogamer og karsporeplanter," Meddel. om Grønland **3**: 645–749. 1892, is often cited separately.

Lauridsen, Peter. Bibliographia groenlandica, eller fortegnelse paa vaerker, afhandlinger og danske manuskripter, der handle om Grønland indtil aaret 1880 incl. Meddel. om Grønland 13. 247 p. 1890.

Includes (p. 83–93) unannotated list of 134 botanical titles, arranged alphabetically by author.

Ostenfeld, C. H., ed. Flora arctica containing descriptions of the flowering plants and ferns, found in the Arctic regions, with their distribution in these countries. Pt. I. Pteridophyta, Gymnospermae, and Monocotyledones. xi, 134 p. 95 fig., map (in text). Copenhagen, 1902.

Bibliography; descriptive flora with keys, references, local and general range (212 species). The area covered includes all Greenland and Baffin Land, the extreme northern part of Canada, Alaska, Siberia and Russia, Nova Zembla, etc. Pt. 1 by O. Gelert and C. H. Ostenfeld. No more published.

——— The flora of Greenland and its origin. Biol. Meddel. K. Danske Videnskab. Selsk. v. 6, no. 3. 71 p. illus. (maps). 1926.

General features of flora; plants brought by Norsemen; phytogeography; endemic species (8); tabular list of 390 vascular plants, with distribution indicated by districts; bibliography.

Porsild, M. P. Alien plants and apophytes of Greenland. Meddel. om Grønland **92**: 1–85. 2 maps. 1932.

Annotated list of 91 introduced angiosperms, with extralimital range; apophytes (annotated list of 63 species), hemerophytes, discussion of plant introduction by Norsemen, bibliography.

Local

Abromeit, Johannes. Samenpflanzen (phanerogamen) aus dem Umanaksund Ritenbenks-distrikt. *In* Botanische ergebnisse der von der Gesellschaft für erdkunde zu Berlin unter leitung Dr. v. Drygalski's ausgesandten Grönlandsexpedition nach Dr. Vanhöffen's sammlungen bearbeitet. B. 105 p. pl. 2–5 (3 col.). 31 cm. Stuttgart, 1899. (Bibliotheca botanica h. 42, [lfg. 2].)

Annotated list of 142 phanerogams from northwest Greenland collected in 1891–93. The vascular cryptogams are listed by the same author in pt. A, p. 69–75, 1897, of the same work.

Böcher, T. W. Biological distributional types in the flora of Greenland. A study on the flora and plant-geography of south Greenland and Eastgreenland between Cape Farewell and Scoresby Sound. Meddel. om Grønland v. 106, no. 2. 339 p. 147 fig. (incl. maps), 2 pl. 1938. (6. og 7. Thule-expedition til Sydøstgrønland 1931–33.)

Table of collectors with references to publications, climate, geology, topography, tabular list of vascular plants collected; annotated list of 391 vascular plants known from the region, with references; biological types, phytogeography, bibliography. The numerous maps of ranges show the known stations for all of Greenland.

——— Phytogeographical studies of the Greenland flora based upon investigations of the coast between Scoresby Sound and Angmagssalik. Meddel. om Grønland v. 104, no. 3. 56 p. 13 fig. (incl. maps), pl. (map). 1933.

Includes tabular list of about 150 vascular plants known from the region indicated (65°55′–69°55′ N. Lat.), with indication of range; bibliography.

Devold, J., and **Scholander, P. F.** Flowering plants and ferns of southeast Greenland. Skrift. Svalbard og Ishavet (Oslo) no. 56. 209 p. (incl. 46 fig., 7 pl. (maps)), fold. map. 1933.

Botanical explorations, list of localities; annotated list of 233 vascular plants; plant formations; bibliography. Covers area in southeast Greenland "south of Angmagssalik, and in Kangerdlugsuak lat. 68°15′ N."

Dusén, Per. Zur kenntnis der gefässpflanzen Ostgrönlands. Bihang K. Svenska Vetensk.-Akad. Handl. v. 27, afd. 3, no. 3. 70 p. 2 fig., 6 pl. (incl. map). 1901.

Botanical explorations, localities, plant formations, with footnote references; annotated list of vascular plants collected by the Swedish expeditions of 1899 and 1900. (East coast, 70°–75° N. Lat.)

Gelting, Paul. Studies on the vascular plants of East Greenland between Franz Joseph Fjord and Dove Bay (Lat. 73°15′–76°20′ N.). Meddel. om Grønland v. 101, no. 2. 340 p. 47 fig. (incl. maps), 4 pl. 1934.

Botanical explorations, descriptive list of localities; annotated list of 163 vascular plants known from region, with localities and collectors; climate, phytogeography, life forms, etc.; bibliography.

Hartz, Nikolaj. Fanerogamer og karkryptogamer fra nordøst-Grønland, c. 75°–70° N. Br., og Angmagsalik, c. 65°40′ N. Br. Meddel. om Grønland **18:** 315–393. 1 fig. 1895.

Botanical collections; annotated list of 165 vascular plants from northeast Greenland with localities; table showing local distribution of 248 vascular plants in northeast and northwest Greenland; table showing local distribution of the 165 vascular plants listed above; phytogeography; list of 64 vascular plants collected by E. Bay at Angmagsalik.

Kruuse, Christian. List of the phanerogams and vascular cryptogams found in the Angmagsalik District on the east coast of Greenland between 65°30′ and 66°20′ lat. N. Meddel. om Grønland **30:** 209–287. 1906.

Localities, botanical explorations, bibliography; annotated list of 183 vascular plants.

——— List of phanerogams and vascular cryptogams found on the coast 75°–66°20′ lat. N. of East Greenland. Meddel. om Grønland **30:** 143–208. 1905.

Localities; annotated list of 178 vascular plants.

Lundager, Andreas. Some notes concerning the vegetation of Germania Land, north-east Greenland. Meddel. om Grønland **43:** 347–414. 18 fig., pl. 17 (map). 1917 (1912).

Mainly ecological; includes tabular list of 48 phanerogams.

Ostenfeld, C. H. Flowering plants and ferns from north-western Greenland collected during the Jubilee expedition 1920–22 and some remarks on the phytogeography of north-Greenland. Meddel. om Grønland **68:** 1–42. 1 fig. (map). 1925. (Reprinted as Arb. Bot. Have København nr. 105. 1925.)

Botanical collections, localities; annotated list of 97 vascular plants; tabular list showing distribution of 125 vascular plants found in Greenland north of 76° N. Lat., with discussion; bibliography.

——— Flowering plants and ferns from Wolstenholme Sound, N. W. Greenland. Meddel. om Grønland **64:** 189–206. 1927 (1923). (Den II. Thule ekspedition til Grønlands nordkyst 1916–18, nr. 6.)

List of 95 vascular plants, with localities and occasional annotations.

——— Plants collected during the first Thule expedition to northernmost Greenland. Meddel. om Grønland **51:** 371–381. 1915.

Localities visited; annotated list of 45 vascular plants (also mosses), including 28 not before reported from the northern part of northwestern Greenland; list of 10 additional species from that region not collected on this expedition.

——— Two plant lists from Inglefield Gulf and Inglefield Land (77°28′ and 79°10′ N. Lat.), N. W. Greenland. Meddel. om Grønland **64:** 207–214. 1927 (1923). (Den II. Thule ekspedition til Grønlands nordkyst 1916–18, nr. 7.)

Includes lists of vascular plants from Inglefield Gulf at Kangerdlugssuak (40 species), and from Cape Agassiz, Inglefield Land (44 species); also list of 6 species from Cape Leiper, Rensselaer Bay, 78°44′ N. Lat.

——— The vegetation of the north-coast of Greenland based upon the late Dr. Th. Wulff's collections and observations. Meddel. om Grønland **64:** 221–268. 5 fig. (incl. map), 5 pl. 1927 (1923). (Den II. Thule ekspedition til Grønlands nordkyst 1916–18, nr. 9.)

Table of vascular plants, showing localities; annotated list of 70 vascular plants; general features of vegetation and climate, biological features, plant formations. (Northern coast from Low Point, 83°6′ N. Lat., to Midgaardsormen, 81°15′ N. Lat.)

——— and **Lundager, Andreas.** List of vascular plants from north-east Greenland (N. of 76° N. lat.) collected by the Danmark-expedition 1906–1908. Meddel. om Grønland **43:** 1–32. 3 fig., 6 pl. 1910.

Localities visited, bibliography; annotated list of 92 vascular plants.

Porsild, A. E. Contributions to the flora of West Greenland at 70°–71°45′ N. lat. Meddel. om Grønland **58:** 157–196. 2 fig. 1926. (Reprinted in Arb. Danske Arkt. Sta. Disko nr. 12. 1926.)

Localities visited by author in 1921, with notes on plants of each; list of 114 vascular plants from Pâtût: annotated list of 203 vascular plants from the Nordost Bugt and the Waygat coast of Nûgssuaq Peninsula; discussion of affinity between flora of Nordost Bugt and that of Scoresby Sound; bibliography.

Porsild, M. P. List of vascular plants collected by Dr. N. C. Engell in the vicinity of the great glacier of Jakobshavn, about 69° lat. n. Meddel. om Grønland **34:** 243–251. 1910.

Partly annotated list of vascular plants, with localities.

——— The plant-life of Hare Island off the coast of West Greenland. Meddel. om Grønland **47:** 249–274. 10 fig. (incl. map). 1911 (1910). (Arb. Danske Arkt. Sta. Disko nr. 3.)

Botanical explorations, geology, climate, plant formations; annotated list of 112 vascular plants; phytogeography, bibliography.

——— and **Porsild, A. E.** The flora of Disko Island and the adjacent coast of West Greenland from 66°–71° N. lat. With remarks on phytogeography, ecology, flowering, fructification and hibernation. Part 1. Meddel. om Grønland **58:** 1–155. 1926. (Preprinted in Arb. Danske Arkt. Sta. Disko nr. 11. 1920.)

History of botanical investigations, annotated list of important localities; annotated list of 285 species and varieties of vascular plants.

Rikli, Martin. Die den 80°n erreichenden oder überschreitenden gefässpflanzen. (Vorläufige mitteilung.) Vierteljahrsschr. Naturf. Gesell. Zürich **62:** 169–193. 1917.

Annotated tabular list of 112 vascular plants, with highest latitude record, localities, etc.; bibliography, phytogeography. Covers northern Greenland, Ellesmere Land, Spitzbergen, Franz Joseph Land.

Seidenfaden, Gunnar. The vascular plants of south-east Greenland 60°04′ to 64°30′ N. lat. Meddel. om Grønland v. 106, no. 3. 129 p. 23 fig., map. 1933.

Botanical explorations, descriptive list of localities; annotated list of 206 vascular plants known from area, with localities and collectors; phytogeography, etc.; bibliography.

——— and **Sørensen, Thorvald.** The vascular plants of northeast Greenland from 74°30′ to 79°00′ N. Lat. and a summary of all species found in East Greenland. Meddel. om Grønland v. 101, no. 4. 215 p. 59 fig. (incl. maps), 4 pl. 1937.

Botanical explorations, with annotated list of localities; annotated list of 128 vascular plants; ecology, etc.; tabular list of 321 vascular plants of East Greenland (60°04′–79°30′ N. Lat.), showing range by districts, etc., with critical notes on many species; phytogeography; bibliography.

Simmons, H. G. A revised list of the flowering plants and ferns of north western Greenland with some short notes about the affinities of the flora. Report of the 2d Norwegian Arctic expedition in the *Fram* 1898–1902. no. 16. 111 p. 1 fig. (map). Kristiania, 1909.

Geographical and botanical explorations, gazetteer of localities, phytogeography; annotated list of 122 species; bibliography. Covers area north of Cape York (76° N. Lat.).

Sørensen, Thorvald. The vascular plants of East Greenland from 70°00′ to 73°30′ N. Lat. Meddel. om Grønland v. 101, no. 3. 177 p. 7 fig. (incl. maps), 20 pl. 1933.

Climate, gazetteer; annotated list of 172 species of vascular plants; local distribution, bibliography.

Vaage, Jakob. Vascular plants from Eirik Raude's Land (East Greenland 71°30′–75°40′ lat. N.). Skrift. Svalbard og Ishavet (Oslo) no. 48. 87 p. (incl. map). 3 pl. 1932.

Botanical explorations, list of localities, general features of flora; annotated list of 183 vascular plants; tabular list of species, showing localities and collectors; bibliography.

UNITED STATES

GENERAL

Atwood, A. C. State and local floras . . . Annotations by S. F. Blake. Bul. Wild Flow. Preserv. Soc. no. 1. 16 p. 1930.

Briefly annotated list, geographically arranged, of principal regional and local floras of the United States; includes popular floras, but comparatively few strictly local lists.

Bergen, F. D Popular American plant names. Bot. Gaz. **17:** 363–380. 1892; **18:** 420–427. 1893; **19:** 429–444. 1894; **22:** 473–487. 1896; **26:** 247–252, 253–

258. 1898. (Reprinted from Jour. Amer. Folk-Lore v. 5–7, 9–10. 1892–1894, 1896–1898.)

Lists of vernacular names, with localities where used, systematically arranged; mostly from northeastern United States, some from Southern and Western States.

Brendel, Frederick. Historical sketch of the science of botany in North America from 1635 to 1840 [and to 1858]. Amer. Nat. **13**: 754–771. 1879; **14**: 25–38. 1880. (Reprinted, 31 p.—Additions . . . l. c. **14**: 443. 1880.)

Sketch of early collectors and botanists (in region north of Mexico).

Britton, N. L. A list of State and local floras of the United States and British America. Annals New York Acad. Sci. **5**: 237–300. 1890. (Reprinted as Contrib. Columb. Univ. no. 14.)

List of 791 floras and floristic papers, geographically arranged.

——— and **Shafer, J. A.** North American trees; being descriptions and illustrations of the trees growing independently of cultivation in North America, north of Mexico and the West Indies. x, 894 p. 781 fig. 26 cm. New York, 1908.

Full descriptions of all species growing without cultivation in the United States and Canada; illustration of each species.

Bruhin, T. A. Prodromus florae adventicae boreali-americanae. Vorläufer einer flora der in Nordamerika eingewanderten freiwachsenden oder im grossen cultivirten pflanzen. Verhandl. K. K. Zool.-Bot. Gesell. Wien **35**: 387–450. 1886.

Annotated systematic list of 623 escaped or cultivated introduced species, with localities and native habitat.

Carter, J. M. G. A synopsis of the medical botany of the United States. x, 176 p. 23.5 cm. St. Louis, 1888.

Bibliography; compiled systematic list, with medicinal properties and dosage; indices.

Gray, Asa. Synoptical flora of North America . . . Continued and ed. by **B. L. Robinson.** v. 1–2, pt. 1. 27 cm. New York, 1878–97 (v. 1, pt. 1, 1895–97; v. 1, pt. 2, 1884; v. 2, pt. 1, 1878).

Descriptive flora, with synoptical keys to genera and partial keys to species, of the flowering plants of the North American continent north of the Mexican border, including Greenland. Incomplete: v. 1, pt. 1, Ranunculaceae—Polygalaceae; v. 1, pt. 2, Caprifoliaceae—Compositae; v. 2, pt. 1, Goodeniaceae—Plantaginaceae. A supplement to v. 1, pt. 2, issued in 1886, containing additions (including new species), corrections, and index consists of p. 445–480, and is to be carefully distinguished from the same pages of the first edition. There are other irregularities in paging in reissues. Vol. 1, pt. 2 and v. 2, pt. 1, containing all the Gamopetalae, were reissued in 1886 by the Smithsonian Institution.

Heller, A. A. Catalogue of North American plants north of Mexico, exclusive of the lower cryptogams. 2d ed. 8 p., 9–236 numb. l., 237–252 p. 23 cm. [n. p.,] 1900. (1st ed., 1898.)

Systematic name-list of 16,673 species and varieties of vascular plants (14,534 in ed. 1). A third edition, issued 1912–14, was never completed (p. 13–276). It covers *Ophioglossum* to *Uva-ursi* (*Arctostaphylos*) part, and reaches no. 15319.

Henkel, Alice. Wild medicinal plants of the United States. Bul. Bur. Plant Ind. U. S. Dept. Agr. no. 89. 76 p. 1906.

Briefly annotated alphabetical list of botanical and vernacular names of medicinal plants of commercial value, compiled from trade lists of drug dealers.—See also her Weeds used in medicine. Farmers' Bul. U. S. Dept. Agr. no. 188. 1904; American root drugs. Bul. Bur. Plant Ind. U. S. Dept. Agr. no. 107. 1907; American medicinal barks. l. c. no. 139. 1909; American medicinal leaves and herbs, l. c. no. 219. 1911; American medicinal flowers, fruits, and seeds. Bul. U. S. Dept. Agr. no. 26. 1913.—The material in all these publications, with some omissions and additions, is summarized and condensed in **Sievers, A. F.** American medicinal plants of commercial importance. Misc. Pub. U. S. Dept. Agr. no. 77. 74 p. 128 fig. 1930.

Johnson, Laurence. A manual of the medical botany of North America. xi, 292 p. 160 fig., 9 col. pl. 23.5 cm. New York, 1884.

Systematically arranged account; descriptions, habitat, part used, constituents, preparation, medical properties, and uses. Covers the United States (mainly eastern) and Canada only.

Medsger, O. P. Edible wild plants. xv, 323 p. illus., plates. 21 cm. New York, 1939.

Descriptive account of edible plants of the United States and Canada, the plants grouped according to the parts used; synoptical annotated lists of the same, divided geographically.

Meisel, Max. A bibliography of American natural history; the pioneer century, 1769–1865. The rôle played by the scientific societies; scientific journals, natural history museums and botanic gardens; State geological and natural history surveys; Federal exploring expeditions in the rise and progress of American botany, geology, mineralogy, palaeontology and zoology. 3 v. 23.5 cm. New York, 1924–29.

Vol. 1: Annotated bibliography of publications relating to history, biography, and bibliography of American [i. e. the United States] natural history from 1769 to 1865; selected list of biographies and bibliographies of principal United States naturalists up to 1865. Vol. 2: History and bibliography of principal institutions, societies, surveys, and periodicals concerned with natural history, organized up to 1844. Vol. 3: Same continued to 1865; annotated bibliography of books, pamphlets, and miscellaneous articles on natural history, 1590–1865, chronologically arranged; chronological table of publications of the different institutions; indices of authors and institutions; appendices.

Millspaugh, C. F. American medicinal plants; an illustrated and descriptive guide to the American plants used as homoeopathic remedies . . . 2 v. 180 col. pl. 29 cm. New York, 1884–87. (Republished as: Medicinal plants . . . 1892.)

Treats of 180 plants, native (128 species), naturalized, or cultivated, each illustrated by a colored plate showing floral details; descriptions, history and habitat, part used, preparation, chemical constituents, physiological action; bibliographies (general and by species).

Muenscher, W. C. Poisonous plants of the United States. xvii, 266 p. 75 fig. 22 cm. New York, 1939.

General considerations, lists of special groups of poisonous plants, etc.; systematic list of over 400 vascular plants, with description, distribution, poisonous properties, symptoms, treatment, references; bibliography.

Pammel, L. H. A manual of poisonous plants chiefly of eastern North America, with brief notes on economic and medical plants. 2 v. (viii, 977 p.). 458 fig., plates. 23.5 cm. Cedar Rapids, 1910–11. (Also issued as 1 vol., 1911.)

General consideration of plant poisons; annotated systematic list of the more important poisonous plants of the United States and Canada, with bibliography; chemistry of alkaloids; annotated systematic list of poisonous plants of the United States and Canada (including bacteria and cellular cryptogams); tabular systematic catalog of poisonous plants of world, showing properties and locality; bibliography (by H. S. Kellogg) of 1,097 titles.

Sargent, C. S. Manual of the trees of North America (exclusive of Mexico). 2d ed. xxvi, 910 p. 783 fig., map. 22 cm. Boston, New York, 1922. (1st ed. 1905.)—Corrections and emendations of the second edition . . . Jour. Arnold Arboretum **7**: 1–21. 1926.

Descriptive flora, with keys. Includes 717 species with additional varieties; figure of each species; glossary; map showing tree regions.

——— The silva of North America: a description of the trees which grow naturally in North America exclusive of Mexico. 14 v. 740 pl. 37 cm. Boston, 1891–1902.

Systematic treatment of trees of North America north of Mexico; full synonymy, descriptions, range, wood, history, cultivation, with frequent biographical footnotes. The beautiful plates are by C. E. Faxon.

Saunders, C. F. Useful wild plants of the United States and Canada. 3d and rev. ed. 275 p. 20.5 cm. New York, 1934. (1st ed., 1920.)

Popular treatment, the plants classified by uses; classified list, geographically arranged.

Sudworth, G. B. Check list of the forest trees of the United States, their names and ranges. Misc. Cir. U. S. Dept. Agr. no. 92. 295 p. [1927.]

Systematic list of native and escaped species and varieties of trees (also cultivated varieties), with detailed range and vernacular names (with localities where used). Revision of his earlier publication of same title. Bul. Div. Forestry U. S. Dept. Agr. no. 17. 1898.

Van Dersal, W. R. Native woody plants of the United States, their erosion-control and wildlife values. Misc. Pub. U. S. Dept. Agr. no. 303. 362 p. 44 pl., 3 maps. 1938.

Map of zones of plant growth, with discussion, etc.; briefly annotated alphabetical list of woody plants, with vernacular names, notes on viability, value for wild and domestic animals; bibliography.

Watson, Sereno. Bibliographical index to North American botany; or citations of authorities for all the recorded indigenous and naturalized species of the flora of North America, with a chronological arrangement of the synonymy. Pt. I. Polypetalae. Smithsn. Misc. Collect. no. 258. vi, 476 p. 1878.

A laborious compilation of synonymy, never completed (Ranunculaceae—Cornaceae). Covers North American continent north of Mexico, and Greenland.

Yanovsky, Elias. Food plants of the North American Indians. Misc. Pub. U. S. Dept. Agr. no. 237. 83 p. 1936.

Systematic list of 1,112 species, with brief annotations and references; bibliography. Intended to summarize all previous publications on the subject. (United States and Canada.)

Youngken, H. W. The drugs of the North American Indian. Amer. Jour. Pharm. **96**: 485–502. 1924; **97**: 158–185, 257–271. 1925.

Annotated list of 75 plant drugs (without botanical names), also animal and mineral drugs; annotated list of plant drugs alphabeted by scientific names; bibliography. (United States, principally.)

NORTHEASTERN

Britton, N. L. Manual of the flora of the northern states and Canada. 3d ed., rev. and enl. xxiv, 1,112 p. 20 cm. New York, 1907. (1st ed., 1901; 2d ed., 1905.)

Descriptive flora of vascular plants, with keys, based on the text of Britton and Brown's Illustrated Flora, and covering the same area.

——— and **Brown, Addison.** An illustrated flora of the northern United States, Canada and the British possessions, from Newfoundland to the parallel of the southern boundary of Virginia, and from the Atlantic Ocean westward to the 102d meridian. 2d ed., rev. and enl. 3 v. 4,666 fig. 27.5 cm. New York, 1913. (1st ed., 1896–98.)

Descriptive flora of vascular plants, with keys and partial synonymy; figure of each species (4,666 species); extensive index of vernacular names.

Chamberlain, L. S. Plants used by the Indians of eastern North America. Amer. Nat. **35**: 1–10. 1901.

Lists of useful plants of the tribes belonging to the Algonquian and Iroquoian families, with uses and references; bibliography.

Clute, W. N. A dictionary of American plant names. [2d ed.] 248 p. 20.5 cm. Joliet, Ill., 1923 [193?]. (1st ed., 1923.)

List of wild plants of northeastern North America (Gray's Manual range), with vernacular names, the families systematically arranged, the genera and species alphabetically; index. Cover title: American plant names. 2d ed.

Day, M. A. The local floras of New England. Rhodora **1**: 111–120, 138–142, 158, 174–178, 194–196, 208–211. 1899.—Addenda. l. c. **2**: 73–74. 1900. (Reprinted. 28 p. 1899.)

Briefly annotated list of 258 State and local floras.

Gray, Asa. Gray's new manual of botany (7th ed. illustrated). A handbook of the flowering plants and ferns of the central and northeastern United States and adjacent Canada, rearranged and extensively revised by **B. L. Robinson** and **M. L. Fernald.** 926 p. 1,036 fig. 21.5 cm. New York, 1908. (1st ed., 1848, 2d ed., 1856, 3d ed., 1859, 4th ed., 1863, 5th ed., 1867, 6th ed., rev. and extended westward to the 100th meridian by Sereno Watson and J. M. Coulter [1890], 2d issue [with additions and corrections. 1891].)

Descriptive flora, with keys, of the vascular plants (4,079 species and 806 varieties and forms, of which 3,413 and 766, respectively, are native); glossary. Covers area from Nova Scotia, Prince Edward Island, and New Brunswick and the greater part of Quebec and Ontario west to Minnesota and south to eastern Kansas, Missouri, Kentucky, and Virginia. Earlier editions had title, Manual of the botany of the northern United States . . . —See also **Robinson, B. L.**, and Fernald, M. L. Emendations of the seventh edition of Gray's Manual. Rhodora 11:33–61. 1909.

Hough, R. B. Handbook of the trees of the Northern States and Canada east of the Rocky Mountains. x, 470 p. 498 fig. 24.5 cm. Lowville, N. Y., 1907.

Descriptive account, with excellent photographs of each species showing leaves, fruit, buds, trunk, and usually cross section of wood, and small maps showing range; wood, uses; systematic synopsis with keys.

Klincksieck, Paul. Les plantes d'Europe adventices ou naturalisées aux États-Unis d'Amérique, constatées à deux intervalles: 1832 et 1896. Bul. Soc. Bot. France 54 (Sess. Extr.): xxx–xlii. 1908.

List of 501 species extracted from Britton and Brown's Illustrated flora, with indication of those (116 species) recorded by Schweinitz in 1836.

New England Botanical Club. Committee on floral areas. Preliminary lists of New England plants I–XXXIII. *In* Rhodora v. 1–11, 18–23, 27–29, 31, 38. 1899–1909, 1916–21, 1925–27, 1929, 1936.

Tabular lists of different families, showing ranges in the New England States, with annotations on rarer species. Not published in systematic order; unfinished.

Toussaint, Anatole, *abbé.* Europe et Amérique (nord-est). Flores comparées comprenant tous les genres européens et américains, les espèces communes aux deux contrées, naturalisées et cultivées. Bul. Soc. Amis Sci. Nat. Rouen **45**: 109–441. 1910: **46**: 93–414. 1911. (Reprinted 650 p. 1912.)

Systematic list of genera of vascular plants in Europe and the eastern United States and Canada, with number of species in each region, mention of species common to the two areas (with range), cultivated species, etc.; statistics.

SOUTHEASTERN

Chapman, A. W. Flora of the southern United States: containing an abridged description of the flowering plants and ferns of Tennessee, North and South Carolina, Georgia, Alabama, Mississippi, and Florida . . . 3d ed. xxxix, 655 p. 23 cm. Cambridge, Mass., 1897. (1st ed., 1860; 2d ed., 1883.)

Descriptive flora, now nearly obsolete, lacking the numerous discoveries of the last 40 years. Reissue of ed. 2 [1892] contains 2 supplements, paged continuously (N. Y. Bot. Garden).

Coker, W. C., and Totten, H. R. Trees of the southeastern states, including Virginia, North Carolina, South Carolina, Tennessee, Georgia, and northern Florida. 2d ed. vii, 417 p. illus., plates. 20.5 cm. Chapel Hill, 1937. (1st ed., 1934.)

Briefly descriptive flora, with keys, range, variations; bibliography, glossary.

Porcher, F. P. Resources of the southern fields and forests, medical, economical, and agricultural; being also a medical botany of the Southern States; with practical information on the useful properties of the trees, plants and shrubs. New ed., rev. and largely augmented. xv, 733 p. 23.5 cm. Charleston, S. C., 1869. (1st ed., 1863.)

Annotated, systematically arranged list of useful plants of the Confederate States, wild and cultivated.

Small, J. K. Flora of the southeastern United States: being descriptions of the seed-plants, ferns, and fern-allies growing naturally in North Carolina, South Carolina, Georgia, Florida, Tennessee, Alabama, Mississippi, Arkansas, Louisiana, and in Oklahoma and Texas east of the one hundredth meridian. 2d ed. xii, 1,394 p. 25 cm. New York, 1913. (1st ed., 1903.)

Descriptive flora, with keys.

——— Manual of the southeastern flora; being descriptions of the seed plants growing naturally in Florida, Alabama, Mississippi, eastern Louisiana, Tennessee, North Carolina, South Carolina and Georgia. xxii, 1,554 p. illus. 23.5 cm. New York, 1933.

Physiographic provinces, etc.; descriptive flora of 5,557 spermatophytes; text figure of floral characters for each genus. The Pteridophyta of the same region are described in his Ferns of the Southeastern States . . . 517 p. illus. (incl. map). 1938.

WESTERN

Abrams, LeRoy. An illustrated flora of the Pacific States, Washington, Oregon, and California. v. 1 (xi, 557 p. 1,299 fig.). 27.5 cm. Stanford University, 1923.

Descriptive flora of vascular plants with keys, partial synonymy, and figure of each species. To be completed in 4 volumes; v. 1, Filicales—Aristolochiaceae.

Coulter, J. M. New manual of botany of the central Rocky Mountains (vascular plants) . . . rev. by **Aven Nelson.** 646 p. 22 cm. New York, [1909]. (1st ed., 1885.)

Descriptive flora of vascular plants, with keys, covering Colorado, Wyoming, Black Hills of South Dakota, northern half of New Mexico, adjacent Arizona, eastern Utah, most of Montana, southern Idaho; glossary. Includes 2,733 accepted species.

Frye, T. C., and **Rigg, G. B.** Northwest flora. 453 p. 25.5 cm. Seattle, [1912].

Keys to the flowering plants of Oregon, Washington, Idaho, and southwest British Columbia, with descriptions of genera and higher groups. Remarkable for the ingenuous ingenuity of its explanations of generic names.

Howell, Thomas. A flora of northwest America. Containing brief descriptions of all the known indigenous and naturalized plants growing without cultivation north of California, west of Utah, and south of British Columbia. Vol. 1. Phanerogamae. 792, [24] p. 24 cm. Portland, Oreg., 1897–[1903].

Descriptive flora of Oregon, Washington, and Idaho, with partial keys. No more published. Dates of issue: p. 1–112, Mar. 15, 1897; 113–274, Apr. 1, 1898; 275–386, Aug. 21, 1900; 387–474, Mar. 21, 1901; 475–562, Nov. 20, 1901; 563–666, July 20, 1902; 667–792, Aug. 10, 1903. The additional 24 pages, not found in all copies, are index.

Kirkwood, J. E. Northern Rocky Mountain trees and shrubs. xvii, 340 p. 87 fig., 35 pl. 23 cm. Stanford University, 1930.

Annotated descriptive flora (248 species), with keys. Includes region from Yellowstone Park, Wyoming, "north and northwest through Montana and Idaho and the Canadian Rockies, and covers the various ranges and the immediately adjacent plains."

McMinn, H. E., and **Maino, Evelyn.** An illustrated manual of Pacific coast trees . . . xii, 409 p. 415 fig., col. pl. 20 cm. Berkeley, 1935.

Descriptive flora of native trees growing from British Columbia to California, inclusive, and about 400 introduced species and varieties; bibliography; list of trees recommended for planting (by H. W. Shepherd).

Palmer, Edward. Food products of the North American Indians. Rpt. U. S. Commr. Agr. **1870**: 404–428. pl. 19–28. 1871.—Plants used by the Indians of the United States. Amer. Nat. **12**: 593–606, 645–655. 1878. (Also in Amer. Jour. Pharm. **50**: 539–548, 586–592. 1878.)

Annotated list, the plants classified according to parts utilized. Refers mainly to western United States. The second part includes edible and some other useful plants.

Rydberg, P. A. Flora of the prairies and plains of central North America. vii, 969 p. 600 fig. 23.5 cm. New York, 1932.

Descriptive flora of 3,988 vascular plants, with keys; covers Kansas, Nebraska, Iowa, Minnesota, North Dakota, South Dakota, southern Manitoba, and southeastern Saskatchewan.

——— Flora of the Rocky Mountains and adjacent plains, Colorado, Utah, Wyoming, Idaho, Montana, Saskatchewan, Alberta, and neighboring parts of Nebraska, South Dakota, North Dakota, and British Columbia. 2d ed. xii, 1,144 p. 23 cm. New York, 1922. (1st ed., 1917.)

Descriptive flora of vascular plants, with keys; brief sketch of life zones; glossary. Includes 6,029 species. The only essential change in the 2d edition is the addition of p. 1111–1144 containing "Additions and corrections."

Stuhr, E. T. Manual of Pacific coast drug plants . . . 189 p. map. 23.5 cm. [n. p.,] 1933.

Annotated list of 1,160 wild and cultivated plants, the families arranged alphabetically; bibliography, glossary.

Sudworth, G. B. Forest trees of the Pacific slope. 441 p. 207 fig. (partly on pl.), 2 maps. 23 cm. Washington, 1908.

Description and figure of each species, with detailed account of habit and bark, size and age, range and occurrence. Publication of U. S. Department of Agriculture, Forest Service.

ALABAMA

General

Cary, C. A., Miller, E. R., and Johnstone, G. R. Poisonous plants of Alabama. 42 p. 40 fig. 23 cm. [Auburn, 1924?] (Also as Alabama Polytech. Inst. Ext. Serv. Cir. 71.)

Annotated list.

Harper, R. M. Economic botany of Alabama, pt. 2. Catalogue of the trees, shrubs and vines of Alabama, with their economic properties and local distribution. 357 p. (incl. 66 fig., 23 maps). 23 cm. University, Ala., 1928. (Geol. Survey Alabama, Monogr. 9.)

Annotated bibliography; climate, soils, etc.; annotated list of woody plants (including cultivated ones) with uses. Pt. 1 is his Geographical report on forests (Monograph 8, 1913).

Mohr, C. T. The medicinal plants of Alabama. Systematic list of the medicinal plants, occurring within the limits of the state, with notes on their distribution and proper time of collecting the parts used. 17 p. 21.5 cm. Mobile, [1890].

Annotated systematic list, without mention of uses.

——— Plant life of Alabama. Contrib. U. S. Natl. Herb. v. 6. 921 p. 13 pl. (incl. map). 1901. (Reprinted as "Alabama edition," publication Alabama Geological Survey.)

Annotated systematic list of the plants (Myxomycetes—Carduaceae); physiography, plant associations; tables of geographic distribution. Total number of species and varieties 4,473, of which 2,476 are spermatophytes.

Local

Bates, F. A. Indigenous botany of Perry County. Proc. Med. Assoc. State Alabama **6**: 58–68. 1853.

Annotated list of medicinal plants.

Cocks, R. S. Catalogue of trees growing naturally in the vicinity of Sardis, Dallas County, Alabama. Jour. Arnold Arboretum **6**: 189–195. 1925.

Annotated list of about 130 species and varieties, nearly all growing on about 300 acres of land along the Alabama River.

Denny, Andrew. Report on the indigenous botany of Clarke County. Proc. Med. Assoc. State Alabama **5**: 41–69. 1852; **6**: 30–33. 1853.

Annotated list of medicinal plants, with properties and uses.

Earle, F. S. The flora of the metamorphic region of Alabama. Bul. Alabama Agr. Expt. Sta. 119. p. 41–120. 1902.

Ecology; annotated list of 1,146 species and varieties of vascular plants. The area covered extends from Lee County north to Cleburne County, then southwest to Chilton County, then southeast to Lee County, and includes the southernmost extension of the Appalachian mountain system.

ARIZONA

See also New Mexico (Castetter; Castetter and Opler).

General

Ewan, Joseph. Bibliography of the botany of Arizona. Amer. Midl. Nat. **17**: 430–454. 1936.

Annotated chronological list; geographic and author indices.

Local

Britton, N. L. A list of plants collected at Fort Verde and vicinity and in the Mogollon and San Francisco Mountains, Arizona, 1884–1888, by Dr. E. A. Mearns, U. S. A. Trans. New York Acad. Sci. **8**: 61–76. 1889. (Reprinted as Contrib. Herb. Columbia Col. no. 9.)

List of plants, with localities and collecting numbers.

Britton, N. L., and Kearney, T. H. An enumeration of the plants collected by Dr. Timothy E. Wilcox, U. S. A., and others in southeastern Arizona during the years 1892–1894. Trans. New York Acad. Sci. **14**: 21–44. 1894.

List of species, mostly from Fort Huachuca and Fort Apache.

Castetter, E. F., and Underhill, R. M. The ethnobiology of the Papago Indians. Univ. New Mexico Bul. 275 (Biol. Ser. v. 4, no. 3). 84 p. 1935. (Ethnobiological studies in the American southwest II.)

General features of Papago life; account of useful plants and animals, classified by uses, with lists of species and vernacular names; bibliography. The region covered is a narrow strip along the Mexican border between the Gila River in southern Arizona and the Altar in northern Sonora.

Davidson, Anstruther. Flora of Clifton district, Arizona. Bul. So. Calif. Acad. Sci. **3**: 110–111. 1904; **4**: 18–19, 35–36. 1905; **5**: 67–70. 1906; **6**: 34–36. 1907.

Annotated list of dicotyledons collected by the author. (Greenlee County.)

Hough, Walter. The Hopi in relation to their plant environment. Amer. Anthropol. **10**: 33–44. 1897.

Includes list of 144 useful plants, with Hopi and botanical names and uses, grouped by uses. (Northeastern Arizona.)

Matthews, Washington. Navajo names for plants. Amer. Nat. **20**: 767–777. 1886.

Systematically arranged list of vernacular names, with botanical identifications, translations, and notes on uses. (Northeastern Arizona.)

Nelson, Aven. Flora of the Navajo Reservation. Amer. Bot. **26**: 48–56, 87–89. 1920; **28**: 20–25. 1922.

Annotated list of 152 vascular plants collected by W. N. Clute in 1919; not in systematic order. (Northeastern Arizona.)

Patraw, P. M. Check-list of plants of Grand Canyon National Park. Nat. Hist. Bul. Grand Canyon Nat. Hist. Assoc. no. 6. ix, 75 p. map. 1936.

Annotated list of cellular and vascular plants. (Processed publication.)—See also **Dodge, N. N.** Trees of Grand Canyon National Park. l. c. no. 3. 69 p. illus. 1936. (Processed publication.) (Coconino County.)

Read, A. D. The flora of the Williams division of the Tusayan National Forest, Arizona. Plant World **18**: 112–123. 2 fig. 1915.

Plant zones; unannotated list of vascular plants. (Coconino County.)

Reagan, A. B. Plants used by the White Mountain Apache Indians of Arizona. Wisconsin Archeol. **8**: 143–161. 1929.

Plant zones, agriculture, running account of useful plants; annotated alphabetical list of scientific names; no vernacular names given. (Apache County.)

Russell, Frank. The Pima Indians. Ann. Rpt. Bur. Amer. Ethnol. **26**: 3–389. 100 fig., 47 pl. 1908.

Includes (p. 69–80) annotated list of wild edible and medicinal plants, with vernacular names. These Indians inhabit the Gila River and Salt River Valleys, in Pinal and Maricopa Counties.

Thornber, J. J. Vegetation groups of the Desert Laboratory domain. *In* **Spalding, V. M.** Distribution and movements of desert plants. p. 103–112. Washington, D. C., 1909. (Carnegie Institution of Washington Pub. no. 113.)

Includes list of 449 species (442 vascular) from an area in the vicinity of Tucson (Pima County).

Whiting, A. F. Ethnobotany of the Hopi. Bul. Mus. North. Arizona 15. viii, 120 p. 1939.

Includes (p. 61–100) annotated systematic list of useful plants, with vernacular names and uses; bibliography. (Northeastern Arizona.)

ARKANSAS

General

Branner, J. C., and Coville, F. V. A list of the plants of Arkansas. Ann. Rpt. Geol. Survey Arkansas **1888** (4): 155–242. 1891.

Botanical explorations, bibliography; list of 1,610 vascular plants (plus a very few cellular) with localities for scarcer species.—See also **Buchholz, J. T., and Palmer, E. J.** Supplement to the catalogue of Arkansas plants. Trans. Acad. Sci. St. Louis **25**: 91–155. pl. 6–13. 1926. (Includes bibliography.)

Harvey, F. L. The arboreal flora of Arkansas. Amer. Jour. Forestry **1:** 413–424, 451–458. 1883. (Reprinted 20 p.)

Geology, general features of forest distribution; annotated systematic list of trees, with vernacular names and local distribution.—A more popular recent work, with description and figure of most of the species, is **Turner, L. M.** Trees of Arkansas. Ext. Cir. Col. Agr. Arkansas no. 180. 112 p. illus. 1937. (Revision of Buchholz, J. T., and Mattoon, W. R. Common forest trees of Arkansas . . . 1924.)

Local

Evans, J. P. Medicinal plants of the Cherokees. Proc. Amer. Pharm. Assoc. **8:** 391–397. 1859.

Annotated list, classified by uses. Region referred to is in northern Arkansas and Oklahoma.

Palmer, E. J. The ligneous flora of Hot Springs National Park and vicinity. Jour. Arnold Arboretum **7:** 104–135. 1926.

History, botanical explorations, geology, physiography, plant associations; annotated list of woody plants. (Garland County.)

——— The ligneous flora of Rich Mountain, Arkansas and Oklahoma. Jour. Arnold Arboretum **5:** 108–134. 1924.

Geology, general features of flora; annotated list of woody plants collected by author. (Le Flore County, Oklahoma, and Polk County, Arkansas.)

——— The Red River forest at Fulton, Arkansas. Jour. Arnold Arboretum **4:** 8–33. 1923.

General features of flora; annotated list of woody plants. Additions are given by author, l. c. **7:** 140–141. 1926. (Hempstead County.)

CALIFORNIA

General

Brewer, W. H., Watson, Sereno, and Gray, Asa. Botany [of California]. 2 v. 27 cm. Cambridge, Mass., 1876–80.

The classical work on the flora of California, although long out of date. Full descriptions, pertinent synonymy, and localities; general range stated; glossary; historical sketch of botanical collections (**2:** 553–559). Gamopetalae by A. Gray, various families by other authors.

Hall, H. M., and Grinnell, Joseph. Life-zone indicators in California. Proc. Calif. Acad. Sci. IV, **9:** 37–67. 1919.

Discussion of life zone concept; lists of characteristic plants and animals; bibliography.

Jepson, W. L. A flora of California. v. 1, pt. 1–7; v. 2, pt. 1–3; v. 3, pt. 1. illus., plates. 27 cm. San Francisco, Berkeley, 1909–39.

Descriptive flora of spermatophytes, including numerous varieties and forms, with localities, exsiccatae, and references. The 11 parts so far issued cover Pinaceae-Convolvulaceae. The part numbers on the cover titles indicate the order of appearance of the text and do not correspond with the consecutive pagination.

——— A manual of the flowering plants of California. 1,238 p. 1,023 fig., map. 22.5 cm. Berkeley, c1923–25.

Keys to and descriptions of the vascular plants, including Pteridophyta; local range and often extralimital range given; short discussion of plant geography and life zones. Enumerates 4,019 species, of which 3,727 are native and 1,416 endemic. Treatment of various groups contributed by specialists.

——— The silva of California. 480 p. 11 fig., 85 pl., 3 maps. 32.5 cm. Berkeley, 1910. (Memoirs of the University of California, v. 2.)

Descriptive flora with keys, synonymy and references; details of distribution, vernacular names and uses, forest provinces, life zones, dendrological characteristics, bibliography.

——— The trees of California. 2d ed. 240 p. 124 fig. 20.5 cm. Berkeley, 1923. (1st ed., [1909].)

Descriptive flora with keys; illustration of nearly every species; uses, vernacular names, forest provinces; relation to fires, endemic species, etc.

McMinn, H. E. An illustrated manual of California shrubs . . . xi, 689 p. 775 fig., col. pl. 24.5 cm. San Francisco, 1939.

Descriptive flora including about 800 species and 200 varieties of native shrubs, undershrubs, and woody vines; bibliography, horticultural uses.

Schneider, Albert. The medicinal plants of the California Indians. Merck's Report **15**: 63–66, 95–96, 127–128. 1906.

Annotated list, the species alphabetically arranged by botanical names; bibliography.

——— Pharmacal plants and their culture. Bul. State Bd. Forestry Calif. no. 2. 175 p. 1912.

Includes bibliography and chapter, "The native and introduced medicinal and poisonous plants of California, with index to common names," containing annotated list of 869 plants, arranged alphabetically.

Local

Abrams, LeRoy. Flora of Los Angeles and vicinity. x, 432 p. 18 cm. Stanford University, 1917. (Earlier eds. 1904, 1911.)

Descriptive flora with keys of the Spermatophyta of the coast area of Los Angeles and Orange Counties.

——— A phytogeographic and taxonomic study of the southern California trees and shrubs. Bul. N. Y. Bot. Gard. **6**: 300–485. pl. A–J. 1910.

Annotated catalog with keys to genera and species; type locality, local range, exsiccatae; physiography, climatology, life zones. Covers area south of 35° 45′ N. Lat.—See also **Parish, S. B.** Additions and emendations. Muhlenbergia **7**: 73–81. fig. 12. 1911.

Applegate, E. I. Plants of the Lava Beds National Monument, California. Amer. Midl. Nat. **19**: 334–368. 6 fig. 1938.

Geology, climate, geography, relationships and general features of flora; annotated list of about 200 plants. The Lava Beds National Monument lies in northeastern Siskiyou and northwestern Modoc Counties.

Barrett, S. A., and **Gifford, E. W.** Miwok material culture. Bul. Pub. Mus. Milwaukee **2**: 117–376 (incl. 2 maps). fig. 23–87, pl. 28–76. 1933.

Includes (p. 140–178) annotated account of plant food, medicines, etc. The Indians lived in the Sierra Nevada and in part of the Sacramento-San Joaquin Valley.

Barrows, D. P. The ethnobotany of the Coahuilla Indians of southern California. 82 p. 24 cm. Chicago, 1900.

Includes account of useful plants. The Cahuilla, Coahuilla, or Coahuila Indians live in western Riverside and southwestern San Bernardino Counties.

Blankinship, J. W. On the natural history of the Farallon Islands. Zoe **3**: 144–165. 1892.

Geology; list of vascular and cellular plants including those in gardens.

Brandegee, Mrs. M. K. Curran. Catalogue of the flowering plants and ferns growing spontaneously in the city of San Francisco. Zoe **2**: 334–386. pl. 17 (map). 1892. (Reprinted.)—Additions . . . l. c. **3**: 49–50. 1892.

General features of flora, list of plants flowering in January; annotated list of 578 vascular plants and 42 mosses.

——— Flora of Bouldin Island. Zoe **4**: 211–218. 1893.

General features of flora; partly annotated list of vascular plants. Bouldin Island is a small island at the junction of the San Joaquin and the Mokelumne Rivers (San Joaquin County.)

——— List of the plants described in California, principally in the Proc. of the Cal. Acad. of Sciences, by Dr. Albert Kellogg, Dr. H. H. Behr, and Mr. H. N. Bolander; with an attempt at their identification. Bul. Calif. Acad. Sci. **1**: 128–151. 1885.

The scope of the paper is indicated by its title.

Brandegee, T. S. Flora of the Californian islands. Zoe **1**: 129–148. 1890.

Tabular list giving distribution of 512 vascular plants by islands, with footnote annotations; brief bibliography. Includes San Clemente, Santa Catalina, Santa Cruz, Santa Rosa, and San Miguel Islands.—For additions see **Clokey, I. W.** Notes on the flora of Santa Cruz Islands [Island]. Bul. So. Calif. Acad. Sci.

30: 60–61. 1931.—Also **Hoffman, R.** Notes on the flora of the Channel Islands off Santa Barbara, California. Bul. So. Calif. Acad. Sci. 31: 46–60, 101–120. 1932.

Brandegee, T. S. Flora of the Providence Mountains. Zoe 5: 147–153. 1903.
Physiography, etc.; partial list of spermatophytes collected by the author. (In the Mohave Desert, in San Bernardino County.)

Chesnut, V. K. Plants used by the Indians of Mendocino County, California. Contrib. U. S. Natl. Herb. 7: 295–408. fig. 66–78, pl. 10–21. 1902.
Annotated systematic list of cellular and vascular plants used by various tribes of Indians; list of species classified by uses; index of Indian names.

Congdon, J. W. Mariposa County as a botanical district. Zoe 2: 234–236. 1891; 3: 25–43, 125–131. 1892.
Plant zones; partly annotated lists of foothill plants totaling 508 species, and list of 122 species of the coniferous belt.

Constance, Lincoln. Flora of Redwood Peak, Oakland Hills, California. Res. Studies State Col. Washington 3: 12–24. 1935.
Geology, associations, etc.; unannotated list of vascular plants. (Alameda County.)

Coville, F. V. Botany of the Death Valley expedition. Contrib. U. S. Natl. Herb. v. 4. 363 p. 21 pl., map. 1893.
Itinerary, plant distribution (general and local), characteristics and adaptations of desert flora; annotated systematic list of vascular and cellular plants (1,261 species and varieties); catalog of specimens collected, numerically arranged; annotated bibliography; index to map. (Inyo County.)

Davidson, Anstruther, and Moxley, G. L. Flora of southern California. 452 p. 19.5 cm. Los Angeles, 1923.
Keys to all groups down to and including species, with descriptions of families and genera; local range. Includes Santa Barbara, San Bernardino, Riverside, Imperial, San Diego, Orange, and Los Angeles Counties.

Eastwood, Alice. A flora of the south fork of Kings River from Millwood to the head waters of Bubbs Creek. Pub. Sierra Club no. 27. 96 p. 9 fig. 1902.
Botanical explorations; descriptive flora of vascular plants, with keys to families and genera. (In Fresno and Tulare Counties, in the southern Sierra Nevada.)

——— Notes on the plants of San Nicolas Island. *In* her Studies in the herbarium and field no. 2. Proc. Calif. Acad. Sci. III, Botany 1: 89–120. pl. 8. 1898.
Geology, etc.; annotated list of 80 vascular plants collected in 1897 by Mrs. Blanche Trask.—For additions see **Howell, J. T.** The vascular plants from San Nicolas Island . . . l. c. IV, 21: 277–284. 1935.

——— The plant inhabitants of Nob Hill, San Francisco. Erythea 6: 61–67. 1898.
General features of flora; list of 64 vascular plants, of which 55 are introduced. (San Francisco County.)

Fawcett, R. A. Flora of Riverside and vicinity. Western Riverside County, to the east edge of the Coachella Valley, and a contiguous portion of San Bernardino County to the north foot of the San Bernardino Mountains and keys for determining the names of native trees, shrubs, herbs, ferns, and most of the weed plants of cultivated fields (the grasses omitted). Occas. Papers Riverside Junior Col. v. 9, no. 1. 172 p. illus., map. 1939.
Unannotated flora in form of keys, the authorities for botanical names omitted; glossary.

Greene, E. L. Flora franciscana. An attempt to classify and describe the vascular plants of middle California. 480 p. 24.5 cm. San Francisco, 1891–97.
Incomplete (Ranunculaceae–Cynarocephalae [Compositae]). Pt. 4 (Cynarocephalae, p. 353–480) is rare, most of the stock having been destroyed by fire before distribution. Notable for its attribution of generic names to Dioscorides, Catullus, Lobelius, and the like.

——— Manual of the botany of the region of San Francisco Bay, being a systematic arrangement of the higher plants growing spontaneously in the counties of Marin, Sonoma, Napa, Solano, Contra Costa, Alameda, Santa Clara, San Mateo, and San Francisco. xiii, 342 p. 24.5 cm. San Francisco, 1894.

Descriptive flora. Omits sedges, grasses, ferns, and some minor groups.—See also **Burtt Davy, J.** Additions . . . Erythea 4: 90, 145, 152. 1896.

Greene, E. L. Notes on the botany of Santa Cruz Island. Bul. Calif. Acad. Sci. 2: 377–416. 1887.

Botanical explorations, etc.; annotated list of 321 vascular plants.—For additions, see **Brandegee, T. S.**, Proc. Calif. Acad. Sci. II, 1: 201–205. 1888.

——— The vegetation of the summit of Mount Diablo. Erythea 1: 166–179. 1893.

Topography, climate, botanical explorations; annotated list of 156 spermatophytes. (Contra Costa County.)

——— The vegetation of the summit of Mount Hamilton. Erythea 1: 77–97. 1893.

List of 212 vascular plants. (Santa Clara County.)

Hall, H. M. A botanical survey of San Jacinto Mountain. Univ. Calif. Pub. Bot. 1: 1–140. 14 pl. (incl. 2 maps). 1902.

Botanical collections; topography, life zones, relationships of flora; annotated list of spermatophytes collected in or above the yellow pine belt (excludes the foothill flora). (Riverside County.)

——— and **Hall, C. C.** A Yosemite flora; a descriptive account of the ferns and flowering plants, including trees, of the Yosemite National Park. vii, 282 p. 170 fig., 11 pl. 19 cm. San Francisco, 1912.

Popular descriptive flora of the vascular plants; the grasses, sedges, and rushes are omitted. (In Tuolumne, Mariposa, and Madera Counties.)

Jepson, W. L. Botany of the Marysville Buttes. Bul. Torrey Club 18: 317–327. 1891.

General features of flora; annotated list of vascular plants collected by the author in April 1891. The Marysville Buttes are an isolated group of mountains in the level plain between the upper Sacramento and Feather Rivers. (Sutter County.)

——— A flora of western middle California. 2d ed. 515 p. 20.5 cm. San Francisco, [1911]. (1st ed., 1901.)

Descriptive flora, with keys, of the Spermatophyta. Includes area from San Francisco Bay north to southern boundary of Mendocino, Lake, and Colusa Counties, east to the Sacramento and San Joaquin Rivers, south to the Pajaro River and Pacheco Pass (essentially the southern boundary of Santa Clara and Santa Cruz Counties).

——— Vegetation of the summit of Mt. St. Helena. Erythea 7: 105–113. pl. 1899.

Topography, general features of flora; annotated list of 36 species found above 2,400 feet elevation. (Napa County.)

Johnston, I. M. The flora of the pine belt of the San Antonio Mountains of southern California. Plant World 22: 71–90, 105–122. 2 fig. 1919.

Geography, collectors, life zones; annotated list of 315 native species and varieties of vascular plants. (On the Los Angeles-San Bernardino County boundary line.)

McClatchie, A. J. Flora of Pasadena and vicinity. *In* **Reid, H. A.** History of Pasadena . . . p. 605–649. illus. Pasadena, 1895. (Reprinted.)

Topography, etc.; annotated list of cellular and vascular plants. (Los Angeles County.)

Merriam, C. H. Results of a biological survey of Mount Shasta, northern California. North Amer. Fauna no. 16. 179 p. illus., 4 pl. 1899.

Includes (p. 135–169) an annotated list of plants collected on Mount Shasta. (Siskiyou County.)

Millspaugh, C. F., and **Nuttall, L. W.** Flora of Santa Catalina Island (California). Pub. Field Mus. Nat. Hist. 212, Bot. Ser. v. 5. 413 p. 14 pl., map. 1923.

Descriptive flora of vascular plants, with keys, synonymy, and citation of exsiccatae; enumeration of non-vascular plants, with occasional descriptions; description of island and of principal collecting stations; annotated list of collectors. Includes 882 species, of which 467 are vascular plants.

Munz, P. A. A manual of southern California botany. xxxix, 642 p. 310 fig. 22.5 cm. Claremont, 1935.

Geology, physiography, life zones, phytogeography, endemism; briefly descriptive flora of vascular plants; list of collectors for whom species have been named, glossaries. Covers Los Angeles, San Diego, Orange, Riverside, San Bernardino, and Ventura Counties, and part of Santa Barbara, Kern, and Inyo Counties.

Parish, S. B. A bibliography of the southern California flora. Bul. So. Calif. Acad. Sci. **8**: 71–75. 1909; **9**: 57–62. 1910; **19**: 24–29. 1920.
Unannotated list of papers, chronologically arranged under authors' names, referring to Los Angeles, San Diego, San Bernardino, Riverside, and Orange Counties. In the 1920 supplement the area is extended to include Santa Barbara, Ventura, Kern, and Inyo Counties.

——— An enumeration of the pteridophytes and spermatophytes of the San Bernardino Mountains, California. Plant World **20**: 163–178, 208–223, 245–259. 3 fig. 1917; **21**: 220–221. 1918.
Geography, climate, etc.; annotated list of vascular plants. (San Bernardino County.)

——— The immigrant plants of southern California. Bul. So. Calif. Acad. Sci. **19** (4): 3–30. 1920.
General history of introduction of weeds into California; bibliography; annotated list of 281 species and varieties of introduced plants.

——— Plant ecology and floristics of Salton Sink. *In* **MacDougal, D. T.** The Salton Sea. (Carnegie Institution of Washington Pub. no. 193.) p. 85–114. Washington, 1914. (List preprinted, 11 p., 1913, with title: Catalogue of plants collected in the Salton Sink.)
Includes (p. 104–114) annotated list of cellular and vascular plants. (At the junction of San Diego, Imperial, and Riverside Counties.)

Peirson, F. W. Plants of Rock Creek Lake basin, Inyo County, California. A check list. 16 p. 23 cm. Los Angeles, 1938.
Annotated list of vascular plants collected by the author at 10,500 feet elevation and higher.

Smiley, F. J. A report upon the boreal flora of the Sierra Nevada of California. Univ. Calif. Pub. Bot. v. 9. 423 p. 7 pl. 1921.
Keyed list of the vascular plants of the higher Sierra Nevada (mainly above 6,500 feet); type locality, general range, life zone, local exsiccatae, with numerous critical notes; geography, physiography, climatology, life zones, statistics of flora. Includes 582 species and varieties.

Sparkman, P. S. The culture of the Luiseño Indians. Univ. Calif. Pub. Amer. Archaeol. and Ethnol. **8**: 187–234. pl. 20. 1908.
Includes (p. 228–234) annotated systematic list of useful plants, with vernacular names. The paper relates to the region of Rincon, San Diego County.

Voegelin, E. W. Tübatulabal ethnography. Anthropol. Rec. **2**: 1–84. 16 fig. (incl. map.), 6 pl. 1938.
Includes account of useful plants (p. 14–21), with Indian vernacular names. (Processed publication.) (Upper part of the Kern River valley.)

Yates, L. G. Insular floras. Ann. Rpt. State Mineralogist Calif. **9**: 179–188. 1890. (Reprinted in his Channel Islands [1890], p. 11–20.)
Includes tabular list of vascular plants from San Miguel, Santa Rosa, Santa Cruz, and Anacapas (Anacapa) Islands of the Santa Barbara group. The flora of Anacapa Island has apparently not been listed elsewhere.—For additions see **Hoffmann, R.** Notes on the flora of the Channel Islands. Bul. So. Calif. Acad. Sci. **31**: 46–60, 101–120. 1932.

COLORADO

See also Utah (Graham).

General

Allison, E. M. Bibliography and history of Colorado botany. Univ. Colorado Studies **6**: 51–76. 1908.
Partly annotated list referring principally to Spermatophyta, arranged alphabetically by authors, their papers arranged chronologically; biographical notes on botanists who have worked in Colorado.

Cary, Merritt. Principal trees and shrubs of Colorado. In his Biological survey of Colorado. North Amer. Fauna **33**: 212–246. fig. 30–39 (incl. map). 1911.
Annotated list, with details of local distribution.

Durrell, L. W., and Newsom, I. E. Colorado's poisonous and injurious plants. Bul. Colorado Expt. Sta. 455. 71 p. 92 fig. 1930.
Annotated list, alphabetically arranged by vernacular names.

Ramaley, Francis. Colorado plant life. viii, 299 p. illus., 10 pl. (3 col.). 23.5 cm. Boulder, 1927. (University of Colorado semicentennial series, v. 2.)
General popular account of life zones, associations, etc.; no list of species; bibliography.

——— European plants growing without cultivation in Colorado. Annales Jard. Bot. Buitenz. Suppl. **3** (2) : 493–504. 1 fig. (map). 1910.
Topography, climate, sources of introduction; annotated list of 95 angiosperms; local distribution, etc.

Rydberg, P. A. Flora of Colorado. Bul. Colorado Agr. Expt. Sta. **100**. xxii, 448 p. 1906.
Keys to the orders, families, genera, and species of vascular plants (2,912 species) ; general range, altitudinal range, and Colorado localities given.

Local

Ashton, R. E. Plants of Rocky Mountain National Park. iv, 157 p. illus., col. pl., fold. tab. 23 cm. Washington, 1933.
A popular flora, with partial keys, but including all the species (about 700) of vascular plants known to occur in the Park; bibliography. Published by the National Park Service. (North central Colorado, northwest of Boulder, in Larimer, Grand, and Boulder Counties.)

Cockerell, T. D. A. The alpine flora of Colorado. Amer. Nat. **40**: 861–873. 1906.
Lists of species classified by geographical and altitudinal range, based on Rydberg's Flora of Colorado.

Daniels, F. P. The flora of Boulder, Colorado, and vicinity. Univ. Missouri Studies Sci. Ser. v. 2, no. 2. 311 p. 1911.
Physiography, plant zones, bibliographies; annotated list of 1,225 vascular plants.—See also **Cockerell, T. D. A.** Notes on the flora of Boulder County, Colorado. Torreya **8**: 177–183. 1918. (Covers Boulder County.)

Eastwood, Alice. A popular flora of Denver, Colorado. 57 p. 23 cm. San Francisco, [1893].
Annotated list of 487 vascular plants. (Denver County.)

Holm, Theodor. The vegetation of the alpine region of the Rocky Mountains in Colorado. Mem. Natl. Acad. Sci. v. 19, no. 3. 45 p. 7 pl. 1923.
Annotated list of about 170 plants collected by the author; phytogeography, morphology.

Ramaley, Francis. Botany of the San Luis Valley in Colorado, I. Univ. Colorado Studies **17**: 27–44. 4 fig. (incl. maps), 2 pl. 1929.
Geography, soils, climate, history, etc.; unannotated list of vascular plants; vegetation of sand dunes and sand hills; bibliography. (Saguache County.)

——— Vascular plants of the Tolland region in Colorado. Univ. Colorado Studies **12**: 27–51. 1917.
List of 722 vascular plants, annotated by symbols. Tolland is in Boulder Park, in Gilpin County, at 8,889 feet elevation.

——— Woody plants of Boulder County. His The silva of Colorado III. Univ. Colorado Studies **5**: 47–63. 2 fig. (incl. map). 1907.
Botanical explorations, economic plants, geography, plant zones; annotated list of 112 woody plants.

Schmoll, H. M. Vegetation of the Chimney Rock area, Pagosa-Piedra region, Colorado . . . iii, 58 p. 5 fig. (incl. map). Chicago, 1935. (Dissertation — Chicago.)
Mainly ecological; includes tabular list of cellular and vascular plants, with indication of their plant association; bibliography. The Chimney Rock area is in Archuleta County. (Processed publication.)

Soth, Blanche. List of plants collected above timber line on Pikes Peak, with altitudinal extensions and notes. Bul. Torrey Club **38**: 237–242. 1911.
Alphabetical list of 157 spermatophytes, with list of altitudes for some of these; notes on various species. (On boundary line between Teller and El Paso Counties.)

CONNECTICUT

See also New York (Taylor, Flora of the vicinity of New York).

General

Graves, C. B., Eames, E. H., Bissell, C. H., Andrews, Luman, Harger, E. B., and Weatherby, C. A. Catalogue of the flowering plants and ferns of Connecticut growing without cultivation. Bul. Conn. Geol. and Nat. Hist. Survey 14. 569 p. 1910.—First supplement. Additions to the flora of Connecticut. l. c. Bul. 48. 94 p. 1930 [1931].

Annotated systematic list of 2,228 species, varieties, and named forms (1,942 species, of which 1,481 are native), plus 169 fugitives. Supplement includes "Additions" published by Harger and others in Rhodora v. 19 and 24. Harger main author of supplement; R. W. Woodward and G. H. Bartlett replaced Bissell and Andrews as joint authors of supplement.—See also **Eames, E. H.** Further additions to the Connecticut flora. Rhodora **33**: 167–170. 1931.

Local

Andrews, Luman. A list of the flowering plants and higher cryptogams growing upon the summit of Meriden Mountain, Conn. [15 p.] 20.5 cm. Southington, Conn., 1900. (Reprinted as Conn. School Document 195 (1901, no. 3). 1901.)

Soil, general features of flora; annotated list of 287 vascular plants. (In town of Meriden, New Haven County.)

Berzelius Society. A catalogue of the flowering plants and higher cryptogams growing without cultivation within thirty miles of Yale College. vii, 71 p. map. 23 cm. New Haven, 1878.

Notes on rarer species, etc.; list of vascular plants and bryophytes, unannotated except for localities. Includes over 100 species from the northern shore of Long Island, not found in Connecticut.—See also **Eaton, D. C.** Plants new to the Connecticut flora. Bul. Torrey Club **10**: 102. 1883.—**Miller, E. S.** A few additions to the Berzelius catalogue. l. c. **10**: 120–121. 1883.

Bissell, C. H., and Andrews, Luman. Flora of the town of Southington and vicinity. A list of the fern and seed plants growing without cultivation. Rpt. Bd. Educ. Conn. [**1900/01**] (Pub. Doc. no. **8**) : 391–506. map. 1902.

Physiography, etc.; annotated list of 1,201 vascular plants (including 990 native species.) (Hartford County.)

Blewitt, A. E. Flora of Waterbury, Conn., and vicinity. 160 p. 23.5 cm. [Lancaster, Pa.] 1926.

Physiography, etc.; annotated list of 1,271 vascular plants, of which 953 are native. (New Haven County.)

Brace, J. P. List of plants growing spontaneously in Litchfield and its vicinity. Amer. Jour. Sci. **4**: 69–86, 292–309. 1822.

Partly annotated list of vascular plants, on the Linnaean system. Brace's herbarium is still preserved at Williams College, Williamstown, Mass. (fide C. A. Weatherby). (Litchfield County.)

[**Holcomb, Irving.**] Native trees, shrubs and woody vines growing in the vicinity of Granby, Conn. 1 leaf. 27 cm. [Granby, 1901.]

Unannotated list of 144 woody plants. (Hartford County.)

Leonard, E. J. Catalogue of the phaenogamous and vascular cryptogamous plants, found growing in Meriden, Conn. ("Incompleted.") Trans. Scient. Assoc. Meriden, Conn. **1**: 1–40. 1885.

Sparsely annotated list of 749 vascular plants.—See also **Kendrick, Mrs. E. B.** Additional plants . . . found growing in Meriden . . . since issue of Catalogue . . . l. c. **2**: 54–57. 1887.—Also **Davis, C. H. S.** A list of the forest trees and shrubs to be found in Meriden, Conn. l. c. **3**: 46–78. 1889. (New Haven County.)

Rogers, Mrs. E. E. Flora of Norwich and vicinity. Rpt. Bd. Educ. Conn. [**1900/01**] (Pub. Doc. no. **8**) : 342–372. 1902.

Annotated list of 833 vascular plants. (New London County.)

DELAWARE

See also Pennsylvania (Keller and Brown; Pennell).

Local

Tatnall, Edward. Catalogue of the phaenogamous and filicoid plants of Newcastle County, Delaware. Arranged according to the natural system, as recently revised by Prof. A. Gray and others. With the synonyms of modern authors. 112 p. 22.5 cm. Wilmington, 1860.
Annotated list of 1,106 species and varieties of vascular plants; unannotated list of diatoms and desmids, by C. Febiger.

Williamson, C. S. Notes on the flora of central and southern Delaware. Torreya **9**: 160–166. 1909.
List of miscellaneous species with localities.

DISTRICT OF COLUMBIA

General

Hitchcock, A. S., and Standley, P. C. Flora of the District of Columbia and vicinity. Contrib. U. S. Natl. Herb. v. 21. 329 p. 42 pl. 1919.
Bibliography, general features of flora; annotated systematic list of vascular plants with keys. Lists formally 1,630 species (of which 287 are introduced); 108 others, occurring only as waifs, incidentally mentioned. Covers the area within 15 miles of the Capitol, including parts of adjacent Virginia and Maryland. Replaces Ward, L. F. Guide to the flora of Washington and vicinity. Bul. U. S. Natl. Mus. no. 22. 264 p. 1881, and 6 lists of additions, 1884–1901 (2d by F. H. Knowlton, 3d–5th by T. Holm, 6th by E. S. Steele).—See also **McAtee, W. L.** Seventh supplement to the flora of the District of Columbia and vicinity. Proc. Biol. Soc. Washington **43**: 21–54. 1930.

Ricker, P. L. A sketch of botanical activity in the District of Columbia and vicinity. Jour. Washington Acad. Sci. **8**: 487–498, 516–521. 1918.
Botanical collectors and societies; annotated bibliography.

FLORIDA

General

Small, J. K. Florida trees; a handbook of the native and naturalized trees . . . ix, 107 p. 23.5 cm. New York, 1913.
Descriptive flora, with keys; general range briefly indicated; wood characters very briefly described.

——— Shrubs of Florida; a handbook of the native and naturalized shrubs. . . x, 140 p. 23.5 cm. New York, 1913.
Descriptive flora, with keys; general range briefly indicated.

Stuhr, E. T. Medicinal plants of Florida. Jour. Amer. Pharm. Assoc. **17**: 761–766. 1928.
Unannotated systematic list.

Local

Baerecke, J. F. Analytical key to the ferns and flowering plants in the Atlantic section of middle Florida (wild and cultivated). 2d ed. 177 p. 20 cm. Deland, 1914. (1st ed., 1906.)
Annotated flora, in form of keys; authorities for botanical names omitted.

Bowman, H. H. M. Botanical ecology of the Dry Tortugas. Papers Dept. Mar. Biol. Carnegie Inst. Washington **12**: 109–138. 7 fig. (maps), 6 pl. 1918.
Geology, climate, general features of vegetation, discussion of each key with list of species and map showing location of each species; ecology; list of marine plants.

Chapman, A. W. A list of plants growing spontaneously in the vicinity of Quincy, Florida. West. Jour. Med. and Surg. (Louisville, Ky.) n. s., **3**: 461–483. 1845. (Reprinted, 23 p.)
Unannotated list of vascular plants. (Gadsden County.)

Harper, R. M. Geography and vegetation of northern Florida. Ann. Rpt. Florida State Geol. Survey 6: 163–437 (incl. map, plates). 1914. (Reprinted.)

Includes lists of principal species, divided according to habit, from 20 local regions north of the southern boundaries of Lafayette, Alachua, Putnam, and St. Johns Counties, with account of geology, soils, topography, vegetation types, and economic features of each region; annotated bibliography.

——— Geography of central Florida. Ann. Rpt. Florida State Geol. Survey 13: 71–288 fig. 2–43 (incl. maps). 1921.

Includes (p. 84–153) brief lists of the commonest plants in 10 regions from Levy, Marion, and Volusia to Hillsborough, Polk, Osceola, and Brevard Counties, with notes on topography, geology, soils, vegetation, agriculture, etc., of each region.

Hitchcock, A. S. A list of plants collected in Lee County, Florida. Proc. Iowa Acad. Sci. 9 (1901): 189–225. 1902.

General features of flora; annotated list of 549 vascular plants collected by the author in 1900.

McAtee, W. L. A list of plants collected on St. Vincent Island, Florida. Proc. Biol. Soc. Washington 26: 39–51. 1913.

List of 261 vascular plants collected in late autumn and winter, with notes on fruiting and flowering. (Near Apalachicola, in Franklin County.)

Melvill, J. C. List of the phanerogams of Key West, south Florida, mostly observed there in March, 1872. Mem. Manchester Lit. and Phil. Soc. III, 8: 138–154. 1884.

Topography; annotated list of 168 vascular plants, plus a few sedges, grasses, and ferns. (Monroe County.)

Millspaugh, C. F. Flora of the sand keys of Florida. Pub. Field Mus. Nat. Hist. 118, Bot. Ser. 2: 191–245. illus. (maps). 1907.

List of vascular plants of each of the sand keys to the westward of Key West, with map of each key, showing graphically the distribution of each species; table summarizing local distribution.

Murrill, W. A. Flora of Alachua County, Florida. Preliminary check list. 35 p. 28 cm. Gainesville, 1937.

Soils, etc.: list of vascular plants, annotated as to abundance, divided into vines, shrubs, trees, Compositae, etc.; the grasses, sedges, and rushes "reserved for a separate publication." (Processed publication.)

Small, J. K. Flora of Miami; being descriptions of the seed-plants growing naturally on the Everglade keys and in the adjacent Everglades, southern peninsular Florida. xii, 206 p. 23.5 cm. New York, 1913.

Descriptive flora, with keys; general range stated. (Dade County.)

——— Flora of the Florida Keys; being descriptions of the seed-plants growing naturally on the islands of the Florida reef from Virginia Key to Dry Tortugas. xii, 162 p. 23.5 cm. New York, 1913.

Descriptive flora, with keys; extralimital range given.

GEORGIA

General

Anonymous. Catalogue of the flora in Georgia, arranged according to the Linnean and natural system. *In* White, George. Statistics of Georgia. suppl. p. 43–77. Savannah, 1849.

Unannotated alphabetical list of vascular plants, of uncertain authorship.

Harper, R. M. Georgia's forest resources. South. Woodlands 1 (3): 4–23, (4): 1–19, (5): 3–19, (6): 15–32. illus. (maps), 3 pl., map. 1907–08.

Botanical explorations and publications, phytogeographical regions; annotated list of 134 trees, in reverse of the usual systematic order, with mention of about 20 others of probable occurrence.

Justice, R. S. Some medicinal and poisonous plants of Georgia. Bul. Univ. Georgia v. 39, no. 9. iii, 49 p. 2 fig. (maps). 1939.

Includes annotated alphabetical lists of medicinal and poisonous plants.

Local

Feay, W. T. Catalogue of phaenogamous plants, growing spontaneously within thirty miles of Savannah, Geo., with their periods of inflorescence, heights, stations, etc., arranged alphabetically for convenience of reference, and numbered consecutively for facilitating exchanges. Oglethorpe Med. and Surg. Jour. **3**: 167–217. 1860.

Annotated alphabetical tabular list of 794 plants.

Harper, R. M. Notes on the flora of middle Georgia. Bul. Torrey Club **27**: 320–341. pl. 22. 1900.

Topography, etc.; lists of species belonging to different floral areas; records of rare species. (In Clarke, Jackson, Oconee, and Walton Counties.)

——— Notes on the flora of south Georgia. Bul. Torrey Club **27**: 413–436. 1900.

Lists of species of different habitats; annotated list of rarer vascular plants collected by the author in 1895–97, mostly in vicinity of Americus, Sumpter County.

——— A phytogeographical sketch of the Altamaha grit region of the coastal plain of Georgia. Annals N. Y. Acad. Sci. **17**: 1–357. 28 pl., map. 1906.

Topography, climatology, habitat groups, etc.; annotated list of plants (739 vascular); general features of distribution; bibliography. (A roughly diamond-shaped region in southern Georgia, south of 32°50′ N. Lat., extending from the southwest corner of the State northeast to Screven County.)

Wright, A. H., and **Wright, A. A.** The habitats and composition of the vegetation of Okefinokee Swamp, Georgia. Ecol. Monog. **2**: 109–232. 75 fig. (incl. maps). 1932.

Ecological; includes lists of vascular plants of different habitats. (The area is mostly in Charlton County.)

IDAHO

See also Oregon (Hemenway); Washington (St. John, Flora of southeastern Washington; St. John and Warren, Preliminary list of the plants of the Kaniksu National Forest).

Local

Holzinger, J. M. Report on a collection of plants made by J. H. Sandberg and assistants in northern Idaho in the year 1892. Contrib. U. S. Natl. Herb. **3**: 205–287. pl. 3–4. 1895.

List of vascular and cellular plants, with data of collections of each. A critical review by A. A. Heller (Bul. Torrey Club **23**: 155–157. 1896) should be consulted in connection with this list. (Principally in Nez Perce, Latah, and Kootenai Counties.)

Spinden, H. J. The Nez Percé Indians. Mem. Amer. Anthropol. Assoc. **2**: 165–274. fig. 2–6, pl. 6–10. 1908.

Includes (p 200–205) an account of plants used as food. These Indians inhabited Idaho, Oregon, and Washington from the Bitterroot Mountains to the Blue Mountains.

ILLINOIS

See also Indiana (Schneck); Missouri (Engelmann Botanical Club).

General

Carter, J. M. G. A synopsis of the medical botany of Illinois. 45 p. 24.5 cm. Chicago, 1884.

Briefly annotated systematic list, including cultivated plants.

Miller, R. B., and **Tehon, L. R.** The native and naturalized trees of Illinois. Bul. Illinois Nat. Hist. Survey **18**: 1–339 (incl. 149 fig., 98 pl.). 1929.

Keys; list, with descriptions, distribution, uses; forest reserves, etc.

Patterson, H. N. Catalogue of the phaenogamous and vascular cryptogamous plants of Illinois, native and introduced. 54 p. 21.5 cm. Oquawka, 1876.

Annotated list of 1,542 vascular plants.

Local

Brendel, Frederick. Flora Peoriana; the vegetation in the climate of middle Illinois. 89 p. 26 cm. Peoria, 1887.

Topography, climate, habitats, etc.; list of vascular and cellular plants (835 vascular), with very brief annotations. (Peoria County.) An enlarged and corrected issue of his Flora Peoriana originally published in Természetrajzi füzetek v. 5, pt. 2–4, 1882. (Reprinted, 107 p.)

Gates, F. C. Contribution to the flora of Cass County, Illinois. Trans. Illinois State Acad. Sci. **15**: 165–170. 1922.

Unannotated list of 393 vascular plants from vicinity of Virginia, Cass County.

——— Contributions to the flora of Hancock County, Illinois. Trans. Illinois State Acad. Sci. **18**: 225–234. 1925.

Topography, etc.; unannotated list of over 650 vascular plants.

——— The vegetation of the beach area in northeastern Illinois and southeastern Wisconsin. Bul. Illinois State Lab. Nat. Hist. **9**: 255–372. pl. 37–56. 1912.

Mainly ecological; contains annotated list of plants; bibliography.

Gleason, H. A. The vegetation of the inland sand deposits of Illinois. Bul. Illinois State Lab. Nat. Hist. **9**: 23–174. 20 pl. 1910.

Mainly ecological; contains annotated list of vascular plants; bibliography. (Includes several areas north of 40° N. Lat.)

Henderson, L. B. The plants of Castle Rock. A preliminary report. Trans. Illinois State Acad. Sci. **21**: 144–151. 1929.

Geology, general features of flora; list of cellular and vascular plants, ecologically arranged. (Ogle County.)

Huett, J. W. Essay toward a natural history of La Salle County, Ill. Pt. I—Botany. Flora La Sallensis. 136 p. 23.5 cm. Ottawa, Ill., 1897.

Botanical explorations, topography, geology, climate, general features of flora, etc.; annotated list of vascular plants.

Patterson, H. N. A list of plants collected in the vicinity of Oquawka, Henderson County, Ills. 18 p. 23 cm. Oquawka, 1874.

List of 709 vascular plants (also mosses), with flowering dates and occasional annotations. Has only cover title; caption title reads: Catalogue.

Pepoon, H. S. An annotated flora of the Chicago area. xxii, 554 p. illus. (incl. maps), pl. 23.5 cm. Chicago, 1927. (Bulletin 8, Natural history survey [Chicago Academy of Sciences].)

Geology, floral regions; annotated list of 1,903 vascular plants. Replaces the earlier "Flora of Cook County, Illinois, and a part of Lake County, Indiana" of W. K. Higley and C. S. Raddin, 1891.—See also **Buhl, C. A.** Supplement . . . Bul. Chicago Acad. Sci. **5**: 5–12. 1934. (Region in vicinity of Lake Michigan from north of Waukegan, Illinois, nearly to Michigan City, Indiana, including Cook, Du Page, and part of Will and Lake Counties, Ill., and northern part of Lake and Porter Counties, Ind.)

Raddin, C. S. Catalogue of the phaenogamous plants of Evanston and vicinity, for 1883. 26 p. 19 cm. Evanston, 1883.

Unannotated list of 706 species. (Cook County.)

Ridgway, Robert. Notes on the native trees of the lower Wabash and White River valleys, in Illinois and Indiana. Proc. U. S. Natl. Mus. **5**: 49–88. 1883 (1882).—Additions and corrections to the list of native trees of the lower Wabash. Bot. Gaz. **8**: 345–352. 1883.—Additional notes . . . Proc. U. S. Natl. Mus. **17**: 409–421. pl. 10–15. 1895.

Local distribution, dimensions of large trees; annotated list of 92 trees, with special mention of large specimens. (Southeastern Illinois and adjacent Indiana from about 37°50′ to 38°50′ N. Lat., with especial reference to vicinity of Mount Carmel, Wabash County, Illinois.)

Sampson, H. C. An ecological survey of the prairie vegetation of Illinois. Bul. Illinois State Nat. Hist. Survey **13**: 523–577. illus. (maps), pl. 48–77. 1921.

Includes (p. 559–568) tabular list of vascular plants, with indication of their plant associations.

Thone, Frank. Preliminary check list of the vascular plants of the Illinois State Park at Starved Rock, LaSalle County. Trans. Illinois State Acad. Sci. **17**: 100–106. 1925.

Botanical explorations; unannotated list of vascular plants; bibliography.—See also **Ries, D. T.** Additions to the flora of Starved Rock State Park. l. c. **32**: 89–90. 1939.

INDIANA

See also Illinois (Pepoon; Ridgway).

General

Coulter, J. M. A century of botany in Indiana. Proc. Indiana Acad. Sci. **1916**: 236–260. 1917.

Brief summary of local botanical history; mostly unannotated bibliography, arranged alphabetically by authors. Incorporates and replaces, except for a few annotations, Underwood, L. M. Bibliography of Indiana botany. l. c. **1893**: 20–30. 1894.

Coulter, Stanley. A catalogue of the flowering plants and of the ferns and their allies indigenous to Indiana. Ann. Rpt. Indiana Dept. Geol. and Nat. Res. **24**: 553–1002, 1019–1074. 1900.—Additions to the flora of Indiana. Proc. Indiana Acad. Sci. **1900**: 136–143. 1901.—Contributions . . . l. c. **1901**: 297–303. 1902.

Geography, physiography, plant societies, poisonous plants, etc.; bibliography; annotated list of vascular plants.—See also **Deam, C. C.** Plants new to Indiana I–XX. *In* Proc. Indiana Acad. Sci. 1904–06, 1909, 1911, 1914–16, 1918, 1920–23, 35 (1925), 37 (1927), 39 (1929)–44 (1934), passim. 1905–35. Title varies.

Deam, C. C. Flora of Indiana: on the distribution of the ferns, fern allies and flowering plants. Proc. Indiana Acad. Sci. **34**: 39–53. 1925.

Summary of floral investigations; topography, climate, life zones, botanical habitats, bibliography. Total flora (vascular plants) 1,980 species.

——— Shrubs of Indiana. 2d ed. Pub. Dept. Conserv. Indiana 44. 380 p. (incl. 153 pl., maps). 1932. (1st ed., 1924.)

Annotated descriptive flora of shrubby plants; list of excluded species.

——— Trees of Indiana. 2d rev. ed. Pub. Dept. Conserv. Indiana 13. 326 p. (incl. 140 pl., maps). 1932. (Other eds., 1912, 1918, 1919, 1921.)

Descriptive flora, with keys; list of excluded species; tables of measurements of large trees and of specific gravity of woods.

McDonald, E. S., *sister.* The life-forms of the flowering plants of Indiana. Amer. Midl. Nat. **18**: 687–773 (incl. 7 pl.). 1937.

Includes systematic list of 2,109 species of spermatophytes of the State, furnished by C. C. Deam, with ecological annotations by symbols, but no data on distribution.

Local

Baird, J. F., and **Taylor, J. L.** Catalogue of phaenogamous and vascular cryptogamous plants of Clark County, Indiana. *In* Manual of the public schools of Clark County, Ind., for 1878–79. p. 46–65. 1878.

List of vascular plants, with flowering dates.

Barnes, C. R. Catalogue of phaenogamous and vascular cryptogamous plants found growing wild in Jefferson County, Indiana . . . To which is added a list of plants growing in Clark County, but not found in Jefferson. By J. F. Baird. 9 p. 24 cm. Madison, 1878.—Addenda. Bot. Gaz. **3**: 13. 1878.

Unannotated list of vascular plants.

Blatchley, W. S. A catalogue of the uncultivated ferns and fern allies (Pteridophyta) and the flowering plants (Spermatophyta) of Vigo County, Indiana. Ann. Rpt. Indiana Dept. Geol. and Nat. Res. **21**: 577–708. 1897.

Topography, soils, etc.; annotated list of 853 vascular plants.

Bradner, E. A partial catalogue of the flora of Steuben County. Ann. Rpt. Indiana Dept. Geol. and Nat. Res. **17** (1891): 135–159. 1892.

Partly annotated list of 729 vascular plants.

Clark, H. W. Flora of Eagle Lake and vicinity. Proc. Indiana Acad. Sci. **1901**: 128–192. 7 fig. 1902.

Physiography, general features of flora, plant formations; annotated list of 578 vascular plants. Eagle or Winona Lake is in Kosciusko County.

Coulter, J. M. A partial list of the flora of Jefferson County, Indiana. Ann. Rpt. Geol. Survey Indiana **6** (1874) : 230–277. 1875.—Some plants new to Jefferson County. Bot. Bul. (i. e. Bot. Gaz.) **1**: 34–35. 1876.

Annotated list of 721 plants, nearly all vascular.

Erlanson, E. W. List of Indiana plants, chiefly from Putnam County, collected 1910–1915 by Earl J. Grimes. Proc. Indiana Acad. Sci. **33** (1923) : 123–162. 1924.

List of vascular plants, with data.

Evermann, B. W., and **Clark, H. W.** The flora of Lake Maxinkuckee and vicinity. *In* their Lake Maxinkuckee . . . Pub. Dept. Conserv. Indiana no. 7, v. 2, p. 117–447. 1920.

Aquatic flora, land flora, etc.; annotated lists of algae, aquatic plants, and vascular plants (838 species). (Marshall County.)

Heimlich, L. F. Native plants of White County III–VI. Proc. Indiana Acad. Sci. **1920**: 219–224. 1921; **1921**: 117–119. 1922; **1922**: 281–289. 4 fig. 1923; **1923** (v. **33**) : 225–231. 5 fig. 1924.

Partly annotated list of 440 vascular plants. Titles vary slightly. The first two papers of the series were his The primrose-leaved violet in White County. l. c. **1914**: 213–217. 1915.—The trees of White County, Indiana, with some reference to those of the State. l. c. **1917**: 387–471. 32 fig. (incl. maps). 1918.

Lyon, M. W., jr. List of flowering plants and ferns in the Dunes State Park and vicinity, Porter County, Indiana. Amer. Midl. Nat. **10**: 245–295. 1927.—Supplement. l. c. **12**: 33–43. 1930.

Habitats, brief bibliography: annotated list of 825 species and varieties of vascular plants (including supplement).

McDonald, E. S., *sister*. The ferns and the flowering plants of Saint Joseph County, Indiana. Amer. Midl. Nat. **15**: 203–240. 1 fig. (map). 1934.

List of vascular plants, with localities; bibliography.

Meyncke, O. M. The flora of Franklin County . . . Bul. Brookville [Ind.] Soc. Nat. Hist. **1**: 13–38. 1885; **2**: 45–49. 1886.

Annotated list of spermatophytes.

Peattie, D. C. Flora of the Indiana dunes. A handbook of the flowering plants and ferns of the Lake Michigan coast of Indiana and of the Calumet district. 432 p. 38 fig., map. 19.5 cm. Chicago, 1930.

Annotated descriptive flora of over 1,300 vascular plants, with keys. (In Lake, Porter, and La Porte Counties.)—See also **Buhl, C. A.** Notes on the flora of the Indiana dunes. Amer. Midl. Nat. **16**: 248–253. 1935.

Phinney, A. J. Catalogue of the flora of central-eastern Indiana. (Alpine, or elevated, district of the State.) Ann. Rpt. Indiana Dept. Geol. and Nat. Hist. **12**: 196–236. 1883.

General features of flora; annotated list of 789 vascular plants from Delaware, Wayne, Randolph, and Jay Counties, mostly (720) from Delaware County.

Price, Gladys, and **Welch, W. H.** Enumeration of the vascular flora of a limestone area of the Bloomington quadrangle, Monroe County, Indiana. Proc. Indiana Acad. Sci. **39** (1929) : 127–131. 1930.

Unannotated list of 399 species and varieties of vascular plants in an area of 4 square miles northeast of Bloomington.

Schneck, Jacob. Catalogue of the flora of the Wabash Valley below the mouth of White River, and observations thereon. Ann. Rpt. Geol. Survey Indiana **7** (1875) : 504–579. 1876.

Topography, general features of flora, measurements of large trees; annotated list of 867 vascular plants and Characeae. The region covered is in Gibson and Posey Counties, Ind., and Wabash, Edwards, White, and Gallatin Counties, Ill.—For additions, see the author's Some plants of the lower Wabash. Bot. Gaz. **2**: 83. 1877.—Also **Deam, C. C.** Additions . . . Proc. Indiana Acad. Sci. **1911**: 365–369. 1912.

Smith, C. P. Report upon a botanical survey of the Indiana State forest reservation. Ann. Rpt. State Bd. Forestry Indiana **3**: 73–[161] (incl. 45 pl.). map. 1904.

Soil, etc.; annotated lists of trees and of other vascular plants. (Clark County.)

Van Gorder, W. B. Flora of Noble County. Ann. Rpt. Dept. Geol. and Nat. Res. Indiana **18** (1893) : 33–71. 1894.
Annotated list of 724 vascular plants.

Welch, W. H. Enumeration of the vascular flora of Jasper County, Indiana. Proc. Indiana Acad. Sci. **36**: 213–220. 1927.—Additions . . . I–II. l. c. **40**: 119–121. 1931; **45**: 77. 1936.
Unannotated list of 662 species and varieties.

Wilson, Guy. Flora of Hamilton and Marion Counties, Indiana. Proc. Indiana Acad. Sci. **1894**: 156–176. 1895.
Annotated list of cellular and vascular plants.—See also **Douglass, B. W.** Additions to the flora of Marion County . . . l. c. **1904**: 223–224. 1905.

Young, A. H. Manual of the botany of Jefferson Co. Rpt. Geol. Survey Indiana **2**: 241–292. 1871.—Notes on some interesting plants found in Jefferson County. Bot. Bul. (i. e. Bot. Gaz.) **1**: 6–8. 1875.
Annotated list of 609 angiosperms; medicinal uses indicated.

IOWA

See also Minnesota (MacMillan); Wisconsin (Smith, Ethnobotany of the Meskwaki Indians).

General

Bode, I. T., and MacDonald, G. B. A handbook of the native trees of Iowa. 95 p. illus. 19 cm. Ames, [1928?].
Annotated descriptive list, with figure of each species; popular treatment.

Cratty, R. I. The immigrant flora of Iowa. Iowa State Col. Jour. Sci. **3**: 247–269. 1929.—Notes on the immigrant flora . . . I. l. c. **8**: 309–311. 1934.
Annotated list of 263 vascular plants, with date of first record.

——— The Iowa flora. An annotated list of the ferns, fern allies and the native and introduced flowering plants of the State represented in the Iowa State College herbarium. Iowa State Col. Jour. Sci. **7**: 177–252. 1933.
Annotated list of 1,608 vascular plants, of which 1,263 are native.

Fitzpatrick, T. J. Manual of the flowering plants of Iowa. 2 pts. (143 p.) 22.5 cm. [Lamoni, Iowa,] 1899.
Descriptive flora; distribution given in some detail. Pt. 1, Polypetalae; pt. 2, Gamopetalae; no more published.

Greene, Wesley. Plants of Iowa. A preliminary list of the native and introduced plants of the State, not under cultivation. 264 p. 22.5 cm. Des Moines, 1907. (Bulletin of the State Horticultural Society.)
Annotated list of 3,115 species, of which about 1,600 are vascular plants.

Macbride, T. H. Twenty-five years of botany in Iowa. Proc. Iowa Acad. Sci. **19**: 43–63. 1912. (Reprinted and rev. 1914.)
Sketch of botanical work; unannotated bibliography for 1888–1912.

Pammel, L. H. A century of botany in Iowa. Proc. Iowa Acad. Sci. **31** (1924) : 45–68. [1925?]
Biological organizations, explorations, running account of botanical activities in various fields.

——— Native shrubs of Iowa. Rpt. Iowa State Hort. Soc. **49**: 34–43. 2 pl. 1915.
Annotated list.

——— and **Fogel, E. D.** A catalogue of the poisonous plants of Iowa. Proc. Iowa Acad. Sci. **14**: 147–176 (incl. 4 maps). 1907.
Bibliography; annotated systematic list of 275 plants.

——— **King, C. M.**, and others. The weed flora of Iowa. rev. ed. Iowa Geol. Survey Bul. no. 4. vi, 715 p. 512 fig. 1926. (1st ed. 1913.)
Systematic list, with brief descriptions, range, and method of treatment; chapters on seed characters, weed migrations, etc.; bibliography.

Rauch, J. H. Report [on the medical and economical botany of Iowa]. Proc. Iowa State Med. and Chirurg. Soc. **2**: 11–52. 1851.
Annotated list of medicinal and other useful plants of Iowa.

Local

Barnes, W. D., Reppert, Fred, and Miller, A. A. The flora of Scott and Muscatine Counties. Proc. Davenport Acad. Sci. **8**: 199–287 (incl. 2 pl.). 1901.
Annotated list of 1,068 vascular plants.

Boot, D. H. Plant studies in Lyon County, Iowa. Proc. Iowa Acad. Sci. **24** (1917): 393–414 (incl. fig. 80, pl. 11–12). [1918?]
Mainly ecological; includes tabular list of 162 vascular plants.

Burk, M. M. J. Flowering plants of Black Hawk County, Iowa. Proc. Iowa Acad. Sci. **33** (1931): 135–149. [1932?]
General features of flora; unannotated list of angiosperms.

Conard, H. S. Plants of central Iowa, being a third edition of the Grinnell flora . . . 99 p. illus. 22 cm. Grinnell, [1929?] (1st ed. [1923], 2d ed. [1925].)
Unannotated flora in form of keys, without authorities for the botanical names. A 5th ed., 1939 (not seen), has title: Plants of Iowa.

Diehl, W. W. The flora of The Ledges region of Boone County, Iowa. Proc. Iowa Acad. Sci. **22**: 77–104. pl. 18–19. 1915. (Reprinted as Contrib. Bot. Dept. Iowa State Col. no. 61.)
Ecology; annotated list of vascular plants; bibliography.

Fink, Bruce. Spermaphyta of the flora of Fayette, Iowa. Proc. Iowa Acad. Sci. **4**: 81–107. 1897.
Topography; annotated list of over 700 seed plants from within 5 miles of Fayette, Fayette County.

Fitzpatrick, T. J. Forest trees and shrubs of Decatur County. *In* **Bain, H. F.** Geology of Decatur County. Ann. Rpt. Iowa Geol. Survey **8**: 309–314. 1898.
Annotated list.

——— Notes on the flora of northeastern Iowa. Proc. Iowa Acad. Sci. **5**: 107–133. 1898.
Annotated list of vascular plants and a few cellular plants collected by the author in Winneshiek, Allamakee, Clayton, Dubuque, Jackson, Clinton, and Scott Counties in 1895, with localities.

——— and **Fitzpatrick, M. F. L.** Flora of southern Iowa. Proc. Iowa Acad. Sci. **5**: 134–173. 1898; **6**: 173–202. 1899.
Annotated list of vascular plants and a few cellular plants, mostly collected by the authors, with localities.—See also **Anderson, J. P.** Plants new to the flora of Decatur County, Iowa. l. c. **12**: 133–138. 1905.—Also his Additions . . . Iowa Nat. **2**: 7–8. 1906.

Fults, J. L. A botanical survey of Lee County, Iowa. Iowa State Col. Jour. Sci. **8**: 251–293. 1934.
Physiography, climate, soils, botanical explorations; annotated list of about 645 vascular plants.

Geiser, S. W. The flowering plants of a typical locality in northeastern Iowa. 19 p. 23 cm. Fayette, Iowa, 1918. (Bulletin of the Upper Iowa University, v. 20, no. 5, Biol. Ser. no. 2.)
List of spermatophytes of region around Fayette, with annotative information conveyed by a number system. (Fayette County.)

Gow, J. E. Preliminary list of the flowering plants of Adair County. Proc. Iowa Acad. Sci. **8**: 152–159. 1901. (Reprinted in Ann. Rpt. Iowa Geol. Survey **27**: 336–344. [1917?].)
Partly annotated list of phanerogams.—See also his Forest trees of Adair County, Iowa. Proc. Iowa Acad. Sci. **6**: 56–63. map. 1899.

Hitchcock, A. S. A catalogue of the Anthophyta and Pteridophyta of Ames, Iowa. Trans. Acad. Sci. St. Louis **5**: 477–532. 1891.
Topography, etc.; annotated list of vascular plants.—See also **Lindly, J. M.** Flowers of Story County. Trans. Illinois State Acad. Sci. **18**: 19–24. 1911. (Refers almost entirely to Ames.) (Story County.)

McDonald, Malcolm. The woody plants of southeastern Iowa. Proc. Iowa Acad. Sci. **44** (1937): 65–74. [1938?]
Tabular list of 168 species and varieties, showing distribution by counties; bibliography.

Mueller, H. A. A preliminary list of the flowering plants of Madison County. Proc. Iowa Acad. Sci. **11**: 261–279. 1904.
Topography; annotated list of spermatophytes, except grasses and sedges.

Nagel, J. J., and Haupt, J. G. List of phaenogamous plants, collected in the vicinity of Davenport, Iowa . . . during the years 1870 to 1875, inclusive. Proc. Davenport Acad. Nat. Sci. **1**: 153–164. 1876.

List of spermatophytes, with flowering dates. (Scott County.)

Oleson, O. M., and Somes, M. P. A flora of Webster County, Iowa. Proc. Iowa Acad. Sci. **13**: 25–58. 1906.

Topography, geology; annotated list of 852 vascular plants.

Pammel, L. H. Flora of northern Iowa peat bogs. Ann. Rpt. Iowa Geol. Survey **19**: 735–777. fig. 106-117. 1909. (Reprinted as Contrib. Bot. Dept. Iowa State Col. no. 40.)

General botanical features of peat bogs; tabular list of cellular and vascular plants of peat bogs in Iowa, Wisconsin, Michigan, and Virginia; bibliography.

——— The flora of Pine Hollow, Dubuque County, Iowa. Proc. Iowa Acad. Sci. **30** (1923): 263–277. 14 fig. [1924?]

General features of flora; unannotated list of vascular plants.

——— **MacDonald, G. B., and Clark, H. B.** The native and cultivated forest trees and shrubs of the Missouri River basin. Proc. Iowa Acad. Sci. **22**: 23–56. pl. 4–15. 1915.

Bibliography, topography, economic uses, distribution, etc.; annotated list with localities. Covers principally the area between Sioux City and Hamburg.

Peck, M. E. The flowering plants of Hardin County. Proc. Iowa Acad. Sci. **12**: 193–241. 1905.

Annotated list of 704 spermatophytes.

Rigg, G. B. Notes on the flora of Calhoun County, Iowa. 37, [3] p. map. 19.5 cm. Iowa City, 1896.

Physiography; annotated list of 173 vascular plants; lists of species grouped by habitats.

Schadt, Conrad. Flora of Iowa County. Ann. Rpt. Iowa Geol. Survey **20**: 186–198. 1910.

Unannotated list of vascular plants.

Shimek, Bohumil. The botanical manuals and the Iowa flora. Univ. Iowa Studies Nat. Hist. **14** (2): 17–37. 1931.

Corrections in habitats and ranges, as relating to Iowa plants, in Gray's Manual and Britton's Manual.

——— Botanical report. The prairies. *In* his Geology of Harrison and Monona Counties. Ann. Rpt. Iowa Geol. Survey **20**: 426–483 (incl. pl. 33–41). 1910.

General features of prairie flora, etc.; alphabetical lists of vascular plants, ecologically arranged; climate, etc.; annotated list of weeds.

——— Early Iowa locality records. Proc. Iowa Acad. Sci. **22**: 105–119. 1915 (1916).

Discussion of important localities in earlier publications, particularly Council Bluffs.

——— The flora of Lyon County. Ann. Rpt. Iowa Geol. Survey **10**: 157–184. 1900.—Addenda . . . Bul. Lab. Nat. Hist. Univ. Iowa **5**: 215–216. 1901.

Soil, general features of flora; annotated list of vascular plants, ecologically divided.

——— The plant geography of the Lake Okoboji region. Bul. Lab. Nat. Hist. Univ. Iowa v. 7, no. 2. 90 p. 1 fig. (map), 8 pl., map. 1915.—Additional notes. l. c. v. 7, no. 3 bis, p. 3–5. 1917.

Includes tabular list of vascular and cellular plants. (Dickinson County.)

——— The plants of Winneshiek County. Ann. Rpt. Iowa Geol. Survey **16**: 147–211. 1906. (Reprinted with title: Flora of Winneshiek County.)

Collectors, etc.; annotated list of vascular plants; bibliographical footnotes.

——— The prairies. Bul. Lab. Nat. Hist. Univ. Iowa **6** (2): 169–240. 13 pl., map. 1911.

Mainly ecological; includes alphabetical list of typical prairie plants of Iowa, with indication of ecological habitat; bibliography. The list of prairie plants is nearly the same as that in his Botanical report. The prairies. 1910.

——— The sand-flora of Iowa. Bul. Lab. Nat. Hist. Univ. Iowa **7** (3, i. e. 4): 6–24. 5 pl. 1917.

Ecology; tabular list of vascular plants, with habitat indicated.

Somes, M. P. Notes on the flora of Johnson County, Iowa. Proc. Iowa Acad. Sci. **20**: 27–101. 1913.

Topography; annotated list of 1,005 vascular plants.

Tolstead, W. L. A flora of Winneshiek and Allamakee Counties and Clayton County in the vicinity of McGregor. Iowa State Col. Jour. Sci. **12**: 321–384. 2 pl. 1938.

Physiography, geology, soils, climate, botanical explorations; annotated list of bryophytes and vascular plants; plant formations, bibliography.

Tuttle, F. M. Flora of Mitchell County. Proc. Iowa Acad. Sci. **26** (1919): 269–299. fig. 92–95. [1920?]

General features of flora; annotated list of vascular plants.

Verink, E. D. A preliminary report on the flora of Linn County. Proc. Iowa Acad. Sci. **21**: 77–99. 1914.

Annotated list of vascular plants, the families arranged alphabetically.

Wolden, B. O. The plants of Emmet County, Iowa. Proc. Iowa Acad. Sci. **39** (1932): 89–132. [1933?]

General features of region, bibliography; annotated list of 930 species and varieties of vascular plants. Replaces Cratty, R. I. Flora of Emmet County, Iowa. Proc. Iowa Acad. Sci. **11**: 201–251. 1904; Iowa Nat. **1**: 36–38. 1905.

KANSAS

General

Gates, F. C. Handbook of Kansas trees. Rpt. Kansas State Bd. Agr. **47** (186–A: Trees in Kansas): 148–372. fig. 117–209. 1928.

Descriptive flora of native and principal cultivated species; each species illustrated by figures of the leaves, fruit, and details; bibliography; keys.

——— Principal poisonous plants in Kansas. Tech. Bul. Kansas Agr. Expt. Sta. 25. 67 p. 64 fig. 1930.

General considerations; annotated systematic list of poisonous vascular plants.

——— Woody plants, native and naturalized in Kansas. Trans. Kansas Acad. Sci. **41**: 99–118 (incl. maps). 1938.

Briefly annotated list of 208 trees, shrubs, and woody vines, with the range of each species shown on a small map of the State.

Hitchcock, A. S. Flora of Kansas. A series of maps illustrating the distribution of flowering plants by counties . . . Maps prepared by J. B. Norton and J. M. Westgate. 2 l. 20 (i. e. 22) maps. 23 cm. Manhattan, 1899. (Partly reprinted from Industrialist, v. 24, 1898.)

Consists of a foreword and about 1,350 tiny maps showing distribution of each vascular plant by counties.

Kellerman, W. A., and **Kellerman, S. D.** Kansas school botany consisting of the analytical flora of Kansas, to which is prefixed an introductory botany and glossary. 58, 197 p. 109 fig. 19 cm. Manhattan, 1888.

Keys to all vascular plants with brief descriptions, unannotated. The illustrations belong to introductory botany.

Smyth, B. B. Check list of the plants of Kansas. 2d ed. 36 p. 23.5 cm. Topeka, 1892 (1893). (1st ed., 1892.)

List of 2,037 plants (of which 1,933 are vascular), with local distribution indicated by districts. The 2d ed. is merely the first ed., with additions on p. 34–36, dated Mar. 1893.

——— Preliminary list of medicinal and economic Kansas plants, with their reputed therapeutic value. Trans. Kansas Acad. Sci. **18**: 191–209. 23 fig. 1903.

Very briefly annotated list of 307 plants, alphabetically arranged.

——— and **Smyth, L. C. R.** Catalogue of the flora of Kansas I–II. Trans. Kansas Acad. Sci. **23/24**: 273–295. 1911; **25**: 63–128. 1913.

Annotated list of plants from Bryophyta through Monocotyledones (Engler and Prantl system); 800 species, of which 668 are vascular; includes cultivated plants. Never completed.

Welch, J. M. The medical flora of Kansas; or the medicinal plants indigenous in that State. Trans. Natl. Eclect. Med. Assoc. **10**: 458–472. 1883.

Annotated list.

Local

Anderson, J. P. Annotated list of plants collected in Greeley and Wallace Counties, Kansas. Iowa Nat. **3**: 2–20. 1917. (Reprinted.)
Topography; annotated list of 226 vascular plants.

Castle, W. E. A list of flowering plants and ferns. Collected in Franklin County, Kansas, during the months of April, May, September, and October, 1890–92. Trans. Kansas Acad. Sci. **13**: 80–89. 1893.
List of 452 vascular plants collected by author in vicinity of Ottawa.

Doell, J. H. The flora of the sand hills of Harvey County, Kansas. Kansas Univ. Sci. Bul. **25**: 113–147 (incl. 4 fig., pl. 13–18). 1938 [1939].
Ecological; includes unannotated list of vascular plants; bibliography.

Fraser, S. V. The flora of Cloud County. Trans. Kansas Acad. Sci. **34**: 220–230. [1931].
General features of flora; unannotated list of slightly over 600 vascular plants.

Hancin, John. Flora of Saline County: ferns and flowering plants. Trans. Kansas Acad. Sci. **42**: 139–149. 1939.
Soil, effects of cultivation on flora, etc.; essentially unannotated list of vascular plants.

Harnly, H. J. A provisional list of the flowering plants of McPherson County. Trans. Kansas Acad. Sci. **15**: 75–78. 1898.
Unannotated list of 183 angiosperms.

Maus, P. M. Flora of Wabaunsee County, Kansas. Trans. Kansas Acad. Sci. **32**: 88–104. [1930?]
Topography, general features of flora; annotated list of 495 vascular plants, the families arranged alphabetically.

Neher, S. J. The flora of Osborne County, Kansas. Trans. Kansas Acad. Sci. **37**: 77–82. 1934.
Physiography, general features of flora; unannotated list of vascular plants, the families arranged alphabetically.

Ross, Bennington, and **Hall, H. H.** A preliminary survey of the flora of Crawford County State Park, Kansas. Trans. Kansas Acad. Sci. **42**: 157–164 (incl. map). 1939.
Geology, soil, topography, climate; annotated tabular list of spermatophytes. (Near Pittsburg.)

Stevens, O. A. Plants of Manhattan and Blue Rapids, Kansas, with dates of flowering. Amer. Midl. Nat. **5**: 71–87, 98–112, 113–129, 201–204. 1917.
Annotated list of vascular plants. The localities are in Riley and Marshall Counties, respectively.

Stieferman, M. A., *sister.* The seed plants of Sedgwick County, Kansas. Trans. Kansas Acad. Sci. **42**: 179–195 (incl. map). 1939.
Soil, topography, etc.; unannotated list of angiosperms.

Weber, Clement. The flora of Clay County, Kansas. Trans. Kansas Acad. Sci. **33**: 87–106. [1931].
History; mostly unannotated list of about 600 vascular plants.

——— The flora of Sheridan County, Kansas. Trans. Kansas Acad. Sci. **35**: 161–178. [1932].
Physiography, etc.; unannotated list of 476 vascular plants.

——— Plants that grow in Ellsworth County. Trans. Kansas Acad. Sci. **32**: 68–79. [1930?].
Topography, soil, etc.; unannotated list of 443 vascular plants.

KENTUCKY

See also Ohio (Braun).

General

Frazee, L. J. Report on the indigenous botany of Kentucky. Proc. Kentucky State Med. Soc. **14**: 47–62. 1869.
Sketch of botanical explorations and publications; unannotated alphabetical list of medicinal plants.

Garman, Harrison. The woody plants of Kentucky. Bul. Kentucky Agr. Expt. Sta 169. 62 p. 20 fig. (in text and on 10 pl., incl. 2 ports.). 1913.
Annotated list, the local distribution given in detail.

Spilman, C. H. Catalogue of Kentucky plants. Trans. Kentucky State Med. Soc. **1852**: 306–318. 1853.
Unannotated alphabetical list of vascular plants.

Local

Greenwell, R. A., *sister.* A flora of Nelson County, Kentucky, with a selected list of economically important plants. Biol. Ser. Catholic Univ. Amer. no. 20. xiv, 204 p. 1935.
Topography, geology, botanical explorations; annotated list of 863 vascular plants, with keys.

Hussey, John. Report on the botany of Barren and Edmonson Counties. Rpt. Prog. Geol. Survey Kentucky II, v. 1, pt. 2. 32 p. 1876. (Reprinted in Geological survey of Kentucky. Timber and botany, B. [no. 2]. 1884.)
Unannotated list of vascular plants.

Kearney, T. H. Notes on the flora of southeastern Kentucky, with a list of plants collected in Harlan and Bell Counties in 1893. Bul. Torrey Club **20**: 474–485. 1893. (Reprinted as Contrib. Herb. Columbia Col. no. 43.)
Rarer plants of special localities; unannotated list of vascular plants collected.

Linney, W. M. Report on the botany of Madison, Lincoln, Garrard, Washington, and Marion Counties, Kentucky. 57 p. 26.5 cm. Frankfort, Ky. [1882]. (Publication of Geological survey of Kentucky.)
Unannotated list of 682 vascular plants.

——— Report on the timbers of Boyle and Mercer Counties. Rpt. Prog. Geol. Survey Kentucky II, v. 5, pt. 11. 36 p. 1880.
Includes list of vascular plants.

Nelson, J. C. Plants of Boone County, Kentucky. Proc. Indiana Acad. Sci. **1918**: 125–143. 1919.
Annotated list of vascular plants.

Price, S. F. Flora of Warren County, Kentucky. 31 p. 20 cm. New London, Wis., 1893.
Unannotated list of 714 vascular plants. The United States Department of Agriculture copy contains numerous corrections and additions in Miss Price's hand.

LOUISIANA

General

Cocks, R. S. A list of the shrubs of Louisiana. Jour. Arnold Arboretum **3**: 173–182. 1922.
Annotated list, with flowering and fruiting dates.

——— A list of the trees of Louisiana. Jour. Arnold Arboretum **2**: 204–216. 1921.
Annotated list; dimensions of various large trees.

Hale, Josiah. Report on the medical botany of the State of Louisiana. New Orleans Med. and Surg. Jour. **9**: 152–173, 287–313. 1852–53.
Annotated systematic list.

Riddell, J. L. Catalogus florae ludovicianae. New Orleans Med. and Surg. Jour. **8**: 743–764. 1852. (Reprinted.)
Unannotated list of vascular plants based on studies of J. L. Riddell, Josiah Hale, and W. M. Carpenter. Refers to a much more extensive "Plants of Louisiana" sent by the author to the Smithsonian Institution in 1851, but never published.

Local

Brown, C. A. Plants observed on an excursion to Grand Isle, Louisiana. Bul. Torrey Club **57**: 509–513. 1 fig. (map). 1930.
General features of flora; list of vascular plants, alphabetically arranged under three habitat groups. (Jefferson Parish.)

Cocks, R. S. The flora of the Gulf Biologic Station. Bul. Gulf Biol. Sta. Cameron, La. no. 7. 42 p. 1907.

History of botanical work in Louisiana; annotated list of plants collected near Cameron, Cameron Parish; notes on rare plants from prairie region of southwestern Louisiana.

Dodson, W. R. The principal plants of economic value in the Florida parishes and southwest Louisiana. *In* **Clendenin, W. W.** A preliminary report upon the Florida parishes of east Louisiana . . . Geol. and Agr. State Exp. Sta. [Louisiana] pt. 3. p. 249–256. [1896.]

Systematic list, classified by uses, mostly unannotated.

Featherman, Americus. Report of botanical survey of southern and central Louisiana, made during the year 1870. 131 p. 23 cm. New Orleans, 1871.

Contains (p. 77–129) a list of cellular and vascular plants, with localities.

——— Third annual report of botanical survey of southwest and northwest Louisiana made during the year 1871. Ann. Rpt. Louisiana State Univ. **1871:** 101–161. 1872.

Geology, general botanical features, author's itineraries, economic plants, etc.; list of vascular and cellular plants, with localities, under title Flora ludoviciana (p. 143–160).

Langlois, A. B. Catalogue provisoire de plantes phanérogames et cryptogames de la Basse-Louisiane, États-Unis d'Amérique. 35 p. 24 cm. Saint-Étienne, [1887].

Unannotated list of 1,214 vascular plants, also cellular plants. Refers particularly to Plaquemines Parish.

Lindly, J. M. Some of the flowering plants of Calcasieu Parish, Louisiana. Proc. Iowa Acad. Sci. **13:** 161–166. 1906.

List of seed plants observed by author in the town of Welsh in winter of 1893–94, the families arranged alphabetically. (Jefferson Davis Parish.)

Lloyd, F. E., and Tracy, S. M. The insular flora of Mississippi and Louisiana. Bul. Torrey Club **28:** 61–101. pl. 8–11. 1901. (Reprinted as Contrib. Dept. Bot. Columbia Univ. no. 174.)

Climate, physiography, plant formations; annotated list of vascular plants of Mississippi Sound islands and delta; bibliography.

MAINE

General

Fernald, M. L. The Portland catalogue of Maine plants. 2d ed. Proc. Portland Soc. Nat. Hist. **2:** 41–72. 1892. (Reprinted.)—[1st]–2d supplement . . . l. c. **2:** 73–96, 123–137. 1895–97. (1st ed., by G. L. Goodale and Joseph Blake, 1868.)

List of 1,410 species of vascular plants, with occasional notations (by signs) concerning distribution. The supplements contain many additional species, with annotations.

Maine forest service. Forest trees of Maine. [6th ed.] rev. by H. B. Peirson. 86 p. illus. 23.5 cm. Augusta, 1938. (1st ed., 1907.)

Annotated descriptive list, with illustration of nearly every species.

Perkins, A. E. Common plants as domestic remedies in Maine. Torreya **30:** 63–68. 1930.

Running account of native medicinal plants.

Scribner, F. L. Ornamental and useful plants of Maine. Ann. Rpt. Maine Bd. Agr. **19:** 157–237. illus., 2 pl. 1874. (Reprinted 1875.)

Annotated systematic list, including medicinal plants.

Local

Chamberlain, E. B. List of plants [in the vicinity of Oxford]. Bul. Josselyn Bot. Soc. Maine **1:** 15–23. 1907.

Tabular list of vascular plants collected in Oxford, Otisfield, and Norway in July 1907 by members of the Josselyn Botanical Society. (Oxford and Cumberland Counties.)

Coburn, L. H. Flora of Birch Island in Attean Pond. Rhodora **22**: 129–138. 1920.

Geology, general features of flora; list of 255 vascular plants, divided according to habitat. Attean Pond is one of the Moose River chain of lakes in northern Somerset County.

Cook, M. P. A list of plants seen on the island of Monhegan, Maine, June 20–25, 1900. Rhodora **3**: 187–190. 1901.

Unannotated list of vascular plants. (Lincoln County.)

Fassett, N. C. The Hypocrites. Maine Nat. **2**: 141–143. 1922.

Geology, topography; annotated list of 27 vascular plants. (Three small islands off the coast of Boothbay, Lincoln County.)

Fernald, M. L. The vascular plants of Mount Katahdin. Rhodora **3**: 166–177. pl. 32. 1901.

Botanical explorations; annotated list of vascular plants.—See also **Stebbins, G. L.** Some interesting plants . . . l. c. **31**: 142–143. 1929.—**Ewer, S. J.** Notes on Katahdin plants. l. c. **32**: 259–261. 1930.—**Norton, A. H.** Plants apparently new . . . l. c. **37**: 455. 1934. (Piscataquis and Penobscot Counties.)

Graves, A. H. Woody plants of Brooklin, Maine. Rhodora **12**: 173–184. 1 fig. (map). 1910.

Annotated list of 100 species and varieties. (Hancock County.)

Harvey, L. H. The floristic composition of the vascular flora of Mount Ktaadn, Maine. Rpt. Mich. Acad. Sci. **11**: 37–47. 1909.

Phytogeographic; includes tabular list of 120 arctic-alpine species; bibliography.

Hill, A. F. The vascular flora of the eastern Penobscot Bay region, Maine. Proc. Portland Soc. Nat. Hist. **3**: 199–304. 4 fig. (maps), map. 1919.

Topography, geology, etc.; annotated list of 747 vascular plants; phytogeography. (Townships of Brooklin, Deer Isle, Stonington, and Swan Island in Hancock County, and Isle au Haut, Knox County.)

Jewell, H. W. Franklin County flora. Pt. 1 (15 p.). 19.5 cm. Farmington, Maine, 1898.

Annotated list of 223 vascular plants, arranged in order of flowering.

Knowlton, C. H. Flora of Mt. Saddleback, Franklin County, Maine. Rhodora **5**: 35–38. 1903.

Geology; partly annotated list of vascular plants.

——— Plants and plant societies at Roque Bluffs, Maine. Rhodora **17**: 145–155. 1915.

Geology, etc.; lists of vascular plants, divided according to habitat; bibliography. (Washington County.)

McAtee, W. L. Plants collected on Matinicus Island, Maine, in late fall, 1915. Rhodora **18**: 29–45. 1916.

General features of flora; annotated list of 217 vascular plants; bibliography. Matinicus Island is 18 miles offshore from Rockland, in Knox County.

Morrell, J. M. H. Some Maine plants and their uses, "wise and otherwise." Rhodora **3**: 129–132. 1901.

Annotated list consisting mostly of edible and medicinal plants.

Ogden, E. C. The herbaceous flowering plants (exclusive of grasses and sedges) growing without cultivation in the vicinity of Orono, Maine. Univ. Maine Studies II, 34. 77 p. illus. (map). 1935. (Maine Bul. v. 37, no. 8.)

Botanical explorations, topography; annotated list of plants; bibliography. (Penobscot County.)

Rand, E. L., and Redfield, J. H. Flora of Mount Desert Island, Maine. A preliminary catalogue of the plants growing on Mount Desert and the adjacent islands. 286 p. map in pocket. 23 cm. Cambridge, 1894.

Geology (by W. M. Davis), etc.; annotated list of vascular and cellular plants (1,286 species, of which 716 are vascular).—See also **Rand, E. L.** Additions to the plants of Mount Desert Island. Rhodora **10**: 145. 1908; **Fellows, D. W.**, and others. Plant list. Bul. Josselyn Bot. Soc. Maine **2**: 16–23. 1908 (list of vascular plants observed in 1908); **Taylor, W. R.** Additions . . . Rhodora **23**: 65–68. 1921; **Stebbins, G. L., jr.** Further additions . . . l. c. **31**: 81–87. pl. 185–186. 1929. (Hancock County.)

Redfield, J. H. Insular vegetation [Great Duck Island]. Bul. Torrey Club **12:** 103. 1885; **20:** 409–410. 1893.

List of vascular plants observed by the author in 1885 and 1893.—See also **Rand, E. L.** Plants from the Duck Islands, Maine. Rhodora **2:** 207–209. 1900. (About 10 miles south of Mount Desert, in Hancock County.)

St. John, Harold. Plants of the headwaters of the St. John River, Maine. Res. Studies State Col. Washington **1:** 28–58. 1 fig. (map), pl. 2–3. 1929.

Botanical explorations, geology, physiography; list of vascular plants, with localities and collectors (indicated by symbols), of the valley of the St. John River from near mouth of Allagash River to the headwaters. (Aroostook and Somerset Counties.)

MARYLAND

See also District of Columbia (Hitchcock and Standley).

General

Shreve, Forrest, Chrysler, M. A., Blodgett, F. H., and Besley, F. W. The plant life of Maryland. 533 p. 15 fig., 39 pl. (incl. map). 26 cm. Baltimore, 1910. (Maryland Weather Service. Special publication v. 3.)

Physiography, climatology, plant geography, ecology, etc.: annotated list of 1,400 species and varieties of vascular plants; bibliography (p. 94–97).

Local

Killip, E. P., and Blake, S. F. Flowering plants and ferns. (Natural history of Plummers Island, Maryland. II.) Proc. Biol. Soc. Washington **48:** 118–134. 1935.

Mostly unannotated list of 720 species and 9 varieties of vascular plants. Plummers Island, in Montgomery County, is the property of the Washington Biologists' Field Club.

Redmond, P. J. A flora of Worcester County, Maryland. Contrib. Biol. Lab. Catholic Univ. Amer. no. 11. 104 p. 1932.

Physiography, geology, climate; annotated list of vascular plants, with keys.

Shriver, Howard. List of wild flowers and trees in vicinity of Cumberland, Maryland. With time of blooming, and locality, where found, and references to Gray's and Wood's botanies for descriptions. 38 p. 19.5 cm. Cumberland, 1901.

List of 848 vascular plants, with localities, arranged by flowering dates. (Allegany County.)

Sollers, Basil. Check list of plants, compiled for the vicinity of Baltimore. 72 p. 20 cm. Baltimore, 1888.

Unannotated list of vascular plants.

MASSACHUSETTS

General

Emerson, G. B. A report on the trees and shrubs growing naturally in the forests of Massachusetts. 2d ed. 2 v. 121 pl. 23.5 cm. Boston, 1875. (1st ed., 1846.)

Descriptive flora, with account of uses, etc. The so-called 3d–5th editions (3d, 1878, 4th, 1887, and 5th, 1894 and later dates) seem to be the same as the 2d ed.

Williams, S. W. Report on the indigenous medical botany of Massachusetts. Trans. Amer. Med. Assoc. **2:** 863–927. 1849.

Annotated list of 319 species.

Local

Andrews, Luman. Catalogue of the flowering plants and ferns of Springfield, Massachusetts, growing without cultivation. Bul. Mus. Nat. Hist. Springfield no. 3. 221 p. (incl. 10 pl., map). 1924.

Annotated list of vascular plants.—See also **Seymour, F. C.** Additions to the flora of Springfield. Rhodora **29:** 241–246. 1927. (Hampden County.)

Bicknell, E. P. The ferns and flowering plants of Nantucket. Bul. Torrey Club **35**: 49–62, 181–201, 471–498. 1908; **36**: 1–29, 441–456. 1909; **37**: 51–72. 1910; **38**: 103–133, 447–460. 1911; **39**: 69–80, 415–428. 1912; **40**: 605–624. 1913; **41**: 71–87, 411–427. 1914; **42**: 27–47, 331–349, 549–570. 1915; **43**: 265–276. 1916; **44**: 369–387. 1917; **45**: 365–383. 1918; **46**: 423–440. 1919.

Amply annotated list of vascular plants, in 20 parts, the last two of which are supplements. Includes over 950 species. Replaces Owen, M. L. A catalogue of plants growing without cultivation in the county of Nantucket, Mass. 1888. (Nantucket County.)

Bouvé, T. T. The botany of Hingham. *In* History of the town of Hingham, Massachusetts. v. 1, p. 87–138. Cambridge, 1893.

Botanical explorations, etc.; unannotated list of vascular plants.—See also his The trees and shrubs of Hingham. l. c. 139–156. 3 pl.—Also **Knowlton, C. H.** Notes on the plants of Hingham, Massachusetts. Rhodora **26**: 175–177. 1924. (Plymouth County.)

Churchill, J. R. List of trees and plants growing naturally in Milton, Mass. *In* **Teele, A. K.** The history of Milton, Mass., 1640–1887. p. 600–613. [Boston, 1888?] (Reprinted 16 p. [1888?])

Annotated list of plants. (Norfolk County.)

Cobb, N. A. A list of plants found growing wild within thirty miles of Amherst. 51 p. 24 cm. Northampton, Mass., 1887.

Unannotated list of vascular and cellular plants, with occasional localities. Replaces Tuckerman, Edward, and Frost, C. C. A catalogue of plants growing without cultivation within thirty miles of Amherst College. 1875.—For additions see **Clark, H. L.** Additions to the flora of Amherst . . . Rhodora **1**: 164–165. 1899.—**Harper, R. M.** Further additions . . . l. c. **2**: 68–70. 1900. (Hampshire County and much of Franklin, Hampden, and Worcester Counties.)

Dame, L. L., and Collins, F. S. Flora of Middlesex County, Massachusetts. 201 p. map. 23.5 cm. Malden, 1888.

Geology, etc.; collectors; annotated list of vascular (1,544 species and varieties) and cellular plants.—See also **Cook, M. P.** Some additions to the "Flora of Middlesex County . . ." Rhodora **1**: 80–82. 1899.—**Smith, E. C.** Further additions . . . l. c. **1**: 97–98. 1899.—**Hosmer, A. W.** Further additions . . . l. c. **1**: 223–224. 1899.

Deane, Walter. Flora of the Blue Hills, Middlesex Fells, Stony Brook and Beaver Brook Reservations, of the Metropolitan Park Commission, Massachusetts. viii, 144 p. 22.5 cm. Boston, 1896.

Annotated list of vascular and cellular plants. The reservations are in Norfolk (Blue Hills), Middlesex (Middlesex Fells, Beaver Brook), and Suffolk (Stony Brook) Counties.—See also **Clark, Arthur.** A few plants of the Blue Hills Reservation. Rhodora **4**: 74–76. 1902.

Fogg, J. M. The flora of the Elizabeth Islands, Massachusetts. Rhodora **32**: 119–132, 147–161, 167–180, 208–221, 226–258, 263–281. 1 fig. (map). 1930. (Contrib. Gray Herb. no. 91.)

Geology, history, botanical explorations, topography, plant associations, changes in flora, origin of flora, phytogeography; annotated list of vascular plants (558 indigenous species). (Dukes County.)

Gardner, G. B. The Nantucket flora. *In* **Douglas-Lithgow, R. A.** Nantucket, a history. p. 245–268. New York, 1914.

General features of flora; unannotated compiled list of vascular plants. (Nantucket County.)

Goodale, A. S., compiler. A check list of Pteridophyta and Spermatophyta (fern-allies and seed-plants) occurring in the Connecticut River watershed in Massachusetts. 72, 5 p. 29 cm. Amherst, Mass., 1929.

Unannotated list of vascular plants occurring in Hampshire and parts of Franklin, Hampden, Worcester, and Berkshire Counties. (Processed publication.)

Hedge, C. E. Wild flowers of Plymouth and vicinity, 1804–1904. 46 p. 20.5 cm. Boston, 1904.

Unannotated list of vascular plants and a few fungi. (Plymouth County.)

Hervey, E. W. Flora of New Bedford and the shores of Buzzards Bay. rev. ed. 137 p. 23 cm. New Bedford, 1911. (1st ed., 1860: A catalogue of the plants . . .)

Unannotated list of vascular and cellular plants (1,524 species) preceded by a "Procession of the flowers," consisting of an annotated list of plants arranged in groups according to flowering dates. (Bristol County.)

Hill, E. S. Flora. *In* **Green, S. A.** The natural history and the topography of Groton, Massachusetts. [v. 1,] p. 1–26. Groton, 1912.

List of vascular plants, with localities and flowering dates; sedges and grasses omitted. (Middlesex County.)—See also **Knowlton, C. H.** Rare plants in Groton . . . Rhodora **14:** 234–235. 1912.

Hoffmann, Ralph. Flora of Berkshire County, Massachusetts. Proc. Boston Soc. Nat. Hist. **36:** 171–382. 1922.

History of botanical collecting, physiography, plant associations; annotated list of 1,656 vascular plants (1,377 species); lists of fugitive and excluded species.—See also **Wallace, G. J.** Some recent additions . . . Rhodora **41:** 128–130. 1939.

Jackson, Joseph. A catalogue of the flowering plants and ferns of Worcester County, Massachusetts. 3d ed., rev. and enl. v, 102 p. 9 pl. 22 cm. Worcester, 1909. (1st ed., 1883; 2d ed., 1894, with title: Flora of Worcester County. A catalogue of the phaenogamous and vascular cryptogamous plants . . . [Also issued as Appendix A to his Through glade and mead.])—Additions . . . 49 p. Worcester, 1927.

Briefly annotated list of 1,661 species and varieties of vascular plants (including additions).—See also **Potter, D.,** and **Woodward, N. P.** Notes on the flora of Worcester County, Massachusetts. Rhodora **37:** 80–88. 1935.

Lewis, I. F. The flora of Penikese, fifty years after. Rhodora **26:** 181–195, 211–219, 222–229. pl. 146–147. 1924–1925.

Contains (p. 219, 222–229) annotated list of vascular plants, with asterisk indicating those not recorded in D. S. Jordan's list (The flora of Penikese Island. Amer. Nat. **8:** 193–197. 1874). Pteridophyta by Alma G. Stokey; Spermatophyta by John M. Fogg. (Dukes County.)

McAtee, W. L. The winter flora of Muskeget Island, Massachusetts. Rhodora **18:** 93–99. 1916.

Topography, general features of flora; annotated list of 54 vascular plants. (Nantucket County.)

New England Botanical Club. Reports on the flora of the Boston district. *In* Rhodora v. 9–18, 20–26. 1907–16, 1918–24.

Annotated list of vascular plants (2,030 species, 421 varieties, forms, and hybrids, of which 1,242 species and 301 varieties and forms are native) of most of eastern Massachusetts (essentially the area from the New Hampshire line south to about 42° N. Lat. and west to eastern border of Worcester County), excluding Cape Cod. An index to families, by C. H. Knowlton, in Rhodora **31:** 41–43. 1929. Greater part of list by C. H. Knowlton and Walter Deane.

Palmer, E. J. The spontaneous flora of the Arnold Arboretum. Jour. Arnold Arboretum **11:** 63–119. 1930.—Supplement . . . l. c. **16:** 81–97. 1935.

Physiography, geology, general features of flora; annotated list of vascular plants. (In Jamaica Plain, Norfolk County.)

Robinson, John. The flora of Essex County, Massachusetts. 200 p. 26 cm. Salem, 1880.

Plant geography, biographical sketches of early botanists, bibliography; annotated list of vascular and cellular plants.

Simonds, A. B., Whittemore, G. F., Farrar, W. G., and **Hartwell, E. A.** Catalogue of the phaenogamous and vascular cryptogamous plants of Fitchburg and vicinity. 39 p. 23 cm. Fitchburg, 1885.

Unannotated list of 816 vascular plants. (Worcester County.)

Stone, G. E. A list of plants growing without cultivation in Franklin, Hampshire and Hampden Counties, Massachusetts. vii, 72 p. 21.5 cm. Amherst, 1913.

Botanical explorations, etc.; partly annotated list of vascular plants (1,493 species).

Tracy, C. M. Studies of the Essex flora: an enumeration of all plants found growing naturally within the limits of Lynn, Mass., and towns adjoining, with notes as to localities and habits. 99 p. 20 cm. Lynn, 1892. (1st ed., 1858.)

General features of flora; annotated list of spermatophytes. (Essex County.)

Waltham. Botany Club. A partial list of the native flora of Waltham, Mass. 12 p., 36 leaves. 14.5 cm. Waltham, 1883.

Unannotated list of vascular plants, mosses, and lichens, with flowering dates. (Middlesex County.)

Young, H. A. Catalogue of the flora of "Oak Island," Revere, Massachusetts, with notes. Bul. Essex Inst. **14**: 141–157. map. 1883. (Reprinted, 19 p.)

General features and statistics of flora; annotated list of 362 vascular plants, mosses, and hepatics. The introduction is by John Robinson.—For additions see **Rich, W. P.** Oak Island and its flora. Rhodora **4**: 87–94. 1902. (Suffolk County.)

MICHIGAN

General

Beal, W. J. Michigan flora. Ann. Rpt. Mich. Acad. Sci. **5**: 1–147. 1904. (Reprinted with subtitle: A list of the fern and seed plants growing without cultivation.)—Additions . . . l. c. **10**: 85–89. 1908.

Climate, distribution, etc.; annotated list of 2,365 species (including the "Additions") of vascular plants; bibliography. Replaces Beal and Wheeler, C. F. Michigan flora. 1892.—See also **Hermann, F. J.** Notes on the flora of Michigan I. Rhodora **38**: 362–367. 1936.—**Hanes, C. R.** Additions to the flora of Michigan. Papers Mich. Acad. Sci. **23**: 135–139. 1938.—**Hanes, C. R.** Plants new or rare in Michigan records. l. c. **24**: 3–7. 1939.—See also **Farwell, O. A.** Contributions to the botany of Michigan no. 14. Rpt. Mich. Acad. Sci. **17**: 167–182. 1916. Contributions . . . no. 15–16. Amer. Midl. Nat. **11**: 72–85, 261–269. 1928–29.—Also his Notes on the Michigan flora I–III. Rpt. Mich. Acad. Sci. **20**: 161–195. 1918 (1919); **21**: 345–371. 1920; **22**: 177–185. 1921. Notes . . . IV–VII. Papers Mich. Acad. Sci. **1**: 85–100. 1923; **2**: 11–46. 1923; **3**: 87–109. 1924; **23**: 123–134. 1938.—Also his Botanical gleanings in Michigan [I]–VII. Amer. Midl. Nat. **8**: 263–280. 1923; **9**: 259–282. 1925; **10**: 19–46, 199–219. 1926–27; **11**: 41–71. 2 fig. 1928; **12**: 44–78, 113–134. 4 fig. 1930.

Gleason, H. A. The plants of Michigan. Simple keys for the identification of the native seed plants of the State. xlvii, 158 p. 19 cm. Ann Arbor, 1918.

Unannotated keys.

Lyons, A. B. The indigenous medical plants of Michigan. Detroit Lancet **1**: 107–118, 197–209. 1878. (Reprinted 25 p., with cover title: Medicinal plants indigenous in Michigan.)

Running account of medicinal plants, in systematic order.

Otis, C. H. Michigan trees. A handbook of the native and most important introduced species. 6th ed., rev. 247 p. illus. (incl. map), pl. 19 cm. Ann Arbor, 1923. (University bulletin n.s., v. 25, no. 5.) (1st issue, 1913.)

Annotated descriptive list, with keys; figure of each species.

Spalding, V. M. Native medicinal plants of Michigan. Proc. Mich. Pharm. Assoc. **4**: 18–27. 1877.

List of vascular and cellular plants, annotated only as to abundance.

Stearns, Frederick. The medicinal plants of Michigan. Proc. Amer. Pharm. Assoc. **7**: 236–285. 1858.

Annotated alphabetical list of nearly 600 wild medicinal plants.

Woodcock, E. F. Observations on the poisonous plants of Michigan. Amer. Jour. Bot. **12**: 116–131. 1925.

Annotated list of 156 vascular plants; bibliography.

Local

Allabach, L. F. Plants collected in Les Cheneaux Islands, Michigan. Trillia **6**: 33–40. 1921.

Geology, topography, general features of flora; mostly unannotated list of vascular plants. Les Cheneaux Islands are in northern Lake Huron, 14 miles northeast of Mackinac Island. (Mackinac County.)

Allmendinger, E. C. Flora of Ann Arbor and vicinity. Proc. Ann. Arbor Scient. Assoc. **1876**: 85–116. 1876.

Mostly unannotated list of 848 vascular plants, found within a 4-mile radius. (Washtenaw County.)

Billington, Cecil. The flowering plants and ferns of Warren Woods, Berrien County, Michigan. Papers Mich. Acad. Sci. **4** (1): 81–110. 1 fig. (map), pl. 9–11. 1925.

Plant associations; annotated list of 358 vascular plants.

Brown, C. A. Ferns and flowering plants of Isle Royale, Michigan. 90 p. 21 fig. 23.5 cm. [Washington, D. C., 1937.]

Physiography, climate, botanical explorations, general features of flora, changes in vegetation; annotated list of 671 species and varieties of vascular plants; bibliography. A publication of the National Park Service. (Keweenaw County.)

Cole, E. J. Grand Rapids flora: a catalogue of the flowering plants and ferns growing without cultivation in the vicinity of Grand Rapids, Michigan. xxii, 170 p. 2 pl., map. 25.5 cm. Grand Rapids, 1901.

Geology, topography, climate, etc.; annotated list of 1,290 vascular plants. (Kent County.)

Coleman, Nathan. Catalogue of the flowering plants of the southern peninsula of Michigan, with a few of the Cryptogamia. 49 p. 21 cm. Grand Rapids, 1874. (Kent Scientific Institute. Miscellaneous publications no. 2.)

Mostly unannotated list of vascular plants.

Cooper, W. S. A catalogue of the flora of Isle Royale, Lake Superior. Rpt. Mich. Acad. Sci. **16:** 109–131. 1914.

Annotated list of 519 vascular plants; bibliography. (Keweenaw County.)

Dachnowski, Alfred. Flora of the Marquette quadrangle. Rpt. Mich. Acad. Sci. **9:** 88–103. 1907.

Botanical explorations, topography, botanical distribution, bibliography; tabular list of vascular plants, with local habitats indicated. (Marquette County.)

Daniels, F. P. Ecology of the flora of Sturgis, Mich., and vicinity. Rpt. Mich. Acad. Sci. **4:** 145–159. 1904.

General features of flora, plant formations; list of 670 vascular plants, ecologically arranged. (St. Joseph County.)

——— The flora of the vicinity of Manistee, Mich. Rpt. Mich. Acad. Sci. **4:** 125–144. 1904.

Topography, plant associations; list of 750 vascular plants, divided ecologically. (Manistee County.)

Darlington, H. T. Contributions to the flora of Gogebic County, Michigan. Ann. Rpt. Mich. Acad. Sci. **22:** 147–176. 1921.—Papers Mich. Acad. Sci. **1:** 74–82. 1923.

Topography, general features of flora; annotated list of vascular plants.

——— Vegetation of the Porcupine Mountains, northern Michigan. Pt. II. Floristics. Papers Mich. Acad. Sci. **22:** 33–68. pl. 5–6. 1937.

Annotated list of vascular plants; bibliography. (Ontonagon County.)

Davis, C. A. The native vegetation of Tuscola County. Notes on the factors affecting plant distribution. *In* his Report on the geology of Tuscola County, Michigan. Rpt. State Bd. Geol. Survey Mich. **1908:** 290–346. 1909.

Temperature, soils, plant associations; annotated list of vascular plants.

Dodge, C. K. A botanical trip to Thunder Bay Island. Rpt. Mich. Acad. Sci. **10:** 40–41. 1908.

Includes running account of plants observed. (In Lake Huron, Alpena County.)

——— Catalog of plants. *In* **Ruthven, A. G.** A biological survey of the sand dune region on the south shore of Saginaw Bay, Michigan. Pub. Mich. Geol. and Biol. Survey **4** (Biol. Ser. **2**): 65–120. 1911.

Annotated list of 886 vascular plants; general features of vegetation. (Huron County.)

——— Flora of St. Clair County, Michigan, and the western part of Lambton County, Ontario. Ann. Rpt. State Hort. Soc. Mich. **29:** 231–313. 1900.

Annotated list of 1,112 vascular plants.

——— The flowering plants, ferns and their allies of Mackinac Island. Rpt. Mich. Acad. Sci. **15:** 218–237. 1913.

Topography, general features of flora; annotated list of over 400 vascular plants. Two unnumbered pages of Addenda are included. (Mackinac County.)

——— Observations on the flowering plants, ferns and fern allies growing wild in Marquette County, Michigan, in 1916 and 1917, especially in the vicinity of the Huron Mountain Club. (Contributions to the botany of Michigan II.) Misc. Pub. Mus. Zool. Univ. Mich. no. 5. 44 p. map. 1918.

Annotated list of vascular plants.

Dodge, C. K. Observations on the flowering plants, ferns and fern allies growing wild in Schoolcraft County and vicinity in the Upper Peninsula of Michigan in 1915. Pub. Mich. Geol. and Biol. Survey **31** (Biol. Ser. **6**) : 75–123. 1921.

Physiography, etc.; annotated list of vascular plants.

——— Observations on the flowering plants, ferns and fern allies growing without cultivation in Tuscola County, Michigan. Pub. Mich. Geol. and Biol. Survey **31** (Biol. Ser. **6**) : 165–222. 1921.

Physiography, etc.; annotated list of vascular plants.

——— Observations on the flowering plants, ferns, and fern allies on and near the shore of Lake Huron from Linwood Park near Bay City, Bay County, to Mackinaw City, Cheboygan County, including the vicinity of St. Ignace, Mackinac, and Bois Blanc Islands, Mackinac County, Michigan. Pub. Mich. Geol. and Biol. Survey **31** (Biol. Ser. **6**) : 15–74. map. 1921.

Geography, physiography, etc.; annotated list of vascular plants.

——— Observations on the wild plants at Whitefish Point and Vermilion, near the south shore of Lake Superior, and other parts of Chippewa County, Michigan, in 1914. Pub. Mich. Geol. and Biol. Survey **31** (Biol. Ser. **6**) : 125–164. 1921.

Physiography, habitats, etc.; annotated list of vascular plants.—See also **Hermann, F. J.** Additions to the flora of Chippewa County . . . Papers Mich. Acad. Sci. **22**: 85–89. 1937.

——— Results of the Mershon expedition to the Charity Islands, Lake Huron. Rpt. Mich. Acad. Sci. **13**: 173–190. 1911.

Annotated list of 372 vascular plants. The Charity Islands are in Saginaw Bay in Arenac County.

Farwell, O. A. A catalogue of the flora of Detroit. Rpt. Mich. Acad. Sci. **2**: 31–68. 1901. (Reprinted with additions from 11th Ann. Rpt. Commr. Parks and Boulevards, Detroit, 1900.)—Additions and corrections to the flora of Detroit. l. c. **6**: 209–214. 1904. (*In* his Contributions to the botany of Michigan no. 8.)

Annotated list of 885 species and minor forms of vascular plants. This forms one [no. 7a] of his Contributions to the botany of Michigan. (Wayne County.)

——— The flora of Parkedale Farm, with special reference to Stony Creek Valley. Rpt. Mich. Acad. Sci. **15**: 150–192. 1 fig. (map). 1913. (His Contributions to the botany of Michigan no. 9.)

Topography, plant associations; annotated list of 479 vascular plants. (Oakland County.)

Gates, F. C., and Ehlers, J. H. An annotated list of the higher plants of the region of Douglas Lake, Michigan. Papers Mich. Acad. Sci. **4**: 183–284. 1925.—Additions . . . l. c. **8**: 111–120. 1928; **13**: 67–88. 1931.

List of 1,026 vascular plants; bibliography. The "Additions" bring the total to 1,177 species. (In Emmet and Cheboygan Counties.)

Gilmore, M. R. Some Chippewa uses of plants. Papers Mich. Acad. Sci. **17**: 119–143. 1933.

Annotated systematic list, with Chippewa names and uses.

Hebert, P. E. Ferns and flowering plants of Berrien County, Michigan. Amer. Midl. Nat. **15**: 328–342. 1934.

Unannotated list of 895 vascular plants.

Kellogg, H. S. The flora of the Rainy River region. Proc. Iowa Acad. Sci. **22**: 60–75. pl. 16–17. 1915. (Reprinted in Contrib. Bot. Dept. Iowa State Col. no. 60.)

List of cellular and vascular plants mostly collected by the author in 1914. (Cheboygan and adjacent counties.)

Pammel, L. H., and Cratty, R. I. Some notes on plants of Whitehall, Michigan. Proc. Iowa Acad. Sci. **30** (1923) : 279–285. 7 fig. [1924.]

General features of flora; unannotated list of vascular plants. (Muskegon County.)

Sutton, J. M. Flora of the Detroit Zoological Tract. Ann. Rpt. Mich. Acad. Sci. **19**: 263–271. fig. 13 (map). 1917.

Unannotated list of vascular plants. (Oakland County.)

Walpole, B. A. Flora of Washtenaw County, Michigan. 80 p. (incl. map). 23 cm. Ypsilanti, [1924].

Collecting localities; annotated list of 1,202 vascular plants; bibliography (p. 5).—See also **Hermann, F. J.** Additions to the flora . . . Papers Mich. Acad. Sci. **21**: 81–87. 1936; **22**: 91–94. 1937; **24**: 17–23. 1939.

MINNESOTA

See also Wisconsin (Cheney, A contribution to the flora; Pammel; Smith, Ethnobotany of the Ojibwe Indians).

General

Rosendahl, C. O., and Butters, F. K. Trees and shrubs of Minnesota. vii, 385 p. illus. (incl. map). 26.5 cm. Minneapolis, 1928.

Descriptive flora of native and cultivated woody plants, with keys; general features of vegetation; glossary. Replaces Minnesota trees and shrubs, by F. E. Clements, C. O. Rosendahl, and F. K. Butters, 1912, though not considered a second edition.

Upham, Warren. Catalogue of the flora of Minnesota, including its phaenogamous and vascular cryptogamous plants, indigenous, naturalized, and adventive. 193 p. 24 cm. Minneapolis, 1884. (Ann. Rpt. Geol. and Nat. Hist. Survey Minn. 1883, pt. 6.)—Supplement to the flora of Minnesota. Bul. Geol. and Nat. Hist. Survey Minn. **3**: 46–54. 1887.

History of botanical exploration, including bibliography, topography, etc.; annotated list of vascular plants.

Weschcke, Carl. An enumeration of the medicinal plants of the State of Minnesota. Pharm. Rundschau (N. Y.) **8**: 155–157. 1890.

Unannotated systematic list of vascular plants.

Local

Arthur, J. C., Bailey, L. H., and Holway, E. W. D. Plants collected between Lake Superior and the International Boundary, July, 1886. Bul. Geol. and Nat. Hist. Survey Minn. **3**: 10–43. 1887.

List of vascular and cellular plants, with localities. (St. Louis and Lake Counties.)

Bailey, L. H. Chippeway plant names. Bot. Gaz. **12**: 37–39. 1887.

List of Indian names obtained in northern Minnesota, with botanical equivalents.

Densmore, Frances. Uses of plants by the Chippewa Indians. Ann. Rpt. Bur. Amer. Ethnol. **44**: 275–397. pl. 28–63. 1928 (1929).

Table of plants used, with scientific and vernacular names and translations of latter, uses, and references; sections on medicinal, food, dye, charm, and decorative plants; references. The data were gathered in Minnesota, Wisconsin, and Ontario.

MacMillan, Conway. The Metaspermae of the Minnesota valley. A list of the higher seed-producing plants indigenous to the drainage basin of the Minnesota River. xiii, 826 p. 2 maps. 25.5 cm. Minneapolis, 1892. (Rpt. Geol. and Nat. Hist. Survey Minn. Bot. Ser. 1.)

Annotated list of 1,174 Angiospermae, with rather extensive synonymy and references; general features of plant distribution, statistical tables, etc; bibliography. Covers most of southern and southwestern Minnesota and small adjacent areas in South Dakota and Iowa (see p. 571).

Pammel, L. H. The flora of Star Island and vicinity. Ames Forester **3**: 119–141. 3 fig. (maps), 7 pl. 1915.

Bibliography, geology, topography, plant formations, etc.; list of cellular plants, annotated list of vascular plants. (In Cass Lake, Cass County.)

Reagan, A. B. Plants used by the Bois Fort Chippewa (Ojibwa) Indians of Minnesota. Wisconsin Archeol. n. s., **7**: 230–248. 1928.

Annotated systematic list. (Koochiching County.)

Rosendahl, C. O., and Butters, F. K. Reputed Minnesota plants which probably do not occur in the State. Minn. Bot. Studies **4**: 461–473. 1916.

Bibliography; list of vascular plants incorrectly reported or not authenticated by specimens.

Wheeler, W. A. A contribution to the knowledge of the flora of southeastern Minnesota. Minn. Bot. Studies **2**: 353–416. pl. 21–27. 1900.
Topography, plant associations; annotated list of 578 vascular plants collected in the summer of 1899, in the vicinity of Winnebago, Faribault County.—See also **Rosendahl, C. O.** An addition . . . l. c. **3**: 257–269. 1903.

——— A contribution to the knowledge of the flora of the Red River Valley in Minnesota. Minn. Bot. Studies **2**: 569–600. pl. 34–41. 1901.
Soil, etc.; list of 325 angiosperms collected in August 1900, with localities and collectors' names. The Red River of the North constitutes the boundary between Minnesota and North Dakota.

MISSISSIPPI

See also Louisiana (Lloyd and Tracy).

General

Lowe, E. N. Plants of Mississippi. A list of flowering plants and ferns. 24 cm. Bul. Miss. State Geol. Survey 17. 292 p. (incl. map.) 1921.
Topography, floral regions; annotated list of vascular plants.

Phares, D. L. Synopsis of the medical flora of the State of Mississippi. Ann. Rpt. Miss. State Bd. Health **1**: 139–180. 1877.
Annotated systematic list.

Local

Brown, C. S. Tishomingo State Park, botany. Bul. Miss. State Geol. Survey **32**: 37–77. 1936.
Annotated list of vascular plants. (Tishomingo County.)

Lauderburn, D. E. Trees of the Yazoo-Mississippi Delta and their economic values. Ext. Bul. Miss. State Col. **67**. 32 p. illus. 1933.
Annotated list, with uses. Adapted from Putnam, J. A., and Bull, Henry. Trees of the bottomlands of the Mississippi River Delta region. Occas. Paper South. Forestry Expt. Sta. no. 27. 1932. (Region from Tennessee line to Vicksburg, extending east for 10 to 70 miles to the loessial bluffs.)

MISSOURI

General

Bush, B. F. The trees, shrubs and vines of Missouri. Ann. Rpt. State Hort. Soc. Missouri **37**: 353–393, i–v. 1895.
Annotated list of 294 species.—See also his Notes on trees and shrubs of Missouri. Amer. Midl. Nat. **10**: 133–158. 1926. (Supplementary notes (on 97 species) to Tracy's Flora (1886); most of the species mentioned are excluded from the flora of the State.)

Palmer, E. J., and Steyermark, J. A. An annotated catalogue of the flowering plants of Missouri. Annals Missouri Bot. Gard. **22**: 375–758. 6 fig. (maps), pl. 15–21. 1935.
Botanical explorations, physiography, climate, geology, botanical regions, bibliography; annotated list of spermatophytes (2,281 species plus 209 varieties and forms). This figure includes the Pteridophyta, not treated in this paper; for the latter see their The ferns and fern allies of Missouri. Amer. Fern Jour. **22**: 105–122. 2 fig. 1932 (1933); also **Pinkerton, M. E.** Ferns and fern allies of Missouri. Annals Missouri Bot. Gard. **20**: 45–77. fig. 1 (map), pl. 5, fold. tab. 1933.—See also **Steyermark, J. A.** Plants new to Missouri. Rhodora **40**: 249–255. 1938.—**Palmer, E. J., and Steyermark, J. A.** Additions, corrections, and revisions . . . Annals Missouri Bot. Gard. **25**: 775–794. 1938.—Replaces Tracy, S. M. Flora of Missouri. Rpt. Missouri State Hort. Soc. 28 (1885), app. 106 p. 1886. (Also reprinted with title: Catalogue of the phaenogamous and vascular cryptogamous plants of Missouri.)

Pammel, L. H. Poisonous and medicinal plants of Missouri. Bul. Missouri State Bd. Hort. no. 14 46 p. 26 fig. [1910?]
Local laws relating to poisonous plants, etc.; running account of principal species.

Local

Bush, B. F. Notes on the mound flora of Atchison County, Missouri. Ann. Rpt. Missouri Bot. Gard. **6**: 121–134. 1895.

Topography, general features of flora; annotated list of angiosperms collected by author in 1893.

Daniels, F. P. The flora of Columbia, Missouri, and vicinity. Univ. Missouri Studies Sci. Ser. v. 1, no. 2. ix, 319 p. map. 1907. (Has also continuous paging (143–461) of volume.)

Annotated list of native and introduced vascular plants (1,058 species and 47 varieties); ecology, bibliography.—For corrections and additions see **Bush, B. F.,** Amer. Midl. Nat. **12**: 185–194, 343–362. 1931; **16**: 83–93. 1935.—See also **Rickett, H. W.** Flora of Columbia, Missouri. 1931. (Boone County.)

Engelmann Botanical Club. A preliminary check list of the cryptogams and phanerogams in the vicinity of Saint Louis, Missouri. [9], 63 p. 21 cm. Saint Louis, 1911.

Unannotated list of cellular and vascular plants, alphabetically arranged, with localities indicated. The area covered includes a small part of adjacent Illinois. Replaces Eggert, Henry (i. e. Heinrich). Catalogue of the phaenogamous and vascular cryptogamous plants in the vicinity of St. Louis, Mo. 16 p. 23 cm. St. Louis, [1891].—See also **Steyermark, J. A.** Recent additions to the flora of St. Louis County, Missouri. Rhodora **36**: 375–376. 1934.

Mackenzie, K. K., Bush, B. F., and others. Manual of the flora of Jackson County, Missouri. xix, 242 p. 24.5 cm. Kansas City, Mo., 1902.

Physiography, etc.; briefly descriptive annotated flora of vascular plants (1,141 species and 51 varieties).

Missouri Botanical Garden. Check-list of the ferns and flowering plants indigenous to the Missouri Botanical Garden extension, Gray Summit, Missouri. Missouri Bot. Gard. Bul. **16**: 65–74, 83–91. 1928.

Unannotated list of vascular plants. (Franklin County.)

——— Medicinal plants of our local flora. Missouri Bot. Gard. Bul. **10**: 118–123, 137–142, 153–157, 166–169. 1922.

Annotated list, the species alphabetically arranged and divided into herbaceous and woody groups. Refers to the vicinity of St. Louis.

——— Plants of our local flora which have been used by man as food. Missouri Bot. Gard. Bul. **10**: 102–110. 1922.

Annotated list, classified by parts used.

Palmer, E. J. Catalogue of the plants of Jasper County, Missouri (fernworts and flowering plants). Annals Missouri Bot. Gard. **3**: 345–401. 1916.

Physiography, geology, floral areas; list of vascular plants, with collecting numbers of the author.

——— Flora of the Grand Falls chert barrens. Trans. Acad. Sci. St. Louis **19**: 97–112. 1910.

Geology, topography, general features of flora; annotated list of vascular plants. (Newton and Jasper Counties.)

Rickett, H. W. Flora of Columbia, Missouri. Univ. Missouri Studies v. 6, no. 1. 84 p. map. 1931.

Essentially a revision of Daniels' Flora (q. v.) and supplementary to it; list, with occasional annotations (1,005 species).—For corrections and additions see **Drouet, F.** Notes on the flora . . . Rhodora **35**: 359–364. 1933; **37**: 189–196. 1935. (Boone County.)

——— A list of plants from the Missouri Ozarks. Amer. Midl. Nat. **11**: 243–254. 1928.

Itinerary; list of vascular plants collected by the author in 1927, with localities. (South-central Missouri.)

MONTANA

General

Blankinship, J. W. A century of botanical exploration in Montana, 1805–1905: collectors, herbaria and bibliography. Montana Agr. Col. Sci. Studies **1**: 3–31. 1904.

Annotated list of collectors; annotated bibliography.

Blankinship, J. W. Native economic plants of Montana. Bul. Montana Agr. Expt. Sta. 56. 36 p. 1905.

Annotated alphabetical list of wild economic plants; bibliography; list of species classified by uses.

——— and **Henshall, H. F.** Common names of Montana plants. Montana Agr. Col. Sci. Studies **1**: 113–139. col. pl. 1905.

Alphabetical lists of vernacular and botanical names, with equivalents.

Rydberg, P. A. Catalogue of the flora of Montana and the Yellowstone National Park. Mem. New York Bot. Gard. v. 1. xi, 492 p. map. New York, 1900.

List of collectors, with dates and location of specimens; annotated list of 1,976 vascular plants, with references and citation of exsiccatae.—See also **Blankinship, J. W.** Supplement to the flora of Montana: additions and corrections. Montana Agr. Col. Sci. Studies **1**: 33–109. 6 fig. 1905.

Local

Grinnell, G. B. The Cheyenne Indians, their history and ways of life. 2 v. plates, map. 24.5 cm. New Haven, 1923.

Includes (v. 2, p. 166–191) annotated systematic list of 94 useful plants, with botanical and vernacular names.

Jones, M. E. Montana botany notes, containing descriptions of new species, list of plants not heretofore recorded from the State, and notes on disputed species. Bul. Univ. Montana no. 61, Biol. Ser. no. 15. 74 p. (incl. 2 fig., 5 pl.). 1910.

Partly annotated list of about 1,700 vascular and cellular plants, with localities, constituting essentially a flora of the Flathead Lake region, Flathead County, but including species from other localities; gazetteer.

Standley, P. C. Flora of Glacier National Park, Montana. Contrib. U. S. Natl. Herb. **22**: 235–438. pl. 33–52. 1921.

Briefly annotated list, with keys and partial descriptions, of the vascular plants (955 species); physiography, life zones, botanical explorations, annotated bibliography. A more popular account of the same flora is his "Plants of Glacier National Park." U. S. Dept. Interior, National Park Service. 110 p. 150 fig., 5 col. pl. 1926.—For additions, see **Graff, P. W.** Unreported plants from Glacier National Park. Bul. Torrey Club **49**: 175–181. 1922.—**Maguire, Bassett.** Distribution notes concerning certain plants of Glacier National Park, Montana. Rhodora **36**: 305–308. 1934; **41**: 504–508. 1939.—**McLaughlin, W. T.** Notes on the flora of Glacier National Park, Montana. Rhodora **37**: 362–365. 1935. (Includes bibliography.) (Flathead and Glacier Counties.)

NEBRASKA

General

Clements, F. E. A preliminary list of the botanical expeditions in Nebraska, 1803–1893. Rpt. Bot. Survey Nebraska **3**: 39–42. 1894.

Chronological list of 18 expeditions with localities covered.

Petersen, N. F. Flora of Nebraska. A list of the ferns, conifers and flowering plants of the State with keys for their determination. 3d ed. 220 p. 21 cm. Lincoln, 1923. (1st ed., 1912.)

Annotated list of vascular plants, with keys.

Pool, R. J. Handbook of Nebraska trees. A guide to the native and most important introduced species. Bul. Nebraska Conserv. and Soil Survey 7. 2d ed. 179 p. illus. (incl. maps). 1929. (Contribution from the Botanical survey of Nebraska n. s., no. 3.) (1st ed., 1919.)

Annotated descriptive list, with figure of each species, and maps showing local range of most species.

Pound, Roscoe. Bibliography of the flora of Nebraska. Rpt. Bot. Survey Nebraska **3**: 43–48. 1894.

Unannotated list of papers, arranged by authors; critical notes on Aughey's Catalogue of the flora of Nebraska. Republished with additions in Pound, R., and Clements, F. C. The phytogeography of Nebraska. 2d ed. *In* University of Nebraska. Botanical survey of Nebraska **8**: 25–30. 1900.

Winter, J. M. An analysis of the flowering plants of Nebraska, with keys to the families, genera, and species, and with notes concerning their occurrence, range, and frequency within the State. Bul. Conserv. Dept. Conserv. and Surv. Div. Univ. Nebraska no. 13. iii, 203 p. 23 cm. [Lincoln,] 1936. (Contribution from the Botanical survey of Nebraska n. s., no. 10.)

Annotated list of angiosperms, with keys.

Local

Bessey, C. E. The forest trees of eastern Nebraska. Proc. Iowa Acad. Sci. **13**: 75–87 (incl. maps). 1906.

Annotated list of 67 species, each with a small map showing local range.

Gilmore, M. R. Some native Nebraska plants with their uses by the Dakota. Collect. Nebraska State Hist. Soc. **17**: 358–370. 1913.

Annotated systematic list, with vernacular names.

——— A study in the ethnobotany of the Omaha Indians. Collect. Nebraska State Hist. Soc. **17**: 314–357. 1913.

Running account of useful plants grouped by uses; annotated systematic list, with vernacular names; list of plants grouped by uses; bibliography. (Eastern Nebraska.)

——— Uses of plants by the Indians of the Missouri River region. Ann. Rpt. Bur. Amer. Ethnol. **33**: 43–154. 30 pl. 1919.

General considerations: annotated systematic list of plants, with vernacular names and uses; lists of plants discussed, with names in different Indian languages; bibliography. Refers mainly to Nebraska.

Rydberg, P. A. Flora of the sand hills of Nebraska. Contrib. U. S. Natl. Herb. **3**: 133–200. pl. 2. 1895.

Author's itinerary in 1893, topography, floral districts, etc.; annotated list of vascular plants collected; numerical list of specimens. The sand hill region extends from about 98°–103° W. Long., with the Niobrara and North Platte Rivers as north and south boundaries. The author collected chiefly in Thomas, Hooker, and Grant Counties.

Smith, J. G., and **Pound, Roscoe.** Flora of the sand hill region of Sheridan and Cherry Counties and list of plants collected on a journey through the sand hills in July and August, 1892. *In* University of Nebraska. Botanical survey of Nebraska **2**: 5–30. 1893.

Topography, general features of flora: list of vascular and cellular plants collected, with localities.

Tillotson, C. R. The trees of Lincoln and vicinity. Ann. Rpt. Nebraska State Bd. Agr. **1906/07**: 213–236. 1907.

Annotated list, including cultivated species. (Lancaster County.)

NEVADA

General

Tidestrom, Ivar. Flora of Utah and Nevada. Contrib. U. S. Natl. Herb. v. 25. 665 p. 15 pl., map. 1925.

Physiography, life zones; plant communities (by H. L. Shantz), the foothill-montane-alpine flora and its environment (by A. W. Sampson); systematic list of vascular plants (about 3,700 species), with keys, references, local habitat, and general range. Treatments of several families contributed by various authors.—For additions (relating principally to Utah) see **Maguire, Bassett.** Distributional notes on plants of the Great Basin region I–II. Leaflets West. Bot. **1**: 183–188. 1935; **2**: 23–26. 1937.

NEW HAMPSHIRE

General

Flint, W. F. The distribution of plants in New Hampshire. *In* **Hitchcock, C. H.** The geology of New Hampshire. p. 381–415. pl., map. Concord, 1874.

General features of flora; list of vascular plants (and lichens) with local range indicated. The latest State list, but long out of date.

Flint, W. F. The trees and shrubs composing the New Hampshire forests. Their distribution, relative abundance, and utility. Rpt. Col. Agr. and Mech. Arts New Hampshire **19**: 60–99. 1890. (Reprinted with some changes from Rpt. Forestry Comn. New Hampshire 1st, 1885.—Reprinted 42 p. 1891.)

Annotated list.—See also **Foster, J. H.** Trees and shrubs of New Hampshire. Bienn. Rpt. Forestry Comn. New Hampshire 1930. [pt. 2.] 131 p. illus. 1931. (Popular account.)

Local

Batchelder, F. W. Preliminary list of plants growing without cultivation in the vicinity of Manchester, New Hampshire. rev. and enl. Proc. Manchester Inst. Arts and Sci. v. 4, pt. 2. 59 p. 1909. (1st ed., 1900.)

List of 1,063 species and 360 additional forms of vascular plants, annotated as to abundance; additional notes on 26 species. The list of 1900 contained 817 names. (Hillsboro County.)

Beetle, A. A. Flowering plants and ferns of the Fox research forest, Hillsboro, New Hampshire. Bul. C. A. Fox Res. and Demonstr. Forest no. 9. 40 p. 1938.

Annotated list of vascular plants. The list contains rather numerous errors; for corrections of the most important ones see **Weatherby, C. A.**, and **Blake, S. F.** On certain plant records from Hillsboro, New Hampshire. Rhodora **41**: 34–36. 1939.—Also **Weatherby, C. A.** A correction. l. c. 138–139. 1939. (Hillsboro County.)

Harris, W. S. The flora of Windham, N. H. *In* **Morrison, L. A.** Supplement to the history of Windham in New Hampshire. p. 69–97. Boston, 1892. (Reprinted 31 p. 1892.)

Contains list of 606 vascular plants. (Rockingham County.)

Jesup, H. G. A catalogue of the flowering plants and higher cryptogams, both native and introduced, found within thirty miles of Hanover, N. H., including a few cultivated species . . . vii, 91 p. map. 23 cm. [Hanover,] 1891.

Physiography, etc.; annotated list of 1,161 vascular plants. Hanover is in Grafton County; parts of Sullivan, Merrimack, and Belknap Counties in New Hampshire, and of Orange, Windsor, and Rutland Counties, Vermont, are included.—See also **Eggleston, W. W.** Flora of Mt. Moosilauke. Rhodora **2**: 97–99. 1900.

Pease, A. S. List of plants on Three Mile Island. Pteridophyta and Spermatophyta. Appalachia **12**: 266–276. 1911.

Botanical explorations; unannotated list of 218 vascular plants. (In Lake Winnipesaukee, near Center Harbor, Belknap County.)

——— Notes on the botanical exploration of the White Mountains. Appalachia **14**: 157–178. 1917.

History of earlier botanical collecting, with bibliographical footnotes; list of collectors with dates.

——— Vascular flora of Coös County, New Hampshire. Proc. Boston Soc. Nat. Hist. **37**: 39–388. pl. 5–11. 1924.—Plants new to Coos County, N. H. Rhodora **30**: 104–105. 1928.

Physiography, soils, climate, general features of flora, history of botanical investigations, bibliography, etc.; annotated list of 1,297 vascular plants.

NEW JERSEY

See also New York (Taylor, Flora of the vicinity of New York); Pennsylvania (Keller and Brown; Smith).

General

Britton, N. L. Catalogue of plants found in New Jersey. Final Rpt. Geol. Survey New Jersey **2**: 27–642. 1889 [1890]. (Reprinted.)

Physiography, bibliography; annotated list of vascular and cellular plants (5,641 species and varieties, of which 1,995 are vascular plants).

Halsted, B. D. Forest trees of New Jersey. Bul. New Jersey Agr. Expt. Sta. 202. 52 p. 25 fig. 1907.

Annotated compiled list; bibliography.

——— The poisonous plants of New Jersey. Bul. New Jersey Agr. Expt. Sta. 135. 28 p. 10 fig. 1899.

Annotated account, the plants grouped according to poisonous character.

Local

Fender, F. S. The flora of Seven Mile Beach, New Jersey. Bartonia **19:** 23–41. 1938.

Botanical explorations, habitats, affinities of flora; annotated list of 416 vascular plants; bibliography. Seven Mile Beach is an island off the shore of Cape May County.

Harshberger, J. W. An ecological study of the New Jersey strand flora. Proc. Acad. Nat. Sci. Phila. [**52**] (1900): 623–671. 1901.—Additional observations on the strand flora of New Jersey. l. c. **54** (1902): 642–669. 6 fig. 1903.

Includes (p. 661–671; v. 54, p. 665–669) lists of plants known to occur on the beaches and salt marshes.

La Wall, C. H. Botanical notes from Longport, New Jersey. Bartonia **3:** 12–21. 1911.

Annotated list of 160 vascular plants. (Atlantic County.)

Moldenke, H. N. A list of plants observed at and near The Pines, Branchville, New Jersey, during the Branchville nature conference, May 24–26, 1935. 9 p. [n. p., 1935?]

Unannotated list of 371 vascular plants. (Processed publication.) (Sussex County.)

——— The observed flora of Watchung, N. J., and its immediate vicinity, together with a schematic classification of the plant kingdom to orders. 3d ed. 59 p. 23.5 cm. Watchung, 1934.

Unannotated list of cellular and vascular plants (1,170 wild species and varieties); statistics of the plant kingdom, giving the number of genera and species in the world by orders. The two previous editions of the list were issued in 1933 (hectographed). Ten multigraphed lists of additions have been issued (1 p., [n. d.]: 2 p., [n. d.]; 2 p., 1934; 5 p., 1935; 7 p., 1937; 1 p., 1937; 2 p., 1938; 3, 2, 3 p., 1939).—For general features of flora, see his The flora of the Watchung Mountains. Torreya **36:** 57–61, 88–93. 2 fig. (incl. map). 1936. (Somerset County.)

Stone, Witmer. The plants of southern New Jersey, with especial reference to the flora of the pine barrens and the geographic distribution of the species. Ann. Rpt. New Jersey State Mus. **1910:** 23–828. 5 fig. 129 pl. 1911 [1912]. (Reprinted.)

Local herbaria, phytogeography, floral districts, etc.; annotated list of vascular plants, with keys; bibliography, gazetteer. (Burlington and Monmouth Counties and southward.)

Walker, H. G., and **Broadhurst, Jean.** Flora. Englewood and vicinity. 45 p. 16.5 cm. [n. p., 1901.]

Unannotated list of vascular plants, without authorities for botanical names, covering area in Bergen County along Hudson River between Edgewater and Closter west to the Hackensack marshes.

NEW MEXICO

General

Castetter, E. F. Uncultivated native plants used as sources of food. Univ. New Mexico Bul. 266 (Biol. Ser. v. 4, no. 1). 62 p. 1935. (Ethnobiological studies in the American Southwest I.)

Annotated list, alphabetically arranged by botanical names, of native plants eaten by Indians of New Mexico and Arizona; bibliography; index of vernacular names.

Standley, P. C. A bibliography of New Mexican botany. Contrib. U. S. Natl. Herb. **13:** 229–246. 1910.

Annotated list.

——— Some useful native plants of New Mexico. Ann. Rpt. Smithsn. Inst. **1911:** 447–462. 13 pl. 1912.

Running account, the plants grouped by uses.

Standley, P. C. The type localities of plants first described from New Mexico. Contrib. U. S. Natl. Herb. **13**: 143–227. pl. 21, map. 1910.

Itineraries of Wislizenus, Wright, Whipple's expedition (Bigelow), Pope's expedition, Parke's expedition; descriptive list of type localities, with lists of species; annotated systematic list of species described from New Mexico.

Wooton, E. O. Native ornamental plants of New Mexico. Bul. New Mexico Agr. Expt. Sta. 51. 40 p. plates. 1904.

Running account of native plants of possible ornamental value.

——— Trees and shrubs of New Mexico. Bul. New Mexico Agr. Expt. Sta. 87. 159 p. illus. 1913.

Annotated treatment, with keys.

——— and **Standley, P. C.** Flora of New Mexico. Contrib. U. S. Natl. Herb. v. 19. 794 p. 1915.

Annotated list of 2,903 species of vascular plants, with keys; gazetteer.

Local

Castetter, E. F., and **Opler, M. E.** The ethnobiology of the Chiricahua and Mescalero Apache. A. The use of plants for foods, beverages and narcotics. Univ. New Mexico Bul. 297 (Biol. Ser. v. 4, no. 5). 63 p. 1936. (Ethnobiological studies in the American southwest III.)

Includes running account of plants used by these Indians in New Mexico and adjacent regions. The Mescalero lived mainly in New Mexico, west to the Rio Grande, east to northwestern Texas, south into northern Mexico; the Chiricahua west of the Rio Grande into southeastern Arizona and south into Chihuahua and Sonora.

Emerson, F. W. An ecological reconnaissance in the White Sands, New Mexico. Ecology **16**: 226–233. 4 fig. (incl. map). 1935.

Includes list of 62 known vascular plants; bibliography. The area, occupying about 270 square miles, is covered with dunes composed of loose crystals of gypsum. (Otero County.)

Robbins, W. W., Harrington, J. P., and **Freire-Marreco, Barbara.** Ethnobotany of the Tewa Indians. Bul. Bur. Amer. Ethnol. 55. xii, 124 p. 7 fig., 8 pl., map. 1916.

Tewa classification and nomenclature of plants; glossary of Tewa names for plant organs, etc.; annnotated list of wild and cultivated plants, with Tewa names and uses; bibliography. The Tewa are a tribe in the upper Rio Grande Valley, New Mexico.

Standley, P. C. Notes on the flora of the Pecos River National Forest. Muhlenbergia **5**: 17–30. 1909.

Topography, botanical explorations, general features of flora; list of more interesting plants collected by the author in 1908, with localities. The Pecos River National Forest, now a part of the Santa Fe National Forest, was in San Miguel, Mora, and Santa Fe Counties.

Stevenson, M. C. Ethnobotany of the Zuñi Indians. Ann. Rpt. Bur. Amer. Ethnol. **30**: 31–102. 3 pl. 1915.

Annotated lists of useful plants, classified by uses, with vernacular names. The Zuñi Indian Reservation is in western McKinley and Valencia Counties.

Watson, J. R. Manual of the more common flowering plants growing without cultivation in Bernalillo County, New Mexico. Univ. New Mexico Bul. 49 (Biol. Ser. v. 3, no. 1). xii, 107 p. 1908.

Annotated flora with keys and brief descriptions; the grasses and sedges omitted.

NEW YORK

General

Brown, H. P. Trees of New York State, native and naturalized. Tech. Pub. New York State Col. Forestry 15. 401 p. illus., map. 1922.

Description and full-page illustration of each species; keys; life zones. The "glossary of derivations" is a remarkable production. Revised and enlarged edition issued as Trees of northeastern United States, native and naturalized. 490 p. illus. 24 cm. Boston, 1938.

House, H. D. Annotated list of the ferns and flowering plants of New York State. New York State Mus. Bul. **254**: 5–759. 1924.

Annotated list of about 2,876 species and many varieties, with synonymy; list of principal local floras.

——— A bibliography of the botany of New York State. New York State Mus. Bul. **188** (Rpt. State Bot. 1915): 66–105. 1916. [1917].

Mostly unannotated list of floristic publications, geographically arranged.

——— Wild flowers of New York. 2 v. 264 col. pl. 30 cm. Albany, 1918 [1920]. (New York State Museum Memoir 15.) (Reissued, 1934, in 1 vol. as Wild flowers.)

Popular descriptive flora of herbaceous plants, noteworthy for its excellent color photographs.

Lee, C. A. A catalogue of the medicinal plants, indigenous and exotic, growing in the State of New-York, with a brief account of their composition and medical properties. iv, 64 p. 23 cm. New York, 1848.

Annotated list. Reprinted from New York Jour. Med. **9**: 151–166, 298–313. 1847; **10**: 1–27. 1848.

Peck, C. H. Edible wild fruits of New York. Trans. Albany Inst. **12**: 83–102. 1893.

Running account of edible wild fruits.

Torrey, John. A flora of the State of New-York, comprising full descriptions of all the indigenous and naturalized plants hitherto discovered in the State; with remarks on their economical and medicinal properties. 2 v. 161 (i. e. 162) pl. 29 cm. Albany, 1843. (Natural history of New York [div. 2].)

The classical flora of the State, still of value for its descriptions and illustrations. Occurs with both plain and hand-colored plates.

Local

Bausor, S. C. A review of some medicinal plants. Pt. 2. Medicinal plants of our local flora. Torreya **37**: 45–54. 1937.

Annotated systematic list of 100 more common medicinal plants in vicinity of New York City.

Beauchamp, W. M. Onondaga plant names. Bul. Torrey Club **15**: 214, 262–266. 1888.—Onondaga Indian names of plants. l. c. **16**: 54–55. 1889.

Running account of vernacular names, with translations and English names.—See also his more extensive paper with same title in Jour. Amer. Folklore **15**: 91–103. 1902. (Refers to the Indians of the Onondaga Valley.)

Beckwith, Florence, Macauley, M. E., and **Fuller, J. B.** Plants of Monroe County, New York, and adjacent territory. Proc. Rochester Acad. Sci. **3**: 1–150. pl. 1–2 (maps). 1896.—Supplementary list. l. c. **5**: 1–38. 1910.—Second supplementary list. l. c. **5**: 59–99. 1917.

Collections, special localities, geology, climate, etc.: annotated list of 1,761 vascular plants (including supplements); annotated bibliography. M. S. Baxter is joint author in place of J. B. Fuller, in suppl. 1–2.

Bradford, G. W. A catalogue of plants, growing spontaneously in the vicinity of Cortland Academy, Homer, Cortland County, New-York. Ann. Rpt. Regents Univ. State N. Y. [**1832**]: 66–71. 1833.

Unannotated alphabetical list of 563 vascular plants.

Burgess, E. S. The Chautauqua flora: a catalogue of the plants of Chautauqua County, New York, native or naturalized; extending through the cryptogamous plants, to the end of the Hepaticae. 37 p. map. 23 cm. Clinton, N. Y., 1877.

Annotated list of 880 plants (761 vascular). (Processed publication.)

Burnham, S. H., and **Latham, R. A.** The flora of the town of Southold, Long Island and Gardiner's Island. Torreya **14**: 201–225, 229–254. 1914.—1st–5th supplementary list. l. c. **17**: 111–122. 1917; **21**: 1–11, 28–33. 1921; **23**: 3–9, 25–31. 1923; **24**: 22–32. 1924; **25**: 71–83. 1925.

Topography, general features of flora; partly annotated list of insect galls, cellular and vascular plants. (Suffolk County.)

Clute, W. N. The flora of the upper Susquehanna and its tributaries. xix, 142, x p. map. 21 cm. Binghamton, 1898.—1st supplement. 16 p. Binghamton, 1901.

Physiography, climate, general feature of flora, etc.: annotated list of 1,105 vascular plants. The area covered includes Broome, Chemung, Chenango, Cortland, Otsego, Steuben, and Tioga, and parts of Allegany, Delaware, Madison, and Schuyler Counties, New York, and parts of Bradford, Susquehanna, Tioga, and Wayne Counties, Pennsylvania.

Dallas, E. M. Minnewaska's flora. 13 p. 21 cm. [n. p.,] 1896.

Unannotated list of vascular plants. (Ulster County.)

Fenno, F. E. Plants of the Susquehanna valley and adjacent hills of Tioga County. New York State Mus. Bul. **67** (Rpt. State Bot. 1902) : 47–160. 1903.—[1st]–2d supplementary list . . . l. c. **75** (Rpt. State Bot. 1903) : 57–60. 1904; **219/220**: 247–250. 1920.

Topography; annotated list of vascular plants.

Ferguson, W. C. Ferns and flowering plants of the Hempstead Plains, Long Island, New York. Torreya **25**: 109–113. 1925 (1926).

List of native vascular plants. (Nassau County.)

——— Some interesting plants from Long Island, N. Y. Torreya **22**: 43–49. 1922.—Contributions to the flora of Long Island, N. Y. 2d paper. Bul. Torrey Club **51**: 177–201. 1924.—Addenda . . . Torreya **24**: **88**. 1924.—Contributions . . . 3d–[4th] paper. Bul. Torrey Club **52**: 133–136. 1925; **53**: 303–308. 1926.—Contributions . . . 5th–[6th] paper. Torreya **28**: 45–51. 1928; **30**: 9–14. 1930.

Lists of rarer species collected by the author, with localities and frequent annotations. To be taken as supplementary to Jelliffe's Flora of Long Island (q. v.).

Goodrich, Mrs. L. L. H. Flora of Onondaga County as collected by the members of the Syracuse Botanical Club. 210 p. port. 23.5 cm. Syracuse, N. Y., 1912.

Annotated list of vascular plants; list of mushrooms.

Gordinier, H. C., and Howe, E. C. The flora of Rensselaer County, New York. A record of the phenogams and vascular cryptogams growing . . . independent of cultivation. 39 p. 22.5 cm. Troy, 1894.

List of 1,345 vascular plants with localities of rarer species.

Gray, Asa. A catalogue of the indigenous flowering and filicoid plants growing within 20 miles of Bridgewater, (Oneida Co.) N. Y. Ann. Rpt. Regents Univ. State N. Y. [1832] : 57–65. 1833.

Unannotated list of vascular plants. This is Gray's first published paper; it was omitted from his bibliography in Amer. Jour. Sci. III, **36**: app. (42, 25 p.). 1888.

Grier, N. M. The native flora of the vicinity of Cold Spring Harbor, N. Y. Amer. Midl. Nat. **9**: 245–256, 283–318, 384–437. 1924–25.—The geology of Long Island with especial reference to the Cold Spring Harbor region and its flora. l. c. 531–533. 1925. (Reprinted 265 p. [including blank pages for memoranda].)

List of 1,865 cellular and vascular plants (1,059 vascular) with localities and collectors' names; geology, soils, climate, botanical explorations; bibliography.—See critical review by N. Taylor, Rhodora **27**: 213–215. 1926, and reply by Grier, l. c. **28**: 242–245. 1926. (Suffolk County, on Long Island.)

Haberer, J. V. A list of plants in the vicinity of Utica, for April, May and a portion of June. 20 p. 23 cm. Utica, 1888.

Annotated list of vascular plants. (Oneida County.)

Hollick, C. A., and Britton, N. L. The flora of Richmond County, New York. A catalogue of the phaenogamous and vascular cryptogamous plants . . . growing in Richmond Co., independent of cultivation. 36 p. 24.5 cm. Staten Island, 1879.

Briefly annotated list of vascular plants.—Additions published as follows: Bul. Torrey Club **7**: 11–12. 1880; **8**: 48. 1881; **9**: 149–151. 1882; **12**: 38–40. 1885; **13**: 83–84. 1886; **16**: 132–134. 1889; **18**: 213–214. 1891; **22**: 460–462. 1895.—Numerous minor additional notes are cited by J. H. Barnhart, Mem. Torrey Club **5**: 42. 1915.

Hotchkiss, Neil. A botanical survey of the Tug Hill plateau. New York State Mus. Bul. 287. 123 p. 7 fig. (maps), 4 pl., map. 1932.

Botanical investigations, ecology, etc.; annotated list of vascular plants; bibliography. The Tug Hill plateau is in Lewis, Jefferson, Oswego, and Oneida Counties.

Hough, F. B. A catalogue of the indigenous, naturalized, and filicoid plants, of Lewis County. Arranged according to the natural method adopted by Professor Torrey, in the State catalogue. Ann. Rpt. Regents Univ. State N. Y. **59**: 249–283. 1846.

Annotated list of vascular plants.

House, H. D. The vegetation of the eastern end of Oneida Lake. New York State Mus. Bul. **197** (Rpt. State Bot. 1916): 61–110. pl. 5–12. 1918; **266** (Rpt. State Bot. 1924): 28–40. 1925.

Geology, climate, life zones, plant associations; partly annotated list of vascular plants. (Chiefly Oneida and Madison Counties.)

——— and **Alexander, W. P.** Flora of the Allegany State Park region. New York State Mus. Handbook 2. 225 p. 33 fig., map. 1927.

Annotated list of vascular plants; lists of species of special localities; bibliography. (Cattaraugus County.)

Howell, G. R., and **Tenney, Jonathan.** Botany. In their Bi-centennial history of Albany . . . p. 21–29. New York, 1886.

Unannotated list of vascular plants (sedges and grasses mostly omitted) of Albany County. This has been attributed to J. S. Markle and C. H. Peck.

Hoysradt, L. H. Catalogue of the phaeogamous [sic] and acrogenous plants growing without cultivation within five miles of Pine Plains, Duchess [sic] Co., N. Y. Bul. Torrey Club v. 6, suppl. xxxii p. [1875–79.]

Annotated list of spermatophytes, the acrogens (pteridophytes) omitted.

Jelliffe, S. E. The flora of Long Island. 160 p. 24.5 cm. Lancaster, Pa., 1899.

Geology, soils, etc.; bibliography; annotated list of cellular and vascular plants (2,238 species, of which 1,383 are vascular plants).—See also **Grout, A. J.** Additions to the recorded flora . . . Torreya **2**: 49–53. 1902.—**Jelliffe, S. E.** Additions . . . l. c. **4**: 97–100. 1904. (Kings, Queens, Nassau, and Suffolk Counties.)

Latham, Roy Flora of the State Park, Orient, Long Island, N. Y. Torreya **34**: 139–149. 1934.

General features of flora; annotated list of vascular plants. (Suffolk County.)

Mead, S. B. A catalogue of plants growing spontaneously in the vicinity of North-Salem Academy. Ann. Rpt. Regents Univ. State N. Y. [**1830**]: 89–97. 1831.

Alphabetical list of 745 spermatophytes, with occasional annotations. (Westchester County.)

Miller, E. S., and **Young, H. W.** Catalogue of the phaenogamous and acrogenous plants of Suffolk County, Long Island. 15 p. 23 cm. Port Jefferson, 1874.

Unannotated list of 871 vascular plants.—See also **Young, H. W.** Additions to Suffolk County catalogue. Bul. Torrey Club **5**: 34. 1874.—**Miller, E. S.** Suffolk County notes (and varying titles). l. c. **6**: 155–157, 171–172, 258–259. 1877–78; **7**: 17–18. 1880.

Paige, E. W. Catalogue of the flowering plants of Schenectady County. 48 p. 24 cm. Albany, 1864.

Physiography; annotated list of phanerogams.

Paine, J. A. Catalogue of plants found in Oneida County and vicinity. Ann. Rpt. State Cabinet Nat. Hist. N. Y. **18**: 53–192. 1865. (Reprinted.)

Annotated list of vascular plants; includes numerous records from outside Oneida County.—See also **Haberer, J. V.** Plants of Oneida County, New York, and vicinity, I. Rhodora **7**: 92–97, 106–110. 1905.

Peck, C. H. Plants of North Elba, Essex County, N. Y. New York State Mus. Bul. **6**: 65–266. map. 1899.

Physiography, etc.; annotated list of vascular (492 species) and cellular plants of North Elba, a town in the Adirondack Mountains.

——— Plants of the summit of Mt. Marcy. New York State Mus. Bul. **5** (Rpt. State Bot. 1898): 657–673. 1899.

Physiography, soil, general features of flora, list of 206 vascular and cellular plants, with remarks on some of the species. Mount Marcy (altitude 5,344 ft.) is the highest mountain in New York.—See also **Adams, C. C.,** and others. Flowering plants from the Alpine zone above timber-line on Mt. Marcy. Ecology **1**: 226–229. 1920. (Essex County.)

Raup, H. M. Botanical studies in the Black Rock forest. Black Rock Forest Bul. 7. 161 p. (incl. 9 fig., 11 pl.), fold. map. 1938.

Ecology; includes (p. 109–158) annotated list of vascular plants collected by the author in 1936–37. Black Rock Forest is a tract of privately owned land, approximately 3,200 acres in extent, lying west of the Hudson River in Orange County, N. Y.

Sartwell, H. P. Catalogue of plants, growing without cultivation in the vicinity of Seneca and Crooked Lakes, in western New-York. Ann. Rpt. Regents Univ. State N. Y. **58**: 273–290. 1845.

Unannotated list of vascular plants. (**Yates County.**)

Smith, A. M. Corrected and enlarged list of plants found on the Adirondack League Club tract. Adirondack League Club Yearbook **1904**: 43–61. [1904.] (1st ed., List of plants . . . l. c. 1898.)

Unannotated list of lichens, hepatics (by Caroline C. Haynes), mosses and vascular plants. (**Herkimer County.**)

Stearns, W. A. List of plants of Fishkill, N. Y. and vicinity. 23 p. 18.5 cm. [Fishkill, 1880?]

"Partial and incomplete" annotated list of vascular plants. (**Dutchess County.**)

Stetson, Sereno. The flora of Copake Falls, N. Y. Torreya **13**: 121–133. 4 fig. 1913.

Topography, etc.; annotated list of angiosperms (ferns in supplement).—See also **Burnham, S. H.** A supplementary list of plants of Copake Falls . . . l. c. 217–219.—**Stetson, S.** 1913 notes on the flora of Copake Falls . . . l. c. **14**: 42–45. 1 fig. 1914. (**Columbia County.**)

Stewart, P. A., and **Merrell, W. D.** The Bergen Swamp: an ecological study. Proc. Rochester Acad. Sci. **8**: 209–262. 4 fig. (maps), 2 pl. 1937.

Includes unannotated list of vascular plants. Bergen Swamp is in the northeastern part of **Genesee County.**

Taylor, Norman. Flora of the vicinity of New York. Mem. N. Y. Bot. Gard. v. 5. vi, 683 p. 9 maps. 1915.

Geology, plant geography; bibliography (by J. H. Barnhart); annotated list of 2,651 species of vascular plants (of which 2,038 are native) with keys. The area covered includes all of Connecticut and New Jersey, Long Island, southeastern New York, and eastern Pennsylvania. See critical review by **Fernald, M. L.,** Rhodora **17**: 62–70. 1915.

——— Native trees of the Hudson River valley. Bul. N. Y. Bot. Gard. **7**: 90–147. pl. 131–150. 1909.

Annotated list, with short descriptions.

——— The vegetation of Long Island. Pt. I. The vegetation of Montauk. A study of grassland and forest. Mem. Brooklyn Bot. Gard. **2**: 1–107. 30 fig. (incl. diagrs., map). 1923.

Mainly ecological; includes (p. 85–107) list of 495 species of vascular plants, with localities. (**Suffolk County.**)

Wiegand, K. M., and **Eames, A. J.** The flora of the Cayuga Lake basin, New York. Vascular plants. Mem. Cornell Agr. Expt. Sta. 92. 491 p. map. 1926.

Physiography, soils, explorations, composition of flora; annotated list with keys of 1,637 vascular plants, of which 1,146 are native species. Replaces **Dudley, W. R.** The Cayuga flora. 1886. (Mostly in **Seneca, Cayuga,** and **Tompkins Counties.**)

——— **Eames, A. J.,** and others. Vascular plants (Pteridophyta and Spermatophyta). *In* A preliminary biological survey of the Lloyd-Cornell Reservation . . . Bul. Lloyd Libr. Bot., Pharm. and Mat. Med. **27** (Entom. Ser. no. 5): 25–52. 1926.

General features of flora; annotated list of vascular plants. This is followed by lists of cellular plants by other authors. The region covered is 15 miles northeast of Ithaca and 1 mile east of McLean, in **Tompkins County.**

Willis, O. R. Report of the flora of Westchester County. *In* **Bolton, Robert.** The history . . . of the county of Westchester . . . Revised. v. 1, p. 771–825. New York, 1881. (Reprinted.)

Geology, etc.; annotated list of 1,188 vascular plants.—See also **Britton, E. G.** Additions to the Westchester County flora. Bul. Torrey Club **13**: 6–7. 1886.—**Day, E. H.** Additions . . . l. c. **13**: 94–95. 1886.—**Martens, J. W., jr.** [Additions.] l. c. **16**: 123–124. 1889.

Winchell, Alexander. Catalogue of plants found growing without cultivation in the vicinity of Amenia Seminary, Dutchess County, N. Y. Ann. Rpt. Regents Univ. State. N. Y. **64**: 256–279. 1851.

Unannotated list of vascular plants and Characeae.

Wright, John, and **Hall, James.** A catalogue of plants, growing without cultivation, in the vicinity of Troy. 42 p. 23 cm. Troy, 1836.
Annotated alphabetical list of vascular plants. (Rensselaer County.)

Zabriskie, J. B. A catalogue of plants indigenous and cultivated, found in the vicinity of Erasmus Hall. Ann. Rpt. Regents Univ. State N. Y. [**1834**]: 176–181. 1835.
Unannotated alphabetical list of vascular plants and a few fungi. (Kings County.)

Zenkert, C. A. The flora of the Niagara frontier region. Ferns and flowering plants of Buffalo, N. Y., and vicinity. Bul. Buffalo Soc. Nat. Sci. v. 16. x, 328 p. illus., map. 1934.
Botanical investigations, topography, geology, climate, life zones, human changes; annotated list of 1,587 species and 115 varieties; ecology, plant societies, bibliography. Covers a radius of about 50 miles around Buffalo, including a part of the Niagara Peninsula, Ontario. Replaces Day, D. F. The plants of Buffalo . . . l. c. **4**: 65–279. 1882–84 (reprinted as A catalogue of the native and naturalized plants of the city of Buffalo and its vicinity . . . 1884); **5**: 85–96. 1886, and also his Catalogue of the Niagara flora. Ann. Rpt. Commrs. State Reserv. Niagara **4**: 67–133. 1888. (Reprinted 67 p.)

NORTH CAROLINA

See also Tennessee (Cain); Virginia (Kearney).

General

Coker, W. C., and **Totten, H. R.** The trees of North Carolina. 106 p. 20.5 cm. Chapel Hill, 1916.
Annotated list of 167 trees, with key to species.—See also **Ashe, W. W.** Additions to the arborescent flora of North Carolina. Jour. Elisha Mitchell Sci. Soc. **34**: 130–140. 1918.

Hyams, C. W. The flora of North Carolina from Ranunculaceae to Salviniaceae. Bul. North Carolina Agr. Expt. Sta. 164. p. 287–365. 1899.
List of 2,685 species of vascular plants, their distribution within the State indicated by letters. Replaces the "Catalogue of the indigenous and naturalized plants of the State," by M. A. Curtis, in Agriculture of North Carolina, pt. 3. 156 p., 1867, and "A preliminary list of additions to Curtis' Catalogue . . . flowering plants," by M. E. Hyams in Jour. Elisha Mitchell Sci. Soc. **1884/85**: 72–76. 1885.

——— Medicinal plants which have been collected and used in North Carolina. Bul. North Carolina Agr. Expt. Sta. 150. p. 327–409. 1898.
Briefly annotated systematic list.

Local

Blomquist, H. L., and **Oosting, H. J.** A guide to the spring and early summer flora of the Piedmont, North Carolina. 2d ed. xviii, 138 p. 23 cm. Durham, 1936. (1st ed., A guide to the spring flora of the Lower Piedmont, North Carolina, 1934.)
Annotated list of vascular plants, with keys; glossary.

Coker, W. C. A visit to Smith Island. Jour. Elisha Mitchell Sci. Soc. **34**: 150–153. pl. 10–16. 1918.
Includes list of woody plants observed by the author in April. (Off the coast of Brunswick County.)

——— The woody plants of Chapel Hill, North Carolina. Jour. Elisha Mitchell Sci. Soc. **19**: 42–49. 1903. (Reprint 8 p.)
List of 133 native trees, shrubs, and woody vines, with date of first flowering. (Orange County.)

Croom, H. B. A catalogue of plants, native or naturalized, in the vicinity of New Bern, North Carolina; with remarks and synonyms. [2d ed.] x, 52 p. 23.5 cm. New York, 1837. (1st ed., 1833.)
Unannotated list of vascular plants, with occasional critical notes in an appendix. (Craven County.)

Ewing, C. O., and Stanford, E. E. Botanicals of the Blue Ridge. Jour. Amer. Pharm. Assoc. **8**: 16–26. illus. 1919.

Account of general features of region and of method of collecting crude drugs, with enumeration of the more important species. The Blue Ridge, in western North Carolina and Virginia, is the chief source of native botanic drugs in the United States.

House, H. D. Woody plants of western North Carolina. 34 p. 22.5 cm. Darmstadt, 1913.

Plant zones, list of endemic species; annotated systematic list of trees, shrubs, and woody vines.

Kearney, T. H. The plant covering of Ocracoke Island; a study in the ecology of the North Carolina strand vegetation. Contrib. U. S. Natl. Herb. **5**: 261–319. fig. 33–50. 1900.

Mainly ecological, but includes list of cellular and vascular plants; bibliography. (Hyde County.)

Lewis, I. F. The vegetation of Shackleford Bank. Econ. Paper North Carolina Geol. and Econ. Survey no. 46. 32 p. 11 pl. (incl. map). 1917.

Topography, climate, plant formations, general features of flora, soil-binders, etc.; list of vascular plants. (A sand reef on coast between Beaufort Inlet and Cape Lookout, Carteret County.)

McAtee, W. L. Notes on the flora of Church's Island, North Carolina. Jour. Elisha Mitchell Sci. Soc. **35**: 61–75. 1919.

General features of flora; annotated list of vascular plants collected by the author. (Currituck County.)

Memminger, E. R. A list of plants growing spontaneously in Henderson County, N. C. Jour. Elisha Mitchell Sci. Soc. **30**: 126–149. 1915.

Unannotated list of vascular plants.

Peattie, D. C. Flora of the Tryon region of North and South Carolina. An annotated list of plants growing spontaneously in Polk County, North Carolina, and adjacent parts of South Carolina, in Greenville and Spartanburg Counties. Jour. Elisha Mitchell Sci. Soc. **44**: 95–125, 141–229. 1928–29; **45**: 59–100, 245–290. 1929–30; **46**: 129–160. 1931.—Additions, corrections and deletions . . . l. c. **53**: 311–323. 1937.

Botanical explorations, physiography, soils, climate; annotated list of 1,155 vascular plants.

Small, J. K., and Heller, A. A. Flora of western North Carolina and contiguous territory. Mem. Torrey Club **3**: 1–36. 1892.

Includes list of species collected by authors. (Mainly in Blue Ridge at Grandfather Mountain, Table Rock, Blowing Rock, Stone Mountain, etc.)

Wood, T. F., and McCarthy, Gerald. Wilmington flora; a list of plants growing about Wilmington, N. C., with date of flowering. Jour. Elisha Mitchell Sci. Soc. **3**: 77–141. map. 1886. (Reprinted 69 p.)

Partly annotated list of 1,217 vascular plants. (New Hanover County.)

NORTH DAKOTA

General

Bergman, H. F. Flora of North Dakota. Bien. Rpt. Agr. Col. Survey North Dakota **6** (1911–12): 151–372. [1918.] (Reprinted.)

Botanical explorations, physiography, types of vegetation, etc.; annotated list of vascular plants, with keys. The 15-page index follows the index of the whole volume, also two unnumbered pages of corrections.—For additions, see **Metcalf, F. P.** Notes on North Dakota plants. Jour. Washington Acad. Sci. **10**: 188–198. 1920.—Also **Stevens, O. A.** New records and other notes on North Dakota plants. Bul. Torrey Club **49**: 93–105. 1922.

Lunell, Joel. Enumerantur plantae Dakotae Septentrionalis vasculares, I–XIV. Amer. Midl. Nat. **4**: 152–165, 211–228, 229–244, 297–310, 355–366, 409–418, 419–431, 467–487, 503–517. 1915–16; **5**: 1–13, 31–46, 55–71, 93–98, 233–241. 1917–18.

Annotated list of 1,311 species and varieties. Remarkable for its nomenclature, which recognizes no dividing line in 1753, but takes generic names from such authors as Dioscorides, and binomials from Fuchs and others of like date. Many of the largest Linnaean genera appear under unfamiliar and botanically indefensible names, as *Tessenia* for *Erigeron*, *Cunigunda* for *Eupatorium*, and *Eupatorium* for *Agrimonia*.

Local

Bell, W. B. The plants of the Williston area. Ann. Rpt. Agr. Expt. Sta. North Dakota **18** (1) : 53–75. 1908. (Also in Bien. Rpt. Agr. Col. Survey North Dakota **4** (1907–08) : 28–51. 1910.)

Annotated list of spermatophytes, the families arranged alphabetically. (Williams County.)

——— Report on the plant survey of the Morton area in Morton County, North Dakota. Bien. Rpt. Agr. Col. Survey North Dakota **5** (1909–10) : 34–60. 1910.

Annotated list of vascular plants.

Bergman, H. F. A report on the plant survey of Barnes County, N. Dak. Bien. Rpt. Agr. Col. Survey North Dakota 6 (1911–12) : 121–150. 3 pl. [1918.]

Topography, climate, plant formations; annotated list of vascular plants.

Stevens, O. A. Plants of Fargo, North Dakota, with dates of flowering. Amer. Midl. Nat. **7**: 54–62, 79–100, 135–156. 1921.

Physiography, soil, etc.; annotated list of flowering plants; list of flowers arranged according to flowering date. (Cass County.)

OHIO

General

Bigelow, J. M. To the medical profession of Ohio. Ohio Med. and Surg. Jour. **2**: 97–143. 1849. (Reprinted, 47 p., with title: A list of the medicinal plants of Ohio.)

Annotated list of 202 medicinal plants of Ohio.

Detmers, Frederica. Medicinal plants of Ohio. Ohio Nat. **10**: 55–60, 73–85. 1910.

Briefly annotated list of wild and commonly cultivated cellular and vascular plants used in medicine.

Kellerman, W. A. Bibliography of Ohio botany. Bul. Ohio Agr. Expt. Sta. Tech. Ser. **1**: 180–201. 1893.—Additions . . . Ann. Rpt. Ohio Acad. Sci. **4**: 5–18. 1896.

Annotated list of works referring particularly to Ohio plants, chronologically arranged. The additions carry this list through 1895.—Essentially the same material is given in the prefatory matter of **Kellerman, W. A.**, and **Werner, W. C.** Catalogue of Ohio plants. Rpt. Geol. Survey Ohio **7** (2) : 58–79. 1893.

Miller, E. M. Bibliography of Ohio botany. Bul. Ohio Biol. Survey **27** (v. 5, no. 4) : 281–376. 1932. (Ohio State University Bulletin v. 37, no. 5.)

Mostly unannotated list of publications (1755–1931), arranged alphabetically by authors.

Schaffner, J. H. Field manual of the flora of Ohio and adjacent territory. 638 p. 17.5 cm. Columbus, 1928.

Bibliography; flora of vascular plants in form of keys; glossary.

——— Poisonous and other injurious plants of Ohio. Ohio Nat. **4**: 16–19, 32–35, 69–73. 1904.

Briefly annotated systematic list of 229 cellular and vascular plants.

——— Revised catalog of Ohio vascular plants, arranged according to the phyletic system of classification; with notes on the geographic distribution in the State, based mainly on specimens in the Ohio State herbarium of the Ohio State University. Bul. Ohio Biol. Survey **25** (v. 5, no. 2) : 87–215. 3 maps. 1932. (Ohio State University Bulletin v. 36, no. 9).)—Additions I–VII. Ohio Jour. Sci. **33**: 288–294. 1933; **34**: 165–174. 1934; **35**: 297–303. 1935; **36**: 195–203. 1936; **37**: 260–265. 1937; **38**: 211–216. 1938; **39**: 232–234. 1939.

Bibliography, phytogeographical regions; annotated list of 2,309 species (plus varieties) of vascular plants. Replaces Kellerman, W. A. The fourth State catalogue of Ohio plants . . . (1899) and its 3 supplements. The 7th list of additions is by Schaffner and C. H. Jones.

——— Trees of Ohio and surrounding territory; including the area westward to the limits of the prairie and south to the thirty-seventh parallel. Proc. Ohio Acad. Sci. **5**: 71–191. 1909. (Spec. Paper Ohio Acad. Sci. no. 15.)

Annotated list, including many cultivated species, with keys.—See also his Check list of Ohio shrubs. Ohio Nat. **8**: 205–209. 1907.

Local

Aiken, W. H. Catalogue of the ferns and flowering plants of Cincinnati, Ohio, and vicinity. Bul. Lloyd Libr. 15 (Bot. Ser. 1). 57 p. 1911.

Botanical collectors and publications; annotated list of vascular plants. (Hamilton County.)

——— Check list of Hamilton County, Ohio, plants exclusive of the lower cryptogams. Jour. Cincinnati Soc. Nat. Hist. **20:** 199–230. 1904.

Climate, bibliography, etc.; unannotated list of vascular plants.

——— List of medicinal plants, wild or cultivated, growing in the vicinity of Cincinnati, Ohio, with notes as to the parts used for medicinal purposes. Jour. Cincinnati Soc. Nat. Hist. **20:** 231–234. 1904.

Systematic list. (Hamilton County.)

[**Bigelow, J. M.**, and **Hor, Asa.**] Florula lancastriensis; or a catalogue comprising nearly all the flowering and filicoid plants, growing naturally within the limits of Fairfield County, with notes of such as are medicinal. Proc. Med. Conv. Ohio **1841:** 49–79. 1841. (Reprinted, 33 p. Lancaster, 1841.)

Unannotated list of vascular plants; annotated list of 190 medicinal plants. Another issue of same date, 22 pages, contains only the unannotated list.

Braun, E. L. The Lea herbarium and the flora of Cincinnati. Amer. Midl. Nat. **15:** 4–75. map. 1934.

Plant communities, changes in flora, bibliography; list of 1,302 species and varieties of vascular plants growing within a radius of 20 miles around Cincinnati, with occasional annotations. The herbarium of Thomas G. Lea, collected in 1834–44 and now at the Academy of Natural Sciences of Philadelphia, includes 714 species from the region. (Hamilton and Clermont Counties, Ohio, and Boone, Kenton, and Campbell Counties, Ky.)

Craig, Moses. A catalogue of the uncultivated flowering plants growing in the Ohio State University grounds. Bul. Ohio Agr. Expt. Sta. Tech. Ser. **1:** 49–110. map. 1890.

Geology, phenology, etc.; annotated list of 468 angiosperms.—See also **Kellerman, W. A.**, and **Werner, W. C.** Corrections and additions . . . l. c. **1:** 224–232. 1893. (Franklin County.)

Foltz, K. O. The phaenogamic flora of Summit County, —Pt. I. Ann. Rpt. Ohio State Acad. Sci. **2:** 21–31. [1894.]

Unannotated list of spermatophytes. No more published.

Griggs, R. F. A botanical survey of the Sugar Grove region. Bul. Ohio Biol. Survey **1** (3) : 245bis–340 (incl. 29 fig., pl., map). 1914. (Ohio State University Bulletin v. 28, no. 25.)

Geology, ecology, etc.; partly annotated list of 972 vascular plants. The area covered is in Fairfield and Hocking Counties.

Jones, H. L. Catalogue of the phanerogams and ferns of Licking County, Ohio. Bul. Scient. Lab. Denison Univ. 7. 102 p. map. 1892.—Additions and corrections to the flora. l. c. **8** (2) : 3–6. 1894.

Geology, climate, special localities; annotated list of 945 vascular plants.

Kellerman, W. A., and **Jennings, O. E.** Flora of Cedar Point. Ohio Nat. **4:** 186–190. 1904.—Additions . . . I. (**Kellerman** and **H. H. York**) l. c. **6:** 540. 1906.—Additions . . II. (**O. E. Jennings**) l. c. **6:** 544–545. 1906.—Additions . . . III. (**M. E. Stickney, J. H. Schaffner**, and **C. A. Davies**) l. c. **10:** 61–63. 1910.

Unannotated list of vascular plants.—See also Additions . . . of 1911–12, by **E. L. Fullmer.** l. c. **12:** 473. 1912; **13:** 78. 1913. (Erie County.)

Moore, D. M. A botanical survey of the campus of Denison University. Jour. Scient. Lab. Denison Univ. **20:** 131–153. 7 fig., pl. 17–19 (incl. map). 1923.

Physiography, general features of flora; list of 321 vascular plants. (In Granville, Licking County.)

Morgan, A. P. Flora of the Miami Valley, Ohio. 60 p. 17 cm. Dayton, Ohio, 1878.—Supplement . . . for 1878. p. 61–68. [n. d.]

Unannotated list of vascular and cellular plants. The area covered is not further defined than in the title.

Moseley, E. L. Flora of the oak openings west of Toledo. Proc. Ohio Acad. Sci. **8:** 79–134. map. 1928. (Spec. Paper Ohio Acad. Sci. no. 20.)

241306°—42——14

Origin of "oak openings," a sandy tract in Lucas, Fulton, and Henry Counties, about 130 square miles in area, soil, changes in flora, etc.; annotated list of vascular plants.

Moseley, E. L. Sandusky flora. A catalogue of the flowering plants and ferns growing without cultivation, in Erie County, Ohio, and the peninsula and islands of Ottawa County. Spec. Papers Ohio State Acad. Sci. no. 1. 167 p. map. 1899.
Climate, geology, etc.; annotated list of over 1,100 vascular plants.—See also **Griggs, R. F.** Additions to the Sandusky flora. Ohio Nat. **1**: 97–98. 1901.

Pieters, A. J. The plants of western Lake Erie, with observations on their distribution. Bul. U. S. Fish Comn. **21**: 57–79. pl. 11–20. 1901. (Reprinted.)
Mainly ecological; includes alphabetical list (p. 77–79) of cellular and vascular plants.

Roth, Conrad. Vascular plants of Scioto County. *In* **Bannon, H. T.** Stories old and often told, being chronicles of Scioto County, Ohio. p. 245–266. Baltimore, 1927.
Unannotated list.

Selby, A. D., and Craig, Moses. A preliminary list of the plants of Franklin County, Ohio. Ann. Rpt. Columbus Hort. Soc. 1890, pt. 2. 19 p. 1891.—Additions . . . l. c. 1891. 1 leaf. 1892.
Unannotated list of 1,069 vascular plants.

Wetzstein, Albert. Pteridophyta and Spermatophyta found in Auglaize County, Ohio. Proc. Ohio State Acad. Sci. **4** (7, Ann. Rpt. 14): 353–364. 1906.
Barely annotated list of vascular plants.

Wright, A. A. Preliminary list of the flowering and fern plants of Lorain County, Ohio. 30 p. map. 23 cm. Oberlin, 1889.—Additions . . . 11 p. map. 23 cm. Oberlin, 1893. [Oberlin College Laboratory Bulletin no. 1 and Suppl.]
Unannotated list of vascular plants. The annotated list of additions brings the total to 975 forms (931 species).

OKLAHOMA

General

See also Arkansas (Evans; Palmer, The ligneous flora of Rich Mountain); Texas (Tharp).

Bogue, E. E. An annotated catalog of the ferns and flowering plants of Oklahoma. Bul. Oklahoma Agr. Expt. Sta. 45. 48 p. 1900.
Annotated list of about 750 vascular plants.

Jeffs, R. E., and Little, E. L. A preliminary list of ferns and seed plants of Oklahoma. Pub. Univ. Oklahoma Biol. Survey **2**: 39–101. 1930. (Contr. Dept. Bot. Univ. Oklahoma n. s. no. 1.)
Physiography, floral areas, botanical explorations; unannotated list of vascular plants; bibliography.—For additions see **Little, E. L., and Olmsted, C. E.** New records of flowering plants from southeastern Oklahoma. Proc. Oklahoma Acad. Sci. **15**: 44–49. 1935. See also for key to species included in the 1930 publication (excluding grasses, sedges, etc.), **Jeffs, R. E.** A key to the ferns and seed plants of Oklahoma. 147 p. 27.5 cm. Norman, 1931. (Processed publication.)

Shannon, C. W. The trees and shrubs of Oklahoma. Cir. Oklahoma Geol. Survey 4. 41 p. 1913.
Annotated list.—See also **Phillips, G. R., Gibbs, F. J., and Mattoon, W. R.** Forest trees of Oklahoma . . . ed. 5, rev. Pub. Oklahoma Div. Forestry no. 1. 135 p. illus. (incl. map). 1939. (1st ed., 1927, by Mattoon and Phillips.)

Stemen, T. R., and Myers, W. S. Oklahoma flora . . . xxix, 706 p. 494 fig. 20 cm. Oklahoma City, 1937.
Annotated descriptive flora of vascular plants, with keys; the grasses, sedges, and rushes omitted; lists of hay-fever, poisonous, drug, and edible plants, etc.—For a critical review, see Fernald, M. L. A flora of Oklahoma. Rhodora **40**: 179–183. 1938.

Local

Eskew, C. T. The flowering plants of the Wichita Mountains Wildlife Refuge. Amer. Midl. Nat. **20:** 695–703. 1938.

Topography, botanical explorations, climate, ecology; unannotated list of 447 spermatophytes; bibliography. (Comanche County.)

Little, E. L. Flora of Muskogee County, Okahoma. Amer. Midl. Nat. **19:** 369–389. 1938.

Botanical explorations, relationships of flora, biological spectrum of flora; briefly annotated list of 829 vascular plants; bibliography.

——— and **Olmsted, C. E.** Trees and shrubs of the southeastern Oklahoma protective unit. Proc. Oklahoma Acad. Sci. **16:** 52–61. 1936.

Topography, climate, botanical explorations, ecology; annotated list of 152 woody plants; bibliography. (Northern McCurtain, southern Le Flore, and eastern Pushmataha Counties.)

Vestal, P. A., and **Schultes, R. E.** The economic botany of the Kiowa Indians as it relates to the history of the tribe . . . xiii, 110 p. (incl. 1 fig. (map), 4 pl.). 27 cm. Cambridge, Mass., 1939.

History of tribe; annotated list of cellular and vascular plants used, with vernacular names, referring principally to Oklahoma; list of plants classified by uses, etc.; bibliography.

OREGON

See also Idaho (Spinden); Washington (Piper and Beattie); British Columbia (Brown).

General

Hemenway, A. F. Botanists of the Oregon country. Quart. Oregon Hist. Soc. **5:** 207–214. 1904.

Account of botanical collectors in Oregon, Washington, Idaho, and adjacent States, from Haenke to Howell.

Sheldon, E. P. The forest wealth of Oregon. 32 p. 4 pl. 20 cm. Portland, 1904.

History, economic value; annotated list of trees, with uses.

Local

Applegate, E. I. Plants of Crater Lake National Park. Amer. Midl. Nat. **22:** 225–314 (incl. 3 pl.). 1939.

Topography, general features of flora, botanical explorations; annotated list of about 570 species and forms of vascular plants, with keys and brief descriptions. (Klamath County.)

Benson, G. T. The trees and shrubs of western Oregon. Contrib. Dudley Herb. Stanford Univ. v. 2. 170 p. map. 1930.

Plant zones; annotated list with synonymy, type locality, local range, and citation of specimens. Refers primarily to the region west of the Cascade Mountains.

Coville, F. V. Notes on the plants used by the Klamath Indians of Oregon. Contrib. U. S. Natl. Herb. **5:** 87–108. 1897.

Annotated systematic list of useful plants; list of Indian names, with botanical equivalents. (Klamath County.)

Gilkey, H. M., and **Powell, G. M.** Handbook of northwest flowering plants. vi, 407 p. illus. 20 cm. Portland, Oreg., 1936.

A popular work with brief descriptions, covering the region in Oregon and Washington from the summit of the Cascade Mountains to the coast, south to Lane County, Oregon, omitting northern Washington. The grasses, sedges, rushes, and willows are omitted.

Gorman, M. W. The flora of Mount Hood. Oregon Out of Doors **1** (1): 64–96. 1920.

Physiography, etc.; annotated list of 283 vascular plants found above 2,000 feet elevation. Replaces Howell, Thomas. The flora of Mount Hood. Mazama **1:** 28–48. 1896. Mount Hood is in Hood River, Multnomah, Clackamas, and Wasco Counties.

Gorman, M. W. List of plants in the vicinity of Portland, Oregon. Muhlenbergia **2**: 351–416. 1916–17.

Annotated list of "835" vascular plants, never published in full. Reaches Malva (Engler and Prantl system), p. 416. (Multnomah County.)

House, H. D. Vegetation of the Coos Bay region, Oregon. Muhlenbergia **9**: 81–100. 1914.

General features of flora; list of about 210 vascular plants collected by author in 1911–12, with data. (Coos County.)

Leach, Lilla. Wildflowering in the Wallowas. Mazama **13**: 50–60. illus. 1931.

Includes unannotated list of vascular plants collected by Lilla and John R. Leach. The Wallowa Mountains are in southern Wallowa and Union and northern Baker Counties.

Nelson, J. C. A list of Oregon plants not mentioned in the local manuals. Muhlenbergia **2**: 345–350. 1916.

List of 63 species, mostly introduced plants, with localities.

Peck, M. E. Flora of the east slope of the Cascade Mountains, in Crook County, Oregon. Proc. Iowa Acad. Sci. **22**: 143–164. 1915 (1916).

General features of flora; annotated list of vascular plants.

——— A section of Upper Sonoran flora in northern Oregon. Proc. Iowa Acad. Sci. **23** (1916): 317–334. [1917?]

Physiography, plant associations; annotated list of angiosperms collected by the author in 1915 near Umatilla, Umatilla County, on the south bank of the Columbia River.

Wynd, F. L. The flora of Crater Lake National Park. Amer. Midl. Nat. **17**: 881–949. 1936.

Bibliography; unannotated list of vascular plants, with keys. (Klamath County.)

PENNSYLVANIA

See also New York (Clute; Taylor, Flora of the vicinity of New York).

General

Illick, J. S. Pennsylvania trees. Bul. Penn. Dept. Forestry 11. 5th ed. 237 p. plates. 1925. (Reprinted, 1928. 1st ed., 1914.)

Includes systematic descriptive list of trees with plate of each species.

Porter, T. C. Flora of Pennsylvania . . . Edited with the addition of analytical keys by J. K. Small. xv, 362 p. map. 24 cm. Boston, 1903.

Annotated list of 2,201 species of spermatophytes, with keys. For the vascular cryptogams, see his Catalogue of the Bryophyta . . . and Pteridophyta . . . found in Pennsylvania. 66 p. Boston, 1904.

Local

Baron, J. J. Floral life of western Pennsylvania: a bibliography. Trillia **8**: 45–57. 1925.

Partly annotated list of botanical publications, alphabetically arranged by authors.

Benner, W. M. The flora of Bucks County, Pennsylvania. 331 p. 24 cm. Philadelphia, 1932.—Additions . . . Bartonia **15**: 32–34. 1933.

Topography, geology, soils, climate, phytogeography, botanical history, etc.; annotated list of vascular plants; bibliography.

[**Christy, C. W.**] Preliminary check-list of the flora of Crawford County, Pa. [12] p. 23 cm. [n. p., ante 1891.]

Unannotated list of vascular plants.

Claypole, E. W. Catalogue of the names of plants found in Perry County during the summer of 1882 and 1883. *In* his Preliminary report on the palaeontology of Perry County . . . Rpt. Prog. 2d Geol. Survey Penn. F^2: 113–145. 1885.

Unannotated list of vascular plants.

Darlington, William. Flora cestrica. An attempt to enumerate and describe the flowering and filicoid plants of Chester County, in the State of Pennsylvania.

With brief notices of their properties, and uses, in medicine, domestic and rural economy, and the arts. xxiii, 640 p. map. 19 cm. West-Chester, Pa., 1837.

Descriptive flora of 1,073 vascular plants, on the Linnaean system.—For additions, see **Stone, H. E.** A centennial survey of the Chester County flora. Bartonia **11**: 36–48. 1929.—The "third edition" (1853) of Darlington's work has less detailed descriptions.

Dolph, J. M. [A list of the plants of Wayne County, Pa.] *In* **White, I. C.** The geology of Susquehanna County and Wayne County. Rpt. Prog. 2d Geol. Survey Penn. **G**[5]: 31–37. 1881.

Unannotated list of vascular plants.

Fussell, Linnaeus. List of Delaware County plants. Proc. Delaware Co. Inst. Sci. **1**: 49–76. 1906.

Unannotated list of "something less than 1400 species" of vascular plants.—See also **Pennell, F. W.** Some additions . . . l. c. **4**: 68–74. 1909.

Heller, A. A. Preliminary report on the flora of Luzerne County, Penn. Bul. Torrey Club **20**: 55–67. 1893.

General features of flora; list of 325 vascular plants, with localities.

Illick, J. S. Woody plants of Franklin County, Pennsylvania. Bul. Penn. Dept. Forestry 21. 42 p. 1921.

Annotated list of 247 native and cultivated woody plants. Revision and enlargement of his Check-list of the woody plants of a portion of the South Mountains near Mont Alto . . . 1913.

Jennings, O. E. A botanical survey of Presque Isle, Erie County, Pennsylvania. Annals Carnegie Mus. (Pittsburgh) **5**: 289–421. 4 fig. (maps), pl. 22–51 (incl. tab. and map). 1909.

Physiography, ecology, with footnote references; unannotated list of 439 vascular plants.

Keller, I. A., and **Brown, Stewardson.** Handbook of the flora of Philadelphia and vicinity, containing data relating to the plants within the following radius: eastern Pennsylvania, north to the Blue Mountains, and west to the Susquehanna; all of New Jersey except the northern counties; and New Castle County, Delaware. With keys for identification of species. viii, 360 p. 23 cm. Philadelphia, 1905.

Annotated list of vascular plants, with keys.—For additions (primarily supplementing the presentation of the Philadelphia flora given in Taylor, N., Flora of the vicinity of New York) see **Long, Bayard.** Some changes in the aspect of the list of the Philadelphia flora. Bartonia **8**: 12–32. 1924; also his Some noteworthy indigenous species new to the Philadelphia area. l. c. **10**: 30–52. 1929.

Kelly, J. P. The ferns and flowering plants of central Pennsylvania. 120 p. illus. (incl. map). 19 cm. State College, Pa., 1937. (Pennsylvania state college studies, v. 2, no. 1.)

General ecological features; unannotated list of 1,166 vascular plants in form of keys, without authorities for botanical names, covering region within 35 miles of State College, Centre County, including parts of adjacent counties.

King, W. L. The flora of Northampton County, Pennsylvania. Torreya **12**: 97–107, 124–132, 165–173, 183–189, 208–215. 1912.

Geography, geology, climate; annotated list of 1,304 vascular plants.—For additions see **Rau, E. A.** Notes on the flora of Northampton County . . . l. c. **12**: 287–289. 1912.

Kline, W. A., Brendle, T. R., and **Mumbauer, J. R.** A preliminary list of the flora of the Perkiomen region . . . 41 p. 24 cm. [n. p.,] 1924.

Bibliography; annotated list of vascular plants collected by the authors. (Montgomery County.)

Lick, D. E., and **Brendle, T. R.** Plant names and plant lore among the Pennsylvania Germans. 300 p. illus. 25 cm. Lancaster, Pa., 1927.

Annotated list of wild and cultivated plants, with vernacular names and much interesting and curious information about uses; bibliography. Originally published as Pennsylvania German Society. Proceedings and addresses . . . v. 33, pt. 3. 1923.

Lutz, H. J. The vegetation of Heart's Content, a virgin forest in northwestern Pennsylvania. Ecology **11**: 1–29. 9 fig. (incl. maps), pl. 1. 1930.—Additions . . . l. c. **15**: 295–297. 1934.

Ecological: includes (p. 11–13) tabular list of vascular plants. (In Watson Township, Warren County.)

Miller, John. Flora of Erie County, Pennsylvania. 47 p. port. 23 cm. [Erie, 1923.]
Annotated list of 1,184 species and 36 varieties of vascular plants.

Pennell, F. W. Flora of the Conowingo Barrens of southeastern Pennsylvania. Proc. Acad. Nat. Sci. Phila. **62:** 541–584. 1910–11.—Further notes . . . l. c. **64:** 520–539. 1913.
Climate, physiography, soil, composition of flora; annotated list of 217 vascular plants, with supplementary list of "occasional species" (77) and "prominent introduced species" (7). The area covered is the "Conowingo or Serpentine Barrens in Chester and Delaware counties." The supplement contains also records from the serpentine barrens of Centerville, New Castle County, Del.

Shafer, J. A. A preliminary list of the vascular flora of Allegheny County, Pennsylvania. Annals Carnegie Mus. (Pittsburgh) **1:** 14–127. 1901.
List of vascular plants, with few annotations.—See also **Jennings, O. E.** Additions and corrections . . . l. c. **3:** 473–479. 1906, the latter bringing the total to about 991 species.

Small, J. K., and Carter, J. J. Flora of Lancaster County: being descriptions of the seed-plants growing naturally in Lancaster County, Pennsylvania. xvi, 336 p. 23.5 cm. New York, 1913.
Physiography, etc.; descriptive flora of spermatophytes, with keys.

Smith, A. H. On colonies of plants observed near Philadelphia. Proc. Acad. Nat. Sci. Phila. [19]: 15–24. 1867.
Annotated list of 106 vascular plants occurring on waste ground along the Delaware River in Pennsylvania and New Jersey, in the vicinity of Philadelphia.—See also **Burk, Isaac.** List of plants recently collected on ships' ballast in the neighborhood of Philadelphia. l. c. **29:** 105–109. 1877.—Also **Martindale, I. C.** More about ballast plants. Bot. Gaz. **2:** 127–128. 1877.

Twining, Alfred. Flora of northeastern Pennsylvania. 89 p. (incl. port.). 23 cm. Scranton, Pa., 1917.
Annotated list of 1,534 vascular plants from the vicinity of the Lackawanna and Wyoming valleys. Replaces "A catalogue of the flowering plants and vascular cryptogams found in and near Lackawanna and Wyoming valleys," by W. R. Dudley and C. O. Thurston. 1892, and "Additions. . . ." 1895.

RHODE ISLAND

General

Providence Franklin Society. The ferns, fern allies and flowering plants of Rhode Island. A revision of the first fifty-eight pages of James L. Bennett's "Plants of Rhode Island" published by the Providence Franklin Society in 1888. 78 p. Providence, 1920.
Unannotated list, with reference to authority for inclusion of each species; bibliography. Unreliable; see critical review by M. L. Fernald, Rhodora **24:** 96–100. 1922.

Local

Bailey, W. W. Notes on the flora of Block Island. Bul. Torrey Club **20:** 221–239. 1893.
Soil, general features of flora; mostly unannotated list of 294 vascular plants collected by the author and J. F. Collins in 1892.—For additions see **Hollick, A.** Annals N. Y. Acad. Sci. **11:** 63–70. 1898, also Torreya **6:** 190. 1906. (Newport County.)

Davis, L. D. The trees and plants of Conanicut Island. Proc. Newport Nat. Hist. Soc. **8:** 32–42. 1892.
General features of flora; unannotated list of angiosperms, the families arranged in a quite unintelligible order. (Newport County, in Narragansett Bay.)

Reynolds, E. S. The flora of the Great Swamp of Rhode Island. Rhodora **9:** 117–122. 1907.
Topography, botanical explorations, etc.; partial list of vascular plants collected. (Washington County.)

Rives, W. C., and Bailey, W. W. Native plants of the island of Rhode Island. Proc. Newport Nat. Hist. Soc. **3**: 87–89. 1885; **4**: 13–15. 1886; **5**: 32–35. 1887; **6**: 24. 1888.

Four unannotated lists of vascular plants of towns of Newport, Middletown, Portsmouth, etc. (Newport County.)

SOUTH CAROLINA

See also North Carolina (Peattie).

General

Elliott, Stephen. A sketch of the botany of South-Carolina and Georgia. 2 v. 12 pl. 22.5 cm. Charleston, 1821–24 (1816–24).

A classic among the earlier descriptive floras of the United States; now long out of date, but still the only available local treatise on the flora of the State. For dates of issue see Barnhart, J. H., Bul. Torrey Club **28**: 680–688. 1901.

Porcher, F. P. Report on the indigenous medicinal plants of South Carolina. Trans. Amer. Med. Assoc. **2**: 677–862. 1849.

List of about 410 plants, annotated as to uses.

Ravenel, H. W. A list of the more common native and naturalized plants of South Carolina. *In* South Carolina. Published by the State Board of Agriculture. p. 312–359. Charleston, 1883.

Nominal list, including only species with popular names; total number of vascular plants estimated at 1,810 species; bibliography.

Local

Bachman, John. Catalogue of phaenogamous plants and ferns, native or naturalized, found growing in the vicinity of Charleston (S. C.). South. Agr. **8**: 189–196, 286–291. 1835. (Preprinted 1834. 15 p.)

Unannotated list of vascular plants, alphabetically arranged. (Charleston County.)

Coker, W. C. Observations on the flora of the Isle of Palms, Charleston, S. C. Torreya **5**: 135–145. 3 fig. 1905.

Descriptive account of vegetation, with included enumeration of species. (Charleston County.)

——— The plant life of Hartsville, S. C. 129 p. 15 pl. 27 cm. Columbia, 1912.

Climate, geology, general features of flora, etc.; key to trees of Hartsville; annotated list of 628 vascular plants. (Darlington County.)

Gibbes, L. R. Botany of Edings' Bay. Proc. Elliott Soc. Nat. Hist. Charleston, S. C. **1**: 241–248. 1859. (Reprint 8 p.)

Geology, etc.; unannotated list of 70 vascular plants; general features of flora. (Small island between north and south mouths of Edisto River, Charleston County.)

——— A catalogue of the phoenogamous [sic] plants of Columbia, S. C. and its vicinity. 13 p. 20.5 cm. Columbia, 1835.

Unannotated tabular list on the Linnaean system, with notes on a few species. (Richland County.)

Ives, S. A., and Smith, X. A. The vascular plants of Horry County, South Carolina. Furman Bul. **20** (9): 25–40. 1938.

List of 600 vascular plants, annotated as to habitat.

McGregor, E. A. List of plants from Batesburg, S. C., and vicinity. Jour. Elisha Mitchell Sci. Soc. **33**: 133–145. 1917.

Unannotated list of vascular plants collected by the author and now in the herbarium of Stanford University. (Lexington County.)

Porcher, F. P. A medico-botanical catalogue of the plants and ferns of St. John's, Berkley, South-Carolina. South. Jour. Med. and Pharm. **2**: 255–286, 397–417. 1847. (Reprinted as thesis, 55 p.)

Annotated alphabetical list of native medicinal plants, with bibliography; alphabetical tabular list of vascular plants. (Berkeley County.)

Weatherby, C. A., and Griscom, Ludlow. Notes on the spring flora of the coastal plain of South Carolina north of Georgetown. Rhodora **36**: 28–55. 1 fig. 1934.

General features of flora; list of vascular plants collected by the authors in 1932, with data; short supplementary lists from Orangeburg, Lexington, and Saluda Counties.

SOUTH DAKOTA

See also Minnesota (MacMillan).

General

Over, W. H. Flora of South Dakota. An illustrated check-list of flowering plants, shrubs and trees of South Dakota. 161 p. illus., pl. 23 cm. Vermillion, 1932.

Annotated list of vascular plants.

——— Trees and shrubs of South Dakota. Cir. So. Dakota Geol. and Nat. Hist. Survey 11. 13 p. 1923.

Annotated systematic list.

Saunders, DeAlton. Ferns and flowering plants of South Dakota. Bul. So. Dakota Agr. Expt. Sta. **64**: 99–227. 1899.

Annotated list.—See also **Visher, S. S.** Additions to the flora of South Dakota. Muhlenbergia **9**: 45–52, 69–77. 1913.

Local

Jones, S. R. Preliminary report on the flora and fauna of the eastern part of the Rosebud Reservation, now known as Gregory County. Bul. So. Dakota Geol. Survey **4**: 123–142. 1908.

Plant regions, general features of flora; annotated list of vascular plants, also list of birds.

McIntosh, A. C. A botanical survey of the Black Hills of South Dakota. Black Hills Engineer **19**: 157–276. 82 fig., map. 1931.

Topography, geology, soils, climate, botanical explorations, fossil plants, sources of flora, ecology; annotated list of 1,300 vascular plants; bibliography. The Black Hills are in Lawrence, Meade, Pennington, Custer, and Fall River Counties.

Reagan, A. B. Notes on the flora of the Rosebud Indian Reservation, South Dakota. Trans. Kansas Acad. Sci. **20**: 191–196. 1906.

Soils, general features of flora; partly annotated list of vascular plants and 2 fungi. The same list of plants, omitting the fungi, is reprinted in Bul. So. Dakota Geol. Survey **4**: 143–153. 1908. Also issued in German in Ber. Deut. Bot. Gesell. **25**: 342–348. 1907. The Rosebud Indian Reservation includes Mellette, Todd, and Tripp Counties, and the eastern edge of Washabaugh and Bennett Counties.

Rydberg, P. A. Flora of the Black Hills of South Dakota. Contrib. U. S. Natl. Herb. **3**: 463–536. pl. 17–20 (incl. map). 1896.

Geography, topography, floral areas, etc.; annotated list of vascular plants collected by the author.—See also **Visher, S. S.** Additions to the flora of the Black Hills. Torreya **9**: 186–188. 1909; Muhlenbergia **8**: 135–137. 1913; **9**: 33–39. 1913.—Also **McIntosh, A. C.** Additions . . . Black Hills Engineer **16**: 160–167. 2 fig. 1928. (Includes bibliography.)

Visher, S. S. List of plants of Harding County, northwestern South Dakota . . . determined by Aven Nelson. Bul. So. Dakota Geol. Survey **6**: 32–68. pl. 2–4. 1914.

Annotated list of 468 vascular plants, also some lichens and mosses; bibliography.

——— Plants of the Pine Ridge Reservation. Collected by S. S. Visher, determined by Dr. P. A. Rydberg. Bul. State Geol. and Biol. Survey So. Dakota **5**: 84–108. 1912.

Annotated list of vascular plants. The reservation includes Washington and Shannon and nearly all of Washabaugh and Bennett Counties.

TENNESSEE

General

Gattinger, Augustin. The flora of Tennessee and a philosophy of botany. 296 p. illus., port. 22 cm. Nashville, 1901.

General features of flora; annotated catalog of 2,218 vascular plants. The abbreviation "O. S.," not explained in the text, means "over the whole State."

——— The medicinal plants of Tennessee . . . xxxi, 128 p. 22.5 cm. Nashville, 1894.

Descriptive flora of medicinal plants.

Local

Cain, S. A. Certain floristic affinities of the trees and shrubs of the Great Smoky Mountains and vicinity. Butler Univ. Bot. Studies **1:** 129–150 (incl. 2 maps). 1930.

Phytogeographic; includes list of 248 woody plants, divided into groups according to geographical affinities; bibliography. (Eastern Tennessee.)

——— An ecological study of the heath balds of the Great Smoky Mountains. Butler Univ. Bot. Studies **1:** 177–208. 4 fig. 1930.

Includes lists of species. The area covered is along the Tennessee-North Carolina State line.

Freeman, C. P. Ecology of the cedar glade vegetation near Nashville, Tennessee. Jour. Tenn. Acad. Sci. **8:** 141–228. 22 fig. 1933.

Includes (p. 198–205) list of vascular plants collected near La Vergne, Rutherford County.

Gattinger, Augustin. The Tennessee flora; with special reference to the flora of Nashville. Phaenogams and vascular cryptogams. 109 p. 22 cm. Nashville, 1887.

General features of flora; annotated list of 1,708 vascular plants, of which 1,251 occur within 30 miles of Nashville.

TEXAS

General

Cory, V. L., and **Parks, H. B.** Catalogue of the flora of Texas. Bul. Texas Agr. Expt. Sta. 550. 130 p. map. 1937 [1938].

Unannotated list of 5,099 species and varieties of vascular plants, their distribution indicated by districts. The paper is dated July 1937, but was actually issued, according to information furnished by Mr. Cory, on January 18, 1938.

Geiser, S. W. A century of scientific exploration in Texas, pt. 1: 1820–1880. Field and Lab. So. Methodist Univ. **4:** 41–55. 1936; **7:** 29–52. 1939.

Briefly annotated alphabetical list of collectors and explorers, with biographical references. The 1936 list reprinted with additions, in his Naturalists of the frontier, p. 317–336. [Dallas,] 1937.

Lewis, I. M. The trees of Texas. An illustrated manual of the native and introduced trees of the State. Univ. Texas Bul. 1915, no. 22. vi, 169 p. 48 fig. 1915.

Annotated descriptive treatment of the trees, of a popular nature.

Parks, H. B. Valuable plants native to Texas. Bul. Texas Agr. Expt. Sta. 551. 173 p. map. 1937 [1938].

Annotated list, with brief descriptions; distribution indicated by districts. The "value" of most of the species included is hypothetical. The paper is dated August 1937, but was actually issued, according to information furnished by Mr. Cory, on January 18, 1938.

Winkler, C. H. The botany of Texas. An account of botanical investigations in Texas and adjoining territory. Univ. Texas Bul. 1915, no. 18. 27 p. port. 1915.

Botanical collectors and authors; annotated bibliography of 121 titles.

Local

Blankinship, J. W. Plantae Lindheimerianae. Pt. III. Rpt. Missouri Bot. Gard. **18:** 123–223. pl., port. 1907.

Intended to supplement and complete Parts I–II of "Plantae Lindheimerianae," by G. Engelmann and A. Gray (1845–50). Numbering of Lindheimer's specimens; biographical sketch; list of Lindheimer's nos. 449–573 of fascicle III (1845–46), nos. 652–754 of fascicle IV (1847–48), with names, data, and indication of those which were types of new species; similar list of nos. 652–1282 from Comal County and vicinity (1849–51); bibliography of Texas botany.

Clover, E. U. Vegetational survey of the lower Rio Grande valley, Texas. Madroño **4:** 41–66, 77–100. illus. (incl. maps). 1937.

Botanical explorations, climate, topography, geology, drainage, local ecological nomenclature, plant associations; list of bryophytes and vascular plants collected by the author, with localities; bibliography. Covers the area along the Rio Grande from Zapata County to the coast.

Coulter, J. M. Botany of western Texas. A manual of the phanerogams and pteridophytes of western Texas. Contrib. U. S. Natl. Herb. v. 2. 588 p. 3 pl. 1891–94.

Descriptive flora of the vascular plants, intended to include all species west of the 97th meridian.

Fletcher, H. T. Notes on the vegetation of the Green Valley region. Pub. West Texas Hist. and Sci. Soc. **2**: 6–50. 1928.

Physiography, climate, etc.; partly annotated list of vascular plants, with notes on uses. The region covered is in Brewster and Presidio Counties.

Hartman, Carl. A list of trees and shrubs occurring in the vicinity of Huntsville, Texas. Trans. Texas Acad. Sci. **12** (2): 66–90. 1913.

Soil, etc.; annotated list of 106 species. (Walker County.) Also in second issue of v. 12, pt. 2, p. 32–56. 1931.

Havard, Valery. Report on the flora of western and southern Texas. Proc. U. S. Natl. Mus. **8**: 449–533. 1885.

Climate, topography, etc., with lists of species of different areas; annotated list of native economic plants.

Heller, A. A. Botanical explorations in southern Texas during the season of 1894. Contrib. Herb. Frankl. and Marshall Col. no. 1. 116 p. 9 pl. 1895.

Botanical regions, author's itinerary; annotated list of vascular plants collected mainly around Corpus Christi, Kerrville, and San Antonio.

Mackensen, Bernard. The trees and shrubs of San Antonio and vicinity. A handbook of the woody plants growing naturally in and about San Antonio, Texas. 51 p. illus. 19 cm. San Antonio, 1909.

Climate, soils, plant formations, general features of flora; annotated list of woody plants, with brief descriptions. (Bexar County.)

Metz, Mary Clare, *sister*. A flora of Bexar County, Texas. Contrib. Biol. Lab. Catholic Univ. Amer. no. 16. xiii, 214 p. map. 1934.

Botanical explorations, physiography, geology; annotated list of nearly 900 vascular plants, with keys.

Muller, C. H. Vegetation in Chisos Mountains, Texas. Trans. Texas Acad. Sci. **20**: 3–31. 1937.

Topography, climate, general features of flora; annotated list of vascular plants collected by the author in 1931–32; bibliography. (Brewster County.)

Pace, Lula. McLennan County plants. *In* **Adkins, W. S.** Geology and mineral resources of McLennan County. Univ. Texas Bul. **2340**: 183–197. 1923.

Annotated list of vascular plants.

Palmer, E. J. The ligneous flora of the Davis Mountains, Texas. Jour. Arnold Arboretum **10**: 8–45. 1929.

Physiography, climate, plant associations, botanical explorations, etc.; annotated list of woody plants. (Jeff Davis County.)

——— The ligneous flora of the Staked Plains of Texas. Jour. Arnold Arboretum **2**: 90–105. 1920.

Physiography, general features of flora; annotated list of woody plants collected by the author in Randall and Armstrong Counties on the Llano Estacado.

Parks, H. B., Cory, V. L., and others. The fauna and flora of the Big Thicket area. 51 p. illus. (incl. map). 23 cm. [n. p.,] 1936. (Reprinted, 1938.)

Includes unannotated list of vascular plants (p. 27–51). Cover title reads: Biological survey of the east Texas Big Thicket area. The area covered extends from southern Newton County north into Sabine County and west to Grimes County.

Sperry, O. E. A check list of the ferns, gymnosperms, and flowering plants of the proposed Big Bend National Park of Texas. Sul Ross State Teach. Col. Bul. **19** (4): 9–98. 2 fig. (incl. map). 1938.

Bibliography, gazetteer; list of vascular plants collected by the author and others, with localities. (Brewster County.)

Stillwell, Norma. Key and guide to native trees, shrubs and woody vines of Dallas County. 48 p. illus. 21.5 cm. Dallas, 1939.

Annotated systematic list of 86 species, with key and popular descriptions, the species of Cacti not enumerated; bibliography.

Tharp, B. C. Ecologic investigations in the Red River valley. *In* Sellards, E. H., and others. Investigations on the Red River . . . Univ. Texas Bul. 2327: 89–155. 1923.

Includes lists of spermatophytes observed in the area, chiefly in Wichita County, Texas, and adjacent Oklahoma.

Young, M. S. The seed plants, ferns, and fern allies of the Austin region. Univ. Texas Bul. 2065. 98 p. 1920.

Annotated list of vascular plants. (Travis County.)

UTAH

See also Nevada (Tidestrom).

Local

Chamberlin, R. V. The ethnobotany of the Gosiute Indians of Utah. Mem. Amer. Anthropol. Assoc. 2: 329–405. 1911. (Also in Proc. Acad. Nat. Sci. Phila. 63: 24–99. 1911.)

Running account of useful plants; annotated list of scientific names, list of vernacular names. These Indians inhabit the desert country bordering on Great Salt Lake.

Eastwood, Alice. Report on a collection of plants from San Juan County, in southeastern Utah. Proc. Calif. Acad. Sci. II, 6: 270–329. pl. 44–46. 1897 (1896).

General features of flora; annotated list of 161 vascular plants (and Characeae) collected by the author in 1895.

Garrett, A. O. Spring flora of the Wasatch region. 5th ed. vi, 240 p. 20 cm. Salt Lake City, 1936. (1st ed., 1911.)

Keys and descriptions of spring flowering plants of eastern edge of Great Basin as far south as Manti.

Graham, E. H. Botanical studies in the Uinta Basin of Utah and Colorado. Annals Carnegie Mus. (Pittsburgh) v. 26. 432 p. 13 pl. (incl. map), fold. tab. 1937.

Botanical explorations, physiography, geology, climate, zones of vegetation, plant communities, phytogeography; bibliography; annotated list of 1,104 species and varieties of vascular plants. The region covered is in northeastern Utah and northwestern Colorado.

Presnall, C. C., and Patraw, P. M. Plants of Zion National Park. Zion-Bryce Mus. Bul. no. 1. 69 p. illus. 1937.

Annotated list of vascular plants, with brief popular descriptions of many species; authorities for botanical names not given. Processed publication of National Park Service. (Washington and Kane Counties.)

VERMONT

See also New Hampshire (Jesup).

General

Barnhart, J. H. The local floras of Vermont. Bul. Vermont Bot. Club. 2: 11–16. 1907.

Chronological list of 64 titles.

Burns, G. P., and Otis, C. H. The trees of Vermont. Bul. Vermont Agr. Expt. Sta. 194. 244 p. illus. 1916.

Annotated descriptive account, with figure of each species; wood anatomy of more important species.—See also **Jones, L. R., and Rand, F. V.** Vermont shrubs and woody vines. Bul. Vermont Agr. Expt. Sta. 145. p. 49–119. illus. 1909. (Popular account.)

Dole, E. J., ed. The flora of Vermont. 3d rev. ed. An annotated list of the ferns and seed plants of the State of Vermont. Compiled by a committee of the Vermont Botanical Club. xiv, 353 p. port. 22.5 cm. [Burlington ?], 1937.

Botanical explorations, geology, statistics; annotated list of 2,572 vascular plants (1,861 species). Replaces the Vermont Botanical Club's Flora of Vermont, published as Bul. Vermont Agr. Expt. Sta. 187. p. 137–258. 1915, which in turn replaced the Flora of Vermont, prepared by a committee consisting of E. Brainerd, W. W. Eggleston, and L. R. Jones and published as Contributions to the botany of Vermont no. 8. 1900.

Local

Flynn, N. F. Flora of Burlington and vicinity. A list of the fern and seed plants growing without cultivation. ix, 124 p. 23 cm. Burlington, 1911. (Contributions to the botany of Vermont, 9.)—Additions . . . Bul. Vermont Bot. Club **7**: 16–17. 1912.

Annotated list of 1,240 vascular plants.—For additions, see **Blake, S. F.** Six weeks' botanizing in Vermont, I. Notes on the plants of the Burlington region. Rhodora **15**: 153–168. 1913. (Chittenden County.)

Kennedy, G. G. Flora of Willoughby, Vermont. Rhodora **6**: 93–134. pl. 54–56, map. 1904. (Reprinted.)—Additional notes from Willoughby. l. c. **6**: 148. 1904.

Geography, soils, botanical collectors, etc.; annotated list of 690 vascular plants. (Orleans County.)

Kittredge, E. M. Ferns and flowering plants of Woodstock, Vermont. 57 p. 16.5 cm. Woodstock, 1931.—Supplement . . . [7] p. 1936.—Grasses and sedges of Woodstock, Vermont. 27 p. Woodstock, 1939.

Annotated list of vascular plants. Supplements not seen. (Windsor County.)

VIRGINIA

See also District of Columbia (Hitchcock and Standley); Iowa (Pammel, Flora of northern Iowa peat bogs); North Carolina (Ewing and Stanford).

Local

Artz, Lena. Plants of the shale banks of the Massanutten Mountains of Virginia. Claytonia **3**: 45–50. 1937; **4**: 10–15. 1937.

Geology, characters and range of the most characteristic species; list of vascular plants; bibliography. The Massanutten Mountains lie along the boundary line between Shenandoah, Rockingham, Page, and Warren Counties.

Erlanson, E. W. The flora of the peninsula of Virginia. Papers Mich. Acad. Sci. **4**: 115–182. 1925.

General features of flora; list of 1,052 vascular plants collected by E. J. Grimes and the author in 1920–21, with localities; bibliography. The "peninsula of Virginia" lies between the York and the James Rivers.

Fernald, M. L. Last survivors in the flora of Tidewater Virginia. Rhodora **41**: 465–504, 529–559, 564–574. illus. (maps), pl. 570–583. 1939. (Contrib. Gray Herb. 128.)

Author's itinerary in 1938 and 1939, with notes on plants observed; annotated list of noteworthy species. (Southeastern Virginia.)

——— Local plants of the inner coastal plain of southeastern Virginia. Rhodora **39**: 321–366, 379–415, 433–459, 465–491. pl. 474–487. 1937. (Contrib. Gray Herb. 120.)

Soils, running account of itinerary and collecting; critical notes on numerous vascular plants, phytogeography.

——— Noteworthy plants of southeastern Virginia. Rhodora **40**: 364–424, 434–459, 467–485. pl. 509–535. 1938. (Contrib. Gray Herb. 123.)

Author's itinerary in 1937; annotated list of rarer species collected.

——— Plants from the outer coastal plain of Virginia. Rhodora **38**: 376–404, 414–452. pl. 440–452. 1936. (Contrib. Gray Herb. 115.)

Phytogeography; critical notes on plants collected in 1935.

——— and **Griscom, Ludlow.** Three days of botanizing in southeastern Virginia. Rhodora **37**: 129–157, 167–189. pl. 332–351. 1935. (Contrib. Gray Herb. 107.)

Critical notes on plants collected by the authors in 1933 in Princess Anne and Norfolk Counties.—See also **Fernald, M. L.** Midsummer vascular plants of southeastern Virginia. l. c. **37**: 378–413, 423–454. pl. 384–405. 1935. (Contrib. Gray Herb. 109.)

Kearney, T. H. Report on a botanical survey of the Dismal Swamp region. Contrib. U. S. Natl. Herb. **5**: 321–550. fig. 51–90, pl. 65–76, 2 maps. 1901.

Mainly ecological; includes annotated list of species; bibliography. The area covered includes Princess Anne, Norfolk, and part of Nansemond Counties, Va., and Currituck, Camden, Pasquotank, and Perquimans Counties, N. C.

Merriman, P. R. Flora of Richmond and vicinity (exclusive of grasses, sedges and trees). 353 p. (incl. 53 pl.). 19 cm. Richmond, Va., 1930.

Popular descriptive flora. (Henrico County.)

Murrill, W. A. The natural history of Staunton, Virginia. xiii, 216 p. 4 pl. 20 cm. New York, 1919.

Contains alphabetical lists of vascular plants collected (p. 110–182). (Augusta County.)

Small, J. K., and **Vail, A. M.** Report of the botanical exploration of southwestern Virginia during the season of 1892. Mem. Torrey Club **4**: 92–201. pl. 75–82. 1893–94.

Physiography, etc.; annotated list of vascular and cellular plants collected mostly within 20 miles of Marion. (Marion is in Smyth County.)

WASHINGTON

See also British Columbia (Brown); Idaho (Spinden); Oregon (Gilkey and Powell; Hemenway).

General

Cooper, J. G. Report on the medical flora of Washington Territory. Trans. Amer. Med. Assoc. **10**: 221–237. 1857.

Annotated list of wild medicinal plants.

Piper, C. V. Flora of the State of Washington. Contrib. U. S. Natl. Herb. v. 11. 637 p. 22 pl., map. 1906.

List of vascular plants, with keys and citation of specimens; physiography, plant associations, collectors, bibliography; gazetteer.

Rigg, G. B. Notes on the history of botany in the State of Washington. Washington Hist. Quart. **20**: 163–173. 1929.

Botanical explorations and collectors, with especial reference to the work of institutions.

Local

Densmore, Frances. Plants used by the Makah. *In* her Nootka and Quileute music. Bul. Bur. Amer. Ethnol. **124**: 307–321. 1939.

Tabular list of plants used by the Makah Tribe; annotated list of medicinal and other useful plants, classified by uses. (Neah Bay, Clallam County.)

Gorman, M. W. Flora of Hamilton Mountain, Washington. Mazama **6**: 62–77. 1920.

Physiography, etc.; annotated list of 198 vascular plants. Hamilton Mountain is a small volcanic peak 2,432 feet high, in Skamania County.

Jones, G. N. A botanical survey of the Olympic Peninsula, Washington. Univ. Washington Pub. Biol. v. 5. 286 p. 9 pl. (incl. map). 1936.—Supplementary notes on the flora of the Olympic Peninsula. Leaflets West. Bot. **2**: 105–108. 1938.

Physiography, geology, climate, plant zones, phytogeography, introduced plants, life forms, botanical explorations, bibliography; annotated list of "over 1,000 species and varieties" of vascular plants, with keys and citation of specimens. The region is bounded by the Pacific Ocean, the Strait of Juan de Fuca, the valley of the Chehalis River, and (on the east) Puget Sound, Admiralty Inlet, and Hood Canal.

——— The flowering plants and ferns of Mount Rainier. Univ. Washington Pub. Biol. v. 7. 192 p. 9 pl. 1938.

Botanical explorations, plant zones, biological spectrum; annotated descriptive flora of vascular plants (729 species) of Mount Rainier National Park. (Pierce and Lewis Counties.)

Muenscher, W. L. C. Flora of Whatcom County, Washington. Muhlenbergia **9**: 101–116, 117–132. 1914.

Topography, etc.; annotated list of 462 vascular plants.—See also his Some changes in the weed flora of Whatcom County, Washington. Torreya **30**: 130–135. 1930.

Piper, C. V., and **Beattie, R. K.** Flora of the northwest coast, including the area west of the summit of the Cascade Mountains, from the forty-ninth parallel south to the Calapooia Mountains on the south border of Lane County, Oregon. xiii, 418 p. 23.5 cm. Lancaster, Pa., 1915.

Descriptive flora of the vascular plants (1,617 species and subspecies) with keys; glossary.—See also **Nelson, J. C.** Additions to the flora of western Oregon. Torreya **18**: 21–35, 220–226. 1918; **20**: 37–45. 1920; **21**: 24–28. 1921; **22**: 98–102. 1922; **23**: 63–67. 1923; also his The introduction of foreign weeds in ballast as illustrated by ballast-plants at Linton, Oregon. l. c. **17**: 151–160. 1917.

Reagan, A. B. Plants used by the Hoh and Quileute Indians. Trans. Kansas Acad. Sci. **37**: 55–70. 1934.

List of Indian plant names; annotated systematic list of vascular plants, with uses. The Hoh and Quileute Indians live on the Pacific coast of the Olympic Peninsula, about due west of Seattle, Washington.

St. John, Harold. Flora of southeastern Washington and of adjacent Idaho. xxv, 531 p. 11 fig., map. 21.5 cm. Pullman, 1937.

Physiography, climate, life zones; descriptive flora of vascular plants. The region covered extends from Spokane to Walla Walla Counties, Washington, with a strip about 15 miles wide in adjacent Idaho. Replaces Piper, C. V., and Beattie, R. K. Flora of southeastern Washington and adjacent Idaho. 1914, which replaced the same authors' The flora of the Palouse region . . . 1901.—See also **Daubenmire, R. F.** Range extensions for southeastern Washington and adjacent Idaho. Leaflets West. Bot. **2**: 199–200. 1939.

——— and **Courtney, W. D.** The flora of Epsom Lake. Amer. Jour. Bot. **11**: 100–107. 1 fig. 1924.

List of vascular plants found in the drainage basin of the lake, which is a "small lake of nearly pure magnesium sulphate," near Oroville, Okanogan County.

——— and **Hardin, Edith.** Flora of Mt. Baker. Mazama **11**: 52–102. illus. 1929. (Reprinted.)

Plant zones, botanical explorations; keyed flora of 333 vascular plants, with brief descriptions.—See also **Muenscher, W. C.** Additions to our knowledge of the flora of Mount Baker . . . Madroño **4**: 263–270. 1938. (Adds 228 species; bibliography.) (Whatcom County.)

——— and **Jones, G. N.** An annotated catalogue of the vascular plants of Benton County, Washington. Northwest Science **2**: 73–93. 4 fig. (incl. map). 1928. (Contrib. Dept. Bot. State Col. Washington no. 9.)

Topography, ecology, etc.; annotated list of 403 species of vascular plants.—See also **Powell, G. M.** Additions to the annotated catalogue . . . Northwest Science **13**: 72. 1939.

——— and **Warren, F. A.** The plants of Mount Rainier National Park, Washington. Amer. Midl. Nat. **18**: 952–985. 1937.—Additions . . . l. c. **20**: 243–244. 1938.

Geology, life zones, botanical explorations, bibliography; annotated list of 695 spermatophytes. Replaces Piper, C. V. The flora of Mount Rainier. Mazama **2**: 93–117. 1901; Additions and corrections. l. c. **2**: 270–271. 1905. (Pierce and Lewis Counties.)

——— and **Warren, F. A.** Preliminary list of the plants of the Kaniksu National Forest, Idaho and Washington. 36 p. 28 cm. [Pullman], 1925. (Contrib. Bot. Dept. State Col. Washington no. 2.)

Annotated list of 451 vascular plants, with a single key to the species. (Pend Oreille County, Washington, and Bonner County, Idaho.) (Processed publication.)

Suksdorf, W. N., and Howell, Thomas. The flora of Mount Adams. Mazama **1**: 68–97. 1896.

Unannotated list of 480 vascular plants. Mount Adams is in the southwestern corner of Yakima County, extending into Skamania County.

Wylie, R. B. The flora of Iowa Rock, a small rocky island in Puget Sound. Proc. Iowa Acad. Sci. **16**: 99–101. pl. 1909.

Topography, etc.; unannotated list of cellular and vascular plants. (Small island near Friday Harbor, San Juan County.)

WEST VIRGINIA

General

Brooks, A. B. West Virginia trees. Bul. West Virginia Agr. Expt. Sta. 175. 242 p. illus. 1920.

Description and full-page illustration of each species; list of shrubs and shrubby vines; glossary.

Core, E. L. The botanical exploration of West Virginia. Proc. West Virginia Acad. Sci. **10** (1936): 46–64. 1937.

Annotated chronological list of 161 collectors, with index.

Fontaine, W. M. Forest trees, shrubs, and medicinal plants. *In* **Maury, M. F.,** and **Fontaine, W. M.** Resources of West Virginia. p. 111–141. Wheeling, 1876.

Annotated list of trees, partial list of shrubs; unannotated list of 76 medicinal plants.

Millspaugh, C. F. The living flora of West Virginia. West Virginia Geol. Survey [Rpt. **5** (A)]: 1–389, 454–487. 1913.

Topography; annotated list of collectors; annotated list of 3,411 plants, of which 1,648 are vascular. Incorporates and replaces the two following earlier floras of the State: Millspaugh, C. F. Preliminary catalogue of the flora of West Virginia. Bul. West Virginia Agr. Expt. Sta. 24. p. 311–537. 1892. Millspaugh, C. F., and Nuttall, L. W. Flora of West Virginia. Pub. Field Columbian Mus. no. **9**, Bot. Ser. **1**: 69–276. pl. 5–7. 1896.—For additions, see **Strausbaugh, P. D.,** and **Core, E. L.** Some additions to the Millspaugh check list of West Virginia spermatophytes. Proc. West Virginia Acad. Sci. **4** (1930): 38–48. 1931. Additions . . . l. c. **9**: 29–31. 1936.

West Virginia. Biological Survey. A preliminary bibliography of West Virginia biology. I. West Virginia Univ. Bul. ser. 37, no. 5–II. 20 p. 1936. (West Virginia Univ. Studies II.)

Unannotated list of botanical and zoological publications.

Local

Core, E. L. [Plant relations on Spruce Knob.] *In* **Tilton, J. L.,** and others. West Virginia geological survey. Pendleton County. p. 311–313. Wheeling, 1927.

Plant associations, etc., with partial list of species. Spruce Knob is the highest mountain in the State. An additional list by **A. B. Brooks** is given on p. 310.

Frye, W. M. The flora of Hampshire County, West Virginia. Proc. West Virginia Acad. Sci. 8 (West Virginia Univ. Bul. ser. 35, no. 15): 59–82. 1 fig. (map). 1935.

Physiography, soil, climate, rare plants; unannotated list of about 939 vascular plants.

Hill, C. N. A botanical survey of Marion County, West Virginia. Proc. West Virginia Acad. Sci. 8 (West Virginia Univ. Bul. ser. 35, no. 15): 99–110. 1935

Physiography, soil, general features of flora, plant societies; list of genera of vascular plants, with indication of number of species found; bibliography.

WISCONSIN

See also Illinois (Gates, The vegetation of the beach area . . .); Minnesota (Densmore).

General

Bruhin, T. A. Vergleichende flora Wisconsins. Verhandl. K. K. Zool.-Bot. Gesell. Wien **26**: 229–286. 1877.—Nachträge und berichtigungen . . . l. c. **27**: 859–866. 1878.—Zweiter nachtrag . . . . l. c. **28**: 633–644. 1879.

Collectors and collections; comparison of European and Wisconsin floras; list of 1,300 vascular plants, with localities; bibliography.

Cheney, L. S. An historical review of the work done on the flora of the territory now included within the limits of Wisconsin. Pharm. Rev. **18**: 557–565. 1900; **19**: 2–15. 1901.

Account of botanists and their explorations and publications, with bibliographical footnotes.

[Clohisy, Matt.] Wisconsin trees. 148 p. illus. 20 cm. Milwaukee, 1927. (Reprinted from The Milwaukee Journal.)
Annotated descriptive list, with notes on uses; popular treatment.

Denniston, R. H., and **Kremers, R. E.** Medicinal plants of Wisconsin. Bul. Univ. Wisconsin **738** (Gen. Ser. 542) : 22–31. 1914.
Tabular list of 259 native vascular plants, with indication of part used and reference to standard dispensatories.

Fassett, N. C., and others. Preliminary reports on the flora of Wisconsin I–XXVI. Trans. Wisconsin Acad. Sci., Arts and Lett. **24**: 249–268, 357–361. illus. (maps). 1929; **25**: 169–214. illus. (maps). 1930; **26**: 263–288. illus. (incl. maps). 1931; **27**: 207–249. illus. (maps). 1932; **28**: 171–196. illus. (maps). 1933; **29**: 299–318. illus. (maps). 1935; **30**: 17–25. illus. (maps). 1937.
List of species, with ranges, of various families and orders; not published in systematic order.

Swezey, G. D. Catalogue of the phaenogamous and vascular cryptogamous plants of Wisconsin. Geol. Survey Wisconsin, Survey of 1873–79. **1**: 375–395. 1883.
Essentially unannotated list of vascular plants.

Local

Cheney, L. S. A contribution to the flora of the Lake Superior region. Trans. Wisconsin Acad. Sci., Arts and Lett. **9**: 233–254. 1893.
List of 345 vascular and cellular plants (233 vascular), with localities. The region covered is in northern Wisconsin, along the north shore of Lake Superior, and the Minnesota-Ontario boundary.

——— and **True, R. H.** On the flora of Madison and vicinity, a preliminary paper on the flora of Dane County, Wisconsin. Trans. Wisconsin Acad. Sci., Arts and Lett. **9**: 45–135. pl. 2 (map). 1893.
Physiography, etc.; annotated list of 900 plants (of which 751 are vascular).

Fuller, A. M. A botanist afield on Washington Island. Year Book Pub. Mus. Milwaukee **6** (1926) : 66–78. fig. 61–64. 1927.
General features of flora, etc.; annotated alphabetical list of vascular plants. (Door County.)

Lueders, H. F. The vegetation of the town Prairie du Sac. Trans. Wisconsin Acad. Sci., Arts and Lett. **10**: 510–524. pl. 17 (map). 1895.
Physiography, general features of flora; unannotated list of vascular plants. (Sauk County.)

Marshall, Ruth. The vegetation of Twin Island. Trans. Wisconsin Acad. Sci., Arts and Lett. **16** (2) : 773–797. pl. 62–63 (maps). 1910.
Plant formations, etc.; list of vascular plants. (In Lake Spooner, Washburn County.)

Pammel, L. H. Woody plants of western Wisconsin. Proc. Iowa Acad. Sci. **1** (2) : 76–80. 1892.
Topography, geology; unannotated list of 115 trees, shrubs, and woody vines. (La Crosse County, Wis., and Houston and Fillmore Counties, Minn.)

Russel, Howland. Check list of the flora of Milwaukee County. Bul. Wisconsin Nat. Hist. Soc. n. s., **5**: 167–250. 1907.
Annotated list of vascular plants.

Smith, H. H. Ethnobotany of the Forest Potawatomi Indians. Bul. Pub. Mus. Milwaukee **7**: 1–230. pl. 1–38, port. 1933.
History of tribe, etc.; annotated classified lists of useful plants; bibliography; alphabetical list of Indian vernacular names. (Forest County.)

——— Ethnobotany of the Menomini Indians. Bul. Pub. Mus. Milwaukee **4**: 1–174. 36 pl. 1923.
General account of the Menomini tribe, who live in Shawano County; annotated lists of medicinal, edible, and other useful plants, with vernacular names.—See also **Judson, J. E.** Wild flowers used by the Indians for medical purposes. Proc. West Virginia Acad. Sci. **11** (1937) : 165–168. 1938.

Smith, H. H. Ethnobotany of the Meskwaki Indians. Bul. Pub. Mus. Milwaukee **4**: 175–326. pl. 37–45. 1928.
Annotated list of plants used by the Meskwaki or Fox Indians of Wisconsin, with native names. These Indians now live on the Meskwaki Reservation at Tama, Tama County, Iowa, where most of the data were obtained.

——— Ethnobotany of the Ojibwe Indians. Bul. Pub. Mus. Milwaukee **4**: 327–525. pl. 46–77. 1932.
History of tribe; annotated classified lists of useful plants; alphabetical list of Indian vernacular names. These Indians live in the forest and lake region of northern Wisconsin (in Vilas, Bayfield, Iron, and Clark Counties, etc.), also on Bear Island, Leech Lake, Cass County, Minnesota.

Tracy, C. T. Catalogue of plants growing without cultivation in Ripon and the near vicinity. 26 p. 17 cm. Ripon, Wis., [1889].
Unannotated list of vascular plants. (Fond du Lac County.)

Wadmond, S. C. Flora of Racine and Kenosha Counties, Wisconsin: a list of the fern and seed plants growing without cultivation. Trans. Wisconsin Acad. Sci., Arts and Lett. **16**: 798–888. 1910.
Physiography, etc.; annotated list of 901 vascular plants.

Wheeler, W. M. The flora of Milwaukee County. Proc. Nat. Hist. Soc. Wisconsin **1888**: 154–190. [1888?]—1st supplement. l. c. **1889**: 229–230. [1889?]
Geology, soils, climate, bibliography; annotated list of 691 vascular plants, raised in the supplements to 921. First supplement not seen (cf. Bul. Torrey Club **17**: 78–79. 1890).—For 2d and 3d supplements see **Bennetts, W. J.** Additions . . . Bul. Wisconsin Nat. Hist. Soc. **1**: 161–166. 1900; **2**: 39–44. 1902.

WYOMING

See also Montana (Rydberg).

General

Nelson, Aven. First report on the flora of Wyoming. Bul. Wyoming Agr. Expt. Sta. 28. p. 45–218. 2 fig., map. 1896.
Annotated list of vascular (and a few cellular) plants, chiefly collected by the author in 1894–95; lists of species reported by other botanists; total (phanerogams) 1,295 species and varieties.

—— The trees of Wyoming and how to know them. Bul. Wyoming Agr. Expt. Sta. 40. p. 57–110. illus. 1899.
Annotated list of native trees, with brief descriptions.—See also **Cary, Merritt.** Notes on the distribution of conspicuous trees and shrubs in Wyoming. *In* his Life zone investigations in Wyoming. North Amer. Fauna **42**: 55–81. fig. 3–17. 1917.

Nelson, E. E. The shrubs of Wyoming. Bul. Wyoming Agr. Expt. Sta. 54. 47 p. illus. 5 pl. 1902.
Annotated list of native shrubs.

Local

McDougall, W. B., and **Baggley, H. A.** Plants of Yellowstone National Park. iv, 160 p. illus., col. pl. 23.5 cm. Washington, D. C., 1936.
A popular treatment, with keys and without authorities for botanical names, apparently intended to list all vascular plants known from the region. Issued by National Park Service. The park is in the northwestern corner of Wyoming, slightly overlapping the Montana and Idaho lines.

Russell, C. P. A concise history of scientists and scientific investigations in Yellowstone National Park. With a bibliography of the results of research and travel in the park area. 144 p. 26.5 cm. [Washington, D. C., 1934.]
Botanical bibliography, p. 37–39. Published by National Park Service. (Processed.)

WEST INDIES

For Bermuda Islands, see Insular Floras; Trinidad and Tobago, see Venezuela.

Descourtilz, M. E. Flore pittoresque et médicale des Antilles, ou histoire naturelle des plantes usuelles des colonies françaises, anglaises, espagnoles et portugaises . . . 2. éd. 8 v. 600 col. pl. 21 cm. Paris, 1833. (1st ed., 1821–29.)

Descriptive account, the plants classified by uses, with vernacular names, description, chemical and medicinal properties, method of use. The second edition is not essentially different from the first.

Grisebach, A. H. R. Flora of the British West Indian islands. xvi, 789 p. 23 cm. London, 1864 [1859–64].

Still the only general flora of the West Indies, from the Bahamas to Trinidad (not including Cuba, Hispaniola, and Puerto Rico). Descriptive flora with synonymy, local and extralimital range (the latter notations now often untrustworthy); separate index of vernacular names. Dates of publication (see Urban, Symb. Antill. **1**: 56): p. 1–192. 1859; 193–315. 1860; 315–506. 1861; 507–789, preface. 1864. Grisebach's paper "Die geographische verbreitung der pflanzen Westindiens," Abhandl. K. Gesell. Wiss. Göttingen **12**: 3–80. 1865 (separate, 80 p.) gives an account of the plant geography based on his Flora; 4,401 species, of which 2,155 are endemic. For statistics of the number of species recorded from different islands, see Urban, Symb. Antill. **1**: 57. 1898.

Grosourdy, René de. El médico botánico criollo. 4 v. in 2. 24 cm. Paris, 1864.

Includes annotated lists of timbers, medicinal plants, poisonous plants, etc. Covers the West Indies and adjacent America.

Urban, Ignatz. Biblographia Indiae occidentalis botanica. In his Symbolae antillanae . . . **1**: 3–195. 1898.—Continuatio I–III. l. c. **2**: 1–7. 1900; **3**: 1–13. 1902; **5**: 1–16. 1904.

Critically annotated list, arranged by authors, with notes referring to their travels in the West Indies. The titles in the first paper are classified in a conspectus.

——— Notae biographicae peregrinatorum Indiae occidentalis botanicorum. In his Symbolae antillanae . . . **3**: 14–158. 1902.

Bibliography; alphabetical list of collectors, with brief biography and itinerary, statement of location of collections, reference to biographies; geographical conspectus, listing collectors and their dates.

BAHAMA ISLANDS

General

Britton, N. L., and Millspaugh, C. F. The Bahama flora. viii, 695 p. 23.5 cm. New York, 1920.

Complete flora, with keys, descriptions, extralimital range and pertinent synonymy of the native and naturalized plants; brief description of region; list of collectors and collections; annotated bibliography. Musci by E. G. Britton, Hepaticae by A. W. Evans, Lichenes by L. W. Riddle, Algae by M. A. Howe, Myxophyceae by F. S. Collins, Diatomeae by C. S. Boyer, Fungi by F. J. Seaver. Lists 1,982 species (of which 1,028 are vascular plants) of which 185 (133 vascular) are endemic.

Local

Northrop, A. R. Flora of New Providence and Andros (Bahama Islands). With an enumeration of the plants collected by John I. Northrup and Alice R. Northrup, in 1890. Mem. Torrey Club **12**: 1–98. 19 pl., map. 1902. (Reprinted in Osborn, H. F. A naturalist in the Bahamas: John I. Northrop. 1910.)

Physiography, botanical regions, botanical explorations; list of 542 species (461 spermatophytes), with localities and collector's numbers; relationship of flora; table of local and extralimital distribution.

CUBA

General

Fors, A. J. Las maderas cubanas . . . [2. ed.?] 106 p. 22.5 cm. Habana, 1937. (1st ed., 1929.)

Annotated list of trees, alphabetically arranged by vernacular names; list of vernacular names arranged by families. Publication of the Dirección de montes y minas. Secretaría de agricultura, comercio y trabajo, Cuba.

Gómez de la Maza, Manuel. Ensayo de farmacofitología cubana. Resumen de las propiedades medicinales, con especialidad las recientemente estudiadas, de muchas plantas indígenas ó de cultivo . . . 112 p. 19.5 cm. Habana, 1889.

Briefly annotated list of 179 cellular and vascular plants, with vernacular names and uses.

——— and **Roig y Mesa, J. T.** Flora de Cuba. Bol. Estac. Expt. Agron. Santiago de las Vegas 22. 178 p. 33 pl., 4 port., map. 1914.

Of value chiefly for its vernacular names (separately indexed) and notes on economic plants, these classed according to uses. Contains also botanical statistics, list of fossil plants, and short notes on Cuban herbaria, collectors, and botanical works. Not a flora.

Grisebach, A. H. R. Catalogus plantarum cubensium exhibens collectionem Wrightianam aliasque minores ex insula Cuba missas. iv, 301 (i. e. 303) p. 22 cm. Lipsiae, 1866.

The fundamental list of Cuban vascular plants. Includes citation of exsiccatae, but not specific localities; separate index of vernacular names. Contains 3,263 species (2,984 phanerogams), of which 939 (929) are endemic.

Léon, J. S. S., *hermano.* Las exploraciones botánicas de Cuba. Mem. Soc. Cubana Hist. Nat. "Felipe Poey" **3**: 178–224. port., map. 1918. (Reprinted.)

General review of botanical explorations in Cuba from earliest times to 1917; bibliographies. The map shows the principal localities at which collections have been made.

Montagne, Camille, and **Richard, Achille.** Flora cubana, ò descripcion botánica usos y aplicaciones de las plantas reunidas en la isla de Cuba por D. Ramon de la Sagra. 3 v. and atlas of 122 pl. (20 col.). 40 cm. Paris, 1845–55(56). (Historia física, política y natural de la isla de Cuba, por D. Ramon de la Sagra. Segunda parte, Historia natural, tomo IX–XII, Botánica.)

Vol. I: Climate, calendar of flowering, etc.; systematic list of vascular and cellular plants, etc. (the cryptogams by Montagne). Vol. 2–3: Descriptive flora of 1,248 phanerogams, with vernacular names, by Richard. There is also an incomplete French edition (see Urban, Symb. Antill. **1**: 146. 1898).

Roig y Mesa, J. T. Diccionario botánico de nombres vulgares cubanos. Bol. Estac. Expt. Agron. Santiago de las Vegas 54. viii, 897 p. 48 pl. 1928.

Bibliography; annotated list of vernacular names of wild and cultivated plants with botanical equivalents; brief descriptions, local occurrence, uses.

Sauvalle, F. A. Flora cubana. Enumeratio nova plantarum cubensium vel revisio catalogi Grisebachiani, exhibens descriptiones generum specierumque novarum Caroli Wright, (Cantabrigiae) et Francisci Sauvalle, synonymis nominibusque vulgaribus cubensis (sic) adjectis. 324 p. 25.5 cm. Havanae, 1873.

Systematic list of 3,350 vascular plants mostly collected by Charles Wright, with vernacular names and some synonyms; Wright's collecting numbers cited only from p. 49 on. Vernacular names separately indexed. Originally appeared in Anales R. Acad. Cien. Habana v. 5–9. 1868–1873. For information about Wright's work in Cuba, see Underwood, L. M. A summary of Charles Wright's explorations in Cuba. Bul. Torrey Club **32**: 291–300. map. 1905. (Reprinted as Contrib. Dept. Bot. Columbia Univ. no. 217.)

Seguí, D. H. Ojeada sobre la flora médica y tóxica de Cuba. 93 p. Habana, 1900.

Annotated list of medicinal plants, with vernacular names; bibliography. Not seen; data taken from Urban, Symb. Antill. **3**: 11. 1902.

Local

Combs, Robert. Plants collected in the district of Cienfuegos, province of Santa Clara, Cuba, in 1895–1896. Trans. Acad. Sci. St. Louis **7**: 393–491. pl. 30–39, map. 1897. (Reprinted as Contrib. Bot. Dept. Iowa State Coll. no. 7.)

Phytogeography, plant habitats; annotated list of vascular plants, with vernacular names, based on the author's collections.

——— Some Cuban medical plants. Pharm. Rev. **15**: 87–91, 109–112, 136. 1897.

Annotated systematic list of plants used as domestic remedies, with vernacular names and uses.

Fernow, B. E. List of trees on the Sierra Maestra. *In* his The high Sierra Maestra. Forestry Quart. **4**: 259–269. 1906.

Annotated list, alphabetically arranged by vernacular names, with brief description of wood and uses.

Gómez de la Maza, Manuel. Flora habanera. Fanerógamas. 597 p. 19 cm. Habana, 1897.

Keys to and descriptions of the phanerogams of Havana, with keys to all the Cuban genera of each family. Contains many new names omitted from Index Kewensis.

Jennings, O. E. A contribution to the botany of the Isle of Pines, Cuba, based upon the specimens of plants from that island contained in the herbarium of the Carnegie Museum under date of October, 1916. Annals Carnegie Mus. (Pittsburgh) **11**: 19–290. pl. 5–28. 1917.

Botanical collections (1831–1916), bibliography, physiography, plant associations; annotated list of 757 species and varieties (731 vascular plants) with synonymy, general range, exsiccatae; keys to species under each family (in Spermatophyta). Includes all recorded species as well as those collected by the author.—See also **Roig, J. T.** Reseña sobre una excursión botánica a Isla de Pinos. Mem. Soc. Cubana Hist. Nat. "Felipe Poey" **5**: 72–76. 1923.

HISPANIOLA

General

Urban, Ignatz. Flora domingensis. *In* his Symbolae antillanae, v. 8. 860 p. 1920–21.

Systematic list of Spermatophyta of Santo Domingo and Haiti, with synonymy, citation of exsiccatae, and statement of extralimital range; vernacular names (separately indexed). Lists 3,000 species (p. 762), of which 983 are endemic. The Pteridophyta are treated in his Symbolae antillanae **9**: 273–397. 1925; the collectors, plant formations, and plant geography in **9**: 1–54. 1923. In the latter contribution the total number of phanerogams is raised to 3,088, of which 1,048 are endemic.—See also his Plantae haitienses novae vel rariores a cl. Er. L. Ekman 1917–[26] lectae I–IV. Arkiv Bot. v. 17, no. 7. 72 p. 1 fig. 1921; v. 20A, no. 5. 65 p. 2 fig. (maps), 3 pl., no. 15. 94 p. 2 fig., pl. 1926; v. 21A, no. 5. 97 p. 1927. Plantae haitienses et domingenses . . . V–[X] a cl. E. L. Ekman 1924–[30] lectae. l.c. v. 22A, no. 8. 98 p. 4 pl., no. 10. 108 p. pl., no. 17. 115 p. 1 fig. 1929; v. 23A, no. 5. 107 p. 5 pl. 1930; v. 23A, no. 11. 103 p. 1 fig., 5 pl. 1931; v. 24A, no. 4. 54 p. 3 pl. 1932.

Local

Barker, H. D., and others. Identification des plantes d'Haïti par leurs noms créoles. Bul. Serv. Tech. Dept. Agr. Haïti 18. 23 p. 1930.

Alphabetical lists of Creole vernacular names, with botanical equivalents.

——— and **Dardeau, W. S.** Flore d' Haïti. Clé et description des ordres, familles et genres des spermatophytes d'Haïti avec la liste de la plupart des espèces. viii, 456 p. 24 cm. Port-au-Prince, 1930.

Keys to families and genera of Spermatophyta with brief descriptions; enumeration of species under each genus, with vernacular names; glossary.

Ekman, E. L. A list of plants from the island of Gonave, Haiti. Arkiv Bot. v. 23A, no. 6. 73 p. 1930.

General features of flora, botanical explorations, etc.; annotated list of vascular plants collected by the author, with vernacular names; distribution in Haiti given.

——— Plants observed on Tortue Island, Haiti. Arkiv Bot. v. 22A, no. 9. 61 p. 1929.

Physiography, geology, climate; annotated list of 889 vascular plants, with Haitian range; phytogeography.

——— Plants of Navassa Island, West Indies. Arkiv Bot. v. 22A, no. 16. 12 p. 2 pl. 1929.

Physiography, geology, climate, general features of flora; list of 102 vascular plants, with data.

Häussler, F. Haitianische pflanzen, die von den bewohnern der insel als heilpflanzen geschätzt und verwendet werden. Schweiz. Apoth. Zeit. **52**: 261–264, 275–278. 1914.

Annotated compiled list, with vernacular names and uses; bibliography.

Ostenfeld, C. H., and others. Plants from Beata Island, St. Domingo collected by C. H. Ostenfeld. (Botanical results of the Dana-expedition 1921–22, no. 1) . . . Dansk Bot. Arkiv v. 4, no. 7. 36 p. 17 fig., 3 pl. 1924.

General features of flora; list of vascular plants (by I. Urban), lists of cellular plants by other authors.

Schiffino, José. Riqueza forestal dominicana. 125 p. plates. 23 cm. Santo Domingo, 1927.

Annotated list of trees of Dominican Republic, alphabetically arranged by vernacular names, with names in other countries of tropical America. Work of same title, 1939, 105 p., cited in Taxon. Index v. 2, entry 634. 1939.

Urban, Ignatz. Zur hochgebirgsflora von Sto. Domingo. *In* his Symbolae antillanae . . . **6**: 280–292. 1909.

Running account of vegetation; list of spermatophytes occurring at 2,000 meters elevation or more, divided according to geographical affinities, with general range indicated.

JAMAICA

General

Beckwith, M. W. Notes on Jamaican ethnobotany. 47 p. 2 pl. 23 cm. Poughkeepsie, 1927.

Bibliography; alphabetical list of 134 medicinal plants, with uses, and general account of native plant medicines; general account of food plants.

Fawcett, William. Economic plants. An index to economic products of the vegetable kingdom in Jamaica. 78 p. 21.5 cm. Kingston, 1891.

Annotated alphabetical list of native and introduced plants of economic importance.

——— A provisional list of the indigenous and naturalized flowering plants of Jamaica. 57 p. 21.5 cm. Kingston, 1893.

Unannotated systematic list of 2,412 species.

——— and **Rendle, A. B.** Flora of Jamaica, containing descriptions of the flowering plants known from the island. v. 1, 3–5, 7. illus., plates. 22 cm. London, 1910–36.

Descriptive flora of vascular plants, with keys, synonymy, exsiccatae, vernacular names, uses, extralimital range; bibliographies; lists of collectors, with dates (in prefaces of several volumes). Not yet completed. Vol. 7 is by S. LeM. Moore and A. B. Rendle.

Local

Hitchcock, A. S. List of plants collected in the Bahamas, Jamaica and Grand Cayman. Ann. Rpt. Missouri Bot. Gard. **4**: 47–179. pl. 11–14. 1893.

Systematic list of vascular plants based mainly on the writer's collections, with localities; relationships of Bahama flora; tables and statistics of distribution. Superseded by other floras except as to records of plants of Grand Cayman.—See also **Fawcett, William.** Plants collected in the Cayman Islands. Bul. Bot. Dept. Jamaica **11**: 6–7. 1889.

LESSER ANTILLES

See also French Guiana (Lanessan, Les plantes utiles . . . for account of useful plants of Guadeloupe (p. 172–187, 490–527) and Martinique (p. 153–171, 421–489)); Puerto Rico (Britton and Wilson, Botany of Porto Rico and the Virgin Islands).

General

Grisebach, A. H. R. Systematische untersuchungen über die vegetation der Karaiben, insbesondere der insel Guadeloupe nach der sammlungen Duchassaing's. Abhandl. K. Gesell. Wiss. Göttingen **7**: 151–286. 1857. (Reprinted, 138 p.)

Systematic list of vascular plants of the Lesser Antilles, with citation of islands on which they are found and of the collectors or authors reporting them; 1,486 species, of which 1,325 are Spermatophyta.

Local

Anonymous. Flora of St. Vincent and adjacent islets. Bul. Misc. Inform. Kew **1893**: 231–296. 1893.

Systematic list of the vascular plants of St. Vincent and the Grenadines (Bequia, Cannouan, Mustique, Union) prepared by the botanists at Kew, with citation of exsiccatae and statement of extralimital range; brief account of physical geography; statistical plant geography. Enumerates about 1,313 species including cultivated plants.

Anonymous. Timbers of Dominica. West Indian Bul. **9**: 329–345. 1909.

Annotated list of 176 timber trees, alphabetically arranged by vernacular names, with botanical names where known; list classified by uses. Revision of **Imray, John.** The useful woods of the island of Dominica. Technologist **2**: 382–391. 1862.

Alexander, W. H. The flora of St. Christopher. Bul. Amer. Geog. Soc. **33**: 207–219. 1901.

Tabular list of principal economic plants, native and introduced, with vernacular names and uses.

Ballet, Jules. Description des plantes. *In* his La Guadeloupe. Renseignements sur l'histoire, la flore, la faune, la géologie, la minéralogie, l'agriculture . . . v. 1, [pt. 1,] p. 176–527. Basse-Terre, 1890 (1894).

Running account with descriptions, uses, vernacular names, the plants grouped principally by habit, the botanical names not given; tabular systematic list (p. 368–521), based on Grisebach, of vascular plants (Filices by Mazé) and bryophytes, with vernacular names that serve to identify the species described in the first section; systematic list of cultivated plants.

Boldingh, Isaac. A contribution to the knowledge of the flora of Anguilla (B. W. I.). Recueil Trav. Bot. Néerland. **6**: 1–36. 1909.

List of 150 vascular plants; general range given.

——— The flora of the Dutch West Indian islands. 2 v. 9 pl., 4 maps. 25.5 cm. Leyden, 1909–14.

Vol. 1, The flora of St. Eustatius, Saba, and St. Martin; vol. 2, The flora of Curaçao, Aruba, and Bonaire. Systematic lists of vascular plants with citation of exsiccatae and statement of extralimital range; historical account of collectors, with detailed notes on some of the larger collections; phytogeographical data; bibliographies; separate indices to vernacular names. Vol. 1 contains 806 species, of which 674 are wild; vol. 2 contains 394 wild species.

———. Flora voor de Nederlandsch West-Indische eilanden. xx, 450 p. 23.5 cm. Amsterdam, 1913.

Keys to and descriptions of the plants listed in his English flora of the same islands, with vernacular names (separately indexed) and mention of the islands on which the plants occur.

——— Lijst van planten die door de bewoners van de drie Nederlandsche Antillen, St. Eustatius, Saba en St. Martin als geneeskrachtig worden beschouwd tevens een vergelijkend oversicht van het medicinaal gebruik dat bij verschillende schrijvers over die planten wordt gevonden. Bull. Kolon. Mus. Haarlem **38**: 93–112. 1907.

Annotated systematic list of medicinal plants, with vernacular names, uses and references.

Box, H. E. A note on the vegetation of Redonda, B. W. I. Jour. Bot. **77:** 311–313. 1939.

Geology, etc.; list of plants collected or observed by author on July 18, 1938. (About 15 miles northwest of Montserrat, 16°55′ N. Lat., 62°16′ W. Long.)

Britton, N. L. The flora of the American Virgin Islands. Mem. Brooklyn Bot. Gard. **1:** 19–118. 1918.

Annotated systematic list of the plants (including principal cultivated species) of St. Thomas, St. Jan, and St. Croix, formerly the Danish West Indies; physical geography; annotated bibliography; list of collectors. Musci by E. G. Britton, Hepaticae by A. W. Evans, Lichenes by L. W. Riddle; references to literature on Fungi and Algae. Enumerates 1,052 native species (excluding fungi and algae), of which 27 are endemic.

——— The vegetation of Anegada. Mem. N. Y. Bot. Gard. **6:** 565–580. 1916.

Physiography, etc.; annotated list of 123 vascular and cellular plants, with general range given; the lichens determined by L. W. Riddle.

Duss, Antoine, *pére.* Flore phanérogamique des Antilles françaises (Guadeloupe et Martinique) . . . avec annotations du Professeur Dr. Édouard Heckel sur l'emploi de ces plantes. Annales Mus. Colon. Marseille v. 3. xxvii, 656 p. 1897.

Systematic list, including cultivated species, of the phanerogams of Guadeloupe, Martinique, and neighboring small islands; brief descriptions, uses, local occurrence, vernacular names; sketch of physical geography, climate, and plant associations.—For additions see **Privault, Daniel.** Contribution à l'étude de la flore de la Martinique. Bul. Soc. Bot. France **84:** 289–294. 1937.

Grébert, René. Les forêts de la Guadeloupe. Bul. Agence Écon. Colon. Auton. et Terr. Afr. (France) **27:** 639–702, 765–875, 941–1015. 1934.

Geography, climate, soil, destruction of forests, phytogeography, zones of vegetation, composition of principal forests; systematic list of timbers, with vernacular name, habitat, brief description, character of wood, use.

Kervégant, D. Les plantes utiles et ornementales de la Martinique. Bul. Agr. Martinique **5:** 1–94. 1936; **6:** 1–142. 1937.

Annotated classified lists of wild and cultivated useful plants, arranged alphabetically by vernacular names. Unfinished; pt. 1, pot-herbs and vegetables; pt. 2, fruits. List consists mostly of cultivated plants.

Maycock, J. D. Flora barbadensis: a catalogue of plants indigenous, naturalized, and cultivated, in Barbados . . . xx, 446 p. 2 maps. London, 1830

Arranged according to Linnaean system, with Latin descriptions and synonymy; separate index of vernacular names.

Mazé, Hippolyte. Contribution à la flore de la Guadeloupe. 190 p. 25 cm. Basse-Terre, 1892.

Unannotated list of wild and cultivated spermatophytes, with vernacular names.

Millspaugh, C. F. Flora of the island of St. Croix. Pub. Field Columbian Mus. 68, Bot. Ser **1:** 441–546. map. 1902.

The species listed (1,029) are incorporated in Britton's Flora of the American Virgin Islands. Biographical sketches of collectors, with bibliographical references; account of cultivated plants.

Schomburgk, R. H. The history of Barbados; comprising a geographical and statistical description of the island; a sketch of the historical events since the settlement; and an account of its geology and natural productions. xx, 722 p. illus., plates, map, table. 25.5 cm. London, 1848.

Contains (p. 582–633) a list of wild and cultivated vascular plants (896 phanerogams), with indices of vernacular names.

Stehlé, Henri, Stehlé, M., and **Quentin, L.** Flore de la Guadeloupe et dépendances. v. 1–2, pt. 1. illus., plates, port., 4 maps, table. 23 cm. Basse-Terre, 1935–37.

Vol. 1, Essai d'écologie et de géographie botanique, by H. Stehlé. Vol. 2, pt. 1, Catalogue des phanérogames et fougères avec contribution à la flore de la Martinique, by H. and M. Stehlé and Quentin: general features of flora, list of endemic phanerogams of Guadeloupe and dependencies, statistics, bibliography; annotated list of vascular plants, with references, synonymy, vernacular names, local and extralimital range; briefly annotated alphabetical list of collectors, Vol. 2, pt. 1 covers part of the Polypetalae and Apetalae.

Wheeler, L. R. The botany of Antigua. Jour. Bot. **54:** 41–52. 1916.
Account of physical geography and climate; general features of flora; plant associations. No systematic list of species.

Wikström, J. E. Öfversigt af ön Sanct Barthelemi's flora. K. [Svenska] Vetensk.-Akad. Handl. **1825:** 411–433. 1826.
Bibliography (in footnotes), botanical explorations; list of 301 vascular plants, with occasional critical annotations.

PUERTO RICO

General

Asenjo, C. F. Apuntes acerca de las plantas medicinales de Puerto Rico. Revista Agr. Puerto Rico **28:** 460–469, 741–748. 1937; **29:** 136–141. 1938.
Extensively annotated systematic list of the principal medicinal plants, with popular and official uses and chemical composition. The parts listed include Typhaceae–Orchidaceae.

Barrett, O. W. The food plants of Porto Rico. Jour. Dept. Agr. Porto Rico **9:** 61–208. 1925.
Annotated alphabetical list of vernacular names of wild and cultivated food plants and corresponding index of scientific names.

Britton, N. L., and Wilson, Percy. Botany of Porto Rico and the Virgin Islands. *In* **New York Academy of Sciences.** Scientific survey of Porto Rico . . . v. 5–6. 626, 663 p. 1923–30.
Full descriptive flora of vascular plants, with keys and synonymy; extra-limital range given; vernacular names, annotated bibliography. The treatment of the Pteridophyta is by W. R. Maxon. Covers area from Mona Island to Anegada and St. Croix.

Brush, W. D., Murphy, L. S., and Mell, C. D. Trees of Porto Rico. *In* **Murphy, L. S.** Forests of Porto Rico . . . Bul. U. S. Dept. Agr. **354:** 56–97. 1916.
Systematic list of 172 species, with size, vernacular names, uses, and brief description of wood.

Cook, O. F., and Collins, G. N. Economic plants of Porto Rico. Contrib. U. S. Natl. Herb. **8:** 57–269. fig. 12–13, pl. 13–60. 1903.
Alphabetical catalog of vernacular and scientific names, with brief descriptive notes and account of uses.

Otero, J. I., and Toro, R. A. Catálogo de los nombres vulgares y científicos de algunas plantas puertorriqueñas. Bol. Estac. Expt. Insul. Rio Piedras, P. R. no. 37. 248 p. 1931.
Botanical explorations; systematic list of 2,063 vernacular names, with equivalent botanical names; indices, including references to Britton and Wilson's Botany of Porto Rico and the Virgin Islands; alphabetical list of English vernacular names, with Spanish and botanical equivalents; bibliography.

Urban, Ignatz. Flora portoricensis. *In* his Symbolae antillanae, v. 4. 771 p. port. 1903–11.
Systematic list with synonymy, citation of specimens, and statement of extra-limital range; history of botanical exploration, plant geography; separate index of vernacular names. Enumerates 2,056 species, of which 271 are endemic.

Local

Britton, N. L. La vegetación de Cayo Icacos. Revista Agr. Puerto Rico **12:** 91–96. 1924.
Unannotated list of spermatophytes collected by the author and others on 30 January 1923. (East of Cape San Juan.)

——— The vegetation of Mona Island. Annals Missouri Bot. Gard. **2:** 33–58. pl. 1–2. 1915.
Physiography, soil, general features of flora; annotated list of about 292 vascular and cellular plants, with general range indicated.

Gifford, J. C., and Barrett, O. W. Trees of the Luquillo region. *In* **Gifford, J. C.** The Luquillo Forest Reserve, Porto Rico. Bul. Bur. Forestry U. S. Dept. Agr. **54:** 35–46. 1905.
Systematic list, with vernacular names and notes on height, wood, and uses.

Wilson, Percy. The vegetation of Vieques Island. Bul. N. Y. Bot. Gard. **8**: 379–410. 1917.

Physiography, botanical explorations; annotated list of 579 plants (555 vascular), with range in adjacent islands.

South America

ARGENTINA

See also Chile (Neger, Pflanzengeographisches aus den südlichen Anden; Philippi, Sertum mendocinum); Paraguay (Morong and Britton; Parodi); Patagonia; Uruguay (Paccard).

The following publications have not been available:

Roibon, Federico. Plantas indígenas medicinales de la Provincia de Corrientes para la Exposición de Filadelfía. 2. ed. Corrientes, 1903. (Cited from Kurtz, F. Bol. Acad. Nac. Cien. Córdoba (Rep. Argentina) **19**: 339. 1913.)

Rojas Acosta, Nicolás. Catálogo de las plantas medicinales del Chaco austral. Revista Farm. (Buenos Aires) **47**: 214–225, 247–258, 284–298, 315–327. 1907; **52**(?): 73–95, 120–143, 151–163, 1912. (Cited from Hicken, C. M. Darwiniana 1: 408. 1929.)

——— Plantas medicinales de Corrientes. Revista Farm. (Buenos Aires) **45**: 115–145. 1905. (Cited from Kurtz, F. Bol. Acad. Nac. Cien. Córdoba (Rep. Argentina) **19**: 226. 1913.)

Sparn, Enrique. Bibliografía de la flora fanerogámica de la Provincia de Córdoba (Argentina). Revista Mus. Prov. Cien. Nat. Córdoba 1938, no. 3, 108 p. (Cited from Bot. Zentralbl. n. f., **32** (v. **174**, Lit.): 60. 1939.)

General

Blondel, Raoul. Las plantas medicinales en la pabellón de la República Argentina. *In* La República Argentina en la Exposición universal de Paris de 1889. **2**: 103–126. Paris, 1890.

Running account of medicinal plants, grouped by uses, with vernacular names.

Castro, E. B. Las maderas argentinas. Su importancia industrial. 344 p. table. 25 cm. Rosario, 1918.

Vernacular names, brief description of tree and wood, uses; no index, no obvious order of arrangement of species.

Domínguez, J. A. Contribuciones a la materia médica argentina. (Primera contribución.) Trab. Inst. Bot. y Farm. Fac. Cien. Méd. Buenos Aires [no. 44]. xxiii, 433 p. 1928.

Biographical accounts of workers on medicinal plants of South America, from Cristóbal Colón to Glaziou, with list of publications; early American medicine; annotated systematic list of medicinal plants of Argentina, with vernacular names; tabular alphabetical list of 524+111 plants, showing results of phytochemical investigations; annotated systematic list (Piperaceae-Lauraceae) of materia médica argentina, with descriptions, pharmacological information, and vernacular names (indexed).

——— Datos para la materia médica argentina. Trab. Mus. Farm. Fac. Cien. Méd. Buenos Aires no. 1. xxix, 278 p. 1903; no. 25. x, 141 p. 1910.

Systematic list, with descriptions, vernacular names, chemical composition, and medicinal uses.

——— Synopsis de la matière médicale argentine. Trab. Mus. Farm. Fac. Cien. Méd. Buenos Aires no. 7. 24 p. 1905.

Botanical explorations, work on medicinal plants; annotated list of native medicinal plants, with vernacular names and localities.

Hauman, Lucien. Les modifications de la flore argentine sous l'action de la civilisation. (Essai de géobotanique humaine.) Mém. Acad. Roy. Belg. Cl. Sci. Collect. in-4°. II, v. 9, fasc. 3. 99 p. 1928.

Includes (p. 79–94) list of adventive phanerogams annotated by means of symbols.

——— Les phanérogames adventices de la flore argentine. Anales Mus. Nac. Hist. Nat. Bern. Rivad. Buenos Aires **33**: 319–345. 1925.

Annotated list of 376 species; supplementary list of cosmopolitan species.—For supplementary lists see **Parodi, L. R.**, and **Molfino, J. F.** Nuevas adiciones a la flora fanerogámica adventicia . . . Revista Fac. Agron. Univ. La Plata **18**: 139–156. 4 fig. 1928.—Also **Molfino, J. F.** Adiciones . . . Anales Mus. Nac. Hist. Nat. Bern. Rivad. Buenos Aires **34**: 89–119. 10 pl. 1928.

——— **Vanderveken, G.**, and **Irigoyen, L. H.** Catalogue des phanérogames de l'Argentine. Anales Mus. Nac. Hist. Nat. Buenos Aires **29**: 1–351. 1917; **32**: 1–314. 1923.

Incomplete; list of 2,392 phanerogams (Gymnosperms, Monocotyledons, Dicotyledons through Droseraceae) with references pertaining to local distribution; bibliography. Pt. 1 by Hauman and Vanderveken, pt. 2 by Hauman and Irigoyen.

Hieronymus, Georg. Plantae diaphoricae florae argentinae ó revista sistemática de las plantas medicinales, alimenticias ó de alguna otra utilidad y de las venenosas, que son indígenas de la República Argentina ó que, originarias de otros países se cultivan ó se crian espontáneamente en ella. Bol. Acad. Nac. Cien. Córdoba (Rep. Argentina) **4**: 199–598. 1882. (Reprinted.)

Annotated systematic list of useful plants, with vernacular names and local range. Said to have been reprinted in 1930. 421 p. (cf. Millán, R. Bibliografía, p. 268. 1935.)

Holmberg, E. L. La flora de la República Argentina. *In* Segundo censo de la República Argentina Mayo 10 de 1895 . . . v. 1, p. 383–474. 38 fig., 13 pl., map. Buenos Aires, 1898.

Botanical explorations, local herbaria, plant formations, statistics of flora.—See also **Stuckert, Teodoro.** Observaciones al capítulo La flora argentina por el Doctor E. L. Holmberg . . . Anales Soc. Cien. Argentina **48**: 67–105. 1899. (Reprinted, 41 p.)

Kurtz, Fritz. Essai d'une bibliographie botanique de l'Argentine. 2. ed. (1912). Bol. Acad. Nac. Cien. Córdoba (Rep. Argentina) **19**: 221–376. 1913; **20**: 369–467. 1915.

List of works relating to the flora of Argentina (principally) and other South American countries (Colombia, Ecuador, Venezuela, Peru, Bolivia, southern Brazil, Paraguay, Uruguay, Chile), arranged alphabetically by authors, followed by classified lists of the publications. Continued by L. Hauman and A. Castellanos in Physis **5**: 263–291. 1922 (for the years 1914–21) and **8**: 436–467. 1927 (for 1922–26).—See also **Hicken, C. M.** Contribución al conocimiento de la bibliografía botánica argentina. Darwiniana **1**: 191–318. 1927. Segunda contribución . . . l. c. 319–430. 1929. Tercera contribución . . . l. c. 431–539. 1930. (Intended to complement Kurtz's "Essai" (1912); includes brief list of herbaria and exsiccatae.)

Latzina, Eduardo. Index de la flora dendrológica argentina. Lilloa **1**: 95–211. 14 pl. 1937.

Index of vernacular names; briefly annotated systematic list of 839 species, with data on physical properties of the woods; bibliography. Replaces his paper of same title published in Publ. Com. Honor. Reducc. Indios **3**: 3–117. 1935. (Also issued as Trab. Inst. Bot. y Farm. Cien. Méd. Buenos Aires no. 54.)

Millán, Roberto. Bibliografía agrícola argentina hasta 1930. Bol. Min. Agr. Rep. Argentina v. 36, suppl. 302 p. 1935.

List of periodicals cited; alphabetical list of vernacular names with cross references to botanical names and citation of publications upon them (almost entirely restricted to useful or injurious plants, including cultivated species); bibliography of papers on various agricultural subjects, including medicinal and other useful plants, also of papers arranged by Provinces. The papers cited are practically all in Argentine periodicals and include few that are strictly botanical.

Molfino, J. F. Plantas usuales de la flora argentina. Almanaque Min. Agr. Rep. Argentina **9**: 457–466. 1934.

Systematic list of medicinal plants for which authentic chemical and physiological data exist, with vernacular names and note on properties.

Niederlein, Gustav. Produits des exploitations et des industries forestières. *In* Exposition universelle internationale de 1889 à Paris. Catalogue spécial officiel . . . de la République Argentine. p. 203–264. Paris, 1889.

List of 1,638 specimens of woods, medicinal and other useful plants, classified by uses and then arranged under geographical divisions, with vernacular names and brief annotations.

——— La riqueza florestal de la República Argentina . . . *In* La República Argentina en la Exposición universel de Paris de 1889. **2**: 1–101. Paris, 1890.

Annotated tabular list of trees (502 numbers), alphabetically arranged by vernacular names, with botanical names, locality, dimensions, characters of wood, uses; alphabetical annotated lists of woods used for tanning and dyeing, textiles, medicinal plants.

Parodi, L. R. Las plantas indígenas no alimenticias cultivadas en la Argentina. Revista Argentina Agron. **1:** 165–212. 1 fig. (map). 1912.

Includes annotated systematic list of about 200 spermatophytes, with vernacular names.—See also his Contribution à l'étude des plantes alimentaires indigènes cultivées en Argentine. Revue. Bot. Appl. et Agr. Trop. **16:** 177–189. fig. 4–6. 1936.

Paz Soldán, M. F. Diccionario geográfico-estadístico nacional argentino. x, 485 p. maps. 27 cm. Buenos Aires, 1885.

Includes (p. 46–58) list of medicinal and other useful plants, alphabetically arranged by vernacular names, with botanical names and Provinces.

Spegazzini, Carlos, and **Girola, C. D.** [Catálogo de las maderas.] *In* Catálogo descriptivo de las collectiones de productos naturales, agrícolas, animales é industriales . . . para la Exposición internacional de agricultura de 1910 . . . Pub. Mus. Agr. Soc. Rural Argentina **1:** 322–413. 1911. (Reprinted with index, xv p., and title: Catálogo descriptivo de las maderas que se exhibieron en la Exposición internacional de agricultura de 1910, l. c., no. 1, app.)

List of 554 specimens (including duplications of species), with vernacular names, brief description of tree and wood, note on uses.

Storni, J. S. Vegetales que utilizaban nuestros indígenas para su alimentación. 102 p. map. 26 cm. [Tucumán?], 1937 (1938).

General considerations, geoethnographic regions; annotated list of wild and cultivated edible plants, arranged in no obvious order, with vernacular names and uses.

Local

Anonymous. Flore de Tucumán. Herbier de M. Lillo. Collection de MM. Schickendantz et Lillo, Tucumán. *In* Exposition universelle internationale de 1889 à Paris. Catalogue spécial officiel . . . de la République Argentine. p. 341–356. Paris, 1889.

Tabular list of about 600 vascular plants, with occasional vernacular names.

Alvárez, Antenor. Flora y fauna de la provincia de Santiago del Estero. 176 p. illus. 23 cm. Santiago del Estero, 1919.

Systematic list of vascular plants (p. 43–116) ; classified lists of useful plants (p. 119–126) ; bibliography.

Ball, John. Contributions to the flora of north Patagonia and the adjoining territory. Jour. Linn. Soc. Bot. **21:** 203–240. 1884.

General features of geography, flora, etc.; annotated list of vascular plants collected in northern "Patagonia" (chiefly from region of Bahía Blanca) and southern Buenos Aires Province by G. Claraz, with vernacular names and uses.

Berg, Cárlos. Enumeracion de las plantas européas que se hallan como silvestres en la provincia de Buenos Aires y en Patagonia. Anales Soc. Cien. Argentina **3:** 183–206. 1877. (Reprinted.)

Annotated list of 154 plants, of which 148 are vascular.

Bettfreund, Carlos. Flora argentina. Recolección y descripción de plantas vivas . . . 3 v. 156 col. pl. 27 cm. Buenos Aires, [1898–1901].

Descriptions and rather crude colored plates (by F. Burmeister) of 156 phanerogams.

Burkill, I. H. Notes on some plants collected in the Aconcagua valleys by Philip Gosse. *In* **FitzGerald, E. A.** The highest Andes . . . p. 361–376. illus. London, 1899.

General features of flora; also annotated list of plants collected by Philip Gosse in the Las Cuevas and Horcones Valleys, mostly at 10,000 feet or higher.

Chodat, Robert, and **Wilczek, Ernest.** Contributions à la flore de la République Argentine. Énumération critique des plantes récoltées par M. E. Wilczek à Saint-Raphaël et dans la vallée de l'Atuel. Bul. Herb. Boissier II, **2:** 281–296, 475–490, 521–544. 27 fig. 1902.

Itinerary, general features of flora; list of vascular plants, with references and data. (Province of Mendoza.)

Fries, R. E. Zur kenntnis des alpinen flora im nördlichen Argentinien. Nova Acta R. Soc. Scient. Upsal. ser. IV, v. 1, no. 1. 205 p. 2 fig., 9 pl., map. 1905.

Enumeration of the vascular plants hitherto known from the alpine region of northern Argentina (Prov. Jujuy, northern Salta, part of Los Andes), with synonymy, citation of exsiccatae, and statement of general range. The climate, plant formations, and plant geography are discussed; bibliography.

Gancedo, Alejandro. Flora arbórea del territorio nacional del Chaco y proyecto de ley. 237 p. illus., map. 28 cm. Buenos Aires, 1916.
Brief description of tree, wood, uses; photographs of nearly every species; list of species classified by uses, etc.

Grisebach, A. H. R. Plantae lorentzianae. Bearbeitung der ersten und zweiten sammlung argentinischer pflanzen des Professor Lorentz. Abhandl. K. Gesell. Wiss. Göttingen **19:** 49–280. 2 pl. 1874. (Reprinted 231 p., 1874.)
Phytogeography, etc.; list of 927 vascular plants, with localities, occasional vernacular names, and frequent annotations.

——— Symbolae ad floram argentinam. Zweite bearbeitung argentinischer pflanzen. Abhandl. K. Gesell. Wiss. Göttingen **24:** 3–345. 1879. (Reprinted 1879.)
Phytogeography; list of 2,265 vascular plants, with localities, occasional vernacular names, and frequent annotations. Based on collections of Lorentz, Hieronymus, and Schickendantz; includes all species listed in his Plantae Lorentzianae.

Hassler, Emil. Contribuciones à la flora del Chaco argentino-paraguayo. 1. parte. Florula pilcomayensis. Trab. Mus. Farm. Fac. Cien. Méd. Buenos Aires no. **21.** 154 p. 1909.
Botanical explorations; bibliography; annotated list of 701 vascular plants known from the region of the lower Pilcomayo River.

Hauman, Lucien. Étude phytogéographique de la région du Río Negro inférieur (République Argentine). Anales Mus. Nac. Hist. Nat. Buenos Aires **24:** 289–443. 19 fig. 1913.
Bibliography, geology, climate, plant associations, distribution; annotated list of 486 vascular plants. The area covered lies between the Río Negro and the Río Colorado, reaching about 100 km. inland.

——— La végétation de l'île de Martin García dans le Río de La Plata. Pub. Inst. Invest. Geog. Fac. Filos. y Letras Univ. Buenos Aires no. 10. 38 p. 3 fig. (incl. maps), 8 pl. 1925.
Physiography, climate, plant associations, etc.; annotated list of vascular plants. The area is noteworthy for the fact that its flora is essentially in its primitive condition.

——— La végétation des hautes cordillères de Mendoza (République Argentine). Anales Soc. Cien. Argentina **86:** 121–188, 225–348. 7 fig. (incl. map), pl. 7–25. 1918.
Physiography, plant associations, statistics of flora, ecology, phytogeography; annotated list of 417 vascular plants; bibliography.

Hicken, C. M. Canistellum Neuqueni. Plantas recogidas en las cordilleras del Neuquén por el Sr. Franco Pastore . . . Physis **1:** 116–133. 1912.
Geography, etc.; list of 129 vascular plants with locality, general and local range.

——— Chloris platensis argentina. 292 p. 26.5 cm. Buenos Aires, 1910. (Apuntes de historia natural, v. 2.)
A flora of the city of Buenos Aires and vicinity, being an annotated list of 1,261 vascular plants, with references and general range; bibliography.

——— Plantae fischerianae. Contribución al conocimiento de la flora del Río Negro. Physis **2:** 1–18, 101–122. 1915–16.
Annotated list of 284 vascular plants collected by Walter Fischer in the vicinity of Río Negro, a station on the railroad from Bahía Blanca to Neuquén.

——— Plantae flossdorfianae in monte Famatina lectae . . . Darwiniana **1:** 22–42. 1922.
Annotated list of 86 vascular plants collected at 3,000 to 4,000 meters altitude, with extralimital distribution. (Province of San Juan.)

——— Plantae Vattuonei. Darwiniana **1:** 95–153. 2 fig. 1924.
Botanical exploration, general features of flora; bibliography; list of 187 vascular plants collected by Dr. Ildefonso Vattuone in 1923, with localities, local and general range, and frequent annotations. (Provinces of Jujuy and Salta.)

——— Primitiae sanzinianae. Las primeras plantas recogidas en Mendoza por Renato Sanzin. Darwiniana **2:** 37–57. 1930.
List of about 300 vascular plants, with localities.

Hicken, C. M. Sertularium andinum. Plantae a cl. Gerth lectae . . . Darwiniana **1**: 47–78. 1923.
Localities; annotated list of 117 spermatophytes, with localities. (Province of Mendoza, region of Río Atuel and Río Diamante.)

Hieronymus, Georg. Sertum sanjuaninum ó descripciones y determinaciones de plantas fanerógamas y criptógamas vasculares recolectadas por el Dr. D. Saile Echegaray en la provincia San Juan. Bol. Acad. Nac. Cien. Córdoba (Rep. Argentina) **4**: 3–73. plate. 1881.
Annotated list of 191 vascular plants.

Hosseus, C. C. Flora argentina. Estudios comparativos sobre la vegetación de las provincias de La Rioja y de San Juan. Bol. Acad. Nac. Cien. Córdoba (Rep. Argentina) **26**: 5–160. 28 fig. 1921.
Itineraries of botanical collectors; annotated list of vascular plants (reaches Leguminosae, part).

Johnston, I. M. A collection of plants from the high cordilleras of northwestern San Juan. Physis **9**: 297–326. 1929.
Author's itinerary; annotated list of 89 vascular plants, with references.

Kerr, J. G. The botany of the Pilcomayo expedition; being a list of plants collected during the Argentine expedition of 1890–91 to the Rio Pilcomayo. The identifications and the description of new species by Mr. N. E. Brown. Trans. and Proc. Bot. Soc. Edinb. **20**: 44–78. 1894.
List of about 200 vascular plants (and 1 lichen) from the vicinity of Fortin Page, on the Río Pilcomayo, in the Gran Chaco.

Kurtz, Fritz. Dos viajes botánicos al Río Salado superior (Cordillera de Mendoza). Bol. Acad. Nac. Cien. Córdoba (Rep. Argentina) **13**: 171–210. 1893.
Botanical explorations, general features of flora and annotated lists of plants of different botanical regions in the area; bibliography. German version (Bericht über zwei reisen zum gebiet des oberen Rio Salado . . .) in Verh. Bot. Ver. Prov. Brandenb. **35**: 95–120. 1895.

——— Enumeracion de las plantas recogidas por G. Bodenbender en la precordillera de Mendoza (Octubre de 1896). Bol. Acad. Nac. Cien. Córdoba (Rep. Argentina) **15**: 502–522. 1897.
General features of flora, phytogeography; list of phanerogams, with locality and general range, collected between Chilcas and Challao on the north and Cacheuta and Puente del Ferro-carril Andino on the south.

Lillo, Miguel. Contribución al conocimiento de los arboles de la Argentina segun colecciones y observaciones de Santiago Venturi . . . vi, 127 p. 28.5 cm. Buenos Aires, 1910.—Segunda contribución . . . Notas sobre el herbario Venturi . . . 69 p. 22 cm. Tucumán, 1917.
Annotated list of 358 species (347 botanically identified), with vernacular names (indexed), localities, notes on size and use. The material is chiefly from northern and central Argentina.

——— Estudio preliminar de una colección de plantas procedentes de Tartagal (departamento de Orán, Salta). [Pub.] Mus. Cien. Nat. Univ. Nac. Tucumán [6]: 3–14. 1925.
Annotated list of vascular plants, the families alphabetically arranged.

——— Reseña fitogeográfica de la provincia de Tucumán. *In* Reunión nacional de la Sociedad argentina de ciencias naturales, 1st, Tucumán, 1916. p. 210–232. pl. 15–30 (incl. map). Buenos Aires, 1919. (Reprinted.)
Includes lists of vascular plants from different phytogeographic regions.

Lorentz, P. G. La vegetacion del nordeste de la provincia de Entre-Rios. 179 p. 2 maps. 27 cm. Buenos Aires, 1878.
General features of flora; briefly annotated list of species observed ("Florula entreriana"), including fungi (by F. de Thümen); phytogeography. (Not seen; reference and annotation furnished by P. C. Standley.) Includes 620 vascular plants and 32 fungi, according to review in Bot. Jahresber. Just **6** (2): 1092–1094. 1882.

——— and **Niederlein, Gustav.** Botánica. Enumeracion sistemática de las plantas collectadas durante la expedicion. *In* Informe oficial de la Comision científica . . . de la expedicion al Rio Negro (Patagonia) realizada en los meses de Abril, Mayo y Junio de 1879 . . . Entrega II, Botánica. p. [171]–294. 12 pl. Buenos Aires, 1881.

Annotated list of 337 vascular plants from the southern pampas and from region of Río Colorado, Río Negro, and Río Neuquén. Essentially the same list of species, unannotated but with vernacular names added, with title: Flora de la pampa de Buenos-Ayres, de Patagonie et des Andes australes, appeared in Exposition universelle internationale de 1889 à Paris. Catalogue spécial officiel . . . de la République Argentine. p. 357–364. Paris, 1889.

Molfino, J. F. Contribución a la flora de la región de Bahía Blanca. Physis **5**: 1–27. 6 fig. 1921.
Botanical explorations, topography, soils, plant associations; annotated list of 265 phanerogams.

——— La flora espontánea del campo "Las Hermanas" en Gutiérrez (F. C. S.). Anales Soc. Rural Argentina **55**: 279–287. 1921.
List of 441 vascular plants, with vernacular names. The locality is about 25 km. from Buenos Aires.

Monticelli, J. V. Anotaciones fitogeográficas de la Pampa Central. Lilloa **3**: 251–382. 26 fig. (incl. maps), 16 pl. 1938.
Geological history, ecology, bibliography, etc.; annotated list of 311 plants, all but 1 vascular, with vernacular names. The area covered is essentially the Territory of La Pampa.

Niederlein, Gustav, and Hieronymus, Georg. Résultats de l'exploration du territoire de Misiones de 1883 à 1888. *In* Exposition universelle internationale de 1889 à Paris. Catalogue spécial officiel . . . de la République Argentine. p. 270-340. Paris, 1889.
Tabular list of 1,553 vascular plants and lichens, with occasional vernacular names, statistics, ecological groupings, etc. Apparently the same material with Niederlein as author and with title: Resultados botánicos de exploraciones hechas en Misiones, Corrientes y países limítrofes desde 1883 hasta 1888, in Bol. Mens. Mus. Prod. Argent. **3**: 273–347. 1890.

Parodi, L. R. Ensayo fitogeográfico sobre el partido de Pergamino. Estudio de la pradera pampeana en el norte de la provincia de Buenos Aires. Revista Fac. Agron. y Vet. Univ. Buenos Aires **7**: 65–271. 23 fig. (incl. maps), 16 pl. 1930. (Reprinted.)
Physiography, soil, climate, plant associations, phenology, dissemination, modifications in flora caused by civilization, etc.; annotated list of 488 vascular plants; bibliography.

Pilar Rodrigo, América del. Una excursión botánica a la provincia de Corrientes. Revista Mus. La Plata n. s., **1937** (Secc. Of.): 100–110. 5 fig. (incl. map). 1938.
Running account, including mention of a large number of species with vernacular and botanical names.

Ringuelet, E. J. Estudio fitogeográfico del Rincón de Viedma (Bahía de Samborombón). Revista Fac. Agron. Univ. La Plata **21**: 15–186. 8 fig. (incl. map), 18 pl. 1938.
Mainly ecological; includes annotated list of 173 vascular plants. His Flora del Rincón de Viedma . . . (resumen), Physis **15**: 491–493. 1939, is an abstract of the above. (East coast of the Province of Buenos Aires.)

Schickendantz, Friedrich. Catálogo razonado de las plantas medicinales de Catamarca. Anales Circulo Med. Argentino **5**: 52–53, 119–126, 247–255. 1881–82.
Annotated systematic list, with vernacular names.

Seckt, Hans. Flora bonariensis. Clave para la determinación de las familias y de los géneros de las plantas. 302 p. 19.5 cm. Buenos Aires, 1918.
Keys to families and genera of region of Buenos Aires, with list of more conspicuous species; glossary.

——— Flora cordobensis. Clave para la determinación de las familias y de los géneros de las plantas que se encuentran silvestres ó cultivadas en la provincia de Córdoba. 632 p. 22 pl. 24.5 cm. Córdoba (Rep. Arg.), 1929–30. (Reprinted from Revista Univ. Nac. Córdoba v. 16 and 17.)
Descriptive keys to families and genera of spermatophytes; annotated alphabetical list of species under each family, with vernacular names; glossary.

Spegazzini, Carlos. Contribución al estudio de la flora de la Sierra de la Ventana. 86 p. 27 cm. La Plata, 1896.
Annotated list of 351 vascular plants (also cellular plants) of the Sierra de la Ventana, Province of Buenos Aires; estimated by the author to represent at least two-thirds of the flora. Published by the "Ministerio de obras públicas de la

provincia de Buenos Aires."—See also his Algunas observaciones sobre la flora de la Ventana. Revista Fac. Agron. y Vet. La Plata 2: 347–349. 1896.

Spegazzini, Carlos. Contribución al estudio de la flora del Tandil. 60 p. 26.5 cm. La Plata, Buenos Aires, 1901.
Annotated list of 370 vascular (and a few cellular) plants of the Sierras de Tandil, Province of Buenos Aires.

——— Flora de la provincia de Buenos Aires. v. 1 (xiv, 161 p.). illus. 27 cm. Buenos Aires, 1905. (At head of title: Anales del Ministerio de agricultura República Argentina. Secc. de biología vegetal.)
Plant formations; descriptive flora of vascular plants, with keys, vernacular names, uses. Incomplete (only vol. 1 published); includes Ranunculaceae–Anacardiaceae.

——— Flórula de la ciudad de La Plata y su partido. Bol. Oficina Agr. Ganad. Prov. Buenos Aires 1: 101–105, 159–170, 197–208, 235–245, 265–275, 299–310, 331–341, 365–371. 1901; 2: 392–402, 485–498. 1902.
Annotated descriptive flora, with keys, vernacular names, and uses. Incomplete; Ranunculaceae–Leguminosae (part).

BOLIVIA

See also Argentina (Kurtz); Colombia (Weddell).
The following publications have not been available:
Díaz Romero, Belisario. Farmacopea callaguaya. Enumeración de las plantas medicinales y productos naturales que emplean los Callaguayas. 13 p. La Paz, 1904. [Cited from Kurtz, Bol. Acad. Nac. Cien. Córdoba (Rep. Argentina) 20: 376. 1915.)
Peña, Rafael. Flora cruceña. 287 p. 8°. Sucre, 1901. (Cited from Kurtz, Bol. Acad. Nac. Cien. Córdoba (Rep. Argentina) 20: 393. 1915.)—For extract from this work, see Serrano, Honorio. Flora boliviana. Geo: Agr., Colon. y Ramas Anex. Bolivia epoca 4, no. 17 (unpaged). 1938. (Annotated list of medicinal plants, alphabetically arranged by vernacular names.)

General

Granado, J. T. del. Plantas bolivianas. 284 p. illus. (incl. maps), 5 col. pl. 22.5 cm. La Paz, 1931.
Treats of useful plants, cultivated and wild.

Herzog, Theodor. Die pflanzenwelt der bolivischen Anden und ihres östlichen vorlandes. viii, 258 p. 25 fig., 3 maps. 25.5 cm. Leipzig, 1923. (Die vegetation der erde, v. 15.)
Botanical explorations, bibliography (95 titles); phytogeography, plant formations, useful plants, list of vernacular names, etc.

Local

Buchtien, Otto. Contribuciones a la flora del Bolivia, I. parte. 197 p. 22 cm. La Paz, 1910.
List only, with descriptions of new species and citation of localities, apparently confined to the author's collections. No citation of exsiccatae.

Hemsley, W. B., and Pearson, H. H. W. On a small collection of dried plants obtained by Sir Martin Conway in the Bolivian Andes. Jour. Linn. Soc. Bot. 35: 78–90. 1901.
Botanical explorations, general features of flora; annotated list of 46 angiosperms collected at localities up to 18,700 feet, 38 of the species being from 12,000 feet or above.

Herzog, Theodor. Die von Dr. Th. Herzog auf seiner zweiten reise durch Bolivien in den jahren 1910 und 1911 gesammelten pflanzen I–VI. Meded. Rijks Herb. Leyden no. 19. 84 p. 1913; no. 27. 90 p. pl. 1915; no. 29. 94 p. 1916; no. 33. 19 p. 1918; no. 40. 77 p. 1921; no. 46. 31 p. 1922.
Treatments of various families by specialists. A list of the families published through 1921 (pt. 1–5) is contained in Herzog's Die pflanzenwelt der bolivischen Anden . . . Vegetation der erde 15: 19–22. 1923.

——— Pflanzenformationen Ost-Bolivias. Bot. Jahrb. Engler 44: 346–405. pl. 3 (map). 1910.
Ecological; includes (p. 399–404) alphabetical list of vernacular names, with botanical equivalents; bibliography.

Rusby, H. H. New species from Bolivia collected by R. S. Williams. Bul. N. Y. Bot. Gard. **6**: 487–517. 1910; **8**: 89–135. 1912.

Descriptions of new species only, systematically arranged; the ferns, grasses, and orchids omitted. A few of the species are from Peru.

——— On the collections of Mr. Miguel Bang in Bolivia. Mem. Torrey Club v. 3, no. 3. 67 p. 1893; **4**: 203–274. 1895; **6**: 1–130. 1896. Bul. N. Y. Bot. Gard. **4**: 309–470. 1907.

List of species collected, from 1885 on, with localities and collecting numbers; complete except for Gramineae. Specimens of Rusby and R. Pearce are also cited.

——— **and Britton, N. L.** An enumeration of the plants collected by Dr. H. H. Rusby in South America, 1885–1886. Bul. Torrey Club **15**: 177–184. 1888; **23**: 471–499. 1896; **15**: 247–253. 1888; **16**: 13–20, 61–64, 153–160, 189–192, 259–262, 324–327. 1889; **17**: 9–12, 53–60, 91–94, 211–214, 281–284. 1890; **18**: 35–38, 107–110, 261–264, 331–334. 1891; **19**: 1–4, 148–151, 263–266, 371–374. 1892; **20**: 137–140. 1893; **25**: 495–500, 542–545. 1898; **26**: 142–152, 189–200. 1899; **27**: 22–31, 69–84, 124–137. 1900; **28**: 301–313. 1901; **29**: 694–704. 1902. (Reprinted as Contrib. Dept. Bot. Columbia Univ. no. 6, 181, 200. 229 p.)

Systematic list of cellular and vascular plants with localities, collecting numbers, and occasional annotations. The plants are mostly from Bolivia. Treatment of cryptogams by E. G. Britton and others. Pt. II (Mosses) has title: An enumeration of the plants collected by H. H. Rusby, in Bolivia, 1885–86.

BRAZIL

See also Argentina (Kurtz).

General

Almeida Pinto, Joaquim de. Diccionario de botanica brasileira ou compendio dos vegetaes do Brasil, tanto indigenas como acclimados . . . Coordenado e redigido em grande parte sobre os manuscriptos do Dr. Arruda Camara . . . xix, 433 p. 16 pl. 27 cm. Rio de Janeiro, 1873.

Annotated list, arranged alphabetically by vernacular names, with botanical equivalents and uses; no index of scientific names.

Corrêa, M. P. Diccionario das plantas uteis do Brasil e das exoticas cultivadas. v. 1–2. illus., plates. 32.5 cm. Rio de Janeiro, 1926–31.

Annotated list, alphabetically arranged by vernacular names, with brief descriptions, uses, range, vernacular names, frequent chemical analyses; index of botanical and vernacular names. Incomplete (A–E).

——— Flora do Brazil. Algumas plantas uteis, suas applicações e distribuição geographica. xi, 154 p. 24 cm. Rio de Janeiro, 1909.

Briefly annotated lists of woods and useful plants, alphabetically arranged by vernacular names; index of botanical names.

Fonseca, Eurico Teixeira da. Indicador de madeiras e plantas uteis do Brasil. v, 343 p. 23 cm. Rio de Janeiro, 1922.

Annotated list of woods and useful plants, alphabetically arranged by vernacular names; no index of botanical names.

Freise, F. W. Plantas medicinaes brasileiras. Bol. Agr. São Paulo **34**: 252–494. 1933. (Reprinted.)

Annotated list, arranged alphabetically by vernacular names, with brief descriptions, uses, dosage; list of botanical names, with vernacular equivalents.

Hoehne, F. C. Plantas e substâncias vegetais tóxicas e medicinais. 355 p. 252 fig., 26 col. pl. 32.5 cm. S. Paulo–Rio, 1939. (Originally published in "O Estado de S. Paulo," 1934–38.)

Running account in systematic order, with vernacular names; not confined to plants of Brazil.

Löfgren, Alberto. Manual das familias naturaes phanerogamas. Com chaves dichotomicas das familias e dos generos brasileiros. xviii, 611 p. 24.5 cm. Rio de Janeiro, 1917.

Keys to families and genera, with family descriptions; brief notes on uses (including extralimital forms); some vernacular names.

Martius, K. F. P. von. Systema materiae medicae vegetabilis brasiliensis. xxvi, 155 p. 22 cm. Lipsiae, Vindobonae, 1843.

Annotated lists, classified by uses; vernacular names.

Martius, K. F. P. von. Ueber die pflanzen-namen in der Tupi-sprache. 18 p. 25 cm. München, 1858. (Reprint from Bul. K. Bayer. Akad. Wiss. 1858, no. 1–6.—Also in his Beiträge zur ethnographie und sprachenkunde Amerika's, zumal Brasiliens. v. 2, p. 371–427. Leipzig, 1867.)

Notes on the language; alphabetical list of vernacular names, with botanical equivalents and usually indication of locality.

——— **Eichler, A. W.**, and **Urban, Ignatz**, eds. Flora brasiliensis . . . 15 v. in 40. 3805 pl., 2 maps. 47.5 cm. Monachii, Lipsiae, 1840–1906.

Descriptive flora of vascular plants and mosses, by many collaborators; includes some extralimital species of bordering regions. The greatest of floras, containing 20,733 pages (i. e., columns; and excluding indices), with descriptions of 22,767 species (5,689 described as new), of which 6,246 are illustrated. Vol. 1, pt. 1 contains biographical sketches of principal Brazilian collectors, with their itineraries (those of the earlier collectors shown on a map), biographical sketches of the collaborators, chronological list of fascicles, systematic list of contents, index of families, and various statistics; also an illustrated account of plant formations. Urban gives the number of plates as 3,811.

Matta, A. A. da. Flora medica brasiliense. 318 p. 19 cm. Manaos, 1913.

Annotated list, the species arranged alphabetically by vernacular names (indexed); bibliography.

Mello Moraes, A. J. de. Phytographia ou botanica brasileira applicada á medicina, ás artes e á industria. Seguida de um supplemento de materia medica, inclusive as plantas conhecidas e applicadas pelos indios em suas enfermidades. lxxix, 464 p. port. 24 cm. Rio de Janeiro, 1881.

Annotated list of useful plants, alphabetically arranged by vernacular names; no index of botanical names. The appendix includes accounts of the medicinal plants of Espirito Santo and Pernambuco, etc., by other authors.

Moreira, N. J. Diccionario de plantas medicinaes brasileiras. 144 p. 22 cm. Rio de Janeiro, 1862.—Supplemento . . . 57 p. Rio de Janeiro, 1871.

Alphabetical list of vernacular names, with botanical equivalents and uses.

Peckolt, Theodor. Brasilianische nutz- und heilpflanzen. Pharm. Rundsch. (N. Y.) **11**: 133–136, 181–184, 257, 287–289. 1893; **12**: 35–39, 87–88, 109–112, 165–169, 187–189, 240–242, 285–287. 1894; **13**: 33–36, 89–91, 215–217, 237–240. 1895. Pharm. Rev. **14**: 51–54, 80–84, 154–156, 225–227, 246–250. illus. 1896; **15**: 25–27, 44–46, 70–72, 131–135, 154–155, 236–237. 1897. Pharm. Arch. **1**: 71–72, 85–92, 127–136, 167–168, 185–188, 200–208, 229–236. 1898; **2**: 92–100. 1899; **3**: 141–151, 161–163. 1900; **4**: 41–55. 1901. Pharm. Rev. **20**: 51–54, 110–112, 492–497. 1902.

Brief descriptions, vernacular names and uses; families not in systematic order. Title varies; paper partly in German, partly in German with English translation. Pharmaceutical Review is a continuation of Pharmaceutische Rundschau. Title of parts in Pharm. Rev. is Medicinal plants of Brazil. Incomplete; chiefly Monocotyledones and Apetalae.

——— Heil-und nutzpflanzen Brasiliens. *In* Ber. Deut. Pharm. Gesell. v. 6–16, 19–22. 1897–1906, 1909–12.

Annotated list with vernacular names, uses, and often chemical analyses; species arranged by families. Title varies; is sometimes "Nutz- und heilpflanzen." Incomplete; chiefly Polypetalae and Gamopetalae.

——— Historia das plantas alimentares e de gozo do Brasil contendo generalidades sobre a agricultura brasileira, a cultura, uso e composição chimica de cada uma dellas. 5 pt. 9 tab. 21 cm. Rio de Janeiro, 1871–84.

Geology, climate, agriculture, useful plants in general; edible plants discussed in detail (alphabetically arranged by vernacular names; incomplete; reaches cacaozeiro only), also mihlo (*Zea mays*), mandioca, coffee.

——— Die nahrungs- und genussmittel Brasiliens. Zeitschr. Allg. Oesterr. Apoth.-Ver. **20**: 425–430, 442–446, 459–464, 478–484, 496–501, 510–515. 1882.—Verzeichniss von volksnamen und botanischen benennungen . . . l. c. **21**: 182–186, 197–200, 214–216. 1883.

General account of plant formations, soils, agriculture, edible plants; alphabetical index of vernacular names mentioned, with botanical equivalents.

——— Volksbenennungen der brasilianischen pflanzen und produkte derselben in brasilianischer (portugiesischer) und der von der Tupisprache adoptirten namen. Pharm. Arch. **1**: 14–17, 97–102, 149–161, 216–228, 237–248. 1898; **2**: 12–20, 26–40, 50–60, 78–80, 100, 117–120, 301–312, 328–332, 372. 1899; **3**: 18–20, 41–47, 61–68, 181–189. 1900; **4**: 28–36, 96–100, 116–120, 137–140, 173–180,

192–200. 1901; **5:** 37–40, 76–80, 93–100, 112–116, 125–132. 1902–03; **6:** 14–16, 22–23, 64, 79–80, 93–96, 124–128. 1903. Pharm. Rev. **23:** 76–80, 236–238, 369–375. 1905; **24:** 17–24, 33–38, 65–68, 161–165, 193–199, 225–227. 1906. (Reprinted as Pharmaceutical science series. Monographs no. 15. 1907.)

Extensive list of vernacular names with botanical equivalents.

Peckolt, Theodor, and **Peckolt, Gustavo.** Historia das plantas medicinaes e uteis do Brazil contendo a descripção botanica, cultura, partes usadas, composição chimica, seu emprego em diversas molestias, doses, usos industriaes, etc. 8 v. 21.5 cm. Rio de Janeiro, 1888–1914.

Annotated systematic list with brief description, uses, and often chemical analyses. Vol. 1, 230 p., vol. 2–7, 1369 p., vol. 8, 137 p. (the last not seen; cf. Revista Flora Med. **2:** 531).—For alphabetical index to the botanical and vernacular names see **Costa, O. de A.** Indice analytico . . . *In* Revista Flora Med. (Rio de Janeiro) v. 2, no. 9–v. 4, no. 11. 1936–38.

Sampaio, A. J. de. Bibliographia botanica, relativa á flora brasileira, com inclusão dos trabalhos indispensaveis aos estudos botanicos no Brasil. Bol. Mus. Nac. Rio de Janeiro **1:** 111–125, 225–245. 1924; **2** (3): 35–61, (5): 19–38. 1926; **3** (1): 37–45. 1927; **4** (3): 97–119. 1928.

A series of lists, alphabetically arranged by authors, of works published since 1840. Very incomplete, and including many works not relating to the flora of Brazil.—See also his Apontamentos para a revisão da Flora brasiliensis de Martius. Lavoura (Rio de Janeiro) **17:** 19–53. 1913.

Local

Barbosa Rodrigues, João. Mbaé kaá tapyiyetá enoyndaua, ou a botanica e a nomenclatura indigena . . . vi, 87 p. tab. 25 cm. Rio de Janeiro, 1905.

Includes (p. 50–68) list of Tupy vernacular names of plants, in part systematically arranged, with derivations and botanical equivalents.

Bornmüller, Joseph. Florula riograndensis. Bearbeitung der von Alfred Bornmueller in den jahren 1903 bis 1907 in Rio Grande do Sul gesammelten pflanzen. Revista Sudamer. Bot. **1:** 129–148, 161–168. 1934; **2:** 1–10, 33–48. 1935.

List of vascular plants, with localities and collector's numbers.

Dusén, Per. Sur la flore de la Serra do Itatiaya au Brésil. Arch. Mus. Nac. Rio de Janeiro **13:** 1–119. 1905.

Botanical explorations, topography, geology; two annotated lists of vascular plants (and hepatics), one of the species of higher elevations, the other of those of lower elevations.—Supplementary notes published by author under title: Beiträge zur flora des Itatiaia. Arkiv Bot. v. 8, no. 7. 26 p. 10 fig., 5 pl. 1909; v. 9, no. 5. 50 p. 5 fig., pl. 1910.—Itatiaya, in the State of Rio de Janeiro, is generally regarded as the highest elevation in Brazil.

Edwall, Gustavo. Indice das plantas do herbario da Commissão geographica e geologica de S. Paulo. Bol. Commis. Geog. e Geol. São Paulo **11:** 49–215. 1896.

List of vascular and cellular plants, with localities and dates; the families alphabetically arranged.

Emrich, Karl. Os nomes populares das plantas do Rio G. do Sul . . . 76 p. 22.5 cm. Porto Alegre, 1935.

Lists of vernacular and botanical names, with equivalents.

Freise, F. W. Namensverzeichnis brasilianischer werkhölzer. Zeitschr. Weltforstw. **2:** 563–576. 1935.—Nachträge. l. c. **4:** 704–708. 1937.

Alphabetical lists of vernacular names of trees furnishing commercial woods, with botanical name and locality where vernacular name is used.

Garcia, Rodolpho. Diccionario de brasileirismos (peculiaridades pernambucanas). xviii, 291 p. 25 cm. Rio de Janeiro, 1915.

Bibliography; alphabetical list of words including many vernacular names of plants, with etymology and botanical equivalents; no index of botanical names. Appeared also in Revista do Instituto historico e geographico brasileiro, and as a reprint from the same (p. 637–947. 1915).

Glaziou, A. F. M. Plantae Brasiliae centralis a Glaziou lectae. (Liste des plantes du Brésil central recueillies en 1861–1895.) Mém. Soc. Bot. France 3. 661 p. port. 1905–13.

Systematic list, with data and collector's numbers, of the dicotyledons only out of a total of more than 12,000 species collected by the author in Rio de Janeiro, São Paulo, Espirito Santo, Minas Geraes, and Goyaz; brief account of climate, topography, soils, etc.

Hoehne, F. C. Botanica [Matto Grosso ao Amazonas] pt. 1–12. 216 pl. 32 cm. (pt. 7, 23 cm.) Rio de Janeiro, 1910–22. (Commissão de linhas telegraphicas e estrategicas de Matto Grosso ao Amazonas. Annexo 5, Historia natural.)

Annotated list of species of various families collected by the author in Matto Grosso and Amazonas, partly contributed by other authors. Plates 1–63 are in atlas to pt. 1; pts. 7, 10, and 11 have plates numbered separately (5, 13, 6); the remainder of the plates continue the numbering of pt. 1.

Huber, Jacques. Lista das plantas colligidas na Ilha de Marajó no anno de 1896 (Materiaes para a flora amazonica I). Bol. Mus. Paraense Hist. Nat. e Ethnogr. **2:** 288–321. 2 pl. 1898.

Plant formations, etc.; list of 245 vascular plants, with vernacular names, locality, and general range. (In mouth of Amazon.)—See also, for corrections and account of plant formations, his Sur la végétation du Cap Magoary et de la côte atlantique de l'île de Marajó (Amazone). Bul. Herb. Boissier II, **1:** 86–107. 1901.—See also **Miranda, V. C. de.** Os campos de Marajó e a sua flora considerados sob o ponto de vista pastoril . . . Tres capitulos extrahidos d'uma obra posthuma . . . publicados e annotados pelo Dr. J. Huber. Bol. Mus. Goeldi Hist. Nat. e Ethnogr. **5:** 96–151. 1908. (Plant associations, annotated lists of forage and nonforage plants, alphabetically arranged by vernacular names; alphabetical list of vernacular names with botanical equivalents.)

——— Plantae cearenses. Listes des plantes phanérogames récoltées dans l'état brésilien de Ceará en septembre et octobre 1897. Bul. Herb. Boissier II, **1:** 290–329. 1901

Plant zones, etc.; list of phanerogams collected by author, with data and general range.

——— Plantae duckeanae austro-guyanenses. Enumeração das plantas siphonogamas colleccionadas de 1902 a 1907 na Guyana brasileira pelo Sr. Adolphe Ducke . . . (Materiaes para a flora amazonica VII.) Bol. Mus. Goeldi Hist. Nat. e Ethnogr. **5:** 294–436. map. 1909.—Plantae duckeanae austro-guyanenses. Bul. Soc. Bot. Genève II, **6:** 179–212, 215–216. 17 fig. 1914.

Itinerary, general features of flora of various localities; list of phanerogams collected in Pará, with data and general range. Incomplete: pt. 1 covers Alismaceae–Meliaceae; pt. 2 contains only descriptions of new species.

——— Plantas vasculares colligidas ou observadas na região dos Furos de Breves em 1900 e 1901. (Materiaes para a flora amazonica V.) Bol. Mus. Paraense Hist. Nat. e Ethnogr. **3:** 400–446. 1902.

List of 189 vascular plants, with vernacular names, locality, and general range. The specimens come mostly from Rio Aramá and Boa Vista do Macujubim, at the mouth of the Amazon between the Amazon and the estuary of the Pará.

Le Cointe, Paul. A Amazonia brasileira III. Arvores e plantas uteis (indigenas e acclimadas) . . . viii, 486 p. plates. 24 cm. Belem-Pará, 1934.

List of plants, alphabetically arranged by vernacular names, with brief account of uses; index of generic names. This is vol. 3 of his L'Amazonie brésilienne.

Löfgren, Alberto. Ensaio para uma synonimia dos nomes populares das plantas indigenas do estado de S. Paulo. 115 p. 22.5 cm. S. Paulo, 1895.—2. parte por **Gustavo Edwall.** 70 p. 22.5 cm. São Paulo, 1906. (Boletim da Commissão geographica e geologica de São Paulo no. 10, 16.)

Alphabetical lists of vernacular names, with brief description and account of uses; index. Includes introduced species.

——— and **Edwall, Gustavo.** Flora Paulista I–IV. 4 v. 22 cm. São Paulo, 1897–1905. (Boletim da Commissão geographica e geologica de São Paulo no. 12–15.)

Keys and descriptions (in Portuguese), based on Flora brasiliensis and later monographs, of the plants of São Paulo. Parts 1 and 3 by Löfgren, 2 and 4 by Edwall. Incomplete; covers (v. 1) Compositae; (2) Solanaceae, Scrophulariaceae; (3) Campanulaceae, Cucurbitaceae, Calyceraceae, Valerianaceae; (4) Myrsinaceae.

Luederwaldt, Hermann, and Fonseca, José Pinto da. Lista das plantas colecionadas e observadas. *In* their A Ihla dos Alcatrazes. Revista Mus. Paulista **13**: 500–509. 1923.

List of about 170 cellular and vascular plants collected in 1920, with some vernacular names. (Off coast of São Paulo.)

Luetzelburg, Philipp von. Estudo botanico do nordéste. Pub. Inspect. Fed. Obras Contra Secc. (Brasil) no. 57, Sér. I, A. 3 v. plates (incl. map) and atlas of fold. maps. [1922–23.]

Includes (v. 3, p. 95–261) list of vascular and cellular plants collected by the author in northeastern Brazil, with localities, the families arranged alphabetically; (p. 263–271) alphabetical list of vernacular names. Refers principally to region from Bahia to Piauhy and Ceará, but includes some States south of this area.

Lutz, Bertha. The flora of the Serra da Bocaina. Proc. Amer. Phil. Soc. **65** (Suppl.): 27–43. 1926.

General features of flora; list of vascular plants, chiefly collected by Dr. Adolphe Lutz. The Serra de Bocaina is part of the Serra do Mar, in São Paulo, and about 50 km. from Itatiaya.

Malme, G. O. A. Zur kenntnis der phanerogamenflora des sandgebietes im süden von Rio Grande do Sul. Svensk Bot. Tidskr. **30**: 1–29. 1936.

Botanical explorations, general features of flora; list of spermatophytes, with localities and occasional annotations.

Moore, S. LeM. The phanerogamic botany of the Matto Grosso expedition, 1891–92. Trans. Linn. Soc. Bot. II, **4**: 265–516. pl. 21–39, map. 1895.

Annotated list of phanerogams collected by the author in the Province of Matto Grosso; general features of vegetation, climate, analysis of flora.

Navarro de Andrade, Edmundo, and Vecchi, Octavio. Les bois indigènes de São Paulo. v, 376 p. illus. 22.5 cm. São Paulo, 1916.

Vernacular names, brief descriptions, uses, illustration of fruit of most species; bibliography. Species alphabetically arranged by vernacular names, no botanical index.

Novaes, José de Campos. Index florae campinensis. Revista Centro Sci., Letr. e Art. Campinas **6**: 57–76. 1904.

Unannotated list of 731 phanerogams of Campinas, São Paulo, the families arranged alphabetically.

Pereira, Huascar. Pequena contribuição para um diccionario das plantas uteis do estado de S. Paulo (indigenas e aclimadas). 779 p. illus. 23 cm. São Paulo, 1929.

Annotated list of wild and cultivated useful plants, alphabetically arranged by vernacular names; bibliography, glossary; no index of scientific names.

Pickel, Bento. Catálogo do herbário da Escola Superior de Agricultura em Tapera (Pernambuco). (Herbarium Pickel.) Bol. Mus. Nac. Rio de Janeiro **13**: 63–132. 1938.

General features of flora; list of vascular plants collected by the author in the States of Pernambuco and Paraíba, with data. The part cited includes Ophioglossaceae–Cyperaceae.

Pilger, Robert. Beitrag zur flora von Mattogrosso. Botanischer bericht über die expedition von Dr. Herrmann Meyer nach Central-Brasilien 1899. Bot. Jahrb. Engler **30**: 127–238. 1 fig. (map). 1901.

Botanical explorations and publications; list of angiosperms collected, with data; ecology.

Pinheiro, E. C. As florestas e as essencias florestaes do estado do Pará. Bol. Min. Agr., Ind. e Comm. (Brasil) **19**: 552–585, 709–738. illus. 1930.

Includes (p. 566–575) list of 423 timbers alphabetically arranged by vernacular names, with botanical equivalents.

Rocha, F. D. da. Botanica medica cearense. 144 p. 17 cm. Ceará, 1919.

Annotated list of medicinal plants, alphabetically arranged by vernacular names; systematic list of botanical names, with vernacular equivalents.

Sampaio, A. J. de. Contribuição ao estudo da flora do estado de Minas Geraes (Brasil). Arch. Mus. Nac. Rio de Janeiro **18**: 1–38. 1916.

General features of flora; list of plants collected between Palmyra and Queluz de Minas in 1905–06, with localities and vernacular names (indexed).

Sampaio, A. J. de. A flora de Matto Grosso. Memoria em homenagem aos trabalhos botanicos da Commissão Rondon. Arch. Mus. Nac. Rio de Janeiro **19:** 1–125. 11 maps. 1916.
Botanical explorations; compiled list of cellular and vascular plants, with localities and collectors' names; bibliography.

——— Nomes vulgares de plantas de Amazonia (com anotações de **A. Ducke**). Bol. Mus. Nac. Rio de Janeiro **10:** 3–69. 1934.
Alphabetical list of vernacular names with botanical equivalents.

——— Nomes vulgares de plantas do Distrito Federal e Estado do Rio. Bol. Mus. Nac. Rio de Janeiro **13:** 161–293. 1938.
Alphabetical list of vernacular names, with botanical equivalents.

Smith, L. B. Relação das plantas vasculares colhidas no norte e nordeste do Brasil. Bol. Inspet. Fed. Obras Contra Sêc. (Brasil) **10:** 123–156. 1938.
List of vascular plants, with data, collected by Dr. Francis Drouet in 1935 in the States of Pará and Ceará.

Tastevin, Constantino. Nomes de plantas e animaes em lingua Tupy. Revista Mus. Paulista **13:** 687–763. 1923. (Reprinted, 75 p.)
Alphabetical list of vernacular names, somewhat annotated, and with scientific names where known.

Usteri, Alfred. Flora der umgebung der stadt São Paulo in Brasilien. 271 p. 72 fig., 1 pl., map. 25.5 cm. Jena, 1911.
History, botanical explorations, climate, geology, plant formations, phenology, bibliographies; key (in Latin) to families, genera, and species of vascular plants.

Warming, Eugenius. Lagoa Santa. Et bidrag til den biologiske plantegeografi. K. Danske Vidensk. Selsk. Skr. Naturv. og Math. Afd. VI, **6:** 153–488. 43 fig., pl. 1892.
Ecological; includes (p. 414–436) unannotated list of cellular and vascular plants; French résumé, bibliography.—Portuguese translation by Alberto Löfgren. 282 p. Bello Horizonte, 1909. (Minas Geraes.)

Wettstein, Richard, and **Schiffner, Viktor.** Ergebnisse der botanischen expedition der Kaiserlichen akademie der wissenschaften nach Südbrasilien 1901. 2 v. in 3. Denkschr. Math.-Naturw. Kl. K. Akad. Wiss. Wien v. 79, pt. I–II. vi, 454 p. 23 fig. (incl. map), 41 pl. (pt. col.). 1908–31; v. 83. 358 p. 15 fig., 24 pl. 1927.
Itinerary; list of plants collected, with data and frequent annotations. Incomplete; includes pteridophytes, gymnosperms, and various families of monocotyledons, dicotyledons, and cellular plants. (São Paulo, with a few plants from Paraná and Rio de Janeiro.)

BRITISH GUIANA

General

Dalton, H. G. Natural history of British Guiana. The vegetable kingdom. *In* his The history of British Guiana . . . v. 2, p. 169–264. London, 1855.
Includes notes on cultivated and wild economic plants; an unannotated alphabetical list of vernacular names of "some of the most common trees and plants," with botanical names; a similar annotated list without botanical names, of timber trees; and an unannotated systematic list of cellular and vascular plants, without authorities, compiled from various sources and of little use as a flora.

Harrison, J. B., and **Bancroft, C. K.** Food plants of British Guiana. Jour. Bd. Agr. Brit. Guiana **10:** 143–177. 1917. (Reprinted l. c. v. 19, no. 1, 1926.)
Briefly annotated lists of wild and cultivated food plants, alphabetically arranged by vernacular names.

Hohenkerk, L. S. Botanical identifications of British Guiana trees and plants. Jour. Bd. Agr. Brit. Guiana **11:** 98–106, 178–185. 1918; **15:** 196–199. 1922.
List of miscellaneous plants, with vernacular (mostly Arawak) names, the families arranged alphabetically.

——— British Guiana timbers. Jour. Bd. Agr. Brit. Guiana **12:** 152–185. 1919.
List of about 60 important timber trees, alphabetically arranged by vernacular names; properties of each wood.

——— A list of British Guiana woods. Timehri III, **4:** 244–293. 1917.
Bibliography; annotated alphabetical list of vernacular and botanical names of trees about 250 species).

Rodway, James. On some of the domestic medicines used in Guiana. Timehri **4**: 294–300. 1885.

Running account of wild medicinal plants.

Schomburgk, Richard. Versuch einer fauna und flora von Britisch Guiana. (Flora.) *In* his Reisen in Britisch-Guiana in den jahren 1840–44. v. 3, p. 787–1212, 1226–1260. Leipzig, 1848.

Brief account of plant geography; annotated list of cellular and vascular plants, divided into 4 groups according to habitat and based on his own collections and those of Robert Schomburgk and earlier botanists.

Stone, Herbert, and Freeman, W. G. The timbers of British Guiana. A report upon a collection made by the Hon. A. G. Bell . . . 110 p. pl. 22 cm. London, 1914.

List of 97 trees, arranged alphabetically by vernacular names, with specific names when known, description of wood, bark, uses; bibliography.

Local

Bentham, George. Contributions towards a flora of South America. Enumeration of plants collected by Mr. Schomburgk in British Guiana. Jour. Bot. Hook. **2**: 38–103, 127–146, 210–223, 286–324. 1840; **3**: 212–250. 1841; **4**: 99–133, 321–323. 1842. London Jour. Bot. **1**: 193–203. 1842; **2**: 42–52, 359–378, 670–674. 1843; **4**: 622–637. 1845; **5**: 351–365. 1846.

List of 997 ("912") vascular plants, with localities and frequent annotations, collected by Robert Schomburgk. Includes also some plants from French Guiana by other collectors.

Brown, N. E., and others. Report on two botanical collections made by Messrs. F. V. McConnell and J. J. Quelch at Mount Roraima in British Guiana. Trans. Linn. Soc. Bot. II, **6**: 1–107. 14 pl. 1901.

Physiography, phytogeography, etc.; list of known flora (above 5,000 feet), with extralimital range (401 species, of which 327 are vascular plants); bibliography; partly annotated list of plants collected (by N. E. Brown, C. H. Wright, et al.).

Davis, T. A. W., and Richards, P. W. The vegetation of Moraballi Creek, British Guiana: an ecological study of a limited area of tropical rain forest. Pt. II. Jour. Ecology **22**: 106–155. 2 fig., pl. 12–14. 1934.

Includes (p. 134–155) lists of plants grouped by association and size and alphabetically arranged by vernacular names, with botanical equivalents. (Near junction of Moraballi Creek and Essequibo River, on east side of latter, 6°11'' N. Lat.)

Gleason, H. A. Studies on the flora of northern South America XIII. The Tate collection from Mount Roraima and vicinity. Bul. Torrey Club **56**: 391–408. 1929.

Botanical investigations; descriptions of new species, etc.; list of 50 vascular plants from above 5,000 feet on Mount Roraima additional to N. E. Brown's list (1901).

Graham, E. H. Flora of the Kartabo region, British Guiana. Annals Carnegie Mus. (Pittsburgh) **22**: 17–292. 2 fig. (maps), pl. 3–18. 1934.

History, physical geography, botanical explorations, plants of Georgetown, geology, climate, plant associations; systematic list of 624 vascular plants, with keys, brief descriptions, citation of exsiccatae, general range; bibliography.

Oliver, Daniel. The botany of the Roraima Expedition of 1884: being notes on the plants observed, by Everhard F. im Thurn; with a list of the species collected, and determinations of those that are new. Trans. Linn. Soc. Bot. II, **2**: 249–300. 3 fig., pl. 37–56. 1887.

Botanical collections, physiography, general features of flora (by im Thurn); annotated list of vascular and cellular plants, by Oliver, E. Marchal, H. N. Ridley, J. G. Baker, and W. Mitten.

Smith, A. C. Notes on a collection of plants from British Guiana. Lloydia **2**: 161–218. 1 fig. 1939.

Includes 50 new species and records of 62 others new to British Guiana, collected by the author in southern part of British Guiana drained by Rupununi and upper Essequibo Rivers.

CHILE

See also Argentina (Kurtz); Colombia (Weddell); Patagonia. For Juan Fernandez and Masafuera, San Ambrosio and San Felix Islands, see under Insular Floras.

The following publications have not been available:

Guajardo, Amador. Botánica médica nacional ó sea plantas medicinales de Chile. Santiago, 1892. (Cited from Reiche, K. Grundzüge der pflanzenverbreitung in Chile, p. 32. 1907.)

Gunckel L. Hugo. Enumeración de plantas colectadas en la Isla de Quiriquina. Revista Universitaria (Santiago de Chile) **12:** 159–174. 1927. (Cited from Looser, G. Bibliografía botánica chilena. Revista Bibl. Chilena **1927:** 230. 1927.)

Maldonado, Ernesto. Tratado de arboricultura forestal y adorno. 2 v. 8°. Santiago de Chile, 1926. (Cited from Porter, C. Reseña histórica y bibliografía razonada de las ciencias naturales en Chile. v. 14, p. 91. 1929.) (Includes descriptions, range, and uses of important wild and cultivated trees; information furnished by Carlos Muñoz.)

Rivera, M. J. Apuntes sobre la vegetación de la costa de Curicó. 50 p. Santiago de Chile, 1903. (Cited from Porter, C. Reseña histórica y bibliografía razonada de las ciencias naturales en Chile. v. 14, p. 130. 1929.)

General

Baeza R., V. M. Los nombres vulgares de las plantas silvestres de Chile y su concordancia con los nombres científicos y observaciones sobre la aplicación técnica y medicinal de algunas especies. 2. ed. aumentada . . . 270 p. 26 cm. Santiago [de Chile], 1930. (1st ed., 1921.)

Bibliography; alphabetical list of vernacular names, with explanation of derivation and botanical equivalents; alphabetical list of botanical names with vernacular equivalents.

Fuentes, Francisco. Algunas plantas venenosas de la flora chilena. Revista Agron. **1:** 68–72. 1922.

Running account of about 80 poisonous plants. Not seen; cited from Porter, C. Reseña histórica y bibliografía razonada de las ciencias naturales en Chile. v. 14, p. 60–61. 1929.

Gay, Claude. Historia física y política de Chile . . . Botánica. 8 v. 22.5 cm. and atlas of 103 col. pl. 37.5 cm. Paris, Santiago, 1845–54.

Descriptive flora of vascular and cellular plants, in Spanish except for the Latin diagnoses; Compuestas by J. Remy, Plantas celulares (v. 7–8) by C. Montagne, some other groups contributed by various authors. The caption title, Flora Chilena, is ordinarily used in citations. The plates are divided into two numberings, 83 and 16 with 4 extra.

Latcham, R. E. La agricultura precolombiana en Chile y los países vecinos. viii, 336 p. 18.5 cm. [Santiago,] 1936.

Account of Indian tribes at time of the Spanish conquest; running account of wild and cultivated edible plants; method of cultivation, bibliography.

Looser, Gualterio. Bibliografía botánica chilena. Revista Bibl. Chilena **1927:** 212–230, 364–390. 1927.

Titles additional to those in Reiche's bibliography (1907), alphabetically arranged by authors. Includes general works referring to Chile, as well as those restricted to that country.

——— Catálogo de plantas vasculares nuevas de Chile. Revista Universitaria (Santiago de Chile) **23:** 215–275. 1938.

List of about 500 species of vascular plants not given in the floras of Gay and Reiche and described principally since 1918, with references; bibliography.

Medina, J. T. Voces chilenas de los reinos animal y vegetal que pudieran incluirse en el diccionario de la lengua castellana . . . 149 p. 27 cm. Santiago de Chile, 1917.

Annotated alphabetical list of various vernacular names of plants and animals, with notes on etymology.

Murillo, Adolfo. Plantes médicinales du Chili. xii, 234 p. 28 cm. Paris, 1889. (Exposition universelle de Paris, 1889. Section chilienne.)

Bibliography; annotated systematic list, with vernacular names, references, brief description, medicinal uses; native plants only.

Philippi, Federico. Catalogus plantarum vascularium chilensium adhuc descriptarum. viii, 377 p. 25.5 cm. Santiago de Chile, 1881. (From Anales Univ. Chile, 1881.)

List only, with references to original publications and to Gay's Flora. The number of species is given as 5,358.

Philippi, R. A. Elementos de botánica para el uso de los estudiantes de medicina i farmacia en Chile. ix, 571 p. 24.5 cm. Santiago de Chile, 1869.

Introduction to botany; systematic descriptive account of plant families, with mention of native and principal exotic economic and medicinal species, with vernacular names and uses.

——— Sobre las maderas de Chile. Anales Univ. Chile **29**: 282–295. 1867.

Annotated systematic list of 83 woody plants, with vernacular names.

Porter, C. E. Reseña histórica y bibliografía razonada de las ciencias naturales en Chile. Tomo 14, Botánica agrícola e industrial. 167 p. 23 cm. Santiago (Chile), 1929.

Annotated list, alphabetically arranged by authors, of publications relating to economic and agricultural botany. Vol. 11 (Criptógamas), 12 (Fanerógamas), 13 (Fitogeografía, etc.), and 15 (Botánica médica) are announced but not yet published.

Reiche, K. F. Flora de Chile. v. 1–6, pt. 1. 24 cm. Santiago de Chile, 1896–1911. (Reprinted, at least in part, from Anales Univ. Chile 1894–1909.)

Descriptive flora, with keys; occasional vernacular names. Incomplete: contains Polypetalae, except Cactaceae; Gamopetalae; lacks Apetalae (except Nyctaginaceae, Amaranthaceae, Phytolaccaceae and Chenopodiaceae), Monocotyledones, Gymnospermae.

——— Grundzüge der pflanzenverbreitung in Chile. xiv, 374 p. 55 fig., 33 pl., 2 maps. 26 cm. Leipzig, 1907. (Die vegetation der erde, v. 8.)

Botanical explorations, ecology, and phytogeography. Includes (p. 27–45, 372) bibliography of 550 titles (plus additions) on the Chilean flora. There is a Spanish translation, Geografía botánica de Chile, by Gualterio Looser (2 v., 1934–37 [1938]).

——— Los productos vejetales indígenas de Chile. Bol. Soc. Fom. Fabril Santiago (Chile) **32**: 481–486, 679–684, 776–784. 1915.

Annotated classified list of useful plants. Essentially the same material in his earlier work of the same title (28 p. 1901; 2. ed. 30 p. 1910).

Turrill, W. B. Botanical exploration in Chile and Argentina. Bul. Misc. Inform. Kew. **1920**: 57–66, 223–224. 1920.

Historical sketch, with references to principal publications; account of regions relatively unexplored botanically; bibliography.

Local

Castillo, Luis, and Dey J., J. Jeografía vejetal del Río Valdivia i sus inmediaciones. 2. ed. corr. i aum. 120 p. 64 fig. 24.5 cm. Santiago de Chile, 1908. (1st ed. by Castillo appeared in Anales Agron. Min. Industr. i Obras Púb. (Chile) **2**: 418–506. 46 fig. 1907, with title: La jeografía botánica del curso inferior del Río Valdivia . . .)

Plant formations, list of trees and shrubs, with short descriptions and uses; unannotated list of principal cellular and vascular plants; list of vernacular names with botanical equivalents. Second edition varies only slightly from first edition.

Espinosa Bustos, M. R. Enumeración de plantas colectadas en "Los Bronces." Revista Chilena Hist. Nat. **28**: 88–97. fig. 11. 1924.

Mostly unannotated list of cellular and vascular plants collected by the author in 1923. (Department of Cachapoal.)

——— Plantas de Aysen. Bol. Mus. Nac. Chile **14**: 65–82. 1935.

List of vascular plants collected in the Province of Aysen, on the Latcham-Macqueen expedition in 1934.

Grandjot, F. de, and Grandjot, Karl. Der Potrero Grande in der Kordillere von Santiago. Verhandl. Deut. Wiss. Ver. Santiago de Chile n. f., **3**: 30–66. 6 fig. (pt. on pl. and incl. map). 1936. (Reprinted.)

Geology, climate, general features of flora; annotated list of plants collected by the authors.

Jaffuel, Félix. Florula de los alrededores de las Termas de Chillán. Revista Chilena Hist. Nat. **42**: 76–90. 1939.

General features of flora, etc.; mostly unannotated list of 258 vascular plants, with some vernacular names.

Jaffuel, Félix, and **Pirion, Anastasio.** Plantas fanerógamas del valle de Marga-Marga. Revista Chilena Hist. Nat. **25**: 350–405. 1923.
Climate, etc.; annotated list of phanerogams, with vernacular names and uses. Marga-marga is near Quilpué, in the vicinity of Valparaiso.

Johnston, I. M. Papers on the flora of northern Chile. 1. The coastal flora of the departments of Chañaral and Taltal. Contrib. Gray Herb. **85**: 1–138. pl. 1 (map). 1929.
Physiography, climate, general features of flora, botanical explorations; annotated list of 394 species and varieties of vascular plants. The area covered extends approximately from lat. 24°30′ to 26°30′ S., west of long. 70°20′ W.

——— Papers on the flora of northern Chile. 2. The flora of the nitrate coast. Contrib. Gray Herb. **85**: 138–163. 1929.—New records for the flora of the nitrate coast. Revista Chilena Hist. Nat. **36**: 4–8. 1932.
Physiography, climate, general features of vegetation, botanical explorations, bibliography (in text); annotated list of 117 vascular plants. Covers region from 20° to 24° S. Lat.

Looser, Gualterio. Excursión botánica a la Alta Cordillera de Las Condes (Andes cerca de Santiago de Chile). *In* Ostenia. Collección de trabajos botánicos dedicados a Don Cornelio Osten . . . p. 217–246. 2 pl. Montevideo, 1933.
General features of flora; annotated list of cellular and vascular plants collected in 1930.

Meigen, Fritz. Skizze der vegetationsverhältnisse von Santiago in Chile. Bot. Jahrb. Engler **17**: 199–294. 1893.
General features of flora; list of vascular plants, with localities, containing all species known from the area between Río Maipu and Río Aconcagua, except the coast and the accompanying mountains.

Moore, Eduardo. Contribución a la flórula de Paredones. Revista Chilena Hist. Nat. **30**: 384–400. 1926.
Soil, botanical explorations; unannotated list of cellular and vascular plants. (On the coast of the Department of Vichuquén, Province of Curicó.)

Neger, F. W. Informe sobre las observaciones botánicas efectuadas en la cordillera de Villarrica en el verano 1896–97. Anales Univ. Chile **103**: 903–967. map. 1899.
Ecology, phytogeography; list of 540 hepatics, mosses, and vascular plants, with indication of ecological habitat; annotated list of lichens. (Not far north of the city of Valdivia.)

——— Introduccion a la flora de los alrededores de Concepción. Anales Univ. Chile **98**: 209–251. 1897.
Climate, plant formations, plant geography; unannotated list of 567 vascular plants.

——— Pflanzengeographisches aus den südlichen Anden und Patagonien. Bot. Jahrb. Engler **28**: 231–258. 1900.
Plant formations, phytogeography; list of 540 plants (436 vascular) with general range indicated by symbols. (Upper course of Río Tolten and Río Maitchin, and region of Lakes Rucachoroi, Huihui, Quillen, and Trommen.)

——— Die vegetationsverhältnisse im nördlichen Araucanien (flussgebiet des Rio Biobio). Bot. Jahrb. Engler **23**: 382–411. 1896.
Ecological; includes (p. 402–411) list of vascular plants of the upper Biobio region.

Philippi, Federico. A visit to the northernmost forest of Chile. Jour. Bot. **22**: 201–211. 1884.
Running account of the vegetation of Fray Jorge, Province of Coquimbo.—Spanish translation by Francisco Fuentes, with revised list of botanical names, under title: Una visita al bosque más boreal de Chile. Bol. Mus. Nac. Chile **13**: 96–109. 1930.

——— and **Philippi, R. A.** Excursion botánica en Valdivia desde Los Cuncos en el departamento de La Union, a traves de la Cordillera de La Costa, hasta la mar . . . Anales Univ. Chile **27**: 289–324. 1865.
General features of flora; annotated list of vascular plants, with descriptions of new species.

Philippi, R. A. Catalogus praevius plantarum in itinere ad Tarapaca a Friderico Philippi lectarum. Anales Mus. Nac. Chile 2. secc. Bot. [no. 8.] viii, 96 p. 2 pl. 1891.

Itinerary; list of 417 plants (414 vascular) with localities and occasional vernacular names, collected on journey from Copiapó into Tarapacá in 1884–85. Also published under title: Verzeichniss der von Friedrich Philippi auf der hochebene der provinzen Antofagasta und Tarapacá gesammelten pflanzen.

Philippi, R. A. Florula atacamensis seu enumeratio plantarum in itinere per desertum atacamense observatarum. 62 p. 6 pl. 34 cm. Halle, 1860. (With his Reise durch die wueste Atacama . . . in sommer 1853–54.)

General features of flora; list of 419 plants (417 vascular) with localities and occasional vernacular names. The "Reise" also issued in Spanish, the Florula being included in the "Viage" as p. 175–236.

——— Sertum mendocinum. Catálogo de las plantas recojidas cerca de Mendoza . . . Anales Univ. Chile **21**: 381–407. 1862. —Sertum mendocinum alterum . . . l. c. **36**: 159–212. 1870.

Lists of 109 and 255 vascular plants, respectively, collected around the city of Mendoza and along route to Chile, with localities and occasional annotations.

Pöhlmann, Robert, and **Reiche, Karl.** Beiträge zur kenntnis der flora der flussthäler Camerones und Vitor und ihres zwischenlandes (19° s. Br.). Verhandl. Deut. Wiss. Ver. Santiago de Chile **4**: 263–305. map. 1900.

Itinerary of Dr. Pöhlmann, climate, altitudes, general features of flora; list of vascular plants, with localities.

Reiche, K. F. Cuadro de la flora de la Isla de la Mocha. *In* his La Isla de la Mocha. Anales Mus. Nac. Chile **16**: 64–90. 1903.

Annotated list of vascular and cellular plants.—For additions, see **Schneider, C. O.** Anotaciones relativa a la flora de la Isla Mocha. Comun. Mus. Concepción **1**: 14–18. 1936. (Off the coast, 38°18′ S. Lat.)

——— Die vegetations-verhältnisse am unterlaufe des Rio Maule (Chile). Bot. Jahrb. Engler **21**: 1–52. 1895.

Topography, climate, ecology; list of 664 vascular plants; statistics of flora.

Santa Cruz, Alcibíades. Plantas medicinales de la región de Concepción. Revista Chilena Hist. Nat. **25**: 241–252. 1923.

Annotated list, with vernacular names and uses.

Valenzuela, P. A. Glosario etimológico de nombres de hombres, animales, plantas, ríos y lugares, y de vocablos incorporados en el lenguaje vulgar, aborígenes de Chile, y de algún otro país americano. 2 v. 23 cm. Santiago de Chile, 1918–19.

Includes some vernacular names of plants, with botanical names and derivations.

Vidal, Roberto, and **Silva, Rigoberto.** Escursión botánica a la cordillera andina. Anales Univ. Chile II, **3**: 279–316. 1925.

Botanical explorations in the Andes of the Province of Talca; list of 186 vascular plants collected by the authors on the Volcán Descabezado Grande, Cerro Azul and vicinity, Province of Talca, with notes on some species.

COLOMBIA

See also Argentina (Kurtz); Mexico (Bukasov); Panama (Seemann).

The following publications have not been available:

Londoño, J. B. Plantas medicinales de Antioquia. Medellin, [n. d.] (Cited from Pérez Arbeláez. Plantas medicinales y venenosas de Colombia, p. [3]. 1937.)

Uribe, J. A. Flora sonsonesa. Sonsón, [n. d.]. (Cited from Pérez Arbeláez, Plantas medicinales y venenosas de Colombia, p. [3]. 1937.)

General

Apolinar María, *hermano.* Vocabulario de terminos vulgares en historia natural colombiana. Revista Acad. Colomb. Cien. Exact., Fís. y Nat. **1**: 196–203, 349–361. 2 col. pl. 1937–[38]; **2**: 256–263, 358–365, 543–556. 6 col. pl. 1938–39; **3**: 76–86. 1939.

Annotated alphabetical list of botanical and zoological vernacular names, with scientific equivalents. Reaches (p. 86, 1939) no. 500, Arra-u. A republication, with corrections and additions, of his unfinished work with similar title in Bol. Soc. Colomb. Cien. Nat. v. 16–19. 1927–30.

Cortés, Santiago. Flora de Colombia. 2. ed. 311 p. 8 fig., 7 col. pl. 24 cm. Bogotá, [1919]. (1st ed., 1897.)

Not a flora. Contains remarks on medicinal and other useful plants arranged by families, and list of Colombian vernacular names with botanical equivalents. The second edition contains sketch of geographical botany of Colombia.

Duque Jaramillo, J. M. Manual de bosques y de maderas tropicales. 2 v. (227 p.). 141 fig. 31 cm. Manizales, 1931.

"Tomo II" (i. e., p. 119–227, index) includes a brief descriptive account of the more important native and exotic timbers, with vernacular name and character of wood.

Pérez Arbeláez, Enrique. Plantas medicinales y venenosas de Colombia. Estudio botánico, étnico, farmacéutico, veterinario y forense. 295 p. 122 fig. 24 cm. Bogotá, 1937.

General considerations; systematic list of medicinal and poisonous plants, excluding the barbascos, with vernacular names; annotated account of the same, with additional vernacular names, uses, etc. Includes the information in his Plantas medicinales más usadas en Bogotá.

——— Plantas útiles de Colombia. v. 1 (172 p.). 116 fig. 24.5 cm. Bogotá, 1935 (1936).

Bibliography of the Colombian flora; systematic list of all useful cellular and vascular plants, with vernacular names and uses; more detailed account of the cellular and vascular plants (except dicotyledons).

Robledo, Emilio. Lecciones de botánica médica, industrial y agrícola. 2. ed. notablemente reformada. 586 p. 182 fig. 22.5 cm. Medellín, 1937. (1st ed., [1924].)

A textbook, cited here only for its Spanish vernacular names of native and exotic plants, including the principal useful plants of the world.

Triana, J. J., and **Planchon, J. E.** Prodromus florae novo-granatensis. 2 v. 4 pl. 24.5 cm. Paris, 1862–67.

List, with citation of exsiccatae, vernacular names, and descriptions of new plants and critical notes. Incomplete: [v. 1], 1862, Ranunculaceae–Sapindaceae (De Candollean order); [v. 2], 1863–7, Cryptogams by various authors. Reprinted with changed pagination from Ann. Sci. Nat. The treatment of phanerogams was continued after publication of these volumes, the whole series referring to phanerogams being as follows in Ann. Sci. Nat. IV, Bot. **17:** 5–190. 1862 (Ranunculaceae–Malvaceae); **17:** 319–382. 1862 (Sterculiaceae–Ternstroemiaceae, part); **18:** 258–381. 1862 (Ternstroemiaceae, part–Sapindaceae); V, **14:** 286–325. 1872 (Terebinthaceae, Rutaceae); **15:** 352–382. 1872 (Terebinthaceae–Olacineae); **16:** 361–382. 1872 (Coriarieae–Rhamneae); **17:** 111–194. 1873 (Geraniaceae–Papayaceae). The series containing the Cryptogams was as follows: IV, **19:** 286–382. 1863; **20:** 228–300. 1863; V, **1:** 95–198. 1864; **2:** 193–271. 1864 (Filices, by G. Mettenius); **3:** 270–311. 1865 (vascular cryptogams except Filices, by A. Braun); **3:** 337–376. 1865; **4:** 324–378. 1865; **5:** 301–342. 1866; **7:** 301–354. 1867.

Local

Cuatrecasas, José. Resumen de mi actuación en Colombia con motivo del II centenario del nacimiento de Mutis. Trab. Mus. Nac. Cien. Nat. y Jard. Bot. Madrid, Ser. Bot. no. 33. 158 p. 8 fig., 8 pl. 1936.

Includes (p. 23–148) systematic list of cellular and vascular plants collected by the author in the central and eastern Cordilleras, with data.

Curran, H. M. The lands of Loba, Colombia. Trop. Woods, Yale Univ. School Forestry **19:** 11–38. 1929.

General features of region and forests; list of trees collected, with vernacular names and notes on wood, the families arranged alphabetically; alphabetical list of vernacular names; physical properties of woods collected. (In Department of Bolívar, between Magdalena and Cauca Rivers, about 8°15′ to 9°15′ N. Lat.)

Espina, Ramón, and **Giacometto, Juan.** Trees of the Sierra Nevada de Santa Marta. Trop. Woods, Yale Univ. School Forestry **30:** 17–37. 1932.

Plant zones, collecting localities; list of trees collected, with vernacular names, the families arranged alphabetically; alphabetical list of vernacular names.

Pérez Arbeláez, Enrique. Frutas de Cundinamarca. Bol. Agr. Colombia Supl. no. 22. 103 p. 63 fig. 1933. (Reprinted from Bol. Agr. Colombia, v. 4, 6. 1931–33.)
Bibliography, systematic list; annotated systematic list, with vernacular names.

——— Plantae medicinales más usadas en Bogotá. Bol. Agr. Colombia Supl. no. 32. 112 p. 74 fig. 1934.
Systematic list, with vernacular names, followed by a brief account of the species and their uses.

Record, S. J., and Kuylen, Henry. Santa Marta Valley, Colombia. Trop. Woods, Yale Univ. School Forestry **23**: 9–23. 1930.
Topography, general features of flora; annotated list of trees, shrubs, and lianas collected by the authors in January 1930, with vernacular names, the families arranged alphabetically; alphabetical list of vernacular names with botanical equivalents.

Seifriz, William. Die höhenstuden der vegetation in der Sierra Nevada von Santa Marta (Colombia). Bot. Jahrb. Engler **68**: 107–112, 113–125. 1 fig. (map), 8 pl. (incl. map). 1936–37.
Includes list of species collected by the author [in 1932] at different altitudes; bibliography.

Toro, R. A. Contribución a la flora de Antioquia. Revista Soc. Colomb. Cien. Nat. **20**: 20–32, 57–64. 1931.
List of plants collected by the author, with vernacular names, localities, and collector's numbers; the families arranged alphabetically. Incomplete (Acanthaceae–Mimosaceae); includes cultivated plants.

——— Una contribución a nuestro conocimiento de la flora silvestre y cultivada de San Andrés. Revista Soc. Colomb. Cien. Nat. **18**: 201–207. 1929.—Una contribución a nuestro conocimiento de la flora de San Andrés y Providencia. l. c. **19**: 56–58. 1930.
Unannotated list of 136 vascular plants, with occasional (English) vernacular names. (Two small islands in Caribbean Sea, off coast of Nicaragua; San Andrés (St. Andrews Island) in 12°31′ N. Lat., 81°43′ W. Long., Providencia (Old Providence Island) in 13°18′ N. Lat., 81°18′ W. Long.)

Weddell, H. A. Chloris andina. Essai d'une flore de la région alpine des cordillères de l'Amérique du Sud. 2 v. 90 pl. 32.5 cm. Paris, 1855–57 (–1861). (Expédition dans les parties centrales de l'Amérique du Sud . . . sous la direction du comte F. de Castelnau. 6. ptie.)
Alpine flora of the Andes, from Colombia and Venezuela to Chile, with descriptions and citation of exsiccatae. Incomplete: vol. 1, Compositae; vol. 2, Calycereae–Frankeniaceae (43 families).

ECUADOR

See also Argentina (Kurtz); Colombia (Weddell). For Galapagos Islands, see under Insular Floras.

General

Diels, Ludwig. Beiträge zur kenntnis der vegetation und flora von Ecuador. Bibliotheca Bot. no. 116. 190 p. 2 fig., map, tables. 1937.
Ecology; alphabetical list of botanical collectors in Ecuador, with brief biographical sketches, itineraries, location of collections; annotated list of vascular plants collected by the author in 1933; bibliography.—Spanish translation by Reinaldo Espinosa, Contribuciones al conocimiento de la vegetación . . . 364 p. map. Quito, 1938.

Jameson, William. Synopsis plantarum aequatoriensium . . . viribus medicatis et usibus oeconomicis plurimarum adjectis. 3 v. 16.5 cm. Quito, 1865.
Descriptive flora, in Latin, with remarks in Spanish; notes on useful plants at end of each family. Incomplete: v. 1, Ranunculaceae–Columelliaceae; v. 2, Caprifoliaceae–Labiatae; v. 3 (136 p., unfinished), Verbenaceae–Plantago.

Popenoe, Wilson. Economic fruit-bearing plants of Ecuador. Contrib. U. S. Natl. Herb. **24**: 101–134. pl. 34–49. 1924.
Systematic list of wild and cultivated fruits, with annotations.

Rimbach, August. The forests of Ecuador. Trop. Woods, Yale Univ. School Forestry **31**: 1–9. 1932.

Botanical regions, etc.; list of vernacular names of trees, with botanical equivalents.

Local

Cordero, Luis. Enumeración botánica de las principales plantas, así útiles como nocivas, indígenas ó aclimatadas, que se dan en las provincias del Azuay y de Cañar de la República del Ecuador. 304 p. col. pl. 23.5 cm. Cuenca, 1911.

Useful plants and weeds, systematically arranged, with vernacular names and uses.

Meyer, Hans. In den Hoch-Anden von Ecuador: Chimborazo, Cotopaxi, etc. . . . 14, 551 p. 37 pl., 3 maps. 25.5 cm. Berlin, 1907.

Includes (p. 512–527) tabular list of cellular and vascular plants collected by the author on Chimborazo, Altar, Antisana, Cotopaxi, and Quilindaña, with locality and altitude indicated. The ferns were named by Hieronymus, the phanerogams by J. Bornmüller, Hieronymus, and Pilger, the cellular plants by other botanists.

FRENCH GUIANA

General

Benoist, Raymond. Les bois de la Guyane française. Arch. de Bot. (Caen) Mém. v. 5, no. 1. 291 p. 10 fig., 58 pl. 1931.

General features of forests, wood structure; systematic list, with keys, vernacular names, brief descriptions, wood structure, uses.

——— La végétation de la Guyane française. Bul. Soc. Bot. France **71**: 1169–1177. 1925; **72**: 1066–1078. 1926.

Physiography, climate, botanical regions with lists of characteristic species.

Bertin, André, Bettenfeld, Marcel, and Benoist, Raymond. Les bois de la Guyane française et du Brésil. vi, 318 p. illus. (incl. 3 maps), 2 pl. 25 cm. Paris, 1920. (Mission d'études forestières envoyée dans les colonies françaises par les Ministères de la guerre, de l'armement et des colonies. v. 5.)

Mainly of interest to foresters; short descriptions, habit drawings, and vernacular names given.

Devez, G. Les plantes utiles et les bois industriels de la Guyane. vi, 90 p. 25 cm. Paris, 1932.

List of plants, classified by uses; annotated list, alphabetically arranged by vernacular names; index of botanical names; bibliography.

Greshoff, Maurits. De nuttige planten van Fransch Guyana in verband met Suriname beschouwd. Bul. Kolon. Mus. Haarlem **25**: 23–45. 1901.

Annotated list of 180 plants, alphabetically arranged by vernacular names, with brief account of uses and Surinam vernacular names.

Heckel, Édouard. Les plantes médicinales et toxiques de la Guyane française (catalogue raisonné et alphabétique). 93 p. 25 cm. Mâcon, 1897. (Also issued as Annales Mus. Colon. Marseille v. 4, p. 67–159. 1898.)

Annotated list, alphabetically arranged by vernacular names; systematic index.

Lanessan, J. L. de. Guyane. *In* his Les plantes utiles des colonies françaises. p. 125–152, 350–420. Paris, 1886. (Annexe aux notices coloniales publiées à l'occasion de l'Exposition universelle d'Anvers en 1885.)

Annotated systematic lists of woods and of medicinal and other useful plants, with brief descriptions and vernacular names. Contains similar sections on other French colonial possessions.

Sagot, Paul. Catalogue des plantes phanérogames et cryptogames vasculaires de la Guyane française. Annales Sci. Nat. VI, Bot. **10**: 361–382. 1880; **11**: 134–180. 1881; **12**: 177–211. 1881; **13**: 283–336. 1882; **15**: 303–336. 1883; **20**: 181–216. 1885.

Botanical explorations; annotated list of vascular plants (incomplete: Dilleniaceae–Myrtaceae).

Stone, Herbert. Les bois utiles de la Guyane française. Annales Mus. Colon. Marseille v. 24 (s. 3, v. 4), fasc. 2, p. 39–135. 2 pl. 1917; v. 25 (s. 3, v. 5),

fasc. 3. 159 p. pl. 3–7. 1917; v. 26 (s. 3, v. 6), fasc. 1. 68 p. 1918; v. 28 (s. 3, v. 8), fasc. 2. 98 p. 1922; v. 28, suppl. 37 p. 1924.
Descriptions of wood anatomy of each species, with vernacular names.

Local

Benoist, Raymond. Contribution à l'étude de la flore des Guyanes. Plantes recoltées en Guyane française en 1913 et 1914. Bul. Soc. Bot. France **66:** 261–266, 317–328, 357–370, 381–398. 1920; **68:** 134–144, 311–323, 484–490. 1921–22; **69:** 51–57, 104–110. 1922.
Botanical regions; list of phanerogams collected by the author.

Cadet de Gassicourt, C. L. Matière médicale des Galibis et des Guaripons, peuples naturels de la Guiane. Jour. de Pharm. et Chim. II, **2:** 310–326. 1816.
Annotated list of 157 plants, classified by uses, with scientific and French vernacular names.

PARAGUAY

See also Argentina (Kurtz; Hassler; Kerr).

General

Bertoni, M. S. Las plantas usuales del Paraguay y paises limítrofes . . . I. Introducción, nomenclatura y diccionario de los géneros botánicos latino-guaraní. 78, 2 p. 23.5 cm. Asunción, [1914]. (Descripción física y económica del Paraguay 31.)
Pt. I (all published) treats of the Guaraní language and includes botanical generic names with Guaraní equivalents. Pages 1–47 are a reprint, with some changes, of p. 1–39 of his Plantas usuales del Paraguay, Alto Paraná y Misiones . . . Anales Cient. Paraguayos sér. I, no. 2. 1901.

Chodat, Robert. Plantae hasslerianae soit énumération des plantes récoltées au Paraguay par le Dr. Emile Hassler . . . de 1885–1895. Bul. Herb. Boissier v. 6, app. 1. p. 1–42. 1898; v. 7, app. 1. p. 43–87. 1899.—Plantae hasslerianae . . . de 1885–1895 et de 1898–1900. l. c. II, **1:** 395–442. 1901; **2:** 297–312, 382–403, 733–747, 811–824. 1902.—Plantae hasslerianae . . . de 1885 à 1902. 2. ptie. l. c. II, **3:** 50–66, 239–255, 342–355, 387–421, 538–552, 612–641, 701–732, 780–811, 906–941, 1007–1039, 1097–1127. 1903; **4:** 61–92, 169–196, 257–292, 475–489, 548–563, 688–693, 824–839, 879–909, 1051–1068, 1155–1172, 1273–1288. 1904; **5:** 65–90, 288–305, 481–506, 603–613, 671–699. 1905; **7:** 279–296, 597–624, 665–681, 795–826. 1907. (Reprinted.)
Plant formations (II, **3:** 50–54); lists of species collected, with data, arranged by families (not in systematic order), each (in part 2) with a synopsis of the distribution of the species. Enumerates 3,889 species of vascular plants. The second part, beginning in 1903, does not in general include the collections listed in the earlier papers of the series. E. Hassler is joint author of ptie. 2.—See also **Hassler, E.** Addenda ad Plantas hasslerianas . . . 20 p. 23 cm. Genève, 1917.

——— and **Vischer, Wilhelm.** La végétation du Paraguay. Résultats scientifiques d'une mission botanique suisse au Paraguay. I–XIV. Bul. Soc. Bot. Genève II, **8:** 83–160, 186–264. 123 fig., 3 col. pl. 1916–17; **9:** 55–107, 165–244. fig. 124–227, col. pl. 4–7. 1917; **11:** 211–299. fig. 228–280. 1920; **12:** 25–54, 157–218. fig. 281–327. 1921; **17:** 127–164. 52 fig. 1926; **18:** 246–294. 38 fig. 1926 [1927]. (Reprinted, 509, 49 p.)
Climate and physical geography; chapters on various families of flowering plants, considered mainly in reference to anatomy, ecology, and biology. Vischer was joint author through no. XI, L. Rehfous for no. XIII–XIV.

Fiebrig-Gertz, Carlos. Guarany names of Paraguayan plants and animals. Revista Jard. Bot. y Mus. Hist. Nat. Paraguay **2:** 99–149. 1923 (1930).
Etymology; systematic list of plants, with Guarany names; Guarany names of animals, etc.

Hassler, Emil. Enumeración preliminar de las plantas usuales del Paraguay. Revista Inst. Paraguayo **3:** 161–170. 1901; **4:** 204–206. 1901.
Unannotated list of vernacular (and botanical) names alphabetically arranged under the families.

Parodi, Domingo. Catálogo alfabético de algunas plantas usuales del Paraguay, de Corrientes y de Misiones. Revista Farm. (Buenos Aires) **26**: 93–100, 211–216 234–252, 280–288, 320–324, 351–360, 393–396, 423–432. 1887; **27**: 32–36, 65–72, 99–108, 137–144, 177–180, 214–218, 246–256, 273–294. 1888.

Annotated, essentially alphabetical list of useful plants identified at Kew. (Vol. 26 not seen, data from Darwiniana **1**: 397.)

——— Notas sobre algunas plantas usuales del Paraguay, de Corrientes y de Misiones. Anales Soc. Cient. Argentina **4**: 80–86, 124–135, 212–217, 243–251, 300–315. 1877; **5**: 33–45. 1877. (Reprinted 61 p. 1877; 2 ed. xxvii, 123 p. 1886 (not seen).)

Annotated, essentially alphabetical list of Guaraní (and Spanish) vernacular names with botanical equivalents; no index to botanical names. The identifications are frequently incorrect; see Hassler's "Enumeración" (1901). A less complete list (Catálogo alfabético Guaraní-Castellano de las plantas indígenas ó natura i zadas en el Paraguay . . .) was published in Revista farmacéutica de Buenos Aires 1860–61, and an English version of the latter in Pharm. Jour. ser. 2, v. 2–4, 1861–62. According to E. Hassler, Bul. Herb. Boissier II, **8**: 979–980, 985–986. 1908, Parodi's first edition is a plagiarism from Martius' "Systema materiae medicae vegetabilis brasiliensis" and his "Pflanzen-namen in der Tupisprache"; in the 2d edition he copied also from Hieronymus' "Plantae diaphoricae."

Rodríguez, P. M. Plantas medicinales del Paraguay. Redactado en presencia de libros y revistas. Aumentado con observaciones y datos farmacológicos. 141 p. 18.5 cm. Asunción, 1915.

Annotated list, alphabetically arranged by vernacular names, with botanical names, properties, uses, etc.

Rojas, Teodoro. Herbario del Jardin botánico del Paraguay (especies determinadas hasta la fecha). Revista Jard. Bot. y Mus. Hist. Nat. Paraguay **1**: 145–163. 1921 (1922); **2**: 154–192. 1923 (1930).

Systematic list of vascular plants, with collectors' numbers and data: Hymenophyllaceae—Orchidaceae (Engler and Prantl system).

Local

Fiebrig-Gertz, Carlos, and Rojas, Teodoro. Ensayo fitogeográfico sobre el Chaco boreal. Revista Jard. Bot. y Mus. Hist. Nat. Paraguay **3**: 1–87 (incl. plates). 1933.

Includes running account of vascular plants, with enumeration of species; bibliography.

Morong, Thomas, and Britton, N. L. An enumeration of the plants collected by Dr. Thomas Morong in Paraguay, 1888–1890. Annals N. Y. Acad. Sci. **7**: 45–280. 1892–93.

Annotated list of vascular plants (and Musci), mainly within a radius of 100 miles of Asuncion. Brief descriptions of a considerable proportion of the species are given.

PATAGONIA

See also Argentina (general works); Chile (general works).

The term Patagonia is variously used. It is here taken to include the portion of Argentina south of the Río Negro and its tributary, the Río Limay (including Lake Nahuel-Haupi); that is, the Territories of Río Negro, Chubut, and Santa Cruz; and in Chile, the island of Chiloë and all the area south of it; that is, the Provinces of Chiloë and Magallanes. Tierra del Fuego is combined with Patagonia.

General

Hauman, Lucien. Étude phytogéographique de la Patagonie. Bul. Soc. Roy. Bot. Belg. **58**: 105–179. 7 fig., 13 pl., map (in text). 1926.

Botanical regions, with lists of species; bibliography.

Macloskie, George, and others. Reports of the Princeton University expeditions to Patagonia, 1896–1899. v. 8, Botany. 3 v. 106 fig., 35 pl. 33.5 cm. Princeton, 1903–14.

Besides sections on the general features of the vegetation and the Hepaticae and Bryophyta, by other authors, the work includes the following by Macloskie: list of vascular cryptogams with general range and brief descriptions; "Flora patagonica," an annotated list of the flowering plants known from Patagonia, with keys, brief descriptions, local and extralimital range; key to families; botanical exploration, bibliography, topography, general characters and origin of flora. Pt. 3 (Supplement) is a "Revision of Flora patagonica" by Macloskie and Per Dusén and consists of additions, deletions, and critical notes. In this work Patagonia is limited essentially as above defined.

Local

Al'bov, N. M. Contributions à la flore de la Terre de Feu I–II. Revista Mus. La Plata **7**: 277–308, 353–402. 12 pl. 1896.

General account of explorations, climate, vegetation, etc., of the "Canal de Beagle"; (pt. 2) "Énumération des plantes . . .", a list of 225 vascular plants, with data, based on Al'bov's collection in 1896; the "Cyperaceae and Gramineae" (p. 393–402) by F. Kurtz.

——— Essai de flore raisonnée de la Terre de Feu. vi, 85, xxiii p. port. 37 cm. La Plata, 1897 (1903). (Anales del Museo de La Plata . . . Sección botánica I.)

Mainly phytogeographical; includes unannotated systematic list of 615 species and varieties of vascular plants. The author includes not only the archipelago of Tierra del Fuego but also the greater part of the north coast of the Strait of Magellan, the west slope of the Andes north to about 44° S. Lat., and the islands off this coast. The title page is dated 1897, but the preface (dated 1903) states that this work is posthumous, Al'bov having died in December 1897. Said to have been published in Russian in 1901 (cf. Darwiniana **1**: 193. 1927).

Autran, Eugène. Florule du Lac Nahuel Haupí et de ses environs. *In* his Les parcs nationaux argentins. Bol. Min. Agr. Rep. Argentina **7**: 14–39. 2 pl. 1907. (Reprinted in Trab. Mus. Farm. Fac. Cien. Méd. Buenos Aires no. **13**.)

List of 376 vascular plants, with localities.

Ball, John. Further contributions to the flora of Patagonia. Jour. Linn. Soc. Bot. **27**: 471–500. 1891.

Geography, etc.; annotated list of 92 vascular plants collected by W. Andrews in the interior of Patagonia, on a trip from Port Desire to the Río Negro and on an expedition up the valley of the Río Limay to the vicinity of Lake Nahuel-Haupí.

Donat, Arturo. Contribuciones al conocimiento de la flora de la Patagonia oriental: la flora de la región costanera. Darwiniana **2**: 58–71. 1930.

General features of flora, plant associations; list of spermatophytes collected by the author in 1928, with their plant associations. (Coastal region of Santa Cruz, between the Pampa de Castillo and the Río Deseado.)

Dusén, Per. Die gefässpflanzen der Magellansländer nebst einem beitrage zur flora der ostküste von Patagonien. *In* Wissenschaftliche ergebnisse der Schwedischen expedition nach dem Magellansländern. v. 3, Botanik. p. 77–266. illus., pl. 4–14. Stockholm, 1900. (Reprinted.)

Botanical explorations, bibliography; annotated list of vascular plants collected by the author, with synonymy, references, and localities.

——— Zur kenntnis der gefässpflanzen des südlichen Patagoniens. Öfvers. K. Vetensk.-Akad. Förhandl. **58**: 229–263. 5 fig. 1902.

Annotated list of vascular plants collected in south and west Patagonia by Nordenskiöld, Borge, and the author.

Espinosa Bustos, M. R. Los alerzales de Piuchué. Bol. Mus. Nac. Chile **10**: 36–93 (incl. fig. 21–58, pl. 2–9). 1917.

General features of flora; unannotated list of cellular and vascular plants; bibliography. (On island of Chiloë.)

——— [Escursión a la Península de Taitao. Lista de las plantas colectadas.] Bol. Mus. Nac. Chile **9**: 154–162. 1916 [1917].

Partial list of vascular and cellular plants collected in 1916–17, with vernacular names and some notes on uses. In author's report as chief of the section of botany. Taitao (Taytao) is on south-central coast of Chile, about 46°20′ S. Lat.

Franchet, A. R. Phanérogamie. *In* Mission scientifique du Cap Horn, 1882–83. v. 5, Botanique. p. 313–400. 12 pl. (pt. col.). Paris, 1889.

Botanical explorations; partly annotated list of 216 vascular plants collected by the expedition south of the Strait of Magellan, with localities; critical notes on species collected by Commerson in 1767.

Hambleton, Santiago. La vegetación del Canal y Río Baker (Patagonia occidental). Revista Argentina Agron. **3**: 159–173. 1936.

General features of flora; list of 223 vascular plants, with localities. The localities covered are: Chonos or Guaitecas Islands, San Quintín, Canal Baker and another fjord of the Gulf of Penas, Río Baker, Cordón Atravesado, and the region of the lakes.

Hieronymus, Georg. Sertum patagonicum ó determinaciones y descripciones de plantas fanerógamas y criptógamas vasculares recogidas por el Dr. Cárlos Berg en las costas de Patagonia. Bol. Acad. Nac. Cien. Córdoba (Rep. Argentina) **3**: 327–385. 1881.

Annotated list of 176 phanerogams, with data and some vernacular names, collected in 1874 near the mouth of the Río Negro, at San Blas Bay, and at the mouth of the Río Santa Cruz.

Hosseus, C. C. Apuntes sobre la vegetación del Lago Argentino y del Río Santa Cruz. Trab. Inst. Bot. y Farm. Fac. Cien. Méd. Buenos Aires **37**: 3–22. 1917.

List of vascular plants collected in 1904–05 by J. Koslowsky, with data, brief descriptions, local and general distribution.

——— La vegetación del Lago Nahuel Huapí y sus montañas. Trab. Inst. Bot. y Farm. Fac. Cien. Méd. Buenos Aires no. 33. 102 p. 1915.

Author's itinerary; annotated list of vascular plants collected by him in the territories of Río Negro and Neuquén in 1913, with data, vernacular names, soil, plant formations, general range, and uses.

Reiche, Karl. Informe. *In* Krüger, P. Memoria jeneral sobre la espedicion esploradora del Río Palena, Diciembre 1893–Marzo 1894. Anales Univ. Chile **90**: 715–747. 1895.

General features of flora, etc.; unannotated list of 197 vascular and cellular plants. (On the mainland, not far south of the island of Chiloë.)

——— La jeografía botánica de la region esplorada del Río Manso. Anales Univ. Chile **101**: 436–465. 1898.

Botanical explorations, plant zones; list of 317 plants (276 vascular) with indication of zones.

Rendle, A. B. Mr. Hesketh Prichard's Patagonian plants. Jour. Bot. **42**: 321–334, 367–378. pl. 465. 1904.

General features of flora; list of Angiospermae collected at western end of Lake Argentina, lat. 50° S., with localities and general range.

Roivainen, H. Contribuciones á la flora de Isla Elisabeth, Rio de las Minas y Puerto San Isidor de Prov. de Magallanes, de Puerto Barroso de Prov. de Chiloë y de los alrededores de Termas de Chillan de Prov. de Ñuble, Chile. Annales Bot. Soc. Zool.-Bot. Fenn. Vanamo v. 4, no. 8. 22 p. 1933.

Alphabetical list of vascular plants collected by the author in 1928–29, with localities.

Skottsberg, Carl. Die vegetationsverhältnisse längs der Cordillera de los Andes S. von 41° S. Br. Ein beitrag zur kenntnis der vegetation in Chiloé, West-Patagonien, dem andinen Patagonien und Feuerland. (Botanische ergebnisse der Schwedischen expedition nach Patagonien und dem Feuerlande 1907–09. V.) K. Svenska Vetensk.-Akad. Handl. v. 56, no. 5. 366 p. 24 fig. (incl. maps), 23 pl. 1916.

Climate, ecology, plant formations, etc.; annotated list of vascular plants, with localities and extralimital range, based chiefly on the collections of this expedition; general features of flora, bibliography.

——— Zur flora des Feuerlandes. Floristische beobachtungen über gefässpflanzen, gesammelt in den jahren 1902 und 1903. *In* Wissenschaftliche ergebnisse der Schwedischen südpolarexpedition 1901–1903. v. 4, pt. 4. 41 p. 2 pl., map. Stockholm, 1916.

Annotated list of vascular plants collected by the author.

——— Zur gefässpflanzenflora Westpatagoniens. Göteborgs K. Vetensk.- o. Vitterh.-Samh. Handl. IV, v. 28, [no. 3]. 29 p. 7 fig. 1924.

Annotated list of vascular plants collected on Nordenskjöld's 1920–21 expedition to Kelly-Fjord and the San Tadeo Glacier; general features of the flora.

Spegazzini, Carlos. Nova addenda ad floram patagonicam. Anales Soc. Cient. Argentina **47**: 161–177, 224–239, 274–290. 1899; **48**: 44–59, 172–190, 239–

242, 329–332. 1899; **53**: 13–34, 66–80, 136–144, 166–185, 242–251, 275 292. 1902. Anales Mus. Nac. Buenos Aires **7** (II, 4) : 135–308. 1902.

Additions to the flora of Patagonia, new species. and critical notes.

Spegazzini, Carlos. Plantae Patagoniae australis. Revista Fac. Agron. y Vet. La Plata **3**: 485–589. 1897.

List of 441 vascular plants from the area between the Strait of Magellan and about 45° S. Lat., with localities and occasional annotations.

——— Primitiae florae chubutensis. Revista Fac. Agron. y Vet. La Plata **3**: 591–633. 1897.

Plant zones; list of 229 vascular plants, with localities, from the Territory of Chubut.—See also **Hosseus, C. C.** Algunas plantas de Cabo Raso (Chubut). Physis **1**: 534–540. 3 fig. 1915.

Wildeman, Émile de. Les phanérogames des terres Magellaniques. 222 p. 23 pl. 36 cm. Anvers, 1905. (Commission de la *Belgica*. Expédition antarctique belge. Rapports scientifiques . . . Botanique.)

List of species collected on the voyage of the *Belgica;* systematic enumeration of the phanerogams of the Austro-Antarctic American flora, with references and local distribution; statistical table of the same, showing distribution by islands, etc. Covers area from Wellington Island to Staten Island, Tierra del Fuego, and adjacent mainland; lists 539 species and varieties from the islands (in addition to those confined to the mainland).

PERU

See also Argentina (Kurtz) ; Colombia (Weddell).

General

Herrera, F. L. Catálogo alfabético de los nombres vulgares y científicos de plantas que existen en el Perú. vii, 363 p. 17.5 cm. Lima, 1939.

Consists of two sections: Exploraciones botánicas en el Perú (reprinted from Revista Mus. Nac. Lima 6: 296–358. 1937), an annotated list of collectors, with account of regions visited and location of collections; Nombres vulgares con sus respectivas sinonimias científicas, an alphabetical list of vernacular names with botanical equivalents; also brief bibliography.

Macbride, J. F. Flora of Peru pt. I–II, VI. Pub. Field Mus. Nat. Hist. Bot. Ser. v. **13**. map. 1936–38. (Pt. I. no. [1], 3; pt. II, no. 1–3; pt. VI, no. 1–2. Pub. Field Mus. 351, 357, 363, 367, 369, 393, 428.)

Recent botanical explorations, phytogeography (by A. Weberbauer); annotated list of spermatophytes, with keys, brief descriptions, citation of exsiccatae with localities, occasional vernacular names. Treatment of several families contributed by other authors, especially by P. C. Standley.

Sievers, Wilhelm. Die nutzpflanzen, ihr anbau und dessen geographische verbreitung. *In* his Reise in Peru and Ecuador . . . 1909. Wiss. Veröffentl. Gesell. Erdk. Leipzig **8**: 302–316. map 4–5. 1914.

Running account of principal edible plants of Peru, wild and cultivated.

Weberbauer, August. Die pflanzenwelt der peruanischen Anden in ihren grundzügen dargestellt. xii, 355 p. 63 fig., 40 pl., 2 maps. 25.5 cm. Leipzig, 1911. (Die vegetation der erde, v. 12.)

Botanical explorations, bibliography, physical geography, geology, climate, botanical regions, plant formations, cultivated plants, phytogeography.

Local

Ball, John. Contributions to the flora of the Peruvian Andes, with remarks on the history and origin of the Andean flora. Jour. Linn. Soc. Bot. **22**: 1–64. 1885.

General features of flora, tables of zonal distribution, phytogeography; list of vascular plants collected by the author in upper valley of Rimac River in the Peruvian Andes in 1882, with frequent annotations.

Bruns, Ferdinand. Beiträge zur kenntnis der vegetation des peruanischen küstengebietes. Mitt. Inst. Allg. Bot. Hamburg **8**: 1–85. 11 fig. 1929.

Botanical investigations, physiography, geology, climate, statistics of flora, plant formations, phytogeography; list of 399 vascular plants (357 native) known from the region, with references, localities, and collectors' names, and general range; bibliography.

Escomel, Edmundo. Flora y fauna de Arequipa. 103 p. 24.5 cm. Lima, 1936.

Includes (p. 3–29) notes on useful plants, wild and cultivated, systematically arranged, with vernacular names; (p. 73–88) alphabetical lists of botanical and vernacular names mentioned in the previous section, with equivalents and page references. Publication of the Dirección general de fomento.

García y Merino, Manuel. Los nombres vulgares de nuestras plantas. Bol. Soc. Geog. Lima 5: 294–301. 1895.

Explanation of some of the most common root words in Spanish vernacular names used in Peru, with examples.

Harms, Hermann. Uebersicht der bisher in altperuanischen gräbern gefundenen pflanzenreste. *In* Festschrift Eduard Seler . . . herausgegeben von W. Lehmann. p. 157–186. pl. 2. Stuttgart, 1922.

Annotated bibliography; annotated systematic list of species, including those previously reported.

Herrera, F. L. Contribución a la flora del departamento del Cuzco. 2. ed. primera parte. 241 p. 21 cm. Cuzco, 1921. (First ed., 1919, in Revista Universitaria Cuzco, v. 8, no. 28.)

List of 656 native and cultivated plants, with remarks and often descriptions; bibliography.

——— Filología Quechua. Etimologías de algunos nombres vernaculares de plantas indígenas en el departamento del Cuzco. Revista Mus. Nac. Lima 8: 81–98. 1939.

Includes alphabetical list of Quechua vernacular names with botanical equivalents.

——— Plantarum cuzcorum herrerarianum. Estudios sobre la flora del departamento del Cuzco. 2 v. plates. 21 cm. Lima, 1930–33.—[I]–III. suplemento. Revista Universitaria Cuzco 2, epoca, 14 (1): 149–160. 1930; 23 (1): 254–262. 1931; 23 (2): 111–147. 1934.

Botanical explorations, bibliography; list of 1,023 cellular and vascular plants (including supplements) collected by the author, with vernacular names (indexed), general range, and data for specimens collected; statistics. Replaces his Chloris Cuzcoensis (1926). The second volume includes species 733–1017 of the flora, as well as notes on various useful plants, some local lists, a list of vernacular names, etc. The sketch of botanical explorations in vol. 1 is reproduced from Revista Universitaria Cuzco 2. epoca, 13: 205–223. 1929.—See also his La flora en el departamento del Cuzco. Revista Mus. Nac. Lima 4: 119–133. 1935. (Adds 77 species.)

——— Plantas que curan y plantas que matan de la flora del Cuzco. (Estudio folklórico.) Revista Universitaria Cuzco 27 (75): 4–76. 1938.

List of vernacular names of medicinal and poisonous plants, classified by uses; annotated list, alphabetically arranged by vernacular names, with habit, properties, and citation of exsiccatae.

——— Sinonimia científica y vulgar de algunas plantas indígenas en el departamento del Cuzco. Revista Universitaria Cuzco 13: 365–395. 1929.

Reciprocal alphabetical lists of botanical and vernacular names.

——— La vegetación de la Quebrada del Urubamba. Revista Chilena Hist. Nat. 36: 122–135. illus. 1932.

Includes lists of 170 species, by localities and formations. (Department of Cuzco.)

Huber, Jacques. Plantas vasculares colligidas e observadas no baixo Ucayali e no Pampa del Sacramento, nos mezes de outubro a dezembro de 1898. (Materiaes para a flora amazonica VI.) Bol. Mus. Goeldi Hist. Nat. e Ethnogr. 4: 510–619. 7 fig. 1906.

Author's itinerary; list of vascular plants, with vernacular names, data, range, and frequent annotations. (Northeastern Peru, in the Amazon drainage.)

Johnston, I. M. The vascular flora of the guano islands of Peru. Contrib. Gray Herb. 95: 26–35. pl. 6–7. 1931.

Geology, climate, etc.; list of vascular plants collected by Dr. R. C. Murphy, 1919–20. Of the several scores of islands apparently only 4 (Lobos de Tierra, San Lorenzo, San Gallan, and Viejas, between 6°26′ and 14°14′ S. Lat.) support any vascular plants.

Málaga Santolalla, Fermín. Flora. *In* his Monografía del departamento de Cajamarca. Bol. Soc. Geog. Lima **20**: 94–107. 1906.

Briefly annotated list of useful and poisonous plants.

Williams, Llewelyn. Woods of northeastern Peru. Pub. Field Mus. Nat. Hist. 377, Bot. Ser. v. 15. 587 p. 18 fig., 2 maps. 1936.

Explorations of author, plant formations, climate, forest products, etc.; systematic list, with brief descriptions of each tree and its wood, and citation of author's collections; tables of anatomical characters; alphabetical list of vernacular names, with botanical equivalents; bibliography. (Departments of Loreto, Amazonas, and San Martín.)

Yacovleff, Eugenio, and Herrera, F. L. El mundo vegetal de los antiguos peruanos. Revista Mus. Nac. Lima **3**: 241–322. 39 fig. 1935; **4**: 29–102. fig. 48–65. 1935. (Reprinted, 102 p. 1935.)

Extensively annotated list of useful plants, the species arranged chronologically according to their first mention in the older writers (beginning with Pizarro, 1533); systematic list, index of botanical names; bibliography.

SURINAM (DUTCH GUIANA)

See also Dutch East Indies (Sirks); French Guiana (Greshoff).

General

Hering, C. J. Overzicht van de cultuurgewassen en boschproducten in verband met nijverheid en handel in de kolonie Suriname, Nederlandsch Guiana, Z. A. xiii, 296, xii p. 24.5 cm. Paramaribo, 1902–03.

Discussion of economic plants, both wild and cultivated, classified by uses; vernacular names. The palms are dealt with in a separate paper: Nuttige palmen der kolonie Suriname. 42, iv p. Paramaribo, 1903.

Pfeiffer, J. P. De houtsoorten van Suriname. 2 v. illus., plates. 22.5 cm. Amsterdam, 1926–27. (K. Ver. Koloniaal instituut te Amsterdam. Mededeeling no. 22. Afd. Handelsmuseum no. 6.)

Part 1, systematic list of trees, describing wood anatomy, with vernacular names and uses; bibliography; part 2, study of technical properties of some of the woods.

Pulle, A. A. An enumeration of the vascular plants known from Surinam. 8, 555 p. 17 pl., map. 24.5 cm. Leiden, 1906.

List with synonymy, citation of exsiccatae, vernacular names, and statement of general range; botanical explorations, statistics of distribution (2,101 species, of which 293 are endemic), phytogeographical remarks; bibliography.

———, ed. Flora of Surinam (Dutch Guyana). v. 1–4, pt. 2. 22.5 cm. Amsterdam, 1932–39. (K. Ver. Koloniaal instituut te Amsterdam. Mededeeling no. 30. Afd. Handelsmuseum no. 11.)

Principal botanical collections; descriptive flora of phanerogams, with general and local distribution, citation of exsiccatae, synonymy, vernacular names. The families are not in systematic order. Not yet completed.—See also Lanjouw, J. Additions to Pulle's Flora of Surinam I. Plants collected by J. Lanjouw in 1933. Recueil Trav. Bot. Néerl. **32**: 215–261. 4 fig. (incl. maps), pl. 2. 1935.

——— Zakflora voor Suriname. Deel 1. 2. uitgave. Bul. Kolon. Mus. Haarlem no. 47. 194 p. 1911.

Keys to the families and genera of wild and cultivated plants; index of vernacular names.

Sack, Johannes. Lijst van eenige planten, die in Suriname als geneeskrachtig worden beschouwd met opgave van de ziekten tegen welke zij worden aangewend. Bul. Inspect. Landb. West-Indië **13**: 1–23. 1908.

Systematically arranged annotated list of medicinal plants, wild and cultivated, with Dutch and "neger-engelsche" names. French résumé by É. de Wildeman in Bul. Sci. Pharm. **16**: 160–164, 204–209. 1909, under title, Plantes médicinales des Guyanes.

Westeroüen van Meeteren, [J. M.] Surinaamsche planten en cultuurgewassen, boomen en houtsoorten. Korte opsomming der tot heden bekende en gebruikt wordende soorten. 52 p. 26 cm. [Haarlem, 1883.]

Briefly annotated alphabetical list of vernacular and scientific names of cultivated plants, native and introduced, also woody plants.

URUGUAY

See also Argentina (Kurtz).

General

Arechavaleta, José. Flora uruguaya. Enumeración y descripción breve de las plantas conocidas hasta hoy y de algunas nuevas que nacen espontáneamente y viven en la República Oriental del Uruguay. vol. 1–4, pt. 3. Anales Mus. Nac. Montevideo v. 3. xxi, 492 p. 1898–1901; v. 5. xlviii, 375 p. illus., pl. 1903–05; v. 6. 502 p. illus., plates. 1906–08; v. 7. 224 p. illus., plates. 1909–11.

Descriptive flora, with occasional vernacular names and uses. Incomplete; reaches Cuscuta (Bentham and Hooker system). Vol. 2, p. xxxvi–xliii, contains biographical sketches of botanists who have collected in the region.

Berro, Mariano B. La vegetación uruguaya. Plantas que se hacen distinguir por alguna propiedad útil ó perjudicial. Anales Mus. Nac. Montevideo **2**: 91–196. 1899.

Bibliography; annotated systematic list of useful and harmful plants, with vernacular names (not indexed).

González, Matías, Coppetti, Víctor, and Lombardo, Atilio. Plantae diaphoricae. Florae uruguayensis. Anales Univ. Uruguay **37**: 3–154 (incl. plates). 1928; 1936 (137): 119–377 (incl. plates). 1936.

Annotated descriptive list of medicinal plants, each illustrated by a plate, with vernacular names and uses.

Herter, Wilhelm. Enumeratio plantarum vascularium sponte nascentium Republica orientali uruguayensi adjectis plantis adventiciis, plantis cultis principalibus, nominibus vernaculis, distributione in Republica, numeris collectionum Gibert et Herter. 191 p. plates (4 col.), map. 19 cm. Montevideo, 1930. (Estudios botánicos en la región uruguaya IV. Florula uruguayensis II.)—Additamenta ad floram uruguayensem. [I]–III. Revista Sudamer. Bot. **2**: 111–128. 2 fig., col. pl. i. 1935; **3**: 146–178. col. pl. iv. 1936; **4**: 179–232. col. pl. vi. 1937.

Botanical explorations; list of vascular plants known from Uruguay, with vernacular names and range by Provinces; 2,998 species including additions.—See also his Plantae uruguayenses novae vel criticae I. Revista Sudamer. Bot. **5**: 13–36. 7 fig. (incl. map). 1937.—Also his Flora ilustrada del Uruguay. 1. lfg. Beih. Repert. Spec. Nov. Fedde 118, no. 1. 16 pl. 1939.

Osten, Cornelius, and Herter, Wilhelm. Plantae uruguayenses. Anales Mus. Nac. Montevideo II, **1**: 325–404. pl. 25–29. 1925; **2**: 103–127, 301–319. 1925–27; **3**: 43–56. 1929. (Reprinted, at least in part.)

Partly annotated list of vascular plants, with synonymy, references, general range, vernacular names, exsiccatae with their localities, index to exsiccatae, keys to genera and species, bibliographies. Intended as a critical revision of the flora of Uruguay; reaches Arundinella (Engler and Prantl system).

Paccard, Ernesto. Lista de algunas plantas medicinales de las Repúblicas Oriental y Argentina. 77 p. col. plates. 22.5 cm. Montevideo, 1905.

Briefly annotated list, arranged alphabetically by vernacular names, with botanical equivalents; list classified by medicinal uses. No index of botanical names.

Sobron, F. C. y. Plantas medicinales de la República del Uruguay y noticias sobre los indígenas. 44 p. 26.5 cm. Madrid, 1874.

Annotated list of native and cultivated species, arranged alphabetically by vernacular names; no index.

Local

Gibert, Ernest. Enumeratio plantarum sponte nascentium agro montevidensi cum synonimis selectis. 146 p. 21 cm. Montevideo, 1873.

Unannotated systematic list, with occasional vernacular names.

Herter, Wilhelm. Apuntes sobre la flora del Palmar de Castillos, Departamento de Rocha, República Oriental del Uruguay. *In* Ostenia. Colección de trabajos botánicos dedicados a Don Cornelio Osten . . . p. 193–204. Montevideo, 1933.

General features of region; unannotated list of cellular and vascular plants.

Herter, Wilhelm. Zur kenntnis der flora von Osturuguay. Beih. Bot. Centralbl. Abt. B, **59**: 245–286. 14 fig. 1939.

Unannotated list of about 1,000 cellular and vascular plants of the departments of Rocha and Treinta y Tres; general features of flora, bibliography.

VENEZUELA

See also Argentina (Kurtz); Colombia (Weddell); West Indies (Grisebach); Lesser Antilles (Boldingh [for Curaçao, Aruba, and Bonaire]).

General

Ernst, Adolf. La exposicion nacional de Venezuela en 1883 . . . 704 p. plan, pl. 32 cm. Caracas, 1884.

Includes (p. 134–286, 361–518) chapters on woods and other useful plants, with annotated lists of species, and including cultivated plants.

Knuth, Reinhard. Initia florae venezuelensis. Beih. Repert. Spec. Nov. Fedde v. 43. 768 p. 1926–28.

List of vascular plants known from Venezuela, with synonymy, vernacular names, and citation of exsiccatae; annotated list of collectors, with brief itineraries and partial bibliography.—For additions, see **Suessenguth, K.** Für Venezuela neue pflanzen der sammlung Vogl. Revista Sudamer. Bot. **1**: 81–86. 1934.—Also **Suessenguth, K.**, and **Beyerle, R.** Ergänzungen zu den "Initia florae venezuelensis" von R. Knuth. Bot. Arch. **39**: 373–381. 1939.

Pittier, H. F. El estado actuel de nuestros conocimientos acerca de la flora de Venezuela. 20 p. 23 cm. Caracas, 1931.

Includes briefly annotated list of collectors; account of the author's own collecting; statistics of flora by families.

——— El estudio de los productos forestales en Venezuela. Bol. Cient. y Técn. Mus. Com. Venezuela **1**: 3–44. 1927.

Includes systematic list of 603 woody plants, with vernacular names (indexed).—See also **Williams, Llewelyn.** Maderas económicas de Venezuela. Bol. Técn. Min. Agr. y Cría (Venezuela) no. 2. 97 p. 24 fig. 1939.

——— Flora venezolana: plantas medicinales. Mem. Cong. Venezol. Med. 4th, [192?] **2**: 230–241. 1925.

"An attempt at a classification of the medicinal plants of Venezuela according to their real or supposed virtues"; vernacular names given. (Not seen; notice by author in Bot. Abstr. **15**: 793. 1926.)

——— Genera plantarum venezuelensium. Clave analítica de los géneros de plantas hasta hoy conocidos en Venezuela. 354 p. 23.5 cm. Caracas, 1939.

Key to the genera of vascular plants.

——— Manual de las plantas usuales de Venezuela. xvi, 458 p. 42 pl. (incl. 6 port.) 23.5 cm. Caracas, 1926. — Suplemento . . . viii, 129 p. 24.5 cm. Caracas, 1939.

Alphabetical list of vernacular names of the commoner and the economic plants, with uses and cross index by scientific names; botanical investigations, plant formations, principal economic plants classified l uses; bibliography. The illustrations are mostly photographs of trees or habitats. The supplement includes a list of woody plants, with vernacular names, and gives the number of vascular and cellular plants known from Venezuela in 1931 as 8,783.

Local

De Verteuil, L. A. A. G. Native medicinal plants. Agr. Record Trinidad **1**: 17–24. 1890.

Annotated list of 65 species, with vernacular names, classified by uses. (Trinidad.)

Ernst, Adolf. Florula chelonesiaca; or, a list of plants collected in January, 1874, in the island Tortuga, Venezuela. Jour. Bot. **14**: 176–179. 1876.

Physiography, etc.; annotated list of 69 vascular and cellular plants, with vernacular names. Tortuga is about 50 miles north of Venezuela and 52 miles northwest of Margarita Island.

——— Plants used medicinally at Caracas, Venezuela, South America, and their vernacular names. Jour. Bot. **3**: 143–150, 277–284, 306–322. 1865.

Annotated list, arranged alphabetically by vernacular names, with botanical equivalents and uses.

Ernst, Adolf. Sertulum naiguatense; notes on a small collection of alpine plants from the summit of Naiguatá, in the mountains of Caracas. Jour. Bot. **10**: 261–264. 1872.

List of 32 cellular and vascular plants collected on April 23, 1872, by James M. Spence.—See also his Enumeracion de las plantas más notables que fueron observadas en la excursion á Naiguatá. Repertorio Caraqueño **1**: 141–146. 1879 (abstract in Bot. Centralbl. 1880(3): 1179), for additional species collected by himself on Aug. 25, 1879.

——— Verzeichniss der auf der venezuelanischen inselgruppe Los Roques im September 1871 beobachteten pflanzen. Bot. Zeit. **30**: 539–541. 1872.

Geology, etc.; annotated list of 26 vascular and cellular plants, with vernacular names. The island group Los Roques lies about 20 miles north of Caracas.

Freeman, W. G., and Williams, R. O. The useful and ornamental plants of Trinidad and Tobago. 2d ed., rev. Mem. Dept. Agr. Trinidad and Tobago no. 4. 192 p. 1928. (1st ed., 1927.)

Alphabetical list of vernacular and scientific names, with cross-references; brief descriptions, uses; lists of genera classified by uses; ornamental plants; systematic list of genera included.

Gleason, H. A. Botanical results of the Tyler-Duida expedition. Bul. Torrey Club **58**: 277–506. pl. 17–45, map. 1931. (Reprinted with addition of index.)

Itinerary (by G. H. H. Tate), geology (by C. B. Hitchcock), plant associations (by Tate); list of vascular plants and Musci, with data and general range, by Gleason and numerous collaborators; table of temperatures and weather. Refers primarily to Mount Duida, in southern Venezuela near the village of Esmeralda. Mount Duida previously had been entirely unknown botanically; 14 genera and nearly 200 species were new.

——— and **Killip, E. P.** The flora of Mount Auyan-tepui, Venezuela. Brittonia **3**: 141–204. 5 fig. 1939.

Topography, geology, general features of flora; annotated list of 244 vascular plants (including 41 new species), collected [in 1937–38] by an expedition from the American Museum of Natural History headed by G. H. H. Tate. Mount Auyan-tepui, the flora of which was previously unknown, is a part of the Roraima-Duida mountain complex, lying about 150 miles northwest of Roraima and 250 miles northeast of Duida.

Jahn, Alfredo. Los páramos venezolanos. Sus aspectos físicos y su vegetación. Bol. Soc. Venezol. Cien. Nat. **1**: 93–127. 1931.

General features of vegetation, list of paramos with altitudes, area, etc., climate; partial list of plants found at 3,500 to 4,000 meters altitude.

Johnston, J. R. Flora of the islands of Margarita and Coche, Venezuela. Proc. Boston Soc. Nat. Hist. **34**: 163–312. pl. 23–30 (incl. 2 maps). 1909. (Contrib. Gray Herb. 37.)

Botanical explorations, physiography, list of vascular plants of Margarita, with synonymy, citation of exsiccatae, and general range; annotated list of useful plants, classified by uses; general features of Coche, with list of vascular plants; phytogeography, bibliography.

Marshall, R. C. Trees of Trinidad and Tobago. 101, viii p. 20 pl 21.5 cm. Trinidad, 1934.

List of wild and cultivated trees, with vernacular names; brief descriptive flora of the native species.

Sandwith, N. Y. Contributions to the flora of tropical America. XXXVII. Notes on the flora of Tobago. Bul. Misc. Inform. Kew **1938**: 353–384. 1938.

General features of flora; annotated partial list of vascular plants collected by the author in 1937, including many additions to the published parts of Williams and Cheesman, Flora of Trinidad and Tobago, and new records for the island in other families.

Williams, R. O. Notes on the plants of Patos. Bul. Misc. Inform. Kew **1924**: 273–280. pl. 1–4. 1924.

Brief sketch of vegetation, with especial reference to cacti, and list of 27 species of flowering plants. Patos is a small island 24 miles from Port-of-Spain and 2¾ miles from the coast of Venezuela.

——— and **Cheesman, E. E.** Flora of Trinidad and Tobago. v. 1, pt. 1–6; v. 2, pt. 1. 21.5 cm. Trinidad, 1928–34.

Descriptive flora with keys, brief synonymy, general range. Not yet completed. E. E. Cheesman is coauthor of v. 2, pt. 1, and of part of v. 1, pt. 2.

ABBREVIATIONS OF PERIODICALS CITED

Abhandl. Gebiet Auslandsk. Hamburg. Univ. Hamburgische universität. Abhandlungen aus dem gebiet der auslandskunde.
Abhandl. K. Akad. Wiss. Berlin. Abhandlungen der Königlichen akademie der wissenschaften zu Berlin.
Abhandl. K. Gesell. Wiss. Göttingen. Abhandlungen der K. gesellschaften der wissenschaften zu Göttingen.
Abhandl. Naturw. Ver. Bremen. Abhandlungen hrsg. vom Naturwissenschaftlicher verein zu Bremen.
Acta Bot. Fenn. Societas pro fauna et flora fennica. Acta botanica fennica.
Acta Horti Gothoburg. Acta horti gothoburgensis. (Meddelanden från Göteborgs botaniska trädgård.)
Acta Soc. Scient. Fenn. Acta Societatis scientiarum fennicae.
Actes Cong. Internatl. Bot. Paris. Actes du 1er Congrès international de botanique tenu à Paris . . . 1900.
Actes Soc. Linn. Bordeaux. Actes de la Société linnéenne de Bordeaux.
Agr. Cir. Dept. Agr. Fiji. Agricultural circular issued by the Department of agriculture, Fiji.
Agr. Gaz. N. S. Wales. Agricultural gazette of New South Wales.
Agr. Notes Agr. Ext. Serv. Univ. Hawaii. University of Hawaii. Agricultural extension service. Agricultural notes.
Agr. Prat. Pays Chauds. L'agriculture pratique des pays chauds.
Agr. Rec. Trinidad. The agricultural record. Official organ of the Central agricultural board of Trinidad.
Agricolt. Colon. Agricoltura coloniale (Istituto agricolo coloniale italiano, Firenze).
Alabama Polytech. Inst. Ext. Serv. Cir. Alabama polytechnic institute. Extension service. Circular.
Almanaque Min. Agr. Rep. Argentina. República argentina. Almanaque del Ministerio de agricultura de la nación.
Amer. Anthropol. American anthropologist.
Amer. Bot. American botanist.
Amer. Fern Jour. American fern journal.
Amer. Jour. Bot. American journal of botany.
Amer. Jour. Forestry. American journal of forestry.
Amer. Jour. Pharm. American journal of pharmacy.
Amer. Jour. Sci. American journal of science.
Amer. Midl. Nat. American midland naturalist.
Amer. Nat. American naturalist.
Ames Forester. The Ames forester.
Anales Agron. Min. Industr. i Obras Púb. (Chile). Ministerio de industria i obras públicas. Anales agronómicos.
Anales Cient. Paraguayos. Anales científicos paraguayos.
Anales Circulo Méd. Argentino. Anales del Circulo médico argentino.
Anales Inst. Biol. México. Anales del Instituto de biología de la Universidad nacional de México.
Anales Inst. Fís.-Geog. Costa Rica. Anales del Instituto físico-geográfico de Costa Rica.
Anales Mus. Nac. Chile. Anales del Museo nacional de Chile.
Anales Mus. Nac. Costa Rica. Anales del Museo nacional. República de Costa Rica. (Later included in Instituto físico-geográfico de Costa Rica.)
Anales Mus. Nac. Hist. Nat. Bern. Rivad. Buenos Aires. Anales del Museo nacional de historia natural Bernardino Rivadavia, Buenos Aires.
Anales Mus. Nac. Hist. Nat. Buenos Aires. Anales del Museo nacional de historia natural. (Later changed to preceding title.)
Anales Mus. Nac. México. Anales del Museo nacional de México.
Anales Mus. Nac. Montevideo. Anales del Museo nacional de Montevideo.
Anales R. Acad. Cien. Habana. Anales de la Real academia de ciencias médicas, físicas y naturales de la Habana.

Anales Soc. Cient. Argentina. Anales de la Sociedad científica argentina.
Anales Soc. Españ. Hist. Nat. Anales de la Sociedad española de historia natural.
Anales Soc. Rural Argentina. Anales de la Sociedad rural argentina.
Anales Univ. Chile. Anales de la Universidad de Chile.
Anales Univ. Uruguay. República oriental del Uruguay. Anales de la Universidad.
Ann. Rpt. Agr. Expt. Sta. North Dakota. Annual repor of the North Dakota agricultural experiment station.
Ann. Rpt. Bot. Off. Brit. Columbia. Annual report of the Botanical office of the province of British Columbia.
Ann. Rpt. Bur. Amer. Ethnol. Annual report of the Bureau of American ethnology.
Ann. Rpt. Colon. Herb. (Natal). Colonial herbarium. Annual report (sometimes Report). *See also* Rpt. Natal Bot. Gard. (Rpt. Colon. Herb.).
Ann. Rpt. Columbus Hort. Soc. Annual report of the Columbus horticultural society . . . including . . . quarterly journal.
Ann. Rpt. Commr. Parks and Boulevards, Detroit. Annual report of the Commissioner of parks and boulevards, city of Detroit.
Ann. Rpt. Commrs. State Reserv. Niagara. Annual report of the Commissioners of the State reservation at Niagara.
Ann. Rpt. Florida State Geol. Survey. Florida state geological survey. Annual report.
Ann. Rpt. Geol. and Nat. Hist. Survey Canada; Ann. Rpt. Geol. Survey Canada. Geological and natural history survey (*later* Geological survey) of Canada. Annual report (*earlier* Report of progress, q. v.)
Ann. Rpt. Geol. and Nat. Hist. Survey Minn. The geological and natural history survey of Minnesota. Annual report.
Ann. Rpt. Geol. Survey Arkansas. Annual report of the Geological survey of Arkansas.
Ann. Rpt. Geol. Survey Indiana; Ann. Rpt. Indiana Dept. Geol. and Nat. Hist. (*or* **Res.**). Annual report of the Geological survey of Indiana. *Later* Indiana. Department of geology and natural history (*later* natural resources). Annual report.
Ann. Rpt. Iowa Geol. Survey. Iowa geological survey. Annual report.
Ann. Rpt. Louisiana State Univ. Annual report of . . . Louisiana state university.
Ann. Rpt. Maine Bd. Agr. Annual report of the secretary of the Maine board of agriculture.
Ann. Rpt. Mich. Acad. Sci. (See Rpt. Mich. Acad. Sci.)
Ann. Rpt. Miss. State Bd. Health. Annual report of the Mississippi state board of health.
Ann. Rpt. Missouri Bot. Gard. Annual report of the Missouri botanical garden.
Ann. Rpt. Nebraska State Bd. Agr. Annual report of the Nebraska state board of agriculture.
Ann. Rpt. New Jersey State Mus. Annual report of the New Jersey state museum.
Ann. Rpt. Ohio Acad. Sci. Annual report of the Ohio academy of science.
Ann. Rpt. Quebec Soc. Protect. Plants. Annual report of the Quebec society for the protection of plants.
Ann. Rpt. Regents Univ. State N. Y. Annual report of the regents of the University of the State of New York.
Ann. Rpt. Smithsn. Inst. Annual report of the Smithsonian institution.
Ann. Rpt. State Bd. Forestry Indiana. Annual report . . . of the State board of forestry [of Indiana].
Ann. Rpt. State Cabinet Nat. Hist. N. Y. Annual report of the State cabinet of natural history of New-York.
Ann. Rpt. State Hort. Soc. Mich. Annual report of the secretary of the State horticultural society of Michigan.
Ann. Rpt. State Hort. Soc. Missouri. Annual report of the State horticultural society of Missouri.
Ann. Rpt. State Mineralogist Calif. Annual report of the State mineralogist [of California].
Annalen K. K. Naturhist. Hofmus. Wien. Annalen des K. K. Naturhistorischen hofmuseums in Wien.

Annalen Wiener Mus. Naturgesch. Annalen des Wiener museums der naturgeschichte.
Annales Bot. Soc. Zool.-Bot. Fenn. Vanamo. Annales botanici Societatis zoologicae-botanicae fennicae Vanamo. (Suomalaisen eläin-ja kasvitieteellisen suran Vanamom Kasvitieteellisia julkaisuja.)
Annales Jard. Bot. Buitenz. Annales du Jardin botanique de Buitenzorg.
Annales Mus. Colon. Marseille. Annales du Musée colonial de Marseille.
Annales Mus. Congo Bot. Annales du Musée du Congo. Botanique.
Annales Sci. Nat. Bot. Annales des sciences naturelles . . . Botanique.
Annales Serv. Bot. Tunisie. Annales du Service botanique de la Direction générale de l'agriculture, du commerce et de la colonisation de la régence de Tunisie.
Annales Soc. Bot. Lyon. Annales de la Société botanique de Lyon.
Annales Soc. Scient. Bruxelles. Annales de la Société scientifique de Bruxelles.
Annali Bot. Annali di botanica.
Annali R. Scuola Sup. Agr. Portici. Annali della Regia scuola superiore di agricoltura in Portici.
Annals Bolus Herb. Annals of the Bolus herbarium.
Annals Bot. Annals of botany.
Annals Bot. Soc. Canada. Annals of the Botanical society of Canada.
Annals Carnegie Mus. (Pittsburgh). Annals of the Carnegie museum.
Annals Missouri Bot. Gard. Annals of the Missouri botanical garden.
Annals N. Y. Acad. Sci. Annals of the New York academy of sciences.
Annals Nat. Hist. Annals of natural history (*later* Annals and magazine . . .).
Annals Natal Govt. Mus.; Annals Natal Mus. Annals of the Natal government museum (*later* Annals of the Natal museum).
Annals Roy. Bot. Gard. Peradeniya. Annals of the Royal botanic gardens, Peradeniya.
Annals So. African Mus. Annals of the South African museum.
Annals Transvaal Mus. Annals of the Transvaal museum.
Annals Univ. Stellenbosch. Annals of the University of Stellenbosch.
Annuaire Conserv. et Jard. Bot. Genève. Annuaire du Conservatoire et du Jardin botaniques de Genève.
Annuario R. Ist. Bot. Roma. Annuario del R. Istituto botanico di Roma.
Anthropol. Records. Anthropological records (University of California).
Appalachia. Appalachia. The journal of the Appalachian mountain club.
Apuntes de historia natural. Apuntes de historia natural. Buenos Aires.
Arb. Bot. Have København. Arbedjer fra den Botaniska have i København.
Arb. Danske Arkt. Sta. Disko. Arbejder fra den Danske Arktiske station paa Disko.
Arb. Deut. Landw.-Gesell. Arbeiten der Deutschen landwirtschafts-gesellschaft.
Arch. Bot. (Forlì). Archivio botanico per la sistematica, fitogeografia e genetica . . . Forlì.
Arch. de Bot. (Caen) Bul. Mens.; Arch. de Bot. (Caen) Mém. Archives de botanique fondées par René Viguier . . . Bulletin mensuel, *and* Mémoires.
Arch. Méd. et Pharm. Mil. Archives de médicine et de pharmacie militaires.
Arch. Méd. Navale. Archives de médicine navale.
Arch. Mus. Nac. Rio de Janeiro. Archivos do Museu nacional do Rio de Janeiro.
Arch. Pharm. Archiv der pharmazie.
Arkiv Bot. Arkiv för botanik. Utgifvet af K. Svenska vetenskapsakademien.
Arq. Univ. Lisboa. Arquivos da Universidade de Lisboa.
Atti Cong. Bot. Internaz. Genova. Atti del Congresso botanico internazionale di Genova.
Atti Cong. Studi Colon. 1., Firenze, 1931. Atti del primo congresso di studi coloniali Firenze, 8–12 aprile 1931.
Atti Ist. Bot. Giov. Briosi R. Univ. Pavia. Atti dell'Istituto botanico "Giovanni Briosi" e Laboratorio crittogamico italiano della R. Università di Pavia.
Atti R. Accad. Sci. Fis. e Mat. (Napoli). Atti della Reale accademia delle scienze fisiche e matematiche.
Atti Soc. Ligust. Sci. Nat. e Geog. Atti della Società ligustica di scienze naturali e geografiche.
Atti Soc. Nat. e Mat. Modena. Atti della Società dei naturalisti e matematici di Modena.
Austral. Nat. The Australian naturalist. Journal and magazine of the Naturalists' society of New South Wales.
Austral. Zool. Australian zoologist.
Bartonia. Bartonia. Proceedings of the Philadelphia botanical club.

Beih. Bot. Centralbl. Beihefte zum Botanischen centralblatt (B. B. C. Beihefte, etc.).
Beih. Repert. Spec. Nov. Fedde. Repertorium specierum novarum regni vegetabilis. Hrsg. von Prof. Dr. Friedrich Fedde. Beihefte.
Beih. Tropenpfl. Beihefte zum Tropenpflanzer.
Ber. Deut. Bot. Gesell. Berichte der Deutschen botanischen gesellschaft.
Ber. Deut. Pharm. Gesell. Berichte der Deutschen pharmaceutischen gesellschaft.
Ber. St. Gall. Naturw. Gesell. Bericht über die thätigkeit der St. Gallischen naturwissenschaftlichen gesellschaft.
Bibl. Contrib. Libr. Harvard Univ. Library of Harvard university. Bibliographical contributions.
Bibliotheca Bot. Bibliotheca botanica.
Bienn. Rpt. Agr. Col. Survey North Dakota. Biennial report of the director of the Agricultural college survey of North Dakota.
Bienn. Rpt. Forestry Comn. New Hampshire. State of New Hampshire. Biennial report of the Forestry commission.
Bihang K. Svenska Vetensk.-Akad. Handl. Bihang till Kongl. Svenska vetenskaps-akademiens handlingar.
Biol. Meddel. K. Danske Vidensk. Selsk. Biologiske meddelelser udgivne af det Kgl. Danske videnskabernes selskab.
Biol. Ser. Catholic Univ. Amer. The Catholic university of America. Biological series. (*Sometimes* Contributions from the Biological laboratory . . .)
Black Hills Engineer. [Published quarterly by the South Dakota state school of mines.]
Black Rock Forest Bul. Black Rock forest bulletin. Cornwall-on-the-Hudson, N. Y.
Blumea. Blumea. Tijdschrift voor de systematik en de geographie der planten. Uitgegeven door het Rijksherbarium te Leiden.
Bol. Acad. Nac. Cien. Córdoba (Rep. Argentina). Boletín de la Academia nacional de ciencias en Córdoba (República argentina).
Bol. Agência Geral Colón. (Lisboa). Boletim da agência geral das colónias.
Bol. Agr. Colombia; Bol. Agr. Colombia Supl. República de Colombia. Ministerio de industrias. Boletín de agricultura *and* Suplemento al Boletín de agricultura.
Bol. Agr. e Pec. Moçambique. Boletim agrícola e pecuário. Colónia de Moçambique.
Bol. Agr. São Paulo. Boletim de agricultura. Secretaria de agricultura, industria e comercio do estado de São Paulo.
Bol. Cient. y Técn. Mus. Com. Venezuela. Boletín científico y técnico del Museo comercial de Venezuela.
Bol. Commis. Geog. e Geol. São Paulo. Boletim da Commisão geographica e geologica de São Paulo.
Bol. Estac. Expt. Agron. Santiago de las Vegas. Estación experimental agronómica, Santiago de las Vegas. Boletín.
Bol. Estac. Expt. Insul. Rio Piedras, P. R. Estación experimental insular Rio Piedras, P. R. Boletín.
Bol. Inspet. Fed. Obras Contra Sêc. (Brasil). Ministério da viação e obras públicas. Inspetoria federal de obras contra as sêcas. Boletim.
Bol. Mens. Mus. Prod. Argent. Boletín mensual del Museo de productos argentinos.
Bol. Min. Agr., Ind. e Comm. (Brasil). Boletim do Ministerio da agricultura, industria e commercio.
Bol. Min. Agr. Rep. Argentina. República argentina. Boletín del Ministerio de agricultura. (*Later* Boletín del Ministerio de agricultura de la nación.)
Bol. Mus. Goeldi Hist. Nat. e Ethnogr.; Bol. Mus. Paraense Hist. Nat. e Ethnogr. Boletim do Museu Goeldi de historia natural e ethnographia. (*Previously* Boletim do Museu paraense . . .).
Bol. Mus. Nac. Chile. Boletín del Museo nacional de Chile.
Bol. Mus. Nac. Rio de Janeiro. Boletim do Museu nacional. Rio de Janeiro.
Bol. Oficina Agr. Ganad. Prov. Buenos Aires. República argentina. Provincia de Buenos Aires. Boletín de la Oficina agrícola ganadera.
Bol. Pro-Cult. Region. S. C. L. Mazatlán, Sinaloa. Boletín de Pro-cultura regional, S. C. L., Mazatlán, Sin.
Bol. R. Orto Bot. e Giard. Colon. Palermo. Bollettino del R. orto botanico e giardino coloniale di Palermo.
Bol. R. Soc. Geog. (Madrid). Boletín de la Real sociedad geográfica.

Bol. Serv. Agr. e Com., Colon. e Flor. Angola; Bol. Serv. Agr. e Com. Angola. República portuguesa. Colónia Angola. Boletim dos Serviços de agricultura e comércio, colonização e florestas. (*Later* Boletim da Direcção dos Serviços de agricultura e comércio.)
Bol. Soc. Africana Italia. Bollettino della Società africana d'Italia.
Bol. Soc. Broteriana. Boletim da Sociedade broteriana.
Bol. Soc. Colomb. Cien. Nat. Boletín de la Sociedad colombiana de ciencias naturales. (*Later* Revista . . . q. v.)
Bol. Soc. Fom. Fabril Santiago (Chile). Boletín de la Sociedad de fomento fabril.
Bol. Soc. Geog. Ital. Bollettino della Società geografica italiana.
Bol. Soc. Geog. Lima. Boletín de la Sociedad geográfica de Lima.
Bol. Soc. Geog. Lisboa. Boletim de Sociedade de geografia de Lisboa.
Bol. Soc. Ibérica Cien. Nat. Boletín de la Sociedad ibérica de ciencias naturales.
Bol. Soc. Venezol. Cien. Nat. Boletín de la Sociedad venezolana de ciencias naturales.
Bol. Técn. Min. Agr. y Cría (Venezuela). Ministerio de agricultura y cría. Boletín técnico.
Bot. Abstr. Botanical abstracts.
Bot. and Zool. See Syokubutu oyobi dôbutu.
Bot. Arch. Botanisches archiv. Zeitschrift für die gesamte botanik und ihre grenzgebiete.
Bot. Bul. Bd. Agr. and Forestry Hawaii. Botanical bulletin . . . of Hawaii. Board of agriculture and forestry.
Bot. Bul. Dept. Agr. and Stock Queensland. Queensland. Department of agriculture and stock. Botany bulletin.
Bot. Centralbl. (Zentralbl.) Botanisches centralblatt, *later* zentralblatt.
Bot. Gaz. Botanical gazette. Chicago. (Vol. 1, Botanical bulletin.)
Bot. Jahrb. Engler. Botanische jahrbücher für systematik, pflanzengeschichte und pflanzengeographie hrsg. von A. Engler.
Bot. Jahresber. Just. Botanischer jahresbericht. (*Later* Just's botanischer jahresbericht.)
Bot. Mag. (Tokyo). Botanical magazine, published by the Tokyo botanical society (*later* Botanical society of Japan).
Bot. Notiser. Botaniska notiser.
Bot. Zeit. Botanische zeitung.
Bothalia. Bothalia. A record of contributions from the National herbarium, Union of South Africa.
Brittonia. Brittonia. A series of botanical papers. Published by the New York botanical garden.
Brotéria Sér. Bot.; Brotéria Cien. Nat. Brotéria série botanica; Brotéria série trimestral: ciências naturais.
Bul. Acad. Hippone. Bulletin de l'Académie d'Hippone. Bône.
Bul. Acad. Internatl. Géog. Bot. Bulletin de l'Académie international de géographie botanique.
Bul. Agence Écon. Colon. Auton. et Terr. Afr. (France). Bulletin de l'Agence économique des colonies autonomes et des territoires africains sous mandat (*formerly* Bulletin de l'agence générale des colonies).
Bul. Agr. Congo Belge. Bulletin agricole du Congo Belge . . . publié par la Direction de l'agriculture.
Bul. Agr. Martinique. Bulletin agricole. Service de l'agriculture de la Martinique.
Bul. Alabama Agr. Expt. Sta. Bulletin. Alabama agricultural experiment station.
Bul. Amer. Geog. Soc. Bulletin of the American geographical society.
Bul. Bern. P. Bishop Mus. Bulletin of the Bernice P. Bishop museum
Bul. Bio-geog. Soc. Japan. Bulletin of the Bio-geographical society of Japan.
Bul. Bot. Dept. Jamaica. Bulletin of the Botanical department, Jamaica.
Bul. Brookville [Ind.] Soc. Nat. Hist. Bulletin of the Brookville society of natural history.
Bul. Buffalo Soc. Nat. Sci. Bulletin of the Buffalo society of natural sciences.
Bul. Bur. Agr. Philippine Isl. The government of the Philippine Islands. Department of the interior. Bureau of agriculture. Bulletin.
Bul. Bur. Amer. Ethnol. Smithsonian institution. Bureau of American ethnology. Bulletin.
Bul. Bur. Forestry Philippine Isl. Department of agriculture and natural resources. Bureau of forestry. Bulletin.
Bul. Bur. Forestry U. S. Dept. Agr. United States department of agriculture. Bureau of forestry. Bulletin.
Bul. Bur. Plant Ind. U. S. Dept. Agr. United States department of agriculture. Bureau of plant industry. Bulletin.

Bul. C. A. Fox Res. and Demonstr. Forest. Caroline A. Fox research and demonstration forest. Bulletin. (Concord, N. H.)
Bul. Calif. Acad. Sci. Bulletin of the California academy of sciences.
Bul. Chicago Acad. Sci. Bulletin of the Chicago academy of sciences.
Bul. Cl. Sci. Acad. Roy. Belg. Académie royale de Belgique. Bulletin de la classe des sciences.
Bul. Col. Hawaii. College of Hawaii publications. Bulletin.
Bul. Colorado Agr. Expt. Sta.; Bul. Colorado Expt. Sta. Bulletin. Colorado agricultural experiment station. (*Later* Colorado experiment station.)
Bul. Conn. Geol. and Nat. Hist. Survey. Connecticut. State geological and natural history survey. Bulletin.
Bul. Conserv. Dept. Conserv. and Survey Div. Univ. Nebraska. University of Nebraska. Conservation and survey division. Conservation department. Bulletin.
Bul. Dept. Agr. and Forestry, Union So. Africa (Plant Ind. Ser.). Bulletin. Union of South Africa. Department of agriculture and forestry (Plant industry series).
Bul. Dept. Agr. Fiji. Department of agriculture, Fiji. Bulletin.
Bul. Dépt. Agr. Indes Néerland. Bulletin du Département de l'agriculture aux Indes Néerlandaises.
Bul. Dept. Scient. and Ind. Res. New Zeal. Department of scientific and industrial research. Bulletin.
Bul. Div. Forestry U. S. Dept. Agr. Bulletin. United States department of agriculture. Division of forestry.
Bul. Dom. Expt. Farms Canada. Dominion of Canada. Department of agriculture. Dominion experimental farms. Bulletin.
Bul. Écon. Madagascar. Madagascar et dépendances. Bulletin économique.
Bul. Essex Inst. Bulletin of the Essex institute.
Bul. Forest Branch Dept. Interior Canada. Department of the interior, Canada. Forest branch. Bulletin.
Bul. Forest Dept. Union So. Africa. Union of South Africa. Forest department. Bulletin.
Bul. Geog. Soc. Phila. Bulletin of the Geographical society of Philadelphia.
Bul. Geol. and Nat. Hist. Survey Minn. Geological and natural history survey of Minnesota. Bulletin.
Bul. Gulf Biol. Sta. Cameron, La. Gulf biologic station, Cameron, La. Bulletin.
Bul. Herb. Boissier. Bulletin de l'Herbier Boissier.
Bul. Illinois State Lab. Nat. Hist.; Bul. Illinois State Nat. Hist. Survey. Bulletin of the Illinois state laboratory of natural history. (*Later* Bulletin of the Illinois state natural history survey.)
Bul. Imp. Inst. Bulletin of the Imperial institute.
Bul. Inspect. Landb. West-Indië. Bulletin. Inspectie van de landbouw in West-Indië (Paramaribo).
Bul. Jard. Bot. Buitenz. Bulletin du Jardin botanique de Buitenzorg.
Bul. Jard. Bot. État Bruxelles. Bulletin du Jardin botanique de l'état à Bruxelles.
Bul. Josselyn Bot. Soc. Maine. Bulletin of the Josselyn botanical society of Maine.
Bul. K. Bayer. Akad. Wiss. Bulletin der K. Bayerische akademie der wissenschaften.
Bul. Kansas Agr. Expt. Sta. Bulletin. Agricultural experiment station. Kansas state college of agriculture.
Bul. Kentucky Agr. Expt. Sta. Kentucky agricultural experiment station. Bulletin.
Bul. Kolon. Mus. Haarlem. Bulletin van het Koloniaal museum te Haarlem.
Bul. Lab. Nat. Hist. Univ. Iowa. Bulletins from the laboratories of natural history of the State university of Iowa. Continued as University of Iowa studies in natural history.
Bul. Libr. Harvard Univ. Bulletin of the Library of Harvard university.
Bul. Lloyd Libr. Bot. Pharm. and Mat. Med.; Bul. Lloyd Libr. Bot. Ser. Bulletin of the Lloyd library of botany, pharmacy and materia medica, *and* Botanical series.
Bul. Misc. Inform. Kew. Bulletin of miscellaneous information. Royal botanic gardens, Kew.
Bul. Miss. State Geol. Survey. Mississippi state geological survey. Bulletin.
Bul. Missouri State Bd. Hort. Missouri state board of horticulture. Bulletin.

Bul. Montana Agr. Expt. Sta. Bulletin. Montana state college of agriculture Experiment station.
Bul. Mus. Hist. Nat. Paris. Bulletin du Muséum (national) d'histoire naturelle.
Bul. Mus. Nat. Hist. Springfield. Museum of natural history, Springfield. Bulletin.
Bul. Mus. North. Arizona. Museum of northern Arizona. Bulletin.
Bul. N. Y. Bot. Gard. Bulletin of the New York botanical garden.
Bul. Nat. Hist. Soc. New Brunsw. Bulletin of the Natural history society of New Brunswick.
Bul. Natl. Mus. Canada Biol. Ser.; Bul. Mus. Natl. Canada Sér. Biol. National museum of Canada. Bulletin. Biological series. (*French ed.* Musée national du Canada. Série biologique.)
Bul. Nebraska Conserv. and Soil Survey. The Nebraska conservation and soil survey. Bulletin.
Bul. New Jersey Agr. Expt. Sta. New Jersey agricultural experiment stations. Bulletin.
Bul. New Mexico Agr. Expt. Sta. Bulletin. New Mexico college . . . Agricultural experiment station.
Bul. North Carolina Agr. Expt. Sta. North Carolina agricultural experiment station. Bulletin.
Bul. Ohio Agr. Exp. Sta. Tech. Ser. Ohio agricultural experiment station. Technical series. Bulletin.
Bul. Ohio Biol. Survey. Ohio biological survey. Bulletin.
Bul. Oklahoma Agr. Expt. Sta. Oklahoma agricultural experiment station. Bulletin.
Bul. Orto Bot. R. Univ. Napoli. Bullettino dell'Orto botanico della R. università di Napoli.
Bul. Penn. Dept. Forestry. Pennsylvania department of forestry. Bulletin. (*Later* Pennsylvania department of forests and waters. Bulletin.)
Bul. Pictou Acad. Scient. Assoc. The bulletin of the Pictou academy scientific association.
Bul. Pub. Mus. Milwaukee. Bulletin of the Public museum of the city of Milwaukee.
Bul. Sci. Pharm. Bulletin des sciences pharmacologiques.
Bul. Scient. Lab. Denison Univ. Bulletin of the scientific laboratories of Denison university (*later* Journal . . .).
Bul. Serv. Tech. Dept. Agr. Haiti. République d'Haïti. Service technique du Départment de l'agriculture. Bulletin.
Bul. So. Calif. Acad. Sci. Bulletin of the Southern California academy of sciences.
Bul. So. Dakota Agr. Expt. Sta. Bulletin. United States experiment station South Dakota (*later* Agricultural experiment station South Dakota . . .).
Bul. Soc. Acclim. France; Bul. Soc. Natl. Acclim. France. Bulletin de la Société (nationale) d'acclimatation de France. (1888-95, Revue des sciences naturelles appliquées. Bulletin mensuel . . .).
Bul. Soc. Amis Sci. Nat. Rouen. Bulletin de la Société des amis des sciences naturelles, Rouen.
Bul. Soc. Bot. France. Bulletin de la Société botanique de France.
Bul. Soc. Bot. Genève. Bulletin de la Société botanique de Genève.
Bul. Soc. Bot. Ital. Bulletino della Società botanica italiana.
Bul. Soc. Géog. Com. Bordeaux. Société de géographie commerciale de Bordeaux. Bulletin.
Bul. Soc. Géog. de l'Est. Bulletin de la Société de géographie de l'Est.
Bul. Soc. Géog. et d'Arch. Oran. Bulletin de la Société de géographie et d'archéologie d'Oran.
Bul. Soc. Hist. Nat. Afrique Nord. Bulletin de la Société d'histoire naturelle de l'Afrique du Nord.
Bul. Soc. Hist. Nat. Autun. Bulletin de la Société d'histoire naturelle. Autun.
Bul. Soc. Hist. Nat. Toulouse. Bulletin de la Société d'histoire naturelle de Toulouse.
Bul. Soc. Linn. Normandie. Bulletin de la Société linnéenne de Normandie.
Bul. Soc. Roy. Bot. Belg. Bulletin de la Société royale de botanique de Belgique.
Bul. Soc. Sci. Nat. Maroc. Bulletin de la Société des sciences naturelles du Maroc.
Bul. State Bd. Forestry Calif. California. State board of forestry. Bulletin.

Bul. State Geol. and Biol. Survey So. Dakota; Bul. So. Dakota Geol. Survey. State geological and biological survey South Dakota. Bulletin, *also* South Dakota geological survey. Bulletin.
Bul. Tech. and Sci. Serv. Min. Agr. Egypt. Ministry of agriculture, Egypt. Technical and scientific service. Bulletin.
Bul. Texas Agr. Expt. Sta. Texas agricultural experiment station. Bulletin.
Bul. Torrey Club. Bulletin of the Torrey botanical club.
Bul. U. S. Dept. Agr. United States department of agriculture. Bulletin.
Bul. U. S. Fish Comn. Bulletin of the United States fish commission.
Bul. U. S. Geol. Survey. Bulletin of the United States geological survey.
Bul. U. S. Natl. Mus. United States national museum. Bulletin.
Bul. Univ. Georgia. Bulletin of the University of Georgia.
Bul. Univ. Montana Biol. Ser. Bulletin, University of Montana. Biological series.
Bul. Univ. New Mexico Biol. Ser. Bulletin University of New Mexico. Biological series.
Bul. Univ. Wisconsin. Bulletin of the University of Wisconsin.
Bul. Vermont Agr. Expt. Sta. Bulletin Vermont agricultural experiment station.
Bul. Vermont Bot. Club. Bulletin. Vermont botanical club.
Bul. West Virginia Agr. Expt. Sta. Bulletin. Agricultural experiment station. College of agriculture, West Virginia university.
Bul. Wild Flow. Preserv. Soc. Bulletin. Wild flower preservation society.
Bul. Wisconsin Nat. Hist. Soc. Bulletin of Wisconsin natural history society.
Bul. Wyoming Agr. Expt. Sta. Wyoming agricultural experiment station. Bulletin. *Later* Bulletin. University of Wyoming. Agricultural experiment station.
Bul. Yale Univ. School Forestry. Yale university. School of forestry. Bulletin.
Bult. Sci. Fak. Terkult. Kjuŝu Imp. Univ. Fukuoka Japan. Bulteno scienca de la Fakultato terkultura Kjuŝu imperia universitato Fukuoka, Japanujo.
Butler Univ. Bot. Studies. Butler university botanical studies.
Canad. Alpine Jour. Canadian alpine journal.
Canad. Field-Nat. Canadian field-naturalist (Ottawa field-naturalists' club).
Canad. Jour. Sci., Lit. and Hist. The Canadian journal of science, literature and history.
Canad. Nat. and Geol.; Canad. Nat. Canadian naturalist and geologist. *Later* Canadian naturalist.
Canad. Rec. Sci. Canadian record of science.
Candollea. Candollea. Organe du Conservatoire et du Jardin botaniques de la ville de Genève.
Cavanillesia. Cavanillesia. Rerum botanicarum acta (Barcinone).
Cir. Oklahoma Geol. Survey. Oklahoma geological survey. Circular.
Cir. So. Dakota Geol. and Nat. Hist. Survey. South Dakota geological and natural history survey. Circular.
Claytonia. Claytonia. Published by the Committee on state flora of the Virginia academy of science.
Collect. Nebraska State Hist. Soc. Collections of the Nebraska state historical society.
Colon. Rpt. Misc. Gt. Brit. Colonial reports, miscellaneous.
Compt. Rend. Acad. Sci. Paris. Comptes rendus hebdomadaires des séances de l'Académie des sciences.
Compt. Rend. Assoc. Franç. Avanc. Sci. Association française pour l'avancement des sciences . . . Compte rendu.
Comun. Mus. Concepción. Comunicaciones del Museo de Concepción.
Contrib. Arnold Arboretum. Contributions from the Arnold arboretum.
Contrib. Biol. Lab. Catholic Univ. Amer. Contribution from the Biological laboratory of the Catholic university of America.
Contrib. Bot. Dept. Iowa State Col. Agr. Contributions from the Botanical department of the Iowa state college of agriculture and mechanic arts.
Contrib. Bot. Lab. Univ. Penn. Contributions from the Botanical laboratory of the University of Pennsylvania.
Contrib. Canad. Biol. Mar. Biol. Sta. Canada. Contributions to Canadian biology, being studies from the Marine biological stations of Canada (Supplement to the Annual report of the Department of marine and fisheries, Fisheries branch). (*Later* Further contributions . . .)

Contrib. Dept. Bot. Columbia Univ. Contributions from the Department of botany of Columbia university. (*Earlier* Contributions from the herbarium of Columbia college.)
Contrib. Dept. Bot. State Col. Washington. Contributions from the Department of botany of the State college of Washington.
Contrib. Dudley Herb. Stanford Univ. Contributions from the Dudley herbarium of Stanford university.
Contrib. Gray Herb. Contributions from the Gray herbarium of Harvard university.
Contrib. Herb. Columbia Col. Contributions from the herbarium of Columbia college. (*Later* Contributions from the Department of botany of Columbia university.)
Contrib. Herb. Frankl. and Marshall Col. Contributions from the herbarium of Franklin and Marshall college, Lancaster, Pa.
Contrib. Lab. Bot. Univ. Montreal. Contributions du Laboratoire de botanique de l'Université de Montreal.
Contrib. N. S. Wales Natl. Herb. Contributions from the New South Wales national herbarium.
Contrib. U. S. Natl. Herb. Contributions from the United States national herbarium.
Cowichan Field Nat. Club Bul. Cowichan field naturalists' club. Bulletin.
Dansk Bot. Arkiv. Dansk botanisk arkiv. Udgivet af Dansk botanisk forening.
Darwiniana. Darwiniana. Carpeta del "Darwinion," laboratorio particular del doctor C. M. Hicken.
Denkschr. Math.-Naturw. Kl. K. Akad. Wiss. Wien. Denkschriften der Kaiserliche akademie der wissenschaften. Mathematisch-wissenschaftliche klasse.
Detroit Lancet.
Ecol. Monog. Ecological monographs.
Ecology. Ecology (*continuation of* Plant world).
Econ. Paper North Carolina Geol. and Econ. Survey. North Carolina geological and economic survey. Economic paper.
Edinb. New Phil. Jour. The Edinburgh new philosophical journal . . .
Erythea. Erythea. A journal of botany, west American and general.
Ext. Bul. Miss. State Col. Extension bulletin. Extension department of Mississippi state college.
Farmers' Bul. U. S. Dept. Agr. United States department of agriculture. Farmers' bulletin.
Field and Lab. So. Methodist Univ. Field and laboratory, contributions from the science departments of Southern Methodist university.
Final Rpt. Geol. Survey New Jersey. Geological survey of New Jersey. Final report of the state geologist.
Flora. Flora; oder allgemeine botanische zeitung. Jena.
Forestry Quart. Forestry quarterly.
Forests Dept. Bul. West. Austral. Forests department bulletin. Western Australia.
Fortschr. Bot. Fortschritte der botanik . . . Berlin.
Furman Bul. The Furman bulletin (running title: Bulletin of Furman university). Greenville, S. C.
Further Contrib. Canad. Biol. Mar. Biol. Sta. Canada. Further contributions to Canadian biology, being studies from the Marine biological station of Canada (Supplement to the Annual report of the Department of marine and fisheries, Fisheries branch). (*Earlier* Contributions . . .)
Gard. Bul. Straits Settlem. The gardens' bulletin, Straits Settlements.
Gard. Chron. Gardeners' chronicle.
Gartenfl. Gartenflora.
Geelong Nat. Geelong naturalist.
Geneesk. Tijdschr. Nederland.-Indië. Geneeskundig tijdschrift voor Nederlandsch-Indië.
Geo: Agr., Colon. y Ramas Anex. Bolivia. Geo: agricultura, colonización y ramas anexas, Bolivia.
Geol. and Agr. State Expt. Sta. [Louisiana]. Geology and agriculture . . . under direction of state experiment stations, Baton Rouge.
Geol. Survey Wisconsin Survey. Geological survey of Wisconsin. Survey of . . .
Göteborgs K. Vetensk.- o. Vitterh.-Samh. Handl. Göteborgs Kungl. vetenskaps- och vitterhets-samhälles handlingar.

Hawaiian Forester and Agr. Hawaiian forester and agriculturist.
Hawaiian Plant. Rec. Hawaiian planters record.
Heil- u. Gewürz-Pflanz. Heil- und gewürz-pflanzen.
Inst. Paper Imp. Forestry Inst. Oxford. Imperial forestry institute. University of Oxford. Institute paper.
Iowa Geol. Survey Bul. Iowa geological survey bulletin.
Iowa Nat. Iowa naturalist.
Iowa State Col. Jour. Sci. Iowa state college journal of science.
Jahrb. Hamburg. Wiss. Anst. Jahrbuch der Hamburgischen wissenschaftlichen anstalten.
Jahresber. Schles. Gesell. Vaterl. Cult. Jahresbericht der Schlesischen gesellschaft für vaterländische cultur.
Jahresber. Ver. Erdk. Dresden. Jahresberichte des Vereins für erdkunde zu Dresden.
Joint Bul. Vermont Bot. and Bird Clubs. Joint bulletin. Vermont botanical and bird clubs.
Jor. Sci. Mat., Fís. e Nat. Lisboa. Jornal de sciências matemáticas, físicas e naturaes publicado da Academia das sciências de Lisboa.
Jour. Amer. Folk-lore. Journal of American folk-lore.
Jour. Amer. Pharm. Assoc. Journal of the American pharmaceutical association.
Jour. and Proc. Hamilton Assoc. Journal and proceedings of the Hamilton association of Nova Scotia.
Jour. and Proc. Mueller Bot. Soc. West. Austr. Journal and proceedings of the Mueller botanic society of Western Australia.
Jour. and Proc. Roy. Soc. N. S. Wales. Journal and proceedings of the Royal society of New South Wales.
Jour. and Proc. Roy. Soc. West. Austr. Journal and proceedings of the Royal society of Western Australia.
Jour. Arnold Arboretum. Journal of the Arnold arboretum.
Jour. Asiatic Soc. Bengal. Journal of the Asiatic society of Bengal.
Jour. Bd. Agr. Brit. Guiana. Journal of the Board of agriculture of British Guiana.
Jour. Bombay Nat. Hist. Soc. Journal of the Bombay natural history society.
Jour. Bot. Journal of botany, British and foreign.
Jour. Bot. Hook. The journal of botany, being a second series of the Botanical miscellany . . . By William Jackson Hooker.
Jour. Cincinnati Soc. Nat. Hist. Journal of the Cincinnati society of natural history.
Jour. de Bot. Journal de botanique.
Jour. de Pharm. et Chim. Journal de pharmacie et de chimie.
Jour. Dept. Agr. Kyushu Imp. Univ. Journal of the Department of agriculture, Kyushu imperial university, Fukuoka, Japan.
Jour. Dept. Agr. Porto Rico. Journal of the Department of agriculture, Porto Rico.
Jour. Dept. Agr. Victoria. Journal of the Department of agriculture of Victoria.
Jour. Dept. Agr. West. Austr. Journal of the Department of agriculture, Western Australia.
Jour. East Africa and Uganda Nat. Hist. Soc. Journal of the East Africa and Uganda natural history society.
Jour. Ecology. Journal of ecology.
Jour. Elisha Mitch. Sci. Soc. Journal of the Elisha Mitchell scientific society.
Jour. Fac. Agr. Hokkaido Imp. Univ. Journal of the Faculty of agriculture, Hokkaido imperial university.
Jour. Fed. Malay States Mus. Journal of the Federated Malay States museums.
Jour. Jap. Bot. Journal of Japanese botany.
Jour. Linn. Soc. Bot. Journal of the Linnean society of London. Botany.
Jour. Malay. Branch Roy. Asiatic Soc. Malayan branch Royal Asiatic society. Journal.
Jour. Polynesian Soc. Journal of the Polynesian society (New Plymouth, N. Z.).
Jour. Scient. Lab. Denison Univ. Journal of the Scientific laboratories of Denison university.
Jour. So. African Bot. The journal of South African botany.
Jour. Soc. Bibl. Nat. Hist. Journal of the Society for the bibliography of natural history. London.

Jour. Soc. Trop. Agr. (Formosa). Journal of the Society of tropical agriculture (Nettai nôgaku kwaishi, Taiwan).
Jour. Straits Branch Roy. Asiatic Soc. Journal of the Straits branch of the Royal Asiatic society.
Jour. Tenn. Acad. Sci. Journal of the Tennessee academy of science.
Jour. Trans. Victoria Inst. (London). Journal of the transactions of the Victoria institute, or Philosophical society of Great Britain.
Jour. Trav. and Nat. Hist. Journal of travel and natural history. Edinburgh.
Jour. Washington Acad. Sci. Journal of the Washington academy of sciences.
Jour. West Austral. Nat. Hist. Soc. Journal of the West Australian natural history society.
K. Danske Vidensk. Selsk. Skr., Naturv. og Math. Afd. Kongelige Danske videnskabernes selskabs skrifter. Naturvidenskabelig og mathematisk afdeling.
K. Svenska Vetensk.-Akad. Handl. K. Svenska vetenskaps-akademien handlingar.
Kansas Univ. Sci. Bul. Kansas university science bulletin (caption title reads: University of Kansas. Science bulletin).
Kudoa. Kudoa (Taihoku, Formosa).
Lav. Ist. Bot. R. Univ. Modena Mem. Lavori dell' Istituto botanico della R. università de Modena. Memorie.
Lav. R. Ist. Bot. Catania. Lavori eseguiti presso il R. istituto botanico di Catania.
Lavoura. A lavoura. Boletim da Sociedad nacional de agricultura. Rio de Janeiro.
Leaflets West. Bot. Leaflets of western botany. San Francisco.
Lilloa. Lilloa. Revista de botanica (Universidad nacional de Tucumán. Instituto Miguel Lillo).
Linnaea. Linnaea. Ein journal für die botanik in ihrem ganzen umfange.
Lloydia. Lloydia, a quarterly journal of biological science. Cincinnati, Ohio.
London Jour. Bot. The London journal of botany.
Lunds Univ. Årsskr. Lunds universitets årsskrift. (Acta universitatis lundensis.)
Madroño. Madroño. A west American journal of botany.
Maine Bul. Maine bulletin.
Maine Nat. The Maine naturalist, journal of the Knox academy of arts and sciences . . .
Malpighia. Malpighia, rassegna mensile di botanica.
Mazama. Mazama. Published by the Mazamas, Portland, Oregon.
Meddel. Göteborgs Bot. Trädgård. *See* Acta Horti Gothoburg.
Meddel. om Grønland. Meddelelser om Grønland udgivne af Kommissionen for ledelsen af de geologiske og geografiske undersøgelser i Grønland.
Meded. Dept. Landb. [Nederland.-Indië]. Mededeelingen uitgaande de Department van landbouw Batavia.
Meded. Kolon. Inst. Amsterdam Afd. Handelsmus. Koloniaal instituut te Amsterdam. Mededeeling. Afdeeling handelsmuseum.
Meded. Lands Plant. Buitenz. Mededeelingen uit 's Lands plantentuin Buitenzorg.
Meded. Rijks Herb. Leiden. Mededeelingen van 's Rijks herbarium Leiden.
Mem. Acad. Cien. y Artes Barcelona. Memorias de la (Real) Academia de ciencias y artes de Barcelona.
Mém. Acad. Malgache. Mémoires de l'Académie malgache.
Mém. Acad. Roy. Belge Cl. Sci. Collect. in-4°. Académie royale de Belgique. Classe des sciences. Mémoires. Collection in-4°.
Mém. Acad. Sci. Inst. France. Mémoires de l'Académie des sciences de l'Institut de France.
Mém. Acad. Sci. St. Pétersb. VI. Sci. Math., Phys. et Nat. Mémoires de l'Académie des sciences de St. Pétersbourg. 6. sér., Sciences mathématiques, physiques et naturelles.
Mem. Amer. Anthropol. Assoc. Memoirs of the American anthropological association.
Mem. Austral. Mus. Sydney. Australian museum, Sydney. Memoir.
Mem. Bot. Survey So. Africa. Botanical survey of South Africa. Memoir.
Mem. Brooklyn Bot. Gard. Memoirs of the Brooklyn botanic garden.
Mem. Canada Geol. Survey Biol. Ser. Canada. Department of mines. Geological survey. Memoir. Biological series (title varies).

Mem. Carnegie Mus. (**Pittsburgh**). Memoirs of the Carnegie museum. Pittsburgh.
Mém. Cl. Sci. Acad. Imp. Sci., Belles-Lettr. et Arts Lyon. Mémoires de l'Académie impériale des sciences, belles-lettres et arts de Lyon. Classe des sciences.
Mem. Cong. Venezol. Med. Memoria del IV Congreso venezolana de medicinas.
Mem. Cornell Agr. Expt. Sta. Memoir. Cornell university. Agricultural experiment station.
Mem. Dept. Agr. Trinidad and Tobago. Memoirs of the Department of agriculture, Trinidad and Tobago.
Mém. Herb Boissier. Mémoires de l'Herbier Boissier.
Mém. Inst. Égyptien; Mém. Inst. Égypte. Mémoires de l'Institut égyptien. (*Later* Mémoires à l'Institut d'Égypte.)
Mem. Lit. and Phil. Soc. Manchester; Mem. Manchester Lit. and Phil. Soc. Memoirs of the literary and philosophical society of Manchester. (*Later* Memoirs of the Manchester literary and philosophical society.)
Mem. N. Y. Bot. Gard. Memoirs of the New York botanical garden.
Mem. Natl. Acad. Sci. Memoirs of the National academy of sciences. Washington, D. C.
Mem. R. Accad. Sci., Lett. ed Arti Modena. Memorie della Regia accademia di scienze, lettere ed arti in Modena.
Mem. R. Soc. Españ. Hist. Nat. Memorias de la Real sociedad española de historia natural.
Mém. Sect. Sci. Nat. et Méd. Inst. Roy. Colon. Belge Collect. in-8°. Institut royal colonial belge. Section des sciences naturelles et médicales. Mémoires. Collection in-8°.
Mém. Soc. Bot. France. Mémoires publiées par la Société botanique de France.
Mem. Soc. Cubana Hist. Nat. "Felipe Poey." Memorias de la Sociedad cubana de historia natural "Felipe Poey."
Mém. Soc. Hist. Nat. Afrique Nord. Mémoires de la Société d'histoire naturelle de l'Afrique du Nord.
Mém. Soc. Imp. (Natl.) Sci. Nat. Cherbourg. Mémoires de la Société impériale [*later* nationale] des sciences naturelles de Cherbourg.
Mém. Soc. Linn. Normandie. Mémoires de la Société linnéenne de Normandie.
Mém. Soc. Sci. Nat. Maroc. Mémoires de la Société des sciences naturelles du Maroc.
Mem. Torrey Club. Memoirs of the Torrey botanical club.
Merck's Report. Merck's report: a practical journal of pharmacy.
México antiguo. El México antiguo.
Minnesota Bot. Studies. Minnesota botanical studies (Geological and natural history survey. Reports of the survey . . . Botanical series).
Misc. Cir. U. S. Dept. Agr. United States department of agriculture. Miscellaneous circular.
Misc. Pub. Dept. Agr. N. S. Wales. Department of agriculture, Sydney, New South Wales. Miscellaneous publication.
Misc. Pub. Mus. Zool. Univ. Mich. University of Michigan. Museum of zoology. Miscellaneous publications.
Misc. Pub. U. S. Dept. Agr. United States department of agriculture. Miscellaneous publication.
Missouri Bot. Gard. Bul. Missouri botanical garden bulletin.
Mitt. Deut. Schutzgeb. Wiss. Beih. Deut. Kolonialbl.; . . . Ergänzungsh. Mitteilungen aus den Deutschen schutzgebieten. Wissenschaftliche beihefte zum Deutschen kolonialblatte; . . . Ergänzungshefte.
Mitt. Forschungsreis. u. Gelehr. Deut. Schutzgeb. Mittheilungen von forschungsreisenden und gelehrten aus den Deutschen schutzgebieten. *Later* Mitteilungen aus den Deutschen schutzgebieten . .
Mitt. Geog. Gesell. Hamburg. Mitteilungen der Geographischen gesellschaft in Hamburg.
Mitt. Inst. Allg. Bot. Hamburg. Mitteilungen aus dem Institut für allgemeine botanik in Hamburg.
Mitt. Seminars Orient. Sprach. Univ. Berlin. Mittheilungen des Seminars für orientalische sprachen an der Königlichen Friedrich Wilhelms-Universität zu Berlin.
Mitt. Ver. Erdk. Dresden. Mitteilungen des Vereins für erdkunde zu Dresden.
Monde des Plant. Le monde des plantes.

Monog. e Rap. Colon. Min. Colon. (Italy). Ministero delle colonie. Direzione centrale degli affari coloniali. Ufficio di studi coloniali. Monografie e rapporti coloniali.
Montana Agr. Col. Sci. Studies. Montana agricultural college science studies.
Muhlenbergia. Muhlenbergia. An illustrated monthly journal of botany. Reno, Nevada [etc.].
Nat. and Appl. Sci. Bul. Univ. Philippines. University of the Philippines. Natural and applied science bulletin.
Nat. Hist. Bul. Grand Canyon Nat. Hist. Assoc. Natural history bulletin. Grand Canyon natural history association.
Naturaliste. Le naturaliste; revue illustreé des sciences naturelles.
Naturaliste Canad. Le naturaliste canadien.
Natuurk. Tijdschr. Nederland.-Indië. Natuurkundige tijdschrift voor Nederlandsch-Indië.
New Orleans Med. and Surg. Jour. New Orleans medical and surgical journal.
New York Jour. Med. The New York journal of medicine.
New York State Mus. Bul. (Rpt. State Bot.) New York state museum bulletin (Report of the botanist).
New York State Mus. Handbook. New York state museum handbook.
New Zeal. Jour. Sci. and Technol. New Zealand journal of science and technology.
North Amer. Fauna. United States department of agriculture. Bureau of biological survey. North American fauna.
Northwest science. Northwest science. Published by the Northwest scientific association, Cheney, Washington.
Notizbl. K. Bot. Gart. Berlin. Notizblatt des Königl. botanischen gartens und museums zu Berlin. (*Later* Notizblatt des botanischen gartens . . .)
Notulae Syst. Mus. Hist. Nat. Paris. Notulae systematicae, Muséum national d'histoire naturelle, Paris.
Nouv. Annales Mus. Hist. Nat. Paris. Nouvelles annales du Muséum d'histoire naturelle, Paris.
Nouv. Arch. Miss. Scient. et Litt. (France). Nouvelles archives des missions scientifiques et littéraires.
Nouv. Arch. Mus. Hist. Nat. Paris. Nouvelles archives du Muséum d'histoire naturelle.
Nova Acta R. Soc. Scient. Upsal. Nova acta Regiae societatis scientiarum upsaliensis.
Nuovo Gior. Bot. Ital. Nuovo giornale botanico italiano.
Nyt Mag. Naturvidensk. Nyt magazin for naturvidenskaberne. Oslo.
Occas. Paper South. Forest Expt. Sta. Occasional paper. Southern forest experiment station.
Occas. Papers Bern. P. Bishop Mus. Bernice P. Bishop museum. Occasional papers. (Vol. 1-8 have two pagings; at top of page for individual number, at bottom of page for whole volume.)
Occas. Papers Riverside Junior Col. Occasional papers of Riverside junior college.
Occas. Papers Univ. Hawaii. University of Hawaii. Occasional papers.
Öfvers. K. Vetensk.-Akad. Förhandl. Öfversigt af Kongl. vetenskaps-akademiens förhandlingar. Stockholm.
Oesterr. Bot. Zeitschr. Oesterreichische botanische zeitschrift.
Official Year Book Union So. Africa. Official year book of the Union of South Africa and of Basutoland, Bechuanaland Protectorate, and Swaziland.
Oglethorpe Med. and Surg. Journ. Oglethorpe medical and surgical journal.
Ohio Jour. Sci. Ohio journal of science.
Ohio Med. and Surg. Jour. Ohio medical and surgical journal.
Ohio Nat. The Ohio naturalist.
Oregon out of doors. Published by the Mazamas, Portland, Oregon.
Ottawa Nat. Ottawa naturalist.
Papers and Proc. Roy. Soc. Tasmania. Papers and proceedings of the Royal society of Tasmania.
Papers Dept. Mar. Biol. Carnegie Inst. Washington. Papers from the Department of marine biology of the Carnegie institution of Washington.
Papers Mich. Acad. Sci. Papers of the Michigan academy of science, arts and letters.
Pharm. Arch. Pharmaceutical archives.
Pharm. Jour. and Trans. Pharmaceutical journal and transactions.
Pharm. Rev. Pharmaceutical review. Continuation of Pharmaceutische Rundschau.

Pharm. Rundschau (N. Y.). Pharmaceutische rundschau; eine monatschrift für die wissenschaftlichen und gewerblichen interessen der pharmacie . . . in den Vereinigten Staaten.
Pharm. Weekbl. Nederl. Pharmaceutisch weekblad voor Nederland.
Phil. Trans. Roy. Soc. Philosophical transactions of the Royal society.
Philippine Agr. Rev. The Philippine agricultural review.
Philippine Jour. Sci. Philippine journal of science.
Physis. Physis. Revista de la Sociedad argentina de ciencias naturales.
Pittonia. Pittonia. A series of papers relating to botany. By E. L. Greene.
Plant World. Plant world; a monthly journal of botany. Binghamton, N. Y. [etc.].
Pop. Bul. Bur. Sci. Philippine Isl. The government of the Philippine Islands. Department of agriculture and natural resources. Bureau of science, Manila. Popular bulletin.
Postelsia. Postelsia. The yearbook of the Minnesota seaside station.
Prilozh. Trudy Prikl. Bot., Gen. i Selek. (Suppl. Bull. Appl. Bot., Gen. and Plant-Breed. Leningr.) Prilozhenie k Trudam po prikladnoĭ botanike, genetike i selektsii (Supplement to the Bulletin of applied botany, of genetics and plant-breeding). Leningrad.
Proc. Acad. Nat. Sci. Phila. Proceedings of the Academy of natural sciences of Philadelphia.
Proc. Amer. Acad. Arts and Sci. Proceedings of the American academy of arts and sciences.
Proc. Amer. Pharm. Assoc. Proceedings of the American pharmaceutical association.
Proc. Amer. Phil. Soc. Proceedings of the American philosophical society.
Proc. and Trans. Nova Scot. Inst. Sci. Proceedings and transactions of the Nova Scotian institute of natural science (*after* v. 7, of science).
Proc. and Trans. Roy. Soc. Canada. Proceedings and transactions of the Royal society of Canada.
Proc. Ann Arbor Scient. Assoc. Proceedings of the Ann Arbor scientific association.
Proc. Asiatic Soc. Bengal. Proceedings of the Asiatic society of Bengal.
Proc. Biol. Soc. Washington. Proceedings of the Biological society of Washington.
Proc. Boston Soc. Nat. Hist. Proceedings of the Boston society of natural history.
Proc. Calif. Acad. Sci. Proceedings of the California academy of sciences.
Proc. Cambridge Phil. Soc. Proceedings of the Cambridge philosophical society.
Proc. Davenport Acad. Sci. Proceedings of the Davenport academy of natural sciences.
Proc. Delaware Co. Inst. Sci. Proceedings of the Delaware county institute of science.
Proc. Elliott Soc. Nat. Hist. Charleston, S. C. Proceedings of the Elliott society of natural history of Charleston, South-Carolina.
Proc. Hawaiian Acad. Sci. (Spec. Pub. Bern. P. Bishop Mus.). Proceedings Hawaiian academy of science. Bernice P. Bishop museum special publication.
Proc. Indiana Acad. Sci. Proceedings of the Indiana academy of science.
Proc. Iowa Acad. Sci. Proceedings of the Iowa academy of science.
Proc. Iowa State Med. and Chirurg. Soc. Proceedings of the Iowa state medical and chirurgical society.
Proc. Kentucky State Med. Soc. Proceedings of the Kentucky state medical society.
Proc. Linn. Soc. N. S. Wales. Proceedings of the Linnean society of New South Wales.
Proc. Manchester Inst. Arts and Sci. Proceedings of the Manchester institute of arts and sciences (New Hampshire).
Proc. Med. Assoc. State Alabama. Proceedings of the Medical association of the state of Alabama.
Proc. Med. Conv. Ohio. Proceedings of the Medical convention of Ohio.
Proc. Mich. Pharm. Assoc. Proceedings of the Michigan pharmaceutical association.
Proc. Nat. Hist. Assoc. Miramichi. Proceedings of the Natural history association. Miramichi, Chatham, New Brunswick.
Proc. Nat. Hist. Soc. Wisconsin. Proceedings of the Natural history society of Wisconsin.
Proc. Newport Nat. Hist. Soc. Proceedings of the Newport natural history society. Newport, R. I.

Proc. Ohio Acad. Sci. Ohio academy of science. Proceedings.
Proc. Oklahoma Acad. Sci. Proceedings of the Oklahoma academy of science.
Proc. Pan-Pacific Scient. Conf. Proceedings of the first Pan-Pacific scientific conference (Bernice P. Bishop museum special publication).
Proc. Portland Soc. Nat. Hist. Proceedings of the Portland society of natural history. Portland, Maine.
Proc. Rhodesia Scient. Assoc. Proceedings of the Rhodesia scientific association.
Proc. Rochester Acad. Sci. Proceedings of the Rochester academy of science. Rochester, N. Y.
Proc. Roy. Soc. Queensland. Proceedings of the Royal society of Queensland.
Proc. Roy. Soc. Victoria. Proceedings of the Royal society of Victoria (Transactions and proceedings 1861–87).
Proc. U. S. Natl. Mus. Proceedings of the United States national museum.
Proc. Washington Acad. Sci. Proceedings of the Washington academy of sciences. Washington, D. C.
Proc. West Virginia Acad. Sci. The West Virginia academy of science. Proceedings.
Prog. Rei Bot. Progressus rei botanicae.
[Pub.] Bur. Govt. Lab. Philippine Isl. Department of the interior. Bureau of government laboratories. [Publications.]
Pub. Com. Études Hist. et Scient. Afrique Occid. Franç. Publications du Comité d'études historiques et scientifiques de l'Afrique occidentale française.
Pub. Com. Honor. Reducc. Indios. Ministerio del interior. Comisión honoraria de reducciones de Indios. Publicación. (Buenos Aires.)
Pub. Dept. Conserv. Indiana. State of Indiana. Department of conservation Publication.
Pub. Field Columbian Mus. Bot. Ser.; Pub. Field Mus. Nat. Hist. Bot. Ser. Publication of the Field Columbian museum, *later* Field museum of natural history. Publication. Botanical series.
Pub. Inspect. Fed. Obras Contra Secc. (Brasil). Publicação. Ministerio da viação e obras publicas. Inspectoria federal de obras contra as seccas.
Pub. Inst. Invest. Geog. Fac. Filos. y Letras Univ. Buenos Aires. Facultad de filosofía y letras de la Universidad nacional de Buenos Aires. Publicaciones del Instituto de investigaciones geográficas.
Pub. Inst. Natl. Étude Agron. Congo Belge. Publications de l'Institut national pour l'étude agronomique du Congo Belge.
Pub. Mich. Geol. and Biol. Survey (Biol. Ser.). Michigan geological and biological survey. Publication. Biological series.
Pub. Middle Amer. Res. Ser. Tulane Univ. Louisiana. The Tulane university of Louisiana. Middle American research series.
Pub. Mus. Agr. Soc. Rural Argentina. Museo agrícola de la Sociedad rural argentina. Publicación.
[Pub.] Mus. Cien. Nat. Univ. Nac. Tucumán. Universidad nacional de Tucumán. Museo de ciencias naturales (*later* Museo de historia natural).
Pub. Oklahoma Div. Forestry. Division of forestry. Oklahoma planning and resources board. Publication.
Pub. Sierra Club. Publication of the Sierra Club.
Pub. Univ. Oklahoma Biol. Survey. Publications of the University of Oklahoma. Biological survey.
Pub. West Texas Hist. and Sci. Soc. West Texas historical and scientific society. Publications.
Quart. Oregon Hist. Soc. Quarterly of the Oregon historical society.
Queensland Agr. Jour. Queensland agricultural journal.
Queensland Nat. Queensland naturalist.
Rassegna Econ. Colon. (Italy). Rassegna economica delle colonie.
Rec. Albany Mus. Records of the Albany museum, Grahamstown, South Africa.
Rec. Auckland Inst. and Mus. Records of the Auckland institute and museum.
Recueil Trav. Bot. Néerland. Recueil des travaux botaniques néerlandais.
Relaz. e Monog. Agr.-Colon. Ist. Agr. Colon. Ital. Relazioni e monografie agrario-coloniali. Istituto agricolo coloniale italiano. Firenze.
Rend. Sem. Fac. Sci. R. Univ. Cagliari. Rendiconti del Seminario della facoltà di scienze della R. Università di Cagliari.
Repert. Spec. Nov. Fedde. Repertorium specierum novarum regni vegetabilis hrsg. von F. Fedde.
Res. Pub. Univ. Hawaii. University of Hawaii. Research publication.
Res. Studies State Col. Washington. Research studies of the State college of Washington.
Revista Acad. Colomb. Cien. Exact., Fís. y Nat. Revista de la Academia vclombiana de ciencias exactas, físicas y naturales.
Reiosta Agr. (Guatemala). Revista agrícola.

Revista Agr. Puerto Rico. Revista de agricultura de Puerto Rico.
Revista Agron. Revista agronómica. (Not seen.)
Revista Argentina Agron. Revista argentina de agronomía.
Revista Bibl. Chilena. Revista de bibliografía chilena.
Revista Centro Sci., Letr. e Art. Campinas. Revista do Centro de sciencias, letras e artes de Campinas.
Revista Chilena Hist. Nat. Revista chilena de historia natural.
Revista Fac. Agron. Univ. La Plata; Revista Fac. Agron. y Vet. La Plata. Revista de la Facultad de agronomía. Universidad nacional de La Plata. *Earlier* Revista de la Facultad de agronomía y veterinaria.
Revista Fac. Agron. y Vet. Univ. Buenos Aires. Universidad de Buenos Aires. Revista de la Facultad de agronomía y veterinaria.
Revista Farm. (Buenos Aires.) Revista farmaceutica.
Revista Flora Med. (Rio de Janeiro.) Revista da flora medica.
Revista Inst. Paraguayo. Revista del Instituto paraguayo.
Revista Jard. Bot. y Mus. Hist. Nat. Paraguay. Revista del Jardin botánico y Museo de historia natural del Paraguay.
Revista Mus. La Plata. Revista del Museo de La Plata. Secc. Oficial.
Revista Mus. Nac. Lima. Revista del Museo nacional. Lima, Perú.
Revista Mus. Paulista. Revista do Museu Paulista.
Revista Mus. Prov. Cien. Nat. Córdoba. (Not seen.)
Revista Soc. Colomb. Cien. Nat. Revista de la Sociedad colombiana de ciencias naturales (*earlier* Boletín . . . q. v.)
Revista Sudamer. Bot. Revista sudamericana de botánica.
Revista Univ. Nac. Córdoba. Revista de la Universidad nacional de Córdoba.
Revista Universitaria Cuzco. Revista universitaria. Organo de la Universidad nacional del Cuzco.
Revista Universitaria (Santiago de Chile). Revista universitaria. Universidad católica. Santiago de Chile.
Revue Bot. Appl. et Agr. Trop. Revue de botanique appliquée et d'agriculture tropicale.
Revue Bot. (Toulouse). Revue de botanique; bulletin mensuel de la Société française de botanique. Toulouse.
Revue Cult. Colon. Revue des cultures coloniales.
Revue de Madagascar.
Revue Gén. Bot. Revue général de botanique.
Revue Géog. Revue de géographie.
Revue Hort. Algérie. Revue horticole de l'Algérie.
Revue Sci. Nat. Appl. Soc. Natl. Acclim. France. Revue des sciences naturelles appliquées. Bulletin mensuel de la Société nationale d'acclimatation de France.
Revue Tunis. Inst. Carthage. Revue tunisienne. Organe de l'Institut de Carthage.
Revue Zool. et Bot. Afric. Revue de zoologie et de botanique africaines.
Rhodora. Rhodora. Journal of the New England botanical club.
Riv. Ital. Essen., Profumi e Piante Off. Rivista italiana delle essenze, dei profumi e delle piante officinali.
Rpt. Austral. and New Zeal. Assoc. Adv. Sci.; Rpt. Australas. Assoc. Adv. Sci. Report of the Australian and New Zealand association for the advancement of science (*formerly* Australasian association).
Rpt. Bd. Educ. Conn. Report of the Board of education of the state of Connecticut.
Rpt. Bot. Survey Nebraska. University of Nebraska. Botanical survey of Nebraska. Report.
Rpt. Col. Agr. and Mech. Arts New Hampshire. Report of the Board of trustees of the College of agriculture and mechanic arts to the New Hampshire legislature.
Rpt. Colon. Herb. (Natal). *See* Ann. Rpt. Colon. Herb. (Natal).
Rpt. Forestry Comn. New Hampshire. Report of the Forestry commission of New Hampshire.
Rpt. Geol. and Nat. Hist. Survey Minn. Report on the Geological and natural history survey of Minnesota.
Rpt. Geol. Survey Ohio. Report of the Geological survey of Ohio.
Rpt. Govt. Bur. Microbiol. N. S. Wales. New South Wales. Report of the Government bureau of microbiology.
Rpt. Great Barrier Reef Comm. Reports of the Great Barrier reef committee.
Rpt. Iowa State Hort. Soc. Report of the Iowa state horticultural society (cover and running title: Transactions).

Rpt. Kansas State Bd. Agr. Report of the Kansas state board of agriculture.

Rpt. (Ann. Rpt.) Mich. Acad. Sci. Report of the Michigan academy of science (*sometimes* Annual report).

Rpt. Missouri State Hort. Soc. Report of the Missouri state horticultural society.

Rpt. Natal Bot. Gard. (Rpt. Colon. Herb.). Report on Natal botanic gardens Colonial herbarium. Report. *See also* Ann. Rpt. Colon. Herb. (Natal).

Rpt. Prog. Geol. Survey Canada. Geological survey of Canada. Report of progress (*later* Annual report, q. v.).

Rpt. Prog. Geol. Survey Kentucky. Geological survey of Kentucky. Reports of progress.

Rpt. Prog. 2d Geol. Survey Penn. Second geological survey of Pennsylvania. Report of progress.

Rpt. Prov. Mus. Nat. Hist. Brit. Columbia. Report of the Provincial museum of natural history (British Columbia).

Rpt. So. African Assoc. Adv. Sci. Report of the South African association for the advancement of science. (*Later* South African journal of science.)

Rpt. State Bd. Geol. Survey Mich. Report of the State board of geological survey of Michigan.

Rpt. Transvaal Dept. Agr. Report of the Transvaal department of agriculture.

Rpt. U. S. Commr. Agr. Report of the commissioner of agriculture. Washington, D. C.

[Rpt.] West Virginia Geol. Survey. West Virginia geological survey. [Report.]

Sarawak Mus. Jour. Sarawak museum journal.

Schweiz. Apoth. Zeit. Schweizerische apotheker-zeitung. Journal suisse de pharmacie.

Sci. Bul. Mus. Brooklyn Inst. Arts and Sci. The museum of the Brooklyn institute of arts and sciences. Science bulletin.

Scient. Contrib. Trop. Plant Res. Found. Tropical plant research foundation. Scientific contributions.

Sitzungsber. K. Preuss. Akad. Wiss. Sitzungsberichte der Königlich preussischen akademie der wissenschaften.

Sitzungsber. Math.-Naturw. Cl. K. Akad. Wiss. Wien. Akademie der wissenschaften in Wien. Mathematisch-naturwissenschaftliche klasse. Sitzungsberichte.

Sitzungs-Ber. Naturw. Gesell. Isis Dresden. Sitzungs-berichte der naturwissenschaftlichen gesellschaft Isis in Dresden.

Skr. Svalbard og Ishavet (Oslo). Skrifter om Svalbard og Ishavet.

Skr. Vidensk.-Selsk. Christiania I, Math.-Naturv. Kl. Skrifter udgivne af Videnskabs-selskabet i Christiania I, Mathematisk-naturvidenskabelig klasse (*later* Norske videnskaps akademi).

Smithsn. Misc. Collect. Smithsonian miscellaneous collections.

So. African Jour. Nat. Hist. South African journal of natural history.

So. African Jour. Sci. South African journal of science.

So. Austral. Nat. The South Australian naturalist.

South. Agr. Southern agriculturist.

South. Jour. Med. and Pharm. Southern journal of medicine and pharmacy.

South. Woodlands. Southern woodlands.

Spec. Papers Ohio State Acad. Sci. Ohio state academy of science. Special papers (later are part of Proceedings).

Spec. Pub. Amer. Geog. Soc. American geographical society. Special publication.

Spec. Pub. Bern. P. Bishop Mus. Bernice P. Bishop museum. Special publication.

St. Paul Med. Jour. St. Paul medical journal.

Sul Ross State Teach. Col. Bul. Sul Ross state teachers college bulletin. Alpine, Texas.

Svensk Bot. Tidskr. Svensk botanisk tidskrift.

Syokubutu oyobi dôbutu (Botany and Zoology).

Taxon. Index. The taxonomic index. Issued by the New York botanical garden for the American society of plant taxonomists.

Tech. Bul. Dept. Agr. and Comm. Philippines. Commonwealth of the Philippines. Department of agriculture and commerce. Technical bulletin.

Tech. Bul. Kansas Agr. Expt. Sta. Technical bulletin. Agricultural experiment station. Kansas agricultural college.

Tech. Pub. New York State Col. Forestry. Technical publication of the New York state college of forestry.
Technologist. The technologist . . . ed. by P. L. Simmonds.
Teysmannia.
Therapeutic Gaz. (Detroit). The therapeutic gazette. A monthly journal devoted to therapeutics . . .
Tijdschr. Indische Taal-, Land- en Volkenk. Tijdschrift voor Indische taal-, land-, en volkenkunde (Batavia).
Timehri. Timehri. The journal of the Royal agricultural and commercial society of British Guiana.
Torreya.
Trab. Inst. Bot. y Farm. Fac. Cien. Méd. Buenos Aires; Trab. Mus. Farm. Fac. Cien. Méd. Buenos Aires. Trabajos del Instituto de botánica y farmacología . . . (Facultad de ciencias médicas de Buenos Aires; *earlier* Trabajos del Museo de farmacología . . .)
Trab. Mus. Nac. Cien. Nat. Madrid Ser. Bot.; Trab. Mus. Nac. Cien. Nat. y Jard. Bot. Madrid Ser. Bot. Trabajos del Museo nacional de ciencias naturales (*later* y jardin botánico). Serie botánica.
Trans. Acad. Sci. St. Louis. Transactions of the Academy of science of St. Louis.
Trans. Albany Inst. Transactions of the Albany institute.
Trans. Amer. Med. Assoc. Transactions of the American medical association.
Trans. and Proc. Bot. Soc. Edinb. Transactions and proceedings of the Botanical society of Edinburgh. (*Earlier* Transactions . . .)
Trans. and Proc. New Zeal. Inst. Transactions and proceedings of the New Zealand institute. (*Later* Transactions and proceedings of the Royal society of New Zealand.)
Trans. and Proc. Phil. Soc. Adelaide. Transactions and proceedings of the Philosophical society (1877–79); *later* . . . Royal society of South Australia.
Trans. and Proc. Roy. Soc. New Zeal. Transactions and proceedings of the Royal society of New Zealand. (*Earlier* Transactions and proceedings of the New Zealand Institute.)
Trans. and Proc. Roy. Soc. So. Austral. Transactions and proceedings of the Royal society of South Australia. (*Earlier* Transactions and proceedings of the philosophical society.)
Trans. and Proc. Roy. Soc. Victoria. Transactions and proceedings of the Royal society of Victoria.
Trans. Bot. Soc. Edinb. Transactions of the Botanical society of Edinburgh. (*Later* Transactions and proceedings . . .)
Trans. Hort. Soc. London. Transactions of the Horticultural society of London.
Trans. Illinois State Acad. Sci. Transactions of the Illinois state academy of sciences.
Trans. Internatl. Cong. Trop. Agr. 3d, London, 1914. Transactions of the third international congress of tropical agriculture . . . London, 1914.
Trans. Kansas Acad. Sci. Transactions of the Kansas academy of science.
Trans. Kentucky State Med. Soc. Transactions of the Kentucky state medical society.
Trans. Linn. Soc. Bot. Transactions of the Linnean society. Botany.
Trans. Linn. Soc. Zool. Transactions of the Linnean society. Zoology.
Trans. Natl. Eclect. Med. Assoc. Transactions of the National eclectic medical association.
Trans. Nat. Hist. Soc. Formosa. Transactions of the Natural history society of Formosa.
Trans. New York Acad. Sci. Transactions of the New York academy of science.
Trans. Ottawa Field Nat. Club. Transactions of the Ottawa field naturalists' club.
Trans. Phil. Inst. Victoria. Transactions of the Philosophical institute of Victoria.
Trans. Roy. Canad. Inst. Transactions of the Royal Canadian institute.
Trans. Roy. Irish Acad. Transactions of the Royal Irish academy.
Trans. Roy. Soc. Canada. *See* Proc. and Trans. Roy. Soc. Canada.
Trans. Roy. Soc. Edinb. Transactions of the Royal society of Edinburgh.
Trans. Roy. Soc. Mauritius. Transactions of the Royal society of arts and sciences of Mauritius (part of the time under French title).

Trans. Roy. Soc. So. Africa. Transactions of the Royal society of South Africa.
Trans. Roy. Soc. So. Austral. Transactions of the Royal society of South Australia. (*Formerly* Transactions and proceedings. . .)
Trans. Scient. Assoc. Meriden, Conn. Transactions of the Scientific association, Meriden, Conn. (*Later* Proceedings and transactions, Meriden scientific association.)
Trans. So. African Phil. Soc. Transactions of the South African philosophical society.
Trans. Texas Acad. Sci. Transactions of the Texas academy of science.
Trans. Wisconsin Acad. Sci., Arts and Lett. Transactions of the Wisconsin academy of sciences, arts and letters.
Transvaal Agr. Jour. Transvaal agricultural journal.
Trav. Off. Natl. Mat. Prem. Vég. (Paris) Notice. Travaux de l'Office national des matières premières végétales pour la droguerie, la pharmacie, la distillerie et la parfumerie. Notice.
Trillia. Trillia. Proceedings of the Botanical society of western Pennsylvania.
Trop. Natuur. De tropische natuur. Orgaan van de Nederlandsch-Indische natuurhistorische vereeniging.
Trop. Woods, Yale Univ. School Forestry. Tropical woods, Yale university school of forestry.
Tropenpflanzer. Der tropenpflanzer. Zeitschrift für tropische landwirtschaft. Organ des Kolonial-wirtschaftlichen komitees.
U. S. Dept. Agr. Yearbook. United States department of agriculture. Yearbook.
U. S. Geol. Survey Bul. United States department of the interior. Geological survey. Bulletin.
U. S. Geol. Survey Prof. Paper. United States department of the interior. Geological survey. Professional paper.
Univ. Calif. Pub. Amer. Archaeol. and Ethnol. University of California publications in American archaeology and ethnology.
Univ. Calif. Pub. Bot. University of California publication in botany.
Univ. Colorado Studies. University of Colorado studies.
Univ. Iowa Studies Nat. Hist. University of Iowa studies in natural history. (Continuation of Bulletin from laboratories of natural history.)
Univ. Maine Studies. University of Maine studies.
Univ. Missouri Studies Sci. Ser. Science series. University of Missouri studies.
Univ. New Mexico Bul. University of New Mexico bulletin.
Univ. Texas Bul. University of Texas bulletin.
Univ. Washington Pub. Biol. University of Washington publications in biology.
Verhandl. Bot. Ver. Prov. Brandenb. Verhandlungen des Botanischen vereins der provinz Brandenburg.
Verhandl. Deut. Wiss. Ver. Santiago de Chile. Verhandlungen des Deutscher wissenschaftlicher verein zu Santiago de Chile.
Verhandl. K. Akad. Wetensch. Amsterdam Afd. Natuurk. Verhandelingen der Koninklijke akademie van wetenschappen te Amsterdam, afdeeling natuurkunde.
Verhandl. K. K. Zool.-Bot. Gesell. Wien. Verhandlungen der (K. K.) zoologisch-botanischen gesellschaft in Wien.
Veröffentl. Gesell. Palästina-Forsch. Gesellschaft für Palästina-forschung. Veröffentlichung.
Věstník K. České Společn. Náuk Třída Mat.-Přír. (Mem. Soc. Roy. Sci. Bohême Cl. Sci.). Věstník Královské České společnosti náuk v Praze. Třída matematicko-přírodovecká (Mémoires de la Société royale des sciences de Bohême).
Victorian Nat. Victorian naturalist.
Vierteljahrsschr. Naturf. Gesell. Zürich; . . . Beibl. Vierteljahrsschrift der Naturforschenden gesellschaft in Zürich; *also* Beiblatt zur . . .
Washington Hist. Quart. Washington historical quarterly.
Webbia. Webbia. Raccolta di scritti botanici, edita da Ugolino Martelli.
West Amer. Sci. The West American scientist.
West. Austral. Yearbk. Western Australian year-book.
West Indian Bul. West Indian bulletin.
West. Jour. Med. and Surg. (Louisville, Ky.) Western journal of medicine and surgery.
West Virginia Univ. Bul. West Virginia university bulletin.
Wisconsin Archeol. Wisconsin archeologist.
Wiss. Veröffentl. Gesell. Erdk. Leipzig. Wissenschaftliche veröffentlichungen der Gesellschaft für erdkunde zu Leipzig.

Year Book Pub. Mus. Milwaukee. Year book of the Public museum of the city of Milwaukee.
Ymer. Ymer. Tidskrift utgifven af Svenska sällskapet för antropologi och geografi.
Zeitschr. Allg. Oesterr. Apoth.-Ver. Zeitschrift des Allgemeinen osterreichischen apotheker-vereines.
Zeitschr. Deut. Morgenländ. Gesell. Zeitschrift der Deutschen morgenländischen gesellschaft.
Zeitschr. Ethnol. Zeitschrift für ethnologie. Organ der Berliner gesellschaft für anthropologie, ethnologie und urgeschichte.
Zeitschr. Gesell. Erdk. Berlin. Zeitschrift der Gesellschaft für erdkunde zu Berlin.
Zeitschr. Weltforstw. Zeitschrift für weltforstwirtschaft.
Zion-Bryce Mus. Bul. Zion-Bryce museum bulletin. Zion-Bryce natural history association. National park service.
Zoe.

AUTHOR INDEX

GEOGRAPHICAL INDEX

This index includes the names of all countries and all geographical divisions used as headings; all States of the United States and Provinces of Canada; all separate islands and island groups listed in the titles or annotations, except those in inland waters; some current synonyms of the primary geographical names adopted (which follow the decisions of the United States Board of Geographic Names); and some names of regions, principally in Africa, which do not have definite boundaries. Such names as Ile de Dissée, Ilha de Marajó, Isle of Pines, etc., are indexed under Dissée, Marajó, Pines. Page numbers in the case of names which are not themselves used as headings refer to the page on which the geographic division containing the reference begins, except in the case of islands, where the page number is that of the actual entry.

GEOGRAPHICAL GUIDE to

FLORAS OF THE WORLD

An annotated list with special reference to useful plants and common plant names

---

PART II

WESTERN EUROPE

Finland, Sweden, Norway, Denmark, Iceland, Great Britain with Ireland, Netherlands, Belgium, Luxembourg, France, Spain, Portugal, Andorra, Monaco, Italy, San Marino, and Switzerland

---

By

S. F. BLAKE

Senior Botanist

Crops Research Division
Agricultural Research Service

UNITED STATES DEPARTMENT OF AGRICULTURE

MISCELLANEOUS PUBLICATION No. 797

Washington, D. C. February, 1961

For sale by the Superintendent of Documents, Government Printing Office, Washington 25, D. C. Price $2.75

FLORAS OF THE WORLD

An annotated selected list of floras and floristic works relating to vascular plants, including bibliographies and publications dealing with useful plants and vernacular names.

Part II

Western Europe: Finland, Sweden, Norway, Denmark, Iceland, Great Britain with Ireland, Netherlands, Belgium, Luxembourg, France, Spain, Portugal, Andorra, Monaco, Italy, San Marino, and Switzerland

By S. F. BLAKE,[1] formerly *senior botanist, Crops Research Division, Agricultural Research Service*

CONTENTS

INTRODUCTION

The first part of this work, dealing with Africa, Australasia, North America, South America, and the islands of the Atlantic, Indian, and Pacific Oceans, was published in 1942.[2] It contained about 2,597 primary and 428 subsidiary titles, a total of 3,025, and covered approximately 82,252,955 square kilometers (31,757,-897 square miles)[3] of land surface.

The present part, dealing with western Europe from Finland to Spain, Italy, and Switzerland (thus covering Europe with the omission of Germany, the old empire of Austria-Hungary, the Balkans region, Poland, and European Russia as now constituted)

---

[1] Sidney F. Blake, world-known botanist and expert bibliographer, completed verifying the bibliographical data contained in Part II prior to his death on December 31, 1959. Alice Cary Atwood, coauthor of Part I, also an expert bibliographer, continued to assist Dr. Blake in the preparation of the manuscript from her retirement in 1942 until only a few weeks before her death in 1947.

Grateful acknowledgment is made to Bernice G. Schubert, who assisted in the final checking of this work.

[2] U.S. Department of Agriculture Miscellaneous Publication 401. pt. 1, 336 p. 1942. (Dated June, but first copies received at U.S. Department of Agriculture on 31 July, 1942.)

[3] Figures of area, here as elsewhere in this introduction, are derived from The Columbia Lippincott Gazetteer of the World (1952). The regions listed under Finland that are now or always have been Russian or largely Russian (with a total of 54 primary and 26 secondary titles) are not incorporated in the figures for area.

contains approximately 3,757 primary and 3,084 subsidiary titles, a total of 6,841, and treats an area of about 3,044,174 square kilometers (1,175,357 square miles). Thus, it has one primary title for each 810 square kilometers (313 square miles), while the first part had one for every 31,672 square kilometers (12,228 square miles), or a ratio of 39 to 1 in favor of the present part; if the subsidiary titles are included in both cases, the ratio rises to 61 to 1.

The purpose of this publication is essentially, as in the previously published part, to furnish an annotated, geographically and alphabetically arranged catalog of all the now useful floras and floristic works, including those in periodical or serial literature, that list or describe the complete vascular flora (or the phanerogams or pteridophytes only) of any region or locality. The work also includes all essentially complete publications on useful, weedy, poisonous, or woody plants, vernacular names, and botanical bibliography; and for each country (but mostly not for parts of a country) one or more works on ecology and phytogeography, when these are available. For the most part only the later works relating to a given area or subject are treated in full, although earlier publications of the same scope are nearly always mentioned under the titles of works that have replaced them. In general, all available local floras (such as the floras of towns or equivalent districts), regardless of age, are entered even though there may be later works (such as county floras) of wider scope that cover their territory. A considerable proportion not only of the very local floras but also of those of counties, provinces, and similar regions are old and by modern standards inadequate, but they are included because they will be of use to the local student and because they are needed to give a complete picture of our present knowledge. Publications dealing with only a part of the phanerogamic flora, such as a single family or larger division, or a single group of useful plants, such as those producing dyes or resins, are excluded, except for complete papers on edible or medicinal plants. No attempt has been made to evaluate the titles listed, but the presence of an item may be regarded as an indication that it has some current utility. Publications that are primarily ecological or phytogeographical have been excluded unless they contain fairly complete lists of the plants of the area discussed. Works of the "popular" type, intended for beginners in botany, have been omitted except for a few that provide a sufficient number of illustrations to be of value to the scientific worker.

Except for unintentional omissions, this work contains all the available nominally complete publications on the vascular flora and the useful (including cultivated) and injurious plants, vernacular names, and botanical bibliography of the countries treated, as well as a considerable number of incomplete publications on the same subjects, which for one reason or another have seemed worthy of inclusion. It differs in scope from the first part principally by including all complete works on pteridophytes, weeds, and poisonous and cultivated plants; by giving a few general works on ecology, phytogeography, and related subjects; and by indicating the libraries in which all works not available in Washington were examined or the name of the correspondent who supplied information about them. Parenthetically, it may be stated that only about 64 of the works given in part I were not then available in Washington, and that some of them have since been added to libraries here.

Like part I, this list is fundamentally based on the subject card catalog of botany in the library of the United States Department of Agriculture, begun in 1906 under the direction of the late Dr. Frederick V. Coville by Majorie F. Warner and Alice C. Atwood, and continued by them and their assistants and successors until July 1952.[4] However, the smaller and the more local publications on European botany are so much less completely represented in the catalog, and so many of them are not available in this country, that much more extensive research has been necessary both in literature and in American and foreign libraries to attain the essential completeness of the first part. As before, not only the Washington libraries (Department of Agriculture, Library of Congress, Smithsonian Institution, National Library of Medicine (formerly Surgeon-General's Library), and Geological Survey), but those of the Gray Herbarium and the Arnold Arboretum (now in process of combination into one) and the New York Botanical Garden have been examined. In addition, in the summer of 1950, the author examined the libraries at the Muséum National d'Histoire Naturelle in Paris; the Botaniska Avdelningen of the Naturhistoriska Riksmuseum in Stockholm; the Universitetets Botaniske Museum, Copenhagen; the Botanisch Museum en Herbarium, Utrecht; the Rijksherbarium and the collection of the Koninklijke Nederlandse Botanische Vereniging, Leiden; the British Museum (Natural History), South Kensington; the Royal Botanic Gardens, Kew; and the private libraries of N. Douglas Simpson and George Taylor, in England. The authorities of the Lloyd Library, Cincinnati, have lent many volumes and afforded help in other ways, and Rogers McVaugh of the University of Michigan has looked up many French publications in the University library.

In nearly every country the author has benefited by the assistance of one or more botanists, without whose help the work would be much less complete. The following merit particular mention in this connection: In Belgium, Walter Robyns; in Denmark, J. B. Hansen; in Finland, Hans Luther; in France, Georges Dillemann of Paris, Bernard de Retz of Versailles, and Paul Jovet of the Muséum National; in Great Britain, D. H. Kent of London, H. S. Marshall, E. J. Salisbury and George Taylor of Kew, and N. Douglas Simpson of Bournemouth; in Iceland, Áskell Löve, now of Montreal, Canada; in Italy, Luigi Fenaroli, Valerio Giacomini, Rodolfo Pichi-Sermolli, and the late Francesco Sappa; in Luxembourg, F. L. Lefort, now of Montreal, Canada; in Nether-

[4] Current publications are listed in the monthly Bibliography of Agriculture, published by the Department of Agriculture library and now (1959) in its 23d volume, but no additions are made to the botany card catalog. The botany subject catalog is now published in book form.

lands, S. J. van Ooststroom; in Norway, Johannes Lid; in Portugal, A. R. Pinto da Silva; in Spain, Antonio de Bolós, the late Arturo Caballero, and E. F. Galiano; in Sweden, Eric Hultén and the late Gustaf Haglund; in Switzerland, Alfred Becherer.

The principal bibliographical sources utilized in the preparation of this work have been The Bradley Bibliography (covering works through 1900); the Bibliographical Contributions from the Lloyd Library; the bibliographies of Saelán for Finland; Krok for Sweden; A. Blytt and others for Norway; Christensen for Denmark; Löve's manuscript bibliography for Iceland; Druce for Great Britain; De Wildeman and Durand for Belgium; Béguinot, Saccardo, and the Bullettino bibliografico della botanica italiana for Italy; Eduard Fischer, the Schweizerische Botanische Gesellschaft, and the Schweizerische Landesbibliotek for Switzerland; and the following serial publications: Biological Abstracts (v. 1–34 (4)), Bohnensieg's Repertorium annuum literaturae botanicae periodicae (v. 1–8), Botanical Abstracts (v. 1–15), Botanisches Centralblatt (from v. 174, Zentralblatt; v. 1–179 (n.s., 37), h. 1–8) and its Beihefte (v. 1–9), Bulletin de la Société Botanique de France (v. 1–106(1/2)), Engler's Botanische Jahrbücher (v. 1–78(4)), Fortschritte der Botanik (v. 1–20), International Catalogue of Scientific Literature (v. 1–14), Journal of Botany, British and Foreign (v. 1–80(956)), and Just's Botanischer Jahresbericht (v. 1–63, which cover literature through 1935).

The arrangement adopted is an alphabetical-geographical one, summarized in the table of contents. Except in a few cases, each publication has been listed in full only once, but all necessary cross-references have been supplied. Under each geographical heading (except Scandinavia, the provinces of France, and a few other special cases), works covering the whole region are listed first (under the heading *General*), followed by the local and partial titles (under the heading *Local*); works relating to a large part of a country not readily placed under its geographical subdivisions have been either combined with the general titles or separated (under the heading *Partial*). Wherever desirable, the geographical information given in the title has been supplemented in the annotation; in Sweden the name of the län is given wherever it does not coincide with that of a landskap, in Denmark the name of the amt (county), and in Italy and Spain, where the arrangement is by regions, the name of the province is added in each case. Figures of longitude always refer to the Greenwich meridian.

Authors' given names, when more than one, are cited only as initials, but the full names and dates, whenever ascertainable, have been given in the index; for this information the writer is under great obligations to many of the cooperating botanists listed in a paragraph above (especially, for Finland, H. Luther; Sweden, E. Hultén, T. E. Hasselrot; Norway, J. Lid; Denmark, J. B. Hansen; Iceland, Á. Löve; Great Britain, D. H. Kent, N. D. Simpson, G. Taylor; Netherlands, S. J. van Ooststroom; Belgium, W. Robyns; Spain, E. F. Galiano; Portugal, A. R. Pinto da Silva; Switzerland, A. Becherer; for France and Italy no source of such information was found). Spanish names are given in the usual form; Portuguese names are alphabetized under the final one, but the preceding family name or names are also written out in full. Every work cited has been examined by the writer in Washington libraries, except those for which the name of another library is given at the end of the title (all of which were examined personally) or those annotated "Not seen," for each of which the name of the correspondent supplying the data is quoted.

The examination and annotation of so many thousand publications has brought to light so many common faults and omissions that some comment will not be out of place. The essential features of even the shortest list of plants include a title that accurately describes its contents; the most definite possible statement of the geographical area covered (at the very least the name of the province or county in which a given locality is placed) with reference, in the case of small or obscure places, to the distance and direction from some locality that can be found on any reasonably detailed map, supplemented, if desirable, by details of latitude and longitude; and a statement of the material on which the list is based, whether the author's collections and observations alone or all the published and unpublished information. Any list that attempts to represent the known flora of a region should also include a statement of the number of species included (the significant feature is the number of species, not that of the taxa, although the number of infraspecific forms and hybrids may be added if desired), preferably divided among the vascular cryptogams, gymnosperms, monocotyledons, and dicotyledons, and a bibliography in which the titles and references are given accurately and in full, not merely to the volume of a work or periodical or the first page of a paper. The alphabetical arrangement of all the genera of vascular plants without regard to family or higher divisions, frequently used in Swedish and Dutch local works, fails to give any idea of the composition of the flora and is not to be commended, however convenient it may be from the viewpoint of those who regularly employ it.

Additional items that add greatly to the value of a flora include an account of the topography, hydrography, geology, soils, and climate; botanical explorations and list of collectors, with full names and some biographical information; list of herbaria in which specimens are deposited; general and special features of vegetation, forest conditions, ecology, phytogeography, life zones, origin of flora, endemic species; notes on native and cultivated useful plants; local vernacular names; lists of doubtful and excluded species, with references; a gazetteer, including particularly localities not to be found in ordinary atlases, supplemented, if possible, by a map; a list of botanical names first published in the work; and an index. The inclusion or omission of keys, descriptions, references, and synonymy depends on the purpose for which the publication is intended and on the availability of such information in other works. Authors of supplements to published material should give a full reference to the title and place of publication of the original paper.

In the introduction to the first part of this work twenty titles dealing with the useful and the woody plants of the world were listed with annotations. It seems worth while to present the list again here, with a few deletions and a greater number of titles covering some additional subjects. They represent for the most part only a selection from the great array of publications available. For useful plants in general, Buschan's Vorgeschichtliche Botanik der Cultur- und Nutzpflanzen,[5] de Candolle's Origine des Plantes Cultivées,[6] Cazzuola's Dizionario di Botanica Applicata,[7] Clute's Useful Plants of the World,[8] Freeman and Chandler's The World's Commercial Products,[9] Good's Plants and Human Economics,[10] Gram, Jensen, and Mentz's Nytteplanter,[11] Guillaumin's Les Plantes Cultivées[12] and his Les Plantes Sauvages,[13] Heuzé's Les Plantes Industrielles,[14] Hill's Economic Botany,[15] Holland's Overseas Plant Products,[16] Perrot's Matières

---

[5] BUSCHAN, GEORG. VORGESCHICHTLICHE BOTANIK DER CULTUR- UND NUTZPFLANZEN DER ALTEN WELT AUF GRUND PRÄHISTORISCHER FUNDE. xii, 268 p. 23.5 cm. Breslau, 1895.—Account of early history of cultivated and useful plants in Egypt, Greece and Rome, etc., and in prehistoric times; list of prehistoric localities with mention of plant remains found and references; bibliography.

[6] CANDOLLE, ALPHONSE DE. ORIGINE DES PLANTES CULTIVÉES. 3. éd. iv, 385 p. 21 cm. Paris, 1886. (Bibliothèque scientifique internationale 43.) (1st ed. 1883.)—General considerations; discussion of the principal and many minor cultivated plants, with consideration of uses, vernacular names, and history, with references, the species grouped according to the parts used; list of species, grouped geographically and according to parts used, with indication of approximate age in cultivation and name of native country; general observations and conclusions. The classical work on this subject; there are various editions in English from 1884 to 1914.

[7] CAZZUOLA, FERDINANDO. DIZIONARIO DI BOTANICA APPLICATA ALLA MEDICINA, ALLA FARMACIA, ALLA VETERINARIA, ALL'ORTICOLTURA, ALL'AGRICOLTURA, ALL'INDUSTRIA E AL COMMERCIO. 720 p. 8vo. Pisa, 1876.—Dictionary of about 20,000 useful plants of the world, with botanical and vernacular names, native country, uses, cultivation, etc. (Not seen; title and annotation from notice in Nuovo Gior. Bot. Ital. 9: 168–169. 1877.)

[8] CLUTE, W. N. THE USEFUL PLANTS OF THE WORLD. 3. ed. 4 p.l., 219 p. 22 cm. Indianapolis, 1943. (1st ed., 1928, reprinted from Amer. Bot. v. 33–34, suppl. 1927–28.)—Popular treatment, with fairly extensive lists of principal species grouped by uses.

[9] FREEMAN, W. G., and CHANDLER, S. E. THE WORLD'S COMMERCIAL PRODUCTS. A DESCRIPTIVE ACCOUNT OF THE ECONOMIC PLANTS OF THE WORLD AND OF THEIR COMMERCIAL USES. 4 p.l., viii, 391 p. illus. (incl. maps), col. plates. 28 cm. London (also a Boston ed.), 1907.—Account of principal food plants and other plants of commercial importance grouped by uses; relates mostly to cultivated plants.

[10] GOOD, R. D. PLANTS AND HUMAN ECONOMICS. xii, 202 p. 8 maps. 19 cm. Cambridge, 1933.—General considerations, discussion of different groups of useful plants with mention of principal species; short bibliography, systematic list of species mentioned.

[11] GRAM, KAJ, JENSEN, HJALMAR, and MENTZ, AUGUST. NYTTEPLANTER. 4 p.l., 503 p. 300 fig. 25 cm. København, 1937.—Semipopular treatment of useful plants of the world; bibliography.

[12] GUILLAUMIN, ANDRÉ. LES PLANTES CULTIVÉES: HISTOIRE, ÉCONOMIE. 352 p. 16 pl. (on 8). 23 cm. Paris, 1946. (Bibliothèque scientifique.)—General considerations; running account of cultivated useful (including some ornamental) plants, grouped by uses; new cultural procedures, etc.

[13] GUILLAUMIN, ANDRÉ. LES PLANTES SAUVAGES: BIOLOGIE ET UTILISATION. 219 p. 8 pl. (on 4). 22.5 cm. Paris, 1948. (Bibliothèque scientifique.)—Edaphic factors, plant formations, botanical features of different regions, etc.; running account of principal useful wild plants, grouped by uses; mention of various plants characterized by peculiar form, fruit, biology, etc.

[14] HEUZÉ, GUSTAVE. LES PLANTES INDUSTRIELLES. 3. éd. 4 v. illus. 18.5 cm. Paris, 1893–95. (1st ed. 1859–60.)—Treatment of principal industrial plants (including some medicinal plants, but no food plants except condiments and spices) with account of history, cultivation, properties, uses, etc.; deals mainly with cultivated species.

[15] HILL, A. F. ECONOMIC BOTANY. A TEXTBOOK OF USEFUL PLANTS AND PLANT PRODUCTS. 2 ed. xii. 560 p. 250 fig. 23.5 cm. New York, 1952. (1st ed. 1937.)—Discussion of useful plants, grouped by uses; systematic list of species; bibliography, composed entirely of publications in English.

[16] HOLLAND, J. H. OVERSEAS PLANT PRODUCTS. vii, 279 p. 18.5 cm. London, 1937.—Briefly annotated alphabetical list of trade names, with botanical names and localities, of "all the natural products of vegetable origin, imported on a commercial scale into the docks under the control of the Port of London Authority and into other ports, for landing and delivery to the consignees in the markets of the United Kingdom. Others of economic value in the countries of production, including many that have been sent to Kew from time to time for identification, are also included." Has bibliography.

Premières Usuelles du Règne Végétal,[17] Regel's Pflanzen in Europa Liefern Rohstoffe,[18] Rosenthal's Synopsis Plantarum Diaphoricarum,[19] Ulbricht's Grundriss der Pflanzenlichen Rohstoffkunde,[20] Wehmer's Die Pflanzenstoffe,[21] and von Wiesner's Die Rohstoffe des Pflanzenreichs;[22] for edible plants, Bois' Les Plantes Alimentaires,[23] Holland's Food and Fodder Plants,[24] Maurizio's Die Geschichte unserer Pflanzennährung,[25] and Sturtevant's Notes on Edible Plants;[26] for medicinal plants, Baillon's Traité de Botanique Médicale Phanérogamique,[27] Bruntz and Jaloux's Plantes Officinales et Plantes à Drogues Médicamenteuses,[28] Dragendorff's

---

[17] PERROT, ÉMILE, and others. MATIÈRES PREMIÈRES USUELLES DU RÈGNE VÉGÉTAL. THÉRAPEUTIQUE–HYGIÈNE—INDUSTRIE. 2 v. (xxviii, 2343 p.). illus. (incl. maps). 24.5 cm. Paris, 1943–44.—Pharmacological classification of drugs, chemistry of plant products, classification of plant products (with lists of species), phytogeography of plant products (with 8 maps), bibliography; systematic account of useful (including medicinal) cellular and vascular plants, with discussion of chemistry, distribution, cultivation, uses, etc., and references.

[18] REGEL, KONSTANTIN VON. PFLANZEN IN EUROPA LIEFERN ROHSTOFFE. viii, 294 p. 22 cm. Stuttgart, 1944.—Discussion of European "Sammelpflanzen" (wild plants collected for commercial use); origin of cultivated plants, etc.; running account of wild plants of Europe (principally Germany) yielding industrial raw materials, grouped according to uses (fibers, wood, cork, oils, dyes, etc.), with references; general discussion. Does not include medicinal or edible plants (except fodder plants).

[19] ROSENTHAL, D. A. SYNOPSIS PLANTARUM DIAPHORICARUM. SYSTEMATISCHE UEBERSICHT DER HEIL-, NUTZ- UND GIFTPFLANZEN ALLER LÄNDER. xxvi, 1359 p. 22.5 cm. Erlangen, 1861.—Systematic list of useful, medicinal, and poisonous plants, with German vernacular names, officinal names, range, brief account of uses.

[20] ULBRICHT, HERBERT. GRUNDRISS DER PFLANZENLICHEN ROHSTOFFKUNDE. EIN LEHR- UND NACHSLAGEBUCH. viii, 147 p. 20 fig. 23.5 cm. Leipzig, 1952 (c. 1951).—Brief treatment of the principal botanical raw materials, grouped under such headings as carbohydrates, fats and oils, tanning materials, glycosides, dyes, fibers, woods, etc. with short bibliographies (restricted to German works).

[21] WEHMER, CARL. DIE PFLANZENSTOFFE, BOTANISCH-SYSTEMATISCH BEARBEITET. BESTANDTEILE UND ZUSAMMENSETZUNG DER EINZELNEN PFLANZEN UND DEREN PRODUKTE. PHANEROGAMEN. 2., neubearb. und verm. aufl. 2 v. (1511 p.). 25.5 cm. Jena, 1929–31. (1st ed. 1911.)—Ergänzungsband zur zweiten auflage: nachträge aus den jahren 1930–1934, mitbearb. von Magdalene Hadders. Aus dem nachlass hrsg. von Hans Amelung . . . iv, 244 p. port. 25.5 cm. Jena, 1935.—Systematic list of 4493 spermatophytes, with habitat, products, chemical properties, references.

[22] WIESNER, JULIUS VON. DIE ROHSTOFFE DES PFLANZENREICHS. 4. aufl. . . . Herausgegeben von Paul Krais und Wilhelm von Brehmer. 2 v. (2253 p.). illus. 25 cm. Leipzig, 1927–28. (1st ed. 1873.)—Encyclopedic work with especial emphasis on chemical features, the plants grouped under such headings as Alkaloide, Farbstoffe, Hölzer, etc., with extensive bibliographies and annotated systematic lists of species.

[23] BOIS, DÉSIRÉ. LES PLANTES ALIMENTAIRES CHEZ TOUS LES PEUPLES ET À TRAVERS LES AGES. HISTOIRE, UTILISATION, CULTURE. 4 v. 698 fig. 25.5 cm. Paris, 1927–37.—Descriptive account of wild and cultivated edible plants (except cereals), with history, uses, vernacular names, and references. Vol. 1, vegetables; v. 2, fruits; v. 3, spices and condiments; v. 4, sources of drinks. "May be regarded as the fourth edition of A. Paillieux and D. Bois, Le potager d'un curieux . . ." 1885.

[24] HOLLAND, J. H. FOOD AND FODDER PLANTS. Bul. Misc. Inform. Kew 1919: 1–84. 1919.—Annotated systematic list of more important food and fodder plants, with mention of uses, amount imported into Britain, etc., and briefer treatment of some of the less important ones; bibliographies.

[25] MAURIZIO, ADAM. DIE GESCHICHTE UNSERER PFLANZENNÄHRUNG VON DEN URZEITEN BIS ZUR GEGENWART. xx, 480 p. 90 fig., pl. (p. 323). 25 cm. Berlin, 1927.—Fully documented study of vegetable foods and drinks used by man before the beginning of agriculture, or still used by the more primitive peoples or by others in time of famine; includes tabular systematic list (p. 444–466) of over 700 vascular and cellular plants of North Temperate Zone with indication of parts eaten and whether used by primitive peoples, as survivals among civilized races, in prehistoric times, or in war famine years (1914/18), with references. Originally published, in shorter form, as Pożywienie róslinne i rolnictwo w roswoju dziejowym. xx, 409 p. 59 fig. 24.5 cm. Warszawa, 1926. There is also a French edition, Histoire de l'alimentation végétale depuis la préhistorie jusqu'à nos jours. 633 p. Paris, 1932. (Not seen.)

[26] STURTEVANT, E. L. STURTEVANT'S NOTES ON EDIBLE PLANTS. Edited by U. P. Hedrick. vii, 686 p. port. 31 cm. Albany, 1919. (New York Dept. of Agriculture. 27th Annual report. v. 2, pt. 2; also as Report of the New York agricultural experiment station 1919, II.)—Alphabetical list of 2897 plants, with habitat, uses, principal English names and references; bibliography. Edible fungi are not included.

[27] BAILLON, H. E. TRAITÉ DE BOTANIQUE MÉDICALE PHANÉROGAMIQUE. 1,499 p. 3,487 fig. 24 cm. Paris, 1883–84.—Systematic account of medicinal plants, with vernacular names and uses, brief descriptions of families and genera and the more important species.

[28] BRUNTZ, LOUIS, and JALOUX, MARCEL. PLANTES OFFICINALES ET PLANTES À DROGUES MÉDICAMENTEUSES. NOMENCLATURE MÉTHODIQUE DRESSÉE D'APRÈS TOUTES LES ÉDITIONS DES CODES OFFICIELS DES MÉDICAMENTS DES DIVERS ÉTATS DU MONDE PRÉCEDÉE DE CONSIDÉRATIONS GÉNÉRALES SUR LES PHARMACOPÉES. PRÉFACE DE . . . GUIGNARD. xvi, 260 p. table. 25 cm. Paris, 1918.—Historical account of the pharmacopoeias of each country of the world, with citation of titles and editions; systematic list of medicinal plants, with citation of the pharmacopoeias in which their different parts and products are listed.

Die Heilpflanzen,[29] Dujardin-Beaumetz and Égasse's Les Plantes Médicinales,[30] Dupuis and Réveil's Flore Médicale,[31] Héraud's Nouveau Dictionnaire des Plantes Médicinales,[32] Hocking's Dictionary of Terms Used in Pharmacognosy,[33] von Hovorka and Kronfeld's Vergleichende Volksmedizin,[34] Lloyd's Origin and History of All the Pharmacopeial Vegetable Drugs,[35] Mitlacher's Die Offizinellen Pflanzen und Drogen,[36] Steinmetz's Codex Vegetabilis[37] and his Materia Medica Vegetabilis,[38] and Youngken's Textbook of Pharmacognosy;[39] for poisonous plants, Bernhard-Smith's

---

[29] DRAGENDORFF, GEORG. DIE HEILPFLANZEN DER VERSCHIEDENEN VÖLKER UND ZEITEN. IHRE ANWENDUNG, WESENTLICHEN BESTANDTHEILE UND GESCHICHTE . . . vi, 884 p. 24.5 cm. Stuttgart, 1898.—Briefly annotated systematic list, without references and essentially without vernacular names.

[30] DUJARDIN-BEAUMETZ, GEORGES, and ÉGASSE, ED. LES PLANTES MÉDICINALES INDIGÈNES ET EXOTIQUES, LEURS USAGES THÉRAPEUTIQUES, PHARMACEUTIQUES ET INDUSTRIELS. vii, 845 p. 1,034 fig., 40 col. pl. 28.5 cm. Paris, 1889.—Descriptive account of more significant medicinal plants of the world, alphabetically arranged, with account of properties and uses; index of botanical and vernacular names, list of species classified by uses.

[31] DUPUIS, ARISTIDE, and RÉVEIL, O. FLORE MÉDICALE USUELLE ET INDUSTRIELLE DU XIXE SIÈCLE . . . NOUVELLE ÉD. COMPLÈTEMENT REFONDUE ET AUGMENTÉE D'IMPORTANTES SUPPLÉMENTS PAR M. J. L. DE LANESSAN. 3 v. 27.5 cm. Paris, n.d. (Lloyd Library.)—ATLAS ICONOGRAPHIQUE . . . 3 v. 150 pl. [Paris, 1887?] (Not seen.)—Account of medicinal and a few other useful plants, alphabetically arranged by French vernacular names, with description, habitat, culture, parts used, method of collection, chemical composition, and uses.

[32] HÉRAUD, AUGUSTE. NOUVEAU DICTIONNAIRE DES PLANTES MÉDICINALES. DESCRIPTION, HABITAT ET CULTURE, RÉCOLTE, CONSERVATION, PARTIE USITÉE, COMPOSITION CHIMIQUE, FORMES PHARMACEUTIQUES ET DOSES, ACTION PHYSIOLOGIQUE, USAGES DANS LE TRAITEMENT DES MALADIES, MEMORIAL THÉRAPEUTIQUE. PRECÉDE D'UNE ÉTUDE GENERALE SUR LES PLANTES MÉDICINALES. 7. éd. viii, 657 p. 292 fig. 19 cm. Paris, 1949. (1st ed. 1875.)—Generalities, including alphabetical list of families with their medicinal properties, and list of species classified by their medicinal properties; list of medicinal plants, alphabetically arranged by French vernacular names, with description, chemical composition, uses, etc.; list of diseases with corresponding remedies.

[33] HOCKING, G. M. A DICTIONARY OF TERMS USED IN PHARMACOGNOSY AND OTHER DIVISIONS OF ECONOMIC BOTANY. xxv, 284 p. illus., front. 24.5 cm. Springfield, Illinois, 1955.—Alphabetical (multilingual) list of vernacular and scientific names of plants (also animals and minerals) used in pharmacy and medicine generally, as well as in agriculture and commerce, and of technical terms and names of pharmacological preparations, with brief but informative annotations; bibliography, list of pharmacological serials, list of selected periodicals on some important industrial plants, dictionary of descriptive terms relating to drugs.

[34] HOVORKA, OSKAR VON, and KRONFELD, A. VERGLEICHENDE VOLKSMEDIZIN. EINE DARSTELLUNG VOLKSMEDIZINISCHER SITTEN UND GEBRÄUCHE, ANSCHAUUNGEN UND HEILFAKTOREN DES ABERGLAUBENS UND DER ZAUBERMEDIZIN. 2 v. (459,960 p.). 383 fig., 28 pl. 26 cm. Stuttgart, 1908–09. (Harvard College library.)—Vol. 1 includes treatment of plants used in popular and scientific medicine (also zoological items, etc.), alphabetically arranged under vernacular names, with scientific and vernacular names and uses, discussion of history, etc.

[35] LLOYD, J. U. ORIGIN AND HISTORY OF ALL THE PHARMACOPEIAL VEGETABLE DRUGS, CHEMICALS AND PREPARATIONS WITH BIBLIOGRAPHY. vol. 1. Vegetable drugs. 8th and 9th decennial revisions (botanical descriptions omitted). xiv, 449 p. 18 pl. 20.5 cm. Cincinnati, 1921.—Alphabetical list of botanical drugs in the U. S. Pharmacopeia, with their history in that pharmacopeia and in the world at large; extensive bibliography.

[36] MITLACHER, WILHELM. DIE OFFIZINELLEN PFLANZEN UND DROGEN. EIN SYSTEMATISCHE ÜBERSICHT ÜBER DIE IN SÄMTLICHEN STAATEN EUROPAS SOWIE IN JAPAN UND DEN VEREINIGTEN STAATEN VON AMERIKA OFFIZINELLEN PFLANZEN UND DROGEN MIT KURZEN ERLÄUTERNDEN BEMERKUNGEN. vii, 136 p. 25.5 cm. Wien und Leipzig, 1912. (Kew library.)—Systematically arranged list of officinal plants (cellular and vascular) of the countries mentioned, with officinal names and countries where used, and brief statement of chemical properties and uses.

[37] STEINMETZ, E. F. CODEX VEGETABILIS. 2. ed. var. pag. 27 cm. Amsterdam, (pref. 1957). (1st ed., (pref. 1947).)—List of 1216 vegetable drugs, alphabetically arranged by scientific names, with names in English, Dutch, German, French, Italian and various other European and Asiatic languages, habitat, parts used, active constituents, and uses; index of vernacular names.

[38] STEINMETZ, E. F. MATERIA MEDICA VEGETABILIS. 3 pt. (pt. 1–2, 479 p.; pt. 3 (indices), 1 pl., 110, 4 p.). 33.5 cm. Amsterdam, (pref. 1954). (Processed publication.)—Annotated list of the vegetable materia medica of the world, alphabetically arranged by the second components (usually) in the officinal names, with brief account of part used and uses in English, Dutch, German, and French, 1 or 2 vernacular or trade names in each language, and indication of poisonous species; indices to Latin names (officinal and scientific) and to each of the languages used.

[39] YOUNGKEN, H. W. TEXTBOOK OF PHARMACOGNOSY. 6. ed. xvii, 1063 p. 521 fig., front. 23.5 cm. New York [etc.], 1948. (Reprinted with corrections 1950. 1st ed. 1921.)—1950 supplement to Sixth edition . . . 1 p.l., 14 p. [1950.]—General considerations, bibliography (general works), morphological classification of crude drugs, therapeutic classes; descriptive treatment of vegetable (p. 69–909) and animal (p. 910–971) drugs (also antibiotics and bacterial biological products (p. 972–990), further bibliography, and microanalytical methods), systematically arranged, with synonyms, brief history, botanical origin, parts used, production and commerce, description, histology, constitution, uses, adulterants, etc.

Poisonous Plants of All Countries,[40] Lewin's Traité de Toxicologie,[41] and Pammel's Manual of Poisonous Plants;[42] for woody plants, Bärner and Müller's Die Nutzhölzer der Welt,[43] Howard's Manual of the Timbers of the World,[44] Meyer's Buch der Holznamen,[45] and Rehder's The Bradley Bibliography;[46] for plants connected with mythology, magic, and folklore, Aigremont's Volkserotik und Pflanzenwelt,[47] Delatte's Herbarius,[48] Folkard's Plant Lore, Legends, and Lyrics,[49] Friend's Flowers and Flower Lore,[50]

---

[40] BERNHARD-SMITH, ARTHUR. POISONOUS PLANTS OF ALL COUNTRIES. 2. ed. xii, 112 p. 152 + 33 fig. 19 cm. London, 1923. (1st ed. 1905.)—List of plants grouped by poisonous principles; list of poisonous plants (152 spermatophytes, also list of mushrooms) grouped by properties, with brief description and names of toxic principles; lists of poisonous microfungi and bacteria, alphabetical list of reputed poisonous plants. Most of the species treated are British, and the work is very far from complete.

[41] LEWIN, LOUIS. TRAITÉ DE TOXICOLOGIE . . . TRADUIT ET ANNOTÉ PAR G. POUCHET. vi, 1120 p. col. pl. 25 cm. Paris, 1903.—Includes Poisons végétaux (p. 556–938), a systematic treatment of vascular and cellular poisonous plants of the world, with properties, symptoms, etc.

[42] PAMMEL, L. H. A MANUAL OF POISONOUS PLANTS CHIEFLY OF EASTERN NORTH AMERICA, WITH BRIEF NOTES ON ECONOMIC AND MEDICAL PLANTS. 2 v. (viii, 977 p.) 458 fig., plates. 23.5 cm. Cedar Rapids, 1910–11. (Also issued as 1 vol., 1911.)—General consideration of plant poisons; annotated systematic list of more important poisonous plants of United States and Canada, with bibliography; chemistry of alkaloids; annotated systematic list of poisonous plants of United States and Canada (including bacteria and cellular cryptogams); tabular systematic catalog of poisonous plants of world, showing properties and locality; bibliography (by H. S. Kellogg) of 1097 titles.

[43] BÄRNER, JOHANNES, and MÜLLER, J. F. DIE NUTZHÖLZER DER WELT. 4 v. (687, 780, 803, 1196 p.). 25.5 cm. Neudamm, 1942–43.—List of over 12,000 species of woody plants (mostly trees), with synonyms, concise statement of habit, range, vernacular and trade names in different countries (including those where the trees are cultivated), properties and uses, wood characteristics, and references; the families arranged systematically, the genera and species alphabetically. Of v. 4 (Registerband, 1943) only an imperfect photostatic copy (p. 1–175, 1137–1183) has been seen. Dr. Bärner writes me that 2 printings of this volume made in 1943 were destroyed by war and post-war action except for (so far as he knows) 2 copies. The contents are as follows: p. 1–175, botanische namen; p. 177–1136, handels- und eingeborenennamen; p. 1137–1183, literatur; p. 1185–1196, autorenkürzungen.

[44] HOWARD, A. L. A MANUAL OF THE TIMBERS OF THE WORLD. THEIR CHARACTERISTICS AND USES. 3. ed. xxiii, 751 p. illus., pls. 22.5 cm. London, 1948. (1st ed. 1920.)—Annotated list, arranged alphabetically (partly by vernacular, partly by scientific names), with range, vernacular names, character of wood, and uses; short bibliography, list of timbers grouped geographically, indices of botanical and vernacular names.

[45] MEYER, HANS. BUCH DER HOLZNAMEN . . . xviii, 564 p. 25.5 cm. Hannover, 1933–36.—Alphabetical list of vernacular names of trees in all languages, with botanical equivalent, family name and habitat; no index to scientific names.

[46] REHDER, ALFRED. THE BRADLEY BIBLIOGRAPHY. A GUIDE TO THE LITERATURE OF THE WOODY PLANTS OF THE WORLD PUBLISHED BEFORE THE BEGINNING OF THE TWENTIETH CENTURY. 5 v. 30 cm. Cambridge, [Mass.], 1911–1918. (Pub. Arnold Arboretum no. 3.)—Very comprehensive bibliography of works on woody plants from earliest times through 1900, divided as follows: vol. 1, general works, morphology, anatomy, physiology, pathology, ecology, floras, ethnobotany; v. 2, works dealing with individual species, genera, and families, arranged systematically; v. 3, collections, arboriculture, diseases, economic uses (the last arranged both geographically and systematically); v. 4, forestry; v. 5, index.

[47] AIGREMONT, DR. [pseudonym]. VOLKSEROTIK UND PFLANZENWELT. EINE DARSTELLUNG ALTER WIE MODERNER EROTISCHER UND SEXUELLER GEBRÄUCHE, VERGLEICHE, SPRICHWÖRTER, REDEWENDUNGEN, RÄTSEL, VOLKSLIEDER EROTISCHEN ZAUBERS UND ABERGLAUBENS, SEXUELLER HEILKUNDE, DIE SICH AUF PFLANZEN BEZIECHEN. 2. aufl. 2 v. (165, 121 p.). 22.5 cm. Leipzig, 1910. (Harvard College library.) (1st ed. 1908–09.)—Discussion of various cellular and vascular plants in their connection with sexual practices, beliefs, and resemblances.

[48] DELATTE, ARMAND. HERBARIUS. RECHERCHES SUR LE CÉRÉMONIAL USITÉ CHEZ LES ANCIENS POUR LA CUEILETTE DES SIMPLES ET DES PLANTES MAGIQUES. 2. éd., revue et augmentée. 3 p. l., 176 p. 25 cm. Liége, Paris, 1938. (At head of title page: Bibliothèque de la Faculté de philosophie et lettres de l'Université de Liége. Fasc. 81.) (1st ed. 1936.)—Origin of the customs of ancient herborists, etc.; detailed account of proper times for collecting herbs, rites, songs, sacrifices, methods of collecting, treatment of plants after collecting, etc., with full references; indices of authors cited, plant names, etc.

[49] FOLKARD, RICHARD. PLANT LORE, LEGENDS, AND LYRICS. EMBRACING THE MYTHS, TRADITIONS, SUPERSTITIONS, AND FOLK-LORE OF THE PLANT KINGDOM. xxiv, 610 p. illus., 6 pl. (incl. front.). 23 cm. London, 1884.—Bibliography; collected treatment of plant lore, grouped into chapters on world-trees and sacred plants of the ancients, plants of the Christian church, of the fairies, of the devil, plant symbolism, plants and the planets, etc., with abundant quotations from literature; alphabetical list of plants with the beliefs, fables, etc., connected with them.

[50] FRIEND, HILDERIC. FLOWERS AND FLOWER LORE. 2 v. (xvi, 704 p.). illus. 22 cm. London, 1884.—Bibliography; running account of myths and legends, beliefs and superstitions connected with a great number of plants, grouped in chapters entitled the fairy garland, bridal wreaths and bouquets, flowers for heroes, saints, and gods, the magic wand, superstitions about flowers, the language of flowers, witches and their flower-lore, flowers and the dead, etc.; collected critical notes and references dealing with each chapter.

Gessmann's Die Pflanzen im Zauberglauben,[51] de Gubernatis' La Mythologie des Plantes,[52] Teirlinck's Flora Diabolica,[53] his Flora Magica,[54] and his Plantenkultus,[55] and Thiselton-Dyer's The Folk-lore of Plants;[56] and for titles on geology, zoology, ethnology, and so on, Bourlière's Éléments d'un Guide Bibliographique du Naturaliste.[57] The most complete dictionary of botanical terms in any language is Font Quer's Diccionario de Botánica;[58] the most compact catalog of generic and family names and definition of ordinary descriptive terms is given in Willis' Dictionary of Flowering Plants and Ferns.[59]

There is no work that covers the ornamental plants of the world as a whole, but those of temperate regions are described in the following books: Bailey's Standard Cyclopedia of Horticulture[60]

---

[51] GESSMANN, G. W. DIE PFLANZEN IM ZAUBERGLAUBEN. EIN KATECHISMUS DER ZAUBERBOTANIK. MIT EINEM ABHANGE ÜBER PFLANZE-SYMBOLIK. 2 p.l., 252 p. 12 fig. 19.5 cm. Wien [etc.], (1899).—General considerations; annotated list of 120 plants connected with witchcraft, alphabetically arranged by vernacular names, with botanical name, characteristics, uses, etc., and index; position of plants in astrology, magic salves and smoking materials, magical acceleration of growth, reincarnation of plants, etc.; bibliography; alphabetical list of German vernacular names of plants, with their symbolical significance (no botanical names).

[52] GUBERNATIS, ANGELO DE. LA MYTHOLOGIE DES PLANTES OU LES LÉGENDES DU RÈGNE VÉGÉTAL. 2 v. (xxxvi, 295; 2 p.l., 374 p.). 22.5 cm. Paris, 1878-82.—Vol. 1, Botanique générale: short or long annotations on such subjects as Achille, Adam (arbre d'), Antoine (herbes de saint-), bernacles (arbre aux), feuille, jalousie, lune, magiques (plantes), moly, sacrés (plantes et herbes), Yggdrasill, etc., the whole arranged alphabetically; vol. 2, Botanique spéciale: alphabetical list of vernacular names in various languages, with account of myths and legends connected with them. No index of botanical names.

[53] TEIRLINCK, ISIDOOR. FLORA DIABOLICA. DE PLANT IN DE DEMONOLOGIE. 322 p. 24 cm. Antwerpen, [1924]. (Harvard College library.)—Running account of diabolical and anti-diabolical plants, grouped by uses, etc., with Flemish and other vernacular names, botanical names, references, superstitions, etc.; systematic list of species, bibliography, index.

[54] TEIRLINCK, ISIDOOR. FLORA MAGICA. DE PLANT IN DE TOOVERWERELD. 388 p. 24 cm. Antwerpen, 1930. (Harvard College Library.)—Running account of plants associated with magic, grouped by uses, etc., with Flemish and other vernacular names, botanical names, references, superstitions, etc.

[55] TEIRLINCK, ISIDOOR. PLANTENKULTUS. EEN TWEEDE HOOFDSTUK VAN DE PLANTLORE. 425 p. Antwerpen, 1904-12.—A treatise dealing with popular religious beliefs connected with plants, for example plant names which are derived from sacred names. (Not seen; title and annotation mostly from notice in Bot. Jahresber. Just 43 (1): 417, 1922.)

[56] THISELTON-DYER, T. F. THE FOLK-LORE OF PLANTS. 3 p. l., 328 p. 19.5 cm. London (also New York edition), 1889.—Popular beliefs, myths, fables, and uses of a large number of plants, grouped into chapters on plant worship, plants in witchcraft, plants in fairylore, dream plants, plants and the weather, doctrine of signatures, plants in folk-medicine, etc.

[57] BOURLIÈRE, FRANÇOIS. ÉLÉMENTS D'UN GUIDE BIBLIOGRAPHIQUE DU NATURALISTE. ix, 302 p. 28.5 cm. Macon, 1940.—Suppléments I et II. 303-368 p. Paris, 1941.—Unannotated, geographically and systematically arranged list of publications (6,357 titles including supplements) including general (world) and regional (including country) but very few local works on natural history in general, zoology, botany, geology, and ethnology; briefly annotated, geographically arranged list (p. 335-353) of principal zoological expeditions, with indication of animal groups included. The botanical sections, at least, although helpful have numerous important omissions.

[58] FONT QUER, PÍO, and others. DICCIONARIO DE BOTÁNICA . . . SEGUIDO DE UN VOCABULARIO IDEOLÓGICO EN EL QUE SE ORDENAN CONCEPTUALMENTE LAS VOCES DEL DICCIONARIO. xxxix, 1244 p. illus. 23 cm. Barcelona [etc.], 1953.—Etymology, etc.; dictionary of descriptive terms and family and higher group names (nearly all in Spanish orthography) in all branches of botany, with extensive classified list of terms in index; bibliography, including botanical dictionaries in other languages.

[59] WILLIS, J. C. A DICTIONARY OF THE FLOWERING PLANTS AND FERNS. 6. ed. rev. xii, 752, liv p. illus. 19 cm. Cambridge, 1931. (Several reissues, the last in 1955. 1st ed. 1897.)—Alphabetical list of generic, family, and group names and botanical terms, with brief definitions of the terms; mention of the family name, number of species, general range, and frequent notes on uses, peculiarities of structure, and features of biology under the genera; characters of the families and of their principal divisions, uses, features of biology, etc., under the family entries. Pages 705-752 constitute a supplement; p. i-xlix a key to families based on Engler and Prantl; p. l-liv a synopsis of the Bentham and Hooker system.

[60] BAILEY, L. H. THE STANDARD CYCLOPEDIA OF HORTICULTURE . . . 6 v. xx, 3639 p. 4,056 fig., 120 pl. (pt. col.). 27 cm. New York, London, 1914-1917. (Corrected reissue [very little changed] 1922, reissued as "new ed." in 3 v. 1925; various reissues.)—Abbreviations of authors' names, synopsis of plant kingdom (by K. M. Wiegand), key to families and genera, explanation of specific epithets; encyclopedic treatment of cultivated ornamental, fancy, and edible plants of United States and Canada (20,602 species), alphabetically arranged by genera, with keys, descriptions, notes on cultivation, etc.; by Bailey, with many collaborators. Includes general articles on horticultural subjects, biographies of leading horticulturists, etc.

and his Manual of Cultivated Plants,[61] Bonstedt's Pareys Blumengärtnerei,[62] Chittenden's Royal Horticultural Society Dictionary of Gardening (see beyond, under Great Britain), Fournier's Arbres, Arbustes et Fleurs de Pleine Terre (beyond, under France), Rehder's Manual of Cultivated Trees and Shrubs,[63] and Schneider's Illustriertes Handbuch der Laubholzkunde.[64] The most comprehensive work on ornamental (and other useful) plants of tropical regions is Macmillan's Tropical Gardening and Planting;[65] and the most extensive series of illustrations of plants grown in greenhouses in temperate regions is contained in Graf's Exotica.[66]

The available general works on floristics, useful plants, and so on of each country, as well as the floral statistics and the number of publications cited, are discussed in the following pages.

ANDORRA.—The little state of Andorra, with an area of 495 square kilometers (191 square miles), on the Pyrenean border between France and Spain, has a flora of about 1,150 species (Losa España and Montserrat, 1951). Three floras have been published (in 1913, 1919, and 1951), the latest by Losa España and Montserrat; unfortunately only the earliest is in a readily accessible journal. There are 3 primary titles cited, of which 2 are general and 1 local, and 3 subsidiary titles.

BELGIUM.—The kingdom of Belgium, with an area of 30,507 square kilometers (11,799 square miles), has a vascular flora of 1,710 species, of which 1,367 are native (Hauman and Balle, 1934). It is well provided with manuals and descriptive floras, of which Crépin's Manuel (not revised since 1884) and Goffart's Nouveau manuel (based on Crépin and brought up to date) are standard for the French-speaking part of the country, and

---

[61] BAILEY, L. H. MANUAL OF CULTIVATED PLANTS MOSTLY COMMONLY GROWN IN THE CONTINENTAL UNITED STATES AND CANADA. rev. ed. 1,116 p. 195 fig. 22 cm. New York, 1949. (1st ed. 1924.)—Glossary of specific epithets, list of abbreviations of authors' names; keyed, descriptive manual of cultivated vascular plants (5,347 species).

[62] BONSTEDT, CARL, *ed.* PAREYS BLUMENGÄRTNEREI. BESCHREIBUNG, KULTUR UND VERWENDUNG DER GESAMTEN GÄRTNERISCHEN SCHMUCKPFLANZEN. UNTER MITWIRKUNG VON O. KRAUSS [etc.]. 2 v. (1 p.l., viii, 940; 4 p.l., 792 p.). 1,004 fig., 48 col. pl. 28.5 cm. Berlin, 1931–32.—Keyed, systematically arranged, descriptive treatment of hardy and greenhouse ornamental vascular plants (including woody plants) cultivated or capable of cultivation in Germany. Successor to Vilmorins Blumengärtnerei, which was originally based on Vilmorin-Andrieux et Cie. Les fleurs de pleine terre. The 2. ausg., by Fritz Encke and collaborators, is in process of publication (v. 1. xv, 941 p. 572 fig. 22 col. pl. 27.5 cm. Berlin, 1958).

[63] REHDER, ALFRED. MANUAL OF CULTIVATED TREES AND SHRUBS HARDY IN NORTH AMERICA EXCLUSIVE OF THE TROPICAL AND WARMER TEMPERATE REGIONS. 2. ed. rev. and enl. xxx, 996 p. front. (map). 22 cm. New York, 1940. (1st ed. 1927.)—List of references for illustrations; keyed, descriptive manual of woody plants (about 2,500 species fully described, many more briefly and unsatisfactorily mentioned with briefer descriptions), with indication of winter hardiness and references to illustrations; list of abbreviations of authors' names. Covers Canada and most of the United States except extreme southern states and most of California.

[64] SCHNEIDER, C. K. ILLUSTRIERTES HANDBUCH DER LAUBHOLZKUNDE. CHARACTERISTIK DER IN MITTELEUROPA HEIMISCHEN UND IM FREIEN ANGEPFLANZTEN ANGIOSPERMEN GEHÖLZ-ARTEN UND FORMEN MIT AUSSCHLUSS DER BAMBUSEEN UND KAKTEEN. 2 v. (810, 1,070 p.). 460, 628 fig. 25 cm. Jena, 1904-12.—Register . . . vii, 136 p. 25 cm. Jena, 1912.—Keyed, descriptive treatment of wild and hardy cultivated woody angiosperms of Middle Europe.

[65] MACMILLAN, H. F. TROPICAL GARDENING AND PLANTING WITH SPECIAL REFERENCE TO CEYLON. 5. ed. x, 560 p. illus., 4 pl. (3 col. incl. front.) 23 cm. London, 1943. (Reprinted 1946. 1st ed. 1910, with title: A handbook of tropical gardening and planting . . .)—General considerations (soils, horticulture, etc.); chapters on ornamental trees, shrubs, climbers, herbs, pot plants, etc., for lower and higher elevations, sand-binding plants, shade trees, fruits, vegetables, spices, medicinal and poisonous plants, oils, rubber, fibers, dye plants, fodder plants, perfume and honey plants, insect and fungous enemies, etc., with classified lists of species, short descriptions, notes on cultivation, etc.

[66] GRAF, A. B. EXOTICA. PICTORIAL CYCLOPEDIA OF INDOOR PLANTS. 643 p. incl. 4,000 fig. 4 col. pl., (map and illustrated glossary on lining papers). 28.5 cm. Rutherford, N. J., 1957.—General considerations on cultivation, etc.; series of about 4,000 photographs of mostly greenhouse plants (including pteridophytes), the families arranged alphabetically, followed by brief descriptions; popular account of plant geography of the world.

Macleod and Staes' Geïllustreerde flora for the Flemish part. A detailed descriptive flora of vascular and cellular plants including full consideration of vernacular names, uses, folklore, and biology is being prosecuted under the direction of Robyns with the vascular plants, by Lawalrée, already about a quarter published. A catalog of vascular and cellular plants with detailed localities was published by De Wildeman and Durand in 1898-1907, with subsequent additions, and a more modern listing of vascular plants by Hauman and Balle, showing distribution by districts, came out in 1934; the two works include an essentially complete bibliography through 1934. In addition, there is a continuous bibliography by Van Aerdschot for the years 1902-1931, while Robyns (1951) has given a list of the more general and comprehensive works. Two works are cited on native and cultivated trees, by Aigret and Wesmael respectively, and two historical works on cultivated plants, by De Wildeman and Morren; the cultivated as well as the wild plants are described in several of the older floras. There are a number of works on medicinal or poisonous plants, of which that by Chalon (1912) is the most recent and the most readily accessible. There is an old list of naturalized and introduced plants by Devos (1870), with some recent additions by Lawalrée and others. The Walloon vernacular names are listed by Boxus, Feller, and Lawalrée, the Flemish ones by Paque and Lawalrée. There are works on botanical history, plant associations and phytosociology, phytogeography, and on the history of rare plants over the years.

Of the nine provinces, all but two (Flandre Orientale, Luxembourg) possess their own floras, five of them dating from 1845 to 1878, with a joint flora of Luxembourg and Namur in 1902. The number of local publications per province ranges from 1 (Flandre Occidentale, Limbourg) to 22 (Liège). Of 106 primary titles cited, 43 are general or partial, 8 provincial, and 55 local; and there are 65 subsidiary titles, making a total of 171. Most of the provincial and local titles are floras or florules, but there is a work on the useful, medicinal, and poisonous plants of Louvain, a flora of the poisonous plants of Liège Province, and several local works on Walloon names in the same province.

DENMARK.—The kingdom of Denmark, with an area (excluding the Faroes) of 42,936 square kilometers (16,576 square miles), has a flora of about 1,500 vascular plants, including naturalized species (Rostrup, 1953). About 206 additional microspecies of *Rubus, Hieracium,* and *Taraxacum* are given by Raunkiær (1950). It is well provided with general works and possesses an appropriate number of special publications and local floras, some of them rather old. There are two standard, pocket-size modern floras, by Raunkiær and Rostrup respectively, the former containing many introduced species and microspecies not given in Rostrup, the latter with more detailed statement of local distribution; also a flora of the Faroes (in Faroese) as well as an annotated catalog (in English). There are some excellent popular floras, including those by Gram and Jessen and by Hvass, and several works on woody plants, including the cultivated species.

A detailed account of plant distribution, with maps showing the range of most of the species, is being actively published family by family. Colored illustrations of practically every species are available in the various iconographies cited. Christensen's publications on botanical history and bibliography, extending through the year 1939 and continued to date in less detailed form by other botanists, constitute the only complete botanical bibliography of any European country, or indeed of any country in the world, with a comparable literature. There is no single comprehensive list of Danish vernacular names, but they are included in Jenssen-Tusch's Scandinavian work of 1867 and in Lyttkens' Swedish nomenclator of 1904-15, and a very extensive compilation of vernacular names prepared by Johan Lange from Axel Lange's manuscript collections is now in course of publication; and there is an extensive list of Faroese vernacular names. More or less adequate works are available on plant associations, weeds, cultivated plants, poisonous plants, edible, medicinal, and other wild economic plants, and on various minor subjects.

The 22 counties (amter) of Denmark are too small and too numerous to afford a satisfactory basis for the subdivision of the literature and, moreover, have very seldom been the subject of special floras, and the 50-odd botanical districts employed in current distributional studies are obviously still less adapted for that purpose. The local works have consequently been grouped under the headings of Jylland (Engl. Jutland, the mainland with some associated islands) and the island groups of Bornholm, Faroes, Fyn, and Sjælland (Zealand). In each case the name of the county (amt) to which a given flora relates is given in parentheses if not included in the title. Among the principal local floras, most of them old, are those of Bornholm (1956), the Faroes (1952), Falster (1863), Nordfyn (1910) and Sydfyn (1857), Haderslev (1873), Horsens (1915), Lolland (1864), Roskilde (1874), Nordostsjælland (1872) and Sydvestsjælland (1873-74), and Tondern (1877). There are also floras, mostly modern, of numerous medium-sized to small islands, including Ærø, Als, Anholt, Brandsø, Endelave, Fanø, Hesselø, Hjelm, Jordsand, Læsø, Langli, Manø, Rømø, Samsø, Seirø (Sejerø), Tunø, and Vorsø, as well as of a few smaller ones. Of 99 primary titles cited, 32 are general and 67 insular or local; and there are 44 subsidiary titles, making a total of 143.

FINLAND.—The republic of Finland, with a land area of 305,379 square kilometers (117,907 square miles), has a flora of 1,175 indigenous species (Ulvinen, 1948; microspecies not included). The 9 floras listed here, of which 6 are fairly modern, are mostly of the pocket manual or field guide type and are all in Finnish except for those in Swedish by Alcenius and Brenner; the standard work is Hiitonen's Suomen kasvio (1933), but more recent data on distribution is available in Hiitonen and Poijärvi's school flora (1958). The fundamental catalog of the species and their detailed distribution is Hjelt's Conspectus (1888–1926), in Latin and Swedish. In 1889 Saelán and others published a series of over 1400 diagrams showing occurrence of each species by botanical provinces. The standard name list of the flora is by Hiitonen (1934), but more recent data on distribution and a more up-to-date nomenclature are to be found in Hylander's Förteckning (1955; see under Scandinavia) of the plants of northwest Europe. There is a full, briefly annotated bibliography through 1900 (Saelán), supplemented by annual lists from 1950 on, and a very extensive catalog of vernacular names (Suhonen, 1936), as well as a list of vernacular names of garden plants (1951). Useful wild plants of all kinds are treated by Rautavaara (1942-43), and there are works on medicinal plants by three other authors. The series of publications on phenology covers two centuries (1750-1955), but has never been digested. An annotated catalog of cultivated plants (1897) is available, as well as a general account of vegetation (in English) and some other general works.

The distribution of plants and animals in Finland has for many years been expressed in terms of biogeographical districts having for the most part no connection with the 10 present administrative districts (lääni). A double system of terminology and corresponding abbreviations is in use for them, the older Latin one and the modern Finnish; the former, which is still used by many Finnish botanists, has been followed in the text on the basis of its wider intelligibility, but the Finnish names are entered as synonyms and in the index. Of the 28 districts as here taken,[67] including 13 which are partly (4) or wholly (9) Russian, 5 have separate full catalogs or name lists of the species (although only 4 of these, for Alandia, Isthmus karelicus, Karelia pomorica occidentalis, and Lapponia rossica, are now of any importance), and about a dozen others are covered by floras relating to more than one district, most of which are a century old or more. The distribution of each species by phytogeographical districts is available in Hiitonen and Poijärvi's flora (1958) except that citations of the Soviet districts have been omitted and must be sought in Hiitonen's earlier Suomen kasvio (1933). From 1 to 26 local floras are cited for each of the districts except Lapponia rossica, for which only 4 general works have been given. Of 238 primary titles cited (including 54 relating to areas which are now or always have been Russian or largely Russian), 35 are general, 19 provincial, and 184 local; and there are 188 subsidiary titles (of which 26 relate to largely Russian areas), making a total of 426.

FRANCE.—The republic of France, with an area of 550,786 square kilometers (212,659 square miles), has a flora, including Corsica, of 4,354 species (Coste, 1900–06). The standard modern descriptive flora is the 3-volume work by Coste (1900–06, reissued 1937), giving a figure of each species and little or no attention to synonymy and infraspecific taxa. A more popular but very useful work is Bonnier's Flore complète illustrée, in 12 volumes, with a good colored figure of each species and a rather detailed account of local range, variations, vernacular names, and uses. The standard detailed flora is the 14-volume Flore de France by Rouy

[67] Including Lapponia rossica, treated as a unit consisting of Lapponia Imandrae, Lapponia murmanica, Lapponia ponojensis, Lapponia tulomensis, and Lapponia Varsugae, and corresponding essentially to the Kola Peninsula.

and some collaborators (1893–1913), whose infraspecific categories are unfortunately not those in use by most other botanists. The century-old flora of Grenier and Godron may still be regarded as a standard descriptive work. There are good pocket manuals by Acloque, Bonnier, and Fournier. There is an old (1868) enumeration of species (4,294) by Bautier, including a reciprocal list of localities with their rarer species, a catalog by Camus (1888), a list of species and races by Léveillé (1916), and two lists of hybrids, neither of which is detailed enough to be of much value. Under iconographies, besides the perennially useful illustrations in Coste and Bonnier, mention should be made of the works of Baillon (500 colored plates), Cusin and Ansberque (4,759 plates, black impressions of the plant with small colored details), and Jaume St. Hilaire (1,000 colored plates). The most extensive bibliographies are those of Holden (to about 1910) and Rehder (to 1900). Dilleman published serially a useful list of departmental and local floristic works, unfortunately suspended when only about a third of the departments had been treated. The standard work on vernacular names is Rolland's Flore populaire, the greatest collection of this sort in existence; very much shorter lists are given by Gatin and Guyettant, and there are numerous local catalogs of patois names.

The leading descriptive manual of trees and woody plants is that by Mathieu (4th ed. 1897); Gayffier gives 200 excellent habit photographs, Guinier began but never completed a series of colored plates and habit photographs, and Hickel describes and figures the seeds, fruit, and seedlings of wild and cultivated woody plants. No less than 18 works (including the partial works) describe in greater or less detail the medicinal plants, which are also noted in many of the floristic publications; the most comprehensive treatment is the recent 3-volume treatise by Fournier. There are several works on edible plants, of which the most complete appears to be that by Ducomet, one on forage plants, three on useful plants in general, and three on weeds. Fournier has published a 3-volume work on hardy cultivated ferns and flowering plants, apparently referring to western Europe in general. Among miscellaneous works are publications on ecology, botanical regions, folklore, toponymy, extinct and disappearing species, a guide for botanizing, and other subjects.

Of the 90 departments (including Belfort and Corse), 6 (Ariège, Aude, Haute-Garonne, Loire, Lozère, Nièvre), have no separate departmental flora of their own and are not included in any joint departmental or provincial flora, although each of them (except Lozère, which has but 1) possesses several local floras; 13 (Bas-Rhin, Haut-Rhin, Haute-Savoie, Hautes-Alpes, Indres, Isère, Nord, Puy-de-Dôme, Savoie, Seine, Seine-et-Marne, Seine-et-Oise, Vaucluse) do not have separate departmental floras but are included in joint departmental, provincial, or regional floras; the remaining 71 departments have separate floras, but 14 of them are better covered by joint floras, all but one of which are later than the individual floras concerned. Sixty-three departments are dealt with by floras published (or begun) between 1825 and 1900, only 20 by floras since 1900. The statistics, by decades, are as follows:[68] 1825-40, 3 floras (plus 1 now replaced by a later joint flora); 1841–50, 6 (plus 3 now similarly replaced); 1851–60, 4 (plus 3); 1861–70, 11 (plus 1); 1871–80, 7 (plus 2 now replaced by later joint floras, and 1 largely replaced by a flora not yet completed); 1881–90, 17 (plus 3 now replaced by joint floras); 1891-1900, 15; 1901–10, 5; 1911–20, 6; 1921–30, 4; 1931–40, 2; 1941–50, 0; 1951–58, 3. For each department except 4 (Ardèche, Ardennes, Charente, and Indre-et-Loire, all of which possess departmental floras), one or more local publications are listed here; the only departments with 10 or more are Ariège (12), Aude (10), Bouches-du-Rhône (19), Haute-Savoie (13), Hérault (14), Savoie (11), and Var (11). Of 719 primary titles cited, 70 are general, 39 partial, 81 provincial (including Paris region), 142 departmental, and 387 local; and there are 680 subsidiary titles, making a total of 1,399.

GREAT BRITAIN WITH IRELAND.—The United Kingdom of Great Britain and Northern Ireland and the Republic of Ireland (Eire), treated together for phytogeographical and historical reasons—that is, the British Isles—have an area of 313,700 square kilometers (121,120 square miles) and a flora of 2,822 species and microspecies (Dandy, 1958; Druce, 1932, gave 2,669; both include naturalized species). The standard older manuals are by Babington (10th ed. 1922) and Bentham (7th ed. rev., 1924), with Hayward's Botanist's pocket-book (19th ed. 1930) filling a particular function; a more modern work in date and in point of view is the flora by Clapham and others. There are several catalogs, in part annotated, the most useful being Dandy's List of British vascular plants, Druce's Comital flora (complete list with distribution by vice-counties), and Watson's Topographical botany. The standard sets of illustrations are Fitch's Illustrations, Ross-Craig's Drawings (in progress), and the colored plates in Sowerby's English Botany. Druce has provided a geographically arranged list of local floras (1933) and the bibliographies by Holden (to about 1910) and Rehder (through 1900) are helpful. The standard works on vernacular names are Britten and Holland, Fisher, and Prior; more local works on the same subject will be found listed under England (General), Devon, Lincolnshire, Somerset, Ireland, Scotland, and Wales. Elwes and Henry's Trees is a large and elaborate treatment of both the native and the cultivated species, but Makins' and Step's works are of more utility for identification. There are several books on medicinal plants, of which Barton and Castle is perhaps the most scientific and Culpeper certainly the most popular; four on poisonous plants, two on edible plants, three dealing with useful plants in general, and three on weeds or adventive plants. Cultivated plants are described by Bean (trees and shrubs), Chittenden (encyclopedic), Edlin, Kirk, and Makins. Among the miscellaneous works are publications on biography, first records, herbals, hybrids, extinct

---

[68] The figures here given refer to the number of departments covered by separate or joint floras, not to the number of separate publications.

and disappearing species, ecology, phytogeography, origin of flora, plant formations, and other subjects.

For Great Britain as a whole, with Ireland, 703 primary works are cited, of which 80 are general, 44 at the level of major political or historical divisions, 130 county (comital), and 449 local; and there are 722 subsidiary titles, making a total of 1,425.

Channel Islands.—The Channel Islands, with an area of 194 square kilometers (75 square miles), have a flora of 690 species (Lester-Garland, 1903). There is no general flora but all the islands are covered by partial floras, and there is a list of the plants of the whole archipelago showing their local distribution. Six local works are cited, including one on Guernsey vernacular names, and there are 14 subsidiary titles, making a total of 20.

England.—England, with an area of 130,346 square kilometers (50,327 square miles), has a flora of 1,687 species (author's count, based on Druce's Comital flora). It has no general flora beyond those for the British Isles, but 5 general titles are cited, mostly on Saxon and Old English vernacular names. Of the 40 counties (plus the London area) all but 2 possess one or more separate county floras; one of these, that for Lancashire, is incomplete, lacking the "commonest species." The only two counties without individual floras, Leicestershire and Rutland, are covered by a joint flora. Three others, Cumberland, Durham, and Northumberland, each of which is the subject of a much older flora of its own, are comprised in a joint flora in 1939; and Yorkshire, whose only comprehensive flora dates from 1840, is fully covered by three partial floras dating from 1888 to 1906. Of these 41 floras, 4 were published from 1840 to 1869, 8 from 1887 to 1899, 12 from 1903 to 1914, 11 from 1926 to 1939, and 6 from 1948 to 1957. From 1 to 33 (York) local publications are cited for each county except 2 (Huntingtonshire, Rutland). Of 365 primary titles dealing with England, 5 are general, 78 county, and 282 local; and there are 355 subsidiary titles, giving a total of 720.

Ireland.—The island of Ireland, or Eire, including the Republic of Ireland and Northern Ireland, with an area of 83,851 square kilometers (32,375 square miles), has a flora of 1,133 species (Webb, 1953). All the species are described in the general floras of Great Britain; the only complete descriptive flora of Ireland is that of Webb (2d ed. 1953), but there is also a semipopular flora by Adams (1931). The distribution of the species by districts is stated in Praeger's The botanist in Ireland, which contains a great deal of general information on the flora. The detailed distribution in Ireland was covered in More's Contributions towards a Cybele hibernica (2d ed. 1898), and increases in knowledge were periodically compiled by Praeger through 1946. There is a list of Gaelic names of medicinal plants with notes on uses, and another of Gaelic names provided only with English (not scientific) equivalents, a list of botanists with biographical information, by Praeger, and a few minor works.

Of the 32 counties, only 7 are provided with individual floras, dating from 1878 to 1950; one of these, that for Westmeath, is incomplete, omitting "all the commoner species." On the remaining 25, 12 are completely covered by the regional floras by Dickie, Praeger, and Stewart and Corry, and 2 more are comprehended in Dickie's work except for their extreme southern part. Eleven counties (Carlow, Kildare, Kilkenny, Laoighis (Queen's), Longford, Louth, Meath, Offaly (King's), Tipperary, Waterford, and Wexford) have no county or regional floras, but all the known species of each are recorded by Praeger (1934). Of the 32 counties, local publications in one form or another are cited for all but 10 (Carlow, Kilkenny, Laoighis, Londonderry, Louth, Meath, Monaghan, Offaly, Tyrone, and Wicklow) ; 4 of these are covered by county or regional floras, 6 (Carlow, Kilkenny, Laoighis, Louth, Meath, Offaly), are not. Of 66 primary titles cited, 20 are general (including partial), 9 county, and 37 local; and there are 67 subsidiary titles, making a total of 133.

Isle of Man.—The Isle of Man, with an area of 569 square kilometers (220 square miles) has a flora of 721 species (Paton, 1933). Of 3 primary titles cited, 2 are general and 1 local; and there are 6 subsidiary titles, making a total of 9.

Scotland.—Scotland, with an area of 78,749 square kilometers (30,405 square miles), has a flora of 1,451 species (author's count, based on Druce's Comital flora; includes the bracketed records). There is no modern flora of the country, which is completely covered by the general British floras. There are general works on Gaelic and old Scottish vernacular names, two long series of papers on local distribution, some partial floras, and a few minor works. Of the 35 geographical divisions here adopted (not precisely corresponding to the number of counties (33), since the Inner and Outer Hebrides, which are made up of parts of three counties are treatd as two entities), there are individual county floras for 15, 9 others are included in joint floras dealing with more than one county; and 1 (Sutherland) is fairly covered by floras of its East and West vice-counties. Of the 10 remaining divisions, 3 (East, Mid, and West Lothian) are treated in Martin's Field-Club flora of the Lothians, but not as county units; the 7 remaining divisions (Argyll, Bute, Dunbarton, Inner Hebrides, Inverness, Nairn, Ross and Cromarty) are not dealt with in detail in any county flora, but most of them are included in Ewing's Glasgow catalogue or Lee's Flora of the Clyde area. Of the 25 possessing county floras (of which 9 are covered by 5 joint floras), 6 date from 1831 to 1864, 10 from 1879 to 1900, 5 from 1912 to 1925, and 4 from 1936 to 1941. Of 132 primary titles cited, 13 are general or partial, 30 county, and 89 local; and there are 143 subsidiary titles, making a total of 275.

Wales.—Wales, with an area of 20,761 square kilometers (8,016 square miles), has a flora of 1,606 species (Hyde and Wade, 1934 and 1954, plus the fern allies). There is no modern descriptive flora except for the ferns, but there are good lists with distribution data of both seed plants and fern (the fern allies not covered), and an old local flora with Welsh vernacular names (another list of Welsh names will be found under Diamond in Shropshire), as well as a work on native and cultivated trees. Of the 12 counties, all but 4 (Flint, Merioneth, Montgomery, and

Radnor) have comital floras (in the case of Anglesey and Caernarvon a joint flora, in the case of Denbigh not a complete flora). Of these, 4 date from 1895 to 1911, 4 from 1926 to 1950. From 1 to 7 local publications are cited for all counties except Breconshire. Of 51 primary titles, 4 are general, 13 county, and 34 local; and there are 80 subsidiary titles, making a total of 131.

ICELAND.—Iceland, an island republic with a surface of about 102,923 square kilometers (39,700 square miles), about 83 percent of which is destitute of vegetation, has a flora of 662 species[69] (Gröntved, 1942), of which 224 belong to the genera *Hieracium* and *Taraxacum*. It possesses an excellent detailed catalog by Gröntved (1942), with a considerable series of subsequent additions, and three modern descriptive floras, one in English by Ostenfeld and Gröntved (1934), two in Icelandic by Löve (1945) and Stefánsson (revised by Steindórsson, 1948), respectively, both the latter well illustrated. There are no separate works on vernacular names or useful plants except Halldórsson's volume on useful plants in 1783, but the vernacular names are more or less fully given in the floras. There are modern treatments of ornamental plants and of topography, vegetation, and plant formations. An extensive bibliography, lacking a few floristic titles, is given by Gröntved, and Dr. Áskell Löve has prepared an essentially complete, briefly annotated typescript catalog of the whole botanical literature of which he has kindly given the author a copy.

Of the 58 primary titles given, 9 are general and 49 are local floras, or in most cases no more than local lists based on one or a very few season's observations; and there are 35 subsidiary titles, making a total of 93. Aside from the lists for some islands, which may well be essentially complete, the papers most meriting the title of floras are those on Eyjafjörður (Eyja Fjord; Óskarsson, 1949), Reykjanes (peninsula; Hadač, 1949), Snæfellsnes (peninsula; Jónsson, 1899), and Vestfirðir (Vestfjarda Peninsula; Steindórsson, 1946).

The local floras here listed are not divided geographically, but in each case the position is indicated according to an arbitrary division of the island into 6 regions based on lines of latitude and longitude, as explained in the text.

ITALY.—The republic of Italy, with an area of 301,020 square kilometers (116,224 square miles), has a flora of 3,877 species, including the commonly cultivated plants (Fiori, 1923; the species taken in a broad sense). Arcangeli (1894), whose specific concept was probably nearer that of the average modern taxonomist, had 4,932 species. The standard flora is Fiori's Nuova flora analitica (1923–29), a 2-volume work; Baroni's Guida botanica d'Italia (3d ed. 1956) is the standard pocket manual. There are several older and much larger works. A name list of native, adventive, and outdoor-cultivated spermatophytes by Ciferri and Giacomini is in course of publication. There have been three editions of an Iconographia florae italicae intended to accompany

[69] Or 428 species, excluding *Hieracium* and *Taraxacum*, according to Steindórsson in Stefánsson, 1948; 535 (of which 382 are indigenous) on the same basis according to Löve and Löve, 1956.

Fiori's floras, with small but excellent figures which have been widely used, and there is a folio work by Perini with colored plates of 400 plants of northern Italy as well as Tenore's rare Flora napolitana with 250 colored plates. There is a list of Italian botanists by Saccardo with biographical notes, lists of publications, and much supplementary information, two rather extensive selected bibliographies by Béguinot (1920, 1938), and several regional bibliographies (Liguria, Sardinia, Sicily, Veneto, and southern Italy).

The standard collection of vernacular and patois names is by Penzig; there are some partial lists (Jaberg, a graphic representation of selected species with the patois names printed on large maps in the appropriate places; D. Saccardo, medicinal plants), an extensive list of the names of woody plants, and a number of local catalogs. There are several works on trees, the most outstanding being a series of papers published by the Touring Club Italiano. There are cited half a dozen works on medicinal plants, one on edible plants (there is another on Piedmonte), one on forage plants (there are others for Bergamo and Parma), one on ornamentals, one on useful and injurious plants in general, three on weeds, and one on adventive plants. Saccardo has published a list of cultivated plants giving the date of introduction of each into Italy, as well as a larger work presenting a list of Italian vascular plants, including naturalized and adventive species, with the date of the earliest records and the names of the recorders. There are publications on physiography in relation to plant and animal life, plant zones, phytogeography, botanical history, review of floristic work, locality names derived from plant names (with several local works on the same subject), plants of the small islands, plants of Virgil and the other classical writers, and on various other subjects. Among the local or regional publications some are of general interest, such as Bonacelli's La natura e gli Etruschi.

Of the 18 regioni (formerly called compartimenti) into which Italy is here divided (Val d'Aosta being included in Piemonte), there are floras of all but 5 (Abruzzi e Molise, Calabria, Campania, Emilia, Puglie). Of these 13 floras (including 2 which take in some adjacent areas, and 1 which is a joint flora), 5 date from 1844 to 1869, 5 from 1884 to 1913, and 3 from 1932 to 1950. Of the 91 provinces, 38 have floras (including 3 provinces into which 1 province has since been split, and 3 others covered by joint floras). Of these, 13 date from 1802 to 1863, 10 from 1870 to 1883, 12 from 1893 to 1921, and 3 from 1942 to 1954. Of 655 primary titles, 61 are general, 68 regional, 84 provincial, and 442 local; and there are 489 subsidiary titles, giving a total of 1,144.

LUXEMBOURG.—The grand-duchy of Luxembourg, with an area of 2,586 square kilometers (999 square miles) has a flora of about 1,150 species (Lefort in litt., 1957). It has one old descriptive flora (1875) and two catalogs (one annotated) of about the same date, but the present state of knowledge of its flora is better represented by Goffart's Nouveau manuel (see under Belgium). Extensive lists of additional records have been published from

1950 on as a result of renewed interest in the flora and in preparation for the issue of a phytogeographic chart of the country. There is a short modern separate bibliography, but the floristic titles are given much more fully in Lefort's historical treatment (1950). There is a work on medicinal plants by Feltgen (1903), one on native and cultivated trees by Koltz (1875), one on introduced plants by Fischer (1872) to which numerous additions have been made in recent years, and works on vernacular names by Klein and Weber. There is an old list of localities with their characteristic species (1877), and a critical account of botanical history, relating to both vascular and cellular plants, by Lefort (1950). Of 12 primary titles cited, 11 are general and 1 local; and there are 7 subsidiary titles, giving a total of 19.

MONACO.—The tiny principality of Monaco, with an area of 1.5 square kilometers (0.58 square miles), surrounded on the landward side by the French department of Alpes-Maritimes, has no published flora that is restricted to it but is best covered by Ardoino's Catalogue of 1862 which reports 1000 species from Monaco and the adjacent town of Menton, formerly included in the principality.

NETHERLANDS.—The kingdom of Netherlands, with an area of 40,892 square kilometers (15,765 square miles, including inland water areas, much of which is in process of being converted to land), has a flora of about 1,350 species (excluding adventives and "run wild" species). It is well provided with general floras and other general works, and has a large number of local lists, about half of which proceed from excursions of the Nederlandsche Botanische Vereeniging (now the Koninklijke Nederlandse Botanische Vereniging). The standard smaller floras are by Heimans, Heinsius, and Thijsse (19th ed. 1956), Heukels and van Ooststroom (14th ed. 1956), and Heukels and Wachter (8th ed. 1952); all describe the plant associations and botanical districts, and include cultivated as well as wild species. The latest essentially full treatment of infraspecific categories and adventive plants is contained in Heukels and van Ooststroom, Flora van Nederland (14th ed. 1956). Complete older floras on a larger scale are Oudemans' De flora van Nederland (2d ed., 3 vol., 1872–74) and Heukels' work in three volumes (1909–11) with the same title. The Koninklijke Nederlandse Botanische Vereniging is now publishing a cooperative work, the Flora Neerlandica, with full treatment of species and lower entities and detailed local range but without information on popular uses and vernacular names, of which five parts have appeared, extending through the Orchidaceae.

The standard detailed catalog with localities and citation of specimens is the Nederlandsche Botanische Vereeniging's Prodromus florae batavae (2. ed., 1901–16); the yearly crop of new records and new localities has been published, first by Heukels and later by Kloos and others, in a series that began in 1896 and still continues. Figures of most of the species are given in Heukels' De flora van Nederland and in his Flora von Nederland (14th ed., 1956), and colored plates for both vascular and cellular plants in the Flora batava by Kops and others (1800–1934). Individual maps showing the range of about 772 species are available. There is a geographically arranged list of publications on the local flora up to 1940, as well as a bibliography of phytosociology. There is an extensive list of vernacular names by Heukels (1907) and another by Gerth van Wijk, and an extremely detailed study of the patois names of 8 common species in south Netherlands and adjacent Belgium by Pauwels and Grootaers; also, among the provincial and local floras, a list of standardized Frisian vernacular names and a partial list of South Limburg patois names. The wild and cultivated useful plants are described by van Hall, Heukels (Landbouwflora), van der Trappen, and Willinge Prins, and the poisonous plants by Miquel; there are about 8 publications on one aspect or another of weeds, and they are discussed in other works on useful and cultivated plants. Cultivated plants in general are described by Bisschop van Tuinen, van Hall, and Heukels, herbaceous plants by Boom and Ruys and by Sipkes, and trees by Boom. There are works on ecology, phytogeography, botanical districts, and plant associations; on botanical history (to 1873), on botanical institutions and herbaria, and a list of Dutch biologists with dates and references to biographical notices; and three glossaries (Dutch-English or Latin-Dutch) of scientific terms. There is a seed-atlas with illustrations of the seeds of nearly all the native as well as many cultivated or naturalized species.

Of the 11 provinces (plus Noordzee-eilanden, here separated for convenience) only 5 possess provincial floras, all of which are old (Friesland, 1840; Groningen, 1860; Limburg, 1887; Noordzee-eilanden, 1898; Zeeland, 1878), but this is of no great importance in view of the general physiographic uniformity of the country and the abundant information on distribution available in other works. From 1 (Groningen) to 17 (Noord-Holland) local floras are cited for the different provinces; with only 2 exceptions, each province has 4 or more. Of 161 primary titles cited, 52 are general, 10 provincial, and 98 local; of the local floras, some 46 are lists based on excursions of the Nederlandsche Botanische Vereeniging; and there are 80 subsidiary titles, making a total of 241.

NORWAY.—The kingdom of Norway, with an area of about 323,916 square kilometers (125,182 square miles, including water areas) has a flora, including naturalized plants, of 2,341 species (Lid, 1952) of which 477 are species and microspecies of *Hieracium* and *Taraxacum*. It has two modern floras, one of which (by Lid) gives more detailed ranges while the other (by Nordhagen) includes the principal cultivated plants, and there are a number of older or popular floras, including two with abundant colored plates (see Bolin and Lindman, under Scandinavia). Illustrations of practically all the species except, presumably, many of the microspecies are available, and a systematic iconography (Nordhagen) is in process of publication. The bibliographical data restricted to Norway alone cover only the years 1859–1905, but those up to 1814 and for 1901–14 are available in other works. The useful plants, particularly the edible and medicinal species, are the subject of several publications, and there are more or less satisfactory works on vernacular names, weeds, cultivated plants, plant geography, and some minor subjects.

The 20 provinces (fylker) into which continental Norway is divided are usually combined for phytogeographic purposes into 18, 2 of them being very small and enclosed by others. Of these 2 (Finnmark (1934) and Rogaland (1892)) possess full separate floras, only the former of which is adequate by modern standards, and three others (Nord-Trøndelag, Sør-Trøndelag, and Troms) as well as parts of two more (Møre og Romsdal, Nordland) are covered by floras that include the plants of two or more provinces; 11 have no provincial floras. From 1 to 13 local floras are cited for each of the 18 fylker except Finnmark, which however, possesses the most detailed of the provincial floras. Svalbard, with 25 titles, not included in the above figures, has a flora (1927) covering the two principal islands or island-groups Spitsbergen and Bear Island; the last complete flora of the third island administratively included in Svalbard, Jan Mayen, published in 1902, contains barely two-thirds (39) of the vascular plants now known, many of which have not yet been reported in print. Of 172 primary titles cited (including Svalbard), 29 are general, 7 provincial, and 136 local; and there are 64 subsidiary titles, making a total of 236.

PORTUGAL.—The republic of Portugal, with an area (continental) of 88,607 square kilometers (34,216 square miles) has a flora of approximately 3,000 species, including introduced plants (Pinto da Silva in litt., 1957; Coutinho, 1939, had 2,845, including some cultivated species). It has two good floras (Coutinho, 1939, and Sampaio, 1947), the first with a long series of additional notes and records. There is a rather old (1913) unannotated catalog of the species by Sampaio. There is no separate bibliography, but Portuguese works up to about 1858 are given by Colmeiro along with Spanish titles, and Bellot Rodríguez has published a supplement carrying the Portuguese titles through 1935. There is an alphabetical list of vernacular names with scientific equivalents and some notes on uses (Vasconcelos, 1915); vernacular names are also given in Coutinho's and Sampaio's floras and in Colmeiro's Diccionario (see Spain). There is a work on native and cultivated trees, several on medicinal plants, and one on poisonous plants, but only partial works on useful plants in general and on weeds. There are several works on plant zones and botanical regions, an old work on the history of botany, and some miscellaneous minor publications.

Of 11 provinces, only 1 (Trás-os-Montes e Alto Douro) has a flora of its own (1944). The number of local floras of the provinces ranges from 1 (Beira Baixa, Ribatejo) to 13 (Estremadura). Of 68 primary titles, 24 are general, 1 provincial, and 43 local; and there are 54 subsidiary titles, making a total of 122.

SAN MARINO.—The little republic of San Marino, lying on the border between the regioni of Emilia and Marche in northeastern

Italy, has an area of 61 square kilometers (23 square miles) and a flora of 714 species (Pampanini, 1930). The only flora is the list by Pampanini.

SCANDINAVIA.—Under this heading are placed 22 primary and 33 subsidiary titles relating to two or more of the countries Denmark, Finland, Iceland, Norway, and Sweden, or to large parts thereof (Lapland). These works include 8 floras (of which 4 are scientific, 4 popular or semipopular with hundreds of colored illustrations), an iconography, a bibliography, 3 lists of species with indication of range by countries or regions, an atlas of distribution maps, 3 works on vernacular names, 1 on weeds, and a few on miscellaneous subjects.

SPAIN.—The republic of Spain, with an area of 492,921 square kilometers (191,424 square miles) has a flora of 6,204 vascular plants, including cultivated species (Lázaro é Ibiza, 1920–21). There are several complete floras (Amo y Mora, 1871–73; Willkomm and Lange, 1861–80, with supplement 1893; Lázaro é Ibiza, 1920–21), none of which can be regarded as satisfactory at the present time. The latest flora, by Caballero (1940), includes only the commoner or more interesting species. An entirely uncritical enumeration of the plants of Spain and Portugal with detailed local distribution by Colmeiro, in 5 volumes (1885–89), contains 9,791 species of which 6,064 are phanerogams. Publications by Willkomm and Caballero contain illustrations of about 300 characteristic, mostly or all endemic species. Colmeiro gave a bibliography in 1858 and a fuller but unannotated one in 1885, Willkomm a better one in 1896, and Rehder a pretty complete classified list of publications through 1900. Colmeiro's Diccionario gives a very extensive list of Spanish and Portuguese vernacular names of the Old and New Worlds and lists of Arabic names used in Spain and of old Spanish names of the 14th and 15th centuries, Asín Palacios has a study of Romance names of the 11th or 12th century, Más y Guindal a catalog of vernacular and pharmaceutical names of medicinal substances, and there are vernacular names in numerous floras and other works. There is an extensive treatment of native and naturalized woody plants by Laguna and Ávila (1883–90), and an anonymous work on woody plants of mountainous areas (1888). A catalog of cultivated plants grouped by uses has been provided by Dantín Cereceda, who has also published a list of 1,165 weedy plants, both of these with ample vernacular names. There are half a dozen works on native and cultivated medicinal plants. The wild and cultivated useful plants are treated by de la Puerta (1891) and in a less detailed way in other works. The ecology and phytogeography are treated by Willkomm (1896), and again recently by Font Quer, and Reyes Prósper has written on the steppe areas scattered over Spain. Willkomm gives a good account of botanical explorations up to about 1895.

Of the 14 regions (including, as here taken, the Baleares) into which the 48 provinces are grouped, 9 possess floras (4 regions being covered by 2 joint floras), of which 4 date from 1795 to 1867 and 5 from 1905 to 1937. The 5 regions lacking floras are

Andalucía, Asturias, Estremadura, León, and Murcia. Of the 48 provinces, 11 have floras, of which 3 date from 1839 to 1861, 3 from 1882 to 1899, and 5 from 1913 to 1953. Of 202 primary titles, 31 are general, 20 regional, 14 provincial, and 137 local; and there are 153 subsidiary titles, making a total of 355.

SWEDEN.—The kingdom of Sweden, with an area of 449,165 square kilometers (173,423 square miles, including water areas) has a flora, including naturalized plants, of about 2,104 species (Lindman, 1926, including 206 species of *Taraxacum* and *Hieracium* and some introduced species; Löve and Löve (1948) give 1,645 native or thoroughly naturalized species, not including micro-species or any *Hieracium* or *Taraxacum*). The standard modern flora, by Lindman (1926), is now out of date, but its place will be taken by a new Scandinavian flora by Hylander (see under Scandinavia) which, although covering a much larger area, provides fuller descriptions and more detailed ranges than were given by Lindman. There are four popular or semipopular and incomplete floras (Bolin, Lagerberg, Lindman, and Ursing, the three first cited under Scandinavia but referring primarily to Sweden), illustrated with colored plates, and several for the most part older works that are complete and more technical. Colored illustrations of more than half the species are available in Swedish works. Krok's bibliography catalogs the botanical literature very fully through 1918 and gives appropriate biographical information, but its usefulness is severely restricted by lack of a subject index; the years from 1945 on are covered by a current series of lists.

There are a number of works on vernacular names, mostly lists of would-be standard names selected or composed by individual authors but including two catalogs of Middle Age names and two extensive compilations of vernacular and patois names, of which that by Lyttkens is much the larger and better documented. There are several works on medicinal plants, nearly all popular or semipopular in character, and some similar works on other useful plants, particularly those that are edible, of which three go back to Linnaeus. There is one work on weeds (1885) and several on cultivated plants, all of which except that by Andersson (1867) are mere name lists or descriptive treatises without mention of uses, although one of them is restricted to ornamental species. There is a compilation of first records for Swedish plants, a work on agricultural climatology, a descriptive treatment of dicotyledonous seedlings, and some works on miscellaneous subjects including phytogeographic studies of north Sweden and south Sweden but not of middle Sweden. The range of practically all the species is shown on the maps in Hultén's Atlas (see under Scandinavia).

In their statement of local distribution, Swedish botanists employ the names and boundaries of the old provinces (landskap), disregarding the more modern administrative division into län, and this practice has been followed in the present work, but the names of the respective län have been added in parentheses when, as is usually the case, they do not coincide with the landskap.

Of the 29 provinces or equivalent divisions recognized (including 5 divisions of Lapland), 21 possess individual floras, of which 7 date from 1851 to 1884, 6 from 1897 to 1909, and 8 from 1924 to 1952. Of the 8 that lack individual floras, 5 (Åsele, Lule, and Lycksele Lappmark, Norrbotten, Västerbotten) are included in a joint regional flora of 1878; one (Gästrikland) is fairly covered by a partial flora of 1863 with extensive subsequent additions embracing the whole area; one (Södermanland) by a joint flora of 1871; and only Ångermanland completely lacks a provincial flora. There is an abundance of local floras, a great many of them relatively modern. From 3 to 32 (Västergötland) local works are cited for each province except 4, of which 2 (Dalsland, Öland) have 1 each and 2 (Gästrikland, Hälsingland) have none; of these 4, only 1 (Öland) has a modern flora. Of 315 primary titles, 47 are general, 7 partial, 27 provincial, and 234 local; and there are 357 subsidiary titles, making a total of 672.

SWITZERLAND.—The republic of Switzerland, officially termed La Confédération Suisse, with an area of 41,295 square kilometers (15,944 square miles), has a flora of 2,578 species (Binz, 1941). There are two good modern pocket manuals, Schinz and Keller (1923) and Binz (1957), the latter also with a French edition, and several less useful older floras. An oldtime favorite manual, that of Gremli, is now out of date (last edition 1901). There is an old list (1897), showing the distribution of each plant in the cantons, and from 1892 to date the most systematic and complete listing of new records that is available for any country in the world. There are several lists of alpine plants, and an extensive catalog of alleged hybrids (Brügger). There is a collection of small but excellent colored plates by Hegetschweiler (the drawings by Labram), usually listed as numbering 480 but actually amounting to 882 in some sets, and two duodecimo atlases of alpine plants with colored habit figures of 500 and 400 species, respectively, as well as some smaller ones; also a set of 255 excellent heliotype plates by Marret. Thommen has published a pocket volume giving a small figure of each species, mostly with enlarged details, intended for field use. There is an excellent classified bibliography by Fischer going through 1900, a less complete classified catalog by Rehder ending with the same year, and a current listing of new publications from 1890 to date, containing abstracts up to the year 1915. Durheim (1856) provided the most extensive collection available of French, German, Italian, and Romansch vernacular names, Kohler gave a list of Germanic names, Savoy and Empeyta the Romand names and Coaz the names of woody plants in the four languages; and there are two works on locality names derived from Germanic vernacular names and one on the French names. Brunies began a list of Romansch names used in Graubünden, abruptly terminated by death when he had reached only the letter F.

There is a popular manual of wild and cultivated woody plants by Kienli, a descriptive flora of native and naturalized woody plants by Empeyta, and a table of the altitudinal limits of trees by Bühler. There are four works on medicinal plants, one on poisonous plants, an old work on edible plants, and a work on useful alpine plants. There are four publications on weeds and adventive plants, none of them complete. There is no treatment of cultivated plants as such, although the cultivated woody plants are partly covered in one popular work. The general phytogeography has been described by Christ in a classical work, that of alpine plants by Jerosch, and Furrer has given an account of plant formations, ecology, history of vegetation, agriculture, and kindred subjects. There is an old work on the plants of the ancient lake dwellings, and a monograph on plant remains from the Paleolithic on. Bruhin gave a sketch of botanical history from Gesner to de Candolle, and a list of identifications of the plants described by Gesner and his contemporaries. Two important works relating to all Central Europe have been listed under Switzerland, which they cover: Hegi's Illustrierte Flora von Mittel-Europa, an unmatched collection of information on many subjects besides description and identification, and Kirchner's Lebensgeschichte der Blütenpflanzen Mitteleuropas, dealing with biology and ecology; and other important Germanic publications that refer also to Switzerland are mentioned by name on p. 28.

Of the 22 cantons here recognized (25 if the halb-cantons were treated separately), 14 are the subject of separate floras, 7 are covered by 3 joint floras, and only 1 (Bern) lacks a flora, but for it there is available a modern flora covering the greater part of the canton. Of the 21 effective canton floras, 4 date from 1839 to 1866, 9 from 1881 to 1896, and 8 from 1910 to 1949. Four of the cantons (Appenzell, Fribourg, Uri, and Zug) have no local floras or other local publications cited, and for 4 others only 1 each is listed; 12 of the other cantons have from 4 to 13 each, Valais has 24, and Graubünden 27. Of 234 primary titles listed, 56 are general (a few of these partial), 42 cantonal, and 136 local; and there are 191 subsidiary titles, making a total of 425.

In the first part of this work a list of 100 selected floras and floristic works was presented, considered to include the items of greatest general utility. A similar list is given here for the countries included in the present part, but the number has been reduced to 67, including only the major publications, because of the impossibility of choosing among the very numerous county and provincial floras. For Andorra: Losa España and Montserrat, Aportación al conocimiento de la flora de Andorra; for Belgium: Crépin, Manuel de la flore de Belgique; De Wildeman and Durand, Prodrome de la flore belge; Goffart, Nouveau manuel de la flore de Belgique et des régions limitrophes; Hauman and Balle, Catalogue des ptéridophytes et phanérogames de la flore belge; Lawalrée, Flore générale de Belgique; for Denmark (see also Scandinavia): Christensen, Den danske botaniks historie med tilhørende bibliografi; Dansk Botanisk Forening, Danmarks topografisk-botaniske undersøgelse; Raunkiær, Dansk ekskursions-flora; Rostrup, Den danske flora; Rasmussen, Føroya flora (see also Ostenfeld and Gröntved, under Iceland); for Finland (see also Scandinavia): Alcenius, Finlands kärlväxter; Hiitonen, Suomen kasvio; Hiitonen and Poijärvi, Koulu- ja retkeilykasvio; Hjelt, Conspectus florae fennicae; Saelán, Finlands botaniska litteratur; for France: Bonnier, Flore complète illustrée en couleurs de France, Suisse et Belgique; Coste, Flore descriptive et illustrée de la France; Fournier, Le livre des plantes médicinales et vénéneuses de France, and his Les quatre flores de France; Rolland, Flore populaire; Rouy, Flore de France; for Great Britain with Ireland: Bentham, Handbook of the British flora; Britten and Boulger, A biographical index of deceased British and Irish botanists; Britten and Holland, A dictionary of English plant-names; Clapham and others, Flora of the British Isles; Dandy, List of British vascular plants; Druce, The comital flora of the British Isles; Fitch, Illustrations of the British flora; Ross-Craig, Drawings of British plants; Sowerby and Smith, English botany; Praeger, The botanist in Ireland; Webb, An Irish flora; for Iceland: Gröntved, The Pteridophyta and Spermatophyta of Iceland; Löve, Íslenzkar jurtir; Ostenfeld and Gröntved, The flora of Iceland and the Faroes; Stefánsson, Flóra Íslands; for Italy: Arcangeli, Compendio della flora italiana; Baroni, Guida botanica d'Italia; Ciferri and Giacomini, Nomenclator florae italicae; Fenaroli, Flora delle Alpi; Fiori, Nuova flora analitica d'Italia; Fiori and Paoletti, Iconographia florae italicae; Penzig, Flora popolare italiana; for Luxembourg: Koltz, Prodrome de la flore du grand-duché de Luxembourg (and see Goffart under Belgium); for Netherlands: Gerth van Wijk, A dictionary of plant names; Heukels, De flora van Nederland; Heukels and van Ooststroom, Flora van Nederland; Koninklijke Nederlandse Botanische Vereniging, Flora Neerlandica, and its (as Nederlandsche Botanische Vereeniging) Prodromus florae batavae; for Norway (see also Scandinavia): Lid, Norsk flora; Nordhagen, Norsk flora; for Portugal: Coutinho, Flora de Portugal; Sampaio, Flora portuguesa; for San Marino: Pampanini, Flora della repubblica di San Marino; for Scandinavia: Hultén, Atlas över växternas utbredning i Norden; Hylander, Förteckning över Nordens växter, and his Nordisk kärlväxtflora; for Spain, Lázaro é Ibiza, Botánica descriptiva; Caballero, Flora analítica de España; Willkomm and Lange, Prodromus florae hispanicae; for Sweden (see also Scandinavia): Krok, Bibliotheca botanica suecana; Lindman, Svensk fanerogamflora; Lyttkens, Svenska växtnamn; for Switzerland: Binz and Thommen, Flore de la Suisse; Fischer, Flora helvetica; Hegi, Illustrierte flora von Mitteleuropa; Schinz and Keller, Flora der Schweiz; Thommen, Taschenatlas der schweizer flora.

ANNOTATED LIST OF FLORAS AND FLORISTIC WORKS

Andorra

General

Barnola, J. M. de. Flora vascular del principado de Andorra. Mem. Soc. Ibérica Cien. Nat. no. 1. 90 p. 1 fig., map. 1919. (Library of Academy of Natural Sciences, Philadelphia.)

Hydrography, geology, etc.; unannotated list of known vascular plants [1034 species], with localities and vernacular names. Replaces **Cousturier, Paul,** and **Gandoger, Michel.** Florule de la république d'Andorre (Pyrénées espagnoles). Bul. Soc. Bot. France **60:** 524-531, 550-557. 1913. (Bibliography, notes on distribution, list of localities with altitudes; list of over 900 species of vascular plants, with localities, based mostly on the collections of Cousturier in 1913.)—See also **Marcet, A. F.** Contribución a la flora de Andorra. Bol. R. Soc. Españ. Hist. Nat. 44: 77-86. 1946. (Extensive list of vascular plants, with localities, collected by author and E. R. Svensson-Sventenius in September 1942, including 10 additional genera and 91 species and varieties.)

Losa España, Mariano, and Montserrat, Pedro. Aportación al conocimiento de la flora de Andorra. Primo Congreso Internacional del Pirineo del Instituto de Estudios Pirenaicos [Monog.] no. general 53 (Botánica 6). 184 p. illus., 6 pl. (on 3), maps (distr. maps). 24.5 cm. Zaragoza, 1950 (cover date 1951).

Botanical explorations, geology, general features of flora, plant associations; annotated list of known vascular plants (about 1150 species); list of fungi, hepatics, and mosses, with localities; short list of vernacular names. At head of title page: Consejo superior de investigaciones científicas. Also printed in Actas del Primo congreso internacional del Pirineo [etc.] v. 1. San Sebastian, 1950. (This issue not seen.)—See also **Losa España, M.** La vegetación dominante en Andorra. 35 p. 5 pl., map. 22 cm. Barcelona, 1954. (At head of title page: Sociedad farmacéutica del Mediterráneo latino. Grupo español.) (Geology, climate, topography, general features of vegetation and plant zones; in Spanish and French.)

Local

Marcailhou-d'Ayméric, Hippolyte. Contribution à la flore de l'Andorre. Ascensions au puig de Coma Pedrosa (2946m.) et au puig dels Pessons (2865m.). Explor. Pyrén. Bul. Soc. Ramond **33** (II, **2** [i.e. 3]): 28-58. 1898.

Itinerary, with lists of species collected at different localities.

Belgium (Belgique; België)

See also France, General (Bonnier, Flore complète illustrée; Bonnier et Layens, Flore complète portative; Camus, Catalogue des plantes; Coste, Flore descriptive; Dubois, Matière médicale indigène; Rolland, Flore populaire); also France, Partial (Bonnier and Layens, Nouvelle flore); Netherlands, General (Pauwels and Grootaers, Enkele bloemnamen); Switzerland, General (Probst, Wolladventivenflora Mitteleuropas).

General

Includes some partial works.

The general works here listed may be classified as follows: *Floras and manuals:* Bertrand, Cogniaux, Crépin (Manuel), Delogne, Devos (Flore), Enckels, Goffart, Lawalrée, MacLeod, Mathieu, Paque (Guide). *Catalogs:* De Wildeman and Durand, Hauman and Balle, Massart (two on littoral and alluvial flora). *Bibliographies:* (Devos), (De Wildeman and Durand), Duvigneaud, Georlette (phytosociology), Gravis, (Hauman and Balle), Rehder, Robyns, Van Aerdschot. *Vernacular names:* Boxus (Walloon), Feller (Walloon), (Lawalrée), Paque (De Vlaamsche volksnamen: Flemish), Vandenbussche. *Trees:* Aigret, Wesmael. *Useful plants* (all mostly or entirely medicinal except Hermans and Lawalrée): Aigret and François, Boxus (Walloon), Chalon, Hermans, (Lawalrée), Rottenburg, Sonet, Thielens, Vandenbussche, Van Heurck. *Weeds and adventive plants:* Devos (Les plantes naturalisées), Höck. *Cultivated plants:* Aigret (trees), Delogne, De Wildeman (historical), Morren (horticultural history). *Miscellaneous:* Chalon (rare plants), Crépin (Guide: botanical history), Devos (Coup d'oeil: biographical and bibliographical), Duvigneaud (historical and bibliographical, 1940–1944), Georlette (phytosociology), Gravis (history and bibliography), Lebrun (plant associations), Massart (Esquisse: plant associations, phytogeography).

Aigret, Clément. Flore analytique et descriptive des plantations le long des routes de l'État en Belgique. Annales Trav. Pub. Belg. II, **10:** 473–526, 779–800, 997–1042, 1191–1286. fold. pl. 21, 27–29, 33–38. 1905. (Reprinted 217 p., 10 pl. Original in Arnold Arboretum library; reprint not seen.)

Keyed, descriptive flora of the native and exotic trees planted along the state roads of Belgium, with notes on cultivation, injurious insects, wood, etc.—See also Wesmael, below.

——— **and François, Vital.** Flore de la Belgique. Plantes médicinales et traité de médécine familière . . . xvi, 199, 12 p. 20.5 cm. Olloy-lez-Mariembourg, 1886.

Includes annotated, classified lists of native and principal exotic medicinal plants, with vernacular names, uses, etc. Botanical names (except family names) are not given. For other works on medicinal plants see Boxus, Chalon, Rottenburg, Sonet, Thielens, Vandenbussche, and Van Heurck below.

Bertrand, Marcel. Synopsis de la flore de Belgique. Nouvelle édition du Synopsis de M. C. Alexis. 236 p. illus. (113 plates of fig.). 18 cm. Gembloux, 1928.

Annotated flora of vascular plants in form of keys. (Not seen; title and annotation supplied by W. Robyns.)

Boxus, Robert. La flore médicale wallonne. 179 p. 24 cm. Huy, 1939.

Alphabetical list of 1847 Walloon names of native and cultivated plants, with short descriptions, range, flowering dates, medicinal properties and uses. (Not seen; title and annotation supplied by W. Robyns.)—See also (for Walloon names) Feller, below, and Bastin, Boxus, Dahmen, Desfresne, and Germay, under Liège; and, for works on medicinal plants, Aigret and François, above, and additional references there given.

Chalon, Jean. Les plantes médicinales et vénéneuses de la flore belge. Bul. Soc. Roy. Bot. Belg. 48: 107–194. 1912.

Discussion of earlier works; annotated list of native and cultivated medicinal and poisonous plants, vascular and cellular, systematically arranged, with uses, etc.—See also Aigret and François, above, and additional references there given.

——— and others. Que sont devenues nos plantes rares de 1862? Bul. Soc. Roy. Bot. Belg. 52: 18–93. 1913.

Annotated lists, by numerous authors, of rarer vascular plants and Characeae of various areas in Belgium; index.—See also his Un sujet intéressant d'observation. Bul. Soc. Roy. Bot. Belg. 48: 74–77. 1912. (Unannotated list of vascular plants of Belgium indicated as very rare by Crépin in the first edition of his Manuel in 1862 [actually 1860].)

Cogniaux, Alfred. Petite flore de Belgique à l'usage des écoles. 4. éd. 349 p. 138 fig. 16 cm. Bruxelles, 1910. (1st ed. 1883.)

Annotated flora of vascular plants and Characeae in form of very briefly annotated keys. (Not seen; title and annotation supplied by W. Robyns.)

Crépin, François. Guide du botaniste en Belgique (plantes vivantes et fossiles). vii, 495 p. 19 cm. Bruxelles, Paris, 1878. (Gray Herbarium library.)

History of botany in Belgium, botanical gardens, libraries, botanical regions, list of fossil flora, lists of rarer plants of various regions, bibliographies.

——— Manuel de la flore de Belgique. 5. éd. lxviii, 495 p. 634 fig., map. 17 cm. Bruxelles, 1884. (Numerous reissues, sometimes numbered as new editions. 1st ed. 1860.)

Glossary (terms in French and Flemish), key to families; annotated, keyed flora of vascular plants and Characeae (1,437 species, of which 1251 are indigenous), with descriptions of families and genera (and keys to the species) and French and Flemish vernacular names. Still the standard manual of the Belgian flora.

Delogne, C. H. Flore analytique de la Belgique. Plantes indigènes et cultivées. xv, 655 p. 20.5 cm. Namur [pref. 1888]. (Gray Herbarium library.)

Botanical zones, etc.; briefly annotated flora of wild and cultivated spermatophytes [4,240 species] in form of keys, with vernacular names; glossary.

Devos, André. Coup d'oeil sur l'histoire de la flore belge. Bul. Soc. Roy. Bot. Belg. 27 (1): 15–70. 1888.

Critical account of botanists and botanical publications; list of old herbaria; bibliography, chronologically arranged, containing separately published works only.—See also (for bibliographies) De Wildeman and Durand, Duvigneaud, Georlette, Gravis, Hauman and Balle, Rehder, Robyns, and Van Aerdschot, below.

——— Flore complète de la Belgique. Espèces indigènes et plantes cultivées sans abri. xxiii, 739 p. 17 cm. Mons, 1885.

Annotated flora of wild and hardy cultivated vascular plants with keys and usually brief diagnoses, and notes on uses; bibliography.

——— Les plantes naturalisées et introduites en Belgique. Bul. Soc. Roy. Bot. Belg. 9: 5–122. 1870.

General remarks, statistics; annotated list of 512 phanerogams, including cultivated plants.—See also Höck, below.

De Wildeman, Émile. Notes pour l'histoire de la botanique et de l'horticulture en Belgique. Mém. Acad. Roy. Belg. Cl. Sci. Collect. in-8° v. 25. 2 p.l., 832 p. 1950.

Discussion of early horticulturists in Belgium; alphabetical list of the Latin polynomial names of plants in J. Hermann, Recensio plantarum in horto magistri Joannis Hermanni, pharmacopoei Bruxellensis, excultarum, 1652, and its Appendix, 1653 (the latter on p. 600–638), with discussion of their identity and citation of synonyms from other, mostly pre-Linnaean authors; tabular systematic list of "la plupart" of the plants cultivated in Belgium toward the middle of the 17th century, with indication of their presence in the gardens of Hermann, P. Ricart, and B. Wynhouts; alphabetical list of polynomial names of plants cultivated in garden of P. Ricart at Lille in 1644; discussion of early gardens of Belgium and of general problems of plant introduction and naturalization.—See also Morren, below.

——— and **Durand, Théophile.** Prodrome de la flore belge. 3 v. (543, 530, 1112 p.). 25 cm. Bruxelles, 1898–1907.

Statistics of flora, increase in knowledge of phanerogams by years (1854–1893), list of naturalized species (with dates), botanical regions, bibliography through 1905 (605 titles); annotated list of cellular and vascular plants (8,896 species, of which 1,258 are native phanerogams), with localities. Introduction and treatment of phanerogams by Durand, remainder by De Wildeman.—See also **Magnel, Louis.** Variétés observées en Belgique. Bul. Soc. Roy. Bot. Belg. 57: 59–74. 1924. (Extensive list of varieties and minor forms not listed in the Prodrome.)—Also his Annotations au Prodrome . . . l.c. 57: 130–138. 1925. (Lists of native and introduced plants, with localities.)—Also **Isaacson, Adolphe.** Annotations au Prodrome . . . l.c. 59: 32–33. 1926. (List with localities.)—Also **Matagne, H.** Quelques phanérogames nouvelles pour la flore belge. l.c. 71: 68–72. 1938; 75: 123–131. 1943. (Annotated lists of spermatophytes new to the Belgian flora.)—Also **Lawalrée, André, and Vanden Berghen, C.** Notes sur quelques phanérogames de la flore belge. l.c. 78: 100–105. 1946. (Annotated list of 26 angiosperms, some new to Belgium.)—Also **Lawalrée, André.** Adventices intéressantes de notre pays. l.c. 81: 40–44. 1949. (Account of 10 adventive species not previously recorded.)—Also his Contribution à l'étude de la flore adventice de la Belgique. l.c. 86: 137–143. 1 fig. 1953. (List of 21 plants, some new to Belgium.)—Also **Druet, J., and Duvigneaud, J.** Note sur quelques phanérogames adventices. Lejeunia 15: 59–62. 1955. (Includes some species new to Belgium.)—Also **Lambinon, J.** Contribution à l'étude de la flore adventice de la Belgique. 1. Adventices rare ou nouvelles pour la Belgique. Bul. Soc. Roy. Bot. Belg. 89: 85–100. 1957. (Data on 30 species.)

Duvigneaud, Paul. Botany. The study of the flora of Belgium. *In* Hacquaert, Armand, *ed.* Natural sciences in Belgium during the war. A series of articles on scientific progress. p. 65–91. 2 fig. (ports.). 26.5 cm. Ghent, 1946.

Classified account of ecological and floristic work on cellular and vascular plants (1940–44), with references and discussion.—See also Devos (Coup d'oeil), above, and additional references there given.

Enckels, R., Gielen, G., and Michiels, N. Geïllustreerde flora voor Zuid-Nederland. 3 p.l., xxi, 100 p. 213 fig. 19.5 cm. Antwerpen, 1953. (Biokosmos-collectie no. 2–5.)

Elementary flora of wild and cultivated vascular plants in form of keys, with localities for scarcer species. Refers only to Flemish Belgium.

Feller, Jules. Flore populaire wallone. Bul. Folklore (Bruxelles) 1: 154–172, 203–224, 277–293. 1891–92; 2: 101–122, 207–218, 293–309. 1893–95.

Bibliography; systematic list of phanerogams, with Walloon vernacular names, localities where used, discussion of meaning, etc. Incomplete; covers 34 families (Ranunculaceae-Amygdalaceae).—See also Boxus, above, and additional references there given.

Georlette, René. Matériaux pour une bibliographie phytosociologique de la Belgique. Annales Gembloux 54: 38–259. 1948.

Sketch of previous work on plant sociology in Belgium, arranged geographically; bibliography.—See also Devos (Coup d'oeil), above, and additional references there given.

Goffart, Jules. Nouveau manuel de la flore de Belgique et des régions limitrophes . . . Ancien Manuel de la flore de Belgique par François Crépin, remanié, mis à jour et étendu au nord de la France, au G.-D. de Luxembourg, aux provinces rhénanes et au sud de la Hollande. 3. éd. augmentée d'environ 200 espèces étrangères, introduites. 42, 525 p. illus., map, port. 19.5 cm. Liège [pref. 1945]. (1st ed. 1935.)

Topography and physical regions (by O. Tulippe) phytogeographic regions and habitats (with lists of species), glossary, keys to families and to some special groups; briefly annotated flora of vascular plants in form of keys, with localities for the rarer species.

Gravis, Auguste. La botanique en Belgique de 1830 à 1905. 33 p. 26 cm. Bruxelles, 1914. (Extrait du Mouvement scientifique en Belgique.)

List of principal Belgian publications in each field of botany, with mention of principal investigators.—See also Devos (Coup d'oeil), above, and additional references there given.

Hauman, Lucien, and **Balle, Simone.** Catalogue des ptéridophytes et phanérogames de la flore belge. Énumération mise à jour et révisée au point de vue systématique et synonymique des espèces et de leurs variétés. 126 p. 24.5 cm. Gembloux, 1934. (Suppl. au t. LXVI du Bulletin de la Société royale de botanique de Belgique.)

Systematic list of vascular plants (1,710 species of which 1,367 are definitely native) with indication of range by districts; statistics, bibliography of publications since 1903.

Hermans, R. Wilde planten op het woudloopersmenu. 65 p. illus. 12 × 15.5 cm. Brussel, 1945.

List of species, grouped by uses; popular account of edible wild vascular plants, alphabetically arranged by vernacular names.

Höck, Fernando. Ankömmlinge in der pflanzenwelt Mitteleuropas während des letzen halben jahrhunderts. Bot. Centralbl. Beih. **9:** 241–255, 321–333, 401–417. 1900; **10:** 284–300. 1901; **11:** 261–281. 1902; **12:** 44–54. 1902; **13:** 211–234. 1902; **15:** 387–407. 1903; **17:** 195–210. 1904; 2. Abt., **18:** 79–112. 1904.—Neue ankömmlinge . . . 2. Abt., **26:** 391–433. 1910.—Ergänzungen . . . 2. Abt., **32:** 71–110. 1914.

List of 627 vascular plants that have become widely adventive or naturalized in "Mitteleuropa" (Germany, Austria-Hungary, Belgium, Netherlands, Luxembourg) since about 1850, with extralimital range (including other regions where introduced), and local records with references. The original list ends in vol. 17, and each part after the first contains additions to the preceding parts. Vol. 18 contains a summary list of 665 species, with references to the earlier parts and brief indication of native range and local occurrence; also discussion of sources of origin, etc. The "Neue ankömmlinge" contains a list of 358 additions, with localities and references. The "Ergänzungen" contains a similar list of 178 additions, a list of genera with number of species mentioned and references to earlier parts, and brief general considerations.—See also Devos (Les plantes naturalisées), above.

Lawalrée, André. Flore générale de Belgique . . . Ptéridophytes. 2 p. l., iv, 195 p. 62 fig. (incl. distr. maps). 23.5 cm. Bruxelles, 1950.—Spermatophytes. v. 1–3, fasc. 2., illus. (incl. distr. maps). 24.5 cm. Bruxelles, 1952–59. (At head of title page: Ministère de l'agriculture. Jardin botanique de l'État.)

Keyed, descriptive flora of vascular plants, with detailed citation of localities, general range, habitat, vernacular names, uses, folklore, ethnology, etc. Vol. 3, fasc. 2 reaches *Potentilla.* The whole flora, to be published under the direction of Walter Robyns, will include volumes on bryophytes, algae, fungi, and lichens.

Lebrun, Jean; Noirfalise, Albert; Heinemann, Paul; and **Vanden Berghen, Constant.** Les associations végétales de Belgique. Bul. Soc. Roy. Bot. Belg. **82:** 105-207. 1949.

General considerations; systematic treatment of the associations, subassociations, and higher groups in the phytosociology of Belgium, with short descriptions; bibliography, supplementing that in the following work, which it otherwise replaces: **Louis, Jean,** and **Lebrun, Jean.** Première aperçu sur les groupements végétaux en Belgique. Bul. Inst. Agr. Sta. Rech. État Gembloux 11: 1–86. 20 fig. on 10 pl. (on 5). 1942.—See also Massart (Esquisse), below.

MacLeod, Julius, and **Staes, G.** Geïllustreerde flora voor België . . . 8. uitgave. Opnieuw herzien en bijgewerkt door Paul van Oye. xxviii, 257 p. 272 fig., map. 16 cm. Antwerpen, 1947. (1st ed. 1892.)
Key to families; pocket flora of vascular plants in form of very briefly annotated keys.

Massart, Jean. La 50e herborisation générale de la Société royale de botanique de Belgique. Sur le littoral belge. Bul. Soc. Roy. Bot. Belg. **51:** 69–187. plates (incl. maps), 3 maps. 1913.
Includes (p. 152–185) "Flore du littoral belge entre Westende et la frontière française," with indication of ecological habitats.—See also the second title below.

——— Esquisse de la géographie botanique de la Belgique. Recueil Inst. Bot. Léo Errera v. 7 bis. xi, 322 p. 101 fig. incl. maps. 1910; and Annexe. xiii p. 466 i.e. 462 fig. on pls., 9 maps, 2 diagrams. 1910.
Geological history, climate, soils, general features of plant associations, geobotanical districts and their plant associations, bibliography.—See also Lebrun, above.

——— Liste géographique des ptéridophytes et des phanérogames qui habitent les districts littoraux et alluviaux de la Belgique . . . Recueil Inst. Bot. Léo Errera v. 7, Annexe. 121 p. plates, charts, maps. 1908.
Includes (p. 4–35) tabular list of vascular plants, with indication of geographical and ecological distribution within the region, and of general distribution outside it; also ecological lists. Is the floristic and tabular supplement to his ecological study entitled "Essai de géographie botanique des districts littoraux et alluviaux de la Belgique." l.c. **7:** 167–584. 23 fig. (incl. maps). 1908 (also published in Bul. Soc. Roy. Bot. Belg. v. 44–46. 1907–09).—See also the second title above.

Mathieu, Charles. Flore générale de Belgique, contenant le description de toutes les plantes qui croissent dans ce pays. 2 v. (655, 561 p.). 23 cm. Bruxelles, 1853 (cover date, 1854).—Supplément . . . ii, 43 p. 22.5 cm. Bruxelles [etc.], 1855. (Arnold Arboretum library.)
Briefly annotated descriptive flora of vascular (1882 species, including some cultivated plants) and cellular plants. Vol. 1, phanerograms; v. 2, cryptogams.

Morren, Édouard, and **Devos, André.** Index bibliographique de l'hortus belgicus. Catalogue méthodique des plantes ornementales qui ont été décrites, figurées ou introduites en Belgique de 1830 à 1880. xxiv, 617 p. 24.5 cm. [Bruxelles] 1887. (Arnold Arboretum library.)
History of horticulture in Belgium; bibliography of botanical gardens, societies, private collections, periodicals, etc.; systematic list of species, with date of introduction and reference to Belgian illustrations. Pages v-xxiv originally appeared in Belg. Hort. **26:** 235–253. 1876; pages i-xxiv, 1–264 as supplement to Bul. Féd. Soc. Hort. Belg., vol. for 1883/85, issued 1887, and pages 265–617 with the Bulletin for 1886, issued 1888.—See also De Wildeman (Notes), above.

Paque, Égide. Guide de l'herborisateur en Belgique. (Plantes phanérogames et cryptogames spontanées ou fréquemment cultivées.) Nouv. éd. entièrement remaniée et complétée. 2 p. l., 117 p. 18.5 cm. Namur, 1900. (Lloyd Library.)
Scarcely annotated and very condensed synoptical flora of cellular and vascular plants, the cellular plants very incompletely treated. The author intended the work as a new edition or successor of **Bellynck, Auguste.** Catalogue des plantes soit spontanées, soit cultivées en grand, observées en Belgique, à l'usage des herborisations. 80 p. 17 cm. Namur, 1876. (List of wild and cultivated vascular plants (and in very condensed form the cellular plants), with brief diagnostic characters and indication of frequency of occurrence. In Arnold Arboretum library.)

——— De Vlaamsche volksnamen der planten van België, Fransch-Vlaanderen en Zuid-Nederland met aanduiding der toepassingen en der genezende eigenschappen der planten. 569 p. 675 fig. 22 cm. Namen, 1896. (Gray Herbarium library.)—De Vlaamsche volksnamen der planten van België, Fransch-Vlaanderen, Noord-Brabant, Hollandsch-Limburg, enz. met aanduiding der toepassingen en der genezende eigenschappen der planten. Bijvoegsel . . . 156 p. 45 fig. 25 cm. Brussel, 1912 (cover date, 1913). (Gray Herbarium library.)

Brief bibliography; alphabetical list of Flemish vernacular and patois names of vascular and cellular plants with botanical equivalents, localities where used, and derivation; list of Belgian (French) vernacular names, with botanical equivalents; list of botanical names, with vernacular equivalents and notes on uses. There is no index of scientific names in the Bijvoegsel.—See also, for other works on vernacular names, Boxus, above, and additional references there given; also Vandenbussche, below.

Rehder, Alfred. (Phytography.) Belgium and Luxembourg. *In* his The Bradley Bibliography. A guide to the literature of the woody plants of the world published before the beginning of the twentieth century. Compiled at the Arnold Arboretum of Harvard University under the direction of Charles Sprague Sargent. v. 1, p. 391–393. 29.5 cm. Cambridge, Mass., 1911.—Additions and corrections. l.c. **1**: 532. 1911; **5**: xx. 1918.

Essentially unannotated list of works (including those in periodical and serial literature) on "Dendrography" and on "General phytography," chronologically arranged.—See also l.c. **1**: 15–16, for alphabetical list of periodicals and serials relating to botany; **1**: 49–50, botanic gardens; **1**: 73, vernacular names; **1**: 80, botanical history; **1**: 226, phenology; **1**: 296, individual noteworthy trees; **3**: 15. 1915, alphabetical list of periodicals and serials relating to arboriculture and gardening; **3**: 28, collections, gardens, arboretums, museums and exhibitions (not herbaria); **3**: 33, history of gardening; **3**: 77–78, dendrological, pomological, economical, and miscellaneous works dealing with horticulture, fruit trees, medicinal and other useful or ornamental plants; **3**: 106, arboriculture and general horticulture; **3**: 250, pharmacopoeias and dispensatories; **3**: 270–271, economic products and useful and medicinal plants; **4**: 12. 1914, periodicals on forestry and woods; **4**: 19, exhibitions of forests products; **4**: 27, history of forests; **4**: 84, dendrography; **4**: 357, forests. Various smaller sections of no floristic significance have been omitted from the above summary.—See also Devos (Coup d'oeil), above, and additional references there cited.

Robyns, Walter. Une flore générale de Belgique. Bul. Jard. Bot. État Bruxelles **21**: 1–10. 1951.

Sketch of comprehensive works on the vascular and cellular plants of Belgium (since 1876), with bibliography covering general works on all groups of Belgian plants (from 1781 on). Does not include local floras or works on single genera.—See also Devos (Coup d'oeil), above, and additional references there given.

Rottenburg, V. H. Flore médicale belge ou nomenclature des plantes médicales qui croissent naturellement en Belgique ou qui y sont cultivées. 79 p. 24 cm. Bruxelles, 1881.

List of native and cultivated medicinal plants, arranged by families, with properties, uses, and doses. (Not seen; title and annotation supplied by W. Robyns.)—See also Aigret and François, above, and additional references there given.

Sonet, Ernest. Flore médicale belge. viii, 421 p. illus. 22 cm. Schaerbeek, 1899.

Systematic treatment of over 600 native and cultivated medicinal plants, with descriptions, properties, and uses, and figure of nearly every species; dictionary of medical terms. (Not seen; title and annotation supplied by W. Robyns.)—See also Aigret and François, above, and additional references there given.

Thielens, Armand. Flore médicale belge. iii, 335 p. 19.5 cm. Bruxelles et Leipzig, Paris, 1862.

Annotated list of principal native and cultivated medicinal plants, systematically arranged, with synonyms, European vernacular names, description, dose, etc.—See also Aigret and François, above, and additional references there given.

Van Aerdschot, Paul. Travaux botaniques publiés en Belgique ou par des botanistes belges [en 1902 et 1903—1930 et 1931]. [I]-XIV. Bul. Soc. Roy. Bot. Belg. 41 (2): 240–243. 1904; **42** (2): 242–251. 1906; **44**: 363–376. 1908; **46**: 387–413. 1910; **48**: 326–361. 1912; **52**: 294–311. 1913 (1914); **54**: 224–245. 1921; **55**: 205–224. 1923; **56**: 191–205. 1924; **57**: 199–210. 1925; **58**: 247–261. 1926; **61**: 182–207. 1929; **63**: 55–80. 1930; **65**: 35–76. 1932.

Unannotated lists of publications arranged alphabetically by authors and mostly by 2-year periods. The first 4 parts were unnumbered, and the first 2 had no title; the third was "Rapport sur les travaux botaniques . . ."—See also Devos (Coup d'oeil), above, and additional references there given.

Vandenbussche, L. Onze volkstaal voor kruiden en artsenijen. xxiv, 650 p. Menin, 1955.

List of 2683 names of herbs and medicaments, arranged alphabetically by scientific names, with Flemish, French, German and English equivalents. (Not seen: cited from Nat. Mosana **10** (1/2): B94. 1957.)—See also Aigret and François, above, and additional references there given; also Paque (De Vlaamsche volksnamen), above.

Van Heurck, Henri, and **Guibert, Victor.** Flore médicale belge. xii, 455 p. 21 cm. Louvain, Bruxelles, 1864.

Annotated systematic list of native and cultivated medicinal vascular plants, with brief descriptions, French, Walloon, and Flemish vernacular names, properties and uses.—See also Aigret and François, above, and additional references there given.

Wesmael, Alfred. Flore forestière de Belgique, ou description et histoire des végétaux ligneux qui croissent spontanément en Belgique ou qui sont cultivés dans les forêts. Bul. Féd. Soc. Hort. Belg. **1865**: 337–508. 1866. (Reprinted, 174 p. In Arnold Arboretum library.)

Systematic list of 113 trees, shrubs, and woody vines; list of cultivated forms of native species; descriptive flora, with keys, vernacular names, local range, uses.—See also Aigret (Flore analytique), above.

ANVERS (ANTWERPEN; ANTWERP)

See also Brabant (Piré and Muller, Flore analytique du centre de la Belgique); also Limbourg (Vandenborn, Catalogue des plantes croissant spontanément aux environs de Saint-Trond . . .).

General

Van Heurck, Henri, and **De Beucker, J. I.** Antwerpsche analytische flora inhoudende de beschryving van alle planten, welke in de provincie Antwerpen, in de vrye natuer, groeijen of er algemeen gekweekt worden. Pt. 1. xxxvi, 192 p. 23 cm. Antwerpen, 1861. (Gray Herbarium library.)

Bibliography, annotated list of botanists; annotated, keyed list of phanerogams, with localities and vernacular names. Incomplete (Ranunculaceae-Caprifoliaceae): no index.

Local

Paque, Égide. Catalogue des plantes plus ou moins rares observées aux environs de Turnhout. Bul. Soc. Roy. Bot. Belg. **19** (1): 7–25. 1880.

Topography, soils, etc.; annotated list of vascular plants and Characeae within a radius of 4 leagues of Turnhout.—See also his Herborisations de 1881. l.c. **21** (2): 22–28. 1882.—Also his Nouvelles recherches sur la flore belge. l.c. **22** (1): 29–43. 1883.—Also **Adriaensen, J.,** and **Haeck, P.** Lijst van meer of minder zeldzame planten uit de omstreken van Turnhout . . . Bot. Jaarb. Kruidk. Genootsch. Dodonaea **4**: 240–250. 1892.

Traets, J. De vegetatie van het Kalmthoutse Reservaat. Meded. Plantensociol. Karter. Gembloers 26. 43 p. 1 fig., map. 1955-56. ("Overdruk van Calmpthoutania 1955–56 no. 4.")

Includes (p. 39–42) alphabetical lists of vascular plants and of bryophytes and lichens.

Van den Broeck, Henri. Catalogue des plantes observées aux environs d'Anvers. Bul. Soc. Roy. Bot. Belg. **22** (1): 112–173. 1883.—1er supplément.

l.c. **23** (2): 142–158. 1884. (Reprinted 64 p., 17 p.)—2e [–4e] supplément. l.c. **27** (2): 7–14. 1888; **42** (2): 13–22. 1905; **56**: 28–32. 1923.

Geography, plant zones; annotated list of vascular plants, Characeae, mosses, and hepatics, with indication of range by zones. The last two supplements contain bryophytes only. Replaces **Van Haesendonck, Constant.** Prodrome de la flore des environs d'Anvers et d'une partie de la Campine. 65 p. 20.5 cm. Bruxelles, 1841. (Annotated list of vascular plants (1,010 species), with habitat and local distribution. Not seen; title and annotation supplied by W. Robyns.)—See also **Hennen, Joseph.** Plantes subspontanées observées dans les environs immédiats d'Anvers. Bul. Soc. Roy. Bot. Belg. **57**: 28–30. 1924. (List of introduced angiosperms.)

Van Haesendonck, Constant. Florule des environs de Westerloo (partie méridionale de la Campine anversoise). Bul. Soc. Roy. Bot. Belg. **7**: 275–311. 1868.

Annotated list of vascular plants and Characeae.

Van Rompaey, Émile. Ingebrachte planten in de omgeving van Antwerpen. Natuurwetensch. Tijdschr. **3**: 34–37, 134–138. 1921. (Library of Museum of Comparative Zoology, Harvard University.)

List of introduced angiosperms, with localities and brief notes on flowering, fruiting, etc.

Verbist, Aloïs. Florule des environs de Hoogstraeten. Bul. Soc. Roy. Bot. Belg. **40** (2): 32–45. 1903.

Lists of common and rarer vascular plants, the latter annotated.

BRABANT

See also Flandre Orientale (Roucel).

General

Piré, Louis, and **Muller, Félix.** Flore analytique du centre de la Belgique. x, 299 p. 18 cm. Bruxelles, 1866.

Bibliography; annotated, keyed list of vascular plants and Characeae, with localities. Includes all of Brabant, part of Antwerp, part of East Flanders, a small part of Hainaut.—See also **Anonymous.** Liste des plants nouvelles ou des stations nouvelles . . . découvertes . . . en 1874. Bul. Soc. Roy. Linn. Bruxelles **4**: 32. 1875.

Van Heurck, Henri, and **Wesmael, Alfred.** Prodrome de la flore du Brabant, ou catalogue raisonné des plantes qui croissent spontanément dans cette province, et de celles qui y sont généralement cultivées. 96 p. 18 cm. Louvain, 1861. (Gray Herbarium library.)

Geology, annotated list of local botanists, etc.; annotated list of vascular plants (870 species), with localities.—See also **Thielens, Armand,** and **Wesmael, A.** Annotations à la flore de la partie septentrionale du Brabant. Bul. Soc. Roy. Bot. Belg. **1**: 201–208. 1862.—Also **Thielens, A.** Nouvelles annotations . . . l.c. **3**: 141–149. 1864.—Also **Baguet, Charles.** Annotations nouvelles à la flore de la province de Brabant. l.c. **15**: 118–143. 1876.

Local

Antoine, Joseph. Liste de quelques plantes plus ou moins rares des environs de Jodoigne. Bul. Soc. Roy. Bot. Belg. **8**: 215–218. 1869.

List of scarcer vascular plants, with localities.—See also **Thys, Joseph.** Notes sur la florule des environs de Jodoigne. l.c. **8**: 460–465. 1869. (Additional list.)

Baguet, Charles. Flore de Louvain (intra muros). Bul. Soc. Roy. Bot. Belg. **41** (2): 157–165. 1904.

Unannotated list of 512 vascular plants.

Carron, G., and **Zwendelaar, H.** Florule des environs de Bruxelles. Bul. Soc. Roy. Linn. Bruxelles **12**: 9–18, 41–50, 101–108. 1884–85; **13**: 64–68, 106–112, 199–202, 226–235. 1885–87; **14**: 33–39, 63–73, 110–120, 127–134, 159–168, 173–177. 1877–88; **15**: 16–24, 42–52, 77–87, 104–117, 139–152. 1888–89. (Reprinted, 152 p. 1889; reprint in Gray Herbarium library.)

Annotated list of vascular plants and Characeae, with vernacular names (not indexed) and localities; covers a radius of 3 leagues around Brussels. Replaces **Dekin, Adrien,** and **Passy, A. F.** Florula bruxellensis seu catalogus plantarum circa Bruxellas sponte nascentium. 72 p. 21 cm. Bruxellis, 1814. (Previous publications, topography, soils, etc.; briefly annotated flora of vascular and cellular plants (1,175 species, of which 840 are vascular) on the Linnaean system, mostly without special localities. Covers region within about 7 leagues radius of Brussels.)

Dandois, Honoré. Florule nivelloise. Bul. Soc. Roy. Bot. Belg. **2:** 58–66. 1863.—Additions . . . l.c. **4:** 139–154. 1865.—Nouvelles annotations . . . l.c. **5:** 256–258. 1866.

Lists of vascular plants and Characeae, with localities, from cantons of Nivelle and Genappe.

Duvigneaud, Paul, Vanden Berghen, C., and **Heinemann, Paul.** À propos de la disparition d'un site naturel: le marais de Bergh et sa flore. Bul. Soc. Roy. Bot. Belg. **74:** 139–153. 2 fig. 1942.

History of the Marais de Bergh, a marsh about 15 km. from Brussels, recently nearly destroyed; plant associations; annotated list of cellular and vascular plants; bibliography.—See also **Matagne, Henri,** and **De Lange, J. E.** Compléments à la florule de Bergh. l.c. **75:** 132–136. 1943. (List of additional vascular plants, bringing their total number to 320 species from an area of less than 1 sq. km.).

Ghysebrechts, L., *abbé.* Additions à la florule des environs de Diest. Bul. Soc. Roy. Bot. Belg. **24** (1): 351–362. 1885; **24** (2): 38–55. 1885; **27** (2): 14–22. 1888; **39** (1): 37–45. 1900; **41** (2): 8–11, 102–105. 1903; **42** (2): 163–169. 1905; **44:** 131–146. 1907; **47:** 55–181. 1910.

Lists of vascular and some cellular plants. Titles vary.

Kickx, Jean, *fil.* Accurata descriptio plantarum officinalium et venenatarum tum phanerogamarum tum cryptogamarum, in agro lovaniensi sponte crescentium, addita earum historia . . . Annales Acad. Lovan. v. 9, pt. 2. 2 p. l., xiii, 348 p. 1827.

Account of medicinal and otherwise useful vascular plants and a few lichens and fungi of the Louvain area, also of poisonous species, with descriptions, notes on variations, biology, habitat, uses or injurious properties, etc. Half-title: Plantae officinales et venenatae agri lovaniensis.)

Lecoyer, C. J. Notice sur la flore de Wavre et de ses environs. Bul. Soc. Roy. Bot. Belg. **11:** 275–282. 1872.—Liste de plantes observées dans les environs de Wavre. l. c. **18** (2): 50–51. 1879.

Topography, etc.; annotated list of rarer vascular plants.

FLANDRE OCCIDENTALE (WEST-VLAANDEREN; WEST FLANDERS)

General

Van de Vyvere, Ernest. Flore de la Flandre-occidentale, ou catalogue des plantes phanérogames indigènes et cultivées dans cette province, avec leur usage, le temps de leur floraison, les lieux où on les trouve le plus communément, et les synonymes des anciens botanistes belges. Annales Soc. Méd.-Chirurg. Bruges **10** (1849): 95–116, 185–226, 336–363, 501–556. 1849. (Reprinted vi, 152 p. 24 cm. Bruges, 1850; reprint in Gray Herbarium library.)

Annotated list of wild and cultivated spermatophytes (1,152 species), with notes on uses.

Local

Anonymous. Flore de l'arrondissement de Furnes et d'une partie de celui d'Ypres, avec description géologique, accompagnée d'une liste zoologique et détails sur quelques animaux et insectes de ce pays. Par un pharmacien. xxix, 66 p. 25 cm. Ypres, (n.d.). (Arnold Arboretum library.)

Includes (p. 13–20) unannotated list of vascular plants, on the Linnaean system. Has been attributed to De Tollenaere, Aimé. (See **De Wildeman** and **Durand,** Prodrome de la flore belge 1: 57. 1907.)

FLANDRE ORIENTALE (OOST-VLAANDEREN; EAST FLANDERS)

See also Brabant (Piré and Muller, Flore analytique du centre de la Belgique).

Local

Roucel, F. A. Traité des plantes les moins fréquentes, qui croissent naturellement dans les environs des villes de Gand, d'Alost, de Termonde et Bruxelles . . . 2 p.l., xxix, 118 p. 20 cm. Bruxelles et Paris, 1792. (Arnold Arboretum library.)

Glossary; descriptive list of scarcer vascular and cellular plants (131 numbered species), with French and Flemish vernacular names and uses.

Vander Meersch, Émile. Notice sur la florule du Kraene-Poel. Bul. Soc. Roy. Bot. Belg. **13**: 224–241. 1874.

General features of flora; annotated list of vascular plants.

Van Langendonck, H. J. Étude sur la flore et la végétation des environs de Gand. Bul. Soc. Roy. Bot. Belg. **68**: 117–180. 9 fig. (incl. map). 1935.

Mainly phytogeographical and ecological; includes list of noteworthy plants, with indication of those (all of infraspecific rank) not listed in Hauman and Balle's Catalogue, partial lists of adventive and extirpated species, and bibliography.

Van Rompaey, Émile. Compte rendu de l'herborisation générale du 28 juin 1942 dans la région de Rupelmonde. I. Plantes vasculaires. Bul. Soc. Roy. Bot. Belg. **75**: 106–117. 1 fig. (map). 1943.

Includes tabular list of 333 vascular plants, alphabetically arranged by botanical names. The following work should perhaps be inserted here, but its nature is uncertain. **Beeldsnijder, —.** Catalogue des plantes, qui se trouvent à Rupelmonde, province d'Utrecht. 2. éd. iv, 36 p. n.p., n.d. (1st ed., 1823.) (Not seen; cited from Pritzel, Thesaurus literaturae botanicae p. 19. 1872.)

HAINAUT (HENEGOUWEN)

See also Brabant (Piré and Muller, Flore analytique du centre de la Belgique).

General

Michot, N. L., *abbé.* Flore du Hainaut. 3 p.l., xxxii, 421 p. 23 cm. Mons, 1845.

Annotated descriptive flora of vascular and cellular plants on the Linnaean system, with localities for scarcer species and notes on uses. Replaces **Hécart, G. A. J.** Florula hannoniensis. Mém. Soc. Agr., Sci. et Arts Valenciennes **2**: 153–208. 1836. (List of 1,441 taxa (1,152 vascular), on the Linnaean system, including cultivated plants, with localities or habitats.)

Local

Bernimolin, H. Catalogue des plantes spontanées et cultivées du Tournaisis, avec indication des localités où on les rencontre. vi, 133 p. 18 cm. Tournai, 1883. (Lloyd Library.)

Annotated list of wild and cultivated vascular plants and Characeae, with vernacular names and localities.

Hardy, Apollon, and Lebrun, Arthur. Florule des environs de Beaumont et de Montbliart. Bul. Soc. Roy. Bot. Belg. **5**: 247–256. 1866.

List of less common vascular plants and Characeae, with localities.

Hocquart, Léopold. Flore du département de Jemmape [sic], ou définitions des plantes qui y croissent spontanément, faites d'après le système de Linnée [sic], a l'usage des élèves en botanique. viii, 303 p. 16 cm. Mons, 1814. (New York Botanical Garden library.)

Briefly annotated, keyed flora of vascular and cellular plants (the vascular plants with short descriptions), with localities (1,514 species, of which 1,062 are vascular); list of communes covered by the flora. Covers the region of Mons, Tournay, Chimay, and Charleroy.

Marissal, F. V. Catalogue des phanérogames observées depuis 1842 dans les environs de Tournay, disposé suivant le système de M. B-C. Dumortier.

1 p.l., 91 p. 22.5 cm. Tournay, 1846. (Extrait des Annales de la Société de littérature, des sciences et des beaux-arts de Tournay.) (Lloyd Library.)
Annotated list of spermatophytes (803 species).

Riomet, Bienaimé. Florule d'Angre et des environs . . . suivie de la florule des muscinées et lichens par G. Lochenies. Mém. et Pub. Soc. Sci., Arts et Let. Hainaut V, **4**: 195–309. 1891 (1892). (Reprinted, 116 p. Mons, 1892.)
Bibliography, geology (by Lochenies); annotated list of vascular plants (550 species); followed (p. 277–309) by list of bryophytes and lichens by G. Lochenies.

LIÈGE (LUIK)

See also Luxembourg (Crépin, L'Ardenne sous le rapport de sa végétation).

General

Derive, Théodore. Flore vénéneuse de la province de Liège, ou description des plantes nuisibles ou suspectes qui croissent spontanément dans cette partie du royaume. 128 p. 12 pl. 21 cm. Verviers, 1839.
Descriptions of the poisonous vascular plants. (Not seen; annotation supplied by W. Robyns.)

Durand, Théophile. Catalogue de la flore liégeoise. Bul. Féd. Soc. Hort. Belg. 1877, suppl. 80 p. 1878.—Additions . . . Bul. Soc. Roy. Bot. Belg. **19** (2): 56–65. 1880.—Annotations . . . l.c. **20** (2): 102–108, 145–149. 1881.
Annotated list of vascular plants and Characeae (1,202 species, 297 races, and 117 varieties in the original Catalogue, of which 1,012, 291, and 114 respectively are native), with localities and indication of soil preference. Is a corrected summary, with omission of many minor locality records, of the following earlier work: **Durand, T.**, and **Donckier, Henri.** Matériaux pour servir à la flore de la province de Liège. l.c. **12**: 370–419. 1873 [1874]; **13**: 486–542. 1874 [1875]; **14**: 270–327. 1875 [1876]. (Botanical explorations, with titles of publications; list of vascular plants, with detailed localities. Each part covers all the vascular plants.)—See also **Devos, André.** Note sur quelques plantes rares trouvées de 1871 à 1881, principalement dans la province de Liège. l.c. **21** (2): 135–140. 1882.

Local

Bastin, Joseph, *abbé*. Les plantes dans le parler, l'histoire et les usages de la Wallonie malmédienne. Collection "Nos Dialectes" no. 8. 260 p. map. 23 cm. Liège, 1939.
Vernacular names and folklore of indigenous and cultivated plants of the Malmédy region, grouped by trees, shrubs, food plants, medicinal plants, ornamental plants, etc. (Not seen; title and annotation supplied by W. Robyns.)—See also Boxus, Dahmen, Desfresne, and Germay, below, and Boxus and Feller under Belgium, General, above.

Beaufays, Gustave. Flore verviétoise contenant la description de toutes les plantes qui croissent spontanément dans les environs de Verviers. viii, 145 p. 18.5 cm. Verviers, 1857.
Descriptive flora of vascular plants, on the Linnaean system. (Not seen; annotation supplied by W. Robyns.) An issue in 1872 is listed (as not seen) in Rehder, Bradley Bibliography **1**: 392. 1911.—See also Fonsny and Collard, Germay, Michel, and Roncart, below.

Boxus, Robert. Dictionnaire Wallon Hutois des noms de plantes indigènes et cultivées. 93 p. 20.5 cm. Huy, [n.d. 1949?].
Annotated alphabetical list of 943 Walloon names used in Huy and vicinity; indices of French and scientific names.—See also Bastin, above, and additional references there given.

Cluysenaar, P. G. Florule de Huy. Plantes phanérogames et plantes cryptogames vasculaires observées à Huy et aux environs de cette ville. Bul. Cercle Nat. Hutois **3/4**: 81–201. 1890.
List of vascular plants, with localities and a few annotations. (Not seen; title and annotation supplied by W. Robyns.)

Cogniaux, Alfred. Coup d'oeil sur la végétation des environs de Visé. Bul. Soc. Roy. Bot. Belg. **3:** 81–92. 1864.

Topography, etc.; list of 100 vascular plants, with localities.—See also **Hardy, Apollon.** Coup d'oeil rapide sur l'histoire de la botanique au pays de Visé. 15 p. 21.5 cm. Visé, 1922. (Historical notes, with lists of rare or interesting vascular plants. Not seen; title and annotation supplied by W. Robyns.)—Also **Bouillenne, Raymond.** Relevé des plantes qui furent determinées, en septembre 1920, sur les ruines de Visé. Bul. Soc. Roy. Bot. Belg. **55:** 19–27. 1922. (Annotated tabular list of 74 vascular plants found on the ruins of Visé, destroyed in August 1914; discussion of methods of introduction, etc.)

Dahmen, M. Contribution à l'histoire de la botanique en Pays Wallon. Les noms wallons des plantes. Bul. Bot. Liégeois **1:** 4–49. 1937. (With Arch. Inst. Bot. Univ. Liège v. 13.)

Systematic lists, with French and Walloon vernacular names, of native and non-native vascular plants; not intended to be complete. Refers mainly if not entirely to Province of Liège. (Processed publication.)—See also Bastin, above, and additional references there given.

Desfresne, Jules. Vocabulaire du règne végétal à Coo et aux environs. Bul. Soc. Liégeoise Lit. Wallone **49:** 169–194. 1907. (Harvard College library.)

Alphabetical list of Walloon vernacular names, with French vernacular names, botanical names, and some annotations; includes also terms relating to parts of plants, etc.—See also Bastin, above, and additional references there given.

Donckier, Auguste. Catalogue des plantes des environs de Goé classées d'après leurs altitudes. Bul. Soc. Roy. Bot. Belg. **10:** 67–75. 1871.

Briefly annotated list of vascular plants divided into groups according to altitudinal range.—See also **Thielens, Armand.** Trois jours d'herborisation aux environs de Goé, Welkenraedt et la forêt d'Hertogenwald. l.c. **9:** 406–418. 1871. (Account of different localities and habitats with short lists of species.)

Fonsny, Henri, and **Collard, Ferdinand.** Florule de Verviers et des ses environs comprenant les espèces indigènes et subspontanées ainsi que la flore exotique des graviers. 2 p. l., xliii, 402 p. 15.5 cm. Verviers, 1885. (British Museum (Natural History) library.)

Annotated, keyed flora of vascular plants and Characeae, with localities for scarcer species.—See also Beaufays, above, and additional references there given.

Frédéricq, Léon. En nouvelle Belgique. Guide du promeneur et du naturaliste dans le district de Malmédy. 2. éd. 115 p. map. 19 cm. Bruxelles, 1924.

General description of the country and vegetation with annotated list of principal vascular plants. (Not seen; title and annotation supplied by W. Robyns.)—See also Siegers, below.

Germay, R. L'herboriste verviétois. Recueil de 100 plantes médicinales qui croissent dans l'arrondissement. 47 p. 18.5 cm. Verviers, 1889.

List of 100 plants, alphabetically arranged by French vernacular names, with Walloon names, brief description, and uses. Botanical names are not given.—See also Bastin and Beaufays, above, and additional references there given.

Lebrun, François. Florule des environs de Spa. Bul. Soc. Bot. France **20** (Sess. Extraord.): cxxii-cxlvi. pl. 3 (map). 1873.

Geology, with lists of localized species; annotated list of vascular plants and Characeae.—See also Lejeune, below.

Lejeune, A. L. S. Flore des environs de Spa, ou distribution selon le système de Linnaeus, des plantes qui croissent spontanément dans le département de l'Ourte et dans les départemens circonvoisons . . . 2 v. (254; 350, viii p.). 21 cm. Liège, 1811–1813. (Gray Herbarium library.)

Annotated, briefly descriptive flora of vascular plants.—See also his Revue de la flore des environs de Spa. Contenant l'énumération de toutes les plantes y décrites, avec les observations, les descriptions, les additions et les corrections nécessaires pour la mettre le plus possible à la hauteur de la science. viii, 263 p. 21.5 cm. Liège, 1824. (List of 1,496 species of vascular plants, including some from Luxembourg only, with references to his "Flore," and frequent annotations. In Gray Herbarium library.)—Also Lebrun, above.

Ley, Wilhelm. Florule des environs d'Eupen. Bul. Soc. Roy. Bot. Belg. **10**: 75–85. 1871.
Unannotated list of spermatophytes.

Marchal, Élie, and Hardy, Apollon. Catalogue des plantes plus ou moins rares de la vallée de la Meuse, de Liège à Maestricht. Bul. Soc. Roy. Bot. Belg. **7**: 240–273. 1868.
Annotated list of scarcer vascular plants, with localities.

Maréchal, Arthur. Coup d'oeil écologique, phytosociologique et floristique sur la pointe extrême du Condroz, située aux portes de Liège. Lejeunia **3**: 29–56. pl. 4 (map). 1939.
Includes (p. 39–44) list of principal vascular plants, with localities, followed by list of *Rubus* (p. 45–49); also bryophytes, lichens, and bibliography. Covers area of 21 square kilometers between the rivers Meuse and Ourthe, from Angleur south to Ougrée, Nomont, and Mery.

——— Florule de Liège (intra muros). Lejeunia **2**: 17–49. 1938.
Geography, geology, adventive plants, etc.; annotated list of native and naturalized vascular plants (about 503 species) found in the commune of Liège and in part in the adjacent communes of Ougrée and Grivegnée.—See also his Florule de la banlieue liégeoise. Coup d'oeil retrospectif depuis 1890 ou disparition de nombreuses stations naturelles d'espèces plus remarquables. Bul. Bot. Liégeois [Lejeunia] **1**: 67–73. 1937. (Running account of species that have disappeared from different localities in the suburbs of Liège.) (Processed.)—Also his Plantes rares observées autrefois dans la region de Liège. Lejeunia **10**: 35–41. 1950. (Discussion of ecological requirements as a guide in the search for rare plants, followed by a short annotated list.)

——— La montagne Saint-Pierre. Îlot biologique de plantes remarquables et rares. Lejeunia **5**: 37–39. 1941.
Topography, geology, general features of flora; list of the more remarkable vascular plants, with localities; origin of flora, bibliography. Refers only to the Belgian part of the region.—See also Netherlands, Limburg (Dumoulin, Guide du botaniste dans les environs de Maestricht; de Wever, Planten van den St. Pietersberg).

Michel, Mathieu. Flore de Fraipont, Nessonvaux et leurs environs, y compris toute la vallée de la Vesdre, depuis Limbourg jusqu'à Chénée . . . 389 p. 16.5 cm. Verviers, 1877. (Gray Herbarium library.)
Briefly descriptive flora of vascular plants.—See also **Michel, M., and Remacle, N.** Additions à la flore de Fraipont et Nessonvaux. Bul. Soc. Roy. Bot. Belg. **18** (2): 98–103. 1879.—Also **Michel, M.** Note sur les plantes naturalisées ou introduites dans la vallée de la Vesdre. l.c. **21** (2): 34–42. 1882.—Also **Durand, Théophile.** Étude comparative sur la végétation de la vallée de la Vesdre avant et après 1840. l.c. **21** (2): 59–66. 1882.—Also his Note sur les récentes découvertes botaniques dans le bassin de la Vesdre. l.c. **24** (2): 142–146. 1885.— Also **Goffart, Jules.** Compte rendu de l'herborisation annuelle de la Société royale de botanique de Belgique, dans la vallée de la Vesdre (les 28 et 29 juin 1936). l.c. **69**: 181–187. 1937.—Also Beaufays, above.

Roncart, Robert. Quelques particularités de la flore des environs de Verviers. Mém. Soc. Roy. Sci. Liège III, v. 16, art. 15. 28 p. 1 fig. 1931. (Also as Arch. Inst. Bot. Univ. Liège v. 9, no. 3. 1932).
Bibliography; list of 221 adventive plants of the gravels of the Vesdre River; discussion of persistence of adventive plants, etc.—See also **Visé, Auguste.** La flore adventice de la région de Verviers. Lejeunia **6**: 99–119. 1942. (Origin of wool manufactured at Verviers, factors in introduction, botanical explorations, etc.; tabular list of 232 adventive angiosperms found by author in vicinity of Verviers, annotated as to abundance.)—Also **Roncart, R.** Sur la flore adventice des graviers de la Vesdre. Herborisation de la Société botanique de Liège (20 octobre 1947). l.c. **11**: 131–133. 1947. (List of angiosperms collected.)—Also **Visé, A.** Flore adventice de la region

de la Vesdre. Revue Verviét. Hist. Nat. **12**: 95–97. 1955; **13**: 2–9. 1956. (Not seen.)—Also his Florule adventice de la vallée de la Vesdre. Bul. Soc. Roy. Bot. Belg. **90**: 287–305. 1958. (Corrected list of all species recorded from the area, unannotated.)—Also Beaufays, above.

Siegers,—. Zusammenstellung der bei Malmédy vorkommenden phanerogamen und gefässkryptogamen mit ihren standorten. Beilage zu Programm des Progymnasiums zu Malmédy **405**: 1–32. 1885.

Annotated list of vascular plants [638 species], with habitat and local distribution. (Not seen; title and annotation supplied by W. Robyns.)—See also Frédéricq, above.

Strail, C. A. Florule de Chaudfontaine et de Magnée (Liège) ou catalogue des plantes qu'on rencontre dans une circonférence d'un rayon de une lieue à une lieue et demie, en prenant cette dernière commune pour centre. Bul. Soc. Roy. Bot. Belg. **2**: 284–322. 1863.

Geology, etc.; annotated list of vascular plants.

LIMBOURG (LIMBURG)

See also Netherlands, Limburg (Foerster, Flora excursoria des regierungsbezirkes Aachen).

Local

Vandenborn, Henri. Catalogue des plantes croissant spontanément aux environs de Saint-Trond, de Cortessem, dans les bruyères de Diepenbeek, Genck, Niel, Beverloo, Ostham, et sur les bords de la Meuse, à Mechelen et Boorsheim. Bul. Soc. Roy. Bot. Belg. **4**: 4–36. 1865.

Soils; annotated list of vascular plants and Characeae.

LUXEMBOURG (LUXEMBURG)

See also Namur (Paque, Flore analytique et descriptive des provinces de Namur et Luxembourg).

Local

Aigret, Clément. Coup d'oeil sur la florule des environs de Villance. Bul. Soc. Roy. Bot. Belg. **40** (2): 48–62. 1901.

List of rarer vascular plants (also mosses and lichens), with reference to Crépin's Manuel for the common species; based on researches of Augustin Dolisy.

Ansembourg, Victor d', *comte*. Flore adventice du secteur ardennais Assenois-Neufchâteau. Lejeunia **11**: 119–130. 1947.

General considerations, list of some characteristic indigenous species of different habitats; list of introduced species, grouped by habitats, with discussion.

Crépin, François. L'Ardenne sous le rapport de sa végétation. Bul. Féd. Soc. Hort. Belg. **1862**: 313–366. 1863. (Reprinted, 60 p., in Arnold Arboretum library.)

Botanical explorations, physiography, etc.; annotated list of vascular plants (660 species) and 3 Characeae. Covers most of Luxembourg except the extreme southern part, as well as the southeastern margins of Liège and Namur.—See also **Van Bastelaer, D. A.** Herborisation dans un coin des Ardennes belges. Bul. Soc. Roy. Bot. Belg. **3**: 228–262. 1864. (Running account of author's and Dr. J. Tosquinet's explorations, with mention of plants observed.)

Gilson, V. Aide mémoire de botanique descriptive indiquant les principales espèces de plantes vasculaires qui croissent dans la partie de la région jurassique belge circonscrite par les limites de l'arrondissement de Virton. 30 p. 21 cm. Bruxelles, 1884.

Systematic list of vascular plants, unannotated but with localities. (Not seen; title and annotation supplied by W. Robyns.) The following papers relate to Virton and other localities in the Jurassic region of southeastern Luxembourg. **Simon, F. J.** Quelques plantes nouvelles pour la région jurassique. Bul. Soc. Roy. Bot. Belg. **28** (2): 89. 1889. (List of 9 species, with localities.)—Also **Even, Charles.** Liste des plantes vasculaires observées dans les terrains jurassiques de la province de Luxembourg. l.c. 35 (2): 34–36. 1896. (List with localities.)—Also his Plantes vasculaires observées dans les terrains jurassiques de Luxembourg. l.c. **41** (2): 12–14. 1903. (List with localities.)—Also **Verhulst, Antoine.** Remarques sur la florule de Virton et des environs. l.c. **46**: 88–100. 1909. (List with localities.)—Also his Plantes du Jurassique belge non renseignées dans le "Catalogue" de Montmédy, avec l'indication de leurs principales stations. Bul. Soc. Nat. et Archéol. du Nord de la Meuse **21** (Sci. Nat.): 3–31. 1909. (List of about 100 species, with localities. Intended as supplement to Pierrot and others, Catalogue des plantes vasculaires de l'arrondissement de Montmédy. [See under France, Meuse.] Not seen; title and annotation supplied by R. McVaugh.)—Also his Essai de phytostatique en Jurassique belge. VI. Bul. Soc. Roy. Bot. Belg. 57: 147–155. 1925. (Botanical explorations; lists of species that have not been seen since 1836 or for 25 or 30 years, and of those discovered since 1864. The earlier parts in this series are physiographic and ecological; for general account of region see pt. I. l.c. 54: 7–48. 1921.)

Legrain, Joseph, *abbé.* Notice sur la flore de Mirwart. Lejeunia **11**: 81–112. pl. 17 (map). 1947.

Geology, changes in flora, phytosociology; annotated list of vascular plants (625 species) of an area of 3.5 kilometers radius around Mirwart.

NAMUR (NAMEN)

See also Luxembourg (Crépin, L'Ardenne sous le rapport de sa végétation).

General

Paque, Égide. Flore analytique et descriptive des provinces de Namur et Luxembourg (plantes indigènes et cultivées) accompagnée d'une carte botanique, des étymologies des noms, des propriétés des plantes, etc. 2 p.l., xxxii, 594 p. 341 fig., map. 23 cm. Namur, 1902. (Arnold Arboretum library.)

Botanical regions, geology, glossary, bibliography, etc.; annotated, keyed descriptive flora of native and hardy cultivated vascular plants, with localities, vernacular names, and notes on uses. Replaces **Bellynck, Auguste.** Flore de Namur, ou description des plantes spontanées et cultivées en grand dans la province de Namur, observées depuis 1850 . . . xxxii, 353 p. 22 cm. Namur, Bruxelles, 1885. (Keys, glossary, etc.; annotated descriptive flora of wild and hardy cultivated vascular plants (1,117 species); list of excluded species. The work is dedicated to the Virgin Mary, patron saint of Namur, as is also Paque's Flore.)—See also **Paque, Égide.** Note sur quelques trouvailles intéressantes. Bul. Soc. Roy. Bot. Belg. 42 (2): 97–98. 1904. (Notes on 5 plants.)

Local

Crépin, François. Un coup d'oeil sur la florule des environs de Han-sur-Lesse. Bul. Soc. Roy. Bot. Belg. 1: 41–69. 1862.

Topography, geology, florulas of different habitats, list of rarer species, excluded species, ecology.—See also his Florule des environs de Han-sur-Lesse. 16 p. 21.5 cm. Bruxelles, 1873. (A condensed version of his earlier paper, with some changes in identifications; both omit a large number of the commoner plants. In Gray Herbarium library.)

Determe, Clément. Catalogue des plantes rares croissant aux environs de Mariembourg. Bul. Soc. Roy. Bot. Belg. 2: 156–166. 1863.

Geography, soils, etc.; annotated list of rarer vascular plants.—See also **Determe, Staneslas.** Liste de quelques plantes trouvées aux environs de Mariembourg. Bul. Soc. Roy. Bot. Belg. **22** (2): 49. 1883.

Devos, André. Florule de Marche-les-Dames. Bul. Soc. Roy. Bot. Belg. **22** (1): 7–28. 1883.—Supplément . . . l.c. **23** (2): 164–165. 1884.

Geology, botanical explorations, etc.; annotated list of the rarer vascular and cellular plants.—See also **Delhaise, Hubert,** and **Simon, F. J.** Florule . . . Second supplément. Bul. Soc. Roy. Bot. Belg. **28** (2): 86. 1889.

Denmark (Danmark)

See also Scandinavia; Sweden, General (Lyttkens).

General

The general works here listed may be classified as follows: *Floras and manuals:* (Gram and Jessen: see Lagerberg under Scandinavia), Hvass (popular: see Bolin under Scandinavia), Lange (Haandbog), (Mentz and Ostenfeld: see Lindman under Scandinavia), Raunkiær (Dansk ekskursions-flora), Rostrup. *Catalogs:* Dansk Botanisk Forening (Danmarks topografisk-botaniske undersøgelse: plant distribution), Jessen (Oversigt), Mortensen (synonymy). *Iconographies:* (Gram and Jessen: see Lagerberg under Scandinavia), Hagerup, (Hvass: see Bolin under Scandinavia), (Mentz and Ostenfeld: see Lindman under Scandinavia), (Oeder: see under Scandinavia). *Bibliograhies:* Christensen, Olsen (excursions of Dansk Botanisk Forening). *Vernacular names:* Jenssen, (Jenssen-Tusch: see under Scandinavia), Lange (Plantenavne: cultivated plants). *Trees:* Gram and Jessen, Kiær, Møller, O. G. Petersen, Vedel, Warming. *Useful and poisonous plants:* Frederiksen (Botanik for landmænd), Grønlund (poisonous plants), Heger, Hornemann, Jacobsen (medicinal), Lind (Middle Age medicinal plants), H. Petersen (edible). *Weeds:* Frederiksen (2), Hansen, Jessen and Lind, Lange (Oversigt: adventive plants). *Cultivated plants:* Frederiksen (Botanik for landmænd), Gram and Jessen (woody plants), Kiær, Lange (Plantenavne: nomenclature), (Møller: woody plants), (O. G. Petersen: woody plants), Vedel (woody plants). *Miscellaneous:* Balslev (plant associations), (Christensen: biographical), Dansk Botanisk Forening (5te beretning: botanical districts), (Jenssen: glossary of scientific names and terms), Olsen (excursions of Dansk Botanisk Forening), Ostenfeld (plant distribution), Raunkiær (De danske blomsterplanters naturhistorie: morphology, anatomy, plant biology), Warming (ecology).

Balslev, Vilhelm, and **Simonsen, Kristen.** Danske plantesamfund. Et grundlag for botaniske udflugter. 2. opl. 117 p. incl. illus., 2 col. pl. 19.5 cm. København, 1945. (1st ed. 1932. 3. ændrede udg. ved Bodil Lange. 114 p. 1957; not seen.)

Semipopular but comprehensive description of plant associations and habitats, including keys to woody plants; table showing percentage of total area in crop land, forest, bog, etc.; bibliography. Only vernacular names are given; no index.—See also Warming, below.

Baltzer, Vilhelm, and **Lange, Morten.** Illustreret flora på grundlag af Björn Ursing Fältflora. 225 p. 3 fig. (glossary), col. pls. (incl. pl. 1–4 on lining papers). 20 cm. København, 1957.

Briefly descriptive, popular flora of spermatophytes, with keys to families and genera, and colored figure of each species.

Christensen, Carl. Den danske botaniks historie med tilhørende bibliografi. 2 v. (884, 680 p.). 235 fig. (mostly ports.). 25 cm. København, 1924–26.

Vol. 1, chronologically arranged botanical history, mainly in the form of rather extensive biographies of Danish botanists; list of botanists at various institutions, from 1639 to present time; subject and personal indices; vol. 2, list of periodicals; list of authors, chronologically arranged by date of first publication, with brief biographical sketches and unannotated chronological lists of their writings; classified index to papers listed (including those in his Den danske botaniske litteratur 1880–1911), species and author indices. Includes (vol. 2) all Danish literature through 1879 (and all Norwegian down to 1814), as well as all later publications of Danish authors who began publishing before 1880; list (p. 511–518) of works by foreign authors published in Denmark. Replaces **Warming, Eugenius.** Den danske botaniske literatur fra de aeldste tider til 1880. Bot. Tidsskr. **12:** 42–131, 158–247. 1880–81.—For continuation, see **Christensen, C.** Den danske botaniske litteratur 1880–1911. xxi, 279 p. 70 port. 23.5 cm. København, 1913. (List of joint works, lexicons, Danish journals, etc.; list of authors, chronologically arranged by date of their first publication (on p. 1–80, those authors who had already appeared in Warming's bibliography, as a continuation of which Christensen's was originally published), with biographical notices and chronological lists of publications; list of school books, textbooks, etc.; list of reports on excursions and minor publications; list of translations and longer reviews; index of authors. Includes about 3000 titles.)—Continued by his Den danske botaniske litteratur. Bibliographia botanica danica 1912–1939. 4 p. l., 350 p. 78 port. 23.5 cm. København, 1940. (Additions to Den danske botaniks historie, v. 2; list of Danish authors who published both before and after 1 Jan. 1912 and who were listed in Den danske botaniske litteratur 1880–1911, with supplementary biographical notes and continuation of their lists of titles (p. 4–152); list of Danish authors who published first after 1911, with biographical notices and lists of publications (p. 153–292); classified list of publications by authors of only one or very few papers, without biographical notices; list of works published in Denmark by foreign authors, alphabetically arranged by author; index to Danish authors. Includes about 5,200 titles by 1,100 authors.)—See also **Grøntved, Julius.** Dansk botanisk litteratur i 1940, 1941, 1942 og 1943 (–1944, 1945, 1946 og 1947). Bot. Tidsskr. **46:** 185–238. 1944; **48:** 254–293. 1949. (Unannotated lists, alphabetically arranged by authors, the Danish and foreign authors separated.)—Also **Hansen, Alfred.** Dansk botanisk litteratur i 1948, 1949 og 1950 (–1954, 1955 og 1956). l.c. **49:** 57–88. 1952; **50:** 279–316. 1955; **53:** 383–429. 1957. (Similar lists.)—Also Olsen, below.

Dansk Botanisk Forening. Danmarks topografisk-botaniske undersøgelse. 1–22. *In* Bot. Tidsskr. v. 43–45, 47, 49–50, 53. 1935–1956.

A series of papers, not in systematic order, on the distribution in Denmark of the vascular plants; the treatment of each species giving references, extralimital and local range, methods of dispersal, prehistoric migrations and fossil occurrence, followed by a synopsis (in Danish and English) of the Danish distribution of all the species treated, and bibliography; each paper or group of papers (up to no. 9) preceded by 2 maps showing the botanical districts and the soils of Denmark. Really a continuation of an earlier series, The distribution within Denmark of the higher plants. Results of the topographic-botanical investigation. I-II, for which see under Ostenfeld below. The individual papers are as follows: **Jessen, Knud.** Nr. 1. Liliiflorernes [Liliales] udbredelse i Danmark. l. c. **43:** 71–132. fig. 3–34 (maps). 1935; **Sørensen, Thorvald.** No. 2. Primulaceernes [Primulaceae] udbredelse . . . l.c. **43:** 133–172. 15 fig. (maps). 1935; **Bøcher, T. W.** Nr. 3. Udbredelsen af Ericaceae, Vacciniaceae og Empetraceae i Danmark. l.c. **44:** 5–40 incl. fig. 3–12 (maps). 1937; **Wiinstedt, Knud.** Nr. 4. Juncaceernes udbredelse i Danmark. l.c. **44:** 41–126 incl. 28 fig. (maps). 1937; **Grøntved, Julius.** Nr. 5. Polygonaceernes udbredelse . . . l.c. **45:** 9–58. 24 fig. (maps). 1939; **Køie, Aase.** Nr. 6. Hypericaceernes udbredelse . . . l.c. **45:** 59–72. 7 fig. (maps). 1939; **Køie, Aase, and Køie, Mogens.** Nr. 7. Udbredelsen af Geraniaceae, Araceae, Lemnaceae og Droseraceae i Danmark. l.c. **45:** 73–97. 23 fig. (maps). 1939; **Wiinstedt, K.** Nr. 8. Hieraciumarternes udbredelse . . . l.c. **45:** 101–130. 28 fig. (maps). 1939; **Wiinstedt, K.** Nr. 9. Cyperaceernes udbredelse . . . I. Scirpoideae. l.c. **47:** 3–64. 27 fig. (maps). 1943; **Mikkelsen, V. M.** Nr. 10. Udbredelsen af Juncaginaceae, Alismataceae og Hydrocharitaceae i Danmark. l.c. **47:** 65–93. 14 fig. (maps). 1943; **Hoff, Mogens.** Nr. 11. Crassulaceernes og Saxifragaceernes udbredelse . . . l.c. **47:** 95–121. 10 fig. (maps). 1943; **Andersen, Alfred.** Nr. 12. Pyrolaceernes og Plumbaginaceernes udbredelse . . . l.c. **47:** 123–142. 11 fig. (maps). 1943; **Wiinstedt, K.** Nr. 13. Cyperaceernes udbredelse . . . II. Caricoideae. l.c. **47:** 143–244. 55 fig. (maps). 1945; **Hansen, Alfred.** Nr. 14. Campanulaceernes og Lobeliaceernes udbredelse . . . l.c. **47:** 245–276 incl. 13 fig. (maps). 1948; **Grøntved, Julius.** Nr. 15. Orchidéernes udbredelse . . . l.c. **47:** 277–370 incl. 37 fig. (maps). 1948; **Egholm, Bjarne.** No. 16. Umbelliferernes udbredelse . . . l.c. **47:** 373–480 incl. 49 fig. (maps). 1951; **Hansen, Alfred.** Nr. 17. Udbredelsen af Caprifoliaceae, Adoxaceae, Dipsacaceae og Cucurbitaceae i Danmark. l.c. **47:** 481–509. incl. 14 fig. (maps). 1951; **Wiinstedt, K.** Nr. 18. Pteridofyternes udbredelse . . . l.c. **49:** 305–388 incl. 46 fig. (maps). 1953; **Gröntved, Johannes.** Nr. 19. Typhaceernes og Spaganiaceernes udbredelse . . . l.c. **50:** 209–238 incl. 11 fig. (maps). 1954; **Rasmussen, S.M.** Nr. 20. Euphorbiaceernes, Malvaceernes og Violaceernes udbredelse . . . l.c. **50:** 239–278 incl. 24 fig. (maps). 1954; **Pedersen, Anfred.** Nr. 21. Rubiaceernes, Polygalaceernes, Linaceernes, Oxalidaceernes og Balsaminaceernes udbredelse . . . l.c. **53:** 139–196 incl. 27 fig. (maps). 1956; **Larsen, Kai.** Nr. 22. Ranunculaceernes udbredelse . . . l.c. **53:** 197–252 incl. 35 fig. (maps). 1956; **Pedersen, A.** Nr. 23. Cruciferernes udbredelse . . . l.c. **54:** 192–304 incl. 52 fig. (maps). 1958; **Hansen, A.** Nr. 24. Gentianaceernes, Menyanthaceernes; Asclepiadaceernes, og Apocynaceernes udbredelse . . . l.c. **54:** 305–332 incl. 12 fig. (maps). 1958.—See also Jessen (Oversigt), below, and Ostenfeld, below.

——— 5te beretning fra Komiteen for den topografisk-botaniske undersögelse af Danmark. Bot. Tidsskr. **29:** 84–90. 1908.

List of the 47 botanical districts recognized, with names of local investigators and citation of published and unpublished local floras.—See also Ostenfeld, below.

Frederiksen, Henry. Botanik for landmænd. 2. opl. 255 p. 139 fig., 16 col. pl. (on 8). 20.5 cm. København, 1948. (1st ed. 1946.)

Includes briefly descriptive semipopular account of cultivated useful plants, fodder plants, weeds (grouped by color), and useful native plants, annotated lists of poisonous plants and bee plants, etc.—See also (for useful or poisonous plants) Grønlund, Heger, Hornemann, Jacobsen, Lind, and H. Petersen, below; and (for weeds) Frederiksen and Petersen, Hansen, Jessen and Lind, and Lange (Oversigt), below.

——— **Grøntved, Peter, and Petersen, H. I.** Ukrudt og ukrudtbekæmpelse. 320 p. 113 i.e. 118 fig. (incl. maps), 16 col. pl. (on 8). 22.5 cm. København, 1950.

Includes (p. 18–105) descriptive account of weeds, alphabetically arranged by vernacular names, with methods of control; also key (p. 123–141; includes dicotyledons only) based on cotyledons and early leaves, with figures of 96 species.—See also Frederiksen, above, and additional references there given.

Gram, Kaj, and Jessen, Knud. Træer og buske i vintertilstand. 99 p. incl. 12 pl. on 6. 33 fig. 21 cm. København, 1945.

Discussion of characters of twigs, buds, etc.; treatment of native and cultivated shrubs and trees in form of key based on winter characters; short bibliography. Intended as a second edition of their Nøgle til at bestemme de i Danmark vildtvoksende og hyppigst forvildede, løvfældende løvtræer og buske i vintertilstand. Bot. Tidsskr. **43:** 484–500. illus. 1936.—See also B. Møller, Kiær, O. G. Petersen, and Warming, below.

——— and **Jessen, Knud.** Vilde planter i Norden.

See Lagerberg under Scandinavia.

Grønlund, Christian. Danske giftplanter. *In* Smaastykker ved udvalget for folkeoplysnings fremme **6:** 305–358. 38 fig. 15 cm. 1874. (Reissued 1887.)

Popular treatment of Danish poisonous plants (Not seen; annotation supplied by J. B. Hansen.)—See also Frederiksen, above, and additional references there given.

Hagerup, Olaf, and **Petersson, Vagn.** Botanisk atlas. Danmarks dækfrøede planter. 550 p. incl. [515] pl. 29 cm. København, 1956.

Habit figures, often with details (in some cases the details only), of essentially all native (and some adventive) angiosperms except the microspecies; no text except for some family or specific descriptions. A supplement containing gymnosperms, vascular cryptogams, and genera of mosses is in preparation.

Hansen, Kristian. Markens vigtigste ukrudtsplanter samt nogle angreb af snyltesvampe og skadedyr paa markens afgrøder. 3. gjennemsete oplag. 41 p. 20.5 cm. Kongens Lyngby, 1906. (1st ed. 1903.)

List of most important weeds, with brief statement of manner of spreading, methods of control, etc.; also a few fungi and insects. The 4. ed. (1909) does not have scientific names.—See also Frederiksen, above, and additional references there given.

[Heger, J. S.] Afbildinger af danske oeconomiske planter, med beskrivelser over deres egenskaber og anvendelse . . . 3 v. (288 p.). 288 col. pl. 20 cm. Kjøbenhavn, 1826–35.

Account of useful and medicinal plants, with descriptions and uses. (Not seen; annotation supplied by J. B. Hansen.)—See also Frederiksen, above, and additional references there.

Hornemann, J. W. Forsøg til en dansk oeconomisk plantelære. 3., forøgede oplag. 2 v. (xii, 1042 p.; 2 p.l., 990 p.). 2 pl. 20 cm. Kjøbenhavn, 1821–37. (Gray Herbarium library.) (1st ed. 1806.)

Descriptive flora of vascular and cellular plants on the Linnaean system, with vernacular names in various Scandinavian languages, and uses. Title of vol. 2 reads: Dansk oeconomisk plantelære.—See also Frederiksen, above, and additional references there given.

Hvass, Hans. Flora i farver.

See Bolin under Scandinavia.

Jacobsen, Edward, and Petersen-Koch, P. Danske medicinalplanter. En populær illustreret vejledning i indsamling og behandling af vore medicinske planter. 51 p. 46 fig. 8vo. Randers, 1918.

Popular account of the medicinal plants occurring in Denmark with descriptions, directions for collecting, and uses. (Not seen; title and annotation from notice in Bot. Jahresber. Just **46** (2): 374. 1928.)—See also Frederiksen, above, and additional references there given.

Jenssen, Jens. Ordbog for gartnere og botanikere. Med vejledning til forstaaelse af de ved planternes betegnelse forekommende græske og latinske slægts–, arts– og varietetsnavne samt botaniske kunstudtryk. 5. opl. gennemset af J. B. Petersen. 3 p. l., 278 p. 23.5 cm. København, 1956. (1st ed. 1886.)

Alphabetical list of Latin and Greek generic and specific epithets and botanical terms, with Danish equivalents and etymology; includes native and exotic plants. Ed. 5 is an unchanged reproduction of ed. 4.—See also Lange (Plantenavne), below.

Jessen, Knud. Oversigt over karplanternes udbredelse i Danmark. Udarbejdet paa grundlag af den Topografisk-botaniske undersøgelses materiale. Bot. Tidsskr. **39**: 137–210. pl. 2 (map). 1926.

Historical review, bibliography, list of 57 topographical districts with lists of collectors; alphabetical list of native and cultivated vascular plants with range indicated by districts.—For a condensation of the historical account and description of the topographical districts in English see Ostenfeld, below. Ostenfeld's map shows the geographical names, not given in Jessen.

——— and **Lind, Jens.** Det danske markukrudts historie. K. Danske Vidensk. Selsk. Skrift. Naturv. og Math. Afd. VIII, v. 8. 496 p. table. 1922-23.

Discussion of prehistoric occurrence, early records, methods of distribution, apophytes, anciently cultivated plants, weeds associated with cultivation of clover and grass, anthropochorous plants (with annotated tabular list); extensively annotated list of weedy spermatophytes, alphabetically arranged by scientific names, with discussion of vernacular names, general distribution, local occurrence, oldest extralimital and local records, biology, uses and harmful properties, methods of introduction; bibliography, index to vernacular names.—See also **Wiinstedt, Knud.** Danske jernbaneplanter. Bot. Tidsskr. **45**: 195–199. 1945. (Discussion of numerous species occurring along railroads.)—Also **Helbæk, Hans.** Prehistoric food plants and weeds in Denmark. A survey of archaeobotanical research 1923–1954. Danmarks Geol. Undersøg. II, **80**: 250–261. pl. 9. 1954. (General discussion, description of localities; tabular list of angiosperms, intended to summarize work accomplished since 1922; bibliography, Danish summary.)—Also **Pedersen, Anfred.** Indslæbte planter ved jernbanerne. Flora og Fauna **61**: 81–109. 20 fig. (incl. maps). 1955. (Discussion of the occurrence and spread of plants along Danish railroads, with detailed treatment of 7 species; bibliography, English summary.)—Also Frederiksen, above, and additional references there given.

Kiær, Egil. Havens blomster i farver. 192 p. incl. 128 col. pl. on 64. 18.5 cm. København, 1955. (Politikens håndbøger no. 88.)

Good colored figures of 508 herbaceous garden plants (including many horticultural varieties), alphabetically arranged by scientific names, with short descriptions. The colored figures in this and in the following work are by Verner Hancke.—See also his Havens buske og træer. 192 p. incl. 128 col. pl. on 64. 18.5 cm. København, 1956. (Politikens håndbøger no. 125.) (Popular handbook of cultivated trees, shrubs, and woody vines, arranged alphabetically by scientific names, with 357 colored figures and brief descriptions.)—Also Gram and Jessen, above, and additional references there given.

Lange, Johan. Haandbog i den danske flora. 4. omarbejdede og forøgede udgave. clxxxviii, 925 p. 21 cm. Kjøbenhavn, 1886–88. (1st. ed. 1851.)—Rettelser og tilføjelser . . . 40 p. 21 cm. Kjøbenhavn, 1897.

Glossary; descriptive flora of vascular plants, with keys to families and genera, and vernacular names.—See also **Møller, Otto, and Ostenfeld, C. H.**

De i de senere aar i Danmark iagttagne findesteder for mindre almindelige karplanter. I. Pteridophyta og Monocotyledones. Bot. Tidsskr. **24**: 377–409. 1902. (List of collaborators; list of rarer species with localities. Intended to supplement Lange's Haandbog, but does not include the records already given in his Rettelser og tilføjelser. Covers only pteridophytes and monocotyledons; no more published.)

——— Oversigt over de i nyere tid til Danmark indvandrede planter med særligt hensyn til tiden for deres indvandring. Bot. Tidsskr. **20**: 240–287. 1896.

General remarks about introduced plants, biliography; annotated list of adventive and naturalized phanerogams, with localities and dates.—See also **Møller, Otto.** Oversigt over de siden 1894 i Danmark indslæbte planter. l.c. **22**: 115–130. 1898. (List of additional species, with localities and short descriptions.)—See also Frederiksen, above, and additional references there given.

Lange, Johan (horticulturist). Plantenavne. 84 p. 21 cm. København, 1949. (Almindelig dansk gartnerforening håndbogsserie 24.)

Alphabetical list of scientific names of plants cultivated in Denmark, each with a standardized Danish name; alphabetical list of Danish names with scientific equivalents; comments on various vernacular names.—See also Jenssen, above.

Lind, Jens. Om lægeplanter i danske klosterhaver og klosterbøger. 115 p. 19.5 cm. København, 1918.

Account of Danish medicinal plants used in the Middle Ages, with discussion of book sources and plants found around ruins and sites of old monasteries and castles; treatment of individual medicinal plants, with history, distribution, and use in medicine. (Not seen; cited from review in Bot. Tidsskr. **36**: 193-194. 1918.)—See also Frederiksen, above, and additional references there given.

Mentz, August, and **Ostenfeld, C. H.** Billeder af Nordens flora.

See Lindman under Scandinavia.

Møller, Birgitte. Træer og buske i Danmarks skove. En botanisk, forstlig og kulturhistorisk skildring af de i vore skove voksende træer og buske. 315 p. illus. 21.5 cm. København, 1916.

Popular treatment of native and cultivated trees and shrubs, with notes on biology, large specimens, uses and folklore.—See also Gram and Jessen, above, and additional references there given.

Mortensen, M. L., and **Ostenfeld, C. H.** Alfabetisk fortegnelse over Danmarks karplanter med synonymer. Udgivet af den af "Botanisk Forening" nedsatte Komité for den topografisk-botaniske undersøgelse af Danmark. 96 p. 19.5 cm. København, 1905.

Alphabetical list of valid scientific names and synonyms of native and introduced vascular plants (including 1,272 native or thoroughly naturalized species), covering all the names used in Danish floras and many of those used in other Scandinavian and north German floras.

Olsen, Sigurd. Oversigt over Dansk Botanisk Forening's ekskursioner. Bot. Tidsskr. **45**: 274–325. map. 1940.

List of excursions of the Society grouped under the 53 botanical districts adopted by the Topografisk Botanisk Undersøgelse, with localities visited, dates, and references to published reports; also list of extralimital excursions. Includes many partial lists for localities not covered by individual floras in the present bibliography.—See also Christensen, above.

Ostenfeld, C. H. The distribution within Denmark of the higher plants. Results of the topographic-botanical investigation. I. A brief historical survey of the investigation. K. Danske Vidensk. Selsk. Skrift. Naturv. og Math. Afd. IX, **3**: 1–15. 1 fig. (port.), pl. 1 (map). 1931.

History of the project, begun by the Botaniske Forening i Köbenhavn (later Dansk Botanisk Forening) in 1904; list of the 53 (i.e. 57) topographical districts, with brief account of geography, soils, and collaborators, and map showing districts and towns.—See also **Jessen, Knud.** The distribution . . . II. The distribution of the Papilionaceae within Denmark. l.c. IX,**3**: 17–97. 36 fig. (maps) on pl. 2–10. 1931. (Alphabetical list of species, with references, extralimital and local range, habitat, methods of dispersal, prehistoric migrations and fossil occurrence; general summary of distribution, bibliography.)—No more of this series published; for continuation see Dansk Botanisk Forening (above).—Also Dansk Botanisk Forening (5te beretning), above.

Petersen, Herluf. Danmarks spiselige vilde planter. 6. udg. 96 p. illus. 21 cm. København, 1941. (1st ed. 1940.)

Guide to collection and identification of edible wild plants (70 species), together with recipes for use; also recipes for teas. (Not seen; title and annotation supplied by J. B. Hansen.)—See also Frederiksen, above, and additional references there given.

Petersen, O. G. Træer og buske. Diagnoser til dansk frilands-trævækst. 2 p. l., ix, 517 p. 248 fig., col. pl. (front.). 24 cm. Kjøbenhavn og Kristiania, 1916. (John Crerar Library.)

Descriptive account of trees and shrubs growing in the open air in Denmark; the great majority are cultivated species.—See also Gram and Jessen, above, and additional references there given.

Raunkiær, Christen. De danske blomsterplanters naturhistorie. Vol. 1. Enkimbladede. lxix, 724 p. 293 fig. 26 cm. Kjøbenhavn, 1895–99.

Systematically arranged account of plant biology treating of such subjects as anatomy and development of leaves, flowers, and fruit, overwintering methods, etc. Covers monocotyledons only; no more published.

——— Dansk ekskursions-flora. 7. udg. ved K. Wiinstedt. xxxi, 380 p. 19 cm. København, 1950. (1st ed. 1890.)

Pocket flora of vascular plants in form of briefly annotated keys, with indication of life-form and brief statement of local range; includes some cultivated species.

Rostrup, Emil. Den danske flora. En populær vejledning til at lære de danske planter at kende. 18 omarbejdede udg. ved C. A. Jørgensen. lxiv, 527 p. 154 fig. (fig. 10, map), front. 19 cm. København, 1953. (1st ed. 1860, with title Vejledning i den danske flora . . .)

Glossary, list of legally protected species with localities, key to genera on Linnaean system, keys to woody plants and water plants; keyed flora of vascular plants (about 1500 species), with short descriptions and statement of range in terms of Topografisk-botaniske Undersøgelse districts; glossary of specific names. Naturalized species are included but not cultivated plants. The treatment of *Rubus, Taraxacum,* and *Hieracium* includes only the principal species, with reference to Raunkiær's Dansk ekskursions-flora for keys to microspecies.

Vedel, Helge and Lange, Johan. Træer og buske i skov og hegn. 224 p. illus., col. pls. 19 cm. København, 1958.

Descriptive treatment of wild and cultivated trees and shrubs. (Not seen; title and annotation supplied by J. B. Hansen.)

Warming, Eugenius. Dansk plantevækst. 2 v. (vi. 325; 3 p.l., 376 p.) illus. 25.5 cm. København og Kristiania, 1906–1909.

Monographic treatment of ecology and plant formations, with bibliographies. Vol. 1, Strandvegetation; v. 2, Klitterne[dunes].—3. Skovene [woods]. Bot. Tidsskr. v. 35. 635 p. 283 fig. incl. maps. 1916–19. (An ecological and biological monograph of the forests of Denmark. Chapter 3 (p. 33–145, fig. 24–89), Danmarks ved- og halvvedplanter, is an account of the individual species of trees and shrubs discussing geological history, manner of occurrence, range, size, biology, variation, etc., with references; briefer notices of the cultivated species are included. The treatment of such groups as *Rosa* and *Rubus* is very summary. Also published separately.) —See also Balslev, above, and (for trees) Gram and Jessen, above, and additional references there given.

BORNHOLM

General

Larsen, Arne. Bornholms flora. Bot. Tidsskr. 52: 189–316. 10 i.e. 11 fig. (incl. map). 1956.

Topography, climate, plant associations, changes in flora, phytogeography; annotated list of vascular plants, with localities; bibliography, gazetteer.

Includes records from Christiansø. Replaces **Bergstedt, N. H.** Bornholms flora. Bot. Tidsskr. **13**: 133–198. map. 1883. (Botanical explorations; annotated list of Characeae and vascular plants (1043 species). Reprinted 68 p. 1883; reprint in Riksmuseet library, Stockholm.) (Bornholm amt.)—See also **Olsen, S.E.S.**, and **Mikkelsen, V. M.** Højsommer-ekskursion til Bornholm 2.–4. august 1957. Bot. Tidsskr. **54**: 80–84. 4 fig. 1958. (Includes supplement to Larsen's flora.)

FAERÖES (FAROES)

See also Iceland, General (Ostenfeld and Gröntved, The flora of Iceland and the Færoes).

General

Ostenfeld, C. H. Phanerogamae and Pteridophyta. *In* Warming, E., *ed.* Botany of the Faeröes. v. 1, p. 41–99. 25.5 cm. Copenhagen, 1901.

Annotated list of vascular plants (285 native or naturalized species.) Sections by the same and other authors on geography, phytogeography, climate, cellular plants, etc., are included in the same volume. The same author's Additions and corrections (l. c. **3**: 835–863. 1908) is a partly annotated, corrected list of 298 native or naturalized vascular plants.—See also **Nordhagen, Rolf,** and **Omang, S. O. F.** Et bidrag til Faerøenes flora. Planter samlet på Faerøene av Olaf Hanssen i 1926. Bergens Mus. Årb. **1929**, Naturv. Rekke no. 7. 8 p. 1930. (Includes additional species.)—For sketch of ecological conditions, plant formations, and bibliography, see **Ostenfeld.** The land-vegetation of the Faeröes, with special reference to the higher plants. *In* Warming, E., *ed.* Botany of the Faeröes. v. 3, p. 867–1026. fig. 165–195. Copenhagen and Christiania, 1908. (Revised from his Plantevæxten paa Færøerne med særlig hensyntagen til blomsterplanterne. Bot. Tidsskr. **28**: 1–142. 29 fig. 1907. Preprinted 1906; preprint not seen.)

Patursson, Gazet. List of popular plant names from the Faeröes. *In* Warming, Eugenius, *ed.* Botany of the Faeröes. v. 3, p. 864–866. 25.5 cm. Copenhagen and Christiania, 1908.

Alphabetical list of vernacular names, with botanical equivalents.

Rasmussen, Rasmus. Føroya flora. 2. útg. xxviii, 231 p. 108 fig. 20.5 cm. Tórshavn, 1952. (1st ed. 1936).

Briefly annotated flora of vascular plants in form of descriptive keys; illustrated glossary, list of specific epithets with Faroese equivalents. Written in the Faroese language.—See also **Oskársson, Ingimar.** Viđbot til Hieracium-grođurin i Føroyum. (Supplement to the Hieracium-flora of the Faeroes.) Frođskapparit (Annales Soc. Scient. Faroensis) **3**: 115–127. 7 fig. 1954. (Not seen.)

——— Føroysk plantunøvn. 207 p. 67 fig. 22 cm. Tórshavn, 1950.—Uppískoyti til Føroysk plantunøvn. 209–255 p. 22 cm. Tórshavn, 1951.

Previous publications on Faroese and some other Scandinavian vernacular names; alphabetical list of Faroese vernacular names of vascular and some cellular plants, with citation of names from previous writers and discussion (often including uses); bibliography, index of scientific names. The supplement includes an annotated list of additional vernacular names and a systematic list of vascular plants of the Faroes, each with a standardized "vernacular name," many of which do not occur in the list proper.

Warming, Eugenius. Historical notes on the botanical investigation of the Faeröes. *In* Warming, E., *ed.* Botany of the Faeröes. v. 1, p. 1–5. 25.5 cm. Copenhagen, 1901.

Brief sketch, with bibliography.

Local

Rasmussen, Rasmus. Vegetationen i de faerøske fuglebjærge og deres nærmeste omgivelser. Bot. Tidsskr. **48**: 46–70. 2 fig. 1946.

Running account of vegetation on the "fuglebjærge", scattered areas inhabited by breeding colonies of sea birds, with lists of species observed at various localities and on various small islands.

FYN (FYEN; FUNEN)

Includes islands of Ærø, Brandsø, Fyn, Langeland, and numerous smaller islands.

General

Lange, Axel. Fynske plantenavne og folkeminder om planter. 2 p. l., 150 p. 24 fig. 22.5 cm. Espe, 1932. (Reprinted from Fynsk Hjemstavn v. 2–5. 1929–32.)—Nogle noter til "Fynske plantenavne . . ." Fynsk Hjemstavn **12:** 22–29, 42–45. 1939. (Both in Botaniske Museum library, Copenhagen.)

Treatment of vernacular names and uses, alphabetically arranged by vernacular names.

Local

Andersen, Anton. Nordfyns flora. Bot. Tidsskr. **30:** 329–454. 1910.

Botanical investigations, topography, general features of flora, plant formations, bibliography; annotated list of vascular plants (1,212 species). Covers northern part of Odense and Assens amter, north of a line from near Kerteminde running a little south of Odense (city) through Vissenbjerg and Aarup to Føns.

Gröntved, Johannes. Ærøs vegetation og flora. Bot. Tidsskr. **49:** 117–159. 4 fig. (fig. 1, map). 1952.

Topography, geology, climate, general features of vegetation, botanical explorations; annotated list of vascular plants, including some cultivated species; bibliography. (In Svendborg amt.)

Grüner, Johanne. Om floraen paa halvøen Østerøen ved Nyborg. Bot. Tidsskr. **43:** 40–49. 2 fig. (incl. map). 1934.

Topography, general features of flora, with partial lists of species of different localities, including many new to the district. (In Svendborg amt.)

Hansen, Alfred. Om floraen på øen Nyland i det sydfynske øhav. Flora og Fauna **62:** 45–49. 1956.

History, topography, etc.; alphabetical list of angiosperms (37 species and 1 hybrid) of a small island which has been formed within the last century. (In Svendborg amt.)

——— Om vegetationen på øen Vresen i Storebælt. Flora og Fauna **65:** 1–6. 3 fig. (incl. map.) 1959.

History of the island; lists of species found by author and earlier visitors (total about 128 species); general features of flora.

Lange, M. T. Den sydfyenske ögaards vegetation, en floristisk skitse. Vidensk. Meddel. Naturh. For. Kjöbenhavn 1857: 199–272. 1857.

Geography, physiography, general features of flora, plant formations, statistics; annotated tabular list of Characeae, mosses, and vascular plants (882 i.e. 385 species) of south Fyn and adjacent islands and islets, including Langeland, Ærö, and others. (In Svendborg amt.)—See also **Mortensen, M. L.** Ekskursionen til det sydlige Langeland den 18.–21. Juli 1903. Bot. Tidsskr. **26** (Meddel. Bot. For. København): vi–x. 1904. (Itinerary, with mention of numerous species observed, including about 100 species not recorded from Langeland in Lange's paper.)—Also (for Ærø) Gröntved, above.

Wiinstedt, Knud. Karplantevegetationen paa Brandsø. Bot. Tidsskr. **40:** 204–219. 1 fig. (map). 1928.

Topography, botanical explorations, plant associations, etc.; barely annotated, alphabetically arranged list of wild and cultivated vascular plants (338 species). (Island in Odense amt.)

JYLLAND (JUTLAND)

See also Sjælland (Thomsen, Sams-øgruppens plantevæxt, which includes list of plants of Endelave).

Includes the whole Danish mainland and the Danish North Frisian Islands (principally Fanø, Jordsand, Manø, Rømø) on the west coast, and the islands of Als, Anholt, Endelave, Læsø, Tunø, and some smaller ones off the east coast.

Local

Andersen, Svend. En adventivflora paa den nedrevne spritfabriks tomt i Aalborg. Bot. Tidsskr. **43:** 51–56. 1934.—Et supplement. l.c. **43:** 236–238. 1935.

Includes list of about 270 adventive plants found around a burned-down alcohol factory in city of Aalborg. The supplement adds 24 species. (In Aalborg amt.)

Böcher, T. W. Vegetationen paa Randbøl Hede med særlig hensyntagen til det fredede areal. K. Danske Vidensk. Selsk. Biol. Skr. v. 1. no. 3. 234 p. 57 fig. (incl. maps), 13 pl. on 7 (pl. 1, map). 1941.

Ecological; includes alphabetical lists of vascular and cellular plants (p. 208–217) and bibliography; German summary. (In Ribe amt.)

Børgesen, Frederik, and **Jensen, Christian.** Utoft hedeplantage. En floristisk undersøgelse af et stykke hede i Vestjylland. Bot. Tidsskr. **26:** 177–221. 15 fig. (incl. map). 1904.

Topography, geology, climate, plant associations, list of species of meadow land along Grinsted-Aa, list of species of cultivated ground, changes in flora due to cultivation and to fires, list of species found in heath land, etc.; biliography. Utoft Plantation is mostly in Grinsted sogn. (In Ribe amt.)

Christiansen, Albertus. Verzeichnis der pflanzen-standorte in Schleswig-Holstein und den eingeschlossenen gebieten Oldenburgs, Hamburgs und Lübecks nach Schmeil-Fitschen, Flora von Deutschland. v, 62 p. 2 fig. (maps). 18 cm. Leipzig, 1913.

Briefly annotated list of vascular plants, with localities for rarer species. Includes North Schleswig (present Danish provinces of Aabenraa, Haderslev, Sønderborg, and Tønder) which from 1864 to 1920 was part of Germany. The 2. aufl. (1949) does not cover North Schleswig. Replaces **Prahl, Peter.** Kritische flora der provinz Schleswig-Holstein, des angrenzenden gebiets [sic] der Hansestädte Hamburg und Lübeck und des fürstentums Lübeck. Unter mitwerkung von . . . R. von Fischer-Benzon und . . . E. H. L. Krause. II. teil, 2 (i.e. v. 2, pt. 2). Kritische aufzählung und besprechnung der im gebiete beobachteten oder aus demselben angegebenen gefässpflanzen und ihrer formen. ix, 345 p. 23 cm. Kiel, 1889–90. (Annotated list of vascular plants, (1,134 species, the merely adventive or escaped species not numbered), with habitats, localities, descriptions of varieties, and some critical notes. In Gray Herbarium library.) There are later editions of the taxonomic part of Prahl's flora, the latest being: Flora der provinz Schleswig-Holstein, des angrenzenden gebietes der Hanse-städte Hamburg und Lübeck und des fürstentums Lübeck. 5. verm, aufl. des 1. teiles der Kritische flora der provinz Schleswig-Holstein. Bearb. von Paul Junge. ix, 357 p. 18.5 cm. Kiel, 1913. (Flora of vascular plants in form of briefly annotated keys. In Gray Herbarium library. 1st ed., Kritische flora der provinz Schleswig-Holstein, I. teil. 1888.)

Christiansen, Werner, and **Christiansen, Willi,** *eds.* Das botanische schrifttum von Schleswig-Holstein, Hamburg und Lübeck. 2 p. l., 323 p. 22.5 cm. Kiel, 1932–36.

Annotated, chronologically arranged bibliography covering all phases of botany, including geography, floristics, and sociology of vascular plants (1148 titles, p. 1–123), folk botany, biographies, etc., with subject, locality, scientific name (generic), and author indices. Covers North Schleswig (present Danish provinces of Aabenraa, Haderslev, Sønderborg, and Tønder) as well as German Schleswig. Publication of Arbeitsgemeinschaft für floristik in der N.S.-Kulturgemeinde, Gaudienstelle Schleswig-Holstein. Replaces **Fischer-Benzon, Rudolf von.** Geschichte der floristischen erforschung des gebietes. *In* Prahl, Peter. Kritische flora der provinz Schleswig-Holstein . . . II. teil, 1 (i.e., v. 2, pt. 1). 63 p. 23 cm. Kiel, 1890. (Brief sketch of earlier botanical investigations; alphabetical list of botanists connected with the flora of the region, with titles of publications and biographical and critical notes. In Gray Herbarium library); also **Fischer-Benzon, R. von.** Zur botanischen litteratur Schleswig-Holsteins, der angrenzenden gebiete und Helgolands. Ein nachtrag zu Prahl, Kritische flora . . . teil II. 1890. 12 p. 22 cm. Kiel, 1891. (Not seen; title supplied by J. B. Hansen.)—The Christiansens' bibliography is continued by **Christiansen, Willi.** Botanik. Zeitschr. Gesell. Schleswig-Holsteinische Gesch. Ergänzungsband. Bibliographie zur schleswig-holsteinischen geschichte und landeskunde **2:** 20–29, 124–129, 221–226, 330–334, 454–458. 1937–43 (covers 1934–38); **3:** 86–91. 1951 (covers 1939–44); **4:** 64–72. 1954 (covers 1945–50); **5:** 86–99. 1956 (covers 1951–54). (Vol. 2 in Harvard College library, others in Library of Congress.)—See also Knuth (Geschichte), below.

Edelberg, Lennart. Jordsands vegetation. Bot. Tidsskr. **48**: 91–96. 3 fig. (fig. 1, map). 1946.

Topography, zones of vegetation; alphabetical list of angiosperms; brief bibliography. (The southernmost of the Danish North Frisian Islands, in Tønder amt.)

Fischer-Benzon, Rudolf von, and **Steinvorth, Johannes.** Ueber die flora der umgegend von Hadersleben. Programm der Lateinischen Schule in Hadersleben für das schuljahr **1872/73**: 3–36. map. 25 cm. 1873.—Nachtrag. Programm . . . **1873/74**: 1. 1874.

General features of flora; list of vascular plants (809 species), with localities and partial keys. The Nachtrag adds 13 species, and is followed (p. 2–38) by a key to the genera. (Haderslev, in Haderslev amt.) (Not seen; titles and annotations supplied by J. B. Hansen.)

Gabrielsen, E. K., and **Iversen, Johannes.** Die flora von Skallingen. (Die vegetation der halbinsel Skallingen. I.) Bot. Tidsskr. **42**: 355–383. 12 fig. (incl. map). 1933.

Topography, effects of human culture, botanical explorations, general features of flora; alphabetical tabular list of vascular plants [212 species] from the peninsula of Skallingen and the adjacent island of Langli. (In Ribe amt.)

Galløe, Olaf, and **Jensen, Christian.** Plantevæksten paa Borris Hede. Bot. Tidsskr. **27**: 249–275. 2 fig. (incl. map). 1906.

Plant formations, damage due to fires; list of lichens, mosses, and vascular plants. (In Ringkøbing amt.)

Gröntved, Johannes, and others. Hammer Bakker. En botanisk undersøgelse, iværksat af Dansk botanisk forening. Bot. Tidsskr. **39**: 239–298. 21 fig., map. 1926.

Ecology, etc.; annotated list of cellular and vascular plants, by various authors (vascular plants p. 291–298, by Grøntved). (A nature reserve of 76 hectares, 10 kilometers northeast of Aalborg, in Aalborg amt.)

Hansen, H. M. Nørholm hede. En formationsstatistisk vegetationsmonografi. K. Danske Vidensk. Selsk. Skr. Naturv. og Math. Afd. IX, **3**: 99–195. 23 fig. (incl. maps), map. 1932.

Ecological; includes (p. 118–123) alphabetical lists of vascular plants, lichens, and mosses; English summary. (In Ribe amt, 10 kilometers northeast of Varde.)

Jacobsen, J. P. Fortegnelse over de paa læsø og Anholt i 1870 fundne planter. Bot. Tidsskr. **11**: 88–113. 1879–80.

Includes slightly annotated list of cellular and vascular plants from Anholt (p. 103–113). The list for Læsø (p. 88–103) is replaced by Wiinstedt (1932), below. (Læsø in Hjørring amt; Anholt, in Randers amt.)—See also **Paulsen, Ove.** Om vegetationen paa Anholt. Bot. Tidsskr. **21**: 264–286. 3 fig. (fig. 1, map). 1898. (Topography, plant formations, etc.; additions and corrections to Jacobsen's list.)—Also **Gröntved, Johannes.** Bidrag til Anholts flora. l.c. **41**: 310–316. 2 fig. 1931. (Changes in vegetation; list of additional species and localities.)—Also **Hansen, Alfred,** and **Olsen, S.E.S.** Nye planter for Anholt 1931–1957. l.c. **54**: 65–68. 2 fig. 1958. (Raises number of taxa to 430.)—Also Wiinstedt, Karplantevegetationen paa Læsø, below.

Knuth, Paul. Flora der nordfriesischen inseln. viii, 163 p. 17.5 cm. Kiel und Leipzig, 1895. (Gray Herbarium library.)

Bibliography, plant formations, relationship between plants and insects; annotated, keyed, descriptive flora of vascular plants (491 i.e. 492 species), with localities. Covers the now again Danish islands in Tønder amt (Rømø or Röm and Jordsand) as well as the German islands.—See also Edelberg (Jordsands vegetation), above, and Wiinstedt (Rømø's vegetation og flora), below.

——— Geschichte der botanik in Schleswig-Holstein. 216 p. 3 pl. 23 cm. Kiel und Leipzig, 1890–92.

Account of botanists and their work, with brief biographical notes and titles of publications (the principal ones annotated); divided into pre-

Linnaean and post-Linnaean periods, the latter grouped by subject matter, Includes North Schleswig (the present four southern provinces of Jylland, —Aabenraa, Haderslev, Sønderborg, and Tønder), at that time part of Germany.—See also W. Christiansen, above.

Lange, Axel. Vegetationen paa Tunø og Hjelm. Bot. Tidsskr. **37:** 1–22. 1920.

Topography, geology, general features of flora, etc., of Tunø, with alphabetical list of 412 vascular plants (383 native) with habitats; topography, etc., of Hjelm with unannotated alphabetical list of 151 vascular plants. (Tunø, in Aarhus amt; Hjelm off Ebeltoft (Æbeltoft) peninsula, in Randers amt.)

Mathiesen, Hans, and Nielsen, Johannes. Botaniske undersøgelser i Randers Fjord og Grund Fjord. Bot. Tidsskr. **53:** 1–34. 12 fig. (incl. maps). 1956.

Includes alphabetical tabular list (p. 12–14) of vascular plants and Characeae, followed by notes on various species. (Randers amt.)

Mentz, August. Store Vildmoses plantevækst. *In* Vildmosearbejdet. Udg. af K. Danske Landhusholdningsselskab. p. 54–66 incl. illus. 25.5 cm. København, 1945.

Description of the vegetation of Store Vildmose, the second largest sphagnum bog in Denmark, now nearly destroyed, with mention of all the characteristic species. (In Aalborg amt.) (Not seen; title and annotation supplied by J. B. Hansen.)

Ostenfeld, C. H. De smaa holme i Kattegat. *In* his Træk af vegetationen i omegnen af Frederikshavn. Bot. Tidsskr. **25:** 103–108. fig. 2. 1903.

General features of flora of Deget, Kjølpen, and Hirsholmene, small islands in the Kattegat east of Frederikshavn; list of 156 angiosperms and 17 mosses, showing occurrence by islands. (In Hjørring amt.)

Pedersen, Anfred. Floraen på Fanø og Manø. Bot. Tidsskr. **50:** 1–34. 5 fig. (incl. 2 maps). 1953.

Topography, geology, climate, botanical explorations, plant associations; annotated list of vascular plants (709 species from Fanø, 380 from Manø), with localities and indication of respective plant associations; bibliography. (In Ribe amt.)

Pedersen, Kristen. Findesteder for sjældnere forekommende, vildtvoksende planter i Viborgegnen. *In* Foreningen for naturkundskab, Viborg, 1912–1937. p. 17–37. 2 fig. 24 cm. Viborg, 1937. (Botaniske Museum library, Copenhagen.)

General features of vegetation; extensive list of noteworthy vascular plants, with localities. Refers mostly to Viborg amt, but in part to adjacent amter (Aalborg, Randers, Ringkøbing.)

Petit, Emil. Udkast til en floristisk beskrivelse af Als. Bot. Tidsskr. **12:** 13–41. 1880.

Topography, general features of flora, etc.; annotated list of vascular plants (690 species). (Island in Sønderborg amt.)—See also his Supplement til "En floristisk beskrivelse af Als." Bot. Tidsskr. **18:** 6–11. 1892. (Brings total to 735 species.)—Also **Petersen, Hans.** Beitrag zur flora von Alsen. 50 p. 21 cm. Sonderburg, 1891. (Beilage zum Programm des Königl. realprogymnasiums zu Sonderburg.) (Botanical explorations, description of localities and habitats; briefly annotated list of vascular plants; list of vascular plants in Petit's list not found by author. In British Museum (Natural History) library.)

Rahn, Knud, and Andersen, H. B. Kongenshus Hede—en botanisk oversigt. Hedeselsk. Tidsskr. **77:** 287–296, 311–325. 10 fig. (fig. 1, map). 1956.

Topography, history, general features of vegetation, etc.; annotated alphabetical list of vascular plants, also mosses and lichens; bibliography. (In Viborg amt.)

Raunkiær, Christen. Floraen paa Holmslands Klit og Tipperne. *In* his Botaniske studier **1:** 31–54. 5 fig. 26 cm. 1934. (Botaniske Museum library, Copenhagen.)

Ecological; includes (p. 45–50) alphabetical list of 245 vascular plants of a dune area in West Jutland. (In Ringkøbing amt.)

Ravn, F. K. Fortegnelse over karplanter, fundne paa Jyllands nordspids. Samt: Sæbys flora ved E. Rostrup. Bot. Tidsskr. **21**: 113–138. 1897.

Botanical explorations; list of vascular plants, with localities, known in Jylland north of a line drawn from Aalbæk through Gaardbogaard to Tværsted Plantage, those known from Sæby but not from Skagen being distinguished typographically; statistics. (In Hjørring amt.)

Stoltenberg, Nicolaus. Beitrag zur kenntniss der flora Tonderns . . . 41, 46 p. 22 cm. Tondern, 1877. (Library of Botaniske Museum, Copenhagen.)

Account of special localities, phenology, etc.; annotated list of angiosperms (about 550 species). (In Tønder amt.)

Tholle, Johannes. Flora og folketro i Sønderjylland. Sønderjysk Månedsskrift **31**: 145–152, 169–176, 177–192. 8 fig. 1955.

Account of popular beliefs and uses, with mention of 152 plants. (Not seen; title and annotation supplied by J. B. Hansen.)

——— Flora og folketro især i Ringköbing amt. Hardsyssels Aarbog **28**: 28–64. 2 pl. 1934.

Account of popular beliefs and uses, with mention of 177 plants. (Ringköbing amt.) (Not seen; title and annotation supplied by J. B. Hansen.)

——— Flora og folketro, især i Vendsyssel. Vendsyssels Aarbøger **10**: 57–96. 1934.

Account of popular beliefs and uses, with mention of 129 plants, summary, and references. (Vendsyssel amt.) (Not seen; title and annotation supplied by J. B. Hansen.)

Wiinstedt, Knud. Horsensegnens flora (Pteridophyta og Phanerogamae). Bot. Tidsskr. **34**: 89–160. 2 fig. (incl. map). 1915.

Geology, botanical explorations, bibliography, botanical formations; annotated list of vascular plants. Covers radius of about 2 Danish miles (about 15 kilometers) around Horsens. (In Aarhus, Skanderborg, and Vejle amter.)

——— Karplantevegetationen paa Læsø. Bot. Tidsskr. **42**: 89–181. illus., map. 1932.

Geography, climate, geology, topography, botanical explorations, plant associations and habitats, biological spectrum, list of plants from Nordre Rønner (a group of small islands 7 km. north of Læsø); annotated alphabetical list of vascular plants of Læsø, including cultivated species; list of common Danish species not found on Læsø; bibliography. (In Hjørring amt.) —See also Jacobsen, above.

——— Rømø's vegetation og flora. Bot. Tidsskr. **46**: 303–346. 1 fig. (map). 1946.

Topography, climate, botanical explorations, geology, plant associations, phytogeography, list of common Danish species not known from Rømø; briefly annotated alphabetical list of vascular plants [490 species], including cultivated species; bibliography. (One of the North Friesian Islands, in Tønder amt.)—See also **Pedersen, Anfred.** Floraen på Fanø og Manø. Bot. Tidsskr. **50**: 9, footnote. 1953. (Adds 14 species.)

——— Vegetationen paa Reservatet Vorsø i Horsens Fjord. Bot. Tidsskr. **44**: 260–306. fig. 2–20, map. 1938.

Geography, geology, zones of vegetation, plant formations, etc.; alphabetical list of spermatophytes, with localities. (Island in Aarhus amt; also spelled Vaarsø.)

SJÆLLAND (SEELAND; ZEALAND)

Includes islands of Amager, Falster, Hesselø, Laaland (Lolland), Møn (Moen), Samsø, Sejrø, Sjælland (Zealand), and numerous smaller islands.

Local

Anderson, Svend. Møens Klint. 1. Om akklimatiserede, fremmede planter i Møens Klinteområde. Flora og Fauna **56**: 73–80. 1950.

Botanical explorations, general features of flora with running account of naturalized species. (In Præstø amt.)—See also his Møens Klint. 2. Agerukrudtsfloraen paa Høje Møen. Flora og Fauna **57**: 65–69. 1951. (Account of field weeds with alphabetical list of species.)

Böcher, T. W. Vegetationsstudier paa halvøen Ulvshale. Bot. Tidsskr. **46:** 1–42. 10 fig. (incl. map). 1942.

Ecological; includes (p. 35–38) partly annotated list of vascular plants; bibliography, German summary. (On island of Møn, in Præstø amt.)

Christensen, Carl. Vegetationen paa øerne i Smaalandshavet. Bot. Tidsskr. **26:** 321–342. 1905.

General features of flora, ecology; tabular list of vascular plants, showing occurence on the different islands. (Small islands of Askø, Fæmø, Fejø, Lilleø, Raagø, Raagø Kalv, Skalø, and Vejrø, directly north of Laaland, in Maribo amt.)

Ferdinandsen, Carl, and **Winge, Øjvind.** Kobberdammene i Aldershvile skov med Bagsværd. En topografisk-botanisk undersøgelse som grundlag for studiet af tilgroningen. Bot. Tidsskr. **33:** 1–44. 15 fig., 4 maps. 1912.

Mainly ecological; includes (p. 10–14) list of vascular plants and larger fungi. (In København amt.)

Grøntved, Peter. Om floraen i nogle nordvestsjællandske skove. Bot. Tidsskr. **43:** 325–356. 1 fig. (map). 1935.

Geology, general features of flora; tabular list of vascular plants, with indication of life form and localities; bibliography. Relates to seven forests in the Tissø-Skarridsø region in Holbæk amt.

——— Om plantevæksten i og ved Tissø. Bot. Tidsskr. **43:** 197–219. 1 fig. (map). 1935.

Geology, general features of flora, etc.; tabular list of vascular plants, with indication of life form and occurrence; bibliography. (A lake in Holbæk amt, near Gørlev.)

Koch, H. P. G. Om Falsters vegetation. Vidensk. Meddel. Naturh. For. Kjöbenhavn II, 4 (1862): 79–152. 1862 [1863].

Topography, general features of flora; annotated list of cellular and vascular plants (916 vascular species) of Falster and adjacent small islands. (In Maribo amt.)—See also his Tillæg til . . . Falsters vegetation. Vidensk. Meddel. Naturh. For. Kjøbenh. IV, 3 (1881): 77–88. 1881. (Brings total to 968 vascular plants.)—Also **Kring, Laurids.** Forandringer i plantevæksten på Falster. Flora og Fauna **59:** 85–92. 1953. (Includes lists of species that have changed their status and of additional species, with discussion. Essentially no scientific names are given.)

Mortensen, Hans. Nordostsjællands flora. Bot. Tidsskr. II, **1** [v.5]: 8–168. map. 1872.

Bibliography, topography, general features of flora, floras of special localities, etc.; annotated list of lichens, hepatics, mosses, Characeae, and vascular plants (1167 species of vascular plants), with localities. Covers Fredriksborg amt and part of København and Roskilde amter, north of a line a little south of København and Roskilde, and includes the islands of Amager and Saltholm on the east and Ovrø (Orø) and Eskildsø on the west.

Nielsen, Peter. Sydvestsjællands vegetation. Bot. Tidsskr. II, **2** [v. **6**]: 261–403. map. 1873–74.

Botanical explorations, plant associations, etc.; annotated list of Characeae and vascular plants, with localities for scarcer species; French summary (p. 389–403). Covers southwestern part of Sorø amt, bounded essentially by the Susaa, Sorø, and Slagelse.

Petersen, O. G. En excursion til Hesseløen. Bot. Tidsskr. III, **2** [v. **10**]: 48–51. 1877.

Author's explorations; briefly annotated list of angiosperms. (In Holbæk amt.)—See also his Bemækninger til et manuskript af H. C. Lyngbye om Hesseløens flora. Bot. Tidsskr. **13:** 79–82. 1882. (List of species found by Lyngbye, with comments.)

Raunkiær, Christen. Allindelille Fredskov. Statistical investigations of the plant formations. *In* his Botaniske studier. v.1. p. 165–226. 6 fig. (incl. map). 26.5 cm. København, 1935. (Arbejder fra den Botaniske have i København no. 132.) (Arnold Arboretum library.)

Mainly ecological; includes list of 362 vascular plants. Allindelille Fredskov is a small forest about 8 kilometers north of Ringsted, in Sorø amt.

Rostrup, Emil. Gammelmose. Beskrivelse af en staten tilhørende tørvemose i Vangede. Bot. Tidsskr. 27: 319–359. map. 1906.

History, changes in vegetation; tabular list of vascular plants, showing occurrence in different years from 1844–1905; list of mosses, fungi, and lichens. (A peat bog north of Vangede, in København amt.)

——— Lollands vegetationsforhold. Vidensk. Meddel. Naturh. For. Kjöbenhavn II, 6 (1864): 37–119. 1864.

Topography, botanical explorations, phytogeography, general features of flora, plant formations, phenology; annotated list of 927 vascular plants and Characeae. (In Maribo amt; the name often spelled Laaland.)

Terslin, H. C. Strandvegetation ved Gilleleje. Gilleleje strandklinters blomsterplanter. 20 p. 4 fig. 23 cm. Gilleleje, 1939. (Reprinted from Fra det gamle Gilleleje. 1938. In Botaniske Museum library, Copenhagen.)

Includes (p. 11–16) tabular list of 161 vascular plants. (In Frederiksborg amt.)

Tholle, Johannes. Flora og folketro især i Nordsjælland. Frederiksborg Amts Aarbøger **1935:** 115–151. 6 fig. 1935.

Account of popular beliefs and uses of 105 Danish plants, mostly from Nordsjælland; index of species according to uses; bibliography. (Not seen; title and annotation supplied by J. B. Hansen.)

——— Flora og folketro især paa Lolland-Falster. Aarbog Lolland-Falsters Hist. Samfund **24:** 128–149. 1936.

Account of popular beliefs and uses, with mention of 123 plants; no references. (Not seen; title and annotation supplied by J. B. Hansen.)

Thomsen, Christen. Roskilde-egnens flora. I. Karkryptogamer og blomsterplanter. iv, 92 p. 20 cm. Roskilde, 1874. (Reprinted from Indbydelsesskrift til afgangsprøven og aarsprøverne i Roskilde Kathedralskole i Juni og Juli 1874, no. 1.) (Gray Herbarium library.)

Topography, etc.; annotated list of vascular plants (934 species), with localities. (In København amt.)

——— Sams-øgruppens plantevæxt. Bot. Tidsskr. II, 4 [v. 8]: 86–142. 1875.

Topography, general features of flora, etc.; annotated list of vascular plants of Samsø group (667 species); partly annotated list of vascular plants from Endelave Island, with account of topography and general features of flora. (Samsø group, in Holbæk amt; Endelave, in Aarhus amt.)—See also **Mortensen, Hans.** Exkursion till Samsø d. 26 til 29 Juli 1887. Meddel. Bot. For. Kjøbenhavn **2** (3): 38–42. 1888. (Account of excursion with mention of species observed and list of new records.)—Also **Ostenfeld, C. H.** Ekskursionen til Syd-Samsø den 22.–23. Juni 1918. Bot. Tidsskr. **36:** 307–310. 1919. (Includes mention of numerous species.)—Also **Böcher, T. W.** Højsommerekskursionen til Samsø . . . l.c. **44:** 223–230. 3 fig. 1937. (Includes lists of species from different localities.)—Also his Flora och vegetation paa Samsø. *In* Rasmussen, Bent, *ed.* Samsø, Danmarks midtpunkt. p. 41–46. 28 cm. København, 1948. (General features of vegetation, with mention of various species.)—Also **Olsen, S.E.S.,** and **Hansen, Alfred.** Forsommerekskursionen til Samsø 29–30. Juni 1957. Bot. Tidsskr. **54:** 77–80. 1 fig. 1958. (Includes mention of numerous species.)

Wiinstedt, Knud. Charlottenlund Skov. En botanisk-historisk studie. Bot. Tidsskr. **38:** 340–363. 1925.

History; alphabetical list of species observed by author in 1922, followed by notes on various species, general considerations, and bibliography. (In København amt.)

——— Nogle indslæbte planter i Pedersborg ved Sorø. Bot. Tidsskr. **51:** 371–376. 1954.

Previous work on introduced plants in Denmark, etc.; list of spermatophytes found around a factory in Pedersborg manufacturing cork products; bibliography. (In Sorø amt.)

——— Om vegetationen paa Sejrø. Bot. Tidsskr. **37:** 83–106. 1920.

Soils, botanical investigations, general features of flora; annotated list of about 400 vascular plants. (In Holbæk amt.)

FINLAND (SUOMI)

See also Scandinavia (all titles except Hartman, Jenssen-Tusch, Nordhagen, Oeder, Rehder).

All titles in Finnish have been provided with a translation. In all cases where the translated title, in English or German, is given within parentheses or without either parentheses or brackets, it will be understood that it was given in the original and that an abstract or summary in the same language is given in the publication. When the translation of the Finnish title is given in brackets, it will be understood that it is added here and that no abstract or summary is given unless its presence is specifically referred to.

The biogeographical districts are arranged alphabetically by their names in the older (Latin) nomenclature; the more modern Finnish names, with their abbreviations, are given in parentheses.

General

The general works here listed may be grouped as follows: *Floras and manuals:* Alcenius, Brenner, Hiitonen (Suomen kasvio), Hiitonen and Poijärvi, Jalas, Kärki (see Bolin under Scandinavia), Kivirikko, Lagerberg (see under Scandinavia), Lönnrot and Saelán, Mela, Olsoni (see Bolin under Scandinavia), Pulkkinen. *Catalogs:* Hiitonen (Suomen putkilokasvit), Hjelt (Conspectus), Lindberg, Saelán and others. *Bibliography:* Saelán. *Vernacular names:* Haapanen (garden plants), Suhonen. *Trees:* Aro, Hjelt (Utbredningen), Hult. *Medicinal (and poisonous) plants:* Cantell, Eneberg, Ervast, (Lönnrot and Saelán), (Rautavaara), Wirzén. *Other useful plants:* (Lönnrot and Saelán), Rautavaara. *Cultivated plants:* Elfving, Haapanen (vernacular names). *Miscellaneous:* Aro (forestry dictionary), Cajander (phytogeography), Erkamo (botanical investigations), Häyrén (floristic regions), Kalela (phytogeography), Kujala (vegetation), Lönnrot (botanical dictionary), Luther (influence of war on flora), Mali (agricultural dictionary), Moberg (phenology), Nuttonson (crop geography), Ulvinen (statistics).

Alcenius, Otto. Finlands kärlväxter: de vilt växande och allmännast odlade. 12. uppl. 428 p. 18.5 cm. Helsingfors, 1953. (1st ed. 1863.)

Key to genera and some families on Linnaean system; keyed, briefly descriptive pocket manual of vascular plants, with rather detailed local distribution; includes cultivated species. Revised (8. ed. and later) by Åke Nordström. Does not cover Russian Karelia or the Kola peninsula.

Aro, Paavo, and others, *eds.* Suomalais-ruotsalais-saksalais-englantilainen metsäsanakirja . . . Finnish, Swedish, German, English forest dictionary. Various pagings. 25 cm. Helsinki, 1944.

Alphabetical list of Finnish names and terms with equivalents in the three other languages; alphabetical list of words in each of the other languages with references to the first section. Includes some zoological words but not the terms of descriptive botany.—See also Lönnrot, below.

Brenner, Magnus. Floristisk handbok, innefattande i Finland vildt växande samt förvildade och allmännare odlade fröväxter och högre sporväxter, för läroverken i Finland. 2 p.l., 260 p. 22.5 cm. Helsingfors, 1886.

Key to genera on Linnaean system; briefly annotated manual of wild and cultivated vascular plants and Characeae in form of key, with notes on uses.

Cajander, A. K. Kasvien vaellusteistä Suomeen. [On the migration routes of plants into Finland.] Lännetär n.s., 2: 12–45. 1914.

Discussion of the migration of plants into Finland in the Postglacial period. (Not seen; title and annotation supplied by H. Luther.) German summary with title: Zur kenntnis der einwanderungswege der pflanzenarten nach Finnland. Acta Forest. Fenn. v. 21, [no. 2]. 16 p. 1921.—See also **Linkola, Kaarlo.** Suomen kasviston historia. [History of the flora of Finland.] *In* Oma Maa 5: 599–622. 1924. (Description of the migration of plants to Finland in Postglacial time. In University of Minnesota library.) —Also **Kalela, Aarno.** Mistä ja milloin Suomi on saanut kasvistonsa? [Whence and when has Finland received its flora?] Suomen Luonto 8: 9–30. 1945. (Not seen; title supplied by H. Luther.)—Also Kalela, below.

Cantell, Sulo, and **Saarnio, Väinö.** Suomen myrkylliset ja lääkekasvit. [Poisonous and medicinal plants of Finland.] 436 p. 18.5 cm. Hämeenlinna, 1936.

Descriptions, uses, and cultivation of the wild and cultivated medicinal and poisonous plants. (Not seen; title and annotation supplied by H. Luther.) —See also Eneberg, Ervast, Lönnrot and Saelán, Rautavaara, and Wirzén, below.

Elfving, Fredrik. Anteckningar om kulturväxterna i Finland. Acta Soc. Fauna et Flora Fenn. v. 14, no. 2. 116 p. 2 maps. 1897.

General considerations, discussion of localities and regions; annotated list of cultivated plants giving range and for the more important species extensive data; German summary, including annotated list of the more important species.—See also Haapanen, below.

Eneberg, E. E. Korta anvisningar för insamling af inhemska växtdroger. Agr.-Ekon. Försöksanst. Landtmannaskr. 4. 3 p.l., 41 p. 35 fig. (partly on pls.). 1915.

Includes descriptive account of principal native and cultivated vascular and cellular medicinal plants, alphabetically arranged by scientific names, with Swedish and Finnish names and directions for collecting, but no account of uses.—See also Cantell, above, and additional references there given.

Erkamo, Viljo. Suomen kasvistollisesti tutkitut seudut. [Regions of Finland floristically investigated.] Luonnon Tutkija 52: 86–87 incl. fig. (map). 1948.

Map showing the localities for which local floras are published or available in manuscript, indicating by symbols their degree of completeness, but without titles or names of localities or authors. Covers present day Finland, Petsamo region, Karelia ladogensis, and Isthmus karelicus, but not the rest of Russian Karelia or Russian Lapland.—See also his Kansallisherbaariostamme ja kasviemme levinneisyyden tuntemuksen historiasta. [On our national herbarium and the history of the knowledge of the distribution of our plants.] Luonnon Tutkija 53: 67–72. 5 fig. (maps). 1949. (Sketch of the growth of the national herbarium (Herbarium Musei Fennici) and of increase in the knowledge of plant distribution. One of the maps shows the approximate number of species from each parish represented in the national herbarium.)

Ervast, Petter. Akademisk afhandling om Finlands inhemska läkemedel. vi, 70 p. 19.5 cm. Helsingfors, 1840.

Account of indigenous medicinal plants, with vernacular names and uses. (Not seen; title and annotation supplied by H. Luther.)—See also Cantell, above, and additional references there given.

Haapanen, Arvo. Yleisimpien puutarhakasvien suomenkielinen nimistö. [The Finnish names of the commonest garden plants.] Maataloustiet. Aikak. 23: liite [suppl.] 1. 14 p. 1951.

Alphabetical list of scientific names with equivalent standardized Finnish names, followed by Finnish-Latin list.—See also Elfving, above, and Suhonen, below.

Häyrén, Ernst. Om växtgeografiska gränslinjer i Finland. (Über phytogeographische grenzlinien.) Meddel. Geog. För. Finland v. 10, [no. 1]. 23, 8 p. 3 fig. 1915.

Sketch of the floristic, phytophysiognomic, phytotopographic, and agricultural regions of Finland.—See also Kalela, Kujala, and Nuttonson, below.

Hiitonen, Ilmari. Suomen kasvio. [Flora of Finland.] 771 p. 437 fig., map. 22.5 cm. Helsingissä, 1933. (Vanamon kirjoja no. 32.)

Glossary, key to genera on Linnaean system; annotated, keyed, descriptive flora of vascular plants (wild and cultivated), the range given by botanical provinces (including those in Russian Karelia and Russian Lapland). *Taraxacum* and *Hieracium* are treated briefly. This is the standard flora of Finland.

——— Suomen putkilokasvit. Luettelo Suomessa luonnonvaraisina, villiytyneinä ja yleisimmin viljeltyinä kasvavista putkilokasveista. [Vascular plants of Finland. List of the indigenous, naturalized, and most commonly cultivated vascular plants in Finland.] 160 p. 1 fig. (map). 22.5 cm. Helsingissä, 1934.

Unannotated systematic list of wild and cultivated vascular plants of Finland, intended for use in making exchanges; alphabetical list of parishes with indication of the corresponding provinces (lääni) and phytogeographical regions (the latter shown on the map). In *Taraxacum* and *Hieracium* only the species collectivae are listed.

——— and **Poijärvi, Arvi.** Koulu- ja retkeilykasvio. [School and excursion flora.] kahdeksas, tarkisettu painos. [8. rev. ed.] 472 p. 311 fig., map. 20 cm. Helsingissä, 1955. (1st ed. 1932, by Hidén [= Hiitonen] and Poijärvi; 9. ed. 1958, not seen.)

Glossary, keys to genera and some families on Linnaean and natural systems; annotated, keyed, briefly descriptive flora of vascular plants (wild and cultivated), showing range by botanical provinces. A pocket manual corresponding to Hiitonen's Suomen kasvio, with distribution brought up to date (but distribution in Russian Karelia and Russian Lapland not given.)

Hjelt, Hjalmar. Conspectus florae fennicae. (7 vol.) Acta Soc. Fauna et Flora Fenn. v. 5, no. 1 (in 3 pts.). 562 p. 2 maps. 1888–95; v. 21, no. 1. 1 p. l., 261 p. 1902; v. 30, no. 1. 2 p. l., 410 p. 1906; v. 35, no. 1. 3 p. l., 411 p. 1911; v. 41, no. 1. 1 p. l., 502 p. 1919; v. 51, no. 1. 1 p. l., 450 p. 1923; v. 54. 2 p.l., 397 p. port. 1926.

Annotated list of vascular plants (v. 1, pteridophytes, gymnosperms, monocotyledons; v. 2–7, dicotyledons) with general statement of local range and detailed local distribution in Latin, and critical notes in Swedish. Each volume contains bibliography and index. *Rosa, Erigeron,* and *Hieracium* are omitted; adventive and cultivated plants are included. The ranges given include Russian Lapland and Russian Karelia.—See also his Notae Conspectus florae fennicae. 20, [4] p. 23 cm. Helsingforsiae, 1888. (Bibliography, explanation of symbols, etc. Issued with first installment of Conspectus.)—Also his Kännedomen om växternas utbredning i Finland med särskilt afseende å fanerogamer och ormbunkar. Acta Soc. Fauna et Flora Fenn. v. 5, no. 2. 152 p. 3 tab. 1891. (Also published as thesis; 2 p.l., 150 p. 3 tab. 22.5 cm. Helsingfors, 1891.) (History of botanical investigations with included bibliography; list of first records for each species in Finland.)

——— Utbredningen af Finlands träd, buskar och ris med särskilt afseende å deras gränser. Meddel. Finska Forstför. **14**: 134–176. 2 maps. 1898.

General features of forest regions; annotated descriptive account of the distribution of woody plants in Finland. (Not seen; title and annotation supplied by H. Luther.)—See also his Plantes ligneuses. *In* Atlas de Finlande 1910. Fennia 30 (1), Carte 20, p. 50–59. 1910/11. (Discussion of ranges, with references to maps in the Atlas. There are also issues in Finnish and Swedish.)—Also Hult, below.

Hult, Ragnar. Vedväxternas utbredning i Finland. (Die verbreitung der holzpflanzen in Finland.) Vetensk. Meddel. Geog. För. Finland **3**: 1–63. 2 maps. 1896.

General features of distribution of woody plants (93 species) in Finland and special features of distribution in 20 sub-areas; statistics.—See also Hjelt, Utbredningen, above.

Jalas, Jaakko, *ed.* Suuri kasvikirja. [Great plantbook.] v. 1. 851 p. 274 fig. (incl. maps), 32 col. pl. 24.5 cm. Helsingissä, 1958.

Flora of vascular plants, without keys or descriptions but with rather detailed account of local and extralimital distribution. Vol. 1, Pteridophyta through Monocotyledones.

Kärki, Eino. Värikuvakasvio.

See Bolin under Scandinavia.

Kalela, Aarno. Die ostgrenze Fennoskandiens in pflanzengeographischer beziehung. Veröffentl. Geobot. Inst. Rübel Zürich v. 20. 68 p. 4 fig. (maps). 1943.

Discussion of the geology, climate, and phytogeography of Russian Karelia and Russian Lapland; bibliography.—See also Cajander and Häyrén, above.

Kivirikko, K. E. Koulukasvio. [School flora.] toinen painos [2. ed.]. 318 p. map. 19 cm. Porvoo, 1917. (1st ed. 1912.)

Key to genera on Linnaean system; keyed, briefly annotated pocket manual of wild and cultivated vascular plants, with short descriptions; *Taraxacum* and *Hieracium* treated very briefly. (Not seen; title and annotation supplied by H. Luther.)

Kujala, Viljo. Vegetation. *In* Suomi. A general handbook on the geography of Finland. Fennia **72**: 209–234. 14 fig. (incl. maps). 1952.

Vegetation zones, vegetation of various habitats, vegetation districts, short bibliography.—See also Häyrén, above, and additional references there given.

Lagerberg, Torsten; Linkola, Kaarlo; and Väänänen, Heikki. Pohjolan luonnonkasvit. [Wild plants of the North.]
See Lagerberg under Scandinavia.

Lindberg, Harald. Enumeratio plantarum in Fennoscandia orientali sponte et subsponte nascentium. Förteckning öfver ormbunkar och fröväxter vildtväxande och förvildade i Finland och angränsande delar af Ryssland. Luettelo saniaisista ja siemenkasveista jotka kasvavat villeinä tai metsistyneinä Suomessa ja siihen rajoittuvissa osissa Venäjää. vii, 79 p. map. 23 cm. Helsingfors, 1901.
Name list of 1,159 species (plus 887 forms of lower rank) of vascular plants. Replaced by Hiitonen's Suomen putkilokasvit (above) except for the list of *Hieracium* (529 species and subspecies).

Lönnrot, Elias. Kasvikon oppisanoja. [Botanical terminology.] Suomi 18: 1–108. 1859.
Classified list of Finnish botanical terms with definitions and Latin and Swedish equivalents; full index.—See also Aro, above.

——— **and Saelán, Thiodolf.** Flora fennica. Suomen kasvio. uusi parannettu laitos. [new rev. ed.] xx, 426 p. 3 tab. 19.5 cm. Helsingissä, 1866. (1st ed. by E. L[önnrot]., 1860.)
Key to genera on Linnaean system; briefly annotated pocket manual of wild growing vascular plants with short descriptions, vernacular names, and notes on uses. Completely replaced as a flora; cited here for its notes on uses of plants and vernacular names. (Not seen; title and annotation supplied by H. Luther.)—See also Cantell, above, and additional references there given.

Luther, Hans. Krigets spår i Finlands flora. (Traces of the war in the flora of Finland.) Memo. Soc. Fauna et Flora Fenn. 24: 138–160. 1948.
Discussion of the influence of war and invasion on the migration of plants, especially weeds, with list (p. 140) of species known only from the parts of Finland now surrendered to Russia, list (p. 147) of species found only in the area of Russian deployment in East Karelia, running account of weeds in the area of the Russian garrison in the Hangö peninsula, etc.; extensive bibliography.

Mali, Liisa. Maatalouden sanakirja . . . Agricultural dictionary. xxviii p., 596 col., 177 p. 25 cm. Helsinki, 1958.
Alphabetical list of Finnish names and terms used in agriculture, with Swedish, German, and English (including U.S.A.) equivalents; table of measurements, Swedish, German, and English indices, index of scientific terms. Includes the commoner vernacular as well as scientific names of agricultural plants and a few descriptive terms of botany.

Mela, A. J. Suomen kasvio. [Flora of Finland.] viides painos toimittanut A. K. Cajander. [5. ed. edited by A. K. C.] x, 763 p. map. 22.5 cm. Helsingissä, 1906. (Suomalaisen kirjallisuuden seuran toimituksia. 53. osa. III.) (Library of Riksmuseet, Stockholm.) (1st ed. 1877, with title Lyhykäinen kasvioppi ja kasvio.)
Key to genera on Linnaean system; annotated, keyed, descriptive flora of vascular plants. Replaced by Hiitonen's Flora except for the treatment of *Hieracium* by J. P. Norrlin, which includes more than 250 species and is the last comprehensive treatment for Finland, although now out of date.

Moberg, Adolf. Naturalhistoriska daganteckningar gjorda i Finland år 1750–1845. Notiser Sällsk. Fauna et Flora Fenn. Förhandl. 3: 95–250. 1857.
Discussion of sources, bibliography; tabular systematic list of flowering plants and one *Equisetum* (p. 113–190), including some cultivated species, with dates for flowering, fruiting, and for some woody plants leafing and leaf-fall as observed at different localities in different years; also zoological and climatic data. The data are mostly rather scattered.—See also his Fenologiska iakttagelser i Finland åren 1750–1845. Bidr. Känn. Finlands Natur och Folk v. 55. xi, 165 p. 1894. (Supplement to the first paper; tabular list of dates for plants (p. 1–107), arranged by species; also zoological and climatic data.) —Also his Klimatologiska iakttagelser i Finland föranstaltade och utgifna af Finska Vetenskaps-Societeten. Första delen. År 1846–1855. I. Naturalhistoriska anteckningar 1846–55. Bidr. Finlands Naturkänn., Etnog. och Statis. v. 7. xviii, 361 p. 7 tab. on 4, map. 1860. (Data on plants, p. 1–250, giv-

ing dates of flowering and in some cases of fruiting, leaf-unfolding, and leaf-fall.)—Also his Klimatologiska anteckningar i Finland . . . Andra delen. År 1856–1875. I. Fenologiska anteckningar. Bidr. Känn. Finlands Natur och Folk v. 41. xi, 318 p. 1885. (Similar data on plants, p. 1–226.)—Also his Sammandrag af de klimatologiska anteckningarna i Finland 1859 [–1893]. (Title varies somewhat in earlier years; the first one is Sammandrag af klimatologiska observationerna i Finland för år 1859.) *In* Öfvers. Finska Vetensk.-Soc. Förhandl. v. 5–36. 1863–93. (Yearly reports including dates of flowering, fruiting, and leaf-unfolding (and for some cereals date of sowing) at different stations, as well as zoological and climatic data. The information is scanty in earlier years, and until v. 21 (data for 1878) only vernacular names were given; beginning in that year scientific names were added and the list was put in tabular form.)—Also **Kihlman, A. O.** Beobachtungen über die periodischen erscheinungen des pflanzenlebens in Finnland 1883. xxxii, 97 p. 28 cm. Helsingfors, 1886. (Description of localities and their vegetation; systematic list of spermatophytes and 1 *Equisetum*, with dates of flowering and fruiting and for some woody plants also leafing and leaf-fall, at different localities; tabular summary. The data are presented in more detail than in any other report in the whole series. Publication of the Societas pro fauna et flora fennica.)—Also his Sammandrag af de klimatologiska anteckningar i Finland år 1894. Öfvers. Finska Vetensk.-Soc. Förhandl. **37:** 245–270. 1895. (Data on plants p. 254–267.)—Also his Pflanzenphänologische beobachtungen in Finnland 1895. Bidr. Känn. Finlands Natur och Folk 58: 1–45. 1900. (Data arranged by localities, the species alphabeted under each.)—Also **Reuter, Märta.** Pflanzenphänologische beobachtungen in Finnland 1896, 1897, 1898 und 1899 [–1900, 1901 und 1902]. l.c. v. 85, no. 4. 107 p. 1 fig. (map). 1936; v. 85, no. 5. 76 p. 1 fig. (map). 1937. (List of stations and observers; tabular list of plants showing dates of flowering, fruiting, leaf-unfolding, and autumnal change in color (also sowing, heading, and harvest in the case of some cultivated species) at different stations, with supplementary lists of species having more scanty data.)—Also **Brotherus, V. F.** Pflanzenphänologische beobachtungen in Finnland 1903 [–1917]. l.c. 64: 141–171 (for 1903), 289–317 (for 1904), 427–457 (for 1905). 1907; **66:** 241–269. 1908 (for 1906); v. 71, no. 1. 44 p. 1911 (for 1907); v. 76, no. 2. 38 p. 1914 (for 1908); v. 76, no. 3. 37 p. 1914 (for 1909); v. 77, no. 6. 44 p. 1919 (for 1910); v. 77, no. 7. 35 p. 1919 (for 1911); v. 78, no. 4. 45 p. 1919 (for 1912; data erroneous, replaced by a corrected ed. by Märta Reuter with same series no. 33 p. 1935); v. 78, no. 5. 30 p. 1920 (for 1913); v. 80, no. 1. 33 p. 1921 (for 1914); v. 80, no. 2. 32 p. 1921 (for 1915); v. 80, no. 4. 30 p. 1925 (for 1916); v. 80, no. 5. 27 p. 1925 (for 1917). (List of localities, geographically arranged, with alphabetical lists of species showing flowering and in some cases fruiting, leafing, and autumnal leaf-coloring dates under each locality.)—Also **Reuter, Märta** (in v. 80, no. 7 and 8, as Pipping, Märta). Pflanzenphänologische beobachtungen in Finnland 1918, 1919 und 1920 [–1951–1955]. l.c. v. 80, no. 7. 77 p. 1 fig. (map). 1927 (for 1918–20); v. 80, no. 8. 57 p. 1 fig. (map). 1927 (for 1921-23); v. 80, no. 9. 46 p. 1 fig. (map). 1928 (for 1924–26); v. 85, no. 3. 65 p. 1 fig. (map). 1935 (for 1927–30); v. 87, no. 4. 97 p. 1 fig. (map). 1941 (for 1931–35); v. 89, no. 1. 118 p. 1 fig. (map). 1942 (for 1936–40); v. 92, no. 1. 79 p. 1 fig. (map). 1948 (for 1941–45); v. 92, no. 3. 64 p. 1 fig. (map). 1952 (for 1946–50); v. 100, no. 1. 63 p. 1 fig. (map). 1957 (for 1951–55). (Data arranged as in her reports for 1896–1902; see above.)—Also **Johansson, O. V.** Die phänologie in Finnland 1750–1950. Comment. Biol. Soc. Scient. Fenn. v. 11, no. 1. 55 p. 1951. (Historical sketch of the investigations and mention of their applications.)—Also his Nachträge zu dem phänologischen Bande XI . . . l.c. v. 11, no. 11. 13 p. 1954. (Includes list of localities with different Finnish and Swedish names, and bibliography.)

Nuttonson, M. Y. Ecological crop geography of Finland and its agro-climatic analogues in North America. Amer. Inst. Crop Ecol. Internatl. Agro-Climat. Ser. Study 10. 35 p. incl. 1 fig. (map), tab. 1950.

Geography and topography, climate, forests, soils, farm crops, comparison with localities in North America, bibliography.—See also Häyrén, above.

Olsoni, Börje. Floran i färg.
See Bolin under Scandinavia.

Pulkkinen, Asko, and **Hagfors, E. A. M.** Koulukasvio. [School flora.] kolmas, täydennetty painos. [3. enlarged ed.] 542 p. 349, i.e. 350 fig. (incl. map). 19 cm. Porvoo, Helsinki, 1941. (Library of Riksmuseet, Stockholm.) (1st ed. 1932.)

Botanical regions, key to families and genera; briefly annotated, keyed, briefly descriptive flora of vascular plants; key to genera on Linnaean system.

Rautavaara, Toivo. Mihin kasvimme kelpaavat. Luonnonkasviemme käyttömahdollisuudet leivän lisänä ja ruoan aineksina, mausteina, kahvin ja teen korvikkeina, lääkkeinä, rehuna sekä teknillisiin tarkoituksiin. [For what our plants are useful. Possible uses of our wild plants as additions to bread and as food materials, spices, coffee and tea substitutes, medicines, fodder, and for technical purposes.] 2 pt. (363 p.). 181 fig. 18.5 cm. Porvoo, Helsinki, 1942–43. (At head of title page: Kansanhuoltoministeriön julkaisuja.)

Popular treatment of useful plants, grouped according to season (pt. 1. spring and early summer plants; pt. 2, summer and autumn plants), with brief descriptions, methods of collecting, uses, recipes, etc.; classified lists of species, bibliographies.—See also Cantell, above, and additional references there given.

Saelán, Thiodolf. Finlands botaniska litteratur till och med år 1900. Acta Soc. Fauna et Flora Fenn. v. 43. xi, 633 p. 1916.

Alphabetical list of Finnish authors (with full names but not dates) with list of their publications, the latter arranged chronologically, with subject of each work briefly indicated as well as references to reviews; similar list of works by non-Finnish authors referring to Finland; short list of exsiccatae, classified index, lists of Finnish and foreign periodicals cited.—See also (for a chronologically arranged list of titles) his Öfversigt af Finlands botaniska litteratur. Notiser Sällsk. Fauna et Flora Fenn. Förhandl. 7 (n.s., 4): 33–131. 1867. (Chronological list of publications, 1635–1865, those printed outside Finland in a separate list; index of authors.)—Also **Pyykkö, Maire.** Suomen kasvitieteellinen kirjallisuus v. 1950 [–1955]. The Finnish botanical literature in 1590 [i.e. 1950] [–1956]. Arch. Soc. Zool. Bot. Fenn. Vanamo **5**: 138–149. 1951; **7**: 5–16. 1952; **8**: 5–16. 1953; **9**: 5–21. 1954; **10**: 5–21. 1956; **11**: 205–227. 1957.—Also **Sorsa, Pentti.** Suomen . . . 1956. l.c. **13**: 149–166. 1959. (The last not seen.) (Unannotated, classified lists of publications by Finnish authors, as well as papers by non-Finnish authors on Finland; alphabetical index of authors, brief systematic index. Purely Finnish titles are provided with an English translation.)

——— **Kihlman, A. O.,** and **Hjelt, Hjalmar.** Herbarium Musei fennici. Enumeratio plantarum Musei fennici qvam edidit Societas pro fauna et flora fennica. 2. ed. I. Plantae vasculares. xix, 156 p. 2 maps. 1889. (1st ed. by William Nylander and T. Saelán, 1859.)

History of Herbarium Musei Fennici, list of additions to flora since first edition, etc.; outline diagrams (about 1,429, including 1,047 species and subspecies) showing occurrence of each species and many varieties and hybrids in each botanical province; list of localities on margin of the range of a large number of species, as well as records of various minor forms and adventive plants.

Suhonen, Pentti. Suomalaiset kasvinnimet. (Finnische pflanzennamen.) Annales Bot. Soc. Zool.–Bot. Fenn. Vanamo v. 7, no. 1. vi, 465 p. 1 fig. (map). 1936.

Account of previous Finnish botanical vocabularies, list of cooperators and localities, bibliography; alphabetical list of scientific names of vascular and cellular native and cultivated plants, with corresponding Finnish names with localities and references; index of vernacular names, brief German summary. Replaces **Reinholm, H. A.** Suomalaisia kasvu-nimejä. [Finnish plant names.] Suomi **10**: 159–301. 1851. (Previous work on Finnish vernacular names; alphabetical list of vernacular names, with botanical equivalents and localities where used, the various names also collected under a single selected name for each species; list of generic and specific names with single selected vernacular equivalents.)—See also **Erkamo, Viljo.** Lisätietoja kansanomaisista kasvinnimistämme. (Neue beiträge zur kenntnis der finnischen pflanzennamen.) Arch. Soc. Zool. Bot. Fenn. Vanamo **4**: 127–130. 1950.

(Alphabetical list of scientific names and localities; short bibliography.) —Also Haapanen, above.

Ulvinen, Arvi. Maamme luonnonvaraisten putkilokasvien lajilukumäärä v:n 1944 tilaston mukaan. (The numbers of indigenous Finnish vascular plants according to statistics in 1944.) Arch. Soc. Zool. Bot. Fenn. Vanamo 1: 85–86. 1948.

List of botanical districts recognized in Finland with the number of species of vascular plants known in each according to Hiitonen and Poijärvi, Koulu-ja retkeilykasvio (1944); total number inside pre-war boundaries of Finland was 1175.—See also his Maamme kasvien yleisyyden arvioinnista. (Zur häufigkeitsbestimmung der pflanzenarten in Finnland.) Arch. Soc. Zool. Bot. Fenn. Vanamo **3**: 131–142. 5 fig. (incl. maps). 1949. (Discussion of methods of evaluating and representing the frequency of occurrence of plant species in Finland; bibliography.)

Wirzén, J. E. A. Dissertatio academica enumerationem plantarum officinalium Fenniam sponte inhabitantium sistens. 1 p. l., 90 p. 20.5 cm. Helsingforsiae, 1837. (Thesis.)

List of 355 officinal plants (312 vascular), on the Linnaean system, with local habitat and range but without notes on uses.—See also Cantell, above, and additional references there given.

ABOËNSIS, REGIO (Ab). (VARSINAIS-SUOMI. V)

See also Alandia (Eklund, titles published in 1932, 1934, and 1935, under Bergroth).

The Regio aboënsis consists of the southwestern part of the Finnish mainland and the adjoining islands east of about 21° E. Lat., bordered on the north by Satakunta and Tavastia australis, on the east by Nylandia, on the south by Nylandia and the Baltic, on the west by the island province of Alandia and the Gulf of Bothnia. It is equivalent essentially to the southern part of Turun ja Porin lääni (north to nearly 61° N. Lat.) and the adjacent westernmost part of Uudenmaan lääni.

Local

Åberg, Gunnar. Förteckning över anmärkningsvärda växter från norra Nagu. Memo. Soc. Fauna et Fl. Fenn. **8**: 274–282. 1933.

List of noteworthy vascular plants, with localities. (Nagu, Finnish Nauvo.) —See also his Anmärkingsvärda växter från Ab Nagu. Memo. Soc. Fauna et Fl. Fenn. **10**: 239–242. 1934. (Additional list.)—Also **Eklund, Ole.** Viktigare växtfynd i Nagu sydskärgård (Ab) sommaren 1927. l. c. **4**: 20–22. 1928. (List with localities.)—Also **Auer, A. V.** Putkilokasvihavaintoja Nauvosta. [Observations on vascular plants from Nauvo.] l. c. **17**: 195–200. 1942. (List with localities.)—Also Pettersson under Eklund (Über die ursachen), below.

Auer, A. V. Havaintoja Turun kaupungin kasvistosta. (Observations on the flora of the town Turku.) Turun Ylioppilas **4**: 140–183. 1955.

Annotated list of 247 species, mostly weeds and anthropochores; bibliography. (Not seen; title and annotation supplied by H. Luther.)

——— Muhkurin kasvisto. Die flora des eichenwaldes von Muhkuri. Silva Fenn. 41. 36 p. 2 fig. (maps). 1937.

Topography, geology, general features of flora; tabular list of vascular plants (198 species and forms), bryophytes, and lichens, and annotated list of fungi.

——— Putkilokasvihavaintoja. [Observations on vascular plants.] Memo. Soc. Fauna et Flora Fenn. **19**: 57–77. 1944.

Extensive list of noteworthy vascular plants, with localities, mostly from vicinity of Turku (Åbo), but partly from Isthmus karelicus, Ostrobottnia media, and Satakunta.

Bergroth, Ossian. Anteckningar om vegetationen i gränstrakterna mellan Åland och Åbo-området. Acta Soc. Fauna et Flora Fenn. v. 11, no. 3. 78 p. map. 1894.

Topography, geology, general features of flora; annotated list of vascular plants (523 species). Refers to parishes of Houtskär, Iniö, Kustavi (Gustavs), and Taivassalo (Töfsala) in Regio aboënsis and to two parishes in Alandia (see Bergroth under Alandia for additional titles relating to that region).—See also **Eklund, Ole.** Ergebnisse einer botanischen reise in den kirchspielen Houtskär und Iniö (Südwestfinnland) im jahre 1928. Memo.

Soc. Fauna et Flora. Fenn. **5**: 28–64. 2 fig. (maps). 1929. (General features of flora, etc.; briefly annotated list of vascular plants, with localities; bibliography.)—Also his Botaniska resor i Åbo skärgård sommaren 1930. l. c. **7**: 31–36. 1931. (List of 18 noteworthy species.)—Also his Merkittäviä kasvilöytöjä Lounais-Suomen saaristosta. [Noteworthy plant finds from the archipelago of southwest Finland.] Luonnon Ystävä **36**: 123–128. 1932. (Additional notes from Åbo region and Alandia.)—Also Eklund (Über die ursachen) below.

Cajander, A. K. Kasvistollisia tutkimuksia Mynämäen, Mietoisten ja Karjalan kunnissa. [Floristic investigations in the parishes of Mynämäki, Mietoinen, and Karjala.] Acta Soc. Fauna et Flora Fenn. v. 23, no. 2. 146 p. 4 fig., map. 1902.
Topography, geology, plant associations, etc.; annotated list of vascular plants; German summary.

Eklund, Ole. Anteckningar om växtvärlden i Korpo västra skärgård (Ab). Meddel. Soc. Fauna et Flora Fenn. **49**: 76–108. 1925.
Topography, general features of flora: tabular list of vascular plants (altogether 443 species) of 12 islands in the group including Finnö, Kalö, and Lempersjö.

——— Botaniska anteckningar från Österskär (Ab, Korpo). Meddel. Soc. Fauna et Flora Fenn. **49**: 20–42. 4 fig. 1925.
Topography, general features of flora; tabular list of vascular plants of Alskär, Bredskär, Österskär, and Sundskär (altogether 287 species).

——— Botaniska anteckningar från Utö i Korpo skärgård. Meddel. Soc. Fauna et Flora Fenn. **45**: 99–106. 1 fig. (map). 1920.
General features of flora; tabular list of vasuclar plants of Utö and the adjacent islet Enskär, with indication of frequency.

——— Die gefässpflanzenflora beiderseits Skiftet im schärenarchipel Südwestfinnlands. Kirchspiele Korpo, Houtskär, Nagu, Iniö, Brändö, Kumlinge, Sottunga und Kökar. Bidr. Känn. Finlands Natur och Folk 101. 324 p. 2+ 84 fig. (maps), map. 1958.
Topography, list of localities (1965), etc.; annotated list of vascular plants with detailed localities. (First four localities in Prov. Aboensis, last four in Alandia.)—See also **Skult, Henrik.** Skogsbotaniska studier i skärgårdshavet med speciell hänsyn till förhållandena i Korpo utskär. Acta Bot. Fenn. 57. 244 p. incl. 8 pl. (on 4). 19 fig. (maps), 12 diagr. 1958. (Mostly ecological; includes (Skogsfloran, p. 191–198) list of 305 vascular plants of woodland habitats, also list of cellular plants; extensive bibliography.)

——— Über die ursachen der regionalen verteilung der schärenflora Südwest-Finnlands. Eine kausalitätsanalyse mit besonderer berücksichtigung der pflanzenwelt in den kirchspielen Korpo und Houtskär. Acta Bot. Fenn. 8. 133 p. 45 fig. (maps). 1931.
Mainly phytogeographical; includes (p. 16–21) unannotated list of the known vascular plants (699 species and subspecies), of the parishes Korpo and Houtskär, with indication of those known only from one or from both parishes.—See also **Pettersson, Bror.** Beiträge zur kenntnis der Taraxacumflora des südfinnländschen schärenarchipels. Memo. Soc. Fauna et Flora Fenn. **8**: 303–316. 1933. (Records of 42 species from the islands of Nagu, Korpo, and Houtskär (all in Regio aboensis) and from Kökar in Alandia.)

——— Vegetationen å Vidskär och Jurmo (Ab, Korpo). Meddel. Soc. Fauna et Flora Fenn. **47**: 178–215. 3 fig. (fig. 1, map). 1921.
General features of flora, plant formations; tabular list of vascular plants from Vidskär, Jurmo, Huvudskär, Grundvikharu, Sand-Örskär, and Stor-Örskär.—See also his Märkliga växtfynd i Ab, Korpo. Meddel. Soc. Fauna et Flora Fenn. **47**: 62–67. 1921. (Detailed records of some noteworthy species.)—Also his Botaniskt från Ab Korpo 1924. Memo. Soc. Fauna et Flora Fenn. **1**: 13–16. 1927. (Includes correction to list, on p. 16.)

Erkamo, Viljo. Tietoja Piikkiön-Kuusiston seudun kasvistosta (V). (Beiträge zur kenntnis der flora in der gegend von Piikkiö und Kuusisto in Südwestfinnland.) Arch. Soc. Zool. Bot. Fenn. Vanamo **3**: 155–164. 1949.
Botanical explorations, geology, etc.; list of known vascular plants (409 species), followed by annotated list of rarer species; discussion of synanthropous species, bibliography.

Flinck, J. A. Viktis sockens kärlväxter. Acta Soc. Fauna et Flora Fenn. v. 19, no. 6. 54 p. 1900.

Topography, geology, ecology, botanical explorations, etc.; annotated list of vascular plants of Viktis (Vihti) parish.—See also **Brotherus, V. E.** Vähäsen Luonais-Vihdin kasvikunnasta. A little about the plant world in SW. Vihti. Luonnon Ystävä **10:** 199–208. 1906. (General account of vegetation with mention of numerous species.)

Jalas, Jaakko. Laitilan seudun putkilokasviston pääpiirteet. (Kasvistollisia tutkimuksia Lounais-Suomessa. I.) (Floristic studies in southwest Finland. I. Main features of the vascular plant flora in the Laitila area.) Arch. Soc. Zool. Bot. Fenn. Vanamo **6:** 29–43. 1951.

Full list, divided into distributional groups, of vascular plants with localities, mostly from Laitila but partly from Kalanti and Kodisjoki (in Regio aboensis) and Hinnerjoki (in Satakunta), with a few records from Pyhäranta (Regio aboensis) and Rauma (Satakunta).

——— and **Vuorisalo, Aarne.** Tietoja Lokalahden pitäjän putkilokasvistosta. (Kasvistollisia tutkimuksia Luonais-Suomessa. III.) (Floristische untersuchungen in Südwestfinnland. III. Materialen zu einer gefässpflanzenflora von Lokalahti.) Arch. Soc. Zool. Bot. Fenn. Vanamo **10:** 125–145. 2 fig. (maps). 1956.

General features of flora, itinerary; annotated list of vascular plants (529 species and subspecies); bibliography.

Koskimies, A. E. Lisätietoja Sauvon ja Karunan pitäjien kasvistosta (V). (Additional notes on the flora of the parishes Sauvo and Karuna in southwest Finland.) Arch. Soc. Zool. Bot. Fenn. Vanamo **2:** 106–117. 1949.

Unannotated list of common species; annotated list of scarcer vascular plants, with localities (the two lists together including all species known from the region); bibliography.

Lindgren, Leo. Ruissalon putkilokasvisto. (The vascular plant flora of Ruissalo.) Turun Ylioppilas **3:** 145–181. 1954.

The island Ruissalo (Runsala) off Turku (Åbo) and 7 smaller islands are divided into 120 squares, the flora of which is listed (531 species); main features of vegetation, bibliography. (Not seen; title and annotation supplied by H. Luther.)

Luotola, V. L. Tutkimuksia Kustavin kasvillisuudesta ja kasvistosta. (Untersuchungen über die schärenvegetation und -flora im kirchspiel Kustavi.) Annales Soc. Zool.-Bot. Fenn. Vanamo **15:** 152–248. 1931.

Botanical explorations, topography, geology, climate, general features of vegetation of different habitats; annotated list of vascular plants (571 species); bibliography.

Malmio, Bruno, and **Erkamo, Viljo.** Tietoja Uudenkaupungin seudun ja Pyhämaan kasvistosta. (Beiträge zur kenntnis der flora in der gegend von Uusikaupunki und Pyhämaa in Südwestfinnland.) Arch. Soc. Zool. Bot. Fenn. Vanamo **5:** 105–126. 1951.

Botanical explorations; list of known vascular plants (681 species, subspecies, and hybrids), list of rarer species with localities; phytogeography, changes in flora, list of vernacular names for various species, bibliography.—See also **Laine, Tapio, and Laine, Unto.** Täydentävia tietoja eräiden putkilokasvien esiintymisestä ja levinneisyydestä Uudenkaupungin seudulla. (Ergänzende beiträge zur kenntnis des auftretens und der verbreitung einiger gefässpflanzenarten in der gegend von Uusikaupunki in Südwestfinnland.) Arch. Soc. Zool. Bot. Fenn. Vanamo **11:** 17–33. 5 fig. (maps). 1956. (Adds 101 species.)—Also **Malmio, Bruno.** Tietoja Uudenkaupungin seudun Hieraciumlajeista (Mitteilungen über die Hieracium-arten der gegend von Uusikaupunki.) l. c. **11:** 33–48. 5 fig. 1956. (List of 57 species.)

Nikoskelainen, Ritva. Katariinanlaakson ja sen lähiympäristön putkilokasvisto. (The vascular flora of the region of Katariinanlaakso.) Turun Ylioppilas **4:** 229–248. 1955.

Briefly annotated list of vascular plants (364 species) of an oakwood area south of Turku (Åbo). (Not seen; title and annotation supplied by H. Luther.)

Nylander, A. A. Stirpes cotyledoneæ paroeciæ Pojo, periculum botanicum. 1 p. l., 22 p. 19 cm. Helsingforsiae, 1844. (Thesis, J. M. Tengström praes.) (British Museum (Natural History) library.)

Topography, plant regions, etc.; list of vascular plants and Characeae (443 species), with indication of frequency, and localities for rarer species. (Pojo, Finnish Pohja.)

Olsson, P. H. Bidrag till kännedomen om floran i Kimito skärgård. Acta Soc. Fauna et Flora Fenn. v. 11, no. 11. 50 p. 1895.

Topography, geology, general features of flora, climate, statistics, etc.; briefly annotated list of vascular plants (617 species), often with localities, also list of bryophytes. Covers area from 59° 42' to 60° 16' N. Lat., about 21° to 22° E. Long. Olsson's records are not regarded as reliable. (Kimito, Finnish Kemiö.)—See also **Olsoni, Börje.** Växtfynd i Ab Kimito-Hitis sommaren 1926. Memo. Soc. Fauna et Flora Fenn. 3: 38–39. 1927. (List of some noteworthy species.)—Also his Växtfynd i Ab Kimito-Hitis sommaren 1927. l.c. 4: 60–62. 1928.—Also his Nya växtfynd in Ab Dragsfjärd och Hitis. l.c. 7: 28–29. 1931.—Also his Anmärkningsvärda växter från Ab Kimito-Hitis sommaren 1931. l. c. 8: 77–78. 1932.—Also his Botaniskt från Ab Kimito-Hitis 1932–35. l. c. 12: 142–146. 1937.—Also his Utricularia neglecta Lehm. i Hitis. l. c. 15: 91–92. 1939.—Also his Botaniskt från Ab Kimito-Dragsfjärd-Hitis 1936–38. l. c. 15: 92–94. 1939.—Also his Botaniskt från Ab Kimito-Hitis 1939–44. l. c. 21: 193–196. 1946.

——— Svenska växtnamn i sydvästra Finland. Bot. Notiser **1896: 1–10.** 1896.

Alphabetical list of botanical names, with Swedish vernacular equivalents, collected by author in three summers' work in southeastern coastal and insular Regio aboënsis and in Alandia.

Pesola, V. A. Paraisten ja Lohjan kalkkilouhosten kasvillisuudesta. (On the vegetation in the limestone quarries of Parainen and Lohja, SW Finland.) Arch. Soc. Zool. Bot. Fenn. Vanamo 7: 57–77. 1952.

Mainly ecological; includes tabular list of vascular plants (309 species and subspecies); bibliography.

Renvall, Thorsten. Muurilan kappelin luonto ja erittäin sen putkilokasvisto. [Natural features and especially the vascular flora of Muurila chapel.] Länsi-Suomi **6** (1): i–iv, 1–100. 1897.

Topography, general features of flora; partly annotated list of vascular plants (486 species) of Muurila (now written Muurla). (Not seen; title and annotation supplied by H. Luther.)

Saelán, Thiodolf. Fröväxter från barlastplatsen invid Åbo slott, samlade af John Lindén och Enzio Reuter. Meddel. Soc. Fauna et Flora Fenn. **11:** 212–216. 1885.

Annotated list of phanerogams found as weeds around Åbo castle.—See also **Pesola, V. A.** Turun linnan-kentän nykyinen kasvisto. [Present flora of Abo castle grounds.] Luonnon Ystävä **21:** 39–47. 1917. (Includes list of 238 species and forms of vascular plants.)

Selin, Gustaf. Ett bidrag till nordvestra Nylands flora. Notiser Sällsk. Fauna et Flora Fenn. Förhandl. **6** (n.s., 3): 123–144. 1861.

Geology, general features of flora, botanical explorations, etc.; annotated list of vascular plants (476 species) and Characeae from Karislojo (Karjalohja) and adjacent parts of Kisko and Lojo (Lohja) parishes.

Vieras, Irja. Pähkinäisten saariryhmän kasvillisuus ja kasvisto. (Die vegetation und flora der inselgruppe Pähkinäinen.) Acta Soc. Fauna et Flora Fenn. v. 58, no. 3. 22 p. 1 fig. (map). 1935.

Topography, geology, general features of flora; briefly annotated list of vascular plants (352 species); bibliography. (Along the boundary between Rymättylä and Korppoo parishes.)

Zetterman, A. J., and Brander, A. E. Bidrag till sydvestra Finlands flora. Notiser Sällsk. Fauna et Flora Fenn. Förhandl. **7** (n. s., 4): 1–29. 1867.

Topography, etc.; annotated list of vascular plants and Characeae (611 species), also Musci (154 species). Relates to the parishes of Korpo (Korppoo), Nagu (Nauvo), Pargas (Parainen), and Rimito (Rymättylä) and the coastal tracts of Pikkis (Piikkiö), Reso (Raisio), St. Karins (Kaarina), and St. Marie (Maaria) parishes.—Also **Reuter, Enzio.** Floristiska notiser från Ab Pargas. Memo. Soc. Fauna et Flora Fenn. **21**: 123–134. 1945. (List of noteworthy vascular plants, intended to supplement the records from Pargas in Hjelt's Conspectus; bibliography.)

ÅLANDIA (Al). (AHVENANMAA. A)

See also Aboënsis, Regio (Eklund, Die gefässpflanzen beiderseits Skiftet; Olsson, Svenska växtnamn).

The phytogeographical province of Alandia, coextensive with the lääni of Ahvenanmaa or Åland, and consisting of the archipelago off the southwest corner of Finland west of about 21° E. Long.

General

Bergstrand, C. E. Naturhistoriska anteckningar om Åland. 24 p. 22.5 cm. Stockholm, 1851. (Reprinted, Bot. Notiser **1852**: 1–11, 23–26, 35–44. 1852. Original not seen; data supplied by H. Luther.)

Topography, general features of flora, etc.; briefly annotated list of known vascular plants (652 species).—See also his Tillägg till Ålands naturalhistoria. Bot. Notiser **1852**: 129–141. 1852. (Further notes on topography, climate, statistics, etc.; additional records bringing total to 685 species.)

Palmgren, Alvar. Die artenzahl als pflanzengeographischer charakter sowie der zufall und die säkulare landhebung als pflanzengeographische faktoren. Ein pflanzengeographischer entwurf, basiert auf material aus dem Åländischen Schärenarchipel. Acta Bot. Fenn. v. 1. 142 p. 2 maps. 1925. (Published also as Fennia v. 46, no. 2. 1925.)

Includes (p. 46–51) unannotated list headed "Die ursprünglichen arten Ålands" (650 species of vascular plants), divided into trees and shrubs, grass-like plants, and herbs; *Taraxacum* and *Hieracium* are represented by only 3 specific names. The author regards as "original species" only those which occur entirely independent of man or of his activity in modifying habitats; about 100 species which are neither cultivated nor ruderal or ballast plants are consequently omitted from the list. The same species names incorporated into a single list and provided with botanical authorities are given (p. 21–28) in his Die einwanderungswege der flora nach den Ålandsinseln. Acta Bot. Fenn. **2**: 1–199. 1927.—See also **Brenner, Widar.** Några anmärkningsvärdare växtfynd från den öståländska skärgården sommaren 1931. Memo. Soc. Fauna et Flora Fenn. **8**: 19–20. 1932. (Records of several noteworthy species.)—Also **Palmgren, Alvar.** Tillägg till Dr. Brenners meddelande: Några anmärkningsvärdare växtfynd . . . l. c. **8**: 22–24. 1932. (Further notes on distribution of some species mentioned by Brenner.)—Also his Kompletterande fyndorter och synpunkter til Ålands flora 1. l. c. **10**: 398–434. 1935. (Mention of author's previous publications on the distribution of plants in Åland; additional records for numerous species; additional records for Brändö, Kumlinge, Kökar, and Föglö parishes.)

Local

Bergroth, Ossian. Anteckningar om vegetationen i gränstrakterna mellan Åland och Åbo-området. Acta Soc. Fauna et Flora Fenn. v. 11, no. 3. 78 p. map. 1894.

Topography, geology, general features of flora; annotated list of vascular plants (523 species) from parishes of Brändö and Kumlinge in Alandia and from four parishes in Regio aboensis (see Bergroth under latter for additional titles relating to that region).—See also **Eklund, Ole.** Anmärkningsvärdare växtfynd i öståländska skärgården. Memo. Soc. Fauna et Flora Fenn. **7**: 36–38. 1931. (Additions to Bergroth's list and Palmgren's 1921 and 1927 lists.)—Also his Merkittävia kasvilöytöja Lounais-Suomen saaristosta. [Noteworthy plant finds from the archipelago of southwest Finland.] Luonnon Ystävä **36**: 123–128. 1932. (Additional notes.)—Also his Viktigare växtfynd i SW-Finland 1933. Memo. Soc. Fauna et Flora Fenn. **10**: 31–35. 1934. (Additional records.)—Also his Eine pflanzengeographische neueinteilung Südwest-Finnlands. l. c. **10**: 272–288. 6 fig. (maps). 1934. (Proposal of a new division, Regio insularis, Ri).—Also **Palmgren, Alvar.** Kompletterande fyndorter och synpunkter till Ålands flora 1. l.c. **10:** 398–434. 1935. (Includes (p. 426–429) additional records for Brändö and Kumlinge.) —Also **Eklund, Ole.** Wichtigere gefässpflanzenfunde aus SW-Finnland 1934. l. c. **11:** 4–8. 1935. (Further records.)—Also his Botaniskt från SW-Finland 1935. l. c. **12:** 24–25. 1936. (Additional records.)

Lindén, John. Bidrag till kännedomen om Sunds sockens flora på Åland. Primula **1887:** 61–76. 1887.

Annotated list of 140 noteworthy species of Sund parish. (Not seen; title and annotation supplied by H. Luther.)

Liro, J. I. Åländska växtnamn. Acta Soc. Fauna et Flora Fenn. v. 40, no. 4. 40 p. 1915.

Formation of vernacular names, etc.; alphabetical list of about 280 vascular and cellular plants, with vernacular names and localities where used and notes on uses. Refers mostly to Eckerö and Hammarland parishes.

Palmgren, Alvar. Studier öfver löfängsområdena på Åland. Ett bidrag till kännedomen om vegetationen och floran på torr och på frisk kalkhaltig grund. II. Floran. Acta Soc. Fauna et Flora Fenn. **42** (1): 171–174. 1915.

Annotated list of vascular plants (366 species) growing in hardwood meadows (laubwiesen), with detailed localities. Part I (169 p. 8 tab. 1915) has title Vegetationen; pt. III (475–633 p. 2 pl., 2 maps, 8 tab. 1916 (1917)), Statistisk undersökning af floran.—See also **Samuelsson, Gunnar.** Några bidrag till floran på Åland. Svensk Bot. Tidskr. **11:** 130–133. 1917. (List with localities, mostly from parish Föglö.)—Also **Palmgren, A.** Die entfernung als pflanzengeographischer faktor. Acta Soc. Fauna et Flora Fenn. v. 49, no. 1. 113 p. map. 1921. (Includes (p. 104–113) tabular list of vascular plants (Die verteilung der die laubwiesenvegetation bildenden arten auf Åland) showing occurrence by parishes.)—Also his Zur kenntnis des florencharakters des nadelwaldes. Eine pflanzengeographische studie aus dem gebiete Ålands. Acta Forest. Fenn. v. 22, [no. 2]. 114 p. map. 1922. (Includes (p. 81–109) annotated list of vascular plants with detailed localities.) Apparently all the species are included in his 1925 and 1927 lists (above under General).—Also **Cedercreutz, Carl.** Die gefässpflanzenvegetation der seen auf Åland. Acta Bot. Fenn. 38. 77 p. 8 pl. on 4. 1947. (Includes (p. 39–62) list of aquatic vascular plants (73 species) with very detailed localities.)—Also **Jaatinen, Stig.** Bidrag till kännedomen om de åländska sjöarnas strandvegetation. l. c. 45. 354 p. 34 fig., 8 pl. on 4, map. 1950. (Includes (p. 281–306) annotated list of mosses and vascular plants of lakeshores.)

Widlund, Uno. Förteckning över kärlväxter på Sälskär och Norrskär i Ålands nordvästra skärgård. Memo. Soc. Fauna et Flora Fenn. **26:** 100–107. 1951.

General features of flora; annotated list of vascular plants (172 species).

ISTHMUS KARELICUS (Ik). (KARJALAN KANNAS. Kk)

See also Karelia australis (Nylander).

The Isthmus karelicus, the upper part of the isthmus between Lake Ladoga and the Gulf of Finland, is now Russian.

General

Hiitonen, Ilmari. Karjalan kannas kasvien vaellustienä lajien nykylevinneisyyden valossa. (Die karelische landenge als einwanderungsweg der pflanzenarten im lichte ihrer heutigen verbreitung. Annales Bot. Soc. Zool. Bot. Fenn. Vanamo v. 22, no. 1. vi, 206 p. 132 fig. (maps). 1946.

Discussion of the Karelian Isthmus as a migration route for plants; includes (p. 197–206) list of native and naturalized vascular plants (814 species and subspecies, excluding *Hieracium* and *Taraxacum*) of the region, annotated by symbols; bibliography (p. 146–154), extensive German summary. Covers the Isthmus karelicus and parts of Karelia australis, Karelia ladogensis, and Savonia australis.

Malmberg, A. J. Förteckning öfver Karelska näsets kärlväxter. Notiser Sällsk. Fauna et Flora Fenn. Förhandl. 9 (n.s. 6): 301–326. 1868.

General features of flora; list of vascular plants (638 species), annotated as to abundance, with localities for scarcer species.—See also **Lindberg,**

Harald. Fanerogamer från Karelska näset. Meddel. Soc. Fauna et Flora Fenn. 22: 4–7. 1896. (List of new records.)—Also his Anmärkningsvärda växtfynd på Karelska näset. l. c. 24: 81–84. 1900. (List of 15 species.)

Local

Fagerström, Lars. Ett bidrag till kännedomen om vegetation och flora i Terijoki socken på Karelska näset. Memo. Soc. Fauna et Flora Fenn. **15:** 94–140. 8 fig. (fig. 1, map), 2 pl. (on 1). 1939–40.

Botanical explorations, topography, climate, geology, plant formations; annotated list of vascular plants of Terijoki parish (706 species, of which about 250 are anthropochorous, plus 59 Taraxaca. *Hieracium* is not listed); bibliography.—See also his Anteckningar om floran och vegetationen i Terijoki socken. I. Memo. Soc. Fauna et Flora Fenn. **17:** 121–142. 4 fig. 1941. (Ecological and phytogeographic; extensive bibliography.)—Also his Några tillägg till floran i Terijoki socken. l. c. **21:** 165–166. 1945. (Additional records bringing total to 713 species plus 61 Taraxaca.)—Also **Ulvinen, Arvi.** Merenrantakasveja Suomenlahden itäosasta. (Meeresstrandpflanzen aus dem östlichen teil des Finnischen Meerbusens.) Annales Bot. Soc. Zool. Bot. Fenn. Vanamo **20** (Notul. 15): 32–35. 1944. (Includes records from Terijoki, Kuolemajärvi, and Uusikirkko in Isthmus karelicus and from Johannes and Koivisto in Karelia australis.)

Hidén [= Hiitonen], Ilmari. Tietoja Sakkulan pitäjän kasvistosta. [Notes on the flora of Sakkula parish.] Meddel. Soc. Fauna et Flora Fenn. **49:** 124–131. 1925.

General features of flora; annotated list of noteworthy vascular plants, list of formae.

Paasio, Ilmari. Korpikylän kruununpuiston putkilokasvistosta. [On the vascular flora of Korpikylä forest reserve.] Luonnon Ystävä **32:** 211–218. 1928.

Topography, etc.; briefly annotated list of vascular plants (220 species) of the forest reserve Korpikylä in Kivennapa parish.

KARELIA AUSTRALIS (Ka). (ETALÄ-KARJALA. EK)

See also Isthmus karelicus (Hiitonen; Malmberg; Ulvinen under Fagerström); Savonia australis (Ulvinen).

Karelia australis is on the extreme southeastern border of present Finland, on north side of the Gulf of Finland, and now for the greater part Russian; bordered on north by Savonia australis, on northeast by Karelia ladogensis, on southeast by Isthmus karelicus, on south by Gulf of Finland, on west by Nylandia.

General

Nylander, William. Collectanea in floram karelicam. Notiser Sällsk. Fauna et Flora Fenn. Förhandl. **2:** 109–201. 1852.

Botanical investigations, topography, botanical regions, general features of vegetation; briefly annotated list of vascular plants (655 species), mosses (202 species), and lichens (175 species). Covers Karelia in a broad sense, between 60° 10′ and 65° N. Lat., 27° and 36° E. Long.

Local

Brenner, Magnus. Bidrag till kännedom af Finska vikens övegetation. Notiser Sällsk. Fauna et Flora Fenn. Förhandl. **11** (n. s., 8): 1–38. 1871.—Ytterligare bidrag till kännedomen . . . l. c. **11:** 445–448. 1871.—Bidrag till kännedomen . . . III. tillägg till Hoglands fanerogamflora. Meddel. Soc. Fauna and Flora Fenn. **11:** 33–40. 1885.

Topography, geology, general features of flora, botanical explorations; annotated list of vascular plants and Characeae (412 species, of which 411 are vascular); list of doubtful records. Covers islands of Hogland (Suursaari), Tytärsaari, and Lavansaari. The third paper brings the total to 422 species (421 vascular).—See also **Saelán, Thiodolf.** Några anteckningar om floran på Hogland. Meddel. Soc. Fauna et Flora Fenn. **25:** 73–81. 1900. (Additional species and new records for Hogland; list of vascular plants observed on Haapasaari (Aspö), an island 2.5 Swedish miles northeast of Hogland.)—Also **Olsoni, Börje.** Växtvärlden på Tytärsaari och

Säyvi. Memo. Soc. Fauna et Flora Fenn. 2: 48–56. 2 fig. 1927. (General features of flora; list of vascular plants showing occurrence on each island (254 species).)—Also **Häyrén, Ernst.** Antropokorer och apofyter i Suurkylä på Hogland sommaren 1939. l. c. 18: 65–67. 1943. (List of 76 species and 1 hybrid, with notes on changes in the weed flora; bibliography.)

Erkamo, Viljo. Bolshevikkiajan merkeistä Viipurin kasvistossa. (Über die spuren der Bolschewikenherrschaft in der flora der stadt Viipuri.) Annales Bot. Soc. Zool.-Bot. Fenn. Vanamo v. 18, no. 3. 1 p. l., 24 p. 1943.

Sources of weedy plants, plants of ruined situations, changes in abundance, etc.; includes tabular lists of weeds connected with Russian occupation and weeds of house ruins.

——— Ruutumenetelmästä kasvistollisessa tutkimustyössä. [The square-method in floristic research.] Luonnon Ystävä 43: 62–68, 97–107. 1 fig. (map). 1939.

Discussion of method, topography, general features of flora; tabular list of vascular plants of the Pihkalanjärvi area in Nuijamaa, Jääski, and Viipuri parishes; bibliography.

Jalas, Jaakko. Kasvistollisia tietoja Seiskarista (Ka). (Floristic notes from the Seiskari Island (Ka).) Arch. Soc. Zool. Bot. Fenn. Vanamo 1: 110–116. 1 fig. (map). 1948.

Previous explorations, with mention of species reported; list of additional species found by author, with notes on certain species; bibliography.

Krohn, Väinö. Piirteitä Säkkijärven ja sen ympäristön kasvistosta vuosina 1913 ja 1923. (Zur flora des kirchspiels Säkkijärvi nebst umgegend in den jahren 1913 und 1923.) Annales Soc. Zool.-Bot. Fenn. Vanamo 3: 1–104. 1924.

Geology, climate, influence of man, plant formations, etc.; annotated list of vascular plants; ecology, unannotated list of cellular plants (not of all groups).

——— Über die vegetation und flora des äussersten schärengürtels Süd-west-Kareliens. Annales Bot. Soc. Zool.-Bot. Fenn. Vanamo v. 15, Beih. 307 p. map. 1931.

Geography, climate, human population, plant formations, influence of man; annotated alphabetical list of vascular plants (457 species); list of rarer species with localities; phytogeography, etc. Relates to the archipelago between 60° 15′ and 60° 23′ N. Lat., 27° and 27° 35′ E. Long.—See also **Fagerström, Lars.** Växtfynd i Aspö (Haapasaari) skärgård i Karelia australis. I. Memo. Soc. Fauna et Flora Fenn. 28: 60–66. 1953. (List of 120 vascular plants; bibliography.)

Lindén, John. Anteckningar om växtligheten i södra Karelen. Meddel. Soc. Fauna et Flora Fenn. 16: 115–186. 1891.

Botanical investigations, topography, geology, general features of flora, plant distribution, statistics; annotated list of vascular plants (530 species); phenology. Relates to area between 60° 43′ and 61° 13′ N. Lat. and about 28° 50′ and 30° E. Long.

Porkka, O. H. Satunnaiskasveja Antrean pitäjästä. (Zufällige pflanzen aus dem kirchspiel Antrea.) Annales Soc. Zool.-Bot. Fenn. Vanamo 3: 233–238. 1925.

Annotated list of 48 weedy plants from Antrea parish.

Toivari, Lauri. Salo-Urpalan kruununpuiston kasvillisuudesta ja kasvistosta v. 1934–36. (Über die vegetation und flora des Staatsforstes von Salo-Urpala in Südostfinnland in den jahren 1934–36.) Arch. Soc. Zool. Bot. Fenn. Vanamo 4: 34–53. 4 fig. (incl. map). 1949.

Topography, plant formations; annotated list of vascular plants; bibliography. (In Virolahti parish.)

Ulvinen, Arvi. Kotkan sataman ja rata-alueen kasvisto vv. 1914–1924. (Über die flora im hafen und im bahngebiet der stadt Kotka 1914–1924.) Annales Soc. Zool.-Bot. Fenn. Vanamo 5: 332–345. 1927.

Annotated alphabetical list of 217 adventive plants and other plants found with them on commercial areas in Kotka.

Valle, K. J. Täydentäviä tietoja Jääsken putkilokasvistosta. [Supplementary notes on the vascular flora of Jääski.] Acta Soc. Fauna et Flora Fenn. v. 46, no. 7. 40 p. 1919.

Botanical explorations; annotated list of 378 vascular plants, bringing total for the parish (with those recorded by Linden, 1891; see above) to 578.

KARELIA BOREALIS (Kb). (POHJOIS-KARJALA. PK)

See also Karelia australis (Nylander); Karelia ladogensis (Chydenius and Furuhjelm; Hjelt; Linkola); Kuusamoensis, Regio (Wainio).

Karelia borealis is on south-central part of border with Russia, its southeastern part now Russian; between about 61°50′ and 63°50′ N. Lat., about 28° and 32°15′ E. Long.; bounded on north by Ostrobottnia kajanensis, on east by Karelia pomorica occidentalis and Karelia onegensis, on south by Karelia olonetsensis and Karelia ladogensis, on west by Savonia australis and Savonia borealis.

Local

Axelson, W. M. Putkilokasvio Pielisen ja Höytiäisen välisellä kannaksella. [Vascular flora of the isthmus between lakes Pielinen and Hoytiäinen.] Acta Soc. Fauna et Flora Fenn. v. 23, no. 4. 78 p. 1902.

Topography, geology, ecology, statistics; annotated list of vascular plants (422 species).

Koskimies, A. E. Havaintoja Keski-Karjalan kasvistosta. (Beobachtungen über die flora in Mittel-Karelien.) Annales Bot. Soc. Zool.-Bot. Fenn. Vanamo v. 5, no. 2. iv, 68 p. 7 fig. (maps). 1934.

Topography, geology, general features of flora; annotated list of known vascular plants (522 species), with localities. Refers to the parishes of Kesälahti, Kitee, Rääkkylä, NW-Tohmajärvi, Pyhäselkä, Kiihtelysvaara, Tuupovaara, and S.-Ilomantsi in Karelia borealis and Uukuniemi (except its eastern part) in Karelia ladogensis.—See also **Tuomikoski, Risto.** Muistiinpanoja Tuupovarran pitäjän ja sen lähiseutujen putkilokasveista. [Notes on the vascular plants of Tuupovaara parish and vicinity.] Memo. Soc. Fauna et Flora Fenn. **17:** 32–43. 1 fig. (map). 1941. (Extensive list of vascular plants from Tuupovaara parish.)—Also **Kallio, Paavo,** and **Nikoskelainen, Ritva.** Kasvistollisia havaintoja Pohjois-Karjalasta. (Floristic notes from North Karelia.) Arch. Soc. Zool. Bot. Fenn. Vanamo **9:** 135–145. 3 fig. (fig. 1, map). 1955. (List of noteworthy species from an area extending north to northern Ilomantsi and Juuka; list of 213 species from Kolvanan uuro; bibliography.)

Tuomikoski, Risto. Suojärven Paperon kasvisto. [Flora of Papero, Suojärvi.] Luonnon Ystävä **37:** 90–97. 1933.

Topography, general features of flora, etc.; briefly annotated list of vascular plants.—See also **Saarnijoki, Sakari.** Satunnaiskasveja Kollaan taistelupaikoilta. [Adventive plants from the Kollaa battleground.] Luonnon Ystävä **46:** 69. 1942. (List of weeds from the 1939 war front in Kollaa.)

Uutela, Elvi. Enon ja Länsi-Ilomantsin kasvisto. (Die flora des kirchspiels Eno und des westlichen teiles von Ilomantsi in Ostfinnland.) Annales Bot. Soc. Zool.-Bot. Fenn. Vanamo v. 10, no. 5. iv, 54 p. 6 fig. (maps). 1939.

Topography, geology, climate, cultivation, plant formations; annotated list of vascular plants (401 species).

KARELIA KERETINA (Kk). (KIERETIN KARJALA. KK; FORMERLY POHJOIS-VIENA. PV)

See also Karelia olonetsensis (Linkola); Kuusamoensis, Regio (Borg; Lindberg; Wainio); Lapponia rossica (Akademiiâ Nauk SSSR).

Karelia keretina is a region (always Russian) extending from the northern half of the White Sea westward to the ante-1944 Finnish border, from about 65° to 67° N. Lat., bordered on the north by Lapponia Imandrae, on the east by the White Sea (Beloye More), on the south by Karelia pomorica occidentalis, on the west by Regio kuusamoensis.

Local

Herlin, Nils. Beitrag zur kenntnis der flora in der Provinz Karelia keretina in Ostkarelien. Memo. Soc. Fauna et Flora Fenn. **19:** 91–103. 1 fig. (map). 1944.

Botanical explorations, general features of flora, notes on weeds; annotated list of vascular plants (about 270 species, of which about 130 are ruderal plants) of vicinity of Kiestinki.

Söyrinki, Niilo. Havaintoja kyläkasvistosta Vienan Karjalassa sotakesänä v. 1941. [Observations on the village flora in Viena Karelia in the summer campaign of 1941.] Luonnon Ystävä **45**: 150–164. 1941.

General features of weed flora; annotated list of weeds from western part of province.—See also his Oikaisuja ja lisäyksia Kiestingin kasviluetteloon. [Corrections and additions to the plant list of Kiestinki.] Luonnon Ystävä **46**: 70. 1942. (Additional notes and corrections.)

——— Kasvistosta Oulankajoen-Pääjärven alueella Kieretin Karjalassa. (Über die flora im gebiet von Oulankajoki-Pääjärvi, Karelia Keretina, Ostkarelien.) Annales Bot. Soc. Zool.-Bot. Fenn. Vanamo v. 27, no. 2. iv, 118 p. 12 fig. (fig. 1, map). 1956.

General features of vegetation, introduced plants, floral districts, etc.; annotated list of vascular plants (423 species).

KARELIA LADOGENSIS (Kl). (LAATOKAN KARJALA. LK)

See also Isthmus karelicus (Hiitonen; Malmberg); Karelia australis (Lindén; Nylander); Karelia borealis (Koskimies).

Karelia ladogensis is on northern end of Lake Ladoga; now nearly all Russian. Bordered on the north by Karelia borealis, on the east by Karelia olonetsensis, on the south by Lake Ladoga and Isthmus karelicus, on the west by Karelia australis and Savonia australis.

Local

Chydenius, J. J., and **Furuhjelm, J. E.** Berättelse öfver en naturhistorisk resa i Karelen, företagen på Sällskapets pro Fauna et Flora Fennica bekostnad. Notiser Sällsk. Fauna et Flora Fenn. Förhandl. **4** (n.s., **1**): 79-118. 1859. (Also as Bidrag Finlands Naturkänn., Etnog. och Statis. **3**: 79-118. 1859.)

Authors' itinerary, topography, general features of flora, etc.; list of noteworthy vascular plants, with localities; phenology, lists of insects, etc. Covers nearly all the coastal region of northern Lake Ladoga that was included in Karelia ladogensis, from Kexholm (Käkisalmi, now Priozersk) to Salmis (Salmi), with the Walamo (Valaam) island group.

Hjelt, Hjalmar. Anteckningar från en botanisk resa i Karelen sommaren 1876. Meddel. Soc. Fauna et Flora Fenn. **6**: 19–69. 1881.

Author's itinerary, topography, general features of flora, plant formations; list of noteworthy vascular plants, with localities, from Impilahti, Pälkjärvi, Ruskeala, and Suistamo parishes in Karelia ladogensis, as well as Tohmajärvi in Karelia borealis.

Huuskonen, A. J. Lisätietoja Pälkjärven pitäjän kasvistosta. (Beiträge zur kenntnis der flora des kirchspiels Pälkjärvi.) Annales Bot. Soc. Zool.-Bot. Fenn. Vanamo v. 21, no. 2. ii, 25 p. 2 fig. (incl. map). 1945.

Botanical explorations, plant regions, changes in vegetation; annotated list of vascular plants (541 species, not counting the cultivated and ornamental plants.)

Linkola, Kaarlo. Studien über den einfluss der kultur auf die flora in den gegenden nördlich vom Ladogasee. Acta Soc. Fauna et Flora Fenn. v. 45. 2 v. (vii, 432; 491 p. tables, maps). 1916–21.

Physiography, plant formations, discussion of plants of the region in relation to human culture; (pt. 2, p. 143–483) account of botanical explorations, annotated list of vascular plants (734 species) with detailed distribution. The area covered extends from 61° 15′ to 62° 25′ N. Lat., 30° 15′ to 30° 50′ E. Long., and includes parts of Karelia ladogensis, Karelia borealis, Karelia olonetsensis, and Karelia onegensis.—See also **Pankakoski, Antero.** Sortavalan kaupungin satunnaiskasvistosta vv. 1925–1930. [On the adventive flora of Sortavala city in 1925–1930.] Luonnon Ystävä **35**: 168–171. 1931. (Additional notes on weeds.)—Also **Huuskonen, A. J.** Lisiä Etelä-Salmin kasvistoon. [Additions to the flora of South-Salmi.] l. c. **36**: 148–150. 1932. (Additional notes from southern Salmi parish.)—Also his Lisätietoja Suistamon ja Soanlahden kuntien kasvistosta. [Additional notes on the flora of Suistamo and Soanlahti parishes.] l. c. **43**: 217–220. 1939. (Extensive list of additional records from Suistamo and Soanlahti.)

Norrlin, J. P. Symbolae ad floram ladogensi-karelicam. Meddel. Soc. Fauna et Flora Fenn. **2**: 1–33. 1878.
Botanical explorations, topography, general features of flora; annotated list of vascular plants, bryophytes, and lichens. Relates to the northwestern shore region of Lake Ladoga (Impilaks, Ruskeala, Sortavala, etc.).

Pankakoski, Antero. Ekologis-kasvistollisia tutkimuksia Hiisjärven luonnonpuistossa. (Ökologisch-floristische untersuchungen im Naturpark von Hiisjärvi in Südostfinnland.) Annales Bot. Soc. Zool.-Bot. Fenn. Vanamo v. 10, no. 3. 2 p.l., 154 p. 30 fig. (incl. maps), map. 1939.
Topography, general features of flora, soil reactions; annotated list of vascular plants (263 species), with list (p. 132-134) of 109 additional species from adjacent territory; table giving pH relations of species investigated, bibliography. (In Salmi parish.)

Räsänen, Veli. Kurkijoen ja sen naapuripitäjien putkilokasvisto. [Vascular flora of Kurkijoki and neighboring parishes.] Kuopion Luonnon Ystäväin Yhdistyksen Julkaisuja ser. B, v. 2, no. 2. 117 p. 24 cm. 1944. (Library of Riksmuseet, Stockholm.)
Annotated list of vascular plants (801 species, including 81 cultivated ones, some of which have escaped); bibliography.

KARELIA OLONETSENSIS (Kol). (AUNUKSEN KARJALA. AK)

See also Karelia australis (Nylander); Karelia ladogensis (Linkola); Karelia onegensis (Fagerström, under Norrlin; Norrlin).
Karelia olonetsensis is a region (always Russian) between Lake Onega and Lake Ladoga, extended in 1944 south to the Swir River.

General

Linkola, Kaarlo. Flora und vegetation. *In* Auer, Väinö, *ed.* Anteil der finnischen forscher an der erforschung von Kola, Ostkarelien und Ingermanland. Fennia **67** (3): 33–46. 1942.
Sketch of natural boundaries, botanical explorations, and vegetational studies; bibliography (almost entirely Finnish).

Local

Elfving, Fredrik. Anteckningar om vegetationen kring floden Svir. Meddel. Soc. Fauna et Flora Fenn. **2**: 113–170. 1878.
Botanical regions, topography, plant formations, etc.; annotated list of vascular plants (488 species and subspecies), mosses, and lichens, with localities.—See also **Cajander, A. K.**, and **Lindroth, J. I.** Matkakertomus kasvitieteellisestä retkestä Aunuksen Karjalaan (Ol.) kesänä 1898. [Itinerary of a botanical trip to Olonets Karelia in the summer of 1898.] Meddel. Soc. Fauna et Flora Fenn. **25**: 22–28. 1900. (Notes on their trip; annotated list of noteworthy plants, including additions to the flora of Karelia olonetsensis and Karelia onegensis.)

Fagerström, Lars, and Luther, Hans. Ett bidrag till kännedomen om floran i Vieljärviområdet i Karelia olonetsensis. Memo. Soc. Fauna et Flora Fenn. **21**: 18–45. 1 fig. (map). 1945.
Botanical explorations, topography, general features of flora, notes on various species; annotated list of vascular plants (452 species and subspecies); bibliography.

Hustich, Ilmari. Eine pflanzengeographische übersicht über das gebiet Kuuttilahti am Sywäri-Swir (Fern-Karelien). Acta Soc. Fauna et Flora Fenn. v. 63, no. 3. 53 p. 2 fig. (fig. 1, map), 2 pl. 1953.
Topography, climate, soils, general features of vegetation, ruderal plants, etc.; annotated list of vascular plants (399 species); bibliography.—See also his Ergänzungen zur flora des gebiets Kuuttilahti am Swir (Fern-Karelien). Memo. Soc. Fauna et Flora Fenn. **20**: 104–107. 1945. (Note on plant associations, etc.; annotated list of noteworthy species.

——— Pflanzengeographische übersicht über das Kuujärvigebiet am mittleren Swir in Fern-Karelien. Memo. Soc. Fauna et Flora Fenn. **20**: 47–77. 3 fig. (fig. 1, map). 1945.

Topography, autumn flowering, soil reactions, general features of vegetation; annotated list of vascular plants (about 550 species); bibliography. (About 61° N. Lat., 34° E. Long. in vicinity of Podporoje (Podporozyhe).)

Palmén, Ernst. Zur kenntnis der flora und vegetation eines uferabschnitts am Laatokkasee nördlich der Syväri-mündung. (Kasvistosta ja kasvillisuudesta eräällä Syvärin suiston pohjoispuolisella Laatokan rannalla.) Annales Bot. Soc. Zool.-Bot. Fenn. Vanamo v. 19, no. 2. 3 p. l., 93 p. 21 fig. 1943.

Topography, geology, ecological conditions, plant zones and habitats; includes (p. 16–25) annotated tabular list of bryophytes and vascular plants (164 vascular species) of an area in vicinity of Gumbaritz extending from about 60° 37′ to 60° 40′ 30″ N. Lat.; bibliography.

Repo, R. Äänislinnan (Kol) satunnaiskasveja. (Adventivpflanzen aus Äänislinna (Petrosawodsk, Aunus-Karelien).) Annales Bot. Soc. Zool.-Bot. Fenn. Vanamo 18 (Notul. Bot. 13): 16–19. 1943.

Briefly annotated list of adventive plants observed by author in 1942.—See also **Fagerström, Lars,** and **Luther, Hans.** Ruderatväxter från Petroskoi bangård sommaren 1943. Memo. Soc. Fauna et Flora Fenn. **21:** 142–150. 1945. (Details for rarer species; briefly annotated list of adventive plants; bibliography.)—Also **Repo, R.** Havaintoja antropokorikasvien esiintymisestä Etelä-Aunukessa v. 1941–44. (Beobachtungen über das auftreten der anthropochoren in Süd-Olonetz in den jahren 1941–44.) Arch. Soc. Zool. Bot. Fenn. Vanamo **3:** 59–77. 1949. (Botanical explorations; annotated tabular list of adventive plants and scarcer ruderal plants from the Olonets Isthmus between Lake Ladoga and Lake Onega, with discussion including short lists from other localities; bibliography.)

KARELIA ONEGENSIS (Kon). (ÄÄNISEN KARJALA. ÄK)

See also Karelia australis (Nylander); Karelia olonetsensis (Cajander, under Elfving; Linkola).

Karelia onegensis is a region (always Russian) northwest of Lake Onega.

General

Norrlin, J. P. Flora Kareliae onegensis. I. Notiser Sällsk. Fauna et Flora Fenn. Förhandl. **13 (n. s., 10):** 1–183. 1871.

Botanical investigations, topography, climate, plant formations, phytogeography, botanical regions, etc. (in Swedish); annotated list of vascular plants (603 species; in Latin). The introductory matter, lacking the list of species, published as dissertation with title Om Onega-Karelens vegetation och Finlands jemte Skandinaviens naturhistoriska gräns i öster. 132 p. 23 cm. Helsingfors, 1871. A German translation of this introduction, with subtitle Ueber die vegetation von Onega-Karelien und die naturhistorische grenze Finnlands sowie Skandinaviens im osten, was published in Acta Forest. Fenn. **23:** 53–161. 1923. The second part of the Flora Kareliae onegensis, published in Meddel. Soc. Fauna et Flora Fenn. **1:** 1–46. 1876, deals entirely with lichens.—See also **Fagerström, Lars.** Växtgeografiska anteckningar från en färd i Fjärr-Karelen sensommaren 1942. Memo. Soc. Fauna et Flora Fenn. **20:** 107–142. 1 fig. (map). 1945. (Author's explorations, general features of vegetation, adventive plants, changes in vegetation; extensive list of noteworthy vascular plants; bibliography.)—Also **Fagerström, L.**, and **Luther, Hans.** En botanisk resa till Schunguhalvön i Karelia onegensis sommaren 1943. l. c. **22:** 84–105. 2 fig. (fig. 1, map). 1946. (Botanical explorations in the Schungu (Sunku or Zaoneshje) Peninsula, general features of vegetation; annotated list of noteworthy vascular plants; bibliography.)

Local

Jalas, Jaakko. Muistiinpanoja antropokoreista Karhumäessä (Kon) syksyllä 1942. (Aufzeichnungen über anthropochoren in Karhumäki, Onega-Karelien, im herbst 1942.) Annales Bot. Soc. Zool.-Bot. Fenn. Vanamo **20** (Notul. Bot. 15): 64–67. 1945.

Annotated list of anthropochorous vascular plants of Karhumäki, on north shore of Lake Onega.—See also **Kytöniemi, Antti.** Kasvitietoja Karhumäestä (Äk). (Botanisches aus Karhumäki, Onega-Karelien.) Annales Bot. Soc. Zool.-Bot. Fenn. Vanamo **20** (Notul. Bot. 15): 35–39. 1944. (List of noteworthy vascular plants including many adventives.)

Sparre, Benkt. Anteckningar om floran på halvön Arsenavolok i Fjärrkarelen. Memo. Soc. Fauna et Flora Fenn. **20:** 17–23. 1 fig. (map). 1945.

Briefly annotated list of vascular plants. (Also called Ashebnavolok and Anschepnavolok; in Sunku parish).

KARELIA POMORICA OCCIDENTALIS (Kpoc). (LÄNSI-VIENA. LV)

See also Karelia australis (Nylander); Karelia olonetsensis (Linkola); Kuusamoensis, Regio (Wainio).

Karelia pomorica occidentalis is an area (always Russian) with northern boundary extending east from the Finnish border at about 65° N. Lat. to the White Sea, the southern boundary extending east at about 63° 20′ to about 34° E. Long., then straight north to about 64° 15′, then northeast to the Gulf of Onega; bounded on the north by Karelia keretina, on the east by Karelia pomorica orientalis, on the south by Karelia onegensis, on the west by Karelia borealis, Ostrobottnia kajanensis, and Regio kuusamoensis.

General

Chernov, V. N. Poleznye rasteniia Karelo-Finskoĭ SSR. [Useful plants of Karelo-Finnish SSR.] 1952.

Not seen; cited from his Vesenniia flora Karelo-Finnskoĭ SSR., p. 154. 1955.

——— Vesenniaia flora Karelo-Finskoĭ SSR. Opredelitel' vesennikh rasteniĭ. [Spring flora of Karelo-Finnish SSR. A key to spring plants.] 154 p. 90 fig. 23 cm. Petrozavodsk, 1955.

Glossary; flora of plants flowering up to middle of June, in form of an ample, annotated key; bibliography. The Karelo-Finskoi SSR. (Karelo-Finskaia Sovetskaia Sotsialisticheskaia Respublika), now the Karel'skaia Avtonomnaia SSR., covers all of Karelia ladogensis, Karelia olonetsensis, Karelia onegensis, Karelia pomorica occidentalis, the western part of Karelia pomorica orientalis and Karelia transonegensis, and most of Karelia keretina.

Erkamo, Viljo. I. O. Bergroths botaniska undersökningar i Karelia pomorica. En sammanställning. Acta Soc. Fauna et Flora Fenn. v. 67, no. 1. 83 p. 1 fig. (map). 1947.

Botanical explorations, Bergroth's itineraries, topography, botanical regions, plant distribution, weedy species, etc.; list of vascular plants collected by Bergroth (513 species and subspecies; *Hieracium* is omitted) with localities; bibliography. In the introduction (p. 17) are listed the 13 species known from the province not collected by Bergroth.

Liro, J. I. Karjalaisia kasvinnimiä. [Karelian names of plants.] Acta Soc. Fauna et Flora Fenn. v. 40, no. 3. 37 p. 1915.

General considerations; alphabetical list of scientific names (including a few cellular plants) with vernacular equivalents and localities where used, and notes on uses; index of vernacular names.

Local

Jalas, Jaakko. Kylien kasvistosta Repolan piirikunnassa Länsi-Pomorian (Kpoc) luonaiskolkassa. (Über die flora der dörfer im kreis Repola in der florenprovinz Karelia pomorica occidentalis (Kpoc), Ostkarelien.) Acta Soc. Fauna et Flora Fenn. v. 66, no. 3. 58 p. incl. 2 pl. 2 fig. (maps). 1948.

Topography, general features of flora, botanical explorations, changes in vegetation, phytogeography, adventive plants; annotated list of vascular plants (304 species); bibliography.

Mannerkorpi, Panu. Uhtuan taistelurintamalle saapuneista tulokaskasveista. (Zur adventivflora an der Uhtua-front.) Annales Bot. Soc. Zool.-Bot. Fenn. Vanamo **20** (Notul. 15): 39–51. 1944.

Includes two lists of adventive plants (altogether 164 species) brought in by war action, with discussion. Relates principally to the stretch of road 11 kilometers long between Pinkojoki and Parakkilampi about 20 kilometers west of Uhtua.

KARELIA POMORICA ORIENTALIS (Kpor). (ITÄ-VIENA. IV)

See also Karelia olonetsensis (Linkola).

Karelia pomorica orientalis is a roughly trapeziform area (always Russian) lying between 63° and 64° 20′ N. Lat., about 33° 50′ and 39° 20′ E. Long., bordered on the north by the Gulf of Onega, on the east by the Onega River, on the south by Karelia transonegensis, on the west by Karelia pomorica occidentalis.

Local

Marmo, Vladi. Kasvistollisia havaintoja Itä-Vienasta. (Observations on the flora of the Eastern Viena district, East Karelia, USSR.) Arch. Sec. Zool. Bot. Fenn. Vanamo **2**: 76–80. 1949.

Topography, geology, mention of noteworthy species of various localities; short bibliography.

KARELIA TRANSONEGENSIS (Kton). (ÄÄNISEN TAKAINEN KARJALA. ÄtK)

See also Karelia olonetsensis (Linkola).

Karelia transonegensis is a trapeziform area (always Russian) bordered on the west by Lake Onega, on the north by Karelia pomorica orientalis, on the east by the Onega River, and on the southeast by the approximate border between the Fennoscandian Archean rock and the Devonian bedrock of northern Russia.

General

Cajander, A. K. Fenno-Scandian kasvitieteellisestä kaakkois-rajasta. [On the southeastern botanical border of Fennoscandia.] Meddel. Soc. Fauna et Flora Fenn. **26**: 172–184. map. 1900.

Topography, geology, distribution boundaries of noteworthy species and of vegetation types.—See also Kalela, Die ostgrenze Fennoskandiens, under Finland, General.

KUUSAMOENSIS, REGIO (Ks). (KUUSAMO. Ks)

See also Lapponia inarensis (Fellman); Lapponia kemensis (Wainio); Ostrobottnia borealis (Auer under Brenner; Kotilainen under Brenner; Rantaniemi; Ostrobottnia kajanensis Parvela).

The Regio kuusamoensis is a region in north-central Finland, bordered on the east by Russia (biogeographical provinces of Lapponia Imandrae, Lapponia tulomensis, Karelia keretina, and Karelia pomorica occidentalis), on the west by Lapponia kemensis and Ostrobottnia borealis, on the south by Ostrobottnia kajanensis. It extends from about 65° 10′ to 67° 40′ N. Lat. and about 27° 25′ to 30° 10′ E. Long. (extreme limits); part of its area, along the eastern border, was transferred to Russia in 1944. Its name, until recently Kuusamo in the older as well as the newer nomenclature, was changed to Regio kuusamoensis because the area, which formerly included only the parish of Kuusamo, was extended in 1927 to cover also Salla, north of Kuusamo.

Local

Backman, A. L. En botanisk resa i Kuusamo sommaren 1902. Meddel. Soc. Fauna et Flora Fenn. **29**: 111–117. 1904 (preprinted 1903).

General features of vegetation with mention of some noteworthy species from region around Paanajärvi, the Kitkajoki, and the upper course of the Oulankajoki.—See also his En botanisk resa i norra Kuusamo sommaren 1908. Meddel. Soc. Fauna et Flora Fenn. **35**: 197–202. 1909. (Notes on the vegetation and flora of the region of the Oulankajoki and the Kitkajoki in northern Kuusamo.)—See also **Brenner, Widar.** En botanisk resa till sydöstra Kuusamo. l. c. **37**: 104–109. 1911. (Account of author's excursion with mention of noteworthy species.)—Also **Backman, A. L.** Floristiska meddelanden från Kuusamo. l. c. **40**: 159–161. 1914. (Short list of noteworthy species, chiefly from the Oulankajoki valley.)—Also **Fortelius, Olof.** Några växtfynd från Kuusamo. l. c. **48**: 129–130. 1925. (Short list of noteworthy species from Kitka, Kyrkobyn, etc.)—Also his Några växtfynd från Kuusamo sommaren 1922. l. c. **49**: 11–14. 1925.—Also **Auer, A. V.** Kasvistollisia havaintoja Pohjois-Suomesta III. [Floristic observations from North Finland III.] Memo. Soc. Fauna et Flora Fenn. **19**: 44–57. 328. 1944. (Extensive list of noteworthy plants, especially from northern Kuusamo and southern Salla.)

Borg, Väinö. Beiträge zur kenntniss der flora und vegetation der finnischen fjelde (alpinen und subalpinen gebirge) I. Acta Soc. Fauna et Flora Fenn. v. 25, no. 7. 170 p. map. 1904.

Geography, topography, geology, climate, botanical regions and zones; annotated tabular list of vascular plants (287 species), showing occurrence in three regions recognized; phytogeography, bibliography. Relates to montane areas in Lapponia inarensis, Lapponia kemensis, Kuusamo, Karelia keretina, and Lapponia Imandrae. Borg uses the term fjelde to cover all mountain areas rising above the coniferous zone.

Herlin, Nils. Ruderaten von Ks, Salla im sommer 1941. Memo. Soc. Fauna et Flora Fenn. 19: 90–91. 1944.

List of weeds from the trainyard at Salla (34 species, including 8 new to the Province).—See also **Pesola, V. A.** Kulttuurikasvupaikkain kasvit Kuusamossa ja Sallassa yli kolme vuosikymmentä sitten. (The plants on the cultural places in Kuusamo and Salla more than thirty years ago.) Suomen Maataloustiet. Aikakausk. **24**: 30–34. 1952. (Includes tabular list of angiosperms observed in ruderal places in 1917.)

Lindberg, Harald. Anmärkningsvärda växtfynd under en resa sommaren 1913 genom Kuolajärvi till Knjäsha vid Hvita hafvet samt vid Kandalakscha. Meddel. Soc. Fauna et Flora Fenn. **40**: 18–28. 1914.

Annotated list of new records from Kuolajärvi region (then in Lapponia kemensis, now in Kuusamo), Karelia keretina, and Lapponia Imandrae.

Pesola, V. A. Huomattavia kasvilöytöja N-Kuusamosta ja Kuolajärveltä. (Bemerkenswerte pflanzenfunde aus N-Kuusamo und Kuolajärvi, zum teil in Lapponia kemensis.) Meddel. Soc. Fauna et Flora Fenn. **44**: 229–246 (German title p. 266). 1918.

List of noteworthy vascular plants from northern Kuusamo and from Kuolajärvi (now Salla).—See also his Über die felsenvegetation in NE-Kuusamo und SE-Kuolajärvi. (NE-Kuusamon ja SE-Kuolajärven kallioiden kasvillisuudesta.) Annales Bot. Soc. Zool.-Bot. Fenn. Vanamo v. 5, no. 7. 18 p. 1934. (Description of numerous localities in the Oulanka (Kuusamo) and Kuolajärvi (Kutsa) region, with table showing occurrence of the plants observed.)—Also **Vaarama, Antero.** Piirteitä Kuolajärven Kutsajoen alueen luonnosta ja listä tietoihin seudun kasvistosta. [Natural features of the Kutsajoki area in Kuolajärvi and additions to the knowledge of the flora of the region.] Luonnon Ystävä **39**: 97–103, 142–145. 4 fig. 1935. (Topography, general features of flora; annotated list of noteworthy vascular plants from the Kutsajoki (river) region in southeastern Salla.)

——— Kansanomaisia kasvinnimiä Kuusamosta ja Kuolajärvelta. [Vernacular names of plants from Kuusamo and Kuolajärvi.] Luonnon Ystävä **22**: 37–48. 1918.

Includes list of vascular plants, alphabetically arranged by scientific names, with vernacular names and notes on uses; also short list of cellular plants with vernacular names.

Wainio, Edvard. Kasvistonsuhteista Pohjais-Suomen ja Venäjän-Karjalan rajaseuduilla. I. [On the floristic relations of the North Finland and Russian Karelian boundary. I.] Meddel. Soc. Fauna et Flora Fenn. v. 4. 1 p. l., 160 p. 1878.

Botanical explorations, topography, climate, plant formations, phytogeography, etc. Title only on preliminary leaf. Reprinted as thesis with addition of lviii p. (23.5 cm. Helsingissä, 1878; in library of Riksmuseet, Stockholm) consisting of a briefly annotated list of vernacular plants (527 species) with localities. Refers to the area between Itä-Repola in Karelia pomorica occidentalis, Pielisjärvi in Karelia borealis, Kuhmo in Ostrobottnia kajanensis, Kuusamo in Kuusamo, and Kiestinki in Karelia keretina.—See also **Linkola, Kaarlo.** Kasviston muutoksia Nurmeksessa. [Changes in the vegetation in Nurmeksen kauppala.] Meddel. Soc. Fauna et Flora Fenn. **45**: 86–88. 1920. (List of about 25 adventive plants not reported by Wainio, and other changes; refers to Karelia borealis.)

LAPPONIA ENONTEKIENSIS (Le). (ENONTEKIÖN LAPPI. EnL)

See also Kuusamoensis, Regio (Borg); Lapponia inarensis (Fellman); Lapponia kemensis (Hustich; Linkola; Montell); Ostrobottnia kajanensis (Parvela).

Lapponia enontekiensis is the narrow northwestern lobe of Finland, north of about 68°5′ N. Lat., bordered on the north and northeast by Norwegian territory, on the southeast by Lapponia inarensis, on the south by Lapponia kemensis, on the west by Swedish territory.

Local

Lindén, John. Bidrag till kännedomen om vegetation och flora inom Enontekis lappmarks björk- och fjällregioner. Acta Soc. Fauna et Flora Fenn. v. 63, no. 1. 82 p. 1943.

Author's itinerary, topography, geology, climate, phenology, tree growth, plant associations, statistics; annotated list of vascular plants (260 species) from the area between 68° 30′ and 69° 19′ N. Lat.—See also **Alm, C. G.**, and **Fries, T. C. E.** Floristiska anteckningar från Karesuando och Enontekis socknar. Svensk Bot. Tidskr. **19**: 250–262. 1925. (Includes additional records.)—Also **Kallio, Paavo.** Eräitä kasvitietoja Kilpisjärven seudulta ja Tornionjokivarrelta. (Some floristic observations from the Kilpisjärvi and Tornionjoki regions of northern Finland.) Arch. Soc. Zool. Bot. Fenn. Vanamo **2**: 51–55. 1949. (List of noteworthy vascular plants, with localities; bibliography.)—Also **Jalas, Jaakko.** Floristisches aus Lapponia enontekiensis, Lapponia tornensis und Troms fylke. l. c. **2**: 90–96. 1 fig. 1949. (Botanical explorations; list of noteworthy vascular plants, with localities; bibliography.)

LAPPONIA INARENSIS (Li). (INARIN LAPPI. InL)

See also Kuusamoensis, Regio (Borg); Lapponia kemensis (Linkola; Wainio); Ostrobottnia kajanensis (Parvela).

Lapponia inarensis is the northeastern lobe of present-day Finland, north of about 68° 15′ N. Lat., bounded on the northwest and northeast by Norway, on the southeast by Lapponia petsamoensis (now Russian), on the south by Lapponia kemensis, on the extreme southwest by Lapponia enontekiensis.

General

Fellman, Jacob. Index plantarum in Lapponia fennica lectarum . . . secundum Georgii Wahlenberg Flora Lappon. Berolini 1812. A. Plantae phanerogamae. Bul. Soc. Imp. Sci. Nat. Moscou **8**: 245–289. 1835.

Briefly annotated list of known spermatophytes (386 species) on the Linnaean system, with localities for rarer species, and vernacular names in Finnish, Esthonian, and Lappish. Relates to Lapponia enontekiensis. Lapponia inarensis, and Lapponia kemensis, as well as Kuusamo. Reprinted with additions and corrections in **Fellman, Jacob.** Anteckningar under min vistelse i Lappmarken. v. 4, p. 167–188. 21.5 cm. Helsingfors, 1906. (Not seen; title supplied by H. Luther.).—See also his B. Plantae cryptogamae. l. c. p. 188–200. 1906. (Annotated list of pteridophytes (25 species), mosses, lichens, algae, and fungi. Not seen; title and annotation supplied by H. Luther.)

Local

Kihlman, Oswald. Anteckningar om floran i Inari Lappmark. Meddel. Soc. Fauna et Flora Fenn. **11**: 45–135. map. 1885.

Author's itinerary, botanical explorations, topography, geology, phenology, botanical regions, statistics; annotated list of vascular plants (329 species). Covers the area of Utsjoki moderförsamling and Inari kapell, about 68° 20′ to 70° 6′ N. Lat., 25° to 29° E. Long.—See also **Laine, Unto, Lindgren, Leo,** and **Mäkinen, Yrjö.** Havaintoja Utsjoen pitäjän länsiosan kasvistosta. (Observations on the flora of western Utsjoki.) Arch. Soc. Zool. Bot. Fenn. Vanamo **9**: 120–135. 6 fig. (fig. 1, map). 1955. (Topography, general features of flora, etc.; list of vascular plants (310 species); bibliography.)

Klockars, Bertel, and **Luther, Hans.** Floristiska iakttagelser i Li, Viibus-Maarestatunturiområdet. Memo. Soc. Fauna et Flora Fenn. **14**: 45–54. 1 fig. (map). 1938.

Topography, general features of vegetation; list of vascular plants, with localities; bibliography.

Parvela, A. A. Inarin viljelyskasvisto ja piirteitä sen historiastra. (Die kulturflora von Inari und züge aus ihrer geschichte.) Acta Agralia Fenn. **31**: 61–83. 1934.

Discussion of the useful and ornamental cultivated plants of Inari, giving length of time species have been cultivated, etc.; bibliography.

LAPPONIA KEMENSIS (Lkem, LATER Lk). (KEMIN LAPPI. KemL)

See also Kuusamoensis, Regio (Borg); Lapponia enontekiensis (Kallio, under Linden); Lapponia inarensis (Fellman); Ostrobottnia borealis (Auer, under Brenner; Cajander; Kalliola; Kotilainen, under Brenner; Rantaniemi); Ostrobottnia kajanensis (Parvela).

Lapponia kemensis is a region in northern Finland, between about 66°50′ and 68°15′ N. Lat., bounded on the north by Lapponia enontekiensis, Lapponia inarensis, and Lapponia petsamoensis (the last now Russian), on the northeast by Russia (Lapponia tulomensis), on the southeast by Regio kuusamoensis, on the south by Ostrobottnia borealis, on the west by Sweden.

Local

Hjelt, Hjalmar, and Hult, Ragnar. Vegetationen och floran i en del af Kemi Lappmark och norra Österbotten. Meddel. Soc. Fauna et Flora Fenn. **12:** 1 p. l., 1–159. 1885.

Authors' itinerary, topography, geology, climate, plant associations, botanical explorations, phytogeography, statistics; annotated list of vascular plants (360 species), with localities. Covers an area between 50 and 70 kilometers wide extending from Karhujupukka (mountain in Ostrobottnia borealis, 66° 28′ N. Lat.) to Saivio (67° 39′ N. Lat.) Title is only on unpaged preliminary leaf.

Hult, Ragnar. Växtgeografiska anteckningar från den finska Lappmarkens skogsregioner. Acta Soc. Fauna et Flora Fenn. v. 16, no. 2. 2 p. l., 200 p. 1898.

List of localities, general features of flora, plant formations; annotated list of vascular plants and bryophytes of northern Sodankylä; list of Musci (bladmossor) of Enare (Inari) and Utsjoki parishes.

Hustich, Ilmari. Floran kring en timmerkoja i Kittilä lappmark. (Botaniska notiser från västra Lappland. 2.) Memo. Soc. Fauna et Flora Fenn. **11:** 162–165. 1936.

Discussion of the flora of a timber camp (Ruottamaa) with lists of species of various habitats.

——— Pallas-Ounastunturin kansallispuisto. [The Pallas-Ounastunturi national park.] Metsätiet. Tutk. Luonnons.-Aluekuv. 1. 60 p. 18 cm. Helsinki, 1938.

Topography, geology, climate, general features of flora; list of vascular plants of the alpine area (180 species). (Not seen; title and annotation supplied by H. Luther.)—See also his Pallas-Ounastunturi nationalpark. Forstl. Forskningsanst. Beskr. Naturskyddsomr. 1. 58 p. 18 cm. Helsingfors, 1939. (The same in Swedish, with slight additions. Not seen; title and annotation supplied by H. Luther.)

——— Pflanzengeographische studien im gebiet der niederen fjelde im westlichen finnischen Lappland. I, II [subtitles omitted]. Acta Bot. Fenn. 19. 156 p. incl. 3 fig. (maps), 6 pl. on 3. 1937; 27. 80 p. 4 fig. (incl. map). 1940.

Phytogeographical and ecological, with special reference to altitudinal range; includes (**27:** 41–62) annotated list of vascular plants from the alpine region, with detailed localities; also (p. 62–73) list of bryophytes from Pallastunturi and Ounastunturi. Covers the mountain areas of the Ylläs-Ounas fjeldgebiet, from 67° 30′ to 68° 25′ N. Lat., and from the Swedish boundary to 25° E. Long.

Linkola, Kaarlo. Lapin tulokaskasvistosta kesällä 1925. [On the adventive flora of Lapland in the summer of 1925.] Luonnon Ystävä **33:** 199–210. 1929.

Author's observations on dispersal of weeds in Lapland; annotated list of adventive plants found by author in Lapponia enontekiensis, Lapponia inarensis, Lapponia kemensis, and Lapponia petsamoensis.

Montell, Justus. Bidrag till kännedomen om adventiv- och ogräsfloran i Muonio socken, Lapponia kemensis. Memo. Soc. Fauna et Flora Fenn. **20:** 82–91. 1945.

Extensive list of adventive plants and weeds, in great part introduced in imports of hay and other horse fodder during the Second World War.

Wainio, E. A. Notes sur la flore de la Laponie finlandaise. Acta Soc. Fauna et Flora Fenn. v. 8, no. 4. 90 p. 1891.

Author's explorations in 1878, with numerous short lists of plants from different localities and habitats; list of known vascular plants, with localities. Relates to parishes of Kuolajärvi (now called Salla in Kuusamo), Kemijärvi (Ostrobottnia borealis), Sodankylä (Lapponia kemensis), and Inari (Lapponia inarensis), extending from about 66° 30′ to 69° 40′ N. Lat., with some slight overlap into Russian territory.

LAPPONIA PETSAMOENSIS (Lps). (PETSAMON LAPPI. PsL.)

See also Lapponia inarensis (Fellman); Lapponia kemensis (Linkola; Wainio); Lapponia rossica; Ostrobottnia kajanensis (Parvela).

Lapponia petsamoensis is an area (Finnish from 1920 to 1940, previously and subsequently Russian) extending northeast from about 68°5′ N. Lat. to the Arctic Ocean at 69°58′ N. Lat., bordered on the east by Lapponia tulomensis, on the south by Lapponia kemensis, on the southwest by Lapponia inarensis, on the northwest by Norway.

General

Parvela, A. A. Petsamon alueen viljelyskasviston historia ja levinneisyys. (Geschichte und verbreitung der kulturpflanzen in Petsamo.) Acta Agralia Fenn. **23:** 33–95. 1 fig. (map). 1931.

Method of investigations, topography, climate, human population; historical and distributional treatment of the various groups of useful and ornamental cultivated plants; bibliography.

Local

Kalliola, Reino. Alpiinisesta kasvillisuudesta Kammikivialueella Petsamon Lapissa. Kasvillisuusmonografia. (Über die alpine vegetation im Kammikivi-gebiet von Petsamo-Lappland. Eine kleine vegetationsmonographie.) Annales Bot. Soc. Zool.-Bot. Fenn. Vanamo v. 2, no. 2. vi, 121 p. 14 fig. (incl. 2 maps). 1932.

Description of plant associations; includes (p. 98–108) annotated list of vascular plants, also (p. 108–110) unannotated list of bryophytes and lichens; bibliography. Relates to an area of about 28 sq. km., about 40 km. south of the Arctic coast, 69° 25′ N. Lat., 30° 20′ E. Long.

Roivainen, Heikki. Tietoja kasvillisuudesta sekä putkilo- ja lehtisammalkasvistosta keskisen Luttojoen seuduilla. (Beobachtungen über die vegetation und die gefässpflanzen- und laubmoosflora in der gegend des flusses Luttojoki in Finnisch-Lappland.) Annales Soc. Zool.-Bot. Fenn. Vanamo **1:** 229–304. 10 fig. 1923.

A vegetational study; includes (p. 276–300) annotated list of vascular plants (236 species) and mosses (104 species).

Söyrinki, Niilo. Studien über die generative und vegetative vermehrung der samenpflanzen in der alpinen vegetation Petsamo-Lapplands. I. Allgemeiner teil. (Tutkimuksia siemenkasvien siemenellisestä ja kasvullisesta lisääntymisestä Petsamon Lapin alpiinisessa kasvillisuudessa. I. Yleinen osa.) Annales Bot. Soc. Zool.-Bot. Fenn. Vanamo v. 11, no. 1. ix, 323 p. 20 fig. (fig. 1, map). 1938.—II. Spezieller teil. l. c. v. 14, no. 1. x, 405 p. map. 1939.

A detailed study of seed and vegetative reproduction; includes (p. 61–68) annotated tabular list of spermatophytes of the Petsamofjelde, about 69° 20′ N. Lat., 30° 30′ E. Long. The second part gives details of phenology, etc., for each species, including habitat and localities.—See also his Beobachtungen über die gefässkryptogamenflora der Petsamofjelde (Lapponia petsamoënsis). l. c. **11** (3): 26–35. 1939. (Annotated list.)

Valle, K. J. Kasvitietoja Petsamosta. [Floristic notes from Petsamo.] Memo. Soc. Fauna et Flora Fenn. **8:** 259–274. 1933.

Extensive list of vascular plants collected by author and others in 1928–1930, with localities.

LAPPONIA ROSSICA (Lr).

See also Lapponia petsamoensis, as well as some of the general works under Finland.

The term Lapponia rossica is used here to cover the phytogeographical provinces of Lapponia Imandrae (Lim), Lapponia murmanica (Lmur or Lm), Lapponia ponojensis (Lp), Lapponia tulomensis (Lt), and Lapponia Varsugae (LV), which in the newer nomenclature have been called Imanteron Lappi (ImL), Muurmannin Lappi (ML or MrL), Ponoin Lappi (PL or PoL), Tuuloman Lappi (TL), and Varsugan Lappi (VL), respectively. With Lapponia petsamoensis, which is here treated separately, they constitute the Kola Peninsula and adjacent Russian Lapland. As they are Russian territory, although regarded phytogeographically as Fennoscandinavian and covered in some of the Finnish publications, only the four following works of general scope are given here. The most complete and detailed treatment of the flora as a whole will be found in Hjelt's Conspectus under the heading Lapponia rossica; ranges are also given by provinces in Hiitonen's Suomen kasvio.

General

Akademiîa Nauk SSSR. Kol'skiĭ Filial Imeni S. M. Kirova. Flora Murmanskoĭ oblasti. v. 1–3. illus. (incl. distr. maps), front. (port.). 25.5 cm. Moskva, Leningrad, 1953–1956.

Keyed, descriptive flora of vascular plants (in Russian) with abundant illustrations, brief notes on uses, and distribution maps for most of the species; B. N. Gorodkov and others, eds. Covers the Kola Peninsula and adjacent Russian Lappland (including Lapponia petsamoensis) and the northernmost part of Karelia keretina south to the Arctic Circle. Vol. 3 goes through Cruciferae (Engler-Prantl system).

Fellman, N. I. Plantae vasculares in Lapponia orientali sponte nascentes. Notiser Sällsk. Fauna et Flora Fenn. Förhandl. 8 (n.s., 5): i–lxx, 1–99. 1882. (Issued separately 1869.)

Botanical investigations (with bibliographical footnotes), physiography, geology, phytogeography, plant zones, statistics, list of excluded species; annotated list of vascular plants (517 species). The introduction is in Swedish, the list in Latin. It covers the 6 provinces of Russian Lapland (Lapponia Imandrae, Lapponia murmanica, Lapponia petsamoensis, Lapponia ponojensis, Lapponia tulomensis, and Lapponia Varsugae), and is still regarded as the most complete list available for Kola-Lapland.

Regel, Constantin. Die vegetationsverhältnisse der halbinsel Kola. (Die pflanzendecke der halbinsel Kola. teil IV.) Beih. Repert. Spec. Nov. Fedde v. 82 (5 lief.). 5–624 p. (unfinished). 31 [i. e. 32] maps. 1935–41.

Phytogeographical and ecological; includes sketch of botanical and other explorations with bibliography (p. 7–41), and other bibliographies, the more important being on p. 505–517 and p. 561–565. On p. 100–105 is given a list of vascular plants (626 species) divided phytogeographically (Die florenelemente). The work relates to the Kola Peninsula proper and adjacent Lapland west to the historical border (including Lapponia Imandrae, Lapponia murmanica, Lapponia petsamoensis, Lapponia ponojensis, Lapponia tulomensis, and Lapponia Varsugae). The earlier parts of the work are ecological, with general descriptions of the vegetation of larger or smaller areas and short lists of species seen at 1235 points where associations were observed; bibliographies. Their general title is Kolos pusiausalio augmenine danga . . . [I]-III dalis. Die pflanzendecke der halbinsel Kola . . . [I.]-III. teil. Lietuvos Univ. Mat. Gamtos Fak. Darbai (Mém. Fac. Sci. Univ. Lithuanie) v. [1] (1922). xxiv, 246 p. incl. 12 fig. on 6 pl. (on 3). 1923 (Lapponia Varsugae); 3 (1924–26): 135–357 incl. fig. 6–7, 3 maps. 1926 (Lapponia ponojensis); 4 (1927–28): 22–206. map. 1928. (Lapponia tulomensis and Lapponia murmanica).

Shlîakov, R. N. Vazhneĭshie dikorastushchie poleznye rasteniîa Murmanskoĭ oblasti. [Most important wild economic plants of Murmansk Region.] 114 p. 30 fig. 20 cm. Moskva, 1956. (At head of title page: Kol'skiĭ filial im. S. M. Kirova A. N. SSSR . . .)

Account of useful plants, grouped by uses, with brief description, habitat, and uses. Covers same area as Akademiîa Nauk SSSR (above).

NYLANDIA (N). (UUSIMAA. U)

See also Savonia australis (Ulvinen).

Nylandia is a region on the southern coast of Finland corresponding in a general way to Uudenmaan lääni and extending from about 22°50′ to about 25°50′ E. Long., north to nearly 61° N. Lat., bounded on the north by Tavastia australis, on the east by Karelia australis, on the south by the Gulf of Finland, on the west by Regio aboënsis.

Local

Brenner, Widar. Växtgeografiska studier i Barösunds skärgård. I. Allmän del och floran. Acta Soc. Fauna et Flora Fenn. v. 49, no. 5. 2 p. l., 151 p. 12 fig. (incl. map), map. 1921.
Topography, geology, climate, ecology, etc.; annotated list of vascular plants (511 species); phytogeography, bibliography. Relates to the Västerskär or Barösunds skärgård, lying west of the open Barösundsfjärd, south of Ingå (Inkoo).

Cedercreutz, Carl. Studien über laubwiesen in den kirchspielen Kyrkslätt und Esbo in Südfinnland mit besonderer berücksichtigung der verbreitung und einwanderung der laubwiesenarten. Acta Bot. Fenn. v. 3. 181 p. incl. illus. (60 [i.e. 61] fig. [maps]). 5 pl. on 3, 4 maps. 1927.
Mainly ecological and phytogeographical; includes (p. 102–138) annotated list with localities of vascular plants (237 species) characteristic of the "laubwiesen," here taken as those formations "die eine Holzvegetation von verschieden Laubbäumen und Sträuchern und eine reiche Gräser- und Kräutervegetation haben;" extensive bibliography.—See also his Vergleichende studien über die laubwiesen im westlichen und östlichen Nyland. Acta Bot. Fenn. v. 10. 63 p. incl. 24 fig. (maps). 1931. (Includes (p. 30–41) list of scarcer vascular plants with localities from "laubwiesen" in eastern Nyland in the area from Orimattila to Elimäki, south to Pyttis (Pyhtää) and Pernå (Pernaja); also (p. 42–49) tabular list of vascular plants (234 species) from much the same area, showing occurrence in special localities; bibliography.)—Also **Gadolin, A. W.** Några kompletterande uppgifter om lövängsvegetationen i Kyrkslätt och Esbo socknar. Memo. Soc. Fauna et Flora Fenn 4: 90–91. 1928. (List with localities.)—Also **Cedercreutz, Carl.** Bidrag till västra Nylands flora. l.c. 4: 172–176. 1928. (List of noteworthy species from Esbo, Kyrkslätt, Sjundeå, and Vihti.)—Also **Brander, Torild.** Om kärlväxtfloran i mellersta Kyrkslätt (N). l.c. 28: 13–24. 1953. (Botanical explorations, etc.; list of noteworthy species; plant distribution, bibliography.)—Also Cedercreutz and Färdig, below.

——— Växtgeografiska anteckningar från Porkala-området. Memo. Soc. Fauna et Flora Fenn. **21**: 170–180. 1 fig. (map). 1945.
General features of vegetation; tabular list of spermatophytes of Prästgårdsholmen, Ekö, and Hooxholmen, three islets in the inmost part of the Tavastfjärd; list of noteworthy vascular plants of the Porkala region (Degerby, Kyrkslätt (Kirkkonummi), and Sjundeå (Siuntio) parishes); bibliography.

Färdig, B. A. Botaniska anteckningar från Nyland: Kyrkslätt, Vitträskholmen. Memo. Soc. Fauna et Flora Fenn. **18**: 115–125. 1 fig. (map). 1943.
Topography, general features of flora; account of plant formations of Vitträskholmen, an island in the Vitträsk in Kyrkslätt (Kirkkonummi) parish, with lists of their vascular plants (total 289 species).

Fagerström, Lars. Floristiskt från Hangö-området. Memo. Soc. Fauna et Flora Fenn. **19**: 36–42. 1944.
Includes two lists of ruderal spermatophytes (118 and 106 species) from Täcktom and vicinity.—See also **Karling, T. G.** Några fynd av adventivväxter i Hangö hamn. Memo. Soc. Fauna et Flora Fenn **19**: 42–43. 1944. (Records of several noteworthy species.)

——— Växtgeografiska studier i Strömfors-Pyttis skärgård i östra Nyland. Med speciellt beaktande av lövängarna, artantalet samt en del arters fördelning och invandring. Acta Bot. Fenn. 54. 296 p. 48 fig. (maps), map. 1954.
Mainly ecological and phytogeographic; includes (p. 40–49) unannotated list of vascular plants (503 species) with remarks on some; bibliography.—See

also his Den botaniska utforskningen av Strömfors-Pyttis skärgård i östra Nyland. Acta Soc. Fauna et Flora Fenn. v. 71, no. 2. 20 p. 1954. (Botanical explorations, bibliography.)

Granö, Olavi. Die flora in ihrer beziehung zur kultur im schärenhof von Porvoo in Südfinnland. Annales Bot. Soc. Zool.-Bot. Fenn. Vanamo v. 25, no. 4. 2 p. l., 47 p. 5 fig. (maps). 1953.
Includes (p. 17–25) annotated list of plants (120 species) characteristic of cultivated habitats, and (p. 40–44) longer tabular list including additional species occurring on such habitats but not confined to them; bibliography.

Häyrén, Ernst. Adventiv-växter från Lappvik i Tenala socken. Meddel. Soc. Fauna et Flora Fenn. 35: 155–163. 1909.
Annotated list of 49 ballast plants and of 6 other adventive plants.—See also **Hidén (later Hiitonen), Ilmari.** Lisätietoja Lappvikin satama-alueen kasvistosta. [Additional notes on the flora of the harbor area of Lappvik.] Meddel. Soc. Fauna et Flora Fenn. 48: 157–162. 1925. (Includes 24 additional species.)—Also **Luther, Hans.** Anteckningar om adventivfloran i Lappvik åren 1925–1939. (Från Tvärminne Zoologiska station.) Memo. Soc. Fauna et Flora Fenn. **16**: 9–15. 1940. (Annotated list of 45 species observed by author, including 14 not before recorded; comparison with earlier lists, etc.)

——— Studier öfver vegetationen på tillandningsområdena i Ekenäs skärgård. Acta Soc. Fauna et Flora Fenn. v. 23, no. 6. 176 p. 4 [i. e. 2] maps. 1902.
An ecological study of the vegetation on newly formed land areas in Ekenäs (Tammisaari) archipelago; includes (p. 161–171) annotated list of the most important algae, mosses, and vascular plants.—See also **Luther, Hans.** Verbreitung und ökologie der höheren wasserpflanzen im brackwasser der Ekenäsgegend in Südfinnland. 2 teile. Acta Bot. Fenn. v. 49. 231 p. incl. 6 fig. (incl. karte 1, map), 4 pl. (on 2), maps 2–100 (on 48). 1951; v. 50. 370 p. 1 fig. (map). 1951. (Teil II contains detailed account of distribution of vascular and cellular water plants, and bibliography.)

——— Über die landvegetation und flora der meeresfelsen von Tvärminne. Ein beitrag zur erforschung der bedeutung des meeres für die landpflanzen. Acta Soc. Fauna et Flora Fenn. v. 39, no. 1. vi, 193 p. 10 fig., 15 pl., map, 3 tab. 1914.
An ecological study; includes (3 folding tables at end) tabular lists of vascular plants (116 species), bryophytes (58 species), and lichens (138 species), showing occurrence on different sea cliffs and skerries in the area, which is in Ekenäs parish.

Hisinger, E. V. E. Flora fagervikiensis eller öfversigt af de vid och omkring Fagervik vexande cotyledoneer och filices. Notiser Sällsk. Fauna et Flora Fenn. Förhandl. **3**: 1–60. 1857. (Issued separately 1855.)
Topography, etc.; briefly annotated list of vascular plants (558 species) from the region of Fagervik mill in the Western part of Ingå (Inkoo) socken, and notes from other parts of westernmost Nylandia. (Fagervik, Finnish Kaunislahti.)—See also **Brenner, Magnus.** Nya växtfynd från västra Nyland, hufvudsakligen från Ingå. Meddel. Soc. Fauna et Flora Fenn. 34: 76–81. 1908. (Extensive list of vascular plants with localities.)

Hustich, Ilmari. Porkkalan kansallispuisto. [Porkkala National Park.] Metsät. Tutk. Luonnons.-aluekur. 3. 48 p. 18 cm. 1939.
A guide to the park; topography, geology, climate, phenology, general features of flora, list of vascular plants (212 species) of the park island Träskö in Porkkala. (Not seen; title and annotation supplied by H. Luther.)—See also his Porkala nationalpark. Forstl. Forskningsanst. Populärvetensk. Beskr. Naturskyddsomr. 3. 48 p. 18 cm. 1939. (Same, in Swedish. Not seen; title and annotation supplied by H. Luther.)

Lemberg, Bertel. En gammal barlastplats. Memo. Soc. Fauna et Flora Fenn. **23**: 94–104. 1947.
Includes list of ballast plants from Lassdal on the western shore of Stor-Pernåvik, with discussion.

Nylander, William. Conspectus florae helsingforsiensis. Notiser Sällsk. Fauna et Flora Fenn. Förhandl. **2**: 9–70. 1852.

Topography, geology, general features of vegetation, climate, phenology; briefly annotated list of vascular plants and Characeae (580 species, of which 574 are vascular), bryophytes, lichens, and a few Hymenomycetes.—See also his Additamentum ad Conspectum . . . Notiser Sällsk. Fauna et Flora Fenn. Förhandl. 2: 203-224. 1852. (Brings total to 584 vascular plants.)—Also **Saelán, Thiodolf.** Några sällsynta växter observerade under en exkursion till Sörnäs lastageplats vid Helsingfors i början af Oktober 1878. Meddel. Soc. Fauna et Flora Fenn. 5: 251–252. 1878. (Annotated list of about 20 weedy plants.)—Also **Brenner, Magnus.** Inom Helsingfors stads område förstörda växtlokaler. l.c. **31**: 12–14. 1906. (Changes in the flora of the Brunnspark in the city.)—Also his Förändringar i Helsingfors stads flora. l.c. **31**: 117–135. 1906. (Changes in environment and flora, especially since 1852.)—Also his Naturskövlingen på Sandviksholmarna vid Helsingfors och dess inverkan på vegetationen. l.c. **47**: 23–30. 1921. (Further changes in vegetation due to destruction of natural habitats.)—Also **Ulvinen, Arvi.** Kasvistollisia tutkimuksia Santahaminassa. [Floristic studies in Santahamina.] l.c. **47**: 40–42. 1921. (Relates mainly to weeds from the Russian army camp on the island.)—Also **Hidén** (later **Hiitonen**), **Ilmari.** Tietoja Suomenlinnan kasvistosta. [Notes on the flora of Suomenlinna (Sveaborg).] l.c. **47**: 67–71. 1921. (Includes mention of numerous species, especially weeds, from the military area Suomenlinna.)—Also **Linkola, Kaarlo.** Zur kenntnis der überwinterung der unkräuter und ruderalpflanzen in der gegend von Helsingfors. (Havaintoja rikkaruohojen ja ruderatikasvien talvehtimisesta Helsingin seudulla.) Annales Soc. Zool.-Bot. Fenn. Vanamo **1**: 91–228. 25 fig. 1922. (Detailed study of the methods of overwintering of weeds and ruderal plants, including (p. 199–221) annotated list of 133 species; bibliography.)—Also **Puolanne, Mielo.** Kasvilöytöjä Helsingin alueelta. [Plant finds in the Helsinki region.]. Memo. Soc. Fauna et Flora Fenn. **1**: 93–94. 1927. (Records of 7 species.)—Also **Pettersson, Bror.** Ferns and flowering plants on erratic blocks with special reference to their modes of dispersal. l.c. **6**: 25–47. 8 fig. 1930. (Discussion of 43 vascular plants found growing on granitic or gneissic erratic blocks in Kallvik (Helsinge [Helsinki] parish), with list of the flora of the neighborhood; bibliography of works on European epiphytes and plants growing on buildings, erratic blocks, etc.)—Also **Puolanne, M.** Helsingin ja sen lähiseudun kasvisto. 1. Taraxacumlajit. (Die flora von Helsinki nebst umgebung. 1. Die Taraxacum-arten.) l.c. **8**: 136–181. 1933. (Descriptive treatment of the 93 species found in the Helsinki region, with local and general range.)—Also **Pettersson, Bror.** A case of long distance dispersal of plants through the import of timber. Acta Phytogeog. Suecica **13**: 96–100. 1940. (List of 163 species found in a sawmill yard at Drumsö, brought in on timber from Poland, with discussion.)—Also **Erkamo, Viljo.** Über die ursprüngliche vegetation im stadtgebiet von Helsinki und ihr verschwinden. (Vorläufige mitteilung.) Annales Bot. Soc. Zool.-Bot. Fenn. Vanamo **19** (Notul. Bot. 14): 27–33. 1 fig. (map). 1944. (Discussion of changes due to cultivation and of the residue of former forest; bibliography.)—Also his Beobachtungen über die mit deutschen truppen im jahre 1944 nach Helsinki eingeschleppten pflanzenarten. l.c. **21** (Notul. Bot. 16): 7–11. 1946. (List of 68 species, with discussion.)—Also his Helsingin seudun kasviston tutkimisesta. (On the investigation concerning the flora of the district of Helsinki.) Arch. Soc. Zool. Bot. Fenn. Vanamo **1**: 123–129. 1 fig. (map). 1948. (Sketch of botanical explorations in vicinity of Helsinki; bibliography.)—Also **Pettersson, Bror.** An alien flora on Drumsö (Helsingfors) introduced by cork bark imported from Morocco and Spain. Memo. Soc. Fauna et Flora Fenn. **27**: 111–117. 1952. (List of 206 species of angiosperms observed in 1940 in a factory yard on Drumsö, Helsingfors, with discussion and bibliography.)

Saelán, Thiodolf. Öfversigt af de i östra Nyland vexande kotyledoneer och ormbunkar. Notiser Sällsk. Fauna et Flora Fenn. Förhandl. **4** (n.s., **1**): 9–77. 1858. (Also as Bidrag Finlands Naturkänn., Etnog. och Statis. **3**: 9–77. 1859.)

Botanical explorations, topography, geology, general features of flora; annotated list of vascular plants and Characeae (648 species, of which 642 are vascular). Covers in general the region from Orimattila, Elimä

(Elimäki), and Sippola south to the coast from Borgå (Porvoo) to Fredrikshamn (Hamina), thus including most of eastern Nylandia and small areas in adjacent Karelia australis and Tavastia australis.—See also his [Beriktiganden och tillägg.]. Notiser Sällsk. Fauna et Flora Fenn. Förhandl. 4 (n.s., 1): 249–253. 1859. (Also as Bidrag Finlands Naturkänn. Etnog. och Statis. 3: 249–253. 1859.) (Corrections and additions.)—Also **Lemberg, Bertel.** Märkligare växtfynd från östra Nyland. Memo. Soc. Fauna et Flora Fenn. **13**: 9–19. 1 fig. 1937. (Records of noteworthy species from Pernå and Stor-Pernåvik, with mention of associated species.)—Also his Tvenne växtfynd från östra Nyland. l.c. **14**: 5–8. 1938. (Notes on various species.) —Also his Märkligare växtfynd från östra Nyland. l.c. **23**: 105–110. 1947. (Further notes.)

Stenroos (later Kivirikko), K. E. Nurmijärven pitäjän siemen- ja saniaiskasvisto. [Seedplant and fern flora of Nurmijärvi parish.] Acta Soc. Fauna et Flora Fenn. v. 9, no. 11. 85 p. 1894.

Topography, general features of flora, plant associations, statistics, etc.; annotated list of vascular plants (486 species).—See also **Aario, Leo.** Vegetation und postglaziale geschichte des Nurmijärvi-sees. Annales Bot. Soc. Zool.-Bot. Fenn. Vanamo v. 3, no. 2. vi, 132 p. 19 fig. (fig. 1, map), map. 1933. (Includes (p. 15–30) annotated list of aquatic vascular plants (119 species) and some mosses and algae; bibliography.)

Tynni, Tellervo. Pukkilan, Askolan ja Porvoon pitäjän pohjoisosan kasvisto. (Die flora der kirchspiele Pukkila und Askola sowie des nördlichen teiles des kirchspiele Porvoo in Südfinnland.) Annales Bot. Soc. Zool.-Bot. Fenn. Vanamo v. 9, no. 6. iv, 102 p. 15 fig. (incl. maps). 1937.

Topography, geology, soils, botanical regions, plant associations, weeds; annotated list of vascular plants (585 species); bibliography.

OSTROBOTTNIA AUSTRALIS (Oa). (ETELÄ-POHJANMAA. EP)

See also Satakunta (Malmgren).

Ostrobottnia australis is a region in west-central Finland, from about 62° to about 63°20′ N. Lat., extending east from the Gulf of Bothnia to about 24° E. Long., bounded on the northeast by Ostrobottnia media, on the southeast by Tavastia borealis, on the southwest by Satakunta.

Local

Ehnholm, Gunnar. Bidrag till kännedomen om kärlväxtfloran i norra Kvarken. Memo. Soc. Fauna et Flora Fenn. **22**: 19–42. 1 fig. (map). 1946.

Geography, climate, topography, origin of flora, plant associations, plant zones, lists of species of several of the islands; annotated list of vascular plants (about 260 species) of the Michelsöarna, a group of small islands between 63° 22′ and 63° 29′ N. Lat., 21° 34′ and 21° 52′ E. Long.; bibliography.

Laurén, Walter. Barlastväxter observerade i Vasa åren 1882–1892. Meddel. Soc. Fauna et Flora Fenn. **22**: 36–44. 1896.

Annotated lists of ballast plants from several localities in Vasa.—See also **Jurvelius, Björn.** Bidrag till kännedomen om adventivfloran i Vasa. Arkiv Svenska Österbotten **2**: 102–112. 1929. (Annotated list of weeds, including comparison with the earlier list. In University of Chicago library.)—Also **Bäck, Ragnar, and Malmberg, Martin.** Adventivväxter i Vasa. Nordenskiöld-samf. Tidskr. **11**: 49–57. 1952. (Mostly adventive plants brought in with horse fodder by the Germans, 1941–44. Not seen; title and annotation supplied by H. Luther.)—Also **Malmberg, M., and Bäck, R.** Nya växtfynd i Vasa. Ett förberedande meddelande beträffande tyska polemokorer och andra adventivväxter. Arkiv Svenska Österbotten **10**: 237–250. 1952. (Includes material not in their other paper. Not seen; title and annotation supplied by H. Luther.)

——— Växtförhållandena i gränstrakterna mellan mellersta och södra Österbotten. Acta Soc. Fauna et Flora Fenn. v. 13, no. 2. 44 p. 1896.

Botanical explorations, topography, soils, general features of flora, statistics; briefly annotated list of vascular plants (499 species, including 30 Hieracia), with localities for scarcer species. Relates mainly to the area between the Lappo and Kyro valleys, from Alavo (Alavus), Kuortane, Nurmo, and Ylistaro north to Nykarleby, as well as the islands south to Bergö.—See also **Widlund, Uno.** Några anmärkningsvärda växter i Gamla Vasa och dess närmaste omgivningar. Arkiv Svenska Österbotten 4: 173–180. 7 fig. 1944. (List of noteworthy vascular plants, with localities. In University of Chicago library.)

Railonsala, Artturi. Kristiinan ja sen ympäristön kasvisto. [Flora of Kristiina and vicinity.] Acta Soc. Fauna et Flora Fenn. v. 58, no. 2. 27 p. 1936.

Topography, general features of flora, botanical explorations; annotated list of vascular plants (607 species, including 33 Taraxaca and 32 Hieracia).—See also his Kriistinan ja sen ympäristön kasvisto. Lisätietoja. [Flora . . . Additional notes.] Luonnon Ystävä 43: 220–222. 1939. (Additions bringing total to 642 species including 39 species of *Taraxacum*, 32 of *Hieracium*, as well as 33 hybrids.)—Also his Kristiinan ja sen ympäristön Taraxacumlajisto. (Die Taraxacum-flora der stadt Kristiina nebst umgebung an der küste des Bottnischen meerbusens.) Annales Bot. Soc. Zool.-Bot. Fenn. Vanamo v. 23, no. 3. 1 p. l., 35 p. 12 fig. (incl. 10 distribution maps). 1949. (Annotated list of 58 species of *Taraxacum.*)—Also **Valovirta, E. J.** Keskieurooppalaista kasvilajistoa Kristiinankaupungin satamassa. (Mitteleuropäisches in der hafenflora der stadt Kristiinankaupunki (Oa). Arch. Soc. Zool. Bot. Fenn. Vanamo 4: 53–60. 1949. (Annotated list of 200 species of adventive plants introduced in fodder, 1941–43, and collected in 1942–48, with discussion; list of additional introductions found in Vaasa and Kaskinen.)

——— Lappfjärdin pitäjän putkilokasvisto. [Vascular flora of Lappfjärd parish.] Memo. Soc. Fauna et Flora Fenn. 7: 92–119. 1931.

Topography, geology, etc.; annotated list of vascular plants (400 species), with localities.—See also his Lappfjärdin Taraxacumlajit. (Die Taraxacumarten von Lappfjärd im westlichen Mittelfinnland.) Annales Bot. Soc. Zool.-Bot. Fenn. Vanamo v. 16, no. 5. 1 p. l., 17 p. 2 fig. 1942. (Annotated list of 46 species.)—Also his Lappfjärdin pitäjän putkilokasvisto.—Uusia löytöjä, täydennkysiä ja oikaisuja. [. . . New discoveries, additions and corrections.] Memo. Soc. Fauna et Flora Fenn. 32: 122–140. 3 fig. (incl. map). 1957. (General features of flora, etc.; extensive list of vascular plants; bibliography.)

——— Niinimaan seudun putkilokasvisto. [Vascular flora of the Niinimaa region.] Memo. Soc. Fauna et Flora Fenn. 8: 199–222. 1932.

Botanical explorations, general features of flora; annotated list of vascular plants (440 species, including 14 species of *Taraxacum* and 39 of *Hieracium*).—See also his Lisätietoja Niinimaan seudun putkilokasvistosta. [Further notes on the vascular flora of the Niinimaa region.] Memo. Soc. Fauna et Flora Fenn. 13: 89–92. 1937. (Additions bringing total to 491 species, including 31 species of *Taraxacum* and 41 of *Hieracium*).

Sarvela, Jaakko. Kasvistollisia tietoja Ilmajoelta ja Kurikasta. [Floristic notes from Ilmajoki and Kurikka.] Memo. Soc. Fauna et Flora Fenn. 12: 65–74. 1936.

Topography, etc.; annotated list of vascular plants (372 species; *Taraxacum* and *Hieracium* are not listed.).

Valovirta, E. J. Über die strandhöhenzonen als hilfsmittel bei der erforschung der schärenflora im landhebungsgebiet. Acta Bot. Fenn. 47. 24 p. 6 fig. (fig. 1, map). 1950.

Topography, etc.; tabular list of vascular plants (146 species) of seven islands (Skötgrund, Långradden, Murgrund Bastuskär, and 4 smaller ones) off the coast near Kristiina, showing occurrence on each island; comparison with Rönnskär archipelago, etc.

——— Untersuchungen über die säkulare landhebung als pflanzengeographischer faktor. Acta Bot. Fenn. 20. 173 p. incl. 61 fig. (incl. maps; fig. 41–61 on 13 pl. [on 7]), tab., map. 1937.

A monographic study of the vegetation of the Rönnskär, Norrskär, and Holmö archipelagos between Vasa and the Swedish coast in relation to the emergence of the islands from the sea; includes annotated lists of vascular plants of Rönnskär archipelago (219 species, p. 47–52) and of the principal islands of the Norrskär archipelago (121 species, p. 77–79).—Also his Muutamia yksityistapauksia kasvillisuuden kehittymisestä maankohoamisalueella. (Einige beispiele der entwicklung der vegetation in einem landhebungsgebiet.) Acta Soc. Fauna et Flora Fenn. v. 72, no. 22. 6 p. 2 fig. (incl. map) 1955. (Account of the vegetation of 3 tiny islets of postglacial origin in the outer part of Rönnskär archipelago, with lists of species (8, 22, and 37 vascular plants respectively).)

OSTROBOTTNIA BOREALIS (Ob). (POHJOIS-POHJANMAA. PP)

See also Lapponia enontekiensis (Kallio, under Lindén); Lapponia inarensis (Fellman); Lapponia kemensis (Hjelt and Hult); Ostrobottnia kajanensis (Parvela; Jalas).

Ostrobottnia borealis is a region in northern Finland between about 64°25′ and 67°10′ N. Lat., bordered on the north by Lapponia kemensis, on the east by Regio kuusamoensis and Ostrobottnia kajanensis, on the south by Ostrobottnia media, on the west by the Gulf of Bothnia and Sweden.

General

Brenner, Magnus. Observationer rörande den nordfinska floran under adertonde och nittonde seklen, särskildt med afseende å kärlväxterna i Öster-Norrbotten, Nord-Österbotten och Kajanien. Acta Soc. Fauna et Flora Fenn. v. 16, no. 4. 307 p. map. 1899.

Chronological account of botanical explorations with lists of plants recorded, topography, geography, general features of flora and plant associations, phytogeography; list of known vascular plants, with detailed localities; tabular list of species showing occurrence in the different parishes; phenology. Covers Ostrobottnia borealis and Ostrobottnia kajanensis and northern part of Ostrobottnia media.—See also his Nya bidrag till den nordfinska floran. Acta Soc. Fauna et Flora Fenn. v. 34, no. 4. 24 p. 1911. (Extensive list of new records and localities, mostly from Ostrobottnia kajanensis.)—Also **Auer, A. V.** Kasvistollisia havaintoja Pohjois-Suomesta. [I]—II. [Floristic observations from northern Finland.] Memo. Soc. Fauna et Flora Fenn. **12:** 131–142. 1937; **14:** 110–123. 1938. (Extensive lists of noteworthy plants, including many records from Ranua, Rovaniemi, Kemijärvi, Alatornio, and Tervola.)—Also **Kotilainen, M. J.** Über die verbreitung der meso-eutrophen moorpflanzen in Nordfinnland. Annales Acad. Scient. Fenn. Ser. A, IV (Biologica), v. 17. 1 p. l., 162 p. incl. 26 fig. (maps), map. 1951. (Includes (p. 73–84) list of 139 vascular moor-plants with detailed localities in Ostrobottnia borealis, Kuusamo, and Lapponia kemensis, followed (p. 84–149) by detailed discussion of the range of each species.)

Local

Cajander, A. K. Die alluvionen der Tornio- und Kemi-thäler. (*His* Beiträge zur kenntniss der vegetation der alluvionen des nördlichen Eurasiens. III.) Acta Soc. Scient. Fenn. v. 37, no. 5. 3 p. l., 223 p. 4 maps. 1909.

A study of plant formations; includes (p. 216–218) unannotated list of vascular plants mentioned from the Tornio and Kemi valleys in Ostrobottnia borealis and Lapponia kemensis.

Huumonen, M. E. Oulun painolastikasvullisuus vv. 1899–1909. [Ballast vegetation of Oulu (Uleåborg) 1899–1909.] Meddel. Soc. Fauna et Flora Fenn. **36:** 84–88. 1910.

List of ballast plants, with dates and some annotations. This and the following papers constitute the longest continuous series of observations on weeds in Finland.—See also his Oulun ympäristön satunnaiset kasvit vv. 1899–1909. [Adventive plants of the vicinity of Oulu 1899–1909.] Meddel. Soc. Fauna et Flora Fenn. **36:** 88–91. 1910. (List with dates and some annotations, consisting mostly of species not in his earlier list.)—Also his Muutamia lisäyksiä ja oikaisuja Oulun seudun putkilokasvistoon. [Some additions and corrections to the vascular flora of the Oulu region.] l.c. **38:** 83–87. 1912. (Includes, p. 85–86, list of additional records.)—Also his Oulun ja sen lähimmän ympäristön painolasti- ja satunnaiskasvisto vv. 1910–12. [Ballast- and adventive-flora of Oulu and its immediate vicinity 1910–12.] l.c. **39:** 175–180. 1913.—Also his Oulun painolastikasvisto vv. 1913–16. [Ballast flora of Oulu 1913–16.] l.c. **43:** 116–118. 1917.—Also **Metsävainio, Kaarlo.** Oulun satunnaiskasvisto vuosina 1917–1925. [Adventive flora of Oulu 1917–1925.] Luonnon Ystävä **30:** 80–86. 1926. (List of harbor and town weeds.)—Also his Satunnaiskasveja Oulusta v. 1926. [Adventive plants from Oulu 1926.] l.c. **31:** 69. 1927—Also his Sattunnaiskasveja Oulusta vv. 1927–1931. l.c. **36:** 28–29. 1932.—Also **Luther, Hans.** Tyska polemokorer och andra adventivväxter från Uleåborg. Memo. Soc. Fauna et Flora Fenn. **24:** 80–86. 1948. (Lists of plants introduced by war (krigsföljeväxter, polemokorer), of mill weeds, and of other weeds, with discussion; bibliography.)—Also Leiviskä, below.

Kalliola, Reino. Pyhätunturin kansallispuiston kasvillisuudesta ja kasvistosta. (Über die vegetation und flora des nationalparks Pyhätunturi.) Silva Fenn. no. 59. 24 p. 13 fig. (fig. 1, map; fig. 6–13 on 2 pl.). 1942.
Botanical explorations, topography, plant zones, general features of flora, plant formations; briefly annotated list of known vascular plants (178 species); bibliography. (An area of about 30 sq. km., including Mount Pyhätunturi with its wooded slopes and the adjacent parts of the moor surrounding it, ca. 67° N. Lat., in Kemijärvi and Pelkosenniemi parishes.)

Keckman, C. E. Anteckningar om floran i Simo och Kemi socknar af norra Österbotten. Acta Soc. Fauna et Flora Fenn. v. 13, no. 4. 66 p. 1896.
Topography, geology, soils, general features of flora, statistics; annotated list of vascular plants (about 572 species). Relates to the coastal region of Kemi and the northwest part of Simo parish.

Leiviskä, Iivari. Oulun kasvisto. [Flora of Oulu.] 19 p. 21.5 cm. Oulussa, 1894.
List of spermatophytes of Oulu (Uleåborg) and of the scarcer species from the vicinity, with very brief annotations. Signed only by author's initials. (Not seen; title and annotation supplied by H. Luther.)—See also his Oulun seudun meranrantojen kasvillisuudesta. [On the vegetation of the seashores of the Oulu region.] Acta Soc. Fauna et Flora Fenn. v. 23, no. 5. 126 p. pl., map. 1902. (Topography, plant zones, general features of flora, etc.)—Also **Huumonen, M. E.** Muutamia lisäyksiä ja oikaisuja Oulun seudun putkilokasvistoon. [Some additions and corrections to the vascular flora of the Oulu region.] Meddel. Soc. Fauna et Flora Fenn. **38:** 83–87. 1912. (Includes list of additional wild plants, p. 83–85, and corrections, p. 87.)—Also Huumonen, above.

Luther, Hans. Botaniska iakttagelser på Hailuoto-Karlö. Memo. Soc. Fauna et Flora Fenn. **24:** 30–49. 8 fig. on 2 pl. (on 1). 1948.
Botanical explorations, topography, general features of flora; annotated list of vascular plants (313 species and subspecies) and 2 Characeae; bibliography.—See also **Widlund, Uno.** Puccinellia phryganodes (Trin.) Scribn. & Merr., ny för Ob jämte några växtuppgifter från Karlö. Memo. Soc. Fauna et Flora Fenn. **25:** 96–100. 1 fig. 1950. (Includes list of noteworthy species.)

Rantaniemi, P. A. Outlines of the botanical investigations in North Ostrobothnia and South Lapland before 1918. Arch. Soc. Zool. Bot. Fenn. Vanamo 8: 31–42. 1 fig. (map). 1953.
Sketch of botanical explorations, especially those of the author, in Kemi Lapland, northern part of North Ostrobothnia down to the Simojoki River, and northern part of Kuusamo; extensive bibliography.

Roivainen, Heikki. Ulkokrunnin (PP, Ii) putkilokasvit. [Vascular plants of Ulkokrunni.] Oulun Luonnon Ystäväin Yhdist. Julkaisuja Sarja [Ser.] A, 2: 3–38. 1954.
Description of the islands of the Ulkokrunni Archipelago, Gulf of Bothnia, and their vegetation; annotated list of vascular plants (223 species and subspecies). (Not seen; title and annotation supplied by H. Luther.)

Teivainen, Lauri. Pisavaaran luonnonpuiston metsäkasvillisuudesta ja kasvitosta. (Zur waldvegetationen und flora des naturschutzgebietes Pisavaara.) Silva Fenn. 65. 34 p. incl. map, 10 fig. on 6 pl. (on 3). 1949.
Topography, geology, climate, plant formations; annotated list of vascular plants (291 species). In Rovaniemi and Tervola parishes.)

Vartiainen, Terttu. Maakrunnin ja Ristikarin putkilokasvit. [Vascular plants of Maakrunni and Ristikari.] Oulun Luonnon Ystäväin Yhdist. Julkaisuja Sarja [Ser.] A, **2:** 38–57. 1954.
Annotated list of vascular plants (207 species) of two islands in Gulf of Bothnia. (Not seen; title and annotation supplied by H. Luther.)

OSTROBOTTNIA KAJANENSIS (Ok). (KAINUU. KN)

See also Ostrobottnia borealis (Brenner).

Ostrobottnia kajanensis is a region in central Finland, extending from about 63°40′ to about 65°30′ N. Lat., bounded on the north by Kuusamo, on the east by Karelia pomorica occidentalis (Russian), on the south by Karelia borealis and Savonia borealis, on the west by Ostrobottnia media and Ostrobottnia borealis.

General

Parvela, A. A. Oulun läänin viljelyskasvit niiden historia ja nykyinen levinneisyys. II. Erikoisosa. (Über die kulturpflanzen im län Oulu (Uleåborg), ihre geschichte und verbreitung. II. Spezieller teil.) Annales Bot. Soc. Zool.-Bot. Fenn. Vanamo v. 2, no. 5. viii, 143 p. 1932.

Alphabetical list of cultivated vascular plants of Oulu län, with localities and indication of number of years they have been cultivated in each area. Covers the whole of Ostrobottnia kajanensis and adjacent parts of Kuusamo, Ostrobottnia borealis, and Ostrobottnia media; also the present Lapin lääni, at that time part of Oulun lääni, and including Kuusamo, Lapponia enontekiensis, inarensis, kemensis, and petsamoensis, and Ostrobottnia borealis and media.

Local

Fagerström, Lars. Anteckningar om floran i några vinterkrigstida mottiområden i Ok Kuhmo, sommaren 1954. Memo. Soc. Fauna et Flora Fenn. **32:** 112–119. 1957.

Includes alphabetical list of weedy plants from Saunajärvi by and Korpisalmi by.

Heikkinen, Lauri. Saksalaiset sotajoukot kasvien levittäjinä. [The German army as a distributor of plants.] Luonnon Tutkija **52:** 25–26. 1948.

List of weeds from a German horse camp of World War II at Hyrynsalmi.

Jalas, Jaakko. Rokua suunitellun kansallipuiston kasvillisuus ja kasvisto. (Vegetation und flora des geplanten Nationalparks von Rokua in Mittelfinnland.) Silva Fenn. 81. 97 p. 18 fig., map. 1953.

A vegetational study; includes (p. 60–83) chapter on flora containing annotated list of vascular plants (252 native species), bryophytes, and lichens; bibliography. (In parish Säräisniemi and vicinity, in Ostrobottnia kajanensis and Ostrobottnia borealis.)

Kyyhkynen, Olli. Huomattavampia putkilokasvi- ja sammallöytöja Kajaanin alueelta ja Pohjois-Savosta vuosina 1916–18. [Noteworthy finds of vascular plants and bryophytes from the Kajaani region and Savonia borealis in 1916–18.] Meddel. Soc. Fauna et Flora Fenn. **45:** 154–165. 1920.

Annotated list of noteworthy vascular plants and mosses, with localities, from Ostrobottnia kajanensis (Sotkamo, etc.) and Savonia borealis (Maaninka, etc.).

——— Suomussalmen kasvisto. [Flora of Suomussalmi.] Acta Soc. Fauna et Flora Fenn. v. 46, no. 1. 139 p. 1919.

Ecology, etc.; annotated list of vascular plants.

Mustonen (i.e. Lönnbohm), O.A.F. Tietoja Kajaanin kihlakunnasta ja etenkin Paltamon pitäjäästä. [Observations on the district of Kajaani and particularly on the parish of Paltamo.] 272 p. 18.5 cm. Hämeenlinnassa, 1885.

Includes chapter "Kasvisto" (p. 39–58) containing briefly annotated list of vascular plants (385 species) of Kajaani county. (Not seen; title and annotation supplied by H. Luther.)—See also **Kyyhkynen, Olli.** Kajaanin kasvistoalueen rajoista ja jaoituksesta. (Über die grenzen und die einteilung der pflanzengeographischen provinz Ostrobottnia kajanensis.) Acta Soc. Fauna et Flora Fenn. v. 49, no. 6. 26 p. 1921. (Discussion of botanical regions in the province, with mention of species from Kajaani and other localities.)

OSTROBOTTNIA MEDIA (Om.). (KESKI-POHJANMAA. KP)

See also Aboënsis, Regio (Auer, 1944); Ostrobottnia australis (Laurén, Växtförhållandena); Ostrobottnia borealis (Brenner); Ostrobottnia kajanensis (Parvela).

Ostrobottnia media is a region on the west coast of central Finland extending from about 62°40′ to about 64°55′ N. Lat. (extreme limits), bordered on the north by Ostrobottnia borealis, on the east by Ostrobottnia kajanensis and Savonia borealis, on the south by Tavastia borealis and Ostrobottnia australis, on the west by the Gulf of Bothnia.

Local

Backman, A. L. Floran i Lappajärvi jämte omnäjd. Acta Soc. Fauna et Flora Fenn. v. 32, no. 3. 138 p. map. 1909.
Botanical explorations, climate, topography, geology, cultivated plants, plant associations; annotated list of vascular plants (470 species), bryophytes, charophytes, and lichens (total 877 species) of the Lappajärvi region; statistics, phytogeography.

Hellström, Fredrik. Förteckning öfver de i Gamlakarleby provinsialläkare-distrikt funna fröväxter och ormbunkar. Meddel. Soc. Fauna et Flora Fenn. 5: 131–159. 1880 (1879).
Topography, soils, general features of flora; annotated list of vascular plants (436 species). Covers Pedersöre härad (Pietarsaaren kihlakunta).—See also **Widlund, Uno.** Adventivväxter i Yxpila hamnområde. Arkiv Svenska Österbotten **6**: 97–103. 1947. (Discussion of plants introduced under war conditions; annotated list of vascular plants from Yxpila harbor brought in with fodder during the Second World War. In University of Chicago library.)

Parvela, A. A. Oulaisten pitäjän kasvisto. [Flora of Oulainen parish.] Acta Soc. Fauna et Flora Fenn. v. 49, no. 3. 78 p. 1921.
Bibliography, climate, ecology, etc.; annotated list of vacular plants (391 species).

Widlund, Uno. Floristiska anteckningar från Gamlakarleby yttre skärgård. Arkiv Svenska Österbotten **4**: 181–196. illus. (incl. map). 1944. (University of Chicago library.)
Brief description of islets visited; tabular list of vascular plants (214 species and forms) showing occurrence on each of 16 islets in the vicinity of Gamlakarleby (Kokkola), from Tankar and Lillgrund in the west to Trullögrund and Lerbådan in the east; lists of species found only on certain islands, etc.

(REGIO ABOËNSIS. See ABOËNSIS, REGIO.)

(REGIO KUUSAMOENSIS. See KUUSAMOENSIS, REGIO.)

SATAKUNTA (St).

See also Aboënsis, Regio (Jalas); Tavastia australis (Mäkelä; Sola).
Satakunta is a region in southwestern Finland (essentially the northern part of Turun ja Porin lääni, south to about 60°50′ N. Lat.), bordered on the north by Ostrobottnia australis, on the east by Tavastia borealis and Tavastia australis, on the south by Regio aboënsis, on the west by the Gulf of Bothnia. Its name and abbreviation are unchanged in the newer nomenclature.

Local

Asp, Gustaf. Hämeenkyrön kasvut, latinaisilla nimillä. [Plants of Hämeenkyrö, by Latin names.] *In* Yrjö-Koskinen, George Z. Kertomus Hämeenkyrön pitäjästä. [History of Hämeenkyrö parish.] Suomi **11**: 103–116. 1852.
Unannotated list of vascular and cellular plants (all groups).

Engman (later Aario), L. E. Parkanon Sydänmaan kasvistosta. [On the flora of Sydänmaa by in Parkano.] Luonnon Ystävä **32**: 91–102. 1928.
Topography, general features of flora, human settlement; briefly annotated list of vascular plants, with localities for scarcer species.

Erämetsä, Heikki. Huittisten pitäjän kasvisto. (Die flora des kirchspiels Huittinen in Südwestfinnland.) Annales Bot. Soc. Zool.-Bot. Fenn. Vanamo v. 25, no. 5. 2 p. l., 26 p. 6 fig. (incl. maps). 1953.
Topography, cultivation, floristic regions, etc.; annotated list of vascular plants (490 species); bibliography.

Ervi, L. O. Köyliön pitäjän putkilokasvisto. (Die flora des kirchspiels Köyliö in der florenprovinz Satakunta in Südwestfinnland.) Arch. Soc. Zool. Bot. Fenn. Vanamo **10**: 64–87. 6 fig. (fig. 1, map). 1956.
Topography, botanical regions, etc.; annotated list of vascular plants (439 species); bibliography.

Häyrén, Ernst. Björneborgstraktens vegetation och kärlväxtflora. Acta Soc. Fauna et Flora Fenn. v. 32, no. 1. 264 p. map. 1909.
Mainly ecological; includes (p. 184–260) annotated list of vascular plants (546 species) and bibliography. Covers Ulfsby (Ulvila) parish, Björneborg (Pori) city, and the rural community Porin maalaiskunta.

Jalas, Jaakko. Rauman-Euran-Luvian alueen putkilokasviston pääpiirteet. (Kasvistollisia tutkimuksia Luonais-Suomessa. II.) (Floristic studies in southwest Finland. II. Main features of the vascular plant flora in the area of Rauma-Eura-Luvia.) Arch. Soc. Zool. Bot. Fenn. Vanamo **8**: 156–191. 1954.
Botanical explorations; annotated list of vascular plants (686 taxa), grouped phytogeographically, with localities; bibliography. Replaces **Lindström, G. I.** Beskrifning öfver Eura socken. Suomi **9**: 93–187. map. 1850. (Includes chapter on Flora och fauna (p. 114–124), with incomplete list of vascular plants (p. 116–119). Also replaces **Vaahtoranta, Eho.** Tietoja Rauman seudun kasvistosta. (Beiträge zur kenntnis der flora der stadt Rauma nebst umgebung.) Arch. Soc. Zool. Bot. Fenn. Vanamo **7**: 52–57. 1952. (Botanical explorations, notes on ballast plants; annotated list of vascular plants, omitting most of the common species; bibliography.)

Malmgren, A. J. Botanisk resa till Satakunta och Södra Österbotten, med understöd af Sällskapet pro fauna et flora fennica, verkställd sommaren 1859 af Th. Simming, P. A. Karsten och A. J. Malmgren. Notiser Sällsk. Fauna et Flora Fenn. Förhandl. **6** (n.s., **3**): 1–42. 1861.
Itineraries, geology, etc.; briefly annotated list of spermatophytes (571 species), showing occurrence in the two provinces, with localities for rarer species; statistics. Relates to region from Loimijoki to Björneborg (Pori), east to Pirkkala, north via Ikalis (Ikaalinen) and Kankaanpää to Kristinestad, Vasa, and Qvarkens skärgård.—See also **Simming, Theodor.** Förteckning öfver de, under en 1859 företagen botanisk resa, i Satakunta och Södra Österbotten observerade ormbunkar, mossor och characeer. Notiser Sällsk. Fauna et Flora Fenn. Förhandl. **6**: 43–61. 1861. (Includes pteridophytes, p. 51–52.)—Also his [Tillägg till förteckningen]. l. c. **6**: 192–193. 1861. (Records for one angiosperm and several bryophytes.)—Also **Saelán, A. T.** Anmärkningar till växtslaget Hieracium. l. c. **6**: 193–195. 1861. (Critical notes.)—Also **Aario, Leo.** Kasvistollisia tietoja Pohjois-Satakunnasta. [Floristic notes from northern Satakunta.] Memo. Soc. Fauna et Flora Fenn. **7**: 267–276. 1932. (List of noteworthy vascular plants from Pori, Reposaari, Merikarvia, and other localities.)

Pohjala, Leo. Kokemäen pitäjän kasvisto. (Die flora des kirchspiels Kokemäki.) Annales Bot. Soc. Zool.-Bot. Fenn. Vanamo v. 5, no. 3. iv, 66 p. 7 fig. (incl. maps). 1934.
Topography, geology, botanical regions, etc.; annotated list of vascular plants, with localities; bibliography.

Såltin, Holger. Säkylän pitäjän kasvisto. (Die flora im kirchspiel Säkylä.) Arch. Soc. Zool. Bot. Fenn. Vanamo **9**: 145–169. 2 fig. (maps). 1955.
Topography, geology, general features of flora, etc.; annotated list of vascular plants (454 species); bibliography.

Sauramo, Matti. Loimaan ja sen lähipitäjien kasvistosta. [On the flora of Loimaa and neighboring parishes.] Satakunta **4**: 200–226. 1916.
Topography, general features of flora; briefly annotated list of vascular plants (446 species) from Loimaa, Mellilä, Alastaro, Metsämaa, and Ypäjä (Not seen; title and annotation supplied by H. Luther.)

SAVONIA AUSTRALIS (Sa). (ETELÄ-SAVO. ES)

See also Isthmus karelicus (Hiitonen).
Savonia australis is a region in southeastern Finland (including most of Mikkelin lääni and a little of Kymen (formerly Viipurin) lääni), bordered on the north by Savonia borealis, on the east by Karelia borealis and Karelia ladogensis, on the south by Karelia australis, on the west by Tavastia australis and Tavastia borealis.

Local

Häyhä, Helka. Taipalsaaren Kylänniemen putkilokasvisto. (The vascular flora of Kylänniemi island, commune of Taipalsaari, South-Savo.) Arch. Soc. Zool. Bot. Fenn. Vanamo **11**: 78–86. 1 fig. (map). 1956.

General features of flora; briefly annotated list of vascular plants (377 species of which 40 are cultivated).

Hult, Ragnar. Bidrag till kännedomen om vegetationen i södra Savolaks. Meddel. Soc. Fauna et Flora Fenn. **3**: 123–163. 1878.

Geology, topography, plant associations, statistics; briefly annotated list of vascular plants (431 species) of the parishes of Ruokolaks (Ruokolahti) and Rautjärvi.

Linkola, Kaarlo. Kasvistollisia pikkutietoja Kangasniemeltä. [Floristic notes from Kangasniemi.] Luonnon Ystävä **43**: 213–217. 1939.

Annotated partial list of vascular plants.

Paasio, Ilmari. Tietoja Savonlinnan seudun putkilokasvistosta. [Notes on the vascular flora of the Savonlinna region.] Luonnon Ystävä **31**: 141–144. 1927.

Annotated list of noteworthy vascular plants.

Ulvinen, Arvi. Kouvolan kasvistosta. I–IV. [On the flora of Kouvola.] Kouvolan Suomalaisen Yhteiskoulun Kertomukset **1928–29**: 1–7. 1929; **1930–31**: 1–11. 1931; **1932–33**: 1–11. 1933; **1934–35**: 1–8. 1935.

Briefly annotated list of weeds and noteworthy other plants from Kouvola (Not seen; title and annotation supplied by H. Luther.)—See also his Kasvilöytöjä Kouvolan seudulta 1 [–3]. (Pflanzenfunde in der gegend von Kouvola.) Memo. Soc. Fauna et Flora Fenn. **6**: 143–145. 1930; **7**: 257–259. 1932; **9**: 177–178. 1934. (Records of noteworthy wild and introduced species.)—Also his Kasvit Kouvolan seutua valloittamassa. [Plants occupying the Kouvola region.] Ankkapurha **2**: 1–31. 1949. (Annotated full list of weeds from Kouvola, divided into groups according to date of introduction. Not seen; title and annotation supplied by H. Luther.)

SAVONIA BOREALIS (Sb). (POHJOIS-SAVO. PS)

See also Ostrobottnia kajanensis (Kyyhkynen, 1920).

Savonia borealis is a region in south-central Finland, extending from about 61°55′ to 64° N. Lat., about 26°20′ to 29°30′ E. Long., bordered on the northwest by Ostrobottnia media, on the northeast by Ostrobottnia kajanensis, on the east by Karelia borealis, on the south by Savonia australis, on the west by Tavastia borealis.

Local

Juva, Ilmari. Kiuruveden pitäjän luonnonvaraiset putkilokasvit. [The spontaneous vascular plants of Kiuruvesi parish.] Kiuruveden Yhteiskoulu Kertomus **1946–47**: 33–39. Iisalmi, 1947.

Briefly annotated list of vascular plants (398 species). (Not seen; title and annotation supplied by H. Luther.)

Kyyhkynen, Olli. Satunnaiskasvistosta muutamilla Savon radan asemilla. [On the adventive flora of some stations along the Savo railway.] Meddel. Soc. Fauna et Flora Fenn. **45**: 150–154. 1920.

List of noteworthy adventive plants from Iisalmi, Kuopio, Varkaus, and other localities (also some from Kajaani in Ostrobottnia kajanensis).

Linkola, Kaarlo. Lisätietoja Kuopion pitäjän kasvistosta. [Additional notes on the flora of Kuopio parish.] Acta Soc. Fauna et Flora Fenn. v. 39, no. 5. 52 p. 1914.

Botanical explorations, etc.; annotated list of vascular plants and mosses.—See also **Hämäläinen, Pentti.** Kuopion kaupungin ja sen lähiympäristön satunnaiskasvistosta. [On the adventive flora of Kuopio city and its immediate vicinity.] Luonnon Ystävä **43**: 154–161. 1930. (Botanical explorations; lists of weeds found by earlier botanists, list of those found by author.)

Pitkänen, Yrjö. Tietoja Pieksämäen putkilokasvistosta. [Contributions to the vascular flora of Pieksämäki.] Kuopion Luonnon Ystäväin Yhdist. Julkaisuja ser. B, v. 1, no. 4. 35 p. 1933.

Annotated list of vascular plants (462 species). (Not seen; title and annotation supplied by H. Luther.)

——— Tuusniemen kasvisto. [Flora of Tuusniemi.] Kuopion Luonnon Ystäväin Yhdist. Julkaisuja ser. B, v. 1, no. 10. 55 p. 1940.

Annotated list of vascular plants (432 species). (Not seen; title and annotation supplied by H. Luther.)

Suhonen, Pentti. Karttulan pitäjän kasvisto. [Flora of Karttula parish.] Kuopion Luonnon Ystäväin Yhdist. Julkaisuja ser. B, v. 1, no. 7. 60 p. 1935.

Topography, general features of flora, annotated list of vascular plants (456 species). (Not seen; title and annotation supplied by H. Luther.)

TAVASTIA AUSTRALIS (Ta). (ETELÄ-HÄME. EH)

See also Satakunta (Sauramo); Savonia australis (Ulvinen).

Tavastia australis is a region in southern Finland, extending from about 60°30′ to 62°15′ N. Lat., and from about 23°10′ to 26°40′ E. Long., bordered on the north by Tavastia borealis, on the east by Savonia australis, on the south by Nylandia and Regio aboënsis, on the west by Satakunta.

Local

Aspelin, E. F., and Thurén, A. Bidrag till Tavastehustraktens flora. Notiser Sällsk. Fauna et Flora Fenn. Förhandl. 7 (n.s., 4): 31–54. 1867.

Topography, climate, phenology, etc.; briefly annotated list of vascular plants (507 species) and bryophytes (174 species). Refers to whole of Wonå (Våna, Vanaja) parish and part of Hattula and Janakkala parishes.

Aulamo, O. I. Riihimäen "myllyalueelta" vuosina 1925-1931 kerättyjä satunnaiskasveja. [Adventive plants collected in a mill region in Riihimäki 1925–31.] Luonnon Ystävä **36**: 146–149. 1932.

Alphabetical list of weeds (157 species) from a mill area in Riihimäki.—See also his Tietoja lastauspaikka- ja satunnaiskasvien esiintymisestä Riihimäellä. [Notes on the appearance of ballast and adventive plants in Riihimäki.] Memo. Soc. Fauna et Flora Fenn. **9**: 161–165. 1934. (Discussion of author's work, with mention of noteworthy species of different groups.)

Björklund, Karin. Padasjoen pitäjän itäpuoliskon kasvisto. (Die flora der östlichen hälfte des kirchspiels Padasjoki in Südfinnland.) Annales Bot. Soc. Zool.-Bot. Fenn. Vanamo v. 10, no. 2. iv, 69 p. 3 fig. (incl. maps). 1938.

Botanical explorations, topography, geology, settlements, botanical regions; annotated list of vascular plants (517 species); bibliography.—See also **Pankakoski, Antero.** Lisiä Padasjoen pitäjän kasvistoon. [Additions to the flora of the parish of Padasjoki.] Luonnon Ystävä **45**: 100. 1941. (Adds 6 species.)—Also **Häyrén, Ernst.** Floristiska iakttagelser i Tavastland. Memo. Soc. Fauna et Flora Fenn. **19**: 144–148. 1944. (List of noteworthy vascular plants from Tavastia australis (Padasjoki, Korpilahti, Kuhmoinen, and other localities on west side of Lake Päijänne), with some from Tavastia borealis (Toivakka, Laukaa, and other localities).)

Bonsdorff, Ernst. Öfversigt af Gustaf Adolfs sockens flora. Notiser Sällsk. Fauna et Flora Fenn. Förhandl. **7** (n.s., **4**): 55–81. 1867.

Topography, climate, general features of vegetation, phenology, etc.; briefly annotated list of vascular plants (427 species), also mosses (135 species) and 3 Characeae. (Parish of Gustaf Adolfs, now Hartola.)

Brander, Torild. Urjalan Kivijärven seudusta. III. Putkilokasvit. [From the region of Kivijärvi in Urjala. III. Vascular plants.] Luonais-Hämeen Kotiseutu- ja Museoyhdistkysen Vuosikirja **20**: 5–47. 8 fig. (incl. map). 1951.

Brief history of botanical explorations in Tavastia australis and Tavastia borealis, etc.; general features of flora of the Kivijärvi area, with bibliography and (p. 42–47) incomplete alphabetical list of vascular plants.—See also **Fagerström, Lars.** Växtfynd i sydvästra Tavastland. I.–III. Memo. Soc. Fauna et Flora Fenn. **28**: 55–60. 1953; **29**: 51–54. 1954; **30**: 34–35. 1955. (Briefly annotated lists of noteworthy species from Urjala, Koijärvi, Humppila, and Ypäjä.)—Also his Luonais-Hämeen kasvimaailman tutkimisesta. (Om sydvästra Tavastlands botaniska utforskning.) I.–II. Luonais-Hämeen Luonto **1**: 45–49. 1956; **2**: 51–54. 1956. (Bibliography of floristic exploration of southwestern Tavastia australis. Not seen; title and annotation supplied by H. Luther.)

Erkamo, Viljo. Lisätietoja Heinolan seudun kasvistosta (EH). (Additional notes on the flora of the Heinola region (EH).) Arch. Soc. Zool. Bot. Fenn. Vanamo **2**: 117–126. 1949.

Botanical explorations, changes in flora, etc.; unannotated list of common species, list of scarcer species with localities (total 457 species of vascular plants); bibliography.

Fagerström, Lars, and **Toivonen, Tauno.** Luonais-Hämeen putkilokasvisto. [Vascular flora of SW-Häme.] Luonais-Hämeen Luonto **3**: 64–81. 1957.

Unannotated list of vascular plants (669 species, not including *Hieracium* and *Taraxacum*), with tabular lists for the individual parishes of Somerniemi, Somero, Ypäjä, Jokioinen, Tammela, Forssa, Humppila, Koijärvi, and Urjala. (Not seen; title and annotation supplied by H. Luther.)

Idman, G. R. Förteckning öfver af undertecknad under en tidrymd af c. 25 år iakttagna kärlväxter i Teisko socken. Memo. Soc. Fauna et Flora Fenn. **3**: 21–31. 1927.

Briefly annotated list of vascular plants, including some cultivated species.

Knabe, C. A. Pflanzenphysiognomische skizzen aus dem südwestlichen Finnland. Allg. Bot. Zeitschr. **2**: 41–44, 64–69, 85–88, 100–101, 128–130. 1896.

Topography, geology, climate, cultivation, account of plant associations with included listing of vascular plants, lichens, and mosses (382 vascular plants) of Kalvola.—See also **Borg, Väinö.**. Kasvillisuus. [The vegetation.] *In* his Fysillis-maantieteellisiä tutkimuksia Kalvolan pitäjässä. [Physicogeographical studies in Kalvola parish.] Tidskr. Geog. För. (Helsingfors) **10**: 428–445. 1898. (General features of vegetation with mention of many species.)

Leopold, Constantin. Anteckningar öfver vegetationen i Sahalahti, Kuhmalahti och Luopiois kapeller af södra Tavastland. Meddel. Soc. Fauna et Flora Fenn. **5**: 81-130. 1880 (1879).

Topography, geology, soils, plant associations; briefly annotated list of vascular plants, bryophytes, and lichens.

Mäkelä, Taimi. Lehdoista ja lehtokasvien leviämisestä Pohjois-Pirkkalan-Tyrvään alueella. (Ueber haine und verbreitung der hainpflanzen im gebiet von Nord-Pirkkala-Tyrvää.) Silva Fenn. 37. 61 p. map. 1936.

Mainly ecological (sylvan plant associations); includes (p. 32–53) annotated list of vascular plants of an area in western Tavastia australis and adjacent Satakunta.—See also **Tapio, Simo.** Tutkimuksia lehtokasvillisuudesta ja lehtokasvien ekologisesta ryhmittymisestä Pirkkalan lehtokeskuksen keskiosassa. (Untersuchungen über die hainvegetation und die ökologische verteilung der hainpflanzenarten im mittleren teil des hainzentrums von Pirkkala in Südfinnland.) Annales Bot. Soc. Zool.-Bot. Fenn. Vanamo v. 25, no. 3. 2 p. l., 57 p. 2 fig. (maps). 1953. (Ecological; includes (p. 50–57) tabular list of vascular plants and bryophytes, not in systematic order.)

Norrlin, J. P. Bidrag till sydöstra Tavastlands flora. Notiser Sällsk. Fauna et Flora Fenn. Förhandl. **11** (n.s., 8): 73–196. 1871. (Preprinted, 73–196 p. Helsingfors, 1870.)

Botanical investigations, topography, plant formations, ecology (tables), etc.; annotated list of 1,225 plants (of which 562 are vascular). Covers the southeast corner of Tavastia australis between about 61° and 61° 17′ N. Lat., 25° and 25° 30′ E. Long. German translation of introductory matter under title: Beiträge zur flora südöstlichen Tavastlands. Acta Forest. Fenn. **23**: 15–52. 1923.—See also **Linkola, Kaarlo.** Über rückgangserscheinungen in der ruderalen beigleitflora der alten kultur in Süd-Häme. Annales Bot. Soc. Zool.-Bot. Fenn. Vanamo 4 (12): 3–7. 1933. (Discussion of changes in frequency of several weedy plants; bibliography.)

Ritala, Hilkka, and **Toivonen, Tauno.** Somerniemen pitäjän kasvisto. (Die flora des kirchspiels Somerniemi.) Arch. Soc. Zool. Bot. Fenn. Vanamo **10**: 95–125. 2 fig. (maps). 1956.

Topography, geology, general features of vegetation; annotated list of vascular plants (531 species); bibliography.

Sola, A. A. Floristisia tutkimuksia Näsijärven länsipuolisissa seuduissa kesällä 1905. [Floristic investigations in the region west of Näsijärvi in the summer of 1905.] Meddel. Soc. Fauna et Flora Fenn. **32**: 82–90. 1906.

Annotated list of noteworthy vascular plants, particularly from the vicinity of Hämeenkyrö.

Soveri, Jorma. Lammin pitäjän kasvisto. (Die flora des kirchspiels Lammi in Süd-Finnland.) Annales Bot. Soc. Zool.-Bot. Fenn. Vanamo v. 4, no. 3. iv, 86 p. 7 fig. (incl. maps). 1933.

Topography, geology, botanical regions, effects of cultivation on flora, etc.; annotated list of vascular plants (546 species), with localities; bibliography. —See also his Lisätietoja Lammin pitäjän kasvistosta. [Further notes on the flora of Lammi parish.] Luonnon Ystävä 38: 100–102. 1934. (List of noteworthy plants, bringing total to 577 species.)

Wainio, E. A. Florula Tavastiae orientalis. Havainnoita Itä-Hämeen kasvistosta. Meddel. Soc. Fauna et Flora Fenn. 3: 1–121. 1 p.l. 1878.

Plant formations, etc.; annotated list of vascular plants (475 species), bryophytes (241 species), and lichens (472 species), with localities. Title appears only on preliminary (unpaged) leaf. The whole paper is in Finnish. Covers all of Luhanka parish, the part of Korpilahti parish on east side of Lake Päijänne, and a small part of Jyväskylä parish (the last in Tavastia borealis).

TAVASTIA BOREALIS (Tb). (POHJOIS-HÄME. PH)

See also Tavastia australis (Häyrén under Björklund; Wainio).

Tavastia borealis is a region in south-central Finland, extending from about 62° to 63°15′ N. Lat., about 23°10′ to 27° E. Long., bordered on the north by Ostrobottnia media, on the east by Savonia borealis, on the south by Savonia australis and Tavastia australis, on the west by Satakunta, Ostrobottnia australis, and Ostrobottnia media.

Local

Brotherus, V. F. Anteckningar till norra Tavastlands flora. Notiser Sällsk. Fauna et Flora Fenn. Förhandl. **13** (n.s., **10**): 185–217. [1872.]

Topography, geology, general features of vegetation; annotated list of vascular plants (416 species), also bryophytes (232 species) and Characeae (3 species). Relates to area between Jyväskylä, Pihtipudas, and Karstula.

Klingstedt, F. W. Om floran i norra Tb. I–II. Acta Soc. Fauna et Flora Fenn. v. 72, no. 13. 21 p. 2 fig. (maps). 1955; v. 74, no. 1. 7 p. 1956.

Topography, botanical explorations; annotated list of vascular plants of the area around the factory villages Äänekoski and Suolahti; bibliography. The second part (not seen) covers *Taraxacum* (45 species) and *Hieracium* (29 species.)

Linkola, Kaarlo. Kasvitietoja Jyväskylästä. [Floristic observations from Jyväskylä.] Meddel. Soc. Fauna et Flora Fenn. **40**: 184–195. 1914.

Changes in the flora; list of noteworthy vascular plants, with localities.—See also his Kasviston muutoksista Jyväskylän kaupungissa ja sen lähistöllä viime vuosikymmenien aikana. [On the changes in the flora of Jyväskylä city and its vicinity during the last decades.] Meddel. Soc. Fauna et Flora Fenn. **43**: 156–171. 1917. (Changes in the flora; extensive annotated list of noteworthy plants (nearly all weeds); bibliography.)—Also **Erkamo, Viljo.** Lisätietoja Jyväskylän kasvistosta. (Beiträge zur kenntnis der flora von Jyväskylä.) Memo. Soc. Fauna et Flora Fenn. **19**: 103–113. 1944. (Discussion of recent botanical explorations and additions to the flora; extensive list of angiosperms (mostly weeds) with localities; bibliography.)—Also **Valovirta, E. J.** Satunnaiskasviuutuuksia Pohjois-Hämeestä. [New adventive plants from North Häme.] Luonnon Tutkija **55**: 101. 1951.

Norrlin, J. P. Några anteckningar till mellersta Finlands (n.v. Tavastlands) flora. Notiser Sällsk. Fauna et Flora Fenn. Förhandl. **13**: 421–436. 1874.

Topography, general features of vegetation; annotated list of commoner species (not complete) of Pihlajavesi kapell (chapel) in Keuru (Keuruu) parish, also annotated list of bryophytes and lichens.

Pynnönen, Alpi. Lehtokasveja Konnevedellä. [Woodland plants from Konnevesi.] Luonnon Ystävä **39**: 129–132. 1935.

Annotated list of the rarer vascular plants.—See also his Muutamia kasvistollisia tietoja Konnevedeltä. [Some floristic contributions from Konnevesi.] Luonnon Ystävä **39**: 181–182. 1935. (Additional records.)—Also his Lisätietoja Konneveden pitäjän kasvistoon. (Beiträge zur kenntnis der flora des kirchspiels Konnevesi.) Annales Bot. Soc. Zool.-Bot. Fenn. Vanamo

18 (Notul. Bot. 13): 1–16. 1 fig. (map). 1943. (Botanical regions, general features of vegetation, etc.; annotated list of the more important vascular plants.)—Also **Häkkinen, Uuno.** Täydennystietoja Konneveden pitäjän putkilokasvistosta. (Ergänzende beiträge zur kenntnis der gefässpflanzenflora des kirchspiels Konnevesi im Mittelfinnland.) l. c. 20 (Notul. Bot.): 67–70. 1945. (Further list of noteworthy species with localities.)—Also his Uusia putkilokasvien kasvupaikkalöytöjä Konnevedeltä. [New vascular plant records from Konnevesi.] Luonnon Tutkija 55: 133. 1951.

Roivanen, Heikki. Tietoja Pihtiputaan ja Kinnulan putkilokasvistosta. [Notes on the vascular flora of Pihtipudas and Kinnula.] Kuopion Luonnon Ystäväin Yhdist. Julk. ser. B. v. 1, no. 1. 48 p. 1927.

Annotated list of known vascular plants (393 species). (Not seen; title and annotation supplied by H. Luther.)

FRANCE

See also Belgium, General (Goffart, Nouveau manuel; Paque, De Vlaamsche volksnamen); Italy, General (Rikli, Das pflanzenkleid der Mittelmeerländer); Netherlands, General (Gerth van Wijk, A dictionary of plant names); Spain, General (Bubani, Flora pyrenaea); Switzerland, General (Bouvier, Flore des Alpes de la Suisse et de la Savoie; Probst, Wolladventiven).

General

See also the classified list of Partial floras (p. 108).

The general titles here listed may be classified as follows: *Floras and manuals:* Acloque, Bonnier (two), Coste, Fournier, Gillet, Grenier, Rouy. *Catalogs:* Bautier (with localities), Camus (two including one of hybrids), Léveillé (two, including one of hybrids). *Iconographies:* Baillon, (Bonnier, Flore complète illustrée), (Coste), Cusin, Jaume St.-Hilaire. *Bibliographies:* Deniker, Dillemann, Holden, Rehder. *Vernacular names:* (Gatin), Guyettant, Kreiter, Rolland. *Pteridophytes:* Rey-Pailhade. *Trees and woody plants in general:* Gayffier Guinier, Hickel (seeds and seedlings), Mathieu. *Medicinal or poisonous plants:* (Ansberque), (Bonnier), Bossu, Bruntz, Cazin, Conte, Cornevin (poisonous plants), Dubois, Fournier, Hariot, P. Hermant, Kosch, Perrot, Rawton, Rodin, Rouget. *Other useful plants:* Ansberque (mostly forage), Bonnier (general), Ducomet (edible), Gibault (edible), C. Hermant (edible), Maniciot (edible), Masclef (general), Rothschild (general). *Weeds:* Fron, Gadeceau, Menault. *Cultivated plants:* Fournier, Gérôme, Guillaumin. *Miscellaneous:* Chevalier (extinct or disappearing species), Davy de Virville (history of botany), Depape (fossil floras), Gatin (dictionary of botanical terms and vernacular names), Gaussen (botanical regions), Gentil (etymology of scientific names), Goujon (climatic influences), Jovet (guide for excursions), Kreiter (etymology), Le Brun (additions to flora), Le Grand (statistics), Meynier (toponymy), Sébillot (folklore), Société botanique de France (reports of excursions), Verlot (guide for botanizing).

Acloque, A. N. C. Flore de France contenant la description de toutes les espèces indigènes disposées en tableaux analytiques . . . 816 p. 2165 fig. 18.5 cm. Paris, 1894. (Reissued 1904.)

Manual of vascular plants (4255 species) in form of keys, with brief indication of range; synopsis of groups of cellular plants, with vernacular names only; illustrated glossary, annotated list of principal medicinal plants, indices. —See also his Flores régionales de la France. Nord.—Ouest.—Sud-ouest et Pyrénées.—Région méditerranéenne.—Sud-est et Alpes.—Nord-est, Vosges et Alsace.—Centre.—Environs de Paris. [431] p. 18.5 cm. Paris, 1904. (Each of the 8 sections (of 44 to 56 pages) contains an annotated list of the vascular plants of the region covered. These lists, bound with the text of his Flore de France, were published in 1904 as separate works under the titles Flore du sud-ouest de la France et des Pyrénées, Flore du sud-est de la France et des Alpes, etc.)

Ansberque, Edme. Flore fourragère de la France. Reproduite par la méthode de compression dite phytoxygraphique . . . 272 pl. on 136. 44 cm. Lyon, 1866.

Nature prints (in sepia) of 708 plants (all but 2 vascular), with brief annotations on plates as to useful or harmful properties; in addition to forage plants, includes medicinal and otherwise useful species, but is not complete for latter. Lithographed index.—See also (for medicinal plants) Bonnier (2), Bossu, Bruntz, Cazin, Conte, Cornevin, Dubois, Fournier, Hariot, P. Hermant, Kosch, Perrot, Rawton, Rodin, and Rouget; and, for other useful plants, Bonnier (2), Ducomet, Gibault, C. Harmant, Manciot, Masclef, and Rothschild; also, under Partial, Bel, Bénézech, Garnaud, and Offner and Pons.

Baillon, H. E. Iconographie de la flore française. 5 v. 500 col. pl. 19 cm. Paris, [1885–94].—Table générale des 500 planches de l'ouvrage. 40 p. 19 cm. [1894.] (With v. 5.)

Indifferent colored plates of 500 species (including some cellular plants), without dissections, accompanied by descriptive text.—See also Bonnier (Flore complète illustrée), Coste, Cusin, and Jaume St. Hilaire, below.

Bautier, Alexandre. Flores partielles de la France comparées. 2 v. (359, 78 p.). 22.5 cm. Paris, 1868. (Arnold Arboretum library.)
Vol. 1, systematic list of vascular plants of France (4294 species), with duration, date of flowering, localities and habitat; v. 2, alphabetical list of localities, with list (by numbers) of rarer plants recorded from them in vol. 1.

Bonnier, Gaston. Flore complète illustrée en couleurs de France, Suisse et Belgique (comprenant le plupart des plantes d'Europe). 12 vol. 721 col. pl. 32 cm. Neuchâtel, Paris, Bruxelles (v. 1–8), Paris, Bruxelles (v. 9), Paris (v. 10–12), [1911–34]. (At head of title page: La végétation de la France, Suisse et Belgique. 2. partie. Ouvrage publié sous les auspices du Ministère de l'instruction publique.)—Tableau générale de la Flore complète . . . 5 p. l., 71, 8 p. 32 cm. Paris [1935].
Full descriptive flora of vascular plants (3346 species), with vernacular names in various European languages (not indexed), uses, local and general range, description of many infraspecific taxa; excellent colored habit figure of each species and subspecies (by Mlle. Julie Poinsot). Corsica is not covered. Vol. 7–12 edited by Robert Douin after Bonnier's death. The volumes, as usual with Bonnier's publications, bear no date; the dates of issue are given in Bibl. Schweiz. Naturw. Lit. **12**: 165–166. 1938, and Jour. Soc. Bibl. Nat. Hist. **2**: 212–215. 1950.

——— Plantes médicinales, plantes mellifères, plantes utiles et nuisibles. 3 p. l., [64 p.]. 64 col. pl. on 32. 18.5 cm. Paris [1920].
Account of 372 useful and harmful plants, systematically arranged, with vernacular names, brief description, uses; colored figure of each species.—See also Ansberque, above, and additional references there given.

——— and **Layens, Georges de.** Flore complète portative de la France et de la Suisse (comprenant aussi toutes les espèces de Belgique, d'Alsace et de Lorraine), pour trouver facilement les noms de plantes sans mots techniques. (At head of title page: La végétation de la France, Suisse et Belgique. 1. ptie.) nouv. éd. rev. et corr. xxvii, 425 p. 5,338 fig., 2 maps. 23 cm. Paris, [19—?] (1st ed. (?). Many reissues, the last seen dated 1948.)
Popular flora in form of illustrated keys, with general range; glossary, etc. Apparently merely a reissue of their Flore complète de la France et de la Suisse . . . [1908], which derives from their Tableaux synoptiques des plantes vasculaires de la flore de la France . . . 2. éd. rev. et corr. xxvii, 412 p. 5,291 fig., map. 22.5 cm. Paris, [1894?]. (Arnold Arboretum library. 1st ed. (?))

Bossu, Antonin. Traité des plantes médicinales indigènes . . . précédé d'un cours élémentaire de botanique . . . 3. éd. corrigée, augmentée, refondue dans plusieurs de ses parties. xxiv, 864 p. 92 fig., and atlas of 60 col. pl. 22.5 cm. Paris, 1872. (Lloyd Library.) (1st ed. 1854.)
Elements of botany, synopsis of classification with summary account of uses under each family; classification of medicinal plants according to uses, with lists of species; treatment of native and cultivated medicinal plants, alphabetically arranged by vernacular names, with additional vernacular names, description, properties, uses, dosage, etc.; list of diseases, with remedies; indices.—See also Ansberque, above, and additional references there given.

Bruntz, Louis, and **Jaloux, M.** Catalogue méthodique des plantes officinales et des plantes à drogues médicamenteuses dressé d'après les éditions de la Pharmacopée française. 2 p. l., 110 p. 25 cm. Dijon, 1919. (Library of Muséum national d'histoire naturelle, Paris.)
Systematic list of the French botanical materia medica, with country of origin, part used, and citation of the different editions of the Pharmacopée française in which the plant was listed.—See also Ansberque, above, and additional references there given.

Camus, E. G. Catalogue des plantes de France, de Suisse et de Belgique. 1 p. l., vii, 325 p. 22 cm. Paris, 1888. (Arnold Arboretum library.)

List of vascular plants and Characeae (4,842 species, of which 4,802 are vascular), with very brief indication of range. Many microspecies are listed but not numbered.

——— Statistique des plantes hybrides signalées dans l'étendue de la flore française. Compt. Rend. Cong. Soc. Sav. Paris et Dépt. Paris, 1898, Sect. Sci. p. 197–212. 1898.

Unannotated list of vascular plant hybrids, without references. Of the many subsequent publications on hybrids, the following is one of the most extensive: **Guétrot,—.** Plantes hybrides de France. I et II (1925–1926)–V–VII (1929–1931). 313 p. illus., 8 pl., ports. 25 cm. Lille, Paris, Gap, 1927–1931. (Author's ideas on nomenclature of hybrids (personal epithets should end in **io**, others in **o**, but he followed different rules in part I/II), etc.; descriptions of numerous hybrids, old and new (numbered 1–69, 100–135 bis), in various genera; biographical notices of several botanists. Nos. V–VII consist mostly of a Monographie des hybrides de saxifrages dactylites dits dactyloides, by Dominique Luizet (1852–1930).)—See also Léveillé (Les hybrides de France), below.

Cazin, F. J. Traité pratique et raisonné des plantes médicinales indigènes. 2. ed. xiv, 1076 p. 24.5 cm. Paris, Boulogne-sur-Mer, 1876, and Atlas. 40 col. pl. 27 cm. (1st ed. 1850.)

Account of native and cultivated medicinal plants, alphabetically arranged by vernacular names, with patois names, descriptions, properties and uses; list of plants, classified therapeutically; calendar for collecting material; no index of scientific names.—See also Ansberque, above, and additional references there given.

Chevalier, Auguste. Les plantes de la flore de France en voie de disparition et leur protection. Bul. Hist. Nat. Appl. 1. ptie., **6**: 40–52, 88–96. 1925.

Annotated list of about 28 vascular plants extinct or apparently so in continental France, list of other species that have become rare; causes of disappearance, methods of protection.—See also **Le Brun, Pierre.** Une enquète sur la flore française. Monde des Plant. IV, **33** (195): 20, (196): 26–28, (197): 35–36. 1932; **34** (199): 3–4, (200): 11–12. 1933. (Annotated list of vascular plants that have disappeared or become rare in special localities or throughout France. Includes Réponses aux enquètes, by various botanists, l.c. p. 11–12. 1933. In Arnold Arboretum library.)—See also his Quelques espèces de la flore de France en voie de disparition. Cong. Internatl. Bot. 8 (Paris 1945), Rapp. et Comm. **21/27**: 161-164. [1954.] (Mention of various species of different localities.)

Conte, E. Les plantes qui guérissent. Les plantes médicinales: caractères botaniques; lieux et époques des récoltes; mode de culture; parties à utiliser, dose et manière de les utiliser . . . 220 p. illus., 12 col. pl. on 6. 17.5 cm. Paris, n.d. [194—?] (Collection Rustica.)

Treatment of wild and cultivated medicinal vascular plants, alphabetically arranged by French names, with vernacular names, brief description, parts used, cultivation, properties, uses; indices, list of diseases with remedies.—See also Ansberque, above, and additional references there given.

Cornevin, Charles. Des plantes vénéneuses et des empoisonnements qu'elles déterminent. xi, 524 p. 52 fig. 21.5 cm. Paris, 1887 (reissue, 1893). (Bibliothèque de l'enseignement agricole publiée sous la direction de M. A. Müntz.) (1st issue in Gray Herbarium library; reissue in U.S. Department of Agriculture library.)

General considerations; systematic account of French poisonous plants, native or cultivated, with poisonous properties, symptoms, references, etc.—See also Ansberque, above, and additional references there given.

Coste, Hippolyte. Flore descriptive et illustrée de la France, de la Corse et des contrées limitrophes . . . Avec une introduction . . . par Ch. Flahault. 3 v. (52, xxxvi, 416; 627; 807 p.). 453 + 4354 fig., map. 25 cm. Paris, 1901 (1900)–06. (Reissued Paris, 1937.)

General features of vegetation and soils, botanical regions and zones (by Flahault); keyed, descriptive flora of vascular plants (4354 species) with local and general range and figure of each species. The standard illustrated flora of France, modeled after Britton and Brown's Illustrated flora of the northern United States. Infraspecific taxa are not mentioned; single authorities (without references to place of publication) are given for all specific names, and few synonyms are listed (these only from Grenier and Godron). The dates of issue of the 16 fascicles of the 1st issue are given in the preface to vol. 3. The colored map seems to be often lacking, at least in the reissue.

Cusin, L. A., and **Ansberque, Edme.** Herbier de la flore française . . . 25 v. [4759] pl. 40 cm. Lyon, 1868 [1867]–1876. (Arnold Arboretum library.)

Habit figures of angiosperms reproduced by "phytoxygraphie" showing a black impression of the plant, with enlarged details or a colored flower; one species to each plate; text little more than a list of species. Vols. 13–25 were by Cusin alone. According to Rehder (Bradley Bibl. 1: 414. 1911) plate 862 (of family 21), 43 (family 39), and 14 (family 78) were never published.—See also **Pouzet, Eugène.** Notes sur l' "Herbier de la flore française" de Cusin et Ansberque. Bul. Mens. Soc. Linn. Lyon n.s., 1: 109–111, 118–119. 1932. (Discussion of the collaboration between Cusin and Ansberque and of the method of printing their plates; he states that Ansberque was actually collaborator only in vol. 1–8, or perhaps only through vol. 7. In W. G. Farlow Library, Cambridge, Mass.)—Also Baillon, above, and additional references there given.

Davy de Virville, Adrien, and others. Histoire de la botanique en France. 394 p. incl. illus. (60 fig., 100 ports.). 16 pl. (on 8). 25.5 cm. Paris, 1954. (Publié par le Comité français du VIIIe Congrès international de botanique, Paris-Nice 1954.)

History of botanical work, including travels and botanical explorations in the world at large by Frenchmen, chronologically arranged and grouped by subjects (floras, agronomy, forestry, anatomy, physiology, mycology, etc.), the first part, covering Middle Ages to 18th century, by Davy de Virville alone, the second by numerous authors; includes mention of a great number of botanists (indexed) with dates and titles of principal works. The chapter on Flore et phytogeographie by Paul Jovet, with collaboration of Gérard Aymonin (p. 243–268), contains a list of deceased botanists with brief biographical notes and titles of works relating to the Paris region, mention of numerous general works, list of authors of regional, departmental, and local floras (with dates but mostly without titles) grouped under 14 regions, etc.

Deniker, Joseph. Bibliographie des travaux scientifiques (sciences mathématiques, physiques et naturelles) publiés par les sociétés savantes de la France depuis l'origine jusqu'en 1888 dressée sous les auspices du Ministère de l'instruction publique. v. 1–2, fasc. 1. 28 cm. Paris, 1895–1922.

A list of all papers relating to mathematical, physical, and "pure" natural science published in French scientific journals from 1700–1888 inclusive, arranged alphabetically by name of department, city or town, society, and journal, then chronologically, essentially unannotated. Vol. 1 (2 p. l., iii, 607 p. 1895–1916), Ain—Orne; v. 2, fasc. 1 (vii, 225 p. 1922), Pas-de-Calais—Sarthe (Le Mans); no more published. Vol. 2, fasc. 1, continued by René Descharmes. Only the first page of each article is given, and for the most part only the initials of given names or the surnames of authors.—See also Dillemann, Holden, and Rehder, below; also, under Partial, Gaussen (Végétation . . . des Pyrénées), Offner and Le Brun.

Depape, Georges. Le monde des plantes à l'apparition de l'homme en Europe occidentale. Flores récentes de France, des Pays-Bas, d'Angleterre. Annales Soc. Scient. Bruxelles sér. B., Sci. Phys. et Nat., 48 (2. pt., Mém. 1, fasc. 1): 39–102. 3 fig. (maps). 1928.

General considerations, sketch of geography of western Europe in Pliocene, running account of Pliocene floras of France, Netherlands, and England with lists of species of different geographical relationships, discussion of fossil floras of various localities, comparison with earlier and later floras, general conclusions, bibliography (arranged geographically).

Dillemann, Georges. Bibliographie des flores régionales de la France. Monde des Plant. V, **40** (237): 22–23, (238): 27–31, (239/240): 36–38. 1939; **41** (241): 3, (242): 6–8. 1940; VI, **42** (244): 12–14, (245): 21–22, (246): 28–29, (247): 36–38. 1947; VI, **43** (249): 3–6, (250/251): 13–15. 1948. (All in Arnold Arboretum library.)

Unannotated list of departmental and local floras and floristic works relating to vascular plants, geographically arranged. The following regions and departments are covered: in no. 237: I. Champagne (Aube, Ardennes); no. 238: (Haute-Marne, Marne); II. Bourgogne (Ain, Côte-d'Or, Nièvre, Saône-et-Loire, Yonne); no. 239/240: III. Lorraine (Meuse, Meurthe-et-Moselle, Moselle); no. 241 (Moselle, continued); no. 242: (Vosges, Massif des Vosges); no. 244: (Massif des Vosges, continued), IV. Alsace (Bas-Rhin, Belfort, Haut-Rhin); no. 245: Franche-Comté (Doubs, Haute-Saône, Jura and Chaîne jurassique); no. 246: région parisienne (general titles); no. 247: région parisienne, general, continued; also Aisne, Brie, Seine-et-Marne, Seine-et-Oise); no. 249: (Aisne, continued, Oise), Normandie (Calvados, Eure, Manche); no. 250/251: (Manche, continued, Orne, Seine-Maritime), région du Nord (in part). No more published.—See also Daniker, above, and additional references there given.

Dubois, François. Matière médicale indigène ou histoire des plantes médicinales qui croissent spontanément en France et en Belgique . . . xiv, 436 p. 22 cm. Tournai, 1848. (Arnold Arboretum library.)

Account of medicinal properties and uses of vascular and cellular plants, the plants grouped (by families) according to use; therapeutic index.—See also Ansberque, above, and additional references there given.

Ducomet, Vital. Les plantes alimentaires sauvages. Ressources de la flore française. 144 p. 18 cm. Paris, 1917.

General considerations, with references; annotated list of edible plants of France, classed by parts used, including wild and cultivated vascular plants and algae, lichens, and mushrooms; no index of scientific names.—See also Ansberque, above, and additional references there given.

Fournier, Paul. Arbres, arbustes et fleurs de pleine terre. (At head of title page: Flore illustrée des jardins et des parcs.) 3 v. (337, 549, 535 p., port.) and atlas. 181 i.e. 182 pl. 25 cm. Paris, 1951–52. (Encyclopédie biologique v. 38, 39, 40, 44.)

Keys to families, genera, and species of hardy cultivated ferns and flowering plants, herbaceous and woody (6,730 species, of which 2,020 are figured) with statement of native country, date of introduction, etc. Refers apparently mainly to western Europe, particularly France.—See also Gérôme and Guillaumin, below.

——— Le livre des plantes médicinales et vénéneuses de France; 1,500 espèces par le texte et par l'image, d'après l'ensemble de nos connaissances actuelles . . . préface par le Dr Henri Leclerc. 3 v. (lxxviii, 447; 504; 636 p.). illus. 23.5 cm. Paris, 1947–48 (1949). (Encyclopédie biologique v. 25, 31, 32.)

Brief history of phytotherapy in the world at large, systematic list of vascular medicinal plants of France (952 genera, in part with mention of species), popular key to species and groups; account of native and cultivated medicinal plants of France, alphabetically arranged by vernacular names, with collected French vernacular names and principal German, English, and Italian names, history, properties, uses, culture, etc.; short bibliography, lists of plants grouped by medical properties and by diseases, indices.—See also Ansberque, above, and additional references there given.

——— Les quatre flores de France, Corse comprise (générale, alpine, méditerranéenne, littorale). Manuel botanique d'excursions de l'étudiant, du touriste, de l'amateur, du spécialiste . . . 2. tirage, xlviii, 1,091 p. illus. (8,075 small fig., incl. map). 20.5 cm. Paris, 1946. (1st printing, Garches [etc.], 1934–40.)

Laws applying to botanical collecting, key to families, etc.; pocket manual of vascular plants (4,217 species) in form of keys, with short statement of range, explanation of generic and specific names, etc., and numerous tiny figures.—For additions and corrections, see Monde des Plant. VI, 42 (247): 33, (248): 43. 1947 (Fournier), (248): 43–46. 1947 (E. Thommen): 45 (265): 11, (266): 21–22. 1950 (M. Breistroffer); 51 (317): 4–6, (318): 11, (319): 15–17. 1956 (A. Berton). (All in Arnold Arboretum library.)

Fron, Georges. Plantes nuisibles à l'agriculture. Caractères botaniques, methodes de destruction. 346 p. 151 fig. 18 cm. Paris, 1917. (Encyclopédie agricole publiée . . . sous la direction de G. Wery.)

Descriptive account of weeds, systematically arranged, with methods of control; annotated tabular list of principal weeds arranged by flowering dates.—See also Gadeceau and Menault, below.

Gadeceau, Émile. Les fleurs des moissons, des cultures, du bord des routes et des décombres (plantes envahissantes). lxviii, 108 p. 39 fig., 100 col. pl. on 50. 16.5 cm. Paris, 1914. (Encyclopédie pratique du naturaliste IV.)

General considerations; semipopular account of principal more or less weedy plants, with description, range, properties, etc.—See also Fron, above, and additional reference there given.

Gatin, C. L. Dictionnaire aide-mémoire de botanique . . . Préface par Ém. Perrot. Révisé et corrigé après la mort de l'auteur par Madame Allorge-Gatin . . . xix, 847 p. illus., port. 18.5 cm. Paris, 1924.

Alphabetical list of botanical abbreviations, vernacular names (French, English, and German), botanical terms (in French, English, German, and Latin, including anatomy, histology, and genetics as well as morphology and taxonomy), and names of families and genera, with definitions or explanations.—See also (for vernacular names) Guyettant, Kreiter, and Rolland, below; also, under Partial, Azais, Bosc, Duboul, Séguy, and Strobl.

Gaussen, Henri. Carte floristique de la France. Annales Géog. 47: 237–256. 1938.

Footnote bibliography; brief treatment of botanical regions, domains, sectors, and districts, with discussion of the area of some species; drawn up with reference to carte 26–1 and planche 25 of the Atlas de France.

Gayffier, Eugène de. Herbier forestier de la France. Reproduction par la photographie d'après nature et de grandeur naturelle des principales plantes ligneuses qui croissent spontanément en forêt. Description botanique, situation, culture, qualités, usages. 2 v. 200 pl. 45 cm. Paris, 1868–73.

Excellent photographs, with descriptive text, illustrating the principal native and naturalized woody plants with notes on uses, etc.—See also Guinier, Hickel, and Mathieu, below.

Gentil, Ambroise. Dictionnaire étymologique de la flore française. xiv, 241 p. 16.5 cm. Paris, 1923. (Encyclopédie pratique du naturaliste XIII.)

List of botanists to whom genera and species of French plants have been dedicated; alphabetical list of generic and specific names of French vascular plants, with derivations.

Gérôme, Joseph. Plantes ornementales herbacées de plein air et rosiers. (Muséum national d'histoire naturelle. (Chaire de culture.) Guides aux collections de plantes vivantes . . . II.) 203 p. 17 cm. Paris, (pref. 1924).

Chronological sketch of introduction of plants into French gardens; list of herbaceous ornamentals and roses, grouped according to duration, season of flowering, etc., with native habitat, date of introduction, frequent remarks on history, horticultural varieties, etc.—See also Fournier (Arbres, arbustes et fleurs), above, and Guillaumin, below.

Gibault, Georges. Les plantes alimentaires indigènes. 27 p. 14.5 cm. Paris, 1904. (Extrait du Petit jardin illustré.) (Oakes Ames Library of economic botany, Harvard University.)

Popular account of principal edible plants.—See also Ansberque, above, and additional references there given.

Gillet, C. C., and Magne, J. H. Nouvelle flore française; descriptions succinctes et rangées par tableaux dichotomiques des plantes qui croissent spontanément en France et de celles qu'on y cultive en grand avec l'indication de leurs propriétés et de leurs usages en médecine, en hygiène vétérinaire, dans les arts et dans l'économie domestique 8. éd. rev. et corr. xxiv, 781 p. illus. 18.5 cm. Paris, 1913. (Arnold Arboretum library.) (1st ed. 1862; ed. 7 and 8 are reissues of ed. 6, 1887.)

Descriptive flora of vascular plants in form of keys, with notes on uses; synonyms indicated in index.

Goujon, Gustave. Les relations entre la végétation française et le climat. xviii, 186, 7 p. 25 fig. (mostly maps) + 33 fig. 32 cm. Paris, 1932. (At head of title page: Ministère de l'air. Mémorial de l'Office national météorologique de France publié sous la direction du Général Dalcambre. no. 23.)

Account of climatic influences (continental or boreal, Atlantic, Mediterranean or southern, and mountain) affecting the French flora, with description of climate, mention of characteristic species, agricultural resources, etc.; bibliography of French works on phytogeography, climate in relation to plant growth, etc.; descriptive treatment of Cupuliferae and Coniferae, taken from Coste's Flore.

Grenier, Charles, and Godron, D. A. Flore de France ou description des plantes qui croissent naturellement en France et en Corse. 3 v. (762, 760, 779 p.). 22 cm. Paris, 1848-56.

Descriptive flora of vascular plants, with localities; old, but still a standard work.

Guillaumin, André. Les fleurs de jardins. 4 v. illus., 256 col. pl., ports. 16 cm. Paris, 1929-36. (Encyclopédie pratique du naturaliste 15, 16, 17, 29.)

Biographical sketches of many French gardeners and horticulturists, bibliography, directions for cultivation, etc.; descriptions and colored plates of French garden plants, grouped according to season (v. 1, spring plants; v. 2–3, summer plants; v. 4, autumn plants); general index in v. 4.—See also Fournier (Arbres, arbustes et fleurs) and Gérôme, above.

Guinier, Philibert. Atlas des arbres, arbustes, arbrisseaux et sous-arbrisseaux croissant spontanément ou naturalisés en France et dans les régions limitrophes. sér. (i.e. no.) 1–9. illus., pls. (pt. col.). 21 cm. Paris, 1912.

Semipopular descriptive account of native or naturalized trees and shrubs, with local and general range, soil and climatic relations, uses, etc.; each of the species is illustrated with a colored plate and a habit photograph. Issued in numbered and separately paged fascicles, not in systematic order; systematic list in Série 1. The work was planned to occupy 28 fascicles, but apparently only 9 were issued.—See also Gayffier, above, and additional references there given.

Guyettant, C. Mémento botanique contenant onze mille trois cents noms vulgaires et tous les noms italiens des plantes utiles ou d'agrément ainsi que leurs noms français et scientifiques. xiii, 436 p. 18 cm. Paris, 1899. (Lloyd Library.)

Alphabetical list of French and Italian vernacular names with botanical equivalents and of botanical names with vernacular equivalents; relates principally to useful or ornamental plants, and includes exotic species.—See also Gatin, above, and additional references there given.

Hariot, Paul. Atlas colorié des plantes médicinales indigènes. x, 221 p. 4 fig., 144 col. pl. on 72. 18.5 cm. Paris, 1900.

Popular treatment of 148 medicinal plants, alphabetically arranged by vernacular names, with vernacular names, uses, etc., and colored plates; briefer account of 216 less important species; list of species, grouped by medicinal uses.—See also Ansberque, above, and additional references there given.

Hermant, Charles, *abbé.* Les plantes alimentaires, condimentaires, saccharifères, oléifères et à boissons de toute la France. 176 p. 100 fig. 8vo. Chalons-sur-Marne, 1938.

Popular account of edible plants, including wild and cultivated species. (Not seen; cited from Monde des Plant. V, 38 (233): 39–40. 1938.)—See also Ansberque, above, and additional references there given.

Hermant, Paul, and Boomans, Denis. La médecine populaire . . . Préface de Albert Marinus. xv, 240 p. illus. 23.5 cm. Bruxelles, Paris, 1929. (Library of Faculté de Pharmacie, Université de Paris.)

Principles of popular medicine (resemblances in color, movement, form, etc.), concept of diseases as material things, animistic concepts, account of several medicinal plants, etc.; the whole with full references.—See also Ansberque, above, and additional references there given.

Hickel, Robert. Graines et plantules des conifères. Bul. Soc. Dendrol. France **1911:** 13–115, 134–204. 93 fig. 1911.—Graines et plantules des angiospermes. l. c. **1912:** 72–141, 147–197. f. 1–23, 2 pl. 1912; **1913:** 7–60, 75–119, 129–178, 185–242. f. 24–85. 1913; **1914:** 31–37. 1914. (Reprinted with title: Graines et plantules des arbres et arbustes indigènes et communément cultivés en France. 2 v. (179, 348 p.). illus. 25 cm. Versailles, 1911–14.)

Descriptive account of the seeds (and in angiosperms also the fruits) of the native and commonly cultivated gymnosperms and woody dicotyledons of France, and separate account of the seedlings; both sections with keys.—See also Gayffier, above, and additional references there given.

Holden, William. Bibliography relating to the flora of France. Embracing botanical section N of the Lloyd Library. Bibl. Contr. Lloyd Libr. v. 1, no. 4. p. 133–186. 23 cm. 1911.

List of books and some separates from journals, alphabetically arranged by authors, with call numbers of those that are in the Lloyd Library; a good working list, although far from complete.—See also Deniker, above, and additional references there given.

Jaume Saint-Hilaire, J. H. Plantes de la France, décrites et peintes d'après nature. 10 vol. 1,000 col. pl., 10 port. 26 cm. Paris, 1808 [1805]–22.

Glossary; semipopular descriptive account of about 1000 wild and cultivated plants, with descriptions, vernacular names in French and various other languages, uses, cultivation.—See also Baillon, above, and additional references there given.

Jovet, Paul, *ed.* Notices botaniques et itinéraires commentés publiés à l'occasion du VIIIe Congrès international de botanique, Paris-Nice, 1954. 48 parts (1,125 p.). illus., maps. 21 cm. Paris, 1954.

Guides for excursions to nearly all parts of France (and to French regions in north Africa), containing account of topography, climate, soils, geobotany, plant associations, general features of flora, etc.; by various authors. Maps and illustrations prepared under direction of H. Gaussen and Paul Rey. Very few parts have been examined.—See also Verlot, below.

Kosch, Alois. Quelle est cette plante médicinale? Les plantes médicinales de France. Leur description et leur activité. 135 p. 49 + 181 fig., 8 col. pl. 20 cm. Paris, 1948.

General considerations; tabular list of 157 medicinal plants, arranged by month of collection, with description, properties, uses, etc.; annotated list of poisonous plants, list of diseases with remedies.—See also Ansberque, above, and additional references there given.

Kreiter, Heinrich. Die von tiernamen abgeleiteten pflanzennamen im französischen. viii, 126 p. Darmstadt, 1912. (Diss.)

Account of French vernacular names of plants derived from animal names. (Not seen; cited from review in Bot. Jahresber. Just **34** (1): 1111. 1914.)—See also Gatin, above, and additional references there given.

LeBrun, Pierre. Principales acquisitions de la flore française depuis 1854. Bul. Soc. Bot. France **101,** Suppl. p. 47–48. 1954.

Running account of about 300 vascular plants added to the flora since the Flore de France of Grenier and Godron (1848–56), arranged geographically and mostly with localities; ***Rosa, Rubus,*** and ***Hieracium*** are omitted.

LeGrand, Antoine. Relevés numériques de quelques flores locales ou régionales de France. Bul. Soc. Bot. France **38**: 190–192. 1891.

List of departmental and regional floras of France, with some critical notes, arranged in order according to number of valid native species included in them (according to Le Grand's concept of species).

Léveillé, Hector. Dictionnaire inventoriel de la flore française. Espèces et races. iv, 44 p. 27.5 cm. Le Mans, 1916. (Arnold Arboretum library.)

Unannotated list of species and races of vascular plants, alphabetically arranged; does not include "micromorphes."

——— Les hybrides de France. Bul. Géog. Bot. **27**: 34–68. 1917.—Correction . . . l. c. **27**: 100. 1917.

List of vascular plants regarded by the author as hybrids, the families arranged alphabetically. The hybrids are listed under the name of only one of the presumed parents (the other not being mentioned), and all are given names ending in "oides," the synonyms being cited. The same new hybrid name often occurs more than once in a genus (e.g., *monspessulanoides* 6 times in *Dianthus*).—See also Camus (Statistique des plantes hybrides), above.

Manciot, André. Alimentation et plantes sauvages. 2 v. (96, 97 p.). illus. 17.5 cm. Paris, n.d. [194–?] (Collections "Toute la nature.") (Arnold Arboretum library.)

Popular account of edible wild plants, grouped by uses; account of poisonous fruits, calendar for collecting, etc. Only vernacular names are given.—See also Ansberque, above, and additional references there given.

Masclef, Amédée, *abbé.* Atlas des plantes de France utiles, nuisibles et ornementales . . . Complément de la "Nouvelle flore" de MM. Gaston Bonnier et Georges de Layens . . . 3 v. 400 col. pl. 24 cm. Paris, 1891. (Reissued 1893.) (Gray Herbarium library.)

Fair to good colored plates, usually with dissections, representing 450 species of vascular plants; explanatory text forms vol. 3.—See also Ansberque, above, and additional references there given.

Mathieu, Auguste. Flore forestière. Description et histoire des végétaux ligneux qui croissent spontanément en France et des essences importantes de l'Algérie. 4. éd. revue par P. Fliche. xxxii, 705 p. 22.5 cm. Paris, Nancy, 1897. (1st ed. 1858.)

Glossary; descriptive flora of native and naturalized woody plants of France (and Corsica) and of the principal ones of Algeria, with keys, botanical description, range, habitat, soil preference, size, rate of growth, reproduction, useful products, injurious insects, etc.—See also Gayffier, above, and additional references there given.

Menault, Ernest, and Rousseau, Henri. Les plantes nuisibles en agriculture et en horticulture et les moyens de les détruire. xvii, 314 p. 80 col. pl. 24 cm. Paris, 1902.

Includes (p. 19–130) account of weedy plants, arranged by flowering dates, with notes on eradication.—See also Fron, above, and additional reference there given.

Meynier, Joseph. Flore. *In* his Les noms de lieu romans en France et a l'étranger. Mém. Soc. Émul. Doubs VII, **3:** 120–167. 1899.

Annotated alphabetical list of Latin plant names and terms, with French locality names derived from them.

Perrot, Émile. Plantes médicinales de France. 4 v. 224 col. pl. 18 cm. Paris, [1928]–43. (Vols. 1–3 are Notice no. 26, 40, 45, Office national des matières premières végétales pour la droguerie . . .)

Bibliography (v. 1), seasons for collecting, cultivated species, etc.; colored plates with descriptive letterpress giving popular account of medicinal uses, methods of cultivation, etc., of 450 species, nearly all French. Name of issuing office varies.—See also Ansberque, above, and additional references there given.

Rawton, Olivier de. Les plantes qui guerissent et les plantes qui tuent. viii, 344 p. 130 fig. 18.5 cm. Paris, 1884. (At head of title page: Bibliotèque instructive.)

Account of French medicinal and poisonous plants, wild and cultivated, arranged by families. There is also an edition in Spanish, apparently a direct translation of the French one, since the ranges given relate primarily to France: Vegetales que curan y vegetales que matan. Tradución de F. G. Brito. viii, 348 p. 130 fig. 18 cm. Paris, Méjico, 1887.—See also Ansberque, above, and additional references there given.

Rehder, Alfred (Phytography.) France. *In* his The Bradley Bibliography. A guide to the literature of the woody plants of the world published before the beginning of the twentieth century. Compiled at the Arnold Arboretum of Harvard University under the direction of Charles Sprague Sargent. v. 1, p. 404–418. 29.5 cm. Cambridge, Mass., 1911. (Publications of the Arnold Arboretum, no. 3.)—Additions and corrections . . . l.c. **1:** 535. 1911; **5:** xxi-xxii. 1918.

Chronological list of floristic publications, with few mainly bibliographical annotations, the titles on dendrography and Corsica separated from the others. It is, for the years covered (1605–1900), the most complete bibliography available. See also the following sections relating to France: Bibliography, general (1: 3); periodicals and serials (1: 20–24); botanic gardens, arboretums, museums, etc. (1: 56–60); dictionaries and lists of vernacular names (1: 74); botanical history (1: 81); phenology (1: 226);

individual noteworthy trees (1: 297); periodicals and serials relating to arboriculture, gardening, and uses of plants (3: 18–23. 1915); works on collections, gardens, arboretums, museums, and exhibitions (not herbaria) (3: 29); history of agriculture and gardening (3: 33); dendrological, pomological, economical, and miscellaneous works dealing with both wild and cultivated plants (3: 83–88); manuals and general treatises on arboriculture and horticulture (3: 110–114); miscellaneous works on arboriculture and horticulture (3: 124–126); pharmacopoeias and dispensatories (3: 255); economic products in general (3: 271); periodicals and serials relating to forestry, timber uses, etc. (4: 12–13. 1914); forestry exhibitions (4: 20); history of forests (4: 27–28); dendrography (4: 85); descriptions of forests (4: 359–362). Various minor sections of no floristic significance have been omitted from this summary.—See also Deniker, above, and additional references there given.

Rey-Pailhade, Constantin de. Les fougères de France. 2 p. l., 133 p. 56 pl. 25 cm. Paris [pref. 1893].

Descriptive account of Filicineae (Hymenophyllaceae, Ophioglossaceae, Osmundaceae, Polypodiaceae), with synonymy, habitat, altitude, and localities, and plate (or part of a plate) of each species; the genera and species arranged alphabetically. A companion work announced as in preparation, Les cryptogames vasculaires de la France (suite aux Fougères), apparently was never published, but Lorenz (Cat. gén. de la librairie française 15: 683) lists his Les selaginelles de France. 8vo. illus. 1899.—See also **Dhien, René,** and **Dubois, Geneviève.** Répartition géographique des fougères françaises. Bul. Soc. Hist. Nat. Toulouse 73: 192–208 incl. illus. (maps). 1939. (List of French Filicineae (Hymenophyllaceae, Ophioglossaceae, Osmundaceae, Polypodiaceae; 56 species) with list of departments in which each is known and small maps showing the range of most of the species; bibliography, including most of the departmental floras and some others.)

Rodin, Hippolyte. Les plantes médicinales et usuelles des champs, jardins, forêts. Description et usages des plantes comestibles, suspectes, vénéneuses . . . 3. éd. xx, 478 p. 200 fig. 16.5 cm. Paris, 1876. (1st ed. 1872.)

General properties (chiefly medicinal) of various plant families; popular account of medicinal plants (mostly French), grouped by uses, short account of poisonous plants. Later editions, including the undated 8th, do not differ from the 3d.—See also Ansberque, above, and additional references there given.

Rolland, Eugène. Flore populaire ou histoire naturelle des plantes dans leurs rapports avec la linguistique et le folklore. 11 v. 23 cm. Paris, 1896–1914.

Bibliography (1: 257–270, with supplements in v. 1–7); systematic list of plants with their vernacular names in different languages (in most European languages in v. 1–3, mostly in French in v. 4–11) with localities and authorities for use, notes on uses and folklore, and lists of personal and locality names derived from them. Covered at first ancient and modern Europe, extreme northern Africa and western Asia, but later consisted mostly of French names. Includes dicotyledons, gymnosperms, Characeae, and some lichens, fungi, and algae; the monocotyledons were omitted, except for Lemnaceae, Potamogetonaceae, and Zosteraceae; presumably they were to be included in v. 12–13, which were never published. Vol. 8–11 edited by Henri Gaidoz (1842–1932); these volumes contain extensive lists of additions and corrections.—See also Gatin, above, and additional references there given.

Rothschild, Jules, and others. Botanique populaire illustrée. Flore pittoresque de la France. Anatomie, physiologie, classification, description des plantes indigènes et cultivées au point de vue de l'agriculture, de l'horticulture et de la sylviculture . . . Avec le concours de MM. Gustave Heuzé, Bouquet de La Grye [etc.] 2. éd. xvi, 473 p. 989 fig., 82 col. pl., map. 28 cm. Paris, [n.d.]. (1st ed., 1885).

Glossary, etc.; popular account of wild and cultivated vascular and cellular plants, with brief descriptions and uses; account of agriculture, horticulture, forests, and fossil flora, with description of regions and zones, mention of principal species, etc.—See also Ansberque, above, and additional references there given.

Rouget, F. Traité pratique des plantes indigènes les plus usitées pour la conservation de la santé. vi, 175 p. 8vo. Toulouse, 1863.

Alphabetical list of indigenous medicinal plants with vernacular names, description, habitat, parts used, doses, properties, etc., followed by list of plants grouped by their medicinal properties, and list of diseases with remedies. (Not seen; title and annotation supplied by G. Dillemann.)—See also Ansberque, above, and additional references there given.

Rouy, Georges, and others. Flore de France ou description des plantes qui croissent spontanément en France, en Corse et en Alsace-Lorraine. 14 v. 22.5 cm. Asnières (Seine), 1893–1913. (Issued as supplements to Annales Soc. Sci. Nat. Charente-Infèr.)— (Supplément.) Conspectus de la flore de France ou catalogue général des espèces, sous-espèces, races, variétés, sous-variétés et formes hybrides contenues dans la "Flore de France." xv, 319 p. port. 23 cm. Paris, 1927.

Bibliography, list of principal exsiccatae, list of French botanists; descriptive flora of vascular plants with keys, references, citation of exsiccatae, details of local range, and extralimital range; special consideration of forms below the specific rank, although the authors' system of nomenclature is peculiar. The standard detailed flora of France, often cited as by Rouy and Foucaud, although Julien Foucaud was co-author only of v. 1–3, as E. G. Camus was of v. 6–7 (and collaborator in early part of v. 8): the other volumes (except treatment of *Rubus*, by J. N. Boulay) were by Rouy alone. The supplement (Conspectus) consists of a systematic list of all valid scientific names in the Flore, with very brief indication of range in France, and an index to families and genera in the Conspectus and the Flore.

Sébillot, Paul. Le folk-lore de France. Vol. 3. La faune et la flore. 1 p. l., ii, 541 p. 25 cm. Paris, 1906.

Livre second, La flore (p. 365–533), discusses, with footnote references, such matters as supernatural origin, planting and cultivation, powers and influences, emblematic significance, relation to health and medicine, use in divination, symbolism and connection with marriage and other customs, connections with legends, etc., of trees and "plantes," considered separately. Only vernacular names are used.

Société botanique de France. Sessions extraordinaires. Bul. Soc. Bot. France v. 2–103, passim. 1855–1956.

Almost from its beginning the Société botanique de France has held summer meetings with accompanying herborizations in different parts of France, rarely in adjacent countries or North Africa. Reports of these meetings with lists of species collected and various other information have been published regularly in the Bulletin of the Society. The more important lists are cited beyond under the respective departments. An alphabetical index to all the reports that seem worth listing is given here, the reference being by department to the volume and year only. The reports are generally to be found in the annual indices under "Session extraordinaire," and for many years were given a separate roman pagination. Ain 23 (1876); Alpes-Maritimes 12 (1865), 30 (1883), 57 (1910), 97 (1950); Alps, from Tyrol to France 98 (1951); Ardennes 32 (1885); Ariège 71 (1924); Aude 9 (1862), 35 (1888); Auvergne 103 (1956); Aveyron 33 (1886); Bas-Rhin 5 (1858); Basses-Alpes 44 (1897); Basses-Pyrénées 15 (1868), 27 (1880), 88 (1941); Cantal 26 (1879), 60 (1913); Charente-Maritime 37 (1890); Cher 17 (1870), 72 (1925); Corse 24 (1877), 48 (1901); Côte-d'Or 17 (1870), 29 (1882); Doubs 16 (1869), 66 (1919); Drôme 59 (1912); Gironde 6 (1859), 49 (1902); Haut-Rhin 5 (1858); Haute-Garonne 11 (1864), 71 (1924); Haute-Savoie 13 (1866); Hautes-Alpes 21 (1874), 69 (1922), 78 (1931), 82 (1935); Hautes-Pyrénées 15 (1868), 54 (1907), 96 (1949); Hérault 4 (1857), 9 (1862), 33 (1886), 40 (1893); Indre-et-Loire 104 (1956); Isère 7 (1860), 59 (1912); Jura 16 (1869), 66 (1919); Landes 88 (1941); Loir-et-Cher 72 (1925); Loire 23 (1876); Loire-Inférieure 8 (1861); Loiret 72 (1925); Lozère 33 (1886); Maine-et-Loire 22 (1875), 104 (1956); Meurthe-et-Moselle 55 (1908); Nièvre 17 (1870); Oise 51 (1904); Paris basin, eastern border, from Seine to Moselle south to Loiret and Haute-Saône, 102 (1955); Puy-de-Dôme 3 (1856), 60 (1913); Pyrénées-Orientales 19 (1872), 35 (1888), 38 (1891); Rhône 23 (1876); Saône-et-Loire 17 (1870); Savoie 10 (1863), 67 (1920); Seine-et-

Marne 28 (1881), 51 (1904); Seine-et-Oise 51 (1904); Var 46 (1899); Vendée 8 (1861), 58 (1911); Vosges 5 (1858), 55 (1908).

Verlot, Bernard. Le guide du botaniste herborisant. Conseils sur la récolte des plantes, la préparation des herbiers, l'exploration des stations de plantes phanérogames et cryptogames, la botanique fossile, et les herborisations aux environs de Paris, dans les Ardennes, la Bourgogne, [etc.]. 3. éd., rev. et augmentée. Avec un introduction par M. Naudin. xvi, 764 p. 34 fig., front. 16 cm. Paris, 1886. (Lloyd Library.) (1st ed. 1865.)

Detailed directions for collecting plants (including fossils), etc.; directions for botanical excursions at numerous localities throughout France and in Corsica, with itineraries and lists of principal species; classified bibliography (p. 736–756).—See also Jovet, above.

Partial

See also Paris (region); also Spain, General (Bubani, Flora pyrenaea).

Under this heading are included works that deal with all or part of two or (usually) more departments but do not correspond to any single one of the old provinces, the principal areas being the following: Alps, Bas-Rhône, Causses Cévenols, Cévennes, Jura Mountains, Loire valley, Lyonnais, Massif Central, Pyrenees, Rhône valley, Riviera, and works dealing with the center, north, south, southwest, and west of France.

Works relating to the old provinces of France in their entirety, such as Alsace, Normandie, and Provence, as well as the Paris region ("région parisienne"), will be found under these provincial names which are alphabeted among the names of the departments.

Cross-references to titles included in the General and Partial sections are not given under the various departments and provinces, but cross-references are given under the departments to the respective provincial floras (such as Alsace, Champagne, Paris region, etc.), which will be found listed in the same alphabet with the departmental floras.

The partial titles here listed may be classified as follows: *Floras:* Babey (Jura), Bonnier (northern France), Boreau (central France), Cariot (basin of the Rhône and the Loire), Écorchard (Paris and western France), Fournier (plaine française), Gandoger (Lyonnais and southeastern France), Grenier (Jura), Guillard (southwest France), Lloyd (western France), Parmentier (Jura), Philippe (Pyrenees), Thompson (Riviera). *Catalogs:* Beille (Massif Central), Braun-Blanquet (high Alps), Fourreau (valley of the Rhône), Gaussen (Pyrenees), Lamotte (central plateau), Magnin (Jura and Lyonnais; Lyonnais), Pantel (Cévennes), Revel (southwest France), Saint-Lager (basin of the Rhône), Zetterstedt (Pyrenees). *Bibliographies:* Gaussen (Végétation . . . des Pyrénées), Offner and Le Brun (Alps). *Vernacular names:* Azais (Languedoc, Provence, etc.) Bosc (Bas-Rhône), Duboul (Languedoc, Gascogne, etc.), Séguy (central Pyrenees), Strobel (toponymy, central southern France). *Medicinal plants:* Bel (southern France), Bénézech (Causses Cévenols; includes also edible plants), Garnaud (Massif Central), Offner and Pons (Alps). *Miscellaneous:* Gaussen (Introduction: list of herbaria, local botanists, etc. of Pyrenees), Lassimonne (botanists, museums, etc., of Massif Central), Miègeville (botanical explorations in Pyrenees), Strobel (toponymy, central southern France), Voigt (vegetation of Riviera).

Azais, Gabriel. Catalogue botanique. Synonymie languedocienne, provençale, gasconne, quercinoise, etc. iv, 185 p. 24.5 cm. Béziers, 1871.

Alphabetical list of French names with corresponding scientific and patois names, alphabetical list of patois names with reference to first list, alphabetical list of patois names of mushrooms with scientific and patois equivalents. (Not seen; annotation supplied by Mme. G. Duprat from microfilm in library of Muséum national d'histoire naturelle, Paris.)—For other lists of vernacular names of the regions mentioned see under Languedoc and Provence; also Bosc and Duboul, below; also under General, above, Gatin, and additional references there given.

Babey, C. M. P. Flore jurassienne ou description des plantes vasculaires croissant naturellement dans les montagnes du Jura et les plaines qui sont au pied . . . 4 v. table. 22 cm. Paris, 1845 (cover date 1846).

Topography, climate, cultivated plants, geology; annotated descriptive flora of vascular plants, with localities and notes on uses. Covers the French and Swiss Jura, from Ain to Doubs and from Genève to Basel.—See also Grenier, Magnin, and Parmentier, below; also Switzerland, General (Friche-Joset; Godet; Thurmann.)

Beille, Lucien. Essai sur les zones de végétation du Massif Central de la France. Bul. Soc. Sci. Phys. et Nat. Toulouse 8: 447–594. map. 1889.

Geography, topography, plant zones, etc.; tabular list of vascular plants (1608 species), with indication of zones in which they occur.—See also **Coste, Hippolyte.** Plantes nouvelles ou récemment découvertes dans les Cévennes et le Massif Central. Monde des Plant. III, **22** (14–129): 5–8, (15–130): 4–7, (17–132): 5–7, (18–133): 6–7. 1921; **23** (19–134): 6–7. 1922. (List with localities.)—Also **Dhien, René.** Les fougères du Morvan. Bul. Scient. Bourgogne **11:** 23–28. 1948. (List of 27 ferns of the Massif du

Morvan, the northeasternmost part of the Massif Central, in the Departments of Cote-d'Ôr, Nièvre, Saône-et-Loire, and Yonne.)—Also his La flore du Morvan. Monde des Plant. 46 (274/5): 7–9. 1 fig. (map). 1951. (List of scarcer angiosperms. In Arnold Arboretum library. For another work on the Morvan, see Gillot under Saône-et-Loire.)—Also Bénézech, Garnaud, Lamotte, Lassimonne, and Pantel, below.

Bel, Jules. Les plantes médicinales du midi de la France. 128 p. 25 cm. Paris, 1897. (Gray Herbarium library.)

Annotated systematic list of wild and cultivated medicinal plants, with uses.—See also Ansberque, under General, above, and additional references there given.

Bénézech, A. Les Causses Cévenols (essai de géographie botanique et d'économie rurale). 4 p.l., 108 p. illus. 25.5 cm. Millau, 1919. (Thèse.) (Gray Herbarium library.)

Plant associations, etc.; includes (p. 56–64) "Les plantes des Causses employées dans l'alimentation ou dans la médecine populaire des campagnes," an annotated list of edible and medicinal plants, with vernacular names and uses. Relates to Aveyron, Gard, Hérault, and Lozère.—See also Beille, above, and additional references there given; also Ansberque under General, above, and additional references there given.

Bonnier, Gaston, and Layens, Georges de. Nouvelle flore du nord de la France et de la Belgique pour la détermination facile des plantes sans mots techniques . . . 8. éd. rev. et corr. 2 p.l., xxxiv, 309 p. 2,282 fig., map. 18 cm. Paris, [1926?] (1st ed. 1887(?).)

Pocket flora of vascular plants in form of illustrated keys; index of botanical names with derivation of generic names, localities for the scarcer species, and brief notes on uses; indices of French and Flemish names, sketch of botanical regions, with map. Covers Belgium and the French departments Pas-de-Calais, Nord, Somme, Oise, Aisne, and Ardennes. The eighth edition differs in no way from the seventh; there are later editions, the last one noted in the literature being the 14. éd. 285 p. 2,713 fig. (date?)

Boreau, Alexandre. Flore du centre de la France et du bassin de la Loire, ou description des plantes qui croissent spontanément ou qui sont cultivées en grand dans les départements arrosés par la Loire et ses affluents, avec l'analyse des genres et des espèces. 3. éd. augmentée. 2 v. (xvi, 356; 1 p.l., 771 p.). 22.5 cm. Paris, 1857. (1st ed. 1840.)

Topography, general features of flora, plant formations, botanical explorations, bibliography; descriptive flora of vascular plants and Characeae (2891 species), with localities. Covers all or part of 29 departments, which are named in the introduction.—See also Cariot, below.

Bosc, J. J. Tableau de quelques végétaux indigènes de la région du Bas-Rhône avec la concordance des noms vulgaires provençaux et languedociens. 33, x p. 24.5 cm. Nîmes, 1893. (Arnold Arboretum library.)

List of vascular (and some cellular) plants, with vernacular names in French, Provençal, and Languedocien.—See also works on vernacular names listed under Languedoc and Provence; also Azais, above, and Duboul, below; also under General, above, Gatin, and additional references there given.

Braun-Blanquet, Josias. La végétation alpine et nivale des Alpes françaises. *In* Chouard, Pierre, *ed.* Étude botanique de l'étage alpin particulièrement en France. p. 27–96. 7 fig., pl. 2–5. 26 cm. Bayeux, 1954.

Includes (p. 88–92) list of 120 angiosperms observed in the Alps of France above 2,900 meters elevation, with localities; bibliography.

Cariot, Antoine, *abbé,* **and Saint-Lager, J. B.** Flore descriptive du bassin moyen du Rhône et de la Loire. 8. éd. rev. et augmentée . . . xxxv, 1,004 p. 18.5 cm. Lyon, 1889. (Cariot's Études des fleurs. Botanique élémentaire, descriptive et usuelle. v. 2.) (Arnold Arboretum library.) (Reissued 1897; 1st ed. 1841, Étude des fleurs . . . by Chirat, Ludovic.)

Descriptive flora of vascular plants and Characeae (2,935 species), with localities. The 1st edition, 1841, was by Chirat; the 3d edition, 1860, although regarded by Cariot as an edition of Chirat, was really a new work. Relates to the departments of Ain, Ardèche, Drôme, Haute-Savoie, Hautes-Alpes, Isère, Loire, Rhône, and Savoie.—See also Boreau, above, and Fourreau, Magnin, and Saint-Lager, below.

Duboul, Axel. Las plantos as camps. Glossoire patois. 2. éd., rev. et corr. 80 p. 24 cm. Toulouse, 1890. (Newberry Library, Chicago.) (1st ed. 1886.)

Alphabetical list of patois names, in some cases with explanation of etymology, and with scientific and French vernacular equivalents; no index. "Les noms patois appartiennent à l'idiome parlé sur les confins de la Gascogne, de la Guyenne, du Languedoc et du comté de Foix;" they were collected in the departments of Ariège, Haute-Garonne, Tarn, and Tarn-et-Garonne.—See also Azais and Bosc, above; also under General, above, Gatin, and additional references there given.

Écorchard, J. M. Flore régionale de toutes les plantes qui croissent spontanément ou qui sont généralement cultivées en pleine terre dans les environs de Paris, et les départements de Seine-Inférieure, Calvados, Eure, Manche, Orne, Maine-et-Loire, Ille-et-Vilaine, Côtes-du-Nord, Finistère, Morbihan, Loire-Inférieure, Vendée, Deux-Sèvres, Charente-Inférieure et Gironde. v. 1. lxxviii, 794 p. 19 cm. Paris, 1877 (cover date, 1878). (Arnold Arboretum library.)

Briefly annotated, keyed, descriptive flora of vascular plants and Characeae, with vernacular names and uses and some localities.—See also his Synopsis de la flore des environs de Paris et des départements maritimes du nord-ouest et du sud-ouest de la France. xliii, 303 p. 17.5 cm. Paris, 1878. (Consists essentially of the descriptions of the families and genera and the keys (but not the descriptions) of the species from his Flore régionale. In Arnold Arboretum library.)

Fournier, Paul, *abbé*. Flore complétive de la plaine française. Genres complexes, espèces collectives, hybrides, classement des sous-espèces et variétés. Région parisienne, ouest, centre, nord, est. xii, 632 p. 565 fig. 18 cm. Paris, 1928.

Brief bibliography; barely annotated keys to the species of a great number of genera of vascular plants, the subspecies and minor variations keyed in footnotes. Also issued as thesis, with title: Étude systématique des genres et espèces complexes de la plaine française. Originally published (1924-27) under the title: Bréviaire du botaniste. Florule de poche des genres et espèces complexes et de leurs hybrides. (Not seen.)

Fourreau, Jules. Catalogue des plantes qui croissent spontanément le long du cours du Rhône. Annales Soc. Linn. Lyon n.s., **16**: 301–304. 1868; **17**: 89–200. 1869. (Reprinted 216 p.)

General features of flora; annotated list of vascular plants, with localities of the rarer species.—See also Cariot, above, and Magnin and Saint-Lager, below.

Gandoger, Michel. Flore lyonnaise et des départements du sudest comprenant l'analyse des plantes spontanées et des plantes cultivées comme industrielles ou ornamentales avec l'indication de quelques-unes de leurs propriétés principales . . . viii, liv, 322 p. 18.5 cm. Paris, 1875.

Flora of vascular plants and Characeae (2,063 species) in form of keys, with local vernacular names, notes on uses, and localities for rarer species; includes cultivated plants.—See also Magnin, below.

Garnaud, —, Huguet, —, and **Perrin, —.** Les principales plantes médicinales du Massif Central. 71 p. illus. 8vo. [Clermont-Ferrand, 1922 ?] (At head of title page: Comité régional des plantes médicinales de Clermont-Ferrand.) (In library of Faculté de Pharmacie, Université de Paris.)

Includes annotated list of medicinal plants, arranged by collecting dates, with vernacular and patois names, botanical characters, properties, uses, etc. There is a condensed ed., 16 p. 1921, without illustrations. (Not seen; annotation supplied by G. Dillemann.)—See also under General, above, Ansberque, and additional references there given.

Gaussen, Henri. Catalogue-flore des Pyrénées. Monde des Plant. **48** (293/297): 11–14. illus. (incl. map), (298/302): 3–7. 1953; **50** (315): 3–5. 1955; **51** (319): 17–18, (320): 25–26. 1956. (Arnold Arboretum library.)

List of vascular plants of the French and Spanish Pyrenees, with distribution indicated by districts. The parts so far published cover pteridophytes, gymnosperms, monocotyledons in part (Alismaceae-Cyperaceae, part, the families arranged alphabetically).—See also his two works below; also

Miègeville, Philippe, Séguy, and Zetterstedt, below; also Spain, General (Bubani, Flora pyrenaea).

——— Introduction à la flore des Pyrénées. Bul. Soc. Hist. Nat. Toulouse 65: 27–48. 1 fig. (map). 1933.

Includes lists of localities (with map), herbaria, local botanists, and botanical authors.—See also his Flora des Pyrénées. Bul. Soc. Hist. Nat. Toulouse 65: 49–119 incl. illus. (incl. maps). 1933. (Detailed taxonomic treatment of the genus *Thalictrum*, by Fernande Flous and H. Gaussen. No more published. This and the preceding paper reprinted as Documents pour la carte des productions végétales. Série: Pyrénées. Tome: Généralités. vol. 4, no. 1. 1 pl., 22, 71 p. illus. (incl. maps). Paris, 1933. In Arnold Arboretum library.)

——— Végétation de la moitié orientale des Pyrénées: sol, climat, végétation. Bul. Soc. Hist. Nat. Toulouse 55: 5–564. 32 pl., 2 maps. 1926. (Reprinted as Documents pour la carte des productions végétales. Série: Pyrénées. Tome: Généralités. v. 1. 559 p.)

Phytogeographical and ecological; cited here for its brief sketch of botanical explorations (p. 14–20) and bibliography (p. 509–553). Relates to all or part of Ariège, Aude, Haute-Garonne, and Pyrénées-Orientales, and in Spain Barcelona, Gerona, Huesca, and Lérida.—See also Deniker under General, above, and additional references there given.

Grenier, Charles. Flore de la chaîne jurassique. Mém. Soc. Émul. Doubs III, v. 10. 1001 p. 1864–69.—Revue de la flore des Monts Jura. l.c. IV, 9: 363–454. 1875. (Reprinted 90 p. 1876.)

Topography, geology, etc.; annotated descriptive flora of vascular plants, with localities for scarcer species. The "Revue" contains the preface to the whole work and additions to p. 1–345 of the "Flore."—See also Babey, above, and additional references there given.

Guillaud, J. A. Flore de Bordeaux et du sud-ouest. Analyse et description sommaire des plantes sauvages ou généralement cultivées dans les parties non montagneuses des bassins de la Garonne, de la Charente et de l'Adour . . . v. 1. Phanérogames. cvii, 217 p. 17.5 cm. Bordeaux, Paris, 1883.

Briefly annotated flora of spermatophytes in form of keys, without localities. No more published. Covers part or all of about 20 departments.—See also Revel, below.

Lamotte, Martial. Prodrome de la flore du plateau central de la France. Mém. Acad. Sci., Belles-Let. et Arts Clermont-Ferrand **17**: 117–226, 285–344, 441–620. 1875; **20**: 455–548. 1878; **21**: 125–190, 491–543. 1879; **22**: 112–174. 1880. (Reprinted with title: Prodrome de la flore du plateau central de la France comprenant l'Auvergne, le Velay, la Lozère, les Cévennes, une partie du Bourbonnais et du Vivarais. 2 v. (628 p.) 23 cm. Paris, 1877–81.)

Botanical explorations, bibliography, geography, topography, geology, general features of flora; annotated list of vascular plants, with localities. Incomplete (Ranunculaceae—Globulariaceae.) Replaces, so far as it goes, **Lecoq, Henri,** and **Lamotte, Martial.** Catalogue raisonné des plantes vasculaires du plateau central de la France. Annales Scient., Litt. et Indus. Auvergne **20**: 101–527. 1847. (Reprinted with title: Catalogue raisonné . . . comprenant l'Auvergne, le Velay, la Lozère, les Cévennes, une partie du Bourbonnais et du Vivarais. 440 p. Paris, 1847. Reprint in Arnold Arboretum library.) (Consists of 2 parts: Considérations générales sur la végétation du plateau central . . . (p. 101–140), with account of topography, general features of flora, etc.; and Catalogue raisonné (p. 141–527), an annotated list of vascular plants and Characeae, with localities.) The following work, despite its title, contains only a partial list of the plants of the central massif and no statement of their local ranges, although it discusses soil preferences: **Lecoq, Henri.** Études sur la géographie botanique de l'Europe et en particulier sur la végétation du plateau central de la France. 9 v. 2 col. pl., table. 24.5 cm. Paris, 1854–58.—See also Beille, above, and additional references there given.

Lassimonne, S. E., and **Lauby, Antoine.** Catalogue des collections botaniques du Massif Central. Revue Scient. Bourbonn. 18. 29–77, 109–167. 1904–05. (Reprinted xxvii, 216 p. map. Moulins (Allier), 1905. Reprint in Gray Herbarium library.)

Geography, geology, topography, climate, general features of flora, etc.; alphabetical list of botanists (with details of herbaria), local publications, societies, and museums, arranged alphabetically by departments. P. 2–206 of reprint printed on recto only.

Lloyd, James. Flore de l'ouest de la France ou description des plantes qui croissent spontanément dans les départements de: Charente-Inférieure, Deux-Sèvres, Vendée, Loire-Inférieure, Morbihan, Finistère, Côtes-du-Nord, Ille-et-Vilaine. 5. éd., publiée par les soins de M. Émile Gadeceau. cxxiv, 458 p. port. 20 cm. Nantes, 1897 (cover date, 1898). (1st ed. 1854.)

Lists of species of different habitats; botanical explorations, physiography, soils, etc. of each department; annotated, briefly descriptive flora of vascular plants and Characeae, with localities.

Magnin, Antoine. Annotations et additions aux flores du Jura et du Lyonnais et remarques sur l'inégale répartition de quelques plantes considérées comme communes. Mém. Soc. Émul. Doubs VI, **9**: 329–412. 1895; **10**: 227–317. 1896.

Bibliography; additional localities for the departments of Doubs, Jura, Ain, Rhône and adjacent Isère, Savoie, and Loire, supplementing those in Cariot and Saint-Lager, Grenier, Parmentier, and Saint-Lager (all given above), and in Rouy (above, under General).—See also **Hétier, François.** Annotations et additions . . . 2. ptie. Contribution à l'étude botanique des bassins lacustres de la chaîne jurassique. Mém. Soc. Émul. Doubs VII, **1**: 85–173. 1897. (Reprinted with the above in their Observations sur la flore du Jura et du Lyonnais. ix, 282 p. 22.5 cm. Besançon, 1894–97.)—Also Archives de la flore jurassienne. 1.–7. année (no. 1–67). 1900 [1899?]–1906, passim, for numerous new records for the flora of the Jura region in the broadest sense. (No. 60–67 not seen.)—Also under Doubs (Magnin, La végétation des monts Jura).—Also Babey, above (for Jura), and additional references there given; also Gandoger, above (for Lyonnais), and reference just below.

——— La végétation de la région lyonnaise et de la partie moyenne du bassin du Rhône ou description topographique, géologique et botanique des regions du Lyonnais, du Beaujolais, de la Dombes et du Bas-Dauphiné . . . i–xvi, 6–513 p. 7 maps. 27 cm. Bâle, Lyon, Genève, 1886. (Arnold Arboretum library.)

Mainly phytogeographical and ecological; includes (p. 302–398) annotated lists of silicicole and calcicole vascular plants with localities; also (p. 457–486) account of naturalized and adventive species and species that have disappeared.—See also Gandoger, above.

Miègeville,—, *abbé.* Histoire de la botanique dans les Pyrénées. Bul. Soc. Bot. France **15** (Sess. Extraord.): xxvii–xxxiv. 1868.

Account of botanical explorations and publications.—See also Gaussen, above, and additional works there cited.

Offner, Jules, and **Le Brun, Pierre.** Un siècle de floristique à travers les Alpes françaises. Bul. Soc. Bot. France **103**: 298–375. 1956.

Changes in flora; condensed account of botanical investigations and publications, region by region; alpine gardens, lists of botanists and locality names, extensive bibliography. Covers region from Haute-Savoie to Alpes-Maritimes, west to Isère, Drome, and Vaucluse.—See also **Le Brun, P.** Un siècle . . . (Additions et corrections). Bul. Soc. Bot. France **104**: 339–351. 1 fig. (map). 1957.—Also Deniker under General, above, and additional references there given.

——— and **Pons, Joseph.** Les plantes médicinales et aromatiques des Alpes françaises. 184 p. 124 fig. 12mo. Gap, 1931.

Popular work containing the descriptions of 124 medicinal plants, alphabetically arranged by vernacular names, with properties, uses, method of harvesting, etc.; calendar of harvesting dates, list of little-used drug plants, etc. (Not seen; cited from Bul. Soc. Bot. France **80**: 251. 1933 and Biol. Abstr. **9**: 2147. 1935.)—See also under General, above, Ansberque, and additional references there given.

Pantel, C. Formation et aspect du relief actuel des Cévennes avec la liste des plantes qui croissent dans ce pays. 77 p. 18.5 cm. Paris, 1885. (Lloyd Library.)

Elementary account of geology and topography; unannotated list of vascular plants, based on collections of author and others, the families arranged alphabetically. (Chain of mountains in Haute-Loire, Ardèche, Lozère, Aveyron, Gard, and Hérault.)—See also Beille, above, and additional references there given.

Parmentier, Paul. Flore nouvelle de la chaîne jurassique et de la Haute-Saône à l'usage du botaniste herborisant. Bul. Soc. Hist. Nat. Autun **7**: 124–431. 1894 [1895]. (Reprinted 307 p. Autun, 1895. In Arnold Arboretum library.)

Bibliography; annotated list of vascular plants, with keys to genera, differential characters under each species and indication of range by districts, followed by key to families.—See also Babey, above, and additional references there given.

Philippe, Xavier. Flore des Pyrénées. 2 v. (605, 502 p.). 22 cm. Bagnères-de-Bigorre, 1859.

Descriptive flora of vascular plants and Characeae, with localities for scarcer species. Covers both French and Spanish slopes. Pritzel gives date as 1859–60, but both volumes are dated 1859.—See also **Loret, Henri.** Notice sur l'herbier et la Flore des Pyrénées de Philippe. Bul. Soc. Bot. France **30**: 50–57. 1883. (Critical notice, pointing out many errors.)—Also **Neyraut, E. J.** Stations de quelques plantes rares ou peu communes des Pyrénées. Compt. Rend. Cong. Soc. Sav. Paris et Dépt. Paris, **1910**, Sect. Sci. p. 114–123. 1911. (List of rare or newly recorded vascular plants collected by author, with localities.)—Also Gaussen, above, and additional references there given.

Revel, Joseph, *abbé.* Essai de la flore du sud-ouest de la France, ou recherches botaniques faites dans cette région. 2 pt. (845 p.). pl. 23 cm. Villefranche, 1885–89; Rodez, 1900. (Publications de la Société des lettres, sciences et arts de l'Aveyron.)

Topography, botanical investigations, special localities with mention of noteworthy species; annotated list of vascular plants (1838 species) with localities. The region covered includes all of Aveyron, Dordogne, Gironde, and Lot, the greater part of Cantal, Corrèze, and Lozère, and parts of Charente, Charente-Inférieure, Gard, Lot-et-Garonne, Tarn, and Tarn-et-Garonne.—See also Guillaud, above.

Saint-Lager, J. B. Catalogue des plantes vasculaires de la flore du bassin du Rhône. viii, 886 p. 27.5 cm. Lyon, 1883. (Issued in 7 pts., 1873–82, with Annales Soc. Bot. Lyon, v. 1–4, 6, 9–10.)

List of vascular plants, with localities. (Southeastern France and adjacent Switzerland.)—See also **Durand, Théophile.** Observations sur le Catalogue . . . Bul. Soc. Roy. Bot. Belg. **21** (2): 7–15. 1882.—Also Cariot, above, and additional references there cited.

Séguy, Jean. Les noms populaires des plantes dans les Pyrénées centrales. Monog. Inst. Estud. Pirenaicos no. gen. 100 (Filología 18). xxix, 444 p. incl. 16 fig. (distr. maps) on 4 pl. (on 2), map. 24 cm. Barcelona, 1953. (At head of title page: Consejo superior de investigaciones científicas. Instituto de estudios pirenaicos.)

Extensive linguistic bibliography, discussion of materials, etc.; systematic list of vascular and a few cellular plants with patois names and localities; discussion of phonetics, origin and formation of names; indices of patois names with abbreviated scientific equivalents, of vernacular and patois names cited in the discussion, grouped by languages, and of scientific and French vernacular names. (In Ariège, Haute-Garonne, and Hautes-Pyrénées.)—See also Gatin under General, above, and additional references there given.

Strobel, Heinrich. Die von pflanzennamen abgeleiteten ortsnamen einiger südfranzösischer departements. 102 p. 21 cm. Tübingen, 1934 [1936]. (Diss.)

List of 100 names of plants and trees, with locality names derived from them and etymological analyses; indices. Relates to departments of Aude, Cantal, Dordogne, Drôme, Gard, and Hérault. (Not seen; title and annotation supplied by G. Rohlfs, Munich.)—See also Gatin under General, above, and additional references there given.

Thompson, H. S. Flowering plants of the Riviera. A descriptive account of 1800 of the more interesting species . . . with an introduction on Riviera vegetation, by A. G. Tansley . . . xxvii, 249 p. 32 pl. (pt. col.). 21.5 cm. London [etc.], 1914.

General features of flora (Tansley); annotated descriptive flora of more important and conspicuous phanerogams; annotated list of vascular cryptogams. Covers French coast of Var and Alpes-Maritimes, from about Hyères eastward, and coast of Liguria, Italy.—See also **Bicknell, Clarence.** Flowering plants and ferns of the Riviera and neighbouring mountains. viii, [166], iv p. 82 col. pl. 29 cm. London, 1885. (Attractive colored plates (two or three species to a plate) of characteristic plants, with appropriate text.)—See also Voigt, below.

Voigt, Alban. Die Riviera. v, 466 p. 6 pl., map. 16.5 cm. Berlin, 1914. (Junk's Natur-führer.)

Account of general features of vegetation, plant formations, etc. (p. 1–283); annotated list of plants cultivated at the La Mortola garden, near Ventimiglia, Italy (p. 288–423).—See also Thompson, above.

Zetterstedt, J. E. Plantes vasculaires des Pyrénées principales. lvii, 330 p. map. 22 cm. Paris, 1857.

Topography, geology, climate, botanical explorations, plant zones, statistics; annotated list of vascular plants (1,477 species), with localities; gazetteer. The area covered is the French slope of the central Pyrenees (in Haute-Garonne and Hautes-Pyrénées) with a small corner of Aragon (Spain).—See also Gaussen, above, and additional references there given.

AIN

See also Bourgogne; Switzerland, General (Binz and Thommen, under Binz).

General

Durafour, Arthur. Fruits sauvages comestibles de l'Ain. Bul. Soc. Nat. Ain **11:** 22–25. 1908.

Descriptive account of 26 wild plants furnishing edible fruits. (Not seen; title and annotation supplied by R. McVaugh from copy in University of Michigan library.)

Huteau, H., and **Sommier, F.** Catalogue des plantes du département de l'Ain. Annales Soc. Émul. (Agr., Sci., Let. et Arts) Ain **27:** 85–292. 1894. (Reprinted, 212 p. Bourg, 1894. In Arnold Arboretum library.)

List of local botanists; annotated list of vascular plants and Characeae, with localities for scarcer species. Replaces **Fray, J. P.** Liste des plantes phanérogames et cryptogames semi-vasculaires du département de l'Ain. 24 p. 24 cm. Bourg, 1878. (Unannotated list of vascular plants and Characeae, without localities. In Gray Herbarium library.) Additions by various authors are given in Bul. Soc. Nat. Ain (later Bul. Soc. Nat. et Archéol. Ain) from no. 5, 1899, on.—See also **Thommen, Édouard.** Contributions à la flore du département de l'Ain. Bul. Soc. Bot. Genève II, **32:** 103–154. 5 fig. (fig. 1, map). 1941. (Collaborators, botanical districts, short bibliography; extensive list of vascular plants, with localities.)

Local

Bouveyron, L. Études floristiques sur le Revermont. Bul. Soc. Nat. et Archéol. Ain **48:** 75–96. 1934.

Description of region; list of about 200 species collected by author, with localities and notes on abundance. (Not seen; title and annotation supplied by R. McVaugh from work in University of Michigan library.)

Calloni, Silvio. Florule des environs de Nantua. Bul. Trimest. Soc. Bot. Lyon II, **3:** 124–137. 1885. (Arnold Arboretum library.)

Botanical explorations, topography, lists of plants of several habitats.—See also **Lingot, Félix.** Les plantes des environs de Nantua . . . Bul. Soc. Nat. Ain **3:** 18–19. 1898; **4:** 12–13. 1899. (List of about 100 species. Not seen; title and annotation supplied by R. McVaugh.)—Also **Roux, Nisius.** Plantes de l'excursion de Nantua. Annales Soc. Bot. Lyon **33:** xl–xli. 1908. (List of noteworthy angiosperms.)—Also his Présentation de plantes de l'excursion de Nantua. l.c. **34:** xviii. 1909. (Short list of species.).

Gillot, Xavier. Rapport . . . sur l'herborisation faite les 29, 30 juin et 1er juillet 1873 dans le Bugey et le Valromey (Tenay, Hauteville, forêt de Mazières, forêt d'Arvières et Colombier du Bugey). Bul. Soc. Bot. France **23** (Sess. Extraord.): civ-cxl. 1876.

Previous publications; itinerary, with mention of plants collected and various critical notes.—See also **Chenevière, —.** Note additionelle sur la flore du Bugey. Bul. Soc. Bot. France **23** (Sess. Extraord.): cxl-cxliii. 1876. (List of noteworthy species from the vallée d'Albarine, with localities.)—Also **Saint-Lager, J. B.** Le "Carex brevicollis" de Tenay et la distribution géographique de cette espèce. Annales Soc. Bot. Lyon **2**: 54–68. pl. 1874; Herborisation à Tenay (suite). l.c. **2**: 88–91. 1875. (Includes mention of various noteworthy species.)—Also **Grenier, Charles.** Herborisations dans les environs de Tenay. l.c. **3**: 41–46. 1876. (Itinerary, with mention of noteworthy species.)—Also **Méhu, Adolphe.** Compte-rendu de l'herborisation faite de Tenay à Hauteville, dans la forêt de Mazières et au Vély. l.c. **3**: 116–127. 1876. (Itinerary, with mention of noteworthy species collected.)—Also **Saint-Lager, J. B.** Notice sur la végétation de la forêt d'Arvières et du Colombier du Bugey. l.c. **3**: 128–141. 1876. (Topography, geology; itinerary, with mention of noteworthy species; phytogeography.)—Also **Meyran, Octave.** Le Colombier-du-Bugey. (Notes botaniques.) Feuille des Jeunes Nat. **14**: 115–117, 130–134. 1884. (Partial lists of vascular plants, arranged by localities and habitats.)—Also **Durafour, Arthur.** Flore du Bugey; observations faites en 1904. Arch. Fl. Jurass. **5** (no. 47/48): 61–62. 1904. (List of noteworthy species, with localities.)—Also **Beauverie, Jean.** Herborisation au Grand Colombier du Bugey. Études Rhodaniennes **5**: 131–144 incl. 4 fig. (incl. 2 maps). 1929. (Itinerary, with mention of various species; bibliography.)

AISNE

See also Picardie.

General

Riomet, L. B. Flore de l'Aisne. Bul. Trimest. Union Soc. Franç. Hist. Nat. no. 12 bis, 13 bis, 14 bis, 16 bis, 20 bis, II. 10 bis. 240 p. (+). illus. (maps). 1952–57.

Annotated list of vascular plants, with localities. Page 240 reaches species 1142 (*Menyanthes*), on Engler-Prantl system. Edited by Marcel Bournérias.

Local

Blin-Paillet, —, and **Blin, Louis.** Statistique botanique de l'arrondissement de Saint-Quentin. Trav. Soc. Acad. Sci. Arts, Belles-Let. et Agr. Saint-Quentin III, **4**: 65–120. 1863.

Unannotated list of spermatophytes, with vernacular names.—See also **Magnier, Charles.** Catalogue des plantes intéressantes de marais de la Somme auprès de Saint-Quentin (Aisne). Revue Bot. Bul. Mens. Soc. Franç. Bot. **2**: 288–293. 1884. (List of 58 noteworthy vascular plants, with localities.)

Jovet, Paul. Le Valois. Phytosociologie et phytogéographie. 389 p. illus., plates. 24.5 cm. Paris, 1949.

Includes (p. 41–54) account of botanical explorations with references. (Mostly in Aisne and Oise.)—See also his Some recent modifications in the flora and the vegetation of the Valois. *In* Lousley, J. E., *ed.* The changing flora of Britain. p. 46–48. 22 cm. Arbroath, 1958. (Changes in flora due to change in habitats.)

La Fons, Alexandre de, *baron de Mélicocq.* Prodrome de la flore des arrondissements de Laon, Vervins, Rocroy et des environs de Noyon. 1 p. l., 68 p. 21 cm. Noyon, 1839. (Arnold Arboretum library.)

Soils; etc., annotated list of spermatophytes.—See also **Copineau, Charles.** Contributions à la flore locale. Environs de Laon. Bul. Soc. Linn. Nord France **7**: 36–43. 1884. (Short lists of species observed by author, arranged by localities.)—Also **Courteville, H.** Note relative à la flore de la région de

Laon. Bul. Soc. Bot. Deux-Sèvres **1927**: 83–90. 1927. List of about 150 phanerogams. In library of Muséum national d'histoire naturelle, Paris.) —Also Riomet, below (first, third, and fourth titles).

Magnier, Charles. Liste des plantes que l'on trouve dans des bois d'Holnon. (Aisne). *In* Poette, Charles. Histoire d'Holnon. p. 24–37. 8vo. Saint-Quentin, 1885.
Annotated list with botanical, French, and patois names. (Not seen; title and annotation supplied by G. Dillemann.)

Riomet, L. B. Essai sur la flore du canton de Rozoi-sur-Serre ou catalogue des plantes qui y croissent spontanément. 39 p. 22 cm. Péronne, 1887. (Library of A. Camus.)
Annotated list of vascular plants, with notes on uses. (In arrondissement of Laon.)

——— Flore de l'arrondissement de Château-Thierry et des environs. 31 l. ca. 15 × 46 cm. n.p., (1922-23). (Library of P. Jovet.)
Annotated list of vascular plants and Characeae, with localities. The work, reprinted from a newspaper, is very rare. The preface (p. 1) is dated 8 février 1922; p. 31 is signed 8 mai 1923. Cover- and caption-title only.

——— Flore de la Thiérache et d'une partie du Laonnois, arrondissement de Vervins et une grande partie de celui de Laon. Revue Bot. Bul. Mens. Soc. Franç. Bot. **9**: 35–163. 1891. (Reprinted with somewhat amplified title, 132 p. Toulouse, 1891. In library of Muséum national d'histoire naturelle, Paris.)
Geography, botanical publications; annotated list of 1,082 vascular plants and Characeae, with localities.—See also La Fons, above.

——— Géographie botanique du canton de Marle. 56 p. 21 cm. Laon, 1899. (Arnold Arboretum library.)
Topography, geology, climate, local distribution of species; annotated list of vascular plants and Characeae. (In arrondissement of Laon.)—See also La Fons, above.

ALLIER

See also Auvergne (Chassagne); Bourbonnais; Saône-et-Loire (Ormezzano).

General

Migout, Abel. Flore du département de l'Allier et des cantons voisins; description des plantes qui y croissent spontanément, suivant la méthode naturelle. 2. éd. 2 p. l., xxxvi, 509 p. 22 cm. Moulins, 1890. (1st ed. 1866.)
Bibliography, list of local botanists, topography, geology, general features of flora, etc.; annotated descriptive flora of over 1,600 species of vascular plants, with keys and localities.—See also **Moriot, J. B.** Contributions à la flore de l'Allier. Revue Scient. Bourbonn. **12**: 165–170. 1899. (List of noteworthy plants from along the Loire River, with localities.)—Also **Lassimonne, S. E.** Documents pour la flore de l'Allier. l.c. **17**: 148–151. 1904. (List of noteworthy plants, with localities.)—Also under Bourbonnais (Lassimonne, Prodromes d'une flore nouvelle des plantes vasculaires du Bourbonnais).

Olivier, Ernest. Flore populaire d'Allier. Noms vulgaires et patois des plantes indigènes et cultivées usités en ce département. 43 p. 21 cm. Moulins, 1886. (Extrait du Bulletin de la Société d'horticulture de l'Allier. Nov. 1886.)
Systematic list of vascular and a few cellular plants, with vernacular and patois names; alphabetical list of vernacular names, with botanical equivalents.

Local

Jourdan, Pascal. Essai phytographique d'une chloris vichyssoise. Flore de Vichy aux naturalistes et aux amis des fleurs . . . Avec une préface de George Sand. vii, 369 p. 12 pl. 18.5 cm. Vichy, 1872. (Gray Herbarium library.)

Phytogeography, etc.; annotated list of the city flora of Vichy (about 70 cryptogams and 400 vascular plants), with notes on vernacular names, symbolism, etc.; list of rare plants of the vicinity of Vichy (by vernacular names only), with localities.

Lassimonne, Étienne. Notes sur la flore du Munet. 16 p. 8vo. Moulins, 1887.

General features of flora; unannotated list of species collected by author. (Not seen; title and annotation supplied by G. Dillemann.)

Olivier, Ernest. Les transformations de la flore aux environs de Moulins (Allier). Revue Gén. Bot. **20:** 151–161. 1908.

Running account of changes in the flora (species that have appeared or disappeared.). Reprinted with some changes from paper of nearly the same title in Revue Scient. Bourbonn. **20:** 5–16. 1907.

Pérard, Alexandre. Catalogue raisonné des plantes croissant naturellement ou soumises à la grande culture dans l'arrondissement de Montluçon (Allier), avec une notice sur la distribution des végétaux dans cette contrée . . . 4 p.l., 248 p. pl. 25 cm. Paris, 1869–71.—Supplément du Catalogue . . . avec une liste de quelques menthes nouvelles ou peu connues. 1 p. l., 29 p. 25 cm. Montluçon, Paris, 1878.

Botanical investigations, topography, geology, plant formations, special localities; annotated list of vascular and cellular plants, including cultivated species. Originally published in Bul. Soc. Bot. France v. 16–18 (1869–71) as a series of papers with various titles.

ALPES-MARITIMES

See also Bouches-du-Rhône (Reynier, Herborisations); Provence (especially Huet); also Monaco (Ardoino, Catalogue des plantes vasculaires qui croissent spontanément aux environs de Menton et de Monaco).

General

Ardoino, H. J. B. Flore analytique du département des Alpes-Maritimes ou description succinte des plantes vasculaires qui croissent spontanément entre le versant est de l'Estérel et la Roïa, les Alpes et la mer. 2. éd. xxx, 470 p. 18.5 cm. Menton, 1879. (1st ed. 1867.)

Topography, botanical investigations; annotated, keyed, briefly descriptive flora of vascular plants.

Burnat, Émile. Botanistes qui ont contribué à faire connaître la flore des Alpes-Maritimes. 2. éd. revue et augmentée par Fr. Cavillier. Boissiera **5:** 3–94. 1941. (At head of title-page: Association des naturalistes de Nice et des Alpes-Maritimes. Mémoire V. Supplément au "Riviera scientifique" (années 1940–1941).) (1st ed., Bul. Soc. Bot. France **30** (Sess. Extraord.): cvii-cxxxiii. 1883 [1885?]. Reprinted, 27 p. [1885.] Reprint in Arnold Arboretum library.)

Alphabetical list of botanists who have collected in or published on the flora of the department and adjacent regions in France and Italy, with biographical data and references, mention of pertinent publications (often with critical annotations), location of herbaria, etc.; includes also titles of some periodicals.

——— and others. Flore des Alpes Maritimes ou catalogue raisonné des plantes qui croissent spontanément dans la chaîne des Alpes Maritimes y compris le département français de ce nom et une partie de la Ligurie occidentale. v. 1–7. map. 25.5 cm. Genève, 1892–1931.

Annotated list of vascular plants, with synonymy, citation of exsiccatae and localities, and frequent critical notes. Incomplete: Renonculacées—Composées Cynaroidées; 1269 species. Vol. 5, pt. 1, by F. Cavillier; v. 5, pt. 2–v. 7, by J. Briquet and F. Cavillier. Covers also a little of Basses-Alpes (canton of Entrevaux) and Var (to Agay), as well as adjacent Liguria (to Albenga) and Piemonte (to Mondovi) in Italy.—See also **Breistroffer, Maurice.** Aperçu sur la flore de Tende-La Brigue (Alpes-Maritimes). Feuille des Nat. n.s., **3:** 77–79. 1948. (Lists of noteworthy species, endemic forms, etc.; includes some species new to the department as a result of change in the frontier.)—Also **Quézel, Pierre.** Contribution à la flore des Alpes-Maritimes. Monde des Plant. **45** (270/1): 67. 1950. (List of noteworthy species. In Arnold Arboretum library.)

Local

Bertoloni, Giuseppe. Notizie intorno a cose naturali osservate nei monti italiani e guidicate una parte utili per se stesse e le altre interessanti agli studiosi della geologica e della botanica. Mem. Accad. Sci. Ist. Bologna II, **3:** 201–244. 1863.

Includes annotated list of vascular plants and lichens collected by author on the Colle di Tenda. The Colle di Tenda, formerly on the boundary line between the Province of Cuneo (Piemonte) and the Department of Alpes-Maritimes, are now in the latter.

Kampmann, Frédéric. Notice sur l'Île Sainte-Marguerite et ses environs. Bul. Soc. Hist. Nat. Colmar **5:** 17–34. 1865.

Climate, plant zones, etc.; unannotated list of vascular and some cellular plants, grouped by geographic affinities; relates also to vicinity of Cannes nearest the islands.—See also his Lettre de M. F.-Kampmann père, à M. Eug. Fournier. Bul. Soc. Bot. France **13:** 44–45. 1866. (List of additions and corrections.)—Also **Reynier, Alfred.** Herborisations aux îles du littoral de la Provence. Bul. Soc. Bot. et Hort. Provence **2:** 197–206. [1881]. (Notes on flora of various islands, including the Îles de Lérins (p. 202–204), with partial lists of species.)

Mader, Fritz. La colline du Château de Nice—sa faune—sa flore. Annales Soc. Let., Sci. et Arts Alpes-Maritimes **22:** 255–291. 1909. (Harvard College library.)

Includes (p. 276–283) list of wild and cultivated vascular plants and Characeae (170 native vascular plants).

Malinvaud, Ernest. Liste méthodique des plantes phanérogames et cryptogames supérieures récoltées pendant la session d'Antibes (mai 1883), suivie d'observations sur quelques espèces. Bul. Soc. Bot. France **30** (Sess. Extraord.): clxxx-cxcvi. 1883.

Extensive list of vascular plants and Characeae (some of the ubiquitous species omitted) collected at Cap d'Antibes, Golfe Jouan, Massif de l'Estérel, Colline de Biot, Îles de Lérins.—See also **Perroud, Louis.** Coup d'oeil sur la flore d'Antibes et de ses environs. Annales Soc. Bot. Lyon **13:** 121–145. 1886. (Includes partial lists of species from different localities and habitats.)

Risso, Antoine. Flore de Nice et des principales plantes exotiques naturalisées dans ses environs. 586 p. 24 pl. 20 cm. Nice, 1844.

Annotated list of vascular and cellular plants, with localities for scarcer species, and in part (Ranunculaceae–Cyperaceae) with descriptions; includes cultivated plants.—See also **Goiran, Agostino.** Un manipolo di piante nizzarde e veronese. Nuovo Gior. Bot. Ital. n.s., **16:** 125–145. 1909. (Records of about 103 noteworthy species and forms of vascular plants from the regions of Nice and of Verona.)—Also Mader, above, and Roubaudi, below.

Roubaudi, Louis. Fin des productions territoriales de Nice. Arbres, arbustes et plantes d'ornement classés par ordre alphabétique . . . *In* his Nice et ses environs . . . p. 282–316. 23.5 cm. Paris, Turin, 1843.

Includes, in addition to a list of cultivated trees, shrubs, and ornamentals, a classified list of medicinal plants (p. 291–293), list of forage plants, lists of wild species of different habitats, and (p. 301–316) a tabular list entitled Catalogue des plantes principales méridionales qui croissent dans les environs de Nice, classées d'après le système de Jussieu, the last based in part on information furnished by abbé J. I. Montolivo. Cavillier (Boissiera **5:** 66. 1941) refers to a Liste de plantes des environs de Nice by Montolivo in an 1883 edition (perhaps error for 1843) of Roubaudi's work (his name misspelled Roubaudy).—See also Mader and Risso, above.

ALSACE (ELSASS)

See also Alsace-Lorraine, which (1871–1919) consisted essentially of Alsace (minus Belfort) and the present Department of Moselle; also Lorraine (Godfrin), and the separate departments that constitute Alsace.

The region and former province of Alsace consists of the Departments of Bas-Rhin and Haut-Rhin and the Territory of Belfort, the latter (constituted in 1871) being that part of Haut-Rhin which was not ceded to Germany in 1871. Only general works on the region are cited under this heading.

Chermezon, Henri, and Hée, Alexandre. Contribution à la flore d'Alsace. Bul. Assoc. Philom. Alsace et Lorraine **7** (1925): 1–9. 1926. (American Museum of Natural History library.)

List of publications on the adventive flora; list of some vascular plants, with localities and annotations.

Kirschleger, Fréderic. Flore vogéso-rhénane ou description des plantes qui croissent naturellement dans les Vosges et dans la vallée du Rhin. 2 v. (vi, 502; 1 p. l., 399 p.). 18.5 cm. Paris, Strasbourg, 1870.

Briefly descriptive flora of vascular plants, with localities; guide for local botanizing. Covers Bas-Rhin, Haut-Rhin (including Belfort), and adjacent Lorraine and Baden. Is essentially a condensed second edition, with shortened descriptions and the omission of many localities, much of the notes on uses, and the bibliographies, of his Flore d'Alsace et des contrées limitrophes. 3 v. 18.5 cm. Strasbourg, Paris, 1852–62. (Annotated descriptive flora of vascular plants, with keys, localities, notes on uses, chemistry, vernacular and pharmaceutical names; vol. 2 includes useful biographical notes on local botanists, and bibliographies; vol. 3: topography, climate, geology, soils, altitudes, general features of vegetation of each region, habitats with lists of species, historical sketch of naturalized plants, cultivated plants, etc.; (p. 189–336) Guide du botaniste herborisateur et touriste à travers les plaines de l'Alsace et les montagnes des Vosges, a detailed account of numerous excursions; bibliography.)—See also **Waldner, Heinrich.** Beiträge zur excursionsflora von Elsass-Lothringen und Flore vogéso-rhénane. 39 p. 22 cm. Heidelberg, 1879. (Refers mostly to Alsace. In Gray Herbarium library.)—Also **Rastetter, V.** Espèces et localités nouvelles pour la flore d'Alsace; années 1953 à 1956. Bul. Soc. Hist. Nat. Colmar **47**: 20–26. 1956. (Records of numerous species. Not seen; cited from Proc. Bot. Soc. Brit. Isles **3**: 83. 1958.)—Also Waldner, Exkursionsflora von Elsass-Lothringen, under Alsace-Lorraine.

Lavialle, M. P. Les plantes médicinales. Leur cueillette et leur culture en Alsace et en Lorraine. 32 p. 22 cm. Mulhouse, 1921.

Includes (p. 13–32) annotated list of 77 important medicinal plants, alphabetically arranged by vernacular names, with figure of each, vernacular names, etc. Publication of Université de Strasbourg, Faculté de pharmacie.

Vosselman, Ph. La flore d'Alsace depuis la mort de Kirschleger. Jour. Pharm. Alsace-Lorraine **6**: 205–208, 224–228. 1879; **7**: 17–19, 43–49, 90–95, 143–148, 201–206. 1880; **8**: 62–67. 1881.

Critical discussion of publications by Waldner and others; extensive list of new records and corrections in vascular plants and mosses. See also reply under same title by H. Waldner, l.c. **8**: 207–209, and response by Vosselman, l.c. 209–212. 1881.

Walter, Émile. La botanique en Alsace et en Lorraine depuis 1870. Bul. Soc. Bot. France **73**: 615–623. 1927.

Sketch of botanical investigations since 1870, with critical mention of publications, but without exact titles.

——— Modifications survenues dans la flore d'Alsace et de Lorraine depuis 1870. Notes rectificatives. Bul. Soc. Bot. France 73 (Sess. Extraord.): 5–61. 1931.

Account of species added since 1870, plants restricted in range or extinct, critical notes on various species, introduced and adventive plants, species erroneously recorded, etc.; bibliography.

ALSACE-LORRAINE (ELSASS-LOTHRINGEN)

See also Alsace and Lorraine, as well as the separate departments concerned.

Alsace-Lorraine, consisting only of the present Departments of Bas-Rhin, Haut-Rhin, and Moselle, with a small part of Vosges, was a Reichsland of Germany from 1871 to 1919. During this period several floras and other works were published which are brought together here since they cannot well be placed under Alsace or Lorraine.

Bossler, Ludwig. Flora der gefässpflanzen in Elsass-Lothringen . . . 385 p. 16.5 cm. Strassburg i. E., 1877. (Gray Herbarium library.)

Synopsis of genera, on Linnaean system; annotated flora, with brief descriptions, and localities for scarcer species. The work was entirely a compilation and condemned by French and German reviewers alike (see Barbiche, Bul. Soc. Hist. Nat. Metz II, **15**: 234. 1880; Flora **61**: 191. 1878), mainly for this reason.

Himpel, J. S. Flora von Elsass-Lothringen. Analytische tabellen zum leichten und sicheren bestimmen der in Elsass-Lothringen einheimischen und häufiger kultivierten gefässpflanzen, namentlich zum gebrauch auf excursionen. 2. aufl. 327 p. 18.5 cm. Strassburg, [1902]. (Gray Herbarium library.) (1st ed. 1891.)

Key to genera, on Linnaean system; briefly annotated flora of vascular plants in form of keys, with range or localities.

Krause, E. H. L. Anmerkungen zum elsass-lothringischen kräuterbuch. ("Florenklein.") Mitt. Philom. Gesell. Elsass-Lothr. **3** (14. jahrgang, 1906): 391–409. 1907; **3** (15. jahrg., 1907): 467–489. 1908; **4** (16. jahrg., 1908): 63–69. 1909; **4** (17. jahrg., 1909): 139–174. illus. 1910; **4** (18. jahrg., 1910): 337–400. 1911; **4** (19. jahrg., 1911): 557–566. 1912; **4** (20. jahrg., 1912): 669–689. 1913. (American Museum of Natural History library.)

Critical notes, new records and localities, etc., for numerous vascular plants.—See also his Die gräser Elsass-Lothringens. Bruchstück einer landesflora. Mitt. Philom. Gesell. Elsass-Lothr. **5** (21. jahrg., 1913): 1–161. 1914. (Annotated list of grasses, including cultivated species. In American Museum of Natural History library.)—Also his Die nelken- und meldenartigen gewächse Elsass-Lothringens. Beih. Bot. Centralbl. **33**, Abt. 2: 441–500. 1915. (Records of 140 apetalous and polypetalous dicots.)—Also his Die nadelhölzer Elsass-Lothringens. Mitt. Gesell. Erdk. u. Kol. Strassburg **5**: 1–22. 1916. (Annotated list of gymnosperms (mostly cultivated). In American Museum of Natural History library.)—Also his Die korb- und röhrenblüter (Syngenesistae und Tubatae) Elsass-Lothringens. Beih. Bot. Centralbl. **35**, Abt. 2: 1–221. 1917. (Records of 352 wild and cultivated Gamopetalae.)

Waldner, Heinrich. Excursionsflora von Elsass-Lothringen. Autorisirte nach Fr. Kirschleger's Guide du botaniste bearbeitete ausgabe. 125 p. map. 15.5 cm. Heidelberg, 1876. (Gray Herbarium library.)

Table of altitudes, climate, agriculture, etc.; guide to excursions, with lists of more important species. Cover-title only. According to Barbiche (Bul. Soc. Hist. Nat. Metz II, **15**: 230. 1880), this is merely an authorized translation of Kirschleger's Guide du botaniste . . . à travers les plaines de l'Alsace et les montagnes des Vosges in his Flore de l'Alsace, and contains almost no data on Lorraine (i.e., Moselle).—See also his Beiträge zur Excursionsflora von Elsass-Lothringen und Flora vogéso-rhénane. 39 p. 22 cm. Heidelberg, 1879. (Refers mostly to Alsace.)

ARDÈCHE

See also Languedoc.

General

Revol, J. Catalogue des plantes vasculaires du département de l'Ardèche. Introduction par M. le Profr. Flahault. Annales Soc. Bot. Lyon **34**: 29–316. map. 1909. (Reprinted incl. Flahault's **Au sujet . . . 316 p. map. Paris,** 1910. In Arnold Arboretum library.)—Supplément au Catalogue . . . l.c. **42**: 51–103. 1922.—[2.] supplément . . . **43**: 75. 1924. (Second supplement not seen.)

Bibliography; annotated list of vascular plants (1,958 species including supplements), with localities.—See also **Flahault, Charles . . . Au sujet de** la géographie botanique de l'Ardèche et du Vivarais. Annales Soc. Bot. Lyon v. 34. [Appendix.] xix p. 1910. (Sketch of soils and phytogeography.—Also **Guffroy, Charles.** Notes sur la flore ardéchoise. Bul. Soc. Bot. France **70**: 12–27. 1923. (List of noteworthy vascular plants, mostly from vicinity of Cheylard, with localities, also some plant galls and mushrooms.) —Also **Breistroffer, Maurice.** Supplément au Catalogue . . . (1.–3. partie.) Bul. Soc. Scient. Dauphiné **63**: 53–56. 1948; Bul. Mens. Soc. Linn. Lyon **23**: 60–64. 1954; **26**: 281–293. 1957. (Extensive lists; part 1 not seen.)— Also **Thiébaut, Joseph.** Complément à la flore de Tournon. l.c. **17**: 61–62. 1948. (List including some species new to department.)—Also **Tallon, Gabriel.** Contribution à la flore du sud de l'Ardèche. l.c. **23**: 169–170. **1954.** (List of noteworthy species.)

ARDENNES

General

Callay, Albert. Catalogue raisonné et descriptif des plantes vasculaires du département des Ardennes . . . Précédé d'une description orographique et géologique . . . et d'une étude de géographique botanique par F. Bestel. xxii, 454 p. port., map. 22.5 cm. Charleville, 1900.

Botanical explorations (by Callay), topography, soils, plant distribution and associations (all by F. Bestel); annotated list of vascular plants, with localities. Replaces **Mailfait, Paul,** and **Cadix, Léon.** Catalogue de la flore du département des Ardennes. Bul. Soc. Hist. Nat. Ardennes **3:** 31–32. 1896; **4:** 33–64. 1897; **5:** 65–112. 1898; **6:** 113–168. 1899. (Annotated list of vascular plants and Characeae, with localities for scarcer species. In American Museum of Natural History library; separate (p. 31–168) in Gray Herbarium library.)—See also **Fliche, Paul.** Note sur la flore du département des Ardennes. Bul. Soc. Bot. France **48:** 376–392. [1902]. (Geology, topography, general features of flora; list of noteworthy vascular plants with localities, chiefly from the cantons of Carignan, Mouzon, and Raucourt.)

ARIÈGE

See also France, Partial (Duboul; Gaussen, and additional floras there cited); Languedoc.

Local

Bourdette, Jean. Additions à la flore du département de l'Ariège. Bul. Soc. Sci. Phys. et Nat. Toulouse **7:** 439–448. 1888.

Annotated list of botanical publications relating to the arrondissement of Foix, list of herbaria; short lists of species observed at 6 localities, including 25 new to the flora.

Claustres, G. Premières données écologiques et floristiques sur la haute vallée de l'Aston. Annales École Natl. Supér. Agron. Toulouse **1:** 133–140. 1 fig. (map.) 1953.

Topography, botanical explorations, geology, climate, etc.; list of noteworthy vascular plants collected between 1800 and 2350 meters elevation, between Port de Fontargente and l'Étang Français de Joucla.—See also his Principaux groupements botaniques subalpins et alpins de la haute vallée de l'Aston. Annales École Natl. Supér. Agron. Toulouse **2:** 3–15. 1954. (Includes supplementary list of species, p. 3–4, and sketch of ecological groups.)

Dupla, Henri. Contribution à la flore de l'Ariège. Quelques plantes de Castillonnais. Monde des Plant. **44** (260/261): 33–36, 45–48, (263) 59–60, 65–66. 1949. (Arnold Arboretum library.)

Extensive list of noteworthy vascular plants of St. Girons region in canton of Castillon.

Giraudias, Ludovic. Notes critiques sur la flore ariègeoise. Bul. Soc. Études Scient. Angers **18:** 37–53. 1889; **19:** 89–116. 1890; **20:** 63–75. 1891. (Reprinted, 59 p.)

Lists of noteworthy vascular plants, with localities and frequent annotations.

Guitard,— Essai sur la flore d'Ussat. Bul. Soc. Hist. Nat. Toulouse **1:** 44–57. 1867.

Botanical publications; unannotated alphabetical list of 607 plants, nearly all vascular.

Jeanbernat, Ernest, and **Timbal-Lagrave, Édouard.** Le massif du Laurenti, ancien Donezan, canton de Quérigut (Ariège). Bul. Soc. Sci. Phys. et Nat. Toulouse **3:** 197–624. 2 pl., map. 1875–76. (Reprinted 434 p. 2 pl., map. Paris, 1879, with title: Le massif du Laurenti, Pyrénées françaises; géographie, géologie, botanique. Original in Harvard College library; reprint in Arnold Arboretum library.)

Botanical explorations, geography, topography, geology, etc.; annotated list of vascular plants and mosses, with localities; list of excluded species.—See also (under Pyrénées-Orientales) their Le Capsir . . . l. c. **6:** 37–283. 1887, for additional records (p. 281–282) for Laurenti.

Lazerges, P. Catalogue des plantes récoltées dans le département de l'Ariège. Bul. Soc. Sci. Phys. et Nat. Toulouse **3**: 103–135. pl. 1–2. 1875–76. (Harvard College library.)
List of angiosperms [560 species], with localities; apparently based on author's collections only.

Marcailhou-d'Ayméric, Hippolyte. Le massif de Tabe . . . (Explorations ariègeoises. [I.]) Explor. Pyrén. Bul. Soc. Ramond **33** (II, **2** [i.e. **3**]): 172–202. 1898.
Topography, etc.; unannotated list of vascular and cellular plants found by author at summit of pic St.-Barthélemy (2100–2350 m. elevation); itineraries, etc.

——— Le Montcalm et le pic d'Estats (3,080 m et 3,141 m) . . . (Explorations ariègeoises. II.) Explor. Pyrén. Bul. Soc. Ramond **35** (II, **4** [i.e. **5**]): 165–185. 1900.
Topography, etc.; lists of vascular and cellular plants at various altitudes up to 3141 meters.

——— Le Mont-Valier (2839 m) . . . (Explorations ariègeoises. III.) Explor. Pyrén. Bul. Soc. Ramond **37** (II, **7**): 156–166, 246–266. 1902.
Topography, etc.; lists of vascular and cellular plants collected at various altitudes from 2180 to 2840 meters.

——— Première ascension du Pic de Serrère (2911 m) (limite de la France et de l'Andorre). Explor. Pyrén. Bul. Soc. Ramond **32** (II, **2**): 5–23. 1897.
Includes lists of species collected at different altitudes up to the summit.

——— and **Marcailhou-d'Ayméric, Alexandre.** Catalogue raisonné des plantes phanérogames et cryptogames indigènes du bassin de la haute Ariège. Bul. Soc. Hist. Nat. Autun **11**: 248–376. map. 1898; **13**: 1–126. 1900; **14**: 1–122. 1901; **15**: 249–413. 1902. Bul. Acad. Internatl. Géog. Bot. **12**: 522–536. 1903; **13**: 89–104, 121–136, 185–216, 289–320, 385–400. 1904; **15**: 137–156, Suppl. p. 1–80. 1905; **16**: Suppl. p. 81–164. 1906; **18**: 1–192. 1908. (Reprinted, 2 v.; incomplete copy seen in Kew library.)—Mém. Acad. Internatl. Géog. Bot. **1**: 151, 82 p. 1911–12. (These 2 parts, forming v. 3 of reprint, not seen; data supplied by Mlle. G. Madier from copy in library of Muséum national d'histoire naturelle, Paris. Second part numbered v. 24 on title page, v. 22 on cover, corresponding to v. 22 of the Bulletin; first part not so numbered.)
Topography, geology, soils, climate, plant zones, botanical explorations, authors' publications; annotated list of spermatophytes (1282 species), with critical annotations, notes on uses and properties. Covers the canton of Ax-les-Thermes (Ariège); also "le cirque formé par les massifs de Puymaurens et de Font-Nègre, les sources de l'Ariège et les crêtes de la Solana d'Andorre."—See also **Claustres, Georges.** Quelques plantes intéressantes de la flore ariégoires [sic]. Monde des Plant. 47 (291/292): 42–44. 1952. (List including some additions. In Arnold Arboretum library.)

AUBE

See also Champagne.

General

Briard, Pierre. Catalogue raisonné des plantes observées jusqu'à ce jour qui croissent naturellement dans le département de l'Aube. Mém. Soc. Acad. Agr., Sci., Arts et Belles-Let. Dépt. Aube 44 (III, **17**): 155–320. 1880; **45** (III, **18**): 5–197. 1881. (Reprinted, 360 p. Troyes, 1881. In Gray Herbarium library.)
Geology, topography, botanical explorations, account of herbaria; annotated list of vascular and cellular plants (1904 species, of which 1326 are vascular), with localities; cultivated and dubious species are included in the numbering. Replaces **Bourguignat, J. R.** Catalogue raisonné des plantes vasculaires du département de l'Aube. v. 1 (viii, 184 p.). 21.5 cm. Paris, 1856. (Annotated list, with references and localities; incomplete (Ranunculaceae—Onagraceae).)—See also **Le Grand, Antoine.** La flore de l'Aube d'après l'ouvrage du Commandant Briard; addition au Catalogue . . . Bul. Soc. Bot. France **28**:

202–207. 1881. (Includes list of additional species and localities.)—Also **Briard.** Florule cryptogamique de l'Aube et Supplément au Catalogue des plantes de ce département. Mém. Soc. Acad. Agr., Sci., Arts et Belles-Let. Dépt. Aube **51** (III, 24): 197–447. 1887; **52** (III, 25): 5–254. 1888; also his Supplément au Catalogue et à la Florule cryptogamique . . . l. c. **55** (III, 28).: 5–148. 1891. (Nearly all cellular plants; vascular plants only on p. 202–204 of 1st pt., p. 230 of 2d pt., p. 8–9 of 3d pt.)—Also **Hariot, Paul.** Sur la flore du département de l'Aube. Compt. Rend. Assoc. Franç. Avanc. Sci. **25** (2): 360–365. 1897. (List of additional species, with localities, and of excluded species.)—Also **Hariot, Paul,** and **Guyot, A.** Contributions à la flore phanérogamique de l'Aube. Additions et rectifications. Mém. Soc. Acad. Agr., Sci., Arts et Belles-Let. Dépt. Aube **66** (III, 39): 5–142. 1902. (List of excluded species, additions to Hariot's La botanique dans le département de l'Aube, etc.; list of additional species and new localities.)—Also **Hemet, L.** Notes de géographie botanique sur l'est du département de l'Aube. 2 pt. (16 p., 3 maps; 52 p., 4 maps). 8vo. Rennes, 1908–09. (Includes annotated lists of 66 and 140 species, respectively, with (in the second part) local florules. Not seen; title and annotation supplied by G. Dillemann.)

Des Étangs, Stanislas. Liste des noms populaires des plantes de l'Aube et des environs de Provins, contenant l'indication des lieux où ils sont usités... Mém. Soc. Agr., Sci., Arts et Belles-Let. Dépt. Aube **12** (no. 91 et 92): 137–246. 1844 [1845].

Annotated alphabetical list of vernacular names, with explanation of origin and citation of localities; list of botanical names, with vernacular equivalents.

Hariot, Paul La botanique dans le département de l'Aube. Mém. Soc. Acad. Agr., Sci., Arts et Belles-Let. Dépt. Aube **55** (III, 28): 165–189. 1891.

Sketch of botanical explorations and publications.

Local

Hariot, Louis, and **Hariot, Paul.** Florule du canton de Méry-sur-Seine. Mém. Soc. Acad. Agr., Sci., Arts et Belles-Let. Dépt. Aube **38** (III, 11): 293–364. [1875.] (Preprinted, 76 p. Troyes, 1874. In Arnold Arboretum library.)

Topography, geology, botanical investigations, annotated bibliography for the whole department, list of herbaria; annotated list of vascular plants and Characeae.

Hariot, Paul. Flore de Pont-sur-Seine. Mém. Soc. Acad. Agr., Sci., Arts et Belles-Let. Dépt. Aube **43** (III, 16): 277–334. 1879.

Soils, plant zones; annotated list of vascular plants and Characeae.

AUDE

See also Languedoc; also Tarn (Doumenjou).

General

Marty, Léonce, and **Sarcos, O.** Les plantes médicinales de l'Aude. 48 p. 24 cm. Carcassonne, 1920. (At head of title page: Comité des plantes médicinales et à essences de la région de Montpellier.)

Includes annotated list of 201 medicinal plants, alphabetically arranged by botanical names, with vernacular names, part used, etc.

Sabarthès, A. Flore. *In* his Bibliographie de l'Aude. p. 63–68. 24.5 cm. Narbonne, 1914. (Extrait du Bulletin de la Commission archéologique de Narbonne.)

Unannotated list of botanical publications relating to Aude, arranged chronologically, including items 558–614; additions (nos. 4609–4610) on p. 536.

Local

Baichère, Édouard, *abbé.* Contributions à la flore du bassin de l'Aude et des Corbières. 1. fasc. Bul. Soc. Études Scient. Aude **2**: 67–110. 1891.

Botanical explorations, etc.; extensive list of vascular and cellular plants, with localities.—See also his Contributions à la flore des Corbières et du bassin de l'Aude. 2. fasc. Mém. Soc. Arts et Sci. Carcassonne **6** (2): 49–90. 1892 (1893). (Similar list.)

——— Étude sur la flore des environs de Carcassonne et liste des plantes vasculaires observées autour de cette ville. 1 p. l., 41 p. 24 cm. Carcassonne, 1891.

Botanical explorations with references to publications, geology, zones of vegetation, itinerary for botanical excursions; annotated tabular list of vascular plants (about 1050 species). Replaces his Note sur la végétation des environs de Carcassonne. Bul. Soc. Bot. France **35** (Sess. Extraord.): xxvii–xxxv. 1889. (Topography, geology, principal localities; list of angiosperms grouped according to abundance, the ubiquitous species omitted.)—See also **Chartier, Louis.** Contributions à la flore de Carcassonne. Bul. Soc. Études Scient. Aude **3**: 203–215. 1892. (List of about 150 vascular plants said to be new to region, with localities.)—Also **Laffage, Achille.** Plantes nouvelles pour la flore des environs de Carcassonne. l. c. **6**: 222–224. 1895. (List of species added to the flora since 1892, with localities and names of collectors.)—Also his Nouvelles plantes et nouvelles stations de plantes rares pour le flore des environs de Carcassonne. l.c. **7**: 156–160. 1896. (List of additional species and new localities.)—Also his Plantes nouvelles pour la flore des environs de Carcassonne. l.c. **8**: 118. 1897. (List of 10 species, with localities.)—Also **Rebelle, Guillaume.** Gerbe de plantes rares ou nouvelles pour la région de Carcassonne. l.c. **21**: 92–95. 1910. (Bibliography; annotated list of angiosperms. In Harvard College library.)—Also his Plantes rares ou nouvelles pour la région de Carcassonne. l.c. **23**: 94–95. 1912. (In Harvard College library.)—See also Clos, below, and Laffage, below.

Calmet, Paul. Florule de Ribaute (Aude). Bul. Soc. Études Scient. Aude **20**: 134–154. 1909.

Topography, etc.; list of 394 plants (367 vascular) with localities and patois names; notes on geology.

Chartier, Louis, and **Respaud, A.** Florule de Caux (Aude). Revue Bot. Bul. Mens. Soc. Franç. Bot. **9**: 163–176. 1891.

Topography; briefly annotated list of vascular plants.

Clos, Dominique. Coup-d'oeil sur la végétation de la partie septentrionale du département de l'Aude. Cong. Scient. France 28. Sess. (Bordeaux, 1861), **3**: 375–402. map. 1863.

Botanical explorations, topography, geology, etc.; list of phanerogams grouped by local range with special reference to Narbonne, etc. Not a full flora.—See also Baichère, Étude sur la flore des environs de Carcassonne, above, and Laffage and Maugeret, below.

Gautier, Gaston. Catalogue de la flore des Corbières . . . mis en ordre par L. Marty. x, 347 p. 23.5 cm. Carcassonne, 1912/13. (Publication de la Société des études scientifiques de l'Aude.) (New York Public Library.)

Geography, etc.; annotated list of vascular plants (2265 species), with localities. Cover date only. The geographical limits are thus defined: on east the Mediterranean, on north and west the Aude River, on south the "thalweg des vallées suivies par le tracé de la grande route de Quillan à Perpignan, depuis le pont sur l'Aude en aval d'Axat, jusqu'à Cases de Pène, et de ce point à la mer par l'Agly." Replaces **Timbal-Lagrave, Édouard.** Florule des Corbières orientales . . . Oeuvre posthume publié par les soins de M. l'abbé Éd. Marçais. Revue Bot. Bul. Mens. Soc. Franç. Bot. **10**: 7–272. 1892. (Topography, etc.; list of angiosperms, with localities and some critical notes. Incomplete; Ranunculaceae to Artemisia (part) only.)

Laffage, Achille. Liste de noms patois usités dans les environs de Carcassonne pour désigner les plantes vulgaires. Bul. Soc. Études Scient. Aude **4**: 111–129. 1893.

Alphabetical list of more than 450 patois names, with standard French vernacular equivalents, and explanation of patois names; very few scientific names given.

Maugeret, Alexander. Liste des plantes que l'on peut récolter à Narbonne pendant le mois de juin (avec indication des localités pour les plus remarquables et pour celles qui ne se trouvent que dans des stations déterminées). Bul. Soc. Bot. France **9**: 638–651. 1862.

Extensive list of vascular plants, in part with localities.—See also **Gautier, Gaston.** Herborisation aux îles de l'étang de Leucate, près de Narbonne, le 16 mai et le 14 juin 1875. Bul. Soc. Bot. France **22**: 300–310. 1875. (Running account of author's visit, with partial lists of phanerogams observed.)—Also his Rapport sur les herborisations dans les environs de Narbonne instituées par le Comice agricole de l'arrondissement de Narbonne. (Cover title. Herborisations dans les environs de Narbonne.) 50 p. 21.5 cm. Narbonne, 1876. (Account of excursions to la Clappe, îles de Planasse and other islands in the étang de Bages, Roquefort and Fraisse-des-Corbières, Pech de l'Agnèle, the islands in the étang de Leucate, island of Sainte-Lucie, and to Roquehaute (in Hérault), with mention of various species. In library of Muséum national d'histoire naturelle, Paris.)—Also his account of numerous excursions in general vicinity of Narbonne. Bul. Soc. Bot. France 35 (Sess. Extraord.): lxxvi–cvi, passim. 1889.—Also **Bru, Fernand.** Notes sur quelques plantes rares ou nouvelles pour la région de Narbonne. Bul. Soc. Études Scient. Aude **21**: 96–97. 1910. (Short list, with localities. In Harvard College library.)—Also **Delpont, J.** Quelques plantes rares ou nouvelles pour la région de Narbonne et des Corbières. Monde des Plant. III, 29 (60): 5–7. 1928. (List with localities. In Arnold Arboretum library.)—Also Clos, above.

Sennen, *frère.* Compte rendu des recherches botaniques faites par les frères des Écoles chrétiennes de La Nouvelle (Aude). Bul. Soc. Bot. France **47**: 424–445. 1 fig. [1901].

Extensive list of vascular plants and Characeae, with localities.—See also his Herborisations aux environs de La Nouvelle (Aude), suite. Bul. Soc. Bot. France **49**: 364–377. pl. 3. 1903. (Another list.)

Timbal-Lagrave, Édouard, and Jeanbernat, Ernest. Rapport sur l'exploration scientifique des environs de Montolieu (Aude). Bul. Soc. Sci. Phys. et Nat. Toulouse **2** (1874): 234–271. [1876?] (Harvard College library.)

Account of authors' explorations, including mention of plants observed [330 species].—See also **Doumergue, François.** Contributions à la flore de Montolieu (Aude) et de ses environs. Revue Bot. Bul. Mens. Soc. Franç. Bot. **4**: 117–148. 1885. (Lists of vascular plants collected on various excursions.—Also **Delpont, L.** Contributions à la flore de Montolieu (complément). Bul. Soc. Études Scient. Aude **6**: 135–138. 1895. (Botanical explorations; list of 101 additional species, with localities.)—Also his Note rectificative et plantes nouvelles pour la flore de Montolieu. l. c. **7**: 161–164. 1896. (Corrections and additions.)—Also his Monographie botanique de Montolieu. l. c. **15**: 107–166. 4 fig. (maps). 1904. (Topography, climate, botanical explorations, short lists of species of various localities and habitats; list (p. 157–164) of vascular plants considered new to the region, with localities.)

AUVERGNE

The region and former province of Auvergne includes the departments of Cantal and Puy-de-Dôme, and a small part of Haute-Loire. Only works relating to the region as a whole are given here.

Aubert, Édouard. Un aperçu sur la botanique en Auvergne et sur les plantes médicinales qui croissent spontanément dans cette région. 153 p. Poitiers, 1929.

Sketch of botanical investigations; descriptive inventory of medicinal plants of the Plateau Central. (Not seen; cited from Bul. Soc. Bot. France **78**: 796. 1932.)

Chassagne, Maurice. Inventaire analytique de la flore d'Auvergne et contrées limitrophes des départements voisins. 2 v. (xl, 358; 542 p.). 24 cm. Paris, 1956–57.

Increase in number of known species, phytogeography, alphabetical list of local botanists with biographical data; annotated list of vascular plants (over 2,000 species and about 2,600 micromorphs and hybrids, plus about 400 taxa from adjoining departments) with geographical origin, biological type, chromosome number, and detailed localities, with special attention to critical groups. Covers central and northern part of the Massif Central, including Cantal and Puy-du-Dôme and adjoining parts of Allier, Aveyron, Corrèze, Haute-Loire, Loire, Lot, Lozère. Replaces Héribaud-Joseph's Flore as an annotated list but the latter contains also keys, descriptions, and notes on uses.

Delhostal, —. Ensag de glosari botanic auvernhat. Oc (Institut d'études occitanes) **11**: (p. ?). 1933. (Reprinted 30 p.)
List of patois names with (in many cases) scientific names and ethnographic notes. (Not seen; title and annotation supplied by J. Séguy.)

Denis, Marcel. État des recherches botaniques en Auvergne. Bul. Soc. Bot. France **75**: 641–652. 1928.
Sketch of botanical work with mention of principal publications on cellular and vascular plants, phytogeography, phytosociology, etc.

Girod, Paul. Le Jardin botanique de Clermont et les botanistes de l'Auvergne. Bul. Soc. Hort. et Vitic. Puy-de-Dôme **11**: 49–83. 1894. (Preprinted 35, xi p. 1893, the added pages containing a list of vascular plants of Auvergne cultivated in the Clermont garden. Preprint in Arnold Arboretum library.)
Includes account of botanical publications relating to flora of Auvergne.

Héribaud-Joseph, *frère.* Flore d'Auvergne . . . avec une introduction et les characées du Plateau Central par l'abbé Hy . . . nouv. [2. ?] éd. xv, 638 p. 16 cm. Paris [etc.], 1915. (Library of W. A. Sledge.) (1st ed. by Gustave, *frère,* and Héribaud-Joseph, *frère,* 1883.)
General features of flora; briefly annotated, keyed flora of vascular plants and Characeae, with short descriptions, localities for rarer species, and notes on uses. Treatment of Characeae contributed by abbé Hy.—See also **Chassagne, Maurice.** Matériaux pour le flore d'Auvergne. Plantes nouvelles et localités de plantes rares de la flore d'Auvergne. Bul. Soc. Bot. France **60** (Sess. Extraord.): x–xlviii. 1914. (Extensive list of noteworthy vascular plants, mostly from Puy-du-Dôme.)

——— Plantes nuisibles aux diverses productions culturales de l'Auvergne. 75 p. 19 cm. Clermont-Ferrand, 1878. (Arnold Arboretum library.)
Annotated systematic lists of weeds and medicinal plants; list of plants suitable for salads, etc.

——— Quelques mots sur la flore d'Auvergne. Monde des Plant. **2**: 37–44, 66–72, 94–101. 1892–93.
List of vascular plants of Puy-de-Dôme not known in Cantal, with localities, and list of plants of Cantal not known in Puy-de-Dôme, with localities and discussion; mention of adventive plants, hybrids, etc. Replaces his Le Puy-de-Dôme et le Cantal ou tableau comparatif des plantes vasculaires et spontanées de ces deux départements. Compt. Rend. Assoc. Franç. Avanc. Sci. **5**: 477–486. 1877.

Luquet, Aimé. Recherches sur la géographie botanique du Massif Central. Les colonies xérothermiques de l'Auvergne. Préface de M. H. Humbert . . . 328 p. 16 pl., 2 maps. 25.5 cm. Aurillac, 1937. (Gray Herbarium library.)
Mainly ecological and phytogeographical; includes (p. 53–172) partial list of species grouped phytogeographically, with local and extralimital range.

Vernière, A. Les botanistes dans l'Auvergne et le Velay. 47 p. 24 cm. Clermont-Ferrand, 1901. (Extrait du Bulletin de la Société des anciens élèves du pensionnat des frères de Clermont-Ferrand no. 41.)
Geology; sketch of botanists and botanical work from the 16th century on, with mention of publications.

AVEYRON

See also Auvergne (Chassagne); Gard (Braun-Blanquet; Martin, Florule); Hérault (Coulouma).

General

Bénézech, A., and **Toulouse, C.** Les plantes médicinales dans le département de l'Aveyron. 1 p. l., 34 p. 24 cm. Millau, 1920. (Library of Muséum national d'histoire naturelle, Paris.)
Includes annotated list of medicinal plants grouped by regions; list of plants to be exploited, calendar for collecting.

Bras, Antoine. Catalogue des plantes vasculaires du département de l'Aveyron. xliv, 533 p. map. 24.5 cm. Villefranche, 1877. (Gray Herbarium library.)

Topography, geology, botanical explorations with biographical data; annotated list of vascular plants, (2040 species), with localities.—See also **Chastaingt, Gabriel.** Additions au tableau de la végétation des environs d'Aubin (Aveyron). Bul. Soc. Bot. France **25**: 100–104. 1878. (List of additional species and localities, with some critical notes.)—Also **Giraudias, Ludovic.** Les plantes rares des environs d'Asprières (Aveyron). Bul. Soc. Études Scient. Angers **11/12**: 147–155. 1882. (Topography, etc.; list of vascular plants, with localities.)—Also **Ivolas, Jean.** Note sur la flore de l'Aveyron. Bul. Soc. Bot. France **32**: 286–292. 1885. (List of 76 additional species, with localities; list of species to be excluded, etc.)—Also **Coste, Hippolyte.** Un ciste hybride nouveau de la science et environ quarante plantes nouvelles pour la flore de l'Aveyron. l. c. **33**: 20–25. 1886. (Includes list of 38 additional species.)—Also his Note sur 150 plantes nouvelles pour l'Aveyron. l. c. 38 (Sess. Extraord.): xlviii-lxx. 1891. (Botanical explorations, list of species to be excluded, bibliography of publications since Bras; list of 150 additional species, with localities.)—Also **Coste, H.,** and **Soulié, J. A.** Note sur 200 plantes nouvelles pour l'Aveyron. l. c. **44** (Sess. Extraord.): lxxxvii–cxxi. [1898?] (List of 200 additional vascular plants and Characeae, with localities and some critical notes.)—Also **Fourès, P.** Note sur quelques plantes nouvelles pour l'Aveyron suivie de nouveaux habitats de plantes rares. Bul. Acad. Internatl. Géog. Bot. **16**: 265–286. 1906. (List of 21 additional vascular plants, with localities; long list of new localities for rarer species.)

Galinat, Maurice. Petite bibliographie du département de l'Aveyron. Monde des Plant. VI, **43** (250/251): 15–16. 1948. (Arnold Arboretum library.)

Mostly unannotated list of floras and floristic works.

Local

Carbonel, Jean. Florule d'Entraygues (Aveyron). 80 p. (place?), 1931.

Geology, climate, etc.; list of plants found by author, with some localities. (Not seen; title and annotation supplied by M. Galinat.)—See also his Les plantes adventices d'Entraygues (Aveyron). Bul. Mens. Soc. Linn. Lyon **10**: 146–150. 1941. (General considerations; list of 66 adventive plants.)

——— Florule de la commune de Saint-Hippolyte. Bul. Assoc. Franç. Bot. **3**: 181-193. 1900.

Soil, topography, general features of flora; annotated list of rarer or more interesting plants.

——— Liste des noms patois de plantes usités dans les cantons d'Entraygues et de Mur-de-Barrez (Aveyron). Bul. Acad. Internatl. Géog. Bot. **13**: 337–352, 401–432. 1904; **15**: 49–56. 1905.

Discussion of patois names, etc.; list of patois names systematically arranged, with derivation and meaning; unannotated list of names of some cultivated plants; index.—See also Luxembourg (Klein).

Coste, Hippolyte. Florule du Larzac, du Causse Noir et du Causse de Saint-Afrique. Bul. Soc. Bot. France **40** (Sess. Extraord.): xci–cxl. 1894.

Geology, topography, general features of flora, lists of species grouped by distribution or ecological preferences; list of rarer vascular plants, with localities, list of common species. The "causses" are limestone plateaus.

——— Mes herborisations dans le bassin du Dourdou. Bul. Soc. Bot. France **35** (Sess. Extraord.): xi–xxviii. 1888.

Topography, geology; extensive list of less common vascular plants.

BAS-RHIN

See also Alsace and Alsace-Lorraine.

Local

Gelshorn, G. Verzeichnis der in der umgegend von Zabern wild wachsenden und häufiger cultivirten gefässpflanzen. iv, 76 p. 24 cm. Zabern,

[1875]. (Beilage zu dem programm des Gymnasiums zu Zabern für das schuljahr 1874–1875.) (Gray Herbarium library.)

Annotated list of vascular plants (1176 species), with localities. Relates to cantons Zabern (Saverne) and Mauersmunster (Marmoutier) and adjacent parts of other cantons.—See also **Walter, Émile.** Die farnpflanzen der umgebung von Zabern. Mitt. Philomath. Gesell. Elsass-Lothr. **3** (Jahrgang 15, 1907): 547–581. illus. 1908. (Annotated list of vascular cryptogams; bibliography. In American Museum of Natural History library.)

Kirschleger, Frédéric. Statistique végétale des environs de Strasbourg. Cong. Scient. France **10** (2, Mém.): 35–61. 1843. (Reprinted, 28 p. n.p., n.d. In New York Botanical Garden library.)

Topography, climate; annotated list of vascular plants (977 species); statistics, general features of flora, classified list of cultivated useful plants, list of cultivated trees and shrubs.—See also Petry, below, and Rehdans, below.

Nicklès, Napoléon. Coup d'oeil sur la végétation de l'arrondissement de Schlestadt. Bul. Soc. Hist. Nat. Colmar **16/17**: 168–239. 1876 (1877).

Topography, regions of vegetation; annotated list of vascular plants (1088 species), with vernacular names. (Also known as Schlettstadt and Sélestat.)

Petry, H. Adventivflora von Strassburg. Mitt. Philom. Gesell. Elsass-Lothr. **1** (2): 38–40. 1893. (American Museum of Natural History library.)

List of adventive angiosperms, with localities.—See also **Ludwig, Alfred.** Beitrag zur adventivflora von Strassburg i. Els. l.c. **2** (Jahrgang 10, 1902): 508–533. 1903.—Also his Neue beiträge . . . l. c. **3** (Jahrgang 12, 1904): 113–125. [1905?] (Extensive lists of adventive vascular plants, with localities and dates. Both in American Museum of Natural History library.)—Also Kirschleger, above, and Rehdans, below.

Petzold, W. Verzeichnis der in der umgegend von Weissenburg im Elsass wildwachsenden und häufiger cultivirten gefässpflanzen. 45 p. 4to. Wissenbourg, 1879. (Schulprogramm.)

Topography, geology, botanical explorations; annotated list of wild and cultivated plants, with localities and German vernacular names. (Not seen; title and annotation supplied by G. Dillemann.)—See also **Recht, H.** Nachtrag zu W. Petzold's Verzeichnis . . . (p?). 1894. (2. beilage zum 23. Jahresber. Gym. Weissenburg.) (Not seen.)

Rehdans, —. Flora der nächsten umgebung Strasburgs. Zum gebräuche fur die schüler bei uebungen im bestimmen der pflanzen und auf excursionen. 56 p. 1884. (Progr. Gymnas. Strasburg in Westpr. 1884.)—2. theil. Die arten der phanerogamen bis zu den labiaten nach natürlichen system. 42 p. 1885. (Wissenschaftliche beilage zum Oster-programm des Königl. gymnasiums in Strasburg.)

List with short diagnoses, and localities for the scarcer species. (Not seen; cited from Bot. Jahresber. Just **12**(2): 250. 1887; **13**(2): 271, 352, 1888.)

Waldner, Heinrich. Ueber die flora von Wasselnheim. Jahresber. Realschule Wasselnheim **1874/75**: 3–9. 1875.

Not seen. Walter (Bul. Soc. Bot. France **73**: 617. 1927) speaks of it as a catalog. (French name Wasselonne.)

BASSES-ALPES

See also Alpes-Maritimes (Burnat); Provence.

General

Laurent, Louis. Catalogue raisonné des plantes vasculaires des Basses-Alpes . . . dressé d'après les documents bibliographiques recueillis par feu L.-A. Dessale . . . complétés et mis en oeuvre par l'auteur. v. 1–2 (393, 357 p.). map (in v. 1). 24.5 cm. Marseille, 1935–40. (Library of Muséum national d'histoire naturelle, Paris.)

Botanical explorations, bibliography, gazetteer, herbaria consulted; annotated list of vascular plants, with Provençal vernacular names, habitat and altitudinal range, detailed localities with names of collectors, and numerous critical notes. Vol. 1 (1935–37) covers Ranunculaceae–Papilionaceae, v. 2 (1939–40) Rosaceae–Compositae (Bidens). The first part of v. 1, going through the Fumariaceae, originally published in Annales Mus. Hist. Nat. Marseille v. 26, Mém. 4. 92 p. map. 1934. The two volumes contain 1202 numbered species, but cultivated, dubious, and even the excluded species are included in the numbering.—See also **Breistroffer, Maurice.** Supplément au Catalogue des plantes vasculaires des Basses-Alpes. (1–4. partie.) Bul. Mens. Soc. Linn. Lyon **15:** 46–48, 49–56. 1946; **17:** 195–198. 1948; **20:** 63–66, 112–115. 1951. (Lists of noteworthy species.)—Also **Mathon, C. C.** Suite au Catalogue de la flore des Basses-Alpes, particulièrement sur la rive droite de la Durance. (1.–3. partie.) Bul. Mus. Natl. Hist. Nat. Paris II, **20:** 536–543. 1948; **21:** 132–139, 371–374, 474–481, 590–592. 1 fig. 1949. (List of vascular plants principally from the right bank of the Durance, with localities, collected mostly by author or Legré, and intended to continue the list of plants of Basses-Alpes beyond the two published volumes of Laurent's Catalogue.)—Also his Seconde addition à la Suite au Catalogue . . . Bul. Mus. Natl. Hist. Nat. Paris II, **22:** 287–290. 1950. (Additional records.)

Local

Lannes, —. Catalogue des plantes les plus remarquables croissant dans le bassin supérieur de l'Ubaye (Basses-Alpes), compris entre Barcelonnette et la frontière de l'Italie d'un côté, et les Hautes-Alpes de l'autre. Altitude minimum 1163 mètres et maximum 3090 mètres. Bul. Soc. Bot. France **26:** 155–175. 1879.

Extensive list of vascular plants, with localities.—See also **Flahault, Charles.** La flore de la vallée de Barcelonnette. *In* Notices publiées par le comité local d'organisation, à l'occasion de la session extraordinaire de la Société botanique de France dans la haute-vallée de l'Ubaye. p. 17–57. illus. 8vo. Montpellier, 1897. (Account of the vegetation and its distribution; unannotated revision of the list published by Lannes in 1879. Not seen; title and annotation supplied by G. Dillemann.)—Also his Liste complémentaire et rectificative des plantes vasculaires observées dans le bassin supérieur de l'Ubaye, de 1135 mètres (altitude de Barcelonnette) à 3400 m. (alt. du sommet des Aiguilles de Chambeyron). Bul. Soc. Bot. France **44** (Sess. Extraord.): cclii–cclxviii. [1898?] (List of vascular plants, with localities or habitats, including corrections and additions.)

BASSES-PYRÉNÉES

General

Bergeret, J. P. Flore des Basses-Pyrénées . . . augmentée par Eugène Bergeret . . . nouvelle éd., complète, publiée avec une préface et des notes par Gaston Bergeret. lxxxvi, 960 p. 25.5 cm. Pau, 1909.

Elementary course in botany; annotated list of wild and cultivated vascular and cellular plants, on the Linnaean system, with localities for scarcer species, notes on uses, and in part with descriptions; indices of French and Béarn vernacular names, etc. In the main a reissue, without any botanical improvements, of Jean Bergeret's Flore des Basses-Pyrénées of 1803; the text of classes 14–24 (Didynamie-Cryptogamie) was first printed in the 1909 volume. Replaced as a flora by Blanchet's Catalogue, below.

Blanchet, Charles. Catalogue des plantes vasculaires du sud-ouest de la France comprenant le département des Landes et celui des Basses-Pyrénées . . . xviii, 172 p. 21.5 cm. Bayonne, 1891. (Gray Herbarium library.)

Annotated list of vascular plants and Characeae [about 2,415 species], with localities.—See also **Parrot, A. G.** Plantes rares ou adventices des Basses-Pyrénées. (1. [–3.] observation.) Bul. Soc. Bot. France **92:** 28–36. 1945; **93:** 254–258. 1947; **94:** 224–226. 1948. (Annotated lists of noteworthy vascular plants from vicinity of Bayonne, Biarritz, and Anglet, with localities; bibliography (in first part).)—Also **Vivant, Jean.** Additions à la flore des plantes vasculaires du Pays basque français. **101:** 193–197. 1954. (Extensive list of species from Haute-Soule, Basses-Pyrénées, not previously recorded from the Pays basque français.)

Local

Ancibure, E., and **Prestat, E.** Catalogue des plantes de la région bayonnaise. 85 p. 25 cm. Bayonne, 1918. (At head of title page: Société bayonnaise d'études régionales.) (Library of Muséum national d'histoire naturelle, Paris.)

Tabular list of vascular plants and cryptogams, with habitat, frequency, vernacular names, and some localities.

Bouillé, Roger de, *comte.* Sur la végétation de quelques-uns des pics les plus elevés des Pyrénées françaises. Bul. Soc. Bot. France **28:** 322–327. 1881.

Lists of vascular plants from Midi d'Ossau, Ar-Sourins, Pic du Ger, Pic du Gabisos, Pic d'Eras taillades, Pic de Bat-Laétouse ou Marmuret (the last in Hautes-Pyrénées), at altitudes up to 3050 meters.—See also Hautes-Pyrénées (Bouillé).

Jovet, Paul. La végétation anthropophile du Pays basque français. Bul. Soc. Bot. France **88:** 254–269. 1941.

Account of adventive vegetation of different habitats, with lists of species; historical account of several introduced species; bibliography.

Richter, J. A. Liste des plantes recueillies de 1870 à 1874 et de 1876 à 1880, dans un rayon de 10 kilomètres, autour de Saint-Jean Pied-de-Port. Bul. Soc. Bot. France **27** (Sess. Extraord.): lxi–lxxi. 1880.

List of vascular plants, annotated only as to abundance, with occasional specific localities.

BEAUJOLAIS

For works on this region and former province, which includes an area in northern Rhône and adjacent Loire, see under Rhône.

BELFORT

See also Alsace.

General

Herbelin, Louis. Histoire de la botanique dans le territoire de Belfort. Bul. Soc. Belfort. Émul. **31:** 197–224. 1912.

Account of botanical explorations and work, with references to publications.

——— Nouvelle flore du territoire de Belfort. Bul. Soc. Belfort. Émul. **42:** 209–257. 1928; **43:** 257–312. 1929; **45:** 205–264. 1931.

Annotated list of vascular plants and Characeae, with localities. Replaces **Parisot, Louis,** and **Pourchot, Louis.** Notice sur la flore des environs de Belfort. [nouvelle édition.] Bul. Soc. Belfort. Émul. **5:** 121–231. 1882. (Physiography, lists of species of different regions, soils, etc.; annotated list of vascular plants and Characeae. (1st ed. in Mém. Soc. Émul. Doubs III, **3:** 57–164. 1859. Parisot was sole author of 1st ed.).)

BERRY

The region and former province of Berry consists essentially of the Departments of Cher and Indre.

Le Grand, Antoine. Flore analytique du Berry contenant toutes les plantes vasculaires des départements du Cher et l'Indre. 2. éd. xxix, 430 p. 18 cm. Bourges, 1894. (Lloyd Library.) (1st ed. 1887.)—Supplément à la Flore du Berry. Mém. Soc. Hist., Lit. et Scient. Cher IV, **15:** 1–82. 1900. (Reprinted, 82 p. 1900. In Gray Herbarium library.)

Topography, key to families; briefly annotated, keyed list of 1448 vascular plants and Characeae (1,430 vascular species), with localities; glossary. Of the 1,407 spontaneous vascular plants, 1,375 are known from Cher. The Supplément contains an extensive list of additions and corrections.—See also **Kersers, L. de.** Localités nouvelles pour la flore du Berry. Bul. Soc. Bot. France **52:** 517–534. 1905. (Extensive list of plants collected by author and others, with localities.)—Also **Lambert, L.** Contribution à la flore du Berry. Monde des Plant. II, **8** (37): 6–7. 1906; (38): 15–16. 1906. (First paper, list of 22 hybrids new to Cher; second, list of newly recorded species. In

New York Botanical Garden library.)—Also **Déribéré-Desgardes, Pierre.** Plantes peu communes en Berry, récoltées dans l'Indre. Bul. Soc. Bot. France **53:** 663–667. 1907. (List of noteworthy plants, with localities.)—Also **Lambert, L.** Liste de plantes nouvelles pour le Berry découvertes en 1907 à Raymond (Cher) ou dans les environs. Bul. Soc. Bot. Deux-Sèvres **19:** 269–270. 1908. (Short list of phanerogams. In Farlow Library, Harvard University.)—Also his Additions à la flore du Berry. l. c. **20:** 59–107. 1909. (Extensive list of vascular plants, with localities. In Farlow Library.)—Also **Kersers, L. de.** Localités nouvelles pour la flore du Berry. Bul. Soc. Bot. France 59 (Sess. Extraord.): xli–xliv. 1915. (List of noteworthy species.)—Also **Touraine, L. L.** Plantes remarquables recoltées dans le département de l'Indre depuis une quinzaine d'années. Bul. Soc. Bot. Deux-Sèvres **1925:** 23–25. 1925. (Mainly a list of introduced weeds from the vicinity of Chateauroux. In library of Muséum national d'histoire naturelle, Paris.)—Also **Rallet, Louis.** Quelques plantes et stations nouvelles pour l'Indre. l. c. **1925:** 30–33. 1925. (List of noteworthy mosses and spermatophytes, with localities. In library of Muséum national d'histoire naturelle, Paris.)

——— Liste des noms populaires des plantes du Berry et principalement du Cher. Mém. Soc. Hist., Lit. (Artist.) et Scient. Cher IV, **13:** 85–126. 1898.

Alphabetical list of vernacular names, with standard French names, scientific names, and localities where used; list of scientific names with vernacular equivalents.

——— Notices biographiques et bibliographiques pour l'histoire de la botanique en Berry. Mém. Soc. Hist., Lit. (Artist.) et Scient. Cher IV, **6** (i.e., **7**): 113–148. 1891. (Reprinted, 38 p. 1891.)

Biographical notices of botanists (alphabetically arranged) connected with the flora of Berry, with mention of publications, local botanical work, etc.

BOUCHES-DU-RHÔNE

See also Provence.

General

Bouisson, A. G. Synopsis analytique des plantes vasculaires du département des Bouches-du-Rhône et éléments du botanique . . . 453 p. illus. 18 cm. Marseille, 1878. (Arnold Arboretum library.)

Includes flora of vascular plants in form of keys; an elementary work.

Castagne, Louis. Catalogue des plantes qui croissent naturellement dans le département des Bouches-du-Rhône . . . Avec une préface . . . par Alphonse Derbès. liv, 203 p. 19 cm. Marseille, 1862. (Gray Herbarium library.)

Biography of Castagne; plant zones, with lists of species; annotated list of vascular plants, with localities.—See also **Blanc, Pierre.** Contribution à la flore des Bouches-du-Rhône. Monde des Plant. III, **24** (29): 6-8. 1923; **25** (31): 5–8, (32): 5–6, (33): 5–6, (34): 7–8, (35): 6–7, (36): 6–7. 1924; **26** (37): 7–8, (39): 7–8, (40): 7–8, (41): 7–8. 1925; **27** (43): 6–8, (47): 3–5. 1926. (List of noteworthy plants, mostly already recorded by others. In Arnold Arboretum library.)

Lions, A. Essai sur les végétaux utiles qui croissent spontanément dans le département des Bouches-du-Rhône, qui y sont cultivés ou qui seraient susceptibles de l'être . . . 2 p. l., 263 p. 22 cm. Marseille, 1863. (Reprinted from Repertoire des travaux de la Société de statistique de Marseille. v. 26. 1863.) (Arnold Arboretum library.)

Annotated tabular list of vascular and cellular plants, alphabetically arranged by French vernacular names, with scientific and Provençal names, habitat or place where cultivated, medical or other economic properties, etc.; no index of scientific names.

Marnac, Émile, and Reynier, Alfred. Préliminaires d'une flore des Bouches-du-Rhône. Plantes adventices, subspontanées, naturalisées, autochtones. Bul. Acad. Internatl. Géog. Bot. **19:** 173–188, 229–260. 1909; **20:** 57–88, 149–200. 1910. (Reprinted 2 p. l., iv, 131 p. Le Mans, 1910, with title, Flore phanérogamique des Bouches-du-Rhône. 1. partie. Préliminaires: plantes subspontanées, adventices, naturalisées.)

Discussion of "spontané" and related terms; critically annotated list of adventive and naturalized vascular plants, as well as some indigenous species whose nativity had been questioned.

Local

Castagne, Louis. Catalogue des plantes qui croissent naturellement aux environs de Marseille. 263 p. 7 pl. 20.5 cm. Aix, 1845.—**Supplément au** Catalogue . . . 125 p. pl. 8–11. 20.5 cm. Aix, 1851.

Annotated list of vascular and cellular plants, with localities for scarcer species. The supplement refers mainly to cellular plants.—See also **Kieffer, F.** Contribution à la flore de Provençe. 8 p. Montpellier, 1907. (List including some species new to Bouches-du-Rhône, as well as new localities. Caption title: Contribution à la flore de Marseille. Not seen; cited from notice in Bot. Centralbl. **111:** 204. 1909.)

Chartrouse, Laugier de. Nomenclature patoise des plantes des environs d'Arles et leur concordance avec les noms français, la synonymie latine des auteurs et les familles naturelles. (p. ?) Arles, 1859.

Not seen; cited from G. Deleuil, Bibliographie botanique provençale. Bul. Mus. Hist. Nat. Marseille **14:** 29. 1954.

Delmas, J. P., *abbé,* **Marnac, Émile, and Reynier, Alfred.** Aperçu sur la montagne Sainte-Victoire près d'Aix-en-Provence. Bul. Acad. Interntl. Géog. Bot. **17:** 17–32. 1907.

Botanical explorations; running account of vascular plants, grouped by localities.—See also **Nègre, Robert.** Les associations végétales du massif de Sainte-Victoire (Provence Occidentale). 85 p. 25 cm. Paris, 1950. (Encyclopédie biogéographique et écologique VII.) (Includes (p. 83–85) list of species not previously recorded.)

Flahault, Charles, and **Combres, Pierre.** Sur la flore de La Camargue et des alluvions du Rhône. Bul. Soc. Bot. France **41:** 37–58. pl. 1–3. 1894.

Botanical explorations, topography, plant habitats, etc.; unannotated list of vascular plants and Characeae collected by authors in spots removed as far as possible from human influence.—See also **Tallon, Gabriel..** Additions à la flore de Camargue. Terre et la Vie **1950:** 265–274. 1950; **1955:** 233–237. 1955; **1957:** 117–119. 1957. (Annotated lists.)

Fonvert, Amédée de, and **Achintre, J. F.** Flore d'Aix-en-Provence. Catalogue des plantes vasculaires qui croissent naturellement dans les environs d'Aix. 2. éd. 2 p. l., iii, 180 p. 18 cm. Aix, 1882. (Library of Muséum national d'histoire naturelle, Paris.) (1st ed., Catalogue des plantes vasculaires . . . 1873; preprinted, 168 p. 1870 (cover date 1871). Preprint in Gray Herbarium library.)

Briefly annotated list of vascular plants (1507 species and varieties), with French and Provençal vernacular names, and localities for the less common species; includes cultivated plants. Replaces **Boyer de Fonscolombe, Étienne.** Calendrier de faune et de flore pour les environs d'Aix ou première apparition des principaux insectes et première floraison des végétaux qui s'y trouvent. Mém. Acad. Sci., Agr., Arts et Belles-Let. Aix **5:** 357–682. 1845. (Reprinted, xi, 14–342 p. 22 cm. Aix, 1845; reprint in Lloyd Library.) (Annotated list of vascular plants, with localities, arranged by date of first flowering (or fruiting in case of ferns), with corresponding list of insects on opposite pages.)—See also **Capoduro, Marius.** Liste de quelques plantes rares des environs d'Aix-en-Provence. Monde des Plant. **4:** 116–118. 1895. (List of noteworthy vascular plants.)

Grenier, Charles. Florula massiliensis advena. Florule exotique des environs de Marseille . . . ou enumération des espèces étrangères introduites autour de Marseille et récoltées par MM. Blaise, Roux, etc. Mém. Soc. Émul. Doubs III, **2:** 387–434. 1858. (Reprinted, 48 p. Besançon, 1857. In Gray Herbarium library.)—Supplément à la Florule exotique . . . l. c. III, **4:** 369–392. 1860. (Reprinted, 24 p. In Gray Herbarium library.)

General considerations; annotated list of adventive phanerogams (about 250 species), with localities and native habitat.—See also **Coste, F.** Localités nouvelles de diverses espèces adventices aux environs de Marseille. Monde des Plant. II, **18** (100): 16–17. (103) 33–34. 1916. (List of adventive angiosperms, with localities. In Arnold Arboretum library.)—Also **Anonymous.** Contribution à la flore des Bouches-du-Rhône. l. c. II, **19** (110): 3–4. 1918. (List of adventive angiosperms collected at Marseille and vicinity. In Arnold Arboretum library.)

Lapraz, Guy. Étude phytogéographique du massif d'Allauch. Annales Fac. Sci. Marseille II, **13:** 101–203. 2 fig. (fig. 1, map). 1940.
Phytogeography of some mountains around Marseille; includes (p. 181–195) list of species new to the flora, and list of new localities.

Laurent, Louis. Carte forestière, agricole et des plantes médicinales de la région nord-est de Marseille. map. ca. 50 × 80 cm. 1923.—Notice . . . 8 p. 12mo. 1923.
The Notice contains a key to the colored map, as well as a list of local species and of medicinal plants. (Not seen; cited from notice in Bul. Soc. Bot. France **71:** 1302. 1925. Published by Service géographique de l'armée.)

——— Une promenade botanique à l'île de Riou. Mém. Acad. Sci., Let. et Beaux-arts Marseille **1925:** 124–141. 1925. (Harvard College library.)
Topography, climate, general features of flora; incomplete alphabetical list of spermatophytes, annotated as to abundance. See also Reynier, Herborisations aux îles du littoral de la Provence, below.

——— and **Deleuil, Gabriel.** La répartition des végétaux dans les îles du Frioul: Pomègues et Ratonneau (rade de Marseille) et la question du reboisement de ces îles. Chêne Soc. Forest. Méditer. et Colon. Bul. **45:** 13–77. 5 fig. (incl. maps). 1938.
Botanical explorations, geography, topography, geology, history, climate, biological environment, etc.; annotated alphabetical list of plants; phytosociology, bibliography, etc. (Library of Muséum national d'histoire naturelle.)—See also Molinier, Le reboisement des îles du Frioul, below, and Reynier, Herborisations aux îles du littoral de la Provence, below.

Lèbre, Émile. Petite histoire et flore des rues d'Aix-en-Provence. Préface de Marcel Provence. 1. partie: La vieille ville. 2 p. l., vi, 151 p. 22 cm. Aix, 1920. (New York Botanical Garden library.)
Historical account of each street, with mention of the adventive plants found on it, with patois names, popular uses, etc.—See also Fonvert and Achintre, above.

Marnac, Émile. Florule de Sainte-Croix (Cassis) (Bouches-du-Rhône). (Contribution à la flore de Provence.) Revue Hort. Bouches-du-Rhône **52:** 28–32, 44–51, 59–61. 1906.
Unannotated list of vascular plants collected by author in 1905. Title varies.

Molinier, René. Une excursion botanique à l'Île Verte (La Ciotat). Chêne Soc. Forest. Méditer. et Colon. Bul. **40:** 182–192. 1935.
Geology; unannotated list of 150 plants observed (p. 184–187); phytosociology. (Not seen; title and annotation supplied by G. Dillemann.)—See also Reynier, Herborisations aux îles du littoral de la Provence, below.

——— Le reboisement des îles du Frioul . . . Chêne Soc. Forest. Méditer. et Colon. Bul. **42:** 23–50. map. 1936.
Includes (p. 30–34) list of vascular plants (about 268 species and varieties), with localities; also account of plant associations, etc. (Islands of Pomègues and Ratonneau.)—See also Laurent and Deleuil, above, and Reynier, Herborisations aux îles du littoral de la Provence, below.

——— and **Tallon, Gabriel.** La flore de la Crau (Bouches-du-Rhône). Historique et acquisitions récentes. Mém. Soc. Bot. France **1950/51:** 95–116. 1951.
Botanical explorations, critical review of erroneously recorded or rare species in earlier publications, list of new localities and newly recorded species.

Pathier, C. X. Florule phanérogamique des environs de Roquevaire. Bul. Soc. Bot. et Hort. Provence **2:** 178–196. [1881]; **3:** 43–61. 1881.
Geology, topography, itineraries for collecting, etc.; list of vascular plants and Characeae, with localities or indication of habitat.

Reynier, Alfred. La botanique à Aix-en-Provence depuis la seconde moitié du XVIe siècle. 10 p. n.p., [1906?]. (From Compte-rendu du Congrès des Sociétés savantes de Provence tenu à Marseille en juillet-août 1906. In library of A. Camus.)
Partial account of botanical explorations, with notes on several botanists.

——— Herborisations aux îles du littoral de la Provence. Bul. Soc. Bot. et Hort. Provence **2:** 197–206. [1881.]

Account of the different islands of the three archipelagos concerned: (1) islands near Marseille—Château d'If, Pomègue et Ratonneau, Île de Maire, Île Plano, Île de Jarro, Île de Riou, Île Verte, Île de Bandol, Île des Imbiers; (2) Îles d'Hyères; (3) archipel de Lérins (Île Sainte-Marguerite, Île Saint-Honorat), with partial lists of the species of each; lists of species found on nearly all the islands, those found only on these islands (in France), etc.

Tallon, Gabriel. La flore des rizières de la région d'Arles et ses répercussions sur la culture du riz. 31 p. 13 fig., tab. 24 cm. [Arles-sur-Rhône, 1950?] (Étude présentée au cours des "Journées du Riz" tenues en Arles-sur-Rhône les 9 et 10 Novembre 1950.) (In Gray Herbarium library. Reprinted, somewhat revised and without illustrations, in Vegetatio **8**: 20–42. tab. 1958.)

Includes tabular list of principal species of rice-fields, grouped by associations, with additional ones listed in text, and discussion.

BOURBONNAIS

The region and former province of Bourbonnais includes Allier and part of Cher.

Lassimonne, S. E. Étude sommaire de la flore du Bourbonnais. Notre Bourbonnais no. 2. (p. ?) 1923.

Phytogeography. plant associations; list of adventive plants. (Not seen, cited from notice in Bul. Soc. Bot. France **70**: 750–751. 1923.)

——— Flore des plantes vasculaires du Bourbonnais. Synopsis des espèces linnéennes ou linnéons. v. 1 (1 p. l., iv, 258 p., 2 l. 3 fig.). 25 cm. Moulins, 1932.

Flora in form of synoptical tables (1,686 species, including cultivated plants); glossary, list of vernacular names with botanical equivalents. No more published.

——— Prodromes d'une flore nouvelle des plantes vasculaires du Bourbonnais. 1. mémoire. Revue Scient. Bourbonn. **1923**: 125–132. 1923; **1924**: 33–64, 145–164. 1924; **1925**: 57–68, 145–163. 1925.—Prodromes . . . 2. mémoire. l. c. **1928**: 41–54. 1928.

Mention of principal floras, changes in topography and vegetation, etc.; list of vascular plants, with localities and some annotations. Not a complete flora, but intended to supplement Migout's Flore du département d'Allier. The deuxième mémoire consists of further additions and corrections.

Pérard, Alexandre. Flore du Bourbonnais comprenant le département de l'Allier et une partie des départements du Cher, de la Creuse, du Puy-du-Dôme et de la Nièvre. 2 v. (112, 48 p.). 24 cm. Montluçon, 1884–86. (Gray Herbarium library.)

Annotated list of vascular plants and Characeae, with localities. The second volume has the subtitle: Matériaux (supplément).

BOURGOGNE (BURGUNDY)

The region and former province of Bourgogne includes the departments of Ain, Côte-d'Or, Saône-et-Loire, and part of Yonne.

Ferrière, Émile. Plantes médicinales de la Bourgogne; emplois et doses. 101 p. 18 cm. Paris, 1892.

List of diseases, with remedies, etc.; list of medicinal plants, alphabetically arranged by botanical names, with uses and dosage.

BRETAGNE (BRITTANY)

The region and former province of Bretagne includes the departments of Côtes-du-Nord, Finistère, Ille-et-Vilaine, Loire-Inférieure, and Morbihan. Only works relating to the region as a whole are given here.

Daniel, Lucien. Les plantes médicinales de Bretagne. 64 p. 67 fig. 18 cm. Rennes, 1924.

Generalities; annotated list of 67 wild medicinal plants, with uses, grouped according to season.

Liégard, Auguste. Flore de Bretagne. xlviii, 405 p. 18.5 cm. Paris, 1879. (Gray Herbarium library.)

Climate, etc.; annotated, keyed, briefly descriptive flora of phanerogams, with notes on uses and habitats but mostly without localities, the families arranged alphabetically; indices of patois and French vernacular names.—See also **Crié, Louis.** La végétation des côtes et des îles bretonnes. Annales Sci. Nat. Bordeaux **6:** 145–164. map. 1887. (Analysis of the elements of the littoral flora with mention of numerous species. Not seen; cited from Bul. Soc. Bot. France **34** (Rev. bibl.): 37. 1887.)—Also **Picquenard, C. A.** Contributions à la flore de Bretagne. Bul. Soc. Sci. Nat. Ouest France **2** (1): 235–244. 1892. (List of noteworthy species.)—Also his Herborisations dans l'Ille-et-Vilaine, le Morbihan et le Finistère d'Octobre 1892 à Octobre 1894. l.c. 5(1): 37–43. 1895. (List with localities.)—Also **Guffroy, Charles.** Notes sur la flore bretonne. [I]–V. Bul. Soc. Bot. France **59:** 316–323, 385–391, 407–413, 495–503. 1912; **74:** 17–39. 1927; **76:** 1064–1076. 1930; **77:** 668–671. 1931; 88: 609–613. 1941. (Lists of noteworthy plants.)—Also **Des Abbayes, Henri.** Quelques phanérogames adventices de Bretagne. [I]–IV. Bul. Soc. Scient. Bretagne (Sci. Math., Phys. et Nat.) **16:** 113–115. 1940; **18:** 28. 1941; **21:** 61–62. 1946; **22:** 73–76. 1948. (Account of various noteworthy adventive plants.)—Also **Corillion, Robert.** Phanérogames intéressantes pour la Bretagne.[I]–VII (title varies). l.c. **22:** 123–125. 1948; **24:** 75–77, 78–80. 1950; **25:** 131–140. 1952; **26:** 107–112. 1953; **27:** 77–84. 1 fig. (map). 1953; **28:** 55–64. 1954. (Records of noteworthy species.)—Also his Notes complétives de floristique armoricaine. l.c. **31:** 81–92. 1958.

Matte, Henri. Flore de Bretagne. Synopsis analytique illustré des plantes vasculaires observées dans le massif amoricain. Bul. Soc. Scient. et Méd. Ouest v. 23–25, 31–32, suppl. 180 p. 113 fig. 1914–16, 1922–23. (P. 1–96 of original not seen. Reprinted in part with title: Flore de Bretagne: synopsis analytique. Trav. Scient. Univ. Rennes v. 17, suppl. 1–128. 87 fig. 1923.)
Bibliography, key to families and genera; annotated list of vascular plants, with brief descriptions and localities. Incomplete (396 species, Ranunculaceae to Paronychiaceae; the part reprinted in 1923 covers Ranunculaceae–Ilex, 224 species).

BRIONNAIS

The region and former province of Brionnais is now part of Saône-et-Loire. For the single flora relating to Brionnais, see Ormezzano under Saône-et-Loire.

CALVADOS

See also Normandie.

General

Hardouin, L., Renou, F., and Le Clerc, E. Catalogue des plantes vasculaires qui croissent spontanément dans le département du Calvados . . . Mém. Soc. Linn. Normandie **8:** 167–293. 1849. (Reprinted xiv, 439 p. 14.5 cm. Caen, 1848 (cover date 1849).)
Topography, geology, botanical investigations; briefly annotated list of vascular plants and Characeae (1,214 species, of which 1,206 are vascular), with localities.—See also **Chevrel, René.** Plantes rares de Calvados et principalement des environs de Caen. Bul. Soc. Linn. Normandie VI, **3:** 10–13. 1911. (List of 24 species.)—Also **Bédel, L.** Plantes rares ou intéressantes des environs de Dozulé (Calvados). l. c. VI, **8:** 90–114. 1916. (Extensive list.)—Also **Meslin, Roger.** Contributions à l'étude de la flore de Normandie. I. Localités nouvelles de plantes du Calvados. l.c. IX, **5:** 62–67. 1948.—Other shorter or longer papers will be found in the same journal.

Local

Dubourg-d'Isigny,—. Aperçus généraux sur la topographie géognostique et botanique de l'arrondissement de Vire. Mém. Soc. Linn. Normandie **6:** 147–170. pl. 10, tab. 1838.
Geology, soils; briefly annotated list of vascular and cellular plants (1318 species, of which 692 are vascular).

Durand-Duquesney,—. Coup-d'oeil sur la végétation des arrondissements de Lisieux et de Pont-l'Évêque, suivi d'un catalogue raisonné des plantes vasculaires de cette contrée. Bul. Trav. Soc. Émul. Lisieux **1**: 73–199. 1846. (Lloyd Library.) ("Reprinted, 127 p. 1847.")

Plant habitats, distribution in relation to soils; annotated list of vascular plants and Characeae.

Poplu, M. C. Flore des rives de la Touque et les falaises de Trouville. vii, 98 p. 18 cm. Pont-l'Évêque, 1873.

Popular account of useful plants, etc., of various families; partly annotated list of vascular plants, evidently very incomplete, with vernacular names.

CANTAL

See also Auvergne.

General

Lauby, Antoine. Botanique du Cantal. Bio-bibliographie analytique suivie d'une liste des végétaux vivants et fossiles nouveaux pour cette région. 2 p. l., 72 p. 25 cm. Paris, 1903. (Extrait de la Revue de la Haute-Auvergne.)

Includes chronological account of students of vascular plants and their publications (p. 29–43), and bibliography (p. 50–65); also list of vascular plants added to the flora since 1883 (p. 70–72), with localities.

Lecoq, Henri. Botanique. *In* Déribier-du-Chatelet, J. B. Dictionnaire statistique, ou histoire, description et statistique du département du Cantal. v. 1, p. 453–510. 25.5 cm. Aurillac, 1852.

Consists of 2 sections: Considérations générales sur la végétation du Cantal (p. 453–464), giving general features of vegetation and descriptions of the different associations; and Catalogue des plantes vasculaires du département du Cantal (p. 465–510), an annotated list of vascular plants and Characeae, with occasional localities.—See also **Personnat, Victor.** Plantes du Cantal manquant au Catalogue publié par M. H. Lecoq, dans le Dictionnaire statistique du départment. Bul. Soc. Bot. France **5**: 595. 1858. (List with localities.)—Also **Gatien,** *frère,* and **Héribaud,** *frère.* Notes sur quelques plantes récemment découvertes dans les montagnes du Cantal. Comp. Rend. Excurs. Soc. Nat. Sci. Nat. Charente-Inf., Soc. Bot. Rochelaise **1**: 109–114. 1878. (Records of 11 species. In Arnold Arboretum library.)—Also **Maranne, Isidore.** Contribution à l'étude de la distribution géographique des végétaux dans le Cantal. Bul. Acad. Internatl. Géog. Bot. **16**: 23–33. 1906. (List of noteworthy plants.)—Also his Deuxième liste de localités nouvelles de plantes rares dans le Cantal. l.c. **17**: 226–229. 1907. (Similar list.)—Also his Quelques localités nouvelles de plantes rares dans le Cantal. l.c. **20**: 146–148. 1910.—Also his Localités nouvelles de plantes rares dans le Cantal. (4.–5. note.) l.c. **22**: 25–27. 1912; **23**: 52–54. 1913.—Also his Quelques localités nouvelles de plantes rares dans le Cantal (6. note). Monde des Plant. II, **15** (85): 95. 1913. (Short list of angiosperms. In Arnold Arboretum library.)—Also **Tallon, Gabriel.** Notes floristiques sur le massif cantalien. Arvernia **2**: 85–92. 1926. (With Revue d'Auvergne v. 41.) (Annotated lists of newly recorded or noteworthy phanerogams; tabular lists of alpine and subalpine phanerogams, with altitudinal range and localities.)

Malvezin, J. E. Aperçu sur l'histoire de la botanique dans le Cantal. Bul. Soc. Bot. France **26** (Sess. Extraord.): xxiii–xl. 1879.

Chronological account of botanists and botanical explorations, with lists of additions made to the flora.

Local

Daude, P. Le touriste au Cantal, contenant notamment l'énumération des principales richesses naturelles en entomologie, zoologie, et botanique. 32 p. 8vo. Saint-Flour, [1902].

Account of various excursions, with lists of noteworthy plants. (Not seen; annotation supplied by G. Dillemann.)

Laronde, A., and **Garnier, R.** Herborisations à Saint-Jacques-des-Blats (Cantal). Revue Scient. Bourbonn. **15**: 16–24, 45–54. 1902.

Topography, geology, plant zones; partial list of vascular and cellular plants, with localities.

Maranne, Isidore. Flore des hauts-plateaux basaltiques du Massif Central (plateau d'Allanche). Bul. Géog. Bot. **26**: 57–99. 1916.

General feature of flora, etc.; annotated list of vascular plants observed by author; list of common plants of Auvergne not found in the region. Covers part of the Massif du Cézallier, including the communes of Allanche, Landeyrat, Pradiers, Vernols, and vicinity.

CHAMPAGNE

The somewhat vaguely defined region of Champagne includes principally the departments of Ardennes, Aube, Haute-Marne, Marne, and part of Yonne. Only works relating to the whole or a large part of the region are listed here.

Laurent, Jules. Catalogue des plantes vasculaires de la Champagne crayeuse. Avec la collaboration de A. Devauversin et Louis Maury. (Études sur la flore et la végétation de la Champagne crayeuse. v. 2.) ix, 269 p. 22 cm. Reims, 1925–[32]. (Library of Muséum national d'histoire naturelle, Paris; p. 65–269 in library of Chicago Natural History Museum.)

Botanical explorations; annotated list of vascular plants (including 1099 native and naturalized species), with localities; gazetteer, short bibliography. Covers all of Marne and part of Ardennes, Aube, and Yonne. According to a note in the copy examined in the Paris Museum, the work was issued in Bul. Soc. Hist. Nat. Reims 1925–32.—Replaces **Fliche, Paul.** La Champagne crayeuse. Étude de géographie botanique. Mém. Soc. Acad. Agr., Sci., Arts et Belles-Let. Dépt. Aube **72** (III, **45**): 171–277. 2 pl. 1908. (Includes (p. 186–201) unannotated list of vascular plants, with indication of those which are introduced.)—See also **Laurent, Jules.** La végétation de la Champagne crayeuse. Étude de géographie botanique. 1 p. l., vii, 355 p. 24 pl., 9 maps. 25 cm. Nemours, 1920. (Études sur la flore et la végétation de la Champagne crayeuse . . . v. 1.) (Ecological; includes (p. 319–342) "Appendice. La florule adventice," containing annotated list of recently naturalized species and list of acclimated perennials which do not spread; also (p. 343–350) bibliography.)—Also **Berton, A.** Quelques plantes de la Champagne crayeuse. Monde des Plant. **50** (315): 6–7. 1955. (List of noteworthy species, mostly from Marne. In Arnold Arboretum library.)

Remy, E. A. Flore de la Champagne, description succincte de toutes les plantes cryptogames et phanérogames des départements de la Marne, des Ardennes, de l'Aube et de la Haute-Marne, leurs propriétés médicales, usages économiques, industriels, et intérêt agricole. xii, 281 p. 19 cm. Reims, 1858.

Briefly descriptive flora of cellular and vascular plants, with some notes on uses but without localities. Hariot (Mém. Soc. Acad. Agr., Sci., Arts et Belles-Let. Dépt. Aube **55**: 179. 1891) states that it is absolutely worthless.

CHARENTE

General

Rochebrune, A. T. de, and Savatier, Alexandre. Catalogue raisonné des plantes phanérogames qui croissent spontanément dans le département de la Charente. xv, 294 p. 22.5 cm. Paris, 1860 (cover date 1861).

Geology, etc.; annotated list of vascular plants and Characeae, with localities for scarcer species; systematic list of specific names, with page references; no alphabetical index.—See also **Duret, V.** Plantes et localités nouvelles de la Charente. Bul. Soc. Bot. Deux-Sèvres **20**: 142–149. 1909. (Extensive list of vascular plants and Characeae, with localities, collected by author in 1877–1880 in the arrondissement of Angoulême. In Farlow Library, Harvard University.)

CHARENTE-MARITIME

Named Charente-Inférieure until 1941.

General

David, P. A., Foucaud, Julien, and Vincent, Pierre. Catalogue des plantes vasculaires qui croissent spontanément dans le département de la Charente-Inférieure . . . 2 p.l., 83 p. map. 25 cm. La Rochelle, 1878. (Bastard-title, Géographie botanique de la Charente-Inférieure.)

Annotated separate lists of common and rare vascular plants and Characeae, with localities for the rare species. Map not seen. Replaces **Faye, Léon.** Catalogue des plantes vasculaires du département de la Charente-Inférieure. viii, 87 p. 18 cm. Sivrai, 1850; Supplément. p. 88–84 (i.e., 94). 18 cm. Sivrai, 1851. (Both in New York Botanical Garden library.)—See also **Foucaud, J.** Herborisations faites dans la Charente-Inférieure en 1878. Bul. Soc. Bot. France **26:** 73–81. 1879. (Extensive list.)—Also **Société botanique rochelaise.** Comptes-rendus des excursions botaniques faites sous les auspices de la Société des sciences naturelles de la Charente-Inférieure. v. 1 (1878)—5 (1882). 1879–1883. (Include many lists of plants collected in Charente-Inférieure, by various authors. In Arnold Arboretum library.)—Also **Giraudias,—.** Herborisations dans la Charente-Inférieure (1881–1885). Revue Bot. Bul. Mens. Soc. Franç. Bot. **4:** 277–288. 1886. (Topography, etc.; extensive list of noteworthy plants, mostly from vicinity of Aulnay.)—Also **Bois, Desiré.** Herborisations dans la région de Royan (Charente-Inférieure). Bul. Soc. Bot. France **64: 145–154. 1917 [1918 ?]. (General features of flora in** vicinity of Saint-Georges-de-Didonne; list of noteworthy species, including list of principal cultivated woody plants.)

Local

Brunaud, Paul. Liste des phanérogames et cryptogames croissant spontanément à Saintes (Charente-Inférieure) et dans les environs. Actes Soc. Linn. Bordeaux **32** (IV, **2**): 116–170. 1878. (Reprinted, 57 p. Bordeaux, 1878. In Arnold Arboretum library.)—Supplément . . . l.c. **34** (IV, **4**): 109–130. 1880. (Reprinted, 26 p. Bordeaux, 1880. In Arnold Arboretum library.)

Bibliography; list of plants (1264 species, varieties, and forms of phanerogams in original list, and 2097 of cryptogams), with some localities and annotations.

Fouillade, A. Introduction à l'étude des modifications de la flore de la Charente-maritime. Annales Soc. Sci. Nat. Charente-Infér. n. s., **1:** 21–90. 1929. (American Museum of Natural History library.)

Discussion of changes in environment, introduced species, earlier floras (Morison, Lesson, Faye, Lloyd, Foucaud), etc. He states that David, Foucaud, and Vincent's Catalogue (1878) gives the same species and localities as ed. 3 of Lloyd (1876).

——— Sur la flore des environs de Tonnay-Charente (Charente-Inférieure). Notes de topographie botanique et observations sur quelques plantes nouvelles, rares ou critiques. Bul. Soc. Bot. Deux-Sèvres **17** (1905): 175–206. 1906. (Farlow Library, Harvard College.)

Botanical explorations, topography, lists of characteristic species of various habitats; list of noteworthy vascular plants, with localities.

Lesson, R. P. Flore rochefortine, ou description des plantes qui croissent spontanément ou qui sont naturalisées aux environs de la ville de Rochefort. 634 p. 21 cm. Rochefort, 1835.

Annotated descriptive flora of vascular and cellular plants (1117 species, of which 909 are vascular), with vernacular names and notes on uses.

Rousseau, Philéas. Catalogue des plantes vasculaires spontanées de l'île de Ré et des plantes qui y sont le plus communément cultivées. Bul. Soc. Sci. Nat. Ouest France **9** (**1**): 147–198. pl. 18 (map). 1899.

History, general features of flora, etc.; annotated list of vascular plants.

CHER

See also all titles under Berry and Bourbonnais; Loir-et-Cher (Segret).

General

Mornet, A. Catalogue des plantes phanérogames du département du Cher d'après la Flore du centre de la France de M. Boreau . . . Mém. Soc. Hist., Lit., Art. et Scient. Cher II, **3:** 51–192. 1876.

List of botanists, etc.; annotated list of vascular plants and Characeae, with vernacular names and localities.—See also **Le Grand, Antoine.** Premier fascicule de plantes nouvelles ou rares pour le département du Cher. Mém.

Soc. Hist., Lit., Artist. et Scient. Cher IV, **1**: 193–202. 1884. (List of 54 species.)—Also **Prudhomme, J.** Adventices et plantes intéressantes du département du Cher. ([I]–II.) Monde des Plant. VI, **43** (252): 18–21. 1948; **44** (255): 8, (256): 11–12. 1949. (Annotated lists of noteworthy vascular plants. In Arnold Arboretum library.)

Local

Duchaussoy, H. Végétation comparée de la Somme et du Cher. Mém. Soc. Linn. Nord France **9**: 1–71. 1898. (Topography, geology, etc.; discussion of species peculiar to each department, account of various localities with mention of their characteristic species, etc.)

CORRÈZE

See also Auvergne (Chassagne); Limousin.

General

Rupin, Ernest. Catalogue des plantes qui croissent dans le département de la Corrèze. Bul. Soc. Scient., Hist. et Archéol. Corrèze **1**: 687–693. 1879; **2**: 247–256, 433–446, 607–621, 817–827, 1879–1880; **3**: 181–196, 355–375, 525–544, 725–747. 1881; **4**: 243–252, 419–433, 587–609. 1882; **5**: 325–363. 1883. (Reprinted, 277 (by error, 377) p. Brive, 1884. In library of Muséum national d'histoire naturelle, **Paris.**)

Brief account of topography and botanical explorations; annotated list of vascular plants (1517 species), with localities; includes some cultivated species.

Local

Crauffon, Emmanuel. La botanique du patois Bas-Limousin. Bul. Soc. Let., Sci. et Arts Corrèze **20** (i.e. **30**): 429–440. 1908.

Includes (p. 435–440) tabular alphabetical list of botanical names, with French and patois equivalents and remarks on derivation, prepared by Oscar Lacombe for the "Annuaire de la Corrèze" of 1853.

Godin de Lépinay, Gaston. Noms patois ou vulgaires des plantes de la Corrèze. Revue Bot. Bul. Mens. Soc. Franç. Bot. **5**: 141–172. 1886.

Pronunciation, previous publications; systematic list of vascular and cellular plants with patois names from Brive, Tulle, Ussel, and Figeac (all in Bas-Limousin).

Laffon, Raoul. Catalogue des plantes qui croissent dans la commune de Saint-Cernin-de-Larche. *In* Rev. Scient. Limousin **9** (année **18**): **177–183.** 1910 to **10** (année 19): 103–106. 1911, passim. maps. ("Reprinted with pref., 83 p.")

Annotated alphabetical list of scientific and vernacular names, with some localities and notes on properties. (Not seen; title and annotation supplied by R. McVaugh from work in University of Michigan library.)

CORSE (CORSICA)

See also Italy, General (Fenaroli).

General

Briquet, John, and **Litardière, René de.** Prodrome de la flore corse. v. 1–3, pt. 2. 25.5 cm. Genève, Bâle, Lyon (v. 1–2, pt. 1), Paris (v. 2, pt. 2–3, pt. 2). 1910–1938.

Botanical explorations, bibliography (with additions in later parts); annotated list of vascular plants, with synonyms, citation of localities and collectors, and numerous critical notes. Vol. 3, pt. 2 reaches Solanaceae (1414 species). Continued by Litardière (v. 2, pt. 2 on) after death of Briquet. The title of v. 1–2, pt. 1 included the words "comprenant les résultats botaniques de sept voyages exécutés en Corse sous les auspices de M. Emile Burnat."

Litardière, René de. La flore adventice de la Corse. Compt. Rend. Cong. Soc. Sav. Paris et Dépt. **61,** Lille 1928 (Sect. Sci.): 112–121. 1929.

General remarks on adventive plants of Corsica; annotated list of 107 phanerogams.

Marsilly, L. J. A. de C. de. Catalogue des plantes vasculaires indigènes, ou généralement cultivées en Corse . . . avec l'indication des stations et des époques de floraison . . . avec le concours de MM. E. Revelière et P. Mabille. 203 p. map. 22 cm. Paris, 1872.

Botanical investigations; annotated list of 1625 vascular plants; gazetteer. —See also **Burnouf, Charles.** Plantes trouvées aux environs de Corte et qui ne figurent pas dans le Catalogue de M. de Marsilly. Bul. Soc. Bot. France **24** (Sess. Extraord.): xxx–xxi. [1878.]—Also **Gillot, F. X.** Rapport sur l'herborisation faite par la Société botanique de France à Erbalunga (Corse), le 29 mai 1877, et sur quelques autres herborisations aux environs de Bastia. l. c. xxxviii–lxii. [1878.] (Itinerary, with short lists of species and some critical notes; followed, p. lxii–ciii, by papers by other authors relating to other localities in Corsica.)—Also **Chabert, Alfred.** Observations sur la flore montagneuse du Cap Corse. l. c. **29** (Sess. Extraord.): l–lvii. 1882. (Annotated list of rare or unrecorded species.)—Also **Petit, Emil.** Additamenta Catalogi plantar. vascul. indig. corsicarum edit. Mr. de Marsilly. Bot. Tidsskr. **14:** 244–248. 1885. (List with localities.)—Also **Vallot, Joseph.** Sur quelques plantes de Corse. Bul. Soc. Bot. France **34:** 131–137. 1887.—Also **Fliche, Paul.** Notes sur la flore de la Corse. l.c. **36:** 356–370. 1 fig. 1889.—Also **Le Grande, Antoine.** Contribution à la flore de la Corse. l.c. **37:** 17–21. 1890.—Also **Debeaux, Odon.** Notes sur plusieurs plantes nouvelles ou peu connues de la région méditerranéenne et principalement [de la Corse et] des Pyrénées-Orientales. Revue Bot. Bul. Mens. Soc. Franç. Bot. **9:** 237–287. 1891; **12:** 177–240. 1894.—Also **Chabert, Alfred.** Contribution à la flore de France et de Corse. Bul. Soc. Bot. France **39:** 66–69. 1892.—Also **Foucaud, Julien,** and **Simon, Eugène.** Trois semaines d'herborisations en Corse. Annales Soc. Sci. Nat. Charente-Infér. **32** (1897): 39–218. 3 pl. 1898. (Reprinted 180 p. 3 pl. La Rochelle, 1898.) (Itinerary, etc.; list of about 860 plants from the littoral region, including 63 new to Corsica.)—Also **Boullu, A. E.,** *abbé.* Herborisations en Corse de MM. Foucaud et Simon. Annales Soc. Bot. Lyon **24:** 63–76. 1899. (Summary of Foucaud and Simon's paper, with list of species (with localities) not found in continental France.)—Also **Foucaud, J.** Additions à la flore de Corse. Bul. Soc. Bot. France **47:** 83–102. pl. 1–5. 1900.—Also **Briquet, John.** Recherches sur la flore des montagnes de la Corse et ses origines. Annuaire Conserv. et Jard. Bot. Genève **5:** 12–119. 3 pl. 1901. (Plant associations, phytogeography; annotated list of newly recorded, rare, or critical vascular plants.)—Also **Lutz, Louis.** Additions à la flore de Corse. Bul. Soc. Bot. France **48:** 49–58. 1901.—Also **Maire, René.** Contributions à l'étude de la flore de la Corse. l.c. **48** (Sess. Extraord.): cxlvi–cxlviii. [1903.]—Also **Lutz, L.** Nouvelles additions . . l. c. **48:** cxlviii–cl. [1903.]—Also **Maire, R.** Remarques sur la flore de la Corse. Revue Bot. Syst. et Géog. Bot. **2:** 21–27, 49–57, 65–73. 1904. (Plant zones, etc.; list of vascular plants new to or rare in Corsica.)—Also **Briquet, J.** Spicilegium corsicum ou catalogue critique des plantes récoltées en Corse du 19 mai au 16 juin 1904, par M. Émile Burnat, accompagné de MM. Jean Burnat, François Cavillier et Émile Abrezol. Annuaire Conserv. et Jard. Bot. Genève **9:** 106–183. 7 fig. 1905. (List of species new to Corsica, itinerary; extensive list of vascular plants, with localities and frequent annotations.)—Also Litardière, below.

Local

Boullu, Antoine, *abbé.* Îles Sanguinaires. *In* his Compte rendu des herborisations d'Ajaccio. Bul. Soc. Bot. France 24 (Sess. Extraord.): lxxxvii–xci. [1878?]—Liste de quelques plantes récoltées aux Îles Sanguinaires. l.c. **26:** 81–82. 1879.

Running account of flora, with short lists of species.

Boyer, Henri. Contribution à la flore de l'extrème sud du Corse ou territoire de Bonifacio. 71 p. illus. incl. maps. 8vo. Montpellier, 1906. (Thèse.)

Geology, geography, climate, vegetation; list of species. Refers to area south of a line connecting the col de Ventilègne with the étang de Balistra. (Not seen; title and annotation from notice in Bot. Centralbl. **105:** 281. 1907.) According to Briquet, Prodrome de la flore corse **1:** xxxvii. 1910, the work contains many errors.

Coste, Hippolyte. Herborisations autour de la ville d'Ajaccio, les 21, 23 et 24 mai. Bul. Soc. Bot. France 48 (Sess. Extraord.): ciii–cvii. 1903.

Three unannotated lists of vascular plants from around Ajaccio, followed (p. cviii–cl) by numerous longer or shorter lists of vascular plants from other localities in Corsica, by Coste and others.

Litardière, René de. Liste des phanérogames et des cryptogames vasculaires observés en Corse en septembre 1902, juillet 1906, juillet 1907 et juillet-août 1908. *In* his Voyage botanique en Corse (juillet-août 1908). Bul. Acad. Internatl. Géog. Bot. **19:** 96–132, 189–211. 1909.

Extensive list of vascular plants, with localities, preceded (p. 37–95) by account of various localities at which author collected.—See also his Contribution à l'étude de la flore de Corse. Bul. Géog. Bot. **24:** 89–108. 1914; **26:** 163–167. 1916. (List of vascular plants from Cap Corse collected by author in May 1913, with localities.)—Also **Litardière, R. de, and Simon, Eugène.** Notice sur les plantes recueillies par M. J. Aylies en Corse durant les années 1917 et 1918. Bul. Soc. Bot. France **68:** 24–41, 86–116. 1921. (List of various vascular plants, chiefly from region of Corte, with localities and notes.)—Also **Litardière, R. de.** Contribution à l'étude de la flore de la Corse. Bul. Soc. Sci. Hist. et Nat. Corse 42 (2): 187–242. 1922. (Itinerary; list of 173 vascular plants, with localities, mostly collected by author in 1919 in the Rotondo massif, upper basin of the Tavignano, in vicinity of Corte.)—Contributions . . . Bul. Soc. Bot. France **71:** 701–713. 1924.—Contributions . . . Notes sur quelques filicinées du Cap, des massifs du Cinto et du San Pietro. Annales Soc. Linn. Lyon **70:** 121–133. 6 fig. 1924.—Nouvelles contributions . . . fasc. 1–4. Arch. Bot. (Caen) v. 2, Mém. no. 1. 44 p. 1928; v. 3, Mém. no. 3. 31 p. 1929; v. 4, Mém. no. 2. 16 p. 1 fig. 1930; no. 3. 10 p. 1930.—Nouvelles contributions . . . fasc. 5–9. Candollea **5:** 153–160. 1932; **7:** 229–241. 1937; **11:** 175–227. 1948; **14:** 121–157. 1953; **15:** 9–14. 1956. (New localities and newly recorded species from various regions and expeditions.)—Also his Notes sur quelques plantes de la Corse orientale (presqu'île cap-corsine, massif du San Pedrone, plaine de la Casinca). Bul. Soc. Bot. France **77:** 452–462. 1930.—Also his Notes sur des ptéridophytes et phanérogames observées en Corse au cours de la session de la Société botanique de France (août 1930). l.c. **79:** 68–77. 1932.

Malcuit, G. Une excursion botanique à l'île Finocchiarola. Bul. Soc. Bot. France **85:** 718–720. 1939.

Topography, geology, general features of flora; list of 17 phanerogams collected by author in July. The archipelago to which the islet belongs, about 3 km. from the east side of Cape Corse, had not previously been explored botanically.

Salis-Marschlins, Ulysses von. Aufzählung der in Korsika und zunächst in der umgebung von Bastia von mir bemerkten cotyledonarpflanzen, nebst angabe ihres standorts, dessen ungefährer höhe über dem meere und dem mehr oder minder häufigen vorkommen einer jeden. Flora **16** (2): 449–461, 465–476, 486–493. 1833; **17** (2), Beibl. 86 p. 1834.

Conditions of travel, etc.; annotated list of vascular plants, with localities. Refers principally to vicinity of Bastia.

CÔTE-D'OR

See also Bourgogne.

General

Royer, C. L. A. Flore de la Côte-d'Or avec déterminations par les parties souterraines. 2 v. (xxvii, 693 p.) 22.5 cm. Paris, 1881-83.

Annotated keyed list of vascular plants, with localities; each genus, in most cases, with two keys to species, one based entirely on subterranean organs, the other on the usual features. Replaces **Lorey, —, [and Duret, L.].**

Flore de la Côte-d'Or, ou description des plantes indigènes et des espèces le plus généralement cultivées et acclimatées, observées jusqu'à ce jour dans ce département, disposées selon la méthode naturelle de Jussieu; suivie d'une table analytique. 2 v. (2 p. l., cli, 1131 p.). 7 pl. 22.5 cm. Dijon, 1831. (Botanical explorations (based on Vallot, Mémoire sur l'histoire de la botanique en Bourgogne. 1828), topography, geology, soils, agriculture, itineraries for collecting; descriptive flora of wild and commonly cultivated vascular plants and Characeae (more than 1500 species), with vernacular names, uses, and localities for scarcer species; key to genera. Duret is often cited as co-author, but his name does not appear on either title page, and the only indication of his share in the work, apparently, is given by some remarks in the preface (p. ii–iii).—See also **Bazot, Louis.** Études de géographie botanique à propos des plantes de la Côte-d'Or. Revue Gén. Bot. **8**: 353–367, 418–425, 451–468, 507–513. 1896. (Phytogeographic, including lists of species of different geographic range, with especial reference to the plateau de Langres.)—Also **Gérard, Frédéric,** *abbé.* Sur quelques plantes rares de la Côte-d'Or et leurs limites géographiques. Compt. Rend. Assoc. Franç. Avanc. Sci. **40**: 457–463. 1912. (Mention of many of the rarer plants and of those which reach the limits of their distribution in the department.) —Also **Genty, P. A.** Suite à la flore du département de la Côte-d'Or de Charles Royer . . . Bul. Scient. Bourgogne **2**: 59–77. 1933. (List of additional species and localities.)—Also his Contributions à la flore de la Côte-d'Or. l. c. **8**: 129–132. 1939. (Short list of additional species and localities.)

Viallanes, J.J.A., and **Arbaumont, Jules d'.** Flore de la Côte-d'Or, contenant la description des plantes vasculaires spontanées ou cultivées en grand dans le département, un aperçu de leurs propriétés médicales et de leurs usages . . . nouv. éd. lxx, 526 p. 16.5 cm. Dijon, 1926. (1st ed. 1889.)

Local distribution; annotated descriptive flora of vascular plants (1350 indigenous species), with keys, localities for scarcer species, and brief notes on uses. The 1926 issue is apparently only a reprint of the "nouv. éd." of 1910 (latter in Gray Herbarium library); it differs from the 1889 issue principally in the addition of an appendix on p. 525–6, listing the species added since 1889.

Local

Genty, P. A. Compte rendu de l'herborisation faite le 6 juin 1932 au parc de "La Trouhaute" et aux environs de Chenôve. Bul. Soc. Bot. France **79**: 811–819. 1933.

Includes (p. 813–819) Florule . . . by P. Genty and P. Guinier, an unannotated list of vascular plants, mosses, and lichens, with the ubiquitous species omitted.

——— L'îlot granitique de Mâlain et sa végétation . . . Bul. Soc. Bot. France **71**: 1069–1084. 1925.

Geology, general features of flora; unannotated list of vascular plants, mosses, lichens, and 1 mushroom (338 species, of which 252 are vascular).—See also his Florules des îlots granitiques des environs de Dijon. Bul. Soc. Bot. France **79**: 667–686. 1932. (Further notes on flora of Mâlain; also lists from Savigny-sous-Mâlain, Mesmont, Remilly-en-Montagne, etc.)

Lachot, Henry. Flore de l'arrondissement de Semur (Côte-d'Or): comprenant toutes les plantes spontanées ou cultivées en grand dans cet arrondissement. Avec des clefs analytiques . . . 714 p. 25.5 cm. Semur, 1885–99. (Gray Herbarium library.)

Geology, soils, etc.; annotated, briefly descriptive flora of vascular plants (1299 species), with localities; list of local botanists. Issued in 2 parts, the second in 2 subparts, all paged continuously.

CÔTES-DU-NORD

See Bretagne; also Ille-et-Vilaine (Mabille, Catalogue des plantes . . . de Dinan et de Saint-Malo).

General

Ferrary, F. Essai sur l'histoire naturelle du département des Côtes-du-Nord. 754, 111 p. 14 cm. Saint Brieuc, 1836. (Library of Muséum national d'histoire naturelle, Paris.)

Annotated list of vascular plants and Characeae, on the Linnaean system, with descriptions, localities, and properties; second part (111 p.), "Application de la méthode naturelle aux plantes composant la flore du département des Côtes-du-Nord," par M. Le Gall, a key to the plants in the first part.

Local

Camus, Fernand. Étude botanique sur l'archipel de Bréhat (Côtes-du-Nord). Compt. Rend. Assoc. Franç. Avanc. Sci. **28** (2): 456–463. 1900.
General features of flora, with mention of numerous species.

Gourio, H., *abbé.* Note sur la flore d'Erquy (Côtes-du-Nord). Trav. Scient. Univ. Rennes **4**: 27–45. 1905.
List of vascular plants, mostly collected by author, with localities.—See also his Note sur la flore d'Erquy (C.-du-N.). Revue Bretonne Bot. **1**: 9–10. 1906. (Mention of various noteworthy plants.)—Also **Daniel, Jean.** Additions à la florule d'Erquy. l.c. **2**: 91. 1907. (Mention of several species.) —Also his Notes sur la flore d'Erquy. l.c. **7**: 158–162. 1912. (Mention of various noteworthy species, color forms, late flowering species, etc.)—Also **Daniel, Lucien.** Recherches sur la flore d'Erquy et l'influence du climat marin sur la végétation. l.c. **11**: 1–160. 10 fig. (incl. maps), pl. 1 (map), 2 ports. 1916; **1919**: 74–159. 1919; **1921**: 53–74. 1921; **1922**: 3–20. 1922; **1923**: 79–85. 1923; **1924**: 87–105. 1 fig. 1925. (Geology, bibliography, account of the flora of different habitats with mention of numerous species (in v. 11); studies in flowering times, color and odor of flowers, etc.)—Also his Sur la répartition anormale de quelques espèces végétales indigènes à Erquy (Côtes-du-Nord). l.c. **1930**: 15–16. 1930. (Note on peculiarities in the distribution of several species.)

CREUSE

See also Limousin.

General

Cessac, Télémaque de, *abbé.* Catalogue des plantes vasculaires de la Creuse. Mém. Soc. Sci. Nat. et Archéol. Creuse **3**: 283–321, 401–451. 1861–62.
Annotated list of vascular plants [1212 species, including many segregates], with localities. Replaces **Pailloux, P. H.** Notes sur les plantes du département de la Creuse. l.c. **1**: 90–106. 1843. (Includes list of vascular plants and Characeae observed by author [720 species], without localities or annotations except for a very few species.)—See also **Martin, Gabriel.** Notes sur la flore de la Creuse. l.c. **5**: 161–183. 1885. (Note on calcicolous plants, etc.; list of newly recorded species and new localities.)—Also his La flore de la Creuse. l.c. **7**: 33–134, 281–290. 1891–92. (Botanical explorations, general features of flora, plants of different soils and habitats, changes in flora; list of new records since 1885.)—Also **Lafay, A.** Additions à la flore de la Creuse. l.c. **20**: 93–98. 1916. (List of additional species and new localities.)—Also **LeGendre, Charles.** Additions à la flore d'une partie de la Creuse. Revue Scient. Limousin **16** (année 34): 148–154. 1930. (List of about 60 species, with localities and notes on abundance. Not seen; title and annotation supplied by R. McVaugh from copy in University of Michigan library.)—Also **Sarrassat, Claude.** Botanique. (Les études scientifiques, archéologiques et historiques dans la Creuse. Histoire naturelle. II.) *In* La centennaire de la Société des sciences naturelles et archéologiques de la Creuse et du Musée de Guéret. 1832–1932. p. 67–85. 25 cm. Guéret, 1932. (Analysis of principal floras; extensive list of additions to flora, bringing total to 1006 species and 363 infraspecific forms; also account of work on cellular plants, botanical collections, etc.)—Also his Supplément à la flore de la Creuse. Mém. Soc. Sci. Nat. et Archéol. Creuse **28**: 138–149. 1943. (Recent botanical explorations; list of newly recorded species and localities, bringing total for Creuse to 1394 taxa.)—Also **Lugagne, René.** Additions à la flore de la Creuse. l.c. **29**: 535–554. 1946. (Extensive list of new localities.)

Local

Frébault, —, and **Jorrand, —.** Catalogue des plantes phanérogames croissant spontanément ou cultivées en grand à Aubusson et dans les environs. Rev. Scient. Limousin 8 (année 15): 61–64, 96–101, 145–146. 1907; 8 (année

16): 162–164, 212–213. 1908.—**Jorrand and Frébault. [1.]–5. supplément . . .** l.c. **9** (année 17): 151–154. 1909; **9** (année 18): 294–295. 1910; **10** (année 19): 150–151. 1911; **10** (année 20): 202. 1912; **11** (année 21): 136–137. 1914.

Lists of phanerogams collected in different years (1904–1911), with some localities and indication of abundance. (Not seen; title and annotation supplied by R. McVaugh from work in University of Michigan library.)

Jourdan, Pascal. Mosaïque de florules rudérales ou examen phytographique des plantes qu'on trouve sur les monuments antiques et les points remarquables du département de la Creuse et les contrées limitrophes. 1. fasc. Mém. Soc. Sci. Nat. et Archéol. Creuse **4**: 157–191. 1873. (Preprinted, 3 p. l., 49 p. 21 cm. Paris, [etc.], 1872, with title Mosaïque de florules rudérales du centre de la France. fasc. 1. In Gray Herbarium library.)

List of vascular and cellular ruderal plants observed by author at the following localities in Creuse and vicinity: Château et église de Guéret, Puy-de-Gaudy, Toulx-Sainte-Croix, les pierres Jomathres de Mont-Barlot, les pierres d'Ep-Nell. No more published.

Pedon, J. B. La vallée de Clairavaux. Revue Scient. Limousin **9** (année 17): 81–85, 97–108. 1909.

Includes (p. 99–108) a "Florule" containing about 400 species, with notes on frequency of occurrence. (Not seen; title and annotation supplied by R. McVaugh from work in University of Michigan library.)

DAUPHINÉ

The region and former province of Dauphiné includes the departments of Drôme, Hautes-Alpes, and Isère. Only works relating to the whole region are given here.

Mutel, Auguste. Flore du Dauphiné ou description succincte des plantes croissant naturellement en Dauphiné ou cultivées pour l'usage de l'homme et des animaux . . . 2. éd. entièrement refondue. 2 v. (viii, 768; 2 p.l., 140 p.). 16.5 cm. Paris, Grenoble, [1848–49]. (1st ed. 1830.)

Annotated descriptive flora of vascular plants, with localities. Vol. 2 is Analyse et tableau des genres par "A^{te} B^{xx}."

Ravaud, —, ***abbé.*** Guide du botaniste dans le Dauhiné. Excursions bryologiques et lichénologiques suivies pour chacune d'herborisations phanérogamiques où il est traité des propriétés et des usages des plantes au point de vue de la médecine, de l'industrie et des arts . . . 1re [–13e] excursion . . . (Bibliothèque du touriste en Dauphiné.) various pagings. 18 cm. Grenoble, [1878–92]. (Lloyd Library.)

Account of various excursions, with itineraries, geology, mention of more important species and their uses, etc.; many of the excursions in two sections, one dealing with mosses and lichens, the other with phanerogams.

Verlot, J. B. Catalogue raisonné des plantes vasculaires du Dauphiné (Isère, Drôme, Hautes-Alpes). Bul. Soc. Stat., Sci. Nat. et Arts Indus. Dépt. Isère ser. 3, v. 3. viii, 408 p. 1872.—Appendice au Catalogue . . . l.c. **11**: 17–66. 1882.

Annotated list of vascular plants and Characeae (2832 species, of which 2823 are vascular), with localities for scarcer species. The appendix ends with no. 918 and was never completed.—See also **Le Brun, Pierre.** Espèces et localités nouvelles pour la flore du Dauphiné. Bul. Soc. Bot. France **64**: 158–161. [1922?]—Also **Offner, Jules.** Observations sur la flore des Alpes du Dauphiné. l.c. **70**: 471–480, 679–689. 1923. (List of noteworthy species.) —Also his Sur quelques plantes nouvellement signalées en Dauphiné ou en Savoie. l.c. **74**: 288–294. 1927.—Also **Breistroffer, Maurice.** Contribution à l'étude des plantes vasculaires du Dauphiné. ([1.]–4. partie.) l.c. **87**: 47–59. 1940; **93**: 327–335. 1947; **94**: 10–15. 1947; **96**: 16–24. 1949. (Lists of noteworthy plants.)—Also **Offner, Jules.** Notes floristiques sur les Alpes du Dauphiné. Bul. Mens. Soc. Linn. Lyon **23**: 201–203. 1954. (List of noteworthy species.)

Villars, Dominique. Catalogue des substances végétales qui peuvent servir à la nourriture de l'homme, et qui se trouvent dans les départemens de l'Isère, la Drôme et les Hautes-Alpes. 1 p. l., 3–48 p. 19 cm. Grenoble, 1793, (Arnold Arboretum library.)

General considerations; account of edible vascular and cellular plants, grouped according to properties and uses.

——— Histoire des plantes de Dauphiné . . . 3 v. 55 pl., tab. 23 cm. Grenoble [etc.], 1786–89.

Out of date as a flora; cited here for its account of uses of vascular and cellular plants.

DEUX-SÈVRES

See also Vienne (Souchet, Flore du Haut Poitou).

General

Sauzé, J. C., and Maillard, P. N. Flore du département des Deux-Sèvres. 3 v. 19 cm. Paris et Niort, 1878–84. (Gray Herbarium library.)

Part 1 (Manuel analytique . . ., in 1 vol., originally published in Mém. Soc. Stat., Sci., Let. et Arts Dépt. Deux-Sèvres s. 2, v. 11. 2 p. l., xxiii, 228 p. 1872; 2. éd. xxxi, 341 p. Paris, 1884): topography, geology, botanical explorations, with included bibliography; keys to families, genera, and species; glossary. Part 2 (Flore descriptive, in 2 vol., Niort, 1878–80 (viii, 501; 2 p.l., 478 p.); reprinted from Mém. Soc. Stat., Sci., Let. et Arts Deux-Sèvres s. 2, v. 15 (4 p. l., 424 p. 1878) and v. 18 (2 p. l., 403 p. 1880): briefly annotated descriptive flora of vascular plants and Characeae, with localities. —Replaces their Catalogue des plantes phanérogames qui croissent spontanément dans le département des Deux-Sèvres. 57 p. 24.5 cm. Niort, 1864. (In Arnold Arboretum library.)

Local

Sauzé, J. C. Exploration botanique de l'arrondissement de Bressuire pendant les années 1864 et 1865. Mém. Soc. Stat., Sci., Let. et Arts Dépt. Deux-Sèvres II, **6**: 285–302. 1866 [1867?]. (Reprinted, 24 p. Niort, 1867. In Lloyd Library.)

Running account of localities with mention of more interesting plants collected.

DORDOGNE

See also Lot-et-Garonne (Queyron).

General

Des Moulins, Charles. Catalogue raisoné des plantes qui croissent spontanément dans le département de la Dordogne. 1. partie. Phanérogames . . . Actes Soc. Linn. Bordeaux **11**: 171–335. 1840. (Reprinted, 165 p. Bordeaux, 1840. In Arnold Arboretum library.)—Supplément . . . 1. fasc. l.c. **14**: 125–191. 1846. (Reprinted, 69 p. In Arnold Arboretum library.)—Additions au premier fascicule du Supplément . . . l.c. **15**: 281–347. 1849.—Supplément au Catalogue . . . 2. fasc. l.c. **15**: 347–456. 1849. (This and the preceding item reprinted, 178 p. In Arnold Arboretum library.)—Catalogue raisonné des phanérogames de la Dordogne . . . Supplément final (1858). l.c. **20**: 451–848. 1859.—Table des matières en forme de catalogue méthodique (pur et simple) servant de rappel aux quatre fascicules successifs dont se compose le Catalogue raisonné. l.c. **20**: 849–903. 1859. (This and the preceding item reprinted, 453 p. 1859. In Arnold Arboretum library.)

The Catalogue (1840) is an annotated list of vascular plants and Characeae, with localities for rarer species. The Table de matières (1859) is a systematic list of the 1321 species recorded, with reference to the parts in which they are treated.—See also **Abzac, J. J. U. d'.** Additions au Catalogue des plantes de la Dordogne de M. Des Moulins. Bul. Soc. Bot. France **26**: 61–62. 1879.—Also **Hoschedé, J. P.** Notes sur quelques plantes récoltées en Dordogne. Bul. Acad. Internatl. Géog. Bot. **12**: 193–206. 1903. (List of noteworthy vascular plants, with localities and some annotations.)

Local

Galinat, Maurice. La flore adventice, sporadique et naturalisée des environs de Périgueux. Monde des Plant. **46** (274/275): 5–6. 1951. (Arnold Arboretum library.)

Annotated list of introduced species.

Virot, Robert. Les modifications récentes et actuelles de la flore et de la végétation du Périgord méridional (Bas-Périgord). Cahiers des Nat. n.s., **13**: 43–61. 1957.

Discussion of changes in vegetation, grouped by habitats; bibliography.

DOUBS

See also Franche-Comté; Haute-Saône (Renauld).

General

Grenier, Charles. Catalogue des plantes phanérogames du département du Doubs. Mém. Soc. Émul. Doubs v. **1**, tome 3: 33–104. 1843. (Reprinted, 72 p. 1843.)

Plant zones, etc.; annotated list of spermatophytes, with localities for scarcer species.

Local

Contejean, C. L. Revue de la flore de Montbéliard. Mém. Soc. Émul. Montbéliard **21**: 47–284. 1892. (Reprinted 238 p. Montbéliard, 1892. In Arnold Arboretum library.)

Botanical investigations, bibliography, topography, geology, climate, plant zones; annotated list of vascular plants. Replaces his Énumération des plantes vasculaires des environs de Montbéliard. Mém. Soc. Émul. Doubs II, **4**: 1–112. 1853; II, **5**: 38–173. 1854. (Botanical explorations, bibliography, topography, geology, climate, plant distribution, naturalized species, chronological table of discovery of noteworthy species; annotated list of vascular plants, with localities; list of noteworthy plants of Doubs including those of Montbéliard.)—See also his Les herbiers et la flore de Montbéliard. Mém. Soc. Émul. Montbéliard **25**: 17–48. 1895. (Account of local herbaria; additions and corrections to his Revue.)—Also his Un dernier mot sur la flore de Montbéliard. l.c. **29**: 44–55. 1902. (Relates principally to naturalized and adventive species, with dates of arrival, lost species, etc.)

Favre, Jules. La flore du cirque de Moron et des Hautes Côtes du Doubs. Etude de géographique botanique. Bul. Soc. Neuchât. Sci. Nat. **49**: 3–130. 5 fig. 1925.

Mainly ecological and phytogeographic; includes sketch of botanical explorations, bibliography, and (p. 18–34) list of rarer vascular plants, with localities. (Refers to area between Les Brenets and Biaufond.)

Magnin, Antoine. Notes sur la botanique à Besançon de 1691 à 1920. Bul. Soc. Hist. Nat. Doubs **33**: 32–105. 1923. (Library of Muséum national d'histoire naturelle, Paris.)

History, publications, etc., of the old Université franc-comtoise, the chair of botany at the Faculté des sciences, the Société d'histoire naturelle du Doubs and the Office mycologique, the Institut botanique, the Société d'émulation du Doubs, the Académie des sciences, belles-lettres et arts de Besançon; list of Besançon botanists with titles of their principal publications, etc.

——— La végétation des Monts Jura notamment des environs de Besançon précedée de la climatologie du département du Doubs. 59 p. map. 22.5 cm. Besançon, 1893. (Library of Muséum national d'histoire naturelle, Paris.)

Climate, zones of vegetation, botanical regions, mention of special localities and habitats, partial bibliography.

DRÔME

See also Dauphiné.

General

Lenoble, Félix. Catalogue raisonné des plantes vasculaires du département de la Drôme. Bul. Soc. Scient. Dauphiné **55** (V, **14**): 1–506. map. 1935.
Botanical explorations, topography, geology, gazetteer, plant zones, short bibliography; list of vascular plants (2239 species), with localities; list of communes.—See also **Breistroffer, Maurice.** Supplément au Catalogue des plantes vasculaires de la Drôme. (1.–4. partie.) Proc.-Verb. Soc. Dauphinoise Études Biol. (Proc.-Verb.) **16**: 49–58. 1937; **17**: 58–68, 121–129. 1938; **20**: 68–84. 1941. (Lists of newly recorded species and new localities. None of these seen; cited from notices in Bul. Soc. Bot. France, passim.)

Local

Chatenier, Constant. Esquisse de la flore de Lus-la-Croix-Haute (Drôme). Bul. Soc. Bot. France 44 (Sess. Extraord.): cxxxv–cxlviii. 1897 [1898].
Topography, etc.; lists of vascular plants of different plant zones; list of rarer species, with localities.

EURE

See also Normandie.

General

Niel, Eugène. Catalogue des plantes phanérogames vasculaires et cryptogames semi-vasculaires croissant spontanément dans le département de l'Eure. Bul. Soc. Amis Sci. Nat. Rouen III, **24**: 225–346. 1888. (Reprinted, 138 p. 1889. In Gray Herbarium library.)
Annotated list of wild and principal cultivated vascular plants and Characeae, with localities for scarcer species. Replaces **Chesnon, C. G.** Catalogue des plantes qui croissent spontanément dans le département de l'Eure. 3 p. l., 60 p. 31 cm. Evreux, 1846. (Statistique du département de l'Eure. Publiée par la Société libre d'agriculture, sciences, arts et belles-lettres du département. 3. partie. 1. section. Botanique.) (Annotated tabular list of over 1800 vascular and cellular plants, with vernacular names, habitat and localities, and medical or other properties. The title above cited, Catalogue des plantes, is really a caption title.)—See also **Fortier, E.** Notes sur quelques nouvelles plantes et raretés du département de l'Eure . . . Bul. Soc. Amis Sci. Nat. Rouen **42**: 210–213. 1907. (List of angiosperms, with localities.)—Also his Notes sur quelques plantes rares ou nouvelles pour le département de l'Eure. l.c. **43**: 108–112. 1908. (List of angiosperms, with localities.)

Local

Duquesne, A. Petit catalogue des plantes phanérogames et cryptogames semi-vasculaires observées, jusqu'à ce jour, autour de Pont-Audemer, dans un cercle déterminé par un rayon de 14 kilomètres environ, à partir de cette ville, et comprenant les cantons de Pont-Audemer et Quillebeuf tout entiers, ainsi qu'une portion des cantons de Beuzeville, de Routot et Montfort-sur-Risle, de Saint-Georges et de Cormeilles . . . Bul. Soc. Amis Sci. Nat. Rouen **20**: 137–200. 1884.
Briefly annotated list of vascular plants and Characeae, with localities.

Toussaint, Anatole, *abbé*, and **Hoschedé, J. P.** Flore de Vernon et de la Roche-Guyon. Bul. Soc. Amis Sci. Nat. Rouen IV, **33**: 103–308. 1898.
Annotated list of about 1700 vascular plants and Characeae. Vernon is in Eure, Roche-Guyon in Seine-et-Oise.—See also their Flore de Vernon et de la Roche-Guyon et plantes rares des Andelys, avec additions comprenant les plantes intéressantes des environs de Louviers et de Pont-de-l'Arche. l.c. **70/71**: 207–353. 1936. (List of vascular plants, with localities. Apparently a fusion of Toussaint, Plantes rares des Andelys l.c. III, **28**: 65–107. map. 1893, and the 1898 paper, cited above, with additions.)

EURE-ET-LOIR

General

Lefèvre, Édouard. Botanique du département d'Eure-et-Loir. viii, 310 p. 22 cm. Chartres, 1866. (Société archéologique d'Eure-et-Loir. Statistique scientifique [ptie.1].)

Botanical explorations, topography, habitats, naturalized plants, etc.; annotated list of vascular and cellular plants (1004 vascular), with localities for scarcer species and vernacular names.

Local

Coudray,—. Essai sur la flore dunoise. Bul. Soc. Dunoise **1:** 79–98. [1867?]

Geography, topography, botanical explorations, etc.; list of 168 rarer vascular plants and Characeae, with localities. Refers to arrondissement of Châteaudun.—See also **Vuez, L.** Notes et matériaux pour servir à la flore des environs de Châteaudun. Bul. Soc. Dunoise **1:** 171–196. [1868?] (Partial lists of vascular and cellular plants, notes on general features of flora, etc.)

FINISTÈRE

See also Bretagne.

General

Picquenard, Charles. Catalogue des plantes vasculaires qui croissent spontanément dans le Finistère. Bul. Soc. Scient. et Méd. Ouest **2:** 125–129, 138–146, 228–254, 342–355. 1893; **3:** 116–128, 129–147, 179–200, 305–318. 1894.

Collaborators, bibliography; annotated list of vascular plants, with localities. Replaces **Crouan, P. L.,** and **Crouan, H. M.** Florule de Finistère. Contenant les descriptions de 360 espèces nouvelles de sporogames, des nombreuses observations et une synonymie des plantes cellulaires et vasculaires qui croissent spontanément dans ce département . . . x, 262 p. 32 pl. 25 cm. Paris [etc.], 1867. (Annotated list of 4188 plants (1131 vascular), with localities for rarer species.)—See also **Langeron, Maurice.** Contribution à l'étude de la flore du Finistère. Bul. Soc. Acad. Brest II, **23:** 131–142. 1898. (List of noteworthy vascular and cellular plants from vicinity of Brest and Le Conquet.)

Local

Guffroy, Charles. Étude floristique de Dibon. Bul. Soc. Bot. France **84:** 545–558. 1938.

Geology, etc.; list of vascular plants, divided into 19 ecological groups.

Miciol, Ernest. Catalogue des plantes des environs de Morlaix. 52 p. 22.5 cm. [Morlaix, 1890–92.] (Issued with Bul. Soc. Études Scient. Finistère v. 10, 12.)

Annotated list of vascular plants, with localities.

Thiébaut, Charles. Une excursion botanique aux îles de Molène, d'Ouessant et de Sein. Bul. Soc. Bot. France **22:** 26–32. 1875.

List of vascular plants collected in September 1874, with localities.

FRANCHE-COMTÉ

The region and former province of Franche-Comté includes the departments of Doubs, Haute-Saône, and Jura. Only works relating to the whole of this region are given here.

Beauquier, Charles. Faune et flore populaire de la Franche-Comté. vol. 2, Flore. 1 p. l., 407 p. 16 cm. Paris, 1910. (Harvard College library.)

General considerations, vernacular names relating to trees and tree growth; alphabetical list of vernacular names with scientific equivalents, numerous patois synonyms, uses, folklore, etc.

Paillot, Justin, and Vendrely, Xavier. Flora Sequaniae exsiccata ou herbier de la flore de Franche-Comté. I–XI. Mém. Soc. Émul. Doubs IV, **6:** 75–100. 1872; **7:** 514–521. 1873; **10:** 477–494. 1876; V, **5:** 12–69. 2 pl. (incl. map). 1881; VI, **4:** 108–115. 1890; **10:** 353–362. 1896; VII, **4:** 380–388. 1900; **7:** 291–309. 1903; **10:** 152–156. 1906.

Lists of vascular and cellular plants distributed in exsiccatae (22 fascicules of vascular plants, totaling 1276 species), with notes on various species. Includes also list of plants transported by the movements of French troops in 1870–71 (IV, **6:** 89–100); "florule du marais du Saône" from Besançon to Morteau (V, **5:** 12–40). Nos. I–II of the series of 11 "livres" are together (IV, **6**); no. VI is lacking, its place being taken by **Renauld, Ferdinand; Flagey, Camille; Vendrely, Xavier;** and **Paillot, Justin.** Contribution à la flore de la Haute-Saône. l.c. V, **7:** 162–200. 1883. Pt. I–II, IV, and VII are by Paillot and Vendrely; pt. III, by Paillot; pt. V, by Paillot, Vendrely, Flagey, and Renauld; pt. VIII–XI, by Vendrely.

Tronchet, Antonin. Aperçu bibliographique sur la floristique et la phytosociologie en Franche-Comté et régions limitrophes. Bul. Soc. Hist. Nat. Doubs **53:** (11 p.). 1949; **54:** (9 p.). 1951. (1st part not seen; 2d reprinted in Annales Scient. Univ. Besançon **6:** 7–15. 1951; only the reprint seen.)

Running account of publications, with bibliography.

GARD

See also Languedoc.

General

France. Comité interministériel des plantes médicinales et à essences. Les plantes médicinales dans le département du Gard. 50 p. 1 fig. 24 cm. Nîmes, 1920.

Sketch of topography, etc.; lists of "espèces se prêtant à une exploitation rémunératrice," grouped geographically, and account of more important ones; calendar for collecting material.

Lombard-Dumas, Armand. La botanique dans le Gard, biographies sommaires des botanistes nés dans le Gard et des auteurs qui ont écrit sur la flore de ce département. 84 p. 8vo. Sommières, 1900.

List of botanists, with biographical notes, account of scientific work, and titles of publications. (Not seen; cited from notice in Bul. Soc. Bot. France **47:** 525–526. 1900.)

Pouzolz, P. C. M. de. Flore du département du Gard ou description des plantes qui croissent naturellement dans ce département. 2 v. (659, 644 p.). 7 pl. 22 cm. Montpellier et Paris, 1856–62. (Gray Herbarium library.)

Annotated descriptive flora of vascular plants, with localities, patois names, and notes on uses. Vol. 2 completed by P. Courcière after the death of Pouzolz. Replaces **Gareiso, Joseph,** *abbé.* Synopsis analytique de la flore du Gard . . . xxxiv, 348 p. 17.5 cm. Nîmes et Paris, 1847. (Unannotated keys to vascular plants, without localities. In Gray Herbarium library.)—See also **Loret, Henri.** Plantes nouvelles pour le Gard avec des observations préliminaires sur la flore de Pouzolz et sur son herbier départemental. 10 p. 25.5 cm. (Nîmes, 1880.) (Corrections of Pouzolz's Flore du Gard, and mention of a few species new to the region. In Gray Herbarium library.)—Also **Martin, Bernardin.** Révision de la flore du Gard comprenant l'énumération des espèces qu'il convient d'en exclure ou de n'y maintenir qu'avec reserve et l'indication des erreurs de diagnose ou de nomenclature qu'il importe d'y rectifier. Mém. Acad. Nîmes VII, **14:** 259–276. 1891 (cover date, 1892). (List of excluded and doubtful species, corrections in identification.)—Also **Cabanès, Gustave.** Catalogue des plantes non mentionées dans l'ouvrage de de Pouzolz et nouvelles pour la flore du département du Gard. (Matériaux pour le prodrome d'histoire naturelle du département du Gard. Partie botanique.) Bul. Soc. Étude Sci. Nat. Nîmes **19:** 8–81. 1891; **20:** 19–46. 1892. (Bibliography; list of about 250 species not recorded by de Pouzolz, with localities and references; long list of critical species, varieties, and forms not described by de Pouzolz, with localities and references.)—Also **Martin, B.** Indication de 250 plantes trouvées dans notre département après la publication de la Flore du Gard et dont l'énumération peut être considérée comme un supplément à la statistique de cette flore. Bul. Soc. Bot. France **40:** 13–23. 1893. (List with localities and names of collectors. Although covering the same ground as the paper by Cabanès (1891–92), the two lists are far from identical.)—Also **Cabanès, G.** Note relative à quelques plantes nouvelles ou rares pour la flore du Gard. Bul. Soc. Étude Sci. Nat. Nimes **22:** 80–92. 1894–95. (Lists of newly recorded species and new localities.)—Also his Note relative à quelques plantes nouvelles pour la flore du Gard. l.c. **23:** 8–13. 1895. (List with localities.)—Also **Barrandon, A.** Observations nouvelles sur les plantes de la région méditerranéenne comprise entre le Rhône, les Cévennes et les Pyrénées, et rectifications à apporter à la flore de France de Grenier et Godron. l.c. **25:** 106–115. 1897. (Extensive list referring to Gard only.)—Also **Cabanès, G.** Notes de botanique. l.c. **28:** 29–40. 1901. (List of newly recorded species, new localities, and corrections.)—Also his Herborisations dans le Gard en 1902. l.c. **30:** 57–77. pl. 1–3. 1902. (List of newly recorded plants and new localities.)

Local

Braun-Blanquet, Josias. Catalogue de la flore du massif de l'Aigoual et des contrées limitrophes. Recherches sur l'exploration floristique, sur la répartition et les conditions de vie des plantes vasculaires. Mém. Soc. Étude Sci. Nat. Nîmes no. 4. 352 p. map. 1933. (Communication de la Station internationale de géobotanique méditerranéenne et alpine, Montpellier, no. 20.)

Botanical explorations, topography, geology, climate, plant zones; annotated list of vascular plants, with localities. The massif de l'Aigoual is mostly in western Gard, overlapping Aveyron and Lozère. Replaces **Martin, Bernardin.** Florule de l'Aigoual et de la contrée avoisinante depuis Valleraugue jusqu'aux environs de Saint-Sauveur-les-Pourcils (Gard). Bul. Soc. Bot. France **42:** 273–285. 1895. (Topography, geology; list of vascular plants, with localities.)—See also **Quézel, Pierre, and Rioux, Jean.** Quelques stations botaniques nouvelles dans les Cévennes (région du Vigan). l.c. **94:** 393–398. 1 fig. (map). 1948. (List of noteworthy vascular plants, with localities.)—Also **Rioux, J.,** and **Quézel, P.** Notes critiques sur quelques espèces du massif de l'Aigoual. l.c. **95:** 134–135. 1948. (Records of 3 species.)—Also **Quézel, P.,** and **Rioux, J.** Plantes adventices de la région de [du] Vigan (Gard). l.c. **97:** 2–4. 1950. (Running account of numerous introduced species, grouped by habitat; bibliography.)

Lagarde, J., and **Cabanès, Gustave.** Les Costières des environs de Nimes. Bul. Soc. Étude Sci. Nat. Nimes **44:** 95–129. 1926.

Includes unannotated list of about 150 vascular plants and 150 fungi. (Not seen; title and annotation supplied by R. McVaugh.)

Lombard-Dumas, Armand, and **Martin, Bernardin.** Florule des causses de Blandas, Rogues et Montdardier (Gard) et des pentes qui les relient aux vallées adjacentes de la Vis, de l'Arre et de l'Hérault. Bul. Soc. Bot. France **38:** 108–110, 142–157. 1891.

Topography, etc.; list of rarer plants, with localities; unannotated list of ubiquitous species.

Martin, Bernardin. Catalogue des plantes vasculaires qui croissent spontanément dans la circonscription de Campestre (Gard). Bul. Soc. Bot. France **22** (Sess. Extraor.): xxxv–li. 1875.—Supplément . . . l.c. **40:** 61–62. 1893.

List of vascular plants, with localities.

——— Florule du cours supérieur de la Dourbie depuis sa source à l'Espérou (Gard), jusqu'au confluent du Trévézel, près de Cantobre (Aveyron). Bul. Soc. Bot. France **37:** 50–67. 1890.—Supplément . . . l.c. **40:** 60–61. 1893.

Topography, geology; list of rarer vascular plants, with localities; unannotated list of common species.

Noël, Henri. Florule du Mont-Duplan. Bul. Soc. Étude Sci. Nat. Nimes **25:** 31–69. 1897. (Reprinted 38 p., with xv p. from Revue du Midi as "Historique et florule du Mont-Duplan." In Arnold Arboretum library.)

History, topography; annotated list of 387 vascular and cellular plants, including many cultivated species. Mont Duplan is a hill near Nîmes.

GASCOGNE (GASCONY)

The region and former province of Gascogne includes the departments of Gers, Hautes-Pyrénées, Landes, and parts of some others.

Levrat, Étienne. La médecine populaire en Gascogne. Revue Pyr. **23:** 260–294. 1911.
Includes annotated list of medicinal plants (p. 287–292), in part with scientific names and uses.

GERS

See also Gascogne; Landes (Ducomet, La botanique populaire dans l'Albret).

General

Dupuy, Dominique, *abbé.* Florule du département du Gers et des contrées voisines . . .3 p.l., ix, 240 i.e. 340 p. (left-hand pages 2–284 blank). 11.5 cm. Auch, 1847. (Gray Herbarium library.)
Brief account of botanical explorations; unannotated list of vascular plants and Characeae, with intermixed keys and diagnostic phrases, without localities.—See also **Duffort, Louis.** Contribution à la flore du département du Gers. [I]-II. Bul. Vulgarisat. Sci. Nat. Gers 1 (1): 14–20. 1901; 1 (3) [by error tome 2]: 3–16. 1902 (reprinted 1922). ([I], list of noteworthy vascular plants and Characeae collected by author, with localities; II, annotated list of local species of *Rosa*, with key.)—Also his Une visite botanique à Concom, juin 1901. l.c. 1 (1): 21–22. 1901. (List of noteworthy plants, including some new to Gers.)—Also **Lemée, Albert.** Contribution à l'étude de la flore du département du Gers. l.c. no. 6 (i.e., v. 1, no. 6): 7–18. 1923 (List with localities.)

Lucante, Angel, *abbé.* Étude sur la flore du département du Gers. Revue Bot. Bul. Mens. Soc. Franç. Bot. 1: 5–8, 37–41, 71–74, 152–155, 189–192, 227–230, 257–260. 1882–83.
Botanical explorations, etc.; unannotated list of 1184 species and varieties of vascular plants known in 1868, based on Dupuy; discussion of method of marking exchange lists of plants. Section III, intended to record the species added since 1868, apparently was never published.

Local

Dader, J., and Rey, P. Les noms gascons des plantes et des formations végétales du Bas-Comminges. (Notes sur la flore gasconne IV.) *In* Documents pour les cartes des productions végétales. Série France-atlantique, tome Aquitaine. Aquitaine centrale, no. 1. Astarac oriental et Bas-Comminges. art. IV. 1 p.l., 13 p. 1945.
Includes lists of scientific names of woody plants, herbs, fruit trees, and cultivated plants with Gascon vernacular names, from the Comminges region in the valley of the Save River.

Dupuy, Dominique, *abbé.* Mémoires d'un botaniste accompagnés de la florule des stations des chemins de fer du Midi dans le Gers. 358 p. 30 fig. 18.5 cm. Paris, 1868.
The "Florule" (p. 117-end) contains a briefly annotated flora in form of key, classified lists of useful and harmful plants, list of rarer species of the principal railway stations, glossary; also (p. 283–312) a set of "couplets à chanter au retour des herborisations," intended to fix in the minds of students the localities at which the rare plants of the region are found.

Lassime, Jeanne. Contribution à l'étude de la végétation du Haut-Armagnac. Le pays lectourois. 102 p. (incl. map). 25 cm. Toulouse, 1932. (On title page: Travail du Laboratoire de botanique et de cryptogamie de la Faculté de médecine et de pharmacie de Toulouse.)
Mainly ecological; includes (p. 71–87) systematic list of medicinal plants of pharmaceutical significance, annotated only as to habitat and abundance.

R., R. de. Notes sur les noms et usages populaires d'un certain nombre de plantes intéressant plus particulièrement la région du Condomois. Bul. Vulgar. Sci. Nat. Gers 3 (i.e. 4): 5–17. 1906.
Alphabetical list of 64 Gascon vernacular names, with botanical equivalents and uses.

GIRONDE

See also Lot-et-Garonne (Queyron).

General

Clavaud, Armand. Flore de la Gironde. Actes Soc. Linn. Bordeaux **35** (IV, 5): 221–446. pl. 9–16. 1881; **38** (IV, 8): 461–608. pl. 31–34. 1884. (Reprinted, iv, 348 p. 12 pl. Paris et Bordeaux, 1882–84. In Gray Herbarium library.)

Descriptive flora of vascular plants, with keys and localities; incomplete (Ranunculaceae to beginning of Amygdalaceae).

Laterrade, J. F. Flore bordelaise et de la Gironde, description caractéristique des plantes qui croissent naturellement dans ce département, avec l'indication de leurs propriétés et leurs usages, precédée de notions élémentaires de botanique. 4. éd. 624 p. 18.5 cm. Bordeaux, 1846. (1st ed. 1811.)—Supplément . . . 625–690 p. 19 cm. Bordeaux, 1857. (Lloyd Library.)

Geography, botanical regions, botanical history; annotated descriptive flora of vascular and cellular plants, with localities for scarcer species and notes on uses.—See also **Delbos, Joseph.** Recherches sur le mode de répartition des végétaux dans le département de la Gironde. Mém. Soc. Phys. et Nat. Bordeaux **1:** 427–369. 1855. (Includes discussion of different habitats with mention of characteristic species.)—Also **Des Moulins, Charles, and Lespinasse, Gustave.** Plantes rares de la Gironde. Cong. Scient. France **28** (Bordeaux, 1861), 3: 405–420. 1863. (List of very local native vascular and cellular plants, with localities; list of naturalized phanerogams, with localities; list of rarer or critical vascular and cellular plants, with localities; list of collecting localities.)—Also **Neyraut, J. E.** Note sur diverses plantes adventives ou nouvelles ou rares pour la Gironde, etc. Actes Soc. Linn. Bordeaux **45** (V, 5) (Compt.-rend.): clxv-clxix. 1892 (cover date 1893). (List of angiosperms, with localities.)—Also **Motelay, L.** Plantes rares ou nouvelles de la Gironde. Compt. Rend. Assoc. Franç. Avanc. Sci. **24** (1): 272–274. 1895. (Mention of various noteworthy plants, especially recently introduced species.)—Also **Labrie, J.** De quelques plantes rares nouvelles pour la flore de la Gironde. Actes Soc. Linn. Bordeaux **59** (VI, 9): 9–20. pl. 1. 1904. (Discussion of several noteworthy species, with descriptions.)—Also **Deysson, J.,** *abbé.* Liste des localités de plantes rares, de formes ou de variétés nouvelles peu répandues dans le département de la Gironde. l.c. **62** (VII, 2): 1–15. pl. 1. 1907. (List of noteworthy vascular plants, with localities and some annotations.)—Also **Henriot, Philippe.** Plantes rares ou nouvelles recueillies aux environs de Sainte-Foy-la-Grande. l.c. **70** (Proc.-Verb.): 106–121. 1918.—Also **Jeanjean, A. F.** Notes sur la flore de la Gironde. l.c. **80:** 73–77. 1928.

Pitard, Joseph. Sur les vicissitudes des espèces rares et adventices du département de la Gironde. Bul. Soc. Bot. France **49** (Sess. Extraord.): cxiii-cxxi. [1903?]

Annotated list of species that are extending their range, species that are disappearing, and principal adventive species.

Local

Chantelat, A. Catalogue des plantes phanérogames et cryptogames qui croissent spontanément aux environs de la Teste-de-Buch. Actes Soc. Linn. Bordeaux **13:** 191–272. 1843 [1844?] (Reprinted 84 p. 1844. In Arnold Arboretum library.)—Supplément et corrections . . . l.c. **17:** 437–458. 1851 [1852].

Annotated list of vascular and cellular plants.

Pitard, Joseph. Liste méthodique des plantes phanérogames, cryptogames supérieures et characées, récoltées pendant la session de Bordeaux (août 1902). Bul. Soc. Bot. France **49** (Sess. Extraord.): c-cxii. [1903?]

Unannotated list of vascular plants collected in coastal region of Gironde.

HAUT POITOU

The region of Haut Poitou includes the departments of Deux-Sèvres and Vienne. The only title relating to Haut Poitou as such (Souché, Flore du Haut Poitou) is given under Vienne.

HAUT-RHIN

See also Alsace; Vosges (Brunotte, Claire); Switzerland, Basel (Binz, Flora von Basel). The 10 departments named Haut-, Haute-, or Hautes- have been alphabeted as though their names consisted of two words, an unorthodox but readily comprehended method.

Local

Issler, Emil. Die gefässpflanzen der umgebung Colmars. Mitt. Philom. Gesell. Elsass-Lothr. 2 (8. jahrgang, 1900): 263–270, (1901): 271–290, 371–395, (1902): 479–507. 1900–02; 3 (1903): 8–30. 1903.—Nachtrag und berichtigungen . . . l.c. 3 (1905): 282–306. 1905. (All in Gray Herbarium library.)

Annotated list of vascular plants, with localities. Covers upper Alsace with the exclusion of the arrondissement (kreis) of Altkirch; that is, the plain between Ensisheim and Schlettstadt (Schlestadt), the Vosges from the Elsässer Belchen to the Weilerthal, and the Kaiserstuhl (in Baden). Replaces **Rosshirt, Karl.** Beiträge zur flora von Kolmar und ergebnisse botanischer ausflüge in die Schweiz. 36 p. 25.5 cm. Colmar, 1888. (Jahresbericht des Lyceums in Colmar . . . 1887–88.) (Localities; list (p. 9–30) of vascular plants of the vicinity of Colmar, based on his own collections; short lists from Switzerland and Italy (p. 31–36). In Gray Herbarium library.)—See also **Issler, E.** Plantes importées par l'industrie lanière. [I]–II. Bul. Soc. Hist. Nat. Colmar n.s., **22**: 1–7. pl. 1–2. 1931; **24**: 151–159. 1935. (Two lists, totaling 102 species, of adventive plants of vicinity of Colmar imported in wool.)

Issler, Émile. Influence de la grande guerre sur la flore des Vosges. Bul. Soc. Hist. Nat. Colmar n.s., **16**: 113–125. 1921; **21**: 159–167. 1929.

General considerations; list of weeds of several altitudes and localities; reforestation. Title of pt. II reads: Essai sur l'influence . . .

——— Die vegetationsverhältnisse der Zentralvogesen mit besonderer berücksichtigung des Hohneckgebietes. Bot. Jahrb. Engler **43**, Beibl. **99**: 6–62. 4 pl. 1909. (Reprinted with title Führer durch die flora der Zentralvogesen . . . 64 p. 4 pl. 1909. In John Crerar Library.)

Topography, geology, climate, plant associations, tree limits, phytogeography; annotated lists of local species, critical species, and hybrids; bibliography.—See also under Vosges, Brunotte and Lemasson, Le botaniste, herborisant au Hohneck et aux environs de Gerardmer.

Kirschleger, Frédéric. Florule des environs de Soultzmatt (Haut-Rhin). *In* Bach, [Marie J. A.]. Des eaux gazeuses alcalines de Soultzmatt . . . p. 261–272. 21 cm. Paris [etc.], 1853.

Soils, etc.; lists of noteworthy plants of three excursions in vicinity of Soultzmatt, with habitats and in some cases localities. Soultzmatt is in the arrondissement of Guebwiller.—See also Marzolf, below.

Marzolf, Georg, jr. Versuch zu einem pflanzen-verzeichnisse von Gebweiler und umgebung. Bestimmt und systematisch aufgestellt nach Kirschleger's Flore d'Alsace etc. (1852–57). Mitt. Philom, Gesell. Elsass-Lothr. 2 (1898): 39–73. [1898.]—Nachtrag . . . l.c. 2 (1900): 261–262. [1900.] (Both in Philadelphia Academy of Natural Sciences library.)

List of wild and cultivated plants, annotated as to abundance, with localities. (French name, Guebwiller.)

Schaeffer, Johannes. Die gefässpflanzen des kreises Altkirch. Ein beitrag zur flora des Reischslande. viii, 73 p. 17 cm. Altkirch, 1895. (Beilage zum Jahresbericht des Gymnasiums Altkirch für das schuljahr 1894–95.)

General considerations; list of about 1200 wild or adventive vascular plants, with localities. (Not seen; title and annotation supplied by B. de Retz.)

HAUTE-GARONNE

See also Languedoc; Tarn (Bel, Nouvelle flore du Tarn et de la région toulousaine); also Spain, Aragón (Soubeiran, Une ascension à la Maladetta); Catalonia (Coste, Florule du Val d'Aran).

General

Timbal-Lagrave, Édouard. Catalogue des plantes spontanées ou cultivées dans le département de la Haute-Garonne, employées en médecine. Annales Soc. Hort. Haute-Garonne [v. ? p. ?]. 1859. (Reprinted, 26 p.)

List of 223 medicinal plants, alphabetically arranged by their French generic names; includes some species not found at all in Haute-Garonne. (Not seen; cited from notice in Bul. Soc. Bot. France **6**: 694. 1859.)

Local

Dop, P. Flore de la région toulousaine. Cong. Assoc. Franç. Avanc. Sci. 39 (Toulouse, 1910), Documents sur Toulouse et sa région **2**: 103–117. 1910. (Reprinted 15 p. Toulouse, 1910. Original not seen; reprint in library of Muséum national d'histoire naturelle, Paris.)

Account of botanical regions; bibliography of floristic works.

Filhol, Édouard, Filhol, Henri, Jeanbernat, Ernest, and **Timbal-Lagrave, Édouard.** Exploration scientifique du massif d'Arbas (Haute-Garonne). Bul. Soc. Sci. Phys. et Nat. Toulouse **2**: 367–478. pl. 2–3. 1875. (Reprinted, 114 p. Toulouse, 1875 (cover date, 1876). Original in Harvard College library; reprint in Arnold Arboretum library.)

Topography, authors' itinerary; list of vascular plants and bryophytes, with localities.

Gadeau de Kerville, Henri. Recherches botaniques et zoologiques effectuées, en 1926 et 1927, dans le cirque d'Espingo et la partie supérieure du val du port de Vénasque (canton de Bagnères-du-Luchon, Garonne). Bul. Soc. Amis Sci. Nat. Rouen **62/63**: 139–203. 4 pl. 1928.

Includes (p. 149–155) unannotated list of vascular plants collected at 1880 to 1990 meters elevation; and (p. 182–187) similar list from upper part of the valley of the port (Pyreneen equivalent of the French term "col") of Vénasque at 2320–2350 meters altitude.

Gaussen, Henri. La flore et son écologie. *In* Henry, S. La forêt de Bouconne. (Étude de géographie historique.) Trav. Lab. Forest. Toulouse tome 1, vol. 3, art. 21, p. 39–57. pl. 3 (map). 1944.

Habitats; annotated partial list of vascular plants and some bryophytes and mushrooms.

Noulet, J. B. Flore analytique de Toulouse et de ses environs. 3. éd. xvi, 376 p. 18.5 cm. Toulouse, 1884. (1st ed. 1855.)

Geology, etc.; annotated list of vascular plants, with localities, followed by keys to genera and species. Replaces **Arrondeau, E. T.** Flore toulousaine, ou catalogue des plantes qui croissent spontanément ou qui sont cultivées en grand aux environs de Toulouse . . . xx, 281 p. table. 18 cm. Toulouse, 1854 (cover date 1855). (Annotated flora of vascular plants in form of keys, with occasional descriptions and localities for scarcer species. In Gray Herbarium library.)—See also Sudre, below.

Sacaze, Julien. Flore populaire du Luchon. Revue Pyr. **2**: 834–841. 1890; **3**: 105–111. 1891.

Tabular list of 304 plants (294 vascular) with scientific, French, and patois names.

Sudre, Henri. Florule toulousaine ou analyse descriptive des plantes qui croissent spontanément ou sont cultivées en grand dans la région sous-pyrénéenne de la Haute-Garonne avec l'indication de leurs propriétés les plus importantes. 239 p. 847 fig. on 20 pl. 21.5 cm. Albi, 1907. (Gray Herbarium library.)

Annotated flora of vascular plants in form of keys, with localities for rarer species. Covers the lower part of Haute-Garonne, that is, the arrondissements of Toulouse. Villefranche, and Muret.—See also **Laredde, Claude.** Sur quelques adventices de la région toulousaine. Bul. Soc. Hist. Nat. Toulouse **80**: 216–220. 1946. (Short list; bibliography of publications on the adventive flora of Toulouse.)—Also Noulet, above.

Walter, Émile. La sapinière de la Cascade d'Enfer, un paradis des fougères. Inventaire de la végétation phanérogamique de cette sapinière. Bul. Soc. Bot. France **83**: 430–434. 1936.

General features of flora, remarks on various ferns; unannotated list of spermatophytes grouped by habit. (In lower cirque of the valley of the Lys, 10 km. from Luchon.)

HAUTE-LOIRE

See also Auvergne; Languedoc.

General

Arnaud, J. A. M. Flore du département de la Haute-Loire, ou tableau des plantes qui y croissent, disposées suivant la méthode naturelle. xix, 108 p. 20 cm. Puy, 1825. (Lloyd Library.)—Supplément . . . 29 p. 20 cm. Puy, 1830. (Extrait des Annales de la Société d'agriculture, sciences, arts et commerce du Puy, pour 1829. In Lloyd Library.)

Topography, etc.; annotated list of cellular and vascular plants (1,208 species, of which 947 are phanerogams), with French and patois names.

Local

Hilaire de Latourette, J. F. B. Flore de l'ancien Velay aujourd'hui partie du département de la Haute-Loire. viii, 775 (by error 175) p. 22 cm. Le Puy, 1848. (Library of Muséum national d'histoire naturelle, Paris.)

Annotated list of spermatophytes, on the Linnaean system, with localities and uses. (Region in central and western Haute-Loire.)

HAUTE-MARNE

See also Champagne.

General

Aubriot, L., and **Daguin, Arthur.** Flore de la Hte-Marne. Catalogue des plantes vasculaires spontanées, subspontanées et de culture générale de ce département. Mém. Soc. Let., Sci., Arts, Agr. et Indus. Saint-Dizier **3:** 7–536. map. 1885. (Reprinted, 2 p.l., 7–536 p. In Arnold Arboretum library.)

Account of various localities, botanical explorations with titles of publications, list of collaborators, topography, etc.; annotated list of vascular plants (1,492 species), with localities, patois names, and notes on uses; includes cultivated plants. The list proper (p. 95–536) has the half-title: Catalogue des plantes vasculaires phanérogames et cryptogames du département de la Hte-Marne. The "nouvelle édition" (Paris, n.d. [1886?]; in Arnold Arboretum library) is precisely the same as to text, but has a long subtitle, and is by "Daguin et Aubriot," not vice versa.—See also **Houdard, Jules,** and **Thomas, C.** Catalogue des plantes vasculaires de la Haute-Marne. Additions et corrections . . . Bul. Soc. Sci. Nat. Haute-Marne v. 9. 202 p. map. 1911–12. (Originally issued in parts with v. 8.) (Botanical regions; annotated list of vascular plants (1494 species, of which 1216 are spontaneous, extensively cultivated, or thoroughly naturalized) with indication of frequency and additional localities; intended as a mise au point of Aubriot and Daguin's flora. The map referred to on p. 12 has not been found.)—See also **Aymonin, Victor.** Observations botaniques de 1914 à 1921. Bul. Soc. Étude Sci. Nat. Haute-Marne **5:** 133–144. 1922. (Alphabetical list of 150 noteworthy species, with localities. Not seen; title and annotation supplied by B. de Retz.)—Also his Observations botaniques en 1922. l.c. **5:** 175–177. 1922. (Not seen.)—Also **Fournier, Paul,** *abbé.* Variétés nouvelles de la flore haut-marnaise. Bul. Soc. Bot. France **70:** 31–34. 1923. (Descriptions of 3 new varieties.)—Also his Espèces et variétés nouvellement reconnues dans la Haute-Marne. l.c. **70:** 84–95. 1923. (List with localities.)—Also his Contribution à la flore de la Haute-Marne. Mém. Acad. Sci., Arts et Belles-Let. Dijon **1923** (II, Bul.): 136–138. 1923. (List of noteworthy plants. In Harvard College library.)—Also **Donnet, E.** Additions à la flore de la Haute-Marne. Bul. Soc. Étude Sci. Nat. Haute-Marne **7:** 282–285. 1924; **8:** 297–301. 1925. (List of additional species, new localities, etc. Not seen; cited from notice in Bot. Jahresber. Just 54 (1): 761–762. 1931.)—Also **Dillemann, Georges.** Suppléments aux Catalogues des plantes vasculaires de la Haute-Marne. Monde des Plant. **45** (267/8): 29–36, 49–52. 1950; **46** (276/77): 32. 1951. (Extensive list of additions and corrections. In Arnold Arboretum library. Also published with same title in Bul. Soc. Étude Sci. Nat. Haute-Marne **16:** 177–206. 1950. In Arnold Arboretum library.)

Local

Demimuid, P., and Demimuid, Charles. Catalogue raisonné des plantes phanérogames croissant spontanément dans l'arrondissement de Wassy. [?,p.] Wassy, 1888. (Reprinted from Bul. Soc. Agr. Arrond. Wassy.)
Not seen; no annotation available.

HAUTE-SAÔNE

See also Franche-Comté; Vosges (Claire).

General

Renauld, Ferdinand. Catalogue raisonné des plantes vasculaires et des mousses qui croissent spontanément dans la Haute-Saône et parties limitrophes du Doubs precédé d'un aperçu sur la géographie botanique ou distribution des plantes dans ce département selon les altitudes et les terrains, avec carte coloriée indiquant les grandes divisions géologiques. 437 p. map. 22.5 cm. Besançon, 1883. (New York Botanical Garden library.)
Geology, with lists of species of different soils, plant zones; annotated list of vascular plants and mosses, with localities and soil relations; discussion of calciphilous species and of distribution as affected by soils. Pages 1–398 are identical, except for the changed title, with his Aperçu phytostatique sur le département de la Haute-Saône suivi d'un catalogue des plantes vasculaires et des mousses. 398 p. map. Paris, 1873. (Another issue, with same paging and date but published at Vesoul and with somewhat different title, in Arnold Arboretum library.) Pages 399–437 form a Supplément . . . by **Renauld, F., Flagey, C., Vendrely, X.,** and **Paillot, J.**; this is a reprint of their Contribution à la flore de la Haute-Saône (caption title, Liste des plantes rares ou nouvelles pour ce département et les parties limitrophes du Doubs). Mém. Soc. Émul. Doubs V, **7**: 162–200. 1882. (Extensive list of vascular plants and mosses, with localities.)—See also **Maire, René.** Contribution à l'étude de la flore de la Haute-Saône. [I.] Bul. Soc. Étude Sci. Nat. Haute-Saône. (I.) **1**: 184–201. 1896. (Not seen; title supplied by G. Dillemann.) II–VII. Bul. Soc. Grayloise Émul. **1**: 158–194. 1898; **2**: 310–323. 1899; **3**: 270–291. 1900; **4**: 341–367. 1901; **6**: 181–220. 1903; **9**: 117–159. 1906. (All in New York Public Library.) (Short bibliography; extensive list of additional species and forms and new localities of vascular plants. All fascicles after the third are preceded by separately paged lists of additions and corrections, included in the pagination given above. The 7th fascicle is the end of the 1st part. Presumably a 2d part, to include cellular plants, was projected but never published.)—Also **Cardot, C.** Plantes et stations botaniques nouvelles de Vosges-Saônoises. (Contributions à la flore de la Haute-Saône.) Bul. Soc. Belfort. Émul. **27**: 1–27. 1908. (Geology, etc.; extensive annotated list of vascular plants, mosses, hepatics, and Characeae, with localities.)

Local

Bonati, Gustave. Sur quelques plantes adventices des gares et voies ferrées de la Haute-Saône. Bul. Soc. Agr., Let., Sci. et Arts Dépt. Haute-Saône **1921**: 85–94. [1922.]

Methods of introduction, etc.; annotated list of naturalized angiosperms; list of adventive plants, with localities.

Humnicki, Valentin. Catalogue des plantes vasculaires des environs de Luxeuil (Haute-Saône). 75 p. 22 cm. Orléans, 1876.—Supplément . . . 77–92 p. Orléans, 1877.—Supplément . . . 93–105 p. Orléans, 1883.

Annotated list of vascular plants growing within a radius of 8 to 10 km. of Luxeuil, based entirely on the author's researches. Second supplement not seen.

Maire, René. Flore grayloise ou catalogue des plantes de l'arrondissement de Gray. 102 p. 22 cm. Gray, 1894. (Arnold Arboretum library.)

Soils, with lists of species; annotated list of vascular plants and Characeae.—See also his Florule adventice de Gray. Feuille des Jeunes Nat. **25**: 155–157. 1895. (Extensive list of adventive plants.)—Also his

Plantes rares ou nouvelles pour la région de Gray observées pendant l'été de 1898 et 1899. (Note préliminaire.) Bul. Soc. Grayloise Émul. **2:** 308–310. 1899. (Annotated list of vascular plants and fungi. In New York Public Library.)

HAUTE-SAVOIE

See also Savoie, General, also Pin and Thompson under Local; also Switzerland, General (Bennett; Binz and Thommen, under Binz; Bouvier; Thommen).

Local

Beauverd, Gustave. Notes floristiques sur le massif de la Fillière (Haute-Savoie). (Contributions à l'étude de la flore des Alpes d'Annecy.) Bul. Trav. Soc. Bot. Genève **10:** 56–97. 1903.

Topography, geology, bibliography; lists of ubiquitous plants, list of rarer or more characteristic plants, with localities (total 979 species of vascular plants).—See also his Circonscription de la Fillière. Bul. Soc. Bot. Genève II, **3:** 21–24. 1911.—Also his L'herborisation du 21 juin au Ballajoux, au Col du Freux et aux alpages de Barme. l.c. **15:** 27–28. 1924.—Also his Résultats de l'herborisation du 1er juillet 1926 dans le massif de la Fillière (Hte Savoie). l.c. **18:** 321–322. 1926.

——— Recherche sur la flore vasculaire du massif de la Tournette (Alpes d'Annecy, Haute-Savoie). (Contribution à la géographie botanique des Alpes de Savoie.) Bul. Soc. Bot. Genève II, **23:** 1–418. 29 fig. (incl. 2 maps), tables. 1931.

Topography, geology, climate, botanical explorations, bibliography, synecology, plant formations, phytogeography, etc.; tabular list of vascular plants (1,352 species, of which 1,094 are native, plus 160 hybrids and infraspecific forms) showing local and extralimital range, etc. The list itself is on unnumbered folded tables bound at the end of the volume.

Briquet, John. Le Mont Vuache. Étude de floristique. Avec la collaboration bryologique de Auguste Guinet. Bul. Trav. Soc. Bot. Genève **7:** 24–146. map. 1894.—Additions et corrections à la monographie du Mont Vuache. l.c. **7:** 232–234. 1894.

Botanical investigations, bibliography, physiography, plant formations, phytogeography; annotated list of 927 vascular plants and mosses (831 vascular plants).—See also **Favre, Jules.** Liste des stations nouvelles de plantes dans les chaînes du Salève et du Vuache. Annuaire Conserv. et Jard. Bot. Genève **18/19:** 193–206. 1915. (Includes lists of new records and additional localities for Mont Vouache, p. 205–206.)—Also **Beauverd, Gustave.** Notes sur la flore du Vuache. Bul. Soc. Bot. Genève II, **18:** 307–309. 1926. (Additional bibliography, summary of additional species and varieties.)

——— Notes floristiques sur les Alpes lémaniennes. Bul. Trav. Soc. Bot. Genève **5:** 191–220. 1889.

General considerations; list of various angiosperms, with localities. (In Haute-Savois and Valais.)—See also his Nouvelles notes . . . Annuaire Conserv. et Jard. Bot. Genève **3:** 46–146. 1899. (Geography, etc.; incomplete list of vascular plants, with localities.)—Also his Les colonies végétales xérothermiques des Alpes lémaniennes . . . Bul. Murithienne **27/28:** 125–212. pl. 1–4 (pl. 4, map). 1900. (Mainly phytogeographic.)

——— Notes sur la flore du Massif de Platé. (Contributions à l'histoire phytogéographique des Alpes occidentales.) Globe **34** (**V, 6**) (**Mém.**)**:** 171–211. 1895. (Reprinted, 53 p. 1895. Reprint not seen.)

Geology, topography, botanical explorations; short lists of species observed in different localities and habitats; discussion of calcicolous, silicicolous, and indifferent species. (Between the valley of the Giffre and that of the Arve.)

Camus, E. G. Note sur la chaîne des Aravis. Revue Savoisienne **43:** 215–242. 1 fig. (map), pl. 1–2. 1902.

Previous publications, topography, geology, list of excursions with mention of principal species; list of vascular plants (apparently intended to be complete), with localities.—See also **Beauverd, Gustave.** Sur la flore des Aravis septentrionaux (Hte-Savoie). Bul. Soc. Bot. Genève II, **1**: 298–316. 1 fig. (map). 1909. (Geology, hydrography, bibliography; annotated list of noteworthy species collected in 1909, mostly from the basin of the Arve; phytogeography.)

Fauconnet, Charles. Herborisations à Salève. 195, liv p. 23.5 cm. Genève et Bâle, 1867.

Topography, geology, general features of flora, account of excursions month by month; annotated list of vascular plants and Characeae, with localities; list of species of the Jura not found on Salève, list of hybrids found by author in vicinity of Geneva, etc.—See also his Promenades botaniques aux Voirons et supplément aux Herborisations à Salève. 62 p. 20.5 cm. Genève, 1868. (General features of flora of Voirons, with lists of noteworthy species to be found on various excursions; supplement to his work of 1867. In Arnold Arboretum library.)—Also **Puget, François,** *abbé.* Sur la végétation du Salève et du territoire qui s'étend de cette montagne au Mont du Vuache. Bul. Soc. Bot. France **13** (Sess. Extraord.): lxi-xci. 1866. (Short lists of species of various localities; unannotated list of vascular plants and lichens.)—Also **Favre, Jules.** Liste de stations nouvelles de plantes dans les chaînes du Salève et du Vuache. Annuaire Conserv. et Jard. Bot. Genève **18/19**: 193–206. 1915. (Includes new records and new localities for Mont Salève, p. 193–204.)

Payot, Vénance. Florule de l'excursionniste aux gorges de la Diozaz. Bul. Soc. Bot. France **23**: 394–397. 1876.

Geology, etc.; unannotated list of vascular plants, mosses, hepatics, and lichens. (Between Saint-Gervais-les-Bains and Chamonix.)

——— Florule du Mont-Blanc. Guide du botaniste et du touriste dans les Alpes pennines. Phanérogames. 291 p. 17.5 cm. Paris, [etc., 1882]. (Arnold Arboretum library.)—2. ptie. Plantes cryptogames vasculaires et cellulaires . . . (fougères ferns). 1 p. l., ii, 22 p. 19 cm. Genève, 1881. (Arnold Arboretum library.)

Annotated list of spermatophytes (1885 species), with localities, within a perimeter of 300 km. around Mont Blanc; 2d part, list of vascular cryptogams, with rather detailed ranges, and Characeae. The second part replaces his Catalogue des fougères, prèles et lycopodiacées des environs du Mont-Blanc ou énumération détaillée des plantes acotylédones vasculaires qui naissent dans les vallées de Sixt, Servoz. Diozaz, Bérard, Valorsine, Trient, Champé, Essert, Ferret, Allée-Blanche, Chapiu, Mont-Joie, comprises dans un rayon de 200 kilomèt. autour de celle de Chamounix suivi d'un catalogue des mousses et des lichens des mêmes localités. 70 p. map. 20 cm. Paris, Genève, 1860. (Topography, gazetteer; annotated list of pteridophytes, with localities, vernacular and patois names, uses, etc.; list of mosses and lichens. In New York Botanical Garden library.)

——— Guide du botaniste au Jardin de la Mer de Glace, ou catalogue des plantes qui croissent à cette limite de la végétation, accompagné d'un aperçu d'une notice sur l'avancement des deux principaux glaciers de la vallée. 13, [2] p. 16 cm. Genève, 1854. (New York Botanical Garden library.)

Topography, etc.; annotated list of plants (110 species, of which 77 are vascular). The Jardin de la Mer de Glace, or Jardin du Courtil, is a sort of rocky island in the middle of the glacier du Talèfre, measuring about 800 by 300 meters, at 2750 meters elevation.—See also **Martins, Charles.** La végétation du Spitzberg comparée à celle des Alpes et des Pyrénées. Bul. Soc. Bot. France **12**: 144–162. 1865. (Includes (p. 156–157) short account of botanical explorations and list of 87 known phanerogams, under title Végétaux phanérogames du Jardin de la Mer de Glace de Chamounix.)—Also **Sommier, Stefano.** Piante del Jardin della Mer de Glace. Nuovo Gior. Bot. Ital. **22**: 251–252. 1890. (Includes 12 phanerogams additional to the 44 listed by Filippo Parlatore in his Viaggio alla catena del Monte Bianco e al Gran San Bernardo . . . p. 138–140. 1850. In Arnold Arboretum library.) —Also Payot, below (Note . . .).

——— Note sur la végétation de la région des neiges ou florule de la vallée de la Mer de Glace au centre du massif du Mont-Blanc. Mém. Acad. Imp. Sci., Belles-Let. et Arts Lyon, Cl. Sci. **17**: 67–85. 1869–70.

Topography, etc.; list of vascular plants, mosses, hepatics, and lichens, with localities (from Entre-la-Porte, Tacul, Leschaux, and Jardin, at about 2,300 to 2,700 meters elevation).—See also his Recherches botaniques dans le massif du Mont-Blanc. Annuaire Club Alpin Franç. **6**: 544–558. 1880. (Includes I, list of mosses of the massif of Mont Blanc (p. 544–549); II, Florule de la base nord de l'Aiguille du Midi entre les altitudes de 2,500 et 3,000 mètres (p. 549–550), with mention of 3 phanerogams and several bryophytes and lichens; III, Flore des îlots du bassin de la Mer du Glace (p. 550–558), including lists of vascular and cellular plants from four botanical "islands" (Entre-la-Porte, Couvercle, Leschaux, Jardin).)—Also his Guide, above.

——— Notice sur la végétation de la région des neiges; ou florule du Jardin de la Mer de Glace, du glacier d'Argentière, au centre du massif de la chaîne du Mont-Blanc. Bul. Soc. Bot. France **37**: 32–34. 1890.

Topography, etc.; unannotated list of vascular plants, mosses, hepatics, and lichens. (A gravelly and rocky area about 100 by 100 meters, on the right side of the glacier d'Argentière and that of la Tour-Noire, at the base of the south-west flank of the Aiguille d'Argentière, at 2684 meters altitude. Needs to be carefully distinguished from the better known Jardin de la Mer de Glace in the Glacier du Talèfre, generally known as that of Chamonix.)

——— Végétation de la région des neiges ou flore des Grands-Mulets (Mont-Blanc). Annales Soc. Sci. Phys. Nat., Agr. et Indus. Lyon III, **6**: 215–222. [1862.]

Topography, etc.; list of vascular (24 species) and cellular plants (58 species).—See also **Martins, Charles.** La végétation du Spitzberg comparée à celle des Alpes et des Pyrénées. Bul. Soc. Bot. France **12**: 144–162. 1865. (Includes (p. 158–159) account of botanical explorations and list of 24 known phanerogams, with title Végétaux phanérogames des Grands-Mulets.)

HAUTE-VIENNE

See also Limousin.

General

Lamy [de la Chapelle], Édouard. Flore de la Haute-Vienne. 65 p. 8vo. Limoges, 1856.

Annotated list, with localities. (Not seen; annotation supplied by G. Dillemann.)—See also his Simple aperçu sur les plantes cryptogames et agames du département de la Haute-Vienne. Cong. Scient. France **26** (1) (Limoges 1859): 459–499. 1860. (Includes 33 vascular cryptogams, with annotations. In University of Michigan library.)—Also **Malinvaud, Ernest.** Catalogue des espèces rares ou critiques qui croissent dans les environs de Limoges, avec la description d'une espèce nouvelle d'Orobanche, découverte à Limoges, et non encore décrite par les auteurs. l.c. **26** (1): 500–510. 1860. (Extensive list of vascular plants, annotated as to abundance. In University of Michigan library.)—Also **Le Gendre, Charles.** Contributions à la flore de la Haute-Vienne. Compt. Rend. Assoc. Franç. Avanc. Sci. **19** (2): 412–415. 1891. (List of noteworthy vascular plants and Characeae, with localities.) —Also his Flore des terrains de serpentine de la Haute-Vienne. Bul. Géog. Bot. **19** (i.e. **29**): 19–34. 1919. (Local occurrence, composition, etc. of serpentine rocks; lists of species occurring on serpentine and other rocks, those occurring also on other rocks but with special forms on serpentine, and those peculiar to serpentine.)—Also his Flore des terrains magnésiens produite par l'érosion des serpentines de la Haute-Vienne. Revue Scient. Limousin **15** (année 32): 21–26, 51–62. 1928. (List of about 110 species, with localities and notes on frequency of occurrence. Not seen; title and annotation supplied by R. McVaugh from work in University of Michigan library.)

Local

Duris, Thomas. Catalogue des plantes recoltées dans les environs d'Eymoutiers (Haute-Vienne). Revue Scient. Limousin **15** (année 32): 142–152, 171–180. 1928.

List of about 430 species, with localities and notes on frequency of occurrence. (Not seen; title and annotation supplied by R. McVaugh from copy in University of Michigan library.)

HAUTES-ALPES

See also Dauphiné; Isère (Offner, La flore du massif des Grandes Russes); Savoie (Thommen, under Perrier de la Bathie); also Italy, Piemonte (Mattirolo, Un'escursione botanica nel gruppo del Viso).

General

Pons, —. Les plantes médicinales des Hautes-Alpes . . . 192 p. 13.5 cm. Gap, 1922. (At head of title-page; Comité interministériel des plantes médicinales et des plantes à essences. Sous-comité départemental des Hautes-Alpes.)

Extensively annotated list of more important wild medicinal plants, alphabetically arranged by vernacular names, with habitat, uses, etc.; briefly annotated list of less important ones, without botanical names; calendar for the collector. Prefatory note by Émile Perrot.

Local

Bonne, Gabrielle. Le Lautaret. Bul. Soc. Hist. Nat. Ardennes **[27]:** 10–32. pl., map. 1932.

List of 479 species, with localities. (Not seen; title and annotation supplied by R. McVaugh from work in University of Michigan library.)—See also **Schoenfeld, W. de.** Rapport . . . sur l'excursion faite du 7 au 10 août au Bourg-d'Oisans, à la Grave, au Lautaret et au Galibier, et dirigée par M. J.-B. Verlot. Bul. Soc. Bot. France **7:** 804–816. 1863. (Itinerary, with partial lists of species observed.)—Also **Brachet, F.** Herborisations au Lautaret, au Galibier (Hautes-Alpes) et dans les environs. Bul. Soc. Études Hautes-Alpes **13:** 158–185. 1894. (Geography, etc.; account of 13 excursions, with mention of principal species; list of principal localities of the upper valley of the Guisanne, with mention of noteworthy species.)—Also **Pellegrin, François.** Quelques observations sur la flore du Lautaret. Bul. Soc. Bot. France **57:** 172–177. 1910. (Partial lists of species from different altitudes, with particular attention to species of the plains occurring at considerable elevations.)—Also **Druce, G. C.** Le Lautaret. Rpt. Bot. Soc. and Exch. Club Brit. Isles **8:** 145–149. 1927. (Includes mention of numerous species.) —Also **Luzzatto, Gina.** Contributo alla flora del Lautaret (Hautes-Alpes). Atti Soc. Ital. Sci. Nat. Milano **70:** 92–119. 1931. (Topography; list of 336 vascular plants collected by author, with localities; bibliography.)

Carret, A. Note sur quelques plantes trouvées au Pic de la Meije. Annales Soc. Bot. Lyon **7:** 171–176. 1880.

List of 3 vascular plants collected at 3,754 meters elevation, with discussion. Same material, with title Plantes récoltées à 3,754 mètres d'altitude, in Feuille des Jeunes Nat. **9:** 48–49. 1879.

Gabriel, Cyprien. Catalogue des plantes croissant dans la région du Délovuy [i.e., Dévoluy]. Bul. Soc. Bot. France **82:** 294–306. 1935.

Geography; mostly unannotated list of vascular plants, mosses, and hepatics collected by author in the regions of Dévoluy and Mont Aurouse.

Heim, Roger. La végétation du bois de la Madeleine et des îlots arbustifs du col du Lautaret. Bul. Soc. Bot. France **69** (Sess. Extraord.): 61–77. 1926.

Topography, phytosociology, etc.; includes (p. 68–72) alphabetical list of vascular plants.

Lannes, —. Catalogue des plantes les plus intéressantes croissant dans la partie supérieure du département des Hautes-Alpes (Briançonnais, le Queyras et le haut du vallon de Vars) . . . Bul. Soc. Études Hautes-Alpes **4:** 145–177, 265–293. 1885.

Extensive list of vascular plants, with localities, collected at 1,100–3,000 meters elevation.—See also **Durenne, —,** and **Petitmengin, Marcel.** Promenade botanique dans les Alpes du Briançonnais (2–9 août 1903). Bul. Acad. Internatl. Géog. Bot. **13:** 253–267 (incl. map). 1904. (Topography, geology; account of excursion in vicinity of Briançon, Cervières, Queyras, Abriès, col de Cellières, etc.)—Also **Brachet, F.** Excursions botaniques de Briançon aux sources de la Durance et de la Clarée (Hautes-Alpes). Petit guide du botaniste herborisant. Bul. Soc. Études Hautes-Alpes **26:** 71–94, 144–166.

map. 1907. (Topography; account of 20 excursions to be made from Briançon, with mention of noteworthy plants.)—Also **Jeanpert, Édouard.** Sur quelques plantes du Queyras. Bul. Soc. Bot. France **59**: 589. 1915. (List of several noteworthy plants, with localities.)—Also his Note sur la flore du Queyras. l.c. **59**: 621–625. 1915. (Itinerary, with mention of noteworthy plants.)—Also **Pons, J.** Note sur quelques plantes nouvelles du Briançonnais. l.c. **69**: 471–480. 1923. (Considerations on introduced species, etc.; list of noteworthy angiosperms from the canton of Briançon, with localities.)—Also his Compte rendu sommaire des herborisations faites par la Société pendant le session de 1922 dans le Briançonnais. l.c. **69** (Sess. Extraord.): 18–61. 1926. (Account of numerous excursions, with mention of noteworthy plants collected.)—Also **Pouzet, Eugène.** Excursion à Saint-Véran en Queyras 16 à 22 août 1932. Annales Soc. Linn. Lyon n.s., **77**: 65–79. 1934. (Includes lists of species from various localities.)

Le Brun, Pierre. L'herborisation au col de la Traversette. Bul. Soc. Bot. France **69**: 481–488. 1923.

Includes lists of vascular plants observed at various altitudes up to the summit (3112 meters).—See also **Soubeiran, Léon,** and **Verlot, Bernard.** Rapport sur l'excursion faite du 2 au 9 août, au Mont-Viso et dans les alpes du Briançonnais . . . Bul. Soc. Bot. France **7**: 783–816. 1860 (see especially p. 793 sqq.).

Magnin, Antoine. Compte-rendu botanique de la session de Gap en juillet et août 1874. Annales Soc. Bot. Lyon **2**: 125–134. 1875.

Topography, geology, plant zones, with partial lists of species of different localities and habitats. Reprinted, with title Aperçu de la végétation des environs de Gap, in Bul. Soc. Bot. France **21** (Sess. Extraord.): xlvi–lii. [1876], under the general title Rapports sur les excursions de la Société . . . , with the following supplementary papers: **Magnin, A.** Rapport . . . sur l'herborisation faite le 24 juillet, au Col de Glaize. l.c. p. lii–lx. (Itinerary, with mention of noteworthy plants collected); **Borel, J.** Enumération . . . des espèces trouvées dans les diverses localités parcourues dans l'excursion au Col de Glaize. l.c. p. lxi–lxxii. (List of vascular plants and Characeae collected, grouped under localities; also list of mosses). Other papers on other localities are also given.—See also **Tillet, Paul.** Observations sur la flore du Laus et des environs de Gap (Hautes-Alpes). Annales Soc. Bot. Lyon **7**: 219–242. 1880. (Includes partial list of plants from numerous localities.)

Reverchon, —. Note sur la flore du Mont Rognouse. Annales Soc. Bot. Lyon **2**: 33–36. 1875.

List of noteworthy species, not in systematic order.

HAUTES-PYRÉNÉES

See also Gascogne (Levrat).

General

Dulac, Joseph, *abbé*. Flore du département des Hautes-Pyrénées (publiée pour le première fois): plantes vasculaires spontanées . . . xii, 641 p. 45 fig., map. 19 cm. Paris, 1867.

Keyed, very briefly annotated flora of vascular plants (1776 i.e. 1778 species), with localities. The index gives the derivation, often fanciful, of all the scientific names. The author, whose classical feeling is indicated by his manufacture of such names as Volataceae (for Aceraceae) and Duotriaceae (for Cistaceae), laid down in his preface 8 selfmade rules which enabled him to rename a considerable number of families, genera, and species. Among these rules were: no personal generic or specific names (or very few); no specific epithets that are substantives, either in apposition or in the genitive, or that represent a relationship with another genus or species (such as bellidifolium, anagalloides); no generic names that are anagrams, or that are modification of other generic names; no family names derived from generic names. Under these rules Dulac was able, for instance, in the small family Amaryllidaceae, to change the names of the family (to **Bractillaceae**), of one of the two genera (*Narcissus* to *Stephanophorum*), and of 5 of the 6 species in the latter genus, all of these being new and completely superfluous names.

Local

Bouillé, Roger de, *comte.* Bat-Laetouse ou Marmuret. Explor. Pyrén. Bul. Soc. Ramond **16**: 107–127. 1881.

Topography, etc.; lists of vascular plants collected at several localities.—See also Basses-Pyrénées (Bouillé).

Chouard, Pierre. Les éléments géobotaniques constituant la flore du massif de Néouvieille et des vallées qui l'encadrent. Bul. Soc. Bot. France **96** (Sess. Extraord.): 84–121. fig. 11–30 on pl. 5–8. 1950.

List of vascular plants (1238 species) of the Néouvieille-Arbizon-Pic de Midi axis of the central Pyrenees between the valleys of the Gave de Pau and the Neste, divided into numerous phytogeographic groups, with indication of localities, plant zones, etc. Accompanied by other papers on physiography, endemism, cellular plants, plant associations, etc.—See also Index bibliographique. l.c. 161–164. 1950. (Bibliography.)

Davy de Virville, Adrien. La flore de deux glaciers inférieurs des Pyrénées. Revue Gén. Bot. 41: 1–23. 1929.

Account of Glacier du Mont-Aigu, on the east slope of Mont Aigu at about 1,400 meters altitude, and of Glacier du Péguère, on the east slope of Mont Péguère, at about the same elevation, with lists of species observed in various habitats.—See also **Neyraut, E. J.** Une excursion botanique au Péguère, dans les Pyrénées françaises. Revue Bot. Syst. et Géog. Bot. **1**: 113–123. 1 fig. (map). 1903. (Topography; itinerary, with lists of species met with at different elevations.)

Debeaux, Odon. Les herborisations des environs de Barèges (Hautes-Pyrénées). Bul. Trav. Soc. Pharm. Bordeaux **4**: 131–142. 1863; **6**: 27–39. 1865. (Reprinted, 26 p. Paris, 1865. In Lloyd Library.)

Topography, etc.; account of Barèges and various adjacent localities, with partial lists of species; general features of flora, etc.

Gerber, Charles, and **Gaussen, Henri.** Deux herborisations au Labassech à cent quinze ans d'intervalle. Bul. Soc. Bot. France **74**: 608–630. 1928.

Botanical explorations, etc.; lists of species observed at various altitudes; bibliography. The mountain called Labassech or Labatsec by the older botanists is the present Céciré.

Ramond de Carbonnières, Louis. État de la végétation au sommet du Pic du Midi de Bagnères. Mém. Mus. Hist. Nat. Paris **13**: 217–281. 1825.

Includes (p. 244–277) annotated list of 133 plants (75 vascular), collected at about 2,600 meters elevation. Republished, with addition of a section on meteorological observations, in Mém. Acad. Roy. Sci. Inst. France [II,] **6**: 81–174. 1827, with title Mémoire sur l'état . . .—See also **Des Moulins, Charles.** État de la végétation sur le Pic du Midi de Bigorre au 17 octobre 1840. Actes Acad. Roy. Sci., Belles-Let. et Arts Bordeaux **6**: 181–292. pl. 1844. (Topography, plant zones, notes on species of thistle and *Parnassia*, etc.; annotated lists of vascular and cellular plants collected at different elevations from 1,800–3,000 meters.)—Also **Martins, Charles.** La végétation du Spitzberg comparée à celle des Alpes et des Pyrénées. Bul. Soc. Bot. France **12**: 144–162. 1865. (Includes (p. 160–162) account of botanical explorations and list of 72 known phanerogams, with title Végétaux phanérogames du sommet du Pic du Midi de Bigorre.)—Also **Joinville, Pierre de.** Récoltes au Pic-du-Midi-de-Bigorre. Monde des Plant. II, 4 (18): 51–52. 1902. (Unannotated list of vascular plants. In Arnold Arboretum library.)—Also **Flahault, Charles.** Compte rendu de l'herborisation de 10–11 août, de Bagnères-de-Bigorre au Pic du Midi; le jardin alpin du Pic du Midi. Bul. Soc. Bot. France **54** (Sess. Extraord.): cxxvi–cxliii. 1909. (Remarks on deforestation, etc.; account of ascent, with lists of species at different elevations; discussion of altitudinal limits, climate, agricultural possibilities, etc.) —Also **Guinet, Camille,** and **Turmel, J. M.** Observations sur la végétation du Pic de Midi de Bigorre. l.c. **96** (Sess. Extraord.): 149–155. 1950. (Sketch of plant associations at different elevations, with ecological data.) —Also **Bouget, Joseph,** and **Bouget, Charles.** Liste floristique du Pic du Midi établie à la date de 1944. Bul. Soc. Hist. Nat. Toulouse **93**: 95–100. 1958. (Unannotated list of 318 known vascular plants growing from the base to the summit (1800-2877 meters), of which 147 were added by the Bougets. Includes additional note by R. Morquer.)

Vallot, Joseph. Flore glaciale des Hautes-Pyrénées. Bul. Soc. Bot. France 32: 133–141. 1885.

Lists of plants collected at 2,600 to 3,352 meters on 11 peaks.

——— Guide du botaniste et du géologue dans la région de Cauterets. xxviii, 329 p. 17 cm. Pau, Paris, 1886. (Lloyd Library.)

Directions for travelling, equipment, etc.; account of 47 excursions to different localities, with itineraries and lists of principal vascular plants; annotated list of known vascular plants, with localities and brief diagnostic characters. Covers a triangular area between Pierrefitte, pic d'Estom-Soubiran, and pic de la Fache.—See also **Mouillard, L.** Contribution à la flore du bassin de Cauterets (Hautes-Pyrénées). Bul. Soc. Bot. France **54** (Sess. Extraord.): xlvi–l. 1908. (Extensive list of species new to the region, with localities.)—Also **Neyraut, E. J.** Rapport sur les herborisations faites aux environs de Cauterets. l.c. 54 (Sess. Extraord.): cii-cxxxv. pl. 1 (map). 1909. (Account of excursion of Société Botanique de France in August 1907, with numerous lists of species observed.)

Ydrac, F. L. Généralités sur quelques plantes de la haute vallée de l'Adour, au point de vue médicinal et toxicologique. Bul. Soc. Bot. France **54** (Sess. Extraord.): xxiv–xxix. 1908.

Annotated list of native medicinal and poisonous phanerogams.

HÉRAULT

See also Languedoc.

General

Juillet, Armand, and **Rodié, Joseph.** Notice sur les plantes médicinales et à essences de l'Hérault. 48 p. 24 cm. Montpellier, 1920. (At head of title-page: Comité des plantes médicinales et à essences de la région de Montpellier.)

Plant zones, etc.; annotated list of more important species, with vernacular names, etc.; calendar for the collector, etc.

Loret, Henri, and **Barrandon, Auguste.** Flore de Montpellier, ou analyse descriptive des plantes vasculaires de l'Hérault. 2. éd., revue et corrigée par H. Loret. 2 p.l., lxxvi, 663 p. 2 maps. 22.5 cm. Montpellier, Paris, 1886. (1st ed., 2 v., 1876; 2d ed. reissued 1888.)

Botanical explorations (with bibliographical footnotes), lists of species of different habitats, glossary, etc.; annotated flora of vascular plants and Characeae [2081 species, 254 varieties and hybrids], with keys or brief descriptions, localities, and uses; list of patois names.—See also **Coste, Hippolyte,** *abbé.* Un bouquet de quarante plantes nouvelles pour la flore d'Hérault. Bul. Soc. Bot. France **40** (Sess. Extraord.): cxliv-cxlviii. 1894. (List with localities.)

Planchon, Louis. Plantes médicinales et toxiques du département de l'Hérault. Mém. Sect. Méd. Acad. Sci. et Let. Montpellier II, 1: 239–337. 1899.

Tabular systematic list of vascular plants, with French and Languedoc names, local distribution, part used, uses.

Thellung, Albert. La flore adventice de Montpellier. Mém. Soc. Natl. Sci. Nat. et Math. Cherbourg 38 (IV, 8): 57–728. 1911–12. (Reprinted as Mitt. Bot. Mus. Univ. Zürich 58. 1912, with same paging.)

List of adventive vascular plants of the Department of Hérault with details of occurrence there and statement of general range and occurrence elsewhere in France; historical review of flora; classification of adventive and naturalized plants, statistics; bibliography. Includes 953 species, subspecies, notable varieties, and hybrids, of which 800 are regarded as truly "adventices." Replaces **Godron, D. A.** Florula juvenalis ou énumération des plantes étrangères qui croissent naturellement au Port Juvénal, près de Montpellier. Mém. Acad. Stanislas 1853: 368–436. 1854. (Reprinted, including his Considérations sur les migrations des végétaux . . . [l.c. 329–367], 116 p. 1856.) (Methods of migration and introduction of plants (his Considérations . . .); list of adventive plants, with name of country of origin. Pritzel (Thesaurus literaturae botanicae, ed. 2, p. 124) cites reprints of 48 and 116 p., dated 1853 and 1854, respectively. Thellung says 1854 reprint includes additions bringing total to 386 species.) Two supplements were published by **Cosson, Ernest.** Appendix Florulae juvenalis . . . Bul. Soc. Bot. France **6:** 605–615. 1859 (brings number of species to 458); Appendix Florulae juvenalis altera, ou deuxième liste de plantes étrangères . . . l.c. **11:** 159–164. 1864; see also **Rioux, Jean,** and **Quézel, Pierre.** La "Flora juvenalis" en 1950. Monde des Plant. 45 (272): 73–74. 1950. (Subsequent history of the locality; only 6 species out of 458 recorded up to 1858 (1859) still persist. In Arnold Arboretum library.)—Also **Pavillard, Jules.** Glanes floristiques autour de Montpellier (3e article). Bul. Soc. Bot. France **102:** 42–43. 1955. (Records of 3 other persistent species.)

Local

Aubouy, Antonin. Catalogue des plantes vasculaires de l'arrondissement de Lodève. (124 p.) 22 cm. Montpellier, 1874.

Annotated list of vascular plants with French and patois names and localities. The pages cited include Ranunculaceae to Scrophulariaceae only; full work not available. Originally published as separately paged supplement in Annales Soc. Hort. et Hist. Nat. Hérault. (Not seen; title and annotation supplied by B. de Retz and G. Dillemann.)

——— Florule de Palavas ou catalogue des plantes vasculaires observées aux environs de cette localité. Revue Bot. Bul. Mens. Soc. Franç. Bot. **3:** 171–206. 1885.

General features of flora; partly annotated list of vascular plants.

——— Herborisations à Murviel-lez-Montpellier (Hérault). Première herborisation, suivie d'une florule du vallon de Fontvalès. (Reprinted, 29 p. 1885, from Annales Soc. Hort. et Hist. Nat. Hérault (vol. ?); original not seen, reprint in library of Muséum national d'histoire naturelle, Paris.)

General features of flora in form of an itinerary; unannotated list of vascular plants [575 species] of the vale of Fontvalès.—See also his Une deuxième herborisation à Murviel-lez-Montpellier suivie d'une florule du vallon de Valcrose. Annales Soc. Hort. et Hist. Nat. Hérault II, **26:** 88–110. 1894. (Includes (p. 101–110) "Liste des plantes observées dans le vallon de Valcrose et sur les coteaux qui l'avoisinent, à Murviel-lez-Montpellier." Reprinted, 27 p.)

Barthés, Melchior. Glossaire botanique languedocien, français, latin de l'arrondissement de Saint-Pons (Hérault) précédé d'une étude du dialecte languedocien. vii, 265 p. 22 cm. Montpellier, 1873. (Harvard College library.)

Includes (p. 39–197) alphabetical list of Languedoc patois names, with French and botanical names, notes on uses, discussion of etymology, etc.; followed by lists of French and botanical names with corresponding equivalents.

Coste, Hippolyte, *abbé,* and **Sennen,** *frère.* Plantes adventices observées dans la vallée de l'Orb à Bédarieux et à Hérépian. Bul. Soc. Bot. France **41:** 98–113. 1894.

Methods of introduction, etc.; annotated list of 98 adventive angiosperms.

Coulouma, Joseph. Flore de Ceilhes et de la haute vallée de l'Orb. Valeur des digitales de Ceilhes et de la région d'après leur teneur en digitaline. 3 p. l., 107 p. 2 maps. 23 cm. Montpellier, 1924. (Université de Montpellier. Faculté de pharmacie. Thèse no. 137.) (New York Botanical Garden library.)

Botanical explorations, climate, geology, ecology, phytogeography, list of new records, list of medicinal plants, etc.; list of vascular plants (p. 88–99) of the upper basin of the Orb from its source to Avène, annotated as to abundance. Covers an area on northern border of Hérault and in adjacent Aveyron. An abstract, with same title, is given by Coulouma in Compt. Rend. Assoc. Franç. Avanc. Sci. 48: 405–410. 1925.

Coustan, Henry. Herborisations autour de Saint-Félix-de-Lodez (Hérault) ou catalogue des plantes vasculaires observées aux environs de cette localité suivi de la liste des oiseaux . . . 64 p. 21 cm. Clermont-l'Hérault, 1897.

Includes unannotated list of vascular plants of the vicinity of Saint-Félix (p. 16–33), also shorter lists from two other localities (p. 34–43) and classified list of local medicinal plants (p. 44–45).

Desjardins, Édouard. Herborisations dans l'Hérault. Catalogue des plantes trouvées à Bessan, Vias, Agde et Roquehaute (Hérault), en mars, août et septembre 1875, 1876 et 1877. Bul. Soc. Hist. Nat. Toulouse **13**: 207–230. 1879.

List of 440 vascular plants, with localities.—See also **Biche, B.** Flore des environs de Roquehaute, communes de Vias et de Portiragnes. Bul. Soc. Étude Sci. Nat. Béziers **5**: 119–144 map. 1881. (Botanical explorations, soils; annotated list of about 400 vascular plants. In Museum of Comparative Zoology library, Harvard University.)

Flahault, Charles. Liste des plantes phanérogames qui pourront être récoltées par la Société botanique de France réunie en session extraordinaire à Montpellier (mai 1893). 28 p. 3 fig. 25 cm. Montpellier, 1893.

Lists of the more interesting species of 15 localities in vicinity of Montpellier.—See also **Mandon, E.** Note sur quelques plantes intéressantes ou nouvelles pour la flore des environs de Montpellier. Bul. Soc. Bot. France **39**: 161–165. 1892. (List of newly recorded species and new localities.) —Also his [Découverte de plantes nouvelles pour notre flore d'Hérault]. l.c. **41**: 463–464. 1894. (Short list.)—Also **Blanc, L.** La végétation aux environs de Montpellier. l.c. **52**: 203–214. 1905. (Topography, geology, discussion of different habitats and their vegetation.)—Also **Pavillard, Jules.** Glanes floristiques autour de Montpellier. ([1.]–2. article.) l.c. **70**: 467–471. 1923: **73**: 732–739. 1927. (Mention of some noteworthy species, with discussion.) —Also **Sennen,** *frère.* La garrigue du littoral, depuis Montpellier jusqu'à Sagunto. l.c. **72**: 92–114. 1925. (Running account of the flora, with mention of 417 species in systematic order, in part with localities.)

Hérail, Joseph. Notes sur la flore des environs immédiats de Béziers. Bul. Soc. Étude Sci. Nat. Béziers **3**: 66–78. 1878. (Museum of Comparative Zoology library, Harvard University.)

Briefly annotated list of phanerogams.—See also his Deuxième note sur la flore . . . Bul. Soc. Étude Sci. Nat. Béziers **4**: 145–155. 1880. (Additional list. In Museum of Comparative Zoology library.)—Also **Sennen,** *frère.* Mes herborisations aux environs de Béziers. Bul. Soc. Bot. France **42**: 181–195. 1895. (Previous botanical explorations; extensive list of vascular plants, with localities.)—Also **Rey-Pailhade, Constantin de.** Liste des plantes adventices de Béziers et des environs (Hérault). Feuille des Jeunes Nat. **31**: 181–185. 1901. (Extensive list, with localities and short annotations.)—Also **Carles, P.** Sur les espèces végétales exotiques des environs immédiats de Béziers (Hérault). Compt. Rend. Acad. Sci. Paris **136**: 1589–1591. 1903. (Discussion of sources of adventive plants, with mention of various species.) —Also **Albaille, S. J.** Notes sur quelques plantes rares ou peu connues des environs de Béziers. Leurs usages et leurs habitats nouveaux. Bul. Soc. Étude Sci. Nat. Béziers **31** (Mém.): 5–18. 1909 (cover date 1910). (List of about 50 vascular plants with localities and notes on uses.)—Also his Sur quelques plantes adventices nouvelles pour les environs de Béziers, Hérault. l.c. **36**: 36–42. 1933. (Not seen; cited from Bot. Jahresber. Just **61/62** (1): 93. 1941.)

Kuhnholtz-Lordat, Georges, and **Blanchet, Georges.** Flore des environs immédiats de Montpellier. v. 2. Les végétaux vasculaires et leurs parasites cryptogames. Préface de Gabriel Arnaud. 457 p. map. 21 cm. Paris, 1948.

List of vascular plants, mostly unannotated, alphabetically arranged by generic names and with keys to the native and adventive species; cultivated species are listed but usually not keyed; families and other groups (with keys), as well as vernacular names, are included in the same alphabet. Under each vascular plant genus the parasites (cellular and vascular) are listed with annotations. Covers a radius of 10 to 12 kilometers around Montpellier. —See also **Rioux, J. A.,** and **Quézel, Pierre.** Note floristique sur la "région" du Montpellier. Monde des Plant. **46** (280/281): 41–43. 1951. (Records of 41 noteworthy plants. In Arnold Arboretum library.)—Also Planchon, below.

Pagès, Eugène. Florule de la vallée supérieure de la Mare et des environs. Bul. Géog. Bot. **22**: 62–147. 1912.

General features of flora, etc.; annotated list of vascular plants and Characeae.

Planchon, Gustave. Des modifications de la flore de Montpellier depuis la XVIe siècle jusqu'à nos jours. 57 p. 27 cm. Paris, Montpellier, 1864. (Arnold Arboretum library.)

History of local botany, bibliography, discussion of changes in flora, geographical origin of added species, etc.; annotated list of species lost, wrongly recorded, adventive, naturalized, etc.

Planchon, Louis. Matériaux pour la flore médicale de Montpellier et des Cévennes d'après Lobel. 44 p. 21.5 cm. Montpellier, 1868. (Extrait du Montpellier médical.)

Work of Lobel; list of vascular and a few cellular wild and cultivated medicinal plants, grouped by families, with notes on some of the species, based mainly on Lobel's Adversaria.

ILLE-ET-VILAINE

See also Bretagne.

General

Picquenard, Charles. Catalogue des plantes vasculaires spontanées du département d'Ille-et-Vilaine. Bul. Soc. Sci. Nat. Ouest France 7 (1): 29–128. 1897.

Geology, soils, botanical explorations, bibliography; annotated list of about 1080 vascular plants, with localities for scarcer species.—See also **Humbert, Henri.** Contributions à la flore de Bretagne. Localités nouvelles pour l'Ille-et-Vilaine. Revue Bretonne Bot. [1]: 23–29. 1906. (Includes list of noteworthy vascular plants (p. 23–27), with localities.)—Also **Nehou, Jean.** Plantes rares et plantes adventices du nord du département d'Ille-et-Vilaine. Bul. Soc. Scient. Bretagne (Sci. Math., Phys. et Nat.) **26:** 93–96. 1953. (Annotated list.)—Also **Dizerbo, A. H.** Plantes nouvelles ou peu communes du sud de l'Ille-et-Vilaine. l.c. **29:** 57–58. 1955. (List with localities.)

St. Gal, Joseph, and **Demarquet, E.** Catalogue raisonné des végétaux spontanés ou cultivés en Ille-et-Vilaine récoltés de 1895 à 1900. xvi, 175 p. 18 cm. Rennes, 1900. (Lloyd Library.)

Brief mention of geology, botanical explorations, etc.; annotated list of wild and cultivated vascular plants, fungi, etc., based on the authors' collections, with localities and vernacular names.

Local

Desmars, J. Catalogue des plantes qui croissent spontanément aux environs de Redon. 75 p. Redon, 1866. (Reprinted from Bul. Soc. Let. et Sci. Redon.)

List of 923 vascular plants found within a radius of six leagues, thus including territory in adjacent Loire-Inférieure and Morbihan. (Not seen; cited from Bul. Soc. Bot. France 14 (Rev. Bibl.): 145. 1867.)

Dubuys, Charles. Catalogue des principales plantes herbacées spontanées aux environs de Rennes (caption title only). *In* École nationale d'agriculture de Rennes. Chaire de botanique. p. 1–12. n.p., n.d. [195–?].

Unannotated list of vascular plants, the weeds and medicinal plants so indicated typographically.

Mabille, Paul. Catalogue des plantes qui croissent autour de Dinan et de Saint-Malo, avec notes et descriptions pour les espèces critiques ou nouvelles. Actes Soc. Linn. Bordeaux **25** (III, 5): 489–644. 1866. (Reprinted, 160 p. Bordeaux, 1866. In Gray Herbarium library.)

Botanical explorations, topography, plant formations, list of naturalized plants; annotated list of vascular and cellular plants (1378 species), with localities. (Dinan in Côtes-du-Nord, Saint-Malo in Ille-et-Vilaine.)—See also **Jeanpert, Édouard.** Localités nouvelles de plantes récoltées aux envir. de Saint-Malo. Bul. Soc. Bot. France **40:** 64–65. 1893.—Also **Thellung, Albert.** [Trouvailles floristiques dans la région de Saint-Malo.] Monde des Plant. 24 (29–144): 5–6. 1923.—Also **Vivien, G.** Contribution à la flore de l'Ille-et-Vilaine. Bul. Soc. Sci. Seine-et-Oise **II, 5:** 82–84. 1924. (Extensive lists of vascular plants from region of St. Malo, Dinard, etc.)—Also **Lemesle, Robert.** Études de quelques phanérogames adventices dans la région nord du département d'Ille-et-Vilaine. Bul. Lab. Marit. Dinard 19: 8–11. 1938. (List of noteworthy adventive plants from vicinity of Saint-Malo and Dinard. In Museum of Comparative Zoology library, Harvard University.) Also **Nehou, Jean.** Les phanérogames adventices dans la région de Saint Malo. Monde des Plant. 44 (262): 56. 1949. (Annotated list; short bibliography. In Arnold Arboretum library.)

INDRE

See also Berry; Indre-et-Loire (Tourlet, Documents).

Local

Chastaingt, Gabriel. Catalogue des plantes vasculaires des environs de La Châtre (Indre). Mém. Soc. Acad. Maine-et-Loire 38: 1–195. 1883. (Preprinted, 2 p.l., 195 p. 1882.)

Topography, geology, general features of flora, etc.; annotated list of vascular plants and Characeae (852 species, of which 846 are vascular).

Rallet, Louis. Étude phytogéographique de la Brenne. Bul. Soc. Sci. Nat. Ouest France V, 5: 1–280. illus. (maps), pl. 1–11, 2 maps. 1936. (Reprinted as thesis.)

Botanical explorations, topography, geology, pedology, climate, list of ponds, etc.; list of 634 vascular plants, with indication of ecological type and localities; lists of cellular plants; detailed account of plant associations, history of vegetation, phytogeography, bibliography. The area is bounded on the north by the Claise River, on the south by the Creuse, on the east by the road from Châteauroux to Limoges, on the west by that from Châtillon to Le Blanc.

INDRE-ET-LOIRE

See also Loire-Inférieure (Nehou).

General

Tourlet, E. H. Catalogue raisonné des plantes vasculaires du département d'Indre-et-Loire. Publié par les soins de J. Ivolas. xxiii, 621 p. port. 25 cm. Paris et Tours, 1908.

Botanical explorations, with bibliographical footnotes; annotated list of vascular plants and Characeae (1,531 species, of which 1,509 are vascular), with detailed localities. Replaces **[Dujardin, Félix].** Flore complète d'Indre et Loire. Publié par la Société d'agriculture, sciences, arts et belles-lettres, et dédiée à M. d'Entraigues, préfet du département . . . 472 p. 3 pl. 21 cm. Tours, 1833. (Annotated, keyed, descriptive flora of vascular and cellular plants [1258 vascular plants and Characeae]; includes cultivated species. The introductory matter is unpaged.) Also replaces **Delaunay, Jules.** Catalogue des plantes vasculaires du département de l'Indre et Loire. 141 p. 8vo. Tours, 1873. (Preface by abbé J. Coqueray; annotated list of vascular plants (1345 species), with localities. Not seen; title and annotation supplied by R. Garestier. Originally published in Bulletin de la Société tourangelle d'horticulture v. 3. 1872, according to critical analysis by Tourlet in Bul. Soc. Bot. France 50: 406. 1903.)

——— Documents pour servir à l'histoire de la botanique en Touraine. 106 p. 25.5 cm. Tours, 1905. (Extrait du Bulletin de la Société pharmaceutique d'Indre-et-Loire.)

Biographical sketches of deceased botanists, list of living botanists native in Touraine or who have collected there, bibliography.—See also his Revision de la flore d'Indre-et-Loire. Bul. Soc. Bot. France 50: 401–428. 1903. (Critical discussion of the floras by Dujardin (1833) and Delaunay (1873); list of additional species, with localities.)

——— Plantes introduites, naturalisées ou adventices, du département d'Indre-et-Loire. Bul. Soc. Bot. France 51: 222–237, 279–287. 1904. (Reprinted with title, Tableau de la flore adventice du département d'Indre-et-Loire. 26 p. Paris, 1904. In library of Muséum national d'histoire naturelle, Paris.)

Annotated list of introduced angiosperms, with discussion of origin, etc.

ISÈRE

See also Dauphiné; Hautes-Alpes (Carret); Savoie (Nétien; Thommen, under Perrier de la Bathie).

General

Gras, Albin. Statistique botanique du département de l'Isère ou guide du botaniste dans ce département. 192 p. 22.5 cm. Grenoble, 1844. (From Statistique générale du département de l'Isère v. 2.) (Arnold Arboretum library.)

Botanical regions and zones, useful plants, trees and shrubs, gazetteer of localities with lists of characteristic species; tabular list of cellular plants, with vernacular names, synonyms, and localities. No formal list of vascular plants.

Local

Cariot, Antoine. Le guide du botaniste à la Grande-Chartreuse et à Chalais ainsi que les localités voisines et sur les montagnes environnantes. 72 p. map. 18 cm. Lyon, 1856.

Advice as to collecting equipment; briefly annotated list of vascular plants; list of species collected and recorded by Villars in this region, arranged by localities; gazetteer.—See also **Cusin, Louis.** Herborisation à la Grande-Chartreuse, les 5 et 6 juillet 1873. Annales Soc. Bot. Lyon **1:** 67–71. 1873. (Itinerary, with mention of noteworthy plants collected.)—Also his Herborisation à la Grande-Chartreuse, les 8 et 9 juillet 1877. l. c. **5:** 195–199. 1878. (List of species not included by Cariot; notes on various other species.) —Also **Emberger, Louis.** À la Grande Chartreuse au début du mois d'août. Bul. Soc. Bot. France **66:** 281–286. 1920. (Account of visit with mention of noteworthy species observed.)

Cuny, Arsène. La flore des montagnes de la Salette (Haut-Dauphiné). Bul. Soc. Scient. Dauphiné **53** (V, **12**): 1–86. 1933. (Preprinted, Grenoble, 1932.)

Soils, discussion of *Hieracium;* account of numerous localities, with lists of characteristic species.

Meyran, Octave. Herborisation à la montagne de Taillefer. Bul. Trimest. Soc. Bot. Lyon II, **3:** 84–92. 1885. (Arnold Arboretum library.)

Topography, itinerary; lists of spermatophytes observed by author from Laffrey to La Morte, then to the Taillefer, and returning from La Morte to Séchilienne.

Nétien, Georges, Mérit, —, and **Perra, —.** Partie botanique. *In* Les grandes excursions de la Société linnéenne. IV. En Oisans. De Vaujany au Col du Glandon par le Col du Sabot et la vallée de l'Eau d' Olle . . . Bul. Mens. Soc. Linn. Lyon **7:** 132–137 incl. 4 fig. 1938.

Itinerary, with mention of characteristic plants; unannotated list of vascular plants collected.—See also other papers referring to various localities in Oisans, with lists, in Annales Soc. Linn. Lyon n.s., **79:** 14–19. 1936; **80:** 20–26. 1937; **81:** (p. ?). 1938; Bul. Mens. Soc. Linn. Lyon **8:** 179–182. 1939.

Offner, Jules. La flore du massif des Grandes Rousses. Revue Gen. Bot. **21:** 257–273. 1909.

Botanical explorations, topography, general features of flora; list of all the alpine plants known from the area.

Tillet, Paul. Florule d'Uriage. (Excursions botaniques en Dauphiné.) 2 p. l., ii, 69 p. 18.5 cm. Paris, 1883. (Lloyd Library.)

Itineraries of eight botanizing trips in vicinity of Uriage, with included partial lists of vascular plants.—See also **Lassimonne, S. É.** Notes de géographie botanique sur Uriage et ses environs. Bul. Soc. Dauphinoise Études Biol. **2:** 136–166. 1910. (In John Crerar Library. Also published in Annales Univ. Grenoble **22:** 349–379. 1910.) (Topography, climate, influence of cultivation; plant associations with lists of characteristic species.)

Verlot, J. B. Les herborisations des environs de Grenoble. Bul. Soc. Bot. France **7**: 634–673. 1860.

Guide for botanical collecting in various localities within 25 to 30 km. of Grenoble, with mention of significant species; systematic list of vascular plants mentioned.—See also **Vidal, Louis,** and **Offner, Jules.** Les colonies des plantes méridionales des environs de Grenoble. Bul. Soc. Stat., Sci. Nat., et Arts Indust. Dépt. Isère **34** (IV, **8**): 505–565 incl. fig. (map). 1905. (Phytogeographic; includes numerous short lists from various localities, and list of 101 southern species with localities.)

JURA

See also Franche-Comté.
For works relating to Jura Mountains in general, see above under France, Partial (Babey, Grenier, Magnin, Parmentier).

General

Michalet, Eugène. Botanique . . . revue et achevée par M. Grenier. *In* Ogérien, *frère*. Histoire naturelle du Jura et des départements voisins. v. 2, p. 1–331. 22.5 cm. Paris [etc.], 1864. (Gray Herbarium library.)

Topography, geology, climate, plant distribution, botanical explorations; annotated list of about 1500 vascular plants, with localities for scarcer species.

Local

Magnin, Antoine. Note sur la flore des environs d'Arbois. 14 p. 8vo. Lyon, 1887. (Library of Muséum national d'histoire naturelle, Paris.)

Account of botanical excursions in vicinity of Arbois, with lists of noteworthy species.

Thevenot, —. Catalogue des plantes vasculaires spontanées observées sur le territoire des cantons de Lons-le-Saunier et de Beaufort et de quelques communes circonvoisines . . . Mém. Soc. Émul. Jura II, **4**: 405–465. 1879. (Reprinted, 61 p. Lons-le-Saunier, 1879. In Arnold Arboretum library.)

Annotated list of vascular plants and Characeae, with localities for rarer species.

LANDES

See also Basses-Pyrénées (Blanchet); Gascogne.

General

Heubert, Camille. Arbres, arbrisseaux et arbustes landais. 96 p. 18.5 cm. Mont-de-Marsan, 1941.

Key to native and cultivated trees and shrubs, followed by annotated list of 127 species, alphabetically arranged by vernacular names, with the scientific names.

Lapeyrère, E. Flore du département des Landes. xxiii, 511 p. 25.5 cm. Dax [1896–1903]. (Issued with Bul. Soc. Borda v. 21–28.)

Bibliography, key to families; briefly annotated, descriptive flora of vascular plants (1673 species and varieties), with partial keys and localities for rarer species; glossary. Preface dated 1892, but according to table of contents on the covers of Bul. Soc. Borda, the work appeared as follows: v. 21, p. i–xxiii, 1–56. 1896; v. 22, p. 57–168. 1897; v. 23, p. 169–264. 1898; v. 24, p. 265–328. 1899; v. 25, p. 329–376. 1900; v. 26, p. 377–424. 1901; v. 27, p. 425–472. 1902; v. 28, p. 473–511. 1903. Replaces **Thore, Jean.** Essai d'une chloris du département des Landes. 2 p. l., xliv, 5–516 p. 18 cm. Dax, 1803. (Annotated list of vascular and cellular plants.) Also replaces **Landry, P.,** and **Behr, J.** Tableau analytique de la flore des Landes; description succincte des plantes croissant spontanément dans le département . . . Bul. Soc. Borda **5**: 273–306. 1880; **6**: 3–25, 245–257, 317–346. 1881; **7**: 71–80. table. 1882. (Key to groups, genera, and species, with habitats but essentially without localities; incomplete (Ranunculaceae-Umbelliferae).)—See also **Heubert, Camille.** Les plantes de la forêt landaise. 61 p. 18 cm. Mont-de-Marsan,

1940. (Popular work, containing the French names of numerous plants (including some new to the Department) grouped by habitat, without scientific names.)

Léon, Jules. Flore landaise et médecine par les plantes vulgaires. xxiii, 269 p. 19.5 cm. Paris, 1876.

Annotated list of vascular plants, with diagnostic characters; supplementary chapters on useful and medicinal plants.

Local

Ducomet, Vital. La botanique populaire dans l'Albret. Essai linguistique. Bul. Acad. Internatl. Géog. Bot. **12:** 127–132, 280–304, 307–310. 1903.

Annotated alphabetical list of patois names of about 300 species, with explanation of meaning. The former duchy of Albret, according to the author, is now included within the department of Gers, Gironde, Landes, and Lot-et-Garonne.

LANGUEDOC

See also France, Partial (Azais); Provence (Capoduro); Spain, Cataluña (Conill).

The region and former province of Languedoc consists of Ardèche, Aude, Gard, Hérault, Lozère, Tarn, and part of Haute-Garonne and Haute-Loire. Only a single work referring to the region as a whole is placed here. See also Azais, under Partial.

Belleval, Charles de. Nomenclateur botanique languedocien. Annuaire Soc. Agr. Hérault **1840:** 67–113. 1840. (Gray Herbarium library.)

Annotated alphabetical list of vernacular names, with botanical equivalents.

LIMOUSIN

The region and former province of Limousin includes the department of Corrèze and part of Haute-Vienne, but Le Gendre in his Catalogue covers also Creuse and a small part of adjacent Charente and Dordogne.

Le Gendre, Charles. Catalogue des plantes du Limousin. Ouvrage edité par la Sociéte botanique et d'études scientifiques du Limousin. 2 v. (1 p. l., 312 p.; 1 p. l., 410 p.). 20 fig. (maps), port. 25.5 cm. Limousin, 1914–1922. (Library of Muséum national d'histoire naturelle, Paris.) (Originally published as supplement to La revue scientifique du Limousin, 1913–22.)

Geography, geology, etc.; annotated list of vascular plants (1570 native species) with rather detailed localities and notes on uses. Covers the departments of Corrèze, Creuse, and Haute-Vienne, and the arrondissements of Confolens (Charente) and Nontron (Dordogne).—See also his Supplément au Catalogue . . . Ouvrage edité [etc.]. 2 p. l., 96 p. 25.5 cm. Limoges, 1926 [1925–26]. (Contains a general list of additions and corrections (68 p.), and another list of species collected principally at Donzenac (Corrèze) by abbé Laubie. In library of Muséum national d'histoire naturelle, Paris. Originally published in La revue scientifique du Limousin, no. 328–334. 1925–26; this not seen.)

Nussac, Louis de. Botanique. *In* his Essai de bibliographie limousine des sciences naturelles. Revue Scient. Limousin **7:** 344–350. 1906. (University of Michigan library.)

Unannotated list of botanical publications. An explanation of the abbreviations used for names of serials is given on p. 331.

LOIR-ET-CHER

See also Loire-Inférieure (Nehou).

General

Franchet, Adrien. Flore de Loir-et-Cher, comprenant la description, les tableaux synoptiques et la distribution géographique des plantes vasculaires qui croissent spontanément ou qui sont généralement cultivées dans le Perche, la Beauce et la Sologne . . . lxxviii, 15, 792 p. 22 cm. Blois, 1885.

Biographical sketches of local botanists, bibliography, plant distribution glossary; annotated descriptive flora of 1289 vascular plants and Characeae with keys, localities, and extralimital distribution.

Local

Franchet, Adrien. Sur une florule adventice observée dans le département de Loir-et-Cher en 1871 et 1872. Bul. Soc. Bot. France **19**: 195-202. 1872.
List of 199 angiosperms found on the sites of army camps at Vendôme, Blois, and Cheverny, with indication of native habitat and discussion.—See also **Nouel, Ernest.** Les plantes de la guerre. Note sur les plantes étrangères observées aux environs de Vendôme à la suite de la guerre de 1870–1871. Bul. Soc. Archéol. Scient. et Litt. Vendômois **12**: 36–56. 1873. (General features of introduction; tabular lists of 148 phanerogams, showing occurrence in 1871 and 1872.)—See also Paris region (Gaudefroy).

Legué, Léon. Catalogue des plantes vasculaires qui croissent naturellement dans le canton de Mondoubleau. x, 106 p. 18.5 cm. Paris-Auteuil, 1891.
Soil, etc.; annotated list of 786 vascular plants.

Martin, Émile. Catalogue des plantes vasculaires et spontanées des environs de Romorantin. [2.éd.] 1 p. l., xi, 533 p. 25.5 cm. Romorantin, 1894. (1st ed. 1875.)
Botanical explorers, etc. (supplementary to the first edition); annotated list of vascular plants and Characeae (1,162 species), with localities.—See also Segret, below.

Segret, L., *abbé.* Catalogue raisonné des plantes vasculaires de la Sologne. Bul. Soc. Bot. France **72** (Sess. Extraord.): 59–122. 1 fig. (map). 1931.
Topography, geography, soils, etc.; annotated list of vascular plants (1,076 species), with localities for rarer species.—See also **Allorge, Pierre,** and **Gaume, Raymond.** Esquisse phytogéographique de la Sologne. Bul. Soc. Bot. France **72** (Sess. Extraord.): 5–59. 1931. (Geography, topography, geology, soils, human history, botanical explorations with references, plant associations, phytogeography, bibliography.)—Also **Segret, L.** La botanique en Sologne, autrefois et aujourd'hui. l.c. **72**: 763–768. 1926. (Changes in topography and vegetation, partial list of species not recorded by Martin from Romorantin, etc.)

LOIRE

See also Auvergne (Chassagne); Lyonnais; also Rhône (Balbis, Gandoger, Magnin, Merit).

Local

Le Grand, Antoine. Statistique botanique du Forez. Annales Soc. Agr., Indus., Sci., Arts et Belles-Let. Dépt. Loire **17**: 43–221, 298–416. 1873.—Supplément . . . l.c. **20**: 273–317. 1876.
Botanical explorations, topography, general features of flora, climate, soils, etc.; annotated list of vascular and cellular plants (1537 species, including supplement, of which 1259 are phanerogams); short list of vernacular names, local herbaria.—See also **Schultz, F. W.** Observations sur la Statistique botanique du Forez de M. A. Legrand. Bul. Soc. Roy. Bot. Belg. **14**: 128–136. 1875. (Critical notes on various species.)—Also **Hervier, Joseph,** *abbé.* Recherches sur la flore de la Loire. 1. fasc. viii, 60 p. 2 pl. (1 col.). 25 cm. St. Étienne, Paris, 1885. (Brief bibliography; list of rarer vascular plants collected by author, with localities and occasional critical notes; includes 104 species and varieties not previously recorded. No more published. In Gray Herbarium library.)

Roux, Claudius. Étude botanique des monts du Forez (Pierre-sur-Haute, Bois-Noirs et Madeleine). Annales Soc. Bot. Lyon **35**: 139–177. 2 port., map. 1911.
Introduction (p. 139–152) is an alphabetical list of botanists, with biographical notes and titles of pertinent publications, by Roux; followed (p. 153–177) by **Alverny, André d'.** Première partie. Géographie botanique des monts du Forez, discussing geography, geology, climate, plant zones. Apparently the projected Liste raisonnée des plantes . . . was never published.

——— Histoire des sciences naturelles et agricoles en Forez (département de la Loire). Annales Soc. Agr., Sci. et Indus. Lyon **1910**: 277–450. ports. 1910 (1911); **1911**: 126–324. ports. 1912.

Alphabetical list of botanists, zoologists, agronomists, engineers, etc., and writers on related subjects, with biographical information and titles of publications; lists of general periodicals, societies with titles of publications, anonymous publications; chronological list of important scientific events, classified list of subjects, list of localities.

LOIRE-INFÉRIEURE

See also Bretagne.

General

Citerne, Paul. Flore de la Loire-Inférieure. Bul. Soc. Sci. Nat. Ouest France **19** (n.s., **9**): 199–222, 274–484. 1909. (Reprinted, 286 p. Nantes, 1909. Reprint not seen.)

Flora of vascular plants in form of briefly annotated keys, essentially without localities; short bibliography. Replaces **Menier, Charles.** Aperçu de la flore de la Loire-Inférieure. *In* Congrès Assoc. Franç. Avanc. Sci. 27, La ville de Nantes et la Loire-Inférieure. v. 2, p. 400–430. Nantes, 1898. (Study of the littoral, the calcareous and the siliceous floras, with partial lists of species; bibliography. Covers phanerogams only. Not seen; cited from notice in Bul. Soc. Sci. Nat. Ouest France 8 (pt. 2, Extraits et analyses): 43. 1898.)

Lloyd, James. Flore de la Loire-Inférieure. 38, xxxii, 335 p. 14 cm. Nantes, 1844.

Botanical explorations, plant formations, etc.; annotated, briefly descriptive flora of vascular plants, with localities. Replaces **Moisan, C. A.** Flore nantaise, ou tableau analytique d'après la méthode de M. Lamark [sic], des plantes naturelles au département de la Loire-Inférieure et de celles qui sont cultivées le plus généralement . . . iii, 725 p. 21 cm. Nantes, 1839. (Annotated descriptive flora of vascular and cellular plants in form of keys, with notes on uses and localities for rarer species; glossary. In Gray Herbarium library.)

Local

Bureau, Édouard. Excursion botanique du Muséum d'histoire naturelle de Paris aux environs de Nantes et sur les bords de l'océan du 5 au 11 août 1892. Bul. Soc. Sci. Nat. Ouest France **3**: 1–36. pl. 1893.

Account of excursion, with mention of noteworthy plants seen.

Gadeceau, Émile. Le Lac de Grand-Lieu. Monographie phytogéographique. Préface par Ch. Flahault. 2 p. l., v, 155 p. 4 fig., 21 pl., map. 25 cm. Nantes, 1909. (Gray Herbarium library.)

Mainly phytogeographic and ecological; includes (p. 46–94) annotated list of vascular plants, grouped according to habitat, and alphabetical list of vernacular names, with explanation of meaning.

——— Liste des plantes observées à l'île Dumet, près Piriac (Loire-Inférieure), le 3 août 1880. Bul. Soc. Sci. Nat. Ouest France **2**: 226–229. 1892.

Botanical explorations, topography, etc.; unannotated list of vascular plants.

——— Promenades botaniques au canal maritime de la Basse-Loire. Bul. Soc. Sci. Nat. Ouest France **2**: 11–20. map. 1892.—Première liste additionelle à la florule du canal maritime . . . l.c. **5**: 27–33. 1895.

Special localities and species, etc.; mostly unannotated lists of 237 species and varieties of angiosperms; the 1895 list includes those listed in 1892. Covers the new lands, roads, ditches, etc., created along the canal running from Martinière to Carnet, parallel to the Loire River.

Nehou, Jean. Les plantes adventices du Val de Loire. Bul. Soc. Scient. Bretagne (Sci. Math., Phys. et Nat.) **20**: 81–99. 4 fig. 1946.

Topography, geology, climate; annotated list of 76 vascular plants introduced from more or less distant sources since the 17th century and capable of maintaining themselves; discussion of degree of naturalization, habitats, etc.; bibliography.—See also his Quelques phanérogames adventices nouvelles dans la région nantaise. Bul. Soc. Bot. France **95**: 193–195. 1948. (Annotated list of additional species, etc.)

St. Gal, Joseph. Flore des environs de Grand-Jouan contenant la description des végétaux vasculaires qui poussent spontanément dans un rayon de 12 à 16 kilomètres autour de l'École d'agriculture de Grand-Jouan et celle des végétaux le plus ordinairement cultivés par l'agriculteur, le forestier et le maraîcher. xlvi, 521 p. 15 cm. Nantes, 1874.—Supplément à la Flore . . . 29 p. 18mo. Nantes, 1885. (Not seen.)

Briefly annotated, descriptive flora of vascular plants and Characeae, mainly in form of keys, with vernacular names, and localities for scarcer species.—See also his Liste des plantes qui croissent spontanément dans le département de la Loire-Inférieure et qui ne sont pas décrites dans la Flore des environs de Grand-Jouan ni dans le Supplément publié en 1885. 48 p. 15 cm. Nantes, 1885. (List of vascular plants and Characeae, with habitats and localities.)

LOIRET

See also Loir-et-Cher (Segret).

General

Jullien-Crosnier, Alexandre. Catalogue des plantes vasculaires du département du Loiret. Bul. Soc. Hort. Orléans et Loiret **14** (III, 2): 188–324. 2 pl. 1889.—[1.]–2. supplément au Catalogue . . . l.c. **17** (III, 5): 153–237. 1898; **19** (III, 7): 477–509. 1905.

Bibliography; annotated list of vascular plants and Characeae (1,720 species). In vol. 14 and 17, the volume numbers printed on title page and cover disagree; those given above are correct.—See also **Benoist, J.,** *chanoine.* Liste des plantes, rares ou assez rares pour le Loiret trouvées en 1912, au cours de ses promenades botaniques. Monde des Plant. II, **15** (81): 61–62. 1913. (List of vascular plants, with localities. In Arnold Arboretum library.)

Local

Benoist, J., *chanoine.* Essai de florule de Pithiviers (Loiret) et des environs. Bul. Acad. Internatl. Géog. Bot. **20** (2): 105–136. 1910.

List of vascular plants and Characeae, with localities and indication of abundance.

Gaume, Raymond. La flore de la forêt d'Orléans aux environs de Lorris (Loiret), excursions des 14, 15 et 16 août 1926. Bul. Assoc. Nat. Vallée du Loing **9**: 101–115. 1926. (Library of Muséum national d'histoire naturelle, Paris.)

Lists of species of different associations; bibliography.

Jullien-Crosnier, Alexandre. Catalogue systématique de quelques plantes nouvelles pour la flore orléanaise. Mém. Soc. Acad. Maine et Loire **12**: 61–80. 1862.

List of 92 vascular plants, with localities, intended to supplement the treatment of the plants of Orléans in Boreau, Flore du centre de la France. éd. 3. 1857.—See also **Humnicki, Valentin.** Catalogue des plantes et des localités nouvelles des environs d'Orléans. 30 p. 21 cm. Orléans, 1876. (List of new records and new localities of vascular plants. In Massachusetts Horticultural Society library.)

Nouel, Ernest. Notice . . . sur un certain nombre de plantes adventices qui ont été recueillies à Orléans, dans l'année 1871. Mém. Soc. Agr. Sci., Belles-Let. et Arts Orléans IV, **14** (v. 45): 35–38. 1872.—Deuxième note sur les plantes adventices importées aux environs d'Orléans par les fourrages de la guerre. l.c. **15** (v. 46): 109–114. 1873.—Cinquième notice sur quelques plantes du département du Loiret. l.c. **16** (v. 47): 5–19. 1874.

The first paper contains a list of 90 angiosperms introduced in the forage supplies of French troops; the second gives a list of 124 such species (including those of first list), showing occurrence in 1871–72; the third contribution (Plantes importées par les fourrages de la guerre, p. 18–19) brings total to 163.—See also Paris region (Gaudefroy).

LORRAINE (LOTHRINGEN)

See also Alsace, Alsace-Lorraine, and the separate departments making up Lorraine. The region and former province of Lorraine consists of the departments of Meurthe-et-Moselle, Meuse, Moselle, and Vosges. The present Department of Moselle, with somewhat smaller territory than it had before 1871, was included in Germany from 1871 to 1919 and constituted German Lorraine or Lothringen. Only works relating to the region as a whole are cited here.

Godfrin, Julien, and **Petitmengin, Marcel.** Flore analytique de poche de la Lorraine et des contrées limitrophes. 1 p.l., viii, 239 p. 18 cm. Paris, 1909.—Atlas de la Flore analytique . . . [par J. Godfrin]. viii, 229 p. incl. 1608 fig. 21 cm. Paris, 1913.

Pocket flora of vascular plants in form of briefly annotated keys, with localities for the rarer species; the atlas gives a figure of each species (mostly taken from Coste's Flore), and a list of synonyms, with equivalents. Includes nearly all the plants of the Vosges and Ardennes Mountains, thus covering parts of Alsace, Ardennes, and Haute-Saône as well as Lorraine.

Godron, D. A. Flore de Lorraine. 3. éd. publiée par MM. Fliche [et] G. Le Monnier. 2 v. (608, 506 p.). 19 cm. Nancy, 1883. (1st ed. 1843–44.)

Descriptive flora of vascular plants, with localities. Keys to families, genera, and species collected at end of vol. 2, otherwise little changed from ed. 2. Covers the old Lorraine, that is, the present Departments of Meurthe-et-Moselle, Meuse, Moselle, and Vosges. The 3. éd. incorporates his Notice sur les explorations botaniques faites en Lorraine de 1857 à 1875 et de leurs résultats. Mém. Acad. Stanislas IV, 7: 126–250. 1875. (Intended to bring ed. 2 of 1857 up to date.)—See also **Fliche, Paul, and Le Monnier, G.,** ***comps.*** Tables dichotomiques de la Flore de Lorraine par D.-A. Godron. 3. éd. 167 p. 18.5 cm. Nancy, 1883. (Consists of the keys to families, genera, and species taken from ed. 3 of Godron's Flore de Lorraine.)

Petitmengin, Marcel. Mise en point sur la flore lorraine. Compt. Rend. Assoc. Franç. Avanc. Sci. **36** (Reims, 1907), Notes et Mém.: 504–519. 1908.

Account of principal discoveries (mostly since 1881), lists of additional species, doubtful species, species to be excluded, etc., grouped by regions; bibliography. Refers to Meurthe-et-Moselle, Meuse, Moselle, Vosges, and adjacent areas including the Vosges Mountains.

Willemet, P. R. Phytographie économique de la Lorraine, ou recherches botaniques sur les plantes utiles dans les arts. 2 p.l., 142 p. pl. 21.5 cm. Nancy, 1780.

Briefly annotated lists of edible, fodder, dye, cosmetic, tannic, ornamental, timber, insecticidal, hygrometric, and otherwise useful plants (but not medicinal plants), grouped by uses.

——— Phytographie encyclopédique, ou flore de l'ancienne Lorraine et des départemens circonvoisins. 3 v. (x, 1394, 94 p.). 21.5 cm. Nancy, 1805. (Lloyd Library; only v. 3 seen.)

Briefly descriptive flora, on the Linnaean system, with vernacular names and mention of uses.

LOT

See also Auvergne (Chassagne).

General

Puel, Timothée. Catalogue des plantes qui croissent dans le département du Lot, classées d'après le système de Linnée, suivi d'une table analytique pour la détermination des genres et des espèces. 248 p. 21 cm. [Cahors, 1845–53.]. (Arnold Arboretum Library.)

Extensively annotated bibliography, list of herbaria; annotated list of vascular plants (1461 species), with localities. P. 237–248 constitute a Supplément; there is no analytical table in the copy seen. Originally published in Annuaire statistique du Lot (not seen) for the years 1845–1852 (Puel himself (p. 237) says published 1845–1853); for full citation of the livraisons see Lucante, Bul. Soc. Études Lit., Scient. et Artist. Lot **11**: 92. 1886.—See also **Personnat, Victor.** Espèces du Lot à ajouter au Catalogue de M. T. Puel (Annuaire du département). Bul. Soc. Bot. France **5**: 595–596. 1858. (List with localities.)—Also his Observations sur quelques plantes du département du Lot. l.c. 7: 22–24. 1860. (Includes several additions.)—Also **Puel, T.** Revue critique de la flore du département du Lot. l.c. 7: 373–375. 1860; 8: 203–210, 291–293, 300–302, 331–333, 445–447, 467–471, 538–541, 584–587, 630–636. 1860–62; 9: 399–400. 1862. (Critical notes on 20 species in v. 7–8; the last part contains a second list of 20 species, annotations on which were never published.)—Also **Malinvaud, Ernest.** Note sur quelques plantes nouvelles ou douteuses pour la flore du département du Lot. Bul. Soc. Linn. Normandie II, 3: 19–25. 1869. (List of additional species and new localities, corrections, etc., mostly from vicinity of Gramat.)—Also his Plantes observées aux environs de Gramat et de La Capelle-Marival (Lot). Bul. Soc. Bot. France 19: 237–239. 1872. (List of newly recorded or rare species for the department.)—Also his Note sur une excursion botanique dans les départements du Lot et de l'Aveyron. Bul. Soc. Linn. Normandie II, 6: 194–202. 1873. (List of noteworthy species from vicinity of Gramat and La Capelle-Marival.)—Also his Herborisations en 1887–88–89 dans le département du Lot, un Alyssum et un Orchis hybride nouveaux pour la flore française. Bul. Soc. Bot. France 36 (Congrès de botanique tenu à Paris . . . 1889): ccxlvi–cclviii. 1890. (Short account of general features of soils, etc., of various localities in the arrondissements of Figeac and Gourdon, with mention of scarcer vascular plants of each.)—Also his Plantes nouvelles pour les départements du Lot et de la Corrèze. l.c. 39: 321–322. 1892. (Short list.)—Also his Traits généraux de la flore du Lot. Faits remarquables de géographie botanique récemment observés dans ce département. Bul. Acad. Internatl. Géog. Bot. 12: 558–562. 1903. (Includes partial list of additions to the flora of the department. Also published in Compt. Rend. Cong. Soc. Sav. Paris et Dépt. 1902 (Sect. Sci.): 135–138. 1903.)—Also his Florulae oltensis additamenta ou nouvelles annotations à la flore du département du Lot. I–IX. Bul. Soc. Bot. France 52: 371–374. 1905; 53: 641–646. 1906; 54: 499–504. 1 fig. 1907; 54: 649–654. 1 fig. 1908; 55: 477–484. 1 fig. 1908; 55: 721–726. 1 fig. 1909; 56: 370–381. 1909; 56: 603–605. 1910; 57: 435–442. 1910; 58: 512–518. 1911; 59: 689–693. 1913. (List of 74 species, with critical notes.)—Also his Revue critique des Crassulacées de la flore du Lot. Compt. Rend. Assoc. Franç. Avanc. Sci. 34 (2, Notes et Mém.): 430–439. 1906. (Bibliography; annotated list.)—Also his Ranunculacées rares ou critiques de la flore du Lot. Compt. Rend. Soc. Sav. Paris et Dépt. 1907 (Sect. Sci.): 145–148. 1908. (Annotated list.)—Also his Crucifères nouvelles pour la flore du Lot. l. c. 1908 (Sect. Sci.): 161–165. 1909. (Annotated list, with localities.)—Also **Lamothe, Camille.** Contributions à la flore du Lot. Herborisations dans la vallée de la Cère. Bul. Soc. Bot. France 56: 246–250. 1909. (Mention of numerous species including additions.)

Local

Galinat, Maurice. Le Quercy, point de rencontre de diverses flores. Bul. Soc. Bot. France 95: 354–363. 1949.

Phytogeographical; cited here for its bibliography (p. 363) of the flora of the Department of Lot.

Giraudias, Ludovic. Énumération des plantes phanérogames et des fougères observées dans le canton de Limogne (Lot). Bul. Soc. Études Scient. Angers 4/5: 152–183. 1876.

Topography, etc.; briefly annotated list of 839 vascular plants, with localities.

Lamothe, Camille. Plantes de la vallée de la Dordogne dans la partie appartenant au département du Lot. Compt. Rend. Cong. Soc. Sav. Paris et Dépt. 1906 (Sect. Sci.): 261–287. 1906.

Topography, geology, plant zones; list of rare or newly recorded vascular plants; unannotated list of more or less interesting plants, the ubiquitous ones omitted; annotated list of introduced, adventive, and naturalized plants.

Lucante, Angel. Catalogue géographique des espèces contenues dans l'herbier de l'abbé J.-P. Bousquet. Bul. Soc. Études Lit., Scient. et Artist. Lot 11: 89–96. 1886; 12: 141–150. 1887; 15: 193–201, 229–237. 1890; 16: 48–58, 99–111, 163–176, 202–211. 1891.

Biographical account of Jean Pierre Bousquet (1844–1879), bibliography; list of wild and cultivated vascular plants (over 1380 species) from the arrondissement of Cahors, with localities. The installments in v. 15–16 were edited by H. Leboeuf after the death of Lucante.

Soulié, A. Catalogue des plantes recueillies à Puycalvel et aux environs et particulièrement dans le canton de Saint-Germain. Bul. Soc. Études Lit., Scient. et Artist. Lot **8**: 208–232, 281–296. 1883; **9**: 18–32, 89–104, 193–204. 1884.

Topography, soils, botanical explorations; annotated list of wild and cultivated spermatophytes (about 900 species, mostly from canton of St.-Germain).

LOT-ET-GARONNE

General

Debeaux, J. O. Revision de la Flore agenaise, suivie de la flore du Lot-et-Garonne. 645 p. port. 24.5 cm. Paris, Agen, 1898. (Arnold Arboretum library.)

Biographical sketches of local botanists, bibliography, additions and corrections to the "Flore agenaise" of Boudon de Saint-Amans (1821) (this portion reprinted from Rev. Bot. Bul. Mens. Soc. Franç. Bot. **13**: 1–340. 1895); annotated list of vascular plants and Characeae (1634 species and 292 varieties) of Lot-et-Garonne, with localities; alphabetical list of Gascon names of commoner wild and cultivated plants, statistics, geology and soils, with lists of characteristic species.—See also **Duffour, Charles,** and **Jeanjean, A. F.** Plantes nouvelles pour le Lot-et-Garonne. Bul. Soc. Bot. Deux-Sèvres **26**: 37–43. 1915. (Remarks on Debeaux's Revision, with unannotated list of additions. In Farlow Library, Harvard University.)

Local

Combes, J. L. Fumel et ses environs. Haut-Agenais. Recherches géologiques et paléontologiques, météorologiques et botaniques. 56 p. 22 cm. Agen, 1855. (New York Botanical Garden library.)

Includes (p. 37–56) annotated alphabetical list of vascular plants and some fungi.

Queyron, Philippe. Catalogue des plantes vasculaires du bassin du Drot (départements de la Dordogne, de Lot-et-Garonne et de la Gironde). Compt. Rend. Cong. Soc. Sav. Paris et Dépt. **1906** (Sect. Sci.): 170–242. map. 1906.

Geography, geology, botanical explorations; annotated list of vascular plants and Characeae, with localities for rarer species. Relates to area from just west of La Réole to Monpazier beyond Villaréal.

LOZÈRE

See also Auvergne (Chassagne); Gard (Braun-Blanquet); Languedoc.

Local

Chauvet, —, and **Broyer, Charles.** La végétation du Mont Lozère. Bul. Soc. Bot. France **86**: 52–59. 1939.

Zones of vegetation, with partial lists of species.—See also **Lenoble, Félix.** Trois semaines d'herborisations en Lozère. Bul. Soc. Bot. France **87**: 16–22. 1940 [?]. (Geology, botanical investigations, account of excursions with mention of various species.)

LYONNAIS

See also France, Partial (Gandoger, Flore lyonnaise).

The region and former province of Lyonnais consists of the departments of Loire and Rhône.

Magnin, Antoine. Observations sur la flore du Lyonnais. Annales Soc. Bot. Lyon **8**: 261–308. 1881; **9**: 201–256. 1882; **10**: 115–168. 1883; **11**: 133–226. 1884; **12**: 25–300. 7 maps. 1885.

Short sketch of botanical explorations, topography and vegetation of different parts of the region with mention of characteristic species, similar account of phytogeographic areas, influence of climate and soil on vegetation with lists of species, plant zones, etc.; annotated lists of vascular plants of siliceous and of calcareous and mixed soils (**12:** 74–170), with discussion of origin of soils, etc.; origin of flora; classified list of introduced species, etc.

——— Prodrome d'une histoire des botanistes lyonnais. Annales Soc. Bot. Lyon **31:** 1–72. 1906; **32:** 1–68. 1907.—Additions et errata . . . **31:** 1 bis-4 bis. 1906.—Additions et corrections . . . **32:** 103–141. 1907; Additions . . . 2e série. **35:** 13–80. 1911.—Rectification . . . **37:** 205. 1912.

Bibliography; list of botanists, horticulturists, etc., grouped more or less chronologically, with biographical notes, account of work and publications, and biographical references, covering also botanical organizations and institutions; classified list of botanical publications (**35:** 57–65), list of botanical artists, list of streets in Lyon dedicated to botanists, list of houses and localities connected with botanists, plants dedicated to Lyonnais botanists, list of portraits, list of families containing several botanists; indices (**32:** 65–68, 140–141; **35:** 78–80).

MAINE-ET-LOIRE

See also Loire-Inférieure (Nehou).

General

Boreau, Alexandre. Catalogue raisonné des plantes phanérogames qui croissent naturellement dans le département de Maine et Loire . . . Mém. Soc. Acad. Maine et Loire **6:** 5–216. 1859. (Reprinted, 216 p. Angers, 1859.)

Botanical explorations (with bibliography); annotated list of vascular plants and Characeae [1810 species], with localities; list of excluded species. Replaces **Guépin, J. P.** Flore de Maine et Loire. 3. éd., revue avec soin et considérablement augmentée. 2 p.l., xcix, 440 p. 18 cm. Angers, Paris, 1845. (1st ed. 1830); Supplément . . . 3 p. l., 51 p. 19 cm. Angers, 1850. (Arnold Arboretum library); Notice sur une flore angevine manuscrite suivie d'un supplément à la Flore de Maine et Loire. Annales Soc. Linn. Dépt. Maine-et-Loire **1:** 53–82. 1854; Additions à la Flore . . . l.c. **2:** 129–142. 1857. (Soils, botanical investigations, etc.; descriptive flora of spermatophytes (1,363 species, including some cultivated ones), with list of cryptogams.—See also **Bouvet, Georges.** Plantes rares ou nouvelles pour de département de Maine-et-Loire. Bul. Soc. Études Scient. Angers **3:** 93–137. 1874. (List of publications since Boreau's Catalogue; extensive list of additional species and localities of vascular plants, Characeae, and mosses.)—Also his Additions à la flore de Maine-et-Loire (observations de l'année 1874). Bul. Soc. Bot. France **22** (Sess. Extraord.): lviii–lxiii. 1875. (Annotated list, with localities.)

Bouvet, Georges. Catalogue raisonné des plantes utiles et nuisibles de la flore de Maine-et-Loire. Bul. Soc. Études Scient. Angers **14:** 137–364. illus. 1885.

Annotated systematic list of useful and harmful vascular and cellular plants, wild and cultivated, with notes on uses; list of species classified by uses.

Corillion, Robert. Abrégé d'histoire de la botanique en Anjou. Bul. Soc. Bot. France **104** (Sess. Extraord.): 38–42. 1957.

Sketch of botanical explorations.

Local

Hy, F. C., *abbé.* Tableaux analytiques de la flore d'Angers. Mém. Soc. Natl. Agr., Sci. et Arts Angers, nouv. pér. (i.e., III), **25:** 193–274. 1884; **26:** 35–128. 1885. (Reprinted with index, 184 p. Angers, 1884 [–85]. In Gray Herbarium library.)

Flora of phanerogams in form of keys and brief diagnostic phrases, scarcely annotated except as to abundance, and without localities.

Préaubert, Ernest. Relevé (Résultats) d'herborisations en Anjou [années] 1901–1902 [–1930–1931]. Bul. Soc. Études Scient. Angers **31:** 157–173. 1902; **35:** 63–79. 1906; **38:** 33–58. 1909; **40:** 15–30. 1911; **42/43:** 1–9. 1914; **49:** 1–14. 1920; **51:** 29–43. 1922; **53:** 51–58. 1924; **55:** 79–92. 1926; **57:** 37–59. 1928; **59:** i–xvii. 1930; **61:** 47–58. 1932.

List of new records of vascular plants, with localities and names of collectors. Title varies; the first reads Relevé d'observations botaniques intéressant la flore vasculaire de l'Anjou.

MANCHE

See also Normandie.

General

Besnou, Léon. La flore de la Manche. Catalogue raisonné des plantes vasculaires et cellulo-vasculaires du département de la Manche. 2 p. l., 380 p. 21 cm. Coutances, 1881. (Gray Herbarium library.)

Annotated list of vascular plants and Characeae, with localities for scarcer species and brief notes on uses.—See also **Bois, Desiré.** Herborisations dans le département de la Manche. Jour. de Bot. 1: 324–330. 1887. (Short bibliography, remarks on cultivated trees, etc.; extensive list of vascular plants, with localities.)—Also **Frémy, P.,** ***abbé.*** Localités nouvelles de plantes rares ou peu communes observées dans le département de la Manche. Bul. Soc. Linn. Normandie VII, **6:** 115–118. 1924. (List with localities.)—Also **Potier de la Varde, Robert.** Contribution à la flore du département de la Manche. l.c. 8 (Proc.-Verb.): 18–20. 1926. (List of 5 noteworthy species.) —Also **Meslin, Roger.** Notes sur la flore du département de la Manche. Not., Mém. et Doc. Soc. Agr., Archéol. et Hist. Nat. Dépt. Manche **39:** 199–224. pl. 1927. (Botanical explorations; annotated list of scarcer vascular plants, mostly from vicinity of Coutainville and Saint-Lô; bibliography.)

Local

Besnou, Léon, and **Lachênée, Bertrand.** Catalogue raisonné des plantes vasculaires de l'arrondissement de Cherbourg. 257 p. 22 cm. Cherbourg, 1862. (Gray Herbarium library.) (Reprinted from Cong. Scient. Franç. Avanc. Sci. 27. sess., v. 2. Original not seen.)

Annotated list of vascular plants, with localities. Replaces **Le Jolis, A. F.** Plantes vasculaires des environs de Cherbourg. Mém. Soc. Imp. Sci. Nat. et Math. Cherbourg **7:** 245–360. 1860. (Reprinted, 120 p. 22 cm. Paris, Cherbourg, 1860. Reprint in Arnold Arboretum library.) (Climate, soils, etc.; annotated list of vascular plants.)—See also **Corbière, Louis.** Herborisations aux environs de Cherbourg. Bul. Soc. Linn. Normandie III, **8:** 358–373. 1884. (Includes (p. 363–368) list of new species and new localities for the region.) —Also his Nouvelles herborisations aux environs de Cherbourg et dans le nord du département de la Manche. l.c. IV, **1:** 97–124. 1888. (Notes on various critical species; list of new records for la Manche, etc.)—Also his Sur l'apparition de quelques plantes étrangères à Cherbourg et à Fècamp. l.c. IV, **1:** 321–330. 1888. (Notes on principal localities; list of noteworthy adventive plants, with localities.)—Also his La flore. *In* Cherbourg et le Cotentin (publié à l'occasion du Congrès de l'AFAS à Cherbourg 1905). p. 93–112. Cherbourg, 1905. (Partial list of phanerogams and cellular plants, arranged by habitats; bibliography. In library of Muséum national d'histoire naturelle, Paris.)

Crié, Louis. Essai sur la végétation de l'archipel Chausey (Manche). Bul. Soc. Linn. Normandie II, **10:** 295–334. 1876.

Geology, climate, general features of the three florules of the area with mention of many species, tabular list of vascular and cellular plants not already mentioned, with localities in the archipel Chausey and in the Channel Islands.—See also **Corbière, Louis.** Compte-rendu des excursions botaniques faites par la Société linnéenne de Normandie aux environs de Granville et aux îles Chaussey les 5, 6 et 7 juin 1891. Bul. Soc. Linn. Normandie IV, **5:** 184–195. 1892. (Itinerary; mention of various noteworthy species, including (p. 190–191) list of nearly 50 vascular plants not recorded by Crié.)—Also **Géneau de Lamarlière, L.** Troisième note sur la flore maritime des côtes de la Manche. Bul. Soc. Bot. France **42:** 39–45. 1895. (Includes (p. 42–44) description of Maîtresse-île, with mention of species of different habitats.)

MARNE

See also Champagne.

General

Brisson, T. P. Catalogue des plantes phanérogames du département de la Marne contenant la description des grandes divisions et celle des familles sous la forme analytique. Plantes médicinales et vénéneuses. 1 p. l., 192 p. 23 cm. Chalons-sur-Marne, 1884. (Gray Herbarium library.)—Supplément au Catalogue . . . 16 p. 23 cm. Chalons, [1885.] (Gray Herbarium library.)

Partial bibliography, key to families, etc.; briefly annotated list of spermatophytes, including cultivated species, with localities; section on medicinal and poisonous plants; classified list of useful and injurious plants, with brief annotations. Pp. 161–192, apparently published in 1884, have title Plantes phanérogames de la Marne. Table méthodique des plantes utiles et nuisibles. Replaces **Lambertye, Léonce,** *comte de.* Catalogue raisonné des plantes vasculaires qui croissent spontanément dans le département de la Marne. xxiv, 207 p. map. 22 cm. Paris, 1846. (Botanical explorations, geology, topography, general features of vegetation, etc.; annotated list of vascular plants and Characeae, with localities.)—See also **Devauversin, A.** Nouvelles localités de plantes rares du département de la Marne. Mém. Soc. Agr., Com., Sci. et Arts Dépt. Marne **1896:** 129–164. 1897. (Extensive list of vascular plants, with localities.)—Also **Géneau de Lamarlière, L.** Sur la flore adventice du département de la Marne. Feuille des Jeunes Nat. **29:** 59–62, 79–83. 1899. (Extensive list of adventive spermatophytes, with localities and names of collectors.)—Also his Contributions à la flore de la Marne. ([1.] –4. note.) Bul. Soc. Bot. France **46:** 272–279. 1900; **47:** 415–422. [1901]; **48:** 39–44 1901; **49:** 345–352. 1903. (Lists of noteworthy vascular plants, with localities and annotations.)—Also **Devauversin, A.** Plantes adventices du département de la Marne. Feuille des Jeunes Nat. **31:** 185–188. 1901. (Extensive list, with localities.)—Also his Contributions à la flore du département de la Marne. Mém. Soc. Agr., Com., Sci. et Arts Dépt. Marne II, **9** (1905–06): 217–311. 1907. (Bibliography for 1896–1906; extensive list of vascular plants and Characeae, with localities.)—Also **Maury, Louis.** Contributions à la flore du département de la Marne. Bul. Soc. Étude Sci. Nat. Reims **19:** 37–87. 1911. (Extensive list of noteworthy vascular and cellular plants, with localities.)—Also **Bourgeois, A.** Documents pour la flore du département de la Marne et spécialement de la région de l'Argonne. l. c. **20:** 19–50. 1912. (Extensive list of noteworthy vascular plants, with localities; also lists of aquatic plants of various habitats.)

Moule, L. Des plantes vénéneuses ou suspectes du département de la Marne. 162 p. 22.5 cm. Châlons-sur-Marne, 1877. (Lloyd Library.)

Account of poisonous plants, arranged by families, with vernacular names, properties, chemistry, effect on man and animals, etc.

Local

Bazot, L. M. Plantes vasculaires de l'arrondissement de Vitry-le-François d'après les herborisations de MM. Thiébaut, Richon, Guillot, L. Bazot. Catalogue avec notes de géographie botanique. [Mém.] Soc. Sci. et Arts Vitry-le-François **17:** 61–329. 1893. (Reprinted, 269 p. 22 cm. Vitry-le-François, 1893. In Arnold Arboretum library.)

Geology, topography, climate, etc.; annotated list of vascular plants (1095 species), with habitat, localities, and in most cases the extreme figures of latitude and longitude for the world-wide range; includes the generally cultivated species. The volume is numbered 16 and dated 1892 on the title-page, but (correctly) 17 and 1893 on cover.—See also **Laurent, Jules.** L'herbier Guillot. Bul. Soc. Étude Sci. Nat. Reims **20:** 3–18. 1912. (Extensive list of noteworthy angiosperms, with localities, from Bassu and other localities in arrondissement of Vitry.)

Plonquet, J. L. Essai sur la topographie médicale du canton d'Ay. 2 v. (126, 119 p.). 8vo. Paris, 1855–56.

Includes (v. 1, p. 27–126), Règne végétal, an annotated alphabetical list of wild and cultivated plants with medicinal characters, habitat, and some localities; (v. 2) 13 tables showing the noteworthy alimentary, medicinal, and poisonous plants of the vicinity of Ay, arranged by natural families. (Not seen; title and annotation supplied by G. Dillemann.)

MAYENNE

See also Sarthe (Deportes).

General

Labbé, E., and **Gentil, Ambroise.** Les plantes médicinales de la région Mayenne-Sarthe. 16 p. 24 cm. Laval, [1921]. (Comité interministériel des plantes médicinales et à essences. Sous-comité Mayenne-Sarthe.)

Soils, etc.; annotated lists of common wild and cultivated medicinal plants, alphabetically arranged by vernacular names, with habitat and part used.

Léveillé, Hector. Catalogue définitif de la flore de la Mayenne. Bul. Géog. Bot. **27**: 8–32, 125–160. 1917.

Collaborators; annotated list of vascular plants, with localities of the rarer species.

——— Petite flore de la Mayenne contenant l'analyse et la description sommaire des plantes vasculaires de la Mayenne. 252 p. 18.5 cm. Laval, 1895.

Bibliography; annotated, keyed, briefly descriptive flora of vascular plants and Characeae, with localities for rarer species. The general distribution in the 6 adjoining departments is also given.—See also his Supplément à la flore . . . Bul. Soc. Agr., Sci. et Arts Sarthe **35** (II, 27): 450–497. 1896; **36** (II, 28): 187–231, 277–318 . 1897–98; **37** (II, 29): 15–50. 1899. (Reprinted, 173 p. Le Mans, 1897 (i.e. 1899). In library of Muséum national d'histoire naturelle, Paris.)—Also his Contributions à la flore de la Mayenne. Bul. Acad. Internatl. Géog. Bot. **8**: 5–12, 43–47, 74–77, 109–110, 120–122, 172–180, 225–229, 264–266. 1899; **9**: 103–109, 142–148. 1900. (Reprinted with title Deuxième supplément à la flore . . . n.p., n.d. In library of Muséum national d'histoire naturelle, Paris.)—Also his Contributions . . . [3. supplément]. l. c. **11**: 92–96, 353–354. 1902; **12**: 7–8, 263–271. 1903; **13**: 156–164, 282–284. 1904.—4.–6. supplement. l. c. **13**: 284–288. 1904; **17** (212): i–xi. 1907; **18**: 495–496, 557–560. 1908; **19**: 25–32, 133–138. 1909.—Also **Chénu, E.** Notes sur la flore de la Mayenne. Revue Bretonne Bot. [1:] 94–96. 1906. (Mention of botanists who have worked on the flora since 1895 and of the species they have added.)

Reverchon, P. Catalogue raisonné des plantes vasculaires du département de la Mayenne. Bul. Soc. Études Sci. Angers **20**: 139–237. 1891; **21**: 105–224. 1892. (Reprinted, 228 p., with title Flore de la Mayenne. Catalogue raisonné . . . Angers, 1892. In Gray Herbarium library.)

Botanical explorations; annotated list of vascular plants and Characeae, with localities.

Local

Barre, J. Géographie botanique du département de la Mayenne. Bul. Mayenne-Sci. **1939**: 32–53. 1939 (?).

Account of 28 communes of the department, with geological data, references to previous catalogs, lists (of rare plants?), etc. (Not seen; cited from notice in Bul. Soc. Sci. Nat. Ouest France V, **9**: 52. 1947.)

MEURTHE-ET-MOSELLE

See also Lorraine; Moselle (Holandre, Nouvelle flore de la Moselle, for records from Briey). Up to 1871 the old Department of Meurthe included the arrondissements of Château-Salins, Sarrebourg, Lunéville, Nancy, and Toul. The two first became German territory in 1871 but are now (since 1927) in Moselle; Lunéville, Nancy, and Toul, with the arrondissement of Briey (detached from the old Moselle) now constitute the Department of Meurthe-et-Moselle.

General

Suard, —. Plantes vasculaires. *In* Statistique du département de la Meurthe. Histoire naturelle. 2. partie. Règne végétal. p. 140–183. 22.5 cm. [Nancy, 1845.] (Excerpt in Arnold Arboretum library.)

Geology, etc.; briefly annotated list of vascular plants, with localities for scarcer species and bibliography. Pritzel (Thesaurus Lit. Bot. 2. ed. 308) cited this work only in the form of a preprint (46 p. 1843), with title Catalogue des plantes vasculaires du dép. de la Meurthe.

Local

Görz, Rudolf. Floristische beobachtungen aus der Lothringer kampfzone. Verhandl. Bot. Ver. Brandenberg **61:** 33–39. 1920.

Topography, cultivation, etc.; running account of about 200 mostly weedy plants observed in an area north of Nomény.

Husson, Camille. Botanique. *In* his Esquisse géologique de l'arrondissement de Toul, suivie d'un aperçu botanique des environs de cette ville. p. 95–105. 8vo. Toul, 1848.

List of scarcer plants, with localities. (Not seen; title and annotation supplied by G. Dillemann.) The three supplements contain no botany, fide Dillemann.

Soyer-Willemet, H. F. Observations sur quelques plantes de France, suivies du catalogue des plantes vasculaires des environs de Nancy. 195 p. 23 cm. Nancy, 1828.

The Catalogue (p. 135–191) is an annotated list of vascular plants, preceded by notes on topography and geology.—See also **Vuillemin, Paul.** Notice sur la flore des environs de Nancy. 33 p. Nancy, 1886. (Phytogeographical and ecological study, with mention of many species. Not seen; cited from review in Bul. Soc. Bot. France **33** (Rev. Bibl.): 183–184. 1886.)—Also **Perroud, Louis.** Aperçu sur la flore des environs de Nancy et de la chaîne des Vosges. Annales Soc. Bot. Lyon **14:** 161–200. 1887. (Account of local ecological florules, with mention of characteristic species, etc.)—Also **Franquet, Robert.** Les environs de Nancy. Bul. Soc. Bot. France **102** (Sess. Extraord.): 11–16. 1955. (Running account of localities visited and noteworthy plants seen; bibliography.)

MEUSE

See also Lorraine.

General

Breton, C. Flore de la Meuse. Plantes vasculaires. Tableaux dichotomiques des familles, des genres et des espèces. 2 p. l., 362 p. 22 cm. Montmédy, 1901 [1890–1901]. (Lloyd Library.)

Unannotated flora in form of keys. Pages 336–359 are a Supplement, containing additions and corrections. On p. 336, the author states that the first fascicle appeared in 1890.

Doisy, Charles. Essai sur l'histoire naturelle du département de la Meuse. Première partie. Flore. xlix, 1108 p. table. 15 cm. Verdun, 1835.

Annotated descriptive flora of vascular plants and Characeae on the Linnaean system, with localities for rarer species and notes on uses; calendar of flowers.—See also **Marjean-Denis,—.** Additions et rectifications à la Flore de la Meuse. Mém. Soc. Philom. Verdun **1:** 205–215. 1840. (List with localities.)—Also **Pierrot, Philogène.** Additions à la Flore de la Meuse. l. c. **8:** 293–327. 1877. (General features of flora; additions to Doisy.)

Local

Pierrot, Philogène. Notice sur les plantes vénéneuses, dangereuses, ou suspectes de l'arrondissement de Montmédy. 2. tirage. 2 p. l., 103 p. 17.5 cm. Montmédy [pref. 1868]. (Gray Herbarium library.) (1st issue, 1868.)

General considerations; descriptive account of 87 poisonous phanerogams.

——— **Cardot, Jules,** and **Vuillaume, A.** Catalogue des plantes vasculaires de l'arrondissement de Montmédy, avec indication de leurs stations, propriétés et usages divers. Bul. Soc. Nat. et Archéol. Nord de la Meuse **3:** 46–80. 1891; **4:** 17–49. 1892; **5:** 48–96. 1893; **6:** 39–87. 1894; **7:** 50–98. 1895; **8:** 50–98. 1896; **9:** 50–114. 1897; **10** (Sci. Nat.) 36–68. 1898; **12** (Sci. Nat.): 69–117. 1900; **13** (Sci. Nat.): 10–59. 1901; **14** (Sci. Nat.): 1–17.

1902; **15** (Sci. Nat.): 1–38. 1903. (Reprinted, 2 p. l., vii, 532 p. 22 cm. Montmédy, 1906. Original not seen, reference supplied by R. McVaugh from copy in University of Michigan library; reprint seen in library of Muséum national d'histoire naturelle, Paris.)

Annotated list of vascular plants, with localities, vernacular names, medical properties and uses, etc.—See also **Evrard, P.** Notes d'herborisations (1911–12). Bul. Soc. Nat. et Archéol. Nord de la Meuse **24** (Sci. Nat.): 29–42. 1912. (Short list of additional species, long list of additional localities. In John Crerar Library.)—Also his Notes d'herborisations 1913. l. c. **25/32** (Sci. Nat.): 20–24. "1913–20." (Processed. In John Crerar Library.)—Also his Notes d'herborisations 1923. l.c. **36** (I. Sci. Nat.): 10–15. 1924. (Processed. In John Crerar Library.)—Also Verhulst, Antoine. Plantes du Jurassique belge non renseignées dans la "Catalogue" de Montmédy (see Belgium, Luxembourg, under Gilson).

MORBIHAN

See Bretagne.

General

Arrondeau, E. T. Botanique. Catalogue des plantes phanérogames observées dans le département. 2 p. l., iv, 119 p. 24 cm. [Vannes] [pref., 1867.] (New York Botanical Garden library.)

Mention of some localities and their characteristic species; briefly annotated list of spermatophytes, with localities. From Histoire naturelle du Morbihan. Catalogues raisonnés des productions des trois règnes de la nature recuillies dans le département. Vannes, 1867. Replaces **LeGall, N. J. M.** Flore de Morbihan. xxiv, 839 p. 19 cm. Vannes, 1852. (General features of flora; annotated descriptive flora of vascular plants [1127 species] with some localities. In Gray Herbarium library. Reissued 1857; reissue not seen.)

Local

Delalande, J. M., *abbé.* Hoedic et Houat, histoire, moeurs, productions naturelles de ces deux îles du Morbihan. Annales Soc. Acad. Nantes et Loire-Infér. **21** (III, 1): 263–380. 1850. (Harvard College library.)

Includes (p. 368–379) partly annotated list of known vascular plants and Characeae, based in part on information furnished by James Lloyd.— See also **Soubeiran, Léon.** Une course aux îles d'Houat et d'Hoedic (Morbihan). Bul. Soc. Bot. France **3**: 553–558. 1856. (Account of author's excursion with mention of numerous species.)

Gadeceau, Émile. Essai de géographie botanique sur Belle-Île-en-Mer. Mém. Soc. Natl. Sci. Nat. et Math. Cherbourg **33** (IV, 3): 177–368. 4 p. l., map. 1903.—Supplément . . . l. c. **35** (IV, 5): 399–414. pl. 1906.—Deuxième supplément . . . l. c. **39** (IV, 9): 333–352. 1923.

Topography, climate, history, cultivated plants; annotated lists of indigenous vascular plants and of naturalized species, with localities for scarcer species; phytogeography, soil, plant formations, etc.; bibliography.

Godron, D. A. Herborisations autour de Lorient, de Port-Louis et à l'île de Groix. Mém. Soc. Natl. Sci. Nat. Cherbourg **19**: 155–210. 1875.

Soils, climate, general features of flora; annotated list of vascular plants, based on the author's collections.

Viaud-Grand-Marais, Ambroise, and **Guyon-Varch,** *abbé.* Catalogue des plantes vasculaires de l'île de Groix (Morbihan). Bul. Soc. Bot. France **30**: 25–35. 1883.

Topography, botanical explorations; annotated list of about 470 vascular plants.

MOSELLE

See also Alsace-Lorraine, Lorraine, and Meurthe-et-Moselle (Suard, Plantes vasculaires).

The Department of Moselle, with somewhat smaller territory than it had before 1871, was included in Germany from 1871 to 1919 and constituted Lothringen or German Lorraine. The arrondissements of Château-Salins and Sarrebourg, formerly in the old Department of Meurthe, have been in Moselle since 1927.

General

Barbiche, R. T., ***abbé.*** Revue critique de la bibliographie botanique locale. Bul. Soc. Hist. Nat. Metz **15**: 229–261. 1880.

Critical discussion of the floras by Waldner and Bassler; extensively annotated chronological list of publications (1806–1878) on flora of Moselle including general works on Lorraine.

Himpel, J. S. Excursions-flora für Lothringen. vi, 222 p. 19.5 cm. Metz, 1888. (Gray Herbarium library.)

Key to genera (on Linnaean system); briefly annotated flora of vascular plants in form of keys, mostly without localities.

Holandre, J. J. J. Nouvelle flore de la Moselle, ou manuel d'herborisation dans les environs de Metz principalement, et les autres parties du département . . . 2. éd. 2 v. (lxxvi, 934 p.). 15 cm. Metz, Paris, 1842. (1st ed. 1829.)

Geology, introduction to botany, synopsis of families; annotated descriptive flora of vascular plants and mosses, with localities.—See also **Friren, Auguste,** *abbé.* [Plantes rares ou nouvelles pour la circonscription de l'ancien département de la Moselle.] Bul. Soc. Hist. Nat. Metz **15**: 147–154. 1878. (Records of various species.)—Also **Barbiche, R. T.,** *abbé.* Contribution à la revision de la flore de la Lorraine allemande. l. c. **23** (II, 11): 81–103. 1904 . (Publications; list of newly recorded, extirpated, and doubtful species of vascular plants, based chiefly on author's collections.)—Also **Friren, A.,** ***abbé.*** Observations sur quelques plantes de la Lorraine. l. c. **26** (III, 2): 97–129. 1909. (Includes 3 sections: I. Localitées à supprimer . . . (p. 97–112), an annotated list of species that have disappeared from specific localities; II. Plantes adventices d'introduction plus ou moins récente (p. 112–128), an annotated list with localities; III. Les plantes adventices du Sablon, près de Metz (p. 128–129), for which see below, under his Flore adventice du Sablon.)—See also **Benoit, Joseph,** *abbé.* Éléments de phytostatique pour le département de la Moselle publiés d'après des notes de feu M. l'abbé T.-R. [sic] Barbiche. Bul. Soc. Hist. Nat. Dépt. Moselle **32** (III, 8): 83–162. 1929. (Extensive lists of newly recorded vascular plants and new localities.)

Local

Barbiche, R. T., ***abbé.*** Florule de l'arrondissement de Thionville. Bul. Soc. Hist. Nat. Dépt. Moselle **12**: 51–90. 1870. (Reprinted, 42 p. In Gray Herbarium library.)

Geology, topography, etc.; annotated list of phanerogams (289 species). (German name, Diedenhofen.)

Brunotte, Camille. Les marais salés de la vallée de la Seille au point de vue botanique. 26 p. map. 8vo. Nancy, 1896. (From Bul. Club Alpin Franç. Sect. Vosgienne.)

General considerations, topography, etc.; list of vascular and cellular plants. (Not seen; cited from notice in Bul. Soc. Bot. France **43**: 73–74. 1896.)

Friren, Auguste, ***abbé.*** Flore adventive du Sablon ou observations sur quelques plantes récemment introduites aux portes de Metz. Bul. Soc. Hist. Nat. Metz II, **15**: 131–154. 1880; **19**: 95–110. 1895.

Geology, etc.; annotated lists of new and rare adventive vascular plants.—See also his Les plantes adventices du Sablon, près de Metz. *In* his Observations sur quelques plantes de la Lorraine. Bul. Soc. Hist. Nat. Metz **26** (III, 2): 128–129. 1909. (Nearly all the species recorded in his earlier papers had disappeared.)

Himpel, J. S. Die flora der umgebung von Metz. iv, 96 p. 19 cm. Metz, 1898. (Beilage zum Jahresber. der Oberrealschule in Metz.) (Gray Herbarium library.)

Annotated list of vascular plants, with localities.

——— Verzeichniss der gefässpflanzen der umgebung Pfalzburgs. Jahresber. Progymnas. Pfalzburg **15** (1885/86): 3–28. 1886.

Not seen; no review available.

Ludwig, Alfred. Die gefässpflanzen von Forbach und umgebung, sowie die darauf beobachteten schmarotzenden pilze, gallen und teratologischen bildungen. I. teil. 42 p. Forbach, 1914. (Beilage zum Jahresber. der Oberrealschule zu Forbach in Lothringen 1914.)

Treatment of pteridophytes, gymnosperms, and monocotyledons, with extensive data on teratological specimens. (Not seen; cited from notice in Bot. Jahresber. Just **44**(1): 977. 1926. No more published?)

NIÈVRE

Local

Gagnepain, François. Contribution à l'étude de la géographie botanique de la France. Topographie botanique des environs de Cercy-la-Tour (Nièvre). Bul. Soc. Hist. Nat. Autun **13**: 127–302. map. 1900.

Botanical explorations, topography, geology, climate, plant associations, statistics, changes in flora; annotated list of vascular plants (1475 taxa, including 1009 native species), with localities; list of adventive plants, grouped according to manner of introduction; alphabetical list of vernacular names, with botanical equivalents; list of locality names derived from vernacular names of plants.—See also **Dhien, René.** Modification d'une flore en cinquante années. Bul. Soc. Bot. France **102**: 242–243. 1955. (List of noteworthy species.)

——— Espèces ou localités nouvelles pour le département de la Nièvre. Bul. Bot. France **42**: 598–613. 1896; **43**: 449–454. 1896; **45**: 129–136. 1898; **47**: 209–221. 1900.

Lists of vascular plants (and higher fungi, in the last installment), with localities. Title varies somewhat.—See also **Delarue, Paul.** Étude sur la flore nivernaise. De quelques plantes rares ou nouvelles pour le département. Mém. Soc. Acad. Nivernais **32**: 43–56. 1930. (Lists of native and adventive plants. In Harvard College library.)—Also his Flore nivernaise. 2. série de plantes rares ou nouvelles. l. c. **34**: 43–62. 1932. (Records of noteworthy species; bibliography. In Harvard College library.)—Also **Dhien, René.** Les fougères de la Nièvre. Bul. Soc. Hist. Nat. Toulouse **80**: 95–99 incl. fig. 1–2. 1945. (List of Filicineae (26 species); bibliography.)—Also his Contribution à la flore de la Nièvre. Bul. Soc. Bot. France **97**: 230–231. 1951. (Additional records.)—Also his La végétation des stations thermales. I. Saint-Honoré-les-Bains (Nièvre). Bul. Soc. Bot. France **101**: 397–400. 1954. [1955]. (Soils; mention of various localities with their noteworthy species.)

——— Essai floristique sur la région de Varzy (Nièvre). Mém. Soc. Acad. Nivernais **23** (**II, 9**): 9–36. 1921. (Harvard College library.)

Geology, botanical explorations; short lists of characteristic species of different habitats; list of rarer phanerogams, with localities.

Joannin-Déponge, —. Flore spontanée (février 1922). Liste des plantes assez rares, rares et très rares de Champvert et ses environs. Mém. Soc. Acad. Nivernais **27**: 119–130. 1925. (Harvard College library.)

Extensive list of vascular plants, with localities, from the vicinity of Decize.

NORD

See also Pas-de-Calais (La Fons; Masclef); also Belgium, General (Paque, De vlaamsche volksnamen der planten).

General

Vandamme, Henri. Flore de l'arrondissement d'Hazebrouck (département du Nord) . . . (cover title, Flore complète . . .). viii, 334 p. 20.5 cm. Paris [etc.], 1850–64. (Arnold Arboretum library.)

Descriptive flora of vascular plants and Characeae, on the Linnaean system, covering the Département du Nord, part of Belgium, Arras (Pas-de-Calais), etc., with localities for rarer species, notes on uses, and index of Flemish names (p. 323–333); includes cultivated plants.

Local

Boulay, J. N., *abbé.* Revision de la flore des départements du nord de la France. 3 v. (63, 46, 58 + 8p.). 17 cm. Lille et Paris, 1878–1880. (Arnold Arboretum library.)

Botanical explorations, bibliography; results of author's collecting, including partial lists of species from various localities, etc. Refers to departments of Nord and Pas-de-Calais. For critical review of fasc. 1 and 2 see **Giard, Antoine.** Bul. Scient. Dépt. Nord **12 (II, 3):** 30–37. 1880.

Bourlet, —, *abbé.* Catalogue des plantes phanérogames qui croissent naturellement dans les fortifications de la ville de Douai. Mém. Soc. Roy. et Cent. Agr., Sci. et Arts Dépt. Nord **12:** 157–246. 1847.

Annotated list of 483 angiosperms.—See also **Gosselin, E.** Contribution à la flore des environs de Douai. Mém. Soc. Agr., Sci. et Arts Centr. Dépt. Nord III, **5:** 59–81. 1895; **7:** 187–212. 1899. (Lists of vascular plants, with localities.)—Also his Contribution . . . l. c. III, **12:** 69–91. 1909–10. (List of adventive plants from various localities; list of noteworthy native plants, with localities.)—Also Froment, below.

Fockeu, Henri. Flore lilloise limitée au périmètre extérieur des glacis. Bul. Univ. Lille II, **5:** 83–121. 1901. (Reprinted 39 p. 1901. Original in library of Faculté de Pharmacie, Université de Paris; reprint in U. S. Dept. of Agriculture Library.)

Brief account of soils, etc.; list of vascular and cellular plants, with localities.

Froment, Pierre, and Froment, *mme.* La flore des ruines de Douai (Nord). Bul. Soc. Bot. France **93:** 393–402. 1946 (1947).

General features of vegetation of bombed areas in Douai and elsewhere; running account of plants observed grouped by habitats, etc.; bibliography.—See also their L'évolution de la flore des ruines de Douai (Nord) durant l'année 1947. l. c. **94:** 410–416. 1948. (Changes in flora of different localities, list of 29 additional species, etc.)—Also Bourlet, above.

Godon, Joseph, *abbé.* Flore du Cambrésis. Mém. Soc. Émul. Cambrai **44:** 209–267. p. l. 1889.

Topography, climate; description of ecological regions in vicinity of Cambrai with lists of species and notes on their local distribution; not a full flora.—On p. 136 of his 1910 paper (see below) the author cites his Catalogue raisonné et descriptif des plantes vasculaires du Cambrésis. Cambrai, Imp. F. Deligne et Cie, 1910. No such work is cited in the annual bibliographies in O. Lorenz, Catalogue général de la librairie française, and in Bibliographie de la France, or in the local bibliographies in Mém. Soc. Émul. Cambrai covering the years 1909–1920, and there is no evidence that it was ever published.

——— Promenades botaniques dans l'Avesnois. Mém. Soc. Émul. Cambrai **64:** 128–245. map. 1910.

Topography, geology, bibliography; account of excursions to various localities in vicinity of Avesnes, with mention of many species, discussion of adventive species; no full list.

NORMANDIE (NORMANDY)

The region of Normandie includes the departments of Calvados, Eure, Manche, Orne, and Seine-Inférieure. Only works relating to the whole of this region are included here.

Corbière, Louis. Nouvelle flore de Normandie contenant la description des plantes qui croissent spontanément ou sont cultivées en grand dans les départements de la Seine-Inférieure, l'Eure, le Calvados, l'Orne et la Manche. xvi, 716 p. 18.5 cm. Caen, 1893 (1894). (Gray Herbarium library.)

Keyed descriptive flora of vascular plants and Characeae, with localities. Replaces **Brébisson, L. A. de.** Flore de la Normandie (phanérogames et cryptogames semi-vasculaires). 5. éd., publiée par J. Morière. xxxv, 518 p. 18.5 cm. Caen, 1879 (cover date, 1880). (1st ed. 1836.) (Keyed descriptive flora of vascular plants and Characeae.)—See also **Corbière, L.** Additions et rectifications à la Nouvelle flore . . . Bul. Soc. Linn. Normandie IV, **9:** 76–116. 1895 (1896).—Also his Deuxième supplément . . . l. c. V, **1:** 150–200. 1897 (1898).—Also **Fortier, E.** Notes sur quelques plantes rares ou nouvelles de la flore de Normandie. Bul. Soc. Amis Sci. Nat. Rouen **45**: 8–13. 1910. (List with localities.)—Also **St. Amand, Hippolyte.** Raretés botaniques observées depuis la publication de la Flore de Normandie, L. Corbière, 1893 et ses Suppléments, 1895 et 1898. Bul. Soc. Étude Sci. Nat. et Mus. Hist. Nat. Elbeuf **31**: 29–33. 1913. (List with localities.)—Other lists of additional or noteworthy species, too numerous to cite here, will be found in Bul. Soc. Linn. Normandie.

Joret, Charles. Flore populaire de la Normandie. lxxxviii, 338 p. 22 cm. Caen et Paris, 1887. (Gray Herbarium library.)

Classical and mythological botany, with references; account of works on popular names in different languages, with footnote references; systematic list of vascular and cellular plants, with vernacular names and localities where used; list of varietal names of apples and pears; index.

Le Héricher, Édouard. Essai sur la flore populaire de Normandie et d'Angleterre. 2 p. l., 103 p. 25 cm. Avranches, Paris, 1857. (Arnold Arboretum library.)—Additions à l'Essai . . . 14 p. 25 cm. [n.p., n.d.] (Arnold Arboretum library.)

Discussion of vernacular nomenclature; running account of vernacular names of plants in Normandy, arranged by families, with English names, discussion of cultivated plants, etc.; index of vernacular names.—See also his Philologie de la flore scientifique et populaire de Normandie et d'Angleterre. viii, 115 p. 22 cm. Coutances, [1883?]. (Systematic list of generic and some specific botanical names of vascular and a very few cellular plants with their French and English vernacular names, with derivation and meaning of all.)

Toussaint, Anatole, *abbè*. **Étude étymologique sur les flores normande et parisienne comprenant les noms scientifiques, français et normands des plantes indigènes et communément cultivées.** Bul. Soc. Amis Sci. Nat. Rouen **41**: 75–333. 1906.

Bibliography; systematic list of vascular plants and Characeae, with explanation of generic names, and under each genus the Norman and French vernacular names with explanation of their meaning and derivation; index.

Viguier, René. La botanique en Normandie et les botanistes normands depuis 1823. Bul. Soc. Linn. Normandie ser. VII, vol. suppl.: 56–90. 1926.

Running account of botanists and their work, with titles of some publications.

OISE

See also Aisne (Jovet); Picardie.

General

Rodin, Hippolyte. Esquisse de la végétation du département de l'Oise. 2. partie. Statistique botanique du département de l'Oise ou catalogue des plantes observées dans l'étendue du département de l'Oise par L. Graves, révisé, annoté et augmenté par H. Rodin. Mém. Soc. Acad. Archéol., Sci. et Arts Dépt. Oise **5**: 661–734. 1864; **6**: 219–283, 527–574, 772–867. 1865–67; **7**: 360–456, 625–715. 1869–70; **8**: 216–255, 713–785. 1871–73; **9**: 211–274, 510–543, 771–842. 1874–76.

Annotated, keyed list of vascular plants and Characeae (1413 species), with detailed localities. Replaces **Graves, Louis.** Catalogue des plantes observées dans l'étendue du département de l'Oise. xv, 302 p. 23 cm. Beauvais, 1857. (Extrait de l'Annuaire du département de l'Oise, 1857.) (Annotated list of 3527 plants (1,333 vascular), with localities.)—See also **Rodin, H.** Esquisse de la végétation du département de l'Oise. Mém. Soc. Acad. Archéol., Sci. et Arts Dépt. Oise **5**: 333–507. 1863. (Reprinted as Esquisse . . . 1. partie. 2 p. l., 156 p. 25 cm. Beauvais, 1864. Reprint in Lloyd Library.) (Geology, table of altitudes, agriculture, climate, forests, flora of walls and ruins, list of hybrids, statistics, weeds, list of albinos, soil relations, lists of purposely and accidentally naturalized plants, fossil plants.)—Also **Camus, E. G.** Quelques plantes nouvelles pour le département de l'Oise. Bul. Soc. Bot. France **48**: 46–47. 1901.

Local

Fauqueux, Charles. Catalogue des plantes du Fayel (Oise). Bul. Soc. Linn. Nord France **24:** 64–92. 1 fig. (map). 1929.

Geology, etc.; annotated list of vascular plants. (A commune in the canton of Estrées-Saint-Denis.)

Molle, J. Catalogue des plantes des environs de Beauvais par un amateur. (Composée pour l'Annuaire de l'Oise de 1892.) 64 p. 22 cm. Beauvais, 1891. (Library ?)

Annotated list of vascular plants and Characeae, with localities. Caption title, Inventaire d'un milier de plantes qu'en rencontre aux environs de Beauvais.

Questier, —, *abbé.* Énumération des plantes intéressantes des cantons de Betz et de Crépy-en-Valois. Mém. Soc. Acad. Archéol., Sci. et Arts Dépt. Oise **1:** 210–217, 325–336, 495–502. 1849–51; **2:** 460–475. 1854.

List of noteworthy vascular plants and Characeae, with localities.

ORNE

See also Normandie.

General

Letacq, A. L., *abbé.* Inventaire des plantes phanérogames et cryptogames vasculaires croissant spontanément ou cultivées en grand dans le département de l'Orne. Bul. Soc. Amis Sci. Nat. Rouen **41:** 369–438. 1906; **42:** 229–292. 1907; **43:** 43–88. 1909.—Supplément à l'Inventaire . . . **44:** 43–208. 1909. (The whole reprinted, 348 p. Reprint not seen.)

Annotated list of vascular plants and Characeae, with localities for scarcer species. The supplement contains additional records, topography, climate, soils and special localities with lists of species, biographical account of botanists with lists of their publications, extensive list of vernacular names with botanical equivalents and localities where used, index.—See also **Lemée, Georges.** Inventaire . . . Second supplément. Bul. Soc. Amis Sci. Nat. Rouen **68/69:** 139–159. 1934. (List of additional species and records, grouped by phytogeographic districts.)—Also his Sur quelques phanérogames nouvelles pour le territoire du Perche. Bul. Soc. Linn. Normandie VIII, **7:** 99–106. 1935. (Topography, etc.; list of noteworthy plants, with localities, mostly from Orne.)

Local

Chevalier, Auguste. Catalogue des plantes vasculaires de l'arrondissement de Domfront avec notes critiques et observations biologiques. Bul. Soc. Linn. Normandie IV, **7:** 98–333. 1893–94.

Botanical explorations, bibliography, soils, etc.; annotated list of vascular plants and Characeae; list of doubtful species, list of vernacular names.—See also his Recherches et observations sur la flore de l'arrondissement de Domfront . . . Bul. Soc. Linn. Normandie V, **1:** 3–56. 1898. (Additional bibliography, phytogeography, botanical districts; list of additional species and localities.)—Also his La flore adventice des ruines du château féodal de Domfront. l. c. V, **1:** 57–78. 1898. (Discussion of adventive species and of their probable dates of introduction.)—Also **Saintange-Savouré, H.** Contribution à la flore du département de l'Orne. l. c. V, **9:** 74–118. 1906. (Geology, bibliography; separate lists of noteworthy species from the arrondissements of Domfront and Argentan.)

Duterte, A. H. Catalogue des plantes phanérogames et cryptogames semivasculaires croissant spontanément à Alençon ou dans un rayon de 20 kilomètres. Bul. Soc. Linn. Normandie III, **8:** 50–158. 1884.

Annotated list of vascular plants and Characeae (1,070 species.)—See also his Additions et rectifications au Catalogue . . . Bul. Soc. Linn. Normandie III, **9:** 72–73. 1885.—Also his Nouvelles additions à faire au Catalogue . . . l. c. IV, **2:** 9–10. 1888.—Also **Letacq, A. L.** Plantes des environs d'Alençon. l. c. VII, **3:** 305–306. 1921. (Notes on several species.)—Also Letacq, below (Catalogue, and Notice).

Frébet, Arsène, *abbé.* Flore du canton de La Ferté-Macé. Annuaire du canton de La Ferté-Macé pour 1883. p. 47–63. 1883(?).
Not seen; no annotation available; title supplied by G. Dillemann.

Letacq, A. L., *abbé.* Catalogue des plantes adventices d'Alençon et du département de l'Orne. Bul. Soc. Hort. Orne **1909:** 42–72. 1909. (Reprinted, 31 p.)
Not seen; cited from Bul. Soc. Amis Sci. Nat. Rouen V, **44:** 139. 1909.

——— Des noms vulgaires de plantes usités dans les cantons de Vimoutiers et de La Ferté-Fresnel (Orne). Bul. Soc. Scient. Flammarion de Marseille **6:** 124–135. 1888.
Systematic list of scientific names with vernacular equivalents, (Not seen; annotation supplied by Mme. G. Duprat, Muséum national d'histoire naturelle, Paris.)

——— Notice sur la flore populaire des environs d'Alençon et de Carrouges (Orne). Annuaire Normand **62:** 290–306. 1895.
Citation of sources, etc.; systematic list of vascular and a few cellular plants, with vernacular names.

PARIS

See also France, Partial (Écorchard); France, General (Jovet, under Davy de Virville).
Essentially all the works cited here relate to a greater or lesser number of departments surrounding Paris, in some cases to a radius of 90 or 100 kilometers. The precise area covered is often stated vaguely in the works themselves, so that it is not possible to annotate all of them definitely. It may be that a few refer mainly to the Départment de la Seine.
For works relating to Paris itself or its immediate vicinity, see under Seine.

Baillon, Henri. Les herborisations parisiennes. Recherche, étude pratique et détermination facile des plantes qui croissent dans les environs de Paris. 2 p. l., 482 p. illus. 15 cm. Paris, 1890.
Popular pocket manual, with keys, localities, etc.

Bautier, Alexandre. Tableau analytique de la flore parisienne d'après la méthode adoptée dans la Flore française de MM. Lamarck et de Candolle . . . Suivi d'un vocabulaire . . . et d'un Guide du botaniste pour les herborisations aux environs de Paris. 22. éd. xlvi, 464 p. 15 cm. Paris, 1891. (Lloyd Library.) (1st ed. 1827.)
Elements of botany, key to genera; briefly annotated flora of vascular plants and Characeae in form of keys, with partial list of localities; alphabetical list of localities with enumeration of their rarer species. Appears to be reissue of 16th ed., 1879.

Bonnet, Edmond. Petite flore parisienne contenant la description des familles, genres, espèces et variétés de toutes les plantes spontanées ou cultivées en grand dans la région parisienne . . . xii, 527 p. 18.5 cm. Paris, 1883. (Gray Herbarium library.)
Briefly annotated descriptive flora of vascular plants and Characeae, with keys and localities for scarcer species. Covers the area within a radius of 95 kilometers around Paris.

Bonnier, Gaston, and **Layens, Georges de.** Nouvelle flore des environs de Paris, de l'Eure, de l'Eure-et-Loir, etc. 2 p. l., xxiv, 271 p. 2,145 fig. 19 cm. Paris, n.d. (Arnold Arboretum library.)
Popular pocket flora in form of illustrated key; index of botanical names with explanations of generic names, localities, and notes on uses; index of vernacular names. Based essentially on their Tableaux synoptiques des plantes vasculaires de la flore de la France. [1894 ?]. Dillemann (Monde des Plant. VI, **42** (247): 36. 1947) states that there have been 13 editions of the Nouvelle flore, but gives no dates.—See also **Bonnier, Gaston.** Album de la Nouvelle flore représentant toutes les espèces de plantes photographiées directement d'après nature en cinquième de leur grandeur naturelle . . . nouv. éd. 183 p. (incl. 169 pl. on 85, with 2,028 fig.). 18 cm. Paris, [1945 ?]. (1st ed. 1906 ?) (Small photographic figures which, although little more than silhouettes, do in general give an idea of the habit, foliage, and inflorescence of the plant; index. Covers all the species within 100 km. of Paris and the common species of the interior.)

[Buc'hoz, P. J.] Flore naturelle et économique des plantes qui croissent aux environs de Paris, au nombre de plus de quatre cents genres et de quatorze cents espèces, contenant l'énumération de ces plantes, rangées suivant le système de Jussieu et par l'ordre alphabétique, les noms triviaux, leur synonimie française, leur description, les endroits où se trouvent les plus rares, leurs propriétés pour les alimens, les médicamens, l'art vétérinaire, les arts et métiers, et l'ornement des jardins . . . 2. éd. 2 v. 24 col. pl. 20.5 cm. Paris, 1803. (Lloyd Library.) (1st ed. 1799.)

The first part (v. 1, p. i–xxxxii, 1–318) is a descriptive flora of cellular and vascular plants; the second (p. 1 bis–176 bis of vol. 1, and the whole (p. 177–659) of vol. 2), Flore économique, is an annotated alphabetical list of vascular and cellular plants with vernacular names and uses.

Camus, E. G. Guide pratique de botanique rurale à l'usage des botanistes, des étudiants en pharmacie . . . 2 p. l., 180 p. 60 pl., map. 22 cm. Paris, 1884. (Library of Muséum national d'histoire naturelle, Paris.)

Calendar of flora listing the species in flower month by month, guide to 78 localities in vicinity of Paris with lists of noteworthy species; series of plates illustrating the species or genera of various families, with accompanying keys on unnumbered pages.

Cosson, Ernest, and Germain de Saint-Pierre, Ernest. Flore des environs de Paris ou description des plantes qui croissent spontanément dans cette région et de celles qui y sont généralement cultivées, accompagnée de tableaux synoptiques . . . 2. éd. liv, 962 p. map. 22 cm. Paris, 1861. (Gray Herbarium library.) (1st ed. 1845, with title: Flore descriptive et analytique des environs de Paris ou description . . .)

List of botanists, bibliography; annotated descriptive flora of vascular plants and Characeae, with localities; index of French and patois names. Covers a radius of 94 kilometers (23.5 leagues) around Paris and is still regarded as the best available flora of the Paris region.—See also their Atlas de la flore des environs de Paris ou illustrations de toutes les espèces des genres difficiles et de la plupart des plantes litigieuses de cette région . . . 2. éd. xiii, [95] p. 47 pl. 24.5 cm. Paris, 1882. (Plates, accompanied by descriptive text, containing illustrations of diagnostic portions of plants. In Gray Herbarium library.) (1st ed. 1845.)

Jeanpert, H. E. Vade-mecum du botaniste dans la région parisienne. Tableaux synoptiques des familles, genres, espèces et variétés . . . 2. éd. rév. et corr. par G. Viennot-Bourgin. vii–xii, 242 p., 241 p. incl. 1634 fig. 21 cm. Paris, 1953. (1st ed. 1911.)

Briefly descriptive flora in form of keys, with figure of each species (1634). Essentially unchanged from 1st ed., except that the list of localities with their characteristic species has been replaced by an index of vernacular names; species naturalized in the area since the first edition have not been inserted. Covers the Paris basin and its periphery (about 10 departments). Localities for the rarer species are given on the illustrations.—See also his Voyage botanique circulaire autour de la flore parisienne. Bul. Soc. Bot. France 51 (Sess. Extraord.): cxxxiii–cxxxvii. 1904 (?). (Mention of some of the noteworthy species of localities in a circle of about 100 kilometers radius around Paris.)

Lanessan, J. L. de. Flore de Paris (phanérogames et cryptogames) contenant la description de toutes les espèces utiles ou nuisibles, avec l'indication de leurs propriétés médicales, industrielles et économiques . . . xlii, 902 p. 693 fig. 18 cm. Paris, 1884. (Gray Herbarium library.)

Glossary, special localities near Paris; keys to all species (vascular and cellular) of the region, with annotated descriptions of the useful and harmful ones, and notes on uses; tabular list of all species, with common names, date of flowering, habitat and localities; includes cultivated species.

——— Flore médicale des environs de Paris. 950 p. 702 fig. 15 cm. Paris, 1880.

Analytical keys to vascular and cellular medicinal plants. (Not seen; title and annotation supplied by P. Fournier.)

Lefébure de Fourcy, Eugène. Vade-mecum des herborisations parisiennes conduisant sans maître aux noms d'ordre, de genre et d'espèces des plantes spontanées ou cultivées en grand dans un rayon de 25 lieues de Paris. 6. éd. 1 p.l., xi, 328 p. 16.5 cm. Paris, 1891. (Lloyd Library.) (1st ed. 1859.)

Briefly annotated flora of vascular and cellular plants, in form of keys; additions by E. G. Camus, p. 311–328. On slip pasted over imprint: Paris, 1896.

Mascré, Marcel, and **Deysson, Guy.** Manuel d'herborisation. Introduction à l'étude de la botanique systématique. [2. éd.] 267 p. 1,040 fig., map. 21.5 cm. Paris, 1953. (1st ed. 1948.)

Popular manual of wild and cultivated vascular plants including most of the species of the Paris region.

Monchot, E. Les plantes spontanées comestibles de la région parisienne. Cong. Naturaliste (Paris 1944) p. 44–71. 26.5 cm. 1947. (Library of P. Jovet.)

Annotated list of edible vascular plants, with brief directions for use; bibliography, index.

Royer, Maurice. Répertoire de bibliographie générale du bassin du Loing. III. Botanique générale. Bul. Mens. Assoc. Nat. Vallée du Loing **7:** 48, 63–66. 1931; **8:** 6–8, 15–16, 42–43, 58–60, 63–64, 70–72. 1932; **9:** 10–12, 30–32, 36, 44, 53–56. 1933; **10:** 16. 1934. (University of Michigan library.)

Unannotated list of floras, works on agriculture and silviculture, and papers on phanerogams, each group chronologically arranged, the last group unfinished.

PAS-DE-CALAIS

See also Nord (Vandamme; Boulay); Picardie.

General

Masclef, Amédée, *abbé.* Catalogue raisonné des plantes vasculaires du département du Pas-de-Calais. lii, 214 p. 24.5 cm. Arras et Paris, 1886. (Arnold Arboretum library.)

Annotated bibliography, collaborating botanists, topography, climate, botanical regions, etc.; annotated list of indigenous or naturalized vascular plants (1,049 species), with localities.—See also **Durand, Théophile.** Quelques considerations sur la flore du département du Pas-de-Calais. Bul. Soc. Roy. Bot. Belg. **26** (2): 23–28. 1889. (Comparison with flora of Belgium.)—Also **Guffroy, Charles.** Notes sur la flore artésienne. Bul. Soc. Bot. France **77:** 629–632. 1931. (List of noteworthy vascular plants, from several localities.)

Local

Acloque, Alexandre. Florule d'Auxi-le-Château (Pas-de-Calais) et des environs. Monde des Plant. **4:** 273–276. 1895.

List of rarer vascular plants, with localities.

Élie, —, and **Lévêque, Édouard.** La flore du Touquet et de Paris-Plage. 225 p. 16 pl. (15 col.). 16 cm. Paris-Plage, 1910.

Topography, climate, soils, etc.; annotated list of vascular plants, evidently incomplete; a "popular" work.

Hill, A. W. The flora of the Somme battlefield. Bul. Misc. Inform. Kew **1917:** 297–300. 1917.

Effects of World War I on topography, vegetation, etc.; list of angiosperms (all weeds) observed on two visits to Arras, Bapaume, etc.

La Fons, Alexandre de, *baron de Mélicocq.* Plantes croissant spontanément dans les environs de Béthune . . . Annuaire du Pas-de-Calais **1848/49** (2. partie): 203–243. 1848–49.

List of rarer species of the swamps along the canal from Béthune (in Pas-de-Calais) to La Bassée (Nord). (Not seen; title and annotation from notice in Masclef, Catalogue raisonné . . . p. xiii.)

Masclef, Amédée, *abbé.* Contributions nouvelles à la flore des collines d'Artois (Cambrésis, Artois, Haut-Boulonnais.) Jour. de Bot. **2:** 271–275, 304–308, 324–328, 342–348, 359–368, 1888.

Botanical explorations (with references); extensive list of the rarer or less known vascular plants, with localities. (In Pas-de-Calais and Nord.)

Rigaux, A. Catalogue des plantes vasculaires et des mousses observées dans les environs de Boulogne-sur-Mer. 38 p. 21.5 cm. Boulogne-sur-Mer, 1877.

List of vascular plants and mosses, with localities, relating principally to the basin of the Liane and the littoral region from Wimereux to Étaples.—See also review by **Antoine Giard,** Bul. Scient. Dépt. Nord **10** (II, 1): 8–11, 50–56. 1878. (List of additional species and localities and corrections. Apparently never quite completed.)—Also his Note sur la flore du Boulonnais. *In* Boulogne-sur-Mer et la région boulonnaise. 1899. Ouvrage offert par la ville de Boulogne-sur-Mer aux membres du XXVIIIe Congrès de l'Association pour l'avancement des sciences . . . v. 1, p. 518–524. 27 cm Boulogne-sur-Mer, 1899. (Chief publications; list of rarer vascular and cellular plants, including some additions to the flora; bibliography.)

PICARDIE (PICARDY)

The region and former province of Picardie includes Somme and parts of Aisne, Oise, and Pas-de-Calais.

Caussin, Octave. Les plantes médicinales de la Picardie. 346 p. 20.5 cm. Paris, 1907. (Arnold Arboretum library.)
Annotated systematic list of wild and cultivated species with vernacular names, brief descriptions, and uses.

PROVENCE

The region and former province of Provence includes the departments of Basses-Alpes, Bouches-du-Rhône, and Var, and part of Alpes-Maritimes and Vaucluse. Only works relating to the whole or a large part of this area are given here.

Bernès, Julien. Les plantes vénéneuses de la Provence. Bul. Soc. Études Scient. et Archéol. Draguignan 16. 72 p. 1947.
"Notes pratiques sur la toxicité des seules Phanérogames et de quelques Champignons parasites des Graminées." (Not seen; cited from Bul. Soc. Bot. France **97**: 34. 1950.)

Capoduro, Marius. Essai sur les noms patois des plantes méridionales les plus vulgaires. Monde des Plant. **6**: 15–16, 79–81, 112–114, 162–163. 1896–1897; **7**: 67–68, 175–176. 1898; **8**: 248–251. 1899.
Annotated list, alphabetically arranged, of patois names in Provence and Languedoc, with discussion of etymology. Incomplete; reaches herbo daï brégant.

Cauvin, G. La médecine populaire en Provence. 90 p. Lyon, 1930. (Thèse.)
Includes account of plants used in popular medicine. (Not seen; cited from Bul. Soc. Bot. France **79**: 957. 1933.)

Charrel, L. Flore de la Provence centrale (Bouches-du-Rhône, Var, Vaucluse) ou catalogue raisonné des plantes vasculaires de cette région. 144 p. 8vo. Toulon, 1912–14.
Annotated list with localities. (Not seen; title and annotation supplied by G. Dillemann.)

Deleuil, Gabriel. Bibliographie botanique provençale (végétaux vasculaires). Bul. Mus. Hist. Nat. Marseille v. 14. 208 p. 1954.
List of 2239 titles, alphabetically arranged by authors, essentially unannotated but with detailed indices. Only the first page of each reference is cited.

France. Comité interministériel des plantes médicinales et à essences. Comité régional de Provence. Notice sur la récolte et la culture des plantes médicinales et à essences en Provence. 68 p. 19.5 cm. Marseille, 1919.
Annotated list of principal wild medicinal plants, with vernacular names, part used, etc.; annotated list of medicinal plants worthy of cultivation.

[Huet, Edmond.] Catalogue des plantes de Provence. Résultat des herborisations faites pendant plus de dix années dans les départements des Bouches-du-Rhône, du Var et des Alpes-Maritimes par MM. R. Shuttleworth, A. Huet et Jacquin, Hanry, complété par les recherches de MM. Thuret, Canut, H. Roux, Blaize, Autheman, Albert, Goaty, Consolat, etc. dans les mèmes départements. 165 p. 22 cm. Pamiers, 1889. (Arnold Arboretum library.)
Unannotated list of vascular plants, with localities and names of collectors. Preface signed E. Huet.

Réguis, Marius. Nomenclature franco-provençale des plantes qui croissent dans notre région. Precédée d'un rapport de M. Achintre. Mém. Acad. Sci., Agr., Arts et Belles-Let. Aix **11**: 1–186. 1878.

Alphabetical list of patois names of wild and cultivated vascular and cellular plants with botanical equivalents; botanical index.

Roux, Honoré. Catalogue des plantes de Provence spontanées ou généralement cultivées . . . avec une préface par . . . A. Derbès. 2 p. l., vii, 654 p. 24.5 cm. Marseille, 1881 [–1891].—Supplément. 657–696 p. 24.5 cm. [1893.]

Annotated list of vascular plants (3172 species in original catalog), including many cultivated species, with localities. The Supplément adds 50 species. Originally published (entirely?) in Bul. Soc. Bot. et Hort. Provence, in which only the installments from **1**: 44–56. [1879?] to **3** [–5]: 113–144. 1883 have been seen, reaching species no. 883.—See also **Legré, Ludovic.** Additions à la flore de la Provence. Bul. Soc. Bot. France **38**: 393–402. 1892; **39**: 402–405. 1893; **44**: 75–79. 1897. (Lists of noteworthy species.)—Also **Le Brun, Pierre.** Contributions à la flore du sud-est de la France. ([1.]–4. note.) l.c. **76**: 1083–1090. 1930; **80**: 770–773. 1934; **82**: 450–453. 1936; **83**: 459–461. 1936. Nouvelles contributions . . . l.c. **93**: 46–49. 1946; **97**: 171–172. 1951; **99**: 7–9. 1952; **101**: 8–11. 1954.—Also **Molinier, René.** Contribution à l'étude de la flore de la Provence occidentale. Chêne Bul. Soc. Forest. Méditerr. et Colon. **38**: 232–247. 1934; **39**: 38–71. 1935. (List of 260 noteworthy species, with localities and phytosociological notes; account of principal plant associations. Not seen; title and annotation supplied by B. de Retz.)—Also **Deleuil, Gabriel.** Contributions à l'étude de la flore provençale. Localités nouvelles de plantes rares et précisions sur certaines localités déjà connues. I–VI, VIII. Annales Fac. Sci. Marseille II, **16**: 87–104, 131–143. 1943–44; Monde des Plant. **45** (266): 23–24. 1950; **46** (282); 51–54. 1951 (two last in Arnold Arboretum library); Annales Fac. Sci. Marseille II, **21**: 245–260. 1952; Bul. Mus. Hist. Nat. Marseille **12**: 87–92. 1953; Monde des Plant. **49** (303/314): 5–7. 1954 (in Arnold Arboretum library). (Lists of noteworthy plants. No. VII not yet published.)—Also **Molinier, René.** Monographies phytosociologiques. Les massifs de l'Étoile et de N.D. des Anges de Mimet (Bouches-du-Rhône). Bul. Mus. Hist. Nat. Marseille **12**: 15–50. 5 fig., map. 1953. (Includes (p. 24–32) list of new localities for Bouches-du-Rhône, and bibliography containing several minor lists of additions to flora of Provence.)

PUY-DE-DÔME

See also Auvergne; Bourbonnais.

General

Huguet, —, and Perrin, —. Rapport à MM. les membres du Conseil général sur les plantes thérapeutiques du Puy-du-Dôme. 35 p. 3 maps. 24.5 cm. Paris, 1919. (At head of title page: Ministère du commerce. Comité interministériel des plantes médicinales.)

Includes account of principal medicinal plants that are commercially exploited in Auvergne and those that are not, and discussion of their occurrence by regions.

Local

Girod, Paul. Florule du Mont-Dore. Revue d'Auvergne **8**: 118–148. 1891.

List of vascular plants and Characeae of the mountain region around Sancy, with localities, based chiefly on the excursion of the Société botanique de France in August 1890.

Luquet, Aimé. Essai sur la géographie botanique de l'Auvergne. Les associations végétales du massif des Monts-Dores. 266 p. 19 pl., map. 25 cm. Saint-Dizier, 1926. (Gray Herbarium library.)

Phytogeographical and ecological; cited here for its sketch of botanical explorations (p. 7–13) and its bibliography.

PYRÉNÉES-ORIENTALES

See also Spain, Catalonia (Sennen, Mes vacances de 1915 en Cerdagne).

General

Gautier, Gaston. Catalogue raisonné de la flore des Pyrénées-Orientales . . . Introduction par C. Flahault. 550 p. 22.5 cm. Perpignan, 1898 [1897]. (Publication de la Société agricole, scientifique et littéraire des Pyrénées-Orientales.)

Plant distribution, topography, geology, plant zones (by Flahault); bibliography; annotated list of 2,672 vascular plants and Characeae, with localities. According to notice in Bul. Soc. Bot. France **44**: 500. 1898, the work was received for review in Dec. 1897, although the only indication of date in it is "1898" on the back cover. Flahault's introduction, according to entry in Bot. Centralbl. **69**: 365. 1897, was separately published at Perpignan (48 p.) in 1896. Replaces **Companyo, Louis.** Histoire naturelle du département des Pyrénées-Orientales. v. 2 (= 3. partie). Règne végétal. 2 p. l., 939 p. 22.5 cm. Perpignan, 1864. (Annotated list of vascular and cellular plants, except algae. A thoroughly unreliable work, according to later authors.)—See also **Le Grand, Antoine.** Notes additionelles au Catalogue de la flore des Pyrénées-Orientales. Bul. Assoc. Franç. Bot. **1**: 62–68. 1898. (List of additional species and localities. Also published in Monde des Plant. II, 7 (100): 93–98. 1898.)—Also **Sennen,** *frère.* Mes herborisations dans les Pyrénées-Orientales. Bul. Soc. Bot. France **46**: 100–116. 1899. (Extensive list of vascular plants, with localities.)—Also **Conill, L.** Additions à la flore des Pyrénées-Orientales. Bul. Assoc. Franç. Bot. **5**: 37–43. 1902. (List with localities.)—Also his Session de l'Académie internationale de géographie botanique dans les Pyrénées-Orientales et à Barcelone du 1er au 11 août 1905. Compte-rendu . . . Bul. Acad. Internatl. Géog. Bot. **16**: 65–93. 1906. (Itinerary, with mention of plants observed.)—Also his Observations sur la flore des Pyrénées-Orientales. Bul. Soc. Hist. Nat. Toulouse **64**: 5–23. 1932; **67**: 129–158. 1935; **72**: 26–62. 1938; **76**: 247–274. 1941. (Lists of newly recorded and critical species, new localities, etc.; bibliographies.)—Also his Contribution à l'étude de la flore du versant septentrional du Mont Canigou (altitude: 430 mètres à 2,785 mètres). Additions au Catalogue raisonné de la flore des Pyrénées-Orientales de G. Gautier. Bul. Soc. Agr., Scient. et Lit. Pyrén.-Orient. **59**: 83–100. 1936. (Account of excursions, with lists of vascular plants including many new records or additions to Gautier's Catalogue. Also published as Documents pour la carte des productions végétales. Ser. Pyrénées, tome Rousillon, v. 2, art. 6. 1936; not seen in this form.)—Also **Horwood, A. R.** The flora of the Pyrénées Orientales. New varieties and additions. Bul. Misc. Inform. Kew **1936**: 95–100. 1936. (Includes (p. 99–100) list of species and varieties collected in the Vernet-les-Bains district by J. W. Wyatt and not given in Gautier's Catalogue.)—Also **Braun-Blanquet, Josias.** Notes critiques sur la flore des Pyrénées-Orientales. Original not seen; reprinted as Commun. Sta. Internatl. Géobot. Medit. et Alp. Montpellier **87**: 219–236. 1945. (Notes on about 80 taxa. Not seen; title and annotation supplied by R. McVaugh.)

Juillet, Armand. Plantes médicinales des Pyrénées-Orientales. 75 p. 8vo. Montpellier, 1920. (Publication of Comité interministériel des plantes médicinales et à essences de la région de Montpellier.)

Introduction by C. Flahault, describing the geographic botany of the region and the general distribution of the medicinal plants; account of medicinal species with description, habitat, and localities; Catalan vernacular names supplied by L. Conill. (Not seen; title and annotation supplied by G. Dillemann.)

Local

Conill, Léon. Aperçu sur la flore de Sournia. Essai de géographie botanique. Bul. Soc. Agr., Scient. et Lit. Pyrén.-Orient. **50**: 111–151. map. 1909.

Geology, plant distribution according to soils, altitude, etc.; unannotated list of spermatophytes, arranged by soils and altitudes; notes on various species.

——— Notes scientifiques sur Vernet-les-Baines et ses environs. Bul. Soc. Agr., Scient. et Lit. Pyrén.-Orient. **58:** 241–358. 1934.

Geology (with bibliography), climate (with bibliography); plant zones, with mention of characteristic species and bibliography; entomology, with lists of species and bibliography.

Jeanbernat, Ernest, and **Timbal-Lagrave, Édouard.** Le Capsir, canton de Montlouis (Pyr.-Orient.). Bul. Soc. Sci. Phys. et Nat. Toulouse **6:** 37–283. 20 pl. [1887.] (Reprinted, with title Le Capsir, canton de Montlouis (Pyrén.-Orient.) (topographie, géologie, botanique)). 255 p. 20 pl. 25 cm. Paris, 1887. Original in Harvard College library; reprint in Arnold Arboretum library.)

Botanical explorations, topography, geology, geography (itinerary); annotated list of vascular plants and mosses, with localities and some critical notes; list of doubtful and excluded species; list (p. 281–282) of new records for their flora of the massif of Laurenti (Ariège).

——— and **Timbal-Lagrave, Édouard.** Quelques jours d'herborisation dans les Albères orientales. Mém. Acad. Sci., Inscript. et Belles-Let. Toulouse VIII, 1 (2): 147–196. 1879. (Reprinted, 52 p. 1879. Reprint in Lloyd Library.)

Topography, etc.; lists of plants observed in various excursions in vicinity of Banyuls, Collioure, Argelès-sur-Mer, etc.; critical notes on some species.

RHÔNE

See also Lyonnais.

General

Anonymous. Flore du département du Rhône. Annales Soc. Linn. Lyon n.s., **1:** 81–128. 1853.

List of vascular plants, with habitat and date of flowering, and localities for scarcer species.

Local

Audin, Marius. Additions à la flore du Haut-Beaujolais. Annales Soc. Bot. Lyon **21:** 57–69. 1896.

Botanical publications; account of topography with mention of numerous species.—See also his Plantes calcicoles du Haut-Beaujolais. Annales Soc. Bot. Lyon **23:** 89–96. 1898. (Geology, distribution of calcicolous species growing in a granitic area.)

Balbis, J. B. Flore lyonnaise, ou description des plantes qui croissent dans les environs de Lyon et sur le Mont-Pilat. 2 v. (xvi, 890, 30; viii, 371 p. Vol. 1 in 2 pts.). 20.5 cm. Lyon, 1827–28.

Annotated descriptive flora of wild and cultivated vascular and cellular plants growing within about 4 leagues of Lyon, as well as those of Mont Pilat (in Loire).—See also [**Roffavier, Georges**]. Supplément à la Flore lyonnaise publiée par le docteur J. B. Balbis en 1827 et 1828; ou description des plantes phanérogames et cryptogames découvertes depuis la publication de cet ouvrage . . . 91 p. pl. 19.5 cm. Lyon, 1835. (Lloyd Library.)—Also **Cusin, Louis.** Notice sur la flore du Pilat. Annales Soc. Bot. Lyon **2:** 118–123. 1875. (Itinerary, with mention of noteworthy plants.)—Also **Saint-Lager, J. B.** Notice géologique sur le Pilat. l.c. **2:** 123–124. 1875.—Also **Lacroix, F.** Rapport de M. F. Lacroix sur l'herborisation faite au Pilat et à Saint-Étienne le 4 et le 5 juillet. Bul. Soc. Bot. France **23** (Sess. Extraord.): clxxii–clxxxii. [1877?] (Botanical explorations; account of excursion with mention of numerous plants.)—Also Coquillat, below.

Coquillat, M. Flore du pavé de Lyon. I. Essai sur une florule du pavé proprement dit. Bul. Mens. Soc. Linn. Lyon **25:** 185–199. 1956.

General considerations; list of 164 species found in the interstices of roads paved with granite blocks, followed by discussion; bibliography.

Gandoger, Michel. La flore du Beaujolais. Bul. Soc. Bot. France **69:** 795–804. 1923.

Note on author's herbarium, etc.; extensive unannotated list of noteworthy vascular plants and Characeae.

Magnin, Antoine. B. Vaivolet et les premiers explorateurs de la flore de Beaujolais. Annales Soc. Bot. Lyon **14**: 37–160. 1887. (Reprinted with title: Énumération des plantes qui croissent dans le Beaujolais . . . 124 p. Bâle, 1887.)

Botanical explorations, Benoît Vaivolet's botanical work, topography and soils, bibliography; compiled list of vascular plants, Characeae, mosses, and a few fungi, with references; list of localities with their principal species, etc.

Merit, J. Autour du Mont Monnet (785 m.), Rhône et Loire. Aperçu de la flore. Bul. Mens. Soc. Linn. Lyon **16**: 173–175, 191–192. 1947.

Geology, etc.; unannotated list of vascular plants, grouped by habitats.

Queney, A., and Coquillat, —. Une station ruderale lyonnaise. Bul. Mens. Soc. Linn. Lyon **14**: 100–103. 1945.

List of adventive plants of an area along the Rhone south of Lyons, with discussion.—See also **Queney, A.** La flore ruderale des bords du Rhône en amont du Pont de la Boucle. Bul. Mens. Soc. Linn. Lyon **19**: 228–232. 1950. (List of 140 species, practically all adventive, of an area north of Lyons, with discussion.)

Sargnon, Louis. Florule de la presqu'île Perrache. Annales Soc. Bot. Lyon **10** (2): 49–67. 1883.

Changes in topography, botanical explorations with lists of plants cited by earlier authors, list of spermatophytes observed by author.

SAÔNE-ET-LOIRE

See also Bourgogne.

General

Carion, J. E. Catalogue raisonné des plantes du département de Saône-et-Loire croissant naturellement ou soumises à la grande culture. Mém. Hist. Nat. Soc. Éduenne **1**: 1–122. 1865. (Harvard College library. Preprinted, 122 p. Autun, 1859. In Gray Herbarium library.)

Annotated list of vascular plants and Characeae, with localities for scarcer species.—See also **Grognot,** (*aîné*). Plantes vasculaires (phanérogames et cryptogames) du département de Saône-et-Loire à ajouter à celles mentionées dans le Catalogue raisonné . . . l.c. **1**: 123–204. 1865. (Extensive list. In Harvard College library.)—Also **Gillot, Xavier.** Plantes rares ou nouvelles du département de Saône-et-Loire. Bul. Trimest. Soc. Bot. Lyon II, **4**: 13–14. 1886. (Mention of several species new to the flora. In Arnold Arboretum library.)—Also **Quincy, Charles.** Deuxième supplément au Catalogue . . . Bul. Soc. Sci. Nat. Saône-et-Loire **37** (n.s., **17**): 50bis-80bis, 81–83, 86–88. 1911. (Extensive list of vascular plants, with localities.)—Also **Chateau, Émile.** Contribution à la flore de Saône-et-Loire. Bul. Soc. Hist. Nat. Autun **26**: 253–302. 1913. (List of collaborators, with biographical data; extensive list of vascular plants and Characeae, with localities.)

Chateau, Émile, and Chassignol, F. Catalogue des plantes de Saône-et-Loire et des cantons limitrophes. Revue Périod. Vulgar. Sci. Nat. et Préhist. de "La physiophile," Soc. Étude Hist. Nat. Montceau-les-Mines, suppl. to v.(?)-12 (no. 14). 2 p.l., 450 p. map. 23.5 cm. 1927–36. (Original not seen; reprint or separate in library of Muséum national d'histoire naturelle, Paris.)

Annotated list of vascular plants and Characeae (1649 species; cultivated and adventive plants and hybrids are given but are not included in this figure), with localities and some notes on uses; bibliography. Covers Saône-et-Loire and bits of at least 6 adjacent departments. According to note in Monde des Plant. V, 40 (238): **30**. 1939, the title of the parts published in 1927–29 was: Flore montcellienne ou catalogue des plantes qui croissent dans un rayon de 80 kms de Montceau-les-Mines.

Quincy, Charles. Notice sur la botanique et les botanistes en Saône-et-Loire. Bul. Soc. Sci. Nat. Saône-et-Loire **37** (n.s., **17**): 21–48, 49bis-80bis, 81–88. 1911.

Account of botanical collectors and their work, with mention of publications. Pages 50bis-80bis, 81–83, 86–88 are Deuxième supplément au Catalogue raisonné des plantes du département de Saone-et-Loire de J. E. Carion (extensive list of vascular plants, with localities).

Local

Bigeard, —. Notes sur la flore des vallées de la Brenne et du Doubs dans le canton de Pierre. Bul. Soc. Sci. Nat. Saône-et-Loire 4: 65–98. 1888.

Topography, soils, general features of flora; annotated list of vascular plants and Characeae (810 species) of the vicinity of Mouthiers-en-Bresse.

Gillot, Xavier. Note sur la flore du Morvan autunois. Bul. Soc. Bot. France **17** (Sess. Extraord.): xlv–l. 1870.

Botanical explorations, etc.; list of 5 species new to Saône-et-Loire, with localities; list of noteworthy vascular plants of the Morvan autunois, with localities; list of species becoming naturalized, etc.—See also his Notice sur les modifications de la flore phanérogamique d'Autun et de ses environs. Cong. Scient. France **42** (tome 1): 343–376. 1877. (Extensive list of naturalized species, grouped by habitats, with discussion of their introduction; lists of species naturalized in the parc de Montjeu and in the vicinity of Autun and the Morvan autunois.)—Also his Note sur la flore du plateau d'Antully. Mém. Soc. Sci. Nat. Saône-et-Loire **2**: 1–19. 1882. (Geology, topography, etc.; annotated list of more interesting vascular plants and Characeae, with localities.)—Also **Chevailler, Jean,** *chanoine.* Quelques plantes qui ont disparu des environs d'Autun. Bul. Soc. Hist. Nat. Autun **29** (Compte rendu des séances et des excursions 1924–25): 57–60. 1926. (Running account of various vascular plants that have disappeared.)

Guillemaut, L. Nomenclature des plantes observées aux environs de Louhans y croissant spontanément ou très répandues dans les jardins ou cultures avec l'indication sommaire de leurs principales propriétés et leurs usages. 158 p. 12mo. Louhans, 1878.

Annotated list with medicinal properties, uses, and habitats, but without localities. (Not seen; title and annotation supplied by G. Dillemann.)—See also his Flore de la Bresse louhannaise. Bresse Louhannaise, Bul. Mens. Soc. Agr. et Hort. Louhans **1889**: 125–130, 139–144. 1889. (List of 240 species, arranged by habitats. Not seen; title and annotation supplied by B. de Retz. Perhaps same material as the next.)—Also his Flore de la Bresse louhannaise. Distribution de la végétation. Plantes les plus communes. *In* Topographie physiologique et médicale de l'arrondissement de Louhans. Notes et remarques sur la Bresse louhannaise. p. 353–372. 12mo. Louhans, 1890. (Unannotated list of plants grouped by habitats. Not seen; title and annotation supplied by G. Dillemann.)

Ormezzano, Quentin, Chateau, Émile, and Gillot, F. X. Florule raisonnée du Brionnais. Bul. Soc. Hist. Nat. Autun **19**: 221–322. map. 1906; **20**: 49–114. 1907; **21**: 53–90. 1908. (Reprinted, 213 p. Autun, 1908; in Gray Herbarium library.)

Botanical explorations, topography, geology, climate, plant formations; annotated list of vascular plants and Characeae, with localities; chapter on introduced plants, with classified lists; annotated list of place names drawn from plants; list of vernacular names. Covers an area of 90,000 hectares, including 59 communes in extreme southern Saône-et-Loire and part of 4 communes in the canton of Donjon in adjacent Allier.

Quincy, Charles. Flore creusotine comprenant, dans un rayon de vingt kilomres., les plantes croissant spontanément, soumises à la grande culture, ou cultivées communément dans les jardins, avec indication de leurs principales propriétés, etc., ouvrage destiné à vulgariser les études botaniques élémentaires de botanique locale, et à classer les herbiers scolaires. (Cover title, Catalogue de la flore creusotine . . .) 1. partie (all publ.). 136 p. (all recto pages from 17 through 127 blank). 21.5 cm. Creusot[sic], 1888 (cover place and date). (Hectographed.)

List of wild and cultivated vascular plants and Characeae (1212 species), with indication of useful and harmful species and occasional localities; annotated list of medicinal plants.—See also his Notes sur la flore du Creusot et de ses environs. Feuille des Jeunes Nat. **11**: 107–111. 1881. (Changes in vegetation, etc.; lists of characteristic native species, grouped by habitats, and of introduced species.)—Also his Florule adventive du Creusot. Revue Bot. Bul. Mens. Soc. Franç. Bot. **3**: 294–304. 1885. (General considerations; annotated list of known adventive plants, divided into those that are native in the central part of France but not in Le Creusot, and those coming from other regions.)—Also his Les plantes adventives du Creusot. Bul. Soc. Sci. Nat. Saône-et-Loire 3: 1–7. 1889(?) (Behavior of adventive plants; list, with localities.)—Also his Additions à la flore adventive du Creusot. Revue Bot. Bul. Mens. Soc. Franç. Bot. 12: 100. 1894. (Adds 3 species.)—Also his Additions à la florule adventice du Creusot. Feuille des Jeunes Nat. 25: 76–77. 1895. (Notes on various adventive species.)

——— Florule des alluvions de la Saône aux environs de Chalon. Bul. Soc. Hist. Nat. Autun 11 (2): 202–216. 1898.

Includes apparently incomplete list of angiosperms, divided by habitats. —See also **Lemosy, E.,** and **Lemosy, Louis.** Notes sur quelques plantes des environs de Chalon. Bul. Soc. Sci. Nat. Saône-et-Loire 28 (n.s., 8): 71–74, 96–99. 1902. (List of noteworthy angiosperms, with localities.)—Also their Notes sur quelques plantes des environs de Chalon recoltées ou observées en 1902. l.c. 29 (n.s., 9): 112–116. 1903. (Additional list.)

——— Simples notes sur les plantes spontanées alimentaires. Bul. Soc. Sci. Nat. Saône-et-Loire 28 (n.s., 8): 100–109. 1902.

Annotated list of edible spermatophytes of vicinity of Chalon, grouped by parts used, and alphabetically arranged by vernacular names.

SARTHE

See also Mayenne (Labbé).

General

Desportes, Narcisse. Flore de la Sarthe et de la Mayenne, disposée d'après la méthode naturelle, avec l'indication des propriétés médicales des plantes et leur usage dans les arts. lx, 528 p. 21.5 cm. Le Mans, Paris, 1838. (Gray Herbarium library.)

Descriptive flora of vascular and cellular plants; out of date as a flora, included here for its notes on medicinal and other uses. Bastard title, Flore du Maine.

Gentil, Ambroise. Bibliographie botanique sarthoise. Bul. Soc. Agr., Sci. et Arts Sarthe 45 (II, 37): 159–173. 1915.

Unannotated list, alphabetically arranged by authors, covering 1786–1915.

——— Botanique sarthoise, origines et progrès. Bul. Soc. Agr., Sci. et Arts Sarthe 48 (II, 40): 213–226. 1922.

Sketch of botanists and botanical explorations, with mention of publications.

——— Glossaire de la flore sarthoise. Bul. Soc. Agr., Sci. et Arts Sarthe 49 (II, 41): 65–166. 1923–24.

Alphabetical list of vernacular names, with explanation and scientific equivalents; list of scientific names with vernacular equivalents.

——— Inventaire général des plantes vasculaires de la Sarthe indigènes ou naturalisées et se reproduisant spontanément. Bul. Soc. Agr., Sci. et Arts Sarthe 33 (II, 25): 365–476. 1892; 34 (II, 26): 59–183, 285–389. 1893–94. (Reprinted 340 p. Le Mans, 1892–94; in Gray Herbarium library.)

Bibliography; annotated list of vascular plants (1192 species), with localities; supplement (34: 367–379), generic index.—See also his Inventaire général . . . 2. supplément. l.c. 44 (II, 36): 233–280. 1913. (Extensive list of additions and corrections.)—Also his Contributions à la flore sarthe. Relevé d'observations faites depuis 1914. l.c. 48 (II, 40): 227–230. 1922.

——— Petite flore mancelle contenant l'analyse et la description sommaire des plantes vasculaires de la Sarthe. 3. éd. 256 p. 18.5 cm. Le Mans, 1898. (Gray Herbarium library.) (1st ed. 1884.)

Annotated, keyed, briefly descriptive flora, with localities for rarer species.

Local

Diard, P. M. Catalogue raisonné des plantes qui croissent naturellement à St.-Calais et dans ses environs . . . revue par Éd. Guéranger. 252 p. 22 cm. St.-Calais, 1852. (Arnold Arboretum library.)

Annotated list of cellular and vascular plants [932 species].—See also **Cottereau, E.,** *abbé.* Recherches sur la flore des environs de Saint-Calais.

Bul. Soc. Agr., Sci. et Arts Sarthe **58** (**III, 9**): 83–128. 1 fig. (map). 1941. (Botanical explorations, geography, geology, origins of flora, plant associations, annotated list of vascular plants that have disappeared since Diard's work, list of new localities, list of species not recorded by Diard, with localities; list of adventive plants of railroads. In New York Public Library.)

[**Maulny,—.**] Plantes observées aux environs de la ville de Mans. 1 p.l., 254 p. 17 cm. Avignon, 1786. (Library of Muséum national d'histoire naturelle, Paris.)

Annotated list of vascular and some cellular plants, alphabetically arranged by vernacular names, with scientific names (not indexed) and localities. —See also **Henry, Adrien.** Flore des environs immédiats du Mans. Bul. Soc. Agr., Sci. et Arts Sarthe **56** (**III, 7**: 35–46. 1937. (List of adventive and naturalized plants, grouped according to means of introduction.)

Roquet,—. Observations botaniques faites aux environs du Lude (1881–85). Bul. Soc. Agr., Sci. et Arts Sarthe **30** (II, **22**): 392–403. 1886.

List of vascular plants, with localities.

SAVOIE

See also Haute-Savoie (Payot, Florule de Mont-Blanc); Switzerland, General (Bennett; Binz and Thommen, under Binz; Bouvier; Thommen).

General

Bouvier, Louis. Histoire de la botanique savoyard. Bul. Soc. Bot. France **10**: 644–667. 1867.

Sketch of botanical explorations, arranged according to nationality of the explorers. Relates to both Savoie and Haute-Savoie.—See also his Notice bibliographique concernant les botanistes dont les voyages ou les ouvrages se rapportent plus ou moins directement aux Alpes savoyardes. l.c. 667–675. 1867. (List of botanists, grouped by nationality, with biographical information and list of pertinent publications.)—Also **Gave, Pierre,** *abbé.* Liste des contributions apportées à la flore de la Savoie depuis 1863 jusqu'en 1905. Cong. Soc. Savantes Savois. **17**: 257–288. 1906. (List of 3 earlier botanists; alphabetical list of 35 foreign botanists who have collected or published on the flora of Savoie, with references to publications; similar list of 69 French botanists. In New York Public Library.)

Chabert, Alfred. De l'emploi populaire des plantes sauvages en Savoie. Bul. Herb. Boissier **3**: 291–301, 334–344. 1895.

Lists of medicinal and other useful plants, classified by uses; poisonous plants; remains of prehistoric plants.—See also his Des plantes sauvages comestibles de la Savoie. Bul. Herb. Boissier **5**: 258–272. 1897. (Edible plants, classified by parts used.)—Also his Noms patois et emploi populaire des plantes de la Savoie. l.c. **5**: 568–578. 1897. (Alphabetical list of vernacular names, with botanical equivalents; various special uses and superstitions.)—Much the same material, with additions and omissions, is given in his paper, De l'emploi populaire des plantes sauvages en Savoie. Bul. Soc. Hist. Nat. Savoie II, **3**: 169–245, 267–276. 1897. Refers to Departments of Savoie and Haute-Savoie.

Constantin, Aimé, and **Gave, Pierre,** *abbé.* Flore populaire de la Savoie. Première partie. Dictionnaire des noms populaires des plantes qui croissent naturellement en Savoie ou qui y sont cultivées en pleine terre . . . 1 p.l., xii, 190 p. 25.5 cm. Annecy, 1908. (Issued as supplement to the Revue savoisienne v. 46–49. 1905–08.)

Alphabetically arranged list of standard French vernacular names (1,339 entries), with botanical name, popular French names in Savoie and Haute-Savoie, patios names, with localities, pharmaceutical names, explanation of vernacular names, notes on uses; bibliography, indices. Includes wild and cultivated vascular and cellular plants.—See also **Gave, P.** Petit supplément . . . Revue Savoisienne v. 52, suppl. 9 p. 1911.

Perrier de la Bâthie, Eugène. Catalogue raisonné des plantes vasculaires de Savoie, départements de la Savoie et de la Haute-Savoie, plateau de Mont-Cenis . . . Avec une préface et un supplément par le Dr. Jules Offner. Mém. Acad. Sci. Belles-Let. et Arts Savoie ser. 5, v. 4. xlv, 433 p. map. 1917; v. 5. 415 p. 1928. (Also issued separately, Paris et Le Carriol, 1917–1928.)

Botanical districts, geology, phytogeography; annotated list of vascular plants, with detailed localities.—See also **Offner, Jules.** Nouvelles acquisitions pour la flore de Savoie. Compt. Rend. Assoc. Franç. Avanc. Sci. **57:** 297–299. 1933. (Running account of newly recorded or noteworthy phanerogams, from various sources.)—Also **Thommen, Édouard,** and **Becherer, Alfred.** Contributions à la flore des départements de la Haute-Savoie, de la Savoie, de l'Isère et des Hautes-Alpes. Bul. Soc. Bot. Genève II, **33:** 109–130. 3 fig. 1942. (List of noteworthy species.)

Local

Beauverd, Gustave. Sur la flore vernale de la Tarentaise (Savoie). Bul. Soc. Bot. Genève II, **4:** 167–216. 15 fig. (fig. 15, map). 1912.

Topography, geology, etc., of various localities, with partial lists of species, critical notes on various species, general considerations on flora, partly annotated bibliography. For additions to bibliography, see Bul. Soc. Bot. Genève II, **24:** 223–224. 1933.—See also **Fournier, Paul.** Plantes de Tarentaise. Monde des Plant. IV, **35** (205): 3–5, (207): 19–20. 1934; **36** (209): 2–3, (212): 12–13, (214): 26. 1935. (List of vascular plants not previously recorded from this area, with localities. In Arnold Arboretum library.)—Also Evrard, below.

Bonjean, J. L. Catalogue des plantes du Mont-Cenis. 37 p. 1846.

Not seen; no review available.—See also **Bouvier, Louis.** Le Mont Cenis: son histoire et sa végétation. 1863. (Not seen; no review available.)—Also **Meyran, Octave.** Une herborisation au Mont-Cenis 12, 13 et 14 juillet 1891. Revue Bot. Bul. Mens. Soc. Franç. Bot. **9:** 329–341. 1891. (Account of excursion with mention of numerous species.)—Also **Durafour, Arthur.** Excursion botanique en Tarentaise, Maurienne et au Mont-Cenis du 3 au 15 août 1904. Bul. Soc. Nat. Ain **10.** année (no. 16): 48–71. 1905. (Itinerary with partial lists of species observed, including Mont-Cenis (p. 66–70). In library of Muséum national d'histoire naturelle, Paris.)—Also **Thompson, H. S.** Liste des phanérogames et cryptogames vasculaires recueillis au-dessus de 8,000 feet (2,440 mètres) dans les districts du Mont-Cenis, de la Savoie, du Dauphiné et des Alpes-Maritimes (Juin-Septembre 1907), avec quelques notes sur les limites altitudinales des plantes alpines. Bul. Acad. Internatl. Géog. Bot. **18:** 194–248 incl. map. 1908; Corrections . . . l.c. **18:** 255–256. 1908. (Plant zones, author's itinerary, statistics of altitudinal range; list of vascular plants and Characeae collected by author in 1907, with localities and altitudes.)—Also **Beauverd, Gustave.** Sur la flore vasculaire des environs de Modane, de Bardonnèche et de Suze (massif du Cenis). Bul. Soc. Bot. Genève II, **13:** 115–184. 1922. (Author's itinerary with mention of numerous species, discussion of various critical plants; statistics, phytogeography, etc., of flora of Susa; bibliography.)—Also **Offner, Jules.** Esquisse botanique de la région du Mont-Cenis. Bul. Soc. Bot. France **67** (Sess. Extraord.): xxv-xxxii. 1923. (Previous publications, topography, plant zones, etc. Followed by lists of noteworthy species by various authors.)

Chabert, Alfred. La flore d'Aix-les-Bains. Bul. Soc. Bot. France **54:** 91–97. 1907.

Discussion of numerous non-native species collected in vicinity of Aix-les-Bains by various botanists and others. "De toutes les flores locales d'Europe, la flore d'Aix-les-Bains est bien celle qui a été la plus embellie par les mystificateurs." Same material in Bul. Soc. Hist. Nat. Savoie II, **13:** 195–200. 1909.

——— Recherches botaniques dans les Alpes de la Maurienne. Bul. Soc. Bot. France **30:** 2–19. 1883.

Topography, etc.; extensive list of vascular plants collected chiefly by author, with localities.—See also **Beauverd, Gustave.** Herborisation de la Société botanique en Maurienne de 13 au 15 avril 1911. Bul. Soc. Bot. Genève II, **3:** 195–232. 6 fig. (incl. map). 1911. (Lists of species of various localities, with notes on general features of vegetation; short list of new records; briefly annotated bibliography.)—Also **Perrier de la Bâthie, Eugène.** Additions à la flore de la Maurienne. l.c. II, **3:** 240–244. 1911.

Chatelain, Maurice. Aperçu de la flore de Faverges et de quelques localités voisines. Revue Savoisienne **42:** 95–116, 210–225, 345–347. 1901.

List of 373 spermatophytes, with localities, collected by author from 1868–1881.

Evrard, F., and Chermezon, Henri. Sur la flore de la Haute-Tarentaise. Bul. Soc. Bot. France **64:** 163–202. 1917.

Botanical explorations (with references); extensive list of vascular plants observed by authors in 1913, with localities. Relates to the region from the Brévières de Tignes to the sources of the Isère, with the neighboring passes and peaks.—See also their La végétation de la Haute-Tarentaise. Bul. Soc. Bot. France **65:** 153–209. 1918. (Topography, geology, climate, plant zones and formations with lists of species.)—Also Beauverd, above.

Menghini, Anna. La flora della Lancebranlette (Piccolo S. Bernardo). Chanousia **2:** 182–220. 1932.

Geology, etc.; annotated list of vascular plants; plant formations, notes on various species, bibliography.

Nétien, Georges. Étude sur la flore du massif des Aiguilles d'Arves. Bul. Mens. Soc. Linn. Lyon **14:** 30–41, 63–64, 65–67, 197. 2 fig. 1945; **16:** 153–157, 167–169. 1 fig. 1947.

Geology, botanical explorations, sketch of flora of different areas with lists of plants observed, plant associations; list **16:** 153–156) of vascular plants (over 300 species and subspecies), with localities for scarcer ones; comparison with other areas, bibliography, etc.

Peyronel, Bruno. Flora del Monte Valaisan al Piccolo S. Bernardo. Chanousia **4:** 136–158. 1940.

List of the vascular plants of Monte Valaisan, with notes on the vegetation. (Not seen; title and annotation supplied mostly by V. Giacomini.)

Pin, C. Catalogue des plantes observées ou citées aux environs d'Aix-les-Bains, Chambéry, Albens, Rumilly, La Chautagne et Les Bauges. 193 p. 19 cm. Aix-les-Bains, 1893.

List of 1765 plants, with localities. (In Savoie and Haute-Savoie. Not seen; title and annotation supplied by B. de Retz.)

Vaccari, Lino, and Pavarino, Luigi. Catalogo delle piante spontanee e coltivate nell'anno 1897 nella Chanousia al Piccolo S. Bernardo. 26 p. 24 cm. Aosta, 1897.

List of the wild and cultivated plants of Chanousia, an alpine garden on the Piccolo San Bernardo. (Not seen; title and annotation supplied by V. Giacomini and F. Sappa.)

SEINE

See also Paris region for works covering Seine and several or many other departments.

The works cited here refer primarily to the city of Paris proper; the secondary references under Gaudefroy bring together papers on the weeds of the environs of Paris from other departments besides Seine.

Local

Gaudefroy, Eugène, and Mouillefarine, Edmond. Notice sur les plantes méridionales observées aux environs de Paris (Florula obsidionalis). Bul. Soc. Bot. France **18:** 246–252. 1871.

Localities, etc.; list of 190 adventive angiosperms, with localities, observed in various parts of Paris and regarded as introduced in the fodder supply of the armies during the two sieges of Paris 1870–71.—See also their La florule obsidionale des environs de Paris en 1872. Bul. Soc. Bot. France **19:** 266–277. 1872. (Describes survival of various species and raises total to 268.) Final notes by **É. Bureau** and **Gaudefroy** (l.c. **21:** 153. 1874) show only 3 species persisting in 1874.—Also **Urban, Ignatz.** [Die durch die heereszüge der jahre 1870 und 71 bewirkte einschleppung zahlreicher fremder pflanzenarten nach Frankreich.] Sitzungs-ber. Gesell. Naturf. Freunde Berlin **1873:** 77–81. 1873 (also in Bot. Zeitung **31:** 678–682. 1873). (Mention of earlier publications on plants introduced as a result of troop movements, with discussion, and particular consideration of species of *Medicago.*) An abstract of these and some other papers referring to the introduction of adventive plants into France during the war of 1870–71 is given in Bot. Jahresber. Just **2** (2): 1105–1106. 1876.—Also **Gagnepain, François.** Quelques plantes rudérales parisiennes. Bul. Soc. Bot. France

47: 337–342. 1901. (Account of the weeds of rue de Tolbiac and vicinity.)—Also **Guffroy, Charles.** Notes sur la flore parisienne. l.c. **58:** 505–512, 583–585. 1911. (List of noteworthy species including many adventives; list of 134 species growing on walls in Garches (Seine-et-Oise) and vicinity, with discussion.)—Also **Jeanpert, H. E.** Localités nouvelles de plantes parisiennes. l.c. **63:** 105–107. 1917; **66:** 235–238. 1919; **67:** 376–378. 1921. (Lists of noteworthy species from Seine and some adjacent departments, including numerous weeds.)—Also **Debray, Marcel,** and **Senay, Pierre.** Sur quelques plantes naturalisées, subspontanées ou adventices, constatées récemment ou retrouvées dans la Seine-Inférieure. Bul. Soc. Linn. Seine-Marit. **9:** 209–217. 1923. (List of about 30 species. Not seen; cited from Bul. Soc. Bot. France **72:** 228. 1925.)—Also **Jovet, Paul.** Florule suburbaine du nord du Paris: Aubervilliers, St-Denis, la Courneuve. l.c. **73:** 999–1006. 1927. (Extensive list consisting essentially of adventive plants.)—Also **Guinet, Camille.** Contribution à l'étude de la flore adventice de la région parisienne. Quelques espèces exotiques naturalisées au Muséum national d'histoire naturelle et susceptible d'être observées dans la région parisienne. l.c. **83:** 225–234. 1936. (Annotated list of some species naturalized on the grounds of the Museum; bibliography.)—Also **Chevalier, Auguste.** Observations sur la flore adventice de Paris. Comp. Rend. Acad. Sci. Paris **211:** 453–457. 1940. (General considerations, discussion of plants persisting from 1870–71.)—Also his Plantes adventices et archéophytes observées à Paris ou aux environs de 1900 à 1940. l.c. **211:** 525–529. 1940. (Mention of numerous species.)—Also **Jovet, P.** Plantes rudérales adventices et naturalisées de Paris et de sa banlieue. Bul. Soc. Bot. France **87:** 286–299. [1941?] (List of about 100 species; bibliography.)—Also his Evolution des groupements rudéraux "parisiens." l.c. **87:** 304–312. [1941?] (Discussion of adventive flora of "décombres enclos.")—Also **Chevalier, A.** Plantes adventices apparues à Paris et aux environs pendant les périodes de guerre. Compt. Rend. Acad. Sci. Paris **218:** 861–863. 1944. (Discussion of 2 species.)—Also his Plantes adventices introduites par la guerre actuelle à Paris et aux environs. Bul. Soc. Bot. France **91:** 102–106 incl. pl. 1–2 (on 1). 1944. (Discussion of 3 species.)—Also **Hibon, Georges.** Remarques sur les fluctuations de quelques plantes adventices dans la région parisienne. l.c. **97:** 18–20. 1950. (Discussion of variation in abundance of various species.)—Also his Végétation de l'emplacement des fortifications de Paris. l.c. **97:** 25–28. 1950; . . . Observations générales. l.c. **97:** 31–33. (History of the fortifications, with systematic list of species observed, and discussion of various species.)

Vallot, Joseph. Essai sur la flore du pavé de Paris limité aux boulevards extérieurs, ou catalogue des plantes qui croissent spontanément dans les rues et sur les quais. Suivi d'une florule des ruines du Conseil d'État. 122 p. 15.5 cm. Paris, 1884. (Gray Herbarium library.)

History, bibliography of similar floras; annotated lists of 209 vascular plants of Paris streets and of 152 species of the Conseil d'État.—See also his Sur les modifications de la flore des ruines du Conseil d'État de 1883 à 1884. Bul. Soc. Bot. France **31:** 321–323. 1884.—Also his Florule du Panthéon. Jour. de Bot. **1:** 52–55. 1887. (List of 32 species, with discussion.)

Weill, Jean. La flore du Bois de Vincennes et son évolution. Bul. Soc. Bot. France **94:** 416–423. 1948; **95:** 12–17. 1948.

Changes due to human influence, etc.; annotated list of noteworthy plants, including species that have disappeared; bibliography.

SEINE-ET-MARNE

See also Aube (Des Étangs); Paris region.

Local

Dalmon, Jacques, and **Dalmon, Henri.** Destruction de la flore des environs de Paris. Variations topographiques de la flore de la forêt de Fontainebleau et de ses alentours. Bul. Assoc. Nat. Vallée du Loing **1:** 35–43. 1913. (Harvard College library.)

Changes in soil, etc., due to civilization; annotated list of some spermatophytes that have disappeared, etc.

Doignon, Pierre. Histoire et bibliographie des recherches phanérogamiques et ptéridologiques dans le massif de Fontainebleau. Cahiers des Nat. n.s., **13**: 35–40. 1957.

Condensed account of botanical explorations with partial references.

Gaume, Raymond. Contribution à l'étude de la flore de la Brie. Bul. Soc. Bot. France **67**: 89–101, 161–169. 1920–21; **69**: 80–84, 608–611. 1922–23.

General features of flora, references to publications; lists of new localities for numerous vascular plants.—See also his Contribution à l'étude de la flore de la Brie. Bul. Soc. Sci. Nat. Seine-et-Oise II, **5**: 66–69. 1924. (List of noteworthy species, with localities.)—Also his Aperçu sur les groupements végétaux du plateau de la Brie. Bul. Soc. Bot. France **72**: 393–416. 1925. (Account of plant associations.)

Laroque, Hippolyte. Indicateur de la flore de Provins et de ses environs précédé d'un aperçu topographique du territoire. 387 p. 16 cm. Provins, 1885 (cover date 1886). (Library of Muséum national d'histoire naturelle, Paris.)

Topography, geology, etc.; "Indicateur de la flore" (p. 115–368), an annotated list of vascular plants and Characeae, with localities.—See also his Botanique. *In* his Géologie descriptive du bassin de la Voulzie suivie de 16 excursions botaniques autour de Provins. p. 267–326. 16mo. Provins, 1891. (Account of botanical excursions to numerous localities, with lists of noteworthy plants. Not seen; title and annotation supplied by G. Dillemann.)

Weil, Lucien. Catalogue des arbres, arbustes et arbrisseaux de la forêt de Fontainebleau. *In* Forêt de Fontainebleau, recherches sur son sol, sa faune, sa flore. fasc. 4, p. 5–20. 1930. (Publication of Association des naturalistes de la vallée du Loing.)

Annotated list of trees, shrubs, and undershrubs, with localities. (Not seen; title and annotation supplied by G. Dillemann.)—See also **Fauvais, Charles,** and **Weil, Lucien.** Sur les fruits comestibles rencontrés en forêt de Fontainebleau. Bul. Assoc. Nat. Vallée du Loing **13**: 43–47. 1930. (Annotated list of 30 edible fruits. In library of Muséum national d'histoire naturelle, Paris.)—Also **Gaume, Raymond.** Rapport sur l'excursion faite par la Société botanique de France en forêt de Fontainebleau le 14 juin 1935. Bul. Soc. Bot. France **83**: 131–144. 1936. (Geology, general features of flora, bibliography.)

SEINE-ET-OISE

See also Eure (Toussaint); Paris region.

General

Guffroy, Charles. L'étude de la flore de Seine-et-Oise au XXme siècle. Bul. Soc. Sci. Seine-et-Oise III, **2**: 49–82, 105–120. 1934.

Annotated bibliography (246 titles) of general works and those referring to Seine-et-Oise; list of species of Seine-et-Oise not figured in Jeanpert's Vade-mecum (1911); index to species listed in the annotated bibliography.—See also his La flore de Seine-et-Oise. Bul. Soc. Sci. Seine-et-Oise II, **1**: 8–9, 16–17, 41–42, 50–52, 63–66. 1919–20. (Includes lists of about 175 very rare species, about 200 rare species, about 145 "Plantes d'origine exotique," etc. Second installment not seen. The next to last item is bv Despaty.)

Local

Allorge, Pierre. Contribution à l'étude du Vexin français. Bul. Soc. Bot. France **60**: 609–612. 1913.

List of noteworthy species from Seine-et-Oise, mostly from the massif du Arthies.—See also his Notes sur quelques plantes intéressantes du Vexin français. Bul. Soc. Bot. France **66** (Sess. Extraord.): xxxvi-xliii. 1922. (Extensive list of noteworthy plants, nearly all from Seine-et-Oise.)

Belèze, Marguerite. Catalogue des plantes nouvelles rares ou intéressantes phanérogames, cryptogames vasculaires et cellulaires ainsi que de quelques hybrides remarquables des environs de Montfort-l'Amaury et de la forêt de Rambouillet (Seine-et-Oise). 78 p. map. 25 cm. Le Mans, 1905. (Arnold Arboretum library.)

Extensive list of rare or noteworthy vascular plants, mosses, hepatics, lichens, mushrooms, and Characeae, with localities.—See also **Guinet, Camille, and Weill, Jean.** Contribution à l'étude de la forêt de Rambouillet: le massif forestier des Yvelines (phanérogames et cryptogames vasculaires). Bul. Soc. Sci. Seine-et-Oise III, **2**: 1–10. 1934. (Topography, etc.; list of noteworthy vascular plants, with localities.)—Also **Guinet, Camille.** Itinéraire botanique en forêt de Rambouillet. Compte-rendu de l'excursion de 18 juin 1935. (Session extraordinaire de Paris, 1935.) Bul. Soc. Bot. France **83**: 190–201. 2 fig. (maps.) 1936. (Topography, etc.; itinerary, with mention of noteworthy species.)

Camus, E. G. Florule du canton de l'Île-Adam (Seine-et-Oise). Bul. Soc. Bot. France **33**: 28–36. 1886.

List of about 200 of the scarcer vascular plants out of about 800 known species, the others being generally distributed in the Paris region.—See also his Supplément à la Florule . . . Bul. Soc. Bot. France **33**: 305–6. 1886.

Ory, R. Florule des environs de Versailles. Bul. Soc. Sci. Seine-et-Oise II, **2**: 4–10. 5 fig. 1921.

List of noteworthy vascular plants, with localities, within a radius of 3 to 4 kilometers.

Réaubourg, Gaston. Note sur les plantes intéressantes indigènes ou adventices des environs de Mantes (Seine-et-Oise). Bul. Soc. Bot. France **54**: 358–368. 1907.

Soils; extensive list of vascular plants, with localities, based mostly on author's work.

Rousse, Émile. Flore de la Roche-Guyon, Seine-et-Oise. 2 p. l., 67 p. 19 cm. Paris, 1893. (Library of Muséum national d'histoire naturelle, Paris.)

Account of excursions; annotated list of spermatophytes, with localities.—See also Eure (Toussaint and Hoschedé, Flore de Vernon et de la Roche-Guyon).

SEINE-INFÉRIEURE

See also Normandie; Paris region.

General

Blanche, Emmanuel, and **Malbranche, A. F.** Description scientifique de la France. Partie botanique. Département de la Seine-Inférieure. Précis Anal. Trav. Acad. Imp. Sci. Belles-Let. et Arts Rouen **1862/63**: 225–387. 1863. (Reprinted with title: Catalogue des plantes cellulaires et vasculaires de la Seine-Inférieure. 166 p. 22 cm. Rouen, 1864. In Gray Herbarium library.)

Annotated list of cellular and vascular plants, with localities for scarcer species.—See also **Chevalier, Joseph.** Notes sur quelques plantes adventices récoltées dans le bassin de la Seine-Inférieure. Bul. Soc. Amis Sci. Nat. Rouen **35**: 49–58. 1900. (List of adventive plants from several departments.)—Also his Contribution à l'étude des plantes adventices en Haute-Normandie. l.c. **60/61**: 13–17. 1926. (Changes in adventive flora.)—Also **Senay, Pierre.** Plantes disparues ou en voie de disparition et plantes nouvelles pour la Seine-Inférieure. Compt. Rend. Assoc. Franç. Avanc. Sci. **53**: 429–431. 1929. (Account of several species that have nearly or quite disappeared and of a few newly recorded ones.)

Local

Coulon, L. Nos excursions. Guide des naturalistes dans les environs immédiats d'Elbeuf. Bul. Soc. Étude Sci. Nat. Elbeuf **28**: 69–128. 1911.

Includes mention of noteworthy plants, with localities, arranged in form of a nature calendar.—See also **Saint-Amand, Hippolyte.** Plantes adventices des environs d'Elbeuf. Bul. Soc. Étude Sci. Nat. Elbeuf **20**: 90–99. 1902. (Annotated list of noteworthy adventive plants; also list of adventive plants collected in Seine-Inférieure and Eure by J. Chevalier, with localities.)—Also his La flore des murs d'Elbeuf. l.c. **30**: 33–38. 1912. (Annotated list of vascular plants.)

Ébran, A. Catalogue des plantes vasculaires du Havre et lieux circonvoisins ou guide des herborisations. 103 p. 22.5 cm. Havre, 1869. (Lloyd Library.)

Annotated list of Characeae and vascular plants (1,162 species), with localities; list of excursions, with enumeration of noteworthy species.—See also **Senay, Pierre.** Contribution à la flore du Havre et des environs. Bul. Mens. Soc. Linn. Seine Maritime **8**: 31–58. 1922. (Annotated list of 98 naturalized plants.)—Also **Debray, Marcel,** and **Senay, Pierre.** Sur quelques plantes naturalisées, subspontanées ou adventices, constatées récemment ou retrouvées dans la Seine-Inférieure. l.c. **9**: 209–217. 1923. (New and additional records of 32 species.)—Also their Étude de la flore adventice du Havre et des environs. l.c. **18**: 3–48. 1932. (Annotated systematic list of 120 species; statistics, bibliography. This and the two above-cited papers not seen; titles and annotations supplied by R. McVaugh from copies in University of Michigan library.)—Also their La flore des ruines du Havre. Bul. Soc. Bot. France **92**: 229–235. 1945 [1946]. (Account of angiosperms found on the bombarded areas of Le Havre, 1944–45; short bibliography.)

Le Turquier de Longchamp, J. A. Flore des environs de Rouen. 2 v. (xxxii, 583 p.). 16.5 cm. Rouen, 1816.—Supplément. iii, 80 p. 22 cm. Rouen, 1825. (Issued as app. to Précis Anal. Trav. Acad. Sci., Belles-Let. et Arts Rouen 1824.)

Annotated descriptive flora of spermatophytes and Characeae, on the Linnaean system, with localities.—See also **Blanche, Emmanuel.** Rectifications et additions à la Flore des environs de Rouen. Précis Anal. Trav. Acad. Sci., Belles-Let. et Arts Rouen **1850/51**: 113–158. 1851. (List of excluded species, list of noteworthy species, list of species to be added.)—Also **Toussaint, Anatole,** *abbé.* Notice sur quelques stations de plantes aux environs de Rouen vers la fin du XVIIIe siècle. Bul. Soc. Amis Sci. Nat. Rouen III, **26**: 81–93. 1890. (Extensive list.)

SOMME

See also Cher (Duchaussoy); Pas-de-Calais (Hill); Picardie.

General

Éloy de Vicq, L. B. Flore du département de la Somme. xxxvi, 562 p. 18.5 cm. Abbeville, 1883.

Topography, climate, biographical sketches of local botanists; annotated descriptive flora of vascular plants and Characeae, with localities. Replaces **Éloy de Vicq, L. B.,** and **Blondin de Brutelette,—.** Catalogue raisonné des plantes vasculaires du département de la Somme. Mém. Soc. Émul. Abbeville [**10**] 1861/65: 295–620. 1865; Supplément au Catalogue . . . l.c. III, **1**: 565–602. 1873. (Annotated list of vascular plants and Characeae, with localities for rarer species. Reprinted, viii, 318 p. 1865; and (Supplément) nouv. éd. 4 p. l., 34 p. 1873; reprints in Arnold Arboretum library.)—See also **Gonse, Ernest.** Supplément à la Flore de la Somme. Mém. Soc. Linn. Nord France **7**: 5–64. 1888; Nouveau supplément . . . l.c. **12**: 6–90. 1908. (Lists of new localities and newly recorded species.)—Also **Brandicourt, Virgile.** [Liste des plantes étrangères au département de la Somme et trouvées dans les cantonnements de 1914 à 1917.] Monde des Plant. II, **19** (115): 24. 1918. (List of 24 adventive plants.)—Also his La dissémination des plantes. Bul. Soc. Linn. Nord France **24**: 22–44. 1929. (Includes (p. 41–44) lists of plants new to the flora of Picardie collected from 1915–28, mostly weeds. Not seen; title and annotation supplied by R. McVaugh.)

Local

Boucher de Crèvecoeur, J. A. G. Flore d'Abbeville et des environs. 3. éd. 116 p. 20.5 cm. Abbeville, 1834. (1st ed. 1803, with title Extrait de la flore d'Abbeville et du département de la Somme.)

Briefly annotated list of wild and cultivated vascular and cellular plants (2331 species) with localities for scarcer species; covers much of the department.—See also **Acloque, Alexandre.** Liste des plantes rares pour le département de la Somme recueillies à Abbeville et dans les environs. Monde des Plant. **4:** 253–256. 1895. (List of rarer vascular plants collected by author, with localities.)

Brandicourt, Virgile. La flore des rues d'Amiens. Bul. Soc. Linn. Nord France **17:** 38–52. 1904.
Discussion of various localities; list of streets, with names of plants found in pavement and on walls in each.

Cauchetier-Chapron, —, and Guffroy, Charles. Catalogue des plantes vasculaires de Montdidier et de ses environs (dans un rayon de 10 kilomètres). Bul. Assoc. Franç. Bot. **5:** 9–16, 65–72, 137–144, 235–238. 1902.
List of communes in area covered; annotated list of angiosperms. (Incomplete; almost entirely Gamopetalae.)—See also their Plantes rares ou nouvelles des environs de Montdidier. Bul. Soc. Bot. France **52:** 39–44. 1905. —Also **Duchaussoy, H.** Flore des environs de Montdidier. Bul. Soc. Linn. Nord France **17:** 243–245. 1905. (References to additional publications relating to the flora.)—Also **Gonse, Ernest.** Contributions à la flore des environs de Montdidier (1883–1899). l.c. **17:** 245–253. 1905. (Extensive list of vascular plants, with localities.)

Caussin, Octave. Flore des tourbières du département de la Somme. 301 p. 2 pl., map. 23 cm. Mayenne, 1912.
General account of peatbogs, description of plant associations in 180 bogs occurring in 19 valleys of the department; list of about 500 vascular plants and 80 bryophytes and Characeae, with localities. (Not seen; title and annotation supplied by B. de Retz.)

——— Flore descriptive du littoral Picard. 1 p. l., 110 p. 18.5 cm. Paris, 1907. (Lloyd Library.)
Annotated, keyed, descriptive flora of spermatophytes and Characeae, with localities. Covers region from the mouth of the Authie to that of the Bresle.

——— Flore du canton de Chaulnes. 1 p.l., xiv, 38 p. 18 cm. Amiens, 1906.
General features of flora, geography, etc.; list of vascular plants and Characeae (about 700 species), with localities for all except the common species.

——— Flore topographique du territoire de Proyart. xi, 35 p. 12mo. Rosières, 1905.
Geography, etc.; annotated list of plants, with localities. (Not seen; title and annotation supplied by G. Dillemann.)—See also his Contributions à la flore locale de la Somme. I. (Proyart.) Bul. Soc. Linn. Nord France **18:** 389–396, 491. 1907. (Extensive list of vascular plants and Characeae, with localities.)

TARN

See also Languedoc.

General

Bel, Jules. Nouvelle flore du Tarn et de la région toulousaine. lix, 371 p. 13 pl. 19 cm. Albi, 1885. (Lloyd Library.)
Botanical explorations, geology, botanical regions, elements of botany, keys to families; annotated, keyed list of vascular plants, with localities and brief notes on medicinal and other uses; glossary, list of Romano-Castraise patois names (not cited in text), indices of French and botanical names. The so-called 2d edition of 1888 is merely a reprint of the 1st with slightly different title (Nouvelle flore du Tarn et de la Haute-Garonne sous-pyrénéenne, fide H. Gaussen in litt.). Replaces **Martrin-Donos, Victor,** *comte de.* Florule du Tarn ou énumération des plantes qui croissent spontanément dans le département du Tarn. v. 1 (xxiii, 872 p. pl.) 21 cm. Paris, 1864. (Topography, geology, etc.; annotated list of vascular plants, with keys, localities, notes on uses (especially on forage value), and frequent descriptions or critical notes. Vol. 2, 1867, Végétaux cellulaires; additions to v. 1 on p. xii–xiii. Both in Gray Herbarium library.)—See also **Caraven-Cachin, Alfred.**

Note sur l'apparition et l'extinction de quelques végétaux dans le département du Tarn. Bul. Comn. Antiq. Castres et Dépt. Tarn **1**: 286–292. 1878. (List of recently naturalized and acclimated angiosperms, with dates, and of a single species that had disappeared.)—Also his Les plantes nouvelles du Tarn (1874–1891). Compt. Rend. Assoc. Franç. Avanc. Sci. 21 (2): 453–456. 1893. (List of adventive plants, arranged chronologically.)—Also **Bel, Jules.** Géographie botanique du département du Tarn. Revue Bot. Bul. Mens. Soc. Franç. Bot. **11**: 432–470. 1893. (Orography, hydrography, geology; list of plants (vascular and cellular) grouped according to soils on which they occur.)—Also **Sudre, Henri.** Notes sur quelques plantes critiques de la flore du Tarn. l.c. **12**: 17–31. pl. 1–2. 1894. (Critical notes on various species.) —Also his Révision des Rubus de l'herbier du Tarn de de Martrin-Donos. Bul. Soc. Bot. France **46**: 81–99. 1899. (Critically annotated list of *Rubus.*)

Caraven-Cachin, Alfred. Catalogue des plantes médicinales, pharmaceutiques, tinctoriales, industrielles et commerciales du département du Tarn. Bul. Comm. Antiq. Castres et Dépt. Tarn **3**: 216–241. 1880; **4**: 133–170. 1881.

Annotated list of 170 medicinal or otherwise useful vascular plants, with patois names, habitat, properties, and uses.

Clos, Dominique. Historique de la flore du département du Tarn. 11 p. 24 cm. Albi, 1898. (Extrait de la Revue du Tarn, 1898.) (Library of Muséum national d'histoire naturelle, Paris.)

Account of botanists and their publications, chronologically arranged.

Local

Clos, Dominique. Phytostatique du Sorézois, bassin méridional du département du Tarn. Mém. Acad. Sci., Inscript. et Belles-Let. Toulouse IX, **7**: 242–301. 1895.

Topography, geology, botanical explorations and publications, partial lists of species of various localities, plant zones, etc.; list of "espèces les plus notables" (p. 270–291), with localities; lists of species of the "causses" (bare limestone plateaus.)

Doumenjou, J. B. Herborisations sur la Montagne-Noire, et les environs de Sorèze et de Castres, suivies du catalogue des plantes phanérogames qui végètent spontanément dans ces localités. 2 p.l., 326 p. 3 col. pl. 22.5 cm. Castres, 1847. (New York Botanical Garden library.)

The Catalogue . . . (p. 141–307) is an annotated list of spermatophytes, with localities. Page 326 is followed by an unpaged leaf of additions and corrections. His Supplément aux Herborisations . . . (61 p. 22.5 cm. Albi, 1851. In New York Botanical Garden library) contains no additions to the list.—See also **Ozanon, Charles.** Note sur les plantes les plus remarquables du versant méridional de la Montagne-Noire, recuillies en juin 1860, dans le canton de Mas-Cabardès, arrondissement de Carcassonne (Aude). Bul. Soc. Bot. France **8**: 119–121, 165–168. 1861. (Botanical explorations, topography, plants of calcareous and granitic soils; list of noteworthy plants.)—Also **Clos, Dominique.** Sur la végétation d'un coin meridional du département du Tarn (Montagne Noire). l.c. **32**: 361–364. 1885. (Mention of noteworthy species.)—Also **Baichère, Édouard,** *abbé.* Herborisations dans le Cabardès et le Minervois (versant meridional de la Montagne Noire) (Aude). l.c. **35** (Sess. Extraord.): l–lii. 1889. (Topography, geology, phytogeography; list of noteworthy plants, grouped by habitats.)—Also **Delpont, J.** Contributions à la flore des Martys (Aude). Bul. Soc. Études Scient. Aude **6**: 139–144. 1895. (Plant habitats, etc.; list of vascular plants grouped by habitats. The locality is part of the Montagne-Noire.)—Also **Gavoy, L.** Dans le Montagne Noire. Liste des plantes. Bul. Soc. Étude Sci. Nat. Carcassonne **22**: 188–193. 1911. (Includes list of about 40 species. In Harvard College library.)—Also Baichère, Étude sur la flore des environs de Carcassonne, under Aude.

Laborie, Eugène. Contribution à la flore du département du Tarn. Bul. Soc. Hist. Nat. Toulouse **23**: 32–49. map. 1889.

Topography, soils, general features of flora, etc., of the Causse de Castres; list of vascular plants, annotated only as to abundance.

TARN-ET-GARONNE

See also Languedoc.

General

Lagrèze-Fossat, Adrien. Flore de Tarn et Garonne, ou description des plantes vasculaires qui croissent spontanément dans ce département. xii, 527 p. 22 cm. Montauban, Moissac, 1847. (Gray Herbarium library.)

Soils, botanical explorations; annotated, keyed, descriptive flora of vascular plants, with localities and French and patois names.

Local

Gaterau, —. Description des plantes qui croissent aux environs de Montauban, ou qu'on cultive dans les jardins . . . 216 p. 20 cm. Montauban, 1789. (Arnold Arboretum library.)

Account of descriptive terms, etc.; annotated list of vascular and cellular plants, on the Linnaean system; glossary, list of French vernacular names with patois equivalents.

VAR

See also Bouches-du-Rhône (Reynier, Herborisations); Provence.

General

Albert, Abel, and **Jahandiez, Émile.** Catalogue des plantes vasculaires qui croissent naturellement dans le département du Var. xliv, 613 p. 16 pl., map. 23 cm. Paris, 1908.

Topography, geology, phytogeography, plant zones (all by Flahault); annotated list of vascular plants (2165 species and 61 hybrids), with localities; bibliography, index of Provençal vernacular names.—See also **Jahandiez, É.** Additions à la flore du Var. Annales Soc. Hist. Nat. Toulon [1] 1910: 75–84. 1910; **7:** 36–74. 2 pl. [1922]; **13:** 42–75. 1927–28. (Extensive lists of additional species and localities. The 1927–28 paper has subtitle Plantes adventices. Vol. 1 in Museum of Comparative Zoology library, Harvard University; vol. 7 and 13 not seen, title and annotation supplied by R. McVaugh from University of Michigan library.)—Also his Plantes nouvelles ou rares pour la flore du Var. l.c. 4 (1913): 92–93. 1913. (Museum of Comparative Zoology library.)—Also **Marnac, Émile,** and **Reynier, Alfred.** Nouvelles contributions à la flore du département du Var. Bul. Soc. Bot. France **70:** 34–46, 94–105, 215–221. 1923. (General considerations on varieties; extensive list of vascular plants (almost all infraspecific forms), with localities.)—Also **Jahandiez, É.** Contribution à l'étude de la flore du département du Var. Annales Soc. Hist. Nat. Toulon **16:** 65–83. [1931.] (List of additional species and localities and of species to be excluded, bringing total to 2234 spontaneous species and 206 adventive or naturalized. In American Museum of Natural History library.)—Also **Bouchard, Jean.** Plantes nouvelles ou peu signalées pour le département du Var. [1.]–[2.] (Supplément au catalogue d'Albert et Jahandiez.) Bul. Soc. Bot. France. **99:** 258–259. 1953; **103:** 619–624. 1957.

Jahandiez, Émile. Bibliographie botanique varoise. Annales Soc. Hist. Nat. Toulon **21:** 62–87. 1937. (Museum of Comparative Zoology library, Harvard University.)

Unannotated chronological list of 325 titles, from 1715 to 1937.

Local

Camus, E. G., and **Camus, Aimée.** Florule de Saint-Tropez et de ses environs immédiats. 38 p. 21.5 cm. Paris, 1912.

Topography, plant associations; annotated list of vascular plants and Characeae of the peninsula of St. Tropez and vicinity.—See also **Camus, E. G.** Additions et corrections à la Florule de Saint-Tropez. Bul. Soc. Étude Flore Franco-Helv. **22:** 26. [1914.] (In Gray Herbarium library.)—Also **Hibon, Georges.** Additions à la Florule . . . Bul. Soc. Bot. France **63:**

207–210. 1919; **68:** 153–156. 1921.—Also **Lemée, Albert.** Notes floristiques sur Saint-Tropez. Monde des Plant. **46**(274/275): 10. 1951. (In Arnold Arboretum library.)

Jahandiez, Émile. Botanique. *In* his Les Îles d'Hyères. Monographie des Îles d'Or: presqu'île de Giens, Porquerolles, Port-Cros, île du Levant; histoire, déscription, géologie, flore, faune. 3. éd. augm. et entièrement refondue. p. 205–366. Toulon, 1929. (Library of Muséum national d'histoire naturelle, Paris.) (1st ed. 1905.)

Annotated list (Florule des Îles d'Hyères et de presqu'île de Giens) of vascular plants (914 species, plus 6 hybrids and 54 adventives) and cellular plants, with localities. The Provençal names given in the 1st ed. are omitted here.—See also **Molinier, René.** Les Îles d'Hyères: étude phytosociologique. Annales Soc. Hist. Nat. Toulon **21:** 91–129. 1937. (Includes (p. 99–103) list of additions. In library of Museum of Comparative Zoology, Harvard University.)

——— **Lantz, L. A., and Parent, H.** Note sur l'histoire naturelle de l'île de Bandol (Var). Annales Soc. Hist. Nat. Toulon **17:** 47–52. 1933. (American Museum of Natural History library.)

Includes unannotated list of spermatophytes (55 species) with remarks on the persistent introduced species, etc.

Laurans, Sylvain. Cotignac et sa flore. Revue Hort. Bouches-du-Rhône **75:** 73–78, 94–97. 1929; **76:** 52–55. 1930; **77:** 35–39, 70–72. 1931.

Topography; annotated list of about 500 phanerogams. (Not seen; cited from review in Bul. Soc. Bot. France **78:** 743. 1932; reference checked by R. E. Woodson in Missouri Botanical Garden library.)

Marnac, Émile. Florule de Pépiole (Pipière). Revue Hort. Bouches-du-Rhône **51:** 34–39, 45–56, 63–69. 1905.—Additions . . . l.c. **52:** 179. 1906.

Topography; briefly annotated list of vascular plants collected by author in 1904. Pépiole (Pipière of H. Roux's Catalogue) is a small hill near the station of Ollioules-Sanary.

——— Florule de Tauroentum. Plage des Lecques, Saint-Cyr (Var). (Contribution à la flore de Provence.) Revue Hort. Bouches-du-Rhône **53:** 102–104, 115–122, 134–142, 152–162, 177–183, 197–203, 220–224, 235–241. 1 fig. (map). 1907.

Topography, geology; annotated list of vascular plants.

Patout, M. R. Abrégé des plantes médicinales croissant dans les environs de Toulon. 85 p. 19 cm. Toulon, 1864. (Arnold Arboretum library.)

Account of plants most used in local popular medicine, alphabetically arranged by vernacular names, with properties, etc.

Perreymond, J. H. Plantes phanérogames qui croissent aux environs de Fréjus, avec leur habitat et l'époque de leur fleuraison. 4 p. l., 90 p. 20.5 cm. Paris, Fréjus, 1833.

Briefly annotated list of vascular plants, alphabetically arranged.

Reynier, Alfred. Aperçu botanique sur la presqu'île Sépat (alias: St. Mandrier), section communale de La Seyne-sur-Mer (Var). Monde des Plant. III, **27** (43): 2–4, (44): 4–6, (45): 2–5, (46): 2–4. 1926. (Arnold Arboretum library.)

Botanical explorations, topography, geology, list of adventive plants; list of angiosperms, with localities.

——— Aperçu botanique sur la Sainte-Baume (Var). Bul. Soc. Bot. et Hort. Provence **1:** 112–123. [1880.]

Geology, topography; partial list of vascular plants belonging to different zones; list of localities with mention of their noteworthy species, list of doubtful or extinct species; shorter treatment of mosses, hepatics, lichens, and mushrooms.—See also **Mader, Fritz.** Le massif de la Sainte-Baume. Une forêt-vierge en Provence. Malpighia **20:** 353–394, 409–455. 1906–07. (Topography, orography; partly annotated lists of woody and herbaceous plants of several localities; discussion of virgin forests of Europe, phytogeography; no comprehensive list of species.)—Also **Reynier, Alfred.** La Sainte-Baume visitée en 1903–1906, par un botaniste suisse. Bul. Soc. Sci. Nat. Provence **2:** 50–57. 1908. (Notes and corrections to Mader's paper.)—Also **Laurent, Louis.** Esquisse de géographie botanique du massif de la Sainte-Baume suivie d'une étude sommaire des plantes médicinales et des cultures de la région . . .

Annales Mus. Hist. Nat. Marseille **19**: 1–96. map, 4 diagr. 1924. (Topography, geology, history, doubtful species, descriptions of various localities with mention of characteristic species, climatic and other factors affecting distribution; annotated lists of medicinal plants, cultivated trees, etc.; bibliography.)—Also **Molinier, René.** Notes sur la flore du massif de la Sainte-Baume. Annales Soc. Hist. Nat. Toulon **20**: 95–124. map. 1936. (Geography, topography, etc.; lists of species to be excluded, of additional species and new localities, etc. In American Museum of Natural History library.)—Also **Molinier, Roger.** La flore de la Forêt domaniale de la Sainte-Baume. Catalogue des espèces présentes dans les limites de la Forêt. Annales Soc. Sci. Nat. Toulon et du Var n.s., **3**: 45–66. 1950. (Not seen; cited mainly from Bul. Mus. Hist. Nat. Marseille **14**: 85. 1954.)—Also **Pialot, Henri.** Note sur la flore de la région centrale du massif de la Sainte-Baume, chaîne principale et plateau du Plant d'Aups. Monde des Plant. **47** (285/286): 3–4. 1952. (List of noteworthy species including numerous additions to the flora. In Arnold Arboretum library.)

Robert, —. Plantes phanérogames qui croissent naturellement aux environs de Toulon. 116 p. map. 20.5 cm. Brignoles, 1838.

Annotated list of localities; annotated alphabetical list of vascular plants and Characeae, with localities for scarcer species.—See also **Meurdefroy, —.** Appendice au Catalogue des plantes phanérogames qui croissent spontanément aux environs de toulon. Bul. Soc. Sci., Belles-Let. et Arts Dépt. Var **10**: 183–199. 1842. (Includes lists of additions and of doubtful species.)—Also **Albert, Abel, and Reynier, Alfred.** Coup d'oeil sur la flore de Toulon et d'Hyères (Var). Bul. Soc. Études Scient. et Archéol. Draguignan **18**: 191–245. 1892. (Geology, topography, general features of flora; unannotated partial lists of southern and nonsouthern vascular plants; list of localities, with their rarer species.)—Also **Arènes, Jean.** Étude sur la végétation des vallées en Provence. IV. Liste méthodique des phanérogames et des cryptogames vasculaires peuplant les vallées toulonaises. Bul. Soc. Bot. France **69**: 730–740. 1923. (Extensive tabular list, with localities.)—Also **Auzende, Joseph.** Guide du botaniste aux environs de Toulon. Manuscript inédit . . . publié et annoté par É. Jahandiez. Annales Soc. Hist. Nat. Toulon **15**: 10–75. [1931.] (Account of excursions in vicinity of Toulon, with lists of noteworthy species. In library of Muséum national d'histoire naturelle, Paris.)—Also **Jahandiez, Émile.** Pertes et acquisitions de la flore toulonnaise depuis un siècle. l.c. **20**: 51–94. 1936. (Discussion of Toulon region as understood by different authors; lists of species that have probably disappeared, of rare species still existing, and of species added (245) since Robert's flora, all with localities and some annotations. In American Museum of Natural History library.)

VAUCLUSE

See also Provence.

Local

Colignon, —. Flore d'Apt. Tableau synoptique de quelques plantes qui croissent aux environs de cette ville. Annales Soc. Lit. Scient. et Artist. d'Apt **2**: 86–160. 1864.

Annotated tabular list of vascular and a few cellular plants, with patois names and localities.

Martins, C. F. Essai sur la topographie botanique du mont Ventoux, en Provence. Annales Sci. Nat. II, Bot., **10**: 129–150, 228–249. pl. 7 (map). 1838.

Topography, climate, geology, botanical explorations, plant zones, with lists of species, etc.; list of known vascular plants.

Palun, Maurice. Catalogue des plantes phanérogames qui croissent spontanément dans le territoire d'Avignon et dans les lieux circonvoisins. 2 p. l., 189 p. 22 cm. Avignon, 1867.

Briefly annotated list of vascular plants and Characeae, with localities for rarer species and patois names.

Pons, Armand. La flore de la chaîne du Grand Lubéron (Vaucluse). Bul. Mus. Hist. Nat. Marseille **11**: 57–82. 1951 (1953).

Annotated list of vascular plants (675 species); remarks on some species, bibliography.

Rouis, E. Notes sur la flore phanérogamique des environs de Carpentras, du Ventoux et des monts de Vaucluse. Mém. Acad. Vaucluse **14**: 52–91, 253–300. 1895. (Reprinted, 1 p. l., 87 p. 24.5 cm. Avignon, 1895. In Lloyd Library.)

Geography, general features of flora (particularly arborescent species), plant zones, phytogeography, etc.; unannotated list of vascular plants and Characeae (1306 species) collected in 1888–1891.

VENDÉE

General

Douteau, J. J. Flore de Vendée. 3 p. l., xxxix, 5–408 p. 14 cm. Paris, 1906. (Arnold Arboretum library.)

Popular flora of vascular plants, in form of keys, briefly annotated but without localities. Reissue with shortened title of his Flore de Vendée. Tableaux dichotomiques des plantes vasculaires recueillies en Vendée jusqu'à nos jours. (Originally published in Revue Sci. Nat. Ouest **6**: 7–24, 40–56, 72–88, 104–120, 136–152, 171–182. 1896.)

Pontarlier, N. C., and **Marichal, H. N.** Catalogue des plantes vasculaires et spontanées du département de la Vendée recueillies par Pontarlier et Marichal, augmenté de la liste des plantes trouvées depuis 1889 jusiq'à ce jour. Revue Sci. Nat. Ouest **4**: 37–64, 107–136. 1894; **5**: 26–45, 102–128. 1895. (Reprinted, 2 p.l., 99 p. Paris, 1895. In Arnold Arboretum library.)

List of vascular plants and Characeae, with localities. A posthumous work, edited by "La revue des sciences naturelles de l'ouest."—See also **Blanchard, Théophile.** Contribution à la flore vendéenne. Bul. Assoc. Franç. Bot. **1** (7): 13–19. 1898. (Extensive list of vascular plants, with localities, including many new to the department. In Arnold Arboretum library.)—Also **Durand, Georges,** and **Charrier, J.** Rapport sur les excursions de la Société botanique de France en Vendée (juin 1911). Bul. Soc. Bot. France **58** (Sess. Extraord.): xcv–cxlii. 1913. (Includes short lists of vascular plants from numerous localities, with some account of botanical regions and ecological habitats.)

Local

Blanchard, Théophile. Liste des noms patois de plantes aux environs de Maillezais (Vendée). Bul. Assoc. Franç. Bot. **3**: 36–38, 53–58, 106–110, 129–131, 193–199. 1900; **4**: 12–15, 99–101, 129–131, 167–174, 201–208. 1901; **5**: 44–52, 113–119, 127–135. 1902.

List of patois names, systematically arranged, with derivation and meaning of names, and uses.

Genevier, Gaston. Extrait de la florule des environs de Mortagne-sur-Sèvre (Vendée). Mém. Soc. Acad. Maine et Loire **20**: 17–51. 1866. (Reprinted, 35 p. Reprint not seen.)

List of vascular plants [about 980 species], with localities and some critical notes. Covers an area in Vendée, Deux-Sèvres, and Maine-et-Loire.

Viaud-Grand-Marais, Ambroise. Catalogue des plantes vasculaires de l'île de Noirmoutier. Bul. Soc. Sci. Nat. Ouest France **2** (1): 161–218. map. 1892. (Reprinted, 64 p. Reprint not seen.)

Topography, geology, botanical explorations; annotated list of vascular plants (about 700 species).

——— and **Ménier, Charles.** Catalogue des plantes vasculaires de l'île d'Yeu. Bul. Soc. Sci. Nat. Ouest France **4** (1): 117–151. 1894.

Geology, topography, botanical explorations and publications, etc.; annotated list of vascular plants. A few additions are included in lists in Bul. Soc. Bot. France **58** (Sess. Extraord.): cxxxv–cxli. 1913.

VIENNE

General

Poirault, J. Catalogue des plantes vasculaires du département de la Vienne. xv, 127 p. 22 cm. Poitiers, 1875. (Gray Herbarium library.)—Supplément au Catalogue . . . 52 p. 22 cm. 1883. (Gray Herbarium library.)

Geography, botanical investigations; annotated list of vascular plants and Characeae, with localities. Replaces **Delastre, C. J. L.** Flore analytique et descriptive du département de la Vienne. xxii, 546 p. 4 pl. 22 cm. Paris, 1842. (Soils, etc.; descriptive flora of phanerogams, with localities; glossary.)—See also **Simon, Eugène.** Additions à la flore de la Vienne. Bul. Soc. Bot. Deux-Sèvres **15**: 179–211. 1904. (Extensive list of phanerogams, with localities and frequent annotations. In Farlow Library.)—Also his Quelques notes floristiques sur la région montmorillonnaise. l.c. **1925**: 16–22. 1925. (List with localities, supplementary to his Additions . . . (1904). Not seen; title and annotation supplied by R. E. Woodson, from copy in Missouri Botanical Garden library.)

Souché, Baptiste. Flore du Haut Poitou, ou analyse des familles, des genres, des espèces et description des plantes qui croissent spontanément ou qui sont l'objet d'une culture en grand dans les départements des Deux-Sèvres et de la Vienne. 2 v. (xl, 322; xx, 283 p.). port. 20 cm. Niort, 1894–1901.

Briefly annotated, descriptive flora of vascular plants and Characeae, with keys. Vol. 2 has subtitle "Matériaux pour une géographie botanique régionale" and contains an alphabetical list of local botanists with dates, and complete list of species, with localities.—See also **Lemesle, Robert.** Quelques phanérogames adventices ou rares dans le Haut-Poitou. Bul. Soc. Bot. France **94**: 284–289. 1948. (Annotated lists of noteworthy adventive and native plants, mainly from vicinity of Poitiers; bibliography.)—Also his Contribution à l'étude des phanérogames adventices du Poitou. Monde des Plant. **45** (273): 79–83. 1950. (Annotated list; bibliography. In Arnold Arboretum library.)

Local

Grelet, L. J. Flore de Châtelleraut et de la Forêt de Châtelleraut (département de la Vienne, 1895–96). Monde des Plant. **6**: 115–119. 1897. (Arnold Arboretum library.)

List of about 200 vascular plants, with localities.

Litardière, René de. La flore des environs de la Station de biologie végétale de Mauroc. Rev. Gén. Bot. **25bis**: 121–132. pl. 6. 1914.

Geology, etc.; sketch of plant formations, with lists of species. (Near Saint-Benoît.)

Richard, O. J. Florule des clochers et des toitures des églises de Poitiers (Vienne). 50 p. 24 cm. Paris, 1888.

General considerations; annotated list of 76 vascular plants found growing on the highest parts of 7 churches in Poitiers in 1886–88; bibliography of works "sur les plantes des villes, des ruines et des monuments publics" (French and foreign).

VOSGES

See also Alsace (Kirschlager); Haut-Rhin (Issler, Influence, and Die vegetationsverhältnisse); Lorraine.

General

Berher, Eugène. La flore des Vosges. Phanérogames, muscinées, lichens. *In* Louis, Léon. Le département des Vosges; description, histoire, statistique. v. 2, p. 9–366. map. 25 cm. Épinal, 1887.

Botanical explorations; annotated list of vascular plants and Characeae (1622 species, plus mosses and lichens), with localities. The other cellular cryptogams, completing the volume, are by other authors. Replaces his Catalogue des plantes vasculaires qui croissent spontanément dans le département des Vosges . . . Annales Soc. Émul. Dépt. Vosges **15** (2): 83–242. 1876; Supplément au Catalogue . . . l.c. 1881: 212–226. 1881; Nouveau supplément . . . l.c. **1883:** 280–296. 1883. Also replaces **Mougeot, J. B.** Considérations générales sur la végétation spontanée du département des Vosges. 356 p. 24 cm. Épinal, 1845. (From Lepage, H., and Charton, Charles. Statistique du département des Vosges.) (Includes (p. 155–355) annotated tabular list of vascular and cellular plants, with localities.)—See also **Gérard, Frédéric.** Notes sur quelques plantes des Vosges. Additions et rectifications. Revue Bot. Bul. Mens. Soc. Franç. Bot. **8:** 51–240, 449–471. 1890. (Reprinted, 216 p. 1890.) (Extensive list of additional species and localities, critical notes, etc.)—Also **Berher, E.** Supplément au catalogue . . . Annales Soc. Émul. Dépt. Vosges **1891:** 157–197. 1891. (Extensive list of additions, corrections, etc.)—Also **Guffroy, Charles.** Notes sur la flore vosgienne. Bul. Soc. Bot. France **59:** 537–545, 599–602. 1912–13. (List of noteworthy plants.)—Also **Lemasson, Constant.** Note sur la flore des Hautes-Vosges. l.c. **60:** 503–504. 1914. (List of noteworthy species.)

Claire, Charles. Les plantes utiles et remarquables de la flore des Vosges. Bul. Soc. Philom. Vosgienne **29:** 203–332. 1904.

Includes annotated systematic lists of poisonous and medicinal plants, etc., with French names and uses, also woody plants.

Haillant, Nicholas. Flore populaire des Vosges. 2 p. l., 220 p. 22.5 cm. Paris, Épinal, 1885-86.

Phonetics, etc.; systematic list of vascular and cellular plants with vernacular and patois names and localities where used. Originally published as supplement to Jour. Soc. Hort. Natl. France s.3, v. 7–8. 1885–86.

Local

Brunotte, Camille, and Lemasson, Constant. Le botaniste herborisant au Hohneck et aux environs de Gérardmer. Bul. Club Alpin Franç. Sect. Vosgienne **12:** 2–7, 18–30, 39–42, 53–60, 66–75. map. 1893. (New York Public Library.) (Republished with title Guide du botaniste au Hohneck et aux environs de Gérardmer. 39 p. map. 1893. In Lloyd Library.)

Sketch of botanical regions, with annotated lists of the rarer or more interesting phanerogams; not a complete flora.—See also **Brunotte, C.** Nouvelles stations de plantes rares dans le massif du Hohneck. Bul. Club Alpin Franç. Sect. Vosgienne **18:** 72–77, 85–91. map. 1899. (Annotated list. In New York Public Library.)—Also **Lemasson, C.** Plantes nouvelles et rares de la flore des Vosges. l.c. **31:** 57–59. 1912; also his Nouvelle contribution à la flore des Vosges. l.c. **32:** 99–103. 1913. (Additional species and localities. Both in New York Public Library.)—Also his Guide du botaniste herborisant au Hohneck et aux environs de Gérardmer. [nouv. éd.] Bul. Soc. Sci. Nancy IV, **1:** 79–109. map. 1921. (Essentially same as 1st ed., with results of subsequent explorations to June 1914 added, and Lemasson given as sole author.)—See also Haut-Rhin (Issler, Die vegetationsverhältnisse der Zentralvogesen mit besonderer berücksichtigung des Hohneckgebietes).

Claire, Charles. Un coin de la flore des Vosges. Plantes des environs de Rambervilliers. Bul. Acad. Internatl. Géog. Bot. **9:** 95–103, 137–141, 234–240. 1900; **10:** 256–260, 283–285. 1901; **11:** 275–276. 1902; **12:** 34, 207–208. 1903.

Annotated list of spermatophytes; incomplete (Ranunculaceae-Rhamnaceae).

Thiriat, Xavier. Catalogue des plantes phanérogames et cryptogames vasculaires qui croissent spontanément dans les vallées de Cleurie et de la Moselotte, entre les villages de Saint-Amé, de Vagney et du Tholy. *In* his La vallée de Cleurie . . . p. 73–115. 18 cm. Mirecourt, Remiremont, 1869. (Bibliotèque municipale d'Épinal.)

Annotated list of vascular plants and Characeae (about 620 species), with patois names (translated in the glossary on p. 416–453). Preceded (p. 63–72) by general features of flora.—See also his Catalogue des plantes phanérogames cultivées ou subspontanées . . . l.c. p. 116–135. 1869. (Annotated list of cultivated or introduced plants (252 species), with patois names. Copy examined borrowed from Bibliotèque municipale d'Épinal, through courtesy of Mme. Duprat, librarian of the Bibliotèque générale du Muséum national d'histoire naturelle, Paris.)

YONNE

See also Bourgogne; Champagne; Seine-et-Marne (Royer).

General

Ravin, Eugène. Flore de l'Yonne: description des plantes croissant naturellement ou soumises à la grande culture dans le département. 3. éd. 460 p. 22 cm. Auxerre, 1883. (1st ed., Catalogue méthodique et raisonné des plantes qui croissent naturellement dans le département de l'Yonne. Bul. Soc. Sci. Hist. et Nat. Yonne 14 (3. trimestre): 40–325. 1861.)

Botanical regions with lists of species, special localities, glossary; annotated, keyed, briefly descriptive flora of vascular plants and Characeae (1448 species, including some cultivated plants), with localities, vernacular names, and indication of uses; list of rare plants of different communes; separate index of vernacular names.—See also **Fliche, Paul.** Notes sur la flore de l'Yonne. Bul. Soc. Bot. France **45**: 40–51. 1898. (Extensive list of vascular plants, with localities.)—Also **Ravin, E.** Découvertes botaniques dans l'Yonne depuis la publication de la Flore de l'Yonne (3. éd. en 1883). Bul. Soc. Sci. Hist. et Nat. Yonne 52 (i.e. 53) (2. partie, Sci. Phys. et Nat.): 59–65. 1900. (List with localities.)

Local

Houlbert, C. V. Flore du Sénonais. Catalogue analytique et descriptif des plantes vasculaires observées dans l'arrondissement de Sens . . . xxxviii, 276 p. 21.5 cm. Sens, 1901. (Arnold Arboretum library.)

Botanical investigations, topography, geology, geography, bibliography, glossary; annotated, keyed, descriptive flora of vascular plants (1197 species), with localities.

Lasnier, —, and Ravin, Eugène. Plantes récoltées dans le département de l'Yonne. Comp. Rend. Cong. Soc. Sav. Paris et Dépt. **1898** (Sect. Sci.): 178–194. 1898.

Sketch of topography and soils; account of the rivers and swamps, with mention of characteristic species; systematic list of aquatic and swamp plants, with localities.

GREAT BRITAIN WITH IRELAND (EIRE)

See also France (General: Depape, Le monde des plantes à l'apparition de l'homme; Normandie: Le Héricher, Essai sur la flore populaire de Normandie et d'Angleterre); Netherlands (Gerth van Wijk, A dictionary of plant-names); Switzerland (Probst, Wolladventivflora Mitteleuropas).

General

Some of the works listed below do not cover Ireland.

The general works here listed may be classified as follows: *Floras and manuals:* Babington, Bentham, Bonnier, Clapham, Hayward, Hooker, Horwood, McClintock, Melderis, Moss, Pratt, Prime, Sowerby, Williams (Prodromus). *Catalogs:* Dandy, Druce (British Plant List; Comital Flora), Linton (hybrids), Watson (Compendium; Cybele; Topographical botany), Williams (alpine flora), Wilson (altitudinal range). *Iconographies:* Boulger, Curtis, Fitch (under Bentham), Hogg, (McClintock), (Moss), Pratt, Roles (under Clapham), Ross-Craig, Sowerby. *Bibliographies:* Druce (Local floras), Holden, Polunin, Rehder, Smart, Trimen. *Vernacular names:* Britten and Holland, Fisher, Grigson, Henslow, Miller, Prior, Rayner; also Cockayne, Earle, and Hoops, under England, General. *Trees and woody plants in general:* Bean, (Chittenden), Elwes, Hadfield, (Kirk), Makins, Step. *Useful and poisonous plants:* Barton (medicinal), Cameron (edible), Culpeper (medicinal), Edlin, Ellis (medicinal and poisonous), Fernie (medicinal), Forsyth (poisonous), Henslow, Grigson, Hill (edible), C. Johnson (poisonous), C. P. Johnson, Pratt (poisonous). *Weeds and adventive plants:* Anonymous, Brenchley, Dunn, Morse. *Cultivated plants:* Bean (woody plants), Chittenden, Elwes (trees), Hadfield (trees), Kirk, Makins (2), Step. *Miscellaneous:* Botanical Society of British Isles (ecology, distribution, changes in flora), Britten and Boulger (biography), Clarke (first records), Druce (Comital flora: detailed distribution; extinct and dubious plants), Egerton-Warburton (synonymy), Gilbert-Carter (glossary of scientific names), Godwin (Quaternary records), Grigson (popular beliefs, vernacular names), Harvey-Gibson (glossary of scientific names), Kent (herbaria and collectors), Linton (hybrids), Macleod (glossary of scientific names), Matthews (phytogeography; origin of flora), Polunin (British floras), Reid (origin of flora), Tansley (plant formations), Turnbull (index to principal floras), Turrill (phytogeography, plant communities), Wilmott (extinct species).

Anonymous. Farm weeds. An aid to their recognition. 2. ed. 80 p. incl. 60 col. pl. (on 30). 21.5 cm. London, (195–?).

Colored plates of 60 common weeds with extensive lists of vernacular names and brief descriptions. The illustrations in nearly all cases include the seedlings. Issued by Shell Chemical Company. Plates by Doris C. R. Thompson, foreword signed by J. L. Hunt.—See also **Brenchley, Dunn,** and Morse below.

Babington, C. C. Manual of British botany, containing the flowering plants and ferns arranged according to the natural orders. . . . 10th ed. with amended nomenclature and an appendix edited by A. J. Wilmott. . . . liv, 612 p. 18.5 cm. London, 1922. (1st ed. 1843.)

Keys to and synopsis of families; briefly descriptive flora of vascular plants and Characeae, with extremely generalized statement of range and partial keys; (appendix) synopsis of British Rubi fruticosi, additional species and notes.

Barton, B. H., and **Castle, Thomas.** The British flora medica; a history of the medicinal plants of Great Britain . . . A new ed., rev., condensed, and partly rewritten, by John R. Jackson. xxii, 447 p. 48 col. pl. 23 cm. London, 1877. (1st ed. 1836–38.)

Description, distribution, properties and uses of 193 vascular plants, with colored figure of each; arranged alphabetically by English names. For other works on medicinal plants, see Culpeper, Ellis, and Fernie, below; for general works on useful (including medicinal) plants, see **Edlin, below, and additional** works there cited; for works on poisonous plants see Forsyth, below, and additional works there cited.

Bean, W. J. Trees and shrubs hardy in the British Isles. 7. ed. 3 v. (xviii, 703; x, 636; viii, 664 p.). illus., 133 pl. (40, 52, 41) on 67. 22.5 cm. London, 1950–51. (1st ed. 1914.)

Bibliography, account of some British and foreign collectors and nurserymen, chapters on propagation and culture, etc.; descriptive account of hardy trees and shrubs, alphabetically arranged by scientific names, with native habitat, date of introduction, notes on cultivation, horticultural varieties, etc. —See also Chittenden, Elwes, Hadfield, Kirk, Makins, and Step, below.

Bentham, George. Handbook of the British flora, a description of the flowering plants and ferns indigenous to, or naturalized in the British Isles. For the use of beginners and amateurs . . . rev. by Sir J. D. Hooker. 7th ed. rev. by A. B. Rendle. lxi, 606 p. 18.5 cm. London, 1924. (Several reissues, the last in 1945. 1st ed. 1858.)

Outline of descriptive botany, etc., key to families; annotated descriptive flora of vascular plants. Noteworthy for its very conservative view of species. There is a previous 7th edition (1900), edited by J. D. Hooker.—Accompanied by **Fitch, W. H., Smith, W. G.,** and others. Illustrations of the British flora . . . 5. ed. rev. xxvii, 338 p. (including 1315 figures on 328 pages). 18.5 cm. London, 1924. (Several reissues, the last in 1944. 1st ed. 1880.) (Excellent small figure of each species, with dissections.)—See also **Butcher, R. W.,** and **Strudwick, F. E.** Further illustrations of British plants. iv, 476 p. 485 fig. 18.5 cm. Ashford, 1930. (Reissued 1944.) (Good drawings of 485 species supplementary to Fitch's "Illustrations," with brief descriptive text.)

Bonnier, Gaston. British flora. Translated and adapted from the French by Ethel Mellor. 2 p. l., xxxiv, 315 p. illus., map. 19 cm. London and Toronto, 1925.

Very briefly annotated pocket manual of vascular plants in form of keys, with glossary, brief account of botanical regions, etc.; brief notes on economic properties given in index. Adapted from Bonnier's La flore du nord de la France et de la Belgique, ed. 16.

Botanical Society of the British Isles. British flowering plants and modern systematic methods. Being the report of the Conference on the study of critical British groups arranged in April 1948, by the Botanical Society . . . edited by A. J. Wilmott. 1 p. l., 102 p. 18 pl. on 10. 21.5 cm. London, 1949.

Papers by various authors on cytology, experimental studies, ecology, and intraspecific variation in their application to the classification of critical groups and on the taxonomic situation in several such groups (water buttercups, *Viola riviniana, Rosa, Rubus, Valeriana officinalis, Galium* and *Jun-*

cus), followed by more or less detailed accounts of various exhibits of these and other genera.—See also their The Study of the distribution of British plants. Being the report of the Conference held in 1950 . . . edited by J. E. Lousley. 1 p.l., 128 p. 28 fig. (mostly maps), 2 pl. (maps). 21.5 cm. Arbroath, 1951. (Discussion of the history and application of the Watsonian vice-county system in mapping the ranges of plants, of different methods of mapping ranges used in Great Britain and other countries, and of various minor points, with bibliographies.)—Also their The changing flora of Britain. Being the report of the Conference held in 1952 . . . edited by J. E. Lousley. 203 p. 24 fig. (mostly maps), 6 pl. on 5. 22 cm. Arbroath, 1953. (Discussion of changes in flora due to natural and human factors, of the Scandinavian, North American, Lusitanian, and Continental elements in the flora, of changes in the weed flora, etc., with bibliographies.)

Boulger, G. S. British flowering plants, illustrated by three hundred full-page coloured plates reproduced from drawings by Mrs. Henry Perrin, with detailed descriptive notes . . . 4 v. 300 col. pl. 32 cm. London, 1914.

Colored plates of selected species, systematically arranged, with popular descriptive text.—See also **Fitch** under **Bentham**, above, and **Roles** under **Clapham, Curtis, Hogg, McClintock, Moss, Pratt, Ross-Craig,** and **Sowerby** below.

Brenchley, W. E. Weeds of farm land. x, 239 p. 41 (i.e. 40) fig., front. 22 cm. London [etc.], 1920.

Mainly concerned with eradication; has chapters on poisonous and injurious weeds (p. 103–116) and uses of weeds (p. 187–205), and alphabetical list of botanical names with corresponding vernacular names.—See also Anonymous, above, and additional references there given.

Britten, James, and Boulger, G. S. A biographical index of deceased British and Irish botanists . . . 2d ed., revised and completed by A. B. Rendle . . . xxii, 342 p. 22.5 cm. London, 1931. (1st ed. 1893.)

Bibliography; alphabetical list of botanists and persons connected with botany, with brief biographical data, titles of principal publications, references to biographies, names of genera or species dedicated to them. For review, with some additions, see Rpt. Bot. Soc. and Exch. Club Brit. Isles **9** (1931): 585–587. 1932.

——— and **Holland, Robert.** A dictionary of English plant-names. xxviii, 618 p. 22.5 cm. London, 1886. (Originally issued, 1878–84, as no. 22, 26, 45 [forming v. 10] of Pub. Engl. Dialect Soc.)

Discussion of vernacular names, bibliography (British and foreign works); alphabetical list of vernacular names with botanical equivalents, localities, references, and frequent quotations from literature; alphabetical list of botanical names, with collected vernacular names. Relates primarily to England but has some Irish and Scottish names. For other works on vernacular names see Fisher, Grigson, Henslow, Miller, Prior, and Rayner, below; also Cockayne, Earle, and Hoops, under England, General.

Cameron, L. C. R. The wild foods of Great Britain, where to find them and how to cook them. xv, 128 p. 25 fig., 2 col. pl. 14 cm. London, 1917. (Gray Herbarium library.)

Popular account of edible wild animals and plants; vascular plants on p. 67–101.—See also Hill, below, and for works on useful plants in general see Edlin, below, and additional references there given.

Chittenden, F. J., *ed.* The Royal Horticultural Society dictionary of gardening. A practical and scientific encyclopaedia of horticulture. 2. ed. [ed.] by P. M. Synge. 4 v. (xvi, 2316 p.). illus. 28 cm. Oxford, 1956. (1st ed. 1951.)—Supplement . . . ed. by P. M. Synge. vii, 334 p. illus. (p. 318–322). 28 cm. Oxford, 1956.

Encyclopedia of horticulture (mainly British), alphabetically arranged, including generic treatments, specific epithets and vernacular names, synonyms, descriptive terms, injurious insects and fungi, garden vegetables, general subjects, etc.; the generic treatments with account of cultivation and list of species, alphabetically arranged, with short descriptions, habitat, etc., and often with keys. Authorities for scientific names are not given. Text unchanged from 1st ed. except for minor corrections and references to supplement. Vol. 1, A-Co; v. 2, Co-Ja; v. 3, Je-Pt; v. 4, Pt-Zy. The Supplement contains (pt. 1) an alphabetical list of genera giving names of recommended varieties with short descriptions or at least indication of color, etc., and (pt. 2) extensive list of additions and corrections.—See also Bean, above, and additional references there given.

Clapham, A. R., Tutin, T. G., and Warburg, E. F. Excursion flora of the British Isles. xxxiii, 579 p. 3 fig. (glossary). 19 cm. Cambridge, 1959.

Keyed pocket flora of vascular plants, with descriptions of the commoner species, the rarer ones given only in the keys; ranges given in rather general terms.

——— **Tutin, T. G., and Warburg, E. F.** Flora of the British Isles. li, 1591 p. 79 fig. 19 cm. Cambridge, 1952.

Keyed, descriptive flora of vascular plants, with brief data on chromosome number, life form, pollination, etc.; short bibliography, glossary. Accompanied by **Roles, S. J.** Flora of the British Isles. Illustrations. pt. 1. 144 p. incl. 552 fig. 29 cm. Cambridge, 1957. (Figures, 4 to a page, covering Pteridophyta to Papilionaceae.)

Clarke, W. A. First records of British flowering plants. 2d ed., rev. and corr. xvi, 194 p. 19 cm. London, 1900. (Arnold Arboretum library.) (1st ed. 1897, reprinted with additions and corrections from Jour. Bot. v. 30–34. 1892–96).—[Additions.] Jour. Bot. **47**: 413–416. 1909.

Bibliography; list of first British records for each species, with date and citation of original notice; summary. The records are brought up-to-date under each species in Druce's Comital flora (1932).

Culpeper, Nicholas. Culpeper's Complete herbal: consisting of a comprehensive description of nearly all herbs with their medicinal properties and directions for compounding the medicines extracted from them. (Title varies.) 430 p. col. pls. 19 cm. London [etc.], [1952]. (1st ed. (later repudiated by author), 1652, with title, The English physitian or an astrologo-physical discourse of the vulgar herbs of this nation . . .)

Extensive list of native and cultivated plants, alphabetically arranged by vernacular names, with brief descriptions, astronomical sign and supposed medicinal virtues, habitat, season; list of diseases with their remedies. The scientific names supplied in the later editions are not always reliable. For partial list of the innumerable editions, see A. Rehder, Bradley Bibliography **3**: 79. 1915; also British Museum catalogue of printed books **46**: 155. 1950. The work has been reprinted far more often and over a much longer period than any other book dealing with British plants, although its actual value is very slight.—See also Barton, above, and additional references there given.

Curtis, William. Flora londinensis: containing a history of the plants indigenous to Great Britain . . . New ed., enl. by George Graves and William Jackson Hooker. 5 v. 647 col. pl. 52 cm. London, 1817–28. (1st ed. 1777 (1775–98); reissue of new ed. in 1835.)

Descriptive flora, with vernacular names in various European languages; now of value chiefly for the excellent colored plates. Issued without pagination or plate numbering. For index, see **B. D. Jackson,** Index to Curtis' Flora londinensis . . . Jour. Bot. **54**: 153–164. 1916; for information on the number of plates, see **F. Cardew,** Jour. Soc. Bibl. Nat. Hist. **2**: 223–224. 1950.—See also Boulger, above, and additional references there given.

Dandy, J. E. List of British vascular plants. Prepared . . . for the British Museum (Natural History) and the Botanical Society of the British Isles. Incorporating the London catalogue of British plants. xvi, 176 p. 22 cm. London, 1958.

List of vascular plants (2137 species, 685 microspecies, and 73 subspecies (total 2895), as well as 538 hybrids), with typographical indication of alien, extinct, and unconfirmed species and of those species that are known only from Channel Islands. Replaces **Britten, James,** and **Rendle, A. B.** List of British seed-plants and ferns . . . 43 p. 23 cm. London, 1907. (Systematic list of specific names with citation of synonyms from Bentham's, Hooker's, and Babington's floras. Only the adopted names with their places of publication are given, not parenthetical authorities or name-bringing synonyms.)—See also their Notes on the "List . . ." Jour. Bot. **45**: 99–108, 433–445. 1907.) Also replaces The London catalogue of British plants . . . 11. ed. 55, [3] p. 22 cm. London, 1925. (1st ed. 1844.) (Systematic list of 2362 species (plus varieties) of vascular plants and Characeae, with indication of number of vice-counties in which each occurs. Edited by J. F. Hanbury and C. E. Salmon; preface signed by Hanbury.)—See also **Salmon, C. E.** References to notes on some additions in The London catalogue, ed. 11, 1925. Jour. Bot. 63: 291–296. 1925.—Also **Druce, G. C.** "Native" British plants. l.c. **64:** 46–48. 1926.—Also **Sprague, T. S.** The nomenclature and orthography of the London catalogue, ed. 11. Rpt. Bot. Soc. and Exch. Club Brit. Isles 7 (1925): 925–933. 1926. Also replaces **British Ecological Society** [i.e., **Clapham, A. R.**]. Check list of British vascular plants. Jour. Ecology **33:** 308–347. 1946. (Unannotated systematic list, including segregate species of doubtful rank but not varieties or forms, with some essential synonyms; intended as a standard for use by contributors to the Biological flora of the British Isles. The two appendices consist of a list of *Rubus* by William Watson, with key to groups, and a list of *Hieracium* by H. W. Pugsley. On the cover of the reprint A. R. Clapham is given as the compiler.)

Druce, G. C. British plant list containing the spermophytes, pteridophytes, and charophytes found either as natives or growing in a wild state in Britain, Ireland, and the Channel Isles. 2. ed. xl, 154 p. 22.5 cm. Arbroath, 1928. (1st ed. 1908: List of British plants . . .)

Composition of flora, extinct species, short bibliography, index to monographs and systematic papers (grouped by genera), geographical divisions with list of vice-counties, etc.; systematic list of 4246 species and subspecies (2141 native) and about 2720 varieties and over 500 hybrids, with indication of native home of introduced plants and mostly very condensed indication of range for the native plants.—See also his Notes on the second edition of the "British plant list". With special reference to the hybrid names and new combinations. Rpt. Bot. Soc. and Exch. Club Brit. Isles 8 (1928): 867–877. 1929.—Also his British plant list (edition II). Additions and corrections. l.c. 8: 878–883. 1929.—Also **Wilmott, A. J.** Nomenclature and corrections to British plant list. [I–] VI. l.c. **12:** 310–318, 515–523, 661–673. 1942–46; **13:** 18–23, 248–255. 1947–48. Watsonia **1:** 114–116. 1949.—Also **Warburg, E. F.** Nomenclature and corrections . . . VII–VIII (title varies). l.c. **1:** 313–316. 1950; **2:** 102–103. 1951.

——— The comital flora of the British Isles . . . being the distribution of British (including a number of non-indigenous) plants throughout the 152 vice-counties of Great Britain, Ireland, and the Channel Islands, with the place of growth, elevation, world-distribution, grade, chief synonyms, and first names by which the plants were recorded as British. xxxii, 407 p. port., map (in pocket). 22.5 cm. Arbroath, 1932.

List of herbaria in author's collection, sketch of H. C. Watson's work on distribution and other principal sources, lists of extinct species and of those confined to different regions and special habitats, list of vice-counties, brief bibliography; list of vascular plants and Characeae, with habitat, local and general range, list of vice-counties in which known, and first British record. Includes 4,468 species, of which 2,270 are given as natives, 106 as colonists and denizens, 293 as established aliens, and 1,799 as adventives.—See also **Druce** and others. New county and other records (from v. 11, Plant records). Rpt. Bot. Soc. and Exch. Club Brit. Isles v. 3 (3)–13, passim. 1913–48. **Wallace, E. C.** Plant records. Watsonia v. 1–2, passim. 1949–53. Proc. Bot. Soc. Brit. Isles v. 1–3, passim. 1954–59. (Annual lists of new records, systematically arranged.)—Also the series entitled Plant notes for 1910, etc. [–1956]. (Title varies.) Rpt. Bot. Soc. and Exch. Club Brit. Isles v. 2 (1910)–13. 1911–48. Watsonia v. 1–2, passim. 1949–53. Proc. Bot. Soc. Brit. Isles v. 1–3, passim. 1954–59.—Also **Hall, P. M.** Corrections and additions to previous reports. Rpt. Bot. Soc. and Exch. Club Brit. Isles **10:** 121–157. 1933; Further corrections . . . l.c. **10:** 552–554. 1934.—Also **Thurston, Edgar.** Comital flora . . . Addenda and corrigenda. l.c. **10:** 555–557. 1934.—Also **Hall, P. M.** A few general additions and corrections . . . l.c. **10:** 558. 1934.—Also his Additions and corrections . . . for Hampshire and the Isle of Wight . . . l.c. **10:** 559–564. 1934.—Also **Baker, F. T.** Additions and corrections . . . for Lincolnshire. l.c. **10:** 564–566. 1934.—Also Jour. Bot. **74:** 201. 1936. (Correction of Angus and East Perth boundary on Druce's map.)—Also **Wilson, Albert.** Druce's Comital flora: corrections and suggestions. Rpt. Bot. Soc. and Exch. Club Brit. Isles **12:** 319–330. 1942.—Also

Wallace, E. C. Additions and corrections . . . for Devon vice-counties 3 and 4. l.c. **12:** 331–337. 1942.—Also **Edees, E. S.** Staffordshire additions . . . l.c. **12:** 532–542. 1944. (Includes bibliography.)—Also **Dony, J. G.** Additions and emendations . . . for v.c. 30 (Bedfordshire). l.c. **12:** 803–814. 1946. (Includes sketch of botanical explorations and bibliography.)—Also **Kent, D. H.** Additions and corrections . . . for Middlesex. [I]–II. Watsonia **1:** 92–101. 1949; **2:** 78–79. 1951.—Also **Webb, D. A.** Irish plant records. l.c. **2:** 217–236. 1952. (Brings Irish records up to date.)

——— The extinct and dubious plants of Britain. Rpt. Bot. Soc. and Exch. Club Brit. Isles **5** (1919): 731–799. 1920.
Annotated lists of extinct vascular plants and those whose British records are questionable.—See also Wilmott, below, and England, General (Salisbury).

——— Local floras. Rpt. Bot. Soc. and Exch. Club Brit. Isles **10** (1932): 399–424. 1933.—Corrections and additions. l.c. 744. 1934.
List of county and local floras of Great Britain and Ireland, geographically arranged (with vice-county numbers), with occasional statistics but otherwise unannotated; includes also numerous short floristic papers.—See also Holden, Polunin, Rehder, Smart, and Trimen, below.

Dunn, S. T. Alien flora of Britain. xvi, 208 p. 22 cm. London, 1905.
Annotated list of 924 introduced vascular plants and Characeae, with localities.—See also **Riddelsdell, H. J.** Notes on Mr. Dunn's "Alien flora," with particular reference to Glamorganshire plants. Jour. Bot. **43:** 89–94. 1905. —Also **Riddelsdell.** l.c. **44:** 138–142. 1906 (review).—Also **Marshall, E. S.** The status of some Britannic plants. l.c. **44:** 207–213. 1906.—Also Anonymous, above, and additional references there given.

Edlin, H. L. British plants and their uses. viii, 152 p. illus., pls., col. front. 22 cm. London [etc.], 1951.
Running account of native and cultivated useful plants (about 450 species) grouped mostly according to uses, partly according to habitat or botanical relationship, with vernacular names, uses, cultivation, bits of history and folklore, etc. Does not include timber trees or edible fungi.—For other general works on useful plants, see Henslow, Grigson, and C. P. Johnson, below; also Barton, above, and additional works there given.

Egerton-Warburton, Geoffrey. Names and synonyms of British plants. Collating the nomenclature of the London Catalogue, English Botany, Babington's Manual, Bentham's Flora, and Hooker's Student's Flora. With an appendix giving other names and their synonyms. And a list of authorities for plant names. xxxvi, 160 p. 16.5 cm. London, 1889. (New York Botanical Garden library.)
List of authorities for botanical names of British vascular plants and Characeae, with titles of the pertinent publications; alphabetical list of botanical names in the works cited (English Botany, ed. 3; London Catalogue, ed. 8; Babington, ed. 8; Bentham, ed. 5; Hooker, ed. 3), with page or plate reference to each work; alphabetical list of synonymous names, with equivalents.—See also Turnbull, below.

Ellis, David. Medicinal herbs and poisonous plants. xi, 179 p. 103 fig. 17 cm. London, Glasgow and Bombay, 1918.
Popular treatment of native and some imported medicinal and poisonous plants, with notes on properties, supply, cultivation, etc.—See also Barton, above, and Forsyth, below, and additional references there given.

Elwes, H. J., and **Henry, Augustine.** The trees of Great Britain and Ireland. 7 v. (2022 p.). 412 pl. (i.e. 417, incl. 5 col. fronts.), port. 32.5 cm. Edinburgh, 1906–13.
Account of native and cultivated trees (not in systematic order), with synonymy, references, description, history, introduction, cultivation, varieties, etc. The illustrations consist principally of excellent habit photographs, but some show leaves and technical details of fruit, twigs, buds, etc.—See also Bean, above, and additional references there given.

Fernie, W. T. Herbal simples approved for modern uses of cure. 3. ed. rev. and enl. xxxi, 596 p. 19 cm. Bristol, London, Toronto, 1914. (Kew library.) (1st ed. 1895?)

Brief history of simpling; list of medicinal plants, alphabetically arranged by vernacular names, with popular uses in medicine, etc.; list of diseases, with remedies; no index of scientific names.—See also Barton, above, and additional references there given.

Fisher, Robert. The English names of our commonest wild flowers. 2 v. (xi, 249; viii, 344 p.). 22 cm. Arbroath, 1932–34.

Pt. 1: alphabetical list of vernacular names of the species of plants that are found in all of the 112 comital districts of England, Scotland, and Wales, with explanation of derivation and meaning and scientific equivalents; all the vernacular names of each species collected under the principal one. Pt. 2: similar list of names of common but less widely distributed species.—See also Britten and Holland, above, and additional references there given.

Forsyth, A. A. British poisonous plants. Bul. Min. Agr. and Fish. [Great Britain] 161. vi, 116 p. 10 pl. (2 col.) (on 5). 1954.

General considerations; annotated list of wild poisonous plants, with account of symptoms and remedies, and references; annotated list of suspected plants, annotated tabular list of plants that affect milk.—See also Ellis, above, and C. Johnson and Pratt (The poisonous . . . plants), below.

Gilbert-Carter, Humphrey. Glossary of the British flora . . . with a preface by Charles E. Raven. 2. ed. rev. and enl. xxiv, 88 p. 20 cm. Cambridge, 1955. (1st ed. 1950.)

Short bibliography; alphabetical list of generic names and specific epithets of British vascular plants, with derivation and meaning, often with notes on pronunciation and name of Greek or Latin classical author who first used the name.—See also Harvey-Gibson and Macleod, below.

Godwin, Harry. The history of the British flora. A factual basis for phytogeography. viii, 384 p. 119 fig., 26 pl. (on 13), tab. 28.5 cm. Cambridge, 1956.

Includes (p. 69–292) list of vascular plants found in Quaternary deposits, with citation of records and discussion; bibliography.—See also Matthews (2), Reid, Tansley, and Turrill, below.

Grigson, Geoffrey. The Englishman's flora. Illustrated with woodcuts from sixteenth-century herbals. 478 p. 44 fig., front. 29 cm. London, 1955.

Discussion of sources, etc.; account of vernacular names, uses, popular beliefs, etc., with frequent quotations from literature, of all the flowering plants "with any considerable tale of names" and of many without such; bibliography, indices.—See also Britten and Holland and Edlin, above, and additional references there given.

Hadfield, Miles. British trees; a guide for everyman. xxv, 458 p. incl. 150 fig. 21.5 cm. London, 1957.

Systematic descriptive treatment of most of the wild and cultivated trees of England, Scotland, and Wales, including cultivated varieties, with details of introduction, cultivation, etc.—See also Bean, above, and additional references there given.

Harvey-Gibson, R. J. British plant names and their derivation. 3 p. l., 50 p. 19 cm. London, 1923.

List of botanical names of genera, species, and families of British vascular plants, with derivation and meaning; alphabetical list of abbreviations of botanical authors' names, with full form of the surname but not the given names, and partial dates.—See also Gilbert-Carter, above, and Macleod, below.

Hayward, W. R. Botanist's pocket-book, containing the chief characteristics of British plants; with botanical name, common name, soil or situation, colour, growth, and time of flowering of every plant, arranged under its own order. 19. ed., rev. by G. C. Druce. xlv, 310 p. 16.5 cm. London, 1930. (Gray Herbarium library.) (1st ed. 1872.)

Tabular systematic list of vascular plants and Characeae, with habitat, color, habit, etc., on verso of pages, and corresponding synoptical key to species on the opposite recto. Ranges are not given except in special cases.

Henslow, George. The uses of British plants traced from antiquity to the present day, together with the derivations of their names. vi, 184 p. 288 fig. 18.5 cm. London, 1905.

Annotated list of spermatophytes, systematically arranged, with explanation of scientific and vernacular names and account of uses.—See also Barton, Britten and Holland, Cameron, and Edlin, above, and additional references there given.

Hill, Jason (*pseudonym*). Wild foods of Britain. 3. ed. 93 p. illus. 18 cm. London, 1944. (1st ed. 1939.)

Popular classified account of edible wild animals and plants (including seaweeds and fungi), with directions for cooking; special recipes, calendar of wild foods, brief bibliography. Refers almost entirely to plants.—See also Cameron, above, and for general works on useful plants see Edlin, above, and additional references there given.

Hogg, Robert, and **Johnson, G. W.** The wild flowers of Great Britain. Botanically and popularly described, with copious notices of their history and uses. 11 v. 924 col. pl. 23 cm. London, 1863–1880. (New York Botanical Garden library.)—Index. 42 p. 8vo. London, 1874–80. (Not seen.)

Good hand-colored plates, without details, with text including account of vernacular names and uses. Plates in v. 1 and in part of v. 2 by Charlotte Gower, the rest by W. G. Smith.—See also Boulger, above, and additional references there given.

Holden, William. Floras of Great Britain. Bibl. Contr. Lloyd Libr. **1** (2): 14–70. 1911.

List of books and some separates from journals, alphabetically arranged by authors, with call numbers of those that are in the Lloyd Library.—See also Druce (Local floras) above and additional references there given.

Hooker, J. D. The student's flora of the British Islands. 3. ed. xxiii, 563 p. 17.5 cm. London, 1884. (Reissued 1897 and 1930. 1st ed. 1870.)

Synopsis of families; briefly descriptive, annotated flora of vascular plants [1413 species], with partial keys, habitat, local and general range; list of excluded species.

Horwood, A. R. A new British flora. British wild flowers in their natural haunts . . . 6 v. illus., pls. (64 col.), 3 maps. 27 cm. London, [1921].

Semipopular treatment of angiosperms, grouped according to habitat, with brief descriptions, habitat, local range, insect and fungous enemies, insect visitors, uses, vernacular names, etc. The sixth volume includes descriptive account of the species not dealt with in detail in v. 1–5. The bibliography (**5**: 223–228) consists mostly of general or topical works, including only a few floras. Some copies are dated 1919, but apparently none were issued until 1921.

Johnson, Charles, and **Johnson, C. P.** British poisonous plants . . . iv, 76 p. 32 col. pl. 19.5 cm. London, 1861. (Gray Herbarium library.) (1st ed. by Charles Johnson, 1856.)

Account of native vascular and cellular plants possessing poisonous properties, systematically arranged.—See also Forsyth, above, and additional references there given.

Johnson, C. P. The useful plants of Great Britain: a treatise upon the principal native vegetables capable of application as food, medicine, or in the arts and manufactures. vi, 324 p. 300 col. fig. on 25 pl. 26 cm. London, 1862 [1861–62].

Local distribution and uses of 300 wild and cultivated vascular and cellular plants, each represented by a colored figure; numerous related useful species mentioned more briefly. The work is sometimes referred to as Sowerby's Useful plants, and one copy examined bore this title on the spine. John E. Sowerby made the illustrations.—See also Barton, Edlin, and Forsyth, above, and additional references there given.

Kent, D. H., Bangerter, E. B., and **Lousley, J. E.** British herbaria. An index to the location of herbaria of British vascular plants with biographical references to their collectors. 101 p. 22 cm. London, 1957 [1958].

Abbreviations of names of institutions, list of herbaria, list of collectors with dates and references to location of their herbaria or exsiccatae, index to herbaria dealing with local floras, list of collections relating to critical genera, bibliography.

Kirk, J. W. C. A British garden flora. A classification and description of the genera of plants, trees and shrubs represented in the gardens of Great Britain, with keys for their identification. x, 584 p. 223 fig. 24 cm. London, 1927.

Briefly descriptive account of genera of spermatophytes cultivated in Great Britain, with mention of principal species; keys to families and genera.—See also Bean, above, and additional references there given.

Linton, E. F. Hybrids among British phanerogams. Jour. Bot. **45**: 268–276, 296–304. 1907.

Annotated list of hybrids among vascular plants.

McClintock, David, Fitter, R. S. R., and Rose, Francis. The pocket guide to wild flowers. 340 p. illus., 112 pl. (64 col.) on 56, map (on lining papers). 20 cm. London, 1956.

Briefly descriptive, partially keyed flora of vascular plants including "all the commoner flowering plants likely to be found wild or looking wild in the British Isles, including the Channel Isles," as well as many rarer species; 1306 species are illustrated. Microspecies are listed in footnotes except in *Rosa, Rubus, and Hieracium.*—See also **McClintock, D.** Supplement to pocket guide to wild flowers. ix, 89 p. (place, date?) (Adds about 450 species. Not seen; cited from notice in Proc. Bot. Soc. Brit. Isles **3**: 251–252. 1959.)

Macdonald, James, and others, *eds.* Exotic forest trees in Great Britain. Bul. Forestry Comn. [Great Britain] 30. vii, 167 p. incl. 2 fig. (maps). 17 i.e., 16 pl. (on 8). 1957.

Includes (pt. II, p. 31–140) Exotic coniferous trees in Great Britain, an extensively annotated alphabetical list of species, with historical notes, cultivation, timber, etc.; (pt. III, p. 141–165) Exotic broadleaved trees in Great Britain, a similar list of dicotyledons.

Macleod, R. D. Key to the names of British plants. vii, 94 p. 22.5 cm. London, 1952.

Bibliography, general considerations on derivation and formation of scientific and vernacular names; alphabetical list of generic names with derivation and meaning, the respective specific names listed under each generic name, with English equivalent and in some cases with derivation and meaning; alphabetical list of specific epithets (except those beginning with a capital), with derivation and meaning; alphabetical list of common names, with scientific equivalent, meaning, and often derivation. The work is essentially restricted to the names given in Bentham's Handbook.—See also Gilbert-Carter and Harvey-Gibson, above.

Makins, F. K. Herbaceous garden flora; a guide to the identification of more than 1,000 species of herbaceous perennials, biennials, and annuals cultivated in British gardens for ornament. x, 373 p. incl. 224 fig. on p. 63–286, plates. 24 cm. London, 1957.

Key to families and genera; figures of 991 herbaceous ornamentals, followed by very brief descriptions (covering also 60 other species.)—See also Chittenden and Kirk, above.

——— The identification of trees and shrubs. How to name any wild or garden tree or shrub likely to be met with in the British Isles. 2. ed. vii, 375 p. incl. 128 fig. 24 cm. New York, 1948. (1st ed. 1936.)

Popular treatment of the wild and (mostly) cultivated trees and shrubs "growing in the open in any part of the British Isles" (1732 species in 1st ed.), with partial keys, very brief descriptions, and in most cases outline drawings of leaves, fruit, etc.—See also his British trees in winter. 56 p. incl. 40 fig. 32 pl. (on 16). 22 cm. London, 1945. (Descriptive account of principal trees in winter condition, alphabetically arranged by vernacular names, with photographs showing bark or habit, key based on twigs, and line drawings of twigs.)—Also Bean, above, and additional references there given.

Matthews, J. R. Geographical relationships of the British flora. Jour. Ecology **25**: 1–90. 17 fig. (maps). 1937.

General considerations; discussion of 11 of the 15 phytogeographical groups recognized in the British flora, with tabular lists of species showing local and general distribution; bibliography.—See also his Plant life in Britain: its origin and distribution. (Masters Memorial Lectures, 1946.) Jour. Roy. Hort. Soc. **71**: 225–239, 259–273. 1946. (Plant distribution in relation to glaciation, introduced species and weeds, flora of various habitats, phytogeography, woodlands, alpine plants.)—Also Godwin, above, and additional references there given.

——— Origin and distribution of the British flora. 173 p. 19 cm. London, 1955. (Hutchinson's university library. Biological sciences.)

General considerations, Pleistocene history, forest development, changes in plant ranges, composition of flora, types of distribution, geographical elements with lists of species.—See also Godwin, above, and additional references there given.

Melderis, Alexander, and **Bangerter, E. B.** A handbook of British flowering plants. 360 p. illus., 24 col. pl. 22 cm. London and Melbourne, 1955.

Semipopular, annotated, descriptive flora of spermatophytes, with notes on biology, properties, uses, vernacular names, etc. Not a complete flora, and authorities for scientific names not given.

Miller, William. A dictionary of English plant names applied in England and among English-speaking people to cultivated and wild plants, trees, and shrubs. viii, 264 p. 23 cm. London, 1884.

Alphabetical lists of vernacular names with botanical equivalents, and of botanical names with vernacular equivalents.—See also Britten and Holland, above, and additional references there given.

Morse, Richard, and **Palmer, Raymond.** British weeds, their identification and control. A practical handbook . . . 207 p. 32 fig., 7 i.e. 8 pl. 22 cm. London, 1925.

General considerations, key; brief account of British weeds, grouped by habitat and alphabetically arranged by vernacular names, with directions for control, notes on uses or poisonous properties; lists of economically useful weeds, poisonous weeds, host plants for insect or fungous pests; list of botanical names, the families arranged alphabetically; indices.—See also Anonymous, above, and additional references there given.

Moss, C. E. The Cambridge British flora. v. 2–3 (in 4 v. incl. 2 v. of plates). 121 fig. (maps), 397 pl., port. 38 cm. Cambridge, 1914–20.

Annotated descriptive flora of vascular plants, with special attention to varieties and hybrid forms; range of each species shown on map, and each illustrated by one or more plates (by E. W. Hunnybun), the details of which, at least in the case of the smaller plants, are frequently imperfectly observed. Incomplete: Salicaceae-Fumariaceae (Engler-Prantl system). No more published.

Polunin, Nicholas. British floras ancient and modern. Rhodora **55:** 209–224. 1953.

Running account of general floras of the British Isles, with especial consideration of that by Clapham (1952).—See also Druce (Local floras), above, and additional references there given.

Pratt, Anne. The flowering plants, grasses, sedges and ferns of Great Britain and their allies the club mosses, horsetails, etc. New ed. rev. by Edward Step . . . 4 v. 4 plain pl. ("A-D"), 319 col. pl. 25 cm. London and New York, 1899–1900. (1st ed. 1855, with title The flowering plants and ferns of Great Britain.)

Account of native vascular plants, with brief descriptions, range, vernacular names, uses, etc.; semipopular, but accurate as far as it goes. The colored plates illustrate about 1500 species. There are various editions and reissues; the first issue of Step's revision was apparently in 1891.—See also Boulger, above, and additional references there given.

——— The poisonous, noxious, and suspected plants of our fields and woods. xii, 208 p. col. pls. 16 cm. London [1857].

Popular account of native and naturalized poisonous plants.—See also Forsyth, above, and additional references there given.

Prime, C. T., and **Deacock, R. J.** The shorter British flora. 2. ed. repr. with minor corrections. xl, 376 p. 232 fig. 19 cm. London, 1957. (1st ed. 1948.)

Popular flora, including about 720 of the commoner species; cited here for its notes on biology, uses, habitats, etc., such as are not found in the ordinary flora.

Prior, R. C. A. On the popular names of British plants, being an explanation of the origin and meaning of the names of our indigenous and most commonly cultivated species. 3. ed. xxvii, 294 p. 20.5 cm. London, 1879. (1st ed. 1863.)

Bibliography; alphabetical list of vernacular names (almost all English), with etymology and explanation of meaning; alphabetical list of botanical names with vernacular equivalents.—See also Britten and Holland, above, and additional references there given.

Rayner, J. F. A standard catalogue of English names of our wild flowers, to which are added the ferns and their allies. 2 p.l., 51, [5] p. 21 cm. Southampton, London [1927].

Systematic list of scientific names of vascular plants and Characeae (1619 species), with "standard" English name for each.—See also Britten and Holland, above, and additional references there given.

Rehder, Alfred. (Phytography.) British Islands. *In* his The Bradley Bibliography. A guide to the literature of the woody plants of the world published before the beginning of the twentieth century. Compiled at the Arnold Arboretum of Harvard University under the direction of Charles Sprague Sargent. v. 1, p. 393–404. 29.5 cm. Cambridge, Mass., 1911. (Publications of the Arnold Arboretum, no. 3.)—Additions and corrections. l.c. **1**: 533–535; **5**: xx–xxi. 1918.

Essentially unannotated list of floristic works (including those in periodical and serial literature) on "Dendrography" and on "General phytography," chronologically arranged, and including practically all the floristic works of the period covered (through 1900) except the most minor ones. For Great Britain, as for all European countries except Belgium, this is for the years covered the most nearly complete floristic bibliography available, and the same is true for many other parts of the world.—See also l.c. **1**: 17–20, for alphabetical list of British periodicals and serials relating to botany; **1**: 51–55, for works relating to botanic gardens; **1**: 73–74, for dictionaries and lists of vernacular names and their etymology; **1**: 80–81, for works on botanical history; **1**: 226, for works on phenology; **1**: 296–297, for noteworthy individual trees and forests; **3**: 16–18. 1915, for alphabetical list of periodicals and serials relating to arboriculture, gardening, and uses of plants (additions, p. 757); **3**: 28–29, for works on collections, gardens, arboretums, museums and exhibitions (not herbaria); **3**: 33, for works on the history of gardening; **3**: 78–83, for dendrological, pomological, economical and miscellaneous works dealing with horticulture, fruit trees, medicinal and other useful or ornamental plants; **3**: 106–110, works on arboriculture and general horticulture; **3**: 250–255, for pharmacopoeias and dispensatories; **3**: 271, for works on economic products; **4**: 12. 1914, for list of periodicals and serials relating to forestry, timber uses, etc.; **4**: 19–20, for works on exhibitions of forest products; **4**: 27, for works on history of forests; **4**: 84–85, for works on dendrography; **4**: 357–359, for works on forests. Various smaller sections of no floristic significance have been omitted from the above summary.—See also Druce (Local floras), above, and additional references there given.

Reid, Clement. The origin of the British flora. vi, 191 p. 22.5 cm. London, 1899.

Origin of flora, means of dispersal, changes in geography and climate, alphabetical list of localities with lists of fossil species found; systematic list of British plants, with account of fossil occurrence, summarized in a table.—See also Godwin, above, and additional references there given.

Ross-Craig, Stella. Drawings of British plants, being illustrations of the species of flowering plants growing naturally in the British Isles . . . with a foreword by Sir Edward Salisbury . . . pt. 1–10. pls. 24 cm. London, 1948–57.

Excellent plates showing habit and details of flower, fruit, and seed, without text. The parts so far issued contain from 22 to 77 plates each (numbered separately in each part), with index; there is separate index covering parts 1–3. Part 10 reaches *Sedum*.—See also Boulger, above, and additional references there given.

Smart, John, and **Taylor, George,** *eds.* Flora. *In* their Bibliography of key works for the identification of the British fauna and flora. 2. ed. p. 94–126. 21.5 cm. London, 1953. (1st ed. 1942.)

The sections on Pteridophyta (p. 111) and Spermatophyta (p. 111–125) include titles of principal general British floras and some foreign works, and (under Spermatophyta) references to taxonomic treatments of various genera arranged by families.—See also Holden, above, and additional references there given.

Sowerby, James, and **Smith, J. E.** English botany; or, coloured figures of British plants. Edited by J. T. Boswell Syme. The popular portion by Mrs. Lankester. The figures by J. Sowerby, J. de C. Sowerby, J. W. Salter, and John Edward Sowerby. Enlarged, re-arranged according to the natural orders, and entirely revised, with descriptions of all the species by the editor. 12 v. 1922 col. pl. 26 cm. London, 1863–86. (Other issues with varying dates. 1st ed. 36 v. 1790–1814; Supplement . . . 5 v. 1831–63.)—Supplement to the third ed. v. 1–4. Compiled by N. E. Brown. viii, 213, viii p. 15 pl. (13 col.). 26 cm. London, New York, [1891]–92.

Descriptive flora of vascular plants and Characeae, with notes on uses, popular beliefs, etc.; the standard illustrated work on British plants. The first edition had 2998 plates, including the supplement, and illustrated cellular as well as vascular plants.—See also **Garry, F. N. A.** Notes on the drawings for "English Botany." Jour. Bot. v. 41–42, suppl. 276 p. 1903–04. (History of the work; list of the 1922 plates of the 3d ed., with references to plate numbers in 1st ed., date of drawings, and notes on habitat, characters, etc., by the artists and others.)—Also Boulger, above, and additional references there given.

Step, Edward. Wayside and woodland trees. A guide to the British sylva. Rev. by A. Keith Jackson . . . and A. Bruce Jackson . . . xxxviii, 186 p. illus., 175 pl. (24 col.). 17.5 cm. London and New York, 1940. (1st ed. 1904.)

Popular but accurate account of native and principal naturalized or widely cultivated trees and larger shrubs; ranges given only in very general terms. —See also Bean, above, and additional references there given.

Tansley, A. G. The British islands and their vegetation. new [2.] ed. 2 v. (xxxviii, 930 p.). 179 fig. (incl. maps), 162 pl. (418 phot.), front. 26 cm. Cambridge, 1949. (1st ed. 1939.)

Physical features, geology, climate, soils, rocks, biotic factors, history of vegetation, description of plant formations and associations, with chapter bibliographies. The 2d ed. is little changed from the first. A condensed, popularized version of the same material is given in his Britain's green mantle, past, present, and future. xii, 294 p. 8 text fig., 142 fig. (phot.) on pls., front. 22 cm. London, 1949. Replaces **Tansley,** and others. Types of British vegetation. 1911.—See also Godwin, above, and additional references there given.

Trimen, Henry. Botanical bibliography of the British counties. Jour. Bot. **12**: 66–73, 108–112, 155–158, 178–183, 233–238. 1874.

Partly annotated bibliography, arranged in the order of Watson's Topographical Botany of county and local floras, with notes on herbaria, etc. Covers England, Wales, and Scotland.—See also Druce (Local floras), above, and additional references there given.

Turnbull, Robert. Index of British plants according to the London Catalogue (eighth edition). Including the synonyms used by the principal authors, an alphabetical list of English names, also references to the illustrations of Syme's English Botany, and Bentham's British Flora. 2 p. l., 98 p. 22 cm. London, 1889. (Gray Herbarium library.)

Alphabetical list of botanical names in the works cited, with page or plate reference to each work (London Catalogue, ed. 8; English Botany, ed. 3; Fitch and Smith's Illustrations of Bentham's British Flora); list of English names, with botanical equivalents.—See also Egerton-Warburton, above.

Turrill, W. B. British plant life. xvii, 315 p. 11 fig. (incl. 9 maps), 72 pl. (48 col.) on 36. 22.5 cm. London, 1948. (The New naturalist series.)

Geological history in reference to plants, statistics of flora, phytogeography, habitats, plant communities, variation, adaptation and heredity, changes in flora. The appendices include list of vice-counties (with map, p. 261) and summary accounts, with short bibliographies, of taxonomic methods, biometrics, ecological methods, cytology, genetics; list of more important county and local floras, general bibliography.—See also Godwin, above, and additional references there given.

Watson, H. C. A compendium of the Cybele britannica . . . 3 pt. (vi, 651 p.). map. 23 cm. Thames Ditton, London, 1868–70.—Supplement . . . iv, 213 p. map. 23 cm. Thames Ditton, London, 1872.

Topographical divisions, plant zones, phytogeography, etc.; pt. 1–2, list of 1428 vascular plants with local distribution by provinces, latitude, and zones, etc., and extralimital distribution; pt. 3, additions and corrections. Does not cover Ireland.

——— Cybele britannica; or British plants and their geographical relations. 4 v. map. 23 cm. London, 1847–59.—Part first of a supplement . . . 119 p. London, 1860.

Vol. 1–3, phytogeography, plant zones, habitats, etc.; list of 1428 vascular plants, with local distribution by provinces, latitude, and zones, vertical range, temperature range, and annotations; v. 4, introduced species, physiography, summaries of distribution, altitudinal range, general features of phytogeography, etc. Does not cover Ireland.

——— Topographical botany: being local and personal records towards shewing the distribution of British plants traced through the 112 counties and vice-counties of England, Wales, and Scotland. 2d. ed., corr. and enl. [by J. G. Baker and W. W. Newbould]. xlvii, 612 p. map. 23 cm. London, 1883. (1st ed. 1873–74.)

Memoir of author, etc.; list of provinces, subprovinces, and vice-counties; list of vascular plants (1428 species), with county and vice-county occurrences and authorities therefor; lists of catalogues, correspondents, etc.; bibliography.—See also **Bennett, Arthur.** Supplement to "Topographical botany," ed. 2. Jour. Bot. v. 43, suppl. 118 p. 1905. (List of local floras cited; list of vascular plants, with additional vice-county records and references, going through 1903.) For *Rubus*, omitted from the Supplement (except for four species), see the following references: **Rogers, W. M.** Handbook of British Rubi. xiv, 111 p. 22 cm. London, 1900. (Appendix, p. 99–107, consists of a list of species with their vice-county records); also his Supplementary records of British Rubi. April, 1900–December, 1908. Jour. Bot. **47:** 310–318, 340–346. 1909. (Similar list; Irish records are omitted); also his Supplementary records . . . (January, 1909–March, 1915.) l.c. **53:** 139–145. 1915. (Similar list.)—Also **Bennett, Arthur, Salmon, C. E.,** and **Matthews, J. R.** Second supplement to Watson's Topographical botany. l.c. v. 67–68, suppl. 99 p. 1929–30. (List going through 1925.)—Also **Druce, G. C.** *In* The comital flora of the British Isles. p. xix–xxi. 1932. (Critical remarks on authorship of records in the two supplements.)

Williams, F. N. The high alpine flora of Britain, being a list of the flowering plants and ferns found at a thousand meters and upwards on the mountains of the British Isles, with authentic references and critical notes. Annals Scott. Nat. Hist. **1908:** 163–169, 242–251. 1908; **1909:** 30–36, 108–114, 164–168, 229–234. 1909; **1910:** 34–39. 1910.

List of mountains over 1000 meters in altitude; list of 164 vascular plants, with localities and altitudinal range. Relates to 64 mountains in Scotland, 3 in Wales, and 2 in Ireland.—See also Wilson, below.

——— Prodromus florae britannicae. xx, 604 p. 24.5 cm. Brentford, 1901–12.

Descriptive flora (in Latin, annotations in English). Incomplete: Sympetalae, some Choripetalae (Rhamnaceae—Droseraceae). Issued in 10 pts.

Wilmott, A. J. Flowering plants. *In* Changes in the British fauna and flora during the past fifty years. Proc. Linn. Soc. **148:** 41–45. 1935.

Brief account of some species that have become extinct, rare, or naturalized, or that have varied in abundance.—See also **Druce (Extinct and dubious plants)** above, and additional references there given.

Wilson, Albert. The altitudinal range of British plants. 2. ed. rev. and enl., 1949. North West. Nat. n. ser., v. 3, no. 1–3/4, Suppl. 117 p. 1955–56. (Reprinted, 1 p.l., 117 p. Arbroath, 1956.) 1st ed., l.c. v. 6, Suppl. 1930–31.)

Climate, bibliography; list of vascular plants, with altitudinal range (in feet) in England, Wales, Scotland and Ireland, localities, and references.—See also **Pickard, J. F.** Altitudinal range of British plants. North West. Nat. n.s. **1** (whole ser., v. **24**): 452, 601–602. 1953–54; **2** (whole ser., v. **25**): 141, 323, 462. 1954. (List of supplementary records to Wilson's 1st ed., mostly made by author.)—Also Williams, The high alpine flora of Britain, above.

CHANNEL ISLANDS (ÎLES NORMANDES)

Local

Baal, H. J. The indigenous trees of Jersey. Bul. Ann. Soc. Jersiaise **15**: 341–346. 1951.

Annotated list of native trees.

Hurst, C. P. The flora of the island of Brechou. Rpt. and Trans. Guernsey Soc. Nat. Sci. 4 (1902): 163–176. 1903.

Topography, general features of flora; annotated list of 205 plants (190 vascular) collected by author in July-August 1902. The flora was previously unknown except for a few lichens.

Lester-Garland, L. V. A flora of the island of Jersey, with a list of the plants of the Channel Islands in general, and remarks upon their distribution and geographical affinities. xv, 205 p. map. 19 cm. London, 1903.

Topography, geology, climate, botanical explorations, bibliography; annotated list of vascular plants and Characeae of Jersey; list of doubtful species; tabular list of vascular plants of Channel Islands with local and general distribution; plant geography, bibliography.—See also **Druce, G. C.** Notes on the flora of the Channel Islands. Jour. Bot. **45**: 395–402, 419–428. 1907. (Extensive list of vascular plants, with localities, from Jersey and other islands.)—Also **Perrédès, P. É. F.** John Piquet (1825–1912). l.c. **50**: 371–374. **1 fig. (port.). 1912.** (Contains list of additional species collected by Piquet.)—Also **Salmon, C. E.** Notes on Jersey plants. l.c. **51**: 16–18. 1913.—Also **Pugsley, H. W.** Notes on Channel Island plants. l.c. **52**: 327–332. 1914. (Records from Jersey and Guernsey.)—Also **Attenborough, T. W.** Botanical report for 1946 [—1950]. Bul. Ann. Soc. Jersiaise v. 14–16, passim. 1949–55.

Marquand, E. D. Flora of Guernsey and the lesser Channel Islands: namely, Alderney, Sark, Herm, Jethou, and the adjacent islets. viii, 501 p. 5 maps. 22.5 cm. London, 1901.

General account of islands; topography, human history, climate (with long list of species in flower in December), geology, botanical explorations, general features of flora, statistics, etc. of Guernsey, followed by annotated list of its vascular and cellular plants (809 vascular) with localities, date of first record (vascular plants), patois names and uses; similar but less detailed treatment of each of the other islands (Alderney, Sark, Herm, Jethou, Lihou, Crevichon, Burhou); list of plants found on Jersey and not on other islands. Includes 2653 species, of which 857 are vascular.—See also **Druce, G. C.** Notes on the flora of the Channel Islands. Jour. Bot. **45**: 395–402, 419–428. 1907. (Extensive list of vascular plants, with localities, from Jersey, Guernsey, Alderney, and Sark.)—Also **Pugsley, H. W.** Notes on Channel Island plants. l.c. **52**: 327–332. 1914. (Records from Jersey and Guernsey.)—Also **McCrea, M. A.** Supplement to Marquand's "Flora of Guernsey . . ." Flowering plants, ferns, mosses, etc., found from 1901 to 1923. Rpt. and Trans. Soc. Guernes. 9 (1923): 167–182. 1924. (List of additions, arranged by islands, with localities, dates, and collectors' names.)—Also her List of new records, 1924–1933 [–1946]. (Title varies.) Rpt. and Trans. Soc. Guernes. **12**: 24, 113–114, 264, 450. 1933–37; **13**: 388–392, 392–393, 393–394. 1946; **14**: 10–11. 1948.—Also **Girard, P. J.** Report of the Botanical Section, 1947 [–1949]. l.c. **14**: 119–120, 234–235, 344–346. 1948–1950.—Also papers by various authors in Jour. Bot. **71**: 106–107. 1933 (Alderney); **75**: 299–300. 1937 (Alderney). Rpt. Bot. Soc. and Exch. Club Brit. Isles **11**: 561–564.

1937 (Alderney). Proc. Bot. Soc. Brit. Isles **1**: 142–151. 1954 (Alderney); **1**: 378. 1955 (Sark). Rpt. and Trans. Soc. Guernes. **15**: 340–342. 1956 (Sark). (Last item not seen.)

——— The Guernsey dialect and its plant names. Rpt. and Trans. Guernsey Soc. Nat. Sci. 5 (1905): 31–47. 1906.

General considerations; alphabetical list of Guernsey patois names, with botanical equivalents and frequent notes on origin of names and uses.

——— The vegetation of small islets. Rpt. and Trans. Guernsey Soc. Nat. Sci. **6** (1909): 38–53. 1910.

Origin of flora, etc.; topography and lists of plants of Houmet Paradis, Houmet Benest, Crevichon, Chapelle Dom Hue, Burhou, Plat Houmet, Houmet Homtolle, Galeux, Longue Pierre, Ecrehos Rocks, and Chausey Islands, each island or group being treated separately.—See also **McCrea, M. A.** Supplement to Marquand's "Flora of Guernsey . . ." Flowering plants, ferns, mosses, etc., found from 1901 to 1923. Rpt. and Trans. Soc. Guernes. 9 (1923): 167–182. 1924.—Also **Tomlin, J. R. le B.** The vegetation of small islets. l.c. 9 (1923): 266–272. 1924. (Lists of species found by author on Lion Rock (Albecq), La Vieille, Lihoumel, Rat Island, and Houmet; and additions to Marquand for Houmet Paradis, Houmet Benest, Crevichon, Chapelle Dom Hue, and Houmet Homtolle.)—Also **Chevalier, Auguste.** Une excursion botanique dans les îles anglo-normandes. Bul. Soc. Linn. Normandie VIII, **6**: 5–8. 1934. (Includes (footnote, p. 6) partial list of angiosperms of Erehou; also notes on species naturalized on some of the other islands.)

ENGLAND

General

Cockayne, Oswald. Saxon names of plants collected. *In* his Leechdoms, wortcunning, and starcraft of early England . . . v. 3, p. 307–350. 25.5 cm. London, 1866. (Rerum britannicarum medii aevi scriptores . . . [no. 35], v. 3.)

Alphabetical list of Saxon names of worts, trees, and some vegetable products, with botanical identifications and references.—See also his A glossary of plants from the Library of the Cathedral, Durham. l.c. p. 297–305. (List of Latin and Greek names, with Saxon vernacular equivalents.)

Earle, John. English plant names from the tenth to the fifteenth century. cxii, 122 p. 15.5 cm. Oxford, 1880.

History of plant names from Theophrastus to modern times, etc.; lists of Anglo-Saxon, mediaeval Latin, and other names from various old vocabularies, etc.; indices.—See also review by **Eugene Fournier,** Jour. des Savants **1881**: 382–383. 1881.

Hoops, Johannes. Über die altenglischen pflanzennamen. 84 p. 21.5 cm. Freiburg i.B., 1889. (Diss.)

Bibliography; running account of numerous Old English plant names, with botanical identifications and pertinent quotations, notes on mythology, etc.; alphabetical list of 500 Old English vernacular names; general discussion of etymology, etc.; detailed treatment of the words beorc, apuldor, box, thung.

Rohde, E. S. The old English herbals. xi, 243 p. pls. (1 col.), ports., facsims. 26 cm. London [etc.], 1922.

Popular account of English herbals from the earliest times, with quotations; bibliography of manuscript herbals, printed English herbals, and principal foreign herbals, with complete titles.

Salisbury, E. J. The waning flora of England. Trans. South-east. Union Scient. Soc. **32**: 35–54. 1927. (American Museum of Natural History library.)

Discussion of vascular plants which have become extinct or appreciably rarer, and alphabetical list of species with names of the counties concerned. —See also Great Britain, General (Druce, The extinct and dubious plants of Britain, and Wilmott).

Bedfordshire

See also Great Britain, General (Dony, Additions and emendations . . . for v. c. 30 (Bedfordshire), under Druce, The comital flora of the British Isles).
Vice-county 30.

General

Dony, J. G. Flora of Bedfordshire. 532 p. 22 fig. (maps), 24 pl. (i.e. 24 fig. on 16 pl. (on 8)), front., map. 22 cm. Luton, 1953.

Botanical explorations (including biographical data), geography, topography, climate, geology, soils, natural regions, botanical districts, bibliography, list of herbaria; annotated list of Charophyta, Bryophyta (by other authors), and vascular plants with detailed localities; list of casual plants and garden escapes, list of fungi (by D. A. Reid), index of botanists, gazetteer. Replaces **Abbot, Charles.** Flora bedfordiensis, comprehending such plants as grow wild in Bedford, arranged according to the system of Linnaeus, with occasional remarks. xii, 351 p. 6 col. pl. 23 cm. Bedford, 1798. (Bibliography; annotated flora of 1225 vascular and cellular plants, on the Linnaean system, with brief diagnoses but essentially without localities.) Also replaces **Saunders, James.** The field flowers of Bedfordshire. vi, 91 p. 21 cm. London and Luton, 1911. (Botanical explorations; annotated list of vascular plants (over 900 species) and Characeae. Although referred to in the preface as a second edition, the work was here first separately published; the list appeared in its original form in the "Bedfordshire Advertiser and Luton Times" in 1898 (not seen).)

Local

Davis, Frederick. Botany of the neighborhood of Luton. *In* his Luton, past and present: its history and antiquities. 2 ed. p. 123–130. 28 cm. Luton, 1874. (Harvard College library.) (1st ed. in his The history of Luton . . . 1855.)

List of spermatophytes, with localities for scarcer species.

Little, J. E. The Ivel district of Bedfordshire. Rpt. Bot. Soc. and Exch. Club Brit. Isles **11** (1935): 50–67. 1936.

Geography, geology, moors and marshes, changes in vegetation, place names derived from plants, changes in woodland, botanical explorations, list of new records, etc.; chronological bibliography.

Berkshire

See also Surrey (Monckton, The flora of the Bagshot district; Whale, British wild flowers . . . growing in and near Egham).
Vice-county 22.

General

Druce, G. C. The flora of Berkshire, being a topographical and historical account of the flowering plants and ferns found in the county, with short biographical notices of the botanists who have contributed to Berkshire botany during the last three centuries. cxcix, 644 p. map (in pocket). 20 cm. Oxford, 1897 [1898].

Topography, climate, geology, general features of flora, botanical districts, history of botanical investigations, bibliography; annotated list of vascular plants and Characeae (1,208 species, of which 893 are native), with localities. The additions of 1918 bring the total to 1,363 species (942 native), with over 500 varieties and 80 hybrids. According to review in Jour. Bot. **36**: 104. 1898, the work, although dated 1897, was not issued until 1898.—See also his Additions to the Berkshire flora. Rpt. Bot. Soc. and Exch. Club Brit. Isles **5** (1917): 443–480. 1918. (Also in Proc. and Rpt. Ashmolean Nat. Hist. Soc. Oxfordshire **1918**: 21–58. 1919. Not seen.)—Also **Brenan, J. P. M.** Notes on the flora of Oxfordshire and Berkshire. l.c. **12** (1943–44): 781–802. **1946.** (Extensive list.)—Also his Records additional to the floras of Oxfordshire and Berkshire. l.c. **13** (1946–47): 227–230. 1948.

Local

Ashbee, C. R., *ed.* Botanical list. For 1879. Flora wellingtonensis. Ann. Rpt. Wellington Col. Nat. Hist. Soc. **10**: 32-57. map. 1880.

Tabular list of vascular plants with localities for scarcer species, earliest flowering date, etc. This appears to be the last edition of this flora; the first issue in this journal (with title Botanical list. Flora wellingtonensis. H. Fraser, *ed.*) appeared l.c. **1**: 51–59. 1869, and was reprinted with additions in v. 3 and v. 4, with different editors' names. A statement in Fraser's introduction indicates that it was separately published earlier.—For additions, see l.c. **22**: 51. 1892; **25**: 49. 1895; **26**: 51. 1896; **27**: 56. 1897; **29**: 45. 1899; **30**: 47. 1900; **31**: 46–47. 1901.

Monckton, H. W. The flora of the district of the Thames Valley drift between Maidenhead and London. 81 p. 16.5 cm. London, 1919. (Gray Herbarium library.)

Surficial geology, short bibliography; list of vascular plants (632 species), with localities, references, and indication of occurrence on alluvium or on gravel.

Reading Natural History Society. A list of the flowering plants, ferns, etc. of the country round Reading . . . xix, 56 p. 18.5 cm. [pref. Reading, 1900.] (Turner Brothers' local series, no. 5.) (Library of N. D. Simpson.)

General features of flora; briefly annotated list of vascular plants.

[Russell, Anna]. A catalogue of plants found in the neighbourhood of Newbury. *In* [Gray, Edward W., *ed.*]. The history and antiquities of Newbury and its environs . . . ; also a catalogue of plants found in the neighbourhood. p. 310–340. 22 cm. Speenhamland, 1839. (Reprinted, 31 p. 1839.)

List of vascular plants, on the Linnaean system, with localities for the scarcer species.—See also her Notes on a list of Newbury plants. Phytologist **3**: 716. 1849. (Minor corrections.)

Stone, F. J. Botany. *In* Field, T. The Radley district; its history, botany, entomology and geology. p. 39–66. 18.5 cm. Oxford, 1912. (Library of N. D. Simpson.)

Running account of vascular and a few cellular plants, grouped by habitats, followed by systematic list of the genera. Druce (Flora of Berkshire, p. cxcv. 1897) cites what is evidently an earlier edition of this work, as follows: "The fauna and flora of Radley and its neighbourhood. Oxford, 1896. The list of plants is edited by F. J. Stone, M. A."

Buckinghamshire

See also London.
Vice-county 24.

General

Druce, G. C. The flora of Buckinghamshire, with biographical notices of those who have contributed to its botany during the last three centuries. cxxvi, 437 p. 3 pl., map (in pocket). 24 cm. Arbroath, 1926.

Topography, geology, botanical districts, history of botanical investigations, table of distribution of plants in 11 inland counties; annotated list of vascular plants and Characeae (1937 taxa, including 939 native species).

Local

Baker, J. G. The botany of Burnham Beeches. Jour. Bot. **55**: 276–278. **1917.**

Unannotated list of vascular plants observed in July and August in a tract of wild forest.

Britten, James. A flora of High Wycombe. Naturalist **2**: 319–325, 348–357. 1866; **3**: 15–19, 33–37, 47–52, 69–77, 116–126, 139–142. 1866–67.

Geology, etc.; annotated list of vascular plants found within radius of 5 miles of High Wycombe. Unfinished; covers Ranunculaceae to Cephalanthera.

Mill, G. G. List of flowering plants found in the neighbourhood of Great Marlow, Bucks, in the early part of the summer of 1843. Phytologist **1**: 983–995. 1844.

Annotated list of angiosperms.

CAMBRIDGESHIRE

Vice-county 29.

General

Evans, A. H. A flora of Cambridgeshire with an appendix on the origin and drainage of the fens. 223 p. map. 19 cm. London, 1939.

Topography, geology, botanical explorations, soils, changes in flora; annotated list of "over 1,000" vascular plants; bibliography, account of the fens; annotated list of Charophyta. Replaces **Babington, C. C.** Flora of Cambridgeshire: or, a catalogue of plants found in the county of Cambridge, with references to former catalogues, and the localities of the rarer species. lvi, 327 p. map. 19 cm. London, 1860. (Botanical history, topography, bibliography, table of local distribution; annotated list of vascular plants and Characeae.) Also replaces **Evans, A. H.** A short flora of Cambridgeshire, chiefly from an ecological standpoint, with a history of its chief botanists.. Proc. Cambridge Phil. Soc. **16:** 197–284. 1911; Additions . . . l.c. **17:** 229–235. 1913. (Account of principal local botanists, list of rare or extinct species; list of vascular plants, with distribution indicated by regions; annotated lists of cellular plants, by other authors.)—See also **Perring, F. H., Sell, P. D.,** and **Walters, S. M.** Notes on the flora of Cambridgeshire (V. C. 29). Proc. Bot. Soc. Brit. Isles **1:** 471–481. 1955. (Changes in vegetation; list of noteworthy species.)—Also **Perring, F. H.,** and **Sell, P. D.** Further notes on the flora of Cambridgeshire. l.c. **3:** 165–171. 1959. (List of noteworthy species.)

Local

Blomefield (formerly **Jenlyns**), **Leonard.** A naturalist's calendar kept at Swaffham Bulbeck, Cambridgeshire . . . Edited by Frances Darwin . . . xviii, 84 p. 19 cm. Cambridge, 1903. (Kew library.)

Mention of other works (mostly British) on same subject; list of dates (mean, earliest, and latest) for first flowering and fruiting of plants, observations of birds, etc., arranged by months. The so-called 2. edition (1922; in library of N. D. Simpson) is merely a re-issue with an unimportant change in the preface.

Evans, A. H. Full list of plants growing in the old fen land at Wicken, omitting obvious intruders. *In* Gardiner, J. S., and Tansley, A. G., *eds.* The natural history of Wicken Fen. p. 50–51. 21.5 cm. Cambridge, 1923.

Unannotated list of vascular plants; see also note on p. 626.

Rishbeth, J. The flora of Cambridge walls. Jour. Ecology **36:** 136–148. pl. 18–19. 1948.

List of 224 plants (186 vascular) found on walls (especially college walls) in Cambridge, with discussion of ecological factors, etc.; bibliography.

CHESHIRE

See also Lancashire (Green, The flora of the Liverpool district).
Vice-county 58.

General

Warren, J. B. L., *3d baron de Tabley.* The flora of Cheshire. Edited by Spencer Moore . . . cxiv, 399 p. port., map. 20.5 cm. London [etc.], 1899. (Gray Herbarium library.)

Biographical sketch of author, botanical districts, biographical list of local botanists, bibliography; annotated list of vascular plants and Characeae. —See also papers by various authors in Naturalist 1899: 353–356. 1899; **1904:** 22–23. 1904; **1926:** 107–112. 1926. Jour. Bot. **38:** 74–76. 1900; **44:** 426. 1906; **45:** 33–34, 163, 212. 1907; **48:** 152–156. 1910; **57:** 91–94, 129–130. 1919; **58:** 137–141. 1920. Lancash. and Chesh. Nat. **4:** 285–297. 1914. (Not seen.) North West. Nat. **17:** 112–113. 1942.—Also **Waterfall, Charles.** Plants from the County of Cheshire, 1910–14. (Not seen; cited from Rpt. Bot. Soc. and Exch. Club Brit. Isles **4:** 228. 1916.)

Local

Baillie, E. J. The city flora. Precursory catalogue of plants within the boundaries of the city of the county of Chester. Proc. Chester Soc. Nat. Sci. **2**: 67–78. 1878. (Reprinted, 12 p.)

Tabular list of vascular plants and Characeae, with localities.—See also **Anonymous.** List of plants, growing within a two mile radius of Chester, found by Miss Cummings, Miss Payne, and Miss A. Payne, and which are not included in the late Mr. E. J. Baillie's City flora, 1875–77. Ann. Rpt. and Proc. Chester Soc. Nat. Sci., Lit. and Art **29**: 31–34. 1900.

Britten, Harry, Newton, P., and **Wild, S. V.** The natural history of the Ballast Hole, Sinderland, Cheshire. North West. Nat. **16**: 40–56. 1 fig. (map), pl. 12. 1941.

Includes (p. 46–52) annotated list of vascular plants and a few fungi. The "Ballast Hole" is an old pit about half a mile long, excavated in 1872–73.

Hitchens, Ernest. The wild plants of the parish. *In* Cartlidge, John E. G. Newbold Astbury and its history . . . p. 191–208. 25.5 cm. Congleton, 1915. (Harvard College library.)

Nature calendar, etc.; unannotated list of vascular plants (461 species).

Robinson, J. F. The flora of Frodsham. Phytologist n.s., **3**: 193–196. 1859.

Mostly unannotated list of vascular plants from which the most common species are omitted.

Sainter, J. D. Botany. *In* his The jottings of some geological, archaeological, botanical, ornithological, and zoological rambles round Macclesfield. p. 83–106. 22.5 cm. Macclesfield, 1878. (Library of N. D. Simpson.)

Tabular list of vascular plants, with habitat, etc. Botany index, p. 149–153.

CORNWALL

See also Devonshire (Martin and Fraser).
Vice-county 1, West Cornwall (with Scilly); vice-county 2, East Cornwall.

General

Davey, F. H. Flora of Cornwall, being an account of the flowering plants and ferns found in the county of Cornwall including the Scilly Isles. lxxxviii, 570 p. 6 ports., map. 23 cm. Penryn, 1909.

Topography, climate, geology, history of botanical investigations, botanical districts, bibliography, phytogeography; annotated list of 1347 vascular plants and Characeae (953 native). Replaces (as to Cornwall) **Keys, I. W. N.** Flora of Devon and Cornwall. Ann. Rpt. and Trans. Plymouth Inst. and Devon and Cornwall Nat. Hist. Soc. **2**: 21–69, 107–181. 1866–67; **3**: 41–83, 99–144, 179–304. 1868–[71]. (Reprinted 330 p.) (Annotated list of vascular plants and Characeae.)—See also **Thurston, Edgar,** and **Vigurs, C. C.** A supplement to F. Hamilton Davey's Flora of Cornwall. Jour. Roy. Inst. Cornwall v. 21, [Suppl.] xx, 172 p. port. 1922. (Corrections and additions, biographical sketch of Davey, etc.; annotated list of additional species and localities.)—Also their Note on the Cornish flora. l.c. v. 21–22, passim. 1923–27.—Also **Thurston, E.** Notes on the Cornish flora (title varies slightly). l.c. v. 22–24, passim. 1928–36.—Also **Rilstone, Francis.** Cornish Rubi. l.c. **22**: 364–365. 1928.—Also his A key to the species of Rubi or brambles recorded from Cornwall. l.c. **23**: 462–476. 1934. (All volumes (21–26) examined in Harvard College library.)—Also his Notes on F. H. Davey's Flora of Cornwall. l.c. **26**: 134–136. 1938. (Notes on several critical genera.)

Local

Allen, C. B. Local botany—being the result of a student's rambles in Penzance and neighbourhood. Ann. Rpt. Roy. Cornwall Polytech. Soc. **40**: 123–146. 1873.

Description of several localities; unannotated list of angiosperms and Equisetaceae, with localities; list of official and officinal medicinal plants included, with uses.

Bastian, H. C. The flora of Falmouth and surrounding parishes. Ann. Rpt. Roy. Cornwall Polytech. Soc. **24:** 83–112. 1856 [1857].

Annotated list of vascular plants.—See also **Bullmore, Ernest.** Flora of Falmouth. Addenda to "The flora of Falmouth . . ." Ann. Rpt. Roy. Cornwall Polytech. Soc. **50:** 261–267. 1882 [1883]. (List of additional species and localities.)—Also **Eager, A. R.** Supplement to Bastian's Falmouth flora. l.c. **56:** 94–96. 1888 [1889]. (Similar list.)—Also Thurston, below.

Couch, Jonathan. Botany. *In* his The history of Polperro, a fishing town of the south coast of Cornwall . . . p. 186–193. 23 cm. Truro, London, 1871.

Briefly annotated list of vascular plants, with localities for some species. Replaces **Couch, T. Q.** The botany of Polperro and its neighbourhood. Ann. Rpt. Roy. Cornwall Polytech. Soc. **16:** 11–21. 1848 [1849]; Additions to the botany . . . l.c. **17:** 29–30. 1849 [1850]. (Annotated list of vascular plants, on the Linnaean system.)

Flower, N. J., Milward, E. M., and **West, R. G.** Flora. *In* The Canterbury Schools, King's School and St. Edmund's. Report on natural history. p. 16–35. 22 cm. [Canterbury, 1945?] (Kew library.)

List of vascular plants and Characeae of the St. Austell district, with localities and habitats.

Montgomery, J. B. List of phaenogamous plants and ferns of western Penwith. Trans. Nat. Hist. and Antiq. Soc. Penzance **2:** 215–222. 1854. (Harvard College library.)

Unannotated list of vascular plants, with localities for some of the rarer species; lists of naturalized species, erroneously recorded or extinct species, and rarer species of neighboring areas.—See also Ralfs, below.

[**Ralfs, John,** and **Penneck, Henry.**] Sketch of the botany of West Penwith. *In* Courtney, John S. A guide to Penzance and its neighbourhood including the Isles of Scilly . . . Appendix, p. 1–23. 20 cm. Penzance, 1845. (Library of N. D. Simpson.)

Description of the flora, p. 1–14, followed by list of additional vascular plants, mosses, hepatics, lichens, and algae, with localities; also a calendar of the flowering of a select number of plants.—See also Montgomery, above.

Stephens, R. D. S. A list of plants found in the parishes of S. Minver, Cornwall, and Bradford Abbas, Dorset. Proc. Dorset Nat. Hist. and Antiq. Field Club **21:** 125–136. 1900.

Note on geology, etc., by J. C. Mansell-Pleydell; list of vascular plants (598 species), divided according to occurrence in one or both parishes.

Thurston, Edgar. The alien and British plants of Par and Charlestown Harbours, Falmouth Docks, and Eastern Green, Penzance. Jour. Roy. Inst. Cornwall **23:** 137–205. 1929. (Harvard College library.)

Sources of alien plants, etc.; annotated list of angiosperms, mostly adventive; table showing local distribution.

Townsend, Frederick. Contributions to a flora of the Scilly Isles. Jour. Bot. **2:** 102–120. 1864.

Topography, general features of flora, soils, etc.; annotated list of vascular plants (348 species) observed by the author in 1862.—See also **Lawson, M. A.** Additions to the flora of the Scilly Islands. Jour. Bot. **8:** 357–358. 1870. —Also **Somerville, Alexander.** Additional records for the Scilly Isles. l.c. **31:** 118–120. 1893.—Also **Marquand, E. D.** Further records for the Scilly Isles. l.c. **31:** 265–267. 1893.—Also **White, J. W.** Plants of Scilly. l.c. **52:** 19. 1914.—Also **Lousley, J. E.** Notes on the flora of the Isles of Scilly. I–II. l.c. **77:** 195–203. 1939; **78:** 153–160. 1940.—Also **Raven, J. E.** Notes on the flora of the Scilly Islands and the Lizard Head. Watsonia **1:** 356–368. 1950.

CUMBERLAND

See also Durham (Upper Teesdale Naturalists' Field Club, The flora of Teesdale). Vice-county 70.

General

Harrison, J. W. H., and **Temperley, G. W.** The flora of the three northern counties. *In* Headlam, Sir Cuthbert, *ed.* The three northern counties of England . . . p. 207–279. 25 cm. Gateshead, 1939.

Topography, general features of flora, etc., of Northumberland, Durham, and Cumberland; annotated list of known vascular plants, with habitat, indication of counties in which each species occurs, and localities in the case of the rarer species; unannotated list of casual and adventive plants.

Hodgson, William. Botany. *In* The Victoria history of the county of Cumberland. v. 1, p. 73–97. map. 31.5 cm. London, 1901.
Botanical explorations, climate, statistics, discussion of botanical districts with account of geology, etc., and lists of rarer species; list of rarer mosses (by C. H. Binstead).

——— Flora of Cumberland, containing a full list of the flowering plants and ferns to be found in the county, according to the latest and most reliable authorities . . . xxxvi, 398 p. 2 maps. 22.5 cm. Carlisle, 1898 (cover date, 1899). (Arnold Arboretum library.)
Soils (by J. G. Goodchild), biographical notices of deceased botanists, plant zones; annotated list of vascular plants and Characeae (1196 species, of which 946 are native).—See also **Bennett, Arthur.** Cumberland plants. Jour. Bot. **37**: 225. 1899.

Local

Baker, J. G. A flora of the English Lake District. viii, 262 p. 22.5 cm. London, 1885. (Gray Herbarium library.)
Phytogeography, botanical history with included bibliography; annotated list of nearly 900 species of vascular plants. Refers to a region about 35 miles long and wide in Cumberland, Westmorland, and northern Lancashire, bounded on the north and east by Allonby, Wigton, Penrith, and Tebay.—See also **Bennett, A. W.** Additional localities for Lake-land plants. Jour. Bot. **23**: 330–331. 1885.—Also **Sidgwick, N. V.** New localities for plants of the Lake District. Naturalist **1894**: 49–51. 1894.—Also **Ley, Augustin,** and **Linton, W. R.** Some plants of the English Lake District. Jour. Bot. **44**: 171–173. 1906.—Also their Correction. l.c. **44**: 216. 1906.—Also **Rogers, W. M.** Plants of the English Lake District. l.c. **45**: 8–12. 1907.—Also **Petty, S. L.** The flora . . . Revision and additions by Mary Thompson [and others] . . . *In* Collingwood, William T., Drury, Agnes, and Hayes, A. G. The Lake counties. new ed. p. 216–227. 24.5 cm. London and New York, 1932. (1st ed. 1902.) (List of scarcer vascular plants, with localities.)

——— On the botany of the Cumberland part of the Pennine Range. Naturalist **1888**: 33–34. 1888.
Topography, geology, agriculture; annotated running list of vascular plants collected in 1887, from headquarters in Alston; altitudinal limits of various species, etc.

Hodgson, William. The botany of the Solway shore. Trans. Cumberl. and Westmorl. Assoc. Advanc. Lit. and Sci. **11**: 114–126. 1886; **14**: 49–72. 1889.
Annotated running account of vascular plants, in systematic order, with mention of localities.

——— Notes on the flora of the Ullswater district. Trans. Cumberl. Assoc. Adv. Lit. and Sci. **7**: 127–154. 1882.
Running account of noteworthy species of each family. Relates to the entire parishes of Patterdale, Martindale, and Watermillock, and part of Barton, Matterdale, and Dacre. Issued as separate (seen in Kew library) with addition of pages 155–173 (not the pages of same number in vol. 7 of the Transactions) containing a list of vascular plants and Characeae (nearly 520 species), with habitats or localities. (In Cumberland and Westmorland.)

Murray, James. Cumbrian plant names. North West. Nat. **12**: 179–182. 1937.
Alphabetical list of vernacular names, with botanical equivalents.

Robson, Joseph. A catalogue of the wild plants of Gosforth (Cumberland) and adjacent district, in 1853. Phytologist **5**: 1–9. 1856.
List of vascular plants and Characeae, with localities, the commonest species omitted.

DERBYSHIRE

Vice-county 57.

General

Linton, W. R. Flora of Derbyshire: flowering plants, higher cryptogams, mosses and hepatics, Characeae. vii, 455 p. pl., 2 maps. 23 cm. London, 1903.

Climate, soils, bibliography; annotated list of vascular plants, bryophytes, and Characeae (1034 species of vascular plants), with localities. Replaces **Painter, W. H.** A contribution to the flora of Derbyshire, being an account of the flowering plants, ferns, and Characeae found in the county. 3 p. l., 156 p. map. 23 cm. London and Derby, 1889. (Arnold Arboretum library.) Notes supplementary to the Flora . . . including the mosses. Naturalist **1899**: 177–208, 241–272. 1899. Additional notes. l.c. **1902**: 5–12. 1902. (The supplements reissued (111 p.) 1902, with addition of index (iv p.) and title A supplement to A contribution to the flora of Derbyshire including a list of mosses found in the county. [In British Museum (Natural History) library.] (Topography, plant zones, bibliography; annotated list of vascular plants (889 species in original) and Characeae; bryophytes added in the supplements.)—See also **Drabble, Eric,** and **Drabble, Hilda.** Notes on the flora of Derbyshire. Jour. Bot. **47**: 199–207. 1909; **49**: 313–317. 1911; **51**: 5–11. 1913; **54**: 133–139. 1916.—Also **Drabble, Eric.** Additions to the flora . . . Jour. Derbysh. Archaeol. and Nat. Hist. Soc. 39: 225–235. 1917.—Also his Notes on the flora . . . l.c. 49 (n.s., 2): 210–219. 1927.—Also **Hollick, K. M.** Botanical records for 1941 [–49]. l.c. 62 (n.s. 15)—69 (n.s., 22), passim. [1942?–50?] (Title varies; first installment covers 1932–41.)

Local

Baker, J. G., and Newbould, W. W. Notes on the flora of Matlock. Jour. Bot. **22**: 334–344. 1884.

Topography, etc.; list of vascular plants collected by authors in August and September.—See also **Hind, W. M.** The flora of Matlock. Phytologist n.s., **2**: 609–612. 1858. (Includes some additional species.)—Also **Purchas, W. H.** Some more notes on Dovedale plants. Jour. Bot. **23**: 181–184, 198–203. 1885.

Garneys, William, and Hagger, John. List of the plants found at and near Repton. *In* Garneys, William, *ed.* Contributions to the flora and fauna of Repton and neighbourhood. (Cover title, Wild flowers of Repton.) p. 1–75. 19 cm. London and Derby, 1881.

List of vascular plants (p. 1–65) and mosses (p. 66–75), with localities. The 1881 work was published as ed. 2 of **Wyatt, William,** and others. Flora repandunensis. The wild flowers of Repton, with a list of I. The birds. II. The moths and butterflies; found in Repton and the neighbourhood. vi, 98 p. front. 19 cm. London, 1866. (List of vascular plants, by W. Wyatt and C. G. Thornton, p. 1–62. In British Museum (Natural History) library.)

Hawkins, Ellen. Catalogue of plants which grow in the neighbourhood of Buxton, with a botanical commentary. *In* Robinson, William H. A handbook to the Peak of Derbyshire, and to the use of the Buxton mineral waters; or, Buxton in 1854 . . . with . . . a botanical appendix by Miss Hawkins . . . p. 205–222. 19 cm. London, 1854. (British Museum (Natural History) library.)

Running account of flora; unannotated list of vascular plants.—See also **Baker, J. G.** On the upland botany of Derbyshire. Jour. Bot. **22**: 6–15. 1884. (List of altitudes, etc.; annotated list of vascular plants collected by author in vicinity of Buxton in August, with altitudes reached.)—See also **West, William.** On the upland botany of Derbyshire. l.c. **22**: 71–74. 1884. (Annotated list)—Also **Rogers, W. M.** Notes on the flora of Buxton. l.c. **23**: 76–80. 1885. (List with localities.)—Also **Hall, F. T., and Hall, R. H.** Notes on the flora of Buxton and district. Rpt. Bot. Soc. and Exch. Club Brit. Isles **12** (1939–40): 338–355. 1942. (Extensive list, with localities.)

Devonshire

See also Great Britain, General (Wallace, Additions and corrections . . . for Devon . . . under Druce, The comital flora of the British Isles); Somerset (Macmillan, Popular names of flowers).

Vice-county 3, South Devon; vice-county 4, North Devon.

General

Friend, Hilderic. A glossary of Devonshire plant names. Rpt. and Trans. Devonsh. Assoc. Adv. Sci., Lit. and Art **14**: 529–591. 1882. (Reprinted, vi, 63 p. 1882, by English Dialect Society. Series C. Original glossaries. Reprint in Kew library.)

Bibliography; annotated list of vernacular names, alphabetically arranged, with scientific equivalents; list of scientific names with vernacular equivalents.

Martin, W. K., and **Fraser, G. T.,** *eds.* Flora of Devon: phanerogams, vascular cryptogams, Charophyta. Promoted by the Devonshire Association. xv, 787 p. 8 pl., 2 maps (1 in pocket). 22.5 cm. Arbroath, 1939.

Botanical divisions, climate, geology, topography, ecology, special floral areas, herbaria, bibliography; annotated list of 1622 species and subspecies (1116 native) and 923 varieties, forms, and hybrids of vascular plants and Characeae, with detailed local distribution and statement of occurrence in Somerset, Dorset, and Cornwall; biographical sketches of local botanists. Replaces (as to Devon) **Keys, I. W. N.** Flora of Devon and Cornwall. Ann. Rpt. and Trans. Plymouth Inst. and Devon and Cornwall Nat. Hist. Soc. **2**: 21–69, 107–181. 1866–67; **3**: 41–83, 99–144, 179–304. 1868–[71]. (Reprinted 330 p., in Arnold Arboretum library.) (Annotated list of vascular plants and Characeae.)—See also **Fraser, G. T.** Some corrections and additions to Flora of Devon, 1939. Rpt. and Trans. Devonsh. Assoc. Adv. Sci., Lit., and Art **73**: 233–274. 1941.—Also 34. [–47.] report on the botany of Devon, by various authors. l.c. v. 74–87, passim. 1942–55.—Also **Patterson, S. C.** Notes on four years' botanical work in S. Devon. Trans. and Proc. Torquay Nat. Hist. Soc. **9**: 119–124. 1946.—Also **Pugsley, H. W.** The flora of Combe Martin. Rpt. and Trans. Devonsh. Assoc. Adv. Sci., Lit. and Art **78**: 193–206. 1946. (Extensive list of vascular plants not recorded from this region. In Harvard College library.)

Local

Briggs, T. R. A. Flora of Plymouth: an account of the flowering plants and ferns found within twelve miles of the town, with brief sketches of the topography, geology, and climate of the area, and history of local botanical investigation . . . xxxv, 432 p. map. 20.5 cm. London, 1880. (Gray Herbarium library.)

Topography, climate, botanical districts, botanical investigations, bibliography; annotated list of vascular plants and Characeae (876 species, of which 874 are vascular); lists of species showing distribution by districts, statistics.—See also **S., H. W.** Records—botany. Trans. Plymouth and Dist. Field Club **1**: 20–21. 1913. (List of newly recorded species and new localities. Not seen; title and annotation supplied by G. Taylor.)—Also **Phillips, E. M.** Flowers. *In* The natural history of the Plymouth rubble heaps. Ann. Rpt. and Trans. Plymouth Inst. and Devon and Cornwall Nat. Hist. Soc. **19**: 15–23. 1947. (Running account by families of plants occurring on bombed areas in Plymouth, followed by tabular list with localities.)

C[aspary], R. The flowering plants and ferns of Sidmouth and its vicinity. 46 p. 10 x 18 cm. London, 1846. (Library of N. D. Simpson.)

List of vascular plants, with localities.—See also Cullen, below.

Chanter, J. R. Botany. Catalogue of plants, etc., growing on Lundy. *In* his Lundy Island . . . p. 163–171. 18 cm. London [1877].

Unannotated list of vascular plants. The list is repeated "with a few additions" in Loyd, L. R. W. Lundy: its history and natural history. p. 224–234. 22.5 cm. London, 1925.—See also **Wright, F. R. E.** Contribution to the flora of Lundy Island. Jour. Bot. v. 71, suppl. [2]. 11 p. 6 pl. 1933. (Geology, general features of flora; annotated list of vascular plants observed by author in July.)—Also his On the origin of the Lundy flora, with some additions. l.c. **73**: 90–95. 1935.—Also his The Lundy Brassica, with some additions. l.c. 74, suppl. I. 8 p. 2 fig., 17 pl. 1936.—Also **Pugsley, H. W.** Flora of Lundy Island. l.c. **72**: 58–59. 1934.—Also **Bowen, H. J. M.** Additions to the flora of Lundy. Watsonia **2**: 303–307. 1953. (General features of flora, etc.; list of noteworthy vascular plants; bibliography.)—Also Green, below.

Cullen, W. H. Flora sidostiensis; or a catalogue of the plants indigenous to the vicinity of Sidmouth. 1 p. l., xix, 57 p. 18 cm. Sidmouth and London, 1849. (Kew library.)

Topography, etc.; briefly annotated list of vascular plants, on the Linnaean system. Replaces **Cresswell, Richard.** The flowering plants and ferns of Sidmouth and its vicinity. 46 p. 10.5 x 18.5 cm. London, Sidmouth, 1846. (Tabular list of 572 vascular plants, with localities for the rarer species. In British Museum (Natural History) library.)—See also Caspary, above.

D'Urban, W. S. M. Botany. *In* his A sketch of the natural history of the neighbourhood of Exeter. [Extracted from Besley's Hand book of Exeter.] p. 105–111. 15 cm. Exeter, [1876]. (Kew library.)

List of the "rarer and more remarkable" vascular plants, with localities.—See also Hiern, under Larter, below.

Evans, H. A. A hand-list of the plants occurring within seven miles of the United Services College, Westward Ho. 1 p.l., 47 p. 22 cm. Bideford, 1881. (Kew library.)

Unannotated list of vascular plants. Printed only on the even-numbered pages.—See also **Saunders, Helen.** Botanical notes. no. III. Rpt. and Trans. Devonsh. Assoc. Adv. Sci., Lit. and Art 38 (II, 8): 491–496. 1906. (Includes (p. 491–492) additions to Evans' Hand-list, and mentions supplements published in 1881, 1883, 1886, and 1894 [not seen].)

Freeman, E. V. List of British plants observed at West Anstey. *In* his Notes on the history of the parish of West Anstey, North Devon, with a list of birds and flora. p. 45–52. 22 cm. n.p., [1920]. (Library of N. D. Simpson.)

Unannotated list of vascular plants, including garden plants.

Green, E. C. Flora. *In* Palmer, Mervyn G., *ed.* The fauna and flora of the Ilfracombe district of North Devon. (Promoted by the Ilfracombe Field Club.) (Cover title, Ilfracombe fauna and flora.) p. 177–262. pl. 7. 22 cm. Exeter, 1946.

Includes briefly annotated list of vascular plants within a radius of 10 miles around Ilfracombe (p. 185–228; 1107 species), also cellular plants (p. 228–262). The species of vascular plants known from Lundy Island are so indicated.

Hannaford, Samuel. Flora tottoniensis; a catalogue of the flowering plants and ferns, growing wild in the vicinity of Totnes. 1 p. l., 38 p. 17.5 cm. London, Totnes, [1851]. (**Arnold Arboretum library.)—Supplement.** (? p.) London, 1852. [Not seen.]

Annotated list of vascular plants growing within 6 miles of Totnes, arranged by the Linnaean system.

Harvey, L. A. List of some plants and animals of Dartmoor. (A.) Vascular plants. *In* Harvey, L. A., and St. Leger-Gordon, D. F. Dartmoor. p. 207–227. 22 cm. London, 1953. (The New naturalist series.)

Briefly annotated list of vascular plants and Charophyta. A tabular list of bryophytes and lichens is given on p. 74–76. Replaces **Brent, Francis.** The botany of Dartmoor and its borders. *In* Rowe, Samuel. A perambulation of the antient [sic] and royal forest of Dartmoor . . . 3d ed. rev. and corr. by J. B. Rowe. p. 350–398. 1 fig. 23 cm. Exeter and London, 1896. (Topography, etc.; list of vascular and cellular plants (about 546 phanerogams), with localities. The 1st ed., 1848, contains Botany of Dartmoor, by **Edward Moore,** p. 217–226. In Kew library.)

Hiern, W. P. Botany of Lynton and neighbourhood. *In* Chanter, John F. A history of the parishes of Lynton and Countisbury . . . p. 134–156. 23 cm. Exeter, 1907.

Topography, bibliography, etc.; unannotated list of vascular plants (558 species; p. 137–149); excluded species, statistics, annotated list of rarer species.

Larter, C. E. Manual of the flora of Torquay. 2 p. l., 83 p. 18.5 cm. Torquay, 1900. (British Museum (Natural History) library.)

Annotated running list of vascular plants. Replaces **Stewart, Robert.** Handbook of the Torquay flora; comprising the flowering plants and ferns growing in and around Torquay, with their respective habitats. iv, 187 p. 17.5 cm. Torquay, London, 1860. (Annotated, briefly descriptive flora of vascular plants. In Gray Herbarium library.)—See also **Hiern, W. P.,** and others, *eds.* 1.—50. report of the botany committee. *In* Rpt. and Trans. Devonsh. Assoc. Adv. Sci., Lit. and Art. v. 41 (III, 1)—90. 1909–1958. (Annual lists of new records and new localities for 6 districts of Devon, namely Torquay, Barnstaple, Torrington, South Molton, Exeter, and Honiton. The series continues.)

Pugsley, H. W. Salcombe plants. Jour. Bot. **73**: 162–166. 1935.

List of vascular plants collected by author in September 1934.

Rogers, W. M. A contribution towards a flora of the Teign Basin, S. Devon. Jour. Bot. **20**: 70–76, 121–124, 133–136, 177–183, 206–209, 239–243, 262–266. 1882.

Annotated list of the rarer vascular plants and Characeae, with localities, terminating with notes on Rosa and a list of the common species not given in the first part of the list.

Saunders, Helen. A list of plants growing wild in the parish of South Molton, and in the neighbouring parishes of North Molton, Filleigh, Chittlehampton, George Nympton, Satterleigh, King's Nympton, and Bishop's Nympton. Rpt. and Trans. Devonsh. Assoc. Adv. Sci., Lit., and Art **26**: 451–466. 1894.

Annotated list of vascular plants and Characeae (535 species), with localities for the scarcer species.—See also her Botanical notes. l.c. **30**: 198–202. 1898. (Additions bringing total to 563 species. In New York Public Library.)—Also her Botanical notes. no. II. l.c. **33** (II, 3): 468–474. 1901. (Includes (p. 474) corrections of errata in the earlier pages.)—Also her Botanical notes. no. III. l.c. **38** (II, 8): 491–496. 1906. (Includes (p. 494–496) additions bringing total to 610 species.)—Also annual lists by Hiern and others cited under Larter, above.

Wright, F. R. E. Braunton. A few nature notes with lists of flora, macro lepidoptera, and birds known to occur in the district. rev. ed. 157 p. illus., pls., map. 19 cm. Barnstaple, 1932. (Library of N. D. Simpson.) (1st ed. 1926.)

Includes 3 chapters of nature notes on local plants, and (p. 111–130) unannotated list of vascular plants and charophytes.

Dorsetshire

See also Cornwall (Stephens, A list of plants found in the parishes of S. Minver, Cornwall, and Bradford Abbas, Dorset); Devonshire (Martin and Fraser, Flora of Devon); Hampshire (Linton, Flora of Bournemouth including the Isle of Purbeck); Somersetshire (Macmillan, Popular names of flowers).

Vice-county 9.

General

Carter, P. W. The botanical exploration of Dorset. Proc. Dorset Nat. Hist. and Archaeol. Soc. 1957; **79**: 73–98. 1958.

Account of botanists, chronologically arranged with references to their writings and discoveries; alphabetical list of botanists with incomplete references to their publications.

Good, R. D'O. A geographical handbook of the Dorset flora . . . including a chapter on the soils of Dorset by K. L. Robinson . . . 39 fig. (maps) 3 maps, front. xi, 225 p. 25.5 cm. Dorchester [pref. 1948].

Brief sketch of botanical explorations, geology, topography, climate, soils, botanical regions, history and constitution of flora, distribution according to climate and soils, comparative abundance of species, brief bibliography; annotated list of vascular plants (1334 species, of which 863 are native), with localities for rarer species. Replaces **Mansel-Pleydell, J. C.** The flora of Dorsetshire, with a sketch of the topography, river system, and geology of the county. 2. ed. xxxviii, 345, xxv p. 2 maps. 23 cm. Dorchester, 1895. (1st ed. 1874.) (Topography, tabular comparison of flora, list of

rare species, drainage, geology; annotated list of vascular plants and Characeae (1173 species), with localities. For principal records of additional species and localities, by various authors, see Jour. Bot. **33:** 121–123, 154–155. 1895; **34:** 432–433. 1896; **42:** 236–239. 1904; **46:** 384–390. 1908; **47:** 73, 378–384, 417–422. 1909; **49:** 363–365. 1911; **50:** 27. 1912; **52:** 40–41. 1914. Proc. Dorset Nat. Hist. and Antiq. Field Club **26:** 75–87. 1905; **29:** 14–29. 1908; **30:** 116–132. 1909.)—See also **Good, R. D'O.** First addendum to the hand-list of the Dorset flora. l.c. **75:** 157–163. 1955. (Additional species and localities. In Harvard College library.)

Local

Barrett, W. B. Notes on the flora of the Chesil Bank and the Fleet. Proc. Dorset Nat. Hist. and Antiq. Field Club **26:** 251–265. 1905.

General features of flora; list of more interesting vascular plants and Characeae, with localities. Covers region from Abbotsbury to the mouth of the Fleet called Smallmouth.—See also his Contributions to a flora of Portland, with special reference to *Limonium recurvum*, **C. E. Salmon. l.c. 33:** 96–143. 1912. (Changes in flora, etc.; annotated list of vascular plants and Characeae, excluding the commonest species, found along coast from Smallmouth south to the Bill of Portland.)

Dale, C. W. Botany of Glanville's Wootton. *In* his The history of Glanville's Wootton, in the county of Dorset, including its zoology and botany. p. 337–387. 19.5 cm. London, 1878. (Arnold Arboretum library.)

Briefly annotated list of 676 plants (431 vascular).

Mansel-Pleydell, J. C., and others. Botany. ***In*** **Braye, John, *ed.* Swanage (Isle of Purbeck). Its history, resources as an invigorating health resort,** botany, and geology. p. 83–101. 22 cm. London, 1890. (Library of N. D. Simpson.)

Page 83–87, list of trees and shrubs and note on December flowers, by Braye; p. 87–101, unannotated list of vascular plants and Characeae found within a radius of 6 miles around Swanage, by Mansel-Pleydell.—See also Linton, Flora of Bournemouth, including the Isle of Purbeck, under Hampshire.

Salter, T. B. Appendix, illustrative of the botany of Poole, and its neighbourhood. *In* Sydenham, John. The history of the town and county of Poole . . . p. 465–491. 22.5 cm. Poole, London, 1839. (Reprinted with title, A short account of the botany of Poole, and its neighbourhood; (being an appendix to Sydenham's History of Poole). [27] p. 1839.)

Geology, topography, general features of flora; alphabetical list of 989 species of vascular plants found within 30 miles of Poole (776 species within 8 miles), annotated as to abundance, with localities for the rarer species.

Wood, H. H. The botany of a Dorset parish. Proc. Dorset Nat. Hist. and Antiq. Field Club **2:** 32–52. 1878.

Notes on introduced plants, etc.; unannotated list of vascular plants (414 species) of the parish of Holwell.

DURHAM

See also Cumberland (Harrison and Temperley, The flora of the three northern counties); also Northumberland (Baker and Tate, A new flora of Northumberland and Durham); also Scotland, Berwickshire (Johnston, The botany of the eastern borders).

Vice-county 66.

General

Potter, M. C. Botany. ***In*** **The Victoria history of the county of Durham.** v. 1, p. 35–81. map. 31.5 cm. London, 1905.

Topography, general features of flora, botanical districts; list of known vascular plants, with indication of range by altitude; also cellular plants. Replaces **Calvert, Robert.** Botany. Flowering plants. *In* his Notes on the geology and natural history of the county of Durham. p. 153–179. 22 cm. Bishop Auckland, 1884. (Briefly annotated list of vascular plants, with localities for scarcer species; the grasses, sedges, and trees, as well as some other groups, omitted. In British Museum (Natural History) library.)—See also **Harrison, J. W. H.** Flora. *In* Scientific survey of north-eastern

England. Prepared for the [British Association] meeting held in Newcastle upon Tyne 31st August to 7th September 1949. p. 49–57. 25 cm. n.p., 1949. (General features of flora of Northumberland and Durham, with mention of numerous species, and notice of changes in flora; short bibliography. In British Museum (Natural History) library.)

Local

Hogg, John. A catalogue of the more rare plants found in the neighbourhood of Stockton. *In* his On the natural history of the vicinity of Stockton-on-Tees. From the History of Stockton by the Rev. John Brewster. p. 39–69. 26.5 cm. Stockton, 1827. (Kew library.)

Annotated list of vascular and cellular plants (341 species, of which 253 are vascular).

——— On the ballast-flora of the coasts of Durham and Northumberland. Annals and Mag. Nat. Hist. III, **19**: 38–43. 1867.

Includes extensive list of ballast plants collected by author or recorded in earlier publications.

Maister, R. H. Weardale flora. Plants found in Weardale 1897–8, by the Rev. R. H. Maister. Trans. Weardale Nat. Field Club **1**: 203–213. 1904. (British Museum (Natural History) library.)

Unannotated list of vascular plants.

Upper Teesdale Naturalists' Field Club. The flora of Teesdale, from Crossfell downwards to the Greta . . . 1 p. l., 26 p. 18.5 cm. Barnard Castle, 1883. (British Museum (Natural History) library.)

Unannotated list of vascular plants. (Along boundary of York, Durham, Westmorland, and Cumberland.)—See also **Salmon, C. E.** Notes upon Teesdale plants. Jour. Bot. **52**: 137–141. 1914.—Also **Marshall, E. S.** Upper Teesdale plants. l.c. **56**: 359–360. 1918.

Essex

See also Hertford (Bishops Stortford and District Natural History Society, List of vascular plants of Bishops Stortford and district); London.

Vice-county 18, South Essex; vice-county 19, North Essex.

General

Boulger, G. S. History of Essex botany. Part I. The botanists of the sixteenth and seventeenth centuries. Essex Nat. **11**: 57–68, 169–184, 229–236. 1899–1900; **13**: 218–229. 1904.

Detailed account of local work of early botanists with quotations from their publications. Unfinished; reaches John Ray.

Brown, G. C. The alien plants of Essex. Essex Nat. **22**: 31–47. 1927.

Extensive list of alien phanerogams collected by author and others during the last 12 years, with localities.—See also his Additional records of Essex plants. l.c. **26**: 184–193, 248–255. 1940.

Gibson, G. S. The flora of Essex, or a list of the flowering plants and ferns found in the county of Essex . . . l, 469 p. 4 col. pl., map. 19.5 cm. London, 1862. (Gray Herbarium library.)

Botanical explorations, climate, topography, general features of flora, list of local botanists, bibliography; annotated list of about 1070 vascular plants and Characeae, with localities; tables showing earliest and latest years of observation of rarer species, comparative abundance, comparison with other counties, etc.—See also **Anonymous.** Flora of Essex. Additional localities for Essex plants, and some observations thereon. Phytologist **6**: 304–308. 1862.—Also **Anonymous.** Extinct species. Remarks on the Flora of Essex. l.c. **6**: 335–342. 1862.—Also **Britten, James.** Some additions to Gibson's "Flora of Essex." Naturalist **1**: 314–315. 1865.—Also **Shenstone, J. C.** A report on the flowering plants growing in the neighborhood of Colchester. Essex Nat. **1**: 22–35. 1887. (List of species not recorded from Colchester region by Gibson, with localities; notes on some of the rarer species recorded by Gibson, with especial notice of Eryngo root.)—Also his Additional notes. l.c. **2**: 34. 1888.—Also his Additions, 1889. l.c. **3**: 222–223. 1889.—Also **Whellan, J. A.** Some Essex plants observed in 1944. l.c. **27**: 284–288. 1945. (In Harvard College library.)

Paulson, Robert. The botanical investigation of Essex. *In* Hutchings, G. E., *ed.* Essex: an outline scientific survey . . . Prepared on the occasion of the Congress of the Southeastern union of scientific societies at Colchester. p. 34–55. 21.5 cm. Colchester, 1926. (Kew library.)
Chronological account of botanists and botanical explorations and publications, including list of noteworthy species recorded during 1903–1925 (p. 48–49), notes on alien plants, principal herbaria, partial bibliography.

Shenstone, J. C. Botany. *In* The Victoria history of the county of Essex. v. 1, p. 31–67. map. 31.5 cm. London, 1903.
Botanical explorations, geology and soils, etc.; list of vascular plants and Characeae, with indication of range by districts; also cellular plants.

Local

Buxton, E. N. Epping Forest, its flowering plants. *In* his Epping Forest. 9. ed., rev. p. 118–129. 3 fig. London, 1923. (1st ed. 1884.)
Annotated list of spermatophytes, prepared by G. S. Boulger.—See also **Powell, J. T.** The existing flowering plants of Epping Forest. Essex Nat. **6:** 1–10. map. 1892. (Boundaries, etc.; annotated list of less common phanerogams.)—Also **Elliott, F. W.** The existing trees and shrubs of Epping Forest. l.c. **10:** 377–387. 1899. (Annotated list.)—Also **Robbins, R. W.** The flora of Epping Forest. Trans. London Nat. Hist. Soc. **1915:** 44–48. 1916. (Habitats, change in flora, notes on rarer species, etc.).—Also **Ross, Joseph.** Some interesting plants of Epping Forest. Essex Nat. **27:** 164–166. 1942. (Mention of various vascular plants.)—Also **Bayne, C. S.** A historical sketch of Epping Forest. *In* The Epping Forest Survey, 2d year. London Nat. **1943:** 32–43. 1944. (Historical account and bibliography; not botanical.)

Gepp, E., and **Manley, F. H.** Botanical list for 1883—Flora felstediensis. Ann. Rpt. Felsted School Nat. Sci. Soc. **2:** 32–43. 1884. (British Museum (Natural History) library.)
List of vascular plants (about 520 species), all found within four miles of Felsted School, except those for which a specific locality is given.—See also **French, John.** On plant distribution in the neighbourhood of Felstead. Essex Nat. **15:** 152–163. 1908. (Sources of flora; unannotated list of flowering plants, grouped according to their supposed invasive character.)

Gibson, G. S. A flora of the neighbourhood of Saffron Walden, Essex. Phytologist **1:** 408–415. 1842.—Supplement . . . l.c. **1:** 838–839. 1844.
Briefly annotated list of vascular plants.

Hope, T. M. The flowers of Hatfield Peverel. *In* his The township of Hatfield Peverel: its history, natural history and inhabitants. p. 238–251. 22.5 cm. Chelmsford, 1930. (Library of N. D. Simpson.)
Unannotated list of angiosperms (378 species).

Lister, Gulielma. List of flowering plants and conifers in Wanstead Park district. Essex Nat. **27:** 128–138. 1942. (Harvard College library.)
List of about 300 spermatophytes, with localities; includes cultivated species. Preceded (p. 121–127. pl.) by her The flora of Wanstead Park district. (Description of localities on the estate.)—See also her Additions to the list of flowering plants found in the Bushwood area of the Wanstead Park district since 1941. Essex Nat. **27:** 293–296. 1946. (Harvard College library.)

——— The present flora of West Ham wastes. Essex Nat. **25:** 171–182. pl. 12. 1937.
History of the area; annotated list of 104 weedy angiosperms.

Warner, Richard. Plantae woodfordienses. A catalogue of the more perfect plants growing spontaneously about Woodford in the county of Essex. vii, 222 p. 18.5 cm. London, 1771.
Annotated list of vascular plants, alphabetically arranged by Ray's Latin names, the Linnaean names also given but not indexed in the original issue. Some copies (e.g., that in Arnold Arboretum library) have bound in p. 223–238, "Index to the Latin names, as given by Linnaeus," issued subsequently, as well as 1 leaf [2 p.] of Errata.—See also [**Forster, T. F.**]. Additions to Warner's Plantae woodfordienses. 1 p.l., 243–255 p. 17.5 cm. n.p., 1784.

(Not seen; title supplied by N. D. Simpson.)—Also **Thompson, Percy.** On an annotated copy of Richard Warner's "Plantae woodfordienses." Essex Nat. **19**: 72–88. pl. 4–7. 1920. (Notes on records and comments in a copy annotated by B. M. Forster.)—Also his On another annotated copy of Warner's "Plantae woodfordienses." l.c. **19**: 221–237. pl. 19–20. 1921. (Notes on a copy annotated by Edward Forster, jr.)—Also his On a third annotated copy of Richard Warner's Plantae woodfordienses. l.c. **20**: 268–276. 2 pl. incl. facsim. 1924. (Notes on a copy annotated by T. F. Forster.)

GLOUCESTERSHIRE

See also Monmouthshire (Shoolbred, The flora of Chepstow); Worcestershire (Lees, The botany of the Malvern hills).

Vice-county 33, East Gloucester; vice-county 34, West Gloucester.

General

Riddelsdell, H. J., Hedley, G. W., and **Price, W. R.,** *eds.* Flora of Gloucestershire. Phanerogams, vascular cryptogams, Charophyta. Edited on behalf of the Cotteswold Naturalists' Field Club . . . clxxxii, 667 p. 43 pl., 6 maps (4 in pocket), col. front. 22.5 cm. Cheltenham, 1948.

Geology, soils, climate, prehistoric vegetation, general features of vegetation, plant associations, changes in weed flora due to war, changes in boundaries, botanical districts, statistics, annotated list of local botanists, etc.; annotated list of vascular plants and Characeae (1,910 species (including microspecies and aliens) of which 1,039 are native, denizen, or colonist) with detailed localities; revised list of *Rubus*, gazetteer, bibliography. Replaces **Strugnell, W. B.** A catalogue of Gloucestershire plants. *In* Witchell, C. A., and Strugnell, W. B. The fauna and flora of Gloucestershire. p. 288–301. 25 cm. Stroud, 1892. (Unannotated list of vascular plants and Characeae, with occasional indication of localities. In Arnold Arboretum library.)—See also review by **Sandwith, C. I.,** and **Sandwith, N. Y.** Proc. Bristol Nat. Soc. **27**: 390–392. 1949.—Also papers by various authors in Proc. Cotteswold Nat. Field Club **30**: 34–39. 1950, and later.

Local

Buckman, James. A botanical guide to the environs of Cheltenham; comprehending a classified arrangement of the indigenous flowers and ferns of the Cotteswold Hills and the vale of Gloucester; with the habitats of the various plants. viii, 59 p. front. 21.5 cm. Cheltenham and London, 1844. (Lloyd Library.)

Topography, geology; annotated list of vascular plants (606 species), on the Linnaean system.—See also **Prentice, Charles.** Addenda et corrigenda to Mr. Buckman's Flora of Cheltenham. Phytologist **2**: 884–885. 1847.—Also **Riddelsdell, H. J.** Buckman's "Botany of the environs of Cheltenham" (1844). Proc. Cotteswold Nat. Field Club **20**: 162–164. 1920. (Critical notes on various doubtful species.)—Also **Sprague, T. A.** The wild flora of Cheltenham. l.c. **30**: 20–28. 1950. (Habitats, weeds, methods of dispersal, etc.; unannotated list of over 200 wild vascular plants found within borough boundaries, list of garden escapes.)—There are numerous additions in various volumes of the Rpt. Cheltenham Col. Nat. Hist. Soc., mostly from 1895 to 1946, and continuing. (Not seen.)

Ellacombe, H. T. Flora of the parish of Bitton. *In* his The History of the parish of Bitton, in the county of Gloucester. p. 241–262. 29 cm. Exeter, 1883.

Unannotated list of vascular plants.

Sandwith, C. I. The adventive flora of the Port of Bristol. Rpt. Bot. Soc. and Exch. Club Brit. Isles **10** (1932): 314–363. 1933.

History of the Port, statistics, sources of origin, etc.; list of 717 adventive angiosperms. Includes the docks at Portishead (Somerset) and Avonmouth (Gloucester), as well as at Bristol.—See also **Abell, R. B., Bannister, C. W.,** and **Townsend, C. C.** Adventive plants in Gloucester (V. cc. 33 & 34), 1952. Proc. Bot. Soc. Brit. Isles **1**: 70. 1954. (Includes list from Sharpness Docks, Bristol, and shorter one from Gloucester Docks.)

Serocold, W. P. List of the flowering plants of the parish of Forthampton and its neighbourhood. 25 p. 21 cm. Tewksbury, 1871. (British Museum (Natural History) library.)

Annotated tabular list of 300 spermatophytes.

White, J. W. The flora of Bristol, being an account of all the flowering plants, ferns, and their allies that have at any time been found in the district of the Bristol coalfields . . . viii, 722 p. 3 pl., map. 22.5 cm. Bristol and London, 1912. (Gray Herbarium library.)

Topography, geology, distribution according to geological formations, climate, history of local botany, biographies, bibliography; extensively annotated list of 1,692 species and varieties of vascular plants and Characeae (1,178 native or naturalized species), with localities. Covers an area of 720 square miles in western Gloucestershire and northern Somerset, from Berkeley and Dunsley south to Frome and Huntspill. Replaces **Swete, E. H.** Flora bristoliensis. xxvi, 138 p. pl., 2 maps. 19.5 cm. London and Bristol, 1854. Also replaces White's own Flora of the Bristol coalfield. 1887.—See also **White, J. W.** Notes supplemental to the flora of Bristol. Jour. Bot. **56:** 11–18, 40–49, 77–87. 1918; Notes on Bristol plants. l.c. **59:** 299. 1921.—Also his The botany of Bristol. Ann. Rpt. and Proc. Bristol Nat. Soc. IV. **7:** 215–211. 1931.—Also series of annual reports by White, later by C. I. Sandwith (later in collaboration with N. Y. Sandwith). l.c. IV, **5:** 88–91. 1920, and continuing.

HAMPSHIRE (AND ISLE OF WIGHT)

See also Great Britain, General (Hall, Additions and corrections . . . for Hampshire and the Isle of Wight, under Druce, The comital flora of the British Isles); Dorsetshire (Salter, Appendix, illustrative of the botany of Poole); Surrey (Monckton, The flora of the Bagshot district).

Vice-county 10, Isle of Wight; vice-county 11, South Hampshire; vice-county 12, North Hampshire.

General

Boulger, G. S. Botanical bibliography of the south-eastern counties. Proc. South East. Union Sci. Soc. **4:** 43–52. 1899.

List of general and local floras and floristic papers of Hampshire, Isle of Wight, Kent, Middlesex, Surrey, and Sussex Counties, in part annotated, with mention of more important herbaria.

Rayner, J. F. The alien or adventive flora of Hampshire and the Isle of Wight. Proc. Isle of Wight Nat. Hist. Soc. **1** (1923): 166–175. 1924.—A list of the alien plants of Hampshire and the Isle of Wight. (Ordinary weeds of cultivation omitted.) l.c. **1** (1924): 229–274. 1925.

Discussion of distinctions between introduced and native species; (second paper) annotated list of nearly 300 alien vascular plants.—See also **Brenan, J. P. M.** A contribution to the alien flora of Southampton. Ann. Rpt. Bot. Soc. and Exch. Club Brit. Isles 13 (1945): 106–112. 1947. (Extensive list of angiosperms, with dates and some annotations.)

Smith, John. The ligneous plants of Hampshire. Trans. Roy. Scott. Arbor. Soc. **12:** 356–370. 1889.

Annotated list of trees, shrubs, and woody vines, with vernacular names.

Townsend, Frederick. Flora of Hampshire including the Isle of Wight; or a list of the flowering plants and ferns found in the county of Southampton, with localities of the less common species . . . new ed., with additions. xxxviii, 658 p. 2 pl. (1 col.), map. 22 cm. London, 1904. (1st ed. 1883.)

Topography, climate, geology, botanical districts, bibliography; annotated list of vascular plants and Characeae (about 1179 species); table showing distribution, statistics, etc.—See also **Rayner, J. F.** A supplement to Frederick Townsend's Flora of Hampshire . . . xix, 132 p. 22 cm. Southampton, 1929. (Changes in flora; memoir of Townsend (by J. Britten, from Journal of Botany); list of additional species and localities.)—Also additional records by various authors, covering the period 1930–1952, in Papers and Proc. Hampsh. Field Club and Archaeol. Soc. **10:** 296–300. 1931; **12:** 82–84, 291–297. 1932–34; **15:** 62–75, 176–177. 1941–42; **16:** 61–63, 194, 290–292. 1944–47; **17:** 81–82, 235, 310–311. 1949–52; **18:** 209–213. 1953. (All in Harvard College library.)

Local

Adamson, R. S. The flora of Ditcham Park, Hampshire. Papers and Proc. Hampsh. Field Club and Archaeol. Soc. 9: 342–360. 1925. (Harvard College library.)

General features of vegetation; alphabetical tabular lists of vascular plants and of bryophytes, annotated as to ecological habitats.

Bell, Thomas. Plants. *In* White, Gilbert. The natural history and antiquities of Selborne . . . Ed. by Thomas Bell. v. 2, p. 369–373. 23 cm. London, 1877.

Alphabetically arranged list of vascular plants, with localities.—See also **Vaughan, John.** Notes on the botany of Selborne. Jour. Bot. 25: 366–370. 1887. (List of rarer vascular plants, with localities.)—Also **Britten, James.** Gilbert White's Selborne plants. l.c. 31: 289–294. 1893. (Unannotated list of 439 plants, mostly vascular, from a list prepared by White himself.)—Also Childs, below.

Bromfield, W. A. Flora vectensis; being a systematic description of the phaenogamous or flowering plants and ferns indigenous to the Isle of Wight. Edited by Sir William Jackson Hooker . . . and Thomas Bell Salter . . . xxxv, 678 p. port. 23 cm. London, 1856.

General features of flora; annotated descriptive flora of vascular plants; gazetteer.—See also **More, A. G.** A supplement to the "Flora vectensis." Jour. Bot. 9: 72–76, 135–145, 167–172, 202–211. 1871. (Reprinted, 30 p.)—Also **Tucker, Robert.** On the flora of the Isle of Wight. l.c. 9: 295–299. 1871.—Also Drabble, O'Brien, and Stratton, below.

Bruce, Hylda. List of wild flowering plants found in Milford parish. Milford-on-Sea Rec. Soc. 1 (2): 9–22. 1910. (British Museum (Natural History) library.)

Unannotated list of spermatophytes (406 species). Preceded by her Some of the wild flowers of Milford. l.c. 1 (2): 5–8. 1910. (Account of some noteworthy species.) Followed by **Melvill, A. H.** Notes on the botany of Milford. l.c. 1 (2): 23–27. 1910. (Account of various noteworthy species. Both in British Museum (Natural History) library.)

Childs, K. A. The Curtis museum. Herbarium list. 24 p. 7 pl. on 4. 22 cm. (Winchester) [pref. 1949]. (Kew library.)

Mention of local botanists, etc.; list of vascular plants, with localities; bibliography. Covers the region in and around Alton, "from Alresford on the south-west to Alice Holt on the north-east, with especial attention to the Gilbert White country."—See also Bell, above.

Clarke, C. B. A list of the flowering plants, ferns, and mosses collected in the immediate neighbourhood of Andover. 1 p. l., 114 p. map. 21.5 cm. Calcutta, 1866. (Lloyd Library.)

Geology, etc.; annotated list of 667 vascular plants, also mosses.

Drabble, Eric, and **Long, J. W.** A list of plants from the Isle of Wight. Rpt. Bot. Soc. and Exch. Club Brit. Isles 9 (1931): 734–757. 1932.

Account of published floras; annotated list of phanerogams observed by the authors from 1916–1931.—See also **Long, J. W.** Some alien plants . . . l.c. 9: 758–760. 1932.—Also Bromfield, above, and additional references there given.

Linton, E. F. Flora of Bournemouth including the Isle of Purbeck and the country within twelve miles of Bournemouth. 2. issue with appendix. viii, 304 p. map. 19.5 cm. Bournemouth, 1919. (British Museum (Natural History) library.) (1st ed. 1900.)

Topography, climate, geology, botanical divisions, etc.; annotated list of vascular plants and Characeae, with localities. The second issue is a reissue of the first, with the addition of an appendix (p. 287 [sic]–304).—See also his Appendix II. p. 304 [sic]–325. 18.5 cm. Edinburgh, [pref. 1925]. (List with localities. In British Museum (Natural History) library.)—Also **Hall, L. B.** Plants of the Bournemouth district. *In* Smith, S. Watson. The book of Bournemouth. Published for the British Medical Association meeting 1934. p. 31–37. 23.5 cm. 1934. (Includes some additions. Not seen; title and annotation supplied by N. D. Simpson.)

Notcutt, W. L. Catalogue of plants observed in the neighborhood of Fareham, Hants. Phytologist **2:** 201–216. 1845.—Remarks on certain plants growing near Fareham. l.c. 490–494. 1846.
Soils, etc.; briefly annotated list of vascular plants (544 species).

O'Brien, Charlotte, and **Parkinson, C.** Wild flowers of the Undercliff, Isle of Wight. vi, 143 p. 8 col. pl. 19 cm. London, 1881.
Briefly descriptive flora of vascular plants, with localities for rarer species, followed by a floral calendar. Relates to the Wilderness at Rookley and St. Helen's Spit as well as to the Undercliff.

Rayner, J. F. The flora of Millbrook station. Papers and Proc. Hampsh. Field Club and Archaeol. Soc. **9:** 99–117. 1921. (Harvard College library.)
Notes on various species; briefly annotated list of 280 plants (mostly vascular) found on reclaimed ground at Millbrook Railway Station from 1915–18.

Reeks, Henry. A list of the flowering plants, ferns, and mosses observed in the parish of East Woodhay . . . Trans. Newbury Distr. Field Club **1870/71:** 60–70. 1871.
Mostly unannotated list of vascular plants and bryophytes (620 species and varieties).

Stratton, Frederic. Flowering plants and ferns and their allies. *In* Morey, Frank, *ed.* A guide to the natural history of the Isle of Wight . . . p. 126–199. 3 pl. 22 cm. Isle of Wight and London, 1909.
Topography, botanical explorations, etc.; annotated list of vascular plants and Characeae (1032 species). The work contains similar lists of cellular plants by other authors. — See also his Isle of Wight plants. Jour. Bot. **51:** 285–294. 1913; **54:** 232–233, 371. 1916.—Also **White, E. H.** Botanical notes. Proc. Isle of Wight Nat. Hist. and Archaeol. Soc. **4:** 145–148. 1951. (Includes account of changes in flora. Not seen; cited from Watsonia **2:** 324. 1953.) — Also Bromfield, above, and additional references there given.

[Warner, F. I.] List of flowering plants, ferns, etc., found within seven miles from Winchester. Rpt. of Proc. Winchester and Hampsh. Scient. and Lit. Soc. **1:** 37–52. 1872.
Annotated list of vascular plants. No author's name is given in the original publication, but according to note in Jour. Bot. **35:** 32. 1897, Warner, at that time secretary of the Society, was the author.

Wise, J. R. The flowering plants of the New Forest district. *In* his The New Forest . . . p. 289–306. 24 cm. London, 1863 [1862].
Annotated list of about 680 species and varieties of vascular plants, divided into three groups according to locality and habitat. There are several later editions. The New Forest is that part of Hampshire between Southampton Water on the east, the Solent on the south, the Avon River on the west, and Wiltshire on the north. — See also **Dyer, W. T.,** and **Trimen, H.** Additions to the lists of Hampshire plants. Jour. Bot. **2:** 316–318. 1864.

Yonge, C. M. John Keble's parishes. A history of Hursley and Otterbourne. xi, 235 p. pls., map. 21.5 cm. London and New York, 1898 (reissued 1899). (Harvard College library.)
The section on "Flowers" (p. 206–235) is a list of vascular plants, with some localities and annotations.

Herefordshire

Vice-county 36.

General

Carter, P. W. Some account of the history of botanical exploration in Herefordshire. Trans. Woolhope Nat. Field Club **34:** 232–267. 1955.
Account of botanists, chronologically arranged with references to their writings and discoveries; alphabetical list of botanists with incomplete references to their publications.

Ley, Augustin. Botany. *In* The Victoria history of the county of Hereford. v. 1, p. 39–54. map. 31.5 cm. London, 1908.
General features of flora, discussion of botanical districts with mention of rarer species; annotated list of vascular cryptogams; no list of spermatophytes. Followed (p. 54–76) by lists of Musci and Fungi by other authors.

Purchas, W. H., and Ley, Augustin. A flora of Herefordshire. xi, xxxvii, 549 p. 3 pl., map. 22 cm. Hereford [pref. 1889]. (Gray Herbarium library.)

Botanical explorations, topography, climate, botanical districts; annotated list of vascular (899 species) and cellular plants, with localities. — See also **Ley, A.** Additions to the Flora . . . Trans. Woolhope Nat. Field Club **1893/4,** suppl. 89 p. 1894. (Library of Academy of Natural Sciences, Philadelphia.)—Also his Additions . . . Jour. Bot. **32:** 207–212. 1894.—Also his Herefordshire Rubi. l.c. 34: 155–162, 217–233. 1896.—Also his Notes additional to the Flora of Herefordshire. Trans. Woolhope Nat. Field Club **1905/07: 69–116.** [1905?] (Extensive list of vascular plants, with localities. Reprinted, 62 p., 1905, including his similar paper with title Mosses. l.c. p. 117–130. Original and reprint in British Museum (Natural History) library.) — Also **Day, F. M.** Plants of the Silurian limestones in the west of the Malvern Hills. Watsonia **2:** 289–296. 1953. (Running account of the flora, with lists of species characteristic of different formations; bibliography. Refers to parishes of Eastnor, Colwall, and Mathon, the last in v.–c.37 [Worcestershire]).

Local

Watkins, B. M. Florula of the Doward Hills. Trans. Woolhope Nat. Field Club **1881:** 53–85. 1881. (Library of Academy of Natural Sciences, Philadelphia.)

Topography, local distribution, etc.; list of vascular plants and Characeae (599 species), almost without annotations. — See also **Ley, Augustin.** Florula of the Doward Hills — mosses. Trans. Woolhope Nat. Field Club **1890/92:** 132–143. 1894 (?). (Includes (p. 143) list of additional vascular plants. In British Museum (Natural History) library.)—Also his List of plants additional to the Florula of the Doward Hills. l.c. **1905/07:** 205–214. 1911. (List of vascular plants and mosses. In British Museum (Natural History) library.)

Hertfordshire

See also London.
Vice-county 20.

General

Hopkinson, John. Botany. *In* The Victoria history of the county of Hertford. v. 1, p. 43–80. map. 31.5 cm. London, 1902.

Botanical explorations, general features of flora, statistics, botanical districts with lists of rarer spermatophytes, running account of pteridophytes; also summary treatment of cellular plants.

Pryor, A. R. A flora of Hertfordshire . . . Edited . . . by Benjamin Daydon Jackson, with an introduction on the geology, climate, botanical history, etc., of the county by John Hopkinson and the editor. lviii, 588 p. 2 maps. 21 cm. London, Hertford, 1887.

Geology, climate, botanical districts, botanical history, list of local botanists, etc.; annotated list of vascular plants (1,116 species); list of cellular plants; tabular list showing distribution. Replaces **Webb, R. H., and Coleman, W. H.** Flora hertfordiensis: or a catalogue of plants found in the county of Hertford, with the stations of the rarer species. 2 p. l., xlvi, 390 p. map. 19 cm. London, Hertford, 1849; Supplement [1]–2 . . . 20, 23 p. 19 cm. 1851-59. (Topography, botanical divisions; annotated list of vascular plants, with localities. All in New York Botanical Garden library.)—See also **Salisbury, E. J.** Botanical observations in Hertfordshire during the year 1909 [–1918]. (Title varies somewhat.) Trans. Hertfordsh. Nat. Hist. Soc. and Field Club **14:** 177–178, 301–302. 1911–12; **15:** 172, 239. 1914–15; **16:** 75–78, 157–160. 1916–17; **17:** 13–15, 143–145, 248–249. 1921. (Records of noteworthy vascular and cellular plants.) — Also **Little, J. E.** Supplementary notes to the Herts flora. Jour. Bot. **55:** 44–52, 74–77. 1917.—Also **Britten, James.** Hertfordshire plants. l.c. **55:** 351–352. 1917.

Salisbury, E. J. Changes in the Hertfordshire flora: a consideration of the influence of man. Trans. Hertfordsh. Nat. Hist. Soc. **18:** 51–68. 1929.

Discussion of species possibly introduced by early settlers of Britain, those becoming more common, those becoming scarcer or now extinct, changes in climate, etc.

Local

Bishops Stortford and District Natural History Society. List of vascular plants of Bishops Stortford and district. Trans. Bishops Stortford and Dist. Nat. Hist. Soc. 1 (2): 3–57. 1952.

List of vascular plants within 6-mile radius around Bishops Stortford. (Not seen; cited mainly from Watsonia 2: 325. 1953.)

Little, J. E. A sketch of the botany of the Ivel district of Hertfordshire. Rpt. Bot. Soc. and Exch. Club Brit. Isles **10** (1932): 375–387. 1933.

Botanical explorations, changes in woodland and in the flora, species added to flora of district since Pryor (1887), bibliography. (Vicinity of Hitchin.) — See also his Botany. *In* Hine, Reginald L. The natural history of the Hitchin region. p. 56–72. 4 pl. 25 cm. Hitchin, 1934. (General features of flora, botanical explorations, list of native species added to flora of Ivel district since Pryor (1887), etc.; bibliography.)

Salisbury, E. J. The ecology of shrub in Hertfordshire: a study in colonization. Trans. Hertfordsh. Nat. Hist. Soc. 17: 53–64. 1918.

Ecological; includes (p. 59–63) tabular list of vascular plants showing occurrence in four areas (Kingsland, Markyate, Bricket, Hertford Heath). — See also his The phanerogamic flora of Bricket Wood. l. c. **21**: 159–160. 1940. (Adds 38 species, bringing total for Bricket Wood to 174 species.) — Also **Benson, R. B.** Some plant records from Bricket Wood. l.c. **21**: 175. 1940.

Wainwright, E. D., *ed.* Flora. *In* Haileybury Natural Science Society. Fauna and flora of Haileybury. 3. issue [i.e., 3. ed.]. p. 34–62. 19.5 cm. Hertford, 1926. (Kew library.) (1st ed. 1888.)

List of vascular plants, with localities.

Whiteman, R. J., *ed.* Flora. *In* Hexton: a parish survey. Made and compiled by the senior students of the Herts. C. C. School, Hexton. p. 19–33. 25 cm. n. p., 1936. (Library of N. D. Simpson.)

Annotated running list of spermatophytes, the "grasses and sedges and lower order plants" omitted.

HUNTINGDONSHIRE

Vice-county 31.

General

Druce, G. C. Botany. *In* The Victoria history of the county of Huntingdon. v.1, p. 29–80. map. 31.5 cm. London, 1926.

Topography, geology, fenland, biographical account of botanists and their work with titles of publications, comparison with botany of adjacent counties; annotated list of known vascular plants and Characeae (890 species). — See also his Huntingdonshire plants. Rpt. Bot. Soc. and Exch. Club Brit. Isles **7** (1925): 949–957. 1926. (List of vascular plants and Characeae, with localities, collected mainly by author in 1876–1925, and mostly not recorded from the county in Watson's Topographical botany.) — Also **Dony, J. G.** A contribution to the flora of Huntingdonshire. Watsonia **1**: 301–307. 1950. (Botanical publications and explorations; list of noteworthy vascular plants.) — Also reports on Flora in Ann. Rpt. Huntingdonsh. Fauna and Flora League for 1949 and later. (Not seen.)

KENT

See also London.
Vice-county 15, East Kent; vice-county 16, West Kent.

General

Hanbury, F. J., and Marshall, E. S. Flora of Kent, being an account of the flowering plants, ferns, etc., with notes on the topography, geology, and meteorology, and a history of the botanical investigation of the county. lxxxiv, 444 p. 2 maps. 19.5 cm. London, 1899. (Arnold Arboretum library.)

Topography, geology, climate, botanical districts, tabular comparison with flora of adjacent counties, history of botanical investigation, bibliography;

annotated list of vascular plants and Characeae. Replaces **Webster, A. D.** The flora of Kent. 34 p. 20 cm. Bromley [1893]. (Geology, etc.; annotated list of vascular plants. In Arnold Arboretum library.) — See also **Bennett, Arthur.** Notes on the "Flora of Kent." Jour. Bot. **37**: 340–343. 1899. — Also **Druce, G. C.** Some Kentish plants. l.c. **40**: 350–353 1902.— Also **Rose, Francis.** Botanical records for metropolitan Kent, 1942–1949. London Nat. **29**: 15–20. 1950. (List of noteworthy vascular plants from the part of Kent that is included in the London area.)

Marshall, E. S. Botany. *In* The Victoria history of the county of Kent. v. 1, p. 45–89. map. 31.5 cm. London, 1908.
Extinct species, botanical regions, topography, geology, botanical explorations; annotated running list of more uncommon species of vascular plants; also lists of cellular plants by E. M. Holmes (except Characeae by Marshall).

Local

Bevis, J. F., and **Griffin, W. H.,** *eds.* Botany. The flora of Woolwich and West Kent. (Districts 1 and 2 of Hanbury and Marshall's Flora.) *In* Grinling, C. H., Ingram, T. A., and Polkinghorne, B. C. A survey and record of Woolwich and West Kent . . . p. 31–230, 497–506. 21.5 cm. Woolwich, 1909. (Gray Herbarium library.)
Annotated list of vascular and cellular plants, based (as to vascular plants) upon Hanbury and Marshall's Flora of Kent (1899).

Cowell, M. H. A floral guide for East Kent, etc., being a record of the habitats of indigenous plants found in the eastern division of the county of Kent, with those of Faversham particularly detailed, and definitely exhibited; together with brief remarks on the uses of the several species in rural or domestic economy, agriculture, medicine, and the manufactures; and on their classical and historical associations; deduced from various authorities. xiv, 98 p. 2 maps. 23 cm. Faversham, 1839. (New York Botanical Garden library.)
Includes (p. 18–54) annotated alphabetical list of vascular plants (542 species) of Faversham, with notes on uses; also similar account of the plants of East Kent; each list preceded by short lists of the noteworthy species of various included localities. Replaces **Jacob, Edward.** Plantae favershamienses. A catalogue of the more perfect plants growing spontaneously about Faversham, in the county of Kent . . . 4 p. l., 146, liii p. pl., front. 18.5 cm. London, 1777. (Includes (p. 1–127) annotated list of vascular plants, alphabetically arranged by Ray's Latin names, the Linnaean names also given in the text and indexed.) — See also Stowell, below.

[Currey, Frederick]. **On the botany of the district lying between the rivers** Cray, Ravensbourne and Thames; being the first report of the Botanical committee of the Greenwich Natural History Club. 24 p. 21.5 cm. London, 1858. (British Museum (Natural History) library.)
General features of flora; "Catalogue of plants growing in the district of Greenwich Natural History Club" (p. 14–24), an unannotated list of vascular plants (833 species), Characeae, and fungi. — See also Webster, below.

Deakin, Richard. The flowering plants of Tunbridge Wells and neighbourhood. v, 371 p. illus. (incl. col. pl.) 21.5 cm. Tunbridge Wells and London, 1871. (Arnold Arboretum library.)
Key to families; annotated descriptive flora of spermatophytes (782 species), with notes on uses, and text figure of nearly every species. Boulger (Proc. South East. Union Sci. Soc. **4**: 48. 1899) refers to a second edition of 1876, which I have not been able to trace. — See also Forster, Jenner, and Roberts, below.

Ewing, J. W. Some materials for a flora of Wrotham and its neighbourhood. 50 p. 18 cm. West Malling, 1883. (Kew library.)
Includes (p. 5–46) briefly annotated list of vascular plants.

Fielding, C. H. A hand-book of Higham: or the curiosities of a country parish. 66 p. 21.5 cm. Rochester and Chatham, 1882. (Library of N. D. Simpson.)
Descriptive account of flora, p. 48–53, and unannotated list of vascular plants (434 species), p. 54–64. According to notice in Jour. Bot. **21**: 188. 1883, the list is far from complete.

Flower, T. B. Flora thanetensis: or, a catalogue of the plants indigenous to the isle of Thanet. vi, 83 p. 19 cm. Ramsgate, 1847. (Arnold Arboretum library.)

Annotated list of vascular plants and algae. — See also Pittock and Wilson, below.

Forster, T. F. Flora tonbrigensis: or, a catalogue of plants growing wild in the neighbourhood of Tonbridge Wells, serving also as a guide to the central flora of Kent and Sussex, arranged according to the Linnaean system . . . With additions by T. Forster. [2. ed.] 56, 216 p. 2 col. pl. 19 cm. Tonbridge Wells, London, 1842. (Lloyd Library.) (1st ed. 1816.)

Additions and corrections to earlier edition; annotated list of vascular and cellular plants found within 15 miles of Tunbridge Wells, on the Linnaean system. The 2. edition is a reprint of the first (as to p. 1–216), with the addition of p. ix–56 of the first section, including biographical sketch of T. F. Forster, a supplementary list of plants (p. 25–43), changes in habitats, account of soils, etc. — See also Deakin, above, and Jenner and Roberts, below.

Haydon, W. T. Catalogue of the flowering plants, etc., found in Dover and its neighbourhood. 35 p. 18.5 cm. Dover, 1890.

List of vascular plants. (Not seen; title and annotation supplied by G. Taylor.) — See also Paley, below.

Hepworth, John. Flora of our ten-mile radius. Rochester Nat. **2:** 353–356, 366–370, 383–387, 389–401, 417–420, 439–441. 1896–97. (Last part seen in library of N. D. Simpson, others in Library of Congress.)

Annotated list of vascular plants of vicinity of Rochester. — See also his Flora of Rochester district. (10 mile radius.) *In* his Rochester and district. A sketch-guide to its geology, flora, and fauna. p. 11–42. 22 cm. Rochester, 1913. (Annotated list of vascular plants and Characeae, with localities for scarcer species. In British Museum (Natural History) library.)

Jacob, J. Wild flowers, grasses and ferns of East Kent. A catalogue compiled . . . 2 p. l., 29 p. 22 cm. Dover, 1936. (Kew library.)

Bibliography, notes on orchids, etc.; partly annotated list of vascular plants and Characeae. — See also Cowell, above.

Jenner, Edward. A flora of Tunbridge Wells, being a list of indigenous plants within a radius of fifteen miles around that place. xx, 134 p. 18 cm. Tunbridge Wells, London [1845].

Annotated list of vascular and cellular plants, on the Linnaean system. The pagination given above is that in the U. S. Department of Agriculture copy. Pp. xv–xx of the introduction constitute a supplement. N. D. Simpson informed the author of two other issues, apparently published in the same year (1845?): one in paper cover, xvi, 134 p., 16 cm.; another in cloth, xxiv, 260 p., 2 col. pl., 18 cm., including supplement on p. xv–xx, the text of the flora printed on one side only. Both, like the one in the U. S. D. A. library, contain a 2-page list of subscribers.—See also Deakin and Forster, above, and Roberts, below.

Kent, D. H. The flora of bombed sites in Canterbury. Watsonia **2:** 12–17. 1951.

Account of bomb damage to Canterbury during World War II; annotated list of vascular plants found in bombed areas.

Lamb, Henry. The flora of Maidstone. A contribution to the natural history of the district. 1 p. l., 15 p. 18.5 cm. Maidstone, 1889. (British Museum (Natural History) library.)

Brief account of habitats; essentially unannotated list of spermatophytes.

Marriott, St. John. British woodlands as illustrated by Lessness Abbey woods. A general survey of the flora and fauna . . . xviii, 81 p. incl. 5 pl. 6 fig., map (in pocket). 21.5 cm. London, Woolwich, 1925. (Kew library.)

Includes (p. 3–23) "Botanical survey," an ecological account; "Flowering plants" (p. 24–31), an unannotated list; also lists of pteridophytes (p. 32), bryophytes, fungi, lichens, and Mycetozoa, as well as of the fauna. Lessness Abbey is in northwestern Kent, 1.5 miles west of Erith.

Morris, W. S. A catalogue of the flowering plants, and ferns, growing in the vicinity of Wye; and the circumjacent country. *In* his The history and topography of Wye. p. 180–189. 22.5 cm. Canterbury [etc.], 1842. (Library of N. D. Simpson.)

Unannotated alphabetical list of vascular plants.

Paley, F. A. The wild flowers of Dover and its neighbourhood, with indices of their English and Latin names. xii, 68 p. 16.5 cm. Dover, [ca. 1850]. (Kew library.)

Running account of vascular plants, with localities.—See also Haydon, above.

Pittock, G. M., and others. Flora of Thanet: a catalogue of the plants indigenous to the island with a few rare aliens. 14 p. 20.5 cm. Margate, 1903. (Gray Herbarium library.)

Unannotated list of 565 phanerogams. "All localities are given in Hanbury and Marshall's Flora of Kent" (ms. note).—See also Flower, above, and Wilson, below.

Roberts, Fred, and **Stebbing, M. A.** Our flowers. *In* Knipe, Henry R., *ed.* Tunbridge Wells and neighbourhood. A chronicle of the town from 1608 to 1915; and papers by various authors relating to the geology, plant and animal life, archaeology, and other matters of the district. p. 87–108. 23 cm. Tunbridge Wells, 1916.

The description of the flora by Roberts ("Our flora") is a general sketch, p. 87–93; followed by a mostly unannotated list ("Our flora II") of the vascular plants with localities for rarer species, by Stebbing, p. 94–108; this is followed by Stebbing, Our trees, p. 109–113, with 1 pl.—See also **Shaw, G. E.** Flowering plants. *In* Given, J. C. M., *ed.* Royal Tunbridge Wells past and present. For the occasion of the Jubilee congress of the South-eastern union of scientific societies July, 1946. p. 99–104. 19 cm. Tunbridge Wells [1946]. (Running account of some of the rarer plants followed by an unannotated list of noteworthy species, etc. In Kew library.) —Also Deakin, Forster, and Jenner, above.

Scargill, J. J. Flowering plants of the Bromley district. Edited for the special committee. (At head of title page: Bromley Naturalists' Society. List IV.) 2 p.l., 32 p. 21.5 cm. Bromley, 1898. (British Museum (Natural History) library.)

Briefly annotated list of spermatophytes (696 species) found mostly within five miles of Bromley Town Hall.

Stowell, H. A. Some observations on the flora of Faversham and its neighbourhood. Phytologist n.s., **1**: 249–256, 375–378. 1856.—Faversham plants. Plantae rariores . . . l.c. **2**: 100–106, 153–156, 180–183. 1857.—Faversham common plants. l.c. 261–262. 1857.

Topography, soils, botanical explorations, changes in flora; list of less common species, with localities; unannotated list of common species.—See also Cowell, above.

Walton, J. W. The wild flowers and ferns of the district round Folkestone . . . Based upon the work of the late Mr. G. C. Walton, F. L. S., as published in 1894. To which is appended a catalogue of mosses of the area, as recorded by E. C. Green . . . (Cover title, The wild flowers, ferns and mosses of the district round Folkestone.) 3 p. l., 40 p. 21.5 cm. Folkestone, 1950 (cover place and date).

Unannotated list of vascular plants and Charophyta (p. 1–33), also mosses, within a radius of about 10 miles around Folkestone, including Dungeness. Replaces **Ullyett, Henry.** Rambles of a naturalist round Folkestone, with occasional papers on the fauna and flora of the district. 3 p.l., 157 p. 19 cm. Folkestone, 1880. (Includes (p. 129–138) unannotated list of vascular plants. In Arnold Arboretum library.) Also replaces **Walton, G. C.** A list of flowering plants and ferns found in the neighbourhood of Folkestone. Revised and brought up to date by G. C. Walton . . . 11 p. 21.5 cm. (Folkestone), 1894. (Unannotated list of vascular plants and Characeae. In British Museum (Natural History) library.)

Webb, E. A. The flora of Chislehurst. *In* Webb, E. A., Miller, G. W., and Beckwith, J. The history of Chislehurst . . . p. 320–335. 29.5 cm. London, 1899. (Library of N. D. Simpson.)

List of vascular plants (about 570 species), with localities for rarer species.

Webster, A. D. The fauna and flora of Greenwich Park. Trans. West Kent Nat. Hist., Micros., and Photog. Soc. **1900/01:** 16–25. 1901. (British Museum (Natural History) library.)

Running account of plants and birds, followed by an unannotated alphabetical list (p. 24–25) of vascular plants, containing some patent errors. The same or essentially the same treatment is given in his The fauna and flora of Greenwich Park. *In* his Greenwich Park: its history and associations. p. 39–53. 23 cm. Greenwich, London, 1902. (In library of N. D. Simpson.)

Wilson, L. W. Thanet flora. Ann. Rpt. Isle of Thanet Field Club **1:** 25–38. 1949.

List of known vascular plants, without localities. (Not seen; title and annotation from Watsonia **2:** 325. 1953.)—See also Notes on species in Thanet and fringing districts. Ann. Rpt. Isle of Thanet Field Club **2:** 48–50. 1950.—Also Additions to Thanet flora. l.c. **3:** 25. 1951.—Also Botanical notes, 1950–1951. l.c. **4:** 36–38. 1952. (These not seen; cited from Watsonia **2:** 325. 1953.)—Also Flower and Pittock, above.

Lancashire

See also Cumberland (Baker, A flora of the English Lake District); Westmorland (Wilson, The flora of Westmorland); York (Stansfield, Flora of Todmorden).

Vice-county 59, South Lancashire; vice-county 60, West (i.e., Mid) Lancashire; vice-county 69, North Lancashire (and Westmorland).

General

Dallman, A. A., and **Wood, M. H.** A biographical list of deceased Lancashire botanists. With a chronological enumeration of their published works and papers. Trans. Liverpool Bot. Soc. **1:** 54–109. 1909.

Alphabetical list of botanists, with brief biographical notes and titles of publications.

Fisher, Harry. Botany. *In* The Victoria history of the county of Lancaster. v. 1, p. 37–86. map. 31.5 cm. London, 1906.

Topography, etc., bibliography; list of vascular and cellular plants ("the commonest . . . not enumerated") with localities.

Local

Alcock, R. H. List of plants noticed by the Bury Natural History Society within fifteen miles of Bury. Rpt. Bury Nat. Hist. Soc. **1868–71:** 14–23. 1872.

Briefly annotated list of vascular plants giving known vernacular names. (Not seen; title and annotation supplied by G. Taylor.)

Bailey, Charles. On the adventitious vegetation of the sandhills of St. Anne's on-the-Sea, North Lancashire (Vice-county 60). Mem. and Proc. Manchester Lit. and Phil. Soc. v. 47, no. 2. 10 p. 3 pl. 1902.—Further notes . . . l.c. v. 51, no. 11. 16 p. 8 pl. 1907.—A third list . . . l.c. v. 54, no. 15. 11 p. 1910.

Annotated lists of over 100 adventive vascular plants.—See also **Whellan, J. A.** The present day flora of the sand-dunes at St. Annes, W. Lanc. V. C. 60. North West. Nat. n.s., **2:** 139–141. 1954. (Notes on numerous species.)

Blackie, J. E. H. The flora of a Manchester housing estate. North West. Nat. **22:** 260–267. 1 fig. (map). 1949.

Topography, etc.; annotated list of angiosperms (261 species) found mostly by writer on the Wythenshawe estate in vicinity of Timperley, Sale, Didsbury, and Gatley. (Includes a small area in adjacent Cheshire.)

Buxton, Richard. A botanical guide to the flowering plants, ferns, mosses, and algae, found indigenous within eighteen miles of Manchester; with some information as to their agricultural, medicinal, and other uses . . . [2. ed.] xxi, 207 p. 17.5 cm. London and Manchester, 1859. (Gray Herbarium library.) (1st ed. 1849.)

Autobiography of author, geology; annotated list of vascular plants, on the Linnaean system, with localities; list of Musci, by Williamson and Sidebotham.—See also **Bailey, Charles.** The botany of the district. *In* Handbook of Manchester. Prepared by the Local committee for the members of the

British Association at the Manchester meeting, 1887. p. 27–42. 16.5 cm. Manchester, 1887. (Changes in flora; general sketch of the botany of the district, with mention of noteworthy species of each family. In library of N. D. Simpson.)—Also Grindon, below.

Dunlop, G. A. An annotated list of the alien plants of the Warrington district. Mem. and Proc. Manchester Lit. and Phil. Soc. v. 52, no. 15. 27 p. 1908.

Sources of introduction, etc.; annotated list of alien vascular plants, some of the widespread species omitted. (Area with radius of 10 miles around Warrington Bridge.)

Dunlop, Margaret. Phytogeography. *In* Grime, Arthur, *ed.* A scientific survey of Blackpool and district. Prepared for the Blackpool meeting of the British Association 1936. p. 20–27. map. 21.5 cm. London, 1936. (Library of N. D. Simpson.)

Partial lists of vascular plants and charophytes, given under habitats.

[Gerard, John]. Flora of the Stonyhurst district. 2. ed. 45 p. 22 cm. Clitheroe, 1891. (1st ed., A preliminary flora . . . 1886. Reprinted from the Stonyhurst Magazine. Both editions in British Museum (Natural History) library.)

Briefly annotated list of vascular plants and Characeae, with localities. Refers to an area in Lancashire and adjacent western Yorkshire.

Green, C. T., *ed.* The flora of the Liverpool district. [rev. ed.] xi, 163 p. 802 fig. on pls., 12 pl. (incl. 2 ports.). 22 cm. Arbroath, 1933. (1st ed. 1902.)

Botanical investigations, etc.; annotated list of vascular plants and Characeae, with 802 figures, illustrating nearly all species; geology (by J. J. Fitzpatrick), meteorology (by J. C. Mitchell), account of special localities with lists of characteristic species, list of contributors; brief bibliography. Covers a radius of 15 miles around Liverpool and 2 miles around Southport, thus including part of Cheshire. Is a revision of the 1902 edition of the same work, which in turn was a revision by a committee of the Liverpool Naturalists' Field Club of its The flora of Liverpool (1872).—See also **Stansfield, Herbert.** Plants of the Liverpool district in the herbarium of John Shepherd (1764–1836). North West. Nat. 13: 10–15. port. 1938. (Extensive list of species collected by Shepherd, with list of new records; brief bibliography.)—Also **Merseyside Naturalists' Association.** Supplement to the Flora of Liverpool. 8 p. 1941. (Not seen; copy in Liverpool Corporation Reference Library, fide Eric Hardy.)—Also **Allen, D. E.** The flora of the Liverpool bombed sites. Rpt. Merseyside Nat. Assoc. 1: 25–27. 1951. (Not seen.)—Also **Hardy, Eric.** The pre-war flora of inner Liverpool. l.c. 1: 28–29. 1951. (Not seen.)—Also **Savidge, J. P.** Botanical records and notes, 1949–53. Proc. Liverpool Nat. Field Club 1953: 12–18. 1954. (Includes many additions, fide Proc. Bot. Soc. Brit. Isles 1: 527. 1955. Not seen.)

Greenlees, Thomas, and **Holden, T. K.** The flora of Bolton. Lancash. and Chesh. Nat. 12: 153–158, 191–195, 238–245, 262–268, 285–292. 1919–20. (Lloyd Library. Reprinted, 1 p.l., 34 p. 1920, in U.S. Dept. Agr. library.)

Annotated list of vascular plants (500 species) found within a radius of 10 miles around Bolton.

Grindon, L. H. The Manchester flora; a descriptive list of the plants growing wild within eighteen miles of Manchester, with notices of the plants commonly cultivated in gardens . . . 2 p. l., viii, 575 p. 232 fig. 21 cm. London, 1859. (Lloyd Library.)

Elements of descriptive botany, keys; annotated flora of vascular and cellular plants, with keys. — See also Buxton, above.

King, F. C. Botany of the district. *In* Smith, Thomas C. A history of Longridge and district. p. 259–271. pl. 21.5 cm. Preston, 1888.

General features of flora; unannotated list of vascular plants (p. 265–269), Characeae and mosses (p. 269–271).

Lacey, W. S. Notes on the flora of the Chorley district of South Lancashire. North West. Nat. n.s., 2: 526–558. pl. 9–10, map. 1955.

General features of flora; annotated list of vascular plants and fungi (492 species); changes in flora, bibliography.

Melvill, J. C. Flora. *In* Nicholls, Walter. History and traditions of Prestwich . . . With . . . The botany of Prestwich by J. Cosmo Melvill . . . p. 164–182. 19 cm. Manchester, 1905.

Principal localities, botanical explorations; annotated running list of vascular plants and a few mushrooms.

——— The flora of Kersal Moor, near Manchester. Jour. Bot. **20:** 211–218. 1882.

Partly annotated list of about 240 vascular plants.

Pearsall, W. H. Florula Furnessiae; its limitations and its lessons. With interspersions of comment by F. Arnold Lees. Lancash. and Chesh. Nat. **7:** 37–40, 65–68, 115–118, 153–156, 165–170, 223–236. 1914. (University of Minnesota library. Reprint, 26 p., in British Museum (Natural History) library.)

Annotated list of noteworthy vascular plants. Furness (or North Lancashire) is the detached portion of Lancashire, north of Morecambe Bay. — See also his Plants of North Lancashire. Lancash. Nat. **5:** 284–286, 324–326, 370–372, 397–400. 1912–13. (List of angiosperms of the Furness district not known in West Lancashire, with localities; list of local plants found in both the Furness region and West Lancashire, with localities. In University of Minnesota library.) — Also his Notes on the flora of North Lancashire. Lancash. and Chesh. Nat. **8:** 57–60. 1915. (List of noteworthy vascular plants observed in 1914, with localities. In University of Minnesota library.) — Also his Notes on the flora of North Lancs. (69b). Rpt. Bot. Soc. and Exch. Club Brit. Isles **4** (1915): 290–294. 1916. (List relating mostly to 1915 collections.) — Also Petty, The constituents of the North Lancashire flora, below.

Petty, S. L. The constituents of the North Lancashire flora, 1597 (?)–1893. 1. [–19.] paper. Naturalist **1894:** 117–124, 149–156, 243–250, 289–296. 1894; **1895:** 53–60, 133–140, 261–268. 1895; **1896:** 101–108, 197–204, 341–348. 1896; **1897:** 85–92, 229–236, 309–316, 325–332. 1897; **1898:** 37–44, 149–156, 309–316, 321–333. 1898; **1900:** 333–335. 1900.

Bibliography; systematic list of vascular plants and Characeae, with chronologically arranged summary of published records for each species, and references. — See also his New plant records for North Lancashire. Naturalist **1907:** 438. 1907. — Also Pearsall, Florula Furnessiae, above.

——— The plants of Leck and neighbourhood, Lancashire. Naturalist **1893:** 91–102. 1893.

Topography, etc.; list of vascular plants and Characeae observed by author, with localities.

——— Some plants of Silverdale, West Lancashire. Naturalist **1902:** 33–54. 1902.

Geology, bibliography; annotated list of known vascular plants and Characeae.

Preston Scientific Society. Flora of Preston and neighbourhood. Compiled by the members of the botanical section of the Preston Scientific Society during the years 1897–1902. 62 p. 16.5 cm. [n. p., 1903.]

Topography, geology, etc.; annotated list of vascular plants (also mosses by Henry Beesley).

Stansfield, Abraham. Observations on the botany of the Forest of Rossendale. *In* Newbigging, Thomas. History of the Forest of Rossendale. p. 285–299. 24 cm. London, 1868 [1867].

History of the forest, etc.; unannotated list of vascular plants and some mosses, hepatics, and lichens. Although the title page date is 1868, some copies of the work were evidently distributed in 1867, since the Library of Congress copy bears a manuscript dedication dated 24 Dec. 1867.

Stansfield, W. H., and Ball, Henry. Botany. *In* Southport: a handbook of the town and surrounding district. Prepared for the meeting of the British Association at Southport, 1903. p. 68–94. 19 cm. Southport, 1903. (British Museum (Natural History) library.)

Notes on various species; briefly annotated list of vascular plants and Characeae, with localities for scarcer species. Followed (p. 94–112) by other botanical papers by other botanists, including treatments of mosses and hepatics. Replaces **McNicoll, D. H.** Plants of Southport. *In* **his Handbook** for Southport, medical and general, with copious notices of the natural history of the district. p. 93–136. col. pl. (front.). 18.5 cm. London [etc.], 1859. (General features of habitats; partly annotated list of vascular plants, not in strictly systematic order. The list is much fuller than that in the third and last ed., the reference for which is as follows: **Grindon, Leo.** Natural history of Southport and its environs . . . The Southport flora. *In* McNicoll, Edward D., *ed.* Handbook for Southport, medical and general . . . 3. ed. p. 64–88. 18.5 cm. Southport, 1883. (Running account of numerous vascular plants, arranged under habitats; not a full flora.) — See also **Stansfield, F. J. S.** Plants of the Southport sand dunes. *In* Watkins, G. D., *ed.* Southport. p. 32–43. 16 cm. Southport, 1936. (Includes mention of various noteworthy species. In library of N. D. Simpson.) — Also **Gresswell, R. K.** Interesting plants found within the county borough of Southport, 1936–38. Compiled by R. Kay Gresswell, from the original notes of W. Waddington. Rpt. Southport Scient. Soc. **5** (34. Rpt.): 12–17. 1946. (List with localities.)

Wheldon, J. A. Some alien plants of the Mersey province. Lancash. Nat. **5**: 167–169, 211–216, 255–257, 337–341, 362–364, 409–411, 425–427. 1912–13; **6**: 32–35, 75–76, 83–86, 134–138, 174–177, 211–212, 248–250, 293–296, 332–335, 371–375, 407–412, 463–464. 1913–14. Lancash. and Chesh. Nat. **7**: 33–36, 83–84, 119–122, 128–130. 1914. (University of Minnesota library.)
Annotated list of noteworthy adventive vascular plants and Characeae, with localities and notes on distinctive characters.

——— **and Wilson, Albert.** The flora of West Lancashire, that portion of the county north of the River Ribble and south of Morecambe Bay. 3 p.l., 511 p. 15 pl., map. 22 cm. Liverpool, 1907. (Gray Herbarium library.)
Topography, botanical districts, discussion of distribution as affected by altitude and by character of soil and rocks, bibliography; annotated list of vascular and cellular plants, with localities.

Whitehead, John, and others. The district flora: as compiled by the Ashton-under-Lyne Linnaean Botanical Society . . . 83 p. 18.5 cm. Ashton-under-Lyne, 1888. (New York Botanical Garden library.)
Annotated list of vascular plants (p. 5–52), headed "The flora as compiled . . . ," followed by a list of mosses and liverworts; p. 81, Supplementary list, and list of casual plants. Covers an area in the southeast corner of Lancashire and adjacent Cheshire (Hyde, Marple), Derby (Glossop), and Yorkshire (Greenfield). — See also **Stubbs, F. J.,** The flora of Ashton-under-Lyne. Lancash. Nat. **6**: 285–287. 1913. (List of noteworthy angiosperms, with localities. Followed (p. 287–290) by additional records by Collier, Fred, and others. In University of Minnesota library.) — Also **[Collier, Fred].** [List of newly recorded species.] l.c. **6**: 252–253. 1913. — Also his Ashton-under-Lyne district natural survey. Lancash. and Chesh. Nat. **7**: 403. 1915. (University of Minnesota library.)

Widnes Secondary School. Field and camera club. Flora of Widnes and district. 1913 (?).
List of plants including 383 native species. (Not seen; cited from notice in Rpt. Bot. Exch. Club and Soc. Brit. Isles **3** (1913): 366. 1914.)

Williamson, Frank. Flora of the Rochdale district. Trans. Rochdale Lit. and Scient. Soc. **11**: 42–71. 1913. (Lloyd Library.)
Topography, geology, general features of flora; annotated list of vascular plants and Characeae (494 species); bibliography. Covers area of 74 square miles in ancient parish of Rochdale and region drained by the river Roch and its tributaries.

LEICESTERSHIRE

Vice-county 55, Leicestershire with Rutland.

General

Horwood, A. R., and Noel, C. W. F. The flora of Leicestershire and Rutland. A topographical, ecological, and historical account with biographies of former botanists (1620–1933). ccxcvii, 687 p. 36 pl. (incl. ports., 5 maps [2 in pocket]). 19.5 cm. London, 1933.
Topography, soils, agriculture, vegetation and altitude, geology, hydrography, botanical districts, woodlands, climate, vegetation types, botanical

explorations (with biographical notices and bibliographical references) for each county separately; herbaria consulted, chronological bibliography; extensively annotated list of about 1400 species and 583 varieties, forms, and hybrids of vascular plants and Characeae, with localities. Replaces Mott, F. T., and others. The flora of Leicestershire, including the cryptogams . . . On the basis of a ms. prepared in 1852 by the late Rev. W. H. Coleman . . . xxvi, 372 p. 2 maps. 22.5 cm. London and Edinburgh, 1886. (Annotated list of botanists, bibliography, botanical divisions; annotated list of vascular (884 species) and cellular plants, with localities; statistics, table showing distribution of each species. In Gray Herbarium library.)

Local

Bemrose, G. J. V. The adventive flora of Leicester and district. Trans. Leicester Lit. and Phil. Soc. **28:** 45–71. map. 1927.

Methods of introduction, etc.; annotated list of alien angiosperms within radius of 5 miles.

Bloxam, Andrew, and Babington, Churchill. Botany of Charnwood Forest. *In* Potter, T. R. The history and antiquities of Charnwood Forest . . . App. p. 35–62. 29 cm. London, 1842.

Annotated list of vascular and cellular plants.—See also **Jackson, A. B.** Charnwood Forest Rubi. Jour. Bot. **44:** 261–266. 1906. (Extensive list.)

Eller, Irvin. The flora of the Vale of Belvoir. *In* his The history of Belvoir Castle, from the Norman Conquest to the nineteenth century . . . p. 391–410. 23 cm. London, Grantham, 1841.

Briefly annotated list of vascular and some cellular plants, the families arranged alphabetically.

Mercer, G. E. The flora of Belgrave and Birstall. Considered from a floristic and ecological aspect. Trans. Leicester Lit. and Phil. Soc. **18:** 76–92. 1 fig. 1914.

Topography, geology, general features of flora; unannotated list of vascular plants (570 species and forms).

Preston, T. A. The flora of the Cropstone Reservoir. Trans. Leicester Lit. and Phil. Soc. **3:** 430–442. 1895.

Running account of changes in water level and plants observed, followed by list of spermatophytes seen. (About 5 miles northwest of city of Leicester.)

LINCOLNSHIRE

See also Great Britain, General (Baker, Additions . . . for Lincolnshire . . . under Druce, Comital flora of the British Isles); Yorkshire (Woodruffe-Peacock, The ecology of Thornton Waste).

Vice-county 53, South Lincolnshire; vice-county 54, North Lincolnshire.

General

Woodruffe-Peacock, E. A. A check-list of Lincolnshire plants. Trans. Lincolnsh. Nat. Union **2** (1909): 1–66. 1909.—Additions and corrections . . . l.c. **2** (1911): 290–299. 1912.

Alphabetical list of vascular plants, with indication of first records and list of divisions of the county in which they occur.—See also annual reports of the Botany section (title varies). Trans. Lincolnsh. Nat. Union from v. 3 (1912–15). 1913–6, on.

——— A critical catalogue of Lincolnshire plants: from all known sources. 1. [–13.] paper. Naturalist **1894:** 85–92, 133–140, 213–220, 257–264, 337–344. 1894; **1895:** 95–102. map. 1895; **1896:** 53–60, 181–188, 245–252, 309–316, 325–332. 1896; **1897:** 133–140, 167–172. 1897.

Sources, etc.; annotated list of known vascular plants and Characeae, showing local distribution. Replaces **Lees, F. A.** Botany and outline flora of Lincolnshire. 29 p. 24.5 cm. Sheffield, 1892. (General features of flora, partly annotated bibliography; list of vascular and cellular plants (1476 species, of which 966 are vascular plants), with date of first record, etc. Reprinted from White, William. White's history, gazetteer, and directory of Lincolnshire. Original not seen; reprint in British Museum (Natural History) library.)—See also **Woodruffe-Peacock, E. A.** The natural history divisions of Lincolnshire. Naturalist **1895:** 289–301. map. 1895. (List of divisions and subdivisions, soils, gazetteer.)

——— Lincolnshire folk names for plants. 30 p. 22.5 cm. n.p., 1894–97. (Issued as supplement to Lincolnshire notes and queries, 1894–97.) (Harvard College library.)

Sources; alphabetical list of vernacular names, with botanical identifications, localities, remarks on uses, etc. Includes Additions and corrections, p. 25–30.

Local

Woodruffe-Peacock, E. A. The East Fen. Trans. Lincolnsh. Nat. Union 3 (1915): 228–236. 1916.

History, etc.; unannotated alphabetical list of vascular plants. The East Fen, containing 29,833 acres, was the last fenland to be drained in Lincolnshire.

London

The City of London (area 1 square mile) is in vice-county 21 (Middlesex).

With the exception of that by Lousley, which refers to the City of London, all the works here listed are general and relate to the "London area" or the "metropolitan area," whose limits have varied somewhat with different botanists but are now generally regarded as bounded by a circle with a radius of 20 miles about St. Paul's Church, thus including the City of London, all of Middlesex County (vice-county 21), and parts of Buckinghamshire (vice-county 24), Essex (vice-counties 18, 19), Hertfordshire (vice-county 20), Kent (vice-county 16, West Kent), and Surrey (vice-county 17)—essentially what may be called "Greater London." Publications on individual localities within this area are placed under the appropriate counties. The principal ones that might be looked for under London are Ardagh (Islington), Hampstead Scientific Society (Hampstead Heath), Warren (Hyde Park), and Webster (Regent's Park), listed under Middlesex; Currey (Greenwich area) and Webster (Greenwich Park), under Kent; Curtis (Flora Londinensis), under Great Britain, General; Groves (South London) under Surrey.

De Crespigny, E. C. A new London flora; or, handbook to the botanical localities of the metropolitan districts . . . xxiv, 179 p. 16.5 cm. London, 1877.

Bibliography, list of localities, etc.; annotated lists of flowering and of cryptogamous plants, alphabetically arranged by scientific names; list of special localities and their characteristic species. Covers region within radius of 30 miles of London. Replaces **Irvine, Alexander.** The London flora; containing a concise description of the phaenogamous British plants, which grow spontaneously in the vicinity of the metropolis, with their localities; arranged in conformity to the natural system . . . xvi, 340 p. 18.5 cm. London, 1838. (Annotated list of vascular plants, with localities and brief descriptions; statistics, list of British plants on the Linnaean system; additional localities, glossary. Covers region east of a line drawn from Southhampton north into Northamptonshire.)

Fitter, R. S. R. The hundred commonest London plants. *In* his London's natural history. p. 256–258. 22 cm. London, 1945.

Unannotated list, based on Bishop, of the most widely distributed plants in the London area.

Kent, D. H., and Lousley, J. E. A hand list of the plants of the London area compiled from the botanical records of the London Natural History Society. Flowering plants, ferns and stoneworts. London Nat. v. 30–36, suppl. 368 p. 10 pl. 1951–57.

Brief account of compilation of list, collaborators, etc.; annotated list of vascular plants and Characeae [1,835 species and 433 infraspecific taxa and hybrids], with localities and names of collectors. P. 334–360 constitute a supplement. "Not intended to be a 'Flora' of the London area, but remains, like its predecessor, primarily a statement of observations made by past and present members of the Society;" but it lists essentially all the species known from the area, except for many casual aliens found in the late 19th and early 20th centuries and apparently not seen since. It covers a radius of 20 miles around St. Paul's Cathedral, including the whole of Middlesex and parts of the counties of Buckingham, Essex, Hertford, Kent, and Surrey, as well as the administrative county of London which includes parts of four of these counties. Replaces **Bishop, E. B., Robbins, R. W.,** and **Spooner, Herman.** Botanical records of the London area, being a list of plants observed within 20 miles of St. Paul's Cathedral. 8 pts. London Nat. 1927–29, 1931–35, suppl. 114 p. 1927–35. (Geographical divisions; list of vascular plants and Characeae, with localities. Most records previous to 1900 (for part north of Thames) or 1914 (for part south of Thames) have been ignored.)—See also **Lousley, J. E.** Botanical records for 1943 [–1957]. London Nat. v. 24 (1944)—37 (1957), passim. 1944 [–58].—Also his Changes in the flora of the London area since 1858. l.c. 37: 34–49. [1958.] (Changes in habitats and flora; bibliography.)

Lousley, J. E. Wild flowers, grasses and ferns of the City of London, 1941–1952. *In* Fitter, R. S. R., and Lousley, J. E. The natural history of the City. App., p. 30–35. 22 cm. London, 1953.

Tabular list of vascular plants (269 species) with dates of occurrence and indication of frequency. Refers to the City proper, only 1 square mile in area.—See also his Wild flowers. l.c. p. 8–19. 6 fig. (Statistics, methods of introduction, garden plants, etc.); also bibliography (p. 36), which has particular reference to publications on the flora of bombed areas.

Melville, Ronald, and **Smith, R. L.** Adventive flora of the metropolitan area. I. Recent adventives on London rubbish. Rpt. Bot. Soc. and Exch. Club Brit. Isles 8 (1927): 444–454. pl. 1928.

Account of localities; list of about 250 angiosperms, with localities, in Essex and Middlesex.—See also Shenstone, below.

Peterken, J. H. G. Habitats of the London area. London Nat. **32:** 2–12. 1953.

Discussion of plant habitats with mention of their characteristic species.

Shenstone, J. C. The flora of London building-sites. Jour. Bot. **50:** 117–124. 1912.

Short lists of vascular plants found at various localities in London, grouped according to source of introduction, with discussion of methods of distribution.—See also **Salisbury, E. J.** The flora of bombed areas. Nature **151:** 462–466. 1943. (Also in North West Nat. **18:** 160–169. 1943.) (General discussion of development of floras in bombed areas, with special mention of various species and their methods of spreading.)—Also **Lousley, J. E. The pioneer flora of bombed sites in central London. Rpt. Bot. Soc. and Exch. Club Brit. Isles 12 (1941–42):** 528–531. 1944. (General considerations; annotated list of 27 angiosperms observed as pioneer colonists on 21 bombed sites in the City of London and vicinity.)—Also **Salisbury, E. J.** List of flowering plants and ferns recorded from bombed sites in London. *In* Fitter, R. S. R. London's natural history. p. 265–268. 22 cm. London, 1945. (List of 126 species, with indication of percentage of total localities the commoner species were found in.)—Also **Lousley, J. E.** The flora of bombed sites in the City of London in 1944. Rpt. Bot. Soc. and Exch. Club Brit. Isles **12** (1943–44): 875–883. 1946. (Annotated list of 113 vascular plants including 1 hybrid.)—Also **Jones, A. W.** The flora of the City of London bombed sites. London Nat. **37:** 189–210. map. [1958.] (Habitats, methods of dispersal, bibliography, etc.; contains lists totaling 342 species.)

Middlesex

See also Great Britain, General (Kent, Additions and corrections . . . under Druce, The comital flora); Hampshire (Boulger, Botanical bibliography of the south-eastern counties); London.

Vice-county 21.

General

Trimen, Henry, and **Dyer, W. T.** Flora of Middlesex: a topographical and historical account of the plants found in the county; with sketches of its physical geography and climate, and of the progress of Middlesex botany during the last three centuries. xli, 428 p. map. 19.5 cm. London, 1869.

Topography, geology, soils, climate, botanical districts, bibliography; annotated list of vascular plants (859 species), with localities; statistics of flora, botanical history, lists of cellular plants. This work provided a model for subsequent county floras of England.—See also the following short papers, by various authors: Jour. Bot. 8: 225. 1870; **9:** 113–114, 271–272, 304, 371. 1871; **11:** 208–209. 1873; **22:** 56, 279–280. 1884; **23:** 36–38, 338–340. 1885; **25:** 14–20, 363–366. 1887; **28:** 120. 1890; **31:** 218. 1893; **48:** 269–278. 1910; **49:** 66. 1911; **52:** 127–131. 1914.—Also **Nicholson, C. S.** The botany of the district. Trans. London Nat. Hist. Soc. **1915:** 40–43. 1916. (Includes lists of probably extinct and not recently recorded species.) —Also **Fitter, R. S. R.** The lost plants of Middlesex. *In* his London's natural history. p. 254–255. 22 cm. London, 1945. (List of 55 plants that have become extinct or nearly so in Middlesex within recent years; based on Salisbury, The waning flora of England, 1927.)—Also **Kent, D. H.** Notes on the flora of Middlesex, 1939–1946. London Nat. **26**: 56–65. 1947. (Extensive list of noteworthy vascular plants.)—Also **Avery, Charles,** and **Watson, W. C. R.** The brambles of Middlesex. l.c. **26**: 66–73. 1947. (Annotated list of species, with some critical notes; tabular list of additional species from nearby counties.)—Also **Kent, D. H.** Middlesex plant records for 1947 [–1948]. l.c. **27**: 29–35. 1948; **28**: 31–36. 1949. (Includes (v. 28) list of extinct or possibly extinct species.)—Also his Additions and corrections to the Comital flora, for Middlesex. [I]–II. Watsonia **1**: 92–101. 1949; **2**: 78–79. 1951. (Intended to bring records in Druce's Comital Flora up to date. The first paper includes bibliography.)

Local

Ardagh, John. Ballard's "Catalogue of Islington plants." (Bibliographical notes XCII.) Jour. Bot. **66**: 185–194. 1928.
List of angiosperms, with localities, from a manuscript prepared by Edward Ballard (1820–1897) in 1837–42.

Cockerell, T. D. A. The flora of Bedford Park, Chiswick. Jour. Bot. **25**: 107–110. 1887.
General features of flora; mostly unannotated list of vascular plants and a few cellular plants. (*In* Chiswick and Acton.)—See also Murray, below.

Hampstead Scientific Society. Hampstead Heath, its geology and natural history. 328 p. pls., map. 22.5 cm. London, 1913. (Kew library.)
Includes, under Plant life, the following chapters: i. **Tansley, A. G.** The vegetation of Hampstead Heath and the neighbouring woods. p. 83–111. pl. 3. (Plant associations, with partial lists of species.) ii. **Watt, H. B.** Hampstead trees and some shrubs. p. 113–143. pl. 4–6. (Descriptions of trees and shrubs, including some cultivated species, with a list alphabetically arranged by vernacular names.) iii. **Whitton, W. A.** Flowering plants. p. 145–174. (Running account of vascular plants; partial list of mosses.) Supplements but scarcely replaces **Wharton, H. T.** The flora of Hampstead. *In* Lobley, J. L. Hampstead Hill; its structure, materials, and sculpturing . . . with The flora of Hampstead . . . p. 73–80. 21.5 cm. London, 1889. (Unannotated list of spermatophytes (455 species). In New York Botanical Garden library.)—See also **Champneys, May.** A flower list of the Hampstead neighbourhood. 32 p. 18 cm. London, 1914. (Annotated list of angiosperms (221 species) the grasses, trees, and shrubs mostly omitted. In library of D. H. Kent.)

Harley, R. M. Flowers and ferns of Harrow. 5 p. l., 36 p. 3 fig., map. 21.5 cm. London, [pref. 1953].
Soils, etc.; annotated list of vascular plants of Harrow-on-the-Hill. Replaces **Horton, W. R. G.** Report on wild flowers. Ann. Rpt. Harrow Nat. Hist. Soc. no. **3**, pt. 1: 3–18. 1948. (Annotated list of angiosperms, intended as a complete flora of Harrow-on-the-Hill (i.e., vicinity of Harrow School). Processed publication; 33 cm. In library of D. H. Kent.)—See also Melvill, below.

Melvill, J. C. The flora of Harrow. new and enl. ed., rev. and edited by . . . W. M. Hind. viii, 176 p. map. 18.5 cm. London and Harrow, 1876. (Gray Herbarium library.) (1st ed. 1864.)
Annotated list of vascular plants, with localities (p. 1–129); also lists of mosses, birds, etc. Covers radius of about 10 miles around Harrow.—See also Harley, above.

Murray, Dion. The flora of Chiswick 1955 as observed by Dion Murray, and listed according to the recently published "Flora of the British Isles" by Messrs. Clapham, Tutin and Warburg. 5 [i.e. 8] p. 18.5 cm. London, 1956.
Systematic list. (Not seen; title and annotation supplied by D. H. Kent.) —See also Cockerell, above.

Warren, J. B. L. The flora of Hyde Park and Kensington Gardens. Jour. Bot. 9: 227–238. 1871.

Annotated list of about 190 vascular plants and Characeae.—See also his Kensington Gardens plants. Jour. Bot. **13:** 336. 1875. (Adds several species.)—Also **Kent, D. H.** Notes on the flora of Kensington Gardens and Hyde Park. Watsonia **1:** 296–300. 1950. (Changes in vegetation, etc.; annotated list of angiosperms observed by author.)

Webster, A. D. The flora of Regent's Park and Primrose Hill. *In* his The Regent's Park and Primrose Hill. History and antiquities. p. 99–102. 22.5 cm. London, 1911. (Yale University library.)
General features of flora, with mention of species that have disappeared and of some introduced ones; short list of trees, shrubs, and herbs.

MONMOUTHSHIRE

Although actually in England, the county is for some administrative purposes counted as in Wales.
Vice-county 35.

General

Clark, J. H. List of the flowering plants of Monmouthshire. Collected by J. H. Clark. *In* his Sketches of Monmouthshire. lxiv p. 17 cm. Usk, 1868. (Library of N. D. Simpson. Reprinted with preface, 2 p.l., lxiv p.; in library of N. D. Simpson.)
Annotated list of vascular plants, with localities for rarer species.—See also **Shoolbred, W. A.** Recent additions to the flora of West Gloucester and Monmouth plants. Jour. Bot. **32:** 263–271. 1894.—Also his West Gloucester and Monmouth plants. l.c. **32:** 311. 1894.—Also his Plants of West Gloucester and Monmouth. l.c. **34:** 29–30. 1896.

Hamilton, S. The flora of Monmouthshire. Being a catalogue of all the flowering plants known to be native to the county; and a guide to localities, together with list of ferns. 2 p. l., 81, xi p. 19 cm. Newport, 1909. (Kew library.)
Geology, etc.; briefly annotated list of vascular plants and Characeae, with localities. According to Hyde and Wade, the list is very incomplete.

Local

Matthews, L. H. Denny Isle. Ann. Rpt. and Proc. Bristol Nat. Soc. IV, 7 (1932): 371–378. pl. 1933.
Includes annotated list of 14 phanerogams; also zoological lists; bibliography. (A limestone rock off Portishead, in Bristol Channel.)

Shoolbred, W. A. The flora of Chepstow. 1 p. l., x, 140 p. map. 22.5 cm. London, 1920. (Gray Herbarium library.)
Topography, geology, climate, botanical explorations, etc.; annotated list of vascular (1004 species) and cellular plants. Covers region in Monmouth bounded by River Wye from Bigsweir Bridge to Beachley and extending 6 or 7 miles west, and adjacent area in Gloucestershire.

NORFOLK

Vice-county 27, East Norfolk; vice-county 28, West Norfolk.

General

Nicholson, W. A., *ed.* A flora of Norfolk, with papers on climate, soils, physiography, and plant distribution, by members of the Norfolk and Norwalk Naturalists' Society. 214 p. 2 maps. 22 cm. London, 1914. (Arnold Arboretum library.)
Climate, soils, topography (by various authors), plant statistics, botanical divisions, list of botanists, brief bibliography; annotated list of vascular plants and Characeae (by Nicholson), with localities; list of mosses and liverworts (by W. H. Burrell). Replaces **Trimmer, Kirby.** Flora of Norfolk: a catalogue of plants found in the county of Norfolk. xxxvi, 195 p. 18.5 cm. London, Norwich, 1866; Supplement . . . 73 p. 19 cm. London, [1885]. (Geology, etc.; annotated list of vascular plants, with localities. Both in Arnold Arboretum library.)—For additional records see papers by various authors, as follows: Jour. Bot. **55:** 191–193. 1917; **57:** 190–192. 1919. Rpt.

Bot. Soc. and Exch. Club Brit. Isles **7** (1923): 252–253. 1924. Trans. Norfolk and Norwich Nat. Soc. **11**: 601–608. 1924; **12**: 477–481, 720–721. 1928–1929.—Also **Clarke, W. G.** Norfolk topography in "The botanist's guide." l.c. **11**: 179–193. 1921. (Changes in topography and vegetation since 1805.) —Also **Nicholson, W. A.,** and **Ellis, E. A.** The botany of Norfolk. Rpt. Brit. Assoc. Adv. Sci. **105** (1935) (App.): 24–34. 1935. (Types of vegetation; bibliography for vascular and cellular plants.)

Swann, E. L. An annotated list of Norfolk vascular plants. Trans. Norfolk and Norwich Nat. Soc. **18**: 3–54. 1955. (Yale University library.)
List of known vascular plants with notes on many of the rarer or more interesting species; list of casual species. "The year 1900 has been taken as an arbitrary starting point for the inclusion of Norfolk plant records" (p. 1).

Local

Babington, C. C. A list of plants observed near Cromer in the months of July, August, and September in 1875 and 1876. Trans. Norfolk and Norwich Nat. Soc. **2**: 380–387. 1878.
Unannotated list of vascular plants and Characeae. — See also **Pigott, B. A. F.** Flowers and ferns of Cromer and its neighbourhood. 1 p.l., 99 p. 18 cm. London [1885]. (Popular account of general features of flora of several localities, followed (p. 83–99) by systematic index of vascular plants and fungi mentioned. (In British Museum (Natural History) library; reissue, 1898, in Gray Herbarium library.)

Burrell, W. H., and **Clarke, W. G.** The fauna and flora of Flordon Common. Trans. Norfolk and Norwich Nat. Soc. **9**: 170–186. 1 fig. (map). 1911.
Includes list of 367 plants, arranged by habitats. Flordon is a village about 7 miles southwest of Norwich.

Deighton, F. C., and **Clapham, A. R.** The vegetation of Scolt Head Island, a preliminary account. Trans. Norfolk and Norwich Nat. Soc. **12**: 86–111. map. 1926.
Geology, plant associations; unannotated list of vascular plants. — See also **Chapman, V. J.** Recent additions to Scolt Head flora. l.c. **13**: 296–302. 1933. —Also his Botany. *In* Steers, J. A. Scolt Head Island. Report for 1933 [–1934]. l.c. **13**: 330–332, 418–419. 1934–35.

Ellis, E. A. Wheatfen Broad, Surlingham. Trans. Norfolk and Norwich Nat. Soc. **13**: 422–451. pl. 2 (map). 1935.
Includes (p. 426–439) tabular list of vascular plants of the Yare Valley swamp.

Galpin, F. W. An account of the flowering plants, ferns and allies of Harleston. With a sketch of the geology, climate, and natural characteristics of the neighbourhood . . . 157 p. front. 22 cm. London and Harleston, 1888. (Gray Herbarium library.)
Geology, botanical investigations, general features of flora, bibliography; annotated list of vascular plants and Characeae. Pages 1–129, 151–157 (index) are devoted to plants, 131–150 to birds (by Charles Candler).

Mann, R. J. The flora of central Norfolk. Mag. Nat. Hist. (ed. Charlesworth) n.s., **4**: 390–407. 1840.
Geology, topography; briefly annotated list of vascular plants and mosses of the region within "an ordinary day's march" of the city of Norwich, with localities for rarer species. — See also **Woodward, S. P.** Flora of central Norfolk. Addenda to Mr. Mann's list of Norwich plants. Annals and Mag. Nat. Hist. **7**: 201–206. 1841. (List of additional vascular plants and bryophytes [bringing total to 708 vascular plants].)

Moxon, J. E. A flora of the neighbourhood of Sandringham, Norfolk. Phytologist **1**: 596–601, 630–635. 1843.
Soils, etc.; briefly annotated list of vascular plants (506 species).

Munford, George. A list of flowering plants found growing wild in western Norfolk. Annals and Mag. Nat. Hist. **8**: 171–191. 1841. (Reprinted 23 p. 1841.)
Geology; annotated list of 722 phanerogams.

Paget, C. J., and Paget, James. Sketch of the natural history of Yarmouth and its neighbourhood, containing catalogues of the species of animals, birds, reptiles, fish, insects, and plants, at present known. 1 p. l., xxxii, 88 p. 22.5 cm. London, 1834. (Kew library.)

Includes (p. 45–88) annotated list of vascular [750 species] and cellular plants. — See also **Lowne, B. T.** List of rare and interesting plants found in the neighbourhood of Great Yarmouth. *In* his A popular natural history of Great Yarmouth and its neighbourhood . . . p. 43–46. 18 cm. Great Yarmouth, London, and Norwich, 1863. (List of vascular plants, with habitats. In Kew library.)

Salisbury, E. J. Vegetation. *In* Oliver, F. W., and Salisbury, E. J. The topography and vegetation of the National Trust Reserve known as Blakeney Point, Norfolk. Trans. Norfolk and Norwich Nat. Soc. **9** (1912–13): 501–542. fig. 9 (map), pl. 1. 1913.

Mainly ecological; includes (p. 509–511) alphabetical list of 60 phanerogams known from Blakeney main bank.—See also in **Oliver, F. W.** Report of Blakeney Point Research Station, 1927–29. l.c. **12** (1928–29): 630–631. 1929. (List of newly recorded plants.)

Simon, P. H. Flora of Holt and district. Rpt. Greshman's School Nat. Hist. Soc. **26**: 11–16. 1948. (Kew library.)

Unannotated list of vascular plants. — See also additional list by **McClelland, G. A. H.** l.c. **26**: 16–17. 1928. — Also **Simon, P. H.** Botanical section. l.c. **27**: 9–13. 1949; **28**: 9–11. 1950. (Lists including additions. In Kew library.)

Northamptonshire

Vice-county 32.

General

Druce, G. C. The flora of Northamptonshire, being a topographical and historical account of the flowering plants and ferns found in the county . . . cxlii, 304 p. 2 pl. 22 cm. Arbroath, 1930.

Topography, geology, botanical districts, biographical notices of local botanists, comparisons with other counties, bibliography; annotated list of vascular plants and Characeae [974 species]. — See also **Bishop, E. B.** A few additions to the flora of Northamptonshire. (In memoriam—George Claridge Druce.) Rpt. Bot. Soc. and Exch. Club Brit. Isles **10** (1933): 571–582. 1934. (Extensive list; refers mostly to vicinity of Bainton.)

Local

Anonymous. List of species in the herbarium of the Oundle School Natural History Society. Ann. Rpt. Oundle School Nat. Hist. Soc. **1946**: 24–59. [1946?]

Annotated list of vascular plants (642 species), mostly from within 5 miles of Oundle, but in part from adjacent Huntingdonshire. Replaces an earlier list, entitled Flora. l.c. **1938**: 15–29. [1938?]—See also **Anonymous.** Plants. l.c. **1948**: 20. [1948?]

Notcutt, W. L. Catalogue of plants observed in the neighbourhood of Daventry, Northamptonshire. Phytologist **1**: 500–508. 1843.

Geology, etc.; annotated list of vascular plants (423 species).

Paley, F. A. A list of four hundred wild flowering plants being a contribution to the flora of Peterborough, with an introduction. xxviii, 26 p. 15.5 cm. Peterborough, 1860. (Gray Herbarium library.)

Includes (p. 1–26) list of spermatophytes (lacking *Carex*, Gramineae, etc.), with occasional localities.

Northumberland

See also Cumberland (Harrison and Temperley, The flora of the three northern counties), Durham (Harrison, Flora, under Potter; Hogg, On the ballast-flora of the coasts of Durham and Northumberland).

Vice-county 67, South Northumberland; vice-county 68, North Northumberland or Cheviotland.

General

Baker, J. G., and Tate, G. R. A new flora of Northumberland and Durham, with sketches of its climate and physical geography . . . Nat. Hist. Trans. Northumberl. and Durham v. 2. 316 p. 2 maps. 1868.

Geology (by George Tate), climate, topography, bibliography; annotated list of vascular plants (1139 species), with localities; list of ballast plants. Replaces **Winch, N. J.** Flora of Northumberland and Durham. Trans. Nat. Hist. Soc. Northumberl., Durham, and Newcastle upon Tyne 2: 1–149. 1838. (Preprinted, 1 p. l., 149 p. Newcastle, 1831 [1832]. In New York Botanical Garden library); Addenda to the Flora . . . p. 151–159. 28.5 cm. [Newcastle-on-Tyne, 1837.] (In New York Botanical Garden library.) (Annotated list of vascular plants, with localities.) — See also **Hull, J. E.** "Baker and Tate." Some corrections, topographical and typographical. Vasculum 22: 131–135. 1936.—Also **Temperley, G. W.** "Baker and Tate." Some further corrections and notes. l.c. 23: 8–13. 1937.

Local

Johnston, George. A flora of Berwick-upon-Tweed. 2 v. (xxiv, 242; 2 p.l., 335 p.). 8 pl. 18 cm. Edinburgh and London, 1829–31.

Topography, geology, etc.; descriptive flora, on the Linnaean system, with localities and notes on uses, popular beliefs, etc. Vol. 1, phaenogamous plants; v. 2, cryptogamous plants. Covers region from northern boundary of Northumberland south to Buddle and Belford, west to Rivers Till and Tweed, and takes in also the plants of Fern (Farne) Islands, Bamborough Castle, and Cheviot.

Lawrence, C. J. Plants of the district. *In* Robertson, W. Ford. Walks from Wooler. p. 49–66. 19 cm. Newcastle-upon-Tyne, 1926. (Library of N. D. Simpson.)

General features of flora with mention of numerous species, grouped by habitats or localities.

Luckley, J. L. Botanical rambles. A flora of Alnwick. 112 p. 22.5 cm. Alnwick, [1893]. (Library of N. D. Simpson.)

Includes systematic list of vascular plants (and a few mosses) with localities, interspersed with accounts of botanical rambles, etc. — See also **Hull, J. E.** Luckley's Flora of Alnwick. Vasculum 15: 20–26. 1929. (Account of Luckley and his Flora of Alnwick, with mention of the noteworthy species.) — Also Tate, Flora of Alnwick, below.

[Richardson, William.] The plants of Holy Island. Phytologist n.s., 6: 10–15. 1862.

Topography, etc.; running account of noteworthy plants collected. The author's name is given as W. R. Alnwick, a pseudonym. — See also **Johnston, George.** Our visit to Holy Island in May, 1854. Hist. Berwicksh. Nat. Club [7] 1873/75: 27–52. 1876. (Includes (p. 49–52) unannotated list of vascular and cellular plants.) — Also **Marsden-Jones, E. M.** Holy Island plants. Jour. Bot. 56: 143–145. 1918. (List of vascular plants observed in vicinity of Trearddur Bay in August 1917.) — Also **Aiken, J.J.M.L.** Appendix. (Meeting at Holy Island — 1st July 1921). Hist. Berwicksh. Nat. Club 24 (1921): 274–275. 1923. (Mention of various species.)

Tate, George. The Farne Islands, with an account of their geology, botany, zoology, and ancient history. Hist. Berwicksh. Nat. Club [3] 1850–56: 222–252. pl. 1857.

Includes (p. 235–236) brief notes on vegetation, and list of 62 angiosperms, with localities. The Farne Islands are a group of 15 to 25 islets 1.5 to 5 miles east of the Northumberland coast, nearest Monkshouse. — See also **Watt, Grace.** Farne Island plants. *In* her The Farne Islands; their history and wild life. Appendix B. p. 200–202. 22 cm. London, 1951. (Unannotated list of present-day Farne Island plants (compiled from specimens collected by T. R. Goddard) and list of plants which have disappeared since 1857 (based on Tate's list).)

Tate, G. R. Flora of Alnwick. Flowering plants, ferns, and their allies. *In* Tate, George. The history of the borough, castle and barony of Alnwick. v. 2, p. 418–435. 26.5 cm. Alnwick, 1868–69.

Annotated list of vascular plants (764 taxa, including 732 species), followed by list of Musci and Hepaticae by Robert Middlemas. — See also Luckley, above.

Nottinghamshire

Vice-county 56.

General

Carr, J. W. Botany. *In* The Victoria history of the county of Nottingham. v. 1, p. 41–74. map. 31.5 cm. London, 1906.

Botanical explorations, general features of flora, botanical districts; list of known vascular plants (854 wild species, plus 130 alien and casual plants), with distribution indicated by districts. — See also his Notes on Nottinghamshire botany. Jour. Bot. **47**: 134–140. 1909. (List of vascular plants not contained in Watson's Topographical botany or in the early county floras, with localities.) — Also his The botany of Nottinghamshire. *In* A scientific survey of Nottingham and district. Rpt. British Assoc. Adv. Sci. **107** (1937) (App.): 66–75. 1937. (Sketch of botanical explorations; lists of extinct, rare, and recently discovered species; botanical districts.)

Howitt, Godfrey. The Nottinghamshire flora; containing the flowering plants, ferns, mosses, hepaticae, lichens, Characeae and algae. 1 p. l., 124 p. map. 17 cm. London [etc.], 1839. (Kew library.)

Annotated list of vascular and cellular plants, with localities for the scarcer species. Replaces **Ordoyno, Thomas.** Flora nottinghamiensis, or a systematic arrangement of the plants, growing naturally in the county of Nottingham . . . 7, v, 344 p. 19.5 cm. Newark, London, 1807. (Annotated, briefly descriptive flora of vascular plants and Characeae, on the Linnaean system.)

Local

Bohler, John. The flora of Sherwood Forest. *In* White, Robert. Worksop, "The Dukery," and Sherwood Forest. p. 303–326. 21 cm. Worksop, 1875. (Library of N. D. Simpson.)

Sketch of Bohler's life; briefly annotated list of vascular plants, bryophytes, lichens, and fungi.

Oxfordshire

Vice-county 23.

General

Druce, G. C. The flora of Oxfordshire, a topographical and historical account of the flowering plants and ferns found in the county . . . 2. ed. (rewritten). cxxxi, 538 p. 20.5 cm. Oxford, 1927. (1st ed. 1886.)

Topography, soils, geology, botanical districts, meteorology (by F. A. Bellamy), sketches of local botanists, bibliography; annotated list of vascular plants and Characeae (2379 forms, of which 1061 are native or naturalized species); general features of flora. — See also **Brenan, J. P. M.** Notes on the flora of Oxfordshire and Berkshire. Rpt. Bot. Soc. and Exch. Club Brit. Isles **12** (1943–44): 781–802. 1946.

Local

Beesley, Thomas. The botany of the neighbourhood of Banbury, with a sketch of the geology. *In* Beesley, Alfred. The history of Banbury . . . p. 571–599. 22 cm. London [1842].

Soils, geology; briefly annotated list of vascular plants (534 species, including those reported by Gulliver; p. 574–592), also cellular plants (p. 592–599). See also Addenda, p. 626–627. Replaces **Gulliver, George.** A catalogue of plants collected in the neighbourhood of Banbury. viii, 37 p. 18 cm. London, 1841. (List of vascular and cellular plants, with localities, arranged mainly by the Linnaean system.)

[Holmes, J. E.] Wild flowers collected in Peppard and the neighbourhood by J. E. Holmes in 1907 and 1908. 17 p. 20 cm. n.p., [1909]. (Kew library.)

List of 474 vascular plants, with localities.

Rutland

See Leicestershire (Horwood and Noel, The flora of Leicestershire and Rutland). Vice-county 55 in part (includes also Leicestershire).

Shropshire

Vice-county 40.

General

Hamilton, W. P., *ed.* Botany. *In* The Victorian history of Shropshire. v. 1, p. 51–98. map. 31.5 cm. London, 1908.

Botanical explorations (with titles of publications), geology, botanical districts; list of known vascular plants (1075 species), with indication of distribution by districts, followed by lists of cellular plants; by various authors. The list of Phanerogamia and Pteridophyta (p. 61–73) is by **R. de G. Benson.**

Leighton, W. A. A flora of Shropshire. xii, 573 p. pls. 22 cm. London, 1841.

Annotated descriptive flora of spermatophytes [876 species], on the Linnaean system, with localities; gazetteer. — See also **Beckwith, W. E.** Notes on Shropshire plants. Jour. Bot. **19**: 48–51, 106–112, 143–149. 1881; **20**: 342–346, 363–368. 1882; **39**: 182–186. 1901. — Also **Bennett, Arthur.** Notes on the flora of Shropshire. l.c. **36**: 380–381. 1898. — Also a long series of records under heading Botany, in Caradoc Rec. of Bare Facts for 1891 [–1893] (v. 1–3). 1892–94; Caradoc and Severn Valley Field Club Rec. of Bare Facts v. 4–52 (except v. 44, 46). 1895–1945. (Not seen; data supplied by G. Taylor.)

Lloyd, L. C., and Rutter, E. M. Handlist of the Shropshire flora. xii, 110 p. map. Shrewsbury, 1957.

Botanical divisions; briefly annotated list of vascular plants, with localities for rarer species.

Local

Benson, R. de G. A flora of the Church Stretton district. Phanerogams. *In* Campbell-Hyslop, C. W., and Cobbold, E. S. Church Stretton. v. 2, p. 53–135. 20 cm. Shrewsbury, 1904.

List of local botanists, geology; annotated list of vascular plants and Characeae.

Diamond, T. P. A flora of Oswestry and district. Containing a list of plants of the neighbourhood, arranged according to their natural orders. With an English-Welsh index. iv, 58 p. 16 cm. Oswestry, 1891. (British Museum (Natural History) library.)

List of vascular plants (about 700 species), with English names and localities; list of English vernacular names, with corresponding Welsh names, the latter (according to letter from the author, in the British Museum copy) collected from the "Llysienaeth Cymraeg" [of Davies] and Dr. Pughe's Dictionary.

Phillips, William. A list of the flowering plants and ferns within a radius of five miles round Shrewsbury. *In* Pidgeon, Henry. An historical and illustrated handbook for the town of Shrewsbury . . . p. 170–193. 16 cm. Shrewsbury, 1878. (Library of N. D. Simpson.)

List of vascular plants, with localities. The same paper (presumably) was republished (with same title) in Phillips, William, and others. A guide to the botany, ornithology and geology of Shrewsbury and its vicinity. p. 5–26. 15 cm. Shrewsbury, 1878. (Not seen; title supplied by George Taylor.)

Somersetshire

See also Devonshire (Martin and Fraser, Flora of Devon); also Gloucester (White, The flora of Bristol).
Vice-county 5, South Somerset; vice-county 6, North Somerset.

General

Macmillan, A. S. Popular names of flowers, fruits, etc., as used in the county of Somerset and the adjacent parts of Devon, Dorset and Wilts . . . Reprinted from the Somerset County Herald. 1 p.l., 297 p. 19 cm. Yeovil, 1922. (Kew library.)

Alphabetical list of vernacular names, with scientific names, notes on uses, popular beliefs, etc.

Murray, R. P. The flora of Somerset. lxi, 437 p. map, front. 23 cm. Taunton, 1896. (Originally published as supplement to Proc. Somersetsh. Archaeol. and Nat. Hist. Soc. v. 39–42. 1893–96, the introduction being the last part published.)

Topography, climate, bibliography, geology, botanical districts, table showing local distribution; annotated list of vascular plants and Characeae (1079 species, of which 962 are native), with localities. — See also **Marshall, E. S.** A supplement to the flora of Somerset. iv, 242 p. 23 cm. Taunton, 1914. (Extensive list of additions and corrections. Published as supplement to Proc. Somersetsh. Archaeol. and Nat. Hist. Soc. v. 59.) — Also his Notes on Somerset plants for 1914 [–18]. Jour. Bot. 53: 122–132. 1915; **54:** 97–105. 1916; 55: 179–191. 1917; **56:** 136–143. 1918; **57:** 147–154, 175–181. 1919. (Title varies.) — Also annual lists of new records and new publications under the heading "The botanical section" in Proc. Somersetsh. Archaeol. and Nat. Hist. Soc. v. 60–88, 91–97. 1915–1943, 1946–1953. — Also **Miller, W. D.** A note on extinct and rare species of the county of Somerset. Rpt. Bot. Soc. and Exch. Club Brit. Isles **10** (1932): 268–276. 1933.

Watson, Walter. The average times of first flowering for Somerset plants. Proc. Somersetsh. Archaeol. and Nat. Hist. Soc. **93:** 108–128. 1949.

Tabular systematic list of spermatophytes showing average date of first flowering, months in which each species flowers, and month of optimum flowering.

Local

Babington, C. C. Flora bathoniensis: or, a catalogue of the plants indigenous to the vicinity of Bath. vi, 62 p. map. 18 cm. Bath, 1834. — Supplement . . . lxiii–105 p. map. 18 cm. [Bath, 1839.] (Gray Herbarium library.)

Briefly annotated list of vascular plants and Characeae; includes some records from adjoining Gloucestershire and Wiltshire. — See also **Jenyns, Leonard.** On the Bath flora. Proc. Bath. Nat. Hist. and Antiq. Field Club **1** (1865–66): 23–60. 1869 (1867?) (Includes (p. 54–60) lists of extinct species and of additions to Babington's Flora.) — Also **Wheatcroft, W. G.** The botany of the Bath district. *In* Morris, J. W., *ed.* Handbook to Bath prepared on the occasion of the visit of the British Association, 1888. p. 241–254. 18 cm. [1888]. (Botanical explorations and publications, etc.; unannotated list of rarer vascular plants; list of additions since Babington's Flora, with localities; annotated list of extinct or erroneously recorded species; the two last lists taken from Jenyns.)

St. Brody, Gustavus. The flora of Weston, and its immediate neighbourhood, including the habitats of Brean-Down, Uphill, Hutton Wood, Worle or Weston Hill, Sand-Point, and Birnbeck Island: with critical remarks on our doubtful species, and botanical memoranda of Steep Holm, Shapwick Moor, and Cheddar Cliffs . . . viii, 174 p. 19 cm. Weston-super-Mare, 1856. (Lloyd Library.)

Briefly descriptive flora of spermatophytes, the species arranged according to months in which they flower; annotated list of ferns. The work contains many errors, according to Murray.

Skene, Macgregor. Botany. *In* A survey of Steep Holm. Proc. Bristol Nat. Soc. IV, 8: 452–459. 1939.

Botanical explorations; list of known vascular plants (254 species and varieties), bryophytes, lichens, and fungi, with references to first records;

general features of flora; bibliography. Replaces **McLean, R. C.,** and **Hyde, H. A.** The vegetation of Steep Holm. Jour. Bot. **62:** 167–175. 1924. (Botanical explorations, geology, general features of flora; unannotated list of vascular plants (220 species) and of bryophytes and lichens.) (Small limestone island in Bristol Channel, off Weston-super-Mare.)

Thompson, H. S. The flora of Steart Island. Jour. Bot. **62:** 1–7. 1924. General account, with included mention of species observed by author. (At mouth of Parrett River.)

Wotton, F. W. A short historical account of the Flat Holme and its natural history. Rpt. and Trans. Cardiff Nat. Soc. **22:** 105–111. 1891. Includes "List of plants collected . . . by T. H. Thomas and John Storrie, June, 1890" (p. 109–111) a barely annotated list of vascular plants, lichens, and fungi. The list is reprinted in Jour. Bot. **29:** 345. 1891, with the omission of some cultivated species. (Island in Bristol Channel off Penarth, near Steep Holm.)

STAFFORDSHIRE

See also Great Britain, General (Edees, Staffordshire additions . . . under Druce, The comital flora of the British Isles); Warwickshire (Bagnall, Flowering plants and ferns indigenous to the neighbourhood around Birmingham).
Vice-county 39.

General

Bagnall, J. E. Botany. *In* The Victoria history of the county of Stafford. v. 1, p. 41–75. map. 31.5 cm. London, 1908.
Typography, general features of flora, botanical regions with lists of rarer species; list of known vascular plants and Characeae (954 species, including addenda), with distribution indicated by districts; also lists of cellular plants.

——— The flora of Staffordshire. Jour. Bot. v. 39, suppl. 74 p. 1901. Botanical districts, bibliography; annotated list of vascular plants and Characeae (1164 taxa, including 882 native species), with localities; sketches of local botanists.—See also **Edees, E. S.** Notes on Staffordshire plants. North West. Nat. **19:** 272–277. 1945. (Critical notes.)

Edees, E. S. The early history of field botany in Staffordshire, 1597–1839. Trans. North Staffordsh. Field Club **82:** 81–110. 1948.
Chronological account of botanical explorations and publications; bibliography.

Local

Anonymous. The flora of Burton-on-Trent and neighbourhood. (Compiled by the Botanical section of the Burton-on-Trent Natural History and Archaeological Society.) Trans. Burton-on-Trent Nat. Hist. and Archaeol. Soc. **3:** 177–190, 269–282. 1896–97; **4:** 75–88, 117–148. 1899–1901.—Errata. l.c. **5:** 67. 1903. (All in British Museum (Natural History) library.)
List of vascular plants and Characeae within 10 miles of Burton-on-Trent, with localities.—See also Brown, below, and Curtis, below.

Brown, Edwin. The flora of the district surrounding Tutbury and Burton-on-Trent . . . *In* Mosley, Oswald. The natural history of Tutbury . . . p. 231–364. col. pl. 27 cm. London, 1863. (Arnold Arboretum library.)
Annotated list of vascular and cellular plants, found mostly within 10 miles of the localities named. See also Addenda et corrigenda, p. 407–408.—Also Anonymous, above, and Curtis, below.

Clifford, Sir Thomas, and **Clifford, Arthur.** Flora tixalliana; or, a catalogue of the most remarkable phaenogamous plants, to be found within a morning's ride of Tixall. *In* their A topographical and historical description of the parish of Tixall, in the county of Stafford. p. 285–308. 26 cm. Paris, 1817 (1818).
Annotated list of spermatophytes on the Linnaean system. According to Pritzel, who quoted from another bibliographer, this was reprinted with title Flora tixalliana. Flore des environs de Tixall. 4to. Paris, 1818.

Curtis, Roger. Adventive flora of Burton-upon-Trent. Rpt. Bot. Soc. and Exch. Club Brit. Isles 9 (1930): 465–469. 1931.

Unannotated list of adventive phanerogams [188 species], with indication of native country.—See also **Burges, R. C. L.** Adventive flora . . . l.c. **12** (1943–44): 815–819. 1946. (Notes on localities, etc.; list of additional species.)

Painter, W. H. List of plants seen within five miles of Biddulph church . . . 1885–92: with which is incorporated a list of plants seen by Mr. Clifford Moore in the neighbourhood of Stafford, King's Bromley, etc. Ann. Rpt. and Trans. North Staffordsh. Nat. Field Club **31**: 74–100. [1897.]

Botanical districts, etc.; briefly annotated list of vascular plants and Characeae (388 species, plus) with localities.

Reader, H. P. The flora of Hawkesyard. Trans. North Staffordsh. Field Club **57**: 105–117. 1923; **60**: 118–132. 1926.

List of vascular and cellular plants observed by author within a radius of 5 miles about Hawkesyard Priory, with localities.

Ridge, W. T. B. The North Staffordshire flora. Trans. North Staffordsh. Field Club v. 56–63, appendix. 173 p. map (p. 158). 1922–29.

Annotated list of vascular plants and Characeae (1105 taxa including 955 species) growing in Staffordshire north of Stafford, with localities and references; list of doubtful species, geology, botanical provinces, climate, bibliography, annotated list of local botanists, index.—See also his Botany. Trans. North Staffordsh. Field Club v. 64–75, passim. 1930–41.—Also **Edees, E. S.** Some additions to the flora of North Staffordshire. l.c. **73**: 49–53. 1939.—Also his Botany. l.c. v. 76–91, passim. 1942–58.

Suffolk

Vice-county 25, East Suffolk; vice-county 26, West Suffolk.

General

Hind, W. M., and **Babington, Churchill.** The flora of Suffolk; a topographical enumeration of the plants of the county, showing the results of former observations and of the most recent researches. 1 p.l., xxxiv, 508 p. map. 19.5 cm. London, 1889.

Topography, geology and climate (all by Wheelton Hind), local distribution, bibliography, list of local botanists; annotated list of vascular plants and Characeae (1339 species in Suffolk and the 3 adjacent counties), with localities; list of fossil plants, tabular view of distribution, history of botanical investigations. Replaces **Henslow, J. S.**, and **Skepper, Edmund.** Flora of Suffolk . . . x, 140 p. 19.5 cm. London, Bury St. Edmund's, [pref. 1860]. (Annotated list of vascular and cellular plants with localities for scarcer species. In Arnold Arboretum library.)—See also **Bennett, Arthur.** Notes on the flora of Suffolk. Jour. Bot. **30**: 8–10. 1892.—Also **Salmon, C. E.** Notes upon Hind's Flora of Suffolk. l.c. **45**: 388–393. 1907.—Also **Simpson, F. W.** The missing, doubtful, new or otherwise interesting flora of the county. Trans. Suffolk Nat. Soc. **3**: 9–15. [1935.] (Lists of extinct, rare, and previously unrecorded species, grouped by habitats.)—Also his Gleanings from a botanist's diary. l.c. **3**: 129–131. [1936.] (Includes some new records.)—Also his Plant records and additions and corrections to the Flora of Suffolk. l.c. **8**: 12–22. [1952.]—Also his New and interesting plants recorded by W. A. Dutt (1870–1939) of Carlton Colville and Lowestoft. l.c. **8**: 136–147. [1954.] (Extensive list of noteworthy vascular plants, mostly from vicinity of Lowestoft and Lothingland.)—Also his Notes and additions to the flora of Suffolk. l.c. **8**: 188–192. [1954.]—Also papers by him, l.c. **9**: 38–43, 249–252. [1955–56]; **10**: 282–284. [1958.]

Salmon, C. E. Botany. *In* The Victoria history of the county of Suffolk. v. 1, p. 47–69, 84. map. 31.5 cm. London, 1911.

Geology, topography, general features of flora, botanical explorations, botanical districts; list of known vascular plants, with partial indication of distribution. Followed (p. 69–84) by lists of cellular plants by other authors.

Local

Atkins, James. Wild flowers of the Ipswich district. Jour. Ipswich and Dist. Field Club **2**: 14–29. 1959. (American Museum of Natural History library.)

Tabular list of angiosperms collected by author, with local vernacular name, date of earliest blooming, habitat, and locality; refers mostly to region within 7 miles of Ipswich. The sedges, grasses, and rushes are omitted. —See also **Batchelder, S. J.** Wild plants growing on waste ground in Ipswich. Trans. Suffolk Nat. Soc. 7: 9–10. 1949. (List of mostly adventive plants.)

Bunbury, C. J. F. List of flowering plants and ferns, found in the parish of Mildenhall, Suffolk. *In* his Arboretum notes. Notes on the trees and shrubs cultivated at Barton. (Cover title, Botanical notes at Barton and Mildenhall, Suffolk.) p. 237-285. 22 cm. Mildenhall, 1889. (Arnold Aboretum library.)
Annoted list of vascular plants (439 species).

——— Notes on wild plants found in the parish of Great Barton. *In* his Arboretum notes. Notes on the trees and shrubs cultivated at Barton. (Cover title, Botanical notes at Barton and Mildenhall, Suffolk.) p. 201–235. 22 cm. Mildenhall, 1889. (Arnold Arboretum library.)
Mention of noteworthy species; annotated list of angiosperms (349 species).

C[rabbe], G[eorge]. A catalogue of plants growing in, and near the parish of Framlingham. *In* Hawes, Robert. The history of Framlingham, in the county of Suffolk . . . p. 449–453. 27 cm. Woodbridge, 1798. (Harvard College library.)
Annotated alphabetical list of vascular plants, the most common species omitted.

Haward, H. List of the flowering plants, except trees, grasses, rushes, and sedges, collected in Wherstead by Mr. H. Haward, and supplied by him. *In* Zincke, Foster B. Wherstead: some materials for its history, territorial, manorial, and during the events between. 2. ed. greatly enl. Appendix D. p. 397–405. 22 cm. London, Ipswich, 1893. (New York Public Library.) (1st ed. 1887.)
Unannotated list of spermatophytes, lacking the groups mentioned in the title.

Mayfield, A. The flora of a Suffolk parish. Jour. Ipswich and Dist. Field Club 3: 18–28. 1911. (American Museum of Natural History library.)
Soils, etc.; list of spermatophytes of Mendelsham parish, with local vernacular names and average date of first flowering.

Mitchell, N. S. P. Flora of three-and-a-half miles of A.45. Trans. Suffolk Nat. Soc. 10: 119–122. 1957.
List of plants observed along road from Stowmarket to Needham Market. (Not seen.)

Surrey

See also London.
Vice-county 17.

General

Salmon, C. E. Flora of Surrey, being an account of the flowering plants, ferns and Characeae, with notes on the topography, climate and geology, and a history of the botanical investigations of the county . . . Edited by William Harrison Pearsall. 688 p. 8 pl., port., map. 22 cm. London, 1931.
Topography, climate, geology, botanical explorations, botanical districts, bibliography; annotated list of vascular plants and Characeae (1259 species, of which 1248 are vascular, the casual plants included but not numbered), with detailed localities.—See also **Druce, G. C.** The flora of Surrey. Rpt. Bot. Soc. and Exch. Club Brit. Isles 9: 680–694. 1932. (Review, with long list of additions.)—Also **Britton, C. E.** Notes on Surrey plants. Jour. Bot. 80: 314–318, 332–337. 1932.

Local

Ager, J. A. M. Flora of St. Thomas's Hospital and bombed sites. St. Thomas's Hospital Gaz. 47: 2–12. 1 fig. (map). 1949.
Account of localities; annotated list of plants (182 species, of which 176 are vascular); bibliography.

Ash, M. A comparative study of the vegetation of Purley Downs golf course and Coulsdon Common. Proc. Croydon Nat. Hist. and Scient. Soc. **11:** 315–329. 4 fig. (incl. 2 maps; 2 on pl.). 1948.
Includes (p. 326–329) tabular list of vascular plants.

Beadall, Arthur. Flora of Warlingham and Chelsham. *In* his Nature notes on Warlingham and Chelsham. p. 8–82. pl. 18.5 cm. Croydon, 1932. (Library of N. D. Simpson.)
Annotated list of vascular plants.

Brewer, J. A. A new flora of the neighbourhood of Reigate, Surrey, containing the flowering plants and ferns of the district, with their localities, times of flowering, etc., and a list of the mosses . . . viii, 194 p. map. 18 cm. London, 1856.
Annotated list of vascular plants (p. 1–139), on the Linnaean system; also the mosses and the fauna.—See also **Mill, J. S.** Plants growing wild in the district of Luxford's "Reigate Flora," omitted both in that work and in the supplementary list by Mr. Holman, published in the old series of the "Phytologist" in September, 1841. Phytologist n.s., **1:** 337–343. 1856. (List with localities.)—Also **Salmon, E. S., and Salmon, C. E.** List of localities. Proc. Holmsdale Nat. Hist. Club **1890/1892:** 8–10, 15–18, 70–71. 1893. (Lists of noteworthy plants with localities; the last by C. E. Salmon alone. In British Museum (Natural History) library.)

Dunn, S. T. Flora of southwest Surrey, including Leatherhead, Dorking, Guildford, Godalming, Farnham and Haslemere. xv, 106 p. 19 cm. London, 1893. (Gray Herbarium library.)
Geology, etc.; annotated list of vascular plants, with localities.—See also Jackson, below.

Groves, Henry, and Groves, James. Phanerogamia, Filicales, and Characeae. Ann. Rpt. South London Micros. and Nat. Hist. Club. **9:** 11–21. 1880. (British Museum (Natural History) library.)
List of vascular plants and Characeae, with distribution indicated by subdistricts. The South London district was bounded on the north by the Thames, on the east by the River Darent, on the west by the western branch of the River Mole, on the south by the junction line of the Chalk and Greensand on the southern slope of the North Downs from the River Mole at Boxhill to the Darent near Otford. Subdistricts 1, 2, and 4 were in Surrey, 3 and 5 in Kent. (In Kent and Surrey Counties.)

Hall, H. R. Flowering plants. *In* his Round the year in Richmond Park. Birds, flowers and trees. p. 38–53. 2 fig. (incl. map). 18.5 cm. London, 1923. (Library of N. D. Simpson.)
General features of flora; unannotated list of vascular plants [about 260 species.]—See also **Collenette, C. L.** The ferns of Richmond Park. London Nat. **27:** 36–37. 1948. (Annotated list of 10 species.)

Jackson, Rose. A list of the flowering plants and ferns occurring within six miles of Haslemere. 2. ed. (revised and extended). Sci. Paper Haslemere Nat. Hist. Soc. no. 4. 33 p. 21 cm. 1924. (British Museum (Natural History) library.) (1st ed. 1909. Supplementary list . . . l.c. no. 4, Suppl. 8 p. [1920].)
Soils, general features of flora; briefly annotated list of vascular plants and Characeae (770 species and 29 varieties and hybrids).

Jones, A. W. The flora of Bookham Common. London Nat. **33:** 25–47. pl., map. 1954.
General features of vegetation, bibliography; annotated list of vascular plants. (Two miles west of Leatherhead.)—See also **Castell, C. P., and Jones, A. W.** Notes on the vegetation of Eastern Plain. London Nat. **37:** 58–65. 2 fig., pl. 2–3. 1958. (Includes list of species.)

Monckton, H. W. The flora of the Bagshot district, the area of the main mass of the Bagshot sands. 32 p. 1 fig. (map). 21.5 cm. London, 1916. (Gray Herbarium library.)
Geology; list of vascular plants and mosses, with localities. Covers region from Bracknell (Berks) south to Aldershot (Hants), and from Esher (Surrey) west to Winchfield (Hants).—See review [by Druce] in Rpt. Bot. Soc. and Exch. Club Brit. Isles 4 (1916): 435–436. 1917, for various additions and corrections.—Also **Monckton, H. W.** The flora of the Bagshot district. Jour.

Bot. 57: 251–257. 1919. (Discussion of occurrence of various species; list of species of *Sphagnum*, with localities.)

Nicholson, George. Flora. 1. Phanerogams and vascular cryptogams. *In* The wild fauna and flora of the Royal Botanic Gardens, Kew. Bul. Misc. Inform. Kew Add. Ser. **5**: 74–91. 1906.

Annotated list of vascular plants.—See also **Jackson, A. B.** Flora. *In* Additions to the wild fauna and flora . . . VII. Bul. Misc. Inform. Kew **1908**: 125–126. 1908.—Also **Rolfe, R. A.,** and **Jackson, A. B.** Phanerogams. *In* Additions . . . X. l.c. **1909**: 369–373. 1909.—Also **Rolfe, R. A.** Flora. *In* Additions . . . XII. l.c. **1911**: 375. 1911.—Also Additions . . . XVII [i.e., XVIII]. l.c. **1936**: 65. 1936.—Also **Airy-Shaw, H. K.,** *ed.* Flora. Phanerogamia. *In* Additions . . . XXI. Kew Bul. **1949**: 236–237. 1949; XXII. l.c. **1952**: 287. 1952.

Parsons, H. F. The flora of the commons near Croydon. Proc. and Trans. Croydon Nat. Hist. and Scient. Soc. **1911–12**: 57–88. 1912.

Plant formations, general features of flora, etc.; tabular list of vascular and cellular plants, showing occurrence on Hayes Common, Keston Common, and other localities, mostly in Surrey, partly in West Kent.—See also **Prime, C. T.** The vegetation of the Blackheath Pebble areas in the Croydon district. Proc. Croydon Nat. Hist. and Scient. Soc. **11**: 269–302. 5 fig. (incl. maps). 1948. (Ecological; includes lists of vascular plants and mosses from Worms Heath, Addington Hills, and Croham Hurst.)

Polunin, Oleg. The flora of Godalming and Charterhouse. Carthusian **21**: 181–184, 253–256. 1954.

Annotated list of spermatophytes. Incomplete; reaches Rosa. (Not seen; annotation supplied by D. H. Kent.)

Pugsley, H. W. Wimbledon wild flowers. Wimbledon and Merton Annual **2**: 35–51. 1904.—List of the flowering plants and vascular cryptogams. l.c. **2**: 150–159. 1904.—Wimbledon wild flowers. [Corrections.] l.c. **3**: 163. 1905.—Additions to the List of flowering plants and vascular cryptogams recorded for Wimbledon. l.c. **4**: 179–186. 1910. (All in library of N. D. Simpson.)

General features of flora; list of vascular plants (572 species), with brief indication of range; extensive list of additional species and new localities, bringing total to 652 species.—See also **Johnson, Walter.** Botany: flowering plants. *In* his Wimbledon Common: its geology, antiquities and natural history. p. 219–264. fig. 19–23 (fig. 19–20, plates). 21 cm. London, Leipsic, 1912. (Discussion of various habitats, with mention of noteworthy species. In library of N. D. Simpson.)—Also **Pearson, A. A.** The flora of Wimbledon Common. An address given to the members of the John Evelyn Club on the 14th Dec. 1917. With list of species recently recorded. 17 p. 22 cm. London, 1918. (General account, followed by unannotated list of recently recorded vascular plants, mosses, hepatics, and fungi. In British Museum (Natural History) library.)—Also **Jones, A. W.** The flora of the golf course on Wimbledon Common. London Nat. **34**: 141–145. 1955. (Habitats; tabular list of vascular plants (83 species).)

Robbins, R. W. The flora of Limpsfield Common. London Nat. **1938**: 62–74. 1939; **1939**: 25–27. 1940.

Tabular list of vascular plants, annotated by symbols. The additions in the second paper bring the total to 369 species.

Robinson, F. E., and **Smith, J. B.** Botanical report for 1878. Flora dulwichiensis. Ann. Rpt. Dulwich Col. Sci. Soc. **1** (1878): 31–45. 1878.

Unannotated list of vascular plants.

Smith-Pearse, T. N. H. A flora of Epsom and its neighbourhood. 107 p. map. 18.5 cm. Epsom, 1917.

Botanical explorations, geology, etc.; annotated list of 760 vascular plants and Characeae.

Whale, William. British wild flowers, etc., growing in and near Egham. 40 p. 13 cm. Egham, [1875]. (Kew library.)

Includes (p. 5–32) "William Whale's list of Egham wild flowers," an alphabetical list of spermatophytes, with localities; also (p. 33–34) "British ferns growing in and near Egham" (alphabetical list with localities).

Covers adjacent parts of Middlesex and Berkshire counties. The copy seen is imperfect, lacking title page and cover.

Whitehouse, Margaret. The present flora of Barnes Common, Surrey. London Nat. **31**: 17–19. 1952.

Running account of the flora of a small area of Surrey heath only 5 miles from Piccadilly Circus.

SUSSEX

See also Hampshire (Boulger, Botanical bibliography of the south-eastern counties); Kent (Forster, Flora tonbridgensis; Jenner, A flora of Tunbridge Wells).

Vice-county 13, West Sussex; vice-county 14, East Sussex.

General

Wolley-Dod, A. H., *ed.* Flora of Sussex. lxxiii, 571 p. 6 pl., 2 maps (1 in pocket). 22.5 cm. Hastings, 1937.

Topography, botanical divisions, geology, chronological list of botanists with account of their work, lists of rare and local plants, of casual aliens, of books and herbaria consulted, and of contributors; annotated list of vascular plants and Characeae (1413 species), with first records and detailed localities. Replaces **Hemsley, W. B.** An outline of the flora of Sussex. Jour. Bot. v. 13, appendix, 34 p. 1875. (Botanical districts, sources; partly annotated list of vascular plants.) Also replaces **Arnold, F. H.** Flora of Sussex, or a list of the flowering plants and ferns found in the county of Sussex, with localities of the less common species. New ed. with numerous additions. xxi, 154 p. 3 pl. (1 col.), map. 19 cm. London [etc.], 1907. (Topography, climate, geology, botanical districts, botanical explorations, bibliography; annotated list of vascular plants and Characeae (1417 species). In Gray Herbarium library. 1st ed. 1887.)—See also Corrigenda . . . Jour. Bot. **76**: 64. 1938.—Also review by **Wallace, E. C.** Rpt. Bot. Soc. and Exch. Club Brit. Isles **11** (1937): 609–614. 1938.

Local

Anonymous. The natural history of Hastings and St. Leonards and the vicinity. 68 p. 18 cm. Hastings, 1878. (Kew library.)

List of vascular (p. 45–53) and cellular (p. 53–60) plants, annotated only as to frequency.—See also **Anonymous.** First supplement. 53 p. 18 cm. Hastings, 1883. (Includes Phanerogamia, p. 46–47, and Filices, p. 48, also cellular plants, p. 48–51; all compiled by E. N. Bloomfield.)—Also **Bloomfield, E. N.,** *ed.* Second supplement. 24 p. 18 cm. St. Leonards-on-Sea, 1888. (Includes Phanerogamia, p. 19–20, Filices, p. 20, cellular plants, p. 21–23.)—Also **Bloomfield, E. N.,** *ed.* Third supplementary list. 32 p. 18 cm. St. Leonards-on-Sea, 1898. (Includes Phanerogamia, p. 24, cellular plants, p. 25–27, statistics, p. 28. Total plants 1,649 species, of which 730 are vascular. All the supplements in Kew library.)—Also **Peatfield, W.** Botany. *In* **Belt, Anthony,** *ed.* Hastings. A survey of times past and present by members of the Hastings Natural History Society and others. p. 190–197. 19 cm. Hastings, 1937. (Partial list of plants, arranged under habitats. In library of N. D. Simpson.)

German, Mrs. P. Flowering plants of Worthing district. *In* Migeod, F. W. H., *ed.* Worthing: a survey of times past and present. p. 69–74. 19 cm. Brighton and Worthing, 1938. (Library of N. D. Simpson.)

Running account of flora, including mention of numerous species grouped by habitats, the widely distributed species omitted.

Mitten, William. Flora. *In* Merrifield, Mary P. A sketch of the natural history of Brighton and its vicinity. p. 193–205. 20.5 cm. Brighton, 1860. (Library of N. D. Simpson. Reissue, 1864, in Library of Congress.)

Unannotated list of vascular plants and hepatics (by Mitten) and lichens (by W. C. Unwin). Preceded by a general sketch of "The botany of the South Downs" (p. 115–143) [by Arthur Wallis].

Reynolds, Bernard. The flora of Christ's Hospital. [2. ed.] 22 p. 21.5 cm. n.p., 1926. (1st ed. 1913.)

List of vascular plants found in vicinity of Christ's Hospital, Horsham. (Not seen; title and annotation supplied by D. H. Kent from copy in British Museum (Natural History) library.)

Roper, F. C. S. Flora of Eastbourne. Being an introduction to the flowering plants, ferns, etc., of the Cuckmere district, East Sussex. xliii, 165 p. map. 18.5 cm. London, 1875. (Gray Herbarium library.)

Topography, geology, climate, botanical explorations, rarer species, statistics, etc.; annotated list of vascular plants and Characeae (700 species) with localities; list of 80 additional species recorded by earlier authors, unannotated lists of mosses, hepatics, and lichens, list of 157 vascular plants reported for East Sussex by Watson but not included in the Eastbourne Flora.—See also his On the additions to the fauna and flora of Eastbourne during 1875. Papers Eastbourne Nat. Hist. Soc. 8th Ann. Rpt. [3 p.] Oct. 15, 1875.—On the additions . . . since 1875. l.c. 10th Ann. Rpt. [5 p.] Oct. 19, 1877.—On the additions to the fauna and flora of the Cuckmere district during the past year. l.c. 11th Ann. Rpt. [5 p.] Oct. 18, 1878; 12th Ann. Rpt. [6 p.] Oct. 17, 1879; 13th Ann. Rpt. [6 p.] Oct. 15, 1880. —On the additions . . . since 1880, with a list of the forms of the genus *Rubus* now known to occur there. Trans. Eastbourne Nat. Hist. Soc. n.s., 2 (1889–90); 178–183. [1890?].—Also **Whitney, N. S.,** and **Milner, B.** The flora of the Eastbourne district. South East. Nat. **1906**: 30–34. 1906. (Account of the rarest species, additions to Roper's Flora, etc. In American Museum of Natural History library.)

Tansley, A. G., and **Adamson, R. S.** A preliminary survey of the chalk grasslands of the Sussex Downs. (Studies of the vegetation of the English chalk IV.) Jour. Ecology **14**: 1–32. pl. 1, map. 1926.

Ecological; includes lists of species, alphabetically arranged.

Weaver, J. Flora. *In* Gordon, Henry D. The history of Harting . . . p. 459–489. 1 fig. 19 cm. London, 1877. (Library of N. D. Simpson.)

Account of trees, etc.; partial list of vascular plants arranged by habitats, followed by list of fungi.

Warwickshire

Vice-county 38.

General

Bagnall, J. E. Botany. *In* The Victoria history of the county of Warwick. v. 1, p. 33–66. map. 31.5 cm. London, 1904.

General features of flora, botanical districts; list of vascular plants and Characeae (905 species) with distribution indicated by districts; also lists of cellular plants.

——— The flora of Warwickshire. The flowering plants, ferns, mosses, and lichens by James E. Bagnall . . . The fungi . . . by W. B. Grove . . . and J. E. Bagnall. xxxiv, 519 p. map (in pocket). 22 cm. London, Birmingham, 1891.

Topography, geology, climate, botanical regions, bibliography; annotated list of vascular and cellular plants (except algae) with localities; tabular list showing distribution (1,073 species and 236 varieties of vascular plants and Characeae, of which 817 species are native); botanical history, with included bibliography.—See also his Notes on the "Flora of Warwickshire." Midl. Nat. **15**: 154–157, 211–214. 1892; **16**: 22–23, 43–47, 69–70, 116–119, 141–143, 161–164, 206–212, 229–234, 251–260. 1893. (Additional records for vascular plants, Characeae, and mosses.)—Also his New localities for rare Warwickshire plants. l.c. **16**: 204–206. 1893.—Also his Notes on rare and local plants . . . l.c. **16**: 260–262. 1893.—Also **Dunn, S. T.** Warwickshire plants. Jour. Bot. **34**: 477. 1896.

Local

Bagnall, J. E. Flowering plants and ferns indigenous to the neighbourhood around Birmingham. Proc. Birmingham Nat. Hist. and Micros. Soc. **1** (1869): 89–97. [1870.]

Briefly annotated tabular list of vascular plants (688 species and varieties). Covers a radius of 10 miles around Birmingham, thus including parts of Staffordshire and Worcestershire as well as of Warwickshire.—See also his Flowering plants and ferns . . . Second list. Proc. Birmingham Nat. Hist. and Micros. Soc. **2** (1870): 73–74. [pref., 1872.] (In New York Botanical Garden library.)—Also his Notes respecting some Birmingham plants. Jour. Bot. **10**: 336–337. 1872.—Also his The flowering plants, ferns, etc. *In* Handbook of Birmingham. Prepared for the members of the British Association, 1886. p. 316–330. 16.5 cm. Birmingham, 1886. (Account of botanical districts with mention of noteworthy species; the total vascular flora is given as 1116 species. Accompanied by papers on cellular plants. In British Museum (Natural History) library.)—Also **Mathews, William.** Botany. Introductory remarks. l.c. p. 313–315. 1886. (Sketch of botanical publications, etc. In British Museum (Natural History) library.)—Also **Grove, W. B.** Flowering plants and ferns. Ecological account. *In* Auden, George A., *ed.* A handbook for Birmingham and the neighbourhood prepared for the 83rd annual meeting of the British Association for the advancement of science. p. 449–452. 19 cm. Birmingham, 1913. (Sketch of flora in relation to the geology. In library of N. D. Simpson.)—Also **Humphreys, John, Cleminshaw, Edward,** and **Grove, W. B.** General account. l.c. p. 452–465. 1913. (Account of several localities with mention of characteristics species.)

——— Notes on Sutton Park: its flowering plants, ferns, and mosses, read at a general meeting of the Birmingham Natural History and Microscopical Society held December 6th, 1876. To which is added The roses and brambles of Warwickshire. 27 p. 21 cm. Birmingham [1877].—Errata. [1 p.] n.p., n.d.—Supplement. [1 p.] n.p., n.d. (All in British Museum (Natural History) library.)

Topography, etc.; annotated list of vascular plants and mosses. The locality is north of Birmingham.

Cumming, Linnaeus. List of the plants found in the country around Rugby. revised and brought up to date. 47 p. 21.5 cm. Rugby, 1923. (Library of N. D. Simpson.)

Annotated list of vascular plants that can be found within walking or cycling distance of the school. At head of title page: The Rugby School natural history society. Replaces **Wait, W. O.** Plants. *In* his Rugby: past and present . . . p. 344–355. 25.5 cm. Rugby, 1893. (Botanical explorations, etc.; unannotated list of vascular plants (about 750 species), Characeae, and fungi. In Harvard College library.)—See also review in Rpt. Bot. Soc. and Exch. Club Brit. Isles **7** (1923): 84–87. 1924.—Also annual reports of the Botanical Section. *In* Rpt. Rugby School Nat. Hist. Soc. v. 57–58, 65–66, 69–73. 1924–40.—Also papers by **D. E. Allen,** with varying titles. l.c. **81**: 18–30. 1948; **82**: 16–21. 1949; **83**: 13–17. 1950; **84**: 28–30. 1951.

King Edward VI Grammar School for Boys Natural History Society. Brown's Coppice; a survey. 79 p. illus. Birmingham, 1954.

Ecological account; includes (p. 26–44) list of vascular plants and cryptogams of a locality near Birmingham. (Not seen; title and annotation supplied by G. Taylor.)

Westmorland

See also Cumberland (Baker, A flora of the English Lake district; Hodgson, Notes on the flora of the Ullswater district); Durham (Upper Teesdale Naturalists' Field Club, A flora of Teesdale); York (Handley, Catalogue of plants growing in the Sedbergh district).

Vice-county 69 (with North Lancashire).

General

Wilson, Albert. The flora of Westmorland. An account of the flowering plants, ferns and their allies, mosses, hepatics and lichens so far known to occur in the county. With particulars as to their localities, usual habitats, grade of citizenship, distribution in the six river basins, altitudinal range, time of flowering, first records, etc. . . . Also a description of the physical geography, topography, geology, climate and botanical features, a bibliography, and a short account of some pioneers of Westmorland field botany. 4 p.l., 413 p. 37 pl., port., map (in pocket). 22 cm. Arbroath, 1938. (Gray Herbarium library.)

Topography, geology, climate, general features of flora, early botanical explorations, bibliography; annotated list of vascular plants (959 species, excluding aliens, etc.), Characeae, bryophytes, and lichens, with localities. Includes some species recorded from North Lancashire but not from Westmorland.

Local

Baker, J. G., and Baker, E. G. The botany of High-cup Nick, Westmoreland. Jour. Bot. 32: 299–307. 1894.
Topography, etc.; annotated list of vascular plants and Characeae.

Gough, Thomas. Plants. *In* Nicholson, Cornelius. The annals of Kendal . . . 2. ed. p. 318–322. 23 cm. London, Kendal, 1861.
List of vascular plants, with localities.

Leach, R. E. The botany of Appleby and district. Naturalist 1894: 27–38. 1894.
General features of flora; list of spermatophytes collected by author, with localities, the Juncaceae, Cyperaceae, and Gramineae omitted.

Wiltshire

See also Somersetshire (Macmillan, Popular names of flowers).
Vice-county 7, North Wilts; vice-county 8, South Wilts.

General

Grose, J. D. The flora of Wiltshire. iv, 824 p. illus. (maps), 11 pl. 22 cm. Devizes, 1957.
Topography, geology, climate, bibliography, biographical notices, list of recorders, botanical districts, list of herbaria; annotated list of vascular plants and Charophyta (1406 species, of which 909 are native); extensive account of habitats and vegetation, gazetteer. Replaces **Preston, T. A.** The flowering plants of Wilts, with sketches of the physical geography and climate of the county. lxix, 436 p. 1 fig., 2 maps. 23 cm. [Devizes] 1888. (Topography, geology, climate, botanical districts, bibliography, etc.; annotated list of spermatophytes (849 species), with detailed local distribution.)

Local

Lewis, W. J. Flora. *In* his Notes on the history of the parish of North Wraxhall, Co. Wilts. With a life of the late rector Francis Harrison . . . p. 157–165. 19.5 cm. London, Brighton, 1913. (Library of N. D. Simpson.)
Notes on various species; tabular list of angiosperms, with habitats or localities, prepared by Miss Lidstone.

Maton, George. The natural history of a part of the county of Wilts, comprehended within the distance of ten miles round the city of Salisbury. 74 p. 21 cm. London, 1843. (Kew Library.)
Botanical explorations; "Vegetabilia" (p. 15–64), an alphabetical list of vascular and cellular plants, with localities; also list of vertebrates.—See also **Campbell, Mrs., and Gullick, Barbara.** Notes on the flora of the Salisbury district. Wiltsh. Archaeol. and Nat. Hist. Mag. 46: 58–63. 1932. (Extensive list.)—Also **Burges, R. C. L.** The Larkhill flora. l.c. 50: 188–190. 1943. (List of noteworthy species.)

Moir, T. R. G., and Peirson, L. G. Handlist of flowering plants of the Marlborough district (10 miles radius). Rpt. Marlborough Col. Nat. Hist. Soc. 83: 46–99. 1935.
Annotated list of vascular plants and Characeae. Additions are recorded in the "Flower list" in each Report of the Society, no. 84–95. 1936–47. Replaces several earlier floras, including **Preston, T. A.** Flora of Marlborough. 2. ed. Rpt. Marlborough Col. Nat. Hist. Soc., App. to no. [12], [15], 20, 23–24. viii, 342 p. pl. 1871–77. (1st ed. 1863.) (Botanical districts, list of collectors, etc.; annotated list of phanerogams, with keys and brief descriptions.) Also replaces **Smith-Pearse, T. N. H.** Hand-list of flowering plants of the district; Marlborough and ten miles round. 32 p. 21.5 cm. [Marlborough] 1919. (Briefly annotated list of phanerogams. Not seen; title and annotation supplied by N. D. Simpson.)

Worcestershire

Vice-county 37.

General

Amphlett, John, Rea, Carlton, and others. The botany of Worcestershire. An account of the flowering plants, ferns, mosses, hepatics, lichens, fungi, and fresh-water algae, which grow or have grown spontaneously in the county of Worcestershire . . . xxxiii, 654 p. map (in pocket). 22.5 cm. Birmingham, 1909.

Botanical districts, explorations (with included bibliography), bibliography; annotated flora of vascular and cellular plants (1185 vascular), with localities and notes on uses. Replaces **Lees, Edwin.** The botany of Worcestershire . . . 2 p. l., xci, 147, 48 p. illus., 3 pl., map. 22 cm. Worcester, 1867. (General features of flora with special attention to introduced plants, topography, climate, botanical explorations, botanical districts, etc.; tabular list of vascular plants and Characeae, showing local distribution; appendix on Worcestershire *Rubus.* In Gray Herbarium library.)—See also **Rea, Carleton.** Appendix to the botany of Worcestershire. Trans. Worcestersh. Nat. Club v. 7/8, Appendix. 139 p. 1920–32. (Extensive list of new records and new localities for vascular and cellular plants (p. 1–135), with supplementary list (p. 136–139) bringing total to 4,567 species (1,285 vascular). In Cornell University library.)—Also **Day, F. M.** Plant records from Worcestershire, 1944–48 [–1949]. l.c. **10** (1948–1949): 193–194, 194–195. 1950. (Author's records. In Cornell University library.)—Also **Hardaker, W. H.** The botany of Worcestershire: appendix II. l.c. **10** (1952–1953): 328–340. 1954. (Author's records based on over 40 years' work. In Cornell University library.)

Humphreys, John. Works relating to the botany of Worcestershire. *In* Bibliography of Worcestershire. Pt. III. 1 p. l., p. 217–247. 1907. (Publication of Worcestershire Historical Society, Oxford.)

Chronologically arranged list of botanical publications, 1549–1906, essentially without annotations. The paper is part III (b) of the Bibliography.

Local

Humphreys, John. A survey of the flora of East Worcestershire. Proc. Birmingham Nat. Hist. and Phil. Soc. **14:** 61–86. 1917.

Geology, sketch of botanical explorations with mention of rarer species recorded by earlier authors, account of various localities with lists of noteworthy species, lists of scarcer plants of different habitats; no full list of species. Relates to the Lickey district, the northern part of the Avon district, and the eastern side of the Severn valley.—See also **Smith, M. M. H.** Ecological survey of the phanerogams of Hartlebury Common, in 1932 and 1933. l.c. **16:** 101–129. 1 fig. (map). 1934. (Account of habitats and plant communities, with included lists of species, complementing Humphreys' list of noteworthy species from that locality given on p. 82–86.)

Lees, Edwin. The botany of the Malvern Hills in the counties of Worcester, Hereford, and Gloucester; with a succinct account of their geology, physical geography, climate, etc., the precise stations of the rarer plants, and the most interesting localities. 3. ed. viii, 166 p. col. pl. 19.5 cm. London, [pref. 1868]. (Gray Herbarium library.) (1st ed. 1843.)

Topography, geology, climate; annotated list of 1844 plants (894 species of vascular plants and Characeae, on the Linnaean system).—See also **Towndrow, R. F.** Worcestershire plants. Jour. Bot. **49:** 98–99. 1911. **51:** 57–58. 1913. (Vicinity of Malvern.)—Also Day, Plants of the Silurian limestones on the west of the Malvern hills, under Herefordshire.—Also Mackie, below.

Mackie, G. E. Catalogue of local flowering plants. *In* his Malvern field handbook and naturalist's calendar . . . p. 14–32. 16.5 cm. Malvern, 1886. (Library of N. D. Simpson.)

Unannotated list of spermatophytes, alphabetically arranged by English names, with habitats or localities; also a "Flower calendar" (p. 5–13) and a "list of local ferns" (p. 33).—See also Lees, above.

Mathews, William. The flora of the Clent and Lickey Hills and neighbouring parts of the county of Worcester. 54 p. 17.5 cm. Stourbridge and London, 1881. (Arnold Arboretum library.)

Geology, general features of flora, botanical investigations; annotated list of vascular plants.—See also **Humphreys, John.** Notes on the flora of the Lickey Hills. Pharm. Jour. 77: 91–93. 1906. (General account of geology, flora, etc., with mention of various species.)

Scott, William. Botany. *In* his Stourbridge and its vicinity . . . p. 539–558. 21.5 cm. Stourbridge, 1832. (Library of N. D. Simpson.)

Alphabetical list of vascular and a few cellular plants, with localities.

Yorkshire

See also Durham (Upper Teesdale Field Naturalists' Club, The flora of Teesdale); Lancashire (Gerard, Flora of the Stonyhurst district).

Vice-county 61, South-east Yorkshire; vice-county 62, North-east Yorkshire; vice-county 63, South-west Yorkshire; vice-county 64, Mid-west Yorkshire; vice-county 65, North-west Yorkshire.

General

Baines, Henry. The flora of Yorkshire. 4 p. l., xvi, 159 p. 2 maps. 22 cm. London, 1840.

Botanical districts, etc.; annotated list of vascular plants (1002 flowering plants) and mosses, with localities. The only general flora of Yorkshire; essentially replaced by the partial floras by Baker, Lees, and Robinson, with the supplement by Lees, listed below.—See also **Baker, J. G.,** and **Nowell, John.** A supplement to Baines' Flora of Yorkshire. Part 1. The flowering plants and ferns: by J. G. Baker . . . ; part 2. The mosses of the county: by John Nowell . . . viii, 188 p. map. 22 cm. London, 1854. (Phytogeography, botanical regions; complete list of vascular plants (p. 37–145); 1208 species, of which 81 are erroneous or dubious), with additional localities.

Baker, J. G. Botany. *In* The Victoria history of the county of York. v. 1, p. 111–172. map. 31.5 cm. London, 1907.

Phytogeography, general features of flora, types of distribution, plant zones, description of different regions (22) with mention of characteristic species; short account of bryophytes, algae, and lichens.

Lees, F. A. A supplement to the Yorkshire floras . . . Ed. by C. A. Cheetham and W. A. Sledge. iv, 134 p. 22.5 cm. London [1942].

Full list of vascular plants and Characeae known from the county, with records additional to those in Baker, North Yorkshire (1906), Lees, The flora of West Yorkshire (1888), and Robinson, The flora of the East Riding of Yorkshire (1902); the species for which no new records are reported being listed by name only.

Woodhead, T. W. Botanical survey and ecology in Yorkshire. Naturalist 1923: 97–128. 7 fig. (ports.). 1923. (Gray Herbarium library.)

Sketch of botanical explorations, work on ecology, history of flora, etc.; bibliography.

Partial

Baker, J. G. North Yorkshire: studies of its botany, geology, climate, and physical geography. 2. ed. Trans. Yorksh. Nat. Union, Bot. Ser. v. 3. xiv, 671 p. 3 maps. 1906. (1st ed. 1863.)

Geology, climate, topography, botanical distribution; annotated list of 1180 vascular plants, also mosses and hepatics, with localities.

Lees, F. A. The flora of West Yorkshire, with a sketch of the climatology and lithology in connection therewith. Trans. Yorksh. Nat. Union, Bot. Ser. v. 2. x, 843 p. map. 1888. (Gray Herbarium library.)

Climate, altitudinal limits of species, geology with lists of species, bibliography; annotated list of vascular (1042 species) and cellular plants, with localities. Replaces **Miall, L. C.,** and **Carrington, Benjamin.** The flora of the West Riding: a list of plants observed in the West Riding of Yorkshire, (exclusive of the Ainsty) with the localities in which they have been observed . . . the phanerogamous portion, by L. C. Miall; the cryptogams, by B. Carrington. xii, 97 p. 2 maps. 19 cm. London, Bradford, 1862. (Briefly annotated list of vascular plants, Characeae, lichens, mosses, and hepatics, with localities. In New York Botanical Garden library.)

Robinson, J. F. The flora of the East Riding of Yorkshire, including a physiographical sketch. Trans. Hull Scient. and Field Nat. Club v. 2. 253 p. map. 1902. (Reprinted, 253 p. map. London, 1902; reprint in British Museum (Natural History) library.)

Botanical investigations with included bibliography, topography, climate, plant associations, list of aliens, etc.; annotated list of vascular plants (1035 species), with localities. A list of mosses and hepatics, by J. J. Marshall, forms an appendix. There are additions by the author and others in Trans. Hull Scient. and Field Nat. Club 3: 98–100, 184–185, 218, 300–302. 1903–07; 4: 103–104, 105–106. 1909.—See also **Flintoff, R. J.** Some plants of the East Riding of Yorkshire. Naturalist **1931:** 229–232, 261–263. 1931.

Wilson, A. K. The adventive flora of the East Riding of Yorkshire. Occas. Papers Hull Scient. and Field Nat. Club 1: 2–28. 1938.

Sources of adventive plants, etc.; list of known adventive angiosperms of East Yorkshire and Hull, with localities.

Local

Balderston, R. R., and Balderston, Margaret. Plants of Ingleton. *In* their Ingleton: bygone and present. p. 163–198. 19 cm. London and Skipton [1888]. (Yale University library.)

List of vascular plants, with habitats and for the rarer species the localities, also partial lists of hepatics, lichens, mosses, etc. (In West Riding.)

Bootham School Natural History Society. Plant list and ferns. *In* Two county parishes. p. 11–29. 1956.

List of vascular plants found in parishes of Overton and Skelton. (In North Riding.) (Not seen; cited from Proc. Bot. Soc. Brit. Isles 3: 78. 1958.)

Crump, W. B. and Crossland, Charles. The flora of the parish of Halifax. Halifax Nat. v. 1–8, suppl. lxxv, 316 p. 1896–1904. (Reprinted 1904; in Arnold Arboretum library.)

Topography, geology, climate, phytogeography, plant distribution, biographies of local botanists, bibliography; annotated list of vascular (663 species) and cellular plants. (In West Riding.)—See also **Moss, C. E.** Changes in the Halifax flora during the last century and a quarter. Naturalist **1900:** 165–172. 1900; **1901:** 99–107. 1901. (Discussion of changes in flora, particularly with reference to species recorded by Bolton in 1775; bibliography.) —Also **Lees, F. A.** In defence of James Bolton, the fungologist. l.c. **1900:** 225–226. 1900.—Also his The volteface of flora: a rejoinder. l.c. **1900:** 229–236. 1900.

Ferguson, D. The natural history of Redcar and its neighbourhood. 128 p. 17.5 cm. London, Redcar, 1860. (Library of N. D. Simpson.)

Includes (p. 119–127) annotated list of vascular plants, with localities; also (p. 117–119, 128) lists of seaweeds and mosses. (In North Riding.)

Fisher, Robert. The flowering plants of Whitby and district with botanical and English names, explanatory preface and index. (Cover title: 12 miles around Whitby. The flowering plants.) 1 p. l., 43 p. 18 cm. Whitby, 1928. (Kew library.) (Reissue, 1929.)

Unannotated list of spermatophytes, by Fisher, and (on p. 43) list of vascular cryptogams by R. J. Flintoff. (In North Riding.)—See also Reynolds, below.

Flintoff, R. J. The natural history of Goathland. III. The flowering plants and ferns. An ecological study. North West. Nat. **16:** 57–75, 164–182, 287–304. map. 1941–42; **17:** 84–100, 229–244, 366–380. 1942–43; **18:** 87–102, 194–204. 1943–44.

Geography, etc.; annotated list of 582 vascular plants, grouped under 31 habitats; index. (In North Riding.)

Gissing, T. W. Materials for a flora of Wakefield and its neighbourhood. Naturalist 2: 6–10, 32–34, 56–60, 80–81, 94–97, 128–131, 147–149, 165–168, 184–186, 212–213, 232–235, 250–252, 269–271, 282–287, 294–295. 1865–66.

(Reprinted, 2 p.l., 59 p. 21 cm. London, Huddersfield, 1867. In Arnold Arboretum library.)

Annotated list of vascular plants and Characeae (606 species). (In West Riding.)—See also **Roberts, George.** A list of the rarer plants found growing on the Carboniferous sandstone in the immediate vicinity of Lofthouse. *In* his Topography and natural history of Lofthouse and its neighbourhood . . . v. [1], p. 389–391. 18 cm. London, Leeds, 1882. (List of noteworthy vascular plants, including cornfield weeds and aliens, with localities. Lofthouse is 3 miles north of Wakefield.)—Also his A list of the rarer flowering plants and ferns of the Lofthouse district. l.c. v. 2, p. 226–236. 18 cm. Leeds, 1885. (Extensive list with localities, bringing total for the Lofthouse-Wakefield region to 697 species.)—Also his Dispersion of plants in Yorkshire: notes on aliens and casuals recently observed in the Wakefield district. Naturalist **1885**: 389–393. 1885. (Methods of introduction; lists of introduced angiosperms, with localities.)—Also **Cryer, John.** Casual and alien plants from Wakefield. Naturalist **1916**: 250–251. 1916.

Handley, John. Catalogue of plants growing in the Sedbergh district, including the Lune basin, from Middleton to Tebay. 48 p. 18 cm. Leeds, 1898. (New York Botanical Garden library.)

List of vascular plants, with localities. (In West Riding and adjacent Westmorland.)

Hobkirk, C. P. Huddersfield: its history and natural history. A descriptive, historical, geological, botanical and zoological sketch of the town and neighbourhood. 2. ed. xi, 256 p. 4 pl. 19.5 cm. Huddersfield and London, 1868. (1st ed. 1859.)

Chapter VI, "Botanical" (p. 173–205), includes annotated list of vascular plants (504 species), also mosses, hepatics, lichens, and some fungi. (In West Riding.)

Holdsworth, Christopher, and others. The botany of Askham Bog. *In* Bootham School. Survey of Askham Bog. p. 13–29 incl. 2 fig. (maps). 24 cm. London, 1946. (Kew library.)

Annotated tabular list of vascular plants, with supplementary list of Juncaceae, Cyperaceae, Gramineae, and Equisetaceae. (In West Riding, 2 miles from the center of the city of York.)

[Howarth, Elijah]. List of plants, collected chiefly in the neighbourhood of Sheffield by Jonathan Salt, and now in the Sheffield Public Museum. vi, 57 p. 21.5 cm. Sheffield, 1889. (British Museum (Natural History) library.)

Biographical notice, index, etc.; tabular list of vascular plants in Salt's herbarium, with data. Publication of Sheffield Literary and Philosophical Society. (In West Riding.)—See also **Innocent, C. F.** A century's changes in the Sheffield district flora. Naturalist **1911**: 216–221. 1911. (Account of vascular plants which have nearly or quite disappeared since being recorded in a manuscript catalogue by Jonathan Salt a century before, and discussion of causes therefor.)—Also Snelgrove, below.

Lee, P. F. The flora of Dewsbury and neighbourhood. Trans. Yorksh. Nat. Union Ser. E. Botany **1885**: 225–248. 1887.—1st supplement . . . l.c. **1887**: 249–264. 1889.—2d supplement . . . Naturalist **1912**: 238–244, 285–289, 306–309. 1912.

Annotated list of vascular plants and Characeae. Second supplement includes extensive list of alien plants. (In West Riding.)

Lees, F. A. The botany of "Nidderdale and the garden of the Nidd." *In* Speight, Harry. Nidderdale . . . a Yorkshire Rhineland. p. 17–63. 23 cm. London, 1894.

Lists of rarer species of special localities; annotated list of vascular plants (722 species); mention of some mosses, hepatics, and lichens. (In West Riding.)

——— Botany of the Ilkley district. *In* Collyer, Robert, and Turner, G. H. Ilkley: ancient and modern. p. xxix–lxi. 22.5 cm. Otley [etc.], 1885. (Harvard College library.)

Soils, etc.; briefly annotated list of vascular (p. xxxvii–liii) and cellular plants, with localities for rarer species. Relates to Wharfedale (between Pool and Addingham), the Washburn and Menston valleys, the Esholt and

Burley woods, and the moorlands from Leathley and Otley Chevin to Addingham and Beamsley. (In West Riding.)

——— The vegetation of Craven in Wharfedale, with its adjacencies in Aire and Ribble: an analysis of its flora in advent and decline. North West. Nat. 12, Suppl. 40 p. 1937–38; 13, Suppl. 41–104 p. 1938–39; 14, Suppl. 105–111 p. 1939.

Annotated list of vascular plants and Charophyta, with special notice of habitats, soil preferences, and vernacular names. Covers region from Skipton to Pennyghent and round by Great Whernside to Greenhow Hill.—See also **Pickard, J. F.** Comments, with addenda and corrigenda . . . l.c. **16: 319–321.** 1942.—See also Rotheray, below, and Windsor, below. (In West Riding.)

Percival, John. The flora of Wensleydale, north-west Yorkshire. Naturalist **1888:** 125–143. 1888.—. . . Additions and corrections. l.c. **1888:** 159. 1888.

Geology, etc.; list of vascular plants (and Characeae, in Additions), with localities for scarcer species and some annotations. Covers main valley of the Yore River between Hawes and Leyburn and the small valleys in this area. (In North Riding.)

Peverett, Frank. Vascular plants of Bradford. A list of 456 species growing within an approximate six-mile radius of the Town Hall. 20 l. 33 cm. Bradford, 1951. (Mimeographed.)

List of vascular plants (456 species), with localities for some of the rarer species. (In West Riding.)—See also **Smith, A. M.,** and **Shaw, G. A.** A supplement to the Peverett list of vascular plants of Bradford. 5 l. 33 cm. (Bradford) 1951. (List with localities adding 138 species, bringing total to 594. Mimeographed; published by the Bradford Naturalists' Society.)—Also Willis, below.

Reynolds, Bernard. Whitby wild flowers. A complete botanic list of the flowers, grasses and ferns of the Whitby district (including Levisham and Scarborough), with notes on their history and habitats. 60 p. 16.5 cm. Whitby, 1915. (Library of N. D. Simpson.)

List of vascular plants (p. 1–22) with localities, followed by notes on various species; list of plants not found recently, bibliography, etc. Also floral notes by Swales, John, p. 49–60. (In North Riding.)—See also Fisher, above.

Rotheray, Lister. Flora of Skipton and district. vii, 133 p. incl. map. 22.5 cm. Skipton, 1900. (New York Botanical Garden library.)

List of vascular and cellular plants, with localities. (In West Riding.) —See also Lees, The vegetation of Craven in Wharfedale, above.

Rowntree, Harold. Vascular plants. *In* Walsh, G. B., and Rimington, F. C., *eds.* The natural history of the Scarborough district. I. Geology and botany. p. 206–285. pl. 6–7. 21.5 cm. Scarborough, 1953.

Annotated list of vascular plants and Characeae. Treatments of the cellular plants are also given (p. 49–205). Publication of Scarborough Field Naturalists' Society. (In North Riding.)

Shuffrey, W. A. The flowering plants and ferns of Littondale. *In* Boyd, William, *archdeacon,* and Shuffrey, W. A. Littondale: past and present. Pt. 2. Halton Gill in the olden time. p. 129–146. 23 cm. Leeds, 1893. (Library of N. D. Simpson.)

Topography, geology; briefly annotated list of vascular plants. (In West Riding.)

Slater, H. H. The flora of Ripon and neighbourhood. Trans. Yorkshire Nat. Union Ser. E. Botany **1882:** 125–152. 1883–84; **1883:** 153–184. 1884.

Annotated list of vascular plants, with localities. (In West Riding.)

Snelgrove, Ernest. Phanerogamia, Pteridophyta, Bryophyta. *In* The botany of the district, subedited by B. H. Bentley. Proc. Sheffield Nat. Club **1:** 49–77. 6 fig., map. 1910. (British Museum (Natural History) Library.)

Topography and geology of the region within 20 miles of Sheffield, subdivided into districts; lists of the characteristic plant associations, including vascular plants, mosses, and hepatics; changes in flora, list of rarer species. (In West Riding.) Same title and treatment in Porter, William S., and Watson, Arnold T., *eds.* Handbook and guide to Sheffield. . . . p. 405–433.

19 cm. Sheffield, 1910. (At head of title page: British Association, Sheffield 1910.) (In library of N. D. Simpson.)—See also Howarth, above.

Soppitt, H. T. Botany. *In* Gray, Johnnie [i.e. Speight, Harry]. Through Airedale from Goole to Malham. p. xiii–xlii. 17.5 cm. Leeds [etc.], 1891.
Compiled list of about 700 vascular plants of Aire valley, with localities. (In West Riding.)

Stansfield, Abraham, *ed.* Flora of Todmorden. Lancashire Nat. **1:** 106–109, 121–123, 135–138, 153–156, 163–166, 179–182. 1907–08; **2:** 53–56, 67–68, 122–127, 158–159, 167–169, 201–204, 249–252, 285–288, 311–314, 347–350, 355–360. 1909–10. (University of Minnesota library.) (Reprinted with addition of A list of the birds of Todmorden by Walter Greaves. 4 p. l., 56, 1A–10A, xii p., 2 l. 21.5 cm. [pref. Manchester, 1911.] (Flora of Todmorden, edited by A. Stansfield. p. 1–56 (vascular plants, p. 1–36). In library of N. D. Simpson.)
Annotated list of vascular plants, mosses, hepatics, and lichens, with localities for scarcer species. The area covered extends into Lancashire. (In West Riding.)

Sturdy, H. H. List of plants. *In* Riley, Frederick. The Settle district and north-west Yorkshire dales. A practical guide book for the visitor and tourist. p. 192–202. 22 cm. Settle, 1923. (Library of N. D. Simpson.)
List of vascular plants, with localities.—See also Windsor, below. (In West Riding.)

Willis, John. On the flora of the environs of Bradford. Jour. Bot. **12** (n.s., 3): 10–18. 1874.
General features of flora; unannotated lists of 607 vascular plants and Characeae found in the Bradford district and on the limestone areas near Bradford. (In West Riding.)—See also **Lees, F. A.** On the flora of the Leeds and Bradford district. Jour. Bot. **12:** 78–83. 1874. (Raises total to 727 species.)—Also **Cryer, John.** Adventive plants on waste ground, Bradford, York, 1919. Rpt. Bot. Soc. and Exch. Club Brit. Isles 5 (1919): 719. 1920.

Windsor, John. Flora cravoniensis, or a flora of the vicinity of Settle in Craven, Yorkshire . . . x, 177 p. 19 cm. Manchester, 1873. (Arnold Arboretum library.)
Topography, geology, botanical explorations, etc.; list of vascular plants, Characeae, mosses, hepatics, and lichens, with localities, from a 15-mile radius around Settle. Reprinted with additions and corrections from papers published in The Phytologist, 1855–58. (In West Riding.)—See also Lees, The vegetation of Craven in Wharfedale, above; also Sturdy, above.

Woodruffe-Peacock, E. A. The ecology of Thornton Waste. Naturalist **1920:** 301–304, 353–356, 381–384. 1920; **1921:** 21–25. 1921.
Includes (p. 353–356, 381–384, 21–24) running account of plants, arranged by habitat. Thornton Moor or Waste was a huge bog, originally about 26 sq. miles in area, between Crowle (Lincolnshire) and Thorne and Goole (Yorkshire). (In West Riding and adjacent Lincolnshire.)—See also **Sledge, W. A.** Thorne Waste plants. Naturalist **1941:** 266–268. 1941. (Notes on various species.)

IRELAND (EIRE)

See also Scotland, General (Cameron, The Gaelic names of plants).
Most of the works under Great Britain, General, refer to Ireland. The principal exceptions are: Britten and Holland, Fisher, Grigson, Miller, Prior, Rayner, Trimen, and Watson.
The vice-county numbers given are those introduced by Praeger (1901), which are independent of the Watsonian series for England, Wales, and Scotland. Another series of numbers uniformly 112 higher, formed by adding the Watsonian total for England, Wales, and Scotland to the Praeger series, is given in W. B. Turrill, British plant life, p. 260. 1948.

General

Includes some partial works.

Adams, John. A student's illustrated Irish flora, being a guide to the indigenous seed-plants of Ireland. viii, 343 p. 578 fig., map. 19 cm. Ashford, Kent, 1931.
Brief account of principal published floras; running account of species (not in strict systematic order) with brief annotations, statement of range by

provinces, keys to genera, and in many cases brief diagnostic characters; poisonous plants, short list of Irish vernacular names, etc. Intended for beginners in botany.

Dickie, George. A flora of Ulster and botanist's guide to the north of Ireland. xix, 176 p. 17 cm. Belfast and London, 1864.

Geology, general features of flora; annotated list of vascular plants (805 species), with localities. Covers "whole of Ulster, excepting the most southern portions of Monaghan and Cavan, comprehending also the northern portions of Leitrim, Sligo, and Mayo belonging to Connaught."

Fitzpatrick, H. M. The trees of Ireland—native and introduced. Scient. Proc. Roy. Dublin Soc. n.s., **20**: 597–656. 1933.

Brief account of introduction of trees into Ireland, list of gardens and other localities; annotated list of native and cultivated trees (nearly all cultivated), with dimensions of large specimens, notes on occurrence of native species, etc.

Hart, H. C. On the range of flowering plants and ferns on the mountains of Ireland. Proc. Roy. Irish Acad. III, **1**: 512–570. table. 1891.

List of mountains; tabular list of 421 vascular plants, showing altitudinal range in 10 different mountain regions of Ireland; discussion of altitudes and ranges; bibliography.

Hogan, Edmund, Hogan, John, and **MacErlean, J. C.** . . . [Gaelic title.] Irish and Scottish Gaelic names of herbs, plants, trees, etc. xii, 137 p. 19 cm. Dublin, 1900. (Gray Herbarium library.)

Manuscript and printed sources; bibliography; alphabetical list of Gaelic vernacular names, with English equivalents (no scientific names); cross-index, English-Gaelic.—See also Moloney and Tidestrom, below, and Cameron under Scotland, General.

Knowles, M. C. A contribution towards the alien flora of Ireland. Irish Nat. **15**: 143–150. 1906.

References to published lists of alien plants in Ireland; annotated list of species collected by author in 1904–1905 in County Kildare, near Lyons.—See also **Johnson, Thomas,** and **Hensman, R.** Agricultural seeds and their weed impurities: a source of Ireland's alien flora. Scient. Proc. Roy. Dublin Soc. n.s., **12**: 446–462. pl. 22–23. 1910. (Includes lists of species introduced in seed of important crop plants, etc.)

Lett, H. W. Botanists in the north of Ireland. Irish Nat. **22**: 21–33. 1913.

List of botanists who worked in northern Ireland or who were natives of Ulster, with biographical notes. Also in Ann. Rpt. and Proc. Belfast Nat. Field Club II, **6**: 615–628. 1913. The material in the two issues of this paper (a presidential address) is not precisely the same, each containing a few names not found in the other.

Mackay, J. T. Flora hibernica, comprising the flowering plants, ferns, Characeae, Musci, Hepaticae, Lichenes and Algae of Ireland. Arranged according to the natural system with a synopsis of the genera according to the Linnaean system. 2 v. (xxxiv, 354; 279 p.). 23 cm. Dublin, 1836.

Botanical investigations, key to genera on Linnaean system; annotated descriptive flora of vascular and cellular plants, with localities and notes on uses. Pt. (i.e., vol.) 1 contains vascular plants and *Chara*, by Mackay; pt. 2 (v. [2], p. 1–156), Musci, Hepaticae, and Lichenes, by Thomas Taylor; pt. 3 (v. [2], p. 157–254), Algae, by W. H. Harvey; p. 255–279, additions, corrections, and index to v. 1 and 2.

Moloney, M. F. . . . [Gaelic title.] Irish ethno-botany and the evolution of medicine in Ireland. 96 p. 19 cm. Dublin, 1919.

Includes (p. 11–50) "The native materia medica of vegetable origin," consisting of "The land flora" (vascular plants, p. 11–48) and "The marine flora" (seaweeds, p. 49–59), a list with Gaelic names and notes on medicinal and other uses.—See also Hogan, above, and additional references there given.

More, A. G. Contributions towards a Cybele hibernica, being outlines of the geographical distribution of plants in Ireland. 2. ed., founded on the papers of the late Alexander Goodman More, by Nathaniel Colgan and Reginald W. Scully. xcvi, 538 p. map. 22.5 cm. Dublin, London, 1898. (1st ed. by David Moore and A. G. More, 1866.)

Extensive bibliography, topography, geology, climate, comparison of Irish and English floras, botanical districts, tabular list showing distribution; annotated list of vascular plants and Characeae, showing local range, latitudinal range, habitat, dates of flowering, vernacular names, etc.—See also **Marshall, E. S.** Remarks on the "Cybele hibernica", ed. 2. Jour. Bot. 37: 269–272. 1899.—Also **Colgan, N.,** and **Scully, R. W.** Remarks . . . A rejoinder. l.c. 37: 315–317. 1899.—Also **Stewart, S. A.** Criticisms of the "Cybele hibernica," ed. 2. l.c. 37: 396–397. 1899.—Also Praeger, Irish topographical botany, below.

O'Connor, Patrick. Liosta de phlanndaí na hÉireann. Hand-list of Irish plants. 4. ed. 32 p. 18.5 cm. Dublin, 1943. (1st ed. 1910, by Johnson, Thomas, and Knowles, M. C., with title Hand list of Irish flowering plants and ferns.)

Name list of vascular plants and Charophyta, with indication of the total number of vice-counties in which each occurs; based on Praeger's The botanist in Ireland. Publication of the National Museum of Ireland.

Praeger, R. L. The botanist in Ireland. xii, 587 p. 29 i.e. 30 fig. (incl. maps), 44 pl. on 22, 6 maps. 19 cm. Dublin, 1934.

Geology, topography, climate, origin of flora, phytogeography, vegetational types, human influences on flora, botanical subdivisions, list of botanists with mention of principal publications and discoveries and references to biographies, systematic list of rare or interesting plants with notes on their distribution, list of records in need of confirmation; topographical and botanical sketch of the counties and noteworthy localities, mountains, islands, etc., with brief account of geology, general features of flora, noteworthy plants, and short bibliographies; list of known vascular plants and Characeae (not including infraspecific taxa), with citation of botanical divisions in which each occurs and summary statement of range; full index. The pagination through 489 is that of the numbers of the sections of the text and not proper page numbers. An admirable work, without exact equivalent in the floristic literature of other countries.

——— Hybrids in the Irish flora: a tentative list. Proc. Roy. Irish Acad. 54 (B): 1–14. 1951.

General considerations, with discussion of certain hybrids; list of known hybrids (over 100), with citation of counties but no references.

——— Irish topographical botany compiled largely from original material. Proc. Roy. Irish Acad. ser. III, v. 7. clxxxviii, 410 p. 6 maps. 1901.

Topography, climate, soils, phytogeography, botanical divisions, botanical investigations, bibliography, tabular list of species showing range; annotated list of vascular plants and Characeae. Intended to complement the 2d ed. of More's Contributions towards a Cybele hibernica.—See also his Irish topographical botany: supplement, 1901–1905. Proc. Roy Irish Acad. 26 (B): 13–45. 1906. (Statistics; list of newly recorded species, principal range extensions, new records arranged by counties; bibliography.)—Also his Phanerogamia and Cryptogamia vascularia. Supplement to "Irish topographical botany," 1906–1928. In his Report on recent additions to the Irish fauna and flora . . . l.c. 39 (B): 57–78. 1929. (Similar treatment.)—Also his A contribution to the flora of Ireland. Including a 3d supplement to "Irish topographical botany." l.c. 42 (B): 55–86. 1 fig. (map). 1934. (List of new records, with conspectus of available records in *Rosa, Hieracium, Euphrasia,* etc.)—Also his A further contribution to the flora of Ireland. l.c. 45 (B): 231–254. 1 fig. (map). 1939. (Similar list.)—Also his Additions to the knowledge of the Irish flora, 1939–1945. l.c. 51 (B): 27–51. 1946. (Similar list.)—Also **Brenan, J. P. M.,** and **Simpson, N. D.** The results of two botanical journeys in Ireland in 1938–9. l.c. 52 (B): 57–84. 1949. (Records of noteworthy vascular plants and Characeae from 27 of the vice-counties.)

——— Some Irish naturalists. A biographical notebook. 208 p. 18 pl. (ports.), front. (port.). 22 cm. Dundalk, 1950.

Author's reminiscences; alphabetical list of botanists, zoologists, and geologists born in Ireland or who have worked there, with biographical information and references; list of societies, institutions, and "team-work," with data.

Praeger, R. L. A tourist's flora of the west of Ireland. xii, 243 p. 17 fig. (incl. maps), 27 pl., 5 maps. 19.5 cm. Dublin, 1909.

Topography, geology, climate, botanical divisions, plant formations, botanical investigations, descriptive list of localities (with in each case the Gaelic name and its meaning) with mention of rarer species and short bibliographies; annotated list of known vascular plants and Characeae, showing occurrence by counties and giving localities and some Gaelic names. The general botanical material, revised and amplified, is incorporated in his The botanist in Ireland, but the Irish locality and vernacular names, as well as details of distribution, are omitted from that work. Covers 8 counties: Clare, Fermanagh, Galway, Leitrim, Limerick, Mayo, Roscommon, and Sligo.

Stewart, S. A. and **Corry, T. H.** A flora of the north-east of Ireland. 2. ed. Flowering plants, vascular cryptogams and charophytes by R. L. Praeger . . . lix, 472 p. 3 maps. 22.5 cm. Belfast, 1938. (1st ed. 1888.)

Botanical explorations, topography, geology, changes in flora, guide for local botanizing, list of botanists; annotated list of vascular plants and Characeae, with localities; list of adventive, casual, and excluded species; annotated list of mosses and liverworts, by W. R. Megaw; gazetteer. Covers Counties Antrim, Derry and Down.—See also **Praeger, R. L.** Additional records for the flora of the northeast of Ireland. Irish Nat. Jour. **8**: 33–37. 1942.—Also **Carrothers, E. N.,** and **Moon, J. M.** Some records for the flora of the north-east of Ireland. l.c. **8**: 214–215. 1944.—Also **Carrothers, E. N., Meikle, R. D.,** and **Moon, J. M.** Records and notes for the flora of the north-east of Ireland. l.c. **9**: 24–28. 1947.—Also **Stelfox, A. W.** Notes on some plants found in the Carnlough district and on the Garron plateau, Co. Antrim. l.c. **9**: 317–326. 1949.—Also **Carrothers, E. N.,** and **Moon, J. M.** Notes on the flora of the north-east of Ireland. l.c. **10**: 284–286. 1952; **12**: 11–14. 1956.—Also **Gunning, B. E. S.,** and **Pate, J. S.** Notes on the County Down flora. l.c. **12**: 162–164. 1957.

Tidestrom, Ivar. The floral alphabet of the Celts. Torreya 23: 41–49. 1923.

Table showing the Gaelic letters with their Irish names (practically all names of trees) and English, Anglo-Saxon, Old Norse, Swedish, and Latin equivalents, and discussion.—See also **Edlin, H. L.** The Gaelic alphabet of tree names. Scott. Forestry **4**: 72–74. 1950. (Discussion of the Gaelic names of the letters and their botanical identification. Drawn up largely from Scottish rather than Irish sources.)—See also Cameron, above, and additional references there given.

Webb, D. A. The flora and vegetation of Ireland. Veröffentl. Geobot. Inst. Rübel Zürich **25**: 46–78. 2 fig. 1952. (Reissued as Notes Bot. Sch. Trinity Col. 5 (4): 17–49. 1953.)

Topography, soils, climate, plant communities, phytogeography, short bibliography. The whole volume consists of papers based on the Ninth International Phytogeographical Excursion to Ireland in 1949.

——— An Irish flora. 2. rev. ed. xxx, 250 p. 160 fig. on 12 pl. (on 6). 19.5 cm. Dundalk, 1953. (1st ed. [1943.].)

Brief bibliography; keyed, descriptive flora of vascular plants (1,133 species), with Irish and English names and local range.

Antrim

See also Ireland, General (Dickie, A flora of Ulster; Lett, Botanists in the north of Ireland; Stewart and Corry, A flora of the north-east of Ireland); Down (Chase, The natural history of Campbell College and Cabin Hill); Sligo (Barrington, Report on the flora of Ben Bulben). Vice-county 39.

Local

Brown, J. J., Kersley, L. W., and **MacDonald, R.** Notes on the flora and fauna of The Maidens. Irish Nat. Jour. **10**: 208–211. 1951.

Includes mention of vascular plants observed. (A group of rocks about 5 miles off the coast near Ballygally.)

Stewart, S. A. Report on the botany of the Island of Rathlin, County of Antrim. Proc. Roy. Irish Acad. II, 4: 82–104. 1884.

Topography, geology, etc.; annotated list of vascular plants, Characeae, and mosses; list of excluded species.—See also **Praeger, R. L.** Three days on Rathlin Island: with notes on its flora and fauna. Ann. Rpt. and Proc. Belfast Nat. Field Club II, 3: 218–222. 1890. (Includes 29 additional species.)—Also his The flora of Rathlin Island. Irish Nat. 2: 53. 1893. (Adds 5 species).)—Also **Bennett, S. A.,** and others. Plant new to Rathlin Island list. Irish Nat. Jour. 1: 10. 1925.

Tate, Ralph. Flora belfastiensis. The plants around Belfast, with their geographical and geological distribution. xix, 92 p. 14.5 cm. Belfast, 1863. (Gray Herbarium library.)

Topography, geology, general features of flora, notes on uses, etc.; briefly annotated list of vascular plants (602 indigenous species) within a radius of 10 to 15 miles around Belfast.—See also Down (Chase, The natural history of Campbell College and Cabin Hill).

Armagh

See also Ireland, General (Dickie, A flora of Ulster . . . ; Lett, Botanists in the north of Ireland).

Vice-county 37.

Local

Praeger, R. L. The flora of County Armagh. Irish Nat. **2**: 11–15, 34–38, 59–62, 91–95, 127–134, 155–159, 182–184, 212–215. pl. 2, 5 (pl. 2, map). 1893. (Reprinted, 37 p.)

Botanical investigations, general features of flora, lists of commoner species; annotated list of rarer species.—See also his Notes on the flora of County Armagh. Jour. Bot. **31**: 238–241. 1893. (Botanical regions, mention of rarer species, etc.; total flora, omitting the doubtfully native species, given as 616 species.)—Also **Carrothers, E. N., Meikle, R. D.,** and **Moon, J. McK.** Botanical notes for the counties of Armagh, Down and Antrim. Irish Nat. Jour. **9**: 304–306. 1949. (List of noteworthy vascular plants, with indication of new records.)

Carlow

Vice-county 13.

Cavan

See Ireland, General (Dickie, A flora of Ulster . . . ; Lett, Botanists in the north of Ireland); also Fermanagh (Stewart, Report on the botany of the mountainous portion of Co. Fermanagh to the west of Lough Erne, and the adjoining district of Co. Cavan).

Vice-county 30.

Clare

See also Ireland, General (Praeger, A tourist's flora of the west of Ireland).

Vice-county 9 (includes also Aran Islands, for which see Galway).

Local

Knowles, M. C., and **O'Brien, R. D.** A botanical tour in the islands of the Fergus estuary and adjacent mainland. Irish Nat. **18**: 57–68. 1909.

Geology, etc.; running account of several islands and adjacent mainland with mention of various plants observed; list of noteworthy species.

Praeger, R. L. Notes on the botany of central Clare. Irish Nat. **14**: 188–193. 1905.

Botanical explorations, general features of flora; list of rarer vascular plants and Characeae, with localities, including 21 species new to the county.

Stewart, S. A. Report on the botany of south Clare and the Shannon. Proc. Roy. Irish Acad. III, **1**: 343–369. 1890.

Topography, geology, general features of flora; annotated list of 579 plants (477 vascular) collected by author.

Cork

Vice-county 3, West Cork; vice-county 4, Mid Cork; vice-county 5, East Cork.

General

Allin, Thomas. The flowering plants and ferns of the County Cork . . . 2 p. l., xiii, 113 p. map. 22 cm. Weston-super-Mare, 1883. (Gray Herbarium library.)

Botanical investigations, climate, general features of flora; annotated list of vascular plants, with localities for scarcer species. Replaces **Power, Thomas.** The botanist's guide for the county of Cork, being a systematic catalogue of the native plants of the county, and more especially of the vicinity of Cork, together with their stations, compiled and arranged according to the natural system of botany. *In* Contributions towards a fauna and flora of the county of Cork, read at the meeting of the British Association held at Cork in the year 1843 . . . v, 130 p. 22 cm. London, Cork, 1845. (Annotated list of 1,809 plants (915 vascular species) with localities.)—See also **Polunin, Oleg.** Notes and additions to the flora of the islands of S. W. Cork. Watsonia **1**: 359–363. 1950. (Annotated list of noteworthy species.) —Also **O'Donovan, J. E., and O'Regan, B.** Notes on some native and alien plants in West Cork. Irish Nat. Jour. **10**: 235–238. 1952. (Remarks on various islands; list of noteworthy vascular plants; bibliography.)

Donegal

See also Ireland, General (Dickie, A flora of Ulster; Lett, Botanists in the north of Ireland); Kerry (Barrington).
Vice-county 34, South Donegal; vice-county 35, North Donegal.

General

Hart, H. C. Flora of the County Donegal, or list of the flowering plants and ferns with their localities and distribution. xxiv, 391 p. map. 21.5 cm. Dublin, London, 1898. (Arnold Arboretum library.)

Topography, geology, phytogeography, altitudinal range, table showing distribution in the region; annotated list of vascular plants and Characeae (737 species, of which 734 are vascular), with localities; climate, list of vernacular names with notes on local plant lore.—See also his Botanical excursions in west Donegal, 1898. Jour. Bot. **37**: 70–76, 125–130, 152–159. 1899. (Includes extensive list of new localities, etc.)—Also **Browning, F. R.** Plants of Donegal, 1926–1927. Rpt. Bot. Soc. and Exch. Club Brit. Isles 8 (1927): 426–427. 1928.

Local

Barrington, R. M. The plants of Tory Island, County Donegal. Jour. Bot. **17**: 263–270. 1870.

Topography, etc.; annotated list of 145 vascular plants.—See also **Praeger, R. L.** Notes on the flora of Tory. Irish Nat. **19**: 189–192. 1910. (Adds 24 species.)

Hart, H. C. Flora of the Croaghgorm Range, Co. Donegal. Jour. Bot. **20**: 198–200. 1882.

Geology, etc.; unannotated list of vascular plants, arranged in groups according to altitude reached. (Croaghgorm or Bluestack Range.)

——— On the flora of Innishowen, Co. Donegal. Jour. Bot. **21**: 23–26, 47–51, 75–80, 150–152, 170–174, 205–209, 275–278, 299–305. 1883.

Geography, geology, topography, botanical explorations, plant geography; annotated list of vascular plants.

——— On the plants of (north) Aran Island, Co. Donegal. Jour. Bot. **19**: 19–23. 1881.

Topography, geology, general features of flora; unannotated list of vascular plants.—See also **Praeger, R. L.** On the flora of Arranmore, Co. Donegal. Irish Nat. Jour. **4**: 50–54. 1932. (Adds 85 species, bringing total to 317.)

Stelfox, A. W. A list of the flowering plants, ferns, etc. (Survey of Inishtrahull. Pt. 4). Irish Nat. Jour. 8: 116–123. 1943.

General features of flora, etc.; annotated list of 119 vascular plants, and some mosses, liverworts, and mushrooms.

Down

See also Ireland, General (Dickie, A flora of Ulster . . . ; Lett, Botanists in the north of Ireland; Stewart and Corry, A flora of the north-east of Ireland).
Vice-county 38.

Local

Chase, C. D. The natural history of Campbell College and Cabin Hill. 64 p. (p. 54–64 blank). front. 24.5 cm. Belfast, 1949.

Includes (chap. V, The wild flowers, p. 30–51) list of about 250 vascular plants with remarks on many of them and supplementary list of extinct species; also lists of trees (p. 15–22), including cultivated species, of cultivated shrubs (p. 23–29), and of mosses (p. 52–53). Relates to an area of 130 acres in the County Down part of the County Borough of Belfast. Replaces **Bennett, S. A., and Chase, C. D.** A list of plants growing in the grounds of Campbell College, Belfast. 13 p. 21.5 cm. n.p., 1925. (Unannotated list of vascular plants and mosses.)—See also Tate, Flora belfastiensis, under Antrim.

Stewart, S. A., and Praeger, R. L. Report on the botany of the Mourne Mountains, County Down. Proc. Roy. Irish Acad. III, 2: 335–380. 1892.

Topography, geology, botanical explorations, altitudinal range, etc.; annotated list of about 593 vascular plants.

Dublin

See also Wicklow (Moore and More, Catalogue of the flowering plants and ferns of Dublin and Wicklow).
Vice-county 21.

General

Colgan, Nathaniel. Flora of the County Dublin: flowering plants, higher cryptogams, and Characeae. lxx, 324 p. map. 23 cm. Dublin, 1904. (Gray Herbarium library.)

Bibliography, botanical investigations, climate, botanical divisions, etc.; annotated list of vascular plants and Characeae [755 species and subspecies] with localities and vernacular names; annotated list of vernacular names.—See also **Brunker, J. P.** Plants of County Dublin. Irish Nat. 31: 94–95. 1922.—Also **Praeger, R. L.** Vegetation-study in the Dublin district. *In* Handbook of the city of Dublin and the surrounding district. Prepared for the meeting of the British Association. September, 1908. p. 72–75. pl. 3. 18.5 cm. Dublin, 1908. (General features of vegetation in Dublin and Wicklow.)—Also **Colgan, Nathaniel.** Phanerogams and vascular cryptogams. l.c. 75–86. pl. 4–5. 1908. (Topography, phytogeography, general features of flora, short bibliography.)—Also **Hudson, H. J.** Changes in Dublin's flora. Irish Nat. Jour. **12**: 190–192. 1957. (Discussion of changes since 1904.)

Rutty, John. An essay towards a natural history of the county of Dublin, accommodated to the noble designs of the Dublin Society; affording a summary view I. Of its vegetables, with their mechanical and œconomical uses, and as food for men and cattle; a catalogue of our vegetable poisons; and a botanical kalendar, exhibiting the respective months in which most of the simples in use are found in flower. II. Of its animals. III. Of its soil . . . 2 v. 4 pl. (pl. 1, map), tables. 20 cm. Dublin, 1772.

The pages treating of plants (v. 1, p. 33–261) include annotated lists of esculent, miscellaneous useful, and dye plants, alphabetically arranged by pre-Linnaean scientific names (p. 33–164), an annotated list of "poysonous [sic] vegetables" (p. 165–177), and a "Kalendarium botanicum hibernicum" (p. 178–261), a listing of most of the indigenous plants and some of the cultivated ones by the months in which they flower or fruit.

Threlkeld, Caleb. Synopsis stirpium hibernicarum alphabetice dispositarum. Sive commentatio de plantis indigenis praesertim dublinensibus instituta. Being a short treatise of native plants, especially such as grow spontaneously in the vicinity of Dublin. . . . With an appendex of observations made upon plants. By Dr. Molyneux . . . (mostly unpaged). 16 cm. Dublin, 1727.

Annotated list of plants, alphabetically arranged by Latin names (mostly from Ray) with English and Gaelic names, notes on uses, etc.; index (p. 23–58) of English and Gaelic names.

Wade, Walter. Catalogus systematicus plantarum indigenarum in comitatu dublinensi inventarum. Pars prima. viii, 275 p., index. 18.5 cm. Dublini, 1794. (New York Botanical Garden library.)

Obsolete as a flora, cited here for its Gaelic vernacular names (indexed); covers the spermatophytes through Polygamia Dioecia, on the Linnaean system.

Local

Hart, H. C. The flora of Howth. With map and an introduction on the geology and other features of the promontory. 2 p.l., 137 p. map. 19 cm. Dublin, 1887. (Gray Herbarium library.)

Topography, geology, soils, botanical investigations, general features of flora; annotated list of vascular plants (520 native species), excluded and introduced species, etc.—See also his Flora of Howth. Jour. Bot. **29**: 377. 1891.

——— Notes on the flora of Lambay Island, County of Dublin. Proc. Roy. Irish Acad. II, **3** (Sci.): 670–693. 1882.

Topography, geology, general features of flora; annotated list of 291 vascular plants.—See also **Praeger, R. L.** Phanerogams and vascular cryptogams. *In* Contributions to the natural history of Lambay, County Dublin. Irish Nat. **16**: 90–99. pl. 20–25 (incl. map). 1907. (Plant associations, notes on rarer species, additions to Hart's list; total number about 322 species and subspecies.)

FERMANAGH

See also Ireland, General (Dickie, A flora of Ulster . . . ; Lett, Botanists in the north of Ireland; Praeger, A tourist's flora of the west of Ireland).

Vice-county 33.

Local

Barrington, R. M. Report on the flora of the shores of Lough Erne. Proc. Roy. Irish Acad. II, **4** (Sci.): 1–24. 1884.

Topography, etc.; annotated list of vascular plants and Characeae (417 species).—See also **Praeger, R. L.** In camp on Lough Erne. Irish Nat. **1**: 110–114. 1892. (Includes list of additional species.)

Stewart, S. A. Report on the botany of the mountainous portion of Co. Fermanagh to the west of Lough Erne, and the adjoining district of Co. Cavan. Proc. Roy. Irish Acad. II, **3** (Sci.): 531–544. 1882.

Topography, general features of flora; annotated list of vascular plants and Characeae collected by author.—See also **Praeger, R. L.** Among the Fermanagh hills. Irish Nat. **13**: 232–241. 1 fig. 1904. (Includes list of additional species and new localities.)—Also **Carrothers, E. N., Meikle, R. D.,** and **Moon, J. McK.** County Fermanagh plants. Irish Nat. Jour. **8**: 397. 1946; **9**: 51–52, 224–228. 1947–49; **10**: 46–50. 1950. (Lists of noteworthy plants. Titles vary slightly.)

GALWAY

See also Ireland, General (Praeger, A tourist's flora of the west of Ireland); Kerry (Barrington); Mayo (Hart, Report on the flora of the mountains of Mayo and Galway).

Vice-county 15, Southeast Galway; vice-county 16, West Galway; vice-county 17, Northeast Galway.

Local

Hart, H. C. A list of plants found in the islands of Aran, Galway Bay. 32 p. map. 20.5 cm. Dublin, 1875. (Gray Herbarium library.)

Geography, topography, geology, botanical explorations, general features of flora; annotated list of vascular plants.—See also **Nowers, J. E.,** and **Wells, J. G.** The plants of the Aran Islands, Galway Bay. Jour. Bot. **30**: 180–183. 1892. (Adds 42 species.)—Also **Colgan, Nathaniel.** Notes on the flora of the Aran Islands. Irish Nat. 2: 75–78, 106–111. 1893. (Brings total to 419 species.)—Also **Praeger, R. L.** Notes on the flora of Aranmore. Irish Nat. 4: 249–252. 1895.

More, A. G. Report on the flora of Inish-Bofin, Galway. Proc. Roy. Irish Acad. II, 2 (Sci.): 553–578. 1876.

General features of flora, comparison with that of Aran; annotated list of 323 vascular plants.—See also **Praeger, R. L.** Notes on the flora of Inishbofin. Irish Nat. 20: 165–172. 1911. (General features of flora; annotated list of additional species (82), etc.)

Kerry

Vice-county 1, South Kerry; vice-county 2, North Kerry.

General

Scully, R. W. Flora of County Kerry including the flowering plants, ferns, Characeae, etc. lxxxi, 406 p. 6 pl., map. 23 cm. Dublin, 1916. (Arnold Arboretum library.)

Botanical investigations, topography, geology, climate, general features of flora, influence of soils, vertical range, botanical districts, bibliography; annotated list of vascular plants and Characeae (840 native or naturalized species and subspecies), with localities.

Local

Barrington, R. M. Report on the flora of the Blasket Islands, Co. Kerry. Proc. Roy. Irish Acad. II, 3 (Sci.): 368–391. 1881.

General features of flora; annotated list of vascular plants of Blasket Islands; tabular list of vascular plants from Tory Island, Co. Donegal; Aran Island, Co. Donegal; Innishbofin Island, Co. Mayo; Arran Islands, Galway Bay; and Blasket Islands.—See also **Praeger, R. L.** Notes on the flora of the Blaskets. Irish Nat. 21: 157–163. 1912. (Brings total to 216 vascular plants for the Blasket Islands.)

Hart, H. C. Report upon the botany of the Macgillicuddy's Reeks, Co. Kerry. Proc. Roy. Irish Acad. II, 3 (Sci.): 573–593. 1882.

Topography, geology, general features of flora; annotated list of vascular plants grouped by altitudinal range.

Kildare

See also Ireland, General (Knowles).
Vice-county 19.

Local

Knowles, M. C. The Douglas collection in the herbarium of the National Museum. Irish Nat. 14: 11–13. 1905.

List of newly recorded or rare vascular plants of County Kildare, with localities.—See also her New plants from Co. Kildare. l.c. 14: 14. 1905.

Kilkenny

Vice-county 11.

Laoighis (Leix; formerly Queen's)

Vice-county 14.

King's

See Offaly.

Leitrim

See also Ireland, General (Dickie, A flora of Ulster . . . ; Praeger, A tourist's flora of the west of Ireland); Antrim (Barrington, Report on the flora of Ben Bulben and the adjoining mountain range in Sligo and Antrim).
Vice-county 29.

Local

Stewart, S. A. Report on the botany of Lough Allen, and the Slieveanierin Mountains. Proc. Roy. Irish Acad. II, 4 (Sci.): 426–442. 1885.

Topography, geology, general features of flora; annotated list of 364 plants (289 vascular) observed by author.

LEIX (QUEEN'S)

See Laoighis.

LIMERICK

See also Ireland, General (Praeger, A tourist's flora of the west of Ireland); also Clare (Stewart, Report on the botany of south Clare and Shannon).
Vice-county 8.

Local

Knowles, M. C., and O'Brien, C. G. The flora of the Barony of Shanid. Irish Nat. **16**: 185–201. 1907.

Geography, geology, general features of flora; list of rarer vascular plants and Characeae, with localities.

LONDONDERRY (DERRY)

See Ireland, Partial (Dickie, A flora of Ulster; Lett, Botanists in the north of Ireland; Stewart and Corry, A flora of the north-east of Ireland).
Vice-county 40.

LONGFORD

See Westmeath (Barrington, Report on the flora of the shores of Lough Ree).
Vice-county 24.

LOUTH

Vice-county 31.

MAYO

See also Ireland, General (Dickie, A flora of Ulster . . . ; Praeger, A tourist's flora of the west of Ireland); Kerry (Barrington).
Vice-county 26, East Mayo; vice-county 27, West Mayo.

Local

Hart, H. C. Report on the flora of the mountains of Mayo and Galway. Proc. Roy. Irish Acad. II, **3** (Sci.): 694–768. 1883.

General features of flora; lists of species from different localities, grouped by altitude.—See also **Colgan, Nathaniel.** Botanical notes on the Galway and Mayo highlands. Irish Nat. **9**: 111–118. 1900.

Marshall, E. S. Plants observed in West Mayo, June, 1899. Jour. Bot. **38**: 184–188. 1900.

List of more interesting vascular plants and Characeae collected in general vicinity of Mallaranway, with localities.

Praeger, R. L. The flora of Achill Island. Irish Nat. **13**: 265–289. 1 fig. (map). 1904.

Botanical explorations, topography, general features of flora, plant associations; list of 416 species and subspecies of vascular plants and Characeae; annotations on the rarer species, statistics.—See also his Achill Island plants. Irish Nat. **14**: 220–221. 1905. (Adds 3 species.)

——— The flora of Inishturk. Irish Nat. **16**: 113–125. 1907.

Topography, general features of flora, botanical explorations; list of vascular plants and Characeae (323 species), with annotations on rarer species; statistics.

——— The flora of the Mullet and Inishkea. Irish Nat. **14**: 229–244. 1905.

Botanical explorations, general features of flora, plant formations; list of vascular plants and Characeae of Inishkea (185 species and subspecies); list of additional species seen on Mullet (165 species). The Mullet is a peninsula, Inishkea two small islands 2 or 3 miles off the Mullet.

——— Phanerogamia and Pteridophyta. Clare Island survey pt. 10. Proc. Roy. Irish Acad. v. 31, pt. 10. 112 p. 6 pl. (incl. map). 1911.

Ecology, etc.; includes list of 393 vascular plants.—See also **Colgan, Nathaniel.** Gaelic plant and animal names, and associated folklore. l.c. v. 31, pt. 4. 30 p. 1911. (General considerations; annotated list of Gaelic names and some unusual English names (also animal names, etc.) from Clare Island and Clew Bay on mainland.)—Also **Praeger, R. L.** The flora of Clare Island. Irish Nat. **12**: 277–294. 3 fig. (incl. map). 1903. (Topography, geology, general features of flora; list of known vascular plants and Characeae, with annotations on various species, etc.)

Meath

Vice-county 22.

Monaghan

See Ireland, General (Dickie, A flora of Ulster . . . ; Lett, Botanists in the north of Ireland).
Vice-county 32

Offaly (King's)

Vice-county 18.

Queen's

See Laoighis.

Roscommon

See Ireland, Partial (Praeger, A tourist's flora of the west of Ireland); Westmeath (Barrington, Report on the flora of the shores of Lough Ree).
Vice-county 25.

Sligo

See also Ireland, General (Dickie, A flora of Ulster . . . ; Praeger, A tourist's flora of the west of Ireland).
Vice-county 28.

Local

Barrington, R. M., and **Vowell, R. P.** Report on the flora of Ben Bulben and the adjoining mountain range in Sligo and Leitrim. Proc. Roy. Irish Acad. II, 4 (Sci.): 493–517. pl. 5 (map). 1885.

Topography, general features of flora; annotated list of vascular plants and Characeae (430 species).—See also **Corry, T. H.** On the heights attained by plants on Ben Bulben. Proc. Roy. Irish Acad. II, 4 (Sci.): 73–77. 1884. (List of vascular plants, with altitudes attained, from 300 feet to summit (1721 feet), with a few from Ben Weisken (to 1963 feet).)—Also **Barrington, R. M.** Sisyrinchium angustifolium on the Ben Bulben range. Irish Nat. **13**: 207–208. 1904—Also **Praeger, R. L.** Plants of the Ben Bulben district. Irish Nat. **14**: 221. 1905. (Notes on 5 species.)—Also **Winder, F. G.,** and **Moore, J. G.** Some notes on the rarer plants of the Ben Bulben range. Irish Nat. Jour. **9**: 68–72. 1947. (Notes on several species.)

Webb, D. A. The vegetation of Carrowkeel, a limestone hill in northwest Ireland. Jour. Ecology **35**: 106–129. 5 fig. (fig. 1, map). 1947.

Ecological; includes lists of plants, ecologically arranged [totaling 248 species].

Tipperary

Vice-county 7, South Tipperary; vice-county 10, North Tipperary.

Local

Hart, H. C. On the botany of the Galtee Mountains, Co. Tipperary. Proc. Roy. Irish Acad. II, 3 (Sci.): 392–402. 1881.

Geology, topography, general features of flora; list of vascular plants, grouped by altitudinal range.—See also **Stelfox, A. W.** Comments on H. C. Hart's "Botany of the Galtee Mountains . . ." Irish Nat. Jour. **8**: 138–142. 1944.—Also his **Further** notes on the flora of the Galtee Mountains and the Glen of Aherlow. l.c. **8**: 327–331. 1946.

Tyrone

See Ireland (Dickie, A flora of Ulster . . . ; Lett, Botanists in the north of Ireland).
Vice-county 36.

Waterford

See Wexford (Hart, Report on the flora of the Wexford and Waterford coasts).
Vice-county 6.

Westmeath

Vice-county 23.

General

Levinge, H. C. The plants of Westmeath. Irish Nat. **3**: 77–80, 96–100, 128–132. 1894 (reprinted 16 p.); **4**: 64. 1895; **5**: 44–47. 1896.

General features of flora, etc.; annotated list of vascular plants and Characeae, "from which all the commoner species have been excluded." Covers County Westmeath.—See also **Linton, E. F.,** and **Linton, W. R.** Westmeath plants. Jour. Bot. **34**: 119–122. 1896. (List of noteworthy species.) —Also **Johnson, Thomas,** and **Knowles, M. C.** The Levinge herbarium. Scient. Proc. Roy. Dublin Soc. n.s., **10**: 122–132. 1903. (List of noteworthy species with data, mostly from Westmeath.)

Local

Barrington, R. M., and **Vowell, R. P.** Report on the flora of the shores of Lough Ree. Proc. Roy. Irish Acad. II, **4** (Sci.): 693–708. 1888.

Topography, notes on distribution, etc.; annotated list of scarcer vascular plants and Characeae, with localities, the species already recorded from all the botanical districts of Ireland in the Cybele Hibernica and its Supplement being omitted. (In Longford, Roscommon, and Westmeath.)—A few additional species are listed by **R. L. Praeger** in Irish Nat. **10**: 39–41. 1901.

Wexford

Vice-county 12.

Local

Hart, H. C. Report on the flora of the Wexford and Waterford coasts. Scient. Proc. Roy. Dublin Soc. n.s., **4**: 117–146. 1883.

General features of flora; annotated list of vascular plants; list of 153 vascular plants of Saltee Islands.—See also **Barrett-Hamilton, Gerald,** and **Glascott, L. S.** Plants found near New Ross, Ireland. Jour. Bot. **27**: 4–8. 1889. (Lists of noteworthy spermatophytes from Counties of Wexford, Kilkenny, and Waterford, with localities.)—Also **Moffat, C. B.** Plants near Ballyhyland, Co. Wexford. l.c. **27**: 105–107. 1889.—Also **Barrett-Hamilton, G.,** and **Glascott, L. S.** Plants found near Kilmanock, Co. Wexford. l.c. **28**: 87–89. 1890.—Also **Barrett-Hamilton, G.,** and **Moffat, C. B.** Notes on Wexford plants. l.c. **30**: 198–200. 1892.—Also their The characteristic plants of Co. Wexford. Irish Nat. **1**: 156–158, 181–183. 1892.—Also **Marshall, E. S.** Some plants observed in Co. Wexford, 1897. Jour. Bot. **36**: 46–51. 1898.

Praeger, R. L. Notes on the flora of the Saltees. I. Phanerogamia. Irish Nat. **22**: 181–191. 1913.

Topography, botanical explorations, changes in flora and fauna due to withdrawal of human influence; list of present flora; comparison with Hart's list of 1883 (in his Report on the flora of the Wexford and Waterford coasts, Scient. Proc. Roy. Dublin Soc. n.s., 4: 117–146. 1883).

Wicklow

Vice-county 20.

General

Brunker, J. P. Flora of the County Wicklow: flowering plants, higher cryptogams and Characeae. [xii] 310 p. map. 22.5 cm. Dundalk, 1950.

Bibliography, botanical explorations, topography, geology, climate, man's influence on flora, calcicole and calcifuge species, vertical distribution, paleobotany, botanical districts; annotated list of vascular plants and Characeae

(804 species), with localities; index to localities. Replaces (as to Wicklow) **Moore, David, and More, A. G.** Catalogue of the flowering plants and ferns of Dublin and Wicklow. Scient. Proc. Roy. Dublin Soc. n.s., **1:** 190–227. 1878. (Reprinted 38 p.) (Annotated list.)—See also **Brunker, J. P.** Flora of Co. Wicklow—new records. Irish Nat. Jour. **11:** 250–251. 1955.

MAN, ISLE OF

See also Scotland, General (Cameron, The Gaelic names of plants).
Vice-county 71.

General

Paton, C. I. A list of flowering plants, ferns, and horsetails of the Isle of Man. North West. Nat. v. 8, suppl. 67 p. 1933.

Botanical investigations (with references), etc.; annotated list of vascular plants and Characeae [721 species], with a few Manx vernacular names; list of escaped plants and doubtful records. Reprinted in Proc. Isle of Man Nat. Hist. and Antiq. Soc. n.s., **3:** 547–619. 1934. (New York Public Library.) Replaces **Kermode, A. S. P.** The Isle of Man. List of flowering plants. 21 p. 21 cm. Douglas, 1900. (Annotated list of angiosperms; list of introduced species. Extracted from Yn Lioar Manninagh **2:** 273–291. 1900. Original not seen; reprint in British Museum (Natural History) library.)—See also **Paton, C. I.** Isle of Man botanical report and revision of catalogues of Manks flowering plants and ferns. 1934–1943. North West. Nat. **19:** 164–170. 1944 [1945]. (Additions and corrections, but no actual revision.)—Also his Botanical notes for 1945. Jour. Manx Mus. **5:** 188. 1946. (List of noteworthy vascular plants, with localities. In British Museum (Natural History) library.)—Also his Botanical report, 1945–1947. Peregrine **1** (5): 19–21. 1948. (Similar list. In British Museum (Natural History) library.)—Also **Allen, D. E.** Recent work on the Manx flora. Proc. Bot. Soc. Brit. Isles **1:** 5–20. 1954. (Phytogeography, etc.; list of noteworthy vascular plants and Characeae, including about 100 additions; list of excluded or doubtful species; bibliography.)

Stenning, E. H. List of the flowering plants, ferns, horse-tails and club-mosses. *In* his Isle of Man. Appendix B. p. 399–416. 22 cm. London, 1950. (The County Books series.)

Unannotated list of vascular plants based on Paton's 1933 list, with some Manx vernacular names.—See also Chap. XVII, The flora of the island. l.c. p. 286–297. 1950. (Popular account with mention of the ecological communities represented.)

Local

Wheldon, J. A. Flora of the Manx curraghs. Lancashire Nat. **2:** 271–274, 301–304. 1909–1910. (University of Minnesota library.)

General features of region; list of principal vascular plants. (An area of flat, depressed land between the Snaefell range and the rising coastline beyond Ballaugh, Jurby and Andreas.)

SCOTLAND

See also Ireland, General (Hogan, Irish and Scottish Gaelic names; Tidestrom, The floral alphabet of the Celts).

General

Includes some partial works.

Bennett, Arthur. Records of Scottish plants for the year 1888 [–1901]; additional to "Topographical botany," ed. 2. Scott. Nat. **10** (n.s. 4): 99–113. 1889; **11** (n.s., 5): 85–88, 137–141, 185–190. 1891. Annals Scott. Nat. Hist. **1892:** 119–127. 1892; **1893:** 95–101. 1893; **1894:** 158–164. 1894; **1895:** 114–118. 1895; **1896:** 113–115. 1896; **1897:** 246–248. 1897; **1898:** 225–229. 1898; **1899:** 92–94. 1898; **1900:** 159–164. 1900; **1901:** 100–102. 1901; **1902:** 32–39, 102–103. 1902.—Additional records to "Topographical botany" . . .

l.c. **1905:** 235–239. 1905; **1906:** 170–171. **1906.**—Some additional botanical records to Scottish counties. l.c. **1908:** 251–252. 1908.

Lists of new vice-county records, with sources but unannotated, arranged under the vice-counties.

Cameron, John. The Gaelic names of plants (Scottish, Irish, and Manx), collected and arranged in scientific order, with notes on their etymology, uses, plant superstitions, etc., among the Celts, with copious Gaelic, English and scientific indices. new and rev. ed. xv, 160 p. port. 22 cm. Glasgow, 1900. (1st ed., 1883, reprinted from Scott. Nat. v. 5–6. 1879–82.)

Annotated, systematic list of scientific names, with Gaelic, Irish, and Manx vernacular equivalents (with translations). According to Hogan (Irish and Scottish Gaelic names . . . p. xi, 1900; see under Ireland) the work contains the "wildest and most absurd attempts at derivation."—See also Hogan under Ireland, General, and Wood, below.

Ewing, Peter. The Glasgow catalogue of native and established plants, being a contribution to the topographical botany of the western and central counties of Scotland. 2. ed. 8, 166 p. 21.5 cm. Glasgow, 1899. (Arnold Arboretum library.) (1st ed. 1892.)

List of vascular and cellular plants, with vice-county records. Covers western Scotland from Ayrshire to western Ross, including the Hebrides. Replaces **Glasgow Society of Field Naturalists.** A contribution towards a complete list of the fauna and flora of Clydesdale and the west of Scotland. 148 p. 19 cm. Glasgow, 1876. (List of vascular plants by Richard McKay, p. 54–84. In Arnold Arboretum library.)

Grierson, Robert. Clyde casuals, 1916–1928. Glasgow Nat. **9:** 5–51. 1931.

Localities, sources of origin, etc.; annotated list of adventive angiosperms (also 2 mosses), with references, summarizing the work of author and others from about 1916 on.

Hooker, W. J. Flora scotica; or a description of Scottish plants, arranged both according to the artificial and natural methods. 2 v. (x, 292; 297 p.). 22 cm. London, 1821.

Keyed, descriptive flora of vascular and cellular plants, with notes on uses. Vol. 1, classes 1–23 of the Linnaean system; vol. 2, cryptogams (pteridophytes p. 152–161) and synoptical descriptive flora of spermatophytes on the natural system with references to fuller treatment in vol. 1.

Lawson, Peter, and son. Synopsis of the vegetable products of Scotland in the museum of the Royal Botanic gardens of Kew. v. p. pl. 23 cm. Edinburgh, 1852.

Classified account of wild and cultivated edible, forage, timber, and otherwise useful plants (but not medicinal plants); refers mostly to cultivated plants. Each section paged separately. At head of title: The Lawsonian collection.

Lee, J. R. The flora of the Clyde area. A handbook of the flowering plants and ferns occurring wild or established within the drainage area of the River and Firth of Clyde. xvi, 391 p. 18 cm. Glasgow, 1933.

Geology, general features of flora, previous publications; annotated, keyed, descriptive flora of vascular plants and Characeae, with indication of vice-county records and some localities. Covers counties of Bute, Dumbarton, Lanark, and Renfrew, large areas of Argyll, Ayrshire, and Stirlingshire, and parts of Kircudbrightshire, Peeblesshire, western Perthshire, West Lothian, and Wigtownshire. Replaces **Hennedy, Roger.** The Clydesdale flora . . . 5. ed. xxxii, 277 p. 8 pl. on 4, port. 17 cm. Glasgow, 1891. (1st ed. 1865.) (Keys to genera on Linnaean system and on natural system; annotated descriptive flora of vascular plants. In Gray Herbarium library.) Also replaces **Elliott, G. F. S., Laurie, Malcolm,** and **Murdoch, J. B.,** *ed.* Fauna, flora and geology of the Clyde area. 567 p. map. 22 cm. Glasgow, 1901. (Includes partly annotated lists, with local distribution indicated, of "Ferns and their allies" by **William Stewart** (p. 106–109) and "Phanerogams" by **Peter Ewing** (p. 110–130); also a table of "Measurements of notable trees" by **John Renwick** and **Richard McKay** (p. 131–147). Relates "to the land draining into the River and Firth of Clyde only.")—See also **Lee, J. R.** Additions to The flora of the Clyde area. Glasgow Nat. **17:** 65–82. 1953. (Extensive list.)

Murray, Alexander. The northern flora; or, a description of the wild plants belonging to the north and east of Scotland with an account of their places of growth and properties. Pt. I (all publ.). xvii, 150, xiv p. 2 facsim. 22.5 cm. Edinburgh, Aberdeen, and London, 1836. (New York Botanical Garden library.)

Descriptive account of angiosperms, on the Linnaean system (Monandria-Pentandria), with localities; included here for its account of plant uses.

Trail, J. W. H. Topographical botany of Scotland. Annals Scott. Nat. Hist. **1898**: 39–46, 98–111, 164–177, 230–235. 1898; **1899**: 95–103, 165–172. 1899; **1900**: 32–42, 103–108. 1900.—Scottish Rubi. l.c. **1902**: 170–176, 233–244. 1902; **1903**: 41–47, 103–107. 1903.—Scottish Hieracia. l.c. **1902**: 244–250. 1902.—Additions and corrections . . . l.c. **1905**: 174–177, 224–235. 1905; **1906**: 33–48, 93–100, 172–184, 230–233. 1906; **1907**: 226–232. 1907; **1908**: 44–48. 1908; **1909**: 178–180. 1909.

List of Watsonian provinces and vice-counties; systematic list of vascular plants and Characeae, with the numbers of the vice-counties in which they are known, and occasional annotations. The "Additions and corrections" and the Supplement (begun 1907: 230) are similarly arranged; the latter is unfinished, ending with Naiadaceae. The Salicaceae, omitted in the first series of papers, are treated l.c. **1906**: 181–184. 1906.

——— Topographical botany of the river-basins Forth and Tweed in Scotland. Trans. and Proc. Bot. Soc. Edinb. **22**: 277–308. 1904.

Geography, botanical explorations (with references to publications); partly annotated list of known vascular plants and Characeae.

West, George. A comparative study of the dominant phanerogamic and higher cryptogamic flora of aquatic habit, in three lake areas of Scotland. Proc. Roy. Soc. Edinb. **25**: 967–1023. 110 fig. on 55 pl. 1905.

General features of lakes in the Loch Ness region (Inverness), the island of Lismore (Argyll), and between Nairn and the Culbin sandhills (Nairn); annotated list of aquatic and semi-aquatic vascular plants and some groups of cellular plants, with localities; descriptive account of the lakes mentioned. —See also his A further contribution to a comparative study of the dominant phanerogamic and higher cryptogamic flora of aquatic habit in Scottish lakes. Proc. Roy. Soc. Edinb. **30**: 65–181. 62 pl. on 31. 1910. (Topography, etc.; annotated list of vascular and cellular plants (p. 71–96); descriptive account of the lakes studied. Refers to areas in Kirkcudbrightshire, Wigtownshire, Fife, and Kinross.)

White, F. B. W. The edible wild fruits of Scotland. Scott. Nat. **3**: 22–28, 73–76. 1875.

Annotated list, with uses.

Wood, John. Scotch names of native wild flowers. Annals Andersonian Nat. Soc. [**1**]: 89–103. 1893.

Running account of old Scotch names of native plants, grouped according to habit of plant or origin of names, with botanical names and notes on etymology and uses. The names are taken from the 1818 edition of John Jamieson, An etymological dictionary of the Scottish language.—See also Cameron, above, and additional reference there given.

ABERDEENSHIRE

See also Banff (Marshall and Shoolbred, Plants observed near Tomintoul, N. B.); also Inverness (Trail, The flowering plants and fern allies of the Cairngorms).

Vice-county 92, South Aberdeen; vice-county 93, North Aberdeen.

General

Dickie, George. The botanist's guide to the counties of Aberdeen, Banff, and Kincardine. xxxii, 344 p. pl., map. 18 cm. Aberdeen [etc.] 1860. (Gray Herbarium library.)

Climate, etc.; topography and geology (by Alexander Cruickshank); annotated list of vascular and cellular plants, with localities.

Local

Macgillivray, P. H. A catalogue of the flowering plants and ferns growing in the neighbourhood of Aberdeen. viii, 44 p. 16 cm. Aberdeen [etc.] 1853. (Gray Herbarium library.)

Topography, general features of flora; annotated list of vascular plants. Covers area from the estuary of the River Ythan south to Garron Point (Kincardine) and west to Banchory Ternan. Replaces **Dickie, George.** Flora abredonensis: comprehending a list of the flowering plants and ferns found in the neighbourhood of Aberdeen; with remarks on the climate, the features of the vegetation, etc., 2 p. l., v, 70 p. 18 cm. Aberdeen (etc.), 1838. (Climate, soils, general features of flora, etc.; annotated list of vascular plants, with localities, on the Linnaean system. In British Museum (Natural History) library.)—See also Trail, Flora of the city parish of Aberdeen, below.

Macgillivray, William. The flora of the Dee, arranged according to the natural system. *In* his The natural history of Dee Side and Braemar. Edited by Edwin Lankester. p. 335–383. 22.5 cm. London, 1855. (Gray Herbarium library.)

Botanical regions; list of vascular plants (p. 337–374) and lichens with habitats and for the scarcer species the localities.—See also his The alpine plants of Braemar. l.c. p. 309–321. (Account of alpine flora in several localities, with included list of species.)—Also his Vegetation of the valley of the Dee. l.c. p. 322–334. (General account of vegetation.)—Also **Barton, J.** Notes on the flora of Braemar, Aberdeenshire. Phytologist n.s., **2:** 281–284, 308–312, 341–346, 417–423. map. 1857–58. (Topography, geology, general features of flora, partial lists of species of different localities.)—Also **Crombie, J. M.** Botany—flowering plants and cryptogamics [sic]. *In* his Braemar: its topography and natural history. p. 54–68. 17 cm. Aberdeen and Edinburgh, 1861. (Mention of some of the rarer plants. In Harvard College library.)—Also **Babington, C. C.** List of plants seen in the valley of Braemar and on Morrone. Scott. Nat. **11:** 81–85, 132–136, 174–184. 1891. (General features of flora, etc.; annotated list of vascular plants and Characeae.)

Roy, John. The flowering plants and ferns of Lochnagar. *In* McConnochie, Alexander I. Lochnagar. p. 161–178. 20 cm. Aberdeen, 1891. (Harvard College library.)

Unannotated list of vascular plants.

Sim, John. On the botany of Scotston Moor and its neighbourhood. A paper read to the Aberdeen Natural History Society . . . January 16, 1866. 30 p. 21 cm. Aberdeen, 1868. (British Museum (Natural History) library.)

General features of flora; annotated list of vascular plants, bryophytes, Characeae, lichens, and diatoms. Scotston Moor is a few miles north of Aberdeen.

Stuart, Peter. Flora of Cairney [sic] parish. *In* Pirie, James. The parish of Cairnie and its early connection with Strathbogie. p. 191–198. 19 cm. Banff, 1906. (Library of N. D. Simpson.)

Unannotated list of phanerogams, excluding Cyperaceae and Gramineae, followed by mention of some noteworthy species.

Trail, J. W. H. The flora of Buchan. Trans. Buchan Field Club **6:** 69–162. map. 1901. (Reprinted 94 p. Peterhead, 1902. In British Museum (Natural History) library.)

Botanical explorations, bibliography; list of vascular plants, summarizing previous records; tabular list of vascular plants and Characeae based on author's work, showing details of local distribution. Covers the region bounded on the north and east by the North Sea, on the south by the River Ythan, on the west by the River Deveron, thus including a small area in Banff.—See also his The flora of Buchan. Annals Scott. Nat. Hist. **1901:** 164–176. 1901. (Annotated list of rare or local species.)—Also his Additions to the flora of Buchan. l.c. **1902:** 45–50, 97–102. 1902.—Also his The flora of Buchan—its distribution, origin, and relations to man. Trans. Buchan Field Club **8:** 2–56. (pref., 1905.) (General features of flora, soils, etc.; annotated list of vascular plants and Characeae (over 700 species); discussion of alien plants, habitats, etc.; additional records for many species. In Harvard College library.)—Also **Matthews, J. R.** The flora of the north-east. *In* Tocher, J. F., *ed.* The book of Buchan (jubilee volume) . . . p. 41–47, 49. pl. 22.5 cm. Aberdeen, 1943. (Sketch of botanical work, pre- and post-glacial flora, etc.; includes (p. 42) partial list of species additional to those recorded by Trail.)

——— Flora of the city parish of Aberdeen. *In* Aberdeen Univ. Studies 91 (James William Helenus Trail: a memorial volume): 57–331. map. 1923.

Changes in topography and flora; annotated list of vascular plants.—See also Macgillivray, above.

——— Florula of a piece of waste ground near Aberdeen. Annals Scott. Nat. Hist. **1896**: 231–245. 1896; **1897**: 24–31, 237–245. 1897; **1899**: 221–230. 1899.

Includes list of 230 angiosperms found (in 1893–96) on 7 or 8 acres of reclaimed ground, with discussion. The third and fourth papers continue the history of the locality for 1897–99.

——— Man's influence on the indigenous flora of Aberdeen. Annals Scott. Nat. Hist. **1911**: 175–180, 232–240. 1911.

Historical sketch of botanical explorations; annotated list of species that have disappeared from the parish of Aberdeen.

Angus (formerly Forfarshire)

Vice-county 90.

General

Gardiner, William. The flora of Forfarshire. xxiv, 306 p. 2 pl. 19 cm. London, 1848.

Topography, botanical districts, list of agricultural weeds; annotated list of vascular and cellular plants, with localities.—See also **Brown, Robert.** The botanical history of Angus. Jour. Bot. 9: 321–327. 1871. (First publication of a manuscript read before the Edinburgh Natural History Society in 1792, containing an annotated list of rarer vascular plants, with localities. French translation (by H. A. Weddell) with title Histoire botanique du comté d'Angus, in Bul. Soc. Bot. France 19: 214–222. 1872.)—Also **Moir, Edward.** Botany. *In* Warden, Alexander J. Angus or Forfarshire, the land and people . . . v. 2, p. 173–193. 24 cm. Dundee, 1881. (General features of flora, with partial lists of species; also partial lists of mosses, hepatics, and lichens.)—Also **Druce, G. C.** Notes on the flora of Forfarshire. Annals Scott. Nat. Hist. **1901**: 104–107. 1901.—Also **Marshall, E. S., and Shoolbred, W. A.** Some Forfarshire plants. Jour. Bot. 43: 114–116. 1905.—Also **Corstophine, R. H.** Forfarshire records. l.c. 46: 299. 1908.—Also **Brebner, James.** The flora of Forfarshire. *In* Paton, A. W., and Miller, A. H. Handbook and guide to Dundee and district . . . p. 597–610. map. 19 cm. Dundee (1912). (At head of title page: British Association, Dundee 1912.) (Account of various localities, with mention of noteworthy species. In British Museum (Natural History) library.)—Also **Marshall, E. S.** Forfarshire plants. Jour. Bot. 54: 211. 1916.—Also **Salmon, C. E.** Notes on Forfarshire plants. l.c. 56: 261–264. 1918.—Also **Graham, R. J. D.** The flora of the Dundee district. *In* Mackie, R. L., *ed.* A scientific survey of Dundee and district . . . Adv. Sci. v. 1, no. 1, suppl. p. 25–29. 1939. (General features of vegetation, botanical regions, botanical publications, list of scientific societies. Covers the whole of Angus, the estuary region of the Tay known as the Carse of Gowrie with the extension of the Sidlaws into Perthshire, and the North of Fife to the East Neuk.)

Local

Arbroath Horticultural and Natural History Association. The flora of Arbroath and its neighbourhood: being a list of flowering plants and ferns, with an appendix of mosses, lichens, and seaweeds. viii, 63 p. 15 cm. Arbroath, 1882. (British Museum (Natural History) library.)

Briefly annotated list of vascular plants (p. 1–45), mosses, lichens, and algae, with additions by W. B. Simson on p. 55–56.

Roger, J. G. The flora of Caenlochan. Trans. and Proc. Bot. Soc. Edinb. 36: 188–194. 1 fig. (map). 1954.

Topography, climate, etc.; lists of species characteristic of different phytogeographical groups, etc.; bibliography.

Willis, J. C., and Burkill, J. H. (i.e., **I. H.**). The phanerogamic flora of the Clova Mountains in special relation to flower biology. Trans. and Proc. Bot. Soc. Edinb. **22:** 109–125. 1902.

Tabular list of known spermatophytes (363 species), with altitude, flowering dates, and notes on floral biology particularly in relation to secretion of honey.

Argyllshire

See also Scotland, General (Grierson, Clyde casuals; Lee, The flora of the Clyde area; West, A comparative study of the dominant . . . flora of aquatic habit); Hebrides, Inner (for islands belonging to Argyll); Inverness (Macvicar, Notes on doubtfully native and introduced plants in Westerness; Ravenshaw, A list of some plants found in the West Highlands); Perthshire (Patton, The vegetation of Beinn Laoigh).

Vice-county 98, Argyll (mainland); vice-county 101, Cantyre. Part of Argyll is also in vice-county 97 (Westerness).

Local

Balfour, J. H. Account of a botanical excursion to the Mull of Cantyre and the Island of Islay, in August, 1844. Phytologist **2:** 291–312, 321–328. 1845.

Itinerary, general features of flora, etc.; list of vascular plants (524 species), with localities.—See also **Somerville, Alexander.** Additions to the known flora of the South Ebudes, v.–c. 102. Jour. Bot. **34:** 433. 1896. (Additional species for Islay.)—Also **Salmon, C. E.** Notes from Cantire. l.c. **36:** 338–340. 1898.—Also his Plant notes from Sutherland and Cantire. l.c. **38:** 299–303. 1900.—Also **Gilmour, J. S. L.** Ten days on West Loch Tarbert. North West. Nat. **20:** 34–37. 1946. (Account of various habitats with mention of noteworthy species.)—The name Cantire or Cantyre is now generally spelled **Kintyre.**

Lee, J. R. The flora of the Arrochar Mountains. Annals Andersonian Nat. Soc. **3:** 80–126. pl. 1–3. 1908. (University of Illinois library.)

Topography, general features of flora; annotated list of vascular plants. Covers the region of Ben Arthur, Ben Narnain, Crois, Ben Ime, Ben Vane, and Ben Vorlich.

Nisbet, Thomas. Phanerogams and ferns of South Ardgoil. Annals Andersonian Nat. Soc. **4:** 1–33. 4 fig. on 2 pl. 1914. (University of Illinois library.)

Topography, plant habitats, etc.; annotated list of vascular plants. Covers region bounded on north by Corrie Odhair and Coilessan Glen, on west by Loch Goil, and on east by Loch Long.

Sprague, Beatrice. Notes on the formation and flora of a shingle island in the River Orchy, Dalmally, Argyll. Trans. Edinb. Field Nat. and Micros. Soc. **5:** 290–315. 3 maps. 1906.

Topography, sources of flora, etc.; includes tabular list of 143 plants (all but 3 vascular).

Ayrshire

See also Scotland, General (Ewing, The Glasgow catalogue . . . ; Grierson, Clyde casuals; Lee, The flora of the Clyde area).

Vice-county 75.

General

Amos, Thomas. Ayrshire plants. Annals Kilmarnock Glenfield Ramblers' Soc. **5:** 32–50. 1907. (British Museum (Natural History) library.)

Unannotated list of known vascular plants (p. 32–47); list of ferns (p. 48–50) by Alexander Laing.

Gilchrist, Andrew. A short sketch of Ayrshire botany and botanists. Annals Kilmarnock Glenfield Ramblers' Soc. **5:** 25–31. 1907. (British Museum (Natural History) library.)

Account of local botanists and their work.

Landsborough, David. The botany of Ayrshire. Published 1872. Republished 1894. Annals Kilmarnock Glenfield Ramblers' Soc. [1] 1893/94: 6–28. port. 1894. (British Museum (Natural History) library.)

List of vascular plants (506 species), with localities and names of collectors. The 1872 publication (not seen) was prepared by Mr. Borland, Mr. Duncan, and David Landsborough, and was based on lists in the volume (v. 5) including Ayrshire in the New statistical account of Scotland. 1845. (Information supplied by G. Taylor.)— See also **Landsborough, David, Jr.**, and **Boyd, D. A.** Supplement to The botany of Ayrshire. Annals Kilmarnock Glenfield Ramblers' Soc. [2] 1898/99: 19–31. 1898. (Extensive list of additional species and localities. In British Museum (Natural History) library.)—Further additions are given in the following series of papers by **Gilchrist, Andrew: Notes on some additions to the botany of Ayrshire.** l.c. **6**: 33–37. 1910; Some old Ayrshire plant records. l.c. **7**: 13–20. 1913; Botany of Ayrshire. Additions to lists already published in the "Annals." l.c. **7**: 64. 1913. (Those in vol. 7 not examined, other in British Museum (Natural History) library.)—Also **Clark, James. Additions to the list of** Ayrshire flowering plants and higher cryptogams already recorded by the Glenfield Ramblers. l.c. **11**: 65–69. 1934. (British Museum (Natural History) library.)

Smith, John. The botany of Ayrshire (by parishes), from original investigation. 107 p. 19 cm. Ardrossan, [pref. 1896]. (Lloyd Library.)
Previous publications; list of known vascular plants, with localities. Printed only on recto of pages.—See also **Rogers, W. M.** Some Clydesdale and s. w. Ayrshire plants. Jour. Bot. **40**: 54–59. 1902.

Local

Balfour, J. H. Account of a botanical excursion to Ailsa Crag, in July, 1844. Phytologist **2**: 257–263. 1845.

Topography, geology, general features of flora, etc.; unannotated list of vascular (86 species) and cellular plants.—See also **Smith, John.** Phaenogamia and filices. *In* The natural history of Ailsa Crag. Annals Andersonian Nat. Soc. **2**: 148–151. 1900. (General features of flora; list of 72 vascular plants observed. The paper also includes a bibliography, p. 152–154.)—Also **Vevers, H. G.** The land vegetation of Ailsa Craig. Jour. Ecology **24**: 424–445. 9 fig. (incl. map), pl. 20–23. 1936. (Includes (p. 444–445) unannotated lists of known angiosperms, ferns, mosses, hepatics, and lichens, arranged alphabetically; brief bibliography.)

Banffshire

See also Aberdeen (Dickie, The botanist's guide to the counties of Aberdeen, Banff, and Kincardine; Trail, The flora of Buchan); Inverness (Trail, The flowering plants and fern allies of the Cairngorms).
Vice-county 94.

General

Craib, W. G. The flora of Banffshire. 1 p. l., 111 p. 21.5 cm. Aberdeen, 1912. (Aberdeen university studies no. 54.) (Also issued as Trans. Banffsh. Field Club, extra number; in Kew library.)

Collaborators, previous publications, etc.; list of vascular plants and Characeae, with localities and authorities for records; tabular list of species, sparingly annotated, showing distribution by parishes.

Local

Craib, W. G. Flora in vicinity of Banff. [Extract from Report of joint meeting of northern literary and scientific societies held at Banff, 1906.] p. 38–68. 15.5 cm. [Banff?, 1906?] (British Museum (Natural History) library.)

List of vascular plants and Characeae, with localities.—See also **Mahood, A. E.** Local flora. *In* Spriggs, Edmund I., *ed.* Banff and district. p. 305–339. 18.5 cm. Banff, 1919 [1920]. (Running account of vascular plants of vicinity of city of Banff, arranged by habitats; bibliography. In Kew library.)

Gordon, J. F. S. Botany. *In* his The book of the chronicles of Keith, Grange, Ruthven, Cairney, and Botriphnie: events, places, and persons. p. 281–301. 26 cm. Glasgow, 1880.

Annotated list of vascular plants (the phanerogams arranged by the Linnaean system) and bryophytes from the parish of Keith and vicinity, with notes on uses; based on data supplied by William Robb and James T. George.

Marshall, E. S., and Shoolbred, W. A. Plants observed near Tomintoul, N. B., July, 1905. Jour. Bot. 44: 154–161. 1906.

List of vascular plants and Characeae collected by authors, with localities. (Mostly in Banffshire (v.c. 94), partly in South Aberdeen (v.c. 92) and East Inverness (v.c. 96).)

BERWICKSHIRE

See also Scotland, General (Trail, Topographical botany of the river-basins Forth and Tweed); Selkirkshire (Hayward and Druce, The adventive flora of Tweedside).
Vice-county 81.

General

Johnston, George. The botany of the eastern borders, with the popular names and uses of the plants . . . (Added title-page: Terra lindisfarnensis. The natural history of the eastern borders. v. 1. Botany.) xii, 336 p. illus., 13 pl. (i.e. 15, incl. map, 1 col.). 22 cm. London, 1853.

Topography soils, climate, hydrography; annotated list of vascular and cellular plants, with localities, vernacular names and uses; fossil flora (by Geo. Tate). The area covered includes the whole of Berwickshire, the Liberties of Berwick, north Durham, and adjacent Northumberland and Roxburghshire. There are many minor additions, particularly in the volumes of the History of the Berwickshire Naturalists' Club.

Local

Anderson, Adam. List of less common plants in the area of the Club, with localities and references. Hist. Berwicksh. Nat. Club **22** (1914): 227–272. 1916 [1914?] (Reprinted, 46 p., [1915]; in British Museum (Natural History) library.)

Systematic list of species recorded in the Proceedings of the Club from 1831 on, with locality, date, and references to Proceedings. Relates to Berwick and various other counties, including Northumberland.—See also **Druce, G. C.** Notes on the flora of Berwickshire. Annals Scott. Nat. Hist. **1907:** 96–101. 1907. (Extensive list.)—Also **Evans, A. H.** Notes on plants found in the district worked by the Berwickshire Naturalists' Club. Hist. Berwicksh. Nat. Club **23** (1916): 217–235. 1917. (List intended to supplement Anderson.)

Kelly, Andrew, and Shaw, William. Botany. *In* Thomson, A. Lauder and Lauderdale. p. 313–341. 23 cm. Galashiels [pref. 1902]. (Harvard College library.)

Annotated list of vascular plants, with localities.

BUTESHIRE

See also Scotland, General (Lee, The flora of the Clyde area).
Vice-county 100 (Clyde Isles).

Local

Bryce, James, [jr.]. The flora of Arran. *In* **Bryce, James,** [sr.]. **The** geology of Arran . . . 4. ed. p. 228–251. 18.5 cm. Glasgow and London, 1872. (1st ed. 1859, with title Geology of Clydesdale and Arran . . .)

General features of flora; annotated list of plants, including for the most part only the rarer species.—See also **Landsborough, David,** and **Landsborough, David,** jr. List of rarer phenogamous plants found in Arran. Reused and enlarged by Prof. Balfour. *In* their Arran: its topography, natural history, and antiquities. [2. ed.] p. 477–482. 19 cm. Ardrossan, London, 1875. (Alphabetical list of of angiosperms and pteridophytes, with localities, followed by shorter lists of cellular plants. In library of George Taylor.)—See also **Travis, W. G.** A contribution to the flora of Arran. Trans. and Proc. Bot. Soc. Edinb. **26:** 120–129. 1913. (Includes list of rare or critical phanerogams and mosses.)—Also **Downie, R. A.** Flowers and plants. *In* his All about Arran. p. 133–142. 17 cm. London and Glasgow, 1933 (reprinted 1935). (General account of flora.)—Also his Appendix III. List of plants. l.c. p. 148–158. (Annotated list of vascular plants, prepared with assistance of Ian W. Tervet; much the same species and localities as given by Bryce, with some corrections in nomenclature.)

CAITHNESS-SHIRE

Vice-county 109.

General

Grant, J. F., and Bennett, Arthur. Contributions towards a flora of Caithness. [I.] Scott. Nat. **9**: 305–307, 357–372. 1888; **10**: 39–43, 77–95. 1889. (Reprinted 41 p. Perth, 1889.)

Bibliography; briefly annotated list of over 600 vascular plants and Characeae, intended to summarize knowledge of the flora.—See also **Bennett, Arthur.** Contributions . . . no. II, IV–V. Annals Scott. Nat. Hist. **1892**: 247–252. 1892; **1904**: 224–233. 1904; **1905**: 36–46. 1905; **1910**: 225–229. 1910; **1911**: 44–51. 1911. (Additional species and localities.)—Also **Grant, J. F.** Flora. *In* Horne, John, *ed.* The county of Caithness. p. 338–356. pl. 22.5 cm. Wick, 1907. (Account of various localities, with mention of characteristic species. In library of George Taylor.)—Also **Crampton, C. B.** The vegetation of Caithness considered in relation to the geology. 132 p. illus. (incl. map). 24.5 cm. Edinburgh, 1911 (cover date). (Ecological; includes (p. 18) list of vascular plants not previously recorded, with localities.)—Also **Bennett, A.** Recent additions to the Caithness flora. Scott. Bot. Rev. **1**: 181–182. 1912. (Also in Trans. and Proc. Bot. Soc. Edinb. **26**: 58–60. 1913.)—Also his Caithness plants. Jour. Bot. **51**: 278–280. 1913.—Also **Marshall, E. S.** Plants of W. Sutherland and Caithness. l.c. **54**: 169–172. 1916.—Also **Bennett, A.** Ulex nanus, Foster in Caithness. Trans. and Proc. Bot. Soc. Edinb. **27**: 135. 1917.—Also his Notes on the flora of Caithness. l.c. **27**: 309–312. 1919.—Also his Notes on Caithness plants. l.c. **29**: 54–56. 1924.

Local

Bennett, Arthur. Contribution towards a flora of Caithness. no. III. Plants of Stroma Isle, Pentland Firth. Annals Scott. Nat. Hist. **1900**: 108–119. 1900.

List of vascular plants and Characeae, with localities, collected by Miss A. M. Geldart in June-July 1899.

CLACKMANNANSHIRE

See Scotland, General (Trail, Topographical botany of the river-basins Forth and Tweed); Perthshire (White, The flora of Perthshire).
Vice-county 87 in part (with South Perthshire and parts of Stirling).

DUMFRIESSHIRE

Vice-county 72.

General

Elliot, G. F. S., and others. The flora of Dumfriesshire, including part of the stewartry of Kirkcudbright. xl, 219 p. map (in pocket). 22 cm. Dumfries, 1896. (Arnold Arboretum library.)

Topography, geology, climate, etc.; annotated list of vascular plants, with localities; index of insects and their host plants. Replaces (as to Dumfries) **M'Andrew James.** A list of the flowering plants of Dumfriesshire and Kirkcudbrightshire . . . 51 p. 21 cm. Dumfries, 1882. (Briefly annotated list of vascular plants, with localities. In Arnold Arboretum library.)

Local

Arnott, Samuel. Popular plant names. Trans. and Jour. Proc. Dumfriessh. and Galloway Nat. Hist. and Antiq. Soc. **17**: 404–410. 1906.—Local plant names. l.c. **23**: 195–201. 1911.—Some local and other plant names. l.c. **24**: 223–228. 1912.

Running account of local vernacular names of many plants.

Corrie, John. Flora. A list of some of the less common plants found in Glencairn. *In* his Glencairn (Dumfriesshire). The annals of an inland parish. Appendix B. p. 186–193. 22 cm. Dumfries, 1910. (Harvard College library.)

Partial list of vascular plants, with localities.

Davidson, Anstruther. The flora of Sanquhar and Kirkconnel parishes. *In* Brown, James. The history of Sanquhar . . . to which is added The flora and fauna of the district by Dr. Anstruther Davidson. [Appendix] p. 1–27. 23 cm. Dumfries [etc.], 1891. (Harvard College library.)

General features of flora; annotated list of vascular plants, Characeae, and mosses (496 species and varieties.) A "2. ed." of the work was issued in the same year.

Johnstone, J. T. Botany. *In* Forrest, William, *pub.* Forrest's guide to Moffat, Moffat district and the Yarrow Locks. 5. ed. p. 107–114. 18 cm. Moffat [pref. 1900]. (Library of George Taylor.)

Sources for records; unannotated list of vascular plants.—See also **Linton, W. R., and Marshall, E. S.** Plants observed near Moffat, Dumfries, July, 1907. Jour. Bot. 46: 212–215. 1908. (List of vascular plants, with localities.)

Dunbartonshire (Dumbartonshire)

See also Scotland, General (Grierson, Clyde casuals; Lee, The flora of the Clyde area); Argyll (Lee, The flora of the Arrochar Mountains); Lanark (Hopkirk, Flora glottiana); Midlothian (general works).

Vice-county 99, partly (an exclave) in vice-county 77.

Local

Lee, J. R. The flora of Easter Dumbartonshire (parishes of Kirkintilloch and Cumbernauld). Glasgow Nat. **15**: 1–18. 1945.

Discussion of vice-counties, topography, etc.; unannotated list of known vascular plants, with indication of occurrence in Clyde and Forth areas. Easter Dumbarton is the exclave to the east of the county proper.

East Lothian (formerly Haddington)

See also Scotland, General (Trail, Topographical botany of the river-basins Forth and Tweed); Midlothian (Balfour and Sadler, Flora of Edinburgh . . . Martin, The Field-club flora of the Lothians; Sonntag, A pocket flora of Edinburgh . . .).

Vice-county 82.

Local

Balfour, J. H. Botany of the Bass. *In* M'Crie, Thomas, and others. The Bass Rock . . . p. 409–431. pl. 21 cm. Edinburgh, 1848.

Remarks on various plants; unannotated list of angiosperms. In some copies the plate is colored and some lack the date on the title page. (Small island at mouth of the Firth of Forth.)—See also his Notes on the flora of the Bass Rock. Proc. Bot. Soc. Edinb. **1855**: 30–32. 1855. (Topography, general features of flora: list of plants observed in 1854 (60 species, of which 39 were vascular plants).)

Edinburgh

See Midlothian.

Elginshire

See Nairnshire and Morayshire.

Fifeshire

See also Scotland, General (Trail, Topographical botany of the river-basins Forth and Tweed in Scotland; West, A comparative study of the dominant . . . flora of aquatic habit); Angus (Graham, The flora of the Dundee district, under Gardiner).

Vice-county 85, in part (includes also Kinross).

General

Young, William. A list of the flowering plants and ferns recorded from Fife and Kinross (V. C. 85). Trans. and Proc. Bot. Soc. Edinb. **32**: 1–173. map. 1936.

Topography, botanical explorations, botanical districts, bibliography; annotated list of known vascular plants and Characeae.—See also review by **Taylor, George.** Jour. Bot. 75: 147–151. 1937.—Also **Macleay, K.N.G.** Additions and emendations to the Comital Flora for Fife and Kinross (v.–c. 85). **Watsonia 2:** 398–410. 1953. (Extensive list; bibliography.)

Local

Borwick, James. Botany of the parish. *In* Campbell, James. Balmerino and its abbey. Appendix III. p. 339–344. 20 cm. Edinburgh, 1867.

List of vascular plants, with localities of rarer species. The same list is given in the new ed. p. 606–610. Edinburgh and London, 1899.

Eggeling, W. J. The flowering plants and ferns of the Isle of May, Firth of Forth, Fife (v.–c. 85). Trans. and Proc. Bot. Soc. Edinburgh **37:** 11–23. 1956.

Botanical explorations, etc.; annotated list of vascular plants; bibliography. Replaces **Sadler, John.** Notes on the flora of the Isle of May, Firth of Forth. Trans. and Proc. Bot. Soc. Edinburgh **11:** 390–392. 1873. (Geology, general features and flora; list of vascular and cellular plants collected by author and others on August 11, 1871.)

Laing, Alexander. List of plants growing in the neighbourhood of Newburgh. *In* his Lindores Abbey and its burgh of Newburgh . . . Appendix XXIII. p. 552–559. 23.5 cm. Edinburgh, 1876.

List of vascular plants, with localities.

Wood, Walter. Botany of the east coast of Fife. List of phenogamous plants, ferns, and mosses. *In* his The East Neuk of Fife. 2. ed. rearranged and enl. Edited . . . by . . . J. Wood Brown . . . p. 525–546. 19.5 cm. Edinburgh, 1887. (1st ed. 1862, p. 351–367.)

List of vascular plants and mosses, with localities. Relates to region between the Firth of Forth on the south and the high ridge of barren ground separating it from the How of Fife on the north, and extends from Elie east to Fifeness and Kingsbarn and west to Scone.

Hebrides, Inner (Ebudes)

See also Scotland, General (Ewing, The Glasgow catalogue); Argyll (Balfour, Account of a botanical excursion to the Mull of Cantyre and the island of Islay).

Vice-county 102, South Ebudes (Islay, Jura, Colonsay, Oransay, etc.); vice-county 103, Mid Ebudes (Mull, Coll, Tiree, etc.); vice-county 104, North Ebudes (Skye, Raasay, Scalpay, Canna, Rhum, Eigg, Muck, etc.). The islands are in the Counties of Argyll and Inverness.

Local

Harrison, J. W. H. The flora of the islands of Rhum, Eigg, Canna, Sanday, Muck, Eilean nan Each, Hyskeir, Soay and Pabbay. Proc. Univ. Durham Phil. Soc. **10:** 87–123. 8 pl. (incl. map). 1939.

Geology; annotated list of vascular plants (650 species, segregates, and hybrids).—See also **Harrison, J. W. H., Clark, W. A., Cooke, R. B.,** and **Harrison, H. H.** Further observations on the flora of the isle of Rhum. Jour. Bot. **79:** 113–116. 1941.—Also **Harrison, J. W. H., Harrison, H. H., Clark, W. A.,** and **Cooke, R. B.** Vascular plants from the Isle of Rhum (V.–C. 104) and the Isle of South Uist (V.–C. 110). l.c. **80:** 113–116. 1942.—Also **Raven, J. E.** Alien plant introductions on the island of Rhum. Nature **163:** 104–105. 1949. (Note on *Polycarpon tetraphyllum, Carex bicolor,* and some others considered to be introduced rather than native.)—Also **Harrison, J. W. H.** Further observations on the vascular plants of the Outer and Inner Hebrides. Trans. and Proc. Bot. Soc. Edinb. **35:** 415–426. 1951. (Extensive list of vascular plants, mostly from Coll and Tiree, Rhum, and Eigg.) —Also **Harrison, J. W. H.,** and **Morton, J. K.** Botanical investigations in the isles of Raasay, Rhum (v.c. 104), Lewis and Harris (v.c. 110) in 1951. Proc. Univ. Durham Phil. Soc. **11:** 12–23. 1951.—Also Macvicar, On the flora of Eigg, below.

———Introduced vascular plants in the Scottish Western Isles. North West. Nat. **23:** 132–135. 1948.

Extensive list of plants, mostly escaped from cultivation or planted, from both Inner and Outer Hebrides, with localities.

Harrison, J. W. H., Harrison, H. W., Clark, W. A., and **Cooke, R. B.** The flora of the Isles of Coll, Tiree and Gunna (V.-C. 110 B). Proc. Univ. Durham Phil. Soc. **10**: 274–308. map. 1941.

Botanical explorations, phytogeographical relationships; annotated list of vascular plants and Characeae (573 species, segregates, and hybrids), with localities.—See also **Harrison, J. W. H.** Further observations on the vascular plants of the Outer and Inner Hebrides. Trans. and Proc. Bot. Soc. Edinb. **35**: 415–426. 1951. (Extensive list of vascular plants, mostly from Coll and Tiree, Rhum, and Eigg.)—Also Macvicar, On the flora of Tiree, below.

———and others. The flora of the Isle of Raasay and of the adjoining islands of South Rona, Scalpay, Fladday and Longay. Proc. Univ. Durham Phil. Soc. **9**: 260–304. 19 fig. on 11 pl. 1937.

Botanical explorations, geology, general features of flora; annotated list of vascular plants and Characeae (603 species, segregates, and hybrids).—See also supplementary papers by Harrison (and others). Rpt. Bot. Soc. and Exch. Club Brit. Isles **11** (1936): 299–303. 1937; **11** (1937): 529–533. 1938. Proc. Univ. Durham Phil. Soc. **11**: 12–23. 1951.

Lawson, M. A. On the flora of Skye. Jour. Bot. **7**: 108–114. 1869.

Phytogeography; unannotated list of vascular plants and Characeae (389 species) collected by author and others on a single visit, with occasional localities.—See also **Linton, W. R.,** and **Linton, E. F.** Additions to the recorded flora of Skye. Jour. Bot. **22**: 367–369. 1884.—Also **Linton, E. F.** and **Linton, W. R.** New county records for Skye, Ross, Sutherland, and Caithness. l.c. **27**: 207–209. 1889. (Records for Skye, p. 207–208.)—Also **Salmon, C. E.** Notes on Scottish plants. l.c. **48**: 225–226. 1910. (List of noteworthy vascular plants collected by A. Wallis, mostly on Skye.)—Also his Notes on plants from Skye. l.c. **54**: 165–169. 1916. (Collected by A. Wallis.)

Loder, J. DeV. Plants. Adapted from Murdoch McNeill's Colonsay. *In* his Colonsay and Oronsay in the isles of Argyll; their history, flora, fauna and topography. Appendix IV. p. 314–385. 1 fig. (map). 25 cm. Edinburgh, London, 1935.

Botanical explorations, geology, soils, local distribution, etc.; annotated list of vascular plants (p. 328–380), bryophytes, charophytes, and algae, with Gaelic vernacular names, localities for rarer species, notes on uses, etc. Replaces **McNeill, Murdoch.** Colonsay, one of the Hebrides . . . vii, 216 p. 19 cm. Edinburgh, 1910. (Includes (p. 86–202) account of botanical explorations, etc.; annotated list of vascular plants and Characeae (580 species and 70 varieties), with vernacular names and uses.

Macvicar, S. M. On the flora of Eigg. Annals Scott. Nat. Hist. **1897**: 34–41, 110–117. 1897.

Geology, etc.; annotated list of vascular plants and Characeae (373 species and varieties).—See also **Spence, D. H. N.** Flora of Eigg: additions and corrections. Trans. and Proc. Bot. Soc. Edinb. **36**: 74–79. 1952. (Additional species and localities; bibliography.)—Also **Davis, P. H.** Additions to the flora of Eigg (V. C. 104). l.c. **37**: 111–113. 1957.—Also Harrison, The flora of the islands of Rhum, Eigg . . . above.

——— On the flora of Tiree. Annals Scott. Nat. Hist. **1898**: 31–38, 81–97. 1898. (Reprinted, 24 p., in British Museum (Natural History) library.)

Topography, woody vegetation, geology, phytogeography, etc.; annotated list of vascular plants and Characeae collected by author in 1896–97.—See also Harrison and others, The flora of the Isles of Coll, Tiree and Gunna, above.

Ritchie, Alec, and **Ritchie, Euphemia.** Botany of Iona and Staffa. *In* their Iona past and present. [3. ed.?] p. 29–32. 22 cm. Edinburgh, 1934. (Library of George Taylor.)

Unannotated list of vascular plants and a few seaweeds. The list was not in earlier editions.

Ross, George. On the flora of Mull. Trans. and Proc. Bot. Soc. Edinb. **13**: 234–242. 1878.

Topography, geology, general features of flora; annotated list of vascular plants collected by author.—See also **Wilmott, A. J.,** *comp.* . . . V.-c. 103. The island of Mull . . . *In* Report on the excursions arranged in 1939. Rpt. Bot. Soc. and Exch. Club. Brit. Isles **12** (1939–40): 236–249. 1942. (Itinerary, botanical explorations, etc.; extensive list of vascular plants.)

Hebrides, Outer

See also Scotland, General (Ewing, The Glasgow catalogue); Hebrides, Inner (Harrison, Introduced vascular plants).
Vice-county 110. In Counties of Inverness and Ross and Cromarty.

General

Harrison, J. W. H. A preliminary flora of the Outer Hebrides. Proc. Univ. Durham Phil. Soc. **10:** 228–273. 7 pl. (incl. map). 1941.

Annotated list of 697 species, segregates, and hybrids known from the group, with localities; bibliography (recent papers only).— See also papers, mostly by Harrison (in part with others) in Proc. Univ. Durham Phil. Soc. **10:** 358–367. 1946; **11:** 1–11, 12–23, 83–90, 135–142. 1951–54; **12:** 29–34, 141–149. 1954–56; **13:** 54–62. 1957. Trans. and Proc. Bot. Soc. Edinb. **35:** 132–156, 415–426. 1950–51.

Local

Atkinson, Robert. Notes on the botany of North Rona and Sula Sgeir. Trans. and Proc. Bot. Soc. Edinb. **33:** 52–60. 2 fig. (maps). 1940.

List of known vascular plants; topography, general features of flora, etc. (North Rona or Ronay is about 38 miles northeast of the Butt of Lewis; Sula Sgeir is 12 miles west of North Rona.)

——— and **Roberts, Brian.** Notes on the islet of Gasker. Scott. Nat. **64:** 129–137. 4 fig. 1952.

Includes (p. 130) list of vascular plants (42 species). (An islet 5½ miles west of Husinish Point on the Atlantic coast of Harris.)

Campbell, M. S. The flora of Uig (Lewis). A botanical exploration. 63 p. 5 (i.e. 6) pl., map. 22.5 cm. Arbroath, 1945.

Botanical explorations, etc.; annotated list of vascular plants; general features of flora, plant associations, and critical notes, by A. J. Wilmott; lists of lichens and algae, by other authors; bibliography.—See also **Wilmott, A. J.** Further botanising in Uig. Scott. Nat. **60:** 82–90. 1948. (List of noteworthy species.)

——— A visit to Scalpay (Harris), V.–C. 110. Rept. Bot. Soc. and Exch. Club Brit. Isles **12** (1941–42): 543–545. 1944.

Includes full list of vascular plants collected on a hasty visit in July 1939 to this island at the entrance to East Loch Tarbert, which had not previously been botanized. There is another island named Scalpay in the Inner Hebrides.

Clark, W. A. The flora of the islands of Mingulay and Berneray. Proc. Univ. Durham Phil. Soc. **10:** 56–70. map. 1938.

Topography, geology, general features of flora, etc.; annotated list of 206 vascular plants.—See also his Noteworthy plants from North Uist, Balesnare, Monach Islands, Harris, Taransay, Mingulay and Berneray, (V. C. 110). l.c. **10:** 124–129. 1939.

Harrison, J. W. H., and **Harrison, J. H.** The vascular plants of Stuley Island, the Isles of Grimsay and Ronay, with some remarks on the flora of Benbecula, South Uist and Barra. Proc. Univ. Durham Phil. Soc. **10:** 499–515. 1950.

Geology, etc.; annotated list of vascular plants of Stuley, Grimsay, Ronay, and Baleshare, with some records from the other islands.

Roberts, Brian, and **Atkinson, Robert.** The Haskeir Rocks, North Uist, Scott. Nat. **67:** 9–18. 3 fig. 1955.

General features of vegetation; list of vascular plants (12 species); zoology, bibliography.

Trail, J. W. H. The plants of the Flannan Islands. Annals Scott. Nat. Hist. **1905:** 187. 1905.

Short annotated list of plants collected by W. Eagle Clarke in 1904, considered to represent almost the entire phanerogamic flora.—See also **Bennett, Arthur.** The plants of the Flannan Islands. l.c. **1907:** 187. 1907. (Adds 9 species.)

Turrill, W. B. The flora of St. Kilda. Rpt. Bot. Soc. and Exch. Club Brit. Isles **8** (1927): 428–444. 1928.

Bibliography; annotated list of vascular plants; analysis of flora.—See also **Petch, C. P.** Additions to the flora of St. Kilda. Jour. Bot. **70:** 169–171. 1932.—Also **Poore, M. E. D.,** and **Robertson, V. C.** The vegetation of St. Kilda in 1948. Jour. Ecology **37:** 82–99. 2 fig. (maps). 1949. (Includes (p. 87, 89-90) notes on changes on vegetation and on introduced plants.)

INVERNESS-SHIRE

See also Scotland, General (Ewing, The Glasgow catalogue; West, A comparative study of the dominant . . . flora of aquatic habit); Banffshire (Marshall and Shoolbred, Plants observed near Tomintoul); Hebrides, Inner and Outer, for islands belong to Inverness.

Vice-county 96, in part (Easterness; includes East Inverness and Nairn); vice-county 97, in part (Westerness; includes West Inverness and part of Argyll).

Inverness, with the addition of Nairn and the northwest part of Argyll, is customarily divided by British botanists into the vice-counties Easterness and Westerness. Easterness includes that part of Inverness east of the watershed between east and west Scotland and also the county of Nairn. Westerness includes that part of Inverness west of this line and also that part of Argyll that is west of Loch Linnhe and northeast of the island of Mull—that is, the districts of Ardgour, Ardnamurchan, and Morven.

Local

Don, John. Flowering plants of Inverness-shire and of some parts of adjoining counties. Part I: Dicotyledons . . . Classified by John Don . . . And collected by the author, with the assistance of Mr. D. M. Shaw and Mr. A. Mackenzie, Inverness, and Mr. John Donald, Inchmore. 1 p. l., 21 p. 18.5 cm. Inverness, [ca. 1898]. (Gray Herbarium library.)

Briefly annotated list of dicotyledons and gymnosperms, with localities. Cover-title only.

Druce, G. C. Notes on the flora of Easterness, Elgin, Banff, and West Ross. Jour. Bot. **26:** 17–26, 116. 1888.

Extensive list of vascular plants and Characeae, with localities and indication of new records.—See also his Plants of Easterness and Elgin. Jour. Bot. **27:** 200–205. 1889. (Running account of author's trip, with mention of noteworthy plants collected.)—Also his Notes on Scotch plants. l.c. **28:** 39–47. 1890. (Extensive list of vascular plants, with localities and some annotations, from Easterness, Perth (Strath Tay), Banff, Elgin, and Ross-shire.) —Also **Marshall, E. S.,** and **Shoolbred, W. A.** Some plants of East Scotland. l.c. **37:** 383–389. 1899. (List of vascular plants from Elgin, East Inverness, Nairn, and East Ross, with localities and some annotations.)—Also **Rogers, W. M.** North-east Highland plants (1903). l.c. **42:** 12–21. 1904. (Lists of plants seen at Tomatin and Dalwhinnie; extensive list of vascular plants from Easterness, Banff, and Elgin.)—Also **Marshall, E. S.,** and **Shoolbred, W. A.** Plants of the Dalwhinnie district, July, 1911. l.c. **51:** 164–167. 1913. (List of noteworthy vascular plants, with localities and some annotations.)

Lang, G. A. A list of wild flowers and ferns of Inverness and district. 33 p. 18 cm. Inverness, [1905?]. (British Museum (Natural History) library.)

List of vascular plants, mostly from vicinity of Inverness, with localities for scarcer species. Cover title only.

Macvicar, S. M. Notes on doubtfully native and introduced plants in Westerness. Annals Scott. Nat. Hist. **1896:** 176–187. 1896.

General considerations; annotated list of doubtful natives, denizens, colonists, aliens, and casuals of the districts of Ardnamurchan and Sunart (in Argyll) and Moidart (in Inverness).

Ravenshaw, T. F. A list of some plants found in the West Highlands, chiefly in the vicinity of Loch Moidart, Inverness-shire, 1856. Phytologist n.s., **2:** 206–208. 1857.

List of vascular plants with localities.—See also **Linton, W. R.,** and **Linton, E. F.** Plants recorded in Westerness additional to "Topographical Botany." Jour. Bot. **22:** 371–373. 1884. (List of vascular plants and Characeae, in part with annotations, collected in an area south of Loch Sunart, in Argyll.) —Also **Linton, E. F.,** and **Linton, W. R.** List of plants observed . . . in V. C. 97 "Westerness" (i.e. the western watershed of Inverness, with that part of Argyleshire N.–W. of Loch Linnhe) in August, 1884. Rpt. Bot. Rec. Club **1884/86:** 153–155. 1887. (List of vascular plants and Characeae, with localities. In Gray Herbarium library.)—Also **Miller, W. F.** New Westerness plants. Jour. Bot. **33**: 345. 1895. (Short list of angiosperms.)—Also **Salmon, C. E.** Notes on Westerness plants. l.c. **41**: 271–275. 1903. (List of vascular plants from vicinity of Corpach, with localities.)—Also **Druce, G. C.** Notes on the flora of Westerness. Annals Scott. Nat. Hist. **1904**: 36–42. 1904. (Extensive list of vascular plants and Characeae from region of Fort William, with localities.)—Also **Marshall, E. S.** Inverness-shire plant notes. Jour. Bot. **55**: 129–134. 1917. (List of noteworthy vascular plants, with localities, from the vicinity of Loch Laggan, in Westerness and Easterness.) —Also **Pigott, Donald.** Some notes on the vegetation and flora of Knoydart. Scott. Nat. **63**: 50–55. 1951. (Includes short lists of vascular and cellular plants from several localities.)

Trail, J. W. H. The flowering plants and fern allies of the Cairngorms. Cairngorm Club Jour. **1**: 197–212. 1895. (Kew library.)

General features of flora, etc.; annotated list of vascular plants, based principally on the work of Dr. John Roy.

Kincardineshire

See also Aberdeen (Dickie, The botanist's guide to the counties of Aberdeen, Banff, and Kincardine; Macgillivray, A catalogue of the flowering plants and ferns growing in the neighbourhood of Aberdeen).

Vice-county 91.

Local

Robertson, E. T., and Gimingham, C. H. Contributions to the maritime ecology of St. Cyrus, Kincardineshire. **Trans. and Proc. Bot. Soc. Edinb. 35**: 370–414. 4 fig. (incl. map). 1951.

Includes (p. 412–414) list of vascular plants (179 species), bryophytes, and lichens.

Watt, V. J. B. The flora of Banchory. *In* his The book of Banchory. p. 183–195. 19 cm. Edinburgh, London, 1947.

List of uncommon species recorded from Banchory Ternan in Dickie's Botanist's guide (1860), followed by notes on some additional species.

Kinross-shire

See Scotland, General (Trail, Topographical botany of the river-basins Forth and Tweed in Scotland; West, A comparative study of the dominant . . . flora of aquatic habit); Fife (Young, A list of the flowering plants and ferns recorded from Fife and Kinross).

Vice-county 85 in part (includes also Fife).

Kirkcudbrightshire

See also Scotland, General (Lee, The flora of the Clyde area; West, A comparative study of the dominant . . . flora of aquatic habit); Dumfriesshire (Elliot, The flora of Dumfriesshire, including part of the stewartry of Kirkcudbright; Arnott, Popular plant names).

Vice-county 73.

General

M'Andrew, James. A list of the flowering plants of Dumfriesshire and Kirkcudbrightshire compiled for the Dumfriesshire and Galloway Natural History and Antiquarian Society. 51 p. 21 cm. Dumfries, 1882. (Arnold Arboretum library.)

Bibliography; annotated list of vascular plants, with localities; list of rarer plants of Wigtownshire.—See also his The botany of Kirkcudbrightshire. *In* Maxwell, J. H. Maxwell's guide book to the stewartry of Kirkcudbright. 6. ed. p. 181–187. 19 cm. Castle-Douglas, 1896. (General features of flora, with mention of noteworthy species of vascular plants, and some cellular plants. In British Museum (Natural History) library.)—Also **Fraser, James.** Some Galloway plants. Trans. and Jour. Proc. Dumfriessh. and Galloway Nat. Hist and Antiq. Soc. III, **2**: 29–34. 1914. (List of newly recorded or rare plants from Kirkcudbright and Wigtown.)

Local

Jeffrey, Mrs. List of wild and naturalized flowers found on St. Mary's Isle, Kirkcudbright. Trans. and Jour. Proc. Dumfriessh. and Galloway Nat. Hist. and Antiq. Soc. **18**: 46–47. 1907.

Unannotated list of angiosperms.

Lanarkshire

See also Scotland, General (Grierson, Clyde casuals; Lee, The flora of the Clyde area); Dunbartonshire (Lee, The flora of Easter Dumbartonshire); Midlothian (Balfour and Sadler, Flora of Edinburgh; Sonntag, A pocket flora of Edinburgh and the surrounding district); Renfrew (Ross, List of flowering plants . . . in the district around Busby).
Vice-county 77.

General

Mackechnie, Robert. Plant recording in Clydesdale. Glasgow Nat. **18:** 3–14. 1958.

Sketch of botanical explorations in Lanark, remarks on introduced plants, bibliography.

Patrick, William. A popular description of the indigenous plants of Lanarkshire . . . xxxiv, 399 p. 16 cm. Edinburgh, 1831. (Gray Herbarium library.)

Topography, geology, climate, etc.; descriptive flora of spermatophytes (604 species), on the Linnaean system, with localities; glossary. The second edition (1 p. l., xxxiv, 399 p. 16 cm. Edinburgh, 1832. In Arnold Arboretum library) is precisely the same as the first as to prefatory matter and text; the dedicatory leaf was replaced by 4 pages of review notices, and a new spine was added to the volume over the original one.

Local

Hopkirk, Thomas. Flora glottiana. A catalogue of the indigenous plants of the banks of the river Clyde, and in the neighbourhood of the city of Glasgow. 1 p. l., iv, 3–170 p. 22 cm. Glasgow [etc.], 1813. (Gray Herbarium library.)

Annotated list of vascular and cellular plants, on the Linnaean system, with localities. Covers in general "the country within a few miles on both sides of the river Clyde, from its Falls, above Lanark, to its junction with the sea some miles below the town of Dumbarton."—See also **Patton, Donald.** The local flora. *In* Kerr, J. G., *ed.* Glasgow. Sketches by various authors. [For the British Association meeting.] p. 298–321. 18.5 cm. Glasgow, 1928. (Brief historical account, followed by description of plant habitats with lists of species; also paragraphs on cellular plants. In British Museum (Natural History) library.)

Patton, Donald. The vegetation of the Tinto Hills. Annals Andersonian Nat. Soc. 4: 84–105. 1925. (University of Illinois library.)

Geology, climate, general features of vegetation; tabular list of vascular plants, with indication of habitats.

——— and **Rennie, William.** The plants of Possil Marsh. Glasgow Nat. **17:** 160–172 incl. fig. (map). 1955.

History, etc.; annotated list of vascular plants and Characeae (216 species); bibliography. (An area of about 70 acres, in Glasgow.)—See also **Ribbons, B. W.** Additions to the flora of Possil Marsh. Glasgow Nat. **17:** 283. 1957. (Adds 3 species.)

Linlithgow

See West Lothian.

Midlothian (formerly Edinburgh)

See also Scotland, General (Trail, Topographical botany of the river-basins Forth and Tweed in Scotland).
Vice-county 83.

General

Balfour, J. H., and Sadler, John. Flora of Edinburgh, being a list of plants found in the vicinity of Edinburgh. vii, 174 p. map. 17.5 cm. Edinburgh, 1863.

Topography, etc.; annotated list of vascular plants (about 1010 species) and of mosses, hepatics, lichens and Characeae, found within 20 or 30 miles of Edinburgh, covering all or part of 11 counties beside Midlothian.

Martin, I. H., *ed.* The Field-Club flora of the Lothians. By the Botanical committee of the Edinburgh Natural History Society . . . [3. ed.] xix, 318 p. incl. 17 plates. front., map. 16 cm. Edinburgh and London, 1934. (1st ed. 1927.)

Keyed, descriptive flora of vascular plants, with some localities but no reference to vice-counties; partial ecological lists of various localities, list of casual plants.

Sonntag, C. O. A pocket flora of Edinburgh and the surrounding district; a collection and full description of all phanerogamic and the principal cryptogamic plants classified after the natural system, with an artificial key and glossary of botanical terms. xii, 246 p. map. 16.5 cm. London and Edinburgh, 1894.

Annotated, briefly descriptive flora of vascular plants, with localities. Covers an area 30 to 40 miles in diameter around Edinburgh.

Local

Fraser, James. Alien plants near Edinburgh. Annals Scott. Nat. Hist. **1904**: 106–113. 1904.—**Evans, William, and Evans, W. E.** Alien plants near Edinburgh. l.c. **1904**: 174–179, 236–240. 1904.—**Fraser, James.** Alien plants near Edinburgh. l.c. **1905**: 96–103. 1905; **1906**: 100–105. 1906.—Alien plants. l.c. **1907**: 37–42. 1907; **1908**: 101–109. pl. 3. 1908; **1909**: 40–44. 1909; **1910**: 43–46. 1910; **1911**: 99–102. 1911.—Alien plants. Scott. Bot. Rev. 1: 39–41. 1912.

Lists of adventive vascular plants and Characeae amounting to about 970 species (through 1911), mostly from Edinburgh or its near vicinity, with localities and some annotations.

Greville, R. K. Flora edinensis; or a description of plants growing near Edinburgh, arranged according to the Linnean system . . . lxxxi, 478 p. 4 pl. 21.5 cm. Edinburgh, 1824.

Annotated descriptive flora of vascular and cellular plants growing within 10 miles of Edinburgh.

Wallace, G. B. The flora of Salisbury Crags. Trans. and Proc. Bot. Soc. Edinb. **29**: 119–126. 1925.

Topography, botanical explorations; tabular list of known angiosperms (178 species); changes in flora, bibliography.—See also **Matthews, J. R.** A note on the flora of Salisbury Crags. l.c. **29**: 226–229. 1926. (Further note on changes in flora.)

MORAYSHIRE (FORMERLY ELGIN)

Vice-county 95.

General

Burgess, J. J., *ed.* **Flora of Moray (vice county no. 95). Flowering and flowerless plants including conifers, ferns, mosses, fungi and algae.** xv, 104 p. port. 22 cm. Elgin, 1935. (British Museum (Natural History) library.)

Botanical explorations, account of some noteworthy species; annotated list of vascular plants and charophytes (p. 1–40), showing distribution by parishes, compiled by the editor; treatment of Coniferae by Peter Leslie (p. 41–50), with paragraphs on biology, uses, afforestation, etc.; also annotated lists of fungi and freshwater algae, by P. Leslie, and of mosses, by George Birnie. Replaces **[Gordon, George]**. Collectanea for a flora of Moray; or, a list of the phaenogamous plants and ferns hitherto found within the province. viii, 40 p. 20.5 cm. Elgin, 1839. (Topography, climate, etc.; list of vascular plants, with localities, on the Linnaean system (724 species); phytogeographic lists, statistics. In British Museum (Natural History) library.)—See also (Field meetings, 1953.) June 27th to July 4th, 1953. Forres. Leader: Miss M. McCallum Webster. Proc. Bot. Soc. Brit. Isles 1: 246–250. 1954. (Includes several additions.)

Local

Stewart, E. J. A., and Patton, Donald. The vegetation of the Culbin Sands. Rpt. Bot. Soc. and Exch. Club Brit. Isles 7 (1923): 253–267. 1924.

History, zones of vegetation; unannotated list of vascular plants. Replaces, so far as listing of species is concerned, their earlier papers: **Patton** and **Stewart.** The flora of the Culbin Sands. Trans. and Proc. Bot. Soc. Edinb. **26:** 345–374. 7 fig. (incl. maps), pl. 57. 1915 (includes bibliography); **Stewart** and **Patton.** Additional notes . . . l.c. **29:** 27–40. 1 fig. (map). 1924. The area, extending west for 6 miles on the shore of Moray Firth (mostly in Elginshire, a little in Nairnshire), was, according to tradition, overwhelmed in 1694 by great accumulations of sand blown in by hurricanes.

Nairnshire

See also Scotland, General (West, A comparative study of the dominant . . . flora of aquatic habit); Inverness (Druce, Notes on the flora of Easterness); Moray (Stewart and Patton, The vegetation of the Culbin sands).

Vice-county 96, in part (Easterness, including also East Inverness).

Local

Thompson, Robert. List of plants found in the parish of Ardclach. *In* his The natural history of a highland parish (Ardclach, Nairnshire). Appendix. p. 265–286. 22.5 cm. Nairn, 1900. (Library of N. D. Simpson.)

Annotated list of fungi, bryophytes, lichens, and vascular plants (vascular plants, p. 268–286).

Orkney

Vice-county 111.

General

Spence, Magnus. Flora orcadensis. Containing the flowering plants arranged according to the natural orders . . . and the mosses by Lieut.-Col. James Grant. xcv, 148 p. 2 maps, 2 ports. 19 cm. Kirkwall, 1914. (Arnold Arboretum library.)

Topography, biographical sketches of local botanists, ecology, geology; annotated list of 565 vascular plants and Characeae, with localities; annotated list of medicinal plants, list of mosses, bibliography. Pages 127–137 constitute a Supplement, based mainly on the work of H. H. Johnston.—See also **Bennett, Arthur.** Notes on the flora of the Orkney Isles. Trans. and Proc. Bot. Soc. Edinb. **27:** 54–59. 1916.—Also **Johnston, H. H.** Observations on "Notes on the flora . . ." l.c. **28:** 43–46. 1920.—Also his Additions to the flora of Orkney, as recorded in Watson's "Topographical Botany" second edition. [1.]–10. paper. l.c. **25:** 173–181. 1895. (Reprinted with corrections, 11 p. 1925; reprint in British Museum (Natural History) library); **26:** 207–226. 1914; **28:** 23–42, 51–66, 98–117, 174–183. 1920–23; **29:** 83–95, 151–170, 297–307, 408–428. 1924–27. Additions . . . 11.–17. paper. 24, 10, 9, 7, 12, 8, 12 p. 22 cm. Edinburgh, 1928–34. (British Museum (Natural History) library.)—Also **Anonymous.** Mr. Magnus Spence's herbarium. Jour. Bot. **65:** 78–79. 1927. (Notice of placing of his herbarium in Stromness Museum.)

Peeblesshire

See also Scotland, General (Lee, The flora of the Clyde area; Trail, Topographical botany of the river-basins Forth and Tweed); Midlothian (Balfour and Sadler, Flora of Edinburgh; Sonntag, A pocket flora of Edinburgh and the surrounding district); Selkirkshire (Hayward and Druce, The adventive flora of Tweedside).

Vice-county 78.

General

Balfour, F. R. S. Botany. *In* Buchan, John W. A history of Peebleshire. v. 1, p. 340–428. 23 cm. Glasgow, 1925. (Reprinted 89 p.)

General features of flora; annotated list of vascular plants and mosses, list of larger fungi; account of trees, with dimensions of some large specimens.

Perthshire

See also Scotland, General (Lee, The flora of the Clyde area; Trail, The topographical botany of the river-basins Forth and Tweed); Angus (Graham, The flora of the Dundee district, under Gardiner).

Vice-county 87 in part, West Perth (87 includes also Clackmannan); vice-county 88, Mid Perth; vice-county 89, East Perth.

General

Barclay, William. [Perthshire botany and botanists.] Proc. Perthsh. Soc. Nat. Sci. 4: clxxxix–cxciv. 1908.
Sketch of local botanists and their work, from Lightfoot to J. Sim (ca. 1863).

White, F. B. W. The flora of Perthshire . . . Edited with an introduction and life of the author, a list of his scientific publications, and an appendix, by J. W. H. Trail . . . lix, 407 p. port., map. 19.5 cm. Edinburgh, 1898.
Origin of flora, memoir of author with bibliography, topography, botanical districts, geology, early history of local botany; annotated list of vascular plants and Characeae. Covers the whole of Clackmannan and part of Stirling, as well as Perth.—See also **Barclay, William.** The additions to the list of Perthshire plants since the publication of Dr. White's Flora. Trans. and Proc. Perthsh. Soc. Nat. Sci. 5: cxlviii–clxv. 1912.—Also **Matthews, J. R.** Recent additions to the list of Perthshire plants. l.c. 8: 184–210. 1927.

Local

Barclay, William. Flora of the right bank of the Tay, between Perth and the mouth of the Earn. Trans. and Proc. Perthsh. Soc. Nat. Sci. 1: 123–134. 1890.
General features of flora; unannotated list of 383 vascular plants.—See also his Flora of the left bank of the Tay between Perth and Glencarse. l.c. 1: 245–255. 1892. (General features of flora; unannotated list of 389 vascular plants.)

——— The flora of the Woody Island. Trans. and Proc. Perthsh. Soc. Nat. Sci. 1: 30–44. 1887.
General features of flora; unannotated list of vascular plants (320 species). (In River Tay, close to Perth.)

Coates, James. The flowering plants. *In* White, F.B.W. The natural history of Kinnoull Hill. Proc. Perthsh. Soc. Nat. Sci. 1: 237–244. 1886.
General features of flora, mention of rarer species; unannotated list of angiosperms.—See also **White, F. B. W.** The ferns, mosses, and fungi. l.c. 244–245, 251–254. 1886. (Includes list of pteridophytes.) Kinnoull Hill is just east of the city of Perth.

Gardiner, William. Contribution towards a flora of the Breadalbane Mountains. Phytologist 1: 468–476. 1843.
Annotated list of vascular and cellular plants collected by author in June-July, 1842.—See also **Balfour, J. B.** (i.e. **J. H.**). Notice of a botanical excursion to the Breadalbane Mountains, in July 1871. **Trans. Bot. Soc. Edinb. 11:** 353–356. 1873. (Includes list of noteworthy species.)—Also **Drummond-Hay, G. W. A.,** and **White, F. B. W.** Notes of a botanical excursion to the Breadalbane Mountains, Perthshire. Scott. Nat. 3: 18–20, 79. 1875. (List of noteworthy species.)—Also **Ewing, Peter.** The summit-flora of the Breadalbane Range. Glasgow Nat. 4: 48–62. 1912. (Ecological conditions; tabular list of 113 vascular plants showing occurrence on the summits of Ben Lawers, Beinn Laoigh, and other mountains; remarks on various species.)—Also **Lee, J. R.** Excursions in Breadalbane (Killin District), July 1915. l.c. 8: 17–22. 1916. (Includes list of rarer vascular plants and bryophytes.)

Patton, Donald. The vegetation of Beinn Laoigh. Rpt. Bot. Soc. and Exch. Club Brit. Isles 7 (1923): 268–319. 1924.
Botanical explorations, topography, geology, climate, general features of flora of region and of special localities, statistics; tabular list of known vascular plants (388 species), with localities indicated; phytogeography, list of mosses and hepatics, bibliography. This mountain, also known as Ben Lui, is 3,708 feet high.

Sadler, John. The wild flowers and ferns of Bridge of Earn including a complete list of the Perthshire ferns. 56 p. 1 fig. 15.5 cm. Edinburgh, [1863?]. (Gray Herbarium library.)
Includes (p. 49–56) incomplete lists of spermatophytes of Moncreiffe and other localities in the vicinity of Bridge-of-Earn.

Renfrewshire

See also Midlothian (Balfour and Sadler, Flora of Edinburgh; Sonntag, A pocket flora of Edinburgh).

Vice-county 76.

General

Ferguson, Daniel. Renfrewshire plants. *In* Hall, Charles A., *ed.* Lists of Renfrewshire plants, macrolepidoptera . . ., with a geological map of the county and introductory notes by the editor. Trans. Paisley Nat. Soc. **2:** 1–40. 1915. (Library of N. D. Simpson.)

Eleven pages of historical and descriptive botany followed by a list of vascular plants and charophytes (740 species) entitled Renfrewshire plants (p. 12–40); also list of excluded species.

Local

Ross, William. List of flowering plants, ferns, and their allies found in the district around Busby. *In* his Busby and its neighbourhood, including the parishes of Carmunnock, East Kilbride, Mearns, and Cathcart . . . Appendix I. p. 121–134. 23 cm. Glasgow, Edinburgh, 1883. (Harvard College library.)

List of vascular plants, with localities, the widely distributed species omitted. (In Renfrew and adjacent Lanarkshire.)

Ross and Cromarty

See also Hebrides, Outer (for works relating to Lewis).

Vice-county 105, West Ross; vice-county 106, East Ross. Vice-county 110 (Hebrides, Outer) includes the Lewis portion, which is in the County of Ross, of the island of Lewis with Harris.

Local

Dixon, J. H. Flowering plants of Gairloch. *In* his Gairloch in north-west Ross-shire . . . p. 256–264. 22.5 cm. Edinburgh, 1886. (Harvard College library.)

Changes in the flora, etc.; unannotated list of flowering plants and pteridophytes.

Druce, G. C. The flora of West Ross with introductory sketches of the topography and climate and also biographical notices of those who have contributed to the flora since 1778. Rpt. Bot. Soc. and Exch. Club Brit. Isles v. 8 (1928), suppl. xxxvii, 112 p. pl. 1929.

Geology, rainfall, mountains and rivers, botanical explorations and bibliography, analysis of flora, plant formations; annotated list of vascular plants and Characeae (809 species of which 636 are native).—See also **Wilmott, A. J.**, and **Campbell, M. S.** Plants collected in West Ross. l.c. **12** (1943–44): 832–833. 1946.

——— Notes on the flora of East Ross. Jour. Bot. **20:** 356–358. 1882.

List of vascular plants and Characeae, with localities, collected in the Black Isle of Cromarty, Dingwall, and in a trip from Strathpeffer to the watershed bounding East Ross. Nearly all are new vice-county records.—See also his Notes on the flora of eastern Ross-shire. Annals Scott. Nat. Hist. **1903:** 226–234. 1903. (List of vascular plants, with localities.)—Also **Marshall, E. S.**, and **Shoolbred, W. A.** Ross-shire plants, 1909. Jour. Bot. **48:** 132–140. 1910. (List of vascular plants and Characeae, with localities, mostly from vicinity of Garve.)

Ribbons, B. W. A contribution to the flora of the Beinn Eighe Nature Reserve. Trans. and Proc. Bot. Soc. Edinb. **36:** 230–265. 1 fig. (map). 1954.

Topography, geology, botanical explorations; annotated list of vascular plants (213 species); bibliography. (In West Ross.)

Roxburghshire

See also Scotland, General (Trail, Topographical botany of the river-basins Forth and Tweed); Berwickshire (Johnston, The botany of the eastern borders); Selkirkshire (Hayward and Druce, The adventive flora of Tweedside).
Vice-county 80.

General

Duncan, James. List of indigenous plants. *In* Jeffrey, Alexander. The history and antiquities of Roxburghshire and adjacent districts . . . v. 4, p. 373–407. 19 cm. Edinburgh and London, 1864.
Annotated list of vascular plants, with localities for scarcer species.

Selkirkshire

See also Scotland, General (Trail, Topographical botany of the river-basins Forth and Tweed).
Vice-county 79.

General

Farquharson, James. List of flowering plants and ferns observed in Selkirkshire. Hist. Berwicksh. Nat. Club 1876/78: 77–88. 1879.
Topography, geology, climate, general features of flora; unannotated list of 484 vascular plants.

Local

Hayward, I. M., and Druce, G. C. The adventive flora of Tweedside. xxxii, 296 p. 79 fig., pl. 24 cm. Arbroath, 1919.
History of local woolen industry, bibliography; annotated list of adventive wool-waste vascular plants (348 species), primarily from vicinity of Galashiels, Selkirk, with some synonymy, general range, description, local occurrence; list of adventive plants not introduced on wool waste (294 species) with records, from Berwick, Peebles, Roxburgh, Selkirk, and Northumberland.

Shetland (Zetland)

Vice-county 112.

General

Druce, G. C. Flora zetlandica. Rpt. Bot. Soc. and Exch. Club Brit. Isles 6 (1921): 457–546. 1922.—Additions . . . l.c. 7 (1924): 628–657. 1925.
Geography, origin of flora, climate, botanical explorations, general features of flora, bibliography; annotated list of 505 species of vascular plants and Characeae (446 native species), including previous records. The Shetland group includes over 100 islands. The supplement increases the flora to 529 species, of which 456 are native. Replaces **Edmonston, Thomas.** Edmonston's Flora of Shetland . . . 2d ed. . . . ed. and rev. by C. F. Argyll Saxby . . . 102 p. pl., port. 17 cm. Edinburgh and London, 1903. (1st ed., Edmonston, Thomas. A flora of Shetland . . . xxvii, 67 p. 18 cm. 1845.)—See also **Johnston, H. H.** Additions to the flora of Shetland. [1. paper.] Trans. and Proc. Bot. Soc. Edinb. **29**: 429–430. 1927; 2.–4. paper. 2, 4, 3 p. 21.5 cm. Edinburgh, 1928–29. (In British Museum (Natural History) library.)—Also **Price, W. R.** Notes on the vegetation of Zetland, 1928. Rpt. Bot. Soc. and Exch. Club Brit. Isles 8 (1928): 770–781. 1929. (Plant associations; notes on general features of flora and on individual species.)

Local

Trail, J. W. H. The flora of Fair Isle. Annals Scott. Nat. Hist. **1906**: 165–170. 1906.
Botanical explorations, topography, etc.; unannotated list of vascular plants; remarks on introduced species, etc.—See also **Pritchard, N. M.** Notes on the flora of Fair Isle. Proc. Bot. Soc. Brit. Isles 2: 218–225. 1957. (Habitats, etc.; list of known vascular plants, with habitat.)—Also **Fitter, R. S. R.** Addenda to the flora of Fair Isle. l.c. **3**: 172–173. 1959.

Turrill, W. B. The flora of Foula. Rpt. Bot. Soc. and Exch. Club Brit. Isles 8 (1928): 838–850. 1929.

Botanical investigations; annotated list of vascular plants; list of vernacular names, analysis of flora, physiography, geology, life-forms.

Stirlingshire

See also Scotland, General (Grierson, Clyde casuals; Lee, The flora of the Clyde area; Trail, Topographical botany of the river-basins Forth and Tweed); Midlothian (Sonntag, A pocket flora of Edinburgh); Perthshire (White, The flora of Perthshire).
Vice-county 86.

General

Stirling, J. S., and **Kidston, Robert.** Notes on the flora of Stirlingshire. [1st]–10th paper. Trans. Stirling Nat. Hist. and Archaeol. Soc. **1890/91:** 88–102. 1891; **1891/92:** 74–102. 1892; **1892/93:** 109–113. 1893; **1893/94:** 88–92. 1894; **1894/95:** 80–86. 1895; **1895/96:** 137–139. 1896; **1896/97:** 103–106. 1897; **1897/98:** 134–135. 1898; **1898/99:** 166. 1899; **1899/1900:** 67. 1900. (Harvard College library.)

The first paper, with title "Notes on the flora of the north-western portion of Stirlingshire", discusses geological districts and general features of flora of the northwestern part of the county and gives a list of vascular plants and Characeae of that part. The second, with title, "Notes on the flora of Stirlingshire, with short geological sketch of the ground" (reprinted, 28 p. 1892), discusses the geology and general features of flora of the whole county, and gives a list of 712 species of vascular plants and Characeae (with 43 varieties) with indication of district in which they occur. The subsequent papers, with title "Notes on the flora of Stirlingshire," contain remarks on excursions and new records, the whole flora amounting to 838 species and 80 varieties.

Local

Croall, Alexander. List of plants found within a radius of 20 miles round Stirling. Trans. Stirling Field Club 1879/80, App. 11 p. 1880. (Harvard College library.)

List of vascular plants and Characeae, annotated only as to abundance.

Sutherlandshire

Vice-county 107, East Sutherland; vice-county 108, West Sutherland.

Local

Bennett, Arthur. Contribution towards a flora of East Sutherland. Annals Scott. Nat. Hist. **1893:** 225–236. 1893; **1894:** 25–29. 1894.

Bibliography; list of vascular plants and Characeae, with localities and names of collectors; apparently intended to summarize available data.—See also **Salmon, C. E.** Plant notes from Sutherland and Cantire. Jour. Bot. **38:** 299–303. 1900. (Includes list of species new to East Sutherland.)—Also **Druce, G. C.** Notes on the flora of East Sutherlandshire. Annals Scott. Nat. Hist. **1903:** 37–41. 1903. (List of vascular plants from vicinity of Golspie.) —Also his East Sutherlandshire notes. l.c. **1903:** 122. 1903.—Also his Plants of Sutherland and Caithness. l.c. **1908:** 39–44, 106–109. 1908. (Extensive list of vascular plants and Characeae; the East Sutherland records relate to the vicinity of Invershin.)—Also **Marshall, E. S.,** and **Shoolbred, W. A.** Some Sutherland plants. Jour. Bot. **47:** 220–223. 1909. (List of vascular plants, with localities; the East Sutherland records relate to vicinity of Oykell Bridge.)

Gray, Archibald. A list of plants observed in West Sutherland (108) . . . with notes on the flora by L. W. Hinxman. Trans. and Proc. Bot. Soc. Edinb. **17:** 220–237. 1888.

General features of flora, etc.; list of vascular plants and Characeae (474 species), with localities for some of them. Refers chiefly to district of Assynt and western part of parish of Tongue, with some species from Durness and Eddrachillis.—See also **Fox, H. E.,** and **Hanbury, F. J.** Botanical notes of a tour in Caithness and Sutherland. Jour. Bot. **23:** 333–338. 1885. (Includes list of species not previously recorded from West Sutherland.)—Also **Salmon, C. E.** Plant notes from Sutherland and Cantire. l.c. **38**: 299–303. 1900. (Includes extensive list of species new to West Sutherland.)—Also **Marshall, E. S.** Plants of W. Sutherland and Caithness. l.c. **54**: 169–172. 1916. (List with localities.)—Also **Wilmott, A. J., and Campbell, M. S.** Autumn botanising at Lochinver (West Sutherland). Rpt. Bot. Soc. and Exch. Club Brit. Isles **12** (1943–44): 820–832. 1946. (Geology, etc.; extensive list of vascular plants and Characeae, with localities.)—Also **Webster, M. M., and Marler, Peter.** A contribution to the flora of West Sutherland. Watsonia **2**: 163–179. 1952. (Topography; extensive list of vascular plants from the Parphe, a region in the vicinity of Kinlochbervie and northward.)

Harrison, J. W. H., and Harrison, H. H. The flora of the island of Handa. Proc. Univ. Durham Phil. Soc. **10**: 1–9. 1938.

Topography, etc.; annotated list of vascular plants.

West Lothian (formerly Linlithgow)

See Scotland, General (Lee, The flora of the Clyde area; Trail, Topographical botany of the river-basins Forth and Tweed); Midlothian (Balfour and Sadler, Flora of Edinburgh; Martin, The Field-club flora of the Lothians; Sonntag, A pocket flora of Edinburgh).

Vice-county 84.

Wigtownshire

See also Scotland, General (Lee, The flora of the Clyde area; West, A further contribution to a comparative study of the dominant phanerogamic and higher cryptogamic flora of aquatic habit in Scottish lakes); also Dumfriesshire (Arnott, Popular plant names).

Vice-county 74.

General

M'Andrew, James. A list of Wigtownshire plants. Trans. and Jour. Proc. Dumfriessh. and Galloway Nat. Hist. and Antiq. Soc. **10** (1893–94): 72–111. 1895.

Botanical explorations, etc.; list of known vascular plants and Characeae, with localities.

M'Candlish, A. C. A list of Wigtownshire plants. 3 p.l., 49 p. 18 cm. (Newton-Stewart, 1931). (Kew library.)

Unannotated list of vascular plants.

Local

Druce, G. C. Wigtownshire. Plants observed in the county . . . in 1883. Rpt. Bot. Rec. Club. **1883**: 65–76. 1884. (Gray Herbarium library.)

Topography, etc.; extensive list of vascular plants and Characeae [552 species], with localities.

Wilson, George. List of plants found in the Glenluce district of Wigtownshire. Trans. and Jour. Proc. Dumfriessh. and Galloway Nat. Hist. and Antiq. Soc. **9** (1892–93): 38–45. 1894.

List of vascular plants, mostly unannotated, and omitting some of the more difficult groups.

WALES

General

Anonymous. The flora of Wales. Jour. Bot. **36**: 10–23. 1898.

Account of botanical explorations, with titles of publications and biographical notes; partly annotated bibliography, arranged by counties. The paper is in great part the work of James Britten, and was originally published in Report of the Royal commission on land in Wales and Monmouthshire. Appendix. p. 140–144. 1896. (Not seen; cf. Jour. Bot. **49**, Suppl. [1]: 4. 1911.)

Davies, Hugh. Welsh botanology . . . xiv, 151, xv, 153–255 p. pl. 22 cm. London, 1813.

Pt. 1 (p. i–xiv, 1–151) is an annotated list of vascular plants of Anglesey, with English and Welsh vernacular names, and list of cellular plants, now replaced by Griffith, The flora of Anglesey and Carnarvonshire; pt. 2 (p. i–xv,

153–255), Llysieuiaith gymreig, yr ail ran . . . An alphabetical catalogue of the Welsh names of vegetables rendered into Latin and English, is a partly annotated list alphabetically arranged by Welsh vernacular names, with scientific and English equivalents, and frequent annotations in Welsh.

Hyde, H. A. Welsh timber trees, native and introduced. 2. ed. viii, 107 p. 24 fig. (fig. 2, map), 25 pl., front. 21.5 cm. Cardiff, 1935. (1st ed. 1931.) (Publication of National Museum of Wales.)

Distribution of Welsh woodlands, etc.; descriptive account of native and cultivated timber trees, with notes on woods, uses, large specimens, English and Welsh names, etc.

——— and **Wade, A. E.** Welsh flowering plants. A handbook to the collection in the Welsh National Herbarium. 2. ed. rev. by A. E. Wade. vi, 209 p. 7 fig. (maps), 4 pl. (ports.) 21.5 cm. Cardiff, 1957. (1st ed. 1934.)

Account of herbarium of National Museum of Wales, composition of flora; discussion of life forms, types of distribution; general features of flora, etc., of each county with references to principal publications; annotated systematic list of known spermatophytes (1343 species) with Raunkiær life forms, habitat, distribution by counties (with citation of localities for rarer species); list of collectors, bibliography. Covers Monmouth as well as the proper Welsh counties.—See also their Welsh ferns; a descriptive handbook. 3. ed. x, 131 p. 67 fig., 10 pl. on 5, front. 21.5 cm. Cardiff, 1954. (1st ed. 1940.) (General features of structure, life history, etc.; keyed descriptive flora of Filicineae (Ophioglossaceae, Marsileaceae, Hymenophyllaceae, Polypodiaceae, Azollaceae) with local, British, and foreign distribution and citation of specimens; list of collectors, bibliography.)

Anglesey

Vice-county 52.

General

Carter, P. W. Some account of the botanical exploration of Anglesey. Trans. Anglesey Antiq. Soc. and Field Club **1952**: 44–69. 1952.

Account of botanists, chronologically arranged with references to their writings and discoveries; alphabetical list of botanists with incomplete references to their publications.

Griffith, J. E. The flora of Anglesey and Carnarvonshire with an account of their flowering plants, ferns and their allies, mosses, marine algae, lichens and Hepaticae. xx, 288 p. map. 22 cm. Bangor, 1895. (Arnold Arboretum library.)

Geology, topography, gazetteer; annotated list of vascular and cellular plants, with localities and English and Welsh vernacular names.—See also **Druce, G. C.** Anglesey and Carnarvonshire plants. Jour. Bot. **40**: 181–188. 1902.—Also **Bickham, S. H., Marshall, E. S.,** and **Shoolbred, W. A.** Carnarvonshire plants. l.c. **51**: 241–244. 1913.—Also **Salmon, C. E.,** and **Travis, W. G.** Notes from Carnarvon and Anglesey. l.c. **55**: 316–323. 1917.—Also **Pugsley, H. W.** Notes on Carnarvonshire plants. l.c. **61**: 19–23. 1923.—Also **Hyde, H. A.** The J. E. Griffith herbarium, etc. l.c. **65**: 127–128. 1927. (Note on acquisition of Griffith's herbarium by the National Museum of Wales.)—Also **Wilson, Albert.** Some plants of Anglesey. North West Nat. **15**: 104–109. 1940.—Also **Conolly, Ann,** and **Smith, P. M.** A contribution to the flora of the west Lleyn Peninsula, Caernarvonshire. Proc. Bot. Soc. Brit. Isles **3**: 151–164. 1 fig. (map). 1959. (Geology, climate, general features of flora; list of about 100 noteworthy species; bibliography.)

Local

Herdman, W. A. First list of plants on Puffin Island. Proc. and Trans. Liverpool Biol. Soc. **4**: 76–79. 1890. (Also as Ann. Rpt. Liverpool Mar. Biol. Sta. **3**: 43–46. 1890.)

Unannotated list of vascular plants.—See also **Phillips, R. W.** The flora of Puffin Island. Rpt. Puffin Island Biol. Sta. **1892/93**: 17–21. 1893(?). (Adds 8 species. Not seen; cited from Jour. Bot. **36**: 20. 1898.)—Also **Wortham, W. H.** The botany of the island. *In* Puffin Island. Proc. Llandudno, Colwyn

Bay and Dist. Field Club 7: 15–18. 1913. (General features of vegetation, with mention of various species. In British Museum (Natural History) library.)—Also **Woodhead, Norman.** The botany. *In* Puffin Island. l.c. **14**: 15–17. 1928. (General features of flora; list of 26 angiosperms collected, with indication of frequency. In British Museum (Natural History) library.) —Also **Lacey, W. S.** The flora of Ynys Seiriol (Puffin Island), with some notes on the effects of myxomatosis. Nature in Wales **3**: 464–470. 1957. (Includes 3 lists of species made at different dates. Not seen; cited from Proc. Bot. Soc. Brit. Isles **3**: 78. 1958.)

BRECKNOCKSHIRE (BRECONSHIRE)

Vice-county 42.

General

Carter, P. W. A history of botanical explorations in Brecknock. Brycheiniog **3**: 157–175. 1957.

Account of botanists, chronologically arranged with references to their writings and discoveries; alphabetical list of botanists with incomplete references to their publications.

Phillips, R. W. The first flora of Breconshire. *In* Phillips, Thomas R., *ed.* The Breconshire border between Wye and Usk, with notes . . . including a First flora of Breconshire. Appendix I, p. i–xxiv. 21.5 cm. Talgarth, 1926. (Kew library.)

Unannotated list of vascular plants, with usually a single cited locality for the rarer species.—See also his The flora of Breconshire. l.c. p. 84–88. (Botanical regions and explorations. In Kew library.) Replaces a number of earlier publications, the most important of which are: **Jones, Theophilus.** List of plants found in the county of Brecon. *In* his A history of the county of Brecknock . . . v. 2, p. 776–785. 30 cm. Brecknock, 1809. (List of vascular plants, a few mosses, and one lichen, on the Linnaean system, with English and Welsh vernacular names, and usually with localities and names of collectors. The same treatment is given on p. 545–548 of the 1-volume reprint of Jones' work published by Edwin Davis at Brecknock, 1898.) **Barrett, W. B.** A contribution towards a flora of Breconshire. Jour. Bot. **23**: 39–44, 83–89, 107–112, 145–149. 1885. (Botanical publications, soils, topography, etc.; annotated list of vascular plants (595 species), with localities; intended to bring together all known records for the county); also **Ley, Augustin.** Recent additions to the flora of Breconshire. l.c. **33**: 135–137. 1895. (List with localities); also **Rogers, W. M.** Radnorshire and Breconshire plants. l.c. **37**: 17–25. 1899. (List with localities, and indication of those that are new county records); also his Radnorshire and Breconshire Rubi. l.c. **37**: 193–198. 1899. (List with localities.)

CAERNARVONSHIRE (CARNARVONSHIRE)

See also Anglesey (Griffith, The flora of Anglesey and Carnarvonshire).
Vice-county 49.

General

Carter, P. W. Some account of the botanical exploration of Caernarvonshire. pt. 1–2. Trans. Caernarvonsh. Hist. Soc. **16**: 52–59. 1955; **17**: 45–61. 1956.

Part 1 not seen; part 2 presents an account of botanists, chronologically arranged with references to their writings and discoveries; alphabetical list of botanists with incomplete references to their publications.

Webster, A. D. The native trees and shrubs of Carnarvonshire. Trans. Roy. Scott. Arbor. Soc. **11**: 481–500. 1887.

Annotated list of 75 woody plants.

Local

Campbell, Bruce. The vascular plants. *In* North, F. J., Campbell, Bruce, and Scott, Richenda. Snowdonia. The national park of North Wales. p. 406–425. 22 cm. London, 1949.

List of vascular plants, with localities for rarer species. Relates to the Caernarvonshire section of the proposed North Wales National Park. (Region of the Snowdon range.)

Inchbald, Peter. Flora of Llandudno. *In* Parry, Richard. Llandudno: its history and natural history. 2. ed. p. 157–169. 19 cm. Llandudno, 1861. (Harvard College library.)

Geology, etc.; annotated list of rarer vascular plants, with localities and Welsh names. The paper (with title Botany) appeared also in Williams, Thomas. Williams's complete guide to Llandudno. The tenth thousand of Llandudno; its history and natural history. p. 189–202. 19 cm. Llandudno, [1864]. (1st issue 1855.)—See also **Robinson, J. T.** The tourist's floral guide. Llandudno. Gard. Chron. n.s., **2:** 9–10. 1874. (Running account of locality with mention of numerous species.)

Wilson, Albert. The botany of Tal-y-Fan. Proc. Llandudno, Colwyn Bay and Dist. Field Club **14:** 82–98. 1928. (British Museum (Natural history) library.)

Climate, geology, edaphic features, etc.; annotated list of phanerogams (p. 86–92) and cryptogams (p. 92–98).

——— The flora of a portion of north-east Caernarvonshire. North West. Nat. **21:** 202–223. 1 fig. (map), pl. 10–11. 1947; **22:** 62–83, 191–211. pl. 6, 10. 1948–49.

Topography, geology, climate, general features of flora, short bibliography; annotated list of vascular plants (667 species), bryophytes, and larger lichens, with localities for scarcer species. Covers the western side of the Conway valley from a little below Bettws-y-Coed to the mouth of the Conway estuary.

——— Notes on the flora of Carnedd Llewelyn. North West. Nat. **4:** 53–56. 1929.

Geology, climate, topography; annotated list of vascular plants collected by author, mostly above 3000 ft. altitude.

Cardiganshire

Vice-county 46.

General

Carter, P. W. Botanical Exploration in Cardiganshire. Ceridigion, Jour. Cardigansh. Antiq. Soc. **1950:** 1–18. 1950.

Account of botanists, chronologically arranged with references to their writings and discoveries; alphabetical list of botanists with incomplete references to their publications.

Salter, J. H. The flowering plants and ferns of Cardiganshire. vi, 182 p. 22 cm. Cardiff, 1935.

Botanical explorations and publications; annotated list of vascular plants and Characeae [including about 890 native or naturalized species], with localities.— See also **Wade, A. E.,** *ed.* A supplement to Dr. J. H. Salter's The flowering plants . . . vi, 45 p. 21.5 cm. Cardiff, 1952. (Extensive list.)

Local

Salter, J. H. The altitudinal range of flowering plants and ferns in mid Wales. North West. Nat. **3:** 131–135, 170–174. 1928.

Geography, climate, etc.; lists of vascular plants reaching their upper limits at various altitudes, from 900–2468 feet. Refers principally to region around Aberystwith, Plynlimmon, and Cader Idris.

——— List of the flowering plants and ferns of Aberystwyth and neighbourhood. 46 p. 20.5 cm. Aberystwyth, [1903?]. (Arnold Arboretum library.)

Botanical investigations; partly annotated list of about 700 vascular plants, with localities for rarer species; intended to include all species from Northern Cardiganshire. Replaces **Morgan, T. O.** Florae Cereticae superioris. A catalogue of plants indigenous in the neighbourhood of Aberystwyth, with their stations and time of flowering; to which is added a list of marine plants found upon the coast. iv, 18 p. 21.5 cm. Aberystwyth, 1849. (Annotated list of vascular plants and marine algae. In Kew library.)

CARMARTHENSHIRE (CAERMARTHENSHIRE)

Vice-county 44.

General

Barker, T. W., The flowering plants of Carmarthenshire. *In* his Handbook to the natural history of Carmarthenshire. p. 42–101. 18 cm. Carmarthen, [pref. 1905].—The ferns, horsetails and club mosses of Carmarthenshire. l.c. p. 101–104. (British Museum (Natural History) library.)

List of spermatophytes (799 species) with localities for rarer species, followed by notes on the more interesting species and list of probable additions to the flora; similar list of pteridophytes (34 species); short list of bryophytes (p. 105–107).—See also **Riddelsdell, H. J.** Caermarthenshire plants. Jour. Bot. **45**: 71. 1907. (List of noteworthy species in herbarium of James Motley.)—Also **Bennett, Arthur.** Carmarthenshire plants. l.c. **46**: 83–90. 1908. (Extensive list of vascular plants, with localities, intended to supplement the records in Watson's Topographical Botany, ed. 2.)—Also **Riddelsdell, H. J.** Caermarthenshire plants. l.c. **46**: 124. 1908. (Comments on species in Bennett's list.)—Also **Hamer, David.** List of Carmarthenshire plants. Encilion **1** (2): 25–29. [1913?] (Extensive list of angiosperms collected by author, with localities; intended as first supplement to Barker's flora. In New York Public Library.)—Also his Carmarthenshire flora. Trans. Carmarthensh. Antiq. Soc. and Field Club **15**: 8–10. 1922. (Extensive list of vascular plants, intended as second supplement to Barker. In New York Public Library.)—Also his Carmarthenshire flora. l.c. **16**: 76–77. 1923. (In New York Public Library.)—Also **Whellan, J. A.** Notes on the flora of south-west Wales. North West. Nat. **17**: 223–227. 1942. (List of rarer vascular plants collected in Carmarthen, Pembroke, and Cardigan.)—Also **Webb, J. A.** Plant records, 1941–44. Proc. Swansea Scient. and Field Nat. Soc. **2**: 188–191. 1944. (Many new records. In Iowa State College library.)

Carter, P. W. Botanical exploration in Carmarthenshire. I. Carmarthen Antiq. **2**: 113–122. 1951; II. Carmarthen Antiq. **2**: 166–177. 1957.

Not seen; cited from Biol. Abst. **31**: 329. 1957; and from author correspondence.

Local

Davies, William. Botany. A catalogue of plants growing in the neighbourhood of Llandeilo. Arranged after the classification of DeCandolle. *In* his Llandeilo-Vawr and its neighbourhood . . . p. 153–164. 18 cm. Llandeilo, 1858. (Library of N. D. Simpson.)

List of vascular plants, with localities.

Innes, (Miss). Botany of the district. *In* Innes, John. "Old Llanelly." p. 148–155. 24.5 cm. Cardiff, 1902.

Alphabetical list of vascular plants, with localities, "founded on the researches of the late Mr. R. V. Innes."

DENBIGHSHIRE

Vice-county 50.

General

Dallman, A. A. Notes on the flora of Denbighshire. Jour. Bot. v. 49, suppl. 50 p. 1911.—Further notes . . . l.c. v. 51, suppl. II. 43 p. 1913.

Botanical investigations, with included bibliography; annotated list of vascular plants with localities, the commoner species omitted.

Local

Rowland, John. A catalogue of plants found growing in the neighbourhood of Wrexham, in Denbighshire. Phytologist **1**: 421–425. 1842.

Unannotated list of vascular plants.

Williams, John. Faunula grustensis: being an outline of the natural contents of the parish of Llanrwst; comprehending some account of its general history, commerce and agriculture; also a triglott catalogue (in Latin, English and Welsh) of the animals and plants found in it, with some notes thereon . . . 148 p. 13.5 cm. Llanrwst, 1830. (Kew library.)

Includes (p. 80–118) "Its botany," a partly annotated list of wild vascular plants, with English and Welsh vernacular names; followed (p. 118–139) by a bare list of garden plants, without Welsh names.

Flintshire

Vice-county 51.

General

Carter, P. W. Notes on the botanical exploration of Flintshire. Publ. Flintsh. Hist. Soc. **16** (Flintshire Miscellany no. 1): 1–21. 1956.

Running account of botanists and their discoveries and publications, arranged chronologically; brief account of geology, soils, etc.; bibliography.

Local

Jones, H. L. A list of plants observed in Flintshire, in 1881 . . . Rpt. Bot. Rec. Club **1880**: 164–166. 1882. (Kew library.)

Unannotated list of 246 species.—See also **Parsons, H. F.** Additional species observed in Flintshire in July and August. Rpt. Bot. Rec. Club **1880**: 166–167. 1882. (Adds 109 species. In Kew library.)—Also **Brown, Robert.** Flintshire plants not recorded in ed. 2 of "Topographical Botany." Jour. Bot. **23**: 357–360. 1885. (List of vascular plants, with localities.)—Also **Dallman, A. A.** Notes on the flora of Flintshire. l.c. **45**: 138–153. 1907. (Extensive list of vascular plants, hepatics, mosses, and lichens.)—Also his Notes on the flora of Flintshire. l.c. **46**: 187–196, 222–230. 1908. (Similar list.)—Also his Notes on the flora of Flintshire. l.c. **48**: 40–53, 73–77, 90–98. 1910. (Extensive list.)—Also **Rogers, W. M.** Notes on the flora of Flintshire. l.c. **48**: 184–185. 1910. (List with localities.)—Also **Dallman, A. A.** Flintshire plants. l.c. **48**: 204–205. 1910. (Notes on several species.)—Also his Further notes on the flora of Flintshire. l.c. **49**: 8–14. 1911. (List with localities.)

Glamorganshire

See also Great Britain, General (Riddelsdell, Notes . . . under Dunn, Alien flora of Britain). Vice-county 41.

General

Carter, P. W. Some account of the history of botanical exploration in Glamorgan. Trans. Cardiff Nat. Soc. **82**: 5–31. 1955.

Account of botanists, chronologically arranged, with references to their writings and discoveries; alphabetical list of botanists with incomplete references to their publications.

Riddelsdell, H. J. A flora of Glamorganshire. Jour. Bot. v. 45, suppl. 88 p. 1907—Further Glamorganshire records. l.c. **47**: 397–412. 1909.

Botanical districts, bibliography; annotated list of phanerogams, with localities.—See also his A flora of Glamorganshire. Rpt. Bot. Soc. and Exch. Club Brit. Isles **10** (1933): 666–669. 1934. (Corrections of data regarding first records for the county in the second supplement to Watson's Topographical botany.)

Trow, A. H., *ed.* The flora of Glamorgan. Including the spermaphytes and vascular cryptogams . . . vol. 1. 2 p.l., 209 p. 22.5 cm. Cardiff, 1911 (1907–12). (Trans. Cardiff Nat. Soc., suppl. to v. 39–44.)

Annotated list of vascular plants, with English and Welsh vernacular names (the Welsh names not indexed) and localities. Indices prepared by H. M. Hallett.

Vachell, Eleanor. A list of Glamorgan plants. Rpt. Bot. Soc. and Exch. Club Brit. Isles **10** (1933): 686–743. 1934.

Bibliography, etc.; briefly annotated list of vascular plants and Characeae, with localities for scarcer species and references for records; similar list of hybrids. Reprinted, with some corrections and omission of list of hybrids, with title List of Glamorgan flowering plants and ferns. *In* Glamorgan

County history. v. 1. Natural history. p. 136–178. pl. 8–14. 31 cm. Cardiff, 1936. Preceded by her flowering plants and ferns. l.c. p. 123–135 1936. (Botanical regions, geology, topography.) Followed by her Glamorgan botanists. l.c. p. 252–254. 1936. (Annotated list of botanists.) Other articles on botany by other botanists are in same volume.—See also her Botanical notes, 1936. **Trans. Cardiff Nat. Soc. 69: 90–91. 1938. (Additional** species and localities.)—Also her Botanical notes. l.c. 71: 29–331. 1941.—Also her Botanical notes, 1939–1946. l.c. **72/78**: 23–26. 1947.—Also **Wade, A. E.** Botanical notes, 1947–8 (–1953–54). l.c. **79**: 52–54. 1950; **80**: 37–38. 1952; **81**: 100–101. 1953; **83**: 25–26. 1956.

Local

Gutch, J. W. G. A list of plants met with the neighbourhood of Swansea, Glamorganshire. Phytologist **1**: 104–108, 118–121, 141–145, 180–187. 1841 –42.—Additions and corrections. l.c. **1**: 380. 1842.

List of vascular and cellular plants, with localities.—See also **Flower, T. B.,** and **Lees, Edwin.** Additions to the List of plants . . . Phytologist **1**: 377–380. 1842.—Also **Dillwyn, L. W.** An alphabetical list of rarer flowering plants and ferns which have been found within twenty miles of Swansea. *In* his Materials for a fauna and flora of Swansea and the neighbourhood. p. 26–44. 25 cm. Swansea, 1848. (Annotated list.)—Also **Ball, John.** Contributions to the flora of south Wales. Bot. Gaz. (London) **1**: 107–109. 1849. (Includes list of noteworthy species from vicinity of Swansea, p. 108–109.)

Harris, Henry. The flora of the Rhondda. 86 p. 19 cm. Bristol, 1905. (Arnold Arboretum library.)

Annotated, briefly descriptive list of vascular plants of the Rhondda Valleys, from Porth to Mardy, and Pontypridd to Blaenrhondda.

Reynolds, L. D. The flowering plants and Pteridophyta of the Caerphilly Basin. North West. Nat. n.s., **3**: 35–57. 3 fig. (maps). 1955.

Topography, geology, botanical divisions; annotated list of vascular plants. (Seven miles north of Cardiff.)

Storrie, John. The flora of Cardiff; a descriptive list of the indigenous plants found in the district of the Cardiff Naturalists' Society, with a list of the other British and exotic species, found on Cardiff ballast hills. v, 129 p. 22 cm. Cardiff, 1886. (Arnold Arboretum library.)

Bibliography; annotated descriptive list of vascular plants, with localities and English and Welsh vernacular names; extinct species, list of ballast plants. Covers the eastern half of Glamorganshire. According to review in Jour. Bot. **25**: 349–351. 1887, the Welsh names given are artificial.—See also **Vachell, Eleanor.** Botany. *In* Hallett, Howard M., *ed.* Handbook to Cardiff and the neighbourhood, prepared for the British Association meeting. p. 218–227. 17 cm. Cardiff, 1920. (Includes lists of vascular plants under localities, with bibliography and list of herbaria. Not seen; title and annotation supplied by N. D. Simpson.)—Also **Hyde, H. A.** Plant distribution in the Cardiff district. *In* Evans, C. J., *ed.* The book of Cardiff. Prepared for the National Association of Head Teachers. p. 28–38. fig. 5–6, pl. 8–9. 23 cm. Oxford, 1937. (Account of plant zones with lists of plants arranged under habitats. In Kew library.)

Sykes, M. H., and **Webb, J. A.** The flora of the bombed areas and slum-clearance sites of Swansea. Proc. Swansea Scient. and Field Nat. Soc. **2**: 291–306. 1947.

List of 242 species. (Not seen; cited from Watsonia **2**: 326. 1953.)

Wade, A. E., and **Smith, R. L.** The adventive flora of the Port of Cardiff. Rpt. Bot. Soc. and Exch. Club Brit. Isles **7** (1925): 999–1027. 1926.—Additions . . . l.c. **8** (1926): 181–183. 1927.

Previous publications, local habitats, etc.; annotated list of adventive angiosperms, principally from the dock areas of Cardiff, Penarth, and Barry, the common weeds of cultivation omitted.—See also **Smith, R. L.,** and **Wade, A. E.** Notes on the adventive flora of the Cardiff district. Rpt. Bot. Soc. and Exch. Club Brit. Isles **12** (1938): 72–83. 1939.

Webb, J. A. The flora of the peninsula of Gower. Proc. Swansea Field Nat. Soc. v. 3, suppl. pt. 1. 25 p. 1956.

Annotated list (Ranunculaceae to Geraniaceae) of vascular plants. (Not seen; cited from Proc. Bot. Soc. Brit. Isles **3**: 77. 1958.)

Merionethshire

Vice-county 48.

General

Carter, P. W. Some account of botanical exploration in Merionethshire. Merioneth Misc. **3**: 1–40. 1955.

Account of botanists, chronologically arranged with references to their writings and discoveries; alphabetical list of botanists with incomplete references to their publications.

Local

Jones, D. A. The flora of Dolgelley and the neighbourhood. *In* Ellis, Thomas P. The story of two parishes, Dolgelley and Llanelltyd. Appendix III. p. 146–155. 22 cm. Newtown, 1928. (Harvard College library.)

General features of flora; unannotated list of vascular plants.—See also Kynoch and Salwey, below.

Kynoch, James. Wild flowers of Barmouth and neighbourhood. 4. ed. . . . and List by the late Rev. T. Salwey . . . 49 p. 17.5 cm. Brighton, 1910. (British Museum (Natural History) library.)

Unannotated list of angiosperms (p. 6–18), followed (p. 18–22) by a list of additions since 1887 (when the second edition was published) and by a reprint (p. 23–46) of Salwey's A list of certain plants . . . (see below) with his notes on the same (p. 47–49). The first edition was by **Himing, F. W.**, and **Brewer, G. S.** Wild flowers of Barmouth found in bloom July 31st to August 16th, 1884. 7 p. 18 cm. Barmouth [pref. 1884]. (Unannotated list of 259 angiosperms. It was published by Kynoch. In Kew library.)—See also Salwey, below.

Ley, Augustin. Montgomery and Merioneth. *In* County catalogues of plants. Rpt. Bot. Rec. Club **1875**: 143–147. 1876. (Kew library.)

List of vascular plants observed by author, mostly in 1875, showing occurrence in one or both counties (285 from Merioneth) and consisting mostly of species not recorded in Watson's Topographical botany; followed by notes on a few of the species.—See also **Rogers, W. M.** Notes on some North Wales plants. Jour. Bot. **24**: 338–343, 363–368. 1886. (List of new county records for Denbighshire, Merionethshire, and Carnarvonshire, followed by extensive list of noteworthy vascular plants, with localities.)—Also **Ley, A.** A second list of species of phanerogams, collected in Merioneth (Co. 48), in Aug. 1886 . . . Rpt. Bot. Rec. Club **1884/86**: 150–151. 1887. (List with localities. In Gray Herbarium library.)—Also **Linton, W. R.** Merionethshire plants. Jour. Bot. **33**: 359–363. 1895. (List of more interesting plants collected by author.)

Salwey, Thomas. A list of certain plants to be met with in the neighbourhoods of Barmouth, Dolgelley, and Harlech, with a few habitats of plants in other parts of Merionethshire. 1 p. l., 40 p. 16 cm. Barmouth, 1863. (Gray Herbarium library.)

List of less common vascular plants (p. 7–24) with localities and some annotations, followed by list of mosses, hepatics, lichens, and algae. Reprinted from **Jones, D.**, *publisher.* The tourist's and visitor's hand-book and guide to Harlech, Barmouth, Dolgelly, Towyn . . . The botany by T. Salwey. 158 p. 16mo. Barmouth, 1863. (Not seen.)—See also Kynoch, above.

Montgomeryshire

Vice-county 47.

General

Carter, P. W. Botanical exploration in Montgomeryshire. The Montgomeryshire collections. Trans. Powys-land Club **49** (2): 1–17. 1946.

Account of botanists, chronologically arranged with brief references to their writings and discoveries.

Local

Ley, Augustin. Montgomery and Merioneth. *In* County catalogues of plants. Rpt. Bot. Rec. Club **1875**: 143–147. 1876. (Kew library.)

List of vascular plants observed by author, mostly in 1875, showing occurrence in one or both counties (254 from Montgomeryshire) and consisting mostly of species not recorded in Watson's Topographical Botany; followed by notes on a few of the species.—See also his A second list of plants . . . observed and collected in County 47 [Montgomeryshire] . . . in Aug. 1886. Rpt. Bot. Rec. Club 1884/86: 149–150. 1887. (List of vascular plants, with localities. In Gray Herbarium library.)—Also **Whitwell, William.** Montgomeryshire notes. Jour. Bot. 34: 172–173. 1896. (List of noteworthy species, with localities.)—Also **Wade, A. E.,** and **Webb, J. A.** Montgomeryshire plant records. North West. Nat. 18: 52–68. 1943. (Geology, topography, etc.; list of vascular plants, with localities, collected from 1939 to 1941, including about 175 species new to the county.)

Morgan, John. The flora of Llanbrynmair. 12 p. 21 cm. Oswestry, 1893. (Reprinted from "Bye-Gones.") (British Museum (Natural History) library.)

List of indigenous and commonly cultivated vascular plants and Characeae, with Welsh vernacular names and occasional notes on medicinal uses and folklore; also short lists of cellular plants.

PEMBROKESHIRE

Vice-county 45.

General

Rees, F. L. A list of Pembrokeshire plants. 53 p. 21 cm. Tenby, 1950.

Annotated list of angiosperms, with localities. Although based only on the work of Mrs. Rees and Bertram Lloyd, it appears to be nearly enough complete as to species to stand as a flora of the county. Published by Tenby Museum. Replaces **Walker, F.** Botany. In Mason, —. Mason's guide to Tenby. 7. ed. p. 258–268. 18.5 cm. Tenby, n.d. (Unannotated list of 750 vascular plants and Characeae of the county, of which 671 grow in Tenby, those not growing there distinguished by an asterisk. In British Museum (Natural History) library.)

Local

Babington, C. C. On the botany of South Pembrokeshire. Jour. Bot. 1: 258–270. 1863.

Topography, general features of flora, botanical explorations (with references); mostly unannotated list of vascular plants and Characeae (about 550 species), mostly found by author in vicinity of Tenby. Covers the southeastern part of Pembrokeshire (containing Tenby and Pembroke), bounded on the north by Milford Haven from its mouth to Carew Castle, then east to East Williamston and to somewhat south of St. Issells. Replaces **F[alconer], R. W.** Contributions towards a catalogue of plants indigenous to the neighbourhood of Tenby. 54 p. 20 cm. London, 1848. (Briefly annotated list of vascular plants [388 species]. In Lloyd Library.)—See also **Trimen, Henry.** Plants of South Pembrokeshire. Jour. Bot. 5: 301–302. 1867. (List of additional species.)—Also **Towndrow, R. F.** Pembroke plants. l.c. 39: 270. 1901. (Short list of new county records, mostly from Tenby.)—Also **Wallis, Anthony.** Pembrokeshire and Carmarthenshire plants. Edited by C. E. Salmon, F. L. S. l.c. 57: 347–350. 1919. (Includes (p. 348–349) list of noteworthy plants from Tenby, Ridgeway, and other localities in Pembrokeshire.)—Also **Pugsley, H. W.** Notes on Pembrokeshire plants. l.c. 62: 102–105. 1924. (List of noteworthy vascular plants, with localities, mainly from Tenby and its vicinity.)—Also **Whellan, J. A.** Notes on the flora of south-west Wales, 1941. North West. Nat. 17: 223–227. 1942. (List of rarer vascular plants collected at Tenby, Nevern, Newport and other localities in Pembrokeshire, also in Cardigan and Carmarthenshire.)—Also Walker, under Rees, above.

Gillham, M. E. An annotated list of the flowering plants and ferns of Skokholm Island, Pembrokeshire. North West. Nat. 24 (n.s., 1): 539–557. 1953.

General features of vegetation; annotated list of vascular plants; bibliography. Replaces **Lockley, R. M.** The flora of Skokholm, 1927–1940. *In* his Dream Island days . . . p. 142–144. 21.5 cm. London, 1943; List of flora recorded at Skokholm 1927–46. *In* his Letters from Skokholm. (Appendix III.) p. 232–236. 21.5 cm. London, 1947. (Mostly unannotated lists.)

——— An ecological account of the vegetation of Grassholm Island, Pembrokeshire. Jour. Ecology **41:** 84–99. 3 fig., map. 1953.
Includes (p. 85) list of 20 angiosperms; also list of algae and lichens.

Hepper, F. N. Flora of Caldey Island, Pembrokeshire. Proc. Bot. Soc. Brit. Isles **1:** 21–36. 1 fig. (map). 1954.
Geology, topography, etc.; annotated list of vascular plants and Characeae (348 species, all but 1 vascular); bibliography.

Sladen, W. J. L. The flora of Skomer. *In* Buxton, John, and Lockley, R. M. Island of Skomer. A preliminary survey of the natural history of Skomer Island, Pembrokeshire, undertaken by the West Wales Field Society . . . Appendix 2. p. 135–141. 22 cm. London, New York, 1950.
Annotated list of vascular plants; also list of lichens, the last determined by A. E. Webb.

Walton, C. L. A contribution to the flora of the St. David's peninsula. 28 p. 21 cm. Haverfordwest, 1951 (cover place and date).
General features of flora, bibliography, etc.; partly annotated list of known vascular plants and Characeae. Published by the West Wales Field Society.

Radnorshire

Vice-county 43.

General

Carter, P. W. The history of botanical exploration in Radnorshire. Trans. Radnorsh. Soc. **20:** 42–58. 1950.
Not seen; cited from Biol. Abst. **31:** 329. 1957.

Local

Ley, Augustin. Radnorshire. *In* Tabular catalogues of common plants for Breconshire, Radnorshire, Selkirkshire, and West Lancashire. Rpt. Bot. Rec. Club **1874:** 80–86. 1875. (Kew library.)
Tabular list of vascular plants from the four counties mentioned, with indication of occurrence in each; includes 303 species from Radnorshire.—See also **Ridley, H. N.** Notes on Radnorshire plants. Jour. Bot. **19:** 170–174. 1881. (Geology, etc.; list of noteworthy vascular plants, with localities.)—Also **Ley, A.** A list of species observed in Radnor in 1881 . . . Rpt. Bot. Rec. Club **1881/82:** 246–247. 1883. (List of vascular plants, with localities. In Gray Herbarium library.)—Also his A list of species observed . . . in Radnorshire (Co. 43) in 1886 . . . l.c. **1884/86:** 144–146. 1887. (List with localities. In Gray Herbarium library.)—Also **Rogers, W. M.** Radnorshire and Breconshire plants. Jour. Bot. **37:** 17–25. 1899. (Extensive list of vascular plants.)—Also his Radnorshire and Breconshire Rubi. l.c. **37:** 193–198. 1899. (Extensive list.)—Also **Wade, A. E.,** and **Webb, J. A.** Radnorshire plant records. North West. Nat. **20:** 156–160. 1946. (Extensive list, with localities and indication of those new to the county.)

——— Notes on the flora of Aberedw. Trans. Woolhope Nat. Field Club **1890/92:** 184–198. 1894[?]. (British Museum (Natural History) library.)
Annotated list of vascular plants and mosses (478 species, of which 358 are vascular) observed by author, mostly in 1886–1890.—See also his First contribution towards a flora of Aberedw, Radnorshire. Woolhope Nat. Field Club **1890/92:** 180–183. 1894[?]. (Topography, general features of flora. In British Museum (Natural History) library.)

Iceland (Ísland)

See also Denmark, General (Hornemann, Forsøg til en dansk oeconomisk plantelære); Scandinavia.

Iceland is now officially divided into 16 districts (sýsla, pl. sýslur). As these are not shown on most maps, the location of the areas covered by local floras has been indicated in the annotations by the terms northeast, north-central, etc. To provide a definite method of indicating position on the map, the parallel of 65° has been taken arbitrarily as dividing north from south Iceland, and the meridians of 16° and 20° as separating east, central, and west Iceland, so that a locality east of 16° and north of 65° is placed in northeast Iceland, one between 16° and 20° and north of 65° in north-central Iceland, one west of 20° and north of 65° in northwest Iceland, and similarly with those south of 65°.

The more common nouns occurring in Icelandic titles are the following: flórunýjungar, floral novelties; fundarstaðir, localities; gróðrarsaga, history of the vegetation; gróður, vegetation; gróðurlýsingar, description of the vegetation; gróðurrannsóknir, investigations of the vegetation; gróðurríki, vegetation, vegetable kingdom; háplöntuflóra, flora of vascular plants; jurtir, plants; plöntuskrá, plant catalog; slæðingar, adventives.

General

Davíðsson, Ingólfur, and **Óskarsson, Ingimar.** Garðagróður aðallega i Reykjavík, Hafnarfirði og á Akureyri. [Garden plants chiefly in R., H., and A.] 450 p. 300 fig., pls. (pt. col.). 24 cm. Reykjavík, 1950.

Lists of species for various purposes, etc.; descriptive treatment of 602 ornamental plants (including woody plants) of which 90 are native, with keys and vernacular names; bibliography.

Gröntved, Johannes. The Pteridophyta and Spermatophyta of Iceland. *In* Gröntved, J., Paulsen, O., and Sørensen, T., *eds.* The botany of Iceland. v. 4, pt. 1. 427 p. 177 fig. (incl. maps). 26.5 cm. Copenhagen, London, 1942.

Botanical explorations, condensed bibliography, topography, geology, soils, climate, vegetational types, altitudinal ranges, botanical districts; annotated list of known vascular plants (622 species, of which 108 are *Hieracium* and 116 *Taraxacum*), with local references, vernacular names (Icelandic, Danish, and English), local and extralimital range, life form, habitat, maximum and average size, notes on uses, and critical notes; bibliography, indices of vernacular and botanical names. Replaces the following earlier floras: **Lindsay, W. L.** The flora of Iceland. Edinburgh New Phil. Jour. n.s., **14:** 64–101. 1861. **Babington, C. C.** A revision of the flora of Iceland. Jour. Linn. Soc. Bot. **11:** 282–348. 1870. **Grønlund, C. H.** Islands flora . . . 2 p.l., 159 p. 20 cm. Kjøbenhavn, 1881. (Gray Herbarium library.) **Strömfelt, H. F. G.** Islands kärlväxter . . . Öfvers. K. Vetensk. Akad. Förhandl. **41** (8): 79–124. 1885.—See also **Steindórsson, Steindór.** Gróðurrannsóknir Íslands II. Náttúrufræðingurinn **13:** 24–48. 4 fig. (maps). 1943. (Includes list of 143 vascular plants entitled Flórunýjungar 1925–1940, and bibliography.)—Also **Óskarsson, Ingimar.** Gróðurrannsóknir. Þrjátíu ára yfirlit. l.c. **13:** 131–152. 1 fig. 1943. (Includes list of localities visited by author and list of noteworthy species.)—Also **Steindórsson, S.** Flórunýjungar 1944. l.c. **14:** 69–74. 1944.—Also his Flórunýjungar 1948. l.c. **19:** 110–121. 1949.—Also **Löve, Áskell.** Some innovations and nomenclatural suggestions in the Icelandic flora. Bot. Notiser **1950:** 24–60. 19 fig. (maps). 1950. (General remarks on the flora; critical notes on 21 genera or species; bibliography.)—Also **Davíðsson, Ingólfur.** Nokkrir fundarstaðir jurta á Austurlandi. Náttúrufræðingurinn **20:** 58–60. 1950. (Notes on 55 species from the east coast.)—Also **Óskarsson, I.** Nýjungar úr gróðurriki Íslands. l.c. **24:** 22–30. 3 fig. 1954; **26:** 102–104. 1956.—Also **Davíðsson, I.** Nokkrir fundarstaðir fremur fágætra jurta. l.c. **24:** 31–35. 1954.—Also **Jónasson, Helgi.** Að austan. l.c. **25:** 36–39. 1955.—Also **Davíðsson, I.** Nokkrir nýir fundarstaðir jurta. l.c. **25:** 39–40. 1955.—Also **Steindórsson, S.** Flórunýjungar 1955. l.c. **26:** 25–31. 1956.—Also records by various authors of single or few species new to Iceland in Náttúrufræðingurinn **17:** 22, 164–174. 1947; **19:** 136, 185–188. 1949; **20:** 187–189. 1950; **21:** 91–93, 173–175. 1951; **22:** 181–182, 183. 1952; **23:** 138–142. 1953; **26:** 26–31, 99–101, 140–141. 1956.

Halldórsson, Björn. Gras-nytiar eða gagn þat, sem hvørr búandi maðr getr haft af þeim ósánum villi-jurtum, sem vaxa i land-eign hanns handa fáfródum búendum og griðmønnum á Islandi skrifat árið 1781. [Useful grasses, or the use every farmer may have of the unsown wild plants which grow on his land—for unlearned peasants and country workers in Iceland written 1781.] 5 p.l., 238 p., indices. 16 cm. Kaupmannahøfn, 1783. (Harvard College library.)

List of vascular and cellular plants, alphabetically arranged by Icelandic vernacular names, with scientific names, vernacular names in Norwegian, Danish, and German, notes on uses, etc. Author's name not on title page, but signed to dedication at end of volume.

Hooker, W. J. List of Icelandic plants. (Appendix E.) *In* his Journal of a tour in Iceland in the summer of 1809. 2. ed. v. 2, p. 309–350. 21.5 cm. London, 1813.

List of vascular and cellular plants, on the Linnaean system, with notes on uses; cited here only for the latter.

Löve, Áskell. Íslenzkar jurtir. 291 p. 127 fig. 20.5 cm. Kaupmannahöfn, 1945.

Elements of descriptive botany, key to families; keyed, descriptive flora of vascular plants, with vernacular names, habitat, and local range. Nearly all the native or naturalized species (552) are figured. Written in Icelandic.

——— and **Löve, Doris.** Cytotaxonomical conspectus of the Icelandic flora. Acta Hort. Gotob. **20:** 65–291. 27 fig. (maps). 1956.

Largely cytotaxonomic and phytogeographic, but includes tabular lists of all known native and naturalized spermatophytes except *Hieracium* and *Taraxacum* (535 species), divided phytogeographically, with critical discussion of numerous species; extensive bibliography.

Ostenfeld, C. H., and Gröntved, Johannes. The flora of Iceland and the Færoes. xxiv, 195 p. 2 maps. 20.5 cm. Copenhagen, 1934.

Brief bibliography; annotated flora of vascular plants in form of keys, with vernacular names; glossary.

Stefánsson, Stefán. Flóra Íslands. 3. útg., aukin. Steindór Steindórsson . . . bjó til prentunar. lviii, 407 p. 253 i.e. 254 fig. (fig. [254], map), front. (port.). 19 cm. Akureyri, 1948. (1st ed. 1901.)

Annotated, keyed, descriptive flora of vascular plants (428 numbered species), with one or more vernacular names for each species; glossary, list of botanists with brief biographical data. The genera *Taraxacum* and *Hieracium* are keyed only to groups, but 108 species of *Taraxacum* (not included in above figure) are listed with localities. Adventive and introduced plants are described but not numbered. Written in Icelandic; ed. 3 edited by S. Steindórsson.

Thoroddsen, Thorvaldur. An account of the physical geography of Iceland with special reference to the plant life. *In* Rosenvinge, L. K., and Warming, E. The botany of Iceland. [v. 1] part 1, [sect.] 2. 1 p.l., 187–343 p. 36 fig. 1914.

Topography, hydrology, geology, climate, general features of vegetation, plant formations; bibliography in form of footnote references.—See also **Hansen, H. M.** Studies on the vegetation of Iceland. *In* Rosenvinge, L. K., and Warming, E. The botany of Iceland. v. 3, pt. 1 (whole no. 10). 2 p.l., 186 p. 12 pl. 1930. (Climate, biological spectra, phytogeography, types of vegetation, detailed discussion of formations in various localities, distribution according to moisture, etc.; bibliography.)—Also **Steindórsson, Steindór.** Studies on the vegetation of the central highland of Iceland. l.c. **3:** 345–547. 20 fig. (fig. 1, map). 1945. (Botanical explorations, topography, geology, climate, description of plant communities, altitudinal range.) —Also **McVean, D. N.** Notes on the vegetation of Iceland. Trans. and Proc. Bot. Soc. Edinburgh **36:** 320–338. 5 fig. (fig. 1, map). 1955. (Climate, snow-cover, list of new locality records, catalog of the species of 9 types of plant communities; short bibliography.)

Local

Áskelsson, Jóhannes. Gróður í Skriðufelli. Árbók Ferðafélags Íslands **1943:** 43–45. 1943. (Cornell University library.)

Includes list of vascular plants (37 species) from the nunatak Skriðufell at Hvítárvatn, between 420 and 450 meters elevation. (South-central Iceland.)

Björnsson, Hálfdán. Gróður í Ingólfshöfða. Náttúrufræðingurinn **20:** 185–186. 1950.

List of vascular plants collected (91 species). (Ingolfshofdi in south-central Iceland.)

Davíðsson, Ingólfur. Gróður á Árskógsströnd. Náttúrfræðingurinn **10:** 72–89. 1940. (Harvard College library.)

Topography, general features of flora, life forms, etc.; list of vascular plants (222 species), annotated as to abundance. (North-central Iceland.) —See also his Notes on the vegetation of Árskógsströnd, North Iceland.

Acta Nat. Island. v. 1, no. 4. 20 p. 1 fig. (map). 1946. (Plant formations, distribution types, biological spectra; annotated list of vascular plants (214 species).)

——— Gróður á Þingvöllum. Náttúrufræðingurinn **16**: 33–35. 1 fig. 1946.
Notes on the vegetation; list of vascular plants (201 species), with vernacular names only. (Thingvellir, in southwest Iceland.)

——— Gróður í Borgarfirði og Njarðvík eystra. Náttúrufræðingurinn **11**: 16–30. 2 fig. 1941.
Topography, general features of vegetation; list of vascular plants (227 species), annotated as to abundance. (Borgarfjörður and eastern Njarðvík, in northeast Iceland.)

——— Gróður í Mýrdal. Náttúrufræðingurinn **9**: 132–137. 1939. (Harvard College library.)
General features of flora; list of vascular plants (198 species). (South-central Iceland.)

——— Gróður í Seyðisfirði. Náttúrufræðingurinn **12**: 24–44. 6 fig. 1942.
Typography, general features of vegetation, phytogeography, life forms; list of vascular plants (231 species), annotated as to abundance. (Seydisfjorður in northeast Iceland.)

——— Gróður í Suðureyjum á Breiðafirði. Skýrsla um hið íslenzk. Náttúrufræðisf. 1941/42: 44–60. 1943. (Cornell University library.)
Topography, statistics of habitat groups, etc.; tabular list of vascular plants (207 species) showing occurrence on different islands in Breida Fjord. (Suðureyjar, northwest Iceland.)

——— Gróður í Videy, Engey og Effersey. Náttúrufræðingurinn **9**: 180–182. 1939. (Harvard College library.)
General features of flora; lists of vascular plants, showing occurrence in these islands. (Southwest Iceland, off Reykjavik.)

——— Gróðurrannsóknir á Hesteyri og í Aðalvík. Náttúrufræðingurinn **7**: 15–24. 1 fig. 1937. (Harvard College library.)
General features of flora; tabular list of vascular plants (193 species), showing occurrence at the two localities mentioned. (Northwest Iceland.)

——— Gróðurrannsóknir á Reykjanesskaga 1937–1938. Skýrsla um hið íslenzk. Náttúrufræðisf. 1937/38: 39–50. 1939. (Botaniske Museum library, Copenhagen.)
General features of vegetation, author's explorations; tabular list of vascular plants (255 species) from Grindavík, Krísuvík, Selsvellir, Seltjarnarnes, and other localities on the southern coast of Reykjanes (peninsula). (Southwest Iceland.)—See also Hadač, below.

——— Nokkrir fundarstaðir jurta. Náttúrufræðingurinn **17**: 120–134. 1947.
Includes (p. 122–131) extensive list of vascular plants collected by author and Johannes Gröntved at Steingrímsfjörður, Bjarnarfjörður, Asparvíkurdalur, and Kaldbaksvík. (Northwest Iceland.)—Also includes (p. 131–134) Gróðurfar í Grímsey á Steingrímsfirði, a list of phanerogams (105 species) grouped by habitat. (An islet off Drangsnes, in Steingrimsfjord, northwest Iceland.)

Fletcher, H. R. Notes on the plants of the north-western peninsula of Iceland. Trans. and Proc. Bot. Soc. Edinb. **32**: 381–395. pl. 28 (map). 1938.
General features of flora; list of vascular plants collected by the St. Andrews University Expedition in 1937. (Southwest and central part of Vestfjarda peninsula, northwest Iceland.)

Guðmundsson, Finnur. List of plants (Pteridophyta and Spermatophyta) found in the Þjórsárver við Hofsjökul (north of Fjórðungssandur), 1951. *In* Scott, Peter, and Fisher, James. A thousand geese. Appendix D. p. 216–218. 1 fig. 22 cm. London, 1953.
Unannotated list of 108 vascular plants of a tundra oasis on the Þjórsá River at the Hofsjökull Glacier, with notes on distribution of some species; also list of 25 species found growing on the roof and walls of a stone and

turf hut at Bólstaður. Reprinted from the 5th Annual report of the Severn Wildfowl Trust. (South-central Iceland.)

Guðmundsson, Þóroddur. Eiðahólmi. Náttúrufræðingurinn **11:** 92–96. 1941.

Topography, etc.; annotated list of vascular plants, alphabetically arranged by vernacular names. (Southeast Iceland.)

Hadač, Emil. The flora of Reykjanes Peninsula, SW-Iceland. *In* Gröntved, Johannes, and Sørensen, Thorvald, *eds.* The botany of Iceland. v. 5, pt. 1. 57 p. map. 26 cm. Copenhagen, London, 1949.

Topography, geology, climate, soils, general features of vegetation, local distribution; annotated list of vascular plants (256 species), with altitudinal range and localities; short bibliography. (Southwest Iceland.)—See also **Davíðsson, Ingólfur.** Slæðingar í Reykjavík og grennd. Náttúrufræðingurinn **19:** 133–135. 2 fig. 1949. (Records of some adventitious species.)—Also his Nýir slæðingar í Reykjavík og ný maríustakkstegund í Mjóafirði. l.c. **21:** 45–46. 1 fig. 1951. (Adds several species.)—Also his Gróðurrannsóknir á Reykjanesskaga, above.

Jóhannsson, Bergþór. Flórulisti af Ströndum. Náttúrufræðingurinn **16:** 175–177. 1946.

List of vascular plants (202 species), with vernacular names only. (Strandir, in northwest Iceland.)—See also his Viðauki við Flórulista af Ströndum. Náttúrufræðingurinn **17:** 175. 1947. (Adds 70 species, with scientific names.)

Johnsen, Baldur. Gróðurríki Öræfa og Suðursveitar í Austur-Skaftafellssýslu. Náttúrufræðingurinn **11:** 54–63. 1941.

Topography, general features of vegetation; unannotated list of vascular plants (220 species). (Öræfi and Suðursveit in southeast and south-central Iceland.)—See also **Steindórsson, Steindór.** Flóra Öræfa og Suðursveitar. Viðauki. Náttúrufræðingurinn **13:** 132–133. 1943. (Adds 15 species.)

——— Gróðurríki Vestmannaeyja. Náttúrufræðingurinn **11:** 47–52. 1941.

General features of vegetation; unannotated list of vascular plants (122 species). (Westman Islands, off southwestern Iceland.) The same or essentially the same material, with same title, by Johnsen *in* Ólafsson, Jóh. G. Vestmannaeyjar. Árbók Ferðafélags Íslands **1948:** 184–190. 2 fig. 1948. (In Cornell University library.)—See also his Observations on the vegetation of the Westman Islands. Vísindafélag Íslendinga Rit 22. 40 p. 1939. (Topography, plant formations, bird colonies, types of vegetation, life forms, biological spectra. Not seen; title and annotation supplied by Á. Löve.)

Jónasson, Helgi. Flórulisti Reykjahverfis, frá Kringluvatni norður að Skógakvísl. Drög til sérflóru 1937. Náttúrufræðingurinn **9:** 89–93. 1939. (Harvard College library.)

Systematic list of vascular plants (203 species), annotated as to abundance. (Reykjahverfi in northeast Iceland.)

——— Gróður í Straumlækjargili. Náttúrufræðingurinn **9:** 101–102. 1939. (Harvard College library.)

Unannotated list of vascular plants (76 species), alphabetically arranged by vernacular names. (Straumlækjargil in northeast Iceland.)

——— Plöntuskrá úr Kelduhverfi. Náttúrufræðingurinn **15:** 178–185. 1945.

Annotated list of vascular plants (225 species). (Northeast Iceland.)

Jónasson, J. N. Gróður í Hegranesi. Náttúrufræðingurinn **11:** 31–40. 1941.

General features of flora; systematic list of vascular plants (181 species), annotated as to abundance (Hegranes in Skagafjörður, North-central Iceland).

Jónsson, Helgi. Bidrag til Øst-Islands flora. Bot. Tidsskr. **20:** 327–357. 1896.

Botanical explorations; annotated list of vascular plants (280 species). (Northeast and southeast Iceland from about 64° 40′ to about 65° 40′ N. Lat.)—See also his Studier over Øst-Islands vegetation. Bot. Tidsskr. **20:** 17–89. 1895. (Topography, geology, general features of flora, plant formations, etc.)

——— Floraen paa Snæfellsnæs og omegn. Bot. Tidsskr. **22**: 169–207. 1899.
General features of flora; annotated lists of lichens and mosses (by other authors) and of 280 vascular plants. (Covers Dalasýsla and Snæfellsnessýsla in northwest and southwest Iceland.)—See also his Vegetationen paa Snæfellsnes. Vidensk. Meddel. Naturh. For. Kjøbenhavn VI, 2 (1900): 15–97. 1901. (Ecological.)—Also **Davíđsson, Ingólfur.** Nokkrir fundarstađir jurta. Náttúrufræđingurinn **17**: 120–134. 1947. (Includes list of additions, p. 121–122.)

——— Vegetationen i Syd-Island. Bot. Tidsskr. **27**: 1–82. 1905.
Bibliography, general features of flora, plant formations; annotated list of 239 vascular plants, mostly collected by author from Reykjanes Peninsula east to Lónsheiđi. (Southwest, south-central, and southeast Iceland.)

Lagerkranz, John. Floristiska anteckningar från en exkursion på Island sommaren 1928. Svensk Bot. Tidsskr. **27**: 273–277. 1933.
Alphabetical list of angiosperms, with localities, collected by author in east and north Iceland, mostly in Eyjafjördur. (Chiefly north-central Iceland.)—See also Óskarsson, Some observations of the vegetation of Eyjafjörđur and Akureyri, below.

Löve, Áskell. Gróđur nyrzt á Hornströndum. Náttúrufræđingurinn **18**: 97–112. 1 fig. (map). 1948.
Author's itinerary, etc.; annotated list of vascular plants (274 species) (Coast and mountains of Horn or North Cape, northwest Iceland.)

Magnússon, Guđbrandur. Gróđur á Holtavörđuheiđi. Náttúrufræđingurinn **4**: 163–165. 1934. (Harvard College library.)
Unannotated list of 82 vascular plants of Holtavarda heath, alphabetically arranged by vernacular names. (Southwest Iceland.)

——— Gróđur í Bæjarhreppi í Strandasýslu. Náttúrufræđingurinn **7**: 27–29. 1937. (Harvard College library.)
Unannotated list of 182 vascular plants, alphabetically arranged by vernacular names. (Northwest Iceland.)

——— Gróđur í Bitru í Strandasýslu. Náttúrufræđingurinn **7**: 140–142. 1937. (Harvard College library.)
Unannotated list of 159 vascular plants, alphabetically arranged by vernacular names. (Northwest Iceland.)

Óskarsson, Ingimar. Botaniske iagttagelser fra Islands nordvestlige halvø, Vestfirđir. Bot. Tidsskr. **39**: 401–445. incl. fig. (map). 1927.
Ecology, etc.; annotated list of 214 vascular plants. (Peninsula between Ísafjörđur and Mjóifjörđur, on south side of Ísafjarđardjúp, in northwest Iceland.)—See also Steindórsson, Vestfirđir, beyond.

——— En botanisk rejse til Øst-Island samt Reyđarfjörđurs karplanteflora. Bot. Tidsskr. **40**: 337–349. 1929.
Annotated list of vascular plants (244 species) from Reyđarfjörđur. (Northeast Iceland.)

——— Gróđur í Öxarfirđi og Núpasveit. Náttúrufræđingurinn **16**: 121–131. 1 fig. 1946.
Plant associations, list of noteworthy species; unannotated list of vascular plants (233 species), with vernacular names only. (Öxarfjörđur and Núpasveit, in northeast Iceland.)

——— Háplöntuflóra hérađanna umhverfis Eyjafjörđ. Eyfirđingarit **1**: 225–250. 1949.
Annotated list of vascular plants (339 species, of which 51 are introduced), with mention of 32 additional Taraxaca and 40 Hieracia reported by other botanists. Covers area on both sides of Eyja Fjord. (North-central Iceland.) (Not seen; title and annotation supplied by Á. Löve.)

——— Some observations of the vegetation of Eyjafjörđur and Akureyri. Vísindafélag Íslendinga (Soc. Scient. Island.) Rit no. 13. 46 p. 6 fig. (incl. map). 1932. (American Museum of Natural History library.)
List of local flora of Iceland; topography, climate, general features of vegetation, cultivated plants; list of indigenous and foreign plants and

vegetables cultivated in gardens, list of acclimated plants; annotated list of 284 wild-growing vascular plants; list of 53 species of adjacent regions. (North-central Iceland.)—See also Lagerkrantz, above.

——— Svarfaðardalur's karplanteflora samt angivelse af arternes højdegrænser over havet. Bot. Tidsskr. **44:** 127–153. 1 fig. (map). 1937.

Topography, climate, general features of flora and plant formations; annotated list of vascular plants (275 species); tables showing altitudinal limits for each species, etc. (North-central Iceland.)

——— The vegetation of the islet Hrísey in Eyjafjörður, North-Iceland. Vísindafélag Íslendinga (Soc. Scient. Island.) Rit no. 8. 18 p. 1 fig. (map). 1930. (American Museum of Natural History library.)

Soil, general features of flora; annotated list of vascular plants (170 species). (North-central Iceland.)

Stefánsson, Stefán. Vatnsdalens vegetation. (Fra Islands væxtrige. II.) Vidensk. Meddel. Naturh. For. Kjøbenhavn V, **6** (1894): 174–212. 1895.

Topography, geology, climate, general features of flora; list of vascular plants (228 species) with localities. (Vatnsdalur, in northwest Iceland.)

Steindórsson, Steindór. Bidrag til Nordvest-Islands flora. Bot. Tidsskr. **45:** 166–179. 2 fig. (maps). 1940.

Geography, botanical explorations, etc.; annotated list of vascular plants (232 species). (Vicinity of Ísafjarðardjúp, mostly its north and east sides, northwest Iceland.)—See also his The Vegetation of Ísafjarðardjúp, North-west Iceland. (Contributions to the plant-geography and flora of Iceland. IV.) Acta Nat. Island. v. 1, no. 3. 32 p. 2 fig. 1946. (Account of plant formations, with lists of species.)—Also Fletcher, above, and Óskarsson (Botaniske iagttagelser), above.

——— The flora at Fjallabaksvegur, South-Iceland. (Contributions to the plant geography and floristic conditions of Iceland. II.) Vísindafélag Íslendinga Greinar **1:** 125–134. 1937.

General features of flora, account of localities, plant formations; alphabetical tabular list of vascular plants (147 species), showing occurrence at 6 different localities; biological spectrum. (In northern Fjallabaksvegur, in region of Landmanna- and Skaptártunguafréttir, south-central Iceland.)

——— Flóra Melrakkasléttu. Náttúrufræðingurinn **11:** 64–74. 1941.

List of vascular plants (216 species), annotated as to abundance; discussion of life forms. (Melrakkaslétta, in northeast Iceland.)

——— Gróður í Ásbyrgi. Náttúrufræðingurinn **2:** 153–155. 1932. (Harvard College library.)

General features of vegetation; unannotated list of vascular plants, alphabetically arranged by vernacular names. (Northeast Iceland.)—See also **Jónasson, Helgi.** Gróður Ásbyrgis. Náttúrufræðingurinn **17:** 116–119. 1947. (Adds 45 species, bringing total to 146.)

——— Gróður í Slúttnesi. Náttúrufræðingurinn **2:** 90–92. 1932. (Harvard College library.)

General features of flora; unannotated list of phanerogams (56 species) observed on 18 June 1931, alphabetically arranged by vernacular names. Preceded by a short paper by **G. G. Bárðarson** (p. 88–90. 1 fig. (map)) describing topography, etc. (Slúttnes, an island in Mývatn, north-central Iceland.)—See also **Jónasson, Helgi.** Gróður í Slúttnesi. Náttúrufræðingurinn **7:** 150. 1937. (Adds 25 species. In Harvard College library.) —Also **Löve, Áskell,** and **Löve, Doris.** Jurtir í Slúttnesi. l.c. **18:** 23–26. 1946. (Botanical explorations; list of known vascular plants (105 species).)

——— Gróðurrannsóknir í Þjórsárdal. I–II. Ársrit Skógræktarfélags Íslands **1941:** 5–35. illus. 1941; **1943:** 76–85. 1943.

General features of vegetation, etc.; annotated list of vascular plants (161 species). Part II adds 12 species, and includes account of wooded areas. (South-central Iceland.)

——— The vegetation and flora of the island Æðey in Ísafjarðardjúp North-west Iceland. (Contributions to the plant-geography and flora of Iceland III.) Vísindafélag Íslendinga Greinar v. 2, no. 1. 26 p. 1 fig., 4 tab. 1942.

Physical features, plant associations; list of 100 species. (Northwest Iceland.) (Not seen; title and annotation supplied by J. B. Hansen.)

——— Vestfirðir. I. Gróður. 92 p. 12 fig. (maps, on 2 pl.), 6 pl. (1 col.). 22 cm. Reykjavík, 1946.
Botanical explorations, general features of flora; list of vascular plants (369 species, including 25 Taraxaca and 25 Hieracia) of the Vestfjarda peninsula with localities for the scarcer species; life forms, bibliography. (Northwest Iceland.)—Also Óskarsson, Botaniske iagttagelser, above.

——— Víðidalur í Lóni. Þáttur úr gróðrarsögu. Náttúrufræðingurinn 8: 168–182. 3 fig. 1938. (Harvard College library.)
Changes in vegetation, etc.; alphabetically arranged tabular list of vascular plants; life forms, general features of flora, etc. (Southeast Iceland.)

Thoroddsen, Thorvaldur. Et besøg paa Grimsø. Geog. Tidsskr. **16:** 204–210. 1 fig. (map). 1902.
Topography, geology, history, etc.; includes brief account of vegetation and list of species collected by author in 1884 (p. 205). (An island about 40 km. north of north-central Iceland.)—See also **Jones, E. W.** The vegetation of Grímsey, Iceland. Jour. Ecology **25:** 222–253. 1 fig. (map), pl. 8–10. 1937. (Topography, climate, plant formations, bibliography; includes mention of all vascular plants seen.)—Also **Steindórsson, Steindór.** Flóra Grímseyjar. Náttúrufræðingurinn **24:** 137–143. 1954. (Botanical explorations, general features of flora; bibliography.)

Italy (Italia)

See also France, General (Guyettant, Memento botanique); Switzerland, General (Dalla Torre, Die alpenpflanzen im wissensschatze der deutschen alpenbewohner; Neuweiler, Die prähistorischen pflanzenreste Mitteleuropas; Probst, Wolladventivflora Mitteleuropas).

General

Includes a few partial works.
The general works here listed may be classified as follows: *Floras and manuals:* Arcangeli, Baroni, Bertoloni, Cesati, Fenaroli (Alps), Fiori, Parlatore, Zenari (northern Italy). *Catalogs:* Ball (south side of Alps), Ciferri. *Iconographies:* Fiori and Paoletti, Negri (medicinal), Perini (northern Italy). *Bibliographies:* Béguinot (La botanica: Lo stato attuale), Béguinot and Landi (island floras), Pasquale (southern Italy), Rehder, Rikli, Saccardo (Cronologia della flora italiana; La botanica), Sarfatti (weeds), Società botanica italiana. *Vernacular names:* Flechia (toponymy), Haussmann (weeds), Italy (Ministero di agricoltura; woody plants), Jaberg, Penzig, D. Saccardo (medicinal plants). *Trees:* Borzì, Italy (Ministero di agricoltura: names of trees), Piccioli, Sartorelli (northern Italy), Touring Club Italiano. *Medicinal and poisonous plants:* Cazzuola, A. Ceruti, O. Ceruti, Chiappini (alpine), Lapierre, Lodi, Mascherpa, Negri, Pagliani, Rovesti, D. Saccardo. *Other useful plants:* Cazzuola, A. Ceruti (forage plants), Lapierre, Riccardo (edible), Silvestri (fodder plants), Terracciano, (ornamental plants), Tenore. *Weeds and adventive plants:* Béguinot and Mazza (adventive plants), Haussmann, Pinolini, Poggi, Saccardo (Cronologia della flora italiana), Sarfatti (bibliography). *Cultivated plants:* (Ciferri), Saccardo (Cronologia delle piante da giardino: list with dates), Terracciano. *Miscellaneous:* Adamović (2: biotic factors, plant zones, etc.; phytogeography), Béguinot (La botanica: history and bibliography; Flora e vegetazione: plant zones, plant formations; Lo stato attuale; historical review with bibliography), Béguinot and Landi (island floras with bibliography), Bilancioni (botanical terms), Bubani (plants of Virgil), Dierbach (mythology), Fraas (plants in the classics), Flechia (toponymy), Giacomini (plant zones, vegetation), Rikli (plant formations, phytogeography), Saccardo (La botanica: list of authors with bibliography, botanic gardens, etc.; Cronologia della flora italiana: list of early records, statistics), Sargeaunt (plants of Virgil), Thiselton-Dyer (plants of Augustan authors).

Adamović, Lujo. Italien. xiii, 671 p. 12 fig. (incl. map), 12 pl. (1–10, 12, maps). 17.5 cm. Berlin, 1930. (Junk's natur-führer.) (John Crerar Library.)
Geography, orography, hydrography, topography, geology, biotic factors, plant formations, plant zones, agriculture, ornamental plants, forestry, animal life (chiefly fishes and invertebrates); separate treatments of the 18 regions of Italy (including Sicily and Sardinia), with account of geography, geology, paleoethnography, list of endemic or characteristic species, gardens and garden plants, zoology.—See also next entry and Béguinot (Flora e vegetazione), Giacomini, and Rikli, below.

——— Die pflanzengeographische stellung und gliederung Italiens. xii, 259 p. 1 fig. (map), 31 maps. 25 cm. Jena, 1933.
General discussion of Mediterranean flora, account of Mediterranean elements in flora of Italy, with tabular lists of species showing range in Italy; treatment of phytogeographical zones, subzones and districts with mention

of characteristic species; extensive bibliography including numerous floras and floristic papers.—See also the entry above and additional references there given.

Arcangeli, Giovanni. Compendio della flora italiana, ossia manuale per la determinazione delle piante che trovansi selvatiche od inselvatichite nell' Italia e nelle isole adiacenti. 2. ed. xix, 836 p. 21 cm. Torino, Roma, 1894. (1st ed. 1882.)

Descriptive flora of vascular plants (6567 taxa, including 4932 species), with keys to families and genera, local and extralimital range; table of altitudes of important localities.

Ball, John. The distribution of plants on the south side of the Alps. Trans. Linn. Soc. II, Bot. **5**: 119–226. 1896.

Statistics; tabular list of spermatophytes, with distribution indicated at 50 localities from Val Tinca and Col di Tenda to Gorizia-Indria and in 10 other mountainous areas of Europe. According to Furrer and Longa (Beih. Bot. Centralbl. 2 abt., **33**: 9. 1915) the data in this work must be used with caution.

Baroni, Eugenio. Guida botanica d'Italia ossia chiavi analitiche per determinare le piante spontanee che vivono principalmente nell'Italia media. xxiv, 574 p. 358 ("360") fig. 18.5 cm. Rocca S. Casciano, 1907. (Gray Herbarium library.)

Pocket manual of vascular plants (3015 i.e. 3017 species), in form of briefly annotated keys, with concise statement of range and vernacular names. Covers the regions of Emilia, Lazio, Marche, Toscana, and Umbria. The 2d and 3d eds. cover all of Italy.—Guida botanica d'Italia ossia chiavi analitiche per determinare le piante spontanee che crescono nella penisola. 3. ed. rived. e corr. da Sara Baroni Zanetti. xxxv, 708 p. 352 fig. 19 cm. Rocca San Casciano, 1955. (Pocket manual of vascular plants in form of briefly annotated keys, with concise statement of range and some vernacular names; includes very numerous varieties. The illustrations are from Fiori and Paoletti's Iconographia.)

Béguinot, Augusto. La botanica. 116 p. 17 cm. Roma, 1920.

Sketch of development of botany in Italy in 19th century; bibliography (p. 45–116) of selected botanical publications, with list of Italian botanical periodicals.—See also his Botanica. 421 p. 21.5 cm. n. p., 1938. (Enciclopedia scientifica monografica italiana del ventesimo secolo. ser. II, no. 1.) (Short sketch of principal Italian botanists (1801–1870) and of the various Italian schools of botany (1871–1900), followed by a more detailed account of Italian researches (after 1901) in the different branches of morphology, physiology, ecology, pathology, genetics, floristics, phytogeography, taxonomy, applied and historical botany; bibliography (p. 339–399) listing principal works of each author.)—Also Béguinot and Landi, Béguinot and others (Stato attuale), **Pasquale, Rehder, Rikli, Saccardo (La botanica; Cronologia della flora italiana), Sarfatti, and Società botanica italiana,** below.

——— Flora e vegetazione. *In* Enciclopedia italiana di scienze, lettere ed arti (Treccani) **19**: 729–736. illus., map. 31 cm. Roma, 1933.

Plant zones, plant formations, geological changes.—See also **Adamović,** above, and additional references there given.

——— and **Landi, Maria.** L'endemismo nelle minori isole italiane ed il suo significato biogeografico. Arch. Bot. (Forlì) **6**: 247–316. 1930; **7**: 39–99. 1931.

Discussion of geography, topography, geology, etc., of each group of islands, with references to botanical publications; critically annotated list of 189 endemic vascular plants, including not only species restricted to the islands discussed but many others found also on the larger islands, the Balearic Islands, or the mainland of Italy. Covers the Ligurian Islands (Bergeggi, Gallinaria, Palmaria, Tinetto, and Tino); the smaller islands around Corsica (very little known; proceeding anti-clockwise around the island, beginning on the northwest side, Île Rousse (Rossa), Pietra, Îles Sanguinaires (Mezzomare, etc.), Lavezzi, Cavallo, Piana, Cerbicali Islands, Gargalo, Giraglia, etc.); those around Sardinia (**Arcipelago della Maddalena,** including the islands Budelli, Caprera, Maddalena, Razzoli, San Stefano,

Santa Maria, Spargi, and the islets and rocks Barettini, Bisce, Corcelli, Monaci, Pecora, Porco, La Presa, and Spargiotto; Tavolara, with the associated islets and rocks Barca Soncia, Cana, Cigni, Figarello, Garofani, Molara, **Piana, Punta; Asinara, Sant'Antioco and San Pietro with adjacent islets);** the Tuscan Archipelago (Capraia, Elba, Giannutri, Giglio, Gorgona, Montecristo, Pianosa, and associated islets); the Ponzian (or Pontine) Islands **(Gavi, Palmarola, Ponza, Santo Stefano, Ventotene, Zannone); the Neapolitan** Islands (also known as Isole Partenopee or Isole Flegree; including Capri, Ischia, Nisida, Procida, Vivara), with "i Galli" or Isole Sirenuse, a group of 3 islets on the western extremity of the Gulf of Salerno; the Eolian or Lipari Islands (Alicuri, Filicuri, Lipari, Panaria, Salina, Stromboli, Vulcano, and numerous associated rocks or islets), with Ustica; the Egadi (Aegadian) Islands (Favignana, Levanzo, Marettimo, and some islets); the Pelagie Islands (Lampedusa, Lampione, and Linosa, with some smaller islets), with Pantelleria; the Maltese Islands (Comino, Gozo, Malta, with some smaller islets and rocks); the islets near Cape Passero, at the southeasternmost point of Sicily (Isola di Capo Passero, I. delle Correnti, Isola Grande and I. Piccola di Marzameni, I. dei Porri, Vindicari); the then Italian islands of the north Adriatic (the Brioni Islands and others along the Istrian coast between Parenzo and Capo Promontore; the then Italian islands in the Gulf of Quarnaro—Cherso, Lussino, Sansego, Canidole, Unie, and the minor islets Levrera and Pregnasig, also the non-Italian Plavnik and Arbe); the then **Italian islands of the south Adriatic (the Tremiti Islands–Capperaia or** Caprara, Cretaccio, S. Domino, S. Nicola; Pianosa; Pelagosa, including Pelagosa Grande and Pelagosa Piccola, with some adjacent rocks; Lagosta and Cazza, off the coast of Dalmatia; Saseno, off Albania); and S. Paulo, S. Pietro, S. Andrea, and S. Nicolicchio in the Gulf of Taranto.

——— and **Mazza, O.** Le avventizie esotiche della flora italiana, e le leggi che ne regolano l'introduzione e la naturalizzazione. Nuovo Gior. Bot. Ital. n.s., **23**: 403–465, 495–540. 1916.

General considerations, bibliography; list of 538 adventive vascular plants with footnote annotations; consideration of associations, distribution, agencies of introduction and dispersal, classification according to extent of naturalization, countries of origin, variation, etc.—See also Haussmann, Pinolini, Poggi, Saccardo (Cronologia della flora italiana), and Sarfatti, below.

——— and others. Lo stato attuale delle conoscenze sulla vegetazione dell'Italia e proposte per la constituzione di un comitato permanente "Pro flora italica" per la regolare sua esplorazione. Atti Soc. Ital. Prog. Sci. 2. riun., Firenze, 1908. p. 563–668. 1909. (Reprinted 107 p. Roma, 1909. Reprint in Gray Herbarium library.)

Geographically arranged review of floristic work on vascular and cellular plants, with citation of numerous publications.

Bertoloni, Antonio. Flora italica sistens plantas in Italia et in insulis circumstantibus sponte nascentes. 10 v. port. 22.5 cm. Bononiae, 1833–54.

The standard older flora of Italy; on the Linnaean system, with synonymy, references, vernacular names, range and localities, detailed descriptions, uses; covers spermatophytes and *Chara* only. Includes 4309 species and 580 varieties (fide Fiori and Paoletti). According to Pampanini (Flora delle Alpi Lepontini occidentali **1**: 24. 1904), the author cited only specimens seen by him (and for the most part preserved in his herbarium). For the Pteridophyta, see his Flora italica cryptogamica. v. 1, p. 5–122. 22 cm. Bononiae, 1858.—See also **Gorini, Gemelli.** Index Bertolonianus . . . 452 p. tab. 19 cm. Mediolani, 1867. (Conspectus of genera, with brief characters and name of natural family; tabular key to families; list of families with their genera; list of genera, species, and varieties in Bertoloni's work, arranged in the same order as the original and without page references.)

Bilancioni, Guglielmo. Dizionario di botanica generale. Istologia–anatomia—morfologia—fisiologia—biologia vegetale. Appendice: biografie di illustri botanici. xx, 926 p. 16 cm. Milano, 1906. (Manuali Hoepli.)

Alphabetical list of botanical terms, with definitions and often discussion; selected list of botanists of world (p. 795–906), with dates and brief account of work, etc.

Borzì, Antonino. Compendio della flora forestale italiana. Prontuario per la sollecita determinazione delle piante forestali indigene all'Italia . . . xliii, 181 p. 17.5 cm. Messina, 1885. (Arnold Arboretum library.)

Pocket manual of woody plants (667 species), in form of keys.—See also Italy (Ministero di agricoltura), Piccioli, Sartorelli, and Touring Club Italiano, below.

Bubani, Pietro. Flora Virgiliana ovvero sulle piante menzionate di Virgilio pareri esposti, considerati, proposti ancora. 135 p. 24.5 cm. Bologna, 1869.—Illustrazioni ulteriori alla Flora Virgiliana. p. 137–144. 24.5 cm. Bologna, 1870.—Ultime note alla Flora Virgiliana. 7, [1] p. 24.5 cm. Bologna, 1876. (All in Arnold Arboretum library.)

Alphabetical list of plant names in Virgil, with pertinent quotations from his works, identification, and when necessary discussion.—See also **Clos, Dominique.** Les plantes de Virgile. 24 p. 21 cm. Toulouse, 1871. (Extrait du Journal d'agriculture pratique et d'économie rurale pour le Midi de la France, Novembre 1871.) (Sources; list of plants mentioned by Virgil, grouped by uses, habit, etc., with references to his works; index of names used by Virgil. Largely a compilation, particularly from Bubani.)—See also Dierbach, Fraas, Sargeaunt, and Thiselton-Dyer, below; also, for additional works on plants in ancient literature, see Rehder, Bradley Bibl. **1:** 514. 1911.

Cazzuola, Ferdinando. Le piante utili e nocive agli uomini e agli animali che crescono spontanee e coltivate in Italia con brevi cenni sopra la coltura, sopra i prodotti e sugli usi che se ne fanno . . . viii, 217 p. 264 fig. 23 cm. Torino e Roma, 1880. (Massachusetts Horticultural Society library.)

Systematic treatment of useful or harmful wild and cultivated vascular and cellular plants (586 species), with vernacular names, general range or origin, and account of uses or injurious properties.—See also (for useful plants in general) A. Ceruti (forage and edible plants), Lapierre, Riccardo (edible), Silvestri (fodder), Terracciano (ornamental), Tenore; and for medicinal and poisonous plants, A. Ceruti, O. Ceruti, Chiappini (alpine), Lapierre, Lodi, Mascherpa, Negri, Pagliani, Rovesti.—The following additional works on medicinal plants have not been available: **Cortesi, Fabrizio.** Botanica farmaceutica. Descrizone delle piante medicinali . . . con cenni descrittivi sulle fanerogame velenose della flora italiana. xii, 208 p. 355 fig. 8vo. Torino, 1910. **Fioravanti, M.** Piante medicinali italiane. Cenni botanici. Coltivazione . . . 132 p. illus. 16mo. Catania, 1925. **Gallo, Giuseppe,** *ed.* Farmacopea italiana ossia dizionario popolare de farmacia e terapeutica . . . Torino, n.d. [1912 or ante]. **Gibelli, G.** Piante medicinali indigene. Genova, 1939. **Grimaldi, L.** Piante medicinali. 3. ristampa. Casalmonferrato, 1922. **Panini, F.** Piante medicinali d'Italia spontanee e coltivate su larga scala. (Manuale Hoepli.) xvi, 350 p. 289 fig. Milano, 1922. **Valenti, F., Cortesi, F.** and **Carlinfanti, E.** Codice per il reconoscimento ed il controllo delle piante medicinali. Parte I. Piante medicinali italiane. 101 p. 40 pl. 8vo. Roma, 1925.

Ceruti, Arturo. Le piante foraggere. 149 p. 93 fig. 25 cm. Torino, 1944.

Account of native and some cultivated forage plants, with short descriptions, uses, and figures of most of the species.—See also Cazzuola, above, and additional references there given.

——— Piante medicinali e alimentari. 2 p. l., 244 p. 97 fig., 16 col. pl., 4 tab. 21.5 cm. Torino, 1957.

Systematic treatment of native and many exotic medicinal plants (and edible plants in so far as they are medicinal) with extensive lists of patois names, brief description, composition, properties, usage, and dosage, and tables showing percentage of vitamins, nutritive substances, and chemical elements in principal food plants and content of amino-acids, proteins, organic acids, etc., in various species; glossary.—See also Cazzuola, above, and additional references there given.

Ceruti, Orazio. Le piante medicinali . . . Riconoscimento, raccolta, conservazione, usi, dosi, ricette, per le malattie dell'uomo e degli animali. 238 p. 43 fig., 32 col. pl. 17.5 cm. Torino, 1945.

Descriptive account of native and cultivated medicinal plants, with vernacular names and uses, grouped by uses.—See also Cazzuola, above, and additional references there given.

Cesati, Vincenzo, Passerini, Giovanni, and Gibelli, Giuseppe. Compendio della flora italiana. 906, lxxii p. and atlas of 137 (i.e. 138) pl. 27 cm. Milano, [1868]–1902. (Gray Herbarium library.)

Flora of vascular plants in form of annotated keys, with full descriptions of families and genera. For dates of issue see notices of publication in Nuovo Gior. Bot. Ital. v. 1–19, 1869–87, and Friedlander's Naturae novitates, 1868–1902; also Chiovenda, Flora delle Alpi Lepontine occidentali **1**: 46–48. 1904, and Nuovo Gior. Bot. Ital. **31**: 284–285. 1924. Text apparently completed 1886; there is also an issue dated 1884 (Arnold Arboretum library).

Chiappini, Vincenzo. Flora alpina officinale. 221 p. 53 fig., 4 col. pl. 20 cm. Rovereto, 1954.

Descriptive account of all the plants of pharmaceutical value growing on the Italian Alps, alphabetically arranged by vernacular names, with properties, etc.; therapeutic index, short bibliography, no index of scientific names. —See also Cazzuola, above, and additional references there given.

Ciferri, Raffaele, and Giacomini, Valerio. Nomenclator florae italicae seu plantae vasculares in Italia sponte nascentes, advenae, aut saepius cultae. Pars 1–2 (1). xii, 362 p. 21.5 cm. Ticini, 1950–54.

Name list of species, subspecies, and hybrids of native, adventive, and cultivated spermatophytes of Italy (not including those grown only in greenhouses), with indication by symbols of the cultivated and adventive species, etc., and citation of name used in Fiori, Nuova flora analitica, when it differs from that used in this work. The parts so far published cover gymnosperms, monocotyledons, and lower dicotyledons.

Dierbach, J. H. Flora mythologica oder pflanzenkunde in bezug auf mythologie und symbolik der Griechen und Römer. Ein beitrag zur ältesten geschichte der botanik, agricultur und medicin [sic]. x, 218 p. 20.5 cm. Frankfurt am Maine, 1833. (Harvard College library.)

Discussion of phenomena of life and growth of plants in relation to the gods; trees and shrubs connected with or sacred to different gods; also edible and ornamental, medicinal and poisonous plants.—See also Bubani, above, and additional references there given.

Fenaroli, Luigi. Flora delle Alpi. Vegetazione e flora delle Alpi e degli altri monti d'Italia. xiv, 369 p. 262 fig., 40 pl. (on 20), 44 col. pl., maps (on lining papers). 23 cm. Milano, 1955.

Short bibliography, climate, edaphic and biotic factors, altitudinal successions, keyed list of plant formations; annotated flora of vascular plants in form of keys, with brief notes on uses. Covers mountains of Italy including Sardinia and Sicily, also Corsica, mainly above 2000 meters elevation. Replaces his Flora delle Alpi e degli altri monti d'Italia. 3. ed. completemente rifatta ed in sostituzione de la "Flora delle Alpi illustrata" del Prof. O. Penzig. xi, 302 p. 141 fig., 30 col. pl. 21.5 cm. Milano, 1932. (1st ed. 1902, Flora delle Alpi illustrata, by O. Penzig.)

Fiori, Adriano. Nuova flora analitica d'Italia contenente la descrizione delle piante vascolari indigene inselvatichite e largamente coltivate in Italia. 2 v. (vii, 944; 1120 p.). 22 fig. 25.5 cm. Firenze, 1923–29.

Flora of vascular plants (3877 species, including the commonly cultivated ones) in form of fairly ample keys, with local range (usually briefly stated), extralimital range, some synonymy, and references to the Iconographia; summary of properties and uses at head of each family. The species are taken in a broad sense; varieties are given rather fully, but formae are omitted, and hybrids and critical species are merely listed. This is the standard modern flora of Italy. Replaces as a working manual **Fiori, A., Paoletti, Giulio,** and **Béguinot, Augusto.** Flora analitica d'Italia, ossia descrizione delle piante vascolari indigene inselvatichite e largamente coltivate in Italia disposte per quadri analitici. 5 v. (v. 4 in 2 pts.). illus. 27 cm. Padova, 1896–1909. This contained some material not in the Nuova flora, namely, descriptions of the genera and of numerous forms and hybrids (with a total of 13,897 taxa, including 3780 native and 357 introduced species), and Fiori's Prodromo di una geografia botanica dell'Italia riguadante la distribuzione delle piante vascolari (1: i–c. 1908), an account of geology, climate, soils, plant associations, botanical divisions and geobotanical bibliography. Vol. 5 was Saccardo, P. A. Cronologia della flora italiana, here treated separately. A. Béguinot was joint author after Paoletti's death (v. 2, pt. 3 to end). For the pteridophytes, see also **Fiori, Adriano.** Filicinae, Equisetinae, Lycopodinae. *In* Società botanica italiana. Flora italica cryptogama. Pars V. Pteridophyta. p. i–v, 1–455, 575–601 (indices). 154 fig. 25.5 cm. Firenze, 1943. (Detailed systematic treatment of ferns and fern allies (84 species and 528 infraspecific forms) with keys, descriptions, synonymy, citation of illustrations, vernacular names, uses, first Italian collections with dates, citation of exsiccatae, habitat, general range and detailed Italian range with citation of localities and collectors; bibliography (including both pteridophytic and general floristic works), list of herbaria, list of exsiccatae cited.) Fiori's systematic treatment is followed by an Appendice: **Giacomini, Valerio.** Saggio fitogeografico sulle pteridofite d'Italia. l.c. p. 457–574. illus. (incl. maps). (Discussion of ecology, altitudinal distribution, position in the composition of the zonal groups and plant formations of Italy, phytogeography.)

——— and **Paoletti, Giulio.** Iconographia florae italicae, ossia flora italiana illustrata con 4400 figure d'assieme e 12973 di analisi rappresentanti tutte le specie di piante vascolari indigene inselvatichite e largamente coltivate finora conosciute in Italia . . . 2. ed. aumentata a cura del A. Fiori. viii, 545 p. 3,878 i.e. 4,445 fig. 27 cm. Sancasciano Val di Pesa, 1921. (1st ed. 1895–1904.)

Contains 4445 small but excellent figures, illustrating each species and many varieties, with abbreviated statement of local range and growth zone; also list of vernacular names with botanical equivalents, and index to figures. The 3. ed. (viii, 549 p. 27 cm. Firenze, 1933; not seen) has 4,419 figures and 13,020 analyses.—See also Perini, below.

Flechia, Giovanni. Nomi locali d' Italia derivati dal nome delle piante. Atti R. Accad. Sci. Torino **15:** 821–842. 1880.

Classification of locality names derived from names of plants; alphabetical list of Latin names of plants (and allied terms) with the Italian locality names derived from them.

Fraas, Carl. Synopsis plantarum florae classicae, oder: uebersichtliche darstellung der in den klassischen schriften der Griechen und Römer vorkommenden pflanzen, nach autoptischer untersuchung im florengebiete entworfen und nach synonymen geordnet. xxxix, 320 p. 21 cm. München, 1845. (2. ausg., Berlin, 1870, essentially unchanged.)

Sketch of botanical knowledge among the early Greeks, author's botanical explorations in Greece, plant regions, etc.; list of vascular and a few cellular plants, systematically arranged, with citation of names from Greek and Latin authors and modern Greek names, annotated as to identity, occurrence, etc., with indices. The 2. ausgabe, Berlin, 1870 (Arnold Arboretum library), differs only in the slightly altered title-page and the omission of 2 leaves of dedication.—See also Bubani, above, and additional references there given.

Giacomini, Valerio, and **Fenaroli, Luigi.** La vegetazione. 272 p. 195 fig. (incl. maps), 118 pl. (on 59; pt. col.; incl. 459 fig.), map (front.) 28 cm. Milano, 1958. (Conosci l'Italia v. 2.)

Botanical regions and zones, plant formations, vegetation of different habitats, etc., of the alpine region, the plain of the Po, Insubria and the Euganean Hills, and the Apennine and Mediterranean region; origin of Italian flora; bibliography of floras and works on vegetation. The authors' names are given only in the introduction.—See also Adamović, above, and additional references there given.

Haussmann, Giovanni, and **Scurti, Iole.** Le piante infestanti. xiv, 305 p. 140 pl. (part. col.). 24 cm. Bologna, 1953.

Generalities on control, etc.; descriptive account of weedy plants, with very numerous patois names and their localities, but not details of control under the individual species. The illustrations include 40 plates of seedlings and seeds.—See also Béguinot and Mazza, above, and additional references there given.

Italy. Ministero di agricoltura, industria e commercio. Nomi volgari adoperati in Italia a designare le principali piante di bosco. Annali Min. Agr., Indus. e Com. v. 60. xii, 227 p. 1873.

Alphabetical list of scientific names of wild and cultivated woody plants, with vernacular names (geographically arranged); index of vernacular names.—See also under Veneto (Goiran, Sopra la pubblicazione . . .).—Also Borzì and Flechia, above, and additional references there given.

Jaberg, Karl, and **Jud, Jakob.** Sprach- und sachatlas Italiens und der Südschweiz. v. 3, karten 530–642. 46 cm. Zofingen, 1930.

Outline maps of Italy, Sicily, Sardinia, and southern Graubünden and Ticino, bearing very numerous vernacular and patois names for the same plant or botanical term printed on the pertinent areas, with reference numbers referring to annotations on the same pages. Karten 530–557 relate to silviculture and forestry, 558–568 to the tree and its parts, 569–598 to individual wild tree species, 599–621 to wild shrubs, wild fruits, etc., 622–642 to wild herbs.—Accompanied by their Der sprachatlas als forschungsinstrument. Kritische grundlegung und einführung in den Sprach- und sachatlas . . . 3 p. l., 243 p. 24.5 cm. Halle (Salle), 1928. (Harvard College library.)—See also Flechia, above, and additional references there given.

Lapierre, Giorgio. Piante aromatiche, medicinali ed industriali. xxiv, 463 p. illus. (incl. 249 numbered fig.). 19.5 cm. Casale Monferrato, 1948.

Includes (p. 32–293) annotated list of wild and cultivated medicinal and other useful plants, alphabetically arranged by vernacular names.—See also Cazzuola, above, and additional references there given.

Lodi, Giuseppe. Piante officinali italiane. vi, 616 p. 328 fig. 24 cm. Bologna, 1941.

General considerations; descriptive list of medicinal plants of Italy, with properties, etc. (Not seen; title and annotation supplied by R. Pichi-Sermolli.)—See also Cazzuola, above, and additional references there given.

Mascherpa, Pietro. Prodotti officinali. *In* Bertuzzi, Alberto, jr., and others. Industrie agrarie, enologiche ed alimentari. p. 511–530. 29.5 cm. Milano, 1949.

Includes lists of native plants that are in the Farmacopea Ufficiale Italiana, of those in other official European pharmacopoeias but not in the Italian, of those not in official pharmacopoeias but studied by Italians, and of plants in popular use and worthy of study; short bibliography.—See also Cazzuola, above, and additional references there given.

Negri, Giovanni. Erbario figurato. Descrizione e proprietà delle piante medicinale e velenose della flora italiana. Con cenni sulle principali specie dell'Africa settentrionale ed orientale, 4. ed. xvi, 459 p. 283 fig., 80 col. pl. (on 40). 32 cm. Milano, 1948. (1st ed. 1904.)

Discussion of early works on medicinal plants, short bibliography; treatment of wild and cultivated cellular and vascular medicinal and poisonous plants, systematically arranged, with description, range, medicinal properties and uses; list of species grouped by medicinal properties. The 4th edition is apparently unchanged from the 3d except for minor corrections in the text; the plates are printed on both sides instead of on recto only.—See also Cazzuola, above, and additional references there given.

Pagliani, Luigi. Le piante medicinali e la loro cultivazione. 2 p.l., 148 p. 135 fig., 7 col. pl. 27.5 cm. Torino, 1928. (Nuova enciclopedia agraria italiana . . . pt. 5.)

Descriptive treatment of wild and cultivated medicinal plants, alphabeted by vernacular names, with properties and uses; bibliography. Contains much the same material as the following work, by the same author, but each includes species not given in the other. Le piante aromatiche e da profumo. 3 p.l., 225 p. 161 fig., 10 pl. (on 8). 27.5 cm. Torino, 1933. (Nuova enciclopedia agraria italiana . . . diretta dai professori Vittorio Alpe [etc.] pt. 5.)—See also Cazzuola, above, and additional references there given.

Parlatore, Filippo, and **Caruel, Teodoro.** Flora italiana, ossia descrizione delle piante che crescono spontanee o vegetano come tali in Italia e nelle isole ad essa aggiacenti . . . 10 v. 23 cm. Firenze, 1848–1894. (Gray

Herbarium library; v. 1–9 in U. S. Dept. Agriculture library.)—Indice generale dei volumini . . . 31 p. Firenze, 1896. (Gray Herbarium library.)

Descriptive flora of phanerogams, with synonymy, iconography, vernacular names, detailed local range (especially in later volumes), extralimital distribution, annotations. Some families (such as Asteraceae) are only synopsized, without mention of species. Vol. 1 to 5 by Parlatore alone; from vol. 6 continued by T. Caruel, after Parlatore's death.

Pasquale, Fortunato. Bibliografia botanica riguardante la flora delle piante vascolari delle provincie meridionali d'Italia. Nuovo Gior. Bot. Ital. n.s., **1**: 259–270. 1894.—1.–[2.] aggiunta . . . Bul. Soc. Bot. Ital. **1897**: 19–22. 1897; **1901**: 238–244. 1901.—3. aggiunta . . . Annali Bot. **4**: 133–137. 1906.—4. aggiunta . . . l.c. 8: 247–251. 1910.

List of principal herbaria; alphabetical list of authors, with their papers. Covers the regions of Abruzzi e Molise, Calabria, Campania, Lucania, and Puglie.—See also Béguinot (La botanica), above, and additional references there given.

Penzig, Otto. Flora popolare italiana. Raccolta dei nomi dialettali delle principali piante indigene e coltivati in Italia. 2 v. (xv, 541; 615 p.). 214 fig. 25 cm. Genova, 1924.

Vol. 1, alphabetical list of scientific names of vascular and cellular plants, with Italian vernacular names and localities where used; bibliography; vol. 2, alphabetical list of vernacular names, with locality and botanical equivalent. Intended to replace all previous local and general lists of Italian vernacular names, but unfortunately did not incorporate the names from many dialect dictionaries; see review by Clemente Merlo in L'Italia Dialettale **1**: 273–275. 1925.—See also Flechia, above, and additional references there given.

Perini, Carlo, and Perini, Agostino. Flora dell'Italia settentrionale e del Tirolo meridionale rappresentata colla fisiotipia. 4 v. 400 col. pl. 34.5 cm. Trento, [1854–65]. (Arnold Arboretum library, v. 1–3.)

Colored plates prepared by the physiotype process, unnumbered but arranged alphabetically in each volume, and bearing brief text indicating habitat and range. Each of the two first volumes contains an alphabetical list of species figured, the third (in copy examined) does not; fourth volume not seen.—See also Fiori and Paoletti, above.

Piccioli, Lodovico. Le piante legnose italiane. vii, 991 p. 150 fig., pl. 23 cm. Firenze, 1890–1903. (Arnold Arboretum library.)

Account of 627 woody plants (native and introduced) with vernacular names, brief descriptions, range, uses, etc.—See also Borzì, above, and additional references there given.

Pinolini, Domenico. Le erbe dannose alle piante coltivate e mezzi per distruggerle. vi, 207 p. 130 fig. 19 cm. Milano, 1924.

Descriptive list of weeds, divided according to occurrence in different crops, with methods of control. (Not seen; title and annotation supplied by R. Pichi-Sermolli.)—See also Béguinot and Mazza, above, and additional references there given.

Poggi, Tito, and Ciferri, Raffaele. Malerbe e lotta. 3. ed. completamente aggiornata. xiv, 428 p. 54 fig., 35 col. pl. 24.5 cm. Casale Monferrato, 1952. (1st ed. 1894, by Poggi, with title Le principali erbe dannose all'agricoltura.)

General considerations; descriptive treatment of 100 principal weeds (and mention of others) with vernacular names and methods of combatting them.—See also Béguinot and Mazza, above, and additional references there given.

Rehder, Alfred. (Phytography.) Italy. *In* his The Bradley Bibliography. A guide to the literature of the woody plants of the world before the beginning of the twentieth century. Compiled at the Arnold Arboretum of Harvard University under the direction of Charles Sprague Sargent. v. 1, p. 423–432. 29.5 cm. Cambridge, Mass., 1911.—Additions and corrections. l.c. **1**: 535. 1911; **5**: xxii. 1918.

Essentially unannotated list of floristic works (including those in periodical and serial literature), chronologically arranged, divided into General and Sicily, the dendrological works in each section separated from the general.—

See also l.c. **1**: 4, for works on botanical bibliography; **1**: 24–26, periodicals and serials relating to botany; **1**: 60–65, botanic gardens; **1**: 74–75, vernacular names; **1**: 81, botanical history; **1**: 226–227, phenology; **1**: 297, individual noteworthy trees; **3**: 23. 1915, alphabetical list of periodicals and serials relating to arboriculture, gardening, and uses of plants; **3**: 29, catalogs of gardens, etc.; **3**: 33, history of gardening; **3**: 88, dendrological, pomological, economical, and miscellaneous works dealing with horticulture, fruit trees, medicinal and other useful or ornamental plants; **3**: 114–115, arboriculture and general horticulture; **3**: 256, pharmacopoeias; **3**: 271, economic products; **4**: 13. 1914, periodicals and serials relating to forestry, timber uses, etc.; **4**: 20, exhibitions of forest products; **4**: 28, history of forests; **4**: 86, dendrography; **4**: 363–364, forests.—See also Béguinot (La botanica), above, and additional references there given.

Riccardo, Salvatore. Le piante spontanee eduli della nostra flora. 52 p. Catania, 1921.

Account of the edible wild plants of the Italian flora. (Not seen; title and annotation from Bot. Jahresber. Just 51 (2): 205. 1929.)—See also Cazzuola, above, and additional references there given.

Rikli, Martin. Das pflanzenkleid der Mittelmeerländer. 3 v. (1418 p.). illus. (incl. maps), plates (12 col.), maps, profiles. 24 cm. 1942–48 (v. 1, 1942–43, reissued 1943).

General account of climate, life-forms, phenology, plant associations, phytogeography, etc.; extensive bibliography (v. 3). Covers Iberian Peninsula, southern France, Italy, southern Switzerland, Balkan region, southeastern Russia, Near East, and northern Africa.—See also Adamović, above, and additional references there given.

Rovesti, Guido. Prima inchiesta sulla produzione italiana delle piante officinali indigene di maggiore importanza per l'erboristeria e per le industrie derivate. xl, 319 p. 24.5 cm. Roma, 1939.

List of the more important native plants used in medicine, etc., alphabetically arranged by vernacular names, with brief account of uses, properties, etc., and list of localities where found and vernacular names with locality where used; statistics, bibliography, indices, pharmaceutical names. Issued by the Commissione consultiva per le piante officinali.—See also Cazzuola, above, and additional references there given.

Saccardo, Domenico. Dizionario dei nomi volgari delle piante medicinali, da essenze, ecc., più in uso e dei corrispondenti scientifici latini. Arch. Farmacog. e Sci. Aff. **6**: 177–234. 1917. (University of Wisconsin library.)

Alphabetical list of about 2500 vernacular names of the more important wild or cultivated Italian medicinal plants, with scientific equivalents and indication of locality where used; alphabetical list of scientific names with collected vernacular names; bibliography.—See also Cazzuola and Flechia, above, and additional references there given.

Saccardo, P. A. La botanica in Italia. Materiali per la storia di questa scienza. Mem. R. Ist. Veneto Sci. Let. ed Arti v. 25, no. 4. 236 p. 1895; v. 26, no. 6. xv, 172 p. 1901.

First part: bibliography; alphabetical list of Italian botanists and writers on Italian flora, with brief biographical details, reference to biographies, list of writings; name list of botanists, grouped geographically by areas they dealt with; annotated list of public and private botanical gardens with brief history, bibliography, list of directors, etc.; chronological list of principal Italian achievements in botany; supplementary biographies. Second part: the last list mentioned, brought up to 1896; list of principal floras of Italy, geographically arranged; supplementary biographies; list of exsiccatae; various supplementary notes; index. The account of botanical gardens and name list of botanists were reprinted in Malpighia **8**: 476–539. 1895, with title Contribuzioni alla storia della botanica italiana.—See also Béguinot (La botanica), above, and additional references there given.

——— Cronologia della flora italiana, ossia repertorio sistematico delle più antiche date ed autori del rinvenimento delle piante (fanerogame e pteridofite) indigene, naturalizzate e avventizie d'Italia e della introduzione di quelle esotiche più comunemente coltivate fra noi. xxxvii, 390 p. 27.5 cm. Padova, 1909. (Flora analitica d'Italia dei professori A. Fiori, G. Paoletti e A. Béguinot v. 5.)

Statistics of introduced species and of increase in number of known species, bibliography; list of Italian vascular plants, with dates of early records and names of recorders; records of additional cultivated species; lists of species known at various periods.—See also (for introduced plant) Béguinot and Mazza, above, and additional references there given.

——— Cronologia delle piante da giardino e da campo coltivate in Italia. Indicatore alfabetico della loro prima introduzione fra noi, dei nomi italiani, della famiglia, della patria d'origine, ecc. Bul. Uffic. Assoc. Ort. Prof. Ital. v. 5–6, suppl. xliv p. 1917–19.

Alphabetical list of plants cultivated in Italy, with habitat and dates of introduction, vernacular names, etc.—See also Terracciano, below.

Sarfatti, Giacomino. Ricerche sulla flora infestante delle culture in Italia. I. Introduzione bibliografica e metodologica. Nuovo Gior. Bot. Ital. n.s., **55**: 527–558. 1 fig. 1949.

General review of the subject of weeds, with bibliography intended to be essentially complete for Italy but excluding works on adventive and ruderal plants, weeds of rice fields, and local weed lists.—See also Béguinot and Mazza, above, and additional references there given.

Sargeaunt, John. The trees, shrubs, and plants of Virgil. 4 p.l., 149 p. 20 cm. Oxford, 1920.

Discussion of descriptive terms used by Virgil, etc.; alphabetical list of plant names used by Virgil, with references to his works, discussion of uses, etc.; list of Italian common names, with Virgilian equivalents, and of botanical names, with equivalents. Some critical notes are given in review by A. W. Pickard-Cambridge. Jour. Roman Stud. **8**: 202–204. 1920.—See also Bubani, above, and additional references there given.

Sartorelli, G. B. Degli alberi indigeni ai boschi dell'Italia superiore. 4 p. l., 454 p. 21.5 cm. Milano, 1816. (Arnold Arboretum library.)

Account of native trees and shrubs of northern Italy, cited here for its account of uses and vernacular names.—See also Borzì, above, and additional references there given.

Silvestri, Antonio de. Le piante pratensi ossia le erbe dei prati e dei pascoli italiani . . . 2. ed. iv, 433 p. illus., port. 24.5 cm. Torino, 1880. (Arnold Arboretum library.) (1st ed. 1879; 6. ed. 1896, not seen.)

Account of the herbs growing in meadows and pastures of Italy, with short descriptions and account of forage value (or lack of it).—See also Cazzuola, above, and additional references there given.

Società botanica italiana. Bullettino bibliografico della botanica italiana. v. [1]–9. 1904–1941.

Classified, mostly annual lists of publications by Italian botanists or on Italian botany, covering years 1904–1938; the entries mostly briefly annotated in the earlier years, not later. Each volume consisted of several fascicles, with annual indices through 1916 and in v. 9, by volume in v. 4–8. Vol. 1–3 edited by G. B. Traverso, v. 4–5 by Cesare Sibilia, v. 6–7 (2) by Albina Messeri, v. 7 (3)–8 by A. Messeri and Rodolfo Pichi-Sermolli, v. 9 by R. Pichi-Sermolli and Rinaldo Corradi.—See also **Giacomini, Valerio.** Rassegna dei lavori italiani di geobotanica, floristica e sistematica per l'anno 1949 [- 1956]. Arch. Bot. (Forlì) **26**: 59–64, 239–249. 1950; **27**: 223–250. 1951; **28**: 48–54, 218–237, 343–351. 1952; **29**: 320–366. 1953; **30**: 155–191. 1954; **31**: 245–273. 1955; **32**: 196–226. 1956; **33**: 261–298. 1957. (Annotated annual lists of systematic, floristic, and ecological publications, alphabetically arranged by authors.)—Also Béguinot (La botanica), above, and additional references there given.

Tenore, Vicenzo, and **Pasquale, G. A.** Compendio di botanica ordinato specialmente alla conoscenza delle piante utili più comuni. 3. ed., riformata ed accresciuta. 4 p. l., 600 p. 20 cm. Napoli, 1870. (New York Botanical Garden library.) (1st ed. 1847.)

Includes (p. 195–580) systematic treatment of cellular and vascular plants with account of useful and harmful species, particularly those of Italy.—See also Cazzuola, above, and additional references there given.

Terracciano, Nicola. Le piante della flora italiana più acconce all'ornamento dei giardini. Atti R. Ist. Incoragg. Napoli V, v. 2, no. 2. 319 p. 1901.

Cultivation and propagation of indigenous plants; annotated list of vascular plants considered suitable for cultivation, with vernacular names, range, date of flowering and fruiting, and methods of propagation.—See also Saccardo (Cronologia delle piante da giardino), above.

Thiselton-Dyer, W. T. Flora. *In* Sandys, John Edwin. A companion to Latin studies. 3. ed. p. 66–89. 23 cm. Cambridge [England], 1921. (Reprinted 1925, 1929, 1938. 1st ed. 1910.)

Classified, briefly annotated summary of plants known to the Augustan poets, with brief pertinent quotations; includes wild and cultivated plants as well as Asiatic plant products used by the Romans; bibliography.—See also Bubani, above, and additional references there given.

[Touring Club Italiano.] [La flora forestale italiana.] Alpe **17**: 337–416. illus. 1930; **18**: 1–96, 289–348, 589–669. illus. 1931; **19**: 105–168, 377–466. illus. 1932; **20**: 161–248, 411–484. illus. 1933.

Treatment of the trees of Italy, not in systematic order, with account of range, botanical characters, wood characters, uses, cultivation, insect and fungous enemies, etc., by numerous contributors. Published as 8 special numbers but within the regular pagination, each number composed of papers by various authors, the contents being as follows (cover-titles): [1] Le querce d'Italia; [2] Le conifere della montagna italiana; [3] Le conifere mediterranee in Italia; [4] Il castagno e il faggio; [5] Le latifoglie a legname pregiato; [6] Gli elementi della macchia mediterranea in Italia; [7] Le latifoglie a legname tenero; [8] Latifoglie varie. The cultivated trees of Italy [La flora forestale esotica] are treated in a supplementary series, as follows: Le conifere esotiche (I. Abeti e pini, II. Specie varie). Alpe **21**: 273–372, 425-508. illus. 1934; Le latifoglie esotiche (I. Specie sempreverdi, II. Specie decidue). l.c. **22**: 161–232, 381–460. illus. 1935.—See also Borzì, above, and additional references there given.

Zenari, Silvia. Flora escursionistica. Chiave botanica analitica per la determinazione delle principali specie vegetali dell'alta Italia. 2 v. (v. 1, xvi, 790 p.; v. 2, 143 p. incl. 56 fig. (in glossary), 51 pl.). 21.5 cm. Padova, 1957. (v. 2 in pocket of v. 1.)

Flora of principal vascular plants of northern Italy in form of keys, with ecological habitat and altitudinal zone but no other indication of range; bibliography. Does not include the species "di minor conto e le rarità." Vol. 2 includes illustrations of chosen species and glossary.

ABRUZZI E MOLISE

See also Campania, General (Tenore, 2); Lazio (Cufodontis, Falqui, Zodda).
The regione of Abruzzi e Molise includes the Provinces of Campobasso, Chieti, L'Aquila, Pescara (formed in 1926), and Teramo.

Local

Abbate, Enrico. Guida dell'Abruzzo. viii, 402, 558 p. 18 cm. 7 maps (in pocket). Roma, 1903. (Harvard College library.)

Includes (parte prima, p. 62–115) account of plant zones, etc., and list of vascular plants, with localities or brief annotations.

Cesati, Vincenzo. Elenco delle piante raccolte de diversi fra li botanichi che presero parte all'escursione sul gruppo della Maiella e del Morrone. Bol. Club Alpino Ital. 7: 161–187. 1873.

List of vascular and cellular plants, with localities, also a few insects and mollusks. (Provinces of L'Aquila, Chieti, Pescara, and Teramo.)

Crugnola, Gaetano. La vegetazione al Gran Sasso d'Italia . . . Contributo alla geografia botanica. vii, 273 p. 23.5 cm. Teramo, 1894. (Arnold Arboretum library.)

Mainly phytogeographical and ecological; includes (p. 39–66) annotated list of spermatophytes (796 species), with localities. The Gran Sasso, whose highest summit is Monte Corno or Monte Cavallo (2914 meters or 9583 feet), is the highest group in the Appennines. (Provinces of L'Aquila, Pescara, and Teramo.) Replaces **Jatta, Antonio.** Ricordo botanico del Gran Sasso d'Italia. Nuovo Gior. Bot. Ital. 9: 197–218. 1877. (General features of flora, plant zones; list of known lichens and vascular plants, with localities and references.)—See also **Crugnola, G.** Un manipolo di piante del Gran

Sasso d'Italia. Riv. Abruzzese Sci., Let. ed Arti **2**: [p?]. 1894. (Not seen.)—Also his Materiali per la flora dell' Abruzzo teramano. Un secondo manipolo . . . Nuovo Gior. Bot. Ital. n.s., **7**: 233–247. 1900. (List of 199 vascular plants from herbarium of Raffaele Petrilli sr., including some additions.)—Also **Fiori, Adriano.** Un manipolo di piante del Gran Sasso . . . Bul. Soc. Bot. Ital. **1907**: 80–83. 1908. (Includes some additions.)—Also **Zodda, Giuseppe.** Nota floristica sul gruppo del Gran Sasso. Nuovo Gior. Bot. Ital. n.s., **49**: 390–398. 1942. (Includes additional species and localities.)

Finamore, Gennaro. Botanica popolare abruzzese. Arch. Stud. Tradiz. Pop. **8**: 29–37, 211–220. 1889.

Alphabetical list of Italian vernacular names, with corresponding scientific names, patois names used in Provinces of L'Aquila and Chieti, and notes on uses.

Grande, Loreto. Primo contributo alla flora di Villavallelonga nella Marsica. Nuovo Gior. Bot. Ital. n.s., **11**: 125–140. 1904.

Physiography; annotated list of 400 vascular plants, with vernacular names. (Prov. L'Aquila.)

Grasso, Vicenzo. Contributo alla flora di alcune località montuose molisane con particolare riguardo alla Montagnola di Civitanova del Sannio (m. 1425) e al commune di Chiauci. Nuovo Gior. Bot. Ital. n.s., **52**: 81–84. 1945.

Localities, etc.; lists of vascular plants, with localities. (Prov. Campobasso.)

Groves, Henry. Flora del Sirente. Nuovo Gior. Bot. Ital. **12**: 51–68. 1880.

General features of flora, with alphabetical lists of species of different botanical regions. (A mountain in Prov. L'Aquila.)

Mauri, Ernesto, Orsini, Antonio, and **Tenore, Michele.** Enumeratio plantarum quas in itinere per Aprutium, vel per Pontificiae ditionis finitimas provincias, aestate anni 1829 collegerunt. Atti Accad. Pontan. (Napoli) **1**: 185–236. 1832.

List of 912 species and varieties of vascular and cellular plants (890 vascular), with localities. (Prov. Campobasso, L'Aquila, and Pescara.) The same list was printed earlier in **Tenore, M.** Succinta relazione del viaggio fatto in Abruzzo ed in alcune parti dello Stato Pontifico nell'està del 1829 . . . 91 p. 25.5 cm. Napoli, 1830. (Not seen; title supplied by R. Pichi-Sermolli.)

Moriniello, Giovanni. Cenno monografico e flora forestale della provincia di Aquila. [17] p. folio. 1875.

Not seen; cited from Rehder, Bradley Bibliography **1**: 423. 1911.

Rovesti, Guido, and **Rovesti, Paolo.** Flora officinale del Parco Nazionale d'Abruzzo e della zona limitrofa. Riv. Ital. Essen., Profumi e Piante Off. **16**: 197–221. illus. (incl. map). 1934.

History, physiography, botanical explorations, etc.; annotated list of medicinal plants, alphabetically arranged by vernacular names, with dialect names, uses, etc.; short bibliography. (In central Appennines, in vicinity of Bisegna, Pescasseroli, and Opi, Prov. L'Aquila.)

Vaccari, Lino, and **Wilczek, Ernest.** Contributo alla conoscenza floristica del Parco Nazionale d'Abruzzo. Chanousia **4**: 179–198. 1940.

Lists from various localities. (Prov. L'Aquila.) (Not seen; cited from Lusina, Nuovo Gior. Bot. Ital. n.s., **60**: 873. 1954.)—See also **Lusina, Giuseppe.** Il Parco Nazionale d'Abruzzo. Nuovo Gior. Bot. Ital., n.s. **60**: 870–873. 1954. (Topography, general features of vegetation, botanical explorations with partial lists of species, bibliography.)—Also p. 882–885 for additional lists of species.—Also **Lusina, G.** Alcune escursionì nel Parco Nazionale d'Abruzzo. l.c. **60**: 890–893. 1954. (Lists from 2 localities.)

Villani, Armando. Primo [-nono] contributo allo studio della flora campobassana. Malpighia **20**: 49–89, 333–338. 1906; **21**: 3–24. 1907; **22**: 25–34. 1908; **23**: 386–409. 1909. Bul. Soc. Bot. Ital. **1910**: 155–159. 1910; **1911**: 41–45. 1911. Nuovo Gior. Bot. Ital. n.s., **19**: 124–152. 1912. Bul. Soc. Bot. Ital. **1914**: 57–64. 1914.

Topography, etc.; lists of vascular plants collected by author or recorded in various publications from different localities in the province; no comprehensive list of species. Title varies; that of 6th is Escursioni botaniche a Capracotta: of 7th, Aggiunte alla flora della provincia di Campobasso; of 8th, Escursioni botaniche a Termoli e a Trivento; of 9th, Escursioni botaniche nel bosco Ramitelli. (Prov. Campobasso.)

——— Primo contributo allo studio della flora della provincia di Chieti. Nuovo Gior. Bot. Ital. n.s., 28: 69–111. 1921.

List of author's papers on flora of Province of Campobasso, lists of species from various localities in Abruzzi and neighboring regions found in an herbarium in Chieti; annotated list of vascular plants of the Province of Chieti, including those found by others, with localities.—See also **Tenore, Michele.** Relazione del viaggio fatto in alcuni luoghi di Abruzzo citeriore nel state del 1831. 132 p. map. 22.5 cm. Napoli, 1832. (Includes (p. 119–130) Index plantarum in Aprutio collectarum, a list of plants on the Linnaean system (199 species, all but one vascular), with localities. In Arnold Arboretum library.)

Zodda, Giuseppe. La flora teramana. Webbia **10**: 1–317. 1954.

Topography, geology, climate, botanical explorations, bibliography; annotated list of vascular plants of the Province of Teramo, with detailed localities. Replaces **Amato, Federico d'.** Contributo alla flora teramana. *In* Monografia della provincia di Teramo. v. 1, p. 191–257. 23.5 cm. Teramo, 1892. (Notes on the vegetation; list of vascular plants (829 species and varieties) with vernacular names, localities, and flowering dates. Not seen; title and annotation supplied by R. Pichi-Sermolli.)—See also **Zodda, G.** Le piante esotiche inselvatiche nel Teramano. (Studi sulla flora teramana. I.) Nuovo Gior. Bot. Ital. n.s., **60**: 840–850. 1954. (Discussion of naturalized species, with list.)—Also his Altre specie nuove della flora teramana. l.c. **61**: 723–724. 1955. (Notes on several species and varieties.)—Also his La flora teramana. Supplemento I. Webbia **12**: 229–270. pl. 24. 1957. (Extensive list of new records.)—Also his Studi sulla flora teramana. II. Lusus e forme rare in piante teramane. Nuovo Gior. Bot. Ital. n.s., **64**: 410–418. 1958. (Includes new records for the Gran Sasso.)

BASILICATA (LUCANIA)

See also Calabria (Terracciano, Synopsis plantarum vascularium Montis Pollini); Campania (Tenore, 2).

The regione of Basilicata, frequently called Lucania, includes the Provinces of Matera and Potenza (the former constituted in 1926 by the separation of eastern Potenza).

General

Gavioli, Orazio. Synopsis florae lucanae. Nuovo Gior. Bot. Ital. n.s., **54**: 10–278. 1948.

Topography, geology, climate, general features of flora, botanical explorations, bibliography; annotated list of vascular plants (3,567 taxa, including 1,899 species), with localities; index. This is the first complete flora of Basilicata; the more important earlier works are: **Barbazita, Francesco.** Saggio della flora lucana, e descrizione di una nuova ombrellifera. Atti R. Ist. Incoragg. Sci. Nat. Napoli 7: 13–95. 1847. (Annotated list of vascular plants and Characeae, on the Linnaean system; refers to vicinity of Balvano and other localities in Basilicata and to adjacent Province of Salerno.) **Terracciano, Achille.** Prodromo della flora lucana ossia catalogo sistematico delle piante spontanee sinora conosciute nella Basilicata. vol. 1 (all publ.). liv, 94 p. 23.5 cm. Caserta, 1893. (Sketch of vegetation; list of Thalamiflorae, with localities. Not seen; title and annotation supplied by R. Pichi-Sermolli.) **Cavara, Fridiano,** and **Grande, Loreto.** Esplorazioni botaniche in Basilicata. Bul. Orto Bot. R. Univ. Napoli **3**: 353–451. 11 fig., pl. 4. 1913. (Earlier botanical explorations, running account of explorations of authors and Lacaita, with many short lists of species from different localities; list (p. 404–451) of vascular plants, lichens, and mosses collected, with localities.)

Local

Gavioli, Orazio. Note sulla flora lucana—Monte Serranetta e suoi contrafforti a sud-est di Potenza. Arch. Bot. (Forlì) **2**: 259–267. pl. 12. 1926; **3**: 175–190. 1927; **4**: 186–201. 1928.

Topography, climate, plant associations; list of 666 vascular plants, with localities (total flora 892 species; 226 common species omitted). (Prov. Potenza.)

——— Note sulla flora lucana—Monti della Maddalena. Nuovo Gior. Bot. Ital. n.s., **38**: 241–294. 1931.

Geography, topography, general features of flora, botanical explorations; annotated list of vascular plants (997 species, varieties, and hybrids). Monti della Maddalena is a name for the mountain group south of Potenza, including M. Pano (1164 meters), Serra della Criva (1360 meters), and M. Arioso (1736 meters). (Prov. Potenza.)

——— Note sulla flora lucana. Primo contributo allo studio della flora del Volturino. Nuovo Gior. Bot. Ital. n.s., **39**: 513–522. 1932.

Topography, geology, etc.; annotated list of 143 vascular plants collected by author in July 1929. (Mountain 1836 meters high, an eastern spur of Monti della Maddalena, in Prov. Potenza.)

Giordano, G. C. Contributo all'illustrazione della flora lucana. Annali R. Ist. Tecn. e Naut. Napoli 3 (1885): 1–47. 1886.

Short account of physiography and geology of Basilicata; list of about 447 vascular plants and 30 mosses. Relates mainly to the Basento valley. (Not seen; title and annotation supplied by F. Sappa.) (Prov. Matera (and Potenza?).)

Terracciano, Nicola. Enumeratio plantarum vascularium in agro murensi sponte nascentium. Nuovo Gior. Bot. Ital. **5**: 5–26, 41–86, 145–156, 225–260. 1873.—Ad Enumerationem . . . addenda. l.c. n.s., **14**: 117–220. 1907.—Ad Enumerationem . . . addenda altera. Bul. Orto Bot. R. Univ. Napoli **3**: 111–192. 1913.

Annotated list of vascular plants (1279 species including first Addenda), with localities and especially in the Addenda with many descriptions or diagnoses. (Vicinity of Muro, in Prov. Potenza.)

——— Florae Vulturis synopsis, exhibens plantas vasculares in Vulture monte ac finitimis locis sponte vegetantes. Atti R. Ist. Incoragg. Sci. Nat Napoli II, **6**: 243–446. 1869. (Reprinted, 206 p. Neapoli, 1869. Reprint in Arnold Arboretum library.)

Topography, general features of flora, statistics; annotated list of vascular plants (997 species), with localities. (Prov. Potenza.)—See also **Poli, Aser.** Contribuzione alla flora del Vulture. Nuovo Gior. Bot. Ital. **17**: 144–146. 1885. (Adds 21 species)—Also **Forti, Achille,** and **Trotter, Alessandro.** Materiali per una monografia limnologica dei laghi craterici del M. Vùlture. Annali Bot. v. 7, suppl. 111 p. 9 fig. (incl. map, profiles), 3 pl. 1908. (Includes Prospetto delle piante superiori vegetanti nei laghi di Monticchio (p. 19–25), by Trotter, an annotated list of Characeae and spermatophytes, followed by remarks on some of them.)—Also **Fiori, Adriano.** Note di floristica calabrese e lucana. 3. Escursione al M. Vulture (Basilicata). Nuovo Gior. Bot. Ital. n.s., **26**: 141. 1919. (New records.)

CALABRIA

See also Campania (Tenore, 2).

The regione of Calabria (Calabrie) includes the Provinces of Catanzaro (formerly Calabria Ulteriore I), Cosenza (formerly Calabria Citeriore), and Reggio Calabria (formerly Calabria Ulteriore II).

General

Svimez. Associazione per lo sviluppo dell'industria nel mezzogiorno. Piante officinali in Calabria. Presupposi e prospettive di industrializzazione. iv, 118 p. 2 maps. 23.5 cm. Roma, 1951.

Includes chapter, Il potenziale erboristico della Calabria (p. 39–60), containing annotated list of 129 native drug plants (species or genera), alphabetically arranged by vernacular names.

Local

Albo, Giacomo. La vegetazione dell'orlo settentrionale della Magna Sila. Nuovo Gior. Bot. Ital. n.s., **33**: 390–437. 1926.

Geology, climate, etc.; list of vascular plants, with localities. Refers to territory of Rossano facing the Gulf of Taranto, from the sea up to about 400 meters altitude. (Prov. Cosenza.)—See also his Notizie ed osservazioni sulla flora del versante settentrionale-orientale della Sila. Nuovo Gior. Bot. Ital. n.s., **42**: 623–664. 8 fig. 1935. (Geology, climate, general features of flora; annotated list of 532 lichens and vascular plants. Refers to coastal zone between the Coriglianeto and Fiumenicà rivers and the slopes of the Sila up to the Monti di Longobucco.)

Bisogni, Carlo. Contributo alla flora d'Hipponium. Riv. Ital. Sci. Nat. **16**: 73–76, 81–86, 97–100. 1896.

Topography, geology, general features of flora; annotated list of phanerogams not previously reported from the area. Hipponium is the ancient name of Monteleone (di Calabria, now Vibo Valentia, in Prov. Catanzaro).

Longo, Biagio. Contribuzione alle flora calabrese. Annuario R. Ist. Bot. Roma **9**: 125–168. 1901.

General features of flora of the Valle del Lao and of the western slope of the Pollino group; unannotated list of vascular plants (575 species). (Prov. Cosenza.)—See also his Contribuzione alla conoscenza della vegetazione del bacino del fiume Lao. Annuario R. Ist. Bot. Roma **9**: 257–276. 1902. (Includes 180 additional species; relates especially to Monte Serramale and the high valley of the Lao, thus entering the Province of Potenza in Lucania.)

——— and **Parisi, R.** Bibliografia botanica silana. *In* Studi silani. p. 295–299. 30 cm. Napoli, 1937.

Unannotated list of 58 titles, alphabetically arranged by authors; most of them are not floristic. Publication of Fondazione politecnica per il mezzogiorno d'Italia, Napoli. (Sila chain in Prov. Cosenza and Prov. Catanzaro.)

Macchiati, Luigi. Catalogo delle piante raccolte nei dintorni di Reggio-Calabria dal settembre 1881 al febraio 1883. Nuovo Gior. Bot. Ital. **16**: 59–100. 1884.

Geology; list of vascular plants, Characeae, and Hepaticae of the vicinity of Reggio, with localities. (Prov. Reggio di Calabria.)

Micheletti, Luigi. Flora di Calabria. 1. [–8.] contribuzione. Bul. Soc. Bot. Ital. **1895**: 169–176. 1895; **1896**: 22–30, 109–116, 141–150, 231–238. 1896; **1897**: 203–208, 256–260. 1897; **1900**: 158. 1900; **1901**: 277–287, 342–349. 1901.

Soils, etc.; lists of mosses, lichens, pteridophytes, and 524 spermatophytes, mostly from vicinity of Catanzaro, with localities. (Prov. Catanzaro.)

Mottareale, Giovanni. Contributo alla flora calabrese. (Erborizzazioni a Laganadi.) Annali R. Scuola Super. Agr. Portici II, v. 4, [art. 8]. 16 p. 1901.

List of spermatophytes collected by author, with localities and some vernacular names. (Prov. Reggio di Calabria.)

Pasquale, Fortunato. Primo contributo alla flora della provincia di Reggio Calabria. Bul. Soc. Bot. Ital. **1897**: 214–224. 1897.—Secondo contributo . . . Riv. Ital. Sci. Nat. **26**: 15–20. 1906.—Terzo [–quinto] contributo . . . Bol. Nat. Suppl. Riv. Ital. Sci. Nat. **26**: 45–52, 105–110. 1906; **27**: 50–51, 77, 86–87. 1907; **28**: 37–44, 49–51. 1908.—Sesto contributo . . . Agricoltura (Napoli) **4**: 49–55. 1913. (Sixth contribution not seen.)

Annotated lists of cellular and vascular plants.

Porta, Pietro. Viaggio botanico intrapreso da Huter, Porta e Rigo in Calabria nel 1877. Nuovo Gior. Bot. Ital. **11**: 224–290. 1879.

Running account of expedition through the three provinces of Calabria (and to Messina region), with mention of numerous species collected, but no proper list.—See also **Solla, R. F.** Osservazioni botaniche durante una escursione in provincia di Cosenza. Malpighia **10**: 168–195, 205–223. 1896. (Botanical explorations (with references), general features of flora; annotated list of vascular plants, mosses, and lichens collected by author at various localities in Cosenza.)—Also **Fiori, Adriano.** Contribuzione alla flora della

Basilicata e Calabria. Nuovo Gior. Bot. Ital. n.s., **7**: 248–271. 1900. (Author's itinerary in 1899; extensive list of noteworthy plants collected on the west coast of Cosenza, in the Sila around Castrovillari, etc.)—Also **Longo, Biagio.** Appunti sulla vegetazione di alcune località di Calabria Citeriore. Annali Bot. **1**: 85–103. 1903. (Localities visited, general features of flora, etc.; list of 108 noteworthy vascular plants, with localities, from northwestern Cosenza, in vicinity of Mormanno and Lungro.)—Also his Nuova contribuzione alla flora calabrese. l.c. **2**: 169–183. 1905. (Similar treatment, referring to northwestern Cosenza from Mormanno south to Monte Montèa.) —Also his Contribuzione alla flora calabrese. Escursione alla Sila. l.c. **3**: 1–12. pl. 1–7. 1905. (Author's itinerary, notes on forests; list of noteworthy vascular plants, with localities, from Cosenza (city) to Timpone della Stella, Miglianò, Macrocioli, etc.)—Also **Trotter, Alessandro.** Notizie ed osservazioni sulla flora montana di Calabria. Nuovo Gior. Bot. Ital. n.s., **18**: 243–278. pl. 8–11. 1911. (General features of flora of Sila, with lists of species of different habitats, etc.; bibliography of Calabrian botany.)—Also **Grande, Loreto.** Note di floristica silana. Bul. R. Orto Bot. Univ. Napoli **3**: 93–110. 1911. (List of noteworthy angiosperms, with localities, collected in 1909-10 by author and others in the Sila Grande and Sila Piccola (the latter the part in Province of Catanzaro).)—Also **Fiori, Adriano.** Note di floristica calabrese e lucana. Nuovo Gior. Bot. Ital. n.s., **26**: 129–141. 1 fig. 1919. (Includes (p. 129–139) notes on author's explorations in 1918 in the Foresta Camere Chiuse della Sila in the comune of Longobucco and in vicinity of Camigliatella, with lists of species of different habitats and list of corrections to names in Solla's paper.)—Also **Sarfatti, Giacomino.** Notizie sulla Sila e la sua vegetazione. l.c. **62**: 504–524. 16 fig. (on 8 pl.) 1956. (Geology, general features of vegetation, with mention of numerous species collected.)—Also **Gramuglio, Gaetano.** Notizie geobotaniche sull'Aspromonte e cenni sulla vegetazione del Monte Basilicò sopra Gambarie. l.c. **62**: 536–544. f. 30–33 on 2 pl. (on 1). 1956. (Topography, geology, etc.; general features of vegetation of M. Basilicò, with mention of numerous species.)

Preda, Agilulfo. Il Monte Cocuzzo e la sua flora vascolare. Nuovo Gior. Bot. Ital. n.s., **7**: 154–174. 1900.

Botanical explorations, topography, geology, general features of flora with lists of species of different areas; annotated list of known vascular plants. (Prov. Cosenza.)

Sarfatti, Giacomino. Ricerche sui pascoli della Sila (Calabria). Webbia **10**: 319–439. 62 fig. (mostly on pls., and including map), tab. 1954.

Discussion of the vegetation of the chestnut, pine, and beech zones, with list of species of each; bibliography. (Prov. Cosenza and Catanzaro.)

Terracciano, Nicola. Synopsis plantarum vascularium Montis Pollini. Annuario R. Ist. Bot. Roma **4**: 1–191. 4 pl. 1891 (1890). (Reprinted 193 p.) —Addenda ad Synopsidem . . . l.c. **9**: 23–88. 1902 (1900).

General features of flora, plant zones; annotated list of vascular plants (1486 species, including addenda); statistics. Monte Pollino is on the boundary between Cosenza and the Province of Potenza (Lucania), the bulk of the massif in Potenza but the highest summit (2270 meters) in Cosenza.—See also his Intorno alla flora del Monte Pollino e della terre adiacenti. Atti R. Accad. Sci. Fis. e Mat. (Napoli) II, v. 8, no. 9. 18 p. pl. 1897 (1896). (Further explorations, with list of 183 additional species, statistics, notes on various species; descriptions of 61 new species, varieties, and forms.)—Also **Gavioli, Orazio.** Contributo allo studio della flora del M. Pollino. Arch. Bot. (Forlì) **8**: 46–64. pl. 1–2. 1932. (Botanical explorations, geography, author's itinerary, etc.; list of 161 vascular plants, with localities. Gavioli states that Terracciano's 1486 species should be reduced by over 500 which came from other parts of Calabria than Monte Pollino.)

Zodda, Giuseppe. Breve visita all'isola di Cirella. Arch. Bot. (Forlì) **30**: 23–31. 2 fig. 1954.

Topography, geology, general features of flora; alphabetical list of 54 vascular plants collected by author on 4 June 1953. (Prov. Cosenza.)

CAMPANIA

See also Lazio (Falqui, Santis).

The regione of Campania includes the Provinces of Avellino (formerly Principato Ultra), Benevento, Caserta (with area somewhat changed from its earlier extent; formerly, with part of Prov. Frosinone in Lazio, called Terra del Lavoro), Napoli, and Salerno (formerly Principato Citra).

General

Tenore, Michele. Flora napolitana, ossia descrizione delle piante indigene del regno di Napoli, e delle più rare specie de piante esotiche coltivate ne' giardini. 5 v. 250 col. pl. 48 cm. Napoli, 1811–38. (v. 1, 2, 5, New York Botanical Garden library; other volumes not seen.)

Descriptive flora of vascular plants and Characeae, on the Linnaean system (3,388 species, including the Appendice); includes some cultivated species. Relates to Abruzzi, Basilicata, Calabria, Campania, and Puglie.

——— Sylloge plantarum vascularium florae neapolitanae hucusque detectarum. vi, 555 p. 23 cm. Neapoli, 1831. (P. 554-555 constitute Appendice [1]. Also issued with addition of 2 more appendices. vi, 639 p. 1831 [–33]. (Latter issue not seen.))—Ad Florae neapolitanae syllogen, Appendix quarta . . . 52 p. 23 cm. n.p., n.d. (Kew library.)—Appendix quinta . . . 56 p. 23 cm. Neapoli, 1842. (Kew library.)

Annotated list with frequent descriptions, of vascular plants and Characeae (3,176 species in original work); index to plates in his Flora napolitana, of which this work is essentially a synopsis.—See also **Giordano, G. C.** Index generalis Syllogis Tenoreanae, appendicumque omnium. Nuovo Gior. Bot. Ital. 7: 163–180. 1875. (Generic index to the Sylloge and its appendices.)

Local

Arata, Maria. Contributo allo studio della flora del Cilento (Salernitano). **Nuovo Gior. Bot. Ital. n.s., 45: 188–211. 1938.—Secondo contributo . . . l.c 45: 410–411. 1938.**

Botanical explorations, geology, plant formations; annotated list of 367 vascular plants. Relates to region between "punta di Ascea e il Capo Palinuro, e nell'interno il paese di Vallo della Lucania e Poderia." (Prov. Salerno.)—See also **Longo, Biagio.** Contribuzione alla flora dei monti del Cilento (Salernitano). Annali Bot. 5: 653–655. 1907. (Relates to Monte Cervate and Monte Alburno.)

Baccarini, Pasquale. Studio comparativo sulla flora vesuviana e sulla etnea. Nuovo Gior. Bot. Ital. 13: 149–205. 1881.

Topography, climate, etc.; includes comparative lists of vascular plants from Vesuvius and Etna, unannotated. (Prov. Napoli and Prov. Catania (Sicilia).)—See also Pasquale, Flora vesuviana, below.

Balsamo, Francesco, and Geremicca, Michele. Botanici e botanofili napoletani. Cenni biografici e storici. **Bul. Orto Bot. R. Univ. Napoli 3: 37–74.** ports. 1913.

Biographical sketches of Neapolitan botanists, with titles of publications. The first part (by Balsamo) is arranged chronologically, the second (by Geremicca) in part alphabetically. (Prov. Napoli.)

Béguinot, Augusto. La vegetazione delle Isole Ponziane e Napoletane. Studio biogeographico e floristico. Annali Bot. 3: 181–453. pl. 8 (map). 1905.

Bibliography, geography, climate, geology, ecology, phytogeography, etc.; annotated list of vascular plants (1063 species), with localities. The Ponzian or Pontine Islands, including Ponza, Zanone, Gavi, Palmarola, Ventotene, and S. Stefano, are in the Province of Latina (Lazio); the principal Neapolitan Islands are Nisida, Ischia, Procida, Vivara, and Capri (Prov. Napoli). Replaced as to Capri by Guadagno, Flora Capraearum nova (see below). For Ischia, see also Gussone, below; for Procida and Vivara, see also Geremicca and Rippa, below.

Calabrese, F. P. Nomenclatura botanica-volgare. Napoli, 1892.

Not seen; no review available.—See also Gusumpaur, below.

Casali, Carlo. Flora irpina. 148 p. 24 cm. Avellino, 1901. (At head of title page: R. scuola di viticoltura ed enologia di Avellino.) (British Museum (Natural History) library.)

Geology, bibliography; annotated list of vascular plants [about 1230 species], with localities; table of altitudes. Covers the Province of Avellino (formerly Principato Ultra); the adjective "irpina" is derived from the name of the Hirpini, an ancient tribe in the region. Replaces **Milani, G. B.** Elenco di piante di Principato Ultra. 20 p. 23 cm. Avellino, 1890. (Unannotated list. Not seen; title and annotation supplied by K. H. Rechinger.) Also replaces **Baccarini, Pasquale.** Materiali per la flora irpina. Nuovo Gior. Bot. Ital. **23:** 47–68. 1891; with additions by **Casali, Carlo,** and **Ferraris, Teodoro.** l.c. n.s., **7:** 215–232. 1900: Bul. Soc. Bot. Ital. **1901:** 86–92. 1901.—See also **Trotter, Alessandro.** Osservazioni ed aggiunte alla flora irpina. l.c. **1905:** 20–29, 32–42, 1905.—Also his Nuove osservazioni . . . l.c. **1906:** 9–24. 1906. —Also **Ferraris, T.** Nuove aggiunte alla flora avellinese. Nuovo Gior. Bot. Ital. n.s., **13;** 59–78. 1906.—Also **Trotter, A.** Nuova serie di osservazioni . . . Malpighia **20:** 305–320. 1906.—Also his Ulteriori osservazioni . . . l.c. **22:** 64–78. 1 fig. 1908.—Also his Nuove osservazioni . . . l.c. **23:** 425–446. 1909. (Brings total to about 1653 species.)

Comes, Orazio. Illustrazione delle piante rappresentate nei dipinti pompeiani. *In* Pompeii e la regione sotterrata dal Vesuvio nell'anno LXXIX. Memorie e notizie pubblicate dall'Ufficio tecnico degli scavi delle province meridionali. p. 177–250. 32.5 cm. Napoli, 1879. (Reprinted 1 p. l., 74 p. Napoli, 1879.)
Annotated list of 50 vascular and cellular plants positively identified in Pompeian wall paintings, alphabetically arranged by scientific names, with quotations from classical authors, etc.; list of 20 doubtful species. There is a German reissue with title: Darstellung der pflanzen in den malereien von Pompeji. Autorisierte . . . uebersetzung. viii, 68 p. 25 cm. Stuttgart, 1895. (Arnold Arboretum library.) (Prov. Napoli.)—See also **Wittmack, Ludwig.** Die in Pompeji gefundenen pflanzlichen reste. Bot. Jahrb. Engler **33** (Beibl. **73**): 38–66. 1903. (List of species identified by Schouw and by Comes in Pompeiian wall-paintings; annotated classified list of plant remains identified by author, etc.; bibliographical footnotes.)

Geremicca, Michele, and **Rippa, Giovanni.** Primo contributo allo studio della flora di Procida e Vivara. Bol. Soc. Nat. Napoli **11:** 18–66. 1897.
Topography, geology, etc.; annotated list of vascular plants (260 species). (Prov. Napoli.)—See also **Béguinot, Augusto.** Contribuzione alla flora di Procida e di Vivara. Bul. Soc. Bot. Ital. **1901:** 386–399. 1901. (Changes in vegetation, etc.; list of plants collected by author, including 126 additional species.)—Also his La vegetazione delle Isole Ponziane e Napoletane, above.

Grande, Loreto. Un'escursione sui Monti di Formicola. Bul. Orto Bot. R. Univ. Napoli **4:** 239–264. 1914.
Botanical explorations; list of species collected, or recorded by others, with localities. (A spur of the Montagna del Matese, in Prov. Caserta.)

Guadagno, Michele. Una escursione botanica a Monte Sacro di Novi in Lucania nel luglio 1907. Bul. Orto Bot. R. Univ. Napoli **2:** 235–250. 1909.
Topography, geology, etc.; annotated list of vascular plants, mostly collected by author. (In the mountain group of Cilento, Prov. Salerno.)

——— Flora Capraearum nova. Flora di Capri. Arch. Bot. (Forlì) **7:** 7–38, 244–275. pl. 1–4. 1931; **8:** 65–80, 143–158, 275–295. 1932.
Bibliography, topography, climate, geology, botanical explorations, plant formations, weeds, etc.; annotated list of vascular plants (842 species) with localities. (Prov. Napoli.) Replaces **Cerio, Ignazio,** and **Bellini, Raffaello.** Flora dell'isola di Capri . . . 95 p. col. pl., map. 19.5 cm. Napoli, 1900. (Topography, geology, climate, bibliography; barely annotated list of about 700 species of vascular plants, with vernacular names in English, French, German, and Italian. In Arnold Arboretum library.)—See also **Cerio, Edwin.** Nota sulla flora caprense. Arch. Bot. (Forlì) **15:** 134–147. pl. 3–7. 1939. (Species mentioned nearly all adventive or escaped.)—Also his Forme macrosomatiche nella flora caprense. l.c. **15:** 219–233. pl. 8–17. 1939. (Account of gigantism in several species.)

——— Prime notizie sulla vegetazione delle Isole Sirenuse. Bul. Orto Bot. R. Univ. Napoli **3:** 75–91. 4 fig. 1913. (Preprinted 17 p. 1911.)
History, geology, general features of flora, climate, etc.; annotated list of 85 phanerogams. The islets, said to be the Isles of the Sirens mentioned by

Homer, are known as "i Galli" by fishermen and include Gallo Lungo, La Rotonda, and La Castellucia. (On south side of Peninsula of Sorrento, in Prov. Salerno.)

——— La vegetazione del Monte Nuovo e le sue origini. Bol. Soc. Nat. Napoli **34** (II, **14**): 238–306. 4 fig., pl. 11. 1923.

History, soils, climate, general features of flora, plant associations; tabular list of vascular plants (355 native); origin of flora, etc. Monte Nuovo, in the Phlegraean Fields, was formed by volcanic action in 1538. (Prov. Napoli.)—See also Terracciano, La flora, below.

——— La vegetazione della Penisola Sorrentina. Bul. Orto Bot. R. Univ. Napoli **5**: 133–178. 1918; **7**: 67–128. 1924; **8**: 239–268. 1926.

Bibliography, topography, geology, climate, botanical explorations; annotated list of vascular plants of the Sorrentine Peninsula, Monti Stabiani, and Capri, with localities. Incomplete; Polypodiaceae to Caryophyllaceae (Engler and Prantl system). (Prov. Napoli and Prov. Salerno.)—See also Lacaita, below.

Gussone, Giovanni. Enumeratio plantarum vascularium in insula Inarime sponte provenientium vel oeconomico usu passim cultarum. xix, 428 p. 20 pl. 23.5 cm. Neapoli, 1854 (–1855).

Topography, geology, climate, general features of flora; annotated list of vascular plants (960 species), with some vernacular names; includes cultivated plants. The volume includes Addenda (p. 401–406), and Addenda II (3 unnumbered pages, dated Junio 1855). Page 400, the last page of text of the main work, is dated Majo 1855. Inarima is an old name for Ischia, also known anciently as Aenaria or Pithecusa. (Prov. Napoli.)—See also **Bolle, Carl.** Petit supplément à la flore de l'île d'Ischia. Bul. Soc. Bot. France **12**: 124–132. 1865.—Also **Migliorato, Erminio.** Seconda nota di osservazioni relative alla flora napoletana. Bul. Soc. Bot. Ital. **1897**: 23–26. 1897. (Includes additions to the flora of Ischia, and notes on changes in the flora around Naples.)—Also his Illustrazione dell'inedita e manoscritta "Flora pithecusana, ossia catalogo alfabetico delle piante vascolari dell'isola d'Ischia" di Giacomo Stefano Chevalley de Rivaz (1834), botanico non conosciuto. Annali Bot. **12**: 177–200. 1914. (Includes annotated alphabetical list prepared by de Rivaz, with list of species not previously recorded.)—Also **Sarfatti, Giacomino.** Su alcune piante rare o nuove dell'isola d'Ischia (Napoli). Nuovo Gior. Bot. Ital. n.s., **64**: 368–380. 10 fig. on 4 pl. (on 2). 1958. (Notes on several species.)—Also Béguinot, above.

Gusumpaur, Federico. Vocabolario botanico napolitano con l'equivalente latino ed italiano. 103 p. 19.5 cm. Napoli, 1887. (Private library of W. T. Swingle.)

Alphabetical list of vernacular names of wild and cultivated plants, with scientific names and frequent annotations; indices of scientific names and of vernacular names, with equivalents. (Prov. Napoli.)—See also Calabrese, above.

Jatta, Antonio. Ricordo botanico del Matese. Bol. Club Alp. Ital. **9** (no. 24): 144–148. 1875.

Topography, etc.; running account of ascent of Monte Miletto, with mention of plants observed. (Prov. Caserta and Prov. Campobasso (latter in Abruzzi e Molise).)

Lacaita, C. C. Catalogo delle piante vascolari dell'ex-Principato Citra. Bul. Orto Bot. R. Univ. Napoli **6**: 101–256. 1921.

Topography, bibliography, botanical explorations, general features of flora; annotated list of vascular plants and Characeae of the present Province of Salerno.—See also **Massalongo, Caro.** Piante provenienti da Amalfi e dintorni, e da altri siti della Penisola Sorrentina. Madonna Verona **14**: 1–32. 1 pl. 1920. (Topography, geology, short lists of species of different habitats; annotated list of 307 vascular plants.)—Also **Chiarugi, Alberto.** La costiera amalfitana. Nuovo Gior. Bot. Ital. n.s., **59**: 441–445. 8 fig. on 2 pl. (on 1). 1953. (Topography, etc., with mention of various species; bibliography.)—Also **Negri, Giovanni,** and **Moggi, Guido.** La vegetazione della pianura costiera del Sele (Salerno—Pesto). l.c. **59**: 461–465. 1953. (Similar lists.)—Also Guadagno, La vegetazione della Penisola Sorrentina, above.

Marcello, Leopoldo. Primo [-terzo] contributo allo studio della flora cavese. Bol. Soc. Nat. Napoli **14**: 53–85. 1901; **16**: 1–15. 1902; **17**: 17–40. 1904.

Topography; annotated lists of vascular and cellular plants (838 species, of which 819 are vascular). (Cava dei Tirreni, Prov. Salerno.)

Moggi, Guido. La flora del Monte Alburno (Appennino lucano). Webbia **10**: 461–645. 29 fig. (partly on 6 pl. [on 3] and incl. maps), map. 1955.

Topography, geology, climate, botanical explorations; annotated list of vascular plants (919 taxa including 903 species) with localities and life form; biological spectrum, phytogeography, short list of mosses and lichens, bibliography. (Prov. Salerno.)

——— La flora officinale del M. Alburno nel Salernitano. Riv. Ital. Essen., Profumi e Piante Off. **36**: 572–576. 1 fig. (port.). 1954.

List of over 200 species of medicinal plants, with habitats but no notes on uses. (Prov. Salerno.)

Pasquale, G. A. Flora vesuviana o catalogo ragionato delle piante del Vesuvio confrontate con quelle dell'isola di Capri e di altri luoghi circonstanti. Atti Accad. Sci. Fis. e Mat. Soc. R. Napoli v. 4, no. 6. 142 p. 1869. (Reprinted, 2 p. l., 142 p. 1869. In Arnold Arboretum library.)

Soils, plant zones, etc.; annotated list of vascular and cellular plants of Mt. Vesuvius region, with some vernacular names, the species of Capri being listed in footnotes. (Prov. Napoli.)—See also **Migliorato, Erminio.** Osservazioni relative alla flora napoletana. Bul. Soc. Bot. Ital. **1896**: 168–171. 1896. (Includes list of additions to the flora of Vesuvius and Capri (p. 168–169).)—Also **Rosa, Francesco de.** La flora vesuviana e l'eruzione dell'aprile 1906. Bol. Soc. Nat. Napoli **20**: 132–153. 1907. (Effects of eruption on different species.)—Also Baccarini, above.

——— and **Avellino, Giulio.** Flora medica della provincia di Napoli ossia descrizione delle piante medicinali che nascono spontaneamente nel perimetro della provincia . . . 200 p. 22 cm. Napoli, 1841. (Kew library.)

Descriptive account of medicinal plants, arranged according to the Linnaean system, with localities and uses.

Pontieri, Lina. La Cerreta di Pietra Palomba sulla riva sinistra dell'Orfanto presso Monteverde. Nuovo Gior. Bot. Ital. n.s., **65**: 425–451. 7 fig. (partly on plates; fig. 1, map), 2 tables. 1959.

Topography, general features of flora, etc.; annotated list of vascular plants (338 taxa); phytogeography, bibliography. (Prov. Avellino.)

Rippa, Anna. Cenni sulla vegetazione del cratere della Solfatara di Pozzuoli. Bul. Orto Bot. R. Univ. Napoli **13**: 11–20. 1 fig. [1935?].

Soils, general features of flora, etc.; unannotated list of spermatophytes collected by author. (A volcano in the Campi Flegrei, still emitting gases, in Prov. Napoli.)

Rippa, Giovanni. Cenni sulla flora di Ariano di Puglia. Bul. Orto Bot. R. Univ. Napoli **11**: 111–134. [1932?].

Soils, climate, plant habitats with lists of species; annotated list of vascular plants collected. (Now Ariano Irpino, in Prov. Avellino.)

Romeo, Antonino. Primo contributo sulla flora del territorio di Pisciotta (Salerno). Annali Fac. Agr. Portici R. Univ. Napoli III, **8**: 160–173. 1937.

Unannotated list of vascular plants. (Prov. Salerno.)

Rosa, Francesco de. Contributo alla flora murale e ruderale di Napoli. Bol. Soc. Nat. Napoli **19**: 219–239. 1906.

Manner of occurrence, etc.; annotated list of 144 vascular plants (139 species) of mural or ruderal habitats in the city of Naples. (Prov. Napoli.)

Tenore, Michele. Flora medica universale, e flora particolare della provincia di Napoli. 2 pt. (620 p.; 178 (i.e. 278) + 44 p.). 21.5 cm. Napoli, 1823. (His Corso delle botaniche lezioni, v. 4, parte 1, sezione 1–2.) (Arnold Arboretum library.)

Annotated descriptive flora of vascular and cellular plants (2091 species), including both the native plants and the medicinal plants of the world in general (the medicinal species marked by an asterisk), with notes on medicinal uses.

——— Saggio sulle qualità medicinali delle piante della flora napolitana, e sulla maniera di servirsene per surrogarle alle droghe esotiche. 1 p. l., 176 p. 20.5 cm. Napoli, 1808. (Arnold Arboretum library.)
Includes classified list of exotic drugs and of native plants which can be substituted for them, followed by discussion of 277 items of local botanical materia medica. There is a 2. ed. (viii, 295 p. Napoli, 1820), not seen (cited in Rehder, Bradley Bibliography **3**: 271. 1915).

Terracciano, Nicola. La flora dei Campi Flegrei. Atti R. Ist. Incoragg. Napoli **61**: 487–822. 1910.
Topography, soils, climate, botanical zones; descriptive list of vascular plants (1,102 species), with localities. The Campi Flegrei (Phlegraean Fields) are a volcanic region west of Naples. (Prov. Napoli.)—See also his Aggiunte alla "Flora" . . . Atti R. Ist. Incoragg. Napoli **68**: 269–454. 4 pl. 1917.—Also his Seconda aggiunta alla Flora . . . l.c. **73**: 1–11. 1921. (Changes in vegetation, lists of plants from various localities, etc., bringing total to 1331 species. In Princeton University library.)—See also Guadagno, La vegetazione del Monte Nuovo, above.

——— I legnami della Terra di Lavoro al concorso agrario regionale del 1879 in Caserta. 155 p. 27 cm. Caserta, 1880. (Arnold Arboretum library.)
Annotated list of 216 woody plants (mostly trees), a large proportion of which are cultivated, with notes on distribution and wood. (Prov. Caserta and part of Prov. Frosinone (in Lazio).)

——— Relazione intorno alle peregrinazioni botaniche fatte per disposizione della deputazione provinciale di Terra di Lavoro in certi luoghi della provincia. 240 p. 8vo. Caserta, 1872.—Seconda [–quarta] relazione intorno alle peregrinazioni botaniche fatte nella provincia di Terra di Lavoro per disposizione della deputazione provinciale. 125; 107; 133, xvii p. 28 cm. Caserta, 1873–78. (Only pt. 4 seen, in library of Muséum national d'histoire naturelle, Paris.)
Geology of various localities, itinerary with mention of plants collected; comprehensive list of species, with localities. (Cited, except for pt. 4, from notices in Nuovo Gior. Bot. Ital. **4**: 301. 1872; **6**: 205. 1874; **11**: 126–127. 1879.) Each part is independent and arranged in same way (no review of pt. 3 seen.) Altogether 2,242 species were recorded, of which 1,840 were vascular. The Terra di Lavoro, later the Province of Caserta, included essentially the present Province of Caserta and part of Prov. Frosinone (in Lazio).—See also his Intorno ad alcune piante della flora di Terra di Lavoro. Atti R. Accad. Sci. Fis. e Mat. Napoli II, v. 4, appendice 2. 10 p. pl. 1891. (List of additional species; list of about 40 angiosperms (many described as new), with critical notes.)

EMILIA-ROMAGNA

The regione of Emilia-Romagna (formerly Emilia) includes the Provinces of Bologna, Ferrara, Forlì, Modena, Parma, Piacenza, Ravenna, and Reggio Emilia. The name Romagna has been used in various ways, in the strictest sense to include only Forlì and Ravenna and part of Ferrara, but usually to include Bologna as well, in a broader sense (as by Zangheri and others) to take in also San Marino and parts of Arezzo and Firenze in Toscana and of Pesara-Urbino in Marche.

Local

Barlotti, Luigi. Le piante medicinali nell'Imolese e dintorni. Rev. Ital. Essen., Profumi e Piante Off. **38**: 60–66. 1956.
Annotated, tabular list of wild and cultivated medicinal plants of the Imola area. (Prov. Bologna.)

Bèguinot, Augusto, and Gabelli, Lucio. La flora alveale del Reno bolognese. Nuovo Gior. Bot. Ital. n.s., **22**: 365–387, 412–474. 1915.
Includes (p. 415–447) annotated tabular list of 520 vascular plants of the channel of the Reno River in Province of Bologna.

Bertolani-Marchetti, Daria. Flora e vegetazione della formazione ofiolitica di Rossena (Reggio Emilia). Nuovo Gior. Bot. Ital. n.s., **55**: 497–517. 1 fig. (map). 1949.
A vegetational study; includes (p. 501–506) list of vascular plants, with indication of life form.

——— Piante officinali spontanee e possibilità erboristiche in particolari ambienti dell'Appennino reggiano. Riv. Ital. Essen., Profumi e Piante Off. **35**: 577–581. 1954.

Topography, etc.; annotated list of medicinal plants. (Prov. Reggio Emilia.)

——— Il popolamento vegetale nelle stazioni salse della Valle Padana. Webbia **9**: 511–621. 16 fig. (mostly on 6 pl. [on 3]), map. 1954.

Geology, climate; list of vascular plants of saline habitats at 24 localities in Emilia (mostly), Lombardia, Piemonte, and Veneto; discussion of individual stations, morphological modifications, bibliography, etc.

Bertoloni, Giuseppe. Vegetazione dei Monti di Porretta e dei suoi prodotti vegetali. *In* Marsili, Carlo. Illustrazione delle terme di Porretta . . . p. 193–332. 24 cm. Bologna, 1867. (Reprinted, 140 p. Bologna, 1867; reprint not seen. Original in Harvard College library.)

Principal cultivated plants; annotated list of vascular plants. Replaces his Notizie intorno alle piante spontanee dei Monti Porrettani. Mem. Accad. Sci. Ist. Bologna II, **4**: 175–242. 1864. (Topography, etc.; annotated list of vascular and cellular plants collected by author.) (Prov. Bologna.)

Bertoni-Campidori, Domenico. La flora di Monte Castellaccio e di Monte Torre, territorio di Faenza. Bul. Soc. Bot. Ital. **1912**: 74–78. 1912.

Topography, etc.; list of vascular plants arranged by habitats. (Prov. Ravenna.)

——— Le piante medicinali della provincia di Ravenna. Romagna Agr. e Zootec. **28**: 62–69. 1934; **29**: 25–32, 81–88, 251–260. 1935.

List of species to be collected, arranged by months; annotated list of wild medicinal plants, with brief description. The installments cited include 65 species (pteridophytes to Viola); apparently no more was published (none in vol. 30–31).

Bolzon, Pio. Flora della provincia di Parma e del confinante Appennino tosco-ligure-piacentino. 120 p. 21 cm. Savona, 1921.

Annotated list of vascular plants (2,764 taxa); bibliography, brief index of local vernacular names. Includes the basins of the following rivers: Enza, Parma, Baganza, Taro, Ceno, Parola, Stirone, Alto Nure, Aveto. (Not seen; title and annotation supplied mostly by F. Sappa.)—See also his Aggiunte alla Flora . . . 3 p. Savona, 1925. (Not seen.)—Also **Lanzoni, Francesco.** Aggiunte alla flora parmense. Arch. Bot. (Forlì) **6**: 189–205. 1930. (Previous publications; annotated list of noteworthy vascular plants, with localities; list of plants growing on the old Campo di Marte southwest of city of Parma.)

——— Sulla flora alveale dei fiumi parmigiani. (Appendice to his Sulla flora alveale della Dora Baltea.) Bul. Soc. Fl. Valdôtaine **12**: 37–55. 1918.

Systematic prospectus of the plants of the channels of the rivers Taro, Baganza, Parma, Enza, and Po. (Prov. Parma.) (Not seen; title and annotation supplied by V. Giacomini.)

Bracciforti, Alberto. Flora piacentina, ossia enumerazione sistematica delle piante della provincia di Piacenza coll'indicazione della loro stazioni, dei caratteri principali di ciascuna famiglia, dei nomi italiani, di quelli del dialetto piacentino, con osservazione e note etimologiche. 384 p. 18.5 cm. Piacenza, 1877. (Gray Herbarium library.)

Annotated list of native and cultivated vascular and cellular plants of the Province of Piacenza (1,837 species, of which 1,670 are vascular plants), with vernacular names but mostly without specific localities, the distribution indicated by habitats.—See also his Dizionarietto botanico piacentino-latino contenente in ordine alfabetico i nomi del dialetto piacentino coi corrispondenti nomi botanici. (Appendice alla Flora piacentina.) 24 p. Piacenza, 1877. (Not seen; no review available. Perhaps includes the collected vernacular names from his Flora.)—Also **Parmigiani, Stefano.** Aggiunta alla "Flora Piacentina" del Prof. Bracciforti. Arch. Bot. (Forlì) **22**: 73–91. 1946; **34**: 263–267. 1958. (List of 355 species, varieties, and forms of vascular plants not included in Bracciforti's Flora, with localities; includes plants previously reported by Pavesi and by Bolzon; (1958) list of 35 taxa.)

Caldesi, Ludovico. Florae faventinae tentamen. Nuovo Gior. Bot. Ital. **11:** 321–347. 1879; **12:** 81–132, 161–196, 257–290. 1880.
List of vascular plants of the mandamento of Faenza (anciently Faventia), with localities. (Prov. Ravenna.)

Casali, Carlo. La flora del Reggiano. (? p.) Avellino, 1899.
List of known vascular plants of the Province of Reggio Emilia, with localities. (Not seen; cited mainly from Negodi, Atti Soc. Nat. e Mat. Modena 75: 5. 1944.)—See also his Aggiunte alla Flora del Reggiano. Nuovo Gior. Bot. Ital. n.s., **6:** 258–283. 1899.—Also his Nuove specie per la flora del Reggiano. Bul. Soc. Bot. Ital. **1900:** 234–236. 1900.—Also his Specie nuove per la flora . . . l.c. **1903:** 73–75. 1903.—Also his Primo [–3.] supplemento alla Flora del Reggiano (?), 73, (?) p. Avellino, 1901; Reggio nell' Emilia, 1903–05. (Not seen.)—Also his 4. [–5.] supplemento . . . Atti e Mem. Soc. Agr. Reggio nell'Emilia n.s., no. 2. (? p.) 1926; no. 9. (? p.) 1928. (Not seen.)

——— I nomi delle piante nel dialetto reggiano. 126 p. Reggio Emilia, 1915. (Estr. da Atti del Consorzio agricolo di Reggio Emilia v. 1.)—Addenda et emendanda. 19 p. Reggio Emilia, 1926. (From l.c. v. 3.)
Rich collection of vernacular names. (Not seen; titles and annotation supplied by F. Sappa.) (Prov. Reggio Emilia.) Replaces his I nomi volgari reggiani delle principali piante nostrali e di quelle esotiche più comunemente coltivate o note. Bol. Consorzio Agr. Regganio 1. 41 p. Reggio Emilia, 1905; [1.] – 2. appendice. 7, 14 p. Reggio Emilia, 1906–11. (Not seen; titles supplied by F. Sappa.)

Castellani, V. Manuale forestale; descrizione delle principali piante allignanti nella provincia di Bologna e loro uso nelle varie construzioni nelle arti e nelle piccole industrie. 146 p. 8vo. Bologna, 1888.
Not seen; no review available.

Centro Erboristico Appenninico Sperimentale, Modena. Piante medicinali spontanee nel territorio modenese. Riv. Ital. Essen., Profumi e Piante Off. **37:** 262–265. 1955.
Systematic list of medicinal plants of Province of Modena, with patois and vernacular names and short bibliography. Compiled largely by F. Panini.

Cicognani, Eug. Flora medica. Cenni sulle piante medicinali delle provincia di Forlì. (? p.) Forlì, 1876.
Not seen; no review available.

Cocconi, Girolamo. Flora dei foraggi che spontanei o coltivati crescono nelle provincie parmensi esposta secondo il metodo naturale. 691 p. 24 cm. Parma, 1856–60. (Arnold Arboretum library.)
Bibliography; annotated list of edible and poisonous plants, wild and cultivated, with brief descriptions, vernacular names, and uses. Relates primarily to forage plants, but includes mention of human uses also. (Prov. Parma.)

——— Flora della provincia di Bologna; vademecum per una facile determinazione delle piante incontrate. xix, 582 p. 17 cm. Bologna, 1883. (Cover title, Flora bolognese.) (Arnold Arboretum library.)
Topography, etc.; flora of over 2,000 species and varieties of vascular plants, in form of annotated keys, with localities.—See also **Mattei, G. E.** Aggiunte alla flora bolognese. 29 p. 23.5 cm. Bologna, 1886. (Annotated list of noteworthy vascular plants. In Arnold Arboretum library.)—Also **Betti, Guiseppe.** Supplemento alla flora bolognese. Riv. Ital. Sci. Nat. **20:** 15–19, 30–32, 56–58, 109–117. 1900. (List of newly recorded species and new localities, intended to summarize all publications since 1883.)— Also **Cobau, Roberto.** Aggiunte alla flora bolognese (entità nuove e località nuove). Arch. Bot. (Forlì) **17:** 3–19. 1941. (List of angiosperms, with localities, intended to include all those added to the flora of Bologna since Betti's supplement.)

Del Testa, Alberto. [Primo] –4. contribuzione alla flora dei dintorni di Cesena. Atti Soc. Toscana Sci. Nat. Pisa Proc. Verb. **7:** 30–43, 204–208. 1890–91; **8:** 60–68. 1892; **9:** 41–43. 1894.
Lists of over 600 vascular plants collected by the author, with localities. Titles vary slightly. (Prov. Forlì.)—See also his Nuova contribuzione alla flora della Romagna. Nuovo Gior. Bot. Ital. n.s., **10:** 234–265. 1903. (Topog-

raphy, etc.; extensive list of vascular plants and Musci from Rimini, Cesena, and other localities.)—Also **Raggi, Luigi.** Contributo alla flora littoranea romagnola. Primo elenco delle piante raccolte lungo il littorale adriatico fra Rimini e Cervia. Riv. Ital. Sci. Nat. **23:** 106–109. 1903. (Short list with localities.)—Also his Florula del bosco dell'Eremo, prov. di Forlì. Bol. Nat. Sup. Riv. Ital. Sci. Nat. **23:** 87–90. 1903. (Unannotated list of vascular plants of a wood just outside the city of Cesena.)—Also his Catalogo delle piante dei contrafforti dell'Appennino romagnolo. Quarta contribuzione alla flora della Romagna. Elenco delle piante raccolte nei dintorni di Mercato Saraceno e di Sarsina. (? p.) Cesena, 1904.—Also his Nuova flora del bosco dell'Eremo (circondario de Cesena). Quinta contribuzione alla flora della Romagna. (? p.) Cesena, 1904.—Also his Secondo catalogo delle piante dei contrafforti dell'Appennino romagnolo. Sesta contribuzione alle flora della Romagna. Elenco delle piante raccolte nei dintorni di Roncofreddo e di Sogiano al Rubicone. (? p.) Cesena, 1904. (The last 3 papers not seen; cited from Nuovo Gior. Bot. Ital. n.s., **11:** 472. 1904.)—Also his Sguardo floristico ai dintorni di Cesena. Nuovo contributo alla flora della Romagna. Nuovo Gior. Bot. Ital. n.s., **11:** 456–472. 1904. (Geography, topography, etc.; lists of vascular plants from different localities and habitats, including cultivated species; sketch of Del Testa's work; bibliography.)

Gabelli, Lucio. Notizie sulla vegetazione ruderale del città di Bologna. Malpighia 8: 41–68. 1894.

Bibliography of works on ruderal flora of cities; annotated list of about 200 vascular plants; statistics, methods of introduction, etc. (Prov. Bologna.)

——— Per un maggior incremento della erboristeria nel territorio bolognese. Arch. Farmacog. e Sci. Aff. **6:** 91–105. 1917. (University of Wisconsin library.)

Includes (p. 95–101) systematic list of 217 wild medicinal plants, with vernacular names and indication of frequency and distribution by zones; also list of 51 cultivated plants used in medicine. (Prov. Bologna.)

Gallizioli, Filippo. Dizionario botanico che comprende i nomi delle piante nelle principal lingue d'Europa oltre quelli di Linneo . . . vii, 371 p. 23 cm. Firenze, 1812. (His Elementi botanico-agrarj vol. [4]; half title, Indice generale degli Elementi agrarj.)

Includes (p. 74–93) Indice II. Che comprende i nomi romagnoli delle piante come si pronunziano a Imola, a Faenza, a Forlì, a Ravenna, e loro territorio, an alphabetical list of vernacular names with localities where used and references to species descriptions in vols. 2 and 3 of his Elementi in which the scientific names are given. (Provinces of Bologna, Forlì, and Ravenna.)

Gibelli, Giuseppe, and **Pirotta, Romualdo.** Flora del Modenese e del Reggiano. Atti Soc. Nat. Modena Mem. III, **1** (v. **16**): 29–220. 1883. (Preprinted 196 p. Modena, 1882. Not seen.)—1. supplemento alla Flora . . . l.c. III, 3 (v. **18**): 1–30. 1884.

Bibliography; annotated list of vascular plants (altogether 1871 species) of the Provinces of Modena and Reggio Emilia, with localities.—See also **Mori, Antonio.** Contribuzione alla flora del Modenese e del Reggiano. Atti Soc. Nat. Modena Mem. III, **5** (v. **20**): 113–126. 1886.—An index (to genera only) to the original list and supplements through 1886 is given l.c. III, **3** (Rend.): 97–105. 1887.—See also **Fiori, Adriano.** Nuove specie e nuove località per la Flora del Modenese e Reggiano. Malpighia **9:** 122–124. 1895.—Also **Vaccari, Antonio.** La vegetazione spontanea della provincia di Modena nei suoi confini politici. Mem. Accad. Sci., Let. ed Arti Modena V, **7:** 151–162. 1947. (Botanical explorations, list of newly recorded plants, partial list of adventives, etc.)—Also Negodi, Flora delle provincie di Modena e Reggio Emilia, below.

Lanzoni, Francesco. La flora del Monte Pelpi. 23 p. Parma, 1942. (Biblioteca della Giovanne Montagna no. 168.)

Includes list of about 130 vascular plants, with vernacular names. (Prov. Parma.) (Not seen; title and annotation supplied by F. Sappa.)

——— La flora del Monte Penna. 15 p. 1 fig. 24 cm. Parma, 1937. (Biblioteca della Giovanne Montagna no. 119.)

Includes list of about 100 vascular plants. (Prov. Parma.) (Not seen; title and annotation supplied by F. Sappa.)

Negodi, Giorgio. Aspetti della flora e della vegetazione del M. Rondinaio (m. 1964 s.m.). (His Studi sulla vegetazione dell'Appennino emiliano e della pianura adiacente. Memoria VI.) Atti Soc. Nat. e Mat. Modena 74 (VI, 22): 1–31. pl. 1–2. 1943.
Mainly ecological and phytogeographical; includes (p. 2–6) list of known vascular plants; bibliography. (Prov. Modena.)

———Flora delle provincie di Modena e Reggio Emilia. (Prospetto delle crittogame vascolari e fanerogame.) Atti Soc. Nat. e Mat. Modena 75 (VI, 22 [bis]): 1–64. 1944.
Botanical explorations, geography; unannotated list of known vascular plants (including cultivated species); bibliography, including separate bibliography of non-vascular cryptogams.—See also Gibelli and Pirotta, above.

——— La flora e la vegetazione dei serpentini di Varana. (His Studi sulla vegetazione dell'Appennino emiliano e della pianura adiacente. II.) Annali Bot. 22 (2): 117–142. 2 fig. 1941.
Ecological; includes (p. 122–124) list of known vascular plants of the serpentine area around Varana; bibliography. (Prov. Modena.)

——— La flora e la vegetazione del M. Cimone. (His Studi sulla vegetazione dell'Appennino emiliano e della pianura adiacente. Memoria IV.) Arch. Bot. (Forlì) 17: 150–195. 3 fig. 1941.
Ecological and phytogeographic; includes (p. 158–165) annotated list of known vascular plants above the tree zone; bibliography. Monte Cimone (2164 meters altitude) is the highest peak in the Modena Apennines. (Prov. Modena.)

Negri, Giovanni, and **Zangheri, Pietro.** Sulla vegetazione del bosco di Ladino (Forlì). Nuovo Gior. Bot. Ital. n.s., 41: 183–188. 1934.
List of 216 vascular plants; biological spectrum, etc.

Nigrisoli, Vittorio, and **Zangheri, Pietro.** Le piante medicinali della Romagna. Guida pratica per riconoscerle e per adoperarle in famiglia. 2. ed. riv. ed aument. xv, 410 p. 15 fig., 36 col. pl. 24.5 cm. Forlì, 1951. (1st ed. 1935.)
Tabular list of medicinal plants, alphabetically arranged by vernacular names, with properties, vernacular names, etc.; annotated catalog of 108 medicinal plants, alphabetically arranged by Italian names, giving description, local distribution, Romagnese patois names, properties and uses.

Panini, F. Piante medicinali spontanee sul territorio modenese e reggiano. Bol. Camera Com. Modena. (? p.) 1924.
Not seen; cited from Bul. Bibl. Bot. Ital. 5 (fasc. 1, 1924): 32. 1927. (Prov. Modena and Reggio Emilia.)

Pasquini, Diego. Flora spontanea delle due stazioni del C.E.A.S. di Pavullo e di Sestola. Riv. Ital. Essen., Profumi e Piante Off. 37: 59–76. illus. incl. port. 1955.
Includes lists of native vascular plants (144 and 292 species respectively) from Pavullo and from Sestola, also lists of medicinal plants; bibliography. (Prov. Modena.)

——— Florula di Montegibbio e Pescaro. Atti Soc. Nat. e Mat. Modena 76 (VI, 23): 103–114. 1945.
Topography, etc.; list of vascular plants with indication of occurrence at each locality; biological spectrum, bibliography. (Prov. Modena.)

——— La vegetazione dei gessi reggiani. Atti Soc. Nat. e Mat. Modena 75 (VI, 22 [bis]): 264–282. 1944.
Mainly ecological; includes (p. 265–269) list of known vascular plants of the gypsum zone which extends from Ventoso (Scandiano) to Vezzano on the Crostolo River, with indication of life forms. (Prov. Reggio Emilia.)

Passerini, Giovanni. Flora dei contorni di Parma esposta in tavole analitiche con alquante nozioni generali intorno alle piante . . . xlviii, 408 p. 19 cm. Parma, 1852. (Arnold Arboretum library.)
Unannotated flora of vascular plants, in form of keys, without localities; index of vernacular names. Includes the commonly cultivated species. Covers the area between the Stirone on the west, the Po on the north, the Enza on the east, and on the south a line "che dal Castello di Guardasone passa successivamente per le alture dello Sporno di Prinzera e di Pellegrino."

(Prov. Parma.)—See also **Avetta, Carlo,** and **Casoni, Vittorio.** Aggiunte alla flora parmense. Malpighia **11:** 209–224. 1897.—Also **Avetta, C.** Annotazione alla flora parmense. l.c. **12:** 164. 1898.

Pavarino, G. L. Intorno alla flora del calcare e del serpentino nell' Appennino bobbiese. (Title varies.) Atti Ist. Bot. R. Univ. Pavia II, **12:** 21–56. 1915 [1907?]; **14:** 19–42. 1914 [1908]; **15:** 89–108. pl. 14. 1918 [1912].
Geology, climate, etc.; tabular lists of vascular plants. (Prov. Piacenza.)

Pavesi, Vittorio. Flora alluvionale della Trebbia e della Nure. (Studio fitogeografico.) Atti Soc. Ital. Sci. Nat. Milano **57:** 189–260. 1919.
Topography, geology, soils, climate, etc.; tabular list of vascular plants (451 species), showing localities; methods of distribution, etc. (The Nure River is in Prov. Piacenza, the Trebbia in Piacenza, Pavia (Piemonte), and for a short distance in Prov. Genova (Liguria); both flow into the Po.)

Picaglia, Luigi. Bibliografia botanica della provincia di Modena. Atti Soc. Nat. Modena Mem. III, **2** (v. 17): 219–230. 1883.—1. supplemento. l.c. III, **11** (v. **26**): 133–156. 1892.
Annotated list of publications, chronologically arranged.

Raggi, Luigi. Flora popolare della Romagna. Contributo allo studio dei nomi volgari delle piante in Romagna. 2. ed. riveduta ed aumentata . . . 57 p. 22 cm. Bologna, 1904. (Lloyd Library.) (1st ed. 1903.)
Bibliography; general features of flora, list of cultivated plants; systematic list of vascular plants (not a complete flora), with Italian names and over 700 patois names. The first edition appeared in Malpighia **17:** 373–388. 1903, with title: Materiali per una flora emiliana. 1. contribuzione. Elenco di 400 voci vernacole romagnole significanti piante della Romagna. The work relates to Bologna, Forlì, and Ravenna, as well as Toscana (Arezzo, Firenze, Pesaro-Urbino) and San Marino.

Revedin, Pietro. Contributo alla flora vascolare della provincia di Ferrara. Nuovo Gior. Bot. Ital. n.s., **16:** 269–334. pl. 1–6 (incl. map). 1909.
Topography, climate, soils, plant zones, bibliography; annotated list of known vascular plants (837 species and varieties).—See also **Béguinot, Augusto.** Una escursione botanica nel littorale della provincia di Ferrara. Bul. Soc. Bot. Ital. **1910:** 125–136. 1910. (Lists of species from various localities, including 28 additional species and 17 subspecies and varieties.)

Soliani, Luigi. Erbario della Pubblica biblioteca Maldotti in Guastalla con cenni illustrativi sulle principali piante che hanno usi medici economici industriali. 149 p. (and index). Guastalla, 1890.
Systematic list of wild and cultivated plants of the region of Guastalla, with notes on uses. (Prov. Reggio Emilia.) (Not seen; title and annotation supplied by F. Sappa.)

Ungarelli, G. Le piante aromatiche e medicinali nei nomi, nell'uso e nella tradizione popolare bolognese. xxxvi, 138 p. 17.5 cm. Bologna, 1921.
Annotated list of medicinal plants, with vernacular names and uses; index of botanical names. (Prov. Bologna.) (Not seen; title and annotation supplied by F. Sappa.)

Vaccari, Antonio. Le piante medicinali spontanee nel Modenese. Atti e Mem. R. Accad. Sci., Let. ed Arti Modena V, **5:** 3–56. 1940–41.
Running account of native medicinal plants, arranged by families, with uses and references; short bibliography. (Prov. Modena.)

Zangheri, Pietro. La flora del circondario di Forlì. Prima contribuzione con cenni sulla storia della botanica e la fitogeografia nel circondario di Forlì. Nuovo Gior. Bot. Ital. n.s., **20:** 45–143. 1 fig. (map). 1913.
Botanical explorations, annotated bibliography, topography, climate, soils, zones of vegetation, origin of flora; annotated list of wild and commonly cultivated vascular plants (875 taxa) of the vicinity of the city of Forlì, with vernacular names (indexed). (Prov. Forlì.)

——— Flora e vegetazione dei calanchi argillosi pliocenici della Romagna e della zona di argille in cui sono distribuiti. Presentazione e prefazione dei Profi. Giuseppe Gola e Giovanni Negri. xv, 279 p. 36 fig., 3 pl. (stereoscopic), map. 24 cm. Faenza, 1942. (His Romagna fitogeografica 2.)
Includes (p. 36–92) list of vascular plants (416 species) with ecological and phytogeographical annotations. The calanchi are systems of small rela-

tively steep and deep valleys which cut the clay slopes of the Subappennine hills. (Prov. of Bologna, Forlì and Ravenna.)

——— Flora e vegetazione dei terreni "ferrettizzati" del preappennino romagnolo. Webbia 7: 1–308. 41 fig., col. front. map. 1950. (His Romagna fitogeografica 3.)

Mainly ecological and phytogeographic; includes (p. 27–105) annotated list of vascular plants (632 species and 199 other taxa) of the "red soils;" bibliography, English summary. (Provinces of Bologna, Forlì, and Ravenna.)

——— Flora e vegetazione delle pinete di Ravenna e dei territori limitrofi fra queste e il mare. Prefazione del Prof. Giovanni Negri . . . 421 p. 30 fig., 4 maps, table. 25.5 cm. Forlì, 1936. (His Romagna fitogeografica 1.) (Kew library.)

Phytogeographical and ecological; includes (p. 71–78) account of botanical explorations and (p. 79–210) annotated list of vascular plants (890 species) with phytogeographical and ecological data. (Prov. Ravenna.) Replaces his Flora di Romagna. La vegetazione delle pinete ravennati. (Nota preliminare e programma di ricerche.) Nuovo Gior. Bot. Ital. n.s., **34**: 1115–1162. 1 fig. (map). 1928. (History, general features of vegetation, botanical explorations; unannotated list of vascular plants (796 taxa).)

——— Sguardo preliminare alla flora e vegetazione dell'alto Appennino romagnolo con particolare riguardo alla Foresta di Campigna. Nuovo Gior. Bot. Ital. n.s. **49**: 59–109. 1942.

Geography, topography, geology, climate; list of about 1000 known vascular plants from above 800 meters elevation, with localities for the rarer species; plant formations, etc. (Mostly in Prov. Forlì, partly in Bologna and Ravenna.)

FRIULI-VENEZIA GIULIA

The regione of Friuli-Venezia Giulia, constituted in 1947, includes the Provinces of Udine (formerly in Veneto) and Gorizia (remnant of the former regione of Venezia Giulia). The rest of historical Friuli is now in Yugoslavia.

General

Gortani, Luigi, and Gortani, Michele. Flora friulana con speciale riguardo alla Carnia. 2 pt. (225, 519 p.) map. 23.5 cm. Udine, 1905–06. (Gray Herbarium library.)

Geography, topography, plant zones, climate, plant habitats of the different zones with lists of species, variation in altitudinal limits, influence of soils on vegetation, phytogeography, botanical explorations, bibliography, notes on excluded species, gazetteer; annotated list of vascular plants [1971 species and 1823 varieties and forms, fide Marchesetti] with localities and Friulian and Carnian vernacular names. Replaces **Pirona, G. A.** Florae forojuliensis syllabus. 170 p. 22 cm. Utini, 1855. (Progr. I. R. Ginnasio liceale di Udine pel 1855.) (Topography, soils, etc.; annotated list of spermatophytes. Covers Friuli and Carnia, that is, the present provinces of Udine and Gorizia and adjacent regions.)

Manzini, V. La virtù delle piante nel Friuli. 28 p. Udine, 1890.

Popular treatment of medicinal plants, grouped under the diseases for which they are used. (Prov. Udine.) (Not seen; title and annotation supplied by F. Sappa. Marchesetti (Atti Mus. Civico Storia Nat. Trieste 11 (III, 1): 294. 1931) cites the same title as published in In Alto 1: 88–90, 111–116. 1890.)

Pirona, G. A. Piante indigene del Friuli utili all'industria. *In* Almanacco del Dott. Vatri per gli anni 1858–59. Udine, 1858–59 (?).

Not seen; cited from Marchesetti, Atti Mus. Civico Storia Nat. Trieste 11 (III, 1): 313. 1931.

——— Vocabolario botanico friulano. Atti della solenna distribuzione dei premj seguita il giorno 6 settembre 1862 all'I.R. Ginnasio liceale di Udine. p. 3–81. 23 cm. Udine, 1862. (University of Michigan library. Reprinted 79 p., 1862; reprint not seen.)

Alphabetical list of local vernacular names, with corresponding Italian names, scientific names, habitat, and often notes on uses or poisonous properties. Reissued (with additions and corrections) as Vocabolario botanico *in* his Vocabolario friulano. p. 481–526. 1871; this reissued by the Società Filologica Friulana in 1928 et seq. (Neither work seen; title and annotation supplied by F. Sappa.)—See also **Pirona, G. A., Carletti, Ercole, and Corgnali, G. B.** Il nuovo Pirona: vocabolario friulano. xxix, 1534 p. 25 cm. Udine, 1935. (Bibliography; vocabulary of the dialect of the general Friuli region (including Udine, Carnia, Gorizia, Tagliamento) with Italian equivalents and examples of usage; "indice dei generi botanici" (p. 1403–1410), an alphabetical list of botanical genera with corresponding Friulian names; under the latter, in text of the dictionary, the vernacular names of the different species are given; also zoological index, etc. Includes many names not in the 1862 publication. In University of Chicago library.)

Rieppi, A. El litum. Erbe mangerecce spontanee del Friuli. 30 p. 8vo. Udine, 1947.

Uses, vernacular and botanical names of a group of edible plants. (Not seen; title and annotation supplied by V. Giacomini.)

Senoner, Adolf. Piante utili all'agricoltura ed alle arti. Catalogo di Adolf Senoner. Annuario Assoc. Agr. Friulana **3**: 25–109. 1860. (Original not seen. Reprinted 85 p. 20 cm. "Estr. dal'Annuario dell'Ass. Agr. fr. Anno III.")

Annotated list of useful plants, including some not known in Friuli, with vernacular names.

Local

Cossettini, G. Piante medicinali della Carnia e del Canal del Ferro. 1925.

Not seen; title taken from Bul. Bibl. Bot. Ital. 5 (2, 1925): 44. 1927. (Prov. Udine.)

Crichiutti, Giovanni. Piante raccolte sul M. Quarnan pr. Gemona (1372 m.). In Alto **9** (1898): 3–5, 16–17, 25. 1898; **10** (1899): 5–6. 1899.

List of 148 species, including many alpines. (Prov. Udine.). (Not seen; title and annotation supplied by F. Sappa.)

Da Santa, Fortunato. Elenco delle piante medicinali comunemente conosciute che crescono nel territorio di Forni di Sopra. (? p.) Udine, 1900.

Not seen; cited from Gortani, Flora friulana **1**: 223. 1905. (Prov. Udine.)

Fornaciari, Giovanni. Elenco delle piante rinvenute nella Valle di Bombaso. Annali Scuola Friulana **1948–49**: (31 p.). 1950.

List of vascular plants collected in Valle di Bombaso above Pontebba between 660 and 1700 meters altitude. (Prov. Udine.) (Not seen; cited from Arch. Bot. (Forlì) **29**: 337. 1953.)

Krasan, Franz. Beiträge zur flora der umgebung von Görz. I–[III]. Oesterr. Bot. Zeitschr. **13**: 345–362, 385–396. 1863; **15**: 101–107. 1865.

Topography, geology, agricultural products, etc.; general features of flora of different botanical regions with mention of numerous species. The third part relates to the Isonzo-thal. Some corrections in scientific names are given by Marchesetti, Atti Mus. Civico Storia Nat. Trieste **9**: 170. 1895.

Leona, Carlo. La flora delle Alpi Giulie nell'Orto botanico alpino "Juliana" di Alberto Bois de Chesne. Atti Mus. Civico Storia Nat. Trieste **18**: 125–264. 2 maps. 1952.

Includes discussion of floristic elements and analysis of the flora of the Julian Alps, with mention of numerous species; gazetteer.

Marchesetti, Carlo de. Bibliografia botanica, ossia catalogo delle publicazioni intorno alla flora del litorale austriaco. Atti Mus. Civico Storia Nat. Trieste **9** (n.s., **3**): 129–210. 1895.—Aggiunte alla bibliografia botanica della Venezia Giulia. l.c. **11** (III, **1**): 217–356. 1931.

Annotated lists containing 1738 titles, alphabetically arranged by authors and referring to all branches of botany. Covers the former Austrian Küstenland, composed of the three crownlands of Görz und Gradisca, Trieste, and Istrien, and the Quarnero (Carnero) Islands; later the greater part of the former Italian compartimento of Venezia Giulia e Zara, including the Provinces of Gorizia, Trieste, Pola, and Fiume, but not including the Province of Zara.

——— Distribuzione geografica delle principali piante medicinali della Venezia Giulia.

A short list of 64 species which might be utilized, divided according to the different provinces. The region covered includes the present Province of Gorizia. (Not seen; cited from Atti Mus. Civico Storia Nat. Trieste **11** (III, 1): 295. 1931.)

——— Una passeggiata alle Alpi Carniche. Bol. Soc. Adriatica Sci. Nat. Trieste **4**: 168–190. 1879.

Includes unannotated list of vascular plants [357 species] collected on the Wishberg (or Vishberg), August 18–19, 1878. (Prov. Udine.)

Minio, Michelangelo. Erborazioni nel bacino medio del Natisone. Contribuzione alla conoscenza botanica delle Prealpi Giulie. Nuovo Gior. Bot. Ital. n.s., **12**: 5–52. 1905.

Topography, climate, plant zones, etc.; partly annotated list of about 800 vascular plants. (Prov. Udine.)

Penzig, Otto. Elenco delle principali piante vascolari trovate nella Valle di Raccolana. *In* Marinelli, Giovanni. Guida del Canale del Ferro . . . p. 62–69. 8vo. Udine, 1894.

List of 555 vascular plants. (Prov. Udine.) (Not seen; cited from Bot. Centralbl. **60**: 278. 1894 and Marchesetti, Atti Mus. Civico Storia Nat. Trieste **11** (III, 1): 312. 1931.)—See also **Crichiutti, Giovanni.** Florula della Valle di Raccolana e del gruppo del Canin. In Alto **16**: 9–12, 20–21, 30–32, 42–43. 1905; **17**: 40–43, 57–58. 1906; **18**: 3–7, 41–44. 1907. (Includes the plants of Monte Canin and Monte Montasio as well as those of the Valle di Raccolana. Not seen; cited mostly from notice in Marchesetti, l.c. 240. 1931.)—Also his Elenco di piante raccolte per la prima volta in Valle di Raccolana e nel gruppo del Monte Canin con cenno sulla distribuzione delle piante arboree. Atti Acad. Sci. Veneto-Trent.-Istriana n.s., **3** (Cl. Sci. Nat., Fis. e Mat.): 104–121. 1906. (Soils, altitudinal limits of trees; list of vascular plants, with localities.)—Also **Paoletti, Giulio.** Florula della Valle di Raccolana e del gruppo del Monte Canin. In Alto **20**: 12–17, 23. 1909. (Cites the principal species of this region. Not seen; cited mostly from notice in Marchesetti, l.c. 308. 1931.)

Pospíchal, Eduard. Flora des oesterreichischen Küstenlandes. 2 v. (xliii, 575; 946 p.). map, 25 tab. (keys). 25 cm. Leipzig und Wien, 1897–99.

Topography, general features of flora; annotated, partially keyed, descriptive flora of vascular plants, with synonymy and localities; includes cultivated species. The area covered is bounded by "die Corada im Coglio, der Mačni vrh bei St. Peter a.K., die Mündung der Arsa und des Quieto in Istrien;" it includes part of the present Province of Gorizia. Marchesetti states that it actually covers only the central part of the Littoral, from the vicinity of Gorizia to the Rio Quieto, omitting the entire alpine zone as well as the district of Parenzo, Rovigno, Pola, and the Quarnero Islands.—See also **Cohrs, Albert.** Beiträge zur flora des nordadriatischen Küstenlandes. Repert. Spec. Nov. Fedde **56**: 66–143. 1953–54. (Extensive list of noteworthy vascular plants collected by author and others in the upper and middle Küstenland.)

Trotter, Alessandro. Un relitto di flora mediterranea nell'alto Friuli occidentale. Atti R. Ist. Veneto Sci., Let. ed Arti **86** (2): 1433–1456. 1 fig. (map), pl. 13–14. 1927.

Topography, geology, etc.; list of vascular plants observed in vicinity of Trasaghis, at south end of Monte Brancot near Gemona, with discussion of several of the noteworthy species; phytogeography. (Prov. Udine.)

Zenari, Silvia. Primo [–secondo] contributo alla flora della Val Cellina (Friuli occidentale). Nuovo. Gior. Bot. Ital. n.s., **27**: 11–37. 1920; **30**: 5–47. 1923.—La flora della Valle Cellina. Aggiunte e correzioni. Arch. Bot. (Forlì) **1**: 51–66. 1925.

Geology, etc.; annotated lists totaling 976 vascular plants (812 species). (Prov. Udine.)—See also her I caratteri della vegetazione in Val Cellina. Arch. Bot. (Forlì) **1**: 101–140, 149–169. pl. 3–5 (3–4, maps). 1925. (Topography, geology, climate, early settlement, plant formations and succession, bibliography.)

LAZIO (LATIUM)

See also Campania (Béguinot; Terracciano (I legnami, and Relazione)).

The regione of Lazio up to 1926 consisted only of the Province of Roma, which, plus a little territory acquired from Campania, is now divided into the Provinces of Frosinone (formed in 1926), Latina (formerly Littoria; formed in 1934), Rieti (1926), Roma, and Viterbo (1926).

General

Maratti, G. F. Flora romana. 2 v. (415, 543 p.) 23 cm. Romae, 1822.

Flora of vascular and cellular plants (2276 species, of which 1962 are vascular; includes some cultivated species), on the Linnaean system, with synonyms (mostly pre-Linnaean), vernacular names, brief annotations, and some localities. Covers the same range as Sanguinetti (see beyond). The work is entirely unreliable.

Pirotta, Romualdo, and **Chiovenda, Emilio.** Flora romana. Annuario R. Ist. Bot. Roma v. 10, fasc. 1–2. 304 p. 1900–01. (Arnold Arboretum library.)

Botanical history, with bibliographical footnotes. Unfinished; reaches Scarpellini (born 1762).—See also **Lusina, Giuseppe.** Bibliografia botanica del Lazio (secc. XIX e XX). Annali Bot. 25: 127–178. 1956. (Annotated list of 339 titles, alphabetically arranged by authors.)

Sanguinetti, Pietro. Florae romanae prodromus alter, exhibens plantas vasculares circa Romam in Cisappenninis Pontificiae dictionis provinciis in Umbria et Piceno sponte venientes secundum sexuale systema dispositas. 3 p. l., 971 p. 8 col. pl. 30 cm. Romae, 1864. (Arnold Arboretum library.) (Reprinted from Atti Accad. Pontif. Nuovi Lincei v. 5–6, 10–12, 15, 17–18, passim. 1853–67.)

Annotated descriptive flora of vascular plants (2,285 species), with vernacular names, uses, and localities. Covered the Province of Rome at its greatest extent, including parts of the present Umbria and Abruzzi e Molise. —See also **Terracciano, Achille.** [Prima] –quarta contribuzione alla flora romana. Nuovo Gior. Bot. Ital. **23**: 495–501. 1891; Bul. Soc. Bot. Ital. **1892**: 113–119, 139–145. 1892; Nuovo Gior. Bot. Ital. n.s., **1**: 129–186. 1894. (Partial lists of vascular plants from various localities.)—Also **Chiovenda, Emilio.** Piante nuove o rare da aggiungersi alla flora romana. Malpighia **11**: 90–113. 1897. (List of 125 spermatophytes, with localities and notes.)—Also his Piante nuove o rare della flora romana. l.c. **12**: 411–420. 1898. (Notes on 38 species.)—Also papers by several authors in Nuovo Gior. Bot. Ital. n.s., **56**: 366–418. 1949; **56**: 650–652. 1950; **57**: 651–655. 1951; **60**: 763–764. 1953; **60**: 897–899. 1954.

Local

Anzalone, Bruno. Flora e vegetazione. *In* Zavattari, Edoardo, and others. Biogeografia dell'isola di Zannone (Arcipelago Pontino). Rend. Accad. Naz. dei XL [Quaranta] IV, **4/5** (whole vol. 76/77): 151–200. fig. 7 (map), pl. 4–10 (pl. 10, map). 1953/54 [1954]. (University of Wisconsin library.)

Annotated list of known vascular plants (338 taxa); list of cellular plants collected; biological spectrum, plant formations, special habitats, etc.; bibliography. (One of the Isole Ponziane.)—See also Campania (Béguinot, La vegetazione delle Isole Ponziane e Napoletane).

——— Flora e vegetazione dei muri di Roma. Annali Bot. **23**: 393–497. pl. 7–12. 1951.

General considerations; list of 385 taxa observed on the walls of Rome, with life form and localities; biological forms, phytogeography, phenology, bibliography. (Prov. Roma.)

Béguinot, Augusto. La flora dei depositi alluvionali del basso corso del fiume Tevere. Studio fitogeografico. Nuovo Gior. Bot. Ital. n.s., **8**: 238–315. 1901.

Habitats, botanical explorations, ecology; unannotated list of vascular plants (664 species) of the lower course of the River Tiber. (Prov. Roma.)

——— Flora e fitogeografia delle Paludi Pontine studiate nelle condizioni anteriori all'attuale bonifica incluso il settore Terracina–Lago di Fondi. Arch. Bot. (Forlì) **10**: 329–382. pl. 4–7. 1934; **11**: 125–168, 275–316. 1935; **12**: 255–316. map. 1936. (Reprinted 210 p. Forlì, 1936; not seen.)

Botanical explorations, botanical regions, etc.; annotated list of 985 species of vascular plants (including additions), with localities; topography, soils, climate, geology, plant associations, phytogeography. Covers an area about 45 km. long and 15 to 22 km. wide from Astura River, Cisterna, and Velletri on the west to Terracina and Lago di Fondi on the east, south of the Monti Lepini and Monti Ausoni, and excluding Cape Circeo. (Prov. Latina and Prov. Roma.)—See also his Appunti sulla flora dell'agro Pontino. Arch. Bot. (Forlì) **13:** 143–147. 1937. (Adds 7 species.)—Also **Ricci, Ignazio.** Osservazioni botaniche nell'Agro Pontino (Lazio). Annali Bot. **25:** 579–588. 1957. (Botanical explorations, changes in vegetation; list of noteworthy species.)

——— Prodromo ad una flora dei bacini Pontino ed Ausonio e dei monti limitrofi. (Herbarium Camillae Doriae II.) Annali Mus. Civico Storia Nat. Genova II, **18** (v. 38): 189–341. 1897. (Reprinted 157 p. Reprint not seen.)
Topography, geology, bibliography, etc.; list of vascular plants (1305 species), with localities. (Prov. Latina and Prov. Roma.)

Bianchini, Arnaldo. Note ed appunti intorno alla "Flora reatina." Riv. Ital. Sci. Nat. **26:** 62–67, 101–113. 1906.
Topography, climate, annotated list of vascular plants grouped by habitats; unannotated list of vascular plants, including cultivated species, of the vicinity of Rieti. (Prov. Rieti.)

Bonaventura, Gustavo. Primo contributo allo studio della flora della costiera di Fregene (Roma). Annali Bot. **25:** 31–53. 1 fig. (map). 1956.
Topography, etc.; list of vascular plants (420 taxa) with indication of biological form; biological spectrum, bibliography.

Cacciato, Alfredo. La florula di uno stagno nella zona archeologica della via Appia Antica, in Roma. Nuovo Gior. Bot. Ital. n.s., **62:** 570–575. 2 fig. 1956.
List of 113 species of a pool along the Appian Way, with discussion. (Prov. Roma.)

——— La vegetazione antropocora dello scalo ferroviario Ostiense di Roma. Nuovo Gior. Bot. Ital. n.s., **59:** 119–143. 1952.
Classification according to method of dissemination, phenology; annotated list of vascular plants (476 species) observed around the railroad yard of Roma Ostiense; discussion, bibliography. (Prov. Roma.)

Cicateri, Manlio. Appunti sulla flora medicinale del Monte Circeo. Annali Bot. **25:** 407–414. 1957.
Unannotated alphabetical list of about 140 medicinal plants; list of the more important species, bibliography. (Prov. Latina.)

Cortesi, Fabrizio. Piante officinali delle provincie di Roma e di Littoria. 91 p. illus. Roma, 1937.
Short descriptions of about 190 wild and a few cultivated medicinal plants. (Prov. Latina and Roma.) (Not seen; cited from Annali Bot. **25:** 147. 1956.)

Cufodontis, Giorgio. La flora vascolare dei Monti Simbruini nel Subappennino laziale. (Herbarium Camillae Doriae III.) Annali Mus. Civico Storia Nat. Giacomo Doria (Genova) **60:** 181–353. 1 fig. (map). 1939.
Geography, geology; list of vascular plants (1063 species), with localities; bibliography, index. Covers essentially the region from Riofreddo and Carsali to Subiaco and Cantari, mostly in Lazio (Prov. Frosinone and Roma), partly in Abruzzi e Molise (Prov. L'Aquila).

Deakin, Richard. Flora of the Colosseum of Rome; or, illustrations and descriptions of four hundred and twenty plants growing spontaneously upon the ruins of the Colosseum of Rome. viii, 237 p. 5 pl. (pt. col.), front. 17.5 cm. London, 1855. (Reissued 1873.)
Annotated descriptive flora of vascular plants (420 species), with notes on uses. (Prov. Roma.)—See also Fiorini-Mazzanti and Rosa, below.

Falqui, Giuseppe. Contributo alla flora del bacino del Liri. Atti R. Accad. Sci. Fis. e Mat. Napoli II, v. 9, no. 11. 51 p. 1899.
List of vascular plants (1280 species), with localities. (Prov. Frosinone and Roma, also Prov. Benevento (Campania) and Prov. L'Aquila (Abruzzi).)

Fiorini-Mazzanti, Elisabetta, *contessa.* Florula del Colosseo. Atti Accad. Pontif. Nuovi Lincei **28:** 8–13, 127–133, 254–261, 305–307, 397–400. 1875; **29:**

8-15, 54-58, 122-125, 236-239, 366-369, 457-466. 1876; **30:** 97-105. 1877; **31:** 155-167. 1878.

Annotated list of vascular and cellular plants [about 340 species, of which 270 are vascular] of the Roman Colosseum. (Prov. Roma.)—See also Deakin, above, and Rosa, below.

Haro,—. Catalogue des plantes phanérogames qui croissent spontanément dans la ville de Rome. Bul. Soc. Sci. Nancy II, **3:** 103-115. 1878.

General considerations; list of over 400 flowering plants, with localities, collected by author within the walls of Rome from 1861-63. (Prov. Roma.)

Macchiati, Luigi. Prima contribuzione alla flora del Viterbese. Atti Soc. Nat. Modena III, 7 (v. 22): 7-61. 1888.

Topography, geology; annotated list of vascular plants (722 species), also mosses and hepatics; includes cultivated plants. Refers to Monte della Pallanzana and neighboring hills, near city of Viterbo. (Prov. Viterbo.)—See also his Note di una escursione botanica alla Pallanzana del gruppo dei Cimini. Nuovo Gior. Bot. Ital. **18:** 157-159. 1886. (Geology, general features of flora.)

Marco, Gennaro de. Flora di Montecassino ossia guida alla determinazione delle piante spontanee di detto luogo . . . 2 pt. (xxxii, 268; 47 p.). 15 pl. 22.5 cm. Montecassino, 1886-87. (Gray Herbarium library.)

Mention of principal trees, description of localities; annotated descriptive flora of vascular and cellular plants (662 vascular species), with vernacular names and keys to families and genera. Perhaps a preprint from his Monte Cassino illustrato nei tre regni della natura . . . v. 1, chap. 4. 1888. (Not seen.) (Hill above Cassino, with a celebrated monastery, formerly in Prov. Caserta, now in Prov. Frosinone.)

Martelli, Ugolini. Una passeggiata sul Monte Velino e Montagne della Duchessa. Bul. Soc. Bot. Ital. **1904:** 110-114. 1904.

Unannotated list of spermatophytes collected by author in August 1903 on Monte Velino (2487 meters high) and Montagne della Duchessa (2266 meters). (Prov. Rieti.)—See also **Anzalone, Bruno.** Contributo alla flora della Montagna della Duchessa (catena di Monte Velino). Annali Bot. **23:** 21-30. 1949. (Includes list of 162 species; bibliography.)—Also **Steinberg, Carlo.** Contributo allo studio floristico e fitogeografico degli alti pascoli della Montagna della Duchessa (Appennino Abruzzese). Nuovo Gior. Bot. Ital. n.s., **59:** 201-251. 13 fig. (fig. 1, map, f. 2-12 on 6 pl. (on 3)). 1953. (Topography, etc.; annotated list of known vascular plants, with indication of life form; plant communities, etc.; bibliography.)

Montelucci, Giuliano. Flora e vegetazione dell Valle dell'Inferno a Roma. (Investigazioni botaniche nel Lazio. V.) Annali Bot. **24:** 241-339, 461-529. 2 fig., pl. 22-23. 1953-54.

Topography, climate, general features of flora; annotated list of vascular plants (681 taxa, including 635 species); biological spectrum, discussion of vegetation, general considerations, bibliography. (Prov. Roma.)

——— La vegetazione dei dintorni di Guidonia. 16 p. 4to. 1941. (Reprinted from Carelia,—. Note geologiche e storiche sul territorio cornicolano. Roma, Minist. Aeronautica.)

Partial lists by localities, with phytogeographical annotations. (Not seen; title and annotation supplied by F. Giacomini.) (Prov. Roma.)

——— La vegetazione del M. Terminillo (Appennino centrale). Webbia **8:** 245-379. 16 fig. (fig. 1, 2, maps; fig. 4-16 on 4 pl. (on 2)). 1952; **9:** 49-359. fig. 17-57 (fig. 17-56 on 12 pl. (on 6)), chart. 1953.

Topography, geology, soils, climate, botanical explorations; annotated list of known vascular plants (859 taxa including 773 species); statistics; (second part) ecology, phytogeography, bibliography. (A mountain formerly in or considered to be in Umbria or Abruzzi, now in Prov. Rieti.) Replaces **Cavara, Fridiano,** and **Grande, Loreto.** Contributo alla flora del Terminillo (Abruzzo). Bul. Orto Bot. R. Univ. Napoli 4: 269-308. pl. 5-8. 1914. (Botanical explorations, general features of flora; annotated list of vascular plants.)

Morgana, Mario. Appunti sulla flora della valle del Sacco. Riv. Ital. Sci. Nat. **24:** 128-132, 147-151. 1904; **25:** 37-42. 1905.

Geology, etc.; mostly unannotated list of 251 vascular plants observed by author in vicinity of Ferentino. (Prov. Frosinone.)

Pirotta, Romualdo. Flora. *In* Abbate, Enrico. Guida della provincia di Roma. p. 171–230. 16.5 cm. Roma, 1890. (Pubblicazione del Club alpino italiano, sezione di Roma.) (Harvard College library.)

Includes lists of species from various localities—Vulcani Laziali, Vulcani Volsinii, Sabatini, Cimini, the Tolfa group, Spratte, Monti Cornicolani, Monti Ernici, Monti Lepini, etc., but no complete flora. (Regione of Lazio.) Republished without change in 2. ed. **1**: 176–235. 17.5 cm. Roma, 1894. (Harvard College library.)

Porchelli, B. M. La flora arpinate. xix, 208 p. 25 cm. Arpino, 1931.

List of vascular plants (921 species) of the vicinity of Arpino, with mention of medicinal and other uses at head of each family. (Formerly in Prov. Caserta, now in Prov. Frosinone.) (Not seen; cited from Bibl. Ital. **1931,** A: 201. 1933.)

Rosa, Pietro. La flora del Colosseo. *In* Colagrossi, P. L'antifeatro Flavio nei sui venti secoli di storia. (Appendici. I.) p. 337–346. 31.5 cm. Firenze, Roma, 1913.

Botanical publications, etc.; unannotated tabular list of vascular plants. (Prov. Roma.)—See also Deakin and Florini-Mazzanti, above.

Santis, Angelo de. Nomi locali di Terra di Lavoro derivati da nomi di piante da colture e vegetazione in genere. Nuovo Gior. Bot. Ital. n.s., **58:** 152–161. 1951.

Alphabetical list of Latin or vernacular names of plants and features of vegetation with corresponding locality names derived from them. (The historical Terra di Lavoro, now partly included in Lazio (Prov. Frosinone and Latina), partly in Campania (Prov. Caserta).)

Sibilia, Cesare. Ricerche floristiche sul territorio di Anagni. Nuovo Gior. Bot. Ital. n.s., **31:** 50–70, 161–176. 1924.

Topography, geology, climate, plant zones and associations; list of vascular plants, with a few annotations; bibliographies. (Prov. Frosinone.)

Steinberg, Carlo. Contributo alla conoscenza della flora dei pascoli erbosi e boschivi di Monterotondo (Roma). Nuovo Gior. Bot. Ital. n.s., **59:** 304–338 incl. 12 fig. (fig. 1, map, fig. 5–12 on 4 pl. (on 2)). 1953.

Soils, climate, etc.; list of vascular plants (p. 310–318), with indication of life form and habitat; plant communities, bibliography. (Prov. Roma.)

Zodda, Giuseppe. Prime notizie sulla flora delle Mainarde. Annali Bot. **19:** 163–201. 1931.

Geography, geology, climate, weeds, general features of flora; annotated list of vascular plants collected by author in August and September 1920. (Mountain group of Mainarde, in vicinity of S. Biagio Saracinesco, formerly in Prov. Caserta, now in Prov. Frosinone; slightly overlaps the Prov. Campobasso (in Abruzzi e Molise).)

LIGURIA

See also Emilia-Romagna (Pavesi); Piemonte (Morteo; Vignolo-Lutati); France, Partial (Thompson; Voigt); Alpes-Maritimes (Burnat).

The regione of Liguria includes the Provinces of Genova, Imperia (formerly Porto Maurizio), La Spezia (formed in 1923 from Genova), and Savona (formed in 1926 from Genova).

General

Frisoni, Antonio. Saggio di una bibliografia scientifica della Liguria. Geografia e storia naturale. G–Flora. Atti Soc. Ligust. Sci. Nat. e Geog. **21**: 159–160. 1910; **22**: 55–67. 1911.

Unannotated list of titles relating to botany (nos. 1400–1695), alphabetically arranged by authors. The whole bibliography reprinted, 157 p. Genova, 1916, with same title. (Not seen; cited from Bul. Bibl. Bot. Ital. 3 (1916): 196. 1916.)

Gismondi, Alfredo. Prospetto della flora ligustica. Guida botanica della Liguria. Descrizione succinta disposta in chiavi analitiche di tutte le piante vascolari che crescono spontaneamente sul suolo ligure, o vi si incontrano subspontanee o naturalizzate, oppure vi sono estesamente coltivate. 913 p. 24.5 cm. Genova, 1950.

General features of flora, plant zones, flora of chestnut woods, list of special localities with noteworthy species, glossary, list of local oil-yielding plants (seeds) with their oil content; briefly annotated flora of vascular plants in form of ample keys, with localities for rarer species and occasional vernacular names and uses. Review in Arch. Bot. (Forlì) **27**: 237. 1951 mentions several omitted species.

Notaris, Giuseppe de. Repertorium florae ligusticae. Mem. R. Accad. Sci. Torino II, **8**: 1–190. 1846; **9**: 125–529. 1848. (Preprinted 495 p. Taurini, 1844. Preprint not seen.)

Bibliography; annotated list of vascular plants (2158 species), with localities; supplementary list of doubtful species.—For supplements, adding 200 species, see **Gennari, Patrizio.** Centuria plantarum Repertorio florae ligusticae addendarum [I]–III [i.e. II]. Mem. R. Accad. Sci. Torino II, **14**: 249–273. 1854; **17**: 453–474. 1858.

Penzig, Otto. Flora popolare ligure. Primo contributo allo studio dei nomi volgari delle piante in Liguria. Atti Soc. Ligust. Sci. Nat. e Geog. **8**: 249–308, 341–379. 1897. (Reprinted 101 p. Genova, 1897. Reprint not seen.)

Systematic list of vascular plants with vernacular names and localities where used; index.—See also **Lagomaggiore, Nicolò,** and **Mezzana, Nicolò.** Contributo allo studio dei nomi volgari delle piante in Liguria. l.c. **12**: 205–278. 1901 [1902]. (Pronunciation, etc.; systematic list of vascular plants with vernacular names and localities.)

——— Florae ligusticae synopsis. (Res Ligusticae, XVII.) Annali Mus. Civico Storia Nat. Genova **38** (II, **18**): 423–531. 1897. (Reprinted 111 p. Genova, 1897. Reprint not seen.)

Topography, plant zones; unannotated list of vascular plants (3,166 species), not including cultivated or naturalized species; bibliography.—See also his Supplemento alla "Synopsis florae ligusticae." Arch. Bot. (Forlì) **1**: 187–204, 246–255. 1925. (Extensive list, including supplement to bibliography.)—Also **Bolzon, Pio.** Ricerche botaniche nella Liguria occidentale. Nota I. [–III.]. Bul. Soc. Bot. Ital. **1925**: 27–36, 77–85. 1925; Arch. Bot. (Forlì) **1**: 256–259. 1925. (Lists of noteworthy vascular plants, with localities, mostly from Prov. Genova.)

Local

Ariello, Giuseppe. Flora delle "Cinque Terre" (Liguria di Levante). (Res Ligusticae XCVIII.) Annali Mus. Civico Storia Nat. Giacomo Doria (Genova) **69**: 101–192. 1957.

Collaborators, etc.; annotated list of vascular plants (802 taxa, including about 671 species). (Prov. Spezia.)

Baglietto, Francesco. Florula della valle del Lagaccio in Genova. 32 p. 24 cm. Genova, 1886. (Arnold Arboretum library.)

List of angiosperms [464 species], with a few annotations. (Prov. Genova.)

Béguinot, Augusto. La vegetazione delle isole liguri di Gallinaria, Bergeggi, Palmaria, Tino e Tinetto. Studio floristico e biogeografico. Annali Mus. Civico Storia Nat. Genova III, **3** (v. 43): 419–469. 1907.

Botanical explorations, topography, geology, bibliography; tabular list of 445 vascular plants, showing distribution by islands; statistics, ecology, etc. (Gallinaria near Albenga, and Bergeggi, near Savona, both in Prov. Savona; Palmaria, Tino, and Tinetto off Portovenere on western side of Gulf of Spezia, in Prov. La Spezia.) Replaces **Barsali, Egidio,** and **Barsanti, L.** Contributo alla flora delle isole del Golfo di Spezia. Atti Soc. Toscana Sci. Nat. Pisa Proc. Verb. **13**: 63–80. 1902. (General features of Palmaria, Tino, and Tinetto; annotated list of 209 known plants (182 vascular).) Also replaces **Preda, Agilulfo.** Materiali per una florula della Palmaria. Nuovo Gior. Bot. Ital. n.s., **10**: 333–355. 1903. (Botanical explorations, topography, etc.; annotated list of vascular plants.)—See also (for Gallinaria) Pollacci, below.

Berger, Alwin. Florula mortolensis. An enumeration of the plants growing wild at La Mortola. 91 p. 19.5 cm. Ventimiglia, 1905.

Annotated list of wild vascular plants. (Prov. Imperia.)—See also France, Partial (Voigt, Die Riviera).

Bertagnon, Esa. Sulla flora medicinale della Liguria. Usi tradizionali dell'Alta Fontanabuona. Atti Accad. Ligure Sci. e Let. **11:** 201–214. 1955.
Annotated list of native medicinal plants of the upper valley of the Fontanabuona (vicinity of Gattorna), with patois names, uses, and discussion of some species; short bibliography. (Prov. Genova.)

Bicknell, Clarence. Flora of Bordighera and San Remo or a catalogue of the wild plants growing in western Liguria in the area bounded by the outer watersheds of the Arma and Nervia torrents. viii, 345 p. map (in pocket). 20.5 cm. Bordighera, 1896.
Annotated list of over 1700 vascular plants, with localities; list of additional species of nearby areas, short index of vernacular names, table of altitudes. Covers region along the coast from Ventimiglia to Arma di Taggia, and inland to Monte Pietravecchia. (Prov. Imperia.)

Nam, Vincenzo. Flora di Alassio e dintorni. Con speciale riguardo alle proprietà medicinali delle piante. xx, 376 p. illus. 8vo. Alassio, 1933.
Descriptive flora, with notes on medicinal properties. (Not seen; cited from notice in Boissiera 5: 66. 1941.) Replaces his Alassio and its flora. 1 p. l., vii, 160 p. 20 cm. Alassio, 1892. (Topography, climate, history, etc.; alphabetical list (p. 65–150) of vascular plants, including some cultivated ones, with localities; short list of seaweeds. In Kew library.) (Prov. Savona.)

Pandiani, Arturo. La vegetazione nel Monte Gazzo (Sestri Ponente). Saggio fitogeografico-floristico. Atti Soc. Ligust. Sci. Nat. e Geog. **23:** 213–285. 1912.
Geology, plant formations, phytogeography, general features of flora; annotated list of vascular plants (533 species); bibliography. (Prov. Genova.)

Penzig, Otto. La flora del Monte di Portofino. Rassegna Mens. Unione Ligure Escursionisti **14:** (p. ?). 1927.
Not seen; cited from E. Vaccari (next ref., p. 264). (Prov. Genova.)—See also **Vaccari, Enrico.** La vegetazione del Promontorio di Portofino (Liguria orientale). Nota preliminare. Arch. Bot. (Forlì) **15:** 258–280. 1939. (Topography, geology, climate, plant associations, etc.)

Pollacci, Gino. L'isola Gallinaria e la sua flora. Atti Ist. Bot. R. Univ. Pavia II, **9:** 107–125. 1911. (Preprinted 19 p. 1905.)
History, topography, geology, fauna, botanical publications, general features of flora; annotated list of vascular plants collected by author. (Prov. Savona.)—See also Béguinot, above.

Puccini, Giuliano. Naturalizzazione e riproduzione spontanea di piante ornamentali nella Liguria occidentale. Nuovo Gior. Bot. Ital. n.s., **57:** 23–33. 1950.
Topography, climate, geology, growth conditions; list of 46 naturalized species of the coast between Capo Verde and Ponte S. Luigi (Prov. Imperia), with discussion; list of ornamental species reproducing spontaneously in gardens, grouped by localities and gardens where found; bibliography.

Ricca, Luigi. Catalogo delle piante vascolari spontanee della zona olearia nelle due Valli di Diano Marina e di Cervo. Atti Soc. Ital. Sci. Nat. Milano **13:** 60–143. 1870.
Topography, general features of flora; annotated list of vascular plants (807 species, 46 varieties, and 2 hybrids). (Prov. Imperia.)

Rovesti, Guido. Le piante aromatiche e medicinali spontanee della provincia di Porto Maurizio. 240 p. 8vo. Porto Maurizio, 1923.
Not seen; cited from Arch. Bot. (Forlì) **2:** 255. 1925. (Now Prov. Imperia.)

Tosco, Uberto. Florula della "Arene Candide" (Finale Ligure). Allionia **1:** 247–256. 1 fig. (map), pl. 1953.
Topography, general features of vegetation, etc.; annotated list of angiosperms (95 taxa); bibliography. (A dune near Finale Ligure Marina, Prov. Savona.)

LOMBARDIA (LOMBARDY)

The regione of Lombardia includes the Provinces of Bergamo, Brescia, Como, Cremona, Mantova (Mantua), Milano (Milan), Pavia, Sondrio, and Varese (the last formed in 1926 from parts of Como and Milano).

See also Emilia-Romagna (Bertolani-Marchetti, Il popolamento); Piemonte (Gabelli); also Switzerland, Graubünden (Braun-Blanquet, Die vegetationsverhältnisse, Die flora des Puschlav; Geiger; Schaad); Ticino (Schröter, 2 titles).

General

Cesati, Vincenzo, *barone.* Flora. *In* Cattaneo, Carlo. Notizie naturali e civili su la Lombardia. v. 1, p. 259–326. pl. [4] (map). 21.5 cm. Milano, 1844. (Reprinted with title: Saggio su la geografia botanica e su la flora della Lombardia. 74 p. map. 22.5 cm. Milano, 1844. In Arnold Arboretum library.)

General features of flora; unannotated list of spermatophytes; statistics. Followed by list of cryptogams (p. 327–348) by Santo Garovaglio and others (ferns, p. 327, by **Santo Garovaglio,–**).—See also his Gestaltung und verhältnisse der pflanzenwelt in der Lombardei. Linnaea **21:** 1–64. 1848. (Topography, geology, climate, general features of flora, etc.; unannotated list of known spermatophytes (2639 species, of which 2568 are native, and 217 subspecies); statistics.)—Also his Die pflanzenwelt in gebiete zwischen dem Tessin, dem Po, der Sesia und den Alpen. Ein nachtrag zu meinem aufsatze: "Gestaltung und verhältnisse . . ." l.c. **32:** 201–262. 2 fig., 2 pl. 1863. (Botanical investigations, general features of flora, etc., of different areas.)—Also **Giacomini, Valerio.** Contributo alla conoscenza della flora lombarda (con osservazioni sistematiche e fitogeografiche). Atti Ist. Bot. Univ. Pavia V, **9:** 127–188. 1950. (Extensive list of vascular plants with localities and some critical annotations.)—Also **Stucchi, Carlo.** Observationi su alcune piante lombarde. Nuovo Gior. Bot. Ital. n.s., **58:** 583–586. 1952. (Various noteworthy species, mostly naturalized.)

Local

Anonymous. Botanica. *In* Cremona e la sua provincia. p. 139–182. 8vo. Cremona, 1863.

A flora of the province of Cremona, with brief indication of localities. (Prov. Cremona.) (Not seen; title and annotation supplied by V. Giacomini.) —See also Sanseverino, below.

Agostini, A. Prospetto delle piante e degli animali più communi e più noti a Castiglione delle Stiviere colle denominazioni rispettive dialettale, italiana e scientifica. 32 p. Castiglione delle Stiviere, 1912.

Contains (p. 1–22) a list of patois names with Italian and scientific equivalents; there are some errors. (Prov. Mantova.) (Not seen; title and annotation mostly supplied by V. Giacomini.)

Ardissone, Francesco. Catalogo delle piante vascolari del Monte Baro. Mem. R. Ist. Lombardo Sci. e Lett., Cl. Sci. Mat. e Nat. **20** (III, **11**): 51–100. 1903.—Aggiunte e correzioni . . . l.c. **20** (III, **11**): 163–164. 1905.

Topography, geology, general features of flora; annotated list of vascular plants [721 species including supplement]. (Prov. Como.)—See also **Cobau, Roberto.** Nuove aggiunte al Catalogo . . . Bul. Soc. Bot. Ital. **1925:** 85–96. 1925. (List of over 80 plants, mostly infraspecific taxa.)

Arietti, Nino. Aspetti floristici della campagna di Montichiari. Comm. Ateneo Sci., Let. ed Arti Brescia **1940/42,** A: 127–188. 4 pl., map. 1943.

Topography, phytogeography, weeds, general features of flora; annotated list of native and commonly cultivated angiosperms of the so-called "brughiera" (heath) of Montichiari. (Prov. Brescia.)

——— La flora della Valle Camonica. Revisione critico-sistematica di un erbario e di note inedite di Ottone Penzig predisposti per un incompiuto catalogo delle piante vascolari della Valle Camonica. Atti Ist. Bot. Univ. Pavia V, **4:** 1–181. 1944.

Previous work, geographical limits, topography, geology, general features of flora, statistics; annotated list of vascular plants (931 species and 370 varieties and forms), with localities and names of collectors. Based entirely on Penzig's herbarium, and so not a complete flora. (Mostly in Prov. **Brescia,** a little in Prov. Bergamo.)—See also Fenaroli, La flora della Conca del Baitone, below.

——— Florula della Lobbia Alta nel gruppo dell'Adamello. Comm. Ateneo Sci., Let. e Arti Brescia **1935**: 257–261. 1936.

Topography, etc.; list of 16 vascular plants observed by author. Lobbia Alta is 3196 meters high. (Prov. Brescia.)

——— La nostra flora nell'economia domestica. 232 p. 120 fig. 8vo. Brescia, 1941.

Treatment of useful plants of Brescia, with Italian names, Brescian patois names, and uses; list classified by uses, list of spontaneous ornamental plants, etc. (Not seen; title and annotation supplied mostly by V. Giacomini.)

Béguinot, Augusto. La vegetazione macrofitica dei Laghi di Mantova. Atti Cong. Limnol. Teor. ed Applic. 4: 173–192. pl. 1–3. 1929.

Includes complete list of the plants of the shores and water of the Laghi di Mantova. (Prov. Mantova.) (Not seen; title and annotation supplied by V. Giacomini.)

Bergamaschi, Giuseppe. Gita botanica agli Appennini Boglelio e Lesima. 14 p. 8vo. Pavia, 1823.—Lettera seconda . . . sopra varie piante degli Appennini, colli oltrepadani e della campagna pavese. 26 p. 8vo. Pavia, 1823.

Lists of plants, particularly of the Apennines of Pavia; the first list in narrative form, the second in systematic order. (Prov. Pavia.) (Not seen; title and annotation supplied by V. Giacomini.)

——— Peregrinazione statistico-fitologica nelle valli Camonica, Seriana, Brembana. 116 p. 8vo. Pavia, 1853.

Systematic treatment of vascular plants collected in the valleys listed, with localities. (Prov. Bergamo and Prov. Brescia.) (Not seen; title and annotation supplied by V. Giacomini.)

Brizi, Ugo, and **Fenaroli, Luigi.** Appunti sulla flora della brughiera lombarda. 32 p. 24.5 cm. Piacenza, 1927. (Estratto dal volume "Le brughiere" pubblicato in onore del Prof. Angelo Menozzi . . .)

Plant formations, lists of mosses; annotated list of vascular plants of the heath area extending from Gallarate to Ticino and Sesto Calende. (Prov. Varese.)

Castelli, Gulielmo. Flora. *In* his La valle di Scalve. Bol. Club Alpino Ital. **30**: 123–125. 1897.

Consists of mention of principal trees, and "Saggio della flora alpina della valle di Scalve" (p. 124–135), a list of vascular plants with localities and indication of abundance, the families arranged alphabetically. (Prov. Bergamo and Prov. Brescia.)

Cesati, Vincenzo. Cenni intorno all'Elenco delle piante spontanee della provincia di Milano del sig. E. con supplemento al medisimo. Biblioteca Ital. **90**: 225–236. 1838; **91**: 330–349. 1838. (Boston Public Library.)

Includes notes on the following publication: Verzeichniss der einheimischen pflanzen in der provinz Mailand, nach Lanfossi, Pollini und eigenem herborisiren, mit einem vorerinnerungen dargestellt. Von E. (Echo, Zeitschrift für Literat. etc. 1837, no. 29 n.f.) [not seen]; followed (second part) by a list of vascular plants, with localities, on the Linnaean system. (Prov. Milano.)

Ciferri, Raffaele, Giacomini, Valerio, and **Poggio, P.** La flora fanerogamica delle risaie dell'Italia transpadana. Atti Ist. Bot. Univ. Pavia ser. 5, Suppl. vol. D. 1 p.l., 26 p. 1949. (Mimeographed.)

List of vascular plants (315 species and 32 subspecies and varieties) of the rice fields north of the Po, with localities; bibliography. (Provinces of Brescia, Cremona, Mantova, and Pavia in Lombardia, also Novara and Vercelli (Piemonte) and Verona (Veneto).)—See also **Koch, Walo.** Zur flora der oberitalienischen reisfelder. Ber. Schweiz. Bot. Gesell. **62**: 628–663. 9 fig. 1952. (Previous investigations; annotated list of weeds collected by author in rice fields in Piemonte (Provinces of Novara and Vercelli) and in Lombardia (Prov. Pavia); bibliography.)—Also **Amato, Francesco d'.** Osservazioni preliminari sulla flora e vegetazione delle risaie e delle paludi a N. del Lago di Massaciuccoli (Versilia). Nuovo Gior. Bot. Ital. n.s., **64**: 153–184. 1 fig. (map), pl. 9–10. 1957. (Includes list of 188 taxa; bibliography.

In Prov. Lucca, in Toscana.)—Also **Pignatti, Sandro.** La vegetazione delle risaie pavesi. (Studio fitosociologico.) Arch. Bot. (Forlì) **33:** 129–193. 2 fig. 1957. (Phytosociological; includes bibliography on rice-field weeds of southern Europe.)

Cobau, Roberto. Flora vascolare spontanea della città di Milano. Nuovo Gior. Bot. Ital. n.s., **23:** 375–402. 1916; **27:** 89–128. 1920; **33:** 39–64. 1926.
Topography, geology, climate, etc., with references; annotated list of vascular plants in the urban flora (549 species including supplements); botanical history, plants of special areas and habitats. (Prov. Milano.)—See also **Stucchi, Carlo.** Sulla attuale presenza e diffusione nel Milanese di alcune piante introdotte. Nuovo Gior. Bot. Ital. n.s., **56:** 294–298. 1949. (Annotated list of various introduced spermatophytes in vicinity of Milan.)

Comolli, Giuseppe. Flora comense disposta secondo il sistema di Linneo. 7 v. 15.5 cm. Como, 1834–36; Pavia, 1846–57.
Geography, topography, geology, climate, table of altitudes; annotated descriptive flora of vascular plants, with vernacular names, localities, notes on uses. (Prov. Como and Prov. Varese, as well as southern part of Canton Ticino in Switzerland.)—See also **Anzi, Martino.** Auctarium ad floram novo-comensem . . . Mem. R. Ist. Lombardo Sci. e Let., Cl. Sci. Mat. e Nat. 14 (III, 5): 177–205. 1879.—Also **Artaria, F. A.** Contribuzione alla flora della Provincia di Como. Rend. R. Ist. Lombardo Sci. e Let. II, **16:** 390–411. 1893; Seconda contribuzione . . . Atti Soc. Ital. Sci. Nat. Milano **35:** 137–164. 1895. (Two lists of vascular plants new to or rare in the Province of Como, with localities.)—Also Scotti, below.

Cornaz, Edouard. Giov. Batt. Patirana et sa flore médicale de Bormio. Bul. Soc. Sci. Nat. Neuchâtel **16:** 119–154. 1888.
Early botanical explorations in district of Bormio; list of medicinal vascular plants in herbarium formed by Patirana (1716–1782), with localities but with Italian vernacular names and medical properties omitted. (Prov. Sondrio.)—See also Peloni, below.

Cozzi, Carlo. La flora urbico-muraria del Gallaratese. Atti Soc. Ital. Sci. Nat. Milano **50:** 300–302. 1912.
Includes list of vascular plants of walls and similar habitats in vicinity of Gallarate. (Prov. Varese.)

——— Florula abbiatense ossia rassegna delle piante vascolari più communi nell'agro di Abbiategrasso. 26 p. 23.5 cm. Abbiategrasso, 1902.
Includes list of 209 vascular plants as well as some fungi and plant galls. (Prov. Milano.) (Not seen; title and annotation supplied by R. Pichi-Sermolli.)—See also his I fiori della Cà di Biss e sue adiacenze. Contributo alla florula abbiatense. Atti Soc. Ital. Sci. Nat. Milano **40:** 19–26. 1901. (List of 100 species.)—Also his I fiori della Cà di Biss. Nuovo manipolo . . . l.c. **41:** 101–107. 1902. (List of 50 species.)—Also his Quarto contributo alla flora del Ticino. l.c. **43:** 7–14. 1904. (Annotated list.)—Also his Ulteriori aggiunte alla florula abbiatense. l.c. **44:** 29–36. 1905. (Annotated list.)—Also his Contribuzione alla flora murale. Bol. Nat. Sup. Riv. Ital. Sci. Nat. **26:** 74–76. 1906. (List of 100 species.)

——— Le piante e i fiori nel vernacolo abbiatense. Nuova ristampa ampliata e corretta. 28 p. 22 cm. Abbiategrasso, 1907. (1st ed. 1904.)
List of native and cultivated plants, mostly phanerogams, arranged alphabetically by local vernacular names. (Prov. Milano.) (Not seen; annotation supplied by F. Sappa, title from Internatl. Cat. Scient. Lit. 8 (M. Bot.): 79. 1910.)

Fenaroli, Luigi. La flora della Conca del Baitone (Val Camonica, gruppo dell'Adamello). Atti Soc. Ital. Sci. Nat. Milano **63:** 221–233. 1 fig. (map). 1924; **65:** 16–24. 1926.
Topography, etc.; annotated list of vascular plants. (Prov. Brescia.)—See also Arietti, La flora della Valle Camonica, above.

——— Note botaniche sul comune di Pisogne. Atti Soc. Ital. Sci. Nat. Milano **62:** 201–208, 209–210. 1923.
Geography, general features of flora, etc.; unannotated list of vascular plants. (Prov. Brescia.)

Fornaciari, Giovanni. Un'isola floristica nelle Alpi Retiche centrali. Atti Accad. Udine VI, **9:** (p.?). 2 pl. 1949. (Repr. 23 p.; only reprint seen.)

Plant zones, general features of flora; tabular list of lichens, bryophytes, and vascular plants of the Refugio Marinelli with indication of duration and zone. (In group of Pizza Bernina, at 2750 to 3100 meters altitude, Prov. Sondrio.)

Furrer, Ernst, and **Longa, Massimo.** Flora von Bormio. Beih. Bot. Centralbl. 2. Abt., **33**: 1–112. map. 1915.

Topography, general features of flora, botanical investigations; annotated list of vascular plants [1,024 species]; bibliography, gazetteer. (Prov. Sondrio.)—See also **Furrer, E.** Vegetationsstudien im Bormiesischen. Vierteljahrsschr. Naturf. Gesell. Zürich **59**: 145–222. 5 fig. (incl. map), pl. 3 (map). 1914. (Mitt. Bot. Mus. Univ. Zürich LVIII.) (Topography, geology, plant associations, etc.)—Also **Fenaroli, Luigi,** and **Longa, M.** Flora bormiese . . . Nuovo Gior. Bot. Ital. n.s., **33**: 153–177. 1926. (Additional records.)—Also **Bertossi, Felice.** Contributo alla conoscenza della flora dei dintorni di Bormio. Atti Ist. Bot. Univ. Pavia V, **9**: 219–225. 1950. (List of noteworthy plants collected by G. Briosi, with localities.)—For the names of 14 additional species, see **Giacomini, V.,** and **Pignatti, S.** Mem. Soc. Ital. Sci. Nat. Milano **11**: 65. 1955.

Geilinger, Gottlieb. Die Grignagruppe am Comersee. Eine pflanzengeographische studie. Beih. Bot. Centralbl. 2. Abt., **24**: 119–420. map. 1909. (Mitt. Bot. Mus. Univ. Zürich XLI.)

Topography, geology, climate, bibliography, etc.; annotated list of Musci and vascular plants; ecology, bibliography, gazetteer. (Prov. Como.)—See also **Rossi, Pietro.** Nuovo contributo alla flora del "Gruppo del Grigne." Nuovo Gior. Bot. Ital. n.s., **32**: 396–441. 1925; **33**: 252–315. 1926. (Topography, botanical explorations, bibliography; list of Bryophyta and vascular plants collected by author, with localities. Incomplete (Bryophyta-Campanulaceae).)—Also **Anonymous.** Escursione sociale in Brianza e alle Grigne 31 Maggio—3 Giugno 1956. l.c. **63**: 426–427. 4 fig. 1957. (Account of excursion of the Società botanica italiana, with mention of numerous species observed, bibliography, etc.)

Giacomini, Valerio. Aspetti scomparsi e relitti della vegetazione padana. Documenti sulla vegetazione recente della "lame" e delle torbiere fra l'Oglio ed il Mincio. Atti Ist. Bot. Univ. Pavia V, **9**: 29–123. f. 8–11 (fig. 8, map). 1946.

Ecological; includes (p. 95–109) tabular list of vascular plants of palustrine formations of the plain between the Oglio and the Mincio; bibliography. (Mostly in Prov. Mantova.)

——— and **Pignatti, S.** Flora e vegetazione dell'Alta Valle del Braulio con speciale riferimento ai pascoli di altitudine. Mem. Soc. Ital. Sci. Nat. Milano **11**: 45–238. 31 fig. (incl. maps), pl. 6–10, map, 26 tab. (partly in text). 1955.

Topography, climate, altitudinal distribution, etc.; annotated list of lichens, bryophytes, and vascular plants (p. 69–109; 595 taxa, including 455 species); plant associations, bibliography. (In Alta Valtellina, in Prov. Bormio.)

Lanfossi, Paolo. Saggio di storia naturale dei contorni di Mantova . . . Parte seconda. Piante. Gior. Fis., Chim., Storia Nat., Med. ed Arti (Milano) II, **10**: 46–57, 138–149, 235–244, 279–298, 370–390, 417–432. 1827.

Annotated list of vascular plants, on the Linnaean system, with localities for rarer species. (Prov. Mantova.)—See also **Barbieri, P.** Aggiunte e osservazioni al Saggio di storia naturale . . . del Dott. Paolo Lanfossi. Poligrafo **13** (1–2): 169–192. 1839; **14** (17–18): 3–52. 1839 [?]. (Not seen; cited from Giacomini, Atti Ist. Bot. Univ. Pavia V, **9**: 119. 1946.)—Also Paglia, Catalogo, below.

Lenticchia, Attilio. Cenni botanichi sulla Punta della Gaeta. (? p.) Menaggio, 1898.

List of plants of the limestone cliff Punta della Gaeta, north of Menaggio, on west shore of Lake Como. (Prov. Como.) (Not seen; cited from Ber. Schweiz. Bot. Gesell. **10**: 88. 1900.)

Longa, Glicerio. Vocabolario botanico-zoologico. I. Flora. *In* his Vocabolario bormino. Studj Romanzi **9**: 279–288. 1912.

Alphabetical list of patois names of wild and cultivated plants in the Bormio area, with Italian vernacular names and scientific equivalents. (Prov. Sondrio.)

Massara, G. F. Prodromo della flora valtellinese ossia catalogo delle piante rinvenute in varie excursioni botaniche nella provincia di Sondrio in cui si distinguono con brevi ragionamenti le specie più rare, i nomi volgari di molte piante, i principali prodotti del suolo e le proprietà che a varie spegie si attribuiscono. xx, 219 p. col. front. 23 cm. Sondrio, 1834. (Arnold Arboretum library.)

Botanical investigations, table of altitudes; annotated list of vascular and some cellular plants of the Province of Sondrio, alphabetically arranged by botanical names, with vernacular names and uses.—See also **Lanfossi, Paolo.** Catalogo delle piante vedute crescere spontaneamente nei territori milanese, valtellinese e bresciano. Poligrafo **1836:** 30–50, 103–134. 1836. (List of vascular plants, bryophytes, fungi, and lichens, on the Linnaean system; refers particularly to Brescia. Not seen; title and annotation supplied by V. Giacomini.)—Also **Anzi, Martino.** Alcune notizie sulla flora valtellinese. *In* Guida alla Valtellina. 2. ed. p. 55–64. 1884. (Not seen).—Also **Longa, Massimo.** Novità per la flora valtellinese. l.c. p. 68–73. 1884. (Not seen.)—Also **Ronchetti, P.** Il mio erbario. Nat. Valtellinese **1885:** 39–41, 55–56, 73–74, 88–90, 119–120, 134–136, 152–155, 166–169, 183–186, 198–210. 1886. (Systematic list of vascular and some cellular plants with localities. Not seen; title and annotation supplied by V. Giacomini.)—Also **Bezzi, Mario.** L'erbario Longa. Rend. R. Ist. Lombardo Sci. e Let. II, **37:** 328–338. 1904. (Includes extensive list of noteworthy plants from Valtellina.)—Also **Craveri, Michele.** Escursione botanica nell'alta Valtellina in tempo di guerra. 15 p. 16mo. Cassino, 1921. (Short list of plants from the pastures of Monte Pedenolo and Monte Braulio. Not seen; title and annotation supplied by V. Giacomini.)—Also **Fornaciari, Giovanni.** Flora e vegetazione delle Valli del Mera e dell'Adda. Contributo: Le felci. Annali Scuola Friulana **1950–51:** 1–74. 1952. (Detailed account of distribution and variation of ferns of Valtellina. Not seen; cited from Arch. Bot. (Forlì) **29:** 336. 1953.)—Also **Sartori-Fanelli, C.** Una relazione inedita statistica e naturalistica sulla Valtellina de Giuseppe Bergamaschi, medico e botanico pavese [1785–1867]. Rass. Econ. Prov. Sondrio **7:** 25–31. 1955. (Bergamaschi's itinerary, etc.; alphabetical lists of species collected by him in 1829 at Sondrio, Tirano, Bormio, Stelvio (in Bolzano), and other localities.)

Moretti Foggia, Amalia. Florula delle piante vascolari del bosco Fontana nei dintorni di Mantova. Atti Soc. Nat. Modena III, **14:** 47–72. 1896.

History, soils; annotated list of vascular plants (532 species). The "bosco Fontana" is an ancient forest covering 228 hectares, 6 km. northeast of Mantova (Mantua) in the comune of Marmirolo. (Prov. Mantova.)—See also **Massalongo, Caro.** Nuovo censimento delle piante fanerogame e crittogame vascolari del bosco Fontana. Nuovi Annali Agr. Min. Econ. Naz. [Italy] **7:** 499–514. 1927. (Plant formations, etc.; unannotated list of 656 species of vascular plants; bibliography.)

Paglia, Enrico. Catalogo dei vegetali spontanei e generalmente coltivati nel Mantovano. *In* his Saggio di studi naturali sul territorio mantovano. p. 447–507. 8vo. Mantova, 1879.

List of wild and cultivated vascular plants (1399 species), and some cellular plants of the Province of Mantova, with localities. (Not seen; title and annotation supplied mostly by V. Giacomini.)—See also Lanfossi, above.

——— Delle erbe nocive ed utili spontanee nei prati mantovani con tavole analitiche per la loro nomenclatura. 40 p. 8vo. Mantova, 1872.

Catalog of harmful and useful plants, with vernacular names and uses. (Prov. Mantova.) (Not seen; title and annotation supplied mostly by V. Giacomini.)

Peloni, Attilio. In herbis salus. Note sulla flora alpina medicinale di Bormio. 71 p. illus. 8vo. Tirano, 1934.

Annotated list of medicinal plants of Bormio. (Prov. Sondrio.) (Not seen; title and annotation supplied by V. Giacomini.)—See also Cornaz, above.

Penzig, Otto. Il Monte Generoso. Schizzo di geografia botanica. Nuovo Gior. Bot. Ital. **11:** 129–147. 1879. (Reprinted 23 p. 1879. Reprint not seen.)

Geology, climate, botanical explorations, plant zones, general features of flora; unannotated alphabetical list of vascular plants, including those recorded by earlier authors. (On border of Prov. Como and Canton Ticino. Called also Monte Calvaggione.)

Piazzoli Perroni, Antonietta. Ricerche sulla flora e vegetazione dei fontanili dell'agro milanese. Nuovo Gior. Bot. Ital., n.s., **63:** 355–410. 6 fig. (fig. 1, map). 1957.
Includes partly annotated list of vascular plants and bryophytes of various spring-heads between the high and low plain of the Po; bibliography. (Prov. Milano.)

Pignatti, Erika, and **Pignatti, Sandro.** Un'escursione al Passo di Gavia. Arch. Bot. (Forlì) **34:** 137–153. 1958.
Topography, botanical explorations; list of known cellular and vascular plants (192 vascular species); discussion of vegetation, bibliography. (On Brescia-Bergamo border.)

Rodegher, Emilio. Elenco delle piante buone e cattive foraggifere dei colli, monti e prealpi della provincia di Bergamo. *In* Pascoli alpini della provincia di Bergamo. Atti della Commissione d'inchiesta sui pascoli alpini. v. 2, Appendice. p. 331–385. 24.5 cm. Milano, 1907. (Publication of Società agraria di Lombardia.)
Topography, plant regions; tabular list of 476 useful and harmful vascular forage plants of the hills and mountains of Bergamo, with localities, and indication of forage value.

——— and **Venanzi, Giuseppe.** Prospetto della flora della provincia di Bergamo. xviii, 146 p. 28 cm. Treviglio, 1894 (1895). (Arnold Arboretum library.)
Table of localities and altitudes; annotated list of vascular plants and Characeae, including cultivated species, with localities for scarcer species. Replaces **Rota, Lorenzo.** Prospetto della flora della provincia di Bergamo. 104 p. 26 cm. Bergamo, 1853. (General features of flora, table of altitudes; annotated tabular list of vascular plants and Characeae, with indication of soil preference and altitudinal limits of range. In Arnold Arboretum library.)—See also **Rodegher, Emilio,** and **Rodegher, Alcide.** Novissimo prospetto della flora della provincia di Bergamo. [I.] [–II.] puntata. Atti Ateneo Sci. Let. ed Arti Bergamo **25** (Atti Accad., Communicazione [4]). 50 p. 1920; 26 (Comm. 4). 64 p. 1921. Continued as Flora della provincia di Bergamo. III. [–VI.] puntata. Bergomum n.s., **3** (2) (Atti Ateneo Sci. Let. ed Arti Bergamo 1927–29, no. 3): 33–48; **3** (3) (Atti . . . 1927–29, no. 4): 56–80; **3** (4) (Atti . . . 1927–29, no. 5): 81–96. 1929; **4** (1) (Atti . . . 1927–29, no. 6): 97–112. 1930. (Systematic list of species and infraspecific entities not included in the Prospetto (1895), with localities and some notes on uses. Not quite completed; covers *Polypodium–Centaurea.*)

Rota, Lorenzo. Prospetto delle piante fanerogame finora ritrovate nella provincia pavese. Gior. Bot. Ital. **2:** 73–82, 247–292. 1847.
General features of flora; list of spermatophytes of the Province of Pavia, with localities for rarer species. Replaces (as to list of phanerogams) **Nocca, Domenico,** and **Balbis, G. B.** Flora ticinensis seu enumeratio plantarum quas in peregrinationibus multiplicibus plures per annos solertissime in papiensi agro peractis observarunt . . . 2 v. (cxxxix, 409; xvi, 393, 39 p.). 27 pl., map. 25.5 cm. Ticini, 1816–21. (Climate, index to map; flora of vascular and cellular plants, with brief descriptions and annotations. Ticinum was the Roman name for the present Pavia.)—See also **Farneti, Rodolfo.** Aggiunte alla flora pavese e ricerche sulla sua origine. Atti Ist. Bot. R. Univ. Pavia II, **6:** 123–164. 1900. (Topography, general features of flora, postglacial history, etc.; list of 355 noteworthy plants, with localities and indication of new records.)—Also **Pavesi, Vittorio.** Elenco di piante dell' alto Apennino pavese. Atti Soc. Ital. Sci. Nat. Milano **45:** 46–54. 1906. (List of 128 vascular plants new to or rare in Province of Pavia.)—Also **Pollacci, Gino.** Aggiunte alla flora ticinese. Atti Ist. Bot. R. Univ. Pavia II, **15:** 53–62. 1918. (Bibliography; list of 53 species and varieties new to the province.)—Also **Fiori, Adriano.** Contribuzione alla flora dei serpentini del Pavese. Bul. Soc. Bot. Ital. **1919:** 39–40. 1919. (Short lists of noteworthy vascular plants collected by Cesare Massa at Zebedassi and Sassi Neri.)—Also **Maffei, Luigi.** Aggiunte alla flora pavese. Atti Ist. Bot. R. Univ. Pavia II, **18:** 137–150. 1921. (Bibliography; 75 additional species listed.)—Also **Segagni, Angela.** Alcune erborizzazioni nella

risaie pavesi. l.c. III, **2**: 117–127. 1925. (Previous work on plants of rice fields in Italy, bibliography; tabular list of 129 vascular plants from various rice fields in vicinity of Pavia.)

Sanseverino, F. Flora cremasca adattata agli usi dell'agricoltura. *In* his Notizie statistiche e agronomiche alla città di Crema e suo territorio. p. 62–82. 16mo. Milano, 1843.

A brief flora of the region of Crema. (Prov. Cremona.) (Not seen; title and annotation supplied by V. Giacomini.)—See also Anonymous, above.

Scotti, Giberto. Flora medica della provincia di Como. xi, 836 p. 21 cm. Como, 1872.

List of cellular and vascular plants, systematically arranged, extensively annotated as to medicinal uses, with vernacular names; includes cultivated plants.—See also Comolli, above.

Spiller, S. Stato attuale delle conoscenze sulla flora mantovana. 8 p. Mantova, 1951.

Historical account, with bibliography, of botanical explorations in the Province of Mantova and of local collections. (Not seen; cited from Arch. Bot. (Forlì) **28**: 346. 1952.)

Traverso, G. B. Flora urbica pavese ossia catalogo delle piante vascolari che crescono spontaneamente nella città di Pavia. Nuovo Gior. Bot. Ital. n.s., **5**: 57–75. 1898.— . . . Centuria seconda. l.c. **6**: 241–257. 1899.

Geology, short bibliography; annotated lists of 200 vascular plants, with vernacular names and table of flowering dates. (Prov. Pavia.)—See also Rota, above.

——— Una salita botanica al Pizzo Arera (Bergamo). Atti Accad. Sci. Veneto-Trent.-Istriana n.s., **5**: 14–26. 1908.

Topography, general features of flora, bibliography; list of 118 vascular plants collected at 1800 meters altitude or above. (Prov. Bergamo.)

Wilczek, Ernest, and **Chenevard, Paul.** Contributions à la flore des Préalpes bergamasques. Annuaire Conserv. et Jard. Bot. Genève **15/16**: 248–287. 1912.

List of vascular plants collected by authors, with localities. (Prov. Bergamo and Prov. Como.)—See also **Chenevard, Paul.** Contributions à la flore des Préalpes bergamasques. l.c. **18/19**: 129–192. 1914. (List of vascular plants, with localities, from Sassina and Varrone valleys, Prov. Como.)—Also **Provasi, Tiziano.** Contributo alla floristica delle valli Sassina e Varrone. Atti Soc. Ital. Sci. Nat. Milano **61**: 179–203. 4 fig. (incl. 2 maps). 1922. (Topography, geology, botanical explorations (with footnote references); list of vascular plants collected by author in July-August, 1920–21, with localities.)

Zersi, Elia. Prospetto delle piante vascolari spontanee o comunemente coltivate nella provincia di Brescia. Aggiunte le esotiche che hanno uso e nome volgare . . . 267 p. 22.5 cm. Brescia, 1871. (Supplement to Commentari dell'Ateneo di Brescia per gli anni 1868, 1869.)

Table of altitudes; briefly annotated list of vascular plants, with vernacular names and localities; includes cultivated species.—See also **Ugolini, Ugolino.** Elenco di piante del Bresciano, aggiunte al Prospetto Zersi, e quadri statistico-tassonomici della flora bresciana. Comm. Ateneo Sci., Let. ed Arti Brescia 1897, App. 56 p. tab. 1897.—Secondo [-settimo] elenco . . . l.c. **1899**: 35–40. 1899; **1900**: 235–241. 1900; **1901**: 235–247. 1901; **1904**: 127–153. 1904; **1907**: 102–142. 1907; **1909**: 122–136. 1909. (Lists of newly recorded species and varieties and new localities. The author extends the Brescian region to include adjacent areas in Bergamo and Mantova as well as in Trento (in Trentino-Alto Adige) and Verona (in Veneto).—Also his Le piante avventizie della flora bresciana. Censimento e dati sulla provenienza, introduzione, diffusione e biologia delle varie specie. Parte I. l.c. **1920**: 62–110. 1921. (General considerations; extensively annotated list of 31 species of gymnosperms and monocotyledons; no more published.)—Also **Arietti, Nino.** Reperti sporadici di flora bresciana. Puntata [1.]–[4]. 3. l.c. **1939**, A: 147–172. 1940; **1942–45**, B: 53–74. 1947; **1948/49**: 203–223. 1950; **1950**: 163–175. 1951. (Records of noteworthy species. Vols. for 1939 and 1950 in Harvard College library.)

MARCHE (MARCHES)

See also Emilia (Raggi).

The regione of Marche includes the Provinces of Ancona, Ascoli-Piceno, Macerata, and Pesaro e Urbino.

General

Paolucci, Luigi. Flora marchigiana, ossia revisione sistematica e descrittiva delle piante fanerogame spontanee finora raccolte nelle regione delle Marche oltre quelle più estesamente coltivate e che talora inselvatichiscono . . . xxv, 656 p. 45 pl. 28 cm. Pesaro, 1890–91 (title page date 1890, cover date 1891). (Gray Herbarium library.)

Botanical publications and collections, topography, geology, climate, plant zones; descriptive flora of phanerogams, with localities.—See also **Paolucci, L., and Cardinali, Federico.** Contributo alla flora marchigiana di piante nuove e di nuove località per alcune sue specie più rare. Malpighia **9:** 125–135. 1895. Secondo contributo . . . Nuovo Gior. Bot. Ital. n.s., **7:** 96–114. 1900.—Also **Bettini, Angelo.** Contributo alla flora marchigiana ossia aggiunta di specie e varietà a quelle elencate e descritte nella Flora marchigiana del Prof. Luigi Paolucci. Arch. Bot. (Forlì) **23:** 39–45, 49–66. 1947. (Annotated list of about 100 species and varieties.)—Also **Brilli-Cattarini, Aldo.** Rinvenimenti floristici marchigiani. (1)–4. serie. Nuovo Gior. Bot. Ital. n.s., **59:** 515–518. 1953; **63:** 117–141. 1956; **64:** 381–409. 1958; **65:** 493–537. 1959. (Lists; bibliography.)—Also **Anzalone, Bruno.** Su alcune piante notevoli o nuove per le Marche o per la provincia di Macerata. l.c. **59:** 549–553. 1953. (Records of 28 species, mostly new to the Province of Macerata.)

Local

Cengia-Sambo, Maria. Contributo alla flora vascolare dell'Urbinate. Nuovo Gior. Bot. Ital. n.s., **35:** 425–506. 1929.

Botanical explorations, topography, climate, list of collectors, list of localities; annotated list of vascular plants; bibliography. The Urbinate is the inner part of the Province of Pesaro e Urbino, west of Pergola, Fossombrone, Trezanni, Pallino, and Schieti.—See also **Luzzatto, Gina.** Risultati di una escursione al Passo del Furlo e revisione di un Contributo alla flora vasculare dell'Urbinate pubblicato nel 1929 dalla Cengia-Sambo. Studi Urbinati **26:** 95–126. 1952. (Includes critical review of Cengia-Sambo's work. Not seen; cited from Arch. Bot. (Forlì) **29:** 345. 1953.)—Also her Secondo contributo di osservazioni sulla flora vascolare dei dintorni di Urbino. l.c. **29** (n.s., 4): 1–41. 1955. (Critical notes and additions. Not seen; cited from Arch. Bot. (Forlì) **33:** 213. 1956.)

——— Le piante medicinali dell'Urbinate. Bul. Soc. Bot. Ital. **1924:** 91–104. 1924.

List of 226 vascular plants, wild and cultivated, with localities and very brief indication of medicinal use. (Same area as her paper cited above.)

Chiosi, Rodolfo. Escursioni botaniche nella Valle del Marecchia. 39 p. 4 fig. 8vo. San Giovanni Valdarno, 1936.

Contains a list, with phytogeographic discussion, and maps showing distribution of some species in the Marchigian Appennines and neighboring mountains. (Not seen; title and annotation supplied mostly by V. Giacomini. In northwestern part of Prov. Pesaro e Urbino.)—See also his Cenni su alcune escursioni botaniche nella Valle del Marecchia. Nuovo Gior. Bot. Ital. n.s., **40:** 536–538. 1933. (Phytogeographic.)

Guidi, Luigi. Catalogo descrittivo dei prodotti della provincia di Pesaro e Urbino inviati all'Esposizione Internazionale di Londra del 1862. 1 p. l., 113 p. 22 cm. Pesaro, 1862. (John Crerar Library.)

Includes (p. 12–96) annotated list of timbers and wild and cultivated vegetable products.

——— Saggio d'una guida per la ricerca e classificazione dei prodotti naturali della provincia di Pesaro e Urbino . . . v. 1 (xlviii, 200 p.). pl. 22.5 cm. Pesaro, 1871.

Evidently issued in 2 parts; only fascicle 1 examined (200 p. n.p., n.d. In New York Botanical Garden library). It bears the cover title: Saggio d'una guida per la raccolta dei prodotti naturali della provincia di Pesaro e Urbino. Parte botanica, and is an annotated list of native and cultivated timbers, fibers, and dye plants (dicotyledons only). The other part, with roman pagination, includes directions for preparing an herbarium, etc. (Giacomini, in litt.). The plate cited in Bradley Bibliography has not been seen. The remaining volume, to cover acotyledons and monocotyledons, evidently was never issued.

Matteucci, Domenico. Il Monte Nerone e la sua flora. Bul. Soc. Bot. Ital. **1893:** 173–180, 244–256, 547–555. 1893.

Topography, geology, etc.; annotated list of vascular and cellular plants. (Prov. Pesaro-Urbino.)

Paolucci, Luigi. Primo elenco delle piante più caratteristiche dei Monti Sibillini: Vettore (m. 2477 s.m.), Priore (m. 2334 s.m.), Sibilla (m. 2213 s.m.). 2 p.l., 46 p. 25.5 cm. Ancona, 1879. (British Museum (Natural History) library.)

Annotated list of 123 angiosperms selected as characteristic of the flora, with localities and brief descriptions. (Mostly in Prov. Ascoli, overlapping Prov. Macerata and Prov. Perugia (the last in Umbria).)

Perri, P. L. Piante officinali della provincia di Macerata. Contributo allo studio della fitoterapia. Riv. Ital. Essen., Profumi e Piante Off. **29:** 432–438. illus. 1947; **30:** 45–49, 89–96, 135–141. illus. 1948.

Botanical regions, list of families with properties, list of plants classified by medicinal uses, list of medicinal plants (mostly wild) alphabetically arranged by vernacular names with habitat, properties, vernacular names, etc.; short bibliography.

Reali, Agostino. Gli alberi e gli arbusti del territorio Camerte; memoria sulla loro utilità e sui loro pregi . . . 3 pt. 536 p. 8vo. Camerino, 1871–76.

Not seen; cited from Rehder, Bradley Bibliography **3:** 88. 1915. (Prov. Macerata.)

PIEMONTE (PIEDMONT)

See also France, Alpes-Maritimes (Bertoloni, Burnat); Haute-Savoie (Payot, Florule du Mont-Blanc); Hautes-Alpes (Lebrun); Savoie (Menghini, Perrier de la Bathie, Peyronel, Vaccari and Pavarino); Switzerland, Ticino (Schröter, Schröter and Rikli); Valais (Favre, Tissière).

The regione of Piemonte includes the Provinces of Alessandria, Asti (formed in 1935), Cuneo, Novara, Torino, and Vercelli (formed in 1926 from the western part of Novara). The Province of Aosta, formed in 1927 from the northern part of the old Province of Torino, became in 1945 a separate regione (Val d'Aosta); at that time about one-third of its area reverted to Torino. Following the practice of Italian botanists, it is here included in Piemonte (as Prov. Aosta).

Owing to the large number of titles included, which is greater than for any other primary division of a country in either part 1 or part 2 of this work, it seems worth while to index the publications on the different provinces by authors, as follows:

Alessandria: Ferraro, (Gola), Morteo, F. Negri, G. Negri (on Crea), (Vignolo-Lutati).

Aosta: Beyer, Bolzon (on Dora Baltea, Ivrea), Briosi, Capra, Luzzatto, Mattirolo (on Gran Paradiso and Valsavaranche), Negri, Vaccaneo (on Lys), Vaccari (8 papers), Valbusa (2 papers).

Asti: Camisola.

Cuneo: Boggiani, Ferraris and Ferro, Gola (2), Ingegnatti, Lorenzo, Mattirolo (on Viso), Sappa (2), Sappa and Piovano, Vignolo-Lutati.

Novara: Ardissone, Armitage and Weiss, Biroli (2), Chiovenda, Chiovenda-Bensi (Piante), Craveri, Franzoni and de Notaris, Gabelli, Rossi (2).

Torino: Bolzon (on Dora Baltea and Ivrea), Charrier, Crosetti, Ferrari, Fontana, Keller, Lanza, Massia, Mattirolo (on Gran Pian and Palazzo Madama), Mussa (4), Negri (on Sabbioni and on Torino), Noelli (3), Pons, Re (2), Santi, Trinchieri, Vaccaneo (on Stupinigi), Vaccari and Wilczek.

Vercelli: Bertolani-Marchetti (2), Bertossi, (Bolzon, Studio), Cesati, Chiovenda-Bensi (Tradizioni), Ferraris (Florula), Negri (on Lucedio), Pellanda, Pomini.

General

Allioni, Carlo. Flora pedemontana sive enumeratio methodica stirpium indigenarum Pedemontii. 3 v. (344, 366 p.; v. 3, plates). 92 pl., front. 40 cm. Augustae Taurinorum, 1785.

List of vascular plants (2800 species, of which 2427 are vascular) with synonymy, localities, notes on uses, and occasional descriptions, arranged by a system of the author.—See also his Auctarium ad Floram pedemontanam cum notis et emendationibus. 2 p.l., 53 p. 2 pl. 26.5 cm. Augustae Taurinorum, 1789. (Additions.)

Colla, Luigi. Herbarium pedemontanum juxta methodum naturalem dispositum additis nonnullis stirpibus exoticis ad universos ejusdem methodi ordines exhibendos. 8 v. 97 (i.e. 98) pl. 22 cm. (v. 8, 27.5 cm.). Augustae Taurinorum, 1833–37.

Annotated flora of vascular and cellular plants with very brief diagnostic characters; index of 2,122 vernacular names (in v. 8). Includes many exotic species; the indigenous ones are distinguished by an asterisk.

Lissone, E. G., and Roberto, Lorenzo. Le piante a fusto legnoso indigene o naturalizzate nel Piemonte e comuni in quasi tutta la penisola e nelle isole. 178 p. 186 illus. 24 cm. Saluzzo, 1926.

Descriptions of 285 woody plants with vernacular names, character of wood, uses, and culture; relates primarily to Piemonte. (Not seen; title and annotation supplied by F. Sappa.)—See also Marchisio, below.

Marchisio, Marco. Vegetazione forestale in Piemonte . . . vii, 258 p. 24 cm. Torino, 1910. (John Crerar Library.)

Capitolo IV, Catalogo delle specie arboree ed arbustacee del Piemonte costituenti selva (p. 32–44), contains annotated lists of the woody plants of the Provinces of Alessandria and Torino, and list of widely distributed species, with Piemonte vernacular names.—See also Lissone and Roberto, above.

Mattirolo, Oreste. "Phytoalimurgia pedemontana" ossia censimento delle specie vegetali alimentari della flora spontanea del Piemonte. Annali R. Accad. Agr. Torino **61:** 107–286. 97 fig. 1919. (Reprinted with title I vegetali alimentari spontanei del Piemonte. (Phytoalimurgia pedemontana.) 2 p.l., 180 p. Torino, Firenze, 1919.)

Annotated account of edible wild plants, grouped according to the parts used, with Italian and Piedmontese vernacular names (indexed); bibliography, chronologically arranged and partly annotated.

Trèves, J. P. Petite flore médicale piémontaise. 411 p. 19 cm. Turin, 1904.

Directions for collecting and preparing medicinal plants, list of most common diseases with indication of plants useful for their treatment; annotated list of about 451 plants with description, habitat, and uses (p. 160–411). (Not seen; title and annotation supplied by F. Sappa.)

Zumaglini, A. M. Flora pedemontana, sive species plantarum phanerogamarum in Pedemonte et Liguria sponte nascentium . . . 2 v. (435, 444 p.). 18.5 cm. Augustae Taurinorum, 1849 (1859?); Bugellae, 1860.

Descriptive flora of vascular plants, with localities for rarer species. Dates of publication uncertain. Although the volumes are dated respectively 1849 and 1860 on their title pages, they may both have been published in 1860, which is the date on the paper cover of each (copy in Gray Herbarium library). In v. 1, p. 435, is the statement "Impressio absoluta . . . mense septembrii 1859", and in v. 2, p. 444, "Impressio absoluta . . . octobr. 1860." The copy in U. S. Department of Agriculture library has title page imprint Bugellae, 1864, in v. 2, and contains iv pages of Curae posteriores . . . ; it lacks the printed cover.—See also **Gola, Giuseppe.** Piante rare o critiche per la flora del Piemonte. Mem. R. Accad. Sci. Torino II, **60:** 193–248. pl. 1910 [1909 ?]. (List of 144 plants new to or rare in the province, with localities; list of recently found adventive plants, with data.)

Local

Ardissone, Francesco. Relazione di alcune escursioni botaniche intraprese a Macugnaga nell'alta Valle Anzasca durante i mesi di luglio, agosto e settembre 1901–1903. Rend. R. Ist. Lombardo Sci. e Let. II, **36:** 986–995. 1903.

General features of flora; unannotated list of vascular plants collected by author or recorded by others. (Prov. Novara.)—See also his Di alcune nuove escursioni botaniche nell'alta Valle Anzasca. Rend. R. Ist. Lombardo Sci. e Let. II, **38:** 483–497. (List of additional species collected; unannotated complete list of 530 species of the area.)—Also **Giacomini, Valerio.** Gli ultimi appunti floristici di Pietro Rossi (1871–1950) in quel di Macugnaga, in Valle Anzasca. Nuovo Gior. Bot. Ital. n.s., **59:** 176–180. 1952. (Unannotated list of vascular plants (about 330 species) collected by Rossi in vicinity of Pecceto.)

Armitage, Eduardo, and Weiss, Enrico. Catalogo nominale delle piante vascolari che crescono spontanee nei dintorni di Pallanza ed Intra. 28 p. 19 cm. Pallanza, 1891. (British Museum (Natural History) library.)

Geography, botanical regions; list of vascular plants [about 792 species], with indication of region and frequency. (Pallanza ed Intra now in comune of Verbania, Prov. Novara.)—See also **Calegari, Matteo.** Contributo alla flora della Zeda (m. 2157) (Lago Maggiore). Malpighia **20:** 219–220. 1906. (Additions and corrections from Pizzo Marona and Pizzo Zeda.)

Bertolani Marchetti, Daria. L'opera e le raccolte dell'Abate Carestia in Valsesia. (Ricerche sulla vegetazione della Valsesia. I.) Nuovo Gior. Bot. Ital. n.s., **61:** 515–578. 19 fig. i.e. 6 pl. (on 3), map. 1955.

Biographical sketch of Abate Antonio Carestia, topography, localities, etc.; list of over 1000 taxa, with localities. (Prov. Vercelli.)

——— La vegetazione della Val Sabbiola. (Ricerche sulla vegetazione della Valsesia II.) Nuovo Gior. Bot. Ital. n.s., **62:** 283–334. 9 fig. (fig. 1, map; fig. 2–8 on 2 pl.). 1956.

Includes (p. 290–310) annotated list of vascular plants (430 taxa) with indication of life form. (Prov. Vercelli.)

Bertossi, Felice. Appunti fitosociologici sulla Val Bagnola (Alta Valsesia). Arch. Bot. (Forlì) **33:** 194–217. 1957.

Includes unannotated list of about 400 taxa found by author. (Prov. Vercelli.)

Beyer, Rudolf. Beiträge zur flora der thäler Grisanche und Rhêmes in den Grajischen Alpen. 30 p. 25.5 cm. Berlin, 1891. (Wissenschaftlige beilage zum programm des Andreas-real-gymnasiums zu Berlin. Ostern 1891. Programm no. 93.) (New York Botanical Garden library.)

Botanical explorations, topography, etc.; list of vascular plants (582 species, including some hybrids) collected by author in 1889, with localities, followed by critical notes on 139 species and forms. (Prov. Aosta.)

Biroli, Giovanni. Flora aconiensis seu plantarum in novariensi provincia sponte nascentium descriptio. 2 v. (xxiv, 218; 260 p.). 20 cm. [Novara] 1808. (Arnold Arboretum library.)

Topography, geology; annotated list of vascular plants (on the Linnaean system), mosses, lichens, algae, and fungi of the Province of Novara, with brief diagnoses, and localities for scarcer species. Biroli's localities are unreliable (see Pampanini, Flora delle Alpi Lepontine occidentali **1:** 32–35. 1904). (Prov. Novara.)—See also **Cesati, Vincenzo.** Syllabus plantarum, quas in ditione novariensi lectas ad Floram aconiensem offert pro appendice prima. Linnaea **11:** 306–312. 1837. (List of vascular and a few cellular plants, with localities.)

——— Flora economica del dipartimento dell'Agogna. 114 p. 21 cm. Vercelli, 1805.

Annotated list of useful plants on the Linnaean system, with brief descriptions and uses; particular mention is made of plants employed for dyeing wool. (Prov. Novara.)

Boggiani, Oliviero. La flore delle Terme di Valdieri. *In* Marchisio, B. Guida alle Terme di Valdieri. p. 145–267. 8vo. Torino, 1898.

Annotated list of 683 vascular plants with localities, altitudinal limits, and some critical notes. (Prov. Cuneo.) (Not seen; cited mainly from Arch. Bot. (Forlì) **11:** 101. 1935.)

Bolzon, Pio. Ricerche botaniche nel bacino della Dora Baltea. Nuovo Gior. Bot. Ital. n.s., **25:** 309–376. 1918.

List of vascular plants, with localities, mostly based on collections of author, from the principal valleys and the lower part of the basin of the Dora Baltea. (Mostly in Prov. Aosta.)

——— Studio fitogeografico sull'antifeatro morenico d'Ivrea. Bul. Soc. Flore Valdôtaine 10, appendice. 117 p. 1915.

Author's itinerary, bibliography, geology, list of species new to Val d'Aosta, list of varieties observed, etc.; annotated list of known vascular plants (639 species), with localities. (Mostly in Prov. Aosta.)—See also his Nuovi materiali per la flora dell'antifeatro morenico d'Ivrea. l.c. **11:** 25–34. 1916. (List with ecological notes; contains list of 100 species not yet known from the region. Not seen; title and annotation supplied by F. Sappa.)

——— Sulla flora alveale della Dora Baltea, con appendice sulla flora alveale dei fiumi parmigiani. Bul. Soc. Flore Valdôtaine **12**: 1–37. 1917.

Includes some partial lists of the florules of the sands of the lower tract of the Piano di Combal with indication of altitudinal limits. (Prov. Aosta.) (Not seen; title and annotation supplied mostly by V. Giacomini.)

Briosi, Giovanni. Alcune erborizzazioni nella Valle di Gressoney. Atti Ist. Bot. R. Univ. Pavia II, **2**: 41–55. 1892.

List of cellular and vascular plants, with localities, collected by author in 1889. The Valle di Gressony (Val de Lys, Valle della Lesa) arises from the glaciers of the Lyskamm and Monte Rosa and enters the Aosta valley near Pont St. Martin. (Prov. Aosta.)—See also Capra, below, and Negri, below.

Camisola, Giuseppe. Flora astese secondo il sistema sessuale di Linneo con cenni sulla virtù di molte piante indigene impiegate in medicina non tanto sull'Astese che su altre provincie del Piemonte nascenti redatta ad uso del popolo. 488 p. 23 cm. Asti, 1854.

Annotated descriptive flora of vascular and some larger cellular plants of the Province of Asti, on the Linnaean system, with vernacular names and uses; section on vernacular names and edibility of fungi; indices of Italian and French vernacular names and names used by Mattioli, with botanical equivalents; similar list of officinal names; list of diseases with remedies, list of dye plants, synopsis of genera, index of botanical names. (Prov. Asti.)

Capra, Giuseppe. Studio botanico-agrario delle Alpi della Valle del Lys (Aosta). Mem. Pontif. Accad. Romana Nuovi Lincei **29**: 81–113. 1911.

Topography, climate, general features of vegetation; annotated tabular list of alpine plants (255 species, mostly forage plants) from Alpi Préal, Val Maronna, Val Bona, and Verhuil; second list, arranged by habitats. His Studio tecnico-economico di alcune Alpi della Valle del Lys. Annali R. Accad. Agr. Torino **53**: 585–715. 1911, includes (p. 696–711) a similar list of 234 species from the same localities. (Prov. Aosta.)—See also Briosi, above, and Negri, below.

Cesati, Vincenzo. Sguardo turistico sulla flora della regione biellese. 14 p. 8vo. Biella, 1882.

Botanical explorations, general features of flora; list of plants of different localities. (Prov. Vercelli.) (Not seen; title and annotation from notice in Nuovo Gior. Bot. Ital. **15**: 308. 1883.)—See also **Negri, Giovanni.** Infiltrazioni illiriche nella flora di un bosco messo-ingrofilo della Valle di Oropa (Monte Rosa). Boissiera **7**: 490–504. 1943. (Includes (p. 496–498) list of herbaceous vascular plants with indication of biological form.)—Also his Appunti informativi per una escursione della Società Botanica Italiana nella Valle di Oropa (Biella), 16–18 giugno 1949. Nuovo Gior. Bot. Ital. n.s., **56**: 452–469. 17 fig. (mostly on 4 pl.). 1949. (Plant zones, lists of species of various associations.)

Charrier, Giovanni. La flora del bacino del Chisola (Alpi Cozie). Arch. Bot. (Forlì) **29**: 40–57, 200–224, 297–307. 1 fig. (map). 1953; **30**: 65–76, 126–137. 1954; **31**: 178–184, 212–244. 1955.

Annotated list of vascular plants (916 taxa), with indication of life form, plant zone, and association. (Prov. Torino.)

Chiovenda, Emilio. Flora delle Alpi Lepontine occidentali ossia catalogo ragionato delle piante crescenti nelle vallate sulla destra del Lago Maggiore. Saggio di flora locale. I. Bibliografia. vi, 151 p. 22 cm. Roma, 1904–06. —Supplemento alla bibliografia. Nuovo Gior. Bot. Ital. n.s., **31**: 275–322. 1924.—II. Pteridophyta. iv, 69 p. 24.5 cm. Catania, 1929. (Arnold Arboretum library.)—III. Gymnospermae. x, 68 p. 27 fig. 25 cm. Modena, 1935.

(I) Alphabetical list of authors, with biographical notes and list of publications, partly annotated. (II) List of pteridophytes, with references and detailed localities. Is "Lavori eseguiti presso il R. Istituto botanico di Catania. II." (III) Annotated list of gymnosperms (10 species), with references, detailed localities, vernacular names, etc. Is "Lavori eseguiti presso l'Istituto botanico dell'Università di Modena. no. 7." (No. III not seen; title and annotation supplied by R. Pichi-Sermolli.) Covers "Valle dell' Ossola, il bacino del Lago d'Orta, le riviere occidentali del Lago Maggiore da Ascona a Sesto Calende, colle valli Intrasca, Intragna e Cannobina." (Prov. Novara, touching Prov. Varese and Canton Ticino (Switzerland).)

Chiovenda-Bensi, Caterina. Piante medicinali nell'uso tradizionale della Valle d'Ossola. Atti Accad. Ligure Sci. e Let. **11:** 32–52. 1955.
Includes annotated lists of native medicinal plants valley by valley, with patois names and uses; notes on various species, bibliography. In this and the following paper the specific identifications cannot always be trusted. (Prov. Novara.)—See also Craveri, below.

——— Tradizioni e usi fitoterapici popolari. La Valsesia. Atti Acad. Ligure Sci. e Let. **13:** 190–205. 1957.
Geography, etc.; annotated list of cellular and vascular plants with vernacular names and notes on medicinal uses; notes on particular species; bibliography. (Prov. Vercelli.)

Craveri, Michele. Piante medicinali ossolane delle antiche farmacopee. Malpighia **25:** 119–137. 1912.
Annotated alphabetical list of 50 plants used in medicine, mentioned in a military pharmaceutical codex of 1832; list of additional species in this codex not given in Rossi's flora (1883). (Prov. Novara.)—See also Chiovenda-Bensi (Piante medicinali), above.

Crosetti, Ettore, and **Fontana, Pietro.** Florula della Rocca di Cavour (Pinerolo). 13 p. 27 cm. Pinerolo, 1911. (Pubblicata per cura del Comizio agrario del circondario di Pinerolo.)
Topography, general features of flora; unannotated list of vascular plants (478 species and 50 varieties and forms). The Rocca di Cavour is a small rocky hill 3 km. in circumference and rising 162 meters above the surrounding plain. (Prov. Torino.)

Ferrari, Enrico. La vegetazione nel territorio di Leynì (Torino) nei rapporti colla coltura agraria. Annali R. Accad. Agr. Torino **55:** 459–515. 1913.
Plant formations, etc.; annotated list of vascular plants (750 species). (Prov. Torino.)

Ferraris, Teodoro. Florula crescentinese e delle colline del Monferrato. (Contribuzioni alla flora del Piemonte I.) Nuovo Gior. Bot. Ital. n.s., **7:** 371–396. 1900; **10:** 531–540. 1903.
Annotated lists totaling 446 vascular plants. (Prov. Vercelli.)

——— and **Ferro, G.** Materiali per una flora del circondario di Alba. Nuovo Gior. Bot. Ital. n.s., **11:** 3–33, 505–520. 1904; **13:** 51–58. 1906.
Geology, topography, etc.; annotated list of vascular plants (758 species and varieties). (Prov. Cuneo.)—See also **Vignolo-Lutati, Ferdinando.** Contributo alla flora del circondario di Alba (title varies). [Nota] 1–6. Nuovo Gior. Bot. Ital. n.s., **27:** 208–222. 1920; Bul. Soc. Bot. Ital. **1922:** 61–68. 1922; **1926:** 11–21. 1926; Nuovo Gior. Bot. Ital. n.s., **34:** 157–167, 1246–1257. 1927–28; **35:** 358–370. 1929. (Topography, etc.; list of additional plants, in part from a wider area.)—See also Vignolo-Lutati, below.

Ferraro, Giuseppe. Botanica popolare di Carpeneto d'Acqui. Arch. Stud. Tradiz. Pop. **4:** 129–137, 165–189, 405–420. 1885.
Annotated list of vascular plants, alphabetically arranged by Italian vernacular names, with scientific names, Carpeneto patois names, uses, popular beliefs, etc. (Prov. Alessandria.)

Fontana, Pietro. Catalogo sistematico delle piante vascolari crescenti nel bacino del Sangone (Alpi Cozie). *In* Studi sulla vegetazione nel Piemonte pubblicati a ricordo del II. centenario della fondazione dell'Orto botanico della R. Università di Torino, 1729–1929. p. 301–360. 25.5 cm. Torino, 1929.
Topography, geology, botanical explorations (by G. Negri); briefly annotated list of vascular plants (1328 taxa), with localities. (Prov. Torino.) —See also **Sappa, Francesco,** and **Charrier, Giovanni.** Aggiunte al "Catalogo . . ." Arch. Bot. (Forlì) **24:** 55–58. 1948. (List of additional species, with localities, bringing total number of taxonomic entities to 1361.)—Also their Saggio sulla vegetazione della Val Sangone (Alpi Cozie). Nuovo Gior. Bot. Ital. n.s. **56:** 106–187. 1 fig., map. 1949. (Geography, topography, geology, climate, plant associations, bibliography; English summary.)—Also **Charrier,**

G. Seconde [-4.] aggiunte al "Catalogo sistematico delle piante vascolari crescenti nel bacino del Sangone (Alpi Cozie)" di Pietro Fontana. l.c. **56:** 476–477. 1949; **57:** 688–690. 1 fig. 1951; **58:** 634–635. 1 fig. 1952.

Franzoni, Alberto, and **Notaris, Giuseppe de.** Elenco delle piante del Margozzolo e sue adiacenze, compilato su note fornite dall'Avv. Alberto Franzoni di Locarno . . . e del Senatore Prof. Giuseppe de Notaris. *In* De-Vit, Vincenzo. Il Lago Maggiore, Stresa e le Isole Borromee. Notizie storiche. v. 1, Appendice. p. 517–530. 24.5 cm. Prato, 1875. (University of Illinois library.)

Unannotated list of spermatophytes. Relates to the area between Lago d'Orta and Lago Maggiore, essentially Monte Mottarone (Montarone) and its region. (Prov. Novara.)—See also **Nobili, Giuseppe.** Note sulla flora del Monte Mottarone. Nuovo Gior. Bot. Ital. n.s., **2:** 102–108. 1895. (Geology, topography, botanical explorations; annotated list of vascular plants. Incomplete; 100 species, Ranunculaceae–Aceraceae.)

Gabelli, Lucio. Alcune notizie sulla vegetazione ruderale delle adiacenze del Lago Maggiore. Mem. Pontif. Accad. Romana Nuovi Lincei **32:** 235–271. 1914.

General considerations; partly annotated list of 198 ruderal vascular plants, with lists of species grouped by habitat; urban flora of Arona (133 vascular plants); various shorter lists of species, conclusions. (Prov. Novara, also Prov. Varese (in Lombardia).)

Gola, Giuseppe. Le piante vascolari della Val Maira (Alpi Cozie). Atti R. Ist. Veneto Sci., Let. ed Arti **92** (2): 1283–1335. 1933; **93** (2): 527–566. 1934.

Table of altitudes, etc.; list of vascular plants, with localities. (Prov. Cuneo.)—See also his I caratteri della vegetazione della Valle Maira (Alpe Cozie). *In* Studi sulla vegetazione nel Piemonte pubblicati a ricordo del II. centenario della fondazione dell'Orto botanico della R. Università di Torino, 1729–1929. p. 1–50. 1929. (Topography, geology, climate, general account of vegetation of different altitudes and habitats.)

——— La vegetazione dell'Appennino piemontese. Annali Bot. **10:** 189–338. 1912.

Geography, topography, climate, geology, ecology; list of vascular plants, with localities. (In Provinces of Cuneo and Alessandria (Piemonte) and Savona (Liguria).)

Ingegnatti, Annibale. Catalogo delle principali specie vegetali che crescono spontanee nel circondario di Mondovì. 76 p. 27.5 cm. Mondovì-Breo, 1877. (Memoria desunta dalla Cronaca del R. Liceo G. B. Beccaria per l'anno scolastico 1875–76.)

Annotated alphabetical list of vascular and cellular plants. (Prov. Cuneo.)

Keller, Robert. Vegetationsskizzen aus den Grajischen Alpen. 152 p. 22 cm. Winterthur, 1904. (Wissenschaftliche beilage zum Programm des Gymnasiums und der industrieschule Winterthur. Schuljahr 1904/05.)

Annotated list of plants of the Bardonecchia basin (p. 7–44), followed (p. 45–149) by general features of vegetation. (Prov. Torino.) (Not seen; title and annotation supplied by F. Sappa.)

Lanza, Miranda. La flora della Valle di Valprato. Bul. Soc. Flore Valdôtaine 14, appendix. 64 p. 1920.

Geology, climate, general features of flora; annotated list of vascular plants; list of plants characteristic of different soils, bibliography. (Prov. Torino.) (Not seen; cited from Revista Club Alp. Ital. **40:** 49. 1921, in Harvard College library.)

Lorenzo, Roberto. Le piante a fusto legnoso indigene o naturalizatte della provincia di Cuneo. 71 p. 4to. Alba, 1889.

Not seen; no review available.

Luzzatto, Gina. La vegetazione della Touriasse (2440 m.) nella regione del Piccolo S. Bernardo (Alpi Graie). Chanousia **2:** 77–113. 1 fig. 1932. (Brooklyn Botanic Garden library.)

Climate, general feature of flora, plant associations, biological spectrum; annotated list of vascular plants (309 species and infraspecific forms); critical notes on various species, bibliography. (Prov. Aosta.)

Massia, Pietro. Nomi locali canavesani da nomi di piante. Malpighia **27**: 101–118. 1915.
Bibliography; annotated alphabetical list of Latin or latinized names of plants, with geographical names derived from them. (Prov. Torino.)

Mattirolo, Oreste. Bibliografia botanica del Parco Nazionale del Gran Paradiso. *In* Il Parco Nazionale del Gran Paradiso. v. 2, p. 55–59. 1928. (Reprinted, [5] p.; only the reprint seen, in library of Joseph Ewan.)
Unannotated, chronologically arranged bibliography. (Prov. Aosta.)

——— Un'escursione botanica nel gruppo del Viso. Bol. Club Alp. Ital. **21**: 172–179. 1888.
Bibliography; unannotated lists of vascular plants collected by author at several localities in July 1887. (Prov. Cuneo.)—See also **Gandoger, Michel.** Voyage botanique au mont Viso. Bul. Soc. Bot. France **36**: 437–446. 1889. (Lists of species collected at various altitudes up to 2700 meters.)—Also **Petitmengin, Marcel.** Considérations botaniques sur le massif du mont Viso. Bul. Soc. Sci. Nancy III, **6**: 100–128. 3 pl. 1906. (Short account of topography, geology, general features of flora; annotated list of some rare or characteristic species, with localities.)

——— Spigolature botaniche nella regione del "Gran Pian" sopra Noasca. *In* Il Parco Nazionale del Gran Paradiso. v. 3, p. 65–79. 27.5 cm. Torino, 1932.
Alphabetical lists of vascular plants from several localities. (Prov. Torino.)

——— Spigolature botaniche nella Valsavaranche. *In* Il Parco Nazionale del Gran Paradiso. v. 2, p. 39–54. 6 fig. 1928. (Reprinted [16] p.; only the reprint seen, in library of Joseph Ewan.)
Alphabetical lists of vascular plants from 9 localities; list of fungi. (Prov. Aosta.)

——— La vegetazione del fossato di Palazzo Madama in Torino. Mem. Mat. e Sci. Fis. e Nat. Soc. Ital. Sci. III, **22**: 157–182. 4 fig. 1924.
History, etc.; annotated list of 141 vascular plants and mosses found in the ditch of the Palazzo Madama in the center of Torino from 1886 to 1922. (Prov. Torino.)—See also Noelli, Flora ruderale torinese, below.

Morteo, E. Florula alluvionale di un tratto del torrente Orba negli anni 1904, 1905, 1906. Malpighia **20**: 487–511. 1906.
Factors of dissemination, soil preferences, etc.; annotated list of vascular plants found by author along the Orba River from Predosa to the river Bormida. (Prov. Alessandria, also Prov. Savona in Liguria.)

Mussa, Enrico. Appunti sulla flora cacuminale del Rocciamelone. Malpighia **23**: 167–170. 1909.
Plant zones; list of 26 vascular plants found at 2400–3450 meters altitude. (A mountain 3537 meters high, the highest in the Val di Susa, Prov. Torino.)

——— Note floristiche delle Prealpi torinesi fra la Dora Riparia e la Stura di Lanzo (zona delle pietre verdi). Atti Soc. Ital. Sci. Nat. Milano **47**: 139–157. 1908.
Topography, general features of flora, geology, etc.; list of vascular plants of Monte Musinè, with localities. (Prov. Torino.)—See also his Sguardo alla vegetazione del M. Musinè (Valle di Susa). Nuovo Gior. Bot. Ital. n.s., **44**: 715–730. 1938. (Geology, general features of flora with lists of species of different localities and habitats, lists of woody and of adventive plants, notes on phenology.)—Also his Florula del M. Musinè (Valle di Susa). l.c. **47**: 705–714. 1941. (List of known vascular plants, with indication of local range and habitat.)—Also **Piovano, Giovanni.** Spigolature di floristica pedemontana. I. l.c. **58**: 627–629. 1952. (Includes 3 additions.)

——— Saggio d'un censimento floristico dell'agro torinese. Nuovo Gior. Bot. Ital. n.s., **38**: 605–623. 1931.
Botanical explorations; mostly unannotated list of known vascular plants from area "fra la Chisola e l'Orco, le montagne de San Giorgio di Piossasco a Castellamonte e la collina di Torino da Moncalieri a Chivasso." (Prov. Torino.)—See also his Calendario floristico torinese. Rass. Mens. Municip. Torino **12** (9): 45–53. 1932. (Popular account of the seasonal succession, especially the flowering, of the phanerogams in the local flora of Torino;

statistics. Not seen; cited from Biol. Abst. 8: 29. 1934.)—Also Re, Flora torinese, below.

——— Vegetazione e florula del Monte Pirchiriano ("Sagra di San Michele," Val di Susa). Nuovo Gior. Bot. Ital. n.s., **46**: 654–665. 1939.
Topography, geology, general features of flora; list of vascular plants. (Prov. Torino.)

Negri, Francesco. Elenco delle piante più notevoli del Monte di Crea e regioni vicine. *In* Onorato, Corrado. Notizie storiche del Santuario di Nostra Signora di Crea . . . p. 418–446. 24.5 cm. Casale, 1889.
Briefly annotated list of vascular plants. (In Casale Monferrato, Prov. Alessandria.)—See also Negri, G., La vegetazione delle colline di Crea, below.

Negri, Giovanni. La vegetazione dei "sabbioni" dell'alta pianura padana. *In* Studi sulla vegetazione nel Piemonte pubblicati a ricordo del II. centenario della fondazione dell'Orto botanico della R. Università di Torino, 1729–1929. p. 621–673. 25 cm. Torino, 1929.
Ecological; includes (p. 630–636) annotated tabular list of 295 vascular plants of various sandy areas. (Prov. Pavia (Lombardia) and Prov. Torino.)—See also **Tosco, Uberto.** Decadimento floristico dei "sabbioni" di Grugliasco (Torino). Nuovo Gior. Bot. Ital. n.s. **58**: 27–59. 1 fig. (map), 2 pl. 1951. (Includes list (p. 38–45) of plants, not in systematic order.)

——— La vegetazione del bosco Lucedio (Trino Vercellese). Contributo allo studio fitogeografico dell'alta pianura padana. Mem. R. Accad. Sci. Torino II, **62** (Sci. Fis., Mat. e Nat.): 387–448. 1912.
Ecological and phytogeographic; includes (p. 415–423) list of 428 vascular plants; also (p. 445–448) unannotated list of vascular plants of alluvial woods along the Po between Crescentino and Trino, and similar list from the heath between Arborio and Gattinara; bibliographical footnotes. (Prov. Vercelli.)

——— La vegetazione della collina di Torino. Mem. R. Accad. Sci. Torino II, **55** (Sci. Fis., Mat. e Nat.): 113–188. 1905.
Mainly ecological; includes (p. 157–188) list of native and naturalized plants; bibliographical footnotes. Relates to the hills southeast of the city, from Moncalieri to Lavriano. (Prov. Torino.)

——— La vegetazione delle colline di Crea. Mem. R. Accad. Sci. Torino II, **56** (Sci. Fis., Mat. e Nat.): 387–437. 1906.
Ecological and phytogeographic; includes (p. 412–437) list of native and naturalized vascular plants; bibliographical footnotes. (Prov. Alessandria.) —See also Negri, F., above.

Noelli, Alberto. La flora dei massi erratici dell'anfiteatro morenico di Rivoli. Nuovo Gior. Bot. Ital. n.s., **33**: 208–218. 1926.
Topography, general features of flora; list of about 300 vascular plants, grouped according to frequency of occurrence; bibliographical footnotes. (Prov. Torino.)

——— Flora ruderale torinese. Nuovo Gior. Bot. Ital. n.s., **20**: 546–558. 1913.
Bibliography of works on ruderal, especially urban, floras; classification of ruderal plants; annotated tabular list of vascular plants of the city limits of Torino [about 272 species]. (Prov. Torino.)—See also Mattirolo, La vegetazione del fossato di Palazzo Madama in Torino, above.

——— La vegetazione del terrazzo diluviale di Rondissone (Torino). Nuovo Gior. Bot. Ital. n.s., **20**: 531–545. 1913.
Effects of cultivation, etc.; tabular list of vascular plants. (Prov. Torino.)

Pellanda, Giuseppe. La flora estiva dei Monti d'Oropa. vii, 682 p. illus. 19.5 cm. Biella, 1904. (John Crerar Library.)
Table of altitudes, glossary, etc.; annotated descriptive flora of vascular plants, with Italian and Piedmontese vernacular names, range, properties and uses, and notes on propagation and cultivation; key to families. Relates primarily to plants flowering or fruiting from June to September. Cover dated 1906, but 1904 is actual date of issue (fide F. Sappa). (Prov. Vercelli.)

Pomini, Luigi. Le piante officinali e del sottobosco spontanee o coltivate della provincia di Vercelli . . . 388 p. illus. 16.5 cm. Vercelli [1937].

Briefly annotated list consisting mostly of medicinal plants, arranged alphabetically by vernacular names, with habitat, uses, etc.; list classified by uses.

Pons, Giovanni. Primo contributo alla flora popolare valdese. Bul. Soc. Bot. Ital. **1900:** 101–108. 1900.—Flora popolare valdese. Secondo contributo. l.c. **1900:** 216–222. 1900.

General considerations; list of plants, with vernacular names (from the Val S. Martino) and uses. Unfinished; covers Ranunculaceae–Pomaceae. (Prov. Torino.)

Re, G. F. La flora segusina . . . riprodotta nel metodo naturale di De Candolle e comentata da Beniamino Caso . . . xxx, 406 p. 19 cm. Torino, 1881. (Arnold Arboretum library.)

Annotated list of vascular plants [1682 species] of the old province of Susa (anciently Segusa) based on Re's Flora segusiensis of 1805, with vernacular names, uses, and localities for scarcer species. (Prov. Torino.)—See also **Caso, Beniamino.** Aggiunte e correzione alla Flora segusina . . . 15 p. 19 cm. Torino, 1882. (Arnold Arboretum library.)—Also **Mattirolo, Oreste.** La flora segusina dopo gli studii di G. F. Re (Flora segusiensis, 1805—Flora segusina, Re-Caso, 1881–82). Saggio storico-bibliografico-botanico. Mem. R. Accad. Sci. Torino II, 58 (Cl. Sci. Fis., Mat. e Nat.): 217–300. 1908. (List of botanists, with biographical information and titles of publications, statistics, bibliography; list of 695 vascular plants, with localities, bringing the total for the region to 2213 species and 279 varieties. Preprinted 1907.)—Also **Beauverd, Gustave.** Sur la flore vernale de la vallée de Suse (Piémont). Bul. Soc. Bot. Genève II, **6:** 159–162. 1914. (Running account of noteworthy species, including new records.)—Also his Sur la flore vasculaire des environs de Mondane, de Bardonnèche et de Suze. l.c. II, **13:** 115–184. 1922. (Includes lists of species from several localities and bibliographies.)

——— Flora torinese. 2 v. (372, 340 p.; v. 2 in 2 fasc., these paged cont.). 19.5 cm. Torino, 1825–27. (Gray Herbarium library.)

Annotated descriptive flora, on the Linnaean system, with localities and some vernacular names. Covers a radius of 5 Piedmont miles (about 10 kilometers) around Torino. (Prov. Torino.) Replaces **Balbis, G. B.** Flora taurinensis, sive enumeratio plantarum circa taurinensem urbem nascentium. xvi, 224 p. 18 cm. Taurini, 1806. (List of vascular and cellular plants, on the Linnaean system, with brief descriptions and annotations, covering region within a radius of 1.5 kilometers around Torino.)—See also **Mussa, Enrico.** La flora dell'agro torinese dopo i lavori di G. Balbis e de G. F. Re e considerazioni sopra l'indigenato di talune specie. Mem. R. Accad. Sci. Torino II, v. **65** (Cl. Sci. Fis., Mat. e Nat.), no. 2. 41 p. 1916. (Review of some recent publications, changes in natural conditions, annotated bibliography; annotated list of 320 vascular plants (mostly exotic) additional to Re's Flora.)—Also Mussa, Saggio d'un censimento floristico dell'agro torinese, above.

Rossi, Stefano. Flora del Monte Calvario. 15 p. 27.5 cm. Domodossola, 1883. (On cover, Estratto dalla Cronaca della Fondazione Galletti—anno 1883.) (Arnold Arboretum library.)

General features of flora; unannotated list of vascular plants [725 species]. (Prov. Novara.)

——— Studi sulla flora ossolana. 112 p. 26 cm. Domodossola, 1883.

List of wild and cultivated vascular plants (1675 species), with localities. Covers the valleys of the Toce (Formazza), Anzasca, and Vedro in the vicinity of Domodossola. (Prov. Novara.) (Not seen; title and annotation supplied by F. Sappa.) Chiovenda (Flora delle Alpi Lepontine occidentali **1:** 73. 1904) intimates that there are rather numerous errors.—See also **Rossi, S.** Nuove piante trovate in Val d'Ossola. Mem. Pontif. Accad. Nuovi Lincei **6:** 63–66. pl. 1–4 (i.e., 2 fold. pl.). 1890. (Descriptions and illustrations of 4 angiosperms.)—Also his Alcune forme vegetali e varietà nuove raccolte nella Valle Ossolana. l.c. **7:** 81–88. pl. 1–4 (i.e., 2 fold. pl.). 1891. (Descriptions of 4 minor forms or varieties.)—Also **Craveri, Michele.** Alcune specie vegetali dell'Ossola ritenute rare per la flora del Piemonte. Malpighia 24: 163–172. 1911. (List of 25 noteworthy species.)—Also Schröter and Rikli, under Switzerland, Ticino.

Santi, Flavio. Appunti sulla flora delle valli di Lanzo. *In* Club Alpino Italiano. Le valli di Lanzo. p. 475–490. 27 cm. Torino, 1904.

Includes brief lists of the more interesting species of the valleys of Lanzo. (Prov. Torino.) (Not seen; title and annotation supplied by F. Sappa.)

Sappa, Francesco. Ricerche sulla vegetazione boschiva a Pinus silvestris L. nella regione di Canale e considerazioni ecologiche ed agricole di tali formazioni. Annali R. Accad. Agr. Torino **82:** 61–145. 1940.

Includes (p. 123–132) tabular list of vascular plants, with indication of localities; bibliography (p. 143–145). (Region of Canale d'Alba "tra i colli e la pianura torinese, il Monferrato propriamente detto e la Langhe," in Prov. Cuneo.)

——— Un settore del fronte occidentale italiano (Alpi Marittime) nel suo paesaggio botanico. Annali R. Accad. Agr. Torino **85:** 1–50. 1 fig. (map). 1942.

Mainly ecological and phytogeographical; includes (p. 6–15) tabular list of vascular plants from the crest between the summit of Belletz and Monte Girauda, with indication of life-form and habitat. (Prov. Cuneo.)

——— and **Piovano, Giovanni.** La Val Pesio e la sua vegetazione (Alpi Marittime). [pt. 2.] La flora. Webbia **7:** 353–458. 1 fig. (map). 1950.

Topography, geology, climate; annotated list of vascular plants (1347 taxa) with biological form, plant zone, localities, altitudinal range; bibliography. (Prov. Cuneo.)—See also their La Valle Pesio e la sua vegetazione (Alpi Marittime). pt. 1. Lav. Bot. Ist. Bot. Univ. Torino **8:** 201–222. 1947. (Botanical explorations, topography, geology, climate, bibliography.)

Trinchieri, Giulio. Osservazioni su la flora spontanea e avventizia dell' Orto botanico di Torino. Malpighia **19:** 3–44. 1905.

Annotated list of vascular plants (261 species); statistics, habitats, etc.; bibliography. (Prov. Torino.)

Vaccaneo, Roberto. Ricerche sulla vegetazione dei boschi di Stupinigi. *In* Studi sulla vegetazione nel Piemonte pubblicati a ricordo del II. centenario della fondazione dell'Orto botanico della R. Università di Torino, 1729–1929. p. 361–408. map. 25.5 cm. Torino, 1929.

Ecological; includes (p. 397–405) list of vascular plants (547 species and varieties); bibliography. (Prov. Torino.)—See also **Tosco, Uberto.** *Gentiana Pneumonanthe* L. ed altri nuovi reperti per i boschi di Stupinigi (Torino). Nuovo Gior. Bot. Ital. n.s., **59:** 568–571. fig. 1 (map). 1953. (Includes several additions.)

——— La vegetazione delle morene del ghiacciaio del Lys (Monte Rosa). Bol. Com. Glaciologico Ital. **14:** 105–172. 1934.

Phytogeographic; includes list of 274 species. (Prov. Aosta.) (Not seen; title and annotation supplied by L. Fenaroli.)—See also Briosi and Capra, above, and Vaccari (Contributo), below.

Vaccari, Lino. Catalogue raisonné des plantes vasculaires de la vallée d'Aoste. v. 1 (viii, 635 p.). 24.5 cm. Aoste, 1904–11. (Arnold Arboretum library.)

Botanical regions, etc.; annotated list of species, with detailed localities, citation of collectors, and critical notes. Vol. 1, Thalamiflores et Calyciflores (Ranunculaceae to Anacardiaceae); no more published (?). At head of title page: Société de la flore valdôtaine. Aoste (Italie). The records are given under 5 botanical regions; (1) the lower part of the valley of the Dora Baltea (Doire) from the morainic amphitheater of Ivrea up to Pré-St.-Didier; (2) the eastern Graian Alps from Ivrea to the massif of Rutor; (3) the region from the Rutor massif to Mont Blanc; (4) the western Pennine valleys; (5) the central Pennine valleys. Bolzon (Bul. Soc. Flore Valdôtaine 10, app. p. 7. 1915) states that the first pages of vol. 2 were published, but the writer has not been able to see them. (Prov. Aosta.)

——— Un coin ignoré dans les Alpes Graies. Herborisations dans la vallée de Champorcher (vallée d'Aoste). Bul. Soc. Flore Valdôtaine **1:** 31–52. 1902.

General considerations, bibliography; unannotated lists of vascular plants of different localities and altitudes. (Prov. Aosta.)—See also his La flore de la serpentine, du calcaire et du gneiss dans les Alpes Graies orientales. Notes de géographie botanique. Herborisations faites les jours 11–18 août 1902. Bul. Soc. Flore Valdôtaine **2**: 52–75. 1903. (Account of various localities, with partial lists of species. Relates to the area between the Dora Baltea, the valley of Fénis, and Mont Mars.)—Also **Guyot, Henry.** Contribution sur la phytogéographie des Alpes Graies orientales. l.c. **18**: 42–58. 1925. (Contains list of noteworthy plants found on an excursion in the Champorcher and Soana valleys. Not seen; title and annotation mostly from Bot. Jahresber. Just **54** (1): 628. 1931.)—Also **Vaccari, L.** Le escursioni botaniche in Val d'Aosta compiute dalla Società Botanica Italiana in occasione del 40. anniversario della "Chanousia," giardino mauriziano alpino al Piccolo S. Bernardo (25–27 agosto 1937). Nouvo Gior. Bot. Ital. n.s., **44**: 568–580. 1937. (Phytogeography; itineraries with mention of noteworthy species (from Piccolo San Bernardo, Courmayeur, and Cogne); alphabetical list of the more noteworthy alpine species of the area.

——— Complément à l'exploration floristique du Val d'Ollomont. Bul. Murithienne **32**: 61–71. 1903.

Geology; list of vascular plants observed in July 1902 at various localities and altitudes up to 3448 meters. (Prov. Aosta.)—See also **Guyot, Henry.** Contribution phytogéographique sur le versant méridional des Alpes Pennines. Bul. Soc. Bot. Genève II, **13**: 185–216. 1921. (Consists mostly of lists of plants from various localities and altitudes in the "haute vallée d'Ollomont" and the "prairies des environs des Chalets de By;" bibliography.)—Also his Ueber einige seltene pflanzen der südseite der Penninischen Alpen. Verhandl. Naturf. Gesell. Basel **56** (2): 111–118. 1945. (List of some noteworthy species from Ollomont, Aosta Valley, etc.)

——— Contributo alla flora del "Naso." Chanousia **4**: 34–57. 1940.

Lists of species (total 59) collected by several botanists on the Naso, an island in the Lys Glacier of the Monte Rosa range. (Prov. Aosta.) (Not seen; title and annotation supplied by A. Becherer.)

——— Flora cacuminale delle valle d'Aosta. Nuovo Gior. Bot. Ital. n.s., **8**: 416–439, 527–542. 1901.

Description of localities; tabular lists of vascular plants (202 species and 16 varieties) found at 2600–3142 meters and at 3200–4200 meters, showing localities; discussion, statistics. (Prov. Aosta.)

——— La flora delle Colle di Gigante. Riv. Mens. Club Alp. Ital. **18**: 347–349. 1899.

Notes on the flora of several localities; lists of vascular plants observed at various elevations from 2800 to 3350 meters. (Prov. Aosta.)

——— Flora nivale del Monte Rosa. Bul. Soc. Flore Valdôtaine **7**: 17–79. 1911. (Republished with some additions in Atti dei Laboratori scientifici A. Mosso **3**: 173–206. 1912; this not seen.)

Botanical explorations (with bibliography); tabular list of vascular plants (225 species) known from 2676 to 3200 meters altitude, with localities and altitudinal range, also cellular plants; general features of flora, phytogeography. (Prov. Aosta.)—See also **Fenaroli, Luigi.** Risultati botanici di un soggiorno al Monte Rosa. Bul. Soc. Bot. Ital. **1926**: 6–11. 1926. (List with localities.)—Also **Luzzatto, Gina.** L'estremo limite altitudinale di alcune specie di piante sul Monte Rosa: nuovi reperti. Nuovo Gior. Bot. Ital. n.s., **42**: 674–677. 1935. (List with discussion.)—Also her Flora subnivale al Corno del Camoscio sopra il Col d'Olen. Atti Soc. Ital. Sci. Nat. Milano **76**: 70–87. 1937. (Botanical explorations, etc.; list of 194 species and varieties of vascular plants known from the area (at 2800 meters and above), with localities and references; supplementary list of 14 species from the Col d'Olen at other localities.)

——— La vegetazione della Grivola (3969 m.) nel gruppo del Gran Paradiso. Riv. Mens. Club Alp. Ital. **25**: 212–217. 1906. (Harvard College library.)

Includes lists of vascular and cellular plants collected by author in August 1904 at various altitudes from 3400 to 3969 meters; also list of 25 vascular plants, showing extreme altitudes reached on Grivola and in other localities. (Prov. Aosta.)

——— and **Wilczek, Ernest.** La vegetazione del versante meridionale delle Alpe Graie orientali (Valchiusella, Val Campiglia e Val di Ceresole). 1. contribuzione: In Valchiusella e Val Campiglia. Nuovo Gior. Bot. Ital. n.s., **16:** 179–231. 1909.

Botanical explorations (with bibliographic references); partial lists of vascular plants from different localities; climate, general features of flora, altitudinal limits, phytogeography. (Prov. Torino.)

Valbusa, Ubaldo. La florula di due isole glaciali del Rutor con appendice morenica epiglaciale. Nuovo Gior. Bot. Ital. n.s., **44:** 705–714. 1938.

Description of the Rutor glacier, etc.; annotated list of 20 phanerogams observed at 2776–2805 meters elevation. (Prov. Aosta.)

——— L'isola glaciale nella florula della "Brenva" e del "M. Bianco." *In* Studi sulla vegetazione nel Piemonte pubblicati a ricordo del II. centenario della fondazione dell'Orto botanico della R. Università di Torino, 1729–1929. p. 697–711. 25.5 cm. Torino, 1929.

Topography, glacial conditions, etc.; annotated list of 19 angiosperms observed at 3000 to 3300 meters elevation in the Brenva Glacier, on the east slope of Mont Blanc. (Prov. Aosta.)

Vignolo-Lutati, Ferdinando. Le Langhe e la loro vegetazione. *In* Studi sulla vegetazione nel Piemonte pubblicati a ricordo del II. centenario della fondazione dell'Orto botanico della R. Università di Torino, 1729–1929. p. 95–285. map. 25.5 cm. Torino, 1929.

Botanical explorations, bibliography, topography, geology, climate, general features of flora, table of altitudes; annotated list of vascular plants (1555 taxa including 1,230 species). The area covered, taken in the original list to include only the country between the rivers Tanaro and Bormida di Millesimo, was extended eastward in the appendices to the river Bormida di Spingo. (Mostly in Prov. Cuneo, extending also into Prov. Alessandria (Piemonte) and Prov. Savona (Liguria).)—See also his Le Langhe . . . Aggiunte e considerazioni sui loro confini orientali. Nuovo Gior. Bot. Ital. n.s., **39:** 523–537. 1932.—Le Langhe . . . Appendice 2. [–3.]. l.c. **44:** 652–666. 1937; **51:** 30–38. 1944. (The third appendix brings the total to 1,647 taxa, including 1,296 species); Appendice 4. Lavori Bot. Ist. Bot. Univ. Torino **8:** 267–275. 1947. (Brings total to 1,663 taxa. Not seen; cited from Biol. Abst. **22:** 1632. 1948.)

PUGLIE (APULIA)

See also Campania (Tenore, 2).

The regione of Puglie or Puglia includes the Provinces of Bari, Brindisi (constituted in 1927 from part of the Province of Lecce), Foggia, Lecce (formerly, long before its subdivision, called Terra d'Otranto), and Taranto (formed in 1923 from part of the Province of Lecce, and at first called Ionio).

General

Amico, Antonio. Cenni storici sulla attività botanica nelle Puglie. Nuovo Gior. Bot. Ital. n.s, **56:** 266–279. 1949.

Brief account of botanical work, and list of local botanists with some biographical data.

Local

Baselice, Luigi. Flora biccarese. Botaniche peregrinazioni nell'agro biccarese per la primevera del 1841. 67 p. 8vo. Campobasso, 1842.

Not seen; no review available. (Biccari, in Prov. Foggia.)—See also **Villani, Armando.** L'erbario di Luigi Baselice. Bul. Ort. Bot. R. Univ. Napoli **3:** 227–231. 1913. (Note on his herbarium, preserved at Biccari.)—Also his Le piante di Biccari conservate nell'erbario Baselice nell'erbario Ziccardi. Nuovo Gior. Bot. Ital. n.s., **20:** 394–416. 1913. (Includes list of plants of Biccari in the Ziccardi herbarium, and list of plants of Biccari in the Baselice herbarium not recorded in Baselice's publication.)

Béguinot, Augusto. La vegetazione delle Isole Tremiti e dell'isola di Pelagosa. Studio fitogeografico. Mem. Mat. e Fis. Soc. Ital. Sci. III, **16:** 155–220. map. 1910.

Botanical explorations (with references), topography, etc., of each island, ecology, origin of flora, etc.; tabular list of vascular plants (444 species), showing distribution by islands. The Tremiti Islands (in Prov. Foggia) include Capperaia (Caprara), Cretaccio, Pianosa, San Domino, and San Nicola. The Pelagosa (Pelagruža) Islands, consisting of Pelagosa Grande and Pelagosa Piccola, are about 45 kilometers northeast of Pianosa; they now belong to Yugoslavia.—See also **Cortesi, Fabrizio.** Nuova contribuzione alla flora delle Isole Tremiti. Annali Bot. **8**: 239–240. 1910.—Also **Pampanini, Renato.** Una collezione botanica delle Isole Tremiti. Bul. Soc. Bot. Ital. **1916**: 29–35. 1916. (List of plants collected by A. Gurgo in 1886, including new records.)—For later works on Pelagosa Islands, see **Baldacci, Antonio.** La flora delle isole Pelagose. Mem. R. Accad. Sci. Ist. Bologna VI, **8**: 53–61. 1911. (Botanical explorations, topography, general features of flora; list of 160 vascular plants.)—Also **Hirc, Dragutin.** Florula Palagruških otoka. [Florula of the island of Pelagosa grande.] Glasnik Hrvatsk. Prirod. Društva **23** (1/2): 86–103. 1911. (Geography, geology, climate, botanical explorations, general features of flora, etc.; partly annotated list of angiosperms.)—Also **Béguinot, A.** Recenti contributi alla flora ed alla ecologia dell'isola di Pelagosa. Bul. Soc. Bot. Ital. **1911**: 242–249. 1911. (Review of 4 papers by Baldacci, Ginzberger, and Hirc published in 1911.)

Bisceglia, Vito. Sulla flora della provincia di Bari. Atti R. Ist. Incoragg. Sci. Nat. Napoli **1**: 63–103. 1811.

Account of useful plants (those containing soda, dye-plants, and medicinal plants).—See also de Michele, Palanza, and Zodda, below.

Bruni, Achille. Descrizione botanica delle campagne di Barletta. 212 p. pl. 22.5 cm. Napoli, 1857.

Geology, soils, climate, etc.; list of vascular plants and a few fungi (758 species, of which 753 are vascular), with flowering dates, the families arranged alphabetically; short list of vernacular names with botanical equivalents, statistics, notes on many of the species listed. (Prov. Bari.) Replaces his Enumeratio plantarum in agro baruletano provenientium. Annali Accad. Aspir. Nat. Napoli **1**: 343–384. 1843. (Briefly annotated list of vascular plants, with localities, on the Linnaean system. Not seen; title and annotation supplied by V. Giacomini.)

Corti, Roberto. Erborizzazioni nelle leccete del territorio di Arneo (fra Taranto e Gallipoli). Nuovo Gior. Bot. Ital. n.s., **57**: 34–56. 1 fig. (map). 1950.

Vegetational study, including (p. 48–53) list of 195 angiosperms, with indication of life form. (Prov. Taranto and Lecce.)—See also **Ferente, Adriana.** Note floristiche e fenologiche nel territorio di Arneo (fra Taranto e Gallipoli). Nuovo Gior. Bot. Ital. n.s., **59**: 287–303. 2 fig. (incl. map), pl. 4. 1953. (Includes (p. 290–295) list of spermatophytes collected at Mandra Torriglione, with vernacular names, also phenological table; bibliography.)

Giorgi, C. de. Flora spontanea e coltivata di Terra d'Otranto. (7 p.) Palermo, 1873.

(Cited by Migliorato, Annali Bot. **9**: 162. 1911, as not seen by him.) (Present Provinces of Brindisi, Lecce, and Taranto.)—See also Marinosci, below.

Grimaldi, Valeria. La vegetazione della Gravina della Madonna della Scala presso Massafra (Taranto). Nuovo Gior. Bot. Ital., n.s., **63**: 163–185. 9 fig. (fig. 1, map). 1957.

Topography, geology, climate, general features of vegetation; annotated list of vascular plants (226 species); bibliography.

Groves, Henry. Flora della costa meridionale della Terra d'Otranto. Nuovo Gior. Bot. Ital. **19**: 110 (bis)– 219. pl. 2–6. 1887.

General features of flora, etc.; list of vascular plants collected by author and his associates, with localities. Refers mostly to coastal region between Otranto and Taranto. (Prov. Lecce and Prov. Taranto.) Replaces his Contribuzione alla flora della Terra d'Otranto. Nuovo Gior. Bot. Ital. **9**: 49–74. 1877. (Includes botanical regions, geology, botanical explorations, as well as list.)—See also his The coast flora of Japygia, S. Italy. Jour. Linn. Soc.

Bot. 21: 523–537. 1885. (Notes on various rarer species, useful plants, phytogeography, etc.)—Also **Personè, F.** Prima contribuzione alla flora di Terra d'Otranto. Bul. Soc. Bot. Ital. **1910:** 13. 1910. (List of additional species.)

Longo, A. Primo contributo alla conoscenza scientifica dei termini dialettali, usati nel Leccese per indicare le piante indigene spontanee e coltivate ed i prodotti più usati derivati da piante esotiche. (? p.) Bologna, 1931.

Not seen; cited from Nuovo Gior. Bot. Ital. n.s., **59:** 303. 1953. (Prov. Lecce.)

Marinosci, Martino. Flora salentina. 2 v. (v, 294; 2 p.l., 271, xl p.). 19.5 cm. Lecce, 1870 [–71?]. (Collana di opere scelte edite ed inedite di scrittori di Terra d'Otranto diretta da Salvatore Grande. v. 13–14.)

Briefly descriptive flora of the then Province of Lecce (the present Provinces of Brindisi, Lecce, and Taranto), with vernacular names and uses, on the Linnaean system (1684 species, of which 1629 are vascular; includes cultivated plants). Biography of author at end of v. 2. Cover title, La flora salentina; cover of v. 2 dated 1871.—See also de Giorgi, above.

Michele, Gabrielle de. Flora bitontina e della provincia di Bari. 142 p. 21 cm. Trani, 1894. (Lloyd Library.)

Annotated descriptive flora of angiosperms, with vernacular names and notes on uses. Relates mostly to the area around Bitonto and is evidently by no means complete. (Prov. Bari.)—See also Bisceglia, above, and Palanza and Zodda, below.

Migliorato, Erminio. Contributo alla flora vascolare delle isole Cesaree e di Gallipoli (Golfo di Taranto). Annali Bot. **9:** 163–167. 1911.

Topography; tabular list of 57 vascular plants, alphabetically arranged, with localities. (Prov. Lecce.)

Nannizzi, Arturo. Contributo alla conoscenza della flora dell'isolotto di "S. Nicolicchio" nel Mar Grande di Taranto. Atti R. Accad. Fisiocr. Siena X, **1:** 497–507. 1927.

Includes list of 23 vascular plants collected by author on August 26, 1926, with notes on many of them; also list of fungi; footnote references to earlier publications. Publication of Istituto botanico della R. Università di Siena. (Prov. Taranto.)—See also **Cataldi, Olinda.** La flora dell'isolotto di "S. Nicolicchio" (Taranto). Nuovo Gior. Bot. Ital. n.s., **58:** 591–593. 1952. (Includes list of species collected by author, mostly additional to those reported by Nannizzi.)

Palanza, Alfonso. Flora della Terra di Bari. Pubblicata dopo la morte dell'autore a cura di A. Jatta. 90 p. 25.5 cm. Trani, 1900. (From "La Terra di Bari," vol. 3, p. 155–244. 1900.) (Gray Herbarium library.)

Bibliography, general features of flora; annotated list of vascular plants (total 1160 species) of the Province of Bari, with localities. The species recorded by others but not found by the author are given in small type at the end of each family.—See also Bisceglia and de Michele, above, and Zodda, below.

Pasquale, G. A., and Licopoli, Gaetano. Di un viaggio botanico al Gargano. Atti Accad. Sci. Fis. e Mat. Soc. R. Napoli v. 5, no. 18. 31 p. 1873.

Itinerary, general features of flora, etc.; annotated list of spermatophytes, mosses, and lichens observed by authors in April 1871. (Prov. Foggia.)—See also **Béguinot, Augusto.** Ricordi di una escursione botanica del versante orientale del Gargano. Nuovo Gior. Bot. Ital. n.s., **16:** 97–123. 1903. (Botanical explorations, with references; lists of species collected by author, with localities.)—Also **Trotter, Alessandro.** A traverso il Gargano. Notizie ed osservazioni botaniche. Bul. Ort. Bot. R. Univ. Napoli **3:** 233–249. 4 fig. 1913. (Botanical explorations; author's itinerary, with mention of numerous species; phytogeography, list of noteworthy micromycetes.)—Also **Fiori, Adriano.** Flora nemorale e boschi del Gargano. Annali R. Ist. Super. Forest. Naz. Firenze **1:** 133–157. 1 fig., 3 pl. 1916. (Includes list of vascular plants of montane and submontane zones, etc.)—Also **Messeri, Albina.** Erborizzazioni in Val Carbonara fra S. Marco in Lamis e S. Giovanni Rotondo (Gargano) nel giugno 1952. Nuovo Gior. Bot. Ital. n.s., **59:** 525–531. pl. 10. 1953. (General features of flora; list of vascular plants observed.)—Also **Corti, Roberto.** Leccete diradate e garighe a *Sideritis sicula* Ucria sul

Gargano, sul fianco sinistro della Valle Carbonara. l.c. **59**: 578–584. pl. 12. 1953. (Includes two lists of plants observed.)

Rodio, Gaetano. Contributo allo studio della flora pugliese. Bul. Orto Bot. Univ. Napoli **15** (1): 27–79. pl. 3–7. 1941. (Arnold Arboretum library.)

Topography, climate, soils, principal cultivated plants, phytogeography, plant formations; annotated list of vascular plants collected in the territory of Locorotondo and Fasano; bibliography. (Prov. Bari.)

Scaramuzzi, Franco. La flora infestante di alcune colture presso Bari. (Ricerche sulla flora infestante delle colture in Italia. III.) Nuovo Gior. Bot. Ital. n.s. **56**: 58–105. 9 diagr. 1949.

Includes (p. 67–73) tabular list of 162 weedy angiosperms; bibliography, English summary. (Prov. Bari.)

Selvaggi, Eugenio. Vocabolario botanico martinese. 249 p. 21.5 cm. Bari, 1950. . . . Appendici [1–2]. 30 p. 21.5 cm. Bari, 1950. (All in library of Raphael G. Urciolo and Max Leopold Wagner, Washington, D. C.)

Annotated alphabetical list of wild and cultivated plants of the Martina Franca region, with scientific names, uses, popular traditions, etc.; no index of scientific names. The Appendici include (p. 5–12) section on "Flora ruderale." (Prov. Taranto.)

Trotter, Alessandro, and **Romano, Michele.** Sulla flora di M. Crispiniano in Puglia. (At head of title: Comitato "Pro flora italica" VI.) Nuovo Gior. Bot. Ital. n.s., **21**: 398–435. pl. 9. 1914.

Topography, botanical regions, description of various localities with lists of characteristic species; list of cellular and vascular plants observed by authors, with localities. (Prov. Foggia.)

Tuseo, Chiara. Appunti sulla flora dei dintorni di Ginosa (Prov. di Taranto). Nuovo Gior. Bot. Ital., n.s., **60**: 903–912. pl. 18. 1954.

General features of vegetation; list of recorded species, with indication of biological form.

Viesti, Maria. Appunti sulla flora stradale e ruderale dei dintorni di Molfetta (Bari). Nuovo Gior. Bot. Ital. n.s., **63**: 487–495. 1957.

Climate, etc.; annotated list of about 140 angiosperms found in country lanes around Molfetta; list of species by habitats. (Prov. Bari.)

Villani, Armando. Primo contributo allo studio della flora lucerina. Nuovo Gior. Bot. Ital. n.s., **22**: 175–235. 1915.

Topography, botanical explorations, bibliography; annotated list of vascular plants, including those recorded by others, with localities. Refers to Lucera and other parts of the Capitanata. (Prov. Foggia.)

Zodda, Giuseppe. Flora del territorio di Bari. (Nota floristica.) Arch. Bot. (Forlì) **18**: 1–26. 1942.

Topography, geology; list of known vascular plants of the Province of Bari (those observed by author, with his localities, and those recorded by Palanza but not seen by author, with localities.)—The following primarily ecological papers by various authors relate to scattered localities in the Province of Bari and contain lists of species too short for separate citation: Nuovo Gior. Bot. Ital. n.s., **55**: 251–275. 1948; **56**: 345–356. 1949; **58**: 594–596, 606–613. 1952; **59**: 144–156, 361–367, 378–387, 468–474. 1952–53; **60**: 592–603, 731–739. 1954; **62**: 478–485. 1 fig., pl. 46. 1956; **63**: 688–699. 2 fig. 1957.—See also Bisceglia, de Michele, and Palanza, above.

SARDEGNA (SARDINIA)

See also Spain, Andalucía (Lindberg, Itinera mediterranea).

The regione of Sardegna includes the Provinces of Cagliari, Nuoro (constituted in 1927 from parts of both the other provinces), and Sassari.

General

See also Gandoger and Terracciano, below, under Local.

Barbey, William. Florae sardoae compendium. Catalogue raisonné des végétaux observés dans l'île de Sardaigne. 263 p. 7 pl., port. 33 cm. Lausanne, 1884–[85].

List of endemic species of Sardinia and Corsica; list of 2,550 known species (1,656 vascular) of Sardinia, with reference to Moris' "Flora sardoa" and frequent annotations; journal of Schweinfurth's journey across Sardinia; supplement, by Ascherson and Levier; addenda; total number of species listed, 2856. According to Th. Herzog (Bot. Jahrb. Engler **42**: 342. 1909), the list is very uncritically compiled.—See also **Nicotra, Leopoldo.** Prime note sopra alcune piante di Sardegna. Malpighia **9**: 240–250. 1895. Ulteriori note . . . l.c. **9**: 364–369. 1895. Ultime note . . . l.c. **10**: 328–348. 1896.—Also **Cavara, Fridiano.** Addenda ad Floram sardoam. Bul. Soc. Bot. Ital. **1900**: 263–267. 1900.—Also **Fiori, Adriano.** Erborizzazioni primaverili in Sardegna. Nuovo Gior. Bot. Ital. n.s., **20**: 144–154. 1913. (At head of title, Comitato "Pro flora italica." IV.)—Also **Negodi, Giorgio.** Contributo alla flora della Sardegna . . . l.c. n.s., **38**: 449–462. 1931.—Also **Schmid, Emil.** Beiträge zur flora der insel Sardinien. Vierteljahrsschr. Naturf. Gesell. Zürich **78**: 232–255. 1933. (Mitt. Bot. Mus. Univ. Zürich 146.) (Phytogeography; annotated list of noteworthy spermatophytes collected by author in 1923 and 1932.)—Also **Landi, Maria.** Contributo alla flora della Sardegna. Arch. Bot. (Forlì) **10**: 52–70. 1934.

Béguinot, Augusto. Bibliografia botanica della Sardegna. Bul. Ist. Bot. Univ. Sassari v. 2, mem. 7. 31 p. 1922.—Aggiunte alla Bibliografia . . . l.c. v. 1 (fasc. 5), mem. 9. 4 p. 1923.

Partly annotated list of 349 (plus 31) titles, covering floristics (vascular and cellular plants), agriculture and useful plants, fossil plants, etc., as well as general works on the island.

Cara, Alberto. Vocabolarietto botanico sardo-italiano. 4 p. l., 94 p. 23.5 cm. Cagliari, 1889. (In library of R. G. Urciolo and M. L. Wagner, Washington, D. C.)

Alphabetical list of vernacular names (often with localities) with Italian or scientific equivalents and some annotations. Replaces his Vocabolarietto botanico relativo alla Sardegna (italiano-latino scientifico-sardo). 64 p. Cagliari, 1887. (Not seen; title supplied by G. Martinoli.)—Additional names and terms are given in his Secondo saggio di un vocabolario etimologico sardo. 126 p. 23.5 cm. Cagliari, 1894. (In library of R. G. Urciolo and M. L. Wagner.)

Cortesi, Fabrizio. Piante medicinali ed aromatiche della Sardegna. Riv. Ital. Essen., Profumi e Piante Off. **13**: 257–262. illus. 1931.

Previous publications on this subject; annotated list of about 40 more important medicinal and aromatic plants, with vernacular names and local occurrence.

Cugusi-Persi, Ignazio. Repertorio alfabetico dei nomi degli alberi, arboscelli, frutteci, radici, corteccie, fiori, semi, erbe e piante sarde in italiano-sardo e sardo-italiano coll'indicazione terapeutica delle piante medicamentose e repertorio alfabetico dei nomi dei pesci in italiano-sardo-francese e sardo-italiano-francese. 79 p. 18 cm. Cagliari, 1879.

Includes (p. 1–38, 39–62) reciprocal Italian-Sardinian and Sardinian-Italian vernacular names of plants, with brief notes on uses; no scientific names. (Copy lacking title page seen in library of R. G. Urciolo and M. L. Wagner, Washington; title supplied by R. McVaugh from copy in University of Michigan library.)

Moris, G. G. Flora sardoa seu historia plantarum in Sardinia et adjacentibus insulis vel sponte nascentium vel ad utilitatem latius excultarum. 3 v. 111 (i.e. 114) pl., map. 30.5 cm. Taurini, 1837–59.

Plant zones, etc.; annotated descriptive flora of Dicotyledones and Gymnospermae (1141 species).—See also **Mattirolo, Oreste.** Reliquiae Morisianae ossia elenco di piante e località nuove per la flora di Sardegna recentemente scoperte nell'erbario di G. G. Moris. Atti Cong. Bot. Internatl. Genova, 1892. p. 374–413. 1893. (List including 30 species and varieties new to Sardegna.)—Also **Martelli, Ugolino.** Monocotyledones sardoae, sive ad Floram sardoam . . . continuatio. 3 fasc. (viii, 152 p.). 10 pl. 32 cm. Firenze [etc.]. 1896–1904. (An unfinished attempt to treat the Monocotyledons, not reached by Moris; covers Orchidaceae, Iridaceae, Amaryllidaceae, Dioscoreaceae, Liliaceae in part.)

Mundula, Antonio. Le piante medicinali che crescono spontanee e coltivate nella Sardegna. Vol. 1. Crittogame (vascolari), gimnosperme, angiosperme (monocotiledoni). 171 p. 19 cm. Sassari, 1910. (John Crerar Library.)
Account of medicinal plants with brief descriptions, habitat, uses, vernacular names, references, etc.; tabular key to plants included, indices.

Rolla, Pietro. Flora popolare sarda; miscellanea di toponimia e dialettelogia italiana. 151 p. 8vo. Casale, 1896.
Contains names of plants on p. 7–55. (Not seen; title and annotation supplied mostly by F. Sappa, who adds "lavoro non riuscito.")

——— Toponimia sarda. 1 p.l., 43 p. 24 cm. Cagliari, 1893. (Private library of R. G. Urciolo and M. L. Wagner, Washington, D. C.)
Includes (p. 3–26), Nomi locali di Sardegna derivati dal nome delle piante, an alphabetical list of scientific and some vernacular names with names of localities derived from them and etymological notes.

Sanna, A. Piante officinali della Sardegna. pt. 1. 233 p. Sassari, 1935.
Not seen; cited from Bul. Bibl. Bot. Ital. **9** (1935): 56. 1939.

Seghetti, G. Appunti sulla flora medica sarda. (? p.) Arezzo, 1918.
List of 180 medicinal plants. (Not seen; cited from mention in Riv. Ital. Essen., Profumi e Piante Off. **13**: 258. 1931.)—See also **Mameli, Efisio.** Elenco, località e diffusione delle piante medicinali e aromatiche sarde. Bol. Comit. "Pro piante medicinali" di Sassari **1922**: 37–52. 1922. (List of 80 species. Not seen; cited from notice mentioned above.)

Local

Béguinot, Augusto. Rilievo floristico e fitogeografico di alcune piccole isole della Sardegna nord-orientale. Arch. Bot. (Forlì) **5**: 79–93. 1929.
Separate lists of vascular plants of the islands and islets Figarello, Barca Sconcia, Garofani, Cigni, Piana, Punta, Cana, and Molara, with account of topography, geology, etc. (Prov. Sassari.)

——— and **Vaccari, Antonio.** Le piante vascolari sinora note per l'isola Tavolara e considerazioni fitogeografiche sulle stesse. Arch. Bot. (Forlì) **3**: 269–290. pl. 10–11. 1927; **5**: 46–78. 1929.
Botanical explorations; annotated tabular lists of 318 vascular plants; phytogeography, etc. (Prov. Sassari.)

Casu, Angelo. Contribuzione allo studio della flora delle saline di Cagliari. Pt. III. Resistenza fisiologica della flora delle saline all'azione del sale marino. Annali Bot. **5**: 273–355. 1 fig. (profile). 1907.
Includes (p. 323–336) tabular list of 331 spermatophytes common along the whole littoral of Cagliari Province.

——— Lo stagno di Sta. Gilla (Cagliari) e la sua vegetazione. Parte 2. Constituzione ed ecologia della flora. Mem. R. Accad. Sci. Torino II, **62** (Cl. Sci. Fis., Mat. e Nat.): 295–333. map. 1912.
Includes (p. 299–306) list of algae and vascular plants. (Prov. Cagliari.)

Desole, Luigi. Studio floristico e fitogeografico delle piccole isole della Sardegna nord-occidentale. 1. nota. L'Isola Piana. Studi Sassaresi **23**: 94–109. 1945. (Original not seen; reissued as Bul. Ist. Bot. Univ. Sassari **3**: 94–109. 6 fig. (fig. 2, map). n.d.)
Topography, climate, plant associations; list of vascular plants (82 species). The Isola Piana or Isola delle Gessiere is about 18 miles from Alghero. (Prov. Sassari.)—See also his Studio floristico . . . 2. nota. Isola Rossa (Aggius); Isola dei Porri (Stintino); Isola Foradada (Alghero). Nuovo Gior. Bot. Ital. n.s., **61**: 290–326. 18 fig. (mostly on 6 pl. [on 3]; fig. 1, map). 1955. (Similar treatment with lists of species collected (30, 35, and 71 respectively). Reissued as Bul. Ist. Bot. Univ. Sassari v. 4, fasc. 3, no. 2. p. 290–326. 1954 [1955?].) (Prov. Sassari.)

Falqui, Giuseppe. Contributo alla flora della Sardegna. 45 p. 24.5 cm. Cagliari, 1905.
List of 648 species of vascular plants and mosses. (Prov. Cagliari and Nuoro.) (Not seen; title and annotation supplied by G. Martinoli.)

——— Un'escursione botanica sul Gennargentu. 43 p. 23.5 cm. Cagliari, 1907.

Plant formations; list of 326 species collected by author in 1903, with localities. Gennargentu (altitude 1,832 meters) is the highest mountain in Sardinia. (Prov. Nuoro.) (Not seen; title and annotation supplied by G. Martinoli.).

Gandoger, Michel. Catalogue des plantes récoltées en Sardaigne. 1 p. l., 16 p. 25 cm. Paris, 1917.

Brief bibliography; unannotated list of vascular plants and Characeae, with localities; apparently based mainly on his own collections made during more than 20 years and on those of Martelli and Nicotra. Reproduced from manuscript.

Gennari, Patrizio. Florula di Palabanda. Nuovo Gior. Bot. Ital. **21**: 28–34. 1889.

Unannotated list of nearly 400 vascular plants of the valley of Palabanda southeast of Cagliari. (Prov. Cagliari.)

Marcialis, Efisio. Piccola flora spontanea dei dintorni di Cagliari. 66 p. 8vo. Cagliari, 1889.

List of vascular and some cellular plants, with vernacular names and localities; obviously far from complete. (Prov. Cagliari.) (Not seen; title and annotation from Bot. Jahresber. Just **17** (2): 291–292. 1892.)—See also **Gennari, Patrizio.** Repertorium florae calaritanae ex horto sicco academico depromptum. 180 p. 20 cm. Calaris, 1890. (Briefly annotated and unimportant list of several hundred bacteria, algae, fungi, lichens, and vascular plants. Not seen; title and annotation supplied by F. Sappa.)

Martinoli, Giuseppe. Il colle di Acquafredda (Siliqua: Sardegna mer.) e la sua vegetazione. Nuovo Gior. Bot. Ital. n.s., **49**: 476–482. 1942.

Topography, plant formations; list of 151 known vascular plants. (A hill 253 meters high, 5 kilometers south of Siliqua, in Prov. Cagliari.)

——— La flora e la vegetazione dell Capo S. Elia (Sardegna meridionale). Nuovo Gior. Bot. Ital. n.s., **57**: 57–148 incl. 23 fig. (mostly on 6 pl.). 1950.

Mainly a study of types of vegetation; includes (p. 64–79) sketch of botanical explorations and list of known vascular plants with indication of biological form; bibliography. (Prov. Cagliari.)

——— La flora e la vegetazione dell'isola Rossa (Golfo di Teulada, Sardegna meridionale). Annali Univ. Ferrara sez. IV, Bot. **1**: 63–73. 1 fig. (map), pl. 1955.

Topography, etc.; list of vascular plants; plant formations, bibliography. (Prov. Cagliari.)

Mola, Pasquale. Flora fanerogamica della massiccio plutonico nord-occidentale di Sardegna. 31 p. 25 cm. Sassari, 1921.

Incomplete and unannotated list of plants, referring particularly to the region of Bosa. (Prov. Nuoro.) Replaces his Catalogo della raccolta delle piante e degli animali esistenti nella regione di Bosa. 45 p. 24 cm. Cagliari, 1916. Also replaces his Flora e lepidotterofauna sarda (regione di Bosa). 69 p. 25 cm. Sassari, 1919. (None of these 3 works seen; titles and annotations supplied by G. Martinoli.)

Nannetti, Alfonso. La flora di Osilo. Bul. Ist. Bot. Univ. Sassari v. 2, mem. 1. 83 p. pl. 1914.

Topography, geology, climate, general features of vegetation, plant formations; annotated list of vascular plants (513 species). According to an introductory note by Béguinot in v. 2, fasc. 1, of the Bullettino, the paper was distributed as a preprint in 1914, although the fascicle as a whole was not issued until 1922. (Prov. Sassari.)—See also **Desole, Luigi.** Addenda a "La flora di Osilo." Nuovo Gior. Bot. Ital. n.s., 54: 364. 1948; Seconda addenda alla "Flora di Osilo," con nuova stazione di Melissa officinalis L. l.c. **62**: 373–379. 2 fig. (fig. 2, map). 1956. (Second paper also issued as Bul. Ist. Bot. Univ. Sassari v. 4, fasc. 4, no. 3. 1 p. l., 6 p. 1955 [1956?].

Negodi, Giorgio. La flora dell'isola Asinara. Arch. Bot. (Forlì) **2**: 35–44, 107–119. 1926; **3**: 71–82. pl. 3–4. 1927.

Botanical explorations, topography, geology, plant formations, bibliography, etc.; annotated list of vascular plants (431 species). (Prov. Sassari.)

Pampanini, Renato, and Martinoli, Giuseppe. Gli isolotti il Toro e la Vacca (Sardegna sudoccidentale) e la loro flora. Rend. Semin. Fac. Sci. Univ. Cagliari **16** (1946): 123–138. 1948. (Harvard College library.)

Topography, geology, botanical explorations; lists of known vascular plants collected on each island with notes on some of them and descriptions of 5 new varieties and forms. (Prov. Cagliari.)

Rovinetti, Fernando. Profilo fitogeografico del Monte Ortobene (Nuoro–Sardegna). Nuovo Gior. Bot. Ital. n.s., **64**: 431–454. 14 fig. (incl. 2 maps; in part on 4 pl. [on 2]). 1958.

Includes (p. 435–438) list of vascular plants with indication of life form; bibliography. (Prov. Nuoro.)

Terracciano, Achille. La "Flora sardoa" di Michele Antonio Plazza da Villafranca redatta con i suoi manoscritti. I–III. Mem. R. Accad. Sci. Torino II, v. 64 (Cl. Sci. Fis., Mat. e Nat.), no. 15. 54 p. 1914; v. 65, no. 13. 53 p. 1916 [1915]; v. 67, no. 1. 78 p. 1933 [1930].

Account of Plazza's [1748–1791] manuscript, statistics, comparison with Moris' flora, etc.; annotated list of cellular and vascular plants (818 species, with Sardinian references, data, vernacular names, and some critical notes).

Vacca-Concas, S. Manuale della fauna e della flora popolare sarda indigena e naturalizzata (in dialetto del Campidano di Cagliari) con la aggiunta dei nomi delle piante e degli animali esotici più conosciuti. 230 p. 16mo. Cagliari, 1916.

Includes (p. 113–230) alphabetical list of vernacular names of plants with notes on distribution and uses. (Prov. Cagliari.) (Not seen; title from Béguinot, annotation supplied by V. Giacomini.)

Vaccari, Antonio. Flora dell'Arcipelago di Maddalena (Sardegna). Malpighia 8: 227–277. pl. 5 (map). 1894.—Supplemento alla Flora . . . [1]—2. l.c. **10**: 521–534. 1896; **13**: 200–210. 1899.—Aggiunte . . . l.c. **22**: 15–24. 1908.—Nuove aggiunte alla Flora dell'Arcipelago di Maddalena e contributo alla flora di alcune isole adiacenti alla Sardegna. Atti Soc. Nat. e Mat. Modena 59 (VI, 7): 31–46. 1928.

Botanical investigations; annotated list of vascular plants (743 species including all supplements); bibliography. Index, l.c. **13**: 203–210. Covers the islands of Budelli, Caprera, Maddalena, Razzoli, San Stefano, Santa Maria, and Spargi, the islets and rocks Barettini, Bisce, Corcelli, Monaci, Pecora, Porco, La Pressa, and Spargiotto, and the adjacent coast from Cape Ferro to the mouth of Fiume Liscia. (Prov. Sassari.) The 1928 paper includes sketch of publications on the flora, and list of new or noteworthy records; also records from Isolotto Mortorio (on coast a little south of Cape Ferro) and Isolotto Figarello (at entrance to Golfo Aranci), with sketch of geology; account of San Antioco island, off southwest Sardinia (Prov. Cagliari), with list of 62 known spermatophytes; list of exsiccatae from the Maddalena Arcipelago; supplement to index in vol. 13. Replaces **Gennari, Patrizio.** Florula di Caprera. Nuovo Gior. Bot. Ital. **2**: 90–145. 1870. (Includes list of 461 vascular plants.)—See also **Vaccari, A.** Osservazioni ecologiche sulla flora dell'Arcipelago di Maddalena. Malpighia **22**: 101–172. 1908. (History, geology, topography, climate, soils, plant associations, phenology, distribution by islands, phytogeography, origin of flora.)

SICILIA (SICILY)

See also Campania (Baccarini); Spain, Andalucía (Lindberg).

The regione of Sicilia includes the Provinces of Agrigento (formerly Girgenti), Caltanissetta, Catania, Enna (formerly Castrogiovanni; formed in 1927), Messina, Palermo, Ragusa (formed in 1926), Siracusa, and Trapani.

The island of Malta, a dependency of Great Britain, is included here for phytogeographic reasons.

General

Calcara, Pietro. Florula medica siciliana o esposizione delle piante indigene medicinali. 1 p. l., viii, 220 p. 17 cm. Palermo, 1851. (Arnold Arboretum library.)

Annotated systematic list of medicinal plants, with vernacular names, brief descriptions, and uses.

Cannarella, Pietro. Notizie sulla flora ruderale della Sicilia. Bol. R. Orto Bot. (e. Giard. Colon.) Palermo **9**, App.: 51–146. 1910.

Classification of ruderal plants; annotated list of 590 vascular plants growing on walls, along roads, etc., with citation of records; includes native as well as introduced species.

——— Saggio di bibliografia floristica della Sicilia e delle isole adiacenti. Nuovo Gior. Bot. Ital. n.s., **15**: 93–178. 1908; **17**: 529–562. 1910.

List of periodicals and bibliographical sources; unannotated list of publications on the botany, agriculture, botanic gardens, etc., of Sicily, arranged alphabetically by authors, and shorter lists for the Eolian (Lipari) Islands, Ustica, Pantelleria, Pelagic Islands, and Malta; chronological list of authors. First part, phanerogams (710 titles); second part, cryptogams (387 titles).

Farina, Vicenzo. La flora sicula ossia manuale delle piante che vegetano nella Sicilia preceduto da un breve saggio su la botanica generale. 370 p. 19 cm. Sciacca, 1874. (Arnold Arboretum library.)

The second part (p. 111–370) is an annotated list of 401 plants, alphabetically arranged by Italian vernacular names, with Sicilian and scientific names, native habitat and local occurrence, and in many cases medicinal uses; list of Italian and Sicilian vernacular names, with equivalents; no botanical index.

Gussone, Giovanni. Florae siculae synopsis exhibens plantas vasculares in Sicilia insulisque adjacentibus huc usque detectas secundum systema Linneanum dispositas. 2 v. in 3 (1 p. l., v, 582; 920 p. Vol. 2 in 2 parts) 22 cm. Neapoli, 1842 [–45].

Descriptive flora of vascular plants (2,586 species and 638 varieties), on the Linnaean system, with references and localities.

Lojacono-Pojero, Michele. Flora sicula o descrizione delle piante vascolari spontanee o indigenate in Sicilia. 3 v. in 5 (vol. 1–2 each in 2 pt.). 101 pl. 32–35 cm. Palermo, 1888 (1886?)–1909. (Gray Herbarium library.)

General features of vegetation, list of endemic species; flora of vascular plants, with full descriptions, synonymy, and iconography, and localities.—See also his Schizzo orografico della Sicilia o itinerario botanico dell'isola. Nuovo Riv. Forest. **13**: 181–215. 1890. (Account of geography, topography, and general botanical features of island.)—Also his Addenda et emendanda ad Floram siculam. Malpighia **20**: 37–48, 95–119, 180–218, 290–300. 1906. (Additional species and localities, based on the material of Thalamiflorae and Calyciflorae Dialypetalae in the Todaro herbarium. The paper apparently was never finished.)

Mattei, G. E. Piante utili indigene di Sicilia. Riv. Ital. Essen., Profumi e Piante Off. **21**: 284–285. 1939.

Brief account of more important medicinal and otherwise useful plants.—See also his Piante medicinali, da essenze, ed altre utilizzabili della flora siciliana. l.c. **20**: 161–164. 1938. (List of native medicinal and otherwise useful plants, alphabetically arranged by botanical names, with notes on occurrence and uses. Incomplete, letters A–C only.)

Nicotra, Leopoldo. Syllabus florae siculae, sive vascularium plantarum Siciliam sponte incolentium enumeratio ex gravioribus notitiis hactenus congestis. Atti e Rend. Accad. Sci., Let. ed Arti Zelanti Acireale n.s., 7 (Cl. Sci.): 73–140. 1896.

Unannotated list of vascular plants, with a short appendix of critical notes.—See also **Nicotra, L.,** and **Campagna, Giovanni.** Addenda ad floram siculam nonnulla. Malpighia **22**: 3–14, 537–549. 1908. (List with localities.) —Also **Zodda, Giuseppe.** Entità nuove o importanti della flora sicula osservate dal . . . Rend. e Mem. R. Accad. Sci., Let. ed Arte Zelanti Acireale Mem. Cl. Sci. III, 5: 99–162. 1909. (Extensive list of cellular and vascular plants, with localities. In New York Public Library.)—Also **Nicotra, A.,** and **Campagna, G.** Addenda ad floram siculam nonnulla. 16 p. 24 cm. Messina, 1910. (Not seen; cited from Internatl. Cat. Scient. Lit. M. Bot. **10**: 234 1913.)

Piazza, Giuseppe. Vocabolario botanico italiano-siciliano e viceversa . . . 59 p. 20.5 cm. Nicosia, 1902. (John Crerar Library.)

Alphabetical list of 888 Italian common names, with scientific and Sicilian vernacular equivalents; list of 615 Sicilian names, with Italian equivalents; classified list of 160 useful species.

Riccobono, Antonio. Le piante della flora siciliana più acconce all'ornamento dei giardini. 2 v. 17 cm. 1906–25.

Key to species, list of scientific and vernacular names; descriptive account of plants considered suitable for gardens. (Not seen; title and annotation supplied by V. Camarrone.)

Tornabene, Francesco. Flora sicula viva et exsiccata, seu collectio plantarum in Sicilia sponte nascentium hucusque cognitarum juxta methodum naturalem vegetabilium exposita in horto botanico Regiae universitatis studiorum Catinae. 687 p. 4 pl. 27 cm. Catinae. 1887.

History, climate, topography, phytogeography, botanical investigations, references, fossil flora, geology, discussion of doubtfully native species; annotated list of vascular plants, Characeae, and lichens (2,488 species, of which 2,412 are vascular), with localities; statistics.

———Quadro storico della botanica in Sicilia che serve di prolusione all'anno scolastico 1846 e 1847 nella Regia università degli studi in Catania. 70 p. 20 cm. Catania, 1847. (Arnold Arboretum library.)

Chronological account of botanists and botanical work in Sicily from the earliest times, with titles and references.

Local

Albo, Giacomo. La flora dei Monti Madonie. Nuovo Gior. Bot. Ital. n.s., **12**: 217–260. 1905.

Annotated list of cellular and vascular plants (about 530 vascular plants). The Monti Madonie (Madonian Mountains) make up the western extremity of the Nebrodes. (Prov. Palermo.)—See also Strobl, Flora der Nebroden, below.

——— La flora di Donna Lucata. Nuovo Gior. Bot. Ital. n.s., **23**: 327–374, 541–558. 1 fig. (map). 1916; **24**: 5–16. pl. 1–5. 1917.

General features of flora, topography, geology, climate, bibliography, etc.; annotated list of vascular plants (548 species). (Prov. Ragusa.)

——— L'isola di Capo Passaro e la sua vegetazione. Naturalista Siciliano 23 (n.s., 3): 179–201. 6 fig. 1919.

Topography, geology, soils, general features of flora (with especial reference to dwarfed and reduced forms), bibliography; annotated list of angiosperms (166 species). (Small island only 300 meters off the tip of Capo Passaro, in Prov. Siracusa.)

——— La vita delle piante vascolari nella Sicilia meridionale-orientale. Pt. II. Flora. vi, 308 p. 24.5 cm. Ragusa, 1919.

Annotated list of 1432 vascular plants, with local range and Sicilian vernacular names. (Prov. Catania, Ragusa, and Siracusa.) (Not seen; title and annotation supplied by F. Sappa.)

Assenza, Vincenzo. Dizionario botanico dialettale-italiano-scientifico della maggior parte delle piante spontanee coltivate ed ornamentali della Contea di Modica. 223 p. 25 cm. Modica, 1923. (1st ed., Dizionarietto vernacolo-italiano ed italiano-vernacolo-botanico della maggior parte delle piante spontanee coltivate ed ornamentali che crescono nelle campagne di Modica e di Scicli (Siracusa), 1894.)

Alphabetical lists of patois names, Italian names, and scientific names, the last with collected common names and notes on the distinctive characters of each plant but no notes on uses. (Prov. Ragusa.) (Not seen; title and annotation supplied by V. Camarrone.)

Bianca, Giuseppe. Flora dei dintorni di Avola. Atti Accad. Gioenia Sci. Nat. Catania **15**: 295–348. 1839; **19**: 51–132. 1842; **20**: 259–322. 1843 [1844 ?]; II, **1**: 251–363. 1844 [1845 ?]; **2**: 81–161. 1845; **3**: 33–107, 255–345. 1846 [–1847 ?]; **6**: 217–378. 1850 [1851 ?]; **8**: 21–110. 1853; **13**: 157–252. 1857; **14**: 23–86. 1859.

Topography, climate, general features of flora; briefly annotated descriptive flora of angiosperms, on the Linnaean system, with Italian, French, and Sicilian vernacular names; incomplete (Monandria-Diadelphia; 638 species). (Prov. Siracusa.)

Borg, John. Descriptive flora of the Maltese Islands including the ferns and flowering plants. 846 p. 22.5 cm. Malta, 1927.

Botanical investigations (with included bibliography), geology, climate, general features of flora; descriptive flora of vascular plants, with localities, extralimital range, and English, Italian, and Maltese vernacular names. (To Great Britain.)—See also Gulia, Lanfranco, and Sommier and Gatto, below.

Cacciato, Alfredo. Erborizzazioni sul Monte S. Anna presso Caltanissetta. Nuovo Gior. Bot. Ital. n.s., **57**: 382–390. 3 fig. 1951.

Geology, general features of flora; list of vascular plants (209 taxa) with indication of life form; short bibliography. (Prov. Caltanissetta.)

Campoccia, Gesualdo. Flora nissena [primo elenco]. 36 p. Caltanissetta, 1902.

List of 623 vascular plants collected by author in vicinity of Nissena (city of Caltanissetta), without indication of localities. (Prov. Caltanissetta.) (Not seen; cited from Bot. Centralbl. **93**: 91. 1903.)

Cannarella, Pietro. Flora urbica palermitana. Bul. Soc. Bot. Ital. **1909**: 73–81, 172–183. 1909; **1912**: 23–31. 1912.

Annotated list of 262 vascular plants of the city of Palermo. (Prov. Palermo.)

Cosentini, Ferdinando. Saggio di topografia botanica della campagna detta l'Arena di Catania col catalogo delle piante che spontaneamente vi nascono. Atti Accad. Gioenia Sci. Nat. Catania **1**: 147–170. 1825.

General features of flora, etc.; unannotated alphabetical list of spermatophytes. Relates principally to the area from the point where the eruption of Etna of 1669 stopped to the fiume Simeto. (Prov. Catania.)

Di Mattei, Vincenzo. Le piante arboree in provincia di Siracusa. 99 p. 26 cm. Siracusa, 1913.

Not seen; cited from International Catalogue of Scientific Literature **13**: 191. 1916.

Fanales, Filippo. Contributo alla conoscenza della flora delle sciare di Marsala. Bol. R. Orto Bot. Palermo **3**: 3–65. 1899.

Geology, general features of flora; annotated list of phanerogams. Sciara (pl. sciare) is a local name applied to dry shrubby areas where the shelly surficial limestone is little broken up. (Prov. Trapani.)

Francini, Eleonora, and Messeri, Albina. L'isola di Marettimo nell'arcipelago delle Egadi e la sua vegetazione. Webbia **11**: 607–846. 154 fig. (incl. maps; partly on pl. 18–52), map 6. 1956.

Topography, climate, soils, etc.; annotated list of vascular plants, with indication of life form, and a few mosses and lichens (altogether 515 taxa); general features of vegetation, bibliography, English summary. (Prov. Trapani.)

Gulia, Gavino. Repertorio botanico maltese contenente i nomi scientifici colle correspondenti voci italiane ed inglesi delle piante le quali sono in Malta conosciute sotto uno populare denominazione colle indicazione dei loro usi, delle epoche delle respettive fasi vegetative, del metodo di perpetuarle, con alcuni altre instruzioni, aggiuntevi la patria loro originaria e la referenza del sistema Linneano. 4 p.l., 68 p. 21.5 cm. Malta, 1855–56. (On cover: Repertorio botanico preceduto da una prefazione bibliografico-critica. fasc. 1.) (Kew library.)

Botanical explorations; alphabetical list of Maltese vernacular names, with scientific, Italian and English names, and uses. (To Great Britain.)—See also Borg, above, and additional references there cited.

Lanfranco, G. G. Guide to the flora of Malta. 66 p. 33 pl. (300 fig.). 20.5 cm. Valetta, Malta, 1955.

List of 300 selected species grouped by color, with notes on habitat and uses, figure of each and Maltese vernacular names; systematic list of the species. (To Great Britain.)—See also Borg, above, and additional references there cited.

Lojacono-Pojero, Michele. Le Isole Eolie e la loro vegetazione con enumerazione delle piante spontanee vascolari. Atti Soc. Acclim. e Agr. Sicilia **17**: 177–328. 1878. (Reprinted 140 p. 23.5 cm. Palermo, 1878. Reprint in Gray Herbarium library; original not seen, cited from Cannarella.)

Topography, geography, climate, cultivated plants, general features of flora, etc.; annotated list of vascular plants (499 species), with localities. The

Isole Eolie or Liparee (Aeolian or Lipari Islands) include Alicudi (Alicuri), Filicudi (Filicuri), Lipari, Panaria (Panarea), Salina, Stromboli, Vulcano, and some islets, all in the Province of Messina.—See also [**Ludwig Salvator,** *erzherzog*]. Filicuri. *In* his Die Liparischen Inseln. 5. heft. viii, 37 p. illus., pls., map. 41 cm. Prag, 1895. (Includes (p. 2–3) alphabetical list of vascular plants collected by author.)—Also his Alicuri. *In* his Die Liparischen Inseln. 6 heft. viii, 29 p. illus., pls., map. 41 cm. Prag. 1896. (Includes (p. 2) alphabetical list of vascular plants collected by author.)—Also **Zodda, Giuseppe.** Une gita alle Isole Eolie. Atti R. Accad. Peloritana **19** (1): 73–108. 1904. (List of hepatics and lichens collected by author, with localities; list of newly recorded vascular plants, species that have apparently disappeared, etc. In library of Academy of Natural Sciences, Philadelphia.)—Also **Traverso, G. B.** Elenco della piante determinate nel R. Istituto botanico di Roma sopra exemplari raccolti nelle Isole Eolie dal 9 al 13 aprile 1900. 2 p. Padova, 1904. (List of about 70 species of phanerogams, lichens, mosses, etc. Not seen; cited from notice in Bot. Jahresber. Just **32** (2): 1248. 1906.)

Lopriore, Giuseppe. Studi comparativi sulla flora lacustre della Sicilia. 116 p. 4to. Catania, 1901.

Includes description of the vegetation of different lakes and swamp formations, list of over 1,000 vascular plants, and phytogeographical conclusions. (Not seen; title and annotation supplied by V. Giacomini.)

[**Ludwig Salvator,** *erzherzog.*] Ustica. xii, 132 p. illus., pls., maps. 41 cm. Prag, 1898.

Includes (p. 8–13) unannotated list of known vascular plants furnished by Prof. N. Borzì. (Prov. Palermo.)

Maravigna, Carmelo. Saggio di una flora medica catanese, ossia catalogo delle principali piante medicinali, che spontaneamente crescono in Catania e ne'suoi contorni, con la indicazione delle loro mediche azioni. Atti Accad. Gioen. Sci. Nat. Catania **2**: 67–120. 1827; **3**: 77–124. 1829. (Reprinted 104 p. 4 pl., according to Pritzel, Thesaurus ed. 2, p. 202. 1872.)

Annotated list of medicinal plants, alphabetically arranged by vernacular names. (Prov. Catania.)

Nicotra, Leopoldo. Introduzione allo studio della flora medica di Messina. Atti R. Accad. Peloritana **2**: 269–280. 1880. (American Museum of Natural History library.)

Short classified account of medicinal plants. (Prov. Messina.)

——— Prodromus florae messanensis plantas exhibens phanerogamas sponte virentes juxta methodum naturalem digestas. 460 p. 19 cm. Messanae, 1878–83. (Gray Herbarium library.)

Annotated, briefly descriptive flora of spermatophytes (1416 species) of the Province of Messina, with localities.—See also his Excerpta ex schedulis ad meum Prodromum florae messanensis augendum et expoliendum. Revue Bot. Bul. Mens. Soc. Franç. Bot. **4**: 260–268. 1885. (List with localities.)—Also his Pteridophytarum messanensium conspectus. Malpighia **7**: 91–96. 1893. (Annotated list of Pteridophyta (26 species).)—Also **Zodda, Giuseppe.** Contribuzione allo studio della flora peloritana. Riv. Ital. Sci. Nat. **18**: 100–104, 128–130. 1898; **19**: 12–16, 37–40. 1899. (Topography, general features of flora, etc.; list of vascular plants from new localities, or at unusual flowering dates, etc. Refers to region of Punta del Faro.)—Also his Osservazioni critiche e geografiche sulla flora vascolare del Peloro. l.c. **20**: 139–144. 1900; **21**: 19–20. 1901. (Extensive list of vascular plants, with localities and some critical notes.)—Also **Nicotra, L.** Variazioni recenti nella flora messinese. Nuovo Gior. Bot. Ital. n.s., **11**: 34–47. 1904. (Changes in abundance of numerous species since Gussone's time, with lists.)—Also **Gramuglio, Gaetano.** Note erboristiche sul territorio di Mistretta. l.c. **61**: 397–402. 1955. (Notes on the vegetation with 3 lists of species observed in a trip through the Caronie Mountains.)—Also Zodda, below.

Parlatore, Filippo. Flora palermitana ossia descrizione delle piante che crescono spontanee nella valle di Palermo. v. 1 (xxii, 442 p.). 22.5 cm. Firenze, 1845 [–47 ?].

Topography, geology, localities; annotated descriptive flora of vascular plants, in Latin and Italian, with vernacular names. Incomplete (267 species,

Gramineae-Liliaceae); no more published. Originally published in Giornale botanico italiano v. 1–2 (1844–1847), there including 269 species. (Prov. Palermo.)—See also **Ferruzza Loncao, Gaetano.** Flora palermitana ossia descrizione delle piante spontanee od inselvatichite che crescono nella Conca d'Oro. fasc. 1. Talamiflore. 92 p. 25 cm. Palermo, 1902. (Descriptive flora, covering Thalamiflorae; no more published. Covers the area between Capo Gallo and Capo Zafferano surrounded by Monti Pellegrino, Gallo, Cuccio, and Catalfano. Not seen; title and annotation supplied by V. Camarrone.) His La flora della Conca d'Oro. Studio bio-geografico. (30 p. 25 cm. Palermo, 1902) is entirely phytogeographic and not floristic. (Not seen; title and annotation supplied by V. Camarrone.)

Ponzo, Antonino. Escursioni nei dintorni di Licata. Malpighia **16:** 227–260. 1902.

Topography, geology, general features of flora; annotated list of vascular plants collected by author. (Prov. Agrigento.)

——— La flora trapanese. 140 p. Palermo, 1900.

Topography, climate, geology, cultivation, botanical work, plant regions, special localities with lists of characteristic species, statistics, endemic species, phytogeography; annotated list of vascular plants of the vicinity of Trapani. (Prov. Trapani.) (Not seen; review in Nuovo Gior. Bot. Ital. n.s., 8: 352–353. 1901.) Replaces **Cassisa, Salvatore.** Flora trapanese. (? p.) Trapani, 1888. (Not seen.)—See also **Ponzo, A.** Aggiunte alla flora trapanese. Bul. Soc. Bot. Ital. **1901:** 370–381. 1901. (List with localities.)

——— La florula dei dintorni di Alcamo. Bul. Soc. Bot. Ital. **1903:** 200–212, 318–330. 1903; **1904:** 262–270. 1904.

Two lists (the first annotated) of vascular plants collected by author (467 species); lists of species recorded by other botanists; plant associations. First part has title, La flora . . . (Prov. Trapani.)—See also his Appunti sulla vegetazione dei dintorni di Alcamo. Nuovo Gior. Bot. Ital. n.s., **11:** 356–385. 1904. (Geology, general features of flora, weeds of cultivation, plant associations.)

——— Le plantule della flora trapanese. 1 [–4.] contributo. Nuovo Gior. Bot. Ital. n.s., **33:** 341–389. 62 fig. 1926; **34:** 546–592. 55 fig. 1927; **35:** 169–213. 46 fig. 1928; **47:** 579–590. 11 fig. 1940.

General considerations; descriptions of the seedlings of 330 species, not in systematic order; bibliographies. (Prov. Trapani.)

Ross, Hermann. Beiträge zur flora von Sicilien. I. –II. teil. Erläuterungen und kritische bemerkungen zum Herbarium siculum I. [–II.] centurie. Bul. Herb. Boissier **7:** 262–299. 4 fig. 1899; II, **1:** 1201–1232. 1901.

Phytogeography, etc.; list of 200 vascular plants, with data from his exsiccatae and notes on distribution, etc.

Silipranti, Giovanni. Contribuzione alla flora dei dintorni di Noto. Atti Soc. Nat. Modena III, **6:** 22–44. 1887.

Annotated list of 200 vascular plants observed by the author in 1882–83. (Prov. Siracusa.)

Sommier, Stefano. Flora dell'isola di Pantelleria. 110 p. port. 26 cm. Firenze, 1922. (On cover: R. Istituto botanico di Firenze n. 1.)

Bibliography; annotated list of known vascular plants (483 species), bryophytes (75 species), lichens (47 species), and fungi (2 species). Replaces the tabular listing of species in his Le Isole Pelagie (see below). (Prov. Trapani.)—For account of topography, geology, and general features of flora, see **Ross, Hermann.** Contribuzione alla conoscenza della flora sicula. II. Isola di Pantelleria. Bul. Soc. Bot. Ital. **1906:** 38–45. 1906.

——— Le Isole Pelagie, Lampedusa, Linosa, Lampione e la loro flora. 344 p. 4 fig. 24.5 cm. [Palermo, 1906–08.] (Issued as Appendice. Bol. R. Orto Bot. [from v. 6, Bol. R. Orto Bot. e Giard. Colon.] Palermo v. 5, no. 1/2—7, no. 1/3.)

Botanical explorations, annotated bibliography, history, geology; annotated list of vascular and cellular plants of each island, considered separately; table showing occurrence of the vascular plants (530 species) on Lampedusa, Linosa, Pantelleria, Malta, etc.; list of 155 additional species from Pantelleria not found in the Pelagic Islands; phytogeography, statistics, general features of flora, etc. (The Isole Pelagie are in Prov. Agrigento.)

Sommier, Stefano and **Caruana Gatto, Alfredo.** Flora melitensis nova. 4 p.l., 502 p. pl. 24.5 cm. [Palermo, 1913–15.] (Issued as Appendice [II]. Bol. R. Orto Bot. e Giard. Colon. Palermo v. 11, no. 4. 1912 [1913] (p. 1–64); App. I. Bol. R. Orto Bot. Palermo n.s., v. 1. 1914–15 (p. 65–502).)

History, Maltese language, topography, geology, climate, cultivation, general features of flora, annotated list of botanical writers and collectors, annotated bibliography; annotated list of vascular plants (913 species), with pertinent references, Maltese vernacular names, localities, and notes on uses, also similar list of cellular plants; tabular list of vascular plants showing local and extralimital occurrence; separate lists of species from the islets of Cominotto and Filfola. Covers the islands Comino, Gozo, and Malta, the islets Cominotto and Filfola, and some smaller ones. (To Great Britain.)—See also **Sommier, S.** Osservazioni sulla flora maltese. Nuovo Gior. Bot. Ital. n.s., **23**: 295–326. 1916. (Phytogeography; tabular list of Pteridophyta and Bryophyta of the Maltese Islands, Lampedusa, Linosa, and Pantelleria.) —Also Borg, above, and additional references there cited.

Strobl, Gabriel. Flora der Nebroden, mit bezug auf die flora ganz Siciliens. *In* Flora v. 61–70. 1878–87. (For page references to the very numerous installments, see A. Rehder, Bradley Bibliography **1**: 431. 1911.)—Die Dialypetalen der Nebroden Siziliens. Verhandl. K. K. Zool.-Bot. Gesell. Wien **53**: 434–458. 1903.

Bibliography, list of herbaria, topography, geology, climate, plant formations; annotated list of vascular plants (about 1550 species and 430 varieties). The final part (1903) contains a supplementary bibliography, the treatment of Dialypetalae, and a list of mosses and lichens. (In Prov. Messina, Catania, Enna, and Palermo.)—See also Albo, La flora dei Monti Madonie, above.

——— Flora des Etna. *In* Oesterr. Bot. Zeitschr. v. 30–38. 1880–88. (For page references to the very numerous installments see Rehder, Bradley Bibliography **1**: 432. 1911.)

Bibliography, herbaria, explorations; annotated list of vascular and cellular plants. (Prov. Catania.)—See also Tornabene, below.

Taranto, Emmanuele, and **Gerbino, Xaverio.** Catalogus plantarum in agro calato-hieronensi collectarum. 51 p. 29 cm. Catania, 1845.

List of vascular plants of Caltagirone (752 species), with habitats, localities, flowering dates, and some annotations. (Prov. Catania.) (Not seen; title and annotation supplied by R. Pichi-Sermolli.)

Tornabene, Francesco. Flora aetnea, seu descriptio plantarum in Monte Aetna sponte nascentium. 4 v. pl. 24.5 cm. Catinae, 1889–92.

Composition of lava, climate, plant zones, botanical history; flora of vascular plants and lichens with descriptions, local range, iconography, vernacular names (not indexed), uses; synopsis of fossil flora. (Prov. Catania.)—See also **Casalaina, Gaetana.** Su alcune piante della flora etnea. Arch. Bot. (Forlì) **14**: 316–322. 1938. (Critical notes on 27 species recorded by Tornabene, which the author considers should be excluded.)—Also Strobl (Flora des Etna), above.

Ventimigla, L. Nomenclatura siciliana dei vegetali erbacei più comuni. 62 p. 8vo. Palermo, 1903.

Gives names and short descriptions of 225 species. (Not seen; cited from Bul. Bibl. Bot. Ital. **1**: 31. 1904. A 2. ed. in 1906 is listed.)

Virga, Carmelo. Notizie storiche e topografiche d'Isnello e del suo territorio. 130 p. 8vo. Palermo, 1878.

Contains (p. 119–130) an apparently very incomplete list of wild plants, on the Linnaean system, with Sicilian vernacular names. (Prov. Palermo.) (Not seen; title and annotation from notice in Bot. Jahresber. Just **6** (2): 736. 1882.)

Zodda, Giuseppe. Sulla vegetazione del Messinese. Saggio di geografia botanica. Rend. e Mem. R. Accad. Sci., Let. ed Arti Zelanti Acireale ser. 3, v. 3, Mem. Cl. Sci. IV. 100, xxvii p. 1905.

Mainly ecological and phytogeographical; includes tabular, ecologically annotated list of vascular plants (p. i–xxvii), geographically divided. (Prov. Messina.)—See also Nicotra, Prodromus florae messanensis, above.

TOSCANA (TUSCANY)

See also Emilia (Raggi): Lombardia (Ciferri).
The regione of Toscana includes the Provinces of Arezzo, Firenze (Florence), Grosseto, Livorno (Leghorn), Lucca, Massa e Carrara (formerly called Apuania), Pisa, Pistoia (formed in 1926 from part of Firenze), and Siena.

General

Bonacelli, Bendetto. La natura e gli Etruschi. *In* Comitato permanente per l'Etruria. Studi etruschi. v. 2, p. 427–569. 27 cm. Firenze, 1928.
An attempt to summarize knowledge of the agriculture and wild and domesticated plants and animals utilized by the ancient Etruscans; not restricted to Etruria proper.

Caruel, Teodoro. Di alcuni cambiamenti avvenuti nella flora della Toscana in questi ultimi tre secoli. Atti Soc. Ital. Sci. Nat. Milano 9: 439–477. 1866.
Sources of information; annotated alphabetical list of introduced and established, adventive, and extinct species.

——— L'erborista toscano. Chiave analitica per trovare sollecitamente il nome delle piante che nascono selvatiche in Toscana. 304 p. 15.5 cm. Firenze, 1876. (Arnold Arboretum library.)
Continuous brief keys including 2,415 species of vascular plants, unannotated.

——— Prodromo della flora toscana, ossia catalogo metodico delle piante che nascono salvatiche in Toscana e nelle sue isole, o che vi sono estesamente coltivate, con la indicazione dei luoghi nei quali si trovano, del tempo della loro fioritura e fruttificazione, dei loro nomi volgari ed usi. xxii, 767 p. 23 cm. Firenze, 1860–64.—Supplemento al Prodromo della flora Toscana. Atti Soc. Ital. Sci. Nat. Milano 8: 429–479. 1865. (Reprinted 51 p. 1866.)—Secondo supplemento . . . Nuovo Gior. Bot. Ital. 2: 252–297. 1870. (Reprinted 48 p. Reprints of both supplements in Arnold Arboretum library.)
Topography, general features of flora, bibliography, etc.; annotated list of vascular plants, with localities (2,415 species, including supplements; the vascular cryptogams first listed in second supplement).—See also **Baroni, Eugenio.** Supplemento generale al "Prodromo della flora toscana di T. Caruel." 638 p. 25 cm. Firenze, 1897–1908. (Includes all the species listed by Caruel with new localities, corrections, and additional species, bringing total number to 2,624. In Gray Herbarium library.)

——— Statistica botanica della Toscana ossia saggio di studi sulla distribuzione geografica delle piante toscane. 3 p.l., 374 p. map. 25 cm. Firenze, 1871.
Topography, geology, botanical explorations, tabular list of species showing local distribution by regions, description of botanical regions, tabular list showing occurrence on mainland and the islands, floral calendar, etc. Supplements his Prodromo (above).

Savi, Gaetano. Materia medica vegetabile toscana. 2 p.l., 56 p. 60 pl. 39 cm. Firenze, 1805.
Account of native medicinal plants, grouped according to properties, with descriptions, vernacular names, and uses.—See also **Groves, Henry.** Notes on some indigenous Tuscan remedies. Pharm. Jour. and Trans. III, 5: 230–233. 1874. (Running account of native medicinal plants arranged by families, with uses but mostly without vernacular names.)—Also Targioni-Tozzetti, below.

Targioni-Tozzetti, Ottaviano. Catalogus plantarum medicinalium in Etruria sponte nascentium, systemates [sic] Linneano distributus. Trans. Roy. Med.-Bot. Soc. London **1827/29** [v. 2]: 40–60. [1829?]
List of 332 medicinal plants (all vascular except 2 lichens), with vernacular and officinal names and habitat but without uses.—See also Savi, above.

——— Dizionario botanico italiano che comprende i nomi volgari italiani specialmente toscani e vernacoli delle piante raccolti de diversi autori e dalla gente di campagna, col corrispondente latino botanico. 2. ed. 2 v. (xii, 308; 248 p.). 22 cm. Firenze, 1825. (1st ed. 1809.)
Alphabetical list of vernacular names with authorities or localities and scientific equivalents; list of scientific names with collected vernacular names and uses or properties. Replaced as a general work on Italian vernacular

names by Penzig, Flora popolare italiana; cited here because of its special significance for Tuscan vernacular names.

Local

Arcangeli, Giovanni. Le piante fino ad ora raccolte in Gorgona. Ricerche e Lav. Ist. Bot. R. Univ. Pisa **2:** 109–144. 1888.

Bibliography, geology, general features of flora; list of vascular and cellular plants (349 vascular species), with habitats and localities. (Prov. Livorno.)—See also **Sommier, Stefano.** Piante raccolte durante la gita sociale alla Gorgona. Bul. Soc. Bot. Ital. **1899:** 117–126. 1899. (Brings total to 427 species.)—Also his La flora dell'Arcipelago Toscano. Nuovo Gior. Bot. Ital. n.s., **9:** 319–354. 1902. (Includes, p. 334–338, list of plants from Gorgona including 44 new to the island.)—Also **Arcangeli, Giovanni.** Sopra varie piante ed alcuni minerali raccolti di recente. Atti Soc. Toscana **Sci. Nat. Pisa (Proc. Verb.) 14:** 4–7. 1903. (Includes short list of vascular plants, with localities, collected by author on Gorgona in August 1902; also list of species collected on the Scogli Neri along the shore between Chioma and Castiglioncello not far from Fortullino.)

Archbald, A. B. Flora dell'alto Serchio e del Lima, ossia catalogo delle piante della regione appennina lucchese . . . pubblicata per cura di A. Carina. 82 p. map. 18 cm. Lucca, 1874. (Gray Herbarium library.)

Geography, topography, plant zones, etc.; annotated tabular list of vascular plants (over 1200 species), with localities. (Mostly in Prov. Lucca, partly in Prov. Massa e Carrara and Prov. Pistoia.)—See also **Andreucci, Arnolfo.** Contributo alla conoscenza della flora dell'alta valle del Serchio (Garfagnana). Malpighia **27:** 260–275, 407–438, 503–524. 1916. (Bibliography, topography, climate, geology, general features of flora, etc.; tabular list of 186 vascular plants, showing plant zones and flowering period; list of the same, with localities and some annotations.)—Also **Arrigoni, P. V.** Sui limiti altimetrici dei consorzi rupestri di leccio in Garfagnana. Nuovo Gior. Bot. Ital. n.s., **63:** 531–590. 9 fig. (on 4 pl. [on 2]), 2 maps. 1957. (Includes (p. 558–566) list of 140 vascular plants.)—Also **Padula, Michele.** Contributo allo studio dei limiti altimetrici del faggio in Garfagnana. l.c. **63:** 591–678. 21 fig. (on 6 pl. [on 3]), maps 3–4. 1957. (Includes (p. 615–625) list of 222 vascular plants.)—Also Santarelli, below.

Bicchi, Cesare. La flora lucchese di fronte alla flora generale d'Italia ed alle flore speciali della Toscana e della Sicilia. 76 p. Lucca, 1877. (In Cronaca annuale del R. Liceo Macchiavelli in Lucca nell'anno scolastico 1876–1877.)

Includes unannotated list of known phanerogams (1,802 species) of the Province of Lucca. (Not seen; title and annotation mostly from notice in Bot. Jahresber. Just **6** (2): 733. 1882.)—See also Paoli, below.

Bonaventura, Gustavo. Primo [–terzo] contributo allo studio della flora del M. Fumaiolo. Bol. R. Ist. Sup. Agr. Pisa **8:** 671–709. 3 fig. (maps). 1932; **9:** 392–401. 1933; **10:** 1–7. 1934.—Quarto contributo . . . Annali Facoltà Agr. R. Univ. Pisa n.s., **1:** 215–225. 1938.

Topography, geology, references to previous publications; lists of vascular plants collected by author or previously recorded, with localities. (Prov. Arezzo, also Prov. Forlì in Emilia.)

Caruel, Teodoro. Florula di Montecristo. Atti Soc. Ital. Sci. Nat. Milano **6:** 74–109. 1864. (Reprinted 38 p., in Gray Herbarium library.)

Geography, botanical explorations, topography, soils; briefly annotated list of vascular plants (343 species). (Island in Tuscan Archipelago, in Prov. Grosseto.)

——— and **Levier, Emilio.** Saggio di un calendario florale per Firenze. Nuovo Gior. Bot. Ital. **3:** 121–165. 1871.

List of plants in flower during each 10-day period throughout the year, the names given in capitals at date of first flowering, in italics at date of last flowering; statistics. Refers to the valley in which Florence lies and the neighboring hills. (Prov. Firenze.)

Corradi, Rinaldo. La flora spontanea del parco delle Cascine (Firenze). Nuovo Gior. Bot. Ital. n.s., **53:** 336–342. 1946.

History, botanical explorations; list of about 700 vascular plants, with indication of habitat and life-form. (An area of 118 hectares northeast of Firenze, Prov. Firenze.)

Corti, Roberto. Aspetti geobotanici della selva costiera. La selva pisana a S. Rossore e l'importanza di questa formazione relitta per la storia della vegetazione mediterranea. (Ricerche sulla vegetazione dell'Etruria. X.) Nuovo Gior. Bot. Ital. n.s., **62**: 75–262. 7 fig. (maps), pl. 5–30, map. 1956.
Includes (p. 85–195) annotated list of vascular plants and a few bryophytes and lichens (over 700 taxa); English summary. (Prov. Pisa.)

——— Rilievi nelle pinete delle colline a sud-ovest di Firenze. (Ricerche sulla vegetazione dei dintorni di Firenze 3.) Nuovo Gior. Bot. Ital. n.s., **41**: 25–120. 22 tab. (1, 4–22 in text). 1934.
Includes (p. 28–34) list of vascular plants collected by author, with indication of their biological form (Raunkiær). (Prov. Firenze.)

Ferrarini, Erminio. Le erbe infestanti di un terreno dell'Alta Lunigiana (Massa Carrara). (Ricerche sulla flora infestante delle colture in Italia.) Nuovo Gior. Bot. Ital. n.s., **61**: 133–181. 11 fig. (fig. 1, map). 1954 (1955).
Includes list of weedy vascular plants (262 species). (Prov. Massa e Carrara.)

——— Flora e vegetazione dei castagneti e dei cerreti dell'Alta Valle del Taverone (Alta Lunigiana). Nuovo Gior. Bot. Ital. n.s., **64**: 485–640. 52 fig. (incl. maps; partly on plates), map. 1958.
Includes (p. 493–522) annotated list of vascular plants (543 taxa, including 526 species); bibliography, English summary. (Prov. Massa e Carrara.)

Fiori, Adriano. Confronto tra la flora del M. Ferrato (serpentino) a quella della Calvana (calcare alberese). (Comitato "Pro flora italica" II. La flora dei serpentini della Toscana II.) Nuovo Gior. Bot. Ital. n.s., **21**: 216–240. 1914.
Contrast between floras of Monte Ferrato (serpentine) and Monte Calvana (calcareous), with lists, bibliography; annotated list of certain vascular plants. (Prov. Firenze.)—See also **Messeri, Albina.** La vegetazione delle roccie ofiolitiche di Monte Ferrato (presso Prato). (Ricerche sulla vegetazione dei dintorni di Firenze 4.) Nuovo Gior. Bot. Ital. n.s., **43**: 277–372. 20 fig. (incl. map). 1936. (Includes list (p. 283–301) of all known vascular plants of Monte Ferrato, with indication of their biological form.)

——— Rilievi geografici e forestali sulla flora del bacino della Cecina e località finitime. Annali R. Inst. Super. Forest. Naz. Firenze **5**: 149–186. 3 pl. (incl. 2 maps). 1920.
Includes lists of species grouped by habitat, but no complete flora. (In Provinces of Grosseto, Pisa, and Siena.)—See also **Vergnano, Ornella.** Erborizzazioni su alcune serpentine della Val di Cecina. Nuovo Gior. Bot. Ital. n.s., **60**: 330–332. 1953. (Three lists, including some species not given by Fiori.)

Francini, Eleonora. La vegetazione del Laghetto di Sibolla (Valdarno inferiore). (Ricerche sulla vegetazione dell'Etruria marittima. II.) Nuovo Gior. Bot. Ital. n.s., **43**: 62–130. 9 fig. (incl. map). 1936.
Ecological; includes (p. 68–72) unannotated list of vascular plants. (A lake, also called Lago di Sibolla, 2.5 kilometers northeast of Altopascio, in Prov. Lucca.)

——— La vegetazione di Monte Ceceri. (Ricerche sulla vegetazione dei dintorni di Firenze. 1.) Nuovo Gior. Bot. Ital. n.s., **39**: 631–644. 1932.
Geology, etc.; list of vascular plants collected by author, with indication of life-form. (Prov. Firenze.)

Gabelli, Lucio. Contributo alla flora murale e ruderale del Senese. Atti Pont. Accad. Romana Nuovi Lincei **68**: 137–146. 1915.
General considerations, localities, etc.; list of 127 ruderal vascular plants, with localities, mostly from an area southeast of Siena. (Prov. Siena.)

Giuli, Giuseppe. Piante che spontaneamente vivono dentro Siena e suoi contorni. (? p.) Siena, 1862.
Not seen; no review available. (Prov. Siena.)

Marcucci, Emilio. Piante del Casentino. *In* Beni, Carlo. Guida illustrata del Casentino. 2. ed. p. 153–190. 16 cm. Firenze, 1889. (1st ed. 1881.)

List of vascular plants (839 species), with localities. (Prov. Arezzo.) (Not seen; title and annotation supplied by R. Pichi-Sermolli.)

Marsilii, Hiram. La flora infestante di un podere della regione collinare della Val d'Elsa. (Ricerche sulla flora infestante delle colture in Italia. V.) Nuovo Gior. Bot. Ital. n.s., **60**: 615–659. 5 fig., tab. 1953.

Includes list of weeds (208 species) from a farm near Castelfiorentino. (Prov. Firenze.)—See also Sarfatti, below.

Maugini, Elena. La vegetazione del bosco di Poggio a Caiano (Firenze). Nuovo Gior. Bot. Ital. n.s., **53**: 331–336. 1946.

Climate, woody plants, etc.; list of vascular plants with indication of locality, life-form, etc. (Prov. Firenze.)

Messeri, Albina. Rilievi a Poggio Pratone. (Ricerche sulla vegetazione dei dintorni di Firenze. 2.) Nuovo Gior. Bot. Ital., n.s., **39**: 645–658. 1932.

Geology, etc.; list of vascular plants collected by author in July, with indication of life-form. (Prov. Firenze.)

Montelucci, Giuliano. Contributo alla flora del Valdarno superiore. Piante raccolte nei dintorni di Rignano sull'Arno. Nuovo Gior. Bot. Ital. n.s., **40**: 479–530. 1 fig. (map). 1933.

Botanical explorations, topography, geology, plant zones and habitats, climate; annotated list of vascular plants (569 species) collected by author in 1912-1932. (Prov. Firenze.)—See also his Rilievi sulla vegetazione del Valdarno superiore (Rignano sull'Arno). II. contributo. l.c. n.s., **50**: 35–74. 6 fig. 1943. (Adds 107 species; biological spectrum, phytocenoses, bibliography.)

Moris, G. G., and Notaris, Giuseppe de. Florula Caprariae sive enumeratio plantarum in insula Capraria vel sponte nascentium vel ad utilitatem latius excultarum. Mem. R. Accad. Sci. Torino II, **2** (Sci. Fis. e Mat.): 59–300. 6 pl. 1840. (Preprinted 244 p. Taurini, 1839; in Arnold Arboretum library.)

Topography, geology, agriculture, climate; briefly annotated list of vascular and cellular plants (701 species, of which 470 are vascular). Capraria is the ancient name of the present Capraia (or Capraja), east of the north tip of Corsica, formerly in Liguria (Prov. Genova), now in Prov. Livorno.—See also **Sommier, Stefano.** Aggiunte alla flora di Capraia. Nuovo Gior. Bot. Ital. n.s., 5: 106–139. 1898. (Botanical explorations, etc.; annotated list of 134 vascular plants not in Moris and de Notaris' Florula, with notes on others, as well as on some cellular plants.)—Also his Capraia. l.c. n.s., 9: 339–342. 1902. (List of species collected by author and others, including some **additions.)**

Nannizzi, Arturo. Le piante medicinali ed aromatiche che crescono spontanee nella provincia di Siena. (Manuale pratico per uso degli allievi del corso di erboristeria.) 205 p. 137 illus. 18 cm. Sienna, 1934.

Directions for collecting, etc.; descriptive list of about 150 medicinal plants native in the Province of Siena. (Not seen; title and annotation supplied by F. Sappa.)

Neri, Francisco. Contribuzione alla flora toscana. La flora del Volterrano. Atti Soc. Toscana Sci. Nat. Pisa (Proc. Verb.) **9**: 45–59. 1894.

Botanical explorations, list of localities with altitudes, etc.; list of vascular plants (408 species), with localities, of the vicinity of Volterra. (Prov. Pisa.) —See also his Contribuzione alla flora toscana. "Ancora la flora del Volterrano." l.c. **9**: 213–216. 1895. (List of vascular and cellular plants, with localities.)—Also his Ancora dell'erbario Amidei. l.c. **8**: 223–225. 1893; **9**: 216–220. 1895. (Includes list of 99 additional species, with localities.)

Paoli, H. De lucensibus artis plantarum doctoribus commentariolum et synopsis plantarum in agro lucensi additamenta. 38 p. Lucae, 1905.

Historical sketch of the botanists of the Province of Lucca of the 16th to 18th centuries; list of plants new to the province; list of exotic plants cultivated in the botanical garden of Lucca. (Not seen; cited from Bot. Centralbl. **102**: 352. 1906.)—See also Bicchi, above.

Pellegrini, Pietro. Flora della provincia di Apuania. 450 p. 8vo. Apuania, 1942.

List of vascular and cellular plants (2350 species, of which 1741 are phanerogams), with detailed localities and bibliography; includes the commonly cultivated plants. (Now Prov. Massa e Carrara.) (Not seen; title and annotation from notice in Arch. Bot. (Forlì) **18**: 183–184. 1942, and from V. Giacomini.)

Piccioli, Lodovico. Guida alle escursioni botaniche nei dintorni di Vallombrosa con chiavi analitiche per determinare i nomi delle piante che vi crescono. 297 p. 23.5 cm. Firenze, 1888. (Arnold Arboretum library.)

Flora of vascular plants in form of annotated keys, with vernacular names and localities. (Prov. Firenze.)—See also **Solla, R. F.** Caratteri propri della flora di Vallombrosa. Bul. Soc. Bot. Ital. **1893**: 52–60, 124–128, 197–207, 276–285, 381–393. 1893. (Topography, plant associations, altitudinal range of various species, phytogeography, etc.)

Pichi-Sermolli, Rodolfo. Flora e vegetazione delle serpentine e delle altre ofioliti dell'alta valle del Tevere (Toscana). Webbia **6**: 1–378. 22 fig. (mostly on pls.), 2 maps. 1948.

Mainly ecological and phytosociological; includes (p. 37–76) sketch of botanical explorations and list of vascular plants of the ophiolitic soils in upper valley of the Tiber (in Tuscany), with localities, indication of biological form, and frequent annotations; bibliography, English summary. (Prov. Arezzo.)

Preda, Agilulfo. Contributo alla flora vascolare del territorio livornese. [1.]–4. centuria. Nuovo Gior. Bot. Ital. n.s., **2**: 108–118, 217–222. 1895. Bul. Soc. Bot. Ital. **1896**: 6–11, 190–194. 1896.

Topography, botanical regions, climate; list of 400 vascular plants, with localities, of which many are recorded as new to the Province of Livorno.—See also **Barsali, Egidio.** Aggiunte alla flora livornese. Bul. Soc. Bot. Ital. **1904**: 202–207. 1904. (Short list of phanerogams and lichens collected by Dr. Danielli.)

Puccinelli, Benedetto. Synopsis plantarum in agro lucensi sponte nascentium. Atti R. Accad. Lucchese Sci., Let. ed Arti **11**: 313–392. 4 pl. 1841 (Monandria-Tetrandria); **12**: 1 p.l., 1–144 p. 5 pl. 1843 (Tetrandria-Enneandria); **13**: 3 p.l., 1–96 p., 3 l. 2 pl. 1845 (Decandria-Didynamia); **14**: 91–297. 4 pl. [1848? t. p. date, 1852] (Tetradymia-Polygamia Monoecia). (Reprinted 2 v.; xlix, 326; 327–531 p. 1841–48. Only v. 1 seen, in library of Muséum national d'histoire naturelle, Paris; data for v. 2, which is the installment in vol. 14 of the Atti, supplied by R. Pichi-Sermolli.)

Botanical explorations; descriptive flora of phanerogams on the Linnaean system, with localities. The roman pagination is a synopsis of families on the natural system, not in the original work. (Prov. Lucca.)—See also his Additamentum ad Synopsin plantarum . . . Gior. Bot. Ital. anno 1, tomo 1, [parte 1]: 118–123. 1844. (Covers Monandria-Decandria.)—Also **Bicchi, Cesare.** Aggiunta alla Flora lucchese del Prof. Benedetto Puccinelli. 26 p. 27 cm. Lucca, 1860. (List of phanerogams. Not seen; title and annotation supplied by R. Pichi-Sermolli.)

Raimondi, Carlo. Le piante medicinali spontanee e coltivabili nel territorio senese e grossetano. Atti Accad. Fisiocr. Siena VIII, **10**: 129–140. 1918. (Library of Academy of Natural Sciences of Philadelphia.)

Annotated list of wild medicinal plants, and notes on cultivation of others. (Prov. Grosseto and Prov. Siena.)

Sandri, Giulio, and **Fantozzi, Pietro.** Contribuzione alla flora di Valdinievole. Nuovo Gior. Bot. Ital. n.s., **2**: 129–180, 289–333. 1895.

Topography, geology, botanical regions, etc.; annotated list of vascular plants (1,165 species). (Prov. Lucca and Prov. Pistoia.)

Santarelli, Enrico. Contribuzione alla flora alveale del Serchio. Atti Soc. Toscana Sci. Nat. Pisa Mem. **34**: 3–45. 1922.

General considerations, plant associations, etc.; list of 409 vascular plants, with localities; bibliography. (Prov. Lucca.)—See also Archbald, above.

Sarfatti, Giacomino. La flora infestante del podere "Cascine" (Firenze). (Ricerche sulla flora infestante delle colture in Italia. II.) Nuovo Gior. Bot. Ital. n.s., **56**: 21–57. 1 fig. (map). 1949.

Includes (p. 24–28) tabular list of 198 weedy vascular plants of a small farm near Florence, on the Arno; English summary. (Prov. Firenze.)—See also Marsilii, above.

Savi, Gaetano. Flora pisana. 2 v. (xiii, 485; 500 p.). 2 pl. 22.5 cm. Pisa, 1798.

Descriptive flora of vascular and cellular plants, on Linnaean system, with localities for the scarcer species. Covers a great part of the Province of Pisa.

Simi, Emilio. Flora Alpium versiliensium exhibens plantas in illis apuanarum Alpium regionibus sponte crescentes secundum systema sexuale Linnaeanum et systema naturale De-Candolleanum digestas. 274 p. 22 cm. Massae, 1851. (Kew library.)

Descriptive flora of vascular and cellular plants [over 500 species, of which 463 are vascular], on the Linnaean system, with localities; list of the species, arranged by the Candollean system. (Prov. Lucca and Prov. Massa e Carrara.)—See also **Milani, G. B.** Supplemento alla flora delle Alpi versiliensi. 36 p. 8vo. Salerno, 1885. (List including over 200 additional species as well as new localities. Not seen; title and annotation mostly from notice in Nuovo Gior. Bot. Ital. **18**: 62. 1886.)—Also **Rossetti, Corrado.** Contribuzione alla flora della Versilia. Atti Soc. Toscana Sci. Nat. Pisa Mem. **9**: 384–426. 1888. (Geography, plant zones, topography, botanical explorations; partial list of vascular plants (465 species, of which over 420 are first recorded here), with localities.)—Also his Seconda contribuzione . . . l.c. Proc. Verb. **8**: 120–143. 1892. (List of mountains explored by author, with altitudes; list of additional species and new localities.)

Sommier, Stefano. Una cima vergine nelle Alpi apuane. Nuovo Gior. Bot. Ital. n.s., **1**: 11–34. pl. 1–3. 1894.

Topography, general features of flora, etc.; annotated list of 140 vascular plants found on the summit of Procinto, and list of 43 additional plants of the adjacent Cintura del Procinto. Procinto is an almost inaccessible summit with an altitude of 1,177 meters, not before visited by a botanist. (Prov. Lucca.)

——— La flora dell'Arcipelago Toscano. Nuovo Gior. Bot. Ital. n.s., **9**: 319–354. 1902; **10**: 133–200. 1903.

Botanical explorations, with bibliographical footnotes; separate annotated lists of vascular plants of each island supplementary to Caruel's treatment in his "Statistica . . ." (1871); (v. 10) tabular list of vascular plants (1,411 species; also 247 Bryophyta) showing distribution by islands. (Prov. Grosseto and Prov. Livorno.) Covers the following islands: Monte Argentaro (p. 322–325); Elba (325–332); Giannutri (332–333); Giglio (333–334); Gorgona (334–338); Pianosa (338–339); Capraia (339–342); Montecristo (342–346); Palmaiola and Cerboli (346–347); Topi (347–349); La Scola di Pianosa (349–350); La Praiola (350–351); Isolotto della Cappa (350–351); Argentarola (351); Pan di Zucchero (352–353); Troia (353); Formiche di Grosseto, Formica di Burano or di Ansedonia, Formica di Montecristo or Scoglio d'Africa (354).—For Argentarola and Scoglio Rosso, see also his Aggiunte alla flora del Monte Argentaro . . . Bul. Soc. Bot. Ital. **1903**: 232–236. 1903 (see p. 234–236).—For Capraia, see also Moris and de Notaris, Florula Caprariae, above.—For Elba, see also his [Alcune piante . . .] Bul. Soc. Bot. Ital. **1904**: 304–305. 1904. Also **Corti, Roberto.** Appunti sulla vegetazione dell'isola d'Elba. I. Una gita a M. Orello e ai monti tra Rio Alto e Portolongone, con osservazioni sui distretti ofiolitici dell'isola. Nuovo Gior. Bot. Ital. n.s., **47**: 494–504. 9 fig. 1940. (Includes short lists of species from various localities.) Also **Negri, Giovanni.** Excursione della Società Botanica Italiana all'isola d'Elba (Aprile 1950). l.c. **57**: 276–293 incl. 18 fig. (mostly on 4 pl.). 1950. (Topography, climate; description of different habitats with extensive lists of species.)—For Giannutri, see also Tanfani, below. —For Giglio, see also Sommier, L'isola del Giglio, below.—For Gorgona, see also Arcangeli, above.—For Monte Argentaro, see also **Sommier, S.** Aggiunte alla flora del Monte Argentaro . . . Bul. Soc. Bot. Ital. **1903**: 232–236. 1903 (see p. 232–234). Nuovi materiale per la flora del Monte Argentaro. l.c. **1911**: 37–41. 1911. Ulteriore contributo alla flora del Argentaro. Nuovo Gior. Bot. Ital. n.s., **19**: 116–123. pl. 9. 1912. (Adds 63 species to the 952 recorded by him through 1911.)—For Montecristo, see also Caruel, Florula di Montecristo, above.—For Pianosa, see also Sommier, La flora dell'isola di Pianosa, below.

——— La flora dell'isola di Pianosa nel Mar Tirreno. Nuovo Gior. Bot. Ital. n.s., **16**: 357–438. 4 fig. 1909; **17**: 123–164. 1910.

Botanical explorations, general features of flora, list of plants of adjacent islets Scola and Scarpa; annotated list of vascular plants (510 species), also list of cellular plants; statistics, phytogeography. (Prov. Livorno.)

——— L'isola del Giglio e la sua flora. Con notizie geologiche del Prof. C. de Stefani. clxxii, 164 p. 10 fig., 5 pl., map (in pocket). 28 cm. Torino, 1900.

History, fauna, geology, annotated bibliography; annotated list of 1048 plants (674 vascular species). (Prov. Grosseto.)—See also his La flora dell'Arcipelago Toscano, above (p. 333–334).

Tanfani, Enrico. Florula di Giannutri. Nuovo Gior. Bot. Ital. **22**: 153–216. 1890.

History of island, topography, geology, general features of flora, botanical explorations, bibliography; annotated list of vascular and cellular plants (173 species, of which 127 are spermatophytes.) (Prov. Grosseto.)—See also **Sommier, Stefano.** Piante vascolari nuove raccolte a Giannutri dal 3 al 7 marzo 1897. Bul. Soc. Bot. Ital. **1897**: 129–136. 1897. (Annotated list of 61 species.)—Also his La flora dell'arcipelago toscano, above (p. 332–333).—Also his Nuove aggiunte alla florula di Giannutri. Bul. Soc. Bot. Ital. **1903**: 228–232. 1903.

Tassi, Attilio. Cenno sulla botanica agraria, medica, economica ed industriale della provincia di Siena. 130 p. pl. 29 cm. Siena, 1865. (Estratto dall'Annuario corografico-economico-amministrativo della provincia di Siena.) (New York Botanical Garden library.)

Annotated list of economically significant vascular and a few cellular plants, with vernacular names and uses.

——— Sulla flora della provincia senese e Maremma toscana. 63 p. pl. 22.5 cm. Siena, 1862. (From "Guida di Siena pubblicata in occasione del X. Congresso degli scienziati italiani.") (Arnold Arboretum library.)

List of botanists, bibliography, botanical explorations, topography, general features of flora, useful plants, etc.; list of wild and cultivated vascular and cellular plants (1,556 species), mostly unannotated and without localities but with indication of introduced species, cultivated species, and those collected by author near Siena. (Prov. Siena and part of Grosseto and Livorno.)—See also **Sommier, Stefano.** Una gita in Maremma. Bul. Soc. Bot. Ital. **1892**: 314–329. 1892. (General features of flora; list of rarer species collected by author in April 1892, with localities. Apparently the paper was never completed.)—Also **Tassi, Flaminio.** Contribuzione alla flora senese. Prima. Orchidaceae . . . Atti R. Accad. Fisiocr. Siena IV, **6** (Suppl.): 121 (i.e. 119 bis)–128. 1895. (List of additional species and localities.)—Also **Nannizzi, Arturo.** Contribuzione alla flora vascolare della provincia di Siena. l.c. X, **3**: 601–609. 1929. (List of 50 vascular plants not previously recorded, with data. In library of Academy of Natural Sciences, Philadelphia.)

Tassi, Flaminio. La botanica nel Senese. Notizie storiche. (Contributo alla storia della botanica in Italia.) Bul. Lab. ed Orto Bot. R. Univ. Siena **7**: 3–56. 2 fig., port. 1905.

Chronological account of botanists connected with the flora of the Province of Siena, with mention of publications, herbaria, etc., and bibliographical footnotes.

Vannelli, Siro. La flora infestante di un podere presso Montecatini Terme. (Ricerche sulla flora infestante delle colture in Italia. IV.) Nuovo Gior. Bot. Ital. n.s., **59**: 388–435. 12 fig. 1953.

Includes list of weeds (241 species) found on a small farm; bibliography. (Prov. Pistoia.)

TRENTINO-ALTO ADIGE

See also Veneto (Dalla Torre; Pollini); Switzerland, General (Dalla Torre).

Essentially equivalent to the former South Tirol, or Tirol south of the Alps. Acquired from Austria in 1919; from about 1927 to 1947 known as Venezia Tridentina. Includes the Provinces of Bolzano (German Bozen, also called Alto Adige; formed in 1927 from northern part of Trento, enlarged in 1948 by transfer of German-speaking communes from Trento), and Trento. The name Trentino has generally been used as equivalent to the Province of Trento.

General

Bonapace, Benedetto. La florula arvense del Trentino. Studi Trent. Sci. Nat. 25: 158–170. 1948. (Library of New York State College of Agriculture, Ithaca.)

Tabular list of spontaneous and naturalized agrophilous plants (154 species), showing distribution, etc.; essentially a list of weedy plants.

Da Grauno, Atanasio. Piante ed erbe medicinali della regione tridentina. 3. ed. migl., ampl. ed illus. 253 p. pls. (12 col.), front. (port.). 19.5 cm. Trento, 1937.

Bibliography; list of medicinal plants (all or mostly native) with patois names, description, habitat, uses, etc., alphabetically arranged by Italian vernacular names; list of diseases with remedies.

Dalla Fior, Giuseppe. La nostra flora. Guida alla conoscenza della flora della regione tridentina. 582 p. 21 + 827 fig. 17.5 cm. Trento, 1926.

Botanical explorations; flora of vascular plants in form of keys; alphabetical list of patois names, glossary of specific epithets, list of local geographical names derived from names of plants. Ranges are given only in general terms.

Dalla Torre, K. W. von, and Sarnthein, Ludwig, *graf von.* **Flora der** gefürsteten grafschaft Tirol, des landes Vorarlberg, und des fürstenthumes Liechtenstein . . . 6 v. port., maps. 22.5 cm. Innsbruck, 1900–1913.

Vol. 1, annotated bibliography, alphabetically arranged by authors, with brief biographical notices and index; v. 2–5, annotated list of cellular plants; v. 6 (in 4 parts, 1909–1913, paged separately), annotated list of vascular plants, with detailed localities; botanical explorations; list of botanists additional to those listed in the bibliography and the account of botanical explorations, with biographical notes; supplement to bibliography (thru 1907). Area covered included the then Austrian Tyrol (including the present Italian regione of Trentino-Alto Adige). Replaces **Hausmann, Franz von.** Flora von Tirol. Ein verzeichniss der in Tirol und Vorarlberg wild wachsenden und häufiger gebauten gefässpflanzen . . . 3 v. (4 p.l., xiv, 1614 p.). 18 cm. Innsbruck, 1851–54. (Annotated, keyed, briefly descriptive flora, list of vascular plants (2,313 species), with localities; bibliography, etc.) Also replaces **Ambrosi, Francesco.** Flora del Tirolo meridionale ossia descrizone delle specie fanerogame che crescono spontanee sopra il suolo trentino e nelle terre adjacenti comprese fra la catena delle Alpi Retiche sino ai confini del Lombardo-Veneto, loro proprietà, ec. . . Flora Tiroliae australis (on extra title page) . . . v. 1–2, pt. 1 (xx, 964; 820 p.). 23 cm. Padova, 1854–57. (General features of flora, botanical explorations, bibliography, etc.; annotated descriptive flora of vascular plants, with localities, vernacular names in Italian, French, and German, and uses; includes cultivated species. Incomplete: v. 1, Monocotyledones; v. 2, Coniferae, Monochlamydeae, Dichlamydeae Gamopetalae (part). In Gray Herbarium library.)—Also replaces **Facchini, Francesco.** Flora von Südtirol. Mit einem vorworte und anmerkungen von Fr. B. v. Hausmann. (Zur flora Tirols. 1. heft.) Zeitschr. Ferdinandeums Tirol u. Vorarlberg ser. III, heft 5, Naturw. Abt. viii, 151 p. 1855. (Title of list proper, Elenchus plantarum quas in Tirolia cisalpina collegit Dr. Facchini. Annotated list of spermatophytes on the Linnaean system, with localities for rarer species; Anhang (p. 129–151), a series of notes by the editor.)—See also **Murr, Josef.** Le mie scoperte botaniche nel Trentino dal 1897 al 1906, con alcune aggiunte. Studi Trent. Sci. Nat. 1: 230–246 incl. 4 pl. 1920. (List of noteworthy vascular plants and mosses, with references to original publications of the records. In Harvard College library.)—Also **Pfaff, Wilhelm.** Südtirolische kriegsbotanik. Schlern 4: 15–21. 1923. (Sources of introduction, etc.; list of introduced or adventive plants, with localities. In University of Chicago library.)—Also his Nachträgliches zur kriegsbotanik. l.c. 5: 72–76. 1924. (Includes additional list, bringing total to 181 species, and discussion. In University of Chicago library.)—Also **Murr, J.** Aggiunte alla flora delle provincie di Bolzano e Trento e del territorio confinante. Studi Trent. 8, Cl. II. Sci. Nat. ed Econ.: 101–106. 1927. (List of vascular plants, with localities. In library of Academy of Natural Sciences, Philadelphia.)—Also **Dalla Torre, K. W. von.** Beiträge zur flora von Tirol und Vorarlberg. Nach dem herbarium und dem handschriftlichen nachlasse des hauptmann-auditors Friedrich Beer bearbeitet. Veröffentlich. Mus. Ferdinand. Innsbruck **7:** 1–120. 1929. (Extensive list of vascular plants, with localities.)—Also **Gams, Helmut.** Die fortschritte in der erforschung der flora und vegetation von Tirol in den letzten jahren. Ber. Naturw.-Med. Ver. Innsbruck **42:** 185–194. 1931. (Brief summary of work (from 1927 on) on the flora, plant sociology, and history of the vegetation, with lists of publications.)—Also **Handel-Mazzetti, Hermann.** Zur floristischen erforschung des ehemaligen landes Tirol und Vorarlberg. Ber. Bayer. Bot. Gesell. **26:** 56–80. 1943. (Bibliography; extensive list of vascular plants, with localities and references, intended to include all significant records from the former Tirol and Vorarlberg since Dalla Torre and Sarnthein's flora of Tirol, Vorarlberg and Liechtenstein (1909–1913) and Murr's flora of Vorarlberg and Liechtenstein (1923–27). Minor forms, as well as microspecies in such genera as *Rosa* and *Hieracium,* are omitted.)—Also his Zur floristischen erforschung von Tirol und Vorarlberg. l.c. **27:** 175–185. 1947; Österr. Bot. Zeitschr. **96:** 83–108. 1949; Verhandl. Zool.-Bot. Gesell. Wien **93:** 81–99. 1953; **94:** 114–137. 1954; **95:** 155–167. 1955. (Similar lists.)—Also **Bonapace, Benedetto.** Entità floristiche rare o endemiche della regione Trentino-Alto Adige. Studi Trent. Sci. Nat. **30:** 184–200. 1953. (List of 153 taxa, with localities.)—Also **Handel-Mazzetti, Hermann.** Angaben aus Prof. Dr. Kotulas pflanzenkatalogen, als beitrag zur floristischen erforschung von Tirol und Vorarlberg. Verhandl. Zool.-Bot. Gesell. Wien **95:** 114–154. 1955. (Extensive list.)

Gams, Helmut. Ueberblick über die floren- und vegetationsgeschichte Tirols. I. teil. Schlern **23:** 435–438, 491–493. 1949; **24:** 15–17, 67–72, 125–130, 325–327, 450–454. 1950; **25:** 26–29. 1951.

Includes in first three installments discussion of floristic and geobotanical investigations with special reference to history of vegetation, plant zones, action of man, etc. (Not seen; cited from Arch. Bot. (Forlì) **27:** 231. 1951.)

Unterforcher, August. Rätoromanische ortsnamen aus pflanzennamen. Beitrag zur tirolischen namenforschung. Zeitschr. Ferdinandeums Tirol u. Vorarlberg III, **36:** 371–399. 1892.

Alphabetical list of botanical names (mostly Latin), with locality names derived from them. Relates primarily to German-speaking part of Tirol (and to Vorarlberg), but cites names from Italian part also. The author's etymology and botany have been severely criticized; see Dalla Torre and Sarnthein, Flora der gefürsteten grafschaft Tirol **1:** 312 and xxiv. 1900.

Local

Artzt, F. F. A. Botanisches von Bad Ratzes. ***In*** **Prossliner, Karl. Das "Bad Ratzes" in Südtirol. Ein topographisch-kunstgeschichtlich-naturwissenschaftliche lokalskizze.** 2. aufl. p. 70–101. 8vo. Bilin, 1895. (1st ed. 1883.)

List of vascular plants (728 species, the common ones omitted), also some mosses and lichens, with localities. (Prov. Bolzano.) (Not seen; cited mainly from Dalla Torre and Sarnthein, Flora der gefürsteten grafschaft Tirol . . . 1: 8. 1900.)—See also **Rottenbach, Heinrich.** Zur flora der umgebung von Ratzes in Südtirol. Deut. Bot. Monatsschr. **18:** 160–163. 1900. (List of additions)—Also **Junge, Paul.** Beitrag zur kenntnis der flora der umgebung von Ratzes in Südtirol. l.c. **21:** 19–21. 1903. (List including additions.)

Bolzon, Pio. Contributo alla flora dell'Alto Adige. Nota [I]–X. Bul. Soc. Bot. Ital. **1926:** 22–32. 1926. Nuovo Gior. Bot. Ital. n.s., **34:** 192–200. 1927; **35:** 97–111, 371–375. 1928; **36:** 647–660. 1930; **38:** 295–307, 690–697. 1931–32; **40:** 118–126, 553–559. 1933–34; **42:** 227–232. 1935.

Lists of vascular plants and some mosses collected by author at various localities and altitudes, with indication of new records. (Prov. Bolzano.)

——— Flora del Monte Marmolada (Dolomiti Agordino-Fassane) con osservazioni sopra talune associazoni. Nuovo Gior. Bot. Ital. n.s., **21:** 143–215. pl. 2 (map). 1914.

Geography, altitudes, bibliography, plants of special localities, etc.; list of vascular plants (425 species), with localities. Monte Marmolada is the highest summit in the Dolomites. (Prov. Trento, and Prov. Belluno in Veneto.)—See also his Piante silicicole del Monte Marmolada (Dolomiti Agordino-Fassane). **Bul. Soc. Bot. Ital. 1915: 6–11. 1915. (Mention of** various species from siliceous habitats.)

Bonomi, Lino. Naturalisti, medici e tecnici trentini. Contributo alla storia della scienza in Italia. iii, 222, [12] p. illus., 24 pl. 25 cm. Trento, 1930.
Short biographies of dead and living scientists and technical men, including 27 botanists. Published by the Museo di storia naturale della Venezia Tridentina in Trento. (Prov. Trento.) (Not seen; cited mostly from Bibl. Ital. **1930,** A: 277. 1932.)

Chiarugi, Alberto. Erborizzazioni nella Val Gardena (Alpi Veneto-Tridentine). Bul. Soc. Ital. **1924:** 106–120. 1924.
List of vascular plants collected by author in July-August 1923, with localities. (Prov. Bolzano.)

Cobau, Roberto. Erborizzazioni sul Pasubio e nella Valle di Canale (1929 e 1930). Arch. Bot. (Forlì) **8:** 119–132. 1932.
Topography, etc.; list of vascular plants observed by author, with localities; weeds introduced during war, etc. (Prov. Trento and Prov. Vicenza, the latter in Veneto.)

Cobelli, Giovanni de. Contribuzione alla flora dei contorni di Rovereto. Pub. Mus. Civico Rovereto 18. 80 p. 1890. (Estratto dai Programmi dell' I. R. Scuola reale superiore Elizabettina de Rovereto 30 (1888–89), 31 (1889–90).) (Gray Herbarium library.)
Annotated list of vascular plants. (Prov. Trento.)—See also his Calendario della flora roveretana. Pub. Mus. Civico Rovereto 37. 74 p. 1900. (Climate, statistics; list of 992 phanerogams arranged according to flowering dates, with indication of the months in which each flowers; index.)—Also Stefani, below.

Cobelli, Ruggero de. La florula di Serrada. Pub. Mus. Civico Rovereto 30. 30 p. 1896.
Annotated list of vascular plants (568 species); includes some cultivated species. Summarizes his three earlier papers in Nuovo Gior. Bot. Ital. v. 25 and n.s., v. 1 and 2. 1893–95. (Prov. Trento.)—See also his Materiali per la fauna e la flora di Serrada e florula della cima di Monte Maggio. Pub. Mus. Civico Rovereto 35. 41 p. 1899. (Includes list of additional species for Serrada (p. 31–32), and Florula della cima di Monte Maggio (p. 35–41), a list of 97 spermatophytes collected in a small area on the summit of Monte Maggio at 1857 meters elevation.)

Diettrich-Kalkhoff, Emil. Flora von Arco und des unteren Sarca-tales (Südtirol). xix, 150 p. 1 col. pl. 23 cm. Innsbruck, 1916. (Gray Herbarium library.)
Topography, geology, climate, general features of flora; annotated list of cellular and vascular plants (2048 species and varieties, including 1282 vascular plants), with localities. (Prov. Trento.)

Entleutner, A. F. Flora von Meran in Tirol. Deutsche Bot. Monatsschr. **1:** 179–181 (163 [bis] –165 [bis] in orig.). 1883; **2:** 25–27, 67–68, 89–90, 97–99, 123–124, 135–138, 150–153, 165–166, 180–184. 1884; **3:** 10–15, 41–44, 52–55, 105–108, 135–138, 164–168. 1885; **4:** 11–14, 27–30, 100–102, 117–120. 1886. (Reprinted 59 p.; reprint not seen.)
Topography, etc.; annotated list of vascular plants (1144 species). (Prov. Bolzano.)—See also **Ladurner, Arthur.** Zur flora von Meran in Südtirol. l.c. **19:** 140–142. 1901. (Alphabetical list of additional species.)—Also his Beiträge zur flora von Meran. [II]– (III). Oesterr. Bot. Zeitschr. **54:** 410–412. 1904; **55:** 397–399. 1905. (Additional alphabetical lists, bringing total to nearly 1,500 species, the limits of the area being somewhat enlarged.)

Foletto, Angelo. La valle di Ledro. Cenni geografici, statistici e storici con guida e carta corografica. 136 p. 8vo. Riva, 1901.
Includes remarks on the flora of the valley and its investigation (p. 15–16), notes on the forest trees (p. 46–47), and (p. 16–22) Prospetto d'alcune più rare e critiche specie della regione che crescono in Val di Ledro, a list of 180 noteworthy species with localities. (Prov. Trento.) (Not seen; title and annotation from Dalla Torre and Sarnthein, Flora der gefürsteten grafschaft Tirol **6** (4): 139. 1913.)—See also **Luzzani, Filiberto.** Aggiunte alla flora della Val del Chiese e dintorni. Studi Trentini **13:** 3–25. 1932. (Extensive list of mostly infraspecific forms from Val del Chiese, Val di Ledro, and Val Vestino. In library of Academy of Natural Sciences, Philadelphia.)

Gambillo, Carlo. Note botaniche. *In* his La valle di Rendena. Annuario Soc. Alp. Trident. **8:** 116–121. 1882.

This chapter consists mostly of a list of spermatophytes, in part with localities, prepared by Michele de Sardagna. (Prov. Trento.)

Gelmi, Enrico. Il Monte Bondone di Trento con ispeciale riguardo alla sua flora. Bul. Soc. Veneto-Trent. Sci. Nat. Padova **1:** 62–76. 1880.

Topography, etc.; extensive list of spermatophytes, grouped by botanical regions and habitats. (Prov. Trento.)—See also Pedrotti, below.

——— Prospetto della flora trentina. 1 p. l., vi, 197 p. 18 cm. Trento, 1893. (Arnold Arboretum library.)

Annotated list of vascular plants of the Province of Trento, with localities for scarcer species.—See also his Aggiunte alla flora trentina. Prima lista. Atti R. Accad. Sci., Let. ed Arti Rovereto III, **2:** 227–238. 1896.—Also **Evers, Georg.** Beiträge zur flora des Trentino, mit rücksicht auf Gelmi's Prospetto . . . Verhandl. K. K. Zool.-Bot. Gesell. Wien **46:** 55–89. 1896. (Extensive list, with localities and annotations.)—Also **Briquet, John.** Quelques notes d'herborisations dans le Tyrol méridional. Bul. Herb. Boissier **5:** 469–484. 1897. (Extensive list of noteworthy angiosperms, with localities. Reprinted in Bul. Lab. Bot. Gén. Univ. Genève **1:** 280–295. 1897.)—Also **Gelmi, E.** Aggiunte alla flora trentina. Nuovo Gior. Bot. Ital. n.s., **5:** 304–321. 1898. (List with localities.)—Also his Nuove aggiunte . . . Bul. Soc. Bot. Ital. **1900:** 68–76. 1900. (List with localities.)—Also **Biasoni, Luigi.** Di alcune piante trovate durante il periodo bellico e postbellico. Studi Trent. **3:** 33–40. 1922; **5:** 145–154. 1924. (Extensive lists of vascular plants, including numerous additional species. In Harvard College library.)—Also **Dalla Fior, Giuseppe.** Contributi alla conoscenza della flora spontanea e avventizia del Trentino. l.c. **6:** 57–77. 1925. (Lists of newly recorded species, native and introduced. In American Museum of Natural History library.)—Also **Luzzani, Filiberto.** Aggiunte alla flora della Val del Chiese e dintorni. Studi Trent. Sci. Nat. **13:** 3–25. 1932. (Extensive list of vascular plants (mostly infraspecific taxa) from Val del Chiese, Val di Ledro, and Val Vestino. In library of Academy of Natural Sciences, Philadelphia.)—Also **Dalla Fior, G.** Bricciche floristiche. l.c. **16:** 171–178. 1935. (Annotated list of newly recorded species and new localities, new altitudinal records, and noteworthy adventive plants.)—Also **Biasoni, L.** Nuove contribuzioni alla flora del Trentino. l.c. **16:** 194–205. 1935. (List of noteworthy vascular plants with localities.)

Graziadei, Damiano. Vegetabili selvatici mangerecci del Trentino. Annuario Soc. Alp. Trident. **10:** 113–122. 1884.

Running account of wild and cultivated edible plants, with vernacular names. (Prov. Trento.)

Heimerl, Anton. Flora von Brixen a. E. Ein mit standorts- und höhenangaben versehenes verzeichnis der im weiteren gebiete von Brixen a. E. (Südtirol) beobachteten wildwachsenden höheren sporen- und samenpflanzen, der nutzgewächse und ziergehölze. xxi, 321 p. 23.5 cm. Wien und Leipzig, 1911. (Gray Herbarium library.)

Bibliography, general features of flora, phytogeography; annotated flora of vascular plants, with localities and vernacular names. (Brixen, Ital. Bressanone, in Prov. Bolzano.) Replaces **Bachlechner, Gregor.** Verzeichniss der phanerogamen pflanzen, welche in der gegend von Brixen wild wachsen, mit angabe einiger fundorte und der blüthezeit u. s. w. Progr. K. K. Gymnas. Brixen **9:** 1–94. 1859; Beiträge zur flora von Brixen. l.c. **15:** I–V, 6–30. 1865. (Not seen; cited from Dalla Torre and Sarthein, Flora der gefürsteten grafschaft Tirol . . . **1:** 12. 1900.)—See also **Hellweger, M.** Aggiunte alla flora del circondario di Bressanone. Studi Trent. **8,** Cl. II, Sci. Nat. ed Econ.: 106–111. 1927. (List of vascular plants, with localities. In library Academy of Natural Sciences, Philadephia.)—Also Klebelsberg, below.

Huber, Gottfried. Monographische studien im gebiete der Montigglerseen (Südtirol) mit besonderer berücksichtigung ihrer biologie. viii, 180 p. 4 fig. (incl. 3 maps). 24 cm. Stuttgart, 1905. (New York Botanical Garden library.)

Includes (Anhang, p. 163–173) unannotated list of fungi, lichens, mosses, vascular plants, and cultivated plants; also bibliography. Reissued (or originally issued ?), without the Anhang, in Arch. Hydrobiol. u. Planktonkunde **1:** 1–81, 123–210. 1905. (Prov. Bolzano.)

Huter, Rupert. Flora der gefässpflanzen von Höhlenstein und der nächsten umgebung. 63 p. 23 cm. Sexten, 1872. ("Als manuscript gedruckt.")
Annotated list of vascular plants (1080 species), with localities, of region around Höhlenstein (Landro) near Toblach (Dobbiaco), including Sexten (Sesto) and Gsies (Valle de Casies) in Prov. Bolzano, also Ampezzo, Buchenstein, Enneberg, etc., in Prov. Belluno (Veneto). (Not seen; title and annotation supplied mostly by K. H. Rechinger.)—See also Pampanini and Zardini, Flora di Cortina d'Ampezzo, under Veneto.

Kirchlechner, Josef. La flora forestale colla geografia botanica delle Alpi Tridentine. xxxii, 46 p. 6 pl., map. 22 cm. Trento, 1900. (According to Dalla Torre and Sarnthein, Flora der gefürsteten graftschaft Tirol **6** (4): 160. 1913 and **1**: 347. 1900, it was originally published in Almanacco agrario Trento for 1895, 1896, 1897, and 1898, under title I principali alberi, arbusti, e frutici indigeni dei nostri boschi.)
Phytogeography; annotated list of trees, shrubs, and suffrutescent plants (164 species), with notes on uses. (Mostly in Prov. Bolzano.) (Not seen; title and annotation supplied by K. H. Rechinger.)

Klebelsberg, Raimund von. Die alpine flora des Plose-gebirges (2561 m.) bei Brixen a. E. (Südtirol). Ber. Ver. Schutze und Pflege Alpenpfl. **4**: 61–88. 1904.—. . . Nachtrag: Flechten und moose. l.c. **5**: 74–80. 1905.
Geology, changes in forest limits, botanical publications; list of known vascular plants with localities. The Nachtrag contains a list of lichens and mosses extracted from Dalla Torre and Sarnthein's Flora, and corrections of two erroneous identifications in his first paper. (Prov. Bolzano.)

Leybold, Friedrich. Ein botanischer ausflug auf den Gaukofel in Südtyrol. Flora **38**: 305–316. 1855.
Includes lists of species from various habitats. (Prov. Bolzano.)

———Der Schleern bei Botzen in Südtyrol. Flora **37**: 433–444, 449–456. 1854.
Includes extensive lists of vascular plants from different altitudes. (Prov. Bolzano.)

Luzzatto, Gina. Contributo alla flora dell'alta valle di Sulden. Atti Soc. Ital. Sci. Nat. Milano **67**: 71–92. 1928.
Topography, geology, botanical explorations; list of 345 vascular plants collected or observed by author, with localities; bibliography. (Prov. Bolzano.)

Orsi, Oswaldo. Piante sospette e venefiche. Almanacco Agr. Trento **1891**: 203–222. pl. 3–10. 1891.
List of suspected and known poisonous plants, with localities. (Prov. Trento.) (Not seen; title and annotation supplied by V. Giacomini.)

Ostermaier, Josef. Pflanzenvorkommnisse in der umgebung der Franz-Schlüterhütte im Villnöstale (Südtirol). Ber. Ver. Schutze u. Pflege Alpenpfl. **3**: 70–71. 1903.
Topography, general features of flora. Followed (p. 72–73) by an unannotated alphabetical list of spermatophytes, apparently by same author, with title Flora der umgebung der Franz Schlüterhütte und der angrenzenden gebirge. The shelterhut referred to is on the Kreuzkofeljoch at 2300 meters altitude. (Prov. Bolzano.) According to Heimerl (Flora von Brixen, p. v. 1911) the whole was reprinted with some changes in Das Villnöstal und seine umgebung. p. 94–101. (Roths Illustrierte führer, no. 9.)—See also **Naumann, Arno.** Die botanischen ergebnisse eines dreitägigen sammelausfluges in die umgebung der Franz-Schlüterhütte (D.–Oe. A.–V.). Sitzungsber. u. Abhandl. Naturw. Gesell. Isis Dresden **1909** (Abhandl.): 86–102. 1910. (Account of plant formations, with annotated lists of plants observed; phytogeography.)

Paoletti, Giulio. Contribuzione alla flora del bacino di Primiero (Trentino). Atti Soc. Veneto-Trent. Sci. Nat. Padova II, **1**: 3–28. 1893.
Geology, bibliographical footnotes, etc.; lists of vascular plants collected by author at various localities; complete list of species mentioned (316). (Prov. Trento.)—See also his A proposito della mia "Contribuzione . . ." Nota. Bul. Soc. Veneto-Trent. Sci. Nat. Padova **5**: 132–134. 1893. (Justification of some records that had been questioned.)

Pedrotti, Giovanni. The alpine flora of Molveno in the Brenta Dolomites. 11 l. incl. 8 col. pl. 11.5 × 16.5 cm. (Trento) [1931 ?]. (British Museum (Natural History) library.)

Alphabetical tabular list of angiosperms with habitats or localities; climate, etc. (Prov. Trento.) (An advertising booklet of Mr. and Mrs. Bettega, proprietors of the Hotel Molveno.)—See also his The high alpine flora of Bondone (June–October) and the marvellous alpine flora of Molveno on the lake–Dolomites (March–November). 31 p. Trento, 1935. (Not seen; cited from Bul. Bibl. Bot. Ital. 9 (fasc. 1, 1936): 38. 1939.)—See also Gelmi, Il Monte Bondone, above.

——— and **Bertoldi, Vittorio. Nomi dialettali delle piante indigine del** Trentino e delle Ladinia Dolomitica presi in esame dal punto di vista della botanica, della linguistica, et del folclore [sic]. xii, 588 p. 100 pl. on 50, 11 maps on 6. 25 cm. Trento, (pref. 1930). (Harvard College library.)

Bibliography; list of native vascular plants, alphabetically arranged by scientific names, with Italian and dialect names (with localities), etymological notes, notes on distribution and uses, list of locality names derived from vernacular names; indices. (Eastern part of Provinces of Bolzano and Trento.)—See also **Pedrotti, G.** Le piante coltivate del Trentino e i loro nomi dialettali. Studi Trent. Sci. Nat. 17: 30–63, 180–214. 1936. (Annotated, classified list of cultivated plants, with vernacular names and uses.)

Pfaff, Wilhelm. Die eislöcher in Ueberetsch, ihre vegetationsverhältnisse und ihre flora. 72 p. 31 pl. 23.5 cm. Innsbruck, 1933. (Klebelsberg, R. von, *ed.* Schlernschriften. Veröffentlichungen zur landeskunde von Südtirol. no. 24.)

Includes Verzeichnis der bisher in den Eislöchern von Ueberetsch und in ihrer nächsten Umgebung beobachteten Pflanzen (p. 52–72), an annotated list of known cellular and vascular plants, with localities. The "Eislöcher" are holes or small caves between truncated blocks of porphyry from which cold air flows out, especially in summer, in the region of the Eppaner Gand. (Ueberetsch is Upper Adige, Prov. Bolzano.)

Sardagna, Michele de. La flora alpina del Trentino nella sua estesione geografica. Annuario Soc. Alp. Trident. 11: 197–231. pl. 1885.

General discussion of alpine floras; list of 446 species, with extralimital distribution; indication on the plate of the altitudinal distribution of 150 species. (Prov. Trento.)

——— **Florula di Predazzo, Paneveggio e suoi dintorni.** *In* **Riccabona,** V. Il gruppo della Pale di S. Martino. Annuario Soc. Alp. Trident. 9: 85–89. 1883.

Unannotated list of spermatophytes. (Prov. Trento.)

Schmolz, Karl. Die flora der Sellagruppe. *In* Bindel, Karl. Die Sellagruppe. Zeitschr. Deut. u. Österr. Alpenver. 30: 366–369. 1899.

General features of flora; unannotated list of vascular plants [162 species] observed by writer. (On boundary of Bolzano, Trento, and Belluno Provinces.)

Stefani, Attilio. Sommario analitico della flora di Rovereto. Atti I.R. Accad. Sci., Let. ed Arti Rovereto III, 5: 253–473. 1899.

Geography, bibliography, etc.; key to genera, on Linnaean system; unannotated key to species of spermatophytes, including some cultivated species; key to families, index. (Prov. Trento.)—See also de Cobelli, G., above.

Zaniboni, Silvio. Cenni sinottici della materia medica spontanea del Trentino. 86 p. 8vo. Brescia, 1867.

Includes (p. 8–49) treatment of the medicinally significant species with their uses. (Prov. Trento.) (Not seen; title and annotation from Dalla Torre and Sarnthein, Flora der gefürsteten grafschaft Tirol 6 (4): 211. 1913.)

Zenari, Silvia. Piante critiche delle Alpi venete. Parte I–III. Nuovo Gior. Bot. Ital. n.s., 55: 68–81. 1948; 56: 198–222. 1949; 58: 97–147. 1951.

Records of mostly minor forms of vascular plants, with some critical notes. Relates to Prov. Bolzano (especially Alpi Passirie and Brevnie) and to Cadore (Prov. Belluno in Veneto).

UMBRIA

The regione of Umbria includes the Provinces of Perugia and Terni (the latter constituted in 1926 from part of Perugia).

General

Barsali, Egidio. Prodromo della flora umbra. Nuovo Gior. Bot. Ital. n.s., **36**: 548–623. 1930; **38**: 624–689. 1931; **39**: 346–415, 549–602. 1932.—Aggiunte al Prodromo . . . II. l.c. **40**: 338–341. 1933.

Botanical explorations, list of authors with biographical details and bibliography; annotated list of vascular plants (1,573 species), with localities. The last part in 1932 includes Aggiunte [I]. Replaces **Batelli, Andrea.** Contribuzione allo studio della flora umbra. 3 pt. (56, 115, 20 p.). 25 cm. Perugia, 1885–88. (Geography, topography; list of vascular plants, with localities, observed mostly by author but in part by others and including records from publications. The second part contains a brief account of botanical explorations. Titles varies somewhat in each part; that given is the cover title to the first part. In Lloyd Library.)—See also **Di Benedetto, M.** Contributo alla flora di Gubbio (Umbria). Annali Bot. **20**: 364–366. 1934. (List of species not recorded from Gubbio, including 3 new to Umbria.)—Also **Lusina, Giuseppe.** Contributo alla flora umbra. l.c. **21**: 534–541. 1938. (List of additional species and localities.)

Local

Barsali, Egidio. Sulla flora ruderale di Perugio. Bul. Soc. Bot. Ital. **1913**: 168–173. 1913.

Annotated tabular list of vascular plants of streets, walls, and waste places of city of Perugia. (Prov. Perugia.)

Bolli, Mario. Flora e vegetazione del M. Subasio. II. contributo. Annali Fac. Agr. Univ. Perugia **6**: 179–228. 10 fig. on 5 pl. (on 3), 2 tab. 1949.

Topography, geology, botanical explorations; annotated list of mosses and vascular plants (560 species, of which 531 are vascular) with indication of biological form; biological spectrum, bibliography. (Prov. Perugia.)

Calestani, Vittorio. La vegetazione nei dintorni d'Orvieto. Nuovo Gior. Bot. Ital. n.s., **14**: 546–574. 1907.

Topography, geology, soils, etc.; list of vascular plants (741 species), grouped under habitats; annotated list of the more interesting species. (Prov. Terni.)

Corazza, Giovanni. Contribuzione alla flora dei dintorni di Spoleto. 184 p. 22 cm. Spoleto, 1889. (Arnold Arboretum library.)

Topography, geology, climate, general features of flora; annotated list of known vascular plants, with vernacular names. Publication of Accademia Spoletina. (Prov. Perugia.)—See also **Faggioli, Fausto.** Registrazione di alcune piante non ancora indicate nella flora di Spoleto. Atti. Accad. Spoletina **1917**: 18. 1917. (Not seen; cited from Bul. Bibl. Bot. Ital. 4 (1917–1923): 17. 1925.)

Faggioli, Fausto. Relazione sulle piante medicinali spontanee o coltivabili nel circondario di Spoleto. Arch. Farmacog. e Sci. Aff. **6**: 241–255. 1917. (University of Wisconsin library.)

Includes lists of cultivated and of spontaneous medicinal plants, alphabetically arranged by vernacular names; also similar lists derived from a manuscript of Pietro Fontana and from Silvestri's Contributo. (Prov. Perugia.)

Silvestri, Filippo. Contributo allo studio della flora mevanate. Appunti. 40 p. 15 cm. Perugia, 1891. (Harvard College library.)

Annotated list of vascular plants collected by author in vicinity of Bevagna (the ancient Mevania). (Prov. Perugia.)

Terracciano, Achille. Le piante spontanee dell'Isola Minore nel Lago Trasimeno. Nuovo Gior. Bot. Ital. **21**: 146–155. 1889.

Topography, geology, etc.; annotated list of vascular and cellular plants. (Prov. Perugia.)

Terrenzi, Giuseppe. Contribuzione allo studio della flora narnese: appunti e note. 64 p. 8vo. Terni, 1890.

Not seen, no review available; cited from Bot. Centralbl. **44:** 303. 1890. (Narni, in Prov. Terni.)

VAL D'AOSTA

Val d'Aosta, formerly a part of the Province of Torino, became a separate province (Aosta) in 1926 and a separate regione in 1945, at which time the southern third of its area reverted to Torino. In accordance with the practice of Italian botanists, it is here treated under the regione of Piemonte (as Prov. Aosta).

VENETO

See also Emilia-Romagna (Bertolani-Marchetti, Il popolamento); Friuli-Venezia Giulia (Zenari); Lombardia (Ciferri); Trentino-Alto Adige (Bolzon (Flora del Monte Marmolada); Huter; Schmolz; Zenari).

The regione of Veneto (for a time called Venezia Euganea; the Venetia of the older writers) includes the Provinces of Belluno, Padova, Rovigo, Treviso, Venezia, Verona, and Vicenza. The Province of Udine, formerly in Veneto, has been in Friuli-Venezia Giulia since 1947.

General

Saccardo, P. A. Della storia e letteratura della flora veneta sommario. x, 208 p. 24 cm. Milano, 1869.

List of authors, arranged more or less chronologically, with biographical sketches and annotated lists of publications.

Visiani, Roberto de, and **Saccardo, P. A.** Catalogo delle piante vascolari del Veneto e di quelle più estesamente coltivate. Atti R. Ist. Veneto Sci., Let. ed Arti III, **14:** 71–111, 303–349, 477–519, 703–737, 1091–1139, 1503–1545, 1735–1776. 1868–69. (Reprinted 1 p.l., 292 p. Venezia, 1869. In Arnold Arboretum library.)

Annotated tabular list of vascular plants and Characeae (2,953 species). Covers also the Province of Mantova in Lombardia, and the Province of Udine which is now in the regione of Friuli-Venezia Giulia.—See also **Bolzon, Pio.** Supplemento generale al Catalogo . . . l.c. VII, **9** (vol. **56**): 431–509. 1898. (Covers all records additional to the Catalogo, bringing total (fide Marchesetti) to 3,130 species and 710 varieties; includes bibliography.)—Also his Contribuzione alla flora veneta. Nota 4. [–12.] Bul. Soc. Bot. Ital. **1899:** 134–139. 1899; **1900:** 87–95, 274–283, 332–338. 1900; **1901:** 73–83. 1901; **1902:** 25–29. 1902; **1903:** 33–39. 1903; **1904:** 32–34. 1904; **1905:** 60–64. 1905. (Lists of newly recorded species and forms and new localities. Bonis, Antonio de, is coauthor of Note 5 and 8.)—Also his Nuove aggiunte alla flora veneta. l.c. **1910:** 69–77. 1910. (New records, arranged by provinces.)

Local

Béguinot, Augusto. La flora delle mura e delle vie di Padova. Studio biogeografico. Malpighia **24:** 413–428. 1912; **25:** 61–84. 1912; **27:** 244–259, 439–454, 547–582. 1915–16.

Annotated list of ruderal floras of Italy, with partial list of those of other parts of Europe (additions **27:** 244), sketch of botanical investigations; annotated list of 327 vascular plants of the city of Padua; floristic elements, ecology, etc. (Prov. Padova.)

——— Flora padovana, ossia prospetto floristico e fitogeografico delle piante vascolari indigene inselvatichite o largamente coltivate crescenti nella provincia di Padova con notizie storico-bibliografiche sulle fonti della flora. 3 pt. (764 p.). 20 pl. 23 cm. Padova, 1909–14. (Gray Herbarium library.)

Pt. 1 (103 p.), chronological account of botanists from 1561 on, with lists of species recorded by the earlier ones, and annotated titles of publications; pt. 2 (p. 105–607), annotated list of vascular plants (1335 species), with localities; pt. 3 (p. 609–764), plant geography, plant associations, etc.—See also his Contributo alla Flora padovana. Bul. Ist. Bot. Univ. Sassari v. 2, Mem. 4. 10 p. 1922. (List of 77 plants.)—Also his Aggiunte alla Flora padovana. Arch. Bot. (Forlì) **7:** 324–329. 1931. **(List of 23 plants.)**

——— L'industria della erboristeria nella provincia di Padova, suo presente e suo avvenire. Arch. Farmacog. e Sci. Aff. **6:** 297–331. 1917. (University of Wisconsin library.)

Includes list of 78 medicinal plants of commercial significance, with vernacular names and brief notes on uses; annotated list of 65 species not handled commercially; annotated list of 44 cultivated species.

——— Risultati principali di una campagna botanica sui Colli Berici. Bul. Soc. Bot. Ital. **1904**: 381–396. 1905.

Topography, geology, soils, climate, human influence, etc., with lists of species characteristic of various soils or habitats; not a complete flora. (Prov. Vicenza.)

——— Saggio sulla flora e sulla fitogeografia dei Colli Euganei. Mem. Soc. Geog. Ital. v. 11. 192 p. map. 1904.

Botanical explorations, topography, geology, climate; annotated list of vascular plants, the species grouped under each family according to soil preferences, with localities for the scarcer species; discussion of habitats, plant associations, origin and development of flora, etc. (Prov. Padova.) Replaces **Trevisan, Vittore.** Prospetto della flora euganea. 67 p. 21.5 cm. Padova, 1842. (Unannotated list of 2,340 species and 187 varieties of vascular and cellular plants (1,370 vascular species). In Arnold Arboretum library.)—See also **Béguinot, A.** Prospetto delle piante vascolari finora indicate per i Colli Euganei e per la Pianura Padovana. 38 p. 23.5 cm. Padova, 1905. (Alphabetical list of vascular plants, with references to his Saggio and to other publications. In British Museum (Natural History) library.)—Also **Panebianco, H.** Osservazioni sulla flora marnicola delle colline di Teolo negli Euganei. Atti Accad. Scient. Veneto-Trent.-Istriano n.s., 5 (Cl. Sci. Nat., Fis. e Mat.): 30–43. 1908. (Ecological; includes (p. 33–36) list of 203 spermatophytes.)

——— La vita delle piante superiori nella Laguna di Venezia e nei territori ad essa circonstanti. Studio biologico e fitogeografico. xv, 348 p. 75 pl. (pl. 1, map). 26 cm. Venezia, 1913. (Pubblicazione n. 54 dell'Ufficio idrografico del R. Magistrato alle acque. In New York Botanical Garden library.)

Botanical explorations (with references), geography, soils, chemical composition of water, climate, plant associations, variations in duration and size, periodical phenomena of vegetation, ecological adaptations, polymorphism and variability, origin and development of flora; annotated tabular list of native and extensively naturalized vascular plants (1,082 species); list of doubtful species, bibliography. Covers the region from the mouth of the Tagliamento to that of the Adige, including besides the Laguna di Venezia as commonly understood, that of Caorle; an area about 100 km. long and 7 to 15 km. wide, nearly all in Prov. Venezia, a little in Prov. Padova.

Cobau, Roberto. Le erbe infestanti dei campi coltivati a tabacco nel Canale di Brenta. Atti Soc. Ital. Sci. Nat. Milano **51**: 247–264. 1913.

Climate, order of appearance of various weeds, etc.; annotated list of 137 weeds of tobacco fields in northern part of the Brenta valley, from the then Austrian border nearly to Bassano. (Prov. Vicenza.)—See also his Risultati di alcune erborizzazioni invernali eseguite nel Canale di Brenta. Malpighia **25**: 501–507. 1913. (Alphabetical list of 105 species found in flower in late December in vicinity of San Nazario.)

——— Le piante avventizie esotiche osservate nel Vicentino. Arch. Bot. (Forlì) **4**: 20–35, 97–114. 1928.

Annotated list of 115 adventive angiosperms, with discussion of occurrence; bibliography. (Prov. Vicenza.)—See also his Di alcune altre piante avventizie esotiche osservate nel Vicentino. l.c. **18**: 83–86. 1942. (Notes on several species.)

Dalla Torre, K. W. von. Die geschichte der floristischen erforschung des Monte Baldo. *In* Festschrift zur feier des siebzigsten geburtstages des Herrn Professor Dr. Paul Ascherson (4. Juni 1904). p. 1–17. 24 cm. Leipzig, 1904.

Chronological account of botanical explorations, with references to publications. Monte Baldo is more or less on the border between the Province of Verona and the Province of Trento (in Trentino-Alto Adige).—See also Pollini, below.

Errico, Pasquale d'. Flora e boschi dell'altopiano di Asiago. Italia Forest. e Mont. **2**: 219–238. 4 fig. 1947.

Includes (p. 222–228) tabular list of 278 vascular plants of the plateau of Asiago or Sette Communi, with indication of life form and ecological habitat; bibliography. (Prov. Vicenza.)

Falda, Leop. Florula del Monte Summano (Provincia di Vicenza). xii, 34 p. 8vo. Vicenza, 1899.

Not seen; cited from Bibl. Ital. **33**: 331. 1899.

Goiran, Agostino. Erborizzazioni estive ed autunnali attraverso i Monti Lessini veronesi. Bul. Soc. Bot. Ital. **1892**: 151–155, 250–254, 269–275, 306–310, 361–369, 411–417, 445–452. 1892; **1893**: 14–21, 88–93, 184–189, 261–270, 295–305, 344–349, 433–438, 497–501, 539–547. 1893.

Annotated list of 799 dicotyledons. (Prov. Verona.)

——— Erborizzazioni recenti (aprile, maggio 1895) in una stazione veronese inondata dell'Adige nel settembre 1882. Bul. Soc. Bot. Ital. **1895**: 224–232. 1895.

List of vascular plants observed on an area along the Adige near Verona which was flooded in 1882 and not yet returned to cultivation. (Prov. Verona.)— See also his Seconda contribuzione alla flore atesina a proposito di due specie nuove pel Veronese . . . Bul. Soc. Bot. Ital. **1896**: 254–255. 1896. (Adds several species.)—Also his Nuove specie da aggiungersi alla flora atesina. l.c. **1898**: 57: 1898. (Additional species.)

——— Flora veronensis (Phanerogamae.) Le piante fanerogame dell' agro veronese. Censimento. 2 pt. (261, 695 p.). 19 cm. Verona, 1897–1904.

Annotated list of spermatophytes of the Province of Verona (2,342 species and 2,355 infraspecific entities), with localities and zonal distribution, early synonymy, vernacular names and uses, etc. (Not seen; cited mainly from review in Nuovo Gior. Bot. Ital. n.s., **12**: 382–385. 1905. According to A. Forti, Madonna Verona **14**: 118. 1920, the work was published as vol. 2 (1897, 1900) of Sormani Moretti, Monografia statistico-economica della provincia di Verona.) Replaces **Pollini, Ciro.** Florae veronensis quam in prodromum florae Italiae septentrionalis exhibit. 3 v. (xxx, 535; 754; 898 p.). 12 pl. 23 cm. Veronae, 1822–24. (Topography, geology, altitudinal zones, botanical explorations, etc.; annotated descriptive flora of vascular and cellular plants (2,701 species), on the Linnaean system. Covers Provinces of Padova, Verona, and Vicenza, Italian Tirol, and adjacent regions.) Also replaces **Goiran, Agostino.** Prodromus florae veronensis. Nuovo Gior. Bot. Ital. **14**: 17–53, 75–97. 1882; **15**: 5–68. 1883; **16**: 105–167. 1884; **17**: 5–26. 1885; **18**: 169–217. 1886. (Extensively annotated list with localities. Incomplete; vascular cryptogams, gymnosperms, Viscaceae, Loranthaceae, Monocotyledones.)—See also his Pteridophytae [sic] (agri veronensis). Atti del Congresso dei naturalisti italiani promosso dalla Società italiana di scienze naturali, Milano, 15–19 settembre 1906. p. 482–513. 25 cm. Milano, 1907. (Annotated list of 42 species and subspecies and 111 varieties, with vernacular names and detailed localities.)—Also his Un manipolo di piante nizzarde e veronesi. Nuovo Gior. Bot. Ital. n.s., **16**: 125–145. 1909. (Includes new records for the province.)—Also **Forti, Achille.** Di alcune entità da confermare o da aggiungere alla flora veronese. Bul. Soc. Bot. Ital. **1911**: 249–253. 1911. (Discussion of 6 species and varieties.)—Also his Nuove entità da confermare o da aggiungere alla flora veronese. Atti e Mem. Accad. Agr., Sci. e Let. Verona **96** (IV, 21): 309–318. 1920. (Discussion of 3 species.)

——— Sopra la pubblicazione del R. Ministero di agricoltura, industria e commercio avente per titolo Nomi volgari adoperati in Italia a designare le principali piante di bosco. Mem. Accad. Agr., Arti e Com. Verona **54** (II, 12): 271–378. 1877. (Preprinted as Erbario forestale veronese. 108 p. 1876; preprint in Arnold Arboretum library.)

Consists principally of "Erbario forestale veronese" (p. 279–378), an annotated list of 155 trees and shrubs, alphabetically arranged by botanical names, with Italian and Veronese vernacular names, local occurrence, uses, etc. (Prov. Verona.)

Grigolato, Gaetano. Flora medica del Polesine, ovvero descrizione delle piante medicinali che nascono nella provincia di Rovigo. fasc. 1–5. 224 p. 11 col. pl. folio. Rovigo, 1843–47.

Account of medicinal plants, extending to the Didynamia, which the author intended to complete in 4 more fascicles. (Not seen; cited from Saccardo, Della storia e letteratura della flora veneta sommario, p. 144. 1869.)

——— Illustrazione alle piante vascolari crescenti spontanee nel Polesine di Rovigo raccolte, classificate ed offerte fino dal MDCCCXLV per l'erbario secco di Monza . . . xi, 82 p. 28 cm. Rovigo, 1854. (Lloyd Library.)

Soils, geography, etc.; annotated list of 807 Characeae and vascular plants, including cultivated species, with vernacular names. Cover title, Illustrazione delle piante vascolari . . . (Prov. Rovigo.) Replaces his Elenco delle piante vascolari che nascono nel Polesine. Rovigo, 1847. (Not seen.)—See also **Terracciano, Achille.** Le piante dei dintorni di Rovigo. Centuria I. [–III]. Nuovo Gior. Bot. Ital. **22:** 414–419. 1890; **23:** 287–295. 1891. (Annotated list of 300 vascular plants.)—Also **Bonis, Antonio de.** Le piante del Polesine. l.c. **24:** 202–208. 1892; Bul. Soc. Bot. Ital. **1893:** 271–276. 1893. (Lists of 200 native species and 40 naturalized species, with localities.)—Also **Béguinot, Augusto.** Contributo alla conoscenza della flora littoranea del Polesine (prov. di Rovigo). l.c. **1911:** 232–242. 1911. (Lists of species collected at various localities centering around Porto di Levante.)

Marignoni, G. B. Cenni storici e bibliografici sulla flora vicentina. 32 p. 8vo. Schio, 1913.

Geology, botanical explorations, bibliography (74 titles), herbaria. (Prov. Vicenza.) (Not seen; cited from Bot. Jahresber. Just 44 (1): 1280. 1926.)

Marzari-Pencati, Giuseppe. Elenco delle piante spontanee fino ad ora osservate nel territorio di Vicenza. 58 p. 8vo. Milano, 1802.

Geography, etc.; list of over 1,000 plants (about 953 vascular) on the Linnaean system. (Prov. Vicenza.) (Not seen; title and annotation supplied mostly by F. Sappa.) According to Saccardo (Della storia e letteratura della flora veneta sommario, p. 87. 1869), the list, rearranged alphabetically, was reissued without name of author as Flora vicentina ossia elenco delle piante che spontaneamente crescono nella provincia fino ad ora osservate dai botanici. *In* Forti, Luigi. Statistica generale della provincia di Vicenza . . . v. 2, p. 64–113. 8vo. Bassano, 1828–29. (Not seen.)—See also **Baldini, Luigi.** Aggiunta all'Elenco stampato in Milano delle piante spontanee fino ad ora osservate nel territorio vicentino. 8 p. 12mo. [Vicenza, 1802.] (Note on several vascular cryptogams and some mosses and hepatics. Not seen; cited from Saccardo, l.c. p. 87.)—Also **Moretti, Giuseppe.** Appendice all'Elenco delle piante spontanee del Vicentino. Gior. Fis., Chim., Stor. Nat., Med. ed Arti (Milano) **8:** 121–136. pl. 4. 1815; II, **3:** 16–23, 252–258. 1820. (List of 96 species with localities. According to Saccardo, l.c. 99, both parts were reissued with different title and the number of species increased to 106.)—Also **Romano, Girolamo.** Notizia sopra diverse piante da aggiungersi alla flora vicentina. Gior. Ital. Let. **24** (II, 8): 269–308. 1821. (Adds about 111 species. Not seen; cited from Saccardo, l.c. 80.)—Also **Beggiato, F. S.** De studio botanicae cum nonnullarum plantarum (prov. vicentinae) enumeratione. 16 p. 8vo. Patavii, 1830. (Thesis. Not seen; according to Saccardo, l.c. p. 152, it includes more than a hundred additions.)—Also **Cobau, Roberto.** Notizie su le piante rare del Vicentino. Nota 1. [–2]. Nuovo Gior. Bot. Ital. n.s., **47:** 662–674. 1940; **49:** 398–414. 1942. (Bibliography (42 titles); lists of newly recorded or noteworthy species, with localities and annotations.)

Minio, Michelangelo. Elementi per un calendario florale di Belluno. Atti Accad. Sci. Veneto-Trent.-Istriana III, **10:** 29–56. 2 fig. 1919.

Includes tabular list of phanerogams, showing flowering dates. (Prov. Belluno.)

——— La flora urbica di Venezia (Notizia storica, dal sec. XVII al sec. XIX.) Nuovo Gior. Bot. Ital. n.s. **34:** 1300–1317. 1928.

Chronological account of publications on the urban flora of Venice, with lists of species recorded or added by various authors up to the middle of the 19th century. (Prov. Venezia.)

——— Un lembo meridionale della flora del Bellunese; escursioni nella valle del Tegorzo. Atti R. Ist. Veneto Sci., Let. ed Arti 78 (2): 679–701. 1919.

Topography, climate, general features of flora, phenology; list of vascular plants collected, in part with localities. (Prov. Belluno.)

——— Sulla vegetazione della conca dolomitica di Caiada (Belluno). Nuovo Gior. Bot. Ital. n.s., 25: 377–404. 1918.

Topography, geology, plant formations; annotated list of vascular plants collected by author. (Prov. Belluno.)

Monti, Lorenzo. Dizionario botanico veronese che comprende i nomi volgari veronesi delle piante da giardino col corrispondente latino Linneano, cui aggiungonsi altre specie indigene e i nomi italiani. pt. I–II. 159 p. 16mo. Verona, 1817.

Alphabetical list of scientific names of cultivated and wild plants, with patois equivalents. (Prov. Verona.) (Not seen; annotation supplied by V. Giacomini.)

Moricand, M. E. Flora veneta, seu enumeratio plantarum circa Venetiam nascentium, secundum methodum Linnaeanam disposita. v. 1. x, 439 p. 20.5 cm. Genevae, 1820.

Flora covering the spermatophytes of the vicinity of Venice [751 species], on the Linnaean system, with localities and brief descriptions. No more published. (Prov. Venezia.)

Naccari, F. L. Flora veneta o descrizione delle piante che nascono nella provincia di Venezia disposta secondo il sistema Linneano e colla indicazione al metodo di Jussieu modificato del de-Candolle arrichita di osservazioni medico-economiche. 6 v. 1 pl. 26 cm. Venezia, 1826–28. (Arnold Arboretum library.)

Flora of vascular and cellular plants (1,342 species of which 949 are vascular), with synonymy, descriptions, localities and uses. (Prov. Venezia.)—See also **Chiamenti, A.** Manipolo di piante clodiensi. Atti Soc. Veneto-Trent. Sci. Nat. Padova 5: 176–183. 1878. (List of angiosperms not included in Naccari's Flora from vicinity of Chioggia, with localities.)—Also **Pignatti, Sandro.** Contributo alla flora della provincia di Venezia. Atti R. Ist. Veneto Sci., Let. ed Arti (Cl. Sci., Mat. e Nat.) **109**: 305–326. 1951. (List of noteworthy species.)

Pampanini, Renato. Contributi alla storia dell'esplorazione floristica del Cadore dal 1838 al 1931. Atti R. Ist. Veneto Sci., Let. ed Arti **91** (2): 1631–1721. 1932.

Alphabetical list of vascular plants new to or rare in the region; annotated chronological list of collectors, with lists of species collected in different localities (with dates), and bibliographical footnotes. (Prov. Belluno.)—See also his Altri contribuzioni alla storia della esplorazione floristica del Cadore. Arch. Bot. (Forlì) **24**: 205–235. 1948. (Additions to bibliography and to list of new or rare plants.)—Also Zenari (2), below.

——— Contributo alla conoscenza della flora del Cadore (Alpi Orientali). [I]–VII. Arch. Bot. (Forlì) **15**: 90–99. 1939; **16**: 133–165. 1940; **18**: 49–71. 1942; **19**: 61–79. 1943; **20**: 1–10. 1946; **21**: 44–55. 1946; **22**: 52–72. 1946.

Systematic list of vascular plants (mostly infraspecific forms) with localities, including 841 taxa (87 species) new to the flora of the Cadore. The Cadore is an area in the eastern part of the Dolomites, including the upper valley of the Piave and the neighboring rivers Ansiei and Boite. (Prov. Belluno.)—See also Zenari (2), below.

——— and **Zardini, Rinaldo.** Flora di Cortina d'Ampezzo. Arch. Bot. (Forlì) **23**: 109–156. 1947; **24**: 1–20, 65–88, 129–168. 1948.

Botanical explorations: list of known vascular plants (1907 taxa, including 1060 species) with localities and names of collectors (by Pampanini); general features of flora, with mention of special localities and lists of characteristic species of various habitats (by Zardini). Covers the whole commune, that is, essentially the upper valley of the Boite. (Prov. Belluno.) Replaces **Zardini, R.** La flora montana ed alpina di Cortina d'Ampezzo. Milano, 1939. (Not seen.)—See also Huter, under Trentino.

Pollini, Ciro. Viaggio al Lago di Garda e al Monte Baldo in cui si ragiona delle cose naturali di quei luoghi . . . 152 p. pl. 17.5 cm. Verona, 1816. (Arnold Arboretum library.)

Includes alphabetical list, with some annotations, of the plants of the Lago di Garda (p. 14-20), and similar lists (p. 79–129) of those of Monte Baldo and vicinity, arranged by localities. (On border of Prov. Verona and Prov. Trento.)—See also **Béguinot, Augusto.** Contributo alla flora del Lago di Garda e di regioni finitimi. 32 p. 27.5 cm. Messina, 1924. (Critical notes on about 72 species. In British Museum (Natural History) library.)—Also his Appunti fitogeografici su alcuni laghi della regione del Garda e del Trentino occidentale. Arch. Bot. (Forlì) **7**: 296–323. pl. 5–7. 1931. (Includes lists of plants from the shores of various lakes.)

Re, Filippo. Florae atestinae prodromus. 136 p. 22.5 cm. Mutinae, 1816. (Gray Herbarium library.)

Briefly annotated list of vascular plants, on the Linnaean system, with short descriptions; also some fungi, lichens, and mosses. Ateste is the ancient name of Este. (Prov. Padova.)

Rovesti, Paolo. Le piante officinali della farmacopea estense. Riv. Ital. Essen. Profumi e Piante Off. **37**: 421–428. illus. incl. port. 1955.

Includes list of 144 native medicinal plants with uses, alphabetically arranged by vernacular names but mostly without scientific names, extracted from the Farmacopea per gli stati estensi, by Antonio Goldoni and others, published at Modena in 1839. (Prov. Padova.)

Saccardo, Domenico. Le piante sponantee nel Regio orto botanico di Padova. Atti Soc. Veneto-Trent. Sci. Nat. Padova II, **2**: 452–470. pl. 3. 1896.

List of 994 species (378 vascular) growing without cultivation in the Royal Botanic Garden of Padua; only the Algae and Fungi annotated. (Prov. Padova.) Replaces **Saccardo, P. A.** Florula spontanea Horti botanici patavini . . . Nuovo Gior. Bot. Ital. **4**: 212–220. 1872.

Saccardo, Francesco. Florula del Montello (provincia di Treviso). 2. ed. riveduta e aumentada da P. A. Saccardo e A. Trotter. 26 p. Treviso, 1920. (1st ed., Bul. Soc. Veneto-Trent. Sci. Nat. Padova **6**: 5–18. 1895.)

List of vascular plants (684 species in 1st ed.), without detailed localities. (Not seen; title and annotation supplied by F. Sappa.)

Saccardo, P. A. Flora tarvisina renovata. Enumerazione critica della piante vascolari finora note nella provincia di Treviso. Atti R. Ist. Veneto Sci., Let. ed Arti **76** (2) (IX, **1**, pt. 2): 1237–1545. 1917.

Botanical explorations, bibliography, list of collectors; annotated list of vascular plants (1,717 species, of which 1,422 are indigenous). Replaces his Prospetto della flora trevigiana. l.c. III, **8**: 1087–1132. 1863; **9**: 427–445, 481–497, 605–638, 837–877. 1863–64. (Reprinted 156 p., 1864; reprint not seen); Aggiunte alla Flora trevigiana per cura del s.c. P. A. Saccardo e G. Bizzozero. l.c. V, **6**: 681–719. 1880. (Topography; annotated list of vascular plants (1,556 species, including additions of 1880).)—See also **Fiori, Adriano.** Piante da aggiungersi alla flora del bosco Cansiglio e del M. Cavallo nel Trevigiano. Bul. Soc. Bot. Ital. **1918**: 35–41. 1918. (List of vascular plants not recorded from these localities by Saccardo.)

——— La flora trevigiana. Notizie storiche e bibliografiche. Atti Ateneo Treviso (v.? p.?) 1910. (Reprinted 28 p. Treviso, 1910.)

History of botanical explorations in Province of Treviso, and bibliography. (Not seen; title and annotation supplied by F. Sappa.)

——— Florula di Recoaro e dei monti circonstanti. *In* Club Alpino Italiano. Guida alpina di Recoaro. p. 27–34. 1883.

List of phanerogams and cryptogams. (Prov. Vicenza.) (Not seen; title and annotation supplied by F. Sappa.)

——— and **Traverso, G. B.** La flora delle vette di Feltre. Atti R. Ist. Veneto Sci. Let. ed Arti **64** (2): 833–908. 1905.

Botanical explorations, general features of flora; annotated list of vascular plants. (Prov. Belluno.)

Sandi, A. F. Enumeratio stirpium plantarum phanerogamarum agri belunensis, quas hucusque patrio municipio cohortante conlegit. 30 p. 8vo. Belluni, 1837.

Topography, etc.; unannotated list of vascular plants. (Prov. Belluno.) (Not seen; title and annotation taken mostly from Saccardo, Della storia e letteratura della flora veneta sommario, p. 111. 1869. According to Minio (Nuovo Gior. Bot. Ital. n.s., **19**: 353. 1912) the rarity of the work is reported to be due to the author's attempt to suppress it soon after publication.)—See also **Bizzozero, Giacomo.** Contribuzioni alla flora veneta. III. Escursione nelle Alpi bellunesi. Atti R. Ist. Veneto Sci., Let. ed Arti VI, **1**: 575–586. 1883. (List of 35 species and varieties new or rare in Belluno, also list of other species collected.)—Also **Toni, Ettore de.** Note sulla flora del Bellunese. Nuovo Gior. Bot. Ital. **21**: 55–76. 1889. (Botanical explorations, list of species not given by Sandi, etc.; annotated list of noteworthy vascular plants; annotated list of Algae.)—Also **Bolzon, Pio.** Sulla flora della Dolomiti bellunesi. Bul. Soc. Bot. Ital. **1907**: 7–14. 1907. (Partial lists from Monte Civetta and Monte Pelmo.)—Also **Pampanini, Renato.** Materiali per una flora della provincia di Belluno. I [–III]. l.c. **1908**: 32–38, 123–126. 1908; **1909**: 56–61. 1909. (Lists of noteworthy vascular plants, with localities.)—Also **Minio, Michelangelo.** Contributo alla flora del Bellunese. Nota [1.] –9. l.c. **1909**: 47–51. 1909; **1911**: 294–296. 1911; **1913**: 62–68, 145–150. 1913; **1914**: 118–123. 1914; **1916**: 18–28. 6 fig. 1916; Nuovo Gior. Bot. Ital. n.s., **24**: 291–296. 1917; Bul. Soc. Bot. Ital. **1921**: 14–18. 1921; **1922**: 23–28. 1922. (List of noteworthy vascular plants, with localities; the 1916 and 1921 papers concern teratological specimens.)—Also his L'erbario di A. F. Sandi e il suo valore per la flora vascolare del Bellunese. Nuovo Gior. Bot. Ital. n.s., **19**: 349–388. 1912. (Discussion of Sandi and his work; list of Belluno plants in the herbarium, with some localities, etc.)—Also **Bolzon, P.** Nuove ricerche sulla flora bellunese. Bul. Soc. Bot. Ital. **1918**: 8–12. 1918. (Partial bibliography; notes on various species.)—Also his Nuove ricerche botaniche nelle Alpi bellunesi. Bul. Soc. Bot. Ital. **1922**: 70–73. 1922. (List of noteworthy plants from various localities.)—Also his Ricerche botaniche in prov. di Belluno e nell'Alto Adige. l.c. **1924**: 24–34. 1924. (Partial lists of species from various localities.)

Scapaccino, M. Studio sulla flora medicinale-aromatica dell'altipiano dei Sette Comuni. 85 p. 4to. Venezia, 1931.

Not seen; title taken from Bul. Bibl. Bot. Ital. 7 (1932): 40. 1933. (Prov. Vicenza.)

Soravia, Pietro. Tecnologia botanico-forestale della provincia di Belluno. 319 p. 19 cm. Belluno, 1877. (Arnold Arboretum library.)

Account of native and cultivated woody plants, alphabetically arranged by botanical names, with vernacular names, local distribution, woods, uses, etc.

Toni, Ettore de. Sui nomi vernacoli di piante nel Bellunese. Serie [1.]–2. Atti R. Ist. Veneto Sci., Let. ed Arti **56** (VII, 9): 195–206. 1898; **57** (VII, **10**): 177–186. 1899.

Alphabetical lists of scientific and vernacular names, with respective equivalents, and some critical notes on origin of names, etc. (Prov. Belluno.) Some additional names are given in the footnotes in **Minio, M.** L'erbario di A. F. Sandi . . . Nuovo Gior. Bot. Ital. n.s., **19**: 349–388. 1912.

Vaccari, Lino. Florula della cima del Monte Grappa (Prov. di Venezia). Bol. Ann. Club Alp. Bassanese **2**: 56–72. 1895. (Reprinted 21 p.)

Notes on the vegetation, list of collectors with biographical data; list of vascular plants (442 species and varieties) with names of collectors. (Prov. Vicenza.) (Not seen; title and annotation supplied mostly by R. Pichi-Sermolli.)—See also **Bolzon, Pio.** Sulla flora castrense del Monte Grappa e dell'Alto Cadore. [Nota I]–II. (title varies.) Nuovo Gior. Bot. Ital, n.s., **29**: 194–204. 1919; Bul. Soc. Bot. Ital. **1920**: 37–43. 1920. (Lists of noteworthy species.)

Vasolin, Egidio. Flora dei monti Toraro e Campomolon e dei loro confrafforti. (? p.) Firenze, 1900.

Not seen; no review available. Cited from Nuovo Gior. Bot. Ital. n.s., **34**: 1425. 1928. (Prov. Vicenza.)

——— Florula dell' altipiano di Tonezza. Atti R. Ist. Veneto Sci., Let. ed Arti 58 (2): 71–96. 1899.

Topography, general features of flora; list of vascular plants, with localities. (Prov. Vicenza.)

Zenari, Silvia. Contributo alla conoscenza della flora del Comelico (Alto Cadore). (Note di critica sistematica. 1. contributo.) Mem. R. Accad. Sci., Let. ed Arti Padova n.s., **58:** 81–109. 1942.—Piante critiche delle Alpi orientali. (Note . . . 2. contributo.) Arch. Bot. (Forlì) **18:** 127–163. 1942. —Un manipolo di piante critiche delle Alpi venete. (Note . . . 3. contributo.) Atti R. Ist. Veneto Sci., Let. ed Arti **101** (2), Cl. Sci. Mat. e Nat.: 559–576. 1942.—Scrofulariacee delle Alpi cadorine (Note . . . 4. contributo.) Mem. R. Accad. Sci., Let. ed Arti Padova n.s., **59:** 27–44. 1943.—Contributo alla conoscenza della flora delle Alpi cadorine. (Note . . . 5. contributo.) Atti R. Ist. Veneto Sci., Let. ed Arti **102** (2), Cl. Sci. Mat. e Nat.: 647–664. 1943.—Campanulaceae delle Alpi venete. (Note . . . 6. contributo.) Nuovo Gior. Bot. Ital. n.s., **50:** 252–266. 1944.—Composite delle Alpi venete. (Note . . . 7. contributo.) Arch. Bot. (Forlì) **19:** 109–144. 1943.

Records of new or critical forms; bibliographies. (Prov. Belluno.)—See also her Piante critiche delle Alpi venete, under Trentino.

——— La distribuzione stazionale delle entità floristiche del Cadore. Parte I–V. Arch. Bot. (Forlì) **22:** 17–31. 1946; **24:** 41–54, 111–118. 1948; **25:** 83–94. 1949; **26:** 46–58, 222–232. 1950; **27:** 102–111, 207–215. 1951; **28:** 205–217. 1952; **29:** 6–39, 114–123, 188–199, 308–319. 1953; **30:** 42–57, 85–96, 138–149. 1954; **31:** 34–45, 117–124. 1955; **32:** 22–33, 71–86. 1956–57.

List of vascular plants (1,048 species) collected by author with data on life form, grade of hygrophily, altitudinal range, localities, etc.; bibliography. Includes a great number of infraspecific taxa. (Prov. Belluno.)

Luxembourg (Grand-Duchy)

See also Belgium, General (Goffart, Nouveau manuel; Höck, Ankömmlinge; Rehder).

General

Anonymous. Notes bibliographiques. I. Botanique. *In* Nature et tourisme en Luxembourg. p. 108–109. 29.5 cm. Luxembourg, 1949–50 (1949).

Unannotated list of publications on various phases of botany, including forestry.

Feltgen, Ernest. Die einheimischen heilpflanzen, medicinisch-pharmaceutische flora des luxemburger landes. 250 p. illus., 2 pl. 23 cm. Luxemburg, 1903.

Short bibliography; list of wild and cultivated medicinal plants, alphabetically arranged by scientific names, with German, French, and Luxemburger vernacular names, officinal names, uses, chemical principles, etc.; alphabetical list of the species showing local occurrence.

Fischer, Eugène. Les plantes subspontanées et naturalisées de la flore du grand-duché de Luxembourg. Pub. Inst. Roy. Grand-ducal Luxemb. Sect. Sci. Nat. et Math. 12: 1–115. 1872. (Preprinted with index, 126 p. 1871; in Gray Herbarium library.)

Source of introduced plants, etc.; annotated list of 318 phanerogams.—See also Robert, under Local.

Klein, E. J. Eigentümliche übereinstimmung von luxemburgischen und südfranzösischen pflanzennamen. Ein vergleichende studie. 18 p. 23 cm. Luxemburg, 1913.

Discussion of resemblances in meaning but not in spelling between patois names of plants in Luxembourg and in Aveyron (see Carbonel), the plants taken up in order by families.

Koltz, J. P. J. Catalogue des plantes vasculaires de la flore du grand-duché de Luxembourg. Recueil Mém. et Trav. Soc. Bot. Grand-Duché Luxemb. **1:** 46–78. 1874.

Unannotated list of vascular plants and Characeae, the cultivated or introduced species distinguished by italic type.

——— Dendrologie luxembourgeoise. Catalogue des arbres, arbrisseaux et arbustes spontanés, subspontanés ou introduits dans les cultures du grand-duché de Luxembourg. 2 p. l., 217 p. 23 cm. Luxembourg, 1875. (Arnold Arboretum library.)

Annotated list of 1424 wild and cultivated woody plants, with indication of range and, in the case of the cultivated species which constitute the great bulk of the work, of the date and place of introduction into Luxembourg.

——— Guide du botaniste dans ses recherches des plantes rares ou peu répandues du grand-duché de Luxembourg. Recueil Mém. et Trav. Soc. Bot. Grand-Duché Luxemb. **2/3**: 65–111. 1877.

Alphabetical list of localities, with lists of noteworthy vascular plants of each and their flowering dates.

——— Prodrome de la flore du grand-duché de Luxembourg. Pub. Inst. Roy. Grand-ducal Luxemb. Sect. Sci. Nat. et Math. **13**: 91–359. 1873.—ptie. 2. Recueil Mém. et Trav. Soc. Bot. Grand-Duché Luxemb. **4/5**: 175–426. 1880; **6/8**: 38–95. 1882; **13**: 91–349. 1897. (Reprinted, 2 v. (v. 2 in 2 pts.), Luxembourg, 1873–97; in Arnold Arboretum library.)

Botanical investigations, with bibliography; annotated list of wild and cultivated spermatophytes (1511 species); annotated, keyed, descriptive flora of cryptogams (except algae and fungi). The 1873 part contains the spermatophytes; 1880, pteridophytes, Characeae, and mosses; 1882, hepatics; 1897, lichens. According to Lefort (see his Contribution [p. 125], below), about 400 of the 1511 spermatophytes recorded by Koltz are not indigenous.—See also **Koltz, J. P. J.** Plantes phanérogames découvertes dans le grand-duché de Luxembourg depuis la publication de la Flore luxembourgeoise de Tinant (1836). Recueil Mém. et Trav. Soc. Bot. Grand-Duché Luxemb. **1**: 12–39. 1874; **2/3**: 54–57. 1877. (Lists amounting to 226 species, with localities and references.)—Also **Fischer, Eugène.** Plantes phanérogames nouvelles ou rares de la flore luxembourgeoise. l.c. **6/8**: 116–124. 1882; **11**: 50–68. 1886. (Annotated list, with localities.)—Also **Robert, Joseph.** Standorte seltener pflanzenarten. (Beiträge zur flora des grossherzogtums Luxemburg. I.) Arch. Trim. Sect. Sci. Nat. Phys. et Math. Inst. Grand-Ducal Luxemb. n.s., **5**: 1–12. 1910. (Extensive list of vascular plants, with localities; bibliography of principal publications on the botany of Luxembourg on p. 19–20.)—Also papers by various authors in Bul. Soc. Nat. Luxemb. **54** (n.s., **43**): 161–208. 2 fig., 7 pl. 1950; **55**: 121–174. 10 pl. on 5. [1951]; **56**: 60–66. 4 pl. on 2, 67–88. 5 pl. on 3, 89–90. 1952; **57**: 155-182. 8 pl. on 4. 1953; **58**: 76–134. 10 pl. on 5. 1954; **59** (n.s., **48**): 57–88. 8 fig., 8 pl. on 4, 89–99. 2 pl. on 1. 1955. (Reissued as Mus. Hist. Nat. Luxemb. Serv. Carte Group. Vég. [Études] 2 (pt.), 3 (pt.) 10, 11, 13, 19, 23, 24.)—Also **Lawalrée, André.** Les ptéridophytes du grand-duché de Luxembourg. Bul. Soc. Roy. Bot. Belg. **83**: 225–240. pl. 1951. (Annotated list of 44 species and hybrids.)

Krombach, J. H. G. Flore du grand-duché de Luxembourg. Plantes phanérogames. 3 p. l., 564 p. 25.5 cm. Luxembourg, 1875.

Bibliography, geology, hydrography, climate; annotated descriptive flora of 1,353 spermatophytes (including some cultivated species), with notes on uses. Replaces **Tinant, F. A.** Flore luxembourgeoise, ou description des plantes phanérogames, recueillies et observées dans le grand-duché de Luxembourg, classées d'après le système sexuel de Linnée [sic]. 512 p. 22.5 cm. Luxembourg, 1836. (Annotated descriptive flora of vascular plants and Characeae (1,462 species) on the Linnaean system. The 2. éd. (512 p. Luxembourg, 1855, in Arnold Arboretum library) appears to be merely a reissue). Also replaces (as to Luxembourg) **Löhr, M. J.** Taschenbuch der flora von Trier und Luxemburg mit berücksichtigung der nahe- und glan-gegenden. lxvi, 318 p. 17 cm. Trier, 1844. (Topography, geology, climate, key to genera on Linnaean system; briefly descriptive flora of 1,561 phanerogams, with localities for scarcer species. In Gray Herbarium library.)

Lefort, F. L. Contribution à l'histoire botanique du Luxembourg. Bul. Mens. Soc. Nat. Luxemb. 54 (n.s., **43**): 31–160. 18 pl. on 9 (incl. 10 port., 8 facsim.) 1950.

Critical history of botanical work in Luxembourg and adjacent areas, with lists of species added by various botanists and included bibliography; relates to both vascular and cellular plants.

Weber, Josef. Lezeburjesch-latein-fransesch-deitschen dixionèr fun de planzen. Recueil Mém. et Trav. Soc. Bot. Grand-Duché Luxemb. **12**: 43–144. 1890.

Alphabetical list of Luxembourg patois names (also horticultural terms) with indication of gender and scientific, French, and German equivalents; index of botanical names.

Local

Robert, Joseph. Die adventivflora von Diekirch. (Beiträge zur flora des grossherzogtums Luxemburg. II.) Arch. Trim. Sect. Sci. Nat., Phys. et Math. Inst. Grand-Ducal Luxemb. n.s., 5(1): 13–20. 1910.

Manner of occurrence, etc.; annotated alphabetical list of adventive plants; bibliography.

MONACO

See also general works under France, Alpes-Maritimes. (Monaco is an enclave in this department.)

All the species are described in the general floras of France and Italy, and all should be covered by the floras of Alpes-Maritimes by Ardoino and Burnat.

General

Ardoino, H. J. B. Catalogue des plantes vasculaires qui croissent spontanément aux environs de Menton et de Monaco avec l'indication des principales espèces de Nice, Sospel, Vintimille, S. Remo, etc. xiii, 46 p. 23.5 cm. Turin, 1862.

Topography, statistics; annotated list of vascular plants (1,000 species from Menton and Monaco, with additional species from the other localities mentioned, in smaller type), in part with localities. Adventive plants are not included. The work is cited here rather than under France because it is apparently the only flora covering at all closely the Principality of Monaco. Menton, adjoining Monaco on the east, was included in the Principality until 1848.

NETHERLANDS (HOLLAND; NEDERLAND; PAYS-BAS)

See also Belgium, General (Goffart, Nouveau manuel; Höck, Ankömmlinge; Paque, De Vlaamsche volksnamen; Vandenbussche, Onze volkstaal); France, General (Depape, Le monde de plantes à l'apparition de l'homme); Switzerland, General (Fischer, Unsere heilpflanzen; Probst, Wolladventivflora Mitteleuropas).

The local titles are arranged under the names of the provinces, except that works on the West Frisian (North Sea) Islands, which belong to the Provinces of Friesland, Groningen, and Noord-Holland, are placed under the heading Noordzee-eilanden.

General

See also Noordzee-eilanden, below (van Eeden, Lijst der planten die in de Nederlandsche duinstreken gevonden zijn; Vuyck, De plantengroei der duinen).

Includes a few partial works.

The general works here listed may be classified as follows: *Floras and manuals*: Buekers, van Hall (Flora Belgii septentrionalis), Heimans, Heukels (De flora van Nederland), Heukels and van Ooststroom, Heukels and Wachter, Koninklijke Nederlandse Botanische Vereniging, A. C. Oudemans, C. A. J. A. Oudemans (De flora van Nederland), Suringar. *Catalogs:* Heukels (Bekende groeiplaatsen), Nederlandsche Botanische Vereeniging, Sloff (fluviatile district). *Iconographies:* (Heukels, De flora van Nederland), (Heukels and van Ooststroom), Kops. *Bibliographies*: van Dieren (phytosociology), Jansen and Wachter (floristic), Rehder. *Vernacular names*: (Blöte-Obbes), Commissie voor de wetenschappelijke benaming (ornamentals), Gerth van Wijk, Heukels (Lijst; Woordenboek), Nijdam, Pauwels. *Trees*: Blöte-Obbes, Boom (cultivated), Hendriks. *Useful and poisonous plants:* van Hall (Neêrlands plantenschat), Miquel (poisonous plants), van der Trappen, Willinge Prins. *Weeds:* Bisschop van Tuinen, van Fraassen (weed seedlings), van Hall (Neêrlands plantenschat), Henrard, Heukels (Landbouwflora), Kloos (woolwaste), Kruijne, Sissingh, Thijsse. *Cultivated plants:* Bisschop van Tuinen, Blöte-Obbes (woody plants), Boom (trees), Boom and Ruys (herbaceous plants), Commissie voor de wetenschappelijke benaming (ornamentals), van Hall (Neêrlands plantenschat), Heukels (Landbouwflora), Hendriks (woody plants), Sipkes (herbaceous plants). *Miscellaneous:* Backer (2: Dutch-English botanical terms, glossary of generic and specific names), Beijerinck (seed atlas), Fiet (list of Latin terms and Dutch equivalents), Goethart (distribution maps), Hoogenraad (phytogeography, botanical districts, plant associations), Jansen, Wachter, and Sloff (biographical data), Jongkindt Coninck (Latin and Greek terms with equivalents), de Leeuw (ecology, floral districts), C. A. J. A. Oudemans (De ontwikkeling: historical and bibliographical), Sirks (herbaria), van Soest (botanical districts), Westhoff (plant associations).

Backer, C. A. Dutch-English taxonomic-botanical vocabulary. (Unpaged.) 16 × 20.5 cm. Leiden, 1949. (Processed publication. 2. ed. Bogor, 1956; not seen.)

Alphabetical list of Dutch botanical terms with English equivalents.—See also Fiet, below, and Jongkindt Coninck, below.

——— Verklarend woordenboek der wetenschappelijke namen van de in Nederland en Nederlandsch-Indië in het wild groeiende en in tuinen en parken gekweekte varens en hoogere planten. xii, 664 p. 27.5 cm. Groningen, Batavia, 1936.

Pronunciation and accentuation of botanical names, list of abbreviations; alphabetical list of generic and specific names of wild and cultivated vascular plants of Netherlands and of Dutch East Indies, with derivation and meaning, in the case of personal names with biographical sketches; list of abbreviations of names of botanists and persons for whom plants were named.

Beijerinck, Willem. Zadenatlas der Nederlandsche flora ten behoeve van de botanie, palaeontologie, bodemkultuur en warenkennis, omvattende, naast de inheemsche flora, onze belangrijkste cultuurgewassen en verschillende adventiefsoorten. 316 p. incl. 140 pl. 27 cm. Wageningen, 1947. (Mededeeling no. 30 van het Biologisch station te Wijster, Dr.)

Definitions, key to genera based on seeds or fruits; drawings of the seeds or achenes of wild-growing Dutch spermatophytes (totaling about 1375 species), with very brief description of color and surface characters.

Bisschop van Tuinen, Klaas. Nederlandsche planten. Beschrijving onzer akkerbouw-, tuinbouw-, boschbouwgewassen en onkruiden, benevens een kort overzicht der voor de planten schadelijke en nuttige dieren. 4 p.l., 171 p. 19.5 cm. Amsterdam, 1874. (Library of Koninklijke Nederlandse Botanische Vereniging, Leiden.)

Description of agricultural, horticultural, and silvicultural cultivated plants and weeds (342 species), with mention of their injurious insects.—See also (for cultivated plants) Blöte-Obbes, Boom, Boom and Ruys, Commissie voor de wetenschappelijke benaming van cultuurgewassen, van Hall (Neêrlands plantenschat), Hendriks, Heukels (Landbouwflora), Sipkes, van der Trappen, Willinge Prins; and (for weeds) van Fraassen, van Hall, Henrard, Heukels, Kloos, Kruijne, Sissingh, Thijsse.

Blöte-Obbes, M. C. Boom en struik in bos en veld. Over aard, gebruik en folklore van onze bomen en struiken, met inbegrip van onze gekweekte vruchtbomen. 4 p. l., 272 p. illus., pls. (part. col.), col. front. 25 cm. Utrecht, 1953.

Historical account of trees in relation to man from earliest times in world at large; popular treatment of native and cultivated trees and shrubs, with history, folklore, uses, and extensive lists of Dutch, Old Dutch, and Flemish names.—See also Bisschop van Tuinen, above, and Boom (Nederlandse dendrologie), below, and additional titles there given.

Boom, B. K. Nederlandse dendrologie. Geïllustreerde handleiding bij het bepalen van de in Nederland voorkomende soorten en variëteiten der gekweekte houtgewassen. 3. druk. 444 p. 134 fig. 23.5 cm. Wageningen, 1949. (Flora der cultuurgewassen van Nederland. deel 1.) (1st ed. 1933; 4. druk. 480 p. Wageningen, 1959; not seen.)

Glossary, keys to genera in summer and winter condition; flora of cultivated woody plants in form of keys, with short descriptions and statement of native range; date of introduction into cultivation (in world at large), etc.—See also Bisschop van Tuinen and Blöte-Obbes, above, and additional titles there given.

——— and **Ruys, J. D.** Flora der gekweekte, kruidachtige gewassen. Handleiding bij het bepalen van de in Nederland voorkomende soorten en variëteiten van gekweekte, winterharde vaste planten, twee- en eenjarige kruiden alsmede bol- en knolgewassen. 450 p. 23 cm. Wageningen, 1950. (Boom, Flora der cultuurgewassen van Nederland. deel 2.)

Glossary, key to families; annotated, briefly descriptive flora of hardy cultivated herbaceous plants in form of keys, with statement of native range and date of introduction into cultivation (in world at large).—See also Bisschop van Tuinen, above, and additional references there given.

Buekers, P. G. Zakflora. Lijsten voor het bepalen van alle in het wild groeiende, en van veel gekweekte planten van Nederland. Vrij bewerkt naar Dr. O. Schmeil en J. Fitschen. iv, 229 p. 217 fig. 19 cm. Zutphen [pref. 1904].

Flora of vascular plants in form of keys, with vernacular names; glossary. (Not seen; annotation furnished by S. J. van Ooststroom.)

Commissie voor de wetenschappelijke benaming van cultuurgewassen. Naamlijst van siergewassen waarin opgenomen bloemisterijgewassen, kruidachtige sierplanten voor de tuin en het in de bloemsierkunde gebruikte plantenmateriaal. 90 p. 26 cm. 's-Gravenhage, 1948.

Annotated, alphabetical, tabular list of scientific names of ornamental plants, with mention of a large number of horticultural varieties; incomplete lists of scientific names with Dutch equivalents and of Dutch names with scientific equivalents. At head of cover: Ministerie van landbouw, visserij en voedselvoorziening. Directie van de landbouw.—See also (for cultivated plants) Bisschop van Tuinen, above, and additional titles there given; and (for vernacular names) Blöte-Obbes, above, and Gerth van Wijk, Heukels (Lijst; Woordenboek), Nijdam, and Pauwels, below.

Dieren, J. W. van, and **Scheygrond, Arie.** Bibliographia phytosociologica neerlandica I, 1922–1934. Nederl. Kruidk. Arch. 45: 218–223 1935.—**Scheygrond, Arie,** and **Vlieger, Jan.** Bibliographia . . . II, 1935–1939. l.c. **50:** 355–364. 1940.
Unannotated chronological list of publications on the phytosociological aspect of ecology in Netherlands and its colonies, as well as those by Dutch authors for other regions.—See also Jansen and Wachter, and Rehder, below.

Fiet [Fiets], Albert. Plantenterminologie, of alphabetische verzameling van kunstwoorden de planten betreffende, met hunne vertaling, ten dienste van tuinlieden, bloemisten, en bloemenvrieden. Opnieuw bewerkt door W. F. A. Grimme, Sr. 8. geheel omgewerkte druk. 204 p. 17.5 cm. Assen, [pref. 1946]. (1st ed. 1885.)
Alphabetical list of specific epithets and descriptive terms (in Latin) with Dutch equivalents; alphabetical list of the generic names of wild and cultivated plants with indication of accent.—See also Backer (Dutch-English botanical vocabulary), above, and additional references there given.

Fraassen, A. M. van. De kiemplanten der tweezaadlobbige akkeronkruiden (1.). Meded. Direct. Landb. Landbouwvoorlichtingsdienst 67. 34 p. incl. 40 fig. on 7 pl. (on 4)., illus. (in glossary). 1952.
Keyed, descriptive treatment of 40 dicotyledonous weed seedlings with illustrations of each.—See also Bisschop van Tuinen, above, and additional references there given.

Gerth van Wijk, H. L. A dictionary of plant names. 2 v. (xxiv, v, 1444; 1 p.l., xxxiii, 1696 p.) 29.5 cm. The Hague, 1911–16.
Vol. 1, bibliography; alphabetical list of scientific names with extensive lists of corresponding English, French, German, and Dutch vernacular names; v. 2, index to vernacular names.—See also Commissie voor de wetenschappelijke benaming, above, and additional references there given.

Goethart, J. W. C., and **Jongmans, W. J.** Planten-kaartjes voor Nederland . . . Bewerkt naar: den Prodromus Florae Batavae, de gegevens van H. Heukels en die van verschillende andere floristen. afl. 1–25. 16 p. 512 maps. 30 cm. Leiden, 1902–1907. (Arnold Arboretum library.)
Range of each species in Netherlands shown in red on individual maps. Issued in 25 fascicles, the species in each fascicle, but not in the whole work as issued, in systematic order. No index.—See also **Goethart** and **Jongmans.** Plantenkaartjes voor Nederland. Nederl. Kruidk. Arch. III, **2:** 1068–1073. 4 maps. 1903. (Discussion of methods used.)—For continuation, see Instituut voor het vegetatieonderzoek van Nederland (I.V.O.N.). Plant-maps for the Netherlands . . . Blumea **2:** 1–23 (incl. 21 maps). 1935. **Sloff, J. G.** Plantenkaartjes voor Nederland. Nederl. Kruidk. Arch. **46:** 558–575 (incl. 18 maps), 657–689 (incl. 32 maps). 1936; **47:** 195–227 (incl. 32 maps). 1937; **48:** 250–265 (incl. 16 maps). 1938; **49:** 307–326 (incl. 20 maps), 387–398 (incl. 12 maps). 1939; **50:** 272–284 (incl. 13 maps), 398–416 (incl. 17 maps, and note on contents of earlier parts). 1940. Plantenkaartjes. **51:** 446–447 (incl. 2 maps), 451–482 (incl. 32 maps). 1941; **52:** 203–210 (incl. 6 maps), 416–438 (incl. 23 maps). 1942; **58:** 94–109 incl. 16 maps. 1951.—Also 11 maps on grasses (17 species) in **Tjeenk Willink, W. E. J.** *ed.* Grassen en granen. Zwolle, 1951. (Not seen; title and annotation supplied by S. J. van Ooststroom.)

Hall, H. C. van. Flora Belgii septentrionalis, sive index plantarum indigenarum, quae hucusque in VII provinciis foederatis repertae sunt. Flora van Noord-Nederland . . . 2 v. (861, 477 p.). 21.5 cm. Amsterdam, 1825–1840. (Arnold Arboretum library.)
Annotated, partly keyed, descriptive flora of vascular and cellular plants, on the Linnaean system. Text in Latin and Dutch, on opposite pages. Out

of date as a flora, still of value for its detailed localities.—See also his [Eerste] [–zevende] nalezing op de Flora belgii septentrionalis. Bijd. Nat. Wetensch. **1:** 135–150. 1826; **2:** 110–124. 1827; **3:** 47–62. 1828; **5:** 78–87. 1830; **6:** 159–170. 1831; **7:** 26–38, 333–347. 1832. (Lists of additional species and localities.)—Also his Nieuwe bijdragen tot de Nederlandsche flora. pt. 1 (no more published). Tijdschr. Nat. Geschied. en Phys. **8:** 203–259. 1841. (Another list.)

——— Neêrlands plantenschat, of landhuishoudkundige flora, behelzende eene beschrijving der onkruiden, vergiftige en nuttige inlandsche planten en der in onzen landbouw gekweekte gewassen. 2 p. l., iii, 332 p. 288 fig. 24.5 cm. Leeuwarden, 1854.

Systematic list, with vernacular names, brief descriptions, and uses, of useful and injurious native plants and of cultivated plants.—See also Bisschop van Tuinen, above, and additional references there given.

Heimans, Elie, Heinsius, H. W., and Thijsse, J. P. Geïllustreerde flora van Nederland. Handleiding voor het bepalen van de naam der in Nederland in het wild groeiende en verbouwde gewassen en van een groot aantal sierplanten. 18. druk bewerkt door J. Heimans, met medewerking van J. H. Kern, G. Kruseman Jr., en T. J. Reichgelt. viii, 1182 p. illus. (incl. 4 maps). 14.5 × 18.5 cm. Amsterdam, 1953. (1st ed. 1899; 19. ed. 1956, unchanged from 18th (not seen).)

Briefly annotated flora of wild and cultivated vascular plants (and a few cellular) in form of illustrated keys, mostly without details of range; discussion of pollination, plant galls, life forms, ecology, treatment of plant associations in form of key, botanical districts (by J. L. van Soest), glossary of scientific names, explanation of abbreviations of authors' names.

Hendriks, W. J. Onze loofhoutgewassen. 2. druk, herzien en uitgebreid. 4 p. l., 615 p. illus. 25 cm. Wageningen [pref. Dec. 1957].

Descriptive treatment of wild and (mostly) cultivated woody angiosperms (including vines), with keys to genera but not to species; includes numerous horticultural varieties.

Henrard, J. T. Bijdrage tot de kennis der Nederlandsche adventiefflora. [1.]–2 mededeeling. Nederl. Kruidk. Arch. **1915:** 94–102. 1915 [1916]; **1916:** 177–205. 1916 [1917].

Annotated lists of adventive plants, with critical notes. The 1st part consists mostly of Cruciferae, the 2d mostly of Gramineae.—See also Bisschop van Tuinen, above, and additional references there given.

Heukels, Hendrik. Bekende groeiplaatsen der tot de bijgenoemde familiën behoorende planten. *In* Levende Nat. v. 1–4, passim. 1896–1899.—Tot dusverre bekende groeiplaatsen der tot de bijgenoemde familiën behoorende planten. l.c. v. 4–7, passim. 1899–1903.

These papers contain lists of all localities of plants known from the Netherlands. (Not seen; titles and annotation supplied by S. J. van Ooststroom.)—See also his Nieuwe plantensoorten en variëteiten gevonden in Nederland [in 1896] [–1931]; Nieuwe vindplaatsen in Nederland van zeldzame plantensoorten [in 1896] [–1931]. (Title varies in earlier years.) *In* Levende Nat. v. 2–36, passim. 1897–1932. (Yearly lists of newly recorded plants and new localities, sometimes separate and sometimes combined, the titles varying considerably in earlier years. In library of Rijksherbarium, Leiden.)—Continued with same or similar title by **Kloos, A. W.,** and **Wachter, W. H.** (for 1932–42). l.c. v. 37–47. 1933–43; by **Kloos** (for 1943–47). l.c. v. 51. 1948; by **Kloos** and **S. J. van Ooststroom** (for 1948–51). l.c. v. 52–55. 1949–52; by **van Ooststroom** (for 1952). l.c. v. 56. 1953; and by **van Ooststroom** and **T. J. Reichgelt** (for 1953–58). l.c. v. 57–62. 1954–59. (Vol. 37–47 seen in library of Rijksherbarium, Leiden; other vols. (except 57) in library of U. S. Dept. of Agriculture.)

——— De flora van Nederland. 3 v. (xxxv, 629 p. map; xxiii, 695 p.; xv, 511 p.). 2,047 fig., map. 25 cm. Leiden en Groningen, 1909–1911.

Corrections to v. 1–3 (1: ix–xxiv), glossary, plant formations, alphabetical list of genera and species with Dutch, German, French, and English vernacular names; keyed descriptive flora of vascular plants with European and local range, biological notes, and vernacular names, and figure of each species.

Heukels, Hendrik. Landbouwflora voor bouw- en weiland. Handleiding tot het bepalen van den naam van de gekweekte planten en van de in bouw- en weiland in het wild groeiende planten. 4 p. l., 187 p. 20 cm. Amsterdam, 1894. (Library of Koninklijke Nederlandse Botanische Vereniging, Leiden.)

Flora of cultivated plants and weeds of arable land and pastures, in form of keys, briefly annotated, with vernacular names.—See also Bisschop van Tuinen, above, and additional references there given.

——— Lijst van Nederlandsche, Hoogduitsche, Fransche en Engelsche namen van planten. vii, 98 p. 24.5 cm. Leiden, Groningen, 1910. (Arnold Arboretum library.)

Alphabetical list of scientific names of Dutch vascular plants with Dutch, German, French, and English equivalents, extracted from the author's De flora van Nederland; index of vernacular names.—See also Commissie voor de wetenschappelijke benaming, above, and additional references there given.

——— Woordenboek der Nederlandsche volksnamen van planten. Uit de gegevens, verzameld door de Commissie voor Nederlandsche plantennamen, bewerkt door H. Heukels, Secretaris dier Commissie. viii, 332 p. 25 cm. [Amsterdam?], 1907. (Arnold Arboretum library.)

Short bibliography, list of regions and cooperators; list of native and some cultivated vascular plants alphabetically arranged by scientific names, with modern Dutch vernacular names (with localities), old Dutch vernacular names (from early herbals, etc.), and Flemish names, with authorities for the two latter groups. Publication of the Nederlandsche Natuurhistorische Vereeniging.—Replaces **Nederlandsche Natuurhistorische Vereeniging.** Voorloopige lijst van Nederlandsche volksnamen van planten. 95 p. 24 cm. Amsterdam, 1904. (Tabular systematic list of Dutch vascular plants, with collected vernacular names. Gray Herbarium library.)—See also **Nederlandsche Natuurhistorische Vereeniging.** Nederlandsche plantennamen, voor algemeen gebruik gekozen door de Commissie voor Nederlandsche plantennamen. 2 p. l., 79 p. 17 cm. Epe (pref. 1906). (Alphabetical list of scientific names of wild-growing vascular plants, each with a "standard" Dutch vernacular name (these indexed). In Kew library.)—Also Commissie voor de wetenschappelijke benaming, above, and additional references there given.

——— and **Ooststroom, S. J. van.** Flora van Nederland. 14. druk. 890 p. 1038 fig. 18 cm. Groningen, 1956. (1st ed. 1900, by Heukels, with title Geïllustreerde schoolflora voor Nederland; all eds. through the 13th had same title.)

Flora of vascular plants in form of ample keys, containing all the wild species and subspecies and many cultivated ones; lists of adventive and hybrid plants, notes on phytogeographical districts, system of plant associations, etc. (Not seen; title and annotation supplied by S. J. van Ooststroom.)

——— and **Wachter, W. H.** Beknopte schoolflora voor Nederland. 8. druk bewerkt door Dr. S. J. van Ooststroom. xii, 414 p. 1 fig. (map). 18.5 cm. Groningen, 1952. (1st ed. 1932; 9. druk. xii, 424 p. 17.5 cm. Groningen, 1957. Not seen.)

Phytogeographical districts, list of plant associations; pocket flora of vascular plants in form of annotated keys, with localities for the rarer species; includes cultivated plants. The standard pocket flora for Netherlands.

Hoogenraad, H. R. Biogeographie. *In* Mulder, G. J. A. Handboek der geographie van Nederland. v. 2, p. 1–129. fig. 1–28 (maps). 26 cm. Zwolle, 1951.

Includes general material (with map of phytogeographic districts, fig. 10); discussion of botanical districts under heading Flora (p. 31–75); short history of the flora and fauna since the beginning of the Tertiary (p. 108–123); and bibliography (p. 123–129; includes very few local floras). Followed by **Adriani, M. J.,** and **Vlieger, Jan.** Plantensociologie. l.c. p. 130–164. fig. 29–35. 1951. (General account of plant associations; bibliography.)—See also de Leeuw, Sissingh, van Soest, and Westhoff, below.

Jansen, Pieter, and **Wachter, W. H.** Lijst der plaatsen, omtrent welker plantengroei gegevens bestaan. (Floristische aanteekeningen 27, 31, 33.) Nederl. Kruidk. Arch. **1930:** 258–287. 1930; **45:** 252–263. 1935; **50:** 169–181. 1940.

Unannotated, geographically arranged lists of 721 publications on the local flora of Netherlands; general works on the whole country are not given. The introduction includes list of localities visited by the Nederlandsche Botanische Vereeniging from 1847 on.—See also van Dieren, above, and additional references there given.

——— **Wachter, W. H. and Sloff, J. G.,** and others. Personalia [I]–V. Nederl. Kruidk. Arch. **49:** 437–452. 1939; **50:** 182–205. 1940; **51:** 340–379. 1941; **52:** 371–415. 1942; **53:** 232–257. 1943.

Alphabetical lists of Dutch botanists and other biologists, with dates and references to biographical notices. H. Engel, joint author, no. IV–V, and W. van Dijk in no. V.

Jongkindt Coninck, A. M. C. Dictionnaire latin-grec-français-anglais-allemand-hollandais, des principaux termes employés en botanique et en horticulture, ainsi qu'une liste de genres de plantes, dérivés de mots, dont l'explication se trouve dans ce dictionnaire. 2. éd. améliorée. 3 p. l., 78 p. 24 cm. Bussum, 1907. (1st ed. [1894].)

Alphabetical list of Latin and Greek descriptive terms (including many specific names and a few generic) and radical words, with equivalents in four modern languages; list of some generic names, with their roots.—See also Backer (Dutch-English botanical terminology), above, and additional references there given.

Kloos, A. W. Nederlandse wol-adventieven. Levende Nat. **43:** 275–279, 338–342, 372–376. 12 fig. 1939. (Lloyd Library.)

Includes (p. 373–376) list of all wool-adventive plants known in Netherlands, with indication of localities.—See also his Nieuwe vondsten van wol-adventieven in Nederland in 1939 [–1950]. Levende Nat. **44:** 375–376. 1940; **45:** 209–210. 1941; **46:** 220. 1942; **47:** 190. 1943; **51:** 46–47. 1948 (for 1943–47); **52:** 114. 1949; **53:** 136. 1950; **54:** 135. 1951. (Vols. 44–47 not seen; titles supplied by S. J. van Ooststroom.)—See also Bisschop van Tuinen, above, and additional references there given.

Koninklijke Nederlandse Botanische Vereniging. Flora neerlandica. Flora van Nederland. deel 1, aflev. [1]–5. illus. 26 cm. Amsterdam, 1948–58.

Flora of vascular plants, with keys, rather detailed descriptions, local and general range, full treatment of infraspecific taxa. Edited by Theodorus Weevers et al. [Deel 1, aflev. 1] includes Pteridophyta (mostly by S. J. van Ooststroom) and Gymnospermae (B. H. Danser); deel 1, aflev. 2 is Gramineae (P. Jansen); aflev. 3–4, Cyperaceae (J. H. Kern and T. J. Reichgelt); aflev. 5, Orchidaceae (P. Vermeulen).

Kops, Jan, and others. Flora batava. Afbeelding en beschrijving der Nederlandsche gewassen. 28 v. 2,240 col. pl. 24.5 × 33 cm. Amsterdam [etc.], 1800–1906; 's-Gravenhage, 1911–1934.

Colored plates of vascular and cellular plants, with descriptive text in Dutch and French; not arranged in systematic order. Begun by Kops, continued by F. W. van Eeden, then by L. Vuyck. For indices, see Naamlijst der Nederlandsche gewassen afgebeeld en beschreven in deel i–xxv . . . door **L. Vuyck en H. C. van de Pavord Smits.** 131 p. 19 cm. 's-Gravenhage, 1920; and Naamlijst . . . deel xxvi-xxvii . . . 16 p. 19.5 cm. 's-Gravenhage, 1930 (latter in Arnold Arboretum library.) The 28th volume, by L. Vuyck, W. J. Lütjeharms, and A. de Wever, contains its own index.—See also Heukels (De flora van Nederland) and Heukels and van Ooststroom, above.

Kruijne, A. A. Vegetatieve herkenning van onze akkeronkruiden. Meded. Direct. Landb. Landbouwvoorlichtingsdienst 65. 55 p. incl. 172 fig. on 12 pl. (on 7). 1952.

Key to weeds, intended for popular use.—See also Bisschop van Tuinen, above, and additional references there given.

Leeuw, W. C. de, *comp.* The Netherlands as an environment for plant life. Presented to the members of the Sixth International Botanical Congress by the Netherlands Botanical Society. 3 p. l., 19 p. 46 fig. (mostly maps). 22.5 cm. Leiden, 1935.

Account of edaphic, climatic, and biotic factors; brief treatment of floristic districts, etc.—See also Hoogenraad, above, and additional references there given.

Miquel, F. A. W. De Noord-Nederlandsche vergiftige gewassen. 2. vermeerderde uitgave. 220 p. 30 col. pl. 22 cm. Amsterdam, 1838. (1st ed. 1836.)

Descriptive account of poisonous cellular and vascular plants, with vernacular names and properties.—See also Bisschop van Tuinen, above, and additional references there given.

Nederlandsche Botanische Vereeniging. Prodromus florae batavae. Vol. 1 (Phanerogamae et cryptogamae vasculares.) Editio altera. Nieuwe lijst der Nederlandsche phanerogamen en vaatkryptogamen . . . pt. 1–4 (xxx, 2451 p.). Nijmegen, 1901–1904; Groningen, 1916. (1st ed. 1850.)

List of collectors; list of vascular plants of the Netherlands, with citation of specimens in herbarium of the Nederlandsche Botanische Vereeniging and in private herbaria (with locality, date, and collector's name) and of additional published records. Laurens Vuyck was compiler of vol. 1. Incomplete; covers dicotyledons and monocotyledons (1659 species, including many adventives and escapes from cultivation). The Pinaceae and Pteridophyta were not treated in the 2d ed.; for the former, see ed. 1, **1**: 251. 1850, for the latter l.c. 331–341. 1850. Vol. 2, pt. 1–2 of ed. 2 (Plantae cellulares . . . Nijmegen, 1893–98; in New York Botanical Garden library) contains bryophytes and lichens; no more published.—See also **Tombe, F. A. des.** Verzeichniss der neuen und bemerkenswerthen gefässpflanzen welche in den Niederlanden 1901–1910 gefunden wurden. I. Meded. Rijks Herb. Leiden no. 8. 61 p. 1912. (Collaborators, bibliography; annotated list of species, the families arranged alphabetically.)—Also **Heukels, Hendrik.** Voor Nederland nieuwe plantensoorten en nieuwe vindplaatsen van zeldzame planten. [I]–IV. Nederland. Kruidk. Arch. **1913**: 150–164. 1913 [1914]; **1915**: 125–138. 1915 [1916]; **1917**: 129–135. 1917 [1918]; **1919**: 105–123. 1920. (Annotated list of newly recorded vascular plants and new localities. Covers Pteridophyta-Umbelliferae; no more published.)—Also his Toevoegingen . . . l.c. **1919**: 123–125. 1920.—Also **Kloos, A. W.** Mededeeling van den heer A. W. Kloos, Jr. op de vergadering der Commissie voor het flor. onderzoek v. Nederland op 29 Dec. 1913. l.c. **1913**: 55–61. 1913 [1914]. (List of 32 noteworthy vascular plants, with localities.)—Also his Aanwinsten van de Nederlandse (*earlier* Nederlandsche) flora in 1914 [–50]. l.c. 1914–16, 1919, 1921–23, 1926–27, 1929–32, passim. 1914 [1915]–32; v. 43 (1933) –54, 56–58, passim. 1933–51. Acta Bot. Neerl. **1**: 122–156. 6 fig. 1952. (Annotated, mostly yearly lists of vascular plants new to the Netherlands, mostly adventive species or infraspecific taxa.) Continued by **Ooststroom, S. J. van, and Reichgelt, T. J.** Floristische notities 1–18 [–35–58]. l.c. **5**: 102–114. 4 fig. (incl. map), 322–334. 5 fig. (incl. map). 1956; **7**: 33–52. 5 fig. 1958.—Also Heukels, Bekende groeiplaatsen, above.

Nijdam, J. Woordenlijst in vier talen op het gebied van tuinbouw en plantkunde. 97 p. 24 cm. 's-Gravenhage, 1945.

Reciprocal lists in Dutch, German, English, and French of names of many common plants and botanical agricultural terms.—See also Commissie voor de wetenschappelijke benaming, above, and additional references there given.

Oudemans, A. C. Onze flora. Beschrijving van de familiën, voornaamste geslachten en soorten der in Nederland in het wild groeiende, verwilderde, verbouwde en aangeplante gewassen, alsmede van eenige fraaie en nuttige Midden-Europeesche planten. Naar de 2. uitgave van Carl Hoffman's Botanischer bilderatlas. 4 p.l., xlii, 191, xxii p. 364 fig., 80 col. pl. 30 cm. Zutphen, 1900.

Popular descriptive flora, with floral calendar (p. xxxi–xxxv) and vernacular names. Many of the less conspicuous species omitted.

Oudemans, C. A. J. A. De flora van Nederland. 2. en verm. druk. 3 v. 352 fig. 22.5 cm. and atlas of 91 col. pl. 24 × 29 cm. Amsterdam, 1872–74. (1st ed. 1859–62.)

Flora containing 1,483 species of vascular plants, with running descriptions, keys, and vernacular names. The plates illustrate 462 species. Formerly the standard larger flora of the Netherlands, now replaced by Heukels, De flora van Nederland, and by Flora Neerlandica.

——— De ontwikkeling onzer kennis aangaande de flora van Nederland uit de bronnen geschetst en kritisch toegelicht. I–IV. Nederl. Kruidk. Arch. II, **2**: 214–277, 279–391. 1877; **3**: 1–75, 325–398. 1878–81.

(I): Chronological list of general works on the flora of Netherlands, 1554–1873, with table of contents of each; list of papers in scientific periodicals, arranged by periodicals; (II–IV): critical account of the work on Dutch plants of Dodoens, Lobel, Junius, de l'Ecluse, Pelletier, etc., with lists of species.

Pauwels, J. L., and Grootaers, Ludovic. Enkele bloemnamen in de Zuidnederlandsche dialecten. viii, 321 p. 7 maps. 25.5 cm. 's-Gravenhage, 1933. (Noord- en Zuidnederlandsche dialectbiblioteek deel 5.)

Sources, bibliography, etc.; very detailed account of the vernacular names of *Convolvulus sepium, Papaver rhoeas, Taraxacum officinale, Paeonia officinalis, Syringa vulgaris, Dianthus caryophyllus* and *D. plumarius,* and *Cheiranthus cheiri,* with etymological discussion. Relates to Provinces of Limburg, Noord-Brabant, and Zeeland, as well as to northern Belgium.—See also Commissie voor de wetenschappelijke benaming, above, and additional references there given.

Rehder, Alfred. (Phytography.) Holland. *In* his The Bradley Bibliography. A guide to the literature of the woody plants of the world published before the beginning of the twentieth century. Compiled at the Arnold Arboretum of Harvard University under the direction of Charles Sprague Sargent. v. 1, p. 390–391. 29.5 cm. Cambridge, Mass., 1911.—Additions and corrections. l.c. 1: 532. 1911; 5: xx. 1918.

Essentially unannotated list of floristic works (including those in periodical and serial literature), chronologically arranged.—See also l.c. 1: 3, for works on botanical bibliography; 1: 15–16, alphabetical list of periodicals and serials relating to botany; 1: 50–51, botanical gardens; 1: 73, vernacular names; 1: 80, botanical history; 1: 296, noteworthy individual trees; 3: 14–15, (757). 1915, alphabetical list of periodicals and serials relating to arboriculture, gardening, and uses of plants; 3: 28, collections, gardens, museums, and exhibitions (not herbaria); 3: 33, history of gardening; 3: 75–77, dendrological, pomological, economical, and miscellaneous works dealing with horticulture, fruit trees, medicinal and other useful plants; 3: 104–106, arboriculture and general horticulture; 3: 249–250, pharmacopoeias and dispensatories; 3: 270, economic products; 4: 19. 1914, exhibitions of forest products; 4: 27, history of forests; 4: 84, dendrography; 4: 356–357, forests. Various smaller sections of no floristic significance have been omitted from the above summary.—See also van Dieren, above, and additional references there given.

Sipkes, Cornelis. Flora van onze gekweekte kruidachtige vollegrondsgewassen. 303 p. 18.5 cm. Zutphen, 1922.

Descriptive flora of cultivated herbaceous plants with keys to the families and genera. (Not seen; title and annotation furnished by S. J. van Ooststroom.)—See also Bisschop van Tuinen, above, and additional references there given.

Sirks, M. J. Botany in the Netherlands. Edited for the organizing committee of the Sixth International Botanical Congress. vii, 140 p. illus., ports. 23 cm. Leiden, 1935.

Includes sketches of the history of botanical institutions and herbaria, with mention of the principal collections included and more important publications. Covers prewar Indonesia as well as the Netherlands proper.

Sissingh, Gerard. Onkruid-associaties in Nederland. Een sociologisch-systematische beschrijving van de klasse Ruderato-secalinetea Br.-Bl. 1936. Avec un résumé: Les associations messicoles et ruderales des Pays-Bas. Verslag. Landbouwk. Onderzoek. 56 (15). iii, 224 p. 18 fig. (mostly maps), 10 pl. (on 5). 1950.

Detailed description of the alliances and associations of weeds, with extensive bibliography and alphabetical tabular list (p. 217–224) of species showing life-form, method of spreading, and sociological classification.—See also Bisschop van Tuinen and Hoogenraad, above, and additional references there given.

Sloff, J. G., and Soest, J. L. van. Het fluviatiele district in Nederland en zijn flora. Nederl. Kruidk. Arch. 48: 199–249. 1 fig. (map). 1938; 49: 268–306. 1939.

Account of previous publications, with mention of species listed, discussion of terminology (fluviatiele); annotated list of vascular plants, with references and range in Netherlands.

Soest, J. L. van. Plantengeografische districten in Nederland. Levende Nat. **33:** 311–318. 1 fig. (map). 1929. (American Museum of Natural History library.)

Discussion of botanical districts; short bibliography. The same treatment, somewhat condensed, is given in the works by Heimans, Heukels and van Ooststroom, and Heukels and Wachter cited above.—See also Hoogenraad, above, and additional references there given.

Suringar, W. F. R. Geïllustreerde zakflora. Handleiding tot het bepalen van de in Nederland wildgroeiende planten, in aansluiting met de werken der Nederlandsche Botanische Vereeniging. 13. druk, bewerkt door Dr. L. Vuyck. vii, 714 p. 558 fig. 18.5 cm. Groningen, 1950. (Library of Koninklijke Nederlandse Botanische Vereniging, Leiden.) (1st ed. 1870, with title Handleiding tot het bepalen . . .)

Briefly annotated pocket flora of vascular plants in form of keys, without localities. At one time the standard pocket flora, now replaced by Heukels and Wachter's Beknopte schoolflora (see above).

Thijsse, J. P. Onkruiden. 128 p. 96 col. pl. on 48. 16 cm. Amersfoort, [193–]. (Natuur-wetenschaplijke zakboeken, v. 3.) (Rijksherbarium library, Leiden.)

Popular treatment of weedy plants.—See also Bisschop van Tuinen, above, and additional references there given.

Trappen, J. E. van der. Herbarium vivum, of verzameling van gedroogde voorbeelden van nuttige gewassen, vooral van dezulke, die in Nederland inlandsch zijn of geteeld worden, en wier kennis voor landbouw, handel en fabrijken belangrijk is, met derzelver beschrijving. 2 v. (747, 955 p.). 23 cm. Haarlem, 1839–43.

Account of useful plants, native and introduced, arranged by the Linnaean system, with vernacular names in Dutch, German, French, English, and Italian, description, uses, etc.—See also Bisschop van Tuinen, above, and additional references there given.

Westhoff, Victor, Dijk, J. W., and **Passchier, H.** Overzicht der plantengemeenschappen in Nederland. 2. druk bewerkt door V. Westhoff, met medewerking van Ir. G. Sissingh. 118 p. 19 cm. Amsterdam, 1946. (Bibliotheek van de Nederlandsche Natuurhistorische Vereeniging nr. 7.) (Arnold Arboretum library.) (1st ed. 1942, by Dijk, J. W., Westhoff, V., and Vlieger, Jan.)

Classification of plant communities, with diagnoses; bibliography.—See also Hoogenraad, above, and additional references there given.

Willinge Prins, J. A., and **Bavelaar, Hen.** De plant als genees-, genot- en voedingsmiddel. Overzicht van in Nederland bekende of gebruikte gewassen, voor ieder die belang stelt in geneeskunde, botanie en voedingsleer. 347 p. 134 fig., 4 col. pl. 24.5 cm. Amsterdam (pref. 1952).

General considerations; treatment of native, cultivated, and exotic (if used in Netherlands or Indonesia) medicinal and edible plants and "genusspflanzen," arranged by families, with Dutch names and uses; partial list of diseases, with remedies.—See also Bisschop van Tuinen, above, and additional references there given.

DRENTE

Local

Kloos, A. W. Verslag van de excursie der Kon. Ned. Bot. Ver. te Coevorden en omgeving op (21) 23–27 (29) Augustus 1947. Nederl. Kruidk. Arch. **57:** 67–73. 1950.

Itinerary, etc.; alphabetical list of vascular plants observed, with greatly abbreviated scientific names.

——— Verslag van de excursie gehouden te Hoogeveen en omgeving op 21 Augustus 1933 en volgende dagen. Nederl. Kruidk. Arch. **44:** 44–63. 1934.

Itinerary, etc.; alphabetical list of vascular plants (517 species and varieties); alphabetical list of additional species from Deventer (Prov. Overijssel) and Gorselse heath (Prov. Gelderland); plant associations.

——— Verslag van de excursie te Borger en omgeving op 22–26 Aug. 1938. Nederl. Kruidk. Arch. **49**: 21–32. 1939.
Itinerary, etc.; alphabetical list of 532 species and forms of vascular plants.

Sande Lacoste, C. M. van der, and **Suringar, W. F. R.** Phanerogamen en vaatkryptogamen, in het oostelijk en zuidelijk deel van Drenthe waargenomen . . . 9–18 Juli 1859 [1860]. Nederl. Kruidk. Arch. 5 (2. stuk): 242–261. 1861. (Gray Herbarium library.)
List of vascular plants, with localities. (From Ane, Assen, Gramsbergen, Koevorden, Valthe, etc.)

Vuyck, Laurens. Verslag der excursie, gehouden te Meppel op 28 Juli 1917 en volgende dagen. Nederl. Kruidk. Arch. **1917**: 32–40. 1917.
Itinerary, etc.; alphabetical list of vascular plants collected.

——— Verslag der excursie, gehouden te Roden, op 24 Juli 1923 en volgende dagen. Nederl. Kruidk. Arch. **1923**: 24–35. pl. 1924.
Alphabetical list of vascular plants collected.

FRIESLAND

See also Noord-Holland (van Goor, De halophyten en de submerse phanerogamen (der Zuiderzee)).
For the coastal islands, see Noordzee-eilanden (West Frisian Islands).

General

Bruinsma, J. J. Flora frisica, of naamlijst en kenmerken der zigtbaarbloeijende planten van de provincie Friesland . . . vi, 187 p. 22.5 cm. Leeuwarden, 1840.
Descriptive flora of spermatophytes (721 species), on the Linnaean system; lists of medicinal, poisonous, and useful plants, etc.—See also **Boer, Petrus de.** [Over de flora van Friesland.] Nederl. Kruidk. Arch. II, **3**: 228–230. 1879.—Also **Koopmans-Forstmann, Dora,** and **Koopmans, A. N.** Bijdrage tot de kennis van de flora van Friesland. l.c. 1928: 59–74. 1928. (Annotated list of rare or newly recorded vascular plants.)—Also their Fluviatiele invloeden aan de Friese kust. (Mededeeling van de Zuiderzee commissie no. 11.) l.c. **1930**: 228–320. 1 fig. 1930. (Includes account of localities and discussion of occurrence of various species.)

Franke, Douwe, and **Ploeg, D. T. E. van der.** List fan offisiële Fryske plantenammen. 56 p. 24 cm. Drachten, 1951.
Systematic list of wild and cultivated vascular plants, with standardized Frisian and Dutch equivalents. Publication of the Fryske Akademy. The annotation (n.) in the list indicates that a name is neuter; (pl.) that it occurs only in the plural.

Local

Feekes, Willem. De Piamer Kooiwaard en Makkumerwaard. Nederl. Kruidk. Arch. **53**: 288–330. 5 fig. (incl. 3 maps), pl. 1943.
Changes in vegetation of the sandbanks due to flooding, plant associations, etc.; alphabetical tabular list of phanerogams and a few mosses.

Kloos, A. W. Verslag der excursie, gehouden te Beetsterzwaag en omgeving op 25 Augustus 1927 en volgende dagen. Nederl. Kruidk. Arch. **1927**: 28–45. 1928.
Alphabetical list of 410 vascular plants; comparison with other localities.

Koopmans-Forstmann, Dora, and **Koopmans, A. N.** De spontane vegetatie van het Kornwerderzand en van de verbindingsdijk met de Friese kust. (Mededeeling van de Zuiderzee-commissie no. 20.) Nederl. Kruidk. Arch. **43** (1933): 361–396. 1933.
Annotated lists of phanerogams, with date of first occurrence. The Kornwerderzand is a pumping station connected with the mainland by a dyke, on the coast between Harlingen and Makkum.

Leeuw, W. C. de. Gaasterland I. (Mededeeling van de Zuiderzee-commissie no. 8.) Nederl. Kruidk. Arch. **1930:** 119–128. 1930.

Botanical investigations, topography, general features of flora; alphabetical tabular list of 256 vascular plants.

Nederlandsche Botanische Vereeniging. Phanerogamen en cryptogamen waargenomen to Wolvega, Scheene, Terissert, Mildam, Knijpe en Heerenveen, den 31. Juli 1881. Nederl. Kruidk. Arch. II, 3: 421–425. 1882.

List of vascular plants and bryophytes.—See also **Kloos, A. W.** Verslag van de excursie gehouden te Wolvega en omgeving op 27 Augustus 1934 en volgende dagen. Nederl. Kruidk. Arch. **45:** 184–195. 1935. (Itinerary, etc.; alphabetical list of vascular plants.)

GELDERLAND

See also Noord-Holland (van Goor, De halophyten en de submerse phanerogamen (der Zuiderzee)).

Local

Anonymous. Lijst der planten, waargenomen en verzameld op de excursie van 23 tot 27 Julij 1860 [1859]. Nederl. Kruidk. Arch. 5 (2. stuk): 218–234. 1861. (Gray Herbarium library.)

List of vascular plants, with localities, collected at Zevenaar, Winterswijk, Groenlo, etc.—See also Sloff, below.

Abeleven, T. H. A. J. Flora van Nijmegen. Nederl. Kruidk. Arch. II, **5:** 251–340, 552–596. 1888–89. (Reprinted 90 and 46 p. In Botanical Museum library, Utrecht.)

Annotated list of 1391 plants (938 vascular) of the region around Nijmegen. Replaces **Gevers Deynoot, P. M. E.,** and **Abeleven, T. H. A. J.** Flora noviomagensis, sive enumeratio plantarum circa Noviomagum sponte crescentium (plant. phanerog. et cryptog. contin.) . . . Flora van Nijmegen . . . 3 p.l., ii, 169, ix p. 18.5 cm. Nijmegen, 1848. (Annotated list of 1085 plants (798 vascular), with localities. Gray Herbarium library.)—See also additions in Nederl. Kruidk. Arch. III, **1:** 298–300. 1898; **2:** 380. 1901.—Also **Vuyck, Laurens.** Verslag van de zomerexcursie 1919. l.c. **1919:** 21–33. 1920. (Includes alphabetical list of vascular plants from vicinity of Nijmegen.)

Bondam, Rutger. Overzicht der flora van Harderwijk. Nederl. Kruidk. Arch. II, **5:** 177–230. 1888. (Reprinted 54 p. In Botanical Museum library, Utrecht.)

Annotated list of vascular plants.

Gorter, David de. Flora zutphanica. 2 p. l., 88 p., index. 17.5 cm. Zutphaniae [pref. 1781]. (New York Botanical Garden library.)

List of 560 plants (493 vascular) of the vicinity of Zutphen, with vernacular names and localities, on the Linnaean system.—See also **Nederlandsche Botanische Vereeniging.** Phanerogamae en cryptogamae vasculares waargenomen op de excursiën . . . in de omstreken van Zutphen, op den 26 en 27 Augustus 1897. Nederl. Kruidk. Arch. III, **1:** 611–626. 1899. (Extensive list of vascular plants, with localities.)

Kloos, A. W. Verslag der excursie, gehouden te Eibergen en omgeving op 22 Augustus 1936 en volgende dagen. Nederl. Kruidk. Arch. **47:** 35–48. 1937.

Itinerary, etc.; alphabetical list of vascular plants collected; lists of additional species from nearby localities.

Kobus, J. D. Flora van Wageningen en aangrenzende gemeenten. 44 p. 17 cm. Wageningen, (pref. 1886). Library of Koninklijke Nederlandse Botanische Vereniging, Leiden.)

List of vascular plants (695), annotated as to abundance, with localities for rarer species.

Kok Ankersmit, H. J. Naamlijst van planten binnen de gemeente Apeldoorn, tusschen de jaren 1850 en 1878, waargenomen. Nederl. Kruidk. Arch. II, **3:** 175–213, 223–228. 1879.

Annotated list of vascular plants. For additions, see Kok Ankersmit, l.c. II, **3:** 284–296, 418–419. 1881–82; II, **4:** 35–36, 156, 157. 1883–84; III, **2:** 57–58, 607. 1900–02.

Koopmans, A. N. De flora van Berg en Bos en van de Gelderse Tuin. 1 p.l., 28 p. 21 cm. Apeldoorn, [194–?]. (Rijksherbarium library, Leiden.)

Annotated lists of vascular plants of these two localities in vicinity of Apeldoorn, including some cultivated trees and shrubs.

Nederlandsche Botanische Vereeniging. Phanerogamae et cryptogamae vasculares waargenomen te Terborg (Wisch en Sillevorde) en Doetinchem den 25 en 26 Juli 1885 . . . Nederl. Kruidk. Arch. II, **4:** 402–407. 1886.

List of vascular plants, with localities.—See also **Vuyck, Laurens.** Verslag van de zomerexcursie 1915 in de omstreken van Doetinchem. Nederl. Kruidk. Arch. **1915:** 20–33 incl. fig. (map). 1915 (1916). (Includes alphabetical list of vascular plants, bringing total for Doetinchem to 506 species.)

——— Phanerogamae en cryptogamae vasculares waargenomen op de excursie . . . op 9 en 10 Augustus 1890 . . . naar Vaassen, Epe, Wissel en Hoenderloo. Nederl. Kruidk. Arch. II, **6:** 95–100. 1892.

List of vascular plants, with localities.

Sloff, J. G. Lijst van phanerogamen en vaatkryptogamen, die thans te boek staan voor Winterswijk en omgeving. Nederl. Kruidk. Arch. **50:** 31–50. 1940.

Alphabetical list of known vascular plants (784 species). Preceded (p. 23–28) by an account of botanical explorations, by same author. Replaces **Vuyck, Laurens.** Verslag der excursie, gehouden te Winterswijk, op 24 Juli 1925 en volgende dagen. Nederl. Kruidk. Arch. **1925:** 8–23. 1926. (Alphabetical list of 501 spermatophytes; comparison with earlier observations in same region.)

Soest, J. L. van. Flora van Arnhem I–VIII. Nederl. Kruidk. Arch. **1922:** 65–96. fig. (map). 1923; **1923:** 68–115. 2 fig. (1 map). 1924; **1924:** 91–133. fig. (map). 1925; **1925:** 97–137. 1926; **1926:** 216–242. 1927; **1928:** 3–25. 1928; **1929:** 405–414. 1930; **43 (1933):** 259–271. 1 fig. (map). 1933.

Annotated systematic list of 980 vascular plants with frequent critical notes; botanical districts.

GRONINGEN

For the coastal islands, see Noordzee-eilanden.

General

Hall, H. C. van. De planten der provincie Groningen. *In* Bijdragen tot de kennis van den tegenwoordigen staat der provincie Groningen. Uitgegeven door de Commissie voor de statistieke beschrijving der provincie Groningen. v. 1, p. 355–376. 22.5 cm. Groningen, 1860.

Descriptive list of vascular plants, mostly arranged by habitats. (Not seen; title and annotation furnished by S. J. van Ooststroom.)

Stratingh, G. A., Westerhoff, R., and **Bosman Tresling, J.** Initia florae groninganae of proeve van eene naamlijst der planten welke in de provincie Groningen gevonden worden. vi, 66 p. 21 cm. Groningen, 1825. (Library of Koninklijke Nederlandse Botanische Vereniging, Leiden.)

Annotated list of spermatophytes (430 species) on the Linnaean system. Authors' names do not appear on title page but are given in introduction. On title page: Uitgegeven door het Genootschap ter bevordering der natuurlijke historie te Groningen.

Local

Vuyck, Laurens. Verslag van de zomerexcursie 1920. Nederl. Kruidk. Arch. **1920:** 20–31. 1921.

List of vascular plants collected in vicinity of Ter Apel.

LIMBURG

See also Belgium, General (Paque, De Vlaamsche volksnamen der planten); Liège (Marchal and Hardy, Catalogue des plantes plus ou moins rares de la vallée de la Meuse, de Liège à Maestricht).

General

Nederlandsche Botanische Vereeniging. Phanerogamae et cryptogamae vasculares waargenomen in de provincie Limburg door de leden der

Nederlandsche Botanische Vereeniging van 1861 tot 1886. Nederl. Kruidk. Arch. II, **5**: 1–36. 1887.

List of vascular plants, with localities. Replaces the earlier list of essentially same title (but "van 1861 tot 1883") l.c. II, **4**: 304–334. 1885.—For additions, see l.c. II, **6**: 210, 527. 1893–95.

Local

Dumoulin, L. J. G. Guide du botaniste dans les environs de Maestricht, ou indication des phanérogames et des cryptogames vasculaires croissant spontanément dans ces environs. 176 p. 19.5 cm. Maestricht, 1868. (Arnold Arboretum library.)

Annotated alphabetical list of vascular plants (978 species, of which 904 are indigenous), with localities and vernacular names.—See also **Haas, E. J. M. de,** and **Hoevenaars, J. J.** Aanwinsten voor de flora van Maastricht en omstreken . . . 1890–1894. Nederl. Kruidk. Arch. II, **6**: 565–567. 1895. —Also Belgium, Liège (Maréchal, La montagne Saint-Pierre).—Also de Wever, Planten van de St. Pietersberg, below.

Foerster, Arnold. Flora excursoria des regierungsbezirkes Aachen sowie der angrenzenden gebiete der belgischen und holländischen provinz Limburg. Phanerogamen und gefässkryptogamen. Nebst uebersicht der geognostischen, der oro- und hydrographischen verhältnisse dieses florengebietes. xxx, 468 p. 19 cm. Aachen, 1878. (Gray Herbarium library.)

Topography, geology, plant formations; annotated flora of vascular plants in form of keys, with localities. Relates to region of Aachen (Aix-la-Chapelle) in western Germany and the adjacent Province of Limburg (Netherlands) as well as the Province of Limbourg (Belgium).—See review by T. Durand, Bul. Soc. Roy. Bot. Belg. **18** (2) (Compt. Rend.): 77–83. 1879.

Garjeanne, A. J. M. De plantengroei om Venlo. Levende Nat. **22**: 173–185. 6 fig. 1917. (Lloyd Library.)

Topography, etc.; list of known vascular plants, with some annotations. —See also Nederlandsche Botanische Vereeniging, below (second entry).

Heimans, Jacob. Verslag. Phanerogame planten. *In* Verslag van het biologisch onderzoek van de Maas en hare oevers. Jaarb. Natuurh. Genootsch. Limburg **1918**: 113–124. 1918.

Lists of vascular plants, in part arranged by habitats.—See also Belgium, Liège (Marchal and Hardy, Catalogue des plantes plus ou moins rares de la vallée de la Meuse, de Liège à Maestricht).

Kloos, A. W. Beknopt verslag van de Pinksterexcursie op 12–15 Juni 1943 te Plasmolen en omgeving. Nederl. Kruidk. Arch. **54**: 130–134. 1947.

Itinerary; unannotated alphabetical list of vascular plants (506 species).

——— Verslag der excursie, gehouden te Eysden, op 29 Juli 1924 en volgende dagen. Nederl. Kruidk. Arch. **1924**: 22–32. 2 fig. 1925.

Alphabetical list of vascular plants collected.

Nederlandsche Botanische Vereeniging. Phanerogamae et cryptogamae vasculares, na afloop der zomervergadering, verzameld en waargenomen op de excursiën te Roermond en omstreken op den 16 en 17 Augustus 1900. Nederl. Kruidk. Arch. III, **2**: 623–629. 1902.

List of vascular plants, with localities.

——— Phanerogamae en cryptogamae vasculares waargenomen op de excursie . . . op den 26 en 27 Augustus 1893, van Venlo narr Blerik, Baarlo, Steil, Velden, Arcen, Hamert, Venraai, Oostrum, Wansum, Wel en De Welsche Hut. Nederl. Kruidk. Arch. II, **6**: 568–578. 1895.

List of vascular plants, with localities.—See also short lists by **Garjeanne, A. J. M.,** and **Rieter, L.** Nederl. Kruidk. Arch. III, **2**: 15–16. 1900.—Also **Garjeanne, A. J. M.** Een verarming van de Venlosche flora. Jaarb. Natuurh. Genootsch. Limburg **1914**: 133–140. [1914.] (Includes mention of numerous introduced plants.)—Also Garjeanne, above.

Sloff, J. G. Verslag van de excursie rond Sittard, Juli 1930. Nederl. Kruidk. Arch. **1931**: 41–62. 1931.

Itinerary, general features of flora, etc.; alphabetical list of 697 vascular plants.

Vuyck, Laurens. Verslag van de zomerexcursie, gehouden te Arcen van 25–30 Juli 1921. Nederl. Kruidk. Arch. **1921**: 20–33. 1922.

Alphabetical list of 471 vascular plants; comparison with the flora as observed in 1893.

Wever, August de. Eenige Z.-Limburgsche plantnamen. Natuurh. Maandbl. Limburg **22**: 27–29, 41–43. 1933.

Annotated list of patois names, with botanical equivalents.

——— Lijst der wildgroeiende en eenige gekweekte planten in Z.-Limburg. [I]–X. Meded. Natuurh. Genootsch. Limburg **1911**: 29–41. 1911; Jaarb. Natuurh. Genootsch. Limburg **1912**: 123–160. 2 pl. 1912; **1913**: 43–115. pl. 1913; **1914**: 9–103. pl. 1914; **1915**: 5–92. 1915; **1916**: 5–96. 1916; **1917**: 3–52. 1917; **1918**: 3–47, 49–90. 1918; **1919**: 3–34. 1919; **1929/23**: 3–48. 1923. (Vol. 1916–1920/23 seen in Rijksherbarium library, Leiden.)

Botanical explorations, etc.; list of cooperating botanists (1914 volume); annotated list of wild and cultivated plants, with localities and frequent critical notes. Title varies in first two parts; that of first part is Wildgroeiende planten in Zuid-Limburg.—See also **Kloos, A. W.** Verslag van de excursie in Zuid-Limburg, Pinksteren 1916. (Henrard, Jansen, Kloos, Wachter.) Nederl. Kruidk. Arch. **1916**: 68–77. 1916 (1917). (List of noteworthy plants collected at various localities.)—Also **Wever, A. de.** De Zuid-Limburgsche flora. Winst en verlies over 1922–1932. Natuurh. Maandbl. Limburg **21**: 58–60, 68–69, 76–77. 1932. (Lists of additional species, new localities, extinct or now very rare species, etc.)—Also his Zuid-Limburgsche flora. Aanwinsten over 1932–1934. l.c. **24**: 103–104. 1935.—Also his Zuid-Limburgsche flora (1935–1944). l.c. **32**: 100–101. 1943. (Not seen, title and annotation supplied by S. J. van Ooststroom.)

——— Planten van de St. Pietersberg. *In* **Schaik, D. C. van.** De Sint Pietersberg. p. 187–257. 24.5 cm. Maastricht, 1938. (Library of Koninklijke Nederlandse Botanische Vereniging, Leiden.)

Survey of vascular plants, with notes on localities and habitats.

Willemse, A. De flora van Noord-Limburg. Jaarb. Natuurh. Genootsch. Limburg **1918**: 165–177. 1918; **1919**: 153–162. 1919. (American Museum of Natural History library.)

Annotated list of vascular plants of Noord-Limburg, with vernacular names and localities. Incomplete: Choripetalae and Sympetalae only.—See also **Jansen, Johannes.** Plantenlijst Noord-Limburg. Natuurh. Maandbl. Limburg **18**: 2–3. 1929. (List of scarcer vascular plants, with localities.)

NOORD-BRABANT

See also Noordzee-eilanden (van Eeden); Belgium, General (Paque, De vlaamsche volksnamen der planten).

Local

Butaye, R., and Haas, Emmanuel de. Lijst der planten te Oudenbosch en omstreken waargenomen door R. Butaye en E. de Haas. 1885–1887. Nederl. Kruidk. Arch. II, **5**: 542–551. 1889.

Barely annotated list of vascular plants.

Hoven, F. J. J. Slingsby van. Flora van 's Hertogenbosch of naamlijst van de planten welke in de omstreken van 's Hertogenbosch gevonden worden, met aanwijzing van hare groeiplaatsen. 2. veel vermeerderde en verbeterde uitgaaf. 193 p. 17 cm. Leeuwarden, 1879. (Library of Koninklijke Nederlandse Botanische Vereniging, Leiden.) (1st ed. 1848.)

Annotated list of vascular and cellular plants. 's Hertogenbosch is also known as Bois-le-Duc.—See also **Vuyck, Laurens.** Verslag der excursie, gehouden te 's-Hertogenbosch op 26 Juli 1918 en volgende dagen. Nederl. Kruidk. Arch. **1918**: 19–30. 1919. (Includes alphabetical list of 367 vascular plants.)

Kloos, A. W. Verslag van de excursie gehouden te Heusden en omgeving op 22 Augustus 1932 en volgende dagen. Nederl. Kruidk. Arch. **43**: 68–81. 1933.

Itinerary, etc.; alphabetical list of vascular plants (588 species), with excessively abbreviated scientific names; notes on various species.

Kuijper van Wäschpenning, J. A. B. Eerste naamlijst van zigtbaarbloeijende planten, welke in de omstreken van Breda gevonden worden. 27 p. 21 cm. Breda, 1826.—Tweede naamlijst . . . 15 p. 22.5 cm. Breda, 1828. (Both in library of Rijksherbarium, Leiden.)

Unannotated lists of vascular plants (213 and 116 species), on the Linnaean system.—See also **Oudemans, C. A. J. A.** Nederl. Kruidk. Arch. II, **2**: 143–146. 1877. (Notes of N. van Aken on numerous species.)—Also **Vuyck, Laurens.** Verslag van de zomerexkursie 1916 in de omstreken van Breda. l.c. **1916**: 19–30. pl. (map). 1916. (Alphabetical list of vascular plants (455 phanerogams); sketch of botanical explorations, with references.)

Nederlandsche Botanische Vereeniging. Phanerogamae en cryptogamae vasculares waargenomen op de excursiën . . . in de omstreken van Bergen-op-Zoom, op den 22 en 23 Augustus 1896. Nederl. Kruidk. Arch. III, **1**: 347–357. 1898.

List of vascular plants and fungi, with localities.

——— Verslag der in Augustus 1911 gehouden excursies. Nederl. Kruidk. Arch. **1912**: 56–66. 1 fig. (map.) 1912.

Itinerary, notes on particular species; alphabetical list of vascular plants, with localities. Relates to Biesbosch, Raamsdonk, Oosterhout, and Dongen.

Tombe, F. A. des. Lijst der planten, waargenomen in de omstreken van Tilburg van 14–15 Mei (vóórexcursie) en 17–21 Juli 1910. Nederl. Kruidk. Arch. **1911**: 47–62. 1911.

List of vascular plants.

NOORD-HOLLAND

See also Noordzee-eilanden (van Eeden); Overijssel (Feekes, De flora van Schokland en Urk).

Local

Eeden, F. W. van. De duinen en bosschen van Kennemerland. Bijdragen tot de kennis van het landschap en den plantengroei in een gedeelte der provincie Noord-Holland. 2 p. l., 107 p. 23 cm. Groningen, 1868. (Arnold Arboretum library.)

General account of region, with mention of numerous species; annotated list of vascular plants (p. 91–107). Kennemerland is the region between Haarlem and Alkmaar.

Feekes, Willem. De ontwikkeling van de natuurlijke vegetatie in de Wieringermeer-polder, de eerste groote droogmakerij van de Zuiderzee. Nederl. Kruidk. Arch. **46**: 1–295 (incl. illus., maps). Plates. 1936. (Also issued as thesis, 295 p. 1936.)

Ecological; includes (p. 260–280) annotated tabular list of vascular plants now found in the drained area connecting the former island of Wieringen with the mainland.—See also van Soest and Venema, below.

Goor, A. C. J. van. De halophyten en de submerse phanerogamen. *In* Redeke, H. C., *ed.* Flora en fauna der Zuiderzee. Monografie van een brakwatergebied. p. 47–53. 25 cm. Helder, 1922.

Tabular list of halophytes and submersed plants (vascular), showing occurrence in different localities, with discussion; bibliography. Refers to littoral region of Noord-Holland, Utrecht, Gelderland, Overijssel, and Friesland.

Hoek, Julie, and Redeke, H. C. Flora van Helder; handleiding tot het bestemmen der in en om Helder, Huisduinen en het Koegras wildgroeiende en op openbare plaatsen algemeen aangeplante kruiden, heesters en boomen. 4 p. l., 216 p. 20 cm. Helder, 1904. (Arnold Arboretum library.)

Annotated descriptive flora of wild and commonly cultivated vascular plants with keys to families and genera.

Kloos, A. W., and Leeuw, W. C. de. De spontane vegetatie van den proefpolder te Andijk in 1928. (Mededeeling no. 3 van de Zuiderzee-Commissie.) Nederl. Kruidk. Arch. **1928**: 149–161 incl. 1 fig. 1929.

Topography, general features of flora; alphabetical tabular list of angiosperms with indication of localities.—See also their De vegetatie van den proefpolder te Andijk in 1929. (Mededeeling . . . no. 7.) Nederl. Kruidk. Arch. **1930**: 113–118. 1930. (Changes in vegetation.)

Koning, C. J. "Gooiland." Een geognostiche biologische studie. 46 p. 28.5 cm. Bussum, 1900. (Library of Koninklijke Nederlandse Botanische Vereniging, Leiden.)

Contains (p. 35–46) briefly annotated list of fungi, lichens, bryophytes, and vascular plants of "Het Gooi" (southeastern part of Noord-Holland), with localities. Reprinted from De Natuur, 1899–1900.—See also Nederlandsche Botanische Vereeniging, Planten waargenomen te Hilversum, below.

Kruseman, Gideon. Flora van Marken. (Mededeeling no. 26 van de Zuiderzee-commissie.) Nederl. Kruidk. Arch. 45: 1–9. 1935.

Botanical explorations, etc.; alphabetical list of phanerogams, with indication of locality and collector, followed by list of algae by other authors. Marken is an island in the former Zuiderzee now in IJsselmeer. Replaces **Boerlage, J. G.** De flora van Marken. Nederl. Kruidk. Arch. II, 5: 56–60. 1887.

Loosjes, Adriaan. Flora harlemica, of lijst der planten rondom Haarlem in het wild groeijende volgens het zaamenstel van Linnaeus. viii, 53 p. 21 cm. Haarlem, 1779.

Includes annotated list of vascular plants (252 species) of region around Haarlem. (Not seen; title and annotation furnished by S. J. van Ooststroom.)

Nederlandsche Botanische Vereeniging. Phanerogamae en cryptogamae vasculares waargenomen op de excursie . . . op 30 en 31 Augustus 1891 . . . van Alkmaar naar Heilo, Kallantsoog, Petten, het Zwanenwater, Bergen en Schoorl. Nederl. Kruidk. Arch. II, 6: 234–240. 1893.

List of vascular plants, with localities.—See also second title below.

——— Phanerogamae et cryptogamae vasculares, na afloop der zomervergadering, verzameld en waargenomen op de excursiën in de omstreken van Muiderberg op den 17 en 18 Augustus 1901. Nederl. Kruidk. Arch. III, 2: 629–632. 1902.

List of vascular plants and fungi, with localities.

——— Planten waargenomen te Alkmaar, in 't Alkmaarsche bosch en te Bergen bij Alkmaar op den 20. Augustus 1871. Nederl. Kruidk. Arch. II, 1: 271–273. 1873.

List of vascular plants and Characeae.—See also second title above.

——— Planten waargenomen te Hilversum, Huissen en Blaricum . . . op den 6. Juli 1872. Nederl. Kruidk. Arch. II, 1: 304–307. 1874.

List of vascular plants—See also **Vries, Hugo de.** Bijdrage tot de flora van het Gooi. Nederl. Kruidk. Arch. II, 5: 524–532. 1889. (List of additional species.)—Also Koning, above.

Rombouts, J. G. H., and Merkus Doornik, J. J. F. H. T. Flora amstelaedamensis, plantarum quae prope et circa Amstelaedamum sponte nascuntur enumeratio et descriptio. Pars prima, continens plantas phanerogamas et cryptogamas vasculares. lx, 136 p. 23 cm. Trajecti ad Rhenum, Amstelaedami, 1852.

Key to genera on the Linnaean system, synopsis of families on the natural system; annotated descriptive flora of vascular plants (452 species).—See also **Sande Lacoste, C. M.** Bijdrage tot de kennis der flora van Amsterdam. Nederl. Kruidk. Arch. I, 4: 243–244. 1858. (Includes additional species.)

Soest, J. L. van. De Pteridophyta en Phanerogamae van Wieringen. *In* **De flora van Wieringen.** (Mededeeling no. 14 der Zuiderzee-commissie.) Nederl. Kruidk. Arch. 1931: 308–355. 1931.

Annotated list of vascular plants, the families arranged alphabetically. Preceded and followed by other papers on the topography, botanical explorations, and bibliography (p. 245–249), phytogeography, ecology, etc., of the same area. Wieringen was an island in the former Zuiderzee. Replaces, as to Wieringen, **Harting, Pieter.** Naamlijst van planten op de eilanden Texel en Wieringen . . . Verslag. en Meded. K. Akad. Wetensch. Amsterdam Afd. Natuurk. 7: 257–266. 1858. (History, etc.; list of vascular plants, with indication of occurrence on the two islands.)—See also Feekes, above, and Venema, below.

Venema, H. J. De houtige gewassen der eilanden van de voormalige Zuiderzee. Jaarb. Nederl. Dendrol. Ver. 11: 72–74. [1936].

Annotated list of woody plants, including cultivated species, from the islands Wieringen, Marken, Urk, and Schokland in the former Zuiderzee.—See also Feekes, above, and van Soest, above.

Vlieger, Jan. Pteridophyta en Anthophyta. Varens en zaadplanten. *In* Zinderen Bakker, E. M. van. Het Naardermeer. Een geologische, historische en botanische landschapsbeschrijving van Nederlands oudste natuurmonument. p. 193–217. 24 cm. Amsterdam, 1942. (Library of Rijksherbarium, Leiden.)

Annotated alphabetical list of vascular plants.

Westhoff, Victor, *ed.* Landschap, flora en vegetatie van de Botshol nabij Abcoude. 102 p. 13 fig. (incl. maps). 22.5 cm. Baambrugge, 1949. (Publication of Commissie voor de Vecht en het Oostelijk en Westelijk Plassengebied.) (Botanical Museum library, Utrecht.)

Includes (p. 14–27), account of botanical explorations, alphabetical tabular list of vascular plants, and discussion of some noteworthy species, all by Westhoff.

NOORDZEE-EILANDEN (WEST FRISIAN ISLANDS)

See also Friesland, General, and Groningen, General.
For the North Frisian Islands, see Denmark (Jylland).

General

Eeden, F. W. van. Lijst der planten die in de Nederlandsche duinstreken gevonden zijn (Flore des dunes maritimes de la Neêrlande). **Nederl.** Kruidk. Arch. II, **1:** 360–451. 1874.

Bibliography; list of vascular and cellular plants of the Dutch islands and duneland, with localities. The list is reprinted in Jour. Bot. **13:** 142–154. 1875.—See also **Witt Hamer, H. M. de.** Supplement op de Lijst . . . l.c. II, **2:** 118–125. 1876.—Also Vuyck, below.

Holkema, Franciscus. De plantengroei der Nederlandsche Noordzee-eilanden: Texel, Vlieland, Terschelling, Ameland, Schiermonnikoog en Rottum. viii, 268 p. 23.5 cm. Amsterdam, 1870.

Botanical explorations, topography, climate, etc.; annotated list of vascular (1331 species) and cellular plants; ecology, etc.; table of distribution of the species by islands.—See also **Sande Lacoste, C. M. van der.** Nederl. Kruidk. Arch. II, **1:** 105–106. 1871. (Records of 12 additional species from Ameland, Terschelling, Texel, and Vlieland.)—Also **Westhoff, Victor.** The vegetation of dunes and salt marshes on the Dutch islands of Terschelling, Vlieland and Texel. Summary of the author's work "De vegetatie der duin- en wadgebieden van Terschelling, Vlieland en Texel." 5 p. l., 3–131 p. 24 cm. 's-Gravenhage, 1947. (Proefschrift.) (Mainly ecological and phytosociological; includes account of changes in vegetation since 1870 (p. 40–43).)—Also, for Terschelling, short lists of additions by **Eeden, F. W. van.** Nederl. Kruidk. Arch. II, **4:** 397–398. 1886, and by **Vuyck, Laurens,** and others. l.c. III, **1:** 154–155. 1897; also Nederlandsche Botanische Vereeniging. Lijst der planten waargenomen op Terschelling . . . 1888 (entered separately below); also **Dieren, J. W. van.** Organogene dünenbildung. Eine geomorphologische analyse der dünenlandschaft der west-friesischen insel Terschelling mit pflanzensoziologischen methoden. xix, 304 p. 44 fig. (incl. 32 fig. on 16 pl. (on 8)), 2 maps. 24 cm. 's-Gravenhage, 1934. (Proefschrift.) (Includes (p. 272–287) lists of the species belonging to different plant associations.)—Also, for Texel, a short list of additions by **Groll, H. W.** l.c. II, **4:** 36. 1883; also **Hoogenraad, H. R., and Visser Smits, Dirk de.** Bijdrage tot de kennis der flora van Texel. l.c. III, **2:** 1039–1046. 1903. (Notes on 66 species, including 8 new to Texel.)—Also, for Vlieland, a short list of additions by **Groll, H. W.** l.c. II, **4:** 145. 1884.—There are more recent separate comprehensive lists for Ameland (see Kloos, 1936, below); Rottum (see Schipper, 1898, below); and Schiermonnikoog (see Kloos, 1927, below).

Vuyck, Laurens. De plantengroei der duinen. 5 p. l., 368 p. 24.5 cm. Leiden, 1898. (Proefschrift.) (Arnold Arboretum library.)

Mainly historical and ecological; includes lists of vascular plants of the coastal dunes recorded by previous authors and (p. 132–140) those added during the last 50 years; also (p. 311–363): "Bijlage. De flora der Noordzee-eilanden . . .," a tabular list of known vascular plants of Texel, Vlieland, Terschelling, Ameland, Schiermonnikoog, Rottum, Borkum, Juist, Norderney, Baltrum, Langeroog, Spiekerooge, Wangerooge, Nordstrand, Pellworm, Hooge, Langenes, Oland, Apelland, Föhr, Amrum, Sylt, and Röm, indicating occurrence on each island. The islands listed, from Borkum to Sylt, inclusive, including the East Frisian and part of the North Frisian Islands, are German; Röm (Romo) is Danish.—See also van Eeden, above.

Local

Kloos, A. W. Verslag van de excursie gehouden op Ameland op 26–31 Augustus 1935. Nederl. Kruidk. Arch. **46**: 328–339. 1936.

Itinerary, etc.; alphabetical list of vascular plants; lists of additional species recorded by earlier investigators (total 527 taxa).—See also **Koopmans-Forstmann, Dora,** and **Koopmans, A. N.** Amelandse plantennamen. Nederl. Kruidk. Arch. **46**: 355–358. 1936. (List of local vernacular names.)

——— Verslag der excursie, gehouden op Schiermonnikoog op 28 Augustus 1926 en volgende dagen. Nederl. Kruidk. Arch. **1926**: 7–17. 1927.

Alphabetical list of 309 vascular plants; comparisons with earlier investigations of same island, with references to earlier publications, and list of additional species recorded.

Nederlandsche Botanische Vereeniging. Lijst der planten waargenomen op Terschelling . . . van 5–10 Augustus 1886. Nederl. Kruidk. Arch. II, **5**: 103–117. 1888.

List of vascular and cellular plants, with localities.

Schipper, W. W. De flora van het eiland Rottum. Nederl. Kruidk. Arch. III, **1**: 358–375. 1898.

Corrections of his preliminary paper with same title (l.c. 209–215. 1897); list of species that have disappeared since 1870, list of additions; unannotated list of vascular plants now occurring on Rottum (142 species and varieties); discussion of habitats, duration, etc.

OVERIJSSEL

See also Noord-Holland (van Goor, De halophyten en de submerse phanerogamen (der Zuiderzee); Venema, De houtige gewassen der eilanden van de voormalige Zuiderzee).

Local

Feekes, Willem. De flora van Schokland en Urk. Nederl. Kruidk. Arch. **52**: 79–112. 2 fig. (maps). 1942.

Plant associations, changes in flora, brief bibliography, etc.; tabular lists of known vascular plants of Schokland and Urk (the latter in Noord-Holland), showing occurrence as recorded by various botanists at different dates. Both Schokland and Urk were formerly islands in the Zuiderzee, now parts of the Noordoost-polder. Replaces **Swart, J. J.** Over de flora van Urk en Schokland. (Mededeeling no. 1 van der Zuiderzeecommissie.) Nederl. Kruidk. Arch. **1929**: 425–457. 1 fig. (map). 1930. (Botanical explorations, etc.; annotated list of phanerogams.)—See also **Bakker, D., and Zweep, W. van der.** De plantengroei van de Staart van Urk. Levende Nat. **53**: 1–9. 5 fig. (fig. 1, map). 1950. (Popular account; includes (p. 4) several species not previously reported.)—Also Venema, De houtige gewassen der eilanden van de voormalige Zuiderzee, under Noord-Holland.

——— and **Bakker, D.** De ontwikkeling van de natuurlijke vegetatie in de Noordoostpolder. Van Zee tot Land no. 6. 92 p. 14 fig., 6 maps. 1954.

Includes (p. 34–42) incomplete tabular list of vascular plants occurring in the Noordoostpolder and the bordering region, divided into 5 groups.

Leeuw, W. C. de. Vollenhove. Verslag van de excursie der Ned. Botanische Vereeniging op 21 Juli 1928 en volgende dagen opgesteld in samenwerking met de subcommissie voor het onderzoek van de flora der Zuiderzee. Nederl. Kruidk. Arch. **1929**: 458–578. 3 fig. (maps), 2 pl. 1930.

Mainly ecological; includes annotated tabular list of vascular plants (p. 476–497), alphabetically arranged.

Nederlandsche Botanische Vereeniging. Lijst der phanerogamae en cryptogamae vasculares waargenomen van het station Heino naar Wijhe op den 28 Juli 1888. Nederl. Kruidk. Arch. II, **5**: 404–418. 1889.
List of vascular plants, with localities.

——— Phanerogamae en cryptogamae vasculares waargenomen op de excursie . . . op den 27 en 28 Augustus 1892, te Steenwijk, Steenwijkerwold, Kallenkote, Eeze, Eesveen, Oldemarkt en Giethoorn. Nederl. Kruidk. Arch. II, **6**: 344–355. 1894.
List of vascular plants, with localities.

——— Phanerogamae en cryptogamae vasculares waargenomen op de excursiën . . . in de gemeenten Hengelo (O.), Losser, Weerselo en het Ambt-Delden van den 21–26 Augustus 1895. Nederl. Kruidk. Arch. III, **1**: 216–231. 1897.
List of vascular plants, with localities.

——— Phanerogamen en cryptogamen waargenomen te Almelo, in Albergen (gemeente Tubbergen) en in 't Vriezeveen, den 8 en 9 Juli 1870. Nederl. Kruidk. Arch. II, **1**: 194–197. 1872.
Unannotated list of vascular and cellular plants.

——— Phanerogamen en cryptogamen waargenomen te Delden en in 't Twikkelerbosch op den 10 en 11 Mei 1870. Nederl. Kruidk. Arch. II, **1**: 198–202. 1872.
Unannotated list of vascular and cellular plants.

Sloff, J. G. Verslag van de excursie der K. N. B. V. naar Ootmarsum en omgeving van 1 tot 9 Augustus 1946. Nederl. Kruidk. Arch. **57**: 7–20. 1950.
Itinerary, etc.; alphabetical list of vascular plants collected (600 species and varieties) with local distribution indicated.

Soest, J. L. van. De flora van Blokzijl. (Mededeeling no. 38 van de Zuiderzeecommissie.) Nederl. Kruidk. Arch. **47**: 354–373 (incl. maps). 1937.
Geology, botanical explorations, etc.; alphabetical list of vascular plants, with abbreviated botanical names; cellular plants, ecology.

——— De flora van de IJseldelta. Verslag van de excursie der Nederlandsche Botanische Vereeniging naar Kampen 1931. (Mededeeling van de Zuiderzeecommissie no. 21.) Nederl. Kruidk. Arch. **43** (1933): 421–482. 13 fig. (maps). 1933.
Topography, geology, botanical explorations, bibliography; annotated alphabetical list of 754 vascular plants,—See also Top, below.

Top, W. G. Flora campensis. Naamlijst der zichtbaar-bloeiende en van een gedeelte der bedekt-bloeiende planten, welke in en in de omstreken van Kampen in het wild gevonden worden. 2 p.l., 62 p. 20 cm. Kampen, 1890. (Botanical Museum library, Utrecht.)
Annotated list of vascular plants, bryophytes, and lichens (724 species, of which 576 are vascular). Replaces **Bondam, Rutger.** Flora campensis. Naamlijst der zigtbaar bloeijende planten, welke in de omstreken van Kampen, in het wild groeijende, gevonden worden . . . 3 p. l., 39 p. 23 cm. Kampen, 1845. (Briefly annotated list of 442 phanerogams. In Gray Herbarium library.) Also replaces **Bondam, R., and Top, W. G. Flora campensis** . . . 54 p. 21 cm. Kampen, 1849. (Annotated list of vascular and cellular plants (635 species, of which 532 are vascular). In Botanical Museum library, Utrecht.)—See also van Soest, above.

UTRECHT

See also Noord-Holland (van Goor, De halophyten en de submerse phanerogamen (der Zuiderzee)).

Local

Gevers Deynoot, P. M. E. Flora rheno-trajectina, seu enumeratio plantarum Trajecti ad Rhenum sponte crescentium (Plant. phaner. et cryptog. contin.) Flora van Utrecht . . . xii, 180 p. 17.5 cm. Utrecht, 1843. (Arnold Arboretum library.) (Reissued 1847.)
Briefly annotated list of vascular and cellular plants of the vicinity of Utrecht; no index. Text in Dutch.

Steenis, C. G. G. J. van. De flora van het Soesterveen. Nederl. Kruidk. Arch. **1926:** 285–316. map. 1927.

Plant formations, etc.; partly annotated list of vascular plants of the moor at Soest.

ZEELAND

General

Verdoorn, Frans. Historisch overzicht der Zeeuwsche floristiek. Arch. Zeeuwsch Genootsch. Wetensch. **1927:** 69–84. 1927.

Chronological sketch of botanical work, with citation of principal publications.

Walraven, A. Lijst van openbare en bedektbloeiende vaatplanten in Zeeland. Nederl. Kruidk. Arch. II, **3:** 108–141. 1878.

Number of species on different islands, etc.; tabular list of 886 vascular plants (741 native), with local distribution indicated. Replaces **Bosch, R. B. van den.** Enumeratio plantarum Zeelandiae Belgicae indigenarum [prima]—quarta. Tijdschr. Nat. Gesch. en Phys. **8:** 1–55. 1841; **9:** 245–265. 1842; **12:** 1–22. 1845; Nederl. Kruidk. Arch. **1:** 84–115. 1848. (List of vascular and cellular plants, mostly unannotated.)—See also **Walraven, A.,** and **Lako, Daniël.** Tweede lijst van openbare- en bedektbloeiende vaatplanten in Zeeland. Nederl. Kruidk. Arch. II, **6:** 101–135. 1892. (Raises total to 948 species of which 751 are indigenous.)

Local

Kloos, A. W. Verslag van de excursie, gehouden te Neuzen en omgeving op 21 Augustus 1937 en volgende dagen. Nederl. Kruidk. Arch. **48:** 18–29. 1938.

Itinerary, etc.; alphabetical list of vascular plants (607 species and forms); list of 31 additional species previously found in region.

——— and **Sloff, J. G.** Schouwen 1929. Nederl. Kruidk. Arch. **1930:** 129–146. 1930.

Previous botanical explorations, itinerary, etc.; alphabetical list of vascular plants (438 phanerogams) with localities; short list of mosses and algae. Relates to Schouwen and Duiveland.—See also van Vloten and Vuyck, below.

Lako, Daniël. Oostelijk Zeeuwsch Vlaanderen. Excursie 6–13 Aug. 1909. Nederl. Kruidk. Arch. **1910:** 109–119. 1910.

Systematic list of vascular plants, with localities. Preceded (p. 108. pl. 1 (map)) by short account of the excursion and map of region visited (vicinity of Hulst).—See also Walraven, below.

——— Verslag van de in Juli/Aug. 1913 gemaakte excursies. Nederl. Kruidk. Arch. **1913:** 17–30. 1 fig. (map). 1913.

Itinerary, etc.; systematic list of 424 vascular plants, annotated as to abundance, with localities. Relates to Oostburg and vicinity.

Nederlandsche Botanische Vereeniging. Phanerogamae et cryptogamae vasculares, waargenomen op de excursiën op het eiland Walcheren (na afloop der zomervergadering te Middelburg) op 25 Juli 1902 en volgende dagen. Nederl. Kruidk. Arch. III, **2:** 1027–1038. 1903.

List of vascular plants, with localities.—See also **Sipkes, Cornelis.** Botanische indrukken van een gemobiliseerde in Walcheren. Levende Nat. **22:** 121–126, 204–211, 273–278. illus. 1917. (Remarks on various species, etc.; alphabetical list (p. 276–278) of vascular plants observed on Walcheren and in West-Zeeuwsch-Vlaanderen, with localities. In Lloyd Library.)

Schipper, W. W. Naamlijst van phanerogamen en vaatkryptogamen op Noord-Beveland met vermelding van groeiplaatsen. Nederl. Kruidk. Arch. **1931:** 544–574. 1 fig. (map). 1932.

Soils, etc.; alphabetical list of 467 vascular plants, with localities.

Vloten, G. van, and Vuyck, Laurens. Phanerogamae en cryptogamae vasculares verzameld op den 24 Augustus 1894, van Zierikzee over Nieuwerkerk, langs Sir Jansland, door de Reigersweide, naar Bruinisse en station Zype door de leden der Nederlandsche Botanische Vereeniging . . . Nederl. Kruidk. Arch. III, **1:** 47–48. 1896.—Phanerogamae en cryptogamae vasculares verzameld en waargenomen op de excursiën der Nederlandsche Botanische Vereeniging op 23 Augustus 1894, van Zierikzee over Rengerskerke oostelijk van Serooskerke door Schuddebeurs naar Zierikzee . . . en op den 25 Augustus 1894 van Haamstede door de duinen (H) en terug over Renesse (R) . . . l.c. 48–52. 1896.

List of vascular plants (and 2 fungi), with localities, from region of Schouwen and Duiveland.—See also Kloos and Sloff, above.

Vuyck, Laurens. Verslag der excursie, gehouden to Goes, op 24 Juli 1922 en volgende dagen. Nederl. Kruidk. Arch. **1922:** 23–33. 1923.

Alphabetical list of vascular plants collected.

Walraven, A. Overzigt van de flora van oostelijk Zeeuwsch-Vlaanderen. Nederl. Kruidk. Arch. **4:** 252–312, 495–515. 1858–59.

Topography, geology, soils, etc.; annotated list of spermatophytes (558 species); list arranged by habitats, statistics. Refers to region of Hulst and Axel.—See also **Walraven** (no title). l.c. **5:** 64–65. 1860. (Additional species. In Gray Herbarium library.)—Also Lako, Oostelijk Zeeuwsch Vlaanderen, above.

ZUID-HOLLAND

See also Noordzee-eilanden (van Eeden).

Local

Hoogenraad, H. R., and Iterson, F. K. van. Flora van de omstreken van 's-Gravenhage. 194 p. 18 cm. 's-Gravenhage, 1906. (Arnold Arboretum library.)

Annotated list of vascular plants. Replaces **Anonymous.** Flora hagana. Proeve eener lijst van naam en vindplaats der wildgroeiende planten van 's-Gravenhage. 45 p. 20 cm. 's-Gravenhage, 1895. (List of vascular plants (811 species) with a few notes on habitats or localities. In Botanical Museum library, Utrecht.) ('s-Gravenhage = The Hague.)

Jaspers, G. P. J., and Bruijn, J. G. de. Nieuwe chorologie der Pterido- en Spermatophyta. (Het plantendek van de Krimpenerwaard V.) Nederl. Kruidk. Arch. **57:** 396–416. 1950.

Previous publications, etc.; list of known vascular plants, with localities or indication of previous records; bibliography. Replaces **Peeters, M. J. J., Scheygrond, Arie, and Vries, D. M. de.** Chorologie der Pterido- en Anthophyta. (Het plantendek van de Krimpenerwaard II.) Nederl. Kruidk. Arch. **1926:** 325–350. map. 1927. (Includes list of vascular plants (p. 336–349).) The Krimpenerwaard is the region between Rotterdam, Schoonhoven, and Gouda.

Molkenboer, J. H., and Kerbert, Coenraad. Flora leidensis. Sive elenchus plantarum spontanearum phanerogamicarum, quae hucusque prope Lugd. Batavorum repertae sunt, secundum ordines naturales digestus. xxxii, 389 p. 19.5 cm. Lugduni Batavorum, 1840.

Descriptive flora of vicinity of Leiden (phanerogams only, 726 species); descriptions in Latin, annotations in Dutch. Replaces **Mulder, Nicolaus** (i.e., **Claas**). Elenchus sive catalogus earum plantarum, quae prope urbem Leidam, in duarum horarum circuitu, nascuntur . . . *In* Annales Acad. Lugd.-Bat. 1816–17. 127 p. 1817. (Annotated list of vascular and cellular plants on the Linnaean system, with vernacular names.)

Nederlandsche Botanische Vereeniging. Lijst der phanerogamae en cryptogamae vasculares waargenomen in de omstreken van Brielle (Rockanje en Oost-Voorne) . . . op den 12 en 13 Augustus 1887. Nederl. Kruidk. Arch. II, **5:** 398–403. 1889.

Unannotated list of vascular plants.—See also Wiegand Bruss, below.

Trappen, J. E. van der. Bijdrage tot de kennis aan Neêrlands flora. 2 p. l., 52 p. 22 cm. 's-Gravenhage, (pref. 1852). (Library of Koninklijke Nederlandse Botanische Vereniging, Leiden.)

Includes unannotated list of vascular and cellular plants (p. 1–21) of the "Westland," between The Hague and Hoek van Holland, followed by notes on various species (p. 22–25); also list of plants collected by author in other parts of Netherlands.

Voo, B. P. van der. In het Polderland. Schetsen van het landschap en den plantengroei in de omstreken van Rotterdam . . . 142 p. plates. 19.5 cm. Amsterdam, 1898. (Rijksherbarium library, Leiden.)

Includes Naamlijst der planten uit de omstreken van Rotterdam (p. 117–131), a briefly annotated list of vascular plants, with localities for the scarcer species, followed (p. 132–142) by notes on various species.

Weevers, Theodorus. De plantengroei van het eiland Goeree in verband met zijn bodem en geschiedenis. Nederl. Kruidk. Arch. **1920:** 80–139. map. 1921.

History, ecology, etc.; list of about 600 vascular plants. Relates only to the former island Goeree, not to Overflakkee, with which it is now connected.

Wiegand Bruss, C. J. E. Eenige floristische gegevens omtrent het eiland Voorne. (Voorne. Contribution to a biological landscape study on the Island of Voorne. no. 5a.) Nederl. Kruidk. Arch. **45:** 264–288. 1 fig. (map). 1935.

Tabular list of 424 plants (402 vascular) collected in June 1934 in western part of island, with Rockanje as a center.—See also **Kloos, A. W.** Verslag van de excursie, gehouden op Voorne 24–29 Juli 1948. Nederl. Kruidk. Arch. **57:** 115–118. 1950. (Itinerary, etc.; alphabetical list of vascular plants, the names excessively abbreviated.)—Also Nederlandsche Botanische Vereeniging, above.

Witt Hamer, H. M. de. Flora delfensis. 61 p. 19 cm. Delft, 1868. (Library of Koninklijke Nederlandse Botanische Vereniging, Leiden.)

Annotated list of vascular plants, bryophytes, and lichens (553 species, of which 482 are vascular) of the region around Delft, with vernacular names.—See also his [Eene bijdrage over de flora van Delft]. Nederl. Kruidk. Arch. II, **2:** 30–32. 1875. (Erroneously recorded species, etc.)

NORWAY (NORGE)

See also Denmark, General (Hornemann); Scandinavia; Sweden, General (Lyttkens).

General

The general titles here given may be classified as follows: *Floras and manuals*: (Berg: popular, see Bolin under Scandinavia), A. Blytt (Haandbog), M. N. Blytt, Gjærevoll (popular, alpine), Hoffstad, (Lagerberg, Holmboe, and Nordhagen: see Lagerberg under Scandinavia), Lid, Nordhagen, Sørensen (popular). *Iconographies:* (Berg: see Bolin under Scandinavia), (Lagerberg: see under Scandinavia), (Mentz: see Lindman under Scandinavia), Nordhagen. *Bibliographies*: A. Blytt (Norsk botanisk litteratur), Hauge, Kleppa. *Vernacular names*: Aasen, Høeg, Schübeler (Viridarium norvegicum), Thorsrud. *Trees*: Resvoll (in winter condition). *Useful plants:* Christophersen (medicinal), Frølich (medicinal), Holmboe (edible: Gamle norske matplanter, Gratis mat av ville planter; Viking plants: Nytteplanter), Reichborn-Kjennerud (medicinal), Tonning. *Weeds*: Holmboe (Nogle ugræsplanters invandring: history of introduction), (Korsmo: see under Scandinavia). *Cultivated plants:* Nuttonson (crop geography), Schübeler (2). *Miscellaneous*: E. Dahl (plant geography), K. Dahl (biogeography), Faegri (geographical districts), Hauge (freshwater plants), Holmboe (Einige grundzüge: plant geography).

Aasen, Ivar. Norske plantenavne. 54 p. 21 cm. [Christiania, 1860.] (Reprinted from Budstikken **1860:** 9–37. 1860. Original not seen; reprint in Gray Herbarium library.)

List of scientific names of Norwegian plants, on the Linnaean system, with vernacular equivalents and localities where used; index.—See also Høeg, Schübeler (Viridarium norvegicum), and Thorsrud, below.

Berg, G. A. Floraen i farger.

See under Scandinavia (Bolin).

Blytt, Axel. Haandbog i Norges flora . . . Efter forfatterens død afsluttet og udgivet ved Ove Dahl. xi, 780 p. 661 fig., port. 19 cm. Kristiania, 1902–06. (Gray Herbarium library.)

Keyed, descriptive flora of vascular plants, with special attention to infraspecific taxa. There is a photolithographic reissue in 1926 (not seen).

——— Norsk botanisk litteratur 1859–1864 [–1868–1870]. Bot. Notiser **1865:** 103. 1865 (for 1859–64); **1866:** 81. 1866 (for 1865); **1867:** 124. 1867 (for 1866); **1868:** 144. 1868 (for 1867); **1871:** 101–102. 1871 (for 1868–70).—**Wille, Nordal.** Norsk botanisk litteratur 1871–79 [–1891–1900]. l.c. **1881:** 28–32 (for 1871–79), 102. 1881 (for 1880); **1883:** 59–62. 1883 (for

1881–82); **1892:** 72–86. 1892 (for 1883–1891); **1902:** 49–74. 1902 (for 1891–1900).—**Holmboe, Jens.** Norsk botanisk litteratur 1901–1905. l.c. **1906:** 257–275. 1906.

Unannotated lists of titles, alphabetically arranged by authors. All Norwegian botanical literature down to 1814 is covered by C. Christensen, Den danske botaniks historie (see under Denmark); that for 1901–14 is partly covered by the International Catalogue of Scientific Literature, M. Botany v. 1–14 (see section 8,000da in its index). Articles by Norwegian authors published in Sweden are included in Krok, Bibliotheca botanica suecana (see under Sweden).—See also Hauge and Kleppa, below.

Blytt, M. N. Norges flora eller beskrivelser over [in v.3, af] de i Norge vildtvoxende karplanter . . . 3 v. (1348 p.). 23.5 cm. Christiania, 1861–76.

Descriptive flora of vascular plants. Vol. 2 and 3 edited by A. Blytt. Pages 1229–1348 constitute a Tillægshefte, containing additions and index.—See also **Blytt, Axel.** Nye bidrag til kundskaben om karplanternes udbredelse i Norge. Forhandl. Vidensk.-Selsk. Christiania 1882, no. 1. 26 p. 1882; 1886, no. 7. 33 p. 1886; 1892, no. 3. 73 p. 1892; 1897, no. 2. 42 p. 1897. (Extensive lists of vascular plants, with localities and names of collectors.)

Christophersen, Erling. Norske medisinplanter. 69 p. illus. 19 cm. Oslo, 1943.

Popular treatment of the more important wild medicinal plants, with patois names, description, distribution, collection, and uses.—See also Frølich, Reichborn-Kjennerud, and Tonning, below.

Dahl, Eilif. Forelesinger over norsk plantegeografi. 2 p.l., 114 p. 47 fig. (maps, diagrs., profiles). 29.5 cm. Oslo, (pref. 1950). (Processed.)

Includes chapters on general considerations, history of plant life in Norway, ecological conditions, floral elements, tree distribution, mountain flora, water plants, and man's influence on flora, each with bibliography.—See also K. Dahl and Holmboe (Einige grundzüge), below.

Dahl, Knut, Lid, Johannes, and **Munster, Thomas.** A division of Norway into bio-geographical sectional areas agreed upon by botanists and zoologists. Skr. Vidensk.-Selsk. Kristiania I. Mat.-Naturv. Kl. **1924, no. 7.** 18 p. map. 1924.

List of 41 biogeographical regions, with geographical limits.—See also E. Dahl, above, and Holmboe (Einige grundzüge), below.

Faegri, Knut. Norges planter.

See Christiansen under Scandinavia.

——— Praktische richtlinien bei dem etikettieren von botanischen sammlungen aus Norwegen. Svensk Bot. Tidskr. **27:** 202–210. 1933.

List of old and new geographical districts, general geographical terms, system of citing localities, list of principal maps; bibliography.

Frølich, Kristina. Viltvoksende medisinske planter. Med. forord av Prof. Dr. N. Wille. 3. oplag omarbeitet og øket utgave. 38 p. 32 fig. 19 cm. Kristiana [1921]. (1st ed. 1908.)

Treatment of 32 wild medicinal plants, all but two vascular, with vernacular names and directions for collecting but no account of uses.—See also Christophersen, above, and additional references there given.

Gjærevoll, Olav, and **Jørgensen, Reidar.** Fjellflora. 160 p. incl. 150 col. fig., 5 fig. (incl. maps). 10.5 × 15 cm. Trondhjem, (pref. 1952).

Popular pocket book with colored figures and descriptions of 150 mountain plants and short descriptions of 88 others; also sketch of plant formations, etc.

Hauge, H. V. Botanisk litteratur om norske innsjøer. Nytt Mag. Naturv. **82:** 68–82. 1941. (Also as Meddel. Norsk Bot. For. **1940:** 69–82. 1941.)

Running account of investigations on cellular and vascular flora of Norwegian inland waters; geographically arranged list of localities, with references; bibliography.

Høeg, O. A. On Norwegian plant names, with some examples of word maps. Nytt Mag. Bot. **3:** 95–101. 5 maps (in pocket). 1954.

General considerations, some examples; cited mainly for its bibliography of titles on Norwegian vernacular names. Essentially the same material, in

Norwegian, is given in his Litt om norske plantenavn . . . Blyttia 13: 101–108. 5 maps (in pocket). 1955.—See also Aasen, above, and additional references there given.

Hoffstad, O. A. Norsk flora. 10. utg. ved Erling Christophersen. 507 p. 1,140 + 31 fig. 19 cm. Oslo, 1957. (1st ed. 1891.)
Keyed flora of vascular plants, with very brief descriptions and brief statement of range; glossary of specific epithets. Authorities for scientific names are not given.

Holmboe, Jens. Einige grundzüge von der pflanzengeographie Norwegens. Bergens Mus. Aarb. 1924/25, Naturv. Raekke, no. 3. 54 p. 1925.
Sketch of topography, climate, geology, botanical regions, changes through cultivation; bibliography.—See also E. Dahl and K. Dahl, above.

——— Gamle norske matplanter. Avhandl. Norske Vidensk. Akad. Oslo I. Mat.-Naturv. Kl. 1929, no. 2. 36 p. 2 fig. (incl. map). 1929.
General considerations; account of prehistoric food plants, grouped by parts used, with references; bibliography.—See also his Gratis mat and Nytteplanter, below, and Tonning, below.

——— Gratis mat av ville planter. Med oppskrifter av Juliane Solbraa-Bay. 144 p. 54 fig. 20.5 cm. Oslo, 1941.
Account of about 150 edible wild plants (vascular and cellular), with instructions for cooking. (Not seen; annotation supplied by J. Lid.) The work has also been translated into Swedish (see Holmboe under Sweden).—See also the title above and additional references there given.

——— Nogle ugræsplanters indvandring i Norge. Nyt Mag. Naturv. 38: 129–262. 3 fig. (maps). 1900.
Sources, general considerations; treatment of 19 weeds, with detailed history of their occurrence in Norway and less detailed account for Denmark and Sweden; list of the same species, with quotation of printed and herbarium data, arranged chronologically; bibliography.

——— Nytteplanter og ugræs i Osebergfundet. *In* Brøgger, Anton W., *ed.* Osebergfundet. v. 5, p. 1–78. 32 fig. 36 cm. Oslo, 1927. (Harvard College library.) (Preprinted 1921, fide J. Lid.)
List of plant remains (mostly useful plants and weeds) found in the ship-burial of a Viking queen (ca. 850 A. D.) on the Oseberg farm in Slagen parish, near Sem, Vestfold, with discussion; general account of the useful plants and weeds of Norway in Viking times; bibliography. The Oseberg find is the richest ever made in Norway.—See also his Gamle norske mat-planter, above, and additional references there given.

Kleppa, Peter. Norske floraer. Norwegian floras. Blyttia 13: 113–117. 1955.
Unannotated list of floras containing keys or descriptions, mostly general but a few local, alphabetically arranged by authors' names, with bibliographical details.—See also A. Blytt (Norsk botanisk litteratur), above, and additional reference there given.

Korsmo, Emil. Ugras i nåtidens jordbruk.
See under Scandinavia.

Lagerberg, Torsten, Holmboe, Jens, and **Nordhagen, Rolf.** Våre ville planter.
See under Scandinavia (Lagerberg).

Lid, Johannes. Norsk flora. 2. utg. 771 p. 390 fig. (incl. map, fig. 390). 21 cm. Oslo, 1952. (1st ed. 1944.)
Keyed flora of vascular plants [2341 species (including *Taraxacum* and *Hieracium*), 100 infraspecific taxa, and 308 hybrids], with short descriptions and local range; alphabetical list of herreder and byer with indication of the fylke in which each is located; glossary, list of botanical authors' names with dates and indication of author's occupation and geographical location, glossary of specific and generic names, indices. Most of the species are represented in the figures, at least by the critical parts. The microspecies of *Taraxacum* (235), being all those known from Norway, and *Hieracium* (242), being only a part of the recorded forms, are listed with localities but not described; a similar list of microspecies of *Rosa* in the 1st edition (p. 325-329) is omitted from 2d edition. The work is written in nynorsk (Landsmål, New

Norse).—See also his Nye plantefunn 1952–1954 (–1955–1957). Blyttia **13:** 33–49. 1955; **15:** 109–127. 2 fig. 1957. (Alphabetical lists of noteworthy records; bibliography.)

Mentz, August, and Ostenfeld, C. H. Billeder af Nordens flora.
See under Scandinavia (Lindman).

Nordhagen, Rolf. Norsk flora. Med kort omtale av innførte treslag, pryd- og nytteplanter. 2 p.l., xxiii, 766 p. 19 cm. Oslo, 1940.—Illustrasjonsbind. 1.–2 hefte. 380, xxiii p. incl. 501 fig. on 376 pl. (on 188). 21.5 cm. Oslo, 1944–48.

Keyed flora of vascular plants (including cultivated woody plants and the more important ornamental and useful plants) with short description, habitat, brief statement of range, standardized Norwegian vernacular name, biological notes and uses. The treatment of *Rosa, Hieracium,* and *Taraxacum* is not complete. Most of the species will be illustrated.

Nuttonson, M. Y. Ecological crop geography of Norway and its agro-climatic analogues in North America. Amer. Inst. Crop Ecol. Internatl. Agro-Climat. Ser. Study 12. 31 p. incl. 1 fig. (map), tab. 1950.

Principal features of geography and topography, climate, forests, soils, and farm crops, comparison with localities in North America, bibliography. —See also Schübeler (2), below.

Reichborn-Kjennerud, Ingjald. Våre folkemedisinske lægeurter. (Bidrag til norsk folkemedisin III.) 109 p. 21 cm. Kristiania, 1922. (Tillegg til Tidsskrift for Den norske lægeforening 1922. Den norske lægeforenings smaaskrifter no. 32.) (University of Chicago library.)

General account of Norwegian popular medicine, origin of vernacular names, etc.; systematic list of cellular and vascular medicinal plants, with vernacular names and uses (both usually with localities and references); bibliography (p. 101–105; including other countries, but mostly Scandinavian).—See also Christophersen, above, and additional references there given.

Resvoll, T. R. Vinterflora. Vore vildtvoksende løvtrær og buske i vinterdragt. iv, 83 p. 52 fig., 14 pl. 19.5 cm. Kristiania, 1911.

Popular account of native broad-leaved trees and shrubs in winter condition, with keys for identification. (Not seen; cited mainly from review in Nyt Mag. Naturv. **49:** I. 1911.)

Schübeler, F. C. Die culturpflanzen Norwegens . . . Mit einem anhange über die altnorwegische landwirtschaft . . . vi, 197 p. 17 fig., 24 pl., map. 29 cm. Christiania, 1862.

Discussion of effect of sunlight and temperature on limits of growth of plants, etc.; annotated list of cultivated plants, with special reference to growth conditions and acclimatization; discussion of ancient Norwegian agriculture; brief French summary. There is a condensed English version by M. R. Barnard, entitled Synopsis of the vegetable products of Norway . . . 31 p. pl., map. 28.5 cm. Christiania, 1862. (Includes annotated, classified lists of cultivated plants and forest trees, alphabetically arranged by English names; no index.)—See also Nuttonson, above, and Schübeler, below.

——— Viridarium norvegicum. Norges væxtrige. Et bidrag til Nord-Europas natur- og culturhistorie. 3 v. (610, 687, 679 p.). 167 fig., 4 maps, port. 29 cm. Christiania, 1885–89.—Tillæg til Viridarium norvegicum. Nyt. Mag. Naturv. **32:** 141–242. 1891. (Reprinted 102 p.)

Topography, climate, phenology, etc.; annotated list of principal cellular and vascular plants (wild and cultivated), with account of occurrence in Norway and other regions, behavior under cultivation, uses, flowering and fruiting, and extensive lists of Norwegian, Swedish, Danish, Icelandic, German, English, French, and Italian vernacular names; table showing northern limit of each species. Replaces his Die pflanzenwelt Norwegens. Ein beitrag zur natur- and culturgeschichte Nord-Europas. 2 v. (468 p.). 79 fig. (incl. maps), 15 maps. 28.5 cm. Christiania, 1873–75.—See also Aasen, above, and additional references there given (for vernacular names); also Nuttonson and Schübeler, above.

Sørensen, H. L. Norsk skoleflora til bruk ved undervisning og botaniske utferder. 19. oppl. ved Rolf Nordhagen. 2 p. l., xxxvi, 303 p. 24+ 53 fig. (in key and glossary). 19 cm. Oslo, 1951. (1st ed. 1873, with title **Norsk flora for skoler.**)

Popular manual of vascular plants with keys (including one on Linnaean system), short descriptions, and brief annotations; illustrated glossary, list of wild medicinal plants.

Thorsrud, Arne, and **Reisæter, Oddvin.** Norske plantenavn. xviii, 276 p. 21 cm. Oslo, 1948.

Discussion of scientific and popular nomenclature; alphabetical list of scientific names of wild and cultivated vascular and cellular plants, with selected Norwegian vernacular equivalents; alphabetical list of vernacular names with botanical equivalents, the vernacular names of each species collected under the one that is selected as standard; list of abbreviations of authors' names. The authors have adopted with only few changes the vernacular names used in the floras of Nordhagen (1940) and Lid (1944). Publication of Instituttet for blomsterdyrking ved Norges Landbrukshøgskole. —See also Aasen, above, and additional references there given.

Tonning, Henrik. Norsk medicinsk og oeconomisk flora, indeholdende adskillige planter, som fornemmelig ere samlede i Tronhiems stift . . . samt oplyses med fornødne beskrivelser, og anmærkninger udi læge-kunsten, landhuusholdningen, farverier m. m. . . . v. 1. [16] 187 p. 21 cm. Kiøbenhavn, 1773.

Account of 240 medicinal and agricultural plants, with vernacular names and uses. Vol. 1 ends with Didynamia; no more published. (Not seen; annotation supplied mostly by J. Lid.)—See also Christophersen (Norske medisinplanter) and Holmboe (Gamle norske matplanter); above, and additional references there given.

AKERSHUS

Akershus fylke was formerly Akershus amt. It is here taken, as in Lid's Norsk flora, to include the enclaved Oslo fylke (formerly Kristiania amt).

Local

Blytt, Axel. Christiania omegns phanerogamer og bregner med angivelse af deres udbredelse samt en indledning om vegetationens afhængighed af underlaget. viii, 103 p. 23 cm. Christiania, 1870.

Geology, soils, phytogeography, etc. (with abstract in French); annotated list of 951 vascular plants (and 12 doubtful species). Replaces **Blytt, M. N.** Enumeratio plantarum vascularium, qvae circa Christianiam sponte nascuntur. 76 p. 23 cm. Christianiae, 1844. Also replaces **Lund, Nikolai.** Haandbog i Christianias phanerogame flora. 4 p. l., xvii, 334 p. table. 17.5 cm. Christiania, 1846. (Gray Herbarium library.)

Poulsson, Edvard. Fortegnelse over Sjursøens bregner og blomsterplanter. Nyt Mag. Naturv. **62:** 52–59. 1925.

Geology, general features of flora; unannotated list of vascular plants. (Small island 2.5 kilometers south of Oslo.)

Størmer, Per. Vegetationsstudien auf der insel Håøya im Oslofjord unter besonderer berücksichtigung der gefässpflanzen und moose. Skr. Norske Vidensk.-Akad. Oslo I. Mat.-Naturv. Kl. 1938, no. 9. 155 p. 42 fig. (incl. map). 1938.

Botanical explorations, geology, climate, general features of flora, plant associations, ecology; annotated alphabetical lists of known vascular plants (578 species), mosses, lichens, fungi, and myxomycetes; bibliography. (An island in Oslofjord, about 30 kilometers south of Oslo.)

AUST-AGDER

Aust-Agder fylke was formerly Nedenes amt.

Local

Blytt, M. N. Fortegnelse over phanerogame planter og brægner, samlede ved Arendal 1838. Bot. Notiser **1840:** 1–8. 1840.

List of vascular plants [520 species], mostly without specific localities.

Engelbrethsen, Peder. Havskjærenes flora. Naturen (Bergen) **29** (III, 9): 138–144. map. 1905.

General features of flora of small islands and rocks in the Skagerak, off Borøen at mouth of the Oksefjord, with partial lists of species. Relates to Store and Lille Buskjærene, Sandskjærene, Kollholmen, etc.

Lynge, Bernt. Vegetationsbilleder fra Sørlandets skjærgaard. Nyt Mag. Naturv. **50:** 53–88. 1 fig., 2 pl. 1912.
Botanical explorations, climate, life forms, general features of flora, plant formations; annotated alphabetical list of known vascular plants from Dybvaag sogn.

BERGEN

See Hordaland.

BUSKERUD

See also Hordaland (Lid); Vestfold (Dyring, Holmestrandsfjordens fanerogamer og karkryptogamer).
Buskerud fylke was formerly Buskeruds amt.

Local

Baardseth, Egil. A study of the vegetation of Steinsfjord, Ringerike. Nytt Mag. Naturv. **83:** 9–47. 11 fig. (incl. map). 1942. (Reprinted in Meddel. Norsk Bot. For. **1941:** 9–47. 1942.)
Topography, hydrography, habitats and formations; briefly annotated, alphabetical list of angiosperms of the lake and its shores; bibliography. (In Norderhov and Hole herreder, 60°5′ N. Lat., 10°20′ E. Long.)

Dahl, Ove. Botaniske undersøgelser fornemmelig i Hallingdal. Forhandl. Vidensk.-Selsk. Christiana 1908, no. 4. 42 p. 1908.
Topography, geology, soils, climate, forest history, botanical explorations with mention of species observed; author's explorations, with partial lists of species from numerous localities. Relates mainly to western Buskerud.

Hagen, Asbjørn. Mølens flora. 69 p. 8 fig. (fig. 1, map). 22.5 cm. Oslo, 1950.
Topography, general features of vegetation (by K. G. Gleditsch), botanical investigations, plant associations, epiphytic plants, rarer species, etc.; alphabetical list of vascular plants (p. 40–46; 296 species and 5 hybrids), with vernacular names, and notes on a few of the species; lists of mosses, lichens, Mycetozoa, fungi, and algae; bibliography. Mølen is an islet in Oslofjord 4 kilometers south of the southern point of Hurumlandet. Publication of Landsforbundet for naturfredning i Norge.

Høeg, O. A. G. T. Holms liste over plantenavn fra Numedal og Sandsvær i 1750-årene. Nytt Mag. Naturv. **80:** 89–107. 1939. (Reprinted in Meddel. Norsk Bot. For. **1938:** 89–107. 1939.)
Includes Holms' list of plants with localities, vernacular names, and modern botanical names, and some observations on medicinal uses, followed by editorial notes on various species in the list.

Lid, Johannes. Tofteholmens plantevekst. I. Karplantene. *In* Resvoll-Holmsen, Hanna. Tofteholmen i Hurum. Norsk Geog. Tidsskr. **2:** 485–491. 1929.
Botanical explorations, general features of flora; unannotated alphabetical list of known vascular plants.—See also **Resvoll-Holmsen, Hanna.** Om flora og vegetasjon. l.c. **2:** 491–499. fig. 6–10. 1929. (General features of flora, phytogeography.)

Poulsson, Edvard. Fortegnelse over Kongsbergs og omegns vildtvoxende fanerogamer og karkryptogamer. Nyt Mag. Naturv. **31:** 340–379. 1890.
Topography, geology, general features of flora; briefly annotated list of vascular plants (over 600 species).

Resvoll-Holmsen, Hanna. Om planteveksten i grensetrakter mellem Hallingdal og Valdres. Skr. Norske Vidensk.-Akad. Oslo I. Mat.-Naturv. Kl. 1931, no. 9. 50 p. (incl. 2 maps). 10 pl. on 5. 1932.
Author's explorations, geology, topography, climate, general features of flora of different formations; list of vascular plants, with localities; altitudinal limits of some lowland species; bibliography. Covers an area along the boundary of Buskerud and Oppland fylker.

Rui, Halfdan. Planteliste fra Skrim i Øvre Sandsvær, Buskerud. Blyttia 9: 66–69. 1951.
List of localities; list of vascular plants (406 species).

FINNMARK

See also Nordland (Qvigstad, Navne paa dyr og planter); Troms (Norman, Norges arktiske flora; Rønning, under Norman).
Finnmark fylke was formerly Finmarkens amt.

General

Dahl, Ove. Floraen i Finnmark fylke. Nyt Mag. Naturv. v. **69.** vii, 430 p. illus. (ports.), 17 pl. (incl. map), port. 1934.
Detailed account of botanical explorations with included bibliography, phytogeography, discussion of special habitats; annotated list of known vascular plants, with detailed distribution. Replaces **Lund, Nikolai.** Forelöbig beretning om en botanisk reise i Østfinmarken i sommeren 1842. Bot. Notiser **1846:** 33–48, 65–95. 1846. (Includes list of known spermatophytes of Finnmark.)—See also **Hedberg, Olle,** and **Nygren, Axel.** Some species introduced by the Germans in the province of Finnmark in northern Norway. Bot. Notiser **1948:** 272. 1948. (List of 9 adventive angiosperms not previously recorded from Finnmark.)—Also **Nissen, Kristian.** Noen gledelige botaniske iakttagelser i Finnmark sommeren 1950. Blyttia **9:** 23–25. 1951. (Notes on 2 species and on change in climate in last 25 years.)—Also **Rune, Olof,** and **Rønning, O. I.** Noen plantefunn i Finnmark 1953. l.c. **11:** 1–4. 1955. (List with localities; bibliography.)

HEDMARK

See also Oppland (Blytt, Botanisk reise i sommeren 1836).
Hedmark fylke was formerly Hedemarkens amt.

Local

Holmboe, Jens. Vegetationen paa Hamar domkirkes ruiner. *In* Gade, F. G., and Holmboe, Jens. Aarsberetning for det Biologiske selskab i Kristiania 1903. Nyt Mag. Naturv. **42:** 49–53. 1904.
Annotated list of 44 phanerogams found on the ruins of Hamar domkirke with discussion and comparison with other floras of similar localities.

Nyhuus, Olinus. Floraen i Trysil. Nytt Mag. Naturv. **76:** 21–72. 1 fig. (port.). 1936.
Botanical explorations, biographical sketch of Nyhuus; annotated list of known vascular plants with vernacular names and detailed localities. Prepared by Jens Holmboe from author's manuscript and notes. Replaces **Sørensen, H. L.** Beretning om en botanisk reise i omegnen af Fæmundsøen og i Trysil. Nyt Mag. Naturv. **15:** 185–240. 1868.

Resvoll-Holmsen, Hanna. Om fjeldvegetationen i det østenfjeldske Norge. Arch. Math. og Naturv. v. 37, no. 1 (i.e. 2). 266 p. 1 fig. (map), 14 pl. on 8, map. 1920.
Mainly ecological; includes (chap. IX, p. 227–252) "Iagttagne karplanter," a list of vascular plants, unannotated but with detailed localities, and bibliography (p. 261–264). Covers the upper part of Østerdalen, Gudbrandsdalen, and Valdresdalene and their principal side valleys, as well as the Fæmundsand Aursundstrakterne, or between about 61° and 62°45′ N. Lat. and about 9°30′ E. Long. and the Swedish boundary.

——— Statistiske vegetationsundersøgelser fra Foldalsfjeldene. Skr. Videnskapsselsk. Kristiana I. Math.-Naturv. Kl. 1914, no. 7. 75 p. 4 pl., map. 1914.
Mostly ecological; includes (p. 69–74) briefly annotated list of vascular plants. Relates to the Fadfjeld and vicinity, north of Grimsbo.

Rud, Jon. Mjøs-egnens flora. Fortegnelse over de i Mjøs-trakterne forekommende, vildtvoxende karplanter. *In* Indbydelsesskrift til eksamen ved Hamar offentlige skole. p. 3–32. 20 cm. Hamar, 1884.
Annotated list of 613 vascular plants. Covers an area on both sides of the lake Mjøsen, in both Hedmark and Oppland. (Not seen; title and annotation supplied by J. Lid.)

HORDALAND

See also Sogn og Fjordane (Knaben).
Hordaland fylke was formerly Søndre Bergenhus amt. It is here taken, as in Lid's Norsk flora, to include the enclaved Bergen fylke (formerly Bergen amt).

Local

Blytt, M. N. Indberetning om en botanisk reise i sommeren 1833. Mag. for Naturv. **12**: 1–76. 1836. (Library of Museum of Comparative Zoology, Harvard University.)
Account of author's excursion in southern and western Norway with list (p. 41–73) of vascular plants (370 species) observed at Voss (in Hordaland) and mention of noteworthy species seen at other localities, especially Kristiansand (Vest-Agder), Egersund, Stavanger, and Finnøy (Rogaland), and Stord and Bergen (Hordaland).

Hanssen, Olaf, and Nordhagen, Rolf. Die adventivflora einiger westnorwegischen kornmühlen in den jahren 1927–1928. Bergens Mus. Årb. 1930, Naturv. Rekke no. 4. 10 p. 1931.
Localities, etc.; alphabetical list of adventive plants, with localities, mostly from Osterfjord and Sörfjord in northern Hordaland and from near Stavanger in Rogaland.

Havås, Johan. Om vegetasjonen på toppen av Hårteigen. Bergens Mus. Årb. **1927**, Naturv. Rekke no. 3. 15 p. 1928.
Includes alphabetical lists of vascular plants (42 species), mosses, lichens, and a rust from the summit of Hårteigen; bibliography.

Holmboe, Jens. Den botaniske ekskursion i Bergens skjærgaard efter det 16de Skandinaviske naturforskermøte 17de og 18de juli 1916. Bergens Mus. Aarb. 1918/1919 (i.e., 1917/18), Naturv. Raekke no. 16. 31 p. 1920.
Includes alphabetical lists of vascular plants of the islands Lille Turø, Møgster, and Anuglen, with notes on geology, general features of flora, etc.

Lid, Johannes. The vascular plants of Hardangervidda, a mountain plateau of southern Norway. Nytt Mag. Bot. **7**: 61–128. 52 i.e. 53 fig. (maps). 1959.
Topography, geology, botanical explorations, etc.; annotated list of vascular plants (455 species and 29 hybrids), with detailed localities; vegetation tables, bibliography, index. (In Hordaland, Telemark, and Buskerud fylker.)—See also Selland, below.

Lie-Pettersen, O. J. Fortegnelse over i skjærgaarden ved Feiefjorden af forfatteren noterede karplanter. *In* his Entomologiske bidrag til skjærgaardsfaunaen i det vestlige Norge. Bergens Mus. Aarb. **1904** (11): 22–25. 1905.
Annotated list of 159 vascular plants. Relates to Radøen and vicinity, north of Bergen.

Lillefosse, Torkel. Strandebarms flora. Nytt Mag. Naturv. **81**: 15–42. 1940. (Reprinted in Meddel. Norsk Bot. For. **1939**: 15–42. 1940.)
Climate, soil, botanical explorations, etc.; annotated list of vascular plants [743 species].

Naustdal, Jakob. Karplantefloraen på Gullfjellet i Fana. (Vascular plants from Gullfjellet near Bergen, W. Norway.) Blyttia **9**: 73–105. 2 fig. (fig. 1, map). 1951.
Botanical explorations, geology, climate, plant zones, plant associations; annotated alphabetical list of vascular plants; bibliography.

Norman, J. M. Botanisk reise i et strög af kysten mellem Stavanger og Bergen fra c. 59°12′ n. B. til c. 60°8′ n. B. Nyt Mag. Naturv. **8**: 249–335. 2 tab. 1855.
Includes (p. 290–324) annotated list of vascular plants (320 species), from some of the islands between 59°12′ and 60°8′ N. Lat., namely Karmöen, Fosenö, Bukken, Bömmelen, Fæö, Mosterö, Storöen, and Kvalöerne, along the coast of Hordaland and Rogaland.

Samuelsson, Gunnar. Studien über die vegetation bei Finse im inneren Hardanger. Nyt Mag. Naturv. **55**: 1–108. 1 fig., 7 pl. (incl. map). 1917 **(1916)**.

Topography, geology, climate, botanical regions, plant associations; alphabetical lists of vascular plants, mosses, and lichens, with localities; bibliography.—See also **Faegri, Knut.** Floristiske notater fra Finse. Floristic notes from Finse. Blyttia **8:** 70–74. 1950. (Alphabetical list of noteworthy vascular plants, with localities; discussion of possible changes in flora; English abstract.)—Also **Lid, Johannes.** Litt om floraen på Finse. Notes of the flora of Finse. l.c. **12:** 11–17. 1954. (Additions bringing total to 286 vascular plants.)—Also **Danielsen A., Saebø, S., and Wendelbo, P.** Nytt fra Finse. New plant records from Finse. l.c. **12:** 97–99. 1954. (List and bibliography.)

Selland, S. K. Hardangeromraadets flora. Karplantefloraen ved Hardangerfjorden og paa Hardangervidden. Bergens Mus. Aarb. 1919/20, Naturv. Raekke no. 10. 205 p. 3 fig., pl., port. 1921.

History of botanical investigations (by Jens Holmboe); annotated list of known vascular plants; bibliography.—See also **Apold, Johannes,** and **Nordhagen, Rolf.** Neue oder seltene adventivpflanzen aus Hardanger. l.c. 1930, no. 3. 6 p. 1931.—Also Lid, above.

——— Om vegetationen i Granvin. Nyt Mag. Naturv. **42:** 183–215. 1904.

Physiography, climate, botanical explorations, general features of flora; annotated list of 557 vascular plants.

——— Om vegetationen paa Voss og Vossestranden. Nyt Mag. Naturv. **44:** 159–200. pl. 4 (map). 1906.

Botanical explorations, physiography, climate, agriculture, etc.; annotated list of 572 vascular plants.

MØRE OG ROMSDAL

See also Hedmark (Resvoll-Holmsen, Om fjeldvegetationen i det østenfjeldske Norge); Nord-Trøndelag (Sundfær, Floraen i Nidaros bispedømme); Oppland (Blytt, Botanisk reise i sommeren 1836.)

Møre og Romsdal fylke was formerly Romsdals amt.

Local

Bjørlykke, Bjørn. Vegetasjonen på olivinsten på Sunnmøre. Undersøkelser foretatt somrene 1934–1936. Nytt Mag. Naturv. **79:** 51–126. 10 fig. 1939.

Mainly ecological; includes (p. 84–97) tabular alphabetical list of vascular plants growing on olivenite at various localities; bibliography.

Dahl, Ove. Botaniske undersøgelser i Romsdals amt med tilstødende fjeldtrakter. 1893. Forhandl. Vidensk.-Selsk. Christiania 1893, no. 21. 32 p. 1893.

Essentially an itinerary of his 1893 trip with mention of noteworthy plants collected at each of many localities scattered over Romsdals amt (now Møre og Romsdal fylke) and adjacent part of Kristians amt (now Oppland).

——— Kystvegetationen i Romsdal, Nord- og Søndfjord. Forhandl. Vidensk.-Selsk. Christiania 1896, no. 3. 76 p. 1896.

Author's itinerary, botanical explorations, topography, account of localities with short lists of vascular plants of various habitats; running lists of arctic, subarctic, boreal, and Atlantic species, with localities. Covers region from Kvitholmen fyr (lighthouse) on border between Nordmøre and Romsdal fogderier, south to Vaerø in Askevold prestegjeld.

——— Plantegeografiske undersøgelser i det indre af Romsdals amt med tilstødende fjeldtrakter. K. Norske Vidensk. Selsk. Skr. **1893:** 77–113. 1894; **1894:** 1–28. 1895.

Topography, etc.; account of author's work in 1890, 1892, and 1893 in various localities, with mention of numerous species. (Eastern Møre og Romsdal and adjacent herred of Lesja in Oppland.)

——— Plantegeografiske undersøgelser i ydre Søndmøre, 1894. Forhandl. Vidensk.-Selsk. Christiania 1894, no. 11. 44 p. 1895.

Author's itinerary, geology, topography, incomplete annotated lists of species of different habitats and localities, general features of flora. Relates chiefly to the islands along the coast between Molde and Stat (Stad).

Dahl, Ove. Vegetationen i Troldheimen (Surendals-Sundalsfjeldene). Forhandl. Vidensk.-Selsk. Christiania 1891, no. 4. 21 p. 1891.
Topography, geology, general features of flora with mention of noteworthy species observed at numerous localities, phytogeography.—See also his Nye bidrag til kundskaben om vegetationen i Troldheimen og fjeldpartiet mellem Sundalen og Lesje. Forhandl. Vidensk.-Selsk. Christiania 1892, no. 11. 33 p. 1892. (Similar account, with longer list of species.)—Also **Gjærevoll, Olav.** Trollheimens planteverden. Populær oversikt. 30 p. illus. (incl. maps). 21 cm. Trondhjem, 1950. (Popular treatment with mention of many species; bibliography. Publication of Trondhjems Turistforening.)

Goksøyr, Harald. Das pflanzenleben auf Rundøy, Sunnmøre in Norwegen. 184 p. (incl. 27 fig., map). 27.5 cm. Oslo, 1938.
Topography, geology, animal life, climate, botanical explorations, plant associations, cultivated plants, weeds, phytogeography, etc.; annotated list of vascular plants, also list of bryophytes; bibliography, tables showing composition of plant associations. Rundøy is an island along the coast 30 km. west of Ålesund. Publication of Det norske videnskabs-akademi i Oslo.

——— Svinøy, ei norsk utøy ved Stad. Nyt Mag. Naturv. **70:** 27–32. 1 fig. 1932.
Topography, animal life, etc.; running account of vegetation, including mention of plants observed. (An islet about 16 kilometers north of Stad.)

Høeg, O. A. Norske plantenavn hos nordmørspresten Eilert Hagerup Kjempe. Blyttia **8:** 99–114. 1950.
Notes on Kempe's manuscript (ca. 1768–1771); list of 103 cellular and vascular plants with vernacular names (mostly from Tingvoll in Nordmøre, partly from Grytten in Romsdal) and annotations; bibliography.

Larsen, F. G., and **Greve, Henrik.** Fortegnelse over de i Kristiansund og nærmeste omegn voxende karplanter (med undtagelse af mosserne). *In* Program fra Kristiansunds Skole 6. Indbydelsesskrift til hovedexamen i Juli 1870 ved Kristiansunds lærde- og realskole. p. 71–84. 17.5 cm. Kristiansund, 1870.
Unannotated list of vascular plants; separate list of ballast plants. (Not seen; title and annotation supplied by J. Lid.)

Lid, Johannes. Ivar Aasens herbarium. Nytt Mag. Naturv. **81:** 57–80. 1940. (Reprinted in Meddel. Norsk Bot. For. **1939:** 57–80. 1940.)
Historical notes; list of 509 plants (397 vascular), collected by Aasen in 1837–39, with vernacular names and often localities, from Skodje and Ørskougs sogne in Søndmør. Aasen was a celebrated Norwegian philologist, founder of the literary language Nynorsk (Landsmål) and author of Norske plantenavne (see under Norway, General).

Nordhagen, Rolf. En botanisk ekskursjon i Eikisdalen. Bergens Mus. Årb. 1930, Naturv. Rekke no. 8. 35 p. 6 fig. (incl. 2 maps). 1931.
Botanical explorations, plant associations, general features of flora; annotated alphabetical list of known vascular plants; bibliography.

Ranes, Olav. Planteliste frå Grip på Nordmøre. Nytt Mag. Naturv. **77:** 90–91. 1937. (Reprinted in Meddel. Norsk Bot. For. **1937:** 90–91. 1937.)
Geology, botanical explorations, etc.; alphabetical list of known vascular plants (34 species). Grip is a group of islets and skerries 14 kilometers northwest of Kristiansund.—See also **Gjærevoll, Olav.** Et tillegg til floraen på Grip, Nordmøre. Blyttia **7:** 15. 1949. (List of 14 species new to the archipelago.)

Wille, Nordal. Karplanter og ferskvandsalger fra øerne Husøy, Ona og Røsholmen paa Nordmør. Nyt Mag. Naturv. **61:** 53–89. 2 pl. 1924 (1922).
Includes (p. 72–74) alphabetical list of vascular plants.

NORDLAND

See also Troms (Norman, Norges arktiske flora).
Nordland fylke was formerly Nordlands amt.

Local

Blytt, A. G. Bidrag til kundskaben om vegetationen i den lidt sydfor og under Polarkredsen liggende del af Norge. Efter undersøgelser anstillede

i Ranen i sommeren 1870 i selskab med student W. Arnell. Forhandl. Vidensk. Selsk. Christiania **1871**: 125–181. 1872. (Reprinted.)
Physiography, botanical investigations, altitudinal limits of various belts, phytogeography, plant formations; annotated list of vascular plants (558 species), also Characeae and mosses. Covers area along the Ranenfjord including the præstegjeld (parishes) of Nesne, Hemnæs, and Mo.

Dahl, Ove. Botaniske undersøkelser i Helgeland. I–II. Skr. Videnskapsselsk. Kristiania I. Mat.-Naturv. Kl. 1911, no. 6. 2 p. l., 221 p. 1912; 1914, no. 4. 2 p. l., 178 p. 1 fig., 2 pl. 1915.
Botanical explorations (with references), topography, geology, general features of flora, etc.; annotated lists of vascular plants, with localities.—See also **Mejland, Yngvar.** Om floraen nord for Røsvatn. Blyttia **1**: 124–126. 1943. (List of noteworthy vascular plants, with localities, from region of Korgen and Hattfjelldal.)

Dyring, Johan. Junkersdalen og dens flora. Et bidrag til kundskaben om de indre dele af Salten. Nyt Mag. Naturv. **37**: 255–307. 1 fig. (map). 1900.
Geography, topography, geology, climate, animal and plant life, botanical explorations, human population, etc.; annotated list of known vascular plants. Covers also Tjøris- and Skaitidal, Balvand, and Graddis.

Grønlie, A. M. The ornithocoprophilous vegetation of the bird-cliffs of Røst in the Lofoten Islands, northern Norway. Nytt Mag. Naturv. **86**: 117–243. 17 fig. (fig. 1, map). 1948.
Ecological and phytosociological; includes (p. 199–200) alphabetical tabular list of vascular plants showing their occurrence on different islands, and (p. 201–214) annotated list of vascular plants found on Vedøy and Trenyken; also lists of bryophytes, lichens, and a few algae; bibliography. Røst is the most southwestern group in the Lofoten Islands.

Neuman, L. M. Bidrag till kännedomen af floran vid Saltenfjord och på Sulitälma-området i Norge. Bot. Notiser **1905**: 251–282, 323–327. 3 fig. 1905.
Author's itinerary, topography, geology, general features of flora; partly annotated list of more interesting vascular plants, with localities.

Norman, J. M. Yderligere bidrag til kundskaben om karplanternes udbredning i det nordenfjeldske Norge søndenfor Polarkredsen. Arch. Math. og Naturv. **8**: 1–186. 1883.
List of 542 vascular plants and Characeae, with detailed localities and phenological data, from Nord-Trøndelag, Sør-Trøndelag, and southern Nordland.

Qvigstad, Just. Navne paa dyr og planter i nordnorske stedsnavne. II. Tromso Mus. Årsh. v. 46, no. 1. 72 p. 1923.
Includes (p. 26–72), E. Planter, a list of plants, alphabetically arranged by vernacular names, with lists of locality names derived from them and reference to maps. The abbreviations for localities are explained in the first part of the paper, l.c. **45** (1): 3–6. 1922. Covers the fylker of Nordland, Troms, and Finnmark.

Sommerfelt, S. C. Physisk-oeconomisk beskrivelse over Saltdalen i Nordlandene. K. Norske Vidensk.-Selsk. Skr. v. 2 (2), pt. 2, p. 1–148. 1827.
Includes chapter "Om Planteriget" (p. 38–81), consisting largely of a partly annotated list of cellular and vascular plants (vascular plants, p. 69–79).

Strompdal, Knut. Planteliste frå Velfjord i Nordland. Nytt Mag. Naturv. **80**: 49–80. 1939. (Reprinted in Meddel. Norsk Bot. For. **1938**: 49–80. 1939.)
Geography, geology, botanical explorations (with references), etc.; annotated list of vascular plants, with localities, the genera *Rosa*, *Taraxacum*, and *Hieracium* omitted.

NORD-TRØNDELAG

See also Nordland (Norman, Yderligere bidrag); Sør-Trøndelag (Bryn, Indberetning . . . om en botanisk reise i det trondhjemske; Hoffstad, Vegetationen og floraen paa kysten af Trondhjems stift nordenfor Trondhjemsfjorden).
Nord-Trøndelag fylke was formerly Nordre Trondhjems amt.

General

Sundfær, John. Floraen i Nidaros bispedømme. Praktisk handbok for skoler og ved botaniske utferder. xxi, 234 p. 18 cm. Trondhjem, 1923. (Riksmuseet library, Stockholm.)

Bibliography; briefly annotated flora of vascular plants with keys, short descriptions, and localities. The bishopric of Nidaros covers Nord- and Sør-Trøndelag and the northern half of Møre og Romsdal.—See also **Gjærevoll, Olav.** Frå floraen i Trøndelag. I–IV. Årbok (earlier Årsberetn.) K. Norske Vidensk. Selsk. Mus. **1948**: 78–82. 2 fig. (maps). 1949; **1950**: 62–71. 3 fig. (incl. maps). 1951; **1951**: 95–99. 1 fig. (map). 1952; **1954**: 69–75. 3 fig. (maps). 1955. (Records of noteworthy species.)

Local

Notø, Andreas. Beitstadens flora. K. Norske Vidensk. Selsk. Skr. 1915, no. 3. 22 p. 1916.

Geography, topography, geology, general features of flora; annotated list of vascular plants. Covers the old herred of Beitstad, now forming the herreder of Beitstad and Malm.

——— Meraker flora. K. Norske Vidensk. Selsk. Skr. 1920, no. 6. 54 p. 1921.

Geography, topography, geology, general features of flora, botanical explorations; annotated list of known vascular plants.

——— Spredte bemerkinger om floraen i Nordre Trondhjems amt. K. Norske Vidensk. Selsk. Skr. 1913, no. 3. 10 p. 1913.

List of noteworthy vascular plants, with localities, collected by author in southern Nord-Trøndelag.

ØSTFOLD

Østfold fylke was formerly Smaalenenes amt.

Local

Breien, Karen. Vegetasjonen på skjellsandbanker i indre Østfold. Nyt Mag. Naturv. **72**: 131–282. illus. (incl. maps). 1933.

Mainly ecological; includes (p. 182–194) alphabetical tabular list of vascular plants from shellbank areas, mostly in vicinity of Kolbjørnvik.

Collett, Robert. Zoologisk botaniske observationer fra Hvaløerne. Nyt Mag. Naturv. **15**: 1–83. 1866.

Topography, general features of fauna and flora, zoological list, botanical explorations; annotated list of vascular plants (579 species), also Characeae. —See also **Hauge, Nils.** Plantefunn fra Fredrikstad omegn og Hvaler. Blyttia **6**: 51–52. 1948. (List of noteworthy species from the vicinity of Fredrikstad and the Hvaler Archipelago.)

OPPLAND

See also Buskerud (Resvoll-Holmsen, Om planteveksten i grensetrakter mellem Hallingdal og Valdres); Hedmark (Resvoll-Holmsen, Om fjeldvegetationen i det østenfjeldske Norge; Rud, Mjøs-egnens flora).

Oppland fylke was formerly Kristians amt.

Local

Blytt, Axel. Beretning om en i sommeren 1863 foretagen botanisk reise til Valders og de tilgrændsende tracter. Nyt Mag. Naturv. **13**: 1–149. map. 1864. (Reprinted with title: Botanisk reise i Valders og de tilgrændense egne.)

Author's itinerary, account of localities and plants observed, plant zones; annotated list of vascular plants (676 species), with localities and indication of zones.—See also **Printz, H. C.** Beretning om en i sommeren 1864 foretagen botanisk reise i Valders. Nyt Mag. Naturv. **14**: 51–96. 1865. (Mostly phenological; includes (p. 91–96) list of rarer phanerogams, with localities, including some new to Valders.)

Blytt, M. N. Botanisk reise i sommeren 1836. Nyt Mag. Naturv. **1**: 257–356. 1838.

Itinerary, with notes on species observed; annotated list of 439 vascular plants of the Dovrefjeld.—See also his Plantegeografisk sammenligning imellem Dovrefjeld og Val d'Eynes med omliggende fjelde i Östpyrenæerne. Forhandl. Skand. Naturforsk. Möde 7, Bilag Q: 3–29. 1857. (Includes (p. 7–29) unannotated list of known spermatophytes (475 species) of Dovre and Drivdal.)—Also **Hoch, Frederik.** Supplementer til Dovres flora. Nyt Mag. Naturv. **12**: 341–355. 1863. (Includes list of noteworthy plants from Dovre.)—Also **Blytt, Axel.** Nogle bemærkninger i anledning af Hr. stud. med. F. Hochs "Supplementer til Dovres flora." l.c. **14**: 46–50. 1866. (Reply to Hoch.)—Also **Barnard, M. R.** Botanical rambles on the Dovre Fjeld. *In* his Sport in Norway . . . p. 276–320. 19 cm. London, 1864. (Annotated list of vascular plants (the pteridophytes including all the species known from Norway), with running title: Alpine flora and ferns of Norway; is essentially a translation of Blytt's 1838 paper. Followed by "Hints to botanists on the Dovre Fjeld" (p. 321–330), an account of botanical explorations and collecting localities.)—Also **Nordhagen, Rolf.** Planteveksten langs Dovrebanen. Norske Turistfor. Aarb. **1923**: 10–48. 34 fig. (incl. ports., maps). 1923. (Semipopular account of general features of flora, plant immigration, etc., with mention of many species. Refers to the Dovrefjeld, Sundalsfjeldene, and Troldheimen. In University of Wisconsin library.)

Collett, Robert. Zoologisk-botaniske observationer fra Gudbrandsdalen og Dovre. Nyt Mag. Naturv. **14**: 97–160. 1866.

Includes (p. 110–122) annotated list of vascular plants from Lillehammer.

Jørgensen, Reidar. Karplantenes høidegrenser i Jotunheimen. Nyt Mag. Naturv. **72**: 1–130. 10 fig. (incl. map). 1932.

Physiography, etc.; previous studies in altitudinal range of plants in Norway; account of localities studied, with tabular lists of species showing altitudinal range; annotated alphabetical list of species; list of species, showing highest altitude reached in the region and in northern Norway and northern Sweden, etc.; bibliography.—See also **Dahl, Eilif,** and **Hygen, Georg.** Nye høydegrenser på Surtningssuen i Jotunheimen. Fra Oslostudentenes sommerekskursjon. (New altitudinal limits on Surtningssuen in Jotunheimen.) Blyttia **9**: 106–110. 1 fig. 1951. (Itinerary, etc.; list of noteworthy altitudinal records from Rauhamrane and Surtningssuen in central Jotunheimen.)

Mork, Elias, and **Heiberg, H. H. H.** Om vegetasjonen i Hirkjølen forsøksområde. Meddel. Norske Skogforsøksvesen **5** (hefte 4; whole no. 19): 617–684. 11 fig. (fig. 1, map), map, tab. 1937. (Library of New York State College of Forestry, Syracuse.)

Includes (tab. 8) tabular list of vascular plants [about 300 species], mosses, hepatics, and lichens of the Hirkjølen experimental area in the district of Ringebu, 23 km. from Atna, showing ecological relationships of each species; English abstract, bibliography.

Nordhagen, Rolf. Sikilsdalen og Norges fjellbeiter. En plantesosiologisk monografi. Bergens Mus. Skrift. no. 22. xv, 607 p. 210 fig. incl. maps. 1943.

Ecological; includes bibliography, illustrations (line drawings) of a large percentage of the species cited, and alphabetical list (index) of vascular plants mentioned [381 species] which, except for *Hieracium* and *Taraxacum*, constitutes a complete list of the known species (the species marked with an asterisk are not known from the region); also partial lists of bryophytes and lichens. (In the Jotunheimen region, at 1000 meters and above.)

Norman, J. M. Beretning om en i Gudbrandsdalen fortagen botanisk reise. Nyt Mag. Naturv. **6**: 212–291. 1851.

Running account of itinerary and plants seen, topography, climate, zones of vegetation; annotated list of vascular plants (445 species), with indication of plant zones inhabited by each. Relates to the præstegjeld Vaage and Lom.—See also **Kleiven, Magne.** Studies on the xerophilous vegetation in northern Gudbrandsdalen, Norway. Nytt Mag. Bot. **7**: 1–60. 8 fig. (fig. 1, map). 1959. (Mainly ecological; includes (p. 20–24) alphabetical list of vascular plants (113 species); bibliography. In Dovre, Lom, Sel, and (mostly) Vågå.)

Resvoll-Holmsen, Hanna. Om vegetationen ved Tessevand i Lom. Skr. Videnskapsselsk. Kristiania I. Mat.-Naturv. Kl. 1912, v. 2, no. 16. 50 p. 3 pl. 1912.

Ecological; includes (p. 40–49) list of vascular plants, with localities for the noteworthy species.

OSLO

See Akershus.

ROGALAND

See also Hordaland (Blytt, Indberetning; Hanssen, Die adventivflora; Norman, Botanisk reise).

Rogaland fylke was formerly Stavanger amt.

General

Hoffstad, O. A. Stavanger amts flora. Fortegnelse over de i Stavanger amt hidtil bemærkede fanerogamer og karkryptogamer tilligemed angivelse af deres omtrentlige udbredelse inden amtet. Stavanger Mus. Aarsber. **1891:** 23–56. 1892.

Author's explorations, etc.; briefly annotated list of known vascular plants, with some localities.—See also his Nogle nye voksesteder for fanerogamer og karkryptogamer i Stavanger amt. Stavanger Mus. Aarsber. **1894:** 45–52. 1895. (List of additional species and localities.)—Also **Røskeland, Askell.** Undersøgelser over karplanternes udbredelse i Stavanger amt. l.c. **13** (1902): 31–42. 1903. (List of additional species and localities.)

Local

Bryhn, Niels. Bidrag til Jæderens flora. Nyt Mag. Naturv. **22:** 245–320. 1876.

Botanical explorations, topography, geology, general features of flora, dunes and dune plants; annotated list of Characeae and vascular plants (690 species), mostly collected by author in 1875.—See also **Nedkvitne, Knut.** Litt om strandvegetasjonen på Jæren. On the beach vegetation of Jæren, SW. Norway. Blyttia **15:** 81–89. 1 fig. 1957. (Includes lists.)

Dahl, Ove. Botaniske undersøgelser i indre Ryfylke. I–II. Forhandl. Vidensk.-Selsk. Christiania 1906, no. 3. 36 p. 1906; 1907, no. 4. 58 p. pl. 1907.

(1906) Geography, topography, geology, climate, botanical explorations, general features of vegetation; (1907) author's itinerary with short lists of species of different localities; annotated list of vascular plants of inner Ryfylke.

Dyring, Johan. Planteliste fra Sogndal. Et bidrag til kundskapen om vegetationen i Dalene. Nyt Mag. Naturv. **52:** 217–284. 1914.—Tillegg . . . l.c. **59:** 183. 1921.

Botanical explorations, topography, geology, climate, general features of vegetation, plant formations; annotated list of vascular plants. (Name now Sokndal.)

Moe, Asche. Dates of flowering for native and garden plants at Stavanger 1897–1926. Skr. Norske Vidensk.-Akad. Oslo I, Mat.-Naturv. Kl. 1928, no. 3. 49 p. 1928.

Alphabetical tabular list of wild and cultivated spermatophytes with flowering dates.

Nordhagen, Rolf. Vegetationsstudien auf der insel Utsire im westlichen Norwegen. Bergens Mus. Aarb. 1920/21, Naturv. Raekke no. 1. 149 p. 36 fig., map. 1922.

Mainly ecological; includes annotated list of vascular plants (264 species), bryophytes, and lichens; bibliography.

SØR-TRØNDELAG

See also Hedmark (Resvoll-Holmsen, Om fjeldvegetationen i det østenfjeldske Norge); Møre og Romsdal (Dahl, Vegetationen i Troldheimen); Nordland (Norman, Yderligere bidrag); Nord-Trøndelag (Sundfær, Floraen i Nidaros bispedømme); Oppland (Blytt, Botanisk reise i sommeren 1836); also Sweden, Jämtland (Kilander).

Sør-Trøndelag fylke was formerly Søndre Trondhjems amt.

Local

Barth, J. B. Knudshø eller fjeldfloraen, en botanisk (plantegeographisk) skitse. 75 p. 18 cm. Kristiania, 1880.

Popular account including running list of the flora of Knudshø, a locality in the Dovrefjeld; includes Tillæg (p. 65–67) by J. E. Zetterstedt.

Bryn, Halfdan. Indberetning til det Kongelige Norske Videnskabers Selskab om en botanisk reise i det trondhjemske sommeren 1886. K. Norske Vidensk. Selsk. Skr. **1886/87:** 1–20. 1888.

Topography, geology; annotated list of vascular plants from region bounded by Ritsen (Rissa) and Skougdalen on the west and Lexvigen (Leksvik) on the east and north.

Fondal, Einar. Floraen i Brekken herred i Sør-Trøndelag. K. Norske Vidensk. Selsk. Skr. 1955, no. 3. 44 p. 8 fig. (fig. 1–2, maps). 1955.

Botanical explorations, geology, climate, general features of flora; annotated list of vascular plants; gazetteer, bibliography.

Haugen, J. E. Frå floraen i Oppdal serleg Storlidalen. K. Norske Vidensk. Selsk. Skr. 1948, no. 2. 22 p. 1950.

Author's explorations; briefly annotated list of vascular plants.—See also **Gjærevoll, Olav,** and **Sørensen, N. A.** Plantegeografiske problemer i Oppdalsfjellene. (Phytogeographical problems in the mountains of the Oppdal district, southern Norway.) Blyttia **12:** 117–152. 5 fig. (incl. maps). 1954. (Discussion of various noteworthy species; bibliography.)

Hoffstad, O. A. Vegetationen og floraen paa kysten af Trondhjems stift nordenfor Trondhjemsfjorden. Nyt Mag. Naturv. **37:** 1–39. 1900.

Author's itinerary in 1896 and 1897, plant formations; list of vascular plants observed, with localities.

Lyche, R. T. Adventivfloraen i Buvik. K. Norske Vidensk. Selsk. Skr. 1931, no. 2. 14 p. 1931.

Annotated alphabetical list of adventive plants collected in 1918–30.—See also his Bidrag til Norges adventivflora. K. Norske Vidensk. Selsk. Skr. 1934, no. 5. 16 p. 1934. (Systematic list of spermatophytes, with dates and localities; intended as a supplement to his 1931 paper, but including also plants from various localities in different parts of Norway.)—Also his Bidrag til Norges adventivflora 1934–1937. l.c. 1938, no. 3. 18 p. 1938. (List of species collected in different years (1918–1937), etc.; systematic list of adventives collected at Buvik, 1934–1937.)—Also his Adventivfloraen i Buvik 1938. K. Norske Vidensk. Selsk. Forhandl. **12** (B. Meddel.): 37–40. 1940. (List of species from Piene's mill, Buvik, in 1938. This is the richest locality for adventive plants in the northern mountain region.)—Also **Fondal, Einar.** Floristiske notater fra mølleområdet i Buvik. I. K. Norske Vidensk. Selsk. Mus. Årsberet. **1950:** 72–73. 1951. (List of noteworthy adventive plants; bibliography.)

Nordhagen, Rolf. Planteveksten paa Froøene og naerliggende øer. Bidrag til kundskapen om naturforholdene i Norges skjærgaard. K. Norske Vidensk. Selsk. Skr. 1916, no. 8. 151 p. 12 fig. (incl. maps), 5 pl. 1917.

Topography, animal life, climate, geology, plant associations, phytogeography, etc.; list of known vascular plants (237 species) of island groups of Froan, Mausund, Sulen, and Halten, with some additional species from the northern tip of Frøya.

Ouren, Tore. Floraen i Budal herred i Sør-Trøndelag. K. Norske Vidensk. Selsk. Skr. 1952, no. 1. 101 p. 25 fig. (mostly distr. maps; fig. 1, map). 1952.

Botanical explorations, topography, climate, soils, plant distribution, alphabetical list of plants with local vernacular names; list of known vascular plants, with localities; list of localities, bibliography.

Ouren, Tore. Floraen i Soknedal herred i Sør-Trøndelag. K. Norske Vidensk. Selsk. Årbok **1959:** 71–121. 23 fig. 1959.
Topography, geography, etc.; annotated list of vascular plants (496 species); bibliography. (Not seen; title and annotation supplied by J. Lid.)

Pettersson, Bror. Botaniska anteckningar från Dyröya och några angränsande öar vid norska västkusten. Acta Soc. Fauna et Fl. Fenn. v. 62, no. 5. 36 p. 1 fig. (map). 1939.
Topography, geology, climate, human culture, plant associations; annotated list of vascular plants, bryophytes, and lichens; bibliography. The Dyröya are a group of islets off the north coast of Fröya (island).

Storm, Vilhelm. Vejledning i Throndhjems omegns flora med en kortfattet form- og systemlære till skolebrug og selvstudium. xliii, 128 p. 17 cm. Throndhjem, 1869.
Annotated list of 560 vascular plants from the vicinity of Trondheim. (Not seen; annotation supplied by J. Lid.) There is an 2. forögede oplag. xliv, 131 p. Throndhjem, 1882. (Not seen; presumably little different from 1st ed.)—See also his Notitser til Throndhjems omegns flora. I–IV. K. Norske Vidensk. Selsk. Skr. **1885:** 1–36. 1886: **1886/87:** 21–39, 40–51. 1888; **1888/90:** 17–31. map. 1892. (Bibliography; additions and critical notes on various species.)

SOGN OG FJORDANE

See also Møre og Romsdal (Dahl, Kystvegetationen i Romsdal, Nord- og Søndfjord); also Oppland (Jørgensen, Karplantenes høidegrenser i Jotunheimen).
Sogn og Fjordane fylke was formerly Nordre Bergenhus amt.

Local

Blytt, Axel. Botaniske observationer fra Sogn. Nyt Mag. Naturv. **16:** 81–304. map. 1869. (Reprinted, viii, 223 p. 1869, with title: Om vegetationsforholdene ved Sognefjorden . . .)
Botanical explorations, topography, climate, general features of flora, plant zones, phytogeography, etc.; annotated list of known vascular plants (737 species) of a rather narrow strip along Sogne Fjord, with localities; table of altitudes. Includes (Tillæg. p. 267–304) list of mosses (by N. Wulfsberg), and meteorological data (by C. Stabell), and index to map. The reprint contains a French résumé of the prefatory matter.—See also **Wendelbo, Per.** Bidrag til Sogns flora. A contribution to the knowledge of the flora of Sogn, W. Norway. Blyttia **15:** 136–143. 2 fig. 1957. (General features of flora; alphabetical list of noteworthy species.)

Dahl, Ove. Botaniske undersögelser i Søndfjords og Nordfjords fjorddistrikter i 1896–97. Forhandl. Vidensk.-Selsk. Christiania 1898, no. 3. 71 p. 1898.
Botanical explorations, geology, climate, general features of vegetation; annotated list of vascular plants observed by author.

Faegri, Knut. Ueber die längenvariationen einiger gletscher des Jostedalsbre und die dadurch bedingten pflanzensukzessionen. Bergens Mus. Årb. 1933, Naturv. Rekke no. 7. 225 p. 47 fig. (incl. map). 1934.
Ecological; includes tabular lists of vascular plants of Nigardsbre, Bersetbre, Åbrekkebre, and Bøyabre glaciers; bibliography.

Knaben, Gunvor. Botanical investigations in the middle districts of western Norway. Univ. Bergen Årb. 1950, Naturvitensk. Rekke 8. 117 p. 20 fig. (fig. 1, map), 2 pl. (incl. front.), map. 1952.
Mainly ecological and phytogeographic; includes (p. 88–109) annotated list of vascular plants known from the mica schist mountains south of Arnafjord and Vik (i.e., in Vik herred in Sogn), with some records from the adjacent northernmost part of Hordaland.

Melheim, Anders. Om floraen på hustak i Hornindal. The flora on turf-covered roofs in Hornindal, W. Norway. Blyttia **11:** 33–61. 1 fig. 1953.
Includes list of 125 species of vascular plants found on roofs, with discussion; bibliography.

Nordhagen, Rolf. Botaniske streiftog i Ytre Nordfjord. (Floristiske undersøkelser på Vestlandet. 1.) Univ. Bergen Årb. 1953, Naturvitensk. Rekke 1. 39 p. 1954.

Botanical explorations, notes on flora of special localities, etc.; annotated list of vascular plants collected chiefly in the Hyen district; English summary, bibliography.

Samuelsson, Gunnar. Om floran i Nordfjord. [I]–II. Nytt Mag. Naturv. **78**: 37–65. 1938; **83**: 49–62. 1942. (Reprinted in Meddel. Norsk Bot. For. **1937**: 37–65. 1938; **1941**: 49–62. 1942.)

General features of flora in vicinity of Randabygda, with annotated list of known vascular plants; also list of scarcer plants found by author in Hennebygda and at other localities. The second part includes additional species and localities for Randabygda, bringing total to 452 species; also extensive list of vascular plants from Nordfjord, with localities.—See also his Nordfjords Hieraciumflora. Publicerad efter hans död av S. O. F. Omang. Arkiv Bot. II, v. 1, no. 1. 270 p. 1949. (List with descriptions of many new species; bibliography.)

SVALBARD

(Spitsbergen and associated islands, including Jan Mayen.)

General

Ostenfeld, C. H., *ed.* Flora arctica containing descriptions of the plants and ferns, found in the Arctic regions, with their distribution in these countries. Pt. I. Pteridophyta, Gymnospermae, and Monocotyledones. xi, 134 p. 95 fig., map (in text). Copenhagen, 1902.

Bibliography; descriptive flora with keys, references, local and general range (212 species). The region covered includes Svalbard as well as areas in North America and Asia. Pt. 1 by Otto Gelert and C. H. Ostenfeld; no more published.

Polunin, Nicholas. Circumboreal Arctic flora. xxviii, 514 p. illus. incl. map. 24 cm. Oxford, 1959.

Botanical regions (10), key to families; partially keyed, descriptive flora of vascular plants (892 species), mostly with only generalized statement of range; figure of each species, the authorities for botanical names given only under the illustrations. A large number of species are given as "agg.;" no infraspecific taxa mentioned. Covers northern Russia, northern Alaska, northern Canada, and Greenland; in our range only Spitsbergen, Bear Island, and Jan Mayen.

BJØRNØYA (BEAR ISLAND; BEEREN-EILAND)

See also Spitsbergen (Andersson and Hesselman, Bidrag till kännedomen om Spetsbergens och Beeren Eilands kärlväxtflora; Resvoll-Holmsen, Svalbards flora).

General

Hanssen, Olaf, and **Holmboe, Jens.** The vascular plants of Bear Island. Nyt Mag. Naturv. **62**: 210–235 (incl. map). 1925.

Botanical investigations; annotated list of 55 vascular plants known from the island; bibliography. Replaces **Fries, T. M.** Om Beeren-Islands fanerogam-vegetation. Öfvers. K. Vetensk.-Akad. Förhandl. **26** (1869): 145–156. 1870. (History with included bibliography, general features of flora; annotated list of 38 vascular plants.)

JAN MAYEN

General

Kruuse, Christian. Jan Mayens karplanter. Bot. Tidsskr. **24**: 297–302. 1902.

Annotated list of 39 vernacular plants known from the island.

Ostenfeld, C. H. Contribution à la flore de l'île Jan-Mayen. Bot. Tidsskr. **21**: 18–32. 1897.—Note corrective. l.c. **21**: 220. 1897.

Botanical investigations, with bibliographical footnotes; annotated list of cellular (by other authors) and vascular plants known from the island; 178 species, of which 28 are vascular plants. Replaces **Reichardt, H. W.**

Flora der insel Jan Mayen. *In* Die internationale polarforschung 1882–83. Die österreichische polarstation Jan Mayen . . . Beobachtungs-ergebnisse. v. 3, VII. theil, Botanik. 16 p. Wien, 1886. (Botanical explorations; annotated list of 70 plants, of which 28 are vascular, with references, localities, and general range.)

Russell, R. S. and **Wellington, P. S.** The vegetation of Jan Mayen Island. (Physiological and ecological studies on an Arctic vegetation. I.) Jour. Ecology **28**: 153–179. fig. 1–2 (fig. 1, map), pl. 10–11. 1940.
Geography, geology, climate, description of plant communities, bibliography. The authors give the number of known species of vascular plants as 58, including 15 added by their expedition (but not listed) and refer to a paper by themselves and D. Westwood, The flora of Jan Mayen Island, as in the press; this paper was never published.

Local

Bird, C. G. List of plants collected in 1934 in Jan Mayen Island. Jour. Bot. **73**: 121–124. 1935.—Correction. l.c. **74**: 64. 1936.
Bibliography; list of vascular plants, mosses, hepatics, lichens, and algae collected, with localities; includes 2 (i.e., 1) phanerogams new to the island.

Dusén, Per. Beiträge zur flora der insel Jan Mayen. Bihang K. Svenska Vetensk.-Akad. Handl. v. 26, afd. III, no. 13. 16 p. 1 fig., 1 pl. 1900.
Botanical explorations, topography, etc.; annotated list of 24 vascular plants collected by author; also list of mosses.

Gandrup, Johannes. A botanical trip to Jan Mayen. Dansk Bot. Arkiv v. 4, no. 5. 34, [1] p. 8 fig. 1924.
General features of vegetation; annotated list of 28 vascular plants collected by author (by Gandrup and M. P. Porsild); also lists of cellular plants, by other authors; bibliography.

Steindórsson, Steindór. Jan Mayen. Náttúrufræðingurinn **28**: 57–89. 12 fig. (fig. 1, map). 1958.
Topography, general features of vegetation, plant formations; alphabetical list of 32 spermatophytes observed by author; bibliography, English summary.

Spitsbergen (Spetsbergen; Spitzbergen)

General

Hadač, Emil. The introduced flora of Spitsbergen. Meddel. Norges Svalbard- og Ishavs-undersøk. **49**: 13–16. 1941.
List of localities and years in which adventive plants have been collected; alphabetical list of 52 known adventive plants, with localities and dates; short bibliography.—See also Dahl and Hadač, below, and Høeg and Lid, below.

Nathorst, A. G. Nya bidrag till kännedomen om Spetzbergens kärlväxter, och dess växtgeografiska förhållanden. K. Svenska Vetensk.-Akad. Handl. n. f., v. 20, no. 6. 88 p. 2 maps. 1883.
Botanical explorations, with bibliographical footnotes; annotated list of 123 vascular plants, with localities; phytogeography, with table showing localities in the area for each species; table showing worldwide range of each species; temperature records.—A condensed version in German, including list of species without localities or annotations, is given in his Studien über die flora Spitzbergens. Bot. Jahrb. Engler **4**: 432–448. 1 fig. (map). 1883.

Resvoll-Holmsen, Hanna. Observations botaniques. *In* Albert I, *prince of Monaco*. Résultats des campagnes scientifiques . . . fasc. 44. (Exploration du nord-ouest du Spitsberg . . . par la Mission Isachsen. 5. partie.) 81 p. 9 pl. 35.5 cm. Monaco, 1913.
Botanical explorations of Spitsbergen, with references, general features of vegetation of west coast, discussion of autumn coloration, etc.; list of known vascular plants (124 species), with citation of known localities; list of cellular plants new to Spitsbergen.

——— Svalbards flora med endel [sic] om dens plantevekst i nutid og fortid. 56 p. illus. (incl. map). 19 cm. Oslo, 1927. (Arnold Arboretum library.)

General features of vegetation and paleobotany; annotated, briefly descriptive flora of Spitsbergen and Bear Island (135 vascular plants, of which 133 are known from Spitsbergen and 54 from Bear Island), with localities.

Local

Andersson, Gunnar, and **Hesselman, Henrik.** Bidrag till kännedomen om Spetsbergens och Beeren Eilands kärlväxtflora grundade på iakttagelser under 1898 års svenska polarexpedition. Bihang K. Svenska Vetensk.-Akad. Handl. v. 26, Afd. III, no. 1. 88 p. 29 fig., 4 pl. 1900.

Botanical investigations, bibliography; annotated list of vascular plants.

——— and **Hesselman, Henrik.** Verzeichnis der in König Karls Land während der schwedischen Polarexpedition 1898 gefundenen phanerogamen. Öfvers. K. Vetensk.-Akad. Förhandl. **55:** 555–557. 1898.

Annotated list of 26 vascular plants.

Asplund, Erik. Beiträge zur kenntnis der flora des Eisfjordgebietes. Arkiv Bot. v. 15, no. 14. 40 p. 2 fig. 1919.

Botanical explorations, etc.; annotated list of vascular plants; bibliography. Includes (p. 2) list of 5 species additional to Nathorst's list of 1883; also (p. 6) list of phanerogams observed on the largest of the Edinburgh Islands off Prince Charles Foreland.—See also Hadač, Die gefässpflanzen des "Sassengebietes" Vestspitsbergen, below.

Brown, R. N. R. The flora of Prince Charles Foreland, Spitsbergen. Trans. and Proc. Bot. Soc. Edinb. **23:** 313–320. 1908.

Topography, short bibliography, etc.; annotated list of vascular plants collected in 1906 and 1907 by Dr. W. S. Bruce (55 species; 3 additional species are mentioned in the introduction).

Dahl, Eilif. On the vascular plants of eastern Svalbard. Chiefly based on material brought home from the "Heimland" expedition 1936. Skr. Svalbard og Ishavet 75. 50 p. incl. 3 fig. (maps). 1937.

Botanical explorations; annotated list of vascular plants, with localities and frequent critical notes; list of new records, bibliography. Relates to the west coast of the Storrfjord north to the northern parallel of Barents Island, also Edge Island, Barents Island, King Karls Land, and the northern coast of Northeast Land west to about 20° E. Long., with Storøya and Sjuøyane (including Phillipsøya).

——— and **Hadač, Emil.** Et bidrag till Spitsbergens flora. Meddel. Norges Svalbard- og Ishavs-Undersøk. 63. 15 p. 1946.

Localities visited, etc.; list of vascular plants collected by authors at various localities on the west and north coast of West Spitsbergen and in North East Land; short bibliography.

Hadač, Emil. Die gefässpflanzen des "Sassengebietes" Vestspitsbergen. Skr. Norges Svalbard- og Ishavs-undersøk. 87. 71 p. 24 fig. (incl. maps), 24 pl. (incl. 56 distribution maps). 1944.

Botanical explorations, geology, climate; annotated list of known vascular plants (124 species, of which 114 are native), with localities and frequent critical notes; phytogeography, bibliography. Relates to region at inner end of Eisfjord (about 15°35′ E. Long., 78°10′—78°22′ N. Lat.), bounded by the Eisfjord on the north and by the Sassenelv, Eskerelv, Brentpass, and Adventelv.

Hagen, Asbjørn. Plants collected in Vestspitsbergen in the summer of 1933. Meddel. Norsk Polarinst. 70. 14 p. pl. 1952.

Alphabetical lists of vascular plants from Kapp Linné, on the southern entrance of Isfjorden, and from Russekeila; also list of bryophytes, lichens, and fungi from Kapp Linné.

Høeg, O. A., and **Lid, Johannes.** Adventive plants in Spitsbergen. K. Norske Vidensk. Selsk. Förhandl. **1:** 176–178. 1929.

Conditions of propagation, etc.; annotated alphabetical list of 26 angiosperms collected in 1928; references to earlier publications.—See also Hadač, The introduced flora of Spitsbergen, above.

Lagerkrantz, John. Några botaniska iakttagelser sammanställda under trenne sommarresor till Västspetsbergen inom Svalbards polararkipelag. Del 1. 71 p. incl. 7 pl. on 4. 25 cm. Stockholm, 1948. (Riksmuseet library, Stockholm.)

Includes sketch of botanical explorations with titles of publications (p. 7–23); list (p. 44–46) of vascular plants (73 species) collected by author around Sveagruvan, preceded by account of topography and general features of flora; account of author's visit to Isfjorden, with lists of vascular plants from Advent-Bay region, Advent Point, and other localities; comparative list of species from Van Mijen Bay and Isfjorden (p. 57–62).

Lid, Johannes. Flowering plants. *In* Iversen, Thor. Hopen (Hope Island), Svalbard. Results of a reconnaissance in the summer 1924. Skr. Svalbard og Ishavet **10:** 28–29. 1926.

Botanical explorations, list of 12 previously known phanerogams; annotated list of phanerogams collected in 1924 by Iversen and Koefoed, including 8 additional species. Followed by lists of mosses, lichens, and fossil plants by other authors.—See also **Turnbull, Robert.** First record of plants from Hope Island, Barentz Sea. Collected by W. S. Bruce, F. R. S. G. S. Trans. and Proc. Bot. Soc. Edinb. **21:** 166–168. 1900. (General features of flora; list of lichens, etc., and 8 flowering plants collected in 1898, some of which are not reported by Lid.)

Michelmore, A. P. G. Botany of the Cambridge expedition to Edge Island, S. E. Spitsbergen, in 1927. Pt. I. Bul. Misc. Inform. Kew **1934:** 30–39. 1 fig. (map). 1934.

Author's itinerary, etc.; annotated list of vascular plants collected on Edge Island and Barents Island (60 species and 4 hybrids), plus a few from West Spitsbergen; short list of fungi; bibliography.

Polunin, Nicholas. Plant life in Kongsfjord, West Spitsbergen. Jour. Ecology **33:** 82–108. 1 fig. (map), pl. 6–9 (on 2). 1945.

Topography, climate, plant associations, plant succession, etc.; list of vascular plants collected or observed by author in 1933, also mosses, etc.; bibliography.

Scholander, P. F. Vascular plants from northern Svalbard, with remarks on the vegetation in North-East Land. Skr. Svalbard og Ishavet no. 62. 153 p. 61 fig., 5 pl. (maps), 2 maps. 1934.

Botanical explorations, localities; annotated list of 91 vascular plants; phytogeography, ecology, bibliography.

Triloff, E. G. Verbreitung und ökologie der gefässpflanzen im gebiete des Hornsundes; ein beitrag zur vegetationskunde Spitzbergens. Bot. Jahrb. Engler **73:** 259–360. map. 1943.

Phytogeographical and ecological; includes (p. 277–279) unannotated list of 65 vascular plants found by author in the Hornsund region and vicinity in southwestern Spitsbergen.

TELEMARK

See also Hordaland (Lid).
Telemark fylke was formerly Bratsbergs amt.

Local

Blytt, M. N. Fortegnelse over phanerogame planter og brægner bemærkede i sommeren 1838 i Skiensfjordens omgivelse. Bot. Notiser **1840:** 25–32, 46–48, 103–108. 1840.

List of vascular plants (662 species), with localities for the scarcer species; statistics.

Dyring, Johan. Flora grenmarensis. Et bidrag til kundskaben om vegetationen ved Langesundsfjorden. Nyt Mag. Naturv. **49:** 99–276. 1911.

Geography, topography, geology, plant formations, botanical explorations (with references); annotated list of known vascular plants. Grenmar is the old Norsk name for the region around the Langesundsfjord.

Wille, Nordal. Vegetationen i Seljord i Telemarken efter 100 aars forløb, Nyt Mag. Naturv. **40:** 65–98. 1902.

Discussion of H. J. Wille's list of plants of Seljord (Sillejord) of 1786, later botanical explorations, discussion of various species; comparative alphabetical lists of vascular plants listed by Wille and those now known.

TROMS

See also Nordland (Qvigstad, Navne paa dyr og planter).
Troms fylke was formerly Tromsø amt.

General

Benum, Peter. The flora of Troms fylke. A floristic and phytogeographical survey of the vascular flora of Troms fylke in northern Norway. Skr. Tromsø Mus. v. 6. 403 p. 31 fig. (incl. maps)+546 fig. (distr. maps) on 91 pl. (on 46). 1958.

Topography, geology, climate, botanical explorations, annotated list of localities, etc.; annotated list of vascular plants [733 species and 93 hybrids] with habitats, altitude, detailed localities, Fennoscandian and general range; bibliography. *Hieracium* is omitted.

Norman, J. M. Norges arktiske flora. 2 v. (1487, 623 p.). map. 24.5 cm. Kristiania, 1894–1901. (Botaniske Museum library, Copenhagen, complete copy. Incomplete copies in Library of Congress and Gray Herbarium.)

Vol. 1, Special plantetopografi: geography, floral districts with their included localities, discussion of tree limits and growth limits with table of localities, general list of localities with altitudes, short bibliography, list of abbreviations (p. 44–45); systematic list of vascular plants with detailed localities, frequency, flowering and fruiting dates, etc.; v. 2, Oversigtlig fremstilling af karplanternes udbredning, forhold til omgivelserne m.m.: similar list with horizontal and vertical distribution, topographical statistics, environmental relations, methods of spreading, etc. Covers Norway above the Arctic Circle, that is, northern Norland and all of Troms and Finnmark. Replaces his earlier series of papers on the same subject: Index supplementarius locorum natalium specialium plantarum nonnullarum vascularium in provincia arctica Norvegiae sponte nascentium, quas observavit . . . K. Norske Vidensk.-Selsk. Skr. 5: 1–58. 1865; Specialia loca natalia plantarum nonnullarum vascularium et Characearum et lichenum in agro arctico Norvegiae confinisque sponte nascentium observavit . . . l.c. 5: 241–378. col. pl. 1868; Notationes summatim conceptae observationum florae arcticae Norvegiae posteriore tempore effectarum, correctione addita latitudinis geographicae terminorum polarium prius indicatorum. Arch. Math. og Naturv. 5: 436–517. 1881.—See also **Jalas, Jaakko.** Floristisches aus Lapponia enontekiensis, Lapponia tornensis und Troms fylke. Arch. Soc. Zool. Bot. Fenn. Vanamo **2**: 90–96. fig. 1. 1949. (Botanical explorations; list of noteworthy vascular plants, with localities; bibliography.)—Also **Reiersen, Johannes.** Nyere plantefunn fra Lofoten—Vesterålen—Hinnøy. Blyttia **7**: 44–46. 1949. (List of 17 vascular plants not previously recorded from the archipelago.)—Also **Benum, Peter.** Nyare plantefunn i Nord-Noreg. New finds of vascular plants in northern Norway. Blyttia **8**: 1–11. 1950. (Alphabetical list of noteworthy species, mostly from Troms fylke.)—Also **Rønning, O. I.** Some new plant finds from Arctic Norway. Acta Borealia, A. Sci. 7. 20 p. 1954. (Includes (p. 4–10) list of noteworthy plants from Nordland (mostly) and Finnmark; those from Troms will be reported separately.)

Local

Benum, Peter. The flora of the western parts of Senja and Kvaløy in Troms fylke, northern Norway. Tromsø Mus. Årsh. 56, no. 1 (Naturhist. Avd. no. 12). 117 p. 1 fig. (map), 4 pl. 1935.

Topography, geology, climate, list of localities, plant associations; annotated list of known vascular plants, with localities; bibliography. (About 69°2′ to 69°55′ N. Lat.)

——— Floraen på Karlsøy i Troms fylke. (The flora of Karlsøy Island in Arctic Norway.) Nytt Mag. Naturv. **77**: 39–80. 1 fig. (map.) 1937. (Reprinted in Meddel. Norsk Bot. For. **1937**: 39–80. 1937.)

Topography, geology, climate, botanical explorations, zones of vegetation; annotated list of about 330 known vascular plants.

Devold, Joakim. Fra Balsfjords flora. Nytt Mag. Naturv. **80**: 1–39. 1939. (Reprinted in Meddel. Norsk Bot. For. **1938**: 1–39. 1939.)

Botanical explorations, with list of localities and collectors; annotated alphabetical list of vascular plants; bibliography.

Heintze, August. Växtgeografiska anteckningar från ett par färder genom Skibottendalen i Tromsö amt. Arkiv Bot. v. 7, no. 11. 71 p. 1908.

Plant formations, etc.; annotated list of vascular plants of the pine and birch zones; bibliography.

Jørgensen, Eugen. Lidt om vegetationen ved Kaafjorden i Lyngen. Nyt Mag. Naturv. **34:** 77–101. 1893.

Topography, etc.; annotated alphabetical list of vascular plants (268 species), also list of bryophytes.

——— Om floraen i Nord-Reisen og tilstødende dele af Lyngen. Forhandl. Vidensk.-Selsk. Christiania 1894, no. 8. 104 p. 1894.

Botanical explorations, topography, geology, general features of vegetation, author's itinerary; annotated list of vascular plants and bryophytes.—See also **Fridtz, Randor.** Undersøgelser over karplanternes udbredelse i Nord-Reisen. Nyt Mag. Naturv. **37:** 230–254. 1900. (Botanical explorations, general features of flora; list of newly recorded species and new localities for rarer species.)

Jørgensen, Reidar. Die höhengrenzen der gefässpflanzen in Troms fylke. K. Norske Vidensk. Selsk. Skr. 1936, no. 8. 106 p. 10 fig. (incl. map). 1937.

Author's itinerary, botanical explorations, topography (with list of vascular plants from each mountain, showing altitudinal range); alphabetical list of 225 vascular plants, showing highest altitudinal range and locality, with additional records from outside the area; general considerations, bibliography.

Mejland, Yngvar. Om floraen på Javreoaivve i Nordreisa. (On the flora of Javreoaivve in Nordreisa, Troms district, northern Norway.) Nytt Mag. Naturv. **79:** 165–191. 1 fig. (map). 1939.

Geography, topography, geology, general features of flora, botanical explorations; annotated list of known vascular plants; bibliography. The flora is said to include a greater number of the rarer Arctic species than any other Scandinavian locality of similar size.—See also Notø, Fjeldfloraen, below.

Notø, Andreas. Fjeldfloraen mellem Altevand og Kirkesdalen. Tromsø Mus. Aarsh. **27** (1904): 1–19. 1905.

Geography, topography; alphabetical list of widespread vascular plants found by author in 1902, and annotated alphabetical list of scarcer species; list of scarcer arctic species known from Javreoaivve and Kirkestinden.—See also Mejland, above.

——— Florula tromsøensis. Ed. nova. Tromsø Mus. Aarsh. **23** (1900): 113–180. 1901.

Phytogeography, etc.; annotated list of vascular plants (406 species, plus 72 hybrids and subspecies, of which 337 species and 51 hybrids and subspecies are from Tromsø proper). Replaces **Norman, J. M.** Florula tromsøensis. Tromsø Mus. Aarsh. **15:** 157–173. 1893, and is regarded by the author as a new edition of that work.

——— Indre- og mellem-Kvænangens karplanter. Nyt Mag. Naturv. **40:** 293–362. incl. fig. (map). 1902.

Topography, botanical explorations, etc.; annotated list of vascular plants.

VEST-AGDER

See also Hordaland (Blytt, Indberetning).

Vest-Agder fylke was formerly Lister og Mandals amt.

Local

Fridtz, R. E. Undersøgelser over floraen paa kysten af Lister og Mandals amt. Skr. Vidensk.-Selsk. Christiania I. Math.-Naturv. Kl. 1903, no. 3. 219 p. 1903.

Short bibliography, botanical explorations, topography, climate, general features of flora of different formations and localities; annotated list of vascular plants (974 species); tabular list, showing occurrence by localities; statistics.

VESTFOLD

Vestfold fylke was formerly Jarlsberg og Larviks amt.

Local

Dyring, Johan. Holmestrandsfjordens fanerogamer og karkryptogamer. Et bidrag til kunnskapen om vegetasjonen i den nordlige del av Vestfold og de tilstøtende strøk av Buskerud fylke. Nyt Mag. Naturv. **59:** 45–184. 1921 (1922).

Geography, geology, climate, botanical explorations, etc.; annotated list of vascular plants. Covers an area mostly in Vestfold but partly in Buskerud, extending about 20 kilometers northwest to the vicinity of Sande from a base line between Horten and Hurumlandt.

Nordhagen, Rolf. Planteveksten. *In* Berg, Lorens. Tjømø. En bygdebok. p. 23–50. 30 fig. 23 cm. Kristiania, 1920.

Survey of the vegetation, with incomplete list of vascular plants. (Not seen; title and annotation supplied by J. Lid.)

——— Planteveksten i Nøtterø herred. *In* Berg, Lorens. Nøtterø, en bygdebok. p. 28–59. fig. 11–42. 23 cm. Kristiania, 1922. (Riksmuseet library, Stockholm.)

Survey of the vegetation, including mention of numerous vascular plants.

PORTUGAL

See also Trás-os-Montes e Alto Douro (Rozeira, for account of phytogeography and plant zones); also Spain, General (Amo y Mora, Flora fanerogámica; Colmeiro, La botánica, Diccionario, and Enumeración; Font Quer, Geografía botánica de la península ibérica; Gandoger, Catalogue des plantes recoltées, and Notes sur la flore espagnole; Lázaro é Ibiza, Regiones botánicas; Más y Guindal, Memorandum de sínonimias; Rehder, (Phytography); Texidor y Cos, Flora farmacéutica; Willkomm, Grundzüge der pflanzenverbreitung).

The continental European part ("continente") of the republic of Portugal is divided into 18 administrative districts which are grouped, in many cases fractionally, into 11 provinces. The latter, although in most cases having the same or nearly the same names as the 6 (later 8) earlier provinces (existing through 1936) have different boundaries in every case but two (Algarve and Minho), although the names and boundaries of the administrative districts have not changed. The local floras here listed have been grouped under the present provinces, whose correlation with the old provinces and the districts is described under the respective provincial headings.

The Atlantic island groups ("ilhas adjacentes") of Azores (Açores) and Madeira have already been treated in part 1 (p. 86, 90) of this work.

General

The general works here listed may be classified as follows: *Floras*: Coutinho (Flora de Portugal), Sampaio (Flora portuguesa). *Catalog*: Sampaio (Lista). *Bibliographies*: Bellot Rodríguez, Brito. *Vernacular names*: (Cardoso), (Coutinho, Flora), (Sampaio, Flora), Vasconcelos. *Trees and woody plants:* Coutinho (Esbôço). *Useful and poisonous plants:* Cardoso (some groups of useful plants), Costa (medicinal), Figueiredo (medicinal and edible), Moller (Catalogo), Vasconcellos (Plantas fanerogâmicas venenosas, and Plantas medicinais), (Vasconcelos). *Weeds*: Silva, Vasconcellos (Plantas vasculares infestantes). *Cultivated plants*: (Cardoso), (Costa), Coutinho (Esbôço, trees; Flora de Portugal). (Figueiredo), (Lautensach), (Sampaio, Flora portuguesa), (Vasconcellos and Feio). *Miscellaneous:* Albuquerque (2: plant zones and natural regions), Girão (vegetation), Brito (history of botany), Daveau (plant associations, phytogeography), Fernandes (additions to flora). Ficalho (flora of the Lusiad), Lautensach (botanical regions, plant associations), Moller (As regiões botanicas), Rothmaler (phytogeography).

Albuquerque, J. P. Manique e. Carta ecológica de Portugal. 58 p. 2 fig. (maps) and folding map in cover. 30.5 cm. Lisboa, 1954. (At head of title page: Ministério da economia. Direcção geral dos serviços agrícolas.)

General considerations, phytogeographical zones, ecological zones, agroclimatic regions, etc.; French summary.—See also the following entry, also Daveau, Girão, Lautensach, Moller (As regiões botanicas), and Rothmaler, below; also Trás-os-Montes e Alto Douro (Rozeira).

——— Zonas fito-climáticas e regiões naturais do continente português. Bol. Soc. Broteriana II, **19:** 569–591. map. 1945.

Short account of plant zones and natural regions.—See also the entry above, and additional references there given.

Bellot Rodríguez, Francisco. Notas bibliográficas sobre la botánica portuguesa. Anales R. Acad. Farm. Madrid 2. época, año **6:** 217–277. 1940.

Sketch of development of botany in Portugal; alphabetical list of about

150 authors, with titles of their works (including works by Portuguese authors or by outside authors on Portugal or Portuguese possessions; goes through 1935). Intended to supplement Colmeiro's La botánica y los botánicos de la península hispano-lusitana (see under Spain, General.)—See also Brito, below.

Brito, Anthero de. Historia da botanica em Portugal. 2. ed. 63 p. 16.5 cm. Lisboa, 1888. (Bibliotheca do povo e das escolas no. 65.) (1st ed. 1883.)

A series of biographical notices on native and foreign botanists who worked on the flora of Portugal from the 16th century to the middle of the 19th, with lists of their botanical publications. The so-called 2d edition is merely a reissue.—See also Bellot Rodríguez, above.

Cardoso, O. M. Pereira. Plantas tintoriais portuguesas. 61 p. pl. 19 cm. Porto, 1927.

Descriptive account of cellular and vascular plants used as dyes and mordants. (Not seen; title and annotation supplied by A. R. Pinto da Silva.) There being no general work on the useful plants of Portugal, this title and the following are cited to supplement the works on medicinal plants by Costa, Figueiredo, and Moller.—**Palhinha, Ruy Telles.** Plantas aromáticas de Portugal. Lista das plantas aromáticas espontâneas, subespontâneas e cultivadas que se encontram em Portugal. Brotéria Sér. Trimest. Ciên. Nat. **15:** 97–113. 1946. (Running account of 233 aromatic or odoriferous vascular plants, systematically arranged, with vernacular names and brief account of properties.)

Costa, Manuel dos Santos. Historia das plantas medicinais portuguesas. x, 424 p. 19 cm. Lisboa, 1899.

List of wild and cultivated medicinal plants, alphabetically arranged by vernacular names, with descriptions, uses, directions for collecting, etc. (Not seen; title and annotation supplied by A. R. Pinto da Silva.)—See also Palhinha under Cardoso, above, and Figueiredo, Moller (Catalogo das plantas medicinaes), and Vasconcellos and Feio, below.

Coutinho, A. X. Pereira. Esbôço de uma flora lenhosa portuguesa. 2. ed. Pub. Dir. Geral Serv. Flor. e Aqüíc. [Portugal] v. 3, tomo 1. 371 p. 50, 14 fig. 1936. (1st ed. 1887, as vol. 2 of his Curso de silvicultura.)

Annotated descriptive flora of woody plants (including cultivated species) in form of keys, with localities and vernacular names; glossary; alphabetical list of botanical names (468 species and hybrids); account of some cultivated trees, etc.—See also **Morais, A. Taborda de.** As árvores notáveis de Portugal. [I]–IV. Annuário Soc. Broteriana **2:** 27–45. illus. 1936; **3:** 11–47. illus. 1937; **4:** 37–48. illus. 1938; **5:** 15–32. illus. 1939. (Account of large specimens of various trees.)—Also **Rothmaler, Werner.** Árvores de Portugal. Bol. Soc. Broteriana II, **15:** 133–148. 1941. (Critical notes on various trees, from *Taxus* to *Quercus*.)

——— Flora de Portugal (plantas vasculares) disposta em chaves dicotómicas. 2. ed. dirigida pelo Dr. Ruy Telles Palhinha. 2 p.l., 938 p. 25 cm. Lisboa, 1939. (1st ed. 1913.)

Briefly annotated flora of vascular plants in form of keys, including cultivated species (total 2,845 species); glossary, list of vernacular names with scientific equivalents. The first edition was followed by his Notas da flora de Portugal I–VII. 25 cm. Paris [etc.], 1914–30 (nos. I, II, V, VI seen in library of Mme. Valentin Allorge, Paris) and a Suplemento . . . Bol. Soc. Broteriana II, **10:** 43–194. 1935, the last bringing together in systematic form the information given in the Notas, as well as new material.—See also his Breves considerações estatísticas acêrca da flora portuguesa. l.c. **28:** 95–121. 1920. (Statistics, phytogeography, list of endemic species and varieties, etc., based on his 1st edition.)—Also **Morais, Artur Taborda de.** Notas sôbre a flora portuguesa. l.c. II, **11:** 153–168. 2 fig. (incl. map), pl. 1936. (New localities and records or critical notes for 30 species.)—Also **Vasconcellos, J. de Carvalho e.** Plantas recentemente herborizadas em Portugal. l.c. II, **13:** 41–43. 1938. (Critical notes on 5 angiosperms.)—Also **Rothmaler, Werner.** Species novae vel criticae florae lusitanicae. l.c. II, **13:** 273–282. 1938. (New records, critical notes, etc., for various vascular plants.)—Also **Rothmaler, Werner, and Silva, A. R. Pinto da.** Florae Lusitaniae emendationes. I–III. Agron. Lusitana 1: 236–254, 373–393. 1939; 2: 75–90. 1940. (Critical notes on synonymy, occurrence, etc., of vascular plants, intended to supplement the second edition of Coutinho's Flora.)—Also **Morais, A. T. de.** Novas áreas da fitogeografia portuguesa. Bol. Soc. Broteriana II, 14: 97–138. 8 fig. 1940. (New localities and records or critical notes for 40 species.)—Also **Rothmaler, W.** Sôbre algumas plantas críticas. Brotéria Sér. Trimest. Ciên. Nat. 9: 5–17. 2 fig. (maps). 1940. (Critical notes, new records, etc., for about a dozen angiosperms; German summary.)—Also **Silva, A. R. Pinto da.** Notas soltas sôbre a flora portuguesa. I. Agron. Lusitana 2: 225–231. 1940. (New records, critical notes, etc., concerning 30 vascular plants.)—Also **Vasconcellos, J. de Carvalho e.** Anotações do herbário do Instituto Superior de Agronomia. Anais Inst. Super. Agron. (Lisboa) 11: 7–17. 1940; 12: 171–180. 1941. (New records, critical notes, etc., for 55 vascular plants.)—Also **Garcia, J. G.** Plantas vasculares raras ou novas para a flora portuguesa. Bol. Soc. Broteriana II, 19: 1–26 (incl. 4 pl.). 1944. (Critical notes on 4 angiosperms.)—Also his Estudos sobre a flora de Portugal. I. l.c. II, 20: 43–92 (incl. 5 pl.). 1946. (List of species and localities additional to Coutinho's Flora, with discussion; French summary.)—Also his Algumas novidades para a flora ibérica. Anales Jard. Bot. Madrid 6 (2): 421–438 (incl. 6 pl. on 3). 1946. (Records of 10 angiosperms, mostly new to the flora of Portugal.)—Also **Silva, A. R. Pinto da,** *ed.* De flora lusitana commentarii ad normam herbarii Stationis agronomicae nationalis. I–XI. Agron. Lusitana 8: 5–18. 1946; 9: 5–41. 3 fig. (incl. maps), 2 pl. 1947; 10: 5–23. 6 fig. (incl. maps). 1948; 10: 79–121. 3 fig. (incl. maps). 1948; 11: 5–37 incl. 4 fig. (i.e. 4 pl., incl. 3 maps). 1949; 13: 19–87. fig. 1–86 (incl. maps), pl. 1–3. 1951; 14: 5–58. fig. 1–2 (incl. map), pl. 1–6. 1952; 15: 95–151. illus. (incl. maps), pl. 1. 1953; 18: 5–48. pl. 1–2. 1956; 18: 83–108. 6 fig. 1956; 19: 5–22. 1957. (A series of critical notes, new records, revisionary studies, etc., by various authors, intended to supplement the second edition of Coutinho's Flora; index in part X.)—Also **Fernandes, Rosette.** Notas sobre a flora de Portugal. I–VII. Bol. Soc. Broteriana II, 23: 119–157. 10 fig. (i.e., 10 pl.). 1949; Mem. Soc. Broteriana 6: 5–61. 18 fig. (i.e. 2 fig., 16 pl.). 1950; Anuário Soc. Broteriana 18: 9–29. 1 fig. (i.e. 1 pl.). 1952; Mem. Soc. Broteriana 9: 85–102. 1953; Bol. Soc. Broteriana II, 28: 131–176 incl. 8 pl. 1954; l.c. II, 30: 121–139. 1956; 31: 183–217. 11 fig. (i.e., 11 pl.). 1957. (Range extensions, new records, and critical notes concerning numerous species; bibliographies.)—Also **Beliz, J. Malato,** and others. Notas de floristica. I–VI. Mem. Soc. Broteriana 6: 63–77. 9 fig. 1950; 7: 5–14. 2 fig. (i.e., 2 pl.) 1951; Anuário Soc. Broteriana 18: 31–48. 3 fig. (i.e. 3 pl.). 1952; Mem. Soc. Broteriana 10: 11–27. 1 fig., 4 pl. 1955; Bul. Soc. Broteriana II, 31: 15–35. 1 fig. (map), 4 pl. (on 2). 1957; II, 32: 253–264. 3 pl. 1958. (Range extensions, new records, and critical notes concerning various species; bibliographies.)—Also **Mendes, E. J.,** and **Romariz, C.** Anotações e aditamentos à flora lusitânica. Bol. Soc. Portug. Ciênc. Nat. 20 (II, 5): 1–18. 3 fig. (maps), 3 pl. (on 2). 1954. (Records of noteworthy species.)—Also Fernandes, below.

Daveau, Jules. Géographie botanique du Portugal. [I]–III. Bul. Herb. Boissier 4: 209–228, 281–313. 1896 (reprinted in Bol. Soc. Broteriana 14: 3–54. 1897); Bol. Soc. Broteriana 19: 3–140. 1903; 21: 16–85. 1906.

Topography, habitats, plant zones and associations, phytogeography, etc., with included lists of species. Pt. I, La flore littorale du Portugal (1896); pt. II, La flore des plaines et collines voisines du littoral (l.c. 19: 3–93); pt. III, Les stations de la zone des plaines et collines (l.c. 19: 93–140; 21: 16–85). The treatment of the two other zones (montane and subalpine) was never published.—See also Albuquerque, above, and additional references there given.

Fernandes, Abílio. Progrès récente dans l'étude de la flore vasculaire du Portugal. Anuário Soc. Broteriana 21: 6–25. 1955.

Lists of spontaneous and subspontaneous vascular plants added to the flora of Portugal since 1947, and in part since 1939, without annotations or references.

Ficalho, F. M. C. de M., *conde de.* Flora dos Lusiadas. 99 p. 22.5 cm. Lisboa, 1880. (Gray Herbarium library.)

Discussion and identification of plants mentioned in Luiz de Camões' poem, grouped geographically; index of scientific names.

Figueiredo, J. J. de. Flora pharmaceutica e alimentar portugueza, ou tractado daquelles vegetaes indigenas de Portugal, e outros nelle cultivados, cujos productos são usados, ou susceptiveis de se usar como remedios e alimentos, distribuidos segundo o systema Linneano . . . 2 p. l., 600 p. 20.5 cm. Lisboa, 1825.
Account of 531 medicinal or edible plants (519 vascular) with vernacular names, description, part used, habitat, taste, etc. but in most cases without specific account of medicinal use.—See also Costa, above, and additional references there given.

Girão, Aristides de Amorim. Aspectos dominantes da vegetação. *In* his Geografia de Portugal. p. 194–212. fig. 92–95 (incl. maps), pl. 25–28, map. 30.5 cm. Porto, 1941.
General considerations, distribution of important trees, climatic adaptations, phytoclimatic zones, natural regions; bibliography.—See also Albuquerque (Carta), above, and additional reference there given.

Lautensach, Hermann. Das spontane und subspontane pflanzenkleid. *In* his Portugal auf grunde einiger reisen und der literatur. I. teil: Das land als ganzes. Petermanns Mitt. Ergänzungsh. **213:** 100–138. 1932.
Botanical investigations, general features of flora, climate, botanical regions, geological history, adventive plants, forests, plant associations, fruit trees, agriculture.—See also Albuquerque (Carta), above, and additional references there given.

Moller, A. F. Catalogo das plantas medicinaes que habitam o continente portuguez. Instituto v. 25–29, passim. 1878–82. (Reprinted 180 p. 23 cm. Coimbra, 1882. Reprint not seen.)
Annotated list of vascular and cellular medicinal plants, with vernacular names, distribution, flowering dates, uses, etc.—See also Costa, above, and additional reference there given.

——— As regiões botanicas de Portugal. Bol. Soc. Broteriana **17:** 89–154. 1900–01.
Statistics, limits of distribution of some trees, phytogeography, botanical regions and zones. Is a selection and translation of material relating to Portugal in Willkomm, Grundzüge der pflanzenverbreitung auf der Iberischen Halbinsel (see under Spain). The paper is often attributed to J. A. Henriques, who wrote only the brief preface. It is now rather out of date. —See also Albuquerque (Carta), above, and additional references there given.

Rothmaler, Werner. Importância da fitogeografia nos estudos agronómicos. Palestras Agron. **2** (1): 49–60. 13 i.e. 14 fig. (maps) on 10 pl. (on 6). 1939.
Condensed account of phytogeographical groups, origin of flora, forest types, areas of endemism, centers of origin.—See also Albuquerque (Carta), above, and additional references there given.

Sampaio, Gonçalo. Flora portuguesa. 2. ed. dirigida por Américo Pires de Lima. xliii, 792 p. 850 fig., 13 pl. 20.5 cm. Porto, 1946 (1947). (1st ed. (incomplete), 1909–14, with title Manual da flora portugueza.)
Key to families; flora of vascular plants (2,378 [i.e. 2,387] species) in form of keys, with flowering dates, local range, and vernacular names; glossary, index of vernacular names with scientific equivalents, list of botanical authors cited. Garden plants are mentioned under their genera but not described or numbered. Title page date 1946, but printer's note at end shows that printing was completed in May 1947. The first ed. (416 p. incl. 741 fig. on numerous pls. 18.5 cm. n.p., n.d. [1909–14]. [no title page]) is a flora of vascular plants in form of very briefly annotated keys, and goes through family 115 (Plantaginaceae) and species 1,809 (*Litorella uniflora*). For dates of publication see Bol. Soc. Broteriana II, **12:** 307–308. 1937.—See also his Iconografia selecta da flora portuguesa. 3 l., 5 l. 150 pl., port. 40 cm. Lisboa, 1949 (1950). (Plates of 150 species, without descriptions and mostly without details; partial bibliography, list of species figured with their synonyms, index. At head of title page: Ministério da educação nacional. Instituto para a alta cultura.)—Also Fernandes, above.

——— Lista das espécies representadas no Herbário português. Pteridófitas e spermáfitas. 148 p. 19 cm. Pôrto, 1913 (cover date).

Systematic list of Portuguese vascular plants, unannotated but with numerous footnotes dealing with points of nomenclature. Was an essentially complete list of the flora of Portugal, the species not represented in the Herbário português of the Faculdade de sciências do Pôrto being distinguished by an asterisk. At head of title page: Universidade do Pôrto. Faculdade de sciências.—See also his Apêndice à Lista das espécies representadas no Herbário português. 10, [3] p. 19 cm. (Pôrto), 1914 (cover date); Segundo apêndice . . . 9, [2] p. 18 cm. (n.p.), 1914 (cover date); Terceiro apêndice . . . 15, [4] p. 19 cm. (n.p.), 1914 (cover date). (Lists of changes in nomenclature, with synonymy.)

Silva, A. R. Pinto da. Elementos para o estudo da determinacão de origem dos trigos pelas sementes estranhas. Revista Agron. **25:** 165–224. 1937.

Includes (p. 172–179, with supplementary list on p. 180) list of about 135 weed seeds found in wheat (*Triticum*) of Portuguese origin. The work is cited here as providing, in conjunction with the following titles, the principal information available on Portuguese weeds.—See also **Santos, P. E. Cavique.** Subsídios para a análise botanica dos fenos do Gerez. Pub. Dir. Geral Serv. Flor. e Aqüíc. **9** (2): 219–234. map. 1942. (Includes (p. 222–224) lists of weeds found in hay from 5 different "currais" (small pens where cattle are enclosed for the night) and surrounding pastures.)—Also **Rothmaler, Werner.** Sobre a sistemática e a sociologia dos linhos de Portugal. Agron. Lusitana **6:** 253–279. 1944. (Includes (p. 259–272) list of weeds found in flax seed.)—Also **Teixeira, J. Brito.** As impurezas do centeio produzido em Portugal. (Ensaio de caracterização.) Agron. Lusitana **10:** 123–157. 4 fig., 1 pl., 3 tab. 1948. (Includes (p. 131–134) list of about 162 weed seeds found in domestic seed of rye, *Secale cereale.*)—Also **Vasconcellos, J. de C. e.** Ervas infestantes das searas de trigo. 404 p. illus., 100 fig. on 50 pl. (on 25). 23.5 cm. Lisboa, 1958. (Methods of control, etc.; systematic treatment of weeds found in wheat, with description, local and general range, methods of propagation and control; bibliography. Published by Federação nacional dos produtores de trigo. The figures illustrate the ligules and other details of grasses and the fruits of other important weeds.) —Also Vasconcellos (Plantas vasculares infestantes), below.

Vasconcellos, João de Carvalho e. Plantas fanerogâmicas venenosas da flora portuguesa. (Preâmbulo para o seu estudo). Vinha Portug. **38** (449/450): 20–23. 3 fig. 1929.

Mostly unannotated list of known or suspected poisonous spermatophytes, with vernacular names.

——— Plantas vasculares infestantes dos arrozais. 3 p. l., 188 p. illus. 23.5 cm. Lisboa, 1954. (At head of title page: Ministério de economia. Comissão reguladora do comércio de arroz. [Publicações 22.])

General considerations; descriptive account of weeds of Portuguese rice fields, with general and local distribution, methods of control, etc.; bibliography.—See also his Sementes estranhas do arroz. Pub. Comis. Regulad. Com. Arroz no. 27. 26 p. illus. Lisboa, 1958. (Descriptive account of fruiting parts of weeds occurring in rice seed; distribution by regions, bibliography.)—Also Silva, above.

——— and **Feio, F. M.** Plantas medicinais e aromáticas. (Elementos para o seu estudo.) Min. Econ. Dir. Geral Serv. Agr. Sér. Estud. e Inform. Téc. 34. 200 p. illus., 2 tab. 22 cm. 1949.

General considerations; systematic treatment of a number of native or cultivated medicinal plants with description, distribution, uses, cultivation, etc.; short bibliography.—See also Costa, above, and additional references there given.

Vasconcelos, Augusto de. Dicionário das plantas de Portugal (espontâneas e subespontâneas) que teem nome popular com o rigoroso binome scientifico, as suas propriedades, usos e muitas indicações úteis. Com uma apreciacão de Gonçalo Sampaio . . . [8], 91, [1, 2] p. 17.5 cm. Porto, 1915.

Alphabetical list of vernacular names, with scientific equivalents and some notes on uses. (Photostatic copy in U. S. Dept. Agriculture Library.)—See also Cardoso, Coutinho (Flora), and Sampaio (Flora), above.

ALGARVE

The present Province of Algarve has the same boundaries as the old province of the same name and consists of the district of Faro.

Local

Daveau, Jules. Aperçu sur la végétation de l'Alemtejo et de l'Algarve. Jor. Sci. Math., Phys. e Nat. Lisboa **8:** 235–280. 1882.

General features of flora, with lists of species observed at different localities; systematic list of vascular plants observed, with localities. The author traveled hastily from Pinal Novo (Estremadura) to Beja and Mértola (Baixo Alentejo), and more leisurely in Algarve from Vila Real de Santo Antonio via Tavira, Faro, and Portimão to Lagos.—See also **Silva, A. R. Pinto da,** and **Fontes, F. C.** Plantas raras e plantas novas para o Algarve. Agron. Lusitana **13:** 77–87. 1951. (List of noteworthy species; bibliography.)

Primo, S. da Costa. Quelques observations sur la végétation de Sagres et du Cap de S. Vicente. Bul. Soc. Portug. Sci. Nat. **12:** 125–134. pl. 20–22. 1936.

Plant zones, geology, climate; unannotated lists of spermatophytes observed by author; general features of flora.

Veiga, S. P. M. Estacio da. Plantas da Serra de Monchique observadas em 1866. Jor. Sci. Math., Phys. e Nat. Lisboa **2:** 120–130, 192–213. 1870 [1869?].

Mostly unannotated list of vascular plants, mosses, and lichens (260 spermatophytes), with vernacular names.

ALTO ALENTEJO

See also Algarve (Daveau).

The Province of Alto Alentejo consists of nearly all the northern part of the old Province of Alemtejo (district of Evora and most of the district of Portalegre, the western part of which is now in Ribatejo).

Local

Fernandes, Abílio, and **Fernandes, Rosette.** Herborizações nos domínios da Fundação da Casa de Bragança. II. Vila Viçosa. Bol. Soc. Broteriana II, **22:** 17–96. 50 fig. (on 33 pl.), map. 1948.

Authors' explorations with mention of species observed; list of vascular plants collected, with localities and some annotations.

——— **Garcia, José,** and **Fernandes, Rosette.** Herborizações nos domínios da Fundação da Casa de Bragança. I. Vendas Novas. Mem. Soc. Broteriana **4:** 5–89. 74 fig. (mostly on pls.), 5 pl. (pl. 5, map). 1948.

Authors' excursions in vicinity of Vendas Novas, with mention of plants observed; list of vascular plants, with localities, and descriptions of the chromosomes of some monocotyledons.—See also **Fernandes, Abílio,** and **Fernandes, Rosette.** Herborizações nos domínios . . . III. Vendas Novas (2. lista). Anuário Soc. Broteriana **15:** 7–34. 5 fig. (i.e. 5 pl.) 1949. (Additions.)

BAIXO ALENTEJO

See also Algarve (Daveau).

The Province of Baixo Alentejo consists of the southern part of the old Province of Alemtejo (district of Beja) and the southern part of the old Province of Estremadura (southern part of district of Setúbal).

Local

Daveau, Jules. Excursions botaniques. Bas Alemtejo. Bol. Soc. Broteriana **5:** 148–158. 1887.

Account of author's explorations in general vicinity of Serpa, with mention of characteristic species seen.

Galvão, J. Mira. Os barros de Beja. Biblioteca do Agricultor Alentejano no. 2. 61 p. 22 cm. Beja, 1941.

Includes an appendix, "Principais plantas da flora espontanea da região dos barros de Beja" (p. 57–61), a list with vernacular names and notes on uses. (Not seen; title and annotation supplied by A. R. Pinto da Silva.)

Sampaio, Gonçalo. Flora vascular de Odemira. Bol. Soc. Broteriana **24:** 7–132. 1909.

Topography, general features of flora, botanical explorations; annotated list of spontaneous vascular plants (891 species), with localities; the principal cultivated species are listed at the end of each family.

——— Um passeio botanico ao Torrão. Bol. Soc. Broteriana 18: 47–79. 1901 (1902).

General features of flora; annotated list of 222 vascular plants collected by author in March 1899.

BEIRA ALTA

See also Beira Litoral (Henriques).

The Province of Beira Alta includes most of the old province of the same name (district of Vizeu, except for the northern strip), the northern half of the old Province of Beira Baixa (consisting of most of the district of Guarda), and a small area from the district of Coimbra in the old Province of Douro.

Local

Henriques, J. A. De Macieira até Castro Daire. Bol. Soc. Broteriana **18:** 163–172. 1901 (1902).

Geology, topography, general features of flora; list of vascular plants collected by author in September 1901, with localities.

——— Uma excursão botanica na Serra do Caramullo. Bol. Soc. Broteriana **4:** 113–123. 1886.

Topography, geology, general features of flora; list of 127 vascular plants observed by author in June 1884, with localities.

Sampaio, Gonçalo. Flora vascular de Trancoso. Anais Fac. Ciên. Pôrto **20:** 129–154, 209–223. 1936.

Briefly annotated list of vascular plants (405 species) collected by author in July 1908, with vernacular names.

BEIRA BAIXA

The Province of Beira Baixa includes most of the southern part of the old province of the same name (the whole district of Castelo Branco), part of the old Province of Douro (part of the district of Coimbra), and part of the old Province of Estremadura (part of the district of Santarém).

Local

Henriques, J. A. Expedição scientifica á Serra da Estrella em 1881. Secção de botanica. 133 p. 2 maps. 32.5 cm. Lisboa, 1883.

Botanical explorations, botanical regions, statistics; annotated list of 716 plants (600 vascular) known from above 400 meters elevation. A few corrections in identification are given in review in Bul. Soc. Bot. France **31** (Rev. Bibl.): 157–159. 1884.—See also **Sampaio, Gonçalo.** Flora vascular das lagoas. *In* Notas sobre a analyse bacteriologica e chimica e sobre a flora e a fauna das aguas das lagoas de Serra da Estrella. Bol. Direcç. Geral Agr. (Lisboa) **9** (7): 5. 1910. (List of 5 vascular plants of the lakes in the Serra da Estrêla, with localities.)—Also **Barros, J. J. de.** Serras de Portugal. A. Contribuição para o estudo floristico e fitosociológico de "Serra da Estrêla." Arquivo Univ. Lisboa **15:** 187–203. 4 pl. 1934. (List of localities, general features of flora; annotated alphabetical lists of spermatophytes, ferns, Characeae, mosses, hepatics, fungi, and lichens, collected by author. In Harvard College library.)

BEIRA LITORAL

The Province of Beira Litoral includes most of the old Province of Douro (the greater part of the districts of Aveiro and Coimbra) and the northern part of the old Province of Estremadura (the greater part of the district of Leiria and a small part of the district of Santarém).

Local

Baptista, M. D. Ensaio de huma descripção, fizica e economica de Coimbra, e seus arredores. Mem. Econ. Acad. R. Sci. Lisboa **1:** 254–298. 1789.

Includes (p. 258–268), "Florae conimbricensis specimen," an unannotated list of vascular and a few cellular plants, on the Linnaean system, with indication of those that are cultivated.

Henriques, J. A. Esboço da flora da bacia do Mondego. Bol. Soc. Broteriana **22:** 21–113. 1906; **23:** 200–215. 1907; **24:** 214–239. 1909; **25:** 191–221. 1910; **26:** 85–177, 210–327. 1911. (Reprinted xvi, 438 p. 19.5 cm. Coimbra, 1913. [Fide A. R. Pinto da Silva.])

Topography, geology, climate, plant zones; annotated list of vascular plants, with keys. The basin of the Mondego is bounded on the north by the serras of Caramullo and Bussaco (Buçaco), the hills of Murtede, Cantanhede, and Azarede, and the Serra de Buarcos; on the south by the serras of Estrella, Açor, Louzã, and Sicó, and the hills of Albergaria up to Lavos.

——— A vegetação espontanea do Bussaco. Bol. Soc. Broteriana **3:** 109–123. 1885.

Geology, etc.; unannotated list of 407 cellular and vascular plants of the Forest of Buçaco.—See also **Dalgado, D. G.** A list of the flora of Bussaco. *In* his Bussaco: its monastery, battle, and woods, and its uses as a health resort. Appendix II. p. 86–110. 22 cm. Lisbon, 1916. (Unannotated alphabetical list of vascular plants, wild and cultivated, with indication of those that are cultivated or naturalized. Based on the work of Henriques and on information supplied by Ernesto de Lacerda. In Harvard College library.)—Also **Figueiredo, J. de Melo de.** Subsídios para o estudo da flora lenhosa e herbácea da Mata do Buçaco. Bol. Min. Agr. (Lisboa) **12** (1/6): 21–56. 1930. (Geology, history, etc.; unannotated list of vascular plants, including numerous cultivated species.)—Also **Franco, João do Amaral.** Notas sobre a flora lenhosa da Mata do Buçaco. Bol. Soc. Broteriana II, **25:** 197–248. 1951. (Annotated systematic list of 124 woody plants (nearly all exotic), with descriptions and some critical notes on nomenclature.)

DOURO LITORAL

The Province of Douro Litoral consists of the northern part of the old Province of Douro (district of Porto and northern strip of district of Aveiro) and a small portion of the old Province of Beira Alta (part of the district of Vizeu).

Local

Johnston, E. J. Breves apontamentos para a flora phanerogamica do Porto. Revista Soc. Instruc. Porto **1:** 10–16, 57–63, 121–125, 167–171, 232–235, 302–306, 390–394. 1881; **2:** 248–252. 1882.

General features of flora; briefly annotated list of angiosperms. Apparently never finished, at least in this journal; covers Ranunculaceae—Amaryllidaceae.—See also **Luso, Augusto.** Herbarium cryptogamicum do Porto e seus arredores. Jor. Hort. Prat. **3:** 73–74, 85–87, 122–124, 223–225. 1872; **4:** 30–31, 54. 1873. (Annotated list of pteridophytes and cellular cryptogams, with localities.)—Also **Johnston, E. J.** Esboço d'um calendario da flora dos arredores do Porto. Annaes Sci. Nat. **1:** 5–16, 84–90, 127–134, 181–186. 1894; **2:** 30–34, 147–154, 209–214. 1895; **3:** 33–38, 129–136, 203–210. 1896. Suplemento . . . l.c. **8:** 123–139. 1903. (List of angiosperms observed by author in vicinity of Oporto, arranged by date of flowering, with localities. The last installment printed in 1896 is index.)—Also **Freyn, Josef.** Contribuição para a flora do Porto. Bol. Soc. Broteriana **16:** 216–218. 1899 (1900). (List of angiosperms collected by Otto Buchtien in 1890–91 at Porto.)—Also **Fonseca, J. L. Lebois.** Lista de espécias herborizadas pelo sócio Sr. Júlio L. Lebois Fonseca, no Porto e arredores. Anuário Soc. Broteriana **7:** 11–15. 1941. (List of noteworthy vascular plants.)—Also Sampaio, below.

Sampaio, Gonçalo. Estudos de flora local. I. Vasculares do Porto. Revista Sci. Nat. e Soc. **4:** 150–158, 195–202. 1896; **5:** 26–42, 122–138. 1897.

List of botanical collectors with some biographical information, topography, principal crops, general features of flora; annotated list of vascular plants, with localities and some vernacular names. Incomplete, covering Ranunculaceae–Valerianaceae.—See also his Estudos sobre a flora dos arredores do Porto. I. Primulaceae. Annaes Sci. Nat. **6:** 51–62. 1900. (Keyed descriptive treatment.)—Also his Estudos sobre a flora dos arredores do Porto. Annuario Acad. Polytech. Porto **27:** 147–171. 1904. (Monograph of local forms of *Spergularia*.)—Also Johnston, above.

Silva, J. Gomes da. Um capitulo de therapeutica indigena. Flora medica portuense. 4 p. l., 168 p. map. 24.5 cm. Porto, 1881. (Arnold Arboretum library.)

Includes bibliographies and systematic list of 186 native medicinal plants, with vernacular names in various languages, uses, etc.

ESTREMADURA

The Province of Estremadura consists of a small part of the old province of the same name, including almost all of the district of Lisboa, the southern part of the district of Leiria, and the northern part of the district of Setúbal.

Local

Almeida, J. L. Ferreira de. Flórula da Mata Nacional das Mestras. Subsídios para o seu estudo. Bol. Soc. Broteriana II, **23:** 89–104. pl. 1. 1949.

History, soils, etc.; unannotated list of vascular plants (168 species.)

Brandão, J. I. Seixas. Memorias dos annos de 1775 a 1780 para servirem de historia á analysi, e virtudes das agoas thermaes da villa das Caldas da Rainha. xxxi, xiv, 281 p. 19.5 cm. Lisboa, 1781.

Includes (p. 258–263) "Catalogo das plantas, que nesta primavera de 1780 observamos no sitio dos banhos das Caldas da Rainha, e sua immediação," an unannotated alphabetical list with some vernacular names and indication of the medicinal species.—See also **Lopes, J. Pinto.** Plantas colhidas em Santa Catarina (Caldas da Rainha). Anuário Soc. Broteriana **9:** 32–38. 1943. (Unannotated list of vascular plants; short list of vernacular names with botanical equivalents.)

Daveau, Jules. Excursion botanique aux îles Berlengas et Farilhões. Bol. Soc. Broteriana **2:** 13–31. 1884.

General features of flora, bibliography; annotated list of vascular plants (112 species), also hepatics, algae and lichens.

Felgueiras, Guilherme. Contribuição para o estudo da flora e da fauna da Mata de Leiria. Bol. Sec. Est. Agr. (Lisboa) **1:** 113–155. 1919. (Reprinted, 45 p.)

General features of vegetation; list of vascular plants collected by author, with occasional annotations and vernacular names. The Mata de Leiria (or Pinhal do Rei) is a large pine forest near the town of Leiria, but in the Province of Estremadura. A. R. Pinto da Silva informs me that the paper, rearranged by A. Taborda de Morais, and with an annex on the medicinal uses of some species, was republished in **Pinto, A.** Arala. O Pinhal do Rei. v. 2, p. 166–194. 1939.—See also **Silva, A. R. Pinto da.** Duas herborizações. Revista Agron. **28:** 132–140. 2 fig. 1940. (Includes (p. 133–137) list of plants from São Pedro de Moel, at the southeast end of the Pinhal do Rei, including additions to the flora. In Iowa State College library.)

Lisbon. Instituto superior de agronomia. Secção de botânica. Elenco florístico da Tapada da Ajuda. Agros **39:** 148–205. 1956.

Unannotated lists, with vernacular names, of "Plantas vasculares espontâneas e subespontâneas" (p. 148–166), "Plantas ornamentais e florestais cultivadas" (p. 166–188), and "Plantas agrícolas e hortícolas" (p. 188–205). Replaces **Vasconcellos, J. de Carvalho e.** Herborizações na Tapada da Ajuda. Anais Inst. Super. Agron. **3:** 176–190. 1930; Herborizações . . . Apêndice à "Relação das plantas herborizadas" publicada nos "Anais do Instituto Superior de Agronomia" vol. III-1929. l.c. **6** (2): 74–77. 1934. (Briefly annotated list of vascular plants (369 species and varieties, including appendix) found in the Park of Ajuda (Lisbon), with vernacular names.)

Luisier, Alphonse. Apontamentos sobre a flora da região de Setúbal. Bol. Soc. Broteriana **19:** 172–274. 1903.

Author's explorations; list of vascular plants (1,004 species) of the vicinity of Setúbal and the Serra d'Arràbida, with localities and some annotations; includes cultivated plants.

Markham, Lady, and others. List of plants collected round Estoril during three seasons. 38 p. 22 cm. Lisboa, 1910.

List, containing many errors. (Not seen; title and annotation supplied by A. R. Pinto da Silva.)

Myre, Mário. Contribuição para o estudo de algumas communidades vegetais da classe Rudereto-secalinetales Br.-Bl. dos arredores de Lisboa. Bol. Soc. Broteriana II, **19**: 699–727. 3 fig. 1945.

Includes (p. 714–718) tabular list of weedy plants, not in systematic order.

Palhinha, R. Telles. Monsanto, arboreto de Lisboa. Ideal de um botânico. 28 p. 26 cm. Lisboa, 1943. (Publicações culturais da Câmara municipal de Lisboa.)

Includes (p. 19–28) a list of herbaceous and suffrutescent plants collected in the Serra de Monsanto. (Not seen; title and annotation supplied by A. R. Pinto da Silva.)

Pereira, J. F. Ensaio d'uma flora do concelho dos Olivais ou enumeração das plantas, que espontaneamente vegetão neste concelho, e das que são mais cultivadas em seus campos, hortas, pomares, jardins, etc. 96 (+) p. 26 cm. n.p. [pref. 1880].

List with local distribution. (Not seen; title and annotation supplied by A. R. Pinto da Silva from imperfect copy in library of Instituto Botânico, Universidade de Lisbôa. The pages cited cover Ranunculaceae-Labiatae.)

Rozeira, Arnaldo. Apontamentos para o estudo da flora vascular de Alcobaça. Anais Fac. Ciên. Pôrto **31**: 44–64. 1946. (Reprinted as Pub. Inst. Bot. Gonçalo Sampaio Faculd. Ciên. Univ. Porto no. 16. 25 p. 1946. Original not seen.)

List of 250 vascular plants collected in June 1944, principally in general vicinity of Alcobaça.

Vasconcellos, J. de Carvalho e. Herborizações na Cêrca dos Jerónimos (Estação Agrária Central). Revista Agron. **24**: 164–175. 1936. (Iowa State College library.)

List of angiosperms collected, with vernacular names.

Withering, William. Florae ulyssipponensis specimen: with some characteristic remarks, and an index. *In* The miscellaneous tracts of the late William Withering . . . v. 1, p. 249–297. 21.5 cm. London, 1822.

List of vascular and cellular plants (421 species), with localities and some annotations, arranged according to date of collection; includes cultivated species. Olisipo is the ancient name of Lisbon.

MINHO

The Province of Minho has the same limits as the old province of the same name (districts of Viana do Castelo and Braga).

Local

Caldas, M. P. Flora medica da Ribeira de Visella. xxxii, 94, iv p. 8vo. Porto, 1901.

Not seen; annotation not available. Title supplied by A. R. Pinto da Silva.

Lima, A. Pires de. As plantas medicinais do Gerês. Anais Faculd. Farm. Porto **10**: 41–132. 1950. (Repr., 96 p., as Pub. Inst. Bot. Gonçalo Sampaio Faculd. Ciên. Univ. Porto no. 49. Original not seen.)

Annotated list of medicinal plants [103 species] of the Serra do Gerês, with vernacular names and uses; bibliography.

Pereira, C. L. Flora da bacia do Minho. Anais Fac. Ciên. Porto **17**: 129–162, 204–239. 1932.

Topography, etc.; list of vascular plants (773 species), with localities.

——— A flora do concelho de Paredes de Coura. Bol. Soc. Broteriana **28**: 33–70. 1920.

Topography, general features of flora, agriculture, etc.; list of vascular plants (432 species), including cultivated plants, with vernacular names, some localities, and indication of frequency of occurrence. Followed (p. 71–94) by papers on the bryophytes and lichens by other authors.

Sampaio, Gonçalo. Flora de Ponte do Lima. Almanaque de Ponte do Lima **5**: 256–265. 1923.

List of vascular and some cellular plants, grouped more or less ecologically, with vernacular names. (Not seen; title and annotation supplied by A. R. Pinto da Silva.)

——— Flora vascular de Caldelas. Anais Fac. Ciên. Pôrto **19:** 65–96, 177–182. 1934–35.

General features of flora; annotated list of vascular plants (330 species) collected by author, with vernacular names.

Silva, A. R. Pinto da, and **Sobrinho, L. G.,** *eds.* Flora vascular da Serra do Gerês. Agron. Lusitana **12:** 233–380. 1950 [1951].

Geography, etc.; list of known vascular plants, with citation of collections and some critical annotations especially on nomenclature. The whole of vol. 12 of Agronomia Lusitana is devoted to articles on the flora and vegetation of the Serra do Gerês by various authors. Replaces **Henriques, J. A.** A vegetação da Serra do Gerez. Bol. Soc. Broteriana **3:** 155–225. 1885. (Includes annotated list of 483 plants (343 vascular).)—See also Emendas e aditamentos à "Flora vascular da Serra do Gerês." Agron. Lusitana **12:** 633–637. 1950 [1951]. (Additions and corrections.)

RIBATEJO

The Province of Ribatejo includes most of the old Province of Estremadura (nearly all the district of Santarém and a narrow eastern strip of the district of Lisboa) and a small part of the old Province of Alemtejo (western part of district of Portalegre).

Local

Frazão, E. A. Mendes, and **Possolo, D. F.** Reconhecimento agrológico do Ribatejo. Bol. Estaç. Agr. Cent. (Lisboa) sér. A, no. 22. 131 p. illus. incl. tables, map. 1936.

Includes (p. 34–88, passim) partial lists of vascular plants of different soils and habitats.

TRÁS-OS-MONTES E ALTO DOURO

This province includes the old Province of Trás-os-Montes (districts of Vila Real and Bragança) and a narrow strip of the old Provinces of Beira Alta and Beira Baixa (northern part of the districts of Vizeu and Guarda).

General

Rozeira, Arnaldo. A flora da província de Trás-os-Montes e Alto Douro. Estudo de distribuição geografica. Mem. Soc. Broteriana v. 3. 203 p. 21 fig. (maps). 1944. (Also issued as thesis.)

Botanical explorations, phytogeography and plant zones of Portugal (with review of the classifications proposed) and of Trás-os-Montes, climate, ecology; list of known vascular plants (1424 species), with localities and collectors' names; French summary, bibliography. Replaces **Coutinho, A. X. P.** Apontamentos para o estudo da flora transmontana. Bol. Soc. Broteriana **2:** 129–163. 1884; Emendas e additamentos á lista das plantas transmontanas . . . l.c. **3:** 48–49. 1885. (Plant zones, botanical explorations; annotated list of 735 vascular plants.)—See also [**Carneiro, A. de Barros**]. Plantas herborizadas em Trás-os-Montes e no Minho . . . (1. lista). Anuário Soc. Broteriana **9:** 18–30. 1943. (Extensive list of vascular plants, with localities, including some species not given by Rozeira in his Flora.)—Also **Rozeira, A.** Plantas novas ou pouco citadas para Trás-os-Montes e Alto Douro III. Bol. Soc. Broteriana II, **19:** 489–492. 1945.

Local

Henriques, J. A. Apontamentos para o estudo da flora transmontana. Vegetação da Serra do Marão. Bol. Soc. Broteriana **3:** 38–47. 1885.

Topography, general features of flora; list of vascular plants collected by author in August and September 1882 between Moledo and Ermida (up to 1415 meters altitude).

Lopes, J. M. Miranda. A flora do concelho de Vimioso. Bol. Soc. Broteriana II, **4:** 130–154. 1927; **5:** 234–255. 1928; **6:** 266–278. 1930; **8:** 176–189. map. 1933.

Geography, topography, general features of flora, climate, geology, agriculture, botanical explorations (in pt. 3); mostly unannotated lists of vascular plants, with vernacular names, and (in pts. 2–4) localities.

Mariz, Joaquim de. Uma excursão botanica em Traz os Montes. Bol. Soc. Broteriana **7:** 3–34. 1889.

Account of author's trip from Bragança to Miranda do Douro, Vimioso, and back to Bragança, with mention of noteworthy species seen.—See also his Outra excursão botanica na mesma provincia. l.c. **7:** 35–76. 1889. (Similar account of vicinity of Tôrre de Moncorvo and Freixo; full list of lichens and vascular plants collected on both excursions, with localities. This paper and the earlier one republished, 78 p. 1889, with title Duas excursães botanicas na provincia de Traz os Montes [fide A. R. Pinto da Silva].)

Mendonça, F. A., and **Vasconcellos, J. de Carvalho e.** Contribuïções para a topografia florística da região duriense. [I]–II. Anais Inst. Vinho Porto **3:** 297–300. 1942 (repr. 38 p.; only reprint seen); **5:** 119–200. 1944.

Two lists of vascular plants, with data (157 and 294 taxa respectively, including some duplications), referring to region along the Douro River and its tributaries from a little west of Pêso da Régua to near Barca de Alva. Part II published in collaboration with L. A. G. Barbosa and F. A. Garcia.—See also **Vasconcellos, J. de Carvalho e.** Breve reconhecimento da flora da região duriense. Anais Inst. Vinho Pôrto **1:** 65–74. 1940. (Includes mention of numerous species from different habitats.)—Also **Mendonça, F. A.,** and **Vasconcellos, J. de C. e.** Estudo fitogeografico da região duriense. l.c. **15.** 52 p. maps, diagrs. 1954. (Includes (p. 35–52) Relação geral das plantas herborizadas [1st part], an annotated list of 33 pteridophytes.)

San Marino

See also Italy, Emilia (Raggi, and others referring to Romagna).
For descriptions of species, see the general floras of Italy.

General

Pampanini, Renato. Flora della repubblica di San Marino. 1 p. l., 228 p. 25.5 cm. San Marino, 1930.

Botanical explorations, list of botanists with biographical data and lists of species they added to the flora, bibliography; list of cellular and vascular plants (714 vascular species), with localities.

Scandinavia

Includes works relating to Denmark, Finland, Iceland, Norway, and Sweden as a whole or to two or more of these countries or large parts thereof (particularly Lapland).

The general works here listed may be classified as follows: *Floras and manuals*: Bolin (popular), Christiansen (popular), Hartman, Hermann, Hylander (Nordisk kärlväxtflora), Lagerberg, Lindman, Wahlenberg (Lapland). *Catalogs*: Berlin, Hylander (Förteckning), Löve. *Iconographies*: Christiansen, Lagerberg, Lindman, Oeder. *Bibliography*: Rehder. *Vernacular names*: Jenssen-Tusch, Nordhagen, Qvigstad (Lapland). *Weeds*: Korsmo. *Miscellaneous*: Andersson (botanical regions of Lapland), Du Rietz (floristic regions), Hultén (atlas of distribution maps), Polunin (list of botanists), Sjörs (phytogeography, ecology).

Andersson, N. J. Conspectus vegetationis Lapponiae . . . 2 pt. (2 p. l., 39, x p.). 20 cm. Upsaliae, 1946. (Diss.)

Account of botanical regions with lists of characteristic species, grouped geographically, and with Swedish, Norwegian, and Russian Lapland treated separately; general considerations.—See also his Anteckningar under en resa i Umeå, Piteå och Luleå Lappmarker sommaren 1845. Bot. Notiser **1846:** 1–30. 1846. (Account of author's trip with mention of noteworthy plants from different localities.)

Berlin, August. Den geografiska utbredningen af skandinaviska halföns fanerogamer och ormbunkar i tabellarisk öfversigt uppstäld. 150 p. 23.5 cm. Stockholm, 1875 (cover date, 1876). (John Crerar Library.)

Definition of geographical regions used; tabular list of vascular plants (1,780 species) showing occurrence in 22 regions of Sweden, Norway, and Lapland.

Bolin, Lorentz, and **Post, L. O. A. von.** Floran i färg. 4. uppl. 212 p. incl. illus. (in glossary) and 128 col. pl. (on 64). 18.5 cm. Stockholm, 1953. (1st ed. 1950.)

Recognizable colored figures of 564 common plants of Sweden and Scandinavia in general, with illustrated glossary, brief notes on habitat, etc., and partial list of species arranged by habitats. The earlier editions differ little in text and not at all in plates. There are also Danish, Finnish, and Norwegian adaptations. For Denmark: **Hvass, Hans.** Flora i farver. 192 p. incl. 128 col. pl. (on 64). 18.5 cm. København, 1952. For Finland: **Kärki, Eino.** Värikuvakasvio. [Flora in colors.] 4. painos. 198 p. incl. 131 pl. (128 col.), map. 18.5 cm. Porvoo, 1955. (1st ed. 1951); Swedish ed. by **Olsoni, Börje.** Floran i färg. Redigerad för Finland . . . 3. uppl. 211 p. incl. 131 pl. (128 col.), map. 18.5 cm. Helsingfors, 1955. (1st ed. 1951.) (Finnish eds. not seen; data supplied by H. Luther.) For Norway: **Berg, G. A.** Floraen i farger. xxii, 179 p. incl. illus. (in glossary) and 128 col. pl. (on 64). 18.5 cm. Oslo, 1951.

Christiansen, M. S. Danmarks vilde planter. fasc. 1–10. 28 cm. København, 1958–1959. (Not seen.)

A popular work intended to consist of 20 fascicles in 2 volumes, amounting to over 600 pages of text, containing descriptions of about 1,200 species, with illustrations of about 900 in colors and 300 in black and white; edited by T. W. Böcher, drawings by Henning Anthon. There are also editions in Swedish and Norwegian, as follows: **Hultén, Eric,** *ed.* Vår svenska flora i färg; **Faegri, Knut,** *ed.* Norges planter.

Du Rietz, G. E. Die regionale gliederung der skandinavischen vegetation. Führer für die Vierte I. P. E. Svenska Växtsociol. Sällsk. Handl. 8. 60 p. 32 pl. on 16 (incl. map). 1925.

Sketch of floristic and vegetational regions and zones in Sweden, Norway, Denmark, and Finland.

Hartman, C. J. Handbok i Skandinaviens flora, innefattande Sveriges och Norges växter till och med mossorna . . . 11., helt och hållet omarbetade upplagan utgifven af Carl Hartman. Förra delen: Fanerogamer och ormbunkar. 2 p.l., lxxxiii, 616 p. 22.5 cm. Stockholm, 1879. (Gray Herbarium library.) (1st ed. 1820.)

Bibliography, key to genera on Linnaean system; annotated descriptive flora of vascular plants. The first part of a 12th ed., edited by T. O. B. N. Krok, with area enlarged to include Finland and Denmark, was published in 1889 (2 p.l., 128 p. 22.5 cm. Stockholm, 1889). (In Gray Herbarium library.) It covered pteridophytes, gymnosperms, and part of the monocotyledons.— See also Holmberg, under Hylander, below.

Hermann, Friedrich. Flora von Nord- und Mitteleuropa. xi, 1154 p. 20.5 cm. Stuttgart, 1956.

Flora of vascular plants in form of keys, with fairly detailed statement of distribution. Covers most of Europe except western France, Iberian Peninsula, and most of Italy. Replaces his Flora von Deutschland und Fennoskandinavien sowie von Island und Spitzbergen. 524 p. 22 cm. Leipzig. 1912.

Hultén, Eric. Atlas över växternas utbredning i Norden. Fanerogamer och ormbunksväxter. (Atlas of the distribution of vascular plants in NW. Europe.) 119 p. (including 108 maps on 33 p.), 512 p. (including 1,847 i.e. 1,894 maps on 474 p.). 22 cm. Stockholm, 1950.

Sources of author's data, discussion of phytogeographic groups and migrational groups (with English summary, p. 98*–110*), alphabetical list of principal Swedish words with English equivalents (p. 114*–119*; including both botanical and nonbotanical terms); series of 1920 maps (the last 26 blank) showing the range of about 1,847 vascular plants with brief annotations on habitat, altitudinal range, etc. Covers Denmark, Norway, Sweden, Finland, and (for general distribution) adjacent Russia.

Hylander, Nils. Förteckning över Nordens växter utgiven av Lunds botaniska förening. 1. Kärlväxter. (Additional title-page: List of the plants of N.W. Europe (Sweden, Norway, Denmark, East Fennoscandia, Iceland and the Faeroes) . . . 1. Vascular plants.) 4. utvidgade uppl. ix, 175 p. 22.5 cm. Lund, 1955. (1st ed. 1907, by Lunds Botaniska Förening, with title Förteckning över Skandinaviens växter.)

Systematic list of vascular plants of Denmark, Finland (plus Russian Karelia and Russian Lapland), Norway, Sweden, and Iceland, with indication of countries in which each occurs (except for the merely casual species); list of plants officially protected, with localities; alphabetical list of synonyms, with equivalents. Infraspecific forms and hybrids are listed; *Hieracium* and *Taraxacum* are omitted, and only the principal species of *Rosa* are given. Although published primarily as an exchange list (with unit values for each plant), it is widely used in Scandinavian countries as a standard of nomenclature. The mosses, algae, and lichens are similarly listed in another volume by other authors (2. Mossor, 3. Alger, 4. Lavar. 1936–37). The list of abbreviations of authors' names with full names, dates, and brief biographical information that was given in the 3d ed. has been omitted.—See also his Nomenklatorische und systematische studien über nordische gefässpflanzen. Årsskr. Uppsala Univ. 1945, no. 7. 337 p. 1945. (Discussion of the principles of nomenclature, nomenclature and segregation of certain families and genera, and the nomenclature, infraspecific forms, etc., of a long list of vascular plants. Intended as a commentary on ed. 3 of the Förteckning.)—Also his Tillägg och rättelser till Förteckning över Nordens växter. 1. Kärlväxter (Lund 1955). Bot. Notiser **112:** 90–100. 1959. (Corrections, with references.)

——— Nordisk kärlväxtflora omfattande Sveriges, Norges, Danmarks, Östfennoskandias, Islands och Färöarnas kärlkryptogamer och fanerogamer. v. 1 (xv, 392 p.). 54 fig., map. 24.5 cm. Stockholm, 1953.

Discussion of boundaries, etc.; annotated, keyed, descriptive flora of wild and cultivated vascular plants, with concise but rather detailed statement of range in each country concerned (including Finland, and western Russia from the region of Lake Ladoga and Lake Onega northward, extending east to 40° E. Long., thus including the biological provinces of Karelia pomorica orientalis and Karelia transonegensis as well as the Kola Peninsula), references to distribution maps, chromosome number, mention of medicinal and poisonous plants, occasional critical notes, and usually one vernacular name in each language. Vol. 1 includes vascular cryptogams, gymnosperms, and monocotyledons except Cyperaceae and Orchidaceae. Replaces **Holmberg, O. R.** Skandinaviens flora. v. 1–1b, häfte 1. map, diagr. 23.5 cm. Stockholm, 1922–31. (Annotated descriptive flora of vascular plants with keys, synonymy, range in each country and general range, and references to some illustrations and exsiccatae. Incomplete; Pteridophyta-Salicales only (Engler-Prantl system). Covers Sweden, Norway, Denmark, Finland, and adjacent Russia east to Lake Onega and the White Sea, including the Kola Peninsula. Vol. 1, häfte 1, has title: Hartmans Handbok i Skandinaviens flora. Vol. 2a (1928) contains Levermossor (Hepaticae), by H. W. Arnell.)—See also **Hylander, N.** Taxa et nomina nova in opere meo: Nordisk kärlväxtflora I (1953) inclusa. Bot. Notiser **1953:** 352–359. 1953. (List of new names and new taxa, with synonymy.)

Jenssen-Tusch, H. C. L. Plantenavne i forskellige evropaeiske sprog. 1. afd. Nordiske plantenavne. xviii, 276 p. 23.5 cm. Kjøbenhavn, 1867.—Tillæg til Nordiske plantenavne. 1 p. l., 277–368 p. 23.5 cm. Kjøbenhavn, 1870.—Navnefortegnelse til Nordiske plantenavne. 122 p. 23.5 cm. Kjøbenhavn, 1871. (All in Gray Herbarium library.)

Bibliography; alphabetical list of scientific names of vascular and cellular plants with Scandinavian (Iceland, Faroes, Sweden, Norway, Denmark) vernacular names and authorities therefor and localities; supplement, index.

Korsmo, Emil. Ugras i nåtidens jordbruk. [2.utg.] xii, 635 p. 494 fig. 25 cm. Oslo, 1954. (1st ed. 1925 as Ugress i nutidens jordbruk. Biologiske og praktiske undersøkelser.)

Includes (avsnitt V, De enkelte ugrasarter, p. 74–479) account of 200 weeds, arranged according to duration and method of dispersal, with descriptions of plant and seed (including number per plant, weight, etc.), germination, range, method of dispersal, control, etc. Relates primarily to Scandinavia and northern Europe. The bibliography and the vernacular names in other languages than Norwegian, given in the first edition, are omitted. Of the 1st ed. there were adaptations in Russian (not seen), Swedish (Ogräs. Ogräsartenas liv och kampen mot dem i nutidens jordbruk . . . Översättning från norskan av L. Bergendal. 6 p.l., 410 p. 393 (i.e. 364) fig. 25.5 cm. Stockholm, 1926), and German, the last (Unkräuter in ackerbau der neuzeit . . . Berlin, 1930) edited by H. W. Wollenweber and with considerably enlarged bibliography.—See also his Ugressfrø. Unkrautsamen. Weed seeds. 175 p. 34 col. pl. 35 cm. Oslo, 1935. (Descriptions in Norwegian, German, and English of the fruits and seeds of 306 weeds, mainly of northern and central Europe, in part from North and South America, involving many which occur as impurities in commercial seed of grains, grasses, and legumes, with colored figures of the seed of each species; tabular index of scientific names with corresponding vernacular names in "American," Danish, English, French, Dutch, Italian, "Canadian," Norwegian, Russian, Swedish, and German. Includes all the species in the German edition of his earlier work (1930) and 101 others. The species are arranged according to duration, etc., not systematically.)—Also his Anatomy of weeds. Anatomical descriptions of 95 weed species with 2,050 original drawings. 4 p. l., 413 p. incl. 193 fig., 91 pl. Oslo, 1954. (Descriptive account of anatomy of stem, leaf, and underground parts of 95 common and widespread weeds, amply illustrated.)

Lagerberg, Torsten. Vilda växter i Norden. 2. omarb. och utök. uppl. 4 v. (1851 p.). 957 fig., 1 [50] pl. (on 26; black, showing microscopic wood sections), 919 col. pl. (on 460). 24 cm. Stockholm, 1947–49. (1st ed. 1937–39. 3. omarb. uppl. 4 v. (1874 p.). 961 fig., 50 pl. (black), 919 col. pl. Stockholm, 1956–58. Not seen.)

Flora of vascular plants (860 species), with non-technical descriptions, good colored photograph of each species in its habitat, local and extralimital range, notes on biology, etc., and Swedish, Danish, Norwegian, and Finnish vernacular names (usually only 1 name in each language). The work, although accurate, is not complete as a flora and omits many or most of the species in such genera as *Carex, Hieracium,* etc. Covers Sweden, Norway, Denmark, and Finland. The cellular plants are treated in a similar way in a supplementary, unnumbered volume by **Nannfeldt, J. A.,** and **Du Rietz, G. E.** Vilda växter i Norden. Mossor, lavar, svampar, alger. 2. revid. och kompl. uppl. xlix, 450 p. 361 fig., 212 col. pl. (on 106). 24 cm. Stockholm, 1952. (1st ed. 1945). There are also Danish, Finnish, and Norwegian adaptations of this work. For Denmark: **Gram, Kaj,** and **Jessen, Knud.** Vilde planter i Norden. 2. udvid. og forbed. udg. 4 v. (1520 p.). 1044 fig., 919 col. pl. (on 460). 24 cm. København, 1949–51. (1st ed. 1937–40, by **Jessen, Knud,** and **Mentz, August.** 3. udvid. og forbed. udg. v. 1–3. 1957–58.) The cellular plants are treated in a supplementary volume, numbered 5 on the spine, by **Gelting, Paul,** and others, based on the 1st ed. of the similar volume in the Swedish edition. For Finland: **Lagerberg, Torsten, Linkola, Kaarlo,** and **Väänänen, Heikki.** Pohjolan luonnonkasvit. [Wild plants of the north.] . . . 3 v. (1558 p.). 924 fig., 804 col. pl. (on 402). 24 cm. Porvoo, Helsinki, 1938–40. (Riksmuseet library, Stockholm.) For Norway: **Lagerberg, Torsten,** and **Holmboe, Jens.** Våre ville planter. 6 v. 737 fig., 804 col. pl. (on 402). 24 cm. Oslo, 1937–40. (Arnold Arboretum library); rev. og forøk. utg. by **Lagerberg, Holmboe,** and **Nordhagen, Rolf.** v. 1–4. 1950–55; with v. 7–8 (1952–54) on cellular plants, by **Du Rietz** and others. (Vol. 4 not seen.)

Lindman, C. A. M. Bilder ur Nordens flora. 2. och 3. uppl. 3 v. (2 p.l., 635 p.). 663 col. pl. 24.5 cm. Stockholm, 1917–1926. (1st ed. 1901–05.)

Semipopular descriptive flora of vascular plants with notes on biology, uses, etc., and delicately drawn colored plate of each species (based on the plates in Palmstruch, Svensk botanik). Not a complete flora, but includes the common or characteristic species of Sweden, Norway, and Finland; ranges are given only in general terms. Plates 520–663 with corresponding text constitute a supplement (to the contents of the 1st ed.), and were also issued separately; there are separate indices, 3: 421–445, 629–635. A 1932 reissue is listed in a catalog. There is also a Danish-Norwegian adaptation: **Mentz, August,** and **Ostenfeld, C. H.** Billeder af Nordens flora. 2. forøgede udg. 3 v. 34 fig., 519 col. pl. 24.5 cm. København, 1917–23. (v. 1–2, plates; v. 3, tekst–xi, 472 p.)—v. 4. Tillægsbind. x, 176 p. col. pl. 520–663. 24.5 cm. København, 1924–27. (All in John Crerar Library.) (1st ed. 1901–07.)

Löve, Áskell, and **Löve, Doris.** Chromosome numbers of northern plant species. Atvinnudeild Háskólans (Reykjavík) Rit Landbúnaðardeildar ser. B, no. 3. 131 p. 1948.

Statistics of number of native or thoroughly naturalized species of vascular plants in Denmark, Faroes, Finland, Iceland, Norway, and Sweden (1859 in whole area); systematic list of species with indication of the Scandinavian countries in which they are found and their chromosome number, with references for latter; bibliography (titles of the papers cited are not given), generic index. Microspecies are not given, and no named species in *Taraxacum* and *Hieracium*. The nomenclature and range are based essentially on Hylander's Förteckning, ed. 3. The authors give the following number of species for the different regions (excluding microspecies, *Hieracium*, and *Taraxacum*): Denmark 1306, Faroes 324, Finland 1,284, Iceland 440, Norway 1,351, Sweden 1645, total 1859.

Nordhagen, Rolf. Studier over gamle plantenavn. I. Motiver i nordiske navn på skinntryter og blåbær (Vaccinium uliginosum og V. myrtillus). Bergens Mus. Årb. **1945** i.e. **1944,** Naturvitensk. Rekke no. 10. 1 p.l., 144 p. 9 fig. (incl. maps). 1946.

Detailed discussion of Norwegian, Swedish, and other vernacular names of *Vaccinium uliginosum* and *V. myrtillus;* cited here for its bibliography of works on vernacular names and etymological works in general.

Oeder, G. C., and others. Icones plantarum sponte nascentium in regnis Daniae et Norvegiae, in ducatibus Slesvici et Holsatiae, et in comitatibus Oldenburgi et Delmenhorstiae: ad illustrandum opus de iisdem plantis, regio jussu exarandum, Florae Danicae nomine inscriptum. 17 v. 3060 col. (or black) pl. 39 cm. Hafniae, 1766 [1761]–1883.—Icones plantarum sponte nascentium in regnis Sueciae et Norvegiae. Supplementum . . . editum a F. M. Liebmann et Joh. Lange. v. 1 (unicum). 180 col. pl. 39 cm. Hauniae, 1853–1874.

Colored (or plain) plates of Scandinavian vascular and cellular plants, not in systematic order; the text in earlier volumes consisting principally of synonymy and statement of source of the specimens figured, in later volumes including descriptions. The work is almost universally cited by botanists under the title Flora danica. The geographical details in the title were changed several times; the form cited above was used in v. 1–4. There were also issues with Danish and German text (not seen), the latter of 15 volumes only; see **C. Christensen,** Den danske botaniks historie **2:** 558–559 (under Denmark, General) for dates of volumes, names of later editors, and other details.—See also **Lange, Johan.** Nomenclator "Florae danicae" sive index systematicus et alphabeticus operis, quod "Icones florae danicae" inscribitur cum enumeratione tabularum ordinem temporum habente, adiectis notis criticis. viii, 354 p. 27 cm. Hauniae, 1887. (History of "Flora danica"; chronological list of plates, with corrected nomenclature, and separate list of critical annotations; tabular systematic list of species with reference to plates and indication of local distribution; alphabetical index to plates.)

Polunin, Nicholas. Botanical research in Scandinavia: 375 contemporary workers and their special interests. Canad. Field.-Nat. **64:** 105–119. 1950.

Alphabetical list of botanists in Finland, Sweden, Norway, Iceland, and Denmark, with their addresses and special fields of work.

Qvigstad, Just. Lappiske plantenavne. Nyt Mag. Naturv. **39:** 303–326. 1901.

Bibliography; list of 350 vernacular names, with scientific equivalents, localities, and references. Covers Norwegian, Swedish, Finnish, and Russian Lapland. The author, Rektor Just Knud Qvigstad (4 April 1853–15 March 1957), apparently reached the greatest age ever attained by any one who has contributed to the literature of botany.

Rehder, Alfred. (Phytography.) Scandinavia. *In* his The Bradley Bibliography. A guide to the literature of the woody plants of the world published before the beginning of the twentieth century. Compiled at the Arnold Arboretum of Harvard University under the direction of Charles Sprague Sargent. v. 1, p. 350–358. 29.5 cm. Cambridge, Mass., 1911.—Additions and corrections. l.c. **1:** 531. 1911; **5:** xviii–xix. 1918.

Essentially unannotated list of floristic works (including those in periodical and serial literature) on "Dendrography" and "General phytography," chronologically arranged; covers Denmark, Iceland, Norway, and Sweden.—See also l.c. **1:** 3, for works on botanical bibliography; **1:** 8–9, (519), alphabetical lists of periodicals and serials relating to botany, arranged by countries; **1:** 40–41, (521), botanical gardens; **1:** 73 vernacular names; **1:** 80, botanical history; **1:** 222–223, phenology; **1:** 296, noteworthy individual trees; **3:** 10–11. 1915, alphabetical list of periodicals and serials relating to arboriculture, gardening, and uses of plants; 3: 33, history of gardening; 3: 66–67, dendrological, pomological, economical, and miscellaneous works relating to arboriculture, fruit trees, horticulture, and cultivated useful or ornamental plants; 3: 100, arboriculture and general horticulture; 3: 247, pharmacopoeias; 3: 269–270, economic products; 4: 5, (539). 1914, list of periodicals and serials relating to forestry, timber uses, etc.; 4: 24, history of forests; 4: 79, dendrography; 4: 342–344, forests (general and by countries). Several smaller sections of no floristic significance have been omitted from the above summary.—Also Hjelt, under Sweden, and additional references there cited.

Sjörs, Hugo. Nordisk växtgeografi. 4 p. l., 229 p. 108 fig. (incl. maps, diagrs.). 22.5 cm. Stockholm, 1956.

Textbook of Scandinavian (particularly Swedish) phytogeography, discussing floral provinces, floristic elements, autecology, synecology, history of flora, vegetation of marine and fresh water habitats, general vegetational regions, heath, steppe, meadow, and bog vegetation up to the tree limit, bare mountain vegetation, phytogeography; bibliography.

Wahlenberg, Georg. Flora lapponica exhibens plantas geographice et botanice consideratas, in lapponiis svecicis scilicet Umensi, Pitensi, Lulensi, Tornensi et Kemensi nec non lapponiis norvegicis scilicet Norlandia et Finmarkia utraque indigenas, et itineribus annorum 1800, 1802, 1807 et 1810 denuo investigatas. lxvi, 550 p. 30 pl., map, 2 tab. 23.5 cm. Berolini, 1812.

Phytogeography, botanical investigations, plant zones, climate, etc.; annotated flora of vascular (529 species) and cellular plants, on the Linnaean system, with diagnoses, localities, and frequent critical notes.—See also **Sommerfelt, S. C.** Supplementum Florae lapponicae . . . xii, 331 p. 3 col. pl. 20 cm. Christianiae. 1826. (Additional species and localities, and critical notes.)—Also his Bemærkinger ved Supplementum Florae lapponicae. Mag. for Naturv. 8: 163–166. 1828. (A few additional records, all but one referring to cellular plants. In library of Museum of Comparative Zoology, Harvard University.)

SPAIN (ESPAÑA)

See also France, Partial (Gaussen, Miègeville, Philippe, and Zetterstedt, all on the Pyrenees); Italy, General (Rikli, Das pflanzenkleid der Mittelmeerländer).
The original spelling and accentuation of titles have been followed.

General

Includes a few partial works.
The general titles here listed may be classified as follows: *Floras and manuals*: Amo y Mora, Bubani (Pyrenees), Caballero (Flora), Lázaro é Ibiza (Botánica descriptiva), Willkomm and Lange. *Catalogs and lists*: Colmeiro (Enumeración), Gandoger (2), Reyes Prósper (steppes), Vicioso. *Iconographies*: Caballero (Illustraciones), Willkomm (Illustrationes, under Willkomm and Lange). *Bibliographies*: Colmeiro (La botánica), Rehder, (Willkomm). *Vernacular names*: Anonymous (woody plants), Asín Palacios (Romance names), Colmeiro (Diccionario), (Dantín Cereceda, 2 titles), (Fleury de la Roche: medicinal), (Juscafresa: medicinal), (Laguna: woody plants), (Lázaro é Ibiza), (Mantero: medicinal), Más y Guindal (Memorandum: medicinal), (Puerta: medicinal), (Texidor y Cos: medicinal), (Willkomm and Lange). *Trees and woody plants in general*: Anonymous, Dantín Cereceda (Catálogo), Laguna. *Medicinal plants*: Bassagaña, Fleury de la Roche, Juscafresa, (Lázaro é Ibiza), Mantero, Más y Guindal (2), Puerta, Texidor y Cos. *Other useful plants*: (Anonymous: woody plants), (Colmeiro, Diccionario), (Laguna: woody plants), (Lázaro é Ibiza), Puerta, (Reyes Prósper). *Weeds*: Dantín Cereceda (Ensayo). *Cultivated plants*: Dantín Cereceda (Catálogo), Fleury de la Roche (medicinal), (Juscafresa: medicinal), (Mantero: medicinal), (Más y Guindal: Memorandum: medicinal). *Miscellaneous*: Colmeiro (Enumeración: botanical history and vernacular names), Font Quer (vegetation), Lázaro é Ibiza (Regiones botánicas), Reyes Prósper (ecology of steppes), Rivas Goday (2: plant zones, plant associations), Willkomm (phytogeography, botanical explorations).

Anonymous. Catálogo razonado de los objetos expuestos por el Cuerpo de ingenieros de montes. 194 p. 3 maps. 25 cm. Madrid, 1888. (Esposición universal de Barcelona. 1888. Issued with Revista Mont. (Madrid). v. 12. 1888.) (Arnold Arboretum library.)

Includes (p. 48–80) annotated list of 140 native and naturalized woody plants of Spain, with vernacular names, range, and uses; also lists of trees yielding charcoal, wood ashes, alkali, tannin, etc.—See also (for woody plants) Dantín Cereceda (Catálogo) and Laguna, below, and (for useful plants in general), Bassagaña, below, and additional references there given.

Amo y Mora, Mariano del. Flora fanerogámica de la península ibérica, ó descripcion de las plantas cotyledóneas que crecen en España y Portugal. 6 v. 22 cm. Granada, 1871–73.

Annotated descriptive flora of spermatophytes, with localities.—See also his Flora cryptogámica de la península ibérica, que contiene la descripcion de las plantas acotyledóneas que crecen en España y Portugal . . . viii, 849 p. 22 cm. Grenada, 1870. (Includes treatment of pteridophytes, p. 791–857. In Arnold Arboretum library.)

Asín Palacios, Miguel. Glosario de voces romances registradas por un botánico anónimo hispano-musulmán (siglos XI–XII). liv, 420 p. 2 pl. 25.5 cm. Madrid, 1943.

Discussion of the manuscript, short bibliography; alphabetical list of 726 Romance names of plants with Spanish vernacular names, citation of references in original text with Spanish translation, and discussion of etymology, botanical identification, etc.; indices of botanical, Spanish, Arabic, and Greek names.—See also **Alvarez López, Enrique.** Commentarios históricos y botánicos con motivo de un "Glosario" hispano-musulmán de los siglos XI al XII. Anales Jard. Bot. Madrid **7:** 5–175. 1947 (1949). (Critical discussion of the manuscript and of the commentary by Asín Palacios.)—Also **Font Quer, Pío.** Commentarios al "Glosario de voces romances" del Dr. Asín Palacios. Mem. R. Acad. Cien. y Art. Barcelona **30:** 213–234. 2 fig. 1950. (Also paged as v. 30, no. 9. 24 p.) (Critical discussion of various names.)—Also Colmeiro (Diccionario), below, and additional references there cited.

Bassagaña, Pedro. Flora médico-farmacéutica abreviada, ó sea descripcion de las plantas medicinales, indígenas y exóticas mas usadas y algunas otras notables que no lo son, con indicacion de su duracion, del país en que habitan, de la época en que florecen las que crecen en España, de usos farmacéuticos, sus virtudes medicinales, etc. 612 p. 25 cm. Barcelona, 1859 (cover date, 1860).

Briefly descriptive flora of native, cultivated, and exotic medicinal plants, with uses and vernacular names. For other works on medicinal and other useful plants see Anonymous, above, and Dantín Cereceda (Catálogo), Fleury de la Roche, Juscafresa, Lázaro é Ibiza (Botánica descriptiva), Mantero, Más y Guindal (2), Puerta, and Texidor y Cos, below.

Bubani, Pietro, Flora pyrenaea per ordines naturales gradatim digesta. Opus posthumum editum curante O. Penzig. 4 v. 27.5 cm. Mediolani, 1897–1901.

Annotated flora of vascular plants, with synonymy, citation of localities (mostly based on specimens collected or examined by author), and in most cases descriptions or critical notes; refers to the Pyrenees on both sides of the French-Spanish border but chiefly to Spanish territory. The author's nomenclature is capricious, accepted names of genera and species being changed on preposterous grounds; the date 1753, which marks the beginning of binomial nomenclature, is ignored.—For discussion of Bubani's Flora, with rectification of his nomenclature, see **Malinvaud, Ernest.** Un coup d'oeil sommaire sur la littérature botanique pyrénéenne, Bubani et son Flora pyrenaea. Bul. Soc. Bot. France **54** (Sess. Extr.): l–liii. 1908; **55** (Sess. Extr.): xliv–liv. 1908; **56** (Sess. Extr.): xxii–xxx. 1910; **57** (Sess. Extr.): xxv–xxix. 1911.—For other floras of the Pyrenees (French side) see Gaussen, Miègeville, Philippe, and Zetterstedt, all under France, Partial.

Caballero, Arturo. Flora analítica de España. xiv, 617 p. 268 fig. 21.5 cm. Madrid, 1940.

Unannotated key to families, genera, and species of vascular plants; glossary, list of generic synonyms; alphabetical list of species included, with habitat, indication of altitudinal range but not details of distribution, and vernacular names. Not a complete flora; includes the commoner or more interesting species (about 3,000).

——— Illustraciones de la flora endémica española. Anales Jard. Bot. Madrid **1:** 201–220. pl. 1–13. 1942; **2:** 266–347 incl. 22 pl. 1942; **3:** 328–381 incl. 16 pl. 1943; **4:** 459–491 incl. 9 pl. 1944; **5:** 523–557 incl. 9 pl. 1945; **6** (2): 549–591 incl. 12 pl. 1946; **7:** 655–691 incl. 10 pl. 1947 (1948); **8:** 523–579 incl. 15 pl. 1948.

Descriptions, illustrations, and discussion of endemic or essentially endemic Spanish spermatophytes, not in systematic order.—See also Willkomm (Illustrationes), below, under Willkomm and Lange.

Colmeiro, Miguel. La botánica y los botánicos de la península hispano-lusitana. Estudios bibliográficos y biográficos. x, 216 p. 28 cm. Madrid, 1858.

Annotated, classified list of 932 works and periodicals, with index, followed by biographical sketches, with index; list of genera dedicated to Spanish and Portuguese botanists, with index.—See also Rehder and Willkomm (2), below; also Bellot Rodríguez under Portugal, General.

——— Diccionario de los diversos nombres vulgares de muchas plantas usuales ó notables del antiguo y nuevo mundo, con la correspondencia científica y la indicacion abbreviada de los usos é igualmente de la familia á que pertenece cada planta. 235 p. 23 cm. Madrid, 1871.

Alphabetical list of vernacular names (mostly Spanish, some Brazilian and American Indian), with botanical equivalents and indication of uses; lists of Arabic names used in Spain, with bibliography; lists of names used in Spanish works of 14th and 15th centuries.—See also Anonymous and Asín Palacios, above, and Dantín Cereceda (2), Fleury de la Roche, Juscafresa, Laguna, Lázaro é Ibiza (Botánica descriptiva), Mantero, Más y Guindal (Memorandum), Puerta, Texidor y Cos, and Willkomm and Lange, below.

——— Enumeración y revisión de las plantas de la península hispano-lusitana é islas Baleares, con la distribución geográfica de las especies, y sus nombres vulgares, tanto nacionales como provinciales. 5 v. 27 cm. Madrid, 1885–89.

Botanical history from earliest times, with lists of Arabic names (with bibliography), list of species known to Arabs, Middle Age vernacular names, list of botanical authors and collectors, with bibliography; list of vascular and cellular plants, with detailed local distribution and vernacular names (not indexed). Includes 9,791 species (6,064 phanerogams) from Spain, Portugal, and the Balearic Islands. Unfortunately the work is entirely uncritical (Willkomm, Vegetation der erde **1**: 23. 1896).

Dantín Cereceda, Juan. Catálogo metódico de las plantas cultivadas en España. 2. ed. 187 p. 46 fig. 21.5 cm. [Madrid?], [pref. 1943]. (1st ed. 1920.)

List of 582 cultivated plants and wild woody plants, grouped by uses and in part by habitat, arranged in systematic order in each group, with collected vernacular names (often with localities); bibliography, indices. Includes extensive lists of horticultural varieties of some important fruits. Publication of Ministerio de agricultura. Sección de publicaciones, prensa y propaganda. —See also Bassagaña, above, and additional references there given.

———Ensayo de un catálogo metódico de las plantas fanerógamas dañinas o nocivas a los cultivos en España con la correspondencia entre los nombres científicos y los vulgares y seguido de índices alfabéticos. 95 p. 23 cm. Madrid [pref. 1916].

Unannotated systematic list of 1,165 more or less weedy phanerogams, with ample vernacular names. Publication of Servicio de publicaciones agrícolas, Ministerio de fomento.

Fleury de la Roche, A. Las plantas bienhechoras consideradas en sus aplicaciones a la medicina, la agricultura, la industria y la economía doméstica. Traducción directa del francés y adaptación por Francisco Salas. 339 p. illus., 12 col. pl. 22 cm. Barcelona, 1931. (French ed. 1906.)

Descriptive account, with uses, of principal wild and cultivated medicinal plants, arranged alphabetically by vernacular names; list of diseases with remedies.—See also Bassagaña, above, and additional references there cited.

Font Quer, Pío. Geografía botánica de la península ibérica. *In* Vidal de La Blache, Paul M. J. Geografía universal. v. 10, p. 143–271. fig. 65–91 (mostly maps and profiles), pl. 19–46 (on 14). 28.5 cm. Barcelona, 1953. (Only the excerpt seen, in library of J. Cuatrecasas.)

General account of climate, soils, topography, and human influence, followed by chapters discussing the vegetation from the point of view of climate, soils combined with climate, and soils alone, also the mountain region and human influence.—See also Lázaro é Ibiza (Regiones botánicas), Reyes Prósper, Rivas Goday (2), and Willkomm, below.

Gandoger, Michel. Catalogue des plantas récoltées en Espagne et en Portugal pendant mes voyages de 1894 à 1912. 2 p.l., 378 p. 25 cm. Paris, 1917. (Arnold Arboretum library.)

Unannotated list of vascular plants and Characeae collected by author, with localities. Lithographed from manuscript.

——— Notes sur la flore espagnole [I]–X. Bul. Soc. Bot. France **45:** 17–29, 588–604. 1898–99; **47:** 120–124, 132–143. 1900; **48:** 405–418. 1902; **52:** 438–462. 1905; **55:** 154–161, 220–224. 1908; **56:** 104–111, 132–138. 1909; **57:** 54–62, 94–100, 501–508, 564–568. 1910–11.

Includes incomplete lists of vascular plants collected by author at numerous localities in different parts of Spain, with indication of new records. Nos. VIII–IX are his 3.[–4.]voyage en Portugal.

Juscafresa, Baudilio. Las plantas medicinales de nuestra flora nacional. Almanaque Agr. Ceres **12:** 317–333, 335–347, 349–367, 369–378. 1953.

General considerations; annotated list of native and cultivated medicinal plants, alphabetically arranged by vernacular names, with descriptions, uses, and additional vernacular names; no index.—See also Bassagaña, above, and additional references there given.

Laguna, Máximo, and **Ávila, Pedro de.** Flora forestal española, que comprende la descripcion de los árboles, arbustos y matas que se crían silvestres ó asilvestrados en España . . . 2 v. (372, 459 p.). 26 cm., and atlas 2 v. (80 col. pl.). 50 cm. Madrid, 1883–90.

Descriptive flora of 553 species of native and naturalized trees, shrubs, and woody vines, with references, vernacular names (and localities), local and general range, and, in case of more important species, cultivation and uses.—See also Anonymous, above, and additional references there given.

Lázaro é Ibiza, Blas. Botánica descriptiva. Compendio de la flora española. Estudio de las plantas que viven espontáneamente en España y de las más frecuentemente cultivadas que tienen aplicaciones en medicina, agricultura, industria y horticultura. 3. ed. corregida y aumentada. 3 v. 1,000 fig., map. 22 cm. Madrid, 1920–21. (Arnold Arboretum library.) (1st ed. 1896.)

Sketch of principal botanical systems, brief bibliography, glossary; partly keyed flora of cellular and vascular plants, with brief descriptions, brief statement of range, vernacular names, and notes on uses; brief sketch of botanical regions. Contains 9,014 species (6,204 vascular, of which 5,948 are spontaneous phanerogams), including cultivated species.

——— Regiones botánicas de la Península Ibérica. Anales Soc. Españ. Hist. Nat. **24:** 161–207. map. 1895.

Floristic elements, botanico-geographical regions with mention of characteristic species.—See also Font Quer, above, and additional references there given.

Mantero, Antonio. Flora española medicinal y aromática. (Guia del recolector). 1 p. l., lii, 243 p., 3 l. 43 pl. (37 col.) on 22. 20.5 cm. León, 1944.

Popular treatment of 200 wild medicinal plants, with vernacular and patois names, habitat, description, part used, directions for collection. Scientific names given only in unpaged index at end.—See also Bassagaña, above, and additional references there cited.

Más y Guindal, Joaquín. Fitoterapia vulgar hispana. Concepto de su flora. Las investigaciones fitoquímicas. Anales R. Acad. Farm. (Madrid) 2. época, año **7:** 417–445. 1941 (1942).

Includes running account of various Spanish medicinal plants, classified by uses; discussion of phytochemical investigations in general, etc.; many footnote references.—See also Bassagaña, above, and additional references there cited.

——— Memorandum de sínonimias, precedencias, nombres científicos y vulgares de los productos químicos, vegetales ó animales, plantas y sus partes y preparaciones de aplicación á la farmacia . . . xvi, 830 p. 23.5 cm. Madrid, 1901. (John Crerar Library.)

Bibliography; alphabetical list of Spanish and Portuguese vernacular and pharmaceutical names (often with indication of locality where used) with botanical equivalents, and of botanical names with collected vernacular names. Includes exotic substances and chemical as well as plant and animal products.—See also Bassagaña and Colmeiro (Diccionario), above, and additional references there cited.

Puerta, Gabriel de la. Botánica descriptiva y determinacion de las plantas indígenas y cultivadas en España de uso medicinal, alimenticio é industrial. 2. ed. corregida y considerablemente aumentada. 1 p. l., 669 p. 169 fig. 21.5 cm. Madrid, 1891. (1st ed. 1877, with title Tratado práctico de determinacion de las plantas indígenas . . .)

Descriptive account of useful wild and cultivated vascular and cellular plants, systematically arranged, with vernacular names, local range, and uses; key to families, synopsis of genera, glossary, etc.—See also Bassagaña, above, and additional references there cited.

Rehder, Alfred. (Phytography.) Iberian Peninsula. *In* his The Bradley Bibliography. A guide to the literature of the woody plants of the world published before the beginning of the twentieth century. Compiled at the Arnold Arboretum of Harvard University under the direction of Charles Sprague Sargent. v. 1, p. 419–423. 29.5 cm. Cambridge, Mass., 1911.—Additions and corrections. l.c. **1**: 535. 1911; **5**: xxii. 1918.

Essentially unannotated list of floristic works (including those in periodical and serial literature) on "Dendrography" and "General phytography," chronologically arranged, divided into General (including Spain and Portugal) and Balearic Islands, the former with the dendrological works separated from the general.—See also l.c. **1**: 3, for works on botanical bibliography; **1**: 24, for alphabetical list of periodicals and serials relating to botany; **1**: 60, botanical gardens; **1**: 74, vernacular names; **1**: 81, botanical history; **1**: 226, phenology; **1**: 297, noteworthy individual trees; **3**: 23. 1915, alphabetical list of periodicals and serials relating to arboriculture, gardening, and uses of plants; **3**: 29, botanic gardens; **3**: 88–89, fruit trees, gardening, and useful or ornamental plants; **3**: 114, arboriculture and general horticulture: **3**: 255–256, pharmacopoeias; **3**: 271, economic products; **4**: 13. 1914, periodicals and serials relating to forestry, timber uses, etc.; **4**: 20, exhibitions of forest products; **4**: 85–86, dendrography; **4**: 362–363, forests.—See also Colmeiro (La botánica), above, and additional references there cited.

Reyes Prósper, Eduardo. Las estepas de España y su vegetación. 304 p. 28 fig., map. 24.5 cm. Madrid, 1915.

Mainly geographical and ecological; includes (p. 157–214) unannotated list of charophytes and vascular plants, with vernacular names (not indexed); also (p. 251–300) account of forage plants (in part with analyses), useful, medicinal, and ornamental plants. Covers portions of all the regions except Asturias, Estremadura, and Provincias Vascongadas. The author's understanding of the distribution of steppes in Spain is not agreed to by all botanists.—See also Font Quer, above, and additional references there given.

Rivas Goday, Salvador. Los grados de vegetación de la Península Ibérica (con sus especies indicadoras). Anales Inst. Bot. A. J. Cavanilles **13**: 269–331. 1 fig. 1955.

Discussion of plant zones (grados de vegetación, cinturones or cingulus de vegetación), with mention of characteristic species; bibliography.—See also his **Übersicht uber die vegetationsgürtel der Iberischen Halbinsel. Kennzeichende arten und gesellschaften. Veröff. Geobot. Inst. Rübel 31**: 32–69. map. 1956. (Condensation of his earlier paper.)—See also Font Quer, above, and additional references there given.

——— and others. **Aportaciones a la fitosociología hispánica (proyectos de comunidades hispánicas). Nota I. Anales Inst. Bot. A. J. Cavanilles 13**: 333–422. 3 tables. 1955.

Systematic description of classes, orders, alliances and associations, with citation of characteristic species; bibliographies.—See also Font Quer, above, and additional references there given.

Texidor y Cos, Juan. Flora farmacéutica de España y Portugal. xxiv, 1248 p. 222 fig. 20 cm. Madrid, 1871.

Descriptive account of the wild and cultivated medicinal plants of the Iberian Peninsula, on the Linnaean system, with vernacular names, uses, and localities. (Not seen; title and annotation supplied by E. F. Galiano.)—See also Bassagaña, above, and additional references there given.

Vicioso, Carlos. Notas sobre la flora española. Anales Jard. Bot. Madrid **6** (2): 5–92 incl. pl. 1–4. 1946.

Extensive list of spermatophytes collected by author in different parts of Spain, with localities and occasional critical notes.

Willkomm, Moritz. Grundzüge der pflanzenverbreitung auf der Iberischen Halbinsel. xiv, 395 p. 21 fig., 2 pl., 2 maps. 25.5 cm. Leipzig, 1896. (Die vegetation der erde . . . hrsg. von A. Engler und O. Drude. v. 1.)

Botanical explorations, bibliography, topography, geology, climate, distribution of some important species; detailed account of phytogeographical regions and plant formations; list of principal cultivated plants.—See also Font Quer, above, and additional references there given.

——— and **Lange, Johan.** Prodromus florae hispanicae, seu synopsis methodica omnium plantarum in Hispania sponte nascentium vel frequentius cultarum quae innotuerunt. 3 v. 24 cm. Stuttgartiae, 1870 (1861)–80.—**Willkomm, Moritz.** Supplementum sive enumeratio et descriptio omnium plantarum inde ab anno 1862 usque ad annum 1893 in Hispania detectarum . . . ix, 370 p. 23 cm. Stuttgartiae, 1893.

Bibliography, list of collectors; annotated, briefly descriptive flora of 5089 vascular plants, with partial keys, synonymy, local and extralimital range, and vernacular names (separately indexed); cultivated species are included. The Supplement, by Willkomm alone, brings the total number of species (including hybrids, introduced plants, and widely cultivated plants) to 5438. For dates of publication, see **Wiltshear, F. G.** Jour. Bot. **53**: 371. 1915.—See also **Willkomm, M.** Illustrationes florae hispaniae insularumque Balearium. Figures de plantes nouvelles ou rares décrites dans le Prodromus florae hispanicae ou récemment découvertes en Espagne et aux îles Baléares . . . 2 v. 183 col. pl. 35.5 cm. Stuttgart, 1881–92. (Descriptions, with good colored figures, of over 200 new or rare species, not in systematic order. Also issued with Spanish subtitle. For dates of publication, see **Wiltshear, F. G.** Jour. Bot. **53**: 372. 1915.)

ANDALUCÍA (ANDALUSIA)

See also Murcia (Rouy).
The region of Andalucía includes the Provinces of Almería, Cádiz, Córdoba, Granada, Huelva, Jaén, Málaga, and Sevilla.

Local

Barras, Francisco de las. Datos para la flórula sevillana. Actas Soc. Españ. Hist. Nat. **26** (II, 6): 187–191, 219–222, 260–263. 1897; **27** (II,7): 59–62, 63–64, 79–80, 83, 93–96, 109–112, 143–144, 154–160. 1898; **28** (II, 8): 35–36, 37–38, 75–76, 116–117, 154–156, 166–168, 255–261. 1899.

List of vascular and cellular plants, with localities, including cultivated species. Covers the Province of Sevilla.

Boissier, Edmond. Voyage botanique dans le midi de l'Espagne pendant l'année 1837. 2 v. (x, 248; 757 p.). 181 (i.e. 206) col. pl., col. diagr. 35.5 cm. Paris, 1839–45.

Vol. 1, author's itinerary, with description of localities and mention of many plants, phytogeography, plates; v. 2, annotated list of 2015 species (1,900 vascular) known from the Province of Málaga and the southern part of the Province of Granada (including a few from outside this area) with citation of localities and collectors, vernacular names, and general range.

Ceballos, Luis, and **Martín Bolaños, Manuel.** Estudio sobre la vegetación forestal de la provincia de Cádiz. xviii, 353 p. plates, chart, 2 maps. 27.5 cm. Madrid, 1930.—Mapa forestal . . . 4 maps. 27.5 cm. Madrid, 1931.

Geography, hydrography, geology, climate, plant formations; annotated list of native and naturalized woody plants (299 species), with vernacular names and references; account of exotic species, principal forests. Publication of Instituto forestal de investigaciones y experiencias. La Moncloa, Madrid.

——— and **Vicioso, Carlos.** Estudio sobre la vegetación y la flora forestal de la provincia de Málaga. xix, 285 p. plates, chart, map. 27 cm. Madrid, 1933.—Mapa forestal . . . 4 maps. 27 cm. Madrid, 1933.

Topography, geology, soils, climate, plant associations and formations; annotated list of wild or naturalized woody plants (354 species), with synonyms, vernacular names, references, notes on local distribution. Publication of Instituto forestal de investigaciones y experiencias. La Moncloa, Madrid.

Cuatrecasas, José. Estudios sobre la flora y la vegetación del Macizo de Mágina. Trab. Mus. Cien. Nat. Barcelona v. 12. 510 p. 25 fig. (incl. maps), 8 pl., map. 1929.—Adiciones y correcciones . . . Cavanillesia **3**: 8–19. 1930.

Topography, geology, plant associations, cultivated plants, altitudinal zones, phytogeography; annotated list of vascular plants (996 species and 118 varieties); list of medicinal species; bibliography. (Massif of Mágina, Province of Jaén.)

Ghersi y Vila,—. Catálogo de las plantas que crecen en Cádiz y su provincia. Actes 3. Cong. Internatl. Bot. Bruxelles, 1910, **2**: 161–183. [1912.]

Unannotated list of vascular plants.—See also Pérez Lara, below.

Hervier, Joseph. Excursions botaniques de M. Elisée Reverchon dans le massif de La Sagra, et à Velez-Rubio (Espagne) de 1899 à 1903 [–1904-1905]. Bul. Acad. Internatl. Géog. Bot. **15**: 1–32, 57–72, 89–120, 157–170. 1905; **16**: 201–221. 1906; **17**: 33–64, 193–208, 230–231. 1907.

Localities visited, bibliography; lists of plants collected, with localities and critical notes. (Massif de La Sagra, on border of Grenada, Jaén, and Murcia; Vélez-Rubio, in Province of Almería.)—See also **Pau, Carlos.** Carta a un botánico. 1.–4. (8, 11, 8, 10 p.) Segorbe, 1904–07. (Corrections and criticisms. Not seen; cf. Bot. Jahresber. Just **35** (3): 473–474. 1911, and Bul. Brit. Mus. Nat. Hist. Bot. **1**: 120–121. 1954.)—Also **Lacaita, C. C.** La Sierra de Cazorla et les excursions d'Elisée Reverchon. Bul. Soc. Bot. Genève II, **21**: 120–134. 3 fig. (incl. map). 1929. (Topography, geography, critical notes, list of 19 additional species.)—Also **Borja Carbonell, José.** Une excursión a la Sierra de La Sagra (Granada). Anales Inst. Bot. A. J. Cavanilles **13**: 455–468 incl. 8 fig. on 5 pl. 1955. (Includes lists of species from various localities.)

Laza Palacios, Modesto. Estudios sobre la flora y la vegetación de las Sierras Tejeda y Almijara. Anales Jard. Bot. Madrid **6** (2): 217–370. 2 fig., 2 tab. 1946.

Geology, climate, plant associations; annotated list of vascular plants, with localities; bibliography. (Province of Málaga.)

Linberg, Harald. Itinera mediterranea. Ein beitrag zur kenntnis der Westmediterranen flora auf grand eines materials von gefässplflanzen, gesammelt in Tunesien und Sizilien im jahre 1924 und in Spanien und Marokko im jahre 1926. Acta Soc. Scient. Finn. n. s. B, v. 1, no. 2. 1 p. l., 178 p. 50 fig. i.e, 32 pl. on 16. 1932.

Author's itinenary; list of vascular plants (2,100 species), with localities. Includes numerous records from Spain, mostly from Andalucía (Provinces of Cádiz, Córdoba, Granada, Jaén, and Málaga, as well as Gibraltar), some from vicinity of Madrid.

Pau, Carlos. Contribución a la flora española. Plants de Almería. Mem. Mus. Cien. Nat. Barcelona, Ser. Bot. v. 1, no. 3. 34 p. 4 pl. 1925.

List of 346 vascular plants collected by Enrique Gros in 1921, with localities and some critical notes. (Province of Almería.)—See also his Plantas de Almería. Cavanillesia **2**: 162–163. 1930. (Mention of several noteworthy species from the Sierra de Abrocena, between the Sierra de Baza and Sierra Nevada.)

——— Contribución al estudio de la flora de Granada. Treballs Inst. Catalana Hist. Nat. [**2**:] 195–227. 1 fig., pl. 10–13. 1916.

List of 243 vascular plants collected by Enrique Gros at various localities, with some critical notes. (Province of Granada.)—See also his Nueva contribución al estudio de la flora de Granada. Mem. Mus. Cien. Nat. Barcelona, Sèr. Bot. v. 1, no. 1. 74 p. 9 pl. 1922. (Itinerary of E. Gros; list of vascular plants (including 640 numbered species) with localities and frequent critical notes; brings total number of species to 883. In Kew library.)

——— Diez días en Sierra Morena. *In* Tomo Extraord. R. Soc. Españ. Hist. Nat. p. 287–298. 1921.

Notes on critical species, etc.; list of species collected, the common ones omitted. (Vicinity of Villaviciosa and Ovejo, Province of Córdoba.)

Pérez Lara, J. M. Flórula gaditana seu recensio celer omnium plantarum in provincia gaditana hucusque notarum. Anales Soc. Españ. Hist. Nat. **15**: 349–475. 1886; **16**: 273–372. 1887; **18**: 35–143. 1889; **20**: 23–94. 1891; **21**: 191–280. 1892; **24**: 279–335. 1895; **25**: 173–222. 1896; **27**: 21–92. 1898. (Reprinted, 682 p. 24 cm. Madrid, 1886 [–1898]. In Arnold Arboretum library.)—Flórula gaditana. Addenda et emendanda. Mem. R. Soc. Españ. Hist. Nat. **2**: 5–62. 1903.

Botanical explorations, general features of flora, bibliography; annotated list of vascular plants [1,905 species (including supplement), according to Wolley-Dod], with localities. (Province of Cádiz.)—See also **Font Quer, Pío.** Notas sobre la flora gaditana. Bol. R. Soc. Españ. Hist. Nat. **27**: 39–46. 3 fig. 1927. (Notes on several species.)—Also Ghersi y Vila, above.

Rivas Goday, Salvador. Contribución al estudio de la vegetación y flora de la provincia de Granada. Excursión botánica a Sierra Baza y Zugar. Anales R. Acad. Farm. (Madrid) 2. época, año **7**: 58–129. 6 fig. (incl. map). 1941.

Botanical explorations, author's itinerary with notes on plants observed, soils, plant associations and zones; list of 284 vascular plants, with localities and some annotations; bibliography. (In comarca of Baza.)

——— and **Bellot Rodríguez, Francisco.** Estudios sobre la vegetación y flora de la comarca Despeñaperros-Santa Elena. Anales Jard. Bot. Madrid **5**: 377–503 incl. 19 fig. (partly on pls., and incl. maps). 1945; **6** (2): 93-215 incl. fig. 24–34 (partly on pls.). 1946.

Geography, topography, botanical explorations, plant zones and plant formations, ecology; annotated list of vascular plants (702 species), with localities; analysis of flora, statistics, annotated list of medicinal plants, itineraries for excursions, bibliography. (In eastern part of Cordillera Mariánica (Sierra Morena), between Almuradiel (Prov. Ciudad Real) and Santa Elena (Prov. Jaén).)—See also **Galiano, E. F.** Anotaciones a la flora de Sierra Morena. Plantas de Aldeaquemada. Anales Inst. Bot. A. J. Cavanilles **12** (1): 501–508. 1 fig. (map). 1954. (Botanical explorations, general features of flora; list of noteworthy plants; bibliography. In Province of Jaén.)

Romero de la Cruz, José. Datos para la flora del litoral granadino. 40 p. 21 cm. Granada, 1927. (Thesis.)

List of species collected in the littoral zones of the Province of Granada. (Not seen; title and annotation supplied by E. F. Galiano.)

Vicioso, Benito. Plantas de Andalucía. Bol. Soc. Aragon. Cien. Nat. **7**: 71–81. 1908.

List of vascular plants observed by author in vicinity of Algeciras, Almuñécar, Motril, etc., with localities. (Provinces of Cádiz and Granada.) —See also **Pau, Carlos.** Plantas de Andalucía. Cavanillesia **4**: 49–57. 1931. (Account of some noteworthy species from various localities collected by Luis Ceballos and Carlos Vicioso.)—Also his Sobre plantas de Andalucía. l.c. **5**: 41–44. 1932. (Similar list.)

Wolley-Dod, A. H. A flora of Gibraltar and the neighbourhood. Jour. Bot. 52, Suppl. xxvi, 131 p. 1914.

Geology, topography, climate, botanical explorations and publications, botanical districts, phytogeography; annotated list of 1,462 species of vascular plants (1,307 native, of which 587 are native on Gibraltar proper). Covers Gibraltar and adjacent part of Province of Cádiz. Replaces **Debeaux, Odon.** Synopsis de la flore de Gibraltar. Actes Soc. Linn. Bordeaux 42 (**V, 2**): 121–378. pl. 10–11 (pl. 10, map). 1888. (Reprinted 261 p., 1889. In Arnold Arboretum library.) (Botanical explorations, etc.; annotated list of 845 i.e., 1,005 species [according to Wolley-Dod 1,037 species] of vascular plants; general features of flora.) Also replaces **Frere, B. H. T.** A guide to the flora of Gibraltar and the neighbourhood. 2 p.l., 159, vi p. 22 cm. Gibraltar, 1910. (Glossary; briefly annotated descriptive flora of phanerogams (except sedges, rushes, and grasses), with descriptions of non-British species.)—See also **Wolley-Dod, A. H.** Gibraltar plants. Jour. Bot. **52**: 10–15. 1914. (Descriptions of several new plants. See also his Euphorbia gibraltarica. l.c. p. 47.)—Also his Flora calpensis; a list of plants recorded from Gibraltar and the Campo district of Spain. v, 100, A–K, xviii p. 31.5 cm. Gibraltar, 1949. (Annotated list of vascular plants, with localities, a posthumous publication based on his manuscript in the Gibraltar Museum. Not complete; lacks, e.g., Gramineae. Mimeographed publication, prepared at the American Consulate, Gibraltar.)

ARAGÓN

See also Valencia (Pau, Plantas de la Sierra de El Toro).
The region of Aragón includes the Provinces of Huesca, Teruel, and Zaragoza.

General

Ferrando Más, Pedro. Los estudios botánicos en Aragón. Revista Acad. Cien. Zaragoza 20: 84–110. 1937.

Account of botanists and botanical work, arranged chronologically, but far from complete.

Loscos y Bernal, Francisco, and **Pardo y Sastrón, José.** Serie imperfecta de las plantas aragonesas espontáneas, particularmente de las que habitan en la parte meridional. 2. ed., aumentada . . . x, 543 p. 20 cm. Alcañiz, 1866–67. (1st ed.: Series inconfecta plantarum indigenarum Aragoniae . . . e lingua castellana in latinam vertit . . . atque edendam curavit Mauritius Willkomm. Dresdae, 1863.)

Annotated list of 2,624 species (of which 2,316 are vascular plants) of Aragon, with localities and vernacular names. Replaces **Asso, I. J. de.** Synopsis stirpium indigenarum Aragoniae. Auctore C. A. R. Caesaraugustano . . . xiv, 160 [14] p. 9 pl. 21 cm. Massiliae, 1779. (Topography, botanical explorations, etc.; annotated list of 1,057 plants (1,009 vascular) with localities, vernacular names, and uses.) Also replaces his Mantissa stirpium indigenarum Aragoniae. 4 p. l., 159–184, [1] p. pl. 10–11. 22 cm. n.p., 1781. (In Arnold Arboretum library.)—See also **Loscos y Bernal, Francisco.** Tratado de plantas de Aragón. 1.–3. parte. (90, 253, 55 p.) 20.5 cm. 1876–81 (?). Suplemento 5–8. (107 p.) 1883–86(?). (Part 1 (1876–77) contains miscellaneous notes; pt. 2 (1878) contains a list of vernacular names, a catalog of 2,582 phanerogams, completing the Serie imperfecta . . . , and supplements 1–4; pt. 3 contains a list of cryptogams including 29 vascular cryptogams. The supplements bring the total to 2,788 phanerogams. Not seen, except for p. 1–53 of 1st part; title and annotation supplied by A. de Bolós. According to review by Willkomm (Bot. Centralbl. 7: 12–13. 1881) the first two parts originally appeared in the "Semanário farmacéutico" of Madrid and were reprinted from that journal.)—Also **Debeaux, Odon.** Plantes rares ou nouvelles de la province d'Aragon (Espagne) provenant des récoltes de M. Reverchon en 1892–1893 [-189-?]. Revue Bot. Bul. Mens. Soc. Franç. Bot. 12: 31–50. 1894; 13: 337–367. 1895; 15: 129–180. 1897. (Records and critical notes on various vascular plants. Vol. 15 has not been seen.)—Also **Pau, Carlos.** Plantas críticas de Asso. Bol. Soc. Aragon. Cien. Nat. 2: 30–34. 1903. (Critical notes, reductions to synonymy, etc., concerning some 19 species described by Asso.) Also **Vicioso, Carlos.** Plantas aragonesas. l.c. 10: 75–83, 98–104. 1911. (Extensive list of noteworthy spermatophytes collected by author in various localities.)—Also **Escriche, Manuel.** Plantas de Aragón. Revista Acad. Cien. Zaragoza 20: 67–74. 1937. (List of vascular plants collected in Aragón by Custodio Campo García, with some annotations but without localities.)—Also **Cámara Niño, Fernando.** Plantas de los terrenos secos de Aragón. Anales Jard. Bot. Madrid 6: 371–395. 1946. (Mention of some previous botanists, etc.; annotated list of noteworthy species from various localities.)—Also **Font Quer, Pío.** Notas sobre la flora de Aragón. I. Collect. Bot. Barcinon. 3: 345–358. 1953. (List of noteworthy species with localities and some critical notes, from vicinity of Orihuela del Tremedal, Alcalá de la Selva, and Sierra Espadada. Unfinished.)

Local

Echeandia, P. G. Flora cesaraugustana y curso práctico de botánica. Obra póstuma . . . precedida de un discurso . . . por D. Manuel Pardo y Bartolini. 50 [2] p. 21 cm. Madrid, 1861. (Gray Herbarium library.)

Botanical publications, topography, geology, climate, general features on flora, etc. (by Pardo y Bartolini); briefly annotated list of vascular plants and a few bryophytes, on the Linnaean system. (Province of Zaragoza.)—See also **Loscos y Bernal, Francisco.** Comentarios sobre la flora de Zaragoza. 78 p. 23.5 cm. [1876–77?]. (List of 899 phanerogams, with localities and some vernacular names. Only p. 1–36 seen; annotation supplied by A. de Bolós.)—Also **Pau, Carlos.** Relación de plantas zaragozanas. Bol. Soc. Aragon. Cien. Nat. **1**: 129–130. 1902. (Unannotated alphabetical list of plants collected by P. Doménech in vicinity of Zaragosa.)—Also his Plantas de Zaragoza. Cavanillesia **7**: 29–34. 1935. (Critical notes on 11 species.)

Losa España, Mariano. Notas sobre la flora y la vegetación de la Sierra de Guara (Huesca). Collect. Bot. Barcinon. **2**: 65–98. illus. (incl. map). 1948.

Topography, plant zones, notes on various species; annotated list of vascular plants collected; notes on species recorded by others. (Province of Huesca.)

——— and **Montserrat, Pedro.** Aportaciones para el conocimiento de la flora del valle de Ordesa. Collect. Bot. Barcinon. **1**: 127–188. pl. 1–7. 1947.

Botanical explorations, topography, geology, plant associations; annotated list of vascular plants, based on collections made in August 1946; bibliography. (Province of Huesca.)—See also **Cuatrecasas, José.** De flore pyrenaica. I. Ojeada a la cliserie del valle de Ordesa. Cavanillesia **4**: 113–127. pl. 3–6. 1931. (Annotated list of principal trees, lists of the species of various associations.)

Pardo y Sastrón, José. Apendice al catálogo de plantas de Torrecilla de Alcañiz. Datos que podrán servir para escribir el catálogo de plantas de Valdealgorfa. Anales Soc. Españ. Hist. Nat. **30 (II, 10)**: 211–236. 1901.

Partly annotated list of vascular and cellular plants of Valdealgorfa, including cultivated species; list of included species not known from Torrecilla. (Province of Teruel.)

——— Catálogo ó enumeración de las plantas de Torrecilla de Alcañiz así espontáneas como cultivadas. Bol. Soc. Aragon. Cienc. Nat. **1**: 59–64, 121–126, 144–148, 172–179, 204–212. 1902; **2**: 20–25, 46–54, 81–88, 110–117, 139–146, 168–176, 195–197, 214–225, 262–268, 289–297. 1903.

Annotated list of 1,022 species of plants (906 vascular). (Province of Teruel.)

Pau, Carlos. Plantas aragonesas recogidas por D. Benito Vicioso, de Calatayud. Actas Soc. Españ. Hist. Nat. **23**: 124–144. 1894–95.

Extensive list of vascular plants, in part with localities, from Calatayud, Campiel, Moncayo, Saviñan, and other localities in Zaragoza.—See also **Vicioso, Benito.** Plantas de las inmediaciones de Calatayud. l.c. **28**: 219–224. 1899. (List of noteworthy Thalamiflorae, with localities.)

——— Plantas de la provincia de Huesca (6–18 Julio 1903). Bol. Soc. Aragon. Cien. Nat. **4**: 180–187, 288–296, 335–336. 1905; **5**: 173–181. 1906; **7**: 106–119. 1908.

List of angiosperms collected in various localities in Huesca, with annotations on the noteworthy species.

——— Plantas de las cercanias de Teruel, recogidas por D. Juan Benedicto, farmacéutico de Monreal del Campo (1891–93). Actas Soc. Españ. Hist. Nat. **24**: 148–156. 1896.

Unannotated list of vascular plants (not in systematic order), followed by critical notes on a few of them. (Province of Teruel.)—See also **Sennen, *frère*. Plantes observées autour de Teruel pendant los [sic] mois d'août** et de septembre 1909. Bol. Soc. Aragon. Cien. Nat. **9**: 173–184, 226–240, 257–270. 1910. (Topography, etc.; list of vascular plants observed, with localities.)

——— Plantas del Formigal de Sallent (Pirineo aragonés). Actas y Mem. Primer Cong. Nat. Españ. (Zaragoza 1908) 4 (Secc. 3): 243–249. 1909.

General features of the country; annotated list of phanerogams. (Province of Huesca.) **(Not seen; title and annotation mostly supplied by A. de** Bolós.)

——— Plantas recogidas por Don Juan Benedicto, farmacéutico de Monreal del Campo, segun muestras remitidas por el mismo. Actas Soc. Españ. Hist. Nat. **24**: 13–23. 1895.

Lists of vascular plants, not in systematic order, from Guadalajara (with localities) and from Baños de Segura and Monreal del Campo in Teruel, the last quite extensive; annotations on several species.

Rivas Goday, Salvador. De flórula farmacéutica. Importancia farmacobotánica del valle de Tena (Pirineos aragoneses). Farmacognosia **2** (whole no. 2): 65–100. 5 fig. (incl. maps), 10 pl. (on 5). 1943.

Author's itinerary, with mention of plants observed; alphabetical tabular list of plants (113 species) of pharmacological interest. (Province of Huesca.)

Rodríguez y Femenías, J. J. Herborización en Panticosa (Julio y Agosto de 1889). Anales Soc. Españ. Hist. Nat. **19**: 101–106. 1890.

Briefly annotated list of 139 vascular plants collected by author. (Province of Huesca.)

Soubeiran, J. L. Une ascension à la Maladetta (23 août 1862). Annales Soc. Linn. Dépt. Maine-et-Loire **6**: 1–16. 2 pl. 1863.

Account of author's ascent, with lists of plants seen at different points. This mountain, also known as Mont Néthou, is 3,432 meters high. (Province of Huesca, also Haute-Garonne (France).)

Zapater, Bernardo. Flora albarracinense, ó catálogo de las plantas de los alrededores de Albarracín y su sierra. Mem. R. Soc. Españ. Hist. Nat. **2**: 289–338. 1904.

List of vascular plants and Characeae, with localities. (Province of Teruel.)—See also **Pau, Carlos.** Lista de las especias á que pertenecen las plantas recogidas en la Sierra de Albarracín por D. Doroteo Almagro. Actas Soc. Españ. Hist. Nat. **25**: 34–51. 1896. (Extensive list of vascular plants, with localities.)—Also his Mis campañas botánicas. Bol. Soc. Aragon. Cien. Nat. **2**: 11–16. 1903. (Includes (p. 15–16) notes on botanical explorations in the Sierra de Albarracín, with list of noteworthy species collected by author.)—Also his Una visita á San Ginés (Sierra de Albarracín) (8 Junio 1906). l.c. **6**: 55–61. 1906. (Includes mention of various noteworthy species.)—Also his Herborizaciones por la Sierra de Albarracín. l.c. **9**: 57–61. 1910. (Mostly species not recorded by Zapater.)

ASTURIAS

See also Galicia (Buch).
The region of Asturias includes only the Province of Oviedo.

General

Argumosa y Valdés, J. A. Plantas medicinales de Asturias. 97 p. 20 cm. Oviedo, 1951.

Includes obviously incomplete account of wild and cultivated medicinal plants.

——— Sobre investigaciones botánicas en Asturias. Bol. Soc. Españ. Hist. Nat. (Secc. Biol.) **50**: 137–146. 1952.

Sketch of botanical work, from which all modern publications of any floristic significance are omitted.

Local

Barbey-Gampert, M. Esquisse de la flore des Picos de Europa. Bul. Soc. Bot. Genève II, **12**: 219–245. 5 fig. 1921.

Geology, paleontology, climate; unannotated list of known vascular plants (562 species), grouped according to geographical affinity and alphabetically arranged; phytogeography, critical notes on various species. The Picos or Peñas de Europa, the highest portion of the Cantabrian Mountains, are in the Provinces of Oviedo, León, and Santander.—See also **Lacaita, C. C.** Observations sur la flore des Picos de Europa. Bul. Soc. Bot. Genève II, **21**: 135–142. 1929. (Botanical explorations, list of additional species, critical notes on various species.)—Also his Rectification à la flore . . . l.c. II, **21**: 263. 1929.—Also **Lascombes, Georges.** Végétation des Picos de Europa. Les paysages forestiers. Bul. Soc. Hist. Nat. Toulouse **79**: 339–358. 4 pl.

(incl. map), map. 1944. (Topography, geology, climate, botanical explorations, account of forest zones with mention of principal species; bibliography.) —Also **Arrieu, Floréal.** Végétation des Picos de Europa. Les paysages pastoraux. l.c. 359–375. 2 pl. 1944. (Account of subalpine and alpine zones, with partial lists of species from different localities.)—Also **Guinea, Emilio.** De mi primer viaje botánico a Picos de Europa. Anales Jard. Bot. Madrid **7**: 335–356 incl. 1 fig. (map). 8 pl. (on 4). 1947 (1948). (Author's itinerary in 1944; list of noteworthy vascular plants collected, with localities.)

Chermezon, Henri. Contributions à la flore des Asturies. Bul. Soc. Bot. France **66**: 120–130. 1919.

Botanical explorations, bibliography; list of vascular plants not previously recorded from Asturias, collected by author in 1916 in vicinity of Avilés.—See also his Aperçu sur la végétation du littoral asturien. Bul. Soc. Linn. Normandie VII, **3**: 159–213. 1921. (Topography of region around Avilés, plant zones and habitats with numerous lists of species, phytogeography.)

Gay, Jacques. Duriaei iter asturicum botanicum, anno 1835 susceptum. Annales Sci. Nat. Bot. II, **6**: 113–137, 113 (i.e. 213)–225, 340–355. 1836.

Account of Durieu de Maison-Neuve's expedition, with mention of plants collected, grouped by localities; no systematic list. The paper was never completed.—See also **Lacaita, C. C.** Duriaei iter asturicum botanicum. Jour. Bot. **67**: 107–113, 151–154, 207–211, 255–258, 310–318, 324–332. 1929. (Remarks on Gay's paper; publication of his descriptions of or notes on 32 species collected by Durieu.)—Also **Font Quer, Pío, and Rothmaler, Werner.** Sequint les petjades de Durieu . . . (Amb motiu del centenari del seu viatge a Astúries.) Cavanillesia **7**: 173–180. pl. 8–9 (pl. 9, port.). 1936. (Notes on several critical species collected by the authors during a retracing of Durieu's itinerary; German summary.)

Guinea, Emilio. Las plantas medicinales de los Picos de Europa. Agricultura (Madrid) **13**: 684–688. 6 fig. 1944.

Popular account of wild medicinal plants. (In Province of Oviedo, and in León (Prov. León) and Castilla la Vieja (Prov. Santander).)

Stephenson, Thomas. The flora of Asturias. Jour. Bot. **65**: 70–74. 1927.

Topography, etc.; general features of flora in vicinity of Covadonga, with mention of numerous species.

BALEARES (BALEARIC ISLANDS)

The islands of Formentera and Íviza (Íbiza) are sometimes separated as the Pitiusas (Pityusae).

General

Barceló y Combis, Francisco. Flora de las islas Baleares, seguida de un diccionario de los nombres baleares, castellanos y botánicos, de las plantas espontáneas y de las cultivadas. xlviii, 645 p. 22 cm. Palma, 1879–81. (Gray Herbarium library.)

Topography, geology, cultivated plants, climate, statistics of flora, botanical investigations with included bibliography; annotated, keyed, descriptive flora of 1,672 vascular and cellular plants (1,542 native species of which 1,344 are vascular), with localities; alphabetical list of vernacular names, with Castilian and scientific equivalents. Replaced as an enumeration of the species by Knoche's Flora balearica, which does not contain descriptions.

Costa, A. C. La flora de las Baleares y sus exploradores. Especies endémicas. Mem. R. Acad. Cienc. y Art. Barcelona II, **1**: 185–222. 1882.

Botanical explorations, with mention of publications; phytogeography, etc.

Knoche, Herman. Flora balearica. Étude phytogéographique sur les îles Baléares. 4 v. front. (v. 4), illus. (incl. maps), 47 pl., 5 maps, tables. 25 cm. Montpellier, 1921–23.

Vol. 1–2 (constituting 1. partie, Catalogue raisonné de toutes les plantes connues): author's excursions, numerical list of localities cited in text, lists of cultivated plants, subspontaneous plants, dubious and excluded species; list of cellular and vascular plants (1,279 species of vascular plants), with local and extralimital range, the local range often shown on maps; v. 3: history, cultivated plants, climate, topography, geology, sketch of botanical explorations with included bibliography, statistics of flora, phytogeography, plant associations, etc.; v. 4: plates. Replaces **Marès, Paul,** and **Vigineix, Guillaume.** Catalogue raisonné des plantes vasculaires des îles Baleares. 2 p.l., xlvii, 370 p. 9 pl. 25 cm. Paris, 1880. (Botanical explorations with included bibliography, topography, climate, general features of flora, phytogeography, plant zones, table of altitudes; annotated list of vascular plants and Characeae; localities for botanical excursions; list of vernacular names.) —See also **Cañigueral Cid, Juan.** Contribución al estudio de la flora balear. Collect. Bot. Barcinon. **3**: 101–107. 1951. (Records from various islands.) —Also **Palau, Pedro.** Investigaciones botánicas en Baleares. Anales Inst. Bot. A. J. Cavanilles **11** (2): 483–495. 1953. (Mention of various localities and their noteworthy species.)—Also his Nuevas estirpes para la flora de Baleares. l.c. **11** (2): 497–519 incl. pl. 1953. (List of 66 species and varieties.)—Also his Plantas de Baleares. Collect. Bot. Barcinon. **4**: 207–214. 2 fig. (incl. map). 1954. (Records of 3 plants.)

Palau, Pedro. Les plantes medicinals baleàriques. 131 p. 15 cm. Palma de Mallorca, 1954. (Biblioteca "Les illes d'or" 61.)

List of diseases with remedies; annotated list of medicinal plants, alphabetically arranged by vernacular names, with Catalonian, Castilian, and French vernacular names, properties, and uses. Written in Catalonian.

Local

Chodat, Lucie. Contributions à la géo-botanique de Majorque. Bul. Soc. Bot. Genève II, **15**: 153–264. 37 fig. 1924. (Also as thèse, Université de Genève. Institut de botanique. sér. 10. fasc. 6. 116 p. 1924.)

General features of flora with lists of species of different habitats, endemism, annotated list of characteristic and endemic plants, phytogeography, bibliography.—See also **Pau, Carlos.** Una visita a Mallorca. Brotéria Sér. Trimest. Ciên. Nat. **3**: 56–65, 174–179. 1934. (Critical notes on various dicotyledons.)—Also **Rivas Goday, Salvador.** Excursión a Mallorca abril-mayo 1943. Impresión botánica. Farmacognosia 4 (whole no. 6): 139–203. 8 fig. (incl. maps), 24 pl. 1945. (Author's itinerary, with mention of plants observed, plant associations; annotated list of noteworthy species; notes on phytogeography, medicinal plants, bibliography.)—Also **Cañigueral Cid, Juan.** Algunos datos sobre la flora de Mallorca. Collect. Bot. Barcinon. **3**: 310–323. 3 fig. 1953. (Account of excursions to Mola de Planicie and Sa Granja de Esporlas; record of various noteworthy species.) —Also **Garcias y Font, Llorens.** Nueva contribución al estudio de la flora balear. Adiciones y correcciones. Collect. Bot. Barcinon. **3**: 359–366. 1954. (Relates to Mallorca.)—Also **Pignatti, Sandro.** Studi sulla flora e vegetazione dell'isola di Mallorca (Baleari). Arch. Bot. (Forlì) **31**: 46–58, 65–100. 2 fig. 1955. (Annotated list of noteworthy species, bibliography; (p. 65–100) revision of *Limonium* in Balearic Islands.)

Font Quer, Pío. La flora de las Pitiusas y sus afinidades con la de la península ibérica. Mem. R. Acad. Cien. y Art. Barcelona III, **20**: 109–154. 32 fig. (incl. maps). 1927. (Also numbered as v. 20, no. 4. 48 p.)

Phytogeography, botanical explorations, etc.; list of 61 species from Pitiusas (Formentera and Íviza) not yet known in the other Balearic Islands, with details of distribution of several of these; annotated list of endemic species, etc. The author listed the species of some groups in earlier papers, as follows: Pteridofitas de las Pitiusas. Bol. Soc. Españ. Hist. Nat. **19**: 507–511. 1919; Compuestas de las Pitiusas. l.c. **20**: 141–159. 1920; Tubifloras de las Pitiusas. Asociación Española para el progreso de las ciencias, Congreso de Oporto, v. 6 (secc. 4, Ciencias naturales), p. 1–76. Madrid, 1921. (In Farlow Library, Harvard University.)

Garcías y Font, Llorens. Plantes dels voltants d'Artà y Capdepera (Mallorca). Contribució á la flora balear I[–VII.]. Butllet. Inst. Catalana Hist. Nat. **5** (2. época, 2): 40–43. 1905; [**8**] (2. época, 4): 16–19. 1907; [**9**] (2. época, 6): 57–60. 1909; **17** (2. época, 14): 112–120. 1917; **19** (3. época, 2): 110–117. 1919; **22** (2. sèr., 2): 157–158. pl. 10. 1922; **23** (2. sèr., 3): 91–103. 1923.

Lists of vascular plants collected by author in different years, with some vernacular names and (after no. I) brief annotations. No. 7 (misnumbered 8) precedes no. 6 and lacks the subtitle.

Marcos, Antoni. Contribució al coneixement de la flora balear. Flòrula de Cabrera i dels illots pròxims. Cavanillesia 8: 5–52. map. 1936.

Botanical explorations, statistics, general features of flora, geology; annotated list of 284 [286] known vascular plants, based chiefly on the collections of the author and Font Quer.—See also **Palau, Pedro.** Dos excursiones a Cabrera. Collect. Bot. Barcinon. **1:** 315–321. 1947. (Author's explorations, with list of additional species.)—Also his Otra excursión a Cabrera. l.c. **3:** 183–184. 1952. (Mention of several species, one new to the island.)

Pau, Carlos. Relación de plantas ibiceñas. Actas Soc. Españ. Hist. Nat. **29:** 62–69, 231. 1900.

List of vascular plants collected by author on Íbiza in April–May 1899, including many new to the island. A preliminary list based on the same collection was published by Pau, l.c. **28:** 213–216. 1899.—See also **Font Quer, Pío.** Exploració botánica d'Eviça i Formentera. Butllet. Inst. Catalana Hist. Nat. **18** (III, 1): 101–102. 1918. (List of 18 noteworthy vascular plants, with localities.)

Rodríguez y Femenías, J. J. Excursion botánica al Puig de Torrella (Mallorca). Anales Soc. Españ. Hist. Nat. **8:** 39–64. 1879.

Author's itinerary, with mention of noteworthy species; list of vascular plants collected, with localities. The Puig de Torrella is the highest mountain on Majorca (Mallorca).

——— Flórula de Menorca. xv, 198 p. 25 cm. Mahón, 1904. (Arnold Arboretum library.)

Botanical explorations (with references), climate, general features of flora; annotated list of vascular plants, Characeae, mosses, and lichens (973 species, of which 880 are vascular), with localities. Replaces his Catálogo razonado de las plantas vasculares de Menorca. xxx, 116 p. 18 cm. Mahón, 1865–68 (Arnold Arboretum library); Suplemento al Catálogo . . . Anales Soc. Españ. Hist. Nat. 3 (Mem.): 5–68. 1874 (reprinted 64 p.); Additions à la flore de Minorque. Bul. Soc. Bot. France **25:** 238–241. 1878.—See also **Pau, Carlos.** Sobre algunas plantas menorquinas. Butllet. Inst. Catalana Hist. Nat. **14** (II, **11**): 135–142. 1914. (Records and critical notes on various species collected by Font Quer.)—Also **Font Quer, Pío.** Plantes noves per a Menorca. l.c. **17** (II, **14**): 51–52. 1917. (List of 14 plants, with localities.)—Also **Bianor,** *frère.* Quelques jours à Menorca (Îles Baléars). Anuari Junta Cièn. Nat. Barcelona **2:** 595–600. 1917. (Author's itinerary in April 1917, with mention of noteworthy plants observed.)—Also **Font Quer, Pío.** Adiciones a la flora de Menorca. Bol. Soc. Españ. Hist. Nat. **19:** 268–273. 1919. (Additional species.)—Also **Montserrat, Pedro.** Aportación a la flora de Menorca. Collect. Bot. Barcinon. **3:** 399–418. 1953. (Extensive list of noteworthy vascular plants; bibliography.)

CASTILLA LA NUEVA (NEW CASTILE)

See also Andalucía (Lindberg; Rivas Goday and Bellot Rodríguez); Murcia (Rouy).

The region of Castilla la Nueva includes the Provinces of Ciudad Real, Cuenca, Guadalajara, Madrid, and Toledo.

General

Colmeiro, Miguel. Apuntes para la flora de las dos Castillas. 176 p. 20 cm. Madrid, Lima, 1849.

Botanical explorations; annotated list of vascular and cellular plants of Old and New Castile, with localities and vernacular names; no index.

Local

Aterido, Luis. Lista ordenada metódicamente de muchas plantas de la provincia de Cuenca. Actas Soc. Españ. Hist. Nat. **28** (II, 8): 195–202. 1899.

Unannotated list of vascular plants observed or collected by author in the términos of Almonacid del Marquesado and Villarejo de Fuentes.

Bellot Rodríguez, Francisco. Estudios sobre la vegetación y flora de la comarca de Somosierra. Nota primera. (Introducción y Valle alto del Jarama). Anales R. Acad. Farm. (Madrid) **10:** 109–130. 8 fig. on 4 pl. 1944.

Topography, etc.; partial lists of species from various habitats and localities; annotated list of principal medicinal species; bibliography. (Province of Madrid.)

Caballero, Arturo. Apuntes para una flórula de la Serranía de Cuenca. Anales Jard. Bot. Madrid **2**: 236–265. pl. 5, 9, 12. 1942; **4**: 403–457. 2 pl. 1944; **6** (2): 503–547 incl. 3 pl. 1946.

Botanical explorations, etc.; lists of vascular plants (the families not in strict systematic order) with localities and some annotations. Refers to region bounded by El Tobar, Beteta, Hoz de Tragavivos, Cañizares, and Solan de Cabras. (Province of Cuenca.)

Caballero y Villaldea, Sergio. Flórula arriacense. Estudio de las plantas que viven en la provincia de Guadalajara, con aplicación a la medicina, artes, industrias, agricultura y horticultura. v. 1–2 (179, 293 p.). map. 21 cm. Guadalajara, 1924–26.

Vol. 1: Historical sketch of botanical investigations to end of 17th century, with annotated lists of plants known in each century, and account of botanical regions, bibliography, annotated list (400 species) of the pre-Linnaean flora, alphabetical list of pre-Linnaean vernacular names; v. 2: continuation to present time, with annotated list (996 species) of vascular plants known at end of 18th century. The work, if completed as projected, would have been perhaps the most elaborate local flora ever written. Part 1, consisting of v. 1–2, is historical. Part 2, in 4 volumes, was to have been an enumeration and critical revision of the plants hitherto reported from Guadalajara, and part 3, in an undetermined number of volumes, the author's new flora, terminating with a volume on statistics and geographical botany.

Cutanda, Vicente. Flora compendiada de Madrid y su provincia, ó descripcion sucinta de las plantas vasculares que espontáneamente crecen en este territorio. 759 p. table, map. 24.5 cm. Madrid, 1861.

Topography, climate, geology, general features of flora, floral calendar, plant zones, statistics of flora, keys to families and genera (latter on Linnaean system); annotated descriptive flora of vascular plants (1876 species), with localities. (Province of Madrid.)—See also **Pau, Carlos.** Notas sueltas sobre la flora matritense. I–XI. Bol. Soc. Aragon. Cien. Nat. **14**: 204–211, 235–237. 1915; **15**: 63–74, 158–172. 1 fig. 1916; **16**: 61–70. 1917; **17**: 150–156, 190–197, 208–211. 1918. Bol. Soc. Ibérica Cien. Nat. **18**: 80–92. 1919; **20**: 176–188. 1921 (1922); **22**: 87–98. 1923; **23**: 96–107. 1924; **25**: 30–36. 1926. (A series of critical notes and new records.)—Also **Rivas Mateos, Marcelo.** Especies nuevas o raras de la flora de la provincia de Madrid. Bol. Soc. Españ. Hist. Nat. **27**: 184–189. 1927. (Annotated list of several species.)—Also **González-Albo, J.** Datos sobre la flora y fitosociología de la provincia de Madrid. l.c. **38**: 9–18. 2 fig., pl. 1. 1941. (Includes records of several noteworthy species. First printed l.c. **37**: 117–127. 1938, but these pages never distributed, fide E. F. Galiano.)

Más y Guindal, Joaquín. Contribución al estudio de la flora de Guadalajara. Bol. Farm. Militar **5**: 289–293, 390–395. 1927.

List of vascular plants from Congostrina, with some vernacular names, and mention of some other species from Hiendelaencina and some other localities. (Province of Guadalajara.)

——— Datos para el estudio de la flora de la Real Casa de Campo de Madrid. 64 p. 18 cm. Madrid, 1923. (From "Monitor de la farmacia.")

Account of collectors; list of 878 plants (118 cryptogamous), with vernacular names. (Province of Madrid.) (Not seen; title and annotation supplied by E. F. Galiano.)

Pomata, Eladio. Catálogo de plantas recolectadas al estado espontáneo en la provincia de Toledo. Anales Soc. Españ. Hist. Nat. **11**: 241–306. 1882.—Apéndice . . . l.c. **12**: 221–240. 1883.

Annotated list of 816 plants (of which 797 are vascular), including appendix.

Secall, José. Plantas vasculares de San Lorenzo del Escorial y sus alrededores. Catálogo metódico de las que se observan silvestres ó asilvestradas. xiv, 66 p. 24 cm. Madrid, 1889. (Arnold Arboretum library.)

Topography, climate, general features of flora, bibliography; tabular list of vascular plants (1,062 species), with localities. (Province of Madrid.)—

See also his Catálogo metódico de las plantas leñosas silvestres ó asilvestradas que se observan en San Lorenzo del Escorial y sus alrededores. Revista Mont. (Madrid) **12:** 377–382, 401–405. 1888. (List of 91 woody plants, with localities. In Arnold Arboretum library.)

Vélaz de Medrano, Luis, and **Ugarte, Jesús.** Estudio monográfico del Río Manzanares. (Biología de las aguas continentales I.) [Bol.] Inst. Forest. Invest. y Exper. Madrid año 6, no. 11. 68 p. pls. (incl. map). 1933.

Includes (p. 11–14) unannotated, mostly compiled list of vascular plants, with some vernacular names. (Province of Madrid.)

CASTILLA LA VIEJA (OLD CASTILE)

See also Asturias (Barbey-Gampert, and Guinea, on Picos de Europa); Castilla la Nueva (Colmeiro, Apuntes para la flora de las dos Castillas). For titles relating to Province of Palencia, formerly in the region of Castilla la Vieja, see under León.

The region of Castilla la Vieja includes the Provinces of Ávila, Burgos, Logroño, Santander, Segovia, Soria, and Valladolid.

Local

Cámara Niño, Fernando. Estudios sobre la flora de La Rioja baja. 182 p. Madrid, 1940.

Includes (p. 80–179) list of vascular plants (997 species) collected by author in La Rioja baja, principally in vicinity of Corera, Province of Logroño. (Not seen; title and annotation supplied by A. Caballero.)—See also Zubía, below

Font Quer, Pío. Datos para el conocimiento de la flora de Burgos. Trab. Mus. Cien. Nat. Barcelona v. 5 (Ser. Bot.), no. 5. 56 p. 1924.

Annotated list of vascular plants collected by author at various localities in Province of Burgos.—See also his Más plantas de Burgos. Bol. Soc. Ibérica Cien. Nat. **25:** 79–84. 1926. (List of plants from provinces of Logroño, Burgos, and Santander, collected by H. Elías and Mariano Losa, with localities.)

Guinea López, Emilio. Geografía botánica de Santander. 5–408 p. illus., pls. 28 cm. Santander, 1953. (Arnold Arboretum library.)

Discussion of plant regions, weeds, topography, geology, climate, soils, plant formations, etc.; annotated list of vascular plants, alphabetically arranged in a single alphabet. (Province of Santander.)—See also **Dupont, Pierre.** Contributions à la flore du nord-ouest de l'Espagne. I–II. Bul. Soc. Hist. Nat. Toulouse **88:** 120–132. 1953; **90:** 429–440. 1955. (Lists of noteworthy species, including new records from Vizcaya and Santander.)—Also **Laínz, Manuel.** Contribución al catálogo de la flora montañesa. Collect. Bot. Barcinon. **4:** 215–226. 1954. (List of noteworthy species, mostly from Santander.)—Also **Dupont, Pierre,** and **Dupont, Suzanne.** Additions à la flore du nord-ouest de l'Espagne. I. Bul. Soc. Hist. Nat. Toulouse **91:** 313–334. 1956. (Records of noteworthy species from Santander, Asturias, and the Pays Basque; bibliography.)

Gutiérrez Martín, Daniel. Apuntes para la flora del partido judicial de Olmedo. 136 p. Ávila, 1908.

Contains list of plants (including some cryptogams) with localities. (Province of Valladolid.) Not seen; title and annotation mainly from notice in Bot. Jahrb. Engler 41 (Literaturber.): 52. 1908.)

Marcet, A. F. Una excursión á Valvanera. Bol. Soc. Aragon. Cien. Nat. **7:** 133–143. 1908.

Topography, zoology, etc.; annotated list of vascular plants. (Province of Logroño.)

Más y Guindal, Joaquín. Datos para el estudio de la flora de Santoña. Bol. Farm. Militar **2:** 309–314. 1924.

Account of about 150 vascular and a few cellular plants collected on a brief visit, with some vernacular names. (Province of Santander.)

Rivas Mateos, Marcelo. Excursión botánica a Gredos. Bol. Soc. Españ. Hist. Nat. **24:** 379–390. 3 fig. (incl. map). 1924.

Methods of travel, general features of flora, etc.; descriptions of new forms; unannotated list of vascular plants and a few mosses and lichens.

(Province of Ávila.)—See also his Especies botánicas de Gredos. Bol. Soc. Españ. Hist. Nat. **25:** 83–85. 1 fig. 1925. (Critical notes on several species.)—Also **Font Quer, Pío.** Datos acerca de la flora orófila de Gredos. l.c. **25:** 265–270. 1925. (List of several rare species collected by Enrique Gros, with localities and critical notes.)

Vicioso Martínez, Carlos. Materiales para el estudio de la flora soriana. Anales Jard. Bot. Madrid **2:** 188–235. 1942.

List of spermatophytes (731 species) collected by author principally in the oak zones, with localities and some annotations. (Province of Soria.)—See also **Montserrat Recoder, Pedro.** Plantas de los alrededores de Soria. Collect. Bot. Barcinon. **2:** 261–271. 2 fig. 1949. (List with localities, including 46 plants not recorded by Vicioso.)

Zubía, Ildefonso. Flora de La Rioja. 2 v. (72, 213 p.). 19 pl. 21.5 cm. Logroño, 1921.

Geography, geology, climate, ecology (v. 1); annotated list of vascular and cellular plants (including 1679 spontaneous species of vascular plants) in the Province of Logroño and adjacent regions (v. 2). (Not seen; title and annotation supplied by A. Caballero and E. F. Galiano.)—See also **Caballero, Arturo.** Acotaciones a la "Flora de La Rioja" de D. Ildefonso Zubía. Cavanillesia **5:** 24–27. 1932. (Botanical explorations, etc.; list of corrections and additions.)—Also Cámara Niño, above.

CATALONIA (CATALUÑA)

The region of Catalonia includes the Provinces of Barcelona, Gerona, Lérida, and Tarragona.

General

Cadevall y Diars, Juan, Sallent y Gotés, Angel, and Font Quer, Pío. Flora de Catalunya. Enumeració y descripció de les plantes vasculars espontanies de l'antic principat fins avuy conegudes y de les més importants que s'hi cultiven. 6 v. illus. incl. 6 col. pl. 27.5 cm. Barcelona, 1915 (1913–15)–1936 (1937).

Annotated descriptive flora of vascular plants (3,177 species), with keys, vernacular names (Catalonian names in full, also some Castilian and French), localities, and figure of each species; sketch of botanical history and plant zones (v. 1), glossary (v. 6), bibliographical sketch of Cadevall, with bibliography (v. 4). Pío Font Quer was collaborator in v. 4–6, in place of Sallent y Gotés. Publication of l'Institut d'estudis catalans. Secció de ciències. Replaces **Colmeiro, Miguel.** Catálogo metódico de plantas observadas en Cataluña particularmente en las inmediaciones de Barcelona, inclusa una porcion de pirenáicas, con sus nombres botánicos mas usuales, los vulgares catalanes de muchas, y la indicacion de localidades y épocas en que florecen. Seguido de la nomenclatura catalana de las plantas traducida al idioma castellano y al botánico . . . lx, 176, 131 p. 20.5 cm. Madrid, 1846. (Botanical explorations with bibliography, physiography, climate (with included calendar of flowers), general features of flora, plant zones, etc.; annotated list of vascular plants; alphabetical list of Catalan vernacular names, with Spanish and botanical equivalents.) Also replaces **Costa y Cuxart, A. C.** Introduccion á la flora de Cataluña y catálogo razonado de las plantas observadas en esta region. Corregida y aumentada con un suplemento . . . y un apéndice al vocabulario. lxxiii, 356, 96 p. 18.5 cm. Barcelona, 1877. (1st ed. 1864.) (Special herbaria, physiography, plant zones, statistics of flora, etc., bibliography; annotated list of 2414 vascular plants, with vernacular names; alphabetical list of Catalan names, with Spanish and botanical equivalents. In Gray Herbarium library); also his Suplemento al Catálogo . . . 1 p. l., 37 p. 18.5 cm. Barcelona, 1877. (Brings total to 2,671 species. In Gray Herbarium library.) Also replaces **Vayreda y Vila, Estanislao.** Plantas notables por su utilidad ó rareza que crecen espontáneamente en Cataluña, ó sea apuntes para la flora catalana. Anales Soc. Españ. Hist. Nat. **8:** 345–462. pl. 7–10. 1879. (List of noteworthy vascular plants (Ranunculaceae–Boraginaceae), with localities and frequent annotations); also his Nuevos apuntes para la flora catalana. l. c. **11:** 41–151. pl. 2–4. 1882. (Notes on author's excursions, list of localities with altitudes; extensive list of more interesting vascular plants observed, with localities and frequent annotations); also his Plantas de Cataluña. l.c. **30** (II, 10): 491–582. pl. 8–10. 1902. (List of vascular plants (Ranunculaceae–Thymelaeaceae) with localities, collected by author); also his Plantas de Cataluña. Especias y variedades más notables que introducimos en este trabajo. Cavanillesia **4:** 58–62. 1931. (List of new varieties and formae (mostly nomina nuda) and new records, with localities, taken from the lost manuscript of the second part of his Plantas de Cataluña, and covering Thymelaeaceae to Pteridophyta, inclusive.)—See also **Sennen,** *frère.* Flora de Catalogne. Additions et commentaires. Treballs Inst. Catalana Hist. Nat. [**3**]: 55–266. pl. 4–7. 1917. (New localities, critical notes, etc., on 1 or more species each of 413 genera of vascular plants, based on the author's collections.)—Also **Lapraz, Guy.** Contribution à l'étude de la flore de Catalogne. Collect. Bot. Barcinon. **3:** 385–394. 1953; **4:** 41–52, 342–349. 1954. (Extensive list of noteworthy vascular plants, mainly from vicinity of Barcelona, Bagà, Serra del Cadí, and El Miracle (Salsona).)

Calicó, Josep. Apuntes de la flora medicinal de Catalunya. 206 p. 15 cm. Barcelona, 1921.

List of medicinal plants, with vernacular names, descriptions, uses, etc. (Not seen; title and annotation supplied by A. de Bolós.)

Conill, Léon. Botanique catalane pratique. 381 p. Perpignan, 1910.

"Ce volume comprend l'énumération par ordre alphabétique de 563 noms vulgaires de plantes représentant 832 dénominations en langue catalane et 106 en langue languedocien. Pour chaque d'entre elles, l'auteur indique l'origine de nom vulgaire, les noms scientifiques et français, l'habitat, la date de fleuraison et les propriétés." (Not seen; review in Monde des Plant. II, **13:** 4. 1911.)—See also Masclans y Girvès, below.

Font Quer, Pío. Los herbarios de Cataluña y su conservación. Mem. R. Acad. Cien. y Art. Barcelona III, **18:** 431–442. 1925. (Also has the paging v. 18, no. 18, p. 3–14.)

Account of the principal herbaria formed in Catalonia, with particular reference to their present condition.—See also **Bofill y Pichot, J. M.** Discurso de contestación por el académico numerario. l.c. 443–452 (also as p. 15–24). 1925. (Biographical account of various Catalonian botanists; bibliography of Font Quer.)

Masclans i Girvès, Francesc. Els noms vulgars de les plantes a les terres catalanes. Arx. Inst. Estud. Catalans Secc. Ciѐn. v. 23. 251 p. 6 fig. 1954.

Formation of Catalan names, bibliography; alphabetical list of vernacular names, with scientific equivalents, often the locality where used, and some chiefly etymological annotations; list of scientific names, with collected vernacular equivalents.—See also Conill, above.

Local

Barrera y Arenas, Jacinto. Catálogo de la flórula de Teyá y Masnou. Crón. Cient. **7:** 179–184, 200–202, 213–217, 226–230. 1884.

Topography, geology, general features of flora, botanical explorations; annotated list of over 400 vascular plants, with some vernacular names. (Province of Barcelona.)

Batalla, E., and **Masclans, F.** Catálogo de las plantas observadas en la cuenca del Gaià (Tarragona). Collect. Bot. Barcinon. **2:** 343–429. 5 fig. (incl. map). 1950.

Botanical explorations, topography, geology, climate; annotated list of vascular plants (946 species and subspecies).

Bolós, Antonio de. Anotacions a la flora olotina. I. Butllet. Inst. Catalana Hist. Nat. **34:** 131–137. 1934.

Mention of previous floras (all unpublished); list of rarer vascular plants, with localities, of the comarca of Olot, Province of Gerona.

——— Datos botánicos sobre los montes de Falgars (Berga). Collect. Bot. Barcinon. **3:** 325–344. 4 fig. 1953.

Topography; extensive list of noteworthy vascular plants collected in 1945, with localities. (Province of Barcelona.)

——— Vegetación de las comarcas barcelonesas. Descripción geobotánica y catálogo florístico, según estudios efectuados por el proprio autor y por

Oriol de Bolós y Capdevila. 579 p. illus., pls., maps, ports., tables. 27.5 cm. Barcelona, 1950. (At head of title page: Instituto español de estudios mediterráneos. Publicaciones sobre botánica.)

Botanical explorations, plant zones, description of various regions, plant associations, phytogeography; annotated list of vascular plants, with detailed localities; bibliography. (Covers 1,600 square kilometers in Province of Barcelona, centering about the city of Barcelona, between 41° 24′ and 41° 40′ N. Lat., 1°49′ and 2°26′ E. Long.) Replaces **Sennen,** *frère.* Catálogo del herbario barcelonés. xxiv, 71 p. 18 cm. Barcelona, 1918. (Unannotated list of vascular plants of vicinity of Barcelona (1,422 species). Not seen; title and annotation supplied by A. de Bolós.)

Cadevall y Diars, Juan. Flora del Vallés. Mem. R. Acad. Cien. y Art. Barcelona [3. épocal] **2:** 1–138. [1897?]

Topography, geology, climate, general features of flora, agriculture, etc.; briefly annotated list of vascular plants (1,860 species, including cultivated plants). (Province of Barcelona.)

——— Plantes notables dels voltants de Tarrassa. Butllet. Inst. Catalana Hist. Nat. **2:** 81–85, 97–100, 121–123. 1902; **3:** 42–43, 49–51. 1903.

List of noteworthy angiosperms of the vicinity of Tarrassa, with localities and some critical notes. (In Vallés region, Province of Barcelona.)—See also his Una excursió botánica al Ubach (voltants de Tarrassa). Butllet. Inst. Catalana Hist. Nat. [4] (2. época) **1:** 42–44. 1904. (List of vascular plants collected on 7 April 1903.)—Also **Solà, J. S. P.** Notes sobre l'Ubach. Arx. Estud. Centre Excurs. Terrassa 29. 16 p. 1924. (General features of flora; lists of plants, with vernacular names and uses. Not seen; title and annotation supplied by A. de Bolós.)

Centellas, Juan. Física y flórula del término de Tarragona. 137 p. 5 phot., map. 1953.

Geology, climate, etc.; annotated list of plants, with vernacular names. Apparently contains many errors; see review in Collect. Bot. Barcinon. **4** (2): viii. 1954. (Not seen.) (Province of Tarragona.)

Codina, Joaquim. Apuntes para la flora de La Sellera y su comarca. 100 p. 24 cm. Girona, 1908.

List of 835 vascular plants with habitats, vernacular names, and medicinal uses. (Not seen; title and annotation supplied by A. de Bolós.) (In southwestern Gerona.)

Coste, Hippolyte, and **Soulié, J. A.** Florule du Val d'Aran ou catalogue des plantes qui croissent spontanément dans le bassin supérieur de la Garonne, depuis ses sources jusqu'à son confluent avec la Pique. Bul. Géog. Bot. **23:** 91–136. 1913; **24:** 5–47. 1914.

Topography, geography, plant zones, local species, botanical explorations; annotated list of vascular plants (1431 species), with localities. (Province of Lérida, also Haute-Garonne (France).) Followed by **Sudre, Henri.** Les Rubus et les Hieracium récoltés dans la vallée d'Aran. Bul. Géog. Bot. **24:** 47–56. 1914.—See also **Coste, H.** Supplément à la florule du Val d'Aran. Monde des Plant. III, **23** (21–136): 7; (22–137): 6–8; (23–138): 6–8. 1922. (Extensive list of additions. In Arnold Arboretum library.)—Also Llenas y Fernández, below.

Cuní y Martorell, Miguel. Excursion entomológica y botánica á la Cerdaña española (Cataluña). Anales Soc. Españ. Hist. Nat. **10:** 367–389. 1881.

Includes (p. 380–385) mostly unannotated list of vascular plants observed in region of Puigcerdá, Province of Gerona.

——— Excursion entomológica y botánica á San Miguel del Fay, Arbucias y cumbres del Montseny. Anales Soc. Españ. Hist. Nat. **9:** 205–242. 1880.

Includes partly annotated lists of vascular plants from San Miguel del Fay, in Province of Barcelona (p. 211–214), Arbucias, in Province of Gerona (p. 226–231), and Montseny, on border of Gerona and Barcelona (p. 233–235). —See also **Cuixart, S., Suaña, J. M.,** and **Rubió, F. A.** Una visita al Montseny en ple hivern (11–14 de Febrer). Butllet. Inst. Catalana Hist. Nat. 18 (III, 1): 111–118. 3 fig. 1918. (Authors' itinerary, with mention of noteworthy plants observed.)—Also **Font Quer, Pío.** Noves troballes de plantas al Montseny. Butllet. Inst. Catalana Hist. Nat. **20** (3. epoca) **3:** 16–21. 1920. (Discussion of a few species.)

Cuní y Martorell, Miguel. Flora de Calella (Catalunya). Anuari Assoc. Excurs. Catalana 2 (1882): 557–566. 1883.

Mostly unannotated list of vascular plants of the vicinity of Calella, with vernacular names. (Province of Barcelona.) (Not seen; title and annotation supplied by A. de Bolós; volume and page reference supplied by Frick Art Reference Library, New York City.)

——— Resultado de una exploracion entomológica y botánica por el término de la Gerriga (Cataluña). Anales Soc. Españ. Hist. Nat. **12:** 83–101. 1883.

Includes (p. 93–99) mostly unannotated list of vascular plants (and 2 cellular plants). (In El Vallés region, Province of Barcelona.)

Font Quer, Pío. Ensayo fitotopográfico de Bages. lxvi, 159 p. 22 cm. Mahón, [1914]. (Tesis.)

Botanical explorations, geography, topography, geology, climate, general features of vegetation; annotated list of native vascular plants (1,105 species), with localities and list of doubtful species under each family; list of naturalized species. Covers an area of 1,150 square kilometers in the middle valley of the Llobregat and the lower valley of the Cardoner, including Montserrat, Vallés, and other localities. (Province of Barcelona.) Replaces **Sallent, Angel.** Flora del Pla de Bages. 80 p. 1904. (List of 842 species. Not seen; data in part supplied by E. F. Galiano.)—See also **Font Quer, P.** Adicions a la flora de Bages. Butllet. Inst. Catalana Hist. Nat. **17** (2. época, **14**): 51. 1917. (Mentions 3 species.)

——— Los estudios botánicos en la provincia de Lérida. Ilerda **1:** 213–285. pl. 15–20 (ports.). 1943. (Harvard College library.)

Chronological account of botanists and botanical explorations, with references to publications; indices of localities, authors, and species mentioned.

——— Una excursió botánica a la Catalunya transibérica. Treballs Inst. Catalana Hist. Nat. [1]: 9–35. pl. 1–4. 1915.

Author's itinerary, general features of flora; list of 195 vascular plants, with localities and some critical notes. This and the following papers refer mostly or entirely to miscellaneous localities in Province of Tarragona.—See also **Gibert Oliver, A. M.** Notes fitogeográfiques de qualques encontrades de la Catalunya occidental. Butllet. Agrupació Excurs. Reus **3:** 5–21. 1917. (Includes annotated list of 280 vascular plants from western (mostly southwestern) Catalonia. Not seen; title and annotation supplied by A. de Bolós.) Also **Sennen,** *frère.* Une semaine d'herborisation dans les garrigues de Tarragona (du 3 au 11 avril 1917). Butllet. Inst. Catalana Hist. Nat. **17** (2. época, **14**): 62–70. 1917. (Mention of noteworthy species of various localities.)—Also **Font Quer, P.** Contribució al coneixement de la flora catalana occidental. Trab. Mus. Cien. Nat. Barcelona 5 (Ser. Bot. no. 3): 193–233. 2 pl. 1920. (Localities; list of 228 vascular plants, with localities and some annotations.)—Also **Bolós, Antonio de.** De les notes botàniques de D. Ramon de Bolós i Saderra. Butllet. Inst. Catalana Hist. Nat. **21:** 131–133. 1921. (Unannotated short lists of noteworthy phanerogams from different localities in southern Catalonia.)—Also **Sennen,** *frère.* Une seconde semaine d'herborisation sur le littoral de Tarragone entre le Francoli et l'Ebre. Annales Soc. Linn. Lyon n.s., **69:** 102–116. 1923; **70:** 63–70 [bis]. 1924; **72:** 9–16. 1926; **73:** 9–14. 1928. (Annotated lists of noteworthy plants, arranged by localities; list of newly described species and of species new or rare in Catalonia.)

——— Flórula de Cardó. Consideraciones generales sobre las plantas que se crian en su valle en relación con la temperatura, sequedad y pobreza polínica del aire, seguidas del catálogo de dicha flórula. 165 p. illus., 10 pl. on 6, map. 23.5 cm. Barcelona, 1950. (Library of J. Cuatrecasas.)

Botanical explorations, topography, climate, general features of flora, plant associations, discussion of plants and animals in relation to humidity, list of wind-pollinated plants with discussion; annotated list of vascular plants (587 species), with vernacular names and localities. (Province of Tarragona, in vicinity of Tortosa.)

——— Flórula de los valles de Bohí. Ilerda añn 3 (whole no. 9): 7–94. 2 fig. (maps), pl. 1947 [1948?]. (Reprinted, 94 p. 24.5 cm. Lérida, 1948. Original and reprint in library of J. Cuatrecasas.)

Botanical explorations, topography, general features of flora, annotated list of noteworthy vernacular names; annotated list of vascular plants (677 species). (Province of Lérida.)

——— Plantes de Vallferrera. Butllet. Inst. Catalana Hist. Nat. **15:** 51–61. 1915.
Author's itinerary, general features of flora; list of vascular plants collected in 1911 and 1912, with localities. (Province of Lérida.)

Gibert Oliver, A. M. Catàlech de la flora de la ciutat de Tarragona i son terme. 94 p. 21 cm. Tarragona, 1892.
Physiography, climate, general features of flora; annotated list of vascular plants (712 species), with vernacular names. (Not seen; title and annotation supplied by A. de Bolós.) (Province of Tarragona.)—See also **Sennen,** *frère*. Quatre jours d'herborisation aux environs de Tarragona (les 11–12–13 et 14 juin). Bol. Soc. Aragon. Cien. Nat. **8:** 168–175. 1909. (Extensive list of noteworthy angiosperms from vicinity of city of Tarragon, with localities.) —Also Nogués Ferré, below.

——— Flora o catálogo de plantas del término de Vilaseca de Solcina. 54 p. 22 cm. (date?)
Annotated list of vascular (and cellular) plants with vernacular names. (Not seen; title and annotation supplied by A. de Bolós.) (Province of Tarragona.)

Llenas y Fernández, Manuel. Contribución al estudio de la flora del Pirineo Central (Valle de Arán). Mem. Inst. Catalana Hist. Nat. 1. 152 p. 1912. (University of Illinois library.)
Botanical explorations, climate, plant zones; annotated list of vascular plants (945 species). The work is not referred to directly by Coste and Soulié (see above), although a few of Llenas' plants are cited at the close of their list. (Province of Lérida.)

Llensa de Gelcén, S. Inventario razonado de la flora de Hostalrich y su comarca. Anales Escuela Peritos Agr. y Superior Agr. (Barcelona) **5:** 119–290. 13 fig. (incl. map). 1945 [1947].
Topography, soils, general features of flora, botanical explorations; annotated list of vascular plants (1,071 species, of which 959 are native or spontaneous and 112 cultivated). (Province of Gerona.)

Losa España, Mariano. Una excursión botánica al cabo de Creus (Gerona). Comun. Reun. Internac. Soc. Farm. Mediter. Lat. II (?p.). 1956.
General features of vegetation and list of species observed (356). (Not seen; cited from Collect. Bot. Barcinon. **5:** 291. 1956.)

Malga, M. A. Herborisacions per los voltants de Vilafranca del Panadés durant els anys 1902 y 1903. Butllet. Inst. Catalana Hist. Nat. [4] (2. época, **1**): 66–69, 87–93. 1904.
Annotated list of vascular plants. (Province of Barcelona.)

Marcet, A. F. Una excursió al Miracle. Butllet. Inst. Catalana Hist. Nat. [7] (2. época, **4**): 23–32. 1907.
Topography; extensive list of spermatophytes collected. (Along the Río Cardoner in western Barcelona and adjacent Lérida.)

——— Flora montserratina. Bol. Soc. Españ. Hist. Nat. **46:** 332–410, 525–581. illus. 1948–49; **47:** 67–89, 163–170, 307–333, 469–484, 625–651. illus. 1949–50; **48** (Secc. Biol.): 87–108, 327–338. 3 fig. 1950; **49** (Secc. Biol.): 201–264. illus. 1951; **50** (Secc. Biol.): 299–379. illus. 1952; **51** (Secc. Biol.): 51–109. 1954.
Geology, topography, climate, geography, plant zones, botanical explorations, bibliography; annotated flora of vascular plants, with keys, short descriptions, and under each family or genus a list of recorded but doubtful species. (Province of Barcelona.)

Masferrer y Arquimbau, Ramón. Plantas espontáneas en los alrededores de Caldetas. Crón. Cient. **7:** 1–2, 17–19. 1884.
Unannotated list of 182 vascular plants collected by Joaquín Salarich. (Province of Barcelona.)

——— Recuerdos botánicos de Vich, ó sea, apuntes para el estudio de la flora de la comarca de Cataluña llamada Plana de Vich. Anales Soc. Españ. Hist. Nat. **6:** 211–248, 359–398. 1877.

Geography, topography, geology, climate, general features of flora, phenology; barely annotated list of 1020 vascular plants, also a few cellular plants. (Province of Barcelona.)—See also **Sennen,** *frère.* Deux jours d'herborisation dans la plaine de Vich. I. De Manlleu à Ntra. Sra. de la Salut en auto-camion. Butllet. Inst. Catalana Hist. Nat. **20** (3. época, **3**): 191–193. 1920.—II. Aux alentours de Gallissans. l.c. 194–196. 1920. (Mention of some noteworthy species observed.)

Montserrat, Pedro. Flora de la cordillera litoral catalana (porción comprendida entre los ríos Besos y Tordera). Collect. Bot. Barcinon. **4:** 351–398. 1 fig. map. 1955; **5:** 1–86, 297–351. 1957–58.

Botanical explorations, topography, geology, soils, climate, general features of vegetation, principal cultivated plants and forest products; annotated list of vascular plants. Reaches species 736 (Verbena). (Prov. Barcelona.)

Nogués Ferré, Antonio. Apuntes para la flora tarraconense. Bol. Soc. Ibérica Cien. Nat. **22:** 177–218. 1923 (1924).—Nota adicional y fe de erratas . . . l.c. **23:** 51–54. 1924.

Briefly annotated list of 713 vascular plants observed by author in vicinity of city of Tarragon. (Province of Tarragona.)—See also Gibert Oliver, Catàlech de la flora de la ciutat de Tarragona, above.

Pau, Carlos. Plantes observées dans l'Ampourdan (surtout aux environs de Figueras) pendant l'année 1905 par le frère Sennen. Bol. Soc. Aragon. Cien. Nat. **4:** 303–333. 1905.

List of vascular plants, with localities. (Province of Gerona.)—See also **Sennen,** *frère.* 2. note sur la flore des environs de Figueras. Act. y Mem. Primer Cong. Nat. Españ. Zaragoza 1908, secc. 4, p. 274–307. 24 cm. Zaragoza, 1909. (List of vascular plants collected in Catalogne, with localities; author's itinerary with mention of noteworthy species. In library of Muséum national d'histoire naturelle, Paris.)—Also **Queralt, Ramón,** and **Pascual, Ll.** Plantes recollides durant el curs 1915–16 als entorns de Fortianell. Butllet. Inst. Catalana Hist. Nat. **17** (2. época, **14**): 90–96. 1917. (List of noteworthy vascular plants, with localities.)—Also **Esteve Chueca, Fernando.** Reseña de una excursión botánica al Alto Ampurdán: vegetación de la Sierra de Roda y Plana de Castelló (Prov. de Gerona). Anales Inst. Bot. A. J. Cavanilles **14:** 555–596. 1956. (Plant formations, etc.; list of noteworthy species; bibliography.)—Also Vayreda y Vila, Excursió botánica al Baix Ampurdà, below.

Peyre, Armond, Jeanbernat, Ernest, and **Timbal-Lagrave, Édouard.** Botanique. Catalogue des plantes recueillies dans cette excursion. *In* Jeanbernat, Ernest, and others. Une excursion scientifique aux sources de la Garonne et de la Noguéra Pallaresa (Catalogne). Bul. Soc. Sci. Phys. et Nat. Toulouse **1:** 66–85. 1872–73. (Harvard College library.)

List of 491 vascular plants, with localities. (Province of Lérida.)

Rubió i Tudurí, F. A. Notes per a un estudi de la flora del Montsant. Butllet. Inst. Catalana Hist. Nat. **19** (3. época, **2**): 52–54. 1919.

List of vascular plants [144 species], with localities. (Province of Tarragona.)

Salvaña, J. M. Recuerdos botánicos de Igualada y flora aqualatense póstuma de D. José Bausili y Salamanca. Mem. R. Acad. Cien. y Art. Barcelona II, **2:** 59–88. 1885.

Botanical explorations, etc., of Bausili y Salamanca; partly annotated list of wild and cultivated vascular plants (and 3 cellular), with vernacular names, from his manuscript; annotated list of additional spermatophytes collected by Salvaña. (Province of Barcelona.)—See also **Queralt, Ramón.** Plantas de los alrededores de Igualada . . . Bol. Soc. Aragon. Cien. Nat. **15:** 75–77, 91–93. 1916. (Unannotated list of [212] vascular plants.)

Sennen, *frere.* Excursion à la Seo d'Urgel, les 29, 30, 31 août 1927. Bul. Soc. Bot. France **75:** 434–450. 1928.

Climate, etc.; unannotated list of vascular plants collected by author. (Province of Lérida.)

——— Flora de la vall de Núria i dels seus encontorns. *In* Puig i Bosch, Josep. La vall de Núria. p. 51–80. Barcelona, 1929.

Annotated list, with localities and vernacular names. (Not seen; title and annotation supplied by A. de Bolós.) (Province of Gerona.)—See also Vayreda y Vila, Catálech de la flora de la Vall de Nuria, below.

——— Mes vacances de 1915 en Cerdagne (juillet à octobre). Bul. Soc. Bot. France 63: 108–136. 1917.
Author's itinerary, with notes on plants observed; list of rare or newly recorded vascular plants, with localities. (Provinces of Gerona and Lérida, also Pyrénées-Orientales (France).)—See also his Nos découvertes en Cerdagne. l.c. 73: 641–680. 1927. (List of new species and new records, with localities, based primarily on the author's collections distributed in his exsiccatae "Plantes d'Espagne.")—Also his Nombreuses localités de plantes nouvelles pour la Cerdagne, observées durant les vacances des années 1915 à 1926. l.c. 74: 355–410. 1927. (Extensive list of vascular plants, with localities, intended to include plants previously only vaguely or not at all reported from Cerdagne. Refers to the valley of the Segre River, in Province of Lérida.)—Also his Indice de distribution ou aires de dispersion de quelques plantes de la flore de la Cerdagne. l.c. 75: 472–476. 1928. (Notes on distribution of a few species.)

——— Quelques espèces adventices, subspontanées ou cultivées en Espagne et dans le domaine mediterranéen. Cavanillesia 2: 10–42. 1929.
Extensive list of adventive or subspontaneous (or in some cases cultivated) plants, mostly observed by the author in Catalonia, with localities and annotations.

——— Quelques formes nouvelles ou peu connues de la flore de Catalogne, Aragon, Valence. Bol. Soc. Aragon. Cien. Nat. 11: 177–215, 229–251. 1912.
Extensive list of new or noteworthy plants, mostly from Catalonia, distributed in his "Plantes d'Espagne," with localities, critical notes, etc.

——— Une vingtaine de "plantes nouvelles" pour la Catalogne. Plantes non encore signalées aux environs de Tortose. Bol. Soc. Aragon. Cien. Nat. 8: 143–150. 1909.
List of 27 vascular plants new to Catalonia (from vicinity of Tortosa), with localities; extensive list of noteworthy vascular plants of vicinity of Tortosa. (Province of Tarragona.)—See also **Font Quer, Pío.** Vuit dies d'excursió botànica al Ports de Tortosa. Anuari Junta Cièn. Nat. Barcelona 2: 619–632. 1917. (Author's itinerary, with mention of noteworthy plants observed.)

Trémols y Borrell, Federico. Contribución a la flora cataláunica. Catálogo de las plantas observadas en la montaña de Requesens, provincia de Gerona. Mem. R. Acad. Cien. y Art. Barcelona [3. epoca,] 2: 163–176. [1897.]
Topography, geology, general features of flora, climate; unannotated list of vascular plants.

Vayreda y Vila, Estanislao. Catàlech de la flora de la vall de Nuria. xii, 95 p. 21.5 cm. Barcelona, 1882. (Arnold Arboretum library.)
Botanical districts, bibliography; annotated list of vascular and some cellular plants, with vernacular names and notes on uses. At head of title page: Associació d'excursions catalana. (Province of Gerona.)—See also **Cadevall, Juan.** Una excursió á Nuria. Butllet. Inst. Catalana Hist. Nat. [4] (2. época, 1): 93–96, 100–105. 1904. (Includes mention of various noteworthy species.)—Also Sennen, Flora de la vall de Núria, above.

——— Catàleg de la flòrula de "La Mare de Déu del Mont." Treballs Inst. Catalana Hist. Nat. [5]: 359–442. 1920.
General features of flora, plant regions; annotated list of vascular plants. (Province of Gerona.)

——— Excursió botánica al Baix Ampurdà. Anuari Assoc. Excurs. Catalana 1 (1881): 457–467. 1882.
Botanical explorations; lists of vascular plants, with some vernacular names. (Province of Gerona.) (Not seen; title and annotation supplied by A. de Bolós; volume and page reference supplied by Frick Art Reference Library, New York City.)—See also Pau, above.

ESTREMADURA

The region of Estremadura includes the Provinces of Badajoz and Cáceres.

Local

[Caballero, Arturo]. Catálogo de plantas recogidas por D. Arturo Caballero en Guadalupe (Cáceres), 1948–1949. Anales Inst. Bot. A. J. Cavanilles (Anales Jard. Bot. Madrid) 10 (1): 25–73. 1951.

Extensive list of vascular plants, with localities. Prefatory note by Elena Paunero.

Rivas Goday, Salvador, and **Borja Carbonell, José.** Plantas de los riberos del Tajo, en Alconetar (Provincia de Cáceres). Anales Jard. Bot. Madrid **8:** 443–468. 1948.

Botanical explorations; discussion of plant communities, with mention of numerous species; notes on several species; bibliography.

Rivas Mateos, Marcelo. Una excursión á la Sierra de Béjar (provincias de Cáceres, Salamanca y Ávila). Actas Soc. Españ. Hist. Nat. **26** (II, **6**): 204–210. 1897.

Itinerary with mention of various plants and insects observed.—See also **Pau, Carlos.** Ligeras indicaciones sobre un viage botánico. Actas Soc. Españ. Hist. Nat. **29** (II, **9**): 282–288. 1900. (Report of an excursion with mention of numerous species.)—Also **Antigüedad Diez, Felix.** Primera serie incompleta de plantas de tierra de Béjar. 86 p. 20 cm. Béjar, 1932. (List of 300 species (265 phanerogams) collected by author in Sierra de Béjar, Provinces of Salamanca (León) and Cáceres (Estremadura), with vernacular names and uses. Not seen; title and annotation supplied by E. F. Galiano.) Also **Rivas Goday, Salvador.** Algunas plantas críticas de Sierra de Béjar. Bol. R. Soc. Españ. Hist. Nat. **40:** 313–317. pl. 21–22. 1942. (Critical notes on several species.)

——— Flora de la provincia de Cáceres. 307 p. port. 21.5 cm. Serradilla, 1931.

Geography, topography, geology, climate, plant formations; annotated list of cellular and vascular plants, with localities and vernacular names; plant regions, etc.; no index.—See also **Caballero, Arturo.** Dos excursiones botánicas a los alrededores de La Alberca (Salamanca, Cáceres). Anales Jard. Bot. Madrid 7: 645–653 incl. 1 pl. 1947 (1948). (Includes short list of noteworthy plants collected by author.)

GALICIA

The region of Galicia includes the Provinces of La Coruña, Lugo, Orense, and Pontevedra.

General

Merino, Baltasar. Flora descriptiva é ilustrada de Galicia. 3 v. illus. 22.5 cm. Santiago, 1905–09.

Botanical explorations, key to families; annotated descriptive flora of vascular plants (1,854 species including supplement in v. 3), with keys, localities, and vernacular names; geology, plant zones, climate, etc.—See also his Adiciones a la flora de Galicia. 111 i.e. 211 p. 9 pl. Braga, 1917. (Reprinted from Brotéria Sér. Bot. v. 10–14. 1912–16, with index added.)—Also **Ruiz de Azua, A.** Helechos de Galicia. ([1.]–3 serie.) Mem. R. Soc. Españ. Hist. Nat. **15:** 693–697. 2 fig. 1929; Bol. R. Soc. Españ. Hist. Nat. **30:** 129–134. 3 fig. 1930; **31:** 629–631. 1931. (Lists of noteworthy ferns (mostly varieties and forms) with localities.)—Also **Bellot Rodríguez, Francisco.** Contribución a la flora gallega. Adiciones y correcciones. Farmacognosia **4** (whole no. 7): 77–82. 2 fig., 3 pl. 1945. (List of 24 phanerogams.)—Also **Vieitez, Ernesto.** Notas para la flora gallega. Anales Edafol. y Fisiol. Veg. **9:** 431–439. 5 pl. (on 3). 1950. (List of noteworthy species; bibliography.)—Also **Bellot Rodríguez, F.** Mantissa stirpium Gallaeciae. Trab. Jard. Bot. Facult. Farm. Univ. Santiago de Compostela 1. [19] p. 4 fig., port. 1950. (List of noteworthy cellular and vascular plants.)—Also his Adiciones a la flora gallega. Anales Inst. Bot. A. J. Cavanilles (Anales Jard. Bot. Madrid) **10** (1): 383–387. 1951. (List with localities.)—Also his Sinopsis de la vegetación de Galicia. l.c. **10** (1): 389–444 incl. 20 fig. i.e. 20 pl. (on 10). 1951. (Treatment of vegetable communities on the method of Brockman-Jerosch and Rübel; key to plant formations, followed by list of same with mention of principal species; bibliography.)—Also **Laínz, Manuel.** Adiciones inéditas del P. Baltasar Merino a su "Flora de Galicia." Brotéria Sér. Trimest. Ciên. Nat. **22:** 155–168. 1953. (Refers to monocotyledons and pteridophytes.)—Also his Aportaciones al conocimiento de la flora gallega. [I]–II. l.c. **24:** 108–143, 153–160. 1955; Anales Inst. Bot. A. J. Cavanilles **14:** 529–554. 1956 (1957). (Extensive lists.)—Also **Laínz Ribalaygua, J. M.,** and **Laínz, M.** Aportaciones . . . III. Brotéria Sér. Trimest. Ciên. Nat. **26:** 90–97. 1957.

Planellas Giralt, José. Ensayo de una flora fanerogámica gallega ampliada con indicaciones acerca los usos médicos de las especies que se describen. 452 p. 22.5 cm. Santiago, 1852. (Arnold Arboretum library.)

Replaced as a flora by Merino's Flora descriptiva; cited here for its notes on medicinal uses of the plants described.

Local

Areses, Rafael. Nuestros parques y jardines. Contribución al conocimiento de las plantas exóticas cultivadas en España. [Pt. 1.] Galicia. v. 1. Pontevedra. 699 p. 149 fig. 24.5 cm. Madrid, 1953. (At head of title page: Escuela especial de ingenieros de montes.)

Account of 500 plants cultivated in Province of Pontevedra, alphabetically arranged by vernacular names, with vernacular names in various languages, native country, localities where grown, references, notes on cultivation, etc.

Buch, Hans. Über die flora und vegetation Nordwest-Spaniens. Comment. Biol. Soc. Scient. Fenn. v. 10, no. 17. 98 p. 6 fig. (fig. 1–2, maps), 2 pl. (on 1). 1951.

Topography, geology, soils, climate, phytogeography, general features of vegetation; list of known vascular plants (about 1,000 species) of the Province of Pontevedra, plus some from the Provinces of La Coruña, Lugo, and Orense (and District of Viana do Castelo, Portugal), with designation of province and often the exact localities; notes on the vegetation and flora of noncoastal northern Galicia and the mountains of Asturias; bibliography.

Carreira Alvárez, Ezequiel. Contribución al estudio de la flora gallega. Plantas herborizadas en el Valle de Burón (Lugo). Anales Inst. Bot. A. J. Cavanilles **13:** 499–532 incl. 10 fig. 1955.

Geology, topography, general features of vegetation; annotated list of vascular plants, with some vernacular names. (Province of Lugo.)

Losa España, Mariano. Datos para el estudio de la flora gallega. Plantas de las Islas Cíes. Anales Jard. Bot. Madrid **4:** 357–401 incl. 6 pl. (incl. map, pl. 1). 1944.

Topography, geology, general features of flora; annotated alphabetical list of known vascular plants; short list of fungi. (Three islets in Bay of Vigo, in Province of Pontevedra.)

Merino, Baltasar. Contribución á la flora de Galicia. La vegetación espontánea y la temperatura en la cuenca del Miño. xxviii, 320 p. 3 pl., 2 maps. 21.5 cm. Tuy, 1897. (Arnold Arboretum library.)—Suplemento I. 51 p. 5 pl. 22 cm. Tuy, 1898. (Arnold Arboretum library.)—Suplemento II–III. Anales Soc. Españ. Hist. Nat. **28:** 5–30. pl. 1–3. 1899; **30:** 167–199. 1901.

Brief sketch of botanical work in Spain with especial reference to Galicia, by Victor López Seoane (p. vii–xvii), with title Bosquejo histórico de la botánica española (reprinted 15 p. Tuy, 1897; in British Museum (Natural History) library); keys to families; annotated list of vascular plants, arranged by months of flowering; systematic list (898 species, including supplements); discussion of useful plants, arranged by families; list of medicinal plants, grouped by parts used, with commercial value. Refers principally to a stretch of about 15 kilometers, from the mouth of the Miño up to Guyán. (Province of Pontevedra.)

LEÓN

See also Asturias (Barbey-Gampert, Esquisse de la flore des Picos de Europa; Guinea). The region of León includes the Provinces of León, Palencia, Salamanca, and Zamora.

Local

Borja Carbonell, José. Datos para la flora cantábrica. Plantas de Nocedo (León). Anales Inst. Bot. A. J. Cavanilles (Anales Jard. Bot. Madrid) **11** (1): 419–436 incl. 1 fig. 1953.

General features of flora; annotated list of angiosperms (about 300 species) collected in July 1951. (Province of León.)

Casaseca Mena, Bartolomé. Plantas de Zamora (primera lista). Trab. Jard. Bot. Facult. Farm. Univ. Santiago de Compostela **5**: 15–24. 1952.

Extensive list of angiosperms, mostly from vicinity of Corrales, with indication of association to which they belong. (Province of Zamora.)

Hierro, Fibicio. Herborizaciones efectuadas en el partido de Carrión de los Condes (Palencia). Datos para la flórula de Castilla la Vieja. Anales Soc. Españ. Hist. Nat. **30** (II, **10**): 237–252. 1901.

Partly annotated list of vascular plants collected by author. (Province of Palencia.)—See also **Laínz, Manuel.** Datos floristicos sobre la cuenca media del río Carrión. Collect. Bot. Barcinon. **3**: 85–94. 1951. (Topography, climate, with mention of various noteworthy species.)

Laínz, Manuel. Contribución al catálogo de la flora salamantina. Anales Inst. Bot. A. J. Cavanilles **13**: 469–498 incl. pl. 1–2. 1955.

Botanical explorations; extensive list of noteworthy species. (Province of Salamanca.)

Losa España, Mariano. Catálogo de las plantas que se encuentran en los montes palentino-leoneses. Anales Inst. Bot. A. J. Cavanilles **15**: 244–376. 1957 (1958).

Annotated list of vascular plants (1069 species) collected by author and P. Montserrat in the montane region of Palencia and León during 1949–1953.

——— Contribución al estudio de la flora y vegetación de la provincia de Zamora. Plantas de Sierra Segundera, La Puebla de Sanabria, Calabor, etc. 165 p. illus. (incl. map). 24 cm. Barcelona, 1949. (At head of title page: Consejo de investigaciones científicas. Instituto de A. J. de Cavanilles, Sección de Barcelona.)

Botanical explorations, topography, geology, climate, general features of flora, plant formations and associations; annotated list of cellular and vascular plants collected, with frequent critical notes; list of noteworthy vernacular names, account of principal medicinal species, bibliography.—See also his Contribución al estudio de la flora en la provincia de Zamora. (Addenda et corrigenda.) Anales Jard. Bot. Madrid **9**: 485–502 incl. 4 fig. (i.e., 4 pl. on 2). 1950. (Additions and corrections.)

——— Plantas de Alar del Rey, Cervera del Pisuerga y otros puntos de la provincia de Palencia. Bol. Univ. Santiago de Compostela **10** (whole no. **34**): 3–38. 1941.

General features of flora; list of about 600 vascular plants with localities and some annotations, alphabetically arranged by scientific names; bibliography.

——— Plantas de los alrededores de Riaño (León). Anales Jard. Bot. Madrid **2**: 172–187. 1942.

Running account of vegetation with lists of species observed at different localities (not in systematic order); list of notable species observed, with localities and some annotations.—See also **Losa España, M.,** and **Montserrat, Pedro.** Nueva aportación al estudio de la flora de los montes cántabroleoneses. Anales Inst. Bot. A. J. Cavanilles (Anales Jard. Bot. Madrid) **11** (2): 385–462. illus. (incl. maps). 1953. (General features of flora of mountains in vicinity of Riaño, with numerous short lists of species from various localities; annotated list of noteworthy species; bibliography.)

——— and **Montserrat, Pedro.** Aportación al estudio de la flora de los montes cantábricos, Anales Inst. Bot. A. J. Cavanilles (Anales Jard. Bot. Madrid) **10** (2): 413–509 incl. 9 pl. (i.e. 9 fig.). 1952.

Botanical explorations, topography, general features of vegetation; annotated list of noteworthy species collected; bibliography. (Northern part of Province of Palencia, with base at Cervera de Pisuerga, and adjacent Santander.)—See also **Leroy, Eduardo,** and **Laínz, Manuel.** Contribución al catálogo de la flora palentina. Collect. Bot. Barcinon. **4**: 81–123. 4 fig. 1954. (Bibliography; annotated list of noteworthy vascular plants from northern part of Palencia, mostly from Peña Labra.)

Macho Tomè, Aquilino. Reseña de los productos naturales y más especialmente de las plantas medicinales espontáneas en el partido judicial de Saldaña. 141 p. 28 cm. Valladolid, 1893.

Includes (p. 45–126) account of medicinal plants, with vernacular names, habitat, properties, uses, etc. (277 species, of which 276 are vascular). The author's name is wrongly spelled Macho Tonie in the Index-catalogue of the library of the Surgeon-General's office II, **10**: 21. 1905. (Province of Palencia.)

Rojo González, Bernardino. Relación de las especies vegetales que se encuentran en una comarca o partido farmacéutico que comprende el término municipal de Revenga de Campos, y los limitrofes de Población, Lomas, Villasirga, Arconada, Villovieco y Villarmentero, todos de la provincia de Palencia. 184 p. map. 21 cm. 1913.

Incomplete list of cryptogams and phanerogams, with vernacular names and uses. (Province of Palencia.) (Not seen; title and annotations supplied by E. F. Galiano.)

MURCIA

The region of Murcia includes the Provinces of Albacete and Murcia.

Local

Cuatrecasas, José. Excursion botánica a Alcaraz y Riópar. Trab. Mus. Cien. Nat. Barcelona v. 5 (Sér. Bot.), no. 7. 49 p. 1926.

Topography, etc.; list of 432 vascular plants collected by author in June and July, 1923, with localities and occasional critical notes. (Province of Albacete.)

Dantín Cereceda, Juan. Contribución al estudio del carácter de la flora fanerogámica de Albacete. Bol. R. Soc. Españ. Hist. Nat. **12**: 107–121. 1 fig. (map). 1912.

Topography, geology, climate, general features of flora; no full list of species. (Province of Albacete.)

Jiménez Munuera, F. de P. Las plantas de Cartagena. Mem. R. Soc. Españ. Hist. Nat. **2**: 63–118. 1903.

Topography, geology, climate, general features of flora, botanical explorations; annotated list of spermatophytes (about 600 species). (Province of Murcia.)—See also his Plantas de Cartagena. Adiciones y rectificaciones. Suplemento 1. Actas y Mem. Primer Cong. Nat. Españ. (Zaragoza 1908) secc. 4, p. 250–273. 1909. (Additional species and new localities. Title and annotation mostly from notice in Bot. Jahresber. Just **38** (2): 1452–1453. 1915.)—Also **Civit, Emilia.** Catàleg de l'herbari Jiménez Munuera de la Junta de ciències naturals. Anuari Junta Ciên. Nat. Barcelona **2**: 261–276. 1917. (List of 703 spermatophytes from vicinity of Cartagena, with localities.)—Also **Bas Peired, Carlos.** Notas sobre la flora de Cartagena. Collect. Bot. Barcinon. **2**: 25–39. 1948. (General features of flora; annotated list of vascular plants collected.)

Rouy, Georges. Excursions botaniques en Espagne. Orihuela, Murcia, Velez-Rubio, Hellin, Madrid, Irun (Mai 1881–Juin 1882). Revue Sci. Nat. (Montpellier et Paris) III, **2**: 228–256, 557–564. 1883; **3**: 58–81, 229–250. 1884. (Library of Museum of Comparative Zoology, Harvard University. Reprinted, 86 p. Montpellier, 1883; in New York Botanical Garden library.)

Author's itinerary, with list of species from each locality; systematic list of vascular plants collected, with localities; list of new records; critical notes on various species. (Murcia and Hellin are in Province of Murcia; Orihuela is in Valencia, Velez-Rubio in Andalucía, Madrid in Castilla la Nueva, Irun in Provincias Vascongadas.)

NAVARRA (NAVARRE)

See also Provincias Vascongadas (Allorge and Allorge, Plantes . . . principalement du Pays Basque; Gredilla y Guana, Apuntas para la corografía botánica vasco-navarra).

The region of Navarra includes only the Province of Navarra.

Local

Lacoizqueta, J. M. de. Catálogo de las plantas que espontáneamente crecen en el valle de Vertizarana. Anales Soc. Españ. Hist. Nat. **13**: 131–225. 1884; **14**: 185–238. 1885.

Geography, climate, geology, soils, general features of flora; annotated list of vascular plants (834 species) and cellular plants (463 species).

Ruiz Casaviella, Juan. Catálogo metódico de las plantas observadas como espontáneas en Navarra. Anales Soc. Españ. Hist. Nat. **9**: 5–52, 285–307, 371–399. 1880.

Annotated list of vascular plants and Characeae (780 species) collected by author mostly in Caparrosa and vicinity, also very short lists of cellular plants; annotated list of cultivated plants.

PROVINCIAS VASCONGADAS (BASQUE PROVINCES)

See also Murcia (Rouy).
The Provincias Vascongadas include the Provinces of Álava, Guipúzcoa, and Vizcaya.

General

Bouda, Karl, and **Baumgartl, D.** Nombres vascos de las plantas. Acta Salmanticensia Filos. y Let. v. 7, no. 3. 99 p. 1955. (Harvard College library.)

Discussion of sources, bibliography, etc.; running account of Basque patois names, with meaning and derivation, grouped by habit, habitat, properties, etc., the names given in part under the equivalent Spanish vernacular names, with almost no scientific names; indices.

Gredilla y Gauna, A. F. Apuntes para la corografía botánica vasconavarra. 131 p. 7 fig. 26 cm. Barcelona, 1913.

History of botany, physiography, general features of flora; list of 1,975 vascular plants, with localities and vernacular names; bibliography. (Covers Navarra and Provincias Vascongadas.) (Not seen; title and annotation supplied by A. de Bolós.)

Lacoizqueta, J. M. de. Diccionario de los nombres euskaros de las plantas, en correspondencia con los vulgares, castellanos y franceses y científicos latinos. 202 p. 21 cm. Pamplona, 1888.

Systematic list of 833 vascular and 36 cellular plants with Basque, French, and Spanish vernacular names, etymology of Basque names, and uses; index of Basque names. (Not seen; title and annotation supplied by A. de Bolós. Contains many errors, fide Bouda and Baumgartl.)

Local

Allorge, Valia, and **Allorge, Pierre.** Plantes rares ou intéressantes du NW. de l'Espagne, principalement du Pays basque. Bul. Soc. Bot. France **88**: 226–254. 1941.

Botanical explorations; list of noteworthy vascular plants collected by authors in 1926–34, with localities; bibliographical footnotes. (Chiefly from Guipúzcoa, Vizcaya, and Navarra; some also from Asturias, León, and Santander.)

Arízaga, Javier de. Itinerarios botánicos de Don Javier de Arízaga, publicados y anotados por D. A. Federico Gredilla y Gauna. 2 v. (475, 354 p.). 22 cm. Vitoria, 1914–15.

The second volume includes a florula (about 950 species) of Elciego, Navaridas, Leza, and Laguardia. (Prov. Álava.) (Not seen; title and annotation supplied by A. de Bolós and A. Caballero.)—See also **Losa, Mariano.** Algunas comentarios a las listas de plantas que D. Javier de Arizaga recogió en el término de Pipaón (Alava). Bol. R. Soc. Españ. Hist. Nat. **29**: 37–44. 1929. (Topography of Pipaón, account of Javier de Arízaga's work, and discussion of various rare plants cited by him on basis of author's own visit to that place.)—Also his Contribución al estudio de la flora de Álava. Noticia de algunas plantas que viven en la Sierra de Cantabria . . . Anales R. Acad. Farm. (Madrid) 2. época, año **6**: 278–333. 1940. (Topography, general features of vegetation, author's itinerary, etc.; list of vascular plants collected, with localities (mainly from Pipaón and Lagrán).)

Guinea López, Emilio. Vizcaya y su paisaje vegetal. (Geobotánica viscaina.) 4 p. l., 432 p. illus., pls. (some col.), port., maps, tables. 28 cm. Bilbao, 1949. (At head of title page: Junta de cultura de Vizcaya. Recursos del medio biológico.)

Author's autobiography, etc.; annotated list of known vascular plants of Province of Vizcaya with localities, general range, and vernacular names;

topography, climate, edaphic factors, soils, plant associations and botanical regions, discussion of locality names derived from Basque names of plants (by Nestor de Goicoechea), bibliography.—See also Castilla la Vieja (Dupont, under Guinea López).

VALENCIA

See also Murcia (Rouy).
The region of Valencia includes the Provinces of Alicante, Castellón de la Plana, and Valencia.

General

Pau, Carlos. Materiales para la flora valenciana según los datos que suministra Cavanilles en su obra Observaciones sobre el reino de Valencia en los años 1795 (el tomo I) y 1797 (el II). I (no more published). 47 p. 21.5 cm. Valencia, 1905. (Library of Mme. Valentin Allorge, Paris.)
List of vascular plants and 1 Chara (total 776 species), often with localities, and with some critical notes, based on the data in Cavanilles' work. Refers to old kingdom of Valencia, nearly the same as the modern region of that name.

Local

Borja Carbonell, José. Estudio fitográfico de la Sierra de Corbera (Valencia). Anales Jard. Bot. Madrid **9**: 361–483 incl. 5 pl. on 3 (with fig. 1–11), 5 fig. 1950.
Topography geology, botanical explorations; annotated list of vascular plants (961 i.e. 967 species), including some cultivated plants; statistics, plant communities, bibliography. (Province of Valencia.)

Clemente y Rubio, S. de Rojas. Plantas que viven espontáneamente en el término de Titáguas, pueblo de Valencia, enumeradas en forma de índice alfabético. Revista Prog. Cien. Exact., Fís. y Nat. Madrid **14**: 429–445, 484–508, 544–572. 1864. (University of Michigan library.)
List of vascular and cellular plants, with vernacular names and dates of flowering; list of cultivated plants, arranged alphabetically by vernacular names. (Province of Valencia.)

Martínez Martínez, Miguel. Aportaciones a la flora española. Plantas de Alicante. Mem. R. Soc. Españ. Hist. Nat. **14**: 405–480. 8 fig. (incl. maps), pl. 16–27. 1934.
Botanical explorations, author's itineraries with mention of species; list of vascular plants collected by author in different parts of the Province of Alicante, with localities.—See also **Cámara Niño, Fernando.** Alcoy como localidad botánica. Bol. Soc. Españ. Hist. Nat. **36**: 307–314. 1 fig. (map), pl. 40–41. 1936. (Soils, climate, general features of flora; annotated list of several noteworthy species.)—Also his Observaciones botánicas en Alcoy, Sierra Mariola y Sierra Aitana (Prov. de Alicante). l.c. **40**: 329–337. pl. 23. 1942. (General features of flora; list of 46 noteworthy angiosperms, with localities.)—Also Pau, Plantas de la Sierra de Aitana, below.

Pau, Carlos. Herborizaciones por Valldigna, Játiva y Sierra Mariola en los meses de Abril, Mayo y Junio de 1896. Anales Soc. Españ. Hist. Nat. **27**: 411–452. 1899.
Author's itinerary; annotated list of noteworthy vascular plants; unannotated list of common species, in part with localities. (Province of Valencia.)

——— Notas botánicas a la flora española. fasc. 1. 40 p. 8vo. Madrid, 1887.
Includes, in addition to descriptions of 41 new species (nearly all from the Provinces of Valencia and Teruel), a list of species not recorded by Willkomm and Lange from the region (old kingdom) of Valencia. (Not seen; cited from review in Bot. Centralbl. 35: 238. 1888.)

——— Plantas de la Sierra de El Toro (Junio 1903). Bol. Soc. Aragon. Cien. Nat. **2**: 279–289. 1903.
Author's itinerary; list of noteworthy spermatophytes collected, with localities and some annotations. (Provinces of Castellón (Valencia) and Teruel

(Aragón).)—See also his Misceláneas botánicas. I. Butllet. Inst. Catalana Hist. Nat. **20** (3. época, **3**: 197–203. 1920. (Mention of various species from northern part of the Sierra de El Toro.)

——— Plantas de "La Murta" (Alcira). Actas Soc. Españ. Hist. Nat. **23**: 171–175. 1895.
List of noteworthy vascular plants, with localities. (Province of Valencia.)

——— Plantas de la Sierra de Aitana (Alicante). Bol. Soc. Aragon. Cien. Nat. **3**: 279–293. 1903.
List of vascular plants collected by author, with localities, with an appendix describing new species from various parts of Spain.—See also Cámara Niño under Martínez Martínez, above.

——— Una visita á los montes de Ayora (19–23 junio 1907). Butllet. Inst. Catalana Hist. Nat. [7] (2. época, **4**): 85–93. 1907.
Mention of numerous noteworthy angiosperms with localities. (Province of Valencia.)

Rouy, Georges. Denia. *In* his Excursions botaniques en Espagne (mai-juin 1883). Bul. Soc. Bot. France **31**: 269–276. 1884.
List of vascular plants collected by author in 1883, with localities. (Province of Alicante.)

——— Excursions botaniques en Espagne, centre d'herborisations à Jativa (mai 1899 [1879] et juin 1880). Bul. Soc. Bot. France **28**: 153–171. 1881; **29**: 40–47, 108–114, 120–127. 1882.
Author's explorations, with mention of species observed at each locality; list of vascular plants collected, with localities; account of new species, new records, etc. (Province of Valencia.)

Sennen, *frère.* Note sur la flore de Benicarló, Peñíscola, Sta. Magdalena, etc. de la province de Castellón de la Plana. Bol. Soc. Aragon. Cien. Nat. **10**: 131–143, 162–180. 1911.
Topography, botanical explorations; annotated list of vascular plants collected mostly by author.

Vicioso, Carlos. Una visita á los montes de Bicorp (Valencia). Bol. Soc. Españ. Hist. Nat. **15**: 90–94. 1915.
Lists of spermatophytes collected by author, with localities. (Province of Valencia.)—See also his Plantas de Bicorp. l.c. **16**: 135–145. 1916. (Additional list.)

Vilanova y Piera, Juan. Memoria geognóstico-agrícola sobre la provincia de Castellon . . . Mem. R. Acad. Cien. Madrid **4**: 575–803. 6 pl. (incl. 2 maps). 1856 [1859?].
Includes (p. 734–750) list of 272 angiosperms, with vernacular names and localities, alphabetically arranged by vernacular names. (Now Province of Castellón de la Plana.)

SWEDEN (SVERIGE)

See also Scandinavia; Switzerland (Probst, Wolladventiven Mitteleuropas).
For Swedish-English glossary of ecological and other terms, see Hultén under Scandinavia. The letters ä, å, and ö are alphabetized as æ, aa and œ respectively.

General

The general and partial works here listed may be classified as follows: *Floras and manuals*: Backman (Lapland: see beyond, under Partial), (Bolin: see under Scandinavia), Kindberg (Svensk flora), Kindberg (north Sweden: see beyond, under Partial), Krok and Almquist, Lagerberg (mountain flora, popular: see beyond, under Partial), (Lagerberg: see under Scandinavia), Lindman, Neuman, Nyman, Svensson (north Sweeden: see beyond, under Partial), Ursing (popular). *Iconographies*: Andersson (Femhundra afbildningar), (Bolin: see under Scandinavia), (Lagerberg: see under Scandinavia), (Lindman: see under Scandinavia), Palmstruch, (Ursing). *Bibliographies*: Hjelt (botanical theses), Krok. *Vernacular names*: E. M. Fries, R. E. Fries (Svenska namn), T. M. Fries, Hartman, Hylander, Kindberg (Svenska namn), Kungliga Lantbruksstyrelsen, Linné (Lapland; see beyond, under Partial), Lyttkens (Namnlista, Svenska växtnamn), Nathorst. *Useful plants*: Agnér (medicinal), Fröier, Henriksson (3), Hewe, Holmboe (edible) Liljeblad, Lind, Linné (3; see also beyond, under Partial). Lundström (edible), (Nyman), Öberg (medicinal), Retzius, Roselli, Schoug, (medicinal), Tegnér-Högstedt (edible). *Weeds*: Lyttkens (Om svenska ogräs), Witte (adventive plants). *Cultivated plants*: Andersson (Aperçu), R. E. Fries (Svenska namn), Hewe, Hylander (nomenclature), Lilja, Svensson. *Miscellaneous*: G. Andersson (phytogeography of north Sweden; see beyond, under Partial), R. E. Fries (Short history of botany in Sweden), Hård av Segerstad (phytogeography of south Sweden; see beyond, under Partial). Nordstedt (first records), Nuttonson (agricultural climatology), Sylvén (seedlings).

Agnér, Hjalmar. Svenska växters nytta och användning, populär framställning. I. Kärlväxter. 123 p. 13 cm. Stockholm, 1897.
Annotated list of Swedish medicinal plants, alphabetically arranged by scientific names, with brief descriptions, occurrence, uses, etc.; index of Swedish names. (Not seen; annotation supplied by E. Hultén.)—See also Henriksson, Hewe, Liljeblad, Linné (Flora oeconomica, Flora suecica), Nyman, Öberg, Retzius, Roselli, and Schoug, below; also Linné under Partial, beyond.

Andersson, N. J. Aperçu de la végétation et des plantes cultivées de la Suède. 2 p. l., 94 p. 2 maps. 24 cm. Stockholm, 1867. (Arnold Arboretum library.)
Topography, geology, climate, plant formations, phytogeography; account of cultivated plants, classed by uses; lists of cultivated trees, grouped according to their northern limits of growth in Sweden.—See also R. E. Fries (Svenska namn), Hewe, Hylander (**Våra prydnadsväxters namn**), **Lilja, and** Svensson, below.

——— Femhundra afbildningar af mera allmänt förekommande svenska växter för skolan och hemmet. 4 p. l., 95 p. 68 pl. 22 cm. Stockholm, 1870.
Figures of 500 common vascular plants, mostly taken from Bentham's Handbook of the British flora (1865), with descriptive text.—See also Palmstruch and Ursing, below; also Bolin, Lagerberg, and Lindman, under Scandinavia.

Bolin, Lorentz, and **Post, L. O. A. von.** Floran i färg.
See under Scandinavia.

Fries, E. M. Kritisk ordbok öfver svenska växtnammen. xiv, 177 p. 19 cm. Stockholm, 1880.
Bibliography; annotated alphabetical list of vernacular and scientific names; index of scientific names, with vernacular names collected under each.—See also R. E. Fries (Svenska namn), T. M. Fries, Hartman, Hylander (Våra prydnadsväxters namn), Kindberg (Svenska namn), Kungliga Lantbruksstyrelsen, Lyttkens (Namnlista, Svenska växtnamn), and Nathorst, below.

Fries, R. E., and others. A short history of botany in Sweden. 162 p. 9 fig., 2 pl. (maps), front. (ports.). 23 cm. Uppsala, 1950. (At head of title page: Seventh international botanical congress, Stockholm 1950.)
Account of botanists and botanical work in Sweden, arranged chronologically and (post-Linnaean period) by subjects; includes brief treatment of phytogeography (p. 68–78), with very little reference to floristics. The maps show the journeys made by Linnaeus' disciples throughout the world and the areas explored by Swedish botanists in South America and the West Indies.

——— and others. Svenska namn på i Sverige odlade träd och buskar. Förteckning utgiven av Föreningen för dendrologi och parkvård. Lustgården **25/26:** 286–309. 1945.
List of trees and shrubs cultivated in Sweden, alphabetically arranged by scientific names, each with a standardized vernacular name; index.—See also **Hylander, Nils.** Svenska träd- och busknamn. Ett supplement. Lustgården **33/34:** 61–65. 1953.—Also Hylander, Våra prydnadsväxters namn, below.—Also E. M. Fries, above, and additional references there given.

Fries, T. M. Svenska växtnamn. 1. Under medeltiden. Arkiv Bot. v. 3, no. 14. 60 p. 1904.
History, etc.; annotated alphabetical list of Middle Age vernacular names, with botanical equivalents.—See also E. M. Fries, above, and additional references there given.

Fröier, Kåre. Näringsrika vilda växter. 32 p. 19 cm. Stockholm, 1942.
Deals with wild plants rich in nutriment. (Not seen; title and annotation supplied by T. E. Hasselrot.)

Hartman, C. V. Växtnomenclatur, innefattande de i Sverige vilda och odlade växternas släkt- och artnamn samt vanligare botaniska termer, deras betydelse, och betoning jämte förklaring af förkortade autorsnamn. 2. uppl. redigerad av H. Dahlstedt. *In* Handbok i svenska trädgårdsskötseln. 9. afdeln. (Tilläggshäfte). 198 p. 20.5 cm. Norrköping, 1903. (1st ed. 1884.)

Alphabetical list of scientific names of wild and cultivated Swedish plants and of botanical terms, with meaning and accentuation; explanations of abbreviations of authors' names. (Not seen; annotation supplied by E. Hultén, title from Krok.)—See also E. M. Fries, above, and additional references there given.

Henriksson, Johannes. Handbok i odling och fröinsamling af Sveriges medicinalväxter. 79. p. 20 cm. Stockholm, 1898.

Handbook of wild and cultivated medicinal plants of Sweden; cultivation, propagation, preparation.—See also Agnér, above, and additional references there given.

——— Växterna i de gamlas föreställningar, seder och bruk. 143 p. 18 cm. Stockholm, 1911. (Riksmuseet library, Stockholm.)

List of wild and cultivated Swedish plants of use or interest to the people, alphabetically arranged by vernacular names, with account of popular beliefs, sayings, and uses; no index of scientific names.—See also Agnér, above, and Liljeblad, below, and additional references there given.

——— Vartill våra växter duga. 220 p. 19.5 cm. Stockholm, 1923.

Uses of wild and cultivated plants (including some cellular plants). (Not seen; title and annotation supplied by T. E. Hasselrot.)

Hewe, Nils. Nöjsamt, sällsamt, nyttosamt om våra växter. 284 p. 20 cm. Stockholm, 1953.

Popular treatment of various wild and cultivated plants of Sweden, with early history (in world at large), uses (especially in medicine), chemistry, popular beliefs, etc.—See also his Läkeväxter. Övertro och vetande från antiken till nutiden. 187 p. 13 fig. 20 cm. Stockholm, 1952. (Popular account of history (including superstitions) and uses of some of the more common medicinal plants in the world at large and in Sweden, alphabetically arranged by vernacular names.)—Also Agnér, above, and additional references there given.

Hjelt, O. E. A. Sveriges biologiska disputations- och program-litteratur 1700—vårterminen 1910. Bidr. Känn. Finlands Nat. och Folk v. 70, no. 1. 1 p. l., viii, 210 p. 1911.

Unannotated classified list of theses, chronologically arranged in each group. Botanik, p. 135–171.—See also Krok, below.

Holmboe, Jens. Gratis mat av vilda växter. Med recept av Juliane Solbraa-Bay. Översatt och avpassad efter svenska förhållanden av Karl-Gustav Kökeritz. 153 p. 60 fig. Stockholm, 1942.

Not seen; title from Nytt Mag. Naturv. 85: 19. 1945. The original (see under Norway, General) gave an account of about 150 edible wild plants (vascular and cellular), with instructions for cooking.

Hultén, Eric, *ed.* Våra svenska flora i färg.

See Christiansen under Scandinavia.

Hylander, Nils. Våra prydnadsväxters namn på svenska och latin. 196 p. 23 cm. Stockholm, 1948.

Discussion of botanical nomenclature; alphabetical list of scientific names (without authorities) of plants cultivated for ornament in Sweden, with single Swedish vernacular name (mostly artificial) for each; list of these vernacular names with scientific equivalents; list of comments on nomenclature, taxonomy, etc., of various names; bibliography.—See also Andersson (Aperçu) and E. M. Fries, above, and additional references there given.

Kindberg, N. C. Svensk flora. Beskrifning öfver Sveriges fanerogamer och ormbunkar. 2 p.l., 402 p. 20 cm. Linköping, 1877.

Briefly descriptive flora of vascular plants, on the Linnaean system.

——— Svenska namn på våra inhemska kärlväxter. 49 p. 18.5 cm. Stockholm, 1905.

Systematic list of vascular plants with a single selected or manufactured vernacular name for each (or sometimes two); index to vernacular names. —See also E. M. Fries, above, and additional references there given.

Krok, T. O. B. N. Bibliotheca botanica suecana ab antiquissimis temporibus ad finem anni MCMXVIII. Svensk botanisk litteratur från äldsta tider t.o. m. 1918. xvi, 799 p. port. 25.5 cm. Uppsala och Stockholm, 1925.

Alphabetical list of Swedish authors and of non-Swedish authors who have published in Sweden or written on Swedish plants, with lists of their publications and biographical information; includes also exsiccatae, and covers Norwegian literature down to 1814. An admirable work as regards accuracy and completeness, but without subject index (planned but never completed), and the attempt under the more prolific authors to arrange titles by subject matter rather than date makes the finding of individual titles unnecessarily difficult.—See also **Håkansson, Torsten.** Svensk botanisk litteratur 1945 [-1948]. Bot. Notiser **1946**: 557–583. 1946; **1947**: 392–410. 1947; **1948**: 441–460. 1948; **1949**: 452–480. 1949. (Meddelanden från Lunds Botaniska Museum no. 82, 87, 91, 92.) (Unannotated lists of titles, alphabetically arranged by authors, with subject index.)—Also **Hjelmqvist, Hakon.** Svensk botanisk litteratur 1949 [-1957]. l.c. **1950**: 471–496. 1950; **1951**: 408–431. 1951; **1952**: 448–470. 1952; **1953**: 441–461. 1953; **1954**: 434–477. 1954; **108** (1955): 461–485. 1955; **109**: 446–472. 1956; **110**: 448–513. 1957; **111**: 659–684. 1958. (Meddelanden från Lunds Botaniska Museum no. 97, 99, 100, 106, 108, 111, 114, 120, 130.) (Unannotated, classified lists of publications.) —Also Hjelt, above.

——— and **Almquist, Sigfrid.** Svensk flora för skolor . . . I. Fanerogamer och ormbunkväxter. 23. uppl. utg. av Erik Almquist . . . 386 p. 195 fig. 18.5 cm. Stockholm, 1945 (4. tryckningen, 1952, the only issue seen). (1st ed. 1883. 24. uppl., with title: Svensk flora. I. Fanerogamer och ormbunkväxter. 24. uppl. utg. av Erik Almquist. 403 p. 195 fig., 96 col. pl. (also issued without plates). Stockholm, 1956; not seen.)

Pocket manual of wild and commonly cultivated vascular plants, with keys, very brief descriptions, and brief annotations; synopsis of families, glossary of specific epithets, list of synonyms. Authorities for botanical names not given.

Kungliga Lantbruksstyrelsen. Normalförteckning över svenska växtnamn. Ny, reviderad uppl. Meddel. K. Lantbruksstyr. 291. 68 p. 1932. (1st ed. 1894, as no. 17 of this series.)

Systematic list of principal wild and cultivated vascular plants (about 670 wild species in 1st ed.) and some cellular plants, with standard name for each as adopted by the National Agricultural Directorate; indices.—See also E. M. Fries, above, and additional references there given.

Lagerberg, Torsten. Vilda växter i Norden.
See under Scandinavia.

Lilja, Nils. Flora öfver Sveriges odlade vexter, innefattande de flesta på fritt land odlade vexter i Sverige, jemte de allmännare och vackrare fenstervexterna, med kännetecken och kort anvisning om deras odlingssätt. xix, 175 p. 21 cm. Stockholm, 1839. (Gray Herbarium library.)—Första supplementet. xxxv, 83 p. 21 cm. Stockholm, 1840. (Gray Herbarium library.)

Descriptive flora of cultivated plants on the Linnaean system, essentially without notes on uses.—See also his Handbok i de odlade vexternas flora och deras kultur . . . 2 pt. (4 p. l., 276 p.). 22 cm. Örebro, 1842–43. (Descriptive flora of cultivated plants, with directions for cultivation. Unfinished; covers Gramineae-Helichrysum (Endlicher system).)—See also Andersson (Aperçu), above, and Svensson, below.

Liljeblad, Samuel. Utkast till en svensk flora, eller afhandling om svenska växternas väsendteliga kännetecken och nytta. 3. uplagan, med norrska växter tillökt, efter författarens död utgifven. 5 p. l., lviii, 761 p. 2 pl. 18.5 cm. Upsala, 1816. (1st ed. 1792.)

Briefly descriptive flora of vascular and cellular plants on the Linnaean system, with brief account of uses; obsolete as a flora, cited here for its statement of uses, particularly medicinal and forage.—See also (for medicinal plants) Agnér, above, and additional references there given; and (for useful plants in general) Henriksson (Växterna), Hewe, and Holmboe, above, and Linné (3 titles), Lundström, Nyman, and Retzius, below.

Lind, Gustaf, and Verdier, Nils de. Våra medicinalväxter. 4. uppl. 48 p. 50 pl. 21 cm. Stockholm, 1940.

Treatment of medicinal plants. (Not seen; title and annotation supplied by T. E. Hasselrot.)

Lindman, C. A. M. Bilder ur Nordens flora.
See under Scandinavia.

——— Svensk fanerogamflora. 2. uppl. x, 644 p. 329 fig. 22 cm. Stockholm, 1926. (1st ed. 1918.)

Short glossary, key to genera and families on Linnaean system; annotated, keyed, briefly descriptive flora of spermatophytes, including principal cultivated species (2,104 species, including 99 of *Taraxacum* and 107 of *Hieracium*). For many years the standard modern flora of Sweden, now to be replaced by Hylander (see under Scandinavia).—Much the same treatment, slightly condensed, with inclusion of pteridophytes and omission of the detailed account of *Rosa, Taraxacum,* and *Hieracium,* is given in his Illustrerad skol- och exkursionsflora över Sveriges kärlväxter. x, 330 p. 328 fig. 20 cm. Stockholm, 1928.—See also **Hylander, Nils.** Märkligare nytillskott till den svenska kärlväxtfloran sedan 1920. Bot. Notiser **1954**: 132–153. 1954. (Running account of about a score of species added to flora of Sweden, not including "segregate" or "chromosome" species.)

Linné, Carl von. Flora oeconomica. 4 p. l., 30 p. 18.5 cm. Upsaliae, 1748. (Diss., Elias Aspelin respond.) (Reprinted in Linné, Amoenitates Academicae 1. ed. Holm., **1**: 509–539. 1749; 1. ed. Lugd. Bat., **1**: 352–388. 1749; 3. ed. cur. Schreber, **1**: 509–540. 1787.)

List of useful and harmful vascular and cellular plants, with brief notes on uses. The species are listed only by generic name and number, the latter referring to the Flora Suecica. Another issue (83 p. 20 cm. Stockholm, 1749), in Swedish and with vernacular names, some additional species, and indices, has title: Flora oeconomica, eller hushålls-nyttan af de i Sverige, wildt wäxande örter.—See also (for medicinal plants) Agnér, above, and additional references there given; and (for useful plants in general) Liljeblad, above, and additional references there given.

——— Flora suecica, exhibens plantas per regnum Sueciae crescentes, systematice cum differentiis specierum, synonymis autorum, nominibus incolarum, solo locorum, usu oeconomorum, officinalibus pharmacopaeorum. 2. ed. aucta et emendata. 2 p.l., xxxii, 464 p., indices. pl. 20.5 cm. Stockholmiae, 1755. (1st ed. 1745.)

Long out of date as a flora; cited here for its brief account of uses.—See also **Hedbom, Karl.** Linnés inhemska medicinalväxter. Svenska Linné-Sällsk. Årsskr. **2**: 65–110. 1919. (Annotated systematic list of 242 native medicinal plants mentioned by Linné in various publications.)—Also Agnér and Liljeblad, above, and additional references there given.

——— Plantae esculentae patriae. 2 p.l., 29 p. 19 cm. Upsaliae, 1752. (Diss., Johan Hiorth respond.) (Reprinted (with binomial names) in Linné, Amoenitates Academicae **3**: 74–99. 1756; 2. ed. **3**: 74–99. 1764; 2. ed. cur. Schreber, **3**: 74–99. 1787.)

Annotated list of edible vascular plants and a few fungi; list classified by uses. The species are listed with the names and numbers of the Flora Suecica. Another issue (40 p. 19 cm. Stockholm, 1752), in Swedish, with vernacular names, some additional species, and indices, has title: Plantae esculentae patriae, eller wåra inländska äteliga wäxter.—See also Agnér and Liljeblad, above, and additional references there given.

Lundström, Erik. Ätliga vilda växter. En dyrtidsbok. 1 p. l., iv, 54 p. 30 fig. 22.5 cm. Stockholm, 1917.

Annotated list of edible wild cellular and vascular plants (except fungi), grouped by uses; brief bibliography. Krok lists "uppl. 2, betydligt ökad, 1918."—Also his Matnyttiga vilda växter. 32 p. 19 cm. Stockholm, 1942. (Deals with edible wild plants. Not seen; title and annotation supplied by T. E. Hasselrot.)—See also Agnér and Liljeblad, above, and additional references there given.

Lyttkens, August. Namnlista över svenska växtnamn på gagn- och ogräsväxter i överensstämmelse med den av Kungl. lantbruksstyrelsen fastställda Normalförteckningen utgiven med tillägg. 43 p. 22.5 cm. Stockholm [pref. 1907].

Systematic list of useful and harmful plants (including cultivated species), with standard vernacular names from the Kungliga Lantbruksstyrelsen list of 1894 (see above), with addition in footnotes of differing or additional names from the author's own work Svenska växtnamn.—See also **E. M. Fries**, above, and additional references there given.

——— Om svenska ogräs, deras förekomst och utbredning samt intagande af uppgifter om ogräsfrön i fröanalysbevis. 113 p. 20.5 cm. Norrköping, 1885.

List of weeds, divided according to noxious quality and manner of occurrence; annotated list of 322 species (314 vascular), with vernacular names in Swedish, Danish, and German, habitat, and range, and discussion of the more important species; discussion of the spread of weeds, etc.—See also Witte, below, and Scandinavia (Korsmo).

——— Svenska växtnamn. 3 pt. (xix, 1891 p.). 24.5 cm. Stockholm, 1904–15.

Historical sketch of development of vernacular nomenclature of plants; systematic list of native and cultivated vascular plants, with standard names for each, followed under each genus by collected generic and specific vernacular names in German, Danish and Norwegian, and Swedish, with citation of authors using them (with dates), and occasional critical notes; systematic list and index of families, genera, and species, with standard names; indices of Latin generic and all vernacular names; extensive chronological bibliography of works cited (Latin, German, and Dutch; English (only 2 titles, one Britton and Brown's Illustrated flora, the other a trifling work on alfalfa); Danish and Norwegian; Swedish and Finnish), with occasional annotations; list of authors cited, geographically and then alphabetically arranged; list of geographical and other abbreviations. The systematic list and index of botanical and standard vernacular names were reprinted with title Svensk växtnamn-förteckning enligt binärt system (2 p.l., 1705–1741 p. 1914).

The system of forming "standard" vernacular names is the same as that of the Normalförteckning of the Kungl. Lantbruksstyrelsen (see above), and most of the names are identical. The method is essentially binomial in principle but uninomial in practice, a single vernacular name being assigned to each genus and another, adjectival in form, to each species; each specific name as written being composed of the specific component prefixed to the generic so as to constitute a single word. The names of the few infraspecific entities included are formed like those of the species, of two elements only, combined into one. Lyttkens' list includes many more species of both wild and cultivated plants than does the Normalförteckning.—See also E. M. Fries, above, and additional references there given.

Nathorst, A. G. Svenska växtnamn. 1–5. Bihang K. Svenska Vetensk.-Akad. Handl. v. 28, afd. III, no. 9. 72 p. 1903.—Arkiv Bot. 1: 497–513. 1904; v. 2, no. 1. 179 p. 1904; no. 8. 12 p. 1904; no. 9. 31 p. 1904.

Pt. 1: History, critical discussion of lists of vernacular names published by the Lantbruksstyrelsen and in Neuman's Flora; discussion of method approved by author; bibliography; (2) completion of history and discussion, bibliography; (3) author's list of standard vernacular names, with separate list of annotations, index; (4) Linnaeus' position with respect to names; (5) Middle Age literature, with annotated alphabetical list of vernacular names; and literature up to 1792 and later. (For list of reviews and criticisms, see Krok, Bibliotheca Bot. Suec., p. 527.)—See also his Svenska växtnamn. Förteckning på Sveriges viktigaste kärlväxter med svenska namn på arter, släkten, familjer och klasser. Ny reviderad uppl. iv, 112 p. 16.5 cm. Stockholm, 1905. (Systematic list of wild and commonly cultivated vascular plants, with the author's standard name for each.)—See also E. M. Fries, above, and additional references there given.

Neuman, L. M. Sveriges flora (fanerogamerna). . . . med biträde af Fr. Ahlfvengren. xxxvi, 832 p. 18.5 cm. Lund, 1901.

Briefly annotated, keyed, descriptive flora of spermatophytes.—See also his Herbarium suecicum. Förteckning öfver Sveriges fanerogamer i enlighet med Sveriges flora. Med biträde af Fr. Ahlfvengren. 1 p. l., 71 p. 19 cm. Lund, 1901. (List of names from his Sveriges flora. Not seen; annotation supplied by E. Hultén.)

Nordstedt, Otto. Prima loca plantarum suecicarum. Första litteraturuppgift om de i Sverige funna vilda eller förvildade kärlväxterna. Bot. Notiser 1920, Bilaga. iv, 95 p. 1920.—Rättelser till "Prima loca . . ." l.c. **1921**: 282. 1921.

Alphabetical list of vascular plants with references to first records from Sweden, with localities.

Nuttonson, M. Y. Agricultural climatology of Sweden and its agro-climatic analogues in North America. Amer. Inst. Crop Ecol. Internatl. Agro-Climat. Ser. Study 11. 27 p. incl. 2 fig. (maps), tab. 1950.

Geography and topography, climate, comparison with localities in North America, bibliography.

Nyman, C. F. Utkast till svenska växternas naturhistoria eller Sveriges fanerogamer skildrade i korthet med deras växställen och utbredning m. m., deras egenskaper, användning och historia i allmänhet. 2 v. (431, 566 p.). 22.5 cm. Örebro, 1867–68.

Descriptive flora of spermatophytes, with local and extralimital range, vernacular names, first Swedish record for each species, uses, biological notes, etc.—See also (for medicinal plants) Agnér, above, and additional references there given, and (for useful plants in general) Liljeblad, above, and additional references there given.

Öberg, Emil. Sveriges läkedomsväxter, deras uppsökande och användande jämte en kort växtbeskrifning . . . 2. uppl. 2 p.l., 76 p. 18.5 cm. Stockholm, 1885. (Riksmuseet library, Stockholm.) (1st ed. 1868.)

List of wild and cultivated Swedish medicinal plants (trees, bushes, herbs, grasses, ferns, lichens, and fungi), with brief descriptions and notes on uses; index of diseases, index of vernacular names.—See also Agnér, above, and additional references there given.

Palmstruch, J. W., and others, *ed.* Svensk botanik. 11 v. 774 col. pl. 23 cm. Stockholm (later Upsala), 1802–38.—Register till hela arbetet. 24 p. 1843.

Colored plates of 774 vascular plants, not in systematic order. Register not seen. The plates afterwards served as a basis for those in Lindman, Bilder ur Nordens flora (see under Scandinavia). For bibliographical details see Krok, Bibliotheca botanica suecana p. 98–100. 1925.—See also Andersson (Femhundra afbildningar), above, and additional references there given.

Retzius, A. J. Försök til en flora oeconomica Sveciae eller swenska wäxters nytta och skada i hushållningen. 2 v. (vii, 792 p.). 18.5 cm. Lund, 1806. Bihang till Flora . . . 20 p. 20 cm. Lund, 1812. (Bihang in Riksmuseet library, Stockholm.)

List of vascular and cellular plants, alphabetically arranged by scientific names, with vernacular names and uses.—See also Agnér and Liljeblad, above, and additional references there given.

Roselli, C. R. Afhandling om mera än 300 swenska wäxters nytta och anwändbarhet såväl i medicinskt afseende, som i hushållningen och näringarne. 90 p. 17 cm. Stockholm, 1858.

Account of over 300 medicinal, edible, and otherwise useful plants. (Not seen; title and annotation supplied by E. Hultén.) Krok (Bibliotheca botanica suecana p. 596) cites a second ed., with title Swensk örtbok . . . 79 p. 8vo. 1869.—See also Agnér and Liljeblad, above, and additional references there given.

Schoug, Carl. Sveriges apoteksväxter jemte anvisning om deras insamling och förvaring. 24 p. 18 cm. Stockholm, 1881.

List of native medicinal plants, alphabetically arranged by vernacular names, with brief descriptions, notes on collecting and preservation; index of vernacular names, index of species grouped by the parts used. (Not seen; title and annotation supplied by E. Hultén.)—See also Agnér, above, and additional references there given.

Svensson, Per. Flora över Sveriges kulturväxter. cxxvi, 727 p. 22.5 cm. Stockholm, 1893. (John Crerar Library.)

Key to genera, on Linnaean system; descriptive flora of vascular plants cultivated in Sweden (2,853 species), systematically arranged, with mention of country of origin but without annotations or uses; explanation of scientific names.—See also Andersson (Aperçu), Hylander (**Våra prydnadsväxters namn**), and Lilja, above.

Sylvén, Nils. Om de svenska dikotyledonernas första förstärkningsstadium eller utveckling från frö till blomning. I. Speciell del. K. Svenska Vetensk.-Akad. Handl. n.s., v. 40, no. 2. 348 p. 25 pl. 1906.—II. Allmän del. 75 p. 31 cm. Uppsala, 1906. (Akad. afhandl.) (Pt. II, Riksmuseet library, Stockholm.)

Pt. I, descriptive account of dicotyledonous seedlings, with references; bibliography; pt. II, key to the 75 groups recognized; list of these, with brief characterizations and lists of species included; bibliography.

Tegnér-Högstedt, Ida. Matnyttiga vilda växter och deras användning i hushållet utexperimenterad i eget kök. 75 p. Stockholm, 1919.

Treatment of edible wild plants. (Not seen; title and annotation supplied by T. E. Hasselrot.)

Ursing, Björn. Svenska växter i text och bild. Fanerogamer. 398 p. incl., 96 col. pl., illus. (in glossary). 20 cm. Stockholm, 1947.

Glossary, keys to families; keyed, briefly annotated popular flora of spermatophytes, with brief descriptions; list of Latin specific epithets, with meaning; alphabetical index to distribution maps in other publications. Many of the less distinctive species are treated only in the keys, not in the text; critical genera are very incompletely treated. The vascular cryptogams are treated in his second volume, Kryptogamer (on p. 20–45. illus., 9 col. pl. 20 cm. Stockholm, 1947).

Witte, Herrfrid. Några bidrag till kännedomen om vegetationen på våra ruderatplatser. Svensk Bot. Tidskr. **3**: 174–182. 1909.

Extensive alphabetical list of adventive plants, with localities, mostly collected by author in 1906, chiefly in the southeastern part of Sweden.—See also **Aulin, F. R.** Anteckningar till Sveriges adventivflora. Svensk Bot. Tidskr. **8**: 357–387. 1914. (Localities, etc.; extensive alphabetical list of adventive plants collected by author in various parts of Sweden.)—Also his Anteckningar till Sveriges adventivflora. l.c. **10**: 97–100. 1916. (Additional list.)—Also **Pleijel, Carl.** Nya lokaler för adventivväxter. Bot. Notiser **1916**: 277–282. 1916. (List of noteworthy species from Ångermanland, Hälsingland, Småland, and Öland.)—Also **Hylander, Nils.** Die grassameneinkömmlinge schwedischer parke mit besonderer berücksichtigung der Hieracia silvaticiformia. Symb. Bot. Upsal. v. 7, no. 1. xiii, 432 p. 24 pl. (on 12). 1943. (Description of some typical park grasslands; discussion of origin and history of numerous weeds in grass land grouped by country of origin, etc.; monograph of the adventive Hieracia sylvaticiformia (143 species of which 131 are described as new); tables showing occurrence of some of the species listed at different localities, bibliography).—Also **Almquist, Erik.** Järnvägsfloristiska notiser. Ett apropos till järnvägsjubileet. Svensk Bot. Tidskr. **51**: 223–263. 2 fig. (maps). 1957. (Discussion of the flora of railroads, railroad stations, etc., with mention of numerous localities and their species, lists of species introduced by special means, etc.; bibliography.)—Also Nordström, Bidrag till kännedomen om Sveriges ruderatflora, under Skåne.—Also Lyttkens, Om svenska ogräs, above.—Also Blom, under Göteborg.

Partial

Under this heading are grouped works relating to several provinces (landskap, as well as those referring to several or all divisions of Swedish Lapland.

Andersson, Gunnar, and Birger, Selim. Den norrländska florans geografiska fördelning och invandringshistoria med särskild hänsyn till dess sydskandinaviska arter. viii, 416 p. incl. 86 fig. (incl. maps). 26.5 cm. Uppsala och Stockholm, 1912. (Norrländskt handbibliotek. V.) (New York Botanical Garden library.)

Botanical explorations, ecological groups, topography, geology, phytogeography, plant migration, etc.; bibliography. Kartor 1 and 2 show the mountains, passes, and valleys between about 60° and 69° N. Lat. For extensive synopsis, see their Die geographische verteilung und die einwanderungsgeschichte der nordskandinavischen flora. Bot. Jahrb. Engler **51**: 501–593. 14 fig. (incl. maps), pl. 3–4 (maps). 1914.

Backman, C. J., and Holm, V. F. Elementarflora öfver Vesterbottens och Lapplands fanerogamer och bräkenartade växter. 1 p. l., v, 268 p. 22 cm. Luleå, 1878.

Keyed, briefly descriptive flora of vascular plants, on the Linnaean system, with localities for scarcer species. Covers the län of Västerbotten and Norrbotten, including all of Swedish Lapland.—See also **Melander, Carl.** Bidrag till Vesterbottens och Lapplands flora. Bot. Notiser **1883**: 160–162,

205–216. 1883. (List of vascular plants collected by author, from various localities; short list of vernacular names, with botanical equivalents.)—Also **Andersson, N. L.** Bidrag till Vesterbottens växtgeografi. l.c. **1890**: 36–38. 1890. (List of noteworthy species.)—Also **Melander, C.** Anteckningar till Vesterbottens flora. l.c. **1890**: 236–239. 1890. (List of noteworthy vascular plants, mostly from Skellefteå socken.)—Also **Sylvén, Nils.** Bidrag till Västerbottens och Lycksele Lappmarks flora. Svensk Bot. Tidskr. **4**: (137)–(152). 1 fig. 1910. (Mention of numerous noteworthy species, including adventive plants from region of Vännäs, Hällnäs, and Lycksele; bibliography.)—Also **Sörlin, Anton.** Några växtlokaler i Västerbotten. l.c. **14**: 288–291. 1920. (Includes alphabetical list of noteworthy vascular plants from Degerfors and Norsjö socknar.)—Also **Nordenstam, Sten.** Några anmärkningsvärda nya växtfynd inom Västerbottens län. l.c. **34**: 454–463. 2 fig. 1940. (Includes alphabetical list of noteworthy plants from Lycksele socken, and another list from various localities (Malå, Tärna, Vännäs, Vilhelmina, etc.).)

Hård av Segerstad, Fredrik. Sydsvenska florans växtgeografiska huvudgrupper. (Die pflanzengeographischen hauptgruppen der südschwedischen flora.) 1 p.l., 244 p. 409 fig. (maps), table. 30 cm. Malmö, 1924.

Phytogeographic and ecological; has maps showing range of over 400 species; bibliography; German summary. Covers region south of line from Norrköping to Uddevalla. For English summary, see his The main features of the floral plant-geography of southern Sweden. Bot. Notiser **1925**: 222–250. 15 fig. (maps). 1925.—See also his Sydsvenska växtlokaler I. Svensk Bot. Tidskr. **21**: 285–304. 1927. (Extensive alphabetical list of vascular plants, with localities.)

Kindberg, N. C. Sammandrag af norra Sveriges flora (fanerogamer och ormbunkar). 72 p. 19.5 cm. Linköping, 1873. (Riksmuseet library, Stockholm.)

Flora of vascular plants on the Linnaean system, with keys to genera and very brief diagnostic characters of species, and indication of range by provinces for those not of general range. Covers Gestrikland, Dalarne, and all provinces north of them.—See also **Birger, Selim.** Växtlokaler från Norrland och Dalarna. Svensk Bot. Tidskr. **3**: (143)–(158). 1910. (Extensive alphabetical list of vascular plants from various provinces.)—Also **Mörner, C. T.** Botaniska anteckningar från Norrlandsfärder 1916–1919. Bot. Notiser **1920**: 33–40. 1920. (List of noteworthy vascular plants, with localities, from Hälsingland, Härjedalen, and the provinces north of them.) —Also his Ytterligare några norrländska växtlokaler. l.c. **1923**: 133–140. 2 fig. 1923. (Alphabetical list of noteworthy vascular plants, with localities; short bibliography.)—Also his Några bidrag till norrländsk floristik, grundade huvudsakligen på reseanteckningar inom årsföljden 1923–1934. l.c. **1935**: 254–272. 1935. (Alphabetical list of noteworthy vascular plants, with localities; bibliography.)—Also **Arwidsson, Thorsten.** Växtgeografiska notiser från Norrland. I–IV. l.c. **1927**: 337–351. 2 fig. 1927; **1929**: 305–310. 1929; (III–IV) **1931**: 355–374. 1 fig. (map). 1931. (I, extensive alphabetical list of noteworthy vascular plants from various provinces; II, similar list, especially from Härjedalen; III, similar list from coast of Norrbotten and Västerbotten; IV, list of flora of Holmöarna in Norrbotten.)—Also **Grapengiesser, Sten.** Norrländska vegetationsbilder [I]–II. Svensk Bot. Tidskr. **28**: 300–353. 8 fig. 1934; **41**: 469–484. 1 fig. 1947. (Part I includes more or less extensive lists from Ångermanland (Nordingrå socken, Skuleberget in Vibyggerå socken, and other localities), Jämtland (various localities), Norrbotten (Storön in Luleåtrakten), Västerbotten (Bygdeå); bibliography (108 titles), index; II includes extensive list of noteworthy species from various localities, especially in Ångermanland.)

Lagerberg, Torsten. Svenska fjällblommor. 2. tillökade uppl. 185 p. 15 fig. (i.e., 5 fig. [maps] and 10 pl. on 5), 200 pl. on 100. 18 cm. Stockholm, 1940. (Riksmuseet library, Stockholm.) (1st ed. 1931.)

Origin and development of mountain flora, altitudinal limits, plant associations; account of 200 conspicuous or characteristic plants, with description, local and general range, and excellent photograph of each in its native habitat. Refers primarily to Lule and Torne Lappmark.

Linné, Carl von. Flora lapponica, exhibens plantas per Lapponiam crescentes, secundum systema sexuale, collectas in itinere impensis Soc. Reg.

Scient. Upsaliensis, anno 1732 instituto. Additis synonymis, et locis natalibus omnium, descriptionibus et figuris rariorum, viribus medicatis et oeconomicis plurimarum. Ed. altera, aucta et emendata studio et cura Jacobi Edvardi Smith. 2 p. l., L, 390 p., indices. 12 pl., front. 23.5 cm. Londini, 1792. (1st ed. 1737.)

Cited here for its vernacular names (indexed) and notes on uses. Linné, in his Lapland journey of 1732, traveled from Uppsala to Umeå, then up the Ume River to Lycksele, then on to Piteå, to Storbacken on the Lule River and Jockmock, then to Qvickjock and across the Norwegian frontier to Sörfjorden and the Maelström, then back and around near the coast to Luleå and Torneå, then down rapidly to Åbo, and back to Uppsala.

Svensson, Per. Flora öfver Norrlands kärlväxter, till läroverkens tjenst utarbetad. xlviii, 95 p. 21 cm. Hernösand, 1885.

Key to genera on Linnaean system; flora of vascular plants, with very brief diagnoses and indication of range; list of species found in Lapland but not in the rest of Norrland (28 species); list of escaped or adventive plants and commonly cultivated species (72 species). Covers region from Gestrikland and Härjedalen northward.

ÅNGERMANLAND

See also Medelpad (Grevillius, Studier).

The old province (landskap) of Ångermanland constitutes the northern part of Västernorrlands län.

Local

Almquist, Erik. Växtgeografiska bidrag. 6. Ångermanland. Bot. Notiser **1941:** 301–309. 1941.

Botanical explorations, etc.; alphabetical list of noteworthy vascular plants from general vicinity of Örnsköldsvik.

Arnell, H. W. Vårfloran vid Härnösand. Bot. Notiser **1918:** 169–209. 1918.

Mainly phenological; includes (p. 181–198) annotated alphabetical list of wild and more commonly cultivated spring-blooming plants.—See also **Lange, Thorvald.** Några växtlokaler i Härnösandstrakten. Svensk Bot. Tidskr. **20:** 406–413. 1926. (Alphabetical list of vascular plants, with localities; phytogeography.)—Numerous records from Härnösand are also given in **Arnell, H. W.** Anteckningar om södra Ångermanlands kärlväxter. l.c. **19:** 380–401. 1925.

Fristedt, R. F. Växtgeografisk skildring af södra Ångermanland. 1 p. l., 40 p. 21 cm. Upsala, 1857. (Akad. afhandl.) (Gray Herbarium library.)

Geography, topography, plant regions, etc.; briefly annotated list of vascular plants, with localities for the more local species. The author defines southern Ångermanland as the region bounded by a line from the southernmost point of junction of Åsele Lappmark and Jämtland southeast to the coast, with a length of about 25 Swedish miles (about 265 kilometers) and greatest breadth of 10 (about 100 kilometers).—See also his Fortsatta iakttagelser af södra Ångermanlands växtlighet. Bot. Notiser **1858:** 73–87. 1858. (Includes list of additional species and localities.)—Also **Haglund, G. E.** Växtgeografiska bidrag till Ångermanlands flora. l.c. **1923:** 385–400. 1923. (List of localities visited, etc.; alphabetical list of noteworthy vascular plants, with localities; list of mountains visited, with mention of noteworthy species. Refers mostly to vicinity of Sollefteå.)—Also **Haglund, G. E., and Hässler, Arne.** Nya bidrag till Ångermanlands kärlväxtflora. l.c. **1925:** 309–319. 1925. (Similar list; bibliography. Includes also records from northern Ångermanland.)—Also **Arnell, H. W.** Anteckningar om södra Ångermanlands kärlväxter. Svensk Bot. Tidskr. **19:** 380–401. 1925. (Author's explorations from 1862–1921, etc.; extensive alphabetical list of noteworthy vascular plants, with localities; includes some from northern Ångermanland.)—Also **Samuelsson, Gunnar.** Växtlokaler från Ångermanland och Åsele Lappmark. l.c. **21:** 123–138. 1927. (Author's itinerary, with notes on vegetation; list of common species, extensive list of less common vascular plants with localities. Refers to region from Långsele and Sollefteå in Ångermanland to Åsele Lappmark.)—Also **Grapengiesser, Sten.** Norrländska vegetationsbilder. [I]–II. l.c. **28:** 300–353. 8 fig.

1904, **41**: 469–484. 1 fig. 1947. (I: Includes lists of noteworthy species from various localities in Nordingrå socken, Skuleberget in Vibyggerå socken, and other localities; bibliography; II: includes extensive list from various localities, particularly in Ångermanland.)

Holmer, Hilmar. Vegetation och flora inom dämningsområdet för Ramsele kraftverk. K. Svenska Vetenskapsakad. Avhandl. Naturskyddsär. **17.** 88 p. 25 fig. (fig. 1–2, maps). 1959.

Includes (p. 65–85) annotated list of vascular plants (363 species), mosses, hepatics, and Characeae; bibliography, English summary.

BLEKINGE

The old province (landskap) of Blekinge coincides with Blekinge län.

General

Holmgren, Björn. Blekinges flora. 380 p. 142 fig. (maps). 22.5 cm. Karlskrona, 1942.

Botanical explorations, phytogeography, list of localities, bibliography, etc.; briefly annotated list of vascular plants, with detailed localities; list of synonyms, index. Replaces his Blekings [sic] fanerogamer och kärlkryptogamer. 204 p. map. 19 cm. Karlskrona, 1921. Also replaces **Gosselman, C. A.** Blekinges flora . . . 2. uppl. 1 p.l., xii, 217 p. 19.5 cm. Lund, 1865. (Riksmuseet library, Stockholm.) (1st ed., Systematisk förteckning på de i Blekinge vildt växande slägten och arter af fanerogamer och ormbunkar. 2 p.l., 64 p. 20 cm. Karlskrona, 1861. In Riksmuseet library, Stockholm.) Also replaces **Svanlund, Fredrik.** Förteckning öfver Blekings [sic] fanerogamer och ormbunkar . . . 1 p.l., 59 p. 18.5 cm. Lund, 1889. (Riksmuseet library, Stockholm.)—For ecological treatment see **Hult, Ragnar.** Blekinges vegetation. Ett bidrag till växtformationernas utvecklingshistoria. Meddel. Soc. Fauna et Flora Fenn. **12**: 161–252. 1885.

Local

Blom, Carl. Ön Kungsholmens flora. (En blomsteroas i Blekinge skärgård.) Bot. Notiser **1913**: 83–91. 1913.

Annotated list of vascular plants.

Du Rietz, G. E., and Du Rietz, Greta. Floristiska anteckningar från Blekinge skärgård. Bot. Notiser **1925**: 66–76. 1925.

Plant associations, etc.; unannotated alphabetical list of vascular plants, with localities, from islands of Långören, Ungskär, Hästholmen, and Hanö.

Holmgren, Björn. Några anteckningar om Blekinges adventiv-och ruderatflora. Bot. Notiser **1941**: 65–98. 1941.

Annotated lists of adventive plants from vicinity of Karlskrona, Ronneby, Karlshamn and Asarum, Sölvesborg, etc., with dates.—See also **Nordström, K. B.** Bidrag till kännedomen om Sveriges ruderatflora. Bot. Notiser **1903**: 113–122. 1903. (Includes list of ruderal and ballast plants, with data, in part from Karlshamn.)

Westerlund, C. G. Flora. Öfversikt af Ronnebytraktens fanerogamer och ormbunkar. *In* his Bidrag till kännedomen om Ronnebytraktens fauna och flora. p. 103–173. 19.5 cm. Stockholm, 1890.

Annotated list of vascular plants; list of excursions with mention of characteristic species. Replaces **Lindblom, A. E.** Stirpes agri rotnoviensis . . . 5 pt. (84 p.). 19.5, 21 cm. Lundae, 1826–29. (Topography, geology, general features of flora, plant associations; annotated list of spermatophytes (287 species), on the Linnaean system. Incomplete; reaches Decandria Pentagynia *Sagina*). Diss., C. A. Agardh praeses, pt. I. In Lloyd Library.)

DALARNA (DALARNE, DALECARLIA, DALEKARLIA)

The old province (landskap) of Dalarna or Dalecarlia now forms the southern part of Kopparbergs län, except that the district of Orsa Finnmark is in Gävleborgs län.

General

Almquist, Erik. Dalarnes flora. Förteckning över kärlväxterna, grundad på Professor G. Samuelssons samlade material och nyare tillägg, jämte några växtgeografiska synpunkter på Dalafloran samt dess utforskning. xvi, 458 p. 17 fig. (incl. 2 ports.; mostly distribution maps), port. (front.). 21 cm. Stockholm, 1949.

Geology, topography, climate, principal habitats, phytogeography, description of different regions, tabular list of alpine and subalpine species showing local occurrence, types of distribution, synanthropous plants, post-glacial history, relict species, botanical explorations, statistics; annotated list of vascular plants (955 species, of which 835 are regarded as native or naturalized; *Taraxacum* and *Hieracium* are treated incompletely and not included in these figures), with reference to first record, and detailed localities; list of excluded species, bibliography, list of collectors with brief biographical data, partial gazetteer. Replaces **Kröningssvärd, C. G.** Flora dalekarlica. Landskapet Dalarnes indigèna phanerogamer och filices. 66 p. 18 cm. Fahlun, 1843. (In Gray Herbarium library.) Also replaces **Indebetou, Conrad.** Flora dalekarlica. Dalarnes fanerogamer och filices 48 p. 20 cm. Nyköping, 1879. (In Lloyd Library.)

Local

Andersson, Gunnar, and **Hesselman, Henrik.** Vegetation och flora i Hamra kronopark. Ett bidrag till kännedomen om den svenska urskogen och dess omvandling. Skogsvårdsför. Tidskr. 5 (Fackuppsatser): 41–110. 15 fig. (incl. maps). 1907. (Also as Meddel. Stat. Skogsförsöksanst. **4:** 35–102, vii–xii. 15 fig. (incl. maps). 1908.)

Topography, geology, climate, history, plant associations, etc.; annotated list of vascular plants, arranged alphabetically. The issue in Meddel. Stat. Skogsförsöksanst. includes a German summary. (In Gävleborgs län.)—See also Malme and Malme, below.

Arwidsson, Thorsten. Floran inom Töfsingdalens nationalpark. K. Svenska Vetenskapsakad. Skr. Naturskyddsär. no. 25. 47 p. 1 fig. (map). 1934.

Topography, plant regions, plant associations, account of bryophytes and lichens; annotated list of vascular plants (218 species); bibliography. (An area of about 1,365 hectares in Idre parish.)

Cedergren, G. R. Om floran i Norrbärke socken i Dalarnes bergslag och dess förhållande till kalken. Bot. Notiser **1925:** 17–46. 1925.

Botanical explorations, general features of flora, geology, ecology; list of the more noteworthy vascular plants, with localities, list of commoner species, with localities; bibliography.

Lundblad, Karl. Geologi, jordmån och vegetation inom Siljansfors Försökspark i Dalarna. 1 p. l., 112 p. 3 maps. 24 cm. Stockholm, 1927. (Skogsförsöksanst. exkursionsled. 12.)

Topography, climate, geology, plant associations, etc.; tabular alphabetical list of vascular plants, hepatics, and lichens, with indication of plant associations in which they occur; localities for some of the rarer species; bibliography. (Mostly in Mora socken, Kopparbergs län.)

Malme, G. O., and **Malme, Rut.** Floran i Hamra nationalpark. K. Svenska Vetenskapsakad. Skr. Naturskyddsär. no. 23. 25 p. 1932.

Topography, general features of flora; annotated list of vascular plants [98 species] and lichens [131 species]. Hamra nationalpark, a part of Hamra kronopark, is an area of about 23 hectares in Orsa Finnmark (Hamra kapellförsamling), in Los socken in Gävleborgs län.—See also **Persson, Herman.** Mossfloran i Hamra nationalpark. l.c. no. 42. 25 p. 1943. (Adds 2 vascular plants, p. 1, also 12 lichens.)—Also Andersson and Hesselman, above.

Samuelsson, Gunnar. Studien über die vegetation der hochgebirgsgegenden von Dalarne. Nova Acta R. Soc. Sci. Upsal. ser. IV, v. 4, no. 8. 2 p. l., 252 p. 17 fig. (incl. maps), 8 pl. (incl. map). 1917.

Ecological; includes lists totaling 208 vascular plants from above the coniferous zone; bibliography.

Vesterlund, Otto. Floran inom Älvdalens kronopark. Svensk Bot. Tidskr. **20:** 251–271. 1926.

General features of flora; annotated alphabetical list of vascular plants (380 species); list of mosses, comparison with flora of Hamra kronopark.

Wiger, Johan. Floran inom Svärdsjö socken. 1–2. Bot. Notiser **1946:** 507–528. 1946; **1948:** 393–400. 1948.

Topography, geology, climate, general features of flora, botanical explorations, etc.; annotated alphabetical list of vascular plants; bibliography. The second paper contains an extensive list of additional species and localities.

Wiström, P. W. Växtnamn på folkspråket i Dalarne. Landtbrukstidskrift för Dalarne, Aug. 1907, p. 15–16. 1907.

Unannotated list of vernacular names from Sävsnäs and Grangärde socknar, systematically arranged. (Not seen; title and annotation supplied by E. Hultén.)

DALSLAND (DAL)

See also Värmland (Larsson, Flora öfver Wermland och Dal, under Hård av Segerstad, Den värmländska kärlväxtflorans geografi; Myrin, Anmärkningar om Wermlands och Dalslands vegetation).

The old province (landskap) of Dalsland or Dal is the region west of Lake Vänern, forming the northern part of Älvsborgs län.

General

Larsson, L. M. Symbolae ad floram Daliae. 19 p. 21 cm. Carolstadii, 1851. (Diss.) (Gray Herbarium library.)

Botanical regions, topography, geology, general features of flora, etc.; list of noteworthy vascular plants and mosses, with localities.—See also his Flora öfver Wermland och Dal (1868), under Värmland.—Also **Kindberg, N. C.,** and **Sundblad, R. F.** Novitier för Dalslands flora, upptecknade af . . . Bot. Notiser **1854:** 41–43. 1854.

Local

Hård av Segerstad, Fredrik. Pflanzengeographische studien im nordwestlichen teil der eichenregion Schwedens, I und II. Arkiv Bot. v. 27A, no. 1. 405 p. 275 fig. (mostly maps), map. 1935.

Geology, soils, postglacial history and immigration of plants, topography, hydrography, climate, phytogeographical regions, discussion of ecological and phytogeographical relationships of the species (with list, p. 125–129, of the species of 9 islands on the northwest coast of Lake Vänern, grouped according to frequency of occurrence); alphabetical list of known vascular plants (except *Betula alba* group, *Rosa, Hieracium,* and *Taraxacum*), with first record, habitat, detailed local distribution, ecological classification, pH determinations, references to published distributional maps in this and other northern regions; phytogeography, bibliography. Relates to a land surface of about 600 sq. km. on the northwest side of Lake Vänern, including the whole härad of Tössbo in Dalsland, except its southernmost part, and parts of Svanskog socken in the härad of Gillberg and of Kila and Tveta socknar in the härad of Näs, all the latter in Värmland.

GÄSTRIKLAND (GESTRIKLAND)

The old province (landskap) of Gästrikland now forms the southern part of Gävleborgs (Gefleborgs) län.

Local

Hartman, R. W. Gefle-traktens växter, med växtställen för de sällsyntare. 2. uppl. 60 p. 21 cm. Gefle, 1863. (Riksmuseet library, Stockholm.) (1st ed. by Hartman, Carl, and Hartman, R. W. Flora gevaliensis seu enumeratio plantarum circa Gevaliam sponte crescentium. 3 pt. 57 p. 20 cm. Gevaliae, 1847–48. (Diss.))

Topography, phenology; annotated list of vascular plants (693 species), bryophytes, lichens and algae, with localities for rarer species. The 1st edition (in Gray Herbarium library) gave an account of topography, geology, general features of flora, climate, botanical explorations; briefly annotated list of vascular (707 species) and cellular plants (except fungi), with localities for rarer species; list of adventive plants, with dates; statistics. The list of species, with a condensed introduction in Swedish, was reissued with title Gefle-traktens vexter, enligt Prof. Fries' naturliga system. 2 p.l., p.

15–57. 20.5 cm. Gefle, 1848. (In Riksmuseet library, Stockholm.) Both editions (1847–48 and 1863) related only to southeastern Gästrikland (Gävle region and Valbo and Hille parishes.) The following works relate to the province at large: **Dahlstedt, Fredrik.** Bidrag till kännedomen om vegetationen i Gestrikland. Svensk Bot. Tidskr. **10**: 578–608. 1916. (Account of botanical work and publications, phytogeography, numerous lists of species from different habitats and localities, bibliography.)—Also **Almquist, Erik.** Växtgeografiska bidrag. 1. Gästrikland. Bot. Notiser **1919**: 279–283. 1919. (List with localities.)—Also **Arnell, W. H.** Anteckningar om Gästriklands kärlväxter. Svensk Bot. Tidskr. **18**: 85–127. 1924. (Account of botanical explorations, etc.; extensive alphabetical list of vascular plants, with localities, including the discoveries of other botanists.)—Also **Ahlner, Sten.** Anteckningar om adventivfloran vid Gävle. l.c. **23**: 347–355. 1929. (Account of localities, etc.; list of adventive phanerogams collected by author, with localities; short bibliography.)

GÖTEBORGS OCH BOHUS LÄN

Göteborgs och Bohus län is here taken in its modern extent, including the old landskap of Bohuslän and the Göteborg area (extending east to Härryda), which is in the landskap of Västergötland.

General

Fries, Harald, *ed.* Göteborgs och Bohus läns fanerogamer och ormbunkar. Förteckning med fyndorter. 480 p. incl. 30 fig. (mostly maps) + 192 fig. (distribution maps), map. 21 cm. Göteborg, 1945.

Botanical explorations, geography, geology, climate, general features of flora, phytogeography, bibliography; annotated list of vascular plants (1,745 species, of which about 1,150 are native or naturalized) with detailed localities and reference to first published record. The genus *Hieracium* is treated very summarily, and *Taraxacum* is omitted. Covers Göteborgs och Bohus län and Angered and Nödinge socknar in Älvsborgs län. Replaces (as to this area) **Lindeberg, C. G.** Hallands och Bohusläns fanerogamer och ormbunkar . . . Göteborgs K. Vetensk. och Vitterh. Samhäll. Handl. n.f., v. 16 [no. 11]. 27 p. 1878. (Topography, general features of flora, botanical investigations; annotated list of vascular plants.) Also replaces **Palmér, J. E.** Förteckning över Göteborgs och Bohus läns fanerogamer och kärlkryptogamer. 146 p. 21 cm. Uddevalla, 1927. (Annotated list of vascular plants, with localities for scarcer species. In New York Botanical Garden library.)—See also **Fries, Harald.** Tillägg till växtförteckningen i "Göteborgs och Bohus läns fanerogamer och ormbunkar." Acta Hort. Gotoburg. **17**: 77–95. 1947. (Extensive list of additional localities, and a few additional species.)—Also **Woldmar, Stig.** Anteckningar till kärlväxtfloran i mellersta Bohuslän. Svensk Bot. Tidskr. **51**: 106–116. 1957. (Extensive list; bibliography.)—Also **Ivarsson, Reinhold.** Bidrag till Bohusläns flora. l.c. **51**: 558–563. 1957. (List of noteworthy species.)—Also **Ohlander, Manne.** Nyare bidrag till kännedomen om kärlväxtfloran på Orust. l.c. **52**: 52–62. 1958. (General features of flora, etc.; extensive list of noteworthy species; bibliography.)

Local

Areschoug, J. E. Plantae cotyledoneae florae gothoburgensis . . . 2 p. l., viii, 150 p. 21.5 cm. Londini Gothorum, 1836.—Bidrag och anmärkningar till Götheborgs flora. Bot. Notiser **1840**: 17–25. 1840.

Annotated, briefly descriptive flora of spermatophytes (680 species); list of adventive or dubious species. Covers nearly the same area as Wahlberg's Flora, namely Fässberg, Kållered, Råda, Frölunda, Askim, Styrsö, Örgryte, and Partille socknar, part of Hisingen (island), also Hönö, Öckerö, Björkö, etc. Replaces **Wahlberg, P. F.** Flora gothoburgensis. 2 pt. (1 p. l., 117 p.). 19 cm. Upsaliae, 1820–24. (Diss.) (Topography, climate, general features of flora; annotated list of vascular plants (645 species) and mosses. Covers region within 1 Swedish mile (10.7 km.) of the city. In Gray Herbarium library.)—See also **Fries, E. T.** Några växtlokaler i Bohuslän och Göteborgstrakten. Bot. Notiser **1911**: 39–49. 1911. (Includes numerous records from the Göteborg area.)—Also **Fries, Harald.** Några adventivväxter från Göteborg, tagna för 40 år sedan. l.c. **1924**: 93–94. 1924. (List of adventive plants collected by J. A. Hallberg in 1878–1882, with dates.)—Also his Anmärkningsvärda växtlokaler huvudsakligen från Göteborgstrakten och från Tjörn i Bohuslän. l.c. **1924**: 453–456. 1924. (List with localities.)—Also **Nilsson, Gunnar.** Material till Bohusläns och Göteborgstraktens flora. l.c. **1925**: 377–387. 2 fig. 1925. (List of vascular plants, with localities; includes records from the Göteborg region.)—Also **Borgvall, Torsten.** Anmärkningsvärda växtlokaler från Göteborgstrakten. l.c. **1927**: 65–68. 1927. (List with localities.)—Also his Några växtlokaler från Göteborgstrakten samt från Lerums och Skallsjö socknar i Älvsborgs län. Acta Hort. Gothoburg. **3**: 157–162. 1927. (List with localities.)—Also his Växtgeografiska bidrag från Göteborgsområdet samt angränsande delar av Västergötland och Halland. l.c. **4**: 173–178. 1929. (List with localities.)—Also his Växtgeografiska bidrag från Västkustlandskapen. l.c. **6**: 37–46. 1931. (List with localities, mostly from the Göteborg area.)—Also **Borgvall, Torsten, and Haglund, G. E.** The Taraxacum flora of the Gothenburg region. Acta Hort. Gotob. **21**: 1–42. 7 pl. (on 4). 1957. (Botanical explorations; list of 135 species, with localities; bibliography.)—Also Blom, below.

Beurling, P. J. Några ord om vegetationen vid Strömstad i norra Bohuslän. Bot. Notiser **1854**: 129–137. 1854.—Några ord . . . (Tillägg . . .) l.c. **1855**: 97–104. 1855.

Topography; extensive list of vascular plants, arranged by habitats.—See also **Hansson, C. A.** Några nya växtlokaler i Strömstad. Bot. Notiser **1874**: 159. 1874.

Blom, Carl. Bidrag till kännedomen om Sveriges adventivflora. [I]–IV. Acta Hort. Gothoburg. **3**: 143–150. 1927; **8**: 169–188. 1933; **11**: 159–182. 1936; **13**: 385–403. 1940.

Extensive lists of adventive vascular plants, collected mostly by author, with localities; referring in part to Halland, Skåne, Småland, Södermanland, and Västergötland, but mostly (**3**: 148–150, **8**: 169–186, **11**: 159–176, **13**: 385–400) to the Göteborg region.

——— Om floran i Lurs skärgård. Acta Hort. Gothoburg. **4**: 285–300. 3 fig. 1929.

Running account of vegetation in different localities and habitats, with mention of vascular plants observed; refers mostly to Resön.

Fröman, Ingmar. Växtgeografiska anteckningar från Bohuslän. Bot. Notiser **1925**: 170–184. 2 fig. 1925.

Extensive list of vascular plants from various localities in vicinity of Fiskebäckskil, followed by a list of vascular plants from immediate vicinity of Kristineberg (in Fiskebäckskil and Skaftö kommuner).

Gillner, Vilhelm. Strandvegetationen på Stora Amundön i Askims socken, Göteborgs och Bohus län. Acta Hort. Gotoburg. **15**: 243–266. 17 fig. (fig. 1, map, fig. 2–17 on 4 pl.). 1944.

Chiefly ecological; includes unannotated list of known vascular plants; bibliography.

Hård av Segerstad, Fredrik. Om floran i Naverstad. Acta Hort. Gotoburg. **11**: 61–69. 1 fig. 1936.

Geology, topography, etc.; unannotated list of about 450 vascular plants, grouped according to frequency of occurrence; phytogeography, etc.—See also **Lange, Thorvald.** Några växtgeografiska anteckningar från Bohuslän. Svensk Bot. Tidskr. **15**: 302–308. 1922. (Includes some species from Naverstad not in Hård av Segerstad's list.)

Lange, Thorvald. Kärlväxtfloran i Styrsö socken i Göteborgs och Bohus län. Svensk Bot. Tidskr. **6**: 282–311. 1912.

Botanical explorations, general features of flora; annotated list of vascular plants; bibliography.

Lindström, A. A. Marstrandsöns ormbunkar och fanerogamer. Bot. Notiser **1920**: 177–210. 1920.—Tillägg . . . l.c. **1921**: 287–288. 1921.

Botanical explorations; annotated list of vascular plants of Marstrandsön; unannotated list of species collected on several other small islands in the vicinity of Marstrand.—See also his Några växtlokaler för sällsyntare växter i Marstrandstrakten. Bot. Notiser **1913**: 79–80. 1913.

Nilsson, Sture. Floristiska anteckningar från Bohuslän. Acta Hort. Gotoburg. **9:** 23–40. 1934.

Geology, general features of flora, vegetation of shell sand banks, etc.; extensive list of more interesting vascular plants from Svarteborg, Mo, Håby, Foss, and Hede parishes.

Ohlsén, Ragnar. Bidrag till kännedom om kärlväxtfloran i Sörbygdens härad, Bohuslän. [I]–II. Acta Hort. Gothoburg. **2:** 51–69. 1926; **3:** 33–42. 1927.

Topography, geology, general features of flora; annotated list of vascular plants. The first part refers mostly to Hede, Krokstad, Sanna, and Svarteborg socknar, the second part to the northern part of Krokstad socken.

GOTLAND (GOTHLAND)

The old province (landskap) of Gotland has the same boundaries as Gotlands län.

General

Johansson, Karl. Hufvuddragen af Gotlands växttopografi och växtgeografii grundade på en kritisk behandling af dess kärlväxtflora. K. Svenska Vetensk.-Akad. Handl. n.f., v. 29, no. 1. 270 p. map. 1897.

Botanical explorations, bibliography, briefly annotated list of local botanists, climate, plant formations, general features of flora, phytogeography; annotated list of vascular plants (1,003 species). His Förteckning öfver Gotlands fanerogamer och ormbunkar med uppgift på växtlokaler, frekvens m.m. (Stockholm, 1897) is a separate issue of pages 1–20, 109–270 of his Hufvuddragen, containing the introductory matter and the list of species. (In Riksmuseets library, Stockholm.) Replaces **Eisen, Gustaf, and Stuxberg, Anton.** Gotlands fanerogamer och thallogamer, med fyndorter för de sällsyntare. 1 p.l., 76 p. 22 cm. Upsala, 1869. (Annotated list of 957 vascular plants; list of localities with their rarer species. In Gray Herbarium library.)—See also **Johansson, Karl.** Nyare bidrag till kännedomen om Gotlands kärlväxtflora. Bot. Notiser **1910:** 209–258. 1910. (Extensive list, with localities, and additional bibliography.)—Also **Vestergren, Tycho.** Några växtfynd på Gotland. Svensk Bot. Tidskr. **7:** 378–380. 1914. (Notes on 4 species.)—Also **Fries, E. T.** Några floristiska notiser från Gotland. l.c. **8:** 263–266. 1914. (List with localities.)—Also his Några gotländska växtlokaler. l.c. **11:** 134–137. 1917; **14:** 341–344. 1920; **19:** 426–429. 1925. Bot. Notiser **1934:** 261–265. 1934; **1942:** 409–412. 1942. (Lists with localities.)—Also his Några färgvarieteter i Gotlands flora. l.c. **1932:** 101–104. 1932. (Extensive list of color varieties, with localities.)—Also **Ridelius, K. G.** Några för Gotland nya växter. Svensk Bot. Tidskr. **27:** 108–111. 1933. (Adds 4 species.)—Also **Nordin, Ingvar.** Gotländska växtfynd 1956 och 1957. l.c. **52:** 195–201. 4 fig. 1958. (List of noteworthy species; bibliography, including references to earlier papers by him.)

Local

Alroth, Nils. Något om Mästermyrs nutida vegetation. Acta Hort. Gotoburg. **14:** 61–86. 6 fig., map. 1940.

Ecological; includes annotated list of 284 plants (mostly vascular) and list of additional species recorded by J. E. Ljungqvist in 1914 (Mästermyr. En växtekologisk studie. I. v, 57 p. 11 fig., 6 pl. (incl. 2 maps). 26 cm. Karlstad, 1914. (Akad. avhandl.)) but no longer found; bibliography. Mästermyr was a large bog near Hemse, now essentially destroyed by draining.

Arwidsson, Thorsten. Studier över floran och vegetationen på Gotska Sandön med särskild hänsyn till nationalparken. K. Svenska Vetenskapsakad. Avhandl. Naturskyddsär. no. 1. 71 p. 3 fig. (incl. 2 maps), 6 pl. 1938.

Soils, botanical explorations, general features and sources of flora, annotated lists of algae, fungi, lichens, and mosses; annotated list of known vascular plants, with localities; bibliography. Replaces **Holtz, Ludwig.** Flora der insel Gottska-Sandö. Verhandl. Bot. Ver. Prov. Brandenb. **13:** 59–68. 1872; Berichtigung . . . l.c. **14:** 153. 1873. (Soil, climate, etc.; annotated list of 96 vascular and cellular plants.) (Small island north of Gotland.)

Johansson, Karl. Calendarium florae gotlandicae. Blomning och lövspricknning i Visbytrakten. Svensk Bot. Tidskr. **21:** 423–477. 3 fig. 1927.

List of about 700 spermatophytes of vicinity of Visby arranged by date of first flowering; meteorological relations, etc.; list of species arranged by date of leafing; alphabetical list of species, showing dates; bibliography.

Lindman, C. A. M. Kärlväxtfloran på Visby ruiner. Öfvers. K. Vetensk.-Akad. Förhandl. **52:** 519–536. 1895.

List of 98 vascular plants found on the old limestone ruins, with discussion of method of spreading, etc.; German summary.

Wahlin, B. J. O. Vegetationen på Lilla Karlsö. I. Lövskogsfragment. Bot. Notiser **1945:** 81–100. 5 fig. (fig. 1, map). 1945.

Botanical explorations, topography, geology, tree vegetation; incomplete list of vascular plants; bibliography.

HÄLSINGLAND (HELSINGLAND)

The old province (landskap) of Hälsingland forms the northern part of Gävleborgs län.

General

Wiström, P. W. Förteckning öfver Helsinglands fanerogamer och pteridofyter, uppgjord efter J. A. Wiströms efterlämnade anteckningar och med tillägg utgifven. 104, [3] p. 20.5 cm. Wimmerby, 1898. (New York Botanical Garden library.)

Earlier publications, etc.; briefly annotated list of vascular plants, with localities, followed by 3 pages of phenological tables. Replaces **Hartman, R. W.** Helsinglands cotyledoneae och heteronemeae. 35 p. 19 cm. Gefle, 1854. (Acad. afhandl.) (Topography, general features of flora; barely annotated list of 880 vascular and cellular plants (about 600 vascular). In Gray Herbarium library.) Also replaces **Wiström, J. A.** Provinsen Helsinglands fanerogama vexter och ormbunkar. 44 p. 22.5 cm. Gefle, 1867. (Topography, geology, climate, plant formations; briefly annotated list of vascular plants, with localities for scarcer species; list of ballast plants. In Riksmuseet library, Stockholm. Krok cites another issue (as thesis): 44, [3] p. Hudiksvall, 1867.)—See also **Westerlund, C. G.** Bidrag till Hälsinglands flora. Bot. Notiser **1906:** 1–40. 1906; **1909:** 95–96. 1909. (Extensive list, chiefly from Hudiksvall and vicinity; the second part contains *Hieracium* only.)—Also **Birger, Selim.** Bidrag till Hälsinglands flora. l.c. **1906:** 81–84. 1906. (Alphabetical list with localities, from western Hälsingland.)—Also **Johansson, Karl.** Små bidrag till Hälsinglands och Jämtlands flora. l.c. **1911:** 267–273. 1911. (List from Hälsingland, p. 267–269.)—Also **Strömman, P. H.** Bidrag till Helsinglands kärlväxtflora. Svensk Bot. Tidskr. **5:** 359–365. 1911. (Extensive list, with localities.)—Also **Lidman, Gottfrid.** Några anteckningar om Hälsinglands flora. l.c. **19:** 84–97. 1925. (Extensive list, mostly from western Hälsingland, with localities.)—Also **Halden, B. E.** Tre för Hälsingland nya kärlväxter jämte några andra växtfynd från södra Hälsingland. l.c. **20:** 65–67. 1926. (List of noteworthy species, with localities.)

HÄRJEDALEN (HERJEDALEN)

See also Jämtland (Kilander).
The old province (landskap) of Härjedalen forms the southern part of Jämtlands län.

General

Birger, Selim. Härjedalens kärlväxter. 1 p. l., 96 p. 22 cm. Stockholm, 1908.

Geography, list of collectors; annotated list of vascular plants, with localities. Replaces **Thedenius, K. F.** Anmärkningar om Herjedalens vegetation. K. Vetensk.-Akad. Handl. **1838:** 24–76. pl. 1. 1839. (Reprinted, 53 p. 1839; in Riksmuseet library, Stockholm.) Also replaces (as to Härjedalen) **Dusén, K. F.** Bidrag till Härjedalens och Helsinglands flora. Öfvers. K. Vetensk.-Akad. Förhandl. **37** (2): 3–44. 1880. (Botanical explorations, author's itinerary; partly annotated list of 562 vascular plants of Härjedalen.)—For account of botanical explorations, geology, climate, plant formations, phytogeography, etc., with bibliography, see **Birger, S.** Om Härjedalens vegetation. Arkiv Bot. v. 7, no. 13. 136 p. 6 fig., 13 pl. (pl. 1, map). 1908.—See also **Arwidsson, Thorsten.** Växtgeografiska notiser från Norrland. II. Bot. Notiser **1929:** 305–310. 1929. (List of noteworthy vascular plants, especially from Härjedalen.)

Modin, Erik. Växtnamn samt folkliga bruk och föreställningar rörande växter i Härjedalen. *In* Svenska Landsmål [etc.]. Fästskrift till H. F. Feilberg . . . p. 697–737, 802–803. 23.5 cm. Stockholm, 1911.
Includes alphabetical list of scientific names of vascular plants, mosses, lichens, and fungi, with local vernacular names and their explanation and notes on popular uses and beliefs; French abstract.

Local

Arwidsson, Thorsten. Floran inom Sonfjällets nationalpark. K. Svenska Vetenskapsakad. Skr. Naturskyddsär. no. 12. 37 p. 1 fig. (map). 1930.
Topography, plant zones, botanical explorations, gazetteer, etc.; annotated list of vascular plants (250 species); bibliography. (An area of about 2,700 hectares in Hede kronopark south and west of Valmån, including the highest part of Sonfjället.)

Cedergren, G. R. Till kännedomen om floran i norra Härjedalen med särskild hänsyn till Vemdalen. Arkiv Bot. v. 14, no. 17. 72 p. 1916.
Botanical explorations, geology, local distribution, etc.; annotated list of vascular plants; bibliography. For some corrections in locality names, etc., see Svensk Bot. Tidskr. **21:** 18. 1927.—See also **Nordström, K. B.** Växtgeografiska uppgifter från Vemdalstrakten i Härjedalen. Bot. Notiser **1915:** 293–298. 1915. (List with localities.)—Also **Nordin, Ingvar.** Växtanteckningar från trakten kring Vemdalen, Härjedalen sommaren 1950. Bot. Notiser **1954:** 52–53. 1954. (List of noteworthy species.)

Smith, Harald. Vegetationen och dess utvecklingshistoria i det centralsvenska högfjällsområdet. 2 p.l., 238 p. 41 fig. (incl. map), 2 maps. 27 cm. Uppsala, 1920. (Särtryck ur Norrländskt Handbibliotek, IX.)
Topography, geology, climate, plant formations, etc. of the high mountain region of central Sweden; annotated list of vascular plants; bibliography. Covers area in northwestern Härjedalen and southwestern Jämtland between 62°15′ and 63°20′ N. Lat., from about 13° E. Long. west to the Norwegian boundary.—See also **Östergren, Olof.** Några fynd och iakttagelser på det "centralsvenska högfjällsområdet." Svensk Bot. Tidskr. **21:** 18–32. 1927. (Extensive list of vascular plants from vicinity of Fjällnäs in Härjedalen and Enafors in Jämtland, with localities.)—Also **Smith, Harald.** En botanisk undersökning av Neans dalgång. K. Svenska Vetenskapsakad. Avhandl. Naturskyddsär. 16. 21 p. 2 fig. (maps), 3 pl. (on 2). 1957. (Includes some new records.)

HALLAND

See also Skåne (Neuman); Västergötland (Westfeldt, Växtligheten).
The old province (landskap) of Halland has essentially the same boundaries as Hallands län.

General

Ahlfvengren, F. E. Hallands växter. Förteckning över fanerogamer och kärlkryptogamer. xix, 207 p. port., map. 22.5 cm. Lund, 1924.
Botanical explorations, bibliography; annotated list of vascular plants, with localities and references to first records; index to localities. Replaces **Fries, E. M.** Flora hallandica . . . Part. prioris sectio I–X. 159 p. 19.5 cm. Lundae, 1817–19; also (as to Halland) **Lindeberg, C. J.** Hallands och Bohusläns fanerogamer och ormbunkar . . . Göteborgs K. Vetensk.-och Vitterh. Samhäll. Handl. n.s., v. 16, [no. 11]. 27 p. 1878.—See also **Erdtman, Gunnar.** Växtlokaler från Halland och sydvästra Västergötland. Svensk Bot. Tidskr. **19:** 370–379. 1925. (Extensive list of vascular plants, with localities.)—Also **Wiger, Johan.** Adventivfynd och växtlokaler från Halland. Bot. Notiser **1926:** 283–285. 1926. (Alphabetical list of adventive and other plants, mostly from Halmstad.)—Also **Svenson, Sten.** Halländska växtlokaler. Acta Hort. Gothoburg. **4:** 27–45. 1928. (Extensive list of plants, with

new localities; short bibliography of recent papers.)—Also **Blom, Carl.** Halländska växtlokaler. l.c. **5:** 129–134. 1930. (Similar list, mostly from extreme southern Halland.)—Also **Wiger, J.** Nya adventivfynd och växtlokaler från Halland. Bot. Notiser **1931:** 201–206. 1931. (Alphabetical list of plants, mostly from Halmstad and Söndrum.)—Also **Peterson, Bo.** Bidrag till Hallands flora. Acta Hort. Gotoburg. **17:** 345–354. 1947. (List of noteworthy angiosperms, with localities; bibliography supplementary to that in Ahlfvengren.)—Also **Lundegren, Alf.** Nya halländska växtlokaler. Hallands Natur **1949:** 29–32. 1949. (Lists of noteworthy species from different localities. In Riksmuseet library, Stockholm.)—Also **Peterson, Bo.** Glimtar från den pågående inventeringen av Hallandsfloran. l.c. **1949:** 33–39. 2 fig. 1949. (Includes list of noteworthy vascular plants with localities. In Riksmuseet library.)—Also **Wiger, Johan.** Anteckningar om halländska växter. l.c. **1950:** 9–15. 1 fig. 1950. (Alphabetical list of noteworthy vascular plants with localities; includes some new to the province. In Riksmuseet library.)—Also **Nilsson, Örjan.** Några trevliga växtfynd i norra Halland. Svensk Bot. Tidskr. **46:** 493–494. 1952. (Records of several species.)

Local

Höijer, Walter. Förteckning över fanerogamer funna inom strandreservatet i Steninge. Hallands Natur **1947:** 24–26. 1947. (Riksmuseet library, Stockholm.)

Unannotated alphabetical list of spermatophytes.

Malmström, Carl. Tönnersjöhedens försökspark i Halland. Ett bidrag till kännedomen om sydvästra Sveriges skogar, ljunghedar och torvmarker. (Das versuchsrevier Tönnersjöheden in Halland. Ein beitrag zur kenntnis der südwestschwedischen wälder, heiden und torfmoore.) Meddel. Stat. Skogsförsöksanst. **30:** 323–528. 63 fig. (incl. maps), 2 maps. 1937.

Ecological; includes (p. 468–482) alphabetical tabular lists of vascular plants, mosses, lichens, and Hymenomycetes with indication of habitat and notes on some of the rarer species; also bibliography.

Peterson, Bo. Kärlväxtvegetationen på yttermurarna av Varbergs fästning. 4 p. 24 cm. n. p., 1948. (Reprinted from Vår Bygd [Halmstad] **1948:** 30–33. 1948. Original not seen; reprint in library, U.S. Dept. Agriculture.)

Conditions of growth, etc.; unannotated alphabetical list of vascular plants (57 species) collected on the outer walls of the fortifications of Varberg in recent years.

Söderberg, Pontus. Förteckning över de fanerogama växter, som iakttagits inom Halmstad med omnejd. För. Gamla Halmstads Årsbok **6** (1929): 217–229. 1929; 7 (1930): 250–264. 1930.

List of phanerogams, with localities. (Not seen; title and annotation supplied by B. H. Peterson.)—See also Skåne (Neuman, Bidrag til kännedomen af floran på Sveriges syd-vestkust omfattande trakten mellan Halmstad och Engelholm).

Theorin, G. R. A. Växtgeografisk skildring af södra Halland. 1 p. l., xvi, 32 p. 21 cm. Lund, 1865. (Akad. afhandl.) (Lloyd Library.)

Topography, climate, geology, general features of flora, plant associations, botanical explorations; briefly annotated list of vascular plants, with localities. Covers the area south of the Nissan River.

JÄMTLAND (JEMTLAND)

See also Härjedalen (Smith, Vegetationen och dess utvecklingshistoria i det centralsvenska högfjällsområdet); also Medelpad (Falck, Botaniska anteckningar från västra Medelpad och sydöstra Jämtland).

The old province (landskap) of Jämtland or Jemtland forms the northern part of Jämtlands län.

General

Lange, Thorvald. Jämtlands kärlväxtflora. Acta Bot. Fenn. 21. 204 p. (incl. 124 maps). map. 1938.

Botanical explorations, topography, geology, soils, climate, phenology, plant associations; annotated list of vascular plants (756 species and subspecies, plus 88 Taraxaca and 332 Hieracia) with detailed localities; list of collaborators, bibliography. Replaces **Olsson, Peter.** Jemtlands fanerogamer och ormbunkar, upptecknade med angifvande af växtlokaler. Öfvers. K. Vetensk.-Akad. Förhandl. **41** (9): 41–155. 1884; Tillägg . . . l.c. **53**: 101–156. 1896. (Botanical investigations (with bibliographical footnotes), physiography, geology, etc.; annotated list of 828 vascular plants.)—See also **Lange, T.** Några tillägg och rättelser till "Jämtlands kärlväxtflora." Svensk Bot. Tidskr. 34: 47–52. 1940. (Extensive list of new localities, etc.)—Also his Anteckningar till Jämtlands flora. V. Bot. Notiser **1943**: 114–150. 8 fig. (maps). 1943. (Phytogeographical; includes account of original (ursprungliga) species, species found in adjacent districts, immigration and postglacial history, bibliography.)—Also **Hedberg, Olov.** Några växtlokaler från södra Jämtland. Svensk Bot. Tidskr. **39**: 431–432. 1945. (List of vascular plants observed by author on Hottöfjället, in Undersåker; alphabetical list of additional records from various localities.)—Also **Kilander, Sven.** Bidrag till Jämtlands kärlväxtflora. Bot. Notiser **1947**: 171–175. 1947. (List of new locality records; remarks on several species.)—Also **Stenar, Helge.** Några växtlokaler från Jämtland. Svensk Bot. Tidskr. **41**: 295–308. 1 fig. (map). 1947. (Extensive list of vascular plants, with localities; bibliography.)—Also **Danielsson, Bengt.** En exkursion i norra Frostviken. l.c. **42**: 486–492. 1 fig. (map). 1948. (Includes alphabetical list of species from several mountain summits.)—Also **Stenar, Helge.** Floristiska strövtåg kring Bäverloken i Ströms socken. En lokal för vildhumle och glesvia. (Glyceria lithuanica.) Sveriges Natur **40** (no. 1, Årsbok): 34–41. 1 fig. 1949. (List of noteworthy species from region of Bäverloken. Not seen; cited from Sveriges Natur **41** (no. 1A, Årsbok): 164. 1950.)—Also his Ytterligare några växtlokaler från Jämtland. Svensk Bot. Tidskr. **44**: 504–514. 1 fig. (map). 1950. (Extensive list of vascular plants, with localities.)—Also **Danielsson, Bengt.** Bidrag till Jämtlands kärlväxtflora. l.c. **45**: 400–407. 2 fig. 1951. (Alphabetical list of noteworthy vascular plants, largely from Föllinge and Frostviken parishes; bibliography.)—Also **Hakelier, Nils.** Bidrag till Jämtlands kärlväxtflora. Bot. Notiser **1951**: 278–279. 1951. (New localities for various species.)—Also **Flinck, K. E.** Nya växtlokaler för kärlväxter i Jämtland. l.c. **1954**: 53–54. 1954. (Includes 3 additional species.)

Local

Gjærevoll, Olav. Snøleievegetasjonen i Oviksfjellene . . . The snow-bed vegetation of Mts. Oviksfjellen, Jämtland, Sweden. Acta Phytogeog. Suecica v. 25. 106 p. illus. 1949.

Mainly phytosociological; includes (p. 80–90) list of vascular plants, with ecological annotations; English summary.

Henning, Ernst. Agronomiskt-växtfysiognomiska studier i Jemtland. 1 p. l., 34 p. 31 cm. Stockholm, 1889. (Akad. afhandl.)

Includes (p. 31–34) alphabetical list of scientific names of plants mentioned, with Swedish and Jemtland vernacular equivalents.

Kilander, Sven. Kärlväxternas övre gränser på fjäll i sydvästra Jämtland samt angränsande delar av Härjedalen och Norge. (Upper limits of vascular plants on mountains in southwestern Jämtland and adjacent parts of Härjedalen (Sweden) and Norway.) Acta Phytogeog. Suecica v. 35. 198 p. 41 fig. (incl. maps), table. 1955.

Description of localities, etc.; annotated list of vascular plants (p. 92–182) with localities and altitudes reached; bibliography, English summary.

Rune, Olof. Kärlväxtfloran i Offerdals och nordöstra Kalls fjällområde. Arkiv Bot. v. 32A, no. 11. 58 p. incl. 15 fig. (incl. 7 maps). 1946.

Author's explorations, topography, climate, geographical affinities of flora, ecology; annotated list of known vascular plants; bibliography.

LAPPMARK, ÅSELE

Bounded on the northwest by Norway, on the northeast by a line drawn northwest to the Norwegian boundary between the Ume älv and the northernmost branch (Vojmån älv) of the Ångerman älv from a point on the northern boundary of Ångermanland at about 64° 3′ N. Lat. (between Fredrika and Örträsk), on the southeast and southwest by the landskap of Ångermanland and Jämtland.

The boundaries of this and the other Lappmarker are shown on the maps in Holmberg's Skandinaviens flora, in Hylander's Nordisk kärlväxtflora, and at the end of each volume of the Archivum Societatis zoologicae botanicae fennicae Vanamo.

Local

Heintze, August. Växttopografiska undersökningar i Åsele Lappmarks fjälltrakter. I–II. Arkiv Bot. v. 12, no. 11. 196 p. 1913; v. 13, no. 5. 148 p. 1913.

Botanical explorations, distribution on north and south exposures, plant zones and formations; list of vascular plants (345 spontaneous species, excluding *Hieracium*), with detailed localities (including ecological occurrence); bibliography. Second part: plant zones and regions in Swedish Lapland; same list of species with generalized statement of range in Åsele Lappmark, ecological habitats and classification; discussion of vertical distribution (with table), alpine and subalpine species, etc.; bibliography.

Melander, Carl. I Åsele Lappmark sommaren 1880. Bot. Notiser **1881:** 26–28, 50–61, 98–101, 108–114. 1 fig. (map). 1881.

Account of author's trip, with mention of noteworthy plants observed.

Rune, Olof. Några anmärkningsvärda växtfynd i södra Lapplands fjäll. Svensk Bot. Tidskr. **39:** 299–303. 1 fig. 1945.

Short list of noteworthy species collected by author on Lasterfjället and Västra Fjällfjället in the parish of Vilhelmina in northwestern Åsele Lappmark; English summary.

Samuelsson, Gunnar. Växtlokaler från Ångermanland och Åsele Lappmark. Svensk Bot. Tidskr. **21:** 123–138. 1927.

Author's itinerary, with notes on vegetation; list of commoner species, extensive list of less common vascular plants, with localities. Refers to region from Långsele and Sollefteå in Ångermanland to Åsele Lappmark.

LAPPMARK, LULE

Bounded on the northwest by Norway, on the northeast by the Kalixälv to near Nikkaluokta, thence south of Kebnekaise to the reentrant angle in the Swedish-Norwegian boundary, on the southeast by the northern and middle part of the landskap of Västerbotten, and on the southwest by a line drawn NW-SE between the Lilla Lule älv and the Pite älv (nearer the latter, and reaching the Norwegian boundary near Sulitelma).

Local

Åberg, Börje. Kärlväxternas höjdgränser i Lule Lappmark och i Graubünden, en jämförelse. Svensk Bot. Tidskr. **46:** 286–312. 1952.

General considerations; list of vascular plants with citation of altitudinal records from Lule Lappmark and Graubünden, with some from various localities in the Alps, mainly in Switzerland; English summary, bibliography.

Andersson, N. J. Plantae vasculares circa Quickjock Lapponiae lulensis, quarum enumerationem . . . 2 pt. (36 p.). 20 cm. Upsaliae, 1844–45. (Diss.)

Topography, general features of flora, botanical regions; annotated list of vascular plants (356 species) with localities. A slightly condensed German version, by C. T. Beilschmied, is given in Flora **30:** 431–439, 447–454. 1847. —See also his Nya bidrag till Qvickjocks-traktens flora. Bot. Notiser **1866:** 107–110, 119–125. 1866. (List of 31 additional species collected by author and others, with localities; list of new localities for scarcer species; list of botanical names with Lapp equivalents.)—See also Simmons, below.

Björkman, Gunnar. Floran i trakten av Äpartjåkkos magnesitfält. K. Svenska Vetenskapsakad. Skr. Naturskyddsär. no. 33. 36 p. 4 fig., 1 tab. 1937.

Botanical explorations, topography, geology, general features of flora, plant formations, occurrence of *Carex maritima;* annotated alphabetical list of vascular plants; annotated list of mosses and lichens; bibliography.

——— Kärlväxtfloran inom Stora Sjöfallets nationalpark jämte angränsande delar av norra Lule Lappmark. K. Svenska Vetenskapsakad. Avhandl. Naturskyddsär. no. 2. iv, 224 p. 2 maps (1 in pocket). 1939.

Botanical explorations, plant zones, geography and gazetteer; annotated alphabetical lists of indigenous vascular plants (418 species), and anthropochorous (114 species) and apophytic species; discussion of introduced species, plant regions, vertical and ecological distribution, phytogeography, bibliography. (A narrow lake region associated with Stora Lule River, stretching northwest from just below Porjus to boundary with Norway.)

Frödin, John. Växttopografiska anteckningar i Stora Lule älvs källområde. Bot. Notiser **1915:** 113–138. 1915.

Botanical explorations, topography, geology, etc.; alphabetical list of vascular plants, with localities.—See also his Iakttagelser över Porjusområdets adventivflora. Bot. Notiser **1915:** 191–194. 1915. (Alphabetical list of adventive plants, with localities.)—Also his Några märkliga sydberg i Lule Lappmark. Tvänne nya lokaler för Potentilla multifida. Svensk Bot. Tidskr. **9:** 192–219. 2 fig. 1915. (Mainly phytogeographic; includes short lists of species from several localities.)

Selander, Sten. En artförteckning från Luottolako. Svensk Bot. Tidskr. **40:** 305–306. 1946.

Topography, etc.; list of vascular plants collected by author in 1943 (47 species and 1 hybrid). (A plateau 1,300 meters high, in the Sarek Mountains.)

——— Floristic phytogeography of south-western Lule Lappmark (Swedish Lapland). II. Kärlväxtfloran i sydvästra Lule Lappmark. Acta Phytogeog. Suecica v. 28. 152 p. 6 fig. (fig. 1, map), maps. 1950.

Taxonomic notes, explanation of locality names, botanical districts, etc.; annotated list of vascular plants (496 species), with localities; brief English summary, bibliography. The range of about 305 of the species is shown on maps. The region described is situated between those investigated by Björkman (Äpartjåkko and Stora Sjöfallet), Tengwall (Sarek), and Arwidsson (Pite Lappmark). For account of botanical explorations and phytogeography (in English) see pt. I of the same work. l.c. v. 27. 200 p. 33 fig. (incl. maps.), 12 pl. (on 6). 1950.

Simmons, H. G. Några bidrag till Lule Lappmarks flora. Bot. Notiser **1907:** 81–94, 97–112. 1907.

Extensive list of vascular plants collected chiefly by author in 1906 in Jockmock socken (including the region of Kvickjock kapellförsamling); discussion of introduced plants, with lists.—See also **Vesterlund, Otto.** Förteckning över fanerogam- och kärlkryptogamfloran inom Jockmocks och Kvickjocks skogsregion. Svensk Bot. Tidskr. **18:** 292–311. 1924. (Extensive list of vascular plants, with localities.)—See also Andersson, above.

Svensson, N. A. Om den fanerogama och kärlkryptogama vegetationen kring Kaitumsjöarne i Lule Lappmark. Bihang K. Svenska Vetensk.-Akad. Handl. v. 21, afd. 3, no. 1. 46 p. 5 fig. (incl. map). 1895.

Topography, general features of flora, plant formations, etc.; annotated list of 232 vascular plants, also list of mosses.

Sylvén, Nils, and Bågenholm, Gustaf. Ruderatväxter, antecknade från Lule Lappmark sommaren 1901. Bot. Notiser **1902:** 269–270. 1902.

List of ruderal angiosperms observed in 1901 in a field on a farm on the east shore of Lake Skalkan (66° 42′ N. Lat.) which had been planted to oats in 1900.

Tengwall, T. Å. Die gefässpflanzen des Sarekgebietes. *In* Hamberg, Axel, *ed.* Naturwissenschaftliche untersuchungen des Sarekgebirges in Schwedisch-Lappland. v. 3, Botanik, p. 627–674. 1925.

General features of flora, list of localities; list of vascular plants from above the coniferous zone, with detailed localities.—See also **Dahlstedt, Hugo.** Die Hieracien des Sarekgebietes in Lule Lappmark. l.c. 675–693. 1925.—For account of ecology, history of vegetation, bibliography, etc., see **Tengwall, T. Å.** Die vegetation des Sarekgebietes. I–II. l.c. 269–436. 73 fig., pl. 10–11 (maps). 1920; l.c. 703–774. 24 fig. 1925.—Also **Åberg, Börje.** Om kärlväxternas höjdgränser i Sarek. Svensk Bot. Tidskr. **43:** 740–767. 1949. (Botanical explorations, etc.; alphabetical list of known vascular plants, with highest known altitudes for each species and pertinent localities; discussion, bibliography.)

LAPPMARK, LYCKSELE

Bounded on the northwest by Norway, on the northeast by the Skellefte älv from the boundary of the landskap of Västerbotten up to about Slagnäs, then northwest by the Norrbotten-Västerbotten län boundary to the Norwegian boundary, on the southeast by the landskap of Västerbotten and Ångermanland (to about 64° 3′ N. Lat., between Fredrika and Örträsk), on the southwest by a line drawn from the latter point northwest to the Norwegian boundary between the Ume älv and the northernmost branch (Vojmån älv) of the Ångerman älv (nearer the latter).

Local

Gaunitz, D., and **Gaunitz, C. B.** Bidrag till kännedom om kärlväxtfloran i Sorsele socken av Lycksele Lappmark. Svensk Bot. Tidskr. **18**: 128–140. 1924.
Physiography, general features of flora; annotated list of 412 species and varieties of vascular plants.—See also **Gaunitz, D.** Kärlväxtfloran i Sorsele socken, Lycksele Lappmark. Svensk Bot. Tidskr. **33**: 63–84. 1939. (Brings total to 447 plants; includes bibliography.)—Also **Rune, Olof.** Några växtfynd i Lycksele Lappmarks fjällområde. l.c. **42**: 494–497. 1 fig. 1948. (Records of noteworthy species from Sorsele and Tärna socknar.)

Nordenstam, Sten. Några anmärkningsvärda nya växtfynd inom Västerbottens län. Svensk Bot. Tidskr. **34**: 454–463. 2 fig. 1940.
Notes on calciphile species; list of noteworthy vascular plants (particularly *Hieracium* and *Taraxacum*) from Byske, Malå, Tärna, Vilhelmina, and other localities.—See also **Fridén, Axel.** Botaniska notiser från Tärna (Lappland). Bot. Notiser **1951**: 84–87. 1951. (List of noteworthy vascular plants from various localities.)

Vretlind, E. G. Från östra Lappland. Om kärlväxtfloran i Malå socken i lidernas region . . . Svensk Bot. Tidskr. **24**: 58–110. 1930.
Botanical explorations, geography, geology, general features of flora, plant associations, introduced plants, etc.; annotated list of vascular plants (325 species, not counting *Taraxacum* and *Hieracium*); phytogeography, bibliography.—See also his Kärlväxtfloran i Malå, Lappland. Ett tillägg. Svensk Bot. Tidskr. **26**: 461–462. 1932.—Also his Malåfloran (Ö. Lappland) än en gång . . . l.c. **29**: 124–127. 1 fig. 1935.

LAPPMARK, PITE

Bounded on the northwest by Norway, on the northeast by a line drawn not far north of the Pite älv, on the southeast by the middle part of the landskap of Västerbotten, on the southwest by the Skellefte älv up to Slagnäs, then northwest to the Norwegian boundary by the Norrbotten-Västerbotten län boundary.

General

Björnström, F. J. Grunddragen af Piteå Lappmarks växtfysiognomi. 1 p. l., 36 p. 19.5 cm. Upsala, 1856. (Akad. afhandl.) (Riksmuseet library, Stockholm.)
Botanical explorations, topography, botanical regions; briefly annotated list of known vascular plants and Characeae, with localities for scarcer species.

Local

Arwidsson, Thorsten. Floran inom Peljekaise nationalpark. K. Svenska Vetenskapsakad. Skr. Naturskyddsär. no. 5. 34 p. 1 fig. (map). 1926.
Topography, botanical explorations, general features of vegetation, etc.; annotated list of vascular plants (253 species); bibliography. (An area of about 145 sq. km. in Arjeplog socken.)

——— Floristiska notiser från Pite Lappmark, huvudsakligen nordvästligaste delen. Bot. Notiser **1926**: 209–226. 1926.
Species not recorded by Birger (see below), notes on *Calluna vulgaris* and *Draba crassifolia*, etc.; alphabetical list of noteworthy vascular plants, with localities; bibliography.

——— Studien über die gefässpflanzen in den hochgebirgen der Pite Lappmark. Acta Phytogeog. Suecica v. 17. 274 p. 52 (i.e. 53) fig. (incl. maps), 16 (i.e. 17) pl. on 9. 1943.
Topography, geology, climate, botanical explorations, influence of reindeer on flora, altitudinal limits of forest trees and vascular plants in general, list of published maps showing the complete Scandinavian distribution of mountain species, discussion of the problem of the persistence of plants in the northern mountains through the glacial period, general features of vegetation of the Peskehaure delta, short lists of weeds from various localities, list of localities; annotated list of vascular plants, based mostly on the author's work, with frequent critical notes; bibliography. Relates to the western part of Arjeplog socken.

Birger, Selim. Bidrag till Pite Lappmarks flora. Bot. Notiser **1909**: 257–274. 1909.
Sketch of botanical explorations, account of localities visited and their characteristic species; alphabetical list of noteworthy vascular plants, with localities; bibliography.

Johansson, H. E. Växttopografiska anteckningar . . . 1. Lappland jämte övriga delar av Västerbottens och Norrbottens län bearbetade av S. Grapengiesser. Bot. Notiser **1933**: 479–499. 1933.
Itinerary; alphabetical list of vascular plants, with localities, mostly from Arvidsjaur and Arjeplog socknar in Pite Lappmark and Jukkasjärvi socken in Torne Lappmark.

Wistrand, Gunnar. Bidrag till kännedomen om floran i Pite Lappmarks barrskogsregion. (Beiträge zur kenntnis der flora in der nadel-waldregion von Pite Lappmark). Svensk Bot. Tidskr. **28**: 1–33. 5 fig. 1934.
Account of localities, etc.; annotated alphabetical list of vascular plants of the coniferous region, from Arvidsjaur and Arjeplog; bibliography.

LAPPMARK, TORNE

See also Lappmark, Pite (Johansson).
Bounded on the northwest by Norway, on the northeast by Finland, on the southeast by the northern part of the landskap of Norrbotten, on the southwest by the Kalixälv to near Nikkaluokta, thence south of Kebnekaise to the reentrant angle in the Swedish-Norwegian boundary.

General

Laestadius, C. P. Bidrag till kännedomen om växtligheten i Torneå Lappmark. 1 p. l., 46 p. 23 cm. Upsala, 1860. (Gray Herbarium library.)
Geography, botanical explorations, botanical regions; list of known vascular plants and larger fungi, with indication of distribution by regions, and localities for some of the species.—See also **Hägerström, K. P.** Bidrag till Torne Lappmarks och Ofotens flora. Bot. Notiser **1882**: 65–96. 1882. (Localities visited by author and E. V. Ekstrand, with notes on vegetation; list of new records for Tornejärvi region, Torne Lappmark, and Ofoten region, statistics; annotated list of noteworthy vascular plants with localities and critical notes. Relates to Lödingen (on Hindö in the Lofotens), Fagernäs and Bergvik in the Ofoten region, Torneträsk region, etc.)—Also **Lagerberg, Torsten.** Några anmärkningsvärdare växtformer från Torne Lappmark. Svensk Bot. Tidskr. **3**: (19)–(24). 1909. (Annotated list of noteworthy vascular plants from western part of Torne Träsk.)—Also **Sylvén, Nils.** Nya växtlokaler från Torne Lappmark. l.c. **8**: 71–82. 4 fig. 1914. (Botanical publications on the area since 1903, when the railroad through Riksgränsen to Narvik was finished; extensive alphabetical list of vascular plants with localities, from the region of Torneträsk, Vassijaure, and Riksgränsen.)—Also **Frödin, John.** Växttopografiska iakttagelser i mellersta delen af Torne Lappmarks fjällområde. Bot. Notiser **1916**: 25–34. 1916. (Botanical explorations, author's itinerary; alphabetical list of noteworthy vascular plants from various localities, including some south of Torne Träsk.)—Also **Sterner, Ewald.** Floristiska anteckningar från trakten kring Torne träsk och ett par platser i Norges Nordland. Svensk Bot. Tidskr. **10**: 91–96. 1916. (Includes alphabetical list of noteworthy vascular plants from the Torneträsk region, with localities; also shorter list from Andön in the Lofotenöarna and Fagernes-fjället south of Narvik (both in Norway).)—Also **Almquist, Erik.** Växtlokaler från Torne och Lule Lappmarker. l.c. **11**: 125–130. 1917. (Alphabetical lists of noteworthy vascular plants with localities, from the Torneträsk, Abisko-Kebnekaise, and Kebnekaise regions; short list of adventive plants from Malmberget.)—Also **Frödin, John.** Några växtlokaler mellan Kebnekaise och norska gränsen. Bot. Notiser **1918**: 211–213. 1918. (Description of localities; alphabetical list of noteworthy vascular plants from vicinity of Tjäktjavagge.)—Also **Samuelsson, Gunnar.** Anteckningar från Torneträsk-området. l.c. **1920**: 51–61. 1920. (Includes alphabetical lists of noteworthy vascular plants, with localities.)—Also **Alm, C. G.** Floristiska anteckningar från Torneträskområdet. Svensk Bot. Tidskr. **15**: 263–265. 1921. (Alphabetical list of noteworthy angiosperms, with localities, from Abisko, Vaddetjåkko, etc.)—Also **Smith, Harry.** Bidrag till Torne

Lappmarks flora. l.c. **18:** 442–450. 1924. (Extensive list of noteworthy vascular plants, with localities, from the mountains west and northwest of Torneträsk in the north to Sjangeli and Gardetjåkko in the south, with Harrivaara and Tjabrak in the east.)—Also **Nyström, Elis.** Fyra för Torne lappmark nya kärlväxter. l.c. **27:** 111–113. 1933. (Records of 2 species and 2 hybrids.)—Also **Jalas, Jaakko.** Floristisches aus Lapponia enontekiensis, Lapponia tornensis und Troms fylke. Arch. Soc. Zool. Bot. Fenn. Vanamo **2:** 90–96. fig. 1. 1949. (Botanical explorations; list of noteworthy vascular plants, with localities; bibliography.)—Also **Asplund, Erik.** Taxonomical excursion to the subalpine and alpine belts of northern Lappland . . . Seventh International Bot. Cong. Stockholm 1950, Excursions Guides CV. 31 p. 18.5 cm. Uppsala, 1950. (Includes List of the vascular plants of the Torneträsk region (p. 15–31), a briefly annotated, alphabetically arranged list of 492 (i.e. 495) species and subspecies known from the region.) —Also **Nordin, Ingvar.** Växtanteckningar från trakten kring Torne träsk sommaren 1952. Bot. Notiser **1954:** 51–52. 1954. (List of noteworthy species.)

Local

Birger, Selim. Kebnekaisetraktens flora. Ett bidrag till kännedomen om floran i öfversta delen af Kalixälfvens dal. Svensk Bot. Tidskr. **6:** 195–217. 4 fig. (incl. map). 1912.

Botanical explorations, topography, general features of flora, vertical range, etc.; annotated alphabetical list of vascular plants; bibliography.—See also **Almquist, Erik.** Växtlokaler från Torne och Lule Lappmarker. Svensk Bot. Tidskr. **11:** 125–130. 1917. (Localities from Kebnekaise region, p. 129–130.)—Also **Frödin, John.** Iakttagelser i Kebnekaise-områdets sydberg. l.c. **11:** 325–343. 3 fig. (incl. maps). 1918. (Includes additional locality records.)—Also **Alm, C. G.** Bidrag till Kebnekaisetraktens flora. l.c. **15:** 265–268. 1921. (List with localities.)—Also **Du Rietz, G. E.** Bidrag till Kebnekaiseområdets flora. l.c. **20:** 43–51. 1926. (Plant zones; alphabetical list of vascular plants, with localities; bibliography. Relates to Tarfaladalen and Tarfalatjåkko.)—Also **Håkansson, Torsten.** Anteckningar om flora och vegetation i Kebnekaise-området. Bot. Notiser **108:** 276–291. 4 fig. (fig. 1, map). 1955. (Running account of vegetation in various spots mostly near the tourist station, with list of 9 species new to the region and list of mosses, hepatics, and sphagnum; bibliography.)

Fries, O. R. Om trakten mellan Torneå- och Calix-elfvars nedre lopp i växtgeografiskt hänseende. Bot. Notiser **1858:** 153–170. 1858.

General features of flora; briefly annotated list of known vascular plants, based mainly on author's collecting from a center at Björkfors, located at the confluence of the Sangiselv and the Korpikelv.

Fries, T. C. E. Floran inom Abisko nationalpark. Arkiv Bot. v. 16, no. 4. 48 p. 1 fig. (map). 1921 (1919).

General features of flora; annotated list of vascular plants; bibliography. This paper was afterwards considered as K. Svenska Vetenskapsakad. Skr. Naturskyddsär. no. 2. (An area of about 5,000 hectares, along the south side of Torneträsk on both sides of the Abiskojokk, including a part of the Torneträsk itself with the islet Abiskosuolo.)

——— Floran inom Vaddetjåkko nationalpark. K. Svenska Vetenskapsakad. Skr. Naturskyddsär. no. 3. 22 p. 1925.

Geography, topography; annotated list of vascular plants (226 species); bibliography. (An area of about 24.5 square kilometers in **Jukkasjärvi** socken northwest of Torneträsk, 68° 30′ N. Lat.)

——— and **Mårtenson, Samuel.** Floristiska anteckningar från de alpina och subalpina delarna af Karesuando och Jukkasjärvi socknar norr om Torne träsk. Svensk Bot. Tidskr. **4:** (55)–(75). 1910.

Geography, topography; annotated list of vascular plants collected by authors, with localities.—See also **Tengwall, T. Å.**, and **Alm, C. G.** Floristiska bidrag från Karesuando och norra delen av Jukkasjärvi socknar. Svensk Bot. Tidskr. **14:** 232–238. 1920. (Itinerary, notes on various species; alphabetical list of noteworthy vascular plants, with localities.)—Also **Alm, C. G.**, and **Fries, T. C. E.** Floristiska anteckningar från Karesuando och Enontekis socknar. l.c. **19**: 250–262. 1925. (Itinerary, etc.; extensive alphabetical list of vascular plants from Enontekis (in Finland), Jukkasjärvi, and Karesuando socknar.)—Also Sterner, below.

Hedberg, Olov, Mårtensson, Olle, and Rudberg, Sten. Botanical investigations in the Pältsa region of northernmost Sweden with an account of its geology and geomorphology. Bot. Notiser Suppl. v. 3, pt. 2. 209 p. 16 fig. (fig. 1, map). 1952.
Botanical explorations, geology, plant zones and communities; annotated list of vascular plants (Hedberg), annotated list of bryophytes; bibliography. (In Karesuando parish, 69° N. Lat., slightly overlapping the Norwegian border.)

Simmons, H. G. Floran och vegetationen i Kiruna. 1 p. l., 403 p. 22 pl., map. 26 cm. Stockholm, 1910. (Vetenskapliga och praktiska undersökningar i Lappland.)
Study of the flora of Kiruna, with especial reference to the effect of colonization upon it (English summary, p. 386–390); includes bibliography and annotated list of 458 vascular plants.—See also his Die flora und vegetation von Kiruna im schwedischen Lappland . . . Bot. Jahrb. Engler **48**: 1–86. 1 fig., 6 pl. 1912. (Essentially a condensation of his 1910 paper, with additions and corrections; includes (p. 13–19) alphabetical list of vascular plants, and bibliography.)

Sondén, Mårten. Anteckningar om floran inom Tornejavreområdet. Svensk Bot. Tidskr. **1**: 215–242. 1907.
Botanical explorations (with account of botanists' routes), general features of flora; extensive list of vascular plants, with localities, based mostly on author's collecting and fresh material received from others, but including also some species from the publications of Fristedt, Hägerström, and Sylvén. —See also **Sylvén, Nils.** Anteckningar om floran vid Vassijaure-Torne träsk. Svensk Bot. Tidskr. **2**: 12–31. pl. 1–5. 1908. (List of vascular plants, with localities.)—Also many of the papers cited under Laestadius, above; also Sylvén, below.

Sterner, Ewald. Jukkasjärviområdets flora (Torne Lappmark). Arkiv Bot. v. 10, no. 9. 50 pl. 1911.
Topography, history, botanical explorations; annotated list of 365 known vascular plants; discussion of introduced plants, English summary, bibliography.—See also **Johansson, H. E.** Växttopografiska anteckningar . . . 1. Lappland jämte övriga delar av Västerbottens och Norrbottens län bearbetade av S. Grapengiesser. Bot. Notiser **1933**: 479–499. 1933. (Itinerary; alphabetical list of vascular plants, with localities, mostly from Arvidsjaur and Arjeplog socknar in Pite Lappmark and Jukkasjärvi socken in Torne Lappmark.)—Also Fries and Mårtenson, above.

Sylvén, Nils. Torneträskområdets adventivflora. Arkiv Bot. v. 14, no. 11. 57 p. 5 fig. (incl. map). 1915.
Changes in adventive flora since 1903, lists of species from various localities and habitats, list of annual, biennial, and perennial species, etc.; alphabetical tabular list (p. 30–34) of adventive plants from Abisko, Björkliden, Riksgränsen, and Vassijaure, showing localities and occurrence in 1903 and 1913, list of noteworthy species found in 1913, etc. Replaces his Ruderatflora i Torne Lappmark. Bot. Notiser **1904**: 117–128. 1904. (General features of adventive flora; list of 106 species and forms of adventive plants from Abisko, Björkliden, and Vassijaure, with indication of localities.) —See also **Hedberg, Olov.** Bidrag till kännedomen om Torne Lappmarks adventivflora. Svensk Bot. Tidskr. **42**: 77–80. 1948. (List of noteworthy species; bibliography.)—Also Sondén, above.

MEDELPAD

The old province (landskap) of Medelpad forms the southern part of Västernorrlands län.

General

Collinder, Erik. Medelpads flora. Växtgeografisk öfversikt och systematisk förteckning öfver kärlväxterna. viii, 190 p. map. 27.5 cm. Uppsala och Stockholm, 1909. (On cover, Norrländskt handbibliotek. II.)

Botanical explorations, list of botanists who have worked on the flora with dates and very brief biographical notes, topography, geology, climate, plant associations, botanical regions, phytogeography, ecological groups, tabular lists of species showing occurrence in adjacent regions with discussion, origin and history of flora; annotated list of vascular plants (893 species, excluding adventive plants), with localities and names of collectors; bibliography. There is a short list of additions in review by S. **Birger,** Svensk Bot. Tidskr. **3:** (89)–(91). 1909.—See also **Cedergren, G. R.** Bidrag till Medelpads flora jämte några uppgifter från angränsande delar av Jämtland. Bot. Notiser **1923:** 352–366. 2 fig. 1923. (Description of localities, etc.; list of vascular plants from Haverö in Medelpad and adjacent Jämtland.)

Local

Falck, Kurt. Botaniska anteckningar från västra Medelpad och sydöstra Jämtland. Bot. Notiser **1924:** 343–360. 4 fig. 1924.
Includes alphabetical lists of vascular plants from Bergåsen, Lönnån, and Ranklöven near Borgsjö.

Fredrikson, A. T. Ett litet bidrag till kännedomen om västra Medelpads flora. Bot. Notiser **1902:** 235–239. 1902.
List of vascular plants from westernmost part of Hafverö (Haverö) socken.—See also Cedergren under Collinder, above.

Grevillius, A. Y. Bidrag till kännedomen om kärlväxtvegetationen på nephelinsyenitområdet i Alnöns norra del samt på närliggande holmar i Medelpad. Öfvers. K. Vetensk.-Akad. Forhandl. **51** (1894): 215–234. 1 fig. (map). 1894.
Topography, general features of flora, etc., with mention of many plants observed.

——— Studier öfver växtsamhällenas utveckling på holmar i Indals- och Ångermanelfven. Sveriges Geol. Undersök. ser. C. Afhandl. och Uppsats. no. 144 [pt. 1]. 1 p. l., 18 p. 1895.
Account of vegetation of various islands, with mention of numerous plants. Refers mainly to the delta of the Indalsälven in Medelpad (Smackgrundet, island south of Lars-Nils-grundet, Storholmen, Fäbodeholmen, Båthusön) and vicinity of Sollefteå on the Ångermanälven (Remsle-skäret and Hågestaön). All in Västernorrlands län.

NÄRKE (NERIKE)

The old province (landskap) of Närke or Nerike forms the southern part of Örebro län.

General

Hartman, Carl. Landskapet Nerikes flora. För nybörjare utgifven . . . xvi, 260 p. 18 cm. Örebro, 1866. (Gray Herbarium library.)
Topography, general features of flora; keyed descriptive flora of vascular plants (773 species), on the Linnaean system, with localities for the scarcer species.—See also **Gellerstedt, J. D.** Nerikes flora eller kort beskrifning af Nerikes vilda växters kännetecken, till den studerande ungdomens tjenst. 2. uppl. omarbetad och tillökt af C. O. Hamnström. iv, 222 p. 16.5 cm. Örebro, 1852. (1st ed. 1831.) (Briefly descriptive flora of spermatophytes (718 species), on the Linnaean system, with localities. In Riksmuseet library, Stockholm.)—Also Spridda bidrag till Nerikes flora, samlade af Naturvetenskapliga Föreningen "Hedera" i Örebro. Bot. Notiser **1886:** 94–98, 119–122. 1886.—Also **Trolander, A. S.** Växtlokaler i Nerike. l.c. **1888:** 88–93, 116–118. 1888. (Extensive list, with localities.)—Also **Lundelius, Hilding.** Växtgeografiska anteckningar beträffande Närkes fanerogamvegetation. l.c. **1910:** 17–29. 1910. (Localities; extensive list of spermatophytes, with localities.)—Also **Sernander, Rutger.** Närkes flora och dess ställning till den svenska växtvärlden och det nordiska klimatet. *In* Aspling, E., and others. Närke. p. 25–57. 10 fig. 20.5 cm. Upsala, 1922. (General features of flora, plant communities, history of flora. Not seen; title and annotation supplied by E. Hultén.)

Local

Asplund, Erik. Floran inom Garphytte nationalpark. K. Svenska Vetenskapsakad. Skr. Naturskyddsär. no. 4. 35 p. 1925.

General features of flora, etc.; annotated list of vascular plants (348 species). (An area of about 20 hectares on the south slope of the Kilsbergen in Hidinge socken.)

Birger, Selim. De 1882–1886 nybildade Hjälmaröarnas vegetation. Arkiv Bot. v. 5, no. 1. 152 p. 14 fig., 12 pl. (incl. map). 1905.

Account of the development of vegetation on the islands formed in Lake Hjalmar by the lowering of its surface 1.9 meters in 1882–86, with lists of species from different islands, and a complete list of the vascular plants. Replaces **Grevillius, A. Y.** Om vegetationens utveckling på de nybildade Hjelmar-öarne. Bihang K. Svenska Vetensk.-Akad. Handl. v. 18, afd. III, no. 6. 110 p. map. 1893.—See also **Birger S.** Die vegetation einiger 1882–1886 entstandenen schwedischen inseln. Bot. Jahrb. Engler **38**: 212–232. 1 fig. (map), 2 pl. 1906. (Abstract of his longer paper, without full lists of species; bibliography.)

Kjellmert, Gösta. Svennevads kärlväxtflora. Bot. Notiser **1947**: 231–282. 4 fig. (maps). 1947.

Geology, phytogeographic and ecological groups, etc.; annotated list of vascular plants (664 species) with detailed localities; gazetteer, bibliography.

Naturvetenskapliga Föreningen Hedera. Lokalförteckning öfver Örebrotraktens fanerogamer och kärlkryptogamer. 46 p. 17.5 cm. Örebro, 1887. (Riksmuseet library, Stockholm.)

List of vascular plants, with localities.

NORRBOTTEN

See also Sweden, Partial (Backman, Elementarflora).

The old province (landskap) of Norrbotten covers only the coastal part of Norrbottens län. For the remaining (Lapland) part of Norrbottens län see Lappmark, Lule, Lappmark, Pite, and Lappmark, Torne.

Local

Almquist, Erik. Växtgeografiska bidrag. 2. Norrbotten. Bot. Notiser **1920**: 127–133. 1920.

Alphabetical list of vascular plants, with localities, from region of Boden, Lulea, and Torneå; list of adventive plants from Karungi.—See also Svenonius, below.

Arwidsson, Thorsten. Växtgeografiska notiser från Norrland. III. Bot. Notiser **1931**: 355–360. 1931.

List of localities, etc.; alphabetical list of some vascular plants from Haparanda skärgård and Piteå skärgård off the coast of Norrbotten, and Renholmen and some other islands off the northernmost coast of Västerbotten.

Birger, Selim. Vegetationen och floran i Pajala socken med Muonio kapellag i arktiska Norrbotten. Arkiv Bot. v. 3, no. 4. 117 p. 1 fig., 7 pl. (incl. maps). 1904.—Tillägg . . . Bot. Notiser **1907**: 69–73. 1907.

Geology, climate, ecology, plant geography; annotated list of vascular plants.—See also **Samzelius, Hugo.** Vegetationsiakttagelser inom Pajala socken af Norrbottens län. Bot. Notiser **1890**: 173–188. 1890. (Topography, etc.; annotated alphabetical list of vascular plants.)

Heintze, August. Växtgeografiska undersökningar i Råne socken af Norrbottens län. Arkiv Bot. v. 9, no. 8. 63 p. 1909.

Plant formations, etc.; annotated list of vascular plants; bibliography.

Marklund, Erik. Växtlokaler från Norrbotten. Svensk Bot. Tidskr. **11**: 392–400. 1918.

List of localities and corresponding socknar (parishes); extensive alphabetical list of vascular plants, with localities, from Piteå, Neder-Kalix, and Elfsby (Älvsby) socknar.—See also **Alm, C. G.** Bidrag till södra Norrbottens flora. Svensk Bot. Tidskr. **13**: 102–104. 1919. (List of additional species and localities from Älvsby and Storsund.)—Also **Lönnqvist, Oskar.** Floran i två sydbranter i Norrbotten. Bot. Notiser **1947**: 98–100. 1947. (Short lists of vascular plants from the southern precipices of Falkberget and Kvarnberget in Älvsby socken.)

Svenonius, Herman. Luleåtraktens flora. Svensk Bot. Tidskr. **19**: 431–484. 1925.

Geology, flora of various lakes and bogs, sketch of vegetation of Uddskär in Luleå's outer archipelago, with annotated list of 185 species and subspecies; annotated list of vascular plants (541 species and subspecies) of Luleå region; partly annotated bibliography.—See also **Grapengiesser, Sten.** Norrländska vegetationbilder. Svensk Bot. Tidskr. 28: 300–353. 1934. (Includes list of additions to flora of Storön in Kalix archipelago, p. 343–345.)—Also **Svenonius, H.** Växttopografiska anteckningar till Luleåfloran. l.c. **34**: 166–170. 1940. (Annotated list; bibliography.)—Also Almquist, above.

ÖLAND

The island of Öland, which has generally been treated as a phytogeographic unit by Swedish writers, is an old province which is administratively included in Kalmar län.

General

Sterner, Rikard. Flora der insel Öland. Die areale der gefässpflanzen Ölands nebst bemerkungen zu ihrer oekologie und soziologie. Acta Phytogeog. Suecica v. 9. 169 p. 295 fig. (fig. 8–295, distribution maps). 1938.

Botanical explorations, topography, geology, soils, climate, effects of cultivation, plant formations, statistics, bibliography; annotated list of known vascular plants (about 1130 species, of which about 820 are native). For account of ecology and phytogeography see his Ölands växtvärld. xi, 237 p. incl. 113 (i.e. 114) fig. (mostly maps). 25 cm. Kalmar, 1926. (Södra Kalmar län III.) (New York Botanical Garden library.)—Replaces **Sjöstrand, Gustaf.** Enumeratio plantarum in Ölandia sponte nascentium. Nova. Acta. R. Soc. Sci. Upsal. **14**: 455–516. pl. 4. 1850. (Includes annotated list of 862 vascular plants and Characeae.) Also replaces (as to Öland) **Sjöstrand, M. G.** Calmar läns och Ölands flora. 4 p. l., 359, iv p. 19.5 cm. Calmar, 1863. (Annotated, keyed, descriptive flora of vascular plants and Characeae. In New York Botanical Garden library.) Also replaces **Hemmendorff, Ernst.** Om Ölands vegetation, några utvecklingshistoriska bidrag. viii, 52 p. map. 23 cm. Upsala, 1897. (Akad. afhandl.) Includes lists of vascular and cellular plants, ecologically arranged.)—See also **Aulin, F. R.** Botaniska anteckningar från Öland. Svensk Bot. Tidskr. **6**: 373–386. 1912. (Extensive list with localities, including some minor forms not listed by Sterner.)—Also **Segerström, A. L.** Några växtfynd från Öland. l.c. **14**: 350–351. 1920. (Includes numerous varieties and forms not listed by Sterner.)—Also **Almquist, Erik.** Nya fynd av Hieracier och andra kärlväxter på norra Öland. l.c. **43**: 125–129. 1949. (List of noteworthy plants, mostly from Böda.)—Also **Hylander, Nils.** Några öländska växtfynd. l.c. **52**: 348–352. 1958. (Adds 2 species; notes on others.)—Also **Saarsoo, Bernhard.** Några nya och mer anmärkningsvärda växtfynd från Öland. Svensk Bot. Tidskr. **53**: 102–105. 1959.

Local

Ahlquist, Abraham. Flora runsteniensis. 5 pt. (32 p.). 21 cm. Upsaliae, 1815–17.

Geography, soils, topography, etc.; annotated list of vascular plants of Runsten, on a somewhat modified Linnaean system. A thesis, K. P. Thunberg praeses, with five respondents the first of whom, Ahlquist, is considered the author. Also published in Thunberg's Dissertationes academicae no. 171–175.

ÖSTERGÖTLAND

The old province (landskap) of Östergötland has essentially the same boundaries as Östergötlands län.

General

Kindberg, N. C. Östgöta flora. Beskrifning öfver Östergötlands fanerogamer och ormbunkar. 4. uppl. 1 p.l., 326 p. 18.5 cm. Stockholm, 1901. (1st ed. 1861.)

Annotated, keyed, descriptive flora of vascular plants on the Linnaean system, with localities; list of principal cultivated plants with brief descriptions, glossary, botanical regions, etc.—See also **Sörlin, Anton.** Floristiska anteckningar från Östergötland och Upland. Svensk Bot. Tidskr. **10:** 73–75. 1916. (List of noteworthy plants from southeastern Östergötland and northern Roslagen, with localities.)—Also **Malme, G. O.** Strödda bidrag till Östergötlands kärlväxtflora. l.c. **17:** 103–105. 1923. (List with localities, mostly from vicinity of Omberg.)—Also **Hylander, Nils.** Några fynd av sällsyntare växter, huvudsakligen antropokorer, i Östergötland. Bot. Notiser **1932:** 94–100. 1932. (List with localities.)

Local

Elmqvist, C. F. Norrköpings-traktens fanerogamer och ormbunkar, jemte växtställen och värden, sådana de blifvit bestämda af Upsala Botaniska Bytesförening. 2 p. l., 52 p. 17.5 cm. Örebro, 1874. (Riksmuseet library, Stockholm.)

Annotated list of vascular plants, on the Linnaean system, with localities.

Genberg, Erik. Kolmårdens kärlväxter. En inventering och en sammanfatning. Svensk Bot. Tidskr. **52:** 487–541. 4 fig. (fig. 2, map). 1958.

Topography, climate, geology, botanical explorations, general features of vegetation; briefly annotated list of vascular plants (844 species); bibliography. (Krokek, Kvarsebo, Kvillinge (in part), and Simonstorp socknar.) Replaces his Kärlväxterna i Krokek. Ett bidrag till kännedomen om Kolmårdens flora. Svensk Bot. Tidskr. **45:** 121–139. 2 fig. (maps). 1951. (Includes alphabetical list of vascular plants (770 species).)

Gertz, Otto. Dömestorps flora. En hittills okänd växtförteckning från år 1754. Bot. Notiser **1947:** 305–316. 1947.

Account of manuscript flora by Johan Fischerström, with list of vascular plants included; lists of additional species recorded by other authors, with biographical and other notes; bibliography. (Near Linköping.)

——— Vegetationen å ruinerna av Alvastra klosterkyrka. Bot. Notiser **1923:** 457–463. 1923.

Previous publications on the plants of old walls and ruins; list of 61 vascular plants and 9 mosses and lichens found on the ruins of the Middle Age church Alvastra south of Omberg; bibliography.

Hamnström, C. O. Förteckning öfver vexter, insamlade omkring Skeninge åren 1832–37. Bot. Notiser **1843:** 17–24, 33–38. 1843.

Partly annotated list of vascular plants and Characeae, arranged by the Linnaean system.

Hedberg, Olof. Vegetation och flora inom Ombergs skyddsområde. K. Svenska Vetenskapsakad. Avhandl. Naturskyddsär. 5. 64 p. 2 fig. (maps), 10 pl. on 5. 1949.

Mainly ecological and phytohistorical; includes (p. 50–57) annotated list of vascular plants, also list of rusts (p. 58); English summary. Refers to forest reserve on west slope of Mount Omberg, an area of about 30 acres. —See also Hesselman, below.

Hesselman, Bengt. Ombergs kärlväxtflora. Svensk Bot. Tidskr. **32:** 1–88. 2 fig. (maps). 1938.

Botanical explorations, general features of flora, plant geography, etc.; annotated list of known vascular plants; bibliography. Replaces **Holmgren, A. F.** Ombergs phanerogamer och ormbunkar. Bot. Notiser **1851:** 186–187, 193–211, 225–250. 1851. (Includes annotated list of 663 vascular plants. Reprinted 47 p.; reprint in Riksmuseet library, Stockholm.) Also replaces **Dusén, Per.** Ombärgstraktens flora och geologi, till ledning för den Ombärgsbesökande allmänheten framstälda. 2 p.l., 94 p. map. 22 cm. Stockholm, 1888. (Includes annotated list of vascular plants (p. 38–68). In Gray Herbarium library.)—See also Hedberg, above.

Julin, Erik. Floran. *In* his Vessers udde. Mark och vegetation i en igenväxande löväng vid Bjärka-Säby. Acta Phytogeog. Suecica **23:** 145–149. 1948.

List of vascular plants, mosses, and larger fungi, arranged alphabetically. Vessers udde is a small nature park on a peninsula in the lake Stora Rengen in Vist socken.

Olsson, Axel. Harstena flora [I.]. Svensk Bot. Tidskr. **52:** 225–283. 5 fig. (incl. 2 maps). 1958.

Topography, general features of vegetation, ecology; annotated list of vascular plants; bibliography. (Group of islands in Harstena, Gräsmarö, Kråkmarö, and Fångö byar, in Gryt socken.)

Segerström, A. L. Sydvästra Tylöskogens flora. Bot. Notiser **1932:** 191–268. 3 fig. (incl. map). 1932.
Geography, effects of cultivation, botanical explorations (with references), topography, geology, general features of flora, plant associations; annotated list of vascular plants (over 800 species, including 101 Taraxaca and Hieracia). Covers region on east side of Lake Vättern from Motala and Kristberg north to Hammar, Lerbäck, and Lake Åfjärden.

Wessén, C. J. Plantae cotyledoneae in paroecia Ostrogothiae Kärna, quas secundum methodum naturalem celeberr. Friesii disposuit atque congessit. 1 p. l., iv, 62, [4] p. 22 cm. Upsaliae, 1838.
Annotated list of spermatophytes and Characeae (564 species), with localities.

Westerberg, F. O. Bidrag till kännedom om nordöstra Östergötlands kärlkryptogamer och fanerogamer. Svensk Bot. Tidskr. **11:** 226–264. 1 fig. 1917.
Annotated list of vascular plants, mainly from Risinge socken, collected by the author.

SKÅNE (SCANIA)

The old province (landskap) of Skåne includes Kristianstads län and Malmöhus län.

General

Anonymous. Inventeringen av Skånes flora. Bot. Notiser **1947:** 83–88. 2 fig. (maps). 1947.
Contains a list of härader and their included districts.

Areschoug, F. W. C. Skånes flora, innefattande de fanerogama och ormbunkartade växterna. 2. uppl. xxii, 585 p. 24 cm. Lund, 1881. (1st ed. 1866.)
Partly keyed, descriptive flora of vascular plants, with localities. Replaces **Fries, E. M.** Corpus florarum provincialium Sueciae. I. Floram scanicam scripsit . . . xxiv, 394 p. tab. 21.5 cm. Upsaliae, 1835. (Botanical investigations; annotated list of vascular and cellular plants, with short diagnoses.) Also replaces **Lilja, Nils.** Skånes flora, innefattande Skånes vilda och odlade växter . . . ny omarbetad uppl. 2 v. (1018 p.). 17.5 cm. Stockholm, 1870. (1st ed. 1838.) (Geography, climate, botanical investigations, etc.; descriptive flora of vascular plants, wild and cultivated, on the Linnaean system, with localities for scarcer species; list of localities.)—See also **Ljungström, Ernst.** Växtgeografiska bidrag till Skånes flora. Bot. Notiser **1884:** 4–6. 1884. (List of noteworthy species.)—Also **Johansson, Nils.** Bidrag till Skånes flora. l.c. **1891:** 140–141, 157–163. 1891. (List of noteworthy species.)—Also **Lindquist, Bertil.** Nya skånska växtlokaler. l.c. **1930:** 214–224. 4 fig. 1930. (List of noteworthy species, with remarks on distribution.)

Weimarck, Henning. Utforskningen av Skånes flora. (Bidrag till Skånes flora. 52.) Bot. Notiser **111:** 81–122. 1958.
Chronological account of botanical investigations with lists of species recorded or first reported by each author.

Local

Andersen, Svend. Om vegetation og flora paa kalktufskrænten ved Benestad. Skånes Natur **22:** 64–76. 1936. (Riksmuseet library, Stockholm.)
Geology, etc.; alphabetical lists of vascular plants (181 species); comparison with the flora of Stenløse (in Fyn, Denmark). (In Kristianstads län.)

Areschoug, J. E. Stirpes in regione cimbritshamnensi sponte crescentes, quarum enumerationem . . . pt. 1 (all). 2 p. l., 16 p. 21 cm. Londini Gothorum, 1831. (Diss., E. Fries praes.) (Riksmuseet library, Stockholm.)
Annotated list of phanerogams (203 species), on the Linnaean system. Incomplete; reaches Pentandria Digynia (*Chenopodium album*). (Also spelled Cimbrishamn and Simrishamn; in Kristianstads län.)

Asplund, Erik. Stenshuvuds kärlväxtflora. K. Svenska Vetenskapsakad. Skr. Naturskyddsär. 49. 36 p. 3 fig. (fig. 1, map). 1955.
Botanical explorations, topography, general features of vegetation, geology; annotated list of vascular plants; bibliography. (In Kristianstads län.)

Blom, Carl. Ullfloran vid Lackalänga i Skåne. Acta Hort. Gothoburg. 5: 85–96. 1930.
Account of the locality (vicinity of a wool-cleaning establishment); annotated list of adventive angiosperms (106 species); bibliography. (In Malmöhus län.)—See also his Tillägg till ullfloran vid Lackalänga. *In* his Bidrag till Skånes adventiv- och ruderatflora. Bot. Notiser **1933**: 365–366. 1933.—Also his Tillägg (II) till ullfloran vid Lackalänga. *In* his Bidrag till kännedomen om Sveriges adventivflora. III. Acta Hort. Gotoburg. **11**: 181–182. 1936.—Also his Lokaler från Skåne, tillika tillägg (III) till ullfloran vid Lackalänga. *In* his Bidrag till kännedomen om Sveriges adventivflora. IV. l.c. **13**: 402–403. 1940.—Also **Hylmö, Bertil.** Notiser om ullfloran vid Lackalänga. Bot. Notiser **1947**: 81–82. 1947. (List including 8 additional species.)—Also **Lange, Thorvald, Flinck, K. E., and Hylmö, Bertil.** Om ullfloran vid Lackalänga i Skåne. l.c. **1954**: 186–191. 1954. (Earlier work; unannotated list of about 150 angiosperms; bibliography.)

Blomqvist, Theodor. Kullabergs fanerogamarter. Skånes Natur **21**: 40–59. 1934. (Reprinted, 22 p. Original not seen; reprint in Riksmuseet library, Stockholm.)
Botanical explorations; partly annotated list of spermatophytes. (In Malmöhus län.) Replaces **Gyllenstjerna, N. C.** Förteckning på de phanerogama växter, ormbunkar och mossor, hvilka blifvit iakttagne på och omkring Kullaberg i nordvestra Skåne. Bot. Notiser **1851**: 70–84. 1851. (Barely annotated list of vascular plants and mosses; list of localities.)—See also **Andersen, Svend.** Et bidrag til listen over Kullabergstraktens fanerogamer. Skånes Natur **22**: 28–34. 1935. (List with localities. In Riksmuseet library, Stockholm.)—Also **Gertz, Otto.** Kullabergs kärlkryptogamer. l.c. **22**: 35–37. 1935. (List with some localities. In Riksmuseet library, Stockholm.)—Also his Kullaberg. En blick på Kullabygdens botaniska forskningshistoria. l.c. **23**: 56–61. 1936. (Sketch of botanical explorations. In Riksmuseet library, Stockholm.)—Also his Anmärkningar och tillägg. l.c. **23**: 62–63. 1936. (In Riksmuseet library.)—Also **Sylvén, Nils.** Kullabergs kärlväxtflora. Fauna och Flora **49**: 229–251. 5 fig. 1954. (Botanical explorations with lists of species added by each botanist; full bibliography.)—Also his Nya växtfynd från Kullabergsområdet. Skånes Nat. **43**: 68–90 incl. 2 fig. 1956. (List with localities.)—Also his Nya Kullabergsväxter 1956. l.c. **44**: 81–82. 1957.—Also **Rickman, H.** Ytterligare ett bidrag till Kullabergsfloran. l.c. **46**: 111–112. 1959. (Additions and new records.)

Brown, Theodor. Anteckningar till Skånes flora. 2 p. l., v, 30 p. 20.5 cm. Lund, 1870. (Riksmuseet library, Stockholm.)
Topography, etc.; list of scarcer spermatophytes of Widtsköfle (Vittskövle) and Degeberga socknar, with localities; list of common species; list of pteridophytes and mosses. (In Kristianstads län.)

Düben, M. W. von. Enumeratio plantarum in regione landscronensi crescentium. pt. I (all). 2 p. l., 16 p. 20 cm. Lundae, 1835. (Diss., C. A. Agardh praes.) (New York Botanical Garden library.)
Annotated list of phanerogams (262 species), on the Linnaean system. Incomplete; reaches Hexandria Monogynia (*Juncus*). (In Malmöhus län.)—See also **Nilsson, Arvid.** Anteckningar om Landskronatraktens flora. [I] (Bidrag till Skånes flora. 14) –II. Bot. Notiser **1942**: 199–202. 1942; **1952**: 81–96. 1952. (Alphabetical lists of noteworthy species; bibliography.)

Gertz, Otto. Eskilstorpsholmarnas flora. Skånes Natur **22**: 156–163. 1935. (Riksmuseet library, Stockholm.)
Botanical explorations, etc.; alphabetical list of known vascular plants; remarks on various species. *In* his later flora of the island Dynan the author adds three species (Bot. Notiser **1947**: 95. 1947). (In Malmöhus län.)

——— Flora farhultensis. En återfunnen ungdomsskrift af Johan Wilhelm Zetterstedt. (Bidrag till Skånes flora. 26.) Bot. Notiser **1944**: 223–236. 1944.

Includes alphabetical list of 367 plants (337 vascular) from Farhult, with list of additional species from the Höganäs region; also biographical and critical notes, including additional species found by other botanists, and bibliography. (In Malmöhus län.)

——— Floran på Öresundsholmen Dynan i närheten av Klagshamn. Bot. Notiser **1947**: 94–98. 1947.

Geography, topography, etc.; list of 77 known plants, including 1 moss; general features of flora, bibliography. (Small island in southern Öresund, in Malmöhus län.)

——— Måkläppens flora. K. Fysiog. Sällsk. Lund Förhandl. **3**: 61–78. 5 fig. 1934.

Topography, botanical explorations; list of 102 known angiosperms; plant galls, general features of flora, bibliography. (A small island 5 kilometers SSW of Falsterbo, in Öresund, in Malmöhus län.)

Gosselman, C. A. Stirpes rariores territorii ystadiensis. 1 p. l., ii, 18 p. 19.5 cm. Lund, 1851. (Akad. afhandl.)

General features of flora; list of vascular plants of vicinity of Ystad, with localities. (In Malmöhus län.)

Gunnarsson, J. G. Vellingeortens flora innehållande förteckning över de vilda och förvildade växter, som av författaren blivit funna inom ett område beläget mellan Gessie församling i norr och Trälleborgsgränsen i söder, i öster sträckande sig till Trälleborgsjärnvägen, i väster till Öresund. 51 p. 22.5 cm. Malmö, 1932.

Changes in flora, etc.; briefly annotated alphabetical list of vascular plants with localities. (In Malmöhus län.)

Lange, Thorvald. Några anteckningar rörande Kristianstadstraktens flora. Svensk Bot. Tidskr. **21**: 33–44. 1927.

Changes in flora, notes on some rare species, list of adventive plants. Intended to complement the records from the region in Areschoug, Skånes flora. ed. 2. 1881. (In Kristianstads län.)—See also **Tufvesson, Per.** Några adventivfynd från Kristianstad och Åhus. Bot. Notiser **1933**: 367–372. 1933. (Extensive list of adventive plants, with localities and dates.)

Lilja, Nils. Stirpes phanerogamae spontaneae, quas in paroeciis Billinge et Röstånga scanicis observavit . . . Bot. Notiser **1846**: 113–118, 129–137. 1846.

Briefly annotated list of spermatophytes. (In Malmöhus län.)

Lindquist, Bertil. Dalby Söderskog. En skånsk lövskog i forntid och nutid. Acta Phytogeog. Suecica v. 10. 273 p. 99 fig. (incl. maps). 1938.

Ecological; includes tabular list of vascular plants (p. 190–194); bibliography; German summary. (A nature reserve 10 miles from Lund, in Malmöhus län.)

Neuman, L. M. Berättelse om en botanisk resa till Hallands Väderö och närliggande delar af Skånska landet, företagen med understöd af Kong. Vetenskapsakademien år 1882. Öfvers. K. Vetensk.-Akad. Förhandl. **40** (8): 45–85. 1883 [1884.]

Consists of two parts: I, Förteckning öfver kärlväxter, hvilka af mig iakttagits på Hallands Väderö (p. 47–60), an unannotated list of vascular plants [343 species]; II, Anmärkningar rörande för Kullens flora nya eller i andra afseenden märkliga, der förekommande växter (p. 61–83), an annotated list of noteworthy vascular plants (in large part *Rubus*), from Kullaberg, Mölle, Brunnby, etc.; also (p. 83–85) a Tillägg to his Hallands Väderö list. (Part I refers to Kristianstads län, part II to Malmöhus län.)—See also **Vallin, Hervid.** För Hallands Väderö ej förut publicerade fanerogamer och kärlkryptogamer samt i övrigt några för ön intressanta växter. Bot. Notiser **1936**: 519–531. 2 fig. 1936. (Botanical explorations, additions, etc.; bibliography.)—Also his Förteckning över fanerogamer och kärlkryptogamer på Hallands Väderö. Skånes Natur **33**: 69–78. 1946. (Botanical explorations; lists of species reported by earlier writers, with additions, totaling about 600 vascular plants from an area of 310 hectares. In Riksmuseet library, Stockholm.)—Also his Om några på amfibolit förekommande för Hallands Väderö nya fanerogamer och ormbunkar. Bot. Notiser **1950**: 421–422. 1950. (Includes records of additional species.)—Also his Hallands Väderö. Något om öns historik och högre växtliv. Skånes Natur **46**: 7–44.

29 fig. (incl. maps). 1959. (History, general features of vegetation, bibliography.)

——— Bidrag till kännedomen af floran på Sveriges syd-vestkust omfattande trakten mellan Halmstad och Engelholm. Göteborgs K. Vetensk.-och Vitterh. Samhäll. Handl. n.s. v. 19, [no. 2]. 56 p. 1884.
Annotated list of vascular plants; bibliographical footnotes. (Halmstad, in Halland; Engelholm (Ängelholm), in Kristianstads län.)—See also **Lidforss, Bengt.** Några växtlokaler till nordvestra Skånes flora. Bot. Notiser 1885: 177–191. 1885. (List of noteworthy angiosperms, with localities, mostly from region between Skälderviken and the Halland boundary, that is, the area from Engelholm in the south to Båstad and Torekov in the north, also the Kullahalvön west of Skälderviken. In Kristianstads and Malmöhus län.)

Nilsson, Arvid. Borgen, en fornlämning och naturskön bäckravin i Rååns dalgång. Inventering af floran. Skånes Natur **45**: 22–45. 5 fig. (fig. 1, map). 1958.
Includes alphabetical tabular list of vascular plants (318 species, subspecies, and hybrids). (In Fjärestads socken.)

——— Örja sockens vegetation och flora. *In* Boken om Örja p. 149–190. 1952. Landskrona, 1952.
Includes briefly annotated list of all species found by author. (In Malmöhus län.) (Not seen; title and annotation supplied by E. Hultén.)

Nordström, K. B. Bidrag till kännedomen om Sveriges ruderatflora. Bot. Notiser **1903**: 113–122. 1903.
List of ruderal and ballast spermatophytes, with data, mostly from Karlshamn in Blekinge and from Malmö, Hälsingborg, and other localities in Skåne; list of 106 ruderal plants of Sweden based on Neuman's Sveriges flora (1901) and the author's observations, with notes on origin.—See also **Turesson, Göte.** Några adventivväxter från Skåne. Svensk Bot. Tidskr. **6**: 95–96. 1912. (List of adventive plants, nearly all from Malmö, with dates.)—Also **Blom, Carl.** Några anmärkningsvärda adventiv- och ruderatfynd vid Malmö åren 1912–20. Bot. Notiser **1921**: 43–45. 1921. (Alphabetical list of adventive and ruderal plants, with dates.)—Also his Bidrag till kännedomen om Sveriges adventivflora. [I.] Acta Hort. Gothoburg. **3**: 143–150. 1927. (Includes, p. 143–148, list of noteworthy adventive plants from Skåne, mainly from Malmö and Lund.)—Also his Bidrag till Skånes adventiv- och ruderatflora. Bot. Notiser **1933**: 355–366. 1933. (Includes extensive alphabetical list of adventive plants, chiefly from Malmö and Lund, with localities and dates; alphabetical list of escapes from cultivation observed at Malmö, 1912–32; similar list from Lund and other localities.)

Påhlman, Georg. Förteckning öfver ön Hvens fanerogamer och kärlkryptogamer. Bot. Notiser **1912**: 161–182. 1912.
Botanical explorations, topography, geology, cultivation, etc.; annotated list of vascular plants. (An island northwest of Landskrona, in Malmöhus län, the site of Tycho Brahe's observatory.)

Sylvén, Nils. Bidrag till Skånes flora. 1–4. Bot. Notiser **1931**: 55–62, 207–212, 308–312, 383–390. 1931.
The first paper, Nya fyndorter för några inom den spontana Skånefloran sällsynta eller mindre allmänna arter, is an extensive list of vascular plants, with localities, primarily from the region of Svalöf; the second, Nya fyndlokaler för adventivarter inom Skånefloran, lists previous publications on the adventive flora of Skåne and gives an alphabetical list of noteworthy adventive plants, mostly from vicinity of Svalöf; the third, Nya hybridlokaler inom Skånefloran, includes mention of the more important previous publications on hybrid plants of Sweden and alphabetical list of noteworthy hybrids, mostly from vicinity of Svalöf; the fourth, Fyndlokaler för sällsyntare eller mera sällan anmärkta former av några skånska växtarter, includes alphabetical list of noteworthy varieties and forms (mostly color forms) and lists of albinistic forms, with localities. (In Malmöhus län.)

Ulf, Bengt. Kungsmarken. En inventering av kärlväxtfloran jämte kort historik över intressantare arters förekomst inom området. Bot. Notiser **1952**: 425–438. 1952.
General features of flora, historical sketch of area; alphabetical list of vascular plants, with earliest and latest dates (years) of observation; short

bibliography. Kungsmarken, the old Kungsängen Räften, 5 kilometers east of Lund, is an area of 230 hectares which has been meadow and pasture land for several centuries. (In Malmöhus län.)

Weimarck, Henning, and others. Bidrag till Skånes flora. 1–50. Bot. Notiser 1939–56.

A series of partial floras and other papers, forerunners of a flora of Skåne, generally containing account of topography, botanical explorations, plant associations or phytogeography, list of noteworthy species, and bibliography. All are numbered Meddelanden från Lunds Botaniska Museum; nearly all are illustrated. Two are given in full above; all the others that seem significant are listed alphabetically here by the locality name, with the series number and reference. 40. Andrarum. **1949:** 1–48; 41. Eljaröd. **1949:** 49–70; 38. Fjälkinge (etc.). **1948:** 93–102; 47–48. Glimåkra. **1953:** 204–232, 369–398; 33. Hjärsås. **1945:** 425–429; 40. Hörröd. **1949:** 1–48; 29. Kågeröd. **1944:** 401–432: 39. Kvidinge. **1948:** 157–166; 49. Loshult. **1956:** 117–142; 25. Nävlinge. **1944:** 1–62; 8. Oderljunga. **1941:** 133–154; 1, 28. Örkened. **1939:** 357–392, **1944:** 389–400; 50. Rya. **1956:** 325–348; 6. Sjöbo. **1940:** 173–192; 20. Söderåsen. **1943:** 284–294; 42. Stehag. **1949:** 109–126; 46. Tjörnarp. **1953:** 81–96; 12. Vittsjö. **1942:** 153–177.

SMÅLAND

The old province (landskap) of Småland includes the län of Jönköping, Kalmar (excluding the island and old province of Öland, here treated separately), and Kronoberg.

General

Scheutz, N. J. Conspectus florae smolandicae. 1 p.l., xviii, 58 p. 20.5 cm. Upsaliae, 1857.

Botanical explorations with references, general features of vegetation; list of vascular plants and Characeae, with localities.—See also his Några bidrag till Smålands flora. 16 p. 20.5 cm. Vestervik, 1862. (List of vascular plants, with localities. In Riksmuseet library, Stockholm.)—Also his Fortsatta iakttagelser rörande Smålands växtlighet. Bot. Notiser **1871:** 55–57, 82–94, 120–126, 143–148. 1871. (Topography, climate, soils, general features of flora, botanical divisions; extensive list of vascular plants and a few Characeae and bryophytes, with localities.)—Also his Bidrag till Gottlands, Smålands och Blekinges flora. l.c. **1872:** 72–74. 1872. (Includes (p. 72–73) about 10 vascular plants from Småland.)—For his Smålands flora (1864), see below under Local.

Local

Bjurulf, Gunvor. Floran i Månsarps socken med Smålands Taberg. Bot. Notiser **1950:** 387–414. 10 fig. (incl. maps). 1950.

Topography, geology, climate, botanical explorations; discussion of various habitats with mention of numerous species, list of noteworthy species; bibliography, English summary. (In Jönköpings län.)—See also Zetterstedt, Om vegetationen, below.

Blomqvist, S. G. Ballastvegetation vid Kalmar 1912–1914. Svensk Bot. Tidskr. **11:** 289–303. 3 fig. 1918.

Description of ballast localities in and near Kalmar (city); extensive alphabetical list of ballast plants collected by author; list of additional species recorded by earlier writers; notes on some noteworthy native species. (In Kalmar län.)—See also **Blomgren, Nils.** Fynd av adventivväxter vid Kalmar åren 1915–1921. Bot. Notiser **1922:** 77–80. 1922. (List arranged by localities.)—Also **Snell, A. J.** Adventivfloran vid Kalmar åren 1924–1928. Svensk Bot. Tidskr. **23:** 356–365. 1929. (Description of the localities concerned (Tjärhovet, Barlastholmen, Fredriksskans, Ängö, Kalmar västra); list of adventive plants found by author, with dates; changes in flora.)—Also Moqvist, below.

Erikson, Johan. En studie öfver Jungfruns fanerogamvegetation. Arkiv Bot. v. 2, no. 3. 14 p. 1904.—Supplement . . . Bot. Notiser **1915:** 139–140. 1915.

Botanical explorations, plant formations, methods of dispersal, etc., with included mention of 142 phanerogams. (An island in northern mouth of

Kalmarsund, in Kalmar län.)—See also **Du Rietz, G. E.** Die hauptzüge der vegetation der insel Jungfrun. Svensk Bot. Tidskr. **19:** 323–346. 7 fig. (incl. map). 1925. (Geology, plant zones and associations; bibliography.) —Also Sterner. Vegetation och flora i Kalmarsunds skärgård, below.

Fries, E. M. Stirpes agri femsionensis, quarum indicem . . . 7 pt. (100 p.). 19.5 cm. Londini Gothorum, 1825–27.

General features of flora; list of vascular and cellular plants of Femsjö, with indication of frequency of occurrence and occasional annotations; notes on cultivated plants. (In Jönköpings län.)—See also **Fries, T. M.** Botaniska anteckningar rörande Femsjö socken i Småland. Bot. Notiser **1852:** 49–57, 69–78, 86–94. pl. 1852. (General features of flora, account of various excursions, mention of noteworthy plants.)—Also **Hård av Segerstad, Fredrik.** Femsjöfloran och dess förändring under de senaste hundra åren. Svensk Bot. Tidskr. **15:** 174–175. 1921. (Abstract of a lecture, with account of changes in the flora, mention of some added or extirpated species, etc.)

Hässler, Arne. Studier över kärlväxtfloran i Mo härad i nordvästra Småland. Bot. Notiser **1931:** 397–454. 1931.

Topography, geology, botanical explorations, ecology, etc.; list of vascular plants (437 species, not including critical genera), with localities; bibliography, German summary. (In Jönköpings län.)

Hård av Segerstad, Fredrik. Södra Sandsjö sockens fanerogamer. Ett bidrag till kännedomen om de pontiska och nordatlantiska växternas gränslinjer i sydöstra Sverige. Arkiv Bot. v. 11, no. 8. 44 p. map. 1912.

General features of flora; annotated list of 519 vascular plants; bibliography. (In Jönköpings län.)

——— Utkast till en flora över Värnamotrakten; till kännedomen om grönstenarnas inflytande på växternas utbredning. 35 p. incl. map. 23 cm. Värnamo, 1920. (Bihang till Värnamo kommunala mellanskolas årsredogörelse 1919/20. In Riksmuseet library, Stockholm.)

Topography, botanical explorations; annotated list of vascular plants (563 species); alphabetical list of species cultivated in district. (In Jönköpings län.)—See also his Tillägg till Värnamotraktens flora. 2 p. 1923. (Bihang till Värnamo kommunala mellanskolas årsredogörelse 1922/23. Bilaga D. In Riksmuseet library, Stockholm.)

Lund, A. A. W. Wimmerby-florans phanerogamer och ormbunkar. 34 p. 19.5 cm. Upsala, 1863. (Acad. afhandl.) (Gray Herbarium library.)

Physiography, general features of flora, etc.; annotated list of 551 vascular plants. (In Jönköpings län.)

Lundequist, O. F. E. Om Grenna sockens vegetation. Svensk Bot. Tidskr. **23:** 11–43. 1929.

Botanical explorations, topography, geology, climate; annotated list of vascular plants and mosses (about 725 vascular plants, not counting *Rosa, Hieracium,* and *Taraxacum,* and over 200 mosses); bibliography. (In Jönköpings län.)

Lundin, P. E. Bidrag till kännedomen om Västervikstraktens kärlväxter (lokaler och frekvensuppgifter). 2. uppl. 31 p. incl. 4 fig. i.e. map, 3 pl. on 2. Göteborg, 1936. (1st ed. 1933.)

Alphabetical list of vascular plants (including cultivated species), annotated as to abundance. (In Kalmar län.)—See also Witte, below.

Moqvist, C. J. Systematisk förteckning på phanerogama växterna i Calmar stads närmaste omgifning. I. 16 p. 20 cm. Upsala, 1848. (Acad. afhandl.)

List of phanerogams; incomplete, reaching Primulaceae on the Friesian system. (Not seen; title and annotation furnished by E. Hultén and G. Haglund.) (In Kalmar län.)—See also Blomqvist, above.

Porat, C. O. von. Om Jönköpingstraktens flora och fauna. *In* Jönköpings historia. p. 41–59. 26 cm. Jönköping, 1917. (Riksmuseet library, Stockholm.)

Topography, climate, phenology; unannotated list of vascular plants of vicinity of city. Pages 41–49, 59, deal with botany, the rest with zoology. (In Jönköpings län.)

Rosen, Gösta von. Floristiska och växtgeografiska anteckningar från Romanäs-halvön i Säby s:n, Jönköpings län. Bot. Notiser **1934:** 281–309. 5 fig. (incl. maps). 1934.

Geology, general features of vegetation, climate; briefly annotated list of bryophytes, lichens, and vascular plants; bibliography. (In Jönköpings län.)—See also **Hellgren, G.** Sällsynta växter i Säby socken. *In* T. Filén. Boken om Säby. p. 42–47. Tranås, 1951. (Includes about 20 additions. Not seen; title and annotation supplied by E. Hultén.)

Scheutz, N. J. Smålands flora innefattande Kronobergs och Jönköpings läns fanerogamer och ormbunkar. xxiv, 362 p. 17 cm. Wexjö, 1864. (Riksmuseet library, Stockholm.)

Phytogeography; keyed descriptive flora of vascular plants and Characeae, with localities for scarcer species; list of principal cultivated plants.—See also additions listed under his Conspectus florae smolandicae, above.—Also **Zetterstedt, J. E.** Bidrag till Jönköpingstraktens flora. Efter J. E. Zetterstedts anteckningar sammanstälde af Ev. Zetterstedt. Bot. Notiser **1881:** 115–118. 1881. (List of vascular plants, with localities.)—Also **Lindwall, C. W.** Något om fanerogamfloran i Jönköpings län. l.c. **1907:** 75–80. 1907. (List of noteworthy spermatophytes, with localities, from Västbo, Östra and Södra Vedbo härader.)—Also his Ett par små drag af floran i Kronobergs län. l.c. **1908:** 108–111. 1908. (List of noteworthy vascular plants, with localities, from Uppvidinge and Norrvidinge härader.)—Also **Gaunitz, Sven.** Floristiska uppgifter från Småland. l.c. **1928:** 327–330. 1928. (List of noteworthy vascular plants, mostly from Hultsjö (in Jönköpings län) and Ryssby (in Kronobergs län).)—Also **Lindstam, C. S.** Ur Smålands flora. l.c. **1931:** 221–223. 1931. (List of noteworthy spermatophytes, with localities, from Jönköpings and Kronobergs län.)

Sjöstrand, M. G. Calmar läns och Ölands flora. 4 p.l., 359, iv p. 19.5 cm. Calmar, 1863.

Annotated, keyed, descriptive flora of vascular plants and Characeae (1,061 species, of which 1,052 are vascular), with localities.—See also **Scheutz, N. J.** Tillägg till M. G. Sjöstrands Calmarläns flora. Bot. Notiser **1863:** 119–124. 1863. (List with localities.)—Also **Ekstrand, E. V.** Novitier för Kalmar läns flora. l.c. **1866:** 18–24. 1866. (List with localities, intended to include all additions since publication of Sjöstrand's Flora.)

Sterner, Rikard. Flora och vegetation i Allgunnenområdet i östra Småland. Svensk Bot. Tidskr. **48:** 104–116. 1954.

Geology, general features of flora, etc.; extensive alphabetical list of vascular plants. (In Kalmar län.)

——— Om floran och vegetationen i Norra Kvills nationalpark i Kalmar län. K. Svenska Vetenskapsakad. Skr. Naturskyddsär. no. 10. 20 p. 1929.

Topography, plant associations; briefly annotated list of vascular plants (196 species). (An area of about 27 hectares in Norra Kvills kronopark in Rumskulla socken, Kalmar län.)

——— Vegetation och flora i Kalmarsunds skärgård. Acta Hort. Gotoburg. **8:** 189–280. fig. 1–17 (maps), 14 tab., map. 1933.

Topography, geology, botanical explorations, plant formations, etc.; annotated list of vascular plants; bibliography. Covers coastal region from Kalmar north to Figeholm. (In Kalmar län.)—See also Erikson, above.

Vahl, Martin. The vegetation of the Notö. Bot. Tidsskr. **32:** 275–284. 1912.

General features of flora, life forms, with included lists of species. (In Kronobergs län.)

Westerlnud, C. A. Bidrag till kännedomen om Kalmar läns vegetation. 62 p. 20 cm. Kalmar, 1852. (Riksmuseet library, Stockholm.)

Botanical explorations, general features of flora; annotated list of vascular plants and Characeae (848 species); account of excursions.—See also his Botaniska anteckningar sommaren 1853. I. Novitier och nya växtlokaler för Kalmarfloran. Bot. Notiser **1853:** 165–168. 1853. (List with localities.)

Witte, Hernfrid. Om vegetationen på ön Städsholmen i Västerviks skärgård. Bot. Notiser **1930:** 13–34. 3 fig. 1930.

Geography, geology, life forms, general features of flora, cultivated plants; annotated alphabetical list of vascular plants (259 species), mosses, and lichens. (In Kalmar län.)—See also Lundin, above.

Zetterstedt, J. E. Om vegetationen i de högländtaste trakterna af Småland. K. Svenska Vetensk.-Akad. Handl. n.f., v. 6, no. 2. 37 p. 1866.

General features of flora; list of vascular plants, bryophytes, and lichens, with occasional footnotes on distribution (637 species, of which 382 are vascular plants). Relates to the Taberg and to Almesåkra and adjacent socknar, in Jönköpings län.—See also Bjurulf, above.

——— Vegetationen på Visingsö. Bihang K. Svenska Vetensk.-Akad. Handl. v. 5, no. 7. 86 p. 1878.

Plant formations, etc.; annotated list of vascular plants, bryophytes, and lichens (752 species, of which 488 are vascular). (In Jönköpings län.)—See also his Om vegetationen på Visingsö. Bot. Notiser **1879**: 159–160. 1879.—Also **Lundequist, Olof.** Några anmärkningsvärda växter från Gränna och Visingsö. Svensk Bot. Tidskr. **13**: 104–106. 1919. (Includes some records from Visingsö.)

SÖDERMANLAND

See also Närke (Birger); Uppland (Almquist and Asplund; Laurent).

The old province (landskap) of Södermanland includes Södermanlands län, the southern third (Södertörn) of Stockholms län, and the southern half (the part south of the river Norrström) of Stockholms stad.

General

Thedenius, K. F. Flora öfver Uplands och Södermanlands fanerogamer och bräkenartade växter. 1 p. l., 524 p. 22 cm. Stockholm, 1871. (Gray Herbarium library.)

Descriptive flora with keys, on the Linnaean system, with localities for less common species. Replaced, as to Uppland, by Almquist, Upplands vegetation och flora. 1929. Replaces **Hofberg, Herman.** Södermanlands phanerogamer och filices. 87 p. 18.5 cm. Stockholm, 1852. (Topography, geology, bibliography, statistics; annotated list of 884 vascular plants, with localities for less common species. In Gray Herbarium library.)—See also **Samzelius, Hugo.** Några för Södermanland nya växtlokaler. Bot. Notiser. **1884**: 145–148. (List with localities.)—Also **Lindström, A. A.** Bidrag till Södermanlands växtgeografi. l.c. **1888**: 194–198. 1888. (List with localities, chiefly from Nyköping and Bogsta.)—Also **Malme, G. O. A.** Bidrag till sydvestra Södermanlands kärlväxtflora. l.c. **1891**: 97–106. 1891. (Notes on distribution, and lists.)—Also his Ett litet bidrag till Södermanlands flora. Svensk Bot. Tidskr. **19**: 423–425. 1925. (Notes from Bettna, Halla, etc.)—Also **Qvarfort, Sten.** Märkligare växtfynd i västra Södermanland. l.c. **42**: 289–291. 1948. (Alphabetical list of noteworthy vascular plants from Dunker, Gryt, and Hyltinge socknar.)

Local

Blom, Carl. Växtgeografiska anteckningar till Nyköpingstraktens fanerogamflora. Bot. Notiser **1916**: 1–23. 1916.

Botanical explorations, account of species found only by earlier collectors; annotated list of scarcer vascular plants, unannotated list of common species (about 620 native and numerous adventive species). Covers Nyköping city and Nikolai socken with adjacent parts of Helgona and Svärta socknar. (In Södermanlands län.)—See also his Invandrare. l.c. **1912**: 45–47. 1912. (List of noteworthy adventive plants, with localities.)—Also **Arnell, H. W.** Fanerogamfloran i Nyköpingstrakten. l.c. **1917**: 97–110. 1917. (Comparison of Blom's list with author's results in Helgona socken, additions to flora, notes on cultivated plants.)

Carlson, C. A. Flora strengnesensis. 1 p.l., 62 p. 24.5 cm. Upsaliae, 1791. (Diss., C. P. Thunberg praes.)

Annotated list of vascular and cellular plants, on the Linnaean system. (In Södermanlands län.)—See also **Hofberg, Herman.** Upgift på några i och omkring Strengnäs förekommande sälsyntare fanerogamer. Bot. Notiser **1842**: 76–78. 1842; **1843**: 38–48. 1843; Tillägg vid upgifterna öfver Strengnäs-traktens vegetation. l.c. **1843:** 136–137. 1843. (List of 102 noteworthy vascular plants, with localities.)—Also **Clemedson, C. J.** Carolus Axelius Carlsons Flora strengnesensis: en akademisk avhandling om Strängnästraktens flora under 1700-talets senare del. Skr. Fören. Strengnenses no. 1. 35 p. 3 fig., 1 facsim. 21 cm. 1956. (Includes annotated list of vascular and cellular plants, alphabetically arranged.)

Ekström, C. U. Beskrifning öfver Mörkö socken i Södermanland. 218, 27 p. pl., 3 tab. 22 cm. Stockholm, 1828.
Includes (p. 116–137) annotated list of vascular plants. (In Södermanlands län.) (Not seen; title and annotation supplied by E. Hultén and G. Haglund.)

Hellbom, P. J. Förteckning på phanerogamer och ormbunkar i Österåkers socken af Södermanland. 1 p. l., 29 p. 19.5 cm. Upsala, 1851. (Acad. afhandl., Elias Fries praes.) (New York Botanical Garden library.)
General features of flora, etc.; briefly annotated list of 522 vascular plants and Characeae. (In Stockholms län.)

Laurell, J. G. Anmärkningsvärdare fanerogamer och kärlkryptogamer inom Sorunda pastoratsområde af Södertörn uti Södermanland. Bot. Notiser **1898:** 81–92, 97–106. 1898.
Geography, geology, etc.; annotated list of rarer species, list of common species. (In Stockholms län.)

Lindeberg, C. J. Plantarum vascularium in regione Maeleri orientali-boreali sponte crescentium synopsis. Pt. I-II. (33 p.). 21 cm. Upsaliae, 1848. (Diss., Elias Fries praes.)
Topography, climate, general features of flora, etc.; list of vascular plants, annotated as to frequency of occurrence, with occasional localities. Refers to Svartsjölandet, including the islands Färingsön, Munsön, Ekerön, Adelsön, Björkön, Elgarn, and northern boundary of Lake Malar up to Sigtuna. (In Stockholms län.)

Lindström, A. A. Bogsta sockens fanerogamer och ormbunkar. Bot. Notiser **1892:** 261–265. 1892; **1893:** 11-23. 1893.
Annotated list of vascular plants. (In Södermanlands län.)

Lundeqvist, N. W. Flora paroeciae Bränkyrka. 3 pt. (48 p.). 17 cm. Upsaliae, 1827. (Diss., N. J. Sillén praes.) (New York Botanical Garden library.)
Topography, etc.; annotated list of ferns (1 species), mosses, algae, fungi, and spermatophytes (509 species), on the Linnaean system. (In Stockholms stad.)

Malme, G. O. A. Bidrag till Trosatraktens flora. Svensk Bot. Tidskr. **21:** 305–313. 1927.
Geology, general features of flora; alphabetical list of common vascular plants not previously recorded, alphabetical list of noteworthy angiosperms, with localities. Complements the records from Trosa in "Stockholmstraktens växter," 1st ed. (In Södermanlands län.)

Post, Hampus von. Om vextgeografiska skildringar. Bot. Notiser **1851:** 110–127, 161–186. map. 1851.—[Tryckfel och tillägg.] l.c. **1852:** 31–32. 1852.
Includes "Försök till en vext-topografisk skildring af parken vid Ås i Södermanland. 1850," a list of vascular plants arranged by habitats. The map, accidentally omitted from the volume for 1851, was issued with no. 2 for 1852 with instructions to insert it in the earlier volume. (In Södermanlands län.)

Qvarfort, Sten. Några anteckningar om floran i Helgesta socken av Södermanlands län. Bot. Notiser **1926:** 276–282. 1926.—Tillägg. . . . l.c. **1930:** 76–77. 1930.
Geology, geography, general features of flora; list of less common vascular plants collected by author in 1922-25, with localities. (In Södermanlands län.)

Söderberg, D. H. Pandora et flora rybyensis. 23 p. 20 cm. Upsaliae, 1771. (Diss., C. Linné praes.) (Reprinted in Linné, Amoenitates Academicae **8:** 75–106. 1785.)
Includes (p. 16–23 of original, p. 93–106 of reprint) "Flora rybyensis," a list of vascular and a few cellular plants, on the Linnaean system, annotated as to frequency of occurrence. (Locality now spelled Ribby. In Stockholms län.)

Torssander, Axel. Anmärkningsvärdare fanerogamer och kärlkryptogamer i Wårdinge socken (Södermanland). Bot. Notiser **1897:** 157–170. 1897.
Botanical explorations; annotated list of vascular plants based almost entirely on the author's collections. (In Stockholms län.)

UPPLAND

The old province (landskap) of Uppland includes Uppsala län, Stockholms län (except the southern third (Södertörn), which belongs to the landskap of Södermanland, Stockholms stad (except the southern half, which is also in Södermanland), and the eastern part (Fjärdhundra) of Västmanlands län.

General

Almquist, Erik. Träd och buskar i Upplands vegetation. Lustgården **10:** 105–140. 18 fig. (incl. maps). 1929.
List of native and naturalized woody plants (excluding some smaller species), with vernacular names and local range; discussion of distribution, etc.

——— Upplands vegetation och flora. Acta Phytogeog. Suecica v. 1. xii, 622 p. front., 440 fig. (incl. diagrs., maps, 430 distr. maps), maps, table. 1929.
Topography, soils, climate, postglacial evolution, history of civilization, plant associations, ecological distribution of species within district, history of plant immigration; annotated list of known vascular plants (1020 species and subspecies), with localities; bibliography. There are 430 maps showing range. Covers Stockholms stad and län (except the portion south of Lake Malar), Uppsala län, and the eastern part (Fjärdhundra) of Västmanlands län (east of Sala). Replaces **Schagerström, J. A.** Conspectus vegetationis uplandicae. 4 pt. (1 p. l., 83 p.). 20.5 cm. Upsaliae, 1845. (Diss.) (Botanical explorations, topography, geology, climate, phenology, general features of flora, comparison with other regions, plant regions; list of known vascular plants, annotated as to abundance and with localities for the rarer species. In Arnold Arboretum library.) Also replaces (for Uppland) **Thedenius, K. F.** Flora öfver Uplands och Södermanlands fanerogamer och bräkenartade växter. 1871.

Local

Almquist, Erik. Floran inom Ängsö nationalpark. K. Svenska Vetenskapsakad. Skr. Naturskyddsär. 13. 47 p. 1 fig. (map). 1930.
Topography, general features of flora, etc.; annotated list of vascular plants. (An area of about 75 hectares on the island Ängsö in Länna parish, Stockholms län.)

——— and **Asplund, Erik,** *eds.* Stockholmstraktens växter. Förteckning över fanerogamer och kärlkryptogamer med fyndorter och frekvensuppgifter utgiven av Botaniska Sällskapet i Stockholm. 2. uppl. cxxiii, 364 p. (incl. 2 ports., 17 maps). map. 20.5 cm. Stockholm, 1937. (1st ed. 1914, Andersson, Gunnar, and others, *eds.*)
Geography, botanical investigations, geology, Quaternary history, soils, plant associations, introduced plants, disappearing species, climate, phytogeography, statistics, general features of botanical regions, chronological bibliography, list of collaborators; annotated, alphabetically arranged list of vascular plants (1464 species, of which 1008 are native), with detailed localities; list of localities. Covers middle and southern part of Stockholms län (north to Hammarby, Norrtälje, and Rådmansö) and adjacent eastern Södermanland (from Västerljung and Trosa north to Överenhörna), 58° 44′ to 59° 47′ N., and includes the coastal islands. Replaces **Wikström, J. E.** Stockholms flora, eller korrt beskrifning af de vid Stockholm i vildt tillstånd förekommande växter . . . v.1 (viii, vi, 185, 423 p. map). 22 cm. Stockholm, 1840; Tillägg. 27 p. 22 cm. n.p. [1844]. (Includes descriptive flora of spermatophytes (417 species); annotated list of 51 introduced species. Incomplete, reaching *Thalictrum* on the Linnaean system.) Also replaces **Thedenius, K. F.** Stockholmstraktens phanerogamer och ormbunkar . . . Bot. Notiser **1850:** 57–80, 89–125. 1850. (Includes annotated list of vascular plants (886 species), with localities for the scarcer species. Reprinted with index, 64 p. 20 cm. Stockholm, 1850. In Riksmuseet library, Stockholm.)

Tillägg . . . l.c. **1856**: 165–178. 1856. (Reprinted, 16 p. 1857. In Riksmuseet library, Stockholm.)—See also **Almquist, E.**, and **Asplund, E.** Förklaringar och anmärkningar till "Stockholmstraktens växter." Svensk Bot. Tidskr. **33**: 277–302. 1939. (Extensive list of corrections and additional localities.) —Also **Horn af Rantzien, Henning.** Spridda bidrag till kännedomen om Stockholmstraktens flora. l.c. 35: 361–364. 1941. (List of additional localities.)—Also **Köhlin, Per.** Tillägg till "Stockholmstraktens växter." l.c. **36**: 52–56. 1942. (Similar list.)—Also **Arrhenius, O.** Artfördelningen i Stockholmstrakten. l.c. **43**: 215–222. 3 fig. (incl. 2 maps). 1949. (List of socknar with their respective areas and number of indigenous species, and discussion.)—Also **Sandberg, Oswald.** Några bidrag till kännedomen om Stockholmstraktens flora. l.c. **45**: 257. 1951. (Records of several noteworthy species.)—Also Laurent, below.

Eriksson, J. V. Bälinge Mossars utvecklingshistoria och vegetation. Svensk Bot. Tidskr. **6**: 105–194. 42 fig. (incl. map), 4 pl. (incl. map). 1912.
Ecological; includes (p. 178–183) tabular alphabetical list of vascular plants, mosses, and lichens; German summary, bibliography. (In Uppsala län.)

Floderus, Ali. Synopsis plantarum paroeciae Uplandiae Funbo. Pt. 1. 1 p. l., 16 p. 21 cm. Upsaliae, 1853. (Lloyd Library.)
Topography, general features of flora, climate, botanical explorations; briefly annotated list of dicotyledons, Coniferae, and Equisetaceae, with localities for the rarer species. No more published. Dissertation, M. A. Goldschmidt respondent. (In Uppsala län.)

Håkansson, J. W. Floristiska bidrag från södra Lidingö. Svensk Bot. Tidskr. **12**: 373–402. 1918.
Topography, general features of flora, changes in flora, etc.; annotated alphabetical list of vascular plants. (In Stockholms län.)—See also his Lidingö flora. *In* Lidingön och dess natur. p. 103–176. illus. 21.5 cm. Stockholm, 1927. (Topography, general features of flora; annotated alphabetical list of vascular plants (916 species), with localities. In Riksmuseet library, Stockholm.)—Also his Nya floristiska uppgifter från Lidingö. Svensk Bot. Tidskr. **25**: 417–434. 1931. (Extensive alphabetical list of vascular plants, with localities.)—Also **Håkansson, J. W.**, and **Knöppel, J.** Nya tillägg till Lidingö flora. l.c. **39**: 211–241. 1945. (Gazetteer, etc.; extensive list of vascular plants with localities.)

Hamner, J. W. Redogörelse för resultatet av en med understöd ur C. W. Sebardts stipendiefond företagen resa för växtgeografiska studier inom Blidö socken i Stockholms skärgård under sommaren 1908. 17 p. map. 20 cm. Visby, 1909. (Riksmuseet library, Stockholm.)
General features of flora; annotated alphabetical list of vascular plants. (In Stockholms län.)

Hesselman, Henrik. Zur kenntnis des pflanzenlebens schwedischer laubwiesen. Eine physiologisch-biologische und pflanzengeographische studie. Beih. Bot. Centralbl. **17**: 311–460. 29 fig. (incl. maps), pl. 4–8. 1904.
Mainly ecological; includes (p. 342–345) alphabetical list of vascular plants of Skabbholmen (belonging to Lidö, in the parish of Vätö, in Stockholms län).

Horn af Rantzien, Henning. Österskärstraktens flora. En floristisk undersökning av södra delen av Österåker socken, Uppland. Svensk Bot. Tidskr. **40**: 15–30. 6 fig. (maps). 1946.
Botanical explorations, geology, topography, types of vegetation, statistics; list of noteworthy vascular plants, with localities; bibliography. (In Stockholms län.)

Laurent, Vivi. Några studier över Stockholmstraktens adventivflora. Svensk Bot. Tidskr. **13**: 255–294. 1920.
Discussion of methods of introduction, etc., with short lists from various localities; extensive alphabetical list of adventive plants with localities and dates. (In Stockholms län and stad.)—See also **Qvarfort, Sten.** Bidrag till kännedomen om Stockholmstraktens adventivväxter. Svensk Bot. Tidskr. **21**: 139–154. 1927. (Descriptions of various localities, with lists; extensive alphabetical list of adventive plants, with localities and dates.)

Malme, G. O. Ett tillägg till Stockholmstraktens växter. Svensk Bot. Tidskr. 9: 367–371. 1915.

Remarks on flora; extensive alphabetical list of plants not previously recorded from Kummelnäs and Hasseludden in Bo socken, Stockholms län.—See also **Smith, Harry.** Tillägg till Stockholmstraktens växter. Svensk Bot. Tidskr. **10**: 572–577. 1 fig. 1916. (Extensive alphabetical list of vascular plants from the same and other localities in Bo socken.)—Also **Malme, G. O.** Ett litet bidrag till Bo sockens kärlväxtflora. l.c. **17**: 110–111. 1923. (Short list.)—Also **Täckholm, Gunnar.** Floristiska bidrag från Bo socken. l.c. **18**: 514–518. 1924. (Extensive alphabetical list of vascular plants.)

Qvarfort, Sten. Kärlväxtfloran på Runmarö. Svensk Bot. Tidskr. **25**: 36–76. 8 fig. (incl. 2 maps). 1931.

Botanical explorations, topography, general features of flora; annotated list of known vascular plants; bibliography. (In Stockholms län.)—See also **Engeström, Tor,** and **Erhardt, Richard.** Från holmarna kring Runmarö. Bot. Notiser 1931: 229–268. 10 fig. (incl. maps). 1931. (Account of topography, general features of flora of different habitats, etc., of various islands in the archipelago just south of Runmarö, with mention of numerous species.)—Also **Jungstedt, Herbert.** Taraxaca från Runmarö och omnejd. Svensk Bot. Tidskr. 51: 352–353. 1957. (List of 72 species.)

Schagerström, J. A. Plantae cotyledoneae paroeciae Roslagiae Bro. . . . 2 pt. (30 p.). 22.5 cm. Upsaliae, 1839–41. (Diss., E. Fries praes.)

Topography, etc.; list of spermatophytes, Equisetaceae, and Characeae, with localities for the rarer species. (In Stockholms län.)—See also Wahlström, below.

Wahlenberg, Georg. Flora upsaliensis enumerans plantas circa Upsaliam sponte crescentes. 1 p. l., vii, 495 p. fold. map. 21 cm. Upsaliae, 1820.

Descriptive flora of vascular and cellular plants, on the Linnaean system. (In Uppsala län.)—See also **Myrin, C. G.** Corollarium florae upsaliensis. 1 p. l., 123 p. 22 cm. Upsaliae, 1834. (Diss.) (Intended as a supplement to Wahlenberg's Flora; covers vascular plants and mosses. Originally issued in 8 pt., 1833–34.)—Also **Hedbom, Karl.** Strödda anteckningar om några märkligare element i Upsala-floran förr och nu. Svensk Bot. Tidskr. **6**: 459–465. 1912. (List of noteworthy species, with localities.)

Wahlström, J. E. Plantarum vascularium in regione Telgae borealis sponte crescentium synopsis. 3 pt. (2 p. l., 40 p.). 20 cm. Upsaliae, 1847. (Diss., J. A. Schagerström, praes.) (Gray Herbarium library.)

Topography, climate, etc.; briefly annotated list of vascular plants and Characeae. (Covers a considerable area around Norrtälje, in Stockholms län, from Rådmansö north into Väddö socken, southwest into Rö socken.)—See also Schagerström, above.

Witte, Hernfrid. Förteckning över kärlkryptogamer och fanerogamer på Bergshamra i Solna socken, Uppland. Svensk Bot. Tidskr. **36**: 408–428. 3 fig. 1942.

General features of flora, etc.; list of vascular plants (512 species, including hybrids), with localities; bibliography. (In Stockholms län.)

VÄRMLAND (VERMLAND, WERMLAND)

See also Dalsland (Hård av Segerstad, Pflanzengeographische studien im nordwestlichen teil der eichenregion Schwedens).

The old province (landskap) of Värmland or Wermland has essentially the same boundaries as Värmlands län, except that a small area near Karlskoga is in Örebro län.

General

Hård av Segerstad, Fredrik. Den värmländska kärlväxtflorans geografi. Göteborgs K. Vetensk.- och Vitterh.-Samhäll. Handl. VI, ser. B, Mat. och Naturv. Skr. v. 7. 707 p. incl. 39 (i.e. 40) fig. (mostly maps), 270 maps. 1952.

Botanical explorations, climate, recent geological history, plant immigration, geology, soils, topography, recent changes in flora, phytogeography; list of known vascular plants [about 957 species (of which about 802 are native or naturalized) and 100 hybrids, as well as over 100 microspecies of *Taraxacum; Rosa* and *Hieracium* are omitted] with very detailed localities; list of botanists with brief biographical notes, gazetteer, French résumé, bibliography. The figures and maps are nearly all distribution maps. Replaces (as to Värmland) **Larsson, L. M.** Flora öfver Wermland och Dal. 2. uppl. xii, 350 p. 19.5 cm. Carlstad, 1868. (1st ed. 1859.) (Bibliography; keyed, annotated flora of spermatophytes (811 species), on the Linnaean system, with localities.)—See also **Hård av Segerstad, F.** Några anmärkningar till Värmlands flora. Svensk Bot. Tidskr. **51**: 552–558. 1957. (Various additions.)

Local

Arrhenius, Axel. Floristiskt småplock. Spridda småbidrag till kännedomen om Lundsbergs-traktens växtvärld. Lundsbergs Skolas Festskrift **1912**: 193–206. Göteborg, 1912.
Annotated list of vascular plants observed in vicinity of Lundsberg. (Not seen; title and annotation supplied by E. Hultén and G. Haglund.)

Hård av Segerstad, Fredrik. Floran i Norra Finnskoga. Meddel. Värmlands Naturhist. För. 11. 36 p. 3 fig. (fig. 2, map). 1938.
Physiography, botanical explorations, statistical treatment of phanerogams from 10 localities, phytogeography; briefly annotated alphabetical list of known vascular plants; bibliography.

——— Om floran i Rämmen socken. Bot. Notiser **1940**: 9–32. 4 fig. (incl. map). 1940.
Geography, topography, geology, botanical explorations, list of species arranged by frequency of occurrence, etc.; annotated alphabetical list of known vascular plants; bibliography; German summary.

——— Utkast till en flora över trakten kring sjön Kymmen i västra Värmland. Meddel. Värmlands Naturhist. För. 7. 22 p. map. 1934.
Phytogeography, with lists of species; annotated alphabetical list of vascular plants, with localities. (In Gräsmark socken in Fryksdals härad and Gunnarskog socken in Jösse härad.)

Hedlund, Lennart, and **Sundell, Sigurd.** Floran i Munkfors. I. Kärlväxter. Bot. Notiser **1951**: 331–367. 3 fig. (maps). 1951.
Geology, topography, botanical explorations, phytogeography; annotated alphabetical list of vascular plants (578 species, subspecies, and hybrids); bibliography.

Lohammar, Gunnar. Floran på Värmlandsnäs och i sydvästra Värmland. Svensk Bot. Tidskr. **31**: 221–287. 4 fig. (incl. map). 1937.
Botanical explorations, geography, phytogeography, etc.; partial list of vascular plants, with localities; bibliography.

Skårman, J. A. O. Anteckningar om kärlväxtfloran i nordligaste Värmland. Svensk Bot. Tidskr. **6**: 64–91. 1912.
Author's explorations, etc.; list of vascular plants collected, with localities. Relates to area along Klarälven and Fryksdalen from Edebäck and Torsby northward.—See also his Om några förekomster af ädla löfträd i nordligaste Värmland. Svensk Bot. Tidskr. **5**: 393–401. 1912. (On occurrence of *Alnus and Tilia.*)—Also his Bidrag till nordligaste Värmlands flora. l.c. **7**: 367–373. 1 fig. (map). 1914. (Further notes on *Tilia;* additions to his list.)

VÄSTERBOTTEN (VESTERBOTTEN)

See also Norrbotten (Arwidsson, Växtgeografiska notiser från Norrland III [includes list from Renholmen and other islands off Västerbotten]); Sweden, Partial (Backman).

The old province (landskap) of Vesterbotten forms only the coastal part of Västerbottens län. For the remaining (Lapland) part of Västerbottens län, see Lappmark, Åsele, and Lappmark, Lycksele.

Local

Arwidsson, Thorsten. Växtgeografiska notiser från Norrland. IV. Bot. Notiser **1931**: 360–374. 1 fig. (map). 1931.
Botanical explorations on Holmöarna, east of Umeå, with unannotated list of the widely distributed vascular plants and list of scarcer species with localities, the whole amounting to 291 species; bibliography.—See also

Grapengiesser, Sten. En blick på Holmöarnes flora. Bot. Notiser **1922:** 313–316. 1922. (General features of flora, with mention of various species.) —Also **Valovirta, E. J.** Die inseln des Holmö-Archipels. *In* his Untersuchungen über di säkulare landhebung als pflanzengeographischer faktor. Acta Bot. Fenn. **20:** 79–83. 1937. (Includes lists of vascular plants of the islands Lillhällan and Holmögadd.)

Fridner, Gustav. Svenska växtnamn i Degerfors socken och angränsande trakter, samt några uppgifter om en del växters användning m. m. Västerbotten **1926:** 266–277. 8 fig. 1926. (University of Chicago library.)
Alphabetical list of vernacular names with botanical equivalents and some notes on uses.

Grapengiesser, Sten. Bygdeåtraktens flora. Iakttagelser rörande vegetationen i en västerbottnisk kustsocken, med artförteckning. Svensk Bot. Tidskr. **20:** 366–405. 1926.
Geology, ecology, etc.; annotated list of vascular plants; bibliography, English summary.—See also his Norrländska vegetationsbilder. Svensk Bot. Tidskr. **28:** 300–353. 1934. (Includes alphabetical list of additions, p. 345–347, under heading Bygdeå socken i Västerbottens län.)

——— Västerbottniska växtnamn. Västerbotten **1926:** 245–257. 1926. (University of Chicago library.)
Alphabetical list of vernacular names, with botanical equivalents and frequent annotations; refers principally to vicinity of Bygdeå socken.

Laestadius, C. P. Några ord om Umeå-traktens flora. Bot. Notiser **1863:** 105–116. 1863.
General account of vegetation, with mention of numerous species.—See also **Andersson, N. L.** Bidrag till Vesterbottens växtgeografi. Bot. Notiser **1890:** 36–38. 1890. (List of vascular plants, with localities, mostly from Umeå socken.)—Also **Melander, Carl.** Tillägg till sid. 38 i Botaniska Notiser 1890. l.c. **1890:** 158–159. 1890. (Notes on several species.)—Also **Lundqvist, Gösta.** Kärlväxter i Umeåtrakten. Svensk Bot. Tidskr. **11:** 344–354. 1918. (Botanical explorations; annotated alphabetical list of vascular plants collected by author.)—Also **Wahlberg, Lennart.** En ny Calypso-lokal i Västerbotten. l.c. **15:** 261–263. 1921. (Includes short list of plants from a locality about 7 kilometers north of Umeå.)

Lindström, A. A. Bidrag till Norrlands växtgeografi. Bot. Notiser **1910:** 165–172. 1910.
Extensive list of vascular plants, with localities, from vicinity of Byske (including Furuögrund).

Malmström, Carl. Vegetation. The experimental forests of Kulbäcksliden and Svartberget in north Sweden. 2. Skogsförsöksanst. Exkursionsled. **11:** 27–87. fig. 6–27, 2 maps. 1926.
Ecological; includes (p. 76–86) tabular alphabetical lists of spermatophytes, pteridophytes, liverworts, mosses, and lichens, with ecological annotations; bibliography.

Wahlberg, Lennart. Bidrag till kännedomen om hembygdens flora. *In* Västerbotten 1923-27, 1931, 1932, 1941, 1942, 1945, 1947–49, passim.
Annotated lists of vascular plants from different localities in Västerbotten. (Not seen, except for two papers; title and annotation supplied mostly by E. Hultén and G. Haglund.)

VÄSTERGÖTLAND (VESTERGÖTLAND)

The old province (landskap) of Västergötland includes (northern part) Skaraborgs län and (southern part) the greater part of Älvsborgs län. For the Göteborg area see Göteborgs och Bohus län.

General

Rudberg, August. Förteckning öfver Västergötlands fanerogamer och kärlkryptogamer med uppgift om växeställen och frekvens efter kollega Ernst Linnarssons och lektor Bror Forssells med fleres anteckningar . . . xiv, 129 p. 19 cm. Mariestad, 1902.
Botanical explorations, bibliography; annotated list of vascular plants [about 1140 species, not counting the species of *Hieracium* listed on p.

127–129], with localities; alphabetical list of cultivated plants; index, list of additional species and subspecies of *Hieracium*. Plants known only from the northwestern part of Västergötland belonging to the Göteborg floral region (härader of Askim, Säfvedal, and Östra Hising) are not included in the main list but are listed on p. 119–122.—See also **Witte, Hernfrid.** Tillägg till "Rudberg: Växtförteckning öfver Västergötland." Bot. Notiser **1902:** 271–282. 1902. (Extensive list of additions and corrections.)—Also **Rudberg, A.** Några ord om min bok: "Förteckning . . ." l.c. **1903:** 51–53. 1903. (Reply to Witte.)—Also **Westerlund, C. G.** Bidrag till Västergötlands flora. l.c. **1904:** 1–25. 1904. (List of noteworthy angiosperms, with localities, mostly from vicinity of Alingsås; includes a systematic treatment of *Alchemilla* in southern and middle Sweden.)—Also **Rudberg, A.** Exkursioner på området af botanisk litteratur rörande Västergötland. l.c. **1906:** 85–92, 159–165. 1906. (Geography and boundaries, botanical regions, alphabetical list of erroneously spelled locality names with references and corrections, annotated list of erroneously recorded species, list of plants reported from Västergötland without definite localities.)—Also his Nyfunne växter i Västergötland. l.c. **1906:** 249–251. 1906. (List with localities.)—Also **Almquist, Erik.** Växtgeografiska bidrag. 4. Västergötland. l.c. **1921:** 175–180, 221–222. 1921. (List of noteworthy vascular plants, with localities, mainly from region of Karlsborg.)—Also **Nilsson, Gunnar.** Bidrag till Västergötlands kärlväxtflora. Svensk Bot. Tidskr. **19:** 236–249. 1925. (Author's explorations, corrections to Rudberg's Förteckning, etc.; list of vascular plants, with localities, from Älgarås, Borås, Kinnekulle, Lugnås, Mariestad, Skövde, and other localities.)—Also **Erdtman, Gunnar.** Växtlokaler från Halland och sydvästra Västergötland. l.c. **19:** 370–379. 1925. (List of vascular plants observed by author, with localities.)—Also **Hasselrot, Karl.** Några bidrag till kännedomen om Västergötlands flora. l.c. **50:** 217–211. 1956. (List of noteworthy species.)—Also Kjellberg, below.

Local

Albertson, Nils. Floran i Dala socken på nordöstra Falbygden. Svensk Bot. Tidskr. **51:** 287–316. 1957.

Geology, etc.; briefly annotated list of vascular plants; bibliography. (In Skaraborgs län.)

——— Österplana hed. Ett alvarområde på Kinnekulle. Acta Phytogeog. Suecica v. 20. xii, 267 p. 18 fig. (incl. maps), 16 pl. on 8. 1946. (Akad. avhandl., Uppsala.)

Ecological and phytogeographical; includes (p. 165–176) annotated list of 380 vascular plants; extensive bibliography, German summary. (In Skaraborgs län.) The ecological term "alvar" (earlier alfvar), borrowed from the Swedish vernacular, signifies a usually treeless, little or not at all dissected region of horizontal limestone rocks; the typical such area is in southern Öland. For discussion of alvar vegetation in Sweden, with list of characteristic cellular and vascular plants (663 species, of which 434 are vascular) see **Witte, Hernfrid.** De svenska alfvarväxterna. Arkiv Bot. v. 5, no. 8. 94 p. 10 pl. 1906.

——— Varatraktens flora. Svensk Bot. Tidskr. **33:** 205–240. 1939.—Tillägg till "Varatraktens flora." l.c. **37:** 304–306. 1943.

Topography, general features of flora, etc.; annotated list of vascular plants (557 species and 8 hybrids, besides some 50 Taraxaca); bibliography. (In Skaraborgs län.)

Högrell, Bengt. Bergjums fanerogamer i blomningsföljd. Öfvers. K. Vetensk.-Akad. Förhandl. **44:** 595–619. 1887.

Geology, etc.; tabular list of 500 vascular plants, arranged by dates of flowering, with earliest flowering dates for 1880–84. (In Älvsborgs län.)

Kjellberg, Gunnar. Kärlväxternas ståndorter och utbredning i Västergötland. Acta Hort. Gothoburg. **4:** 179–263. 2 fig. (maps). 1929.

Ecological grouping, botanical regions; annotated list of 830 vascular plants, nearly all collected by author, with localities; short bibliography. (In Älvsborgs and Skaraborgs län.)

Linde, Gunnar. Floran inom Leckö eller Sikajocki slottsområde i Kållandsö eller Otterstads socken, Kållands härad av Skaraborgs län . . . 31 p.

incl. 3 fig., pl., map. 23.5 cm. Lidköping, 1939. (Riksmuseet library, Stockholm.)
General features of flora; annotated alphabetical list of vascular plants. (In Skaraborgs län.)

[**Linnarsson, E. J. E.**] Förteckning övfer Sköfde-traktens fanerogamer och ormbunkar. 28 p. 16 cm. Skövde, 1883. (Riksmuseet library, Stockholm.)
Unannotated list of vascular plants. (In Skaraborgs län.)

Nattsén, Theodor. Förteckning öfver fanerogamer och ormbunkar, funna inom Alingsås pastorat, med fyndorter för de ovanligare. Bot. Notiser **1887:** 36–37, 49–60. 1887.
General features of flora; briefly annotated list of vascular plants. (In Älvsborgs län.)—See also **Högman, Samuel.** Växtställen omkring Alingsås. Bot. Notiser **1876:** 99–102. 1876. (List with localities, including some species not in Nattsén's list; not in systematic order.)—Also **Westerlund, C. G.** Bidrag till Västergötlands flora. l.c. **1904:** 1–25. 1904. (List of noteworthy angiosperms, with localities, mostly from vicinity of Alingsås; includes systematic treatment of *Alchemilla* in southern and middle Sweden.)

Neander, Gustaf. Finnerödja sockens kärlväxtflora. Svensk Bot. Tidskr. **33:** 127–187. 16 fig. (incl. port., map). 1939.
Botanical explorations, plant associations, etc.; annotated alphabetical list of about 625 vascular plants (not counting *Hieracium* and *Taraxacum*); bibliography. (In Skaraborgs län.)

Nordstedt, Otto. Sandhems flora. Bot. Notiser **1900:** 75–80, 159–160, 273–278. 1900; **1903:** 35–38, 221–227. 1903.—Tillägg. **1918:** 309. 1918.
Annotated list of vascular plants. (In Skaraborgs län.)

Nordwall, J. F. Kärlväxtfloran i Fägre socken i nordöstra Västergötland. Svensk Bot. Tidskr. **52:** 73–111. 2 fig. (fig. 1, map). 1958.
Topography, general features of flora, etc.; annotated list of vascular plants (553 species); bibliography. (In Skaraborgs län.)

Rudberg, August. Förteckning över Lugnåsbergets fanerogamer och ormbunkar. Bot. Notiser **1887:** 117–125. 1887.
General features of flora; mostly unannotated list of vascular plants. (In Skaraborgs län.)

Sahlén, A. J. Wenersborgs flora eller kort beskrifning på de växter, som förekomma närmast omkring Wenersborg samt på Halle- och Hunneberg, till den studerande ungdomens tjents. 1 p. l., viii, 192 p. 16 cm. Mariestad, 1854. (Gray Herbarium library.)
Briefly annotated list of spermatophytes, on the Linnaean system, with short descriptions. (In Älvsborgs län.)—See also **Kindberg, N. C.** Tillägg till Sahléns Venersborgsflora. Bot. Notiser **1863:** 12–14. 1863. (List with localities.)—See also Svensson, below.

Sandberg, Carl, and **Söderberg, Ivar.** Boråstraktens flora. Svensk Bot. Tidskr. **16:** 221–259. 1922.
Botanical explorations, topography, general features of flora; annotated list of known vascular plants; bibliography. (In Älvsborgs län.) Replaces **N[ordwall], J[ohan].** Förteckning jämte kort beskrifning på vilda och förvildade fanerogamer växande i och omkring Borås (med undantag af gräs och carices). 42 p. 19 cm. Borås, 1867. (Riksmuseet library, Stockholm.) —See also **Westfeldt, G. A.** Bidrag till Boråstraktens flora (Ås härad). Svensk Bot. Tidskr. **20:** 25–42. 1926. (General features of flora; extensive list of vascular plants, with localities.)—Also his Bidrag till Boråstraktens flora. l.c. **23:** 263–269. 1929. (List with localities.)—Also **Sandberg, Carl,** and **Westfeldt, G. A.** Boråstraktens adventivväxter. Acta Hort. Gotoburg. **13:** 17–26. 1939. (Inclusive list, with localities and dates.)—Also Westfeldt (Floran), below.

Skårman, J. A. O. Anteckningar till östra Västergötlands flora. Svensk Bot. Tidskr. **39:** 95–117. 4 fig. 1945.
Botanical explorations, general features of flora; list of vascular plants, with localities, from Hjo, Gredbäck, and Norra Fågelås socknar, the commonest species omitted; bibliography. (In Skaraborgs län.)—See also his Ett ytterligare bidrag till kännedomen om kärlväxtfloran i östra Västergötland, below.

Skårman, J. A. O. Bidrag till nordöstra Västergötlands flora. Svensk Bot. Tidskr. **18**: 204–278. 9 fig. 1924.
General features of flora, comparison with Undenäs-Tived region, etc.; unannotated list of vascular plants (754 species) from Älgarås, Amnehärad, Finnerödja, and Hova; bibliography. (In Skaraborgs län.)

——— Floran i Undenäs och Tived. Ett bidrag till nordöstra Västergötlands växtgeografi. Svensk Bot. Tidskr. **10**: 113–182. 12 fig. (maps). 1916.
Botanical explorations, general features of flora, special localities, phytogeography; annotated list of vascular plants; bibliography. (In Skaraborgs län.)—See also his Ytterligare bidrag till floran i Undenäs och Tived. Svensk Bot. Tidskr. **16**: 417–437. 2 fig. 1922. (Notes on author's explorations, etc.; extensive alphabetical list of vascular plants, bringing total to about 730 species and subspecies.)

——— Floran på Kållandsö och i angränsande delar av Kålland. Svensk Bot. Tidskr. **21**: 171–241. 10 fig. (incl. maps). 1927.
Geography, botanical explorations, general features of flora, plant associations, phytogeography; annotated list of vascular plants (over 650 species, not counting *Hieracium*, *Taraxacum*, and *Rosa*); bibliography. Covers Otterstad, Strö, and Sunnersberg socknar. (In Skaraborgs län.)

——— Floristiska anteckningar från några av öarna i östra Vänern (Torsö, Bromö, Dillö och Onsö). Svensk Bot. Tidskr. **23**: 63–95. 1929.
General features of flora; annotated list of vascular plants; bibliography. (In Skaraborgs län.)

——— Floristiska undersökningar i Ale härad. Svensk Bot. Tidskr. **29**: 413–502. 10 fig. (incl. port., maps). 1935.
Botanical explorations, topography, geology, general features of flora, plant associations, phytogeography, etc.; annotated list of vascular plants; bibliography. (In Älvsborgs län.)

——— Floristiska undersökningar i trakterna av Anten och Mjörn. Svensk Bot. Tidskr. **36**: 312–372. 2 fig. (maps). 1942.
Botanical explorations, topography, general features of flora; annotated list of vascular plants, with localities; bibliography. (In Älvsborgs län.)—See also his Tillägg till "Floristiska undersökningar . . ." Svensk Bot. Tidskr. **37**: 94–99. 1943.—Also **Hasselrot, T. E.** Kompletterande floristiska anteckningar från Anten-Mjörn-området i västra Västergötland. l.c. **47**: 465–487. 1953. (Notes on flora of adjoining areas, etc.; extensive list of vascular plants, bringing total to 674 species; bibliography.)

——— Kinnekulles kärlväxtflora. Svensk Bot. Tidskr. **25**: 293–394. 7 fig. 1931.
Geology, botanical explorations, general features of flora; annotated list of vascular plants (about 850 species and subspecies); bibliography. (In Skaraborgs län.) Replaces **Zetterstedt, J. E.** Kinnekulles phanerogamer och ormbunkar. Bot. Notiser **1851**: 1–42, 49–55. 1851. (Includes list of 632 vascular plants and Characeae.)—See also **Albertson, Nils.** Acta Phytogeog. Suecica **20**: 148. 1946. (Several additional species.)

——— Nya bidrag till floran i trakterna öster om Göta älv. Svensk Bot. Tidskr. **32**: 374–434. 9 fig. (maps). 1938.
Botanical explorations, general features of flora, phytogeography, etc.; annotated list of vascular plants; bibliography. Covers the socknar Fuxerna, Åsbräcka, Fors, and western part of Gärdhem, all south of Trollhättan. (In Älvsborgs län.)—See also **Berggren, August.** Förteckning över några mera anmärkningsvärda växtarter runt Trollhättan. l.c. **23**: 141–145. 1929. (Geology, etc.; list of vascular plants, with localities, collected on both sides of Göta älv from Åkerström to Stallbacka.)

——— Ett ytterligare bidrag till kännedomen om kärlväxtfloran i östra Västergötland. Svensk Bot. Tidskr. **40**: 435–446. 1946.
Botanical explorations, general features of flora, etc.; list of vascular plants collected in the parish of Bredvik on Vättern, north of Gredbäck, the commoner species omitted; bibliography. (In Skaraborgs län.)—See also his Anteckningar till östra Västergötlands flora, above.

Sundin, Torsten. Floran i Billingssocknen Bolum. Svensk Bot. Tidskr. **51**: 264–286. 1 fig. (map). 1957.

Geology, etc.; briefly annotated list of vascular plants; bibliography. (In Skaraborgs län.)

Svensson, G. Bidrag till kännedomen om floran på Halleberg. 30 p. 23 cm. Vänersborg, 1945.

Annotated list of vascular plants; bibliography. (In Skaraborgs län.) (Not seen; title and annotation supplied by E. Hultén and G. Haglund.)—See also Sahlén, above.

[Wästberg, V. A.]. Några iakttagelser om närmaste Mariestads-traktens fanerogamer och ormbunkar, upptecknade till hjelpreda för Mariestads skolungdom vid botaniska exkursioner. 1 p. l., 41 p. 14 cm. Mariestad, 1864. (Riksmuseet library, Stockholm.)

List of vascular plants and Characeae, on the Linnaean system, with localities. (In Skaraborgs län.)

Westfeldt, G. A. Floran i nordöstra delen av Sjuhäradsbygden. Svensk Bot. Tidskr. **48**: 649–770. 7 fig. (maps). 1954.

Topography, geology, botanical explorations, phytogeography, plant habitats; list of vascular plants, with detailed localities; English summary, bibliography. (In Älvsborgs län.)—See also Sandberg, above.

——— Växtligheten på västsvenska kyrkogårdar. Lustgården **37/38**: 187–199. 5 fig. 1957.

Running account of cultivated and wild plants commonly occurring in churchyards in southern Västergötland and northern Halland; English summary. (In Älvsborgs län and Hallands län.)

——— Vegetationen på Kråkeboberg vid Ulricehamn. Svensk Bot. Tidskr. **41**: 159–181. 3 fig. (incl. map). 1947.

Topography, geology, general features of flora, ecological lists of species; tabular partial list of vascular plants and mosses; German abstract, bibliography. The Kråkeboberg (altitude 264 meters) is noted for its rich flora. (In Älvsborgs län.)

Zetterstedt, J. E. Om växtligheten på Vestergötlands siluriska berg med särskild hänsyn till mossvegetationen. Öfvers. K. Vetensk.-Akad. Förhandl. **33** (1): 43–71. 1876.

Account of various mountains, with mention of characteristic vascular plants and mosses; no full list of species. Relates to Kinnekulle, Falbygdens berg, Halleberg, Hunneberg, etc. (In Skaraborgs län.)—See also **Witte, Hernfrid.** Om Falbygdens vegetation. *In* Falköping förr och nu. p. 99–117. 23.5 cm. Falköping, 1910. (Topography, geology, plant formations with mention of numerous species; lists of scarcer plants and of some introduced and cultivated plants. In Riksmuseet library, Stockholm.)—Also **Skårman, J. A. O.** Från exkursioner på silurbergen i östra Falbygden. Svensk Bot. Tidskr. **27**: 365–399. 1 fig. (map). 1933. (Botanical explorations, topography, geology, climate, general features of flora, plant associations, etc.; annotated list of more important vascular plants collected by author in 1932 and 1933 on Mounts Gisseberg, Gerumsberg, Plantaberg, and Varvsberg.)

VÄSTMANLAND (VESTMANLAND, WESTMANLAND)

The old province (landskap) of Västmanland or Westmanland forms the western part of Västmanlands län and the northern part of Örebro län.

General

Iverus, J. E. Beskrivning över Västmanlands fanerogamer och thallogamer. 1 p. l., iv, 326 p. 22 cm. Upsala, 1877.

Topography, etc.; keyed descriptive flora of vascular plants (874 species), with localities for scarcer species. Replaces **Wall, W. A.** Westmanlands flora innefattande provinsen Westmanlands vilda phanerogamer och filices. 2 p. l., 142 p. 20.5 cm. Stockholm, 1852. (Annotated, briefly descriptive flora of vascular plants (756 species) and Characeae, on the Linnaean system, with localities for scarcer species.)—See also **Samuelsson, Gunnar.** Växtlokaler från Västmanland. I-II. Svensk Bot. Tidskr. **17**: 401–448. 1923; **19**: 1–48. 1925. (Botanical publications, list of collectors, etc.; extensive list of scarcer vascular plants, with localities; list of generally distributed species. Intended to supplement Wall's and Iverus' floras.)—Also

Alvén, Edvard. Några växter från Västmanland. Bot. Notiser **1950:** 466–467. 1950. (Records of noteworthy species.)

Local

Binning, Axel. Bidrag till kännedomen om kärlväxtfloran i västra Västmanlands bergslag. Svensk Bot. Tidskr. **15:** 214–242. 1921.

General features of flora, botanical explorations; list of vascular plants, with localities, mainly from Järnboås, Hjulsjö, and Nora socknar. (In Örebro län.)—See also **Elgenstierna, Carl.** Några för Vestmanland nya växtlokaler. Bot. Notiser **1889:** 248–249. 1889. (List of noteworthy vascular plants, nearly all from vicinity of Nora.)—Also **Segerström, A. L.** Ett tillägg till kännedomen om kärlväxtfloran i västra Västmanlands bergslag. Svensk Bot. Tidskr. **17:** 105–108. 1923. (List of noteworthy vascular plants, with localities, from the region of Loka in Grythyttan parish, Västmanland, and from Silverhyttan in Karlskoga parish, Värmland.)

Dahlgren, K. V. O. Salatraktens kärlväxtflora. Svensk Bot. Tidskr. **4:** (107)-(136). 1 fig. 1911.

Geology, plant associations; partly annotated list of known vascular plants (over 600 species); bibliography. (In Västmanlands län.)—See also his Tillägg til Salatraktens kärlväxtflora. Svensk Bot. Tidskr. **17:** 96–102. 1923. (Extensive alphabetical list, with localities.)—Also his Några nya fanerogamfynd från Sala-trakten. l.c. **43:** 121–125. 1949. (Alphabetical list, with localities; short bibliography.)

Hisinger, Wilhelm. Förteckning på växterna i Skinskattebergs socken i Westmanland. 1 p. l., 45 p. 19 cm. Stockholm, 1832. (Riksmuseet library, Stockholm.)

Annotated list of cellular and vascular plants (575 species, including 426 vascular plants). (In Västmanlands län.)—See also **Suneson, Svante.** Bidrag till kännedomen om floran inom Skinnskattebergs och Malingsbo socknar i Västmanland och Dalarna. Bot. Notiser **1929:** 195–204. 1929. (List with localities; short bibliography.)

Post, H. A. von. Westra Mälarstrandens cotyledoner iakttagne och antecknade 1839–43. Bot. Notiser **1844:** 113–142, 145–154. 1844.—Tillägg . . . gjorda åren 1844–45. l.c. **1846:** 49–53. 1846.

Physiography, plant formations; annotated list of 513 phanerogams (including supplement). (In Västmanlands län.)

Troilius, A. M. Om Westeråstrakten i botaniskt afseende. 1 p.l., 43, xl p. 22 cm. Stockholm, 1860. (Akad. afhandl.) (New York Botanical Garden library.)

Botanical explorations, geography, general features of flora; briefly annotated list of vascular plants and Characeae (726 species), with localities for the rarer species. (Name now spelled Västerås, in Västmanlands län.)—See also **Wollert, Arvid.** Några växtfynd inom Västerås hamnområde. Svensk Bot. Tidskr. **24:** 136–137. 1930. (Extensive list of adventive plants, with dates.)—Also **Walldén, N. B.** Västeråstraktens växt-och djurliv. 236 p. 2 pl. 24.5 cm. Västerås, 1955. (Includes list of changes in flora, p. 29–36. Not seen; title and annotation supplied by T. E. Hasselrot.)

SWITZERLAND (SCHWEIZ; SUISSE; SVIZZERA; SVIZZRA OR SVIZRA (ROMANSCH))

See also France, General (Bonnier, Flore complète illustrée; Bonnier and Layens, Flore complète portative; Camus, Catalogue des plantes; Rolland, Flore populaire); France, Partial (Babey, Flore jurassienne; Grenier, Flore de la chaîne jurassique); Italy, General (Rikli, Das pflanzenkleid der Mittelmeerländer).

Works cited under France or Italy that relate also to one or few cantons of Switzerland are referred to under the respective cantons.

Several German floras cover Switzerland also. The only ones here cited in full are Hegi, Illustrierte flora von Mittel-Europa; Kirchner and others, Lebensgeschichte der blütenpflanzen Mitteleuropas. Other important works of this sort are: Ascherson, Paul, and Graebner, Paul, Synopsis der mitteleuropäischen flora (1896-1939); Hochstetter, C. F., Die giftgewächse Deutschlands und der Schweiz (1844); Karsten, Hermann. Flora von Deutschland, Oesterreich und der Schweiz, 2 aufl. (1895); Koch, W. D. J., Synopsis der deutschen und schweizer flora, 3 [i.e. 4.] aufl. (1892-1907); Marzell, Heinrich, Wörterbuch der deutschen pflanzennamen (1937-); Reichenbach, H. L., and others, Icones florae germanicae et helveticae (1834-1915); Thomé, O. W., Flora von Deutschland. Österreich und der Schweiz, 2 aufl., v. 1, Phanerogamen (1903-05).

The cantons are listed alphabetically under their names as written in the language spoken predominantly in each. The entirely or chiefly French-speaking cantons (constituting la Suisse romande) are Fribourg, Genève, Neuchâtel, Valais and Vaud; Ticino is Italian; Bern is largely German, the Jura French-speaking; Graubünden is half German, half Romansch, with two southern valleys Italian; the remaining cantons are German-speaking.

Includes some partial works relating to the Alps, the Jura, French and German Switzerland, etc. Those relating to the Urkantone and the Waldstätte are given under Schwyz.

The general titles here given may be grouped as follows: *Floras or manuals*: Bennett, Binz, Bouvier (Alps), Friche-Joset (Jura), Gaudin, Godet (Jura), Gremli, Hegi, Schinz. *Catalogs*: Brügger (hybrids), Bruhin (plants of Gesner), Christ (alpine), Heer (alpine), Jerosch (alpine), Rhiner, Thurmann (Jura). *Iconographies:* Deutscher und Oesterreichischer Alpenverein (alpine), Hegetschweiler (Sammlung), (Hegi), Marret, Seboth (alpine), Thommen. *Bibliographies*: Bruhin (Uebersicht), Ed. Fischer, Rehder, Schweizerische Botanische Gesellschaft (Referate), Schweizerische Landesbibliotek. *Vernacular names*: Brandstetter (trees, toponymy), Coaz (trees), Correvon (medicinal), Dalla Torre (alpine), Durheim, Empeyta (Romand), (Hegi), H. Jaccard (toponymy), Kohler (Germanic), Oettli (toponymy), Savoy (Romand). *Trees or woody plants in general*: Bühler (altitudinal limits), Coaz (vernacular names), Empeyta, Kienli. *Useful and poisonous plants*: Binz (poisonous), Correvon (medicinal), Dalla Torre (alpine), (Empeyta: woody plants), Eugen Fischer (medicinal), Flück (medicinal), Gesse (edible), Hegetschweiler (Die giftpflanzen), (Hegi), Künzle (medicinal), Morell (lake dwellings), Neuweiler (prehistoric), (Savoy: Romand). *Weeds (and adventive plants in general)*: Buchli, Probst, Stebler, Thellung. *Cultivated plants*: (see under Schaffhausen, Kummer, 3d title). *Miscellaneous*: Becherer (summary of new records), Brandstetter (toponymy), Briquet (biographical), Buchli (ecology of weeds), Christ (phytogeography), Furrer (phytogeography and ecology), H. Jaccard (toponymy), P. Jaccard (zones, formations, etc.), Jerosch (alpine phytogeography), Kirchner (biology), Morell (prehistory), Neuweiler (prehistory), Oettli (toponymy), Schweizerische Botanische Gesellschaft (Fortschritte: lists of new records).

Becherer, Alfred. Ein halbes jahrhundert floristischer neufunde in der Schweiz. Verzeichnis der in den jahren 1901–1950 für die Schweiz neu festgestellten pteridophyten und phanerogamen. Verhandl. Naturf. Gesell. Basel **62**: 224–244. 1951.

Annotated list of principal native and introduced vascular plants first recorded in 1901–50; many critical groups are omitted (such as *Koeleria, Erophila, Rubus, Taraxacum, Hieracium.*).—See also Schweizerische Botanische Gesellschaft (Fortschritte), below.

Bennett, A. W. The flora of the Alps, being a description of all the species of flowering plants indigenous to Switzerland; and of the alpine species of the adjacent mountain districts of France, Italy, and Austria including the Pyrenees. 2 v. 120 col. pl. 20.5 cm. London, 1896. (Various issues, 1891–1900.)

General features of flora, short bibliography; rather popular account of phanerogams, with brief diagnostic characters.

Binz, August. Schul- und exkursionsflora der Schweiz mit berücksichtigung der für Basel in betracht kommenden benachbarten teile Badens und des Elsasses. 6. aufl. xxiii, 438 p. 384 fig. 18.5 cm. Basel, 1949. (1st ed. 1920. 8. aufl. bearb. von A. Becherer. xx, 382 p. 384 fig. Basel, 1957. Not seen.)

Pocket manual of vascular plants in form of briefly annotated keys, with distribution very briefly indicated; list of poisonous plants. The corresponding French edition is: **Binz, August,** and **Thommen, Édouard.** Flore de la Suisse y compris les parties limitrophes de l'Ain et de la Savoie. 2. éd. rev. et augm. xxxvi, 450 p. 418 fig. (incl. map). 19 cm. Lausanne, 1953. (1st ed. 1941.) (Bibliography; pocket manual of vascular plants in form of briefly annotated keys; list of poisonous plants. The number of species in the first edition is given as 2,578.)

Bouvier, Louis. Flore des Alpes de la Suisse et de la Savoie, comprenant: la description des plantes indigènes et des plantes cultivées, les propriétés de chaque famille, de chaque espèce utile ou nuisible, et des données hygiéniques sur tous les fruits de nos climats. 2. éd., augmenté . . . 6 p. l., 228, 812 p. 18.5 cm. Paris, Genève, 1882. (1st ed. 1878.)

Annotated, partially keyed, descriptive flora of spermatophytes, including cultivated plants, with notes on uses.

Brandstetter, J. L. Die namen der bäume und sträuche in ortsnamen der deutschen Schweiz. 86 p. 26.5 cm. Luzern, 1902. (Beilage zum Jahresbericht der Höheren lehranstalt in Luzern für das schuljahr 1901/1902.) (Arnold Arboretum library.)

Short bibliography; list of 99 woody plants (including some repetitions) with discussion of etymology, etc., of vernacular names, and list of locality names derived from them in the German-speaking cantons.—See also

H. Jaccard and Oettli, below; also, for vernacular names in general, Durheim, below, and additional titles there cited.

Briquet, John. Biographies des botanistes à Genève de 1500 à 1931. Ber. Schweiz. Bot. Gesell. v. 50a. ix, 494 p. port. 1940.
Alphabetical list of 286 botanists (Swiss and foreign) who have lived or studied at Geneva, with biographical sketches, references to sources, lists of publications, lists of species, etc., named for them; chronological list of botanists mentioned.

Brügger, C. G. Wildwachsende pflanzenbastarde in der Schweiz und deren nachbarschaft. Jahresber. Naturf. Gesell. Graubündens n.f., **23/24:** 47–123. 1880. (Reprinted with title Beobachtungen über wildwachsende . . .)
General remarks; list of 403 supposed hybrids found by author, with localities, references, and occasional descriptions. Brügger's judgment as to the hybrid nature of his plants has been severely criticized.—See also his Aufzählung neuer pflanzenbastarde der Bündner- und nachbar-floren. (Botanische mitteilungen, I.) Jahresber. Naturf. Gesell. Graubündens n.f., **25:** 54–61. 1882.—Also his Beschreibungen neuer zwischenformen hybriden oder zweifelhaften ursprungs. (Botanische mitteilungen, II.) l.c. **25:** 62–112. 1882. (The two last reprinted with title Mitteilungen über neue pflanzenbastarde der schweizer-flora.)

Bruhin, T. A. Aelteste flora der Schweiz. Aus den werken Conrad Gesner's und seiner zeitgenossen (1516–1565). Ber. St. Gall. Naturw. Gesell. **1864/65:** 18–104. 1865.
Systematic list of vascular plants, with corresponding names from works of Aretius, Fabricius, Gesner, Rhellicanus, and Tragus.

[———] Uebersicht der geschichte und literatur der schweizer-floren. pt. I–II. (30, 40 p.) *In* Jahresbericht über die Erziehungsanstalt des Benediktiner-Stiftes Maria-Einsiedeln im studienjahre 1862/63 und 1863/64. Einsiedeln, 1863–64.
History of Swiss floristic work from Gesner to De Candolle, with references to publications. (Not seen; title and annotation supplied by W. Rytz.)—See also Ed. Fischer, Rehder, Schweizerische Botanische Gesellschaft (Referate), and Schweizerische Landesbibliotek, below.

Buchli, Matheus. Oekologie der ackerunkräuter der Nordostschweiz. Beitr. Geobot. Landesaufn. Pflanzengeog. Kom. Schweiz. Naturf. Gesell. nr. 19. 354 p. (incl. diagrs., maps). 6 pl., table. 1936.
Mainly ecological; includes tabular list of 205 weedy vascular plants (p. 59–64), and bibliography (p. 343–354). Refers to Aargau, Schaffhausen, and Zurich.—See also Probst, Stebler, and Thellung, below.

Bühler, Anton. Studien über die baumgrenze im hochgebirge. Ber. Schweiz. Bot. Gesell. **8:** 19–38. 1898.
Sources; tabular list of trees, showing extreme altitudes reached in 8 different regions in Switzerland; lists from various localities, showing trees observed at different altitudes; general considerations.—See also Coaz, Empeyta, and Kienli, below.

Christ, Hermann. La flore de la Suisse et ses origines. Éd. française traduite par E. Tièche, revue par l'auteur . . . nouv. éd. augmentée d'un Aperçu des récents travaux géobotaniques. xiv, 571, 119 p. 4 pl., 4 maps, tab. 24 cm. Bâle, Genève, Lyon, 1907. (Gray Herbarium library.) (1st ed. 1883, based on his Das pflanzenleben der Schweiz, 1879.)
Account of plant zones, arranged altitudinally and geographically, with description of topography, climate, characteristic species, etc.; origin of flora. The Supplément (Aperçu des récents travaux géobotaniques concernant la Suisse. 119 p.) is mainly a summary of recent investigations. It was also published separately.—See also next title and Furrer, Heer, P. Jaccard, Jerosch, and Thurmann below.

——— Ueber die verbreitung der pflanzen der alpinen region der europäischen Alpenkette. Neue Denkschr. Schweiz. Naturf. Gesell. v. 22, no. 7. 84 p. map. 1867.
Discussion; tabular list of 693 vascular plants, showing distribution in northern hemisphere.—See also title above and additional references there given.

Coaz, Johann. Verzeichnis der in der Schweiz wildwachsenden holzgewächse. (Baume, sträucher und halbsträucher.) 16 p. 20 cm. Bern, 1900.

List of woody plants with German, French, Italian, and Rhaeto-Romanic (Romansch) names, and notes on the occurrence of some of the rarer species. (Not seen; title and annotation mostly from notice in Ber. Schweiz. Bot. Gesell. **11:** 130. 1901.)—See also Brandstetter and Bühler, above, and additional references there given.

Correvon, Henry. Plantes et santé. 331 p. 18.5 cm. Genève, 1917.

Annotated list of wild and cultivated medicinal plants, alphabetically arranged by vernacular names, with uses; account of honey plants, list of diseases, etc. with remedies.—See also (for medicinal or poisonous plants), Binz, above, and Eugen Fischer, Flück, Hegetschweiler (Die giftpflanzen), and Künzle, below; and (for useful plants in general) Dalla Torre, Empeyta, Gesse, Hegi, Morell, Neuweiler, and Savoy, below.

Dalla Torre, K. W. von. Die alpenpflanzen im wissensschatze der deutschen alpenbewohner. 91 p. 1905. (Festschr. herausgegeben anlässlich der V. ordentl. Generalvers. des Vereins zum schutze und pflege der alpenpflanzen [c. V.] zu Bamberg am 24. Juli 1905.)

Bibliography; list of more widespread alpine plants of the Tirol and Switzerland, with Germanic vernacular names and localities where used, and medicinal and other uses. (Not seen; cited from review in Bot. Jahresb. Just **33** (3): 439. 1907.)—See also (for vernacular names) Durheim, below, and additional references there given; and (for useful plants) Correvon, above, and additional references there given.

Deutscher und Oesterreichischer Alpenverein. Atlas der alpenflora. . . 2. neubearb. aufl. Ausführung der farbentafeln nach naturaufnahmen und originalvorlagen von A. Hartinger . . . 5 v. 500 col. pl. 19 cm. Graz, 1897. (1st ed. 1882–84.)

Fair to excellent colored figures, without dissections, illustrating 500 alpine vascular plants of Switzerland, Germany, and Austria. There are later French and German editions, with accompanying single volumes of text; the German by Dalla Torre (München, 1899) has title: Die alpenflora der österreichischen Alpenländer, Südbaierns und der Schweiz (cover title: Handbuch zum Atlas der Alpenflora); the French, by H. Correvon (Genève et Bâle, 1899), has title: Atlas de la flore alpine. The plates in the German edition are arranged by the Engler and Prantl system; those in the French edition, arranged on the Candollean system and so differently numbered, bear the German number in the upper right hand corner. There is at least one reissue (Paris, 1901). —Among the numerous popular works of this sort, the two following may be cited: **Hegi, Gustav.** Alpenflora; die verbreitetsten alpenpflanzen von Bayern, Österreich und der Schweiz. 13. überarb. aufl. Herausgegeb. von . . . Hermann Merxmüller. 96 p. 42 pl. (34 col.), map. 21.5 cm. München, 1956. (1st ed. 1905.) (List of protected species; colored figures of 272 species, and photographs of 34 more, with descriptions.) **Schröter, Carl.** Taschenflora des Alpen-wanderers . . . nach der natur gezeichnet und gemalt von Ludwig Schröter. 27. aufl. bearb. von Dr. W. Lüdi. unpaged text, viii p. (index). 26 pl. (24 col.). 20.5 cm. Zürich, [19–?]. (1st ed. 1889; 25. aufl. 1940.) (Colored figures of 207 species, black figures of 10, with text in German, French, and English.)—See also Hegetschweiler (Sammlung), Hegi, Marret, Seboth, and Thommen, below.

Durheim, C. J. Schweizerisches pflanzen-idiotikon. Ein wörterbuch von pflanzenbenennungen in den verschiedenen mundarten der deutschen, französischen und italienischen Schweiz, nebst deren lateinischen, französischen und deutschen namen . . . viii, 284 p. 23.5 cm. Berne, 1856.

Alphabetical list of scientific names, with principal French and German names and the vernacular and patois names used in Switzerland (French, German, Italian, and Romansch) with localities; separate alphabetical lists of vernacular names in different languages, with scientific equivalents. H. Marzell (Wörterbuch der deutschen pflanzennamen, p. 14. 1937) calls it a rich collection but with many errors, especially in citation of localities. —See also Brandstetter, Coaz, Correvon, and Dalla Torre, above, and Empeyta, Hegi, H. Jaccard, Kohler, Oettli, and Savoy, below.

Empeyta, Eugène. Catalogue descriptif des arbres, arbustes, arbrisseaux et sous-arbrisseaux indigènes ou naturalisés en Suisse suivi d'un dictionnaire des principaux noms vulgaires donnés dans la Suisse romande, à différentes plantes avec leurs synonymes français et latins. 211 p. 27 cm. Genève, 1887. (Arnold Arboretum library.)

Descriptive flora of woody plants, with account of wood and uses; alphabetical list of Romand vernacular names (including herbaceous plants), with French and scientific equivalents.—See also Bühler, above, and additional references there given; also (for useful plants) Correvon, above, and additional references there given; also (for vernacular names) Durheim, above, and additional referencs there given.

Fischer, Eduard. Flora helvetica, 1530–1900. xviii, 341 p. 22 cm. Bern, 1901.—Nachträge. ix, 40 p. 22 cm. Bern, 1922. (Both are: Bibliographie der schweizerischen landeskunde . . . herausgegeben von der Centralkommission für schweizerische landeskunde. Fasc. IV, 5.)

List of bibliographies, periodicals, botanical publications of 16 to 18th centuries and ofthe 19th century, the last classified by botanical groups and regions, the titles arranged chronologically; list of exsiccatae. Includes publications relating to adjacent regions, and occasional very brief annotations. The supplement, also ending with 1900, includes classified list of additions and short list of corrections to the original.—See also Bruhin (Uebersicht), above, and additional references there given.

Fischer, Eugen. Unsere heilpflanzen in neuer wertung und geltung. Praktischer ratgeber für den anbau, das sammeln und die verwendung einheimischer heil- und gewürzpflanzen. 359 p. illus., 80 col. pl. on 40. 25 cm. Zürich, 1941.

Annotated tabular list of the most used Middle European medicinal plants, alphabetically arranged by German vernacular names; extensively annotated list of some 80 species, alphabetically arranged by vernacular names, with other (German language) vernacular names, range, history, directions for collection and cultivation, medicinal and other uses, etc.; table of dates for collecting, list of diseases with plants used therefor, etc.; bibliography.—See also his Heilpflanzen. 64 p. incl. 24 col. pl. 15 cm. Bern, [195–]. (Die Hallwag-taschenbücher 33.) (Popular treatment of about 100 medicinal plants, with uses and dosage. French ed., Plantes médicinales. Lausanne (1952). (Petits atlas de poche Payot 21.) There is also a Dutch edition, Geneeskundige planten. Nederlandse tekst door Drs Roland Claeys. 64 p. illus. Gent, 1955. (Not seen; cited from Bibl. Sci. Helvetica **31**: 173. 1956.))—Also Correvon, above, and additional references there given.

Flück, Hans. Unsere heilpflanzen. Eine gemeinverständliche beschreibung mit angaben über wirkstoffe, wirkung, anwendung, einsammlung und anbau. 2. revid. ausg. xvi, 160 p. 144 col. fig. 17 cm. Thun, 1950. (1st ed. 1941.)

Brief general account of chemical properties, methods of application, calendar for collecting etc.; descriptive account of 144 more important medicinal plants (wild and cultivated, all but 3 vascular), with description, occurrence, chemical properties, uses, and colored figure of each plant; wild substitutes for tea (*Thea*), list of diseases with remedies. There is also a French issue, Nos plantes médicinales . . . xvi, 160 p. 144 col. fig. 17 cm. Lausanne, 1942.—See also Correvon, above, and additional references there given.

Friche-Joset,—,and Montandon, F. J. Synopsis de la flora du Jura septentrional et du Sundgau, contenant un résumé analytique et raisonné des végétaux phanérogames croissant sur les différentes chaines du Jura septentrional . . . Et des végétaux vasculaires du Sundgau . . .par F. J. Montandon. xii, 409 p. pl., tab. 18.5 cm. Mulhouse, 1856. (Arnold Arboretum library.)

Annotated, keyed list of vascular plants, with localities. Reissued, unchanged, with Montandon given as sole author and title Guide du botaniste dans le Sundgau et les lisières du Jura badois, balois, soleurois, bernois et dubisien . . . Mulhause, 1868. (New York Botanical Garden library.) Sundgau is an area in upper Alsace.—See also Godet and Thurmann, below; also France, Partial (Babey, and additional references there given).

Furrer, Ernst. Kleine pflanzengeographie der Schweiz. viii, 331 p. (incl. illus., maps). 16 pl. 19.5 cm. Zürich, 1923.

Discusses geology, climate, agriculture, plant formations, ecology, history of vegetation, etc., with lists of titles. The zweite, umgearbeitete auflage (viii, 127 p. 13 fig., 8 pl. on 4. 19 cm. Zürich, 1942) is considerably condensed.—See also Christ, above, and additional references there given; also Buchli (ecology of weeds).

Gaudin, Jean. Flora helvetica, sive historia stirpium hucusque cognitarum in Helvetia et tractibus conterminis aut sponte nascentium aut in hominis animaliumque usus vulgo cultarum continuata. 7 v. 28 pl. (pt. col.). 20 cm. Turici, 1828–33.

Descriptive flora of phanerogams, on the Linnaean system, with synonymy, references, and detailed descriptions; v. 7 gazetteer, with lists of rarer species of each locality.

Gesse, L. A. Notices sur les plantes qui croissent en Suisse sans culture et qui peuvent servir d'aliment. 11 p. 8vo. Genève, 1817.

Not seen; no review available. Cited from E. Fischer, Flora helvetica p. 74. 1901. There is a translation in Schweizerfreund **3**: 199–201, 209–210, [217–219, suppl.]. 1817. (Not seen; reference supplied by W. Rytz.)—See also (for useful plants in general) Correvon, above, and additional references there given.

Godet, C. H. Flore du Jura ou description des végétaux vasculaires qui croissent spontanèment dans le Jura suisse et français plus spécialement dans le Jura neuchâtelois. xvi, 872 p. 21 cm. Neuchâtel, 1853 (1852–53).

Bibliography, etc.; annotated descriptive flora of vascular plants, with localities for the more local species, those known from the Neuchâtel Jura marked with an asterisk. The list of species, with localities but without descriptions, was published earlier in his Énumération des végétaux vasculaires du Jura suisse et français plus spécialement du canton de Neuchâtel. (vi, 233 p. 21.5 cm. Neuchâtel, 1851.) Relates to part of the cantons of Vaud, Neuchâtel, Bern, Solothurn, and Basel in Switzerland and part of the departments of Ain, Jura, and Doubs in France.—See also his Supplément à la Flore du Jura suisse et français. viii, 220 p. 21 cm. Neuchâtel. 1869. (Additions and corrections, as well as a full list of the species in his Flore.) —Also **Tripet, Fritz.** Modifications apportées à la flore du Jura neuchâtelois . . . Arch. Sci. Phys. et Nat. Genève III, **14**: 281. 1885. (Species lost and added since 1869.)—Also **Lerch, J. F.** Liste de quelques plantes rares et de quelques localités nouvelles pour notre Jura. Rameau du Sapin **27**: 4, 11–12, 13–14, 17–18. 1893. (Extensive list. In Arnold Arboretum library.)—Also **Tripet, F.** Localités et plantes nouvelles pour la flore du Jura (Notes laissées par Chs. Hri. Godet.) l.c. **27**: 24, 25–26, 29–31. 1893. (Extensive list. In Arnold Arboretum library.)—Also **Magnin, Antoine.** Contribution à la connaissance de la flore des lacs du Jura suisse. Bul. Soc. Bot. France 41 (Sess. Extraord.): cviii-cxxviii. 5 fig. (incl. map), pl. 9–10. 1895. (Account of the two lakes in Neuchâtel and three in Vaud, with partial lists of species.) —Also Friche-Joset, above, and additional references there given.

Gremli, August. Exkursionsflora für die Schweiz. Nach der analytischen methode . . . 9, verm. und verb. aufl. xxiv, 472 p. port. 17.5 cm. Aarau, 1901. (1st ed. 1867.)

Briefly annotated flora of vascular plants, in form of keys; a popular flora, now antiquated. According to a prefatory note by the publisher, this edition was revised by Dr. Dill. There is an English translation by L. W. Paitson, 1888, of the 5th German edition, and two French editions (1886, from the 5th German edition and a 2d edition, 1898).

Heer, Oswald. Ueber die nivale flora der Schweiz. Neue Denkschr. Allg. Schweiz. Gesell. Gesam. Naturw. v. 29, abt. 1, [no. 2]. 114 p. 1884.

Includes annotated tabular lists of the high altitude flora of the Rhaetic Alps, of Wallis (Valais) and Chamonix, Zermatt, the Bernese Alps, Switzerland in general, and Swiss plants extending into the Arctic regions.—See also Christ, above, and additional references there given.

Hegetschweiler, Johann. Die giftpflanzen der Schweiz . . . gezeichnet von J. D. Labram. 1 p. l., xxvi, 84 p. 38 col. pl. 21 cm. Zürich, [1824–34]. (Gray Herbarium library.)

List of poisonous plants of Switzerland, list of remedies; colored plates of vascular and cellular poisonous plants, with descriptive text. The following work, by the same author, has not been seen: Flore des plantes vénéneuses de la Suisse, contenant leur description, l'époque de leur floraison, les lieux . . . l'indication de celles qui sont employées en médecine, les symptômes qu'elles produisent . . . et les premiers soins à donner dans les empoisonnements. Destinée à l'usage des écoles et des gens de la campagne. 117 p. 23 pl. 8vo. Payerne, 1849.—See also Correvon, above, and additional references there given.

——— Sammlung von schweizer pflanzen nach der natur und auf stein gezeichnet von J. D. Labram. 80 (+?) hefte. 480 (882) col. pl. 16.5 cm. Basel, [1825–34]. (Gray Herbarium library.)

Small but excellent colored plates, each with a page of descriptive text including notes on uses, the plates unnumbered and not in any order. The number of plates is ordinarily given as 480, but the copy in the British Museum (Natural History) library contains 882, as stated in its Catalogue (G. Taylor, in litt.).—See also Deutscher und Oesterreichischer Alpenverein, above, and additional references there given.

Hegi, Gustav. Illustrierte flora von Mittel-Europa. Mit besonderer berücksichtigung von Deutschland, Oesterreich und der Schweiz. 7 v. in 13. 1271 fig., 280 pl. (mostly col.). 27.5 cm. München, (1906–31).

Descriptive flora of vascular plants, with keys, brief synonymy, vernacular names in German, French, English, and Italian, full descriptions, local and general range, treatment of more common subspecies, varieties, etc., discussion of special features of distribution, habitat, and biology, uses, etymology of generic and specific names, pre-Linnaean names, etc.; includes mention of cultivated species. Summaries of morphological, chemical, and other characteristics are given under most of the families. All the commoner or more widespread species are represented in the colored plates, and practically all the others in text figures. Treatment of numerous families contributed by other authors. Vol. 7 includes key to families, extensive glossary, systematic list of plant groups down to genera, indices of scientific and German vernacular names, classified index of useful plants, and index of plant-names connected with folklore, etc. Altogether, for breadth of treatment and utility of information brought together, the work is without rival among the floras of the world. For dates of publication, given in the original only on the verso of v. 7, p. (vii), see **Becherer, A.** Candollea **5**: 342–344. 1934. Vol. 1–3 (1) of 2. aufl. have been published (München, (pref. 1935)–58); they cover from Pteridophyta through Polygonaceae (848 species), and have more detailed text than the corresponding volumes of 1st edition. Several volumes of 1st ed. have been reprinted.

Jaccard, Henri. Les noms de végétaux dans les noms de lieux de la Suisse française. Bul. Murithienne **32**: 109–172. 1903.—Additions aux noms . . . l.c. **33**: 157–167. 1904 (1905).

General remarks; systematic list of vascular and cellular plants, with lists of locality names derived from vernacular and patois names.—See also Brandstetter and Durheim, above, and additional references there given.

Jaccard, Paul. Flore. *In* Æppli, A., et al. La Suisse; étude géographique, démographique, politique, économique et historique. p. 185–230. illus. 30 cm. Neuchâtel, n.d.

Plant zones, general features of alpine vegetation, account of alpine plants of different regions, similar account of the Jura, zones and formations of the plateau region, forests of montane zone, fossil flora. The copy examined appears to be a (revised?) reissue of a work which came out in 1908, of which only the extract entitled Flora (p. 190–237) from the German edition has been seen. (In Gray Herbarium library.)—See also Christ, above, and additional references there given.

Jerosch, M. C. Geschichte und herkunft der schweizerischen alpenflora. Eine übersicht über den gegenwärtigen stand der frage. vi, 253 p. (incl. 2 tab.). 23.5 cm. Leipzig, 1903.

Mainly phytogeographical; includes bibliography (p. 202–212) and tabular list (p. 228–252) of 420 spermatophytes, with indication of occurrence in western, central, and eastern Alps and in other mountains and other parts of the world, with critical notes on many of the species (p. 117–201).—See also Christ, above, and additional references there given.

Kienli, Walter. Die gehölze der schweizerischen flora und des schweizerischen obstbaues . . . xxviii, 404 p. incl. 199 fig. 17.5 cm. Münsingen, 1948.

Popular pocket manual including native trees and shrubs, all the cultivated fruit trees and some introduced ornamental and useful trees, with illustration of nearly every species.—See also Bühler, above, and additional references there given.

Kirchner, Oskar von, Loew, Ernst, Schröter, Carl, and others. Lebensgeschichte der blütenpflanzen Mitteleuropas. Spezielle ökologie der blütenpflanzen Deutschlands, Oesterreichs und der Schweiz. 4 v. (lfg. 1–62). illus. (incl. maps). 25 cm. Stuttgart, 1908 (1904)–42.

Synopsis and classification of ecological phenomena, bibliography, glossary; systematic account of the biology and ecology of spermatophytes, treating such subjects as habitat, light, soil, and temperature relations, ecological niche, range (including altitudinal range), dissemination, germination and development, rate of growth, size, phyllotaxy, anatomy, flower formation, pollination, seed formation, protective structures, morphology of vegetative and floral parts, special features of life history, etc.; each family provided with separate bibliography. Each volume issued in several abteilungen, few of which have been completed; lfg. 60–62 and some others not seen. Continued by W. Wangerin and C. Schröter after the death of Kirchner and Loew, then (lfg. 61/62) by Theodor Schmucker; treatment of various families contributed by other authors.

Kohler, J. M. Alphabetisch geordnetes namens-verzeichniss der verbreitetsten pflanzenarten in der Schweiz. 16 p. 22.5 cm. Zürich, 1850. (Arnold Arboretum library.)

Tabular alphabetical list of 598 vernacular names with corresponding standard German names and botanical names; includes cultivated plants. No index of scientific names.—See also Brandstetter and Durheim, above, and additional references there given.

Künzle, Johann, and others. Das grosse kräuterheilbuch. Die kunst, gesund zu bleiben nach der giftfreien heilmethode und den originalrezepten von Joh. Künzle Pfr. 12 aufl. 527 p. illus., 108 pl. (106 col.) on 54, col. front. (port.). 25 cm. Olten, copyright 1945. (1st ed. 1945?)

Includes (VI. Kapitel, p. 286–416. 100 col. pl. on 50) Pflanzenbeschreibung, a list of native or cultivated medicinal plants alphabetically arranged by vernacular names, with uses, etc.; this summarized on p. 493–502.—See also Correvon, above, and additional references there given.

Marret, Léon, and others. Icones florae alpinae plantarum. 1.–3. série [=3 v.]. illus. (incl. maps), [255] pl., [55] maps. 27 cm. Paris, (1911)–24.

Excellent heliotype plates of herbarium specimens, as well as text figures and maps, illustrating various groups of plants of the mountains of Europe, particularly Switzerland, France, and former Austria-Hungary; the short text concerned chiefly with range and culture. The system used in numbering the plates is not evident.—See also Deutscher und Oesterreichischer Alpenverein, above, and additional references there given.

Morell, J. R. Flora and fauna of the ancient lake dwellings. I. Plants of the Swiss lake habitations. *In* his Scientific guide to Switzerland. Appendix D. p. 390–392. 19.5 cm. London, 1867.

Unannotated list of vascular plants, grouped by uses, etc.; based on publications by Oswald Heer and Ferdinand Keller.—See also Neuweiler, below, and Wirz under Glarus, Local.

Neuweiler, Ernst. Die prähistorischen pflanzenreste Mitteleuropas mit besonderer berücksichtigung der schweizerischen funde. Vierteljahrsschr. Naturf. Gesell. Zürich **50**: 23–134. 1905. (Reprinted 110 p., without Nachtrag (p. 132–134), as Botanische exkursionen und pflanzen-geographische studien in der Schweiz. 6. heft.)

Previous work, manner of occurrence and localities; annotated systematic list of about 220 cellular and vascular plants (of which over 170 are known from Switzerland), known from the Paleolithic on, with localities and references; list of localities with their species, general considerations, bibliography. —See also his Untersuchungen über die verbreitung prähistorischer hölzer in der Schweiz . . . Vierteljahrsschr. Naturf. Gesell. Zürich **55**: 156–202. 1 fig. 1910. (References, with summaries of some earlier papers, discussion of individual species, etc.; tabular lists showing occurrence of the 37 identified woods from Paleolithic to Roman times.)—Also his Nachträge [I]–II urgeschichtlicher pflanzen. l.c. **80**: 98–122. 1 fig. 1935; **91**: 122–136. 1946. (Lists of additional localities, lists of species with localities and some annotations, lists of localities with their respective species.)—Also Morell, above.

Oettli, Paul Deutschschweizerische ortsnamen. 142 p. 22 cm. Erlanbach-Zürich, [1946?]. (Volksbücher des Deutschschweizerischen sprachvereins. vol. 15.)

Includes (p. 69–91) a chapter, Pflanzen und tiere in ortsnamen, containing a running account of locality names derived from plants (p. 69–86).—See also Brandstetter and Durheim, above, and additional references there given.

Probst, Rudolf. Wolladventivflora Mitteleuropas. Redigiert von Werner Strub. vii, 193 p. port. 23.5 cm. Solothurn, 1949.

List of adventive vascular plants (and one cellular) imported in wool from all parts of the world, with general range, localities (where occurring as woolwaste weeds), dates, and the names of collectors; bibliography. Although nominally referring only to Central Europe (and particularly Switzerland), it actually assembles the records for such weeds from practically all of Europe. Only the names of the actual localities are given, not those of the respective countries.—See also Buchli, above, and additional references there given.

Rehder, Alfred. (Phytography.) Switzerland. *In* his The Bradley Bibliography. A guide to the literature of the woody plants of the world published before the beginning of the twentieth century. Compiled at the Arnold Arboretum of Harvard University under the direction of Charles Sprague Sargent. v. 1, p. 432–438. 29.5 cm. Cambridge, Mass., 1911. (Publications of the Arnold Arboretum, no. 3)—Additions and corrections l.c. **1**: 535. 1911; **5**: xxii. 1918.

Chronological list of floristic publications, divided into "dendrography" and "general phytography," with few, mainly bibliographical annotations. See also the following sections relating to Switzerland: bibliography, general (**1**:4); periodicals and serials (**1**: 26–27); botanical gardens, arboretums, museums, etc. (**1**: 65); dictionaries and lists of vernacular names (**1**: 75); botanical history (**1**:81); phenology (**1**: 227); individual noteworthy trees (**1**: 297–298); periodicals and serials relating to arboriculture, gardening, and agriculture (**3**: 24. 1915); dendrological, pomological, economical, and miscellaneous works dealing with both wild and cultivated plants (**3**: 89); miscellaneous works on arboriculture and horticulture (**3**: 126–127); pharmacopoeias (**3**: 256); economic products in general (**3**: 271); periodicals and serials relating to forestry, etc. (**4**: 13. 1914); history of forests (**4**: 28); dendrography (**4**: 86); description of forests (**4**: 364–367). Various minor sections of no floristic significance have been omitted from this summary.—See also Bruhin (Uebersicht), above, and additional references there given.

Rhiner, Joseph. Tabellarische flora der Schweizer-kantone. 2. aufl. iv, 64 p. 27.5 cm. Schwyz, 1897. (Gray Herbarium library.) (1st ed. 1869.)

Tables showing the distribution of each vascular plant in the cantons; list of latest floras of each canton. Preceded (mostly) by his Abrisse (esquisses complémentaires) zur zweiten Tabellarischen flora der Schweizerkantone. Ber. St. Gall. Naturw. Gesell. **1890/91**: 118–255. 1892; **1894/95**: 173–296. 1896; **1897/98**: 283–332. 1899. (List of principal local floras, without titles (authors' names and dates only); critical notes and abstracts of additions to flora of each canton; list of calciphile and calcifuge plants; list of plants limited to a single canton, and those lacking from a single canton.)

Savoy, Hubert. Essai de flore romande. 209 p. 21 cm. Fribourg, 1900.
Systematic list of scientific names with patois equivalents, etymology, properties, uses, superstitions, etc.; indices. Relates to some of the French-speaking cantons (Fribourg and Vaud) and to French Savoie. (Not seen; title and annotation supplied by A. Becherer.)—See also H. Jaccard and Durheim, above, and additional references there given.

Schinz, Hans, and **Keller, Robert.** Flora der Schweiz. Zum gebrauche auf exkursionen, in schulen und beim selbstunterricht. 3., stark verm. aufl.

2 v. (xxx, 648 p. 37+5+135 fig.; xviii, 582 p.). 19.5 cm. Zürich, 1909–14. (Arnold Arboretum library.) (1st ed. 1900.)

The standard modern descriptive flora of vascular plants (2,587 species in 4th ed.), with keys. Pt. (i.e., vol.) 1 (Exkursionsflora) of each edition (except the first, which was not divided) contains species and occasional subspecies only; pt. (i.e., vol.) 2 (Kritische flora), describes the subspecies, varieties, and formae, and lists the adventive plants. The first volume (Exkursionsflora) of the 4th edition was issued in 1923 (xxxvi, 792 p. 37+135 fig. 19.5 cm. Zürich, 1923). Albert Thellung was "mitarbeiter" of ed. 3, v. 2, and ed. 4, v. 1. There is a French edition of ed. 3, pt. 1, translated by E. Wilczek and H. Schinz: Flore de la Suisse . . . 1. partie. xxii, 690 p. 128 fig. 18.5 cm. Lausanne, 1909 [1908]. (Not seen; title supplied by A. Becherer.)

Schweizerische Botanische Gesellschaft. Fortschritte in der systematik und floristik der Schweizerflora (gefässpflanzen) . . . [Title varies] Ber. Schweiz. Bot. Gesell. **2**: 82–102. 1892; **3**: 121–128. 1893; **6**: 88–100. 1896; **8**: 111–125. 1898; **10**: 123–134. 1900; **11**: 166–204. 2 fig. 1901; **13**: 103–137. 11 fig. 1903; **14**: 114–122. 1904; **15**: 54–71. 1905; **17**: 217–255. 1907; **19**: 125–170. 1910; **20**: 182–246. 1911; **21**: 142–170. 1912; **22**: 110–150. 1913; **23**: 106–133. 1914; **24/25**: 148–252. 1916; **26/29**: 161–267. 1920; **30/31**: 73–108. 1922; **32**: 83–103. 1923; **33**: 83–100. 1924; **34**: 66–89. 1925; **35**: 87–109. 1926; **36**: 71–92. 1927; **37**: 144–180. 1928; **38**: 149–180. 1929; **39**: 79–101. 1930; **41**: 296–334. 1932; **43**: 38–76. 1934; **45**: 248–296. 1936; **48**: 273–317. 1938; **50**: 379–424. 1940; **52**: 476–536. 1 fig. 1942; **54**: 347–398. 1 fig. 1944; **56**: 587–628. 1946; **58**: 131–171. 1948; **60**: 467–515. 1950; **62**: 527–582. 1952; **64**: 355–389. 1954; **66**: 164–193. 1956; **68**: 197–238. 1958.

Annual (later biennial) lists of newly recorded species and forms and new localities, with references. Title varies: in v. 2–11, Fortschritte der schweizerischen floristik im jahr 1891 [–1900]; v. 13–39, Fortschritte der floristik [with numerous subtitles] . . . 1901 [–1929]; v. 41–66, Fortschritte in der systematik und floristik der Schweizerflora (gefässpflanzen) in den jahren 1930 und 1931 [–1956 und 1957]. The lists in v. 2–3 and 8–10 included also cellular plants, the pages given above referring only to the vascular plants; the other lists cover vascular plants only. The compilers were: v. 2–3, J. Jäggi and C. Schröter; v. 6–14, Schröter; v. 15–17, M. Rikli; v. 19–26/29, H. Schinz and A. Thellung; v. 30/31, W. Lüdi and J. Braun-Blanquet; v. 32–35, Lüdi; v. 36–68, A. Becherer.—See also Becherer, above.

——— Referate über die im jahre 1890 [–1929] erschienenen veröffentlichungen, welche auf die schweizerische flora bezug haben. Ber. Schweiz. Bot. Gesell. **1**: 124–155. 1891; **2**: 127–147. 1892; **3**: 108–116. 1893; **4**: 81–109. 1894; **5**: 105–125. 1895; **6**: 65–87. 1896; **7**: 111–131. 1897; **8**: 44–75. 1898; **9**: 85–97. 1899; **10**: 78–105. 1900; **11**: 109–162. 1901; **13**: 33–92. 1903; **14**: 52–109. 1904; **15**: 44–53. 1905; **17**: 121–183. 1908; **19**: 28–56, 56–70, 70–81. 1910; **20**: 48–76, 141–164, 164–181. 1911; **21**: 31–52, 107–118, 119–138. 1912; **22**: 81–92, 93–109, 151–171. 1913; **23**: 54–81, 82–105, 134–162. 1914; **24/25**: 86–116, 117–147, 253–272. 1916; **26/29**: 125–144, 145–160, 268–281. 1920; **30/31**: 64–72, 109–118. 1922; **32**: 70–82, 104–110. 1923; **33**: 74–82, 101–107. 1924; **34**: 58–65, 90–95. 1925; **35**: 76–78, 110–123. 1926; **36**: 62–70, 109–117. 1927; **37**: 125–143, 181–192. 1928; **38**: 127–148, 181–192. 1929; **39**: 64–78, 102–109. 1930.

List of current publications relating to the distribution and taxonomy of Swiss vascular plants, alphabetically arranged by authors and with abstracts or in later volumes brief notices through vol. 24/25, only the titles thereafter. The pages listed above refer only to vascular plants. Title and arrangement of contents vary somewhat. In v. 30/31—36, the inclusive title is Fortschritte in der systematik, floristik und pflanzengeographie der Schweizerflora; in v. 37–39, Bibliographie und fortschritte . . . In v. 19–26/29, the material is divided under the headings Systematik, Floristik, and Pflanzengeographie; in later volumes the first two headings were combined, and with v. 32 "Naturschutz" was added to "Pflanzengeographie." The principal reviewers or compilers in v. 1–17 were C. Schröter and later M. Rikli; in v. 19–26/29, A. Thellung, H. Schinz, and W. Rytz; in v. 32–35, W. Lüdi and Rytz; in v. 36–39, A. Becherer and Rytz. For full list of reviewers, see l.c. **50**: 799–800. 1940. No more published after vol. 39; continued by the

Schweizerische Landesbibliothek (see below)—See also Bruhin (Uebersicht), above, and additional references there given.

Schweizerische Landesbibliothek. Botanik. Bibl. Schweiz. Naturw. Lit. **1:** 92–117. 1927; **2:** 100–130. 1928; **3/4:** 178–255. 1930; **5:** 101–133. 1931; **6/7:** 157–199. 1932; **8:** 112–133. 1934; **9:** 122–146. 1935; **10:** 164–188. 1935; **11:** 139–160. 1936; **12:** 146–174. 1938; **13:** 115–133. 1939; **14:** 129–148. 1940; **15:** 128–154. 1941; **16:** 107–129. 1942; **17:** 117–139. 1943; **18:** 111–131. 1944. Bibl. Schweiz. Naturw. und Geog. Lit. **19:** 117–139. 1944; **20:** 132–148. 1945; **21:** 142–163. 1946; **22:** 187–206. 1947; **23:** 179–201. 1949. Bibl. Sci. Nat. Helvetica **24:** 261–285. 1950; **25:** 272–302. 1950; **26:** 149–169. 1951; **27:** 192–219. 1953; **28:** 180–207. 1953; **29:** 195–221. 1954; **30:** 146–171. 1955; **31:** 152–173. 1956; **32:** 163–186. 1957; **33:** 162–181. 1958 (for 1957).

Unannotated classified lists of publications on Swiss botany, or by Swiss botanists in foreign publications, or by foreign authors in books or journals published in Switzerland. Vol. 1 covered 1925; v. 2, 1926; v. 3/4, 1927–28; v. 5, 1929; v. 6/7, 1930–31; v. 8, 1932; v. 9 and subsequent volumes cover 1 year each. The earlier lists were based on the bibliographies in Berichte der Schweizerischen Botanischen Gesellschaft; beginning with v. 6/7 (for 1930/31), they replace those that formerly appeared in that journal.—See also Bruhin (Uebersicht), above, and additional references there given.

Seboth, Joseph. Die alpenpflanzen nach der natur gemalt . . . Mit text von Ferdinand Graf und einer anleitung zur cultur der alpenpflanzen von Joh. Petrasch. 4 v. 4 col. front., 401 col. pl. 16 cm. Prag, Leipzig, 1879–84.

Cited for its colored plates, showing habit excellently, but without details. There is an English edition (Alpine plants painted from nature . . . ed. by A. W. Bennett. London, n.d.) which, like the German, had one or more little or not at all changed reissues.—See also Deutscher und Oesterreichischer Alpenverein, above, and additional references there given.

Stebler, F. G., and **Schröter, Carl.** Die wichtigsten unkräuter der futterwiesen und ihre bekämpfung. (Beiträge zur kenntniss der matten und weiden der Schweiz. IV.) Landw. Jahrb. Schweiz **5:** 141–225. 20 col. pl. 1891.

Includes tabular list of 79 important weeds of pastures (p. 153–155), annotated by means of symbols, and systematic list of the species with numerous vernacular names, occurrence, methods of eradication, etc.—See also **Stebler, F. G.** Die unkräuter der alpweiden und alpmatten und ihre bekämpfung. (Beiträge zur kenntnis der matten und weiden der Schweiz. XIV.) l.c. **13:** 1–120. 40 fig., 20 col. pl. 1899. (Similar treatment of weeds of alpine pasture lands, with list (p. 3–8) of 154 weeds.)—Also Buchli, above, and additional references there given.

Thellung, Albert. Beiträge zur adventivflora der Schweiz. [I]–III. Vierteljahrsschr. Naturf. Gesell. Zürich **52:** 434–473. 1908; **56:** 269–292. 1912; **64:** 684–815. 1919. (Mitteilungen aus dem Botanischen museum der Universität Zürich. XXXVI, 2; LVI, 2; LXXXIII.)

Short bibliographies; lists of adventive vascular plants with localities and statement of native habitat, including plants new to Switzerland and other noteworthy records. Includes also some species from adjacent areas in France, Germany, Austria, and Italy. Parts I–II are Beiträge zur kenntnis der Schweizerflora VIII, XII.—See also Buchli, above, and additional references there given.

Thommen, Eduard. Taschenatlas der schweizer flora mit berücksichtigung der ausländischen nachbarschaft. 2. verm. aufl. xvi, 309 p. 3,055 fig. 19 cm. Basel, 1951. (1st ed. 1945.)

Small figure of each species of the vascular plants, mostly with enlargements of flower or fruit, and with indication of color of flower; list of more local or dubious species, showing range. Includes the more conspicuous extralimital species of the territory immediately surrounding Switzerland. There is also a French edition, Atlas de poche de la flore suisse comprenant les régions étrangères limitrophes. xiv, 296 p. 3,015 fig. 19 cm. Lausanne, 1945.—See also Deutscher und Oesterreichischer Alpenverein, above, and additional references there given.

Thurmann, Jules. Essai de phytostatique appliqué à la chaîne du Jura et aux contrées voisines, ou étude de la dispersion des plantes vasculaires envisagée principalement quant à l'influence des roches soujacentes. 2 v. (xli, 444; 2 p.l., 373 p.). 7 pl. (incl. maps). 24 cm. Berne, 1849.

Vol. 1: bibliography, topography, climate, soils, general features of flora, etc.; v. 2: full list of vascular plants, with habitat, range, and localities for scarcer species. Includes the Jura in Switzerland, France, and adjacent Germany.—See also Friche-Joset, above, and additional references there given; also (for phytogeography) Christ, above, and additional references there given.

AARGAU (ARGOVIE)

See also Basel (Becherer, Beiträge zur flora des Rheintals zwischen Basel und Schaffhausen): Solothurn, General (Probst).

General

Lüscher, Hermann. Flora des kantons Aargau. xiv, 217 p. 23.5 cm. Aarau, 1918.

Annotated list of vascular plants (1,311 species and 90 hybrids); list of additional species of adjoining regions. (Not seen; title and annotation supplied by A. Becherer.)—See also **Becherer, Alfred.** Die botanische erforschung des Aargaus seit dem erscheinen der Flora von H. Lüscher. Verhandl. Schweiz. Naturf. Gesell. **106** (2): 151–152. 1925. (Incomplete list of additions to Lüscher's Flora and of extirpated species.)

Mühlberg, Friedrich. Die standorte und trivialnamen der gefässpflanzen des Aargau's. Mit benützung eines hinterlassenen manuskripts der Aargauer-flora des Herrn Joseph Fridolin Wieland sel., gewesenen arztes in Schöftland, und mit beiträgen mehrerer botaniker. xxiv, 246 p. 17 cm. Aarau, 1880 [1879]. (New York Botanical Garden library.)

Bibliography; etc.; annotated list of vascular plants, with localities and with special attention to vernacular names.

Local

Binz, August. Über die flora von Rheinfelden-Olsberg. Verhandl. Naturf. Gesell. Basel **44** (1): 285–302. 1933.

Geography, botanical explorations, plant associations, etc., including lists of species added to the flora as it was represented by records in C. F. Hagenbach's Tentamen florae basileensis (1821–1843); bibliography.—See also Bruhin, below.

Boll, Jacob. Verzeichniss der phanerogamen- und kryptogamen-flora von Bremgarten, dem untern Freiamt, Hallwilersee, Limmatthal und den angrenzenden theilen des kantons Zürich. viii, 126 p. 18 cm. Aarau, 1869.

Geology, etc.; annotated list of vascular plants and mosses (1,368 vascular species).

Bruhin, T. A. Synoptische flora des bezirks Rheinfelden und der angrenzenden gebiete zwischen der Sisseln und Ergolz, Kanton Aargau der Schweiz. Deut. Bot. Monatsschr. **11:** 156–157, 173–174. 1893; **12:** 27–28, 106–115. 1894.

Botanical publications, bibliography; annotated list of vascular plants. Incomplete; includes Ranunculaceae–Celastraceae.—See also Binz, above.

Hugentobler, Hans. Wilde heil- und nutzpflanzen der Zofingergegend. Zofinger Neujahrsblatt **34:** 33–68. 1949.

Popular treatment of useful and medicinal plants, grouped by families. (Not seen; title and annotation supplied by W. Rytz.)—See also Lüscher, below.

Lüscher, Hermann. Verzeichnis der gefässpflanzen von Zofingen und umgebung und den angrenzenden theilen der kantone Bern, Luzern, Solothurn und Baselland. Mitt. Aargau. Naturf. Gesell. **4:** 34–135. 1886. (Reprinted 103 p.)

Botanical explorations, geology, bibliography, herbaria, etc.; annotated list of vascular plants (1,145 species); lists of naturalized species, extinct and doubtful species, localities with their principal species; separate lists of German and patois names.—See also his Neue beiträge zur flora der Nordschweiz, mit besonderer berücksichtigung der umgebungen von Zofingen. Deut. Bot. Monatsschr. **9:** 56–60, 84–88, 121–127. 1891. (Extensive list with localities.)—Also Hugentobler, above.

Thurnheer, Anna. Der alte Reusslauf bei Fischbach, ein aargauisches naturschutzgebiet. Pflanzengeographische studie. Mitt. Aargau. Naturf. Gesell. **18:** 1–20. 5 fig., 2 pl. 1928.

Plant associations, etc.; unannotated list of vascular plants.

Zschokke, Eugen. Verzeichniss der in der umgegend von Aarau wildwachsenden phanerogamischen pflanzen, nebst angabe ihrer fundorte. 72 p. 19 cm. Aarau, 1847. (Arnold Arboretum library.)

Annotated list of phanerogams, on the Linnaean system.

APPENZELL

See St. Gallen (Wartmann and Schlatter, Kritische uebersicht über die gefässpflanzen der kantone St. Gallen und Appenzell; Oettli, Beiträge . . . aus dem Curfirsten- und Sentisgebiet).

BASEL (BÂLE)

See also Aargau, Local (Lüscher); Solothurn, general (Probst).

General

Binz, August. Flora von Basel and umgebung. Rheinebene, umgebung von Mülhausen und Altkirch, Jura, Schwarzwald und Vogesen . . . 3. aufl. xliv, 320 p. 18.5 cm. Basel, 1911. (1st ed. 1901.)

Bibliography, botanical investigations, list of local botanists; annotated, keyed flora of vascular plants (about 1,800 species), including the commoner cultivated species, with localities and short descriptions; list of poisonous plants. (Not seen; annotation supplied mostly by A. Becherer.) Replaces **Schneider, Ferdinand.** Taschenbuch der flora von Basel und der angrenzenden gebiete des Jura, des Schwarzwaldes und der Vogesen. 344 p. 19 cm. Basel, 1880. (Key to genera, on Linnaean system; annotated list of vascular plants, with brief diagnoses. In Gray Herbarium library.)—See also **Binz, A.** Neuere ergebnisse der floristischen erforschung der umgebung von Basel. Verhandl. Naturf. Gesell. Basel **21:** 126–144. 1910. (Bibliography; records of noteworthy species.)—Also his Ergänzungen zur Flora von Basel. [I.]–VI. teil. l.c. **26:** 176–221. 1915; **33:** 256–280. 1922; **53:** 83–135. 1942; **56** (2): 60–78. 1945; **62:** 248–266. 1951; **67:** 176–194. 1956. (Bibliographies; extensive lists of noteworthy vascular plants, with localities.)

Local

Becherer, Alfred. Beiträge zur flora des Rheintals zwischen Basel und Schaffhausen. Verhandl. Naturf. Gesell. Basel **32:** 172–200. 1921.

List of rarer vascular plants, mostly collected by author, with localities; bibliography.

——— **Steiger, Emil,** and **Lettau, Georg.** Die flora des Naturschutzreservates an der Rheinhalde oberhalb Basel. Verhandl. Naturf. Gesell. Basel **33:** 127–217. pl. 5. 1922.

Topography, list of lichens, plant associations of mosses with annotated list of species; annotated list of vascular plants (485 species; by Becherer); general features of flora; bibliography.

Binz, August. Die erforschung unserer flora seit Bauhin's zeiten bis zur gegenwart. Verhandl. Naturf. Gesell. Basel **13:** 361–390. 1901.

Chronological account of botanical work in Basel, with references to published floras; lists of extirpated and newly recorded species, adventive plants, etc.—See also **Heinis, Fritz.** Geschichte der botanischen forschung in Baselgebiet 1900–1949. Tätigkeitsber. Naturf. Gesell. Baselland **18** (1948–49): 23–35. 2 fig. (port.). 1950. (Classified account of botanical work, with references.)

Heinis, Fritz. Der Bölchen und seine pflanzenwelt. Tätigkeitsber. Naturf. Gesell. Baselland 8 (1926–30): 55–119. 4 fig., pl. 2–5. 1930. (American Museum of Natural History library.)

Topography, botanical explorations, geology, climate, general features of flora, plant associations, phytogeography; list of vascular plants (199 species), with localities; bibliography. (A mountain 1,102 meters high, a good hour east of Langenbruck, on the Baselland-Solothurn boundary.)

——— Kleine beiträge zur flora von Liestal und umgebung. Tätigkeitsber. Naturf. Gesell. Baselland [2] (1902/03): 48–52. 1904. (American Museum of Natural History library, like all the following supplements.)

List of angiosperms (mostly adventive), with localities.—See also his Floristische beobachtungen im kanton Baselland. l.c. [4] (1907/11): 72–77. 1911. (Extensive list of vascular plants, with localities, mostly from vicinity of Liestal.)—Also his Beiträge zur flora des kantons Baselland. l.c. 7 (1922/25): 57–85. 1926. (Includes (p. 63–74) list of new records, including many from Liestal.)—Also his Eine kolonie adventiver pflanzen im stadtgebiet von Liestal. l.c. **16**: 96–102. 1948. (List of adventive plants.)

——— Theodor Völlmin, ein Gelterkinder botaniker. Tätigkeitsber. Naturf. Gesell. Baselland **6** (1917/21): 6–24. 1923. (American Museum of Natural History library.)

Includes (p. 10–24) extensive list of vascular plants, with localities, mostly from the vicinity of Gelterkinden based on Völlmin's work.

Thommen, J. Volkstümliche pflanzennamen von Baselland. Baselbieter Heimblätter **10:** 393–402, 405–414, 433–436. 1945.

Not seen; no review available. Reference supplied by W. Rytz.

BERN (BERNE)

See also Aargau, Local (Lüscher); Luzern (Lüdi); Solothurn, General (Probst).

General

Züricher, Gertrud. Pflanzennamen im Kt. Bern. Schweiz. Arch. Volksk. **41:** 161–175. 1944.

Extensive alphabetical list of patois names, with scientific names, German vernacular names, and locality.

Local

Bourquin, Jules. Flore de Porrentruy. Actes Soc. Jurass. Émul. II, **37:** 29–208. 1933.

Changes due to agriculture, plant zones, etc.; annotated list of vascular plants. Replaces **Thurmann, Jules.** Enumération des plantes vasculaires du district de Porrentruy. 54 p. 21.5 cm. Porrentruy, 1848. (Physiography, plant formations, etc.; annotated tabular list of about 720 vascular plants. In Gray Herbarium library.)

Fischer, Eduard. Botanik und botaniker in Bern. Verhandl. Schweiz. Naturf. Gesell. **96** (2) (1914, pt. 2): 3–28. [1915].

Sketch of botanical work from the time of Brunfels on, with bibliography (mainly biographical references). Refers to vicinity of city.

Fischer, Ludwig. Verzeichniss der gefässpflanzen des Berner-Oberlandes, mit berücksichtigung der standortsverhältnisse, der horizontalen und verticalen verbreitung . . . Mitt. Naturf. Gesell. Bern **1875** (Abhandl.): 1–196. 1876. (Preprinted 1875.)—Nachtrag . . . l.c. **1882** (Abhandl. 1): 3–17. 1882.—2. nachtrag . . . l.c. **1889:** 109–114. 1889.—3. nachtrag . . . l.c. **1904:** 152–164. 1905.

Geology, etc., bibliography; list of vascular plants, with localities. Intended as a second edition of his Verzeichnis der phanerogamen und gefässkryptogamen des Berner-Oberlandes und der umgebungen von Thun. 128 p. Bern, 1862. (Not seen.)—See also **Briquet, John, and Cavillier, François.** Notes sur quelques phanérogames de l'Oberland bernois. Annuaire Conserv. et Jard. Bot. Genève **20:** 222–261. 1918. (New records and localities for numerous angiosperms.)—Also **Rytz, Walther.** Neue und bemerkenswerte pflanzenfunde aus dem Berner Oberland. Mitt. Naturf. Gesell. Bern **1920:** 156–170. 1921.—Also **Gilomen, Hans.** Beiträge zur flora des Berner Oberlandes und des Wallis. Ber. Schweiz. Bot. Gesell. **51:** 344–347. 1941.—Also **Sulger Büel, Ernest.** Beiträge sur flora des Berner Oberlandes (erhebungen aus den jahren 1926–40). Mitt. Naturf. Gesell. Bern. n. f., **2:** 1–13. 1945. (Principal results of author's collecting, including a number of species new to the region.)

——— and **Fischer, Eduard.** Flora von Bern. Systematische übersicht der im bernischen hügelland zwischen Jura und Alpen wildwachsenden und allgemein kultivierten phanerogamen und pteridophyten. 10. aufl. neu bearb. von Dr. Walther Rytz. xxxvi, 431 p. 18 cm. Bern, 1944. (1st ed. 1855, Taschenbuch der flora von Bern . . .)

Physiography, key to families; annotated, keyed, briefly descriptive flora of vascular plants. Covers the whole hill country between (but not including) the Alps and the Jura; the first eight editions covered a radius of about 3 hours' travel around Bern. (Not seen; title and annotation supplied by A. Becherer.)—See also Rytz, below.

Guthnick, H. J. Die flora. *In* Schweizer, J. J. Das Faulhorn im Grindelwald . . . p. 32–41. 20.5 cm. Bern, 1832.

List of plants from the Faulhorn, the Schwarzhorn, and the Schwabhorn. (Not seen; title and annotation supplied by W. Rytz.)—See also **Martins, Charles.** La végétation du Spitzberg comparée à celle des Alpes et des Pyrénées. Bul. Soc. Bot. France **12:** 144–162. 1865. (Includes (p. 154–155) list of 132 species of phanerogams collected by author and Auguste Bravais in 1841–1846, under title Végétaux phanérogames du sommet du Faulhorn.)

Hess, Emil. Die pflanzengeographischen verhältnisse des Oberhasli. (Das Oberhasli . . . I. teil.) 92 p. 6 fig., 3 pl. 30.5 cm. Bern, 1921. (Erhebungen über die verbreitung der wildwachsenden holzarten in der Schweiz. Bearbeitet und veröffentlicht im auftrage des Eidgen. Departements des innern. lfg. 4.) (Reprinted as Promotionsarbeit, with title: Forstbotanische monographie des Oberhasli von Interlaken bis zur Grimsel.)

Topography, geology, climate; annotated list of 146 wild-growing woody plants, with detailed local range; plant associations.

Lüdi, Werner. Beiträge zur floristik des kantons Bern. Ueberprüfung der verbreitungsangaben aus dem Berner Mittelland und dem Berner Oberland in der Flora der Schweiz von H. Schinz und R. Keller. Ber. Schweiz. Bot. Gesell. **40:** 56–86. 1931.

Discussion of floristic regions in the area, bibliography; extensive list of vascular plants with notes on local range.

——— Pflanzengeographische streifzüge im Hohgantgebiet. Mitt. Naturf. Gesell. Bern **1933:** 135–188. 8 fig. on 4 pl. 1934.

Mainly phytogeographic and ecological; includes (p. 139–144) list of vascular plants of chiefly alpine and subalpine distribution known from the Alps between the Thunersee and the Emme, with indication of localities.

Maurer, Jakob. Verzeichnis der flora des Klusgebietes in der Stockhornkette. *In* Gempeler, D. Heimatkunde des Simmentales. p. 469–503. 21 cm. Bern, 1904.

Short introduction and full catalog of species. (Not seen; title and annotation supplied by W. Lüdi.)

Rytz, Walther. Geschichte der flora des bernischen hügellandes zwischen Alpen und Jura. Mitt. Naturf. Gesell. Bern **1912:** 53–221. 3 fig. (incl. map). 1913.

Geography, geology, topography, climate, botanical explorations, lists of species from various altitudes and habitats, fossil plants, phytogeography, bibliography; tabular list of 1,298 species of vascular plants, with indication of local distribution by districts, etc.—See also Fischer and Fischer, above.

Thiébaud, Maurice. Notes floristiques sur la région biennoise. Bul. Soc. Neuchâtel. Sci. Nat. **76:** 45–58. 3 fig. 1953.

Partial lists of species including adventive plants, from Bienne and other localities on the north shore of the Bielersee.—See also his Sur la flore de la région biennoise et de la chaîne de Chasseral. Bul. Soc. Neuchâtel. Sci. Nat. **78:** 105–130. 1955. (Discussion of stations, records of adventive plants, list of noteworthy species.)

FRIBOURG (FREIBURG)

General

Jaquet, Firmin. Catalogue raisonné des plantes vasculaires du canton de Fribourg et des contrées limitrophes. Mém. Soc. Fribourg. Sci. Nat. v. 5. 381 p. 1930.

Annotated list of vascular plants (1786 species), with localities; gazetteer. Replaces **Cottet, Michel, and Castella, François.** Guide du botaniste dans le canton de Fribourg. Bul. Soc. Fribourg. Sci. Nat. [v. 5] (8.–11. ann.). i–lxii, 1–358 p. 1891. (Annotated list of botanists, gazetteer; annotated list of known vascular plants and Characeae.)—See also **Jaquet, F.** Plantes exotiques de pleine terre introduites, accidentelles ou cultivées dans le canton de Fribourg. Mém. Soc. Fribourg. Sci. Nat. 3: 195–284. 1925. (Annotated list.)

GENÈVE (GENF; GENEVA)

See also France, Haute-Savoie (Fauconnet, Herborisations à Salève).

General

Déséglise, Alfred. Florula genevensis advena. Bul. Soc. Roy. Bot. Belg. **16**: 235–244. 1877 (cover 1878).

Annotated list of adventive phanerogams.—See also his Supplément à la florule exotique de Genève. Bul. Soc. Études Scient. Paris **4**: 18–29. 1881. (Original not seen; reprint 12 p., in Arnold Arboretum library.)—Also his Florula genevensis advena. 3. supplément. Bul. Soc. Roy. Bot. Belg. **22**: 97–111. 1883. (Brings total to 150 species, and includes list of all species found, with dates. Relates mainly to vicinity of city of Geneva but includes records from other localities in the canton. Although called the third supplement it is actually the second.)—Also **Palézieux, Philippe de.** Les plantes adventices des environs de Genève de l'herbier Paiche. Bul. Soc. Bot. Genève II, **36**: 103–109. 1945. (Alphabetical lists.)—Also **Weber, Claude.** Les plantes étrangères dans la florule du canton de Genève. l.c. II, **37**: 165–173. 1946. (New records and localities.)—Also **Duperrex, Claude.** Plantes introduites dans le canton de Genève. Trav. Soc. Bot. Genève **1/2**: 8–12. 1 fig. 1954. (List of localities with mention of species introduced (often intentionally) at each.)

Lendner, Alfred. Répartition des plantes ligneuses croissant spontanément dans le canton de Genève. 2 p. l., 63 p. 2 maps. 32 cm. [Berne, 1906.] (Erhebungen uber die verbreitung der wildwachsenden holzarten in der Schweiz. lfg. 1.) (Arnold Arboretum library.)

Geology, list of trees, shrubs, and undershrubs; annotated list of woody plants, including some cultivated species, with vernacular names, general distribution in area, and lists of localities; general features of forests, notes on some large cultivated trees, etc.; brief bibliography; tables showing occurrence of the different species in the principal oak woods of Geneva.

Reuter, G. F. Catalogue détaillé des plantes vasculaires qui croissent naturellement aux environs de Genève, avec l'indication des localités et de l'époque de la floraison. 2. éd., entièrement refondue et considérablement augmentée . . . xvi, 300 p. 16 cm. Genève, 1861. (1st ed. 1832)

Annotated list of vascular plants and Characeae (about 1,850 species), with localities. Includes a section "Rubi genevenses" by E. Mercier. Covers most of the canton of Genève and adjacent parts of Vaud, Haute-Savoie, and Ain.—See also **Schmidely, Auguste.** Annotation au Catalogue des plantes vasculaires . . . Bul. Trav. Soc. Bot. Genève **3**: 82–155 [160]. 1884 (Extensive list of additions and corrections.)—Also **Beauverd, Gustave.** Contributions à la connaissance de la florule genevoise. Bul. Herb. Boissier II, **6**: 427–428. 1906.—Also **Sulger Büel, Ernst.** [Contribution à la connaissance de la flore aquatique genevoise.] Bul. Soc. Bot. Genève II, **11**: 16–18, (302). 1919.—Also papers on the flora of the canton of Geneva or the Geneva basin, with various titles, by **Thommen, Édouard.** l.c. II, **28**: 208. 1938; **29**: 150–151. 1938; **30**: 277–278, 282, 288. 1940; **31**: 432–434. 1940; **32**: 180–183. 1941; **33**: 131–136. 1942; **34**: 44–50. 1 fig. (map). 1945; **36**: 91–102. 1945; **38**: 45–54. 1948; **39**: 1–7. 1948; **41**: 65–71. 1949; **42/43**: 6–15. 1952.

Trav. Soc. Bot. Genève **1/2:** 1–7. 1954.—Also **Becherer, Alfred.** Altes und neues aus der flora des kantons Genf. Mitt. Naturf. Gesell. Bern n.f., **4:** xxv–xxvi. 1947. (Abstract of a lecture, including list of deceased collectors with dates and mention of some noteworthy recent additions to flora.) —Also his Contributions à la flore de Genève (observations fai tes en 1954 et 1955). Trav. Soc. Bot. Genève **3:** 33–38. 1956.

Local

[**Bouvier, Louis.**] Promenades botaniques. Itinéraire du jeune botaniste dans le canton de Genève et les contrées voisines. 3 p.l., 66 p. 14.5 cm. Genève, Lausanne [pref. 1890]. (New York Botanical Garden library.)

List of collecting localities, with their principal species.

GLARUS (GLARIS)

General

Wirz, Johann. Flora des kantons Glarus. 3 pt. (40, 176, 72 p.). 4 pl. (in pt. 3). 18.5 cm. Glarus, 1893–96. (Kew library; pt. 1 there is a reissue in 1901; original issue of pt. 1 (1893) not seen.)

Flora of vascular plants in form of keys, with habitat and local range. Heft 1, Holzgewächse; h. 2, Kräuter [herbaceous dicotyledons, and most of monocotyledons]; h. 3, Grasartige gewächse und gefässkryptogamen.—See also **Sulger Büel, Ernst.** Bemerkenswerte glarner pflanzen. Ber. Schweiz. Bot. Gesell. **53:** 469. 1943. (List of 17 species.)—Also **Wirz-Luchsinger, Hans.** Ergänzungen zur flora des kantons Glarus. Verhandl. Naturf. Gesell. Basel **56** (2): 79–94. 1945. (History of Wirz's Flora, bibliography; list of vascular plants, with localities, intended to include all those added since 1896.)

Local

Heer, Gottfried. Ueber volkstümliche pflanzennamen des glarnerischen mittel- und unterlandes. 13 p. Glarus, 1893.

"Beispiele von einheimischen Pflanzennamen in verschiedenen Teilen des Kt. Glarus. Beziehungen der Namen zum Sprachgut (Bedeutung, Verwendung, Gebräuche, Standort, Jahreszeit)." (Not seen; title and annotation supplied by W. Rytz.)

Schlittler, Jakob. Vegetationsstudie des Niederurnertales. Mitt. Naturf. Gesell. Glarus **7:** 67–129. 4 maps. 1945.

Geography, geology, climate, general features of vegetation, plant associations; includes list of species arranged by altitudinal zones, and list of introduced species. (Not seen; title and annotation supplied by A. Becherer.)

Wirz, Johann. Die veränderungen in der pflanzenwelt unseres landes unter der einwerkung des menschen. Neujahrsbl. Naturf. Gesell. Kantons Glarus **1:** 1–57. 1898.

Condensed account of geological history in relation to flora; historical account of cultivated plants from the time of the lake-dwellings on; account of weeds and naturalized plants. (Not seen; title and annotation from review in Ber. Schweiz. Bot. Gesell. **8:** 75. 1898.)—See also Switzerland, General (Morell, and additional title there given).

Wirz-Luchsinger, Hans. Beobachtungen über die verbreitung wildwachsender holzarten im kanton Glarus. 130 p. 14 fig., map. 29.5 cm. Bern, 1928. (Erhebungen über die verbreitung der wildwachsenden holzarten in der Schweiz. Bearbeitet und veröffentlicht im auftrage des Eidgen. Departements des innern. lfg. 5.)

Topography, geology, plant regions; annotated list of 150 woody plants, with detailed localities; plant associations.

GRAUBÜNDEN (GRIGIONI; GRISCHUN (ROMANSCH); GRISONS)

See also Ticino (Schröter, Flora des Südens); Italy, General (Jaberg, Sprach- und sachatlas); Lombardia (Fornaciari under Massara; Furrer, Flora von Bormio); Sweden, Lule Lappmark (Åberg, Kärlväxternas höjdgränser).

General

Braun-Blanquet, Josias. Flora raetica advena. Verzeichnis der in Graubünden eingeschleppten und verwilderten gefässpflanzen und wichtigsten kulturgewächse. Commun. Sta. Internatl. Géobot. Méditer. et Alp. Montpellier 110. 111 p. 1 fig. (map). 1951.

Localities; annotated list of introduced and escaped angiosperms, including also some cultivated species; bibliography. Does not include naturalized species, or those which are native in one part of the canton and adventive in another; for these see Braun-Blanquet and Rübel's Flora.

——— and **Rübel, Eduard.** Flora von Graubünden. Vorkommen, verbreitung und ökologisch-soziologisches verhalten der wildwachsenden gefässpflanzen Graubündens und seiner grenzgebiete. Veröffentl. Geobot. Inst. Rübel Zürich 7. heft. 1,695 p. map. 1932–36.

Botanical explorations, botanical regions, bibliography; annotated list of vascular plants, with detailed localities. Replaces **Moritzi, Alexander.** Die pflanzen Graubündens. Ein verzeichniss der bisher in Graubünden gefundenen pflanzen, mit besonderer berücksichtigung ihres vorkommens. (Die gefässpflanzen.) Neue Denkschr. Allg. Schweiz. Gesell. Gesam. Naturw. v. 3, [no. 4]. 158 p. 6 pl. 1839. (Botanical collectors, physiography, botanical regions; annotated list of 1,497 vascular plants; list of vernacular names.)—See also **Becherer, Alfred.** Beiträge zur flora Südbündens. Verhandl. Naturf. Gesell. Basel **68:** 165–193. 1957. (Extensive list.)

Local

Braun (later Braun-Blanquet), **Josias.** Die vegetationsverhältnisse der schneestufe in den Rätisch-Lepontischen Alpen. Ein bild des pflanzenlebens an seinen äussersten grenzen. Neue Denkschr. Schweiz. Naturf. Gesell. v. 48. vii, 347 p. 4 fig., 4 pl., map. 1913.

Topography, geology, altitudinal limits of snow, climate, ecology, plant formations; annotated list of vascular plants (224 species) with detailed localities (above 2,600 meters elevation); history of flora, bibliography. Refers to the region between the Vorderrhein, the Adda, and the Insubrian Lakes (Lago Maggiore, Lago di Lugano, Lago di Como), that is, much of Ticino and Graubünden, and adjacent Italy.—See also **Lüdi, Werner.** Die gipfelflora des Flüela-Schwarzhorns bei Davos. Ber. Geobot. Forschungsinst. Rübel Zürich **1938:** 50–53. 1939. (List of angiosperms found at 3,100–3,150 meters altitude, with discussion and comparison with earlier lists.)

——— Die vegetation des Piz Languard, ein masstab fur klimaänderungen. Svensk Bot. Tidskr. **49:** 1–8. 2 fig. (incl. map). 1955.

Includes list of 78 vascular plants observed at or above 3,000 meters altitude on Piz Languard, in the Upper Engadine, from 1903–1941, with the conclusion that the number of species has increased.

Brockmann-Jerosch, Heinrich. Die flora des Puschlav (bezirk Bernina, kanton Graubünden) und ihre pflanzengesellschaften. xii, 438 p. 5 pl., map. 25.5 cm. Leipzig, 1907. (Die pflanzengesellschaften der Schweizeralpen. I. teil.)

Topography, geology, climate; annotated list of cellular and vascular plants, including the chief cultivated plants; plant associations, altitudinal zones, history of vegetation, list of vernacular names with botanical equivalents and localities; bibliography. (Puschlav or Poschiavo.)—See also **Becherer, Alfred.** Beiträge zur flora des Puschlav. Jahresber. Naturf. Gesell. Graubünden **82:** 131–177. [1950.] (Annotated list of noteworthy vascular plants.)—Also his Neue beiträge zur flora des Puschlav. l.c. **84:** 29–42. 1953.—Also his Beiträge zur flora Südbündens. Verhandl. Naturf. Gesell. Basel **68:** 165–193. 1957. (Extensive list.)—Also Rübel, below.

Brügger, C. G. Gefässpflanzen. *In* Naturgeschichtliche beiträge zur kenntniss der umgebungen von Chur . . . Herausgegeben von der Naturforschenden gesellschaft Graubündens. p. 47–104. 21 cm. Chur, 1874.

List of vascular plants annotated by symbols (see p. v–vi for explanation). —See also Capeder, below.

Brunies, S. E. Die flora des Ofengebietes (Südost-Graubünden). Ein floristischer und pflanzengeographischer beitrag zur erforschung Graubündens. Jahresber. Naturf. Gesell. Graubündens n.f., **48:** 1–326. 6 pl., map. 1906. (Reprinted.)

Topography, geology, climate; annotated list of 825 vascular plants; plant formations, etc.; list of fungi, bibliography.—See also **Braun, Josias.** Bemerkungen zur "Flora des Ofengebiets." Vierteljahrsschr. Naturf. Gesell. Zürich **55**: 287–289. 1910. (Mitteilungen aus dem Botan. museum der Universität Zürich (XLIX), III.) (Includes 31 additional species collected in a single afternoon.)

——— Noms populers da plauntas publichos per adöver impustüt dals magisters rumauntschs. Annalas Soc. Retorumantscha **59**: 180–191. 1945; **60**: 84–121. 1946; **62**: 117–167. 1948. (Vol. 59 in Library of Congress, v. 60 and 62 in Cornell University library.)
Explanation of abbreviations; alphabetical list of scientific names with corresponding Romansch names and localities, etymological notes, uses, folklore, etc. Covers A—Fu only.

Candrian, Moritz. Katalog der Oberengadiner flora (gefässpflanzen). Jahresber. Naturf. Gesell. Graubündens n.f., v. 66, Beilage. xx, 174 p. 1928.
Geography, topography, general features of flora, botanical explorations, bibliography; annotated list of known vascular plants, with localities. Pages v–xiv, 1–37 (to species no. 348) first appeared with title, Flora des Oberengadins. l.c. n.f., **64**: 109–152. 1926.

Capeder, Emil. Excursions- und schul-flora von Chur und umgebung mit berücksichtigung des anschliessenden gebietes von Arosa . . . xliii, 372, 63 p. map. 19 cm. Chur [1904]. (Lloyd Library.)
Key to genera on the Linnaean system; keyed flora of vascular plants, with brief descriptions but without annotations or statement of range; short supplementary list (by Jos. Braun), with localities; list of useful species in the flora, with brief notes on uses. Based principally on Brügger's list of plants of Chur and vicinity (1874), with additions from later publications. The work is very severely criticized in review by M. Rikli in Bot. Centralbl. **96**: 198. 1904.—See also Brügger, above.

Derks, Karsje. Die flora von Klosters. 44 p. 22 cm. Klosters, 1928.
Geography, geology, climate, woody plants, formations, principal localities. (Not seen; title and annotation supplied by A. Becherer.)

Freuler, Bernhard. Die holz- und kulturpflanzen des zentralen Bündner Oberlandes. Jahresber. Naturf. Gesell. Graubündens n.f., **47**: 329–359. 1905 (1906).
Topography, general features of forests, notes on Romansch vernacular names, geology, altitudinal zones of vegetation, tree formations; annotated list of woody plants; tabular list of native and some cultivated woody plants, giving altitudinal range, vernacular names, etc. Relates to the Vorderrheintal.

Geiger, Ernst. Das Bergell. Forstbotanische monographie. Jahresber. Naturf. Gesell. Graubündens n.f., **45**: [1–4], 1–119. 5 (i.e. 6) pl., 2 maps. 1901. (Inaug. diss.)
Includes (p. 17–71) systematic list of 98 wild woody plants, with altitudinal limits, followed by extensively annotated treatment of the same, with vernacular names, discussions of minor forms, local occurrence, uses, etc.

Grisch, Andreas. Beiträge zur kenntnis der pflanzengeographischen verhältnisse der Bergünerstöcke. Beih. Bot. Centralbl. 2. Abt., **22**: 255–316. 1907. (Reprinted as Inaug.-diss. 68 p.)
Topography, geology, climate, general features of flora; annotated list of vascular plants; plant formations, bibliography. Covers the Bergünerstöcke chain as well as the Val Tschitta and Val d'Err and is bounded mainly by the Albula and Julia Rivers.

Hager, Karl. Verbreitung der wildwachsenden holzarten im Vorderrheintal (Kanton Graubünden). 1 p. l., 331 p. 4 pl., 2 maps. 32 cm. 1916. (Erhebungen über die verbreitung der wildwachsenden holzarten in der Schweiz. Bearbeitet und veröffentlicht im auftrage des Schweiz. Departments des innern. lfg. 3.) (Arnold Arboretum library.)
Topography, geology, general features of vegetation, climate, altitudinal limits of agriculture and tree growth, etc.; extensively annotated list of woody plants; plant associations, agricultural conclusions, etc.; bibliography.

Hegi, Gustav. Zur flora der Maloja. Vierteljahrsschr. Naturf. Gesell. Zürich 73, Beibl. **15** (Festschrift Hans Schinz): 233–251. pl. 9. 1928.
General features of flora; list of rarer vascular plants, with localities; plant associations, cultivated plants, etc.

Jäggli, Mario. Flora del S. Bernardino. Parte 1. Notizie introduttive. Censimento delle specie. Bol. Soc. Ticinese Sci. Nat. **35:** 1–203. 4 fig., 9 i.e. 7 pl., map. 1941. (American Museum of Natural History library.)

History, topography, geology, climate, botanical explorations, plant zones, bibliography; annotated list of cellular and vascular plants, with localities (1,333 species and 259 infraspecific forms, including 869 species and 177 infraspecific forms of vascular plants). Covers an area about 14 (north-south) by 10 kilometers, with San Bernardino nearly in the center.

Käser, Friedrich. Die flora des Avers. Jahrb. Schweizer Alpenclub **20:** 364–393. 1 fig. 1885.

Topography, woody plants, general features of flora, account of Hieracia, etc.; unannotated list of about 500 vascular plants (including varieties and hybrids).

——— and **Sulger Büel, Conrad.** Flora von Samnaun. Jahrb. St. Gall. Naturw. Gesell. **54:** 132–208. 2 fig. 1917.

Botanical explorations, geology, topography, cultivated plants, general features of flora; list of vascular plants, with localities.

Killias, Eduard. Eine fahrt auf den Piz Lischanna (3103 m.) am 16. September 1871. 32 p. 14 cm. Chur, 1872.

Account of author's ascent, with list of plants and list of beetles. (Not seen; title and annotation supplied by W. Rytz.)

——— Die flora des Unterengadins mit besonderer berücksichtigung der speciellen standorte und der allgemeinen vegetationsverhältnisse . . . Jahresb. Naturf. Gesell. Graubündens n.f., v. 31, Beilage. lxxv, 266 p. 1887–88.

Bibliography, topography, geology, climate, phenology, plant regions, phytogeography, soils, cultivated plants; annotated list of vascular plants, mosses, hepatics, and lichens (1,657 species, of which 1,082 are vascular plants); list of localities with altitudes.—See also **Braun, Josias.** Zur flora des Unterengadins. Vierteljahrsschr. Naturf. Gesell. Zürich **55:** 290–295. 1910. (List with localities.)—Also **Braun-Blanquet, Josias,** and others. Eine pflanzengeographische exkursion durchs Unterengadin und in den Schweizerischen Nationalpark . . . Beitr. Geobot. Landesaufn. Pflanzengeog. Komm. Schweiz. Naturf. Gesell. 4. 80 p. map. 1918. (Mainly ecological; includes short lists of noteworthy vascular plants from several localities.)—Also **Schröter, Carl.** Über die flora des Nationalparkgebietes im Unterengadin. Jahrb. Schweizer Alpenclub **52:** 170–211. illus., 5 pl., map. 1918. (Popular account including history of the park, plant zones, general features of flora, etc.; bibliography.)

Rübel, Eduard. Pflanzengeographische monographie des Berninagebietes. Bot. Jahrb. Engler **47:** i–vi, 1–616. 20 fig., 36 pl., col. front., tab., map. 1911–12. (Reprinted.)

Mainly ecological and phytogeographic; includes annotated list (p. 297–545) of vascular and cellular plants, the latter by other authors; also alphabetical list of scientific names with Romansch vernacular equivalents (p. 553–563), gazetteer, etc.—See also Brockmann-Jerosch, above.

Schaad, G. I nomi popolari della flora prativa in Val Bregaglia. Vox Romanica **4:** 48–64. 1939. (Harvard College library.)

Alphabetical list of scientific names, with patois names and etymological discussion.

Schibler, Wilhelm. Flora von Davos. Verzeichnis der gefässpflanzen der landschaft Davos und der angrenzenden gebiete. Jahresber. Naturf. Gesell. Graubündens n.f., v. 74, Beilage. xv, 216 p. 1937.

Author's botanical explorations and his bibliography; annotated list of vascular plants, with detailed localities; short list of vernacular names, with botanical equivalents (by Joseph Hartmann). Replaces **Geissler, Otto.** Die flora von Davos mit angabe der fundorte und der zeit der blüthe. 2 p. l., 55 p. 17 cm. Davos, 1882. (Annotated list of phanerogams. In Gray Herbarium library.)—See also **Schibler, W.** Die flora des Davoser landwassertales über 2600 meter. Festschr. 110. Jahresversamml. Schweiz. Naturf. Gesell. Davos p. 93–118. 1929. (Botanical explorations, general considerations, list of mountains visited by author; list of 253 vascular plants of Davos region reaching 2,600 meters elevation or more, with references to localities; statistics.)—Also **Lüdi, Werner.** Die gipfelflora des Flüela-

Schwarzhorns bei Davos, under Braun-Blanquet, Die vegetationsverhältnisse der schneestufe, above.

Schröter, Carl. Das St. Antönierthal im Prättigau in seinen wirtschaftlichen und pflanzengeographischen verhältnissen dargestellt. Landw. Jahrb. Schweiz **9:** 133–272. 32 fig., pl. 2–7, map. 1895.

Includes list of vernacular names (p. 258), lists of vascular plants of several summits between 2,000 and 2,828 meters altitude (p. 270–272), etc.

Seiler, Johann. Bearbeitung der Brüggerschen materialien zur Bündnerflora. Jahresber. Naturf. Gesell. Graubündens n.f., v. 51. 580 p. 7 pl., map. 1909.

Brügger's itineraries in Graubünden, list of his publications and manuscripts, notes on hybrids described by him; list of vascular plants, with localities and occasional critical notes; brief bibliography.—See also **Braun, Josias.** Zu Seilers Bearbeitung . . . Berichtigungen und zusätze. Jahresber. Naturf. Gesell. Graubündens n.f., **52:** 1–51. 1910.—Also **Coaz, Karl.** Berichtigung. l.c. n.f., **53:** 192–194. 1912.—Also **Sprecher, Ferdinand.** Beiträge zur flora Graubündens vorwiegend des Schanfiggs und des Prätigaus. l.c. n.f., **56:** 1–19. 1916.—Also **Thellung, Albert.** Floristische beobachtungen um Vals (Bündner Oberland). l. c. n.f., **61:** 44–54. 1922. (Extensive list.)

Steiger, Emil. Beiträge zur kenntnis der flora der Adula-gebirgsgruppe. Verhandl. Naturf. Gesell. Basel **18:** 131–370, 465–755. 1906.

Botanical investigations, topography, geology, climate; annotated list of known vascular plants, with localities.—See also **Braun, Josias.** Nachtrag zu Steigers verzeichnis der gefässpflanzen des Adulagebietes. Vierteljahrsschr. Naturf. Gesell. Zürich **55:** 296–300. 1910. (Mitt. Bot. Mus. Univ. Zürich XLIX, V.)—Also **Thellung, Albert.** Floristische beobachtungen um Vals (Bündner Oberland). Jahresber. Naturf. Gesell. Graubündens n.f., **61:** 44–54. 1922. (Extensive list of vascular plants.)

Thellung, Albert. Beitrag zur kenntnis der flora von Arosa. *In* Braun, Josias. Neue beiträge zur Bündnerflora. II. Vierteljahrsschr. Naturf. Gesell. Zürich **55:** 267–286. 1910. (*In* Mitteilungen aus dem Botanischen museum der Universität Zürich XLIX.)

Annotated lists of the more interesting native and adventive plants collected by author in June-Oct. 1908, mostly at 1500–2200 meters altitude.—See also his Die flora von Arosa. *In* Jenny, Aug. Arosa und die Chur-Arosa-bahn. Orell Fussli's Wanderbilder no. **372/374:** 47–53. 32 fig. 1914. (Sketch of the vegetation, arranged by habitats. Not seen; title and annotation supplied by W. Rytz.)—Also his Floristische beobachtungen um Arosa Juli-August 1925. Jahresber. Naturf. Gesell. Graubündens n.f., **64:** 347–361. 1926. (Topography, bibliography; annotated list of rarer spermatophytes collected by author.)—Also **Casti, J. B.** Botanisches aus Arosa. Schweiz. Apotheker-zeit. **83:** 525–528. 3 fig. 1945. (Mention of various habitats and their characteristic species.)

Ulrich, August. Beiträge zur bündnerischen volksbotanik. Jahresber. Naturf. Gesell. Graubündens n.f., **39:** 1–23. 1896.

Alphabetical list of botanical names, with vernacular equivalents, and notes on popular uses; list of vernacular names, with botanical equivalents. Relates to Prätigau, especially the vicinity of Schiers. The 2d ed. (75 p. Davos, 1897) has not been seen, but the review, in Ber. Schweiz. Bot. Gesell. **8:** 74–75. 1898, indicates no change in the scope of the work.

LUZERN (LUCERNE)

See also Aargau, Local (Lüscher); Schwyz (Rhiner, Die gefässpflanzen der Urkantone und von Zug; Rhiner, Volkstümliche pflanzennamen der Waldstätten); Solothurn, General (Probst).

General

Schobinger-Pfister, Josef. Taschenbuch für reisende botaniker im kanton Luzern, auf Rigi, Pilatus und umgegend. Nach J. G. Krauer's Prodromus florae lucernensis, vermehrt mit mehr als 500 pflanzenarten. 252 p. 15.5 cm. [Luzern], 1866. (Gray Herbarium library.)

Annotated list of wild and cultivated spermatophytes (1,648 species), on the Linnaean system, with localities.

Steiger von Büron, J. R. Die flora des kantons Luzern, der [des] Rigi und des Pilatus . . . 636 p. 123 fig. 21 cm. Luzern, 1860. (Gray Herbarium library.)

Introduction to botany; annotated descriptive flora of vascular plants [1,684 species, of which 1,414 are wild growing], with localities, on the Linnaean system. Replaces **Krauer, J. G.** Prodromus florae lucernensis, seu stirpium phanerogamarum in agro lucernensi et proximis ejus confiniis sponte nascentium catalogus. xii, 105 p. 14 cm. Lucernae, 1824. (Annotated list of 1,081 spermatophytes, with localities, on the Linnaean system. In New York Botanical Garden library.)

Local

Amberg, Karl. Der Pilatus in seinen pflanzengeographischen und wirtschaftlichen verhältnissen. 267 p. 23 fig. on 7 pl., map. 23 cm. Luzern, 1916. (Inaug.-diss. Eidg. Techn. Hochsch. Zürich.)

Ecology; annotated list of vascular plants; bibliography.

Aregger, Josef. Florenelemente und pflanzenverbreitung im Entlebuch und in den angrenzenden gebieten von Obwalden. Mitt. Naturf. Gesell. Luzern **16**: 3–135. 14 fig. (incl. map and diagr.), 4 pl. on 2. 1950. (American Museum of Natural History library.)

Botanical explorations, topography, geology, climate; discussion of phytogeographic groups, with lists of species and annotations, and of distribution according to altitude and soil reaction; list of 416 alpine and subalpine species of the chain between the Thuner– and Brienzersee and the Vierwaldstättersee, with indication of localities; discussion of the principal plants of the Napf; bibliography.

Lüdi, Werner. Die alpenpflanzenkolonien des Napfgebietes und die geschichte ihrer entstehung. Mitt. Naturf. Gesell. Bern **1927**: 195–265. pl. 10–12 (pl. 10, map). 1928.

Mainly ecological and phytogeographic; includes (p. 199–201) annotated tabular list of alpine plants of the region; bibliography. (In Luzern and Bern.)

Portmann, Hans. Flora des gemeinde Escholzmatt. 1 p. l., 35 p. 23 cm. Winterthur, 1922. (British Museum (Natural History) library.)

Geology, climate, general features of flora, introduced and adventive species; briefly annotated list of about 730 vascular plants.

——— Floristisches über Flühli. 25 p. 22.5 cm. [1937.] (Reprinted from Blätter fur Heimatkunde aus dem Entlebuch **10**: 679–702. 1937. Original not seen.)

Topography, climate, floral areas; list of vascular plants, in part with localities; includes (p. 3–5) list of noteworthy species of the Schimberg area by Dr. Hans Mollet. The copy in U. S. Department of Agriculture library was presented by the author and contains his additions.

Zimmerli, E. Der untergang einer landschaft. Eine skizze des heutigen Wauwilermooses. Leben und Umwelt **7**: 169–178. 4 fig. 1951.

Contains (p. 170–174) account of plant habitats of the ruined bog at Wauwil with lists of species, together constituting a full flora.

NEUCHÂTEL (NEUENBURG)

General

Godet, C. H. Description des plantes vénéneuses du canton de Neuchâtel à l'usage des écoles et des gens de la campagne. 16 col. pl. 23 cm. Neuchâtel, 1845. (Arnold Arboretum library.)

Colored plates of principal poisonous plants, with unpaged descriptive text. Another edition (60 p. 26 pl. Neuchâtel, 1864) is listed in the Bradley Bibliography, with title Les plantes vénéneuses . . . publiée par la Société neuchâteloise d'utilité publique.

——— Énumération des végétaux vasculaires qui croissent dans le canton de Neuchâtel. Mém. Soc. Sci. Nat. Neuchâtel v. 2, [no. 2]. ix, 55, 10 p. 1839.

Botanical explorations, etc.; annotated list of vascular plants (1,335 species, of which 1,186 are definitely known from the canton).—See also his

Flore du Jura (above under Switzerland, General), in which all species known from the canton are marked by an asterisk.—Also **Tripet, Fritz.** [Plantes nouvelles pour le canton.] Bul. Soc. Sci. Nat. Neuchâtel 8: 241–242. 1870. (Adds 6 species.)—Also his Modifications apportée à la flore du Jura neuchâtelois par l'abaissement des lacs. Arch. Sci. Phys. et Nat. Genève III, 14: 281. 1885. (Species lost and added since 1869.)—Also **Dubois, Auguste.** Liste des espèces nouvelles trouvées sur territoire neuchâtelois depuis la publication de la Flore du Jura et du Supplément de Ch. Hri. Godet, soit depuis 1869. Rameau de Sapin **48**: 34–38. 1914. (Annotated list of 46 vascular plants, with references; additional notes on adventive species, lost species, etc. In Arnold Arboretum library. A review, with list of species, is given in Ber. Schweiz. Bot. Gesell. **24/25**: 126. 1916.)—Also **Gaille, Armand.** [Trois graminées nouvelles pour notre canton.] Bul. Soc. Neuchâtel Sci. Nat. **42**: 115. 1918.—Also **Ischer, Adolphe.** Notes floristiques. Bul. Soc. Neuchâtel. Sci. Nat. **66**: 5–11. 2 fig. 1942. (List of noteworthy species.)—Also **Favarger, Claude.** Notes de floristique neuchâteloise. l.c. **75**: 18–19. 1952. (Notes on 9 phanerogams.)

Local

Graber, Aurèle. La flore des gorges de l'Areuse et du Creux-du-Van ainsi que des régions environnantes. Bul. Soc. Neuchâtel. Sci. Nat. **48**: 25–365. fig. 1924. (Mitt. Bot. Mus. Univ. Zürich 99.)

Physiography, geology, climate, phytogeography, plant associations; annotated list of vascular plants, with localities; list of introduced plants; bibliography.

Loeske, Leopold. Beitrag zur flora von Chaux-de-Fonds und umgebung. Deut. Bot. Monatsschr. **22**: 182–186. 1911.

General features of vegetation, etc.; list of vascular plants observed by author in July 1889 up to 1,400 meters elevation, grouped according to abundance and special localities; includes some species from the adjacent French department of Doubs.

Spinner, Henri. La distribution verticale et horizontale des végétaux vasculaires dans le Jura neuchâtelois. Mém. Univ. Neuchâtel. v. 2. 200 p. 7 pl., map. 24 cm. Neuchâtel, 1918.

Bibliography; annotated list of vascular plants, with altitudinal limits of each species; statistics, general consideration.

——— La garide des Valangines. Bul. Soc. Neuchâtel. Sci. Nat. **37**: 132–143. 5 fig., 2 tab. 1911.

Includes tabular list of 226 vascular plants.

Wirth, Carl. Flora des Traverstales und der Chasseronkette. Beih. Bot. Centralbl. 2. abt., **32**: 195–328. 1914. (Also as Inaug.-diss. 143 p. Zürich, 1914.)

Geography, hydrography, geology, climate, general features of vegetation, plant formations, origin of flora; annotated list of vascular plants; bibliography.

ST. GALLEN (ST. GALL)

General

Schlatter, Theodor. Die pflanzenwelt St. Gallens. *In* Die stadt St. Gallen und ihre umgebung. p. 71–147. 23 cm. St. Gallen, 1915.

Topography, climate, plant formations, phytogeography, weeds, cultivated plants, etc. (Not seen; cited principally from Ber. Schweiz. Bot. Gesell. **24/25**: 268. 1916.)

Wartmann, Bernhard. Beiträge zur St. Gallischen volksbotanik. Ber. St. Gall. Naturw. Gesell. **1872/73**: 237–349. 1874.

Partly annotated alphabetical list of scientific names (including cultivated plants) with vernacular equivalents; alphabetical list of about 1,600 vernacular names, with botanical equivalents.—See also **Schlatter, Theodor.** Romanische pflanzennamen im kanton St. Gallen. Jahrb. St. Gall. Naturw. Gesell. **1907**: 89–104. 1908. (List of various plants, with discussion of their Romansch names.)

——— and **Schlatter, Theodor.** Kritische uebersicht über die gefässpflanzen der kantone St. Gallen und Appenzell. Ber. St. Gall. Naturw. Gesell. **1879/80:** 61–238. 1881; **1882/83:** 159–328. 1884; **1886/87:** 247–461. 1888. (Reprinted 558 p.)

Bibliography, etc.; annotated list of vascular plants [about 1,400 species]. Replaces **Wartmann, Jakob.** St. Gallische flora für anfänger und freunde der botanik bearbeitet. iv, 267 p. 19 cm. St. Gallen, 1847. (Gray Herbarium library.)—See also **Vogler, Paul.** Neue ankömmlinger in der flora des kantons St. Gallen. St. Galler Tagblatt 24. Dezember 1903. (List of newly recorded adventive plants. Not seen; cited from reviews in Bot. Centralbl. **97:** 270. 1904 and in Ber. Schweiz. Bot. Gesell. **14:** 108–109. 1904.)—Also **Schmid, Heinrich.** Alpenpflanzen im Gäbrisgebiete und in der umgebung der stadt St. Gallen. Jahrb. St. Gall. Naturw. Gesell. **1904:** 147–200. 1905. (Includes (p. 192–199) list of alpine plants of the area, containing numerous new records and localities.)—Also his Wodurch unterscheidet sich die alpenflora des Kronberggebietes von derjenigen des Gäbrisgebietes? l.c. **1906:** 126–150. 1907. (Comparison of floras of different mountain areas, list of alpine plants of the Kronberg region.)—Also **Heyer, August.** Floristische notizen. l.c. **1906:** 367–376. 1907. (Extensive list.)—Also **Sulger Büel, Ernst.** Beiträge zur flora der kantone St. Gallen und Appenzell aus dem jahren 1890–1908. l.c. **1907:** 76–88. 1908.—Also **Schnyder, Albert.** Beiträge zur flora der kantone St. Gallen und Appenzell (Buchs und umgebung) aus dem jahren **1905–1909** [–1910–1913]. l.c. **1908/09:** 282–294. 1910; **53** (1913): 161–177. 1914.—Also **Schlatter, Theodor.** Beiträge zur flora der kantone St. Gallen und Appenzell. l.c. **1911:** 87–121. 1912.—Also **Kägi, Heinrich.** Neue beiträge zur flora des kantons St. Gallen. l.c. **63** (2): 109–140. 1928.—Also **Heyer, August.** Floristiche mutationen in der umgebung von St. Gallen seit 1910. l.c. **65:** 118–122. 1930. (List of plants that have disappeared, and new records.)

Local

Baumgartner, Gottlieb. Das Curfirstengebiet in seinen pflanzengeographischen und wirtschaftlichen verhältnissen. Ber. St. Gall. Naturw. Gesell. **1899/1900:** 147–390. 1 fig., 14 pl., tab., 3 maps. 1901.

Topography, geology, climate, plant formations, cultivated plants, etc.; annotated list of vascular plants, with altitudinal range and vernacular names; agriculture, etc.—See also his Zauber-, heil-und zierpflanzen unserer einheimischen alpenflora. Jahrb. St. Gall. Naturw. Gesell. **1903:** 96–99. 1904. (Running account of a lecture, with mention of more important species; only vernacular names given.)—Also **Schinz, Hans.** Zur flora des Curfirstengebietes. Vierteljahrsschr. Naturf. Gesell. Zürich **49:** 229–231. 1905.—Zweiter beitrag . . . l.c. **51:** 204–210. 1906.—[Dritter beitrag.] Nordabfall der Curfirsten, von Walenstadt bis Weesen. Bul. Herb. Boissier II, **7:** 654–659. 1907.—Vierter beitrag . . . Vierteljahrsschr. Naturf. Gesell. Zürich **55:** 248–251. 1911.

Kägi, Heinrich. Die alpenpflanzen des Mattstock-Speer-gebietes und ihre verbreitung ins Zürcher Oberland. Jahrb. St. Gall. Naturw. Gesell. **56** (2): 45–254. 1920.

Topography, geology; list of about 323 species (plus *Hieracium*) of alpine and subalpine vascular plants, with detailed localities; phytogeography, altitudinal ranges, ecology; list of 54 rarer nonalpine species, with localities.

Müller, Eduard. Beitrag zur volksbotanik des kantons St. Gallen. Vierteljahrsschr. Naturf. Gesell. Zürich **70:** 104–121. 1925. (Mitteilungen aus dem Botanischen museum der Universität Zürich CX.)

Systematic list of vascular plants (and a few lichens) with vernacular names used in Walenstadt and the Walenstadtberg.

Oettli, Max. Beiträge zur ökologie der felsflora. Untersuchungen aus dem Curfirsten- und Sentisgebiet. Jahrb. St. Gall. Naturw. Gesell. **1903:** 182–352. 4 tab. 1904.

Ecological; includes list of "Felsenpflanzen" (p. 322–333), and 38 "florulae" (p. 333–347) of different spots in the region; bibliography. Refers to an area in St. Gallen and adjacent Appenzell.—See also Baumgartner, above.

Roth, August. Das Murgtal und die Flumseralpen. Eine pflanzengeographische studie. Jahrb. St. Gall. Naturw. Gesell. **52:** 1–283. 8 pl., tab., map. 1913. (Mitteilungen aus dem Botanischen museum der Universität Zürich LXI.)

Topography, hydrography, geology, climate; annotated list of vascular plants, with localities (p. 24–186); plant formations, plant zones, agriculture, bibliography.

Schnyder, Albert. Floristische und vegetationsstudien im Alviergebiet. **Vierteljahrsschr.** Naturf. Gesell. Zürich 75, Beibl. 17. 188 p. 1930. (Mitteilungen aus dem Botanischen museum der Universität Zürich CXXXV.)

Geography, geology, climate, etc.; list of known vascular plants, with localities; plant associations, bibliography. Covers area from the Kreuzegg to Gonzen.—See also his Beobachtungen über pflanzenwanderungen im Alviergebiet. Jahrb. St. Gall. Naturw. Gesell. **60**: 20–47. 1924. (Includes list of adventive, naturalized, and cultivated species.)

Vogt, Margrit. Pflanzengeographische studien im Obertoggenburg. Jahrb. St. Gall. Naturw. Gesell. 57 (2): 170–298. 1921.

Mainly phytogeographical and ecological; includes (p. 183–196) list of known vascular plants of the region grouped by habitats; bibliography. (Refers to valley of Thur River from Wildhaus to Wattwil.)

SCHAFFHAUSEN (SCHAFFHOUSE)

General

Kummer, Georg. Botanik. (Übersicht über die entwicklung der naturwissenschaftlich. forschung im kanton Schaffhausen und chronik der Naturforschenden gesellschaft Schaffhausen. II.) Mitt. Naturf. Gesell. Schaffhausen **2**: 20–46. 1923. (American Museum of Natural History library.)

Chronological account of botanical explorations 1823–1923, with titles of publications and some biographical information; condensed account of botanical explorations in adjacent regions; alphabetical list of additional botanists, with brief biographical data.

——— Die flora des kantons Schaffhausen mit berücksichtigung der grenzgebiete. 7 pts. Mitt. Naturf. Gesell. Schaffhausen **13**: 49–157. pl. 6–9, fold. map. 1937; **15**: 37–201. pl. 1–6. 1939; **17**: 123–260. pl. 3–12. 1941; **18**: 11–110. pl. 2–10. 1943; **19**: 1–130. pl. 1–9. 1944; **20**: 69–208. pl. 5–13. 1946; **21**: 75–194. pl. 12–20. 1947. (Reprinted, 936 p. incl. 2 indices; reprint not seen.)

Topography, geology, hydrography, climate, agriculture, forests, partial bibliography, annotated chronological list of collectors with titles of publications; annotated list of vascular plants (with consideration of neighboring regions—Zürich, Thurgau, and adjacent Germany), with detailed localities; list of ponds and marshes, etc. (Pt. 1, 4, American Museum of Natural History library; pt. 5–7, U. S. Dept. of Agriculture library; pt. 2–3, not seen.) Replaces **Meister, Jakob.** Flora von Schaffhausen. 1 p. l., vii, 202, viii p. 20 cm. Schaffhausen, 1887. (Beilage zum Osterprogramm des Gymnasiums Schaffhausen 1887.) (Key to genera, on the Linnaean system; briefly annotated flora of vascular plants in form of keys. In New York Botanical Garden library.) Also replaces **Kelhofer, Ernst.** Die flora des kantons Schaffhausen. Mitt. Bot. Mus. Univ. Zürich 85. xvi, 298 p. 1920. (Teil II zu E. Kelhofer's Beiträgen zur pflanzengeographie des kantons Schaffhausen.) (Biography and bibliography of author (by A. Barth), botanical investigations (with biographical footnotes); annotated list of vascular plants, with localities; bibliography, list of herbaria and collectors. In Arnold Arboretum library.)

——— Die wildwachsenden pflanzen. (Schaffhauser volksbotanik. 1.) 2. umgearb. und erweit. aufl. Neujahrsbl. Naturf. Gesell. Schaffhausen v. 5 (1953). 130 p. 6 pl. 1952. (1st ed. 1928, as Volkstümliche pflanzennamen und volksbotanische mitteilungen aus dem kanton Schaffhausen.)

Annotated list of wild plants grouped by season, habit, and habitat, with vernacular names and localities where used, popular uses and beliefs, etc.; list of vernacular names grouped according to origin, list of species bearing identical vernacular names; bibliography, indices.—See also his Die kulturpflanzen. (Schaffhauser volksbotanik. II.) 1.–2. teil. Neujahrsbl. Naturf. Gesell. Schaffhausen v. 6 (1954). 142 p. 8 pl. 1953; v. 7 (1955). xxvii, 113 p. 2 fig. (incl. port.), 6 i.e. 8 pl., port. 1954. (Treatment of cultivated plants (omitting ornamentals) classified by uses, with vernacular names, popular uses and beliefs, commercial statistics, lists of horticultural varieties, etc.; bibliographies. Vol. 6 has grains, vegetables, medicinal and miscellaneous cultivated plants; vol. 7, fruit trees and small fruits, as well as a biography and bibliography of the author.)

Local

Kummer, Georg. Die flora von Osterfingen. *In* Osterfingen, ein heimatbuch. Beiträge zur heimatforschung . . . Schaffhausen **3**: 45–64. illus. [1925.]

Geography, climate; list of principal localities with their characteristic species. (Not seen; title and annotation supplied by A. Becherer.)

SCHWYZ (SCHWYTZ)

General

Rhiner, Joseph. Die gefässpflanzen der Urkantone und von Zug. 2. bearb. Ber. St. Gall. Naturw. Gesell. **1891/92**: 147–271. 1893; **1892/93**: 175–260. 1894; **1893/94**: 111–207. 1895. (1st ed. 1870; see below.)

Botanical explorations, topography, geology, gazetteer; annotated list of vascular plants (1,467 species) of Schwyz, Unterwalden, Uri, Zug, and an area in southern Luzern and the St. Gotthard region (Ticino) with localities. Intended as the 2d edition of his Prodrom der waldstätter gefässpflanzen. x, 218 p. 21.5 cm. Schwyz, 1870; and 1. nachtrag . . . 32 p. 21.5 cm. Schwyz, 1872. (Topography, geology, botanical explorations, etc.; annotated list of 1,401 vascular plants; guide to botanical excursions; list of vernacular names, etc. Both in Gray Herbarium library.)

——— Volkstümliche pflanzennamen der Waldstätten nebst gebrauchs- und etymologie-angaben. viii, 104 p. 21 cm. Schwyz, 1866.

Systematic list of scientific names with vernacular equivalents, notes on etymology and uses; alphabetical list of vernacular names. Covers Luzern, Schwyz, Unterwalden, Uri, and Zug. (Not seen; title and annotation supplied by A. Becherer.)—See also his Nachtrag zu den 1866 zu Schwyz erschienenen Volksthümlichen pflanzennamen der Waldstätten. Ber. St. Gall. Naturw. Gesell. **1893/94**: 208–214. 1895. (Alphabetical list of additions.)

Local

Bruhin, T. A. Flora einsidlensis. Systematische aufzählung der in Einsiedeln freiwachsenden und häufiger cultivirten gefässpflanzen. 75 p. 18 cm. Einsiedeln, New York und Cincinnati, 1864.

Topography, botanical explorations; partly annotated list of vascular plants, including cultivated species (984 species, of which 872 are spontaneous); statistics. Originally published (fide W. Rytz) as an anhang to his Uebersicht der geschichte und literatur der schweizerfloren, pt. II. (Original publication not seen. Another edition (Einsiedeln, 1868) is listed by Fischer.) —See also Gander below.

Düggeli, Max. Pflanzengeographische und wirtschaftliche monographie des Sihltales bei Einsiedeln. Vierteljahrsschr. Naturf. Gesell. Zürich **48**: 49–270. illus., 4 pl. (incl. map). 1903. (Arbeiten aus dem Botanischen museum des Eidg. Polytechnikums XI.) (Reprinted as Inaug.-diss. vii, 222 p.)

Geography, topography, gazetteer, geology, moors (with list of plant and animal remains), climate; list of cultivated plants; annotated list of wild cellular and vascular plants (p. 98–145); plant associations, garden plants, origin of flora, historical sketch of agriculture, bibliography.

Gander, Martin. Flora einsidlensis. Tabellen zur bestimmung der Einsiedeln freiwachsenden gefässpflanzen. 136 p. map. 18 cm. Einsiedeln, 1888. (Arnold Arboretum library.)

Key to genera of vascular plants, on the Linnaean system; briefly annotated list of vascular plants, the species provided with brief diagnostic characters. —See also Bruhin, above.

Schröter, Carl. Die flora der Mythen. *In* Müller, Hugo. Die Mythen. Herausgegeben von der Sektion Mythen des Schweizer Alpenclub. p. 24–30. Schwyz, 1919.

List of some scarcer or especially characteristic species, and list of the flora of the summit (63 species), etc. (Not seen; title and annotation supplied by W. Rytz.)

SOLOTHURN (SOLEURE)

See also Aargau, Local (Lüscher).

General

Probst, Rudolf. Verzeichnis der gefässkryptogamen und phanerogamen des kantons Solothurn und der angrenzenden gebiete . . . redigiert von Dr. Max Brosi, Dr. Walter Moser, Dr. Emil Stuber. vii, 587 p. port. 23.5 cm. Solothurn, 1949.

Localities, changes in habitats and in flora, list of botanists, bibliography, list of herbaria; annotated list of vascular plants, with detailed localities and special attention to infraspecific forms (1,299 species [including hybrids] plus about 800 species of adventive plants). Includes the records for indigenous species from an area up to 10 kilometers wide around the canton, thus covering part of Aargau, Luzern, Bern, and Basel, as well as a little of Baden and Alsace. Replaces **Lüscher, Hermann.** Flora des kantons Solothurn. xiii, 238 p. 17.5 cm. Solothurn, 1898 (John Crerar Library); Nachtrag. 24 p. 18.5 cm. Grenschen, 1904 (John Crerar Library); 2. nachtrag . . . Allg. Bot. Zeitschr. **16:** 72–73, 88–90, 122–123, 138–141. 1910. (Bibliography; briefly annotated list of vascular plants, with localities and collectors' names; includes cultivated species.)

Local

Probst, Rudolf. Die adventiv- und ruderalflora von Solothurn und umgebung. Mitt. Naturf. Gesell. Solothurn 5 (17.Ber.): 157–215. 1914.—2. [–5.] beitrag zur adventivflora von Solothurn und umgebung. l.c. **6** (18.Ber.): 11–49. pl. 1920; 8 (20.Ber.): 41–82. 1928; **9** (21.Ber.): 5–50. 1932; **10** (22. Ber.): 87–105. 1934.—6. beitrag zur adventivflora von Solothurn und umgebung, mit berücksichtigung der adventivflora von Olten bis Aarau . . . l.c. **12** (24. Ber.): 3–48. 1938.

Previous publications, principal localities, etc.; annotated lists of adventive plants (about 600 species and varieties of phanerogams in first list); bibliographies. The 1938 paper includes discussion of "wolladventiven" from different countries, extensive list of adventives, and bibliography.—See also his Uebersicht über die adventivflora von Solothurn und umgebung. Ber. Schweiz. Bot. Gesell. **42:** 536–549. 1933. (Summary of his first 4 papers on adventive plants of Solothurn and vicinity, with incomplete lists of species grouped according to occurrence (around wool, cotton, and grain establishments, etc.).—Also his Wolladventivflora Mitteleuropas (above under Switzerland, General).

THURGAU (THURGOVIE)

General

Eberli, Johannes. Beitrag zur thurgauischen volksbotanik. Mitt. Thurgau. Naturf. Gesell. **16:** 129–202. 1904.

General considerations; alphabetical list of scientific names of vascular and cellular plants, with vernacular names (and localities where used) and notes on uses; index of vernacular names.

Hugentobler, Hans. Die adventiv- und ruderalflora des kantons Thurgau. Mitt. Thurgau. Naturf. Gesell. **37:** 72–133. 1954.

General considerations, botanical explorations; annotated list of known adventive and ruderal plants.

Wegelin, Heinrich. Die flora des kantons Thurgau. xi, 195 p. 29 cm. Frauenfeld, 1943.

Botanical explorations, with biographical notes; annotated list of vascular plants. (Not seen; title and annotation supplied by A. Becherer.) Replaces **Naegeli, Otto,** and **Wehrli, Eugen.** Beitrag zu einer flora des kantons Thurgau. Verzeichnis der gefässpflanzen des bezirkes Frauenfeld, des Hinterthurgaus und der Untersees. Mitt. Thurgau. Naturf. Gesell. **9:** 121–178. 1890; Neue beiträge zur flora des kantons Thurgau. l.c. **11:** 27–37. 1894.

(General features of flora; annotated list of vascular plants, with additions in 1894 paper.) Also replaces **Naegeli, O.** Die pflanzengeographische bedeutung der neuentdeckungen in der thurgauischen flora. l.c. **25:** 166–182. 1924. (Account of additions since 1900, grouped by geographical affinities.)—See also **Hugentobler, Hans.** Neue erstfunde von pflanzen im Thurgau und neue standorte seltener pflanzenarten des kantons. l.c. **36:** 132–143. 1953. (Extensive list.)

Local

Baumann, Eugen. Die vegetation des Untersees (Bodensee). Eine floristisch-kritische und biologische studie. 2 p.l., v, 554 p. 31 fig., 15 pl. (incl. maps). 24.5 cm. Stuttgart, 1911. (Archiv für Hydrobiologie. Suppl. Band 1.)

Geology, ecology; annotated list of cellular and vascular plants, with localities and ecological observations; plant associations, phytogeography, bibliography. Covers region immediately surrounding the Untersee from Konstanz west to Stein am Rhein and Radolfzell. Replaces, so far as Switzerland is concerned. **Höfle, M. A.** Die flora der Bodenseegegend . . . viii, 162 p. 25 cm. Erlangen, 1850.—See also **Baumann, E.** Beiträge zur flora des Bodensees und des Rheingebietes. Mitt. Thurgau. Naturf. Gesell. **27:** 145–154. 1928.

Boltshauser, Heinrich. Beitrag zu einer flora des kantons Thurgau. Verzeichniss der in der umgebung von Amriswil wildwachsenden phanerogamen und gefässkryptogamen. Mitt. Thurgau. Naturf. Gesell. **6:** 19–43. 1884; **8:** 72–83. 1888.

Briefly annotated list of vascular plants and mosses.

Brunner, Friedrich. Verzeichniss der wildwachsenden phanerogamen und gefässkryptogamen des thurgauischen bezirks Diessenhofen, des Randens und des Höhgaus. Mitt. Thurgau. Naturf. Gesell. **5:** 11–61. 1882.

General features of region; annotated list of vascular plants [826 species]. —See also **Brunner, Hans.** Beiträge zur kenntnis der flora des bezirks Diessenhofen und seiner umgebung. l.c. **21:** 201–209. 1915.—Also **Naegeli, Otto.** Zur flora von Diessenhofen. . . . l.c. **24:** 107–128. 1922. (List of additional species and localities discovered since 1882, bringing total to "over 1,000" species.)

Josephy, Grete. Die flora des Hudelmooses. Eine oekologische studie. Mitt. Thurgau. Naturf. Gesell. **24:** 129–160. illus. (incl. map). 1922.

Mainly ecological; includes (p. 142–150) annotated list of vascular plants. (Near Amriswil and Zihlschlacht, in upper Thurgau.)

Knecht, S. Beiträge zu einer flora des kantons Thurgau. Mitt. Thurgau. Naturf. Gesell. **4:** 161–206. 1879.

Annotated list of phanerogams collected by author in southern part of "Hinterthurgau" and adjoining portion of kanton Zürich.

TICINO (TESSIN)

See also Graubünden (Braun-Blanquet, Die vegetationsverhältnisse der schneestufe); Italy, General (Jaberg, Sprach- und sachatlas); Lombardia (Comelli, Flora comense; Penzig, Il Monte Generoso); Piemonte (Chiovenda, Flora delle Alpi Lepontine occidentali).

General

Chenevard, Paul. Catalogue des plantes vasculaires du Tessin. Mém. Inst. Natl. Genevois **21:** 3–553. map. 1910.—Additions au Catalogue des plantes vasculaires du Tessin. 11 p. Genève, 1916. (Not seen.)

Botanical investigations, general features of flora, phytogeography, plant zones, geology, climate, bibliography, list of excluded species, table of altitudes; annotated list of vascular plants (1785 species), with localities. Replaces **Franzoni, Alberto.** Le piante fanerogame dell Svizzera insubrica . . . Opera postuma ordinata e annotata dal Dre A. Lenticchia . . . con note ed aggiunte di L. Favrat. Neue Denkschr. Allg. Schweiz. Gesell. Gesam. Naturw. v. 30, abt. 2. 1 p. l., iv, 256 p. 1890. (General features of flora; annotated list of spermatophytes, with vernacular names.) Also replaces **Lenticchia, Attilio.** Le crittogame vascolari della Svizzera insubrica. Malpighia **8:** 305–321. 1894.—See also **Voigt, Alban.** Beiträge zur floristik des

Tessins. Ber. Schweiz. Bot. Gesell **26/29:** 332–357. 1920. (Mitt. Bot. Mus Univ. Zürich 85.) (Adds about 114 species.)—Also his Neue funde in del Schweiz. Allg. Bot. Zeitschr. **24/25:** 23–24. 1922; Weitere funde in del Schweiz. l.c. 45–46. 1922.—Also his Beiträge zur floristik des Tessins. Sitzungsber. u. Abhandl. Naturw. Gesell. Isis Dresden **1920/21** (Abhandl.): 11–19. 1922. (Topography, geology, etc.; extensive list of vascular plants from Sottoceneri region.)—Also his Neue beiträge zur kenntnis der Tessinerflora. *In* Schinz, Hans. Beiträge zur kenntnis der Schweizerflora. XXI, c. Vierteljahrsschr. Naturf. Gesell. Zürich **71:** 160–161. 1926.—Also **Thommen, Édouard.** Observations sur la flore du Tessin (1946) [-(1947)]. Bol. Soc. Ticinese Sci. Nat. **41:** 27–50. 4 fig. 1946; **42:** 111–124. 1947.—Also **Thommen and Dübi, Hans.** Observations . . . (1948 et 1949). l.c. **44:** 52–63. 1 fig. 1950. (Last three references in Lloyd Library.)—Also **Thommen.** Contribution à la flore du Tessin. Bul. Soc. Bot. Genève II, **42/43:** 1–5. 1 fig. 1952.—Also **Dübi, H.** Appunti sulla flora insubrica. Bol. Soc. Ticinese Sci. Nat. **47/48:** 67–102. 1953. (Extensive list of noteworthy species. Not seen.)

Jäggli, Mario. Cenni sulla flora ticinese con tavole per riconoscere arbusti ed alberi. 3. ed. 55 p. 17 fig., 4 pl. 21 cm. Bellinzona, 1953. (1st ed. 1924.)

Botanical explorations with references, botanical regions with lists of characteristic species; annotated key to the spontaneous trees and more important shrubs.

——— Notizie di bibliografia botanica. Bol. Soc. Ticinese Sci. Nat. **1:** 15–21, 34–39. 1905.

Sketch of botanical explorations in Ticino, with references to publications; phytogeography.

Salvioni, Carlo. Nomi locali del cantono Ticino derivati dal nome delle piante. Bol. Storico Svizzera Ital. **11:** 214–218. 1889.

List of Latin (or Italian) names of plants, with the geographical names derived from them; alphabetical list of geographical names mentioned.—See also **Gualzata, Mario.** La flora e la topografia nella toponomastica ticinese. Bol. Soc. Ticinese Sci. Nat. **20:** 39–52. 1925. (General considerations; list of generic names and botanical terms, with geographical names derived from them. In American Museum of Natural History library.)

Local

Bär, Johannes. Die flora des Val Onsernone (bezirk Locarno, kanton Tessin). Floristische und pflanzengeographische studie. Bol. Soc. Ticinese Sci. Nat. v. 11. 413 p. 1915. (Mitt. Bot. Mus. Univ. Zürich 70.) (Gray Herbarium library.)

Annotated list of cellular and vascular plants, with detailed localities. For physiography, ecology, list of ruderal and adventive plants, and bibliography, see the first part of this work, with same title, in Vierteljahrsschr. Naturf. Gesell Zürich **59:** 223–563. 1914 (Mitt. Bot. Mus. Univ. Zürich 69) which was reissued in condensed form as his Die vegetation des Val Onsernone (kanton Tessin). Beitr. Geobot. Landesaufn. Pflanzengeog. Kom. Schweiz. Naturf. Gesell. 5. 80 p. 2 charts, map. 1918.

Bettelini, Arnoldo. La flora legnosa del Sottoceneri (cantone Ticino meridionale). 213 p. 6 pl., 2 maps. 23 cm. Bellinzona, 1904. (Dissertation. Reissued Milano, 1905, and Zurigo, 1905. 1904 issue in Gray Herbarium library; Zurigo reissue in U. S. Dept. Agriculture library.)

Bibliography, geography, geology, climate; annotated list of 166 native and naturalized woody plants (p. 49–150); botanical regions, plant associations, upper limits of vegetation, history of forests, forest economy. Sottoceneri is the name applied to the part of Ticino south of Monte Ceneri.—See also Freuler, below.

Calloni, Silvio. Appunti sulla flora culminicola del Pizzo di Claro. Annali Club Alp. Ticinese **1886:** 137–143. 1887.

List of 40 species, with critical notes on several of them. (Not seen; title and annotation supplied by A. Becherer.)

Chenevard, Paul, and Braun, Josias. Herborisations dans les vallées de Bavona et de Peccia. (*In* Chenevard, P. Contributions à la flore du Tessin. 5. suite.) Annuaire Conserv. et Jard. Bot. Genève **9**: 1–92. 1905.

General features of flora at different localities and altitudes; extensive list of vascular plants, with localities.

Freuler, Bernhard. Prospetto sulla diffusione verticale delle piante legnose spontanee nel Ticino meridionale. Bol. Soc. Ticinese Sci. Nat. **3**: 56–65. 1906.

Alphabetically arranged, tabular list of 98 woody plants, with Italian and Sottocenerino vernacular names and altitudinal limits (with localities).—See also Bettelini, above.

Furrer, Ernst. Botanische skizze von Pizzo Corombe (Colombe), einem dolomitberg im Nordtessin. Ber. Geobot. Forschungsinst. Rübel Zürich **1952**: 54–72. 2 fig. (incl. map). 1953.

Geology, climate, botanical explorations: list of known vascular plants (136 species) from above about 2,300 meters elevation; phytogeography, plant associations.

Jäggli, Mario. Il delta della Maggia e la sua vegetazione. Beitr. Geobot. Landesaufn. Pflanzengeog. Kom. Schweiz. Naturf. Gesell. 10. 174 p. 5 pl., map. 1922.

Mainly ecological; includes annotated list of cellular and vascular plants (p. 103–170).

——— Elenco sistematico della piante del Monte di Caslano. Bol. Soc. Ticinese Sci. Nat. **44**: 24–51. 1 fig. (map). 1950. (Lloyd Library.)

Topography, geology, climate; annotated list of vascular plants (536 species and 25 infraspecific forms and hybrids).

——— Monografia floristica del Monte Camoghè (2,232 m) (presso Bellinzona) e vette circostanti. Bol. Soc. Ticinese Sci. Nat. **4**: 1–249. 6 pl., map. 1907. (Reprinted as Inaug.-diss., Zürich, 1908; also as Mitt. Bot. Mus. Univ. Zürich 37.)

Physiography, hydrography, climate, geology, vertical distribution, plant zones, plant formations, bibliography, etc.; annotated list of vascular plants, with altitudinal range.

——— Sulla florula del colle di Sasso Corbario. Bol. Soc. Ticinese Sci. Nat. **2**: 79–85. 1905.

General features of flora; list of 142 vascular plants of rupestrian situations. (Near Bellinzona.)

Keller, Robert. Beiträge zur kenntnis der flora des Bleniotales. Bul. Herb. Boissier II, **3**: 371–386, 461–487. 1903.

Geography, etc.; annotated list of vascular plants, with localities. Relates to the valley of the Blenio (Brenno, Bregno) and its tributaries, including the whole district of Blenio and the vicinity of Biasca.—See also **Gsell, Rudolf.** Die orchideen des Bleniotales. Bol. Soc. Ticinese Sci. Nat. **38**: 112–123. 1944. (Not seen.)—Also his Les orchidées de Val Blenio (Tessin). Un supplément. l.c. **41**: 51–57. 1946. (Not seen.)

Schröter, Carl. Flora des Südens. Die pflanzenwelt Insubriens (täler zwischen Ortasee und Comersee). 2. aufl. vollständig neu bearb. von E. Schmid. 167 p. 29 fig. (incl. maps; partly on plates), 105 pl. (64 col.). 20.5 cm. Zürich, 1956. (1st ed. 1936.)

Plant zones, gardens and parks; brief account of the 400 species illustrated, many of which are cultivated; bibliography. There is an Italian version of 1st ed. by Mario Jäggli. Flora d'Insubria . . . xii, 151 p. 60 i.e. 62 fig. (incl. map), 72 pl. (part. col.), front. 21 cm. Bellinzona, 1950.

——— and **Rikli, Martin.** Botanische exkursionen in's Bedretto-, Formazza- und Bosco-Tal. Verhandl. Schweiz. Naturf. Gesell. **86**: 305–396. 10 pl. 1904. (Reprinted 92 p., as Botanische exkursionen und pflanzengeographische studien in der Schweiz. heft 1.)

Running account of localities, with mention of occasional species, list of vascular plants new to Tessin (p. 378), bibliography; short lists of vascular plants from various localities in Ticino and Piemonte (Prov. Novara).—See also **Bär, Johannes.** Floristiche beobachtungen im Val di Bosco. (*In* Schinz, Hans. Beiträge zur kenntnis der Schweizerflora IV.) Vierteljahrsschr. Naturf. Gesell. Zürich **49**: 197–229. 1904. (Mitt. Bot. Mus. Univ. Zürich

XXII, 2.) (Running account of flora with mention of numerous species.)—Also **Chiovenda, Emilio.** Contributo alla studio della flora italiana dell'ing. B. D. Biadego di Verona. Nuovo Gior. Bot. Ital. n.s., **30:** 48–72. 1923. (Includes many records from Val Formazza and Valle Antigorio.)—Also **Becherer, Alfred.** Botanische beobachtungen im Val Formazza. Ber. Schweiz. Bot. Gesell. **37:** 1–8. 1928. (Relates to Italian part of area.)—Also Chiovenda under Italy, Piemonte.

UNTERWALDEN (UNTERWALD)

See also Luzern (Aregger, Florenelemente und pflanzenverbreitung im Entlebuch und in den angrenzenden gebieten von Obwalden); Schwyz (Rhiner, Die gefässpflanzen der Urkantone und von Zug; Rhiner, Volkstümliche pflanzennamen der Waldstätten).

Local

Christ, Hermann. Die flora von Engelberg. *In* Fleiner, Albert. Engelberg. p. 87–95. 36 fig. 20.5 cm. Zürich [1890]. (University of California library.)

Lists including about 100 of the more interesting species of various habitats and altitudes, with notes on geology, ethnology, etc.—See also **Lötscher, Conrad.** Die erforschung der flora von Engelberg. Schweiz. Apotheker-Zeit. **66:** 453–457. 1927. (Sketch of botanical explorations.)

URI

See Schwyz (Rhiner, Die gefässpflanzen der Urkantone und von Zug; Rhiner, Volkstümliche pflanzennamen der Waldstätten).

VALAIS (WALLIS)

See also France, Haute-Savoie (Briquet, Notes floristiques sur les Alpes lémaniennes; Payot, Florule de Mont-Blanc); Italy, Piemonte (Vaccari, La flora nivale di Monte Rosa).

General

Jaccard, Henri. Catalogue de la flore valaisanne. Neue Denkschr. Allg. Schweiz. Gesell. Gesam. Naturw. v. 34. lvi, 472 p. 1895.

Account of collectors and their work, botanical districts with list of species, calcicolous and silicicolous species, lists of erroneously recorded and adventive species, bibliography; annotated list of vascular plants [1785 species]; alphabetical list of localities with their altitudes. Appendice (p. 434–460); Observations sur quelques labiées valaisannes, par **M. John Briquet.** Replaces **Angreville, J. E. d'.** La flore vallaisanne [sic]. viii, 218 p. 18.5 cm. Genève, Paris, 1862. (Botanical publications; list of vascular and cellular plants, with localities. In Gray Herbarium library.)—See also **Becherer, Alfred.** Florae vallesiacae supplementum. Supplement zu Henri Jaccards Catalogue de la flore valaisanne. Denkschr. Schweiz. Naturf. Gesell. v. 81. iv, 556 p. 1 fig. (map). 1956. (Bibliography, 1895–1954; critical catalog of all the species with new records and corrections.)

Mariétan, Ignace. Flore pharmaceutique du Valais. Schweiz. Apotheker-Zeit. **88:** 382–389 incl. pl. 1950.

Cultivation of medicinal plants, etc.; list of medicinal plants cited in Flück's Nos plantes médicinales that occur in Valais, with localities.—See also Wolf, below.

Wolf, F. O. Plantes médicinales indigènes ou cultivées en Valais, leurs propriétés et emplois en médecine. vi, 80 p. 20 cm. Sion, 1906.

Treatment of 109 medicinal plants, in alphabetical order, with vernacular names, local range, medicinal properties, and uses. (Not seen; title and annotation supplied by A. Becherer.)—See also Mariétan, above.

Local

Beauverd, Gustave. Aperçu sur la florule du Mont-Gelé de Riddes (Valais). Vallées du Rhône, de Nendaz et d'Isérables. Bul. Murithienne **23/25:** 38–44. 1897.

Topography, etc., account of plant regions, arranged according to altitude, with lists of species. Mont Gelé is 3,208 meters high.

Bielander, Josef. Die pflanzen in Lax (Wallis). Ihre namen und ihrer verwendung. Schweiz. Arch. Volksk. **45**: 81–104. 1948.

General features of local nomenclature, discussion of trees, fruits and vegetables, edible and medicinal plants, popular beliefs and uses, etc.; alphabetical list of vernacular names, with German and scientific equivalents.

Binz, August. Das Binnental und seine flora. 46 p. 1 fig., map. 26.5 cm. Basel, [1908]. (Separatabdruck aus dem Bericht der Realschule zu Basel 1907/08.) (Arnold Arboretum library.)

Topography, geology, etc.; list of excursions with mention of principal plants; plant communities, bibliography.—See also his Les plantes du Binnental. *In* Desbuissons, Léon. La vallée de Binn (Valais) . . . p. 291–310. 3 fig. 24 cm. Lausanne, 1909. (General features of flora; lists of the rarer or more interesting vascular plants from various localities.)—Also **Schmidely, Auguste.** Notes floristiques. Bul. Murithienne **29/30**: 35–39. 1901. (List of noteworthy plants.)—Also next entry.

——— Verbreitung der wildwachsenden holzarten im Binnental (kanton Wallis). 40 p. pl. 32 cm. Bern. 1908. (Erhebungen uber die verbreitung der wildwachsenden holzarten in der Schweiz. lfg. 2.) (Gray Herbarium library.)

Topography, climate, etc.; discussion of altitudinal range, habitat, etc., of each tree and shrub; tree formations, etc.; table showing altitudinal range, soil, etc. for each species.—See also entry above.

Braun-Blanquet, Josias. Le Gornergrat et les Rothörner de Findelen. Aperçu phytosociologique et floristique. (Observations sur la végétation et sur la flore des environs de Zermatt. II.) Bul. Murithienne **41**: 30–55. 1921.

Includes (p. 33–37) list of 107 species and 3 hybrids from the summit of the Gornergrat, altitude 3,100–3,136 meters, and (p. 50–55) lists of species from the summits of the Ober-Rothorn, Unter-Rothorn, and Plattenhorngrat, at 3,000 meters and above.

——— and **Thellung, Albert.** Observations sur la végétation et sur la flore des environs de Zermatt. I. Notes floristiques. Bul. Murithienne **41**: 18–30. 1921.

Extensive list of noteworthy vascular plants, with localities, including many new altitudinal records.—See also **Thellung, A.** Herborisations à Zermatt (Valais) en juillet-août 1922. Monde des Plant. III, **23** (no. 23–138): 4–6, (no. 24–139): 6–7. 1922. (Lists of noteworthy species and altitudinal records.)—Also **Becherer, Alfred.** Floristische beobachtungen im Wallis. Bul. Murithienne **63**: 121–135. [1946.] (Includes I. Zermatt und umgebung (p. 121–128), a list of noteworthy species with references to works on altitudinal distribution of plants especially in the Alps.)—Also his Neue höhenrekorde für blütenpflanzen in den Alpen. Ber. Schweiz. Bot. Gesell. **61**: 273–279. 1951. (Includes new altitudinal records from Zermatt, p. 275–277; bibliography.)—Also his Records d'altitude pour la flore phanérogamique du Valais et des Alpes. Trav. Soc. Bot. Genève **4**: 42–46. 1958. (Includes bibliography.)

Chodat, Robert. La course botanique de la Société "La Murithienne" en 1891 dans la partie supérieure de la vallée de Bagnes. Bul. Murithienne **19/20**: 61–70. 1892.

Includes lists of species observed at Chanrion and in the upper valley of the Bagnes.—See also **Jaccard, Henri.** Flore de Chanrion. Bul. Murithienne **21/22**: 112–113. 1894. (List of additional species, with localities.)—Also **Beauverd, Gustave.** Rapport botanique sur l'excursion de la Société la Murithienne dans les vallées de Bagnes, d'Aoste et du Grand-Saint-Bernard, les 28, 29, 30, 31 juillet et 1er août 1902. l.c. **32**: 13–60. 1903. (Includes short lists of noteworthy species collected at various localities from Châbles to Chanrion and beyond.)—Also **Mariétan, Ignace.** Notes floristiques sur la partie supérieure de la vallée de Bagnes (Fionney). l.c. **46**: 32–51. 1929. (Topography, botanical explorations; account of various localities, with mention of noteworthy species.)—Also **Zwicky, Henry.** Die flora des oberen Bagnes-Tales. Mitt. Naturf. Gesell. Bern n.f., **2**: xxx–xxxii. 1945. (Sketch of botanical explorations, etc.; mention of noteworthy species from various localities, including several new to the region.)—Also his Quelques plantes nouvelles de la vallée de Bagnes. Bul. Murithienne **62**: 199–204. [1945.] (List of noteworthy species.)

Christ, Hermann. Die Visp-taler föhrenregion im Wallis. Bul. Murithienne **40**: 187–273. 1919 (1920).

Mainly phytogeographical; includes (p. 260–273) annotated lists of native vascular plants and of weeds of the "fir" (*Pinus silvestris*) region, which includes the slopes of the Visp valley from Vispach (now Visp: 687 meters elevation) up to the point where the valley divides into two at Stalden (790 meters).

Correvon, Henri. La vallée de Tourtemagne, en Valais. Jahrb. Schweiz. Alpenclub **35**: 187–202. 3 fig., 2 pl. 1900.

Topography, geology, etc.; unannotated systematic list of more characteristic vascular plants.—See also [**Besse, Maurice, and Jaccard, Henri**]. Herborisation dans la vallée de Tourtemagne à l'occasion de la réunion annuelle de la Murithienne les 19, 20 et 21 juillet 1909. Bul. Murithienne **36**: 13–18. 1911.—Also **Beauverd, Gustave.** Nouveautés floristiques de la vallée de Tourtemagne (Valais). Bul. Soc. Bot. Genève II, **15**: 273–288. 1924. (Topography, botanical explorations, plant zones, list of noteworthy vascular plants.)

Farquet, Philippe. Aperçu sur le haut Val Ferret. Les Angroniettes. Bul. Murithienne **44**: 124–129. 1927.

Topography, etc.; lists of species at several localities at 2,300 to 2,550 meters altitude.

———Une florule disparue: les îles de Saxon. Bul. Murithienne **55**: 32–37. 1938.

Topography, history; list of vascular plants, grouped by habitats. (A swampy region between Charrat and Saxon, in the valley of the Rhône.)

——— Les marais et les dunes de la Plaine de Martigny. Esquisse historique et botanico-zoologique. Bul. Murithienne **42**: 113–159. 1925.

Discusses changes due to draining of the marshes, with lists of animals and plants which have disappeared, and lists of species growing at different localities on the dunes.

——— Le Mont Ravoire s/Martigny. Bul. Murithienne **57**: 34–89. [1940].

Geology, hydrography, topography, climate, fauna; lists of vascular plants of different areas; ethnography, bibliography.

Favre, Émile. Guide du botaniste sur le Simplon. Bul. Murithienne v. 5/6, [Suppl. 1]. viii, 318 p. 1875.

Annotated list of vascular plants, with localities; gazetteer. Covers area from Brigue (Brig) to the frontier below Gondo, and includes a few noteworthy plants from somewhat outside this range.—See also papers by **Gandoger, Michel.** Bul. Soc. Bot. France **35**: 185–194. 1888; **Cornaz, Édouard.** Bul. Soc. Sci. Nat. Neuchâtel **22**: 306–307. 1894; and a long series relating mostly to the south side of the Simplon by **Becherer, Alfred.** Verhandl. Naturf. Gesell. Basel **50**: 1–31. 1939; Ber. Schweiz. Bot. Gesell. **50**: 301–320. 1940; Verhandl. Naturf. Gesell. Basel **51** (2): 119–121, 138. 1940; Bul. Murithienne **58**: 113–121. 1941; **60**: 40–58, 58–63. [1943]; **63**: 128–135. [1946]; **64**: 52–55. 1947.—Also Italy, Piemonte (Chiovenda, Flora delle Alpi Lepontine occidentali.)

Gams, Helmut. Von den Follatères zur Dent de Morcles. Vegetationsmonographie aus dem Wallis. Beitr. Geobot. Landesaufn. Pflanzengeog. Kom. Schweiz. Naturf. Gesell. nr. 15. xii, 760 p. 98 fig. (partly on 26 pl.), map. 1927.

Mainly ecological; includes sketch of botanical explorations, etc. (p. 1–7), and annotated list of woody plants (p. 178–184), with vernacular names and altitudinal range.

Jaccard, Henri. Notice botanique sur la vallée du Trient. 22 p. 17.5 cm. Lausanne, 1895.

Geology, etc.; list of generally distributed species, arranged by habitats, and of local species, arranged by localities; the ubiquitous species omitted. (Not seen; title and annotation supplied by A. Becherer.)

Lüdi, Werner. Die pflanzenwelt des Aletschwald-reservates bei Brig (Wallis) . . . Bul. Murithienne **67**: 122–178. 9 fig. (fig. 1, map). [1950.]

Botanical explorations, general features of flora, plant associations; briefly annotated list of vascular plants (299 species), the families arranged alphabetically; descriptions of numerous small areas intended for study of changes in vegetation in future years.

Massart, Jean. La dissémination des plantes alpines. Bul. Soc. Roy. Bot. Belg. 37 (1): 129–150. 1898.

Includes (p. 143–147) list of 87 spermatophytes observed by author on the Triftje and the Schwärze, two rocky "islands" at 2,650–3,000 meters elevation surrounded by glaciers, on the Breithorn and Jumeaux (Castor and Pollux) respectively.

Pannatier, Joseph. La florule du Val des Dix. Bul. Murithienne **31**: 116–149. 1902.

Topography, general features of flora, plant formations; annotated list of 338 vascular plants.—See also his Quelques notes d'herborisations dans le Val d'Hérens. Bul. Murithienne **32**: 190–197. 1903.—Also **Jaccard, Henri.** Notes sur l'herborisation dans les vallées d'Hérens et d'Hérémence. l.c. **33**: 69–76. 1 fig. 1904.—Also **Meylan, Olivier.** Herborisation au Val des Dix et régions voisines du 21 au 23 juillet 1931. l.c. **49**: 51–53. 1932.—Also **Wilczek, Ernest.** Le végétation phanérogamique des environs de la cabane du Val des Dix. l.c. **49**: 81–82. 1932. (List of vascular plants collected at 2,930 meters altitude.)

Stebler, F. G. Flora von Törbel. Bul. Murithienne **45**: 51–93. 1928.

Topography, climate, geology; annotated list of vascular plants, the commoner species omitted; plant associations.

Tissière, P. G. Guide du botaniste sur le Grand St-Bernard. Bul. Murithienne v. 1, [Suppl.]. 117 p. pl. 19 cm. 1868.

Topography, climate, etc.; annotated list of vascular plants, with altitudinal range.—See also **Favre, Émile.** Supplément au Guide . . . Bul. Murithienne 3/4: 41–78. 1876. (Extensive list.)—Also **Gandoger, Michel.** Excursion botanique au Grand Saint-Bernard . . . Bul. Soc. Bot. France **32**: 223–229. 1885.—Also his Deuxième voyage botanique au Grand-Saint-Bernard . . . l.c. **40**: 385–390. 1893.—Also papers by **Wilczek, Ernest.** l.c. 41 (Sess. Extraord.): cclxvi–cclxx. 1895; by **Beauverd, Gustave.** Mém. Herb. Boissier. **21**: 95–96. 1900; Bul. Murithienne **32**: 13–60. 1903; by **Melly, C.** l.c. **35**: 49–54. 1909; by **Chodat, Robert.** Bul. Soc. Bot. Genève II, **19**: 280–281. [1928.]

Whymper, Edward. Scrambles amongst the Alps in the years 1860–69. 5. ed. xviii, 468 p. illus., pls., maps. 22.5 cm. London, 1900. (1st ed. 1871.)

Includes (footnote, p. 98, in chap. 5) list of 9 phanerogams collected by author at 10,500 to nearly 13,000 feet altitude on the Matterhorn, with lists of species from other elevated localities in Europe.—See also **Lindt, Rudolf.** Eine besteigung des Matterhorns. Jahrb. Schweiz. Alpenclub **10**: 250–282. 1 fig. 1875. (Adds, on p. 271–272, several phanerogams and a few lichens to Whymper's list.)

Wilczek, Ernest. Le voyage botanique de MM. E. Burnat, J. Briquet, Fr. Cavillier, E. Wilcez et Abrezol, dans le Valais supérieur, de Brigue à la Furka. Bul. Murithienne **39**: 301–344. 1916.

Includes (p. 317–331) lists of rarer vascular plants collected between Grengiols and Oberwald, in the Rappental, the Blindental, and the Gehrental.

Wurgler, W. La végétation adventice des fraisières valaisannes. Landw. Jahrb. Schweiz **67**: 917–924. 1953.

Extensive tabular list of adventive plants observed in strawberry gardens in the districts of Conches, Sierre, Sion, Cothey, Entremont, Martigny, and Saint-Maurice.

VAUD (WAADT)

See also France, Doubs (Favre).

General

Durand, Théophile, and Pittier, Henri. Catalogue de la flore vaudoise. Bul. Soc. Roy. Bot. Belg. 20 (1): 7–226. 1881; 21 (1): 197–328. 1882. (Reprinted 549 p. Lausanne, 1882.)—[Supplément.] l.c. 25 (1): 187–343. 1886. (Reprinted 250 p. Lausanne, 1887.)

Botanical investigations, bibliography, list of collectors; annotated list of about 1,872 species of vascular plants (1,704 native); phytogeography, botanical regions. Replaces **Rapin, Daniel.** Guide du botaniste dans le canton de Vaud, comprenant en outre le bassin de Genève et le cours inférieur du Rhône en Valais. 2. éd. xxiv, 772 p. 18 cm. Genève et Paris, 1862. (1st ed. 1842.) (Keys to genera on Linnaean system; annotated descriptive flora of vascular plants.)—See also **Jaccard, Henri.** Notes et additions concernant la flore vaudoise. Bul. Murithienne **27/28:** 252–260. 1900.—Also **Beauverd, Gustave.** Additions au Catalogue . . . Bul. Herb. Boissier II, **4:** 1174–1179. 1904.—Also **Jaccard, Henri.** Additions pour la régions des Alpes, et spécialement le bassin Sarinien au Catalogue . . . Bul. Murithienne **33:** 116–146. 1904 (1905).

Local

Aubert, Samuel. La flore de la vallée de Joux . . . Bul. Soc. Vaud. Sci. Nat. **36:** 327–741. 1 fig., pl. 15–20 (incl. map). 1900. (Reprinted 1901.)

Topography, climate, geology, plant formations, phytogeography, etc.; annotated tabular list of vascular plants showing vertical range, localities, habitat, etc.; bibliography.—See also his Notes complémentaires sur La flore . . . Bul. Soc. Vaud. Sci. Nat. **40:** 21–24. 1904; Additions et observations relatives à La flore . . . l.c. **57:** 237–244. 1931.—Also his Plantes étrangères introduites par semis à la vallée de Joux. l.c. **57:** 593–595. 1932. (List of nonindigenous species of which seeds were scattered by an amateur botanist during a period of 40 years.)—Also his Plantes étrangères introduites à la vallée de Joux. Bul. Soc. Bot. Genève II, **38:** 55–62. 1947. (List of principal surviving species.)

Bonner, C. E. B. Étude floristique du Jorat. 128 p. 1 fig. (map). 24.5 cm. London, 1940. (Thèse, Faculté des sciences, Université de Lausanne.)

Geography, topography, climate, forests, plant associations, bibliography, list of collectors; annotated list of vascular plants (*Rubus* omitted), bryophytes, and lichens, with localities. (A plateau northeast of Lausanne.)

Chastain, André. Inventaire floristique de la tourbière de La Vraconnaz (Haut-Jura vaudois). Liste systématique et répartition des espèces. Bul. Soc. Neuchâtel. Sci. Nat. **75:** 45–72. 1952.

Botanical explorations, etc.; tabular list of cellular and vascular plants (530 species and varieties, of which 232 are vascular) showing occurrence in different habitats. The bog is northwest of Sainte-Croix and about 12 kilometers from Fleurier. The same floristic material, presumably, is contained in his Recherches écologiques et floristiques sur le "Pinetum" de la haute-tourbière de La Vraconnaz (Jura helvétique vaudois). Recueil Trav. Inst. Bot. Univ. Montpellier, Suppl. 2. 176 p. illus., pls., map, tab. 1952. (Topography, geology, climate, botanical explorations; list of vascular and cellular plants (530 species, of which 225 are spermatophytes); plant associations, biological spectrum, etc. Not seen; cited from Arch. Bot. (Forlì) **30:** 61. 1954.)

Corboz, François. Flora aclensis, ou catalogue des plantes de la flore suisse qui j'ai trouvées sur le territoire d'Aclens, de 1872 à 1885. Bul. Soc. Vaud. Sci. Nat. II, **22:** 278–301. map. 1887.—Flora aclensis . . . [suppléments]. l. c. **29:** 97–136. 1893; **31:** 227–246. 1895; **35:** 49–60. 1899; **36:** 170–197. 1900. **39:** 211–232. 1903.

Annotated list of vascular plants; the supplements include lists of cellular plants. The 1903 supplement brings the total to 917 species of vascular plants (including cultivated species) and 879 cellular plants, from an area of about 4 sq. km. at 400–518 meters altitude.

Cruchet, Émilie. La flore rudérale et adventice de Lausanne et de ses environs. Mém. Soc. Vaud. Sci. Nat. **4:** 273–328. 1933.

Discussion of areas studied, with classified lists of species; annotated list of species; discussion of sources of adventive plants, bibliography.

Dutoit, Daniel. Contribution a l'étude de la végétation du massif de Naye sur Montreux. Mém. Soc. Vaud. Sci. Nat. **4:** 365–413. 1934.

Mainly ecological; includes sketch of botanical explorations (p. 367–369) and list of 33 species not previously recorded from the area (p. 370–371).

Vetter, J. J. Quelques notes sur la flore des environs d'Orbe. Bul. Soc. Vaud. Sci. Nat. II, **22**: 268–277. 1887.

Includes list of 97 adventive plants, as well as a list of new localities for native species.

Villaret, Pierre. Étude floristique de la vallée d'Anzeindaz. 264 pl. 8 pl., map, tab. Lausanne, 1956. (Thesis).

Annotated list of vascular plants. (Not seen; title and annotation supplied by H. P. Fuchs.)

ZÜRICH (ZURICH)

See also Aargau (Boll).

General

Kölliker, Albert. Verzeichnis der phanerogamischen gewächse des cantons Zürich. xxv, 154 p. 19 cm. Zürich, 1839.

Bibliography, list of herbaria, topography, geology, climate; annotated list of 1,140 phanerogams; statistics, altitudinal limits.—See also **Naegeli, Otto.** Uber die botanische erforschung des kantons Zürich seit der flora von A. Kölliker (1839). Ber. Zürch. Bot. Gesell. **13**: 68–80. 1917. (Running account of additions to the flora (over 250 species and subspecies), grouped by regions. In New York Botanical Garden library.)

Naegeli, Otto. Ueber veränderungen der Zürcher flora im letzten jahrhundert in berücksichtigung der nachbargebiete. Vierteljahrsschr. Naturf. Gesell. Zürich **73** (Beibl. **15**, Festschr. Hans Schinz): 601–641. 1928.

General consideration of the problem of disappearance of species in a region; discussion of particular species and of special habitats and localities. —See also **Schmid, Emil.** Die veränderungen der flora von Zürich binnen der letzen 100 jahre. Jahrb. Zürichsee **14**: 69–85. 6 fig. 1952. (Popular account with mention of species of different habitats that have become rare or extinct.)

——— and **Thellung, Albert.** Die flora des kantons Zürich. 1. teil. Die ruderal- und adventivflora des kantons Zürich. Vierteljahrsschr. Naturf. Gesell. Zürich **50**: 225–305. 1905.

Historical account, classification; annotated list of ruderal and adventive plants (1,049 species and 20 hybrids).—See also **Rikli, Martin.** Die pteridophyten des kantons Zürich. (Flora des kantons Zürich II.) Ber. Zürch. Bot. Gesell. **11**: 14–61. 1912. (List of 40 taxa, with localities and collectors' names. Not seen; title and annotation supplied by A. Becherer.)

Rikli, Martin. Die quellen der Zürcherflora. (Flora des kantons Zürich II, A.) Ber. Zürch. Bot. Gesell. **11**: 3–13. 1912.

Bibliography relating to the flora of Zürich and adjacent regions. (Not seen; title and annotation supplied by A. Becherer.)

Schinz, H. R. Der kanton Zürich in naturgeschichtlicher und landwirtschaftlicher beziehung dargestellt. ii, 401 p. 8vo. Zürich, 1842.

Popular work including account of woody plants, poisonous and medicinal plants, weeds, etc. (Not seen; annotation taken from Ber. Zürch. Bot. Gesell. **10**: 87. 1907.)

Schröter, Carl. Die erforscher der Zürcherflora. I. Die zeit vor Albert Kölliker. Ber. Schweiz. Bot. Gesell. **16** (Ber. Zürch. Bot. Gesell. **10**): 80–93. 1907.

Sketch of botanists from Gesner to K. W. Nägeli, with mention of their principal publications, herbaria, etc.—See also his Vierhundert jahre botanik in Zürich . . . Verhandl. Schweiz. Naturf. Gesell. **99** (1917, 2): 3–28. table. 1918. (Sketch of botanists and botanical work from Gesner on; bibliography (historical and biographical references).)

Local

Hegi, Gustav. Das obere Tösstal und die angrenzenden gebiete, floristisch und pflanzengeographisch dargestellt. Bul. Herb. Boissier II, **1**: 179–212, 337–352, 533–548, 637–652, 689–736, 793–824, 913–944, 1041–1072, 1153–1200, 1233–1300. 1900–01; **2**: 49–108. 1901.—Nachträge zum ersten teil. l. c. **2**: 185–216. 1902.

Bibliography, herbaria; annotated list of 918 vascular plants (including Nachträge); physiography, geology, phytogeography, lists of introduced and cultivated plants, bibliography, list of mosses. (In Cantons of Zürich, St. Gallen, and Thurgau.)

Jäggi, Jacob. Eglisau in botanischer beziehung. *In* Taschenbuch für Eglisau. 50 p. 18.5 cm. Zürich, 1883.
List of less common species, with localities. (Not seen; title and annotation supplied by A. Becherer.)

Keller, Robert. Flora von Winterthur. 2 v. (185, 71 p.). 21.5 cm. Winterthur, 1891–96. (Arnold Arboretum library.)
Pt. 1: bibliography, etc.; annotated list of phanerogams (991 species); pt. 2: phytogeography, especially in relation to glaciation; annotated tabular list of introduced species; bibliography; corrections to pt. 1.

Oberholzer, Ernst. Die montanen und subalpinen pflanzen (mit einschluss einiger nordischer moorpflanzen) des Hohen Ron-gebietes. Ber. Schweiz. Bot. Gesell. **47**: 333–351. 1937.
Brief bibliography; annotated list of vascular plants; comparison with the Zürich Oberland (region between Ricken and Hörnli).

ZUG (ZOUG)

See also Schwyz (Rhiner, Die gefässpflanzen der Urkantone und von Zug; Rhiner, Volkstümliche pflanzennamen der Waldstätten).

General

Ribeaud, Emile. Die zugerischen pflanzennamen. *In* Jahresbericht der Kantonalen industrieschule und des Staedtischen gymnasiums in Zug. Schuljahr 1882–83, Beilage. p. 17–51. 1883.
List of vernacular names of native plants, with etymological annotations and indication of meaning. (Not seen; title and annotation supplied by W. Rytz.)

ABBREVIATIONS OF SERIALS CITED

Aberdeen Univ. Studies. Aberdeen university studies.
Acta Agralia Fenn. Acta agralia fennica. *Preceded on title page by Finnish title:* Suomen Maataloustieteellisen seuran julkaisuja.
Acta Borealia. A, Sci. Acta borealia. A, Scientia.
Acta Bot. Fenn. Societas pro fauna et flora fennica. Acta botanica fennica.
Acta Bot. Neerl. Acta botanica neerlandica.
Acta Forest. Fenn. Acta forestalia fennica . . . Publications of the Society of forestry in Finland. *At head of title page:* Suomen Metsätieteellinen seura—Finska forstsamfundet.
Acta Hort. Gothoburg.; Acta Hort. Gotoburg. Acta horti gothoburgensis (from v. 8 on, . . . gotoburgensis). Meddelanden från Göteborgs botaniska trädgård.
Acta Nat. Island. Acta naturalia islandica. *At head of title page:* Museum rerum naturalium reykjavikense. **Náttúrugripasafnið í Reykjavík.**
Acta Phytogeog. Suecica. Acta phytogeographica suecica edidit Svenska växtgeografiska sällskapet.
Acta Salmanticensia Filos. y Let. Acta salmanticensia iussu Senatus universitatis edita. Filosofía y letras.
Acta Soc. Fauna et Flora Fenn. Acta Societatis pro fauna et flora fennica.
Acta Soc. Scient. Fenn. Acta Societatis scientiarum fennicae.
Actas Soc. Españ. Hist. Nat. Actas de la Sociedad española de historia natural (*with its* Anales).
Actas y Mem. Primer Cong. Nat. Españ. (Zaragoza 1908). Actas y memorias del Primer congreso de naturalistas españoles, celebrado en Zaragoza los dias 7–10 de octubre de 1908. Zaragoza, 1909. (Not seen; title taken from Bot. Jahresber. Just 38 (2): 1452. 1915.)
Actes Acad. Roy. Sci., Belles-Let. et Arts Bordeaux. Actes de l'Académie royales des sciences, belles-lettres et arts de Bordeaux.
Actes Congr. Internatl. Bot. Bruxelles. Actes du 3. Congrès international de botanique tenu à . . . Bruxelles [1912].
Actes Soc. Jurass. Émul. Actes de la Société jurassienne d'émulation.
Actes Soc. Linn. Bordeaux. Actes de la Société linnéenne de Bordeaux.
Adv. Sci. The advancement of science. The report of the British association for the advancement of science.
Agr.-Ekon. Försöksanst. Landtmannaskr. [Finland] Agrikultur-ekonomiska försöksanstaltens landtmannaskrifter. Helsingfors.
Agricoltura (Napoli). Agricoltura. Bollettino mensile della scienza e practica agricola . . . Napoli.
Agricultura. (Madrid). Agricultura. Revista agropecuaria. Madrid.
Agron. Lusitana. Agronomia lusitana. Estação agronómica nacional, Portugal.
Agros. Agros. **Boletim dos estudantes do Instituto superior de agronomia.** Lisboa.
Allg. Bot. Zeitschr. Allgemeine botanische zeitschrift.
Allionia. Allionia.
Almanacco Agr. Trento. Almanacco agrario trento.
Almanaque Agr. Ceres. Almanaque agrícola Ceres.
Almanaque de Ponte do Lima. Almanaque de Ponte do Lima. (Not seen.)
Alpe. L'Alpe. Rivista forestale italiana . . .
Amer. Bot. American botanist.
Amer. Inst. Crop Ecol. Internatl. Agro-Climat. Ser. Study. American institute of crop ecology. International agro-climatological series. Study.
Anais Fac. Ciên. Pôrto. Anais da Faculdade de ciências do Pôrto.
Anais Fac. Farm. Porto. Anais da Faculdade de farmácia do Porto.
Anais Inst. Super. Agron. (Lisboa). Anais do Instituto superior de agronomia. Lisboa.
Anais Inst. Vinho Porto. Anais do Instituto do vinho do Porto. (Not seen.)

Anales Edafol. y Fisiol. Veg. Anales de edafología y fisiología vegetal. Madrid.
Anales Escuela Peritos Agr. y Super. Agr. (Barcelona). Anales de la Escuela de peritos agrícolas y superior de agricultura y de los servicios técnicos de agricultura. Barcelona.
Anales Inst. Bot. A. J. Cavanilles. Anales del Instituto botánico A. J. Cavanilles. *Formerly* (vol. 1–9) Anales del Jardín botánico de Madrid.
Anales Jard. Bot. Madrid. Anales del Jardín botánico de Madrid. *Later* (*from* v. 10) Anales del Instituto botánico A. J. Cavanilles.
Anales R. Acad. Farm. Madrid. Anales de la Real academia de farmacia. Madrid.
Anales Soc. Españ. Hist. Nat. Anales de la Sociedad espanñola de historia natural.
Ankkapurha. Ankkapurha. (Not seen.)
Ann. Rpt. and Proc. Belfast Nat. Field Club. Annual report and proceedings of the Belfast naturalists' field club.
Ann. Rpt. and Proc. Bristol Nat. Soc. Annual report and proceedings of the Bristol naturalists' society; *later* Proc., q. v.
Ann. Rpt. and Proc. Chester Soc. Nat. Sci., Lit. and Art. Chester society of natural science, literature and art. Annual report and proceedings.
Ann. Rpt. and Trans. North Staffordsh. Nat. Field Club. North Staffordshire naturalists' field club and archaeological society. Annual report and transactions.
Ann. Rpt. and Trans. Plymouth Inst. and Devon and Cornwall Nat. Hist. Soc. Annual reports and transactions of the Plymouth institution and Devon and Cornwall natural history society.
Ann. Rpt. Dulwich Col. Sci. Soc. Annual report of the Dulwich college science society.
Ann. Rpt. Felsted School Nat. Sci. Soc. Annual report of the Felsted school natural science society. (Not seen.)
Ann. Rpt. Harrow Nat. Hist. Soc. The annual report of the Harrow natural history society.
Ann. Rpt. Huntingdonsh. Fauna and Flora League. Annual report of the Huntingdonshire fauna and flora league. (Not seen.)
Ann. Rpt. Isle of Thanet Field Club. Annual Report of Isle of Thanet field club. (Not seen.)
Ann. Rpt. Liverpool Mar. Biol. Sta. Annual report of the Liverpool marine biological station on Puffin Island.
Ann. Rpt. Oundle School Nat. Hist. Soc. The annual report of the Oundle school natural history society.
Ann. Rpt. Roy. Cornwall Polytech. Soc. Royal Cornwall polytechnic society. Annual report.
Ann. Rpt. South London Micros. and Nat. Hist. Club. Annual report of the South London microscopical and natural history club.
Ann. Rpt. Wellington Col. Nat. Hist. Soc. Annual report of Wellington college natural history society.
Annaes Sci. Nat. Annaes de sciencias naturaes.
Annalas Soc. Retorumantscha. Annalas de la Società retorumantscha.
Annales Acad. Lovan. Annales Academiae lovaniensis.
Annales Acad. Lugd.-Bat. Annales Academiae lugduno-batavae.
Annales Acad. Scient. Fenn. Annales Academiae scientiarum fennicae. *Preceded on title page by Finnish title:* Suomalaisen tiedeakatemian toimituksia.
Annales Bot. Soc. Zool.-Bot. Fenn. Vanamo. Annales botanici Societatis zoologicae-botanicae fennicae "Vanamo." *Preceded* (v. 1–24) *on title page by Finnish title:* Suomalaisen eläin- ja kasvitieteellisen Seuran Vanamon kasvitieteellisiä julkaisuja.
Annales École Natl. Supér. Agron. Toulouse. Annales de l'École nationale supérieure agronomique. Faculté des sciences de Toulouse.
Annales Fac. Sci. Marseille. Annales de la Faculté des sciences de Marseille. *At head of title page:* Publications de l'Université d'Aix-Marseille.
Annales Gembloux. Annales de Gembloux. Organe trimestriel de l'Association des ingénieurs sortis de l'Institut agronomique de l'État à Gembloux.
Annales Géog. Annales de géographie. Paris.

Annales Mus. Hist. Nat. Marseille. Annales du Musée d'histoire naturelle de Marseille.
Annales Sci. Nat. Bot. Annales des sciences naturelles . . . Botanique.
Annales Scient., Lit. et Indus. Auvergne. Annales scientifiques, littéraires et industrielles de l'Auvergne.
Annales Scient. Univ. Besançon. Annales scientifiques de l'Université de Franche-Comté de Besançon. *Earlier* (v. 1–4) Annales scientifiques.
Annales Soc. Acad. Nantes et Loire-Infér. Annales de la Société académique de Nantes et du département de la Loire-Inférieure.
Annales Soc. Agr., Indus., Sci., Arts et Belles-Let. Dépt. Loire. Annales de la Société d'agriculture, industrie, sciences, arts et belles-lettres du département de la Loire.
Annales Soc. Agr., Sci. et Indus. Lyon. Annales de la Société d'agriculture, sciences et industrie de Lyon.
Annales Soc. Bot. Lyon. Annales de la Société botanique de Lyon.
Annales Soc. Émul. (Agr., Sci. Let. et Arts) Ain. Annales de la Société d'émulation (et d'agriculture, lettres et arts) de l'Ain. *Title varies somewhat.*
Annales Soc. Émul. Dépt. Vosges. Annales de la Société d'émulation du département des Vosges.
Annales Soc. Hist. Nat. Toulon. Annales de la Société d'histoire naturelle de Toulon.
Annales Soc. Hort. et Hist. Nat. Hérault. Annales de la Société d'horticulture et d'histoire naturelle de l'Hérault.
Annales Soc. Hort. Haute-Garonne. Annales de la Société d'horticulture de la Haute-Garonne. (Not seen.)
Annales Soc. Let., Sci. et Arts Alpes-Maritimes. Annales de la Société des lettres, sciences et arts des Alpes-Maritimes.
Annales Soc. Linn. Dépt. Maine-et-Loire. Annales de la Société linnéenne du département de Maine-et-Loire.
Annales Soc. Linn. Lyon. Annales de la Société linnéenne de Lyon.
Annales Soc. Lit., Scient. et Artist. d'Apt. Annales de la Société littéraire, scientifique et artistique d'Apt (Vaucluse).
Annales Soc. Méd.-Chirurg. Bruges. Annales de la Société médico-chirurgicale de Bruges.
Annales Soc. Sci. Nat. Charente-Infér. Annales de la Société des sciences naturelles de la Charente-Inférieure.
Annales Soc. Sci. Nat. Toulon et du Var. Annales de la Société des sciences naturelles de Toulon et du Var. (Not seen.) *Successor to* **Annales de la** Société d'histoire naturelle de Toulon.
Annales Soc. Sci. Phys. Nat. Agr. et Indus. Lyon. Annales des sciences physiques et naturelles, d'agriculture et d'industrie, publiées par la Société impériale d'agriculture, etc., de Lyon.
Annales Soc. Scient. Bruxelles. Annales de la Société scientifique de Bruxelles. Série B, Sciences physiques et naturelles.
Annales Soc. Zool.-Bot. Fenn. Vanamo. Annales Societatis zoologicae-botanicae fennicae Vanamo. *Preceded on title page by Finnish title:* Suomalaisen eläin- ja kasvitieteellisen Seuran Vanamon julkaisuja.
Annales Trav. Pub. Belg. Annales des travaux publics de Belgique.
Annales Univ. Grenoble. Annales de l'Université de Grenoble.
Annali Accad. Aspir. Nat. Napoli. Annali dell'Accademia degli aspiranti naturalisti di Napoli. (Not seen.)
Annali Assoc. Agr. Friulana. Annali dell'Associazione agraria friulana. (Not seen.)
Annali Bot. Annali di botanica.
Annali Club Alp. Ticinese. Annali del Club alpino ticinese. (Not seen.)
Annali Fac. Agr. Portici R. Univ. Napoli. Annali della Facoltà di agraria di Portici della R. università di Napoli.
Annali Fac. Agr. R. Univ. Pisa. Annali (*già* Bollettino) della Facoltà di agraria della R. università. Pisa.
Annali Fac. Agr. Univ. Perugia. Università degli studi di Perugia. Annali della Facoltà di agraria.
Annali Min. Agr., Indus. e Com. Annali del Ministero di agricoltura, industria e commercio (Firenze).

Annali Mus. Civ. Storia Nat. Genova. Annali del Museo civico di storia naturale di Genova. *From* v. 47, Annali del Museo civico di storia naturale Giacomo Doria. Genova.
Annali Mus. Civ. Storia Nat. Giacomo Doria (Genova). Annali del Museo civico di storia naturale Giacomo Doria. Genova.
Annali R. Accad. Agr. Torino. Annali della R. accademia d'agricoltura di Torino.
Annali R. Ist. Super. Forest. Naz. Firenze. Annali del R. istituto superiore forestale nazionale. Firenze.
Annali R. Ist. Tecn. e Naut. Napoli. Annali del R. istituto tecnico e nautico di Napoli. (Not seen.)
Annali R. Scuola Super. Agr. Portici. Annali della Regia scuola superiore di agricoltura di Portici.
Annali Scuola Friulana. Annali della Scuola friulana. (Not seen.)
Annali Univ. Ferrara Bot. Annali dell'Università di Ferrara. Sezione IV. Botanica.
Annals and Mag. Nat. Hist. Annals and magazine of natural history.
Annals Andersonian Nat. Soc. Annals of the Andersonian naturalists' society.
Annals Kilmarnock Glenfield Ramblers' Soc. Annals of the Kilmarnock Glenfield ramblers' society.
Annals Scott. Nat. Hist. Annals of Scottish natural history.
Annuaire Club Alpin Franç. Annuaire du Club alpin français.
Annuaire Conserv. et Jard. Bot. Genève. Annuaire du Conservatoire et du Jardin botaniques de Genève.
Annuaire du Pas-du-Calais. Annuaire du Pas-du-Calais.
Annuaire Normand. Annuaire des cinq départements de la Normandie publié par l'Association normande. *Half-title:* Annuaire normand.
Annuaire Soc. Agr. Hérault. Annuaire de la Société d'agriculture de l'Hérault.
Annuario Acad. Polytech. Porto. Annuario da Academia polytechnica do Porto.
Annuario Assoc. Agr. Friulana. Annuario dell'Associazione agraria friulana. (Not seen.)
Annuario R. Ist. Bot. Roma. Annuario del R. istituto botanico di Roma.
Annuario Soc. Alp. Trident. Annuario della Società degli alpinisti tridentini.
Anuari Assoc. Excurs. Catalana. Anuari de la Associació d'excursions catalana. (Not seen.)
Anuari Junta Cièn. Nat. Barcelona. Junta de ciències naturals. Anuari. Barcelona.
Anuário Soc. Broteriana. Anuário da Sociedade broteriana.
Árbók Ferdafélag Íslands. Ferdafélag Islands. Árbók.
Årbok (*earlier* Årsberet.) K. Norske Vidensk. Selsk. Mus. Det Kongelige norske videnskabers selskab museet. Årbok (*to and including* 1950 Årsberetning).
Arch. Bot. (Caen) Mém. Archives de botanique. Publiées par René Viguier . . . de Caen . . . Mémoires.
Arch. Bot. (Forlì). Archivio botanico per la sistematica, fitogeografia e genetica . . . Forlì. *From* v. 32 (1956), Archivio botanico e biogeografico italiano.
Arch. Farmacog. e Sci. Aff. Archivio di farmacognosia e scienze affini . . . Roma.
Arch. Fl. Jurass. Archives de la flore jurassienne.
Arch. Hydrobiol. u. Planktonkunde. Archiv für hydrobiologie und planktonkunde. (Neue folge d. Forschungsberichte a.d. Biolog. station z. Plön.)
Arch. Inst. Bot. Univ. Liège. Archives de l'Institut botanique de l'Université de Liège. *Later* Université de Liège. Belgique. Archives de l'Institut de botanique.
Arch. Math. og Naturv. Archiv for mathematik og naturvidenskab . . . Kristiania.
Arch. Sci. Phys. et Nat. Genève. Archives des sciences physiques et naturelles. *At head of title page:* Bibliothèque universelle.
Arch. Soc. Zool. Bot. Fenn. Vanamo. Archivum Societatis zoologicae botanicae fennicae "Vanamo." *Preceded by Finnish title:* Suomalaisen eläin- ja kasvitieteellisen Seuran Vanamon tiedonannot ja pöytäkirjat. *Beginning with* v. 6 (1951) *the Latin title comes first and the Finnish title is altered by dropping the words* "ja pöytäkirjat."

Arch. Tradiz. Pop. Archivio delle tradizioni popolari.
Arch. Trim. Sect. Sci. Nat., Phys. et Math. Inst. Grand-ducal Luxemb. Institut grand-ducal de Luxembourg. Section des sciences naturelles, physiques et mathématiques. Archives trimestrielles.
Arch. Zeeuwsch Genootsch. Wetensch. Archief. Vroegere en latere mededeelingen voornamelijk in betrekking tot Zeeland uitgegeven door het Zeeuwsch genootschap der wetenschappen. Middelburg.
Arkiv Bot. Arkiv för botanik. Utgifvet af K. svenska vetenskapsakademien.
Arkiv Svenska Österbotten. Arkiv för Svenska Österbotten.
Arquivo Univ. Lisboa. Arquivo da Universidade de Lisboa.
Ársrit Skógræktarfélags Íslands. Ársrit Skógæktarfélags Íslands.
Årsskr. Uppsala Univ. Uppsala Universitets årsskrift.
Arvernia. Arvernia. Supplément *to* Revue d'Auvergne.
Arx. Centre Excurs. Terrassa. Centre excursionista de Terrassa. Arxiu. (Not seen.)
Arx. Inst. Estud. Catalans Secc. Cièn. Institut d'estudis catalans. Arxius de la Secció de ciències.
Atti Accad. Gioen. Sci. Nat. Catania. Atti dell'Accademia gioenia di scienze naturali in Catania.
Atti Accad. Ligure Sci. e Let. Atti della Accademia ligure di scienze e lettere . . .
Atti Accad. Pontan. (Napoli). Atti dell' Accademia pontaniana (Napoli).
Atti Accad. Pontif. Nuovi Lincei. Atti dell'Accademia pontifica de' nuovi lincei. (Title varies.)
Atti Accad. Sci. Fis. e Mat. Soc. R. Napoli. Società reale di Napoli. Atti dell'Accademia delle scienze fisiche e matematiche.
Atti Accad. Scient. Veneto-Trent.-Istriana. Atti della Accademia scientifica veneto-trentino-istriana.
Atti Accad. Spoletina. Atti della Accademia spoletina.
Atti Accad. Udine. Atti della Accademia di Udine.
Atti Ateneo Sci. Let. ed Arti Bergamo. Atti dell'Ateneo di scienze lettere ed arti. *See* Bergomum.
Atti Ateneo Treviso. Atti dell'Ateneo di Treviso. (Not seen.)
Atti Cong. Bot. Internatl. Genova, 1892. Atti del Congresso botanico internazionale di Genova.
Atti Cong. Limnol. Teor. ed Applic. Atti del IV. Congreso de limnologia teoretica ed applicata. Roma 1929. (Not seen.)
Atti e Mem. Accad. Agr., Sci. e Let. Verona. Atti e memorie dell'Accademia di agricoltura, scienze e lettere di Verona.
Atti e Mem. R. Accad. Sci., Let. ed Arti Modena. R. accademia di scienze, lettere ed arti in Modena. Atti e memorie.
Atti e Mem. Soc. Agrar. Reggio nell'Emilia. Atti e memorie della Società agraria di Reggio nell'Emilia.
Atti e Rend. Accad. Sci. Let. ed Arti Zelanti Acireale. Atti e rendiconti dell' Accademia di scienze, lettere ed arti dei zelanti di Acireale.
Atti. I. R. Accad. Sci., Let. ed Arti Rovereto. Atti della I.R. accademia di scienze, lettere ed arti degli agiati in Rovereto; *earlier* Atti della R. accademia . . . *later* Accademia roveretana . . .
Atti Ist. Bot. (Giov. Briosi) R. Univ. Pavia. Atti dell'Istituto botanico ("Giovanni Briosi") e laboratorio crittogamico italiano della R. università di Pavia.
Atti Mus. Civ. Storia Nat. Trieste. Atti del Museo civico di storia naturale di Trieste.
Atti R. Accad. Econ.-Agr. Georg. Firenze. Atti della Reale accademia economico-agraria dei georgofili di Firenze.
Atti R. Accad. Fisiocr. Siena. Atti della R. accademia dei fisiocriti in Siena.
Atti R. Accad. Lucchese Sci., Let. ed Arti. Atti della Reale accademia lucchese di scienze, lettere ed arti.
Atti R. Accad. Peloritana. Atti della R. accademia peloritana.
Atti R. Accad. Sci. Fis. e Mat. Napoli. Atti della Reale accademia delle scienze fisiche e matematiche. *At head of title*: Società reale di Napoli.
Atti R. Accad. Sci., Let. ed Arti Rovereto. Atti dell' I. R. accademia di scienze, lettere ed arti degli agiati in Rovereto.
Atti R. Accad. Sci. Napoli. Atti della Reale accademia delle scienze, sezione della Società reale borbonica. Napoli.

Atti R. Accad. Sci. Torino. Atti della Reale accademia delle scienze di Torino.

Atti R. Ist. Incoragg. Napoli; Atti R. Ist. Incoragg. Sci. Nat. Napoli. Atti del Reale istituto d'incoraggiamento alle scienze naturali di Napoli; *later omits* alle scienze naturali.

Atti R. Ist. Veneto Sci., Let. ed Arti. Atti del Reale istituto veneto di scienze, lettere ed arti.

Atti Soc. Acclim. e Agr. Sicilia. Atti della Società di acclimazione e di agricoltura in Sicilia.

Atti Soc. Ital. Prog. Sci. Atti della Società italiana per il progresso delle scienze.

Atti Soc. Ital. Sci. Nat. Milano. Atti della Società italiana di scienze naturali e del Museo civico di storia naturale in Milano.

Atti Soc. Ligust. Sci. Nat. e Geog. Atti della Società ligustica di scienze naturali e geografiche.

Atti Soc. Nat. (e Mat.) Modena. Atti della Società dei naturalisti (*later* e matematici) di Modena.

Atti Soc. Toscana Sci. Nat. Pisa. Atti della Società Toscana di scienze naturali residente in Pisa. Memorie; Processi verbali.

Atti Soc. Veneto-Trent. Sci. Nat. Padova. Atti della Società veneto-trentina di scienze naturali residente in Padova.

Atvinnudeild Háskólans (Reykjavík) Rit Landbúnaðardeildar. Atvinnudeild háskólans. Rit landbúnaðardeildar. Reykjavík.

Avernia. Avernia.

Avhandl. Norske Vidensk.-Akad. Oslo I. Mat.-Naturv. Kl. Avhandlinger utgitt av det Norske videnskaps-akademi i Islo I. Matematisk-naturvidenskapelig klasse.

Bedfordshire Nat. The Bedfordshire naturalist. Being the journal of the Bedforshire natural history society and field club. (*Vol.* 1 *as* Journal of the Bedfordshire natural history society and field club.)

Beih. Bot. Centralbl. *See* Bot. Centralbl. Beih.

Beih. Repert. Spec. Nov. Fedde. *See* Repertorium specierum novarum regni vegetabilis hrsg. von F. Fedde. Beihefte.

Beitr. Geobot. Landesaufn. Pflanzengeog. Kom. Schweiz. Naturf. Gesell. Beiträge zur geobotanischen landesaufnahme hrsg. von der Pflanzengeographischen kommission der Schweizerischen naturforschenden gesellschaft.

Belg. Hort. Belgique horticole.

Ber. Bayer. Bot. Gesell. Berichte der Bayerischen botanischen gesellschaft zur erforschung der heimischen flora.

Ber. Geobot. Forschungsinst. Rübel Zürich. Bericht über das Geobotanische forschungsinstitut Rübel in Zürich.

Ber. Naturw.-Med. Ver. Innsbruck. Berichte des Naturwissenschaftlich-medizinischen vereines in Innsbruck.

Ber. Realschule Basel. Bericht der Realschule zu Basel. (Not seen.)

Ber. Schweiz. Bot. Gesell. Berichte der Schweizerischen botanischen gesellschaft.

Ber. St. Gall. Naturw. Gesell. Bericht über die thätigkeit der St. Gallischen naturwissenschaftlichen gesellschaft.

Ber. Ver. Schutze u. Pflege Alpenpfl. Bericht des Vereins zum schutze und zur pflege der alpenpflanzen. (Title varies somewhat.)

Ber. Zürch. Bot. Gesell. Bericht der Zürcherischen botanischen gesellschaft.

Bergens Mus. Aarb. (Årb.) Bergens museums aarbog (*from* 1927 årbok). Naturvidenskelig raekke (*from* 1941 Naturvitenskapelig rekke).

Bergens Mus. Skr. Bergens museums skrifter.

Bergomum. Bollettino della Civica biblioteca. Studi di storia e d'arte . . . Istituto italiano d'arti grafiche, Bergamo. (*Includes, in the numbers examined,* Atti dell'Ateneo di scienze, lettere ed arti, separately paged.)

Bibl. Contr. Lloyd Libr. Bibliographical contributions from the Lloyd library, Cincinnati, Ohio.

Bibl. Ital. Bibliografia italiana.

Bibl. Schweiz. Naturw. Lit. Bibliographie der schweizerischen naturwissenschaftlichen literatur. Herausgegeben von der Schweizer. landesbibliotek. Bibliographie scientifique suisse. Publiée par la Bibliothèque nationale suisse. *Title changed with v.* 19 *to* Bibliographie der schweizerischen naturwissenschaftlichen und geographischen literatur.

Bibl. Schweiz. Naturw. und Geog. Lit. Bibliographie der schweizerischen naturwissenschaftlichen und geographischen literatur. Herausgegeben von der Schweizer. landesbibliotek. (Also has French title.) *Earlier* Bibliographie der schweizerischen naturwissenschaftlichen literatur . . . *With v.* 24, *title changed to* Bibliographia scientiae naturalis helvetica.

Bibl. Sci. Nat. Helvetica. Bibliographia scientiae naturalis helvetica. Herausgegeben von der Schweiserischen landesbibliotek. *Earlier* Bibliographie der schweizerischen naturwissenschaftlichen und geographischen literatur.

Biblioteca Ital. Biblioteca italiana o sia giornale di letteratura, scienze ed arti compilato da varj letterati. Milano.

Bidr. Finlands Naturkänn., Etnog. och Statis. Bidrag till Finlands naturkännedom, etnografi och statistik.

Bidr. Känn. Finlands Natur och Folk. Bidrag till kännedom af Finlands natur och folk.

Bihang K. Svenska Vetensk.-Akad. Handl. Bihang till Kongl. svenska vetenskaps-akademiens handlingar.

Bijd. Nat. Wetensch. Bijdragen tot de natuurkundige wetenschappen.

Biol. Abst. Biological abstracts.

Blyttia. Blyttia. Norsk botanisk forenings tidsskrift.

Boissiera. Boissiera. Mémoires du Conservatoire de botanique et de l'Institut de botanique systématique del l'Universite de Genève . . . (Supplément de Candollea.)

Bol. Ann. Club Alp. Bassanese. Bollettino annuale del Club alpino bassanese. (Not seen.)

Bol. Camera Com. Modena. Bolletino della Camera di comercio di Modena. (Not seen.)

Bol. Club Alp. Ital. Bollettino del Club alpino italiano.

Bol. Comit. Glaciologico Ital. Bollettino del Comitato glaciologico italiano.

Bol. Comit. "Pro piante medicinali" di Sassari. Bolletino del Comitato "Pro piante medicinali" di Sassari. (Not seen.)

Bol. Consorzio Agr. Reggiano. Bollettino del Consorzio agrario reggiano. (Not seen.)

Bol. Direcç. Geral Agr. (Lisboa). [Portugal.] Ministério das obras publicas, commercio e industria. Boletim da Direcção geral da agricultura. Lisboa.

Bol. Estaç. Agr. Cent. (Lisboa). Estação agrária central. Boletim. *Publication of* Direçcão geral dos serviços agricolos. Ministério da agricultura. Lisboa.

Bol. Farm. Militar. Boletín da farmacia militar. Madrid.

[Bol.] Inst. Forest. Invest. y Exper. Madrid. Instituto forestal de investigaciones y experiencias. Madrid [Boletín.]

Bol. Min. Agr. (Lisboa) Boletim do Ministério da agricultura publicado pela Direcção de instrução agrícola. Lisboa. *Earlier (through v.* 1, *no.* 2/4) Boletim de Secretaria de estado de agricultura publicado . . .

Bol. Nat. Sup. Riv. Ital. Sci. Nat. Bollettino del naturalista . . . supplemento mensile alla Rivista italiana di scienze naturali.

Bol. R. Ist. Sup. Agr. Pisa. Bollettino del R. Istituto superiore agrario di Pisa.

Bol. R. Orto Bot. (e. Giard. Colon.) Palermo. Bollettino del R. orto botanico (*later* [*with v.* 6) R. orto botanico e giardino coloniale) di Palermo.

Bol. Sec. Est. Agr. (Lisboa). Boletim da Secretaria de estado da agricultura publicado pela Direcção da instrução agrícola. Lisboa.

Bol. Soc. Adriatica Sci. Nat. Trieste. Bollettino della Società adriatica di scienze naturali in Trieste.

Bol. Soc. Aragon. Cien. Nat. Boletín de la Sociedad aragonesa de ciencias naturales.

Bol. Soc. Broteriana. Boletim da Sociedade broteriana. (*Vol.* 1–3 *as* Sociedade broteriana. Boletim annual.)

Bol. Soc. Españ. Hist. Nat. Boletin de la (*for a time,* Real) Sociedad española de historia natural.

Bol. Soc. Geog. Ital. Bolletino della Società geografica italiana.

Bol. Soc. Geog. Lisboa. Boletim da Sociedade de geografia de Lisboa.

Bol. Soc. Ibérica Cien. Nat. Boletín de la Sociedad ibérica de ciencias naturales. *Earlier* (*through v.* 17) Boletín de la Sociedad aragonesa de ciencias naturales.

Bol. Soc. Nat. Napoli. Bollettino della Società di naturalisti in Napoli.
Bol. Soc. Portug. Ciênc. Nat. Boletim da Sociedade portuguesa de ciências naturais. (Also has French title.)
Bol. Soc. Ticinese Sci. Nat. Bollettino della Società ticinese di scienze naturali.
Bol. Storico Svizzera Ital. Bollettino storico della Svizzera italiana.
Bol. Univ. Santiago de Compostela. Boletín de la Universidad de Santiago de Compostela.
Bot. Centralbl. Botanisches centralblatt; *later* Botanisches zentralblatt.
Bot. Centralbl. Beih. Beihefte zum Botanischen centralblatt.
Bot. Gaz. (London). Botanical gazette, a journal of the progress of British botany . . . London.
Bot. Jaarb. Kruidk. Genootsch. Dodonaea. Botanisch jaarboek uitgegeven door het Kruidkundig genootschap Dodonaea.
Bot. Jahrb. Engler. Botanische jahrbücher für systematik, pflanzengeschichte und pflanzengeographie herausgegeben von A. Engler.
Bot. Jahresber. Just. Botanischer jahresbericht. Systematisch geordnetes repertorium der botanischen literatur aller länder. Herausg. von Dr. Leopold Just. *Later* Just's botanischer jahresbericht . . .
Bot. Notiser. Botaniska notiser. *Vols. for* 1849-56 *entitled* Nya botaniska notiser.
Bot. Notiser Suppl. Botaniska Notiser. Supplement.
Bot. Tidsskr. Botanisk tidsskrift udgivet af Dansk botanisk forening (*v.* [6]–29 *have also French title,* Journal de botanique . . .)
Bot. Zentralbl. *See* Bot. Centralbl.
Bresse Louhannaise, Bul. Mens. Soc. Agr. et Hort. Louhans. La Bresse louhannaise. Bulletin mensuel de la Société d'agriculture et d'horticulture de l'arrondissement de Louhans. (Not seen.)
Brotéria Sér. Bot. Brotéria. [Subtitle varies.] Série botânica.
Brotéria Sér. Trimest. Ciên. Nat. Broteria. Série trimestral. Ciências naturais.
Bul. Acad. Internatl. Géog. Bot. Bulletin de l'Académie internationale de géographie botanique; *later* (*v.* 21 *on*) Bulletin de géographie botanique.
Bul. Ann. Soc. Jersiaise. Société jersiaise. Bulletin annuel.
Bul. Assoc. Franç. Bot. Bulletin de l'Association française de botanique.
Bul. Assoc. Nat. Vallée du Loing. Bulletin de l'Association des naturalistes de la vallée du Loing.
Bul. Assoc. Philom. Alsace et Lorraine. Bulletin de l'Association philomatique d'Alsace et de Lorraine.
Bul. Bibl. Bot. Ital. Bullettino bibliografico della botanica italiana.
Bul. Bot. Liégeois. Bulletin des botanistes liégeois (*With* Archives de l'Institut botanique de l'Université de Liège). *Later* Lejeunia.
Bul. Brit. Mus. Nat. Hist. Bot. Bulletin of the British Museum (Natural History). Botany.
Bul. Cercle Nat. Hutois. Bulletin du Cercle des naturalistes hutois. (Not seen.)
Bul. Comn. Antiq. Castres et Dépt. Tarn. Bulletin de la Commission des antiquités de la ville de Castres et du département du Tarn.
Bul. Féd. Soc. Hort. Belg. Bulletin de la Fédération des sociétés d'horticulture de Belgique.
Bul. Folklore (Bruxelles). Bulletin de folklore (Société du folklore wallon).
Bul. Géog. Bot. Bulletin de géographie botanique; *earlier* Le monde des plantes, *and* Bulletin de l'Académie internationale de géographie botanique.
Bul. Herb. Boissier. Bulletin de l'Herbier Boissier.
Bul. Hist. Nat. Appl. Bulletin de l'histoire naturelle appliquée.
Bul. Inst. Agr. Sta. Rech. Gembloux. Bulletin de l'Institut agronomique et des stations de recherches de Gembloux. *At head of title page:* Institut agronomique de l'État à Gembloux.
Bul. Ist. Bot. Univ. Sassari. Bullettino dell'Istituto botanico della R. università di Sassari.
Bul. Jard. Bot. État Bruxelles. Bulletin du Jardin botanique de l'État à Bruxelles.
Bul. Lab. Bot. Gen. Univ. Genève. Bulletin du Laboratoire de botanique générale de l'Université de Genève.

Bul. Lab. ed Orto Bot. R. Univ. Siena. Bullettino del Laboratorio ed Orto botanico della R. Università di Siena.
Bul. Lab. Marit. Dinard. Bulletin du Laboratoire maritime de Dinard. *At head of cover:* Muséum national d'histoire naturelle.
Bul. Mayenne Sci. Bulletin de Mayenne—sciences. (Not seen.)
Bul. Mens. Assoc. Nat. Vallée du Loing. Bulletin mensuel de l'Association des naturalistes de la vallée du Loing.
Bul. Mens. Soc. Linn. Lyon. Bulletin mensuel de la Société linnéenne de Lyon.
Bul. Mens. Soc. Linn. Seine Maritime. Bulletin mensuel de la Société linnéenne de la Seine Maritime.
Bul. Mens. Soc. Nat. Luxemb. Société des naturalistes luxembourgeois. Bulletins mensuels. (Also has German title.) *Later* (*from n.s.,* 43) Bulletin.
Bul. Misc. Inform. Kew; Add. Ser. Bulletin of miscellaneous information. Royal botanic gardens, Kew; Additional series.
Bul. Murithienne. Bulletin de la Murithienne, Société valaisanne des sciences naturelles. *Earlier volumes* (2–26) *were entitled* Bulletins des travaux de la Société murithienne (du Valais) (title varies somewhat).
Bul. Mus. Hist. Nat. Marseille. Bulletin du Muséum d'histoire naturelle de Marseille.
Bul. Mus. (Natl.) Hist. Nat. Paris. Bulletin du Muséum (national) d'histoire naturelle. Paris.
Bul. Orto Bot. R. Univ. Napoli. Bullettino dell'Orto botanico della R. università di Napoli.
Bul. Sect. Vosgienne Club Alp. Franç. Club alpin français. Bulletin de la Section vosgienne.
Bul. Scient. Bourgogne. Bulletin scientifique de Bourgogne.
Bul. Scient. Dépt. Nord. Bulletin scientifique du département du Nord et des pays voisins (Pas-de-Calais, Somme, Aisne, Ardennes, Belgique) . . . (Title varies somewhat.)
Bul. Soc. Acad. Brest. Bulletin de la Société académique de Brest.
Bul. Soc. Agr. Arrond. Wassy. Bulletin de la Société d'agriculture de l'arrondissement de Wassy. (Not seen.)
Bul. Soc. Agr., Let., Sci. et Arts Dépt. Haute-Saône. Bulletin de la Société d'agriculture, lettres, science et arts du département de l'Haute-Saône.
Bul. Soc. Agr., Sci. et Arts Sarthe. Bulletin de la Société d'agriculture, des sciences et arts de la Sarthe.
[Bul.] Soc. Agr., Scient. et Lit. Pyrén.-Orient. Société agricole, scientifique et littéraire des Pyrénées-orientales. [Bulletin.]
Bul. Soc. Amis Sci. Nat. Rouen. Bulletin de la Société des amis des sciences naturelles de Rouen.
Bul. Soc. Archéol., Scient. et Lit. Vendômois. Bulletin de la Société archéologique, scientifique et littéraire du Vendômois.
Bul. Soc. Belfort. Émul. Bulletin de la Société belfortaine d'émulation.
Bul. Soc. Borda. Bulletin de la Société de Borda.
Bul. Soc. Bot. Deux-Sèvres. Bulletin de la Société botanique des Deux-Sèvres.
Bul. Soc. Bot. et Hort. Provence. Bulletin de la Société botanique et horticole de Provence.
Bul. Soc. Bot. France. Bulletin de la Société botanique de France.
Bul. Soc. Bot. Genève. Bulletin de la Société botanique de Genève. *First series as* Bulletin des travaux . . .
Bul. Soc. Bot. Ital. Bulletino della Società botanica italiana.
Bul. Soc. Dauphinoise Études Biol. Bulletin de la Société dauphinoise d'études biologiques (Bio-club).
Bul. Soc. Dendrol. France. Bulletin de la Société dendrologique de France.
Bul. Soc. Dunoise. Bulletins de la Société dunoise. Archéologie, histoire, science et arts.
Bul. Soc. Étude Flore Franco-Helv. Société pour l'étude de la flore franco-helvétique. Bulletin. *Earlier* (no. 1–2) Société pour l'étude de la flore française.
Bul. Soc. Étude Sci. Nat. Béziers. Bulletin de la Société d'étude des sciences naturelles de Béziers.

Bul. Soc. Étude Sci. Nat. Carcassonne. Bulletin de la Société d'étude des sciences naturelles de Carcassonne.
Bul. Soc. Étude Sci. Nat. Elbeuf. Bulletin de la Société d'étude des sciences naturelles (*from v.* 21, et du Musée d'histoire naturelle) d'Elbeuf.
Bul. Soc. Étude Sci. Nat. Haute-Marne. Bulletin de la Société d'étude des sciences naturelles de la Haute-Marne.
Bul. Soc. Étude Sci. Nat. Haute-Saône. Bulletin de la Société d'étude des sciences naturelles de l'Haute-Saône. (Not seen.)
Bul. Soc. Étude Sci. Nat. Nimes. Bulletin de la Société d'étude des sciences naturelles, Nimes.
Bul. Soc. Étude Sci. Nat. Reims. Bulletin de la Société de l'étude des sciences naturelles de Reims.
Bul. Soc. Études Hautes-Alpes. Bulletin de la Société d'études des Hautes-Alpes.
Bul. Soc. Études Lit., Scient. et Artist. Lot. Bulletin de la Société des études littéraires, scientifiques et artistiques du Lot.
Bul. Soc. Études Scient. Angers. Bulletin de la Société d'etudes scientifiques d'Angers.
Bul. Soc. Études Scient. Aude. Bulletin de la Société d'études scientifiques de l'Aude.
Bul. Soc. Études Scient. et Archéol. Draguignan. Bulletin de la Société d'études scientifiques et archéologiques de la ville de Draguignan.
Bul. Soc. Études Scient. Finistère. Bulletin de la Société d'études scientifiques du Finistère.
Bul. Soc. Études Scient. Paris. Bulletin de la Société d'études scientifiques de Paris.
Bul. Soc. Flore Valdôtaine. Bulletin de la Société de la flore valdôtaine (Aoste).
Bul. Soc. Fribourg. Sci. Nat. Bulletin de la Société fribourgeoise des sciences naturelles.
Bul. Soc. Grayloise Émul. Bulletin de la Société grayloise d'émulation. Gray [France].
Bul. Soc. Hist. et Archéol. Orne. Bulletin de la Société historique et archéologique de l'Orne.
Bul. Soc. Hist. Nat. Ardennes. Bulletin de la Société d'histoire naturelle des Ardennes.
Bul. Soc. Hist. Nat. Autun. Bulletin de la Société d'histoire naturelle. Autun.
Bul. Soc. Hist. Nat. Auvergne. Bulletin de la Société d'histoire naturelle d'Auvergne.
Bul. Soc. Hist. Nat. Colmar. Bulletin de la Société d'histoire naturelle de Colmar.
Bul. Soc. Hist. Nat. Dépt. Moselle. Bulletin de la Société d'histoire naturelle du département de la Moselle; *from* 1871–1913, Bulletin de la Société d'histoire naturelle de Metz.
Bul. Soc. Hist. Nat. Doubs. Bulletin de la Société d'histoire naturelle de Doubs.
Bul. Soc. Hist. Nat. Metz. *See* Bul. Soc. Hist. Nat. Dépt. Moselle.
Bul. Soc. Hist. Nat. Reims. Bulletin de la Société d'histoire naturelle de Reims.
Bul. Soc. Hist. Nat. Savoie. Bulletin de la Société d'histoire naturelle de Savoie.
Bul. Soc. Hist. Nat. Toulouse. Bulletin de la Société d'histoire naturelle de Toulouse.
Bul. Soc. Hort. Allier. Bulletin de la Société d'horticulture de l'Allier.
Bul. Soc. Hort. et Vitic. Puy-de-Dôme. Bulletin de la Société d'horticulture et de viticulture de Puy-de-Dôme.
Bul. Soc. Hort. Orléans et Loiret. Bulletin de la Société d'horticulture d'Orléans et du Loiret.
Bul. Soc. Hort. Orne. Bulletin de la Société d'horticulture de l'Orne. (Not seen.)
Bul. Soc. Imp. Nat. Moscou. Bulletin de la Société impériale des naturalistes de Moscou. Section biologique. (Bı͡ulleten Moskovskogo obshchestva. Ispytateleĭ prirody.)
Bul. Soc. Let. et Sci. Redon. Bulletin de la Société des lettres et des sciences de Redon.

Bul. Soc. Let., Sci. et Arts Corrèze. Bulletin de la Société des lettres, science et arts de la Corrèze.
Bul. Soc. Liégeoise Lit. Wallone. Bulletin de la Société liégeoise de littérature wallone.
Bul. Soc. Linn. Lyon. Bulletin de la Société linnéenne de Lyon. (Not seen.)
Bul. Soc. Linn. Nord France. Bulletin de la Société linnéenne du nord de la France.
Bul. Soc. Linn. Normandie. Bulletin de la Société linnéenne de Normandie.
Bul. Soc. Nat. Ain. See next title.
Bul. Soc. Nat. et Archéol. Ain. Bulletin de la Société des naturalistes de l'Ain (*from no.* 38 (1924) *on*, . . . des naturalistes et des archéologues).
Bul. Soc. Nat. et Archéol. Nord de la Meuse. Bulletin de la Société des naturalistes et archéologues du nord de la Meuse.
Bul. Soc. Nat. Luxemb. Société des naturalistes luxembourgeois. Bulletin. *Earlier* Bulletins mensuels.
Bul. Soc. Neuchâtel. Sci. Nat. Société neuchâteloise des sciences naturelles. Bulletin.
Bul. Soc. Philom. Vosgienne. Bulletin de la Société philomatique vosgienne.
Bul. Soc. Roy. Bot. Belg. Bulletin de la Société royale de botanique de Belgique.
Bul. Soc. Roy. Linn. Bruxelles. Bulletin de la Société royale linnéene de Bruxelles.
Bul. Soc. Sci., Belles-Let. et Arts Dept. Var. Bulletin (trimestriel) de la Société des sciences, belles-lettres et arts du département du Var séant a Toulon.
Bul. Soc. Sci. Hist. et Nat. Corse. Bulletin de la Société des sciences historiques et naturelles de la Corse.
Bul. Soc. Sci. Hist. et Nat. Yonne. Bulletin de la Société des sciences historiques et naturelles de l'Yonne.
Bul. Soc. Sci. Nancy. Bulletin des séances de la Société des sciences de Nancy.
Bul. Soc. Sci. Nat. et Archéol. Creuse. Bulletin de la Société des sciences naturelles et archéologiques de la Creuse.
Bul. Soc. Sci. Nat. Haute-Marne. Bulletin de la Société de sciences naturelles de la Haute-Marne.
Bul. Soc. Sci. Nat. Neuchâtel. Bulletin de la Société des sciences naturelles de Neuchâtel. *Beginning with v.* 26 (1898), Société neuchâteloise des sciences naturelles. Bulletin.
Bul. Soc. Sci. Nat. Ouest France. Bulletin de la Société des sciences naturelles de l'ouest de France.
Bul. Soc. Sci. Nat. Provence. Bulletin de la Société des sciences naturelles de Provence.
Bul. Soc. Sci. Nat. Saône-et-Loire. Bulletin de la Société des sciences naturelles de Saône-et-Loire.
Bul. Soc. Sci. Nat. Seine-et-Oise. Bulletin de la Société des sciences naturelles de Seine-et-Oise. *Earlier* . . . des sciences de Seine-et-Oise, de la Beauce et de la Brie.
Bul. Soc. Sci. Phys. et Nat. Toulouse. Bulletin de la Société des sciences physiques et naturelles de Toulouse.
Bul. Soc. Sci. Seine-et-Oise. Bulletin de la Société des sciences de Seine-et-Oise, de la Beauce et de la Brie. (*Beginning with* ser. III, . . . des sciences naturelles de Seine-et-Oise.)
Bul. Soc. Scient. Bretagne (Sci. Math., Phys. et Nat.). Bulletin de la Société scientifique de Bretagne (Sciences mathématiques, physiques et naturelles).
Bul. Soc. Scient. Dauphiné. Bulletin de la Société scientifique de Dauphiné (ancienne Société de statistique des sciences naturelles et des arts industriels du département de l'Isère).
Bul. Soc. Scient. et Méd. Ouest. Bulletin de la Société scientifique et médicale de l'Ouest.
Bul. Soc. Scient. Flammarion de Marseille. Bulletin de la Société scientifique Flammarion de Marseille.
Bul. Soc. Scient., Hist. et Archéol. Corrèze. Bulletin de la Société scientifique, historique et archéologique de la Corrèze.

Bul. Soc. Stat., Sci. Nat., et Arts Indus. Dépt. Isère. Bulletin de la Société de statistique, des sciences naturelles, et des arts industrielles du département de l'Isère.
Bul. Soc. Vaud. Sci. Nat. Bulletin de la Société vaudoise des sciences naturelles.
Bul. Soc. Veneto-Trent. Sci. Nat. Bullettino de la Società veneto-trentina di scienze naturali.
Bul. Trav. Soc. Bot. Genève. Bulletin des travaux de la Société botanique de Genève.
Bul. Trav. Soc. Émul. Lisieux. Bulletin des travaux de la Société d'émulation de Lisieux.
Bul. Trav. Soc. Pharm. Bordeaux. Bulletin des travaux de la Société de pharmacie de Bordeaux.
Bul. Trimest. Soc. Bot. Lyon. Société botanique de Lyon. Bulletin trimestriel.
Bul. Trimest. Union Soc. Franç. Hist. Nat. Union des Sociétés françaises d'histoire naturelle. Bulletin trimestriel.
Bul. Uffic. Assoc. Ort. Prof. Ital. Bullettino ufficiale della Associazione orticola professionale italiana.
Bul. Univ. Lille. Bulletin de la Université de Lille.
Bul. Vulgarisat. Sci. Nat. Gers. Bulletin de vulgarisation des sciences naturelles, organe de la Société botanique et entomologique du Gers.
Butllet. Agrupació Excurs. Reus. Butlletí de l'Agrupació excursionista de Reus. (Not seen.)
Butllet. Inst. Catalana Hist. Nat. Butlletí de la Institución (Instituciò) catalana d'historia natural.
Cahiers des Nat. Cahiers des naturalistes. Bulletin des naturalistes parisiens. *Earlier* La feuille des naturalistes.
Cairngorm Club Jour. Cairngorm club journal.
Canad. Field-Nat. Canadian field-naturalist.
Candollea. Candollea. Organe du Conservatoire et du Jardin botaniques de la ville de Genève. *Successor to* Annuaire du Conservatoire et du Jardin botaniques de Genève.
Caradoc and Severn Valley Rec. of Bare Facts. Caradoc and Severn Valley field club record of bare facts. (Not seen.)
Caradoc Rec. of Bare Facts. Caradoc record of bare facts. (Not seen.)
Carmarthen Antiq. Carmarthen antiquary. (Not seen.)
Carthusian. The Carthusian. (Not seen.)
Cavanillesia. Cavanillesia. Rerum botanicarum acta.
Ceridigion, Jour. Cardigansh. Antiq. Soc. Ceridigion, journal of the Cardiganshire antiquarian society. (Not seen.)
Chanousia. Chanousia. Giardino botanico alpino dell'Ordine Mauriziano sul colle del Piccolo San Bernardo (metri 2200 s.m.). Annuario.
Chêne Soc. Forest. Méditer. et Colon. Bul. Le chêne. Société forestière méditerranéenne et coloniale. Bulletin.
Collect. Bot. Barcinon. Collectanea botanica a Barcinonensi botanico instituto edita. Barcinone.
Comment. Ateneo Brescia (Sci., Let. ed Arti). Commentari dell' Ateneo di Brescia; *from* 1902 *adds* di scienze, lettere ed arti.
Comment. Biol. Soc. Scient. Fenn. Societas scientiarum fennica. Commentationes biologicae.
Commun. Sta. Internatl. Géobot. Méditer. et Alp. Montpellier. Communication de la Station internationale de géobotanique méditerranéenne et alpine, Montpellier. (Also has German title.)
Compt. Rend. Acad. Sci. Paris. Comptes rendus hebdomaires des séances de l'Académie des sciences. Paris.
Compt. Rend. Assoc. Franç. Avanc. Sci. Association française pour l'avancement des sciences . . . Compte rendu.
Compt. Rend. Cong. Soc. Sav. Paris et Dépt. Comptes rendus du Congrès des Sociétés savantes de Paris et des départements.
Compt. Rend. Excurs. Soc. Sci. Nat. Charente-Inf., Soc. Bot. Rochelaise. Comptes rendus des excursions faites sous les auspices de la Société des sciences naturelles de la Charente-Inférieure, publiés par la Société botanique rochelaise.

Compt.-Rend. Soc. Nat. Luxemb. Société des naturalistes luxembourgeois (Fauna). Verein Luxemburger naturfreunde. Comptes-rendus des séances. Mitteilungen aus den vereinssitzungen. (Order of items in title varies.)
Comun. Reun. Internac. Soc. Farm. Mediter. Lat. II. (Not seen.)
Cong. Assoc. Franç. Avanc. Sci. Association française pour l'avancement des sciences. Congrès. (Not seen.)
Cong. Internac. Estud. Piren. Com. Congreso internacional del Pirineo del Instituto de estudios pirenaicos.
Cong. Internatl. Bot. 8 (Paris, 1954) Rapp. et Commun. Huitième congrès international de botanique Paris 1954. Rapports et communications parvenus avant le Congrès aux sections . . .
Cong. Naturaliste. Congrès naturaliste. Paris, 1944.
Cong. Scient. France. Congrès scientifique de France.
Cong. Soc. Savantes Savois. Congrès des sociétés savantes savoisiennes.
Crón. Cient. Crónica científica. Revista internacional de ciencias. Barcelona.
Danmarks Geol. Undersøg. Danmarks geologiske undersøgelse.
Dansk Bot. Arkiv. Dansk botanisk arkiv. Udgivet af Dansk botanisk forening.
K. Danske Vidensk. Selsk. Biol. Skrift. Kongelige danske videnskabernes selskab. Biologiske skrifter.
K. Danske Vidensk. Selsk. Skrift. Naturv. og Math. Afd. Det K. danske videnskabernes selskabs skrifter. Naturvidenskabelig og mathematisk afdeling.
Denkschr. Schweiz. Naturf. Gesell. Denkschriften der Schweizerischer naturforschenden gesellschaft . . . [*earlier* Neue denkschriften]. Mémoires de la Société helvétique des sciences naturelles.
Deut. Bot. Monatsschr. Deutsche botanische monatsschrift. (Subtitle varies.)
Edinb. New Phil. Jour. The Edinburgh new philosophical journal . . .
Encilion. Yr encilion or the Journal of the Carmarthenshire antiquarian society and field club.
Essex Nat. The Essex naturalist: being the journal of the Essex field club.
Études Rhodaniennes. Les études rhodaniennes.
Explor. Pyrén. Bul. Soc. Ramond. Explorations pyrénéennes . . . Bulletin de la Société Ramond.
Farmacognosia. Farmacognosia. Anales del Instituto de farmacognosia José Celestino Mutis. *At head of title page:* Consejo superior de investigaciones científicas . . .
Fennia. Fennia.
Festschr. 110. Jahresversamm. Schweiz. Naturf. Gesell. Davos. Festschrift für die 110. versammlung der Schweizerischen naturforschenden gesellschaft in Davos.
Feuille des Jeunes Nat. (La) feuille des jeunes naturalistes. *Followed by* La feuille des naturalistes.
Feuille des Nat. La feuille des naturalistes. Revue mensuelle d'histoire naturelle. (Bulletin de la Société des naturalistes parisiens.) *Followed by* Cahiers des naturalistes.
Flintsh. Misc. *See* **Pub. Flintsh. Hist. Soc. (Flintsh. Misc.).**
Flora. Flora; oder Allgemeine botanische zeitung.
Flora og Fauna. Flora og fauna. Utgivet af Naturhistorisk forening for Jylland . . .
För. Gamla Halmstads Årsbok. Föreningen Gamla Halmstads årsbok. (Not seen.)
Forhandl. Skand. Naturforsk. Möde. Forhandlinger ved de Skandinaviske naturforskeres möde. (Title varies slightly.)
Forhandl. Vidensk.-Selsk. Christiania. Forhandlinger i Videnskabs-selskabet i Christiania.
Forstl. Forskningsanst. Populärvetensk. Beskr. Naturskyddsomr. Forstliga forskningsanstaltens populärvetenskapliga beskrivningar av naturskyddsområden.
K. Fysiog. Sällsk. Lund Förhandl. Kungl. fysiografiska sällskapets i Lund förhandlingar.
Geog. Jour. Geographical journal.
Geog. Tidsskr. Geografisk tidsskrift, udgivet af bestyrelsen for det Kongelige danske geografiske selskab.

Géographie. La géographie, bulletin de la Société de géographie.
Gior. Bot. Ital. Giornale botanico italiano.
Gior. Fis., Chim., Stor. Nat., Med. ed Arti (Milano). Giornale di fisica, chimica, storia naturale, medicina ed arti . . . Milano.
Gior. Ital. Let. Giornale dell' italiana letteratura. (Not seen.)
Glasgow Nat. The Glasgow naturalist.
Glasnik Hrvatsk. Prirod. Društva. Glasnik Hrvatskoga prirodoslovnoga društva. *Earlier* Glasnik Hrvatskoga naravoslovnoga družtva. *At head of title page:* Hrvatsko prirodoslovno društvo (Societas scientiarum naturalium croatica.
Globe. Le Globe. Journal géographique. Organe de la Société de géographie de Genève.
Göteborgs K. Vetensk.-och Vitterh.-Samhäll. Handl. (ser. B. Mat. och Naturv. Skr.). Göteborgs kungl. vetenskaps- och vitterhets-samhälles handlingar. (Serien B. Matematiska och naturvetenskapliga skrifter.)
Halifax Nat. The Halifax naturalist and record of the scientific society.
Hallands Natur. Hallands natur. Hallands natursyddsförenings årsskrift.
Hedeselsk. Tidsskr. Hedeselskabets tidsskrift.
Hist. Berwicksh. Nat. Club. History of the Berwickshire naturalists' club.
Ilerda. Ilerda.
In Alto. In alto. (Not seen.)
Instituto. O instituto. (Subtitle varies.) Coimbra.
Internatl. Cat. Scient. Lit. M, Bot. International catalogue of scientific literature. M, Botany.
Irish Nat. Irish naturalist.
Irish Nat. Jour. Irish naturalists' journal.
Italia Forest. e Mont. L'Italia forestale e montana. Rivista bimestrale edita dai professori della Sezione forestale della Facoltà agraria e forestale di Firenze.
Jaarb. Natuurh. Genootsch. Limburg. Jaarboek van het Natuurhistorisch genootschap in Limburg.
Jaarb. Nederl. Dendrol. Ver. Nederlandsche dendrologische vereeniging. Jaarboek.
Jahrb. Schweiz. Alpenclub. Jahrbuch des Schweizer alpenclub.
Jahrb. St. Gall. Naturw. Gesell. Jahrbuch der St. Gallischen naturwissenschaftlichen gesellschaft.
Jahrb. Zürichsee. Jahrbuch vom Zürichsee.
Jahresber. Naturf. Gesell. Graubündens. Jahresbericht der Naturforschenden gesellschaft Graubündens.
Jahresber. Progymnas. Pfalzburg. Jahresbericht des Progymnasiums zu Pfalzburg. (Not seen.)
Jahresber. Realschule Wasselnheim. Jahresbericht der Realschule zu Wasselnheim. (Not seen.)
Jor. Hort. Prat. Jornal de horticultura pratica.
Jor. Sci. Mat. Phys. e Nat. Lisboa. Jornal de sciências matemáticas, physicas e naturaes publicado sob os auspicios da Academia real das sciências de Lisboa.
Jour. Bedfordsh. Nat. Hist. Soc. and Field Club. Journal of the Bedfordshire natural history society and field club. *Later* (with vol. 2) The Bedfordshire naturalist.
Jour. Bot. Journal of botany, British and foreign. London.
Jour. de Bot. Journal de botanique. (Morot), Paris.
Jour. Derbysh. Archaeol. and Nat. Hist. Soc. Journal of the Derbyshire archaeological and natural history society.
Jour. des Savants. Journal des savants.
Jour. Ecology. Journal of ecology.
Jour. Ipswich and Dist. Field Club. The journal of the Ipswich and district field club.
Jour. Linn. Soc. Bot. Journal of the Linnean society of London. Botany.
Jour. Manx Mus. The journal of the Manx museum.
Jour. Northamptonsh. Nat. Hist. Soc. and Field Club. Journal of the Northamptonshire natural history society and field club.
Jour. Pharm. Alsace-Lorraine. Journal de pharmacie d'Alsace-Lorraine.
Jour. Roman Stud. The journal of Roman studies.

Jour. Roy. Hort. Soc. Journal of the Royal Horticultural Society. London.
Jour. Roy. Inst. Cornwall. Journal of the Royal institution of Cornwall.
Jour. Soc. Bibl. Nat. Hist. The journal of the Society for the bibliography of natural history.
Jour. Soc. Natl. Hort. France. Journal de la Société nationale d'horticulture de France. (Earlier titles vary.)
K. (Kaiserliche, Königliche, etc.) Titles beginning with K. filed under word following the K.
Kuopion Luonnon Ystäväin Yhdist. Julkaisuja. Kuopion luonnon ystäväin Yhdistyksen julkaisuja. [Yearbook of the Friends of nature society of Kuopio.]
Lännetär. Lännetär. (Not seen.)
Länsi-Suomi. Länsi-Suomi. (Not seen.)
Lancash. Nat.; Lancash. and Chesh. Nat. The Lancashire naturalist. A monthly journal of natural history for the county of Lancashire and for the adjacent districts. *Later (from v.* 7 *on)* The Lancashire and Cheshire naturalist (with which is incorporated the "Lancashire naturalist").
Landw. Jahrb. Schweiz. Landwirtschaftliches jahrbuch der Schweiz.
Lav. Bot. Ist. Bot. Univ. Torino. Lavori botanichi dell'Istituto botanico dell' Università di Torino.
Leben und Umwelt. Leben und umwelt.
Lejeunia. Lejeunia. Revue de botanique. (Subtitle varies.)
Levende Nat. De levende natuur. Tijdschrift voor natuurvrienden.
Lietuvos Univ. Mat. Gamtos Fak. Darbai. (Mém. Fac. Sci. Univ. Lithuanie.) Lietuvos Universiteto matematikos gamtos fakulteto darbai. Mémoires de la faculté des sciences de l'Université de Lithuanie.
Linnaea. Linnaea. Ein journal für die botanik in ihrem ganzen umfange.
London Nat. London naturalist. The journal of the London natural history society. *Earlier* Transactions of the London natural history society.
Lunds Univ. Årsskrift. Lunds universitats årsskrift. (Acta universitatis lundensis.)
Luonnon Tutkija. Luonnon tutkija. *Earlier* (v. 1–50) Luonnon ystävä.
Luonnon Ystävä. Luonnon ystävä. [The friend of nature.] Eläin- ja kasvitieteellinen aikakauslehti. Julkaisuja. Suomalainen Eläin- ja kasvitieteellinen seura Vanamo. *Later* (v. 51 and after) Luonnon tutkija.
Lustgården. Lustgården. Årsskrift. Föreningen for dendrologi och parkvård. Stockholm.
Maandbl. Natuurh. Genootsch. Limburg. Maandblad van het Natuurhistorisch genootschap in Limburg. *Succeeded* (1912 *and later) by* Jaarboek van het Natuurhistorisch genootschap in Limburg.
Maataloustiet. Aikak. Maataloustieteellienen aikakauskirja. The journal of the Scientific agricultural society of Finland.
Madonna Verona. Madonna Verona. Bollettino del Museo civico di Verona.
Mag. for Naturv. Magazin for naturvidenskaberne.
Mag. Nat. Hist. (ed. Charlesworth). The magazine of natural history . . . conducted by Edward Charlesworth.
Malpighia. Malpighia, rassegna mensile di botanica.
Meddel. Bot. For. Kjøbenhavn. Meddelelser fra den Botaniske forening i Kjøbenhavn.
Meddel. Finska Forstför. Finska forstföreningens meddelanden.
Meddel. Geog. För. Finland. Meddelanden af Geografiska föreningen i Finland.
Meddel. Göteborgs Bot. Trädgård. *See* Acta Hort. Gothoburg.
Meddel. K. Lantbruksstyr. Meddelanden från Kungl. lantbruksstyrelsen.
Meddel. Norges Svalbard- og Ishavs-undersøk. Norges Svalbard- og Ishavsundersøkelser. Meddelelser. *Publication of* Det kongelige departement for handel, sjøfart, industri, håndverk og fiskeri. Olso. *Later* Norsk polarinstitutt. Meddelelser.
Meddel. Norsk Bot. For. Norsk botanisk forening. Meddelelser.
Meddel. Norsk Polarinst. Norsk polarinstitutt (*formerly* Norges Svalbard- og Ishavs-undersøkelser). Meddelelser.
Meddel. Norske Skogforsøksvesen. Meddelelser fra det Norske skogforsøksvesen.

Meddel. Soc. Fauna et Flora Fenn. Meddelanden af Societas pro fauna et flora fennica.
Meddel. Stat. Skogsförsöksanst. Meddelanden från Statens skogsförsöksanstalt. Mitteilungen aus der Forstlichen versuchsanstalt Schwedens.
Meddel. Värmlands Naturhist. För. Meddelanden från Värmlands naturhistoriska förening.
Meded. Centrum Plantensociol. Karter. Gembloers. Centrum voor plantensociologische kartering en centrum voor oekologisch en plantensociologisch onderzoek te Gembloers. Mededeling.
Meded. Dir. Landb. Landbouwvoorlichtingsdienst. Directie van de landbouw. Landbouwvoorlichtingsdienst. Mededeling. Wageningen. *On cover:* [Netherlands.] Ministerie van landbouw, visserij, en voedselvoorziening.
Meded. Natuurh. Genootsch. Limburg. Mededeelingen van het Natuurhistorisch genootschap in Limburg. *Later* (1912 *and subsequently*) Jaarboek van het Natuurhistorisch genootschap in Limburg.
Meded. Rijks Herb. Leiden. Mededeelingen van 's Rijks herbarium Leiden.
Mém. Acad. Imp. Sci., Belles-Let., et Arts Lyon, Cl. Sci. Mémoires de l'Académie impériale des sciences, belles-lettres et arts de Lyon. Classe des sciences.
Mém. Acad. Imp. Sci. St.-Pétersb. Mémoires de l'Académie impériale des sciences de St.-Pétersbourg.
Mém. Acad. Imp. Sci. St.-Pétersb. Divers Savants. Mémoires présentés à l'Académie impériale des sciences de St.-Petersbourg par divers savants.
Mém. Acad. Internatl. Géog. Bot. Mémoires de l'Académie internationale de géographie botanique. (Not seen.)
Mém. Acad. Nîmes. Mémoires de l'Académie de Nîmes.
Mém. Acad. Roy. Belgique Cl. Sci. Collect. in-8°. Académie royale de Belgique. Classes des sciences, Mémoires. Collection in-8°.
Mém. Acad. Roy. Sci. Inst. France. Mémoires de l'Académie royale des sciences de l'Institut de France.
Mém. Acad. Sci., Agr., Arts et Belles-Let. Aix. Mémoires de l'Académie des sciences, agriculture, arts et belles-lettres d'Aix.
Mém. Acad. Sci., Arts et Belles-Let. Dijon. Mémoires de l'Académie des sciences, arts et belles-lettres de Dijon. Bulletin.
Mém. Acad. Sci., Belles-Let. et Arts Clermont-Ferrand. Mémoires de l'Académie des sciences, belles-lettres et arts de Clermont-Ferrand.
Mém. Acad. Sci., Belles-Let. et Arts Savoie. Mémoires de l'Académie des sciences, belles-lettres et arts de Savoie.
Mém. Acad. Sci., Inscript. et Belles-Let. Toulouse. Mémoires de l'Académie des sciences, inscriptions et belles-lettres de Toulouse.
Mém. Acad. Sci., Let. et Beaux-arts Marseille. Mémoires de l'Académie des sciences, lettres et beaux-arts de Marseille.
Mém. Acad. Stanislas. Mémoires de l'Académie de Stanislas.
Mém. Acad. Vaucluse. Mémoires de l'Académie de Vaucluse.
Mem. Accad. Agr., Arti e Com. Verona. Memorie dell'Accademia d'agricoltura, arti e commercio di Verona.
Mem. Accad. Sci. Ist. Bologna. Memorie della Accademia delle scienze dell'Istituto di Bologna.
Mem. Accad. Sci., Let. ed Arti Modena. Accademia di scienze, lettere ed arti in Modena. Memorie.
Mem. and Proc. Manchester Lit. and Phil. Soc. Memoirs and proceedings of the Manchester literary and philosophical society. (Manchester memoirs.) *Running title:* Manchester memoirs.
Mém. Assoc. Nat. Nice et des Alpes-Maritimes. Association des naturalistes de Nice et des Alpes-Maritimes. Mémoire.
Mem. Econ. Acad. R. Sci. Lisboa. Memorias economicas da Academia real das sciencias de Lisboa, para o adiantamento da agricultura, das artes, e da industria em Portugal, e suas conquistas.
Mém. et Pub. Soc. Sci., Arts et Let. Hainaut. Mémoires et publications de la Société des sciences, des arts et des lettres du Hainaut.
Mém. Hist. Nat. Soc. Éduenne. Mémoires d'histoire naturelle (publication de la Société éduenne).
Mem. Inst. Catalana Hist. Nat. Memòries de la Institució catalana d'historia natural. Barcelona.

Mém. Inst. Natl. Genevois. Mémoires de l'Institut national genevois.
Mem. Mat. e Fis. (e Sci. Fis. e Nat.) Soc. Ital. Sci. Memorie di matematica e di fisica della Società italiana delle scienze; *sometimes* Memorie di matematica e di scienze fisiche e naturali . . .
Mem. Mus. Cienc. Nat. Barcelona Ser. Bot. Memorias del Museo de ciencias naturales de Barcelona. Serie botánica.
Mém. Mus. Hist. Nat. Paris. Memoires du Muséum d'histoire naturelle (Paris).
Mem. Pontif. Accad. Nuovi Lincei. Memorie della Pontificia accademia dei nuovi lincei. Serie iniziata per ordine della S. D. N. S. Papa Leone XIII. Roma. (Title varies in later volumes.)
Mem. R. Acad. Cien. Madrid. Memorias de la Real academia de ciencias exactas, físicas y naturales de Madrid.
Mem. R. Acad. Cien. y Art. Barcelona. Memorias de la Real academia di ciencias y artes de Barcelona.
Mem. R. Accad. Sci. Ist. Bologna; Cl. Sci. Fis. Memorie della R. accademia delle scienze dell'Istituto di Bologna; Classe di scienze fisiche.
Mem. R. Accad. Sci. Torino. Memorie della Reale accademia delle scienze di Torino.
Mem. R. Ist. Lombardo Sci. e Lett., Cl. Sci. Mat. e Nat. Memorie del Reale istituto lombardo di scienze e lettere. Classe di scienze matematiche e naturali.
Mem. R. Ist. Veneto Sci., Let. ed Arti. Memorie delle Reale istituto Veneto di scienze, lettere ed arti.
Mem. R. Soc. Españ. Hist. Nat. Memorias de la Real sociedad española de historia natural.
Mém. Sect. Méd. Acad. Sci. et Let. Montpellier. Académie des sciences et lettres de Montpellier. Mémoires de la section de médecine.
Mém. Soc. Acad. Agr., Sci., Arts et Belles-Let. Dépt. Aube; Mém. Soc. Agr. (etc.). Mémoires de la Société académique d'agriculture, des sciences, arts et belles-lettres du département de l'Aube; *earlier* Mémoires de la Société d'agriculture (etc.).
Mém. Soc. Acad. Archéol., Sci. et Arts Dépt. Oise. Mémoires de la Société académique d'archéologie, sciences et arts du département de l'Oise.
Mém. Soc. Acad. Maine et Loire. Mémoires de la Société académique de Maine et Loire.
Mém. Soc. Acad. Nivernais. Mémoires de la Société académique du Nivernais.
Mém. Soc. Agr., Com., Sci. et Arts Dépt. Marne. Mémoires de la Société d'agriculture, commerce, sciences et arts du département de la Marne.
Mém. Soc. Agr., Sci., Arts et Belles-Let. Dépt. Aube. *See* Mém, Soc. Acad. Agr., Sci., Arts et Belles-Let. Dépt. Aube.
Mém. Soc. Agr., Sci., Belles-Let. et Arts Orléans. Mémoires de la Société d'agriculture, sciences, belles-lettres et arts d'Orléans.
Mém. Soc. Agr., Sci. et Arts Centr. Dépt. Nord Douai. Mémoires de la Société d'agriculture, sciences et arts centrale du département du Nord séant à Douai.
Mém. Soc. Agr., Sci. et Arts Valenciennes. Mémoires de la Société d'agriculture, des sciences et des arts, de l'arrondissement de Valenciennes.
Mém. Soc. Arts et Sci. Carcassonne. Mémoires de la Société des arts et des sciences de Carcassonne.
Mem. Soc. Broteriana. Memorias da Sociedade broteriana.
Mém. Soc. Bot. France. Mémoires publiés par la Société botanique de France.
Mém. Soc. Émul. Abbeville. Mémoires de la Société (*in some* royale *or* impériale) d'émulation d'Abbeville.
Mém. Soc. Émul. Cambrai. Mémoires de la Société d'émulation de Cambrai.
Mém. Soc. Émul. Doubs. Mémoires de la Société d'émulation du Doubs.
Mém. Soc. Émul. Jura. Mémoires de la Société d'émulation du Jura.
Mém. Soc. Émul. Montbéliard. Mémoires de la Société d'émulation de Montbéliard.
Mém. Soc. Étude Sci. Nat. Nîmes. Mémoires de la Société d'étude des sciences naturelles de Nîmes.
Mém. Soc. Fribourg. Sci. Nat. Mémoires de la Société fribourgeoise des sciences naturelles.

Mem. Soc. Geog. Ital. Memorie della Società geografica italiana.
Mém. Soc. Hist., Lit. (Artist.) et Scient. Cher. Mémoires de la Société historique, littéraire (artistique), et scientifique du Cher.
Mem. Soc. Ibérica Cien. Nat. Memorias de la Sociedad ibérica de ciencias naturales.
Mém. Soc. Imp. (Natl.) Sci. Nat. et Math. Cherbourg. Mémoires de la Société impériale [*later* nationale] des sciences naturelles et mathématiques de Cherbourg.
Mem. Soc. Ital. Sci. Nat. Milano. Memorie della Società italiana di scienze naturali e del Museo civico di storia naturale di Milano.
Mém. Soc. Let., Sci., Arts, Agr. et Indus. Saint-Dizier. Mémoires de la Société des lettres, des sciences, des arts, de l'agriculture et de l'industrie de Saint-Dizier.
Mém. Soc. Linn. Nord France. Mémoires de la Société linnéenne du nord de la France.
Mém. Soc. Linn. Normandie. Mémoires de la Société linnéenne de Normandie.
Mém. Soc. Linn. Paris. Mémoires de la Société linnéenne de Paris.
Mém. Soc. Natl. Agr., Sci. et Arts Angers. Mémoires de la Société nationale d'agriculture, sciences et arts d'Angers.
Mém. Soc. Natl. Sci. Nat. et Math. Cherbourg. Mémoires de la Société nationale des sciences naturelles et mathématiques de Cherbourg (*earlier omits* et mathématiques).
Mém. Soc. Philom. Verdun. Mémoires de la Société philomathique de Verdun.
Mém. Soc. Phys. et Hist. Nat. Genève. Mémoires de la Société de physique et d'histoire naturelle de Genève.
Mém. Soc. Roy. et Cent. Agr., Sci. et Arts Dépt. Nord. Mémoires de la Société royale et centrale d'agriculture, sciences et arts du département du Nord.
Mém. Soc. Roy. Sci. Liège. Mémoires de la Société royale des sciences de Liège.
[Mém.] Soc. Sci. et Arts Vitry-le-François. Société des sciences et arts de Vitry-le-François. [Mémoires.]
Mém. Soc. Sci. Nat. et Archéol. Creuse. Mémoires de la Société des sciences naturelles et archéologiques de la Creuse.
Mém. Soc. Sci. Nat. Neuchâtel. Mémoires de la Société des sciences naturelles de Neuchâtel.
Mém. Soc. Sci. Nat. Saône-et-Loire. Mémoires de la Société des sciences naturelles de Saône-et-Loire.
Mém. Soc. Sci. Phys. et Nat. Bordeaux. Mémoires de la Société des sciences physiques et naturelles de Bordeaux.
Mém. Soc. Sci. Phys. et Nat. Toulouse. Mémoires de la Société des sciences physiques et naturelles de Toulouse.
Mém. Soc. Stat., Sci., Let. et Arts Dépt. Deux-Sèvres. Mémoires de la Société de statistique, sciences, lettres et arts du département des Deux-Sèvres.
Mém. Soc. Vaud. Sci. Nat. Mémoires de la Société vaudoise des sciences naturelles.
Mém. Univ. Neuchâtel. Mémoires de l'Université de Neuchâtel.
Memo. Soc. Fauna et Fl. Fenn. Memoranda Societatis pro fauna et flora fennica.
Merioneth Misc. Merioneth miscellany.
Metsätiet. Tutk. Lait. Luonnonsaluekuvauksia. Metsätieteellisen tutkimuslaitoksen luonnonsuojelualuekuvauksia. (Not seen.)
Midl. Nat. Midland naturalist.
Milford-on-Sea Rec. Soc. Milford-on-sea record society. An occasional magazine.
Min. Econ. Dir. Geral Serv. Agr. Sér. Estud. e Inform. Téc. Ministério da economia. Direcção geral dos serviços agricolas. Série: Estudos e informação técnica.
Mitt. Aargau. Naturf. Gesell. Mitteilungen der Aargauischen naturforschenden gesellschaft.
Mitt. Bot. Mus. Univ. Zürich. Mitteilungen aus dem Botanischen museum der Universität Zürich.
Mitt. Gesell. Erdk. u. Kol. Strassburg. Mitteilungen der Gesellschaft für erdkunde und kolonialwesen zu Strassburg i. E.
Mitt. Naturf. Gesell. Bern. Mitteilungen der Naturforschenden gesellschaft in Bern.

Mitt. Naturf. Gesell. Glarus. Mitteilungen der Naturforschenden gesellschaft in Glarus. (Not seen.)
Mitt. Naturf. Gesell. Luzern. Mitteilungen der Naturforschenden gesellschaft in Luzern.
Mitt. Naturf. Gesell. Schaffhausen. Mitteilungen der Naturforschenden gesellschaft Schaffhausen.
Mitt. Naturf. Gesell. Solothurn. Mitteilungen der Naturforschenden gesellschaft in Solothurn.
Mitt. Philom. Gesell. Elsass-Lothr. Mitteilungen der Philomathischen gesellschaft in Elsass-Lothringen.
Mitt. Thurgau. Naturf. Gesell. Mitteilungen der Thurgauischen naturforschenden gesellschaft.
Monde des Plant. Le monde des plantes. Intermédiaire des botanistes. Revue internationale de bibliographie . . . (subtitles vary). Directeur H. Léveillé.
Monde des Plant. Le monde des plantes. Revue mensuelle de botanique. Organe de l'Académie internationale de géographie botanique dirigée par M. H. Léveillé. Vol. 1–8 (no. 109). 1891–déc. 1898; with v. 8 (no. 110/111), jan.-févr. 1899, it became Bulletin de l'Académie internationale de géographie botanique (q.v.), a title later (v. 21 to end) changed to Bulletin de géographie botanique. At the time he changed the title of this journal to Bulletin . . ., Léveillé, with characteristic clarity, began the publication of a new journal with the title Le monde des plantes, which still continues; see above.
Monog. Inst. Estud. Pirenaicos. Monografias del Instituto de estudios pirenaicos.
Mus. Hist. Nat. Luxemb. Serv. Carte Group. Vég. [Études.] Grand-ducheé de Luxembourg. Musée d'histoire naturelle. Service de la carte des groupements végétaux. [Études.]
Nat. Hist. Trans. Northumberl. and Durham. Natural history transactions of Northumberland and Durham . . . *Later* Transactions of the Natural history society of Northumberland, Durham, and Newcastle-upon-Tyne.
Nat. Mosana. Natura mosana.
Nat. Siciliano. Il naturalista siciliano, organo della Società di scienze naturali. (Subtitle varies.)
Nat. Valtellinese. Il naturalista valtellinese. (Not seen.)
Náttúrufræðingurinn. Náttúrufræðingurinn. Reykjavík. (Subtitle varies.)
Naturalist. The naturalist, a monthly journal of natural history for the north of England.
Nature. Nature. London.
Naturen (Bergen). Naturen. Illustreret maanedsskrift for populær naturvidenskab udgivet af Bergens Museum.
Natuurh. Maandbl. Limburg. Natuurhistorisch maandblad. Organ van het Natuurhistorisch genootschap in Limburg.
Natuurwetensch. Tijdschr. Natuurwetensschappelijk tijdschrift.
Nederl. Kruidk. Arch. Nederlandsch kruidkundig archief.
Neue Denkschr. Allg. Schweiz. Gesell. Gesam. Naturw. Neue denkschriften der Allgemeinen schweizerischen gesellschaft für die gesammten naturwissenschaften.
Neue Denkschr. Schweiz. Naturf. Gesell. Neue denkschriften der Schweizerischen naturforschenden gesellschaft.
Neujahrsbl. Naturf. Gesell. Kantons Glarus. Neujahrsblatt der Naturforschenden gesellschaft des kantons Glarus. (Not seen.)
Neujahrsbl. Naturf. Gesell. Schaffhausen. Neujahrsblatt herausgegeben von der Naturforschenden gesellschaft Schaffhausen.
Norsk Geog. Tidsskr. Norsk geografisk tidsskrift.
Norske Turistfor. Aarb. Den Norske turistforenings aarbok.
K. Norske Vidensk. Selsk. Forh. Det Kongelige norske videnskabers selskabs forhandlinger.
K. Norske Vidensk. Selsk. Mus. Årsberet. Det Kongelige norske videnskabers selskab. Museet. Årsberetning.
K. Norske Vidensk. Selsk. Skr. Det Kongelige norske videnskabers selskabs skrifter.
North West. Nat. The north western naturalist.

Not., Mém. et Doc. Soc. Agr., Archéol. et Hist. Nat. Dépt. Manche. Notices, mémoires et documents publiés par la Société d'agriculture, d'archéologie et d'histoire naturelle du département de la Manche.

Notes Bot. Sch. Trinity Col. Notes from the Botanical school of Trinity college, Dublin.

Notiser Sällsk. Fauna et Flora Fenn. Förhandl. Notiser ur Sällskapets pro fauna et flora fennica förhandlingar.

Notre Bourbonnais. Notre Bourbonnais. (Not seen.)

Nova Acta R. Soc. Sci. Upsal. Nova acta Regiae societatis scientiarum upsaliensis.

Nuova Riv. Forest. Nuova rivista forestale.

Nuovi Annali Agr. Min. Econ. Naz. [Italy]. Ministerio dell'economia nazionale. Nuovi annali dell'agricoltura.

Nuovo Gior. Bot. Ital. Nuovo giornale botanico italiano.

Nya Bot. Notiser. Nya botaniska notiser. (This title was carried by Botaniska notiser during the years 1849–56.)

Nyt (*later* **Nytt**) **Mag. Naturv.** Nyt magazin (*from v.* 75 Nytt magasin) for naturvidenskaberne. **Oslo.**

Nytt Mag. Bot. Nytt magasin for botanik.

Occas. Papers Hull Scient. and Field Nat. Club. Occasional papers of the Hull scientific and field naturalists' club.

Oesterr. Bot. Zeitschr. Oesterreichische botanische zeitschrift.

Öfvers. Finska Vetensk.-Soc. Förhandl. Öfversigt af Finska vetenskaps-societetens förhandlingar.

Öfvers. K. Vetensk.-Akad. Förhandl. Öfversigt af Kongl. vetenskaps-akademiens förhandlingar.

Oulun Luonnon Ystäväin Yhdist. Julkaisuja. Oulun Luonnon ystäväin Yhdistyksen julkaisuja. [Yearbook of the Friends of nature society of Oulu.] Sarja [series].

Pagine Istriane. Pagine istriane. Periodico mensile scientifico-letterario-artistico con particolare riflesso alla provincia dell'Istria. Capodistria.

Palestras Agron. Palestras agronómicos. Lisboa. *At head of title page*: Ministério da agricultura.

Papers and Proc. Hampsh. Field Club and Archaeol. Soc. Papers and proceedings of the Hampshire field club and archaeological society.

Papers Eastbourne Nat. Hist. Soc. Papers of the Eastbourne natural history society; *later* Transactions.

Peregrine. The peregrine. (Yn shirragh ny ree.) A publication of the Manx field club.

Petermanns Mitt. Ergänzungsh. Dr. A. Petermanns Mitteilungen aus Justus Perthes' Geographischer anstalt. Ergänzungsheft.

Pharm. Jour. Pharmaceutical journal. The official organ of the Pharmaceutical society. London. (Title varies somewhat.)

Pharm. Weekbl. Nederl. Pharmaceutisch weekblad voor Nederland. Amsterdam.

Phytologist. The phytologist: a popular botanical miscellany.

Poligrafo. Poligrafo. (Not seen.)

Précis Anal. Trav. Acad. Imp. Sci., Belles-Let. et Arts Rouen. Précis analytique des travaux de l'Académie impériale des sciences, belles-lettres, et arts de Rouen.

Primula. Primula. (Not seen.)

Proc. and Rpt. Ashmolean Nat. Hist. Soc. Oxfordsh. Ashmolean natural history society of Oxfordshire. Proceedings and report (*earlier* Report).

Proc. and Trans. Croydon Nat. Hist. and Scient. Soc. Proceedings and transactions of the Croydon natural history and scientific society. *Later* Proceedings of the Croydon . . .

Proc. and Trans. Liverpool Biol. Soc. Proceedings and transactions of the Liverpool biological society.

Proc. and Trans. Nat. Hist. Soc. Glasgow. Proceedings and transactions of the Natural history society of Glasgow.

Proc. Bath Nat. Hist. and Antiq. Field Club. Proceedings of the Bath natural history and antiquarian field club.

Proc. Birmingham Nat. Hist. and Micros. Soc. Proceedings of the Birmingham natural history and microscopical society.

Proc. Birmingham Nat. Hist. and Phil. Soc. Proceedings of the Birmingham natural history and philosophical society.
Proc. Bot. Soc. Brit. Isles. Proceedings of the Botanical society of the British Isles.
Proc. Bot. Soc. Edinb. Proceedings of the Botanical society of Edinburgh.
Proc. Bristol Nat. Soc. Proceedings of the Bristol naturalists' society; *earlier* **Annual report and proceedings.**
Proc. Cambridge Phil. Soc. Proceedings of the Cambridge philosophical society.
Proc. Chester Soc. Nat. Sci. Proceedings of the Chester society of natural science.
Proc. Cotteswold Nat. Field Club. Proceedings of the Cotteswold naturalists' field club.
Proc. Croydon Nat. Hist. and Scient. Soc. Proceedings of the Croydon natural history and scientific society; *earlier* Proceedings and transactions . . .
Proc. Dorset Nat. Hist. and Antiq. Field Club. Proceedings of the Dorset natural history and antiquarian field club.
Proc. Holmesdale Nat. Hist. Club. Proceedings of the Holmesdale natural history club.
Proc. Isle of Man Nat. Hist. and Antiq. Soc. The Isle of Man natural history and antiquarian society. Proceedings.
Proc. Isle of Wight Nat. Hist. Soc. Proceedings of the Isle of Wight natural history and archaeological society.
Proc. Linn. Soc. Proceedings of the Linnean society of London.
Proc. Liverpool Lit. and Phil. Soc. (Not seen.)
Proc. Llandudno, Colwyn Bay and Dist. Field Club. Proceedings of the Llandudno, Colwyn Bay and district field club.
Proc. Perthsh. Soc. Nat. Sci. Proceedings of the Perthshire society of natural science. *Replaced by their* Transactions and proceedings.
Proc. Roy. Irish Acad. Proceedings of the Royal Irish academy.
Proc. Roy. Soc. Edinb. Proceedings of the Royal society of Edinburgh.
Proc. Sheffield Nat. Club. Proceedings of the Sheffield naturalists' club. (Not seen.)
Proc. Somersetsh. Arch. and Nat. Hist. Soc. Proceedings of the Somersetshire archaeological and natural history society.
Proc. South East. Union Sci. Soc. South eastern union of scientific societies. Proceedings at the . . . annual congress. (*On cover*: Report and transactions.) [London.]
Proc. Swansea Scient. and Field Nat. Soc. The proceedings of the Swansea scientific and field naturalists' society.
Proc. Univ. Durham Phil. Soc. Proceedings of the University of Durham philosophical society.
Proc.-Verb. Soc. Dauphinoise Études Biol. Procès-verbaux de la Société dauphinoise d'études biologiques. (Not seen.)
Program Realsch. Erst. Ordn. Aachen. Programm der Realschule erster ordnung zu Aachen.
Programm K. K. Gymnas. Brixen. Programm des k.k. gymnasiums in Brixen. (Not seen.)
Programma Imp. R. Scuola Naut. Lussinpicolo. Programma dell'Imperiale reale scuola nautica in Lussinpicolo. (Not seen.)
Pub. Comis. Regulad. Com. Arroz. Publicações de Comissão reguladora do comércio de arroz. Lisboa.
Pub. Dir. Geral Serv. Flor. e Aqüic. [Portugal]. Direcção geral dos serviços florestals e aqüícolas. Publicações.
Pub. Engl. Dialect Soc. English dialect society. Publications.
Pub. Flintsh. Hist. Soc. Flintshire historical society publications.
Pub. Inst. Bot. Gonçalo Sampaio Faculd. Ciên. Univ. Porto. Publicações do Instituto de botânica "Dr. Gonçalo Sampaio" da Faculdade de ciências da Universidade do Porto. (Not seen.)
Pub. Inst. Roy. Grand-ducal Luxemb. Publications de l'Institut royal grand-ducal de Luxembourg.
Pub. Mus. Civico Rovereto. Pubblicazione fatta per cura del Museo civico (*sometimes* Civico museo) di Rovereto.
Rameau du sapin. Le rameau du sapin. Organe du Club jurassien. Neuchâtel. (Lithographed.)

Rass. Econ. Prov. Sondrio. Rassegna economica della provincia di Sondrio.

Rass. Mens. Municip. Torino. Rassegna mensile municipio. Torino.

Rass. Mens. Unione Ligure Escursionisti. Rassegna mensile. Unione ligure escursionisti. (Not seen.)

Recueil Inst. Bot. Léo Errera. Recueil de l'Institut botanique Léo Errera (Université de Bruxelles).

Recueil Mém. et Trav. Soc. Bot. Grand-duché Luxemb. Recueil des mémoires et des travaux publiées par la Société botanique du Grand-duché de Luxembourg.

Recueil Trav. Bot. Néerland. Recueil des travaux botaniques néerlandais.

Recueil Trav. Bot. Univ. Montpellier. Recueil des travaux botaniques de l'Université de Montpellier. (Not seen.)

Rend. Accad. Naz. dei XL [Quaranta]. Rendiconti della Accademia nazionale dei XL [Quaranta]. *Formerly* Società italiana delle scienze detta Accademia dei XL [Quaranta]. Roma.

Rend. e Mem. R. Accad. Sci., Let. ed Arti Zelanti Acireale. Rendiconti e memorie della R. accademia di scienze, lettere ed arti degli zelanti, Acireale.

Rend. R. Ist. Lombardo Sci. e Let. Reale istituto lombardo di scienze e lettere. Rendiconti.

Rend. Semin. Fac. Sci. Univ. Cagliari. Rendiconto del Seminario della Facoltà di scienze della Università di Cagliari.

Repert. Spec. Nov. Fedde. Repertorium specierum novarum regni vegetabilis . . . Hrsg. von Friedrich Fedde.

Repert. Spec. Nov. Fedde Beih. Repertorium specierum novarum regni vegetabilis . . . Hrsg. von . . . Friedrich Fedde. Beihefte.

Revista Acad. Cien. Zaragoza. Revista de la Academia de ciencias exactas, físico-químicas y naturales de Zaragoza.

Revista Agr., Com. y Trab. (Habana). Revista de agricultura, comercio y trabajo (Habana). *Later* Revista de agricultura.

Revista Agron. Revista agronómica. Publicação da Sociedade de sciências agronómicas de Portugal.

Revista Mont. (Madrid). Revista de montes. Madrid.

Revista Prog. Cien. Exact., Fís. y Nat. Madrid. Revista de los progresos de las ciencias exactas, físicas y naturales. Madrid.

Revista Sci. Nat. e Soc. Revista de sciencias naturaes e sociaes. Porto.

Revista Soc. Instruc. Porto. Revista da Sociedade de instrucção do Porto.

Revue Bot. Appl. et Agr. Trop. Revue de botanique appliquée et d'agriculture tropicale.

Revue Bot. Bul. Mens. Soc. Franç. Bot. Revue de botanique. Bulletin mensuel de la Société française de botanique.

Revue Bot. Syst. et Géog. Bot. Revue de botanique systematique et de géographie botanique.

Revue Bretonne Bot. Revue bretonne de botanique pure et appliquées.

Revue d'Auvergne. Revue d'Auvergne. Publiée par la Société des amis de l'Université de Clermont.

Revue des Eaux et Forêts. Revue des eaux et forêts.

Revue Gén. Bot. Revue général de botanique.

Revue Hort. Bouches-du-Rhône. Revue horticole. Journal mensuel des travaux de la Société d'horticulture et de botanique des Bouches-du-Rhône.

Revue Internatl. Bot. Appl. et Agr. Trop. Revue internationale de botanique appliquée et d'agriculture tropicale; *earlier* Revue de botanique appliquée et d'agriculture tropicale.

Revue Périod. Vulgar. Sci. Nat. et Préhist. de "La Physiophilie," Soc. Études Hist. Nat. Montceau-les-Mines. Revue périodique de vulgarization des sciences naturelles et préhistoriques de "La physiophilie," Société d'études d'histoire naturelle, Montceau-les-Mines.

Revue Pyr. Revue des Pyrénées et de la France meridionale.

Revue Quest. Scient. Revue des questions scientifiques.

Revue Savoisienne. Société florimontane d'Annecy. Revue savoisienne; *later* Revue savoisienne, publication périodique de l'Académie florimontane d'Annecy.

Revue Sci. Nat. Appl. Revue des sciences naturelles appliquées publiée par la Société nationale d'acclimatation.

Revue Sci. Nat. (Montpellier et Paris). Revue des sciences naturelles. Montpellier et Paris.

Revue Sci. Nat. Ouest. Revue des sciences naturelles de l'Ouest.

Revue Scient. Bourbonn. Revue scientifique du Bourbonnais et du centre de la France.

Revue Scient. Limousin. La revue scientifique du Limousin . . .

Revue Verviét. Hist. Nat. Revue verviétoise d'histoire naturelle. (Not seen.)

Ricerche e Lav. Ist. Bot. R. Univ. Pisa. Ricerche e lavori eseguiti nell' Istituto botanico della R. università di Pisa.

Riv. Abruzzese Sci., Let. ed Arti. Rivista abruzzese di scienze, lettere ed arti. Teramo.

Riv. Ital. Essen., Profumi e Piante Off. Rivista italiana delle essenze, dei profumi e delle piante officinali; *earlier* Rivista italiana delle essenze e profumi.

Riv. Ital. Sci. Nat. Rivista italiana di scienze naturali.

Riv. Mens. Club Alp. Ital. Rivista mensile del Club alpino italiano.

Rochester Nat. The Rochester naturalist. A quarterly record of the Rochester naturalists' club.

Romagna Agr. e Zootec. La Romagna agricola e zootecnica.

Rpt. and Trans. Cardiff Nat. Soc. Cardiff naturalists' society. Report and transactions.

Rpt. and Trans. Devonsh. Assoc. Adv. Sci., Lit. and Art. Report and transactions of the Devonshire association for the advancement of science, literature and art.

Rpt. and Trans. Soc. Guernes. La Société guernesiaise (*earlier* Guernsey society of natural science and local research). Report and transactions. (Title varies.)

Rpt. Bot. Rec. Club. The Botanical record club. Report . . . *Earlier* (through report for 1878) The Botanical locality record club. (Subtitle varies.)

Rpt. Bot. Soc. and Exch. Club Brit. Isles. The Botanical Society and exchange club of the British Isles. Report.

Rpt. Brit. Assoc. Adv. Sci. Report of the . . . meeting of the British association for the advancement of science.

Rpt. Bury Nat. Hist. Soc. Bury natural history society. Report . . .

Rpt. Cheltenham Col. Nat. Hist. Soc. Cheltenham college natural history society. Report of the proceedings of the Society.

Rpt. Gresham's School Nat. Hist. Soc. Report of Gresham's school natural history society.

Rpt. Marlborough Col. Nat. Hist. Soc. Report of the Marlborough college natural history society.

Rpt. Merseyside Nat. Assoc. Report of the Merseyside naturalists' association. (Not seen.)

Rpt. Proc. Winchester and Hampsh. Scient. and Lit. Soc. Winchester and Hampshire scientific and literary society. Report of proceedings.

Rpt. Puffin Island Biol. Sta. Puffin Island biological station. Report. (Not seen.)

Rpt. Rugby School Nat. Hist. Soc. Report of the Rugby school natural history society.

Rpt. Southport Scient. Soc. Southport scientific society. Report. *Formerly* Southport Society of natural science. Report.

Rpt. Winchester Col. Nat. Hist. Soc. Winchester college natural history society. Report.

Satakunta. Satakunta.

Schlern. Der Schlern. Zeitschrift für heimat- und volkskunde. Bolzano. (Subtitle varies.)

Schweiz. Apotheker-zeit. Schweizerische apotheker-zeitung. Journal suisse de pharmacie. Giornale svizzero di farmacia.

Schweiz. Arch. Volksk. Schweizerisches archiv für volkskunde.

Sci. Paper Haslemere Nat. Hist. Soc. Haslemere natural history society. Science paper.

Scient. Proc. Roy. Dublin Soc. The scientific proceedings of the Royal Dublin society.

Scott. Bot. Rev. The Scottish botanical review. A quarterly magazine including the Transactions of the Botanical society of Edinburgh.

Scott. Forestry. Scottish forestry.
Scott. Nat. Scottish naturalist.
Silva Fenn. Silva fennica. *At head of title page:* Suomen metsätieteellinen Seura . . . Society of forestry in Suomi . . .
Sitzungsber. Gesell. Naturf. Freunde Berlin. Sitzungsberichte der Gesellschaft naturforschender freunde zu Berlin.
Sitzungsber. u. Abhandl. Naturw. Gesell. Isis Dresden. Sitzungsberichte und abhandlungen der Naturwissenschaftlichen gesellschaft Isis in Dresden.
Skånes Natur. Skånes natur. Skånes naturskyddsförenings årsskrift.
Skogsförsöksanst. Exkursionsled. Skogsförsöksanstaltens exkursionsledare.
Skogsvårdsför. Tidskr. Skogsvårdsföreningens tidskrift.
Skr. Fören. Strengnenses. Skrifter utgivna av Föreningen strengnenses.
Skr. Norges Svalbard- og Ishavets-Undersøk. Norges Svalbard- og Ishavets-undersøkelser skrifter. *Earlier* (*through* no. 81) Skrifter om Svalbard og Ishavet.
Skr. Norske Vidensk.-Akad. Oslo. Skrifter utgitt av det Norske videnskaps-akademi i Oslo.
Skr. Svalbard og Ishavet. Skrifter om Svalbard og Ishavet.
Skr. Tromsø Mus. Tromsø museums skrifter.
Skr. Vidensk.-Selsk. Christiania (Kristiania) I, Math.-Naturv. Kl. Skrifter udgivne af Videnskabs-selskabet i Christiania (Kristiania) I, Mathematisk (Matematisk)-naturvidenskabelig klasse (*later* Norske videnskaps akademi).
Skýrsla um Hiđ Íslenzk. Náttúrufraeđisf. Skýrsla um hiđ íslenzka (ílenska) náttúrufræđisfélag (náttúrufræđisfjélag).
South East. Nat. The south eastern naturalist, being the transactions of the South-eastern union of scientific societies.
South-east. Nat. and Antiq. The south-eastern naturalist and antiquary being the volume of transactions of the South-eastern union of scientific societies including the proceedings . . . (title varies somewhat).
St. Thomas's Hospital Gaz. St. Thomas's hospital gazette.
Stavanger Mus. Aarsber. Stavanger museums aarsberetning.
Studi Sassaresi. Studi sassaresi. (Not seen.)
Studi Trent.; Studi Trent. Sci. Nat. Studi trentini; Studi trentini de scienze naturali.
Studi Urbinati. Studi urbinati. (Not seen.)
Studj Romanzi. Studj romanzi. Società filologica romana.
Suomen Luonto. Suomen luonto. (Not seen.)
Suomi. Suomi. Tidskrift i fosterländska ämnen.
Svensk Bot. Tidskr. Svensk botanisk tidskrift.
Svensk Geog. Årsb. Svensk geografisk årsbok. The Swedish geographical yearbook.
Svenska Linné-sällsk. Årsskr. Svenska Linné-sällskapets årsskrift.
Svenska Växtsociol. Sällsk. Handl. Svenska växtsociologiska sällskapets handlingar.
K. Svenska Vetenskapsakad. Avhandl. Naturskyddsär. Kungl. svenska vetenskapsakademiens avhandlingar i naturskyddsärenden.
K. Svenska Vetensk.-Akad. Handl. Kongliga (*later* Kungliga) Svenska vetenskaps-akademiens handlingar.
K. Svenska Vetenskapsakad. Skr. Naturskyddsär. K. Svenska vetenskapsakademiens skrifter i naturskyddsärenden.
Sveriges Geol. Undersök. ser. C. Afhandl. och Uppsats. Sveriges geologiska undersökning. Ser. C. Afhandlingar och uppsatser.
Sveriges natur. Sveriges natur. Svenska naturskyddsföreningens tidskrift. (The Årsbok a separately paged number each year.)
Symb. Bot. Upsal. Symbolae botanicae upsalienses.
Tätigkeitsber. Naturf. Gesell. Baselland. Tätigkeitsbericht der Naturforschenden gesellschaft Baselland.
Terra. Terra . . . Helsingfors.
Terre et la vie. La terre et la vie. Revue d'histoire naturelle.
Tidskr. Geog. För. (Helsingfors). Geografiska föreningens tidskrift. (Also has Finnish title.) Helsingfors.
Tijdschr. Nat. Geschied. en Phys. Tijdschrift voor natuurlijke geschiedenis en physiologie. Leiden.

Tomo Extraord. R. Soc. Españ. Hist. Nat. Real socieded española de historia natural. Tomo extraordinario publicado con motivo del 50. aniversario de su fundación.
Torreya. Torreya.
[Trab.] Inst. Forest. Invest. y Exper. Madrid. *See* [Bol.] Inst. Forest. Invest. y Exper. Madrid.
Trab. Jard. Bot. Facult. Farm. Univ. Santiago de Compostela. Universidad de Santiago [de Compostela]. Facultad de farmacia. Trabajos del Jardín botánico.
Trab. Mus. Cien. Nat. Barcelona. Trabajos del Museo de ciencias naturales de Barcelona. *In part as* Treballs del Museu de ciències naturals de Barcelona. (Série botánica.) *At head of title page (or cover):* Publicaciones de la Junta de ciencias naturales (ciències naturals) de Barcelona.
Trans. and Jour. Proc. Dumfriessh. and Galloway Nat. Hist. and Antiq. Soc. Dumfriesshire and Galloway natural history and antiquarian society. Transactions and journal of proceedings.
Trans. and Proc. Bot. Soc. Edinb. Transactions and proceedings of the Botanical society of Edinburgh. *Earlier* Transactions . . .
Trans. and Proc. Perthsh. Soc. Nat. Sci. Transactions and proceedings of the Perthshire society of natural sciences.
Trans. and Proc. Torquay Nat. Hist. Soc. Torquay natural history society. Transactions and proceedings.
Trans. Anglesey Antiq. Soc. Transactions of the Anglesey antiquarian society and field club.
Trans. Banffsh. Field Club. Transactions of the Banffshire field club and scientific society, Banff, Scotland.
Trans. Bishops Stortford and Dist. Nat. Hist. Soc. Transactions of the Bishops Stortford and district natural history society. (Not seen.)
Trans. Bot. Soc. Edinb. Transactions of the Botanical society of Edinburgh. *Later* Transactions and proceedings . . .
Trans. Buchan Field Club. Transactions of the Buchan field club.
Trans. Burton-on-Trent Nat. Hist. and Arch. Soc. Transactions of the Burton-on-Trent natural history and archaeological society.
Trans. Caernarvonsh. Hist. Soc. Transactions of the Caernarvonshire historical society.
Trans. Cardiff Nat. Soc. Transactions of the Cardiff naturalists' society.
Trans. Carmarthensh. Antiq. Soc. and Field Club. The—year's transactions of the Carmarthenshire antiquarian society and field club.
Trans. Cumberl. (and Westmorl.) Assoc. Advanc. Lit. and Sci. Transactions of the Cumberland (*from v.* 9, Cumberland and Westmorland) association for the advancement of literature and science.
Trans. Eastbourne Nat. Hist. Soc. *See* Papers Eastbourne Nat. Hist. Soc.
Trans. Edinb. Field Nat. and Micros. Soc. Transactions of the Edinburgh field naturalists' and microscopical society.
Trans. Hertfordsh. Nat. Hist. Soc. Transactions of the Hertfordshire natural history society and field club.
Trans. Hull Scient. and Field Nat. Club. Transactions of the Hull scientific and field naturalists' club.
Trans. Leicester Lit. and Phil. Soc. Transactions of the Leicester literary and philosophical society.
Trans. Lincolnsh. Nat. Union. Lincolnshire naturalists' union. Transactions.
Trans. Linn. Soc. Bot. Transactions of the Linnean society. Botany.
Trans. Liverpool Bot. Soc. Transactions of the Liverpool botanical society.
Trans. London Nat. Hist. Soc. Transactions of the London natural history society; *later* London naturalist.
Trans. Nat. Hist. and Antiq. Soc. Penzance. Transactions of the Natural history and antiquarian society of Penzance.
Trans. Nat. Hist. Soc. Northumberl., Durham, and Newcastle upon Tyne. Transactions of the Natural history society of Northumberland, Durham, and Newcastle upon Tyne.
Trans. Newbury Dist. Field Club. Transactions of the Newbury district field club. (Not seen.)
Trans. Norfolk and Norwich Nat. Soc. Transactions of the Norfolk and Norwich naturalists' society.

Trans. North Staffordsh. Field Club. Transactions (and annual report) of the North Staffordshire field club. (Title varies somewhat.)
Trans. Paisley Nat. Soc. Transactions of the Paisley naturalists' society. (Not seen.)
Trans. Plymouth and Dist. Field Club. Plymouth and district field club. Transactions. (Not seen.)
Trans. Powys-land Club. Transactions of the Powys-land club.
Trans. Radnorsh. Soc. Transactions of the Radnorshire Society. (Not seen.)
Trans. Rochdale Lit. and Scient. Soc. Transactions of the Rochdale literary and scientific society.
Trans. Roy. Med.-Bot. Soc. London. Transactions of the Royal medico-botanical society of London.
Trans. Roy. Scott. Arbor. Soc. Transactions of the Royal Scottish arboricultural society.
Trans. Soc. Guernes. *See* Rpt. and Trans. Soc. Guernes.
Trans. South-East. Union Scient. Soc. Transactions of the South-eastern union of scientific societies.
Trans. Stirling Field Club. Transactions of the Stirling field club.
Trans. Stirling Nat. Hist. and Arch. Soc. Stirling natural history and archaeological society. Transactions.
Trans. Suffolk Nat. Soc. Transactions of the Suffolk naturalists' society.
Trans. Weardale Nat. Field Club. Weardale naturalists' field club. Transactions.
Trans. West Kent Nat. Hist., Micros., and Photog. Soc. Transactions of the West Kent natural history, microscopical and photographic society.
Trans. Woolhope Nat. Field Club. Transactions of the Woolhope naturalists' field club.
Trans. Worcestersh. Nat. Club. The transactions of the Worcestershire naturalists' club.
Trans. Yorksh. Nat. Union, Bot. Ser. Transactions of the Yorkshire naturalists' union, Botanical series.
Trav. Lab. Forest. Toulouse. Travaux du Laboratoire forestier de Toulouse.
Trav. Scient. Univ. Rennes. Travaux scientifiques de l'Université de Rennes.
Trav. Soc. Acad. Sci., Arts, Belles-Let. et Agr. Saint-Quentin. Société académique des sciences, arts, belles-lettres et agriculture de Saint-Quentin (Aisne). Travaux . . .
Trav. Soc. Bot. Genève. Travaux de la Société botanique de Genève.
Treballs Inst. Catalana Hist. Nat. Treballs de la Institució catalana d'historia natural.
Treballs Mus. Ciën. Nat. Barcelona. See Trab. Mus. Cien. Nat. Barcelona.
Tromsø Mus. Aarsh. (Årsh.). Tromsø museums aarshefter (årshefter).
Univ. Bergen Årb. Naturvitensk. Rekke. Universitetet i Bergen. Årbok. Naturvitenskapelig rekke.
Universo (Firenze). L'universo (Istituto geografico militare. Firenze).
Västerbotten. Västerbotten. Västerbottens länshembygdsförenings årsbok.
Vasculum. The vasculum. The north country quarterly of science and local history. Newcastle-upon-Tyne. (Subtitle varies.)
Verhandel. Batav. Genootsch. Kunst. en Wetensch. Verhandelingen van het Bataviaasch genootschap van kunsten en wetenschappen.
Verhandl. Bot. Ver. Prov. Brandenb. Verhandlungen des Botanischen vereins der provinz Brandenburg.
Verhandl. (K.K.) Zool.-Bot. Gesell. Wien. Verhandlungen der (K.K.) Zoologisch-botanischen gesellschaft in Wien.
Verhandl. Naturf. Gesell. Basel. Verhandlungen der Naturforschenden gesellschaft in Basel.
Verhandl. Schweiz. Naturf. Gesell. Verhandlungen der Schweizerischen naturforschenden gesellschaft. (Also French and Italian titles.)
Veröffentl. Geobot. Inst. Rübel Zürich. Veröffentlichungen des Geobotanischen institutes Rübel in Zürich.
Veröffentl. Mus. Ferdinand. Innsbruck. Veröffentlichungen des Museum Ferdinandeum in Innsbruck.
Verslag. Landbouwk. Onderzoek. Verslagen van landbouwkundige onderzoekingen. [*Publication of*] Ministerie van landbouw, visserij en voedselvoorziening. Directie van de landbouw. 's-Gravenhage.

Verslag. en Meded. K. Akad. Wetensch. Amsterdam Afd. Natuurk. Verslagen en mededeelingen der Koninklijke akademie van wetenschappen. Afdeeling natuurkunde.

K. Vetensk.-Akad. Handl. Kongl. vetenskaps-akademiens handlingar. *With the* ny följd (v. 1, *for* 1855–56) *this became* Kongliga svenska vetenskaps-akademiens handlingar.

Vetensk. Meddel. Geog. För. Finland. Vetenskapliga meddelanden af Geografiska föreningen i Finland; *later* Meddelanden af Geografiska föreningen i Finland.

Vidensk. Meddel. Naturh. For. Kjøbenhavn. Videnskabelige meddelelser fra den Naturhistoriske forening i Kjøbenhavn (Kjöbenhavn).

Vierteljahrsschr. Naturf. Gesell Zürich; . . . Beibl. Vierteljahrsschrift der Naturforschenden gesellschaft in Zürich; *also* Beiblatt zur . . .

Vinha Portug. A vinha portugueza. Revista geral de agricultura.

Vísindafélag Íslendinga Greinar. Vísindafélag íslendinga (Societas scientiarum islandica). Greinar.

Vísindafélag Íslendinga Rit. Vísindafélag íslendinga. Rit.

Vox Romanica. Vox romanica. Annales helvetici explorandis linguis romanicis destinati.

Watsonia. Watsonia. Journal of the Botanical society of the British Isles. *Replaces* (*from* 1949) The Botanical society and exchange club of the British Isles. Report.

Webbia. Webbia. Raccolta di scritti botanici . . .

Wiltsh. Archaeol. and Nat. Hist. Mag. The Wiltshire archaeological and natural history magazine.

Wimbledon and Merton Annual. Wimbledon and Merton annual.

Zeitschr. Deut. u. Österr. Alpenver. Zeitschrift des Deutschen und österreichischen alpenvereins.

Zeitschr. Ferdinandeums Tirol. u. Vorarlberg. Zeitschrift des Ferdinandeums für Tirol und Vorarlberg.

Zeitschr. Gesell. Schleswig-Holsteinische Gesch. Zeitschrift der Gesellschaft für schleswig-holsteinische geschichte.

AUTHOR INDEX

GEOGRAPHICAL INDEX

This index includes the names of all countries (in English and in the native tongue) and all geographical divisions used as headings. The divisions consist of cantons (Switzerland), counties, departments, landskaper, provinces, and regions. Compound names (mostly hyphenated) are indexed as separate words, alphabetized under the first word; thus, Loir-et-cher precedes Loire, Haut-Rhin precedes Haut-Savoie, and so on. The term "see under" indicates names of provinces that do not appear as headings but which are included in annotations of titles under the region cited.

Regardless of the language, ä, ö, and ü are alphabetized as ae, oe, and ue, å as aa, æ as ae, and ø as oe.

U.S. GOVERNMENT PRINTING OFFICE: 1960-506458